28th Polymer Processing Society Annual Meeting 2012 (PPS-28)

Pattaya, Thailand
11 – 15 December 2012

Volume 1 of 2

ISBN: 978-1-5108-4261-8

Printed from e-media with permission by:

Curran Associates, Inc.
57 Morehouse Lane
Red Hook, NY 12571

Printed by Curran Associates, Inc. (2016)

For permission requests, please contact Polymer Processing Society at the address below.

Polymer Processing Society
810 E. 10th St.
Lawrence, KS 66044
USA

Phone: +1-800-627-0326
+1-785-865-9403
Fax: +1-785-843-6153

pps@allenpress.com

Additional copies of this publication are available from:

Curran Associates, Inc.
57 Morehouse Lane
Red Hook, NY 12571 USA
Phone: 845-758-0400
Fax: 845-758-2633
Email: curran@proceedings.com
Web: www.proceedings.com

TABLE OF CONTENTS

VOLUME 1

EXTRUSION AND EXTRUSION PROCESSES

BLOW MOLDING, ROTOMOLDING AND THERMOFORMING

FIBERS, FILMS AND FOAMS

INJECTION AND MICROINJECTION MOLDING AND MOLDS

MIXING AND COMPOUNDING

REACTIVE PROCESSING, POLYMER BLENDS AND ALLOYS

POLYMER COMPOSITES AND NANOCOMPOSITES

VOLUME 2

RHEOLOGY AND RHEOMETRY

MORPHOLOGY AND STRUCTUR DEVELOPMENT

PROCESS MODELING AND SIMULATION

BIO-BASED & BIODEGRADABLE POLYMERS AND POLYMER RECYCLING

FOOD RHEOLOGY & PROCESSING

NATURAL RUBBER & RUBBER PROCESSING

FRONTIER PROCESSING TECHNOLOGY

SYMPOSIUM HONORING PROF. MUSA KAMAL

MEMORIAL SESSION OF PROF. TOSHIRO MASUDA

PLENARY SESSION

KN-01-01

MICROPELLETIZING USING RAYLEIGH DISTURBANCES

Tim Osswald[1*], William Aquite[1], Natalie Rudolph[2], Daniel Kasperek[3]
[1]Polymer Engineering Center, University of Wisconsin-Madison – osswald@engr.wisc.edu
[2]Fraunhofer-Institut für Chemischexa Technologie ICT – natalie.rudolph@ict.fraunhofer.de
[3]Department and Chair for Medical Engineering, TU Muenchen – daniel.kasperek@pe.mw.tum.de

Abstract - This paper presents a new technique that allows production of micropellets with a relatively spherical shape. This method is an alternative way of producing powders with physical properties demanded in specialized processes such as selective laser sintering processes. The micropellets are achieved by extruding a polymer melt strand through a capillary and then, deforming the strand with a stream of hot air causing breakup. Observed deformation and regimes of breakup are explained considering the molecular structure, viscosity, surface tension and melt strength for different resins and also considering the viscoelastic response of the extruded thread as it influences the breakup process. This paper presents a description of the design of the system, a discussion of the phenomena involved in this process and as well as results that validate each of these phenomena.

Keywords: Rayleigh Disturbance, Micropellets, Powder.

Introduction

Additive manufacturing has grown in importance and gaining more interest for series-production of individual parts featuring highly complex geometric shapes and features. As a consequence there is an increasing demand for micropellets and powders for powder-based polymer processing techniques such as selective laser sintering (SLS) and selective mask sintering (SMS). This represents a challenge for current production techniques and calls for alternative manufacturing methods and the reduction of the cost of current ones [1]. Furthermore these processes are highly influenced by the effects of size, size distribution and shape of the polymer particles.

The standard size of common polymer pellets ranges from 1 mm to 5 mm while micropellets have a size ranging between 50 μm and 1000 μm. To achieve such sizes current processes entail costly cryogenic milling or chemical precipitation processes and are just available for a limited selection of resins. Furthermore cryogenic grinding produces powders with microscopic sharp edges and angles, which lowers bulk density and processability. Cryogenic grinding is also expensive and energy consuming [2]

A new technique that allows the production of micropellets with a controlled size and shape has been developed at the Polymer Engineering Center [3]. This method is an alternative and less energy consuming way of producing polymer pellets with physical properties demanded in specialized processes such as laser sintering were size, geometry, size distribution and morphology are of significant importance. The micropellets are achieved by extruding a polymer melt strand through a capillary and then, deforming the strand with a stream of hot air causing Rayleigh disturbances on the melt and eventual breakup into pellets. The molecular structure, viscosity, surface tension and melt strength of the resin are influential factors on the deformation and breakup of the melt.

Pellets, fibers and threads of different size distributions and shapes obtained under a wide range of process conditions have been physically characterized to proof the different phenomena involved in the pelletization process.

Experimental

The experimental setup for the extrusion of micropellets allows aerodynamic forces to cause perturbations at the interface of a polymer melt with heated air. Polymer chains aligning in the same direction as to which the forces act, leads to stretching of the melt. Then, due to pressure and interfacial velocity fluctuations induced by those forces as well as changes within the fluid, Rayleigh disturbances can occur. Such disturbances create local pressure differences along the interface and hence the interfacial tension minimizes the interface between the two faces to achieve a minimum in surface-to-volume ratio. Disturbances with a wavelength larger than the circumference of the original thread grow and result in a necking down of the fluid thread due to surface tension.

The interfacial tension minimizes the interface between the two faces, as well as the surface-to-volume ratio. The disturbances with a wavelength larger than the circumference of the initial thread grow, decreasing diameter, and increasing deviatoric stresses and capillary forces, break up and the thread disintegrates into droplets [4,5].

Fig. 1 depicts the breakup mechanism as a result of competing surface tension effects and deviatoric stresses.

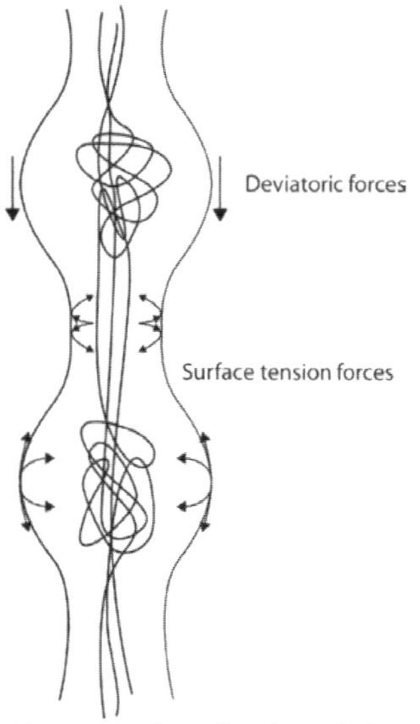

Figure 1. Forces and molecular chain orientation during the process.

The ratio of those competing effects, the deviatoric stresses and surface tension stresses, is defined as the capillary number (Eq. 1),

$$Ca = \frac{\tau R}{\sigma} \qquad (1)$$

where τ is the deviatoric stress, R the strand's characteristic dimension, and σ the surface tension. There is a critical capillary number where break up occur [6]. At this point surface tension pinches off the strand and shapes a droplet. This phenomena is the basic principle used for the design the micropelletizer.

Experimental setup

The experimental setup for the micropelletizer consists of an extruder, an extrusion die, and equipment for collecting the product.

The extruder is a 19 mm (0.75 Inch) Brabender laboratory model 2503, with a Length-to-Diameter-ratio L/D of 25 : 1. It is equipped with three controlled heating bands that use compressed air to cool down. The screw is driven by a 2 KW DC motor which can be varied in screw rotational speed continuously from 0 to 150 min^{-1}. The extrusion die is built up inside a nozzle where hot compressed air flows through. The air is heated up by means of a heat torch with a capacity of 5 KW and able to heat up to 730 cm^3/s (6.5 SCFM) of air to around 350°C.

The nozzle consists of two parts; the mandrel and the collar. The mandrel has a 45° angled conical tip on the outside and a centered melt channel with a diameter of 1 mm. The collar has a 45° angled conical shape on the inside and a centered hole with a diameter of 3 mm to drive the air. All parts are manufactured of stainless steel AISI 304. The nozzle allows a coaxial flow of polymer in the center and hot streaming air around it [7].

Fig 2. presents an schematic drawing of the experimental setup.

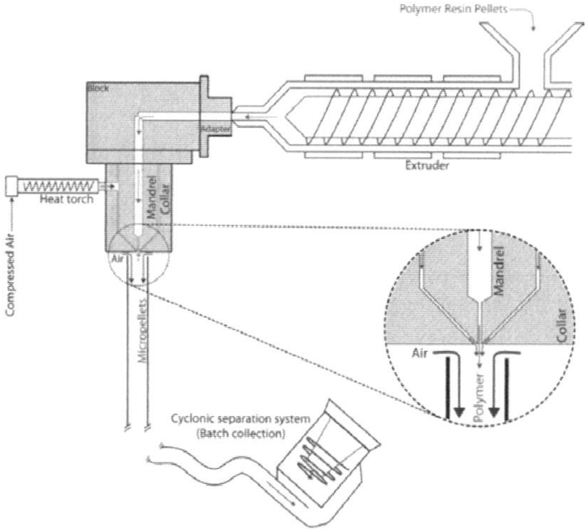

Figure 2. Experimental setup of the experiment.

The break up process is expected to happen where the two phases, air and polymer merge. The extruded polymer melt comes out of the 1 mm capillary and as it exits it, is pulled downwards by the stream of hot air. Then the stretching and breakup process of the thread takes place followed by collection of the particles by a cyclonic separation system.

Resins processed using this device started with polyolefins such as high density polyethylene (PE-HD), low density polyethylene (PE-LD) and Polypropylene (PP). These were chosen to start the study of the influences of the different rheological properties and processing conditions on the development of the breakup. Polylactic acid (PLA), as a resin of interest for selective laser sintering processes, has also been used. Melt fracture index values for resins investigated using this system are shown in Table 1.

	Material	Company	MFI (g/10min, 190°C/2.16kg)
PE-HD	Dowlex IP 40	Dow Chemical	40
	Univel DMDA-6200	Dow Chemical	0.38
PE-LD	EN1807	Westlake Chemical	0.7
PLA	Ingeo 3051D	NatureWorks LLC	17.5
	Ingeo 3251D	NatureWorks LLC	40
PP	Pro-fax 6523	LyondellBasell	4

Table 1. Melt flow index for resins tested

In order to determine the effects of various process conditions and material properties on the formation and detachment of drops, different resins were tested at a range of process conditions. Digital photography have been used to study the process and microscopy to perform size and morphology analysis on collected samples. Modeling and simulation for this process, as a sourcevaluable insight to the process study is discussed in [8].

Results and Discussion

With the described experimental setup, a variation of the controlled process conditions allowed to observe a series of phenomena on the perturbed extrudate. Evidence of the occurrence of Rayleigh disturbances and breakup using the described experimental setup was demonstrated.

Fig. 3 shows peaks and valleys on particles that resulted from the processing of PE-HD. The shape of the particles corresponds with the described phenomena.

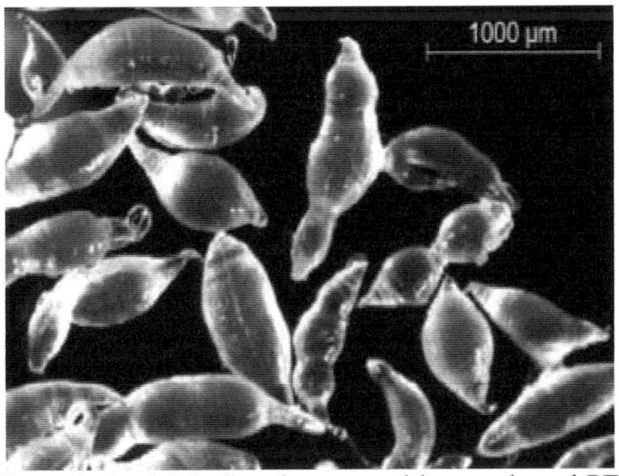

Figure 3. Rayleigh disturbances and lemon shaped PE-HD particles.

Fig. 4 shows PE-HD micropellets obtained under different process conditions where smaller particles are obtained high circularity values.

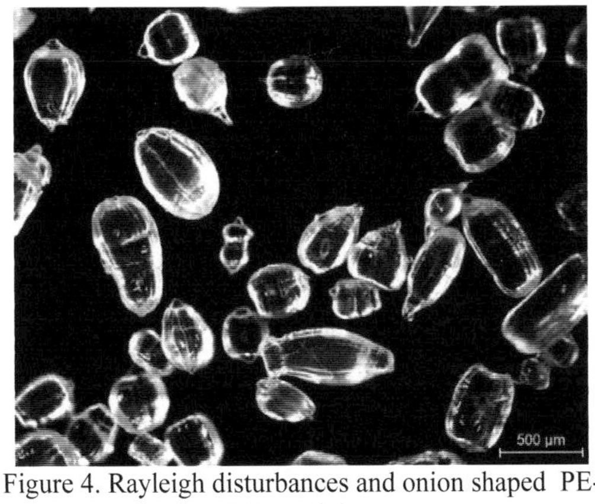

Figure 4. Rayleigh disturbances and onion shaped PE-HD particles.

Fig. 5 and Fig. 6 compare the size distribution and circularity of samples of PE-HD micropellets to commercially available SLS powder. The micropellets sample shows numerous amounts of particles with sizes below 30 µm, proving that it is possible to obtain a high concentration of particles with geometries that are close to spherical at scales not reachable .

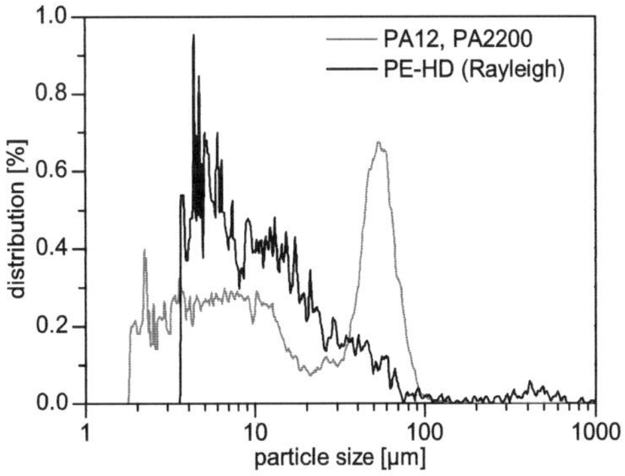

Figure 5. Size distribution of produced PE-HD compared to commercially available PA12, PA2200.

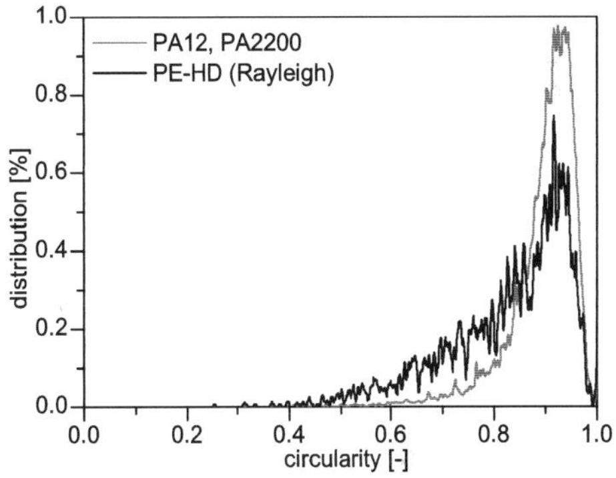

Figure 6. Circularity of produced PE-HD compared to commercially available PA12, PA2200.

Due to the uniformity of the shape of the micropellets obtained using this device, these pellets are expected to show a higher bulk density, improved flowability and a higher surface-area-to-volume ratio in comparison to common polymer pellets [9].

Furthermore, the analysis of samples collected allowed to observe how the deformation and breakup of extruded melt is influenced by the molecular structure and rheological properties of the resin for different process conditions. Hence, there is a limited processing window where Rayleigh disturbances are likely to occur allowing rounded pellets to be produced.

For instance, the interaction of the melt with the air can also result in the fracture of the melt for which the imposed force of the air stream will cause brittle breakup of the polymer thread. The molecular structure of the polymer was found to be of importance for such condition. In extensional flows, as the ones seen in this process, high branching is responsible for increments of strain hardening and viscosity [10,11]. Strain hardening (high melt strength at high extensional rates) correlates with the strength of alignment, defined as the extent to which molecules tend to be aligned along a single axis [12]. Strain hardening is therefore responsible for limited stretching of these types of melts in this process. Fracture is more likely to occur under the deformation imposed by the air considering the higher probability of chain entanglement in highly branched molecular structures. Fig. 6 presents of cylindrical pellets obtained due brittle fracture of a PE-LD melt.

Figure 6. Lentil shaped PE-LD pellets resulting from brittle fracture of the melt.

A similar analysis applies in regard of the value of the viscosity of the resin. Since high viscosity values give a higher density of molecular entanglements, higher melt strength occurs due to enhanced material resistance to extensional flow [13]. Experiments demonstrated that, the lower the viscosity value of a resin, the smaller the pellets were likely to be since the breakup would take place after the melt stretches.

Conclusions

Particles in the micro scale with nearly spherical shape and high circularity were obtained using the micropelletizer, demonstrating that the manufacturing of micropellets using Rayleigh disturbances is feasible. This may open opportunities for a variety of resins to be micropelletized.

For this particular process, melt strength and viscosity are properties that determine the size of the particles after breakup, whereas molecular structure influences the occurrence of breakup. Highly branched polymer resins show strain hardening when processed with the design presented. The higher the viscosity of the melt, the bigger the resulting diameter of the pellets.

The mentioned results are based on a laboratory scale processes but scale-up should be a feasible aim.

References

[1] D. Drummer; D. Rietzel; F. Kuehnlein, in Proceedings of the 3rd International Conference on Additive Technologies; DAAAM Specialized Conference, 2010.

[2] J. Kruth, G. Levy; F. Klocke; T. Childs in CIRP Annuals - Manufacturing Technology, 2007, 56, 730-759.

[3] M. Launhardt, MSc. Thesis, University of Wisconsin: Madison, 2011.

[4] N. Ashgriz; F. Mashayek, Journal of Fluid Mechanics, 1995, 291, 163-190.

[5] T. Funada; D.D. Joseph, International Journal of Multiphase Flow, 2002, 28, 1459-1478.

[6] T. Osswald; G. Menges *in Materials science of polymers for engineers*. Hanser Publishers, Ed; 2003, 202-14.

[7] T. Osswald, Provisional Patent Application P100314US, 2010.

[8] W. Aquite; M. Engerer; M. Launhardt; N. Rudolph; T. Osswald, in ANTEC Annual Technical Conference Proceedings, 2011.

[9] Claude Y. F. Xi; Elizabeth Takacs; Mark Tate; Michael R. Thompson; John Vlachopoulos, in ANTEC Annual Technical Conference Proceedings, 2003.

[10] A.D. Gotsis; B.L.F. Zeevenhoven; C. Tsenoglou,, Journal of Rheology, 2004, 48, 895.

[11] Z. Tadmor; C.G. Gogos *in Principles of polymer processing, Wiley-Interscience,* 2006.

[10] S.A. Khan; R.K. Prud'homme; R.G. Larson, Rheologica Acta, 1987, 26, 144-151.

[13] A.K. Doufas; L. Rice; W. Thurston, Journal of Rheology, 2011, 55, 95

O-01-454

PIEZOELECTRIC FILAMENTS PRODUCED BY COEXTRUSION

R. S. Martins[1], R. Gonçalves[2], J.G. Rocha[3], J.M. Nóbrega[1*], H. Carvalho[4], S. Lanceros-Mendez[2]

[1] *IPC/I3N – Institute for Polymers and Composites, University of Minho, Portugal*
[2] *Centro/Departamento de Física, University of Minho, Portugal*
[3] *Dep. Industrial Electronics, University of Minho, Portugal*
[4] *Centre for Textile Science and Technology University of Minho, Portugal*

Abstract - Considering the significant interest of both the academic and industrial communities in the fields of i/e-textiles (interactive/electronic textiles), the number of developed applications is far below the expected. This fact is mainly related to the difficulties on the development of production methodologies adequate to industrial scale processes. In this areas the integration of piezoelectric materials, that possess sensing/actuating capabilities, such as poly(vinylidene fluoride), PVDF, and can be processed using conventional processing techniques, is very promising and has encouraged a large number of research works. However, until now, most of the developed production methodologies are difficult to adapt to the industrial scale. This work reports recent developments achieved, in the framework of a research project, on the production of piezoelectric filament by coextrusion of PVDF and electrical conductive layers. The developed production methodology involves a conventional coextrusion line, for which a coextrusion die was designed to produce a multilayer filament. This filament comprises an inner layer of an electrically conductive Polypropylene grade and a middle layer of PVDF and is coated with an electrical conductive ink. The piezoelectric response of the produced filaments will be also presented.

Keywords: i/e-textiles ; electroactive filaments; coextrusion.

Introduction

Poly(vinylidene fluoride) (PVDF) motivated the interest of both industrial and academic communities due to its exceptional pyro- and piezoelectric properties [1-3]. The PVDF properties depend both on its degree of crystallinity and orientation of its crystalline phase, which are strongly dependent on the processing conditions employed during production [1,3-6].

A polymeric piezoelectric device comprises at least one piezoelectric layer and two electrically conductive layers, one at each side of the central layer, which are used as electrodes for the connection of electronic conditioning/drive equipment.

The most common form of piezoelectric PVDF based devices are the flat film. Its production starts from the extrusion of a PVDF layer that is subsequently stretched, poled and, finally, coated by metallization. Just a few research works available in the literature focus on the development of filament shaped PVDF sensors. Walter et al. [7] have studied extensively the phase transitions of extruded PVDF monofilaments that were used in the production of composite parts, which exhibited piezoelectric activity. Lund et al. [8] report the production of two-layered filaments, showing that the electroactive phase content is not affected by the conductive inner core and that the β-phase content depends only on the processing temperature and stretch ratio, as happens for the single PVDF filaments. In the works of Mazurek et al [9], that propose a sequential processing methods for the production of piezoelectric cables, it was shown that the cooling at -30°C of PDVF, after stretching, increase the β-phase content, while the coating required for the placement of the outer electrode promotes a significant reduction.

In this work, the development of a new filament shaped PVDF based sensor is proposed, with a coaxial arrangement of the electrode (electrical conductive) and PVDF layers, that is produced by coextrutison. This device is suited for a new range of application, particularly the ones that involve innovative e-textile applications.

Piezoelectric Filament Production

The production line for the piezoelectric filament starts with the coextrusion of the filament inner (electrode) and middle (PVDF) layers, see Fig. 1. For the inner and middle filament layers were used, respectively, an electrically conductive polypropylene (Premix 1396) and an extrusion grade PVDF (Solef Ta-1010).

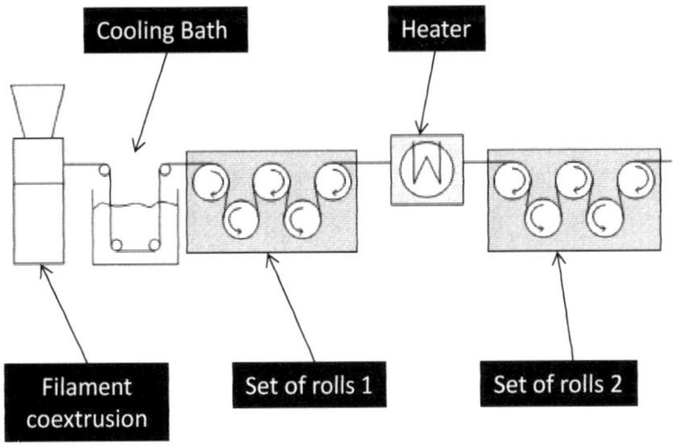

Figure 1 - Extrusion line used to produce the two layered coextruded filaments.

The velocity imposed on the *Set of rolls 1* (Fig. 1), allow to control the linear velocity of the filament at the die exit and, consequently, its diameter. After leaving the die flow channel, the filament is cooled in a water bath to room temperature and is then heated in Heater (Fig. 1), which is a forced convection air temperature controlled oven. Inside the Heater, when it reaches a temperature at which is deformable, the filament is stretched, due to the pulling force promoted by the *Set of rolls 2* (Fig. 1), that should be set to a linear velocity higher than the one at *Set of rolls 1*. The draw ratio imposed to the filament is determined by the relation between the linear velocities V_2 and V_1, respectively, the linear velocity of *Set of rolls 2* and *1*, and is given by $R=V_2/V_1$. The cooling and heating steps are required to assure a better control of the temperature at which the filament draw takes place. At the end of the extrusion line the continuous produced filament is wound.
Afterwards the filament is coated with a silver conducting ink, which forms its third (outer) layer. The two conductive layers are then connected directly to a voltage source and subjected to high voltage, to obtain a poled PVDF layer.

Piezoelectric Filament Characterization

Upon production the piezoelectric filament was tested using two different set-ups, being the generated electrical signals measured with adequate equipment. The two tests performed were: (i) compressing manual and transversely the filament; (ii) submit the filament to tensile loading-unloading cycles in a universal testing machine, with an amplitude of 0.2 mm at a speed of 100 mm/min.
The results obtained for the manual compressional and tensile tests are illustrated Fig. 2 and 3, respectively. In both tests the piezoelectric behavior of the produced filaments is evident.

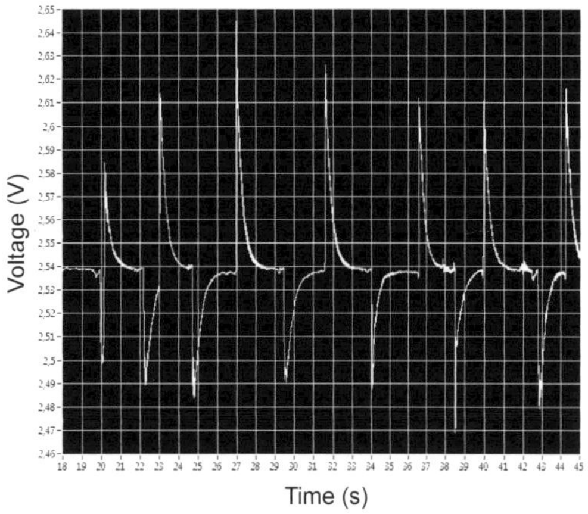

Figure 2. Electrical signal produced in the manual compression test

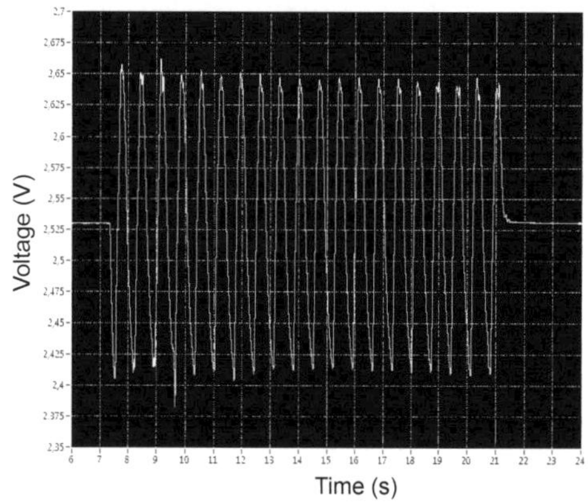

Figure 3. Electrical signal produced in the tensile

Conclusion

The filaments produced in this work, using a coextrusion based process, were shown to generate an electrical signal when stimulated by mechanical stimuli, behaving in a similar way to standard piezoelectric film sensors.

Acknowledgments

The authors gratefully acknowledge funding by Portuguese Foundation for Science and Technology (FCT) for their financial support under POCI, PDTC (PTDC/CTM/108801/2008), Plurianual programmes and PEst-C/CTM/LA0025/2011 (Strategic Project - LA 25 – 2011-2012). They also wish to thank the IN2TEC initiative of the School of Engineering/University of Minho which supported some initial work on piezoelectric filaments.

References

1. J. Lovinger, Developments in crystalline polymers, Elsevier applied science, London (1982).
2. S. Bauer, Journal of Applied Physics 80, p. 5531 (1996).
3. H. S. Nalwa, Ferroelectric Polymers: Chemistry, Physics and Applications Marcel Dekker, Inc, New York (1995)
4. Salimi and A. A. Yousefi, Polymer Testing 22, p. 699 (2003).
5. K. Nakamura, D. Sawai, Y. Watanabe, D. Taguchi, Y. Takahashi, T. Furukawa and T. Kanamoto, Journal of Polymer Science Part B: Polymer Physics 41 (2003), p. 1701.
6. J. Gomes, J. Serrado Nunes, V. Sencadas and S. Lanceros-Mendez, Smart Materials and Structures, 19 (6) (2010) 065010.
7. W. Steinmann, S. Walter, G. Seide, T. Gries, G. Roth, M. Schubnell, Journal of Applied Polymer Science, vol. 120, no. 1, pp. 21-35 (2011)
8. A. Lund, B. Hagström, Journal of Applied Polymer Science,Vol. 120, 1080–1089 (2011)
9. B. Mazurek, S. Różecki, D. Kowalczyk, T. Janiczek, Journal of Electrostatics, Volumes 51-52, Elsevier B.V. pp 180-185 (2001)

O-01-062

IMPROVED PERFORMANCE OF PP/EPR BLENDS BY ADDING NANO-SIO2 PARTICLES: EFFECT OF HIGH-SHEAR PROCESSING ON THE DISPERSION

M. Louizi[1], V. Massardier[1], Ph. Cassagnau[2]
[1,2] Université de Lyon, F-69003 Lyon, France
[1]INSA Lyon, F-69621 Villeurbanne, France
[1,2] CNRS, UMR 5223, Ingénierie des Matériaux Polymères, France
e-mail: molka.louizi1@insa-lyon.fr; philippe.cassagnau@univ-lyon1.fr; valerie.massardier@insa-lyon.fr: corresponding author
URL: www.insa-lyon.fr, www.lmpb.univ-lyon1.fr

Abstract
Fillers-reinforced polymer composites have gained much attention to scientific investigators in recent years, because the addition of fillers can modify the structural, mechanical, thermal and electrical properties of composite materials. The aim of this article is to study new formulations based on polypropylene/ethylene propylene rubber (PP/EPR) used in the manufacture of automobiles resulting from end-of-life-vehicules, in the presence of hydrophobic nano-silica (hydrophobic (SiR805) with the objective of recycling such blends. A new technologically attractive method using high shear processing was performed to obtain nano-composites with high mechanical properties.
High shear processing has proved not only to be an excellent method to obtain better and finer dispersion of 3 wt % of hydrophobic silica (SiR805) in (PP/EPR/ SiR805) nanocomposite upon increasing the screw speed up to 800 rpm but also to enhance the miscibility between PP and EPR by suppressing the coalescence of EPR nodules. An improvement in tensile and impact strength as well as elongation at break was observed.

Keywords: recycling, polyolefins, nanocomposite, nano-silica, core-shell structure, high-shear processing, high-deformation stress

Introduction
Isotactic Polypropylene (iPP) is one of the most important commercial polymers used in the manufacturing of automotive, but it presents poor low-temperature impact strength. Thus, PP is toughened by blending it with olefin elastomers, such as ethylene propylene rubber (EPR), ethylenepropylene-diene monomer (EPDM) or ethylene octene copolymer (EOC) [1,2] to improve its impact strength especially to manufacture bumpers. Consequently, postconsumer recycled PP containing elastomers (EPR, EPDM, EOC…) from automobile bumpers corresponds to a considerable tonnage of waste. The use of inorganic filler can also renforce the stiffness of polymer blends. Fillers-reinforced polymer composites have gained much attention to scientific investigators in recent years, because the addition of fillers can modify the structural, mechanical, thermal and electrical properties of composite materials [3, 4, 5, 6, 7, 8, 9, 10, 11,12,13,14,15, 16]. These properties are affected by the degree of dispersion and amount of filler in the polymer matrix and the interaction between filler and polymer [2]. To overcome the drawback resulted by only adding elastomer or filler, many studies have been done on polymer/elastomer/filler ternary system [17,18,19,20]. The dispersion of fillers in a polymer matrix has been reported to be dependent upon the shear stresses exerted on the fillers during the melt mixing [21,22]. The shear stress can overwhelm the electrostatic and Van der Waals interactions in the fillers and lead to the breakup of the filler agglomerates.
In recent years, it is reported that the use of new techniques such as high-shear processing improve both dispersion of fillers and the distribution of the dispersed phase by introducing strong shear force [23,24,25,26,27,28,29]. This idea is of scientific importance, and corresponds to an important prospect in industrial applications.
In this work, hydrophobic (SiR805) silica is used to prepare PP/EPR/SiO2 ternary composites. The ultimate objective of this study is not only to control the dispersion and distribution of EPR and nano-SiO_2 particles in PP matrix by using a high shear processing technology but also to improve the properties of the PP / EPR in order to improve the performance of bumpers with the perspective of developing a new route for polymer recycling.

Experimental
Materials
(PP/EPR) (78/22) (PP108MF97) supplied by Sabic, contains 22 wt % of EPR (random copolymer containing 50 wt % of ethylene) and 78 wt % of isotactic PP, showing a melt flow index MFI = 10 g/10 min under 2.16 kg at 230°C.

Hydrophobic nanosilica, Aerosil® R805, having a specific surface area of 200 m^2/g and an average diameter of 12 nm, was supplied by Degussa Corp. High-shear processing method, in which 3 wt% of hydrophobic silica were directly melt blended with (PP/EPR) matrix in a high shear extruder Leistritz ZSE 18 HP. The L/D ratio of the screw was about 60. The rotation speed of the screw used in this study was set to be 100rpm, 500 rpm, 800 rpm and 1200 rpm. The melt compounding was carried out at 200°C around 3 min and the blends were fed into the hopper at a flow rate of 4 kg/h.
For the measurements of mechanical properties, all samples were hot pressed at 200°C.

Characterization
Morphologies of the nanocomposites were studied with a Hitachi S800 scanning electron at an accelerating voltage of 30 KV. Prior to observations, samples were cryo-fractured in liquid nitrogen to avoid any plastic deformation and morphology alteration. Some samples were observed with extraction of the dispersed phase. The dispersed phase was selectively extracted in heptane at 24°C for 5 hours.
Tensile tests were carried out on an instron mechanical machine MTS 2/M tester at a speed of 50 mm/min. The mechanical behaviour was investigated on samples obtained by injection moulding at 25° C. A Battendfeld injection moulding machine has been used. The temperature of the injection molding cylinder was kept at 200°C and the injection pressure was maintained at 65 bars.
Impact strength was tested using Zwick D7900 Type 5102-100/00 instrument in compliance with standard ISO 179, the samples were conditioned at -22°C for 48h. All the reported values are the average values of ten experimental results.

Results and discussion
Morphology Analysis
Finer dispersion of fillers and adhesion with the polymer matrix are of crucial importance for improving the mechanical behaviour of nanocomposites. To control the dispersion of nano-silica and the distribution of EPR nodules, the morphology of nanocomposites was investigated.
Fig. 1 illustrates the phase structure of PP/EPR/3 wt % SiR805 nanocomposites prepared under low and high shear processing after selective etching of EPR rubber particles. The solvent used for extraction of EPR nodules is heptane. For PP/EPR/3 wt% SiR805 nanocomposite prepared at 100 rpm, the sizes of EPR nodules detected by SEM were found to range between 0.80 and 1.30 µm. Under the conventionnel shear rate (100 rpm), the fracture surface indicates that EPR nodules are unstable due to the coalescence of nodules in a continuous PP matrix during melt processing and most of them are stacked together. Furthermore, large aggregates of SiO_2 particles with sizes in the range of [0.6-1.0 µm] are observed in the PP matrix.
On the other hand, by processing under high shear rate (800 rpm), the phase structure of PP/EPR/3 wt% SiR805 nanocomposite exhibits a homogenous distribution of the EPR nodules which ranges from 0.2 to 0.5 µm as shown in Fig. 1 (c). It seems to be firmly embedded in the matrix and becomes more uniform. It is apparent that the dispersion of nano-silica is greatly improved with a size of less than 80 nm. This could be tentatively understood as a consequence of the effect of high screw rotation which imparts high shear stresses on the silica agglomerates resulting in the prevention of their agglomeration and the break up of large aggregates into finer ones.

Mechanical properties
Mechanical properties such as modulus, tensile strength and elongation at break are mainly affected by the degree of dispersion of nano-silica as well as the screw rotation speed during processing (Tab. 1). For the PP/EPR/3 wt% SiR805 ternary composite, the increase of shear rate has a positive effect on the elongation at break, a greater increase observed for the nanocomposite processed at 800 rpm reaching up 945% compared to the same nanocomposite processed at 100 rpm (689%). Furthermore, a significant increase in impact strength which reaches 16 $KJ/m^2$is obtained for PP/EPR/3 wt% nanocomposite processed at 800 rpm table 1. PP/EPR/3 wt% SiR805 composites processed at 1200 rpm revealed a noticeable reduction in elongation at break. This finding might be due to the scission of polymer chains.

Figures
Figure 1: SEM images of the cross-sectional fracture at liquid nitrogen temperature for (PP/EPR/3 wt% SiR805) nanocomposites under (a,b) conventionnel shearing (100 rpm) and high shearing (800 rpm) (c) after the extraction of the EPR dispersed phase with Heptane

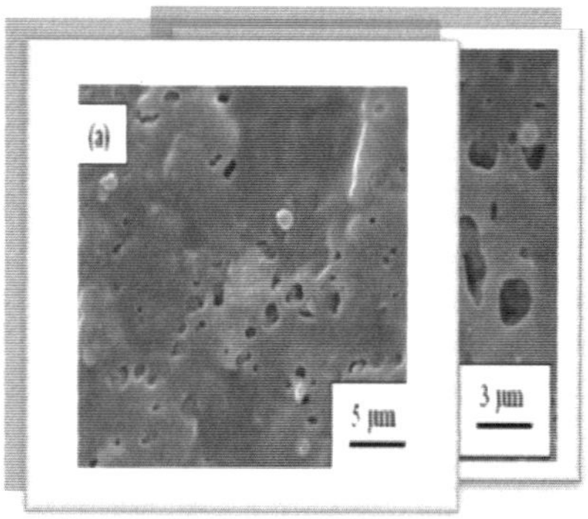

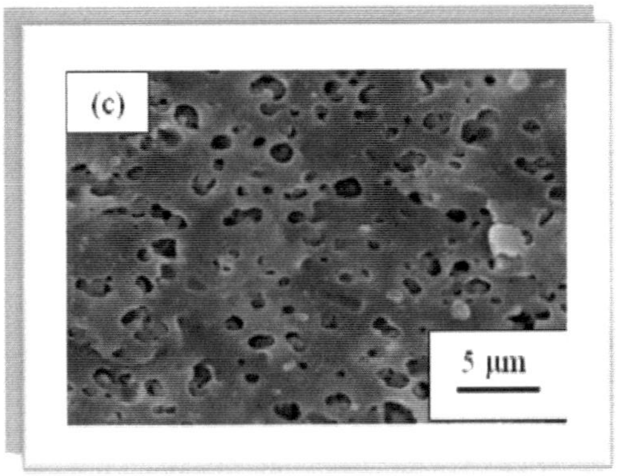

Tables

Table1: Mechanical properties of nanocomposites processed at different shear rates 100, 500, 800 and 1200 rpm

Blends	E (Mpa)	σ_m (Mpa)	E_b (%)	Impact Strength (kJ/m²)
PP/EPR (100 rpm)	430±10	19±1	620±25	12±3
PP/EPR/3 wt% SiR805 (100rpm)	472±5	19±1	689±10	10±3
PP/EPR/3 wt% SiR805 (500rpm)	462±8	21±1	820±05	13±2
PP/EPR/3 wt% SiR805 (800rpm)	443±6	22±1	945±12	17±2
PP/EPR/3 wt% SiR805 (1200rpm)	319±11	19±1	674±40	10±2

Conclusions

The idea of using rigid inorganic nano-particles combined with the effect of high shear processing to improve the compatibility of immiscible polymer blends offers new industrial perspectives.

A homogenous dispersion of hydrophobic nano-silica and the improvement of the distribution of EPR nodules was achieved when the screw rotation speed reaches 800 rpm in (PP/EPR/3% wt) ternary composite. The results indicate that high shear processing is an effective method for improving the performance of PP/EPR/nano-silica ternary composite and offers new perspectives for the "high shear recycling" of polymer composites. So, recycling is important for the plastic industry, energy saving and the environment.

REFERENCES

1. N. Mnif ; V. Massardier ; M. Jaziri *J. Appl, Polym. Sci* 2007, 104,3220.
2. M. Jaziri; N. Mnif; V. Massardier; H. Perier-Camby *Polym.Eng. Sci* 2007,47,1009.
3. Z.Liu; M.J .Gilbert; *J. Appl, Polym* 1996, 59 1087.
4. Y.J. Lee; I. Manas-Zloczower ; D.L .Feke; *Polym eng. sci* 1995,35, 1037.
5. S. H. Mansour; S. L. Abd-El-Messieh *J. Appl, polym. sci* 2002, 83, 1167.
6. A.S Luyt; JA Molefi; H .Krump *Polym.Deg. Stab* 2006,91,1629.
7. Y. Zhou; V. Rangari; H. Mahfuz; S. Jeelani; P.K. Mallick *Mater. Sci. Eng: A* 2005,402,109.
8. Y. Wang ; J. Wang *Polym.Eng. Sci* 1999,39,190.
9. S.G. Lei; S.V. Hoa; M.T .Ton-That *Compos. Sci. Technol* 2006, 66, 1274.
10. S.S. Ray; J. Bandyopadhyay; M. Bousmina; *Macromol. Mater. Eng* 2007, 292, 729.
11. R. Matadi Boumbimba; M. Bouquey; R. Muller; L. Jourdainne; B. Triki; P. Hébraud; P. Pfeiffer *Polym. Test* 2012, 31, 800.
12. I. Hejazi; F. Sharif; H. Garmabi *Mater. Desig* 2011, 32, 3803.
13. S. Zhu; J. Chen; Y. Zuo; H. Li, Y. Cao *Appl. Clay. Sci* 2011, 52, 171.
14. K. Trongtorsak ; P. Supaphol; S. Tantayanon ; *Polym. Test* 2004, 23, 533.
15. C. Chen; J. Samaniuk ; D. G. Baird; G. Devoux; M. Zhang; R. B. Moore; J. P. Quigley *Polym* 2012,53, 1373.
16. Xi. Liu; Y. Wang; W. Yang; Z. Liu; Y. Luo; B. Xie; M. Yang *J. Mater. Sci* 2012, 47, 4620.
17. H.Yang; Q. Zhang; M. Guo; C. Wang; R.N. Du; Q. Fu; *Polym* 2006,47, 2106.
18. S.Q. Chang; T. X. Xie; G. S Yang *J. Appl. Polym. Sci* 2006, 102, 5184.
19. C. Valerio; M. Carla; M. Antonio; F. Giuseppe; F.Angelo; S. Roberta *Nano. Sci. Nanotechnol* 2008, 8, 1823.
20. J. Qian; S. Dang; Z. Huang; Y. Xu *Sci. Plas. Technol* 2011,167, 69.
21. K.K. Chawla; Compos. Mater. Sci. Eng: Springer-Verlag (1987)
22. J.W. Cho; D.R. Paul *Polym* 2001, 42, 1083.
23. Li. Yongjin; Sh. Hiroshi *Polym* 2007,48, 2203.
24. Li. Yongjin; Sh. Hiroshi *Macromol* 2009, 42 2587.
25. Chen. Guang-Xin ; Li. Yongjin; Sh. Hiroshi *Carb* 2007, 45, 2334.
26. Sh. Hiroshi; Li. Yongjin; K. Akira; S. Hironari; *Macromol* 2005, 38, 7880.
27. Li .Yongjin ; Sh. Hiroshi *Europ. Polym. J* 2006, 42, 3202.
28. Sh. Hiroshi; Li. Yongjin; K. Akira; S. Hironari , Nanosci. Nanotechnol 2006, 6, 3923.
29. F. Teyssandier ; P. Cassagnau ; J.F. Gérard; N. Mignard; F. Mélis 2012, 133, 913.

O-01-091

CAPILLARY EXTRUSION AND SWELL OF A HDPE MELT EXHIBITING SLIP

Evan Mitsoulis[1*], Mahmoud Ansari[2], Savvas G. Hatzikiriakos[2]
[1]*School of Mining Engineering & Metallurgy, National Technical University of Athens (NTUA)*
Zografou 157-80, Athens, GREECE, *mitsouli@metal.ntua.gr*
[2]*Dept. Chemical and Biological Engineering, University of British Columbia(UBC), Vancouver, Canada – hatzikir@apsc.ubc.ca*

Abstract - The extrudate (die) swell of a high-density polyethylene (HDPE) melt was studied both experimentally and numerically under slip conditions. The excess pressure drop due to entry (entrance pressure drop), the effect of pressure and temperature on viscosity, and the slip effects on the capillary data analysis have been examined. Using a series of capillary dies having different diameters, *D*, and length-to-diameter *L/D* ratios, a full rheological characterization has been carried out, and the experimental data have been fitted both with a viscous model (Carreau-Yasuda) and a viscoelastic one (the Kaye - Bernstein, Kearsley, Zapas / Papanastasiou, Scriven, Macosko or K-BKZ/PSM model). Particular emphasis has been placed on the effects of wall slip (significant for HDPE). It was found that viscous modelling underestimates the pressures drops (especially at the higher apparent shear rates and *L/D* ratios) and predicts virtually no extrudate swell. On the other hand, the viscoelastic simulations were capable of reproducing the experimental data well, and this was particularly true for the pressure drop. The prediction of viscoelastic extrudate swell presented a problem, since the simulations grossly overpredict it due to the highly elastic nature of the melt. This occurs despite the presence of severe slip at the wall, which brings the swell down considerably. It is argued that other phenomena play role in determining the swelling of HDPE, perhaps associated with its crystallinity and/or sagging and cooling phenomena when simply extruding in the atmosphere.

Keywords: Extrudate (Die) Swell, Capillary Die Flow, HDPE Melt, K-BKZ Constitutive Equation, Slip Effects

Introduction

Capillary rheometry is extensively used in both industry and academia to assess the rheological and processing behaviour of polymer melts at high shear rates before testing their processability in full industrial scale [1]. One important aspect of material performance in processing is "extrudate (die) swell" [2-4]. This is the phenomenon of increasing area or diameter of the extrudate as it comes out from the die, whereupon it suddenly encounters a dramatically different type of flow, i.e., from a constrained flow within the die walls with no-slip or partial slip to a shear-free flow without walls outside the die. As a result of this change in boundary conditions, the polymer swells, sometimes dramatically [2]. The degree of swelling heavily depends on its past deformation history (memory effects), geometric characteristics of the die and viscoelastic properties [2].

The main objective of the study is to measure and predict the extrudate swell of a HDPE melt that exhibits significant slip effects. It is our goal to demonstrate that such slip effects have to be seriously considered before any reliable predictions can be made on macroscopic quantities related to processing of polymers; in this case extrudate swell is the macroscopic quantity which is significant in processes such as extrusion, wire coating, and blow molding, among others.

Experimental

Materials

A HDPE melt was used in this work carefully selected in order to address the effects of slip on extrudate swell. This particular HDPE (m-HDPE) has a molecular weight of about 229,800 g/mol and a polydispersity index of 20 [5]. Its rheological behaviour has been studied previously by Ansari et al. [5].

Rheological Testing

As discussed above, the rheology of this resin was studied by Ansari et al. [5]. The master curves of its linear viscoelastic moduli are also reported here along with predictions of the K-BKZ constitutive equation for several rheological properties at the reference temperature of 190°C.

An Instron capillary rheometer (constant piston speed) was used to determine the extrudate swell and the slip behaviour of this polymer. The viscosity as a function of the wall shear stress, σ_W, and apparent shear rate, $\dot{\gamma}_A = 32Q/\pi D^3$, where Q is the volumetric flow rate and D is the capillary diameter, is also studied as part of the slip study. Three series of dies having various diameters (D=0.079, 0.122 and 0.211 cm) and length-to-diameters ratios (L/D=5, 16, 33) were used (in total 9 dies) to determine directly the viscosity and the slip behaviour through the well known Mooney analysis at 190°C [6].

The extrudate swell measurements were performed by analysing extrudate images (immediately after die exit) taken with a high-resolution Nikon D-90 camera equipped with a Sigma DG Macro 2.8 lens attached to three Kenko extension tubes, which give 1.5X magnification in macro focusing mode. A reference tip with known thickness has been used in order to estimate the extrudate diameters in respect to its size. Such a typical image is depicted in Fig. 1. The reported extrudate swells for each shear rate are the average of analysing at least 5 different images.

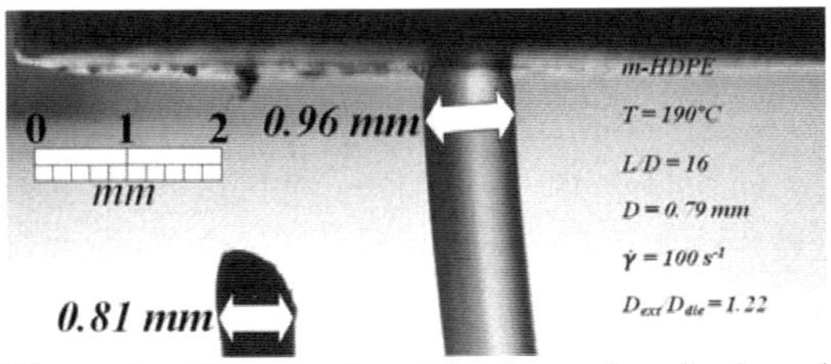

Figure 1 – The experimental setup and method used for extrudate swell measurements. Die dimensions (D=0.79 mm, L/D=16), T=190°C, $\dot{\gamma}_A$=100 s^{-1}.

Governing Equations and Rheological Modeling

We consider the conservation equations of mass, momentum and energy for weakly compressible fluids under non-isothermal, creeping, steady flow conditions. These are written as [1]:

$$\overline{u} \cdot \nabla \rho + \rho\left(\nabla \cdot \overline{u}\right) = 0, \tag{1}$$

$$0 = -\nabla p + \nabla \cdot \overline{\overline{\tau}}, \tag{2}$$

$$\rho C_p \overline{u} \cdot \nabla T = k \nabla^2 T + \overline{\overline{\tau}} : \nabla \overline{u}, \tag{3}$$

where ρ is the density, $\overline{u}$ is the velocity vector, p is the pressure, $\overline{\overline{\tau}}$ is the extra stress tensor, T is the temperature, C_p is the heat capacity, and k is the thermal conductivity. For a *compressible* fluid, pressure and density are connected as a first approximation through a simple linear thermodynamic equation of state [1]:

$$\rho = \rho_0\left(1 + \beta_c p\right), \tag{4}$$

where β_c is the isothermal compressibility with the density to be ρ_0 at a reference pressure p_0 (=0).

The viscous stresses are given for inelastic non-Newtonian compressible fluids by the relation [1]:

$$\overline{\overline{\tau}} = \eta\left(\left|\dot{\gamma}\right|\right)\left(\overline{\overline{\dot{\gamma}}} - \frac{2}{3}\left(\nabla \cdot \overline{u}\right)\overline{\overline{I}}\right), \tag{5}$$

where $\eta\left(\left|\dot{\gamma}\right|\right)$ is the apparent non-Newtonian viscosity, which is a function of the magnitude $\left|\dot{\gamma}\right|$ of the rate-of-strain tensor $\overline{\overline{\dot{\gamma}}} = \nabla \overline{u} + \nabla \overline{u}^T$, which is given by:

$$\left|\dot{\gamma}\right| = \sqrt{\frac{1}{2} II_{\dot{\gamma}}} = \left(\frac{1}{2}\left(\overline{\overline{\dot{\gamma}}} : \overline{\overline{\dot{\gamma}}}\right)\right)^{1/2}, \tag{6}$$

where $II_{\dot{\gamma}}$ is the second invariant of $\overline{\overline{\dot{\gamma}}}$

$$II_{\dot{\gamma}} = \left(\overline{\overline{\dot{\gamma}}} : \overline{\overline{\dot{\gamma}}}\right) = \sum_i \sum_j \dot{\gamma}_{ij} \dot{\gamma}_{ij}, \tag{7}$$

The tensor $\overline{\overline{I}}$ in eq. (5) is the unit tensor.

To evaluate the role of viscoelasticity in the prediction of extrudate swell, it is instructive to consider first purely viscous models in the simulations. Namely, the Carreau-Yasuda model was used to fit the shear viscosity data of the HDPE melt. The Carreau-Yasuda model is written as [1]

$$\eta = \eta_0\left[1 + (\lambda \left|\dot{\gamma}\right|)^{\alpha}\right]^{\frac{n-1}{\alpha}}, \tag{8}$$

where $\eta_{0,C}$ is the Carreau zero-shear-rate viscosity, λ is a time constant, n_C is the Carreau power-law index, and α is the Yasuda exponent (=2 for the simple Carreau model). The fitted viscosity of the HDPE melt by Eq. (8) is plotted in Fig. 2, while the parameters of the model are listed in Table 1.

Table 1: Parameters for the HDPE melt obeying the Carreau-Yasuda model (*Eq. 8*) at 190°C.

Parameter	Value
$\eta_{0,C}$	191,660 Pa·s
λ	2.402 s
n_C	0.149
α	0.550

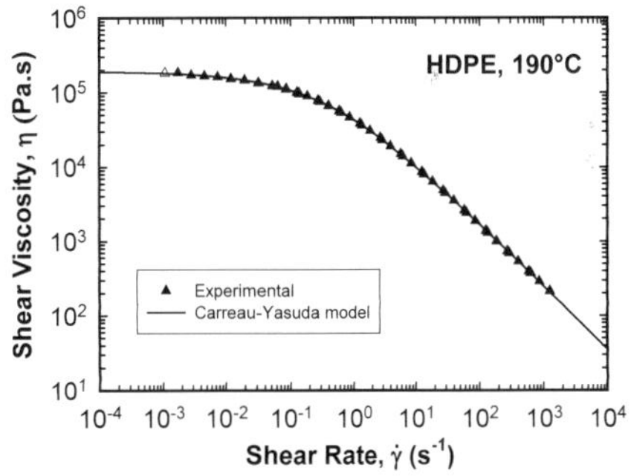

Figure 2 – The shear viscosity of the HDPE melt at 190°C fitted with the Carreau-Yasuda (*Eq. 8*) model using the parameters listed in Table 1.

For the capillary flow simulations the effect of pressure on viscosity should be taken into account as this becomes evident below. This effect is quite significant for LDPE compared to m-LLDPE and less significant for HDPE [7,8]. This effect can be taken into account by multiplying the constitutive relation with a pressure-shift factor, a_p, defined by the Barus equation, that is [8]:

$$a_p \equiv \frac{\eta}{\eta_{p0}} = \exp(\beta_P p), \tag{10}$$

where β_p is the pressure coefficient and p is the absolute pressure, as discussed above.

Viscoelasticity is included in the present work via an appropriate rheological model for the stresses. This is a K-BKZ equation proposed by Papanastasiou et al. [9] and modified by Luo and Tanner [10]. This is written as:

$$\boldsymbol{\tau} = \frac{1}{1-\theta} \int_{-\infty}^{t} \sum_{k=1}^{N} \frac{a_k}{\lambda_k} \exp\left(-\frac{t-t'}{\lambda_k}\right) \frac{\alpha}{(\alpha-3)+\beta I_{C^{-1}}+(1-\beta) I_C} \left[\boldsymbol{C}_t^{-1}(t')+\theta \boldsymbol{C}_t(t')\right] \mathrm{d}t' \tag{11}$$

where λ_k and a_k are the relaxation times and relaxation modulus coefficients, N is the number of relaxation modes, α and β are material constants, and I_C, I_C^{-1} are the first invariants of the Cauchy-Green tensor $\boldsymbol{C}_t$ and its inverse $\boldsymbol{C}_t^{-1}$, the Finger strain tensor. The material constant θ is given by

$$\frac{N_2}{N_1} = \frac{\theta}{1-\theta}, \tag{12}$$

where N_1 and N_2 are the first and second normal stress differences, respectively. It is noted that θ is not zero for polymer melts, which possess a non-zero second normal stress difference. Its usual range is between –0.1 and –0.2 in accordance with experimental findings [1]. The experimental data and their best fit with constants listed in Table 2 are plotted in Fig. 3.

Table 2: Relaxation spectra and material constants for the HDPE melt obeying the K-BKZ model (*Eq. 11*) at 190°C ($\alpha = 11.075$, $\beta = 0.6$, $\theta = 0$, $\overline{\lambda} = 96$ s, η_0 = 190,425 Pa·s).

k	λ_k (s)	a_k (Pa)
1	2.76×10^{-3}	73,646
2	1.98×10^{-4}	292,000
3	2.07×10^{-2}	61,525
4	0.148	59,290
5	1.062	33,818
6	45.67	1,029
7	6.81	9,787
8	514	59.4

In the case of slip effects at the wall, the usual no-slip velocity at the solid boundaries is replaced by a slip law of the following form [1]:

$$a_T u_{sl} = -\beta_{sl} \sigma_w^b, \tag{13}$$

where u_{sl} is the slip velocity, σ_w is the shear stress at the die wall, β_{sl} is the slip coefficient, and b is the slip exponent. It should be noted that the shift-factor a_T is used here to take into account the temperature effects of the slip law at different temperatures.

In 2-D simulations, the above law means that the tangential velocity on the boundary is given by the slip law, while the normal velocity is set to zero, i.e.,

$$\beta_{sl}\left(\overline{t}\overline{n} : \overline{\overline{\tau}}\right)^b = a_T\left(\overline{t} \cdot \overline{u}\right), \qquad \overline{n} \cdot \overline{u} = 0, \tag{14}$$

where $\overline{n}$ is the unit outward normal vector to a surface, $\overline{t}$ is the tangential unit vector in the direction of flow, and the rest of symbols are defined above.

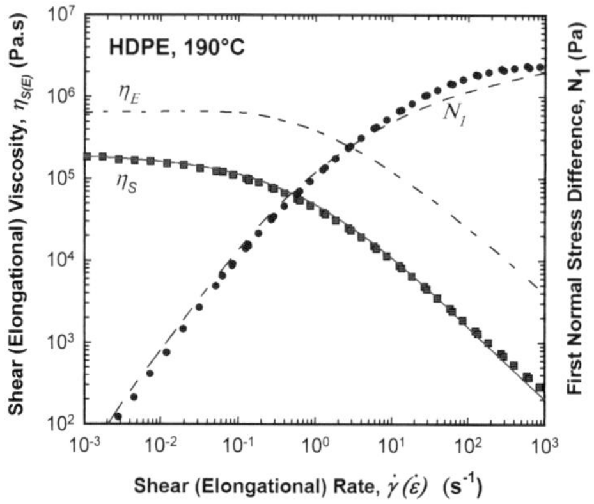

Figure 3 – Experimental data (solid symbols) and model predictions of shear viscosity, η_S, first normal stress difference, N_1, and elongational viscosity, η_E, for the HDPE melt at 190°C using the K-BKZ model (*Eq. 11*) with the parameters listed in Table 2.

Method of Solution

The solution to the above conservation and constitutive equations is carried out with two codes, one for viscous flows (*u-v-p-T-h* formulation) [11] and one for viscoelastic flows [12]. The boundary conditions for the problem at hand are well known and they involve no slip or slip velocity conditions at the solid walls and a constant temperature of 190°C, while at the entry the fully developed velocity profile is given that corresponds to a given flow rate Q, and at the exit zero tractions and zero heat flux are imposed. Symmetry is also imposed at the centerline.

Results and Discussion

Figure 4 shows the results for the extrudate swell for the HDPE melt at T=190°C. The results have been obtained without a reservoir present, because its inclusion gave unrealistically high swell values, meaning the melt did not "forget" its stress state fast enough and wanted to have the same expansion as in the reservoir (12:1 expansion). The so-called "asymptotic" swell values do not differentiate the L/D effect very well, as they give approximately the same swelling for all 3 dies with different L/D lengths. The presence of a second normal stress difference brings the swell down, but at most 10%, which is not appreciable. The same conclusions had been reached by Goublomme et al. [13], who had also studied a HDPE melt, but without adding slip at the wall.

It should be noted that the experimental swell values in this range of apparent shear rates was between 20-70%, having the same trend as in the simulations (upward increase for high rates). Perhaps other phenomena play a role in non-isothermal swelling (or non-annealed swelling) of this polymer melt. Thus here, as in earlier works [13], the correct prediction of the extrudate swell of HDPE melts remains an elusive subject.

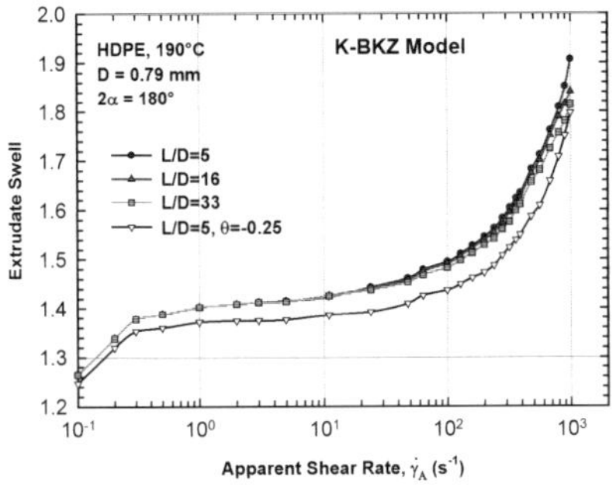

Figure 4 – Asymptotic extrudate swell of the HDPE melt at 190°C as a function of the apparent shear rate for three different L/D values. Simulation results with the K-BKZ model (Eq. 8). Symbols are put to show the continuation steps. The extrudate swell decreases slightly with increasing die length at the higher range of apparent shear rates. Swell values are reported at L_{ext}=16R.

Table 3: Values of the various parameters for the HDPE melt at 190°C.

Parameter	Value
β_c	0.01036 MPa^{-1}
β_p	0.03 MPa^{-1}
β_{sl}	18,800 cm/(s.MPa^b)
b	4.0
ρ	0.762 g/cm^3
k	0.00255 J/(s·cm·K)
C_p	2.721 J/(g·K)
E	28,840 J/mol
R_g	8.3143 J/(mol·K)
T_0	190°C (463 K)

Conclusions

A HDPE has been studied in entry flows through capillary dies with different L/D ratios (=5, 16, 33) with the purpose of predicting the pressure drop in the system and its extrudate swell. The experiments have shown that this particular HDPE slips strongly at the wall. Full rheological characterization was carried out both with a viscous (Carreau-Yasuda) and a viscoelastic (K-BKZ) model. All necessary material properties data were collected for the simulations.

The viscous simulations showed that when all effects are taken into account, virtually no extrudate swell is predicted, a well known deficiency of viscous models.

The viscoelastic simulations with the K-BKZ/PSM model showed that the swell was overpredicted, despite the fact that slip was included. The general trend of the experimental data that showed an exponential increase of extrudate swell, was captured by the model, albeit for the asymptotic swell that does not take into account the presence of the reservoir. An increase in L/D ratio reduces the swelling, and this was more evident experimentally as a long die gives enough time to the material to "forget" its elastic stresses (fading memory).

Acknowledgements

Financial assistance from the Natural Sciences and Engineering Research Council (NSERC) of Canada and the programme "Thales" for basic research from GGET of Greece are gratefully acknowledged.

References

1. Dealy, J.M., Wissbrun, K.F., "Melt Rheology and its Role in Plastics Processing - Theory and Applications", Van Nostrand Reinhold, New York (1990).
2. Tanner, R.I. "Engineering Rheology", 2nd Ed., Oxford University Press, Oxford (2000).
3. Boger, D.V., Walters, K., "Rheological Phenomena in Focus", Rheology Series, Vol. 4. Elsevier, Amsterdam (1993).
4. Bird, R.B., Hassager, O., Armstrong, R.C., Curtiss, C.F., "Dynamics of Polymeric Liquids, Vol. 2: Kinetic Theory", 2nd Edition, Wiley, New York (1987).
5. Ansari, M., Hatzikiriakos, S.G., Sukhadia, A.M., Rohlfing, D.C., "Rheology of Ziegler-Natta and metallocene high-density polyethylenes: broad molecular weight distribution effects", Rheol. Acta, **50**, 17-27 (2011).
6. Mooney, M., "Explicit formulas for slip and fluidity", J. Rheol., **2**, 210-222 (1931).
7. Sedlacek, T., Zatloukal, M., Filip, P., Boltizar, A., Saha, P., "On the Effect of Pressure on the Shear and Elongational Viscosities of Polymer Melts", Polym. Eng. Sci., **44**, 1328-1337 (2004).
8. Carreras, E.S., El Kissi, N., Piau, J.-M., Toussaint, F., Nigen, S., "Pressure Effects on Viscosity and Flow Stability of Polyethylene Melts during Extrusion", Rheol Acta, **45**, 209–222 (2006).
9. Papanastasiou, A.C., Scriven, L.E., Macosco, C.W., "An Integral Constitutive Equation for Mixed Flows: Viscoelastic Characterization", J. Rheol., **27**, 387-410 (1983).
10. Luo, X.-L., Tanner, R.I., "Finite Element Simulation of Long and Short Circular Die Extrusion Experiments Using Integral Models", Int. J. Num. Meth. Eng., **25**, 9-22 (1988).
11. Hannachi, A., Mitsoulis, E., "Sheet Coextrusion of Polymer Solutions and Melts: Comparison between Simulation and Experiments", Adv. Polym. Technol., **12**, 217-231 (1993).
12. Luo, X.-L., Mitsoulis, E., "An Efficient Algorithm for Strain History Tracking in Finite Element Computations of Non-Newtonian Fluids with Integral Constitutive Equations", Int. J. Num. Meth. Fluids, **11**, 1015-1031 (1990).
13. Goublomme, A., Draily, B., Crochet, M.J., "Numerical prediction of extrudate swell of a high-density polyethylene", J. Non-Newtonian Fluid Mech., **44**, 171-195 (1992).

O-01-092

TUBING EXTRUSION OF A FLUOROPOLYMER MELT

E. Mitsoulis[1*] S.G. Hatzikiriakos[2]

[1]*School of Mining Engineering & Metallurgy, National Technical University of Athens (NTUA)*
Zografou 157-80, Athens, GREECE, *mitsouli@metal.ntua.gr*

[2]*Dept. Chemical and Biological Engineering, University of British Columbia(UBC), Vancouver, Canada – hatzikir@apsc.ubc.ca*

Abstract - The tubing extrusion-forming process of a fluoropolymer (FEP) melt was studied both experimentally and numerically. The flow behaviour of a FEP resin was determined by using a tubular die used in industrial-scale operations and these data were compared with simulation results using (i) a viscous model (Cross) and (ii) a viscoelastic one (the Kaye - Bernstein, Kearsley, Zapas / Papanastasiou, Scriven, Macosko or K-BKZ/PSM model) in order to assess the viscoelastic effects. In all simulations, compressibility, thermal and pressure effects on viscosity were taken into account. It was found that the viscoelastic results for the pressure, and hence the stresses at the wall, were always higher than the viscous results. Both were higher than the experimental results. A quadratic slip model plus viscoelasticity was found necessary to reproduce the experiments. The smooth flow curves resulting from this industrial tubular-coating die are a further manifestation that this is an appropriate design for coating fluoropolymers at very high apparent shear rates, exceeding 5000 s^{-1}.

Keywords: Tube forming, non-isothermal die flow, slip-at-the-wall, fluoropolymers, FEP melt, K-BKZ constitutive equation, viscoelasticity.

Introduction

Fluoropolymers are of great commercial and scientific interest due to their unique combination of properties. These include excellent chemical stability and dielectric properties, anti-stick characteristics, mechanical strength, and low flammability [1-2]. Their most important uses are in electronics and electrical applications, especially for wiring insulation, chemical processing equipment, medical devices and laboratory ware and tubing [2-4].

The main focus of this work is on the process of tubing extrusion of FEP copolymers. Its objective is to model the tubing process of melt-processible fluoropolymers. The flow model predictions are performed using both a viscous and a viscoelastic model and these are compared with experimental results obtained from industrial-scale tubing dies for a certain fluoropolymer (FEP). The tubing simulation model includes all important effects in a full numerical scheme, such as pressure- and temperature-dependence of viscosity, compressibility, viscous dissipation, slip-at-the-wall, and viscoelasticity through an appropriate integral constitutive equation.

Experimental

Materials

Tubing extrusion experiments were performed for the FEP-4100 resin (a copolymer of tetrafluoroethylene and hexafluoropropylene) using a tubing extrusion die. This resin has been rheologically characterized previously in detail by Rosenbaum et al. [5]. It has a molecular weight of about 208,000 and a polydispersity index of about 2. The melting point of this resin is around 260°C, determined by differential scanning calorimetry (DSC) analysis. The rheological parameters determined by fitting the experimental results are summarized in Tables 1-3. These include the parameters of the Cross and power-law models for the viscosity of the fluoropolymer melt (Table 1), the parameters for the K-BKZ viscoelastic constitutive model (Table 2), and the physical properties of the FEP-4100 needed for the non-isothermal flow simulations (Table 3).

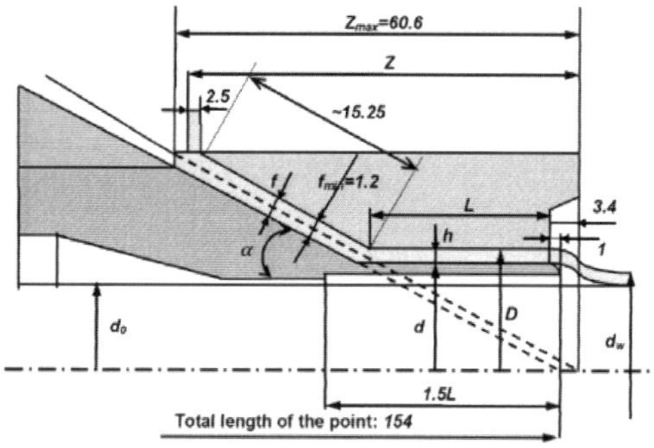

Figure 1 –Nokia Maillefer 4/6 crosshead die design for tubing coating with dimensions in mm (not in scale) [6,7].

A schematic of the tubing extrusion die used in the experiments and the numerical simulations appears in Figure 1 [6,7]. It is a special annular crosshead die attached to the rheometer to mimic the tubing extrusion process. This crosshead die included dies and tips (internal diameter) with equal entry cone angles of $2a$=60° and die land length L of 7.62 mm. In the present work we report data obtained with a die having a diameter D of 3 mm and a tip diameter d of 1.52 mm. The molten polymer enters the tapered die via the uppermost port and is forced around the wire tip guide towards the die orifice. The wire tip guide serves as a mandrel for the molten polymer, giving the extrudate a tubular shape. The die land passage forms the exterior surface of the tubular shape, and the exterior surface of the cylindrical extension forms the interior surface of the tubular shape. This die can also be used for wire-coating. However, in the present study the pressure extrusion makes no use of wire as the die is used in conjunction with a capillary rheometer having a barrel of diameter equal to 25.4 mm.

Governing Equations and Rheological Modeling

We consider the conservation equations of mass, momentum and energy for weakly compressible fluids under non-isothermal, creeping, steady flow conditions. These are written as [4]:

$$\overline{u}\cdot\nabla\rho+\rho\left(\nabla\cdot\overline{u}\right)=0, \quad (1)$$

$$0=-\nabla p+\nabla\cdot\overline{\overline{\tau}}, \quad (2)$$

$$\rho C_p\overline{u}\cdot\nabla T=k\nabla^2T+\overline{\overline{\tau}}:\nabla\overline{u}, \quad (3)$$

where ρ is the density, $\overline{u}$ is the velocity vector, p is the pressure, $\overline{\overline{\tau}}$ is the extra stress tensor, T is the temperature, C_p is the heat capacity, and k is the thermal conductivity. For a *compressible* fluid, pressure and density are connected as a first approximation through a simple linear thermodynamic equation of state [4]:

$$\rho=\rho_0\left(1+\beta_c p\right), \quad (4)$$

where β_c is the isothermal compressibility with the density to be ρ_0 at a reference pressure p_0 (=0).
The viscous stresses are given for inelastic non-Newtonian compressible fluids by the relation [4]:

$$\overline{\overline{\tau}}=\eta\left(|\dot{\gamma}|\right)\left(\overline{\overline{\dot{\gamma}}}-\frac{2}{3}\left(\nabla\cdot\overline{u}\right)\overline{\overline{I}}\right), \quad (5)$$

where $\eta\left(|\dot{\gamma}|\right)$ is the apparent non-Newtonian viscosity, which is a function of the magnitude $|\dot{\gamma}|$ of the rate-of-strain tensor $\overline{\overline{\dot{\gamma}}}=\nabla\overline{u}+\nabla\overline{u}^T$, which is given by:

$$|\dot{\gamma}|=\sqrt{\frac{1}{2}II_{\dot{\gamma}}}=\left(\frac{1}{2}\left(\overline{\overline{\dot{\gamma}}}:\overline{\overline{\dot{\gamma}}}\right)\right)^{1/2}, \quad (6)$$

where $II_{\dot{\gamma}}$ is the second invariant of $\overline{\overline{\dot{\gamma}}}$

$$II_{\dot{\gamma}}=\left(\overline{\overline{\dot{\gamma}}}:\overline{\overline{\dot{\gamma}}}\right)=\sum_i\sum_j\dot{\gamma}_{ij}\dot{\gamma}_{ij}, \quad (7)$$

The tensor $\overline{\overline{I}}$ in eq. (5) is the unit tensor.
To evaluate the role of viscoelasticity in the prediction of Bagley correction, it is instructive to consider first purely viscous models in the simulations. Namely, the Cross model was used to fit the shear viscosity data of the FEP melt. The Cross model is written as [4]

$$\eta=\frac{\eta_{0,C}}{1+\left(\lambda\dot{\gamma}\right)^{1-n_C}}, \quad (8)$$

where $\eta_{0,C}$ is the Cross zero-shear-rate viscosity, λ is a time constant, and n_C is the Cross power-law index. The fitted viscosity of the FEP melt by Eq. (8) is plotted in Fig. 2, while the parameters of the model are listed in Table 1.

Table 1: Parameters for the FEP resin obeying the Cross model (*Eq. 8*) and the power-law model (*Eq. 9*) at 371°C.

Parameter	Value
$\eta_{0,C}$	1542.3 Pa·s
λ	0.0049 s
n_C	0.316
K	11,931 Pa·s^n
n	0.5

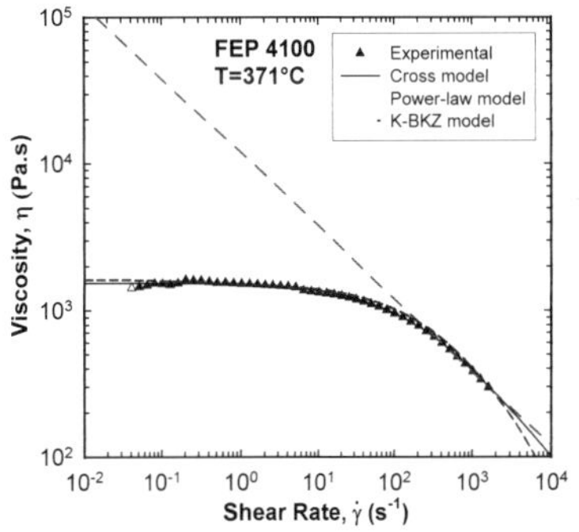

Figure 2 – The shear viscosity of the FEP melt at 371°C fitted with the Cross (*Eq. 8*) and the power-law (*Eq. 9*) models using the parameters listed in Table 1.

For easy checks and simple analytical formulas, the high shear-rate range of the viscosity data can be fitted to the power-law model for the viscosity:

$$\eta=K\,|\dot{\gamma}\,|^{n-1}, \quad (9)$$

where K is the consistency index and n is the power-law index. These values are also given in Table 1.
For the capillary flow simulations the effect of pressure on viscosity should be taken into account as this becomes evident below. This effect is quite significant for LDPE compared to m-LLDPE and less significant for HDPE [8,9]. It is also present for the FEP melt. This effect can be taken into account by multiplying the constitutive relation with a pressure-shift factor, a_p, defined by the Barus equation, that is [9]:

$$a_p\equiv\frac{\eta}{\eta_{p0}}=\exp\left(\beta_p p\right), \quad (10)$$

where β_p is the pressure coefficient and p is the absolute pressure, as discussed above. In our case, β_p shows a power-law dependency on pressure, according to the following equation:

$$\beta_p = mp^{-n_p}, \tag{11}$$

where m=1.39x10^{-4} Pa$^{n_p-1}$ and n_p=0.54, with the pressure p given in Pa.

Viscoelasticity is included in the present work via an appropriate rheological model for the stresses. This is a K-BKZ equation proposed by Papanastasiou et al. [10] and modified by Luo and Tanner [11]. This is written as:

$$\boldsymbol{\tau} = \frac{1}{1-\theta}\int_{-\infty}^{t}\sum_{k=1}^{N}\frac{a_k}{\lambda_k}\exp\left(-\frac{t-t'}{\lambda_k}\right)\frac{\alpha}{(\alpha-3)+\beta I_{C^{-1}}+(1-\beta)I_C}\left[\boldsymbol{C}_t^{-1}(t')+\theta\boldsymbol{C}_t(t')\right]\mathrm{d}t' \tag{12}$$

where λ_k and a_k are the relaxation times and relaxation modulus coefficients, N is the number of relaxation modes, α and β are material constants, and I_C, I_C^{-1} are the first invariants of the Cauchy-Green tensor $\boldsymbol{C_t}$ and its inverse $\boldsymbol{C_t^{-1}}$, the Finger strain tensor. The material constant θ is given by

$$\frac{N_2}{N_1} = \frac{\theta}{1-\theta}, \tag{13}$$

where N_1 and N_2 are the first and second normal stress differences, respectively. It is noted that θ is not zero for polymer melts, which possess a non-zero second normal stress difference. Its usual range is between −0.1 and −0.2 in accordance with experimental findings [4]. The experimental data and their best fit with constants listed in Table 2 are plotted in Fig. 3.

Table 2: Relaxation spectra and material constants for the FEP resin obeying the K-BKZ model (*Eq. 12*) at 371°C (α = 7.174, β = 0.6, θ = −0.11, $\bar{\lambda}$ = 0.76 s, η_0 = 1613 Pa·s).

k	λ_k (s)	a_k (Pa)
1	0.337x10^{-3}	0.49050x10^{6}
2	0.178x10^{-2}	0.22702x10^{6}
3	0.865x10^{-2}	66,224
4	0.562x10^{-1}	4823.1
5	0.488	172.41
6	2.98	18.628
7	16.6	3.6235

In the case of slip effects at the wall, the usual no-slip velocity at the solid boundaries is replaced by a slip law of the following form [4]:

$$a_T u_{sl} = -\beta_{sl}\sigma_w^b, \tag{14}$$

where u_{sl} is the slip velocity, σ_w is the shear stress at the die wall, β_{sl} is the slip coefficient, and b is the slip exponent. It should be noted that the shift-factor a_T is used here to take into account the temperature effects of the slip law at different temperatures.

In 2-D simulations, the above law means that the tangential velocity on the boundary is given by the slip law, while the normal velocity is set to zero, i.e.,

$$\beta_{sl}\left(\bar{t}\bar{n}:\bar{\bar{\tau}}\right)^b = a_T\left(\bar{t}\cdot\bar{u}\right), \qquad \bar{n}\cdot\bar{u} = 0, \tag{15}$$

where $\bar{n}$ is the unit outward normal vector to a surface, $\bar{t}$ is the tangential unit vector in the direction of flow, and the rest of symbols are defined above.

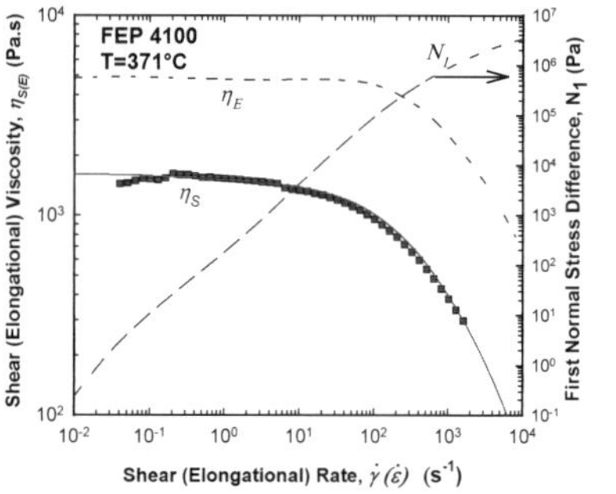

Figure 3 – Experimental data (solid symbols) and model predictions of shear viscosity, η_S, first normal stress difference, N_1, and elongational viscosity, η_E, for the FEP melt at 371°C using the K-BKZ model (*Eq. 12*) with the parameters listed in Table 2.

Method of Solution

The solution to the above conservation and constitutive equations is carried out with two codes, one for viscous flows (*u-v-p-T-h* formulation) [12] and one for viscoelastic flows [13]. The boundary conditions for the problem at hand are well known and they involve no slip or slip velocity conditions at the solid walls and a constant temperature of 371°C, while at the entry the fully developed velocity profile is given that corresponds to a given flow rate Q, and at the exit zero tractions and zero heat flux are imposed.

Results and Discussion

Figure 4 (symbols) shows the apparent flow curves for resin FEP® 4100 obtained experimentally with the tubular die using a d=1.524 mm tip and D=3 mm die at T=371°C. The apparent shear rate was calculated by using the formula which applies to slit dies [4]:

$$\dot{\gamma}_A = \frac{6Q}{0.25(D-d)^2\,0.5\pi(D+d)}, \tag{15}$$

where Q is the volumetric flow rate, d and D is the tip and die diameter, respectively. The apparent wall shear stress was estimated as the average of the stress at the inner and outer walls by using the following formula for a power-law fluid [14]:

$$\tau_{rz} = \frac{\Delta p D}{4L}\left(\frac{2r}{D} - c^2\frac{D}{2r}\right) \tag{16}$$

where τ_{rz} is the shear stress at radius r, Δp is the pressure drop (measured experimentally by means of a load cell that measures the force needed to displace the polymer melt through the reservoir and die), L is the length of the die land, and c is a parameter which depends on the geometry and the power-law index. For a power-law index of n=0.5 and a diameter ratio κ=d/D=0.508, we find from tables c=0.7283 [14].

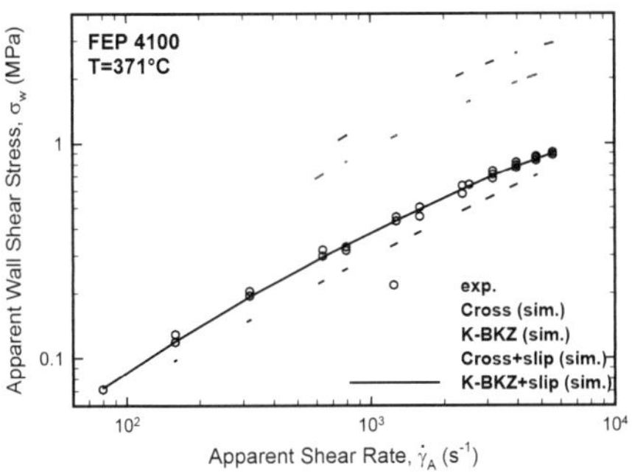

Figure 4 – The apparent flow curve and its predictions for the FEP 4100 melt in the Nokia Maillefer 4/6 crosshead at 371°C. Symbols are experimental data. Lines are simulation results by the Cross and the K-BKZ models with all effects present (influence of slip is evident).

A smooth flow curve is observed in Figure 4 which increases monotonically, an indication of a good design. Together with the experimental data are shown simulation results by the purely viscous Cross model and the viscoelastic K-BKZ model, with no slip and with slip present. The simulations based on the K-BKZ model with slip best describe the experimental data.

Table 3: Values of the various parameters for the FEP resin at 371°C.

Parameter	Value
β_c	0.00095 MPa^{-1}
β_p	0.03 MPa^{-1}
m	1.39×10^{-4} Pa^{n_p-1}
n_p	0.54
β_{sl}	400 cm/(s.MPa^b)
b	2.0
ρ	1.492 g/cm^3
k	0.00255 J/(s·cm·K)
C_p	0.96 J/(g·K)
E	50,000 J/mol
R_g	8.3143 J/(mol·K)
T_0	371°C (644 K)

Conclusions

A FEP melt has been studied in a crosshead die used in tubing extrusion flows. Full rheological characterization was carried out both with a viscous (Cross) and a viscoelastic (K-BKZ) model. All necessary material properties data were collected for the simulations.

Both viscous and viscoelastic models were found to over-predict significantly the experimental apparent flow curve (shear stress vs. shear rate) in a wide range of extrusion rates reaching up to 5600 s^{-1}, where the experiments produced smooth extrudates. Viscoelasticity gave higher results than the purely viscous Cross model.

The smooth experimental curve and the good condition of the extrudates made the addition of slip at the wall a necessity, given the fact that this melt appears to slip in capillary dies. Due to lack of experimental data, a quadratic slip law was assumed, which gave the correct predictions for the experiments by using the viscoelastic K-BKZ model.

Acknowledgements

Financial assistance from the Natural Sciences and Engineering Research Council (NSERC) of Canada and the programme "Thales" for basic research from GGET of Greece are gratefully acknowledged.

References

1. Imbalzano, A.E., Kerbow, D.L., "Advances in Fluoroplastics", Trends Pol. Sci., **2**, 26-30 (1994).
2. Ebnesajjad, S., "Melt Processible Fluoropolymers", Elsevier, New York (2003).
3. Domininghaus, H., "Plastics for Engineers: Materials, Properties, Applications", Hanser, Munich (1993).
4. Dealy, J.M., Wissbrun, K.F., "Melt Rheology and its Role in Plastics Processing - Theory and Applications", Van Nostrand Reinhold, New York (1990).
5. Rosenbaum, E., Hatzikiriakos, S.G., Stewart, C.W., "Rheological Characterization of Well Defined TFE/HFP Copolymers", Rheol. Acta, **37**, 279-288 (1998).
6. Buckmaster, M.D., Henry, D.L., Randa, S.K., "High speed extrusion", U.S. Pat. No. 5,688,457 (1997).
7. Dupont de Nemours, Co., Wilmington, DE, private communication (1998).
8. Sedlacek, T., Zatloukal, M., Filip, P., Boltizar, A., Saha, P., "On the Effect of Pressure on the Shear and Elongational Viscosities of Polymer Melts", Polym. Eng. Sci., **44**, 1328-1337 (2004).
9. Carreras, E.S., El Kissi, N., Piau, J.-M., Toussaint, F., Nigen, S., "Pressure Effects on Viscosity and Flow Stability of Polyethylene Melts during Extrusion", Rheol Acta, **45**, 209–222 (2006).
10. Papanastasiou, A.C., Scriven, L.E., Macosco, C.W., "An Integral Constitutive Equation for Mixed Flows: Viscoelastic Characterization", J. Rheol., **27**, 387-410 (1983).
11. Luo, X.-L., Tanner, R.I., "Finite Element Simulation of Long and Short Circular Die Extrusion Experiments Using Integral Models", Int. J. Num. Meth. Eng., **25**, 9-22 (1988).
12. Hannachi, A., Mitsoulis, E., "Sheet Coextrusion of Polymer Solutions and Melts: Comparison between Simulation and Experiments", Adv. Polym. Technol., **12**, 217-231 (1993).
13. Luo, X.-L., Mitsoulis, E., "An Efficient Algorithm for Strain History Tracking in Finite Element Computations of Non-Newtonian Fluids with Integral Constitutive Equations", Int. J. Num. Meth. Fluids, **11**, 1015-1031 (1990).
14. Hanks, R.W., Larsen, K.M., "The Flow of Power-Law, Non-Newtonian Fluids in Concentric Annuli", Ind. Eng. Chem. Fundam., **18**, 33-35 (1979).

O-01-135

COMPOSITE MODELING VS SCREW PUMPING CHARACTERISTICS FOR SCREW EXTRUSION

K. Wilczynski[1+], A. Nastaj[2], A.Lewandowski[3] and K.J. Wilczynski[4]
[1+] *Faculty of Production Engineering, Warsaw University of Technology, Poland – k.wilczynski@wip.pw.edu.pl*
02-524 Warsaw, Narbutta 85
[2] *Faculty of Production Engineering, Warsaw University of Technology, Poland – anastaj@wip.pw.edu.pl*
[3] *Faculty of Production Engineering, Warsaw University of Technology, Poland – a.lewandowski@wip.pw.edu.pl*
[4] *Faculty of Production Engineering, Warsaw University of Technology, Poland – wilczynski_k@wp.pl*

Abstract - Composite modeling of solids conveying, melting and melt flow in screw extruders, single screw extruders and twin screw extruders, co-rotating and counter-rotating, is based on combining melt conveying models with melting and solid conveying models. The composite models are generally lumped parameters models, and 3-D non-Newtonian FEM computations are not applied in this type of modeling. However, 3D non-Newtonian FEM simulations have been performed to refine the melt flow description in the screw extruders. Based on these simulations screw pumping characteristics may be developed. Simulations performed for different flow rates result in pressure gradient over the screw distance under study and these data may be presented in the dimensionless form. And, these may be expressed by some approximate equations which may be implemented into the composite model of the process. This technique has been applied to develop new composite models of various screw extrusion processes. For example, 3-D FEM Ansys-Polyflow melt flow computations have been carried out to design the screw pumping characteristics which have been implemented into the composite model of the counter-rotating twin screw extrusion process. The model has been validated experimentally using a Leistritz ZSE 27HP modular intermeshing machine.

Keywords: Screw Extrusion, Composite Modeling, FEM Simulations

Introduction

Composite modeling of solids conveying, melting and melt flow in screw extruders is based on combining melt conveying models with melting and solid conveying models. The composite models are lumped parameters models, and 3-D non-Newtonian FEM computations are usually not applied. This approach may be generally sufficient for engineering purposes since the main goal of the engineering designs is to predict the pressure and mean melt temperature profiles along the machine for a given screw and die geometry as a function of the operating parameters.

However, 3-D non-Newtonian FEM simulations may be applied to composite modeling via screw pumping characteristics. FEM simulations performed for different flow rates result in pressure gradient. These data may be presented in the dimensionless form, and this may be expressed by an approximate equation which may be implemented into the composite model.

Screw pumping characteristics for single screw extrusion were modeled numerically, e.g. by Rauwendaal [1], Potente et al. [2], and Syrjälä [3]. Pumping characteristics have been also modeled for twin screw extrusion, for co-rotating extrusion, e.g. by White et al. [2, 4, 5], as well as for counter-rotating extrusion, both tangential by White et al. [6, 7, 8], and intermeshing by Hong and White [9, 10].

In this paper, fully 3-D non-Newtonian FEM Ansys-Polyflow melt flow computations have been carried out to design the screw pumping characteristics which have been implemented into the composite model of the counter-rotating twin screw extrusion process. The model have been validated experimentally using a Leistritz ZSE 27HP modular intermeshing machine.

Experimental

An extensive experimental study has been performed for validation of the model [11]. A Leistritz ZSE 27HP modular intermeshing counter-rotating twin screw extruder was used in the study. Various screw elements were arranged on shafts, thick flighted elements (FD), thin flighted elements (KFD) which have large inter-flight gaps, and shearing elements (ZSS).

The polymers used in the study were (i) a low density polyethylene (Basel Orlen, Purell 3020D), (ii) a polystyrene (BASF, PS Polystyrol 158K), and (iii) a LDPE/PS polyblend of 85/15 composition.

The polymers were fed into the hopper at various rates 4 kg/h, 8 kg/h and at the flood mode. The screw speed was set at 40 rpm and 80 rpm.

FEM Simulation

Ansys-Polyflow program has been applied to model the melt flow in the machine. The problem set-up consists in the definition of the boundary conditions to identify the process. The screws are rotating, and the velocity is assumed to vanish along the barrel. A mass flow rate is prescribed at the entry of the computational domain and it is also an outflow condition at the exit of the domain. Both these conditions assume that the velocity vectors are perpendicular to the boundary.

A flow analysis has been performed for melt flow in the C-chamber, and leakage flows have been identified over the calander gap, the tetrahedron gap, the side gap, and the flight gap. Characteristic pressure profiles have been observed with regions of pressure decrease and increase for thick flighted element, and plateau and linearly declining regions for shearing element Some examples are shown in Figs. 1-2.

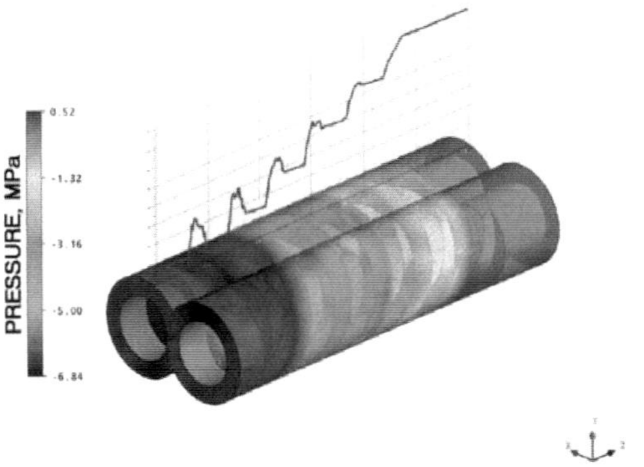

Figure 1. Simulation of pressure distribution for thick flighted screw element

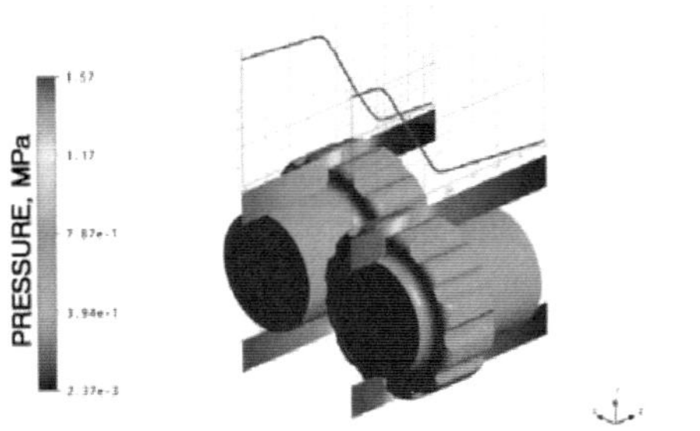

Figure 2. Simulation of pressure distribution for shearing element

Screw Pumping Characteristics

Screw pumping characteristics have been modeled by Hong and White [9, 10] for 2-D non-Newtonian and non-isothermal flow in various types of screw elements of an intermeshing counter-rotating twin screw extruder.

Screw pumping characteristics for counter-rotating twin screw extrusion process are given in terms of the following dimensionless groups:

$$Q^* = f(\Delta p^*) \quad (1)$$

where Q* is the dimensionless flow rate, and Δp* is the dimensionless pressure gradient

$$Q^* = \frac{Q}{2WHL_C N} \quad (2)$$

$$\Delta p^* = \frac{H^{n+1}\cos\varphi}{12k(\pi DN)^n}\frac{\Delta p}{L} \quad (3)$$

where Q is the flow rate, Δp is the pressure, W is the screw channel width, L_C is the C-chamber length, N is the screw speed, H is the screw channel depth, D is the screw diameter, L is the screw length, n is the power law index, k is the consistency.

Screw pumping characteristics have been calculated for thick flighted screw element and for shearing element, for various values of the power law index and for a broad range of flow rates. Some examples of computations are shown in Figs.3-4.

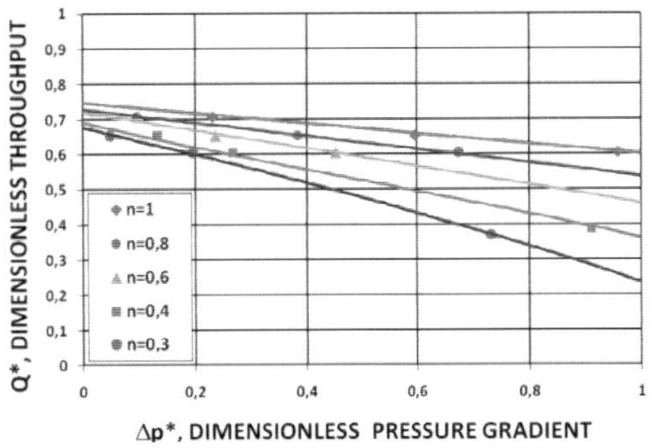

Figure 3. Screw pumping characteristics for thick flighted screw element

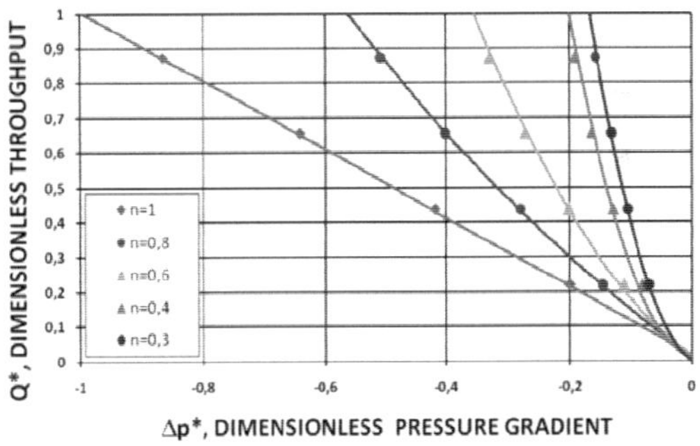

Figure 4. Screw pumping characteristics for shearing element

Thick flighted screw elements have got high pumping capacity which increases with an increase of a power law index. Shearing elements have got no pumping capacity, and a pressure gradient must be applied to force the fluid through an element and this gradient also increases with an increase of a power law index.

It is interesting to note that negative pressure gradient has been observed for thick flighted screw element for some simulation data.

Screw Pumping Characteristics and Composite Modeling

Screw pumping characteristics have been modeled using regression formulas for the thick flighted screw element and for the shearing element. The formulas are the functions of the process parameters and power law index, and they have the following form:

$$Q^* = A - B\,n^C\,\Delta p^* \quad (4)$$

where Q* is the dimensionless flow rate, Δp* is the dimensionless pressure gradient, n is the power law index; A, B, C are the formulas parameters of different

values for the thick flighted screw element and for the shearing element.
The pumping characteristics may be implemented into the composite model of the process.

Composite Modeling for Counter-Rotating Twin Screw Extrusion

Wilczynski et al. [7] have developed the first composite model of metering, melting and solid conveying in a starve fed closely intermeshing counter-rotating twin screw extruder. The model is based on the Hong and White melt flow description [9, 10), and Wilczynski and White melting description [13, 14]. Computations were made for axial fill factor as well as for pressure, temperature and melting profiles, and the results were compared with experiment. Recently, Jiang et al. [15] have presented the global model for closely intermeshing counter-rotating twin screw extruder with flood feeding.
In this paper, a TSEM composite model (Twin Screw Extrusion Model) has been presented. The model is based on the Wilczynski and White melting description [13, 14], and 3-D non-Newtonian FEM computations are appliied to refine the melt flow description.
Screw pumping characteristics defined by Eqs. (1) - (3) have been implemented into the TSEM model. This allows to compute melt conveying flow for composite modeling of the counter-rotating machine. FEM-calculated pressure gradient Δp is available for a given flow rate Q from pumping characteristics defined by Eq.4 at all iterations of computations. Computations are very fast and reliable. An example of computations is shown in Fig.5.

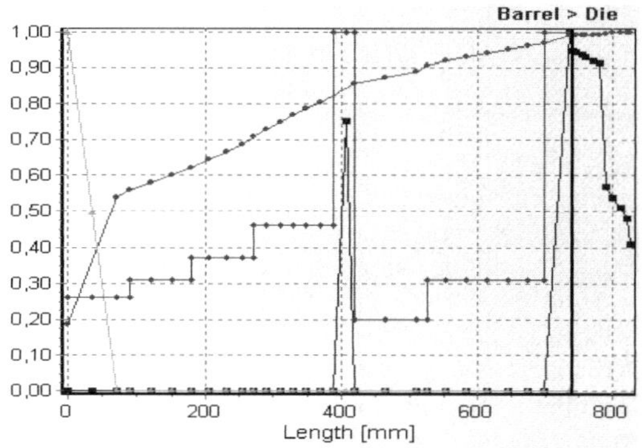

Figure 5. TSEM composite simulation based on the screw pumping characteristics: red line - fill factor, navy line - pressure, green line - temperature, blue line – melting

The TSEM computer model predicts polymer melting behavior, fill factor, and pressure as well as temperature profiles. The model has been validated experimentally for various flow rates and screw speeds [11]. The results of composite simulations were consistent with experimental runs.

It has been observed that when flow rate has been increased, the melting length has been longer, and the filled length has also been increased. However, when screw speed has been increased, the melting length has been shorter, and the filled length has also been decreased.

Conclusions

3-D non-Newtonian screw pumping characteristics have been developed and modeled by some regression formulas for the closely intermeshing counter-rotating twin screw extrusion. The formulas have been implemented into the composite model of the process which have been validated experimentally.
The FEM simulations have confirmed that thick flighted screw elements have got high pumping capacity which increases with an increase of a power law index, and the shearing elements have got no pumping capacity and a pressure gradient must be applied to force the fluid through an element, and this gradient also increases with an increase of a power law index. It is interesting to note that negative pressure gradient has been observed for thick flighted screw element for some simulation data.

References

1. Ch. Rauwendaal, *Polymer Extrusion,* Hanser, Munich 2001.
2. J.L. White, H. Potente, *Screw Extrusion*, Hanser, Munich 2003.
3. S. Syrjala, in *Mixing and Compounding, Ch.11* (ed. I. Manas-Zloczower), Hanser, Munich 2009.
4. Y. Wang, J.L.White, W. Szyd\lowski, *Int. Polym. Process.* 1989, *4*, 262.
5. Y. Wang, J.L.White, W. Szyd\lowski, *J. Non-Newt. Fluid Mech.* 1989, *32*, 19.
6. M.H. Kim, J.L.White, *Int. Polym. Process.* 1990, *3*, 201.
7. D.S. Bang, J.L.White, *Int. Polym. Process.* 1996, *11*, 109.
8. D.S. Bang, J.L.White, *Int. Polym. Process.* 199, *12*, 278.
9. M.H.Hong, J.L.White, *Int. Polym. Process.* 1998, *13*, 342.
10. M.H.Hong, J.L.White, *Int. Polym. Process.* 1999, *14*, 136.
11. K. Wilczynski, A. Lewandowski, K.J. Wilczynski, *Polym. Eng. Sci.* 2012, *52*, 449.
12. K. Wilczynski, Q. Jiang, J.L. White, *Int. Polym. Process.* 2007, *22*, 198.
13. K .Wilczynski, J.L.White, *Int. Polym. Process.* 2001, *16*, 257.
14. K. Wilczynski, J.L.White, *Polym. Eng. Sci.* 2003, *43*, 1715.
15. Q. Jiang, J.L. White, J. Yang, *Int. Polym. Process.* 2010, *25*, 223.

O-01-185

WINDING DEFECTS REFERRED TO SURFACE MODIFICATION OF BLOWN FILMS BY CORONA TREATMENT

F. A. Heinzler[1]*, J. Wortberg[1], F. Hoffmann[2] and T. Kirchhoff[2]

[1] *Institute of Product Engineering, University of Duisburg-Essen, Germany – felix.heinzler@uni-due.de*

[2] *Windmöller & Hölscher KG, Germany – tim_kirchhoff@wuh-group.com*

Abstract – Todays polymer film processing often includes printing and lamination. To enable this further processing, polyethylene-films have to be activated on the surface by corona treatment. Even though the surface treatment is necessary for further processing, some winding defects can be assigned to the correlation between the film surface treatment and winding parameters. Examples for these defects are winding knots which occur as frequent small spots on the roll and film. This paper discusses the effect of corona treatment on the film surface by NC-AFM-measurements and possibilities to improve processing and to reduce defects related to the modification of the surface.

Keywords: winding defects, blown film, corona treatment, AFM-measurement, wound-in air

Introduction

To enable printing or lamination of common films like polyethylene or polypropylene the surface of the film needs a special treatment to increase polarity and surface energy. Even though corona treatment is necessary for further processing, some winding defects can be assigned to the correlation between the film surface treatment and winding parameters. Examples for these defects are winding knots which occur as frequent small spots on the roll. The defect appearance can be compared to foam streak defects at coating [1]. Defects occur at noncritical winding parameter setups when corona treatment is activated, indicating that the treatment is a necessary but critical process step. Within a research project the focus was set on this winding defect and its emergence in the production process of polyethylene-film. The effect of winding knots occurs mandatory in combination with corona treatment. At a closer look the winding knots can be identified as small local compacted air volumes. These spots damage the film layers above.

Surface tension and surface energy

Within this project mono-layer polyethylene-film with a width of 1200 mm and high transparency was used for the experiments. There needs to be a separation between in-line corona treatment directly linked to the extrusion process and the surface treatment of already stored material.

For a common printing or lamination process the surface energy of the film has to be about 42 mN/m (+4 -2 mN/m²). The necessary corona treatment intensity is very much depending on the production machine. Hence in addition the surface energy was measured by a contact angle measuring system with an automated axisymmetric drop shape analysis (ADSA) for comparable databases. As measuring liquids ultrapure water, diiodomethane and ethylene glycol were used. For the interpretation of the contact angle data there are several different methods. A common method for solid polymer surfaces is the method of Owens, Wendt, Rabel and Kaelbe [2]. For solid materials and measuring liquids with known polar and disperse properties the surface tension and energetic polar or disperse parts can be calculated [3].

By using the sessile drop method the following contact angles were measured and surface energies were calculated. The measurements took place within 2 h after the processing.

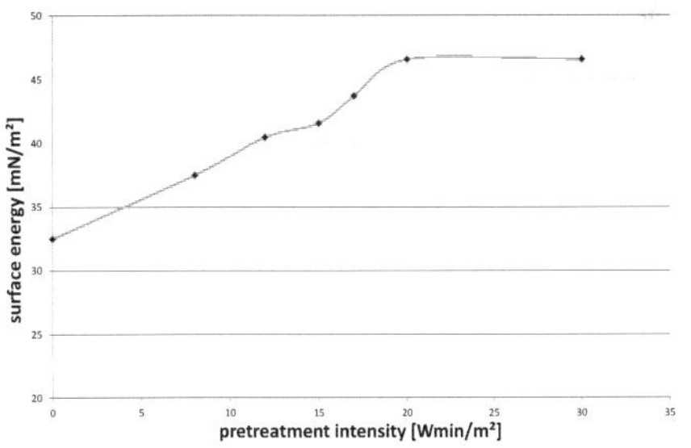

Figure 1 – surface energy according to treatment intensity of stored polyethylene -film

As shown in figure 1, a corona treatment intensity of 15 to 17 Wmin/m² was needed to achieve a proper surface energy.

Above 20 Wmin/m² there was no further increase of the surface energy recorded, but this point will be discussed in the following. Hence even stored PE-films can be activated for further processing like printing or lamination. The next step is the discussion of the effect of corona treatment on the surface and process conditions leading to winding knots at noncritical parameter setups. In order to determine the source for the corona treatment effect, NC-AFM-measurements (no contact - atomic force microscopy) capture the film surface profile in nanometer scale for different intensities.

Film Surface Modification

To get a detailed impression of the polyethylene-film surface, NC-AFM-measurements enable display details of 400 µm², 25 µm² and 4 µm². Every picture has a resolution of 512x512 pixel along the x- and y-range. At this level of details, the modification caused by corona treatment of the film surface can be shown for certain intensities. The recorded images in the following show a color-scaled measurment of the height (left) and amplitude (right) in z-range of the film's surface profile.

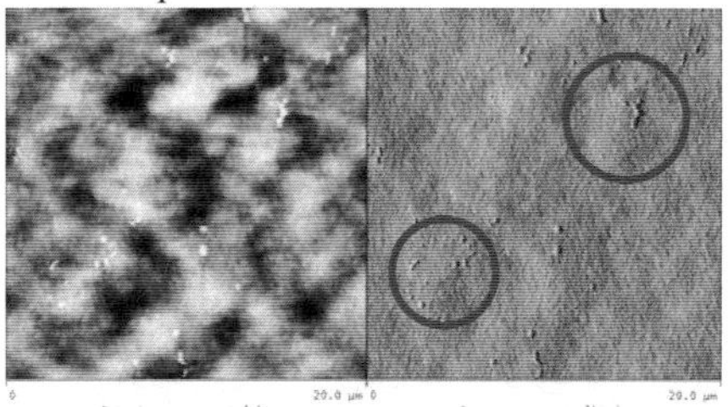

Figure 3 – Film surface without treatment (400µm²)

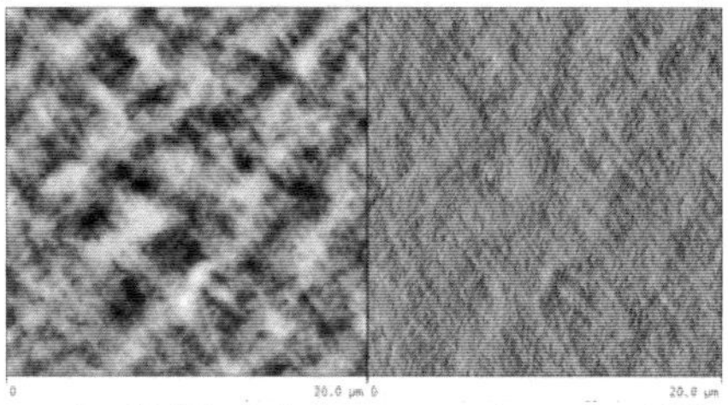

Figure 4 – 10 Wmin/m² treatment intensity (400µm²)

The measurements indicate, that at first a cleaning of the surface takes place. Micro-particles attached to the film are removed by the treatment at low intensity rates. This effect is important for lamination processes and the adhesion of glue or ink without contamination. The next increase of the corona intensity is located within the surface energy area that enables further processing. Unfortunately in this area winding defects have to be considered, so these effects have to be matched with process adaptions in the following to prevent defects.

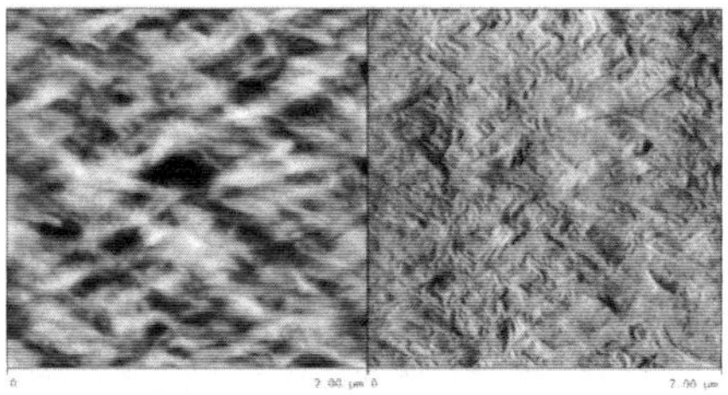

Figure 5 – 8 Wmin/m² treatment intensity (4µm²)

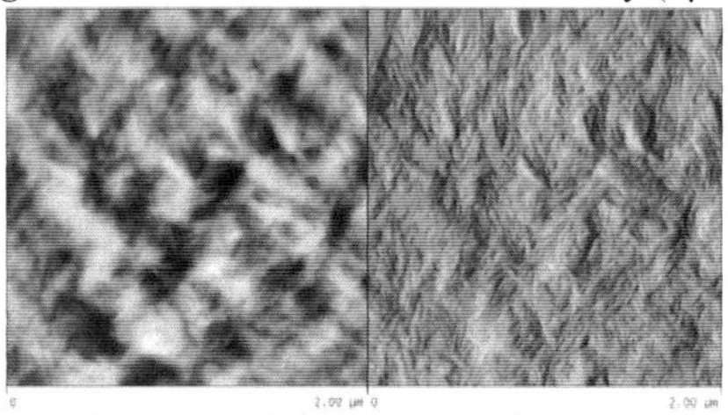

Figure 6 – 17 Wmin/m² treatment intensity (4µm²)

By increasing the corona intensity the film surface profile is smoothed and straightened. The corona treatment levels the film surface and reduces the profile amplitude. In addition to the AFM-images this effect can be shown in a three dimensional plot of the topographic measurement spots.

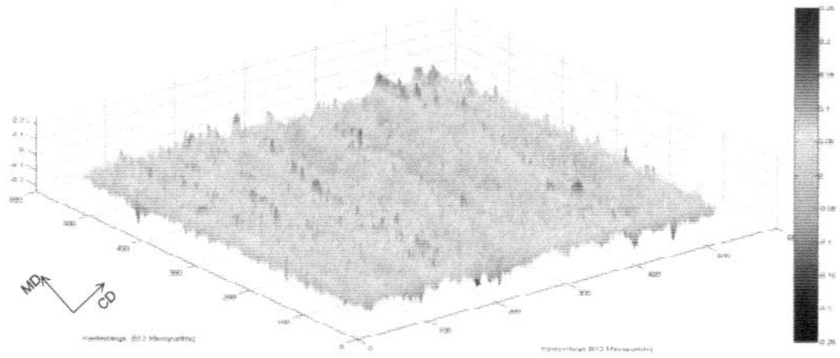

Figure 7 – 12 Wmin/m² treatment intensity (25µm²)

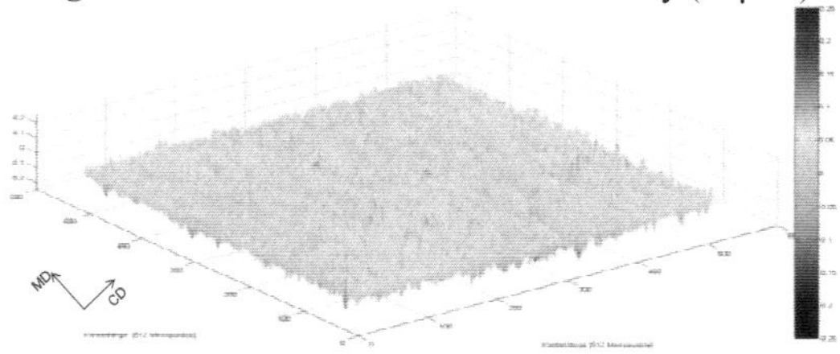

Figure 8 – 17 Wmin/m² treatment intensity (25µm²)

As shown in Figure 8 the roughness profile of the film is straightened and high peaks or amplitudes are leveled. If the intensity is increased further on, this effect is amplified. To testify this, images of pretreated film with intensities up to 60 Wmin/m² were examined.

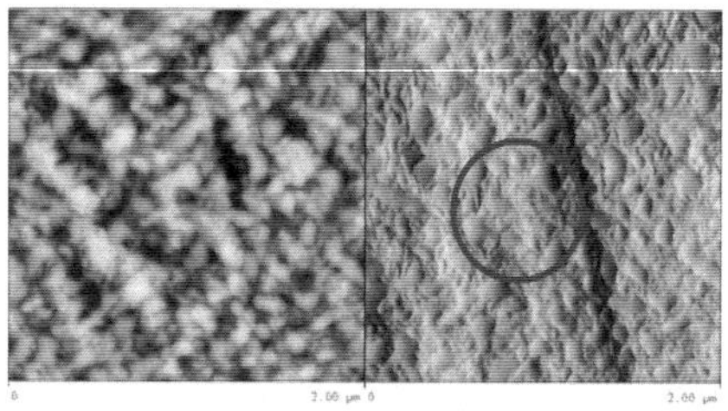

Figure 9 – 60 Wmin/m² treatment intensity (4µm²)

At this high intensity rate not only the edges and amplitude peaks of the roughness profile are smoothed and leveled but also local, plasticized spots occur. The surface energy is no longer increased above a certain level of treatment. Beyond this point, local energetic overload leads to high temperature spots and plasticizing of the film surface.

The corona treatment affects the surface of the film not only by increasing the surface energy or surface tension. The roughness profile is mandatorily influenced as shown in figure 3 to 9. The effect reaches from leveled or smoothed amplitude to local plasticized spots. These effects will influence the contact of film layers inside the roll, the haptic of the film gets more doll, and even the wound-in air will be reduced as the surface roughness is decreased. Before the process conditions are examined, the treatment effect on in-line processes have to be analyzed.

Differences to In-line Corona Treatment

At today's production lines the corona treatment is installed in-line to the extrusion process of the film. To examine differences between the in-line process and the process conditions with stored films, another test-series was setup.

The general effect on the film surface is similar to the stored film test series. Though the effect is similar, it is very much amplified at recently extruded films. The

polymer surface is activated more easily and with less corona intensity.
The following pictures show NC-AFM-measurements of in-line extruded and pretreated polyethylene-films.

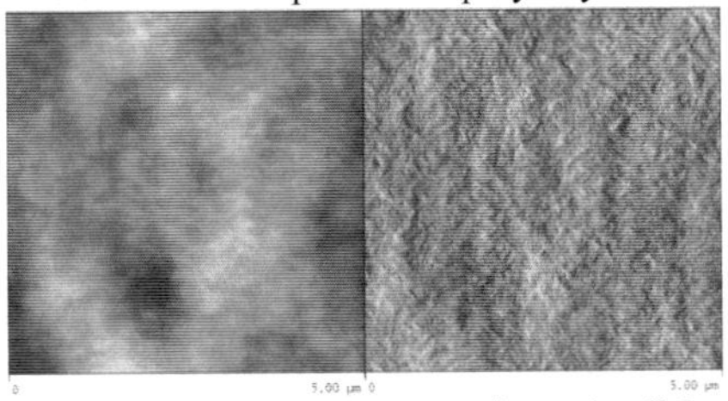

Figure 2 – 8 Wmin/m² treatment intensity (25μm²)

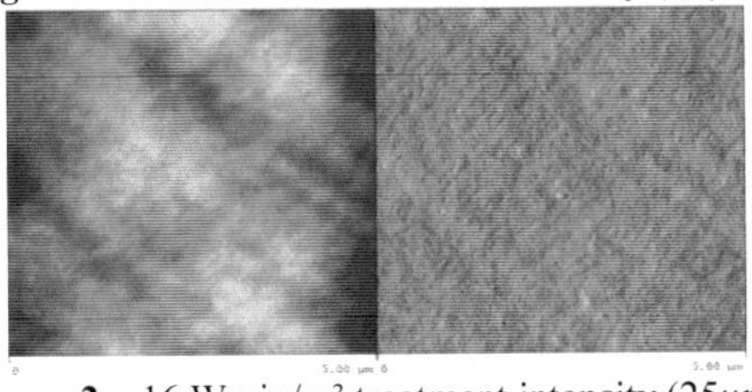

Figure 2 – 16 Wmin/m² treatment intensity (25μm²)

The described effect of leveling the film surface profile can also be seen in these images. But it is much amplified in comparison to the test-series with stored film rolls. Taking a closer look at the surface shows, that the plasticizing effect that occurred at high intensities in-line already takes place at lower intensities. At high intensities i.e. 48 Wmin/m² these effects multiply.

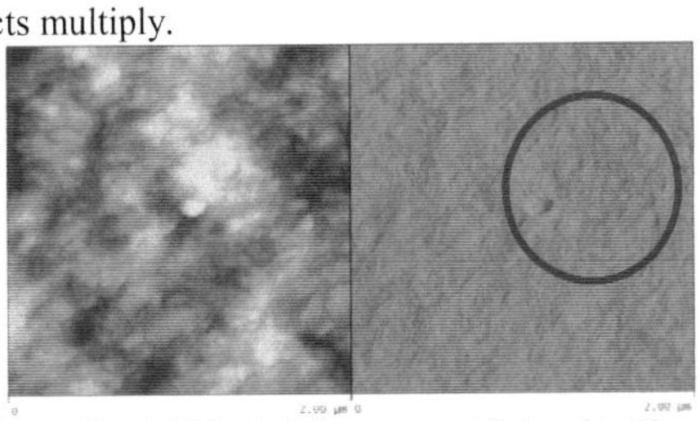

Figure 2 – 16 Wmin/m² treatment intensity (4μm²)

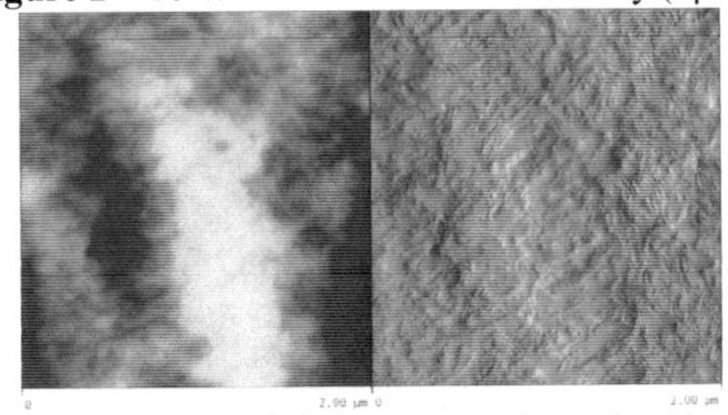

Figure 2 – 48 Wmin/m² treatment intensity (4μm²)

Within both test-series at first the surface was cleaned, smoothed or leveled and at high intensities plasticized spots occur. The still active surface of in-line extruded polyethylene-films amplifies the effect and less treatment intensity is needed to modify the surface.
For the discussion of process conditions and winding defects, the focus has to be set to a reduced roughness profile. The extensive plasticizing takes place at high intensities which are not necessary for further processing. To determine the source of the defects, now the process conditions and influences by process parameters within the build-up process of rolls have to be considered.

Process Conditions

The focused defects only occurred on the roll directly after the winding process. Before winding, the film had no defects and contamination of the film or specks can be excluded. If the layers with defects are cut out of the roll right after the process, the deformation regresses after a small amount of time. If the film is stored the film is permanently damaged. Because there is no contamination as reason for the defects, a correlation between the winding parameters and the effect of corona treatment on the film surface must cause the defects.
Within the first test series the intensity was set to 17 Wmin/m² to guarantee the range between 42-44 mN/m². After that, the winding speed, the web tension and the contact pressure of the contact roll were documented and set in correlation to the amount and size of knots, occurring on the roll surface. The test-setup was a contact winding station with a central winding drive and a contact roll.
The winding station enabled the independent variation of winding speed, web tension or winding tension and pressure of the contact roll. So the different effects of every winding parameter could be validated. Unfortunately the specific parameter range and its influence are very much depending on the winding station, polymer film and roll. At this point the results will be discussed in quality, the impression of the defects on the surface are subjectively rated.
Winding tension and pressure have a common effect on the buildup process of the roll. Winding tension increases the wound-in tension of the film and stabilizes the roll. The pressure of the contact roll increases wound-in tension and reduces air entrainment. If the tension is too low or there is too much air entrainment, the roll is instable and telescopes [4].
On the other hand the error probability of corona pretreated film rolls increases if web tension or pressure is increased. Especially the contact roll pressure has a major effect on roll quality. Because the contact roll's main influence is related to wound-in air, a connection to the surface modifications by corona treatment is indicated. The higher the contact pressure onto the surface of the film roll is, the more air is pressed out during the process. To provoke defects at standard process conditions, the contact pressure was set to 750 N and web tension to 250 N. In case the corona treatment is switched on there is an error probability of nearly 100 %, causing defects to occur immediately on the whole film roll. This parameter reduces the wound-in air to a minimum causing a very hard roll. To increase the air entrainment, pressure and tension were reduced. Reducing the pressure has a very large effect on the defects. The winding tension also influences the quality but the main effect is implemented by pressure. At a contact pressure of

100 N and web tension of about 50 N only occasional isolated spots occur.
The best setup for pretreated film rolls is the usage of an additional contact roll behind the nip. In gap winding mode this additional roll stabilizes the film roll and prevents telescoping. But the air buffer between the film layers is much bigger than in contact winding mode due to a softer coating of the roll. With this setup and a moderate contact pressure of ca. 1 bar, no defects along the roll width occurred at 17 Wmin/m² without telescoping.

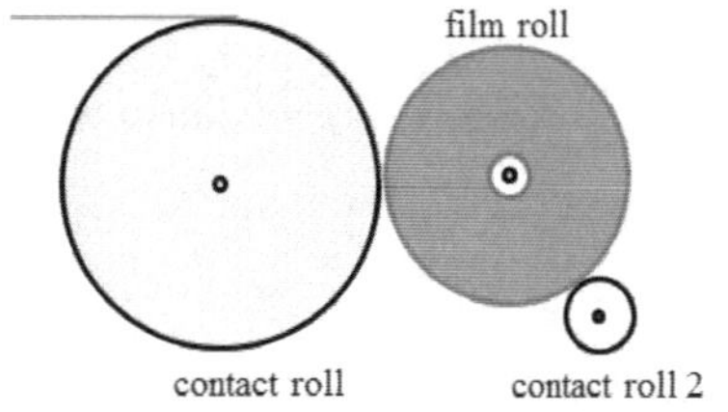

Figure 2 – Roll setup with additional contact roll

Another parameter to influence wound-in air is the winding speed. Increasing the winding speed has a very positive effect on roll quality in relation to the focused defects. By increasing winding speed the defects rated were significant lowered as the air entrainment is increased.
The roughness of the film surface carries a small air coating into the roll. This coating can be influenced by the winding speed and the surface profile. The faster the winding speed is adjusted, the more air is taken into the roll. By this process parameter even high defect rates at high winding tension and contact pressure could be reduced to good quality rolls. This proves the described positive effect of higher amounts of wound-in air and buffers between the film layers. It also links the described effects shown by the AFM-measurements to the process conditions. As already described, air within the film roll is influenced by the winding speed and the film surface. By the leveled surface profile less air is taken into the roll because of the reduced roughness of the surface. At high tension and pressure the film layers get into direct contact and local remaining air is compressed at small spots. Increasing the air entrainment prevents this effect.
To achieve a good roll quality a compromise between tension and pressure for a stable process and winding speed to prevent the appearance of defects has to be adjusted.

Conclusion

Because of the influence on the surface roughness, the connection to the described process conditions and wound-in air is obvious. The analysis of the test series leads to the assumption that the smoothed profile mandatory reduces wound-in air and causes the described effects.
Under high tension and pressure levels, there remains nearly no air buffer between the layers. Because of the smoothed profile, air cannot remain within the roughness-profile and is pressed to local spots. These spots occur as winding knots. By optimized winding parameters the air buffer has to be increased so that the film is not damaged by local remaining air volumes but the roll remains stable. This can be achieved by low but stable adjusted process parameters.

Acknowledgements
This project is supported by

ERDF – European Regional Development Fund
„Investition in unsere Zukunft“
NRW Ziel-2 Programm
CheK.NRW – Chemie und Kunststofftechnik

Title: Maßnahmen zur Erhöhung der Rohstoffeffizienz bei der Kunststofffolienherstellung

References

1. Smith, R. Duane, *Roll and Web Defect Terminology*, TAPPI Press, Atlanta, 1995
2. Mock, U.: PhD Thesis, University of Freiburg, 2004
3. D. K. Owens, R. C. Wendt, *Journal of Applied Polymer Science* 1969, *13*, 1741-&.
4. Roisum, D. R. *Reading a Roll,* Tappi Journal, vol 81, no 4, pp 205-211, Atlanta, 1998

O-01-248

PROTOTYPE SYSTEM AND METHODOLOGY FOR THE CHARACTERIZATION OF THE CONTACT RESISTANCE AT THE POLYMER-CALIBRATOR INTERFACE

O. S. Carneiro*, J. M. Nóbrega, C. Silva, A. Mota and A. M. Ribau

IPC/I3N – Institute for Polymers and Composites, Department of Polymer Engineering, University of Minho, Guimarães, Portugal - olgasc@dep.uminho.pt; mnobrega@dep.uminho.pt; carolina.silva@dep.uminho.pt; a51117@alunos.uminho.pt; angelaribau@dep.uminho.pt

Abstract - In thermoplastic profile extrusion lines the cooling stage usually takes place in a calibrator or in several of these tools in series. This stage is crucial since it determines the produced profile contour and also its level of residual thermal stresses and morphology, affecting, therefore, its behavior in service. One of the most influent parameters affecting the cooling performance of calibrators is the contact resistance at the polymer-calibrator interface, or its inverse, the heat transfer coefficient, $h_{interface}$, taken at the same location. However, the selection of the most appropriate value for this coefficient for a specific set of boundary conditions is still problematic, since it depends on many factors and is difficult to characterize. To overcome this limitation, a prototype calibrator was designed aiming to determine realistic values of the heat transfer coefficient at the polymer-calibrator interface, when using typical and controlled extrusion conditions. The value of $h_{interface}$ is determined by inverse engineering, resorting to an in-house 3D heat transfer modeling code. The prototype was designed in such a way that allows changing easily its construction material, surface finish and length. Also, relevant extrusion parameters can be varied, namely, the extrusion flow rate, extrusion temperature, degree of calibration vacuum and cooling water flow rate and temperature.
In this work the prototype calibrator and the methodology developed to determine $h_{interface}$ are described and a practical example of application is presented.

Keywords: heat transfer coefficient; contact resistance; extrusion cooling; calibrator.

Introduction

Like in all plastics processing technologies, in the extrusion of profiles the cooling stage generally limits the maximum production rate and contributes for the definition of the final dimensions and morphology of the produced part. The low thermal diffusivity of polymers is responsible for the development of considerable thermal gradients during the cooling stage, giving rise to the development of stresses [1-9] that can be frozen in the product (residual thermal stresses) jeopardizing its performance in service [10-14]. In the extrusion of complex thermoplastics profiles the cooling stage takes place in a calibrator (represented in Fig. 1) or in several of these tools in series. During the last decade, this research team developed and verified a methodology for the thermal design of calibrators for thermoplastic profiles that encompasses a non-isothermal 3D code, to model the thermal interchanges during the calibration / cooling stage, geometry and mesh generators and an optimization routine aiming at determining the optimal cooling conditions [11-14]. In previous studies it was concluded that the predicted results (temperature fields developed in the plastic profile during calibration) were highly dependent on the value employed for the heat transfer coefficient at the polymer-calibrator interface, $h_{interface}$ (or its inverse, the thermal contact resistance) [11]. This thermal contact resistance develops at the polymer-calibrator interface, as illustrated in Fig. 1, as a consequence of a partial lack of contact between the two surfaces. In the technical literature one can find values for this parameter ranging from 10 to 10,000 $W.m^2.K^{-1}$ [15]. The lack of systematic studies on this field motivated the development of a prototype system and methodology, for the characterization of the values $h_{interface}$, in a variety of conditions that can be found in extrusion practice. The prototype system and methodology developed are described in the next sections and their use is illustrated with a case study in the subsequent section.

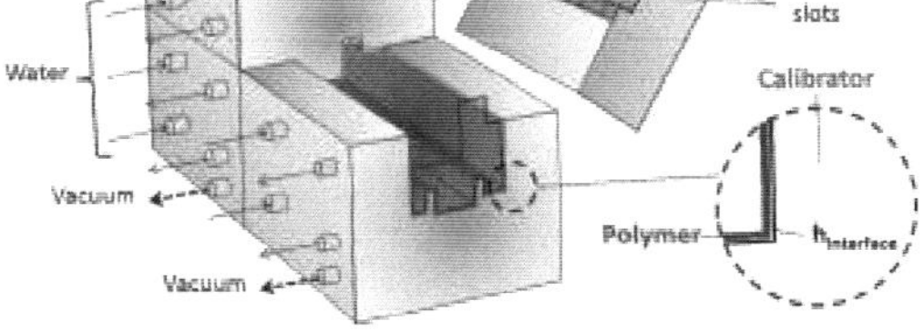

Figure 1 – Calibrator for profiles and thermal contact resistance at the polymer-calibrator interface.

Prototype System

The prototype calibration system consists of a structure used to support the modular prototype calibrator and the monitoring system, as shown in Fig. 2. At the top of the structure, a sliding support is used to mount an infrared thermographic camera that can be axially displaced, which is used to measure the temperatures of the upper plastic surface before entering and after exiting the calibrator.

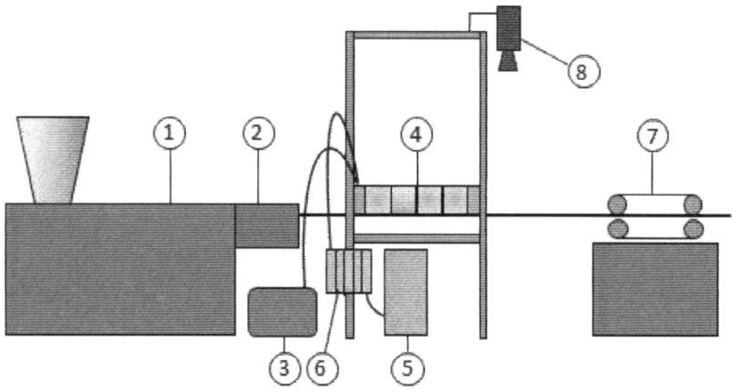

Figure 2. Set-up used to characterize $h_{interface}$: (1) – extruder; (2) die; (3) – vacuum pump; (4) calibrator; (5) thermoregulator; (6) flow meter; (7) pulling rolls.

The calibrator is modular, comprising several building blocks, as illustrated in Fig. 3. Each block contains a main structure and a calibration insert, where the vacuum slots are machined, as can be seen in Fig. 3. Each main block accommodates four independent axial cooling channels, two vacuum channels and five thermocouples (three located at the upper surface and one at each lateral surface). These thermocouples are used to monitor the temperature of the calibration insert surface, which will be used as a boundary condition in the determination of the value of $h_{interface}$, in the numerical modeling code. The decision of using small size thermocouples and to locate them outside the calibration insert was intended to minimize the disturbances induced into the temperature field developed in the vicinity of the interface polymer-calibrator. The calibration insert is screwed to the main block and can be manufactured with different materials and/or surface finishes, and can present a different number, location and dimensions of the vacuum slots. The blocks are fixed together by two clamping screws and the set is fixed to the main structure by two L shaped beams (see Fig. 3).

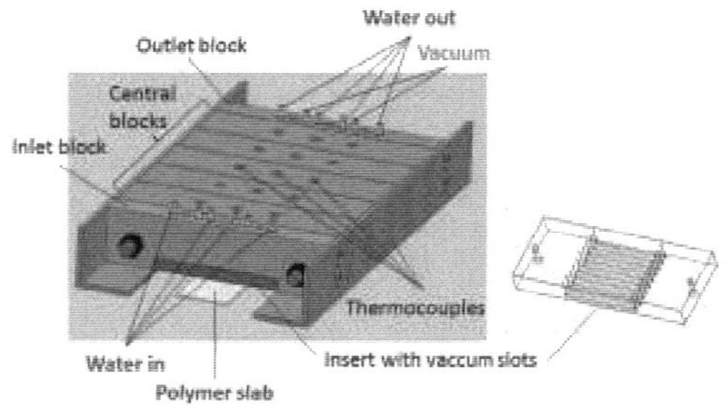

Figure 3. Modular calibrator: building blocks, main components and detailed view of the calibration insert.

In service, the polymer upper surface temperature is measured by an infrared thermographic camera focused at the entrance or exit regions of the calibrator. The temperature at the polymer-calibrator interface cannot be measured, since this would affect significantly the heat transfer conditions in that region.

Methodology

In the extrusion runs needed to determine $h_{interface}$, a polymer slab is extruded and cooled using the modular calibrator, as illustrated in Fig. 2. The temperature of the calibration insert surface is monitored, as already referred, through thermocouples, being the collected data acquired via a data acquisition system. After reaching steady state conditions the infrared thermographic camera is focused on the upper inlet surface of the polymer slab and then on the upper outlet surface of the same slab.

The methodology to follow in the determination of $h_{interface}$ can be summarized as follows. The existing 3D numerical modeling code is used to determine, through an iterative inverse problem solving scheme, the most adequate value of $h_{interface}$ and $h_{polymer}$ (the natural convection heat transfer coefficient, that affects the free surfaces, lower and lateral ones, of the polymer slab); corresponding to a specific extrusion run. The input data needed to solve this 3D heat transfer problem, using the numerical modeling code, include those related with the system geometry, boundary conditions and relevant properties of the polymer and metal of construction of the calibration insert block (more details can be found elsewhere [11]).

Case Study

The system developed was tested using a polystyrene, PS, Edistir N2560 and calibration inserts manufactured in aluminum. The modular calibrator with a length of 300 mm was used. The distance between the die exit and the calibrator entrance and between the calibrator exit and the pulling unit was of 200 mm and 150 mm, respectively. The extruded slab was 15 mm width and 1.2 mm thickness. The relevant properties of the polymer and aluminum, the extrusion conditions and other input data needed to run the numerical modeling code are listed in Table 1.

The temperatures of the insert, read by the 35 thermocouples embedded in the calibrator block, were fitted through a second order polynomial equation. The resulting temperature distribution equation was used in the modeling code as boundary condition at the calibration insert surface.

The temperature field of the polymer slab surface was determined with a thermographic camera ThermaCAM S and corresponding software (ThermaCAM Researcher Professional 2.8 SER-2). With this information, the experimental temperature distribution of the polymer surface center-line was obtained. This temperature distribution was then compared with the corresponding predictions given by the numerical modeling code. Several trial values for $h_{polymer}$ and $h_{interface}$ were tested to obtain an adequate fit between numerical and experimental results. The best guess for these parameters resulted in the following values: $h_{polymer} = 30$ W.m^2.K^{-1} and $h_{interface} = 435$ W.m^2.K^{-1}.

Table 1. Experimental extrusion run data.

Property	**Polystyrene**	**Aluminum**
Density (kg.m^{-3})	1050	2730
Thermal conductivity (W.m^{-1}.K^{-1})	0.17	123
Specific heat (kJ.kg^{-1}.K^{-1})	2.05	0.90
Extrusion conditions		
Extrusion temperature (ºC)	210	
Linear extrusion velocity (m.s^{-1})	0.0263	
Other conditions		
Water temperature (ºC)	30	
Water flow rate (l.min^{-1})	40	
Vacuum pressure (bar)	-0.3	
Room temperature (ºC)	20	

As can be seen in Fig. 4, the temperatures predicted by the code using the above values and the experimental ones (read before and after the calibrator) almost overlap, meaning that the values assume for the two heat transfer coefficients are the adequate (real) ones.

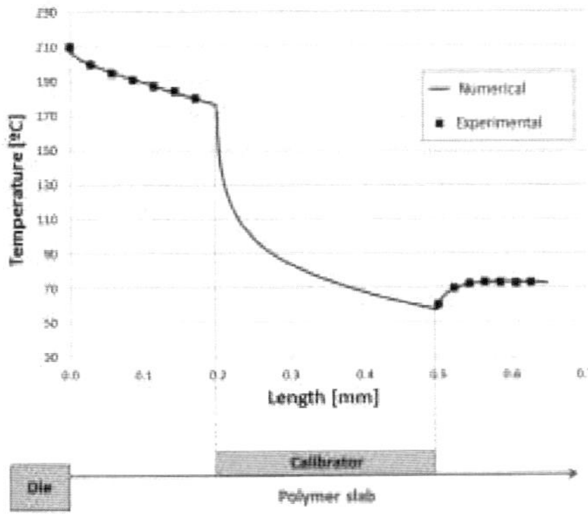

Figure 4. Temperature distribution at the center-line of the polymer slab: experimental measurements and corresponding values predicted by the numerical code for $h_{polymer} = 30$ W.m^2.K^{-1} and $h_{interface} = 435$ W.m^2.K^{-1}.

Conclusion

In this work a prototype calibration system and methodology developed for the characterization of the thermal contact resistance at the polymer-calibrator interface were described and used in a case study. These tools will enable the development of models for the dependence of this parameter on the most significant extrusion/calibration variables. The future implementation of the referred models in the existing 3D numerical modeling code will originate more realistic numerical predictions of the temperature fields developed during the calibration / cooling stage thus leading to better calibrators design.

Acknowledgements

The authors gratefully acknowledge funding from Fundação para a Ciência e Tecnologia, FCT (COMPETE Program) under the project FCOMP-01-0124 - FEDER-010190 (Ref. PTDC / EME - MFE/102729/2008) and FEDER, via FCT, under the PEst-C/CTM/LA0025/2011 (Strategic Project - LA 25 - 2011-2012).

References

1. K.K. Kabanemi; M.J. Crochet *Int. Polym. Proc.* 1992, 7, 60.
2. K.M.B. Jansen *Int. Polym. Proc.* 1994, 9, 82.
3. W.F. Zoetelief; L.F.A. Douven; J. Ingen Housz *Polym. Eng. Sci.* 1996, 36, 1886.
4. H. Ghoneim; C.A. Hieber *Polym. Eng. Sci.* 1997, 37, 219.
5. M.R.Kamal; R.A. Lai-Fook; J.R. Hernandez-Aguilar *Polym. Eng. Sci.* 2002, 42, 1098.
6. E.Q. Clutton; J.G. Williams *Polym. Eng. Sci.* 1995, 35, 1381.
7. A.J. Whittle; R.P. Burford; M.J. Hoffman *Polym. Eng. Sci.* 2000, 40, 2311.
8. Z. Gao; W. Pfandl *Polym. Eng. Sci.* 1996, 36, 2815.
9. J.F.T. Pittman; I. Szarvasy in 3th Esaform Proceedings, Stuttgart, Germany, 2000.
10. J.M. Nóbrega; O.S. Carneiro; F.T. Pinho; P.J. Oliveira *Int. Polym. Proc.* 2004, 19, 225.
11. J.M. Nóbrega; O.S. Carneiro; J.A. Covas; F.T. Pinho; P.J. Oliveira *Polym. Eng. Sci.* 2004, 44, 2216.
12. J.M. Nóbrega and O.S. Carneiro, Plast. Rub. Comp.: Macromol. Eng., 35, 387 (2006).
13. J.M. Nóbrega; O.S. Carneiro; A. Gaspar-Cunha; N.D. Gonçalves *Int. Polym. Proc.* 2008, 23, 331.
14. J.M. Nóbrega; O.S. Carneiro in *Optimization in Polymer Processing*, A. Gaspar-Cunha; J.A. Covas, Eds; Nova Science Publishers, Inc., New York, 2011; 145-167.
15. L. Fradette; P. A. Tanguy; P. Hurez; D. Blouin *Int. J. Numerical Methods for Heat & Fluid Flow* 1996, 6(1), 3.

O-01-061

MODELING THE MOLECULAR WEIGHT LOSS IN SINGLE SCREW EXTRUSION

S. Littek[1]*, V. Schöppner[1], M. Schneider[2] and K. Huber[2]

[1] *Kunststofftechnik Paderborn, University of Paderborn, Warburger Str. 100 33098 Paderborn, Germany – volker.schoeppner@ktp.upb.de; stefan.littek@ktp.upb.de*

[2] *University of Paderborn, Physikalische Chemie, Germany - schneid2@mail.upb.de; klaus.huber@upb.de*

Abstract – Extrusion is the most commonly used process for processing polymers. The high shear rates and temperatures impacting the polymer during extrusion cause thermomechanical degradation processes. With the help of a special test setup, polypropylene (PP) homopolymers with an MFI of 47 g/10 min are subjected to a defined shear rate ($\dot{\gamma}$) and residence time (t_v) at constant temperature. The removed material samples were analyzed and evaluated by gel permeation chromatography (GPC). Based on the experimental results, the influencing parameters were put into a functional relationship and compared with the measurements. A satisfactory correlation was achieved. Furthermore, it was shown with a statistical trend test that there is no significant relationship between the level of stress and the polydispersity. In this paper, the results of the experiments and the functional relationship are presented and discussed.

Keywords: Molecular weight loss, material degradation, polypropylene, chain scission

Introduction

One of the most commonly used techniques for processing, mixing or compounding polymers is the extrusion process [1-3]. Therefore developments in single screw extrusion aim to raise the efficiency of these machines at constant melt quality [4]. An increase in efficiency can be reached by raising the screw speed at constant screw diameter and machine size [5].

The high shear rate and melt temperature which influence the polymer through the extrusion process cause thermomechanical material degradation [3]. The degradation mechanisms differ for every polymer and the following reactions can run individually, successively or in parallel:

- Chain scission (degradation),
- Chain build up (post polymerization),
- Cross-linking or branching.

The degradation reactions in the processing of thermoplastic polymers are based on:

- Thermal effects [6,7],
- Oxidative effects [6,7],
- Water presence (hydrolytic degradation) [6],
- Residence time [7,8],
- Shear deformation [9].

Because of the above-mentioned stresses, the molecular weight and/or molecular weight distribution of the polymer will change in comparison to the original distribution of the raw material. In general, for polypropylene, the material degradation takes place by chain scission and the molecular weight distribution tends to lower molecular weight [10]. Under excessive thermal stress, the chains break and radicals will be formed. Under separation of low molecular constituents and radical transmission, these radicals continue to react [6].

When raising the rotational speed, the focus should also be on melt quality. Due to the higher screw speed, the polymer is exposed to higher mechanical and thermal stress while simultaneously the residence time decreases. The molecular weight loss is thus influenced by pressure, shear rate, melt temperature and residence time. The aim therefore is to generate defined material stresses and to measure the molecular weight loss in comparison to the unstressed raw material. The molecular weight distribution can be analyzed by gel permeation chromatography (GPC or size exclusion chromatography SEC). Because of an interaction of the process parameters in the extruder, the stress on the material has to be exerted by a piece of apparatus where one parameter can be varied and the other remains constant. Commercial systems like the cone-plate rheometer and coaxial rotational rheometer cannot be used because of the high shear rates and flow instabilities. Using a self-designed and self-built cylinder system, it was possible to generate defined material stresses.

Experimental

Test setup

Fig. 1 shows a drawing of the used cylinder system. This system works according to the principle of a coaxial Searle-type rheometer. In the gap between the inner and outer cylinders, energy is fed to the polymer through the rotating movement of the inner cylinder.

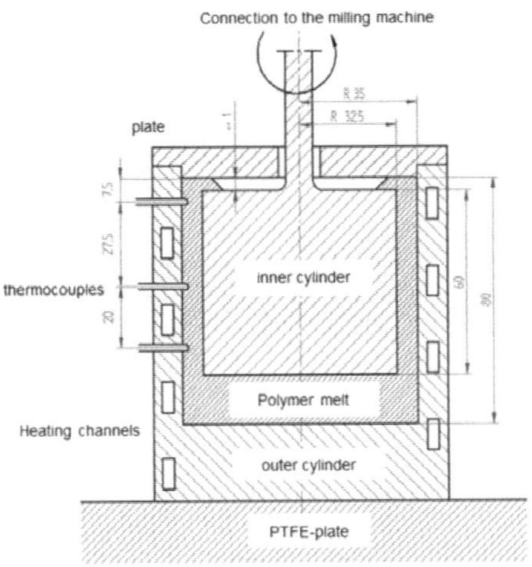

Fig. 1 – Cylinder assembly

The structure consists of a cup-shaped cylinder provided with spiral heating channels, and a second cylinder connected to the drive unit. The setup is also equipped with three thermocouples (Type J, dia. 2 mm) to measure the melt temperature. A cap is mounted to prevent the leakage of polymer melt from the cylinder system at a testing point. The outer cylinder is fixed on a solid, heat-resistant polytetrafluoroethylene plate, which serves as an insulating layer. This outer cylinder is temperature-controlled with a thermal oil, which flows through spiral channels and is heated and conveyed with a heating-cooling device. The inner driven cylinder is heated to the corresponding melt temperature by heater bands in front of each test point. Just before every test point, the heating bands are removed and the inner cylinder lowered into the polymer melt.

Determination of the molecular weight

The molecular weight distributions are determined by the SEC process from GPC Systems Alliance 2000. This is a so-called high-temperature GPC and is linked to a refractive index detector and a viscosity detector. For calibration, eight different narrowly distributed polystyrene standards from the firm PSS Polymer Standards Service are used at 155 °C. As the internal standard, a calibration was made to butylhydroxytoluene (BHT also: 2,6-di-tert-butyl-4-methylphenol) with every measurement. Using the universal calibration relationship, the molecular weight distribution of the examined specimens was determined. From the number average and the weight average molecular weight, the polydispersity was calculated. 1,2,4-trichlorobenzene (TCB) was used as the eluent and a three-step column arrangement with average pore widths of 10^6, 10^5, 10^3 Å was used. This makes it possible to analyze engineering polymers with a broad molecular weight distribution. The injection volume of the polymer was 220 µl, the measuring interval was put at a time of 0.5 s, and a measuring temperature of 155 °C was set. The PP was dissolved in TCB within two hours at 155 °C using BHT.

The reproducibility of the GPC unit was checked with the aid of measurements on four PP specimens. The specimens were the raw PP of a batch, because it was safe to assume that they would have very similarly dispersed molecular weights. The evaluation of the GPC analysis showed a standard deviation of 9 % calculated on the mean value of the four measurements, Table 1.

Table 1 – Monitoring the reproducibility of the GPC unit

	sample			
	1	2	3	4
M [g/mol]	133000	111000	138000	127000
average weight [g/mol]	127250			
standard deviation [%]	9,2			

Because of the good reproducibility of the GPC unit, the validation of the cylinder system was subsequently performed and an examination made as to how precise the stressing of the material is.

Validation

Validation of the measuring unit was carried out by measuring five material samples at a test point and then subjecting them to GPC analysis. The shear rate was 1005 s^{-1}, the melt temperature 230 °C and the residence time 15 s. The purely thermal melting time of the PP was 13 minutes for all five measuring points. A certain amount of polymer is needed for a GPC measurement. For this reason, it was not measuring points but measuring ranges that were defined in order to be able to remove sufficient material. The inner cylinder (Fig. 2, left) was subdivided into three sections in which, over the entire circumferential length, polymer was removed with a spatula for the GPC analysis. In Fig. 2, right, the outer cylinder is shown together with the two areas from which samples were removed. In addition, the inner cylinder, as would be positioned during a test point, is shown as a dotted line.

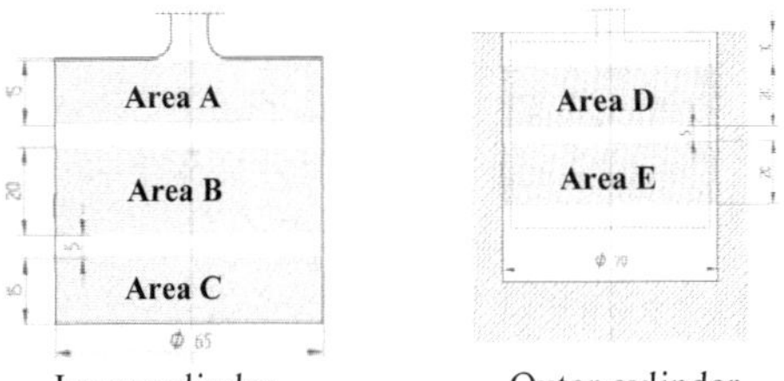

Inner cylinder   Outer cylinder

Fig. 2 – Sampling areas on the inner (left) and outer (right) cylinders

In Fig. 3, the molecular weight distributions of the five test points are shown as examples for the sampling areas B. Here, the standardized molecular weight per class is plotted on the Y axis (frequency).

The molecular weight distributions are fractioned into logarithmized classes and plotted in a logarithmic scale. A look at the individual distributions shows that a logarithmic normal distribution exists.

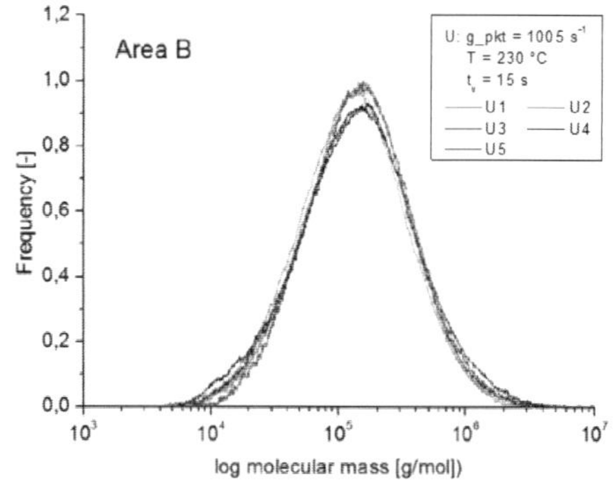

Fig. 3 – Molecular weight distributions of sampling area B

Sampling area B has, compared with the other sampling areas, the lowest deviation and was therefore selected for the subsequent tests [11].

Results and Discussion

The PP 579S from the firm Sabic IP was stressed under the combination of four shear rates and four residence times at constant temperature. The shear rate was in each case increased by the starting shear rate and the residence time was varied over a broad range. The development of the number average (arithmetic mean from two measurements) of the molecular weight distribution vs. the residence time is plotted for different shear rates at constant temperature (230 °C) in Fig. 4. At the time t = 0 s, the value for the raw molten sample is shown as the starting value.

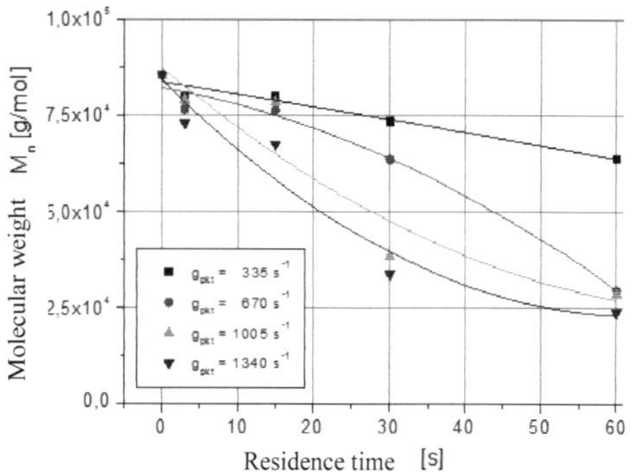

Fig. 4 – Number average of the molecular weight vs. the residence time for various shear rates at a constant temperature (230 °C)

Although the polydispersity fluctuates between 2.15 and 4.83, it should be pointed out that the GPC analysis is flawed by deviations and that the polydispersity reacts very sensitively to fluctuations that occur predominantly to the right of the dispersion curve (bell curve). The question regarding the change and dependence of the polydispersity on the stress was clarified with a statistical trend test. Using the statistical trend test according to von Neumann, an examination was made to see whether a trend could be identified for the obtained values. For this purpose, an evaluation of the stress, oriented to the temperature, shear rate and residence time, was performed. Every test point was evaluated, related to the possible maximum value ($T_2+\dot{\gamma}_4+t_{V4}$ = 1+4+4 = 9). A test point with $\dot{\gamma}$ = 335 s^{-1}, T = 230 °C, t_v = 3 s receives the value (1+1+1)/9 = 1/3. The value 0 represents a low load and 1 reflects a high load.

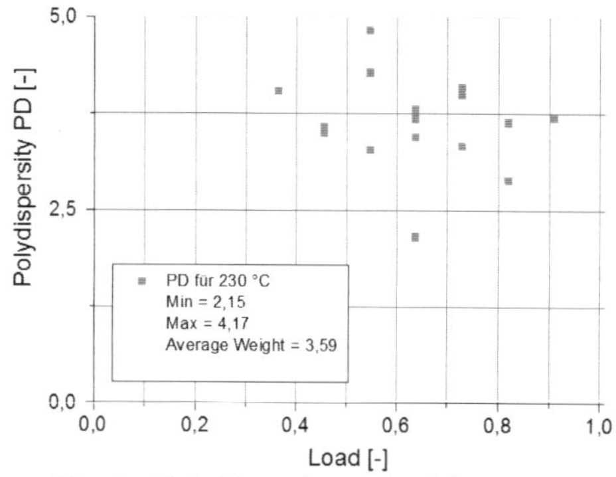

Fig. 5 – Polydispersity vs. load for 230 °C

With the trend test according to von Neumann, no trend exists with a probability of 95 % (Q = 2.11 > P(95%) = 1.223). As shown in Fig. 5, there is thus no relationship between the polydispersity and the load.

The following mathematical relationship was taken as a basis for calculating the material degradation

$$\frac{M_n}{M_{n,0,Auf}} = \frac{1}{\left[\left(\exp\left(T_0\left(\frac{1}{T-T_B}\right)\right)\right) * \left(1+\frac{\dot{\gamma}}{\dot{\gamma}_0}\right) * exp\left(\left(\frac{t_v}{t_{v0}}\right)^{1,25}\right)\right]} \quad \text{Eq. 1}$$

where M_n is the molecular weight to be calculated and $M_{n,0,Auf}$ the unstressed starting molecular weight of the raw material after melting. The prevailing temperature T is standardized with a reference temperature T_B and is included in the denominator of the equation. The shear rate is taken into account by $\dot{\gamma}$ and the residence time by t_v. The sensitivity to temperature is described by T_0, the sensitivity to shear rate by $\dot{\gamma}_0$ and the sensitivity to residence time by t_{v0}.

To describe the results obtained by experiment with the help of Eq. 1, it was necessary to perform a non-linear regression. With the help of the OriginPro software, Eq. 1 was entered, and T_0, $\dot{\gamma}_0$ and t_{v0} were determined. The calculated values were compared with the measurements in Fig. 6. As the starting value, the molecular weight of the melted sample was used. The correlation coefficient R^2 is 0.73. After analysis and evaluation of additional values, a better correlation is targeted.

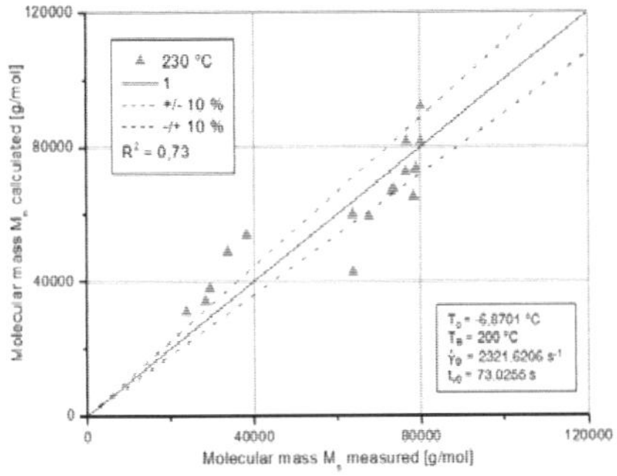

Fig. 6 – Comparison between calculated and measured molecular weight

Experiments were carried out on material degradation using a single screw with a nominal diameter of D = 30 mm and an L/D ratio of 31D:

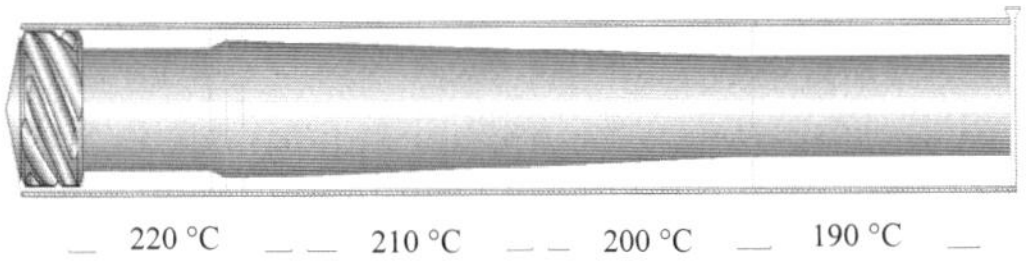

Fig. 7 – Screw used for the experiments

In the tests, only the screw speed was varied between 100 and 2000 min^{-1}, while all other parameters were kept constant. At the die orifice, material samples were removed and evaluated by GPC. Fig. 8 shows the relative molecular weight (related to 100 min^{-1}) plotted against the screw speed.

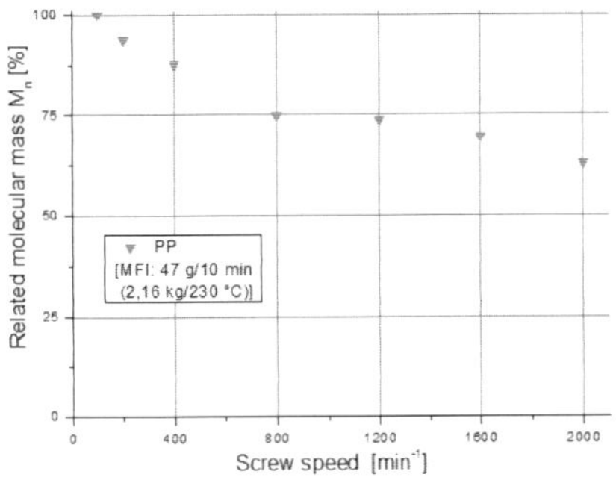

Fig. 8 – Relative molecular weight vs. screw speed

Conclusions

The specific defined stress was applied to the PP satisfactorily with the self-designed unit. Parameterization and modeling of the molecular weight loss was possible with a coefficient of determination of $R^2 = 0.73$. The next step in the project is to examine the influence of temperature on the degradation of PP. For this, the shear rate and residence time will be varied at different temperatures. Furthermore, PP types with different flow behavior than the PP presented here will be examined. Based on these data, a physical-mathematical model will be derived, which describes the material degradation of PP as a function of temperature, shear rate and residence time.

Acknowledgements

We wish to thank the Deutsche Forschungsgemeinschaft, which has made these studies possible by financing the SCHO 551/7-1 project.

References

1. C. Rauwendaal, Polymer Extrusion, Carl Hanser Verlag, 2001
2. C. Rauwendaal, Mixing in polymer processing, Carl Hanser Verlag, 2000
3. L. Reich; S. S. Stivala, Polym. Eng. Sci., 1980, 20, 654
4. J. Wortberg, Kunststoffe 89, 1999, 1, 54
5. J. Wortberg; R. Michels, VDI-Verlag, 1997
6. K. Kircher, Chemische Reaktionen bei der Kunststoffverarbeitung, Carl Hanser Verlag, 1982
7. H. Batzer, Polymerwerkstoffe, Thieme-Verlag
8. G. Menges; E. Haberstroh; W. Michaeli; E. Schmachtenberg, Werkstoffkunde Kunststoffe, Carl Hanser Verlag, 2002
9. A. Casale; R. S. Porter, Poly. Stress Reac., 1979, 2
10. S. V.Canevarolo, Polym. Deg. and Stab., 2000, 709, 71
11. S. Littek; M. Schneider; K. Huber; V. Schöppner, Journal of Plastics Technology, 2012, 8

O-01-340

DETERMINATION OF CORRECTION FACTORS FOR THE NETWORK-THEORY-BASED DESIGN OF EXTRUSION FILM DIES

B. Henke[1*] and V. Schöppner[1]
[1]Kunststofftechnik Paderborn, University of Paderborn, Germany – volker.schoeppner@ktp.upb.de; bianka.henke@ktp.upb.de

Abstract – For manufacturing plastic film and sheet and also for coating sheet material, use is made of dies with a rectangular cross-section, which are known as slot dies. So far, the neck-in effect has been neglected in the calculation for determining the pressure loss in slot dies. This effect causes constriction of the edge zones of the melt in the section between the die and the roll (accounting for a share of approx. 5-10%), which plays a key role in the layout of slot dies. The mathematical formulations adopted for the calculation of the neck-in in these dies take the form of two models for calculating the length of the neck-in and a further numerical model for varying the die geometry. These formulations provide information on the width of the neck-in at the edges, thus permitting an optimum layout of the die and an increase in the cost-efficiency of the production plant. The mathematical formulations of the planar-uniaxial flow model can be implemented in the Excel-based calculation tool, allowing the neck-in length to be calculated. The calculation tool can thus be used to simulate simple flow processes with allowance for the neck-in and hence to design more complex die geometries.

Keywords: film, extrusion die design, neck-in

Introduction

There are many different possibilities for the use of plastics products today. The most important processing method is extrusion, in which film, sheet or webs are produced in a continuous process as so-called endless products [1].

Flat film extrusion is a key production process, in which the melt is pressed through a die with a rectangular cross-section and then taken-off as a flat film. After it has emerged from the die, the melt is cast on a cooling roll [2]. The gap between the die and the roll causes the material to constrict in the edge zone, leading to an undesirable thickening, or beading, in this area [BB98]. This constriction of the edge zones is called neck-in [3].

Slot dies are made wider than the film that is required, since the extent of the neck-in cannot be reliably predicted. During production, technical aids prevent any further propagation of the effect, and the edge zones are cut off, so that the film is in its definitive width when it goes through the subsequent process steps. This method incurs higher die and system costs, thus reducing cost-efficiency [4].

The neck-in effect has been neglected in the die calculation for determining the pressure loss in slot dies. Because of the reduction in the film width, allowance ought to be made for this effect in the layout of slot dies.

Neck-In Effect

A mathematical formulation is to be used in order to permit the neck-in effect to provide information on the extent of the neck-in in the edge zones, thus ensuring that an optimum die layout can be achieved.

The neck-in effect arises on account of the tensile forces that prevail in the film between the die exit and the cooling roll (Fig. 1); it is brought about by the lack of retention forces at the film edges. A lower material viscosity, faster take-off speeds and a greater distance between the slot die and the cooling roll all serve to make the neck-in effect more pronounced [5, 1].

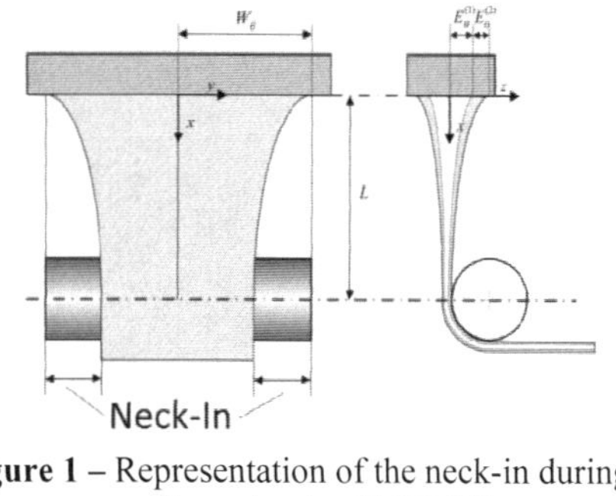

Figure 1 – Representation of the neck-in during flat film extrusion [LSJ+07]

Mathematical description of the neck-in

Dobroth-Erwin model (DEM)

The mathematical description is based on the Dobroth-Erwin model. This formulation describes the neck-in effect in a continuous, stable process in which the disturbance variables are ignored. The mathematical boundaries and flow lines are not displaced in the steady-state melt flow. The flow rate upon output from the die is also identical to the take-off velocity on the cooling roll.

The development of the neck-in is revealed through investigations into the flow lines and the law of conservation of mass. This model works on the assumption that the material in the edge zone expands uniaxially in the take-off direction, while the material in the middle of the film web undergoes uniform planar expansion. In addition, Newtonian, incompressible fluids are assumed in this model.

In this model, the surface tension, the swelling of the extrudate and edge stress effects are given as possible causes of the neck-in effect. The edge stress has the biggest influence and is brought about by the dissimilar forms of strain that prevail in the centre of the film web and at its edges. This effect leads to a thinner film, and hence the stress states and strains in the centre of the film differ clearly from those in the edge zone. A comparison of these states reveals the extent of the beading at the edge. The flow lines in the centre of the film web run in a straight line, parallel to each other. An element flowing between these parallel flow lines has the same width upon reaching the cooling roll as it had when it left the die [6].

Planar-uniaxial flow model (EESM)

The planar-uniaxial flow model is based on the Dobroth-Erwin model. For calculating the neck-in effect, it is assumed that the centre of the film web undergoes uniform planar deformation, while the edge zones deform uniaxially in the take-off direction. In this model, an equation is derived for the edge zone of the film web, and a term for the neck-in length. Figure 2 shows a schematic diagram of the flow model. The edge zones of the film undergo uniaxial deformation, while most of the middle of the film experiences planar deformation. The X-axis is defined as the direction of pull, the Y-axis as the width direction and the Z-axis as the thickness direction [7].

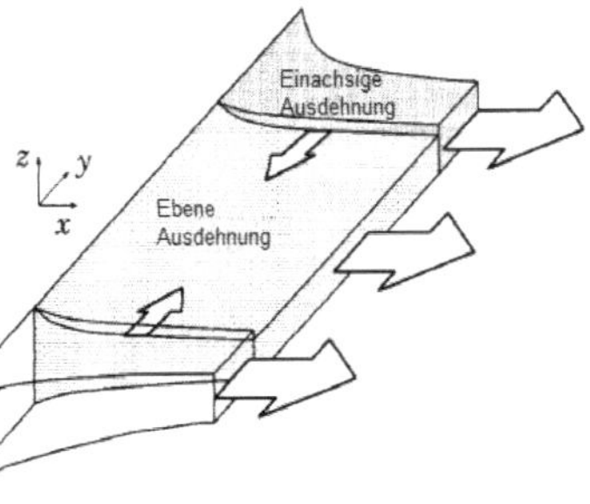

Figure 2 – Schematic diagram of the flow in the flat film [IDI+03]

In the description employed in the model, inertia and gravity are neglected on account of a small Reynold's number. In addition, the surface stress and the swelling of the extrudate are not taken into consideration. The melt and extrudate temperatures are assumed to be constant over the entire width of the film. The temperature drop in the air gap between the die and the cooling roll is ignored [7].

The neck-in effect is brought about by the contracting force of the planar component, which causes the edge zone to be drawn in towards the middle of the film. Via the equilibrium of forces prevailing at the cross-section of the film web with plane x = constant, the edge line comes into contact with the cooling roll in parallel to the direction of pull and can then be defined. The starting point of the edge line is at the point of departure, i.e. at the exit of the die.

For Newtonian fluids, the determination of the neck-in effect can be derived analytically. The following applies for the planar stresses σp_1 and σp_2:

$$\sigma_{p1} = 4\eta_0\dot{\epsilon}_p ; \sigma_{p2} = 2\eta_0\dot{\epsilon}_p \qquad \text{Eq. 1}$$

with the shear viscosity η_0 and the strain rate $\dot{\epsilon}_p = d\epsilon_p(t)/dt$. These conditions then give rise to:

$$y(x) = \sqrt{Lx - \frac{1}{2}x^2} \qquad \text{Eq. 2}$$

for the edge line of the film web. The neck-in length ξ is given as:

$$\xi = \sqrt{\frac{1}{2}L} \qquad \text{Eq. 3}$$

One of the characteristics of non-Newtonian fluids is that they are able to display elastic and viscous behavior at the same time. These properties can be described in a Maxwell model with the aid of the mechanical elements of a spring and a damper [8]. The mathematical formulations for describing the neck-in effect are derived using this model.
In this case, the neck-in length ξ is given as:

$$\xi = \sqrt{\frac{1}{2}\left(1 - 2\dot{\epsilon}_p\tau\right) L} \qquad \text{Eq. 4}$$

On the basis of the mathematical calculation with the aid of the assumptions, the neck-in length will account for a lower percentage share than with the Newtonian fluid due to the positive variables of $\dot{\epsilon}$ (strain rate) and τ (relaxation time). In addition, the neck-in length in non-Newtonian fluids which have a long relaxation time is less pronounced than for shorter relaxation times with the same process parameters [6].

Three-dimensional numerical model

In this model, the thickness and width of the film web is generated for an isothermal film production process with free boundary-value problems. During production, the aim is to achieve a uniform film web that does not require any finishing work. This requirement cannot be achieved with a uniform cross-section for the die. The film web that is produced has necked-in edge zones that have to be removed. The aim of the simulation is to configure a die geometry which will make it possible to produce a film with a uniform web thickness and edge zones that can be used. With the aid of the Lagrange function, adjoint equations were derived for a fixed-width film web, and a minimizing algorithm was compiled via these. The numerical simulation makes it clear that the algorithm converges with the imprecise, approximate adjoint equations. Employing the simulation, an optimum, elliptical die geometry is configured. By making this change, it becomes possible to produce film webs that can be further processed without need for the edge to be trimmed [9].

Comparison of the models

The comparison is intended to permit the rapid selection of a model, making allowance for the neck-in effect in the design of a die, which makes it possible to estimate the extent to which the effect is channeled into the dimensioning of the die. The Dobroth-Erwin model forms the basis for the planar-uniaxial flow model. Thanks to the further development of the model, it is now possible to use the planar-uniaxial flow model to calculate the neck-in length for a Newtonian and a non-Newtonian fluid without any major outlay.

The planar-uniaxial flow model is suitable for implementation in the existing Excel tool. Work is currently underway to incorporate the mathematical description of the neck-in into the Excel tool which is based on the network theory. Verification will then be performed, so that the results will be established in the course of the next few months.

Conclusions

By specifically calculating the neck-in effect, it is possible to reduce the die costs when designing the die and to keep the amount of edge trim during production to a minimum, thus boosting the cost-efficiency of the production plant. The implementation of the planar-uniaxial flow model mathematical formulation in the calculation for determining the pressure loss in slot dies permits a more precise die design for dies with a slot-shaped exit cross-section. In addition, simple flow processes can be simulated and more complex geometries designed, with allowance for the neck-in.

References

1. Prof. Dr.-Ing. V. Schöppner, Institut für Kunststofftechnik Universität Paderborn (KTP): *Skript zur Vorlesung Standardverfahren Extrusion*, Sommersemester
2. H. Greif, A. Limper, G. Fattmann, S. Seibel,.: *Technologie der Extrusion*, Carl Hanser Verlag München Wien, 2004
3. L. Bottenbruch, R. Binsach: *Polyamide*, Carl Hanser Verlag München Wien, 1998
4. Betriebsbesichtigung Firma Wentus Kunststoff GmbH Höxter, Informationsgespräch mit Herrn Pohl und Herrn Warneke, 11.November 2011
5. Gerhard W. Becker: *Die Kunststoffe: Chemie, Physik, Technologie*, Band 1, Carl Hanser Verlag München Wien, 1990
6. T. Dobroth, L. Erwin: *Causes of Edge Beads in Cast Films, Polymer Engineering and Science*, Vol.26 No.7, Ausgabe: 04/1986
7. H. Ito, M. Doi, T. Isaki, M. Takeo: *A Model of Neck-in Phenomenon in Film Casting Process*, Journal of the Society of Rheology, Vol.31 No.3, Ausgabe: 01/2003
8. G. Menges, E. Haberstroh, W. Michaeli, E. Schmachtenberg: *Werkstoffkunde Kunststoffe*, Carl Hanser Verlag München Wien, 2002
9. K. Selvanayagam, T. Götz: *Optimal Control of Film Casting Processes*, AGTM Report 273, 2008

O-01-464

ANALYSIS OF FLOW BEHAVIOR IN DIE EXTRUSION PROCESS FOR VARIOUS INSIDE DIE DESIGNS

M.-Y. Lyu[1*], S. H. Choi[2], and H. J. Kim[2]
[1] *Seoul Nat'l Univ. of Science and Technology, Seoul, S. Korea, *mylyu@seoultech.ac.kr*
[2] *Hankook Tire Central R&D Center, Daejeon, S. Korea, sh.choi@hankooktire.com;187000357@hankooktire.com*

Abstract - Tire for automobile is consisted of several layers and the most outer layer is tread. The tread is manufactured by co-extrusion of different rubber compounds. The die for tread extrusion is composed of three parts, insert die, preformer die, and final die. The final die profile is designed according to required tread profile. Extruded profile of tread is influenced by not only final die profile but insert die and performer die. Computer simulation has been performed to investigate flow behaviors inside die. Velocity distribution, velocity vector, pressure distribution, shear rate, and wall stress for various inside die design have been compared. Velocity profile at the exit of final die was influenced by insert die and preformer die design. The simulation results can be utilized for optimum die design and trouble shooting of tread extrusion.

Keywords: Tire tread, Insert die, Final die, Extrusion, Velocity profile, Pressure distribution

Introduction

Tire for automobile is consisted of several layers and the most outer layer is tread. The tread is manufactured by co-extrusion of different rubber compounds [1,2]. The die for tread extrusion is composed of three parts, insert die, preformer die, and final die. The final die profile is designed according to required tread profile. Extruded profile of tread is influenced by not only final die profile but insert die and performer die [2,3]. Tearing of edge of extruded tread has been reported during tread extrusion process. Computer simulation has been performed to investigate rubber flow behaviors inside die. Velocity, pressure, and wall shear stress have been analyzed to perform trouble shooting.

Computer simulation

Problem in tread extrusion

Fig.1 shows tearing problem occurred during tread extrusion.

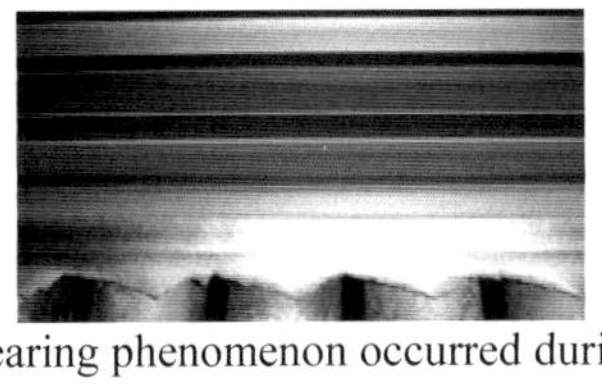

Fig. 1 Tearing phenomenon occurred during tread extrusion

Simulation model

Fig. 2 shows the co-extrusion die consisted of insert die, performer, and final die. Fig. 3 shows modified extrusion die to eliminate tearing problem.

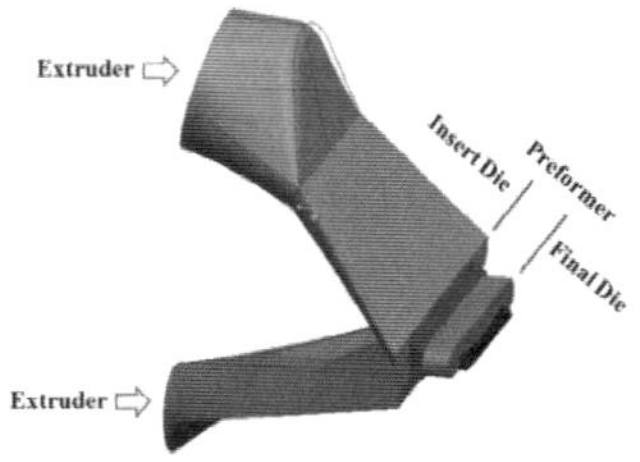

Fig. 2 Co-extrusion die for tread

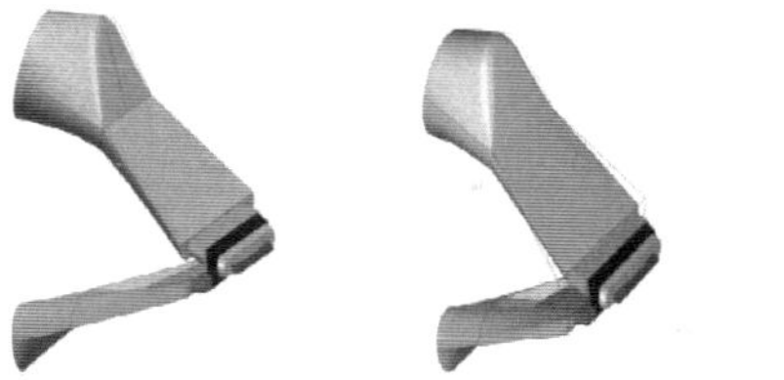

(a) Improved performer (b) Reduced insert channel opening

Fig. 3 Modified co-extrusion dies for tread

Results and discussions

Pressure distribution

Fig. 4 shows pressure distributions for three die models. Original die design, improved performer, and reduced insert channel opening showed 4.68e7 Pa, 4.76e7 Pa, and 4.92e7 Pa of pressure drops respectively from die inlet to die exit.

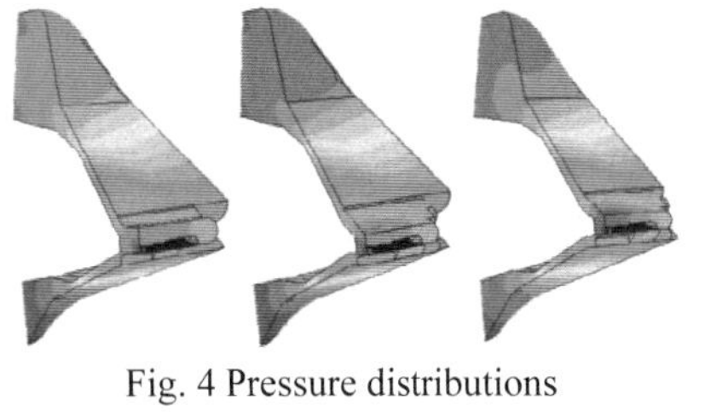

Fig. 4 Pressure distributions

Temperature distribution

Fig. 5 shows temperature distributions for three models. Fig. 6 shows average temperature profile along the flow direction in the die. Original die showed the highest temperature among the tested design. Two modified dies showed similar temperature profiles.

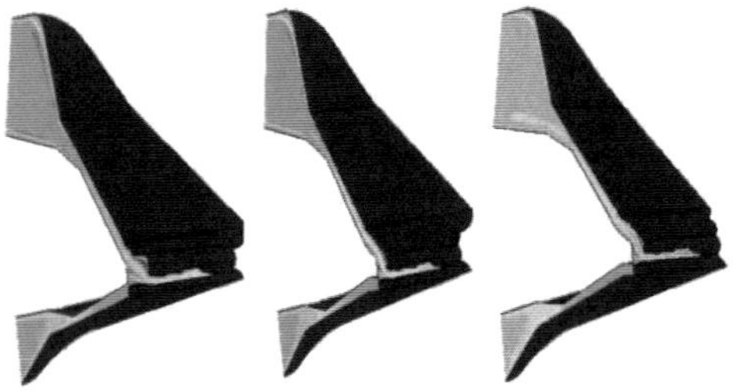

Fig. 5 Temperature distributions

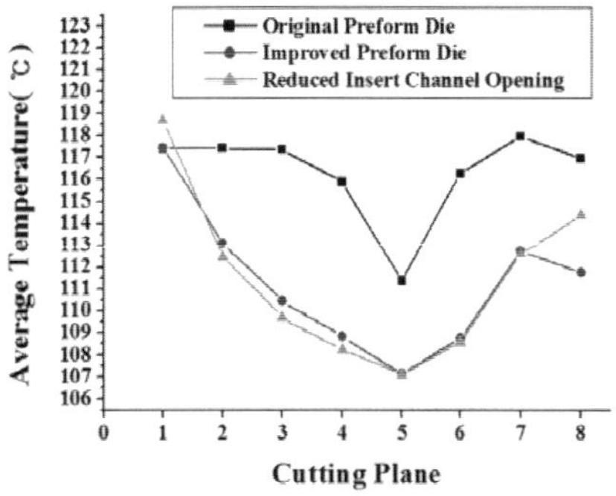

Fig. 6 Average temperature profiles inside die along the flow direction

Velocity at die exit

Fig. 7 shows velocity vector at the die exit for three tested die designs. Original die showed most non-uniform velocity vector and reduced channel opening die showed most uniform velocity vector.

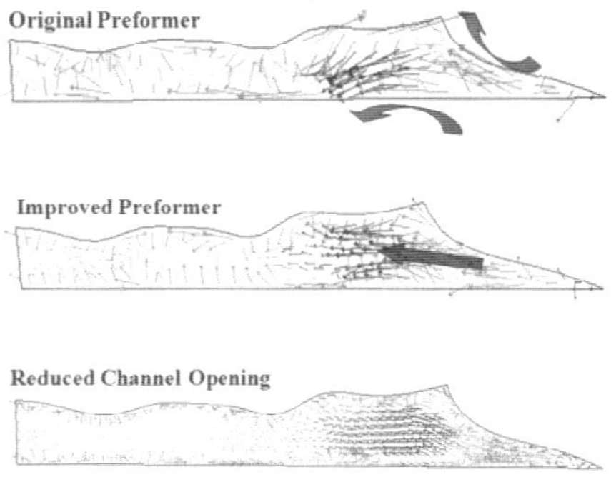

Fig. 7 Velocity vector at the die exit

Wall shear stress at die exit

Fig. 8 shows wall shear stress at the die exit for three tested die designs. Original die showed higher wall shear stress compared with other die designs. Reduced channel opening die showed the smallest and the most uniform wall stress.

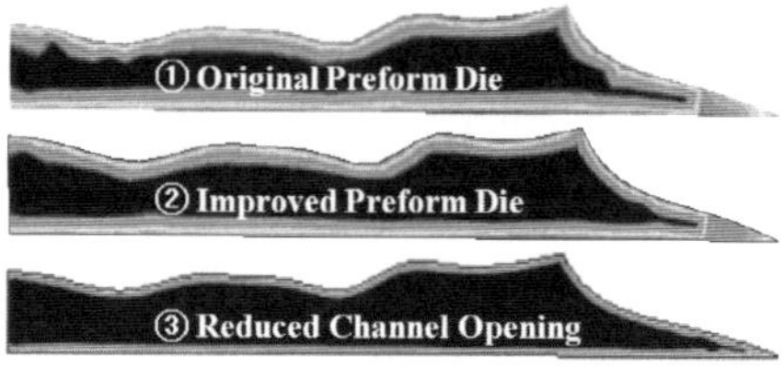

Fig. 8 Wall shear stress at the die exit

Conclusions

Flow behaviors of rubber in the die have been investigated to examine tearing phenomenon of extruded tread. Three die designs have been suggested and comparing flow and thermal characteristics. The most uniform velocity distribution at the die exit showed the lowest wall shear stress and subsequently eliminate tearing problem.

Acknowledgements

This work (research) is financially supported by the Ministry of Knowledge Economy (MKE) and Korea Institute for Advancement in Technology (KIAT) through the Workforce Development Program in Strategic Technology. This work (research) is financially supported by the Ministry of Knowledge Economy (MKE) through New Growth Power Equipment Competitiveness Reinforcement Program.

References

1. J.L. White, Rubber Processing Technology, Materials and Principles, Hanser Publishers, Munich (1995).
2. C.D. Han, Rheology in Polymer Processing, Academic Press, NY (1976).
3. Tim A. Osswald and Georg Menges, Materials Science of Polymers for Engineers, Hanser, New York(1995).

TEMPERATURE EVOLUTION DURING FILAMENT DEPOSITION IN 3D FREE FORM EXTRUSION

F. M. Duarte[1*], S F. Costa[1] and J. A. Covas[1]

[1]I3N/IPC - Institute for Polymers and Composites, Department of Polymer Engineering, University of Minho, Campus Azurém, 4800-058 Guimarães, Portugal – fduarte@dep.uminho.pt; fcosta@gmail.com; jcovas@dep.uminho.pt

Abstract - The evolution in time of the filament temperature during deposition in FFE is measured and compared with computational predictions. The results validate the theoretical model as long as accurate thermal properties are input, particularly the coefficient of heat transfer by convection. A procedure to determine the coefficients of heat transfer by contact and support between filaments was successfully validated.

Keywords: Free Form Extrusion, Process modeling, Rapid prototyping, Extrusion, Heat transfer.

Introduction

The manufacture of 3D parts by extrusion involves the deposition of a semi-molten filament onto successive horizontal planes in the vertical direction, according to a sequence that is dictated by the part geometry. The method is an adaptation of rapid prototyping techniques, where a commercial pre-extruded ABS filament and heater are replaced by a (prototype) mini-extruder capable of processing a wide range of polymer systems. The deposition pathway and operating conditions will determine the mechanical performance of the part: upon contact, adjacent filaments must be sufficiently hot to ensure adhesion, but then should quickly cool down to avoid excessive deformation due to gravity and weight of the filaments above. The cooling stage has been modeled taking into consideration the local heat fluxes with the support, the surrounding environment, and with adjacent filaments, either previously deposited, or deposited later [1-3].
In this work, the evolution of filament temperature during deposition is followed experimentally using an infrared camera and the results are compared with the computational predictions.

Experimental

Material.
An ABS polymer (ABS FA 4475 Bordeaux – Poliversal) with the properties shown in Fig 1 was used. The pellets were ground in a Retsch mill using liquid nitrogen.

Table 1 – Material properties of ABS FA 4475.

Property	Value
Density (kg/m^3)	1050
Thermal conductivity (W/m· ºC)	0.18
Specific Heat (J/kg· ºC)	2200

The emissivity was determined heating a small amount of material in a hot plate to various temperatures. The surface temperature was measured with a fast reading type K thermocouple and by an infrared FLIR SC640 camera. The emissivity of the material was then computed, by adjusting the two measured values using the ThermaCAM Researcher Pro 2.8 SR-2® software. The value of 0.85±0.01 was obtained.

Equipment layout.
The experimental layout is illustrated in Fig. 1. It comprises a prototype mini extruder [4] coupled with a gear pump (Extertal Lenze 32699) and a 1mm die extruding the filament vertically downwards over a computer controlled 3D table (0.05 mm accuracy). The filament temperature evolution is recorded by the infrared camera, positioned above the extrusion head, the data being processed by the CAM Researcher Pro 2.8 SR-2® software.

Figure 1 – Experimental setup.

Heat transfer coefficients.

The **Coefficient of heat transfer by convection, h,** was determined using the following procedure: i) measure the temperature evolution in a portion of filament where only heat exchange by convection exists – a slot on the support table was made, as pictured in Fig. 2); ii) measure the cross section of the filament in the previously select point; iii) compute the coefficient of heat transfer by convection using the correlation of Hilpert [5] considering the air temperature near the point and the measured cross section of the filament; iv) in practice, the environment temperature near to the die is higher than at a distance, hence two values of the coefficient of heat transfer by convection were computed.

In the case of the **Coefficients of heat transfer by contact with the support and between filaments**: i) measure the filament cross-section; ii) using the coefficient of heat transfer by convection determined above, adjust the heat transfer coefficient by contact until the theoretical and experimental temperatures coincide; iii) repeat the procedure for two filaments in contact, in order to determine the heat transfer coefficient by contact.

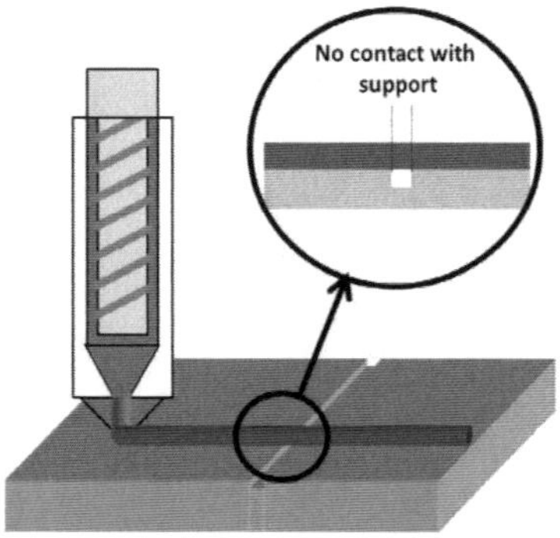

Figure 2 – Slot machined in the table support

Results and Discussion

Fig. 3 shows a typical experimental temperature evolution in time for a filament extruded at 200°C. The thermal images up to 7.2 seconds correspond to the extrusion/deposition stage whereas the others were obtained during cooling.

Figures 4 and 5 show the experimental and predicted temperature evolution for one filament extruded at 200°C and 220°C, respectively, at a point located 60 mm away from the extruder. Two values of h (52 and 62 W/m^{2o}C) were assumed. The agreement between the two types of data is very good, particularly when the highest value of h is taken in (value determined close to the extruder). As expected, the contact area with the support increases with temperature (from 10% to 14% of the perimeter).

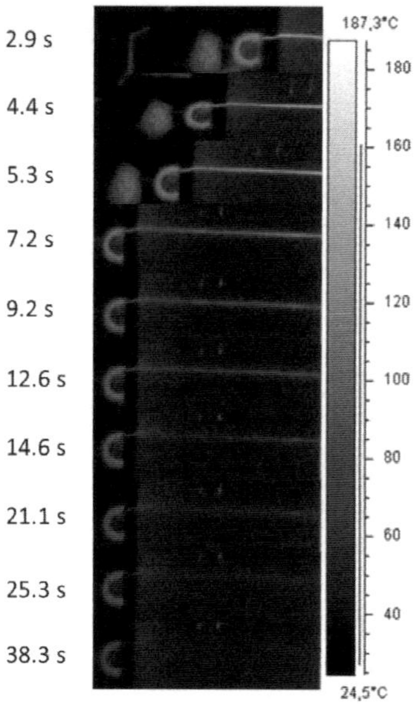

Figure 3 – Thermal images of temperature evolution with time for one filament extruded at 200°C.

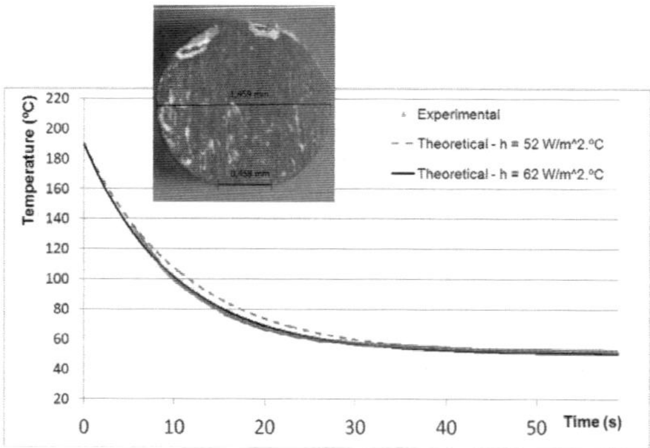

Figure 4 – Temperature evolution at a point located 60 mm away from the extruder (set T=200°C).

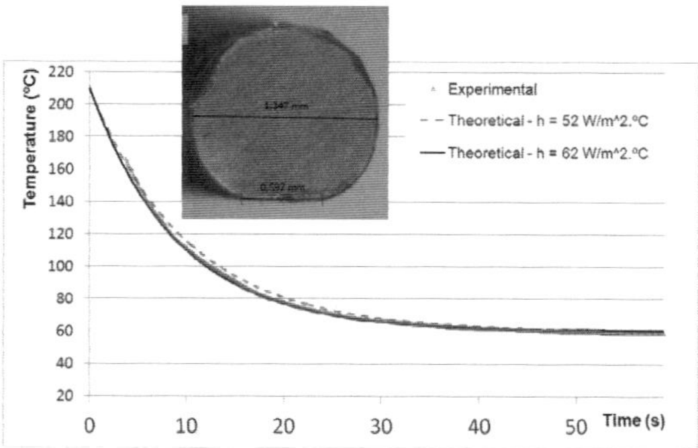

Figure 5 Temperature evolution at a point located 60 mm away from the extruder (set T=220°C).

Figure 6 refers to the deposition of two adjacent filaments and presents the temperature evolution of the second filament at the same location as before, for an extrusion temperature of 200ºC (h=62W/m^{2}ºC). Again, the fit between predictions and experiments is excellent.

Conclusions

A code predicting the temperature evolution of a filament during deposition by Free Form Extrusion was validated by comparing measurements with the corresponding calculations.

The experiments evidenced that the contact area with the support or with other filaments changes with processing conditions. Also, procedures to estimate the values of heat transfer coefficients were developed.

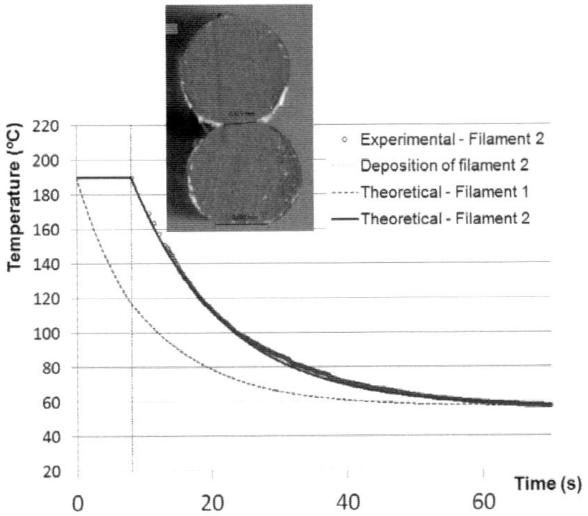

Figure 6 – Temperature evolution of the second filament at a point located 60 mm away from the extruder (set T=200ºC).

References

1. S. F. Costa; F.M. Duarte; J.A. Covas *Materials Science Forum* 2010, *636-637*, 833.
2. S. F. Costa; F.M. Duarte; J.A. Covas *Int J Mater Form* 2008, *1*, 703.
3. S. F. Costa; F.M. Duarte; J.A. Covas, in MATLAB - *A Ubiquitous Tool for the Practical Engineer*, Intech, 2011.
4. J. A. Covas; P. Costa. *Polym Test*, 2004, *23*, 763.
5. J. P. Holman, *Heat transfer*, McGraw-Hill, New York, 1983.

O-01-494

VISCOUS HEATING OF POLYMER MELTS IN PLASTIC PIPE EXTRUSION

Chanyut Kolitawong[1,*], **A. Jeffrey Giacomin**[2] **and Udomkiat Nontakaew**[1]
[1]*Department of Mechanical and Aerospace Engineering, Polymer Research Center, King Mongkut's University of Technology North Bangkok, Thailand – ckw@kmutnb.ac.th, unk@kmutnb.ac.th*
[2]*Department of Mechanical Engineering, Rheology Research Center, University of Wisconsin, Madison, USA – giacomin@wisc.edu*

Abstract - We use bipolar cylindrical coordinates to study the temperature rise from viscous heating in a power law fluid in a pressure-driven axial flow between eccentric cylinders. We begin with the energy equation in bipolar cylindrical coordinates, then obtain a dimensionless algebraic analytic solution for the temperature profile under a steady, laminar, incompressible and fully developed flow for an adiabatic core and an isothermal outer cylinder. We consider this particular pairing of thermal boundary conditions to be closest to the extrusion of plastic pipe from an eccentric annular die. We find that the dimensionless temperature profile depends upon the radius ratio of the inner to outer cylinders, the eccentricity, the angular position, and the power law index *n*. The temperature is a strong function of the gap between the cylinders.

Keywords: Plastic pipe extrusion, plastics tubing, viscous heating, viscous dissipation, sag, pipe extrusion die.

Introduction

In plastic pipe manufacture, if polymer were extruded from a concentric annular die, the pipe wall thickness would not be uniform, and specifically, thicker at the bottom [1, 2]. This is because of the gravity flow of the molten plastic, inside the pipe, occurring during its long residence in the cooling chamber. This is why the die mandrel is normally displaced eccentrically downward, so that the extrudate entering the cooling chamber is thicker on top.

In curved hose manufacture, the hose is suspended freely from the die [3]. Curvature of the host is succeeded by making the die centerpiece eccentric. The extrudate bends away from the thicker side. To obtain a particular hose shape, the die eccentricity must be specifically programmed.

Many researchers have attacked concentric annular die problems analytically and numerically [4]. For instance, Bird *et al.* [5, 6], Michaeli [7],and Baird and Collias [8] extensively review concentric axial annular flow of Newtonian and power-law fluids. Fredrickson and Bird [9] solved analytically for the axial flow of power-law (and Bingham) fluids through concentric annuli. Later, Guckes [10] studied the same fluids numerically through eccentric annuli for large gaps. Shah [11] extensively reported laminar convective heat transfer in concentric and eccentric ducts.

Later, Kolitawong and Giacomin [12] mapped the eccentric cylinder cross section in Figure 1 into bipolar cylindrical coordinates in Fig. 2 [5]. They analytically obtain the dimensionless axial velocity profile between eccentric cylinders for power law liquids.

The temperature rise during plastics pipe and curved hose manufacture may cause plastic degradation. Uneven melt temperatures may also distort the velocity profile, flowing faster than otherwise predicted, where the fluid is hotter. Here, we are interested in viscous heating of a power-law fluid flowing axially, under a pressure gradient, through an eccentric annulus. To study the heat transfer between the eccentric cylinders, the energy equation is first written in bipolar cylindrical coordinates [13]. The temperature distribution in the fluid between the eccentric cylinders is then determined analytically from this energy equation for constant heat fluxes at the inner walls. For this, the temperature of the die mandrel is often uncontrolled, leaving it with zero heat flux at steady state.

Analysis

In bipolar cylindrical coordinates, ξ and θ represent two orthogonal sets of circles. Figs. 1 and 2 show the inner and outer cylinders at constant ξ_1 and ξ_2. The rectangular and the bipolar cylindrical coordinates are related by [5, 14]:

$$x = \frac{a\sinh\xi}{\cosh\xi - \cos\theta} \qquad (1)$$

$$y = \frac{a\sin\theta}{\cosh\xi - \cos\theta} \qquad (2)$$

$$z = \zeta \qquad (3)$$

where $\theta \in [0,\ 2\pi)$, $\xi \in (-\infty,\ \infty)$ and $\zeta \in (-\infty,\ \infty)$. Kolitawong and Giacomin [12] derived the following dimensionless axial velocity profile between eccentric cylinders for power law liquids in the bipolar cylindrical coordinates,

$$v_{\zeta p}(\kappa, \beta) = -\frac{n}{n+1}\left(\frac{d(1-\varepsilon\cos\beta)}{2R_2}\right)^{1+\frac{1}{n}}\left(|\kappa|^{1+\frac{1}{n}} - 1\right) \qquad (4)$$

where n is defined in Eq. (8); $d = R_2 - R_1$, the annular gap; R_2 and R_1, the outer and inner radii of the annulus, respectively. β is the bipolar angular coordinate measured from the x-axis as shown in Fig. 1, and ε, the dimensionless eccentricity:

$$\varepsilon \equiv \frac{e}{d}, \qquad 0 < \varepsilon < 1 \tag{5}$$

They define the dimensionless axial velocity profile such that,

$$v_{\zeta p} \equiv \frac{v_\zeta}{\left[-\frac{R_2^{n+1}}{m}\frac{dP}{d\zeta}\right]^{\frac{1}{n}}} \tag{6}$$

where m is defined in Eq. (8); v_ζ, the dimensional velocity profile of the fluid between the eccentric cylinders; $dP/d\zeta$, the pressure gradient in the axial direction. They also introduce the dimensionless bipolar coordinate,

$$\kappa \equiv \frac{\xi - \xi_0}{\xi_1 - \xi_0} \tag{7}$$

where ξ represents a set of circles located between inner and outer circles, respectively at ξ_1 and ξ_2. The power law viscosity is defined such that:

$$\eta = m\dot{\gamma}^{n-1} \tag{8}$$

where n is the power law index and m is the consistency index. The magnitude of the rate of strain tensor is:

$$\dot{\gamma} \equiv \sqrt{\frac{1}{2}\left(\underline{\underline{\dot{\gamma}}} : \underline{\underline{\dot{\gamma}}}\right)} \tag{9}$$

where the rate of strain tensor is [5]:

$$\underline{\underline{\dot{\gamma}}} = (\nabla \underline{v}) + (\nabla \underline{v})^T \tag{10}$$

From the energy equation written in bipolar cylindrical coordinates [5, 14] for a steady, laminar, incompressible, fully developed flow, we assume that there is only axial velocity in the ζ direction, i.e. $v_\zeta = v_\zeta(\xi,\theta)$, $v_\xi = v_\theta = 0$, and $T(\xi,\theta)$. For this case, the energy equation simplifies to,

$$0 = k\left\{\frac{1}{h}\frac{\partial}{\partial \xi}\left(\frac{1}{h}\frac{\partial T}{\partial \xi}\right) + \frac{1}{h}\frac{\partial}{\partial \theta}\left(\frac{1}{h}\frac{\partial T}{\partial \theta}\right)\right\} - \left\{\tau_{\xi\zeta}\left(\frac{1}{h}\frac{\partial v_\zeta}{\partial \xi}\right) + \tau_{\theta\zeta}\left(\frac{1}{h}\frac{\partial v_\zeta}{\partial \theta}\right)\right\} \tag{11}$$

The shear stress can be written in terms of the velocity gradients such that [12]

$$\tau_{\xi\zeta} = -\eta\frac{1}{h}\frac{\partial v_\zeta}{\partial \xi}, \tag{12}$$

$$\tau_{\theta\zeta} = -\eta\frac{1}{h}\frac{\partial v_\zeta}{\partial \theta} \tag{13}$$

where η is defined in Eq. (8); v_ζ, the velocity profile for the flow between the eccentric cylinders. For small dimensionless eccentricity, $\varepsilon \ll 1$, *i.e.* $\frac{\partial v_\zeta}{\partial \theta} \ll \frac{\partial v_\zeta}{\partial \xi}$, $\frac{\partial T}{\partial \theta} \ll \frac{\partial T}{\partial \xi}$, we substitute Eqs. (12) and (13) into Eq. (11) and rearrange to get,

$$\frac{\partial^2 T}{\partial \xi^2} = -\frac{m}{k}\left(\frac{1}{h}\right)^{n-1}\left|\frac{\partial v_\zeta}{\partial \xi}\right|^{n-1}\left(\frac{\partial v_\zeta}{\partial \xi}\right)^2 \tag{14}$$

Introducing the dimensionless variable from Eq. (**7**), into Eq. (14):

$$\frac{\partial^2 T}{\partial \kappa^2} = -\frac{m}{k}\left(\frac{2}{hc}\right)^{n-1}\left|\frac{\partial v_\zeta}{\partial \kappa}\right|^{n-1}\left(\frac{\partial v_\zeta}{\partial \xi}\right)^2 \tag{15}$$

Rearranging Eq. (6) for dimensional velocity, v_ζ, and substituting into Eq. (15), one gets,

$$\frac{\partial^2 T}{\partial \kappa^2} = -\frac{m}{k}\left(\frac{2}{hc}\right)^{n-1}\left[-\frac{R_2^{n+1}}{m}\frac{dP_\zeta}{d\zeta}\right]^{\frac{n+1}{n}}\left|\frac{\partial v_{\zeta p}}{\partial \kappa}\right|^{n-1}\left(\frac{\partial v_{\zeta p}}{\partial \kappa}\right)^2 \tag{16}$$

Furthermore, we can rearrange Eq. (16) such that,

$$\frac{\partial^2 T^*}{\partial \kappa^2} = -\left|\frac{\partial v_{\zeta p}}{\partial \kappa}\right|^{n-1}\left(\frac{\partial v_{\zeta p}}{\partial \kappa}\right)^2 \tag{17}$$

where the dimensionless temperature profile, T^*, is

$$T^* = \frac{T}{\lambda_0} \quad . \tag{18}$$

where,

$$\lambda_0 \equiv \frac{m}{k}\left(\frac{2}{hc}\right)^{n-1}\left[-\frac{R_2^{n+1}}{m}\frac{dP}{d\zeta}\right]^{\frac{n+1}{n}} \tag{19}$$

We know that $\frac{\partial v_{\zeta p}}{\partial \kappa}$ is positive when $-1 \le \kappa < 0$, and $\frac{\partial v_{\zeta p}}{\partial \kappa}$ is negative when $0 < \kappa \le 1$. Thus, we differentiate Eq. (4) with respect to κ and then substitute into Eq. (17) to get,

$$\frac{\partial^2 T^*}{\partial \kappa^2} = -\lambda_1|\kappa|^{1+\frac{1}{n}} \tag{20}$$

where λ_1 is always real and positive and defined such that,

$$\lambda_1 \equiv \left[\frac{d(1-\varepsilon\cos\beta)}{2R_2}\right]^{\frac{(n+1)^2}{n}} . \tag{21}$$

Assuming that the outer radius R_2 (κ = -1), has a constant temperature T_2, corresponding to $T^* = T_2^*$, and the inner radius R_1 ($\kappa = 1$) has a constant heat flux q_1, *i.e.* heat flowing out at a constant rate.

$$k\left.\frac{\partial T}{\partial \xi}\right|_{\xi=\xi_1} = q_1 \tag{22}$$

Eq. (22) can be written in dimensionless form;

$$\left.\frac{\partial T^*}{\partial \kappa}\right|_{\kappa=1} = \frac{cq_1}{2k\lambda_0} \tag{23}$$

By integrating Eq. (20) twice and by applying these boundary conditions, we can shown that,

$$T^* = \lambda_2\left(1-|\kappa|^{3+\frac{1}{n}}\right)+\lambda_3(1+\kappa)+T_2^* \tag{24}$$

and, by rearranging Eq. (24),

$$\Delta^* = \lambda_2\left(1-|\kappa|^{3+\frac{1}{n}}\right)+\lambda_3(1+\kappa) \qquad \Delta^* = T^* - T_2^* \tag{25}$$

where, Eq. (25) is the main result of this paper, and

$$\lambda_2 = \frac{\lambda_1}{\left(2+\frac{1}{n}\right)\left(3+\frac{1}{n}\right)}, \tag{26}$$

$$\lambda_3 = \frac{\lambda_1}{\left(2+\frac{1}{n}\right)} + \frac{cq_1}{2k\lambda_0} \tag{27}$$

The mandrel temperature is usually uncontrolled due to space limitation. This leads to adiabatic boundary condition after a period of time. Thus, the constant heat flux q_1 can be dropped from Eq. (27).

Results and Discussion

Figs. 3 to 4 show that the dimensionless temperature (Δ^*), in general, increases with the angular distance from the smallest gap, β, and the power law index, n, but decreases with the dimensionless radius ratio, φ. The temperature rises at the core since heat accumulates at the inner wall according to the adiabatic heat flux. Moreover, the more eccentric the die (the higher the ε), the greater the temperature difference between the fluid at the smallest gap ($\beta = 0°$) and the biggest gap ($\beta = 180°$). This is because of the convective flow in the eccentric die. In other words, the dimensionless temperature is proportional to the velocity in the die.

The dimensionless temperatures also decrease by two decades [from $O(10^{-4})$ to $O(10^{-6})$] when the radius ratio (φ) increases from 0.80 to 0.90. When the radius ratio increases, the inner and outer radii approach one another, and so the gap is very small all around the die. In this case, the convective effects in the die diminish, which is why the fluid temperature decreases as the radius ratio increases.

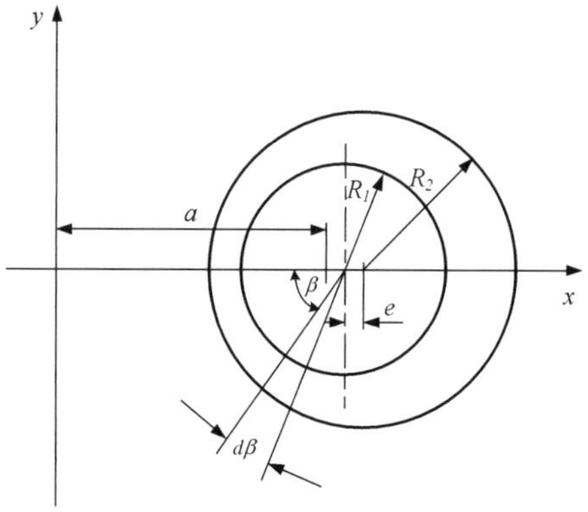

Fig. 1 Cross section of an eccentric annulus. R_1 and R_2 are the inner and outer cylinder radii; and e is its eccentricity.

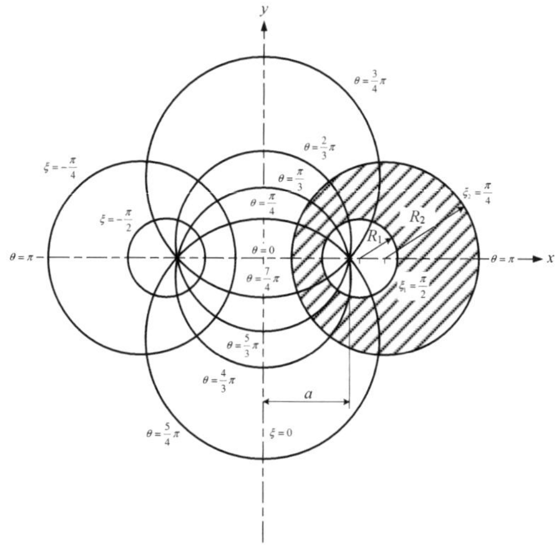

Fig. 2 Bipolar cylindrical coordinates showing the shaded die cross section with the confocal length, a [5].

Conclusion

The solution reveals that the dimensionless temperature distribution is a function of the dimensionless radius ratio, φ, the dimensionless eccentricity, ε, the angular distance from the smallest gap, β, the power law index, n, and is a strong function of, the dimensionless gap, κ. For the adiabatic inner wall, the temperature rises in the gap due to viscous dissipation toward the zero heat flux at the steady state condition. This temperature rise increases if the gap is increased. Moreover, the temperature rise also increases with the power law index, n, to reach the Newtonian temperature profile at $n = 1$.

Acknowledgement

The authors gratefully acknowledge the Thailand Research Fund (TRF) for funding support.

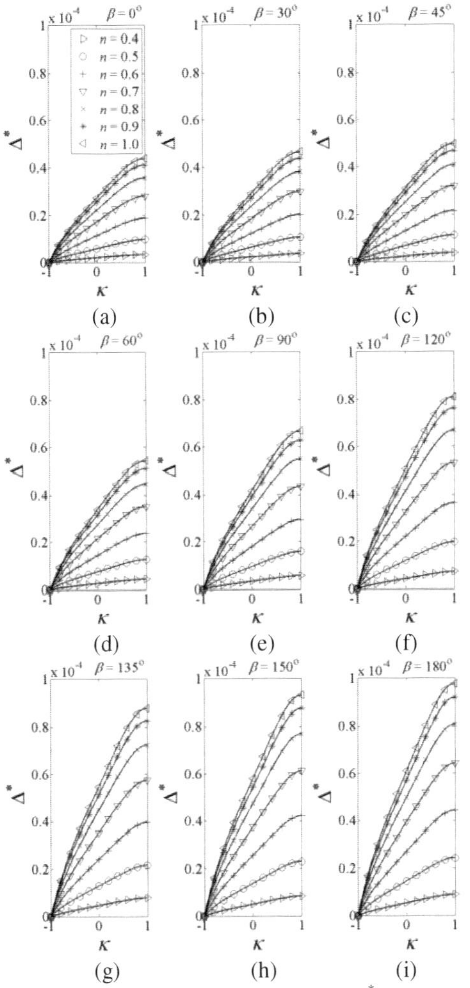

Fig. 3 Dimensionless temperature profiles Δ^* versus the dimensionless gap κ using φ = 0.80 and the dimensionless eccentricity, ε = 0.10, at various angles β and power law indexes n.

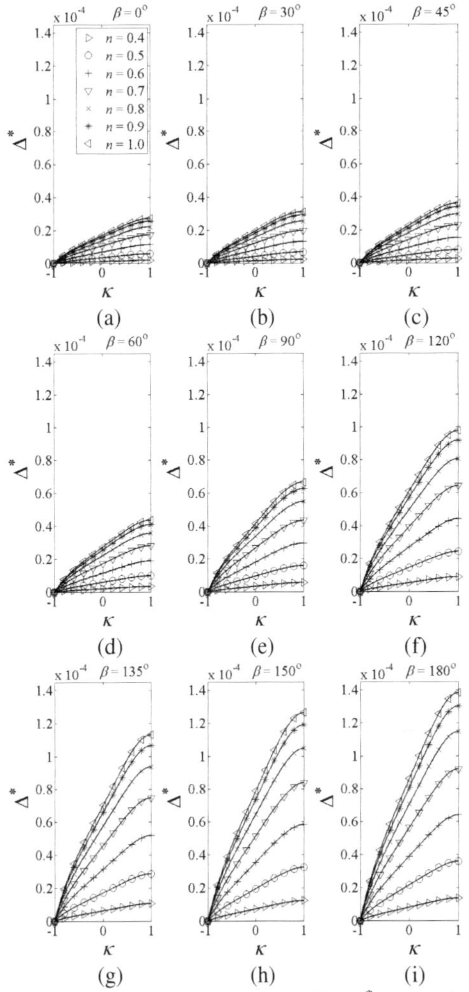

Fig. 4 Dimensionless temperature profiles Δ^* versus the dimensionless gap κ using φ = 0.80 and the dimensionless eccentricity, ε = 0.20, at various angles β and power law indexes n.

References

1. D.N. Githuku, and A.J. Giacomin, *Int. Polym. Proc.*, **VII**(2), 140 (1992).
2. D.N. Githuku and A.J. Giacomin, *J. Eng. Mater.-T. ASME*, **115**(4), 433 (1993).
3. L.A. Goettler, A.J. Lambright, R.I. Leib, and P.J. DiMauro, *Rubber Chem. Technol.*, **54**(2), 277 (1981).
4. W.T. Snyder, *AIChE J.*, **9**(4), 503 (1963).
5. R.B. Bird, R.C. Armstrong and O. Hassager, *Dynamics of Polymeric Liquids. Vol. I: Fluid Mechanics*, 2nd Ed., Wiley, New York (1987), p.181, 595.
6. R.B. Bird, W.E. Stewart, and E.N. Lightfoot, *Transport Phenomena*, 2nd Ed., John Wiley & Sons, Inc., New York (2007), p.53, 64.
7. W. Michaeli, *Extrusion Dies for Plastics and Rubber*, 2nd Ed., Hanser Publishers, New York (1992), p.58.
8. D.G. Baird and D.I. Collias, *Polymer Processing: Principles and Design*, John Wiley & Sons, Inc., New York (1998), p.14.
9. A.G. Fredrickson, and R.B. Bird, *Ind. Eng. Chem.*, **50**, 347 (1958).
10. T.L. Guckes, *Trans. ASME(B)*, **97**, 498 (1975).
11. S. Kakac, R.K. Shah, and W. Aung, Handbook of Single-Phase Convective Heat Transfer, John Wiley & Sons, New York (1987), ch.3.
12.C. Kolitawong and A.J. Giacomin, *Polym.-Plast. Technol. Eng.*, **40**(3), 363 (2001).
13. C. Kolitawong, N. Kananai, A.J. Giacomin, and U. Nontakaew, *J. Non-Newton. Fluid*, **166**, 133 (2011).
14. P. Moon and D.E. Spencer, *Field Theory for Engineers*, D. Van Nostrand Company, Inc., New York (1961), p.361.

O-01-512

REAL TIME OPTICAL METHODS TO MONITOR THE POLYMER MELT FLOW DURING EXTRUSION

S. V. Canevarolo[1]*, P. A. R. Muñoz[2], L. S. Hamester[2] and A. M. C. Santos[2]

[1]*Dep. Engenharia de Materiais, DEMa, Univ. Federal de São Carlos, UFSCar, São Carlos, Brazil - caneva@ufscar.br;*
[2]*Graduate Program in Materials Science and Engineering, Univ. Federal de São Carlos, Brazil – pabloarm@gmail.com, leticia.cosmos@gmail.com, adillysmarcelo@hotmail.com.*

Abstract - This work shows the use of in-line optical detection (turbidity, rheo-polarimetry, colorimetry,..) to quantify in real-time the melt flow behavior of the polymer during extrusion. The optical apparatus, fitted at the extruder exit, consists of an optical detector set on a slit-die, temperature and pressure measurement and control, on-off valve and software that controls all processing variables and collects data from the detectors. The melt turbidity, a variable dependent of the droplets size and shape, is recorded during polymer blend extrusion. It can gives information upon the droplets breakup and/or coalescence and the deformation/recovery hysteresis of the dispersed second phase. In-line rheo-polarimetry, measuring the flow-birefringence directly in the main extrusion stream, can gives in real time the level of the polymer chain orientation via measuring the intensity of the transmitted polarized light. One of the measurement protocols during the blends extrusion is a continuous reduction in the flow pumping pressure allowing the optical measurement to be taken at varying degrees of molecular orientation. Changes in the crossed-polarized transmitted light intensity follow the expected sine-square curve, in which the degree of molecular orientation is the same at the maximum of each curve. Shifting of the peak's maxima along the pressure axis indicates changes in the degree of the polymer matrix chains orientation. More incompatible polymers blends showed a larger deviation as regard to matrix deformation. A similar fact found in the literature in mechanical measurements is the so-called apparent interfacial slip at polymer-polymer interfaces, what reflects in viscosity drop of immiscible blends. In-line colorimetry is also another in-line optical technique that can be measured in real time, for instance to quantify changes in the melt flow color. Preliminary results showed that the in-line colorimetry detector developed can quantify color changes in a discrete manner.

Keywords: Real time monitoring, polymer extrusion, optical properties, rheo-polarimetry.

Introduction

The final properties of a commercial plastic part are intrinsically dependent on the polymer system used including polymers, phase morphology and components content. The morphology is mostly set during the mixing procedure which usually is done in an extruder. So the characterization of the polymer morphology already at the melt stage is of great importance. Better still if it can be done in real time applying in-line characterization techniques.

In the past 15 years our research group has been creating new optical devices to characterize in real time the extrusion process [1-6, 8]. These detectors are made of a slit-die fitted at the extruder exit from where the molten polymer flows, with a pair of transparent windows that allows a light beam to pass orthogonally through the molten flow. Following changes in the transmitted light intensity one is able to quantify in-line the turbidity, which is a function of the type, concentration and particle size and shape of the second phase present in the flow. With the interposition of a pair of crossed polarizers in the optical path, i.e. in-line rheo-polarimetry, is possible to evaluate quantitatively the flow-birefringence, which is a function of the degree of the polymer matrix orientation. Finally using an in-line colorimeter changes in the color due to thermal and mechanical degradation can be used to quantify these properties.

Experimental

Materials

Polypropylene PP HP500N from Braskem, MFR = 11g/10min; High density polyethylene HDPE BT003 from Braskem MFR = 0.33g/10min, Polymethylmethacrylate PMMA from Unigel, MFR = 16.4g/10min, and a general purpose polystyrene PS N1841 from Innova MFR = 11g/10min, all used as received. Yellow acid 49 (MM 426 g/mol) from Quimanil, used as received.

Hardware

The extruder is a Werner & Pfleiderer ZSK 30 twin-screw fitted with a slit-die at its exit, with transparent windows holding an optical detector, which measures in real time the melt turbidity, rheo-polarimetry and colorimetry of the polymer flow. A manual on/off valve is added after the die. An electronic device was fitted inside the extruder panel in order to get input and output signals from the processing parameters. More information can be obtained elsewhere [1-6, 8]. All signals are in real time collected, converted, screen monitored and recorded by the software, developed in

a LabView 8.6 platform, which also controls the die-head pressure via altering the screw speed.
A bench apparatus for in-line color measurement build up in Teflon associated with a sunken pump was used as validation technique. The data were collected in-line using software developed in Lab View 8.6. A RGB LED was used as light source and the light transmission intensity was measured for each color.

Processing operation:

In-line melt turbidity
The three polymer mixtures being PP/PS, HDPE/PS and PS/HDPE, all having 0.1% w/w of the second phase, were processed in the twin-screw extruder at constant operational conditions: temperature profile @ 240°C, screw rotation speed of 70 rpm, feeding rate of 2.0 kg/h and a simple screw profile, built only with conveying screw elements.

In-line melt rheo-polarimetry
Any polymer system submitted to deformation may orient its polymeric chains along the flow what brings about an anisotropic phenomenon, manifested as flow-birefringence [7, 8]. It is well established that the crossed-polarized transmitted light intensity follows a sine-square behavior depending on the degree of chains deformation. The signal intensity in this quadratic relation is measured here in terms of normalized voltage as a function of the optical path difference OPD, also affected by stresses distribution [8]. Herein, this procedure is employed to follow in real time the influence of a disperse phase in incompatible polymer blend systems. Polystyrene resin was used as the vehicle to disperse polymethylmethacrylate, polyethylene and polypropylene at a maximum of 1 w/w% to avoid coalescence effect [9] (PS/PMMA- 0,15; 0,25; 0,50; 0,75 and 1 w/w% of PMMA; PS/HDPE - 0,15 ; 0, 25 ; 0,5 w/w% of HDPE and PS/PP- 0,038; 0,075; 0,15 and 0,25 w/w% of PP) in a twin-screw extruder Werner & Pfleiderer ZSK 30 at steady-state. Screw profile is made up only of a melt seal (2 x KB45/5/42 - Kneading blocks + 20/10 LH) and a distributive mixing zone (2 x KB45/5/28), the remainder formed only by conveying elements. The processing conditions were 250 rpm for screw rotation and temperature kept constant at 260 °C in the whole extruder barrel. Real time monitoring is carried out during the so-called 'stop feeding' extrusion method. Here, the material is fed at a high throughput (10Kg/h), enough to reach high shear stresses, being further ceased so that the extruder 'self-empties' making the shear stress in the slit-die land to reduce and by extension the degree o molecular orientation.

In-line colorimeter:
The in-line colorimeter developed in this work has the radiation source made-up of two bright LED's, one RGB and one white. The white LED is used to measure the turbidity. A bench validation procedure was developed using a water solution of yellow acid 49 (YA49) of increasing concentrations. The water solution is pumped through a Teflon-based slit-die of 1.5 mm thick [6] using a sunken pump, while data are collected in real-time from the moment the signal attains stability. Initially, measurements are carried out with distilled water as the onset base-line. Further, constant volumes of 1 mg/mL YA49 solution are added in time intervals long enough to have a homogeneous solution circulating through the bench system, what is set by the stabilization of the signal.

Results and Discussion

In-line melt turbidity
Figure 1 shows the changes in the in-line melt turbidity during its discharge from the extruder as a function of the die-head pressure. During the discharge the turbidity value changes according to the melt shear viscosity ratio of the polymer mixture. For melt shear viscosity ratio ~1 (PP/PS mixture) the melt turbidity decreases during the increase of the die-head pressure, and so the shear flow rate. Upon reducing the die-head pressure and so the shear flow rate the phenomenon is reversed, indicating the deformation reversibility of the second phase droplets. Increasing the melt shear viscosity ratio to ~4 (PS/HDPE mixture) the second phase droplets have higher viscosity and so are more difficult to be deformed, but still some deformation is present. Finally when the melt shear viscosity ratio is low ~¼ (HDPE/PS mixture) the turbidity signal during the discharge keeps almost constant due to the high viscosity and elasticity of the HDPE matrix, having longer relaxation time then the experiment time scale, and so not having time to the recovery of the droplet's deformation.

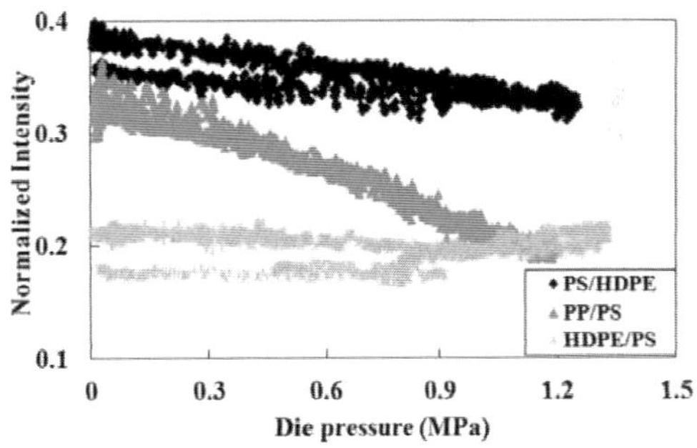

Figure 1 – In-line melt turbidity (normalized intensity) curves as a function of the die-head pressure of various polymer mixtures, as indicated, during their discharge out of an extruder.

In-line melt rheo-polarimetry
As result of decreasing shear stress during 'stop of feeding' procedure, due to the fact that the molten polymer flow rate reduces during the experiment, the crossed-polarized transmitted light intensity displays a sine-square behavior attaining an optical path difference over the half of the first order (275 > OPD >

550nm) for all polymer systems, what is set by the peaks of maximum, shown in Fig. 3 only for PS/PMMA blends.

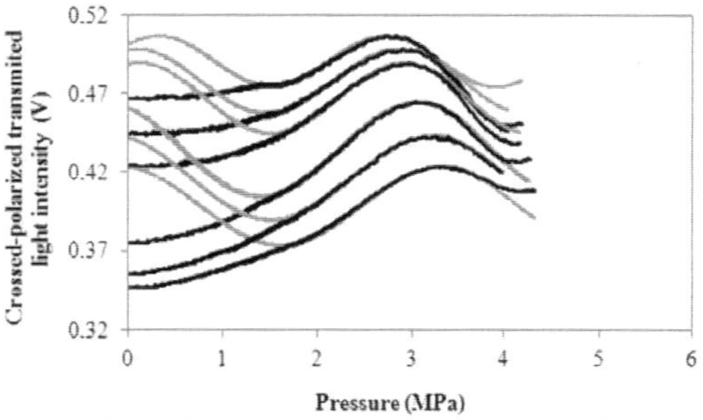

Figure 3 –In-line monitoring results during the 'stop feeding procedure' showing the combined effects of scattering and flow birefringence. From upward to downward: pure PS; 0,15; 0,25; 0,50; 0,75 and 1 w/w% of PMMA.

The peaks, which represent the same degree of chain deformation and so having the same relaxation time, are shifted to higher pressures with increasing disperse phase amount, indicating an anomalous drop in viscosity. This probably is an indication of apparent interfacial slip between blend components, as currently suggested in the literature [10, 11], shown here with an optical device. In order to get the peak's pressure values in the curves at the half of the first order (OPD = 275nm), sine-square fittings were made so that one can evaluate the influence of disperse phase content in the molecular orientation of polystyrene matrix. This can be seen in Fig. 3.

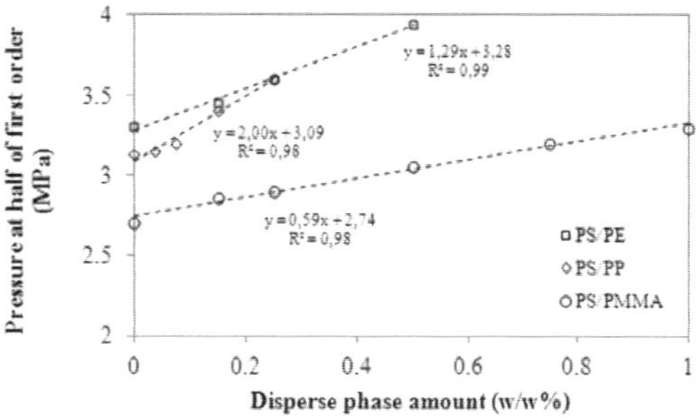

Figure 3 – Pressure values at the half of first order for each immiscible system as function of disperse phase amount.

The most important information in Fig. 3 is the slope of each curve. It is an indication of the influence of unlike structurally polymer pairs on the viscosity drop. According to the literature, less incompatible polymer systems would show a large deviation in viscosity as compared to the blend constituents. Polyolefins are more incompatible with polystyrene matrix [10, 11]. This may explain the larger values in slopes as compared to polystyrene/poly(methyl methacrylate) system. All in all, more force must be applied to dissimilar polymers systems necessary to promote the same degree of chains orientation as that found in pure polystyrene matrix, keeping constant all the other experimental conditions.

In-line colorimetry.

Chemical modifications in polymers which change the polymer color can be monitored by colorimetry. These color changes are caused by formation of chromophore groups such as double bonds, carbonyl groups etc. Thus, in-line measurement of color during processing of polymeric resins is an important tool for quality control, mainly when they undergo degradation, as current found in PVC, PET and polyolefins processing as well as in reactive processing. Hence, an in-line colorimeter device is presented to probe color changes in polymer processing.

YA49 water solutions were used to raise data for the bench validation of the in-line colorimeter device, which are depicted in Fig. 4. It turned out that the relation between absorbance and the analyte concentration just follows Lambert-Beer's law which validates the device for quantitative analysis, as for instance in the determination of chromophore groups during polymer processing. The measured wavelength should be the one corresponding to the complementary color developed; regarding YA49 analyte the blue region is of choice, it is the complementary color of the yellow [12]. The resolution of the device is found to be about 5% in the studied conditions, which means a difference in absorbance of 0.00234 a.u. or 8.2 x 10^{-6} mols of chromophore groups.

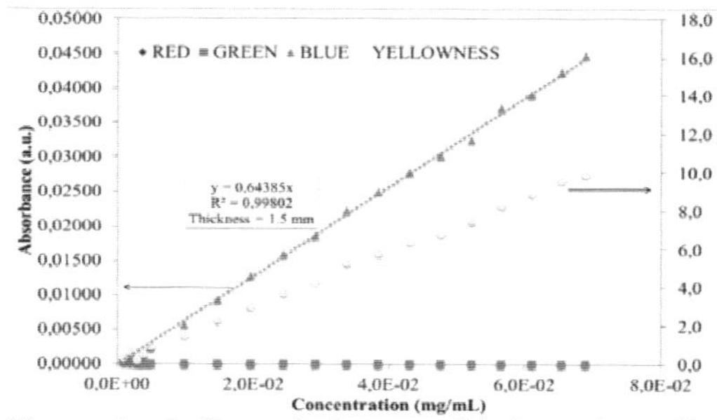

Figure 4 – In-line colorimeter validation using yellow acid 49 solution in flow.

As expected the in-line device has not indicated absorptions for red and green colors, the YA49 presents a single color and the color is stable during dilution. For samples that have changes in more than one color it is possible to measure the yellowness index (YI) and associate it with the color changes in the medium. YI have been used [6, 13] to determinate the kinetics degradation of poliolefins using off-line apparatus, which means further sample preparation and by extension another thermal history, which may affect the final result. Therefore, it is very convenient to use in-line measurements (in this case in-line colorimeter) to get a more precise and fast analysis of structural changes which may lead to polymer color changes.

Conclusions

In this paper we present some real time optical methods to monitor the polymer melt flow during extrusion. First the in-line melt turbidity can quantify the deformation and recovery of a second phase droplets during extrusion/discharge. When the elasticity of the matrix is low the deformation/ recovery cycling of the droplets is reversible. In-line rheo-polarimetry proved to be an attractive technique to follow the degree of molecular orientation when studying diluted blends in real time. It was shown that, compatibility between components plays an important role on the viscosities deviation encountered. Finally preliminary bench results showed that in-line colorimetry can quantify polymer color changes, which can be associated to degradation.

Acknowledgements

To Braskem, INNOVA and Unigel for the donation of the polymers, CNPq for the grant to S.V.C and CAPES/CNPq for the grants to P. A. R. M., L. S. H., and A. M. C. S.

References

1. L.A. Pinheiro, C.S. Bittencourt, L.A. Pessan, S. V.Canevarolo, Macromol Symp, 2006, 347.
2. L. A. Pinheiro, G-H. Hu, L. A. Pessan, S. V. Canevarolo. *Polym Eng Sci.*, 2008, 48, 806.
3. M. K. Bertolino, M. Sc. Thesis, Universidade Federal de São Carlos, 2008.
4. S. V. Canevarolo, M. K. Bertolino, L. A. Pinheiro, V. Palermo, S. Piccarolo, *Macromol Symp*, 2009, 279, 191.
5. L. Zborowski, S. V. Canevarolo, *Polymer Testing* 2011, 31, 254.
6. K. C. Reis, S. V. Canevarolo. *Polym Eng Sci.* 2012, 52, 8, 1784.
7. R. K. N. Janeschitz-Kriegl Hermann, *"Polymer Melt Rheology and Flow Birefringence"*. Springer-Verlag, Berlin, (1983).
8. K. Soares, A. M. C. Santos, S. V. Canevarolo, *Polym Test*, 2011, 30, 848.
9. C. W. Macosko, *Macromol. Symp*. 2000, *149*, 171.
10. L. A. Utracki *Polym. Eng. Sci.* 1983, *23*, 609.
11. R. Zhao, C. W. Macosko *J. Rheol*, 2001, *46(1)*, 167.
12. G. H. Jeffery et al., "Vogel's Textbook of Quantitive Chemical Analysis". Longman Group, UK, (1983).
13. E. Epacher et al. *J. App. Pol. Sci.*, 1999, *74*, 1596

O-01-650

TAIL PELLET MINIMIZATION VIA CONTROL OF SOLIDIFICATION PROCESS AND CUTTING PHENOMENA FOR A HIGH-MFR PP

W. Tanthapanichakoon[1,*], S. Jinarakpong[1], R. Kulwatthanasal[2], S. Kitchoackgaroensakul[1], and S. Lertthanongsak[1]

[1]*Thai Polyethylene co., ltd., 10 i-1 Road Map Ta Phut Industrial Estate, Muang, Rayong, 21150 – winyut@scg.co.th*
[2]*Rayong Engineering & Plant Service Co.,Ltd (REPCO), 10 i-1 Road Map Ta Phut Industrial Estate, Muang, Rayong, 21150*

Abstract
Polypropylene (PP) tail is always in a top list of customer's annoying problems. Compounders do not accept tail pellets due to their unsatisfactory appearance, tendency to cause film breakage, and undesirable problems in feeding and other types of processing. Tail pellets occur mainly with underwater pelletizer, where appropriate solidification of polymer and cutting must be ascertained in the invisible area. Too much solidification may cause hard tails, die freeze-off or severe wear of cutter and die plate while insufficient solidification leads to soft tails and rounded-up fibers around the cutter. Before polymer melt is pelletized, the outer part of polymer should be solidified while inner part remains molten. On the other hand, cutter needs to be sharp and gap between die plate and cutter must always be zero. During operation, grinding cutter is another means to increase cutter sharpness. Zero gap between die and cutter depends significantly on setting up forward and backward pressure of the cutter shaft. It is worth noting that too much forward pressure will reduce the die and cutter lifetime even if it guarantees very low to zero tail pellets. In this case, setting condition at the die exit of a high-MFR PP grade close to T_m was found to minimize tail pellets. Therefore, the appropriate melt temperature, averaged across its cross-sectional area of the die hole exit, is suggested to be between melt temperature (T_m) and crystallization temperature (T_c). Procedure to optimize forward pressure is also provided. TiC128 die plate is recommended above Tungsten carbide due to its much longer lifetime. By these three guidelines, the underwater pelletizer can be optimized for both cost and performance.

Keywords: Tail, Polypropylene, Cutting, Solidification, Pelletizer

Introduction

Polypropylene (PP) production depends on chemical, physical and appearance properties. A significant problem of PP pellets is appearances such as tail pellet, white puffy pellet, black pellet, etc. The most chronic problem is tail pellet, causing a big loss around 100k USD/year. Tail pellet can be specified in two types. One is called "soft tail" similar to hair with length of more than 2 mm, as illustrated in Fig. 1. The other one is called "hard tail", i.e., a rigid tail with length of less than 2 mm, as shown in Fig. 2.

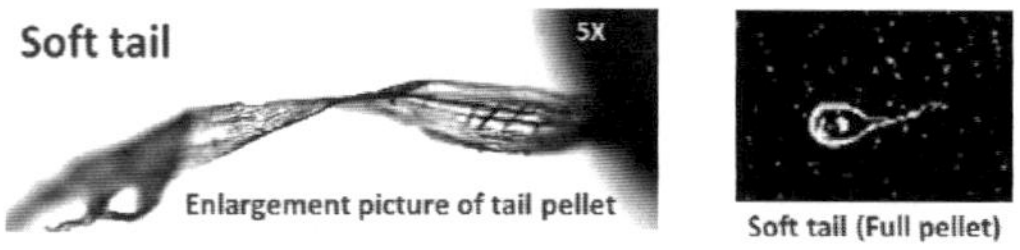

Figure 1: Characteristic of soft tail

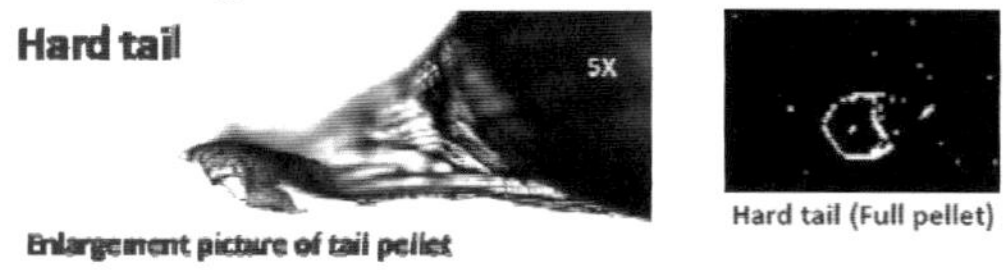

Figure 2: Characteristic of hard tail

50 tail pellets/1 kg or more can affect stretched film application, and break film. Since tail and normal pellets are not simultaneously molten, thickness of stretched film from tail pellets is thinner than that from normal pellets. Thinner part can be easily broken, causing loss of raw material and time. Fig. 3 shows stretched film.

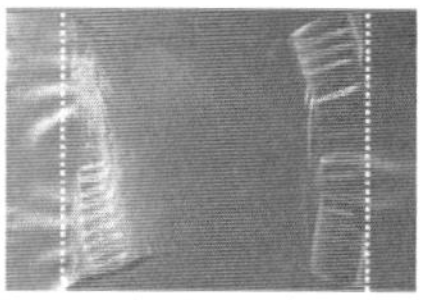

Figure 3: Stretched film

Experimental

1. Tuning die operating conditions

Experiments are conducted with a ZSK-300 extruder with L/D of 19, and screw diameter of 298 mm. The test material is selected to be polypropylene (PP) with MFR of 35 g/10 min at 2.16 kg as the amount of tail pellets varies as seen in Fig. 4. PP is pelletized with underwater pelletizer, which composes of die plate, cutter and water chamber.

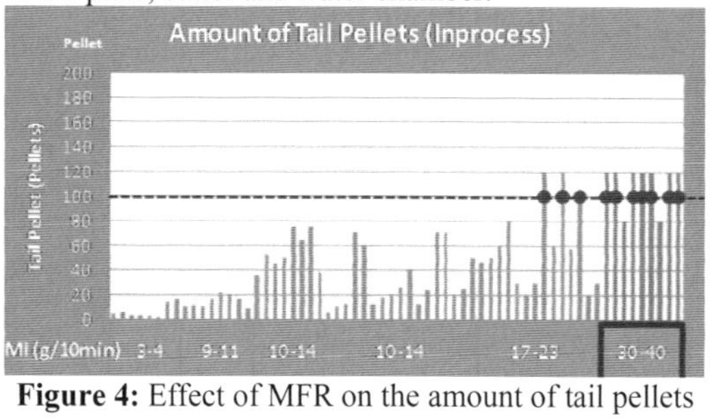

Figure 4: Effect of MFR on the amount of tail pellets

Mass and momentum balances are written in Eq. 1 and 2. [1] Eq. 2 complies with the well known Newton's second law: Mass times Acceleration = Sum of forces acting. [2] The effect of viscosity is present in the term of change in surface forces per unit time and unit volume, which means that, at the same pressure force, less viscosity would induce higher velocity of motion. This affects the amount of heat transfer and viscous heat dissipation, as well as degrees of solidification as shown in the relationship in Eq. 4. Thus, it is reasonable to presume that, at the same die conditions, high MFR PP is more likely to have insufficient time for suitable solidification before being pelletized. This explains why PP tails are found in PP with MFR of 35 more than the rest.

$$\frac{\partial \rho}{\partial \tau} + \nabla \cdot (\rho U) = 0 \qquad \text{- Eq. 1}$$

$$\frac{\partial (\rho U)}{\partial \tau} + \nabla \cdot (\rho U U) = -\nabla p + \mu \nabla^2 U + \frac{(1-\beta^2)}{(\beta^3 + \varepsilon)} A_{mush} (U - U_p) \qquad \text{- Eq. 2}$$

where β is the degree of solidification in Eq. 3.

$$\beta = \begin{cases} 0 & T < T_{solidus} \\ 1 & T > T_{liquidus} \\ \dfrac{T - T_{liquidus}}{T_{liquidus} - T_{solidus}} & T_{solidus} < T < T_{liquidus} \end{cases} \qquad \text{- Eq. 3}$$

$$\frac{\partial (\rho H)}{\partial t} = -\nabla \bar{q} - p(\nabla \bar{v}) - (\nabla \cdot \rho H v) - (\tau : \nabla \bar{v}) + \phi \qquad \text{- Eq. 4}$$

In order to understand more about solidification of polymer, the concept of mushy zone must be well understood. Mushy zone is a porous region, in which polymer is trying to form crystalline and amorphous structures. The mushy zone thickness dramatically reduces with increasing melt velocity [1], indicating low rate of solidification. This explains further why a high-MFR PP tends to have more tail pellets.

Based on Eq. 1, 2 and 3, it can be seen that polymer characteristics that affects solidification process of the polymer includes density of melt and solid, viscosity, heat capacity, latent heat, heat conductivity, rheology, thermal characteristics such as crystallization temperature, and solidus/liquidus temperature. These imply more complication when one attempts to control die condition. In order to control die conditions of PP with MFR of 35, suitable melt solidification must be ascertained. Too much solidification may cause hard tails, die freeze-off, and/or severe wear of cutter and die plate while insufficient solidification leads to soft tails and/or rounded-up fibers around the cutter. Therefore, as the polymer exits the die for pelletization, it should have outer part solidified while the inner part remains unmolten as seen in Fig. 5.

In order to simplify the solidification calculation, temperature is used instead of degrees of solidification. Calculated temperature is averaged across the cross-sectional area of die hole. Based on the previous fact that outer part except inner part becomes solidified, the average temperature across the cross-sectional area can be hypothesized to be between melting temperature (T_m) and crystallization temperature (T_c) of polymer. PP with MFR of 35 has T_m of 150°C and T_c of 118°C. Relationship of hot oil inlet temperature, mass flow rate of polymer and temperature of polymer melt at die exit can be correlated by Eq. 6, which is simplified based on one-dimensional consideration. Parameters related to Eq. 5, 6 and 7 are listed in Table 1.

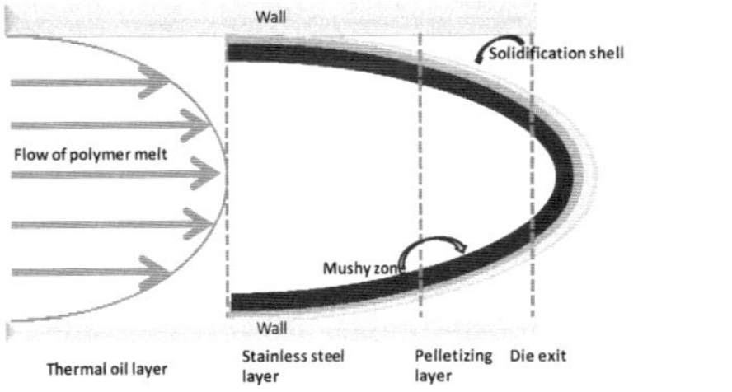

Figure 5: Solidification theme of polymer flow through die hole

$$Q_{PCW} = Q_{hot\ oil} + Q_{polymer} \qquad \text{- Eq. 5}$$

$$U_D A_D (T_{havg} - T_{pcw}) = m_h C_{p,h} (T_{hi} - T_{ho}) + m_m C_{p,m} (T_m - T_{mD}) \qquad \text{- Eq. 6}$$

$$U_D = \frac{1}{\frac{1}{h_0} + \frac{1}{h_w} + \frac{L_{ss}}{k_{ss}} + \frac{L_{ts}}{k_{ts}}} \qquad \text{- Eq. 7}$$

Where

$Q_{polymer}$	= Heat gain of polymer melt (W)
$Q_{hot\ oil}$	= Heat generated from hot oil (W)
Q_{pcw}	= Heat generated from process water (W)
m_m	= Mass flow rate of polymer melt (kg/s)
$C_{p,m}$	= Specific heat of polymer melt (J/kg.K)
T_m	= Temperature of polymer melt at die inlet (K)
T_{mD}	= Temperature of melt at die plate exit (K)
m_h	= Mass flow rate of hot oil (kg/s)
$C_{p,h}$	= Specific heat of hot oil (J/kg.K)
T_{hi}	= Temperature of hot oil inlet (K)
T_{ho}	= Temperature of hot oil outlet (K)
U_D	= Overall heat transfer coefficient of die plate (w/m^2.k)
A_D	= Surface area of die plate (m^2)
T_{pcw}	= Temperature of process water (k)
T_{havg}	= Temperature of hot oil average in die (K)
h_w	= Heat transfer coefficient of water (W/m^{2}K)
h_o	= Heat transfer coefficient of thermal oil
k_{ss}	= Conductivity of die material (W/m.K)
k_{ts}	= Conductivity of pelletizing layer (W/m.K)
L_{ss}	= Distance of die material (m)
L_{ts}	= Distance of pelletizing layer (m)

Our experiment attempts to correlate number of tail pellets with melt temperature of polymer at die exit (T_{mD}), which is calculated from Eq. 6. Experiments vary hot oil inlet temperature and mass flow rate of polymer in order to adjust temperature of polymer melt at die exit toward T_m of this PP. It is worth noting that the calculation assumes melt temperature after screen and before die to be the same as that of die hole inlet.

In order to avoid the effect of regrinding which sharpens the cutter and makes biased experiments, number of tail pellets below 40 pellets (per 5 kg) is not taken into this correlation. This also implies that the optimum operating conditions of die can be more easily found with worn die since newly reconditioned dies are likely providing below 40 pellets per 1 kg almost always.

Table 1: Estimated parameters for calculations of melt temperature of polymer at die exit

No.	Parameter	Value
1	$C_{p,m}$	2,300 J/kg.K
2	m_h	4.608 kg/s
3	$C_{p,h}$	2,520 J/kg.K
4	T_{pcw}	55°C
5	A_D	0.4562 m^2
6	h_w	72,900 $w/m^2.k$ [3]
7	h_o	20,000 $w/m^2.k$ [3]
8	k_{ss}	16.27 W/m.K
9	k_{ts}	42.32 W/m.K
10	L_{ss}	0.004 m
11	L_{ts}	0.04

2. Extension of die plate and cutter lifetime by adjustment of forward pressure and die and cutter material selection

2.1) Forward pressure

Zero gap between die plate and cutter as well as cutter sharpness are the key to eliminate tail pellets by cutter force. Too much cutter force, however, shortens cutters and die plate lifetime. Contact force between die plate and cutter is originated by hydrualic system. To find contact force on die plate, force balance on cutter in Eq.8 needs to be done. Forces acting on a cutter are composed of six forces as shown in Fig. 6.

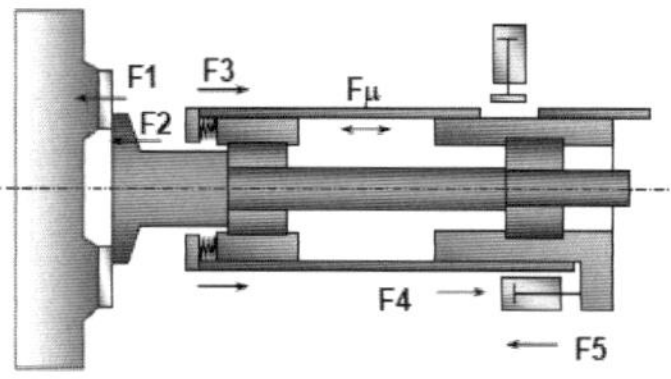

Figure 6: Cutter shaft and forces acting on cutter

$$F = F_1 + F_2 - F_3 - F_4 + F_5 - F_\mu \qquad \text{- Eq.8}$$

Where

F = Force on contact surface between cutters and die plate

F_1 = Propelling force by cutters

F_2 = Pump effects

F_3 = Pre-load spring force

F_4 = Backward force by hydraulic cylinder

F_5 = Forward force by hydraulic cylinder

F_μ = Friction force

Force balance equation (Eq.8) is used to calculate the appropriate forward pressure of cutter by creating cutter pressure profile. Cutter pressure profile is an individual characteristic which comes from experiments. The experiment varies parameters (Propelling force, Forward force and Backward force) by the following step:

1. Run cutter at each speed starting from the minimum operating speed of 400 rpm.
2. Increase forward pressure constantly until cutter moves forward, and record value of forward pressure.
3. Set forward pressure back to zero
4. Increase backward pressure until cutter moves backward, and record value of backward pressure.
5. Repeat Step 1-4 for each cutter speed until the maximum operating speed (1,000 rpm).
6. Use these experimental data to calculate cutter force by force balance equation (Eq. 8).
7. Convert force to pressure, and plot graph between forward pressure and cutter speed by fixing backward pressure at 2 kg/cm^2.

2.2) Die and cutter material selection

Apart from supplying the appropriate cutter force, material of die plate and cutter also significantly affects lifetime of die plate and cutter. Softer die plate and cutter, as well as high cutter sharpness, are prone to achieve very low to zero amount of tail pellets but their lifetime is too short. On the other hand, harder die plate and cutter one, and low cutter sharpness, are likely going to extend their lifetime while increasing probability to generate tail pellets. Finding the optimum point for hardness of die plate and cutter is required.

Results and Discussion

1. Tuning die operating conditions

In Fig. 7, hot oil inlet temperature is linearly proportional to melt temperature of polymer after screen and before die, which is reported on our extruder screen panel. Based on this, it can be seen that hot oil inlet temperature has more effect to the melt temperature more than mass flow rate of polymer

By assumption that melt temperature of polymer after screen and before die is the melt temperature at die hole inlet, melt temperature of polymer at die exit can be calculated by Eq. 6. When this amount is correlated with the number of tail pellets (per 1 kg sample), Fig. 8 can be obtained. Number of tail pellets appears to be reduced as melt temperature at die exit approaches T_m (i.e., 150°C). This goes in accordance with the theory of sufficient solidification for reduction of soft tails. This also implies that Eq. 4, 5 and 6 can be practical for PP tail pellet control in the plant-sized extruder. This also leaves room for further model development into 2D and 3D model, should the geometry of die is well known.

Based on the trend lines of results, the appropriate melt temperature, averaged across its cross-sectional area of the die hole exit, appears to be between melt temperature (T_m) and crystallization temperature (T_c). Too much solidification may cause hard tails, die freeze-off, or severe wear of cutter and die plate.

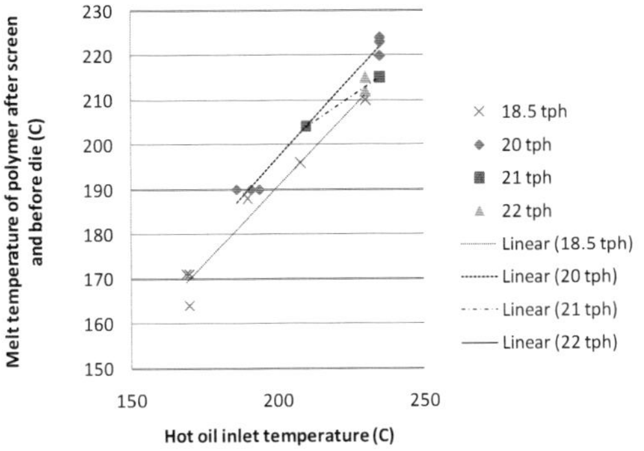

Figure 7: Relationship between hot oil inlet temperature and melt temperature of polymer

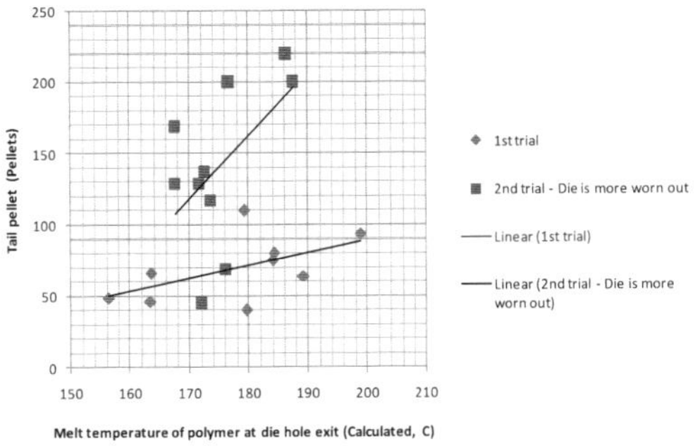

Figure 8: Number of tail pellets at different melt temperatures of polymer at die hole exit.

As the benefit of setting conditions by this guideline, die lifetime can be extended before reconditioning is required. This makes a big difference in terms of cost since each die plate can be several hundred thousands USD for a commercial plant size.

2. Extension of die plate and cutter lifetime by adjustment of forward pressure and die material selection

2.1) Forward pressure

From the experimental with cutter pressures of 5, 10 and 15 kg/cm^2, the result is shown in Fig. 9. Normal contact pressure of die plate is 10 kg/cm^2, but, for harder cutters like TiC143, 15 kg/cm^2 is found to be appropriate. For softer material such as TiC128 and Tool steel, die plate contact pressure of 5 kg/cm^2 is however recommended. Cutter lifetime of TiC128 and tool steel is doubled from 3 weeks to 1.5 months at contact pressure of 5 kg/cm^2. Meanwhile, TiC143 cutter still has the longest lifetime of 3 months despite the use of a higher contact pressure of 15 kg/cm^2.

2.2) Die and cutter material selection

By experimental, lifetime of soft cutters, which include TiC128 and Tool steel, can reach 1.5 months while harder cutters, i.e. TiC143, has a lifetime of 3 months. With TiC128 cutter, die plates made from tungsten carbide (harder) and TiC128 (softer) can last for 12 and 20 months, respectively. This appears to be controversial to general belief that harder material tends to last longer. The reason for better performance of TiC128 die plate is that softer material is easier to regrind and maintain cutter sharpness, thus leading to mush less frequency for cutter grinding. Hypothetically, TiC128 die plate with TiC143 cutter may be the best option that is yet to be verified during the next few years.

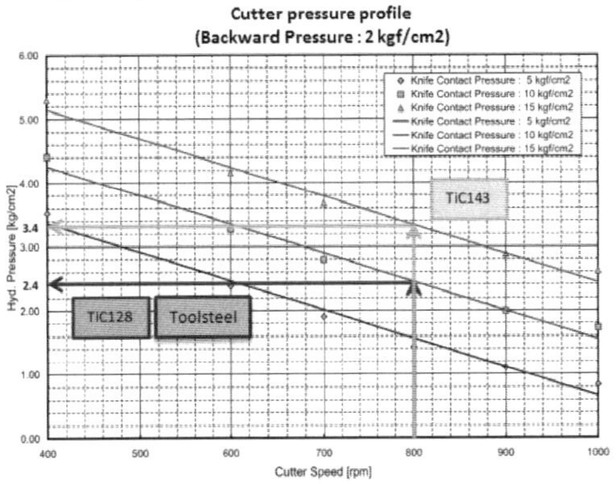

Figure 9: Cutter pressure profile

Conclusions

Even at high MFR which is more susceptible to PP tails, our simplified equations and experimental results point out that, before polymer melt is pelletized, the outer part of polymer should be solidified while inner part remains molten. Appropriate melt temperature, averaged across its cross-sectional area of the die hole exit, is suggested to be between melt temperature (T_m) and crystallization temperature (T_c) by calculations. This optimum condition lengthens the die lifetime before reconditioning, which is needed when die is worn out and too many tail pellets are found. By further selection of die plate and cutter material, lifetime of die plate and cutter can be lengthened. Die plate lifetime can be lengthened from 12 to 20 months, and cutter lifetime can be lengthened from 3 weeks to 3 months without tail pellets and any cutting problem. This saves cost of about 130k USD baht per year.

Acknowledgements

Noppon Summaprasit, Kachamas Tumrongsak, Voravit Varongchayakul, Pramote Komnopad, and Wee Chabthanom are thankfully acknowledged.

References

1. M. Wu; A. Vakhrushev, G. Nummer; C. Pfeiler; A. Kharicha; A. Ludwig *The open transport Phenomena Journal* 2010, 2, 16-23
2. Walter Mitchaeli in *Extrusion dies for plastics and rubber (Design and engineering computations, 3rd edition)*, Hanser, Munich, 2003
3. S. Ren; S. Dang; T. Lu; K. Wang *Advanced Material Research* 2010, 97-101, 2736-2743

*** Formatting Issue - Best Available Paper/Graphic ***

O-01-666

DEVELOPMENT OF CO-ROTATING TWIN SCREW EXTRUDER TO OVERCOME LIMITATIONS OF CONTINUOUS MIXERS AND BATCH MIXERS FOR ADDITIVE MASTERBATCHES

W. Polrut[1*], P. Nutthanara[1], W. Na Ranong[1], S.Treethammakul[1] and W. Tanthapanichakoon[1]
[1]Thai Polyethylene co., ltd., 10 i-1 Road, Map Ta Phut Industrial Estate, Muang, Rayong, 21150, Thailand – wareerop@scg.co.th

Abstract - Additive masterbatches had been widely produced by either continuous mixer or batch mixers. From various screw designs to adjust dispersive and distributive mixing of additives, the co-rotating intermeshing twin screw extruder (c-TSE) was chosen to improve the quality of processing aid masterbatch, slip and antiblock masterbatch, and antimicrobial masterbatch. Due to its continuous operation, it provided higher cost-effectiveness than the batch mixer. The film appearance rating (FAR) was significantly enhanced by more intense mixing screw. Products produced from c-TSE had higher light coefficient (less yellow) than those from batch mixer due to less residence time at high temperature. The potential of free barrel slip is also eliminated by applying c-TSE. From experiments, the efficiency of c-TSEs for these additive masterbatches was improved by reducing the channel depth to provide more friction to drive the material forward. It was found that the screw versatility and line flexibility are main advantages of this extruder. The optimal extrusion temperature was studied from the thermal stability of raw materials to avoid the degradation of masterbatches. Morphological characterization on degrees of dispersion was also useful for screw design adjustment. Theoretical design of screw configuration and in-depth characterization technology, as well as technology to link between formulation-process parameter-properties were the keys for this success.

Keywords: Batch mixer, Continuous mixer, Co-rotating intermeshing twin screw extruder, Additive masterbatches, Light coefficient, Film appearance rating.

Introduction

Continuous mixers and batch mixers have been widely used to produce many kinds of masterbatch for decades. However, the lack of screw versatility to tailor the mixing performance and degassing capabilities are drawbacks of these machines compared to twin screw extruder. From advantages of twin screw extruder on good feeding system, adjustable screw design for homogeneous mixing, devolatilization as well as good self-cleaning action, it becomes competitive choice for additive masterbatches production [1,2,3].

However, in the case of high concentrated slip and antiblock masterbatch, the counter-rotating non-intermeshing type of twin screw extruder is not suitable. The lack of self-wiping efficiency of this extruder resulted in free barrel slip from the additive and difficult to drive the material forward, leading to "no output" problem.

Considering on the banbury mixer, the mixing time, step of material loading and pressure of ram (piston) movement are adjustable before discharging to single screw extruder. The better elongational mixing quality is expected to be achieved at longer mixing time. However, this type of mixer is not applicable to form homogeneous compounded product without pellet tail. Dispersion problem was also encountered in the slip and antiblock masterbatch in the form of film appearance rating (FAR, represents degrees of gel formation) problem, wherein the FAR value at -20 to -40 was lower than the specification of this product (i.e., more than -10). The higher number of material portion loading, the better mixing quality would be expected but the historical experiment showed no improvement; poor dispersion could not be overcame. Generally, we cannot set temperature profile upon banbury mixer because the melt temperature is generated by shear heating only. This banbury mixer is followed by single screw extruder before underwater pelletizing. Simple single screw extruder designs are inefficient dispersing designs. Therefore, single screws are seldom used for compounding, except where special effects of poor dispersion are desired such as streaking and avoiding fiber breakage.

Concerning formulation of additive masterbatches, Lc is another important property. Initial Lc of ingredients which are additive and base resin and also characteristic of extruder to produce this masterbatch are important parameters affecting Lc of additive masterbatches. Many dead zones of counter-rotating, non-intermeshing twin screw extruder could generate more polymer degradation than that of c-TSE. Schematic 1 shows common problem, cause and solution of three kind additive masterbatches.

*** Formatting Issue - Best Available Paper/Graphic ***

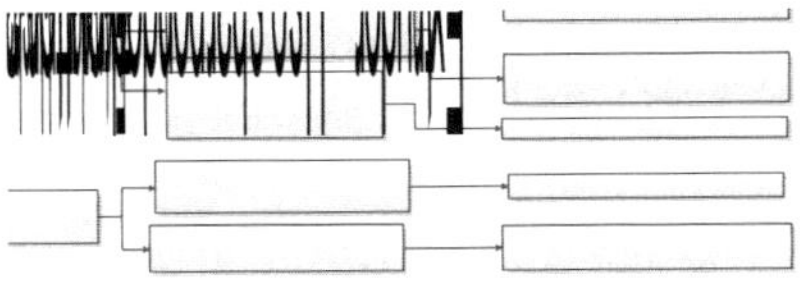

Schematic 1 Additive masterbatches: problem, cause and solution.

From the production problem that showed no output when slip and antiblock masterbatch was run at continuous mixer and product quality problems of additive masterbatches (processing aid masterbatch, slip and antiblock masterbatch, and antimicrobial masterbatch) which are off-spec of film appearance rating (FAR), light coefficient (Lc, represents yellowness) and bad pellet appearance; small tail that occurred at banbury mixer and continuous mixer and attempting to apply theoretical design of screw configuration and in-depth characterization technology, as well as technology to link between formulation-process parameter-properties, a co-rotating intermeshing twin screw extruder was chosen to be a production machine to overcome those obstacles in this study.

Experimental

From the problem analysis shown in Schematic 1, a c-TSE having L/D of 44 and screw diameter of 50 mm was used to produce three types of masterbatch. Commercial LLDPE (MI 1-3 g/10 min) was used as a base resin for slip and antiblock masterbatch, and processing aid masterbatch. Commercial PP (MI of 30-40 g/10min) was used as a base resin for antimicrobial masterbatch. The composition of each masterbatch was illustrated in Table 1.

Table 1 Additive masterbatches composition

Type of masterbatch (M/B)	Composition
Slip and antiblock M/B	LLDPE base resin Slip agent: 5-15 wt% Antiblocking agent 5-15 wt% Antioxidant: 0.1–0.5 wt%
Processing aid M/B	LLDPE base resin Processing aid: 1-5 wt% Antioxidant: 0.01–0.05 wt%
Antimicrobial M/B	PP base resin Antimicrobial: 15–25 wt% Neutralizer: 1–3 wt%

In the production of each masterbatch, additives were premixed using the high speed mixer before transferred into a side feeder, while a base resin was fed into main feeder. Similar screw design was applied for compounding the processing aid and antimicrobial masterbatches, while more intense screw design was used for preparing the slip and antiblock masterbatch to distribute and disperse slip and antiblock agents into the compounded mixture properly. Temperature profiles of slip and antiblock masterbatch and processing aid masterbatch were set based on characterization of color stability of ingredients affecting Lc drop which will be demonstrated in result and discussion part while temperature profile of antimicrobial was not critically affecting Lc of final product. Output and screw speed were optimized for each masterbatch production.

Results and Discussion

c-TSE for additive masterbatches

Based on problems of additive masterbathes production which are Lc, FAR, pellet tail and no output, these could be overcome by using c-TSE. The reasons are that it has shelf-wiping characteristic to overcome free barrel slip causing no output and small channel depth to provide more friction to drive material forward. Adjustable screw configuration to provide more intense screw design to get more dispersive mixing, e.g., in the case of die build-up and pellet tail of processing aid masterbatch as well as FAR problem upon slip and antiblock masterbatch which were morphological analyzed and justified to be caused by inhomogeneous dispersion of additives within the matrix. Furthermore, adjustable screw speed to increase shear mixing and adjustable temperature profile to completely melt the ingredients are also important to make c-TSE becomes another good machine for additive masterbatches production. Moreover, less dead zones of c-TSE than that of counter-rotating, non-intermeshing twin screw extruder gave less Lc drop (higher Lc value) which would be due to less polymer degradation. In addition, less Lc drop could be due to less residence time upon c-TSE compared to that of banbury mixer. Initial Lc of base resin is another important parameter effecting Lc of final product. The initial Lc of the base resin of about 30-40 unit higher than spec, the on-spec of Lc value would be achieved. Therefore, formulation, process parameters and properties should be adjusted properly. The advantages of using this kind of extruder are illustrated in Schematic 2. Moreover, flexibility of production is another advantage of using this type of extruder which is a common type extruder used in industrial.

After optimizing of processing parameters to give good products quality, output of slip and antiblock masterbatch was 50 kg/h while output processing aid and antimicrobial masterbatches was equal at of 80 kg/h.

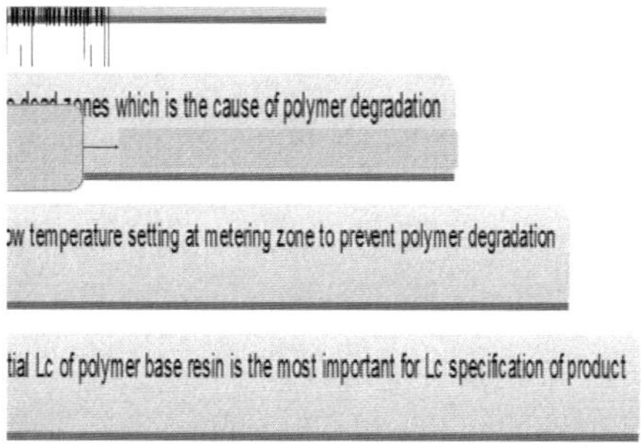

Schematic 2 Capability of c-TSE for additive masterbatches.

Color stability by spectrophotometer

Regarding yellowness of produced masterbatches, a cause of this may come from degradation of raw materials. This characterization part aimed to prioritize the effect of color change of each ingredient during processing which would indicate degradation of substance. Aging experiment was carried out to observe Lc (yellowness) of raw material. The sample was heated from 25 °C to several temperatures which were 100, 150 and 180 °C. At each temperature, the sample was held for 10 min before Lc testing using spectrophotometer. Fig. 1 showed the tendency of Lc value of raw material consisting of slip agent, antioxidant, LLDPE base resin, and slip and antiblock masterbatch. The result showed that slip agent gave the highest drop rate of Lc value indicating that yellowness of masterbatch was mainly caused by slip agent.

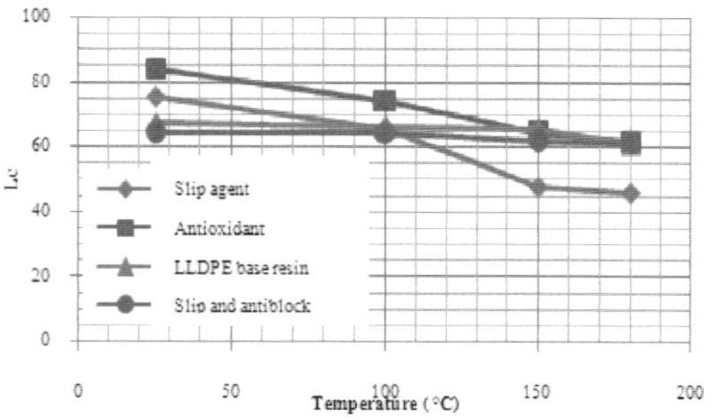

Figure 1 Tendency of Lc value of material after aging at from 25 to 180 °C.

Theoretically [4], the discoloration of compound is caused by thermal or catalytic degradation of slip agent (fatty acid amide) as shown in Schematic 3. From the reaction, fatty acid amide was decomposed to colored alkyl nitrile after heating up to more than 200 °C or catalysed by inorganic substance.

$$\underset{\text{Fatty acid amide}}{\text{R-CONH}_2} \xrightarrow[\text{or catalyst}]{\text{Heat}>200\,^{\circ}\text{C}} \underset{\text{Alkyl nitrile}}{\text{R-C}\equiv\text{N}} + H_2O$$

[R = Fatty acid chain]

Schematic 3 Thermal or catalytic degradation of slip agent (fatty acid amide).

Moreover, TGA results of processing aid masterbatch showed that the processing aid was decomposed at 200 °C. Therefore, melting temperature should be under 200 °C for both slip and antiblock masterbatch and processing aid masterbatch.

Conclusions

With c-TSE, film appearance rating (FAR) of additive masterbatches was significantly enhanced by more intense mixing screw. Products produced from c-TSE had higher light coefficient (less yellow) than those from batch mixer due to less residence time at high temperature. c-TSE can also eliminate the potential of free barrel slip that exists in the case of continuous mixer. With less channel depth to provide more friction to drive the material forward, c-TSE provides more output than both continuous and batch mixer by its continuous mode of operation. In overall, c-TSE has become a more suitable production machine for additive masterbatches when compared with continuous and batch mixers, which have been long used in this industry.

Acknowledgements

We gratefully acknowledge Thai Polyethylene Co., Ltd for financial support.

References

1. R.M. Harris, *Coloring Technology for plastics,* Plastics design library, New York, 1999.
2. Enno E. Agur, *Advances in Polymer Technology*, 1986, *6*, 225-232
3. C. Rauwendaal, *Polymer Extrusion*, 4th Edition, Hanser Publishers, Munich, 2001.
4. J.C. J. Bart, *Polymer Additive Analytics: Industrial practice and case studies*, Firenze: Firenze university press, 2006.

O-01-029

Novel Constructions for Ring-shaped Dies Broaden Processing Capabilities in extrusion

Heinz Gross

Introduction

Ring-shaped die heads serve for the extrusion of various types of semi-finished products such as pipes, foamed films or sheets, blown films or cables. Annular extrusion heads are employed too in the extrusion of hollow bodies by blow molding. For all processes the die has to be re-centered on the die head each time after cleaning.. Retrofitting an ring-shaped head with an elastic tilt joint (Fig. 1) will pay off extremely fast, if dies need to be cleaned frequently, or if dimensions have to be changed at short intervals. It is also worthwhile considering whether to equip ring-shaped heads with a tilt joint, when processing expensive raw materials, or where narrow thickness tolerances must be adhered to.

Fig.1. Three examples of elastic tilt joints inserted into heads of different dimensions and for different processes

Tilting dies in Pipe, Tube and Capillary Production

In order to seal the parting line reliably, the tilting joint must only be compressed by a precisely defined measure. An ideal way of doing this is by using a simple bayonet lock coupling. Figure 2 represents a pipe head with an

Fig. 2. Tilting die with bayonet lock coupling, with a narrow centering fit between head and die, and with fine-pitched thread screws (M6x0.5) designed for fine optimization of the relative position of die and mandrel.

elastic tilt joint and a bayonet lock coupling, thus featuring no conventional centering screws. The bayonet lock coupling reliably generates the forces required for fixing the die and sealing the dividing plane between the head and the die. This way, the operator can easily flangeor remove the die by a spinning motion. The geometry of the bayonet lock coupling automatically makes sure the elastic tilting joint is compressed just as far as is required to seal the parting line between head and die. This also prevents the gasket from being compressed too hard or too little. Additionally, no greasing of flange screws is required and the danger of screw tearing as a result of thread seizing does not exist anymore. Figure 3 shows the bayonet lock coupling of a small head designed as a tilting die right from the start. This head served to produce capillaries from PLA for medical applications, at a mere 0.2 mm outer diameter. Optimizing the relative position of die and mandrel by means of conventional radial centering screws would certainly be impossible to achieve, considering the wall thickness of the capillary of below 0.1 mm

Fig. 3. Bayonet lock coupling of a capillary die with tilting technique

What is more, it is not necessary any more to pre-center the die in a time-consuming process, after fixing it. Tilting dies have narrow fits between head and die, thus merely allowing for the dies to be fixed to the head center. The great benefit here is the fact that the die exactly sits in the same central position each time the process starts, which used to be different with the conventional techniques. The die used to take on a new position then, each time the process was started, and this was mainly determined by the patience and skill of the respective operator.

For the individual process, optimization of the relative position of die and mandrel can most easily be done by special centering screws, as can be seen in Figure 2. While compressing the elastic joint takes only little power, small screws with little thread pitches can be used for tilting the die. In case extraordinarily fine tuning is required, the operator may employ special screws with fine-pitched threads.

Motor-driven Centering of Ring-Shaped Dies

In the long run, however, it is recommended to use inexpensive stepper motors, because once an optimum processing position is found, this position can be stored in the machine control unit, and can always be reproduced exactly at new start-up, for the same pipe geometry. A motor drive for the tilting die can even allow for eccentric differences in thicknesses to be minimized during extrusion by means of a control unit. Figure 4 shows a Flex Ring pipe die head designed for the production of core-foamed PVC pipes.

Fig. 4 Flex Ring die combined with tilting technique which enables to establish a closed-loop control to minimize as well eccentric, as also asymmetric variations in the thickness of extruded pipes

Tilting Die Technique for Extrusion Blow Molding

Integrating elastic tilt joints into blow molding heads will open up entirely new ways of processing. When retrofitting a blow molding head, two low-cost stepper motors, which sit in a 90° angle, can be employed to tilt the die. This way, the die can easily and dynamically be tilted into every needed position, during discharge. Consequently, the die can be centered from the control panel at close to any desired fineness and precision, and the process parameters can even be stored at the end of the production cycle. The tilting technique thus represents the first method ever to enable production at exactly the same relative position of die and mandrel, each time the machine is started up.

Moreover, the relative position of die and mandrel can certainly be changed during discharge of the preform too. This way, the operator can improve, e.g. the wall thickness distribution in crooked curves or in parts with a degree of stretching that varies over the circumference Figure 5. If additionally connecting the tilting technique to the

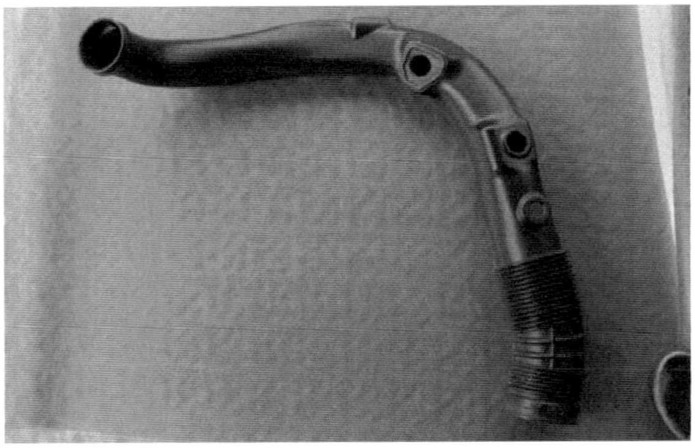

Fig. 5 Extrusion blow molded pipe with complicated shape

GWDS (**G**ross **w**all thickness **d**istribution **s**ystem) technique, which is just as novel, and is currently in the stage of development and testing, it will be possible to include in any usual blow molding machine a dynamic tilting function, and a well-aimed dynamic axial as well as radial closed-loop control of wall thicknesses of the parison. No special die areas will have to be deformed any more by actuators, in order to conduct axial and radial closed- loop control of the wall thickness. Neither will special software have to be implemented into the control unit. At present, this is done by very expensive and sophisticated PWDS (**p**artial **w**all hickness **d**istribution **s**ystems) or Flex Ring tools.

There are in fact certain limitations to the PWDS system in particular, and to the flexring system too, and that is when the degree of deformation in the part does not change gradually, but rather significantly over a very restricted area. In these cases, the GWDS technique can achieve a more homogeneous wall thickness distribution in the part, than PWDS or Flex Ring systems can do (Figure 6).

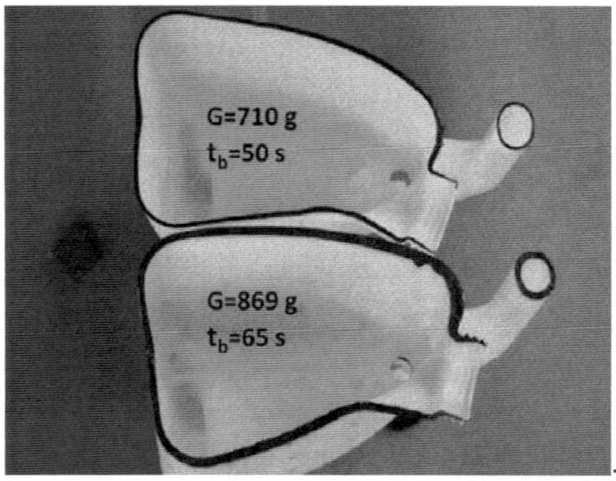

Fig. 6 Comparison of the thickness distribution in a small fuel tank. 3D-tecnology (on top) conventional fabrication method (underneath); G= weight; T_b= cooling time

Furthermore the GWDS technique has overcome the former restriction that a radial wall thickness programming system is limited to a certain die diameter range and can not be applied to small multi-cavity heads. Figure 7 shows a retrofitting unit for a 6-cavity head which is equipped with a tilting joint and a GWDS die in order to produce tubes with crooked curves and with changes in the cross-sectional geometry over their length. The die is dynamically tilted in order to achieve a uniform thickness in the curves and the GWDS technology is uses to profile the thickness distribution of the parison to take into account the cross sectional change of the geometry of the tube.

Fig. 7 Retrofitting unit for a head having 6 profiled cylindrical GWDS-dies (diameter 10 mm) and 12 stepper drives to tilt every die individually during the extraction of the parison

References

1 Deutsche Patentschrift (German patent) DE 10 2009 058 361 B3, published: 01.06.2010

O-01-295

PREDICTING THE MELTING BEHAVIOR WITHIN A SINGLE SCREW EXTRUDER USING 3D FVM SIMULATIONS

Ch. Hopmann, C. Kremer and S. Grammel*

Institute of Plastics Processing at RWTH Aachen University (IKV)
Pontstrasse 49, 52062 Aachen, Germany – grammel@ikv.rwth-aachen.de

Abstract - At the IKV in Aachen we are working on a current research project to develop a finite-volume based numerical simulation method, using 3D flow simulations, allowing us to model the melting process in a single screw extruder. The fundamental equations of fluid dynamics are solved using the software Fluent distributed by ANSYS, Inc. The screw geometry is modeled as a helical shaped channel and the principle of reversed kinematics is applied defining the screw as stationary. The solid bed is modeled as a fluid with a very high viscosity by adjusting the formulation of the Carreau model for temperatures below the melting temperature. The model predicts the melting length and the position of the solid bed. Numerical calculations also predict velocity, temperature and pressure distributions within the screw channel and along the screw length. The simulation results provide a better insight into the flow field and the melting mechanisms within a single screw extruder and promote an understanding of how the design of the screw geometry takes influence.

Keywords: FVM, plasticizing screw, melting zone, flight clearance, helical screw channel

Introduction

The design of an extruder screw has a major influence on the extrusion process and decisively determines whether an extruder functions according to requirements. However, up to now, the influence of the screw geometry on polymer processing is not fully understood and in some cases not even predictable.

As a result the design of single-screw extruders depends very much on trial and error methods. Without expert knowledge and experience the design process is inconceivable and forms a major key element in screw design. Yet any misjudgments or miscalculations by the designer can involve significant costs and shortcomings in quality. Analytical and numerical calculation programs used are based on many simplifications only allowing the screw geometry to be predicted with a certain degree of uncertainty.

Thus, we have reached a point that with current available approaches and methods designed screws are often not suitable and have to be remachined or redesigned, involving high costs.

Up to now, most physical-mathematical models of the melting process are based on experiments made by Maddock [1], Rauwendaal [2] and Potente [3]. The first mathematical model of the melting process was developed by Tadmor [4] in 1966. In his model he assumed the screw and barrel to be unwound, while the barrel is moved at an angle across the screw.

In the past decade more and more 3D numerical models were developed to predict the melting process. A three dimensional simulation of the melting process based on the Finite Volume Method was developed by Chang [5]. He used the unwound approach. The phase change from solid to melt is modeled by introducing a function determining the local liquid fraction. The Cross model is used to describe the shear viscosity of the melt. The viscosity for the solid bed and melting zone is described by an effective viscosity. Numerical calculations of pressure and temperature fields as well as the melting process agree relatively well with experimental data. Chen, Lin and Kuan [6] calculate the melting process three dimensionally using a hybrid FEM/FDM approach. The material and melting mechanism are described similar to that of Chang, using an enthalpy formulation. The pressure field is calculated by means of the Finite Element Method while the energy equation is solved by using the Finite Difference Method. The approach of Chen, Lin and Kuan shows a good agreement with experimental data of Tadmor. Pressure fields, throughputs and local shear rates can be predicted. The local melt pool is precluded. Atakan and Gupta [7,8] simulated the melting process three dimensionally. They used an unwound screw-barrel model. The numerical results were in good agreement with the melting mechanism shown by Maddock.

Although most of the models developed up to now show reasonable results and can be used to predict pressure and temperature fields, a fully developed 3D method to calculate and understand melting in single-screw extruders is not yet developed. Therefore, a 3D calculation method developed at the IKV in Aachen, sets out, in conjunction with flow simulations, to predict the melting process, using a 3D helical-shaped screw channel.

Material

A low density polyethylene (LDPE) was used for the simulation work and is supplied by LyondellBasell. The Carreau model, as shown in Eq.1, is used to describe the melt viscosity.

$$\eta_c = \frac{a_T \cdot A}{(1 + a_T \cdot \dot{\gamma} \cdot B)^C} \quad (1)$$

where η is the shear viscosity (Pas), $\dot{\gamma}$ is the shear rate (1/s), a_T is the shift factor and A, B and C are material specific model constants. These constants are obtained using a capillary rheometer and are A = 4233, B = 0,1333 and C = 0,6697. The solid bed is modeled as a fluid with a very high viscosity by adjusting the formulation of the Carreau model for temperatures below the melting temperature, as shown in Fig. 1.

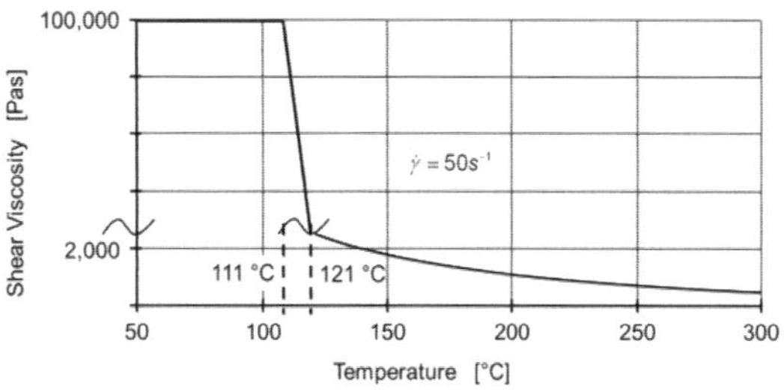

Figure 1: Viscosity for the LDPE.

The temperature shift factor is described by the Williams-Landel-Ferry (WLF) equation (Eq.2) as follows:

$$\log a_T = \frac{8.86 \cdot (T_B - T_{St})}{101.6°C + (T_B - T_{St})} - \frac{8.86 \cdot (T - T_{St})}{101.6°C + (T - T_{St})} \quad (2)$$

where T_B is the reference temperature (°C) and T_{St} is the standard temperature (°C). The density, heat capacity and thermal conductivity of the polymer were modeled temperature dependent.

Model description

In this paper a three-dimensional numerical model is used to analyze the melting process in a single-screw extruder. The conservation equations for mass, momentum and energy of fluid dynamics are solved using the Finite Volume Method. Steady non-isothermal flow simulations are considered for a section of an extruder of variable channel depth and a length of nine times the diameter. The screw geometry is modeled as a helical shaped channel, including the flight clearance. The computational domain is limited by the boundary wall of the screw geometry, as seen in Fig. 2, the barrel, the inlet and the outlet. The computational domain is meshed using hexahedral volumes, with a total of 1,935,000 elements within the model. The operation conditions and screw specifications used in this paper are summarized in Table 1. The barrel is considered to rotate with an angular velocity equal to the screw speed, while the screw is kept stationary. Both the temperatures of the screw and barrel are specified to be 227 °C. A temperature profile is assigned to the inlet of the model as shown in Fig. 3, enabling regions in the vicinity of the screw and barrel to be set with a temperature value above the melting temperature.

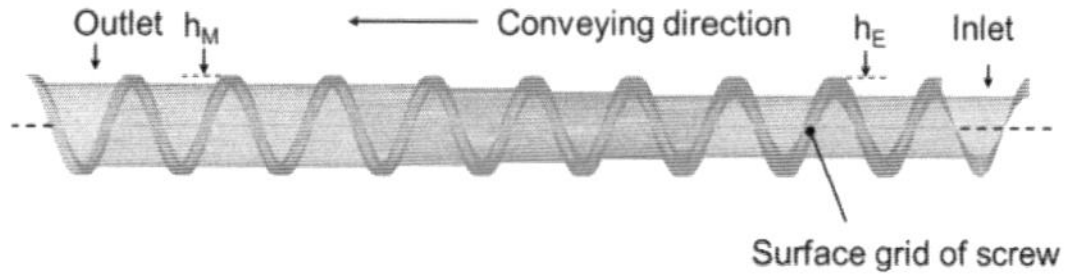

Figure 2: Helical shaped screw channel

Table 1: Model parameters

Model specification	Model 1	Model 2
Channel depth h_E / h_M [mm]	12 / 3	12 / 6
Flight clearance [mm]	0/0.1/0.5/1.0	0.1
Diameter [mm]	60	60
Screw length [D]	9	9
Mass flow rate [kg/h]	86/68/64/43	36
Rotational speed [rpm]	80	40

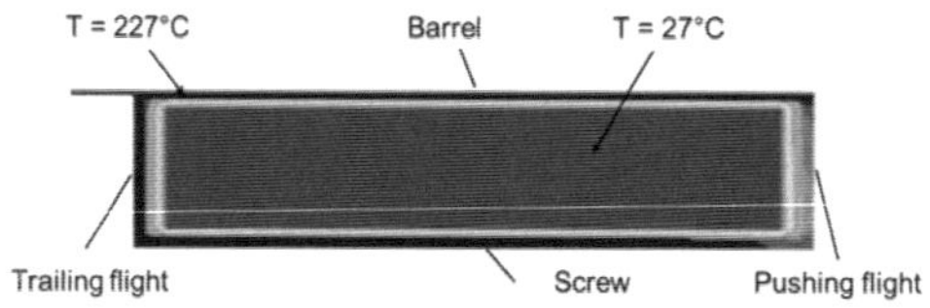

Figure 3:Thermal boundary condition at the inlet

Numerical Simulations

Influence of the viscosity of the solid bed

The influence of the viscosity value assigned to the solid bed is examined in respect to the melting rate, pressure distribution along the screw length and the shape of the solid bed. Four different viscosity values are examined: 50000 Pas, 100000 Pas, 200000 Pas and 300000 Pas. For higher viscosity values than 300000 Pas calculations show no convergence as a result of an increasing discontinuity in the viscosity function. Numerical results show that the viscosity has almost no influence on the melting rate and pressure distribution, but a considerable influence may be observed on the shape of the solid bed. Fig. 4, Fig. 5, Fig. 6 and Fig. 7 display a sequence of cross sectional cuts along the screw channel, parallel to the conveying direction, revealing the solid bed and melt distribution along the screw length. The red regions are regions of high viscosity resembling the solid bed, while the blue regions are regions of low viscosity and resemble the melt. Comparing the shape of the solid bed for the highest and lowest viscosity in Fig. 2 and Fig. 5, it may be observed that the less viscous the fluid of the solid bed is the greater its ease of movement and the less its resistance to be deformed by shear stress. Therefore, the width of the solid bed assigned with the lower viscosity is much greater as its resistance against the axial force exerted on it by the moving barrel is lesser. The higher the viscosity value for the solid bed is the lesser is the deformability and the width of the solid bed. Though the higher viscosity values are more in accordance to the melting model of Maddock, it remains a challenge to choose the corresponding viscosity, relating to the real solid properties of a solid bed.

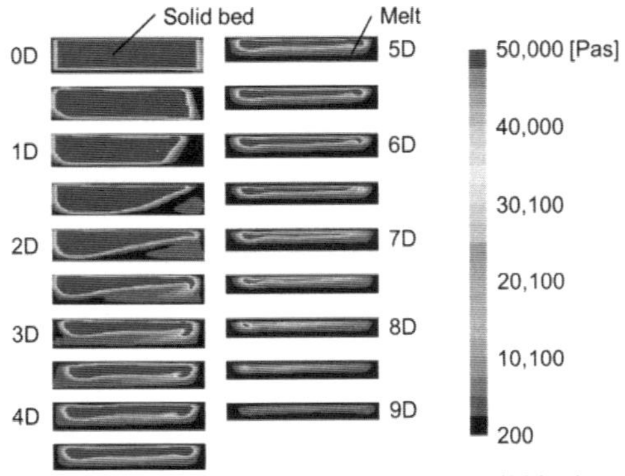

Figure 4: Shape and location of the solid bed (50,000 Pas)

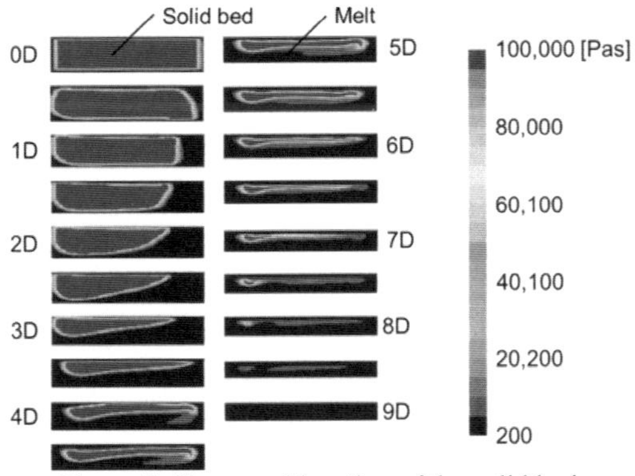

Figure 5: Shape and location of the solid bed (100,000 Pas)

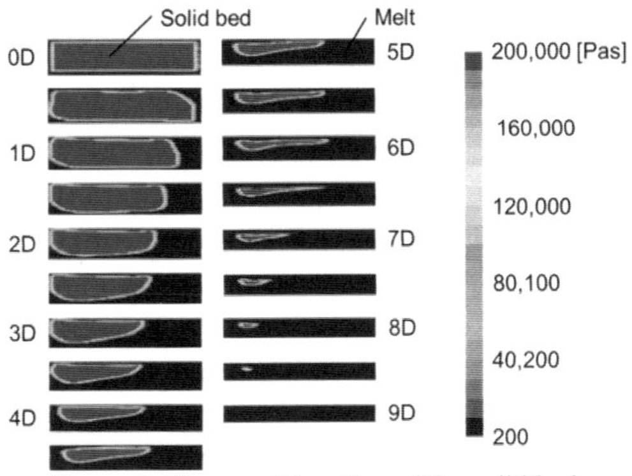

Figure 6: Shape and location of the solid bed (200,000 Pas)

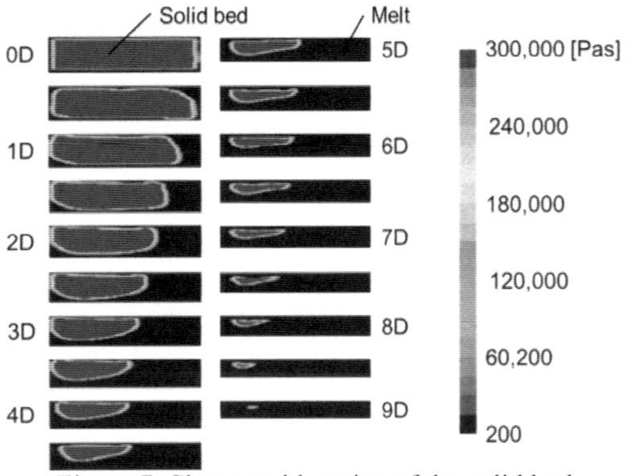

Figure 7: Shape and location of the solid bed (300,000 Pas)

Influence of the flight clearance

The influence of the flight clearance is examined in respect to the melting rate and pressure distribution along the screw length. The flight clearance is varied from 0 mm 0.1 mm, 0.5 mm and 1 mm clearance. Fig. 8 shows a plot of the pressure distribution along the screw length for different flight clearances. Results show that the flight clearance has an effect on the performance of an extruder. The highest pressure build-up of 15 MPa at 0.45 m is achieved by considering no flight clearance. With a flight clearance of 1 mm a pressure build-up of only 10 MPa at 0.54 m is achieved, accounting for a considerable pressure loss of as high as 5 MPa due to leakage flow across the flight clearance.

Fig. 9 shows a plot of how the flight clearance influences the melting performance of an extruder. Results show that when the flight clearance increases from 0.5 mm to 1 mm it reduces the melting length by a considerable amount of almost 2D. As the flight clearance increases, the melt temperatures inside the screw channel may rise considerably, explaining the effect on the melting length. Without any clearance between the flight and the barrel the melting length drops to approximately 7.5D, as the additional heat generated within the clearance space is not taken into account.

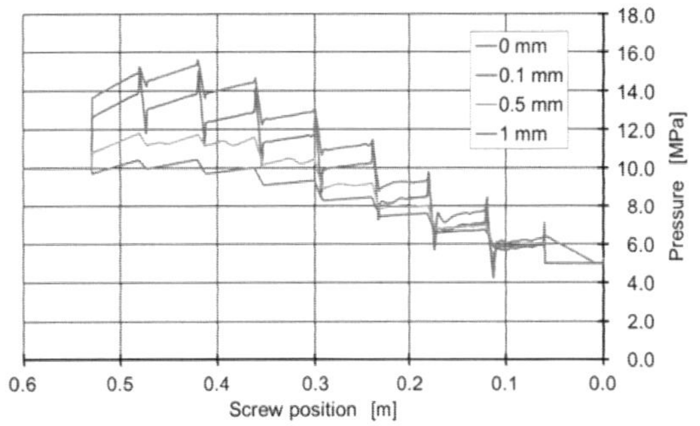

Figure 8: Influence of the flight clearance on the pressure distribution

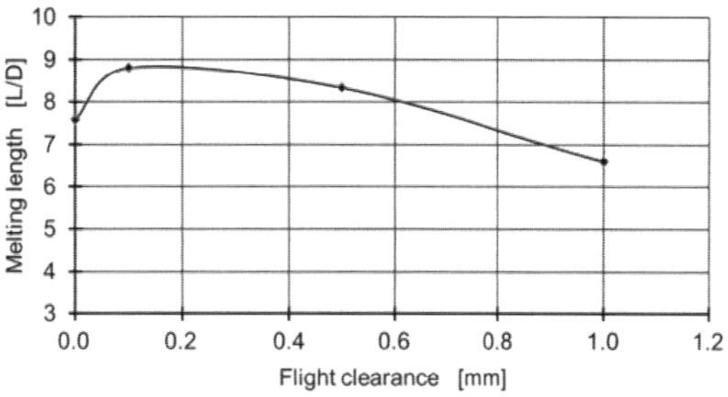

Figure 9: Influence of the flight clearance on the melting length

Conclusions

In this paper a finite-volume-based 3D numerical model to analyze the melting process in a single-screw extruder was presented. The model predicts the melting length and the position and shape of the solid bed within the helical shaped screw channel. Simulations show the influence of the viscosity of the solid bed on the shape and location of the solid bed and processing conditions. Calculations also show how the flight clearance influences the extrusion performance in respect to pressure and melting rate.

Future work

Future work will hold the verification of numerical results with experimental data. Upcoming work will also hold the simulation of the melting process within a barrier screw. Future work will also present the influence of the temperature settings of the barrel and the screw on the melting process.

Acknowledgements

The research project 16513 N of the Forschungsvereinigung Kunststoffverarbeitung has been funded as part of the "Industrielle Gemeinschaftsfoerderung und –entwicklung (IGF)" by the German Bundesministerium für Wirtschaft und Technologie (BMWi) due to an enactment of the German Bundestag through the AiF, to whom we extend our thanks.

References

1. B. Maddock, *SPE ANTEC Tech. Papers*, **15**, 383 (1959).
2. C. Rauwendaal, *Polymer Extrusion*, Hanser, Munich (2000).
3. H. Potente, *Tagungsband zum Seminar Einschneckenplastifizierextruder*, **2**, 2001.
4. Z. Tadmor, *Fundamentals of Plasticating Extrusion*, Polym. Eng. Sci., **6**, 185 (1966).
5. R.Y. Chang, C.W. Hsu and W.Y. Wang, *Three-dimensional numerical analysis of the single screw plasticating extrusion process*, *SPE ANTEC Tech. Papers*, **47**, 735 (2001).
6. C.H. Chen, K.C. Lin, C.F. Kuan and H.C. Kuan, *Hybrid FEM/FDM approximation for single screw extrusion*, SPE ANTEC Tech. Papers, 375 (2007).
7. A. Altinkaynak and M. Gupta, *Investigation of melting in single-screw extruder*, *SPE ANTEC Tech. Papers*, 809 (2006)
8. A. Altinkaynak, M. Gupta, M.A. Spalding and S. L. Crabtree, *Melting in a Single Screw Extruder: Experiments and 3D Finite Element Simulations*, Intern. Polymer Processing XXVI, Carl Hanser Verlag, **2**, 182-196 (2011)

O-02-616

EVALUATION OF PROCESSING CONCEPTS FOR POLYOLEFINS IN ROTATIONAL FOAM MOLDING

Remon Pop-Iliev*

Faculty of Engineering and Applied Science, UOIT-University of Ontario Institute of Technology, Oshawa, Ontario, Canada L1H7K4 – remon.pop-iliev@uoit.ca

Abstract - Several distinct rotational foam molding processing concepts, fundamentally based on the conventional rotational molding process, have emerged in the recent decades and gradually evolved into technologies that advantageously enable the creation of a cellular polymeric structure, i.e., a distinct foamed layer or core within the interior of hollow rotational moldings that can be entirely encapsulated within an integrated layer of solid polymeric skin. However, the rotational molding process inherently suffers from very lengthy and energy-intensive processing cycles, the duration of which can be significantly augmented when processing integral-skin foamed moldings. This is usually attributed to the insulative effects of the thereby formed cellular structure that is creating a delay especially during the mold cooling segment, which combined with the unavoidable thermal gradients across the mold, directly governs the efficiency of the process control. This paper focuses on evaluating the suitability of these technologically-distinctive processing concepts for successfully transforming engineering resins into rotationally molded cellular polymeric composite articles of controllable densities in terms of achieving simultaneous retention of fine-celled foam morphologies while maintaining a uniform thickness of the solid skin and while reducing the processing cycle times and energy consumption. Various polyethylene (PE) and polypropylene (PP) grades were selected to create experimental formulations for processing parts of various shapes consisting of PE based skin and PE based foam, and PP based skin and PP based foam, as well as the combination of the two.

Keywords: rotational molding, polymeric foams, integral skin.

Introduction

The nature of the conventional rotational molding process is cyclic. In its simplest terms, its processing cycle includes subjecting a pre-charged (with plastic), separable (cast or fabricated), vented, metal mold to biaxial rotation (or at least a rocking motion) into a heated area (oven), and subsequently, into a cooled environment (air and/or water sprinklers), after which the mold is opened, the part is removed, and the cycle repeated [1-6]. Large-sized, single-piece, lightweight and durable plastic products that could fulfill challenging demands even under extremely severe conditions of application are permanently needed in the marine, automotive, materials handling, road signage, toys, leisure and other industries [1, 2]. Integral-skin foamed rotational moldings, commonly denoted as cellular composites ideally having a clearly distinct surface layer of solid skin of uniform thickness that is encapsulating a seamlessly coupled fine-celled foamed core or layer of uniform cell density and distribution, seem to be the ideal candidates for accomplishing these objectives. In this context, over recent decades a variety of interrupted and non-interupted rotational foam molding processing concepts have been developed to convert conventional rotational molding into a plastics fabrication method that advantageously allows the creation of a cellular structure, i.e., a distinct foamed layer or core within the interior of hollow rotational moldings that can be encapsulated entirely within a solid skin integrated boundary layer.

Single-Charge Rotational Foam Molding

For a number of years the state-of-the-art for fabricating integral-skin rotationally molded cellular composites was the single-charge rotational foam molding technology. It is based on simultaneously charging the mold with predetermined quantities of non-foamable and foamable resins having a significant particle size difference. The skin-forming non-foamable resins are introduced into the mold in a powder form, whereas the foamable resins intended for forming the foamed core or layer are first extrusion melt-compounded with an adequate quantity of chemical blowing agent (CBA) and then introduced into the mold in a pellet form [7]. The powder particles sinter earlier than the pellets because of their greater total contact area and because their smaller size allows them to migrate towards the internal surface of the rotating heated mold where the temperature is the highest. Thereby, the powder will preferably fuse and form the skin layer before the uninterrupted heating of the mold initiates pellet sintering and subsequently trigger the decomposition of the CBA particles dispersed within the foamable pellets to form the foamed core [8, 9]. This concept may be effectively improved by selecting non-foamable resins that have a lower melting point, density, and viscosity [10], and/or heat capacity [11] than the respective properties pertaining to the foamable resins. In addition, the non-foamable skin-forming resin should preferably comprise of a low and a high zero shear viscosity components [12]. Yet, the mold could be charged with a blend of non-foamable and foamable powders (e.g., reground compounded pellets of a polymer with a CBA) of polymers that have significantly dissimilar melting points and/or significantly dissimilar particle sizes [13]. Also, a more intensive mold rotation and an optimized oven temperature profile are helpful for preventing the premature adherence of the foamable pellets on the skin [14].

Disadvantages of the Single-Charge Method

The production of integral skin cellular polymeric composites using the single-charge rotational foam molding process suffers from undesirably lengthy processing cycle times, high energy-intensiveness, and the lack of efficient means for real-time process control during both the heating and cooling segments of the cycle. This is so because the temperature of the mold and the plastic it is charged with must be first elevated from room temperature to beyond its melting point and then cooled back to room temperature. This disadvantage is augmented even further when processing integral-skin cellular polymeric composites as a result of the extended duration of the mold cooling segment, due to the insulative effect of the developed foam layer or core within the mold. Consequently, it becomes very difficult to effectively control the process so to retain fine-celled foam morphology and skin thickness uniformity. In addition, previously conducted research indicated credible evidence of a strong counter-proportional causality between the duration of the rotational molding cycle and the quality of the obtained cellular morphologies. Moreover, the single-charge process suffers from numerous intrinsic processing contradictions that have to be simultaneously addressed, which is very difficult to achieve because of the quite limited number of controllable processing variables in real time. These include: (i) remove unwanted pin-holes and bubbles formed on and within the solid skin; (ii) ensure completion of the formation of the solid skin layer before the beginning of the activation of the foamable core-forming resin; (iii) prevent CBA-decomposition-evolved blowing gas losses; (iv) prevent foam invasion into the solid skin layer and preserve uniform solid skin thickness, and (v) govern the nucleation, growth, and retention of cells with controlled sizes and desired densities throughout the adjacent foamed core. This makes the process quite difficult to control and not very user friendly. As such, it has not been widely accepted by rotomolders.

Rapid Rotational Foam Molding (RRFM)

Rapid Rotational Foam Molding represents an innovative approach for fabricating integral-skin cellular composites based on rotational molding. It essentially exploits the unique synergistic effects resulting from the deliberate conjunction of extrusion with the rotational molding process. In this context, instead of making decomposition-free foamable solidified pellets in a remote extrusion-based melt compounding operation involving a carrier resin and a chemical blowing agent, the Rapid Rotational Foam Molding processing method offers a design solution that completely decouples the shaping of the part's solid skin from the formation of the skin-encapsulated foamed layer or core. Thereby, improved process controllability, significantly shorter processing cycle duration and reduced energy consumption can be readily achieved. The Rapid Rotational Foam Molding

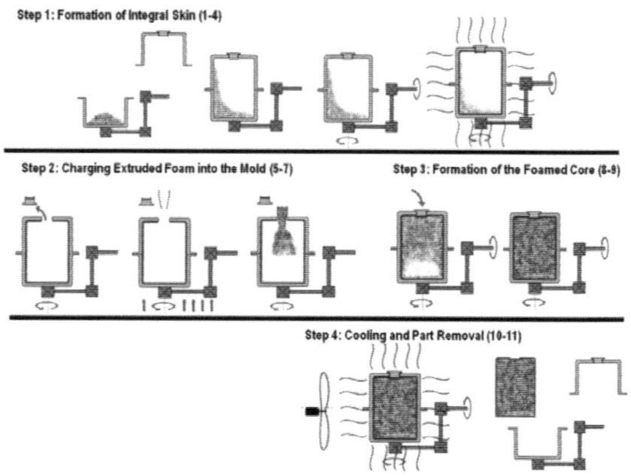

Fig. 1: The 4 Processing Steps of RRFM

processing concept can be summarized into four unique steps depicted in Figure 1.

Step 1: At the inception of the process, the mold is manually charged with a predetermined amount of non-foamable plastic resin in powder form, which will be used to create the part's solid skin. *Step 2:* The arm and the mold assembly are inserted into the oven by means of a carriage and translational mechanism. The mold rotates in a biaxial manner within the oven, utilizing the arm. This step is accomplished at a material-dependent elevated temperature for an accordingly set period of time to accomplish the creation of the skin. Simultaneously, the extruder is charged with predetermined amounts of dry blended polymer resin with a CBA, suitable for achieving polymeric foam that exhibits a desired volume expansion ratio (VER) at a set melt temperature. *Step 3:* The arm and mold assembly are then translated towards the extruder, with the arm rotation switched to a uni-axial mode and the mold aligned with the extrusion port. The extruder subsequently fills the hot mold with extrudate comprised of the desired foam for a period of time dependent on the melt flow rate and screw RPM. The filling process occurs at the mold-extruder interface gate, which has been designed so as to seal the mold during the conventional rotational molding cycle, facilitate the introduction of foam to it during the foam filling stage, and allow the skin to self-heal after the filling process is accomplished thereby allowing the skin to settle back at its original location, undamaged. *Step 4:* The mold is then bi-axially rotated and cooled. Subsequently, the solidified part is removed [15]. In summary, the Rapid Rotational Foam Molding processing concept is distinctively based on directly introducing non-chilled extruded foam on top of the already formed un-foamed hot skin in real time and via a specially designed extruder nozzle-to-mold interface port. The primary purpose of this novel processing approach is to provide the means for achieving: (i) perfect bonding at the interface between the solid skin and the foamed core or layer; (ii) obtain fine cell density in rotationally molded integral skin foams within a significantly wider processing and

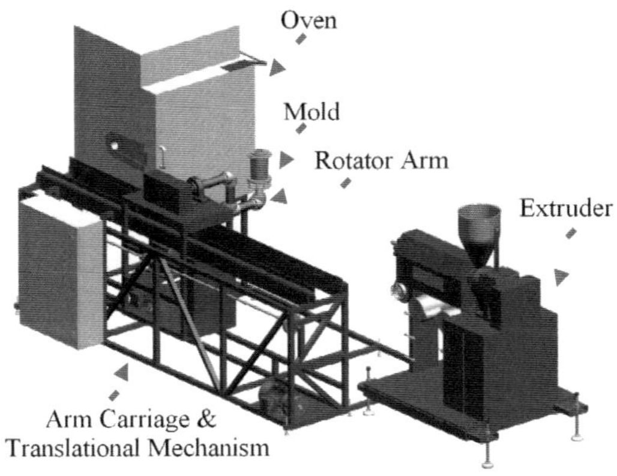

Fig. 2: The Experimental Setup for RRFM

materials window through increased process controllability in comparison with the currently achievable, as well as to (3) dramatically reduce the duration of the processing cycle and consequently increase the efficacy and utility of the process.

Experimental

Multiple batteries of planned Rapid Rotational Foam Molding experimental trials resulting in both polyethylene (PE)- and polypropylene (PP)-based integral-skin rotational moldings were conducted utilizing the custom-built industrial-grade lab-scale Rapid Rotational Foam Molding experimental setup (Fig. 2) and by utilizing three distinct mold shapes. Three typical mold shapes: a cylindrical, a flat plate and a box mold have been subjected to a comparative experimental study to validate mold design solutions through the obtained product quality.

Results and Discussion

The experimental results revealed that Rapid Rotational Foam Molding creates advantageous potentials for the manufacture of advanced multi-layered single-piece cellular composites with a uniform skin wall thicknesses. Fig. 3 depicts typical comparative charts of foamed morphologies

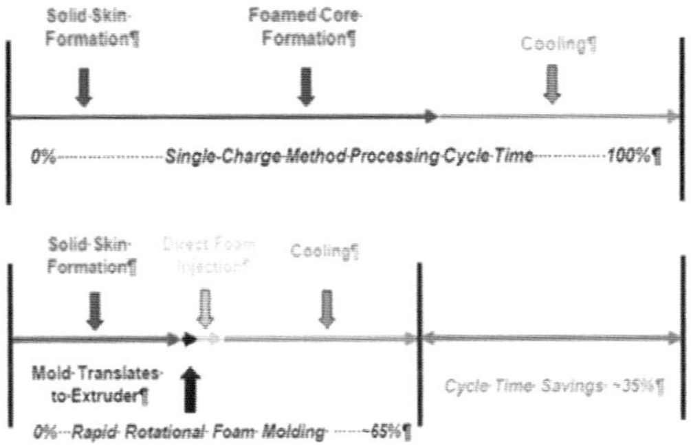

Fig. 4: Single Charge vs. RRFM Cycle Time Comparison

characterized in terms of foam density and average cell size for different polyolefins and volume expansion ratios (VER), respectively. while reducing the processing cycle times and energy consumption to a fraction of previous levels. This is so because, in essence, non-chilled extruded fine-celled foam is directly introduced into the uni-axially rotating mold via a specially designed "self-repairing" interface port to form a foamed layer or core on top of the hot solid skin shell pre-formed using the traditional method. Once the mold is filled-up with foam as desired, the port is closed and the mold is externally cooled to retain the fine-celled foam morphology and solidify the product for removal. In comparison with the single-charge method, Rapid Rotational Foam Molding results in a significant (~35%) decrease in cycle time (Fig. 4), which translates into a noteworthy energy consumption reduction. The experimentally obtained foam morphologies and skin-foam interfaces were satisfactory for both PE and PP grades. Since melt extrusion has been successfully integrated into the rotational foam molding process, it becomes possible to eventually implement physical blowing agents in rotational foam molding for producing a novel class of microcellular polyolefin composites and nano-composites which will possess properties that are currently not achievable. Importantly, this opens a unique opportunity for implementing a wide variety of

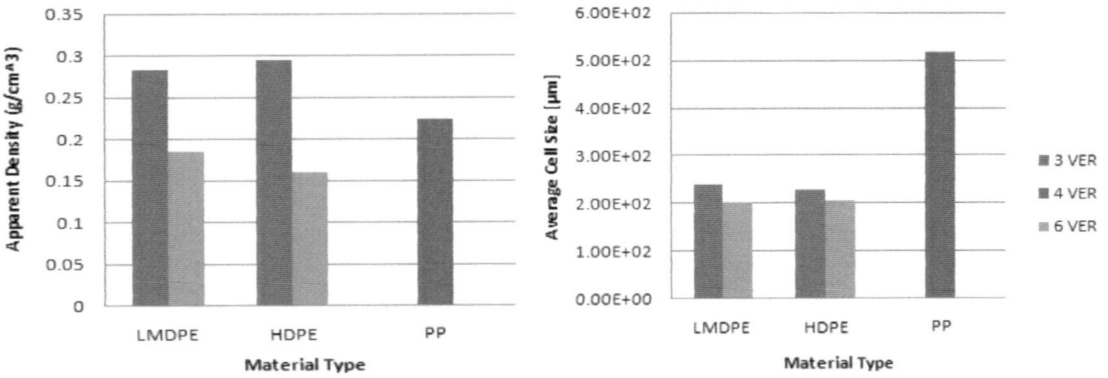

Fig. 3: Foam Density vs. Average Cell Size Comparison

PE and PP grades designed for extrusion foaming for forming the foamed core or layer in Rapid Rotational Foam Molding and thereby compensate for the lack of dedicated PE and PP grades for rotational foam molding.

Conclusions

The Rapid Rotational Foam Molding processing method represents a novel technology intended for manufacturing high-quality lightweight integral-skin cellular rotationally molded composites. Unlike the prior art, the novel Rapid Rotational Foam Molding method is based on intentionally decoupling the traditional rotational molding process for obtaining a hollow article (the solid skin) from the creation of an integral solid skin-surrounded foamed core or layer. This advantageously resulted in devising a less time consuming and less energy intensive technologically-distinctive advanced approach to the processing of polyolefins into rotationally molded cellular polymeric composites of controllable densities and with significantly improved mechanical properties compared to conventionally-processed respective analogs. By exploiting and building on this unique synergy between the traditional rotational molding and the extrusion foaming process it becomes possible to explore and eventually implement physical blowing agents in rotational foam molding for the first time. The successful development of the Rapid Rotational Foam Molding process and its future industrial scale-up will make attainable the manufacture of the next-generation of integral-skin rotationally molded cellular polymeric composites with advanced mechanical, energy-absorption and thermal properties suitable for applications in most challenging environments, not integrally instituted in the past. Such foam-reinforced integral-skin cellular polymeric composites form a very desirable class of products because they can satisfy severe service requirements while demonstrating very high strength-to-weight ratios. Prime applications include floating devices, navigational buoys, and lightweight, ultra-strong and insulative articles demanded by the automotive and the aerospace industry.

Acknowledgements

The author would like to acknowledge the NSERC Discovery Grant and the NSERC/GM Design Chair's funding in support of this project and the contributions of Ms. Kimberly Christian, Mr. Emad Abdalla and Mr. Gregory Eberle, as well as the donation of material and/or services by Equistar Chemical, Crompton Corporation, Lyondell Basel and Ingenia Polymers Inc.

References

[1] R.J. Crawford, J.L. Throne, Rotational Molding Technology, William Andrew Publishing (2002).

[2] R.J. Crawford, M.P. Kearns, Practical Guide To Rotational Moulding, Rapra Technology Limited, (2003).

[3] G.L. Beall, Rotational Molding: Design, Materials, Tooling, and Processing, Hanser/Gardner, Cincinnati, (1998).

[4] G.L. Beall, J.L. Throne, Hollow Plastic Parts: Design and Manufacture, Hanser/Gardner, Cincinnati, (2004).

[5] J.L. Throne, Thermoplastic Foams, Sherwood Techn., Inc., Sherwood Publ., Ohio (1996).

[6] G. Liu, C.B. Park, and J.A. Lefas, Polym. Eng. Sci., 38, 1997 (1998).

[7] R. Pop-Iliev, C.B. Park, Journal of Cellular Plastics, Vol. 39, No. 1, pp. 49-58 (2003).

[8] R. Pop-Iliev, G.M. Rizvi, C.B. Park, Polymer Engineering and Science, Vol. 43, No. 1, pp. 40-54 (2003).

[9] R. Pop-Iliev, D. Xu, C.B. Park, Journal of Cellular Plastics, Vol. 40, No. 1, pp. 13-25 (2004).

[10] C.E. Slade, Canadian Pat., CA983226 (February 10, 1976).

[11] H. Mori, E. Adachi, and Y. Nogichi, U.S. Pat. 3,962,390 (June 8, 1976).

[12] C.C. Lee, U.S. Pat. 5,928,584 (July 27, 1999).

[13] D.G. Needham, U.S. Pat. 5,366,675, (November 22, 1994).

[14] R. I. Kliene, British Pat. GB2,311,245 (September 24, 1997).

[15] R. Pop-Iliev, Acta Physica Polonica A, Vol. 120, No. 2, pp. 292-297 (2011).

O-02-591

THE INFLUENCE OF PLUG MATERIALS ON CONTACT SLIP IN THERMOFORMING

PJ Martin*, HL Choo and CPJ O'Connor
School of Mechanical & Aerospace Engineering
Queen's University Belfast, United Kingdom – p.j.martin@qub.ac.uk

Abstract - Over many years a wide variety of materials have been employed as plugs for thermoforming including metals, woods and various polymer and polymer-based composites. Little agreement currently exists on the optimum choice of these materials as the measurement of friction between soft polymer sheets and plug materials is extremely difficult in practice. This is especially the case when it is considered that slip occurs exclusively through extension of the heated sheet across the contacting surface. It is for this reason that contact slip in thermoforming remains a largely unresolved area of research. In this work the magnitude of slip was experimentally measured for a range of common plug materials. Preform shapes were created by pushing specially designed plugs into heated sheet samples and then freezing the resulting wall thickness distributions. These were then compared in order to provide an indirect method of highlighting the relative magnitudes of slip and thereby friction for the combinations of materials in contact. Four plug materials (acetal (POM), nylon and 2 grades of syntactic foam) were included in the study along with three polymer sheet materials (polypropylene (PP), polystyrene (PS) and aPET). As well as the choice of materials a variety of other processing parameters such as the plug and sheet temperatures and the plug displacement were included in the experiments. The results show that under the same processing conditions, the choice of plug material profoundly changes the magnitude of slip observed. With higher plug temperatures the syntactic foam plugs tended to constrain contact slip more than that of acetal and nylon. However, this pattern was reversed at lower plug temperatures thereby highlighting the strong interdependence of material and temperature in determining the magnitude of slip during contact. These trends were similar for all three sheet materials tested.

Keywords: Thermoforming, Plug, Slip, Friction.

Introduction

Plugs are a common feature of many thermoforming processes and are often essential for ensuring that the wall thickness distribution in a deep-draw thermoformed part is controlled and balanced. Generally they are designed to locally capture the initial sheet thickness and greatly reduce its deformation in key areas prior to the final blowing stage that completes most thermoforming processes. Contact friction between the soft polymer sheet and the rigid plug surface therefore plays a fundamental role in controlling the magnitude of deformation that may occur locally and it is therefore critical in determining the overall wall thickness distribution (WTD). Various experimental studies have been carried on plug-assist thermoforming in order to gain deeper understanding and better control of the process. Typically the process takes place in two stages with the movement of a mechanical plug being used to pre-stretch the heated polymer sheet before forming is completed through the application of air pressure. Previous work has clearly demonstrated the profound effect that the action of the plug has on the final wall thickness distribution produced [1,2]. It is a deliberate feature of the action of plugs that the contacting sheet material is encouraged to slide across its surface during contact and this helps to redistribute thickness [3]. Some general process parameters that have been shown to have a significant effect on the wall thickness distribution include the plug design (shape), polymer sheet properties, plug properties, and plug and sheet temperatures [4]. Measurement of the magnitude of slip during plug contact would be of considerable interest but very little work has been reported. This largely reflects the difficulty in making such measurements, particularly in industry. However, several research groups have attempted to measure contact frictional properties relevant to thermoforming in the laboratory [5,6].

The aim of this study was to further investigate the effects of slip during plug contact through a series of dedicated thermoforming experiments involving different combinations of plug and sheet materials. A key feature of the experiments was their concentration purely on the initial plugging phase of the thermoforming process. This enabled the magnitude of plug slip during the process to be studied in detail. As well as building knowledge of the effects of frictional properties and consequently sheet slip, the results of this work are also of great importance for the development of effective simulations of plug-assisted thermoforming. An ultimate aim of the work was to provide quantifiable data on frictional effects during the process to assist in the development of accurate thermoforming process simulations.

Experimental

Thermoforming of the polymer sheets was performed using the plug assisted thermoforming process. An open ring mould was used so that the final output was the preformed shape created by plug contact alone. After heating the sheet to the required temperature, the plug was pushed into the sheet, and at the end of its movement the sheet was cooled immediately to prevent further deformation and to retain the preform wall thickness distribution. For these tests a special doughnut-shaped plug tip design was created (as shown in Fig. 1) to help to encourage slip during contact with the sheet. The plug materials chosen for this investigation were POM-C, HS Blue Nylon and the Syntactic Foams; Hytac-B1X and Hytac-WFT, all of which are commonly employed as plug materials for thermoforming. The sheet materials investigated were typical thermoforming grades of high impact polystyrene, polypropylene and aPET with starting sheet thicknesses of 0.8, 0.8 and 0.5mm respectively.

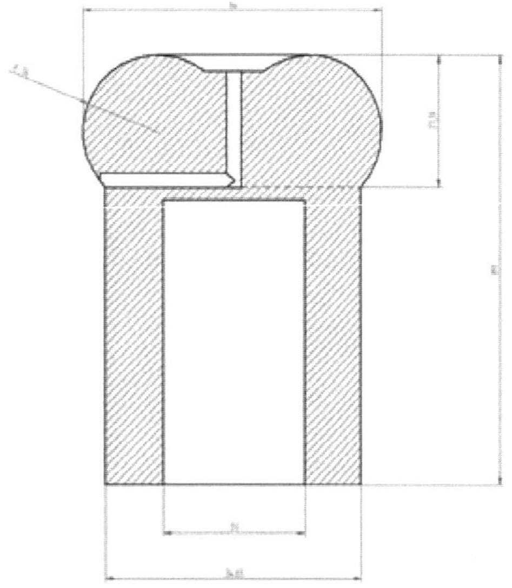

Figure 1. - Doughnut-ended plug design

Table 1 – Plug thermoforming test settings

Test Variable	Settings Used
Plug Material	POM, WFT, B1X
Sheet Material	PP, PS, aPET
Doughnut Radius (mm)	8, 12 & 16
Plug Displacement (mm)	48, 73 & 96
Average Plug Speed (mm/s)	100, 200
Sheet Temperature (°C)	157(PP), 150(PS), 100(PET)
Plug Temperature (°C)	25 & 91 (PP), 25 & 87.5 (PS), 25 & 62.5 (PET)

The other major test variables were: plug displacement; doughnut radius; average plug speed; sheet temperature and plug temperature. These are detailed in Table 1. The sheet temperatures chosen for each sheet material were their normal forming temperatures. The plug temperatures chosen were room temperature (25°C) and the halfway point between this and the chosen sheet forming temperature for the sheet material. In addition, for aPET a plug temperature equal to its forming temperature was also chosen. This was not possible for either PP or HIPS due to risk of melting the plug materials at their much higher forming temperatures.

Results and Discussion

A typical formed part is shown in Fig. 2. The wall thickness was measured along the points shown in the diagram, starting from the centre of the base of the preform.

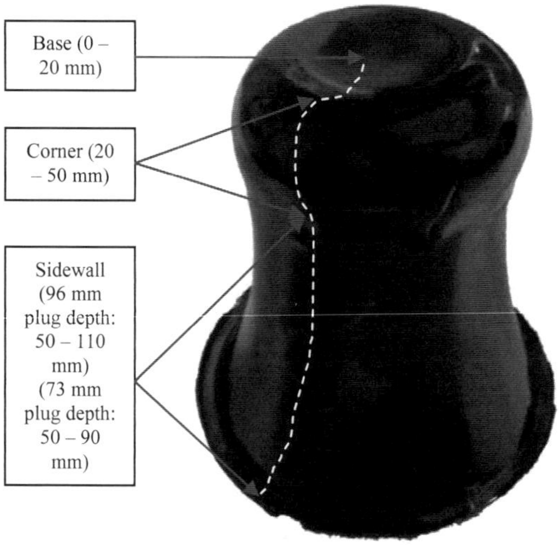

Figure 2. - Typical formed part

Fig. 3 shows how the wall thickness distribution of PP preforms change with plug temperature for the 4 different plug materials: Hytac B1X, Hytac WFT, Nylon and POM. For all plug materials it can be clearly seen that the wall thickness at the base increases with increasing plug temperature. This shows that the ability of the sheet to slide across the plug is reducing with increasing temperature, which confirms that the coefficient of friction is rising with temperature. The effect of sheet cooling due to heat transfer, which might have been expected to counteract this effect, appears to be minimal. Of the plug materials Hytac B1X gives the stickiest contact at the higher plug temperature whereas it is nylon that provides the highest friction at lower temperatures. In all cases the magnitude of slip observed determines the balance in the WTD between the base of the preform and sidewall.

Fig. 4 shows how the wall thickness distribution of PS preforms change with plug temperature for the 4 different plug materials. The overall results are markedly different for those of PP with the material thickness in the base of the preform being much thicker than that in the sidewall regardless of plug temperature.

This suggests a much stickier contact, but it is also likely that the relative ease with which PS may be stretched at forming temperatures also contributes to the large reduction in wall thickness in the sidewall of the preform.

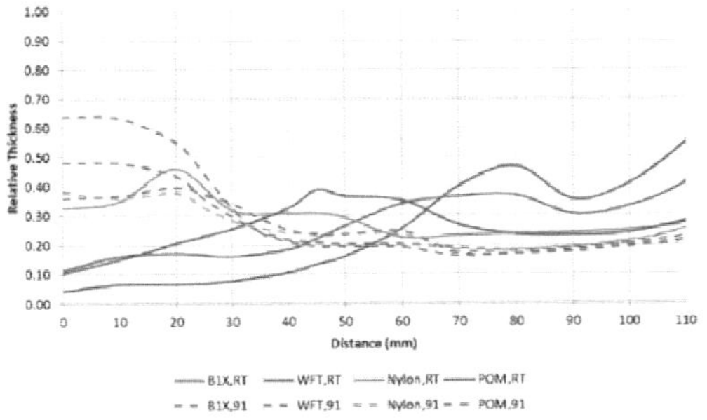

Figure 3 – WTD of PP formed at 157 °C using various plug materials at plug temperature of 25 °C (RT), 91 °C. The average plug speed and plug depth were 200 mm/s and 96 mm respectively.

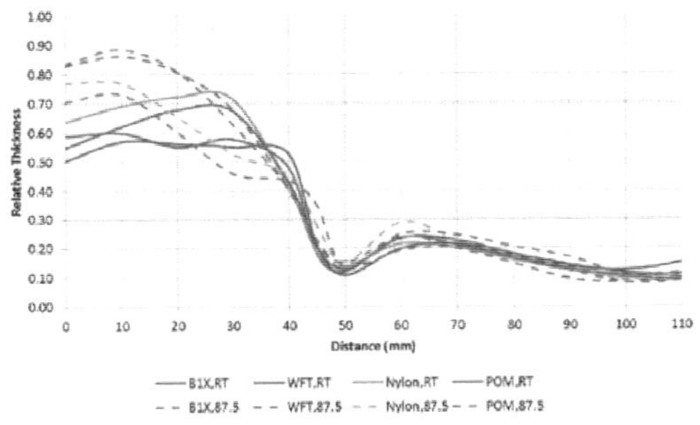

Figure 4 – WTD of PS formed at 150 °C using various plug materials at plug temperature of 25 °C (RT), 87.5 °C. The average plug speed and plug depth were 200 mm/s and 96 mm respectively.

Fig. 5 shows how the wall thickness distribution of aPET preforms change with plug temperature for the 4 different plug materials. These results are broadly similar to those for PP although with aPET the WTD in the preform is close to balance especially at the intermediate plug temperature. The most evenly distributed wall thickness in this example is for a plug temperature of 62.5°C. It is again clear that the magnitude of slip across the surface of the plug reduces as the plug temperature increases. In this case Hytac WFT produces the least slip at the highest plug temperature, with POM the most. However, almost the opposite results are observed at the lowest plug temperature.

For all of the results it is clear that at all plug temperatures there are small but significant differences in the preform WTDs for the different plug/sheet material combinations. This suggests that the selection of an appropriate plug material for a given sheet material in thermoforming is an important decision and likely to significantly change the wall thickness distribution in the production part.

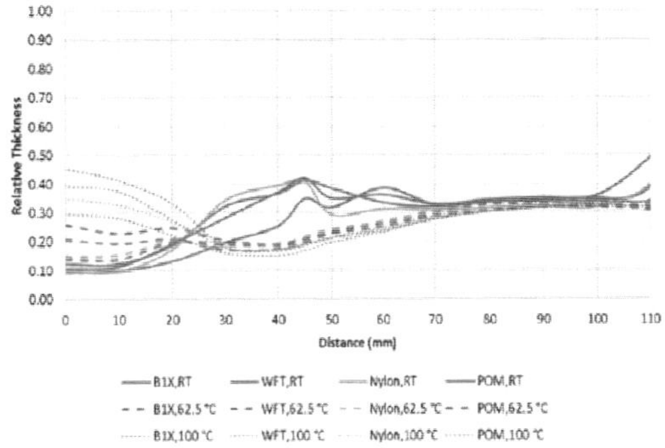

Figure 5 – WTD of aPET formed at 100 °C using various plug materials at plug temperature of 25 °C (RT), 62.5 °C, and 100 °C. The average plug speed and plug depth were 200 mm/s and 96 mm respectively.

Using the results from different plug displacements it was possible to study the progressive development of the WTD in the preform. This is illustrated in Fig. 6 for aPET and the plug material Hytac WFT. In the graph 'distance' refers to the measured distance from the centre of the base of the preform to the clamped edge of the sheet. The 'base', 'corner' and 'sidewall' regions are highlighted. The results confirm that initially all regions are stretched but in the later stages the upper sidewall no longer stretches and the predominant deformation occurs in the base region which suggests that slip occurs at different rates during the process.

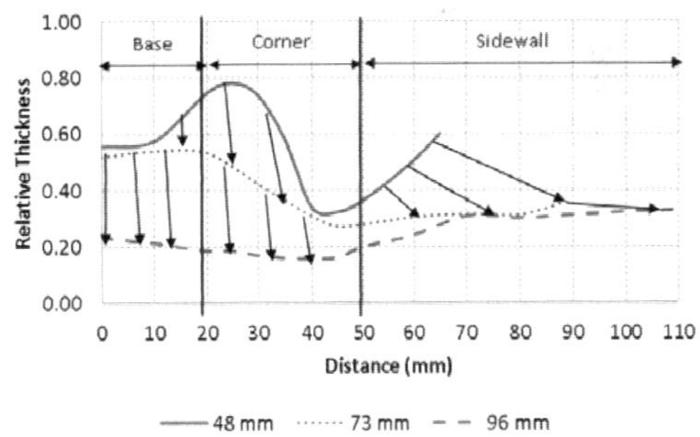

Figure 6 – WTD of aPET formed using WFT at average plug speed of 200 mm/s, plug temperature of 62.5 °C, and plug depths of 48, 73, and 96 mm.

Conclusions

This work has confirmed the importance of plug friction in determining the wall thickness distribution in parts made by thermoforming. Measurements of plug slip show that it occurs progressively with plug displacement, although its effects are most pronounced in the latter stages. It is highest at low plug temperatures and it reduces as the plug temperature is increased. These results suggest that contact friction rises with the temperature at the contacting surface and that interfacial cooling of the sheet has a negligible effect on the magnitude of slip. For the four plug materials investigated there were clear differences in behaviour, and the results show that under the same processing conditions, the choice of plug material profoundly changes the magnitude of slip observed. With higher plug temperatures the syntactic foam plugs tended to constrain contact slip more than that of POM and nylon. However, this pattern was reversed at lower plug temperatures thereby highlighting the strong interdependence of material and temperature in determining the magnitude of slip during contact. These trends were similar for all three sheet materials tested, although the observed differences were even larger with PS giving the lowest slip (highest friction) and aPET the highest slip when tested against Hytac WFT. A finite element based simulation of the plugging process is currently being constructed and it will be used to investigate different friction relationships and compare their performance with the experimental behaviour.

Acknowledgements
The authors would like to thank the European Commission and the members of the PlugIn research project consortium.

References

1. P. Collins; J.F. Lappin; E. Harkin-Jones; P.J. Martin *Plastics Rubber Comps* 2000 *29* 349.
2. R. McCool; P.J. Martin *Polym. Eng. Sci.* 2010 *50* 1923.
3. P. Collins; E.M.A. Harkin-Jones; P.J. Martin. *Intl. Polym. Process.* 2002 *17* 361.
4. P.J. Martin; P. Duncan *Polym. Eng. Sci.* 2007 *47* 804.
5. H. L. Choo; P.J. Martin; R. McCool; E.M Harkin-Jones in Proc. Polymer Processing Society 23rd Annual Meeting, Salvador, Brazil, May 2007.
6. B. Hegemann; P. Eyerer; N. Tessier; K. Kouba; T. Bush in Proc. ANTEC 2003 Conf., Nashville, USA, vol. 1, 2003 791.

O-02-584

LOCAL HEATING USING LASER RADIATION FOR IMPROVED WALL THICKNESS DISTRIBUTION OF THERMOFORMED PARTS

B. Neubig[1]* and C. Bonten[1]

[1] *Institut für Kunststofftechnik, University of Stuttgart, Germany, Böblingerstr. 70, 70199 Stuttgart – benjamin.neubig@ikt.uni-stuttgart.de; christian.bonten@ikt.uni-stuttgart.de*

Abstract - Thermoforming generally describes the shaping of a thermoplastic semifinished part above its softening temperature to a three dimensional part. The process is used for a wide field of applications, such as industrial and packaging goods. However, one of the big challenges is the characteristical nonuniform wall thickness distribution of the shaped product. Hence, in conventional processes pre-blowing or plug assisted prestretching is used to achieve an improved uniformity. The deformability of a thermoplastic sheet also strongly depends on its processing temperature. Taking advantage of this behavior, a novel approach for heating a semifinished part was recently developed at IKT. Therefore, the stage of heating was separated into two technological process steps. At first, the semifinished part is heated by means of a conventional radiant heater from two sides. In a second step, certain areas of the sheet are additionally and locally heated by means of a laser beam. In this manner, a well defined temperature profile was generated that secures a locally different deformability of the sheet.

In this study, the developed heating technique is described, using the example of a plug assisted negative pressure forming of semifinished polystyrene. The test rig is presented and the laser heating process is discussed by means of thermal pictures of the sheet surfaces. In addition, the laser heating process is described mathematically. Finally, experimental and calculated results are compared. Furthermore, a particular generated temperature profile and its effect on the wall thickness distribution for a negative formed part is discussed.

Keywords: Thermoforming, Laser Heating, Temperature Profile, Wall Thickness Distribution

Introduction

In thermoforming the generation of a homogenous or even well defined wall thickness distribution of the finished part is difficult and has ever been a big challenge [1]. For manufacturing large-area products, e.g. bathtubs, it is known that zonal or pattern heating can be applied in order to improve the wall thickness distribution of the final part [2]. Zonal heating can be achieved by a specific temperature adjustment of the heater elements which effects the local energy input to the sheet [1]. Pattern heating acts in the same way by shielding the thermal radiation of the heaters by means of an aluminum plate for instance [2]. Nevertheless, the sharpness of the generated temperature profiles is limited with these methods.

For this reason a novel approach for generating temperature profiles for the manufacturing of small-sized parts has been investigated in this study. Therefore, the stage of heating was separated into two steps. At first, the semifinished part is heated by means of a conventional radiant heater system from two sides. In a second step, a sharp and well defined temperature profile is generated by additional heating certain areas using a laser beam.

The big challenge of this method is to heat a certain area of the sheet quasi-simultaneously without damaging the sheet surface. The energy intensity in the laser focus is very high and the laser focus is very small in comparison to the area that has to be heated. In order to estimate meaningful laser and scanning parameters a mathematical approach for calculating the increase in temperature has been used.

This mathematical approach is based on the heat transfer equation, including internal source term:

$$\frac{\partial}{\partial t}(\rho \cdot \vartheta) = div\left(\frac{\lambda}{c_p} \cdot grad\vartheta\right) + \frac{\dot{s}}{c_p} \qquad \text{Eq. 1.}$$

Herein ρ is the density, λ is the thermal conductivity and c_p is the heat capacity. The source term $\dot{s}$ is in accordance with the internal heat energy input per unit of time and volume. For the local heating with the laser beam, this source term can be derived from the Lambert-Beers' law [3]:

$$I(z) = I_0 \cdot e^{-\alpha \cdot z} \qquad \text{Eq. 2.}$$

I_0 is the intensity absorbed by the material's surface and α is the absorption coefficient of the material at the specific wavelength of the laser beam. For simplicity, an ideal circular laser focus and a constant heat flux distribution in the cross section of the laser beam have been assumed. With this assumption I_0 is given by:

$$I_0 = (1-R) \cdot \frac{P_L}{\pi \cdot r_f^2} \qquad \text{Eq. 3.}$$

Herein R is the reflection coefficient on the sheet surface, P_L is the laser power and r_f is the focal radius. The internal source term as function of depth can be derived as:

$$\dot{s} = \frac{dI(z)}{dz} = -\alpha \cdot (1-R) \cdot \frac{P_L}{\pi \cdot r_f^2} \cdot e^{-\alpha \cdot z} \quad \text{Eq. 4.}$$

Finally, the translational movement of the laser focus can be implemented by using the equation for a circle:

$$(x-x_0)^2 + (y-y_0)^2 \leq r_f^2 \quad \text{Eq. 5.}$$

With that equation the coordinates x_0 and y_0 of the center of the focus and every grid point inside the focus (x and y) can be defined for a certain scanning path and a certain scanning velocity.

Experimental

Machinery

A thermoforming apparatus for laboratory use has been converted to a test rig. This assembly is shown in figure 1 schematically. The test rig is equipped with a radiation heating system consisting of an upper and a lower heating unit each built up of 16 identical discrete ceramic heater elements (Elstein 60 mm x 60 mm; 150 W). Above the upper radiation heating unit, a laser system, including a laser beam source and a beam deflection unit, has been mounted and connected with the control system of the thermoforming machine. In this manner, the whole thermoforming process runs automatically and in a reproducible way. At first, the sheet is moved towards the heating system and evenly heated up on both sides. Then the heating system is pulled over and subsequently, the laser beam generates a specific temperature profile on the sheet surface. That followed, the sheet is retracted to the molding station where shaping takes place.

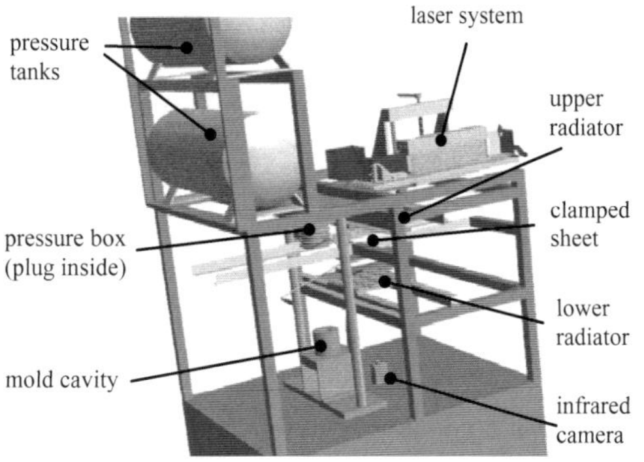

Fig. 1: Design of the test rig

The installed laser system consists of a CO_2-laser beam source, a lens telescope and a galvanometer scanner with integrated plane-field focusing lens. The parameters that can be varied in this laser system are given in table 1.

Tab. 1: Specification of the laser system

laser source	
wavelength	10.6 µm
nominal power	0 to 70 W
operating mode	continuous wave
galvanometer scanner	
scanning velocity	0 to 3.5 m/s
focal distance	435 mm
focal diameter	0.6 to 1.6 mm
area of operations	300 mm x 300 mm

Test Item and Material

In the test rig a female aluminum mold with a single cavity for an axially symmetrical bowl-shaped part is installed. The cavity has an aperture diameter of 70 mm and a bottom diameter of 51 mm. Its maximum draw depth is 75 mm.

For all investigations presented sheets made of impact resistant polystyrene (STYROLUTION: Polystyrol 486 M) with a thickness of 1000 µm have been used.

Experimental Procedure

With the aim to heat a certain area as fast as possible different scanning strategies according to figure 2 have been investigated with regard to temperature increase and scanning time. For these investigations, the largest possible focal diameter and scanning velocity of 1.6 mm and 3.5 m/s have been used whereat the laser power and path distance h have been varied. Adjacent, experimental and calculated results have been compared. Therefore, thermal pictures of the upper and lower sheet surface have been made directly after heating with the laser beam. The installed infrared camera (Cedip Infrared Systems: Jade UC) has been calibrated to the sheet material. However, due to higher accuracy the thermal pictures had to be made without preheating the sheets at room temperature.

For the calculations, the absorption coefficient α at the radiation wavelength of the CO_2-laser ($\lambda = 10.6$ µm) has been determined by means of infrared spectroscopy and using equation 2.

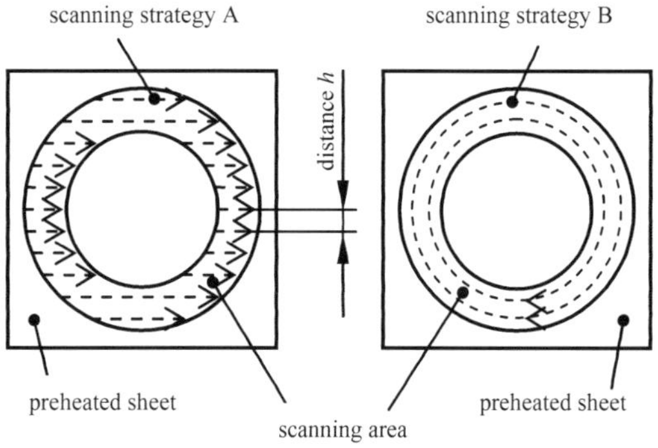

Fig. 2: Different scanning strategies for local heating using laser radiation

In order to improve the wall thickness distribution of the test sample described above, a specific temperature profile has been developed. Appropriate forming parameters for this sample have been determined in preliminary studies. Best results have been achieved by using a plug made out of hardened felt and a forming pressure of 0.5 MPa. Based on these parameters, samples have been manufactured with and without local heating. The samples have been evaluated by measuring and comparing the wall thickness distribution.

Results and Discussion

Absorption Coefficient
With measurements using infrared spectroscopy the absorption coefficient has been determined to be $\alpha \sim 1/100\ \mu m^{-1}$. Due to this value of α most of the laser radiation is absorbed in the surface region of the sheet and thus the laser acts as surface heat source.

Scanning Strategy
First investigations by scanning a single line (length 100 mm) on the sheet surface have shown that a maximal laser power of 40 W can be used until any visible changes on the sheet surface occur. Figure 3 shows the measured temperature distribution of the upper sheet surface for the two scanning strategies according to figure 2. The thermal pictures show that in both cases the ring profile is heated quite homogeneously and nearly up to the same temperature increase. The left picture indicates a temperature increase of more than 8 K and the picture on the right side close to 8 K in the middle of the heated area. However, the overall scanning time differs considerably. With strategy A, the sheet has been heated in an overall scanning time of 0.8 s using a path distance of 1 mm. In comparison, the heating time for strategy B has been 1.3 s, although a path distance of 2 mm has been applied. The difference in scanning time can be explained by the lower number of control steps the galvanometer scanner needs for scanning a line instead of a circle. Consequently, local heating by means of a hatching (strategy A) seems to be the appropriate way.

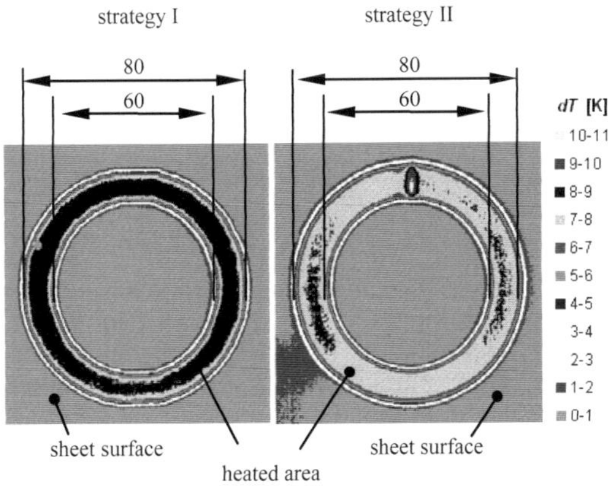

Fig. 3: Thermal pictures of the upper sheet surface after laser heating

The thermal pictures in figure 3 indicate the measured temperature distributions at a cooling time of 3.2 s after the irradiation by the laser beam. In the test rig this is exactly the time that passes until the sheet is moved from the heating station to the forming station. Figure 4 shows the temperature distribution of the lower sheet surface at that time after laser heating by means of strategy A. This temperature distribution is almost homogenous with marginal higher values on the left side. Due to thermal equilibrium, the temperature increase in the middle area of the ring is about 8 K on the average. Hence, directly before forming, the temperature distribution through the thickness can be seen as homogenous, although the laser impinges the sheet only on the upper side.

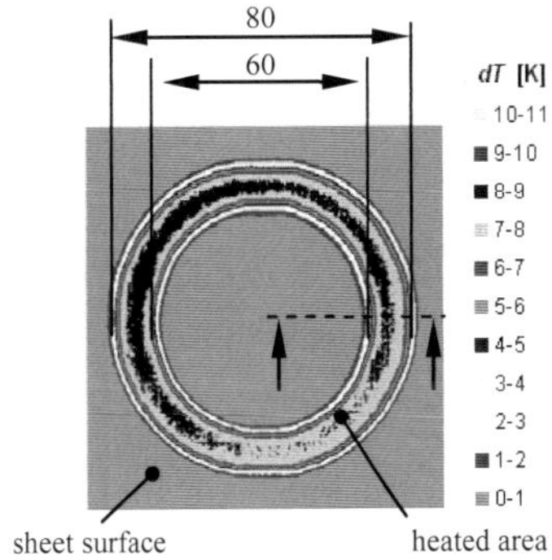

Fig. 4: Thermal picture of the lower sheet surface after laser heating using strategy A

Measurements and Calculations
Figure 5 and figure 6 show measured and calculated values for the temperature increase on the upper and on the lower sheet surface. The path distance has been 1 mm and 2 mm (strategy A) and a cooling time of 3.2 s after irradiation has been applied. The laser power has been set to 40 W. The values are illustrated for the temperature distribution along the dashed line that is indicated in figure 4 for a circular ring with inside diameter of 35 mm and outside diameter of 45 mm. The calculations have been implemented using a measured heat capacity of 1.2 kJ/kg·K, a thermal conductivity of 0.16 W/m·K and a density of 1.04 kg/m³ [4]. The reflection coefficient has been assumed to be 0.05 [5]. Also, cooling due to convection on the sheet surfaces has been considered at which a heat transfer coefficient of 5.7 W/m²·K [1] has been assumed.

The calculations have been carried out by solving the partial differential equation 1 three-dimensionally based on the finite difference method according to Crank and Nicolson [6]. The values show quite good consistencies at the maximum value of the temperature increase in the middle of the heated ring area for both path distances and on both surface sides. Solely, the values of the lateral temperature distribution show small discrepancies. One possible reason for that is, that the Lambert-Beers' law has been implemented only one-dimensionally in the thickness direction of the sheet and thus scattering inside the material has been neglected. However, the measurements and the calculations show that by keeping the laser power, the focal diameter and the scanning velocity constant, different temperature increases can be achieved solely by varying the distance between the scanning paths. Furthermore, the mathematical approach and the calculation method are appropriate to estimate laser and scanning parameters, if a certain temperature increase is required.

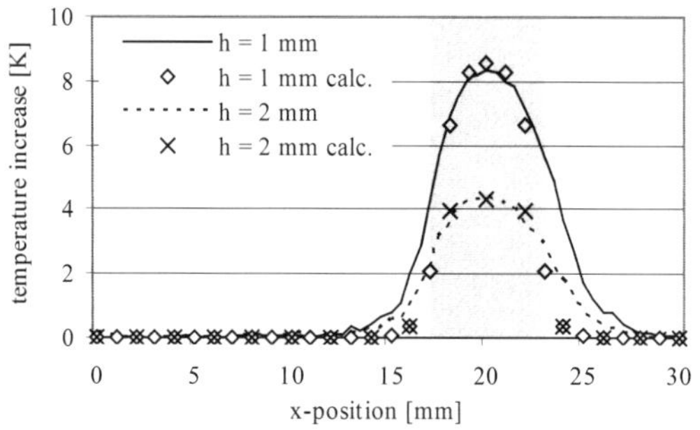

Fig. 5: Measured and calculated temperature increase on the upper sheet surface

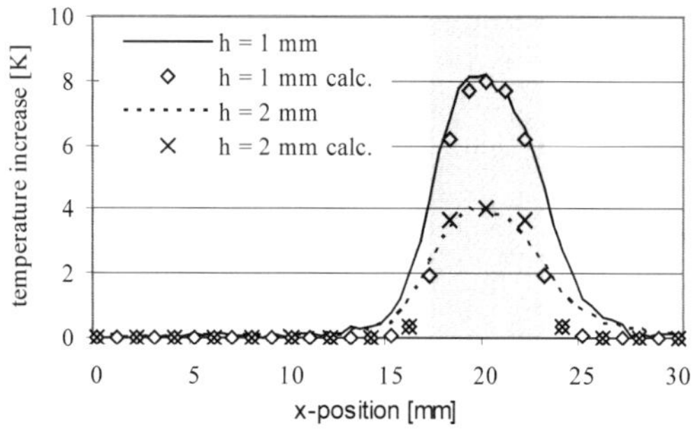

Fig. 6: Measured and calculated temperature increase on the lower sheet surface

Improvement of Wall Thickness Distribution

All parts that have been manufactured without local heating have shown a typical wall thickness distribution. This wall thickness distribution contains three characteristical thick sections as illustrated in figure 7. The first one can be explained by the high energy loss due to heat conduction through the clamping. The other two are mainly related to the high friction between the sheet and the plug. These areas are barely stretched during prestretching and remain as thick sections after pressurized air has been applied. Hence, in order to achieve a more homogenous wall thickness distribution, these areas at the unstretched sheet have been additionally heated by the laser beam.

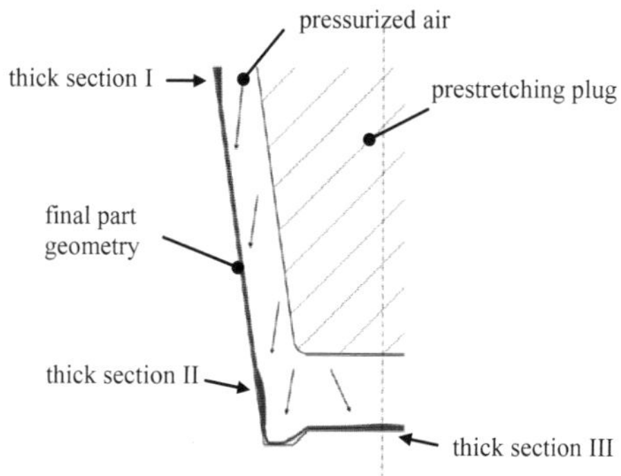

Fig. 7: Typical wall thickness distribution of formed parts without local laser heating

The effect of this local heating is illustrated in figure 8. The figure shows measurements of wall thickness distribution of parts, manufactured with the same forming parameters, but once with local heating and once without. In comparison, applying the local heating all thick sections could be reduced significantly. Altogether the wall thickness along the side wall and along the bottom of the part that has been manufactured by local heating is more homogenous than the one of the other part.

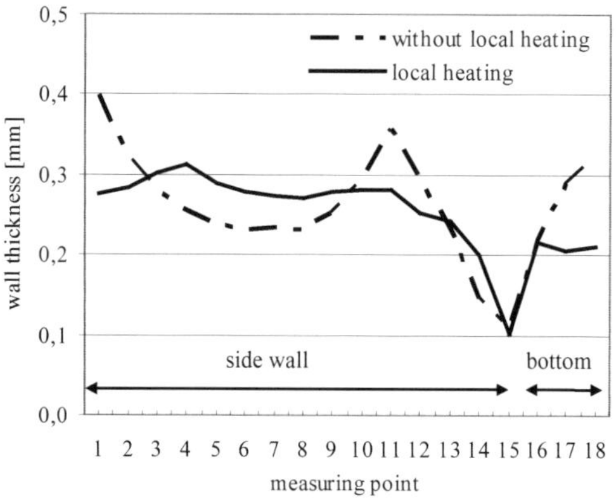

Fig. 8: Comparsion of wall thickness distribution of formed parts with and without local laser heating

Conclusions

This study shows that a laser beam can be an appropriate tool for locally heating a semifinished part in thermoforming applications. Measured and calculated temperature increases for a certain circular ring profile have presented good consistencies. Furthermore, a specific temperature profile generated with the laser beam lead to a more homogenous wall thickness distribution of a thermoformed part.

Acknowledgements

The authors thank Laser Systems GmbH for excellent cooperation and Styrolution Group GmbH for providing material.

References

1. J. L. Throne, *Technology of Thermoforming,* Hanser, New York, 1996
2. A. Illig, *Thermoformen in der Praxis*, Hanser, München, 1997
3. H. Hügel, *Strahlwerkzeug Laser*, B. G. Teubner, Stuttgart, 1992
4. Verein Deutscher Ingenieure, *VDI-Wärmeatlas*, Springer, Berlin, 2006
5. C. Bonten, C. Tüchert, Journal of Reinforced Plastics and Composites, 2002, 21, 699
6. R. Marek, W. Götz, *Numerische Lösung von partiellen Differentialgleichungen mit finiten Differenzen*, Moreno, Buchloe, 1995

O-02-505

FABRICATION OF POLYMERIC MICROFLUIDIC DEVICES BY MICROTHERMOFORMING FOR BIOMEMS APPLICATIONS

Nicole E. Steidle*, Anja Lepple, Marc Schneider, Markus Heilig, Ralf Ahrens, Matthias Worgull and Andreas E. Guber

Institute of Microstructure Technology, Karlsruhe Institute of Technology, Germany – nicole.steidle@kit.edu

Abstract – Tubes with micro- and nanostructures on the inner and outer walls for biological and medical devices have been successfully fabricated. For fabrication a combination of hot embossing and thermoforming was used. The structuring of the surface allows the adjustment of the surface characteristics to an application. Wetting behavior as well as cell behavior can be regulated. This enables improvements in biocompatibility, influences on cell movement, cell alignment and cell adhesion to the surface as well as controlled wetting behavior.

Keywords: Hot Embossing, Thermoforming, Microfluidics

Introduction

Microfluidic devices, like micro catheters, are important in daily medical care. Commonly micro-extrusion is used for the mass production of catheters with diameters typically in the range of 0.3 mm to 10 mm and a wall thickness down to 0.05 mm [1].

To improve biocompatibility, wearing comfort and wearing time of catheters surface modifications are necessary. The influence of micro- and nanostructures on surfaces on the wetting and cell behavior is widely debated. Many examples can be found in nature which show the influence of surface texture on the wetting behavior. The most known example is the lotus leaf. With the combination of hierarchical micro- and nanostructures and wax coatings it is possible to achieve high contact angles with low tilting angles thus enabling the self-cleaning effect of the lotus leaf [2]. Also cell behavior can be influenced by micro- and nanostructures. Yoon et al. discussed the passive control of cell locomotion, the cells movement and location, using micropatterns [3]. Other groups also examined the influence of surface structures to the cell adhesion or alignment [4]. The change of the surface texture will therefore have an effect on the surface properties. For catheters this could improve biocompatibility, wearing comfort as well as wearing time.

To achieve a change of the surface properties surface modifications typically like nanoparticle coatings and physical or chemical surface treatments are used. While using these methods challenging tasks are not to contaminate the system by particles or fractions of the coating [5] as well as to ensure the stability of the new surface characteristics. To assure the stability of the surface treatment and to avoid introducing another material in a medical device a direct structuring of the catheter is preferable to achieve the wanted surface properties.

In micro extrusion structuring of the surface is very difficult. Lee et al. patented an extrusion method with a sacrificial mandrel to fabricate microstructured parts [6]. With this method it is possible to produce micro parts with inside diameters of around some 10's of micrometers. After the extrusion, the mandrel will be etched away.

In this work we introduce a fabrication process which uses hot embossing and thermoforming to produce micro parts for biological and medical applications e.g. micronozzles, catheter tips and microneedles. Hot embossing is a method used for the fabrication of micro- and nanostructures where high-precision manufacturing and precise control is needed [7]. Thermoforming or blow molding is a well-known method for the fabrication of three-dimensional shapes. By heating and introducing a gas pressure thermoplastic polymer foils are formed into a mold. Commercial examples in the macroscopic range reach from coffee cups to car bumpers. In the microscopic scale thermoforming is used for the fabrication of medical devices like miniature balloons for a variety of purposes [8]. The fabrication of structured three dimensional shapes was already discussed by Heilig et al. [9, 10]. He used a combination of hot embossing and thermoforming to fabricate three dimensional shapes with micro or nanostructures on one surface.

This work shows a promising fabrication process to be capable of the mass production of low-cost surface-structured, biocompatible polymeric medical devices, like catheters with micro- and nanostructures on its inner and outer walls. Fig. 1 shows a schematic view of a micro- and nanostructured tube.

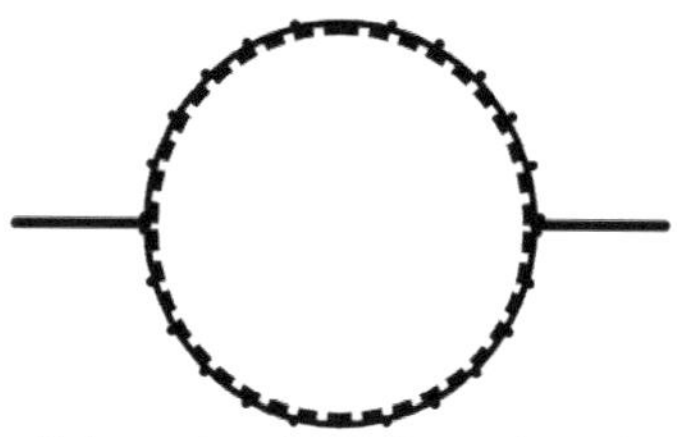

Figure 1: Schematic view of a tube with different structures on the inner and outer wall.

Experimental

Hot Embossing:

Hot embossing is used for high-precision manufacturing of micro- and nanostructures. With hot embossing it is possible to fabricate structured thermoplastic polymer foils which can afterwards be used for thermoforming or twin sheet forming respectively. In this work Polycarbonate (PC) and Polysulfon (PSU) were chosen as materials due to their biocompatibility and therefore high usage in biological and medical applications. To adapt the wetting behavior of a fluid on the outer and inner wall of the tube the foils were structured on boths sides in one process step. A schematic view of the hot embossing process is shown in Fig. 2.

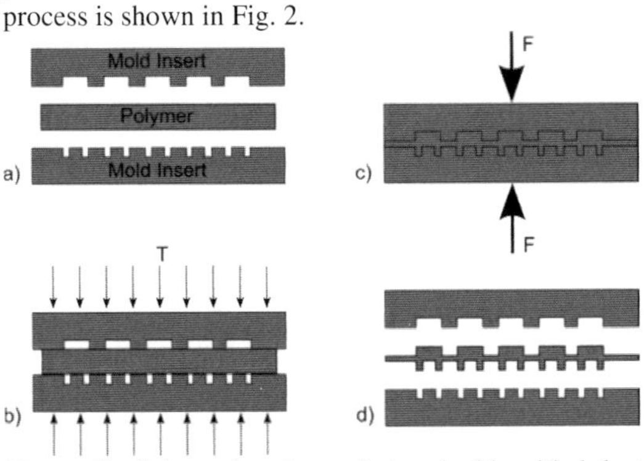

Figure 2: Schematic view of the double sided hot embossing process. a) A thermoplastic polymer is put between to mold inserts which are connected to a heating plate. b) The mold inserts are closed until a touch force is reached and the system is heated to the molding temperature. c) The molding force is applied and d) after cooling the embossed polymer foil can be demolded.

At the beginning one polymer foil is put between two differently structured molds inserts. These mold inserts are connected to heating plates to heat the inserts to the molding temperature. The molding temperature is higly dependent on the material and therefore Differential Scanning Calorimetry (DSC) measurements were executed to get the thermic properties of the polymer foils. The information given in these measurements were also used for the thermoforming process. For PC a molding temperature of 200 °C for PSU a molding temperature of 255 °C was chosen. After inserting the polymer foil the machine will close until a touch force of 500 N is applied. The mold inserts will then be heated to the molding temperature. When the molding temperature is reached a molding force is applied and the polymer foil is isothermally structured. The molding force was chosen to be 75 kN for PC and 50 kN for PSU. The molding force was depending on the size of the structured area. During cooling the molding force is kept until the demolding temperature is reached. After demolding the structured foil can be used for thermoforming. To improve the molding process a the process chamber with mold inserts and polymer foils was evacuated.

Thermoforming:

In this work a special case of thermoforming was used called twin-sheet. Instead of only forming one foil, twin-sheet forming enables the shaping of two foils in one process step. Another benefit is an in process bonding achieving three dimensional shapes like tubes. With a preceding structuring of the foils by hot embossing it is possible to fabricate three-dimensional shapes with micro- and nanostructures on its surface. Fig. 3 shows a schematic view of the twin sheet forming process. Two pre-structured polymer foils are clamped between two mold inserts. The mold inserts are connected to heating plates which heat the inserts and the clamped foil to a temperature below the polymer's glass transition temperature. This low temperature is necessary to prevent the destruction of the pre-embossed micro- and nanostructures during the forming process. The clamping force is around 50 kN to assure the sealing of the mold. After reaching the temperature pressurized gas is inserted between the two foils with a pressure between 0.5 MPa and 5 MPa. Through the gas pressure the polymer foils are pressed into the cavities. The pressure as well as the clamping force is kept constant until the demolding temperature is reached. To improve the forming result an evacuation of the cavities in the mold is done before the forming process.

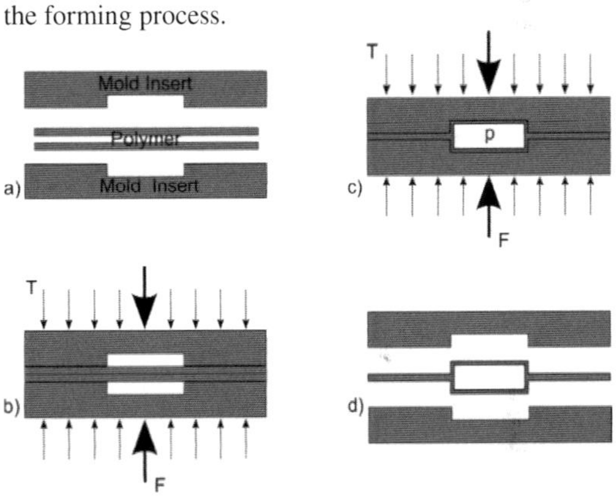

Figure 3: Schematic view of the twin-sheet forming process. a) Two thermoplastic polymer foils are put between two mold inserts. b) The system is closed and heated to the forming temperature. A clamping force is applied. c) After reaching the molding temperature the two foils are pushed into their mold inserts by pressurized gas. The pressure is kept constant until the demolding temperature is reached. d) After depressurizing the three dimensional microstructure can be demolded.

Separation:

Two methods were used for the separation of the thermoformed tubes. First laser cutting with a UV-Laser of a wavelength of 248 nm was used. Different distances to the outer wall of the tube were tested to ensure, that the thermo bonded area was still big enough to keep the two foils together. Fig. 4a shows the cutting area with a distance of 120 µm to the outer wall. The chosen distance was not enough to keep the polymer foils bonded. With a distance of 140 µm, shown in Fig. 4b, the bonded area is sufficient to keep the two foils joined.

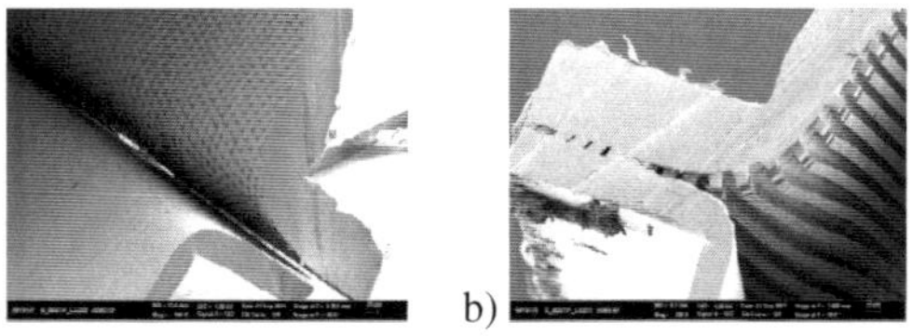
a) b)

Figure 4: Laser cut areas with a distance to the outer wall of a) 120 µm and b) 140 µm.

Second we constructed a hot-cutting method to cut the thermoformed foils with a heated blade. A schematic view of the layout can be found in Fig. 5. Due to the heated blades the polymer is melted during cutting. This improves the bonding strength of the two polymer foils. To achieve a clean cutting line it is important to have a temperature around the glass transition temperature. This assures that the structures on the inner and outer wall are not destroyed during separation due to heat input. Fig. 6 shows the dependency of the temperature and holding time of the blades to the polymer during the cutting process.

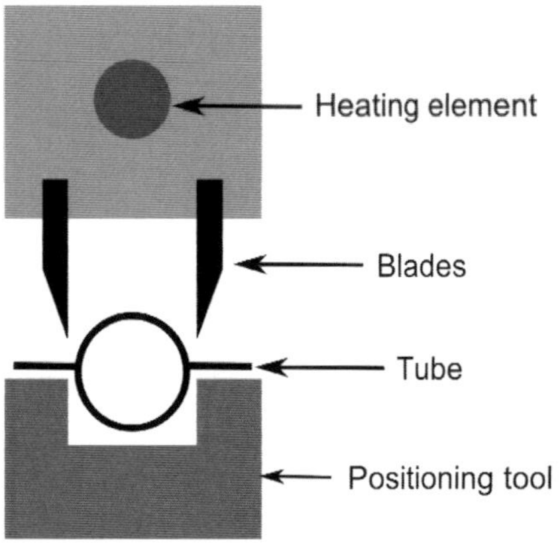

Figure 5: Schematic view of the cutting mechanism. Two heated blades are used to cut through the residual layer of the micro tube. Due to the hot blades the polymer gets melted in the cutting area thus improving the bonding strength.

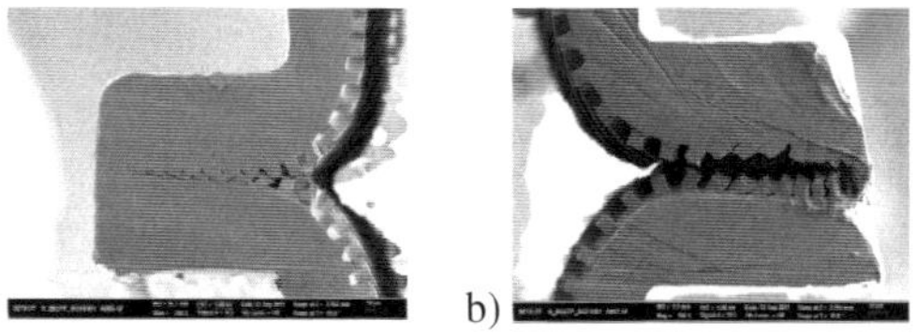
a) b)

Figure 6: Hot-cut residual layers with a cutting-temperature of 145 °C and a) a holding time of 3 minutes and b) without any holding time.

Results and Discussion

Characterization of surface structures:

After hot embossing the structured foils were characterized by dimensions, regularity of structuring and wetting behavior. To characterize the fabricated tubes three methods were used. Scanning Electron Microscopy (SEM) was used to determine distance, diameter and regularity of the micro- and nanostructures. Fig. 7 shows the SEM picture of a hot embossed foil structured with pillars. It also shows the good regularity of the structures over the whole area. To verify the measured dimensions atomic force microscopy (AFM) was used to examine the surface. The AFM measurements were done on a Dimension Icon by Veeco in tapping mode.

a)

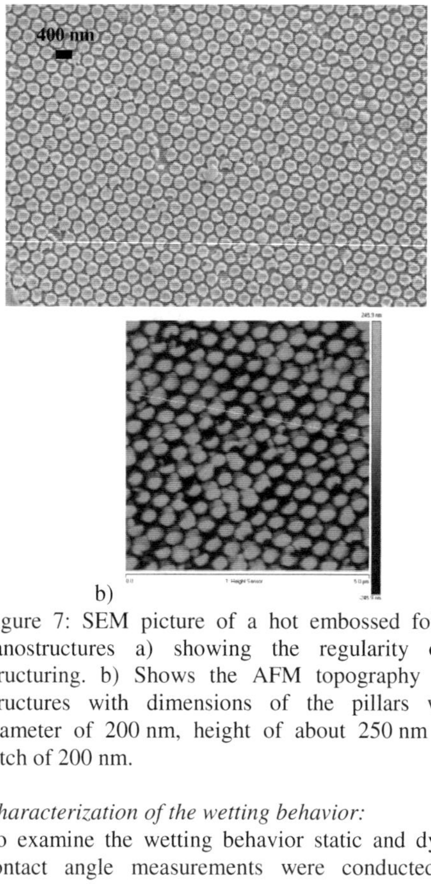

b)

Figure 7: SEM picture of a hot embossed foil with nanostructures a) showing the regularity of the structuring. b) Shows the AFM topography of the structures with dimensions of the pillars with a diameter of 200 nm, height of about 250 nm and a pitch of 200 nm.

Characterization of the wetting behavior:

To examine the wetting behavior static and dynamic contact angle measurements were conducted. The measurements were taken with a Dataphysics OC30 contact angle setup with water as the test fluid. For the static measurements a drop with a volume of 2 µl was dispensed at a temperature of 23 °C and a humidity of 67 %. Depending on the structure and material a change of the contact angle was demonstrated. An increase of the contact angle of PC from the unstructured material from 73° ± 1.2° to a contact angle of 116° ± 3.6° was achieved, thus changing the hydrophilic surface into a hydrophobic surface. Dynamic contact angle measurements were carried out with the sessile drop technique. A small droplet is placed on the sample and the volume is slowly increased until it reaches 10 µl. This volume is halted and when a stable contact angle is reached it is taken as the advancing contact angle. Afterwards the volume is decrease while taking the receding contact before each decrease of the base diameter. The contact angle hysteresis (CAH) is a value for possible contact angles as well as for the adhesion on the surface. High CAH are a sign for high adhesion. For PC a CAH of 32° was measured, with contact angles between 84° ± 4.2° and 116° ± 3.6°.

Results of hot embossing and thermoforming:
Tubes with an outer diameter of 2 mm were successfully fabricated. Fig. 8 shows a formed tube before separation with nanostructures on its inner wall.

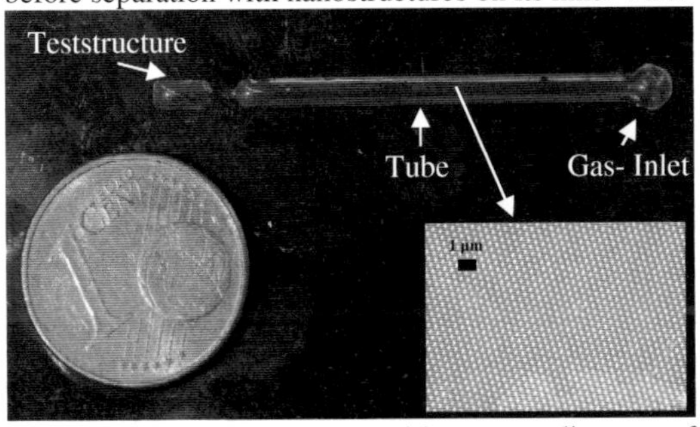

Figure 8: Thermoformed tube with an outer diameter of 2 mm and nanostructures (as seen on small SEM picture) on the outer wall.

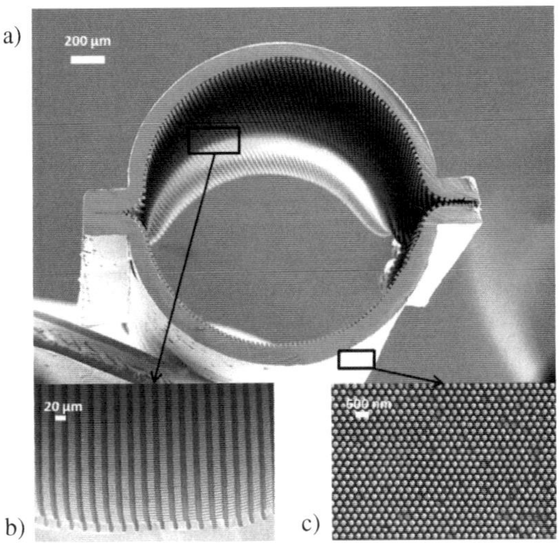

Figure 9: a) shows a separated tube with an outer diameter of 2 mm with b) microstructures on the inside and with c) nanostructures on the outside.

After separation tubes with a length of 35 mm were obtained. Through the structuring of the tubes the wetting behavior was adjusted to achieve a high wetting behavior on one side and a low wetting behavior on the other side. Fig. 9 shows a picture of a tube with microstructures on the inside (Fig 9b) and nanostructures (Fig. 9c) on the outside. Due to the flexibility of thermoforming process it is also possible to produce tubes with changing diameters or other three-dimensional shapes.

Conclusion
A method for the fabrication of tubes for microfluidic applications for biological and medical experiments was introduced. Through the combination of hot embossing and twin sheet forming tubes with micro and nanostructures on the inner and outer wall were fabricated. These structures with diameters as small as 100 nm and a height of 300 nm were achieved. Due to the structuring of the walls the wetting behavior was changed. An increase of the contact angle by 43° was shown, changing the wetting behavior from the hydrophilic regime to a hydrophobic regime. These structured surfaces enable the selective control of fluidic behavior and biocompatibility in a variety of BioMEMS systems. The technology will permit the fabrication of highly specific microfluidic systems such as improved medical equipment e.g. catheters, and specialized research instrumentation such as micropipettes.

Acknowledgements
The authors would like to thank P. Abaffy and G. Papagno for capturing the SEM images. Our thanks goes also to Dr. L. Zimmermann and M. Röhrig for supplying the hot embossing mold inserts. This work is supported by the Bürkert Technology Center (BTC), a cooperation between Bürkert GmbH & Co. KG and the Institute of Microstructure Technology (IMT) at the Karlsruhe Institute of Technology (KIT). This work was carried out with the support of the Karlsruhe Nano Micro Facility (KNMF, www.kit.edu/knmf), a Helmholtz Research Infrastructure at Karlsruhe Institute of Technology (KIT, ww.kit.edu).

References
1. E. Wintermantel, S-W. Ha, *Medizintechnik – Life Science Engineering*, 5th Edition, Springer, Berlin, 2009, 672ff.
2. W. Barthlott, C. Neinhuis, Planta, 1997, 202: 1-8.
3. S.H. Yoon, Y. K. Kim, et al., Lab Chip, 2012, 12, 2391-2402.
4. E. Martinez, E. Engle, et al., Ann Anat. 191, 2009, 126-135.
5. R. S. Lima, B. R. Marple, Journal of Thermal Spray Technology, 16(1), 2007, 40-63.
6. A. P. Lee, U.S. Patent, 5 658 515, 1997.
7. M. Worgull, *Hot Embossing – Theory and Technology of Microreplication*, William Andrew, Oxford, 2009.
8. R. Truckenmüller, S. Giselbrecht, et al., Adv. Mater. 2011, 23, 1311-1329.
9. M. Heilig, M. Schneider, et al., Microsyst. Technol., 2011, 17, 593-600.
10. M. Heilig, S. Giselbrecht, et al., Microsyst. Technol., 2010, 16, 1221-1231.

O-02-380

EPOXY PROCESSABILITY BY REACTIVE ROTATIONAL MOLDING

E. Lecocq[1*], F. Nony[1], L. Létoquart[1], M. Savoye[1], A. Tcharkhtchi[2] and J.-F. Gérard[3]

[1]LSTP, CEA Le Ripault, BP16, 37260 Monts, France ; eva.lecocq@cea.fr

[2] PIMM, ENSAM, 151 boulevard de l'Hopital, 75013 Paris, France,

[3] IMP, INSA de Lyon, 17 avenue Jean Capelle, 69621 Villeurbanne, France,

Abstract – Rotational molding (RM) applications are actually limited by the polyolefin thermomechanical properties. Reactive Rotational Molding (RRM) opens the range of transformable polymers but is industrially quite ignored due to the difficulties to control the conversion and temperature dependent viscosity. This work assesses the potential of polyepoxyde thermoset as a high performance rotomolded material. The effect of processing conditions of epoxy-amine (DGEBA + MXDA) RRM on the end product quality is investigated and correlated with in situ monitoring of the material temperature and ex situ rheological studies. Hydrocysts or frozen pools are obtained if the viscosity rise is too sharp at the solid body rotation time. Lower curing temperatures or thinner walls have to be considered for better thickness uniformity. Processing windows are particularly restricted by the high heat of reaction and the low initial viscosity of the reactive formulation.

Keywords: Reactive Rotational Molding, Processability, Polyepoxydes, Chemorheology

Introduction

Rotational molding is a processing technique used in the manufacturing of hollow plastic products [1]. Polymer powder is placed in a mold which is biaxially rotated and heated. As the powder melts, it begins to adhere to the mold walls. When a homogeneous layer is formed on the inner surface, the mold is cooled and the hollow piece can be released. The low cost of the molds allows efficient manufacturing of small series of large stress-free products with a quite uniform thickness distribution. Nevertheless, this process requires high temperatures and long cycle times to ensure adequate quality of the finished product. The choice of process compatible-materials is limited mainly by the high viscosity of polymers in molten state or their poor temperature resistance [2]. Therefore stabilized polyolefins nowadays represent 95% of the market of rotomolded polymers. However, their technical performances are limited by their low mechanical strength and low heat distortion temperature. So, the application market of rotational molding is actually restricted to storage tanks, kayaks, boat hulls or camper van tops.

Reactive rotational molding (RRM) constitutes an alternative to the classical process limitations [3]. In this technique, liquid monomers are initially injected into the mold. The heating and the rotational motion of the mold allow simultaneous polymerization and spreading of the reactive formulation with increasing viscosity. This process requires significantly lower temperatures and shorter cycle times. Therefore, transformable materials are expanded to PVC plastisol, anionically polymerized lactams and thermosets such as unsaturated polyester, polyurethane [4] or polyepoxyde [5].

These reactive formulations undergo chemical and rheological transformations as they cure. The processability comes from the control of the material viscosity during the rotation. This viscosity is time and temperature dependent as it is affected by many interdependent factors such as conversion of reaction, heat transfer and fluid mechanics. These difficulties result in flow defects on end products such as uncomplete pieces, hydrocysts on inner surface or non-uniform thickness. Efficient processing conditions are actually determined by trial and error with no scientific demarch.

The objective of this project is to assess polyepoxyde as a new material candidate for rotational molding material. This work presents an original investigation of the influence of processing conditions on the processability of a polyepoxyde formulation. Processing windows are identified on the basis of visual examination of the see-through end product and correlated with in situ monitoring of the material temperature and ex situ rheological studies of the reactive liquid.

Experimental

Materials

The liquid epoxy system used was a DGEBA (diglycidyl ether of bisphenol A) DER 332 from Dow Chemicals crosslinked with an aromatic diamine, MXDA provided by Sigma Aldrich. Epoxy resin and hardener were simultaneously mixed at stoechiometric ratio with a dual axisymmetric centrifugal machine (Speedmixer® DAC400, Synergy Devices Limited) at 1500 rpm for 2 minutes at ambient temperature just before the injection.

Methods

Rotational molding were performed on a LAB40® shuttle machine (STP, Canada) in a cubic stainless steel mold with an 21cm-edge. Mold temperature was monitored with a magnetic thermocouple stick to the external surface of the mold. A K-thermocouple (Ø1.5 mm) drilled through the mold to measure the material temperature. The thermocouple protruded from the mold inner surface at a distance chosen to match the final thickness of the epoxy polymer. Temperature data were collected by a thermally-isolated TPaq21® (Datapaq™) telemetry system screwed to the secondary axis. Temperature signal is gathered every 2s by the logger and transmitted in real time by radio waves.

Rheological characterizations were carried out on an ARES® rheometer of TA Instruments using parallel plates (diameter: 25 mm, gap: 1 mm) in multiwave ramp temperature mode (1, 5, 10 and 50 rad/s). Rheological measurements were controlled and treated by Universal Orchestrator® software. Gel time was determined by the crossover of the loss factor at various frequencies.

Results and Discussion

Monitoring of rotational molding

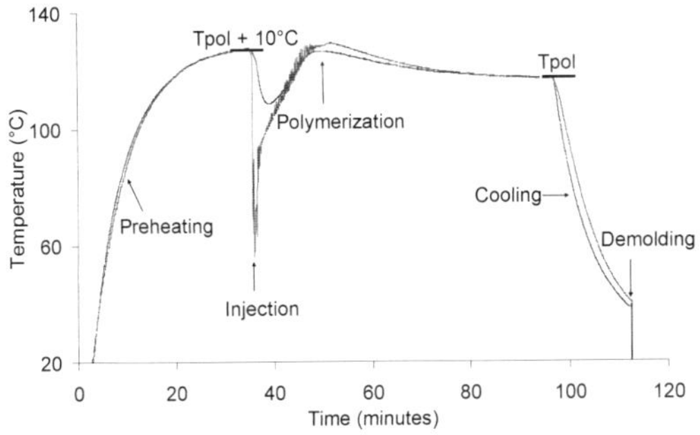

Figure 1 – Temperature monitoring of the mold (blue line) and the formulation (red line) during a reactive rotomolding cycle

The Fig. 1 shows the temperature monitoring of the mold and the material during a rotomolding cycle consisting in 5 steps. During the first step, the biaxially rotated mold is preheated in an electric oven. The efficiency of this convective heat source is low. Furthermore, the stainless steel is not the best suited material for high heat conductivity. These reasons explain the long period needed for this preheating step. When the mold reaches the set point temperature, it is removed from the oven. A controlled amount of the mixed formulation is directly injected onto the K-thermocouple equipped face. The ambient temperature formulation causes an immediate decrease of the material temperature. This injection step could take up to 2 min for large volumes. The mold is closed, put back into biaxial rotation and transferred into the oven at a controlled temperature for the polymerization (T_{pol}). The preheating temperature is set 10°C higher than T_{pol} to prevent the heat loss of the mold during the injection. The axis speeds (v_1/v_2) and the oven temperature (T_{pol}) are controlled and maintained for 1 hour.

The reactive liquid forms a pool at the bottom of the mold due to the low viscosity of the precursors. The pool movement into the rotating mold periodically wets and dewets the material thermocouple. That's why, oscillations can be seen on the monitored material temperature. Indeed, the thermocouple alternatively collects the temperatures of the pool and the internal air cavity. Nevertheless, the rotation of the mold allows a good homogenization of the pool temperature.

The conductive heat transfer from the mold to the liquid activates the epoxy-amine polycondensation. The chemical reaction is highly exothermic and causes a sharp increase of the material and internal cavity temperatures. By inversed heat transfer from the formulation to the mold, this last one is heated beyond the set point temperature. As the viscosity rises, the material is lifted off by the mold walls. When the epoxy thickness reaches the height of the thermocouple, there is no more oscillation. As this height is preset to the final thickness of the polymer, this time corresponds to the complete dispersion of the pool. Thus it can be considered as the solid body rotation (SBR) flow regime.

After an hour of curing, the mold is removed from the oven and a forced cooling step is performed by air fans. When the material temperature reaches 50°C, the epoxy cube is finally demolded. The see-through appearance of amorphous epoxy polymers allows a visual identification of any defect on the end product.

The one hour polymerization step is the studied one. The time when the charged mold is put back into the oven is considered as the time t=0 for the rheological and thermal measurements. The material temperature profile recorded in real time is fitted by 8 consecutive linear ramps transmitted to the ARES® software to measure the real viscosity evolution of the formulation during the process. The difference between in situ temperature (material thermocouple through the polymer) and ex situ temperature (ARES® thermocouple embedded into the lower plate) is lower than 2°C at each point.

Influence of processing conditions on the end product quality

Processing conditions have been investigated varying independently the oven temperature or the injected weight. The epoxy formulation, DGEBA – MXDA has been chosen for its high chemorheological difficulties in link to the process such as its great heat of reaction (510 J/g) and its low initial viscosity (0.8 Pa.s at 25°C and 0.015 Pa.s at 80°C).

A criterion has been needed to determine the quality of the end product. Cubes are considered "without defect" (Fig. 3a) when the polymer is coated over the entire surface with a uniform thickness without any visible flow front. Processability windows are defined by the loss of this state. Three majors flow defects have been identified. The first one is the presence of holes (Fig. 3b) preferably found in the corners or along the edges due to the fact that the liquid experiences the greatest deceleration at the 90° angles as the mold rotates. Another identified defect is the presence of hydrocysts which are multiple flow fronts observed on the entire inner surface (Fig. 3c). Hydrocysts are known as a secondary flow regime occurring before the solid body rotation. They develop from stabilized fingers at the flow front. The last defect called "crystal" (Fig. 3d) comes from the brutal gelation of the pool due to the adiabatic massive exothermy. The frozen pool bounces off the walls with the rotation and wipes off the dissipated first layer.

Figure 3 – Pictures of rotomolded products: (a) no defect, (b) holes, (c) hydrocysts, (d) crystal.

The Fig. 4 shows the material temperature evolution and the corresponding viscosity as a function of the chosen oven temperature T_{pol}. The other processing conditions such as axis speeds (10/7 rpm) and injected weight W_{inj} (270 g) are fixed.

The liquid formulation is injected at ambient temperature in a preheated mold. So, the initial material temperature at t=0, $T_{t=0}$ is lower than the preheating temperature. Nevertheless, higher T_{pol} induces greater $T_{t=0}$. The highest amplitude of the oscillations displayed on the temperature at high T_{pol} comes from the greater difference of temperature between air cavity and material due to the build-up polycondensation exothermy. The higher is the temperature, the faster is the reaction. So the heat is released more quickly leading to a more increasing heating rate. Moreover the heat of reaction released by the high temperature cured formulation is greater than low temperature ones because the reaction is stopped by vitrification at higher conversion. Indeed, DGEBA-MXDA exhibits an infinite glass transition at 126°C.

At the initial time, the reaction conversion can be considered at 0. So, as the initial temperature is high, the corresponding viscosity is low. This order is reversed in a few minutes as viscosity becomes more conversion dependent. The SBR viscosity for which the pool is dispersed, does not vary significantly with the temperature.

The end cube cured at T_{pol}=90°C presents frozen hydrocysts on its inner surface. Indeed, the higher T_{pol} is, the sharper is the viscosity rise. For this cube, the hydrocysts formed at low viscosity had no time to destabilize and for SBR to be achieved before gelation.

Holes can be observed on the edges or corners of the four cubes due to the fact than 270 grams are lower than the minimum amount required to cover the entire surface. Indeed, it exists a minimum viscosity needed to compete with the mold surface tension and avoid dewetting. As every viscosity is associated with a lifted thickness of liquid, a lower thickness cannot be reached.

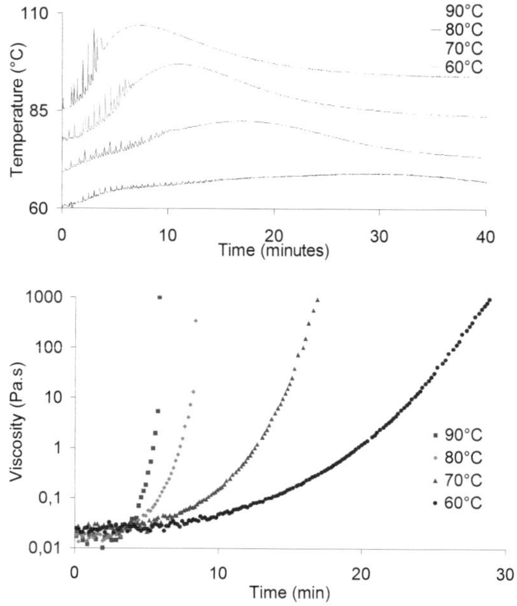

Figure 4 – Influence of the oven temperature on the material thermal profile and corresponding viscosity evolution for the DGEBA-MXDA RRM.

The Fig. 5 shows the material temperature evolution and the corresponding viscosity as a function of the injected weight (W_{inj}). The other processing conditions such as axis speeds (10/7 rpm) and T_{pol} (60°C) are fixed.

At the very beginning, the pool tries to reach slowly the mold temperature. The heating of a larger pool tends to take longer time than a smaller one. When the polycondensation is activated, the heating rate increases sharply due to exothermy build-up. Although the released heat of reaction is proportional to the injected weight, the displayed exotherm area tends to rise exponentially with W_{inj}. This phenomenom can be explained by the fact that a part of the heat of reaction is extracted by the mold walls. This absorption requires a great surface area to volume ratio and is consequently more effective with small pools.

The viscosity evolution is temperature controlled at the beginning then is correlated to the conversion as the reaction is thermally activated. A greater injected weight needs a higher SBR viscosity (η_{SBR}) because the final layer is thicker.

As already discussed, the 270g cube presents holes as the minimal weight to cover the entire surface is 290 g. The 550 g cube presents no flow defect. Nevertheless, at an amount of 770g, the SBR viscosity needed to dissipate the pool is too high and a large pool subsists at the bottom of the mold. A larger pool tends to be more adiabatic because of the low surface on volume ratio causing a exothermy build-up leading to pool gelation (Fig. 3d).

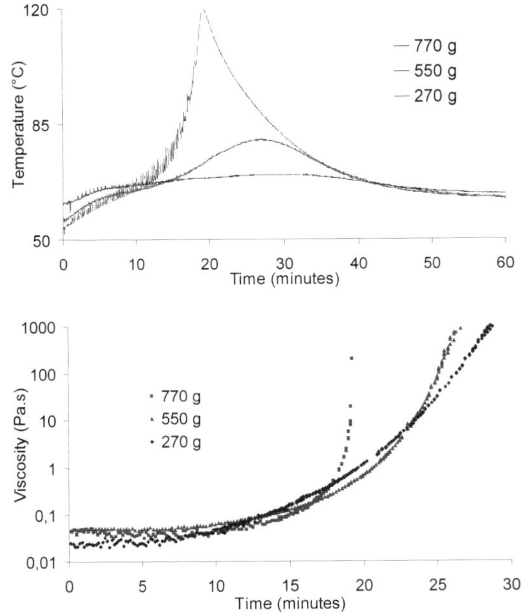

Figure 4 – Influence of the injected weight on the material thermal profile and corresponding viscosity evolution for DGEBA-MXDA RRM.

Determination of processing windows

The Table 1 presents the pool dispersion time t_{SBR}, the gel time t_{gel} and the end cube quality as in Fig. 3a, b, c or d for different processing conditions.

T_{pol}	W_{inj} 270 g	550 g	770 g
90°C	t_{SBR}=4'/t_{gel}=4'/ bc		
80°C	7' / 9' / b		
70°C	12' / 17' / b	12' / 12'/ c	
60°C	17' / 30' / b	18' / 26'/ a	18'/19'/3d
50°C		38' / 50' / a	34'/40'/3a

Table 1 – t_{SBR}, t_{gel} and end product quality for varying injected weights and oven temperatures T_{pol} of DGEBA MXDA rotomolded cubes

Whatever the processing conditions, it can be seen that flow defects such as hydrocysts (c) or crystals (d) appear when t_{SBR} and t_{gel} are too close from each other. The viscosity rise is too sharp for the liquid to go from the pool or the hydrocyst to SBR flow regime before gelation.

A raise of T_{pol} causes a decrease of the initial viscosity and an increase of the reaction rate. So, t_{SBR} and particularly t_{gel} tend to decrease. A raise of W_{inj} causes an increase of the η_{SBR} needed to dissipate the pool and an increase of the reaction heat decreasing t_{gel}. A decrease of T_{pol} or W_{inj} allows a reduction of the viscosity slope tangent to η_{SBR}. These two solutions improve the end product quality to the detriment of cycle time.

Multi-shot injections have to be considered for thick products to reduce the exotherm and decrease η_{SBR}. The formulation can also be modified by additives allowing a dilution of the reaction heat and an increase of the initial viscosity [6].

Conclusions

Influence of the injected weight and oven temperature on the epoxy rotomolding ability has been investigated. A minimum weight corresponding to the critical thickness has to be injected to cover the entire inner surface of the mold. Processing windows can be defined by the time between the SBR viscosity and gelation time. If this time is too short, hydrocysts cannot be destabilized or the pool can froze in mass. To avoid these flow defects, the oven temperature or the injected weight have to be decreased.

Influence of the axis speeds, the formulation viscosity and the mold shape on the end product quality is currently investigated. Rotomolded polyepoxyde has been considered as a potential material for the manufacture of vessels for high pressure gaseous storage.

References

1. R. J. Crawford, J. L. Throne, *Rotational Molding Technology*, Plastics Design Library, 2002
2. M. C. Cramez, M. J. Oliveira, R. J. Crawford, *Polymer Degradation and Stability*, 2002, *75*, 321.
3. H. Harkin Jones, PhD Thesis, Queen's University of Belfast, 1996.
4. J. Viale, F. Nony, P. Mazabraud, *International Journal of Material Forming*, 2008, *1*, 803
5. J.-P. Pascault, J. J. Williams, *Epoxy Polymers: New Materials and Innovations,* Wiley-VCH, Weinheim, 2010.
6. E. Mounif, G. G. Liang, *Polymer International*, 2009, *58*, 954.

TEXTILE REINFORCED MDPE ON ROTATIIONAL MOLDING

N. O-Charoen[1], P. Srinak-Eiang[1], K. Plubthong[1] and H. Hamada[2]
[1]*Department of Materials and Metallurgical Engineering, Rajamangala University of Technology Thanyaburi, Thailand – narongchai.o@en.rmutt.ac.th*
[2]*Department of Advanced Fibro Science, Kyoto Institute of Technology, Kyoto, Japan — hhamada@kit.ac.jp*

Abstract - The objectives of this article is to assess the rotomoldability of the textile insert reinforcement and to investigate how to improve some mechanical properties, the final product quality and develop the process of rotational molding by using the difference of textile types such as materials types and fiber size. However, the textile reinforcement of rotational molding has not been well studied. Rotational molding experiments were carried out in a laboratory scale biaxial machine capable of controlling temperature and time in the oven. Mechanical properties tests, as well as puncture, were performed on the rotationally molded parts. Differential scanning calorimetry and microscopy techniques have also been employed to identify the material and structural parameters.

Keywords: Textile Reinforced, Rotational Molding, Puncture Test

Introduction

Rotational moulding is a mainly technique for the making of hollow plastic parts. More recently, the rotational moulding of high-density polyethylene has gained considerable importance, due to its proven advantages in the making of hollow plastic products. There have been only a few technological developments in rotational molding and the knowledge of the process has essentially been based on trial and error. Not many studies have been carried out to understand the underlying principles that govern this process [1].

In recent years, attention has been paid to their use as a reinforcing material for thermoplastics in the rotational molding. In particular, the automotive industries have shown interest in the advantages that this type offibre - reinforced system can provide. In this work, the rotational moldability were investigated the characterization of textile insert-reinforced and unreinforced polymer samples and measurable some mechanical properties, such as tensile and puncture test. The effect of textile types were also evaluated.

Experimental

Rotational molded specimens were made of LLDPE powder grade M3204RU with an MFI 4 (190°C/2.16kg) supplied by SCG chemical (Thailand). The characteristic of materials and the composites were analyzed using a scanning electron microscope (SEM) (JSM6510, JEOL Co. Ltd.). The textile types are different in two type of materials such as HDPE and Nylon. A biaxially shuttle style rotational molding machine was designed and built in our laboratory (Fig. 1). It consists of a mold rotating arm and a heated oven with a circulation fan. A 3 mm thick sheet of steel rectangular box (Fig. 2) was used for all the molding tests. The oven temperatures ranged between 200°C to 250°C. Mold rotation speeds of 6 rpm (major axis) and 1.5 rpm (minor axis) were used. The description of the textile insert-reinforced rotational molding process, briefly, a two-part steel rectangular box mold is loaded with polymer powder and/or attached the textile with the magnets inside of the mold together. Before loading the polymer or polymer composite, a demoulding agent is applied to the internal surface of the mould. Then the mold is rotated in two axes at relatively low speeds while being heated, so that particle consolidation can take part. Then the mold is cooled down and the product is extracted from it.

Mechanical tests were carried out with the specimens processed at different types of textile. The mechanical tests considered were: tensile strength, impact and puncture test. For the tensile test, the samples used for the tests are followed ASTM D638. The tests were carried out in a Hounsfield tensile tester (model TG18) at room conditions and at a speed of 50 mm/min. The ultimate tensile strength (UTS) was obtained from this test. The puncture tests were carried out using a deep drawing machine fitted with a steel sphere 19×10^{-3}m in diameter (ASTM D3787). The specimens were obtained from the upper side of the rotomoulded rectangular box. Maximal pucture force was measured before the specimen broke.

Fig1. Shows (a) the shuttle style rotational molding machine

Fig 2. A rectangular box mold

Results and Discussion

Tensile strength

As it can be seen in Table 1, the textile (nylon net) insert increase the UTS of the composite materials compared to unreinforced neat LLDPE matrix from 18.81 and 16.80 MPa respectively. This tendency is in agreement with previous results reported for natural fibre-reinforced composites [3,4] and for long glass fibre composites [5,6]. In general terms, the increase in UTS is not modest for both types of reinforcingfibres. In the case of HDPE nets shows that the slighty decreasing in the UTS should be due to the melt of the HDPE fibers to homogenously with the LLDPE matrix (Fig 3). Low mechanical properties at fibhigh contents are mostly associated with the presence of fibre clumps and voids [1]. Reinforcing discrete fibres increase the viscosity of the polymer matrix. This phenomenon has been reported for glasfibres [7] as well as for naturafibres [8]. An increased viscos ity might also contribute to the formation of clumps during melt processing of naturafibre composites. With an adequate fibre treatment, this problem is reduced in the rotational moulding process.

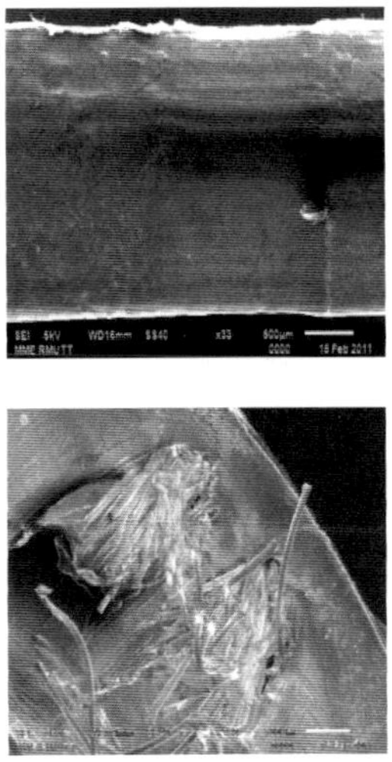

Fig 3. The micrographs shows that (upper) the homogenous surface of LLDPE sample with textile (HDPE) and (lower) the LLDPE sample with textile (nylon) insert

Table 1 show the ultimate tensile strength (UTS) of the rotational molding samples (1) neat LLDPE, (2) LLDPE+HDPE nets and (3) LLDPE+nylon nets

Materials	Thicknes s (mm)	Area (mm^2)	Modulus (MPa)	UTS (MPa)
1. Neat LLDPE	2.78	36.9	722.34	16.80
2. LLDPE+PE	2.90	37.6	772.13	16.71
3. LLDPE+Nylon	2.90	39.0	845.50	18.81

Puncture test

As shown in Fig 4, the nylon nets-reinforced LLDPE specimens show greatest puncture force compared to the HDPE nets-reinforced LLDPE and the neat LLDPE samples at 16.54, 15.88 and 15.42 MPa respectively. It is shown that by using different textile materials reinforced specimens tend to produce the different in the puncture force. This indicates that the coherent fibre network increases the stiffness of the composite. The observed tendencies are in agreement with data from tensile test, which shows that the nylon nets-reinforced specimens with stand higher compression loads at lower deformation levels.

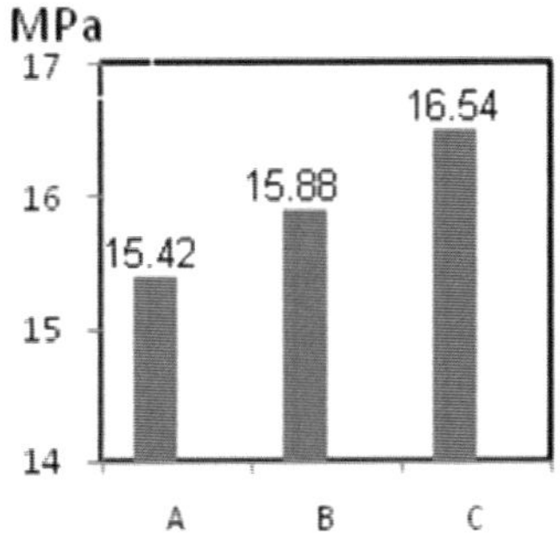

Fig 4. show the results of puncture tests of the rotational molding samples (A) neat LLDPE, (B) LLDPE+HDPE nets and (C) LLDPE+nylon nets

Conclusions

The mechanical properties of textile insert by rotational moulding were characterized using some mechanical tests. The mechanical properties of textile-reinforced composites vary considerably with the types of textile materials. Tensile strength results indicated that, depending on the textile type, tensile strength decreases due to the melt of textile materials.

Puncture behavior was ifluenced by textile type and content. The nylon nets -reinforced composites showed higher impact strengths than the HDPE nets and neat LLDPE matrix. The testing methodology described here can be used to assess the properties of both reinforced and unreinforced rotational moulded products.

Acknowledgements
This research was financially supported by the National Metal and Materials Technology Center, Thailand (MTEC), contract grant no. MT-B-53-MAC-47-188-G

References (Times New Roman 10, bold)

1. F.G. Torres, C.L. Aragon / Polymer Testing 25 (2006) 568–577
2. A.K. Bledzki, S. Reihmane, J. Gassan, Polymer-Plastics Tech. Eng. 37 (4) (1998) 451.
3. R.B. Seymour, Reinforced Plastics Properties & Applications, ASM International, Materials Park, OH, pp. 257 (1991).
4. F.G. Torres, S.F. Bush, Composites A: Appl. Sci. Manuf. 31 (12) (2000) 1421.
5. S.F. Bush, F.G. Torres, E.S. Erdogan, in: Proceedings of the Seventh International Fibre Reinforced Components, FRC98, p. 237 (1998).
6. S.F. Bush, F.G. Torres, J.M. Methven, Composites: Appl. Sci. Manuf. 31 (12) (2000) 1289.
7. M.A. Lopez-Manchado, J. Biagiotti, M. Arroyo, J.M. Kenny, Polym. Eng. Sci. 43 (5) (2003) 1018.

O-02-046

CREEP FORMING OF POLYCARBONATE: EXPERIMENTATION AND MODELLING

S. Zitzlsberger[1]*, C. Maurer[1] and G. Reinhart[2]

[1]*Department of Mechanical, Automotive and Aeronatical Engineering, University of Applied Sciences Munich, Germany – sabine.zitzlsberger@hm.edu; cmaurer@hm.edu*

[2]*Institute for Machine Tools and Industrial Management (iwb), Technische Universität München, Germany – gunther.reinhart@iwb.tum.de*

Abstract - An investigation of creep forming of sheet material from polycarbonate by numerical analysis is presented. The parts are formed by a reconfigurable tooling system consisting of individually height adjustable pins with a flexible silicone layer on top forming the final mold. The required load is applied using a negative form of this system. Investigations include doubly curved glazing components with the ratio of radii being 0.5 and 1 and single curvature. Creep deformation and stress relaxation over process time were studied. The effect of anisotropic creep behavior was observed. The resulting springback was predicted for different forming temperatures and compared to experimental results.

Keywords: Polycarbonate, Creep, Reconfigurable tooling system, Numerical analysis, Springback

Introduction

The process of creep forming is mainly known due to its application in the aircraft industry. The process allows the production of aircraft wing panels made from aluminum alloy with improved physical properties [1]. The deformation by creep takes place as the material with internal stress is exposed to high temperature [2]. On the other hand the process of creep forming is also applied to form for example wind shields from amorphous thermoplastic resins like polycarbonate (PC). Thereby the optical quality of the glazing component is of first importance order which is obtained by the low forming temperature. While creep forming allows producing high quality products on the one hand, high process times and a certain amount of springback on the other hand are disadvantages. The amount of springback depends on the creep behavior of the material at higher temperatures, the sheet thickness and the curvature of the component [2]. This work deals with a sheet thickness of 5 mm as nearly 80 % of applied glazing components are represented thereby.
Creep formed components are in general of low batch sizes and therefore not suited for injection compression molding regarding the mold costs. Molds for thermoforming processes are cheaper than those for injection molding [3]. Nevertheless also these molds are still too cost-intensive for prototypes and small series where batch sizes lie in the order lower than 10. Reconfigurable tooling concepts seem a promising choice regarding cost and time. The system presented in this paper consists of individually height adjustable pins with a flexible silicone layer on top forming the final mold. The same kind of reconfigurable tooling system is also described elsewhere [1, 4]. Therefore the total milling process of the form can be skipped.
The system presented was developed for forming polymethylmethacrylate (PMMA). This material shows a good optical quality also when formed beyond the glass transition temperature. Furthermore the high heat capacity of the material makes it possible to heat it up in an external oven. In contrast creep forming of PC requires a temperature control system within the tool. The temperature control system is implemented by heating wires within the flexible silicone layer in order to provide a constant heat transfer which is necessary for the creep forming process. This integrated temperature control system prevents the reconfigurable tooling system from high temperature and therefore the dearness of the system.
Primarily the paper presents an FE model of the reconfigurable tooling system for simulating creep forming of PC. Furthermore the amount of springback of double curved and single curved components of 5 mm thickness is analyzed. The stress relaxation over process time is compared for different curvatures. The influence of anisotropic creep was studied by the implementation of the Hill potential. The numerical results for different forming temperatures are finally compared to experiments.

Experimental

Numerical procedures for simulating creep forming

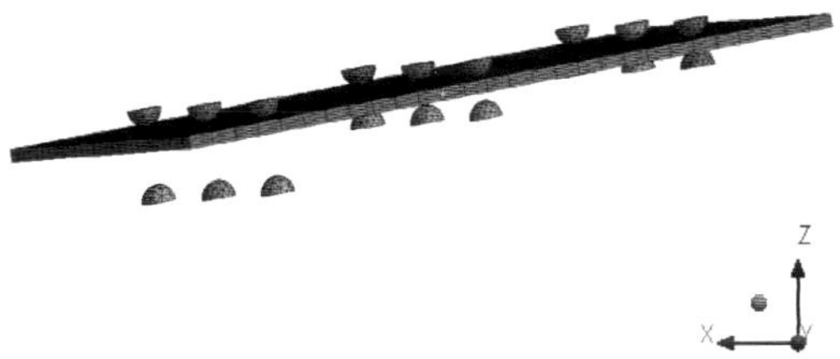

Figure 1 – FE model for the reconfigurable tooling system.

FE simulations of creep forming using ANSYS were carried out. Investigations include single and double curved glazing components. Sheet material from polycarbonate, 5 mm thick, is creep formed at different temperatures and with different curvatures. The used values of the forming parameters are shown in table 1.

The reconfigurable tooling system is represented by hemispheres forming the contact region. To force the sheet to acquire the desired shape the hemispheres on the upper side are moved in negative z-direction. Contact regions are defined between the sheet surface (target) and the hemisphere surface (contact). A friction coefficient of 0.4 therefore is specified. The sheet is meshed with solid quads using four layers to avoid penetration of the hemisphere which can be seen in Fig. 1. The highest point was specified as bonded contact and furthermore the outer edge was fixed to avoid rigid body motion. For analysis of the springback the displacement in z-direction is set to 0 at certain times. The primary creep behavior was modeled by the modified time hardening equation (provided by ANSYS CLASSIC) as follows:

$$\varepsilon_{cr} = C_1 \sigma^{C_2} t^{C_3+1} e^{-\frac{C_4}{T}} / (C_3 + 1) \quad (1)$$

For each temperature (C_4 = 0) individual parameters were found shown in table 2. Furthermore the extruded sheet material exhibits an orientation. Therefore the elastic modulus in both directions was determined for all temperatures. The values are shown in table 3. The anisotropic creep behavior was modeled by the Hill potential. This factor was determined to be 0.96 for creep behavior at 393.15 K.

Table 1 – Values of the forming parameters.

Temperature [K]	Radius [mm]	Ratio of Radii
393.15	959.69	0
401.65	1905.88	0.5
416.15	3805.03	1

Table 2 – Values of the material constants for modified time hardening at different temperatures.

Temperature [K]	C_1	C_2	C_3
393.15	6.9121 E-07	1.5581	-0.18239
401.65	2.4151 E-07	2.8781	-0.03622
416.15	1.0964 E-06	2.0682	0.87867

Table 3 – Values of the elastic moduli at different temperatures.

Temperature [K]	Elastic modulus [N/mm²]	
	x-direction	y-direction
393.15	1671.58	1489.93
401.65	1320.88	1292.64
416.15	499.80	320.79

Creep forming experimentation

Extruded sheet material from polycarbonate (Makrolon AG2677, manufactured by Bayer MaterialScience AG), 5 mm thick, was used for the experiment. Before the forming process each sheet was baked in an oven at 393.15 K for about 10 hours for dehumidification to avoid bubbles when heated up to higher temperatures. The polycarbonate sheet was then clamped within the

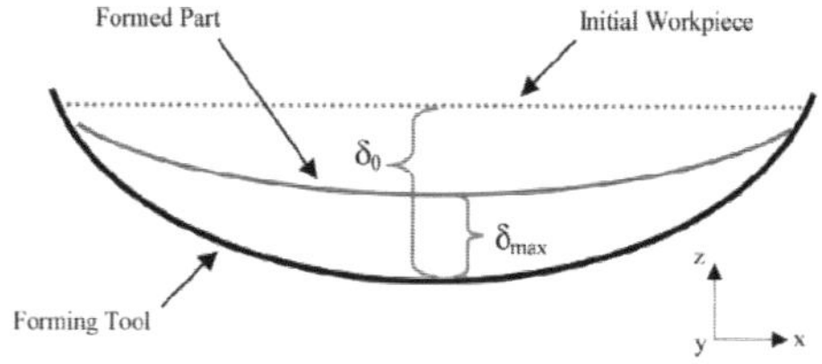

Figure 2 – Springback defined according to [2], where R = 0 represents no springback, R = 1 represents fully springback.

reconfigurable tooling system between two flexible silicone layers having already the desired temperature. The temperature is provided by integrated heating wires (RD 100/0.3). The experimental setup is shown in Fig. 2. The resulting springback was measured by a 3D scanner (Creaform, UNIscan) and compared to the original form using the software CATIA. The springback is defined according to [2] as follows:

$$R = \delta_{max} / \delta_0 \quad (2)$$

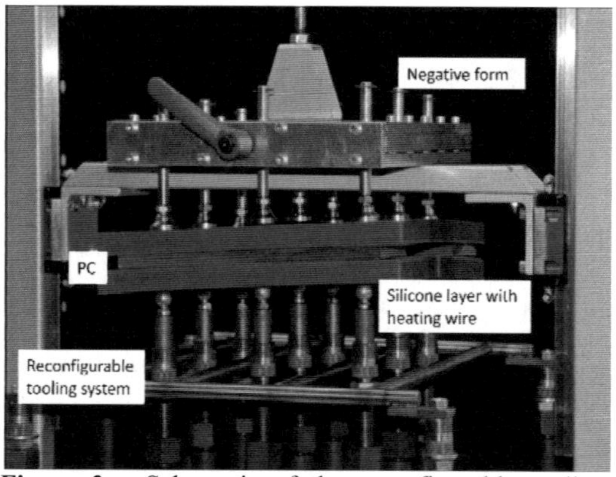

Figure 3 – Schematic of the reconfigurable tooling system.

Results and Discussion

Forming temperature

Fig. 3 shows the resulting springback R over a process time of 100, 1000 and 3000 seconds for different forming temperatures. Solid lines represent the springback in x-direction and dotted lines in y-direction respectively. It is shown that lower forming temperatures provide a lower amount of springback. A forming temperature near the glass transition temperature provides the highest amount of springback. In general low forming temperatures provide higher stress levels due to the higher elastic modulus and therefore cause a higher creep rate. Experimental results are plotted for each temperature at a process time of 3000 seconds. The simulation is found to exhibit good accordance with experimental results for a forming temperature of 401.65 K and 416.15 K. The deviation of the mean value is 2.5 % for 401.65 K and 5.5 % for 416 K respectively. For a forming temperature of 393.15 K a deviation of 20 % was determined. As the material at a temperature of 393.15 K provides a higher elastic modulus than at higher temperatures the material needs to be formed with higher forces. The simulation only represents the hemispheres without the flexible silicone layer. Therefore we propose that due to the compression of the flexible silicone layer at the center hemisphere, the applied stress to the polycarbonate sheet drops to a lower level resulting in a higher amount of springback.

Radius and ratio of radii

The influence of the geometry of the formed glazing component was studied. Therefore two different radii and three different ratios of radii have been studied by FE modeling. The results are shown in Fig. 4. A ratio of radii of 0 represents a single curvature, a ratio of 0.5 represents the double curvature in the right angle direction and a ratio of 1 represents the same curvature in each direction. The influence of the radius as well as the ratio is negligible at the early stage of the forming process. The amount of springback differs more regarding higher process times and smaller radii.

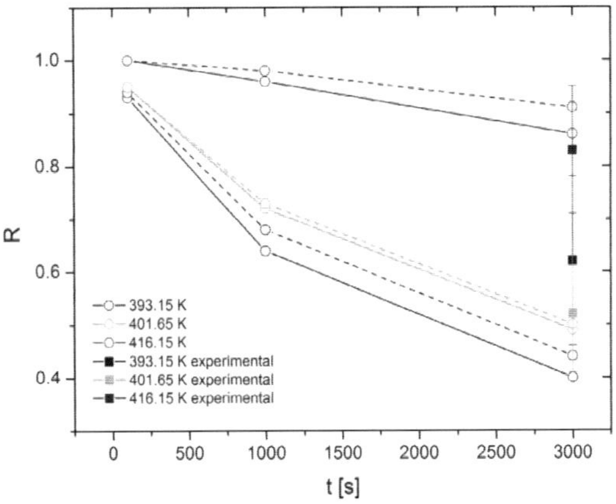

Figure 4 – Springback at various temperatures for x- (solid lines) and y-direction (dotted lines) and experimental results for the process time of 3000 seconds.

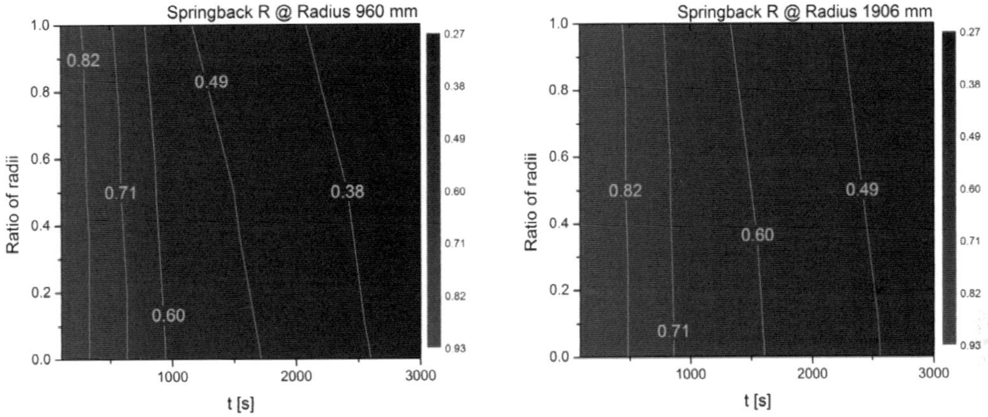

Figure 5 – Variation of the predicted springback for different ratios of radii over process time for a radius of 960 mm and 1906 mm, respectively.

Table 4 – R-values for anisotropic creep behavior.

	Isotropic	Anisotropic
x-direction	0.44	0.32
y-direction	0.40	0.51

Stress relaxation

Fig. 5 shows the simulated maximum (surface) stresses during the creep deformation. The data is plotted for a forming temperature of 393.15 K and a radius of 960 mm. For a ratio of radii of 1 the stress level reaches about 41 MPa. As creep takes place the stress relaxes and therefore the maximum stress is reduced to approximately 4.9 MPa. Springback occurs upon release of the negative form showing residual stresses of 2.4 MPa. This springback is unavoidable in creep forming as the creep rate reduces sharply with the drop of the stress level [1].

Anisotropic creep behavior

Anisotropic creep behavior was modeled by the Hill potential. Therefore creep data was carried out for extrusion in-direction and at right angles. A mean value for the Hill potential of 0.96 was calculated for a temperature of 393.15 K. The deviation between the results of anisotropic and isotropic creep behavior is shown in table 4 for both directions. The deviation in the x-direction is 12 % and in the y-direction 11 % respectively.

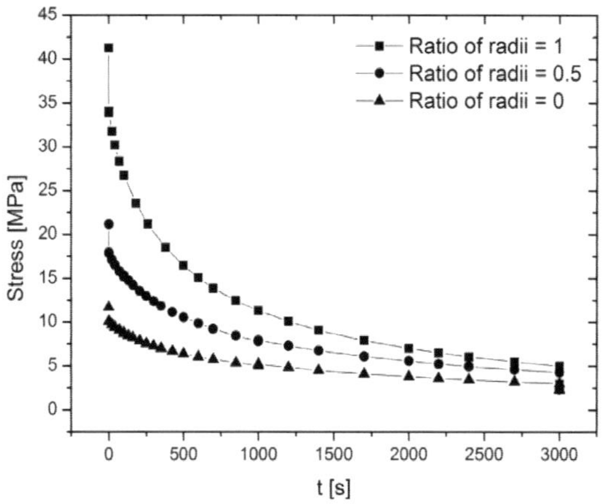

Figure 6 – Stress relaxation over process time for different ratios of radii (Radius = 960 mm, temperature = 393.15 K).

Conclusions

Springback behavior in creep forming sheet material from polycarbonate was observed using FE methods. Analyses were carried out and were compared to experimental results. The following conclusions were obtained.

(1) The FE model was found to be suitable for the prediction of creep forming. Further research has to be done in order to avoid the compression of the silicone layer at the center point to provide full stress levels.

(2) Lower forming temperatures lead to less springback according to higher stress provided.

(3) The ratio of radii is less important for springback behavior especially for big radii.

(4) The anisotropic creep behavior is not negligible. To avoid anisotropic creep tempering of the material should be done before the forming process.

Acknowledgements

This research was supported by the Bavarian State Ministry of Sciences, Research and the Arts under grant D 2-F1116.MÜ/4/2. The authors like to thank *Bayer MaterialScience* for providing creep data at elevated temperatures.

References

1. L. Zhan; J. Lin; T. A. Dean *International Journal of Machine Tools & Manufacture* 2011, *51,* 1.
2. K. C. Ho; J. Lin; T. A. Dean *International Journal of Plasticity* 2004, *20,* 733.
3. P. Schwarzmann in *Thermoforming – A Practical Guide*, A. Illig, Ed.; Carl Hanser, Munich, 2001; 2.
4. X. Bai; Y. Zeng; W. Wu; Z. Li; F. Zou in International Technology and Innovation Conference 2 Proceedings, Hangzhou, 2006, 1202

KN-03-01

MODERN FLAME RETARDANTS FOR POLYMER FOAMS – DEVELOPMENT TRENDS WITH SPECIAL EMPHASIS ON NOVEL POLY(BUTYLENE TEREPHTHALATE) FOAMS

T. Köppl, B. Fischer, V. Altstädt*

Department of Polymer Engineering, University of Bayreuth, Universitätsstr. 30, 95447 Bayreuth, Germany - thomas.koeppl@uni-bayreuth.de; bianca.fischer@uni-bayreuth.de, altstaedt@uni-bayreuth.de

Abstract - Most polymer foams are easily flammable due to their chemical structure and their extensive surface area. For many applications a sufficient flame retardancy is absolutely necessary. Until now, halogen-based additives, for example hexabromocyclododecane (HBCD) for polystyrene, are commonly used as flame retardants. Many groups work on replacing these additives, which are harmful to the human health and environment. Therefore halogen-free flame retardants gain in importance. Chemists and engineers work together to develop alternatives based on inorganic, nitrogen or phosphorous additives. Especially the flame retardancy of polymer foams has been proved to be difficult due to the fact that the additive may not influence significantly the foaming process.

Currently, flame-retarded novel Poly(butylene terephthalate) (PBT) foams are under development. Possible applications of such materials would focus on applications where chemical, heat and abrasion resistance play a decisive role and other foam materials are only usable with restrictions. Some halogen-free approaches for compact PBT are under development and there applicability has to be proved in foamed PBT. The addition of flame retardants shows also an influence on the subsequent material properties and on the formation of the foam morphology.

In this study PBT is processed by foam injection moulding. The influence of the type of the flame retardant on the material characteristics especially mechanical and flame retardance properties were investigated in the foamed PBT. The foam structure depends significantly on the one hand on the type of the flame retardant like particle size or dispersion and on the other hand on the change of the elongational deformation behaviour of the polymer melt, which is important for the cell growth step during the foaming process.

Keywords: polymer foams, flame retardancy, foam injection moulding, MuCell® technology

Introduction

Thermoplastic polyesters like poly(butylene terephthalate) (PBT) are polymers used for injection moulding applications in electrical engineering, communication and transportation. The burning behaviour in these applications plays clearly a decisive role. Applications in mass transport systems are sensitive to fire, smoke and toxicity (FST) aspects in a fire scenario owing to the inherently confined operating conditions. As these materials do not show inherent flame retardancy, the use of materials that delay ignition or rapidly self-extinguish after ignition and produce low toxicity upon burning is therefore required. Nevertheless, other key properties of the polymers should also be taken into consideration, and the addition of flame-retardants should not influence too severely the mechanical stability and processability.

The state of the art constitutes fireproofing by adding halogen-based additives to the polymer, which is mostly not beneficial regarding health and environmental aspects. In the near future this halogen-based compounds (e.g. HBCD) will be banned for the use in polymers (European RoHS law).

Right now foaming of PBT is not yet studied in detail but can be an interesting new foam material. Potential application areas of PBT foams are where other foams are only applicable with restrictions. Hence the chemical, heat and abrasion resistance, the strong toughness and the high heat deflection temperature up uto 150 °C are advantages for such applications.

PBT is not yet commercially available as foamed material. One patent about foam extrusion of modified PBT is available [1] and some rare applications of foam injection moulded parts with the aim to minimize shrinkage and warpage are known [2]. In general there is little scientific consideration on foaming of PBT and all the more not for flame-retarded PBT foams. Structural foams produced by foam injection moulding and moulded parts of bead foams are not known in literature. K. Song et al. investigated for the first time PBT batch foams with a pressure drop method and the blowing liquid chloroform [3]. Processing by foam extrusion with a chemical blowing agent of unmodified and modified PBT is published by Xanthos et al. [4,5]. The repetitive occurring problem is the cell coalescence and the consequent resulting broad cell size distribution.

The main challenge in the foaming process of PBT is a sufficient cell stabilisation during cell growth. Due to the semi-crystallinity no softening range exists, that is unfavourable for foaming and can lead to inhomogeneous macrocellular cells. Usually the cell structure can be set by varying the temperature. The small foam processing window and the additionally low melt strength above the melting point allows indeed no temperature tolerances. Therefore the cell structure depends mainly on the processing conditions and on the melt strength of the polymer that can be optimised by modification of molecule structure or incorporation of additives.

In this study structural PBT foams are processed by foam injection moulding and breathing mould technology. To identify the best foaming conditions the influence of a chemical and a physical blowing agent on the density and foam morphology of the structural foams are investigated. The thickness of the breathing mould is systematically varied to study the foaming behaviour in dependence of the density reduction. Since PBT is easy flammable (UL 94 HB) also the flame retardance of PBT in foamed material is necessary, especially because foams ignite earlier due to a larger surface area and a lower density than the compact material. By means of elongational rheology and microscopy the influence of a halogen-free flame retardant on the foaming behaviour and on the foam morphology is analysed.

Experimental

Foam injection moulding

The injection moulding grade Ultradur® B 4520 (1300 kg/m^3) from BASF SE was used for foam injection moulding. The physical blowing agent N_2 was injected as supercritical fluid by the MuCell®-technique with a concentration of 0.5, 0.7 and 1 %. Compact samples and integral foams with a compact layer and a microcellular foamed core were realised at a Battenfeld BA 1500/630 + 400 BK Sandwich with a melt temperature of 250 °C and a hot runner temperature of 265 °C. To achieve samples with a high expansion ratio and a relative thick foamed core the breathing mould technology is applied. Therefore after melt injection the mould with shut-offs on the parting line opens in the direction of the clamping unit. Breathing of the mould from 2 to maximal 10 mm results in samples with a part thickness up to 10 mm.

Compounding of flame-retarded PBT

To reduce the flammability of PBT the halogen-free flame retardant Exolit® OP 1240 from Clariant Additives, an aluminum phosphinate (abbreviated AlPi), was added with a concentration of 6 and 20 wt.-% by prior melt compounding of PBT with a twin-screw extruder ZSK 26 MCC from Werner & Pfleiderer (L/D = 44, D = 25) at a maximum temperature of 250 °C.

Rheological behaviour

The modified compounds were analysed at 235 °C by elongational rheology by the so-called Rheotens-method. The Rheotens-setup is placed under a capillary rheometer (Rheograph 6000, Goettfert) with a tubular die (diameter 2 mm, length 30 mm). The polymer was extruded with a constant shear rate of 43.2 s^{-1} and the melt strand pass through the counter-rotating draw-off wheels of the Rheotens-setup, which are arranged 95 mm under the die. The wheels have a constant acceleration of 12 mm/s^2 until the rupture of the melt strand. Thereby the force to draw-off the strand is measured in dependence of the draw-off velocity. The characteristic values melt strength (maximum force) and drawability (maximum stretching ratio at break) were investigated.

Foam characterisation

The foam structure was analysed by three-dimensional micro-computertomography (μ-CT) as well as by two-dimensional scanning electron microscopy (SEM). At this the size of the compact layer and foamed core as well as the mean cell size and cell size distribution plays an important role.
The μ-CT-measurements were used for quantitative evaluation of cell sizes and run on a Skyscan 1072 with a cylindrical specimen of 4 mm diameter. The X-ray projections were reconstructed and afterwards evaluated by MAVI-Software from FhG ITWM Kaiserslautern.
By recording SEM micrographs foams with an open cell structure and cells with a diameter down to 500 nm can be by analysed. The SEM-pictures were obtained from a Jeol JSM-IC 848 with an acceleration voltage of 15 kV. The pictures were taken of the surface of cryo-fractured samples. To determine the open cell content a gas pycnometer Ultrafoam 1200e from Quantachrome was used. The final foam density is measured by the buoyancy method according to ISO 1183-1 with cylindrical samples with a diameter of 4 mm.

Results and Discussion

Foamability of PBT

Foaming of PBT could be realised by addition of blowing agents. By using a physical blowing agent, the addition occurs through MuCell® technology directly into the polymer melt. In our case, nitrogen was selected as physical blowing agent and injected into the melt in different concentrations (0.5 / 0.7 / 1 wt.-%). For these experiments, no breathing mould technology was used leading to structural foams with a part thickness of 2.3 mm and 400 μm thick compact layers.
With increasing blowing agent concentration, the foam morphology became finer and more homogeneous. However, the density reduction decreases significantly from 14 % over 9 % to 2 % compared to the compact material.
Higher blowing agent concentrations lead to the formation of a higher amount of interfaces and

nucleating sites as well. This results in foams with high cell density combined with small cell diameters but low density reduction. Lower nitrogen concentrations positively effect the density reduction but suffer in cell morphology, i.d. larger cell sizes and broader cell size distribution.

Applying of breathing mould technology

The nitrogen concentration of 0.5 wt.-% leads to the highest density reduction and that is the reason for fixing this parameter. The breathing mould enables the adjustment of different expansion degrees. By this technology, structural foams with part thicknesses between 2.2 and 10.4 mm could be produced. With increasing breathing run the density reduction increases as well (up to 80 %).

Beside the density reduction, there is also a dependence of mean cell diameter and relative amount of large cells on part thickness.

Increasing the part thickness, the mean cell diameter increases as well. This could be explained by slower cooling rates of the polymer melt in thicker parts leading to delayed stabilisation and more frequently occurring cell coalescence mechanisms. Furthermore with increasing part thickness the polymer melt gets more elongated resulting in thinner cell walls and consequently in lower resistance for cell coalescence. Up to a part thickness of 4.8 mm, which is equal to a density reduction of 45 %, a homogeneous cell structure occurs without the formation of large cells. Above this critical density reduction, the cell sizes are dramatically increased combined with high inhomogeneity.

Concerning the compact layers of the structural foams, it can be mentioned that the amount of compact layer decreases with increasing part thickness from 35 % down to 6 %.

Influence of flame retardant on foaming of PBT

To reduce the flammability of PBT foams the halogen-free flame retardant aluminum phosphinate (abbreviated AlPi) was incorporated by prior melt compounding with 6 and 20 wt.-%. The structural foams were injection-moulded with 0.5 % N_2 using the breathing mould technology with a resulting part thickness of 4.8 mm, where still no cell coalescence and no cells above 500 μm occur. By addition of AlPi the compact layer is increased from 480 μm to 510 μm and 570 μm, respectively, what can be explained by a heterogeneous crystal nucleation and a 6 °C higher crystallisation temperature.

The density is not affected at a constant density reduction of 45 % (700 kg/m^3). The morphology of the foamed core shows different structures (Fig. 1).

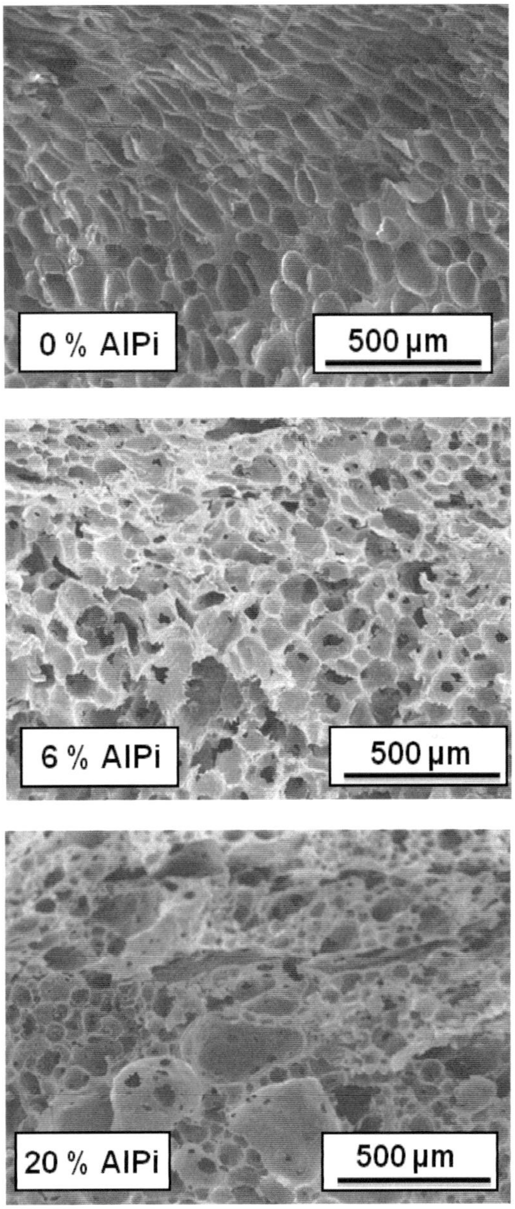

Figure 1 – Structure of foamed cores with different flame retardant concentrations (SEM)

The cell structure is influenced increasingly by the amount of flame retardant. Whereas the neat PBT has a mean cell diameter of 60 μm (± 5 μm, evaluated by SEM) the mean cell diameter increases with 6 wt.-% AlPi slightly to 85 μm and with 20 wt.-% AlPi more than two times to 130 μm.

A possible reason for the inhomogeneous cell structure with 20 wt.-% AlPi is a weak diffusion of blowing agent because of the filler network that results in a inhomogeneous distribution of the blowing agent in the polymer melt.

The flame retardant shows no nucleation effect and on the contrary the cell size distribution becomes broaden. Furthermore, the open cell content is increased from 15 % in the neat PBT foam to 50 % in the flame-retarded foams. The compounded materials were analysed by elongational rheology to understand the change in morphology (Fig. 2).

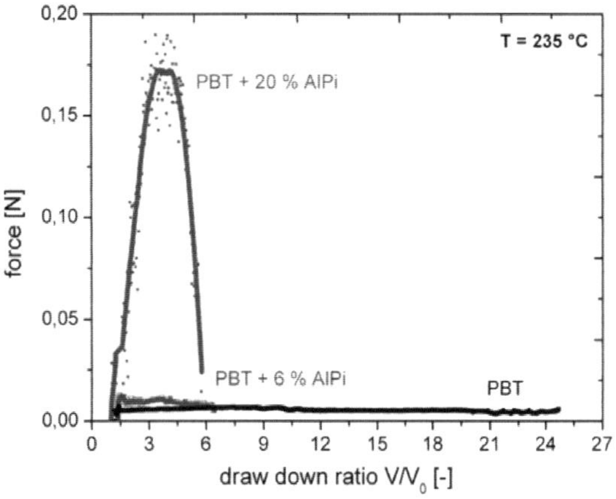

Figure 2 – Melt elongational behaviour of flame-retarded compounds with aluminum phosphinate (AlPi)

The incorporation of AlPi leads at a concentration of 20 wt.-% to a noticeable increase of melt strength because of the formation of a filler network but also to a significant reduction of the drawability because the chain elongation is hindered. The drawability is also strongly decreased with a concentration of 6 wt.-% AlPi. The melt strength is at the same level of the neat PBT. The increased melt strength with 20 wt.-% AlPi shows no positive effect on cell stabilisation since both flame-retarded foams have cells above 500 µm. The strongly reduced drawability can be a reason for the higher open cell content in the flame-retarded foams. Due to the strong elongation of polymer melt during breathing of the mould a rupture of cell struts occur that result in open cells (see Fig. 1).

Conclusions

PBT is a material, that is foamable and that foam structure can be optimised by varying different material and process parameters. Addition of the halogen-free flame retardant AlPi lead to formation of larger cells sizes without a nucleating effect. Nevertheless, foaming of flame-retarded PBT is feasible and is promising for the application in electrical engineering and transportation.
The development of new halogen-free flame retardants open up new possibilities for applications of new engineering thermoplastic foams like PBT foams.

Acknowledgements

The financial support of the German Research Foundation (DFG) in the frame of the research project number Al 474/17-1 is highly acknowledged. PBT was kindly provided by BASF SE and the flame retardant from Clariant Additives. We further acknowledge Anne Lang and Jacqueline Uhm for the experimental support.

References

1. Mitsubishi Eng. Plastic Corp., Japanese Patent JP 002006291063A 257, 2006.
2. GE Plastics, Technical brochure by GE Plastics, Pittsfield, MA, 1999.
3. K. Song, W. Li, J.O. Eckert, D. Wu, R.E. Apfel, J. Mat. Sci. 1999, 34, 387-395.
4. B. Jeong, M. Xanthos, Y. Seo, J. Cell. Plas. 2006, 42, 165-176.
5. M. Xanthos, B.J. Jeong, Polym. Eng. Sci. 2007, 47, 244-253.

KN-03-02

PREPARATION OF MODIFIED UNORIENTED ELECTRICAL GRADE POLYVINYLIDENE FLUORIDE POLYMER FILMS BY CONTINUOUS PROCESS

Vijayakumar R.P.[1], Milind V Mhalgi[2], Devang V. Khakhar[3] and Ashok Misra[4*]

[1] *Visvesvaraya National Institute of Technology, Nagpur, India – vijayakumarrp@che.vnit.ac.in*

[2] *Tata Autocomp Systems Limited-Technical Center –milind.mhalgi@tacogroup.com*

[3] *Indian Institute of Technology Bombay, India – khakhar@iitb.ac.in*

[4] *Intellectual Ventures, Bangalore, India – amisra@iitb.ac.in*

Abstract: The beta (β) phase in polyvinylidene fluoride (PVDF) exhibits piezoelectricity due to its dipolar conformation. It is reported that in the compression molded PVDF films the addition of a small weight percent of onium salts induces beta phase formation and increases the toughness in PVDF films. We have carried out a study of the cast film extrusion of PVDF film with the addition of onium salts. Determination of processing methodology to produce cast film of PVDF added with onium salt is of considerable importance from a practical point of view. Processing of PVDF with onium salts by different extrusion process were carried. In the present work it is observed that the processing in a twin screw extruder attached with ribbon die is a feasible method to prepare PVDF film with the addition of onium salt. WAXD and FTIR studies were carried to determine the formation of betaphase in the prepared films. DSC studies showed a significant increase in the melting temperature compared to pure PVDF film. The study showed a new method to produce highly polar beta phase PVDF films for enhancing electric properties.

Keywords: PVDF, Onium salts, WAXD, FTIR, DSC and SALS.

Introduction

Polyvinylidene fluoride (PVDF) is a meltprocessable fluoropolymer known for its polymorphic phases [1]. Among the different crystal phases of Polyvinylidene fluoride (PVDF), the most useful one is the β phase, which gives piezoelectric and pyroelectric effect because of its dipolar conformation [2]. The highly polar β phase is induced in the melt extruded α phase containing PVDF films by stretching at lower temperatures to draw ratio of ~ 4 or at higher temperatures to higher draw ratios (>4) [3-5].

The production of β phase film by direct extrusion will avoid the additional offline stretching process. It has been reported that by the reinforcement of PVDF by the addition of organically modified layered silicates induces β phase during crystallization. Priya et al [6,7] were the first to study the effect of clay loading on the crystallization behavior of PVDF and they reported that the crystallization rate of PVDF increases in the presence of clay platelets and gives the beta phase, which is stable after annealing. Different groups studied the effect of nanoclay in the phase transformation in PVDF [8,9,10,11]. Patro et al[12] studied the effect of different clay modifiers on the PVDF-clay nanocomposites and reported that the phosphonium modified clay is the most efficient in generating β phase and ammonium and pyridium modified clays improve the toughness of PVDF most significantly among the modifiers studied.

Judovits et al [13] found that the PVDF mixed with different alkyl and aryl based onium salts by means of compression molded films increases the β phase content. However, the phase transformation behavior and nucleation behavior of PVDF with the addition of onium salt was not reported by Judovits et al [13].

Vijayakumar et al [14] found the presence of small fraction of γ phase in the compression molded PVDF films with the addition of Benzyl triphenyl phosphonium chloride and Tetrabutyl ammonium perchlorate. The γ phase has slightly less polarity compared to the β phase, whereas nonpolar α phase is completely absent in all of the produced films. They also found the decrease in the sprehulitic size of PVDF with the addition of onium salts and confirmed the formation of β phase with higher melting point.

The literature reveals that the studies on PVDF with the addition of nanoclay / onium salts were carried only by using the films prepared by solvent casting / compression techniques.

PVDF films with complete β phase by avoiding the γ phase, which forms due to high pressure crystallization, can be produced by direct casting process. The cast film extrusion (continuous process) of PVDF film with the addition of onium salts has not been reported and the main focus of this paper. Determination of processing methodology to produce cast film of PVDF added with onium salt is of considerable importance in industrial point of view.

Experimental

Materials

PVDF, Kynar 720 (Atofina, France) having molecular weight 2.9 x 10^5 g/mol determined using solution viscosity and MFI about 12g/10 min, was used for the study. Two different onium salts, benzyl triphenyl phosphonium chloride (BTPC) and tetrabutyl ammonium perchlorate (TAP) were used. TAP was obtained from Fluka, Sigma Aldrich, Switzerland. BTPC was obtained from two different suppliers, Acros organics, New Jersey, USA and Sigma - Aldrich, St. Louis, USA and refered below as BTPC (A) and BTPC (S) respectively. The onium salts were used as received.

Film preparation and observation

Studies with Varying Cast Roll Temperature.

PVDF film was melt extruded at 230^{0}C by cast film extrusion technique. Film samples were prepared at five different cast roll temperatures namely 75,90,107,120 and 130^{0}C.The cast roll speed was kept at 10 rpm. After this the cast films were uniaxially stretched on a stretching unit at a temperature of 80^{0}C to study the effect of stretching induced phase transformations in these samples.The films were characterized using WAXD &FTIR.The unstretched films were also characterized using SALS based on which the average spherulitic radius was calculated.

Trial I

The composition of 1 & 3 wt% of TAP were manually mixed with PVDF and fed into the barrel through the hopper in a single screw extruder attached with slit die. The length over diameter (L/D) ratio of the screw was 24:1. The raw materials were predried for 1 hour at 110°C before processing. The processing parameters for the cast film extrusion are shown in Table 1. Pure PVDF film was prepared under similar processing conditions.

Table 1- Processing parameter for cast film extrusion - Single screw extruder

Parameter	Value
Extruder – Zone 1	200 °C
Extruder – Zone 2	210 °C
Extruder – Zone 3	220 °C
Die	230 °C
Cast roll	80 °C
Screw speed	19 rpm

Trial II

PVDF and TAP were predried for 3 hours at 110°C. 1 and 3 wt% of TAP were manually mixed with PVDF and fed into the twin screw extruder (APV Baker, 25 L/D). The nozzle temperature was 210°C and the screw speed was 60 rpm. The extrudate was made into small pellets using pelletizer. These pellets were predried at 110°C for 3 hours and film preparation was carried out in the same single screw extruder used in Trial I. The processing parameters for the pure PVDF and the pellets prepared with 1 wt% TAP were same as in Trial I. The film obtained from PVDF + 1wt% TAP had small unmelted material. The temperature at the extruder and die zones was increased by 10°C for processing the pellets prepared with 3 wt% TAP.

Trial III

1 wt % of BTPC (S) was manually mixed with PVDF and fed into the twin screw extruder (APV Baker, 25 L/D). The temperature profile from the feed zone to nozzle was 150, 170, 190, 210 and 230°Cand the screw speed was 60 rpm. The pelletized extrudate was predried at 110 °C for 4 hours in a vacuum oven. Initially the film processing was carried in a single screw extruder with the processing parameters used for PVDF + 3% TAP in Trial II. At this condition the material is not flowing easily. The temperature at the extruder and die zone was increased by 20°C and screw speed was increased to 22 rpm and film preparation was carried out.

Trial IV

1 and 3 wt% of BTPC (S) was manually mixed with PVDF after predrying at 110°C. The film was prepared in a twin screw extruder (Brabender Live station with Gravimetric feeder, Germany) attached with ribbon die (width 150 mm, thickness 0.5 mm). The processing parameter for the cast film extrusion is shown in Table 2. The melt from the die is passed over the electrically operated conveyer. By controlling the speed of the conveyer, the thickness of the film is controlled. Feeder output of 1.5 kg/hr and conveyer speed of 6.12 m/min is used for preparing PVDF films with 1 and 3 wt% of BTPC (A). The other processing conditions remain the same. The film samples of PVDF with 1 and 3 wt% BTPC (A) and BTPC (S) are referred below as P1BT (A), P3BT (A), P1BT (S) and P3BT (S) respectively.

Table 2 - Processing parameter for cast film extrusion - Twin screw extruder

Parameter	Value
Extruder – Zone 1	170 °C
Extruder – Zone 2	170 °C
Extruder – Zone 3	175 °C
Extruder – Zone 4	180 °C
Extruder – Zone 5	190 °C
Extruder – Zone 6	200 °C
Extruder – Zone 7	210 °C
Extruder – Zone 8	220 °C
Extruder – Zone 9	230 °C
Gravimetric feeder output	0.5 kg/hr
Screw speed	35 rpm
Conveyer speed	3.7 m/min

Results and Discussion

Studies with Varying Cast Roll Temperature.

For the unstretched films, the average spherulitic radius increased from 1.8 to 3.7 μm with increase in cast roll temperature. Higher cast roll temperature results in smaller degree of supercooling thereby reducing the nucleation rate. The effect of cast roll temperature on fraction of β-phase is shown in Fig.1. There is a gradual increase in the amount of β-phase till 107^{0}C and after that it decreases substantially at 130^{0}C. Since the crystallinity of all the samples was found to be the same, larger spherulites are obtained with increase in cast roll temperature. Therefore, there is an inverse relation between spherulite size in unstretched film and fraction of β-phase obtained in stretched films.Thus the lower cast roll temperature gives higher fraction of β-phase [15].

Trial I

X-ray patterns for the prepared films are shown in the Fig. 2. Film prepared with pure PVDF showed α phase peaks corresponding to 2θ = 18.4° and 2θ = 19.9°.The α phase peaks are broadened for the 1 and 3 wt% TAP added films. The X-ray patterns show partial shifting of α phase peak towards 2θ = 20.4°, corresponds to β phase, for 3 wt% TAP added film.This implies the formation of some fraction of β phase with the addition of TAP. However, this observation is not in agreement with the results obtained in [14], where the α phase peak completely shifted to $2\theta = 20.4°$ for 0.5 wt% addition of TAP. The observation in the Trial I implies that the added TAP was not mixed uniformly with PVDF. Hence it was decided to prepare the premix in a twin screw extruder to obtain high intensive mixing and then prepare cast film using single screw extruder.

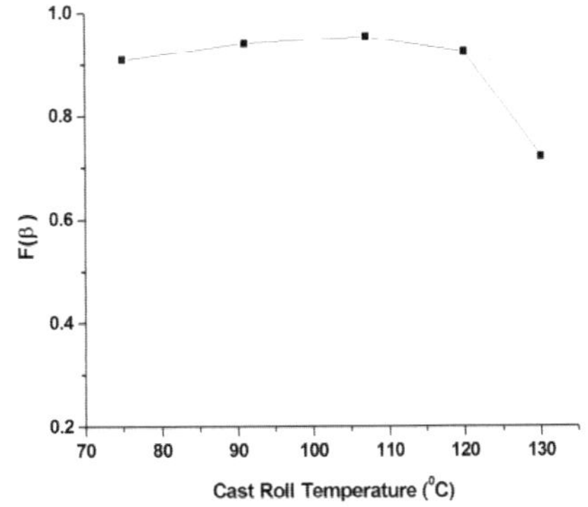

Figure 1- Effect of cast roll temperature on β-phase fraction in stretched PVDF films (stretch ratio 4.2)

Trial II

The FTIR spectra for the films prepared with pure PVDF and with 1 and 3 wt % of TAP (figure not shown) showed that the intensity of α phase peak at 766 cm^{-1} does not decreases with the addition of TAP. However, the intensity of β phase peak at 840 cm^{-1} increases with the addition of TAP. The 811 cm^{-1} peak corresponds to the γ phase is absent for the films prepared with TAP, whereas small peak at 811 cm^{-1} was seen for the films prepared with onium salts in [14]. Thus, the small fraction of γ phase formation in the compression molded films [14] is due to the high pressure applied during processing [1].

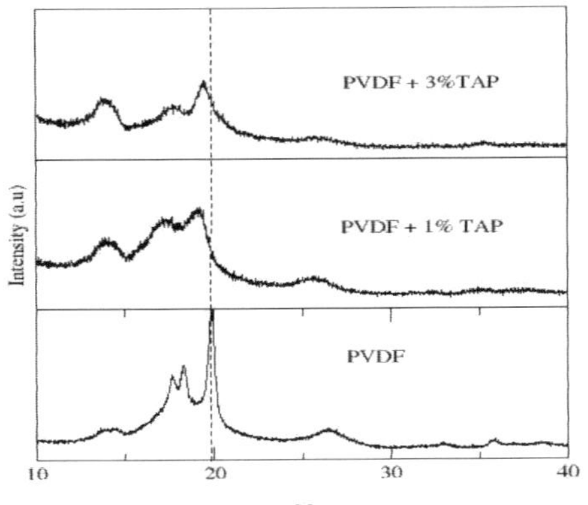

Figure 2 - WAXD pattern for the PVDF film extruded with different weight % of TAP

The complete phase transformation was not obtained due to nonuniform mixing of TAP as the prepared films had unmelted material in both 1 and 3 wt% TAP added films.

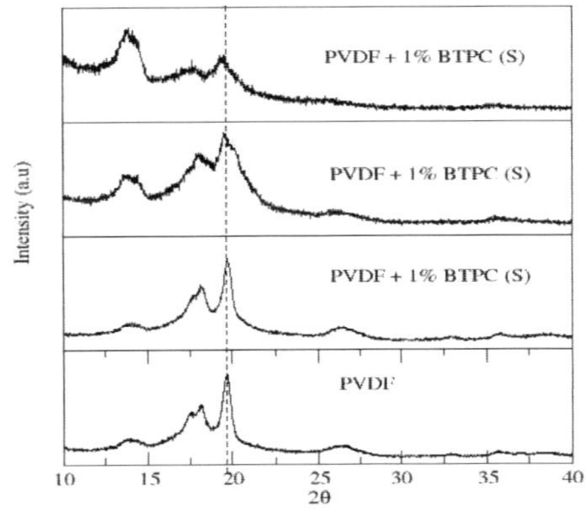

Figure 3 - WAXD pattern for the PVDF film extruded with 1 wt % of BTPC

Trial III

X-ray diffraction traces for the films prepared with pure PVDF and for three film samples of PVDF + 1 wt% of BTPC (S) are shown in the Fig. 3. PVDF films prepared with 1 wt% of BTPC (S) shows superposition of α and β phase peaks. This observation is not in agreement with the results obtained in [14]. Observations in Trial III confirm that the intensive mixing of onium salt with PVDF in a twin screw extruder with L/D ratio of 25:1 and then processing in a single screw extruder with L/D ratio of 24:1 is complex. Hence it was decided to prepare onium salt added PVDF film using twin screw extruder having attached with cast film die.

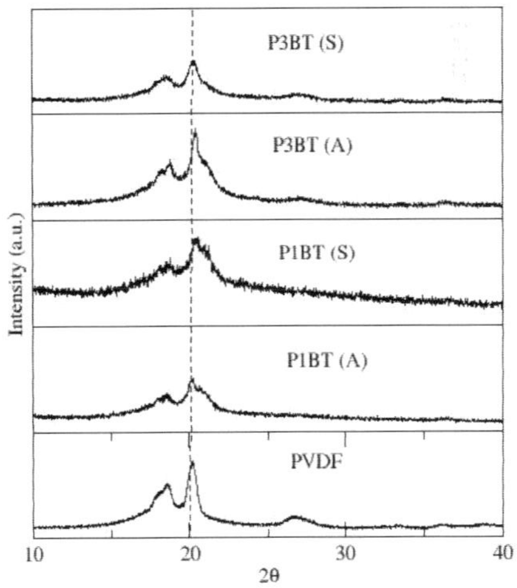

Figure 4- XRD pattern for the PVDF film extruded with different weight % of BTPC

Trial IV

Phase behavior

X-ray diffraction traces for the films prepared with pure PVDF and with 1 and 3 wt% BTPC (A) and BTPC (S) are shown in the Fig. 4. The β phase peak at $2\theta = 20.4°$is seen for the films prepared with different compositions of BTPC, whereas for the pure PVDF only α phase peaks corresponding to $2\theta = 18.4°$ and $2\theta = 19.9°$ are seen. This implies the formation of β phase with the addition of BTPC. This observation is in agreement with the results obtained in [14].

The FTIR spectra for the films prepared with BTPC show a peak at 840 cm^{-1} which corresponds to the β phase, whereas this peak is absent for the pure PVDF film (Fig.5). The γ phase peak at 811 cm^{-1} is absent similar to the observation for the films prepared with TAP in Trial II and confirms that the γ phase formation for the films prepared with onium salts in [14] is due to high pressure applied during processing. The intensity of α phase peak at 766 cm^{-1} decreases with the addition of BTPC. Thus the film prepared with BTPC in Trial IV has β phase

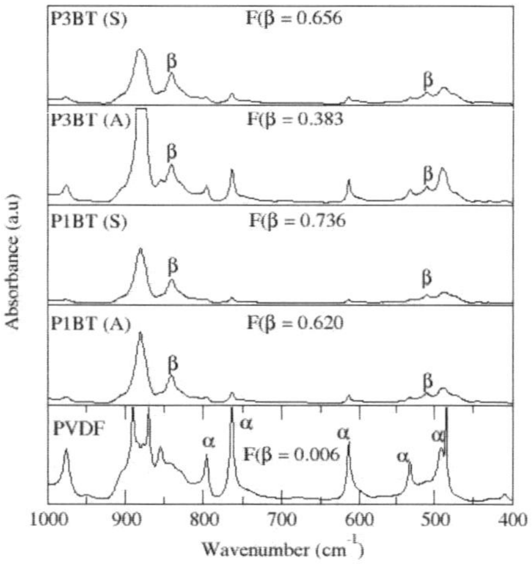

Figure 5 - FTIR spectra for the PVDF film prepared with different weight % of BTPC

Thermal studies

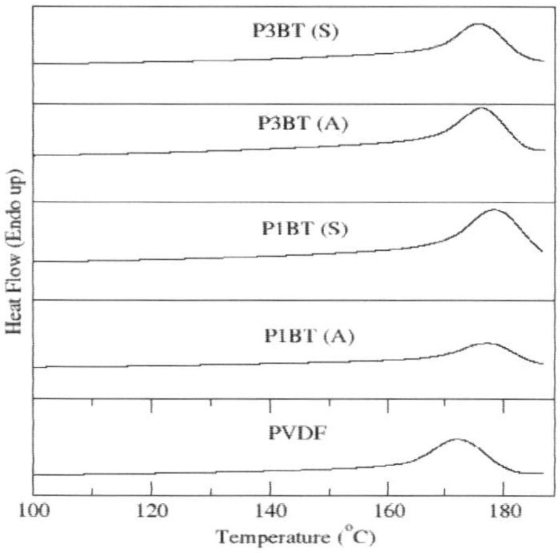

Figure 6- DSC melting curves for the PVDF films prepared with different weight % ofBTPC - *2nd* heating.

The melting curves obtained from differential scanning calorimetry of PVDF with BTPC (A) and BTPC (S) samples are shown in Fig.6. The melting temperature of PVDF increases with the addition of BTPC and confirms the formation of higher melting β phase crystals in the films prepared in Trial IV.

Conclusions

The processing of onium salt added PVDF film was carried in different types of extruder and the prepared films were characterized by different techniques. XRD, FTIR and DSC studies show that the PVDF cast films prepared with the addition of Benzyl Triphenyl Phosphonium Chloride (BTPC) in a twin screw extruder attached with ribbon die havepredominantly β phase crystals. The present work gives a method to produce PVDF films with the addition of BTPC, which have higher β phase content.

Acknowledgements

We thank SAIF, IIT Bombay for providing FTIR facility

References

1. A.J. Lovinger, *Developments in crystalline Polymers-1,* Applied Science Publishers, London, 197 (1982).
2. H .Kawai.,*Jpn. J. Appl. Phys*, 1969, *8*, 1975.
3. K. Matsushige, K. Nagata, S. Imada, and T. Takemura, *Polymer*, 1980, *21*, 1391.
4. J.C.McGrath and I.M.Ward, *Polymer*, 1980, *21*, 855.
5. T. Hsu and P. Geil, *Journal of Materials Science*, 1989, *24*, 1219.
6. L.Priya and J.P.Jog, *Journal of Polymer science part B: Polymer physics*, 2002, *40*, 1682.
7. L.Priya and J.P.Jog, *Journal of Applied Polymer science*,2003, *89*, 2036.
8. KP Pramoda, Ashiq Mohamed, In Yee Phang and Tianxi Liu, *Polymer International*, 2005, *54*, 226-232.
9. Deepak shah, Pralay Maiti, Erica Gunn, Daniel F. Schmidt, David. D. Jiang, Carl A. Batt and Emmanuel P.Giannelis. *Adv. Materials*, 2004, *16*, 1173.
10. Douglas R. Dillon, Kishore K Tenneti, Christopher Y. Li, Frank K Ko, Igore Sics and Benjamin S Hsiao, *Polymer*, 2007, *47*, 1678.
11. Jennifer Buckley, Peggy Cebe, Daniel Cherdack, Jennifer Crawford, B. Seyhan Ince, Matthew Jenkins, Jingjing Pan, Matthew Reveley, Niesha Washington, Natalie Wolchover, *Polymer*, 2006, *47*, 2411.
12. T. U. Patro, M. V. Mhalgi, D. V. Khakhar, A. Misra, *Polymer, 2008, 49*, 3486.
13. L. H. Judovits, N. J. Rodak, S. Zerafati, R. Gupta Ravi, (WO/2007/145668) U.S.Patent, December 21, 2007.
14. R. P. Vijayakumar, D. V. Khakhar, A. Misra, Indian patent application (Application No 1258/MUM/2011) Filed on 18 April 2011
15. Milind V. Mhalgi, Devang V. Khakhar, Ashok Misra, *Polym.Engg. & Sci.*, 2007, **47**, 1992.

O-03-371

EFFECT OF SPINNING CONDITIONS ON FIBER FORMATION BEHAVIOR IN MELT BLOWING PROCESS

R. Ruamsuk, W. Takarada and T. Kikutani*

Department of Organic and Polymeric Materials, Tokyo Institute of Technology, Japan – ruamsuk.r.aa@m.titech.ac.jp; takarada.w.aa@m.titech.ac.jp; kikutani.t.aa@m.titech.ac.jp

Abstract - The effect of melt blowing process parameters on fiber diameters was investigated for two types of polymer, polymethylpentene (PMP) and polypropylene (PP). The results exhibited that the decreasing of the throughput and the increasing of air flow rate reduced the fiber diameter for both polymers, while the decreasing of the die to collector distance (DCD) caused the reduction of fiber diameter only for PP. Moreover, under the similar spinning conditions, diameter of PMP fibers tended to be larger than that of PP fibers. For clarifying this result, on-line measurements were employed to observe the thinning behavior of these two fibers. The measurements revealed that while the air velocity decreases steeply along the spin-line, the fiber velocity measured using a laser doppler velocimeter increased in the region close to the die and then started to decrease gradually with the further increase of the distance where the fiber velocity is almost equal to the air velocity. In most cases, maximum fiber velocity obtained from the on-line measurement was much lower than the fiber velocity calculated from the final mean diameters. The fiber formation near the die was observed by using a high speed video camera. This observation revealed that even though the reducing of throughput and increasing of air flow rate lead to the decrease of the fiber diameters, thinning of the extruded fiber tended to be unstable with the phenomenon similar to the draw resonance in the melt spinning process.

Keywords: Melt blowing process, Fiber diameter, Instability.

Introduction

A nonwoven fabric is broadly defined as a structure of webs or sheets of randomly oriented fibers connected together by physical entanglement or adhesion which means no weaving or knitting procedure. This engineered fabric has been attracting interested by many fields of industries due to its wide range of applications such as filtration, membrane separation, fiber-reinforced plastics, biosensors, and wound dressings [1]. The further growth for nonwoven fabrics industry is expected by reducing of fiber sizes to nanometer range which causes some favorable properties for example, surface-to-volume ratio and variations in the wetting behavior [2]. One of the major processes that produce nonwoven fabrics is melt blowing process which was first invented at the Naval Research Laboratory of USA by Van Wente [3]. The melt blowing process is a single step process which begins with extruding the molten polymer through a die with orifices and extending the extruded fibers with two jets of hot air that rapidly attenuate the extrudates into small diameter fibers. After that the melt-blown fibers mat can be formed on a collector. Due to the demand of the size reduction of fiber diameter, effects of spinning parameters on fiber diameters have been studied. For instance, the path of hot air in the die have been investigated for various shapes such as a square, a rectangular, an annular and, a famous one, a v-slot shape air holes to study the performance of air flows which play an important role in reducing the size of fibers [4,5]. The effect of varied parameters of melt blowing process has been reported. The most influential parameter on fiber diameters is the die temperature, while air flow rate and polymer throughput rate show moderate influence. The parameter that less influence on fiber diameters are DCD and collector speed [6,7].

In this study, the efficiency of the melt blowing apparatus was examined through investigating the effect of different parameter conditions on the properties of resultant fiber web along with the measurement of air and extruded polymer velocities. Moreover, the images of the jet path of molten polymer was captured using a high-speed camera, which has been used by several researchers to investigate a jet path in the electro-spinning process [8,9].

Experimental

Polymethylpentene (PMP) (MFR=3000) and Polypropylenes (PP) (MFR=800 and 1550) supplied by Mitsui Chemical, Inc. were used. Fig. 1 shows a schematic of melt blowing die with 10 orifices used in this study. The die was designed based on the design of a typical commercial melt blowing die. The geometries of the die assembly are: the diameter of the spinneret orifice (D_o) was 0.12 mm, the width of air slot (a) was set to 1 mm for PMP and PP MFR 800 and to 1.5 mm for PP MFR 1550, respectively, and the set back (S) was 1.3 mm for both polymers.

Several on-line measurements were also carried out at various positions below the die; air velocity was measured using a pitot tube (DT-8920, CEM instruments), fiber velocity was measured using a laser doppler velocimetry (LDV; LS50) to study the basic understanding of thinning behavior. Observation of extruded polymer using high speed camera (Vision Research, Inc.) was also performed to explore the thinning behavior of filaments near the spinneret. For this purpose, the set back was set to be 0 mm to expose fibers immediately after extrusion through the spinneret orifice.

The melt-blown fibers were collected on a metal mesh to form a web. Images of the prepared melt-blown webs were recorded by using a scanning electron microscope (SEM; Topcon SM-200). For the analysis of fiber diameter distribution, diameters of 50 fibers on the horizontal line in the middle of SEM image were measured. These webs were also used to investigate the fiber structure by wide-angle X-ray diffraction (WAXD) and differential scanning calorimetry (DSC).

(a)

(b)

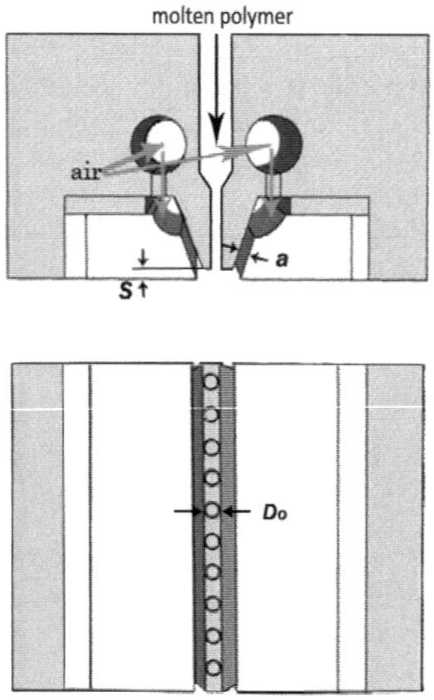

Fig 1 - Schematic of melt blowing die **(a)** side view and **(b)** bottom view. Set back (*S*), diameter of orifice (D_0) and width of air slot (*a*) are indicated in the figure.

Results and Discussion

1. Effect of spinning condition

Table 1 displays mean diameters for PMP and PP under various spinning conditions. For the effect of throughput (Exp. No. 1 and 4) and air flow rate (Exp. No. 2 and 6), both polymers showed the same trend that when the throughput was decreased or the air flow rate was increased, the amount of fine diameter fibers increased and therefore mean diameter decreased. This trend was clearly found in case of PMP. These results suggest that the fiber diameter is efficiently influenced by the air drag force, which is the heart of diameter attenuation in this process. In case of PP, however, even though the mean diameter was decreased as in the case of PMP, it was caused by the increase of the amount of fine diameter fibers without the change of the peak position in the diameter versus frequency diagram.

The DCD influenced fiber diameter of PP (Exp. No.5), in that the mean diameter became 2 μm larger when the DCD was increased from 10 to 20 cm. On the other hand, diameter of PMP fibers was not significantly influenced by the change of DCD (Exp. No. 3), in that the mean diameters remained at around 7 μm for three conditions.

The effect of spinning conditions on fiber structure analyzed by using DSC and WAXD showed insignificant change for both polymers. However, the comparison between the two polymers exhibited the difference in terms of the crystallization behavior in the melt blowing process. In the DSC heating thermogram of PMP shown in Fig.2(a), glass transition temperature (Tg) and melting peak temperature (Tm) were observed at around 35 and 230 °C, respectively, however cold crystallization peak did not appear. Considering that the Tg of PMP is higher than the room temperature, this result suggests that the fiber is crystallized in the spinning process. On the other hand, in the heating thermogram of PP shown in Fig. 2(b), phase transition peak is displayed at around 100℃. This result suggests that no crystallization occurred during the spinning process, and smectic structure was formed probably after the spinning process.

The results from the X-ray measurement were in accordance with the results from DSC. In the WAXD patterns shown in Fig. 3, PMP exhibits the reflection from oriented crystals 200 planes. For PP fiber, the pattern of smectic phase of PP is observed.

Table 1 - Summary of the effect of spinning conditions on fiber diameter.

Exp No.	Polymer	Throughput (g/min)	Air Flow Rate (L/min)	DCD (cm)	Mean Diameter (μm)
1	PMP 3000	0.17	300	10	4.34
		0.50			6.72
2		0.50	100	10	17.12
			200		10.01
			300		6.81
3		0.50	300	10	6.81
				20	6.72
				30	6.93
4	PP800	0.18	350	10	3.13
		0.54			4.49
5		0.54	350	10	4.49
				20	6.19
6	PP1550	0.54	100	10	7.46
			200		3.70
			300		3.38

Note – The numbers after the name of polymers indicate its MFR.

(a.) PMP

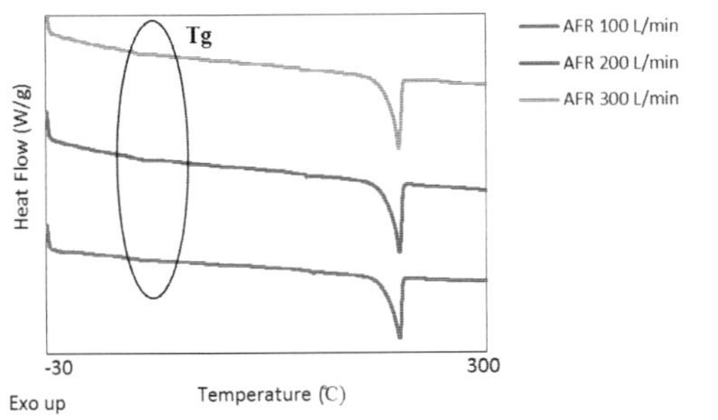

(b.) PP

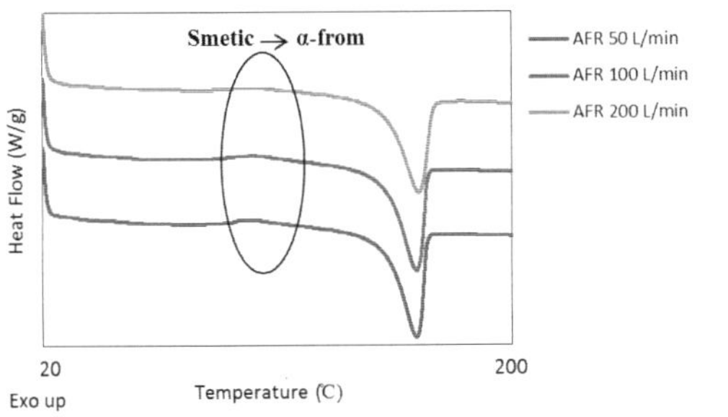

Fig 2 - DSC heating thermograms for (a) PMP and (b) PP web samples melt-blown with three different air flow rates.

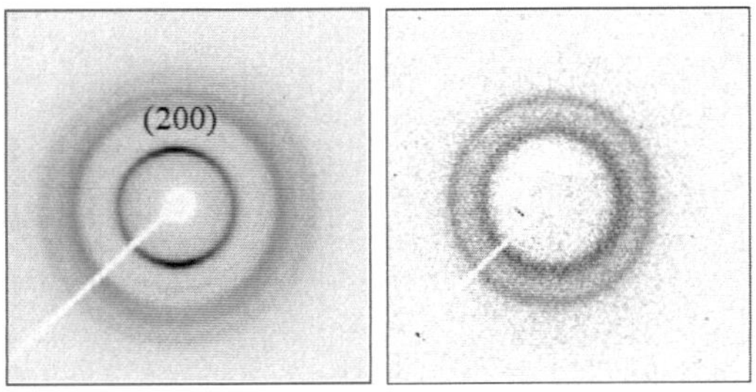

(a.) PMP (b.) PP

Fig 3 - WAXD patterns for the melt-blown fiber web prepared with air flow rate of 100 L/min. (a) PMP, (b) PP. The measurements were performed from the edge direction of fiber web sheet. Direction of sheet is horizontal in the WAXD pattern.

2. Analysis on the fiber formation behavior

Results of air velocity and fiber velocity measurements are shown in Fig. 4. The air velocity profile show rapidly decreases within 5 cm below the spinneret, in which the deceleration was more significant with the increasing of air flow rate. At the distance over 5 cm, the air velocity showed continuous reduction with slower rate. On the other hand, the fiber velocities showed a maximum at about 3 cm for higher air flow rate, 200 and 300 L/min, and at about 5 cm for 100 L/min. These results exhibited that the increasing of the fiber velocity was observed only when air velocity was higher than the fiber velocity. After velocity reached the air velocity, the observed fiber velocity started to decrease along with the reduction of air velocity. It should be noted from Table 2, however, that the maximum fiber velocity from the on-line measurement was lower than the fiber velocity estimated from the mean diameter of final fibers by using the mass conservation equation;

$$V = \frac{tp}{\rho} \Big/ \pi\frac{D^2}{4}$$

where tp is polymer throughput per orifice, ρ is polymer density and D is the mean fiber diameter. These results may suggest that the velocity from the LDV corresponds to the speed of thinning molten polymer itself at the upper part of the spin-line where the shape of the spin-line is almost straight, whereas at the region below the observed maximum fiber velocity, it corresponds to the down-ward speed of fiber with whipping motion, which occurs because of the loosoning of spin-line tension.

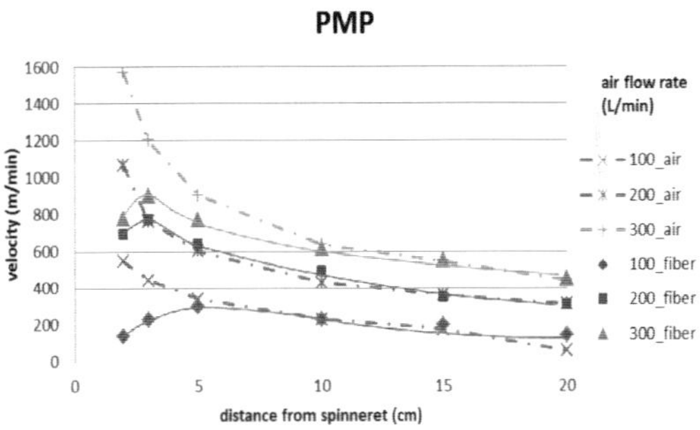

Fig 4 - Variations of air velocity together with PMP fiber velocity with the change of distance from spinneret for three different air flow rates.

Table 2 – Comparison of fiber velocity calculated from mean fiber diameters and the maximum fiber velocity from on-line measurement.

Materials	Air flow rate (L/min)	Fiber velocity from calculation (m/min)	Maximum fiber velocity from on-line measurement (m/min)
PMP	100	297	300
	200	1045	771
	300	1594	902
PP	100	560	327
	200	1712	830

Results of the observation of extruded polymer using a high speed camera for different spinning conditions are displayed in Table 3. The meaning of the marks are; O: continuous and stable formation of fibers, Δ: continuous but unstable formation of fibers, and x: intermittent breakage of fibers. When the spin-line becomes unstable, it shows the periodic oscillation of fiber diameter similar to the one observed in the melt spinning process under the unstable condition defined as draw resonance.

It was found that the extruded fiber attenuates steady and continuously under the conditions of high throughput and low air flow rate, whereas the continuous but unstable formation of fibers and even the breakage of fibers occur under the conditions of low throughput and high air flow rate. The frequency of diameter fluctuation was analyzed from the continuous but unstable fiber formation conditions and the intermittent fiber breakage conditions. The results exhibited that the frequency of fluctuation increases with the increasing of air flow rate. The frequencies are higher for PP than PMP. This instability causes variation of fiber diameter. Accordingly minimum diameter of melt-blown fibers, which appears to be the majority in the web, can be smaller than the diameter expected from the fiber velocity.

Table 3 – Melt blowing conditions for observation using high-speed camera.

Materials	Throughput		Air flow rate (L/min)		
	Volume (cm^3/min)	Mass (g/min)	100	200	300
PMP	0.067	0.055	Δ	Δ	Δ
	0.2	0.166	O	O	O
	0.6	0.498	O	O	O
PP	0.067	0.06	x	x	-
	0.2	0.18	-	-	x
	0.67	0.6	O	O	O

Conclusions

The investigation on the production of nonwoven fabrics by using the melt blowing process was presented. The best condition for producing very fine fiber of two kinds of polymers, polymethylpentene and polypropylene, were investigated. The effect of spinning conditions, i.e. die to collector distance, throughput and air flow rate on the fiber diameters and the fiber structure were different between PMP and PP. Diameter of PMP fiber was influenced by throughput rate and air flow rate but insignificantly by DCD, while DCD is the parameter that also influences the fiber diameter in case of PP fibers. The on-line measurements of air and fiber velocities along the spinning line as well as the observation of spin-line near the nozzle using a high speed camera provided information about the fiber formation behavior and the unstable spinning behavior.

References

1. W. Albrecht; H. Fuchs; W. Kittelmann *Nonwoven Fabrics: Raw Materials, Manufacture, Applications, Characteristics, Testing Processes,* Wiley-Vch, Weinheim, 2003.
2. D.H. Reneker *et al. Advances in Applied Mechanics*. 2007, *41*, 43-195.
3. V.A. Wente *Ind. Eng. Chem.* 1956, *48*, 1342-1346.
4. E. C. A. Schwarz, US Patent 4380570, 1983.
5. M. A. J. Uyttenclaele and R. L. Shambaugh, *Ind. Eng. Chem. Re* . Vol. 28, 1989, 1740-1741.
6. R.R. Bresee; U.A. Qureshi *International Nonwovens Journal*. 2004, 49-55.
7. R.R. Bresee; U.A. Qureshi *Journal of Engineered Fibers and Fabrics*. 2006, *1*, 32-46.
8. M.M. Hohman *et al. Applications. Phy. Fluids.* 2002, *13 (8)*, 2221.
9. D.H. Reneker *et al. J. Appl. Phys*. 2000, *87 (9)*, 4531–4547.

O-03-035

ELECTROSPINNING OF POLYLACTIDE-POLYGLYCOLIDE NANOFIBROUS MEMBRANES WITH ULTRA-LOW DIELECTRIC PROPERTY

Shih-Jung Liu*, Lung-Yi Chiou, Chun-Yi Liaw
Polymer Rheology and Processing Lab
Chang Gung University, Tao-Yuan 333, Taiwan
Email: shihjung@mail.cgu.edu.tw

Abstract - In this study, nanofibrous membranes of polylactide-polyglycolide were fabricated using an electro fiber spinning setup that consisted of a syringe and needle (the internal diameter is 0.42 mm), a ground electrode, an aluminum sheet, and a high voltage supply. The dielectric properties of the electrospun membranes were characterized. The experimental results suggested that the electrospun membranes exhibited ultra-low dielectric behavior. The influences of the nanofibers diameter and the density of the membranes on the dielectric properties were also instigated. It was found that the dielectric constants decreased with the fiber diameter and increased with the density of the nanofibrous membranes. By employing the electrospinning process, one will be able to fabricate polymeric membranes with ultra-low dielectric performance.

Introduction

Polylactide-polyglycolide (PLGA) has been one of the most important biodegradable and biocompatible materials. It is non-toxic and exhibits interesting physical properties. PLGA is the subject of growing scientific and practical interest in the past decades. Depending on the content of lactic acid in the copolymers, PLGA can be completely degraded in a few months. In a low-moisture environment, the degradation of PLGA is further reduced and the total degradation time may take more than a year.

While the dielectric properties of other biodegradable materials such as polylactide (PLA) have been studied by several groups [1-3], no one has investigated the electrical properties of PLGA. In this study, the nanofibrous membranes of polylactide-polyglycolide (PLGA) materials have been prepared using the electro fiber spinning process [4, 5]. Electrospinning is a simple and effective nanofabrication method for preparing nanofibrous membranes with diameters ranging form 5 to 500 nm, which are 102 to 104 times smaller than those prepared by the traditional method of solution or melt spinning. It also has the advantages of fabricating nanofibers from a huge range of materials. Of the major classes of materials, electrospinning is able to produce nanofibers of polymers, composites, semiconductors and ceramics [4, 6, 7]. After fabrication, the dielectric constant (k) of the nanofibrous polylactide-polyglycolide (PLGA) membranes was measured. The influence of the diameters of the nanofibers and the density of the membranes on the dielectric properties of the membrane was also investigated.

Experimental

1,1,1,3,3,3-hexafluoro-2-propanol (HFIP) were purchased from Sigma-Aldrich (Saint Louis, MO, U.S.A.). The poly (D,L)-lactide-co-glycolide (PLGA) used was commercially available materials (Resomer RG 503, Boehringer, Germany) and had a ratio of 50:50 and an intrinsic viscosity of 0.4.The electrospinning setup [5] utilized in this study consisted of a syringe and needle (the internal diameter is 0.42 mm), a ground electrode, an aluminum sheet, and a high voltage supply, as shown schematically in Figure 1. The needle was connected to the high voltage supply, which could generate positive DC voltages and current up to 35 kV and 4.16mA/125W respectively. For the electrospinning of nanofibers, PLGA was first dissolved in HFIP at three different concentrations of 13%, 15% and 17% by weight [8, 9], and was then delivered by a syringe pump with a volumetric flow rate of 3, 4, or 5 ml/hr. The distance between the needle tip and the ground electrode was 8, 12 and 15 cm, and the positive voltage applied to polymer solutions was set to either 15, 18, or 20 kV. All electrospinning experiments were carried out at room temperature [10].

Results and Discussion

Figure 1 shows the SEM micrographs of the electrospun PLGA nanofibers subjected to various processing conditions. All these nanofibrous membranes exhibited microporous structure composed of fully interconnected multi-fibrous layers. The diameters of the spun nanofibers ranged between 427 nm and 1153 nm, and the porosity of the nanofibrous matrix was high.

The dielectric constant (k) of the electrospun nanofibers was also measured. Figure 2 shows the k-values of the nanofibrous membranes as a function of frequency. When the applied electric field is an alternating current field, the frequency of the signal comes into play. The dielectric constants of the polymers can be reduced dramatically by employing the electrospinning process. Furthermore, by adopting the appropriate processing condition, nanofibrous membrane with a k-value of as low as 1.22 could be successfully produced by the electrospinning technique. All k-values of the nanofibrous membranes decreased slowly with the frequency. Furthermore, the experimental result in Figure 2 suggested that the dielectric constant of the membranes decreased with the diameter of the nanofibers.

Figure 1 –SEM images of electrospun nanofibers subjected to various processing conditions.

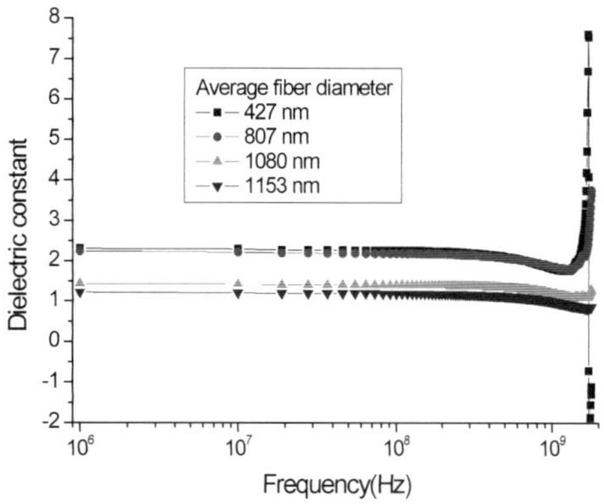

(a)

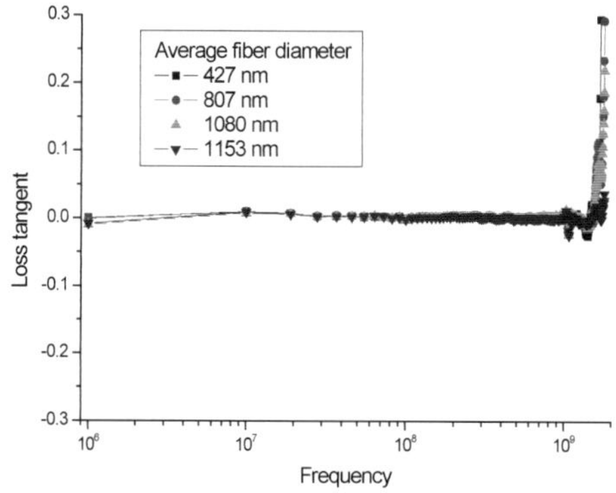

(b)

Figure 2 –Influence of fiber diameter on (a) dielectric constant, and (b) loss tangent of nanofibrous membranes.

Conclusions

In this study, nanofibrous membranes of polylactide-polyglycolide (PLGA) materials have been realized using the electro fiber spinning process. The result of dielectric constant (k) measurements suggests that k-values of the polymers can be reduced dramatically by the electrospinning technique. The influence of the density and the diameters of electrospun fibers on the dielectric properties of the membranes was also investigated. The empirical results indicate that the microporous structures of the membranes play an important role in reducing the k-values of the membranes. By employing the electrospinning process, one will be able to fabricate polymeric membranes with low-dielectric properties.

Acknowledgements

The National Science Council of Taiwan, R.O.C. under the grant NSC98-2221-E-182-003-MY3, has supported this work financially.

References

1. J.K. Jeszka, L. Pietzak, M. Pluta, and G. Bioteux: J. Non-crystal Solids 356 (2010) 818.
2. T. Nakagawa, T. Nakiri, R. Hosoya, and Y. Tajitsu: IEEE Trans. Ind. Appl. 40 (2004) 1020.
3. S. Oh, J.K. Lee, P. Theato, and K. Char: Chem. Mater. 20 (2008) 6974.
4. W.E. Teo, and S. Ramakrishna: Nanotechnology 17 (2006) R89.
5. S.J. Liu, Y.C. Kau, C.Y. Chou, J.K. Chen, R.C. Wu, and W.L. Yeh: J. Membrane Sci. 355 (2010) 53.

6. Z.M. Huang, Y.Z. Zhang, M. Kotaki, and S. Ramakrishna: Comp. Sci. Technol. 63 (2003) 2223.
7. I.S. Chronakis: J. Mater. Process. Technol.: 167 (2005) 283.
8. S.G. Kumbar, S.P. Nukavarapu, R. James, L.S. Nair, and C.T. Laurencin: Biomaterials 29 (2008) 4100.
9. Y. Yang, X. Zhu, W. Cui, X. Li, and Y. Jin: Macromol. Mater. Eng. 294 (2009) 611.
10. S.J. Liu, L.Y. Chiou, J.Y. Liaw, Jpn. J. Appl. Phys. 50 (2011) 10PG05.

O-03-056

PREPARATION OF OPEN-POROUS MICROCELLULAR POLYPROPYLENE FOAMS WITH NANO-FIBRILS BY CORE-BACK FOAM INJECTION MOLDING

Ryoma Miyamoto[1], Masato Fukumori[2], Haruo Shikuma[1] and Masahiro Ohshima[1,*]

[1] *Dept. of Chemical Engineering, Kyoto University, Kyoto, Japan 615-8510*
[2] *West Japan Branch, Showa Denko Gas Products, Osaka, Japan 550-0026*

Abstract – Open-porous microcellular foams with interconnected nano-fibrillated structure were prepared from isotactic polypropylene (i-PP) with a crystalline nucleating agent. 1,3:2,4 bis-O-(4-methylbenzyliden)-D-sorbitol gelling agent (Gel-all MD) was used as the nucleating and gelation agent of PP. The nucleating agent enhances the crystallization rate, provides a gelation of PP with three-dimensional network of highly connected nano-fibrils and gives higher melt-tension to PP. Foaming of the PP in course of crystallization with the help of the nucleating agent induces the bubble nucleation and growth in the inter-lamella region and opens the cell wall with nano-fibers. The core-back foam injection molding was employed to foam PP with the nucleating agent in course of crystallization process and prepare the open-porous microcellular foam with nano-fibrillated structure at an industrial production level. The effects of core-back distance and the nucleating agent, Gel-all MD, on the open pore ratio as well as the cell size, density and crystallinity were experimentally investigated using a 35 tonnage clamping force Mucell® machine with N_2. The uniform microcellular foams, whereas the average cell size was less than 5 μm, the expansion ratio was five-fold and the open pore ratio was over 90 %, could be achieved.

Keywords: Open pore, Microcellular Foam, Foam Injection Molding, Core-back Molding, Polypropylene Foam.

Introduction

Open porous foam has superior properties such as high permeability, selective osmosis, sound adsorption and dampening[1]. Because of these properties, they have been used for many industrial applications: from acoustic insulator, filter, battery electrode supports, separation membrane, battery separator, and tissue attachments and growth supports. However, commercially-available open plastic foams are made of almost polyurethane thermoset foams. There is still very few thermoplastic open porous foam available. The open-porous thermoplastic foams have been produced by either polymer blending or solute leaching.[2,3] Lee et al[2] prepared open porous foam from polyethylene (PE) and polypropylene (PP) blends by an extrusion. They proposed two strategies of opening the cell-wall of thermoplastic foams: 1) preparing polymer blends so that disperse domains can be composed of a low-T_c (soft)polymer and the matrix can be composed of a higher-T_c (hard) polymer, cell-opening can be initiated through soft domains entrapped between growing adjacent cells, 2) preparing polymer blend so that disperse domains can be composed of a higher-T_c (hard) and the matrix can be composed of a lower-T_c (soft) polymer, cell-opening is initiated by the deboning between hard domain and soft matrix. Both strategies exploit a structural in-homogeneity in the polymer blends where the crystallinity or the crystalline temperature is used for creating the inhomogeneity in polymer. Relationship between crystallinity of polymer and its foaming behavior have been also investigated for other polymers either by batch foaming or an extrusion foaming.[4-7] Pilla et al[6] conducted extrusion-foaming of polylactide (PLA) with talc as the nucleating agent and showed that the addition of nucleating agent increased the crystallinity of PLA, deceased the average cell size, and increased the cell density. They also reported the effect of CO_2 foaming agent on the crystallinity and an interesting cavitation phenomenon on the cell wall (cell-opening phenomena). Preparations of open porous foam by injection molding were limited to the cases combining the particulate leaching method. Kramschuster et al[8] conducted foam injection molding to prepare open porous thermoplastic foam from polylactide (PLA) and poly vinyl alcohol (PVOH) with salt as the particulate.

There are a limited number of reports on open porous thermoplastic foam prepared by foam injection molding[9]. One of the reasons why the injection molding could not prepare open porous foam was the conventional foam injection molding could not realize higher expansion ratio. The high expansion ratio is needed for cell-opening. Ishikawa et al [10] and Stumpf et al[11] introduced the core-back foam injection molding technique to foam i-PP with either N_2 or CO_2 as blowing agent. They discussed the similarity and dissimilarity of coreback foam injection molding from batch foaming and indicated the possibility of higher expansion ratio.

In this study, the core-back foam injection molding was employed to prepare open porous PP with a crystalline nucleating agent, Gel-all- MD. The nucleating agent enhances the crystallization during the cooling process of injection molding. The crystalline creates a structural inhomogeneity and open the cell wall with nano-fibrillated structure.

Experimental

Core-back foam injection molding.

A 35 tonnage clamping force foam injection molding machine (J35EL III-F, Japan Steel Work, Ltd.) and a gas injection device (Trexel SCF system SII TRJ-10-A-MPD) were used for the experiments. Nitrogen (N_2) is pressurized to 25MPa and injected into molten polymer in the screw barrel at 0.2wt% of polymer. Mixing at high temperature and pressure by the screw enhances the gas dispersion and dissolution and creates a single-phase N_2/polymer solution. The solution is injected into the mold cavity to foam. The cavity size was 50mm in width, 70mm in length and 2mm in the initial thickness. The core-back foam injection molding is slightly different from the traditional structural foam injection molding process: the polymer fills up the mold and dwelling pressure (40 MPa) is exerted. At the end of dwelling time (6s), a portion of the mold is moved to quickly increase the mold cavity volume, which rapidly reduces the cavity pressure. The sudden pressure drop enhances the bubble nucleation and achieves the fine cell structure of polymer foams. The core-back rate, which is the rate of moving the portion of the mold to open, the core-back distance, which is the distance of moving the portion of the mold from its initial position, can be changed in the rage from 0 to 40 mm/s and from 0 to 8mm, respectively. By manipulating the core-back rate and distance, the expansion ratio of cavity volume and the depression rate of the cavity pressure can be controlled.

Materials

The polymer used in this study was isotactic polypropylene, i-PP (F133A, Prime polymer Ltd). The nucleating agent was 1,3:2,4 bis-O-(4-methylbenzyliden)-D-sorbitol gelling agent (Gel-all MD, New Japan Chemical co., Ltd)(**Fig. 1**). N_2 in 0.999 purity (Izumisangyo, Japan) was used as a physical foaming agent. The barrel temperature of the injection molding machine was set at 210 °C, and the mold temperature was kept at 40 °C.

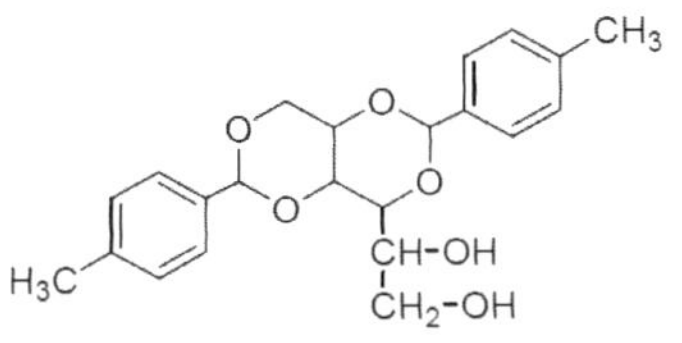

Figure 1 Chemical structure of 1,3:2,4-bis-*O*-(p-methylbenzylidene) - D-sorbitor molecule

Characterization

Differential scanning calorimeter (Pyris, PerkinElmer Ltd) was used to measure the crystallinity of injected PP. The measurement was conducted under N_2 purge. The heating/cooling rates were set to 10 °C /min and the temperature range was 40-200 °C. Crystallinity was calculated using 209J/g as the heat of fusion of the PP α-foam. Gas pycnometer (AccuPyc, Shimadzu Corp.) was used to measure the open pore ratio. The sample volume measured by the gas pycnometer excludes the open pore volume of the sample. Thus, the open pore ratio ε was calculated by Eq. (1).

$$\varepsilon = \frac{V_{app} - V_{mea}}{V_{app}} \times 100 \qquad (1)$$

where V_{app} is the apparent volume, which is geometrically measured by a caliper. V_{mea} is the volume measured by the pycnometer.

Figures 2(a) and (b) shows a schematic view of the core-back foam injection molded sample and SEM image of skin layer and core layer of the sample, respectively. The sample was cut out just after solidifying with liquid N_2. The number of cell was obtained by counting the cells of which size was larger than 5 μm from SEM images and the cell density was calculated by using Eq. 2

$$N_d = \left(\frac{N}{A}\right)^{\frac{3}{2}} \qquad (2)$$

where N is the number of cells observed in one image and A is the area of the image.

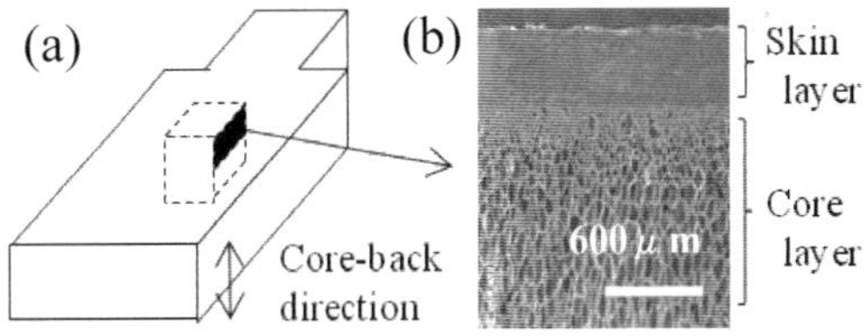

Figure 2 – (a) Schematic view of the core-back foam injection molded sample (b) SEM images of the cross sectional area of the foamed sample cut out in parallel to the core-back direction

In this study, the observed position is the center of the injection molded sample (Fig. 2 a). The sample has a skin and core layers as shown in Fig. 2 b. In the skin layer, no bubble was observed and cellular structure was observed in core layer.

Results and Discussion

Effect of addition of nucleating agent MD on cell morphologies

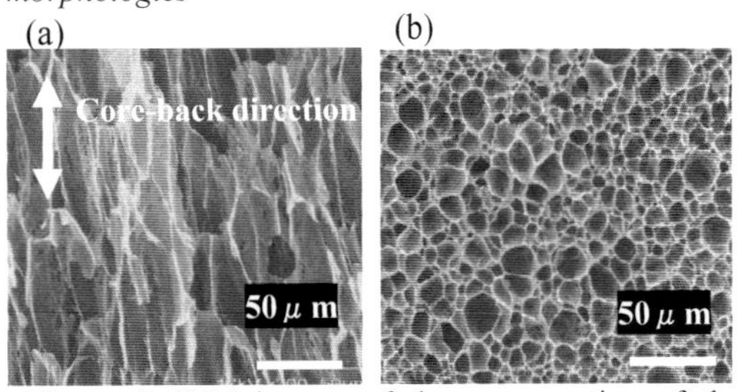

Figure 3 SEM images of the cross section of the foamed sample at core layer (Core-back condition: 8mm/0.4s, MD0.5%);(a) parallel and (b) perpendicular to the coreback direction.

Figure 3 shows that SEM images of the cross sectional areas of foamed sample cut out (a) in parallel to core-back direction and (b) in perpendicular to coreback direction. It is apparent that the cells were elongated to the core-back direction and shaped to be elliptic. On the other hand, the cell morphology observed from the view perpendicular to coreback direction was spherical. In the following discussion, the cell diameter and density were calculated from the SEM images of the area perpendicular to coreback direction assuming spherical cells.

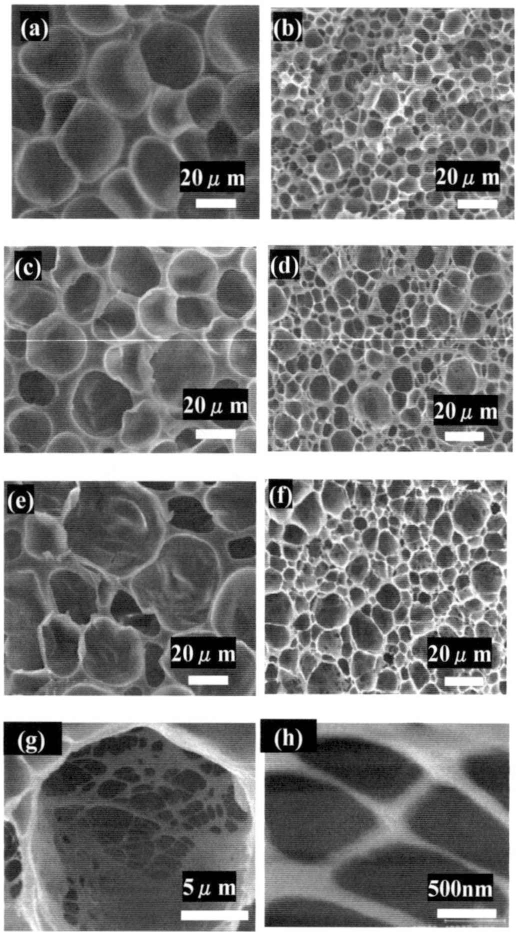

Figure 4 –SEM images of the cross sectional areas of the core layer of samples foamed with and without MD. All images were taken in perpendicular to coreback direction. (a) Core-back condition: 2mm/0.1s, MD0%, (b) 2mm/0.1s, MD0.5%, (c) 4mm/0.2s, MD0%, (d) 4mm/0.2s, MD0.5%, (e) 8mm/0.4s, MD0%, (f) 8mm/0.4s, MD0.5%, (g) magnification of the part (f), (h) higher magnification of the part (g).

Figures 4 a-f show SEM images of the samples with and without nucleating agent. The left column (Figs. 4-a,c,e) shows the images of the sample foamed without nucleating agent at different coreback distances and the right (Figs. 4-b,d,f) is the images of the sample foamed with nucleating agent at different coreback distances. As can be seen, the addition of crystalline nucleating agent, Gel-all, decreased the cell size and increased the cell density. The highest expansion, fivefold of initial thickness, was realized by setting coreback distance at 8 mm and making the sample thickness from 2 to 10 mm (Figs 4 e and f). Figures 4 d, f, g and h clearly show that the cell wall was open and fibrillated by the addition of the nucleating agent.

Figures 5-a and b show the effect of expansion ratio (coreback distance) with the same coreback rate, 20mm/s, on the cell diameter and the cell density of the foamed samples. The cell size and density were observed from the view perpendicular to the coreback direction in Fig. 5. They did not drastically change against core-back distance. However, the cell was elongated to the core-back direction in associated with coreback distance.

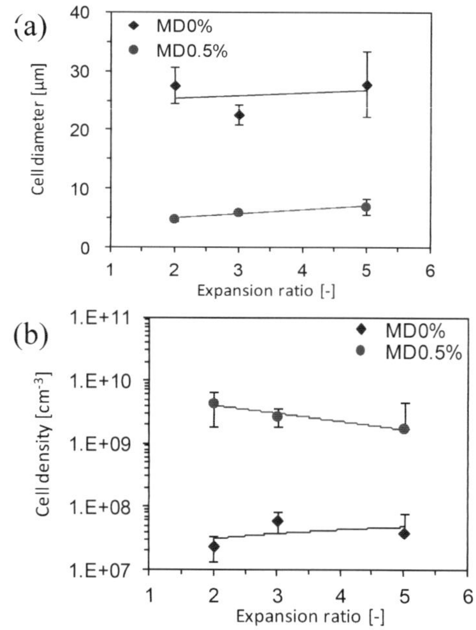

Figure 5 – (a) the cell diameter and (b) the cell density of the foamed sample (error bar indicates min. and max values).

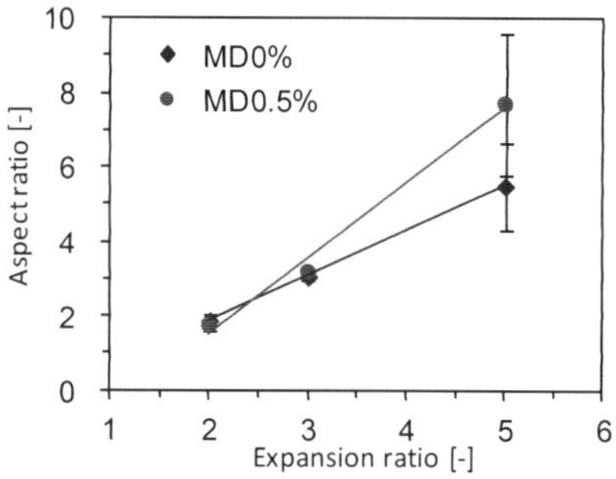

Figure 6 – Aspect ratio of the cell of the foamed sample at core position (error bar indicate min. and max values).

Figure 6 show the effect of expansion ratio on the aspect ratio (i.e., the ratio of cell size measured from the area in perpendicular to coreback direction to the longest length of each cell measured from the area in parallel to the coreback direction). The aspect ratio was changed in proportional to expansion ratio.

As can be seen in Fig. 4, (a)-(f) and Fig.5, Gel-all MD induced the crystallization of the i-PP and enhanced the heterogeneous bubble nucleation. Furthermore, due to gelation effect, the addition of Gel-all MD increased the elasticity at temperature higher than the melting point of iPP. Thus, the nucleating agent could supress bubble growth and reduce cell size.

Since the high expansion ratio of the foam with micro-scale cell size could be realized by core-back injection molding, the thickness in cell wall was decreased and cell-opening was made easier in foam injection molding. In addition, the nucleating agent could enhance crystallization and make the nano-fibril structure when the cell wall opened. **Figure 7** shows the open pore ratio of the foamed samples. The open-pore ratio was increased by addition of nucleating agent, Gel-all- MD and the increase of expansion ratio.

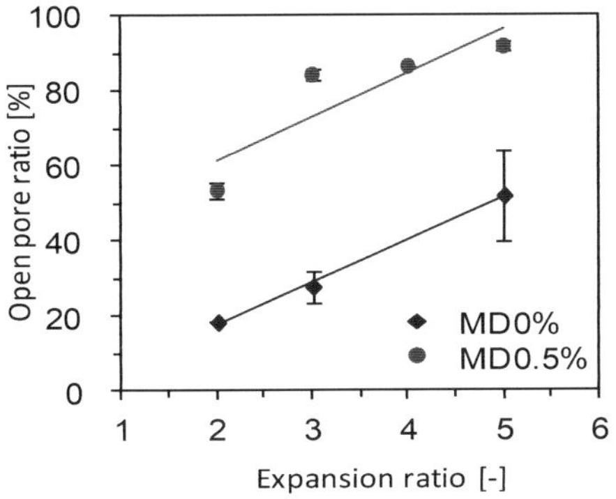

Figure 7 – Open pore ratio of the foamed sample at core position (average value of two experiments, error bar: min. and max value).

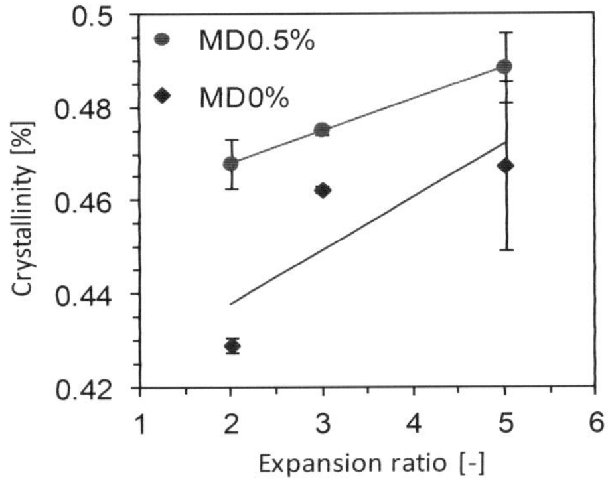

Figure 8 – Crystallinity of the foamed sample (error bar: min and max values).

Figure 8 shows the crystallinity of the samples injection-foamed at different expansion ratios and with different MD concentration but the same coreback rate, 20mm/s. The samples were cut out from the same position that SEM observation was made for. The crystallinity was increased by adding the nucleating agent as well as with the increase of expansion ratio. This indicates the enhancement of crystallization with nucleating agent and the occurrence of oriented or shear-induced crystallization. Controlling the crystallinity by expansion ratio and amount of nucleating agent is the key factor of cell-opening with fibrillated structure in foam injection molding with core-back operation.

Conclusions

Highly Open-porous microcellular poly-propylene foams with nano-fibrils were prepared by core-back foam injection molding. By adding the nucleating agent, the crystallinity was increased, the elasticity was increased and cell walls were fabricated with nano-fibrils. As a result, the cell diameter was decreased, the cell density was increased and the open pore ratio was increased. The high expansion ratio, i.e., 5 times as large as the initial volume with high open pore ratio, higher than 90%, were achieved.

Acknowledgements

The this research was conducted for the NEDO project Development of Innovative Non-Freon used Heat Insulation Technologies

References

1. J. McRae, H. E. Naguib and N. Atalla *Journal of Applied Polymer Sci.* 2010, *116,* 1106-1115.
2. P. C. Lee, J. Wang, and C. B. Park *Ind. Eng. Chem.* 2006, *45,* 175-181.
3. D. Kohlhoff and M. Ohshima *Macromol. Mater. Eng.* 2011, *296,* 770-777.
4. K. Taki, D. Kitano, and M. Ohshima *Ind. Eng. Chem. Res.* 2011, *50,* 3247-3252.
5. M. Mihai, M. A. Hunealt, and B. D. Favis *Journal of Applied Polym. Sci.* 2009, *113,* 2920-2932
6. S. Pilla, S. G. Kim, G. K. Auer, S. Gong, and C. B. Park *Polym. Eng. Sci.* 2009, 49, 8, 1653-1660
7. J. B. Bao, T. Liu, L. Zhao, D Barth, G-H. Hu *Ind. Eng. Chem. Res.* 2011, *50,* 13387-13395
8. A. Kramschuster, L-S Turng *J. Biomed. Mater. Res. Part B* 2010, *92B,* 366-376.
9. H.B Wu, . J Haugen and E. Wintermantel, *Journal of Cellular Plastics 48 (2) 141-159 (2012)*
10. T. Ishikawa, K. Taki, and M/ Ohshima *Polym. Eng. Sci. 52 (4), 2012, pp. 875-883.*
11. M. Stumpf, A. Sporrer, H-W. Schmidt, V. Alstadt, *47(6) 519-534, (2011)*

O-03-069

COMBINING FUNCTIONALITIES IN MULTICOMPONENT SYNTHETIC FIBERS

R. Hufenus[1]*, G. Fortunato[1], L. Gottardo[1], F.A. Reifler[1]

[1] *Empa, Swiss Federal Laboratories for Materials Science and Technology, St. Gallen, Switzerland – rudolf.hufenus@empa.ch, giuseppino.fortunato@empa.ch, laura.gottardo@empa.ch, felix.reifler@empa.ch*

Abstract - Empa's laboratory for Advanced Fibers develops synthetic fibers with distinct functionalities in new combinations. Our custom-made pilot melt-spinning plant enables the prototype production of mono-, bi- and tri-component fibers with various cross-sections and material combinations. Ongoing projects include photocatalytic, drug-release, biodegradable, artificial turf, reinforcing, and rheological core fibers produced on our plant.

Keywords: Synthetic fiber, bicomponent, melt-spinning, reinforcement, biomaterial.

Introduction

Melt spinning is the most commonly used method for manufacturing commercial synthetic fibers. The current trends in polymer melt spinning are manifold. Recent research activities include enhancement of mechanical properties [1, 2], implementation of electric or magnetic functions [3], introduction of biologically active species like drugs or silver composites [4, 5], as well as variation of fiber morphology by bicomponent spinning [6, 7]. Bicomponent fibers are among the most interesting developments in the field of synthetic fibers [8, 9].

Bicomponent or multicomponent fibers are synthetic fibers made from two or more polymers of different chemical and/or physical structure, extruded from a common spinneret to form a single filament [10]. The polymer flows are kept separate up to the spin pack and brought together in or before the spinneret capillary. When the filament leaves the spinneret, it consists of non-mixed components that are fused at the interface. Depending on the characteristics of the different polymers, the multicomponent fiber can provide functional properties such as thermal bonding, self-crimping, unique cross sections, and achieve functionality of special polymers or additives at reduced cost [11]. The three main geometries of multicomponent fibers are side-by-side, core-sheath, and multiple core configurations (Fig. 1).

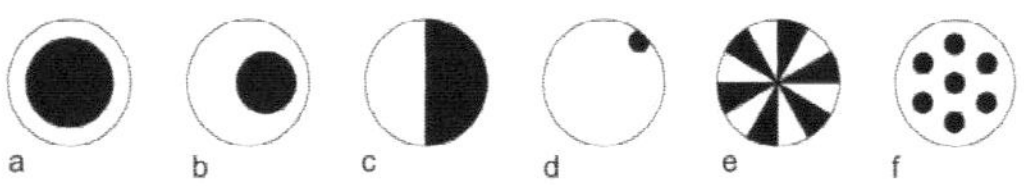

Figure 1 - Typical cross-sections of multicomponent fibers: a) concentric core-sheath, b) eccentric core-sheath, c) 50/50 side-by-side, d) unequal side-by-side, e) segmented pie, f) islands-in-the-sea.

Side-by-side and eccentric core-sheath bicomponent fibers are most commonly used to produce self-crimping yarns applied in voluminous products. By combining two polymers that undergo different shrinkage, the yarn curls up after thermal or other relaxing treatment and develops spiral crimp. Core-sheath types are predominately used as binding fibers for nonwovens, with a standard polymer as core and a low softening-point as sheath. When such fibers are laid in a nonwoven structure and heated to a temperature above the softening point of the sheath polymer, the fibers adhere wherever they cross and touch [12].

Multiple core configurations, like islands-in-the-sea and segmented pie, are mainly applied to produce microfibers with diameters smaller than those obtained via conventional melt spinning [13]. Islands-in-the-sea fibers comprise fibrils dispersed in a dissolvable matrix polymer that will be removed in a follow-up process [12]. In the segmented pie technique, a bicomponent fiber is spun from two incompatible polymers that adhere poorly and split into microfibers when subjected to mechanical stress [11]. Fabrics made from microfibers are very flexible, and the high density of fibers makes them inherently wind- and water-proof, while water vapor from perspiration can evaporate easily.

Experimental

Empa's custom-made pilot melt-spinning plant enables the prototype production of mono-, bi- and tri-component fibers with various cross-sections and material combinations with a throughput of up to several kg/h. A schematic drawing of the melt spinning plant is shown in Fig. 2.

The polymers are melted using two single screw extruders with diameters of 13 mm (1) and 18 mm (2), respectively. The maximum extrusion temperature is 400°C. Spin pumps enable a constant mass flow of 0.5-40 cm^3/min. In order to produce tri-component fibers, an additional piston extruder (3) is installed, with a throughput of 1.5-15 cm^3/min. A special spin pack (4) comprises an oil cooling and heating system (oil temperature 10-360°C) to keep the three polymer melt flows at different temperatures down to the spinneret plate.

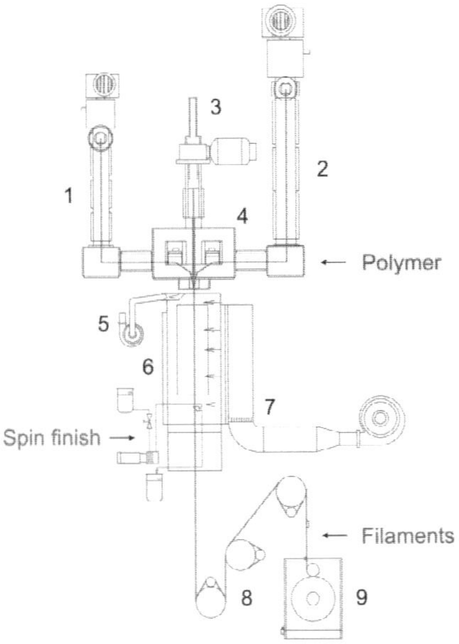

Figure 2 - Schematic assembly of the pilot melt spinning plant (see text for an explanation of the numbered parts).

Evaporating monomers and oligomers are sucked in by an exhaust (5). The extrudate is spun into the 2.8 m free-fall section which is equipped with a removable or extendable chimney with a maximum temperature of 350°C. The quenching chamber (6) has a maximum air flow of 520 m^3/min and a length of 1.4 m; its height within the free-fall section is steplessly variable. The temperature of the quench air (7) can be adjusted between 8°C and 160°C, using either a two-step cooling system or a heater.

After cooling and, where applicable, wetting with a spin finish, the filaments are drawn by three heated godets (8). The maximum temperature of the godets is 210°C, their speed can be varied between 100 rpm and 1800 rpm. The draw ratio, i.e. the ratio of speeds of draw and feed godets, can be chosen accordingly. Finally a winder (9) with a maximum speed of 2000 rpm is used to spool the filaments on a bobbin. Mono- and multifilaments with fineness in the range of 0.15-20 tex (mg/m) per filament can be produced.

Results and Discussion

Photocatalytic fibers for odor control textiles

Many textiles have a distinct ability to adsorb unpleasant odors that can leave the textile through desorption. A smart way to face this problem are odor control textiles that degrade the malodor itself at the surface through e.g. photocatalytic processes on TiO_2 based photocatalysts [14].

We developed PET-based fibers for the degradation of malodor, with highly dispersed photocatalytic nanoparticles incorporated into the fiber [15]. TiO_2 nanoparticles with large surface area and high degree of crystallinity were obtained by flame synthesis (Fig. 3). As-spun fibers were subjected to formaldehyde decomposition tests to evaluate photocatalytic activity. The results indicate that malodor degradation takes place.

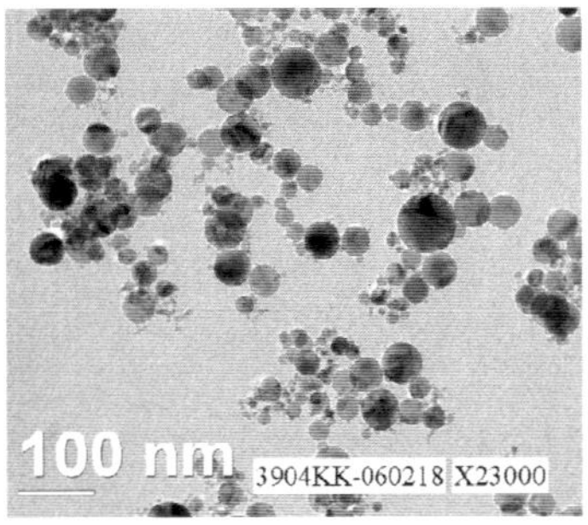

Figure 3 - TEM micrograph of TiO_2 nanoparticles synthesized in an aerosol flame reactor [16].

Controlled-release fibers for antibacterial application

Antibacterial textiles can help preventing nosocomial infections, i.e. infections which are a result of treatment in a hospital. A possibility to achieve an antibacterial effect on textiles is the application of silver based additives, as silver is a well-known fungicide and bacteriostatic agent with low risk of antibiotic resistance [17, 18]. Due to the high surface to bulk ratio and thus ample release of silver ions, silver nanoparticles promise a substantial antibacterial effect [18]. The silver ions act by displacing essential metal ions on the bacteria membrane surface (Fig. 4).

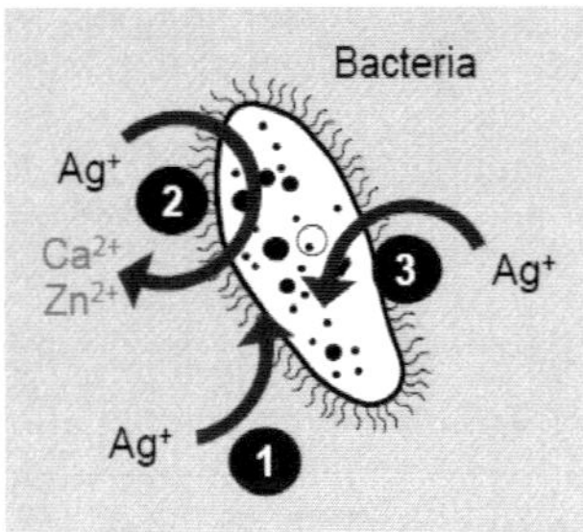

Figure 4 - Ag^+ cations interact with bacteria cells as follows [19]: 1) Damage cell membrane [17]; 2) Displace Ca^{2+} and Zn^{2+} ions; 3) Interact with cell molecules containing sulphur, oxygen or nitrogen [20].

We produced mono- and bicomponent PET fibers loaded with silver-silica particles for antibacterial application. The 5 to 20 nm silver nanoparticles are supported in a porous, amorphous SiO_2 matrix with particle diameters of 0.1 to 1 µm. The silica matrix drags the nanoparticles into the polymer and maintains the predispersed state and the activity of the silver nanoparticles [19]. Silver release was measured by leaching the fibers in aqueous media, concentrations were determined by atomic absorption spectroscopy. With a particle concentration of only 0.2 wt% the silver release is sufficient to render antibacterial effects, as was confirmed by antibacterial tests.

Biodegradable fibers from renewable sources

For temporary textile implants, fibers from biocompatible and biodegradable polymers are preferable. Furthermore, polymers from renewable sources are future-oriented. The commercially available biopolyesters polyhydroxybutyrate (PHB) and polylactide (PLA) combine these aspects. However, inflammatory response to degradation by-products limits the application of PLA as biomaterial, whereas the low crystallization rate of PHB renders melt-spinning difficult.

We produced PHB-PLA core-sheath fibers with a maximum tensile stress of up to 0.34 GPa and a Young's modulus of up to 7.1 GPa [21]. X-ray diffraction measurements revealed that the PLA component alone is responsible for the tensile strength. In vitro biocompatibility studies with human dermal fibroblasts showed that cells adhere on the fibers (Fig. 5), making them good candidates for medical therapeutic approaches.

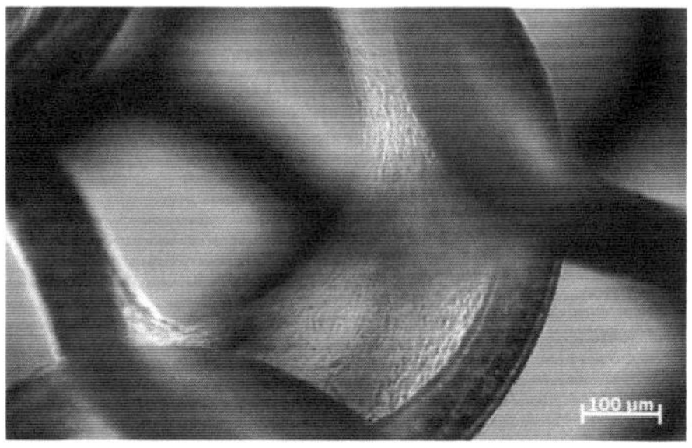

Figure 5 - Human cells on biopolyester fibers show cell nuclei (blue), actin (cytoskeleton, green) and vinculin staining (focal adhesions, red).

Artificial turf fibers for future sports flooring

Artificial turf is a lower maintenance all-weather alternative to natural turf . On one hand, polyamide (PA) carpets have excellent resilience but provoke abrasion injuries (friction burn). On the other hand, polyethylene (PE) monofilaments are skin-friendly but tend to permanent deformation.

PA-PE bicomponent fibers render a robust, skin-friendly artificial turf. The goal of our respective work was to develop a bicomponent monofilament with optimized cross-section and material combination to maximize resilience while minimizing risk of skin abrasion, in order to achieve artificial grass for sports flooring that resembles natural turf with respect to playability and appearance [22]. We succeeded in producing fibers for artificial turf that show a better resilience than up-to-date synthetic grass, without cut-back in skin-friendliness (Fig. 6).

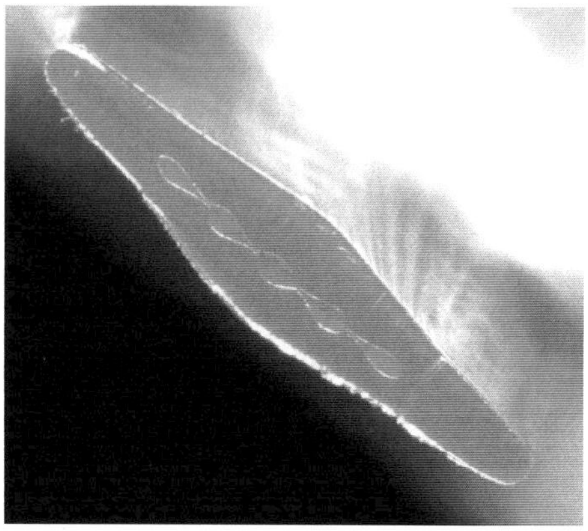

Figure 6 - Cross-section of a PA-PE bicomponent monofilament intended for skin-friendly artificial turf with good resilience.

Bicomponent fibers for the mechanical reinforcement of concrete

The reinforcement of concrete with fibers can be an economical alternative to conventional steel bar reinforcement. We successfully applied polyolefin based bicomponent fibers with high tensile strength and elastic modulus to enhance the mechanical properties of concrete [23]. The effects of an introduction of nanoparticles and other additives, and of a structuring of the fiber surface, were assessed (Fig. 7). The performance of such fiber reinforced concrete was studied with respect to fiber pull-out characteristics and the bending behavior of concrete. Ductile post-peak behavior of fiber reinforced concrete was achieved, making these fibers interesting for the applications in pre-cast elements, industrial floors and earth quake protecting systems.

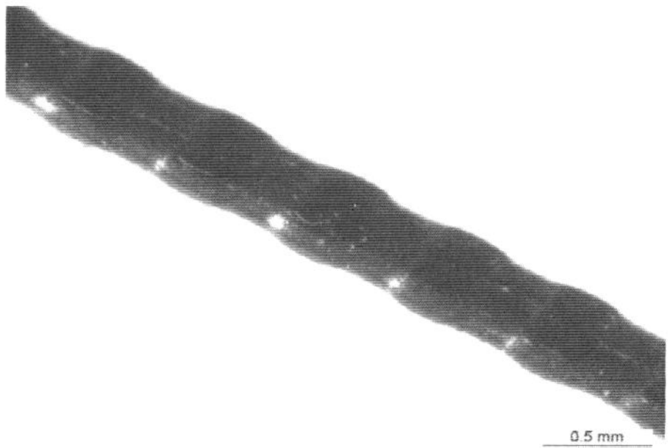

Figure 7 - Polyolefin bicomponent fiber with surface structuring and additives in the sheath that enhance fiber pull-out characteristics and fiber-matrix bond strength.

Fiber with rheological core for flexure-rate responsive damping

Instead of using hard shell components to dissipate the energy of the impact, state-of-the-art concepts of impact protective garments rely on flexible materials [24]. We develop fibers that exhibit rate-dependent viscoelastic properties owing to the presence of a fluid core with distinctive rheological characteristics [25]. The fluid shall be confined inside a suitably structured fiber core that induces fluid flow upon fiber bending, resulting in a rate-adaptive mechanical damping effect (Fig. 8).

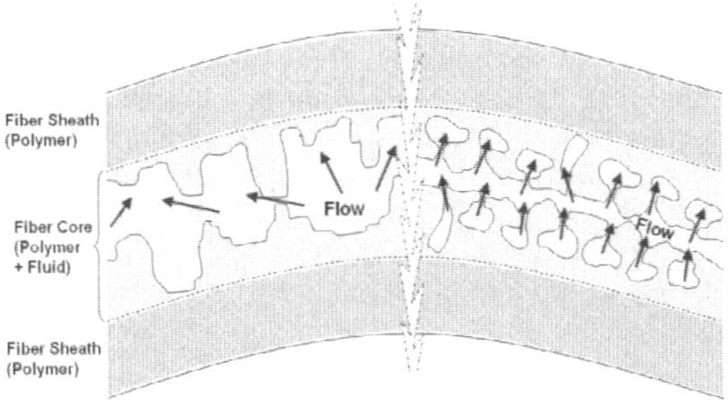

Figure 8 - Illustration of a rheological core fiber with a polymer sheath around a polymer plus fluid mixed core.

The challenges of this project involve the use of non-equilibrium multi-component polymer melt flow behavior during melt-spinning to generate the desired fiber core geometry, filled with a fluid of suitable rheological properties. Fibers with a rheological core are expected to open an entire field of adaptive product applications, notably as protectors against impact, or for rate-adaptive damping in fiber-reinforced high-performance composites.

Conclusions

For textiles to stand their ground in the field of high-tech products, investments in research are highly required. Most effective are product enhancements on the fiber level, the base material of any textile. Here, multicomponent synthetic fibers are among the most interesting developments. Focusing on the prototype production of fibers with innovative cross-sections and new base materials, we believe to contribute to the continuing success of the textile industry.

Acknowledgements

This research was co-funded through grants by the Swiss Innovation Promotion Agency CTI and the Swiss National Science Foundation SNSF. The authors thank the industry partners and collaborators from Empa that contributed to the projects presented.

References

[1] T.D. Fornes, J.W. Baur, Y. Sabba, E.L. Thomas, *Polymer,* 2006, *47,* 1704.

[2] J.K.W. Sandler, S. Pegel, M. Cadek, F. Gojny, M. van Es, J. Lohmar, W.J. Blau, K. Schulte, A.H. Windle, M.S.P. Shaffer, *Polymer,* 2004, *45,* 2001.

[3] Y. Liang, X. Xia, Y. Luo, Z. Jia, *Materials Letters,* 2007, *61,* 3269.

[4] S.Y. Yeo, H.J. Lee, S.H. Jeong, *Journal of Materials Science,* 2003, *38,* 2143.

[5] R. Dastjerdi, M.R.M. Mojtahedi, A.M. Shoshtari, *Fibers and Polymers,* 2008, *9,* 727.

[6] X.Q. Shi, H. Ito, T. Kikutani, *Polymer,* 2006, *47,* 611.

[7] J. Huang, D.G. Baird, A.C. Loos, P. Rangarajan, A. Powell, *Composites Part A: Applied Science and Manufacturing,* 2001, *32,* 1013.

[8] S. Houis, F. Schreiber, T. Gries, *Fibre-Table according to P.-A. Koch: Bicomponent Fibres,* Shaker, Aachen, 2008.

[9] H.J. Koslowski, *Technical Textiles,* 2009, *52,* E202.

[10] F. Fourné, *Synthetic Fibers,* Hanser, 1999.

[11] H.H. Yang in *Handbook of Fiber Chemistry,* M. Lewin, Ed., CRC Press 2007, 31-137.

[12] M. Jaffe, A.J. East in *Handbook of Fiber Chemistry,* M. Lewin, Ed., CRC Press 2007, 1-29.

[13] L.M. Robeson, R.J. Axelrod, M.S. Vratsanos, M.R. Kittek, *Journal of Applied Polymer Science,* 1994, *52,* 1837.

[14] M. Kaneko, I. Okura, *Photocatalysis - Science and Technology,* Springer Verlag, 2002.

[15] F. Reifler, A. Ritter, F. G., A. Vital, K. Akurati, R. Hufenus in Fiber Society Fall Conference, Davis CA, 2007.

[16] K.K. Akurati, A. Vital, F. Reifler, A. Ritter, T. Graule in NSTI Nanotechnology Conference, Santa Clara CA, 2007, 173-176.

[17] I. Sondi, B. Salopek-Sondi, *Journal of Colloid and Interface Science,* 2004, *275,* 177.

[18] C. Damm, H. Münstedt, A. Rösch, *Journal of Materials Science,* 2007, *42,* 6067.

[19] M.J. Height in 46th Man-Made Fibers Conference, Dornbirn, 2007.

[20] D.P. Dowling, K. Donnelly, M.L. McConnell, R. Eloy, M.N. Arnaud, *Thin Solid Films,* 2001, *398-399,* 602.

[21] R. Hufenus, F.A. Reifler, K. Maniura-Weber, A. Spierings, M. Zinn, *Macromolecular Materials and Engineering,* 2012, *297,* 75.

[22] R. Hufenus, C. Affolter, M. Camenzind, F.A. Reifler, *Macromolecular Materials and Engineering,* DOI: 10.1002/mame.201200088

[23] J. Kaufmann, J.F. Lübben, E. Schwitter, *Composites Part A: Applied Science and Manufacturing,* 2007, *38,* 1975.

[24] K. Walker, S. Robson, N. Ryan, L. Mallen, R. Sibbick, G. Budden, A. Mepham, *Adv. Mater. Process.,* 2008, *166,* 36.

[25] L. Gottardo, R. Hufenus, M. Dressler, M. Heuberger in International Symposium on New Frontiers in Fiber Materials Science, Charleston SC, 2011, 30-31.

O-03-125

REINFORCING A PC-FILM BY MEANS OF MONOAXIAL STRETCHING

V. Schöppner[1], A. Wibbeke[1*], B. Scholtes[2], M. Lebsanft[2] and W. Zinn[2]
[1]*Kunststofftechnik Paderborn, University of Paderborn, Germany – volker.schoeppner@ktp.upb.de; andrea.wibbeke@ktp.upb.de*
[2]*Institut für Werkstofftechnik, Metallische Werkstoffe, University of Kassel, Germany – scholtes@uni-kassel.de; m.lebsanft@uni-kassel.de; zinn@uni-kassel.de*

Abstract - When manufacturing uniaxially stretched film made of polycarbonate, elevated strengths are generated in the polycarbonate polymer through the stretching. Previous investigations have demonstrated that uniaxial stretching generates self-reinforcement in polycarbonate. This self-reinforcement is brought about by an increase in strength and Young's modulus through a mutually supporting molecular structure and is a function of the stretching temperature, the stretching ratio and the stretching rate. To measure the forces prevailing during the stretching process, the film is stretched on the tensile testing machine at different stretching temperatures, with different stretching rates and different stretching ratios. Here, the technical strain depends solely on the stretching temperature and the stretching rate. Set against industrial stretched films, the mechanical properties of the stretched polycarbonate film are comparable or even better. Orientating the molecules causes the molecular chains to lie close together, and the stabilizing bonds become active. This change in the polymer's molecular structure can be detected by means of polarized light microscopy, thermomechanical analyses (TMA) and an X-ray diffractometer.

Keywords: monoaxial stretching, orientation, polycarbonate film

Introduction

The monoaxial stretching of polycarbonate film causes the macromolecules to become orientated. This then gives rise to self-reinforcement and anisotropy in the material. The direction-dependence of the mechanical properties is particularly clear: the stiffness and strength display anisotropic behavior. By varying the stretching ratio, stretching rate and temperature profile, it is possible to adjust the mechanical properties on a local basis. A stretching temperature just above the glass transition temperature is necessary to ensure that the macromolecules are able to move and become orientated through the stretching [1]. The challenge lies in determining the degree of orientation that is generated through the stretching. The question then arises as to whether the stretching ratio, stretching rate and stretching temperature have an influence on the orientation.

Forces during deformation

In order to measure the forces that act during stretching, the polycarbonate film is stretched on a universal testing machine in a hot-air oven at different stretching temperatures and with different stretching rates and stretching ratios. The force-deformation curves are then analyzed and the true stress (Eq. 2) plotted over the stretching ratio (Eq. 1). The stretching ratio is defined as follows for this, according to [2]:

$$\lambda = 1 + \varepsilon \tag{1}$$

The true stress is calculated as follows:

$$\sigma = \lambda \times \sigma_N = (1 + \varepsilon) \times \frac{F}{A} \tag{2}$$

Fig. 1 shows the curve for the true stress σ over the stretching ratio λ. It is clear from this that the stretching temperature has the bigger influence on the stress-stretching ratio curve. The stretching rate, by comparison, is seen to have virtually no influence when it is varied slightly between 40 mm/min and 60 mm/min, by about 10 m/min. If the stretching rate is varied to a greater extent between 100 mm/min and 150 mm/min, however, by around 50 m/min, it also has an influence.

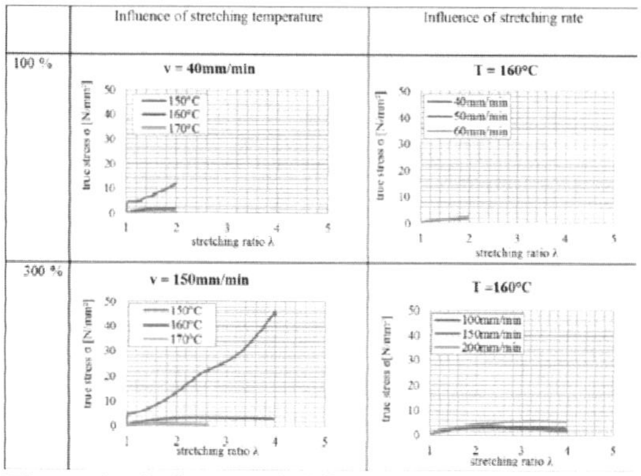

Figure 1 – Influence of different parameter variations on the stress-stretching ratio curve

Producing self-reinforcement in polycarbonate by means of stretching

To produce self-reinforcement in a polycarbonate film, the film is stretched by setting various different process parameters in the narrow gap on an MDO A / MDO B uniaxial stretching unit manufactured by Dr. Collin [3]. This self-reinforcement is a function of the stretching temperature, the degree of stretching and the stretching rate (Fig. 2). The strength is anisotropic as a consequence of the self-reinforcement. In the process, the strength is mainly increased in the machine direction (MD). In the transverse direction (TD), only a small increase in strength is observed. The anisotropy in the mechanical properties set out here provides indirect information on the orientation state of the macromolecules and permits the orientation state to be identified in accordance with [4].

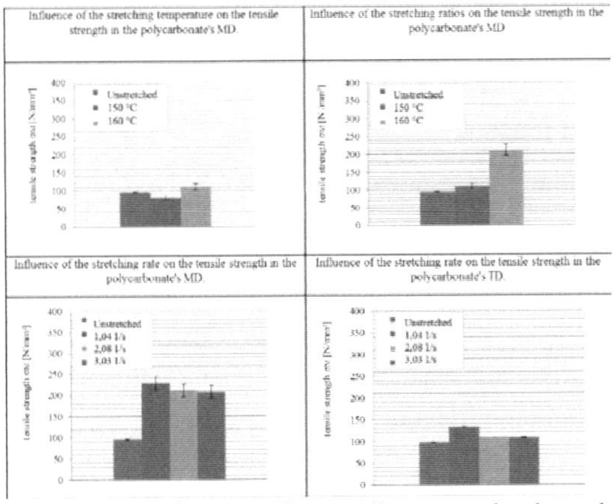

Figure 1 – Increase in tensile strength through stretching the polycarbonate film

Providing proof of orientation in monoaxially stretched polycarbonate film

It was seen in the mechanical analyses that stretching gives rise to anisotropy, which supplies information on the orientation state of the molecules. In addition, orientation can be analyzed by means of polarization microscopy, TMA and measurements in a universal X-ray diffractometer.

Polarization microscope

Fig. 3 shows the micrographs taken through the polarization microscope. The pictures of the non-stretched film show that there is no inherent stress. When the polycarbonate film is stretched, inherent stresses are introduced into it. These are a function of the stretching rate and the temperature profile and are highlighted by the red coloring. The higher the stretching rate, the more inherent stresses will be present in the film. Apart from this, it is shown that the inherent stress can be reduced through annealing at 140 °C and, with a lower stretching rate of 1.04 1/s, more inherent stresses develop than with a high stretching rate of 2.08 1/s. The results measured here tally with the mechanical analyses presented in Fig. 2, which means that polarization microscopy offers a means of depicting the extent of the prevailing inherent stresses.

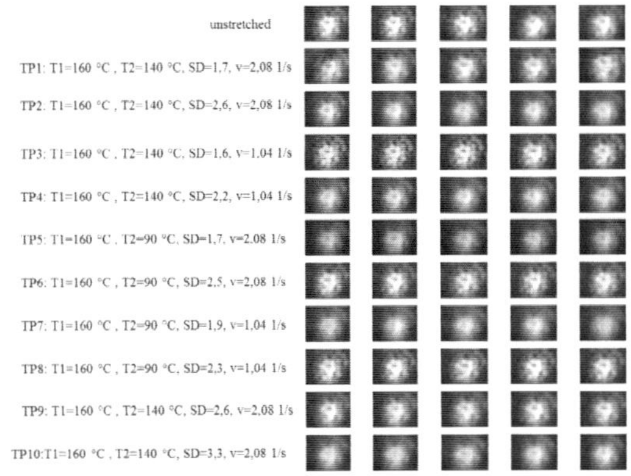

Figure 2 – Micrographs of the non-stretched and stretched polycarbonate films taken through the polarization microscope

Thermomechanical analysis (TMA)

TMA is used to measure dimensional stability in heat by means of penetration, with the specimen undergoing linear heating to this end [5]. The analysis of the coefficients of expansion is important for the evaluation. According to [6], the mean coefficient of expansion α is proportional to the change in length and is defined as follows

$$\alpha = \frac{\Delta L}{\Delta T} \cdot \frac{1}{L_0} \tag{3}$$

where L_0 stands for the starting length, ΔL describes the change in the length of the specimen and ΔT the temperature change. Figure 4 shows the curve of the mean coefficient of expansion over the temperature. Once the glass transition temperature has been exceeded, a deflection in the coefficient of expansion is seen for all the stretched specimens. Test point ten shows the lowest mean coefficient of expansion and hence the greatest change in length during heating. The reason for the high coefficient of expansion is that the molecules have become oriented during stretching and then relax at an elevated temperature. The greatest orientation is thus generated by the highest stretching ratio. It is also seen that a greater degree of orientation is brought about by a reduced stretching rate. TMA allows the degree of orientation to be portrayed and makes it possible to analyze which process settings have an influence on the orientation.

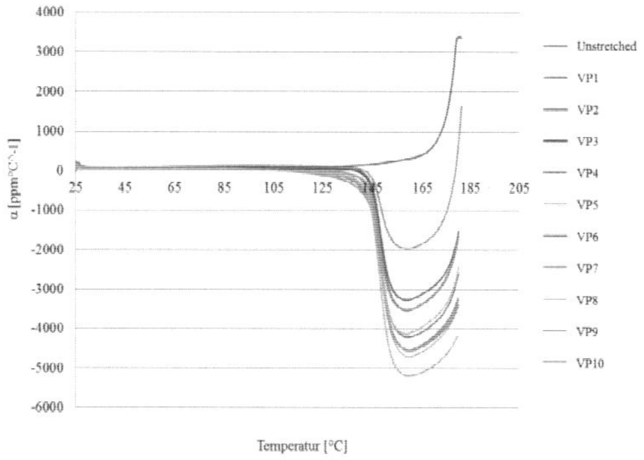

Figure 3 – Profile of the mean coefficient of expansion α

Universal X-ray diffractometry

For semi-crystalline polymers, the X-ray scatter method will show up the distribution of the orientation [7]. It will be investigated, if this method shows a result as well for amorphous polymers. The diffractometer detects the location and intensity of the reflexes that occur in accordance with Bragg's equation. In the following analysis, a polycarbonate with a medium molecular weight was analyzed in the non-stretched and stretched state through X-ray diffraction with Cu-Kα radiation ($\lambda = 0.154056$ nm). In each case, the polycarbonate films were stretched at a stretching temperature of 160 °C and a stretching ratio of 3 (v=2.08 1/s), 4 (v=2.08 1/s), 3 (v=1.04 1/s) and 5 (v=1.04 1/s) and investigated in the MD and TD. Figure 5 shows that there is no direction-dependence of the diffracting structure and also that there is no evidence of the stretching having any influence on this structure. Just the intensity of stretched films is higher compared to unstretched films, because of a higher diffractive volume.

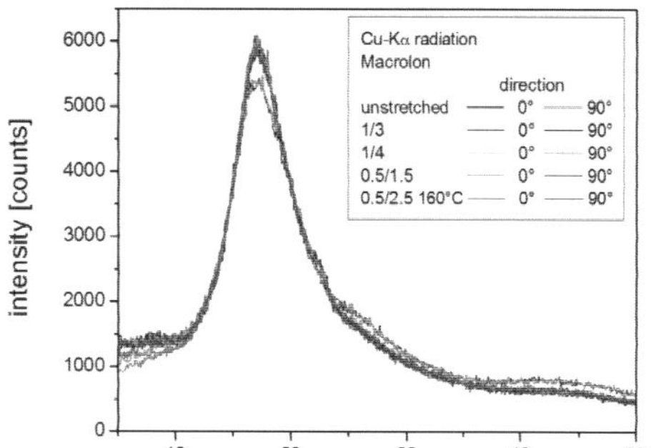

Figure 4 – Comparison of the diffractograms in the MD and TD

Results and Discussion

When stretching the amorphous polycarbonate films, the stretching temperature, stretching ratio and stretching rate have an influence on the stress-stretching ratio curve. This is also shown by the mechanical analyses. The monoaxial stretching generates self-reinforcement in both the MD and the TD, which provides information on the orientation state of the molecules. In addition, orientation has been analyzed through polarization microscopy, TMA and measurements with the universal X-ray diffractometer. Through the polarization microscopy and the TMA, it proved possible to provide evidence of the level of orientation that had been introduced. While the X-ray diffractometer showed that the stretching creates orientation, it was not possible to draw any conclusions about the level of the orientation that is introduced.

Conclusions

Through the orientation of the amorphous molecule structure in the polycarbonate, it is possible to generate self-reinforcement in both the MD and the TD. This is a function of the stretching temperature, stretching rate and level of stretching and can be proven through micrographs taken through the polarization microscope and by means of TMA.

Acknowledgements

The results presented here were obtained within the scope of the DFG's supported collaborative research centre's Transregio 30, for process integrated manufac-ture of functionally graded structures based on thermo-mechanically coupled phenomena. We would like to thank the DFG for their support.

References

1. R. Jürgen; H. Wohlfart: *Technologie der Werkstoffe,* 8. Auflage 2007
2. E., Haberstroh: *Fügen und Umformen von Kunststoffen*, Institut für Kunststoffverarbeitung in Industrie und Handwerk an der RWTH Aachen, 2012
3. V. Schoeppner; A. Wibbeke; M. Sasse: *Manufacture of self-strengthened polycarbonate by extrusion*, Congress of the Polymer Processing Society PPS-27, 2011
4. M. Fleißner: *Makroskopische und molekulare Orientierung in Hochpolymeren*, Farbwerke Hoechst AG, vormals Meister Lucius & Brüning, Frankfurt am Main, 1973
5. G. Menges, E. Haberstroh, W. Michaeli, E. Schmachtenberg: *Werkstoffkunde Kunststoffe*, Institut für Kunststoffverarbeitung, RWTH Aachen, Hanser-Verlag, 2002
6. W. Grellmann; S. Seidler: *Kunststoffprüfung*, Hanser-Verlag, 2005
7. H. Schnell, L. Bottenbruch: *Über Polycarbonate*, Farbenfabriken Bayer AG. Krefeld-Uerdingen, Naturwissenschaften 1967 Heft 12 (1967)

O-03-128

ELECTRICAL CONDUCTIVITY AND ELECTROMAGNETIC INTERFERENCE SHIELDING EFFECTIVENESS OF INJECTION-MOLDED POLYPROPYLENE/CARBON FIBER COMPOSITE FOAMS

A. Ameli, P.U. Jung, C.B. Park*

Microcellular Plastics Manufacturing Laboratory, Department of Mechanical and Industrial Engineering, University of Toronto, 5 King's College Road, Toronto, Ontario, Canada M5S 3G8 – ameli@mie.utoronto.ca; pjung@mie.utoronto.ca; park@mie.utoronto.ca

Abstract – Effort is made to produce cost-effective and lightweight composites with enhanced electrical properties and higher electromagnetic interference shielding effectiveness (EMI SE). Foaming can reduce the product weight and potentially improve the electrical properties. To this end, the through-plane and in-plane electrical conductivities and EMI SE of injection molded solid and foamed polypropylene/carbon fiber (PP-CF) composites were measured at different locations along the injected sample length. The composites microstructure was also characterized in terms of fiber orientation and skin layer size. The through-plane conductivity of the skin and core regions was also measured separately to elaborate the effect of foaming on each region. Compared to the solid composites, in addition to the reduced weight of the composite (~30%), introducing of foaming enhanced the through-plane conductivity in both the core and skin regions (up to 5 orders of magnitude), significantly improved the nonuniformity of the conductivity along the sample length (up to 4 orders of magnitude), and decreased the conductivity anisotropy (up to 4 orders of magnitude). Longitudinal and transversal in-plane conductivity increased slightly with foaming at some locations. These conductivity improvements via foaming resulted in very uniform EMI SE of the foamed composites with a maximum of 30% increase, compared to the solid composites.

Keywords: Polymer composites, Conductivity, EMI Shielding Effectiveness, Foaming, Injection molding, Carbon fiber

Introduction

Conductive polymer composites containing carbon fiber, carbon black and metallic fibers have been extensively investigated in an attempt to replace metal-based EMI shields [1-3]. Conductive composites with micro-sized fillers require high filler loading to achieve the percolation threshold and adequate level of shielding, which adversely affects the weight and economic feasibility of the product. Also, dispersion and distribution of conductive fillers are very important parameters affecting the electrical properties [4-5]. Introduction of foaming (cellular structure) to conductive composites can be an effective way to not only reduce the weight significantly, but also improve the dispersion [6] and distribution [7,8] of the fibers.

In injection molding process, in-plane orientation of high aspect ratio fibers induced by melt flow creates parts with anisotropic electrical conductivity where through-plane (*T-P*) conductivity is several orders of magnitude lower than both longitudinal and transversal in-plane (*I-P*) conductivity [7]. Foaming in injection molding process can change the state of fiber orientation and the size of skin region that has highly unidirectionally oriented fibers and thus enhance the through-plane conductivity. Very little attention however has been paid to this matter. Motlag et al. investigated the conductivity of carbon fiber /carbon black/cyclic olefin copolymer that injection foam molded using chemical blowing agent and reported that, in some cases, foaming increases the through-plane conductivity [7]. They attributed this improvement to both change of fiber orientation and interaction of carbon fiber and carbon black in the foamed samples. In our earlier work, the mechanisms by which foaming contributes to the enhancement of *T-P* conductivity were presented and the effects of processing and foaming variables on the through-plane conductivity were investigated [8].

In this work, the effects of foaming on the uniformity and anisotropy of conductivity were studied by measuring the *T-P* conductivity and longitudinal and transversal *I-P* conductivity of the injected samples at three different locations and in the skin and core regions. EMI SE of the injected solid and foamed composites was also characterized to assess the effect of foaming on the SE.

Experimental

Materials and compounding

The commercial injection-molding grade polypropylene (PP), ZS-761 with a specific gravity of 0.9 $g.cm^{-3}$ and a melt flow index of 21 g/10 min (230°C/2.16 kg) was supplied by SUNOCO Inc., Philadelphia, USA. Chopped carbon fiber with a density of 1.8 $g.cm^{-3}$ was used as the conductive additive. 20 wt. % compound of polypropylene and carbon fiber was prepared at Lubrizol Corp., Wickliffe, Ohio, USA. N_2, as the physical blowing agent was supplied by Linde Gas, Canada.

Experimental setup

A 50-ton Arburg Allrounder 270/320 C injection molding machine (Lossburg, Germany) with a 30-mm diameter screw, and equipped with MuCell® technology (Trexel, Inc., Woburn, Massachusetts) was used. The mold contained a rectangular cavity with a fan gate after the sprue. The cavity dimensions were 152.4×101.6×3.2 mm^3. Table 1 summarizes the processing parameters used in the injection molding of solid and foamed samples.

Microstructure and fiber orientation characterization

The microstructures were examined using a JEOL JSM-6060 scanning electron microscope (SEM). The SEM micrographs were used to measure the skin layer thickness of the samples. To characterize the CF orientation, the length-thickness (*LT*) and length-width (*LW*) planes of composites in both the core and skin regions were metallographically polished with a series of Struer polishers. Digital images of the polished surfaces taken by optical microscopy were quantitatively analyzed to calculate the fiber orientation factors. More details can be found in [8].

Table 1 – Processing parameters used in the injection molding of solid and foamed samples.

Parameter	Solid	Foam
Barrel temperature (ºC)	190	190
Barrel pressure (MPa)	17	17
Screw speed (rpm)	400	400
Metering time (s)	10	10
Injection flow rate (cm^3/s)	100	100
Mold temperature (ºC)	30	30
Pack/hold pressure (MPa)	80	N/A
Pack/hold time (s)	4	N/A
Gas injection pressure (MPa)	N/A	24
N_2 content (wt. %)	N/A	0.3
Set void fraction (%)	N/A	25

Electrical conductivity measurement

For *T-P* conductivity, disk shape samples of 30 mm in diameter and 3.2 mm in thickness were cut from the injected samples near the gate and at the middle and end locations (with equal distances) along the injected sample length. To measure the *T-P* conductivity of the skin and core, the full samples were machined and polished down to 0.3 and 1.8 mm, respectively. Alpha-A high performance conductivity analyzer by Novocontrol Technologies GmbH & Co. KG was used to measure the *T-P* conductivity. To measure the *I-P* conductivity, rectangular samples of 40×10 mm^2 was cut in the length and width directions of the injected sample at similar locations as those for *T-P* samples. HP 4339B resistance meter was used for *I-P* conductivity measurements. In all cases, a voltage of 1.0 volts was applied.

EMI SE measurement

EMI SE in the X-band frequency range (8.2-12.4 GHz) was measured using a coaxial set-up, consisting of a PNA HPE8361A network analyzer and two standard WR-90 coaxial launchers that are connected face to face to guide EM wave, with a copper sample holder in between the launchers. The sample and launcher opening sizes were 25×12.5 mm^2 and 22.5×10 mm^2, respectively. The incident power used was 0 dBm, which corresponds to 1 mW.

Results and Discussion

Through-plane conductivity

Figure 1 depicts the *T-P* electrical conductivity of the skin and core for the solid and foamed PP-CF composites. In all the cases, the *T-P* conductivity of the skin was up to four orders of magnitude lower than the conductivity of its corresponding core region. This is related to the significant in-plane orientation of fibers in the skin region [7], as reflected by the lower through-plane fiber orientation (f_T) in Table 2. In both solid and foamed composites, the *T-P* conductivity of both skin and core increased from near the gate location toward the end location. When foaming was introduced to the composites, the *T-P* conductivity increased up to a maximum of five orders of magnitude in both skin and core regions at all the locations of the injected samples. Overall, these trends correlated very well with the states of fiber orientation, as the through-plane fiber orientation factor (f_T) increased from near the gate location toward the end of the samples in both solid and foamed composites and also when the foaming was introduced to the composites (Table2).

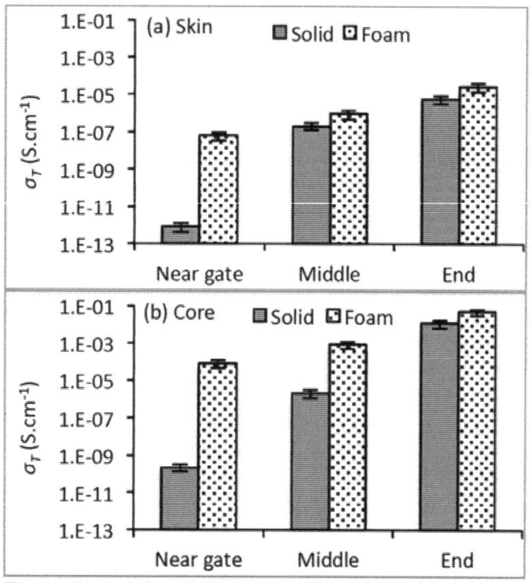

Figure 1 – *T-P* conductivity of (a) skin and (b) core of the solid and foamed PP-CF composites. In all Figures, error bars show ± one standard deviation.

The increase of *T-P* conductivity in the skin region upon the introduction of foaming was attributed to the plasticization effect of the added gas [8]. With the presence of gas mixed in the polymer melt, its viscosity was decreased and resulted in decreased shear stresses and consequently the severity of the fibers one-dimensional orientation in the machine direction was decreased. This resulted in an increased f_T (Table 2) and *T-P* conductivity. The maximum effect of gas was observed near the gate where the fiber orientation factor in the flow direction was the highest due to the longest exposure of the shear flow and the highest speed of the flow at that location.

The improvement of *T-P* conductivity in the core via foaming was attributed to two mechanisms: a) increased fiber-to-fiber contacts around the cells due to the biaxial stretching of the matrix during cell growth and also increased chance of fiber out-of-plane rotation due to the three dimensional growth of the bubbles [8]; and b) decreased in-plane orientation of fibers due to the plasticization effects of the added gas. Compared to the other locations, the greatest improvement in the *T-P* conductivity of the core occurred near the gate where the cell density was the highest and the average cell size was the smallest.

Table 2 – Fiber orientation factors in the longitudinal (f_L), transversal (f_W) and thickness (f_T) directions.

	Skin				Core		
	Thick (mm)	f_L	f_W	f_T	f_L	f_W	f_T
	Solid*						
N. Gate	0.61	0.81	0.18	0.01	0.33	0.63	0.04
Middle	0.43	0.76	0.20	0.03	0.39	0.56	0.05
End	0.33	0.66	0.32	0.04	0.44	0.49	0.07
	Foam*						
N. Gate	0.49	0.74	0.21	0.05	0.27	0.54	0.19
Middle	0.38	0.71	0.33	0.06	0.35	0.44	0.21
End	0.28	0.58	0.35	0.07	0.39	0.41	0.21

*: Standard deviation was less than 17% of the average values in all the cases.

Figure 2 shows the measured *T-P* conductivity of the full solid and foamed composites. The *T-P* conductivity of the foamed composites was always higher than their corresponding solid ones with a maximum difference of about five orders of magnitude near the gate location.

In order to better understand the contribution of skin and core to the overall conductivity, it was calculated using the measured conductivity of the skin and core. The overall resistance of the full sample (R_{eq}) should be governed by the resistance of a series of resistors as the skin (R_s) - core (R_c) - skin (R_s) combination following:

$$R_{eq} = R_s + R_c + R_s = 2R_s + R_c \quad (1)$$

For a resistor, specific resistance (ρ) is:

$$\rho = R.A/t \quad (2)$$

where A and t are the area and thickness of the disk shape sample. A remained the same for the skin, core and full samples and noting that conductivity is the reciprocal of resistivity, Eq. (1) can be rewritten as:

$$\sigma_{eq}^{-1} = 2\alpha\sigma_s^{-1} + (1-\alpha)\sigma_c^{-1} \quad (3)$$

where, σ_{eq}, σ_s and σ_c are the conductivity of the full sample and its skin and core, respectively. α is the ratio between the skin layer thickness (Table 2) and the overall thickness of the sample (3.2 mm).

As Figure 2 shows, the measured and calculated *T-P* conductivity of the full samples agreed reasonably well and the predicted values fell within the scatter range of the measurement replications. According to Eq. (3), the overall conductivity is not only affected by the conductivity values of the skin and core regions, but also by their sizes. Since the *T-P* conductivity of the skin was much lower than that of the core, a thinner skin would result in a higher overall conductivity. As seen in Table 2, foaming decreased the skin layer thickness of the injected samples as well. When gas was added, due to the decreased viscosity of the melt, the injection and filling time was decreased and thus a shorter time was given to the skin to solidify. In addition, the presence of mixed gas and bubbles could have decreased the cooling rate during the cycle time. Consequently, a thicker core was achieved when the sample was foamed. Therefore, in addition to the aforementioned mechanisms, change of the skin size was also another one that contributed to the increase of *T-P* conductivity in the foamed composites.

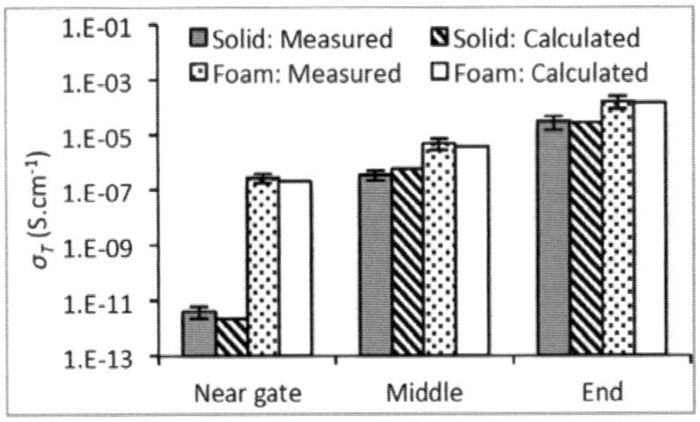

Figure 2 – Measured and calculated overall *T-P* conductivity of solid and foamed PP-CF composites.

In-plane conductivity

Figure 3 shows the in-plane electrical conductivity of the solid and foamed composites in longitudinal and transversal directions. Foaming increased both longitudinal and transversal conductivities of near the gate and middle locations while slightly decreased the conductivity of the end location. The *I-P* conductivity increase was in agreement with the change of fiber orientations after foaming, as the orientation factors indicated a more random distribution of fibers in the foamed cases (Table 2) and it is known that *I-P* conductivity increases by more random distribution of fibers [9]. One reason for the slight decrease of *I-P* conductivity in the end location might be the foaming with larger/more elongated cells and lower cell density, compared to the other locations, as has been explained in [8].

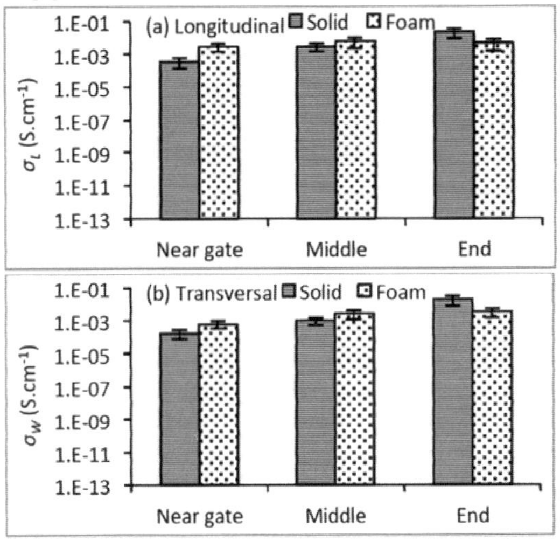

Figure 3 – (a) longitudinal and (b) transversal *I-P* conductivity of solid and foamed composites.

Overall, foaming improved the uniformity of both *T-P* and *I-P* conductivities along the samle length. In the case of *T-P* conductivity, the difference between the conductivity of near the gate and end locations was decreased from seven orders of magnitude in solid composite to only three orders of magnitude when the sample was foamed. The *I-P* conductivity of foamed sample in both longitudinal and transverse directions varied only within the same order of magnitude. Foaming also significantly decreased the anisotropy of the conductivity defined as the ratio of *T-P* and *I-P* conductivities (σ_L/σ_T or σ_W/σ_T). The range of anisotropy decreased from 9×10^7-7×10^2 in the solid composite to only 9×10^3-2×10^1.

Elecromagnetic interference shielding effectiveness

EMI SE is a measure of the material's ability to attenuate EM waves intensity. For an EM radiation, EMI SE is the logarithm of the ratio of incident power (P_i) to transmitted power (P_t), SE = 10 log (P_i/P_t). For example, a SE of 20 corresponds to the blocking of 99% of EM incident wave. The level of shielding required for computer devices is 15-20 dB [10].

Figure 4 shows the EMI SE of solid and foamed PP-CF composites. The data are the average of SE in the X-band frequency range for four replications. The introduction of foaming increased the EMI SE of the composites by aproximately 30% near the gate and 15% in the middle location. The EMI SE remained approximately unchanged in the end location. As Figure 5 shows, a logarithmic relationship was found between the geometric average of conductivity values (σ_T σ_L σ_W) and EMI SE, independent of the sample and location type. The greatest improvement of SE near the gate location was thus attributed to the significant increase of *T-P* conductivity via foaming in this location, which resulted in an inceased average conductivity. In the end location, upon foaming, *T-P* conductivity increased while *I-P* conductivity decreased and the combination resulted in almoct no change in SE. Also, foaming greatly improved the uniformity of the SE of the injected composite along its length. In the solid composites, the SE near the gate was 26% lower than that in the end location and it was reduced to only 6% (close to the standard deviation of the replications) in the foamed samples.

Conclusions

The effects of foaming on the through-plane and in-plane conductivity, and EMI SE of PP-CF injected composites were studied at different sample locations.

By the introduction of foaming, the through-plane conductivity increased in both the core and skin regions up to a maximum of 5 orders of magnitude. The greatest improvements were observed near the gate location. Longitudinal and transversal in-plane conductivity increased slightly with foaming at some locations. Foaming significantly improved the uniformity of both through-plane and in-plane conductivity and decreased the anisotropy of the conductivity.

Compared to the solid composites, in addition to ~30% weight reduction, foamed composites presented very uniform EMI SE values along the sample length with a maximum of 30% increase near the gate location.

The results reveal the potential application of injection foam molding technology to manufacture lightweight conductive products with enhanced EMI SE for applications in the electronics, aerospace and automotive industirs.

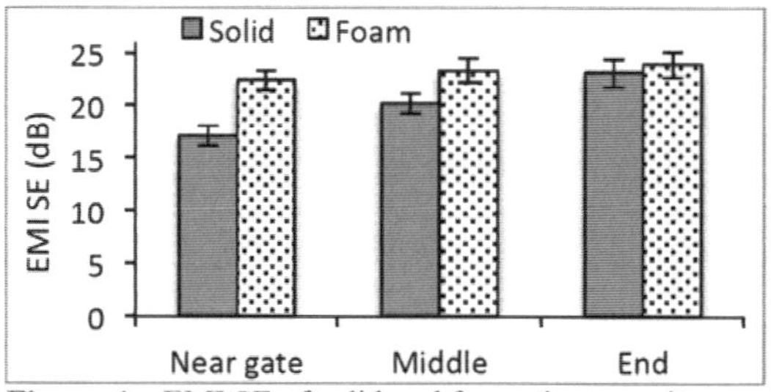

Figure 4 – EMI SE of solid and foamed composites.

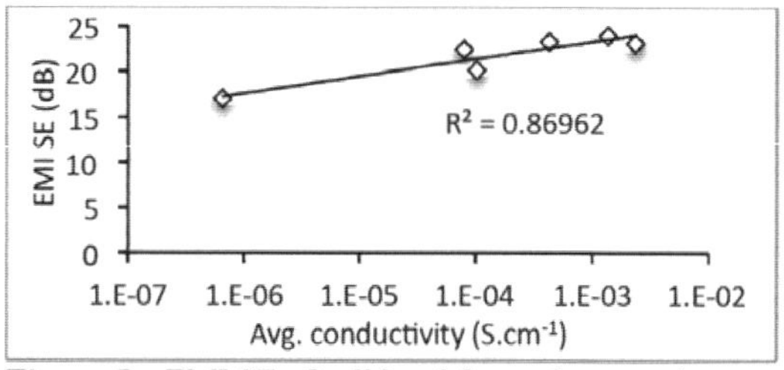

Figure 5 – EMI SE of solid and foamed composites.

Acknowledgements

The work was financially supported by MITACS, Ontario, Canada and the compound was kindly supplied by Dr. Feina Cao of Lubrizol Corp.

References

1. M.H. Al-Saleh; U. Sundararaj, Macromol. Mater. Eng. 2008, 293, 621–630.
2. S.T. Tan, M.Q. Zhang, M.Z. Rong; H.M. Zeng; F.M. Zhao, Polym. Polym. Compos 2001, 9, 257–262.
3. N.C. Das; D. Khastgir; T.K. Chaki; A. Chakraborty, Comp. Part A 2000, 31, 1069–1081.
4. 4. P. Potschke; A.R. Bhattacharyya; A. Janke, Carbon 2004, 42, 965–969.
5. 5. T. Villmow; P. Pötschke; S. Pegel; L. Häussler; B. Kretzschmar, Polymer 2008, 49, 3500-3509.
6. M. Okamoto; P.H. Nam; M. Maiti; T. Kotaka; T. Nakayama; M. Takada; M. Oshima; A. Uskui; N. Hasegawa; H. Okamoto; Nano Lett. 2001, 1, 503.
7. G.H. Motlagh; A.N. Hrymak; M.R. Thompson, Polym. Eng. Sci. 2008, 48, 687-696.
8. A. Ameli; P.U. Jung; CB Park, Comp. Sci. Tech. Submitted.
9. M.T. Kortschot; T.R. Woodhams, Polym. Compos. 1988, 9, 60-71.
10. M.H. Al-Saleh; G.A. Gelves; U-T Sundararaj, Comp. Part A 2011, 42, 92–97.

RAPID PREPARATION OF HIGH-GAS VOLUME FRACTION POROUS POLYIMIDE FILM BY PHASE SEPARATION AND SUBSEQUENT PHOTOPOLYMERIZATION

Kentaro Taki[1]*, Kazunori Hosokawa[1], Shota Takagi[1], Hiroyuki Mabuchi[1], Masahiro Ohshima[1]
[1]*Department of Chemical Engineering, Kyoto University,*
A4-021, Katsura Campus, Nishikyo-ku, Kyoto, 615-8510, Japan – taki@cheme.kyoto-u.ac.jp; takagi@cheme.kyoto-u.ac.jp; oshima@cheme.kyoto-u.ac.jp

Abstract - A porous polymeric film is a promising material for low-dielectric constant film. A high-gas volume fraction porous polyimide thin film, 30 μm in thick was prepared by employing a phase separation of polyimide precursor in *N,N*-dimethylacetamide (solvent) / 2-(diethylamino) ethyl methacrylate / photoinitiator/ polyimide precursor. A wet film of the solution on a substrate was placed between two quarts windows of a high-pressure view cell apparatus. 5 MPa CO_2 gas was filled to the view cell. Immediately, the cast film became unclear because of the phase separation and the solvent drops formed in 60 s. The film was exposed to UV light through the quarts windows to polymerize 2-(diethylamino) ethyl methacrylate with keeping the pressure of CO_2 for 30 s. The cured thin film containing 1 μm in average diameter of numerous pores was treated at 320°C for 1 hour under nitrogen flush. The obtained film was 30 μm in thick, 1 μm in average diameter pores. The minimum relative dielectric constant was 1.536 with 74 % of pore volume at 20 GHz.

Keywords: porous, polyimide, film, photopolymerization, CO_2

Introduction

In electronics industry, the polyimide film has been used for a flexible-printed-cable (FPC) in electronic devices because inherently the polyimide has high-heat resistance, chemically stable and flexible. With the increase of transmission speed in a mobile information devices e.g., cellular phone and smart phone, the polyimide FPC is demanded to reduce its relative dielectric constant to decrease the attenuation of signals. [1] There are two different strategies to reduce the dielectric constant of polyimide. Chemical modifications of polyimide backbone and side chains have been studied as one of the strategies. Another attempt is to involve voids in polyimide films. The pioneering studies to introduce 10 nanometer-order-sized pores in polyimide were reported by Hedrick.[2] They introduced a block polymer which could be pyrolyzed at high temperature to form voids in polyimide matrix. The number of voids was successfully prepared in the polyimide while the void fraction was limited because of small number of pores.

The other technique employed was a physical foaming technique where CO_2-gas dissolved polymer matrix was thermally phase-separated to induce bubble nucleation in rubbery state polymer.[3] The supercritical fluid assisted extraction technique was applied to remove meso-scale domains in polyimide matrix and the rest of the domains became voids.[4] The process is quite sophisticated to create meso-sized pores while it is presumed that the extraction of all domains may not be possible, which might be unfavorable for electronics devices. A water-borne porous polyimide was prepared from the condensation of water vapor onto the polyimide precursor substrate.[5] The micron size water drops formed on the substrate were the porogen. The reduction of dielectric constant of the porous film was succeeded while the production time of the porous film just for forming the water droplets took 48 hours.

In this study, a new concept of production of porous polyimide was developed. The concept involves the solidifying a transient structure of phase separation, which was recently reported in our previous study where UV curable monomers and CO_2 solution was phase-separated with forming a transient porous structure and solidified by a rapid photopolymerization.[6] Moreover, the photopolymerization followed by the unidirectional freezing also enabled us to create unique porous structures from the UV curable monomer/solvent/photoinitiator solid-liquid phase separation.[7]

Here in this study, a mixture of polyimide precursor, solvent and UV curable monomer/ photoinitiator was exposed to the high-pressure CO_2. Immediately, the phase separation occurred and solvent droplets were formed in the mixture. The transient structure of phase separation consisting of the droplets in the wet film was solidified by photopolymerization of the UV curable monomer with UV irradiation. The obtained porous polyimide precursor was transformed to porous polyimide by heat treatment. The relative dielectric constant and loss tangent of the porous polyimide films were measured at 20 GHz. Firstly the porous structure and pore diameters respected to CO_2 pressure and UV intensity, secondary, the relationships among the dielectric constant, loss tangent and void fraction of porous polyimide film are discussed.

Experimental

Preparation. Pyromellitic Dianhydride (B0040) and 4,4'-Diaminodiphenyl Ether (O0088) were purchased from Tokyo Chemical Industry (Tokyo, Japan). Diphenyl(2,4,6-trimethylbenzoyl) phosphine Oxide (photoinitiator, 326-63232) and *N,N*-dimethylacetamide, dehydrated (solvent, 042-25285) were purchased from Wako Pure Chemical (Osaka, Japan). UV curable monomer of 2-(Diethylamino) ethyl methacrylate, which is *tert*-amine group methacrylate was purchased from Tokyo Chemical Industry (Tokyo, Japan). CO_2 gas (99.9% purity) and N_2 gas (99.998% purity) were purchased from Showa Tansan (Kawasaki, Japan). Copper substrate was supplied by the Sumitomo Electric Industries (Osaka, Japan). All the regents were used as received.

Polyimide precursor (PAA) was synthesized in a dried three-neck flask following the previous study. 0.505 g of 4,4'-Diaminodiphenyl Ether was dissolved in 5.5 g of *N,N*-Diethylacetamide in the three-neck flask with a mechanical agitator. Then 0.550 g of Pyromellitic Dianhydride was added to the flask. The polymerization proceeded in 8 hours and yellow-colored viscous liquid was obtained. Typically, 1.58 g of UV curable monomer and 0.201 g of photoinitiator were added to 6.56 g of the precursor solution and stirred until homogeneous solution was obtained.

The solution was cast on a copper substrate (4.5 × 4.5 cm^2) using a wired-bar (bar coater) and automated coater (Imoto Machinery, Kyoto, Japan). The wet film of the solution was place in a high-pressure view cell with two quarts windows pre-heated at 40°C as shown in エラー！参照元が見つかりません。 manufactured by the Tama-Seiki, Kanagawa, Japan. The maximum pressure of CO_2 was 6.5 MPa. The diameter and thickness of quartz window were 65 mm and 25 mm, respectively.

The high-pressure CO_2 gas at a given pressure was filled in the view cell. The wet film became unclear within 30 seconds due to the phase separation. Then the wet film was exposed to UV light (High Pressure 200 Watt Mercury Vapor Short Arc, S2000, ExFo) through the quarts window. The UV curable monomer in the wet film was polymerized for 30 seconds. The CO_2 pressure was released to atmospheric pressure for 60 seconds. After drying the film in room temperature, the porous PAA was formed.

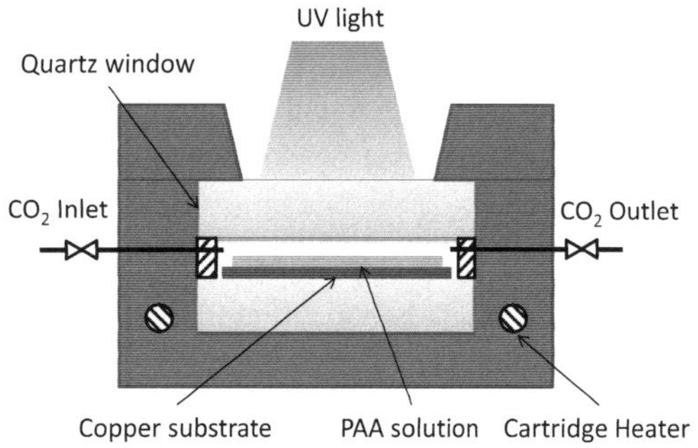

Figure 1 – Scheme of pore formation in polyimide precursor.

The film was transformed to a porous polyimide by heat treatment in nitrogen-flushing oven (DN610I, Yamato Scientific, Tokyo, Japan), where the temperature programed as (1) 3°C/min to 110°C and kept for 60 min, (2) 3°C/min to 320°C and for 60 min under 20 L/min of nitrogen. The copper substrate was removed by immersing into 30 wt.-% of nitric acid aqueous solution, rinsed with distilled water and dried in vacuum chamber.

Characterization. The porous structure of porous PAA and polyimide coated with gold and palladium was observed using a scanning electron microscope (JEM-6700F). The pore diameter of each pore was measured as follows: (1) fit an ellipsoid on pore, (2) calculate a diameter from the area of the ellipsoid assuming a circular pore with the aid of the Image J (National Institute of Health).

The void fraction of porous polyimide was calculated as follows:

$$f = \frac{LA}{w}\frac{1}{\rho_s} \qquad (0.1)$$

where L is thickness, A is area, w is mass of the porous polyimide and ρ_s is the solid density of non-porous polyimide. The thickness was obtained averaging three different points of film. The area was calculated by an image of the film.

The relative dielectric constant and loss tangent of the porous polyimide film were measured using the transmission technique at 20 GHz and room temperature (KEAD, Japan).

Results and Discussion

Micrographs of the cross section of porous PAA film are shown in

Figure 1. In

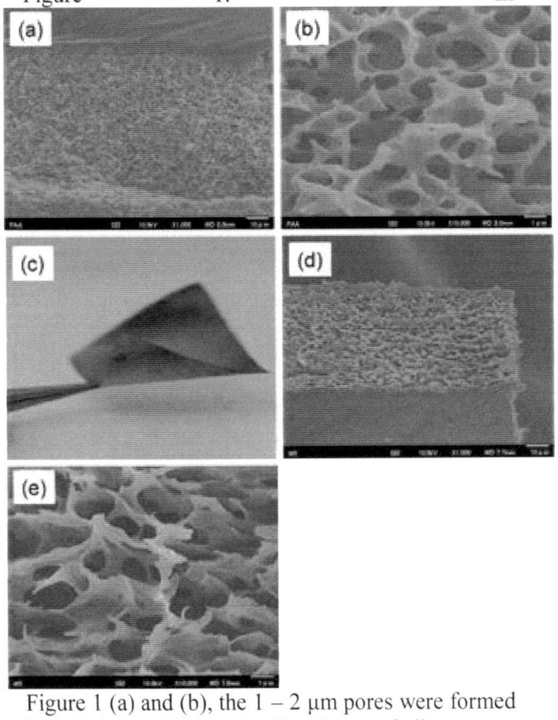

Figure 1 (a) and (b), the 1 – 2 μm pores were formed uniformly in the 50 μm thin film. Most of all pores are inter-connected pores. A 3 μm skin-layer which was exposed to CO_2 atmosphere was formed upper surface of the film.

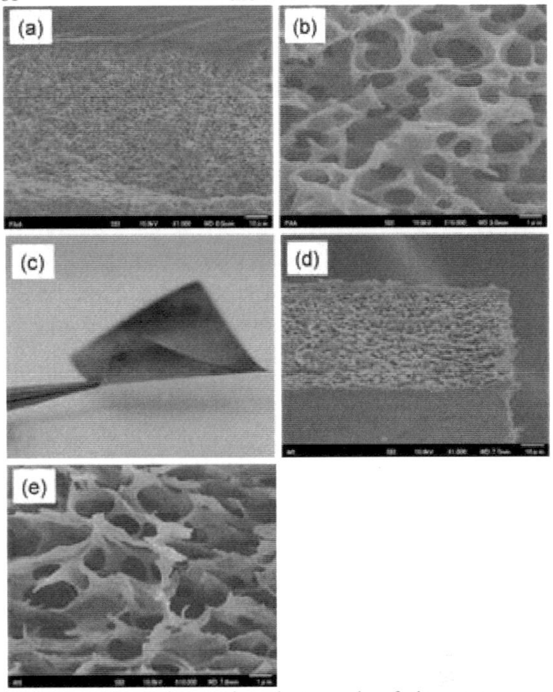

Figure 1 (c) shows the photograph of the porous polyimide thin film clipped with a tweezers. The film is self-standing and flexible.

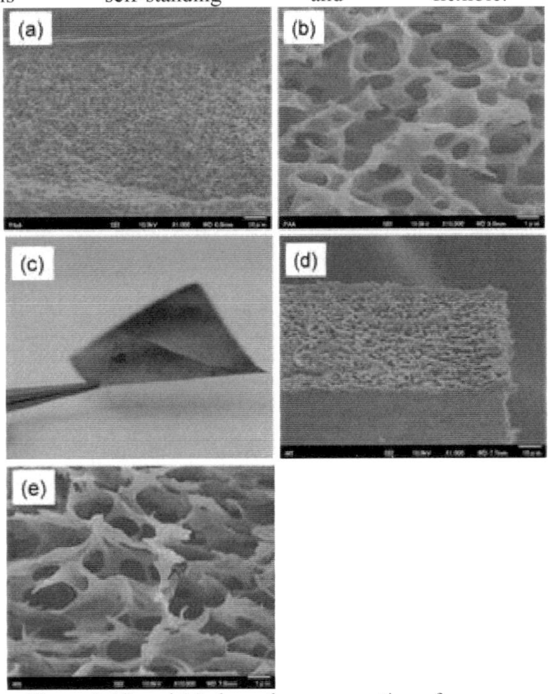

Figure 1 (d) and (e) show the cross-section of porous polyimide film. The thickness decreased to 30 μm while the pore diameter increased at approximately 0.5 μm after the heat treatment. The transformation from PAA to polyimide shrunk the film and increased the open-area of pores.

The effect of CO_2 pressure on pore formation of porous PAA film was investigated with keeping the same UV intensity of 152 mW/cm^2. Figure 2 shows the pore diameters of porous PAA and porous polyimide films. Overall the pore diameters of porous polyimide decreased by the heat treatment of porous PAA. Pore diameter of 0.99 MPa-CO_2 was 0.69 μm to 0.36 μm. It could be the nanoporous film, however, the void fraction of the porous PI was 0.5 %. The volume of the pores was too small for reducing the relative dielectric constant. At 2.2 and 3.0 MPa of CO_2 were also performed to create porous PAA. The porous PAA could not be obtained so that the plots were not shown on the graph. At 3 MPa of CO_2, the porous polyimide was obtained by the heat treatment of the non-porous PAA film subjected to the CO_2 and UV irradiation. The residual solvent vaporized and formed some voids in the film. The pore diameter and void fraction were 0.47 μm and 8%, respectively. The void fraction was still too low to use the low-dielectric constant film.

The CO_2 pressures of 4 to 6.5 MPa, we succeeded to create high void fraction porous films of PAA and polyimide with 1 μm order pores. The void fractions of 4, 5 and 6.5 MPa were 52, 57 and 65 %, respectively. At least 4 MPa of CO_2 was required to create pores in the PAA film at high volume fraction. The pore diameters of porous PAA slightly decreased with the increase of CO_2 pressure in average.

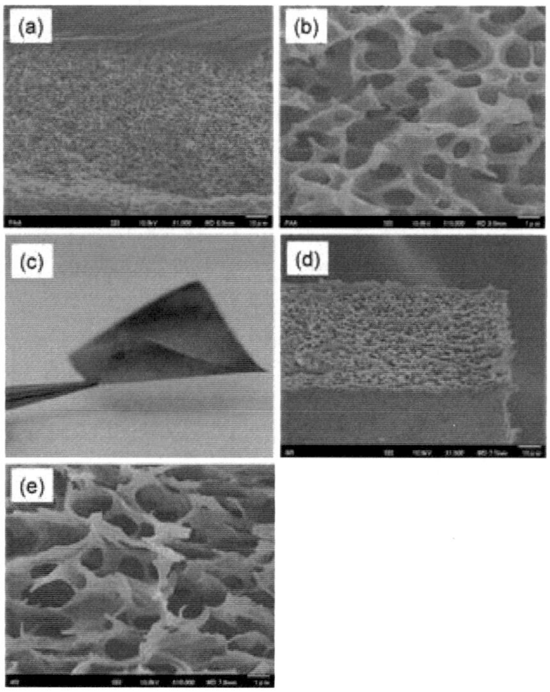

Figure 1 – Porous PAA and polyimide films. (a) cross-section of porous PAA, (b) magnified view of (a), (c) photograph of porous polyimide film, (d) cross-section of porous polyimide film and (e) magnified view of (d).

The effect of UV intensity on porous structure of porous PAA and polyimide was investigated with keeping the same CO_2 pressure level of 6.5 MPa. Overall similar to the effect of CO_2 pressure, the pore diameter of porous polyimide is smaller than that of PAA. The pores in porous PAA shrunk after the heat treatment. At the UV intensity of 0 mW/cm^2, the pore diameter was 0.69 μm, however, the pores were disappeared after the heat treatment. The UV intensity of 38 mW/cm^2 at 365 nm was also performed and the pores could not be kept by the UV irradiation. Higher than 76 mW/cm^2, the porous structures were successfully obtained in PAA and polyimide films. The pore diameters both of PAA and polyimide decreased with the increase of UV intensity.

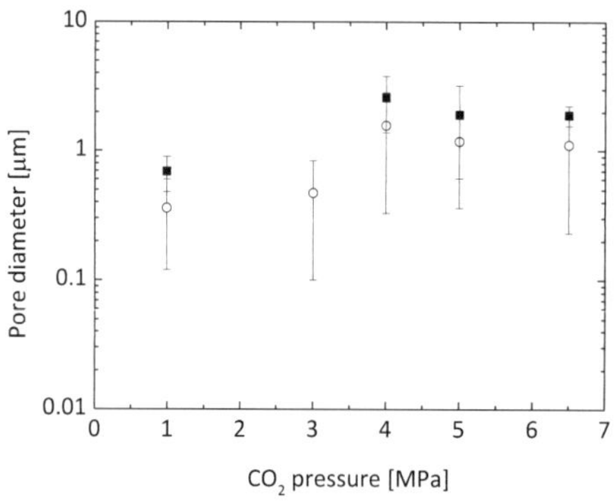

Figure 2 – Effect of CO_2 pressure on pore diameter of porous PAA and polyimide. Solid square: porous PAA. Open square: porous polyimide.

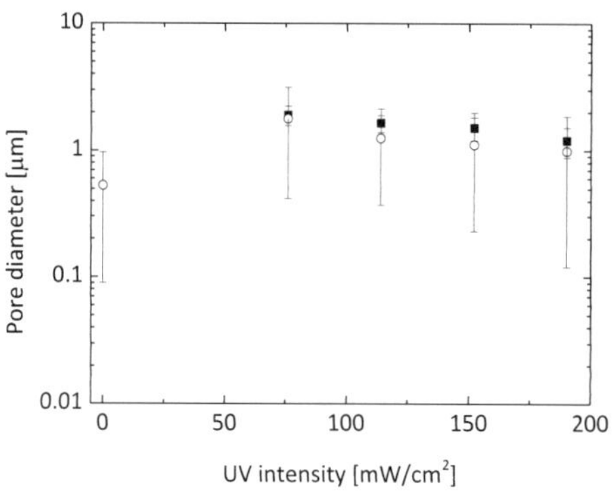

Figure 3 – Effect of UV intensity on pore diameters of porous PAA and polyimide. Solid square: porous PAA. Open square: porous polyimide.

The relative dielectric constant and loss tangent of porous polyimide films were shown in Figure 4. Figure 4 (a) shows the relationship between the relative dielectric constant and void fraction. The dielectric constant decreased with the void fraction proportionally. The line on the graph is the theoretical line from the relative dielectric constant of 3.4 of non-porous polyimide to 1 of the air. The measured relative dielectric constant data are located on the lines.

It is reasonable that the relative dielectric constant of porous polyimide film decreases with the void fraction. The relative dielectric constant achieved lower than 1.5 over 70% of void fraction. These films have uneven surface and large voids in the film, which could not be used for the flexible printed circuit materials.

Figure 4 (b) shows the loss tangent of the porous polyimide film. The horizontal line on the graph is the loss tangent of solid polyimide. The loss tangent data of porous polyimide are larger or same level of the solid polyimide. Although the five data located over 70% of void fraction are lower than the solid polyimide, these films are uneven surface and large voids in the film. The reduction of loss tangent is more important than the relative dielectric constant to reduce the attenuation of the signals in on the electrical circuit. The polymers formed by the photopolymerization of the monomer may inhibit to order the polyimide chain during the heat treatment or the residues of the pyrolized polymer affects the loss tangent as impurity. Further to reduce the loss tangent the temperature profile of heat treatment should be optimized to remove the polymer before the imidization of PAA.

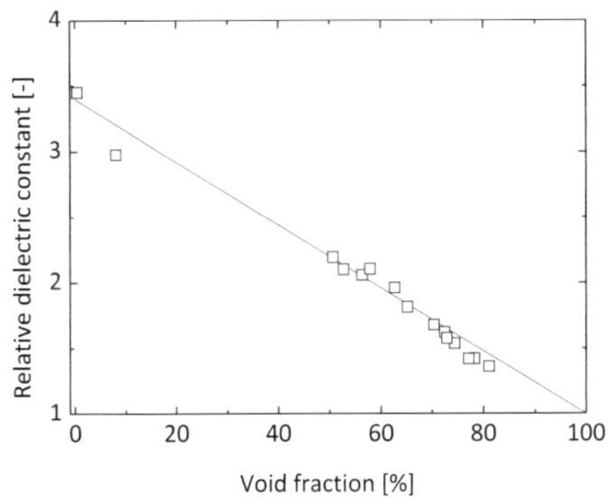

(a)

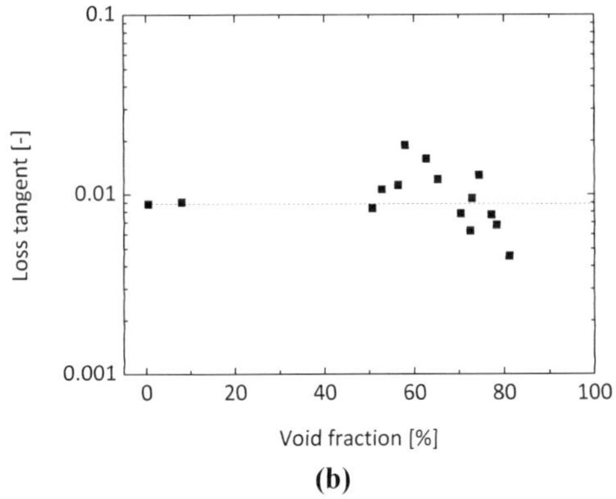

(b)

Figure 4 – Relative dielectric constant (a) and loss tangent (b) of porous polyimide film vs. void fraction.

Conclusions

Porous polyimide film was prepared from the porous polyimide precursor in which pores were formed by means of the CO_2-dissolution induced phase separation and photopolymerization of the tert-amine monomer in the polyimide precursor solution. The process of pore formation in the polyimide precursor took 120 seconds, which was quite fast comparing the previous studies. For the porous polyimide film the gas volume fraction is over than 80 % and the relative dielectric constant decreased to 1.4 % and flexibility. The flexible porous polyimide film is expected to be used a flexible printed cable for high-throughput transmission.

Acknowledgements (Times New Roman 10, bold)

This project was financially supported by the New Energy Development Organization (Grant number: 09A16003d Japan).

References

1. Xiong-Yan, Z. and L. Hong-Jie, *Review of polymer materials with low dielectric constant.* Polymer International, 2010. **59**(5): p. 597-606.
2. Hedrick, J.L., et al., *Nanoporous polyimides*, in *Progress in Polyimide Chemistry Ii*1999. p. 1-43.
3. Krause, B., et al., *Ultralow-k dielectrics made by supercritical foaming of thin polymer films.* Advanced Materials, 2002. **14**(15): p. 1041-+.
4. Mochizuki, A., et al., *Development of photosensitive porous polyimide with low dielectric constant.* Journal of Photopolymer Science and Technology, 2002. **15**(2): p. 159-165.
5. Ren, Y. and D.C.C. Lam, *Properties and Microstructures of Low-Temperature-Processable Ultralow-Dielectric Porous Polyimide Films.* Journal of Electronic Materials, 2008. **37**(7): p. 955-961.
6. Taki, K. and S. Okumura, *Production of Porous Polymeric Films with Unimodal or Bimodal Pore-Size Distributions via Depressurization- and Photopolymerization-Induced Bubble Nucleation in Low-Viscosity UV-Curable Monomer/High-Pressure CO_2 Solutions.* Macromolecules, 2010. **43**(23): p. 9899-9907.
7. Okaji, R., et al., *Preparation of porous honeycomb monolith from UV-curable monomer/dioxane solution via unidirectional freezing and UV irradiation.* Journal of Applied Polymer Science, 2012.

O-03-162

The effect of particle size and coating of $CaCO_3$ on the foaming behavior of a styrene butadiene block copolymer

Bernd Geissler[1]*, Stephan Laske[2], Clemens Holzer[2] and Günter Rüdiger Langecker[2]

[1]*Polymer Competence Center Leoben GmbH., Leoben, Austria* – bernd.geissler@pccl.at
[2] *Department of Polymer Engineering and Science, Chair of Polymer Processing, Montanuniversitaet Leoben, Austria* - stephan.laske@unileoben.ac.at; Clemens.Holzer@unileoben.ac.at; guenter.langecker@unileoben.ac.at

Abstract - In this study the physical foaming of a styrene butadiene block copolymer (SBC) using coated and uncoated calcium carbonate ($CaCO_3$) was examined. The aim was to determine, how the mean particle size and the coating of the $CaCO_3$ influence the foam morphology and the processing of styrene butadiene block copolymer foam. Therefore, several compounds were prepared with a twin screw extruder using $CaCO_3$ grades with distinct mean particle sizes. In the first series $CaCO_3$ grades with the same coating were used. In the next step the influence of the surface coating of the $CaCO_3$ on the foam morphology was investigated. For this purpose, a compound was prepared with one uncoated $CaCO_3$ grade with the same mean particle size as one of the coated ones. All compounds were characterized with a cone plate rheometer to compare the influence of the different $CaCO_3$ grades on the rheological properties. Their foaming behavior was investigated by using a single screw extruder with a screw diameter of 45 mm and supercritical CO_2 as foaming agent. The different calcium $CaCO_3$ grades were compared, due to their effect on the maximum cell density, the cell size and the foam density.

Keywords: styrene butadiene block copolymer, calcium carbonate, CO_2, foaming, d50%

Introduction

Due to their unique properties like high shrinkage and high impact resistance, styrene butadiene block copolymers are often used in the field of plastic packaging. But in recent years increasing raw material prizes and new recycling regulations made it necessary to look for possibilities to reduce the product cost. One possibility is to reduce density via e.g. physical foaming. In literature no information about the foaming behavior of styrene butadiene block copolymers is available, so the first aim was the development of a suitable nucleating agent. Normally nucleating agents are inorganic fillers, foreign polymer phases or chemical blowing agents, which are used to archive small cell sizes, high cell density and uniform cell size distribution [1-2]. According to McClurg [3] ideal nucleants are easily dispersable in the polymer matrix, have a uniform size and shape, a non-wetting surface and are plentiful enough to promote heterogeneous nucleation. Besides talc [4–7] and chemical blowing agents [8], $CaCO_3$ [9–12] is often used as a nucleating agent.

Experimental

Experimental Materials

In this study a SBC-blend and carbon dioxide (CO_2, 99% pure, Linde Gas Austria) as blowing agent were used. Four different types of calcium carbonate ($CaCO_3$), three with different particle size and surface coating and one additional without a surface coating were examined. $CaCO_3$-1 is a precipitated calcium while all others are natural chalk types from the same company.

Table 1 – Physical Properties of the SBC-blend

	SBC-Blend
MVR (200°C/5 kg)	11 g/10 min
Density	1.02 g/cm³

The physical properties of the two polymers and the four $CaCO_3$ grades are summarized in Tables 1 and 2.

Table 2 –Properties of $CaCO_3$ Particles

	d_{50} in µm	d_{98} in µm	Surface coating
$CaCO_3$-1	0.08	-	yes
$CaCO_3$-2	1.8	8	no
$CaCO_3$-3	1.8	8	yes
$CaCO_3$-4	>6	45	yes

Material Preparation

The material preparation was carried out on a co-rotating twin screw extruder with a screw diameter of 25 mm and a screw length of 40 D. For each $CaCO_3$ type we produced a master batch with 20 wt.-% filler amount with a screw speed of 150 rpm and a fixed extrusion temperature of 185°C.

Foaming Experiments

For the foaming experiments a Rosendahl ROEX 45 mm grooved feed single screw extruder with a screw length of 32 D was used. The screw was specially designed for the physical foaming and includes a 4 D long shear element and a 8 D long mixing element. Additionally a static mixer was mounted before the die. For these experiments we used a capillary die with a diameter of 2.5 mm and length of 10.5 mm. The blowing agent was injected by a positive displacement pump (Teledyne Isco syringe pump

260D) above the shear element at a fixed amount of 0.4 wt.-% CO_2 for each formulation. The temperature profile was adjusted to achieve a melt temperature of 180 °C before the die and all experiments were done with a screw speed of 10 rpm and a constant mass flow rate of 6 kg/h. All master batches were dry blended and carefully diluted with the SBC-Blend to archive a filler content of 5 wt.% for each $CaCO_3$ type. Then the mixtures were fed into the barrel through the hopper. After reaching a steady state the melt was collected and 2 mm thick sheets were prepared for the rheological tests with a vacuum hot press (Dr. Collin P 200 PV). Next the volume flow rate was measured and the blowing agent injection was started. A sufficient time was given to stabilize the process until no further change was noticeable. For every setting the whole process was repeated three times.

Rheological Characterization

The rheological properties of the samples were measured with a rotational rheometer (Anton Paar, Physica MCR 501) at 180 °C with cone plate geometry. The cone has an angle of 5° and a diameter of 25 mm. For the characterization of the dynamic rheological properties frequency sweep experiments were performed to measure the complex viscosity, the loss and storage modulus in a frequency range of 0.1 – 500 rad/s. For all samples dynamic stress sweeps were executed to determine the end of the linear viscoelastic regime. Furthermore creep recovery measurements were done to determine the material's elasticity at 180 °C. The creep recovery test consists of a 300 s creep phase with a constant load of 200 Pa followed by a 700 s recovery phase.

Foam Morphology Characterization

For the morphology characterization the samples were fractured in a brittle manner and the microstructure of the fracture surface was examined with an infinite focus microscope (Alicona IFM). For the determination of the mean cell diameter (d_m in μm) and the cell density (N in number of cells per cubic centimeter) the Software ImageJ was used. The cell densities were calculated from the micrographs using Eq. 1 [13]. With N_b as number of cells per area (A in cm^2), $\rho_{SBC\text{-}Blend}$ as density of unfoamed blend in kg/m^3 and ρ_{Foam} as foam density in kg/m^3.

$$N = \left(\frac{N_b}{A}\right)^{\frac{3}{2}} \cdot \frac{\rho_{SBC-Blend}}{\rho_{Foam}} \tag{1}$$

The foam density was measured with a Density Kit (Mettler Toledo) equipped on a precision balance XS205 (Mettler Toledo).

Results and Discussion

Effect of filler size and coating on the rheological properties

Figure 1 shows the influence of the filler on the complex viscosity of the SBC-blend. The complex viscosity increases due to the filler addition, however it is not possible to find a correlation between the filler size and the increase in the complex viscosity. $CaCO_3$-3 shows a slightly higher viscosity than $CaCO_3$-2. Both filler types have the same particle size, but $CaCO_3$-3 is coated. The storage modulus shows hardly any influence of the filler addition (Fig. 2). Similar results are shown in Fig. 3 for the loss modulus.

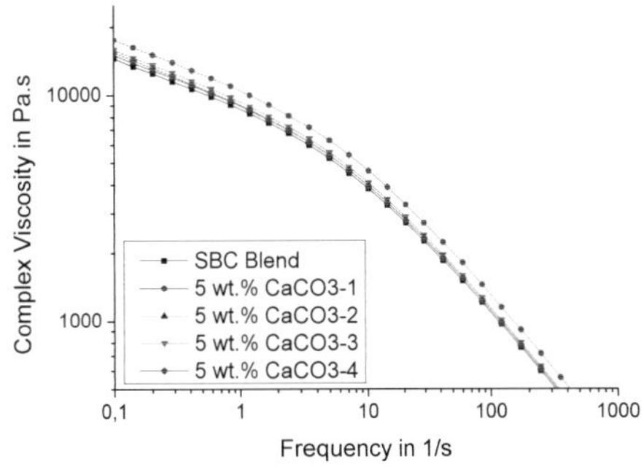

Figure 1- Complex viscosity of SBC-Blend and SBC – $CaCO_3$ compounds

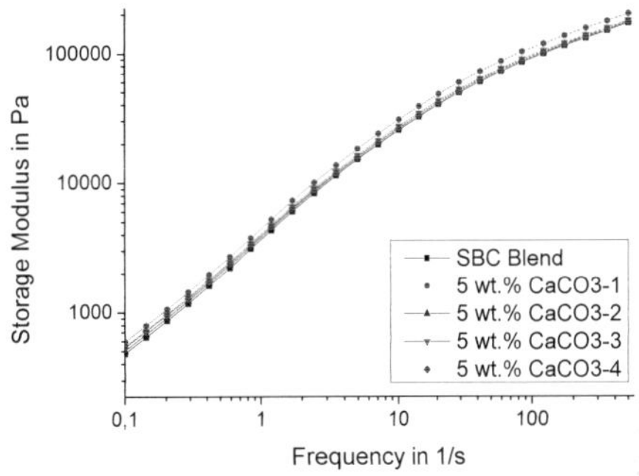

Figure 2- Storage modulus of SBC-Blend and SBC – $CaCO_3$ compounds

Figure 4 represents the time dependent strain of the different formulations during creep and recovery. The results show a dissimilar result compared to the complex viscosity, especially for the creep phase. With increasing particle size, there is an increase in the calculated zero shear viscosity. Similar to the frequency sweep results, $CaCO_3$-3 shows a slightly higher viscosity than $CaCO_3$-2 due to the coating. During the recovery phase all formulations shows a similar behavior. We believe that the recovery phase is dominated by the polymer and the creep phase is more influenced by the filler.

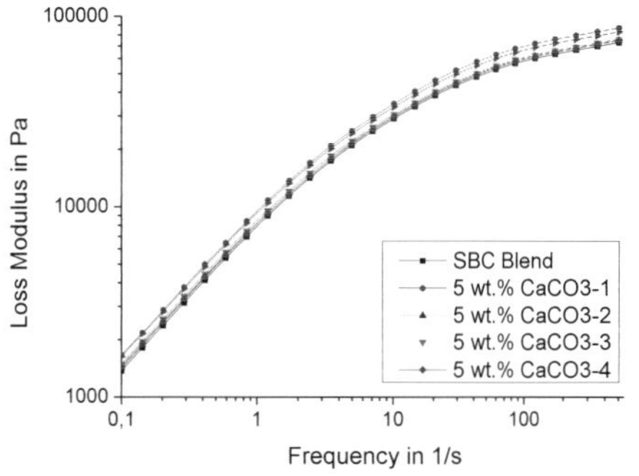

Figure 3- Loss modulus of SBC-Blend and SBC – $CaCO_3$ compounds

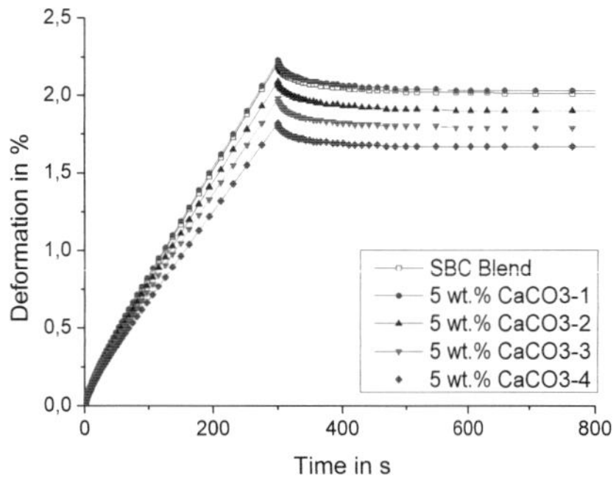

Figure 4- Time dependent strain during creep and recovery

Effect of filler size and coating on the foaming behavior

The SBC blend shows a bad foaming behavior compared to other materials, but it was possible to improve the foam morphology due to the addition of the different $CaCO_3$-grades (Fig. 5). In this study the cell density increases with increasing filler size and slightly increases due the coating. The addition of the filler leads to a slightly lower pressure drop over the capillary die in all cases. The values of the cell densities show the similar tendency, compared to the results from the creep recovery experiments. With increasing zero shear viscosity the number cell density increases.

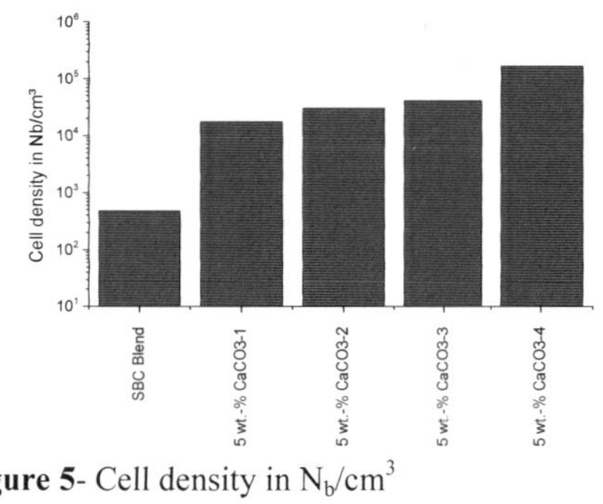

Figure 5- Cell density in N_b/cm^3

We believe that the apparent increasing zero shear viscosity leads to higher stresses in the flow field of the capillary die compared to the pure SBC Blend. According to Leung et. al [14] we believe that the higher stresses at the same deformation rate of the melt, results in a further local pressure reduction and a higher degree of supersaturation around the filler particles.

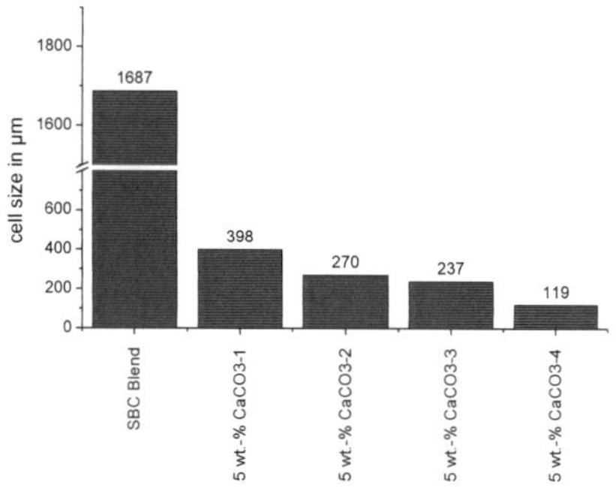

Figure 6- Mean cell size in µm

The same results were observed for the mean cell size (Fig. 6). With increasing particle size of the filler the mean cell size decreased from 398 µm to 119 µm. With the experimental settings we achieved a density reduction of about 40 - 50 % for the different materials (Fig. 7). With these settings, it was not possible to achieve a higher foaming ratio, because of the low blowing agent solubility. The SBC-blend shows the lowest density due to the big cell sizes. Generally the density was strongly influenced by the sample collection and preparation.

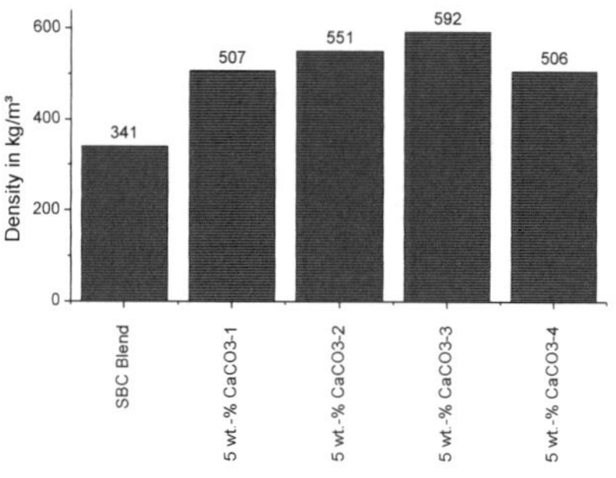

Figure 7- Density

Conclusions

The effect of the size and coating of the $CaCO_3$ on the foaming behavior of a SBC-blend was studied. It was found out that the nucleation rate increases with increasing particle size. The same was observed for the surface coating. The creep recovery experiments correlate with the foaming results. The results showed that the increasing apparent zero shear viscosity due to the addition of the fillers leads to an increase in the cell density and a decrease in the mean cell size. We think that the higher apparent zero shear viscosity leads to higher stresses at the same deformation rate in the flow field of the capillary die, which leads to local pressure fluctuations.

Acknowledgements

The research work of this paper was performed at the Polymer Competence Center Leoben GmbH (PCCL, Austria) within the framework of the COMET-program of the Federal Ministry for Transport, Innovation and Technology and the Federal Ministry of Economy, Family and Youth with contributions by Institute of Plastics Processing, Montanuniversitaet Leoben and Brueckner Maschinenbau GmbH, Gabriel-Chemie GmbH, Greiner Packaging International, Rosendahl Maschinen GmbH and Teufelberger GmbH. The PCCL is funded by the Austrian Government and the State Governments of Styria and Upper Austria.

References

1. C. B. Park, L. K. Cheung, and S. W. Song, *Cellular polymers*, vol. 17, no. 4, pp. 221–251, 1998
2. S. Leung, A. Wong, and C. Park, in ANTEC Tech. Paper, vol. 190, pp. 22–24, 2009.
3. R. McClurg," *Chemical Engineering Science*, vol. 59, no. 24, pp. 5779–5786, Dec. 2004
4. H. E. Naguib, C. B. Park, and P. C. Lee, *Journal of Cellular Plastics*, vol. 39, no. 6, pp. 499–511, Nov. 2003.
5. W. Kaewmesri, *Journal of Cellular Plastics*, vol. 42, no. 5, pp. 405–428, Sep. 2006
6. S. N. Leung, A. Wong, and C. B. Park in Antec, 2009, pp. 22–24.
7. A. Wong and C. B. Park, *Chemical Engineering Science*, vol. 75, pp. 49–62, Jun. 2012
8. A. H. Behravesh, C. B. Park, L. K. Cheung, and R. D. Venter, *Journal of Vinyl and Additive Technology*, vol. 2, no. 4, pp. 349–357, 1996
9. M. Abbasi, S. N. Khorasani, R. Bagheri, and J. M. Esfahani, *Polymer Composites*, vol. 32, no. 11, pp. 1718–1725, Nov. 2011.
10. J. B. Azevedo and M. S. Rabello, in PPS-23, pp. 2–6.
11. H.-X. Huang and J.-K. Wang, *Journal of Applied Polymer Science*, vol. 106, no. 1, pp. 505–513, Oct. 2007
12. F.-C. Chiu, S.-M. Lai, C.-M. Wong, and C. Hui Chang, *Journal of Applied Polymer Science*, vol. 102, no. 3, pp. 2276–2284, Nov. 2006
13. M. Mihai, M. Huneault, and B. Favis, *Polymer Engineering and Science*, vol. 50, no. 3, 2010
14. S. N. Leung, A. Wong, L. C. Wang, and C. B. Park, *The Journal of Supercritical Fluids*, Dec. 2011.

O-03-172

EFFECT OF CRYSTALLINE PHASE SEPARATED HARD SEGMENTS ON PRODUCTION OF MICROCELLULAR THERMOPLASTIC POLYURETHANE FOAMS USING BUTANE

N. J. Hossieny[1], M.R. Barzegari[1], and C. B. Park[1*]
[1]Microcellular Plastics Manufacturing Laboratory
Department of Mechanical & Industrial Engineering, University of Toronto, Toronto, Ontario, Canada M5S 3G8
nemat@mie.utoronto.ca, mrbarzi@mie.utoronto.ca, park@mie.utoronto.ca

Abstract – This study presents a novel technique of utilizing phase separated crystalline and non-crystalline hard segment (HS) domains as effective heterogeneous nucleation sites in production of microcellular thermoplastic polyurethane (TPU) foams. The HSs in TPU are statistically distributed in the soft matrix phase. This distribution can be changed by processing TPU at various conditions such as melt processing or isothermal treatments. In this study, the foaming behavior of "as-received TPU"(AR-TPU) and "processed TPU" (PR-TPU) has been investigated using n-butane as a blowing agent. The isothermal crystallization behavior of AR-TPU and PR-TPU at atmospheric pressure was investigated using regular Differential Scanning Calorimetry (DSC). Furthermore, the effect of butane pressure (55 bar and 103 bar) on the isothermal crystallization of TPU was investigated. Overall, the crystallinity of AR-TPU and PR-TPU increased significantly in presence of butane. However the increase in crystallinity was higher in PR-TPU. It was also observed that T_g of both AR-TPU and PR-TPU decreased in presence of butane. Comparing the foam characteristics of AR-TPU and PR-TPU, it was observed that PR-TPU exhibited remarkable increase in cell density and reduction in cell size over wide temperature ranges of 150 °C and 170°C at a moderate saturation pressure without addition of nucleating agents and/or fillers.

Keywords: Thermoplastic Polyurethane, Microcellular Foams, Heterogeneous Nucleation, Butane, Hard Segments

Introduction

Thermoplastic polyurethane (TPU) elastomer is a block copolymer consisting of alternating soft and hard segments. The soft segment (SS) is a polyol with an ether or ester group in the main chain, which provides rubber-like properties and elasticity of the elastomer. The hard segment (HS) contains urethane linkages generated by the reaction of diisocyanate and short-chain diols and influences the mechanical properties, such as the elasticity, hardness, and tearing strength. The in-favorable thermodynamic immiscibility between HS and SS caused due to the formation of hydrogen bonds in HS drives the system to form a microphase-separated structure at room temperature [1]. TPUs are widely used for different purposes, such as automobile parts, building materials, sports goods, and medical equipments [2]. Recently, expanded TPU bead foams have been developed which can be molded into complex three dimensional products [3]. However one of the drawbacks is the high cost due to high temperature requirement during the sintering of the beads. Production of expanded TPU beads with microcellular foam morphology may significantly reduce the molding cost by improving the inter-bead bonding.

To date, there have been few reports on processing of microcellular TPU foams [4]. Most of the reported microcellular TPU foams in the literature are processed using supercritical CO_2. Butane is a popular physical blowing agent widely used in foam industry due to its highly solubility and low diffusivity. However butane poses significant thermodynamic challenge to achieve controlled foam morphology with fine cells and high density. Up to date, there hasn't been any report of processing microcellular foams in presence of butane. This study presents novel method to produce microcellular TPU foam in presence of butane. Furthermore, the microcellular morphology in TPU is achieved at a mild processing pressure over wide range of temperature's without presence of nucleating agent and/or fillers. Furthermore, this work presents novel technique of melt processing TPU and utilizing the crystalline phase separated HS as a effective heterogeneous nucleating sites to achieve the observed microcellular morphology.

Experimental

Materials

The TPU material used in this study is from BASF with a brand name of Elastollan. The material has a melting point of 171°C, specific density of 1.13 g/cm^3 and hardness of Shore 90A. The HSs are made from MDI (Methylene diphenyl diisocynate) and BDO (Butanediol) as chain extender. The SSs are made from polyether diols and they tend to have high hydrolysis resistance. N-butane was supplied by Linde Canada and used as a blowing agent. The "as-received" TPU material (AR-TPU) was melt compounded in a twin screw DSM Micro-compounder and then extruded to cool in room temperature. The extruded "processed" TPU (PR-TPU) was cut using pelletizer into small rectangular pellets with average dimension of 3mm x 1.5 mm x 1.5 mm.

Foaming Procedure

Both AR-TPU and PR-TPU samples were foamed in an autoclave foaming chamber. The overall setup of the batch foaming process is shown in Fig. 1. Samples were placed in specially designed sample holders and placed in the high pressure vessel. The pressure vessel was vacuumed to remove moisture. Butane was fed to the pressure vessel using a Teledyne ISCO high pressure syringe pump and the sample was saturated at desired pressure for 60 min. After saturation time, pressure was rapidly released, and the foam morphology was fixed by cooling the chamber in the water bath. The foaming experiments were performed over a wide range of temperatures from 150°C up to 170°C and two different pressures of 53 bar and 103 bar respectively.

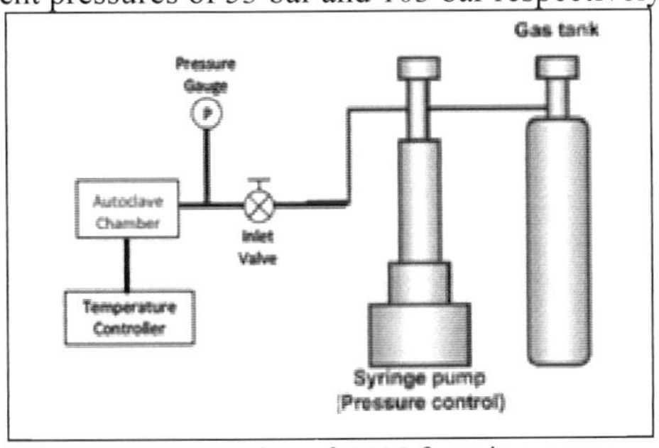

Fig. 1. Schematics of TPU foaming setup.

Foam Characterization

The morphology of the foamed sample was observed with JOEL JSM-6060 scanning electron microscope (SEM). The morphology of the foamed samples was characterized by measuring the expansion ratio ($\rho_{unfoamed}/\rho_{foamed}$) and cell density, that is, the number of cells per volume. Cell density, defined as the number of cells per cubic centimeter of foam (N_f), was obtained by the following equation:

$$N_f = \left(\frac{N}{A}\right)^{3/2} \cdot \frac{\rho_{unfoamed}}{\rho_{foamed}} \qquad \text{Eq.1}$$

where, N is the number of cells that can be seen on the SEM micrograph, and A is the area of the micrograph in unit cm^2. The foam density was measured using water displacement method.

Isothermal Crystallization Analysis of TPU

The effect of annealing on the crystallization behavior of TPU under atmospheric pressure was investigated using DSC (Q2000, TA Instruments). In a typical annealing experiment, approximately 10 mgm sample was equilibrated at 25°C followed by heating to the desired isothermal temperature at a constant heating rate of 20°C/min. Next, the sample was left at isothermal temperature for 60 min. This was followed by cooling at 20°C to -90°C and subsequent heating at 10°C to 230°C.

To investigate the crystallization behavior under butane pressure, the respective TPU sample was saturated in the autoclave foaming chamber. Approximately 15 mgm of sample was loaded into the foaming chamber and ramped at 10°C/min to desired temperature and pressure. This was followed by annealing for 60 min. Since the expansion during foaming may affect the HS crystallization and distribution caused by biaxial stretch, it is important to prevent foaming to occur. Hence after saturation time, the sample was rapidly quenched to room temperature. Butane was released at a very low pressure drop rate avoiding any foaming to take place. The sample was immediately heated in DSC from -90°C to 230°C at heating rate of 10°C/min to investigate the effect of butane pressure on the crystallization behavior of both AR-TPU and PR-TPU.

Results and Discussion

Effect of Thermal Annealing on TPU Crystallization

Fig. 2 shows the DSC endotherm of PR-TPU after the isothermal treatments over range of temperatures. AR-TPU showed similar behavior and hence is not shown. It can be seen that the low temperature melting peak (T_{mlow} = 67.90°C) shifts to higher temperatures (marked with arrow in Fig. 2), generally 15-20 °C above the annealing temperature . The high melting peak (T_{mhigh}= 170°C) remains unaffected. Seymour and Copper [5] have already reported this behavior in their studies and have related the shifting of low melting peak to higher temperature caused by the increase in the length or continuous improvement of the short range ordered HS. This process occurs without affecting any long-range order which may be present in the TPU structure [6]. However after annealing at 180°C as seen in Fig. 3a, PR-TPU shows formation of new multiple peaks at very high temperature's of 197°C and 211°C compared to the single high temperature peak in AR-TPU at 195.2°C.

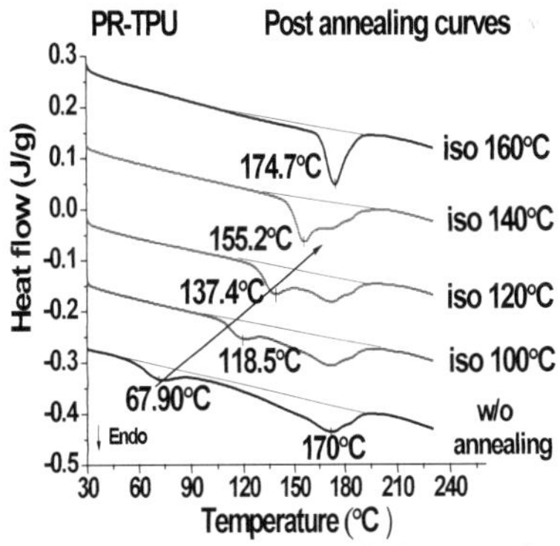

Fig. 2. Post annealing DSC endotherms of PR-TPU.

Fig. 3b shows the plot of Tg of SS after annealing at different temperatures. It can be observed that with increase in annealing temperature the Tg of both AR-TPU and PR-TPU increases. Hesketh et al. [7] reported similar behavior on series of polyurethanes annealed at various temperatures between 120°C and 190°C. They concluded that the fraction of HSs dissolved in SS sections increased by increasing the annealing temperature.

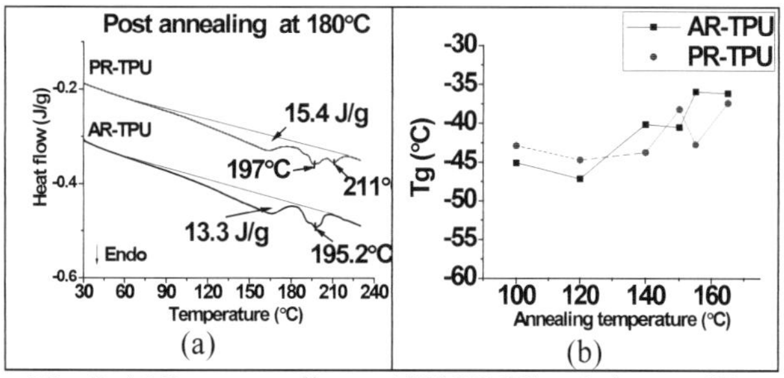

Fig. 3. (a) Post annealing at 180°C DSC endotherm ; (b) Tg after annealing at range of temperature.

Effect of Butane on TPU Crystallization

Fig. 4 compares the melting behavior of AR-TPU and PR-TPU after the samples were isothermally treated at 165°C at atmospheric pressure and with presence of butane at saturation pressure of 55 bar. The T_{mhigh} of AR-TPU decreases from 179.2 °C to 174.2°C and that of PR-TPU decreases from 179.8°C to 173.8°C. This decrease in melting behavior is due to the plasticizing effect of butane. However after saturating with butane, a broad low melting peak is formed in AR-TPU and PR-TPU with significant increase in total heat of fusion (ΔH_f). The ΔH_f of PR-TPU is higher than AR-TPU as seen in Fig. 5a. The presence of butane plasticizes TPU, which may cause decrease in viscosity. The HS dissolved in SS have increased mobility at low viscosity, which may assist in their phase separation. Furthermore, the longer HS length that have phase separated can stack into increased perfection to form crystalline phases. Fig. 5b compares the Tg of AR-TPU and PR-TPU after annealing at atmospheric pressure and in presence of butane at 55 bar. The Tg of both AR-TPU and PR-TPU decreases in presence of butane. This decrease confirms improved purity of SS from dissolved HS and the observed increase in crystallinity.

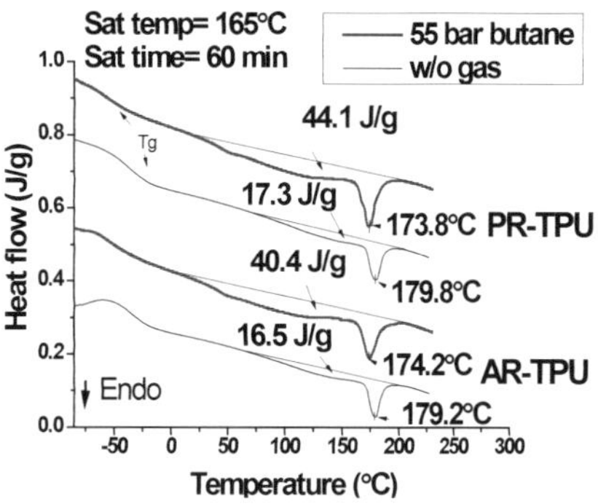

Fig.4. Comparison of DSC melting endotherm of AR-TPU and PR-TPU after annealing at 165°C at atmospheric pressure and in presence of 55 bar butane pressure

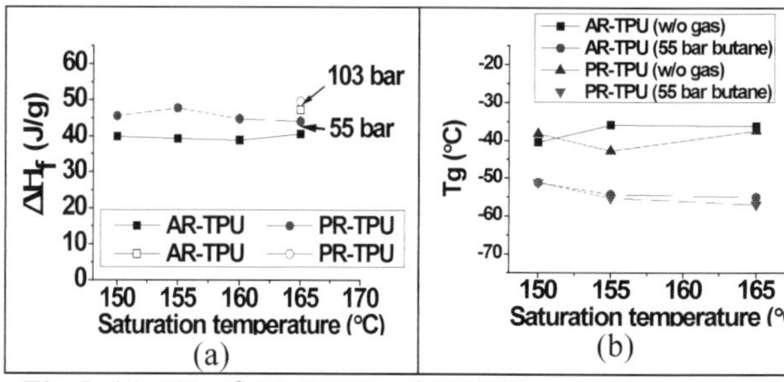

Fig.5. (a) ΔH_f of AR-TPU and PR-TPU at 55 bar and 103 bar butane pressure;(b) Tg after annealing in atmospheric pressure and in presence of butane at 55 bar.

Foaming Behavior of TPU in Presence of Butane

Fig. 6 shows the SEM micrographs of morphology of AR-TPU and PR-TPU foams prepared at temperature of 165°C and saturation pressures of 55 bar and 103 bar respectively. Fig.7 and Fig.8 shows the plot of characterized values of cell densities and average cell size of AR-TPU and PR-TPU over the range of processing temperature's and pressure's. PR-TPU shows microcellular morphology with cell size below 10μm and cell density above 10^9 cells/cm^3 at all processing conditions of temperature and pressure.

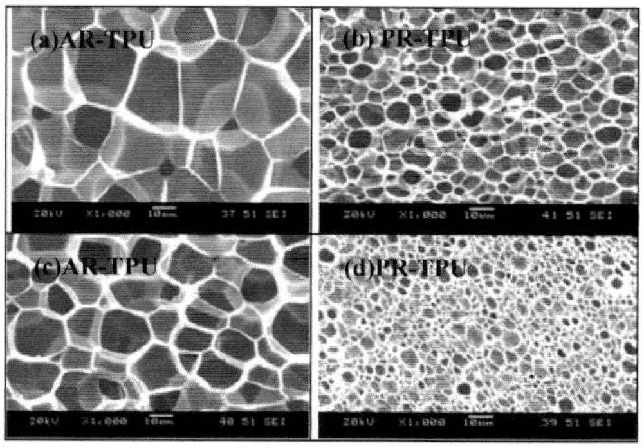

Fig. 6. Foam morphologies of AR-TPU and PR-TPU at 165°C. (a), (b) 55 bar butane; (c), (d) 103 bar butane.

The high cell density, or number of bubbles per unit volume, achieved in microcellular foams is largely determined by nucleation rate. Some of the popular strategies to increase the nucleation rate is by adding particle impurities or nucleating agents. In the presence of particles, bubbles form at polymer-particle interface following heterogeneous nucleation. The Gibbs free energy for heterogeneous nucleation is given by [8],

$$\Delta G^*_{het} = \frac{16\pi\gamma_{lg}^3}{3(P_{bub,cr}-P_{sys})^2} F(\theta_c, \beta) \qquad \text{Eq. 2}$$

$F(\theta_c, \beta)$ is the geometrical factor that relates to the surface geometry of the nucleating agents, θ_c is the contact angle between the bubble surface and the solid surface measured in the liquid phase and β is the semi-conical angle of conical cavities that models the irregular surface of nucleating agents. Hence presence of an interface greatly reduces the free energy for nucleation leading to significant increase in nucleation rate. Recently some researchers have proposed and demonstrated the local system pressure variation to significantly decrease the critical radius for nucleation and increase the nucleation rate. To take into account this pressure fluctuation, the Gibbs free energy for heterogeneous nucleation can be rewritten as [9],

$$\Delta G_{het}^{*} = \frac{16\pi\gamma_{lg}^{3} F(\theta_c,\beta)}{3\left(P_{bub,cr}-(P_{sys}+\Delta P_{local})\right)^2} \quad \text{Eq. 3}$$

Where ΔP_{local} is the difference between P_{sys} in the bulk and the local system pressure. The sorption and diffusion of butane takes place only in the SS regions, the phase separated HS domains and HS crystallites do not take part in this process [10]. The critical point of butane is 152°C and 37.96 bar. At all the investigated foaming conditions, butane is in its supercritical state. The plasticizing effect of butane occurs in pronounced way in its supercritical state. The plasticization of butane is characterized by swelling of TPU and hence increase in SS mobility which may allow the longer HS to move and rearrange into kinetically favored stacking, thus forming crystallites. On the other hand, the isolated HS lying in SS may stack with shorter HS to increase the length of HS and stack into crystallites. The soft segment (SS) phase in vicinity of perfecting HS domains undergo extensional strain causing local system pressure variation and hence significant increase in the nucleation rate. Furthermore, melt processing of TPU causes breaking of HS into broad segment length. Upon cooling the readily crystallizable HS lengths tend to stack to form crystals. However the stacking is not neat giving rise to conformational disorder and formation of crystals with non-uniform thickness [11]. The nucleation rate in PR-TPU may further be significantly increased by reduction in ΔG_{het}^{*} caused by the semi-conical angle β of the HS crystals that have stacked to form irregular crystals from non-uniform thickness. The effect of these crystals as heterogeneous nucleating sites is obvious in foams prepared at 170°C as seen in Fig. 7 and Fig. 8. AR-TPU shows decrease in cell density and increase in cell size. On the other hand, PR-TPU shows microcellular morphology with cell size of approximately 10 μm and cell density of greater than 10^9 cells/cm^3.

Conclusions

This study presents a novel technique of utilizing phase separated crystalline and non-crystalline HSs in TPU microstructure as heterogeneous nucleation sites to produce microcellular foams. Comparing the foam characteristics of AR-TPU and PR-TPU, it was observed that PR-TPU exhibited remarkable increase in cell density and reduction in cell size over wide temperature ranges between 150 °C and 170°C at a moderate pressure of 55 bar without addition of any nucleating agents and/or filler. Finally, microcellular morphology was achieved in presence of butane, which poses significant thermodynamic challenge in processing of such morphology.

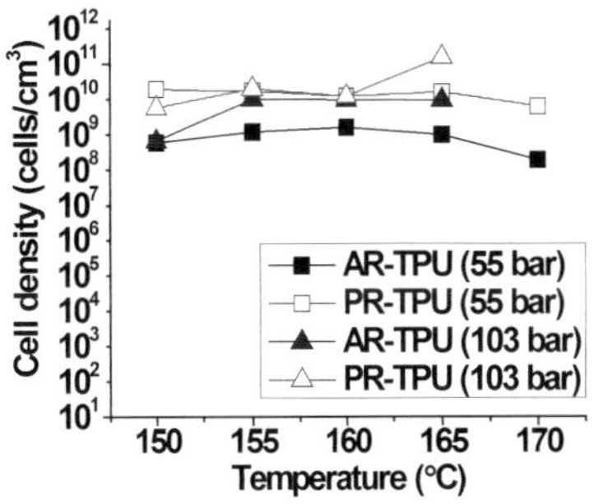

Fig. 7. Cell densities of AR-TPU and PR-TPU foams

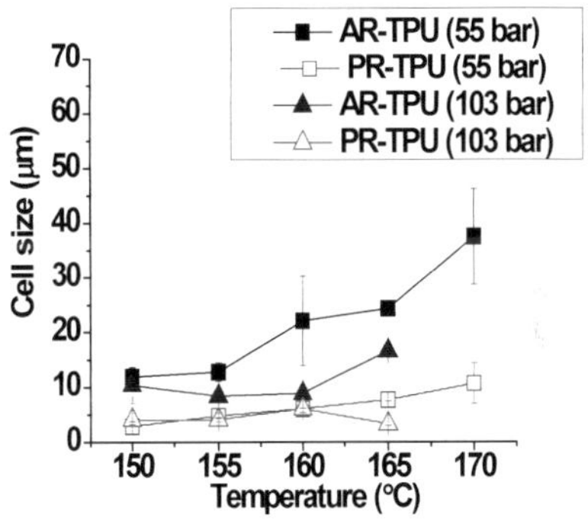

Fig. 8. Average cell size of AR-TPU and PR-TPU foams

References

1. S.L.Cooper; A.V.Tobolsky *J. Appl. Polym. Sci.* 1966, 10, 1837.
2. C.S.Schollenberger *Polyurethane Technol.* 1969, 197.
3. F. Prissok; F. Braun, U.S. Patent 0222442 A1, 2010.
4. X. Dai; Z. Liu; Y. Wang; G. Yang; J. Xu; B. Han *J. Supercrit. Fluids* 2005, 33, 259.
5. R.W. Seymour; S.L.Cooper *Macromolecules* 1973, 6,48.
6. D.I. Collias; D.G. Baird *Polym. Eng. Sci.* 1995, 35(14), 1178.
7. T.R. Hesketh; J.W. Van Bogart; S.L. Cooper *Polym. Eng. Sci.* 1980, 20, 190.
8. S.N. Leung; C.B. Park; H. Li *Plast. Rub. & Comp.: Macromol. Eng.* 2006, 35, 93.
9. A. Wong; R.K.M. Chu; S.N. Leung; C.B. Park; J.H. Zong *Chem. Eng. Sci.* 2011, 66, 55.
10. K. D. Zeigel *J. Macromol. Sci.* 1971, B5, 11.
11. C.D. Eisenbach; H. Nefzger in *Multiphase molecular systems, contemporary topics*, W.M. Cumberston Ed.; Plenium Publ Corp, New York, 1989; Vol. 6, 339-61.

O-03-201

EFFECTS OF EXTENSION PARAMETERS ON THE STRUCTURE OF POLYPROPYLENE FILM

Liang-Yu Chen[1], Chung-Cheng Su[1] and Ren-Haw Chen[1]*

[1] *Department of Mechanical Engineering, National Chiao Tung University, 1001 Ta Hsueh Road, Hsinchu City, Taiwan, R. O. C. – lychen.me99g@nctu.edu.tw; razecities@hotmail.com; chenrh@mail.nctu.edu.tw*

Abstract - Developing the interior structure of a polymer film is vital for creating novel functions of such a film. This study investigates the effects of extension parameters on the super-structure of a polypropylene (PP) film. An extension device is also designed and fabricated to analyze how annealing, extension temperature, extension rate, processing procedures and other parameters influence the super-structures of PP films. Based on super-structure analyses, the relationships between extension parameters and film structures are determined. Experimental results indicate that low extension ratio treatment before total extension process promotes crystallization during processing. In a film with low crystallinity, spindle-shaped microporous structures can be generated by extension on crystalline and amorphous interfaces. As for extension parameters, although a low extension temperature can increase the number of microporous structures generated at a low extension ratio, structure sizes are dispersed. Additionally, although high extension temperatures obtain uniformly sized micropores, films require a high extension ratio to promote micropore formation. Based on the experimental results, this study also describes how parameters influence the micropores of a PP film created during the extension process. Analysis results further demonstrate how extension parameters can be adjusted to improve the fabrication of PP films.

Keywords: Film extension, microporous film, extension parameter

Introduction

Advances in thin film materials have extended thin film applications to a diverse array of industries. In particular, fulfilling the different needs of functional polymer films has received considerable recently. As a major component of liquid crystal display monitors and organic solar cells, must either provide unique optical properties or block specific particles. These usages have also led to increasingly stringent requirements regarding quality and performance of the film, which still fails entirely satisfy requirements of an economic manufacturing method.

As an excellent material for producing films, polypropylene (PP) has a higher melting point, lower density, higher chemical resistance, and better mechanical properties than polyethylene, another widely used semicrystalline polymer. Such advantages make PP an appropriate experimental material.

Conventional film forming processes can be classified into solution casting and extrusion molding. In the former, films are formed with a polymer solution or melted polymer onto a mold. Although the film structure can be determined with mold design, time-related costs and inflexible equipment design are the major obstacles for reducing overhead cost for mass production. Extrusion molding method is characterized by fewer procedures and lower complexity than the former method for continuous production, making it better in terms of lower overhead costs for mass production.

After the extending process, several changes occur on a polymer film, including crystalline properties, optical properties and structure. Extension parameters such as temperature, extension ratio, annealing, strain rate and processing procedures affect these characteristics. Engeln *et al.* [1] investigated the uniaxial extension of polytetrafluoroethene (PTFE). According to their results, a higher temperature decreases Young's modulus. Also, an increasing extension ratio increases implies an increasing birefringence and molecular regularity as well. Yamada and Nonomura considered how preheating, extension, cooling, and thermosetting procedures influence film extension. Their results indicated that additional cooling and thermosetting procedures during the process reduce bowing distortion in the stretching of polyethylene terephthalate (PET) film [2]. Fujita *et al.* analyzed the molecular orientation behaviors of PET film during transverse direction (TD) stretching process. Results involving molecular orientation behavior were also related to transverse direction stretching and a stable heat source [3]. By using small-angle X-ray scattering (SAXS), Ishihara *et al.* observed PET film after stretching. Their results suggested that increased density caused by oriental crystallization is strongly related to the micropores generated by structural changes. Another study found that temperature control and stretching method determine the super-structure of a film [4]. Funaki *et al.* transformed the PP film from smectic structure to α-monoclinic crystal phase by rapid quenching. The density and refractive index differences between spherulite and matrix were decreased, and the transparency was drastically improved [5].

Experimental

Materials

The polypropylene film selected in this study was formed by an inflation method, based on three PP varieties with different molecular weights. The thickness of film is 0.13mm. The film is transparent. According to the thermogravimetric analyzer (TGA), pyrolysis temperature of the film is around 429°C. Based on differential scanning calorimetry (DSC), three glass transition temperatures (T_g) (-3.50°C, -4.73°C, 55.64°C) and three melting points (T_m) (157.46°C, 163.87°C, 164.84°C) were found. Films were cut into rectangular test samples with a length of 500mm and width of 50mm along machine direction. An array of dots with 10mm intervals was marked on the film surface.

Extension device

To simulate commercial production this study analyzed how annealing, extension temperature, extension rate, processing procedures and other parameters influence the super-structures of PP films by designing and fabricating an extension device. Fig. 1 schematically depicts the device design. Samples in this study obtain a uniaxial extension after heat treatment by a heater plate to specific temperature. Distance between the axis of rollers is 370mm. Extension strain rate and temperature could be adjusted according to different requirements.

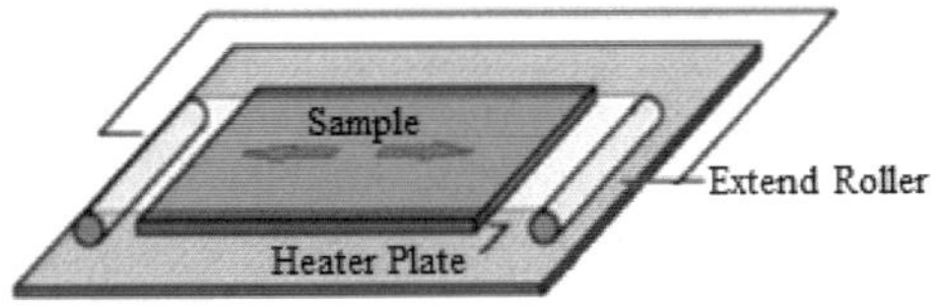

Figure 1 – Schematic design of extension device.

Polarized light examination

An attempt was also made to determine how extension parameters affect the film structure by a ZEISS Axioskop 40 polarized optical microscope (POM) with objective lenses at 5X, 10X, 20X, 50X, 100X magnifications, and ocular lens at 10X magnification. Micrographs were taken with a CCD camera connected to the microscope.

Test projects of extension parameters

This study examined four extension parameters, extension procedure, extension ratio, extension temperature, and extension strain rate. Each operation consisted of 30 min preheatment to form samples at a uniform temperature. Notably, extending direction is the same as original machine direction (MD) of the film.

Extension procedure change: Extension temperature was set to 135°C and the extension strain rate was set at $0.014s^{-1}$. One of the processes was extended to 1.2X extension ratio before extend to 5 times the extension ratio. Another one extended directly to 5X extension ratio.

Extension ratio change: Extension strain rate was set at $0.014s^{-1}$ and the extension temperature was set at 135°C. Extension ratios were set at a range of 3X to 4X, 5X, 6X and 7X.

Extension temperature change: Extension strain rate was set at $0.014s^{-1}$, and the extension ratio was set at 5X. Extension temperatures were set at a range of 55°C to 75°C, 95°C, 115°C and 135°C.

Extension strain rate change: Extension temperature was set at 135°C and extension ratio was set at 5X. Extension strain rates were set at a range of $0.003s^{-1}$ to $0.007s^{-1}$, $0.011s^{-1}$, $0.014s^{-1}$ and $0.018s^{-1}$.

Results and Discussion

Extension procedure change

According to the observation of POM, the film lacks a crystalline structure. However, film with a slight extension has few crystalline structures. When both samples extended to 5X extension ratio, the one with slight extension has a lower transparency appearance, and a more spindle-shaped microporous structure (Fig. 2).

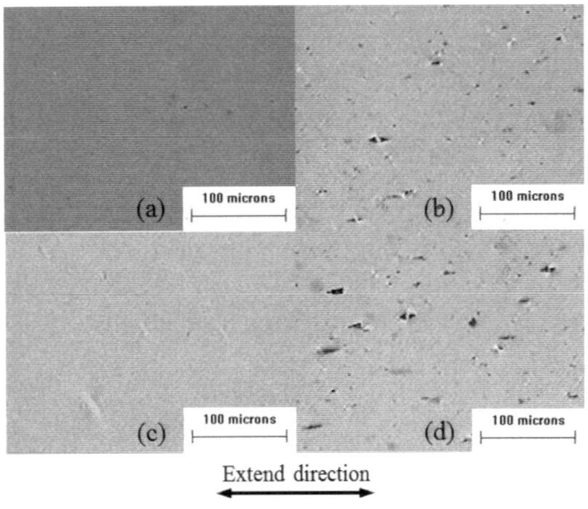

Figure 2 – POM micrographs of extended films through different extension processes. (a) before extension (b) directly extended to 5X (c) extended to 1.2X (d) extended to 5X after extended to 1.2X.

Extension ratio change

During the extension process of the thin films, a higher extension ratio leads to a thinner thickness; in addition, films turn white. POM micrographs indicate that the super-structure of films significantly changes (Fig. 3). According to this figure, quantity of spindle-shaped micropores increased with an increasing higher extension ratio. Major axis lengths of spindle-shaped micropores are around 15~18μm when extension ratio at 3X. As the ratio expands, the average length increases, the crystalline structure in the micropores begins to break, and the film exhibits fibrous characteristics.

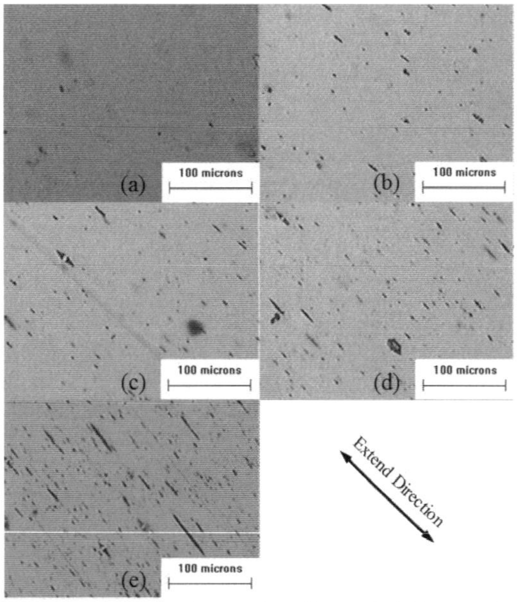

Figure 3 - POM micrographs of extended films under different extension ratios (a)3X (b)4X (c)5X (d)6X (e)7X.

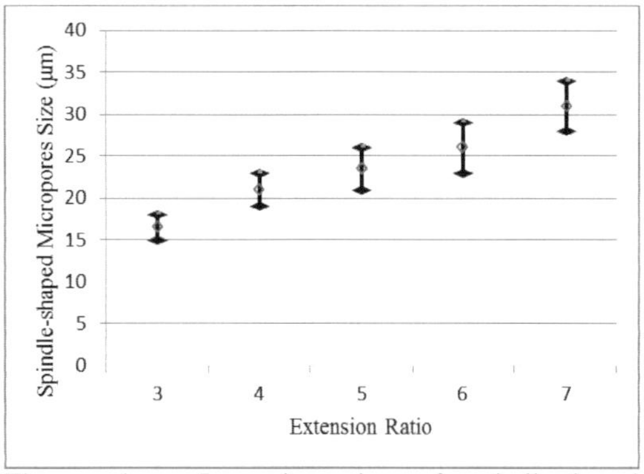

Figure 4 – Increasing size of spindle-shaped micropores as the extension ratio changes

Extension temperature change

Deformation of film extended at a lower temperature is more irregular than a higher temperature. Samples tend to expand at certain point. In contrast, a film extended at higher temperature has a more rubber-like characteristic, in which the film tends to extend uniformly. According to Fig. 5, the major axis lengths of spindle-shaped micropores are around 14~30μm at 55°C. At a higher extension temperature, sizes of micropores are increasingly closer, and fewer broken crystalline structure appear (Fig. 6).

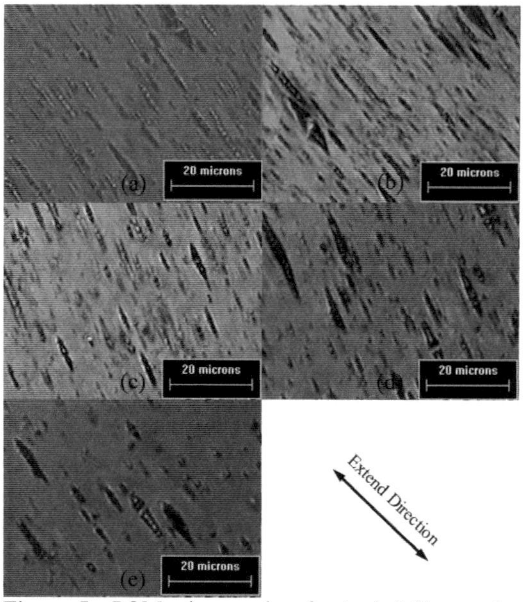

Figure 5 - POM micrographs of extended films under different extension temperatures. (a)55°C (b)75°C (c)95°C (d)115°C (e)135°C.

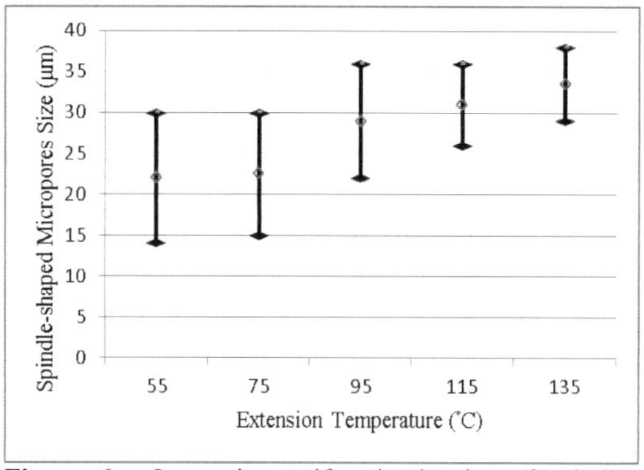

Figure 6 – Increasing uniformity in size of spindle-shaped micropores as the extension temperature increases.

Extension strain rate change

Even under different extension strain rates, thicknesses of samples are all around 61μm, and transparencies resemble each other. This information reveals that films with different extension strain rates do not significantly differ from each other. POM micrographs (Fig. 7) of each sample resemble each other, regardless of the change in extension strain rate. Major axis lengths of spindle-shaped micropores are all around 19~27μm (Fig. 8). Differences between quantities of broken crystalline structure in the same area also resemble each other. This finding reveals that the extension strain rate has a limited influence on the film structure.

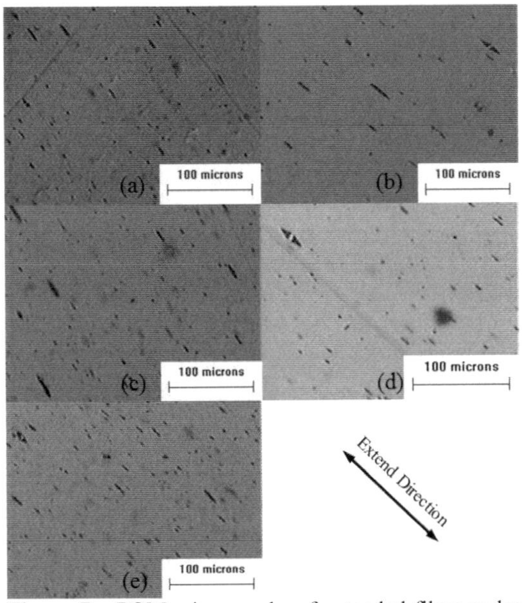

Figure 7 – POM micrographs of extended films under different extension strain rates. (a)$0.003s^{-1}$ (b)$0.007s^{-1}$ (c)$0.011s^{-1}$ (d)$0.014s^{-1}$ (e) $0.018s^{-1}$.

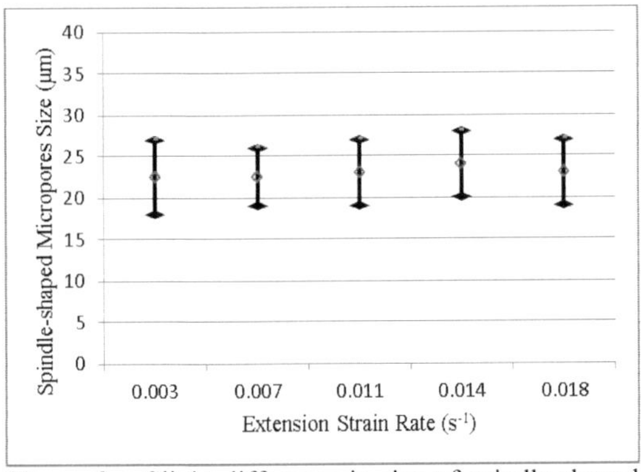

Figure 8 – Slight difference in size of spindle-shaped micropores as the extension strain rate changes.

Conclusions

This study investigates how extension procedure, extension ratio, extension temperature and extension strain rate affect the super-structure of polypropylene film. Based on the experimental results, several trends of extension parameter settings are identified, as summarized in the following:

- Low extension ratio treatment before total extension process promotes crystallization during processing, as well as causes more spindle-shaped microporous structures.
- For a spindle-shaped micropore generated in polypropylene film with low crystallinity, crystal structure can be found at its center, with the amorphous phase surrounded. This finding suggests that spindle-shaped microporous structures can be generated by extension on crystalline and amorphous interfaces.
- Regarding extension parameters, a low extension temperature can increase the number of microporous structures generated at a low extension ratio; however, sizes are dispersed. Although high extension temperatures obtain micropores with a uniform size, films require a high extension ratio to promote micropore formation.

Acknowledgements

The authors would like to thank the National Science Council of the Republic of China, Taiwan, for financially supporting this research under Contract No. NSC 99-2221-E-009-022-MY2. The R&D Center of Membrane Technology, Chung Yuan University is gratefully acknowledged for its information support.

References

1. I. Engeln; R. Hengl; G. Hinrichsen, *Colloid & Polymer Science,* 1984, 262, 780.
2. T. Yamada; C. Nonomura, *Journal of Applied Polymer Science,* 1994, 52, 1393.
3. S. Fujita; C. Nonomura; H. Ishihara; T. Yamada, *Seikei-Kakou,* 1996, 8, 605.
4. H. Ishihara; M. Shibaya; N. Yoshihara, *Journal of the Society of Rheology,* 2005, 33, 161.
5. A. Funaki; T. Kanai; K. Kondo; H. Odaka in PPS 27 Proceedings, Marrakech, 2011, CD.

O-03-034

NANOSTRUCTURED ANTI-REFLECTION FILMS FOR ENHANCEMENT OF SOLAR CELLS PERFORMANCE PREPARED USING PLASMA SPUTTERING AND INFRARED ASSISTED ROLLER EMBOSSING TECHNIQUES

Shih-Jung Liu* and Che-Ting Liao
Department of Mechanical Engineering
Chang Gung University, Tao-Yuan 333, Taiwan
**Corresponding author: shihjung@mail.cgu.edu.tw*

Abstract - This paper reports the continuous fabrication of dual-side nano-structured anti-reflection protective layer for performance enhancement of solar cells using plasma sputtering and infrared assisted roller embossing techniques. Nano-structures were first deposited onto the surface of glass substrates using the plasma sputtering technique. After electroforming, a nickel master mold containing nano-array of 30 nm was obtained. The mold was then attached to the surfaces of the two metallic rollers in an infrared assisted roll-to-roll embossing facility. The embossing facility was used to replicate the nano-structures onto 60 □m thick polyethylene terephthalate (PET) films in the experiments. The embossed films were characterized using UV–vis spectrophotometer, atomic force microscope (AFM), and scanning electron microscope (SEM); its total conversion efficiency for solar cells was also measured by a solar simulator. The experimental results showed that the fabricated films could effectively reduce the reflectance and increase the conversion efficiency of solar cells.

Introduction

Photovoltaic device has gradually become an important device for a new source of power generation. To increase the conversion efficiency of solar energy, the reduction of reflection at the surface of a concentrated photovoltaic device is required [1]. An anti-reflection protective layer, i.e., a coating applied to the surface of a material to reduce light reflection and to increase light transmission, is highly desirable. This study reports a simple and effective infrared assisted hot roller embossing process for continuous fabrication of dual-side nano-structured anti-reflection films for the protective layer of solar cells.

Experimental

The material used in this study was polyethylene terephthalate (PET) films with a thickness of 60 μm.The plasma sputtering experiment was completed on a commercially available sputtering machine (Sputter Demos, Kao Duen Tech., Taiwan) equipped with a gas mass flow control (MFC KD-1000, Kao Duen Tech., Taiwan), a vacuum gauge controller (Terranova Model 934, Duniway Stockroom Corp, Canada), and a DC current plasma generator (MDX 500, Advanced Energy, U.S.A). The sputtering gas used was Argon (Ar), while an aluminum plate was used as the target. During sputtering, the gas ions pass the energy to the target atoms, knocking them out and generating film growth on the glass substrate. Various processing conditions were employed to identify the optimal processing condition. The shape, height, and width of the microstructures on the glass substrates were inspected using scanning electronic microscopy (FEG-SEM 6500F, Jeol Ltd., U.S.A.). The results suggested that glass substrate, which was plasma sputtered by using condition G (i.e., a power of 230 watts, a substrate rotating speed of 6 rpm, an Ar flow rate of 25 sccm, and a sputtering time of 60 mins), could have uniformly distributed nano-structures of 30 nm in diameter. It was then selected for the subsequent procedure. After electroforming, Ni master molds bearing nano-arrays of 30 nm in size were obtained as shown microscopically in Figure 1a. The mold was characterized using an atomic force microscope (AFM 5100, Agilent Tech., U.S.A.), and the measured result in Figure 1b suggests that the surface roughness (Ra) was 21.5 nm.

Results and Discussion

An infrared (IR) assisted roll-to-roll embossing facility, which includes two metallic rollers and two IR heaters with temperature controls, was designed and built in our lab [2]. Figure 2a shows photographically the setup of the IR assisted roll-to-roll embossing facility and the embossing rollers. The power of the IR-heater (Elstein Model FRS/2, Germany) was 500 watts and the highest working temperature of the heater was 750oC. To emboss microstructures on two sides of the plastic films, the Ni mold bearing nanostructures were attached onto the surfaces of the rollers by glues. During roller embossing, the metallic rollers are driven by a servo-motor and heated by the IR heaters before coming into contact with the plastic plate. The distance between the heater and the roller surface (Figure 2b) can be adjusted by the Z-stage, located above and beneath the top and bottom heaters respectively. A few test trials were first carried out to find out the correlation between the emitting temperatures of the IR heater and the temperature at the roller surface. The relevant emitting temperature of the heater was then set to obtain the desired surface temperature of the Ni molds. After the plastic films came into contact with the rollers, the top roller was pushed against the plastic film and the bottom roller by the pneumatic actuator located at top of the embossing facility. The infrared energy was converted into heat at the rollers' surface and was sufficiently high enough to melt thermoplastics at the roller/plastic film interfaces and cause the melt to flow and conform to the microstructures. Anti-reflection films with two-sided nanostructures are thus obtained [3].

To further verify and inspect the optical properties of the fabricated anti-reflection films, an UV/VIS spectrometer with Integration Sphere (V-650, Jasco Co., U.S.A.) was employed, while unpolarized UV-visible light was used as the light source. Figure 3 shows the transmittance and reflectance of the flat and nanostructured PET films. While the solar cell without the nanostructures showed 7.4% of average reflectance in the wavelength range of 300 to 700 nm, the average reflectance of the nanostructrured solar cell decreased to 4.4% in the same range. Furthermore, at the wavelength of 630 nm, the transmittance increased from 89.7% to 93.5% with the nanostructured pattern. The measured result suggests that the overall performance of the films with nanostructures is very much improved when compared to that of the flat films. Nanostructur-patterned films can serve as effective anti-reflective layers over the 300-700 nm spectrum range.

To further inspect the enhancement capacity of the anti-reflection films, photocurrent-voltage measurements were performed using a solar simulator (YSS-50A, Yamashita Denso, Japan) equipped with a xenon lamp. Figure 4 shows the measured I–V characteristic of the solar cells with and without the nanostructured patterns. While the flat PET film shows 1.222V of open circuit voltage (Voc), 65.3% of fill factor (FF), 2.46 mA of short circuit current (Isc), and 6.5% of efficiency, the solar cell with nanostructures shows 1.224V, 71.8%, 2.39 mA, and 7.2% of Voc, FF, Isc and efficiency, respectively. The plastic film with nanostructures displays better energy conversion efficiency than the flat film. The results demonstrate that the fabricated anti-reflection films can effectively enhance the performance of solar cells.

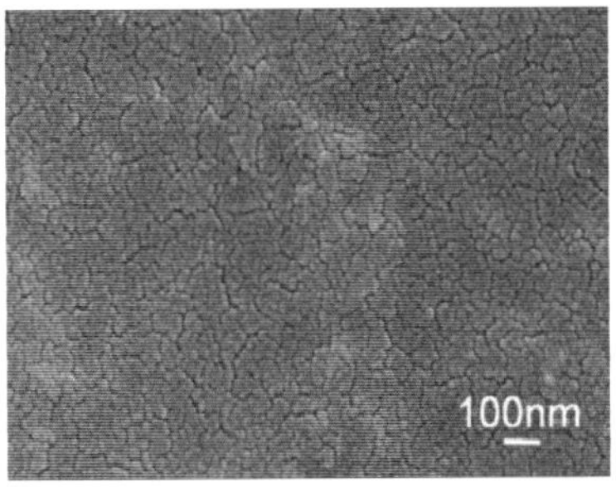

(a)

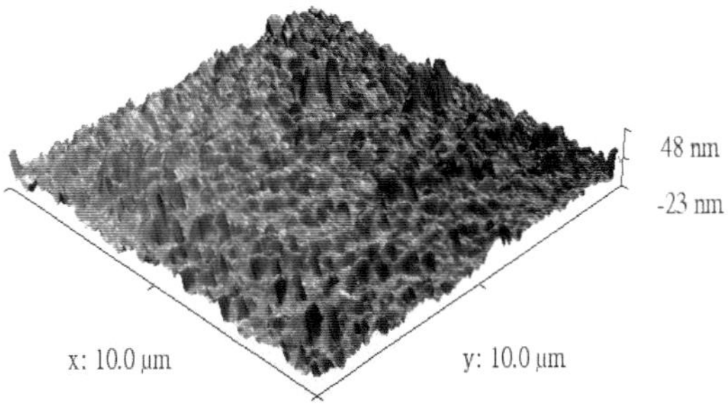

(b)

Figure 1 –(a) SEM, and (b) AFM photos of the Ni mold (Ra=21.5 nm)

Figure 2 –The embossing rollers.

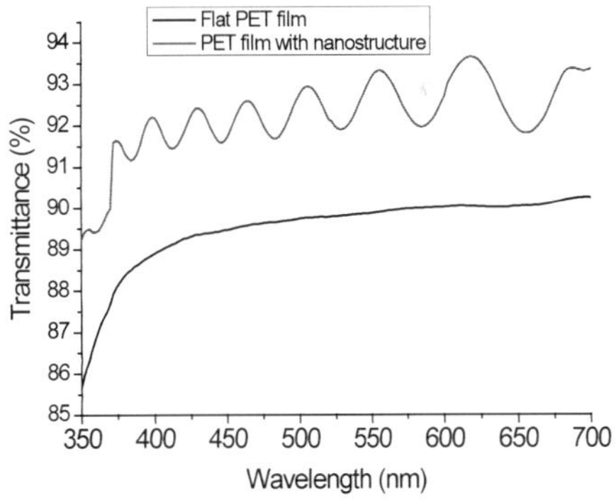

(a)

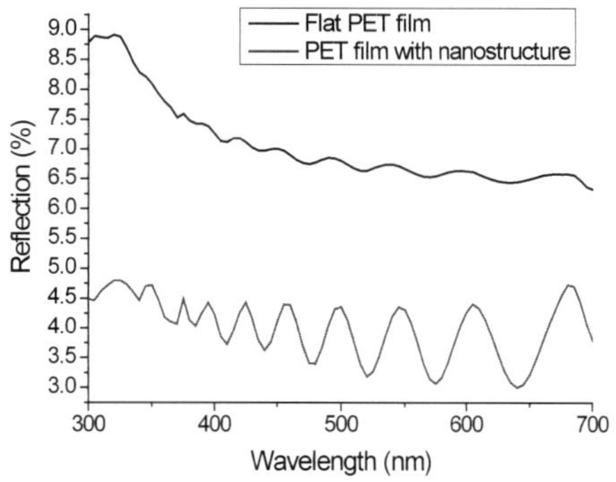

(b)

Figure 3 –Transmittance and reflectance of the solar cell comparing the flat layer and the nanostructured layer.

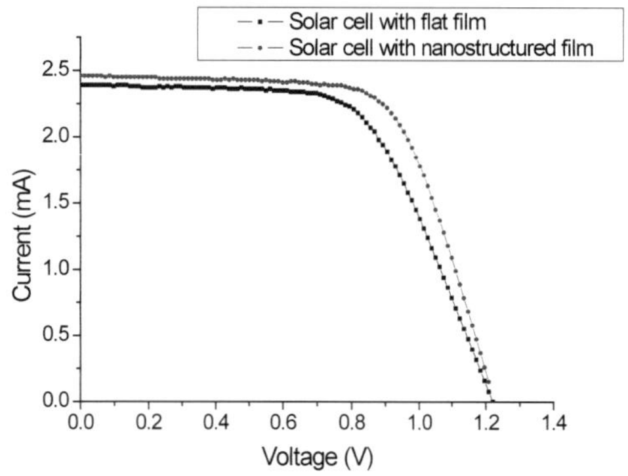

Figure 4 –I-V characteristics of the solar cell with and without nanostructured patterns.

Conclusions

This paper discussed an innovative and effective IR assisted roll-to-roll embossing method for the rapid fabrication of dual-side nanostructured anti-reflection films for performance enhancement of solar cells. The proposed method shows great potential for fast fabrication of the anti-reflection protective layer of solar cells due to its simplicity and versatility.

Acknowledgements

The National Science Council of Taiwan, R.O.C. under the grant NSC100-2221-E-182-020-MY3, has supported this work financially.

References

1. L. El Chaar, L.A. Lamont, N. El Zein, "Review of photovoltaic technologies," Renew. Sustain. Energy Rev. **15**, 2165-2175 (2011)
2. J. Zhao, Martin A. Green, "Optimized antireflection coatings for high-efficiency silicon solar cells," IEEE Trans. Electron Dev. 38, 1925-1934 (1991).
3. S.J. Liu, C.T. Liao, , "Fast fabrication of nanostructured anti-reflection layers for enhancement of solar cells performance using plasma sputtering and infrared assisted roller embossing techniques," Optics Express. 20, 5143-5150 (2012).

O-03-364

ON THE DEVELOPMENT OF NOVEL PIEZOELECTRIC FIBRE SENSOR

Erik Nilsson[1,3*], Anja Lund[1,2], Christian Jonasson[4], Christer Johansson[4], Bengt Hagström[2,3]
[1] *Dept. Materials and Manufacturing Technology, Chalmers University of Technology, SE-412 96 Göteborg, Sweden*
[2] *The Swedish School of Textiles, University of Borås, SE-501 90 Borås, Sweden*
[3] *Swerea IVF, Textiles and Plastics Department, Box 104, SE-431 22 Mölndal, Sweden*
[4] *ACREO, Arvid Hedvalls Backe 4, SE-400 14 Göteborg, Sweden*
***Corresponding Author**: Erik Nilsson erik.nilsson@swerea.se*

Abstract
The fibre is of a bicomponent core/sheath structure, with poly(vinylidene fluoride) (PVDF) as the piezoelectric sheath component and an electrically conductive compound of carbon black (CB) and high density polyethylene (HDPE) as core material. Since late 1960 the piezoelectric properties of PVDF has been known. Today there are a number of commercial products where piezoelectric films is being used as sensor. However little or none knowledge in the field of piezoelectric fibres produced from PVDF. With our concept where fibres is used as a sensor, renders new possibilities to have sensors totally intergraded in the textiles. The development of such fibres is not without difficulties. Formation of piezoelectric β-phase crystallites have been addressed and overcome. Challenges are to induce the orientation of the β-phase crystallites by poling at a high electrical field, and application of electrodes. Different techniques for poling are discussed and characterization of piezoelectric properties is done. Characterization of the fibres in a MTS machine shows that exposed to sinusoidal wave at 0.07% strain a clear response of 4 V is given as output signal from the fibres. In the present study piezoelectric fibres of bicomponent fibres are weaved to a fabric. Such a fabric if strapped around the chest of a person, the fibres can register medical parameters such as pulse and breathing.

Keywords: Piezoelectric, Melt spinning, bi-component fibres

Introduction

The development of lightweight piezoelectric textiles is described. Here we present a novel piezoelectric fibre that can be fully integrated in a textile, either as a tiny single fibre sensor or large area sensor. In previous studies the formation of polar ß-phase crystal and production of the fibres have been described. [1, 2] In the present study methods for poling of piezoelectric fibres is evaluated. Characterization is carried out by measuring the piezoelectric response in axial tension.

PVDF

Its piezoelectric properties were discovered in 1969 [3, 4] and some 20 years later, piezoelectric PVDF films were commercially available. PVDF [-CH_2-CF_2-] can crystallize in at least four different phases known as α, β, γ, and δ. Crystallization from the melt α –phase is formed which has a monoclinic unit cell and the chains are of a TGTG conformation. When been cold drawn α –phase transforms to the polar ß-crystals now in all TTTT conformation which are randomly oriented in the amorphous matrix. To render the piezoelectric properties the ß-crystals has to be oriented in an electrical field in a process called poling.

Poling of PVDF

By applying a sufficiently strong electrical field, the dipoles will orient in the field direction. Extensive research has been done on poling of PVDF films. Up until recently most of the studies on fibres have been concerning electro spun fibres.[5, 6] Melt spun monofilaments in a epoxy matrix has been studied by Walter et al. and Lund et al. have studied the formation of ß-crystals in melt spun fibres. [1, 2, 7] Two different methods are commonly used for poling of PVDF i) In contact poling electrodes are applied on both sides of the PVDF subjecting the PVDF to field. Typical values are approximately 80-300 MV/m.[8] ii) The second poling method is corona poling. Corona discharges occur when a sufficiently high voltage between two asymmetric electrodes, a needle tip and a plate is applied. (Surface of fibre in our case). The discharges will create ions that will charge the surface of the fibres. A field is induced between the surface of the fibre and the core. In corona poling higher field strengths (400-800 MV/m) can be used. [9] When higher field strength is used poling time and temperature can be reduced. [10-12]

To the best of our knowledge, poling of bicomponent PVDF fibres is not previously described in the literature. In this paper, different poling techniques and their effect on the piezoelectric characteristics of our bicomponent yarns are presented. Fibre is evaluated for the application as a sensor in a textile fabric.

Experimental

The PVDF was Solef 1006 from Solvay Solexis (Milan, Italy) According to the supplier, its melting point is 175°C and its density is 1780 kg/m^3. The core electrode was a high density polyethylene (HDPE)

ASPUN 6835A from Dow. MFI = 17 g/10 min at 190°C for 2.16 g, density = 950 kg/m^3 and melting point 129°C. Carbon black (CB) was Ketjenblack EC-600 JD from Akzo Nobel, Netherlands, with a specific gravity of 1800 kg/m^3, bulk density of 100-120 kg/m^3 and BET surface area 1400 m^2/g. All data presented here is according to the supplier.
Compounding of the core material (HDPE+10wt%CB) was done using a Coperion (Stuttgart, Germany) ZSK 26 K 10,6 twin screw extruder.

Fibre spinning
Bicomponent fiber spinning was performed using a melt spinning line from Extrusion Systems Limited (ESL, Leeds, England) equipped with two single screw extruders, one for the core material, and one for the sheath material, see Fig 1 Their temperature settings were identical: extruder zones 1, 2, and 3 were set to 190°C, 220°C, and 230°C. The temperature of the gear pumps and spinneret was set to 230°C. The spinneret used had 24 holes with a diameter of 0.6 mm and a land length of 1.2 mm. The relative rate of metering of polymer to the spinneret controls the relative amounts of core and sheath material in the fiber. Fibres were spun with MDR 4 and solid state draw ratio (SSDR) of 10. The diameter of the produced fibres is controlled by the draw ratio through the entire system.

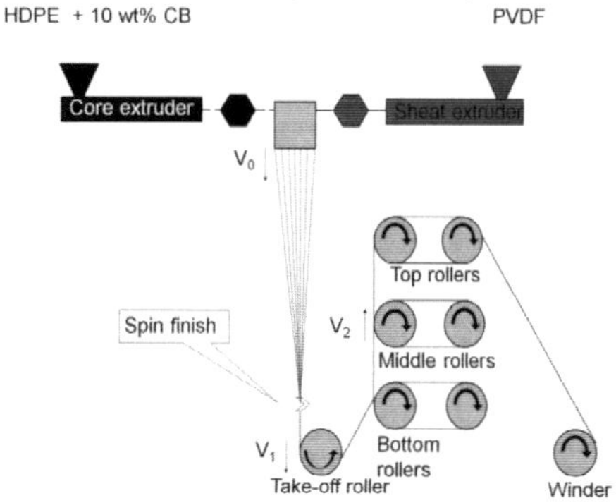

Figure 1 Schematics of the bi-component spinning line.

Contact poling
The sensor constitute of a yarn of 24 fibres, a distance of 25 mm the yarn was embedded in a conductive silver paint matrix, which constitute as outer electrode. Fibre ends were cut off with a razor blade to expose the core of the fibres, which serves as inner electrode. The fibre ends were immersed in conductive silver paint interconnecting all the fibres cores. Poling of the sensors was done altering poling voltage, time and temperature. During heat treatment the specimens were clamped to avoid fibre shrinkage and connected to a high voltage power supply (ES50P-10W) from Gamma High Voltage Research (Ormond Beach, FL). Poling at elevated temperature was done in a heated Heraeus oven. The sample was heated 3 minutes before the voltage was applied and cooled down to room temperature before the voltage was removed.

Corona poling
A point-to-plane corona discharge without a control grid was used to pole the PVDF yarn. Discharges from the needle tip result in charge built up on the PVDF fibre surface, an electrical field between the charged fibre surface and the grounded inner electrode is orienting β-phase crystallites in radial direction. A schematic of corona poling equipment is shown in Fig 2 A Poly(methyl methacrylate) pipe with interconnected needles in a circular pattern surrounding the fibre. A high voltage power supply (ES50R-10W) from Gamma High Voltage Research (Ormond Beach, FL) was connected to the needles and grounded to fibres core electrode. Core of the fibres were interconnected as described above. A high negative corona voltage was applied between tip of the needles and the grounded cores.

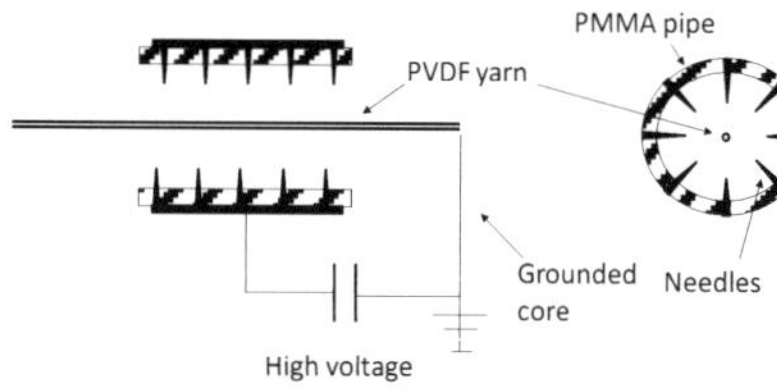

Figure 2 Equipment for corona poling.

Piezoelectric Characterization
For piezoelectric characterization, the sensor-yarn was subjected to a dynamic strain in axial fibre direction using a servo-hydraulic tensile testing machine, Model 66-21B-01 from MTS systems. The bias force was 2N. Samples were tested in oscillating mode, a sinusoidal strain was applied, initially the yarn was exposed to a frequency sweep to find the so called cut-of-frequency, for which the fibres output signal is unaffected by the electrical properties of the fibres and measurement equipment. Finally the output voltage from the fibres was recorded. The sensor electrodes were connected to a data acquisition device, NI DAQPad-6016, from National Instruments (Austin, TX) connected to a computer running a LabVIEW Software, which controlled the measurement. The piezoelectric signals were slightly filtrated to reduce high frequency noise.

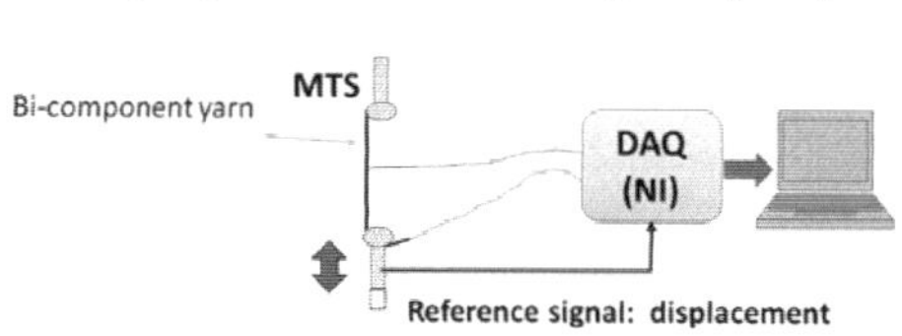

Figure 3 Characterization of fibres, output voltage from the fibres was recorded on a computer.

In addition, an analog signal from the MTS machine proportional to the strain in the sampled. As shown in Fig 3.

Results and Discussion

Fibres used in this study were characterized by XRD in a previous study. [1] In which it was shown that the fibre polar β phase is present after cold drawing.

Piezoelectric properties

In Fig 4 a yarn exposed to a sinusoidal wave at 4 Hz crating a stain in the fibre of 0.07%. Such a strong signal at this low stain is promising for using the fibres to detect even small movements or vibrations. If the amplitude of the strain is reduced 100 times to 0.0007% would create a signal of 0.04 V. So it is believed that small deformation in the human body for examples breathing and even detect the small deformation during a heartbeat could be detected. For comparison reason an un-poled sensor was tested in the MTS machine showing no evidence on piezoelectric response.

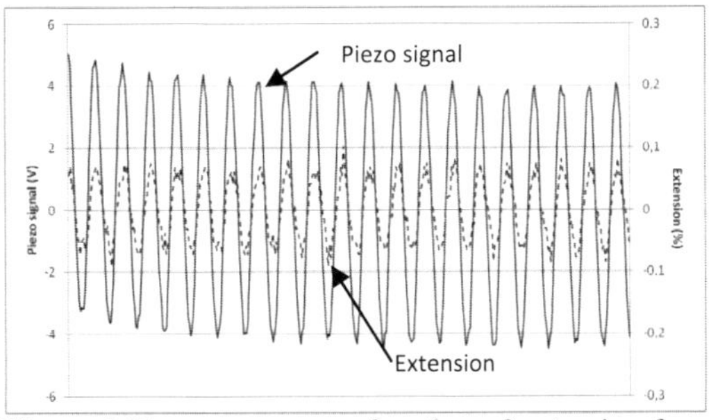

Figure 4 Piezo response as function of extension for a yarn.

The in Fig 4 measured voltage is frequently called open-circuit output voltage V_0 and is given by [13]:

$V_O = g\sigma t$ (Eq. 1)

Where g is the piezoelectric coefficient, σ is the applied stress and t is the PVDF thickness. From Eq. 1 we get:

$g_{31} = VA/Ft$ (Eq. 2)

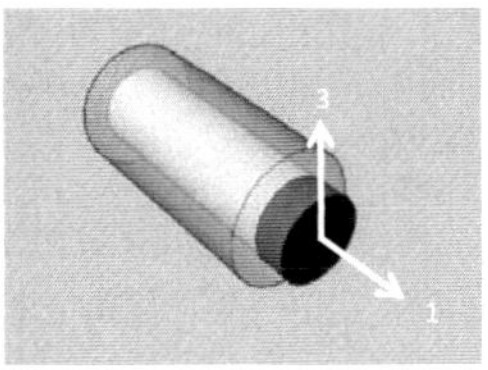

Figure 5 Classification of the axis of the fibre.

Where F is the applied force in axial direction and A is the area of the yarn to which F is applied. The subscripts to the coefficient g indicate that the electrical axis in these measurements is the thickness axis (3), while mechanical deformation was applied along the length axis (1). Consequently, σ will equal F divided by the cross-sectional area of the fibres (assuming that mechanical properties of PVDF are equivalent to that of the core compound, we use the whole cross-sectional area for calculations instead of just the PVDF-area). Each yarn/sensor has 24 fibres with a radius of approximately 30 μm. The output voltage is given from Fig 4. Using Eq. 2 piezoelectric coefficient value g_{31}= 313 $mVmN^{-1}$

Contact poling of fibres

Fibres characterized in Fig 4 were prepared by contact poling at 100 MV/m for 10 minutes at 120°C. Due to the simplicity of the contact poling it was used to study how temperature, time and poling voltage influence the piezoelectric properties. I Fig. 6 it can be seen how these three parameters influence the piezoelectric properties of the fibres.

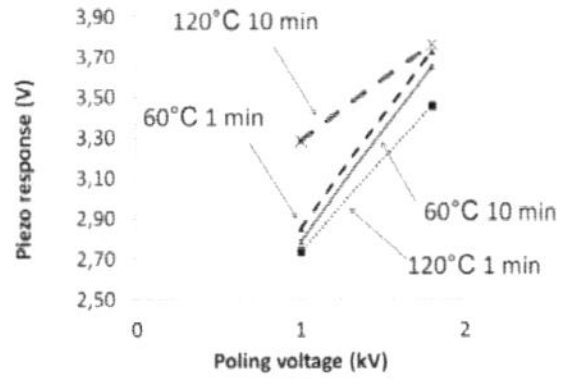

Figure 6 Piezo response as function of poling voltage.

In Fig 6 it can be seen that a high poling voltage enhance the piezoelectric response, no matter if poling is carried out at 60 °C or 120 °C. The thickness of the fibre is 60 μm and the PVDF sheath is 17μm thick, applying a 1 kV voltage to the fibre creates a 60 MV/m field over the sheath material, sufficiently high to permanently orient the dipoles in the material. However, from Fig 6 it can be seen that a 60 MV/m field is not strong enough to fully orient the dipoles. The piezo response is enhanced by raising the temperature from 60°C to 120°C but also extending the time from 1 to 10 minutes. This shows how the time and temperature interacts. When the poling voltage is increased to 1.8 kV time and temperature becomes less important.

We have found that there is some limitation in using contact poling for fibres. As consequence of increasing the number of fibres, in example a textile fabric, the probability of electrical breakdown becomes a problem. Thickness fluctuation of the PVDF sheath caused by carbon black agglomerates in the core material is believed to be the reason. So for poling of many fibres in a textile fabric contact poling is not a suitable method.

Corona poling of fibres

The purpose to find the threshold of the poling voltage sufficient high to orient the polar β phase crystallites. It was shown, with the present equipment setup, that a voltage of -10 kV a measurable output signal of approximately 2 V is recorded from the fibres for poling times as short as 2 seconds. This can be seen in Fig 7 a threshold of the output voltage is reached already after 2 seconds further extending the poling time does not yield a higher piezoelectric effect in the fibres. Poling voltage could be increased up to -15 kV without electrical breakdown when poling time was 2

seconds. However, when the poling time is extended the probability of an electrical breakdown increases. The short poling time and also the fact that we have been able to pole large area of a textile is in favour for using corona poling.
Due to the experimental equipment contact poling of fibres at elevated temperatures has not been done. But for a given poling voltage, increasing the temperature will most likely it will enhance the piezoelectric response.

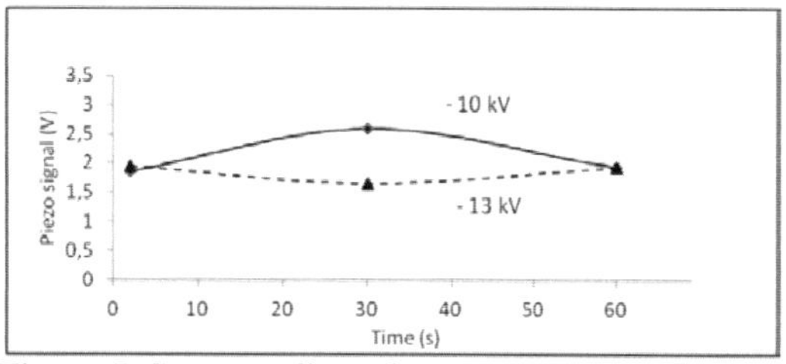

Figure 7 Corona poling of a yarn. Piezo response as function of poling time.

Corona poling of a textile fabric been done and the fabric show a high piezoelectric sensitivity. Strapped around the chest of a person it can detect the inhale and exhale.
It should be noted that the fibres is only sensitive to dynamic forces not static. So for application with low frequencies the signal is slowly diminishing. However, any change in deformation from the static mode will be obtained as a positive or negative voltage peak.

Conclusions

The technology of producing a piezoelectric polymer textile sensor has been demonstrated and evaluated in axial tension. Even deformation as small as 0.07% creates a strong piezoelectric signal of 4 V The sensor can be well suited been used to register heartbeat of a human without any external energy source. For single fibres or a yarn contact poling is a useful method. But for a textile consisting of hundreds or thousands fibres corona poling is proven to be the most suitable method.

Acknowledgements

This work was supported in part by the Swedish Foundation for Strategic Research (SSF) and Vinnova (The Swedish Governmental Agency for Innovation Systems), via the Smart Textiles initiative, for financial support.

References

1. Lund, A., et al., *Piezoelectric polymeric bicomponent fibers produced by melt spinning.* Journal of Applied Polymer Science, 2012: p. n/a-n/a.
2. Lund, A. and B. Hagström, *Melt spinning of beta-phase poly(vinylidene fluoride) yarns with and without a conductive core.* Journal of Applied Polymer Science, 2011. **120**: p. 1080-1089.
3. Fukada, E. and S. Takashita, *Piezoelectric effect in polarized poly(vinylidene fluoride).* Japanese Journal of Applied Physics, 1969. **8**: p. 960.
4. Kawai, H., *The piezoelectricity of poly(vinylidene fluoride).* Japanese Journal of Applied Physics, 1969. **8**: p. 975-976.
5. Chang, S., K. Yong-Tae, and B. Chi-Kyung, *Electrospinning of Poly(Vinylidene Fluoride)/Dimethylformamide Solutions with Carbon Nanotubes.* Journal of Polymer Science: Polymer Physics Edition, 2003: p. 1572-7.
6. Chang, C., et al., *Direct-write piezoelectric polymeric nanogenerator with high energy conversion efficiency.* Nano letters, 2010. **10**: p. 726-731.
7. Walter, S., et al., *Characterisation of piezoelectric PVDF monofilaments.* Materials Technology, 2011. **26**(3): p. 140-145.
8. Wang, T.T. and H. von Seggern, *High electric field poling of electroded poly(vinylidene fluoride) at room temperature.* J. Appl. Phys., 1983. **54**(8): p. 4602-4604.
9. Kumar, A. and M.M. Perlman, *Simultaneous stretching and corona poling of PVDF and P(VDF-TriFE) films. II.* J. Phys. D: Appl. Phys., 1993. **26**: p. 469-473.
10. Day, G.W., et al., *Effects of poling conditions on responsivity and uniformity of polarization of PVF_2 pyroelectric detectors.* Applied Physics Letters, 1974. **24**(10): p. 456-458.
11. Blevin, W.R., *Poling rates for films of polyvinylidene fluoride.* Applied Physics Letters, 1977. **31**(6): p. 6-8.
12. DeRossi, D., et al., *Method of evaluating the thermal stability of the pyroelectric properties of polyvinylidene fluoride: Effects of poling temperature and field.* J. Appl. Phys., 1982. **53**.
13. MeasSpec, *Piezo Film Sensors Technical Manual.* 1999, Measurement Specialties Inc.

O-03-372

DEPOSITION OF HMDSO ON COTTON SURFACE USING A LOW PRESSURE RADIO FREQUENCY GLOW DISCHARGE

A.SHAFIEI[1], S.PILEHVAR[2]*,K.SALEM[1] , H.MORTAZAVI[1], M.GHORANNEVIS[1],P.MANAFI[2], Z.NOBAKHT[3] , Z.FALLAH[2]

[1]Plasma Physics Research Center, Science and Research Branch of Islamic Azad University, Tehran
a.shafiei@mail.com mortazavi19@gmail.com . ghoranneviss@gmail.com k.salem@yahoo.com

[2] Polymer Engineering Group, School of Engineering, Science and Research Branch, Islamic Azad University -
s.pilehvar1988@gmail.com p.manafi@ippi.ac.ir z.fallah@gmail.com

[3]Polymer Engineering Group, South Tehran Branch, Islamic Azad University, Tehran, Iran
Zahraa_6_66@yahoo.com

Abstract

This paper focuses on plasma deposition by an radio frequency plasma (Rf) using hexamethyldisiloxane (HMDSO). HMDSO plasma-polymerized films are deposited onto cotton surfaces using nitrogen/HMDSO mixtures. The chemical and physical properties of the obtained coatings are discussed using contact angle measurements, Fourier transform infrared spectroscopy (FTIR), scanning electron microscopy (SEM). Contact angle results show that the polymerized cotton surfaces become clearly hydrophobic. FTIR results show that if plasma polymerization occurs in HMDSO/N2 mixtures, the deposited film is silica-like containing only few carbon atoms. FTIR results show that Si bonds are produced generally and exhibit high hydrophobosity properties on the surface of samples. SEM results clearly indicate that with increasing plasma treatment time, the deposited films become rougher.

Keywords: HMDSO,PLASMA,POLYMERIZATION

Introduction

Plasma polymerization is a convenient method for the deposition of thin polymer films on a wide variety of substrates [1-2]. Plasma treatment is increasing being used for modifying surface properties of textile such as hydrophilicity and hydrophobicity. Plasma interaction with the surface produces modification of its chemical and morphology. Plasma-polymerized samples have received a large amount of interest due to their unique properties, these samples are generally amorphous, free from pinholes, highly cross-linked, highly resistant to heat and corrosion and very adhesive to different substrates [3-5]. Due to these excellent properties, plasma-polymerized samples can offer many practical applications in the field of mechanics, optics and electronics [6].

Plasma polymerization of silicon compound is an environmentally technology of surface modification for super hydrophilic materials. HMDSO is an organosilicon compound with the formula $O[Si(CH_3)_3]_2$. In our work it is used for surface modification and polymerization. The plasma method for polymerization, has also been used to fabricate thin functional films that facilitate surface modification of materials [7-9]. When a substrate is exposed to plasma, radicals are created. These radicals can initiate polymerization reaction when they are in contact with monomers in a liquid or gaseous phase. Charged particles in plasma with a polymer surface generate radicals at the surface of the polymer though excitation of the polymer molecules and finally a grafting polymer is formed on the surface of the substrate. Among different methods that can be used for super hydrophobic and water repellency treatment, plasma based processes have been widely studied and developed [10-12]. Improving the super hydrophobic and water repellency, coating material must be added to amplify the hydrophobicity. In this work the silicon compounds such as HMDSO and N_2 gas were used. N_2 gas is used as the carrier gas the monomer. HMDSO is one of the most common materials used for coating. In this work we demonstrate a manufacturable method for formation of super hydrophobic and water repellency coating on cotton samples. The water repellency coating was produced by an radio frequency (RF) plasmas process using mixture of N_2 gas and HMDSO liquid. The system is evacuated 10^{-3} Torr. The surface morphologies of plasma treated samples were characterized by scanning electron microscopy (SEM). Wettability has been studied with contact angle measurement. The Fourier transform infrared (FTIR) spectrometry analyzed the chemical composition of our samples. The functionality of the plasma – deposited layer is mainly determined by the nature of the vapor precursors used [13]. These precursors, which lead to the coating formation, are diluted in a main gas, which is usually argon or Nitrogen [14]. Hexamethyldisiloxane (HMDSO) is often preferred as precursor since this compound is generally sufficiently volatile near temperature, relatively non-toxic and non – flammable, cheap and available from commercial sources [15]. Hmdso plasma polymerized thin films can be assayed for a large number of applications in rather different fiels such as protective anti- scratch coatings on plastic substrates, barrier films for food and pharmaceutical packaging, corrosion protection layers, coating for biocompatible materials and low – k dielectric layers for microelectronic applications [15,16]. Plasma deposition from HMDSO mixed with different carrier gases has been extensively studied using low pressure non thermal plasmas.[17,18-20]

Experimental

A schematic diagram of the plasma-polymerization reactor is shown in Figure 1. Two circular stainless steel electrodes are placed within a cylindrical enclosure. The plasma reactor is made of pyrex and the gas gap between the top and middle plates is 2.5 mm.

The power electrode is connected to a 13.56 MHZ, RF supply trough an L-C matching box and the second electrode is grounded.

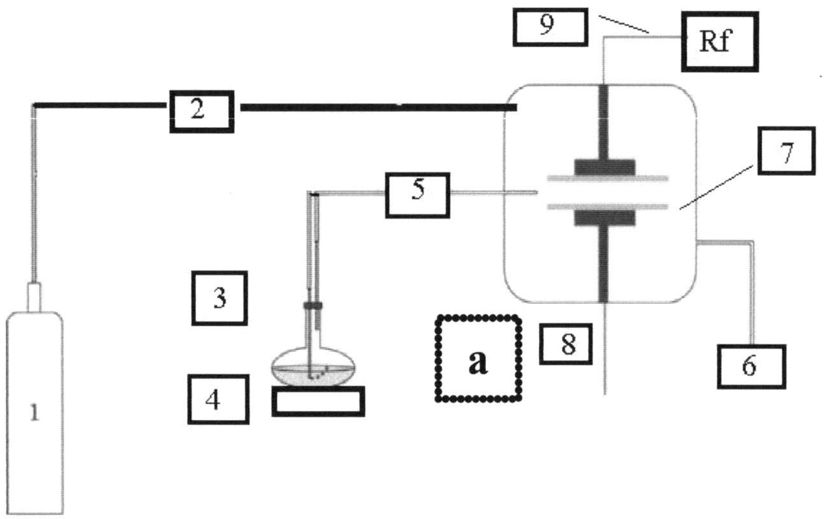

Fig. 1 schematic view of the plasma-polymerization set up Plasma reactor is a Pyrex tube (a), N2 Gas cylinder (1) is connected to reactor trough a needle valve (2), HMDSO is at a balloon and is heated by a oven (3) is connected to reactor by a flow meter, (5), system is connected to Pump (6) sample plate (7) that is grounded (8), Rf power supply (9)

Before starting the polymerization, the Cotton surfaces were cut with an area of 5×5 cm^2 and were placed on the sample plate. After introduction of the substrate into the reactor, the discharge chamber is pumped down below 5×10^{-3} mbar, using a rotary pump and then filled with the working gas/ HMDSO to a pressure of 2×10^{-1} mbar. The liquid HMDSO monomer (Merck>98%) is vaporized by heating to a separated vessel which is connected to the container. HMDSO after changing to gas phase was fed to the chamber and was mixed with carrier gas (N_2). Plasma was operated in a glow discharge made of HMDSO (98%) that was evaporated at high temperature and was mixed with N_2 gas as (30% -70%). In this work, deposition is carried out using fixed percent of HMDSO/N2 mixtures. Some samples were treated at the same plasma power of 30 W in different polymerization times and some of them were polymerized in the same time (2 min) and different plasma's powers e.g. (20, 30 and 50W).

Results and discussion

It is well known that the solid-liquid contact angle is sensitive to the chemical nature of samples and morphology. Wettability of the HMDSO polymerized Cotton samples will depends on changes in surface morphology due to plasma etching action and formation of polar groups on surface. This will also relies on the polar nature of the grafted polymer. As mentioned before, the static contact angles of the treated specimens were evaluated by contact angle measuring system for investigating the wettability. Results have shown that the total surface energy decreases significantly after plasma polymerization. Plasma polymerization is sufficient to change the surface energy and the wettability of surfaces. According to the obtained results the variation rate of water contact angle on the surface is inversely proportional to the variation rate of surface energy. To measuring the water repellency of plasma coating, test was performed according to contact angle measurement system, surface free energy (SFE) of the surfaces have been calculated using the Owens-Wendt (OW) with the sessile drop method. The OW calculations are used to obtain the dispersive (Gama d) and polar (Gama p) components of SFE. It is obvious that the contact angle is increased gradually with increasing treatment time and power of plasma. Surface energy is calculated by measuring the angle of contact of various wetting liquids. Surface energy has two components namely polar component (γ_S^p) and disperse component (γ_S^d). Effect of plasma polymerization time on the contact angle of different liquids and surface energy of polymerized cotton surfaces in different conditions are presented in Table1 and 2. Both components contribute to total surface energy. The polar component is important to understand the hydrophilic surface modification, whereas the disperse component indicates the changes in hydrophobic modification or formation of cross-linked network [21].

The wetting ability is the ability of a liquid to adhere to a solid and spread over its surface. The classical model by Young suggests that equation (1):

$$\gamma_{SV} = \gamma_{LV} \cos\theta + \gamma_{SL} \qquad (1)$$

where θ is the measured contact angle, the surface tension of solid in contact with air, γ_{SV} the surface tension of the liquid in contact with air, and γ_{LV} is the surface tension between the solid and the liquid, Table 1. The work of adhesion W_a between the solid and liquid can be expressed in terms of the equation as follows e.g. equation (2):

$$W_a = \gamma_{SV}(1+\cos\theta) \qquad (2)$$

The solid surface tension can be separated into the intermolecular attraction of polar interaction γ_s^p and dispersion interaction γ_s^d as equation (3):

$$\gamma_S = \gamma_S{}^d + \gamma_S{}^p \qquad (3)$$

Therefore, the work of adhesion can be expressed as the sum of the different intermolecular forces at the interface.

$$W_a = 2(\gamma_{LV}{}^d \gamma_S{}^d)^{1/2} + 2(\gamma_{LV}{}^p \gamma_S{}^p)^{1/2} \qquad (4)$$

then

$$\gamma_{LV}(1+\cos\theta) = 2(\gamma_{LV}{}^d \gamma_S{}^d)^{1/2} + 2(\gamma_{LV}{}^p \gamma_S{}^p)^{1/2} \qquad (5)$$

Liquid	$\gamma_{LV}^p (mJ/m^2)$	$\gamma_{LV}^d (mJ/m^2)$	$\gamma_{LV} (mJ/m^2)$
Water	51.0	21.8	72.8
Diiodomethane	1.3	49.5	50.8

Table 1: values of water and diiodomethane liquid probes

Sample number	Plasma polymerization duration time(min)	Water Contact angle(^{0})	Diiodomethane contact Angle(^{0})	Surface Energy (mJ/m^2)
1	0	0	0	76.35
2	2	85	39	40.49
3	4	95	45	37.24
4	6	104	50	35.05
5	8	116	57	33.26

Table 2. Contact angle of water and diiodomethane liquid probes on polymerized cotton samples in different plasma exposure times with fixed power of plasma (30 W) and 30% HMDSO

Sample number	Power of plasma (W)	Water Contact angle(^{0})	Diiodomethane contact Angle(^{0})	Surface Energy (mJ/m^2)
1	20	100	50	34.45
2	30	85	39	40.49
3	40	80	35	42.77
4	50	74	30	45.97

Table 3 Contact angle of water and diiodomethane liquid probes on polymerized cotton samples in different plasma's power (20-40-50 w) and in a fixed exposure time (2 min)

Liquid	power $\gamma_{LV}^{p}(mJ/m^2)$	$\gamma_{LV}^{d}(mJ/m^2)$	$\gamma_{LV}(mJ/m^2)$
$\gamma_{S}^{p}(mN/m)$	51.0	21.8	72.8
$\gamma_{LV}^{p}(mJ/m^2)$	50.8		

It is obvious that the contact angle is increased gradually with increasing polymerization time and power of plasma and reached to the highest value. The chemical composition of cotton samples were characterized by FTIR spectrum and were recorded on a Thermo Nicolet Nexus 870 FT-IR spectrophotometer, with measurement conditions as follows: mirror velocity of 0.6329, a 45° angle of incidence, and spectral resolution of 4 cm^{-1}. The FTIR results of HMDSO coated cotton surfaces are shown in figure 2 the peaks around 650-700 cm^{-1} are attributed to –CH2-CH=CH-. The small peak of Si-O-Si was assigned at 1000-1150 cm^{-1}. A small peak attributed to Si-CH2-Si bonds appeared at around 1350 cm^{-1}.Also the peak at 1430 cm^{-1} assigned to unsymmetrical rotational of Si-CH3 that after plasma treating has significantly decreased [22,23] more over a CH_x peak is raised in 2900-2960 cm^{-1} according to FTIR it can be found that by increasing treatment time.

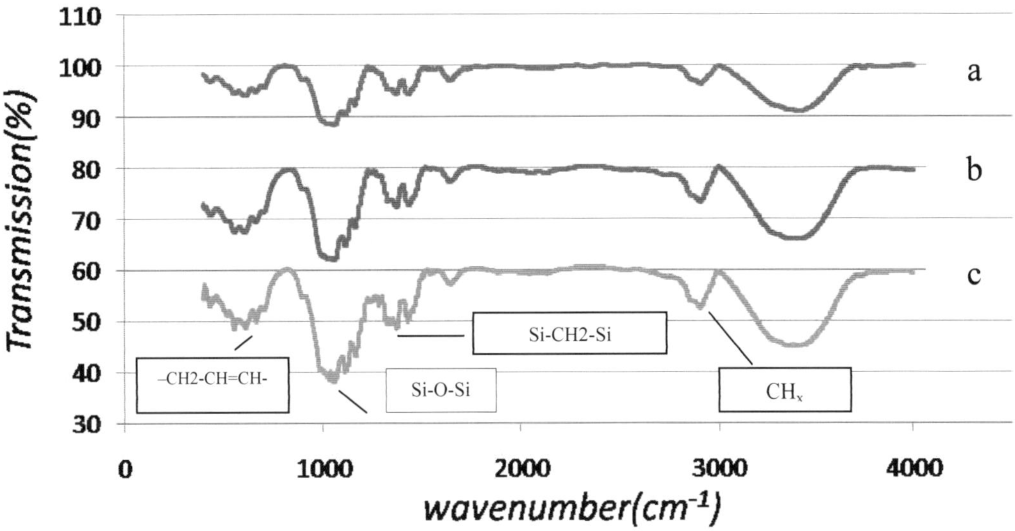

Fig. 2 FTIR spectrum of Cotton polymerized in different times of polymerization (a) 2 min (b) 4 min and (c) 6 min

The SEM images of super-hydrophobic cotton textiles before and after polymerization are shown in Fig.3 It can be observed that the nanostructure on the cotton fiber was damaged to some extent and the surface of the fiber become rougher than the surface before .

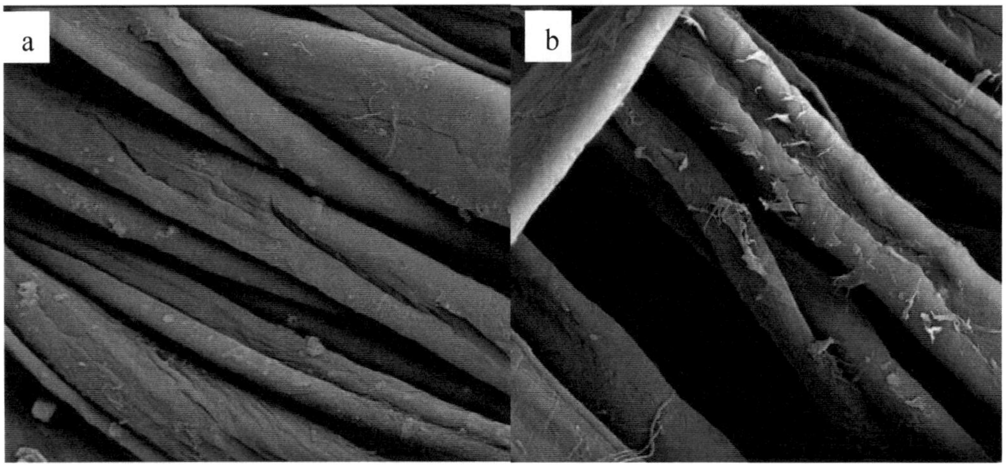

Fig. 3 SEM images of samples origin cotton and (b) 4min polymerized cotton

Conclusions

In this paper, plasma polymerized HMDSO samples were deposited on cotton surfaces using Rf in nitrogen gas at low pressure. SEM results clearly show that the surfaces deposited in N2/HMDSO (glow discharge) were quite rougher than before polymerization, FTIR results show that plasma polymerization leads to formation and increasing intensity of some hydrophobic groups such as Si-O-Si, Si-CH2-Si CH_x and etc. Contact angle data clearly show that after polymerization, cotton samples become hydrophobic remarkable. For example water contact angle of cotton changed from zero to 116° after polymerization, this means by HMDSO polymerization contact angles increased more than 100%.

References

[1] Y.V. Pan, E.Z. Barrios, D.D. Denton, J. Polym. Sci., Part A: Polym. Chem. 36 (1998) 587.
[2] C. Zhang, J. Wyatt, D.H. Weinkauf, Polymer 45 (2004) 7665.
[3] F. Arefi, V. Andre, P. Montazerrahmati, J. Amouroux, Pure Appl. Chem. 64 (1992) 715.
[4] T.P. Kasih, S. Kuroda, H. Kubota, Plasma Processes Polym. 4 (2007) 648.
[5] Y. Sawada, S. Ogawa, M. Kogoma, J. Phys. D: Appl. Phys. 28 (1995) 1661.
[6] A. Hiratsuka A, I. Karube, Electroanalysis 12 (2000) 695.
[7] H. Yasuda, Plasma polymerization, Academic press, London, 1985
[8] R. Szeto, D.W. Hess, J. APPl. Phys. 903 (1981) 152
[9] H. Bieddrman, Y. Osada, Plasma polymerization processes, Elsevier Science Pub, New York, 1992 Chap. 7.
[10] I. Woodward, W.C.E. Schofield, V. Roucoules, J.P.S. Badyal, Languir 19 (2003) 3432.
[11] J.K.Evju, P.B. Howell, L. E. Locasicio, M.J. Tarlov, J.Hickman, J.Appl. Phys. Lett. 84 (2004)
[12] S.M. Mukhopadhayay, P. Joshi, S. Datta, J.Macdaniel, Appl.Surf.Sci.210 (2002) 219.
[13] G. Borcia, N.M. D. Brown, journal of physics D: Applied physics 40 (2007) 1927-1936.
[14]F. Massines, N.Gerardi, A. Fornelli, S. Martin, surface and coating technology 200 (2005) 1855-1861.
[15] S. Kurosawa, H. Harigae, H. Aizawa, K. Terashima, H. Suzuki, J. Photopolym. Sci. Technol. 19
(2006) 253.
[16] L.Zajickova, V. Bursikova, Z. kucerova, J. Franclova, P.Stahel, V. Perina, A. Mackova, journal of physics and chemistry of so;ieds 68 (2007)1255-1259
[17]F. Bnitez, E. Martinez, J. Esteve, thin solid films 377-378 (2000) 109-114.
[18]S. Zanini, C. Riccardi, M. Orlandi, P. Esena, M. Tontini, M. Milani,, V. Cassio, surface and coating technology 200 (2005) 1255-1259
[19]C. Vautrin-Ul, C. Boisse,-Laporte, N. Benissad, A. Chausse, P. Leprince, R. Messina, progress in organic coating 30 (2000) 9-15.
[20] D. Hegemann, H. Brunner, C. Oehr, Surface and coating technology 142-144 (2001) 849-855.

[21] N. V. Bhat, D. J. Upadhyay, Plasmas Polym. **8**, 99 (2003)
[22] R.Morent, N. De Geyter, S.Van Vlierberghe, P. Dubruel, C.Leys, L. Gengember, E.Schacht, E. Payen, progress in organic coating 64 (2009) 304-310
[23] D.L. Pavia, G.M. Lampman, G.S. Kriz, Introduction to spectroscopy, 3rd edn. (Thompson Learning Inc, Washington, 2001

O-03-390

EXTRUSION FOAMED POLY(LACTIC) ACID/POLYPROPYLENE BLENDED

N. Chuapon[1], O. Naithanom[1], S. Areerat[1], P. Piyamanocha[2*], T. Sedláček[2], and P. Sáha[2]

[1]School of Chemical Engineering, Faculty of Engineering, King Mongkut's Institute of Technology Ladkrabang, Bangkok, Thailand, 10520 – kasurat@kmitl.ac.th
[2] Polymer Centre, Faculty of Technology, Tomas Bata University in Zlín, T.G. Masaryka 555, 760 01 Zlín, Czech Republic - sedlacek@ft.utb.cz

Abstract - Poly(lactic) acid (PLA), as an alternative sustainable bio-based polymer, is recently extensively investigated for polymer foaming processes demands. Several methods are utilized for the foam morphology modification, the foaming window expansion, and the physical properties improvement. In present study, various composition of PLA/polypropylene (PP) blends were prepared and subsequently foamed using extrusion foaming process with chemical blowing agent. The foaming process was carried out at selected screws speed and various foaming temperatures. Results revealed the highest expansion ratio for neat PLA in comparison to PLA/PP blends having expansion ratio relatively low. SEM images showed phase separation of PP portion in the polymer matrix acting as nucleating agent but significantly lowering melt strength of PLA. In addition, prepared foam structures were described through the isotropic foam index, while the compatibility of prepared PLA/PP blends was determined via the thermal analysis.

Keywords: Poly(lactic) Acids, Polypropylene, Foam, Isotropic Foam

Introduction

Due to an environmental conservation, utilization of renewable plastic resources or bio-based polymers such as poly(lactic) acid (PLA) is being intense. In contrast, PLA is derived from plants, such as corn, tapioca, and sugarcanes, and degraded by hydrolysis at the ester bond. PLA has been utilized in foaming processes [1-3] and compared with petroleum based polymer such as polyethylene terephthalate (PET). However, PLA has low crystalline content resulting in low foaming ability. Its crystallization can be enhanced by, for example, isothermal annealing, polymer compounding with inorganic fillers, or strain-induced crystallization that could improve final product properties.

In general, foam structure or morphology can be influenced by foaming mechanism; cell nucleation, bubble growth, and stabilization. Cell nucleation could be modified by adding nucleating agents while in bubble growth step, melt strength is an important factor that could indicate final foam morphology and influence on final product for example mechanical and electrical properties [4-6]. In this study, PLA was compounded with high melt strength polypropylene (HMS-PP) and then foamed by extrusion process. The influence of PP on foam morphology and isotropic property was studied.

Experimental

Materials

PLA 2002 biaxial grade supplied by NatureWorks having molecular weight of $1.4x10^5$ g/mol and High melt strength PP $4.3x10^5$ g/mol supplied by Borealis had been chosen in this study. The PLA was dry at 80°C for 4 hrs before performing every experiment. Azodicarbonamide from Sigma-Aldrich was utilized as a blowing agent at constant concentration of 1 phr.

Polymer Compound Preparation

Selected polymers were dry blended with various concentrations as shown in Tab. 1. The compounds were melt mixed in single-screws extruder KETSE 20/40, Brabender GmbH & Co. KG using temperature profile of 175/185/190/190/190°C for the barrel and 190°C for 5 mm diameter extrusion die. The extruded was subsequently cooled down by immerge in water at room temperature. Appropriated take off speed was controlled by means of pelletizer maintaining suitable diameter and size of the extruded. The sketch of experimental apparatus used in this experiment is illustrated in Fig. 1. In addition, pure PLA resin was also processed in the extrusion line in order to perform identical thermal history of all samples. Then, prepared compounds were dried at 80°C for 4 hrs in order to remove all entrapped moisture in the resins during cooling process. Thermal characteristic of tested polymers were examined using differential scanning calorimeter (NETZSCH DSC 204 F1) at temperature range from 25-200°C with heating and cooling rate of 10°C/min.

Table 1 PLA blends composition.

Formula	Weight Fraction (wt%)	
	PLA	PP
PLA	100	0
5%PP	95	5
10%PP	90	10
20%PP	80	20

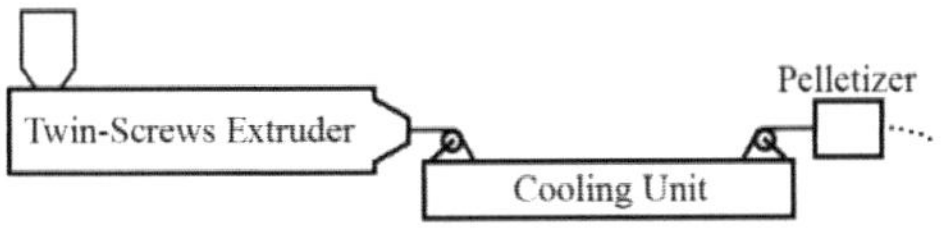

Figure 1 Schematic of extrusion process.

Polymer Foaming and Characterization

The selected blowing agent was dry blended with the compounds and delivered to hopper. The experiments were carried out using single-screws extruder with barrel temperature profile of 175/190/200/200/190°C while extrusion temperature at die was changed to 175, 180, and 190°C. The extrusion slit die had exit dimension of 10x0.5mm. Extrusion rates were controlled by means of screws speeds which were fixed at 20 rpm and the extruded were collected and cooled down at room temperature.

Liquid nitrogen was employed for cryogenic cracking of foam samples then the cross-sectional of cracked surfaces were imaged by the help of SEM (LEO 1455VP). Obtained micrographs were then analyzed by ImageJ computer aid program for bubble diameter and cell density estimation. The uniformity of bubble distribution was described by the isotropic index (η).

Results and Discussion

Thermal Properties

The compound pellets were successfully prepared by the extrusion process whereas the transparency of pure PLA pellets was absolutely dominated by PP. The DSC results of tested polymers are shown in Fig. 2 and thermal properties of tested materials are summarized in Tab. 2. Three major transition regions can be observed from heating scan of pure PLA thermograph referring to glass transition, cold crystallization, and melting temperatures, respectively. The stress store in the resins during quenching from mixing process is clearly observed from the endothermic relaxation peak in the first heating scan where it is diminished in the second heating scan. Cold crystallization, in contract, can be observed from the first heating scan and at the onset of the melting peak while this crystallization behavior is very obvious in the second heating scan resulting from rearrangement of polymer molecules during the first heating cycle. During cooling tests, in addition, crystallization does not occur since PLA is a polymer which is difficult to crystalline [7]. The thermograph of PP indicates unambiguous melting and crystallization temperatures illustrated in both heating and cooling scans while glass transition region including stress relaxation behavior cannot be observed.

Obtained DSC results of all PLA/PP compounds obviously exhibit thermal behavior of both source materials. Stress relaxation peaks of the compounds are smaller than that of pure PLA while cold crystallization is obviously presented in the second heating scans. This cold crystallization behavior of PLA/PP blends, moreover, is exhibited quicker than neat PLA. This

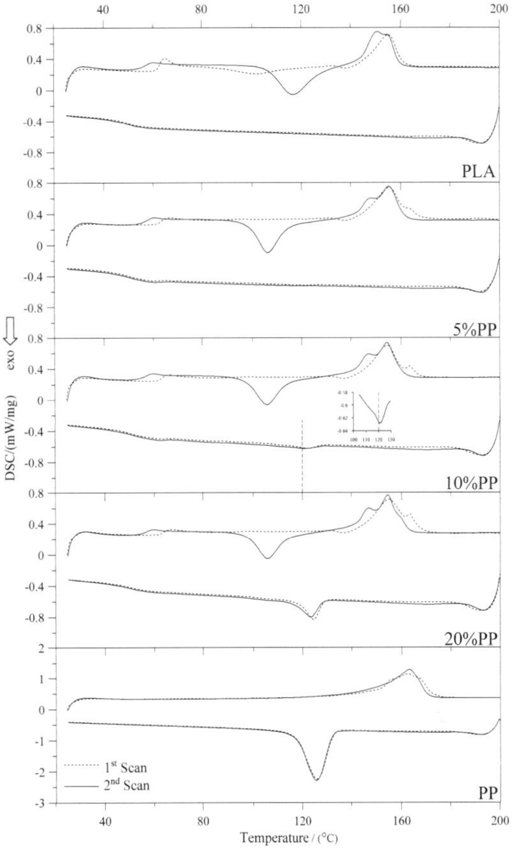

Figure 2 DSC thermographs of tested samples.

could be caused by PP particle induces PLA to crystalline. The overlap peaks in melting region are according to the melting behavior of pure PLA where it could have high polydispersity. The crystallization of PP portion during cooling step, moreover, could be clearly observed from samples containing 20 and 10%wt PP which may leads to phase separation behavior while it cannot be observed from 5%wt PP sample. It is confirmed from SEM images in Fig. 3 where the PP particle in polymer matrix can be indicated.

Table 2 DSC characterization of PLA/PP blends.

Formula	Second Scan					
	T_g (°C)	T_{crys} (°C)	H_{crys} (J/g-PP)	$T_{m,1}$ (°C)	$T_{m,2}$ (°C)	H_f (J/g)
PLA	56.5	-	-	150.7	-	32.8
5%PP	56.6	-	N/A	147.9	155.5	33.6
10%PP	55.8	121.2	19.4	146.9	154.3	33.7
20%PP	55.9	123.6	61.2	146.9	154.4	35.9
PP	-	125.7	94.7	-	163.3	90.9

Remark: $T_{m,1}$ relates to melt temperature of PLA.
$T_{m,2}$ relates to melt temperature of PP.

Foam Morphology

The crossectional surface of extrusion foam samples were analyzed by means of SEM and shown in Fig. 3. The morphological analyzation results are summarized in Fig. 4-5 which obviously show different foam morphologies influenced by PP content and as well as extrusion temperatures. As can be seen from the Fig. 4, bubble sizes are in a range of 40 to 80 micron where the low foaming temperature tends to create larger bubble sizes. Furthermore, tested foam samples have cell density values in the range of 10^6-10^7 cells/cm^3 when varied PP concentrations.

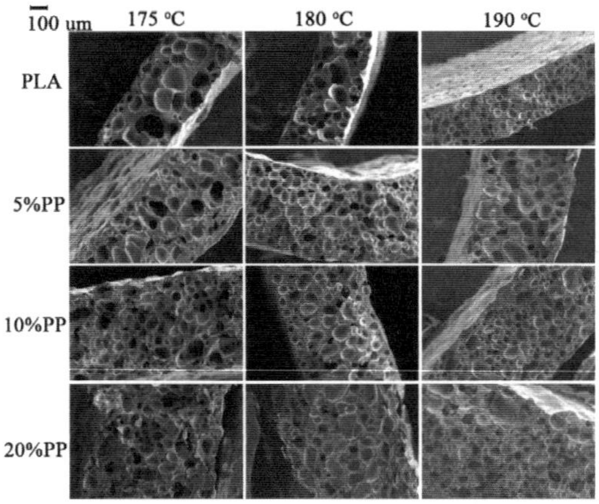

Figure 3 SEM micrographs of PLA/PP blend foams.

In addition, it is seen from Fig. 5 that, all foam samples have significantly identical expansion ratios except pure PLA foam which has a bit higher value. The addition of PP reduces melt strength of PLA matrix through phase separation leading to bubble rupture and coalescence.

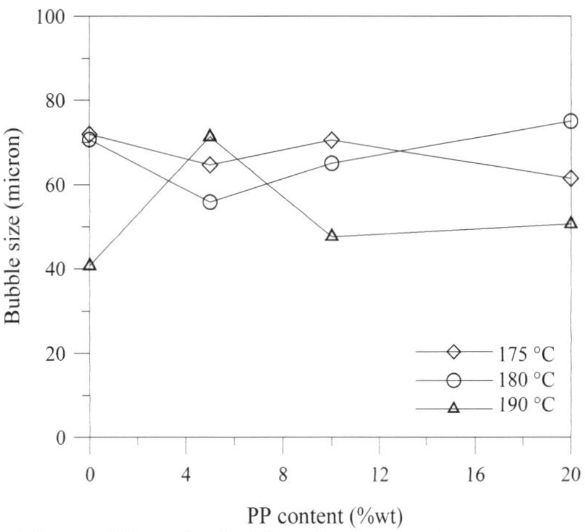

Figure 4 Plotted of bubble diameter vs. PP content.

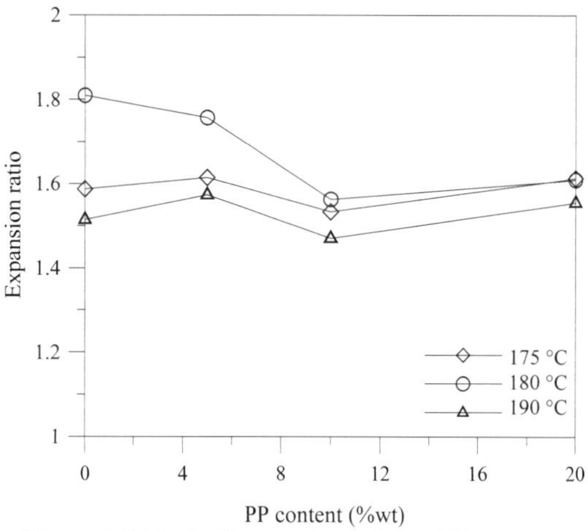

Figure 5 Plotted of expansion ratio vs. PP content.

Isotropic Foam Behavior

Isotropic foam means uniformly of spherical bubble distribution in foam specimen. The isotropic index (η) can be determined through the relationship between surface bubble member (N_S) and isotropic foam cell density (N_F^0) which was proposed by P.Buahom and S.Areerat [8] as the following equation:

$$\eta = \log\left(\frac{N_F}{N_F^0}\right) \quad (1)$$

where

$$N_F = \frac{6}{\pi(d)^3}\left(1 - \frac{1}{\phi}\right) \quad (2)$$

$$N_F^0 = \varphi_0 (N_S)^{3/2} \quad (3)$$

where d and ϕ are average bubble size and foam expansion ratio, respectively. φ_0 is standard isotropic factor having value of 0.1453 [8]. The calculation result is shown in Fig. 6 presenting prepared PLA/PP foams have the index values lie in the isotropic foam region, but, at the upper limit which means that obtained bubble morphologies are nearly non-uniform.

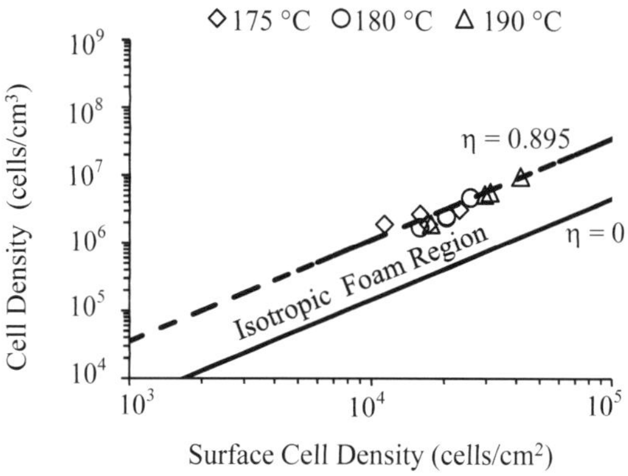

Figure 6 Plotted of bubble density vs. surface bubble density.

In contrast, Fig.7 demonstrates that high expansion PLA and low PP concentration composite foams present non-isotropic foam index while low expansion foams have more uniform bubble shapes.

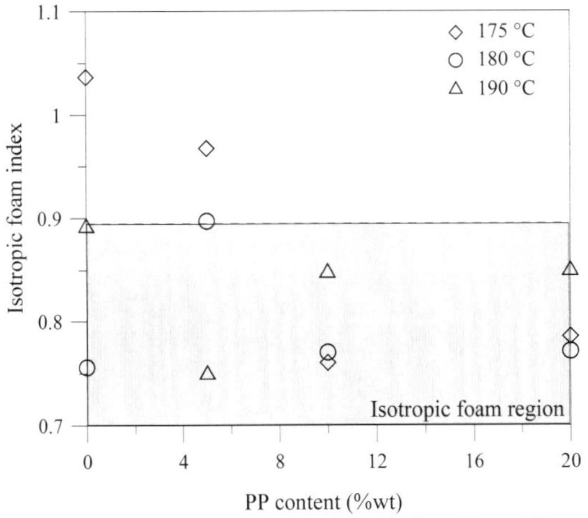

Figure 7 Plotted of isotropic foam index (η) vs. PP content.

Conclusions

The PLA/PP composites were chemically foamed using single-screws extruder with extrusion slit die. The foaming temperatures and PP concentrations were significantly influenced on foam morphologies through phase separation of PP portion and variation of the melt strength. Proper amount of PP is acting as nucleating agent through phase separation behavior. Obtained foam morphologies are in the isotropic region but mostly at the upper limit. It could be concluded that PLA foam morphology can be controlled by adding proper concentration of immiscible pair of polymer such as PP and through the controlling conventional processing conditions such foaming temperature as well.

References

1. D. Li; T. Liu; L. Zhao; X. Lian; W. Yuan. *Ind. Eng. Chem. Res.* 2011, 50, 1997-2007
2. M. Mihai; M.A. Huneault; B.D. Favis, *Polym. Eng. Sci.* 2010, 50, 629-642
3. X. Liao; A.V. Nawaby; H.E. Naguib, *J. Appl. Polym. Sci. 2012*, 124, 585-594
4. D. Garlotta *J. Polym. Environ.* 2001, *9*, 63-83.
5. W. Zhai; Y. Ko; W. Zhu; A. Wong; C.B. Park *Int. J. Mol. Sci.* 2009, 10, 5381-5397.
6. Y.M. Corre; A. Maazouz; J. Duchet; J. Reignier *J. Supercrit Fluids*. 2011, 58, 177-188
7. B. Hararak; N. Kerddonfag; C. Winotapun; P. Kumsang, C. Sripethdee; W. Chinsirikul in AWPP2012 Proceeding, Kyoto, 2012
8. P. Buahom; S. Areerat *J. Cell. Plast.* 2011, 47, 133-152

O-03-402

OBSERVATION OF HIGHER ORDER STRUCTURE AT HEAT SEAL PARTS ON UNIAXIAL DRAWN PP FILM

K. Yamada[1]*, R. Konishi[2], K. Okada[3], T. Tsujii[3], and K. Miyata[4]

[1] *Department of Advanced Fibro-Science, Kyoto Institute of Technology, Japan - kazushi@kit.ac.jp*
[2] *Nakamoto Packs Co. Ltd., Japan - konishi@npacks.co.jp*
[3] *PerkinElmer Japan Co. Ltd., Japan - kiyomi.okada@perkinelmer.com, tetsuya.tsujii@perkinelmer.com*
[4] *Department of Polymer Science and Eng., Yamagata University, Japan - ken@yz.yamagata-u.ac.jp*

Abstract – Heat seal technology is an essential technique for intermediate packaging material for various industries. There are different kinds of sealing technologies, which would be selected for suitable heat sealed properties of the film. Heat sealing technique is conventional technology for heat sealed film by control temperature, pressure and dwell time. The conditions of heat sealed are important for controlled peel strength and heat sealed properties of the film. Furthermore, film strength depends on the drawn ratio and molecular orientation of film. However, the heat sealing ability of higher molecular orientation films are poor due to loss the heat sealing energy for relaxation of oriented molecules in polymer films. Therefore, it is very important to investigate the relationship between heat sealing conditions, molecular orientation, and higher order structure in polymer films on properties of heat sealed film. In this study, cast polypropylene (CPP) film was heat sealed by using heat sealing technique. The heat sealed condition was set at heat sealed time of 0.1 second with pressure of 0.2 MPa at various heat sealed temperature of 145 to 150 °C. The difference of higher order structure of these films was discussed on the basis of results of micro-Raman spectroscopy, FT-IR spectroscopy, DSC and peel test. From the result, it was found that higher order structure of PP films at heat sealed parts depends on heat sealing time, temperature, and drawn ratio of PP films.

Keywords: Heat sealing, CPP film, Raman spectroscopy, microscopic FT-IR, high order structure, crystallinity

Introduction

Heat sealed plastic bags are widely used in food packages, vacuum-seal, and so on. Heat seal technology is an essential technique for intermediate packaging industries. It is very important to understand about heat sealing properties of these packages.[1,2] Basically, there are two typical films, that is, laminated films and homo films. There are different kinds of sealing technologies such as chemical adhesives, ultrasonic welding, heat sealing, impulse heat sealing, and hot air welding. Therefore, heat sealing technique is the most conventional than other methods as mentioned above. These laminated and homo films are then heat sealed for preparing as packaging material applications. Heat seal is made by fusing the polymers to another through the application of heat and pressure. The initial pressure enables intimate contact between the films. Then adhesion is promoted by applied of heat and pressure from the outside of films. The important factors to obtain great heat seal properties include heat seal temperature, pressure, heat seal time and dwell time for cool down after heat sealing. [3-5]

In addition, film strength depends on the drawn ratio and molecular orientation of film. The heat sealing ability of higher molecular orientation films are poor due to loss the heat sealing energy for relaxation of oriented molecules in polymer films. Accordingly, it is very important to investigate the relationship between heat sealing conditions, molecular orientation, and higher order structure in polymer films in order to understand the difference of heat sealed properties at heat sealed parts.

In this research, we used homo-polymer film specimens with the different drawn ratio in order to evaluating the effect of molecular orientation on heat sealed properties. The films were heat sealed with hot plate heat sealing machine at various heat sealing conditions. The difference of higher order structure of the films was discussed on the basis of results of micro-Raman spectroscopy, FT-IR spectroscopy, DSC and tensile test.

Experimental

Materials

In this study, cast polypropylene (CPP) film (TOYOBO, PYLEN FILM-CT, P1011, Japan) was used as a base film. This CPP film was unidirectionally drawn at 0, 1.0, 1.5, 2.0, 2.5 times by uniaxial drawn machine (Nakamoto Packs Co., Ltd., Japan). The drawn film conditions then will be referred as x1.0, x1.5, x2.0 and x2.5, respectively.

Heat Sealing

Figure 1 shows a schematic diagram of heat sealing method in this experiment. The uniaxial drawn and un-drawn films were placed in a bar type heat-sealing machine (Heat seal Tester, Sangyo Co., Ltd, Japan). The heat seal pressure was kept constantly at 0.2 MPa. The heat sealing time was 1.0 second. Heat sealing temperature was varied from 145 to 150 °C.

Peel test

In order to examine the mechanical properties of the heat sealed part, a peel test was carried out with test pieces of 15 mm wide and 100 mm long. A schematic draw of the heat sealed specimen is shown in Figure 2. The peel test was performed at 300 mm/min by using an INSTRON universal testing machine (INSTRON 4466) at 20 °C. The span length was 50 mm.

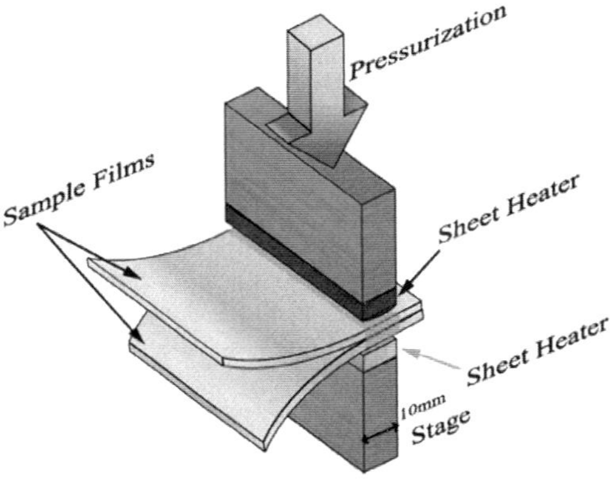

Figure 1 Schematic illustration of heat sealing process.

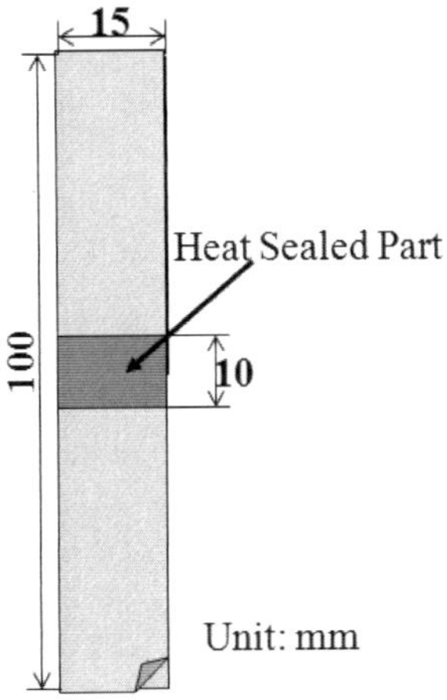

Figure 2 Schematic draw of a specimen for peel test.

DSC Measurement

Thermal properties of heat sealed part of specimens were measured by Differential Scanning Calorimeter (DSC: Perkin Elmer, Pyris1) for investigated the difference of crystallinity with and without heat sealing. The temperature range was set from 40 to 250 °C at a heating rate of 20 °C/min.

Microscopic FT-IR measurement

Heat sealed part of specimens were mapping by microscopic FT-IR (Frontier-Spotlight 400: Perkin Elmer, Japan) to evaluate the crystallinity and the orientation of heat sealed parts with and without heat sealing. Mapping mode was transmittance imaging method. Wave resolution was 4 cm^{-1} with number of scan 2 times. Pixel size was set at 25 μm.

Microscopic Raman spectroscopy

Raman spectra have been measured with the micro-Raman backscattering geometry. The experimental setup as well as the polarization geometries used for obtaining polarized Raman spectra in the 90° scattering geometry has been described previously. The Raman spectra were excited with the linearly polarized 633 nm line of He-Ne laser (LabRAM HR-800: HORIBA Co., Ltd., Japan). The excitation beam was directed to the sample compartment of a properly modulated metallurgical microscope (Olympus BX-41). The microscope was used for the delivery of the excitation laser beam on the sample and the collection of the backscattered light through a beam splitter with objective lens adapted to the aperture of the microscope. Exposed time was set at 1.0 s. The number of scan was set at 5 times.

Results and Discussion

Peel strength for each drawn film

Figure 3 shows the peel strength of drawn film specimens at various stabilization temperatures. Peel strength of cast PP film was about 10 N. Peel strengths for drawn films were lower than cast PP film. In addition, the peel strength decreased with increasing the drawn ratio for PP film. In the case of x2.5 drawn film, the strength was about 1 N/15mm, which was hardly to heat sealed. Therefore, it is considered that there is a linearly relationship between drawn ratio and heat sealed strength, that is, dependence on the orientation degree of PP molecules in film surface area.

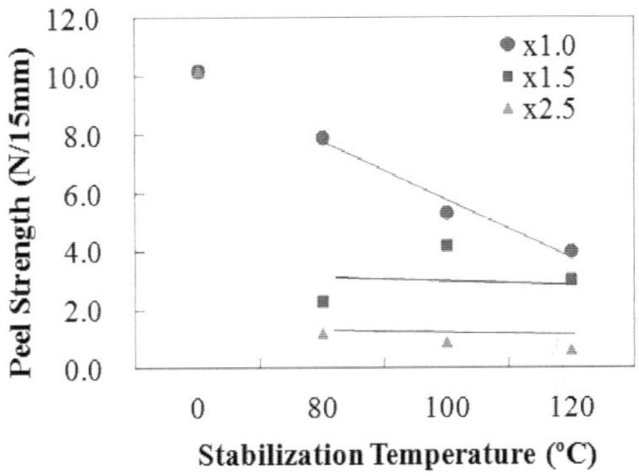

Figure 3 Peel strength of drawn film specimens for various stabilization temperatures at a heat sealed temperature of 145°C.

DSC measurement

We tried to perform DSC measurement in order to evaluate the change in crystallinity of high order structure at heat sealed part of PP drawn films. Figure 4 (a) and (b) show the typically DSC curves of CPP film and x2.5 drawn PP film at 145°C heat sealed temperature, respectively. Melting point and melting enthalpy (ΔH) value of the 1st and the 2nd heating thermogram of CPP film are similar. However, melting point of the 1st heating thermogram for x2.5 drawn film exhibited higher than the 2nd one. This difference depends on the drawn ratio of PP film, which influenced on difference in molecular orientation of PP films. ΔH values of the films and heat sealed films at each heat sealed conditions are shown in Figure 5. It was found that ΔH of non-heat sealed film was lower than heat sealed part due to higher crystallinity of the heat sealed part as compared to the non-heat sealed one, In addition, ΔH of higher drawn ratio was lower than that of lower drawn ratio. It was attributed to the crystallization of the film after heat sealed. It can be considered that ΔH of higher drawn ratio film becomes low due to the energy waste to make it relaxation of molecular orientation in PP films. Accordingly, these results will support the change in secondary order structure of PP films. However, in the results of DSC measurement, it is very difficult to obtain the precise and information of secondary structure change in the edge area of heat sealed part. Hence, we applied the micro FT-IR mapping for heat sealed edge part of PP films for better understanding.

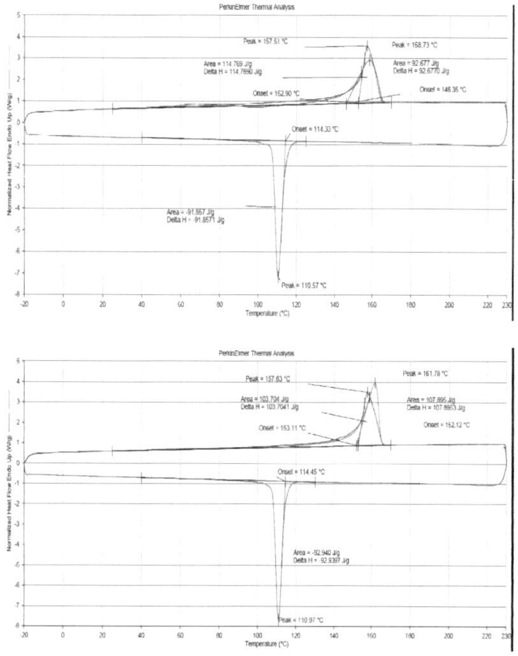

Figure 4 Typically DSC curves of (a) CPP film and (b) x2.5 drawn film at 145 °C heat sealed temperature.

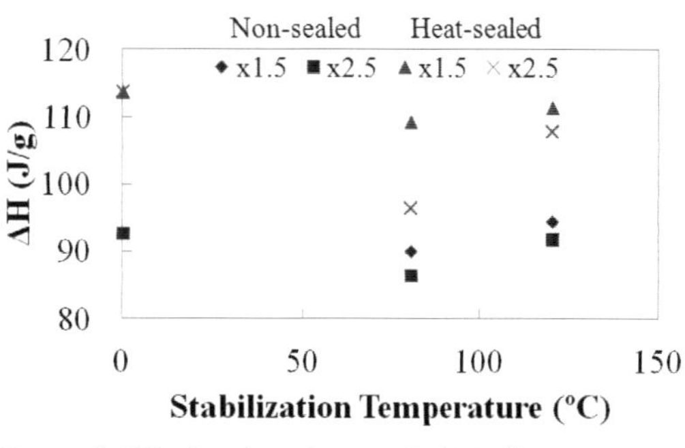

Figure 5 ΔH of various heat sealed PP films at 145°C for each stabilization temperature.

Microscopic FT-IR measurement

Figure 6 illustrates the FT-IR mapping images of PP film of heat sealed edge part. These mapping images were normalized at 972 cm^{-1} and imaged the peak ratio for 998cm^{-1}/972 cm^{-1}. The left area is corresponded to non-heat sealed part and the right area is corresponded to heat sealed part in these images. From the result, light color is corresponded to the higher crystallinity area. In non-heat sealed part, crystallinity was increased with increasing the stabilization temperature so that the color variation is consistent with the molecular orientation of PP film. Because drawn film could raise the creep relaxation without heat stabilization after drawing. Therefore, crystallinity of non-heat sealed part of PP film was decreased with decreasing the stabilization temperature. On the other hand, similar tendency was shown in the heat sealed part of PP films, that is, crystallinity of heat sealed part was increased with increasing the stabilization temperature. However, in this case, it was seen that each crystallinity was lower as compared with non-heat sealed part. Because it is considered that thermal energy of heat sealing is used for the relaxation of molecular orientation in PP films. As a result, crystallinity of heat sealed part is lower than non-heat sealed part. In addition, crystallinity of the edge part of heat sealing indicated the different tendency as compared with heat sealed and non-heat sealed part. It is very interesting and important phenomena for heat sealed films. However, we can't observe the more small area at high spatial resolution in order to have low resolution about 20 μm spatially.

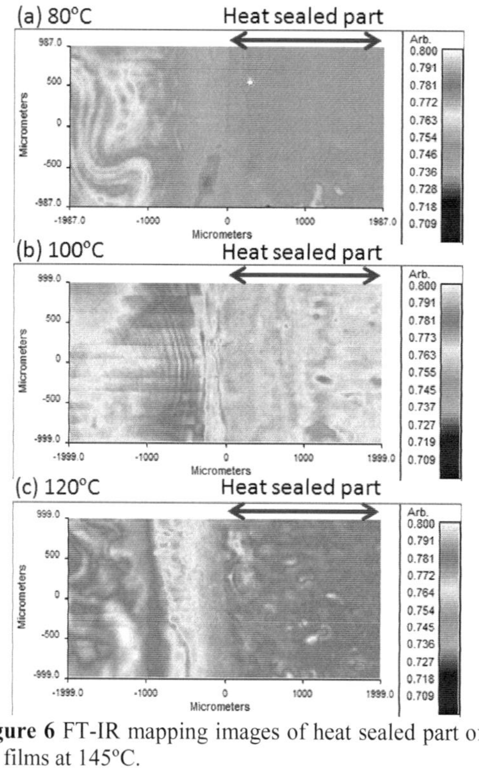

Figure 6 FT-IR mapping images of heat sealed part of PP films at 145°C.

Microscopic Raman spectroscopy

Figure 7 presents the Raman spectra of PP film for each drawn films. After drawing, some peaks were decreased, where is considered that these peaks are corresponding to molecular orientation in PP films. In this study, peak ratio (I_{848}/I_{818}) was calculated in order to evaluate molecular orientation and/or crystallization changes near heat sealed edge part, and then the result was plotted in Figure 8. In Figure 8, peak ratio of neat film part area was higher than heat sealed part. In addition, the slope of strength change was observed at interface area within 200 μm. As the resolution and pitch width are 1 μm and 1 μm respectively, it was found that molecular orientation of PP films indicated the relaxation over an appreciable range of the slope gradually. It should be noted that when this slope is rapidly become steep, stress concentration will arise at this edge part for tensile loading. Therefore, it is very important to control the heat sealing conditions without raising the stress concentration at heat sealed edge part in order to get the better packaging products. In addition, it was indicated that we can analyze of very small area and changes on the basis of the results by using micro-FT-IT and micro-Raman spectroscopy imaging applied for drawn PP films.

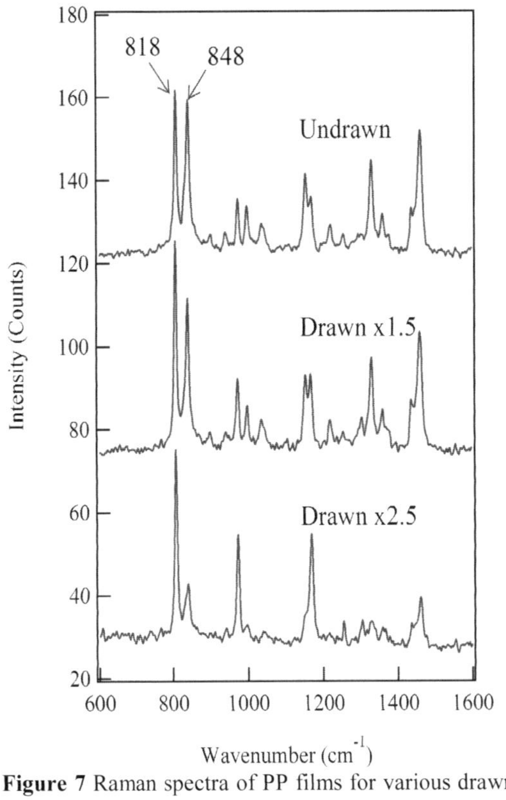

Figure 7 Raman spectra of PP films for various drawn ratio.

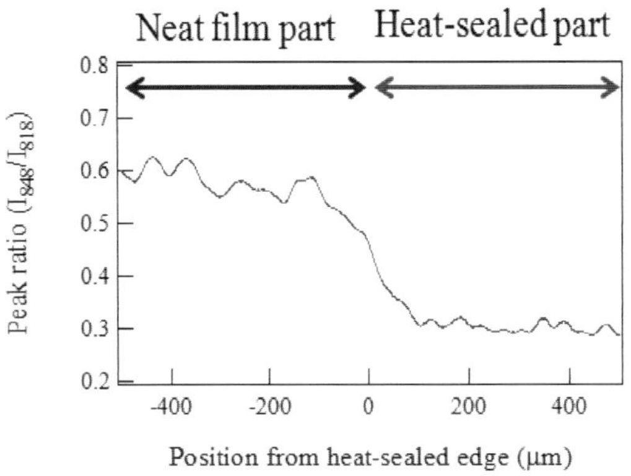

Figure 8 Line mapping data of Raman spectra of drawn PP film around heat sealed edge part at heat sealed of 145°C.

Conclusions

In this investigation, we investigated the heat sealed edge part of drawn PP films by using micro-FT-IR and micro-Raman spectroscopy. From this study, it was indicated that we can analyze the very small area and changes on the basis of the results of heat sealed part by using micro-FT-IT and micro-Raman spectroscopy imaging, which applied for drawn PP films. We can summarize that heat sealed properties were related to crystallinity and molecular orientation in the heat sealed film. Therefore, it is very important to control the heat sealing conditions without arising the stress concentration at heat sealed edge part in order to get the better packaging products.

References

1. S. Farris, C. A. Cozzolino, L. Introzzi and L. Piergiovanni, *Packag. Technol. Sci.* (2009), 22, pp.359-369.
2. Dinesh Aithani, Hugh Lockhart, Rafael Auras and Krittika Tanprasert, *Packag. Technol. Sci.* (2006), 19, pp.245–257.
3. Yasuo Hashimoto, U.S. Ishiaku, Y.W. Leong, Hiroyuki Hamada, Tetsuya Tsujii, *Polymer Engineering and Science* (2006), pp.205-214.
4. Tsujii Tetsuya, Yasuo Hashimoto, U. S. Ishiaku, M. Mizoguchi, Y. W. Leong, Hiroyuki Hamada, *Journal of Applied Polymer Science* (2006), pp.513-519.
5. Tsujii Tetsuya, U. S. Ishiaku, Machiko Mizoguchi, Hiroyuki Hamada, *Journal of Applied Polymer Science* (2005), pp.753-760.

O-03-407

STUDY ON THE APPLICATIONS OF MICROCELLULAR INJECTION MOLDING AND NANOCOMPOSITES TO IMPROVE THE WARPAGE OF THE PLASTIC BOX

Shyh-shin Hwang, Jui-Pin Yang ,Ching-hsin Hu, and Cheng-Chien Li,
Chien-hsin University of Science and Technology, Jhongli, Taiwan. ROC

Abstract
Conventional injection molding parts may cause shrinkage/warpage problem after cooling due to the uneven wall thickness, uneven cooling, or long flow ratio. In this study, both microcellular injection molding and polymer nanocomposites are used to improve the warpage problem over the conventional molding process. Moreover, the polymer nanocomposites foam has smaller cell size and larger cell density compared to those by neat polymer foam. Foam with small cell size would help the warpage problem. So using polymer nanocomposites foam to improve the warpage problem is the goal of this project. A rectangular box is used for this study which has shrinkage/warpage problem along the long side wall. Semi-crystalline polypropylene (PP), amorphous polystyrene (PS), and PP/clay nanocomposites are used in this study. The results show the microcellular injection molding can improve the shrinkage/warpage problem of the box. However, PS foam has less shrinkage/warpage than that of PP. And the addition of clay into PP also helps the shrinkage/warpage problem. The PP/clay foam has the same level of shrinkage/warpage of the box as the PS foam does.

Key Words: Polypropylene, Polystyrene, Nanocomposites, Microcellular, Shrinkage/Warpage.

Introduction

Shrinkage/warpage is a big issue in polymer processing which is caused by the cooling during molding process. Some of the poor mold designs will enhance this problem, like uneven thickness of the parts [1] and unbalanced mold temperature on the core and cavity [2,3]. The shrinkage value for many plastics materials (in order of 0.6%) are high relative to dimensional tolerances. and the shrinkage is not always isotropic in nature. Anisotropic shrinkage behavior can also lead to a degree of warpage (out-of-plane distortion) or internal stresses [4, 5]. Panchal and Kazmer [6] used the shrinkage sensor to measure the in-mold shrinkage of the HIPS and PP. It showed that HIPS behaved relatively isotropic shrinkage. However, semi-crystalline polymers PP demonstrated more anisotropic shrinkage.

Warpage is basically attributed to thermally induced stresses and ''frozen-in'' orientation, associated with large increases in the normal stresses during melt flow of the reinforced composites relative to its matrix [5]. On the other hand, filler reinforcement does not generate large normal stresses and, in some instances, has lower normal stresses than those of the matrix [7]. Kazmer et al. [8] studied the warpage prediction of optical media. The optical media is manufactured by injection-compression molding process which has less residual stress on the parts than that by conventional injection molding. They found that of the process conditions, mold temperature had the greatest effect on warpage.

This problem takes place commonly in conventional injection molding. The most effective way to improve this problem is the addition of filler into the polymer matrix, i.e, composites. Alternatively, it can be improved with the microcellular injection molding [9-10] which creates million of microcells in the polymer. Kramschuster et al [11] studied the effect of the process conditions on the shrinkage/warpage of the plastic box. They found that the gas dosage and injection speed affected the microcellular injection molded part's shrinkage/warpage the most. The author's previous study of microcellular injection molding [12] on the gear shaft showed this process improving the dimensional stability of the gear shaft.

Moreover, polymer nanocomposites also can improve the shrinkage/warpage problem [13]. Prashantha et al. [14] studied the effect of multi-walled carbon nanotube (MWNT) addition on shrinkage and warpage properties of polypropylene (PP) injection moldings before and after annealing. The addition of 2 wt% MWNT into PP can reduce part's shrinkage/warpage up to 48% as compared to neat PP..

Very few papers have published to study the effects of polymer-nanocomposites foaming on the shrinkage/warpage. In this study, polymer/clay nanocomposites and microcellular injection molding are used to study the effect of clay loading on the shrinkage/warpage problem of the plastic box. Materials used are PP, PS, and PP/clay nanocomposites.

Materials and Experimental

PS and PP with melt flow index of 8 and 38 were purchased from Chi-mei Co. and Formosa Chemicals & Fiber Co. respectively. The PP/clay nanocomposites were purchased from masterbatch of P802, nanoMax PP-HiST of Nanocor Co. Ltd. Then the masterbatch (50 wt.% clay) of the nanocomposites were diluted into 1, 3, 5, 7, and 9 wt% of PP/clay nanocomposites. After all, they are molded by conventional and mucell® injection molding processes. The mucell® injection process was performed on a 100 ton ARBURG-420C injecting machine, equipped with microcellular injection molding capability. N_2 was used as the physical blowing agent. Investigating the processing benefits and part's dimentional improvements of combining nanocomposites with the microcellular injecting molding process is the primary purpose of this research.

Results and Discussion

The reason to choose the materials from two different groups is that they behave different characteristics on shrinkage. The semi-crystalline PP has larger shrinkage rate than that of amorphous PS. Figure 1 shows the mold design of the plastic box which can easily show the molding problem caused by shrinkage/warpage. Figure 2 shows the shrinkage/warpage of PP box by conventional molding. The shrinkage/warpage is determined by measuring the distance between the two center points of the long side of the box. It is done by measuring the image of the box.

Shrinkage/warpage of the semi-crystalline PP.

Figure 3 shows the shrinkage of the PP box by conventional and microcellular injection molding. Gas concentration 0 means conventional injection molding. It is seen that the shrinkage/warpage is 6.28 mm for conventional injection molding. If supercritical fluid Nitrogen was introduced into PP and microcellular injection molding was adopted. The shrinkage/warpage is 0.52 mm at the 0.2 wt% of Nitrogen concentration. So the shrinkage reduction is improved dramatically by 92%.

Shrinkage/warpage of the amorphous PS.

Figure 4 shows the shrinkage of the GPPS box by conventional and microcellular injection molding. It is seen that the shrinkage for neat GPPS is 0.3 while it is 6.2 for neat PP on conventional molding. During foaming, the shrinkage is reduced from 0.3 mm to 0.06 mm as the gas content is 0.25 wt%. The shrinkage reduction is improved by 80%. So, for the semi-crystalline and amorphous materials, the shrinkage reduction is better improved by semi-crystalline material than that of amorphous material (Fig. 6). This is caused by the phase change from the amorphous melt to the semi-crystalline solid state accounts for the higher shrinkage of semi-crystalline plastics in comparison to amorphous plastics where the molded shrinkage of the amorphous plastics is due primarily to pressure compensated thermal contraction.

Shrinkage/warpage of the PP/clay nanocomposites.

Figure 6 shows the shrinkage/warpage of the conventional injection molded PP/clay nanocomposites box. By adding 1 wt% of clay into PP, shrinkage/warpage of the nanocomposites can reduce from 6.4 mm to 3.5 mm. By adding more of clay into PP, up to 9 wt%, the shrinkage/warpage of the nanocomposites reduction is improved gradually. The author's previous paper [12] showed the shrinkage for neat PA6, PA6/3 wt.% clay, and PA6/5 wt.% clay at 260 °C/30 MPA is 1.00697, 0.9196, and 0.787 respectively, while it is 0.874, 0.816, and 0.716 at 50 °C/30 MPA, respectively, meaning the shrinkage is 0.1329, 0.103, and 0.071, respectively. The addition of clay apparently reduces shrinkage in PA6/MMT nanocomposites.

Figure 7 shows the shrinkage/warpage of the microcellular injection molded PP/clay nanocomposites box at the gas content of 0.2 wt%. The shrinkage is 0.68 mm for neat PP. By adding 1 wt% clay into PP, the shrinkage can be improved from 0.68 mm to 0.62 mm. The shrinkage reduction is gradually improved with adding more of clay into PP up to 7 wt%.

Figure 8 shows the shrinkage/warpage comparison of the conventional/microcellular injection molded PP/clay nanocomposites box at differing clay contents. It is seen that for the neat PP the shrinkage is 6.4 mm and 0.68 mm for conventional and microcellular molding respectively. The shrinkage improvement is 63%. However, the shrinkage improvement at 1, 3, 5, 7, and 9 wt% clay loading are 82%, 82%, 83%, 96%, and 95% respectively. It is evident that the shrinkage improvement is much better of PP/clay nanocomposites for conventional/microcellular injection molding.

Cell structure.

Figures 9~11 show the cell structures of the microcellular PS, PP, and PP/clay box respectively. They reveal that there are skin, intermediate, and core layers in the thickness direction. It seems that PS has better foaming structure than those of PP and PP/clay boxes which is also related to the shrinkage/warpage behavior.

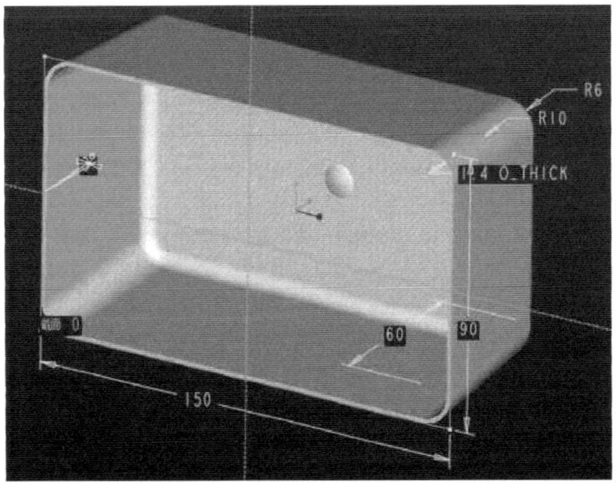

Figure 1. Design of the plastic box.

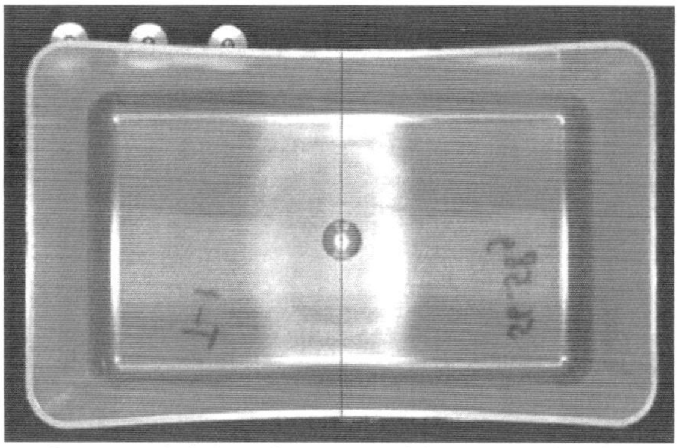

Figure 2. Shrinkage/warpage of the conventional injection molded PP box.

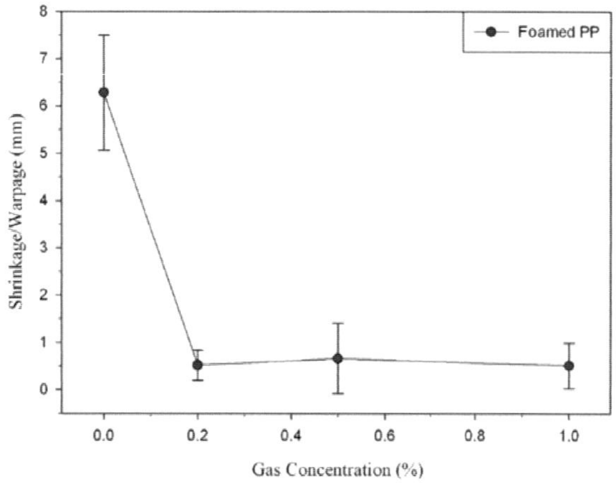

Figure 3. Shrinkage/warpage of the conventional/microcellular injection molded PP box.

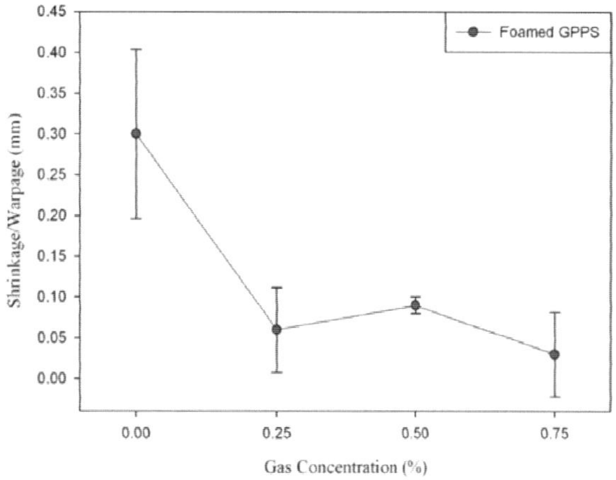

Figure 4. Shrinkage/warpage of the conventional/microcellular injection molded GPPS box.

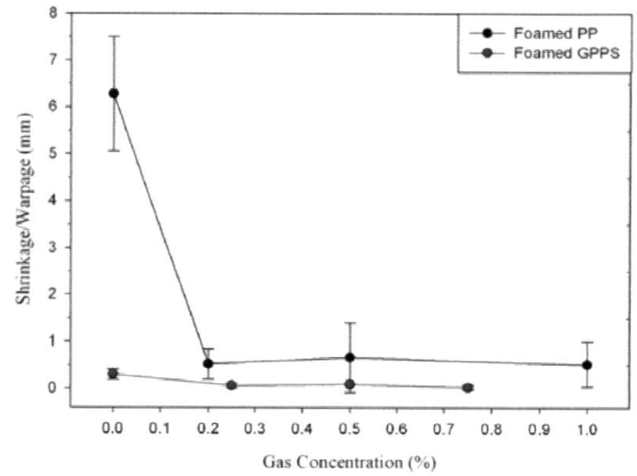

Figure 5. Shrinkage/warpage Comparison of the conventional/microcellular injection molded PP and GPPS box.

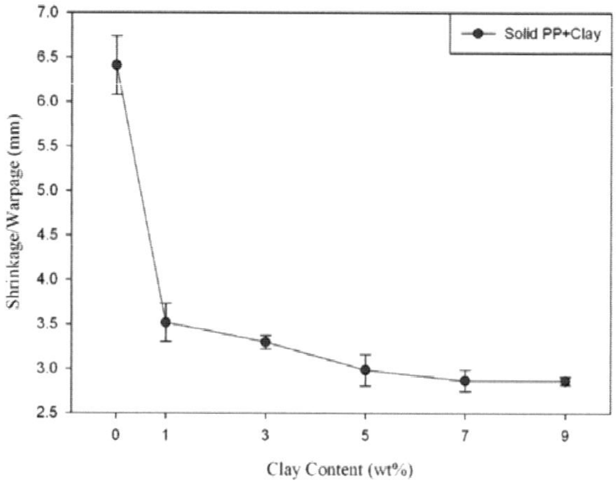

Figure 6. Shrinkage/warpage comparison of the conventional injection molded PP/clay nanocomposites box.

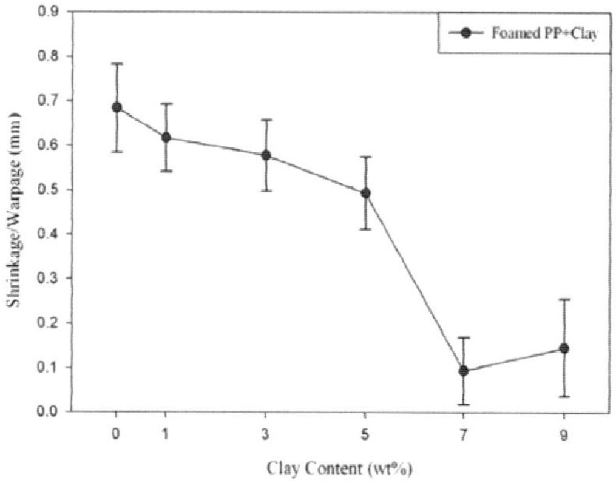

Figure 7. Shrinkage/warpage comparison of the microcellular injection molded PP/clay nanocomposites box at differing clay contents.

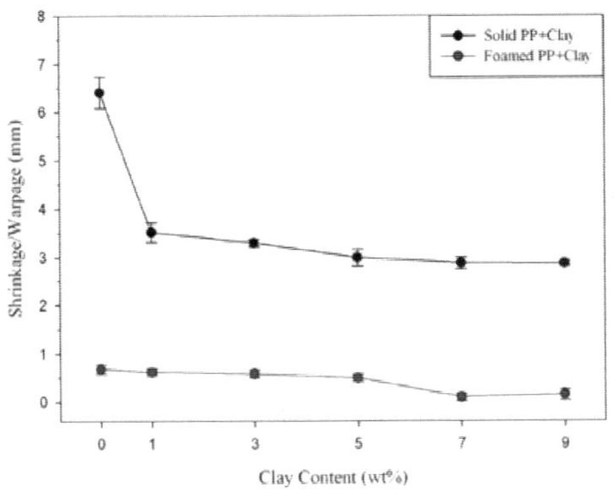

Figure 8. Shrinkage/warpage comparison of the conventional/microcellular injection molded PP/clay nanocomposites box at differing clay contents.

Figure 9. Cell structure of the microcellular GPPS box at the gas content of 0.5 wt% near the gate.

Figure 10. Cell structure of the microcellular PP box at the gas content of 0.5 wt% near the gate.

Figure 11. Cell structure of the microcellular PP/clay box at the clay and gas contents of 1.0 wt%, 0.3 wt% respectively near the gate.

Conclusions

In this study, the combined benefits of shrinkage/warpage improvement of microcellular injection molding and polymer/clay nanocomposites were used to reduce the shrinkage/warpage of the plastic box. Both the semi-crystalline PP and amorphous GPPS were used in this study. Microcellular GPPS box has less shrinkage than that of PP. And microcellular PP/clay nanocomposites box has the same level of shrinkage with that of foamed GPPS.

Acknowledgements

This work was supported by Ministry of Economics under grant 100-EC-17-A-08-S1-143.

References

1. Technical Bulletin, "Designing with Plastics", Hoechst Celanese, Chatham, NJ (1989).
2. M. St. Jacques, *Polymer Engineering & Science,* V22, P. 241-247, 1982.
3. Paro, H. T.; Bosnyak, C. P.; Sehanobish, K. Machine Des 1986, 58, 69.
4. Fischer, J.M. (2003). Handbook of Molded Part Shrinkage and Warpage. William Andrew Publishing/Plastics Design Library
5. C. Shen and H. Li, *Polym. Plast. Technol. Eng.*, 42(5), 971 (2003).
6. R. Panchal and D. Kazmer, ANTEC Technical Paper, 2009.
7. R.M. Pisipati, Ph.D. Dissertation, Virginia Polytechnic Institute and State University, Blacksburg, VA (1983).
8. B. Fan, D. O. Kazmer, W. C. Bushko, R. P. Theriault, and A. J. Poslinski, *Journal of Polymer Science, Part B, Polymer Physics,* vol. 41, pp. 859-872, 2003
9. V. Martini, J. E., Nam. P. Suh, and F. A. Waldman, U.S. Patent No. 4,473,665, 1984.
10. Jonathan S. Colton, Nam P. Suh, *Polymer Engineering and Science* 27 (7) (1987) 500–503.

11. Adam Kramschuster, L-S Turng, *Polymer Engineering and Science*, P.1408~1418, 2005.

12. S. S. Hwang and Z. S. Ke, *International Communications in Heat and Mass Transfer , V.* 35, Issue 3, March, 2008, P. 263-275.

13. S-S Hwang, S-P Liu, Peming P. Hsu , J-M Yeh , Jui-pin Yang, K-C Chang, and S. N Chu, *International Communications in Heat and Mass Transfer*, Volume 38, 2011, May, P. 597-606 .

14. K. Prashantha, J. Soulestin, M. F. Lacrampe, E. Lafranche, P. Krawczak, G. Dupin,, and M. Claes, *eXPRESS Polymer Letters,* V.3,No.10,2009, P.630-638.

STUDY OF THE CRATER FORMATION MECHANISM ON THE SURFACE OF BOPP FILM

Toshitaka Kanai[1*], Satoshi Tamura[2]
[1] *Idemitsu Kosan Co.,Ltd. 1-1, Anesaki-Kaigan, Ichihara-city, Chiba, 299-0193 Japan*
toshitaka.kanai@si.idemitsu.co.jp
[2] *Prime Polymer Co., Ltd . 580-30, Nagaura, Sodegaura-city, Chiba, 299-0265 Japan*
Satoshi1.Tamura@primepolymer.co.jp

Abstract
In an attempt to clarify the formation mechanism of the crater-like surface roughness of BOPP film, various thicknesses of PP sheets were compared in this study. It was found that the thickness of PP sheet has a great effect on the crater shape of BOPP film. In addition, the crystal grains formation in the surface layer which is precursor producing crater-like surface can be predicted from the analysis of the cooling calculation. It was found that the crystal grains in the surface layer are easily created with the increase of the chill-roll temperature. From the analysis of the cooling calculation, it was confirmed that more than 1.5 seconds of the crystallization time is required to produce crystal grains for the same PP. In addition, the crater formation mechanism was researched in terms of the stretching behavior of entire PP sheets. By analyzing the change of the surface roughness with the stretching ratio, the deformation of surface structure starts partially before the craters were formed on the BOPP films surface. New hypothesis of the crater-like film surface roughness formation mechanism was proposed by observing the transformation of crater from sheet to BOPP film and by investigating the relationship between the stress-strain curve and surface roughness change.
Keywords: Film, Surface, Crater-like structure, Polypropylene, Spherulite

1. Introduction

It is a well known fact that there are some crystal types of PP with different crystal systems, and surface roughness of BOPP films has been investigated in association with crystal types [1),2)].
Various studies have reported on the change of the crystal structure during the stretching process [3)-6)]. Although Koike et al. made a report on the change of PP crystal structure during the uniaxial stretching observed by the birefringence technique[7)], no information related to the surface structure was reported. Kanai et al. reported on the PP spherulites deformation and crystalline orientation during stretching in terms of the stretching stress pattern, and on the analytical results of the crystal structure changes observed by birefringence, light scattering and small-angle X-ray scattering (SAXS) [5)]. Kanai et al. also reported that there was a strong relationship between the observation results of crystal structures in PP sheets and the simulation results obtained by cooling calculations conducted with regard to the thermal transfer characteristics of PP sheets using the chill-roll cooling system [8)]. It was concluded in the report that the cooling simulation is an efficient method to control the crystallization of PP sheets. This is an important factor in determining the stretchability of BOPP.
It was found in our previous report [9)] that the crater-like structures on BOPP film surface were related to the morphology of the surface layer of the PP sheets. This report will give some perspective on the formation mechanism of craters with regard to controlling the morphology of the surface layer of PP sheet and also on its connection to changes in the entire structure.

2. Experiments

1) Samples

In order to clarify the mechanism for the formation of the surface craters, PP-A of a high tacticitiy with a meso pentad mmmm value of 96mol% and PP-B of a low tacticity with mmmm value of 91mol% were used. Properties of PP-A and PP-B are shown in Table 1. Melting point of PP-A was 3℃ higher than that of PP-B, because stereoregularity of PP-A is higher than PP-B.

Table1. Properties of PP Resins

Properties	PP-A	PP-B
MFR(g/10min)	3.5	3.0
mmmm (%)	96	91
Tm(℃)	164	161
Mw/Mn (-)	5.2	4.5

PP sheets were produced using a sheet forming machine (GM Engineering) with an extruder with a 35mm diameter. The PP resin was extruded from a T-die with a width of 200mm at 250℃. Then it was cast by a chill-roll with a diameter of 250mm at a speed of 0.8 m/min and PP sheets with various thicknesses were produced at various chill-roll temperatures. In this report, samples were named in order of resin type, chill-roll temperature, and thickness. So a sample 'A80-500' is the sheet made by PP-A at a chill-roll temperature of 80℃ with a thickness of 500μm. The BOPP film names were defined by putting 'f' at the end of PP sheet names, for example 'A80-500f'.

After PP sheets were cut into a square shape of 85×85mm in size, they were stretched to BOPP films by a table tenter (Bruckner KARO Ⅳ) in order to analyze the surface structure. And then after they were preheated at a definite temperature for 1 minute, it was stretched at maximum stretching ratios of 5 times to the machine direction (MD) at first, and then 7 times to the transverse direction (TD). Under other experimental condition, it was also stretched at maximum stretching ratios of 6 times to both the machine direction (MD) and the transverse direction simultaneously.

2) Evaluation of PP sheets and the BOPP films

The sectional structures of PP sheets were observed by an optical microscope (Nikon ECLIPSE-LV100POL). Thermal properties of Tm and ΔHm of PP sheets were measured by DSC (PerkinElmer B014-3018/B014-3003) at the heating rate of 10℃/min.

X-ray diffractions were measured on the PP sheets with a Rigaku Denki RINT-2500 diffractometer with Ni-filtered Cu-Kα radiation. The crystallinity χ_c and the β crystal content named as the K-value of PP sheets were calculated from the X-ray diffraction curves [25,26]. Spherulite size U_{max} of PP sheet was calculated from equation (1) obtained by the Hv scattering pattern which reflects the scattering of the polymer superstructure

$$U_{max} = \frac{4.09}{(4\pi/\lambda)\sin\theta_{max}} \quad \cdots\cdots (1)$$

where λ is the wavelength of He-Ne laser (632.8nm), and sin θ_{max} is the scattering angle which indicates the yield point of scattering intensity to 45°against a plane of polarization, respectively.

The surface structure of BOPP films was observed using a Scanning Electron Microscope (SEM, JOEL JSM5600LV). The shapes of the craters were measured by a highly precise shape measuring machine (Kosaka Laboratory SURFCODER ET4000A) using a diamond head at a head pressure of 70μN. Ten points average roughness Rz defined by JIS B0601 [1994] was used as the parameter of depth of crater on BOPP films surface.

Moreover, in order to investigate the crater formation mechanism at the surface layer of PP sheet, the influence of cooling speed was studied under the sheet forming condition described in section 2.1). The temperature of PP resin starts to drop after PP sheet is extruded from the die and touches the chill-roll. PP crystallization starts when the sheet temperature reaches the crystallization temperature, at 115℃ in the case of PP-A. After sheet temperature is kept at the crystallization temperature until the latent heat ΔH is consumed, the sheet temperature starts to drop again. Crystallization time is defined as the time during which PP sheet temperature is kept at the crystallization temperature, and PP crystallizes during the crystallization time. In this report, it is assumed that heat is conducted by primary thermal conduction expressed in equation (2) and is transferred at the boundary condition expressed in equation (3).

$$\frac{\partial T}{\partial t} = \frac{k}{C_p \cdot \rho} \cdot \frac{\partial^2 T}{\partial x^2} \quad \cdots\cdots (2)$$

$$k\frac{\partial T}{\partial x} = h(T_w - T_\infty) \quad \cdots\cdots (3)$$

where T is sheet temperature (K), t is time (s), k is thermal conductivity (0.28W/m·K), C_p is heat capacity (1.93 J/kg·K), ρ is density of PP, x is sheet position from chill-roll (m), h is heat-transfer coefficient (756W/m^2·K), T_w is surface temperature of sheet (K), and T_∞ is chill-roll temperature (K), respectively. Although the density ρ of PP depends on temperature, the fixed value 890kg/m^3 was used in this study in order to make the calculation simple. Heat-transfer coefficient h was determined by measuring the real temperature of chill-roll and PP sheet using a touched type thermometer.

3. Results and Discussion

1) Relationship between the crystalline structure of sheet and crater shape of BOPP film surface

Properties of PP sheets made by sheet forming machine are shown in Table 2.

Table2. Properties of PP Sheets

Properties	A30-500	A80-500	B30-500	B80-500
PP Resin	A	A	B	B
Chill Roll Temp. (℃)	30	80	30	80
Thickness (μm)	500	500	500	500
Xc (%)	46	65	48	64
Chill Roll Side Rz (μm)	0.21	0.26	0.20	0.21
Opposite Side of Chill Roll Rz (μm)	2.24	2.85	1.63	1.90

It is considered that these crystal grains have some influence on the forming of the crater, because they were not observed in the surface layer on the chill-roll side with the temperature at 80℃ or lower. After measuring the diameter and the depth of these grains in the PP sheet by visual inspection and the diameter and Rz of craters of BOPP films, there seemed to be a good correlation between them (Fig. 1). Therefore, it can be said that the craters were formed from the crystal grains after the PP sheets were stretched to become the BOPP films.

2) Formation behavior of crater

The formation behavior of the crater was observed with SEM using the PP films at various stretching ratios using a table tenter. A lot of apertures appeared on the surface of the opposite side of the chill-roll after being stretched at only 1.5 times in the machine direction. These apertures became precursors at the beginning of stretching process, and these apertures seemed to grow into craters after the stretching ratio got higher. As a result, craters were produced from the apertures created at the beginning of stretching process as the starting point.

SEM images of the cross-sectional view of PP sheets stretched at arbitrary stretching ratios are shown in Fig. 2. Some apertures of around 10-20μm in depth appeared in the early period of the stretching process at 1.5×1.5 times, and the diameter of the aperture increased with increasing stretching ratio. Many thick walls appeared on the stretching ratio at 1.5×1.5 times, and at the end the walls between craters were torn apart into thin walls.

The effect of the total stretching ratio on Rz of the BOPP film which is a good parameter of the crater depth was investigated in Fig.3. Rz showed a maximum value at a stretching ratio of 2.0, and then got lower as the stretching ratio got larger. After the stretching ratio of 2.0, the size of the crater became larger as the wall between craters were torn. The apertures depth that appeared at the beginning of the stretching process got deeper as the stretching temperature was raised from 153 °C to 159 °C. Therefore, it is believed that there is the another crater formation mechanism other than the crystal dislocation system from β crystal to α crystal, because crater shape changed at higher stretching temperature than the β crystal melting point of 148°C.

It is considered that the transformation behavior of this crystal in the surface layer is related to the stretching behavior of the whole film. Fig. 4 shows the changes of stretching force with the stretching ratio. A yield points were observed at a stretching ratio of 1.3 and the stretching force gradually decreases at stretching ratios from 1.3 to 5. It is well known that interlamellar separation in the amorphous phase occurs at the beginning of stretching and the spherulite begins to collapse when the stretching ratio passes the yield point. Therefore, it is presumed that the grains in the surface layer of the sheet begin to collapse similarly to the spherulites in the middle layer of the sheet.

3) The influence of sheet forming conditions on the crater formation of BOPP film

A higher chill-roll temperature is required to produce craters on the BOPP film surface of chill-roll side. Furthermore, crystal grains in PP sheet are needed in order to produce craters on BOPP film.

In order to investigate the influence of the chill-roll temperature on the sheet surface temperature, calculated results of the primary sheet temperatures of the opposite side of chill-roll and chill-roll side are shown in Fig.5 (a) and (b), respectively. The crystallization time got longer as the chill-roll temperature got higher. This graph shows that a crystallization time of more than 1.5 seconds is needed to create the crater on the surface of BOPP film using our experimental method. This result obtained from the PP sheet with the different chill-roll temperatures is similar to PP sheets of different thicknesses. It was confirmed that crater formation can be controlled by the crystallization time which is a function of the thickness of PP sheet and chill-roll temperature. In addition, crater can be formed on the BOPP film surface of the chill-roll side with controlling the crystallization time suitably, while the previous reports discussed only the cases of the crater on the surface of opposite side of chill-roll.

4. Conclusion

In an attempt to clarify the formation mechanism of crater-like surface roughness, different PPs and forming conditions were used in this study. As a result, it was found that chill-roll temperature had a great effect on the shape of the crater and, the crater of the BOPP film stretched from that at 80°C was larger than the PP sheet cast at 30°C. When chill-roll temperature rose, the crystal grain of the PP sheet at surface layer of the opposite side of the chill-roll grew larger. This means that the shape of the crater is closely related to the crystal grain shape.

It was found that the thickness of PP sheet has a great effect on the crater shape of BOPP film because the crystal grains in the surface layer of PP sheet changes accordingly to the thickness of PP sheets. In addition, the crystal grains formation in the surface layer can be predicted from the analysis of the cooling calculation. It was found that over critical crystallization time is required to produce crystal grains in the surface layer of PP sheet with β type trans-crystals.

Furthermore, the relationship between the crater on the BOPP films surface and the chill-roll temperatures at the proceeding of PP sheet was investigated. It was found that the crystal grains in the surface layer are easily created with the increase of the chill-roll temperature. From the analysis of the cooling calculation, it was confirmed that more than 1.5 seconds of the crystallization time is required to produce crystal grains for the same PP.

5. References

1) Fujiyama M., Kawamura Y., Wakino T., Okamoto T., J. Appl. Polym. Sci. 36, 985(1988)
2) Fujiyama M., Kawamura Y., Wakino T., Okamoto T., J. Appl. Polym. Sci. 36, 995(1988)
3)Kanai T., Yonekawa F., and Kuramoto I ,17th Polym. Proc. Society Annual Meeting Abstracts, 17(2001)
4) Kanai T., Seikei-Kakou, 18(1), 53-66(2006)
5) Kanai T., Matsuzawa N., Takebe T., Yamada T., Polymer Processing Society Regional Meeting Europe CD-ROM Abstracts(2007)
6) Kanai T., 24th Polymer Processing Society Annual Meeting CD-ROM Abstracts (2008)
7) Koike Y., and Cakmak M., Polymer 44, 4249-4260 (2003)
8) Kanai T., Takebe T., Matsuzawa N., Yamaguchi H. Asian Workshop on Polymer Processing Plenary Lecture(2009)
9) Tamura S., Ohta K., Kanai T., J. Appl. Polym. Sci. 124,2725-2735(2012)

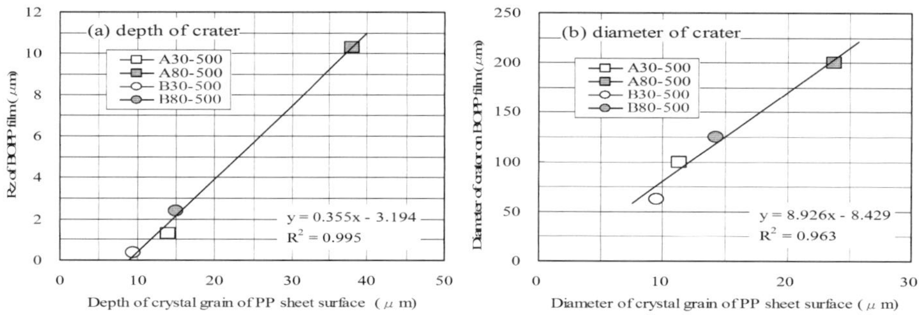

Fig.1 Relationship between the shape of crystal grain on PP sheet surface and shape of craters of BOPP film surface

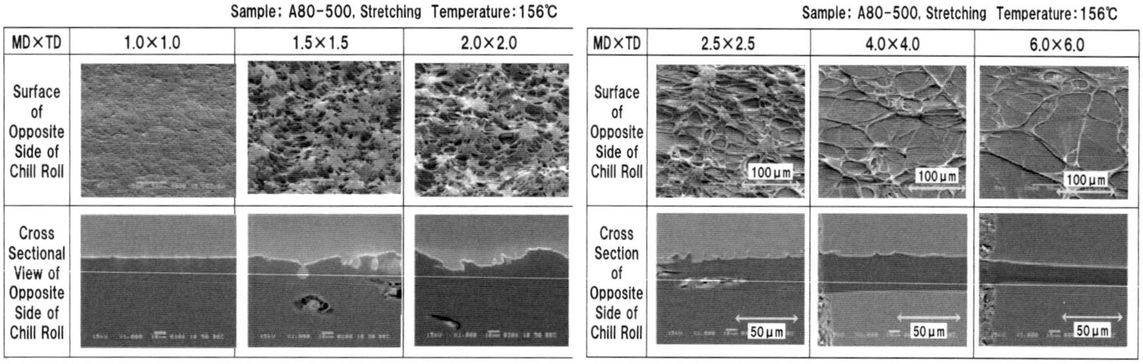

Fig. 2 SEM image of the surface of BOPP film stretched from A80-500 sheet at several Stretching Ratios at 156 ℃

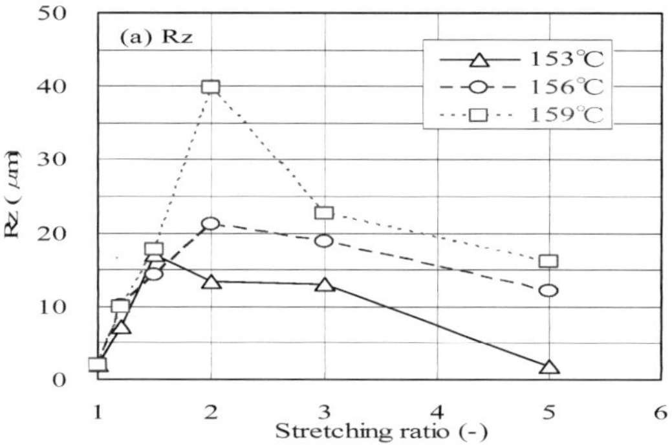

Fig. 3 Dependence of surface roughness on total stretching ratio of A80-500 at different stretching temperatures

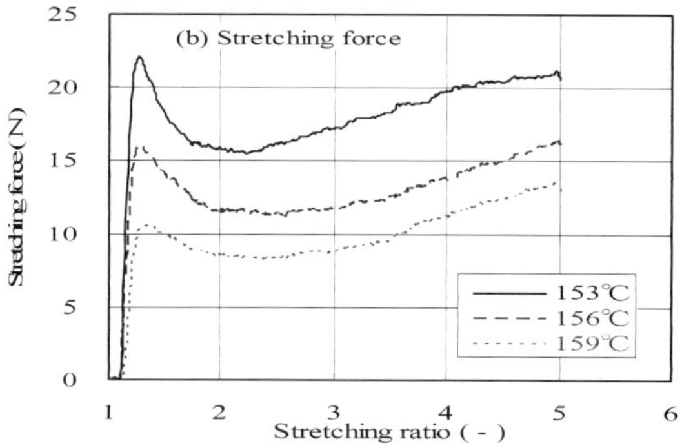

Fig. 4 Dependance of stretching force on stretching ratio of A80-500 at different stretching temperatures

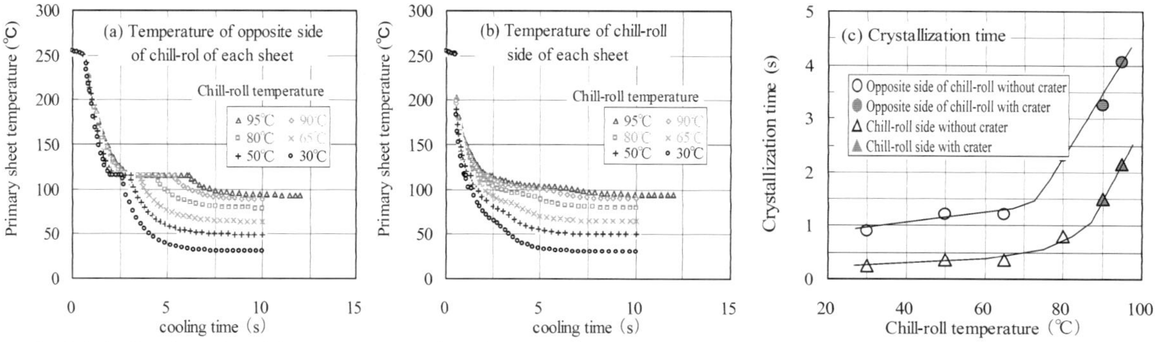

Fig.5 Calculated results of the temperature of PP sheets with different chill roll temperatures

Conditions: Die temperature 250℃, Thickness of sheets 300 μ m, Stretching Temperature 159℃, MD/TD Stretching Ratio 5/7

O-03-507

Fabrication of Carbon-core/TiO_2-Shell Nanofibers and their Application to Dye-sensitized Solar Cell

Asumi Tanaka*, Shinsuke Nagamine and Masahiro Ohshima

Department of Chemical Engineering, Kyoto University, Katsura Campus, Nishikyo-ku, Kyoto 615-8510, Japan,
a.tanaka@cheme.kyoto-u.ac.jp, nagamine@cheme.kyoto-u.ac.jp, oshima@cheme.kyoto-u.ac.jp

Abstract - Carbon-core/TiO_2-shell fibers were fabricated by two methods consisting of electrospinning, sol-gel coating and carbonization. Both methods start with the fabrication of precursor fibers of polyacrylonitrile (PAN) by electrospinning. In the first method (Method 1), the PAN fibers were coated with TiO_2, followed by the carbonization of PAN where crystallization of TiO_2 simultaneously occurs. In the second method (Method 2), the PAN fibers were carbonized prior to the coating with TiO_2. The TiO_2-coated carbon fibers were heated again to crystallize the TiO_2 shells. Both methods could successfully produce the carbon-core/TiO_2-shell fibers. The diameters of the resultant carbon-core/TiO_2-shell fibers were hundreds of nanometers and increased with the concentration of the PAN solution. The main advantage of Method 1 is simplicity, although it has difficulty in controlling the crystal structure of TiO_2 shells. In method 2, the crystal structure of TiO_2 shells could be controlled by selecting the crystallization temperature independently from the carbonization temperature.

Keywords: Electrospinning, Nanofiber, Carbonization, Core-shell structure, Dye-sensitized solar cell

Introduction

Dye-sensitized solar cells (DSSCs) have received intensive attention because of their lower cost, higher efficiency compared to other nonsilicon alternatives and environmental friendliness. Recently, several groups have shown the performance of DSSCs whose anodes were composed of carbon fibers and metallic oxide. Liu *et al.* combined TiO_2 with carbon fibers vertically aligned on the substrate, which complemented the electron conductivity of TiO_2 [1]. Unalan *et al.* conceived that conductive carbon fiber fabric could be an alternative of substrate of transparent conductive oxide (TCO) [2]. Based on this concept, they produced ZnO nanowires on carbon fibers and the flexible composite was employed as the anode of DSSC. The composite served two functions of the electron generation at the ZnO nanowires and their transport through carbon fibers to external circuit.

Recently, we reported the fabrication of carbon-core/TiO_2-shell fibers by carbonizing the composite fibers of poly (vinyl alcohol) (PVA) and TiO_2 prepared by electrospinning and sol-gel coating [3]. In this study, we attempted to fabricate carbon-core/TiO_2-shell fibers using polyacrylonitrile (PAN) as the polymer, which can be easily converted to carbon materials with a high yield. Here we examined two production routes. The first method is similar to our previous approach. The electrospun PAN fibers were coated with TiO_2 shells. The TiO_2-coated PAN fibers were subjected to a thermal treatment to facilitate the carbonization of PAN and the crystallization of TiO_2 (Method 1). The advantage of this method is that thermal treatment is required only once. However, the appropriate temperature ranges of carbonization of PAN and crystallization of TiO_2 do not correspond with each other. In general, the high temperature would promote the carbonization of the core and enhance the electric conductivity. On the other hand, the excessively high temperature results in the crystal transformation of the TiO_2 shells from anatase to rutile, which is less suitable for the application in DSSC. In the second method, PAN fibers fabricated by electrospinning were converted to carbon fibers prior to the coating with TiO_2 shells. After the formation of TiO_2 shells, the composite fibers were heated again to crystallize the TiO_2 shells (Method 2). By this method, the optimal carbonization and crystallization temperatures can be independently selected, although the number of processes is higher than Method 1. The possibility of fabricating carbon-core/TiO_2-shell fibers by these 2 methods was demonstrated and the controllability of the fiber diameter by PAN concentration was discussed.

Experimental

Materials

PAN (M_W = 150,000) was obtained from Sigma-Aldrich. N,N-dimethylformamide (DMF), titanium tetraisopropoxide (TTIP) and ethanol were purchased from Wako Pure Chemical Industries.

Preparation of PAN Fibers

PAN solutions in DMF of various PAN concentrations from 5 to 12 wt% were prepared. This precursor solution was electrospun with a commercial electrospinning device (Kato-Tech, NEU). The applied voltage was 15 kV, the distance from the tip of the needle to the collector was 13.5 cm, and the feed rate of solution was 0.95 mL/h.

Preparation of Carbon/TiO_2 Fibers by Method 1

A nonwoven fabric of PAN fibers was immersed in a TTIP / ethanol solution (1.0 mol/L) for 1 min. After washed with ethanol and dried at room temperature, the fabric was stabilized at 230 °C for 2 h in air and then carbonized at 700 °C for 2 h under continuous N_2

purge. During this process, the crystallization of TiO_2 shells of fibers also proceeded.

Preparation of Carbon/TiO_2 Fibers by Method 2

A nonwoven fabric of PAN fibers was stabilized and carbonized under the same thermal conditions as Method 1. The carbonized fabric was immersed in a TTIP / ethanol solution. After washed with ethanol and dried, the TiO_2 shells of fibers were crystallized at 450 °C for 2 h in N_2.

Characterization of Fabrics

The morphologies of the fibers were studied using a field emission scanning electron microscope (FE-SEM, JEOL, JSM-6340FS, operated at 15 kV) and a transmission electron microscope (TEM, JEOL, JEM-1010, operated at 100 kV). The crystal structures were investigated by an X-ray diffractometer (XRD, Rigaku, Multiflex/CE, Cu Kα radiation, λ=0.154 nm, operated at 40 kV and 40 mA).

Results and Discussion

Morphology of the Resultant Fibers

Fig. 1 presents photographs of carbon/TiO_2 composite fabrics prepared by Method 1 and 2 with PAN concentration of 10 wt%. Both samples are flexible and black. Fig. 2 shows the SEM images of these samples. Both of them are mainly composed of fibers of several hundred nanometers in diameter as shown in Fig. 2 (a) and (b). The magnified images of the fractured cross-section suggest the dense structure of these fibers. It is observed that the shell is partially peeled off from the core in Fig. 2 (c) and (d). The cross-section of the composite fiber fabricated by Method 1 was observed by TEM (Fig. 3). A dark-colored shell is observed around of a light-colored core, and no void is seen between the core and shell.

For ease of observation of TiO_2 shells, we attempted to remove the core material by combustion at 700 °C in air. In case of Method 1, the precursory PAN/TiO_2 fibers were calcined instead of carbonization. The residual TiO_2 was in the shape of hollow fibers with the shell thickness of tens of nanometers (Fig. 4 (a)), demonstrating that the core fibers were completely covered by TiO_2 shells. The similar result was obtained for the carbon/TiO_2 fibers prepared by Method 2 (Fig. 4 (b)).

In these two methods, the formation of TiO_2 shells was conducted by simply immersing the fibers of PAN (Method 1) or carbon (Method 2), followed by drying. The core-shell structure could be formed by both methods, in spite of the difference in surface properties between PAN and carbon fibers. Water is required for the production of TiO_2 from TTIP through hydrolysis and condensation. However no water was added to the TTIP solution in these methods. We speculate that when the fibers were removed from the TTIP solution and dried, the TTIP adhered to the surfaces of the fibers reacted with the atmospheric moisture.

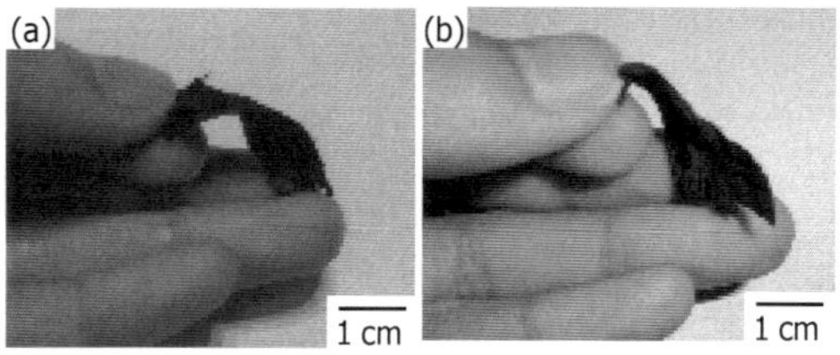

Fig. 1 Photographs of carbon/TiO_2 fabrics prepared by (a) Method 1 and (b) Method 2.

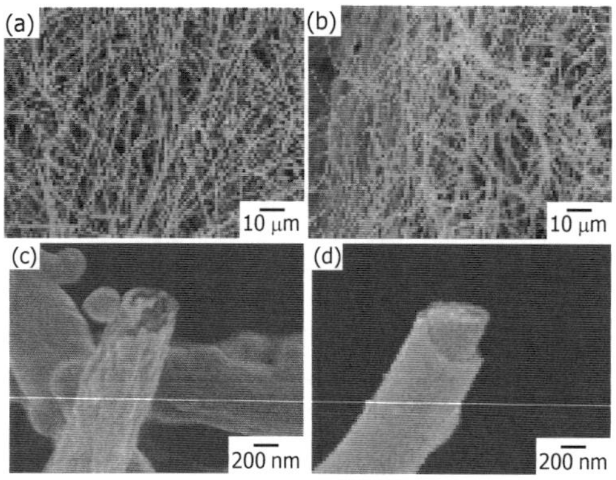

Fig. 2 SEM images of carbon/TiO_2 fibers prepared by (a), (c) Method 1, (b), (d) Method 2.

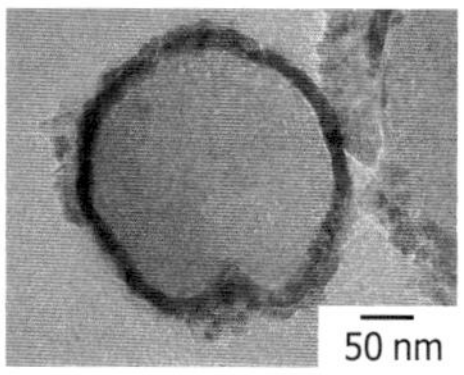

Fig. 3 TEM image of the cross-section of carbon/TiO_2 fiber prepared by Method 1.

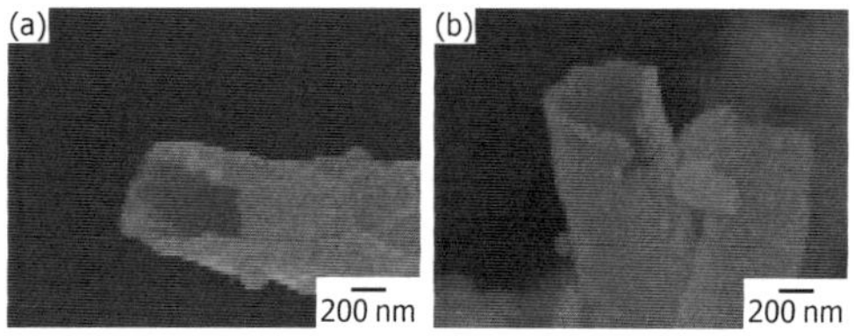

Fig. 4 SEM images of TiO_2 hollow fiber obtained by removing the core of (a) PAN/TiO_2 fiber prepared by Method 1, (b) carbon/TiO_2 fiber prepared by Method 2.

The Crystal Structures of TiO_2 Shells

The crystal structure of TiO_2 shells was investigated by an XRD and the results are presents in Fig. 5. The XRD pattern of the fibers prepared by Method 2 shows the diffraction peaks corresponding to anatase crystal structure of TiO_2. On the other hand, the TiO_2 shells of the fibers prepared by Method 1 are the crystal mixture of anatase and a portion of rutile. The difference in the crystal structure originated from the difference in crystallization temperature. The higher temperature of 700 °C in Method 1 promoted the crystal transformation from anatase to rutile. For control of TiO_2 shell structure, Method 2 was found to be more suitable compared to Method 1.

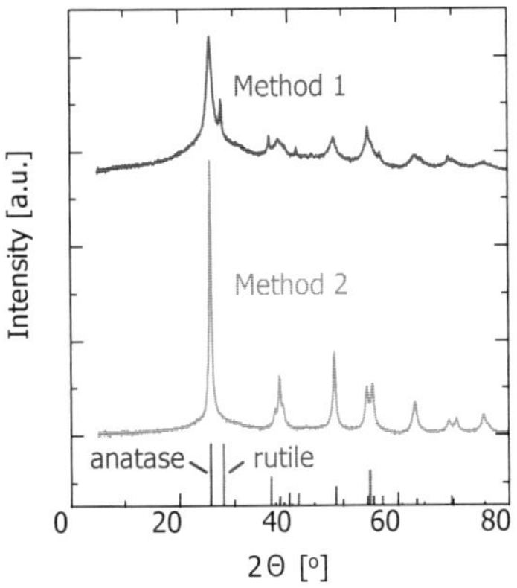

Fig. 5 XRD pattern of carbon/TiO_2 fibers prepared by two methods.

The Diameter of the Fibers

The effects of PAN concentration on the diameters were investigated for the fibers prepared by Method 1. Figs. 6 (a) and (b) present the SEM images of electrospun PAN fibers at different PAN concentrations of 9 wt% and 10 wt%. Some beads are observed in fibers prepared at 9 wt% because of the insufficient viscosity of PAN solution. The PAN concentration higher than 10 wt% is required for the production of bead-free fibers. SEM images of PAN-core/TiO_2-shell fiber and carbon-core/TiO_2-shell fiber prepared at 10 wt% are shown in Fig. 6 (c) and (d). After carbonization, the shrinkage of fiber was observed. The diameters of the fibers were measured from SEM images and are shown as the function of PAN concentration (Fig. 7). The PAN/TiO_2 fibers have the larger diameter than the PAN fibers due to the formation of TiO_2 shells. The diameter of fibers was reduced by carbonization, showing the shrinkage of fibers. The higher concentration of PAN solution led to the larger diameters of the PAN fibers. The dependence on PAN concentration is reflected by the diameters of the finally obtained carbon-core/TiO_2-shell fibers. The diameter of the fiber is controllable by the PAN concentration.

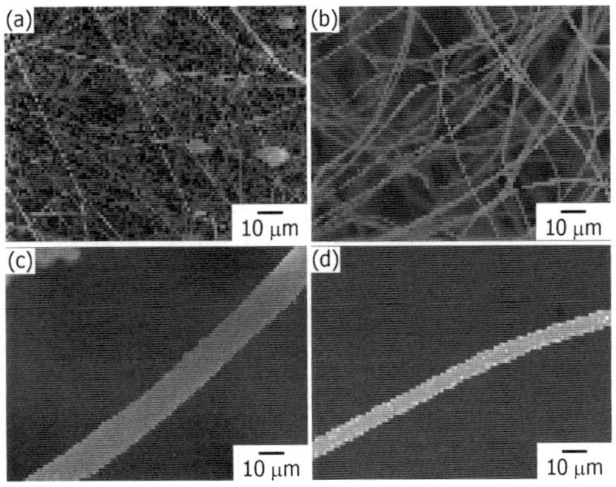

Fig. 6 SEM images of fibers. (a) PAN fibers prepared at 9 wt%. (b) PAN fibers at 10 wt%. (c) PAN /TiO_2 fiber at 10wt%. (d) carbon/TiO_2 fiber at 10wt%.

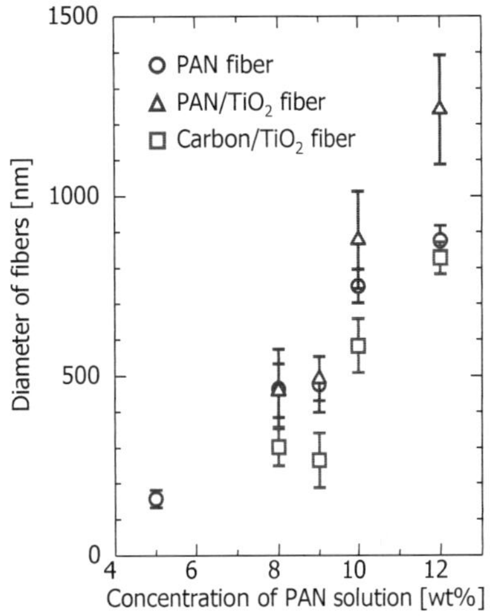

Fig. 7 Diameters of PAN fibers, PAN/TiO_2 fibers and carbon/TiO_2 fibers prepared with PAN concentrations of 5-12 wt%.

Conclusions

In this study, flexible carbon-core/TiO_2-shell fibers were prepared by two methods conducting electrospinning, sol-gel coating and carbonization. In Method 1, TiO_2 shells were formed on PAN fibers and the PAN-core/TiO_2-shell fibers were thermally treated to bring about carbonization and crystallization simultaneously. In Method 2, TiO_2 shells were formed on carbon fibers obtained by carbonization of PAN fibers, followed by the crystallization of TiO_2-shell.

Both methods could produce the carbon-core/TiO_2-shell fibers. Method 1 has difficulty in controlling the carbonization degree of the core and the crystal structure of TiO_2 shells separately, although it is simpler. From this viewpoint, Method 2 is superior to Method 1 because it enables to control the structures and properties of the core and shell by selecting the appropriate temperatures for carbonization and crystallization processes. The diameters of carbon-core/TiO_2-shell fibers were controlled by the concentration of PAN solution. Further study on the application of the carbon-core/TiO_2-shell fibers to the anode of flexible DSSCs is now in progress.

Acknowledgements

The authors appreciate Prof. K. Mae, Prof. T. Maki and Dr. I. Hasegawa in Kyoto University for their help in XRD measurements.

References

1. Liu J *et al.*, Novel dye-sensitized solar cell architecture using TiO_2-coated vertically aligned carbon nanofiber arrays *ACS Appl. Mater. Interfaces* 2009, *1*, 1645.
2. Unalan HE *et al.*, Photoelectrochemical cell using dye sensitized zinc oxide nanowires grown on carbon fibers *Appl. Phys. Lett.* 2008, *93*, 133116.
3. Nagamine S *et al.*, Fabrication of carbon-core/TiO_2-shath nanofibers by carbonization of poly(vinyl alcohol)/TiO_2 composite nanofibers prepared via electrospinning and an interfacial sol-gel reaction *Mater. Lett.* 2011, *65*, 3027.

O-03-557

EFFECT OF D-LACTIDE CONTENT ON STRUCTURE DEVELOPMENT BEHAVIOR OF POLYLACTIC ACID FILMS DURING UNIAXIAL AND SEQUENTIAL BIAXITIAL STRETCHING

Wataru Takarada[1]*, Noppadon Kerddonfag[2], Wannee Chinsirikul[2] and Takeshi Kikutani[1]

[1] *Dept. Organic and Polymeric Materials, Grad. Sch. Science and Engineering, Tokyo Institute of Technology, 2-12-1-S8-32, O-okayama, Meguro-ku, Tokyo, 152-8552,Japan - takarada.w.aa@m.titech.ac.jp; kikutani.t.aa@m.titech.ac.jp*

[2] *National Metal and Materials Technology Center, Thailand – noppadk@mtec.or.th ; wanneec@mtec.or.th*

Abstract - Crystallizability of poly(L-lactic acid) (PLLA) can be controlled by changing the D-lactide content, whereas the crystallizability affects the drawing behavior and structure development during stretching of PLLA films. Two PLLAs were used in this research, i.e. H-PLA with low D-lactide content and high crystallizability, and L-PLA with high D-lactide content and low crystallizability. In the film stretching of H-PLA, crystallization occurred in all the stretching conditions examined, namely the draw ratios of 3.0 and 4.0 in the uniaxial planar stretching and the draw ratios of 3.0 x 3.0 in the sequential biaxial stretching. Crystallization caused stress hardening, and stretching stress increased significantly especially during the TD stretching in the sequential biaxial stretching. On the other hand, in the case of L-PLA, no crystallization occurred in the uniaxial stretching at the draw ratio of 3.0. Accordingly, stress hardening was not observed in the biaxial stretching with the draw ratios of 3.0 x 3.0. It should be noted that spontaneous orientation development was vaguely observed in the uniaxial stretching when the draw ratio was increased to 4.0. It indicated that orientation-induced crystallization started to occur even in L-PLA under highly stretched conditions.

Keywords: polylactic acid, D-lactide content, film stretching, crystallization, birefringence

Introduction

Structure development behavior during the stretching of polymer films has been studied intensely using various crystalline and amorphous polymers [1-3]. In cases of crystalline polymers, crystallization behavior affects the stretching behavior and accordingly leads to the variation of the properties of resultant stretched films. In many cases, crystallization causes the increment of stretching stress and film tenacity. In some cases, however, higher stress hinders stretching up to high draw ratios. Therefore control of crystallization during film stretching is crucial to attain high stretchability as well as good mechanical properties of the stretched films.

Poly(lactic acid) (PLA) is recognized as a sustainable non-petroleum-based aliphatic polyester because PLA can be produced from plants such as corn. The lactide, a monomer of PLA, has two asymmetric carbons in its chemical structure, and therefore it can be classified into L-lactide, D-lactide and DL(meso)-lactide. Basically, lactide is produced by fermentation of glucose and chirality of lactide varies depending on bacterial species and fermentation conditions. In many cases, bacterial species produce L-lactide with optical purity of around 90 to 95 %. Accordingly, commercially available poly(lactic acid) pellets are mainly composed of L-lactide but contains a small amount of D-lactide.

Many researchers reported the effect of D-lactide content on the crystallizability, morphorgy and other properties of PLLA [4-8]. Takasaki et. al. studied the high-speed melt spinning behavior of PLLAs with three different D-lactide contents and reported that the orientation induced crystallization can be enhanced at high take-up velocities, however there was no trace of crystallization from the DSC measurement even at the take-up velocity of 8.0 km/min when D-lactide content was increased to 16.4 % [9].

In this study, on-line birefringence measurement during stretching of PLLA films of two different D-lactide contents was conducted and the effect of D-lactide content on the stretching behavior and crystallization behavior was investigated.

Experimental

Materials

Two commercial grade PLLAs from Nature Works were used. The D-lactide content of PLA4043D (H-PLA) is around 4.5 % and that of PLA4060D (L-PLA) is around 12%. H-PLA is a highly crystalline polymer whereas L-PLA is supplied as an amorphous polymer. Cast sheets were prepared from the individual PLLAs using a laboratory-scale flat film line (Dr. Collin GmbH). The thickness of the PLLA sheets was controlled to 240 μm.

Film stretching

PLLA sheets were stretched using a laboratory scale batch type stretcher (Bruckner GmbH, KARO IV). This apparatus allows us four-sides clamping with the 5-clips configuration on each side. In all experiments, stretching temperature was 80 °C and the sheet was pre-heated for 30 s before the stretching. Uniaxial stretching was conducted along the casting direction (machine direction: MD) with the 4 sides clipping. Strictly speaking, this stretching is defined as "planner stretching" because film width was kept constant. The draw ratios of 3.0 and 4.0, and the strain rates of 0.2, 0.5 and 1.0 s^{-1} were adopted.

Biaxial stretching was also conducted. The sheet was stretched firstly along the MD, and then along the transverse direction (TD). Draw ratios in both MD and TD were 3.0 and the strain rate was 0.5 s^{-1}. This stretching can be called the sequential equi-biaxial stretching. To investigate the effect of structural change between the MD and TD stretchings, TD stretching was started either immediately after or 5 s after the MD stretching..

On-line measurements

Tensile forces in MD and TD and optical retardation were measured during stretching. After the end of stretching, measurements were continued for 60 s to investigate the relaxation behavior.

The tensile force was measured with a tensile force sensor built-in in the center clip. Optical retardation was measured by an optical system consisting of a laser source, a rotating polarizer, a quarter-wavelength plate and two detectors with polarizers. Details of this optical system were reported in our previous paper [10]. Laser was irradiated vertically to the film surface through a hole located on the top cover of the stretching oven. The Laser passed through the film was reflected by a mirror placed below the film and came out from the oven through the same hole after passing through the film again, and then directed to the optical detector using a beam splitter. Variations of film thickness and cross-sectional area during the stretching were estimated assuming the homogeneous deformation and constant volume. Time-course variations of MD and TD stresses were evaluated from the MD and TD forces and cross-sectional area. Time-course variation of birefringence was analyzed from the measured optical retardation and the film thickness.

Results and Discussion

Uniaxial stretching

Changes of MD and TD stresses and MD-TD birefringence during uniaxial stretching and relaxation of L-PLA film conducted at various strain rates are shown in Fig.1. In this case, stretching was stopped at the draw ratio of 3.0. Tensile stress in both MD and TD increased during the stretching and then relaxed during the holding period. As expected, tensile stress in MD was higher than that in MD. The maximum tensile stresses at the end of stretching increased with increasing strain rate. Birefringence showed nearly the same behavior as tensile stress. Through the detailed analysis of these data, linear relations between the stress difference between MD and TD and birefringence was confirmed during the both

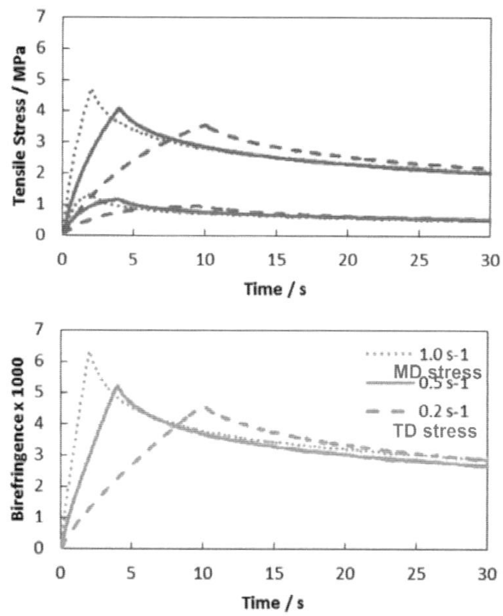

Figure 1 – Changes of MD and TD tensile stresses and MD-TD birefringence during uniaxial stretching and relaxation of L-PLA film. Draw ratio = 3.0. Strain rates are indicated in the figure.

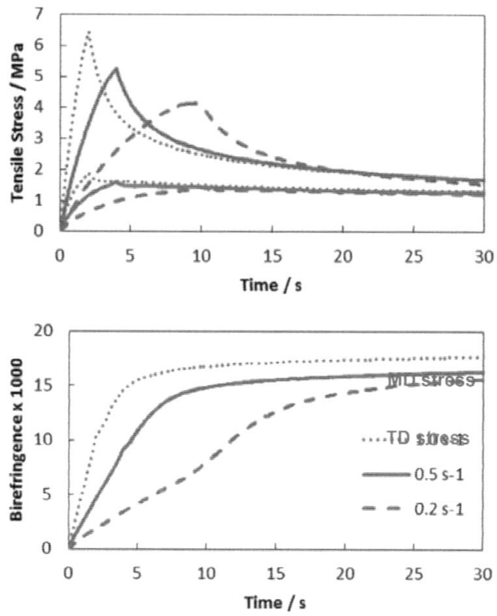

Figure 2 – Changes of MD and TD tensile stresses and MD-TD birefringence during uniaxial stretching and relaxation of H-PLA film. Draw ratio = 3.0. Strain rates are indicated in the figure.

stretching and holding processes. This result indicates that the film was in an amorphous and rubbery state where stress optical rule was applicable.

Changes of MD and TD stresses and birefringence during uniaxial stretching of H-PLA are shown in Fig. 2. In H-PLA, stresses decreased whereas birefringence increased during the holding period. Origin of the spontaneous growth of birefringence is attributable to the orientation induced crystallization. It is well known that molecular orientation enhances crystallization. It also should be noted that the crystallization occurring under the effect of flow/deformation can cause the spontaneous increase of molecular orientation.

Changes of stresses and birefringence during uniaxial stretching and relaxation of L-PLA for the draw ratio of 4.0 are shown in Fig. 3. In comparison with the results for the draw ratio of 3.0, birefringence relaxation during the holding period was less distinct. Even a slight increase of birefringence was observed at the late stage of holding period if the stretching was conducted at a high strain rate. These results indicated that orientation induced crystallization can occur under highly oriented conditions even in L-PLA, which is normally considered as an amorphous polymer,

Changers of stresses and birefringence during uniaxial stretching and relaxation of H-PLA for the draw ratio of 4.0 are shown in Fig. 4. Stress increased steeply at the late stage of stretching. This phenomenon is normally recognized as the "stress-hardening" behavior. It can be seen clearly that the steep increase of stress is accompanied by the steep increase of birefringence. In other words, both stress-hardening and steep birefringence increment were caused by the crystallization under tensile deformation of H-PLA.

Biaxial stretching

Changes of stresses and birefringence during sequential biaxial stretching of L-PLA with and without interval time between MD and TD stretchings are shown in Fig. 5. MD stress was higher than the TD stress during the MD stretching as confirmed in the uniaxial stretching experiments. Both MD and TD stresses as well as birefringence relaxed during the interval between the MD and TD stretchings. During the TD stretching, TD stress increased steeply. On the other hand MD stress decreased if there was no interval between the MD and TD stretchings whereas MD stress increased if TD stretching was applied after the relaxation of 5 seconds. Stresses in MD and TD after the TD stretching were similar irrespective of the interval time. Even though stress difference was close to zero, MD-TD birefringence showed slightly negative values immediately after the TD stretching. After the relaxation birefringence came back to zero. These results indicated that L-PLA appears to be amorphous even after the TD stretching, however a slight memory effect of MD stretching which enhanced the orientation toward TD during the TD stretching was observed. Maximum stresses decreased slightly if there was an interval time because of stress relaxation, however, effect of interval time on the MD-TD birefringence was

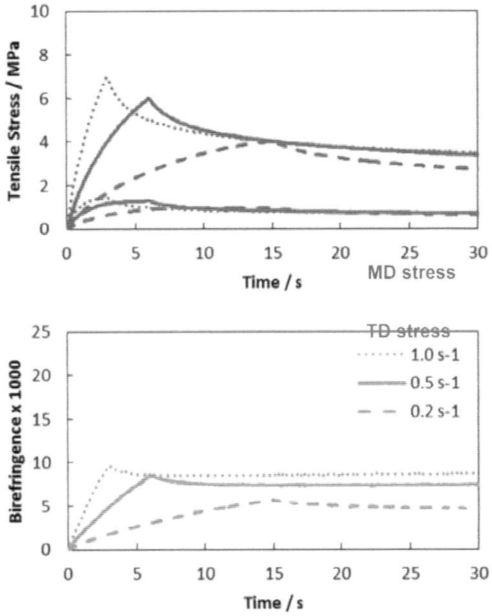

Figure 3 – Changes of MD and TD tensile stresses and MD-TD birefringence during uniaxial stretching and relaxation of L-PLA film. Draw ratio = 4.0. Strain rates are indicated in the figure.

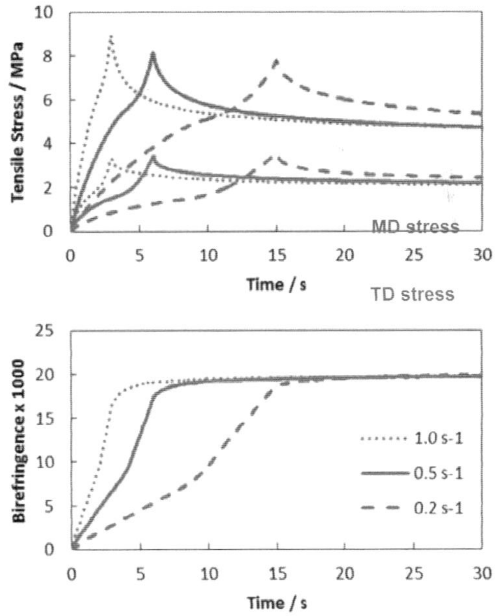

Figure 4 – Changes of MD and TD tensile stresses and MD-TD birefringence during uniaxial stretching and relaxation of H-PLA film. Draw ratio = 4.0. Strain rates are indicated in the figure.

negligible in cases of both immediately after the TD stretching and after the relaxation.

Changed of stresses and birefringence during biaxial stretching of H-PLA are shown in Fig. 6. In the latter stage of TD stretching, both MD and TD stresses increased sharply. The increases were more enhanced if an interval of 5 seconds was applied. At the end of TD stretching, TD stress was much higher than the MD stress. Even though birefringence increased spontaneously during the interval period, it decreased steeply during the TD stretching and showed negative values in both cases of with and without interval time. These phenomena suggested that the development of crystallites during the MD stretching and the holding period enhanced the development of TD stress. It should be noted that the final values of birefringence were similar in the cases of with and without the interval period even though variations of birefringence during the stretching processes were significantly different.

Conclusions

Stretching behavior and structure development of PLLA films of different D-lactide content were investigated. In the case of H-PLA, crystallization started to occur in the late stage of MD stretching, which leads to the spontaneous increase of MD-TD birefringence after the MD stretching and the steep increase of TD stress and sharp decrease of MD-TD birefringence down to negative values during the TD stretching. In the case of L-PLA, the polymer has low crystallizability, and the stretching behavior basically followed the stress-optical rule. However, small spontaneous orientation was found after uniaxial stretching to the draw ratio of 4.0. In case of sequential biaxial stretching, MD-TD birefringence showed negative values immediately after the TD stretching even though the stress difference was negligible. These results suggested that the orientation induced crystallization was occurring even in L-PLA under the effect of stretching.

References

1. J.B.F. Champchesnel, D.I. Bower, I.M. Ward, J.F. Tassin, G. Lorentz, *Polymer*, 1993, 34, 3763
2. S. Murakami, M. Yamakawa, M. Tsuji, S. Kohjiya, *Polymer*, 1996, 37, 3945
3. Y. Koike, M. Cakmak, *Polymer*, 2003, 44, 4249
4. X. Ou, M. Cakmak, *Polymer*, 2008, 49, 5344
5. H. Tsuji, Y. Ikeda, *Polymer*, 1995, 36, 2709.
6. T. Miyata, T. Masuko, *Polymer*, 1997, 38, 4003.
7. K. Jmashidi, S. –H. Hyon, Y. Ikeda, *Polymer*, 1988, 29, 2229.
8. J. J. Cooper-White, M. E. Mackay, J. Polym. Sci. Polym. Phys. B., 1999, 37, 1803.
9. M. Takasaki, H. Itho, T. Kikutani, *J. Macromol. Sci. B. Phys.*, 2003, 42, 57.
10. T. Kikutani, N. Nakao, W. Takarada, *Polym. Eng. Sci.*, 1999, 39, 2349.

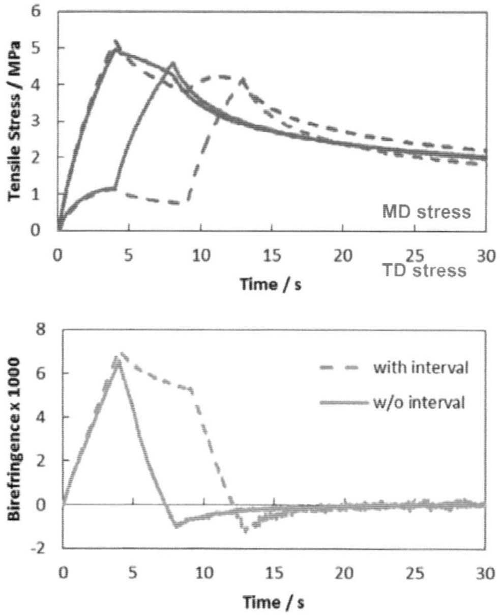

Figure 5 – Changes of MD and TD tensile stresses and MD-TD birefringence during sequential biaxial stretching and relaxation of L-PLA film. MD x TD draw ratio was 3.0 x 3.0. Strain rate was 0.5 s^{-1}. Solid lines and dashed lines are for the conditions of without interval time and with an interval time of 5 s between the MD and TD stretchings, respectively.

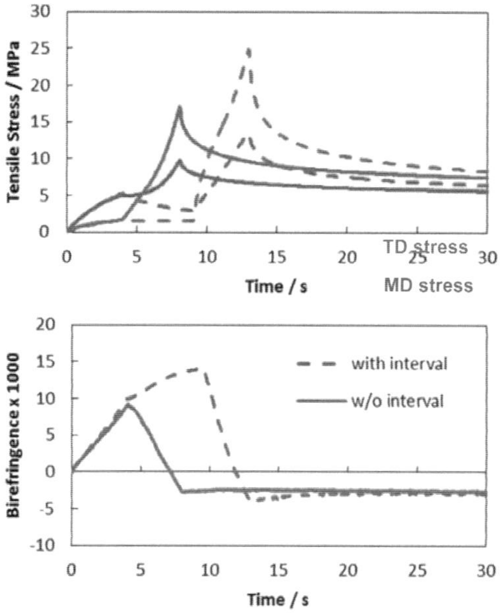

Figure 6 – Changes of MD and TD tensile stresses and MD-TD birefringence during sequential biaxial stretching and relaxation of H-PLA film. MD x TD draw ratio was 3.0 x 3.0. Strain rate was 0.5 s^{-1}. Solid lines and dashed lines are for the conditions of without interval time and with an interval time of 5 s between the MD and TD stretchings, respectively.

O-03-559

DEVELOPMENT OF A GASBAG-ROLLER-ASSISTED HOT EMBOSSING PROCESS WITH ANODIC ALUMINUM OXIDE ROLLER TO FABRICATE NANOSTRUCTURES ON POLYMERIC FILM

C.F. Lu, C.H. Lin and S.Y. Yang*

Department of Mechanical Engineering, National Taiwan University, R.O.C.-syyang@ntu.edu.tw

Abstract To improve the pressure uniformity of roller hot embossing, a gasbag-roller-assisted hot roller embossing process is designed and developed. A gasbag-roller applies the characteristic of fluid isotropic and uniform pressure to enhance the uniformity of the hot roller embossing pressure. Anodic aluminum oxide (AAO) roller mold is employed in this study, which can be fabricated in low cost and large area. A gasbag-roller-assisted hot roller embossing facility is constructed and the experiment is conducted. The experiment demonstrates that the nano-pillars array can be successfully fabricated on PMMA film. The SEM, reflective rate and hydrophobic effect of the PMMA film are also evaluated.

Keywords: Anodic aluminum oxide, hot roller embossing, anti-reflective, hydrophobic, PMMA

Introduction

Recently, anti-reflective and hydrophobic films with nano-pillar array are drawing attention due to their applications in self-cleaning films, display devices, lenses and solar cell. Conventional methods to fabricate nanostructure on the mold are ion beam (FIB) [1], e-beam [2] and photolithography [3]. However, these methods are expensive, time-consuming and of limited area. In 1995, Masuda employed anodic process to fabricate hexagonal nanohole array on the aluminum plate[4]. This method has advantages of low-cost, simple process and large area. Therefore, we employ anodic process to fabricate anodic aluminum oxide on the tube as mold for hot roller embossing.

Experimental

1. Preparation of anodic aluminum oxide (AAO) roller mold

A 99.999% aluminum sheet is bent to a tube and annealed under nitrogen ambient at 470°C for 3 hours. The two-step anodic process is as follows (Fig. 1)

(a)The aluminum is electro-polished in a 1:4 volume mixture of $HClO_4$ and C_2H_5OH under the voltage of 10V for 10 min.

(b)The first anodization is performed in 0.1M phosphoric acid solution at 4°C and 180V for 4 hours.

(c)The oxide layer is etched away by 12wt% H_3PO_4 at 32°C for 4 hours.

(d)The second anodiztion is performed in 0.1M phosphoric acid solution at 4°C and 180V for 10min.

(e)The pore-widening step is performed in 12 wt% H_3PO_4 at 32°C for 10 min.

After the anodic process, a anodic aluminum oxide with 400 nm interpore distance and 250 nm pore diameter is obtained as shown in Fig. 2(a). The SEM image of nanopore is shown in Fig. 2(b). An are of 50mm × 50mm is cut and to be attached on to the surface of a heating roller.

2. A gasbag-roller-assisted hot embossing facility and process

Fig. 3 shows the schematic of gasbag-roller-assisted hot embossing. A gasbag-roller-assisted hot embossing facility is designed and constructed for experiments. As shown in Fig. 4, the facility consists of a gasbag-roller, an AAO roller mold, a driving motors and a heating roller. The gasbag roller is manufactured by wrapping and sealing a silicon tube around an aluminum dumbbell-shaped rod. After being inflated with specific pressure gas (P_{inner}), the gasbag roller can exert pressure over the AAO mold. Above the gasbag roller is a rigid roller, two pneumatic cylinders presses the two ends of the rigid roller to exert specific pressure (the external pressing pressure, P_{outer}). A DC motor drives the gasbag roller, which by friction drives the substrate pass the gasbag roller and AAO tube mold. The moving velocity of the substrate is between 1.4 mm/s ~ 7 mm/s. The AAO roller mold is attached with heat resistant sealant to the heating roller. The power of the heating is 200W and the maximum temperature is 300°C. The substrate is Polymethylmethacrylate (PMMA). The width is 50mm and the thickness is 0.4mm. The glass transition temperature (Tg) is 105°C.

3. Measurement of antireflection and hydrophobic effect

Spectroscope (Hitachi U-3501 UV-VIS-NIR) and the contact angle measurement system are employed in this study for evaluating the reflctive rate and contact angle.

Results and Discussion

1. The fabricated PMMA film with nano-pillar array

Fig. 5 shows 45° view observed under a scanning electron microscope (JOEL JSM-6700F) of the fabricated nano-pillars over a 50mmx50mm PMMA substrate fabricated using the gasbag-roller-assisted hot embossing process. The nano-pillars array is fabricated under the condition of 150°C AAO mold, 1.6 kgf/cm^2 pressure in the gasbag roller and 0.4 kgf/cm^2 pressing pressure upon rigid roller with the driving motor at the speed of 1.47mm/s. Fig. 6 shows the effect of the pressing pressure and motor velocity on nanostructure protrusion under fixed condition of 1.6kgf/cm^2 inner gas pressure of gasbag roller and 150°C AAO mold. The parameters and experimental results are shown in table 1. Five local areas of nano-pillars are randomly selected to measure the average height.

2. Anti-reflection and hydrophobic effects of PMMA films with nano-pillars array

The reflective rate in different wavelength in the PMMA films with nano-pillar array are measured (Fig. 7) . The reflective rate of bare PMMA film without nano-pillar array is about 2.6%~6.4% at the wavelength from 350mm~800mm. The reflective rates of the PMMA films fabricated with pressing pressure of 0.2kgf/cm^2 ~0.4 kgf/cm^2 are 0.1%~0.3%. Moreover, Fig. 8 shows the contact angles of the PMMA with and without nano-pillar array. The contact angle of bare PMMA is 75.9°. With increasing of the pressing pressure, the contact angles of the PMMA film increase. The higher pressing pressure results in deeper protrusion of nano-pillars.

Table 1 –The average heigh of the nano-pillar under various parameters.

Pressing pressure, P_{outer} (kgf cm-2)	Motor speed(mm/s) 1.47	2.21	2.94
0.2	~260	~205	~120
0.3	~290	~265	~150
0.4	~320	~320	~230
0.5	~320	~320	~230

(unit:nm)

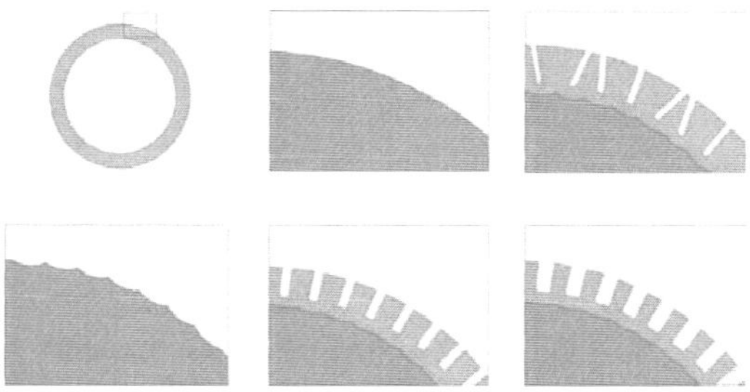

Figure 1 –Two-step anodic process to fabricate anodic aluminum oxide (AAO)

(a)

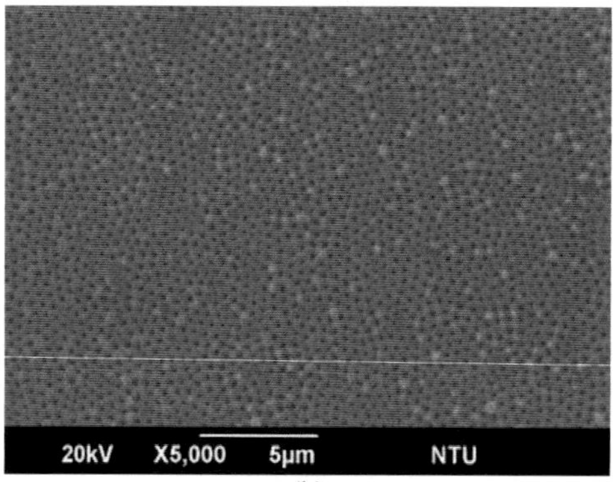

(b)

Figure 2 – (a) AAO tube mold (b) The SEM image of AAO tube mold

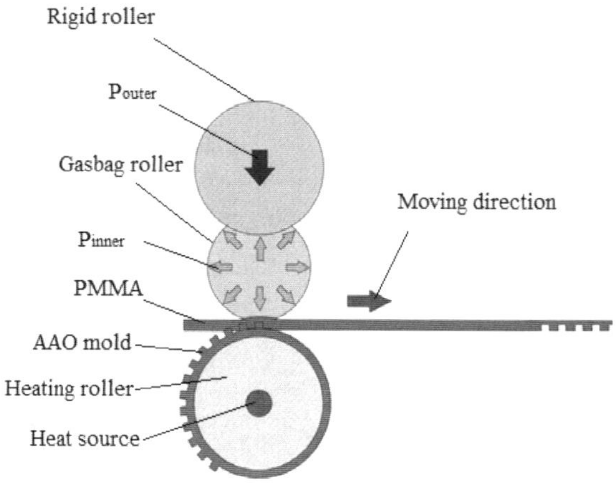

Figure 3–Schematic diagram of gasbag-roller-assisted hot embossing

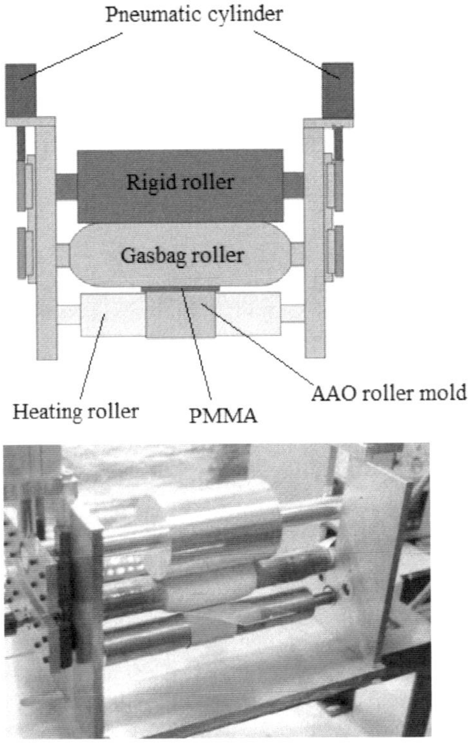

Figure 4 –The sketch and photo of the gasbag-roller-assisted hot embossing facility

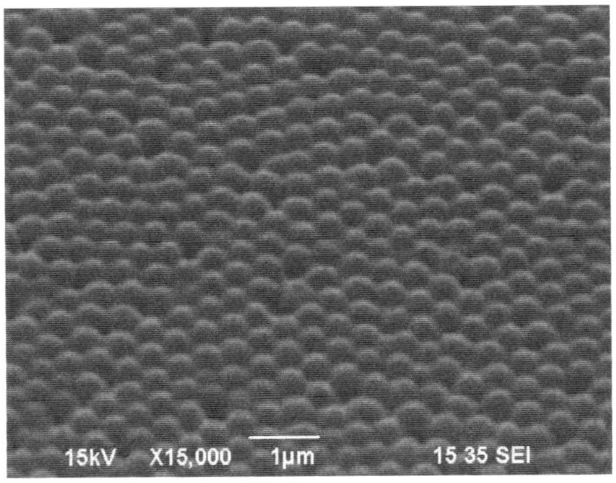

Figure 5–The fabricated nano-pillar array on the PMMA films

Figure 6 –The process window showing the effect of the pressing pressure and the motor speed on the fabrication of nanostructure on the PMMA film

Figure 7 –The reflective rate of the PMMA films without and with the nano-pillars array

Figure 8 –The contact angles of the PMMA films without and with the nano-pillar array

Conclusions

In this study, anodic aluminum oxide (AAO) is employed to produce roller with anodic aluminum oxide nanostructures. In addition, a gasbag-roller-assisted hot roller embossing has been designed and implemented. Nano-pillars structures can be fabricated on PMMA surface by the gasbag-roller-assisted hot roller embossing facility. The hydrophobic and anti-reflective characteristic of PMMA with nanostructures have been verified. The contact angle of PMMA film with nanostructures is 95.65° while the contact angle of a bare PMMA is 75.9°. Nanostructures on the PMMA also decrease reflective rate from 6% with bare PMMA to 0.2%. This study demonstrates an efficient and low-cost method for fabrication of nano-pillars on polymeric films for hydrophobic and anti-reflective function.

References

1. Y. H. Lanyon, G. D. Marzi, Y. E. Watson, A. J. Quinn, J. P. Gleeson, G. Redmond and D. W. M. Arrigan. *Chem.* 2007, *79*, 3048
2. S. Griffith, M. Mondol, D. S. Kong, and J. M. Jacobson *J. Vac. Sci. Technol.* 2002, *B 20*, 2768
3. H. W. Chen, Y. C. Lee, Member, IEEE, F. B. Hsiao 2010 5th IEEE International Conference on Nano/Micro Engineered and Molecular Systems January 20
4. H. Masuda and K. Fukuda *Science* 1995 *268 (5216)*: 1466

O-03-625

SEQUENTIAL BIAXIALLY ORIENTED POLYLACTIC ACID FILMS CONTAINING DIFFERENT D -LACTIDE CONTENTS: DEFORMATION BEHAVIOR AND STRUCTURE DEVELOPMENT

Noppadon Kerddonfag*, Teerapon Yamwong, Bongkot Hararak, Pramote Kumsang, Chao Sripethdee, Charinee Winotapun and Wannee Chinsirikul

National Metal and Materials Technology Center,
National Science and Technology Development Agency,
Ministry of Science and Technology, Pathumthani 12120, THAILAND

Abstract - To investigate the effects of PLA having different D-lactide contents on its biaxial deformation behavior, three grades of NatureWorks LLC's PLA resins were used. These included 4032D, 4043D and 4060D which contained different stereo-regularities and the D-lactide contents of 1.40, 4.25 and 12.0%, respectively. For comparison, the stretching condition was fixed for all samples. The preheating and stretching temperature was conducted at 80°C and the deformation speed was 4,500 mm/min. During the MDO stretching (MD ratio=5.0), all samples showed a similar biaxial deformation behavior in early stage but exhibited different strain hardening behavior when the stretch ratio was higher than 3.0. PLA with lower D-lactide content in particular 4032D showed much more pronounced strain hardening than 4043D whereas 4060D exhibited only a slight increase in draw stress. It was found that the final stresses in MD of 4032D, 4043D and 4060D (at a stretch ratio of 5x1) were about 26.4, 20.1 and 9.6 MPa, respectively. Using the *in-situ* birefringence measurement, during film stretching process, PLA 4032D and 4043D exhibit a stress-birefringence behavior of crystalline development structure while PLA 4060D shows an amorphous-like behavior as agreed with the result from DSC experiment. Moreover, it is found that by incorporating 20wt% of 4060D into either 4032D or 4043D, a remarkable reduction in stretching stress, birefringence and crystallinity were also observed. It is interesting to point out that the D-lactide content in PLA plays a vital role in biaxial deformation behaviors of both neat PLAs and their blends. The influence of regularity of crystallizable PLA's chain segment on structural development under film stretching process is observed not only in MDO stretching but also in TDO stretching.

Keywords: Polylactic acid, Oriented film, Structure and morphology

Introduction

According to increasing oil prices and rising consumer consciousness about global warming, Polylactic acid (PLA) is becoming more popular and attractive biodegradable plastic to replace traditional petroleum based plastics. PLA takes one challenge in packaging applications, especially the biaxially oriented films (BOPLA films) as it has good optical and mechanical properties. Comprehensive studies relating to structural development and crystallization of biaxially stretched PLA films were reported by Ou and Cakmak [1]. Several attempts have been made to extend stretchability of the biaxially oriented biodegradable polylactic acid to the highest possible stretch ratio. One method is incorporating a non-biodegradable component (Ethylene-acrylate copolymer, Biomax), generally known as a toughness modifier, into PLA. However, the disadvantage is that Biomax could reduce optical properties of the films, increasing haze value up to 6-11% [2].

As our previous studies [3,4], one method to enhance the stretchability of BOPLA was incorporating high D-lactide PLA into the lower D-lactide PLA without sacrificing biodegradability. Thus, the objective of this study is to further investigate the effects of PLA having different D-lactide contents (including its blend) on its biaxial deformation behavior related to the structural development. The interesting technique used in this study, to reveal structural development under a rather high deformation rate of about 100%/sec ($1.0 s^{-1}$), is the *in-situ* birefringence measurement that have not been previously reported by other research groups.

Experimental

Materials and Sheet extrusion

The general biaxial film stretching grades of polylactic acids used in this work were NatureWorks LLC-PLA 4032D, 4043D and 4060D containing different D-lactide contents of about 1.40, 4.25 and 12wt%, respectively. In this study, 5 different precursor sheets comprising of neat PLAs and PLA blends (as described in Table 1) were prepared by sheet extrusion. PLA sheet was cast using an extruder with 80 screw rpm, with melting temperature of ~210°C. Cast sheets with thickness of about 0.30 mm were quenched on a water cooled chill roll with controlled temperature of 53°C. Draw speed of the chill roll was kept relatively low in order to minimize orientation in the as-cast sheets.

Table 1 Designation and composition of PLA samples

Sample	PLA 4032D	PLA 4043D	PLA 4060D

4032D	100%	-	-
4032D-H20	80%	-	20%
4043D	-	100%	-
4043D-H20	-	80%	20%
4060D	-	-	100%

Film stretching and birefringence measurement

A sequential biaxial film stretching was conducted on a laboratory stretcher, Brueckner KARO IV. This apparatus allows for four-side clamping with 5-clip configuration on each side. The as-cast sheets, dimension of 75 mm x 75 mm, were first stretched in machine direction (MD) and followed with stretching in the transverse direction (TD). The samples were preheated at 80°C for 45 s and then stretched at 80 °C under a strain rate of 1.0 s^{-1}, corresponding to 4,500 mm/min. After that, the stretched samples were held for 60 s in order to observe the stress relaxation behavior. During stretching process, the stretching forces (MD&TD) and birefringence were simultaneously recorded. The biaxial stresses of MD and TD were calculated from the instantaneous cross sectional area based on the initial sheet's cross sectional area and speed of stretching clip. The birefringence was calculated from the optical retardation obtained from the *in-situ* birefringence measurement apparatus equipped with biaxial film stretching machine.

Thermal analysis by Differential Scanning Calorimetry (DSC)

Perkin Elmer DSC7 was used to determine the transition temperatures and %crystallinity of samples at a heating rate of 10°C/min under nitrogen atmosphere. Samples were enclosed in aluminium pans with crimped-on lids. An empty aluminium pan was used as a reference. Sample crystallinity was calculated by using the heat of fusion of 100% crystalline PLA of 93 J/g [5]. In the case of cold crystallization, enthalpy of cold crystallization must be subtracted from the melting enthalpy.

Results and Discussion

Morphology of precursor sheet by thermal analysis

Fig. 1 shows DSC thermograms of as-cast sheets, including of neat PLAs and PLA blends. All samples show a similar Tg in a range of ~54-55°C but different in details of cold-crystallization and melting behaviors. 4032D and 4043D show a larger cold crystallization and sharp melting peaks compared to those of blend samples. The melting temperature of 4032D (164.3°C), which has the lowest D-lactide content (1.40%), is higher than that of 4043D (150.2°C), as shown in Table2. It was found that the crystallinity (%X_c) of 4032D is about 7.2%. However, there were no traces of crystallization found from 4032D (and also all other samples) during cooling scan in DSC experiment (-10°C/min). It can be suggested that the morphology of 4032D, as well as all cast sheets, formed in the cast sheet extrusion process was likely to be amorphous but it is possible to further crystallize during heating (ΔH_{cc}=-31.0J/g). With regard to PLA 4043D with higher D-lactide content of about 4.25%, the smaller area under cold crystallization peak (ΔH_{cc}= -12.5J/g) and lower melting temperature was clearly observed. This is attributed to the lower degree of chain regularity or crystallizable chain segment of 4043D compared to that of 4032D [6].

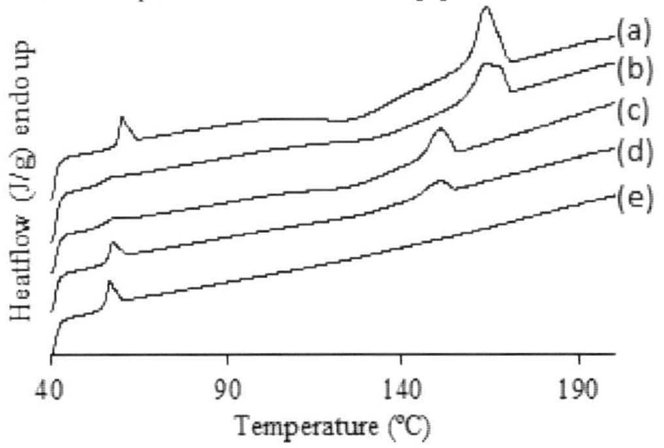

Fig 1 DSC thermograms of cast PLA sheets: (a) 4032D, (b) 4032D-H20, (c) 4043D, (d) 4043D-H20 and (e) 4060D.

On the other hand, 4060D shows neither cold crystallization nor melting peaks, as it contains highest D-lactide content of about 12.0%. It is indicative of amorphous-like PLA owing to the irregularity of its polymer chain. By incorporating 20wt% of high D-lactide PLA (PLA 4060D) into the blends, the exothermic ΔH_{cc} of 4032D-H20 and 4043D-H20 were found decreasing from 36.6 to 22.2 J/g and 12.5 to 5.2 J/g, respectively. In this regard, high D-lactide PLA plays an important role on crystallization of the blends by reducing "cold crystallization ability" of PLA during heating. It is interesting to note that the regularity of chemical structure in PLA's polymer chain can favor or control crystallization occurring in both of heating and also in film stretching process. This will be later discussed.

Table 2 Thermal analysis of PLA sheets.

Sample	T_g (°C)	T_{cc} (°C)	ΔH_{cc} (J/g)	T_m (°C)	ΔH_f (J/g)	%X_c (%)
4032D	55	126.7	-31.0	164.0	38.2	7.2
4032D-H20	55	132.7	-22.2	164.2	23.4	1.3
4043D	55	124.0	-12.5	151.0	13.2	0.8
4043D-H20	55	124.7	-5.2	150.2	6.9	1.8
4060D	54	-	-	-	-	0

MDO stretching and birefringence measurement

According to MDO stretching with deformation speed of about 4,500 mm/min (1.0 s^{-1}), the stretch ratio of MD reached 5 times (MDO=5x) in 4.0 s then held for 60 s to observe the stress relaxation and changing in birefringence. Under this stretching condition, the stretching stresses and birefringences of all samples seem to reach the steady state. Fig. 2 shows a deformation behavior (stretching stresses) and birefringence of each PLA and blended samples as a function of time, including stretching period and holding period for relaxation (see Fig.2 (c)). The plots shown in this paper are the early stage of experiments and the holding period for relaxation is about 10 s after stretching.

During the MDO stretching, all samples showed a similar biaxial deformation behavior in early stage but exhibited different strain hardening behavior when the stretch ratio was higher than 3.0. Of all samples, 4032D exhibits the most pronounced strain hardening with highest stretching stresses (in MD&TD), birefringence and also the crystallinity (see Table 3). The increase of birefringence is an indication of orientation to the direction of stretching force. The birefringence seems to increase to the maximum value as agreed with increasing of MD stress during MDO stretching. Subsequently, it is found that MD stress drastically decreases and finally reaches the plateau in relaxation period while the birefringence reaches the plateau with a small variation. It could be suggested that the "newly formed structure" has been permanently developed during MDO stretching.

4032D-H20, the blended sample of PLA4032D with 20wt% of PLA4060D (D-lactide content=12.0%), exhibits a lower level of birefringence and stretching stresses, in particular, in a range of MD ratio = 3.0 to 5.0 (see Fig. 2 (a)). In the case of 4043D (MDO=5x), which contains D-lactide content=4.25% and its blended sample (4043D-H20) also show a similar tendency of lowering in birefringence, stretching stress and crystallinity like 4032D and 4032D-H20.

According to stress-birefringence relationship as shown in Fig 3, 4060D shows a linear increasing of birefringence which agrees with Stress-Optical Rule (SOR). Under MDO stretching at strain rate of $1.0s^{-1}$, it is found that 4060D exhibits the amorphous orientation behavior, which is similar to A-PET under relatively low strain rate ($0.0025s^{-1}$) and small deformation [7].

It can be seen that the birefringence of 4060D slightly increased (from 0.0134 ± 0.0005 to 0.0143 ± 0.0006) while MD-TD stress decreased (from 8.2 ± 0.3 to 3.5 ± 0.1 MPa) in the period of stress relaxation. It is an indication of permanent structure formed in this interval time. The small endothermic peak, corresponds to melting, was observed from DSC result of this stretched sample at the temperature of about 113°C. The melting temperature (T_m) of MDO stretched 4060D sample is much lower than those of 4032D and 4043D. Nevertheless, this ordered structure or crystal is formed in a certain condition, in particular of adequate time, under strain and elevated temperature.

4032D sample also shows a linear increasing of birefringence at the initial stage of MDO stretching, before deviating from the SOR. The onset of strain-induced crystallization is generally determined from the point that deviates from the SOR line. It is found that at the SOR region the stress optical constants (slopes) of 4032D-H20 and 4043D are relatively the same but that of4043D-H20 is slightly lower. The stress optical constant of 4043D is the highest while 4060D shows the lowest. These birefringence results well agree with the crystallinity of MDO stretched samples (see Table 3).

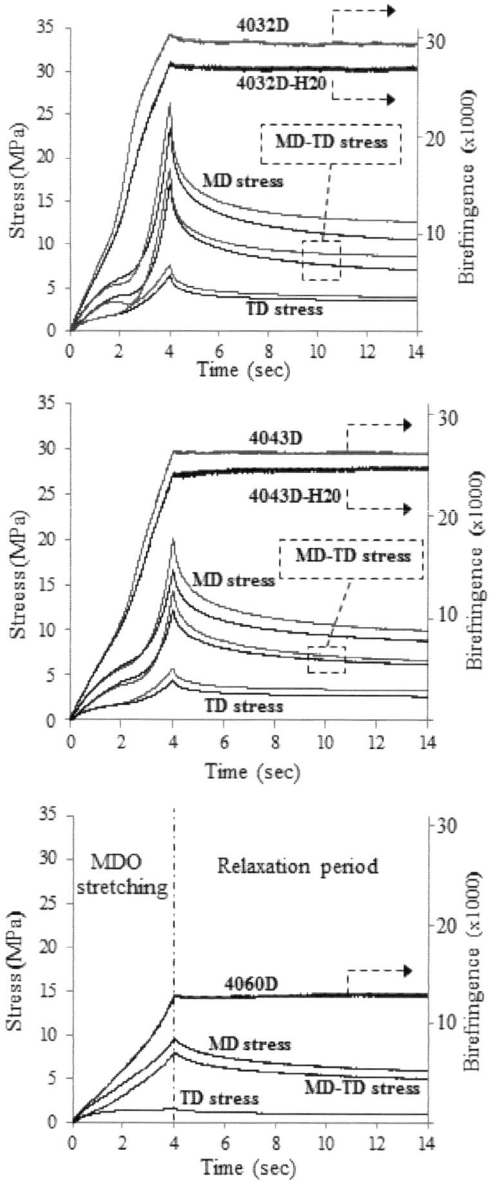

Fig 2 Deformation behaviors (MD&TD stresses) and birefringences of PLA samples (MDO=5x) plotted as a function of time: (a) 4032D and 4032D-H20, (b) 4043D and 4043D-H20 and (c) 4060D.

Table 3 Properties of MDO stretched PLA samples (MDO=5x).

Sample	Max.σ_{MD} *(MPa)*	Max.σ_{TD} *(MPa)*	%A*	Max.Δn *(x1000)*	%X_c *(%)*
4032D	26.4	7.7	100	29.4	43.3
4032D-H20	23.5	6.4	91.5	27.6	37.7
4043D	20.1	5.7	82.5	26.9	31.7

4043D-H20	16.6	4.4	76.4	23.8	27.6
4060D	9.6	1.6	55.8	13.4	2.3**

* *Calculated area under stretching curves as compared to area under stretching curve of 4032D sample.*
** *Small endothermic peak at temperature of about 113°C.*

In this step, it could be concluded that the increasing of D-lactide content in polymer chain of PLA as well as blending with higher D-lactide component has high influence on retarding the structural development, in particular molecular orientation and crystallization, during MDO stretching.

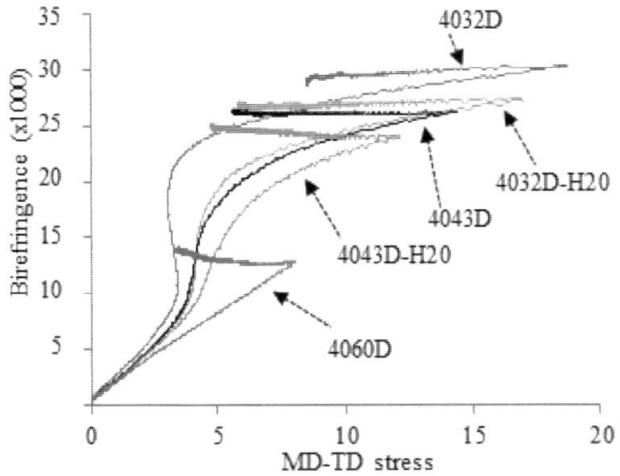

Fig 3 The MD-TD stress-birefringence relationship of PLA samples (MDO=5x) reploted data from Fig.2.

Sequential biaxially film stretching

In order to demonstrate the effect of structure development from the 1st step stretching on the 2nd step stretching, 4043D and 4043D-H20 were selected in this investigation. Fig. 4 presents overall deformation behavior under sequential biaxial stretching, consisting of stretching stresses in MDO and TDO. Refer to MDO stretching, 4043D exhibits higher stretching stresses (MD&TD), birefringence and crystallinity than 4043D-H20 (as depicted in Fig. 2(b) and Table 3). Fig 4 shows that the developed structure in MDO stretching has influence on that in TDO stretching. It seems to be that the more structure in MDO streching is developed, the more stretching energy in TDO step is required. The TD stress of 4043D was found gradually developed in MDO stretching period then abruptly increased in the early stage of TDO stretching and rised up to the higher stress than that of MD stress.

In the case of 4043D-H20, the lower deformation stresses were found along the stretching period, in particular less TD stress at the initial stage of TDO stretching and its shallow slope compared to 4043D. It is possible that low strain hardening or resistance to deformation of 4043D-H20 is attributed to the high D-lactide content of 4060D counterpart that could hinder the structure orientation. The calculated area under stretching curves, corresponding to the stretching energy consumption during film forming, of 4043D-H20 is about 13.2% lower than that of 4043D. The crystallinity of stretched 4043D-H20 film was found to be 12.0% lower than that of 4043D (X_c=25.8% and 29.3%, respectively).

It could be summarized for blending between PLAs containing different D-lactide contents that the higher D-lactide PLA can play an important role on structural development under film stretching process, not only on the MDO stretching but also on the TDO stretching.

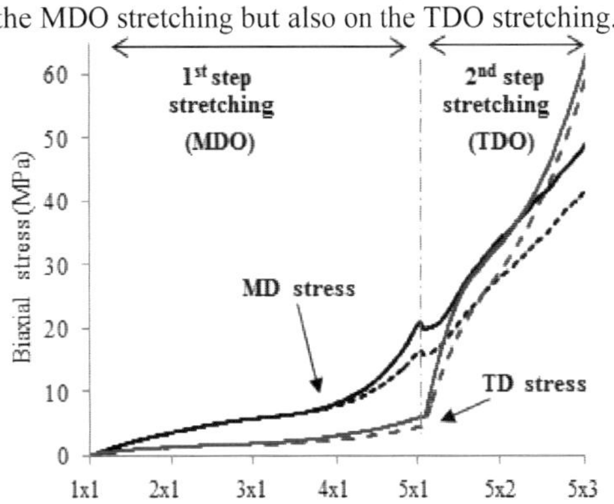

Fig 4 Deformation behaviors of 4043D (solid line) and 4043D-H20 (broken line) under sequential biaxially film stretching with stretch ratio = 5x3.

Conclusions

In this study, the effect of PLA having different stereo-regularities on its biaxial deformation behavior was investigated. The three-grade of PLAs containing various D-lactide contents, PLA 4032D, 4043D and 4060D were used (D-lactide content = 1.40, 4.25 and 12.0%, respectively). Of all samples, 4032D exhibits the most pronounced strain hardening with highest stretching stresses (in MD&TD), birefringence and also the crystallinity. This may be attributed to the highest degree in chain regularity or crystallizable chain segment. On the other hand, the increased D-lactide content has a clear influence on retarding structural development in PLA during MDO stretching, as demonstrated in 4060D. For the blends between PLAs having different D-lactide contents, the higher D-lactide PLA will play an important role on structural development under film stretching process, both MDO and TDO streching. It could be pointed out that the D-lactide content in PLA plays a vital role in biaxial deformation behaviors of both neat PLAs and their blends.

Acknowledgements

The authors would like to acknowledge the National Metal and Materials Technology Center (MTEC) for funding. The authors are grateful to Prof. Kikutani and Dr. Wataru Takarada for valuable advice on setting up a birefringence apparatus and fruitful discussion.

References

1. X. Ou and M. Cakmak *Polymer* 2010, 51, 783.
2. M.S. Lee, U.S.Patent 0148715 A1, 2009
3. W. Chinsirikul, N. Kerddonfag, C. Winotapun and P. Kumsang in PPS 27 Proceeding, Marakech, 2011, CD.
4. B. Hararak, N. Kerddonfag, C. Winotapun, P. Kumsang, C. Sripethdee and W. Chinsirikul, in AWPP-2012 Proceeding, Kyoto, 2012, Proceeding book.
5. E.W. Fisher, H.J. Sterzel and G. Wegner *Kolloid ZZ*

Polymer, 1973, 25, 980.
6. Hongbo Li, Michel A. Huneault *Polymer*, 2007, 48, 6855.
7. D.S. Ryu, T. Inoue and K. Osaki *Polymer*, 1998, 39, 2515.

O-03-626

PROCESSING AND PROPERTY OPTIMIZATION OF POLY(LACTIC ACID)/POLY(BUTYLENE SUCCINATE) BLEND BLOWN FILMS

C. Sripethdee*, B. Hararak, C. Winotapun, N. Kerddonfag, P. Kumsung and W. Chinsirikul

National Metal and Materials Technology Center, National Science and Technology Development Agency Ministry of Science and Technology, Pathumthani, Thailand –chaos@mtec.or.th

Abstract - Poly(butylene succinate) (PBS) and poly(lactic acid) (PLA) are well-received biodegradable polymers. PBS is an interesting material due to its ease of processing, excellent toughness and other physical properties similar to those of PE. However, practical application of PBS has been limited because of its softness. Early studies have been conducted to improve and/or control PBS properties to further expand its potential applications. Those studies mainly involve copolymerization and nanocomposite methods. In this study, the PLA/PBS blend is a system of interest. PLA is a much stiffer material with higher strength, as compared to the PBS. Current attempts are at developing PLA/PBS blends and blown films with synergistic properties. The PLA/PBS blend is immiscible, therefore, the final film performance depends not only on inherent property of each component, but also on the structure or phase morphology of the blends. PLA/PBS blends and blown films were prepared with 20/80 and 40/60 blending ratios. Mixing method of the PLA/PBS blends was another focus under this investigation. Blown films were fabricated using either a compound of the PLA/PBS prepared by intensive mixing, or a dry blend. Interestingly, it was found that blown films of the PLA/PBS blends were stiffer and had higher elongation at break than the PLA and PBS films alone. Phase morphology of the blend films having PLA of 20 wt% showed fiber-like structure of PLA distributed in the PBS matrix. It should be pointed out that blown films prepared from the twin-screw blending appeared to be more uniform with finer fiber diameter (D ~0.4 μm), as compared to those of dry blends (D ~0.7 μm). DSC first scan results revealed that the %crystallinity (%χ_c) of PLA in twin-screw blend films increased with a presence of PBS component.

Keywords: Poly(butylene succinate), Poly(lactic acid), PBS/PLA blends, Blown film.

Introduction

Environmental concerns and a shortage of petroleum resources have driven efforts on bulk production of biodegradable polymers such as aliphatic polyesters, polysaccharides and unsaturated polyesters [1]. These biodegradable materials can be completely degraded into natural ecosystems such as active sludge, natural soil, lake and marine [2]. Accordingly, the biodegradability of biodegradable polymers corresponds to the ability to be chemically transformed by the action of biological enzymes or microorganisms [2,3]. Poly(butylene succinate) (PBS) is one of the commercially used biodegradable polymers with a range of interesting properties including flexible properties, melt processible and compostability [4]. PBS is produced by condensation reaction of 1,4-butanediol and succinic acid [5]. However, PBS is a white crystalline thermoplastic with glass transition temperature of about −35 °C, PBS is very soft at room temperature. These properties can limit PBS in some practical application. Poly(lactic acid) (PLA) is an aliphatic polyester made up of lactic acid (2-hydroxy propionic acid) building blocks. It is also a biodegradable and compostable thermoplastic derived from renewable plant sources, such as starch and sugar [6]. PLA has numerous interesting properties including good mechanical properties (high modulus), thermal stability and low environmental impact [7]. Hence, it was proposed in this study to blend PBS with PLA to exploit advantageous characteristics of each counterpart, namely good toughness and elongation of PBS and high tensile strength and modulus of PLA [8]. Most of early studies on PLA/PBS blend involved investigation of the blend's morphology and thermal property via copolymerization and nanocomposite method. However, limited information was reported on mixing process of PLA/PBS and the resulting blends' or products' properties.

The objective of this research is, thus, to study the blend and blown film from PBS and PLA based on two different mixing methods: (1) dry blending and (2) blend prepared from an intensively mixed PLA/PBS (50/50 wt%) by twin screw extrusion. The PLA/PBS blend is immiscible, therefore, the final film performance depends not only on inherent property of each component, but also on the structure or phase morphology of the blends. Blown film properties of PLA and PBS blends with different compositions were examined.

Experimental

Materials and blends preparation.

Polybutylene succinate (PBS) (GS Pla AZ91TN-grade) with a melt flow rate of 4 g/10 min (190 °C/2.16 kg) and density of 1.26 g/cm^3 was a commercial product of MITSUBISHI CHEMICAL. Poly(lactic acid) (PLA) (4032D-grade, biaxially oriented films) with a melt flow rate of 2.5 g/10 min (190 °C/2.16 kg) and density of 1.24 g/cm^3 was produced by NatureWorks LLC.

PLA and PBS blends of two ratios, 20/80 and 40/60 by weight, were prepared by two methods: (1) dry blending (DB) and (2) twin-screw mixing of PLA/PBS (50/50) and blended with neat PBS, denoted herein as twin-screw blend (TB). Twin-screw extruder used was LabTech, 20 mm screw diameter, 32 L/D ratio. Optimum barrel (zones 1-7), and die set temperatures (zones 8) were ~40, 160, 170, 190, 190, 195, 195 and 200 °C, respectively, and screw speed used was 90 rpm. PLA/PBS blends from the two mixing methods were then fed into a film line for blown film fabrication.

Blown film extrusion.
All blown film samples were produced on a laboratory-scale single screw extruder (HaakePolyLab OS RheoDrive7). The extruder had a 19 mm diameter single screw, 25 L/D ratio, a mixing section at the end, and a compression ratio of 3:1. Optimum barrel (zones 1-4), and die set temperatures (zone 5) were ~40, 225, 225, 225 and 225 °C, respectively. Cooling temperature was ~42 °C. Film samples were produced at a moderate screw speed of 40 rpm and take up speed of 90 m/min. These parameters were kept constant for all films production. Resulting films had thicknesses of ~30±2 μm.

Thermal analysis by differential scanning calorimetry.
Thermal analysis was performed on a Perkin Elmer DSC7. Samples were sealed in aluminum pans and scanned at heating and cooling rates of 10 °C/min under nitrogen atmosphere. Calibration was made based on standard procedure and resulting thermograms were normalized to constant weight and analyzed.

Mechanical properties testing.
Tensile strength, modulus and elongation at break of neat and blended PLA/PBS films were examined using an universal testing machine, Instron 44502 series, in accordance with ASTM D882.

Morphology.
Morphology of final films was observed using a scanning electron microscope (SEM, HITASHI SU-1500) at an accelerating voltage of 7.5 kV. The film samples were fractured in liquid nitrogen. The fracture surface along with machine direction and transverse direction were examined.

Results and Discussion

Thermal properties of the PLA/PBS Blend films.
In order to investigate basic thermal property, Fig. 1 shows the 2nd heating thermograms, of neat PBS and PLA pellets from the same cooling condition of -10 °C/min. PBS exhibits normal DSC thermogram with a cold crystallization temperature of about 95.4 °C and melting temperature of about 109.9 °C. However, PLA show no trace of cold crystallization, only the melting temperature was found at 167.3 °C with very small melting enthalpy. It implied that PLA hardly crystallized under this cooling condition (-10 °C/min).

Fig. 2 shows DSC thermograms of the neat PBS and PLA/PBS blown films from different blending processes, dry blend and twin-screw blend. Like as-received PBS, neat PBS film also reveals a small exothermic peak of cold crystallization and an endothermic melting peak, but at lower temperatures of 90.8 °C and 108.2 °C, respectively. In the case of films produced from PLA/PBS blends, the same melting behavior of PBS was found, independent of either blending composition or mixing technique. There is no indication of co-crystallization between the two phases of PBS and PLA. However, it was found that with increasing PLA content the crystallinity of PBS in the blends decreases, in particular for the 40/60 blend (see Table 1). PLA in the blends seems to have an influence in suppressing the crystallization of PBS matrix in blown films. On the other hand, much higher crystallinities of the PLA counterpart in the blends were found in a range of 34.5-37.9% as compared to the PLA pellet (1.9%) and the PLA film (2.1%). It is clear that PBS is effective in inducing PLA's crystallization in the blend blown films. The changes of crystallinity in the PBS matrix and the PLA phases and these effects on film's mechanical properties will be later discussed.

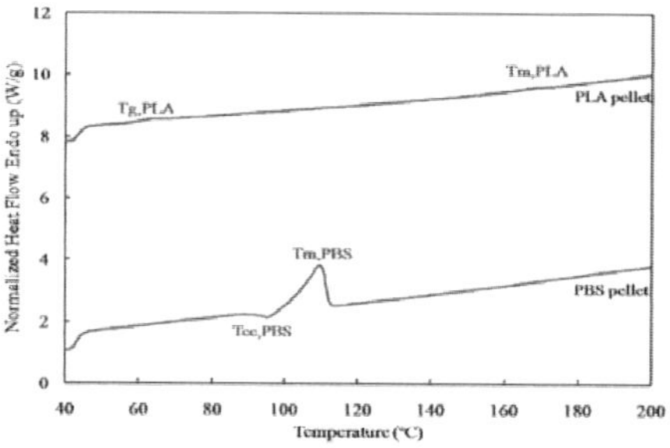

Figure 1 – DSC 2nd heating curves of PBS and PLA pellets, from a cooling rate of 10 °C/min.

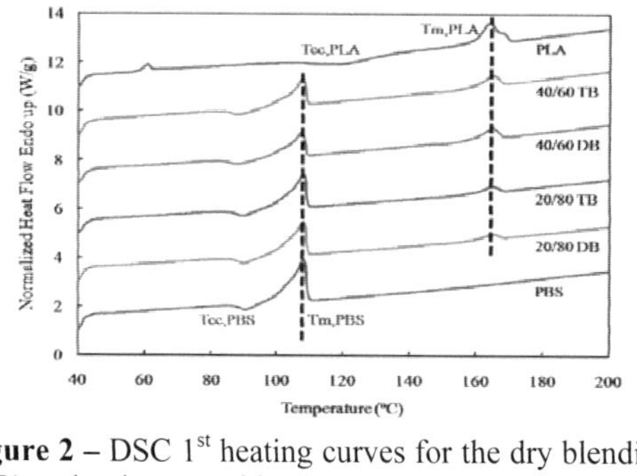

Figure 2 – DSC 1st heating curves for the dry blending (DB) and twin-screw blending (TB) of PLA/PBS blend films with two blend ratios at a heating rate of 10 °C/min.

Phase Morphology of PLA/PBS Blend films.
According to the results obtained from capillary rheometer of PLA and PBS at a temperature of 225 °C (the same temperature with blown film process). Fig. 3 shows calculated viscosity ratios and viscosity ratios with volume fraction between the two phases of PLA and PBS. Under shear rate in a range of 100 to 2,000 s-

[1], general shear rate in extrusion process, the viscosity ratio between PLA and PBS (η_{PLA}/η_{PBS}) was in a range between 0.72 and 1.14. To be more specific, under a film blowing condition used in this study, estimated shear rate was about 250 s^{-1}, and the corresponding-viscosity ratio (η_{PLA}/η_{PBS}) was 0.84. Based on the known principle, processing the blend with a viscosity ratio of lower or close to 1.0, under appropriate elongational flow, can lead to a formation of fiber or fibril structure of the dispersed polymer. This suggests that dispersed PLA can be deformed and present in an elongated or fiber-like structure.

Table 1 – DSC parameters obtained from the first heating scan for the dry blending (DB) and twin-screw blending (TB) of PLA/PBS blend films with two blend ratios.

PLA/PBS ratio	T_{cc}(PBS) (°C)	T_m(PBS) (°C)	$\%\chi_c$(PBS)	T_g(PLA) (°C)	T_{cc}(PLA) (°C)	T_m(PLA) (°C)	$\%\chi_c$(PLA)
pure PBS	90.8 ± 0.1	108.2 ± 0.2	47.7 ± 0.6	-	-	-	-
20/80 DB	90.0 ± 0.2	108.2 ± 0.0	45.6 ± 1.3	57.6 ± 3.1	-	164.9 ± 0.2	34.5 ± 0.6
20/80 TB	89.5 ± 0.2	108.0 ± 0.2	45.8 ± 2.3	57.3 ± 2.2	-	164.9 ± 0.2	37.9 ± 1.1
40/60 DB	88.0 ± 0.3	107.6 ± 0.1	36.4 ± 2.3	58.9 ± 3.1	-	165.0 ± 0.2	34.8 ± 1.5
40/60 TB	87.9 ± 0.4	107.8 ± 0.1	38.6 ± 1.1	57.3 ± 0.0	-	165.0 ± 0.0	35.5 ± 0.6
pure PLA	-	-	-	57.0 ± 0.1	119.3 ± 1.3	164.0 ± 0.2	2.1 ± 0.4

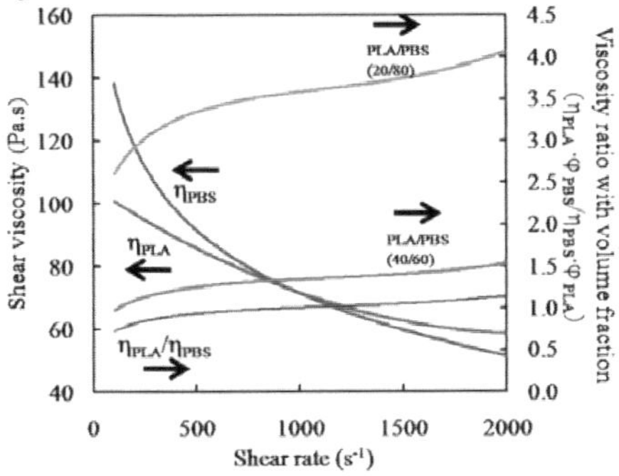

Figure 3 – The rheological properties of PLA and PBS under shear rate in a range of 100 to 2,000 s^{-1}, temperature of 225 °C.

Fig. 4 shows the SEM micrographs of the fracture surface of neat PBS and PLA/PBS blown films. Fractured samples were prepared along MD and TD, in order to investigate the morphology of the blends. It is clear that the neat PBS has a rather smooth surface without yielding (Fig 4(a)). In the case of PLA/PBS blends (Fig 4(b)-(e)), the identifiable two phases of the immiscible blend morphology were observed in all samples. When incorporating 20 wt% of PLA into PBS, Fig 4(b)-DB and (c)-TB, fracture surface in TD, it was found that the size of minor PLA phase from DB is larger than TB. From the images of fracture specimen along MD, the elongated or fiber-like structure of the PLA phase is observed. This morphology can be attributed to the resulting viscosity ratio of lower than 1, as previously discussed. The PLA fibers in both TB and DB were found to be parallel to the film drawing's direction or MD (as indicated by the arrow). Detailed examination on the blends' morphology will be carried out to reveal failure mechanism as well as interfacial adhesion of the PLA fiber and the PBS matrix that can be linked to observed mechanical properties of the blend blown films.

Estimated diameter of PLA fiber in the case of 20/80 twin-screw blend sample (D ~0.4 μm) was almost 50% smaller than that of the dry-blend film (D ~0.7 μm). This can be a result of intensive mixing by twin-screw extruder. With increasing PLA content in the blend from 20 to 40 wt%, fracture surfaces along MD also reveal a preferential orientation of presumably the PLA phase. In fact, we hardly visualize voids or cavity between the PLA and PBS phases, both for 20/80 and 40/60 blend films. Observed fractographs may suggest possible adhesion between the two phases, at least to a certain degree. This hypothesis is a part of our future investigation. It should be noted that the 40/60 blend morphology is different from the 20/80 sample, also indicating an influence of the volume fraction of blend components.

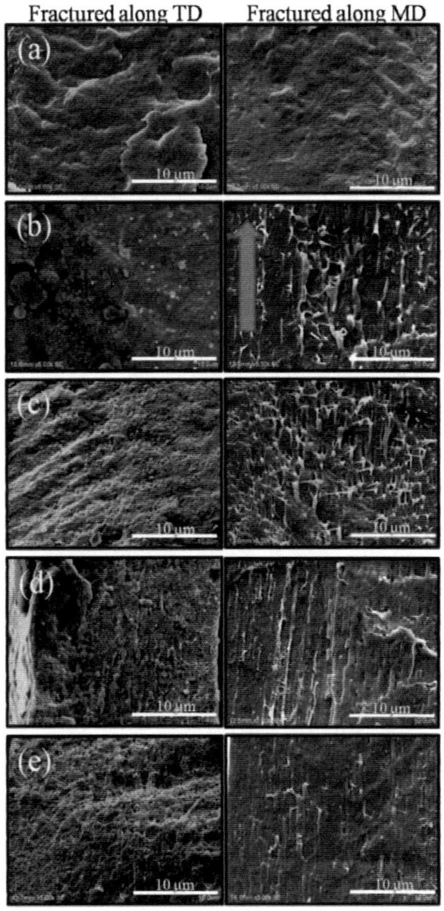

Figure 4 – SEM images of fractured along with TD and MD direction for the dry blending (DB) and twin-

screw blending (TB) of PLA/PBS blend films with two blend ratios, (a) pure PBS, (b) 20/80 DB, (c) 20/80 TB, (d) 40/60 DB and (e) 40/60 TB.

Mechanical properties of PLA/PBS Blend films.

In order to understand effects of compositions and mixing of PLA/PBS blend films on mechanical properties, the stress-strain curves of PLA/PBS blend films in MD are compared in Fig. 5. From this figure the neat PBS film shows ductile failure behavior while the neat PLA exhibits a brittle-like material with high modulus, high tensile strength, low elongation and a sharp break. By incorporating 20 wt% and 40 wt% of PLA in to PBS, the tensile curves of blend films were evidently changed. These reflect, in different yielding characteristics and overall improvement in toughness for the blend samples.

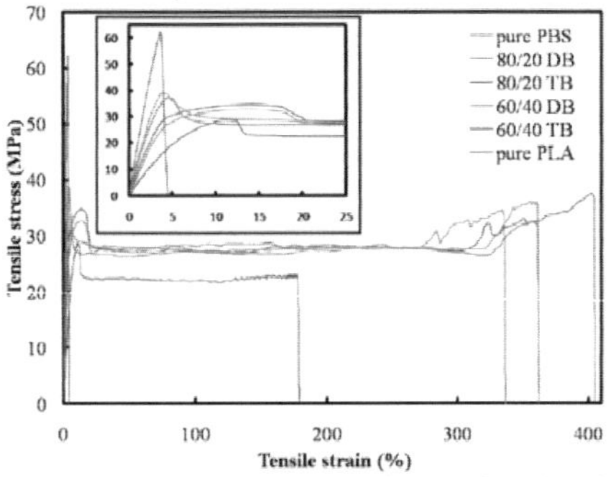

Figure 5 – Stress-strain curve of MD direction for the dry blending (DB) and twin-screw blending (TB) of PLA/PBSblend films with two blend ratios.

Clearly seen in Fig. 5 and in Table 2, tensile strength and modulus of the blend films both dry blend (DB) and twin-screw blend (TB) tend to increase with increasing PLA content In the case of the film containing 20 wt%, PLA, increase of tensile strength and modulus are ~50 and 59% over those of the neat PBS film. Increases in both modulus and tensile strength of these blend blown films nearly follow the rule of mixture. This findings is similar to the results reported by Shibata et al. based on the injection molded PLA/PBS having the same PLA content of 20 and 40 wt% [9].

Interestingly, the elongation at break and toughness of the PLA/PBS blend films were significantly improved over that of each counterpart. The synergistic effect is apparent in the case of 20/80 twin-screw blend (TB) film, as displayed by the highest elongation to break and toughness among all tested samples. Deformation of the 20/80 TB film requires the highest breaking energy, based on the area under the stress-strain curves. Significant improvement of toughness in the 20/80 PLA/PBS twin-screw blend film can be well attributed by the reinforcing effect of the fine PLA fibers developed in the blends. Such morphology was revealed in SEM micrograph (Fig. 4(c)).

However, for the PLA/PBS TB blown film with 40 wt%, elongation at break of this films decreased, compared with 20/80 blend film, but was still higher than that of the neat PBS and PLA films. This may be owing to a higher content of stiff PLA and the resulting structure that differ from the 20/80 PLA/PBS sample. Developed fine fibrous structure of PLA in the PLA/PBS blends can play a vital role in enhancing final film property. This result is achieved via reinforcing mechanism or effective load transfer between the two phases. Effects of structure in the blend of PLA/PBS with similar composition of 20/80 and 40/60 were also reported by Bhatia et al based on compression molding process [8]. It was found that tensile elongation of the blends surprisingly decreased to lower value than that of the neat PBS. Observed structure of dispersed PLA in such blends were largely in droplet form. It is, therefore, important to summarize that, blending composition and resulting morphology of the PLA/PBS blends are vital in affecting final properties of the films. As revealed in this study, the measured tensile modulus and strength of PLA/PBS blown films well conform to the rule of mixture.

Table 2 – Tensile properties of MD direction for the dry blending (DB) and twin-screw blending (TB) of PLA/PBS blend films with two blend ratios

MAT.	E* (MPa)	TSB ** (MPa)	EB*** (%)	Toughness (MPa)
pure PBS	484.5±14.4	22.8±1.0	180.1±46.8	75.6±9.3
20/80 DB	770.2±25.0	34.1±2.5	348.2±18.3	97.8±8.9
20/80 TB	927.9±16.4	38.2±1.1	420.5±10.3	164.4±13.3
40/60 DB	1216.3±126.0	38.6±4.2	348.3±41.2	106.7±10.2
40/60 TB	1366.2±54.2	38.6±3.3	356.6±22.4	111.1±8.9
Pure PLA	2121.5±92.1	62.1±2.7	3.6±0.2	2.7±0.4

* Modulus, ** Tensile strength at break, *** Elongation at break

Conclusions

Based on the known principle, processing the blend with a viscosity ratio of lower or close to 1.0, under appropriate elongational flow, can lead to a formation of fiber or fibril structure of the dispersed polymer. This suggests that dispersed PLA can be deformed and present in an elongated or fiber-like structure. By incorporating 20 and 40 wt% of PLA in to PBS, the tensile curves of blend films were evidently changed. These reflect, in different yielding characteristics and overall improvement in toughness for the blend samples. The synergistic effect is apparent in the case of 20/80 twin-screw blend (TB) film, as displayed by the highest elongation to break and toughness among all tested samples.

Acknowledgements

This research was supported by National Metal and Materials Technology Center (MTEC), NSTDA.

References

1. M. Vert; J. Feijen; A.C. Albertsson; G. Scott; E Chiellini, *Biodegradable Polymer and Plastics*, Royal Society of Chemistry, London, 1992.
2. M. Okada *Prog. Polym. Sci.* 2002, 27, 87.
3. A.K. Mohanty;M. Misra; G. Hinrichsen *Macromol. Mater. Eng.* 2000, 266, 1.
4. S.M. Lee; D. Cho; W.H. Park; S.G. Lee; S.O. Han; L.T. Drzal *Compos. Sci. Technol.* 2005, 65, 647.

5. T. Fujimaki *Polym. Degrad. Stabil.*1998, 59, 209.
6. L.T. Lim; R. Auras; M. Rubino *Prog. Polym. Sci.* 2008, 33, 820.
7. F. Carrasco; P. Pagès; J. Gámez-Pérez; O.O. Santana; M.L. Maspoch *Polym. Degrad. Stabil.* 2010, 95, 116.
8. A. Bhatia; R.K. Gupta; S.N. Bhattacharya; H.J. Choi *Korea-Aust. Rheol. J.* 2007, 19, 125.
9. M. Shibata; Y. Inoue; M. Miyoshi *Polymer.* 2006, 47, 3557.

MICRO-AND NANO-SCALE POLYMER PROCESSING - TOP-DOWN APPROACH AS INJECTION MOLDING AND IMPRINTING TECHNOLOGY

H. Ito[1]*, A. Hariu[1], S. Takahagi[1], T. Takayama[1], and T. Kyotani[2]

[1] *Graduate School of Science and Engineering, Yamagata University, Yamagata 992-8510, Japan*
**Corresponding author e-mail: ihiroshi@yz.yamagata-u.ac.jp*
[2] *Institute of Multidisciplinary Research for Advanced Materials, Tohoku University,Sendai 980-8577, Japan*

Abstract - *Nanopatterning was achieved by application of nanoporous anodic aluminum oxide (AAO) membranes as mold inserts in injection molding and thermal nanoimprinting. The AAO membrane has closely packed nanoholes. Depending on the AAO membrane used, the nanoholes have a diameter of 30–200 nm and a depth that varies from a few hundred nanometers to 60 μm. Effects of structure size on polystyrene and polycarbonate surface properties were investigated. In the case of polystyrene, contact angle measurements with water indicate that the contact angle increased from 85 deg on the smooth surface to 140 deg on the polystyrene nanostructure.*

Keywords: Thermal nano-imprinting, Injection molding, anodic aluminum oxide, nano-rod strucutures

Introduction

Recently, forming a nanoscale polymer material structure to add new functionality to materials has attracted attention. Reportedly, a nanostructure can be added to a polymer surface by filling molten polymer into a substrate with a nanoporous structure fabricated using the anodized aluminum oxide (AAO) method [1–5]. However, because the nanorod and nanotube lengths depend on the AAO film thickness, it is difficult to form and produce a long nanofiber. Moreover it is difficult to fabricate nanostructure with high aspect ratio (length/diameter ratio) by injection molding process.

To clarify the nanorod fabrication mechanism, the behavior of that filling into AAO template of molten polymer was investigated in this study using injection molding and thermal imprinting processes.

Experimental

Thermal nano-imprinting

An AAO membrane (200 nm diameter pores and 60 μm thickness, Anodisc47; Whatman plc.) and polystyrene (PS-679, MFR=18g/10 min; PS Japan Corp.) were used for this study. Figure 1 shows a scanning electron micrograph (SEM) of the AAO membrane surface structure. The AAO membrane was cut into 1×1 cm^2 pieces. Then a 160-μm-thick PS film was fabricated using hot-pressing. Imprinting was conducted using a vacuumed hot-imprinting machine. To evaluate the effects of processing conditions, imprinting pressure, imprinting temperature, and the pressure-keeping-time were varied, respectively, from 0.1 to 3 MPa and from 120°C to 200°C for 5, 10, 15, 30 min. The obtained samples were polished in 4 mol/L $HClO_4$ solution; then the sample rod length was evaluated through observation of cryofracture surfaces using scanning electron microscopy (SEM). To evaluate the surface wettability of the obtained samples, the contact angle against purified water was also measured using surface tension and a contact angle measuring system (DM500; Kyowa Interface Science Co. Ltd.).

Nano-injection molding

In this study, we used two different diameter/depth AAO stampers. A commefcial AAO membrane (200 nm diameter pores and 60 μm thickness, Anodisc47; Whatman plc.) was used. Another AAO sheeet was made in our laboratory. This was made an AAO stamper with a structure of several nanopores for use in this study. Before anodization, all bulk Al substrates (99.999%, 5 mm square, and 0.5 mm thick) were degreased in acetone using ultrasonication. They were electropolished individually to remove large surface irregularities. Electropolishing was conducted by making the substrate an anode in perchloric acid – ethanol solution ($HClO_4$: C_2H_5OH = 2:8 in volumetric ratio) for 10 min. The solution temperature was kept at 5°C and stirred at a constant rate. A flat Ni sheet was used as a cathode for bulk Al electropolishing. A constant voltage of 20 V was applied between the anode and the cathode.

Two-step anodization was then conducted using the electropolished Al substrate to produce an oxide film having hexagonally arranged nanopore structures. The first anodization was performed using the electropolished Al substrate for 1 h. A sulfuric acid solution (0.3 M H_2SO_4; 10°C and 25 V with 1 A) was used. From this anodization, the mean diameter of the nanoporous structure was obtained as 30 nm. The resultant irregular porous thick oxide layer was etched completely by immersing the substrate into a mixture of 1.8wt% chromic acid (H_2CrO_4) and 6 wt% phosphoric acid at 60°C during 5 h. Finally, the desired AAO layer thickness (in this case, 30 μm) was obtained using the subsequent second anodization during 1 h using a sulfuric acid solution (0.3 M H_2SO_4; 10°C and 25 V with 1 A).

For a nanoporous structure with 100 nm diameter, the first anodization was performed using the electropolished Al substrate for 1 h. An oxalic acid solution (1.8 wt% oxalic acid; 10°C and 80 V with 1 A) was used. The second anodization was conducted at 3 h with an oxalic acid solution (1.8 wt% oxalic acid; 10°C and 80 V with 1 A). Finally, the desired AAO layer thickness (in this case, 60 μm) was obtained. Figure 1(b) presents SEM micrographs of surface structure of the AAO stamper obtained using an anodization method. Therefore, a high-aspect-ratio and high-density nanopillar structure surfaces can be produced by molding polymer resins into the nanopores in the AAO mold, thereby generating a strong adhesive force.

The prepared nanoporous structured AAO stamper was applied as mold inserts; they were mounted in the heat insulator, zirconia. They were then set in the mold plate. Recently, a nanopillar array structure for application-patterned media was fabricated using a passive heating system with injection molding [6]. We also performed injection molding for a 200–500 nm height structure [7] and nanopillars (ca. 200 nm) using a heat insulator: zirconia plate [8]. We obtained a nanostructure with a high aspect ratio (over 1.0) using injection molding.

We used dimensions of a square flat mold of 20 × 20 mm. The cavity thickness is 0.5 mm. Nanopatterned polycarbonate (lupilon H7001; Mitsubishi Engineering-Plastics Corp.) samples were prepared using micro-injection molding (modified Elject AU3E; Nissei Plastic Industrial Co. Ltd.). For injection molding, the injection speed, holding pressure, and maximum injection pressure were 120 mm/s, 120 MPa, and 200 MPa, respectively. The injection unit temperature was 350°C; the mold temperatures were 115 and 100°C. We measured the process behavior using a data-logger system inside the molding machine. Before molding, polymer pellets were dried in an oven at 100°C for 24 h.

The molded nanostructure of samples was also observed using SEM. To evaluate the internal molecular orientation of the molded sample, optical retardation was analyzed using a polarizing optical microscope (POM, BX51-P; Olympus Corp.) with a Berek compensator. Furthermore, a skin-shear-core structure of molded parts was analyzed using polarizing microscopy after a thin section was cut parallel to the MD from the molded parts using a microtome (every 30 μm thickness, 2055 Autocut; Leica Microsystems). To analyze the wettability of a surface of obtained samples, the contact angle was also measured using surface tension and a contact angle measuring system (*see above*).

Results and Discussion

Thermal nano-imprinting

Figure 2 shows SEM micrographs of a cross section of nanorod formed at 1 MPa of nanoimprinting pressure, 130°C of nanoimprinting temperature, and 30 min of pressure holding time. The 7-μm-long PS nanorod has 200 nm diameters.

We analyzed the effects of temperature on the rod length. It is readily apparent that the rod length increases when the imprinting temperature increases. The material is completely filled up to 60 mm, which is the AAO film thickness at temperatures greater than 160°C. We also measured the effect of pressure on the PS film rod length. At 130°C of imprinting temperature and 30 min of pressure keeping time, the rod length increases slightly with increased imprinting pressure.

The effect of pressure keeping time on the rod length of the PS film fabricated at 1 MPa of imprinting pressure and 120°C, 140°C, and 160°C imprinting temperature was evaluated. The filling rate of material increases when the imprinting temperature increases. These results indicate that forming of nanorods must increase the imprinting temperature corresponding to reduction of the molten polymer viscosity.

Figure 3 presents the effect of pressure holding time on the rod length of the PS film fabricated at 1 MPa of imprinting pressure and 120°C, 140°C, and 160°C imprinting temperature. The filling rate of material increases when the imprinting temperature increases. These results indicate that forming of nanorods must increase the imprinting temperature corresponding to reduction of the molten polymer viscosity.

Figure 4 portrays the contact angle against purified water of PS film with nanorods fabricated at 1 MPa of imprinting pressure, 130°C of imprinting temperature, and 30 min of pressure keeping time. The contact angle increased from 85° on the smooth surface to 140° on the polystyrene nanostructure, which means that higher water-repellence can be added by forming a nanostructure on the surface.

Nano-injection molding

Figure 5(a) presents a photograph of the thin-wall product with a 30 nm nanopillar array surface structure.

Figure 5(b) also portrays polarized photographs of molded parts. The intrinsic birefringence of PC is large, thereby confirming the interference color. The interference fringe correlates closely with optical retardation and the residual strain. The interference color was observed clearly near the gate; it faded toward the flow end, implying that the residual strain near the gate was greater than at any other position. The interference color became unclear at the higher injection temperature of 280°C.

Figure 6 shows SEM micrographs of a cross section of nanorod of the PC parts fabricated at 120 mm/s of injection speed and 115°C mold temperature. The rod length increased when heat insulator mold used. The length showed 7-μm-long at coventional injection molding. These results indicate that rod length increases 10 times at using heat insultor mold. This results mean that low cooling behavior inside the cavity was important to fabricate nano-structure by injection molding.

We observed SEM images of surface nanopatterns of a molded sample. Figure 7 portrays SEM images of surface nanopatterns of a molded sample. This portion is located near the gate. Using the heat insulator, we observed high replication of surface nano-features: 30 nm rod features. Results show that nano-molding can be performed with the AAO stamper and heat insulator mold. This portion is located near the gate. Using the heat insulator, we observed high replication of surface nano-features: 30 nm rod features. Results show that nano-injection molding can be performed with the AAO stamper and heat insulator mold.

Conclusions

Results of this thermal nanoimprinting study show that molten polymer viscosity reduction is effective for the formation of a nanostructure on a PS film surface. Results indicate that higher water-repellence can be added to PS films by forming a nanostructure on the surface. We also investigated the injection moldability and higher-order structure of thin-wall parts with various nanosurface features. We used a zirconia-insulated mold to increase heating efficiency. The replication accuracy was improved using a heat insulator and a high-flow-grade polymer. Results showed high replication of surface nanofeatures—30 nm rod features—using the injection molding process.

Acknowledgments

This study was partially supported by KAKENHI, a Grant-in-Aid for Scientific Research (C) (project ID: 21560710) from the Japan Society for the Promotion of Science(JSPS).

This work was also supported in part by the Collaborative Research Program (ID: 20100049) of Institute of Multidisciplinary Research for Advanced Materials, Tohoku University.

References

1. J. Tan, X. L. She, F. Yuan, S. and Yang, D. Zhou, J. Porous Matter, Vol. 15, pp. 619-623 (2008).
2. M. Zhang, P. Dobriyal, J-T. Chen, T. P. Russell, J. Olmo, and A. Merry, Nano Letters, Vol. 6, No. 5, pp. 1075-1079 (2006).
3. D. S. Kim, H. S. Lee, J. Lee, S. Kim, K-H. Lee, W. Moon, and T. H. Kwon, Microsyst. Technol., Vol. 13, pp. 601-606 (2007).
4. J. M. Park, N. H. Kim, B-K. Lee, K-H. Lee, and T. H. Kwon, Microsyst. Technol., Vol. 14, pp. 1689-1694 (2008).
5. J. Yeo, J. Ryu, B-K. Lee, E. Byeon, T. H. Kwon, and D. S. Kim, Microsyst. Technol., Vol. 14, pp. 1689-1694 (2010).
6. M. Choi, N. Lee, J. Han, J. Hong, and S. Kang, ANTEC Tech. Paper, pp. 2030-2034 (2007).
7. H. Ito, H. Suzuki, K. Kazama, and T. Kikutani, Current Applied Physics, Vol. 9, pp. e19-e24 (2009).
8. H. Ito, S. Takahagi, K. Zouda, and Y.Kayano, ANTEC 2010 Tech. Paper, pp. 2138-2142 (2010).

Fig. 1 SEM image of the AAO stamper surface structure: a) AAO membrane (200 nm diameter), b) AAO plate (30 nm diameter).

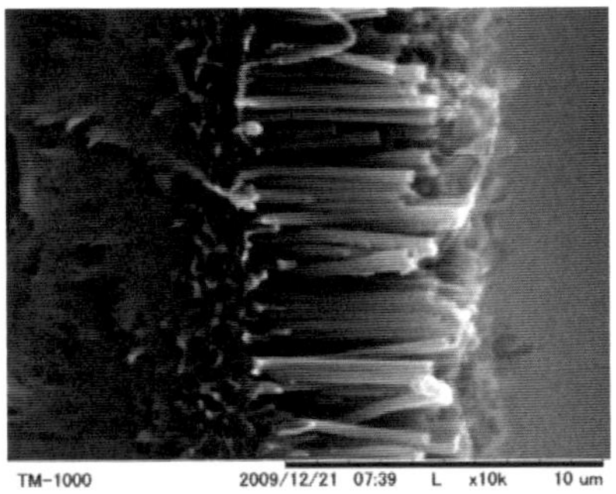

Fig. 2 SEM image of cross-section of PS film after imprinting. Nanorod structure has 200 nm diameter with 7 μm length.

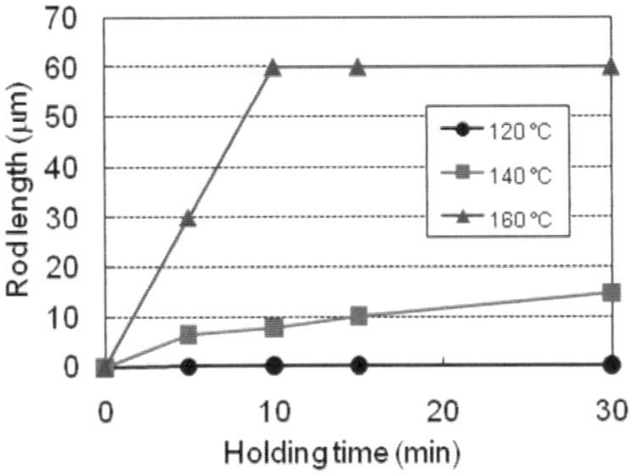

Fig. 3 Dependence of pressure holding time on the rod length.

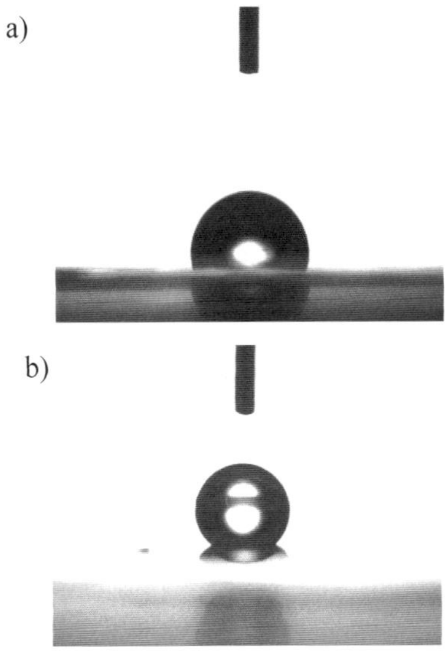

Fig. 4 Image of the water drop used for measuring the contact angle of (a) before and (b) after imprinting.

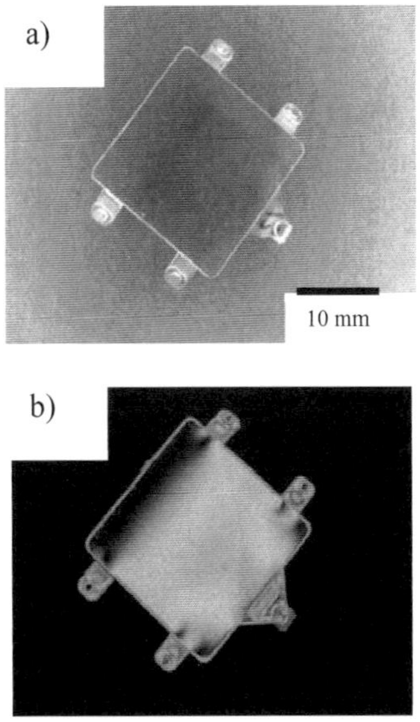

Fig. 5 Optical (a) and polarized (b) photograph of the thin-wall product.

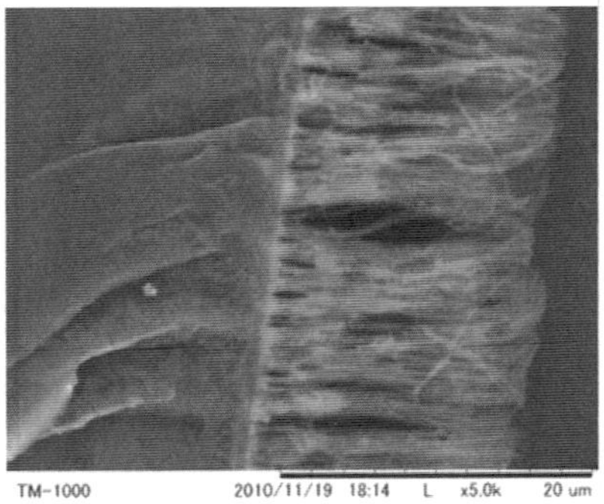

Fig. 6 SEM image of cross-section of PC molded parts after injection process. Nanorod structure has 200 nm diameter with 20 μm length..

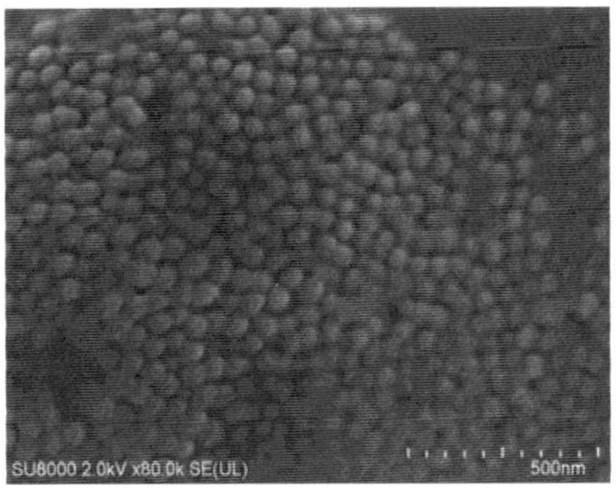

Fig. 7 SEM images of surface nanopatterns of the molded sample (30 nm pillar).

O-04-350

INFLUENCE OF DYNAMIC MOLD TEMPERATURE PROCESSES ON THE WELD LINE STRENGTH OF BI-DIRECTIONAL FILLED TENSILE BARS. A NEW MULTI-FUNCTIONAL TEST INJECTION MOLD TO EXAMINE SURFACE DEFECTS, THIN-WALL PARTS AND FROZEN LAYER FORMATION.

G.R. Berger[1,*], *G.A. Pacher*[2], *W. Friesenbichler*[1]

[1]*Department of Polymer Engineering and Science, Chair of Injection Molding of Polymers, Montanuniversitaet Leoben, Otto Gloeckel Strasse 2, Leoben, Austria – gerald.berger@unileoben.ac.at; walter.friesenbichler@unileoben.ac.at*

[2]*Polymer Competence Center Leoben GmbH, Roseggerstraße 12, Leoben, Austria – gernot.pacher@pccl.at*

Abstract: Dynamic mold surface control continuously gains importance in achieving high grade polymer surface finishes. To examine the influence of dynamic mold surface temperature control on the weld line strength, a new test mold, using two different dynamic mold temperature technologies, was used to manufacture bi-directional filled tensile bars. The mold's nozzle side is equipped with the areal heating and cooling technology BFMOLD™, which tempers a chamber below the cavity by using hot and cold water sequentially. A local electrical ceramic heating element is installed into the ejection side close to the weld line region. The weld-line tensile bars were manufactured in a systematic DoE where mold surface temperature, injection rate and nozzle temperature were systematically varied.

The mold features further a series of exchangeable mold inserts, either for conventional or weld-line tensile bars, thin-thick-thin stripes and complex plate-shaped parts having cut-out, ribs and thin-wall regions. Three new, triple-combined mold cavity sensors were used, each featuring an infrared melt temperature detector, a thermocouple-based mold surface temperature detector, and a piezo-electric cavity pressure detector. The transient mold temperature was acquired either by the IR melt temperature detectors or an IR-thermography camera.

Keywords: variothermal, dynamic mold temperature control, weld line strength, injection molding, surface defects

Introduction

Dynamic mold surface temperature control (also known as rapid heat cycle molding – RHCM) is a variant of the injection molding process where the mold surface is heated before injecting and cooled after filling. Several studies have shown the use of RHCM improves the surface quality of injection-molded parts [1-6]. A very common approach of RHCM is to use water as heating and cooling fluid. Commercial systems consist of separate water-circuits for heating and cooling, which are alternatively pumped trough the mold cooling-channels [7].

A new injection test mold, using two different dynamic mold temperature technologies, was developed. The fixed mold half is equipped with the areal heating and cooling technology BFMOLD™ [9], which controls the temperature of a chamber below the cavity by using hot and cold water sequentially. In the moving mold half an electrical ceramic heating element is installed to heat the weld line region. The mold features further a series of exchangeable mold inserts, either for conventional or weld-line tensile bars, thin-thick-thin stripes and complex plate-shaped parts having cut-outs, ribs and thin-wall regions. The latter is equipped with three new, triple-combined mold cavity sensors to acquire the transient cavity process parameters during RHCM. Each sensor features a piezo-electric cavity pressure detector, a thermocouple mold surface temperature detector, and an infrared melt temperature detector, which can measure both, the melt temperature and prior to the melt front arrival the opposite mold surface temperature.

Response times of both RHCM technologies, surface temperature distribution, and typical measured curves of cavity pressure, melt and mold temperatures will be discussed. Weld-line tensile bars were systematically manufactured based on DoE methods to examine the influence of RHCM processing on tensile strength.

Injection mold and part geometries

Fig. 1 shows a schematic drawing of the ejection-sided mold insert used for the weld line tensile bars. The melt is first injected in the center and then separated twice. Each tensile bar is filled from 2 sides thus a weld line is formed in the middle, where the surrounding cavity is heated by an AlN ceramic heater element 3 mm below the mold surface (Ultramic 600 CER-1-01-00002, 25 mm x 25 mm, [8]).

As mentioned before, the opposite mold insert is planar and heated by the areal heating/cooling technology BFMOLD™ [9], using a heat exchange chamber of

240 mm x 175 mm with maximum temperature of 160°C (Fig. 2). This one is used for all specimen geometries.

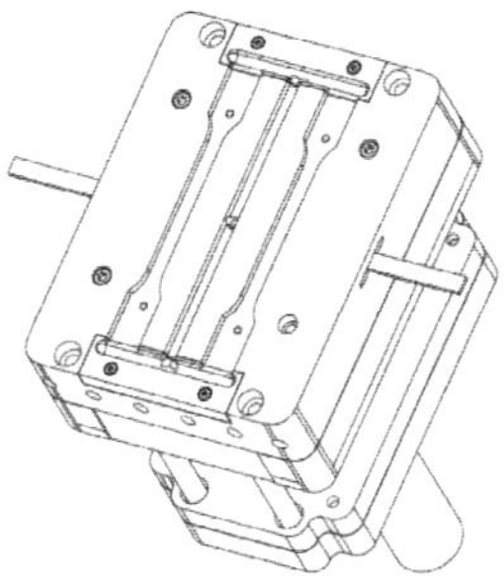

Fig. 1: Schematic of the weld-line tensile bar mold insert with locally underlying ceramic heaters.

Fig. 2: Photography of the mirror-finished, PVD-coated and areal-heated mold half.

Further specimen can be produced by a second ejection-sided mold insert (Fig. 3) where again chosen regions are heated by underlying ceramic heaters. Additionally, three triple-combined FOS MTPS408 cavity sensors (FOS Messtechnik GmbH, front diameters of 4 mm, [10]) may be installed either to measure the opposite mold surface temperature development in between mold closing and melt front arrival or to detect the polymer melt temperature after melt front arrival.

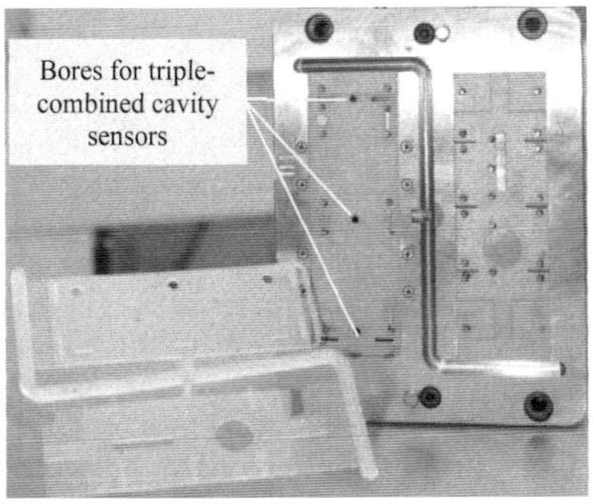

Fig. 3: Plate-shaped specimen to examine surface defects. Locally underlying ceramic heaters and bores for triple-combined cavity sensors.

Experimental

MTPS408 cavity sensor

Fig 4 displays typical measurement curves of a MTPS 408 cavity sensor in a conventional injection molding cycle. Both, the data acquisition and the BFMOLD™ heating started as the mold was closed, where the set cooling temperatures were 60°C. Since the used PC-ABS grade and the mold surface have nearly the same emissivity of 95%, the IR-sensor calibration fits for the monitoring of both materials. Therefore, the IR sensor detected well the rising opposite mold surface temperature until the polymer melt front arrived. The ejection-sided mold surface temperature, which was acquired by the surface-near thermocouple in the MPTS408 sensor, stayed constant until polymer was injected.

As soon as the polymer melt front passed the sensor, both the IR-detector and the cavity pressure reacted very fast, due to their response time of 10 milliseconds, while the type K thermocouple mold temperature sensor was delayed.

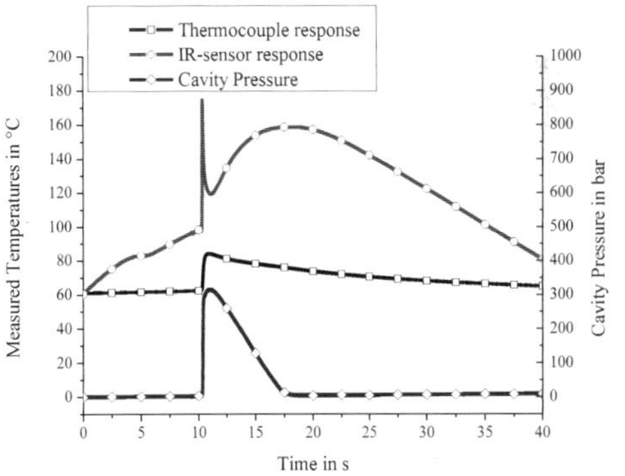

Fig. 4: Typical measurement curves of a FOS MPTS 408 sensor in a variothermal injection molding cycle with black-colored PC-ABS.

The maximum measured temperatures differed. The mold temperature rose only by 15°C due to the several hundred times higher thermal effusivity of the sensor steel compared to the polymer.

The measured melt temperature first increased to nearly the set nozzle temperature of 180°C as the melt reached the IR-sensor. Since the used PC-ABS grade was black-colored its IR-penetration depth was only a few micrometers, thus a limited polymer surface layer contributed to the temperature measurement. This surface layer froze due to cooling by the mold. The second increase was maybe caused either by reheating of the frozen layer by the hot polymer melt core or by a distinct increase of the frozen layer's IR- transmissivity in the lower temperature range, thus the IR-penetration depth was higher. In the remaining cycle time, mold and melt temperatures as well as cavity pressure decreased continuously.

Performance of RHCM technologies

The performance of both RHCM technologies was evaluated first in heating/cooling dry runs with the open mold. Type K thermocouples were applied onto the mold surfaces close to the expected weld-line region. Since the used type K thermocouples were found to be too slow, the monitoring of the surface temperature was done by an IR-thermography camera type Thermacam SC 5000 (FLIR Systems Inc., USA). Fig. 5 displays both, IR-thermography images of the mold surfaces and the rectangular temperature analysis areas marked in black.

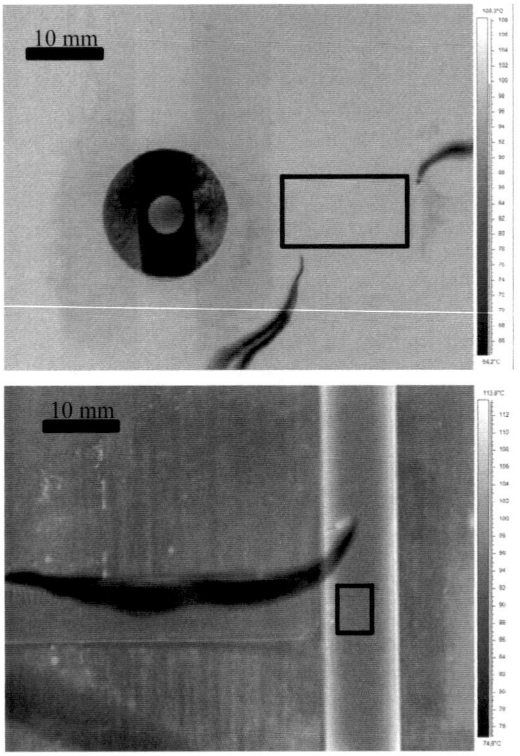

Fig. 5: IR-thermography images of BFMOLD™ (top) and ceramic heater elements (bottom) heated mold surface temperature distributions prior to heating. Shadowed streaks are caused by type K thermocouples; Temperature analysis areas marked by black rectangles.

Fig. 6 shows the averaged IR-thermography-detected surface temperatures during a heating/cooling cycle for both RHCM technologies. The nominal temperature was set to 160°C. While BFMOLD™ generates a global uniform temperature distribution (see Fig. 5 top), its temperature gradient of +1.5 K s^{-1} is low. Ultramic heating showed a temperature gradient of +10 K s^{-1}, however the heated region was small. Both technologies had automatic controller action induced temperature variations, which have to be further reduced.

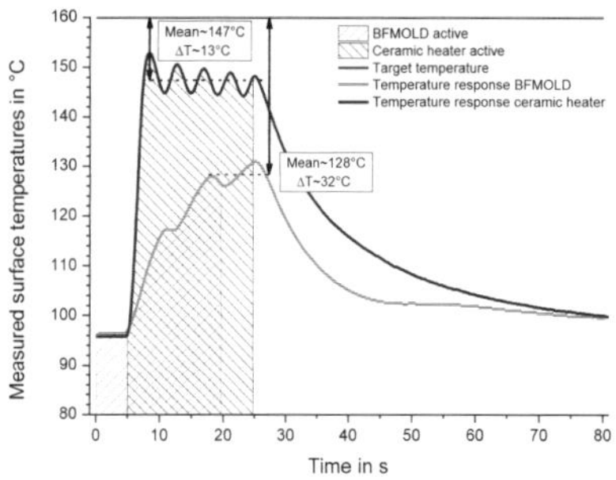

Fig. 6: Comparison of measured surface temperature curves, detected by IR-thermography.

Experimental Design for Tensile Strength evaluation

Natural PP polymer grade Borealis BB412E, Borealis Polyolefins, was chosen for the experiments. Tab. 1 shows the Design of Experiments plan to examine both, the weld line tensile strength and the elongation at break. The required heating times and nominal temperatures for both heating technologies were detected in pre-tests to guarantee the same local mold surface temperatures T_M on both sides of the cavity during the injection phase. Moreover, the switch-over-volume was adapted so that the filling-grade at switch-over-point was equal for all settings. 10 specimens were manufactured and analyzed at each setting.

Tab. 1: 2^3-Design of experiments plan with center point.

Injection rate v_I (cm^3s^{-1})	Nozzle temperature T_N (°C)	Mold Surface Temperature T_M (°C)
60	210	120
60	210	103
60	230	120
60	230	103
30	210	120
30	210	103
30	230	120
30	230	103
45	220	111.5

Results and Discussion

The nozzle temperature, the injection rate and the surface temperature were found to influence the tensile strength significantly; the latter was the most important (Fig. 7). The interactions were not statistically significant.

Fig. 8 displays the main effects plot. As expected, rising nozzle temperature and surface temperature increased the tensile strength. Surprisingly, injection rate showed a negative effect, which may be caused by trapped air at the welding area or by the adapted switch-over-volume from 30 to 31 cm^3. Further experiments will be done to clarify this behavior. Moreover, the center point result differs from the interpolated lines, telling that at least one of the examined parameters has a non-linear influence on the tensile strength. Therefore, a Central Composite

Response Surface Design of Experiment will be added to generate a more suitable mathematical model.

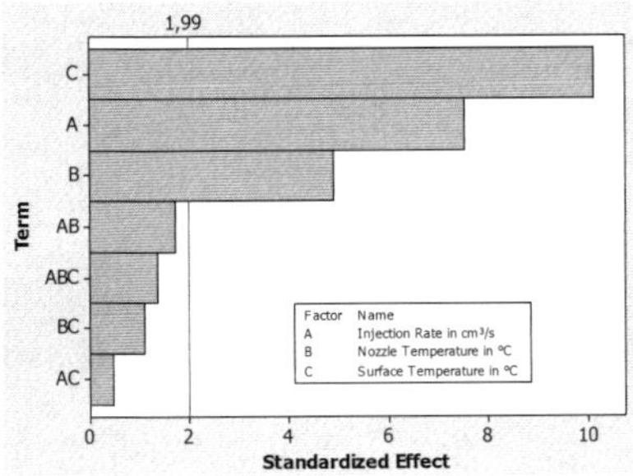

Fig. 7: Pareto Chart of Standardized Effects on the Tensile Strength (in MPa; Alpha = 0.05).

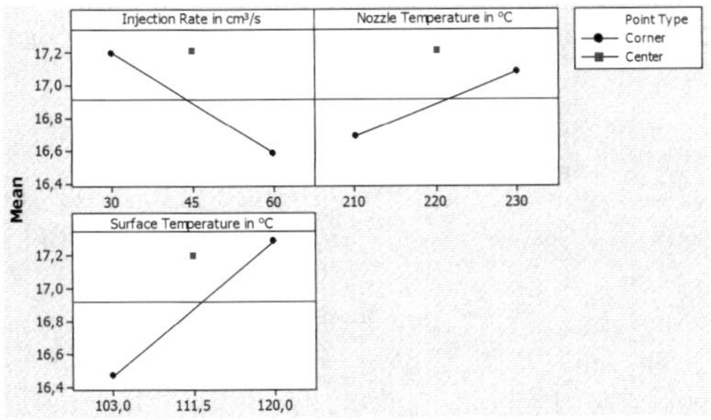

Fig. 8: Main Effects Plot for Tensile Strength (in MPa; Data means).

The influences of surface temperature, injection rate and nozzle temperature on the elongation at break were found to be similar (Tab. 2); however the interaction of injection rate and nozzle temperature was significant since the p-Value < 0.05.

Tab. 2: Estimated effects for elongation at break (significance level = 0.05).

Factors and Interactions	Effect	p-Value
Injection rate v_I (cm^3s^{-1})	-0.4741	0.000
Nozzle temperature T_N (°C)	0.1997	0.050
Mold Surface Temperature T_M (°C)	0.5095	0.000
Interaction $v_I * T_N$	-0.2311	0.025
Other interactions	< 0.11	> 0.050

Conclusions

An injection mold featuring interchangeable inserts for producing tensile bars (with or without weld-lines), thin-thick-thin stripes and complex plate-shaped parts was manufactured. The mold was equipped with novel sensors, allowing the simultaneous measurement of mold temperature, melt temperature and cavity pressure. Furthermore, the mold features two RHCM systems for variothermal processing.

Experiments carried out showed that RHCM has significant influence on mechanical properties. Both, tensile strength and elongation at break were being increased.

Further studies will be carried out to investigate the polymer morphology at the weld-line area, aiming to gain profound knowledge of relationships between RHCM processing, polymer morphology and mechanical properties.

Acknowledgements

The research work of this paper was performed at the Polymer Competence Center Leoben GmbH (PCCL, Austria) within the framework of the COMET-program of the Federal Ministry for Transport, Innovation and Technology and the Federal Ministry of Economy, Family and Youth with contributions by the Chair of Injection Moulding of Polymers, Montanuniversitaet Leoben and Wittmann-Battenfeld GmbH, Schöfer GmbH, Pollmann Austria GmbH, and Ernst Wittner Ges.m.b.H. The PCCL is funded by the Austrian Government and the State Governments of Styria and Upper Austria.

References

1. G. Lucchetta; M. Fiorotto; P.F. Bariani *CIRP Annals – Manuf. Technol. 2012, 61, 539-542.*
2. F. Pisciotti; A. Boldizar; M. Rigdahl; I. Ariño *Polym. Eng. Sci. 2005, 45(12), 1557-1567.*
3. U.A. Theilade; H.N. Hansen *Int. J. Adv. Manuf. Technol. 2007, 33(1-2), 157-166.*
4. G.R. Berger; S. Roock, et al., Improving the polymer surface quality by infrared radiation driven dynamic mold temperature control. *Conference Proceedings of 27th Annual Conference of Polymer Processing Society Polymer Processing Society, Marakech, Morocco, 2011.*
5. G.R. Berger; S. Roock, G. Steinbichler *Kunststoffe International 2011 (7), 26-29.*
6. R.X. Xu; E. Sachs *Polym. Eng. Sci. 2009, 49(2), 305-315.*
7. S.C. Chen; W.R. Jong, J.A. Chang et al. *Int. Polym. Proc. 2006, 46(7), 938-945.*
8. Ultramic® Advanced Ceramic Heaters, *URL: www.watlow.com/products/heaters/ultramic-ceramic-heaters.cfm?famid=10.*
9. Kunststoff - Institut Lüdenscheid für die mittelständische Wirtschaft NRW GmbH: BFMOLD™ - Technologie, URL: *www.kunststoff-institut.de/download/1142.*
10. FOS MTPS 408. Injection Molding Sensor for Pressure and Temperature, *URL: www.fos-messtechnik.de/MTPS_408_2008.pdf.*

O-04-057

ACHIEVING FINE CELL STRUCTURE OF POLYPROPYLENE FOAM WITH HIGH EXPANSION RATIO BY CORE-BACK FOAM INJECTION MOLDING

Shunya Yasuhara[a], Ryoma Miyamoto[a] and Masahiro Ohshima[a,*]
[a] *Dept. of Chemical Engineering, Kyoto University, Kyoto, Japan 615-8510*

Abstract - Microcellular foam injection molding process combined with core-back expansion molding is getting a leading position as a competent technology of producing fine cellular foam. The effects of core-back rate, core-back distance, core-back timing and holding pressure on the cell size and cell density were thoroughly investigated by experiments using a 35 tonnage clamping force Mucell machine (J35ELIII-F, JSW) with nitrogen as a blowing agent for foaming isotactic polypropylene (i-PP). The experimental investigation clarified that the increase in core-back rate increased the depressurization rate of the polymer in mold cavity and thus the higher core-back rate could make the cell size decreased and the cell density increased. Furthermore, manipulation of core-back timing (dwelling time) could control the crystallization and viscosity of polymer in cavity and make the cellular structure changed. Adjusting the processing conditions, i-PP could be foamed with fivefold volume expansion, a density reduction of 80% and fine cell size of 20- 60 μm.

Keywords: polypropylene, injection molding, supercritical fluid, core-back foaming

1. Introduction

Nowadays, plastic foaming technology is indispensable to reduce the weight of plastic products with higher dimensional stability. Especially, the fine cellular foaming technologies with CO_2 or N_2 as a physical blowing agent have been attracting much attention[1-7]. Among these technologies, microcellular foam injection molding process with core-back expansion molding (breathing molding) is getting a leading position as a competent technology of producing fine cellular foams with high expansion ratio [8-11]. The technique slightly differs from conventional microcellular foaming technique: once the polymer fills up the mold cavity, a portion of the mold is moved to expand the cavity volume and reduce the pressure exerting on the polymer in the cavity, which induces bubble nucleation and growth in polymer. By manipulating the core-back rate and distance, the rapid depression of the cavity pressure can be realized. As a result, the foam with high expansion ratio, density reduction of 30 -50% were achieved [10-11].

Over the past few years, the foaming behavior during core-back molding was visually observed through the glass windows of a mold[8,9]. The effects of gas concentration and the core-back rate on bubble nucleation and growth were investigated. It was clarified that the foaming behavior in the cavity of core-back mold expansion resembles the batch pressure quench foaming. But, at the same time, some differences were also pointed out: foaming in coreback expansion molding is conducted in non-isothermal and non-equilibrium state. N_2 provides higher degree of super-saturation and larger cell density with finer cell size than CO_2 in injection molding while in batch foaming the opposite results were obtained[8,9]. However, there has been few study of the effect of operating factors of coreback molding, such as core-back rate, core-back distance and core-back time on cellular structure.

In this study, a 35 tonnage clamping force Mucell machine (J35ELIII-F, JSW) equipped with a core back mold was used to investigate these effects. The machine can change the coreback distance in the range of 2 to 8 mm at the various core-back rate in the range of 0 to 40 mm/s. Adjusting the coreback operating conditions, foamed i-PP was prepared with fivefold volume expansion at maximum, a density reduction of 80% and cell size of 20- 60 μm.

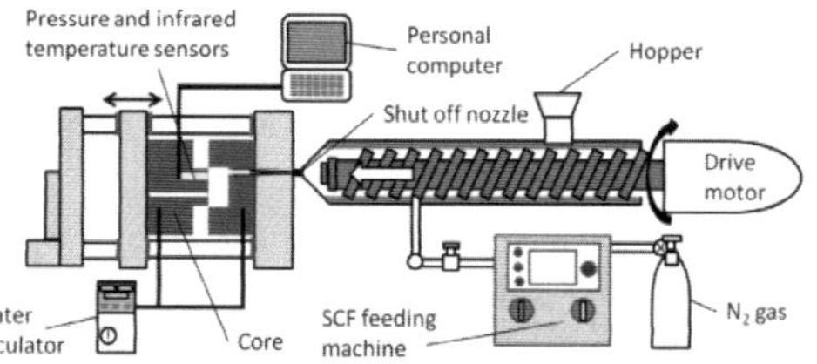

Figure 1— schematic diagram of the injection molding machine and a gas injection device

2. Experimental

2.1 Equipment

Figure 1 illustrates the 35-ton foam injection molding machine (Japan Steel works, J35ELIII-F) equipped with a gas injection device (Trexel SCF system SII TRJ-10-A). The injection device pressurized and injected N_2 into polymer melt in screw barrel and the revolution of the screw created a single phase polymer/N_2 mixture. The core-back mold was a box-type mold having a gate 1.5 mm in diameter and shut-off nozzle. A data logging system (Mold Marshaling system EPD-001, Futaba Corp.) monitored pressure and temperature in the cavity and acquired the data every 20 msec. Ejector-pin-type pressure and infrared temperature sensors were used for pressure and temperature measurements. The core-back distance and rate were monitored by a laser displacement sensor (LB-01/LB-60, Keyence Ltd) and were recorded every 20 msec using an additional data logger (GR-3000, Keyence Ltd). **Figure 2** illustrates a schematic time chart of core-back injection molding (position of movable part of mold, screw position, and clamping force): The core-back distance is a distance of moving the portion of mold part from its initial position (2mm) to open and the core-back rate is the rate of the movement. The core-back expansion was made at the end of dwelling time. Thus, the core-back timing was controlled by dwelling time.

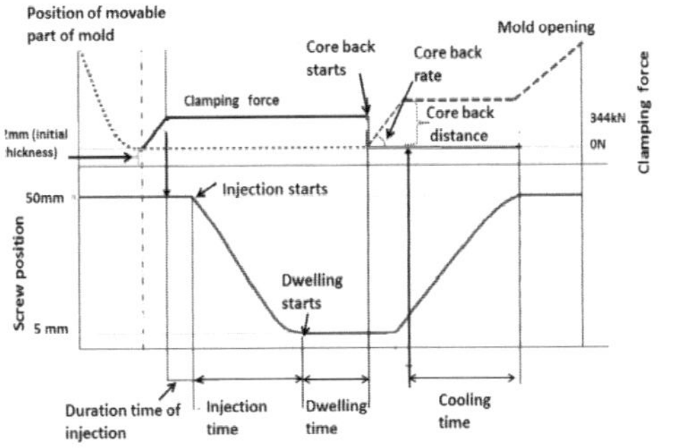

Figure 2 –A schematic time-chart of core back injection molding operation

2.2 Materials

N_2 (Izumi Sangyo) was introduced to injection machine at pressure of 25 MPa so as to make concentration to be 0.2 wt% of polymer weight. The temperatures of 7 zones of the barrel were set at 190, 200, 210, 210, 210, 210 and 210 °C from the bottom of hopper to the tip of barrel respectively. Isotactic Polypropylene (i-PP) (F133A, Prime Polymer, MFR=3 g/10min) was used. The cell morphology was observed with the SEM (Tiny-SEM, Technex) by cutting a piece from the core layer at the center part of sample. The observation was made at areas both in perpendicular and parallel to the coreback direction as shown in **Fig. 3.**

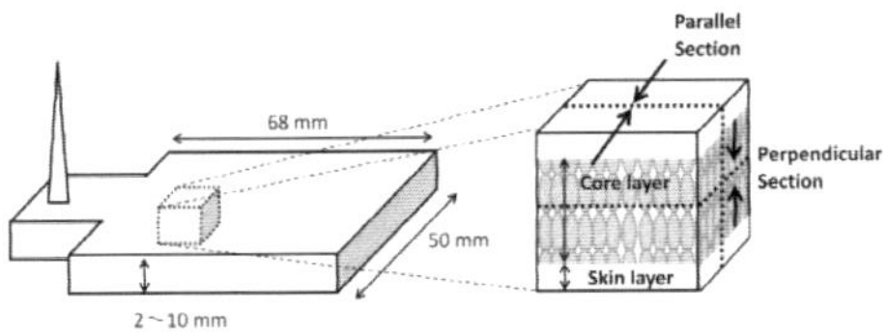

Figure 3 – Sample geometry and position of the sample used for SEM observation

3. Results and Discussion

3.1 Effect of core-back rate

Table 1 shows the condition of experiment to investigate the effect of core-back rate. **Figure 4** shows the SEM images of i-PP foamed at different core-back rates while keeping the core-back distance at 8mm. The SEM images were taken from the area in perpendicular to the core-back direction. The number density per unit area was calculated by Eq.1.

Table 1 – Foaming conditions for investigating the effect of core-back rate at fivefold expansion ratio

Condition	Value
Initial cavity size (mm)	68 × 50 × 2
Final cavity size (mm)	68 × 50 × 10
Injection velocity (mm/s)	40
Holding pressure (MPa)	40
Dwelling time (s)	6
Core-back rate (mm/s)	13, 10, 8, 4

$$N_d = \left(\frac{n}{A}\right) \quad (1)$$

where n is the number of cells observed in a SEM image and A is the area of evaluation. The cells were elongated to the core-back direction and shaped to be elliptic. On the other hand, the cell morphology observed at area perpemdicular to coreback direction was spherical as shown in Fig.4. Thus, N_d was caculated from the SEM image of the area perpemdicular to core-back direction.

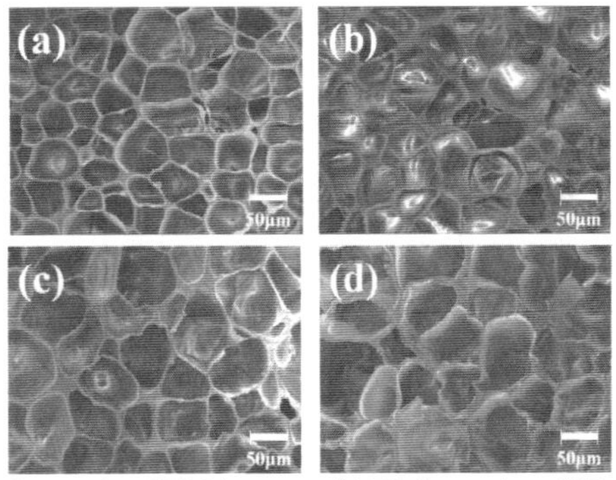

Figure 4 – SEM images of i-PP foamed at different core-back rates at the same core-distance 8 mm (a) 13 mm/s (b) 10 mm/s(c) 8 mm/s, (d) 4 mm/s

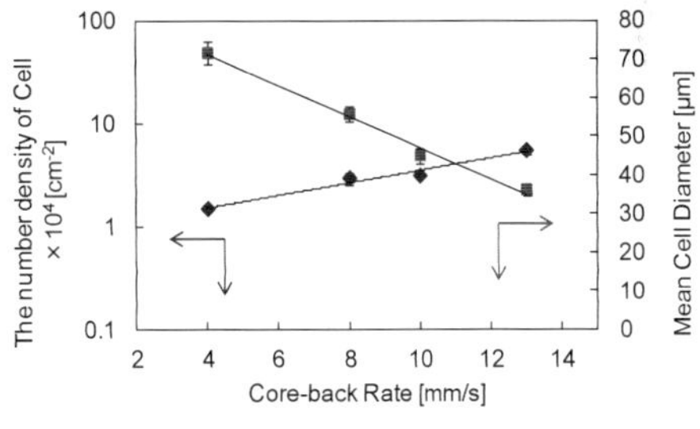

Figure 5 – Cell size and cell density of i-PP foamed at different core-back rates

Figure 5 shows cell size and cell density per unit area of i-PP foamed at different core-back rates. The cell size decreased and the cell density increases as the core-back rate increased. **Figure 6** shows the pressure profiles in mold cavity at different core-back rates. The equilibrium pressure of N_2 in PP was calculated from Henry's constants[14]. The depressurization rates at the equilibrium pressure of N_2 were calculated and listed in **Table 2**. As the core-back rate was increased, the depressurization rate was increased and, consequently, the cell size was decreased and the cell density was increased. These are the same phenomena observed in both extrusion foaming[12] and batch foaming processes[13]; as the higher depressurization rate is made in physical foaming process, the bubble nucleation rate is enhanced and cell size is decreased. Foaming at core-back rate slower than 4 mm/s with 6 s dwelling time could not complete fivefold expansion ratio and the foam density was increased due to the overcooling and solidification (Table 2).

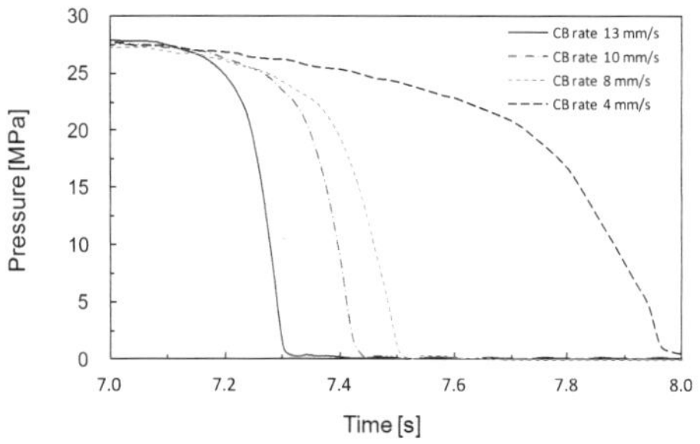

Figure 6 – Pressure profiles at different core-back rates

Table 2 – The effect of core-back rate on foam density and skin layer

Core-back rate [mm/s]	13	10	8	4
Density [g/cm³]	0.213	0.213	0.214	0.281
Skin layer [μm]	458.4	424.1	436.7	468.9
Equilibrium pressure [MPa]	1.79	1.79	1.79	1.79
Depressurization rate [MPa/s]	372.5	300	200	130

Table 3 – Foaming conditions for investigating the effect of core-back distance at fixed core-back rate

Condition	Value
Initial cavity size (mm)	68 × 50 × 2
Final cavity size (mm)	68 × 50 × (4, 6, 10)
Injection velocity (mm/s)	40
Holding pressure (MPa)	40
Dwelling time (s)	6
Core-back rate (mm/s)	20

3.2 Effect of core-back distance

The effect of core-back distance on cell structure was investigated. Experimental conditions were listed in **Table 3**. By changing the core-back distance, the volume of cavity was changed. Thus, foam expansion rate was controlled by the core-back distance. **Figure 7** shows the cell density and cell size at different core-back distances. They were measured from the SEM images of the cross-sectional area perpendicular to the core back direction. **Figure 8** shows the SEM images of the cross sectional area parallel to the core back direction. As can be seen in Fig. 8, the cells were elongated to the core-back direction. **Figure 9** shows the aspect ratio of the cellular size. The ratio was calculated by:

$$\text{Aspect ratio} = \left(\frac{\textit{cell length in long axis}}{\textit{cell size measured at area perpendicular to coreback direction}} \right)$$

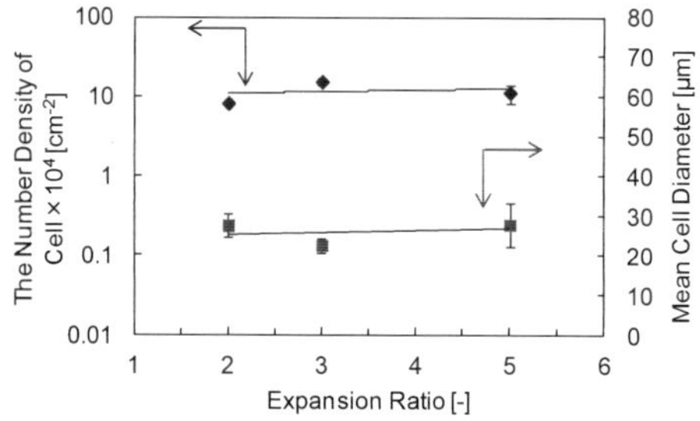

Figure 7– cell size and cell density of i-PP foamed at different core-back distances (i.e, expansion ratios)

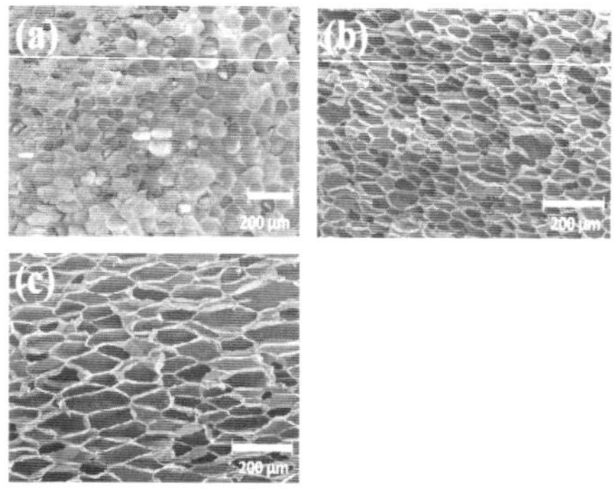

Figure 8 - SEM images of i-PP foamed at different core-back distances. –images were taken at the cross sectional area parallel to the coreback direction. (Expansion ratio /distance; (a) 2 fold/ 2mm, (b) 3 fold/4mm , (c) 5 fold/ 8mm)

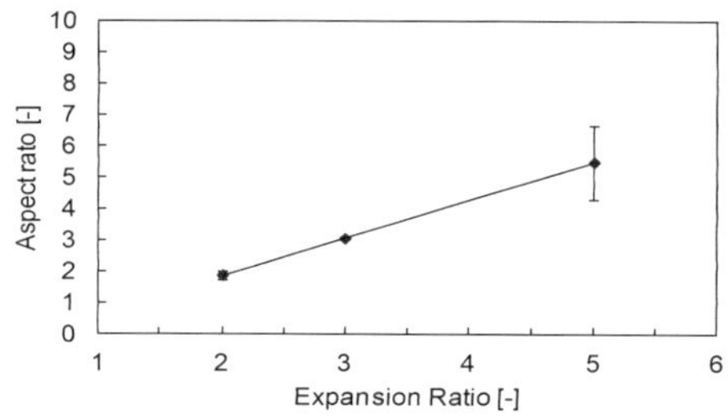

Figure 9 – Aspect ratio of the cell at different core back distances (expansion ratio)

The cell size and cell density observed at the area perpendicular to the core-back direction did not change drastically against the change in core-back distance (expansion ratio) while the cell length in core-back direction was changed in associated with the core-back distance (Figure 9). The bubble nucleation rate was increased with the increase of core-back rate but not affected with the increase of expansion ratio.

Crystallization of i-PP was increased in course of foam expansion and the presence of growing crystals could enhance the bubble nucleation. Therefore, it can be considered that small degree of increase in cell density and the decrease in cell size observed when increasing the core-back distance from 4 to 8mm could be explained by the crystallization-induced bubble nucleation.

3.3 Effect of dwelling time (core-back timing)

As illustrated in Fig.2, the core-back or mold cavity expansion was initiated at the end of dwelling process. Thus the effect of core-back timing was investigated by varying dwelling time as shown in **Table 4**. **Figure 10** shows the thickness of skin layer of the samples foamed at different dwelling time. **Figure 11** shows the cell size and cell density of the foams. The thickness of skin layer increased as the dwelling time increased. After the polymer was injected into the mold, which was temperature-controlled at 40°C, it was cooled down and crystallized from the interface at the mold. Therefore, as the dwelling time increased, the skin layer became thicker. Furthermore, the overall polymer viscosity increased as the dwelling time increased. The increase in polymer viscosity suppresses the bubble growth, reduces the cell size and increases the cell density. However, the longer dwelling time over 10 s could not achieve the specified expansion ratio due to thicker skin layer and higher viscosity. On the other hand, the shorter dwelling time less than 5 s made a hollow structure in the plastic products due to coalescence of bubbles. There existed an appropriate range of core-back time.

Table 4 – Foaming conditions for investigating the effect of core-back timing (dwelling time)

Condition	Value
Initial cavity size (mm)	68 × 50 × 2
Final cavity size (mm)	68 × 50 × 4
Injection velocity (mm/s)	40
Holding pressure (MPa)	40
Dwelling time (s)	6, 8, 10
Core-back rate (mm/s)	10

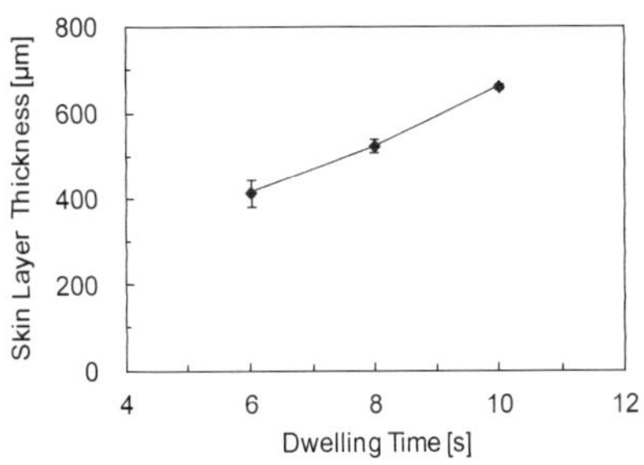

Figure 10 – Thickness of skin layer of the i-PP foamed at different dwelling times.

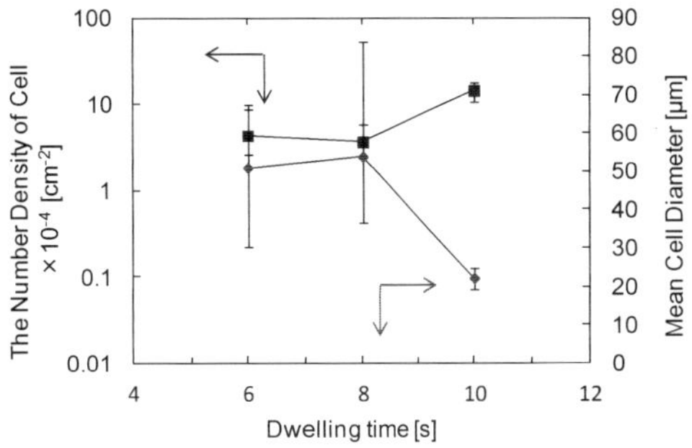

Figure 11 – cell size and cell density of i-PP foamed at different dwelling time.

Conclusions

The core-back foam injection molding was conducted under various conditions. Foaming i-PP with fivefold expansion ratio, density reduction of 80%, and fine cellular structure was achieved. Experimental investigation clarified that the core-back rate changed the depressurization rate of the polymer in mold cavity and the cell size and cell density could be controlled by core-back rate. Furthermore, manipulation of core-back timing (dwelling time) could control the crystallization rate and viscosity of polymer in cavity and could design the cell structure of the foam.

References

1. M. Mahmoodi; A Behravesh; S. Rezavand; A. Pashaei J. Appl. Polym. Sci, 2010,10,1002.
2. S. Chen; W. Liao; R. Chien Elsevier Ltd, 2012, 10, 1016
3. T. Yamada; Y. Murata; H. Yokoi Seikei-Kakou, 2010,10, 4325
4. Y. Moon; K. Lee; S. cha J MECH SCI TECHNOL; 2009, 10,1007
5. M. Stumpf; A. Spörre; H.Schmidt; V. Altstädt J Cell Plast, 2011, 10, 1177.
6. A. Spoerrer; D. Bangarusampath; V. Altstaedt, Ph.D. dissertation , University of Bayreuth, 2008
7. D. Baldwin; N. Suh Polymer Engineering and Science, 1995, 10, 1002
8. T. Ishikawa; M. Ohshima Polymer Engineering and Science. 2011, 51(8),1617-1625,
9. T. Ishikawa, K. Taki, M Ohshima, Polymer Engineering and Science, **2012, 52**,(4), 875–883.
10. T. Kanai; H. Kawato; H. Goda, Proc. of 19th Polymer Processing Society meeting, 2003
11. Mazda,http://www.mazda.com/publicity/release/2008/200809/080909c.html
12. C. Park; L. Cheung Polymer Engineering and Science 1997, 10, 1002
13. K. Taki, Chemical Engineering Science, 2008, 63(14), 3643-3653.
14. G. Li; F. Gunkel; J.Wang; C. Park; V. Altstädt *Journal of Applied Polymer Science*, 2006, *10*, 1002

WIRELESS IN-MOLD SENSOR FOR INJECTION MOLDING: A NEW APPROACH

F. Müller[1*], C. Kukla[2], T. Lucyshyn[1] and C. Holzer[1]

[1] *Department of Polymer Engineering and Science, Montanuniversitaet Leoben, Austria – florian.mueller@unileoben.ac.at; thomas.lucyshyn@unileoben.ac.at; clemens.holzer@unileoben.ac.at*

[2] *Industrial Liaison Department, Montanuniversitaet Leoben, Austria – christian.kukla@unileoben.ac.at*

Abstract: In this paper a novel wireless in-mold sensor is presented which is based on structure borne sound. A distinctive sound is injected in the metal mass of the mold when the melt front passes the sensor head. The sound can be detected by an accelerometer from any outside surface of the mold thus no wires are necessary. The components of the sensor implemented in the mold are purely mechanical. Measurements with a single and a multi cavity mold configuration with two different polymer types are presented. To verify the results obtained by the wireless sensor the results are compared to well-known cavity wall temperature sensors which are suitable for fast melt front recognition. The temperature sensors are located at the same flow path position as the wireless in-mold sensor enabling the comparison of the measured results. In every measurement series the melt front was detected at the same temporal moment by both measurement systems independently showing the independence of the novel wireless sensor from the processed polymer.

Keywords: injection molding, wireless in mold sensor, structure-borne sound, acoustics.

Introduction

The most important process for producing plastic parts is injection molding. Nowadays, the injection molding machine is developed towards high reliability with special focus on repeatability [1, 2]. Opposite to this development, the mold, where the actual part quality arises, is often neglected in terms of non-collecting important process parameters. This emerges from the fact that mainly just two types of in-mold sensors are commercially available [3]. Besides the financial aspect of equipping a mold with in-mold sensors a lot of effort is necessary for planning and drilling cable ducts for the sensor wires. Another important aspect comes up with servicing the mold, where squeezing harms or damages the sensitive sensor wires.

Recently, different wireless in-mold sensors were presented. Fan et al. showed a wireless sensor approach of a self-energized sensor for simultaneously measuring temperature and pressure in a mold [4]. In Müller et al. a purely mechanical in-mold sensor for detecting the melt front was presented [5]. In this paper the sensor concept is shortly described at first. The main contribution of the actual paper is the presentation of measurement results with the novel sensor concept using two additional polymer material families.

Experimental

Principle of Operation

A schematic drawing of the principle of operation of the novel wireless in-mold sensor is shown in Fig. 1. The sensor is shown in initial position shortly before being activated by the filling polymer A. A movable pin C projects inside the cavity B getting depressed by the passing melt front. The accelerated pin impinges on a resonant body D generating a sound which is injected into the metal mass of the mold as structure borne sound. All these parts implemented in the mold are purely mechanical keeping the sensor concept simple in operation.

Since structure borne sound equals a mechanical excitation of the mold the sound can be detected by an accelerometer at any outside surface of the mold. Hence, the sensitive electronic parts are located outside the mold, which decreases mechanical load and simplifies replacement in case of a failure.

After solidification of the polymer the movable pin can be used as an ordinary ejector pin pushing the part out of the mold. When pulling the ejector plate back to its initial position the movable pin will stop again in its initial position (like shown in Fig. 1), assuming not too high decelerations and friction to avoid inertia induced activation. All the components are back in initial position and consequently the cycle can start again. The novel wireless in-mold sensor is capable of detecting the melt front at predefined locations inside the cavity. Hence, the senor can be used as a replacement of cavity wall temperature sensors which are deployed as melt

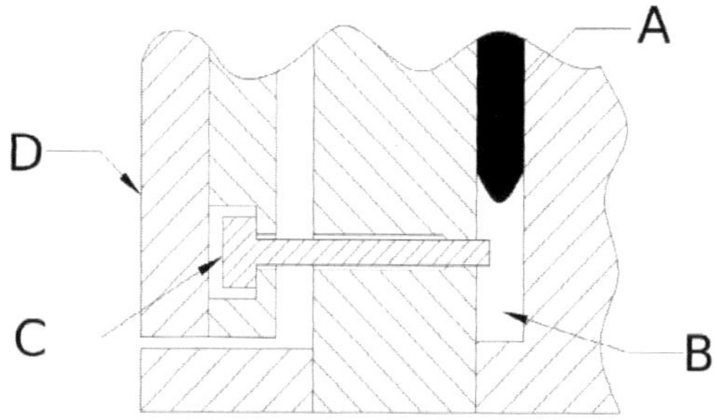

Figure 1: Schematic drawing of the principle of operation of the wireless in-mold sensor. The melt A fills up the cavity B. The movable pin C is depressed by the passing melt front impinging on the resonant body D injecting a sound in the metal mass of the mold.

front detectors. The novel sensor concept yields additional advantages listed in [5].

Experimental Setup

At first a brief instruction on nomenclature for easier understanding is given. The mechanical components implemented in the mold are named acoustic actuator. The accelerometer which is mounted on the outside surface of the mold is called acoustic sensor.
For the measurements a test mold was constructed yielding the possibility of implementing a maximum of two acoustic actuators, one in each of the two fully symmetrical cavities which are gated by a hot runner system. The produced part has a rectangular shape with two different wall thicknesses, i.e. 1.5 mm and 2.0 mm. By replacing the acoustic actuators with blankets the sound generation can be prohibited. For higher flexibility throughout the research project they are constructed as flexible as possible facilitating the easy exchange of components. An acoustic actuator is shown as a rendered section view in Fig. 2. It consists of a sprung movable pin A projecting in the cavity. The spring B ensures the right position of the pin at initial condition. When the pin is activated by a passing melt front it impinges on the exchangeable resonant body C which introduces structure borne sound in the metal mass of the mold.
At the outside surface of the mold an accelerometer (352A60 form PCB Piezotronics) is mounted detecting all the sounds from the mold. After amplification the signal from the acoustic sensor is recorded by a data acquisition system (DAQ) from National Instruments (USB-6366).
To verify the results of the novel wireless in-mold sensor cavity wall temperature sensors (Priamus 4009b with a sensor head diameter of 0.6 mm) were used. Due to their fast response time they are suitable for detecting a passing melt front [6]. To compare the results of both measuring systems to each other, both sensors need to be located at the same flow length position. To realize this requirement they are positioned opposite to each other. The temperature sensors are recorded by the DAQ as well for simple synchronization of the three data channels (one accelerometer and two cavity wall temperature sensors).

Measurements

In Müller et al. [5] it was shown that the novel wireless sensor concept is functional. The results were carried out using a highly fluent Polypropylene (C7069 – 100 NA) from Dow. To verify functionality of the sensor concept with more polymer material families two additional measurement series were performed and the results are shown in this paper. For each of the series at first just one acoustic actuator was implemented in the mold following by measurements with two implemented acoustic actuators. All the measurements were done on a machine from Arburg. As a representative of the

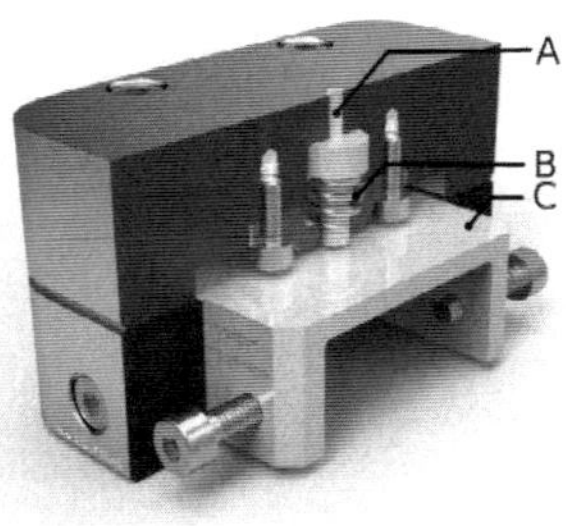

Figure 2: Rendered section view of an acoustic actuator consisting of a sprung movable pin A, a spring B ensuring right position at initial condition and a changeable resonant body C on which the pin impinges.

Polyoxymethylene (POM) material family Ultraform N2320 N 003 from BASF was used. The injection speed was set to 40 cm^3s^{-1} which lead to a filling time of approximately 1.6 s. The mold temperature was set to 70 °C.
The second polymer family was Polybutylenterephthalate (PBT). Ultradur B4300 G6 from BASF was used as a representative. In addition it is a fiber filled material (30 volume percent glass fibers). The injection speed was set to 90 cm^3s^{-1} and for mold safety reasons a maximum filling pressure of 1200 bar was permitted. As a result, the filling time is just slightly shorter than in the measurements with POM. The mold temperature was set to 70 °C.

Results and Discussion

Measurement Results for POM

At first one acoustic actuator was implemented in the mold. For this series the second cavity was neglected. In Fig. 3 an example of the measurements is shown. The red line shows the acoustic sensor signal, the dashed line the recorded temperature of the associated cavity wall temperature sensor. Up to 1.67 s the acoustic signal just contains large levels of noise occurring from different sources, e.g. injection molding machine. At 1.67 s a large deflection (peak A in Fig. 3) is recognized by the acoustic sensor evoked by the impinging movable pin on the resonant body (verification measurements are shown in [5]). At the same temporal moment the detected temperature rises sharply signalizing the melt front passing the temperature sensor. The temporal match between the two different measurement concepts, the wireless acoustic sensor and the cavity wall temperature sensor, is very good.
For the second measurement series with POM the second acoustic actuator was implemented in the second cavity. Now, one (if both actuators are excited at the same moment) or two (if excited successively due to unbalanced filling) deflections are expected. Since small inhomogeneous temperature distributions in hot runner systems cause an unbalanced filling of multi cavity molds two peaks are expected [7]. Despite the implementation of a second acoustic actuator in the

second cavity there is solely one acoustic sensor. However, a channel is added for the second cavity wall temperature sensor. In Fig. 4 an example result of this measurement configuration is shown. Again large levels of noise are detected up to 1.67 s, where the first deflection is detected (peak A in Fig. 4). This deflection corresponds to the already discussed acoustic actuator. At 1.78 s a second deflection (peak B in Fig. 4) is recognized evoked by the second implemented acoustic actuator. For each deflection the associated cavity wall temperature sensor rises at the same temporal moment. In the second measurement series a third peak indicated with E is recognized. This peak occurred sometimes after the second deflection and was generated by an unknown source inside the mold, e.g. a sub-assembly with too big tolerances [5].

Measurement Results for PBT

At first, a series with one implemented acoustic actuator was measured where one example result is shown in Fig. 5. Similar to the measurements with POM large levels of noise were detected during the whole measurement time (red line). At around 1.46 s a deflection A is recognized. This deflection has a good temporal correlation with the steep slope of the temperature signal. Hence the passing melt front was detected by both measurement systems at the same moment.
For the second measurement series the second acoustic actuator was implemented in the second cavity. As before, a second peak is expected generated by the second acoustic actuator. One example measurement is shown in Fig. 6. As expected, two peaks appear marked with A and B at 1.44 s and 1.46 s. Again, the temperature signals rise at the same temporal moments as the deflections appear. In contrast to the measurements with POM the chronological order of the cavity filling has changed, indicated by the temperature sensors. Now the temperature sensor of the second cavity detects the melt front earlier than the temperature sensor of the first cavity. The reason for this chronological swap is unknown but may result from different behavior of the material in combination with small inhomogeneous temperature distributions in the hot runner system and the cavities themselves.

Discussion and Outlook

As it was shown, the novel wireless sensor is capable of generating the same results regarding the detection of a passing melt front as cavity wall temperature sensors. When taking the results of [5] in consideration, the functionality of the wireless sensor concept was shown with three different material types from different material families (PP, POM, PBT).
However, the lack of wires raises issues in terms of which peak is generated by which acoustic actuator. Considering the results shown in Fig. 4 and 6, where the chronological filling order of the cavities swapped

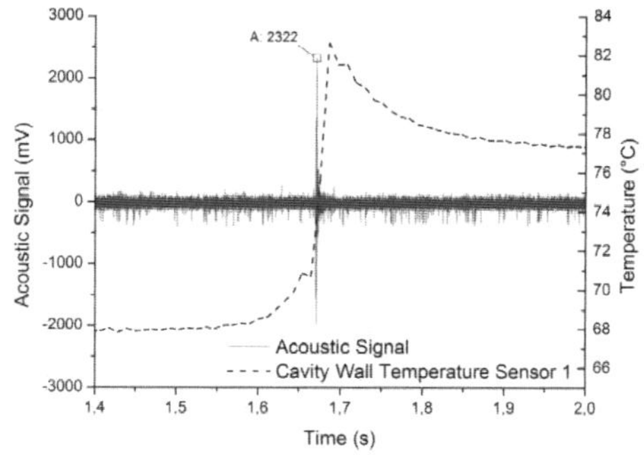

Figure 3: Measurement result for one implemented acoustic actuator with POM. The red line shows the acoustic sensor signal which has a deflection at 1.67 s. At the same moment the temperature detected by the cavity wall temperature sensor (dashed line) rises which indicates a passing melt front.

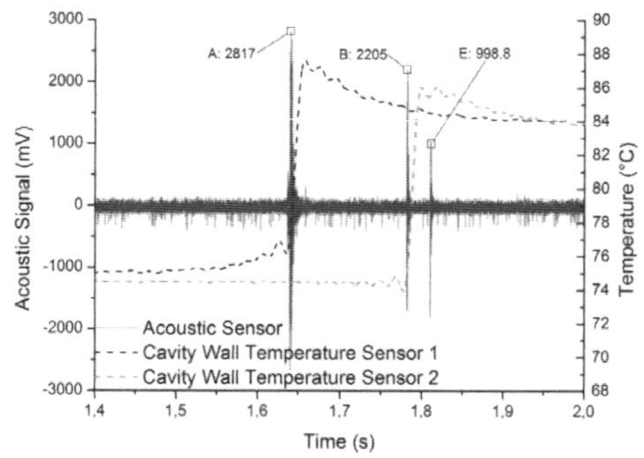

Figure 4: Measurement result for POM with two implemented acoustic actuators. A third channel for the second cavity wall temperature sensor is added. Two deflections generated by the acoustic actuators can be seen at 1.67 s and 1.78 s. For both deflections a good temporal match with the associated temperature sensors is achieved. A third peak E occurs which is declared as upcoming uncontrollable noise.

during the measurements, it would not have been possible detecting this behavior by the simple signal processing method by just looking at the occurring peaks.
Since the acoustic signals are recorded with a high sampling frequency fs = 25 kHz the signal contains further information. In Fig. 7 and 8 the first peak A and the second peak B of the measurements with POM and two implemented acoustic actuators from Fig. 4 are shown in more detail. Comparing both peaks to each other it can be seen that the deflections do have different frequency components as well as a different damping behavior making their shape unique. This is a result of using two different resonant bodies which are designed to generate special frequency components enabling the association of the occurring peak to an acoustic actuator. Thereby the signal processing algorithm has to be robust due to large level of noise.

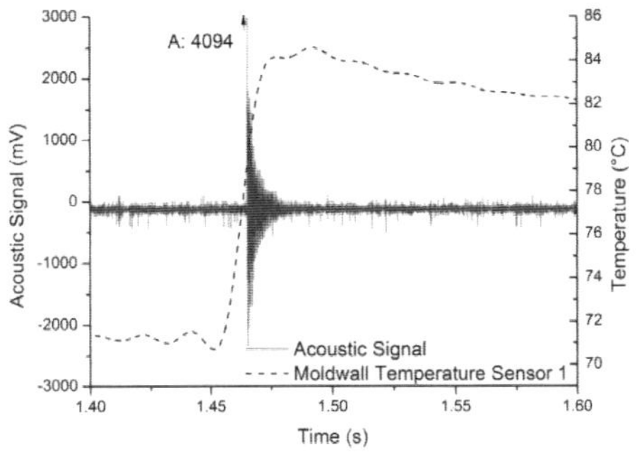

Figure 5: Measurement result for PBT with one implemented acoustic actuator. The red line shows the acoustic sensor signal with a deflection at 1.46 s. At the same temporal moment the temperature signal has a steep slope indicating the passing melt front.

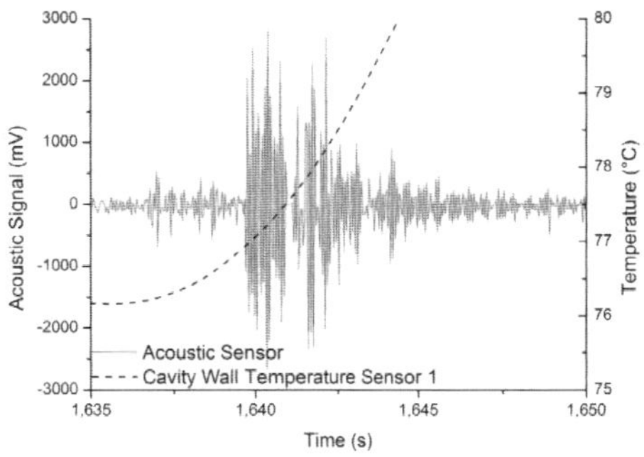

Figure 7: Detailed representation of the peak A (from Fig. 4) of the POM measurements with two implemented acoustic actuators showing the frequency components of the deflection.

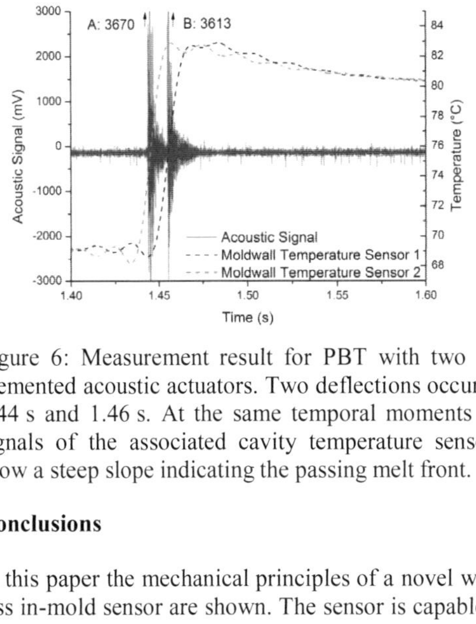

Figure 6: Measurement result for PBT with two implemented acoustic actuators. Two deflections occur at 1.44 s and 1.46 s. At the same temporal moments the signals of the associated cavity temperature sensors show a steep slope indicating the passing melt front.

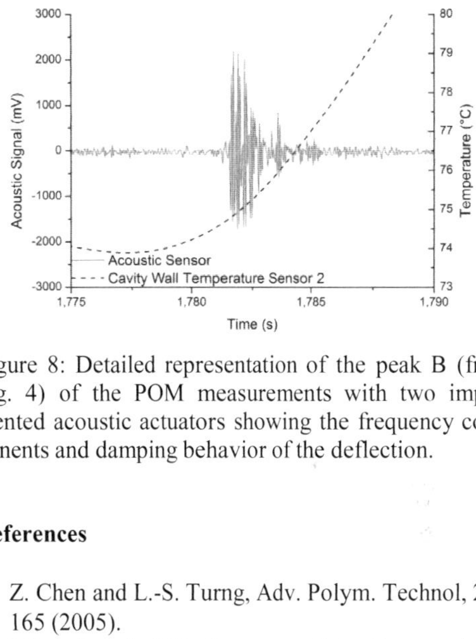

Figure 8: Detailed representation of the peak B (from Fig. 4) of the POM measurements with two implemented acoustic actuators showing the frequency components and damping behavior of the deflection.

Conclusions

In this paper the mechanical principles of a novel wireless in-mold sensor are shown. The sensor is capable of detecting a passing melt front. Measurements with two different polymer types and different number of cavities (which leads to a different number of implemented acoustic actuators) were performed and compared to well-known cavity wall temperature sensors located at the same flow path position as the novel wireless sensor. In every measurement series the melt front was detected at the same temporal moment by both measurement systems independently. These results show the independence of the novel wireless sensor from the processed polymer.

Acknowledgement

The research work was supported by the project 'PolyRegion' which is founded by the European Territorial Cooperation.

References

1. Z. Chen and L.-S. Turng, Adv. Polym. Technol, 24, 165 (2005).
2. M.R. Kamal, W.I. Patterson, D.A. Fara, and A. Haber, Polym. Eng. Sci., 24, 686 (1984).
3. J. Giboz, T. Copponnex, and P. Mélé, J. Micromech. Microeng, 17, 96 (2007).
4. Z. Fan, R.X. Gao, and D.O. Kazmer, "Design of a self-energized wireless sensor for simultaneous pressure and temperature measurement," Advanced Intelligent Mechatronics (AIM), 2010 IEEE/ASME International Conference on Advanced Intelligent.
5. F. Müller, G. Rath, T. Lucyshyn, C. Kukla, M. Burgsteiner, and C. Holzer, J. Appl. Polym. Sci. In Press (2012).
6. D.O. Kazmer, S. Velusamy, S. Westerdale, S. Johnston, and R.X. Gao, Polym. Eng. Sci., 50, 2031 (2010).
7. F. Johannaber and W. Michaeli, Handbuch Spritzgiessen, 2nd edn., Hanser, München (2004).

O-04-145

INFLUENCE OF MASTERBATCHES ON THE BOND STRENGTH OF TWO-SHOT MOLDED PARTS

T. Kisslinger[1]*, T. Lucyshyn[2], G.R. Langecker[2] and C. Holzer[2]

[1] *Polymer Competence Center Leoben GmbH, Leoben, Austria – thomas.kisslinger@pccl.at*

[2] *Department of Polymer Engineering and Science; Chair of Polymer Processing, Montanuniversitaet Leoben, Leoben, Austria – thomas.lucyshyn@unileoben.ac.at; langecker@a1.net; clemens.holzer@unileoben.ac.at*

Abstract - New construction possibilities and the combination of different materials in one processing step without the necessity of further assembling are the outstanding advantages of the multi-component injection molding process. The major difficulty for the injection molding process is to achieve a durable good adhesion between the components. It is known that masterbatches are able to change the mechanical properties of polymers, but their influences on bond strength are still not well explored. The interesting question is whether color masterbatches or other polymers acting as compatibilizers are able to improve the bond strength of two-shot molded parts.
In this work a test mold with core back technology was used to produce two-component injection molded tensile test specimens, which were analyzed regarding their adhesion in a tensile test. The investigations carried out include the influences of the barrel temperature of the second component and the impact of color Masterbatches and other additional polymers on the adhesion of two-shot molded parts.
Different types and amounts of color masterbatches and other polymers were gravimetrically weighed and added to various thermoplastic polymers to investigate their effect on the adhesion.
Based on these studies, it is possible to make recommendations regarding the use of color masterbatches or additional polymers as compatibilizers (including the melt temperature of the second component as well as the material combinations) to achieve a high bond strength.

Keywords: injection molding, multi-component, bond strength, masterbatch.

Introduction

Due to the ambition of time and cost minimization as well as the steadily rising haptic and visual requirements of molded plastic parts, the multi-component injection molding process is becoming increasingly important. Using this process, it is possible to fabricate plastic parts of materials having different optical and mechanical properties.
The dilemma occurs when it comes to the bond strength. The major challenge for the injection molding process is to achieve a durable good adhesion between the components. The material combination, the process parameters and the contact area structure have an effect on the bond strength of multi-component injection molded parts.
The knowledge about adhesion properties has been steadily improved in the past decades. Previously it was assumed that a pure physical adhesion (due to hooking of molecular chains) is responsible for the bond strength. Today various theories of adhesion are distinguished. The most important models to interpret the adhesion are [1]:

1. physical adhesion model (hooking of molecular chains)
2. specific adhesion model
   - 2.1. molecular physics interfacial operations
     - 2.1.1. polarisation theory of De Bryne (dipole character of molecule)
     - 2.1.2. electrostatic theory of Derjagin (transport of electric charge)
     - 2.1.3. diffusion theory of Vojuckij (flexibility of molecule, diffusion of molecule segments)
   - 2.2. thermo-dynamic interfacial operations
     interface thermo-dynamic theory of Zisman, Fowkes, Good and Wu (surface- and interface energy, adsorption, dispersive forces)
   - 2.3. formation of chemical compounds (covalent- and ionic bonds)

So far, there is no recognized theory to describe the processes that take place during the formation of a multi-component plastic part [2].
In quantitative terms, bond strength is frequently characterized by means of a tensile test as a mechanical stress – the so-called bond strength σ_V. This is defined in equation (1) (according to [2]):

$$\sigma_V = \frac{F_a}{A_N} \qquad \text{Eq.1}$$

Fa … external force required to tear the specimen
A_N… molded part area being present prior to fracture

Therefore, to obtain quantitative information about the adhesion of various material combinations, it is unavoidable to produce suitable test specimens and perform strength test measurements. In this study a core-back mold was used to manufacture the two-component tensile bars and a standard tensile testing machine was used for the evaluation of bond strength. At the core-back process the different materials are injected consecutively, so each component runs through the whole injection molding cycle (two-shot process).

Experimental

The core-back mold (Fig. 1), developed at the Polymer Competence Center Leoben GmbH, Leoben, Austria, and manufactured by CAMO Formen- und Werkzeugbau GmbH, Schwanenstadt, Austria, was designed for bond strength investigations of hard/hard and hard/soft two-shot molded parts.

Fig.1: Core-back mold

The used core-back mold was specifically designed for producing several different test specimen geometries. So it is possible to simulate different stress conditions by the usage of a standard tensile testing machine. Possible geometries are tensile, shear stress, puzzle and joint test specimen, which were constructed according to the multipurpose tensile bar of ISO 3167 type A (Fig. 2 and 3). Furthermore, a peel test specimen was designed for clamping the hard component in a sliding carriage to introduce the tensile stress normal to the direction of slide. The purpose of the puzzle and joint test specimen is the testing of material compatibility for in-mold assembly.

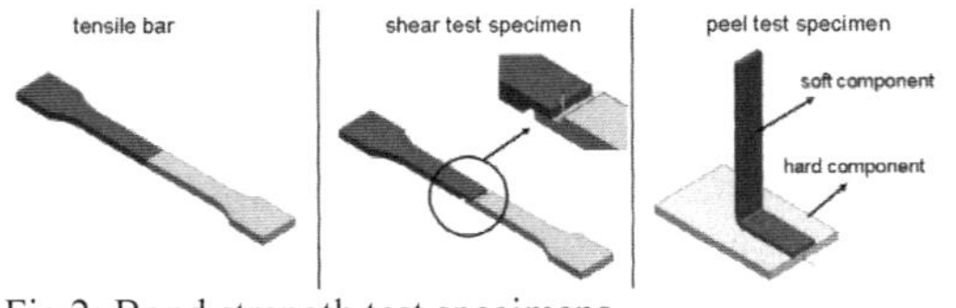

Fig.2: Bond strength test specimens

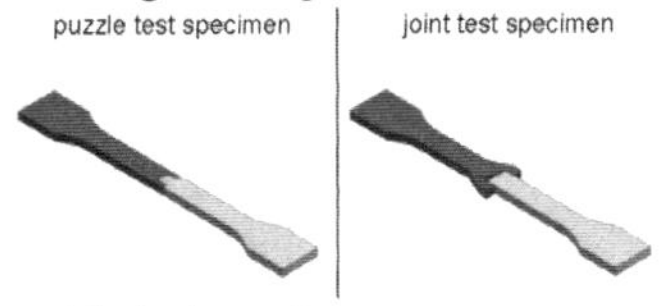

Fig.3: Assembly test specimens

The different test specimens were produced at the Department of Polymer Engineering and Science, Chair of Polymer Processing, Montanuniversitaet Leoben, Austria, using a two-component injection molding machine (HM 110/350H 130V) from Wittmann Battenfeld GmbH, Kottingbrunn, Austria. The subsequent tensile tests were carried out at a tensile testing machine (Zwick/Roell, Zwick GmbH, Ulm, Germany) at a temperature of 23 °C, 50 % atmospheric moisture and a tensile test velocity of 50 mm/min at the Department of Polymer Engineering and Science, Chair of Material Science and Testing of Plastics, Montanuniversitaet Leoben, Austria.

In previous investigations the tensile bar, the shear and the peel test specimens were injection molded by a systematic variation of process parameters. In these analyses the influences of high and low levels of

- melt temperature of the second component,
- mold temperature,
- injection rate of the second component,
- packing pressure of the second component,
- delay time before injection of second component and
- interface roughness

on the bond strength were investigated. The interpretations give approximately equivalent results to values from publications [3, 4 ,5, 6]. It was found that the melt temperature of the second component has the strongest influence on bond strength. The bond strength increased almost always with a higher level of melt temperature of the second component. The increase of adhesion between the components partially amounted to over 100 %.
Based on these results the focus in this work was placed on a more detailed study of the influence of melt temperature and color masterbatches as well as additional polymers on the bond strength. To change the melt temperature the barrel temperature of the second component was successively increased by steps of 10°C.
Two different material combinations were chosen to show the results:

1. PBT – ABS
2. PC– PMMA

The influence of two different color-masterbatches on the bond strength was tested on the material combination 1 and the effect of additional polymers acting as compatibilizers on the bond strength was tested on material combination 2.
All presented adhesion results are average values of 10 measurements based on the tensile bar (thickness 4 mm).
As mentioned before the main aim of these investigations was to achieve an increase on bond strength by the use of color masterbatches as well as additional polymers.

Results and Discussion

Figure 4 shows the bond strength results in dependence of the barrel temperature of the second component from the material combination PBT–ABS. PBT as the first component and a yellow masterbatch were used in two different variations. The dye of the masterbatches was the same, but the first masterbatch was based on LDPE and the second one on PC. The following combinations were invatigated:

a) PBT natural finish (without masterbatch)
b) PBT +2 % masterbatch (based on LDPE) gravimetrically weighed
c) PBT +2 % masterbatch (based on PC) gravimetrically weighed

With PBT natural finish a sharp increase in bond strength occurred between 250 °C and 260 °C barrel temperature of the second component. Both colored PBT showed also a sharp increase in bond strength, but this abrupt rise occurred between 260 °C and 270 °C. This sharp increase in bond strength is most probably caused by melting the crystallites of the first component. If the second component hits the first component with a temperature that melts the crystallites, both components are able to interdiffuse each other in a better way, which is in accordance with the publication of Bottenbruch and Binsack [7]. This improved interdiffusion leads to a significant increase of adhesion. This effect was investigated in detail in prior work [8]. Reasons for the shift of the sharp increase due to the colorations (a↔b,c), have not been explored so far.

Figure 4 also shows the possibility of an increase in adhesion due to the use of masterbatches. From 220 °C to 240 °C it was unfeasible to generate an adhesion between PBT (natural finish) and ABS. If the PBT was colored with masterbatch (PC) an adhesion to ABS was achieved at 230 °C, but at higher temperatures the bond strength was lower than PBT without masterbatch.

On the other hand PBT as the first component was colored with masterbatch based on LDPE. In this case bond strength to ABS could be realized from 220 °C to 280 °C. From 240 °C to 260 °C lower values in bond strength compared with PBT natural finish were measured. From 270 °C the results of adhesion were the same as PBT without masterbatch, thus also the maximum reached adhesion was equal.

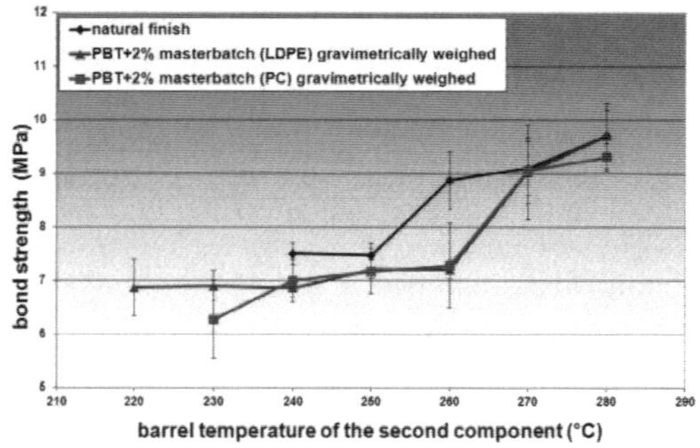

Fig.4: Bond strength results of PBT-ABS; with and without color-masterbatch

Figure 5 shows the bond strength results in dependence of the barrel temperature of the second component from the material combination PC–PMMA. PMMA as the second component and a TPE-S as additional polymer acting as compatibilizer were used in two different variations. At first two percent TPE-S were gravimetrically weighed to PMMA, to make a comparison to these results two percent TPE-S were compounded with PMMA. This compound was manufactured at the Department of Polymer Engineering and Science, Chair of Polymer Processing, Montanuniversitaet of Leoben, Austria, using a counter-rotating twin screw extruder Plasti-Corder PL 2000 from Brabender GmbH & Co. KG, Duisburg, Germany.

d) PMMA natural finish (without masterbatch)
e) PMMA +2 % TPE-S gravimetrically weighed
f) PMMA +2 % TPE-S compounded

It can clearly be seen that TPE-S as additional polymer worked as an adhesion promoter. From 220 °C to 240 °C the gravimetrically weighed TPE-S lead to an average increase of bond strength compared to PMMA natural finish by approximately 30 %. At a barrel temperature (second component) of 250 °C the bond strength results from material combination d and e were quite the same. At 260 °C the TPE-S (gravimetrically weighed) lead to a decrease in adhesion by less than 7 %.

On the other hand the PMMA-TPE-S compound offered an increase in bond strength over the entire measured temperature range. The average increase of bond strength compared to PMMA natural finish was nearly 45 %. Especially the increase of almost 70 % between 220 °C and 240 °C barrel temperature of the second component was an unexpected result.

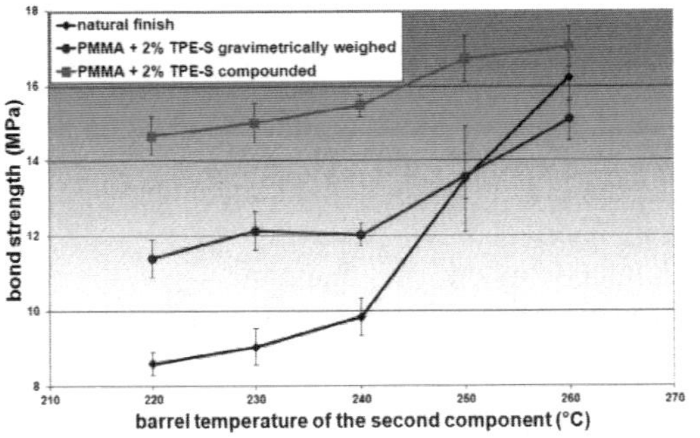

Fig.5: Bond strength results of PC-PMMA; with and without TPE-S

Conclusions

In this study it was shown that for material combinations which offer a fundamental compatibility the bond strength can be strongly influenced by the melt temperature of the second component. For all investigated material combinations the adhesion improved with increasing barrel temperature.

If the first component is a semi-crystalline thermoplastic it is of particular importance to set such a high barrel temperature of the second component that the crystallites of the first component are able to melt. Thus the conditions for diffusion are guaranteed and the bond strength will get a sharp increase.

Also a strong influence on the bond strength could be shown by the addition of color masterbatches and additional polymers acting as compatibilizers. In prior investigated material combinations [8] the adhesion decreased with an addition of masterbatches. The assumption that the adhesion can be influenced positively by using a suitable masterbatch or additional polymer could be confirmed by the investigations described above. Especially at lower barrel temperatures of the second component a considerable increase in the bond strengths compared to the natural finish materials could be achieved. Due to these findings multi-component plastic parts can be manufactured with the required bond strength at lower melt temperatures and thus shorter cycle times. This would result in a significant energy, time and cost reduction.

Acknowledgements

The research work of this paper was performed at the Polymer Competence Center Leoben GmbH (PCCL, Austria) within the framework of the COMET-program of the Federal Ministry for Transport, Innovation and Technology and the Federal Ministry of Economy, Family and Youth with contributions by Department of Polymer Engineering and Science; Chair of Polymer Processing and Chair of Material Science and Testing of Plastics, Montanuniversitaet Leoben and CAMO Formen- und Werkzeugbau GmbH, Schwanenstadt, Austria, Praher Kunststofftechnik GmbH, Schwertberg, Austria, Mahle Filtersysteme Austria GmbH, St.Michael ob Bleiburg, Austria, ZIZALA Lichtsysteme GmbH, Wieselburg, Austria, Simcon kunststofftechnische Software GmbH, Würselen, Germany. The PCCL is funded by the Austrian Government and the State Governments of Styria and Upper Austria.

References

1. W. Brockmann; et al., *Klebtechnik – Klebstoffe, Anwendung und Verfahren*, Wiley-Vch Publisher, Weinheim (2005).
2. H. Potente; R. Kleeschulte in PPS 24th Annual Meeting Proceedings, Italy, 2008, S03-729.
3. I. Kühnert, PhD Thesis, Chemnitz University of Technology, Chemnitz, 2005.
4. C. Jaroschek in *Spritzgießen von Formteilen aus mehreren Komponenten*, 1st Edition, Augustinus Publisher, Aachen, 1994.
5. K. Kuhmann, PhD Thesis, Friedrich-Alexander University of Erlangen-Nürnberg, Erlangen, 1999.
6. S. Brinkmann, PhD Thesis, Faculty of Mechanical Engineering, Aachen, 1996.
7. L. Bottenbruch; R. Binsack *in Polyamide – Kunststoffhandbuch ¾ Technische Thermoplaste*, Hanser Publisher, Munich, 1998, 1st Edition.
8. T. Kisslinger; T. Lucyshyn; M. Doler; C. Holzer; G.R. Langecker in PPS 26 Proceedings, Banff, 2010.

O-04-165

DEMOLDING FORCE MEASUREMENTS OF COATED MICRO-STRUCTURED SURFACES IN A HOT EMBOSSING PROCESS

Tobias Struklec[1]*, Martin Burgsteiner[1], Thomas Lucyshyn[1], Werner Balika[2], Christian Pöschl[2] and Clemens Holzer[1]

[1] *Department of Polymer Engineering and Science, Chair of Polymer Processing, Montanuniversitaet Leoben, Austria – tobias.struklec@unileoben.ac.at*

[2] *Sony DADC Austria AG, 5081 Anif, Austria*

Abstract - This paper is a study of the demolding of micro structured polymer surfaces in hot embossing. One of the major problems of several replication processes is a high number of defective products as a consequence of part damage e.g. ripping or deformation of the micro structure. These defects occur due to a combination of several mechanisms, e.g. adhesion, friction, capillary attraction, chemical bonding, electrostatic and vacuum effects. The influencing aspects, contrary to the several acting mechanisms, are already identified. These four key aspects are the mold (including the mold coatings), the polymer material, the micro structure and the processing parameters. To assess the impact of each of these aspects the demolding force is used as a qualitative summary of the different occurring mechanisms.

In this study the focus was on the effects of different surface coatings on the micro structure in combination with several polymer types. Seven out of eight nickel stampers were coated, while one was left untreated. This way a comparison between the different coatings and the original surface was possible. All nickel stampers had the same micro structure and were prepared in one process (except coating). The simple design of the micro structure was chosen to prevent effects like hindered shrinking inside the mold. Each stamper was used to determine the demolding forces in combination with different polymers like Polypropylene (PP), Cyclo-Olefin-Copolymers (COC), Polycarbonate (PC), Polystyrene (PS) or Poly(methyl methacrylate) (PMMA). While the demolding forces for some polymers were located close to the lower bound of the detection range, others were significantly higher. The coatings showed that most combinations worsened the demoldability while only one coating excelled as a demolding aid.

Keywords: injection molding, micro, microstructures, demolding, molding

Introduction

The trend of modern lab-analysis in life science and medical applications focuses more and more towards "lab-on-a-chip technologies". These life-science applications try to get as many features as possible on an even smaller space [1, 2, 3 and 4]. Therefore, the expectations for these applications increased continuously and are still increasing today. Not only the geometry of the channel (micro-geometry) but also the geometry of the polymer part (macro-geometry) is getting more and more precise, subsequently the dimensional tolerances for the polymer applications are going down. By now tolerances of only a few micrometers are desirable. Additionally, these applications are in medical services which impose many restrictions. These restrictions limit not only the variety of materials to choose from but also affect the processing, since almost no additives or enhancements may be used. The resulting injection molding process is seldom ready for large scale production [5]. This paper focuses on the demolding and demoldability of micro structured surfaces in hot embossing. The interaction of structured polymer surfaces with the surface of the mold has yet to be investigated thoroughly. Still, the interaction leads to different kinds of problems, like structures that may rip in the demolding process. These interactions might also induce bending of the chip, or inhibit demolding altogether.

Many of the mentioned problems only occur in specific setups or for some particular polymers. It can be assumed that the demolding problems are strongly related to the polymer, the geometry of the microstructure, the mold itself and the process settings. While there are studies on the demolding behavior for different geometries [7] as well as certain process settings [6] very little is known about the polymer - mold interaction. In this work a fixed geometry was replicated several times and coated with various coating materials. Using a hot embossing process, the polymer - coating pairings were tested in a regular hot embossing setup detecting the demolding force (F_{dem}) [6, 10] as parameter to describe the demolding quality.

Experimental

All the experiments have been performed at the Paul Scherrer Institute (PSI, Switzerland) using a Jenoptik HEX03 press. The Hex03 is designed like a standard materials testing machine, which allows a time and distance controlled measurement of the force and the displacement during the demolding step. For the highly sensitive demolding step the so-called ''position relative" modus of the HEX03 was used, which had previously been introduced for similar work done at the PSI [7]. This method allows temporary recording with a 20 ms sample time (with a displacement-resolution of 133 nm per measurement step) compared to the regular 1s time steps. This is necessary to represent the demolding with a sufficiently accurate resolution. The demolding happens over a small timeframe in the range of motion of 50 µm.

Setup

The nickel insert with the desired test micro geometry was fastened to the upper Hex03 tool by a steel frame allowing good heat conductivity. At the bottom of the press another arbitrary structured insert was used to keep the polymer substrate at the bottom and ensure the demolding (detaching) takes place on the desired (upper) insert. This is achieved with any insert that has a higher structural density and therefore a larger corresponding demolding force. The polymer substrate was loosely placed on the bottom insert. With a "touch force" of 200 N the polymer substrate was fixed between the two inserts. This ensured an acceptable heat conductivity. After a defined period of time the full molding force was applied and held over at least 5 minutes. Following, the cooling of the tool starts until the temperature falls below the demolding temperature. This triggers the demolding motion and the start of the high resolution measurement.

Parameters

The following parameters were adjusted according to the material and coating setup (Table 1): The molding temperature, the embossing force to ensure a good casting of the microstructure into the polymer substrate and the demolding temperature. Additionally, the demolding took place under vacuum with a constant speed of 0.2 mm/min. All parameters were chosen after initial testing. Both the demolding speed and the demolding temperature were varied before the actual tests to ensure that no unintended influence affects the measurement.

Table 1 – List of the used hot embossing parameters for Polymer 1-4 (P1- P4).

	P1	P2	P3	P4
Molding Temperature (°C)	165	145	135	150
Embossing Force (N)	5000	3000	3000	2000
Demolding Temperature (°C)	80	80	60	85

The demolding speed directly influences the demolding force, so a low velocity was chosen to increase the measurement accuracy. On the contrary the demolding temperature does not affect the demolding force once a 'certain' temperature is reached. Therefore, the demolding temperature was set to a sufficiently low empirical value.

Coatings

The nickel inserts (stampers) were fabricated using a standard lithography and a galvanic process provided by Sony DADC Austria AG. The stamper design was chosen to be a homogeneous pattern consisting of horizontal and vertical micro channels with 50 µm width and depth. All stampers were produced from the same silicon wafer and cut out after galvanizing. This ensured similar surface topology for all inserts.

Two coatings (C1, C2) came from the Leoben Laserzentrum (Austria), four (C3, C4, C5, C6) from Balzers Oerlikon (Liechtenstein) and one coating (C7) from Montanuniversitaet Leoben (Austria). One coating was arbitrarily chosen as the reference coating to compare the coating facilities and procedures. The uncoated nickel was used as reference for the demolding force. Furthermore, all coatings were analyzed regarding their contact angle (Figure 1).

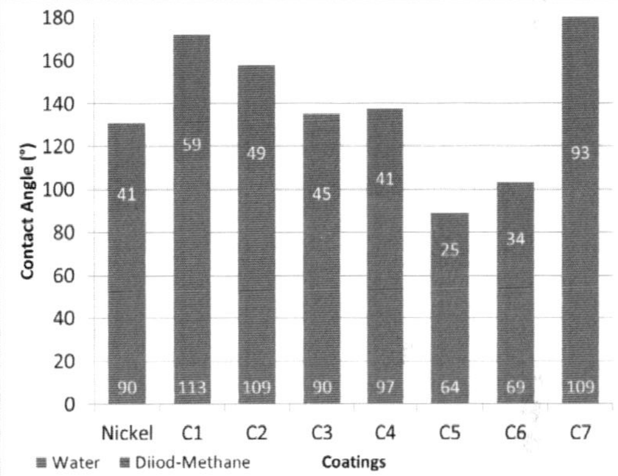

Figure 1 – Measured contact angles of the stampers.

Materials

The following polymers were used for the experiments (Table 1). The material-mix is derived from common material choices in medical application. Additionally, it spans from semi-crystalline polymers to amorphous polymers, which is important because semi-crystalline polymers exhibit different shrinking behaviour and therefore may induce different mechanisms that influence the demolding process. Furthermore, the morphology and material structure may have an effect on the dispersive interaction between the polymer and the insert surface. The polymer substrate for these experiments was a 3 mm thick 100 x 100 mm² wide plate that had previously been manufactured in an injection molding unit.

Results and Discussion

Data Analysis

The sensors in the Hex03 are primarily used to control the hot embossing process. The load cell is usually used in the range of several kN. It was to be expected that measurement resolution in the range of the expected 100 N is low. The raw data acquired in this measurement setup displayed an expected amount of noise. This had to be filtered to reduce the noise to a level that allowed extracting the occurring demolding force from the signal. The filter parameters had to be adjusted several times to increase the detection range without compromising the accuracy of the signal.

Results

Previous work at the PSI [7] measured the force history curves for different structure depths. This way a demolding force (F_{dem}) was defined that could be deduced from discontinuities of the force recording that happen at $F < 0$ at the end of the hot embossing process. The same approach was used to measure the demolding force for different coatings and a selection of polymers (Table 2). The demolding force is represented as the discontinuity from the sudden release of the stamper from the polymer. This leads to a characteristic displacement (leap) in the force curve (Figure 2).

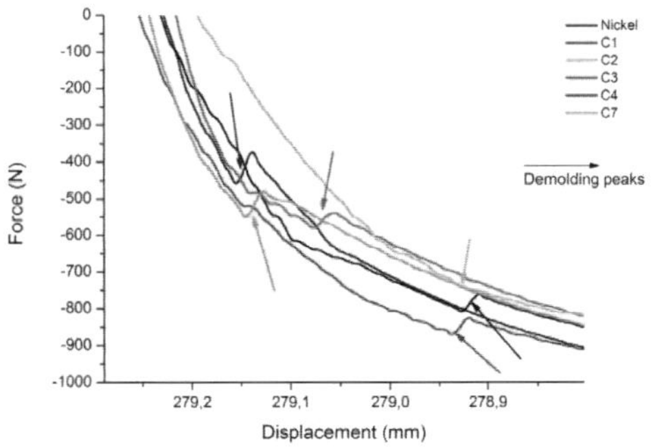

Figure 2 – Measured and smoothed force curves with the associated demolding peak for P2.

In some cases the demolding curves show no such leap, which is an indication that demolding is no sudden release movement but happens as a peeling movement. This peeling movement can occur if the substrate is demolded asymmetrically. For example when the demolding starts at one side of the substrate and moves to the other. In this case if the surface interaction of the polymer and the stamper is high enough, the force measurement can appear locally slightly elevated. The peeling motion may also be in radial direction (inwards or outwards) if a bending of the substrate or stamper occurs. The radial peeling has the same effect as the peeling from side to side. In some of these cases, if the surface interaction is low, no significant change in the force diagram is detectable.

In the experiments of this study radial peeling (bulging) was inhibited for the substrate due to geometrical placement of the anchoring structure beneath the demolding geometry and for the stamper due to a reasonably high stiffness. Minor peeling from one side to another can unfortunately not be controlled and is dependent on the inherent tilt of the system as well as the demolding speed.

Table 2 – Measured demolding forces in Newton for the polymer coating pairing. * could not be demolded. - not measured

	P1	P2	P3	P4
Nickel	0	50	0	0
C1	0	58	0	0
C2	70	70	-	-
C3	73	57	0	-
C4	33	72	-	-
C5	70	*	20	30
C6	63	*	-	0
C7	0	10	0	0

For interpretation of the demolding force, only seemingly instantaneous parallel demolding was used. Nevertheless, tendencies for increased peeling effects were taken into consideration, examining the different force history curves for signs for demolding 'abnormalities'.

To rank applicability of different coatings a relative comparison among them was necessary. It is expected that low surface energy directly affects adhesion [10] thus allowing to rank coatings according to their respective demolding force and surface energy. Furthermore, a distinction between polar and disperse interactions is noted.

Generally, the average demolding force was lower than the expected 200 N found in the previous experiments [8]. Polymer 3 and Polymer 4 even lead to no detectable demolding force for any coating in almost all setups. Polymer 1 and Polymer 2 exhibited distinct force curve leaps in most cases. This corresponds to known demolding difficulties, e.g. sticking effects, for common applications [9]. This steered the primary focus towards Polymer 1 and 2. The findings showed, that for this setup in all cases the C7 coating excelled as demolding aid, reducing the demolding force to almost zero. For an uncoated stamper the demolding force was 0 N for Polymer 1 and 50 N for Polymer 2 (Table 2). All of these force values are an average of at least three measurements. The increasing demolding force for all metal based coatings compared to the uncoated stamper shows the worsening demoldability for Polymer 1 and 2. This deterioration can go up to a degree where a particular coating (C5 and C6) raised the demolding force for Polymer 2 above the level of the auxiliary restrain structure. This led in all cases to a sticking effect of the substrate on the stamper making a regular demolding impossible. In comparison the C7

coating, corresponding to the high contact angle, lowered the demolding force, i.e. in this case to 10 N.
It is to note, that the measurements in some cases varied up to 60 N, which didn't allow an appropriate error calculation. These variations are not uncommon and have been reported in different cases [7] caused by inhomogeneous demolding of the polymer, e.g. peeling.

Discussion
Throughout the measurements several mechanisms showed up in the force diagrams which made the interpretation of some of these curves difficult. The easy to interpret curve showed a distinct force peak that related directly to the demolding (Figure 2). But occasionally the curve took different forms. Among these there are four main errors that occurred and could be seen in the measured force curves in the following way:

- A kink
  Instead of a local force peak, the slope of the force curve changed indicating that the demolding occurred. The reason that the force peak could not be measured is most likely due to a peeling of the polymr substrate off the stamper. This lowered the overall needed force to demold the micro-structure but also explains the change in slope when the demolding started.
- A bulge
  The force curve showed bulge at the demolding position. This indicates most likely a deformation of the polymer which stretches the overall demolding process.
- An elevation
  There is a distinct jump in the demolding force curve, but the level of the measured force remains high, doesn't drop back to the original level and changes its slope.
- A displaced peak.
  In some rare cases a peak similar to the peak being induced from demolding was detected but at an entirely wrong displacement position. The explanation is that this was caused by a relaxation in the measurement equipment.

All of these effects could not be reproduced on demand and were not regarded for the evaluation of the demolding force.

Conclusions
The demolding forces were measured online using the Jenoptik HEX03 embossing machine in a slightly modified setup. The value of F_{dem} corresponds to the height of the peak detected in the recorded force curve, which was confirmed for pillar and hole structures [7 and 10]. The method was sensitive enough to detect differences due to the size of the patterned features or the demolding temperature T_{dem}, or surface coatings. F_{dem} largely depends on the total adhering surface and the corresponding surface properties. This means that for stampers with dense, deep structures the additional area of the sidewalls adds up to the overall surface compared to a non-structured silicon wafer. The measurements clearly showed that the surface coating can have a big impact on the demolding force and the demolding mechanism. This measurement method also makes it possible to detect abnormalities in the demolding process, e.g. peeling due to non-parallel separation. Therefore, it is a good basis for further investigation of different micro structure designs, and can be used as a testing method to determine the suitability of a stamper for easy demoldability.
Among the tested coating the C7 coating excelled all other coatings by far reducing the measured demolding force almost always to 0 N.

Acknowledgements
The research work of this paper was performed in cooperation with Sony DADC (Austria) as a technological partner. The force measurements mentioned in this paper were performed at the Paul Scherrer Institute (PSI, Switzerland) with the help of Mirco Altana.

References
1. Becker, H., et al., 2002, Polymer microfluidic devices, Talanta 56, p. 267-287
2. Gruber, A., E., et al., 2004, Microfluidic lab-on-a-chip systems based on polymers - fabrication and application, Chemical engineering journal 101, p. 447-453
3. Rötting, O., et al., 2002, Polymer microfabrication technologies, Microsystem Technologies 8, p. 32-36
4. Usama, M., et al., 2009, Micro-injection moulding of polymer microfluidic devices, Microfluid Nanofluid 7, p. 1-28
5. Dieudonne, A., M., et al., 2006, Injection molded microfluidic chips featuring integrated interconnections, Lab on a chip 6, p. 1346-1354
6. Trabadelo, V., Schift, H., Merino, S., Bellini, S., & Gobrecht, J., 2008, Measurement of demolding forces in full wafer thermal nanoimprint. Microelectronic Engineering, 85(5-6), 907-909
7. Schmidt, N., 2006, Untersuchung der Haft- und Entformungskräfte beim Heißprägen von Mikrostrukturen, Master Thesis, Paul Scherer Institut
8. Schift, H., 2010, Oral contribution, Paul Scherrer Institut, 5232, Villigen, PSI
9. Pöschl, C., Karl, M., 2010, Oral contribution,Sony DADC, Anif
10. Delaney, K. D., Bissacco, G., Kennedy, D., 2012, A Structured Review and Classification of Demolding Issues and Proven Solutions, Intern. Polymer Processing XXVII 1, p. 77-90

O-04-193

HIGH PRECISION IN SHRINKAGE SIMULATION OF THICK-WALLED PLASTIC PARTS DUE TO IMPROVED MATERIAL DATA

Natalie Rudolph[1,*], Martin Löhner[2], Dietmar Drummer[2] and Tim Osswald[1]
[1] *University of Wisconsin-Madison, Polymer Engineering Center, 1503 University Ave, Madison, WI 53706, USA*
[2] *University of Erlangen-Nuremberg, Institute of Polymer Technology (LKT), Am Weichselgarten 9, 91058 Erlangen-Tennenlohe, Germany*

Injection molding is the most inexpensive and flexible manufacturing process for the production of demanding mass products, such as cell phones, digital cameras, data storage and other optical components that require high dimensional stability. But especially the demand for high precision causes problems in conventional injection molding. The main reason for shrinkage of plastic parts is assumed to be the thermodynamic and thermomechanical behavior of the plastic. One approach is to improve the prediction of the dimensional deviations like sink marks often found in the center of optical parts by means of numerical simulation. Hence, the implemented material data as well as the models incorporating them have to be investigated. This means that the pvT-behavior needs to be known precisely to account for the volumetric changes as a function of temperature, pressure and time, where the last parameter is often neglected. In addition, the mechanical behavior, for example the Young's modulus, has to be determined as a function of temperature to predict the dimensional deviations of plastic parts.
In this work the influence of modifications of the pvT-data as well as the Young's modulus was investigated. In the first approach data from different sources were compared and the impact of the huge deviations between data sets investigated. In a second approach, the impact of temperature dependent modulus data was determined. Therefore, numerical simulations of injection molded thick-walled polycarbonate parts suitable for optical applications were performed using SIGMASOFT®. The results were compared to experimental measurements. As main impact factors the pressure at solidification and the build-up of tensile stresses in the core of the part developing thereafter were identified. It was found that the modulus is a crucial parameter for the shrinkage in the center of the part. In processes with low cooling rates, such as production of thick parts, more attention has to be given to the rapid decrease of the modulus near T_g. In addition, the effect of cooling rate on the pvT-behavior has only a minor effect on the shrinkage of thick lenses. The improvement of the simulation gives a better understanding of the underlying mechanisms and allows the manufacturer to consider these effects already in the design phase.

Keywords: shrinkage simulation, precision, thick-walled plastic parts, Young's modulus, pvT-data.

Introduction

During injection molding, plastics are deformed in the melt state to form the desired geometry. In the subsequent cooling step the melt is cooled down below the glass transition for amorphous polymers to keep this shape. The thermomechanical history during processing is known to affect the shrinkage of polymer parts [1-5]. In case of amorphous materials, especially the cooling rate and pressure at the glass transition are important [6].
The dashed line in Figure 1 shows the change of specific volume during processing in a pvT diagram. Pressure is applied during the injection phase while the temperature stays nearly constant (1). This process is called isothermal injection. During packing more melt is pushed into the part to compensate the volumetric shrinkage that results from the decrease of the temperature (2). Therefore, this process step can be considered isobaric cooling.

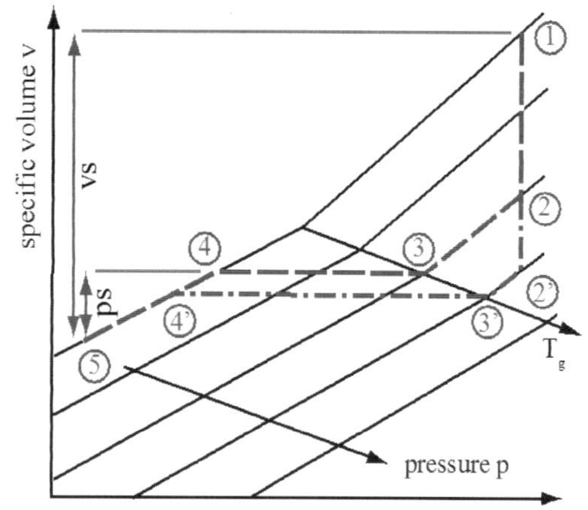

Figure 1 – pvT diagram indicating the change of specific volume during an injection molding process with moderate (dashed) and high (dash dotted) pressure

Once the glass transition is reached at the gate (3), no further melt can flow into the cavity. This point is called sealing time. Now the further decreasing temperature causes shrinkage. The accompanying pressure drop, however, allows expansion of the volume. Because of these counteracting effects, the specific volume stays constant during this so called isochoric cooling until ambient pressure is reached (4). After ambient pressure is reached the volume reduces because of the decreasing temperature till room temperature is reached (5). From point (4) on the part is shrinking away from the walls.

A possible approach to reduce shrinkage is to increase the injection pressure (dash dotted line) and apply more packing pressure on the plastic (2'). Hereby, the part shrinkage (ps) away from the wall can be reduced because the polymer reaches ambient pressure at lower temperatures (4').

The temperature of the melt at sealing time is considered most influential on the shrinkage. No plastic can flow into the part after the gate is frozen. The volumetric shrinkage is not longer compensated by additional melt. The higher the melt temperature at a specific location in the part, the higher is the shrinkage. For example, this effect can be observed in thick parts or at ribbings. If the ribs are too thick in comparison to the part, the shrinkage after sealing time is higher in these areas, which causes sink marks. The pressure and temperature development differs within the part.

For thick-walled parts the cooling time is assumed to be another important parameter because long time periods around the glass transition could lead to relaxation effects or accelerated physical aging [7].

Experimental

In the following the investigated material, the measurement techniques to obtain the pvT-data, the experimental data from injection molding as well as the simulation tools used in this project will be described.

Material

For optical applications amorphous thermoplastics are needed because of their transparency. In this study polycarbonate PC Makrolon LQ2647 from Bayer MaterialScience AG (BMS) is used. Due to its toughness it is suitable for larger parts such as car windows and also thick parts such as automotive headlights. This resin has a solid density of 1.20 g/cm^3 and a glass transition temperature of 150 °C (from pvT-measurements). The pressure dependence of viscosity was implemented into the Cross-WLF model with the parameter D3 = 0.0347 K/bar, which reflects the pressure dependence of the glass transition as proposed by Menges et al. [8]. For the injection molding experiments a PC Makrolon LED 2245 was used, which has the same properties as LQ 2647, but a lower melt viscosity.

In the context of volumetric shrinkage the Young's modulus determines the compliance of the outer layers to the tensile forces acting in the part core. For very long cooling times as found in thick parts, this parameter has to be measured as a function of temperature. Figure 2 shows the Young's modulus as determined by DMA measurements provided by Bayer MaterialScience. Because the data was only measured up to 130 °C, but the glass transition occurs at 150 °C, the modulus data was extrapolated to 150 °C from the last two known temperatures (120 °C and 130 °C). By default SIGMASOFT® would assume the Young's modulus at a constant level above the highest measured temperature (here 130 °C).

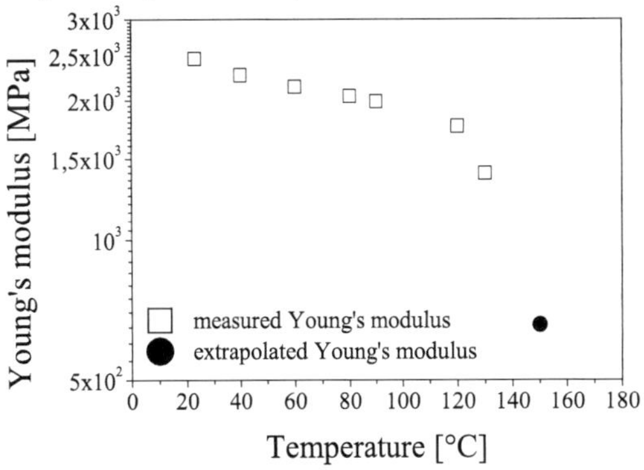

Figure 2 – Young's modulus of polycarbonate measured in a DMA up to 130 °C and extrapolated to 150 °C (black dot)

For pvT measurements of polymers, two different devices can be used; the Piston Die Dilatometer and the Confining Fluid Apparatus. Both techniques were used in this work to obtain pvT-data and the simulation tool SIGMASOFT® was used to calculate the Tait-parameters.

Confined Fluid Apparatus

In the confined fluid technique the specimen is in a closed holding tank surrounded by a fluid as shown in Figure 3.

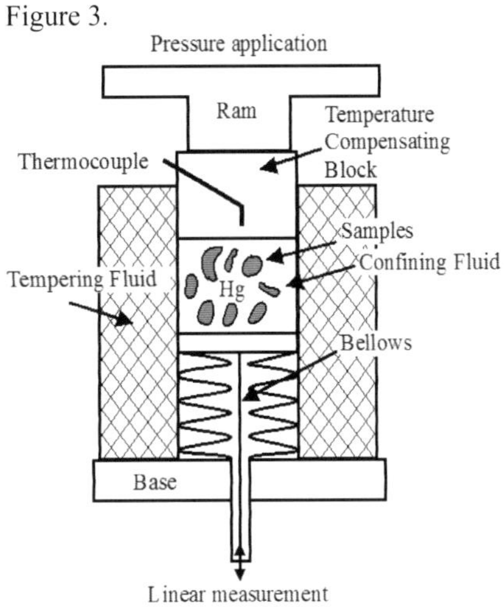

Figure 3 – Schematic of Confined Fluid Apparatus [9]

Mercury is the most common fluid used for this method. The volumetric change of the specimen and the fluid is measured as a function of temperature and pressure. By subtracting the well known volumetric change of the confining fluid the pvT data of the test material can be determined. A density measurement of the specimen is required before the actual test. Together with the recorded volume change the pvT diagram can be generated. The only commercially available apparatus, which is also used for the measurements in this work, is the Gnomix pvT apparatus from Gnomix Inc. The measurements were performed at the Leibniz Institute for Polymer Research (IPF) Dresden. Both isothermal (approx. 750 bar/min) and isobaric (cooling rate of 3 K/min) measurements were performed.

Piston Die Dilatometer
In the piston die technique a cylinder, which is sealed off by two pistons, is filled with polymer pellets, Figure 4. After melting, the pressure is applied with the upper piston. The volume of the polymer is calculated at different temperatures and pressures from the known geometry, initial weight and the measured sample height by detecting the movement of the pistons. In this work, a SWO pvt 100, now Thermo Fischer Scientific, was used for isobaric measurements with a cooling rate of 3 K/min.

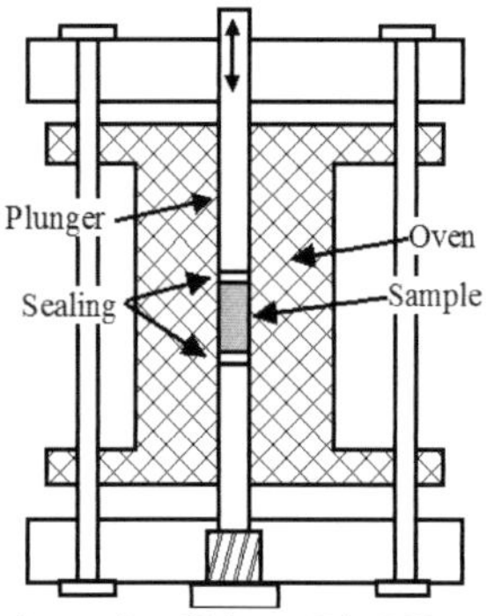

Figure 4 – Schematic of Piston Die Dilatometer [9]

Both techniques have their advantages and disadvantages. The Confined Fluid Method allows both heating and cooling measurements, the measurements are completely hydrostatic and effects of friction, which can occur during the piston movement, are not present. However, the Piston Die Technique does not need a toxic fluid such as mercury and can be used to produce samples, which can be further analyzed.

Tactile Measurements
As a reference to the simulations, the surface topography of thick, coplanar parts was analyzed. The injection molding as well as the tactile measurement was performed by Bayer Material Science [10]. The specimen is a thick quadratic plate with a thickness of 20.1 mm and feed size of 40 mm. Thinner regions are added on two opposite sides of the part to achieve homogeneous filling of the main part. The part was produced at 9 different parameter settings, with a 2^3 DoE and a central point. In this paper the results of the central point will be discussed. In this case, the melt temperature was set to 280 °C, the mold temperature to 117 °C, the packing pressure at 1100 bar and the cycle time was 570 seconds.
The planeness was tested tactile at 60 defined spots on the surface. 30 spots were placed at every coplanar side facing each other. In this work the following locations, Figure 5, will be shown.

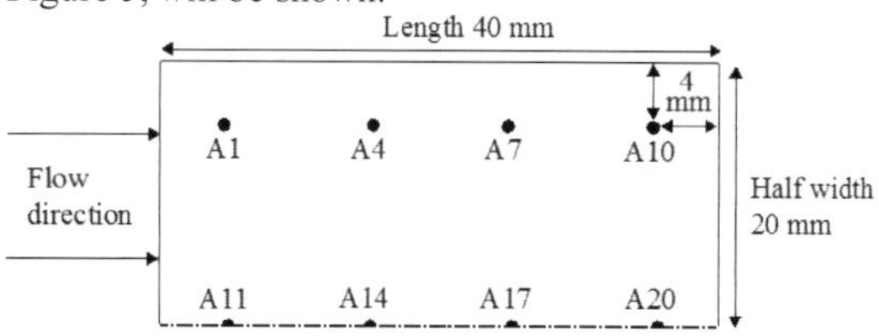

Figure 5 – Part dimensions and arrangement of the used measure points on the coplanar area of the parts

The following changes in geometry were observed on the injection molded parts, Figure 6.

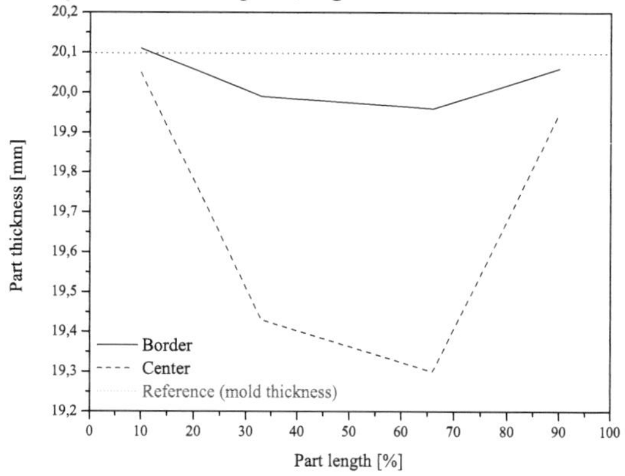

Figure 6 – Part thickness determined by tactile measurements for the central point at the border area (A1 – A10) and in the part center (A11 – A20)

The resulting part thickness is higher at the border areas than in the center. In addition, higher values than the nominal size (20.1 mm) were observed at the border. This effect was more distinctive near the gate than at the end of the flow path. A longer cycle time and a higher mold temperature lead to less shrinkage (not shown). These influences are rather small in comparison to the effect of the packing pressure. A higher packing pressure leads to even thicker parts.

Simulation
The fully 3D injection molding simulation tool SIGMASOFT® was used for the shrinkage prediction of the parts. It can calculate the filling, packing and cooling phase as well as the external cooling of the part which allows a wholesome prediction taking the temperature and pressure distributions in the single phases into account and using them as a starting point for the subsequent step. The calculated shrinkage was determined at the same locations on the coplanar surface as for the tactile measurement, Figure 5.

A screenshot of the simulation results over the cross section at 50 % flow length can be seen in Figure 7, indicating the biggest displacement in the center.

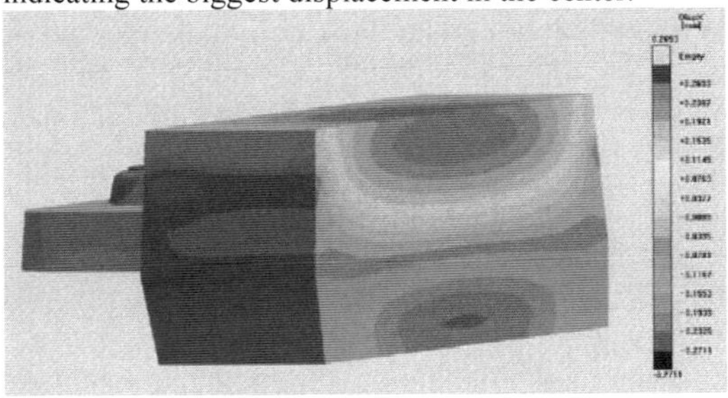

Figure 7 – Exemplified simulation result in SIGMASOFT®

Results and Discussion

Differences in pvT data based on measurement technique

Figure 8 shows the differences between the measurements with the two described techniques as well as data from the Material Data Center® database exemplified with a low and high pressure level.

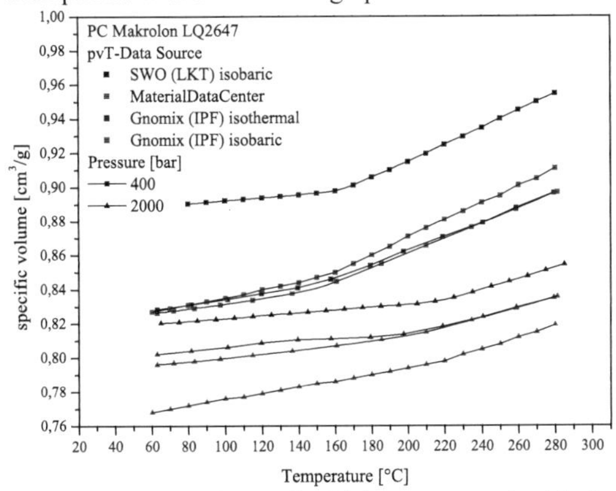

Figure 8 - Comparison of pvT-data measured with the confined fluid vs. piston die method as well as data from the Material Data Center®

The difference between the isobaric and isothermal procedure (IPF) can be seen in a deviation of the curves in the solid state, which is already starting at temperatures above T_g (best observed for 2000 bar). The data from the piston die device (LKT) shows a similar, but slightly smaller slope in the melt and larger deviations in the solid; even compared to the isobaric runs (IPF). Overall, the absolute values of the specific volumes at every pressure level are at significantly higher values. This could indicate that the calculation of specific volume is not accurate enough based on the cylinder geometry and the sample weight. The difference in the slope in the solid state could be related to friction effects in the piston die apparatus. The pvT-data from the database is again different from all the other curves. For low pressures (400 bar) the values are comparable to the Gnomix measurements; however, the overall level of the specific volume is much lower at high pressures of 2000 bar. In total this gives a good idea about the lack of precise data as well as the necessity to further investigate the measuring techniques.

Another effect on the pvT-behavior, which is normally neglected, is the dependence on cooling rate. Due to faster cooling, T_g is shifted to higher temperatures. Therefore, a preliminary study of the effect of cooling rate in a 20 mm thick plastic part was performed. Thus, the temperature distribution over the part thickness and thereof the cooling rate distribution was calculated. The main focus was on the cooling rate of the polymer while passing T_g, because the cooling rate at this particular point determines the shrinkage and the resulting density of the polymer. The temperature decreases faster close to the surface. The cooling rate shows a peak, which decreases with increasing distance to the part surface.

By assuming a constant T_g of 140 °C (neglecting the pressure dependence of T_g) the times at which the single measuring points pass T_g and the corresponding cooling rates can be identified, Column 1, 2 an 3 in Table 1.

Table 1 – Cooling rate and time to reach T_g in dependence of the distance to the part surface while neglecting (column 2+3) or regarding (column 4+5) the change of T_g due to pressure

Distance to part surface [mm]	Time to T_g = 140 °C [s]	Cooling rate @ T_g [K/s]	Time to $T_g(p)$ [s]	Cooling rate @ T_g [K/s] $\left[\frac{K}{s}\right]$
0	115	0.13	**50**	**0.45**
2	400	0.10	**290**	**0.13**
4	520	0.11	520	0.11
6	600	0.12	600	0.12
8	640	0.13	640	0.13
10	650	0.13	650	0.13

Although the solidification times increase with the distance to the surface, the corresponding cooling rates while passing T_g stay nearly constant in the studied process. By taking the higher solidification temperatures due to the pressure dependency of T_g into account, the cooling rates at T_g change at 0 mm and 2 mm distance to the surface, Column 1, 4 and 5 in Table 1. T_g is passed earlier and the resulting cooling rates are higher. For the remaining measuring points the pressure during solidification was low enough to show no effect. Based on this investigation, it was concluded that the influence of cooling rate in the studied part is negligible. What is more, the cooling rates of about 0.13 K/s are in the range of regular pvT-measurements of 0.05 K/s (3 K/min) and therefore, the measured data can be used without modifications or restrictions.

Expansion of border layers

The effect of the expansion at the borders of the coplanar areas shown in Figure 6 is a result of solidification under pressure and relaxation of the over-densified areas during the long cooling time [10]. By analyzing the density distribution of injection molded thin parts solidified under high pressures, over densification was also found by other researchers [3, 4]. Furthermore, this effect was observed by Menges et al. [8] and Rudolph [11] during their exploration of different routes of pvT measurements. However, so far it wasn't observed on a macroscopic scale in the part dimensions. In contrast to earlier works, the thickness of the here presented parts is much higher than that of conventional plastic parts. Therefore, the same percentage of over densification in a 20 mm part results in a much higher absolute expansion volume compared to a 2 mm part.

A layer of 3 mm solidifies at the part surface while packing pressure acts within the part. This happens in the whole part, but especially affects the borders, because the percentage of the affected cross section area is higher compared to the center as shown in Figure 9, top.

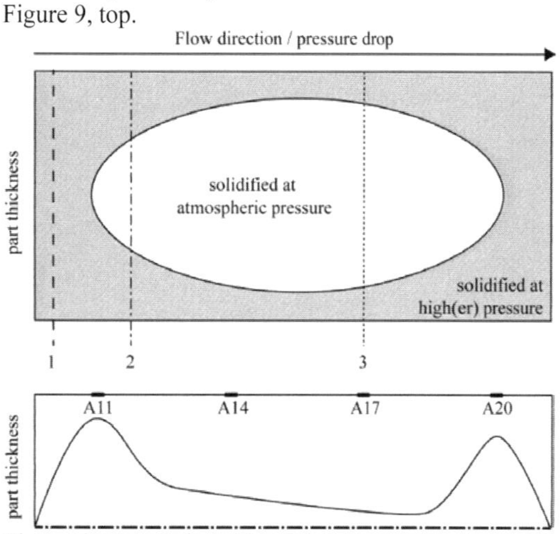

Figure 9 - top: Schematic of the proportion of material solidified at ambient pressure to that at high pressure over the cross section; 1 (dashed line): completely solidified at high pressure, 2 (dash-dotted line): huge part solidified at high pressures, 3 (dotted line): small part solidified at high pressure; bottom: Schematic of the part thickness variation as a result of solidifivation under elevated pressure including the pressure drop effect over the flow length

The density of polymers is increased by solidification under pressure. The resulting tighter package of the molecule chains can relax at temperatures close to T_g or above, but is frozen in upon faster cooling [1, 7, 12-14]. Because of the long cooling time of the studied part, the temperature of the affected areas is close to T_g after demolding. The relaxation of the tight packed areas leads to an increase in volume. The thickness increases especially in the areas where the whole cross section solidified under pressure. The anticipated part thickness related to this effect is shown schematically in Figure 9, bottom. This sketch neglects the overlaying effect of shrinkage, but takes the pressure drop over the flow length into account. It is in good agreement with the observed dimensions shown in Figure 6. A certain expansion can also be expected in the part center because here a small part of the cross section solidified under pressure as well. However, most of the core area solidifies when the pressure has already declined. The change in expansion potential over the flow length is related to the pressure drop. Because the expansion is dependent on the packing pressure, differences in the packing pressure lead to different expansions within the part. Figure 10 shows the pressure during solidification as a function of the part thickness (distance to the surface).

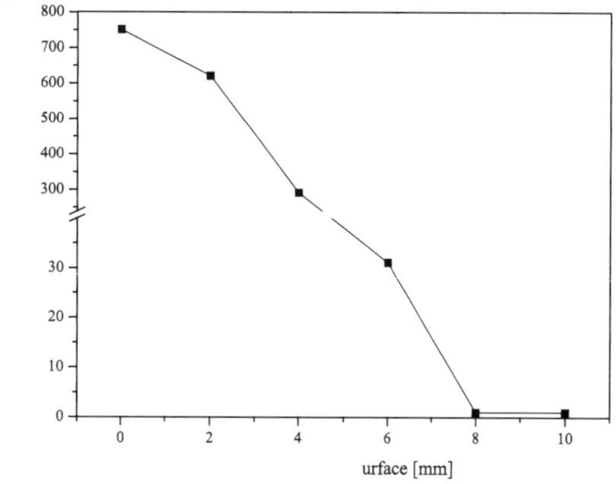

Figure 10 – Pressure at solidification as a function of the distance to the part surface as simulated with SIGMASOFT®

Close to the surface, the solidification pressure is around 750 bar, which should lead to a rather high over densification. With decreasing distance to the wall, the pressure drops until the very core. At 8 and 10 mm distance a tension instead of compression is acting on the material, which is indicated by setting the values to atmospheric pressure in this area. However, even at a distance below 4 mm, the effect of pressure is negligible.

Furthermore, the packing pressure decreases with increasing distance to the gate. Thereby the expansion is higher close to the gate and decreases with increasing distance to the gate.

At the very surface of the part, the polymer solidifies during injection. The pressure in this phase is lower than the packing pressure. Lower densities than in areas solidified under applied packing pressure are anticipated. Wimberger-Friedl and de Bruin [5] measured a similar density distribution over the cross section for thinner parts. Relaxation effects during a long cooling phase lead to an expansion of these areas of higher density.

In comparison to the experiments, the simulations also showed higher part thicknesses at the border of the coplanar area than in the center. However, bigger thicknesses than the nominal size were not observed, because the effect of density differences due to different solidification pressures is not implemented in any simulation tool. The overall thickness differences were much smaller than in the measurements. An effect of the parameter variation on the results was also not observed (not shown). The part thicknesses were nearly constant for all adjustments. A graphic comparison between the measured and simulated results is shown in Figure 11.

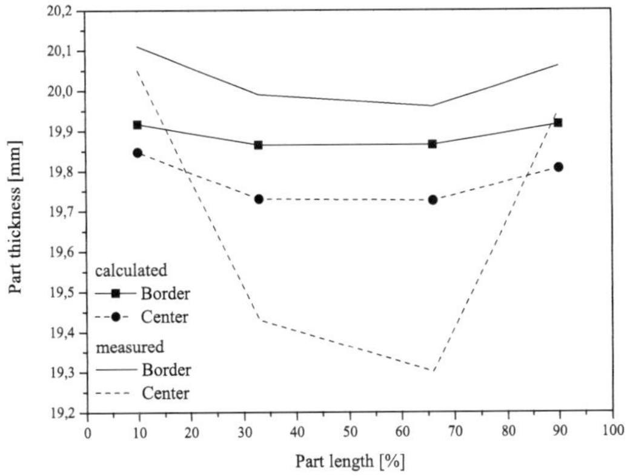

Figure 11 – Part thickness over the flow length at the border and center simulated with SIGMASOFT® using the isobaric pvT-data (IPF) as well as the original, not extrapolated values of the Young's modulus compared to tactile measurements

Influence of Young's Modulus

The low amount of shrinkage and similar simulation results for all parameter variations lead to the conclusion that the modulus might be too high in the simulations. In that case the solidified surface would be hindered from being pulled in by the tension acting in the core of the part. Therefore, the simulations were repeated using the Young's modulus data provided by BMS and extrapolating the values to the glass transition temperature as shown in Figure 2. The results of the simulation are shown in Figure 12 in comparison to the measurement. An improvement of the simulation with the adjusted Young's modulus is clearly observable. The simulation reflects the actual amount of shrinkage in the center of the part very well. The enlargement of the part on the area borders could again not be observed in the simulation results, because it is not incorporated in the models. The expansion of the center of the part is negligible, Figure 9 top, because of the low percentage of pressure solidified cross section. Otherwise the actual shrinkage would still be underestimated. Further variations of the modulus showed that in the part center, the shrinkage depends strongly on the modulus at T_g and is hardly influenced by the values chosen at lower temperatures. At the border, however, where the solidification occurs really fast, also the values at lower temperatures have an impact on the resulting shrinkage (not shown).

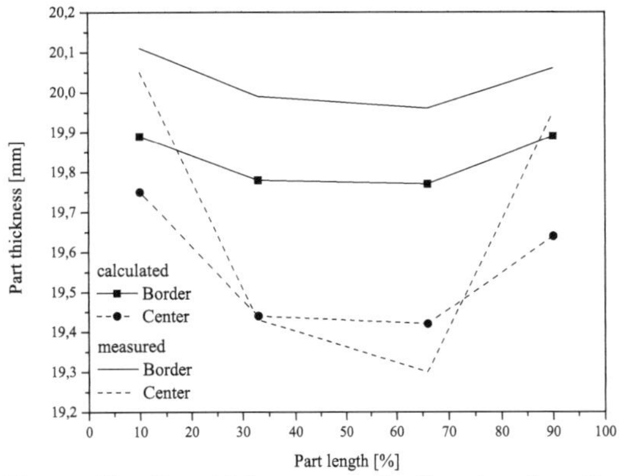

Figure 12 – Part thickness over the flow length at the border and center simulated with SIGMASOFT® using the isobaric pvT-data (IPF) as well as the extrapolated values of the Young's modulus compared to tactile measurements

Influence of pvT data variations

The variation between the isobaric pvT results measured with the Gnomix and the SWO device are quite pronounced, Figure 8. Therefore, simulations were repeated with the isobaric pvT-data from the LKT and the results can be seen in Figure 13.

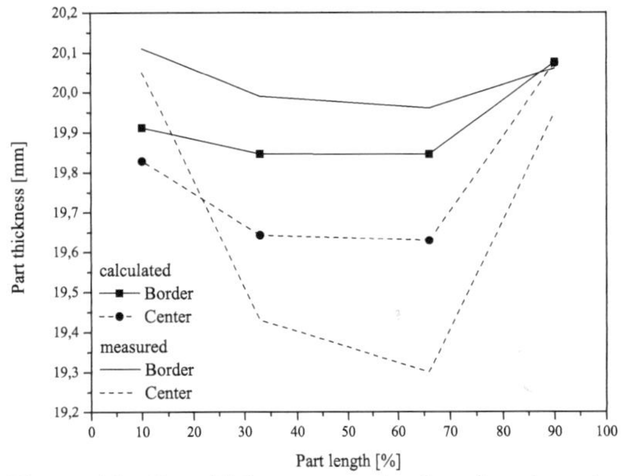

Figure 13 - Part thickness over the flow length at the border and center simulated with SIGMASOFT® using the isobaric pvT-data (LKT) as well as the extrapolated values of the Young's modulus compared to tactile measurements

It can be seen that the predicted shrinkage in the center is neither comparable to the experimental data nor with the good prediction with the IPF data, Figure 12. The reason for this deviation is most likely based on the higher values of specific volume, Figure 8.

In another variation, the pvT-data was adjusted in a way that the thermal expansion in the solid was equally high as the thermal expansion in the melt, which is a very bold overestimation. However, it was found that there was no effect of this modification when using the original modulus data and a significant effect when using the extrapolated modulus data. This emphasizes the significance of the Young's modulus for the shrinkage prediction of thick-walled parts, where the temperature in the majority of the part volume is above T_g for a long time.

Conclusions

By comparing pvT results from different sources, measured with different devices or even different measuring routes, significant deviations in the specific volume can be observed. These deviations further lead to different shrinkage results in numerical simulations. Therefore, it can be concluded that the collection of pvT-data has to be investigated in more depth to improve the shrinkage simulation of thick walled parts. So far, the most promising results were obtained with the data from the Confined Fluid Apparatus.

The simulations show that not only the pvT-data but material data in general have a significant influence on the results. In the presented study of thick walled parts, the impact of the temperature dependence of the Young's modulus, and especially the temperature range around the glass transition, could be seen. By implementing a decrease of the modulus to reasonable values in this temperature range, the experimental results could be closely resembled. However, the accuracy of this modulus prediction should be proven in repeated measurements.

Another effect was observed, which was up until now not discovered in macroscopic properties such as part dimensions. It is the so called over densification, which happens when the solidification occurs during the application of the packing pressure or high pressures in general and if relaxation can occur due to high temperatures.

In future work, the measurement of pvT-data as well as the incorporation of the over densification effect in existing models or the generation of new ones will be investigated in more depth.

Acknowledgements

The authors would like to thank the DFG for the financial support of this work and Dr.-Ing. Olaf Zöllner, Christoph Klinkenberg and Sarah Klönne from Bayer MaterialScience AG for the provided experimental results as well as their helpful input in numerous discussions. In addition, thanks goes to Sigma Engineering GmbH, Aachen, and especially to Dr.-Ing. Marco Thornagel for the support with their simulation tool SIGMASOFT®. We would also like to thank Dr. Jürgen Pionteck from the Leibniz Institute for Polymer Research Dresden for the measurement of pvT data in their Gnomix device as well as the discussion of the results.

References

1. Goldbach, G., Rehage, G., "Die Volumenretardation des Polystyrols nach Druck- und Temperatur- sprüngen", Rheol. Acta, **6**, 30-53 (1967)
2. Kamal, M.R., Kalyon, D.M., Dealy, J.M. "An integrated experimental study of the injection molding behavior of some polyethylene resins", Polym. Eng. Sci., 20, 1117-1125 (1980)
3. Greener, J., "Pressure-Induced Densification In Injection Molding", Polym. Eng. Sci., **26**, 534-542 (1986)
4. Yu, J.S., Lim, M., Kalyon, D.H., "Development of Density Distributions in Injection Molded Amorphous Engineering Parts. Part I", Polym. Eng. Sci., **31**, 145-152 (1991)
5. Wimberger-Friedl, R., de Bruin, J. G., "Gapwise Density Distribution in Injection-Molded Poly(mtehylmethacrylate)", Polym. Eng. Sci., **33**, 383-392 (1993)
6. Simha, R., "Transitions, relaxations, and thermodynamics in the glassy state", Polym. Eng. Sci., 20, 82-86 (1980)
7. Tribone, J.J., O'Reilly, J.M., Greener, J., "Pressure-Jump Volume Relaxation Studies of Polystyrene in the Glass Transition Region", J. Polym. Sci. Polym. Phys., **27**, 837-857 (1989)
8. Menges, G., Thienel, P., Wübken, G., "Eine Abschätzung für die Relaxation von Molekülorientierungen in Kunststoffen", Kunststoffe, **66**, 42-49 (1976)
9. Ehrenstein, G.W., Riedel, G., Trawiel, P.: "Thermal Analysis of Plastics", Hanser Publishers, Munich (2004)
10. Zoellner, O.: Kunststoffoptiken im Mehrschichtspritzguss, PhD thesis, University of Erlangen-Nuremberg (2012)
11. Rudolph, N.: Druckverfestigung amorpher Thermoplaste, PhD thesis, University of Erlangen-Nuremberg (2009)
12. Kimmel, R.M., Uhlmann, D.R., "Effects of High Pressure on Amorphous Polymers. I. Densification of Polymethyl Methacrylate", J. Appl. Phys., **41**, 2917-2927 (1970)
13. Kimmel, R.M., Uhlmann, D.R., "Effects of High Pressure on Amorphous Polymers. II. Annealing of Densified Polymethyl Methacrylate", J. Appl. Phys., **42**, 1892-1896 (1971)
14. Kimmel, R.M., Uhlmann, D.R., "Effects of High Pressure on Amorphous Polymers. III. Thermodynamic Properties of Densified Polymethyl Methacrylate", J. Appl. Phys., **42**, 4917-4925 (1971)

O-04-199

INLINE-COMPOUNDING OF FIBER REINFORCED THERMOPLASTICS ON AN INJECTION MOLDING MACHINE

E. Moritzer[1], R. Kleeschulte[2] and E. Albring[1]*

[1]*Polymer Engineering Paderborn, University of Paderborn, Germany – ellen.albring@ktp.upb.de, elmar.moritzer@ktp.upb.de*

[2] *K-Lab, laboratory for polymers in OWL, University of Paderborn, Germany – rainer.kleeschulte@klab-owl.de*

Abstract – Today, global markets pose great challenges for industrial product development. Complexity, diversity of variants, flexibility and individuality are just a few of the features which the products now have to fulfill. Because of their high capacity for adaptation, polymers are increasingly able to substitute traditional materials such as wood, glass and metals in various fields of application [1].

The traditional development process for new polymer formulations, however, is complex and expensive. Developing polymer formulations which are perfectly suited to their specific applications is therefore out of the question [2]. The integration of the compounding step into the injection molding process would lead to a more efficient and faster development process for new polymer formulations, making it possible to create new product-specific materials. In order to evaluate this new approach to formulation development, the conventional procedure of compounding on a twin-screw extruder with a subsequent injection molding process is compared to inline-compounding on an injection molding machine. Different formulations for fiber reinforced thermoplastics are thus processed in the inline-compounding process on an injection molding machine and, for purposes of comparison, these same formulations are compounded on a twin-screw extruder with subsequent injection molding.

Keywords: inline-compounding, fiber reinforced thermoplastics, co-rotating twin screw extruder

Introduction

Inline-compounding on an injection molding machine is a concept which permits rapid, economic and efficient product-specific material development. In order to achieve this, the compounding operation is integrated in the injection molding process, and the requirements and final formulation are specifically tailored to the corresponding product. It is not a case of developing new standard formulations here, but rather of generating special formulations for defined applications. The inline-compounding injection molding process is thus not in direct competition with conventional compounding on a twin screw, since it pursues a completely different aim. Nevertheless, the new concept naturally has to withstand comparison with the conventional process.

In order to evaluate the new approach to formulation development, the conventional procedure of compounding on a twin-screw extruder with a subsequent injection molding process is compared to the inline-compounding process on an injection molding machine. Different formulations of fiber reinforced thermoplastics (PP and PA6.6) are thus processed in the inline-compounding process (ILC) on an injection molding machine and, by way of comparison, are also compounded on a twin-screw extruder with subsequent injection molding (TSE and IM).

Experimental

System technology and process sequence for inline-compounding on an injection molding machine

The system technology is based on an all-electric injection molding machine without tie bars (type: e-motion 200/100, manufacturer: Engel Austria GmbH, Schwertberg), which can additionally be fitted with a static mixing nozzle (type: SMK-X12/6, manufacturer: Sulzer Chemtec GmbH) to homogenize the melt (Figure 1). A dual multi-purpose specimen to DIN ISO 527 [3] makes it possible to determine all the relevant target variables. The injection molding machine is fed via four loss-in-weight feeders (type: DDW-MD2-DDSR20/28-5, manufacturer: Brabender Technologie KG), which ensure that the raw material is metered in so as to maintain a stable process. A pre-mixing station for the individual components makes sure that the formulation fed in is as homogenous as possible – this is necessary in view of the relatively poor longitudinal mixing action of an injection molding machine. The requirements and the ultimate formulation are specifically tailored to the product in question. The entire process sequence is supported by software from Bayer Technologie, which provides assistance for all the individual steps from data management, via analysis and model compilation, right through to the optimization of the formulation and the design of experiments.

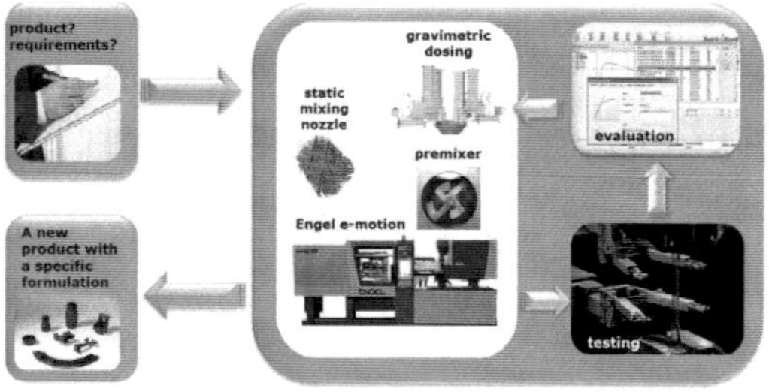

Figure 1 – Schematic representation of the formulation development with the inline-compounding process

System technology of the twin-screw compounding

For compounding the material formulations in the conventional manner, use is made of a ZSK 25 WLE twin-screw extruder (D=25mm) from Coperion Werner & Pfleiderer. The extruder is incorporated in the test setup shown below for this purpose (Figure 2).

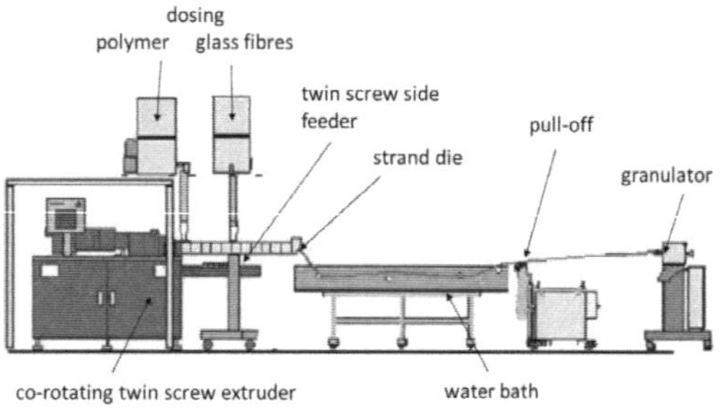

Figure 2 – Schematic diagram of the twin-screw compounding system employed

In order to obtain specimens that can be compared with each other from the compounds produced, the compounds are subsequently processed on the Engel e-motion 200/100 that had already been used for the inline compounding.

Components employed

The matrix material used is a polyamide 6.6 from Radici Plastics. This is mixed with different glass fiber contents (10, 20, 30, 40% by wt.) and glass fiber types from PPG (HP3540, DCS 3540, HP3610). In addition, comparative investigations are conducted on a fiber reinforced polypropylene matrix (Sabic; fiber type: HP3299, coupling agent, dispersion aid) with different contents.

Scanning Electron Microscopy (SEM)

In order to assess fiber/matrix adhesion, REM micrographs were taken of the specimen's fracture surfaces in cooperation with the Chair of Coating Material & Polymers at the University of Paderborn.

The NEON 40 scanning electron microscope from Zeiss was used to take the micrographs. The fracture surfaces of the specimens are observed following the tensile test.

Determining the fiber lengths/fiber contents

In order to measure the distribution of the glass fiber lengths in the test specimen, the fibers must be separated from the matrix material. For this purpose, the individual test specimens are incinerated in a furnace at 650°C as per DIN EN ISO 3453-1 [5]. Following this, it is necessary to deposit the fibers on a specimen slide (5x5cm). For this purpose, the glass fiber specimens are separated in acetone and spread onto the specimen slide. After just a short time, the acetone evaporates and only the glass fibers to be measured remain on the slide. Finally, the specimen slide with the fibers on it is digitized by means of a Slide Scanner from Nikon (resolution 4000 dpi) and evaluated with the aid of the particle analyzing software, PxF Workbench. During the evaluation, pre-classifications must be set by making special adjustments to the reference mark in order to avoid errors. Hence it must be ensured, for instance, that no intersecting fibers, impurities or fibers which are lying on top of each other are wrongly measured together. In order to be able to evaluate a representative quantity of glass fibers, five specimen stages are scanned for each formulation. The number-weighted fiber length l_n was taken as a characteristic parameter [6].

$$l_n = \frac{\sum_{i=0}^{n}(n_i \cdot l_i)}{\sum_{i=0}^{n} n_i} = \sum_{i=0}^{n}(h_{ni} \cdot l_i) \qquad \text{(Eq. 1)}$$

n_i = number of fibers in the respective class
l_i = fiber length of the respective class
h_i= proportion of fibers in the respective class

Determining the mechanical properties

In order to correlate the measured fiber length distributions with the mechanical properties of the molded parts, the tensile strength σ_M, the stiffness E_t (DIN EN ISO 527-1 [3]) and also the impact toughness a_k (DIN EN ISO 179, unnotched [4]) were measured on ten test pieces.

Results and Discussion

Fiber-matrix adhesion (REM)

To permit an assessment of the fiber/matrix adhesion with inline compounding, REM micrographs were taken of the fracture surfaces of a number of specimens. Figure 3 shows excellent fiber/matrix adhesion in a polyamide 6.6 reinforced with 30% by wt. glass fibers.

Figure 3 – REM micrograph of the fracture surface of a polyamide 6.6 matrix with 30% by wt. glass fibers

By comparison, the fiber/matrix adhesion in a non-polar PP matrix is considerably less favorable (Figure 4) and requires the addition of a coupling agent.

Figure 4 – REM micrograph of the fracture surface of a polypropylene matrix with 30% by wt. glass fibers (without addition of a coupling agent)

Comparison of the processing concepts

When the two compounding concepts are compared, it becomes clear that the mechanical performance of the specimens produced by the injection molding compounding process is consistently superior to the mechanical properties of the specimens produced in the conventional manner, i.e. first compounded and subsequently injection molded. An improvement of up to 17% was achieved here. The reason for this is that, with inline-compounding on an injection molding machine, the material and, above all, the fibers are only subject to thermal and mechanical loading once. The fiber lengths in the finished part are thus clearly above the fiber lengths for the conventional process (Figure 7). In addition, Figure 7 shows that the reduction in fiber length caused by the injection molding process itself is very small, since the number-averaged fiber length drops by 20 μm between the TSE granulate and the finished part (TSE and IM). The number-averaged fiber length in the parts produced by inline-compounding injection molding (ILC), by contrast, is up to 50% greater than the fiber lengths in the parts with a conventionally compounded formulation (Figure 7).

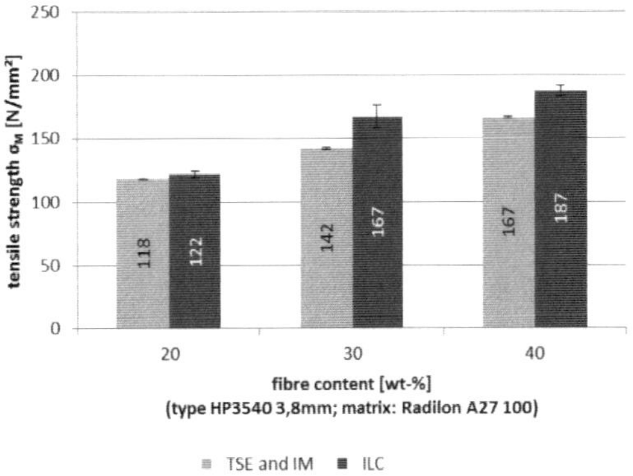

Figure 5 – Comparison of the compounding concepts: influence of the fiber content on the tensile strength with a fiber-reinforced PA6.6 matrix (error indicators show the simple standard deviation)

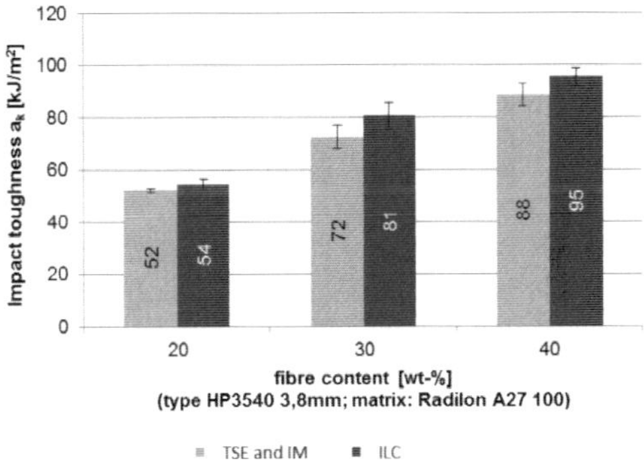

Figure 6 – Comparison of the compounding concepts: influence of the fiber content on the impact strength with a fiber-reinforced PA6.6 matrix (error indicators show simple standard deviation)

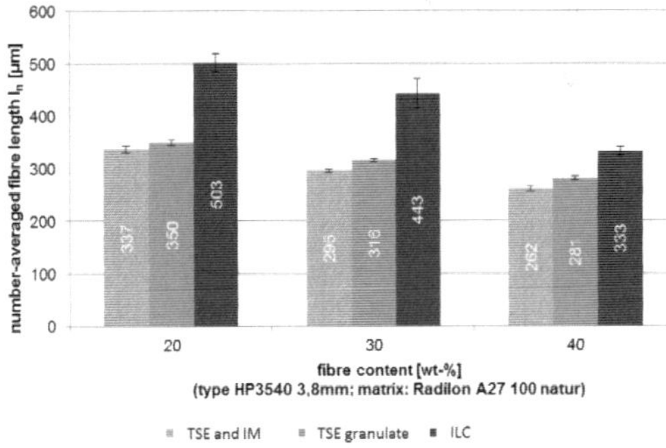

Figure 7 – Comparison of the compounding concepts: influence of the fiber content on the number-averaged fiber length with a fiber-reinforced PA6.6 matrix (error indicators show simple standard deviation)

This state of affairs was also ascertained with a fiber-reinforced PP matrix. Improvements of up to 40% were even achieved here in respect of mechanical properties (Figure 8) [7].

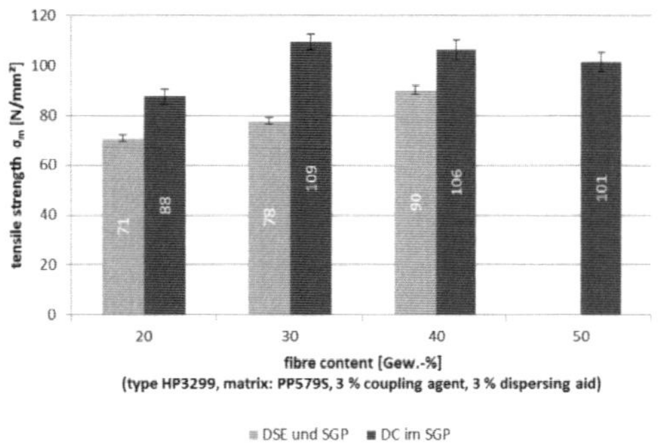

Figure 8 – Comparison of the compounding concepts: influence of the fiber content on the tensile strength with a fiber-reinforced PP matrix (error indicators show the simple standard deviation)

Influence of the fiber type

In addition to this, the influence of different fiber types in the two compounding concepts was investigated. Apart from differences in the starting fiber length, the fiber types essentially differ in respect of their bundle size (Table 1). Fiber type HP3610 is also claimed to have a particularly positive influence on impact strength as well as very good glycol resistance at the same time.

Table 1 – Properties and nature of the glass fiber types investigated

ChopVantage	HP 3540	DCS 3540	HP 3610
< 1.0 mm [%]	30	20	40
1.0 – 2.8 mm [%]	70	80	60
fiber type	E-glass	E-glass	E-glass
diameter[μm]	10	10	10
length [mm]	3.8	3.0	3.8

The results in respect of the influence of the fiber type do not reveal any clear influence of the fiber type on the tensile strength or on the resultant fiber lengths (Figure 8). Here too, however, it is seen that the values of the inline-compounded specimens are above, or at least at the same level as the mechanical properties and fiber lengths of the specimens compounded and injection molded in the conventional manner.

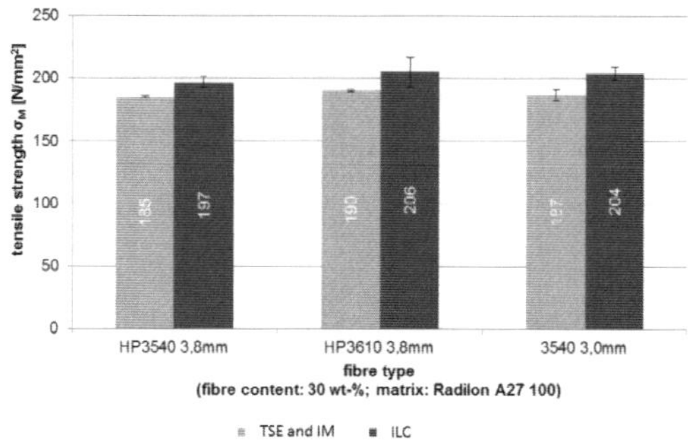

Figure 8 – Comparison of the compounding concepts: influence of the fiber type on tensile strength with a fiber-reinforced PA6.6 matrix (error indicators show simple standard deviation)

Conclusions

To sum up, it can be said that, with the newly developed process of inline-compounding on an injection molding machine, it is possible to achieve better mechanical properties and longer fibers than for the conventional approach to formulation development employing the twin-screw compounding process, while saving on one process step at the same time. In addition, the material experiences less thermal damage, which is to be the subject of further investigations. The use of raw materials also permits cost savings compared with materials that have already been made up beforehand. The new compounding concept thus permits an efficient and economic development of product-specific plastics formulations directly at the processors. A concept is being worked out to facilitate the introduction of inline compounding in companies interested in this process.

Acknowledgements

We would like to extend our thanks at this point to Radici Plastics GmbH, PPG Industries Fiber Glass BV, Sabic Germany GmbH, Addcomp Holland BV and WTH GmbH for their generous supply of materials. Thanks also go to the Chair of Coating Material & Polymers at the University of Paderborn, with Dr. rer. nat. Karl Hiltrop, for the preparation of the REM micrographs. Furthermore the project is being financed with funds from the EU and the Land of North Rhine-Westphalia.

Ziel2.NRW
Regionale Wettbewerbsfähigkeit und Beschäftigung

EUROPÄISCHE UNION
Investition in unsere Zukunft
Europäischer Fonds
für regionale Entwicklung

References

1. J. Stebani, Die Kunststoffwelt im Umbruch. Neue Konzepte, neue Unternehmen, neue Chancen. Kunststoffe 92 (2002) 6, p. 28-31.
2. J. Stebani, G. Maier, E. Bacher, Schneller-umfassender-effizienter. Kunststoffe 97 (2007) 9, p.227-231.
3. DIN ISO 527 T1: Kunststoffe – Bestimmung der Zugeigenschaften, Beuth, Berlin, 1996.
4. DIN EN ISO 179 T1: Kunststoffe – Bestimmung der Charpy-Schlageigenschaften, Beuth, Berlin 2010.
5. DIN EN ISO 3453 T1: Kunststoffe – Bestimmung der Asche, Beuth, Berlin, 2008.
6. H.-P. Thieltges, Faserschädigung beim Spritzgießen verstärkter Kunststoffe, dissertation, RWTH Aachen, 1991.
7. E. Albring, E. Moritzer, R. Kleeschulte, Compoundieren – einfach und direkt. Kunststoffe Heft 11/2012

O-04-205

MECHANICAL PROPERTY OF INJECTION MOLDED UNTWISTED JUTE/PP WITH WELD LINE

Cuntao WANG[1], **YuqiuYANG**[2,*], **Hiroyuki HAMADA**[1]

1 Kyoto Institute of Technology,Sakyo-ku, Matsugasaki, Kyoto,606-8585 Japan.-wangcuntao2005@163.com, hhamada@kit.ac.jp

2 Donghua University, NO.2999 of RD Renmin (N), Songjiang, Shanghai, 201620, P.R. China- amy_yuqiu_yang@hotmail.com

Abstract - Jute fiber is of not only low density and good mechanical property but also natural and degradable property. On the other hand, recyclable Polypropylene (PP) was also good choice for environmental friendly material. Injection molding is one of the most important processes to manufacture plastic composites because it is of high quality products with low cost. However, weld lines are in inevitable when two separate melt fronts rejoin during injection molding. Therefore, in this paper the weld property of injection molded untwisted and twisted jute/PP dumbbell shape specimen with weld line was investigated. Pultrusion technique was adopted to fabricate jute/PP long fiber pellets (LFT) and the re-compound pellets of LFT, i.e. RP was made to improve the fiber distribution. Then LFT, RP, and hybrid mixtures with the hybrid ratios of LFT50:RP50 were used to mold dumbbell shape specimens with or without weld line. In particular, the influence of different pellets and holding pressure on weld line strength of injection molded jute/PP dumbbell shape specimens with weld line was discussed based on tensile test and SEM observation.

Keywords: Long fiber pellets, twists, weld lining strength, hybrid pellets.

Introduction

As well known, most currently marketed composites are made from synthetic fibers and polymers that derive from petroleum which is unsustainable fossil fuel. They are not biodegradable and thus, pose a serious disposal problem at the end of their life [1]. Nowadays, however, public concern about the environment and the availability of limited fossil fuel resources has forced the governments, companies, and scientists to find alternatives to crude oil and to develop sustainable materials from renewable resources.

Natural fibers such as jute fiber are of not only low density and good mechanical property but also excellent degradable property. Further, recyclable Polypropylene (PP) is also good choices for environmental friendly material.

Injection molding is one of the most attractive mass production methods for plastic parts. It is especially used to produce a wide variety of complex geometry articles in one automated production step with low wear of the processing equipment. High production rate, short cycle times and low percentage of scrap are accounting the advantages that make this process very attractive from engineering and economical point of view [2-5]. However, weld line is one of the typical problems seen in injection molding. Although it is not always easy to completely eliminate weld lines, the weld line strength of injection-molded parts could be improved through suitable adjustment of processing parameters such as melt temperature, mold temperature, injection speed and so on [6-16].

In the study pultrusion technique was adopted to fabricate jute/PP long fiber pellets (LFT) and it was found that twisted jute fiber bundles in LFT specimens was not separated and dispersed well. As a result, in this paper untwisted pellets were used and the re-compound pellets of LFT, i.e. RP was made. Then LFT, RP, and hybrid mixtures with the hybrid ratios of LFT50:RP50 were used to mold dumbbell shape specimens with or without weld line. In particular, the influence of different pellets , twists , holding pressure on weld line strength of injection molded jute/PP dumbbell shape specimens with weld line was discussed based on tensile test and SEM observation.

Material and Experimental

The pultrusion technique was adopted to prepare long fiber pellets (Fig. 2a and d) for injection molding. Jute yarns (TESAC) were continuously pulled and coated with PP (Prime Polymer Co., Ltd) resin and after pass through a heated die whereby PP resin impregnates into the jute yarns and sufficient cooling, the impregnated jute yarns were cut into pellets. For twisted pellets, four jute yarns were twisted together when they were simultaneously fed into the pultrusion machine. However, for untwisted pellets, the four jute yarns were parallelly and simultaneously fed into the pultrusion machine. The schematics of untwisted and twisted jute/PP pellets were shown in figure 1, it was worthy to point out that four yarns consisted in the pellet were twisted one.

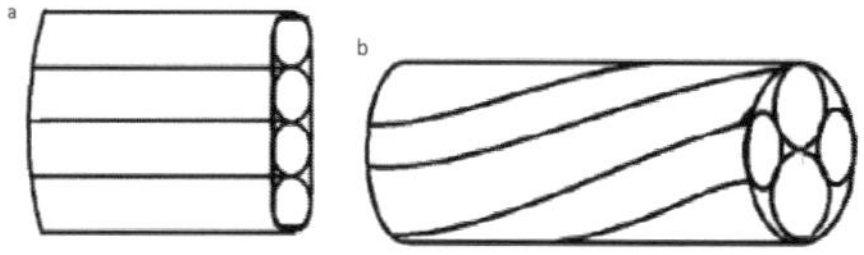

Fig.1 Schematic of untwisted and twisted jute/PP pellets: a) Untwisted pellets; b) Twisted pellets

Table 1 Specification of LFT and RP pellets

Material	LFT (Twisted)	LFT (Untwisted)	RP
Length(mm)	8	5	3
Cross section (mm)	3×3	4×2	4×4
Fiber volume fraction (%)	27	27	27

In order to improve the fiber dispersion the re-compound pellets of LFT i.e. RP (Fig.2b and e) was extruded at a screw speed of 250rpm with barrel temperature 200 ℃ and cut into 3 mm length. Specification of LFT and RP pellets was shown in Table 1. For hybrid LFT/RP (Fig.2c and f), LFT and RP were mixed according to the ratio of LFT 50: RP 50.

These pellets were dried in an oven at 85 ℃ for 16 hours prior to being used for molding to remove the moisture. Dumbbell specimens with 3 mm thickness, 175 mm length and 65 mm gage length (non-weld specimen) and those with weld line by double gate were molded by PO YUEN UW50 injection machine. Dumbbell shaped specimens were prepared at the barrel temperature range from 150 to 200 ℃, mold temperature of 30 ℃, cooling time of 20s.

In order to investigate the effect of holding pressures, in particular, 11, 22, 44 MPa for both RP and hybrid LFT/RP weld specimens and 22, 44, 88 MPa for untwisted jute/PP LFT weld specimens and 44, 88 132 MPa for twisted jute/PP LFT weld specimens was adopted.

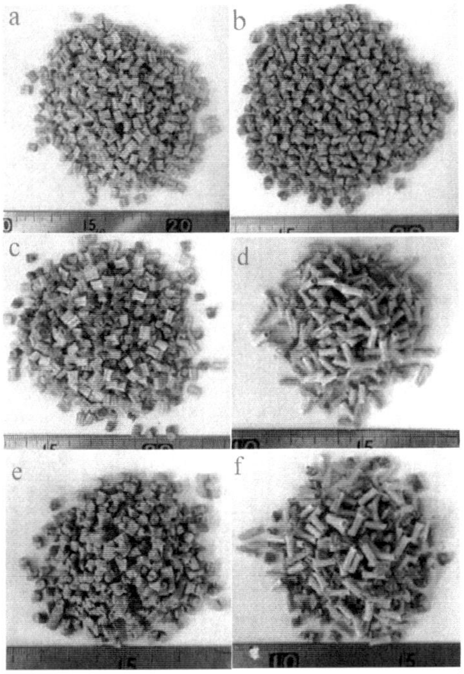

Fig.2 Six kinds of jut/PP pellets: a) Untwisted Jute/PP LFT, b) Untwisted Jute/PP RP, c) Untwisted Jute/PP Hybrid (LFT50:RP50); d) Twisted Jute/PP LFT, e) Twisted Jute/PP RP, f) Twisted Jute/PP Hybrid (LFT50:RP50)

Static tensile tests were conducted for specimens at an Instron type universal testing machine on a constant cross-head speed of 1mm/min. (the room temperature was about 25°C and the humidity was about 65%). Test method was according to JIS K7054-1995 (For fiber reinforced plastics). In particularly, extension meter (Dynamic 2620-601) was used to measure the strain during tensile test.

After tensile test, the part of fracture surface was cut out and coated with a layer of gold. Scanning Electronic Microscope (JSM 5200) was used to observe the fracture surface of specimens after tensile test.

Results and Discussion

Effect of LFT, RP, and Hybrid LFT/RP on tensile property (Non-weld specimens)

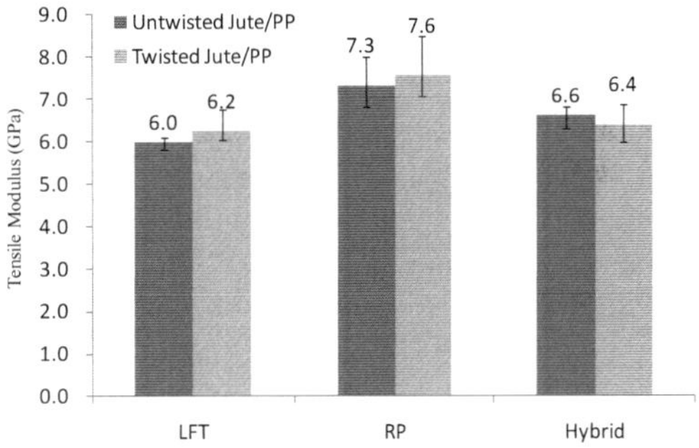

Fig.3 Tensile Modulus of untwisted and twisted jute/PP specimens

The effect of LFT, RP, hybrid LFT/RP on the tensile property was shown in the figures 3~4. It was found that tensile modulus of both RP and hybrid specimens showed the higher value than LFT. In detail, the tensile modulus of untwisted RP and twisted RP was improved by 21.7% and 22.6% than that of LFT specimens, respectively.

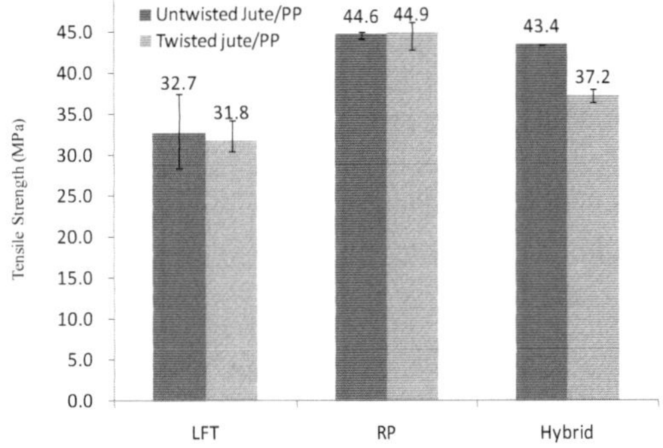

Fig.4 Tensile strength of untwisted and twisted jute/PP specimens

Additionally in figure 4, tensile strength of untwisted RP and untwisted hybrid specimens was improved by 36.4% and 32.7% than that of untwisted LFT, respectively. Tensile strength of twisted RP and twisted hybrid specimens was improve by 41.2% and 17.0% than that of twisted LFT, respectively.

Comparing untwisted and twisted specimens, it was found there was no big difference for both LFT and RP specimens. However, tensile strength of untwisted hybrid specimens increased by 16.7% than that of twisted hybrid specimens. As shown in table 1, size of untwisted LFT and RP pellets was very similar and it was easier to mix and disperse evenly in the hybrid specimens, which could result in the better tensile strength of untwisted hybrid specimens.

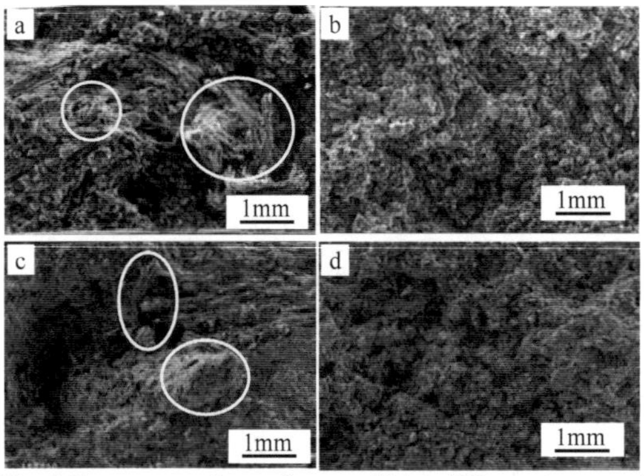

Fig.5 Fracture surface of nonweld jute/PP specimens: a) Twisted LFT: fiber bundles could also be seen in fracture surface; b) Twisted RP: fibers were dispersed more evenly than LFT; c) Untwisted LFT: fiber bundles could be seen in fracture surface; d) Untwisted RP: fibers were dispersed more evenly than LFT

In fracture surface of twisted LFT (Fig 5a) and untwisted LFT (Fig 5c), large and long fiber bundles could be seen. It was thought that fiber bundles were not separated sufficiently and jute fibers were not dispersed evenly. In fracture surface of twisted RP (Fig 5b) and untwisted RP (Fig 5d), large fiber bundles could not be seen and jute fibers dispersed better than LFT which was thought to be helpful to obtain a good interfacial property and led to an increased mechanical property.

Effect of LFT, RP, and Hybrid LFT/RP on weld line strength

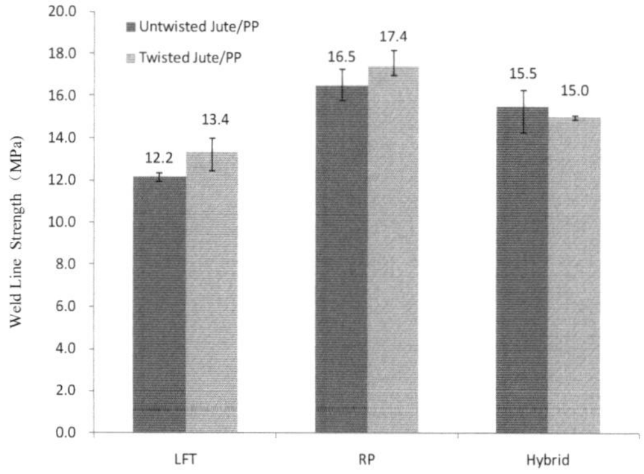

Fig.6 Tensile strength of different weld specimens

In figure 6, for untwisted jute/PP specimens, it was found weld line strength of RP and hybrid specimens was improved by 35.2% and 27.0% than that of LFT, respectively. On the other hand, for twisted jute/PP specimens, weld line strength of RP and hybrid specimens increased by 30.0%, 11.9% than that of LFT, respectively.

Fracture surface of twisted jute/PP with weld line was shown in figure 7. It was found that large fiber bundles existed and most fibers orientated in 90 degree direction to flow direction in LFT specimens. However, fibers were separated and dispersed more uniformly and more fibers orientated in 0 degree direction in RP specimens. Therefore, weld line strength of RP was improved than that of LFT.

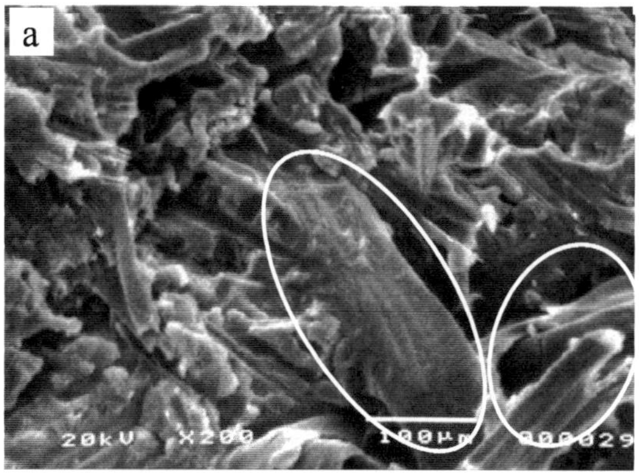

Fig.7 Fracture surface of twisted jute/PP specimens with weld line: a) Twisted LFT: fiber bundles existed and more fibers orientated in 90 degree direction; b) Twisted RP: fibers were dispersed more evenly and more fibers orientated in 0 degree direction.

Effect of holding pressure on weld line strength

As shown in figure 8, it was found that the weld line strength of untwisted LFT increased by 38.3% as holding pressure increased from 22 to 88 MPa. Weld line strength of untwisted hybrid specimens was improved by 14% when holding pressure increased from 11to 44MPa. For untwisted RP, it was thought there was no significant improvement.

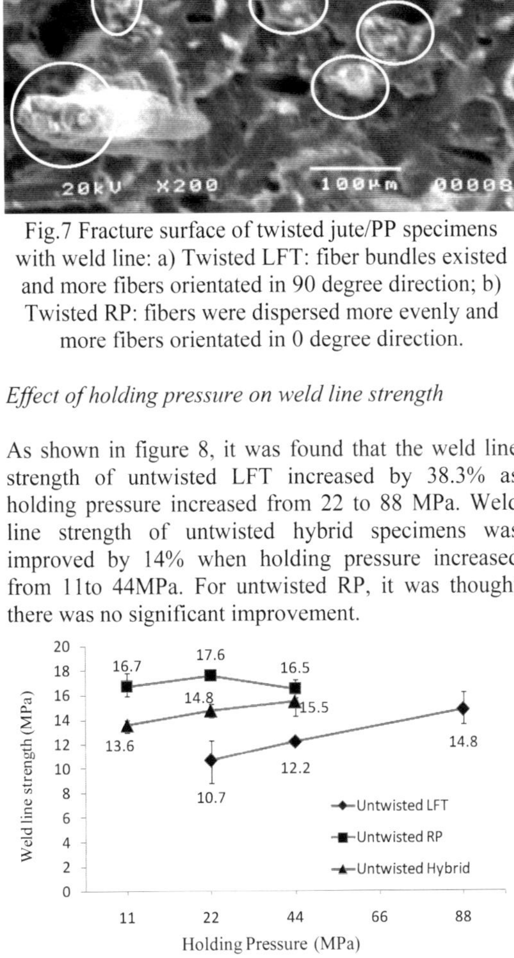

Fig. 8 Weld line strength of untwisted jute/PP at different holding pressure.

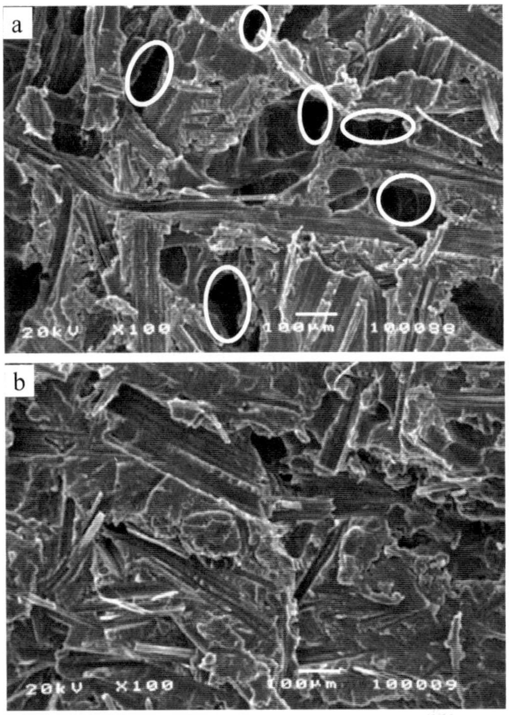

Fig.9 Fracture surface of untwisted LFT at different holding pressure: a) 22MPa; b) 88MPa

Fracture surface of untwisted LFT at different holding pressure was shown in figure 9, it was found that many micro-voids and gaps can be seen in fracture surface of LFT at 22MPa holding pressure (Fig. 9a). However, micro-voids can hardly be seen in fracture surface of LFT at 88MPa holding pressure (Fig. 9b). It was because micro-voids and gaps were reduced a lot when holding pressure increased from 44 to 88 MPa which make the structure more compact and resulted in the improvement of weld line strength.

Conclusions

In this paper, untwisted and twisted jute/PP dumbbell specimens with or without weld line was molded and the effect of different holding pressures and different pellets on the weld line strength was investigated based on the tensile test and SEM observation. The conclusions could be summarized as follow:

1) Tensile strength of RP specimens was improved at least by 35% than that of LFT both for untwisted and twisted jute/PP, because fiber distribution and interfacial property of RP was much better than that of LFT.

2) Weld line strength of RP was improved at least by 30% than that of LFT both for untwisted and twisted jute/PP.

3) Weld line strength of untwisted Jute/PP LFT increased by 38% as holding pressure increased from 22 to 88 MPa because higher holding pressure was more effective to remove the defects such as micro-voids and gaps.

References

[1] Tapasi Mukherjee • Nhol Kao, J Polym Environ 19:715(2011)

[2] F. Corrales, F. Vilaseca, M. Llop, J. Giron`es, J.A. M´endez, P. Mutj, Journal of Hazardous Materials 144 (2007) 730–735

[3] C. Wu, C. Ku, H. Pai, Int J Adv Manuf Technol (2011) 52:131–141

[4] U. S. Ishiaku, O. A. Khondker, S. Baba, A. Nakai,1 and H. Hamada, Journal of Polymers and the Environment, (2005)

[5] C. Wu, W. Liang, Polymer Engineering & Science(2005)

[6] C. Wu, C. Ku, H. Pai, Int. J. Adv. Manuf. Technol. , 52, 131 (2011)

[7] C. Wu, W. Liang, Polym. Eng. Sci. , 1021 (2005)

[8] S. Patcharaphun, P. Jariyatammanukul, POLYM-PLAST TECHNOL, 49, 1305 (2010)

[9] S. Liu, J. Wu, J. Chang, Polym. Eng. Sci. , 1256, 1262 (2000)

[10] L. Turng, H. Kharbas, Polym. Eng. Sci. , 157, 167 (2003)

[11] R. Selden, Polym. Eng. Sci. , 205, 216 (1997)

[12] U. S. Ishiaku, O. A. Khondker, S. Baba, A. Nakai, and H. Hamada, J. Polym. Environ. , 13, 151, 156 (2005)

[13] L. Xie, G. Ziegmann, Int. J. Adv. Manuf. Technol. , 48, 71.81 (2010)

[14] C. Lu, X. Yu, S. Guo, Polym. Eng. Sci. , 1666 (2005)

[15] L. Xie, M. Leester-Scha¨del, G. Ziegmann, S. Bu¨ttgenbach, Int. J. Adv. Manuf. Technol. , 16, 1009 (2010)

[16] J. W. Kim, H. S. Kim and D. G. Lee, Mater. Res. Innovations, 15, 303, (2011)

O-04-217

VISUALIZATION ANALYSIS OF DISCONTINUOUS WRINKLY GROOVES GENERATION PROCESS BEHIND MICRO SAW-TOOTH STEPS PATTERN

Y.Kanetoh[1*] **and H.Yokoi**[2]

[1] *Institute of Industrial Science, The University of Tokyo, Japan –ykaneto@iis.u-tokyo.ac.jp*
[2] *Institute of Industrial Science, The University of Tokyo, Japan –hiyokoi@iis.u-tokyo.ac.jp*

Abstract - In the injection molding for optical products made of polycarbonate, the generation of discontinuous wrinkly grooves in the flow direction can be seen at the downstream of an area of micro saw-tooth patterns located diagonally to the flow direction. In this study, the authors focused on this molded defect and attempted to analyze how these discontinuous wrinkly grooves occur, through dynamic visualization using the glass-inserted mold. The generation process of discontinuous wrinkly grooves was magnified and visualized using a cavity with such a pattern area inclined at 45 degrees along the flow direction. The generation of several peculiar grooves arranged along the flow direction was seen on the surface of molded products, about 7 mm from the downstream of the area of micro saw-tooth patterns. The magnified visualization results showed that air trapped at the inside corner area of the saw-tooth pattern escapes from the edge of the pattern area to the plate area. The escaped air caused the skin layer to partially demold off from the mold surface and reached to the flow front. In the process where the air escape area moves to the downstream according to the movement of the contact area between the flow front and mold surface, the demolded area due to the escaped air is seen to expand and shrink repeatedly. The results clarify that wrinkly grooves are formed sequentially but discontinuously according to the expansion and shrinkage behavior of the air escape area and that an area with low transcription area is also formed inside the wrinkly grooves. Based on these results, a model on the process of molding defect generation was established.

Keywords: Visualization, Micro saw-tooth steps pattern, Wrinkly grooves, Air escape

Introduction

In the injeciton molding of optical parts, the generation of discontinuous wrinkly grooves along the flow direction can be seen at the downstream of an area of micro saw-tooth patterns located diagonally to the flow direction. In this study, the authors aimed to clarify how these discontinuous wrinkly grooves occur through dynamic visualization. This report particularly discusses the visualization results of the generation process of this defect using a cavity with an area of micro saw-tooth patterns located diagonally to the flow direction.

Experimental Method

Cavity with Area of Micro Saw-tooth Steps Pattern

Fig.1 shows the rectangular cavity with an area of micro saw-tooth patterns used in the experiments conducted this time. This area of micro saw-tooth patterns (EFGH) is square, positioned at the center of the cavity, and inclined 45 degrees to the flow direction. In the area, micro saw-tooth patterns of 0.5mm high and 2mm wide are engraved on the movable side mold. This saw-tooth patterns are the concave shape engraved in the E-F(G-H) direction. The cavity can also be depressurized by connecting the vacuum hole at the end of the cavity to a vacuum pump.

Experimental Equipments and Molding Conditions

The molding machine used was the ultra-high speed injection molding machine FN4000-9HTN (NISSEI PLASTIC INDUSTRIAL CO., LTD.) and the resin was PC AD5503 (Teijin Chemicals Ltd.).

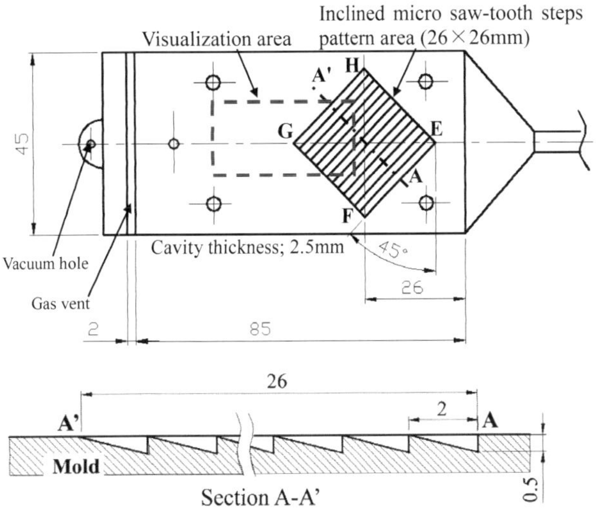

Figure 1 Cavity shape (Unit; mm)

In the visualization experiments, the area enclosed by the dotted lines in Fig.1 was observed using a high speed video camera MEMRECAM fx K4 (NAC Image Technology Inc.). We also magnified and observed the area around the corner G which is positioned at the downstream of the area of micro saw-tooth steps pattern and area of discontinuous wrinkly grooves using a long distance microscope. Experiments were conducted at the following molding conditions; resin temperature of 295°C, mold temperature of 90°C, and injection rate of 40cm^3/s. The process by which these discontinuous wrinkly grooves occur were investigated when the internal pressure of the cavity was reduced by forced exhaust from the vacuum hole, at the injection rate of 16cm^3/s and mold temperature of 115°C.

Results and Discussion

Observation of Discontinuous Wrinkly Grooves at Downstream of Micro Saw-tooth Patterns Area

In the experiments, discontinuous wrinkly grooves were observed about 7 mm downstream of the the area of micro saw-tooth patterns. Fig.2 shows the external view of the molded sample and magnified view of the discontinuous wrinkly grooves. At the part indicated J, discontinuous wrinkly grooves formed along the flow direction can be seen. This defect was also seen on the surface of the same side as the area of micro saw-tooth steps pattern. (2) shows the magnifed image of the part J. In this image, the wrinkly grooves are made up of Groove I which is thick wrinkle. The inside of this defect area is not transcribed onto the mold surface (Low transcription area). Only the Groove I of downstream side is formed of U-shape. In addition, slight wrinkles (Groove II) were also found at the boundary between the low transcription area and other areas. On the inside of Groove I, thinner wrinkles (Groove I') can be observed. At part J, the low transcription area which has a rhomboid shape is made up of discontinuous Groove I and Groove II. Many of these rhomboid areas can be seen lined up along the flow direction.

Visualization of Generation Process of Discontinuous Wrinkly Grooves

Next, magnified observation was carried out on the resin flow behavior from the area of micro saw-tooth patterns to the area of discontinuous wrinkly grooves. Fig.3 shows these images. At (1), the escaped air area M can be observed at the flow front at the downstream of G portion. This area M can be seen to

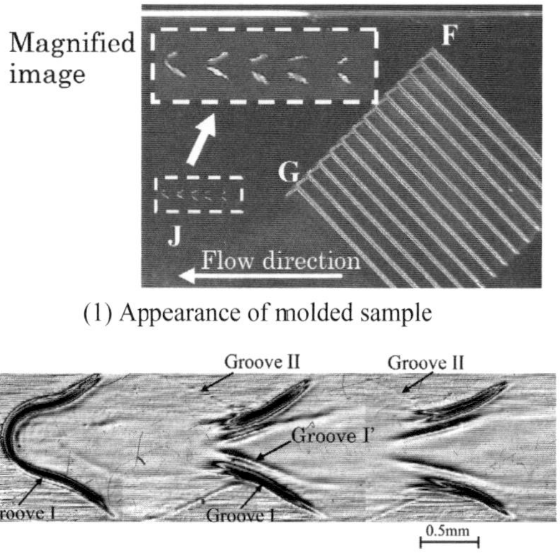

(1) Appearance of molded sample

(2) Magnified observation of "J"

Figure 2 Observation of molded sample

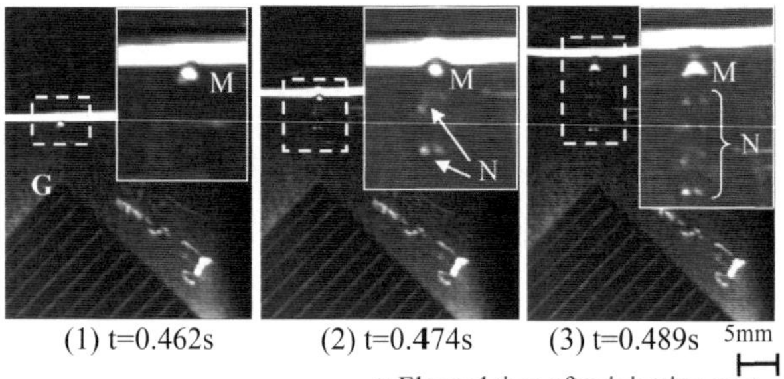

(1) t=0.462s (2) t=0.474s (3) t=0.489s

t; Elapsed time after injection start

Figure 3 Visualization images of flow behavior at the inclined micro-steps pattern area

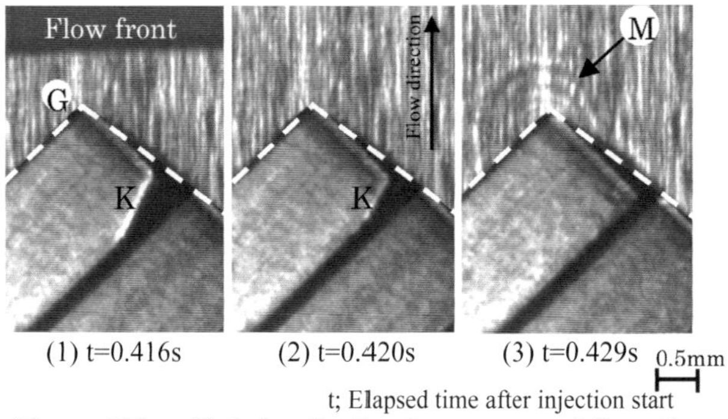

(1) t=0.416s (2) t=0.420s (3) t=0.429s

t; Elapsed time after injection start

Figure 4 Magnified visualization images around G portion

expand and shrink as it moves together with the flow front, and white stripes N are found to form corresponding to area M. The positions of the white stripes N in this figure correspond to the positions of the wrinkly grooves in area **J** in Fig.2.

Fig.4 shows the results of magnifying and visualizing the area around G portion. After the flow front passes G portion, an unfilled area K can be seen at the grooves inside the area of micro saw-tooth patterns. When K is filled, compressed air remaining in the groove moves from G portion to the plate area, thereby forming the escaped air area M as shown by (3). Fig.5 shows the visualized image of M immediately after it reaches the flow front. At (1), the escaped air area M penetrated through the flow front. The width of this deairing route is narrower than width of M. At (2), Groove I are formed at the upstream side of the deairing route passing through the flow front. At (3), the area M widens, forming Groove II at both sides of the widened area M. After this, as shown by (4), other Groove I' are formed on the inside of Groove I. This is thought to decrease the cross-sectional area of P and block the deairing route.

At the downstream of the magnified observed area, the widened area M is seen to shrink, and Groove I are seen to form again during this shrinkage process. When this expansion and shrinkage process of the area of deaired air was repeated, several wrinkles(Groove I, I', and II) were seen to be formed, eventually leading to the formation of the U-shaped Groove I which protrudes out at the downstream.

Effects of Molding Conditions on Discontinuous Wrinkly Grooves

Fig.6 shows the external view of the sample at the injection rate of 16cm^3/s. Unlike Fig.2, an area R of several wrinkles can be seen formed from the F-G side of each micro saw-tooth patterns to the downstream side. When the mold temperature is set high at 115°C or depressurization processing is conducted, no wrinkles are observed in each condition (Fig.7). At high mold temperatures, the adhesion between the mold surface and resin increases, and this minimizes the entry of residual air into the space between the resin surface and mold surface. The effect of depressurization is therefore thought to decrease the pressure of the residual air trapped in the corner of maicro saw-tooth pattern and minimize the demolding of the skin layer from the mold surface.

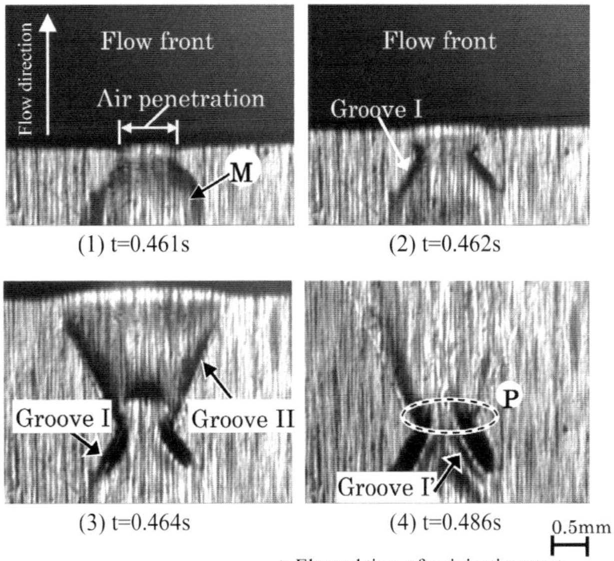

Figure 5 Magnified visualization images of wrinkly grooves generation process near flow front area

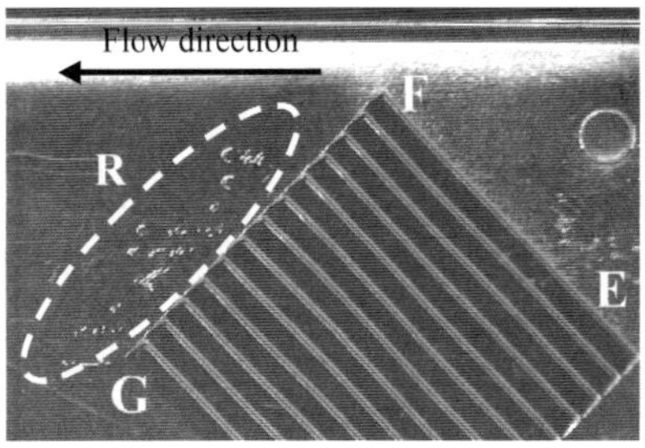

Figure 6 Appearance of molded sample (Injection rate; 16 cm^3/s)

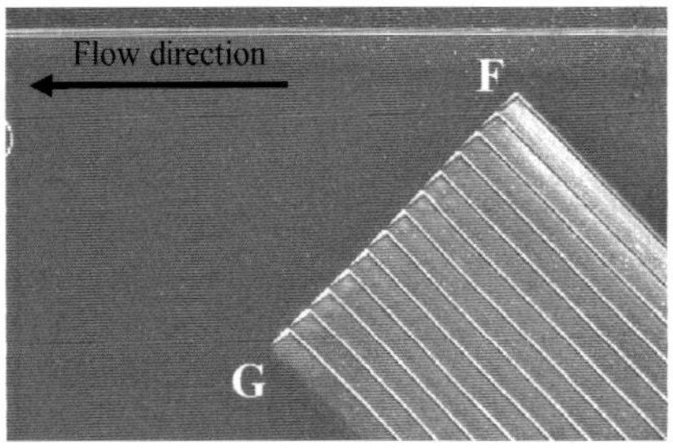

Figure 7 Appearance of molded sample (Mold temp.;115°C or depressurization)

Model of Generation of Discontinuous Wrinkly Grooves

Fig.8 shows the model of the generation of discontinuos wrinkly grooves. When residual air is compressed inside the micro saw-tooth patterns, it enters the plate area and forms the escaped air area. This escaped air area is seen to distribute to the space between the mold surface and skin layer. Resins contacting the escaped air area is melting inside, so when the force of the resin fluidity acts on the curved areas at the upstream of the dome-shaped escaped air, the escaped air area is pushed and moved to the flow front. On reaching the flow front, this area pushes up one part of the contacting area between the flow front and mold surface, thereby securing the deairing route to the atmosphere ((1)-(2)). Here, wrinkles (Groove I) are seen to form in the escaped air area located outside the deairing route. At (3), the escaped air area can be seen to widen from the deairing route as the resins flow,

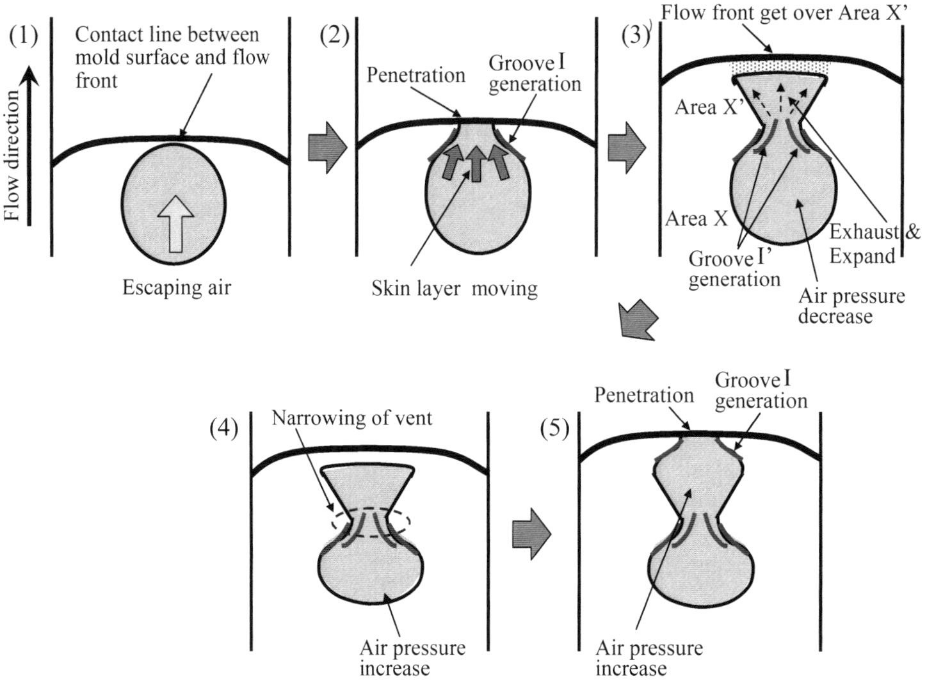

Figure 8 Discontinuous wrinkly grooves generation process behind micro saw-tooth patterns

thereby forming area X'. The formation process of this area X' corresponds to the expansion process of the escaped air area. After this, the flow front rides onto area X' and touches the mold surface, blocking the deairing route at the tip of the escaped air area. In area X at the upstream side of area X', the resin surface raised from the mold surface by the escaped air area derforms largely as it is dragged to the narrow deairing route formed at (2). As a result, new wrinkles (Groove I') are formed on the inside of Groove I. The narrow deairing route formed at (2) is thought to be blocked. At (4), air supply from area X to area X' is stopped, and this stops the expansion of area X'. This correponds to the shrinkage of the escaped air area. Then as resin pressure increases with resin flow, the air pressure in area X exceeds a certain level, and this causes the narrow deairing route to pass area X' again as shown at (5). Air supply to area X' is resumed, and the air pressure in this area also increases. At the flow front, the resins are once again raised from the mold surface, this secures the deairing route and transitions to the expansion process of the escaped air area. When expansion and shrinkage are repeated to the point that the air pressure for raising the resins at the flow front can no longer be secured, the air of deaired air stops growing. Because the deiaring route does not pass through the flow front in this state, the U-shaped Groove I is formed. In addition, as the escaped air area raises the skin layer from the mold surface, the area corresponding to the deformation and movement locus of the escaped air area becomes the low transcription area.[1]

Conclusion

In this study, we visualized the process by which discontinuous wrinkly grooves are formed at the downstream of the area of micro saw-tooth patterns inclined at 45 degrees to the flow direction. In the escaping process of the residual air in the area of micro saw-tooth patterns, the repetition of expansion and shrinkage of the escaped air area at the flow front was found to result in the formation of several discontinuous wrinkly grooves in the flow direction. It was also found that this defect can be effectively reduced by incraesing the mold temperature and depressurizing the internal air pressure of the cavity.

Acknowledgements

The authors would like to thank Teijin Chemicals Ltd. for providing the resin samples. This research was carried out as subproject of U2010 Project held at the Institute of Industrial Science, The University of Tokyo. We thank all participated companies for support and cooperation.

Reference

1. H.Yokoi, T.Otake; Polymer Processing Society Asia/Australia Regional Meeting, 138(1999)

O-04-247

VISUALIZATION ANALYSIS ON FLOW BEHAVIOR OF LONG CARBON FIBER REINFORCED MATERIALS IN INJECTION MOLDING

H.Yokoi[1*], R.Mori[2] and Y.Kanetoh[2]

[1] *Institute of Industrial Science, The University of Tokyo, Japan –hiyokoi@iis.u-tokyo.ac.jp*
[2] *Institute of Industrial Science, The University of Tokyo, Japan*

Abstract - In this study, for the first time we applied the following three visualization methods to the experimental analyses of long carbon fiber orientation process during the cavity filling process in injection molding; (1) glass-inserted mold for observing from the parting face, (2) camera driven system for observing from the flow-front side, and (3) runner exchange system for observing inner material flow assisted by X-ray CT technology. By using the above-mentioned methods, which we have especially applied for analyzing the glass fiber orientation process in previous works, we visually analyzed the cavity filling process and inner material flow in molded samples. As a result, we were able to successfully observe the following phenomena, focusing on the flow front area; deposition of carbon fibers orientated along the flow front's elongational surface, formation of unstable asymmetrical fountain flow along the thickness direction of the sample, alternately changing the diverging location at the flow front along the width direction (asymmetrical zigzag fountain flow phenomenon), and wave-like uneven inner flow caused by stagnant materials at the front along the width of the sample, etc.

Keywords: Injection Molding, Visualization, Long Carbon Fiber Reinforced Plastic, Fiber Orientation, Asymmetrical Fountain Flow

Introduction

Thermoplastics containing carbon fibers or glass fibers are widely used for automobile parts, mechanism and exterior parts, etc, due to their high functionality such as improving mechanical strength, adding conductivity, etc. However, depending on the state of fiber orientation inside the molded product, various defects such as deformation and reduced strength are known to occur.

To date, the authors have been conducting studies on glass fiber orientation behavior including visualization analysis of the fiber orientation process to form seven layers using glass-inserted molds [1], observation of asymmetrical fountain flow using colored marking on a runner exchange system [2], analysis of inner flow behavior and orientation behavior of fibers using colored marking method, X-ray CT technology and three-dimensional image processing software [3]. However, given that there exists no information on the orientation behavior of carbon fibers, in this study, the authors attempted to apply the above visualization analysis technique for glass fibers to carbon fibers in the aim to visualize the resin flow behavior and fiber orientation.

Experimental Method

We prepared two types of materials; (1) Colored material : PP containing 28wt% of long carbon fibers (length; 6.7mm, diameter; ϕ 7μm) and 4wt% titanium oxide, and (2) Normal material: PP containing 28wt% of long carbon fibers. Using a high speed rotary runner exchange system [3], the normal material were switched to the colored material during flow, and molded into a simple rectangular cavity with a thick film-like gate consisting of land and stagnant resin pool as shown in Fig.1. The molding machine used was the ultra-high speed two-color injection molding machine FN4000-9HTN (screw diameter of ϕ 32mm, clamping force of 1750kN). The resin temperature was set at 230°C, mold temperature at 50°C, and injection rate at 20cm^3/s.

To enable observation of flowing resins from outside the mold, two types of visualization mold and unit were utilized ; (1) glass-inserted mold [1] for observing from the parting face, and (2) open end cavity and camera driven system [4] for observing from the flow-front side (See Fig.2), where the distance

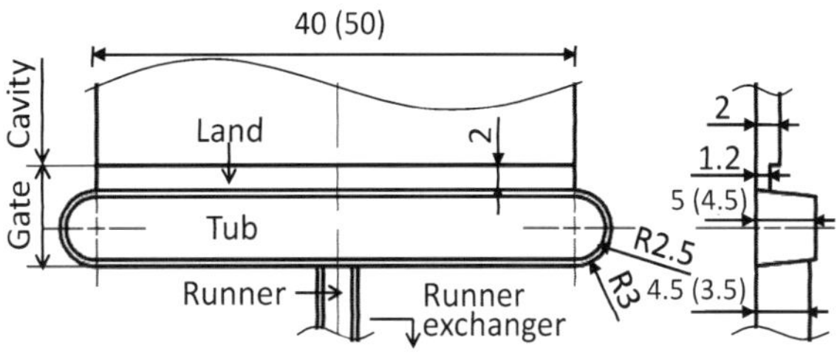

Figure 1 Dimension of the gate and cavity (Unit; mm)

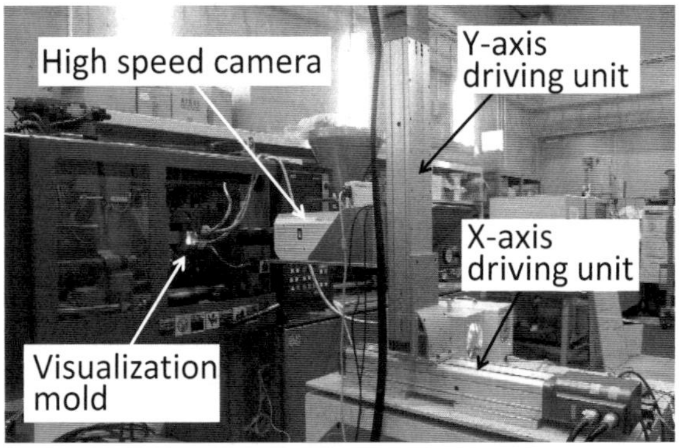

Figure 2 Appearance of the observation system from front direction

between the flow front and camera is kept constant during filling period by moving the camera backward. The resin flow was observed using a high speed video camera MEMRECAM fx K4. As the flow history of the resins becomes frozen after molding, the shape of the boundary between the normal material and colored material inside the molded product was extracted using an X-ray CT device to visualize inner flow. The microfocus X-ray CT device SMX-130CT was used for the observation of inner flow, while the VG Studio MAX was used for three-dimensional image processing.

Experiment Results and Discussion

Dynamic Observation of Flow Behavior from The Parting Face

Fig. 3 shows the observed images of the molding process. First, we observed the resin flow pattern formed by the boundary between the normal and colored materials at the surface of the molded product. At the boundary, colored material can be seen to appear forming a large V-shape (W-shape for some molded samples). The black resin area (A1) stopping at the tip of the flow front breaks up into several parts, some remain at the flow front and move by several mm (A3), and then extend out to the glass cavity surface and adhere to it (A4). At the boundary between the normal and colored materials, a V-shaped concave can be seen at the flow front (B1,B2). One part of the fiber orientation layer along the flow front surface is cracked, peeling into a layer which is touching and adhering to the glass surface (C1,C2) before the flow front. In this experiment, this phenomenon was seen in all molded samples, but no reproducibility in generation area and size was seen.

Observation of Inner Flow

X-ray images were three-dimensionally constructed for the molded sample in which the colored

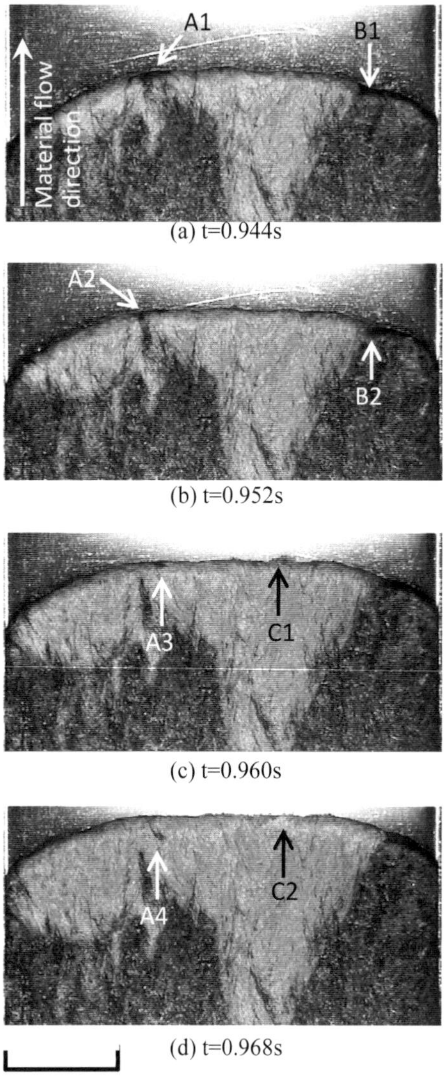

t : Elapsed time after start of injection (s)

Figure 3 Appearance of the observation system from front direction

material was made to flow until they broke through the tip of normal material. Fig.4(a) shows one example of the images, and Fig.4(b) shows the cross-section along the thickness center of the sample and cross-section along the sample width. The dark gray area is the normal material injected in the first stage, while the whitish area is the colored material injected after the runner has been switched.

Fig.4 shows the following unique flow behavior of the fountain flow area; At (i), the inner flow near the front is seen to form a large wave, which is caused by partial stagnation of the flow front material. In addition, fibers inside the colored material are seen to clearly orient along the wave. In the cross-sections along the

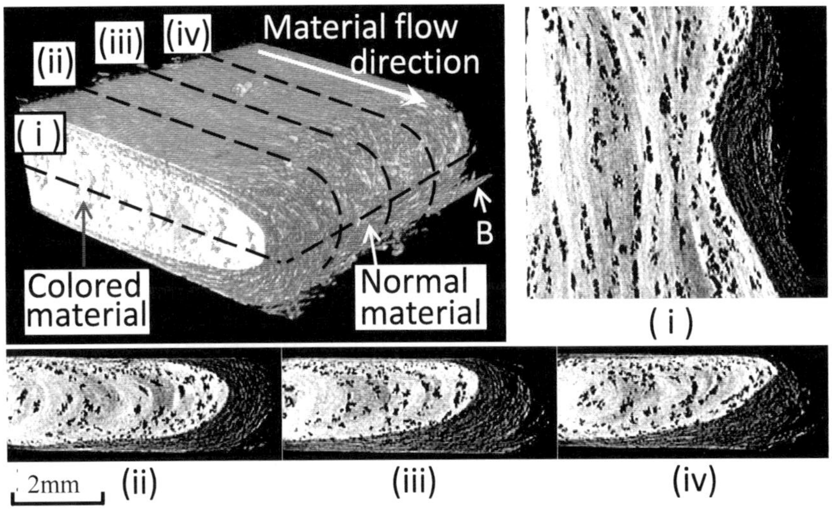

Figure 4 X-ray CT images of the molded sample of CFRTP obtained by runner-exchange operation

thickness direction (ii to iv), the colored material forms an asymmetrical fountain flow leaning towards the upper wall side at the flow front (ii). At (iii) and (iv), the colored material are seen to recover to a symmetrical fountain flow again. As shown by A, at the tip of the flow front, the surface layer peels away from the front surface, and in the inner layer, fibers orient according to the shape of the flow front, and form a new flow front. These phenomena were observed in most of the samples.

Dynamic Observation of Flow Behavior from The Flow-front Direction

Fig.5 shows images of the flow front obtained by front observation (thickness in the horizontal direction and cavity width in the vertical direction) as well as sketches of the resin flow direction. Fig.5(a) shows the complicated flow of the flow front; i.e. how the front material moves towards or from the left and right walls irregularly. From Fig.4(b), it can be seen that Inner layer 1 peels away from the wall while drawing a slanted boundary line and then flows to the left wall. Inside this peeled layer, a convex-shaped flow front is immediately formed, resulting in the resins spurting out from the layers near the cavity walls. It can be seen that the carbon fibers orient and form a layer according to the flow front shape. Although not shown in the figure, the flow front surface layer was also observed to swing in the cavity width direction (up/down direction in the figure), indicating that the carbon fibers orient also along the direction of the sheet width. Generally, resins containing short glass fibers orient in the shape of an arc along the flow front surface layer. These resins are also known to form an assymetrical zig-zag fountain flow in the thickness direction [2]. However, the results of this experiment suggest that resins containing long carbon fibers orienting along the width direction form a three-dimensional assymetrical fountain flow which moves both along the thickness and width directions.

Discussion on Flow Behavior at Front Area

In the flow front of carbon fiber reinforced materials, carbon fibers are seen to orient along the thickness or width direction of the sample, to form laminated layers along the flow front profile. The fibers involved in the outermost layer of the flow front extend to orient according to the flow front profile, making it difficult for further reinforcement of the fibers due to the gradual elongated orientation of fibers in addition to radiative cooling effect of the melt front. For this reason, once the resins near the contact area with cavity wall break up, inner material of the front area tends to spurt out intensively from the breakage area in the form of diverging flow. The diverging flow position is located in the vicinity of the wall. The broken surface layer is pushed and developed by the spurting resins, such that the peeled layer rotates toward the opposite wall side to the broken side, resulting in the occorence of asymmetrical fountain flow. As a result, fine wave-like patterns are observed at the molded surface. At the breakage area, the spurt-out material tends to drag materials on both sides of breakage area along the width direction, due to the parallel orientation of fibers to the flow flont profile along the width direction. Consequently, the opening through which the resins spurt out is widened, enabling the resins to stagnate more easily in the vicinity of the opening (both sides of opening along the width). This causes the resins to spurt out and flow concentratedly at the same position, forming the same large local V-shape (or W-shape) pattern on both sides of the molded surface. The formation of the concave at the flow front is thought to correspond to the boundary between the stagnant layer of old carbon fibers (normal material) and the layer of new carbon fibers (colored material).

Conclusions

In this study, we visualized the cavity filling process and inner flow of carbon fiber reinforced resins using a glass-inserted mold, camera driven system for observing from the flow-front side, and high speed runner exchange system. The following results were obtained.

(1) In the visualization of the cavity filling process, it was found that when resins spurt out once from one part of the flow front, the resins that follow these resins spurt out from the same areas at the back and front in a V-shape (or W-shape) that resembles a tear, and this breakage area starts to widen. No reproducibility was seen in the filling patterns of the flow front.

(2) In the observations using the runner exchange system, a stagnant layer of entangled carbon fibers was found to distribute with an inconsistent thickness at the flow front. Inside this layer, distribution of long fibers orienting orthogonally in the flow direction was observed. Along the thickness direction, resins were found to spurt out from one side of the cavity walls asymmetrically.

(3) In the front visualization of the flow front, the followings were repeatedly observed; fiber orientation along the arc-shaped flow front surface layer, peeling of fiber layer formed on the outermost layer of this surface layer, and the assymetrical spurting of the resins from one side of the cavity walls (zig-zag fountain flow). Also observed was a three-dimensional assymetircal fountain flow in which resins swing in the horizontal direction due to the orientation of long fibers spreading towards the width direction.

(4) Along the cavity width, resins were found to spurt out as if one part of the outermost layer, which is formed along the flow front and thus difficult to deform, had torn. As a result, long fibers orienting along the flow front on both sides of this tear are dragged, pushing and further widening the tear opening. It is thought that this causes the material inside to spurt out from the opening.

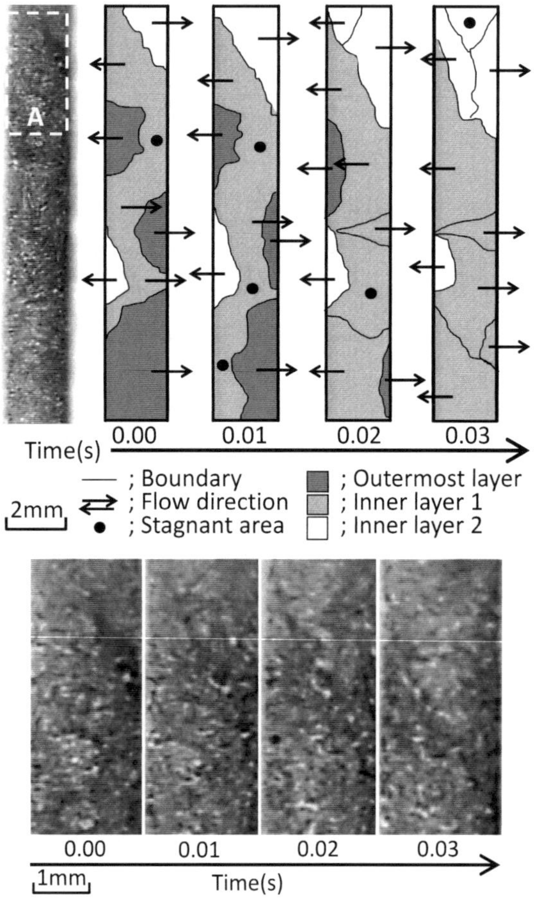

Figure 5 Visualization images of cavity filling process observed from front direction

Acknowledgements

The authors would like to thank Toray for providing the molding materials, as well as all participating companies of this study conducted as part of U2010 and U2011 Project held at the Institute of Industrial Science, The University of Tokyo.

References

1. H. Yokoi, Y. Murata, Y. Nishi; Abstracts for the 9 Annual Meeting of the Polymer Processing Society, 1993, 67.
2. H. Yokoi, Y. Kanetoh; *The journal of the International Polymer Processing*, 2005, **20**, 2, 157.
3. H. Yokoi, Y. Kanetoh; PPS2009 Europe/Africa Regional Meeting Program and Abstracts, 2009, 54.
4. H. Yoki, H. Mitsuhata, N. Masuda; The 3rd Pacific Symposium on Flow Visualization and Image Processing, 2001, 45.

O-04-036

EXPERIMENTAL AND NUMERICAL ANALYSIS OF THE TEMPERATURE DISTRIBUTION OF INJECTION MOLDED PRODUCTS USING PROTRUDING MICROPROBES

Shih-Jung Liu* and Chia-Wei Ho
Polymer Rheology and Processing Lab
Department of Mechanical Engineering, Chang Gung University
Tao-Yuan 333, Taiwan
**Correspondence: 259, Wen-Hwa 1st Road, Kwei-Shan, Tao-Yuan 333, Taiwan.*
Email address: shihjung@mail.cgu.edu.tw

Abstract - This study aims to, experimentally and numerically, examine the three-dimensional temperature distribution along the melt flow path of injection molded parts. A special experimental set-up, which includes an injection mold equipped with protruding microprobes for guiding embedded thermocouples, was designed and built to measure the temperature field along the flow path, i.e., inside the runner and the cavity, of injection molded products. The experimental results suggested that the disturbance induced by the probes remained negligible and precise temperature profiles could be measured at various positions inside the cavity. A significant increase of melt temperature was found to result from the viscous dissipation of the polymeric materials in the runner. Additionally, a commercially available code was employed to simulate and predict the temperature variation in injection molded parts. It was shown that the numerical prediction coincided satisfactorily with the measured temperature data.

Keywords: injection molding, temperature distribution, experimental set-up, numerical simulation

Introduction

In an injection molding process, the polymeric material undergoes a complex history in which it is melted, flows through complex geometries from sprue, runner to cavity, deforms to take the shape of the mold, and finally cools down and solidifies into the final product. Temperature is an important parameter that influences process features such as cycle times, crystallization rates, degree of crystallinity, melt flow properties and molded product qualities. This report aims to, experimentally and numerically, investigate the temperature distribution in injection molded parts. We proposed an experimental device which allows the measurement of three-dimensional temperature field in the depths of the runners and cavities. A plate cavity with a thickness of 3mm equipped with a tubular type device consisting of protruding microprobes guiding embedded thermocouples inside the cavity was designed and built to measure the transient temperature distribution during the molding cycle. In addition, a commercially available code was employed to simulate the transient temperature distributions inside the mold cavity.

Experimental

The resin used in this study was polypropylene (Globalene 6331, LCY Chem., Taiwan). The melt flow index of the polymer was measured at 14.5.

Experiments were conducted on an 80-ton conventional injection-molding machine, with a highest injection rate of 109 cm3/sec. A plate mold with a thickness of 3mm that has tubular needles for guiding embedded thermocouples inside the cavity was built in our lab for the experiments, which made it possible to measure the in situ temperature during molding. The dimension of the runner was 3mmx6mm. Nine orifices of 0.8 mm in diameter were worked through the materials to insert the temperature probes inside the molding cavity. To minimize the flow disturbance as well as the potential inaccuracies due to conduction errors [1-3], the probe was intruded from the far end of the cavity toward filling gate in the direction opposite to that of the melt flow. The temperature probe consists of sheathed K-type thermocouples and tubular needles used to guide the thermocouple in the depths of the cavity. The needles have a length of 80 mm and an internal and external diameter of 0.65mm and 0.8mm respectively, while the diameter of the thermocouple is 0.56mm. For each experiment, nine temperature profiles could be measured (Figure 1). By changing the intrusion lengths of the needles, the temperatures at the same three cross sections (i.e., 25mm, 50mm and 75mm from the gate) along the flow direction of molded parts could be obtained. Furthermore, two couples were inserted into the core of the runner at different locations to monitor the temperature variation along the runner (a1, a2 and a3). The temperature of the mold was regulated by a water-circulating mold temperature control unit [4].

During the injection molding process, the in-runner and in-cavity temperatures were recorded by the thermocouples connected to a personal computer through an RS-232 interface. The response time of the thermocouple was 0.1 second. The transient temperature profiles during the injection molding cycle were recorded by LabVIEW (National Instrument, U.S.A.) data acquisition software.

Results and Discussion

Temperature profiles of injection molding at three different cross sections inside the cavity were measured. Disturbance induced by the probes remain limited and all probes remained stationary during the melt filling process. Good repeatability was obtained for all the measurements. Figure 2 shows the temperature profiles at center, side and corner thermocouples (thermocouples 6, 3 and 1, respectively in Figure 1) of the front section.

All measured temperature profiles in Figures 2 show a slowdown of cooling at a temperature of approximate 130oC, which corresponds to the crystallization temperature of polypropylene materials [5]. In this study, the melt fill temperature used was 185oC. Interestingly, the measured peak temperature at the front section (Figure 2) inside the cavity was approximately 220oC.

The numerical simulation examined the cooling process of injection molding via a three-dimensional filling and cooling code. With constant temperatures (50oC) along the cooling channels, and with an initial melt temperature of 185oC for the polymer melt, the numerical iteration provides the three dimensional temperature distributions inside the molded articles at different elapsed times. The calculated results for injection molding at the front, middle and rear sections are also shown in Figures 2. When compared to the experimental data, the numerical simulation result was in reasonable agreement with the measured temperature profiles of injection molded parts. The results testify to the validity of the heat transfer model for the prediction of the injection molding cooling process.

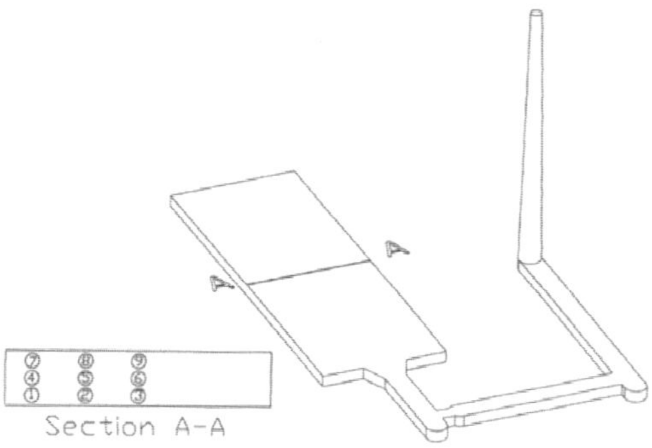

Figure 1 –Positioning of the thermocouples in the devices.

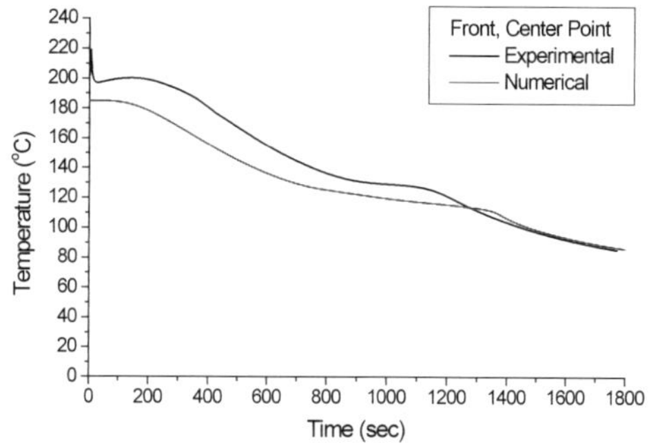

(a)

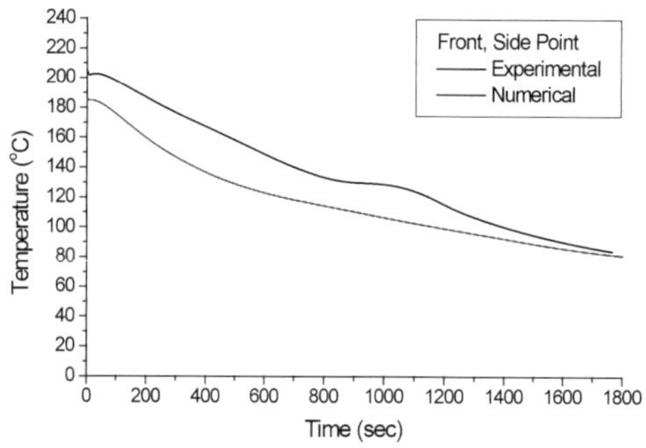

(b)

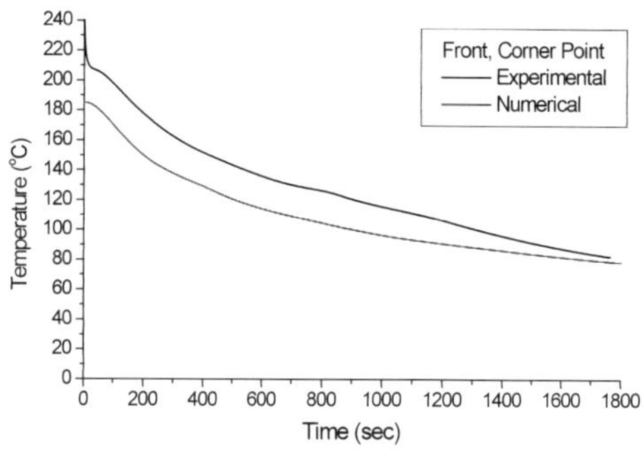

(c)

Figure 2 –Temperature profiles at the front section.

Conclusions

The study in this article has examined the injection molding cooling process, both experimentally and numerically, with the intent of either reaching or understanding a more inclusive or realistic phenomenon. It has been found that the disturbance induced by the probes remains negligible and precise temperature profiles can be measured at various positions inside the cavity. The numerical method utilized a commercially available code to predict and simulate the cooling process in injection molding. Calculated results coincided satisfactorily with the experimental data.

Acknowledgements
The National Science Council of Taiwan, R.O.C. under the grant NSC98-2221-E-182-003-MY3, has supported this work financially.

References

1. R.J. Dickson, Thermal conduction errors of manganin-constantan thermocouple arrays, Phys. Med. Biol. 30 (1985) 445-453.
2. M. Tarnopolsky, I. Seginer, Leaf temperature error from heat conduction along thermocouple wires, Agr. Forest Meteorol. 3 (1999) 185-194.
3. G.J.K. Packer, J.L.B. Gamlen, Calculation of temperature measurement errors in thermocouples in convection heating cans, J. Food Sci. 39 (2007) 739-743.
4. S.J. Liu, C.W. Ho, Experimental and Numerical Analysis of the Temperature Distribution of Injection Molded Products Using Protruding Microprobes, Rev. Sci. Inst., 82 (2011) 055109.
5. S.L. Rosen, Fundamental Principles of Polymeric Materials, Wiley, New York, 1999.

O-04-315

REPLICATION OF MICROSTRUCTURED FREEFORM SURFACE PARTS IN INJECTION MOLDING

M. Burgsteiner[1]*, F. Müller[1], A. Wolfberger[2], T. Lucyshyn[1], C. Kukla[3] and C. Holzer[1]

[1]*Department of Polymer Engineering and Science, Chair of Polymer Processing, Montanuniversitaet Leoben, Austria – martin.burgsteiner@unileoben.ac.at; florian.mueller@unileoben.ac.at; archim.wolfberger@unileoben.ac.at; thomas.lucyshyn@unileoben.ac.at; clemens.holzer@unileoben.ac.at*

[2]*Department of Polymer Engineering and Science, Chair of Chemistry of Polymeric Materials, Montanuniversitaet Leoben, Austria –archim.wolfberger@unileoben.ac.at*

[3]*Industrial Liaison Department, Montanuniversitaet Leoben, Austria – christian.kukla@unileoben.ac.at*

Abstract - Functionalized surfaces equipped with microstructures offer a multitude of possible applications. Due to the established production processes, like lithographic methods (LIGA) or etching, the microstructures are usually replicated on planar polymer parts. Microstructures on freeform surfaces are therefore difficult to realize with common methods. One of the major problems occurring during production (e.g. injection molding) are undercuts in demolding direction. This leads to a high risk of ripping off or deforming of the structures. In this work a novel method for the replication of microstructures on freeform surfaces is presented. A pattern composed of lines was produced on a nickel substrate, using a conventional LIGA process. The nickel substrate was used as a premaster for replication in a bendable stamper material. The bendable material enabled a good fitment to the mold cavity surface, and combines sufficient stiffness for an adequate replication with good flexibility to release undercuts while demolding. Poly(dimethylsiloxane) (PDMS) offers suitable properties as a replication master. For replication testing, a PDMS master was fixed in an injection mold and parts with rounded passages and micro-structured surfaces were molded. To evaluate the quality of replication, optical analyzing methods were used, estimating the geometries of each production step. The results showed very good replication rates of at least 90% of the origin structure shape.

Keywords: freeform surface, injection molding, microstructures, replication, PDMS

Introduction

In the last years a continuous trend pursues the miniaturization of products with the aim to integrate a multitude of functions and thus broaden the field of application without increasing weight or volume of the parts, e.g. using parts with functionalized surfaces via microstructures [1]. Depending on the integrated features these parts are for example able to process complex analyses on a small area ("Lab-on-a-chip", "Micro-Total-Analytical-Systems") [1, 2, 3], have self-cleaning surfaces ("Lotus-effect") [4], or influence the behavior of cell growth via structuring [5, 6].

The trend of surface functionalization by micro-structuring was strongly initialized by the semi-conductor industry, whose production methods like lithography (LIGA) [1, 2, 3], etching [2, 3], laser ablation [1, 2, 3], micromachining [1] or electrical discharge machining (EDM) [1] were adapted for the production of micro-structured master templates for the following replication on polymer parts [2, 3]. Consequently, these masters usually have flat surfaces, and therefore are replicated on planar polymer parts [6].

However, many applications would benefit from 2D as well as 3D microstructures integrated on macro-scale freeform surfaces, for instance medical devices [6], implants with special surface properties for a better acceptance by the human body, but also optical elements, like displays or lenses.

Despite progress was made in machinery [1] and continuous adaption of the established processes the mastering process still suffers from the lack of economical production of microstructures on macro-scaled freeform surfaces on larger areas [1]. Additionally, one of the major problems in the following mass replication (e.g. injection molding) is the demolding of free formed structures yielding an undercut in the demolding direction. This undercut leads to serious problems because of the high risk of ripping off or deformation of the structure [7].

This paper presents a novel method for the replication of structured free form surfaces in injection molding.

Experimental

Principle of Operation

Elsner et al. [8], Gorb et al. [9] and Möllenbeck et al. [10] described a concept where an undercut structure was cast into a flexible poly(dimethylsiloxane) (PDMS) matrix. This concept was developed further in this project with one additional step, so that it is applicable for the injection molding process for common polymers.

Thus the replication process consists mainly of three steps:

I. The positive structure is integrated on the pre-master, produced by lithographic patterning.
II. The 2D premaster is then used to produce the flexible master by casting the premaster structure with a flexible material, e.g. PDMS. This casting process is carried out under vacuum to enhance the quality of the rep-lication, followed by a heat treatment under

elevated temperatures to ensure the complete curing of the master. This PDMS master is demolded from the casting mold.

III. The flexible PDMS master is fixed on a freeform shape of an injection mold cavity. After that the micro-structured freeform surface can be replicated in a conventional injection molding process.

Experimental Setup & Measurements

As a first step a nickel (Ni) substrate was produced using a conventional LIGA process. This nickel premaster included an embossed structure, containing lines with a width of approximately 45 μm and a height of approximately 56 μm. To investigate the dependence of the replication quality on the orientation of the lines, two sets of lines were aligned perpendicular to each other on the substrate, as sketched in Fig. 1. The first type is orientated rectangular to the flow direction, which is later referred to as horizontal orientation (Fig. 1, left); the second one is orientated in flow direction, which is later referred to as vertical orientation (Fig. 1, right).

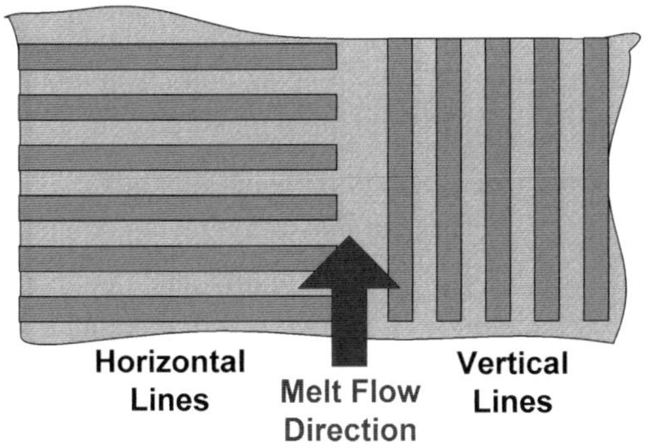

Figure 1 – Sketch of the structure integrated on the nickel substrate, consisting of horizontal lines, shown on the left and vertical lines, sketched on the right.

To investigate the dependence of the replication quality on the orientation of the lines, parts in a flat region of the mold were produced and analyzed. Additional proof-of-concept-measurements were performed in a curved area of the mold, showing the possibility to replicate structures on these types of surfaces.
The flexible master was built up from a two component PDMS system by Wacker Chemie AG, Germany. The mixed components were cast into a mold including the nickel premaster. To enhance the quality of the replication process vacuum was applied. The casting process was followed by a heat treatment at a temperature of about 80 °C to ensure the complete curing before demolding of the PDMS master.
The replication experiments were performed on an ARBURG Allrounder 320C 500-100, using a highly fluent Polypropylene (PP), type C7069 – 100 NA, by DOW Europe GmbH, Switzerland.

To evaluate the quality of the replication, different optical analyzing methods were used, including confocal microscopy (FRT MicroProf), an infinite focus system (Alicona Infinite Focus) as well as scanning electron microscopy (SEM) (ZEISS EVO MA25). The differently oriented lines were evaluated at every step of the mentioned process chain, including the nickel premaster, the PDMS master and the injection molded parts with the replicated microstructure at flat and curved regions.

Results and Discussion

The line geometries of the microstructures were monitored along the complete process chain. For comprehensibility the results are split into the master production results and polymer replication results.

Nickel premaster and PDMS master structure

The geometries of the nickel structures were measured with a confocal microscope, showing structure heights of approximately 56 μm, independent of their orientation. The width of the lines was determined as 45 μm for both types of lines.
Comparing the dimensions of the nickel premaster structures to those of the PDMS master good replication rates were achieved. The height of the PDMS structures was determined as 54 μm, which corresponds to a replication rate of more than 95 %. The replication ratio is defined as the ratio between the replica-dimension to the equivalent master-dimension. The width of the horizontal structures is 46 μm and therefore only slightly different to their counterpart on the nickel premaster, while the width of the vertical lines is equal to those on the nickel substrate.

PP part - flat micro-structured surface investigation

Confocal microscopy was used to evaluate the heights of the structure in the flat region. The gained results show a height of about 56 μm, independent of the orientation. This value is slightly higher than those of the PDMS master, a result of elastic deformations during injection molding, and equal to the nickel premaster, implying a very high grade of replication.
SEM measurements were performed to investigate the width of the structures. Fig. 2 shows a horizontal line, while Fig. 3 displays a vertical one. The structures in these figures appear to be twisted, which is a result of the sample preparation. As shown in Fig. 2, the width of the horizontal structure is approximately 50 μm, which is slightly higher than those of the PDMS master and the nickel premaster. The brighter areas, surrounding the top surface of the structure suggests that the edge of the structure is rounded. Fig. 3 shows the measured width of the vertical structure, which is only 36 μm and therefore significantly lower than the vertical lines in the PDMS master, which had a width of about 45 μm. This result was confirmed by other

measurements made with confocal microscopy, which detected 35 µm. Therefore the width of the structures in flow direction is strongly influenced and is reduced to about 80 % of the master structure, while those of the horizontal lines were raised to 110 % of that of the master.

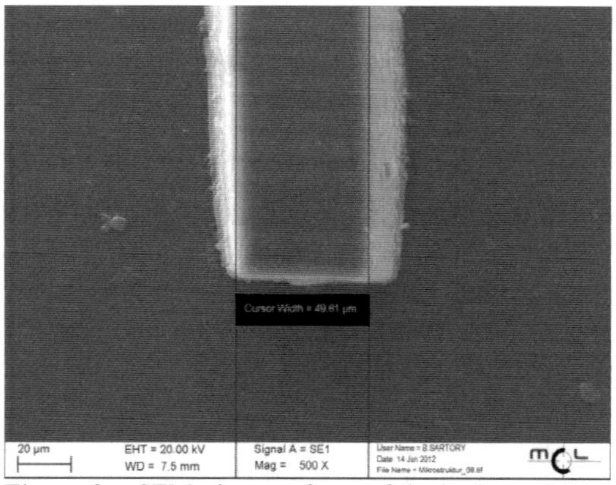

Figure 2 – SEM picture of one of the horizontal lines of the structure on the flat region (rectangular to the flow direction), with a width of about 50 µm, which is slightly higher than those of the master structure with about 46 µm.

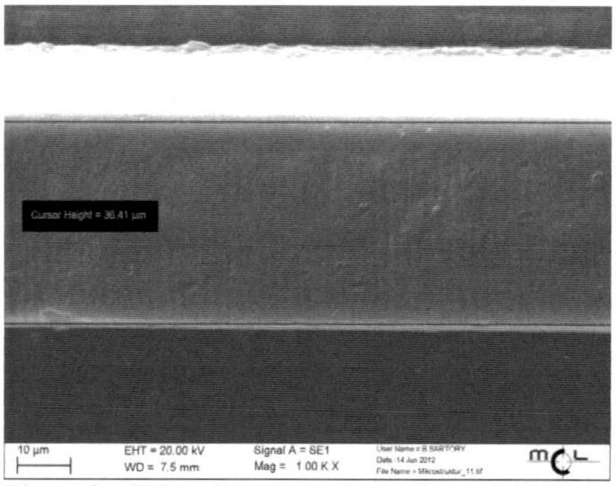

Figure 3 – SEM micrograph of one of the vertical lines of the structure on the flat region (in flow direction), with a width of about 36 µm, which is significant lower than those of the master structure with about 45 µm.

PP part - curved micro-structured surface investigation

The geometries of the structures on the curved area of the molded part were examined by an infinite focus analyzing method with a five-fold magnification. An overview of a small region of the replicated area in the polymer is displayed in Fig. 4, where a slight barreling is visible. The radius on the ground was calculated to be about 6 mm, which is quite close to the value of the cavity. The white spaces in the figure are measurement artifacts and therefore automatically excluded by the analyzing software of the infinite focus system.

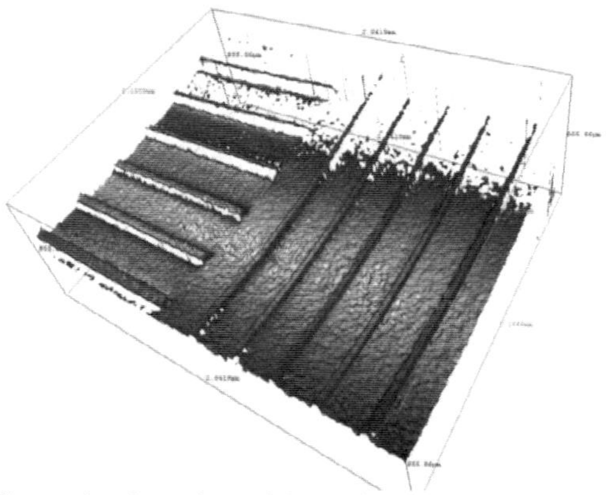

Figure 4 – Overview of the replicated microstructures measured with the infinite focus analyzing method with a five-fold magnification. On the right side the vertical structures are displayed, on the left side the horizontal ones.

Evaluation of the results of the structure height was done separately for the horizontal and the vertical lines. Fig. 5 shows the average height profile of two horizontal lines, calculated from 20 parallel measured profiles after a cylinder subtraction to achieve a flat baseline. As it can be seen in Fig. 4 the height of the structures is about 52 µm, which is close to the dimension of the PDMS master used for the replication. However, these structures are not as high as the structures on the flat areas.

The width of the horizontal structures was determined to be about 48 µm, which is close to the values from the PDMS and nickel structures and also similar to those from the horizontal lines on the flat area.

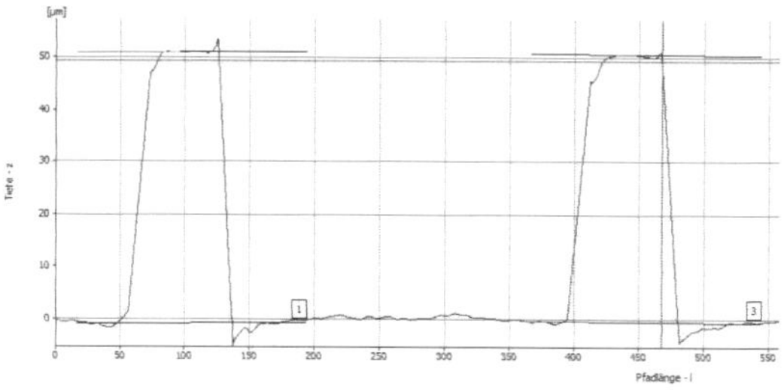

Figure 5 – Average height profile of the horizontal lines of the micro-structured curved surface after a cylinder subtraction for achieving a flat plane for measuring the height.

The average height profile for two vertical lines is shown in Fig. 6, which was also calculated as mentioned before. The height of the vertical structure is measured as approximately 49 µm. This value is thereby significantly lower than that for the PDMS master and lower than those of the horizontal lines and therefore also lower than the heights of the lines on the flat area.

The width of the vertical structures was determined to be about 32 µm. Therefore they yield similar dimensions to the determined width for the vertical line on the flat surface and dimension grades (for the widths) of about 70 % for the vertical and about 110 % for the horizontal lines.

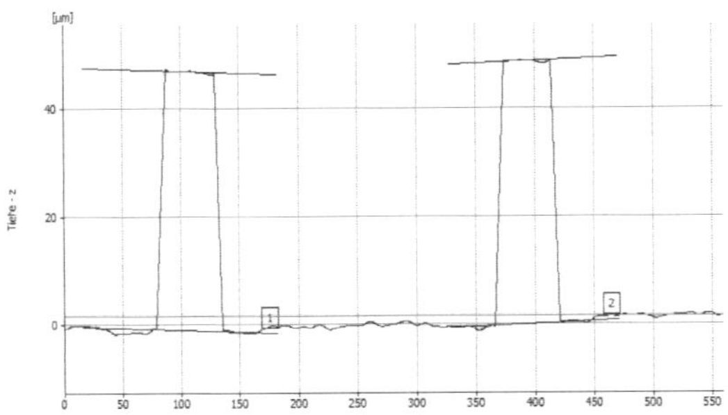

Figure 6 – Average height profile of vertical lines of the micro-structured curved surface after a cylinder subtraction to measure the height.

Discussion and Outlook

The experimental results show a good replication rate of the height of microstructure in every step of the process chain. The reached replication rate of structure height in total was at least about 90 %. While the height in the flat regions is slightly higher, it is reduced slightly in the curved regions compared to the PDMS structure. So it seems that the bending of the PDMS reduces the structural heights, especially in the vertical line region which were oriented in bending direction.
The investigation of the widths shows that the replication of the microstructure from the nickel pre-master in PDMS does not affect the width in a significant manner in contrast to the replication by injection molding. While the width of the horizontal structures, which were oriented rectangular to the flow direction of the polymer melt, was comparable to those in the PDMS, the value of the vertical line is significantly smaller than that of its counterpart. A better filling of the structures perpendicular to flow direction compared to the parallel ones was also observed for cyclic olefin copolymer (COC) by Mönkkönen et al. [11], who investigated the dependence of the filling behavior of linear structures on the melt flow direction among other things. However, the inverse effect was found for other plastic materials. Therefore it was assumed that the low melt viscosity is responsible for that effect, which would be similar to the characteristics of the PP used in that experiment, having a MFR of about 100 g/10 min (230 °C/2.16 kg). Another option is a result of a different filling mechanism of the different oriented structures as described by Lee et al. [12]. However, this has to be clarified in further investigations. In contrast to that, the bending of the PDMS master seems to have no influence on the widths of the structure.

Therefore the next investigations will contain the understanding of the effects mentioned before, the replication of microstructures on real freeform surfaces as well as the development of a mathematical model to predict the needed geometries of microstructures for LIGA process to get a defined microstructure on a freeform surface. Additionally, the stiffness of the PDMS has to be improved to avoid rounding of the edges and barreling of the whole structure, which occurred especially at the horizontal lines, orientated rectangular to the flow direction.

Conclusions

In this paper a novel concept for the replication of microstructures on freeform surfaces was presented and the functionality of the concept was proven. It is shown that a 2D-microstructure produced with a conventional LIGA process can be sufficiently replicated in a PDMS master, which can be used as a structured insert in an injection mold to produce polymer parts with micro-structured freeform surfaces. Additionally, it was found that the achieved replication rates of the structure heights are at least about 90 %. Only the widths of the structures in flow direction are strongly influenced and were reduced to about 80 % of the master structure.

Acknowledgements

This work is dedicated to those people around the work group, who supported the experiments. Without them it would not be possible to realize this project.

References

1. E.B. Brousseau; S.S. Dimov; D.T. Pham *Int. J. Adv. Manuf. Technol.* 2010, 47, 161-180
2. Y. Sun, Y.C. Kwok *Analytica Chemica Acta* 2006, 556, 80-96
3. G.S. Fiorini; D.T. Chiu *BioTechniques* 2005, 38, 429-446
4. X. Kang et al. Applied Surface Science 2007, 253, 8830-8834
5. K. Anselme et al. *Acta Biomaterialia* 2010, 6, 3824 - 3846
6. J.G. Fernandez; C.A. Mills; J. Samitier *Small*, 2009, 6, 614-620
7. Z. Song et al. *J. Vac. Sci. Technol. B* 2008, 26(2), 598-605
8. C. Elsner; J. Zajadacz; K. Zimmer *Microelectronic Engineering* 2011, 88, 60-63
9. S. Möllenbeck et al. *Microelectronic Engineering* 2007, 84, 1007-1010
10. S. Gorb et al. *J. R. Soc. Interface* 2007, 4, 271-275
11. K. Mönkkönen et al., *Polymer Engineering And Science*, 2002, Vol. 42, No. 7, 1600-1608
12. J.G. Lee et al., Polymer Engineering and Science, 2010, Vol. 50, No. 6, 1186-1198

O-04-394

REAL TIME VIS SPECTROSCOPIC MONITORING OF COLOUR DURING INJECTION MOULDING

P Brookbank[1,*] and O Ghita[1]

[1] *University of Exeter, College of Engineering, Mathematics and Physical Sciences, Harrison Building, North Park Road, Exeter, EX4 4QF, UK – P.A.Brookbank@exeter.ac.uk; O.Ghita@exeter.ac.uk*

Abstract

The colormetric analysis of molten polyethylene terathylate (PET) has been performed during injection moulding using a VIS in-line monitoring system. The conversion of spectroscopic data to the CIE LAB colour space has allowed the monitoring of moulded part colour fluctuation shot by shot. The injection moulded parts have subsequently been analysed offline in a standard colourmetric test to establish the CIE LAB values that would be produced by plastics converters. CIE LAB values calculated for on- and offfline parts show similar changes during increasing masterbatch / colour concentration. The CIE LAB values from on- and offline transmission measurements of coloured polymeric parts have shown good comparable results which demostrate the potential offered by the in-line VIS monitoring system.

Keywords: Visible Spectroscopy, Injection Moulding, Quality Control

Introduction

The quality control of colour during plastic injection moulding is a current and ongoing problem for injection moulders across the industry. The method of colour evaluation will influence the control of masterbatch or liquid colour addition and therefore cost. The most common method for colour analysis is by visual comparison of the produced part with a RAL chart. The paint and ink industries use a spectroscopic method for colour determination as they rely on tight reproduction of colour. These industries use the CIE LAB model that conforms to standards such as ISO-12647 and is a way of accurately describing the colour of an object. CIE LAB is a colour space described by the International Commission on Illumination (Commission internationale de l'éclairage) and is used to describe the sensation of the human eye towards colour. The CIE L, A and B values represents colour stimuli as an achromatic signal (L*) and two chromatic channels representing yellow-blue (b*) and red-green (a*) (Figure 1). The colour space is normalised against the illuminant which achieves a colour space that is independent of the colour of the source light.

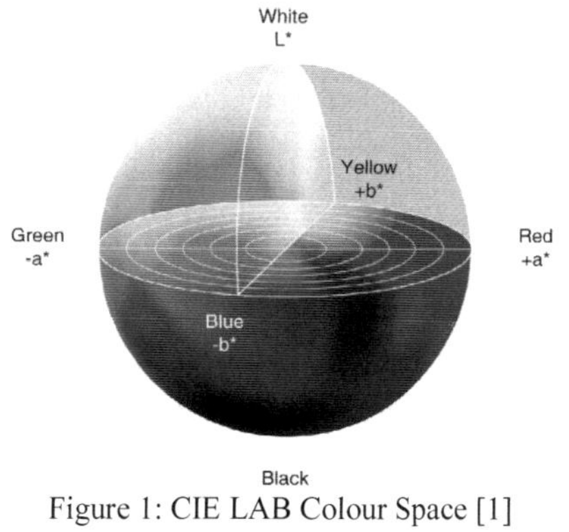

Figure 1: CIE LAB Colour Space [1]

The use of a handheld [2] or automated [3] colorimeter equipment during production is either time-consuming or requires a large capital investment respectively. The introduction of visible colourmetry equipment into the nozzle or end cap of an injection moulder would allow visible colour to be monitored. This may then be used as a quality control tool to detect out of specification parts that may not be detected via visual inspection.

Experimental

A Battenfeld 40 tonne hydraulic injection moulding machine (IM) has been used throughout experimentation. PET (Melinar Plus+) [3] has been used in transmisssion mode for spectral aquisition. PET was processed at 280 C with a mould temperature of 50 C. PET was dried for 3 hours at 121 C in a -40% desicant dryer.

PET was coloured red using a PET specific masterbatch [4] with dosing concentrations of 0.0% w/w to 1.0% w/w in 0.25% w/w increments.

The mould used produced standard tensile test bars and 3 point bend test bars (ISO 527, ISO 178) [7, 8].

The standard nozzle of the IM was replaced with a longer nozzle that allowed the incorporation of two optical probes (Figure 2). The internal diameter of the nozzle was 8mm and represents the pathlength between the optical probes. A high capacity heaterband was used to overcome the additional thermal mass.

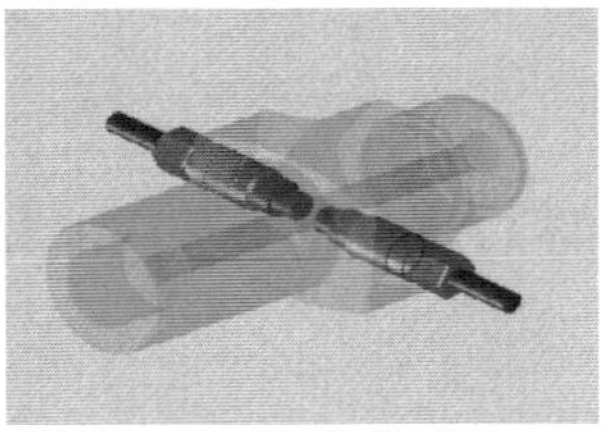

Figure 2: IM Nozzle with Optical Probes Installed

A Ocean Optics DH2000 light source and HR2000+ spectrometer conected to the optical probes via standard SMA connectors were used to carry out visible measurments. Postprocessing and CIE LAB calculation was carried out using the supplied SpectrSuite operating software. Air (empty pathlength) was used to set the spectrometer response to 100% during measurments. Offline measurements on the flexural bars were carried out using the same optical equipment installed in a simple probe holder (Figure 3).

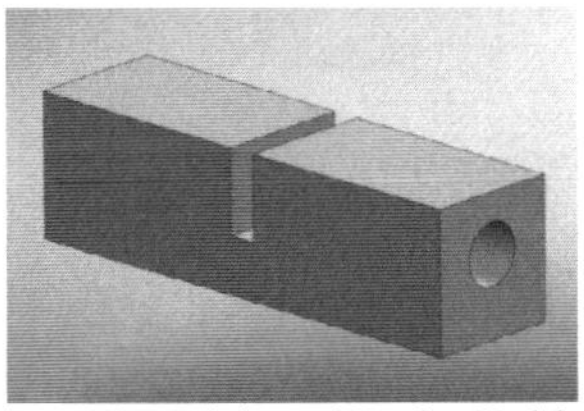

Figure 3: Lab based Probe Holder

Ten repeat measurements have been averaged and the standard deviation is shown as vertical error bars in Figure 4, Figure 5, Figure 6, and Figure 7.

Results and Discussion

The results of transmission measurements taken during the IM processing of PET can be seen in Figure 4. These results show that the addition of red masterbatch prevents the transmission of blue / yellow light in the wavelength range 380-550nm so only red light is transmitted. The increase in masterbatch concentration has the effect of lowering the amount of transmitted light and shifting the 'red peak' further into the red region of the spectra. Natural PET can be seen to interact with the red region of the spectra and will therefore cause the transparent part to appear faintly blue. There is a noticeable amount of noise in the spectra at low (380-450nm) wavelengths which is due to the DH2000 light source not extending far enough into the UV range to produce a stable signal without illumination from a deuterium lamp. The calculated CIE LAB values (Figure 5) reflect the observed changes in the spectra. The reduction in L relating to the increasing darkness of colour, the positive A and B terms relate to a strong red and yellow combination. The gap separating the red and yellow values increases with increasing masterbatch concentrations corresponding to a deepening of the red colour.

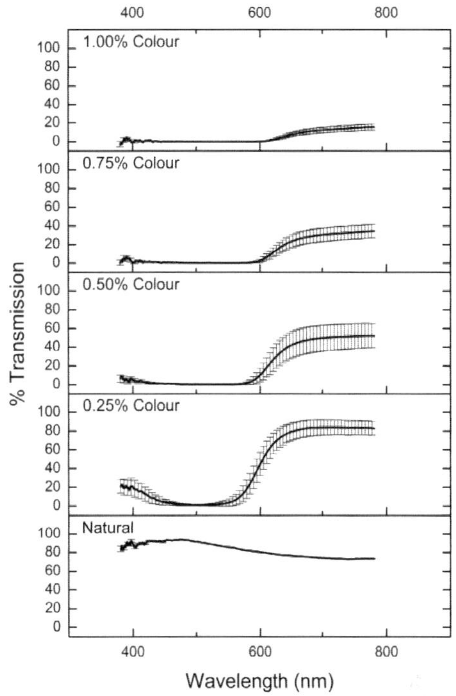

Figure 4: Online Visible Spectra of 0.00% w/w to 1.00% w/w Red Masterbatch PET

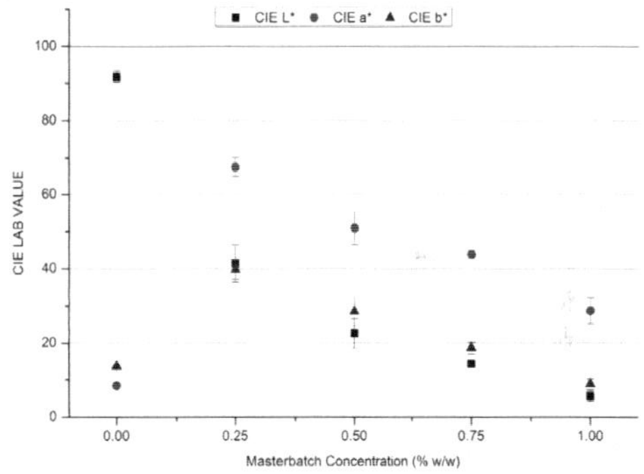

Figure 5: Online - CIE LAB Values Calculated for Red PET

The offline spectra (Figure 6) show the same trends as seen in the online data. The small differences in the intensity of the transmitted light are due either to the decreased molecular mobility [9] of the polymer in the solid state, or to the surface gloss of the part [10]. The reduction in transmitted light reduces the L value. The positive A value of offline data depicts a similar red colour when compared to online data (Table 1). The greatest difference between the on- and offline CIE LAB values can be found in the yellow characteristic in the B value. This is due to the larger amount of light being transmitted in the orange - red wavelength range of 597 – 617nm.

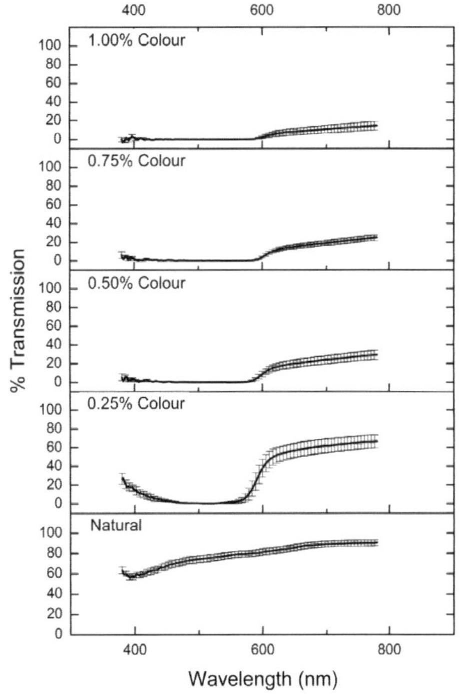

Figure 6: Offline Visible Spectra of 0.00% w/w to 1.00% w/w Red Masterbatch PET

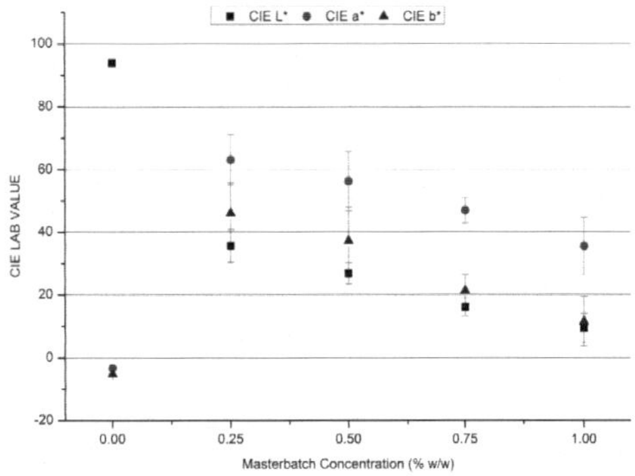

Figure 7: CIE LAB Values Offline PET Transmission

Table 1: CIE LAB Values from all online and offline Analysis

Online PET						
Offline PET						

Conclusions

The inline monitoring of colorimetric information has been successfully carried out using transmission measurements on translucent PET material. The CIE LAB values calculated from offline transmission measurements are equivalent to online results and can be used to provide an absolute colour value for produced parts. Knowing that surfaces interact with light in a complex and varied way that includes the processes of absorption, scattering, refraction and diffraction, but it is the light that is reflected by the surfaces in a scene that we use to identify those surfaces by their colour. The presence of a surface effect may account for the small differences in CIE LAB values when comparing inline results with those obtained using the standard colorimetric measurement. Additionally, the difference in translucence between solid and molten parts is believed to further increase this discrepancy. This technique can be applied to monitor both colour change and tolerance.

Acknowledgements

The authors would like to acknowledge the EU funded project 'OPTIJECT' (Project / Grant N° 262623) for financial support (http://www.optiject.eu/).

References

1. Print Root Available: http://www.printroot.com/forum/f13/lab-l**b*-color-spaces-5728/. Last Accessed 10th October 2012
2. Janssen, A. (2007). *Reflective colour measurement Using the Brontes colorimeter and ALS-NW-45 light source*, 2007
3. Wittmann Group, Press Release - *PowerVision in the UNILOG B6 control* (2011)
4. Distrupol UK – Melinar Plus+ Natuaral PET
5. Spectra Masterbatch UK – PET Specific Red
6. ISO 527 – Plastics: Determination of Tensile Properties (2009)
7. ISO 178 – Plastics: Determination of flexural properties (2010)
8. Ocean Optics - WS-1-SL White Reflectance Standard with Spectralon. Avalable: http://www.oceanoptics.com/Products/ws1diffrefstan.asp.
Last Accessed 10th October 2012
9. B. Smith, *Fundamentals of Fourier Transform Infrared Spectroscopy*, CRC press, 2011, Pg. 39.
10. S. Westland; C. Ripamonti, *Computational Colour Science using MATLAB*, John Wiley & Sons LTD, 2004

O-04-431

INVESTIGATION OF SANDWICH STRUCTURE FROM PP BLENDS SKIN AND XHDPE CORE

Thnatiwat Norraprateep[1, 2*] **and Utai Meekum**[1]

[1] *School of Polymer Engineering, Suranaree University of Technology, Nakhon Ratchasima 30000, Thailand*

[2] *Center of Excellence on Petrochemical and Materials Technology, Chulalongkorn University, Bangkok 10330, Thailand;* thnatiwat@gmail.com

Abstract: The interfacial surface adhesion enhancement of PP/xHDPE/PP sandwich structure derived from the PP blend skins was investigated. EPDM, LDPE, HDPE, and UHMWPE were blended with PP copolymer and used as the structural skins. The structural specimen with various volume fractions of cores were manufactured by co-injection molding. Silane and dicumyl peroxide(DCP) system was employed as crosslinking agent for the xHDPE core. The sauna curing was performed to accelerate the silane/moisture condensation reaction. The mechanical properties by mean of impact, tensile and flexural were obtained in according with the standard testing methods. It was found that HDT and flexural properties of the obtained specimens were directly proportional to the xHDPE core fraction. For the impact and tensile properties, it was observed that only for the PP/xHDPE/PP samples where the tested values were closely and directly related to the xHDPE core fraction. Moreover, rule of mixture models could be well fitted with the mentioned properties. Nevertheless, in the sandwich derived from the PP blend skins; PP/EPDM, PP/LDPE, PP/HDPE and PP/UHMWPE, the impact, tensile and flexural properties were far more correlated to the xHDPE core fraction. Also, the ROM models could not be efficiency explained the outcome. The improvement of the interfacial adhesion through the ethylene fragment in the blended polymers and also the silane/moisture reaction via the sauna curing were suspected and used to explain with the high confidentially hypothesis.

Keywords: Sandwich Structure, PP blend skins, Crosslinked HDPE, and Mechanical properties.

Introduction

The combining multiple materials with different properties are used to produce products with combined features that are not feasible using a single resin system. Many applications are in the areas of combining materials for examples moisture resistance, oxygen barrier, film for toughness, improved heat seal ability, improved appearance. Multiple layer sheets are used in construction, food packaging, and medical applications. Blow molding can be used to produce of two layer gas tanks. Wire coating can be coextruded with a foam layer and a solid insulative layer or two layers of insulation for use in power cables. Generally, since some of the resins systems that provide the unique properties are not compatible, adhesive layers between noncompatible resins are typically employed to adhere the structures together. Several adhesives can be used with different resin systems to provide adequate adhesion between layers. However, adhesives are limited for only a couple of the matrix. It could not be used in the other couples.

The melt state fusion of properties can be normally achieved by co-injection molding technique. Not only the interfacial enhancement but also environment friendly production and cost saving to aesthetics and combinations of engineering properties are succeed[1]. In general, polar resins adhere to polar resins and nonpolar resins adhere to nonpolar resins. Polar and nonpolar resins repel each other, providing no adhesion. Thus, the interfacial adhesion should be considered for the combination of two different materials. Several researchers and industrial applications are interested in the sandwich injected products to achieve the cost saving and create high performance materials from market ready materials [1-7]. In this study, introduction of ethylene segments into PP skins by melt blending with ethylene based polymers were explored. The concept of like-dissolve-like was expected at the interfacial adhesion between PP blends skins; PP/EPDM, PP/LDPE, PP/HDPE and PP/UHMWPE, and xHDPE core. Consequently, the improvement of the interfacial interaction through the partial miscible skin and core would be achieved. By this technique, the physical entanglement, dipole-dipole interaction, or hydrogen force would be evidenced. The perfect combined properties of materials could be initiated. The rule of mixture approaches was attempted to predict properties of the injected sandwich structures[8].

Experimental Procedures

Materials

LDPE, LD1902F, HDPE, H5814J and block copolymer PP, P740J, were supplied from SCG Chemical. Ethylene propylene monomer (EPDM) was obtained from Chemical rubber Co., Ltd. The UHMWPE, UH900, was available from Asahi Chemical Industry. The commercial grade reagents Dicumyl peroxide(DCP) and silane(VTMS), Silquest® A 171, were used as coupling/grafting system.

Samples Preparation

The masterbatch of PP/EPDM, PP/LDPE, PP/HDPE, and PP/UHMWPE at 50:50 by weight were prepared using twin screw extruder at 230°C, respectively. It was used as stock material for the PP blends. The PP blend skin was obtained by premixing virgin PP and pre calculated amount of masterbatch and then melt mixing in the CITSE at the temperature profile of 220°C, 225°C, 225°C, 230°C, and 230°C from feed zone to die head, respectively. Solution of VTMS and DCP at 1 and 0.3 phr was dissolved and used to prepare the xHDPE core. HDPE pellet coated with the silane solution was fed and mixed in twin screw mixer at 160°C, 165°C, 165°C, 170°C, and 170°C from feed to die zones, respectively. The sandwich specimens were fabricated using dual barrels co-injection molding machine at identical barrel temperature profile at 160°C, 165°C, 170°C, 175°C and 180°C from feed to nozzle, respectively. The test specimens were obtained from the artificial balanced family mold. The sandwich sample with the core volume fraction of 0.5, 0.7 and 0.9 were successfully produced. The calculated fraction volume was confirmed by cross section image calculation. The injection molded specimens were divided into 2 sets of samples. One was incubated in the sauna oven at 105°C for at least 12 hours to accelerate the completion of crosslink reaction via silane/water condensation. This process hereby called ***sauna treated*** or ***post cured*** specimen. The rest was allowed to anneal at room temperature and it is named as ***original*** sample.

Performance Testing

Heat distortion temperature(HDT) at 455 kPa, tensile, flexural, and notched impact properties were performed in according to ASTM standards. Cox's model, eq. (1) and the basic rule of mixture(ROM) in the both directions, eq. (2) and (3), were attempted to predict the strength(σ_i) and modulus(E_i) of the injected sandwich composites. For prediction of flexural properties, only transverse model was executed.

$$E_{com} = (1 - v_c)E_s + \left[1 - \frac{\left(tanh\frac{\beta l}{2}\right)}{\frac{\beta l}{2}}\right] v_c E_c \quad (1)$$

$$\text{And } \beta = \sqrt{2\pi \frac{G_s}{E_c} A_c \ln \frac{R}{r}}$$

$$E_{com} = v_c E_c + v_s E_s \quad (2)$$

and

$$\frac{1}{E_{com}} = \frac{v_c}{E_c} + \frac{v_s}{E_s} \quad (3)$$

where E_{com}, E_c and E_s are the modulus of the composite, core, and skin, respectively. v_i is denoted as volume fraction. In Cox's model, r is the radius of the fiber, G_s is the shear modulus of skin, R is the center-to-center distance of the fibers, A_c is the area of the fiber and l is the average fiber length, respectively[10]. The calculated values were compared and discussed with respect to the actual measured ones.

Results and Discussion

Thermal property

The HDT of the sandwich specimens at difference skin and volume fractions are summarized in Table 1. Obviously, the HDT of all sandwich samples are higher when they were underwent sauna cured. Either, contribution of increasing in crystallinity through sample annealing or DCP/silane induced crosslinked is hypothesized. Both phenomena would higher the HDT of the sauna cured samples. For the sandwich, it is observed that HDT is decreased with increasing the low HDT xHDPE fraction for both original and cured. Similarly, at the given volume fraction, it is seen that low HDT of the sandwich is directly caused by the use of PP blends having low HDT. Applying Fox's equation, the outcomes indicate that the model values are lower than the measured ones regardless to the type of samples. It suggests that the reinforcing effect from the interfacial adhesion between skin and core would have the greater effect than the miscibility of the blend.

Table 1 HDT of sandwich specimens with various skins and volume fractions.

Samples	Measured HDT(◦C)		Fox's model	
	Original	Cured	Original	Cured
PP	80±0.9	123.0±0.5	80±0.9	123.0±0.5
PP-EP	79.2±0.8	123.0±1.0	79.2±0.8	123.0±1.0
PP-LD	75.3±0.6	123.5±0.5	75.3±0.6	123.5±0.5
PP-HD	77.3±0.6	118.7±0.8	77.3±0.6	118.7±0.8
PP-UH	72.8±0.3	123.5±1.3	72.8±0.3	123.5±1.3
xHDPE	67.7±1.2	97.7±1.3	67.7±1.2	97.7±1.3
PP/xHDPE50/PP	79.5±0.9	124.3±0.6	73.3±0.9	108.9±0.6
PP/xHDPE70/PP	79.0±1.0	122.7±0.6	70.9±1.0	104.1±0.6
PP/xHDPE90/PP	72.3+1.2	114.0±1.0	68.7+1.2	99.7±1.0
PP-EP/xHDPE50/PP-EP	78.0±0.5	117.0±1.0	73.0±0.5	108.9±1.0
PP-EP/xHDPE70/PP-EP	75.8±0.8	110.2±0.8	70.7±0.8	104.1±0.8
PP-EP/xHDPE90/PP-EP	74.3±0.6	105.5±0.5	68.7±0.6	99.7±0.5
PP-LD/xHDPE50/PP-LD	73.5±0.5	116.8±1.0	71.3±0.5	109.1±1.0
PP-LD/xHDPE70/PP-LD	72.2±0.8	109.7±0.6	69.8±0.8	104.2±0.6
PP-LD/xHDPE90/PP-LD	70.7±0.6	106.0±1.0	68.4±0.6	99.8±1.0
PP-HD/xHDPE50/PP-HD	75.5±0.5	106.2±1.1	72.2±0.5	107.1±1.1
PP-HD/xHDPE70/PP-HD	73.5±0.5	104.2±0.8	70.3±0.5	103.1±0.8
PP-HD/xHDPE90/PP-HD	72.0±1.0	102.7±0.6	68.5±1.0	99.4±0.6
PP-UH/xHDPE50/PP-UH	70.5±0.5	116.2±1.3	70.2±0.5	109.1±1.3
PP-UH/xHDPE70/PP-UH	68.7±0.6	110.3±0.6	69.1±0.6	104.2±0.6
PP-UH/xHDPE90/PP-UH	67.2±0.3	106.3±0.6	68.2±0.3	99.8±0.6

Mechanical properties

Figure 1 illustrates the impact strengths of composites derived from PP blends/xHDPE. It is seen that for the PP/xHDPE/PP structure the impact of the sandwiches are directly proportion to the xHDPE core. The rest of samples using PP blend as skin, the impact strengths do not pronouncedly depend on the core strength. Typically, the strength of the sandwiches at high core fraction is higher than the strength of core, especially for the PP/EPDM and PP/UHMWPE skins. Again, interfacial reinforcing would be suspected for this positive effect.

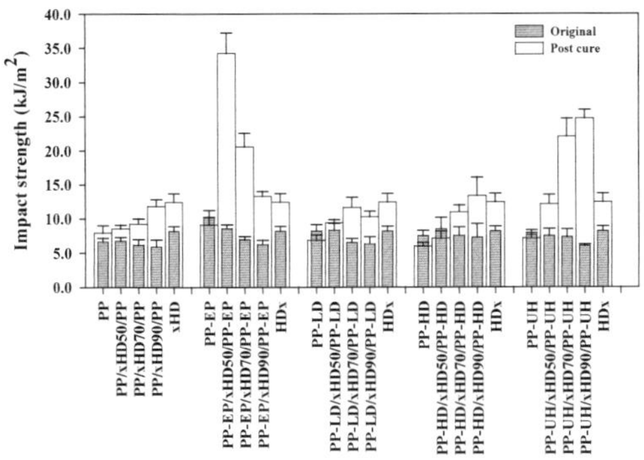

Figure 1 Impact strength of PP/xHDPE/PP and PP blended skins, PP/EPDM, PP/LDPE, PP/HDPE, and PP/UHMPE, sandwich structures; (a) original and (b) sauna cured, respectively

Tensile strength of the sandwich structures with various skins before and after sauna curing is presented in Figure 2(a) and 2(b), respectively. It reveals that the strengths of PP/xHDPE/PP are linearly depended on the higher strength core fractions for both types of sample. Moreover, it is obeyed both ROM and Cox's models. Vice versa, the strength of the sandwich derived from PP blend skins does not closely depend on the faction of core for both original and cured. Also the calculated values are higher than the measured ones especially when the sample was cured. Similarity is found for the tensile modulus shown in Fig. 3. The modulus of the PP/xHDPE/PP is directly proportional to the core and well predicted by ROM for both types of sample. In contrast, the modulus of the sandwiches manufactured from the PP blend skins are not. It reinforces that better adhesion between the PP blends and xHDPE core would have significant contribution to the sandwich composites.

The flexural properties, both strength and modulus, of the sandwich samples are shown in Fig. 4(a) and 4(b), respectively. Generally, the flexural properties are decreased with increasing the xHDPE fraction, regardless to the generic types of skin. Seemingly, the ROM models, both strength and modulus, are well fitted with the tested values for both original and cured. Closer observation, it reviews that the model applied for the sauna cured is better agreement than the original sample. It could state that improving in the surface adhesion through the silane/moisture reaction would be occurred and hence the measurement values are well explained.

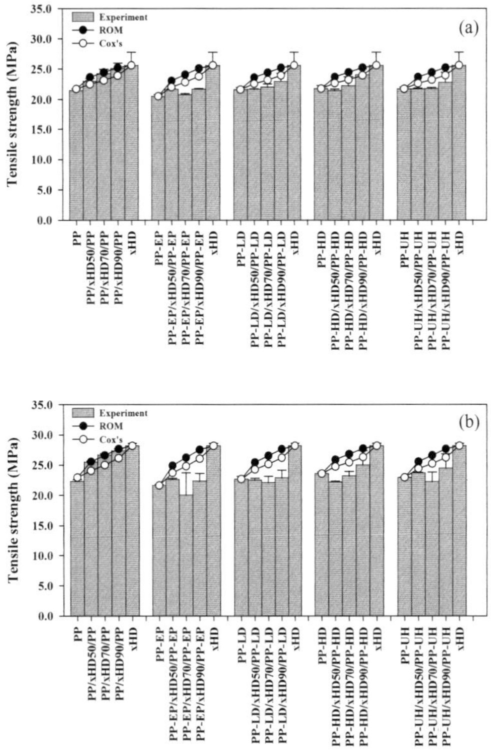

Figure 2 Tensile strength of PP/xHDPE/PP and PP blended skins, PP/EPDM, PP/LDPE, PP/HDPE, and PP/UHMPE, sandwich structures; (a) original and (b) sauna cured, respectively

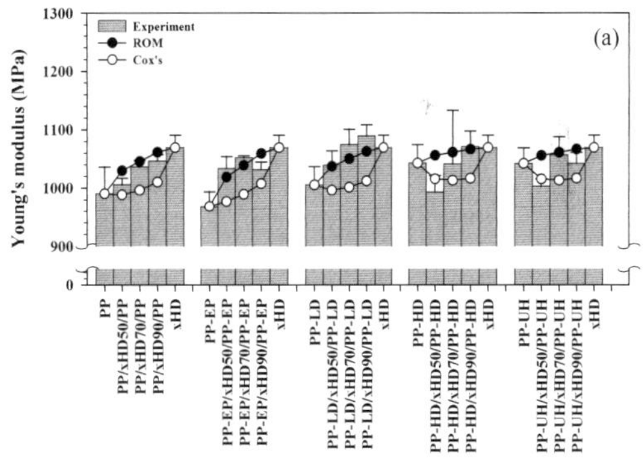

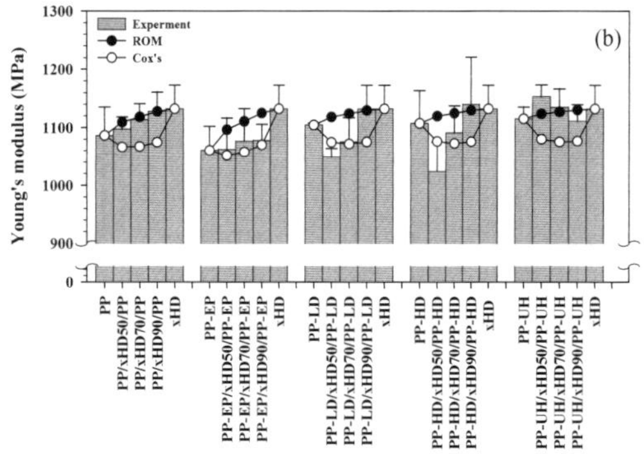

Figure 3 Tensile modulus of PP/xHDPE/PP and PP blended skins, PP/EPDM, PP/LDPE, PP/HDPE, and PP/UHMPE, sandwich structures; (a) original and (b) sauna cured, respectively

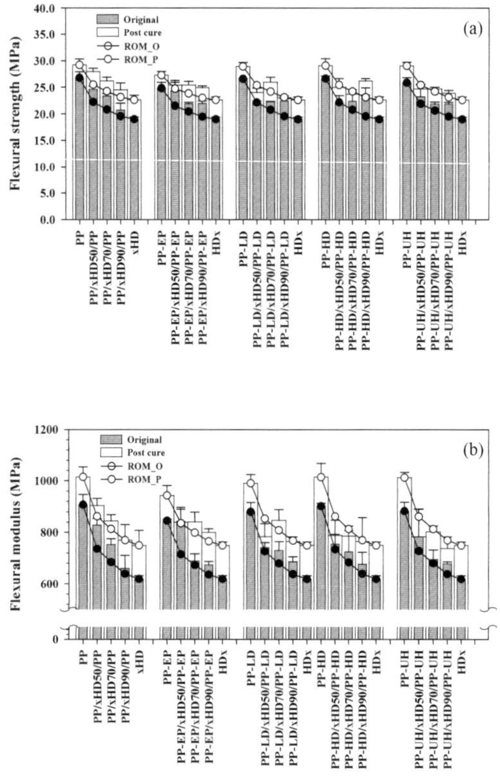

Figure 4 Flexural properties of PP/xHDPE/PP and PP blended skins, PP/EPDM, PP/LDPE, PP/HDPE, and PP/UHMPE, sandwich structures both original and cured; (a) strength and (b) modulus, respectively

Conclusions

The sandwich structures using PP, PP/EPDM, PP/LDPE, PP/HDPE and PP/UHMWPE blends, skins and silane/DCP induced xHDPE core were successfully manufactured by co-injection molding. It was found that HDT and flexural properties of the obtained specimens were directly proportional to the xHDPE core fraction. For the impact and tensile properties, it was seen that only for the PP/xHDPE/PP samples where the tested values were closely related to the core fraction. Moreover, calculation from rule of mixture model was disagreement with the mentioned properties. Nevertheless, in the sandwich derived from the PP blend skins; PP/EPDM, PP/LDPE, PP/HDPE and PP/UHMWPE, the impact, tensile and flexural properties were far more regard to the xHDPE core fraction. Also, the ROM models could not be efficiency explained. The improvement of the interfacial adhesion through the ethylene fragment in the blended polymers and also the silane/moisture reaction via the sauna curing were suspected and used to explain with high confidentially hypothesis. Interfacial investigation by SEM in the coming work is required and it will be used to reinforce the conclusion.

References

1. T. Nagaoka, U.S. Ishiaku, T. Tomari, H. Hamada, and S. Takashima. *Polym. Test.*, 2005, 24, 1062-1070.
2. U.S. Ishiaku, H. Hamada, M. Mizoguchi, W.S. Chow, and Z.A. Mohd Ishak. *Polym. Comp.*, 2005, 26, 52-59.
3. Z.A.M. Ishak, U.S. Ishiaku, and J. Karger-Kocsis, *J. Mater. Sci.*, 1998, 33, 3377.
4. J. Karger-Kocsis, Composites, 1990, 21, 243.
5. J. Karger-Kocsis and K. Friedrich, *Plast. Rubb. Proces. Appl.*, 1989, 12, 63.
6. R. Eggert, R.Walter, H. Hamada, and K. Friedrich, *Advanced Comp. Letters*, 1998, 7, 5.
7. D. Watanabe, U.S. Ishiaku, T. Nagaoka, K. Tomari, and H. Hamada, *Int. Polym. Proces.*, 2003, 18, 199.
8. S. Patcharaphun and G. Mennig. *J. Met., Mat. and Min.*, 2007, 17, 9-16.
9. G. kalaprasad, K. joseph, and S. Thomas. *J. Mat. Sci.*, 1997, 32, 4261-4267.

O-04-442

VISUALIZATION ANALYSIS OF IMBALANCED CAVITY FILLING PHENOMENA IN HOT RUNNER MOLD WITH A VALVE-GATE SYSTEM

R. Takamatsu[1*], Y. Kanetoh[1], H. Yokoi[1]
[1]*Institute of Industrial Science, The University of Tokyo, Japan*
– rtkmt@iis.u-tokyo.ac.jp; ykaneto@iis.u-tokyo.ac.jp; hiyokoi@iis.u-tokyo.ac.jp

Abstract - In hot runner technology, various defects are known to occur, such as stagnation of resins in the manifold and hot runner nozzle, filling imbalance, asymmetrical cavity filling behavior, flow marks, gate marks, etc. However, to date, systematic analysis has rarely been performed on the process and mechanism by which these defects occur. Thus in this study, we visualized uneven flow phenomenon in the cavity of two hot runners with different structures using a visualization mold and a temperature measurement mold that we had designed and developed. In addition, we also clarified the mechanism of void generation by the magnified observation of the area around the valve pin. Using a temperature measurement mold, we measured the temperature distribution of the resin flowing inside the cavity.
Keywords: Hot-runner system, Visualization, Flow behavior, Temperature measurement

Introduction

With hot runner technology, various defects are known to occur, such as stagnation of resins in the manifold and hot runner nozzle, filling imbalance, asymmetrical cavity filling behavior, flow marks, gate marks, etc. However, to date, systematic analysis has rarely been performed on the process and mechanism by which these defects occur. Aiming to clarify the reasons for the generation of various molding defects, in this study we visualized uneven flow phenomenon in the cavity of two hot-runner systems with different structures using a visualization mold and a temperature measurement mold that we had designed and developed. The results are outlined below.

Experimental

Figs.1 and 2 show the visualization mold we designed and developed in this study. In the previous report [1], we discussed and analyzed about the visualization results along the cavity thickness direction using a laser-light-sheet mold with different types of hot-runner system. In the experiments conducted at this time, we installed a prismatic quartz glass at the movable side of the mold to visualize resin flow from the opposite side of the valve gate. At the stationary side of the mold, we installed a hot runner system (TSV4 made by Seiki Corporation) with a 4mm diameter valve gate at two opposing positions along the center line of a 300mm×130mm×2mm rectangular cavity, at a distance of 150mm (Fig.3). We visualized the valve gate I area at the top part of the hot runner system. In this study, we attempted visualization inside the cavity using two hot-runner systems with different resin flow routes (Fig.4). Hot runner (a) is a side-fed type (SF) in which the resin is supplied to the tip of the nozzle from the sprue after crossing the valve pin. Hot runner (b) is of the center-fed type (CF) in which the resin crosses the pin inside the manifold and flows parallel to the pin inside the nozzle. At the movable side of the mold, a pocket is provided at the parting side (indicated by the red box in Fig.2) to enable a visualization or temperature measurement unit to be inserted as required. Fig.6 shows the structure of the temperature measurement unit. As already discussed in the previous report [2], this temperature measurement unit has a cylindrical block installed with sensors which is rotated using a worm gear. The center of the rotated block is aligned to the center of the gate. Small-diameter infrared temperature sensors (EPSSZL, made by Futaba Corporation) are installed 15, 30, 45, 60mm from the center. The injection molding machine used in this study was the ultra-high speed two-color injection molding machine FN4000-9HTN (Nissei Plastic Industrial Co, Ltd.). For the resins, we used PS (PS679 from PS Japan Corporation) and ABS (TOYOLAC 700 from TORAY INDUSTRIES, INC.). The injection rate was set to 40cm^3/s, resin temperature to 220°C, and mold temperature to 45°C.

Results and Discussion

Visualization of the cavity filling phenomena

Using a high speed video camera (MEMRECAM fx K4, made by nac Image Technology Inc.), we visualized the resin flow phenomenon in the cavity. Fig.6 shows the flow pattern inside the visualized area. The elapsed time after the start of injection was set as t[s]. With (a) SF, in the latter half of filling, resin flowing along the center line of the mold becomes slow, and the flow front forms a U-shape. With (b) CF, the flow front shape remains an arc from the start to finish of flow. With SF, the resin meets the pin near the gate. At the start of injection, low temperature resin around the tip of the low temperature pin detours to the back of the pin before spurting out of the gate. This is considered to be the reason for the delayed filling. With CF, the low temperature resin around the tip of the pin flows parallel to the pin and spurts out from the gate. As there are few effects of biased stagnant resin compared to SF, resin flow did not slow down. Next, to clarify the effects of stagnant resin inside the molded product, we conducted injection molding experiments whereby we filled the manifold

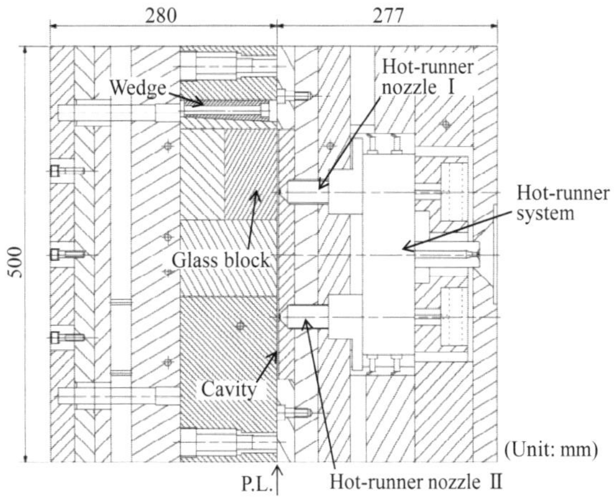

Fig.1 Schematic structure of the experimental mold

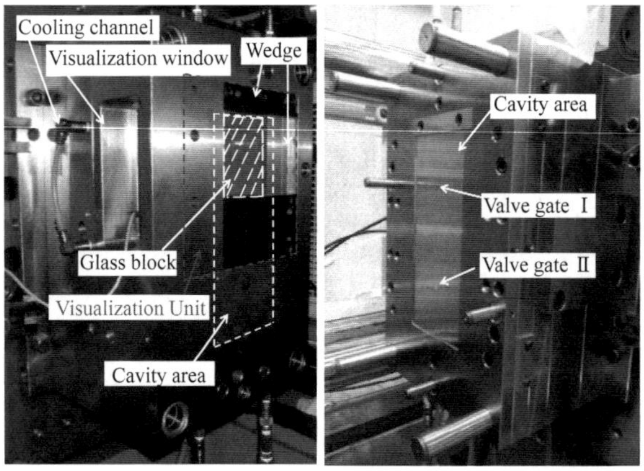

(a) Movable side (b) Stationary side

Fig.2 Appearance of the visualization mold

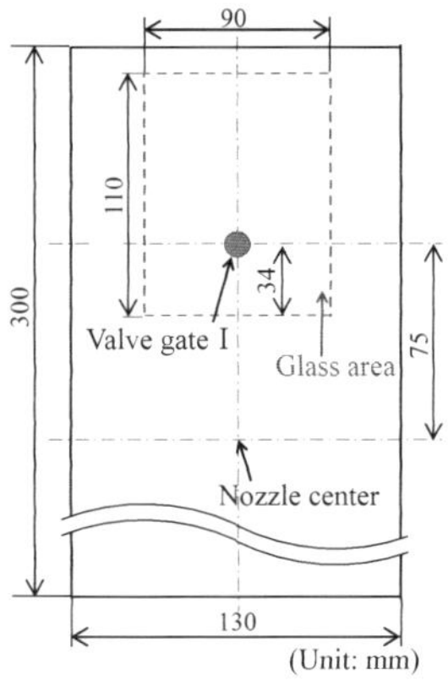

Fig.3 Shape of cavity

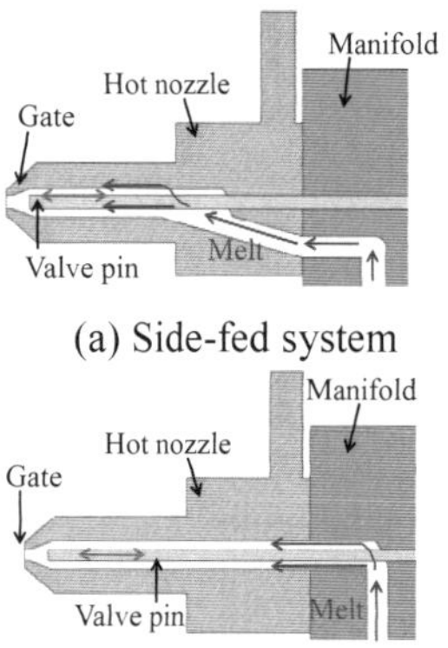

(a) Side-fed system

(b) Center-fed system

Fig.4 Schematic structure of the hot runner

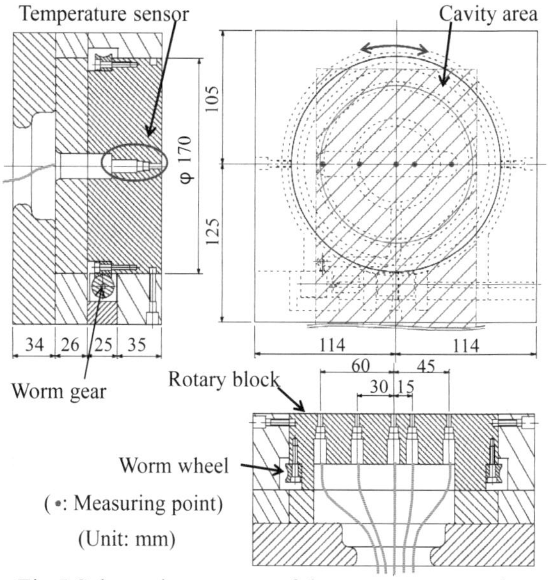

Fig.5 Schematic structure of the measurement unit

with resin, and then switched the resin in the cylinder with resin of different color from the resin in the manifold. Fig.7 shows the process by which resins were switched from transparent to red resins. After moving backwards to the pin, the red resin can be seen to detour to the center of the pin and spurt out from the bottom of the pin. Then this colored resin starts to form a U-shape from both sides of the pin, gradually increasing its area so that the two ends of the U-shape becomes one. The transparent resin (stagnant resin) remains from the pin to the downstream area in the shape of a straight line. Figs.8 and 9 show how stagnant resin (red resin) disappears when the manifold is filled with the red resin and transparent resin continuously injected after this. The samples shown are those of the second to fifth shots after switching. With SF, the stagnant resin forms a vertical line connected to the concave region of the molded product. On the other hand, with CF, immediately after switching at (a), the resin is seen to form a vertical line like SF, but this disappears soon after. Around the gate, the stagnant resin forms a round shape for both SF and CF. This is thought to be due to the stagnant resin present near the tip of the nozzle.

Magnified observation around valve gate

Voids occur around the gate of the molded product as shown in Fig.10, so attempts were made to clarify how these voids are formed by the magnified observation of the area around the gate. The distance between the end of the valve pin and cavity side was set as d[mm] (d=0~11mm). Fig.11 shows how the pin advances after injection. At (a), we can see the singularity that resins are not seen to flow at the center of the gate. However at (b) and (c) where the pin is advancing, several black spots

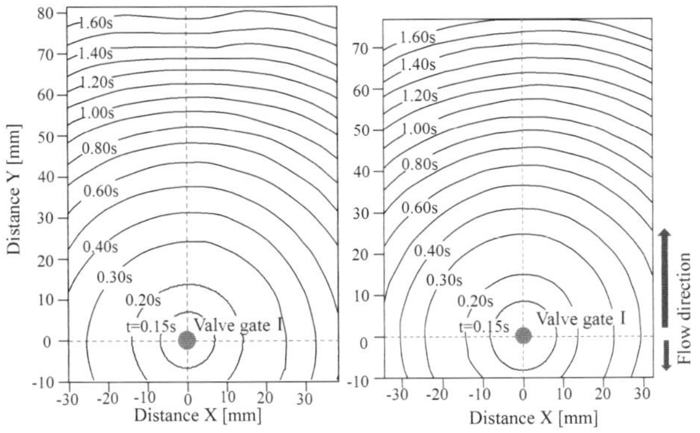

(a) Side-fed system (b) Center-fed system

Fig.6 Cavity filling pattern

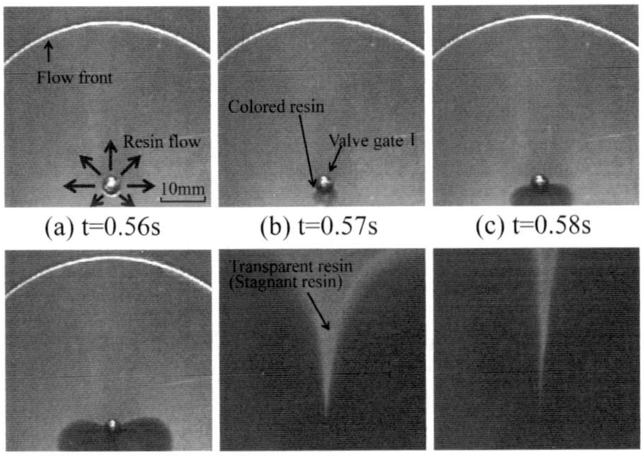

(a) t=0.56s (b) t=0.57s (c) t=0.58s

(d) t=0.60s (e) t=1.00s (f) t=1.40s

Fig.7 Flow behavior around the gate when the resin was changed from transparent resin to colored one

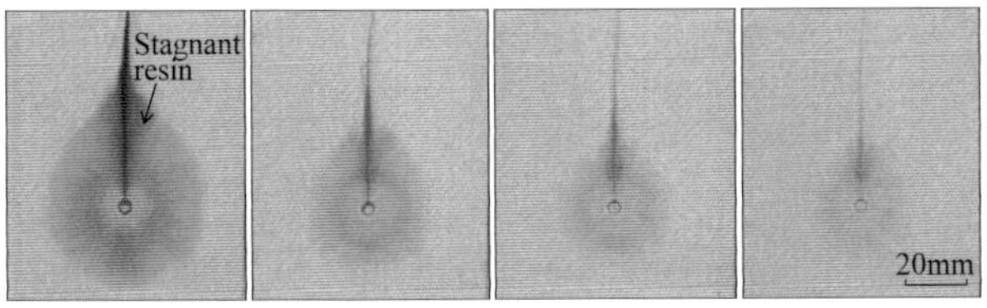

(a) 2nd shot (b) 3rd shot (c) 4th shot (d) 5th shot

Fig.8 Change of colored-resin pattern with shots (Side-fed)

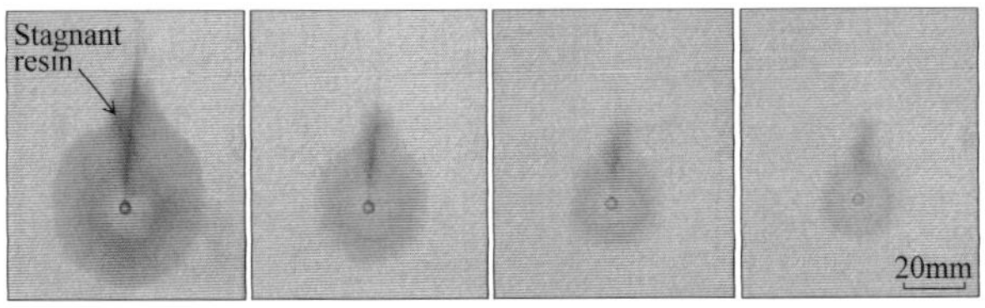

(a) 2nd shot (b) 3rd shot (c) 4th shot (d) 5th shot

Fig.9 Change of colored-resin pattern with shots (Center-fed)

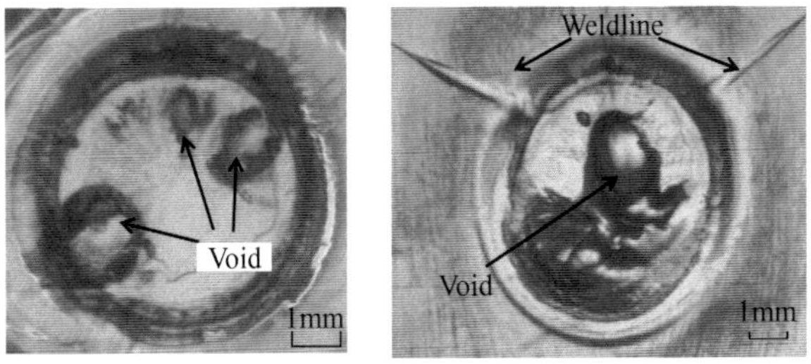

(a) Ordinary condition (b) Lower temperature condition

Fig.10 Magnified observation of molded samples around the valve gate I

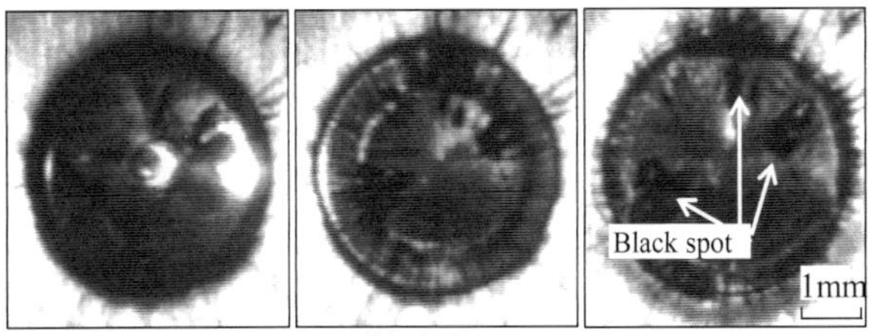

(a) d=11mm (b) d=3mm (c) d=0mm

Fig.11 Visualization images around the valve gate I during the gate closing process

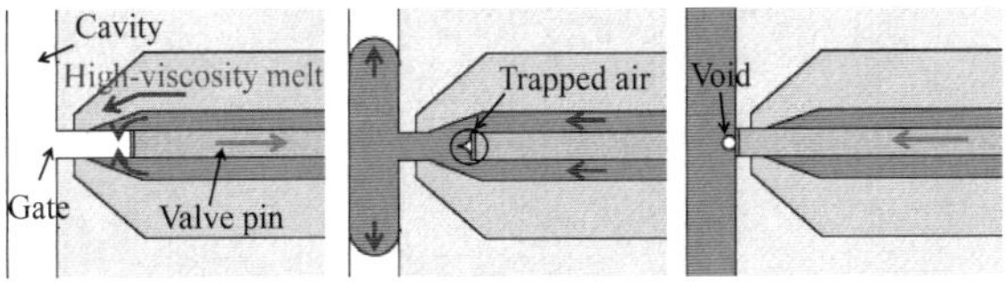

(a) Valve opening (b) During injection (c) Valve closing

Fig.12 Generation mechanism of voids

are seen to be growing. The positions of these black spots correspond to the positions of the voids formed in the molded product. The mechanism by which voids are formed is thought to be as follows; when the valve pin moves backwards, a space is formed at the edge of the pin (Fig.12(a)) . Then the low temperature resin around the exit of the gate flows into this space at the edge of the pin. Because this resin has high viscosity, the space remains and freezes at the edge while trapping air bubbles inside it (b). After injection is completed, these air bubbles are pushed out to the surface of the product together with the resin as the pin moves forward (c), thereby remaining in the product as voids.

Next, at the very low resin temperatures of 190°C, as weldlines are formed as shown in Fig.10(b), magnified observation was carried out under very low temperature conditions. From the flow patterns in Fig.13, as resins spurt out to the top of the gate weakly, resin flow slows down, resulting in the formation of weldlines. Under low resin temperature conditions, even very small temperature changes can result in large resin viscosity

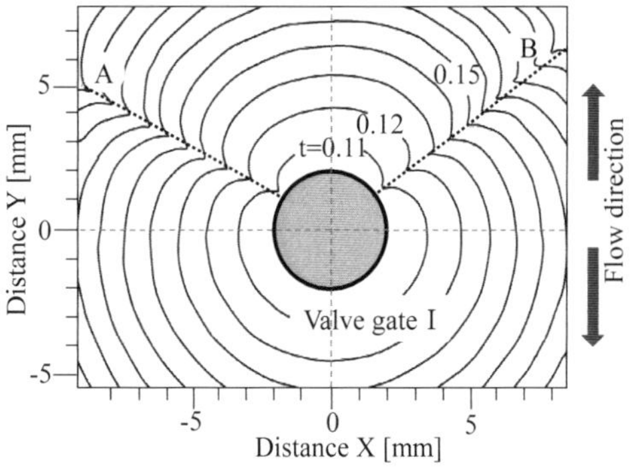

Fig.13 Cavity filling pattern around the valve gate I under a lower temperature condition

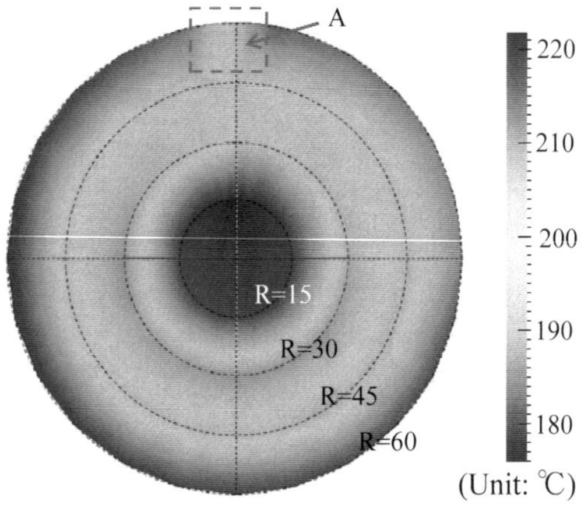

Fig.14 Peak temperature distribution of flowing melt (Resin: ABS)

changes, and this causes the resin flow to slow down or become unstable in the area where the resin spurts out. It is thought that weldlines are formed at the borderline with the resin spurting out from the bottom of the gate as if to surround the stagnant resin. The number of weldlines formed and the direction formed in, depend on how the resin at the tip of the nozzle flows into the cavity, so they differ for each shot.

Measuring temperature distributions inside the cavity

Using the developed temperature measurement mold and black ABS resin, we measured the distribution of temperature at the surface of the resin flowing inside the SF hot runner cavity. Fig.14 shows the mapping of the peak values at each measured point. The resin temperature can be seen to rise as the resin flows from the gate to the outer circumference (R=60mm) of the cavity. This may be because the low temperature stagnant resin near the gate inside the hot nozzle filled with resin before injection started forms a fountain flow near the gate after injection starts, and reaches the surface of the molded product. The temperature measured here was therefore low. After this, the resin at the inner layers of the product flows to the surface layer. However, the longer the flow length, the more shear heating effect will be superimposed, which is thought to be the reason for the higher temperature measured at the downstream area. In areas where the flow was delayed (Fig.14 A), the measured resin temperature is lower than that in the surrounding area. In the case of the SF hot-runner system, the effect of stagnation of low-temperature resin near the gate area becomes more dominant as compared to the effect of temperature rise due to the shear heating when resin flow crosses the pin. As a result, the low-temperature high-viscosity resin is thought to remain locally in the area where the stagnant resin spurts out from the nozzle. An eccentric waving phenomenon of the temperature distribution was also observed in Fig.14, where the temperature at R=45mm was measured lower than those at R=30mm and 60mm. Along the same area where the low-temperature resin appeared on the surface around R=45mm, an eccentric dull flow-mark was generated, suggesting that the above low-temperature melt might cause the dull part of flow-marks during the asymmetric fountain flow process.

Conclusions

(1) In order to clarify the causes of molding defects unique to hot runner molds, we designed and developed an in-cavity visualization mold and a temperature distribution measurement mold.

(2) With the side-fed type hot runner, we observed the formation of the U-shaped flow front. However, because this shape was not seen with the center-fed type hot runner, it is thought to have formed due to the presence of low temperature stagnant resin around the tip of the valve pin.

(3) We clarified the mechanism by which voids are formed at the gate position of molded products through the magnified observation of the valve gate. We also visualized the process by which weldlines are formed at the gate under extremely low temperature conditions.

(4) We measured the temperature distribution of flow resins inside the cavity using the temperature measurement mold, and clarified the relation with the delay in resin flow.

Acknowledgements

The authors would like to thank Seiki Co., Ltd. for providing Hot-runner system. This research was conducted as a subproject of the U'10 & U'11 and U'12 &U'13 project of the Institute of Industrial Science.

References

1. H. Yokoi, K. Nakamichi: JSPP'96 Tech. Papers, p.75~p78, 1996.
2. For proceedings: H. Yokoi, H. Kai in PPS-26 Proceedings, Banff, 2010, File No.G13-332, p7.

O-04-506

INFLUENCE OF HIGH SHEAR RATES ON THE MICROSTRUCTURE OF INJECTION MOLDED POM-PDMS COMPOUNDS

Johanna Aho[1*], Nicolas Javier Alvarez[2], Mogens Papsoee[1], and Ole Hassager[2]

[1] *Novo Nordisk A/S, Brennum Park, 3400 Hilleroed, Denmark – jhaz@novonordisk.com; mpap@novonordisk.com*
[2] *Technical University of Denmark/ Danish Polymer Center, Denmark – nial@dpc.kt.dtu.dk; oh@dpc.kt.dtu.dk*

Abstract

Polyoxymethylene-polydimethylsiloxane (POM-PDMS) compounds can exhibit roughened scale-like topography, and surface delamination can occur when it is subjected to very high shear rates in injection molding. The effect of injection velocity on the final microstructure, surface quality, and mechanical properties of molded POM-PDMS was investigated and compared to a neat POM. Injection molded POM-PDMS compound has a skin with smaller, less oriented PDMS domains and a sub-skin layer with larger, highly extended domains. For the parts molded at high injection velocity, the silicone-rich sub-skin layer is closer to the surface, and can together with the shrinking core pull the layers apart and cause delamination of the skin. The presence of a rough, irregular surface pattern could result from stick-slip behavior on the mold wall due to a high injection velocity. The weld-line strength of neat POM was significantly higher than of POM-PDMS, and for the latter the strength decreased with increasing injection velocity.

Keywords: polyoxymethylene, polydimethylsiloxane, injection molding, microstructure, immiscible blends

Introduction

Low friction between sliding components of drug delivery devices, such as insulin pens, is desired, since it reduces the force needed for dosing the drug [1]. Polyoxymethylene (POM) has good sliding and wear properties, which can, however, further be improved by introducing certain special fillers and additives. Polydimethylsiloxane (PDMS) is commonly added to injection molding thermoplastics in order to reduce the dry friction between parts in the final assembly. However its immiscibility with most thermoplastics poses challenges when compounding, and significantly influences the properties of molded parts. The properties are dictated by the size distribution and shape of the droplets of the dispersed phase in the matrix.

In the compounding and secondary processing, the shear stresses in the continuous phase act to deform the droplets of the dispersed phase, while interfacial tension, σ, acts to keep the droplets spherical, i.e. minimize the surface area. The capillary number, Ca, defined as the ratio of the shear stress τ to the interfacial stress σ/R acting on the spherical droplet (radius R), is the major parameter that quantifies droplet deformation;

$$Ca = \frac{\tau R}{\sigma} \quad (1)$$

the droplets are more likely to be deformed, when the Ca is large.

The critical capillary number Ca_{crit} refers to the point at which deformed droplets break up into smaller ones. When $Ca > Ca_{crit}$, droplets deform affinely but do not break up, since the shear stress dominates the interfacial stress. At the critical capillary number $Ca \approx Ca_{crit}$, the shear and interfacial stresses compete and the elongated threads break up into smaller droplets. When $Ca < Ca_{crit}$, interfacial stress dominates, yielding a relatively stable droplet shape. Ca_{crit} depends on the type of flow (shear or elongation), elasticity of the dispersed and continuous phases, and the ratio between their viscosities, $p = \eta_d/\eta_c$. Droplet break-up predominantly occurs at high strain rates, especially when both shear and extensional deformation are applied ("strong flow"). Another competing process is coalescence, which can occur when small droplets subjected to "weak flow" (soft collisions), have sufficient time to coalesce [2].

Parts molded from POM-PDMS compounds can have impaired visual and mechanical properties when processed under certain conditions. For example, they can exhibit roughened, scale-like topography and delamination of the surface layers. Generally, contamination, moisture, too low mold temperature, incompatible additives, or too high injection rate are suggested as reasons for delamination [3]. For immiscible blends, delamination can occur in the areas subjected to very high shear stress [4], and too high injection velocity has been observed to cause phase coarsening of immiscible components particularly with low viscosity ratio [5].

In this work, injection molding flow simulations, microscopy, rheology and surface science are combined to investigate the microstructure and surface quality of the molded POM-PDMS compound at different injection velocities. In addition, the effect of the injection velocity to the weld-line strength is considered.

Experimental

Materials

A natural-color POM copolymer, polymerized from trioxane and dioxylane, with a solid density of 1.41 g/cm^3 and melt-flow index of 14 g/10 min, melt-blended with a small amount of linear-chain PDMS, was chosen for evaluating the behavior of immiscible blends in injection molding. An unmodified, neat POM copolymer and a PDMS gum (M_w= 550 000 g/ mol; estimated from the zero-shear viscosity) were used for interfacial tension and rheological characterization, and are representative of the individual components in the compound material.

Injection Molding and Simulation

Tensile test specimens and rectangular plaques were injection molded using a 90 ton machine (Husky) with 18 mm screw, and a hot runner mold unit with symmetrical cavity filling through 2 opposing hot runner gates, leading to a welding of the melt fronts in the middle of the part. Injection velocity (V_{inj}) was the only variable with the purpose of producing POM-PDMS parts with remarkable surface distortions (high V_{inj}), subsequently reducing V_{inj} until a good surface was obtained, yet having a proper welding of the melt fronts. As a reference, the same moldings were performed with the neat POM as well.

For the tensile test specimen mold, V_{inj} *(high)* = 60 mm/s and V_{inj} *(low)* is a profiled injection with a switch-over from 40 to 15 mm/s at a screw position of 25 mm. For the plaque mold, V_{inj} *(high)* = 55 mm/s and V_{inj} *(low)* = 16 mm/s.

For estimating the shear rates within the cavity, injection molding simulations were performed using 3D finite element method (Autodesk Simulation Moldflow Insight 2013) with the same process settings and mold geometry as in the molding process described above.

Material Characterization

A parallel-plate oscillating rheometer (AR2000Ex, TA Instruments) under nitrogen purge gas was used to measure the complex viscosity of the POM-PDMS compound, neat POM and PDMS.

Interfacial tension between PDMS and POM was measured by the pendant drop method at 200 °C using a DataPhysics OCA Contact Angle device. The analysis was done using a low Bond number algorithm that is suitable for nearly spherical drop shapes [6].

The PDMS droplet orientation, shape and size were analyzed from the cryo-microtomed cross-sections of the molded plaques at the weld-line by SEM (Inspect S, FEI). Energy-dispersive spectrometry (EDS) mapping was done for tracing the presence of PDMS domains. In addition, molded plaque surfaces were inspected by a Leica M60 stereomicroscope.

The weld-line strength was tested in a tensile tester (Lloyd LRX Plus) at ambient temperature using the cross-head speed of 100 mm/min.

Results and Discussion

Viscosity and viscosity ratio

Complex viscosity of POM-PDMS, of both the components, and their viscosity ratio are shown in Fig. 1. Carreau-Yasuda viscosity fitting is done to extrapolate to higher shear rates. The upwards curvature, seen for the POM-PDMS compound at low frequency is due to the secondary plateau in the elastic modulus G', and originates from the elasticity of the droplet-matrix interface which is well-recognized for immiscible blends [7].

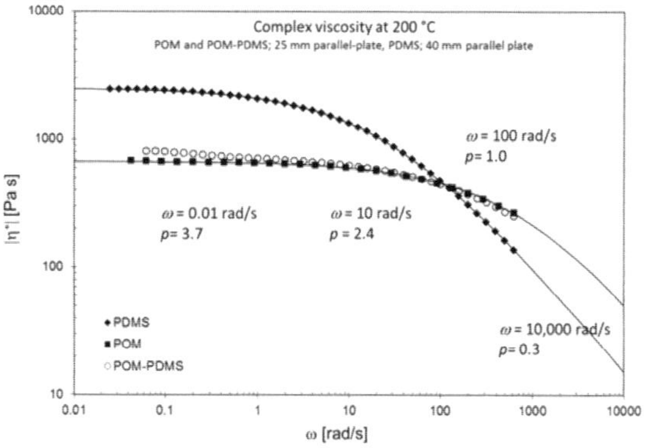

Figure 1. Measured complex viscosity (symbols) and Carreau-Yasuda fit (lines), and the viscosity ratio at different shear rates.

Fig. 1 shows that the PDMS/POM viscosity ratio varies from 0.3 at high frequency to nearly 4 at low frequency, being relatively low especially at the region corresponding to high shear rates typical in injection molding. Note that a low viscosity ratio can promote droplet coalescence [8].

Estimation of the capillary number

The melt experiences the highest shear stress and rate at the gate area when injected into the mold cavity. At the gate, the flow is a mixture of shear and elongation, leading to significant stretching of the PDMS droplets. Therefore, the capillary numbers determined at the gate are compared in order to correlate the role of droplet deformation to the surface roughness and delamination seen in parts injected at high velocity.

Ca was calculated using the gate shear stress values predicted by flow simulations, interfacial tension σ = 12 mN/m, measured as described earlier, and the initial PDMS domain size, estimated from SEM images of cryo-microtomed POM-PDMS granules. The domain radius varied approximately from 0.5 to 5 µm.

The advantage of using the shear stress values predicted by the flow simulations is that the effect of shear heating in the hot runner will be taken into account: At the low injection velocity, V_{inj}= 16 mm/s, the modeled maximum shear rate at the gate, $\dot{\gamma}_{s\,max}$ = 21 228 s^{-1} and at V_{inj}= 55 mm/s, $\dot{\gamma}_{s\,max}$ = 40 132 s^{-1}. The simulations showed a shear-induced maximum increase in the melt temperature of +10.8 °C from the set T=200 °C for V_{inj}= 16 mm/s, and +19.6 °C for V_{inj}= 55 mm/s. Therefore the shear stresses are $\tau_{max} \approx$ 622 000 Pa and 395 000 Pa, respectively. Taking the domain size variation into account, *Ca*(V_{inj}=16 mm/s) ≈ 26…260, and *Ca*(V_{inj}= 55 mm/s) ≈ 16…160. Note that because of the increase in temperature generated by shear-induced heating, *Ca* at the gate is lower for V_{inj}= 55 mm/s. This implies that there is a lower degree of droplet deformation at higher injection velocity. Furthermore, since the *Ca* is directly proportional to the dispersed domain size, the break-up of the droplets in the dispersed phase is highly dependent on the size of the droplets. Since the initial size of the droplets is not controlled, it can vary from compound to compound, and the difference in droplet deformation between the high and low injection velocity is not likely the main reason for the observed surface roughness and delamination. The microstructure analysis shows that coalescence of droplets must also play a role.

Visual observations and microstructure analysis

PDMS can be distinguished as elongated, needle-like domains in cryo-microtomed sections (Figs. 2-3). Figs 2(a) and (b) show a cross-section of the part at a short distance from the weld-line; the frozen-skin layer with oriented, small disperse domains, the sub-skin with highly stretched larger domains, and the less oriented core, as described in [9]. The EDS mapping proved that the lighter patches surrounding the oriented domains are PDMS bled onto the surface after microtoming.

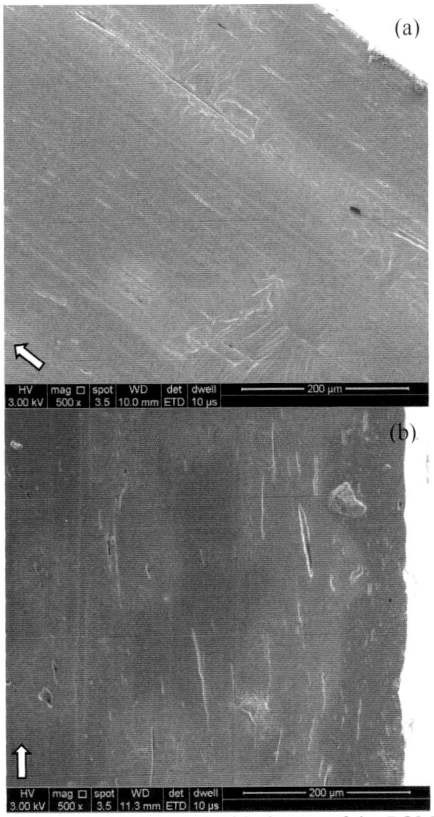

Figure 2. Skin and sub-skin layers of the POM-PDMS cross-section, (a) V_{inj}= 16 mm/s, (b) V_{inj}= 55 mm/s. The arrows indicate the flow direction in the cavity.

The pattern at the weld-line in Fig. 3(a) shows clearly stretched PDMS domains, the largest and most stretched ones being in the sub-skin layer where highest amount of shear deformation is present. According to [9], the skin region is created by the fountain-flow where the extensional deformation dominates, whereas the orientation in the sub-skin layer originates from the shear flow. The observed large PDMS domains suggest that some degree of coalescence has occurred in the sub-skin layer. Due to the strong extensional deformation at the flow front, break-up of the droplets may additionally contribute to the domain size gradient between the skin and the sub-skin layers [9], as well to the coexistence of both small, and very large, highly deformed domains in the sub-skin [10]. Fig. 3(b) shows a long gap formed at the weld-line and a wide irregular blotch of PDMS (shown by EDS) around it. Evidently, the PDMS has spread on the surface from the long, narrow closure that was exposed by microtome cutting. Such a large domain of the dispersed phase could be a result of coalescence at some stage in the cavity. Large amount of PDMS concentrated on the weld-line definitely has an effect on the fusion of the flow fronts, as will be proven later.

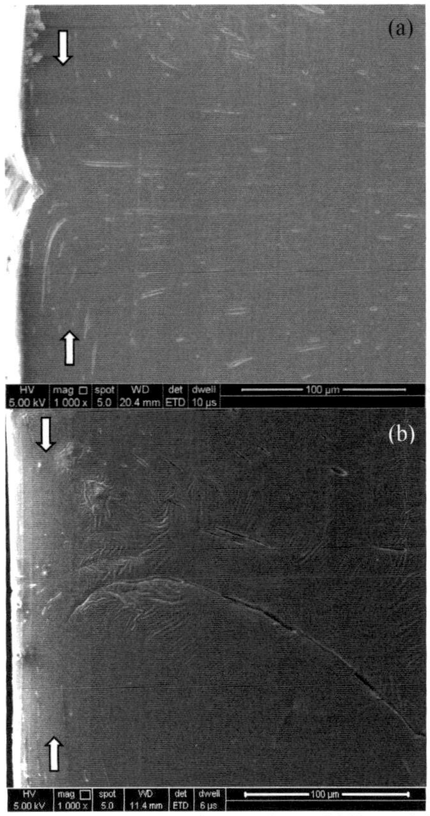

Figure 3. Cross-section at the weld-line of the POM-PDMS (a) V_{inj}=16 mm/s, (b) V_{inj}=55mm/s. The arrows indicate the flow direction in the cavity.

As generally known, with fast mold filling, the frozen skin layer is thinner, meaning in this case that the silicone-rich sub-skin layer will be closer to the surface at the higher injection velocity. Moreover, polymers with high crystallinity such as POM, shrink remarkably while cooling, which can cause the separation of the sub-skin and skin layers, and show as delamination of the surface layers.

Optical stereo microscopy reveals a clear difference in topography between POM-PDMS plaques molded with the low and high injection velocity. For V_{inj} = 16 mm/s, the surface looks relatively smooth to the naked eye, and the microscope shows a rather even pattern of slightly stretched pores (Fig. 4a). However, plaques injected at V_{inj} = 55 mm/s have a very irregular surface pattern: The randomly oriented, rough, turbulent macroscopic domains alternate with smooth surface area (Fig. 4b), which could be a result of alternating occurrence of melt sticking and slipping on the cavity wall, in a similar way as discussed in [11].

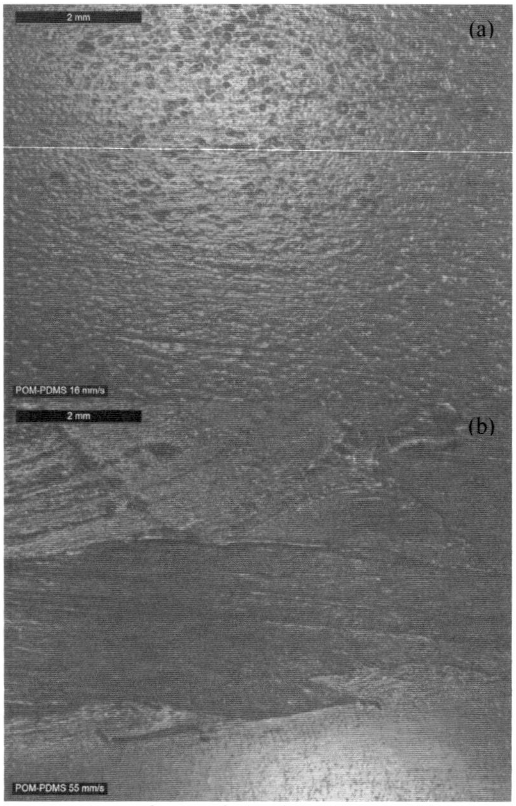

Figure 4. Surfaces of the injection molded POM-PDMS plaques, (a) V_{inj}=16 mm/s, (b) POM, V_{inj}= 55 mm/s

Weld- line strength

The effect of injection velocity on the weld-line strength is material-dependent: For unfilled polymers, higher V_{inj} can improve the weld-line strength due to the higher temperature of the melt flow front which leads to thinner frozen skin layer, and to more efficient fusion of the meeting flow fronts. However, in fiber-reinforced materials increasing the V_{inj} can decrease the weld-line strength due to higher fiber orientation, which both increases the relative bulk strength, and decreases the fusion of the flow fronts [12]. Analogous to this, for immiscible blends with low viscosity ratio the dispersed domains are largely oriented and stretched, thus aligning perpendicularly at the weld-line and leading to weaker flow-front welding [13].

Maximum tensile stress for POM and POM-PDMS with the standard deviation (each with 20 parallel tests) at different injection speeds are shown in the Fig. 5. A clearly weaker weld-line for the high V_{inj} than for low V_{inj} is obtained, even though the PDMS content is very low. Apparently with a dispersed phase of high surface-lubricating capacity – like PDMS – a very low amount is enough to make a difference. The weld-line strength of the neat POM shows the opposite trend with an almost insignificant difference between the velocities. The scatter for the POM-PDMS samples is clearly higher, most likely due to the random distribution of larger extended domains.

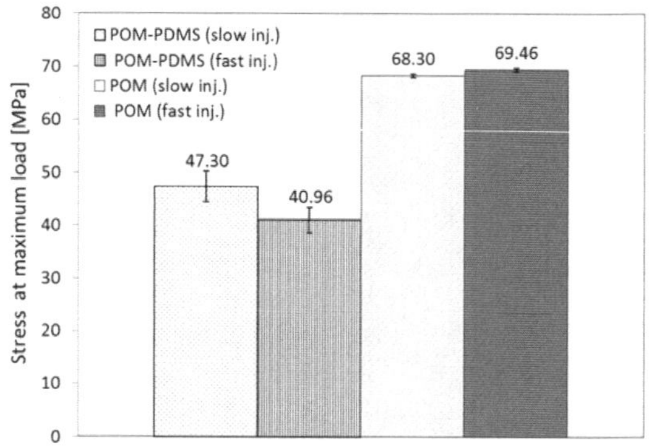

Figure 5. Weld-line strength measured as maximum tensile stress for POM and POM-PDMS molded at high and low injection velocity.

Our results suggest that higher injection velocity can weaken the weld-line, perhaps partly because of large PDMS domains formed by coalescence, partly due to the alignment of the domains subjected to stronger stretching at high injection velocity. However, reaching the best weld-line strength requires balancing between the effects of dispersed domain orientation at high velocity and the cooling flow front/ skin formation at low velocity.

Conclusions

We compared the effect of injection velocity on the final microstructure, surface quality, and mechanical properties of molded POM-PDMS compound and a neat POM. High capillary number at the gate predicts approximately equal extent of PDMS droplet deformation in the POM-PDMS at both studied injection velocities. Also micrographs revealed a skin with smaller, less oriented PDMS domains and larger, highly extended domains in the sub-skin layer at both velocities. For high V_{inj}, the highly oriented, silicone-rich sub-skin layer is closer to the surface due to the thinner frozen skin. Shrinking of the core may pull the layers apart and cause delamination of the skin. The irregular surface pattern of POM-PDMS at high V_{inj} suggests stick-slip behavior on the mold wall. Perpendicular alignment of the PDMS domains at the flow front causes weakening of the weld-line compared to the neat POM. Significantly large extended PDMS domains were found at the weld-line for the high V_{inj}, and its weld-line strength was lower than that for the low V_{inj}.

Acknowledgements

The authors wish to thank the following persons: Regin Jensen, Amine A. Khaled, Rasmus P. Knudsen, Christoffer Raad, Lars Schulte, and Torben Storgaard-Larsen. Novo Nordisk STAR Program and the Danish Research Council are acknowledged for the financial support.

References

1. J.L. Laursen; I.M. Sivebaek; L.W. Christoffersen; M. Papsoee; M.E. Vigild; P. Bronsted; A. Horsewell *Wear* 2009, *267*, 2294.
2. H.E.H. Meijer; M.H.M. Janssen; P.D. Andersson in *Mixing and Compounding of Polymers – Theory and Practice*, I. Manas-Zloczower, ed.; Hanser Publishers, Munich, 2009; 41-177.
3. D. Frenkler; H. Zawistowski. *Hot Runners in Injection Moulds,* RAPRA Technology Ltd., 2001.
4. M. Sepe in *Injection Molding Handbook*, T.A. Osswald; L.-S. Turng; P. Gramann, eds.; Carl Hanser Verlag, Munich, 2008; 645-754.
5. J. Karger-Kocsis; I. Csikai *Polym. Eng. Sci.* 1987, *27*, 241.
6. N.J. Alvarez, L.M. Walker, S.L. Anna. *J. Colloid Interface Sci.* 2009, *333*, 557.
7. M. Gabriele; R. Pasquino; N. Grizzuti. *Macromol. Mat. Eng.* 2011, *296*, 163.
8. A. Maani; M.-C. Heuzey; P.J. Carreau. *Rheol. Acta* 2011, *50*, 881.
9. S. Fellahi; B.D. Favis; B. Fisa. *Polymer* 1996, *37*, 2615.
10. Y. Son; K.H. Ahn; K. Char. *Pol. Eng. Sci.* 2000, *40*, 1376.
11. M.C.O. Chang. *Intern. Polymer Processing* 1996, *XI*, 76.
12. R. Seldén. *Pol. Eng. Sci.* 1997, *37*, 205.
13. S. Guo; A. Ait-Kadi. *J. Appl. Pol. Sci.* 2002, *84*, 1856.

O-04-539

THE EFFECTS OF GAS COUNTER PRESSURE COMBINED WITH DYNAMIC MOLD TEMPERATURE CONTROL ON THE SKIN LAYER THICKNESS AND FOAMING CELL SIZE OF MICROCELLULAR INJECTION MOLDING

Shia-Chung Chen*, Jen-An Chang, Chia-Yen Tseng

Department of Mechanical Engineering, Chung Yuan Christian University, Taiwan, R.O.C.
R & D Center for Membrane Technology, Chung Yuan Christian University, Taiwan, R.O.C.
R&D Center for Mold and Molding Technology, Chung Yuan Christian University, Taiwan, R.O.C.

Abstract - In this study, we developed a foaming control system using the Gas Counter Pressure (GCP) combined with Dynamic mold temperature control (DMTC) during the microcellular injection molding process and investigated its influence on the parts' surface quality, skin layer thickness and foams structures uniformity. The results show that under GCP control, part surface roughness for transparent polystyrene (PS) improved by 90%. When GCP increased, the skin thickness also increased, the weight reduction decreased. Applying mold temperature control alone with temperature in the high temperature near Tg, the surface roughness improved. Increasing mold temperature decreased the skin thickness; however, the cell size distribution became significantly non-uniform. It was found that fast cooling speed is necessary for controlling the cell size and uniformity, alone with 10 °C/sec cooling rate, the average cell size decreased from 192.92 μm to 84.97 μm and uniformity was improved. We conclude that skin thickness, uniform cell size as well as good surface quality could be achieved efficiently by combining of GCP and DMTC.

Keywords: microcellular injection molding; gas counter pressure; mold temperature control; surface roughness; cell size.

***Corresponding Author:** Shia-Chung Chen (E-mail: shiachun@cycu.edu.tw)

Introduction

Microcellular foaming technology of polymers with carbon dioxide or nitrogen used as physical blowing agent was first developed by Martini [1]. A supercritical fluids (SCF) was used in the microcellular injection molding process have characteristics of gas-like diffusivity, low viscosity, and liquid-like density. This implies that microcellular foams offers many advantages, such as warpage minimization, light weight, lower melt viscosity, and elimination of sink marks [2-7]. Despite these advantages, however, the most difficult issue that limits microcellular injection molding wide application is surface defects. Several methods have been proposed to improve the surface roughness, including control of SCF content [8], material modification, In-Mold Decoration (IMD) process, and co-injection molding process [9]. Recently dynamic mold temperature control [10-12] and gas counter pressure (GCP) process [13-18] that have been reported to effectively improve the part surface roughness.

Earlier studies on foaming control follow either by varying temperature alone [7-9] or by varying pressure alone [10-15] independently. Our study proposed that an approach of simultaneous control of both pressure and temperature in foaming process. Therefore, a GCP combined with mold temperature system was established.

Experimental

An Arburg 420C MuCell injection molding machine was used to control the process and the products structures. A GCP gas control unit installed with a high frequency gas control valve connected to the mold was used to adjust and monitor the real time gas pressure inside the mold cavity, and dynamic mold temperature was controlled by two mold temperature control units, as shown in Fig. 1. The specimen dimensions are 100 mm length x 30 mm width x 3 mm thickness. A polystyrene resin (POLYREX PG-33, CHIMEI Chemicals) was used in this study. Nitrogen was used as the SCF blowing agent and GCP gas source. Relevant molding conditions of MuCell with GCP combined with the mold temperature control are presented in Table 1. More detailed system stability, sequence of the process operating and cavity gas pressure monitoring can be seen in our earlier study [18, 19]. Surface roughness was clearly revealed by 3D laser microscope (Keyence VK8550). Morphology of foamed cell were examined by Scanning Electron

Microscope (SEM), Fig. 2, from which solid skin layer thickness, averaged foamed size as well as cell density can be calculated.

Results and Discussion

The comparisons of surface roughness of foaming parts under mold temperature control as well as GCP combined with mold temperature control were shown in Fig. 3. From the results, it can be found that with the condition of mold temperature control alone, the surface roughness can be improved by over 65% from varied in the range of 90~120°C (near Tg). Further, when GCP combined with mold temperature control, the surface roughness can achieve an improving ratio over 90%.

According to the quantitative results and comprehensive comparison of foaming quality such as skin thickness, weight reduction, were shown in Fig. 4 and Fig. 5. The skin thickness increases with the longer GCP holding time, but decreases with higher mold temperature. As a result of skin layer thickness, GCP holding time affects the weight reduction ratio lightly but the mold temperature. On the other hand, the high mold temperature implies that will extend the bubble growing time and cause the cell size non-uniform distributed severely.

In order to improve foam quality, the efficient cooling system was designed. It was found that fast cooling speed is necessary for controlling the cell size and uniformity, as shown in Fig. 6(a) and Fig. 7, alone with 10 °C/sec cooling rate, the average cell size decreased from 192.92 μm to 84.97 μm. The addition of 100 bar GCP and 3 sec holding time decreased the cell size from 73.68 μm to 38.85μm, as shown in Fig. 6(b) and Fig. 8. Combined fast cooling and GCP control improves foam quality efficiently.

Conclusions

In this study, a new method of P-T path control was developed for improving foaming qualities. At first, as the cooling rate is 1°C/s and GCP is 100 bar, effects of various GCP holding time and mold temperature were investigated on their influence on the foaming situation and part surface appearance. It was found that GCP with DMTC can restrict foaming behavior at melt front leading to surface quality improvement, furthermore, increasing mold cooling rate results in smaller foamed cell size and evenly distributed. In a combined GCP and temperature control process, it achieves low skin thickness, more uniform and smaller cell sizes and higher cell density.

Acknowledgements

This research was supported by The Center-of-Excellence Program on Membrane Technology from the Ministry of Education and the R&D Center for Mold and Molding Technology.

Table 1. Parameters of MuCell process

Material	PS
Injection velocity (mm/s)	5
Melt temp. (°C)	210
SCF level (wt%)	0.5
SCF flow rate (Kg/hour)	0.32
Conditions of GCP	
GCP (bar)	100
Gas holding time (s)	0, 3, 5
Conditions of mold temperature control	
Mold temp. (°C)	60, 90, 120
Cooling water temp. (°C)	8
Cooling Rate (°C/s)	1.1, 5.1, 10.9

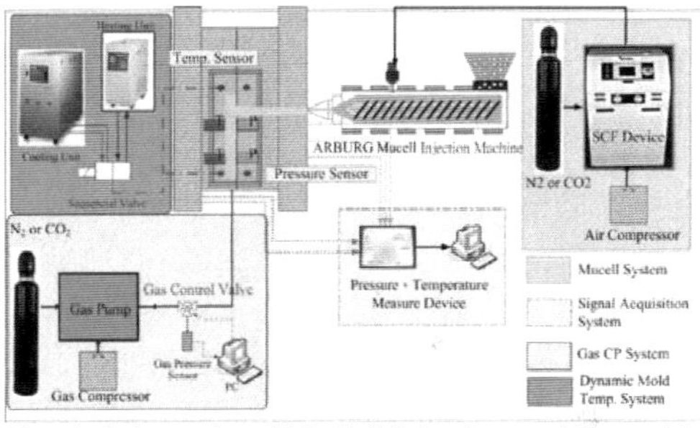

Fig. 1. Schematic of MuCell injection molding with GCP and mold temperature control system.

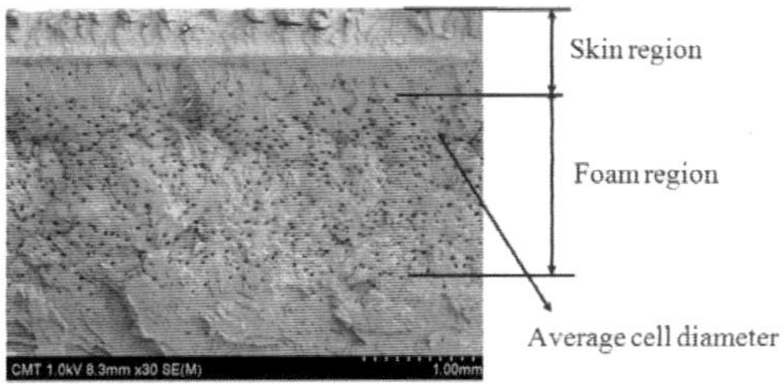

Fig. 2. A typical SEM micrograph of a foamed sample cross section at GCP process

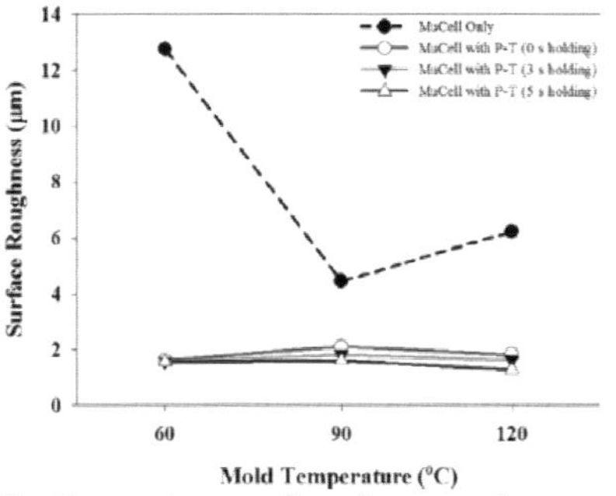

Fig. 3. Comparisons of surface roughness under various GCP holding time

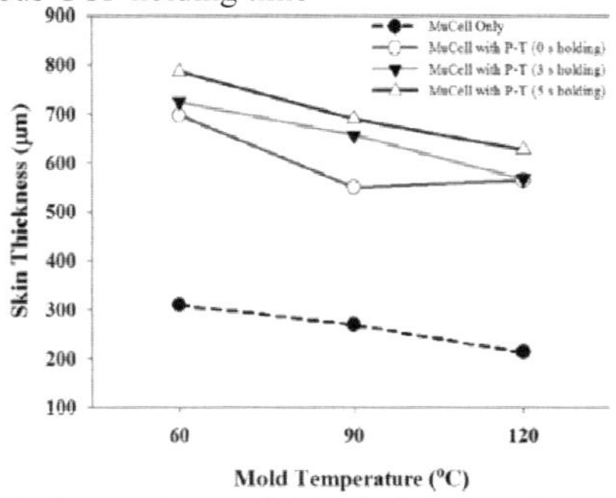

Fig. 4. Comparisons of skin thickness under various GCP holding time

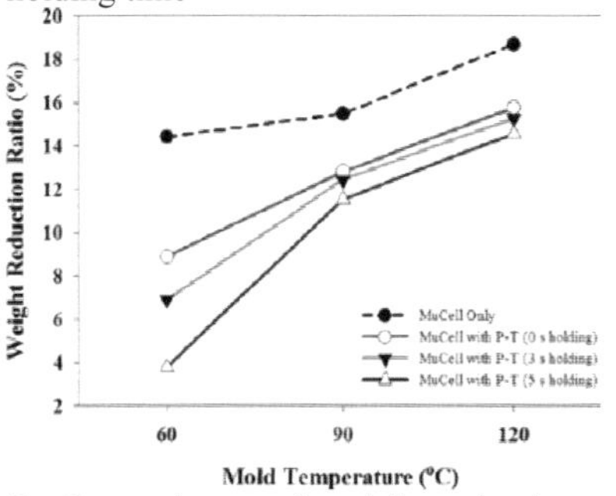

Fig. 5. Comparisons of weight reduction under various GCP holding time

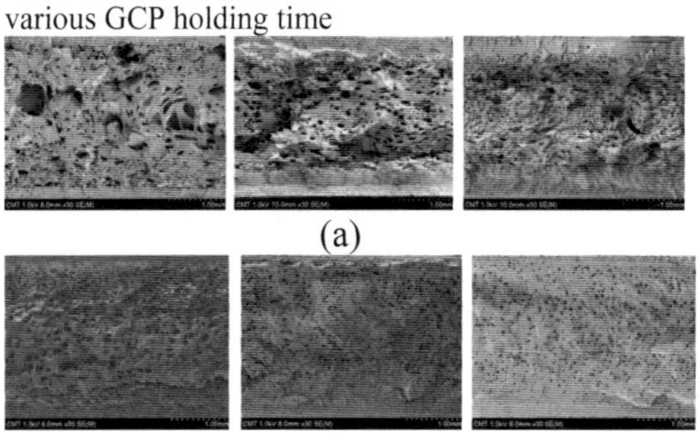

(a)

(b)

Fig. 6. Comparisons of foaming morphology under various cooling speed at 120 °C (left to right is 1.1, 5.1, 10.9 °C/s) (a) without GCP (b) 100 bar GCP

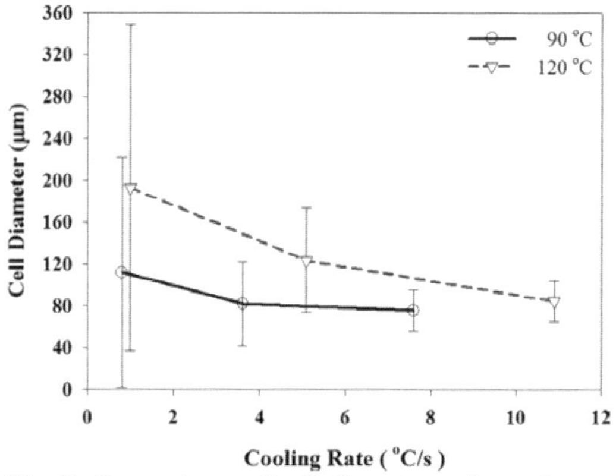

Fig. 7. Comparisons of cell diameter under various cooling speed (without GCP)

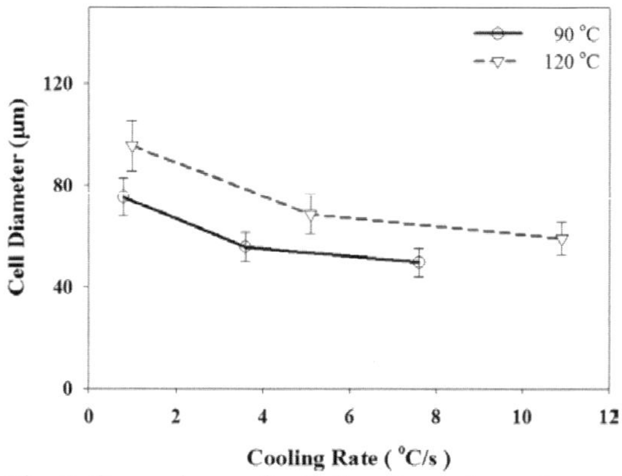

Fig. 8. Comparisons of cell diameter under various cooling speed (with GCP 100 bar)

References

1. J. E. Martini; F. A. Waldman; N. P. Suh *SPE ANTEC Tech. Papers*, 1982, *28*, 674.
2. J. Xu,; D. J Pierick *J. Inject. Mold. Tech.* 2001, *5*, 152.
3. W. Gong; J. Gao; M. Jiang; L. He; J. Yu; J. Zhu *J. Appl. Polym. Sci.* 2011, *122*, 2907.
4. W. Kaewmesri; P. Rachtanapun; J. Pumchusak *J. Appl. Polym. Sci.* 2008, *107*, 63.
5. H. X. Huang; J. K Wang *J. Appl. Polym. Sci.* 2007, *106*, 505.
6. D. I. Goel; D. G. Baird *Polym. Eng. Sci.* 1995, *35*, 1167.
7. D. I. Collias; D. G. Baird *Polym. Eng. Sci.* 1995, *35*, 1167.
8. J. Lee; L. S. Turng *Polym. Eng. Sci.* 2010, *50*, 1281.
9. L. S. Turng; H. Khabas *Int. Polym. Process* 2004, *19*, 77.
10. H. L. Chen; R. D. Chien; S. C. Chen *Int. Commun. Heat Mass Transf.* 2008, *35*, 991.
11. S. C. Chen; Y. W. Lin; R. D. Chien; H. M. Li

Adv. Polym. Technol. 2008, *27*, 224.
12. S. W. Cha; J. D.Yoon; *Polym-Plast Technol. Eng.* 2005, *44, 795.*
13. S. Semerdjiev, *Thermoplastische Strukturschaumstoffe*, VEB Deutscher Verlag für Grundstoffindustrie, Leipzig, 1980.
14. A. K. Bledzki; H. Kirschling; G. Steinbichler; P. Egger in 21st Annual Meeting of the Polymer Processing Society, Leipzig, 2005.
15. W. Michaeli; D. Opfermann; *SPE ANTEC Tech. Papers*, 2006, *64*, 1290.
16. J. D. Yoon; S. K. Hing; J. H. Kim; S. W. Cha *Cellular Polym.* 2004, *23*, 39.
17. A. K. Bledzki; H. Kirschling; G. Steinbichler *P. Egger, J. Cell Plast.* 2004, *40*, 489.
18. S. C. Chen; P. S. Hsu; Y. W. Lin *Int. Polym. Process* 2011, *26*, 275.
19. S. C. Chen; P. S. Hsu; S. S. Hwang *J. Appl. Polym. Sci.* 2012, 1.

STUDY ON RESIN FLOW OF MICRO INJECTION MOLDING MACHINE

Akira Inoue[1], Masayoshi Nomoto[1], Tatsuya Tanaka[1] and Yoshihiko Arao[1].
[1]*Applied Materials Engineering Laboratory, Doshisha University Kyoto Japan*

Abstract
In recent years, micro molding machine becomes to be much in demand because lots of small hospital-use plastic products are made by micro molding with development of micro molding technology.
One of micro molding methods is injection molding. Injection molding method can mold micro component with high precision and it has good production efficiency. However, when micro components are made by current common injection molding machine, it causes disadvantages such as waste of resins, waste of power, etc. Therefore downsized injection molding machine is developed. There is micro injection molding machine (SHINKO SELLBIC CO.LTD) that weighting less than one hundred kilogram. This micro injection molding is plunger injection machine and it can mold product of 0.01 g with general-purpose pellet. This micro injection molding machine has particular kind of screw shape that is circular disk. Screw shape of this micro injection molding machine is different from screw shape of current common injection molding machines. Therefore, resin melting mechanism and resin mixing mechanism for disk type of screw are not well known. This study presents resin flow of this micro injection molding machine when the resins are plasticized. Experiment and computer simulation were performed on this study. From those results, although melt resin flow of this micro injection molding machine is different from melt resin flow of current common injection molding machines, this micro injection molding machine has good kneading action.

Keywords: Micro injection molding, FEM, Static visualization

Introduction

In recent years, micro molding machine becomes to be much in demand because lots of small hospital-use plastic products are made by micro molding with development of micro molding technology.
One of micro molding methods is injection molding. Injection molding method can mold micro component with high precision and it has good production efficiency. However, when micro components are made by current common injection molding machine, it causes disadvantages such as waste of resins, waste of power, etc. Therefore downsized injection molding machine is developed. This micro injection molding machine has particular kind of screw shape that is circular disk. Screw shape of this micro injection molding machine is different from screw shape of current common injection molding machines. Therefore, resin melting mechanism and resin mixing mechanism for disk type of screw are not well known. This study presents resin flow of this micro injection molding machine when the resins are plasticized. Experiment and computer simulation were performed on this study. Flow state of screw of micro injection molding machine was compared flow state of standard screw of current common injection molding machine in computer simulation. In the experiment, Static visualization was performed with colored pellet and polypropylene for research flow state in screw of micro injection molding machine.

Micro injection molding machine

This study used micro injection molding machine (SHINKO SELLBIC CO.LTD) that is shown in Fig.1. And main specification of this micro injection molding machine is shown in Table.1. Breadth of this micro injection molding machine is only 60 centimeter. So this micro injection molding is very small compared to current common injection molding machines. This micro injection molding machine has pre-plunger mechanism and it can mold product of 0.01 g with general-purpose pellet. This micro injection molding machine doesn't waste the resin because the nozzle accepts hot runner mechanism. Micro injection molding machine is better suited for mold of small parts that are like hospital-use. Screw shape of micro injection molding machine (micro screw) is shown in Fig.2 and screw shape of current common injection molding machine (standard screw) is shown in Fig.3. Screw shape of this micro injection molding machine is different from screw shape of current common injection molding machines. This micro injection molding machine has particular kind of screw shape that is circular disk. Therefore injection molding machine could be subminiaturized. Plasticizing unit of micro injection molding machine is shown in Fig.4. Micro screw and Manifold are shown in Fig5. Micro screw and manifold are paired up and screw only rotates. Pellets are moved through channel of screw, melt resin are flowed through channel of manifold to plunger. And then, melt resin are injected to mold. Mold can be easily exchanged because mold is very light. Heater is situated among manifold and nozzle, cooling water flows around micro screw.

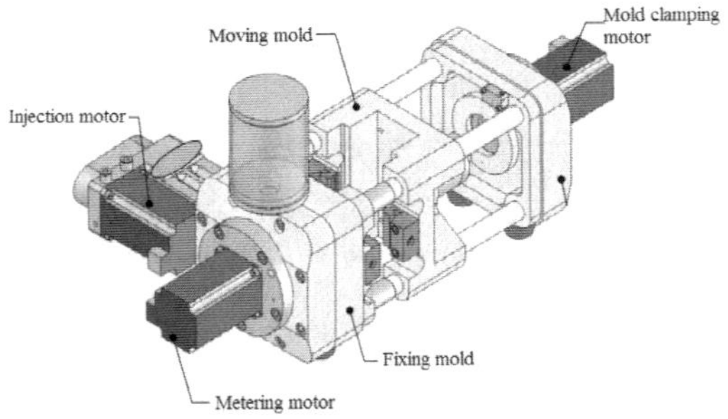

Fig.1 Micro injection molding machine

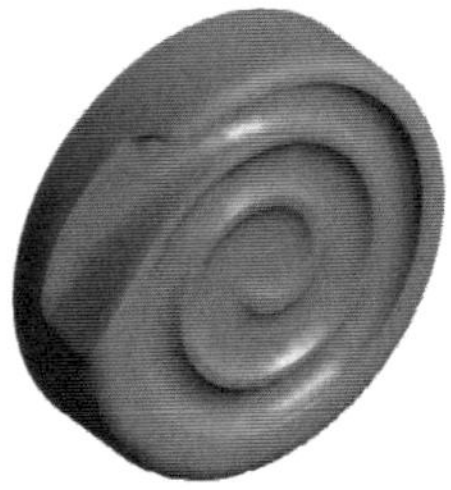
Fig2. Micro screw

Fig.3 Standard screw

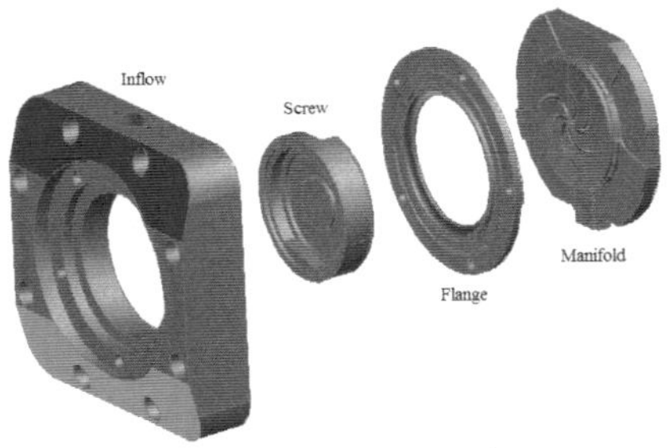

Fig.4 Plasticizing unit

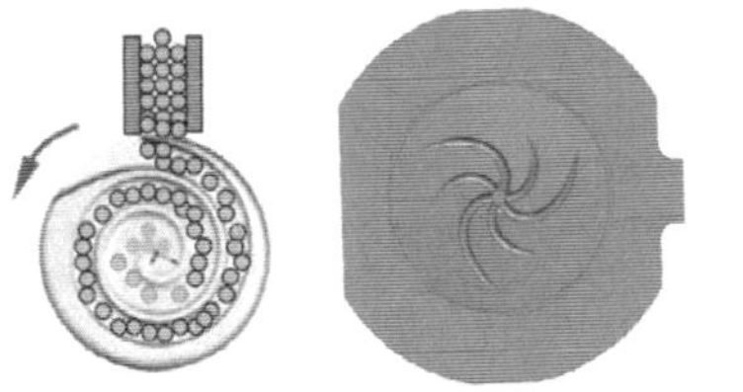
Fig.5 Micro screw and Manifold

Table.1 main specification

Plunger diameter [mm]	φ8	φ10
Plunger stroke [mm]	38	
Injection volume[cm3]	1.9	2.9
Injection average [cm3/s]	5.8	9
Maximal injection speed [mm/s]	115	
Maximal injection pressure [Mpa]	270	173
Maximal pressure keeping [Mpa]	147	94
Plasticizing performance [kg/h]	1.1	
Screw diameter [mm]	φ80	
Mold clamping force [ton]	3	

Computer simulation [1-3]

3DCAD model and mesh are created by GAMBIT (ANSYS). POLYFLOW (ANSYS) was used for this computer simulation and it computes solution by finite element method. Analysis condition is given below.

1. Molten resin is Non-compressible, isothermal and Non-Newtonian fluid.
2. This simulation takes no account of inertia force and gravity force because molten resin is highly viscous fluid.
3. Fluid is fully-melting and fullness in flow channel
4. This simulation takes no account of slip condition on the surface of a solid.
5. Viscosity data of molted resin is Polypropylene of 180 °C.

Base equations of flow analysis are equation of continuity and equation of momentum that are given below.

$$\frac{\partial u_i}{\partial x_i} = 0 \qquad (1)$$

$$\frac{\partial(\rho u_i)}{\partial t} + \frac{\partial(\rho u_i)}{\partial x_i} = \frac{\partial \sigma_{ij}}{\partial x_i} \qquad (2)$$

u_i is the velocity, x_i the coordinate, ρ the density, t the time, σ_{ij} the stress tensor. The constitutive law of material of generalize Newtonian fluid of incompressibility is given below.

$$\sigma_{ij} = -p\delta_{ij} + \sigma'_{ij} = -p\delta_{ij} + \eta\gamma_{ij} \qquad (3)$$

p is the pressure, δ_{ij} Kronecker delta, σ'_{ij} deviator stress tensor, η shear viscosity, γ_{ij} shear rate tensor.

The constitutive equation is assumed to be represented by Carreau model.

$$F(\dot{\gamma}) = \eta_\infty + (\eta_0 + \eta_\infty)(1 + \lambda^2\dot{\gamma}^2)^{\frac{n-1}{2}} \qquad (4)$$

F is viscosity, γ shear rate, η_0 zero shear rate viscosity, η_∞ infinite shear rate viscosity, λ natural time, n power law index.

Rotating speed of screw is 60rpm, simulation time is 1 second because screw goes into a 360-degree roll.

Standard screw (mold clamping force is 40ton, screw length 480millimeter, outside diameter 24millimeter) was performed computer simulation to compared to micro screw.

Particle tracking was performed by use of result of finite element method. From result of particle tracking, mean shear stress of each particles are calculated from the following formula.

$$\tau_{mean} = \frac{\int_0^{t_p} \tau(t)dt}{t_p} \qquad (5)$$

τ_{mean} is mean shear stress, t_p is particle residence time.

Result of computer simulation and Discussion

Internal pressure distribution of micro screw is shown in Fig.6. Micro screw generated very low pressure that compared to standard screw. Specifically, outside of screw has a low pressure. It would appear that fluid constantly can't feed into screw. Histogram of mean shear stress is shown in Fig.7. Vertical axis is frequency and horizontal axis is mean shear stress that each particle was subjected to in Fig.7. From Fig.7, Micro screw generated low mean shear stress that compared to standard screw. Histogram of particle residence time is shown in Fig.8. Vertical axis is frequency and horizontal axis is particle residence time in Fig.8. From Fig.8, micro screw has a longer particle residence time than standard screw. From those results, Micro screw gives pellet low mean shear stress over many generations. It would appear that it takes micro injection molding machine a while to melt pellet because micro screw generated low mean shear stress.

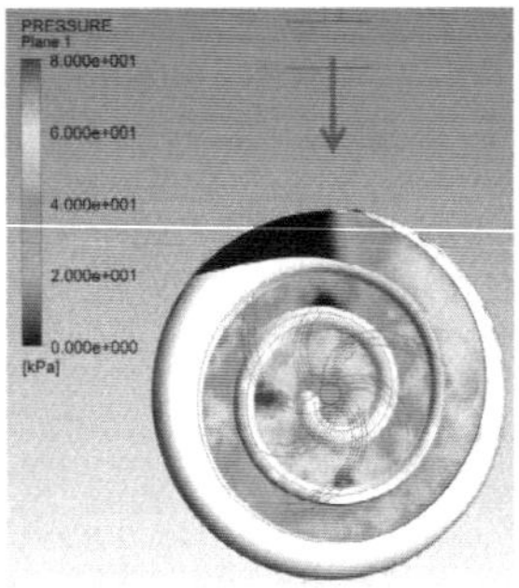

Fig.6 Internal pressure distribution of micro screw

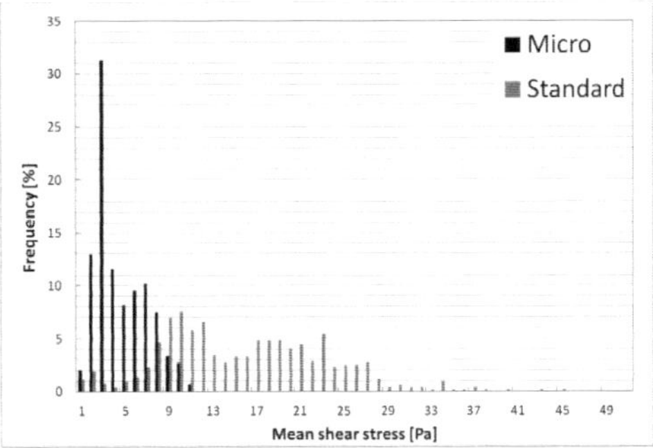

Fig.7 Histogram of mean shear stress

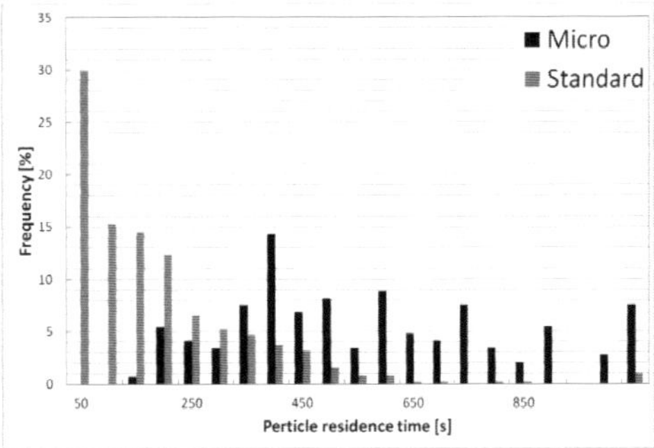

Fig.8 Histogram of particle residence time

Experimental [4-5]

This experiment was used color free polypropylene and colored polypropylene. The pellet put in screw by two ways are given below.

(a) Color free pellet, yellow pellet and blue pellet were put in screw at equal rate to verify mixing performance and test samples were molded by spiral flow mold and evaluated. And screw was pulled out from machine to obtain sample.

(b) To verify flow of resin, yellow pellet was put in screw after color free pellet was put in screw and color free pellet disappeared in hopper. And screw was pulled out from machine to obtain sample.

Manifold temperature was 210 degrees, nozzle temperature was 200 degrees, tool temperature was 60 degrees, injection speed was 3 mm/s and rotating speed was 90 rpm.

Result of experimental

The samples were molded by (a) condition that are shown in Fig.9. From Fig.9, the molding is evenly colored and molding-by-molding don't make no difference. Sample of pulling test is shown in Fig.10. In Fig.10, left sample is screw side, right sample is manifold side. Color of resin is evenly colored at manifold side. At screw side, pellet are left at outside. And color of resin is evenly colored at center. From those results, micro screw has good mixing performance. However, it takes long time to measure resin just like result of computer simulation. It would appear that fluid constantly can't feed into screw. If supply of pellet is increased, measurement time will become fast. It was found that micro injection molding machine has good mixing performance although it takes long time to measurement time.

Sample of pulling test is shown in Fig.11 at (b) condition. In Fig.11, left sample is that was obtained after yellow pellet put in screw and it measured 3 times. Right sample is that was obtained after yellow pellet put in screw and it measured 6 times.

From left sample of Fig.11 pellet are left at outside though yellow pellet was put in screw after color free pellet was put in screw and color free pellet disappeared in hopper. It would appear that pellet is left at outside by centrifugal force before pellet melt. From right sample of Fig.11, resin is concentrated in center though there is no supply of pellet. It would appear that micro screw generates Weissenberg effect. In micro injection molding machine, pellet is left at outside before it melt, and melted resin concentrated in center by Weissenberg effect.

Fig.9 Moldings

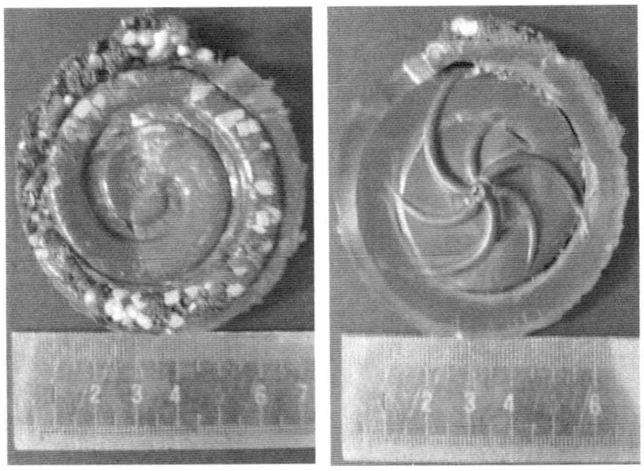

Fig.10 Sample of pulling test at (a) condition

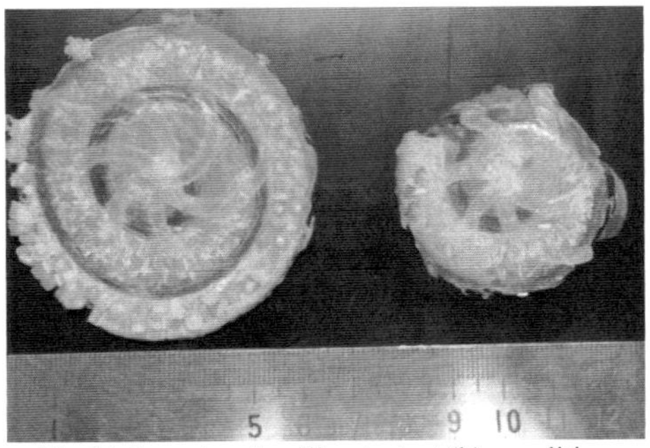

Fig.11 Sample of pulling test at (b) condition

Conclusions (Times New Roman 10, bold)

Moldings of micro screw are evenly colored and micro screw has good mixing performance. However, it takes long time to measurement time. This cause is low pressure in micro screw and pellet constantly can't feed into screw. In micro injection molding machine, pellet is left at outside before it melt, and melted resin concentrated in center by Weissenberg effect.

References

1. Takeshi Ishikawa; Tarou Amano, POLYMER ENGINEERING AND SCIENCE, 2002, Vol.42, No.5
2. Takeshi Ishikawa, POLYMER ENGINEERING AND SCIENCE, 2001, Vol.41, No.5
3. Kazuhisa Fukutani; Sayaka Yamada, KOBE STEEL ENGINEERING REPORTS, 2008, Vol.58, No.2
4. Michihiro Tatsuno; Hidetoshi Yokoi, Institute of Industrial Science, 2002, Vol54, No.3
5. Tomoyuki Akashi; Munetaka Jotaki, The Institute of Electrical Engineers of Japan, 1987, Vol107, No.7

O-04-596

MICROCELLULAR INJECTION MOLDING WITH GAS-LADEN PELLETS: EFFECTS OF CO-BLOWING AGENTS ON MICROSTRUCTURE

Xiaofei Sun[1], Lih-Sheng Turng[1*], Pat J. Gorton[2], Sezen Buell[2], and Pankaj Nigam[2]

[1]*Polymer Engineering Center, Dept. of Mechanical Eng., University of Wisconsin–Madison, Madison, WI 53706, USA*
turng@engr.wisc.edu *(corresponding author)*
[2]*Energizer Personal Care, LLC, Dover, DE 19904, USA*

Abstract. A novel combination approach to producing foamed injection molded parts has been investigated. By combining extruded, gas-laden pellets with microcellular injection molding, the processing benefits and material characteristics of using both nitrogen (N_2) and carbon dioxide (CO_2) as co-blowing agents can be realized, thus yielding features superior to that of using either nitrogen or carbon dioxide alone. Using an optimal content ratio for the co-blowing agents, as well as the proper sequence of introducing the gases, foamed parts with a much better morphology can be produced. In particular, extruding nitrogen gas-laden pellets followed by microcellular injection molding with higher amounts of carbon dioxide produces the finest and densest cell structure. In this paper, the theoritical background is discussed and experimental results show that this combined approach leads to significant improvements in the foam cell morphology for low density polyethylene (LDPE), polypropylene (PP), and high impact polystyrene (HIPS).

Keywords: Microcellular injection molding, foam extrusion, carbon dioxide, nitrogen, co-blowing agents

Introduction

Microcellular injection molding, also known as microcellular foaming, is one of the special injection molding processes. The idea of microcellular foaming was first conceived at MIT in the 1980s. Later, Trexel Inc. combined the idea of microcellular foaming with the injection molding process and commercialized it as the MuCell process [1]. It continues to attract attention because it saves on material costs and energy while improving dimensional stability and production efficiency compared with conventional solid injection molding. Microcellular injection molding requires specially designed supercritical fluid (SCF) delivery and dosing systems to be installed for the delivery of the physical blowing agent (PBA). As such, modifications need to be made on the conventional injection molding machine(s) for the installation of supercritical fluid (SCF) delivery devices.

Alternatively, chemical blowing agents (CBAs) can be used to produce foamed injection molded parts without the need of installing any additional equipment and the dosing process is generally straightforward. However, residuals tend to appear in the parts after the reaction, which leads to the degradation of the polymer matrix and might contaminate the mold [2]. Furthermore, the use of chemical blowing agents does not allow for good control over the foaming process or the porosity and cellular structure of the parts [3, 4].

Another way to achieve foaming using a conventional injection molding machine is to saturate pellets with a physical blowing agent in a high pressure vessel first, and then use them for injection molding [5]. However, the production rate is usually not high enough for continuous mass production of injection molded parts.

Recently, a novel method to produce microcellular injection molded parts was proposed by Lee et al. [6], and is known as supercritical fluid-laden pellet injection molding foaming technology (SIFT). Using this method, gas-laden pellets can be produced by an extruder equipped with a gas injection device. Modification and device additions are only needed for one extruder, and the gas-laden pellets produced can be used by multiple conventional injection molding machines to produce lightweight, foamed parts without any modification or additional equipment for the injection molding machines. Since this process employs commonly used physical blowing agents like nitrogen or carbon dioxide, general concerns surrounding chemical blowing agents are no longer an issue. The feasibility and effectiveness of this method have been verified through preliminary studies and reported in [6].

The two most commonly used physical blowing agents in foaming processes today are nitrogen and carbon dioxide. While both are being widely applied, they have distinct physical properties, as shown in Table 1, which lead to significant differences in the process as well as in the final part quality. Generally speaking, nitrogen provokes stronger nucleation as compared to carbon dioxide, given the same gas content [7], due to its lower solubility, therefore yielding a higher degree of supersaturation. Meanwhile, large amounts of carbon dioxide can be dissolved more readily [8] and yield a much more profound plasticization effect on the polymer melt as compared to nitrogen [9].

Table 1: Key physical properties of N_2 and CO_2.

Properties in LDPE	N_2	CO_2
Density (kg/m^3) at 3 °C and 130 bar	170.8	902.3
Solubility (g/g) at 180 °C and 120 bar [10, 11]	0.011	0.075
Diffusion coefficient (cm^2/sec) at 188 °C and 100 bar [12]	1.5×10^{-4}	1.1×10^{-5}
Diffusion coefficient (cm^2/sec) at ambient conditions [13]	5.1×10^{-7}	5.0×10^{-8}

By combining the SIFT and MuCell processes, the ability to incorporate both nitrogen and carbon dioxide into the foaming process becomes possible. Incorporating both blowing agents via a two-step approach enables the processing benefits and material characteristics of both blowing agents, thus yielding features superior to that of using either nitrogen or carbon dioxide alone.

Some earlier studies suggested that a better foam morphology could be created using a blend of carbon dioxide and nitrogen co-blowing agents. Wong and Park et al. [10] applied a batch foaming blend of CO_2–N_2 gas on polystyrene (PS) film. The 75% CO_2–25% N_2 gas blend appeared to have the best foaming performance and yielded the highest cell densities and cell growth rates. Maio et al. [9] applied a CO_2–N_2 blend on the extrusion foaming of polycaprolactone (PCL). The best results in terms of morphology and density were achieved using a 20/80 weight ratio CO_2–N_2 mixture as the foaming agent in the extrusion. However, the understanding of the foaming mechanism of such a blend is limited, and it has only been applied to batch foaming and extrusion foaming.

By combining the SIFT and MuCell processes, two different blowing agents can incorporated using a two-step process. The first type of blowing agent can be incorporated into the gas-laden pellets using the SIFT process. These extruded, gas-laden pellets can then be used in microcellular injection molding with another blowing agent. In this way, we can create foamed injection molded parts using two different blowing agents at controlled dosages and ratios. A schematic of this approach is shown in Fig. 1.

Theoretical Background

According to classic nucleation theories [15-19], the nucleation rate of the microcellular injection molding foaming process can be expressed as:

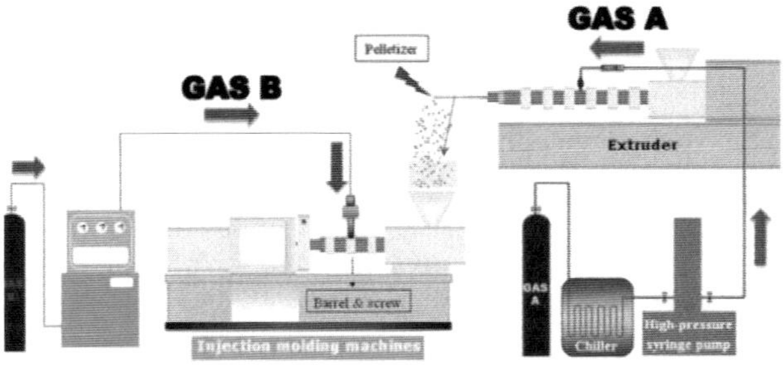

Figure 1: Schematic of the combined SIFT/MuCell process.

$$N = fC \exp\left(-\Delta G^{**}/kT\right) \qquad \text{(Eq. 1)}$$

where N is the nucleation rate, f is the frequency of atomic molecular lattice vibrations, C is the concentration of gas molecules, ΔG^{**} is the activation energy barrier for nucleation, k is the Boltzmann's constant, and T is the absolute temperature. The activation energy barrier can be estimated by:

$$\Delta G^{**} = \frac{16\pi\gamma^3 M^2}{3(RT\rho \ln S)^2} \qquad \text{(Eq. 2)}$$

where γ is the surface energy of the bubble interface, R is the universal gas constant, ρ is the density, and S is the degree of supersaturation. A low ΔG^{**} is favorable for foaming because it will increase the nucleation rate, N, exponentially. Nitrogen has a low solubility in most polymers, and therefore it will create a high degree of supersaturation, S. Carbon dioxide, on the other hand, can substantially reduce the surface energy, γ, due to its plasticizing effect. When nitrogen and carbon dioxide are combined in the same foaming process, both will contribute to a lower ΔG^{**} value and thus a higher nucleation rate. The end result will be to create a finer morphology.

Experimental

A schematic of the experimental setup is shown in Fig. 1. The gas-laden pellets were produced using an Extrudex EDN 45-30 single screw extruder. Gas was injected into the barrel using a TeledyneISCO 260D high precision syringe pump. The injected gas was mixed into the polymer to form a single-phase polymer–gas solution in the extruder. This solution was then extruded through a 12-orifice filament die. Once the strands came out of the die, they went into a cold water bath to be cooled down immediately in an effort to prevent foaming and a loss of gas from the strands. After water cooling, the strands were then pelletized. The pellets were oven dried to remove any moisture and then stored in plastic zipper bags.

When a large amount of gas is introduced, a high degree of supersaturation will inevitably cause the strands to foam, resulting in a loss of gas in the pellets. Hence, the gas content one can introduce in the gas-laden pellets is lower than that in the MuCell process. Nevertheless, it has been reported in [15] that a lower degree of gas content (and thus supersaturation) actually leads to a better surface quality as the foaming tends to occur after the mold cavity is filled and a smooth skin layer is formed.

Once the gas-laden pellets are produced, they can be taken to injection molding machines to produce foamed parts. The injection molding machine used in this study was an Arburg Allrounder 320S equipped with a microcellular injection molding (MuCell) system. Tensile test bar mold (ASTM 638 Type I) was used to mold the parts. The resin used in this study was low-density polyethylene (LDPE, Marlex, Chevron Phillips Chemical Company), high-impact polystyrene (HIPS, Dow Chemical), and polypropylene (PP, Bassel Pro-fax).

Table 2. Key process conditions.

	Extrusion
Temperature	140/150/160/165/170/155/140 °C for LDPE 190/200/210/210/210/200/190 °C for HIPS and PP (sequence from hopper to die)
Screw Speed	20 rpm
	Injection Molding
Nozzle/Runner Temperature	160 °C
Injection Speed	60 cm^3/s
Packing	No
Screw Recovery Speed	15 m/min
Back Pressure	100 bar
Coolant Temperature	15 °C
Cooling Time	25 s

Results and Discussions

Molding Trial 1 was done on LDPE by dosing nitrogen first in the SIFT process to create N_2-laden pellets. Then, in the MuCell process, carbon dioxide was injected into the blend. Figure 2 shows the morphology using this approach.

At a low carbon dioxide MuCell dosage level (0.5%), a large cell size was observed due to the limited amount of nitrogen that was able to be dosed into the polymer using the SIFT process. Still, a clear trend was observed: as the nitrogen level increased in the SIFT pellets, the cell size decreased. Once enough carbon dioxide was introduced (2.5%), along with the small amount of nitrogen in the gas-laden, the morphology improved dramatically. The cell size was reduced from approximately 120 micron on average to about 30 micron.

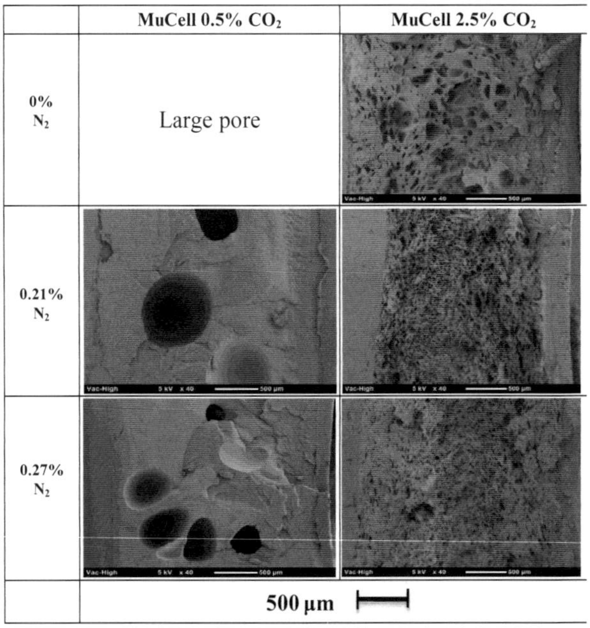

Figure 2: SEM images of Molding Trial 1 (LDPE with N_2 SIFT + CO_2 MuCell). The scale bars are 500 µm.

Molding Trial 2 reversed the sequence of blowing agents. Thus, CO_2-laden SIFT pellets were used with an N_2 MuCell process. Figure 3 shows the outcomes. In this trial, not much difference was observed in the morphology. This was because, in the SIFT process, only a very limited amount of carbon dioxide could be injected (0.55% maximum). This was not sufficient to yield a substantial plasticization effect, and thus could not drive and sustain cell growth. In the MuCell process, even though a larger amount of nitrogen gas was injected (as compared to the SIFT process), the nitrogen had a low solubility in the polymer. Thus, a level of 1.2% nitrogen could not be exceeded, otherwise large voids would appear in the parts due to undissolved nitrogen in the blend (a condition known as "out of solution"), which could serve as pre-existing cell nuclei. However, the number of these pre-existing nuclei from out-of-solution nitrogen was far less than those created by the nucleation of dissolved nitrogen, thus leading to fewer, larger cells.

To take full advantage of the combined effect of nitrogen and carbon dioxide, a large amount of carbon dioxide was needed to lubricate the polymer and to drive and sustain the cell growth. In comparison, only a relatively small dosage of nitrogen was needed to improve cell nucleation. In order to verify if this methodology was applicable to other types of materials, the same process conditions for Molding Trial 1 were applied to HIPS and PP, with three different levels of nitrogen content in the gas-laden SIFT pellets, and 2.5% carbon dioxide in the MuCell process. The SEM images of the morphology are shown in Fig. 4. Both HIPS and PP showed a sharp reduction in cell size and an increase in the cell density once both nitrogen and carbon dioxide were introduced, as compared to using CO_2 MuCell alone. Overall, the cell size in HIPS was finer and more uniform than the cell size in PP.

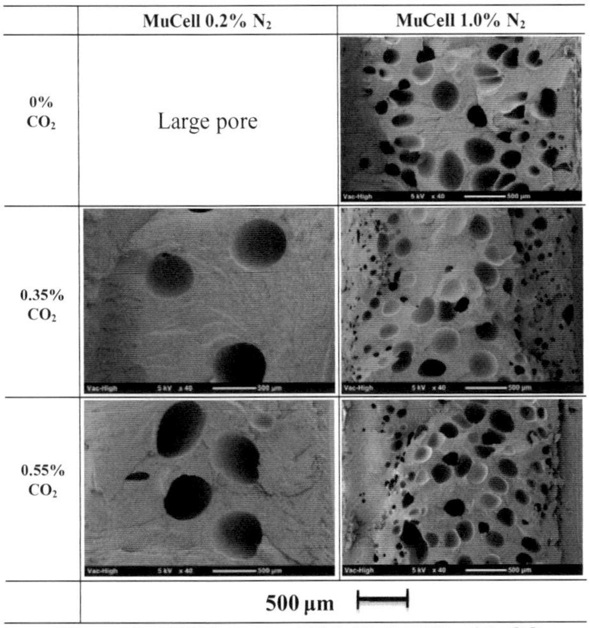

Figure 3: SEM images of Trial 2 (LDPE with CO_2 SIFT + N_2 MuCell). The scale bars are 500 μm.

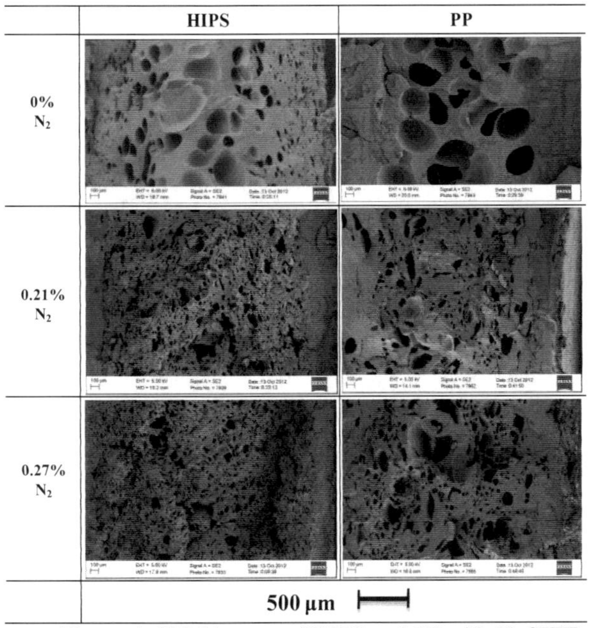

Figure 4: SEM images of HIPS and PP with N_2 SIFT + 2.5% CO_2 MuCell. The scale bars are 500 μm.

Conclusions

In this study, a combined SIFT/MuCell process was successfully carried out using both nitrogen and carbon dioxide as the co-blowing agents. N_2 SIFT + CO_2 MuCell yielded remarkable improvements in foam morphology compared with using either type of blowing agent alone. This methodology has been applied to LDPE, PP, and HIPS, and all three cases showed substantial improvement in the morphology.

Acknowledgements

This research was supported by the Wisconsin Alumni Research Foundation (WARF) Accelerator Program.

References

1. Lee, J. J., Turng, L. S., ANTEC 2009, p. 1662.
2. Krumpholz, T., Michaeli, W., ANTEC 2008, p. 1019.
3. Sauceau, M., Fages, J., *Progress in Polym Sci.*, 2011, **36**, p. 749.
4. Gale, M., *British Plastics and Rubbers*, 2000, p. 4.
5. Pfannschimidt, O., Michaeli, W., ANTEC 1999, p. 2100.
6. Lee, J.J., Turng, L.S., *Polym Eng Sci*, 2011, DOI: 10.1002/pen.22004.
7. S.G. Kim, B.S. Kang and C.B. Park, *Cell. Poly.*, 2006, **25**, p. 19.
8. G. Li, J. Wang, C.B. Park, and R. Simha, *J. Poly. Sci. Part B-Poly. Phy.*, 2007, **45**, p. 2497.
9. E. Di Maio, G. Mensitieri, *Poly. Eng. Sci.*, 2005, **45**, p. 32.
10. Wong, A., Park, C.B., ANTEC 2011.
11. Crank, J., The Mathematics of Diffusion, Oxford University, Press, United Kingdom, 1975.
12. Flumerfelt, R., Lee, J., *J of App Polym Sci*, 1995, **58**, p. 2213.
13. Areerut, S., Ohshima, M., *Polym Eng Sci*, 2004, **10**, p. 1915.
14. Xu, J., Microcellular Injection Molding, Wiley & Sons, Inc., Hoboken, New Jersey, 2010.
15. J.J. Lee, L.S. Turng, E. Dougherty, and P. Gorton, *Polymer*, 2001, **52**, p. 1436.
16. M. Volmer and A. Weber, Z. Phys. *Chem.*, 1926, **119**, p. 277.
17. D.L. Tomasko, A. Burley, L. Feng, S.K. Yeha, and K. Miyazono, *J. Supercrit. Fluids*, 2009, **47**, p. 493.
18. J.S. Colton and N.P. Suh, *Polym. Eng. Sci.*, 1987, **27**, p. 485.
19. J.S. Colton and N.P. Suh, *Polym. Eng. Sci.*, 1987, **27**, p. 493.

O-04-630

OPTIMIZATION PARAMETERS OF INJECTION MOULDING MACHINE FOR REDUCING WARPAGE OF DOG BONE PLASTIC PART

M. Amran*, S. Salmah, M. Zaki, R. Izamshah, M. Hadzley, S. Sivarao and N.I. Syahriah
Faculty of Manufacturing Engineering, Universiti Teknikal Malaysia Melaka, Hang Tuah Jaya 76100 Durian Tunggal, Melaka, Malaysia – mohdamran@utem.edu.my

Abstract - The optimization of processing parameters on warpage of polypropylene (PP) in the application of injection moulding machine was studied. The appropriate parameters were adjusted to reduce the warpage defect on the tensile test specimen of dog bone. The type of injection moulding machine used in this research is Arburg 420C 800-250C. Four parameters that have been investigated; injection pressure, clamping pressure, back pressure and holding pressure. A concept of design of experiment (DOE) has been applied using Taguchi method to determine the suitable parameters. To measure the warpage of the dog bone, digital height gauge was used to measure the flatness of the part surface. According the analysis of variance (ANOVA), the most significant factor that effect the warpage was holding pressure by 57.82%, followed with back pressure by 25.75%, clamping pressure by 16.27% and injection pressure by 0.16%. It is found that the optimum parameters setting that have been obtained were injection pressure at 950 bar, clamping pressure at 600 kN, holding pressure at 700 bar and back pressure at 75 bar. The depreciation value of warpage minimum index in this experiment was decreased by 4.6% after confirmation run.

Keywords: Injection Moulding, Polypropylene, Warpage, Taguchi Method, Injection Parameters.

Introduction

The parameters of injection moulding machine are very important for polymer processing operation in the plastic industry today [1]. The quality of the final product is mainly dependent on the moulding process parameters [2]. The main moulding parameters in injection moulding are; material melt temperature, mold temperature, injection time and required pressure. Some researches choose molding parameters according to the material type and molding geometry and then implement them to assess the molding quality. If there are any variations, iterative corrective actions are performed to reach the quality requirements [3]. However, some defects are still found in the plastics parts. For example, to produce thinner plastic part such as casing and housing of products which have thin shell feature, the significant warpage defect still appeared [4]. In this study, optimization parameter of injection moulding machine for reducing warpage of dog bone plastic part is investigated by using design of experiment (DOE) by Taguchi method.

Experimental

The polymer used in this study was general polypropylene (PP). Arburg 420C 800-250C injection moulding machine was selected. Table 1 shows the general advantages and disadvantages of PP polymer.

Table 1: Advantages and disadvantages of PP

Advantages	Disadvantages
• Low coefficient of friction • Good fatigue resistance • Good impact strength • Excellent flexural strength • Good grade availability	• Poor weatherability • Flammable • Difficult to bond • Broken down by ultraviolet radiation

Almost all the thermoplastics and some thermoset plastics can be injected by injection moulding machine thus adding to the flexibility of the process in this injection moulding machine. Table 2 shows the factors and working level of design factor in Taguchi method. Four factors and three levels were used in this experiment. There are 9 trial processes and each trial are be done for 3 times. This means that 27 injection processes were done to get the most accurate result.

Table 2: Factors and working level of design factor

Symbol	Factor (Pressure)	Level 1	Level 2	Level 3
A	Injection (bar)	900	950	1000
B	Clamping (kN)	550	600	650
C	Holding (bar)	500	600	700
D	Back (bar)	60	75	90

Legend : A for Injection Pressure (bar)
: B for Clamping Pressure (kN)
: C for Holding Pressure (bar)
: D for Back Pressure (bar)

After specimen was produced, the value of the warpage was measured by using the digital height gauge and then the average of the measurement is calculated. Then, by using the results that obtained from the data, analysis was done using the MINITAB software to get the optimum parameter value.

Results and Discussion

Experimental Result

Table 3 shows the summarization of the data that was collected from the experiment. Four factors of pressure were selected for this experiment. There are injection pressure (A), clamping pressure (B), holding pressure (C), and back pressure (D). It is found that the maximum value of deflection (Za) was 0.3276 mm in trial no 1 and minimum value of deflection (Za) was 0.2563 mm in the trial no 3.

Table 3: Summarization of experimental result

No	A (bar)	B (kN)	C (bar)	D (bar)	Average Za (mm)
1	900	550	500	60	0.3276
2	900	600	600	75	0.2914
3	900	650	700	90	0.2563
4	950	550	600	90	0.3239
5	950	600	700	60	0.2722
6	950	650	500	75	0.2721
7	1000	550	700	75	0.2663
8	1000	600	500	90	0.2814
9	1000	650	600	60	0.327

Analysis of Signal to Noise (S/N) Ratio

Taguchi method uses the S/N ratio to measure quality characteristic deviating from the desired value. The aim of this study is to minimize the warpage index within optimal process parameter; the smaller the better quality characteristic was selected in this study. Table 4 shows the S/N ratio value that calculates using MINITAB software. It is found that the highest value of S/N ratio was 11.825 in trial no 3 and the lowest S/N ration value was 9.6931 in trial no 1. According to the value of deflection (Za) as shown in Table 3 and S/N ratio as shown in Table 4, it shows that the different value of the data. This means, the increase value of the deflection will result the decreasing value of the S/N ratio and vice versa. The S/N ratio value also can be calculated using equation 1.

$$-10\log(\mathrm{MSD}) \quad \text{where} \quad \mathrm{MSD} = \frac{1}{n}\sum_{i=1}^{n} y_i^2 \qquad \text{Eq. 1}$$

Where MSD is the mean square deviation, y the observation or data and n is the number of test in a trial.

Table 4: Summarization of S/N ratio

No	A (bar)	B (kN)	C (bar)	D (bar)	S/N
1	900	550	500	60	9.6931
2	900	600	600	75	10.7102
3	900	650	700	90	11.825
4	950	550	600	90	9.7918
5	950	600	700	60	11.3022
6	950	650	500	75	11.3054
7	1000	550	700	75	11.4926
8	1000	600	500	90	11.0135
9	1000	650	600	60	9.709

Determination of S/N Ratio Response Table

The S/N ratio response table for each process parameters at the level of 1, 2 and 3 was created by utilizing the S/N ratio values for warpage in the Table 5. Based on the data from Table 5, the S/N response diagram has been constructed as shown in Fig. 1.

Table 5: The responses table of S/N ratio

Level	A (bar)	B (kN)	C (bar)	D (bar)
Level 1	10.74	10.33	10.67	10.23
Level 2	10.80	11.01	10.07	11.17
Level 3	10.74	10.95	11.54	10.88
Difference	0.06	0.68	1.47	0.93

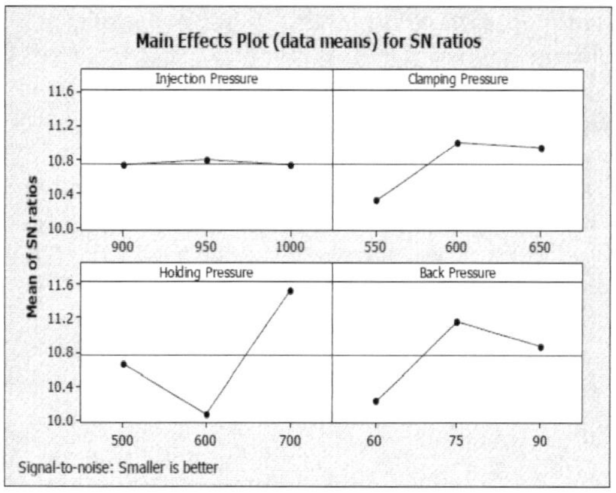

Figure 1: Main effect plot for S/N ratio

Based on the observation on the graphs, in Fig. 1, injection pressure the highest value is at level 2 which is 10.80 of S/N ratio and the pressure is 950 bar. Further, clamping pressure shows an increasing value where the highest value obtained is at level 2, where the S/N ratio is 11.01 in which the pressure is 600 kN. Furthermore, in holding pressure, the maximum value obtained from the graph is at level 3 with the S/N ratio 11.54 and pressure of 700 bar. For back pressure, the maximum value obtained is at level 2, S/N ratio of 11.17 at pressure 75 bar. That means, the most significant factor that affecting warpage are holding pressure and back pressure followed by clamping pressure and injection pressure.

Analysis of Variance (ANOVA)

The analysis of variance (ANOVA) is to be carried out to examine the influence of process parameters on the quality characteristic in this study, the ANOVA will compute the quantities such as degrees of freedom (f), sums of squares (S), variance (V), F-ratio (F), and percentage contribution (P%) as illustrated in Table 6.

Table 6: ANOVA computation using minitab software

Source	f	S	V	F	P (%)
A (bar)	2	0.00001	5.0×10^{-6}	-	0.16
B (kN)	2	0.00103	5.15×10^{-4}	-	16.27
C (bar)	2	0.00366	1.83×10^{-3}	-	57.82
D (bar)	2	0.00163	8.15×10^{-4}	-	25.75
Pooled Error	0	0	0	-	
Total	8	0.00633	3.165×10^{-3}	-	100

According Table 6, the highest value of percentage contribution (P) was in the holding pressure column, it record that 57.82% of P value. This is followed by back pressure at 25.7% of P value. The P value for clamping pressure was 16.27% and the lowest one of P value was 0.16% at injection pressure column. It shows that holding pressure, back pressure and clamping pressure are the significant factor while injection pressure is not much affecting the warpage defect.

Confirmation Run

After the optimal levels of all the control factors were identified, i.e., injection pressure at 950 bar, clamping pressure at 600 kN, holding pressure at 700 bar and back pressure at 75 bar, the last step in Taguchi parameter design is conducting the confirmation run. The combination of the optimal levels of all the factors produces the minimal index of warpage. Table 7 shows the result for non-optimum and optimum process of injection moulding.

Table 7: Result non-optimum and optimum process

	A (bar)	B (kN)	C (bar)	D (bar)	Za (mm)
Non-optimum	900	650	700	90	0.2563
Optimum	950	600	700	75	0.2455

According to the result from Table 7, the percentage of depreciation value of warpage can be determined. The percentage of depreciation of this experiment is 4.6%. The values of warpage after the experiment are decreased because some of the parameters before the experiment are set out of the recommended value of the machine manual.

$$Depreciation\ (\%) = \frac{Before - After}{Before} \times 100$$

$$= \frac{0.2563 - 0.2445}{0.2563} \times 100$$

$$= 4.6\%$$

Conclusions

In this study, the optimization process of injection molding parameters is investigated. Taguchi method was used to find the optimum parameters that influence the warpage defect in dog bone tensile specimen. The S/N ratio and ANOVA analysis were utilized to find the optimum parameters. In conclusion, by using the S/N ratio the optimum parameter that can minimize the warpage defect was injection pressure at 950 bar, clamping pressure at 600 kN, holding pressure at 700 bar and back pressure at 75 bar. From ANOVA analysis, the most significant factor that affects the warpage is holding pressure by 57.82% of contribution, followed with back pressure by 25.75% and clamping pressure by 16.27%. Injection pressure is almost less significant factor which only contributes about 0.16%. The deflection of warpage was decreased to 4.6% after confirmation experiments.

Acknowledgements
The authors gratefully acknowledge Universiti Teknikal Malaysia Melaka (UTeM) for supported the word under Short Term Research Grant (PJP/2012/FKP (25B) S1028).

References

1. J. Avery, *Injection Molding Alternatives; A Guide for Designers and Product Engineers.* Hanser/ Gardner Publications, Inc., Ohio, 1998.
2. B. Ozcelik; T. Erzurumlu *J. Mater. Process. Technol.* 2006, *171*,437.
3. K. S. Nezhad; E. Siores *J. Mater. Process. Technol.* 1997, *63*, 458-462.
4. M.C. Huang; C.C. Tai *J. Mater. Process. Technol.* 2001, *110*, 1.

O-04-313

INJECTION MOULDING OF MICRO STRUCTURES ON COMPLEX PART GEOMETRIES

Prof. Ch. Hopmann, A. Neuss*, R. Schoeldgen and M. Schoengart

Institute of Plastics Processing (IKV) at RWTH Aachen University, Pontstrasse 49, D-52062 Aachen, Germany
zentrale@ikv.rwth-aachen.de; neuss@ikv.rwth-aachen.de

Abstract

Surfaces with functional micro structures play an increasing role in many areas of application. Injection moulding with variothermal temperature control has proven to be a suitable mass production process for the fabrication of plastics parts with micro-structured surfaces. However, technologies available for the fabrication of these functional surfaces are still limited to flat and even shaped part geometries. To transfer the effect to complex part geometries, a technology for the manufacturing of these surface structures on 3D-shaped parts is required. As a first approach, the replication of cone-shaped micro structures with a diameter and height of 30 µm on a complex shaped mould insert is being investigated. The insert has the shape of a ball scraper with a diameter of 23 mm and a curvature radius of 18 mm, on which the dimples are aligned on concentric circles. The variothermal injection mould with an integrated laser-based cavity heating system is adapted to heat 3D-shaped surfaces. It is shown, that an increase in mould temperature from 30 °C to 130 °C leads to a precise and uniform moulding of the microstructures using a polypropylene material. Moreover, the structures are not only moulded completely, but also demoulded free of damage on the entire area. The investigations show that micro structured surfaces on complex part geometries can be produced by injection moulding under sufficient process conditions.

Keywords: injection moulding, micro structures, dynamic mould heating, laser

Introduction

Plastics products with innovative micro- and nano-structured functional surfaces still offer a high potential for new applications, for example in the field of medical- and bio-technology. Examples are selective wetting behaviours in micro dosing equipment or lab-on-a-chip applications [1]. Usually, these types of functional surfaces are created in a multi-step process. After the moulding of the part, several process steps follow to functionalise its surface [2-4]. To succeed in a global market with large differences in production costs, the integration of those functionalities in a product at reasonable costs is one of the driving forces of new production technologies for manufacturing in high-wage countries. In this context there is a growing demand for production technologies, which easily allow geometric shaping of parts and a simultaneous functionalisation within one process step to reduce the complexity of the process chain and subsequently decrease the costs of production. A promising approach to efficiently create these micro-structured surfaces is the replication in the injection moulding process [5, 6]. On the one hand, a precise moulding of surface structures in the µm-range requires elevated mould temperatures during the injection of the polymer melt. On the other hand, a mould temperature low enough for a demoulding without damage and a short cycle-time is required. In order to attain high mould temperatures and ensure a demoulding without damaging the part, variothermal process control is used [7]. Considerable progress has been achieved in the replication of micro-structures by injection moulding over the recent years. In a previous project, it was shown that superhydrophobic plastics surfaces with contact angles of more than 165 ° towards water could be produced in a single process step under certain conditions [8]. However, this technology is limited to parts with a flat surface. To use the effect of functional surfaces on more complex parts, a technology for the manufacturing of these surface structures on 3D-shaped parts is required.

For this purpose, a 3D-shaped mould insert with micro-structured dimples with a width and depth of 30 µm has been investigated. The structured surface of the mould insert has the shape of a ball scraper, which means that the moulded part has the shape of a dish with microstructures on the inner wall. The micro structures were generated by ultra-short pulse laser ablation, imitating the surface structure of the lotus leaf. In this work, the replication and the demoulding of the structures are being investigated without regard to the functionality. However, the size and shape of the structures have shown to be suitable to increase the functionality of the surface regarding the wetting behaviour. In order to rapidly heat the mould surface prior to the injection of the polymer, a mould integrated laser-based heating system was adapted to heat 3D-surfaces. For this purpose, a glass cylinder shaped exactly to the mould insert contour was placed behind the cavity wall. In a first step, the temperature field on the surface of the mould insert was being analysed by means of thermography. In the next step, moulding trials in automatic mode for steady state thermal conditions were performed. The moulded surfaces were then investigated by means of scanning electron microscopy (SEM).

Mould and process technology

Injection mould design

In order to enable the dynamic temperature control of the cavity, an injection mould with an integrated laser heating unit was used [2]. The mould has two cavities and is equipped with interchangeable mould inserts. For the precise centering of the mould halves the guiding pillars are combined with a ball track unit. The laser-heated cavity is equipped with a pressure resistant quartz glass mould insert which was designed to withstand cavity pressures of up to 1000 bar. The laser beam is guided through a flexible beam waveguide into the laser collimation optics that is mounted at the fibre end. The collimation optics, which is widening and collimating the laser beam, is integrated into the moving side of the injection mould. Figure 1 shows a detailed section view of the mould.

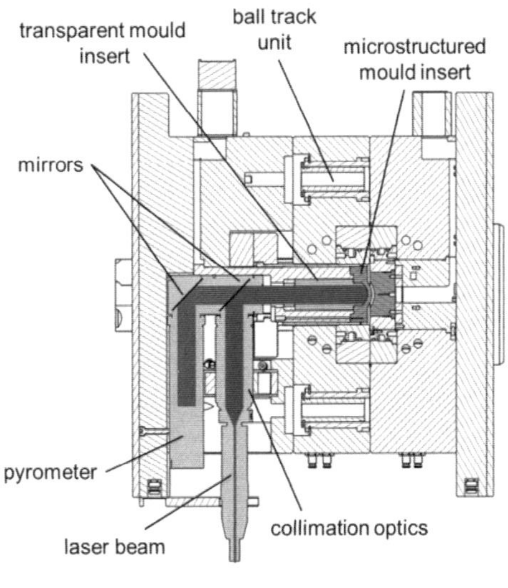

Figure 1: Section view of the injection mould with integrated laser collimation optics

The laser beam passes from the collimation optics through the glass insert. It is absorbed at the backside of the cavity wall on the moving mould half, where the micro-structured mould insert is mounted (Figure 2). The coefficient of emission describing the percentage of the incoming laser power that is being absorbed and transformed into heat was determined to 0.65 by reflectivity measurements.

Next to the collimation optics, a pyrometer is also integrated into the mould. Via semi-transparent mirrors inside the collimation optics, the temperature of the surface within the absorption zone can be measured. This information can then be used for a temperature-based power control of the laser unit.

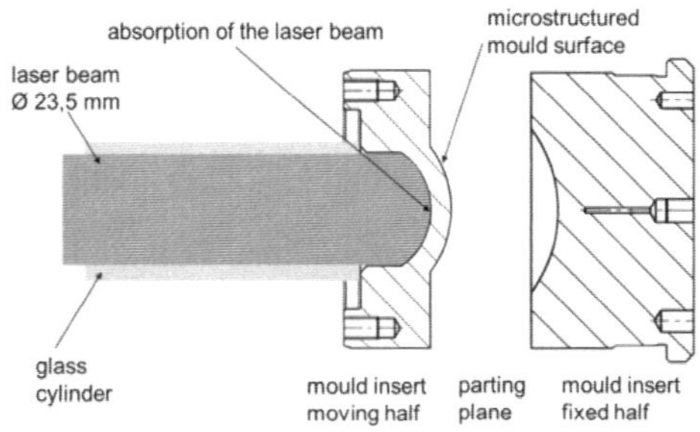

Figure 2: Indirect heating of the cavity surface by the laser radiation

3D-shaped micro-structured mould insert

The micro-structures in the mould insert were created at the Fraunhofer Institute for Laser Technology (ILT), Aachen. To generate the microstructures on the surface of the mould insert, a laser with 10 ps pulse length and a wavelength of 355 nm was used. Figure 3 shows a detail of the mould insert. The structure depth is defined by the number of laser pulses per spot. The ground diameter of the cone-shaped dimples is defined by the average power of the laser and the focal length of the telecentric F-Theta lens used in this experiment.

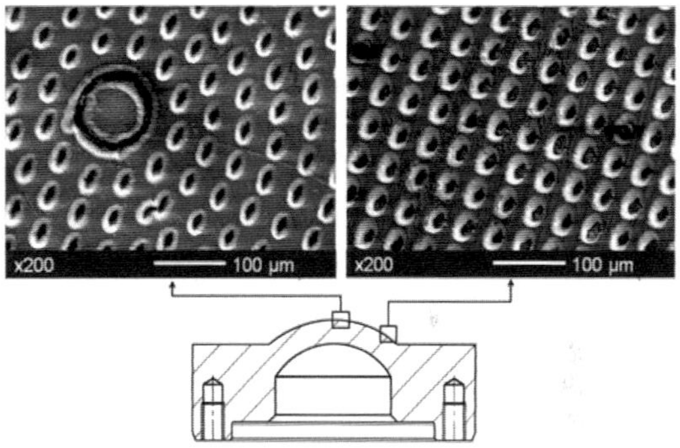

Figure 3: SEM and section view of the mould insert with a micro-structured ball scraper

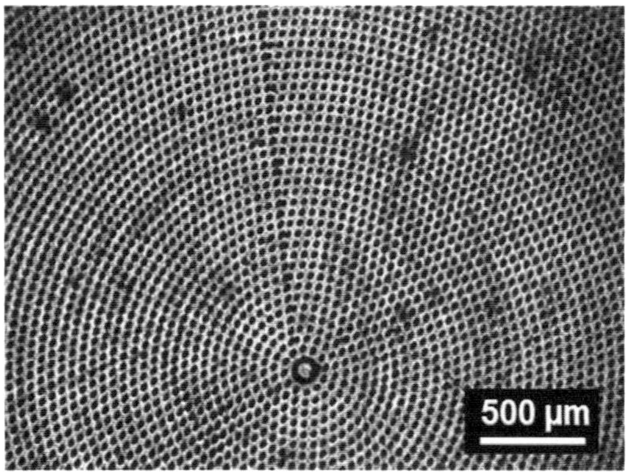

Figure 4: 3D Laser microscopy of the top of the

micro-structured ball scraper

A shorter focal length of the optic results in a shorter diameter of the dimple. In combination with a galvanometer scanner, consisting of a high-performance rotary motor section based on moving magnet technology and a high-precision position detector, a rapid fabrication of dimples in arrays is possible. The spacing of the dimples is arbitrary. Figure 4 shows a 3D laser microscopy image of the mould insert. As it can be seen, the dimples were positioned on concentric circles around the centre. Furthermore, the dimples are aligned parallel to the direction of demoulding, which means that the angle between the moulded cone and the surface of the part varies depending on the position on the ball scraper. A characteristic of structuring with this picosecond-laser system is that no steel melt accumulates at the edge of the microstructures. Because of the very short irradiation pulses, the steel body does not absorb heat. The introduced radiant energy directly evaporates the steel.

Injection moulding process

All moulding experiments were performed on a hydraulic KM CX 160-1000 injection moulding machine, KraussMaffei Technologies GmbH, Munich, Germany, with a 40 mm diameter plasticising unit and a heated extended nozzle. Characteristic injection moulding parameters can be seen in Table 1. As polymer material, a polypropylene PP 513MNK40, SABIC Europe, Sittard, Netherlands, was used.

Parameter	**Unit**	**Value**
Plasticising temperature	° C	200
Maximum injection flow rate	cm^3/s	50
Maximum holding pressure	bar	700
Clamping force	kN	500
Time of holding pressure	s	20
Cooling time	s	20
Mould base temperature	° C	30

Table 1: Injection moulding parameters

These parameters were constant during all experiments. The injection moulding experiments were done in automatic mode to assure that every part is moulded under the same reproducible conditions. The process is controlled by a laser control unit which provides a Linux based process managing software as well as an interface with several digital in- and output connections (LASCON). The control unit is connected both to the diode laser system and the injection moulding machine. At the start of the variothermal injection moulding process, the injection moulding machine closes the mould. When the mould is closed, the injection moulding machine gives a signal to the control unit, which subsequently gives clearance to the laser system to start the heating period. As soon as the desired temperature is reached the heating period is finished and the laser sends a signal to the control unit, which passes a signal to the injection moulding machine to start the injection of the material. After the cavity is filled, holding pressure is applied until the gate freezes, followed by a cooling time. When the temperature of the part is low enough for demoulding free of damage, the mould opens, the part is being ejected and the cycle starts over again. This operation in automatic mode ensures steady state thermal conditions in the injection mould.

Thermography

Before the moulding trials, the heating characteristic of the indirect laser based mould heating system was analysed by means of thermography. An infrared camera Thermovision A40, Flir Systems, Portland, USA, was placed between the halves of the opened injection mould. Figure 5 shows the temperature field on the surface after a heating time of 5 seconds. In previous investigations, a surface temperature of 130 °C was found to be sufficient for a precise moulding of the microstructures with the same polypropylene material. It can be seen, that the temperature-based laser power control unit is able to heat the centre of the surface to the pre-set temperature of 130 °C. However, due to the Gaussian intensity distribution of the laser beam and the high thermal conductivity of the mould steel, the temperature decreases towards the edge of the mould insert.

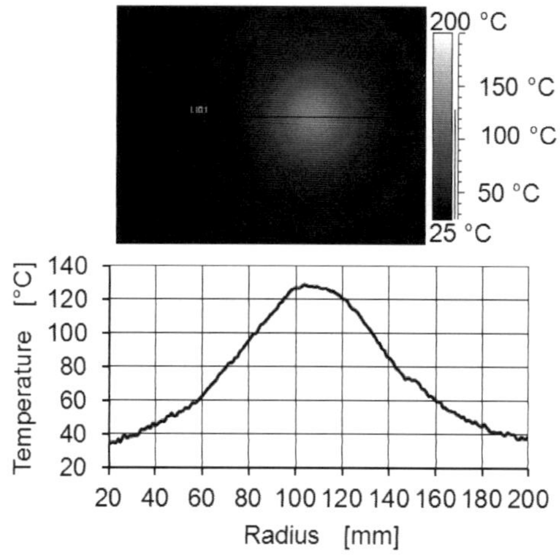

Figure 5: Radial temperature profile of the cavity

Since the pyrometer only measures the temperature of a 2 mm spot in the centre, this temperature drop cannot be detected and compensated by the laser control unit. On this mould insert, micro structures are placed within a radius of 23 mm around the centre. In this area, the maximum temperature difference is 10 K with a minimum temperature of 120 °C.

Moulding results

The moulded surfaces were analysed by means of SEM. Figure 6 shows a SEM picture of a surface created using the micro-structured mould insert in a conventional injection moulding process. The left half shows an area near the centre of the ball scraper, the picture on the right half was taken near the edge. In both cases, it is obvious that the tips of the structures were not filled completely in this process variant. Due to the rapid increase in viscosity as the hot material enters the relatively cold mould, the polymer is not able to fill the ground of the cavity.

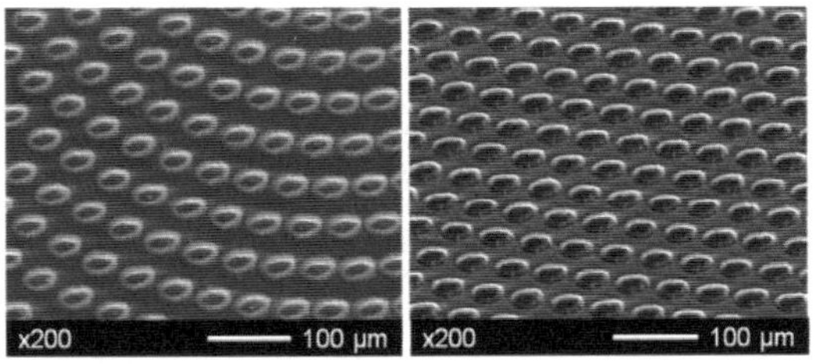

Figure 6: Moulded structures in the center (left) and on the edge (right) of the ball scraper created in the process without additional mould heating

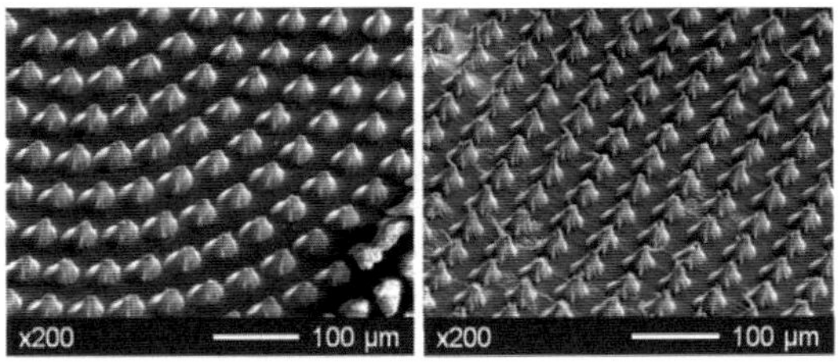

Figure 7: Moulded structures in the center (left) and on the edge (right) of the ball scraper created in the variothermal process (with laser based heating)

Figure 7 shows a surface created in the variothermal moulding process. It can be seen, that in both areas a precise and uniform filling and demoulding of the microstructures was achieved. The increased mould temperature delays the solidification of the polymer melt, leading to a complete filling of the structures. Furthermore, the temperature difference between the center and the edge prior to the injection has no significant effect on the moulding accuracy of the structures.

Conclusion and outlook

In this investigation, a 3D-shaped plastics part with a micro-structured surface was successfully injection moulded and demoulded. Elevated mould temperatures during the melt injection, realised by a laser based mould heating system, are necessary for a complete moulding of the cone shaped dimples. A uniform moulding was achieved on a micro-structured mould insert with the shape of a ball scraper. The focus of this work was the analysis of the moulding and demoulding of the structures regardless of their functionality. However, previous investigations have shown that this type of micro-structured surface shows strongly water-repellent properties. Therefore, the wetting properties of the surfaces will be analysed in future investigations by means of contact angle measurements.

Acknowledgements

The depicted research was funded by the Deutsche Forschungsgemeinschaft (DFG) as part of the program Cluster of Excellence "Integrative Production Technology for High-wage Countries" at RWTH Aachen University. The authors would like to thank KraussMaffei Technologies and SABIC Europe for providing machine technology and polymer materials for the experimens.

References

1. W. Barthlott; C. Neinhuis *International Textile Bulletin* 2001, *1*, 9-12
2. M. Groenendijk *Laser Technology Journal* 2008, *5*, 44-47.
3. V. Zorba; E. Stratakis; M. Barberoglou; E. Spanakis; P. Tzanetakis; C. Fotakis *Applied Physics A* 2008, *93*, 819-825
4. M. Callies; Y. Chen; F. Marty; A. Pépin; D. Quéré *Microelectronic Engineering* 2008, *78-79*, 100-105.
5. W. Michaeli; A. Spennemann *Kunststoffe* 2000, *90*, 52-57.
6. J. Gießauf; G. Pillwein; G. Steinbichler *Kunststoffe* 2008, *98*, 87-92.
7. W. Michaeli; F. Klaiber; S. Scholz in Proceedings of the 4th International Conference on Multi-Material Micro Manufacture (4M), Cardiff, UK, 2008.
8. W. Michaeli; F. Klaiber; M. Schoengart; S. Beckemper in Proceedings of the 7th conference on Multi-Material Micro Manufacture (4M), Oyonnax, France, 2010.

KN-05-01

PROGRESS IN DISPERSIVE MIXING - APPLICATION OF ULTRASONIC TECHNIQUE-

Y.Suetsugu[1*] and A.Sato[1]

[1]*Advanced Technology Research Laboratories, Idemitsu Kosan Co.,Ltd., 1280 Kami-izumi, Sodegaura, Chiba 299-0293 Japan – yoshiyuki.suetsugu@si.idemitsu.co.jp*

Abstract - Mixing and dispersion techniques are in great importance in order to manufacture high quality compounds and composites including nano composites. Progress in dispersive mixing technology will be reviewed in the presentation. A novel dispersion technique for nanoparticles in polymer matrix using ultrasonic wave will then be presented. An ultrasonic unit is attached to a conventional twin screw extruder. It induces a dynamic pressure field with a frequency of 19 kHz in the polymer melt. Dispersion behavior of agglomerate in the ultrasonic zone is visualized with high-speed video camera. Agglomerates of nanoparticles are successfully dispersed in the matrix polymer within 0.5 second of ultrasonic wave application. The dispersion behavior under the ultrasonic dynamic pressure field is very much different from the "Onion dispersion model" and the "Rupture dispersion model" occurring in shear flow in conventional mixers.

Keywords: Mixing, Dispersion, Nanoparticles, Ultrasonic

Introduction

Various types of fillers have been added to polymers or rubbers to improve properties including mechanical properties, thermal properties, electrical properties etc. Manufactures tend to use fillers of smaller size which provide more enhanced properties at low content. Recently many types of nanoparticles are studied and produced.

Properties of the compounds or composites are highly affected by the state of dispersion, as well as the types and content of the fillers. Better dispersion often promises enhanced properties. For instance, tensile and falling weight strengths are greatly improved as the dispersion progresses [1], [2]. Reliability of the electrical properties will be also improved.

Batch and continuous mixers has the history of 100 years. Modular construction of twin screw extrudes for polymer industry has the history of 60 years. There are tremendous efforts to improve the performance of the mixers.

In 1980s, Mannas-Zloczower et al. [3] and Shiga and Furuta [4] presented the dispersion models. Those are the "Rupture model" and "Onion model". Later, Rwei et al. [5] [6] carried out experimental work and stated as follows. At the beginning of the dispersion process, agglomerates rapture takes place under significant stress level. Erosion process like onion peeling becomes dominant later. Agglomerate "gradually" breaks and dispersed in polymers. Based on their observation, rather long residence time at high stress zone would be required in order to get excellent dispersion. Narrowing the gap between tips of rotor or kneading disk and barrel wall, or increasing rotational speed provides higher stress. However, these lower the probability to passing through the high stress zone and elevate the polymer temperature often causing polymer degradation. Dispersing the small or nanoparticles in molten polymer is then not an easy job.

In this presentation, we present and review ultrasonic dispersion technique developed in our laboratory [7], [8], [9]. It provides excellent dispersion and enhanced properties. Agglomerates of nanoparticles can be broken up and dispersed within a second in the polymer matrix.

Experimental

Experimental setup is illustrated in Fig.1. Ultrasonic unit composed of die with flat channel, horn, transducer and ultrasonic generator is constructed. Frequency f and amplitude ξ of the horn are 19 kHz and 7 μ m respectively. Glass windows allow the investigation of dispersion mechanism. The ultrasonic unit is attached to a conventional twin screw extruders. Polymer and filler are melted, mixed and devolatized in the extruder, then flow into the ultrasonic unit.

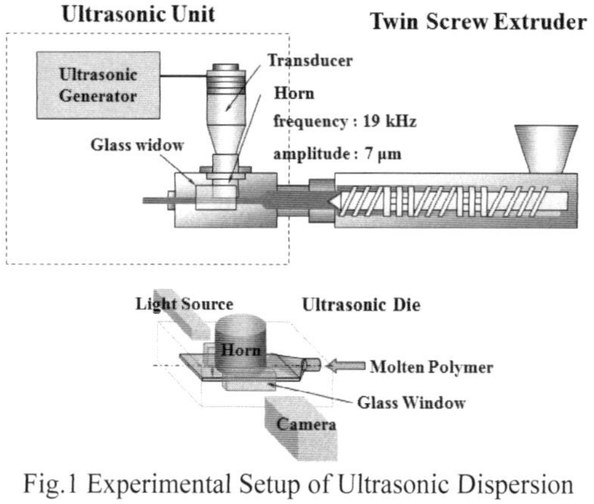

Fig.1 Experimental Setup of Ultrasonic Dispersion

Results and Discussion

Attenuation and reflection of acoustic pressure

Due to the ultrasonic vibration of the horn, compressional wave, with high pressure zone and low pressure zone, is generated and transferred in a media. Wave length λ is given by

$$\lambda = c'/f \quad (1)$$

where c' denotes the velocity in media. Acoustic pressure generated under the horn, P_0, is

$$P_0 = \rho c' \omega \xi \quad (2)$$

ρ the density, ω the angular frequency. The λ and the P_0 in molten polymer are roughly 60 mm and 1 MPa. Attenuation of the pressure wave takes place due to high viscosity of molten polymer.

$$P(x)/P_0 = \exp(-\alpha x) \quad (3)$$

$$\alpha = (4/3)\, \eta^2 \pi^2 f^2 / \rho c'^3 \quad (4)$$

x the distance from the horn, α the attenuation coefficient, η the viscosity of polymer. Attenuation trends are shown in Fig.2. Large reduction of the ultrasonic pressure in high viscous media is expected. Depth of the polymer channel ought to be rather shallow to avoid severe attenuation. In the experiment, the channel depth is 2.5 mm.

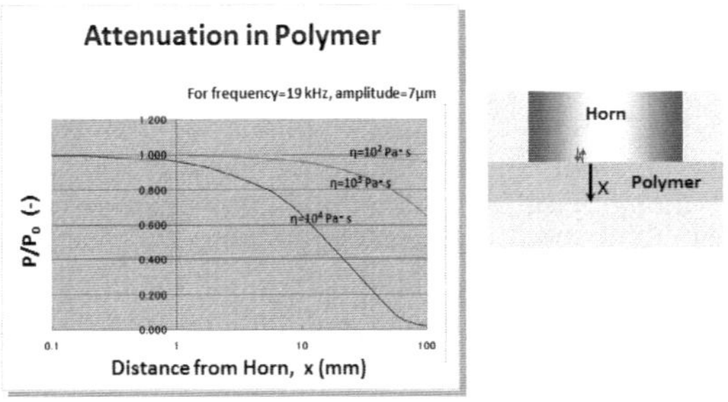

Fig. 2 Attenuation of acoustic pressure in polymer melt

The pressure wave generated at the horn can be reflected at the bottom wall of the channel and the horn. Initial pressure wave can be reflected at the bottom wall of the channel (1st reflected wave). It is then reflected at the horn (2nd reflected wave). Fig.3 illustrates the reflection of the pressure wave for the channel depth=2.5 mm. By summing up 0th wave to 20th wave, the acoustic pressure field of ± 6 MPa with the frequency of 52.6 μ s is predicted.

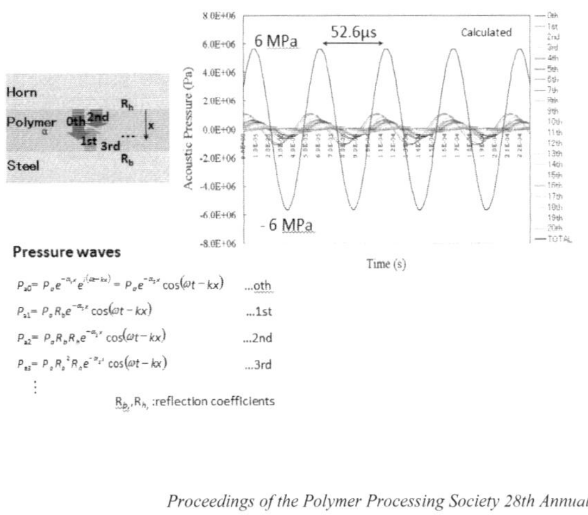

Fig. 3 Acoustic pressure of reflected waves

Investigation of dispersion mechanism

Model agglomerates are inserted into the polymer flow right before the acoustic pressure field zone. Soft agglomerate and compressed agglomerate formed under the pressure of 10 to 100 MPa are used. The agglomerates are successfully dispersed with the application of ultrasonic wave for 0.5 s. Visualization by video cameras reveals what's going on in the acoustic pressure field. Fig.4 shows the photos by high speed video camera. The agglomerate repeats contraction and expansion rapidly with the period of 52.6 μ s, same as the frequency of the acoustic pressure. In the long time scale, 0.2 s, diameter of agglomerate grows 3 times as great as the initial size. Consequently, the agglomerate breakups and is dispersed in the matrix polymer.

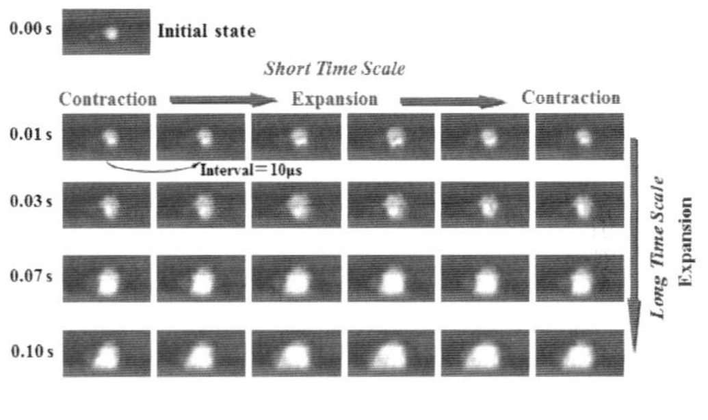

Fig.4 Agglomerate motion in the acoustic pressure field

Fig.5 illustrates the dispersion mechanism presumed under the acoustic pressure field. Inside of the agglomerate should be the particles and air, because highly viscous fluid like polymer melt can't be penetrated into the narrow channels between the particles even after the devolatization. At low pressure period (1 and 3 in Fig.5), the agglomerate rapidly expands with inside air. Agglomerate rapture may occur. At high pressure period (2 in Fig.5), the agglomerate rapidly contracted. Agglomerate may rapture by collision of the fragmented portion of agglomerate. While repeating the "expansion and contraction motion" 9,000 times in 0.5 s, the agglomerate is dispersed and incorporated into the matrix. Size of agglomerate grows in this period by accumulating the volatile components. The mechanism is very much different from the "Rapture model" and "Onion model" in the shear flow.

Virtues and some concerns

As illustrated in Fig.6, the time scale of the ultrasonic dispersion is extremely short. The conventional dispersion, as noted in the introduction, requires considerable period of time. In the ultrasonic dispersion, dispersive mixing is completed in the order of 0.1s. Consequently, the residence time at the ultrasonic zone required is only 1 to 2 s. This promises high throughput.

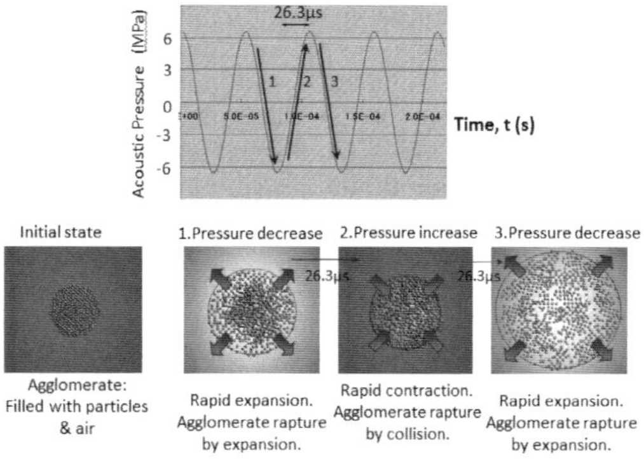

Fig.5 Dispersion model in the acoustic pressure field

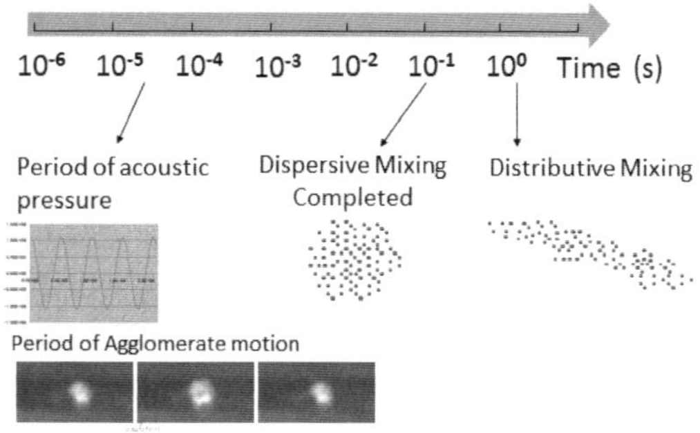

Fig.6 Time scale of the ultrasonic dispersion

In the conventional mixers, the probability of passing through high shear zone is always a serious concern. Major portion of the dispersion would occur at high shear zone like the narrow gap between needing disk tip and barrel wall. Improving both the probability and shear stress leads to low throughput, high power consumption and high polymer temperature often causes degradation.

In the ultrasonic dispersion process, the polymer necessarily flows into the dispersion zone under the horn and agglomerates are dispersed as discussed above, promising high throughput. Rise in temperature in the ultrasonic unit is less than 5℃. No degradation is observed.

It is relevant to control the polymer pressure at the ultrasonic zone to proper range, since it makes use of the pressure of the compressional wave. When the polymer pressure is too high, the "expansion and contraction motion" does not take place and no dispersion occurs. When the polymer pressure is too low, cavitation takes place everywhere in the polymer leading to low dispersion efficiency. Proper range of the polymer pressure is 2 to 7 MPa at ultrasonic unit. Polymer pressure can be controlled by preparing some exits with varying size or number of holes.

Devolatization at the extruder zone leads to better dispersion results at ultrasonic zone. If the polymer possesses many air bobbles or volatile contents, the "expansion and contraction motion" of the agglomerate would be impeded.

Conclusions

Basics, virtures and concerns of the novel technique of ultrasonic dispersion were discussed in comparison with conventional mixers.

1. Ultrasonic unit was developped and mounted to the conventional twin screw extruder.
2. The unit provedes dynamic acoustic pressure wave in polymer melt by utilising reflection of the waves. This induces the agglomerate dispersion with unique "expansion and contraction motion".
3. Dispersion in the unit is completed within 0.5 s and promises high throuput without polymer degradation.
4. Polymer pressure and devolatization are the keys for successful dispersion in the ultrasonic dispersion technique.

References

1. Y.Suetsugu; *International Polymer Processing,* 1990,*3*, 184.
2. Y.Suetsugu; in *Mixing and Compounding of Polymers,* I.Manas-Zloczower and Z.Tadmor, Ed.; HanserPublishers, New York, 1994; Chapter 16.
3. I.Manas-Zloczower; I.A.Nir; Z.Tadmore; *Rubber Chem.Tech.*, 1982, 55, 1250.
4. S.Shiga; M.Furuta; *Nippon Gomu Kyokaishi,* 1982, 55, 491.
5. S.P.Rwei; I. Manas-Zloczower;D.L.Feke; *Polymer Eng.Sci.* 1990, 30,701
6. S.P.Rwei; I. Manas-Zloczower;D.L.Feke; *Polymer Eng.Sci.* 1992, 32,130
7. Y. Suetsugu; A.Sato; *Seikei-kakou,* 2006, 18, 807.
8. Y.Suetsugu; A.Sato; *Japan.Soc.Polym.Process. '07 Sympo. Paper, Yamaguchi, Japan,* 2007, p233-234.
9. Y.Suetsugu; A.Sato; *Japan.Soc.Polym.Process. '06* Sympo. Paper, Gihu, Japan, 2007, p163-164.

O-05-569

A STUDY ON THE MORPHOLOGY AND MECHANICAL PROPERTIES OF TPU/PS BLENDS

M. R. Mansourieh, A. Javadi*, M. Moradi, F. Fallahnejad and M. Bastani
Polymer and Color Engineering Department, AmirKabir University of Technology, P.O. Box 15875/4413, Tehran, Iran; ajavadi@aut.ac.ir; reza.mansourieh@aut.ac.ir; mrzmorady@yahoo.com; f_fallahnejad@yahoo.com

Polystyrene (PS) is a general-purpose glassy polymer with low cost and good processability, which is used extensively in polymeric products. The drawback of this polymer is its brittleness, which in turn limits its applications. In this research, Thermoplastic Elastomers of Polyurethane (TPU) was used to improve the mechanical properties of PS. Blends with different compositions were prepared using a twin-screw extruder and test specimens for impact and tensile characterizations were obtained with an injection-molding machine. The results showed that increasing the amount of TPU leads to higher impact strengths and elongations at break of the blends. It was found that TPU amounts bellow 20 wt. % leads to a rather smooth increase in the mentioned properties whereas in higher TPU contents (30-50 wt. %), the mechanical properties increase dramatically. On the other hand, tensile curves showed that adding only 10 wt. % of TPU changes the behavior of PS from brittle to ductile and therefore yielding occurred in all blends. Scanning Electron Microscopy (SEM) showed a homogeneous dispersion of TPU particles in the PS matrix and the size of TPU domains increased by its higher amounts.
Keywords: PS, TPU, blends, Mechanical properties, Morphology

1. Introduction

Polymeric blends consisting of a glassy matrix and a rubber like polymeric phase are known to exhibit improved impact properties. Polystyrene (PS) is a rigid thermoplastic polymer which has many general applications. It has good process ability and low cost, but its brittleness is caused the restriction of PS applications. Thermoplastic elastomer of polyurethanes (TPUs) are an important class of polymers which have many specialized applications. They are known for their good mechanical strength, wear and tear resistance and low-temperature elasticity. So TPUs are extensively used to improve the different properties of other polymers [1-6]. The improvement of mechanical properties specially impact strength of PS, is one of the aims of researchers to make PS/TPU blends [7-11].

Theocaris and kefalas [7]. Synthesized a polyester based TPU in their lab. and prepared the simple blends of PS/TPU using a brabender. Their results showed that using 5 to 15wt. % of TPU leads to higher impact strength and elongation at break. But in blends containing 20wt. % TPU, these two parameters decreased again. They mentioned that their mixing process was poor and could not make enough dispersion of the rubber phase in PS matrix. So large domains of the TPU did not interact with PS chains. As a result, with increasing TPU in blends, the mechanical properties decreased [3]. Cassu and his coworker prepared the PS/TPU blends using poly(styrene-co-maleic anhydride), SMA, as a compatibilizer. Their results showed that the maximum strain at break is related to blend containing 20 wt. % of TPU and up to 1 wt. % of SMA [8]. Xu and his coworkers investigated the compatibilization effect of poly(styrene-b-4-vinylpyridine) for Syndiotactic Polystyrene (sPS)/TPU blends and found that the added block copolymer reduced the domain size of the dispersed phase in the blends. Also the tensile strength and elongation at break increased, while the dynamic modulus and complex viscosity decreased with the amount of P(S-*b*-4VPy) in the blend. They resulted that the compatibilizing effect of the diblock copolymer is the result of its location at the interface between the sPS and the TPU phases and penetration of the blocks into the corresponding phases, i.e. the polystyrene block enters the non crystalline regions of the sPS, and the poly (4-vinylpyridine) block interacts with TPU through intermolecular hydrogen bonding [11].

2. Experimental

2.1. Materials

The PS used as a matrix was Toporex-525 commercial grade with MFI of 8.5 gr /10 min (200 °C / 5 kg). The TPU was polyester based, Elastollan C70 AWH, producted by BASF company.

2.2. Blend preparation

At first TPU was dried at 70 °C for 3 hours in an oven with air circulation. Then six types of PS/TPU blends, labeled A, B, C, D ,E and F, were made with different amounts of polyurethanes (0, 10, 20, 30, 40 and 50 wt. %, respectively) via melt mixing. A , brabender twin screw extruder

with temperature zones of 150, 160, 170, 175 and 180oC and the rotor's speed of 30 rpm was used to preparing the blends. The standard samples for mechanical properties was prepared by an injection molding machine, Engel made in Austria.

.

2.2. *Mechanical properties*

The Izod Impact tests (with SUN 2500) were performed according to ASTM D256 by using specimens with a 2.5 mm notch depth .

The tensile tests were done by using Cesare Galdabini machine, made in Italy with a strain rate of 5 mm/min according to ASTM D638.

2.3. Morphology

Blends were cryogenically fractured and morphology of the binary blends was analyzed by Scanning Electron Microscopy (SEM) (MIRA II, TESCAN Company).

3. Results and discussion

3.1. Mechanical properties

3.1.1. Impact strength

The Izod impact tests were performed on all samples. At least 10 samples were tested for each blend. The results showed that the impact strength of blends increase with increasing the amount of TPU and this increase is more significant, by using more than 20 % wt. of TPU., table 1 and figure 1.

3.1.2. Tensile strength

The results of tensile tests showed that with increasing TPU content, the modulus and tensile strength of blends decrease but the slope of this decrease is much higher in blends of containing higher amounts of TPU. Also the elongation at break is increased by increasing the amount of TPU and again the slope of this increase is very high in blends of containing high percentage of TPU, table 1and figure 1.

Table 1. The mechanical properties of neat PS and TPU/PS blends.

Samples	Impact strength (j/m)	Tensile strength (MPa)	Elongation at break %	Modulus (MPa)
A	13.71	39.8	0.705	-
B	17.40	35.98	1.08	2945.70
C	18.46	30.35	19.05	2865.23
D	31.15	22.43	24.48	2297.93
E	57.00	17.18	28.06	1638.4
F	164.03	9.19	81.78	698.09

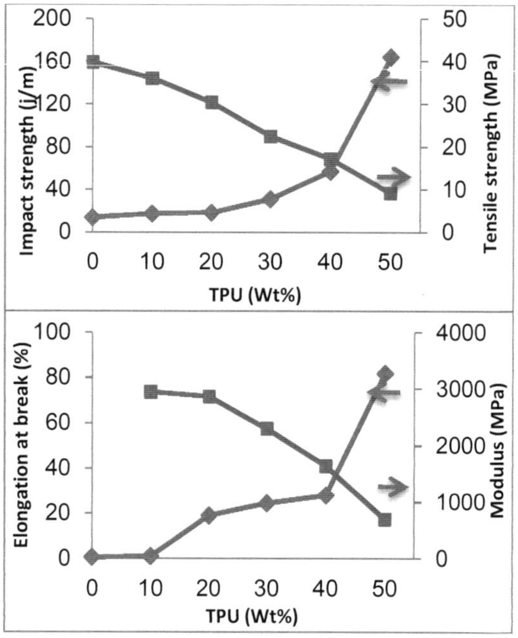

Figure 1. Variation of mechanical properties of neat PS and TPU/PS blends with increase of TPU content.

3.2. Morphology

The microstructure of blends was investigated with Scaning Electron Microscopy (SEM). The micrograph of cryogenically fractured surface of all blends is illustrated in Figure 2. It was found that the morphology of all blends is matrix-disperss and the TPU particles are disperssed homogeniously in PS matrix. The average size of about 100 particles of each blends was determined with Image J software. Table 2 shows the average particle size of TPUs for blends containing 10 to 50 wt. % of TPU. It was found that the particle size of TPU increased by increasing the TPU content. In blends of PS/TPU (90/10), the size of TPUs are in nanometer scale (100-500 nm), figure 3[a], While in 80/20 composition, the droplets were in submicron region, figure 3b. As TPU percentage is added to 30 and 40 wt. %, the dimension of TPU particles becomes biger and reaches to micron size, figure 2c and d and table 2.

Table 2. TPU particle size analysis.

Samples	Min particle size	Max particle size	Mean particle size	Standard deviation
A	-	-	-	-
B	0.115	1.507	0.314	0.215
C	0.273	1.914	0.671	0.252
D	0.280	4.452	1.244	0.768
E	0.443	6.669	1.668	1.001
F	0.303	9.149	2.171	1.632

4. Conclusion

The TPU/PS blends were prepared using a twin screw extruder and the mechanical properties of blends was investigated. It was found that the impact strength and the elongation at break of blends was increased by increasing the TPU content and this increase was much higher in blends containing more than 20 wt. % of TPU. The SEM micrographs showed a homogeneous dispersion of TPU particles in PS matrix and the size of TPU particles was increased with increasing its composition in blends. Despite of smooth surface of TPU domains and also pulling out of many TPU particles, which confirms the poor interaction and weak interface between TPU and PS, it seems that the excellent dispersion of particles makes the improvement of mechanical properties especially in high contents of TPUs.

5. References

1- J. song ; A. Batra ; J. M. Rego , C. W. Macosko , " , *Prog. In Org. Coating*, 2662, 2011.

2-J. H. Wu ; C. H. Li; Y. T. Wu; M.T. Leu; Y. Tsai , *J. composite sci*., 2010. *70*, 1258

3- P.S.O. Patr´ıcio, J.A. de Sales, G.G. Silva, D. Windm oller; J. C. Machado, *J. membrane sci*., 2006, *271*, 177.

4-H. Sheng Wang; X. Dong Chen, M. Qui Zhang, M. Zhi Rong, *Polymer,* 2006, *14*, 1 6.

6- W. Yang, J. Shen, S.H Zhu; C. M. Chan, *J. of app. polym. sci*. 1998 *67*, 2035.

7-P. S. Theocaris; V. Kefalas, *J. of App. Polym. Sci.*, 1991, *42*, 309).

8- S. N. Cassu, M. I. Felisberti, *App. Polym. Sci*., 200493, 2297.

9-P. Potschke;, J. Pionteck ; H. stuz, *Polymer*, 2002, *43*, 6965.

10- J. S. Wang; X.D. Chen; M.Q. Zhang, M. Z. Rong, *Polymer & Polymer Composites*, 2006, *14*, 1.

11- S. Xu; B. Chen; T. Tang, B. Huang, *Polymer*, 1999, *40*, 3399.

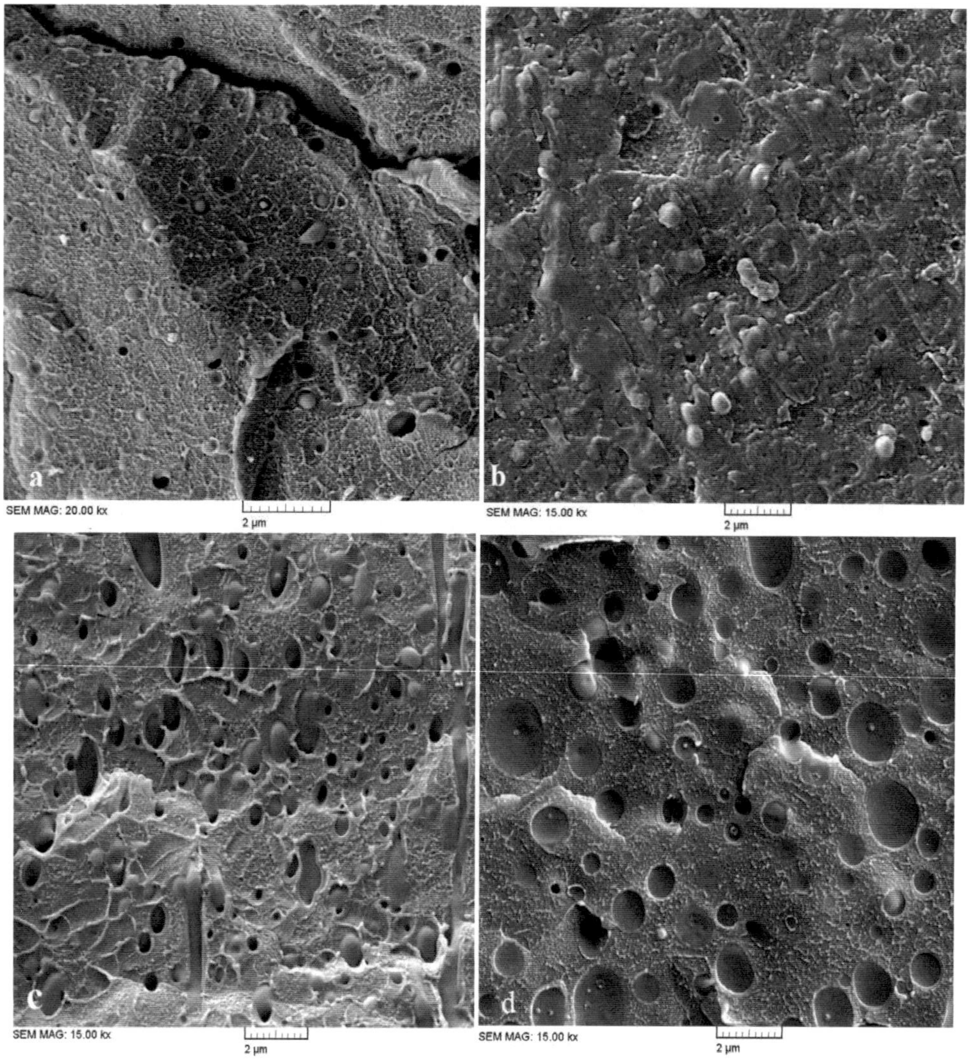

Figure 3. SEM micrographs of the blends with different TPU contents: (a) 10 wt. %, (b) 20 wt. %, (c) 30 wt. % and (d) 40 wt. %.

CHARACTERIZATION OF TRISTIMULUS COLOUR VALUES IN POLYCARBONATE EXTRUSION PROCESS

U. Saeed[1], S. Ahmad[1], J. Alsadi[1], D. Ross[2] and G. Rizvi[1*],

[1] *Faculty of Engineering & Applied Science, University of Ontario & Institute Of Technology, 2000 Simcoe Street North, L1H7K4 Oshawa, Canada*

[2] *Dip SABIC Innovative Plastics, Canada Inc., Cobourg, Ontario, Canada*

Abstract - One of the greatest challenges faced by plastic compounding companies producing small lots, usually for prototyping purposes, is getting the right colour formulation in the first attempt. Any adjustments required, result in significant wastage of material compared to the lot sizes, which are small. In addition, the manufacturing process parameters also affect colour properties, thereby, compounding the problem. Very few studies regarding the effects of processing parameters on color during compounding of polymers have been carried out. In this study, the processing conditions, which can affect the colour tristimulus values, are optimized using the Box Behnken experimental design for a polycarbonate material. The experiment was designed by setting three processing parameters, temperature, speed and feed rate at high and low values. An inter-meshing twin-screw extruder (TSE) was employed and the compounded polycarbonate were molded into flat coupons, which were then analyzed for their CIE L*, a*, and b* values, measured with a spectrophotometer. The obtained data was analyzed using Stat-Ease Design-Expert software for Analysis of Variance (ANOVA). The developed model and experimental data obtained passed all diagnostic tests establishing this model as statistically significant. The polynomial equations represent the quantitative effect of process variables (temperature, speed, and feed rate) and their interactions on the responses. The pigment distribution in the polycarbonate was also examined by means of optical microscopy. The obtained results reveal that the tendency of pigment agglomeration is dependent on treatment levels.

Keywords: Polycarbonate, Pigment, Tristimulus colour values (L a* b*), Experimental design, Optical Microscopy

Introduction

In their natural state, polymer resins meet a variety of performance requirements but are lacking in the color that draws the customer to the end product. Achieving a Marketable color requires the addition of one or more pigments. One of the greatest challenges in the design of an actual product is to obtain realistic manufacturing process parameters prior to plant execution. A few studies regarding the effects of processing parameters on color during compounding of polymers have been carried out by various researchers [1- 2]. It is recommended to use the shortest processing time in order to attain optimal gloss, brilliance, and blend homogeneity. Furthermore, an optimum loading should be used for each specific item; too much pigment is not only costly, but also detrimental as it reduces the impact resistance [3]. The term ''pigment dispersion''in the /PC/ industry describes the relative amount of pigment aggregates and agglomerates that are present in solid media. Several properties of the polymer formulation influence the dispersing process namely the viscosity of the binder, its ability to wet the pigment surface, the attractive forces operating between the pigment primary particles and the production process [4, 5]. Difficulties with dispersion or obtaining a uniform blend can be overcome by reducing the resin viscosity and extending mixing time [4]. There are few quantitative methods for the determination of the state of pigment dispersion; one of these involves direct visual observation using microscopy. Although there are other methods, such as flocculation-gradient techniques or color strength and hiding power determination, during optimization of the production parameters and during development of the formulation, microscopy gives a direct view of the pigment particle size and size distribution.

One of the most significant aspects of color matching is to establish a proper color standard generated by spectrophotometric measurement. Spectrophotometers are particularly helpful quality control tools to measure color and numerically define color differences [5]. For the measurement of color, standard values are used worldwide, for example as determined by an organization called CIE. CIE was founded in 1913 and stands for *Commission Internationale de l'Eclairage*(International Commission on Illumination). The values used by CIE are called L*, a* and b* and the color measurement method is called CIELAB. L* represents the difference between light (where L*= 100) and dark (where L*=0), a* represents the difference between green (-a*) and red (+a*) and b* represents the difference between yellow (+b*) and blue (-b*) [5]. The deviation in L*, a*, b* with respect to target is represented as "dE*, where

$$dE = \sqrt{(\Delta L)^2 + (\Delta a)^2 + (\Delta b)^2} \qquad (1)$$

It represents the magnitude of the difference in color, but does not indicate the direction of color difference. When colored materials are produced they are compared to a standard as a measure of quality control. Instead of using absolute values of color, color differences are used. The total change in color, dE* is

used to represent the color difference in the CIELAB color space [6]
A factorial design, or a fraction of such, can be used to detect which predictor variables, or which two-variable interactions have an effect on the response. These designs are efficient in that they use few experimental observations, but they cannot detect or estimate any quadratic relationships. After the initial exploration of the predictor variables, 3 level full factorial, a central composite design or a Box-Behnken design can be used to investigate the full second order polynomial relationship between the response and the predictor variables. The Box Behnken design is dominant on other response method due to economic point of view [7].

The main goal of this study is to employ the Box-Benhken response surface design to optimize the processing parameters affecting on the color properties of polymer during extrusion process and to characterize the obtained sample using optical microscope in transmission mode. The optimization of the process parameter for color processing will not only improve the quality of the product but reduce the total wastage.

Experimental (Times New Roman 10, bold)
Experimentation was carried out for the investigated grade at SABIC IP Canada Inc. , a blend of two polycarbonate resins was used along with four different pigments, the color formulation of these grades in parts per hundred (PPH) presented in Table 1. The melt flow index (MFI) for Resin 1 was 25 gm/10min and that of Resin 2 was 6.5 gm/10min.

Table 1: Color formulation used for investigated grade

S.No	Type	PPH
1	Resin 1	33
2	Resin 2	67
3	Pigment A	0.20
4	Pigment B	0.05
5	Pigment C	0.0004
6	Pigment D	0.0016
7	Pigemnt E	0.0710

The processing was carried out on twin screw extruder (ZSK26 Coperion Germany) of 25.5 mm, 27 kW with ratios of L/D=37 and D_o/D_i=1.55. The extruder had 10 heating zones, 9 identified on the barrel and 1 at the die. The three process parameters, the temperature of the mixing zone, feed rate to the extruder, and the screw speed were considered, the experimental design level used is shown in table 2. Parameters were varied on 13 different treatments with additional four center point (treatment level 1) for Box Behnken response method to study the effect of variables on color. The additional centre point was added to estimate the experimental error and for the detection of nonlinearity in the responses.

Table 2: Parameters and experimental design level used

Treatment level	Temperature °C	Screw Speed rpm	Feed rate Kg/hr
1	255.00	750.00	19.00
2	255.00	650.00	11.00
3	255.00	850.00	27.00
4	280.00	750.00	11.00
5	255.00	650.00	27.00
6	230.00	750.00	11.00
7	230.00	750.00	27.00
8	230.00	650.00	19.00
9	280.00	750.00	27.00
10	280.00	850.00	19.00
11	255.00	850.00	11.00
12	230.00	850.00	19.00
13	280.00	650.00	19.00

Extruded melt was quenched in cold water and then pelletized. Using injection molding, these pellets were then molded into rectangular chips (3x2x0.1") which were then analyzed for their CIE L*, a*, b* values, measured with a spectrophotometer (CE 7000A, X rite-Inc. USA). The desired color output in terms of CIE L*, a*, b* values was taken as L*= 70.04, a*= 3.42, and b*= 18.09. The Design Expert Software (Version 8, Stat-Ease Inc. USA) was used for the statistical analysis of data in order to evaluate and compare the effects of parameters. The part of sample chips were hot pressed to get thick plates of 70 μm which were then characterize using optical microscope in transmission mode. The degree of pigment dispersion in PC was evaluated by distribution of pigment particle sizes. The micrographs of samples were quantified using Pro image software. The software identifies the pigment particles, calculates their area and generates an area distribution graph. The area distribution was converted to their size distribution.

Results and Discussion
Optical micrographs were obtained at magnification of 100X (Fig.1). The presence of the agglomerates and aggregates in samples is caused by the incomplete pigment dispersion process as seen in these micrographs. The micrographs shown in Fig. 1 (top) were then processed to improve contrast and to convert each micrograph into a binary (negative contrast black/white) micrograph. An example of such a transformation is presented in Fig. 1 (middle). The other micrographs were converted in a similar manner. The Fig. 1 (bottom) shows the results of particle detection using the Image-Pro software.

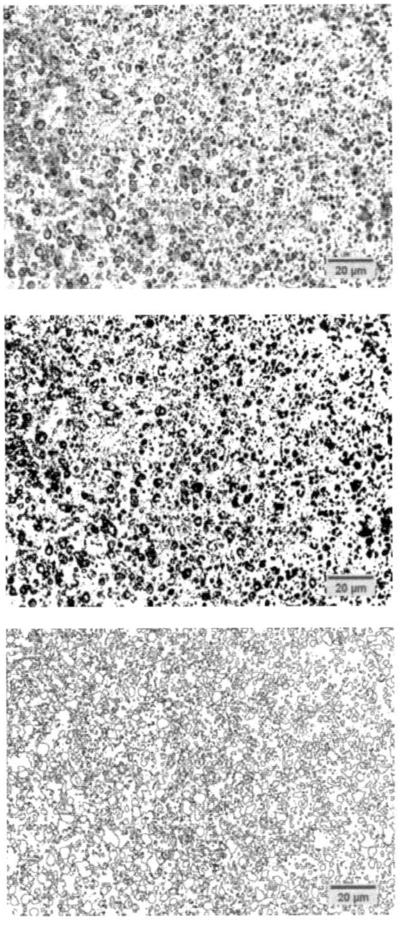

Fig. 1: Micrograph at 100X magnification (a) before-top (b) after the image modification- middle and (c) outlined particles - bottom

As an example, The results of the image analysis of 4 different treatment levels are presented in the Fig. 2, where the size distribution of pigment particles is shown. The quality of the dispersion process can be evaluated from the distribution graph. The smallest particles have a diameter of 0.20 mm and it seems that the peak of treatment 1 is slightly sharper than the other treatment levels in the major distribution region. This implies that treatment 1 have a lower variance of particle density, illustrates better dispersion of the pigments. The results of micrograph can be correlated by calculating the variation (dE) in color tistimulus values with respect to the target by using equation 1. The calculated results demonstrate that the dE for treatment 1 is 0.38 as compared to the other mentioned treatment level which leads the dE equal to 1.18, indication of poor pigment dispersion.

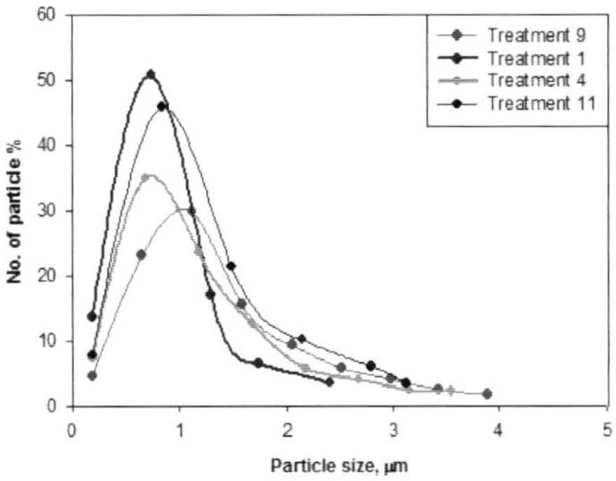

Fig. 2: Particle diameter distribution in samples obtained from treatment level 9,1,4 and 11 presented in table2.

Analysis of variance (ANOVA) was performed to find interactions and to optimize the processing parameters for color properties. The effect of parameters on L*, a* and b* are characterized by using Design Expert software. Sequential F-tests are performed, starting with a linear model and adding terms (quadratic and linear if appropriate) [7]. The F-statistic is considered for each type of model, and the highest order model with significant terms normally would be chosen. The same process was adopted for all tristimulus values and only the significant terms were included based on the F statistics ≤ 0.05 and probability values ≤ 0.1.
Strong interactions were observed between process parameters for L*, a* and b* between the investigated parameters. As an example, Figure 3 presents the contour plot drawn at temperature 273.57°C for response a* with 95% confidence interval, indicates the non linear ship between speed and feed rate. The strong curvature leading to elliptical contours is due to the quadratic terms of all three processing parameter. Therefore, a* value increases with increasing speed and feed rates in two quadrants and decreases in the other two quadrants. Also the interactions between the speeds and feed rates are positive in two quadrants and negative in the other two. It also indicates that the target value is achieved at 738 rpm and 24.06 kg/hr.

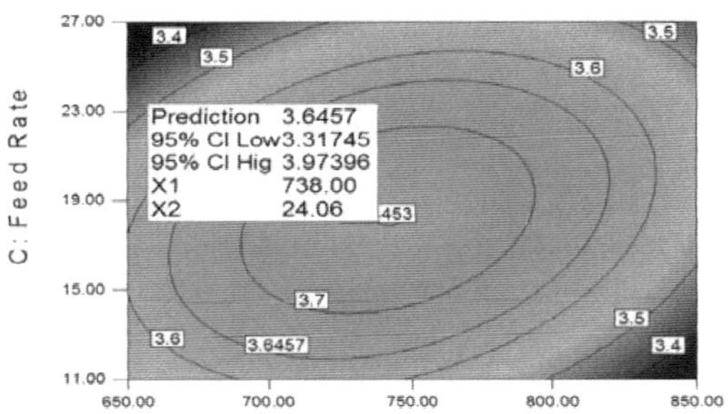

Fig. 3b: Contour plot between speed and feed rate for a*

Due to the results of the ANOVA analysis, multiple linear regression analysis was generated to determine the predicted response for L*, a* and b* function presented in table 3. The parameters and their interactions integrated in the regression model have strong impact on the predicted tristimulus values.

Table 3: Regression model for L*, a* and b*

Response	Regression model
L*	$+70.22-0.29 * A-0.034 * B+0.087 * C+0.25 * B * C+0.084 * B^2$
a*	$+4.01-0.18 * A-0.018 * B+2.500E^{-003} * C-0.043 * A * C+0.097$
b*	$+18.27-0.17 * A-0.021 * B+0.019 * C+0.20 * B * C-0.23 * A^2-0.18 * C^2$

In order to optimize the surface in the feasible region a "Numerical Optimizer" was used. The response optimizer of Design expert calculates a number of local points in the practicable area of the variable. Since factors temperature (A), Speed(B) and Feed rate (C) exert a significant influence on the responses of all color components (i.e. L*, a* & b*), a multi-criteria decision making method was essential and a total desirability function "d" was used to detect the optimum settings of the three parameters with respect to all responses [8]. The desirability function is a measure of overall quality and provides convenient means to compare several responses and to select the optimum with the most desirable properties; it reflects the desirable ranges for each response (d_i). The measured responses are transformed to a dimensionless desirability scale, that ranges between d = 0, for a completely undesired response, to d = 1 for a fully desired response. The value d is the highest at conditions where combination of the different criteria is globally optimal. The predicted values of L*, a* and b* for the investigated grade are presented in desirability figure 4. It was found that the maximum combined desirability of 86% in terms of L* = 70.04, a*=3.64, b*=17.93 achieved at 273.57 °C, 738.00 rpm and 24.06 kg/hr which is in quite good agreement to the preferred target values. The slight variation of the optimized values with respect to the target is in the range of 95% confidence interval. The agreement of optimized values suggests that the model is validating for the prediction of color values.

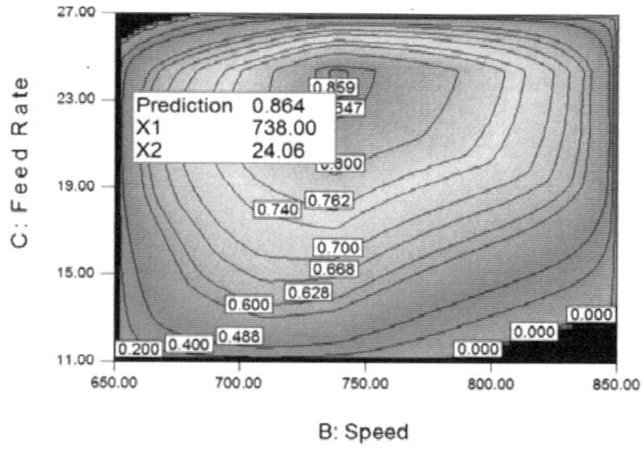

Fig. 4: 2D Desirability curve

Conclusions

In this study, experimental design approach based upon Box Behnken response surface methodology was used to investigate the effects of parameters. The current study indicates the presence of interaction in L* a* and b* present the interaction between the operating parameters. The optimized tristimulus values are closer to target values at lower 273.56 °C, 738.00 rpm and 24.06 kg/hr. The obtained results reveal that the tendency of pigment agglomeration is dependent on treatment levels. The optical method gives quantified information concerning the particle size distribution and, consequently, the degree of the pigment dispersion in the system. Such results are in close relation to the total color difference values. Furthermore, the obtained optimized parameters can be used to achieve the target value for polymer color that would

Acknowledgements

The authors would like to extend their gratitude for the financial support provided by SABIC IP and NSERC as well as the staff of SABIC IP at Cobourg, Canada for their assistance.

References

1. S. P. Rwei, *Polymer Engineering and Science* 45, 1665, (2001)
2. A.Y. Wong in SPE ANTEC 44 Proceeding, China (1998)
3. B. Klein, , Metallic pigments in Coloring of Plastics, Fundamentals, Robert Charvat John, Ed. 2; Wiley & Sons, Vol.1, 1-472, (2004)
4. F.L. Duivenvoorde, C.F. van Nostrum, J. Laven, R. Van der Linde, *J. Coat. Technol*, 72, 145, (2000)
5. ASTM D 2244-93. Standard Test Method for Calculation of Color Differences from Instirnentally Measured Color Coordinates, 222-215.
6. http://itl.nist.gov/div898/handbook.
7. www.weibull.com.
8. M. A. Islam, V. Sakkas, T. A. Albani, *J. Hazard. Mater.* 170, 230, (2009)

O-05-453

ZEOLITE AND TALC – IMPORTANT MINERAL FILLERS FOR ACTIVE PACKAGING APPLICATIONS

C. Holzer[1]*, M. Piontek[1], U. Gallaun[1], S. Laske[1], G. Rieß[2] and H. Flachberger[3]

[1]*Department of Polymer Engineering and Science, Chair of Polymer Processing, Montanuniversitaet Leoben, Austria – clemens.holzer@unileoben.ac.at*

[2]*Department of Polymer Engineering and Science, Chair of Chemistry of Polymeric Materials, Montanuniversitaet Leoben, Austria*

[3]*Department of Mineral Resources, Chair of Mineral Processing, Montanuniversitaet Leoben, Austria*

Abstract

Thermoplastic compounds with mineral fillers have a special position in many applications including automotive, consumer products, packaging materials and myriad of other applications to improve the material properties. Thereby the filler and particle size distributions are one of the most important impacts onto the mechanical properties of the compound. For improved processability of thermoplastic compounds, the used fillers in this project were coated with silane. This modification also improves the bonding between the filler and the polymer matrix.

In this research the influence of coated and uncoated milled zeolite and for comparative reasons talc as well onto the mechanical properties and processability of Polypropylene compounds was investigated. In addition, the rheological properties, which are related to the processability of the filled Polypropylene compounds, were measured. Additionally the zeolite was doped with two different chemicals to realize water absorption for packaging applications. Thereby the water absorption of produced plates was detected by weight measurements. After the measuring time the plates were dried to the releasing rate during the drying process.

Through the use of coated mineral fillers the processability and properties of polymers could be improved and by additional doping of coated zeolite compounds good water absorption was reached. This can be a big advantage in active packaging. For production it is also important that all compounds can be handled without any change of basic process parameters.

Keywords: Zeolite, Active Packaging, Compounding, Coating, Talc.

Introduction

Mineral fillers are the most used fillers in many polymeric applications including automotive, consumer products, packaging materials and many other applications [1, 2, 3]. Since years mineral fillers are not only used for reducing material costs, but also to improve properties such as stiffness, impact strength, hardness, volumetric shrinkage, heat distortion temperature, and electric conductivity [4].

Natural zeolite is a very interesting mineral filler, having a highly regular structure of pores and chambers that allow some molecules to pass through and cause others to be either excluded or degraded [5]. It is in many ways the inorganic equivalent of organic enzymes, as many of them also have specific sized chambers that trap chemicals within our bodies, holding them and either degrade them or react with specific chemicals. Depending on the type this mineral is also able to store up to 40 wt% of its dry weight as water, which is released during heating without destroying the mineral structure [6].

Based on these properties natural zeolite is used for active or intelligent polymer packaging applications to realize interaction between the packaged good and the packaging, for example indication of decomposed vegetables [7, 8].

Experimental

Materials

The mineral fillers used in this study are natural zeolite-clinoptilolite with a high purity level of about 90 % and talc. The zeolite and talc powder was purchased from Paltentaler Minerals GmbH, Austria.

The used base polymer was Polypropylene (MOSTEN GB005) with a melt flow rate (MFR) of 5 g/10 min (230 °C/ 2.16 kg).

For a better interaction between filler and polymer matrix, the fillers were coated with Vinyltrimethoxy-silane and 3-(Trimethoxysilyl)propylmethacrylate.

$CaCl_2$ and Na_2SO_4 were used for doping of the coated zeolite.

Preparation of the compounds

The zeolite and talc powder were incorporated into the base polymer with an amount of 50 wt%. All fillers were coated with the two different coating agents to reach a better processability and interaction between the filler and the polymer. The weight ratio of coating agent and filler was always 1:20.

All of the compounds were manufactured using an intermeshing, co-rotating twin screw extruder Theysohn TSK30/40D (Theysohn Holding, Austria) and water granulation Econ EUP100L (Econ Ltd, Austria). The feed rate was set at 5 kg/h for all polymer-filler-compounds with a screw speed of 100 rpm. To assure an optimal dispergation of the mineral fillers in the polymer matrix a screw geometry with three dispersing sections was used.

The temperature profile can be seen in Table 1.

All compounds were used to produce specimens for testing the viscosity, tensile- and Charpy impact tests and water absorption as well as films for testing the processability.

Table 1. –Temperature profile for the compounding process

Cylinderzones	Temperature PP-Filler Compounds (°C)
hopper	cooling
2	160
3	180
4	200
5	200
6	200
7	200
8	200
die	200

Production of the polymer films

The films were produced by using a single screw extruder E30P (L=30D) (Dr. Collin GmbH, Germany) with a 250 mm flat film die. The thickness of the films was adjusted at 100 µm. The melt temperature was set at 220 °C with a screw speed of 80 rpm.

The temperature profile of the extruder including the temperature of the feed block, the melt tube and the die can be seen in Table 2.

Table 2. –Temperature profile for film extrusion

Cylinderzones	Temperature PP-Filler Compounds (°C)
hopper	170
2	200
3	220
4	220
5	230
melt tube	230
feed block	230
die	230

Production of specimen

The test specimens for tensile tests were produced with an Allrounder 470-1000, 1000 kN closing force, injection molding machine (Arburg, Germany). The melt cylinder temperature was set to 200 °C with a tool temperature of 30 °C.

The temperature profile for the injection molding machine can be seen in Table 3.

The test specimens for viscosity, impact tests and water absorption were produced with a vacuum press P200PV (Dr. Collin GmbH, Germany). The temperature was set to 210 °C and the applied pressure was 16.5 MPa.

Table 3. –Temperature profile for injection molding

Cylinderzones	Temperature PP-Filler Compounds (°C)
Hopper	180
2	185
3	190
4	195
die	200

Complex viscosity

Steady shear measurements, with rate sweep from 10^{-1} to 100 s^{-1}, were performed with the rotational rheometer MCR501 (Anton Paar, Austria). The plate-plate diameter was 25 mm. In order to avoid oxidative degradation of the material during the measurement, the experiments were carried out in nitrogen atmosphere.

The measurements were carried out at 220 °C melt temperature in a frequency range between 0.1 rad/s to 500 rad/s.

Tensile test

A universal tensile testing machine Z010 (Zwick and Co. KG, Germany) was used to carry out tensile tests according to ISO 527-1.

All specimens were tested under standardized conditions (23 °C/50 % r.H.). The data were evaluated using the testXpert II software (Zwick and Co. KG, Germany).

Charpy impact test

A pendulum system (CEAST RESIL 25, Instron/CEAST division, Italy) was used to carry out the Charpy impact tests according to ISO 179. All tests were carried out under standardised conditions (23 °C / 50 % r.H.).

The impact tests were performed using standardised test specimens and were carried out 10 times for each mixture to get a significant average value.

Water absorption

The test specimens were inlayed in a water bowl under standardised conditions (23 °C and 50 % r.H.) and the weight was measured during the test period. After 28 days the specimens were dried and again measured.

Results and Discussion

Processability of the compounds

The processability of all compounds showed no difference with other mineral fillers such as chalk or kaolin. For proper incorporation of the filler in the polymer matrix an optimal screw geometry was chosen. In this project a "sharp" geometry with three dispersing sections was chosen.

In the production of compounds it could be seen that the coating minimizes the bridging of the filler in the side dosage. Since zeolite rapidly absorbs moisture from the surrounding air, it is advisable to get the filler

directly from the dryer to the compounder and use a degassing during the process. Thus it can be said that with these restrictions the used fillers can processed with existing compounding lines or process parameters if a suitable screw is chosen.
The further processing of the pellets with a cast film line proceeded without difficulties. The basic requirement for good processability was sufficient drying of the compound in a desiccant dryer (at least 24 h at 80 °C).
The processing of the compounds in a conventional injection molding machine basically works fine when the same conditions as in the compounding process (adequate drying) are observed. In the machine no agglomeration or segregation occurred. Provided that the compounds are compounded homogeneous, it is possible to use injection molding for processing. Direct processing of fillers is not recommended.
Basically it can be said that zeolite and talc compounds can be well processed with polymer processing machinery considering some procedural necessities.

Complex viscosity
The viscosity curves at a temperature of 220 °C are shown in Figure 1. The viscosity plot shows the viscosity of the polymers in the shear rate range between 0.1 s^{-1} and 500 s^{-1}.
For materials such as high filled polymer melts (PP with 50 wt% zeolite and talc; uncoated and coated), which process a large yield stress, the constant viscosity region may not exist. The viscosity of the high filled zeolite compounds is even higher than the viscosity of the talc compounds. Uncoated zeolite compounds have a viscosity nearly 500 % higher than talc compounds, the difference between the coated compounds is nearly 1700 %.
If the filler is coated, the whole viscosity curve gets to lower values. The coating agents reduce the viscosity of the high filled zeolite/talc compounds for a better processability required in injection molding or film extrusion.

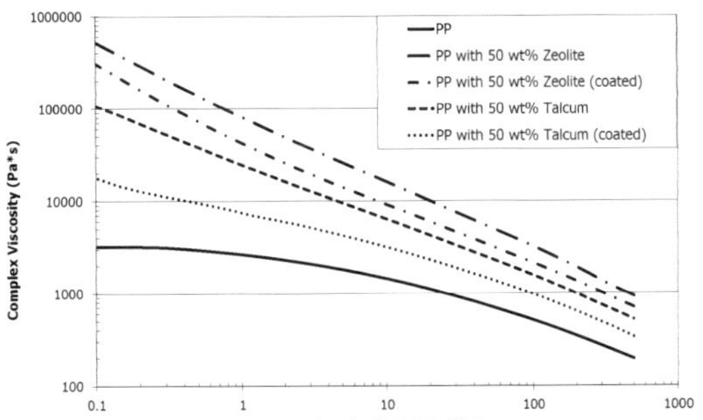

Figure 1: Viscosity curves of PP and PP-zeolite/talc compounds (coated and uncoated)

Tensile test and Charpy impact test
Results for Young's Modulus, tensile strength, elongation and Charpy impact are shown in Table 4.
The Young's Modulus increases for polymers with fillers like zeolite or talc and toughness decreases. Elongation decreases by a decade. Zeolite compounds show an approximately 30 % higher tensile strength and nearly a 200 % higher elongation than talcum compounds. The Young's modulus and toughness instead are about 20 % lower than the talc compounds. The two coating reagents have a small positive influence on Young's modulus und a bigger increasing influence on toughness.
Summing up the results it can be said that the coating of the fillers has a positive influence on the workability, the viscosity and the strength of the zeolite and talc compounds.

Table 4: Mechanical properties of PP and PP-zeolite/talc compounds with a filling degree of 50 wt%

Polymer Compound	Young´s Modulus (MPa)	Tensile strength (MPa)	Elongation (%)	Charpy impact (kJ/m^2)
PP	1580	34	19	3.7
PP-Z	3970	33	1.6	1.2
PP-Z(c)	4200	31.8	1.4	1.6
PP-T	4820	24.1	0.8	1.6
PP-T(c)	4960	25.3	0.9	1.8

Water absorption
The water absorption of the zeolite compounds is shown in Figure 3. Within 4 days 3 wt% of its dry weight can be absorbed by the zeolite compound. During the remaining test period, the moisture content is substantially constant. The drying process can reduce almost all the absorbed moisture. Different zeolite particle sizes were also tested and it can be shown that they have no influence on water absorption. Talc compounds are not able to absorb water as good as zeolite compounds.

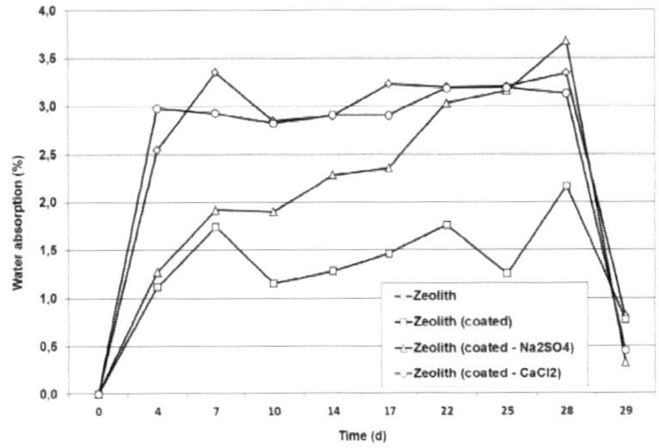

Figure 3: Water absorption of PP-zeolite compounds

Coating of the fillers is needed for better incorporation and mechanical properties but it reduces the water absorption by about 50%. To reach good water absorption and good processability additional doping agents were added (not possible for talc). Doping with $CaCl_2$ showed water absorption like uncoated compounds, with Na_2SO_4 linear water absorption can be achieved throughout the test. This may be of advantage in terms of active packaging for moisture absorption by the packaged product over a certain period. Again, by the drying step almost the entire amount of water is released back.

Conclusions

Comparing the mechanical properties and viscosity curves of uncoated and coated zeolite compounds it can be said that coating of the fillers is needed for better processability and better properties.

By additional doping of the coated zeolite compound with $CaCl_2$ and Na_2SO_4 good water absorption can be reached, which can be a big advantage in active packaging.

In processing the zeolite/talc compounds behave as other compounds with mineral fillers. Since zeolite strongly absorbs moisture, it is very important to dry the compound sufficiently before any processing and use it directly after the drying process. For the incorporation of the powdered mineral filler in the polymer, it is advisable to use an additional degassing, to remove any possible residual moisture.

Basically, all zeolite/talc compounds in extrusion and injection molding process can be handled without any change of basic process parameters. Treatment of the fillers with coating reagents can minimize agglomeration during incorporation.

Through the use of mineral fillers like zeolite and talc the properties of polymers can be improved and an active plastic packaging can be developed that satisfies the individual needs of the consumers.

Acknowledgements

This research work was performed at the Polymer Competence Center Leoben GmbH (PCCL, Austria) within the framework of the K1-program of the Austrian Ministry of Traffic, Innovation and Technology with contributions by the Montanuniversitaet Leoben.

We acknowledge our company partners Paltentaler Minerals and Greiner Packaging. We want to thank also the Chair of Materials Science and Testing of Plastics for the tensile and Charpy impact tests.

References

1. S. Quintavalla; L. Vicini *Meat Science* 2002, *62,* 373-380.
2. S. Kwon; K. J. Kim; H. Kim; P. P. Kundu; T. J. Kim; Y. K. Lee; B. H. Lee; S. Choe *Polymer* 2002, *43,* 6901-6909.
3. Y. Wang; A.J. Easteal; X. D. Chen *Packag. Technol. Sci.* 1998, *11,* 169-178.
4. H. Kim; J. Biswas; S. Choe *Polymer* 2006, *47,* 3981-3992.
5. C. Chuapradit; L. Ruangchuay Wannatong; D. Chotpattananont; Piyanoot Hiamtup; A. Sirivat; J. Schwank *Polymer* 2005, *46,* 947-953.
6. D. Metin; F. Tihminlioglu; D. Balköse; S. Ülkü *Composites* 2004, *Part A 35,* 23-32.
7. S. Nur (Aktas) Dirim; Ali Esin; Alev Bayindirli *Turkish J. Eng. Env. Sci.* 2003, *27,* 1-9.
8. P. Appendini; J.H. Hotchkiss *Innovative Food Science & Emerging Technologies* 2002, *3,* 119-126.

O-05-452

PREPARATION AND PROPERTIES OF PP/VGCF COMPOSITES: EFFECT OF SURFACE MODIFICATION WITH TITANATE COUPLING AGENT

W. Cheewawuttipong, N. Washida, S. Tanoue*, H. Uematsu, and Y. Iemoto
Graduate School of Engineering, University of Fukui, Fukui, Japan- tanoue@matse.u-fukui.ac.jp

Abstract
The purpose of this study is to develop the mechanical properties and the thermal conductivity of polymer composites by twin screw extruder. The vapor grown carbon fiber (VGCF) similar to graphite was used to improve the thermal conductivity of polymer. Two types of polypropylene (PP) with different viscosity were employed as polymer matrix. The thermal conductivity and the mechanical properties increased with content of VGCF. In addition, PP composites showed a ductile failure in tensile testing. The surface modification was used titanate coupling agent for coating to VGCF. The agglomerate of VGCF was observed with content of VGCF. When VGCF treated with titanate coupling agent enhanced to modify interfacial between filler and matrix, well dispersion, and also reduced agglomerate. However, the mechanical properties did not change with titanate coupling agent. On the other hand, thermal conductivity became increased a little by modified VGCF of high content.

Keywords: Polypropylene, Vapor grown carbon fibers, Titanate coupling agent, Twin screw extruder

Introduction

The polymer is achieving the significant economic benefits in comparison with other materials as a result of light weight, lower costs, and excellent electrical insulators with a low thermal conductivity. Consequently, some researchers have worked to develop thermal conductivity of polymer by increasing filler which has high thermal conductivity such as diamond, mica, glass, clay, carbon, boron nitride (BN), aluminium nitride (AlN), silicon carbide (SiC), and alumina (Al_2O_3) [1,2].

Generally, carbon has been used to improve Young's modulus, tensile strength, thermal conductivity, and excellent electrical conductor. Vapor grown carbon fiber (VGCF) is a new type of carbon fibers. The structure of VGCF is similar to carbon nanotubes (CNTs) and also cheaper than those of CNTs. VGCF have benefits in terms of high mechanical and thermal conductivity properties by combining with polymer such as polyamide (PA), polycarbonate (PC), polyethylene (PE), polyether ether ketone (PEEK), polypropylene (PP), polystyrene (PS), polylactic acid (PLA) and Epoxy etc. [3-5].

The titanate coupling agent is usually used in the polymer composites to make molecular bridges at the interface between filler and matrix and modify dispersion of filler such as ceramic powder, carbon black, carbon fiber, calcium carbonate ($CaCO_3$) etc. [6-7]. It clearly demonstrated that the titanate coupling agents (LICA12 and LICA 01) were used in PP/HDPE/$CaCO_3$ with 0.3 – 1 wt.% of filler have effect to modify the mechanical properties of composites [8]. However titanate coupling agent (LICA 12) was used in PP/Talc composites with 0.2 – 1 wt.% of filler. This research illustrated that the melt viscosity of composites decreased and well dispersed. The tensile and flexural properties were decreased whereas impact and elongation properties increased by treated coupling agent [9]. Therefore, it is possible to make well dispersed of filler in matrix and maintain or improve the mechanical properties composites contained VGCF by using surface modification.

The aim of this study was to investigate the VGCF content and surface modification with titanate coupling agent influencing on the thermal conductivity and the mechanical properties of PP/VGCF composites. In this study, PP composites were prepared by twin screw extruder.

Experimental

Material

Polypropylene (PP) was supplied by Sumitomo Chemical Co., Ltd., Japan. Two types of PP, low-viscosity PP (PP-L, MFR 26-29 g/10 min) and high-viscosity PP (PP-H, MFR 7-8 g/10 min), were used in this study. The multiwall vapor grown carbon fiber (VGCF) was produced by Showa Denko K. K. , Co., Japan. Namely VGCF-X; average diameter 15 nm, and average length 3μm was used to adding in PP matrix.

Surface modification

Titanate coupling agent (KR138S) was produced by Ajinomoto Fine-Techno Co., Inc. Japan. It was used by treated VGCF with titanate coupling agent at 0.2 wt.% of filler content. The coupling agent was mixing with isopropyl alcohol to make solutions for filler separately. The solution was stirred for 30 min at room temperature. The treated VGCF were dried to remove solvent at 100 °C for 12 h.

Preparation of composites

The PP/ VGCF composites were mixing by a twin screw extruder at a barrel temperature of 200 °C and the screw rotation speed was 150 rpm. All materials were put in hopper at a feed rate of 0.5 kg/h The machinery was produced by Kurimoto Co,. Ltd, Japan. Materials were dried under vacuum at 80 °C for at least 24 h before compounding.

with VGCF and treated VGCF

Testing

The thermal conductivity was measured with the hot-wire method. The device was produced by Kyoto Electronics Manufacturing Co., Ltd., Japan. The sample size with 17x12x0.05 cm was used. The tensile testing was carried out, according to JIS K7162 test method, using a Shimazu Autograph AGS-J at room temperature. The specimens (JISK7162, Type 1BA) were prepared by injection molding. The dynamic mechanical analysis (DMA) was carried out using the E400-TYPE-DVE (UBM Co., Ltd., Japan). The temperature range was -40 – 165 °C and the heating rate was 2 °C/min in tension mode. Scanning electrical microscopy (SEM) was used. The instrument was produced by Hitachi, Co., Ltd., Japan, to investigate the dispersion behavior of VGCF in composites.

Results and Discussion

Thermal conductivity

The thermal conductivity of PP composites is shown in Fig. 1. The thermal conductivity increased with VGCF content. In addition, the surface modification of VGCF by coupling agent slightly influenced to improve the thermal conductivity at a high content of VGCF in PP matrix.

The dispersion state of VGCF was investigated by SEM photographs as shown in Fig. 2. It is observed that the agglomerate of VGCF was grown up with VGCF content, while the treated VGCF well dispersed and small sizes of agglomerate. It is seen closely in Fig. 2 (b) and (d) found that at 5 wt.% of treated VGCF occurred various sizes of agglomerate and non-uniform dispersion as compare with 10wt.% of treated VGCF was uniformly dispersed in PP matrix. This result finding that a high amount of treated VGCF well dispersed in PP matrix. It is possible to expect the increase of connection frequency between VGCF for increasing thermal conductivity of PP/VGCF composites.

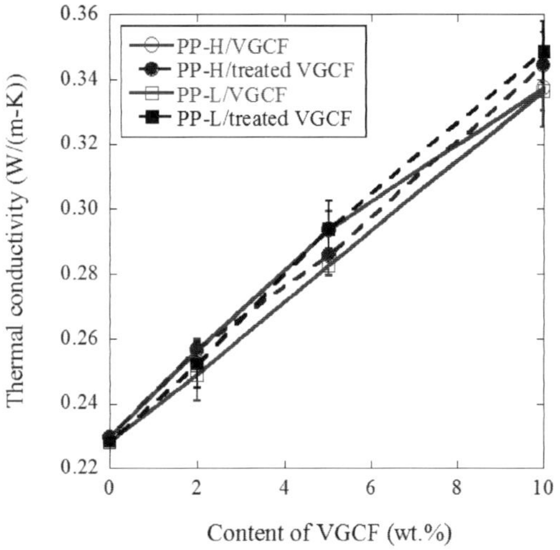

Figure 1 Thermal conductivity of PP/VGCF composites as a function of filler content

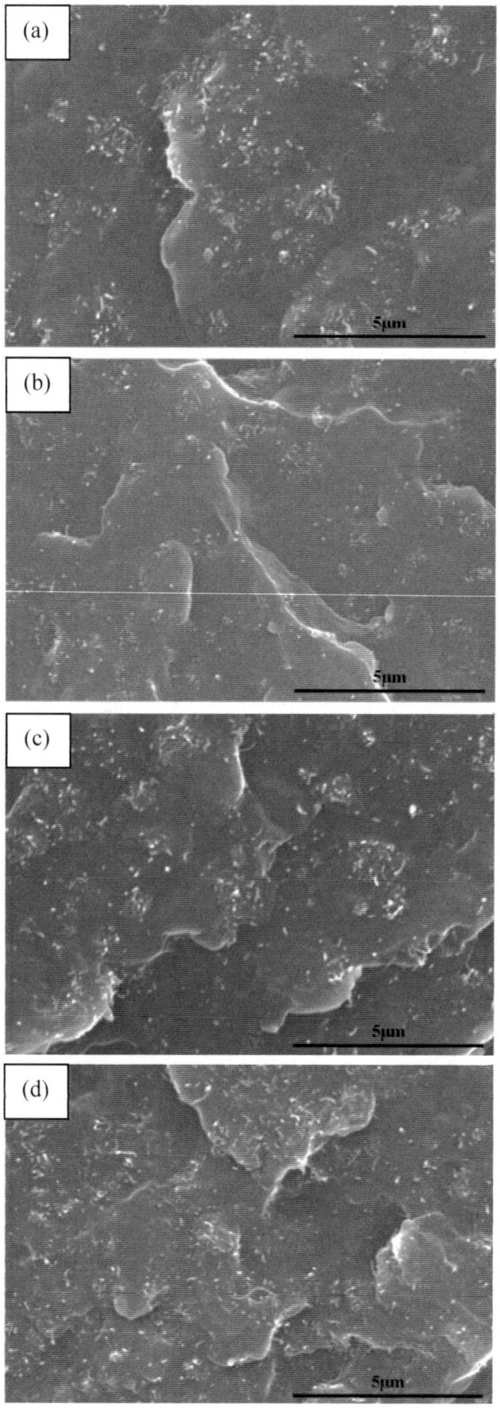

Figure 2 SEM photo on the surface of PP-L composites with VGCF (a) 5 wt.% (c) 10 wt.% and treated VGCF (b) 5 wt.% (d) 10wt.%

Mechanical properties

The stress-strain curves of VGCF and treated VGCF content combined with PP matrix as shown in Fig. 3 and 4. These results showed that the maximum strain to failure was about 630% and 640% for high and low viscosity PP composites, respectively. The strain to failure points of PP composites was alike with PP neat when VGCF content increased. It should be noted that the PP composites showed ductile failure behavior. The mechanical behaviors were almost as same in both matrices. In case of our research, the maximum mechanical behaviors were VGCF 10wt.% of PP composites. It may be explained that mechanical behaviors increased with VGCF content. In addition, this result found that the titanate coupling agent is independent an effect with mechanical properties of PP composites. This result is in contrast to previous studies in the titanate coupling agent is significantly reduced yield and tensile strength properties of polypropylene [9].

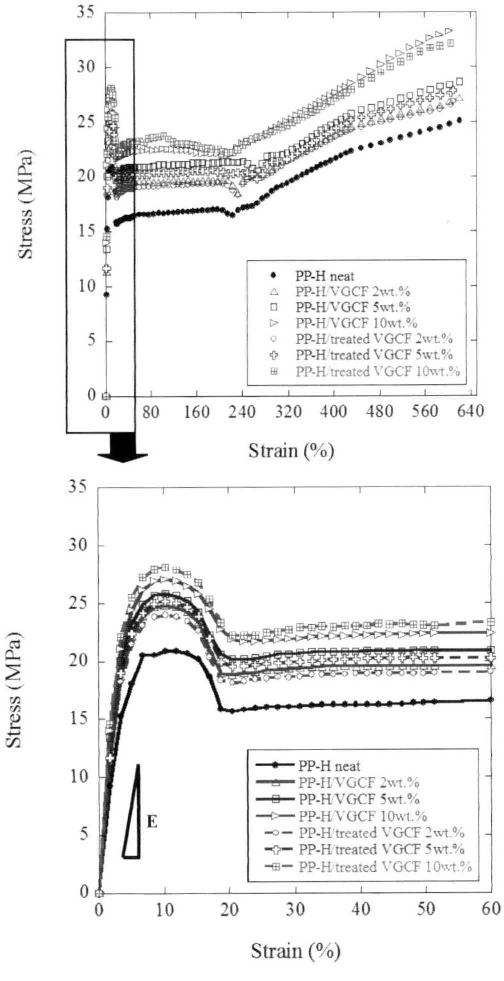

Figure 3 Comparison of stress-strain curve for PP-H/VGCF composites in case using treated VGCF and untreated one.

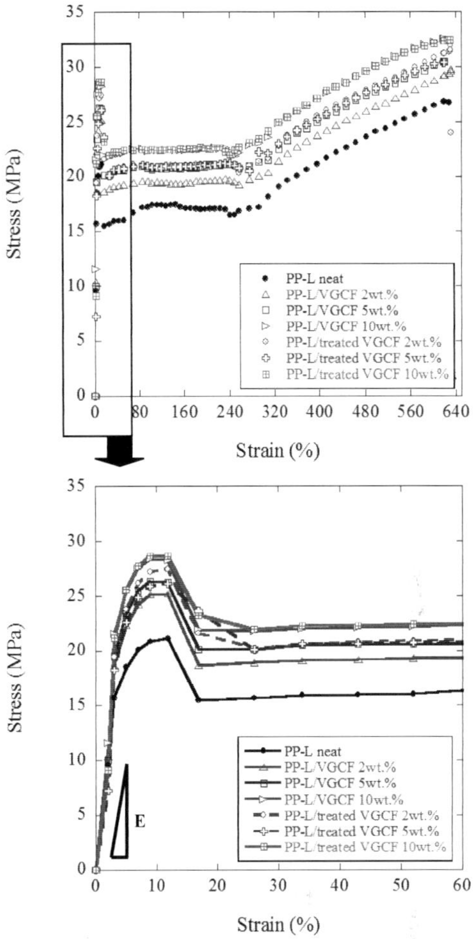

Figure 4 Comparison of stress-strain curve for PP-L/VGCF composites in case using treated VGCF and untreated one.

Dynamic mechanical analysis

The dynamic mechanical analysis data of the composites as a function of filler content with VGCF and treated VGCF shown in Fig. 5 and 6. The storage modulus totally increased with VGCF content. The significant effect of surface treatment on storage modulus can not be seen from the results shown in Fig. 5. Moreover, this result found that the titanate coupling agent has influence on improving loss modulus of PP composites. It could be described that the surface modification is significant to modify the dispersion ability of VGCF in PP matrix and also increase adhesion between filler and matrix by producing molecule bridge between polar molecule (VGCF) and nonpolar molecule (PP) enhanced the ability of PP composites to dissipate energy as shown in Fig. 6. The internal structure should be checked by using X-ray diffraction.

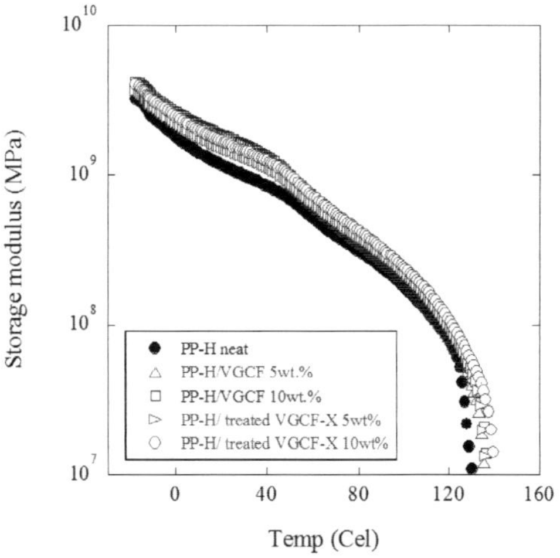

Figure 5 Storage modulus of PP composites as a function with VGCF and treated VGCF content with 5 wt.% and 10 wt.% at frequency 1 Hz

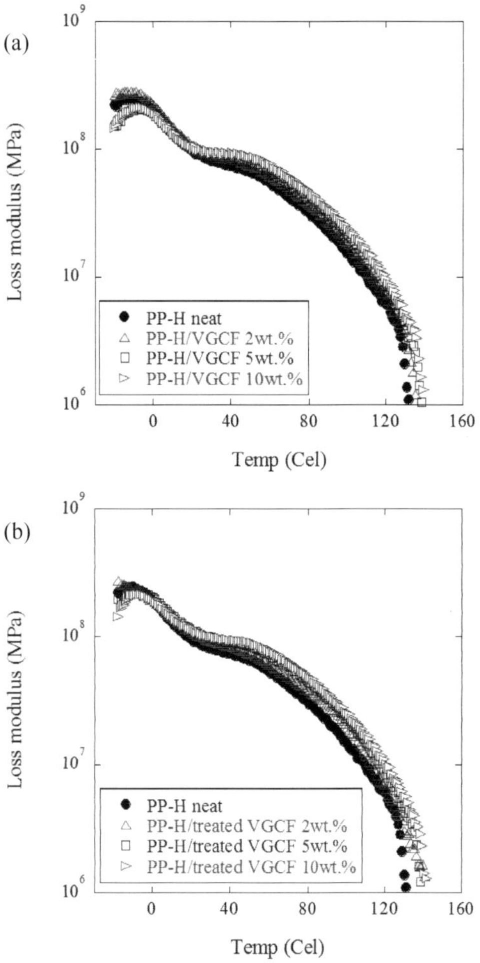

Figure 6 Loss modulus of PP composites as a function with (a) VGCF and (b) treated VGCF content at frequency 1 Hz

Conclusions

The thermal conductivity and the mechanical properties increased with VGCF content and also occurred ductile failure behavior. An agglomerate group of VGCF was growing when increasing in amount of VGCF. Surface modification by treated titanate coupling agent contributes to improve dispersion ability, and reduce agglomerate sizes. In addition, it has an effect to modify adhesion by using chemical reaction lead to increase loss modulus. The mechanical properties were independent of surface modification. In terms of thermal conductivity found that at high content treated VGCF was well dispersion which modified network between filler in PP matrix therefore thermal conductivity increased.

Acknowledgements

We would like to express our thanks to Industrial technology center of fukui prefecture for technical support injection molding.

References

1. B. Weidenfeller, M. Hofer, F. R. Schilling; Composites: Part A 35, 2004, 423-429.
2. B. Weidenfeller, M. Hofer, F. R. Schilling; Composites: Part A 33, 2002, 1041-1053.
3. M. H. Al-Saleh, U. Sundaraj; Carbon 2009, 47, 2-22.
4. I. C. Finegan, G. G. Tibbetts; J. Mater. Res., 2001, Vol. 16, 6.
5. T. Koyama, S. Tanoue, Y. Iemoto; J. Textile Eng. 2009, Vol. 55, 3, 73-83.
6. G. W. Lee, M. Park, J. Kim, J. I. Lee, H. G. Yoon; Composites: Part A 37, 2006, 727-734.
7. N. A. N. Alkadasi, D. G. Hundiwale and U. R. Kapad; J Sci Ind Res, Vol. 64, 2005, 679-685.
8. J. Gonzalez, C. Albano, M. Ichazo, B. Diaz; Eur. Polymer J., 2002, 38, 2465-2475.
9. C. A. Wah, L. Y. Choong, G. S. Neon; Eur. Polymer J., 2000, 36, 789-801.
10. Z. Zhou, S. Wang, L. Lu, Y. Zhang, Y. Zhang; Compos. sci. tech, 2008, 68, 1727-1733.
11. C. Teng, C. M. Ma, K. Chion, T. Lee, Y. Shih; Mater. Chem. Phys., 2011, 26, 722-728.

O-05-445

IBUPROFEN-KOLLIDON® SR MATRICES SOFTENED TOWARDS THE INCREASE OF WEIGHT PERCENT OF IBUPROFEN

C. Wiranidchapong[1]*, N. Ruangpayungsak[1], D. Shuwisitkul[1] and S. Tunvichien[1]
[1]*Faculty of Pharmacy, Srinakharinwirot University, Thailand – chutimav@swu.ac.th*

Abstract - Kollidon® SR has been widely used as a release controlling agent in matrix tablet, produced by direct compression method. It is composed of polyvinyl acetate and polyvinyl pyrrolidone in the ratio of 8: 2. Thermal analysis, using the heating program: heating from 25-100 °C, 10 °C/min; cooling to -60 °C, 10 °C/min; and finally heating to 100 °C, 5 °C/min, revealed the glass transition temperatures (T_g) of Kollidon® SR and ibuprofen, a non steroidal anti-inflammatory drug, around 39.10 °C and -44.10 °C, respectively. Similar analysis was extended to investigate the T_g of matrices containing ibuprofen in Kollidon® SR in the range of 10-90 % w/w. The T_g of the matrices were between 34.09 °C and -43.06 °C. The decrease of the T_g of ibuprofen in Kollidon® SR matrix as the increase of weight percent of ibuprofen could be described by the principle of Gordon-Taylor equation. This finding was in accordance with the ibuprofen-Kollidon® SR matrix softening during keptat 30 and 45 °C. The T_g values were respectively fit to the Gordon-Taylor and Kwei equations, a modified version of Gordon-Taylor equation. Those values were better fit to Kwei equation, indicating an interaction between ibuprofen and Kollidon® SR. This could be confirmed by FTIR analysis revealing an inter-associated hydrogen bonding between ibuprofen and Kollidon® SR.

Keywords: Kollidon SR; Ibuprofen; Glass transition temperature; Gordon-Taylor equation; Kwei equation

Introduction

Monolithic system has widely been used to prolong drug release for several decades. The system is composed of drug dissolved or dispersed in a polymer matrix. Drug release is controlled by the diffusion of dissolved drug through the porosity and the tortuosity of the polymeric matrix. Drug containing in polymer matrix exists in the dissolved state when drug concentration is lower than drug solubility in that polymer. Conversely, non-dissolved drug is dispersed in the polymeric matrix when drug concentration is higher than drug solubility in that polymer [1].

In case of the polymer, used as release controlling agent, exhibiting glass transition temperature (T_g) nearly equal to storage temperature, slightly increase of the temperature may cause the transformation of the polymer into rubbery state. This state of the polymer easily interacts with the drug, dissolved or dispersed the matrix. Wiranidchapong et al. [2-5] revealed an alteration of the T_g of drug-polymer solid dispersion as a function of composition, which could be described by the principle of Gordon-Taylor equation [6-9]. Therefore, drug dissolved in the polymer matrix may decrease the T_g of the matrix system in case of drug exhibiting the T_g lower than that of the polymer [10-14]. This may cause either physical or chemical instability of drug-polymer matrix.

The aim of this study was to investigate the T_g behavior of ibuprofen, Kollidon® SR (KSR), and the mixture of ibuprofen blended with KSR at various weight percent in order to describe the stability problem of ibuprofen in KSR matrix tablet stored at 30 and 45 °C.

Experimental

Materials

Ibuprofen (Lot No. IB1S1461) and Kollidon® SR (Lot No. 14209516K0) were supplied as gifts by BASF, Ludwigshafen, Germany.

Preparation of ibuprofen in KSR mixture

10-90 % w/w ibuprofen in KSR physical mixture were prepared by physically mix with mortar and pestle for 5 min. The mixtures were kept in refrigerated temperature.

Preparation of ibuprofen in KSR matrix tablet

Matrix tablets, 1 cm, i.d., of ibuprofen, KSR, 10, 40, and 70 % w/w ibuprofen in KSR were produced by direct compression. The powder, 320 mg in weight, was compressed by hydraulic press at 100 kg/cm^2. Each kind of tablet was kept at 30 and 45 °C for 3 days. Tablet hardness was measured by Stokes-Monsanto hardness tester at 0, 1, 2, and 3 days. Maximum force applied in the hardness test was 15 kg.

Differential Scanning Calorimetry (DSC) analysis

DSC was carried out using a Mettler Toledo DSC apparatus with a refrigerated cooling system (DSC 823e, Switzerland) and nitrogen as purge gas. The DSC cell was calibrated with indium (melting point 156.9 °C and ΔH = 27.5 J/g). Each sample, i.e., ibuprofen, KSR, and 10-90 % w/w ibuprofen in KSR physical mixtures, was accurately weighed into standard aluminum pan with cover and scanned using the heating program: heating to 100 °C at 10 °C/min; cooling to -60 °C at -10 °C/min; finally heating to 100 °C at 5 °C/min.

Mathematical analysis

T_g of 0-100 % w/w ibuprofen in KSR versus weight fraction of KSR were fitted to Gordon-Taylor and Kwei, a modified version of Gordon-Taylor, equations by nonlinear regression (GraphPad Prism® version 4.0). Eleven experimental data points were used for each fit. The coefficient of determination (R^2), runs test, and the residual plot were used to evaluate the goodness of each fit.

Fourier Transform Infrared Spectroscopy (FTIR) analysis

FTIR spectra of ibuprofen, KSR, and 10-70 % w/w ibuprofen in KSR either physical mixtures or melted mixtures, heated from 25-100 °C at 10 °C/min; cooling to -60 °C at -10 °C/min; and finally heating to 100 °C at 5 °C/min were recorded with a Perkin-Elmer FTIR Spectrum One using potassium bromide disks.

Results and Discussion

Appearance of matrix tablet after hardness test

The produced tablets were white to yellowish white color dependent on weight fraction of KSR in the composition. The yellowish white color was more intense associated with the increase of tablet hardness when weight fraction of KSR was increased. Tablet hardness was measured as illustrated in Table 1.

Table 1: Tablet hardness or appearance of ibuprofen in KSR after applied 15 kg of compression force.

Ibuprofen in KSR (% w/w)	Tablet hardness (kg) or appearance						
	Day 0	Day 1		Day 2		Day 3	
		30°C	45°C	30°C	45°C	30°C	45°C
0	No change	No change	No change	No change	No change	No change	No change
10	Little change	Little change	Little change	Little change	Little change	Little change	Little change
40	Sticky	Sticky	Sticky	Sticky	Sticky	Sticky	Sticky
70	10.5	10	Sticky and cracked	Sticky and cracked	Sticky and cracked	Sticky and cracked	Sticky and cracked
100	4	12	10	10	12	12	10

Hardness of ibuprofen tablet was 4 kg at day 0. The increase of tablet hardness was found when ibuprofen tablet was stored at 30 and 45 °C, while no change was observed in KSR tablet stored at those temperatures for 0-3 days.

At concentration of ibuprofen lower than 50 % w/w, the increase of weight percent of ibuprofen increased elastic manner of the matrix tablet. This indicated that higher concentration of ibuprofen caused a stronger reduction of T_g of KSR.

At 70 % w/w ibuprofen in KSR tablet, weight fraction of ibuprofen was higher than that of KSR, so that, elastic manner according to T_g of KSR lowering was not dominated. However, higher storage temperature made 70 % w/w ibuprofen in KSR matrix tablets more elastic. This might be a result of an interaction between ibuprofen and KSR increased.

DSC analysis

DSC curve of ibuprofen, obtained from heating to 100 °C, 10 °C/min, exhibited melting point around 77.87 °C. T_g of ibuprofen, obtained from cooling the melted ibuprofen to -60 °C, -10 °C/min and then heating to 100 °C, 5 °C/min, was revealed around -44.10 °C as shown in Fig. 1.

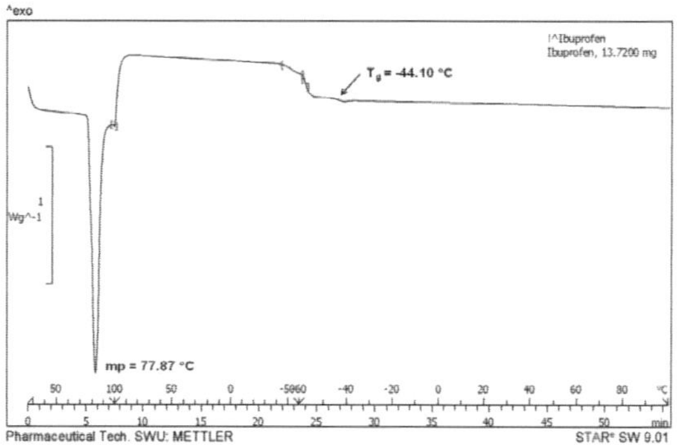

Fig. 1: DSC curve of ibuprofen.

Similar heating program was used to investigate T_g of KSR and the mixture of ibuprofen in KSR at 10-90 % w/w. DSC curve of KSR displayed the T_g around 39.10 °C as shown in Fig. 2.

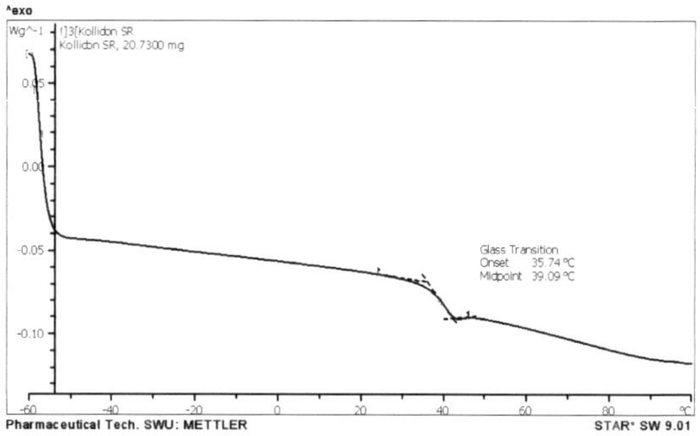

Fig.2: DSC curve of KSR.

T_g of the blends containing ibuprofen in KSR at 10 – 90 % w/w were between 34.09 °C and -43.06 °C. In addition, the T_g of the mixture was decreased when the concentration of ibuprofen was increased as displayed in Fig. 3.

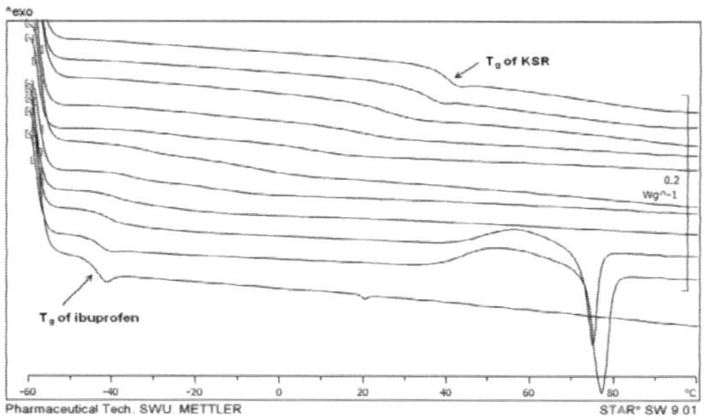

Fig. 3: DSC curve of 0–100 % w/w ibuprofen in KSR (top to bottom).

T_g behavior of ibuprofen in KSR, heated from 25-100 °C at 10 °C/min; cooling to -60 °C at 10 °C/min; and finally heating to 100 °C at 5 °C/min, were in accordance with the principle of Gordon-Taylor equation.

T_g analysis

Fitting T_g values of 0–100 % w/w ibuprofen in KSR to Gordon-Taylor and Kwei equations gave coefficient of determination (R^2) equal to 0.8730 and 0.9026, respectively as shown in Fig. 4.

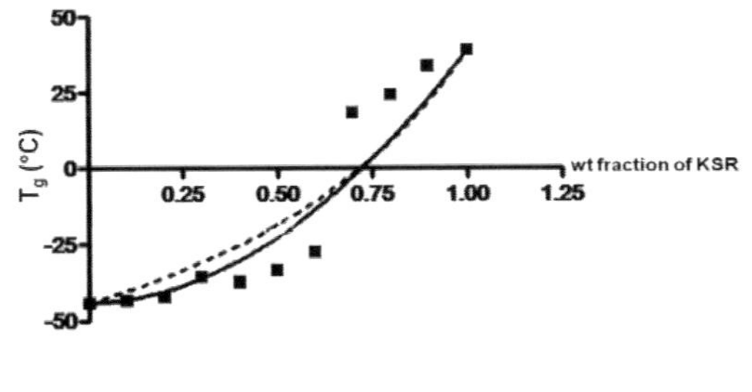

Fig. 4: T_g versus weight fraction of KSR curves based on experimental data (■); Gordon-Taylor equation (- -); Kwei equation (—).

By means of runs test, experimental T_g and estimated T_g obtained from Gordon-Taylor equation were significantly different (P value = 0.0238) but experimental T_g and estimated T_g obtained from Kwei equation were insignificantly different (P value = 0.2619). This indicates that T_g of ibuprofen in KSR is better fit to Kwei equation. However, residuals corresponding to the fits of Gordon-Taylor and Kwei equations were apparently nonrandom as displayed in Fig. 5. The non-randomness of the residuals suggests that even Kwei equation does not completely described T_g behavior of ibuprofen in KSR.

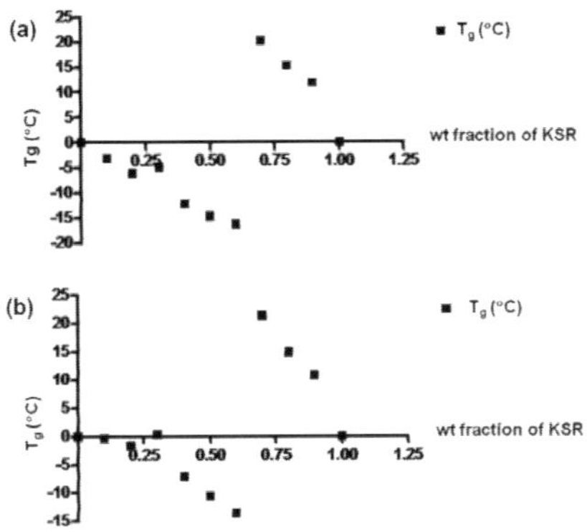

Fig. 5. Residuals corresponding to the Gordon-Taylor fit (a) and the Kwei fit (b).

FTIR analysis

FTIR spectrum of ibuprofen displays peaks around 3000, 1720, and 1230 cm^{-1} corresponding to O-H stretching vibration, asymmetrical carbonyl stretching vibration, and C-O stretching vibration of carboxylic acid, respectively [15]. For KSR, composed of polyvinyl acetate (PVA) and polyvinyl pyrrolidone (PVP) at the ratio of 8: 2, the peak around 1739 cm^{-1} corresponding to the carbonyl stretching vibration of ester group in PVA was observed. In addition, the bands around 3469 and 1663 cm^{-1} corresponding to N-H stretching vibration and carbonyl stretching vibration respectively of amide group in PVP [15-16] were also observed. The N-H stretching band shifted to lower wave number was noticed in both physical mixture and melted mixture of 10–30 % w/w ibuprofen in KSR as seen in Fig. 6-7.

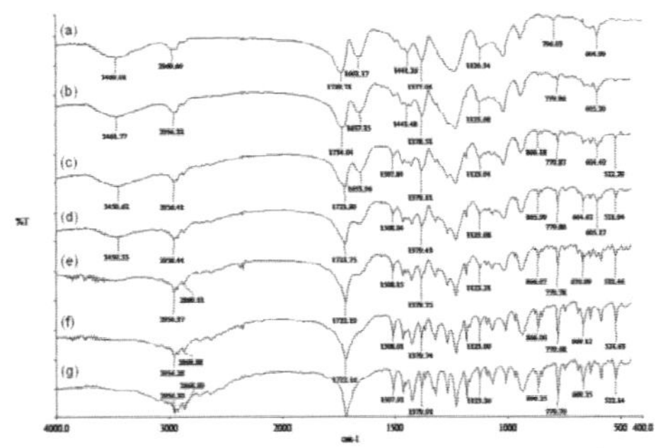

Fig. 6: FTIR spectra recorded at room temperature in the range of 4000-400 cm^{-1}: (a) KSR; physical mixtures of ibuprofen in KSR at (b) 10 % w/w; (c) 20 % w/w; (d) 30 % w/w; (e) 40 % w/w; (f) 70 % w/w; (g) ibuprofen.

Furthermore, the peak around 1663 cm^{-1} shifted to 1658 cm^{-1} was noticed in 30 % w/w ibuprofen in KSR melted mixture. This implied that hydrogen bonding between carbonyl group of PVP and hydroxyl group of ibuprofen might be occurred [16]. This was in agreement with shift of the peak around 1230 cm^{-1} to 1240 cm^{-1} in melted mixture of 10–40 % w/w ibuprofen in KSR as shown in Fig. 8. This indicated that C-O stretching vibration of carboxylic acid group of

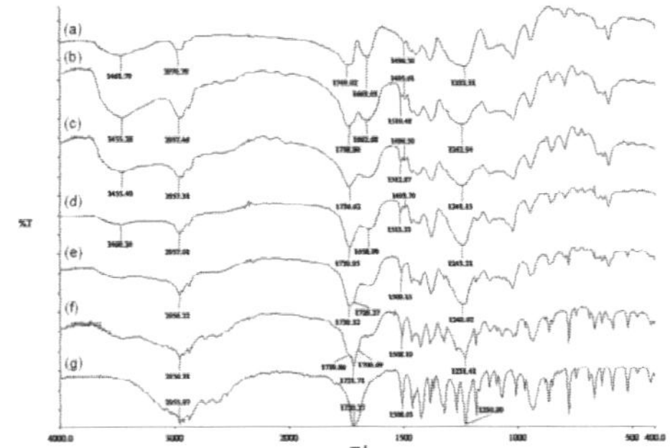

Fig. 7: FTIR spectra of melted mixtures recorded at room temperature in the range of 4000-400 cm^{-1}: (a) KSR; ibuprofen in KSR at (b) 10 % w/w; (c) 20 % w/w; (d) 30 % w/w; (e) 40 % w/w; (f) 70 % w/w; (g) ibuprofen.

ibuprofen was changed to C-O stretching vibration of ester group [15].
The appearance of the shoulder around 1700 cm^{-1} in melted mixture of 70 % w/w ibuprofen in KSR implied the presence crystalline form of ibuprofen in the mixture [16]. This phenomenon revealed rapid re-crystallization of amorphous ibuprofen in 70 % w/w ibuprofen in KSR after DSC analysis [17].

Conclusion

Ibuprofen could interact with KSR, resulting in a reduction of T_g of KSR in either physical mixture or melted mixture. In physical mixture, crystalline ibuprofen partially dissolved in KSR, so that, only dissolved form of ibuprofen made T_g of KSR lowering. In contrast to melted mixture, ibuprofen was completely changed to amorphous form blended with KSR, so that, T_g of melted mixture of ibuprofen in KSR could be described by the principle of Gordon-Taylor equation. This finding supported the phenomenon of ibuprofen in KSR matrix tablet soften.

References

1. C-J. Kim in *Controlled release dosage form design*, C-J. Kim, Ed.; Technomic Publishing, Pennsylvania, 2000, 13-48.
2. C. Wiranidchapong, PhD Thesis, Chulalongkorn University, 2006.
3. C. Wiranidchapong; I.G. Tucker; T. Rades; P. Kulvanich *J. Pharm. Sci.* 2008, 97(11), 4879-4888.
4. C. Wiranidchapong; T. Rades; I.G. Tucker; P. Kulvanich *Thermochimica Acta* 2009, 485, 57-64.
5. C. Wiranidchapong in *Sex hormones*, R.K. Dubey, Ed.; InTech, Rijeka, Croatia, 2012.
6. H.A. Schneider *Makromol. Chem.* 1988, 189, 1941-1955.
7. S.W. Kuo; F.C. Chang *Macromolecules* 2001, 34, 4089-4097.
8. S.W. Kuo; C.F. Huang; F.C. Chang *J. Polym. Sci. Part B: Polym. Phys.* 2001, 39, 1348-1359.
9. M. Maldonado-Santoyo, et al. *J. Polym. Sci. Part B: Polym. Phys.* 2004, 42, 636-645.
10. M.R. Jenquin; S.M. Liebowitz; R.E. Sarabia; J.W. McGinity *J. Pharm. Sci.* 1990, 79, 811-989.
11. M.R. Jenquin; R.E. Sarabia; S.M. Liebowitz; J.W. McGinity *J. Pharm. Sci.* 1992, 81, 983-989.
12. C. Wu; J.W. McGinity *Int. J. Pharm.* 1999, 177, 15-27.
13. C. Wu; J.W. McGinity *Eur. J. Pharm. Biopharm.* 2003, 56, 95-100.
14. S. Hülsmann; T. Backensfeld; R. Bodmeier *Pharmaceut. Dev. Tech.* 2001, 6, 223-229.
15. R.M. Silverstein; G.C. Bessler; T.C. Morrill in *Spectrometric identification of organic compounds*, J. Stiefel, Ed.; John Wiley and Sons, Singapore, 1991, 91-164.
16. S.G. Kazarian; G.G. Martirosyan *Int. J. Pharm.* 2002, 232, 81-90.
17. A.R. Paradkar; M. Maheshwari; A.R. Ketkar; B. Chauhan *Int. J. Pharm.* 2003, 255, 33-42.

PROCESSING OF NANOCOMPOSITES PLA/GRAPHITE USING A NOVEL ELONGATIONAL MIXING DEVICE

René Muller[1], Michel Bouquey[1], Jérôme Rondin[1], Fatima Hassouna[2], Rigoberto Ibarra[3*]

[1]*Laboratoire d'Ingéniérie des Polymères pour les Hautes Technologies (LIPHT), ECPM, Université de Strasbourg – rene.muller@unistra.fr;michel.bouquey@unistra.fr;jerome.rondin@etu.unistra.fr*

[2]*Centre de Recherche Public Henri Tudor – fatima.hassouna@tudor.lu*

[3]*Centro de Investigación en Materiales Avanzados S.C – rigoberto.ibarra@cimav.edu.mx*

Miguel de Cervantes 120, Complejo Industrial Chihuahua, Chihuahuahua, Chih. 31109, México

Abstract

Expanded graphite (EG) was added to Polylactic acid (PLA) and then fully mixed in a novel elongational mixing device (RMX) to obtain PLA/EG nanocomposites. This type of mixer has reported remarkable dispersion efficiency when applied to polymer blends compared to conventional mixing systems. The operation of the mixing device is based on the induced multiple passages of the material (by means of reciprocant pistons), at different flow speeds trough a short capillary die, thus creating high elongational to shear rate ratios. From TEM analysis it could be observed that highly homogeneous materials were obtained at several processing conditions, it means, different combinations of number of cycles and flow speeds at constant set temperature. Some domains of particles ranged from hundred to several hundreds of nanometers across the observed fields. On the other hand, complementing the morphology information, X-rays diffractograms of these samples showed different intensity of the characteristic peak of EG (3 % w/w EG was kept constant), suggesting partial exfoliation of this. Complex viscosity for neat PLA and nanocomposites was evaluated in samples prepared at different mixing conditions. Furthermore, the molecular weight of some neat PLA samples was assessed in order to correlate the PLA degradation process with the operational parameters of the RMX. Viscous heating was found to play a major role throughout the mixing. Final PLA nanocomposites presented a storage modulus, E', between 20-40 % higher than the pristine PLA, being especially remarkable for samples at increasing speed.

Keywords: Mixing, viscosity, PLA degradation, elongational flow

Introduction

Work on polymer/graphite nanocomposites via melt blending has been increasingly gaining attention especially since the "boom" of the graphene in the last decade [1]. In virtue of that, some significant attempts to produce polymer/graphite nanocomposites via melt processing are reported in the literature, i.e. [2, 3, 4]. In these works, different types of graphitic materials like natural, expanded graphite and commercial graphene have been dispersed into a polymer matrix through the use of internal mixers and single and twin-screw compounders. In the case of graphite, variable but rather discreet results have been obtained, i.e. low mechanical properties and large particle size in the final dispersion. Direct addition of graphene nanoplateles [4], on the other hand, has reported much better results but its prohibitive cost makes this method still unreliable. As it is known, high shear contributions to dispersive mixing are common in conventional mixing devices, as those just mentioned. Shear stresses for filler dispersion often involve ranges of high shear rates that considerably reduce the polymer viscosity in the shear thinning zone, which demands high energy inputs to reach reasonable dispersion efficiency. Also, this efficiency is affected by the rotational component of the shear flow that offers no contribution to filler dispersive mixing. In this context, it has been stated that a way to highly improve dispersive mixing is to enhance the contributions of *elongational flow fields* in the overall mixing process [5,6]. Accordingly, greater hydrodynamic stresses are generated in elongational flow and stresses are transferred more efficiently to the agglomerates because rotational motion is absent in pure elongational flow (irrotational flow); as a consequence, the energy consumption is expected to be less in elongational flow relative to shear flow [5]. That is why, in the present work, a new mixer design with predominantly elongational flow contribution (RMX) has been used as a different approach to obtain polymer/graphite nanocomposites. This device is based on similar principles studied in previous works [7,8]. In this novel mixer, the material is induced to pass throughout a central die from opposites cylindrical chambers in a back and forth fashion. An unlimited number of convergent/divergent flows can be produced at different speed, allowing the generation of elongational flow in an efficient manner [9]. Selected polymer matrix for this work was Polylactide acid (PLA), given the great interest at present on biodegradables polymers, as well as on the interest on finding ways to enhance PLA physical properties as to compete with well-established polymers of general use like polyethyleneterephtalate (PET).

Experimental

Amorphous PLA 4042D from NatureWorks® was used as polymer matrix with average molecular weight, Mw, 157,000 g/mol, Mn, 71,000 g/mol and polydispersity 2.21; data were obtained from gel permeation chromatography (GPC). Expanded graphite (EG) Ecophit® from SGL Group was used as a filler; carbon content ≥ 95 %, $D_{50} \sim 5\text{-}7$ μm, $\rho_{powder} \sim 0.115\text{-}0.135$ g/cm^3, $\rho_{real} = 2.25$ g/cm^3, moisture content ≤ 2 (data from technical sheet).

For the mixing operation, before obtaining the definitive PLA/EG blends in the RMX, a pre-mixing step, in order to add the powder, was realized in a Haake Rheomix 600, Thermo Electron at 200° C and 5 rpm. The central mixing operation was carried out in the RMX, which presents the next main features: controlled pistons speed (v) range of 0.18 – 180 mm/s which corresponds, using a 3-cm-diameter pistons, to volumetric flow (Q) values in the range of 0.125 – 125 cm^3/s. A mixing sequence is defined simply by the piston velocity (v or Q) and the number of cycles (N); the pressure in one of the chambers is continuously measured during the mixing sequence by a pressure transducer Fig. 1 shows a general operational scheme:

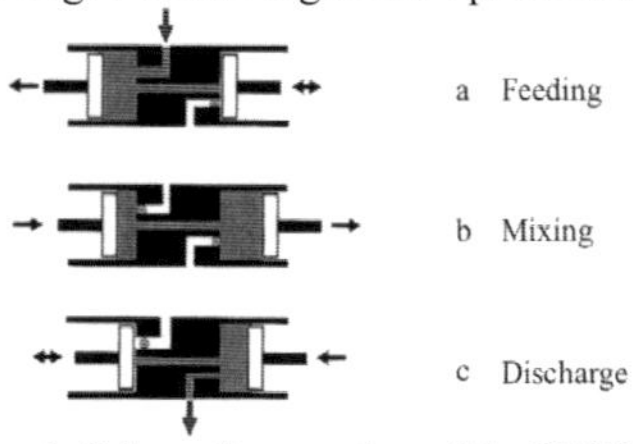

Fig. 1. Schematic operation of the RMX [7].

Mixing sequences (v/N) for this particular study covered: Neat PLA, 10/40, 20/10, 40/10 and 40/20; PLA/EG: 10/10, 10/20, 10/30, 10/40, 20/10 and 40/10. The mixing temperature was 200°C. Concentration of EG was kept constant at 3 % w/w. In general, these conditions were selected to cover in a first approach a relatively low range of speeds and number of cycles with respect to the RMX capacity, preventing large PLA degradation. As mixing element in the RMX a capillary die of ϕ =4 mm and L/D= 7 was used. The shear rate range at the selected operation conditions in the RMX covered from 1800 (10 mm/s) to 7200 s^{-1} (40 mm/s) and it was calculated from $(d\gamma/dt)= [4Q/\pi R^3][(3n+1)/4n]$ (1). A reference sample was prepared at the lowest piston speed, 3 mm/s (540 s^{-1}) and by direct delivery of material from the RMX (half of a cycle). Rheological characterization was carried out in both, Bohlin Inst capillary rheometer HR2000 and Anton Parr MCR301 rheometer. In the first one, shear and elongational viscosities of PLA, in the range of the selected RMX speeds were obtained, as well as global power law index, n. In the second one, by using oscillation mode, the complex viscosity, η*, of neat PLA and PLA/EG samples as a function of frequency (0.1- 600 rad/s) was measured. X-rays characterization was done in a Siemens spectrometer trough a 2Θ sweep from 10 to 30°. Morphology characterization was realized by TEM in an electronic microscope Hitachi 7500. Also, for the assessment of molecular weight of processed neat PLA, GPC measurements were carried out. Dynamical mechanical properties were evaluated in a DMA Nietzsch 202D.

Results and discussion

Rheology of neat PLA samples. Before dealing with mixing of EG filled PLA, a basic understanding of the PLA degradation phenomenon under the RMX mixing conditions was considered necessary. As it is well known, the PLA degrades upon thermal processing due to several undesired reactions, which mainly include hydrolysis of the ester linkage, giving place to molecular weight reduction by chain scission [10]. Concerning that, in Fig. 2, it is possible to appreciate changes in complex viscosity (η*) for neat PLA samples, processed at different RMX mixing conditions. The associated changes in Mw evidence different levels of PLA degradation as a result of possible residence time and thermomechanical effects. To address these factors, in the Fig. 3 it is presented the evaluation of PLA (virgin) stability in terms of η*, as a function of time. Drop of η* with time at an extent depending on temperature is clearly observed. On associating this behaviour to RMX mixing conditions, for instance, the largest mixing sequence (10/40, v=10 mm/s, N=40) takes only about 2.5 min in mixing, but considering the feeding time of the RMX at 200°C

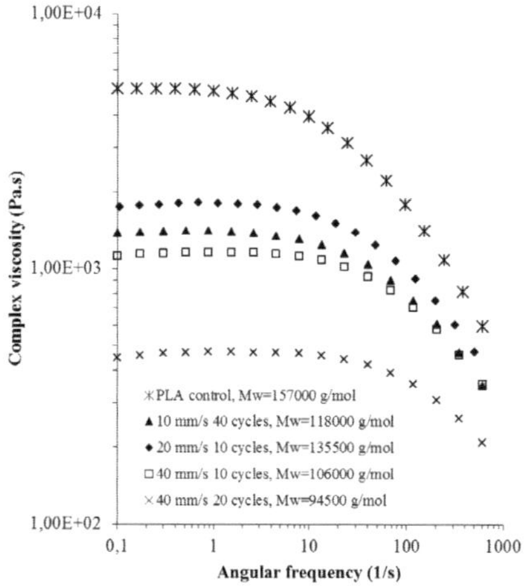

Fig. 2. Complex viscosity as a function of frequency for neat PLA samples obtained at different mixing sequences in the RMX.

(approximately 10 min), the total residence time sums up around 12.5 min (750 s). That is, according to Fig. 3, a decrease of 11 % of the initial value of η*. Of course, this estimation of η* does not reflect what is observed in Fig. 2. So then, there is more involved in PLA large η* drop that simple time effect. Trying to get an insight into the phenomenon, the Fig. 4 presents, for sample PLA 40/10, the curve of the pressure (ΔP) as the piston displaces along the chamber pushing the melt throughout the capillary die. It is evident that, even though the piston speed was set to be constant, ΔP decreases in each cycle during the sequence, what clearly points out to viscosity decrements. In this sense, in the literature [7,8,9] has been referred that at large v and/or N viscous heating become very important, which for the specific case of PLA is crucial in terms of its temperature sensitivity. Therefore, in order to estimate the viscous heating (ΔT) or average adiabatic temperature rise, the next relationship, from ref. [7], was used: $\Delta T = \Delta P/\rho_{PLA}Cp_{PLA}$ (2). Based on this, an accumulated ΔT of about 36-40°C was estimated for sample 40/10 during the mixing sequence. Again from Fig. 3, this would account for a drop of about 77 %

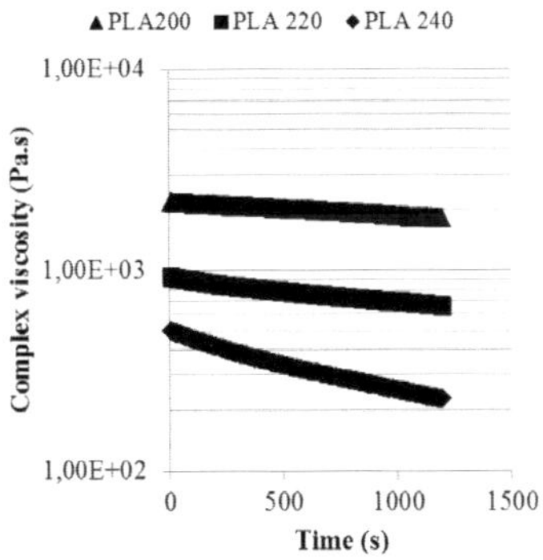

Fig. 3. Complex viscosity of pristine PLA as a function of time at different temperatures.

(only from the jump from 200 to 240°C) in η*, plus the additional drop due to the short time effect at 240°C. These facts are in better agreement with the large drops in η* observed in Fig. 2 respect to pristine PLA. Indeed, the final Mw loss (about 30 % for this sample) evidences that viscous heating affects ΔP not just by a thermally but degradative driven mechanism.

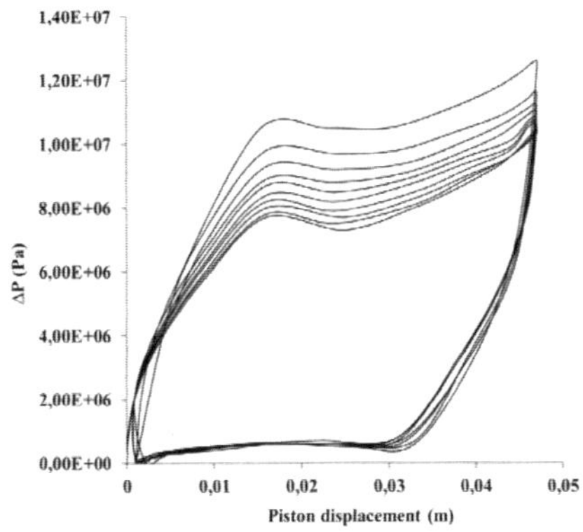

Fig. 4. ΔP behaviour during a mixing sequence of 10 cycles and *v*=40 mm/s, neat PLA.

Rheology of EG filled PLA. Once certain understanding of the relationship rheology/mixing/PLA degradation has been stated, the η* behaviour of EG filled PLA is shown in Fig. 5. In general, samples at increasing *N* (10/20, 10/30 and 10/40) seem to exhibit the net effect of viscous heating, just discussed, and its detrimental effect on η*. In the other hand, the sample at highest *v* (40/10) reports higher η* even though, from Fig. 2, it presented the lowest Mw. A possible explanation to this could lie on the ease of attachment and/or interlocking of even shorter polymer molecules to graphite particles, enhancing hydrodynamic effect on viscosity. Sample 10/10 and 20/10, in turn, processed at the lighter mixing conditions studied, are expected to preserve in a major grade the structural integrity of the polymer matrix and, hence, to show higher η*.

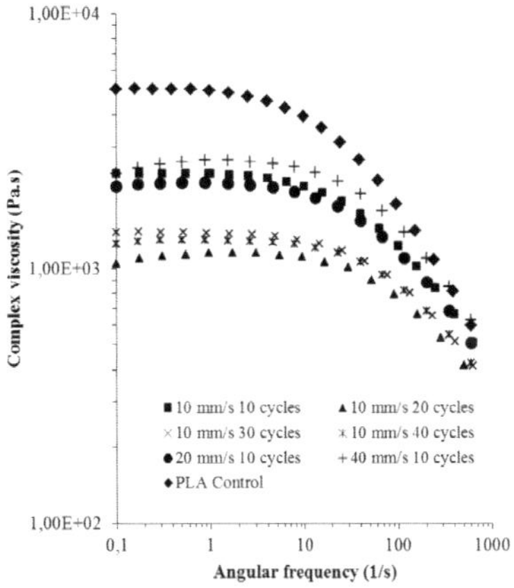

Fig. 5. Complex viscosity as a function of frequency for PLA/EG samples obtained at different mixing sequences in the RMX.

Morphology. Fig.6 presents XRD diffractograms of PLA/EG samples. For the sake of a valid comparison, a reference sample was processed at the lowest RMX specification speed, 3 mm/s, and 0.5 cycles to try to preserve the expanded graphite structure. As expected, the reference sample exhibits the higher intensity of the characteristic peak of graphite (2Θ = 26.38°). Lower peaks intensity for the studied samples with respect to the reference would suggest a partial exfoliation of expanded graphite, even though there is not a clear tendency as a function of mixing conditions. Complementing with TEM, in micrographs of Fig. 7, sample marked as reference and 10/10 showed good particles distribution, a common feature to all samples. Also, the reduction of the global particle size in the sample 10/10 is evident when compared to the reference. In general, most of the samples presented a broad distribution of particle sizes, ranging roughly from 100 nm - 1μm thick. However, in many cases, the identification of particles much thinner than 100 nm could be evident (i.e. sample10/10 at 40000X), giving certain support to X-rays results about partial exfoliation. From Fig. 7, additionally, direct comparison reported bigger particle size for sample 40/10 (lowest Mw) than 10/10. It is important to mention that special difficulty arose to carry out a more reliable assessment of TEM evaluation, since high hardness of the samples, in general, made that thin EG edges have been slightly oriented in the sense of the cut, making not so reliable to measure their thickness in an accurately way. More work will be done on this.

Physical properties. With respect to a basic look at the physical behaviour of PLA/EG blends, table 1 is presented. Samples with increasing *N* (10/20, 10/30, 10/40) reported a similar E' increase over reference neat PLA, whereas samples at higher *v*, 20/10 and 40/10 reported even more remarkable increments. In the other hand, sample 10/10 showed also a significant increase and exhibited the highest glass transition temperature, Tg, (65°C). These results are in agreement with the behaviour of η* observed in Fig.5.

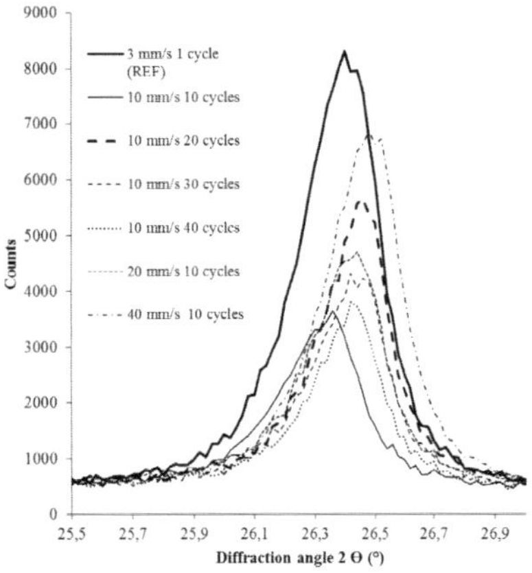

Fig. 6. XRD of PLA/EG samples at different mixing conditions (characteristic EG peak; no peaks for PLA).

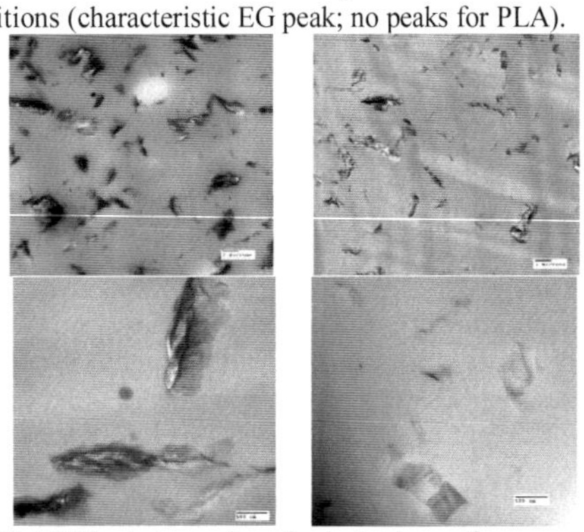

Fig. 7. TEM photographs of, clockwise, reference sample 5000 X (bar=2μm), 10/10 5000 X, 10/10 40000 X, 40/10 40000 X (bar=500 nm).

Sample (*v/N*)	E' (MPa)	% ΔE'	Tg (°C)
PLA ref	2750	-	60
10/10	3600	30.9	65
10/20	3350	21.8	64
10/30	3400	23.6	65
10/40	3300	20	64
20/10	3400	23.6	63
40/10	3800	38.2	61

Table 1 Storage modulus and Tg of PLA/EG blends

It is important to recall that only 3 % w/w of EG has been used, unlike about 10 % w/w of EG added in conventional equipment to reach these values in similar PLA/EG systems [2]. Trying to correlate these results to mixing in the RMX, Fig. 8 shows that at 200°C, and even at 220°C, there is enough margin for elongational flow contributions to the mixing process, since extensional viscosities of PLA remain relatively high (indeed, shear viscosity ratios, η_E/η, increases around three times, 23 to 62) in the shear rate range of interest. Rheological data at 240° C (~T_{max} from viscous heating), would also be important but it was impossible to attain by capillary rheometry given the extremely low viscosity of PLA in such a condition. According to this scenario, it is proposed that samples at higher *v* or *N* follow, predominantly, a mechanism of molecular diffusion or interlocking into EG particles, favouring both reinforcement and partial exfoliation. In these samples, even though the stresses needed for dispersive mixing remains the lowest, because of the highest viscous heating, the multiple exposure of the material to elongational stresses enhances this dual mechanism. In the other hand; samples at the lighter mixing conditions like 10/10 and 20/10, exposed to less viscous heating, are favoured from both, high elongational stresses and major matrix integrity.

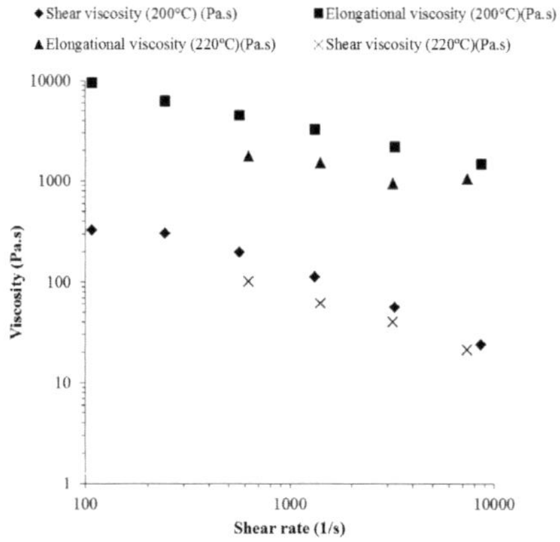

Fig. 8. Shear and extensional viscosity of PLA from capillary measurements covering the RMX range of study.

Conclusions

Despite the extent of PLA degradation, a very acceptable performance of the RMX (good distributive and dispersive mixing) was evidenced, leading to obtain PLA/EG nanocomposites with remarkable reinforcement (high modulus). Certainly, PLA degradation, as promoted by viscous heating, plays a major role during the mixing of PLA/EG nanocomposites in the RMX. It represents, in fact, a basic platform to well define reinforcement and dispersion mechanisms, to be deeper clarified in a future work.

Acknowledgments

Cathy Royer, Neurosciences Centre, University of Strasbourg for all her kindly support on TEM.

References

1. S. Park;R.S. Ruoff, *Nature Nanotech*, 4, 217 (2009).
2. M.Murariu *et al*, *Polymer Deg. and Stab.*,95,5,889 (2010).
3. F. M. Uhl; Q. Yao; C.A. Wilkie, *Polym. Adv. Technol.*, 16, 7, 533 (2005).
4. H. Kim ; C.W. Macosko, Polymer, 50, 3797 (2009).
5. Chris Rauwendall, *Plastic Add. & Comp.*, 21 (1999).
6. Ica Manas-Zloczower; Z. Tadmor (Editor), *Mixing and Compounding of Polymers: Theory and Practice,* 2nd Ed. Carl Hanser Verlag, Munich (2009).
7. M. R. Mackley *et al*, *Journal of Rheology*, 39,1293 (1995).
8. M. Bouquey,; C. Loux; R. Muller; G. Bouchet, *Journal of Appl. Polym. Sci.*,119, 482 (2011).
9. Y. Son, J. *Appl. Polym. Sci.*112, 609 (2009).
10. N. Najafi; M.C. Heuzey; P.J. Carreau, *Composites Science and Technology*, 72, 5, 608 (2012).

O-05-335

THE INFLUENCE OF COMPOUNDING CONDITIONS ON THERMALLY CONDUCTIVE COMPOSITES

Andreas Leuteritz[a], Burak Kutlu[a,b], Ines Kühnert[a,*] and Maximilian Bader[c]

[a] Leibniz-Institut für Polymerforschung Dresden e.V., Hohe Str. 6, 01069 Dresden, Germany
[b] Technische Universität Dresden, Professur für Polymerwerkstoffe und Elastomertechnik, 01062 Dresden, Germany
[c] Robert Bosch GmbH, P.O. 1131, 71332 Waiblingen
**Corresponding Author: Dr. Ines Kühnert, Kuehnert@ipfdd.de*

Abstract

Heat conductivity of polymeric materials is commonly very low, thus limiting the use of such materials in applications where a heat sink is obligatory. To overcome this drawback highly heat conductive fillers like hexagonal boron nitride (hBN) with a very high thermal conductivity are incorporated into resins and thermoplastics. In order to study the influence of compounding conditions on the resulting materials properties hBN was mixed with polyamide at different concentrations. In small scale compounding on a DSM X-plore Microcompounder two step compounding was necessary in order to allow high filler loadings. On the other hand by using a Buss Cokneader MKS 30-18 or Leistritz twin screw extruder ZSE27MAXX the mixing of matrix and filler could be realized at different feeding positions in one compounding step. In order to minimize post-compounding influences specimen from all variants were prepared by hot press and the thermal conductivity was determined using a Netzsch Nanoflash LFA447. Samples prepared with Leistritz extruder showed much higher thermal conductivity owing to a more homogeneous distribution of hBN particles. Depending on the processing conditions during compounding the modulus and strength of injection molded samples did not change very much whereas notched impact strength increases with higher rotational speed. Regarding higher local shear conditions finally the rheological behavior indicates severe polymer degradation. In conclusion, a significant influence of compounding conditions on polymer degradation and thermal conductivity of the used matrix-filler-system could be found. Therefore, the results can be used to optimize the compounding process for such composites.

Keywords: compounding, thermal conductivity, composites, mechanical properties

O-05-243

INFLUENCE OF TWIN-SCREW EXTRUSION CONDITIONS ON CNT DISPERSION AND LENGTH AND THE RESULTING ELECTRICAL AND MECHANICAL PROPERTIES OF POLYCARBONATE BASED NANOCOMPOSITES

Petra Pötschke*, Tobias Villmow, Beate Krause and Regine Boldt
*Leibniz Institute of Polymer Research Dresden, Hohe Str. 6, 01069 Dresden, Germany – *poe@ipfdd.de; villmow@ipfdd.de; krause-beate@ipfdd.de; boldt@ipfdd.de*

Abstract - In this study, the extrusion conditions during melt mixing of multiwalled carbon nanotubes (MWCNTs) with polycarbonate were varied in order to find optimum conditions for electrical and mechanical composite properties.
Using a Berstorff ZE25 extruder, for composites containing 3 wt% Baytubes® C150P the rotation speed was varied between 100 and 1000 rpm and the throughput between 5 and 15 kg/h. In addition, different screw lengths and profiles were used, the temperature profile was changed, and feeding of CNTs was varied between hopper and side feeder. Through these variations, the specific mechanical energy input (SME) was varied between 0.11 and 0.90 kWh/kg. Throughout all mixing conditions, the agglomerate area ratio as determined from light microscopy images on thin sections decreased exponentially with SME. However, the CNT length as determined from TEM investigations on CNTs dissolved from the composites also decreased significantly after the extrusion step and were between 52% and 78% of the initial CNT length (based on x_{50}-values) depending on the extrusion conditions. The mechanical properties consequently varied with the processing. When varying the rotation speed at throughputs of 5 or 10 kg/h, elongation at break as well as the impact strength increased with SME, whereas at 15 kg/h no such increase was observed. Modulus and tensile strength showed the same trends but less intensive. The increase of the properties can be mainly related to the better dispersion which reduces the number of agglomerates which may act as defects in the mechanical tests. The decrease at the screw speed at 15 kg/h may be attributed to the more pronounced nanotube shortening.

Keywords: carbon nanotubes, polycarbonate, melt processing, dispersion, aspect ratio

Introduction

Carbon nanotubes (CNTs) have been shown in the last years as versatile filler to influence several properties favorably. This concerns especially the electrical conductivity of multiwalled CNTs (MWCNTs) which can be transferred to insulating matrices by establishing a percolated filler network. One crucial point in producing such nanocomposites is the filler dispersion, which is difficult to achieve due to strong van-der-Waals forces and entanglements between the tubes. However, especially for mechanical properties, but also for their processability and e.g. surface properties, samples free of primary agglomerates are advantageous. Next to the polymer and nanotubes characteristics and their interactions, the mixing conditions can significantly influence the achievable state of dispersion. This was shown for melt-mixing using small-scale compounders, e. g for PC or PA [1,2]. For laboratory scale investigations, results were presented on PLA and PCL where for twin-screw extrusion the screw configuration, screw speed, throughput and temperature profile were varied [3,4]. On the other hand, McGlory et al. [5] reported about the influence of screw speed, where an optimum concerning the electrical percolation at 100 rpm was achieved.

Characterizing or quantifying the state of dispersion is not often performed. In recent works we introduced to use the fraction of remaining agglomerates as observed in transmission light microscopy as a measure for at least the dispersion on the microscale [1,3,4]. Another measure is the number of remaining agglomerates. Of course, it is also possible to quantify the dispersion from TEM investigations regarding the dispersed nanotubes [6]. Among the indirect methods, melt rheology, DSC, and electrical conductivity are discussed in literature [7].

Not much is known about the shortening of nanotubes during composite processing. Only recently, a methodology was developed to measure nanotube shortening after melt processing using TEM analysis of nanotubes dissolved from the matrix [8,9]. First results on polycarbonate based composites revealed shortening of nanotubes up to 30% of their initial length.

In this contribution, composites based on polycarbonate with 3 wt% MWCNTs were produced using compounding in a laboratory-scale twin-screw extruder by varying certain processing conditions. The composites were injection molded under comparable conditions and investigated concerning their state of dispersion using light microscopy, the CNT length after processing using TEM on dissolved CNTs, and the tensile and impact behavior. Whereas dispersion and CNT length were discussed in context with the specific mechanical energy (SME) input, the effects in mechanical properties were related to the state of dispersion and the CNT length.

Experimental

The employed nanotubes represent industrial multi-walled CNTs (MWCNTs) C150P Baytubes® (Bayer

MaterialScience AG, Leverkusen, Germany; >95% purity C, bulk density 120-170 kg/m^3). As polycarbonate (PC) Makrolon® 2600 (Bayer MaterialScience AG, Leverkusen, Germany) was applied. The polycarbonate based composites filled with 3 wt.% Baytubes® C150P were prepared using a co-rotating twin-screw extruder ZE25 (Berstorff, Germany) at a mean barrel temperature of 260°C, a rotation speed varied between 100 and 1000 rpm, a throughput between 5 and 15 kg/h, and screws with a length-diameter-ratio (L/D) of 36 or 48. PC pellets and the powdery MWCNT material were fed simultaneously into the hopper by gravimetric dosing or CNTs were fed into the sidefeeder. Based on the torque values recorded during the compounding, the specific mechanical energy (SME) was calculated using the following equation:

$$SME = \frac{2\pi N \tau}{m} \quad (1)$$

whereas N is the mixing speed, τ is the torque values, and m is the throughput during the extrusion.

Injection moulding was done using an Ergotech 100/420-310 (Demag, Germany) using a two-cavity mold, fed by a flash gate. Injection velocity was set to 40 mm/s, the mold temperature to 80°C and melt temperature to 300 °C and sample geometries according to the standards for mechanical testing were produced.

Light microscopy (LM) in transmission mode was performed on thin sections with a thickness of 5 µm to quantify the macrodispersion of CNT agglomerates within extruded strands. The agglomerate area ratio A_A was calculated from the ratio of the area of remaining agglomerates A_{Agg} to the total area of the micrograph A_0 (~0.6 mm^2) using the software ImageJ Version 1.43 g. For the determination of the nanotube length distribution after melt compounding a method described by Krause et al. [8] was used by dissolving the composites in chloroform .

Uniaxial tensile tests were performed on injection moulded dogbone type (1A) specimens according to EN ISO 527-2/1A/50 using a Universalprüfmaschine (Zwick GmbH & Co. KG, Germany). Charpy notched impact strength was determined at room temperature according to EN ISO 179/1eA.

Results and Discussion

The composites prepared at different rotation speeds and throughputs using the screw with an L/D ratio of 48 were investigated with regard to the MWCNT macro dispersion as well as the electrical and mechanical properties and for selected composites the nanotubes length distribution was quantified.

It was found that with increasing specific mechanical energy SME the agglomerate area ratio A_A decreased and levelled off at a SME input of around 0.4 kWh/kg (see Figure 1). In addition, the light microscopy images of thin sections in Figure 1 illustrate the state of macro dispersion. Furthermore, the nanotube length distributions illustrated by the x-values were quantified to show the influence of the processing conditions during melt compounding on the CNT length. The results are incorporated in Fig. 1.

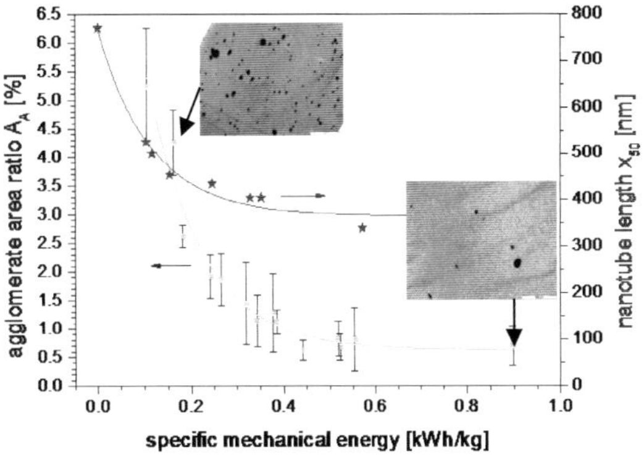

Figure 1 – Agglomerate area ratio A_A and nanotube length x_{50} in dependence on SME for polycarbonate / 3 wt% MWCNT composites including exemplarily light microscopy images on thin sections, variation of rotation speed and throughput for a L/D=48 distributive screw and CNT feeding at hopper.

As expected and described before [9,10], a CNT shortening during the melt compounding took place. If comparing the different mixing speeds (at a throughput of 5 kg/h) a decrease of the nanotube lengths with the rotation speed could be found. The mean value x_{50} and x_{90} decrease from 525 nm and 1155 nm (100 rpm) to 340 nm and 765 nm (1000 rpm) as compared to the initial CNT length values of 770 nm and 2407 nm. Therefore, a significant reduction of the aspect ratio can be concluded.

In addition, also the influence of the throughput on the macrodispersion and (at 300 rpm) on nanotube length was investigated. It was found that with increasing throughput the SME decreased and the values of agglomerate area ratio and CNT length increased. An explanation for this finding is the lower residence time of the material.

The feeding position was varied for composites prepared at 500 rpm and 5 kg/h throughput using the screw with L/D = 36. When the CNTs were fed in the sidefeeder attached at 14L, a higher agglomerate area ratio and longer nanotubes were found in the composite. That feeding Baytubes® C150P in the hopper leads to the better macrodispersion as compared to feeding using a side feeder, was previously described by Müller et al. [11] for polypropylene based composites. The reason for the higher CNT length values for the CNTs fed in the sidefeeder is the shorter residence time for the CNTs in the polymer melt and the lower applied maximal shear stresses in comparison to feeding the CNTs together with the solid PC granules in the hopper.

The comparison between two different screws with the L/D ratio of 36 having either a dispersive profile (with mainly kneading blocks) or a distributive screw profile (having mainly mixing elements) at 100 rpm and 5 kg/h leads to a slightly higher SME for the dispersive screw configuration. This results in slightly better CNT dispersion, however interestingly the CNT shortening is less pronounced when using the dispersive screw configuration.

The mechanical properties were found also to depend on the SME, see figure 2. Introduction of 3 wt% MWCNTs resulted in improved modulus (extruded PC 2340 MPa) and tensile strength (extruded PC 62 MPa); however reduced values of elongation at break (extruded PC 119 %) and notched impact strength (extruded PC 80 kJ/m^2) as compared to extruded PC. In the composites with 3 wt% MWCNTs, when varying the rotation speed at throughputs of 5 or 10 kg/h, elongation at break as well as the impact strength increased with SME, whereas at 15 kg/h much lower increase in both values was observed. Modulus and tensile strength showed the same trends with SME but less intensive. The modulus increased up to 2700 MPa; whereas for the tensiles trength the highest value was 65 % MPa, both values obtained at a medium SME input of 0.55 kWh/kg achieved at 500 rpm and 5 kg/h. The increase of the properties can be mainly related to the better dispersion with SME (see Fig. 1) which reduces the number of agglomerates able to act as defects. The stagnation in increase of mechanical properties with SME despite increasing quality of dispersion at the screw speed of 15 kg/h may be attributed to the more pronounced nanotube shortening.

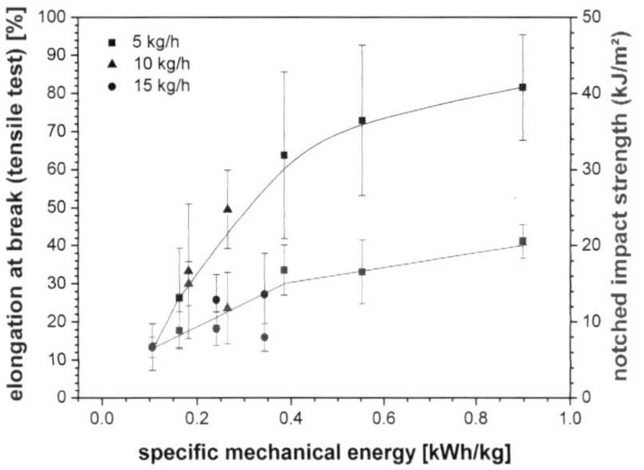

Figure 2 – Elongation at break from tensile tests and notched impact strength of /polycarbonate / 3 wt% MWCNT composites in dependence on SME (L/D=48, distributive screw, variation of throughput and rotation speed, CNT feeding at hopper)

Comparions of the composites produced with the both screw profiles mentioned above leads to slightly better properties for the dispersive screw configuration, which also generated slightly higher SME input.

Conclusions

In our study was found, that for PC/3 wt% MWCNT composites higher specific mechanical energy input achieved by variations in rotation speed and throughput resulted in better CNT macrodispersion. At the same time, the CNTs were shortened more severly with increasing SME. Mechanical properties responded to both, better dispersion leading to improved values, espacially of elongation at break and impact strength, and reduced nanotube length, resulting in negative effects on both properties.

For practical applications, a balance of the counteracting processes of the better dispersion and more pronounced CNT length shortening with increasing SME has to be achieved. It has to be considered that for improved mechanical properties a good CNT dispersion is essential, since the remaining CNT agglomerates lead to defects in the composites. However, for a high electrical conductivity of CNT composites the nanotubes should remain as long as possible which only can be achieved at low SME.

Thus, depending on the aspired property profile and in addition economic aspects suitable energy inputs and mixing conditions have to be selected.

Acknowledgements

We thank the German Federal Ministry of Economics and Technology (project grant ID AiF 336 ZBG) for financial support.

References

1. Kasaliwal G, Göldel G, Pötschke P, *Journal of Applied Polymer Science* **2009,** *112*(6), 3494-3509.
2. Krause B, Pötschke P, Häußler L, *Composites Science and Technology* **2009**, *69* (10), 1505-1515.
3. Villmow T, Pötschke P, Pegel S, Häussler L, Kretzschmar B, , *Polymer* **2008,** *49* (16), 3500-3509
4. Villmow T, Kretzschmar B, Pötschke P., *Composites Science and Technology* **2010,** *70*, 2045-2055.
5. McClory C, Pötschke P, McNally T, *Macromolecular Materials and Engineering* **2011,** *296* (1), 59-69.
6. Pegel S, Pötschke P, Villmow T, Stoyan D, Heinrich G, *Polymer* **2009,** *50* (9), 2123-2132.
7. Logakis E, Pandis, C, Kyritsis, A, Pissis, P, Micusik, M, Omastová, M, Pionteck, J, *Chemical Physics Letters* **2010,** *498,* 125-128.
8. Krause B, Boldt R, Pötschke P, *Carbon* **2011,** *49* (4), 1243 -1247.
9. Krause B, Villmow T, Boldt R, Mende M, Petzold G, Pötschke P, *Composites Science and Technology*, **2011,** *71*(8), 1145-1153.
10. Socher R, Krause B, Müller MT, Boldt R, and Pötschke P. *Polymer* **2012**;*53(2)*, 495-504.
11. Müller MT, Krause B, Kretzschmar B, and Pötschke P. *Composites Science and Technology* **2011**;*71(13)*:1535-1542.

O-05-228

EVALUATION OF MIXING PERFORMANCE OF SPECIAL ROTOR SEGMENTS IN AN INTERMESHING CO-ROTATING TWIN-SCREW EXTRUDER BY APPLYING WEIGHTED INDICES

Sayaka Yamada[a]*, Kazuhisa Fukutani[a], Kazuo Yamaguchi[a], Hideo Funahashi[a]
Katsuhiro Ebata[b], Hideyuki Uematsu[b] and Shuichi Tanoue[b]
[a] *Technical Development Group, KOBE STEEL, LTD., 1-5-5 Takatsukadai, Nishi-ku, Kobe, Hyogo, 651-2271, Japan – yamada.sayaka@kobelco.com; fukutani.kazuhisa@kobelco.com; yamaguchi.kazuo@kobelco.com; funahashi.hideo@kobelco.com*
[b] *Graduate School of Engineering, University of Fukui, 3-9-1 Bunkyo, Fukui, 910-8507, Japan – tanoue@matse.u-fukui.ac.jp;uematsu@matse.u-fukui*

Abstract - An intermeshing co-rotating twin-screw extruder is often used for producing polymer composites. In order to satisfy properties of the polymer composites, it is important to understand mixing performance of segments. In this study, we discussed mixing performance of special rotor segments using a numerical analysis method. These rotor segments have several tip clearances for achievement of both self-cleaning and effective mixing, and these are commercially named VCMT (various clearance mixing technology).The barrel was assumed to be fully filled with isothermal molten polymer. We adopted a particle tracking method in order to obtain the mixing evaluation indices. In addition, we also adopted the following evaluation indices, 1) residence time for the evaluation of uniformity, 2) maximum 1st principal stress for the evaluation of dispersive mixing, 3) probability distributions of each evaluation index. Moreover, we proposed probability distributions of weighted indices as better evaluation. The reciprocal of the residence time of each particle was selected to weight indices. The probability distributions of VCMT were compared with those of two kinds of kneading disk (KD) segments. From the comparison, it was found that the probability of VCMT in large index region is higher than those of KDs. And it was also found that the difference in probability distributions among segments is emphasized by adopting the weighted indices.As a conclusion, it is found that VCMT has better mixing performance than KDs.

Keywords: Melt Mixing, Dispersive Mixing, Twin-Screw Extruder, Special Rotor Segments, Distributive Mixing

1. Introduction

An intermeshing co-rotating twin-screw extruders are capable of melting, melt mixing, reacting, and devolatilization material in the process of transporting and extruding from its die. Featuring a high mixing performance, they have been extensively used for blending and compounding polymer with fillers and/or additives. Lately, its use has been expanded into carbon nano-tubes and nano-composites using clay particles, and there has been an increase in research on the application into recycling environmentally conscious material, such as polylactic acid [1].

In such applications, melt mixing in a twin-screw extruder is one of the important factors that affect the quality of products. While various segments are used in the melt mixing section, including a rotor and a kneading disk (KD), it is important to know the respective mixing performance and use the segments that meet the particular needs.

Advanced technology enables three-dimensional flow analyses of complicated shapes; accordingly, knowledge about the flow around segments in the melt mixing section has been deepened[2)-6)]. In general, many studies on mixing have addressed KDs as segments [2)-4) and 6)], but a few have focused on rotors [7)].

This study was intended to investigate the dispersive mixing, an important mixing mode related to the breakage of aggregates of fillers, of the special rotor segments. These rotor segments have several tip clearances for achievement of both self-cleaning and effective mixing, and these are commercially named VCMT (various clearance mixing technology)[8)]. Characteristics of the residence time distribution of fluid in segments were also studied.

For evaluation, probability distributions of indices were used conventionally. In this study, we also focused on probability distributions of weighted indices for quantitative evaluation.

For comparison, the same evaluation was performed for two kinds of kneading disk segments: N-KD with disk-stagger angle of 90° and R-KD with disk-stagger angle of 45°.

2. Analysis method and conditions

2.1 Properties

The barrel was assumed to be fully filled with isothermal molten polymer.

The molten polymer density of 735 kg/m^3 was used. The Carreau model expressed by eq.1 was applied as a viscosity model, where the zero shear viscosity η_0 of 14527 Pa·s, model parameter λ of 1.18 s, and exponent n of 0.38 were used as properties of polypropylene at 503 K, in reference to the paper by Ishikawa et al. [9)]. Fig. 1 shows the diagram of the shear viscosity η to the shear rate $\dot{\gamma}$.

$$\eta = \frac{\eta_0}{\left[1+\left(\lambda \dot{\gamma}_s\right)^2\right]^{\frac{1-n}{2}}} \qquad \text{eq.1}$$

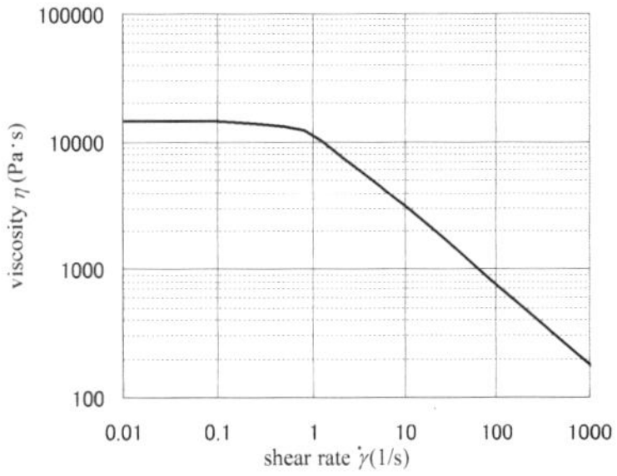

Fig. 1 Viscosity of polypropylene vs. shear rate at 503K

2.2 Analysis conditions

The segment length L of 50.4 mm was selected, and the cross section shape was determined in reference to the paper by Funatsu et al.[3]. Where, the barrel diameter D_1 was 40 mm, and the center distance D_2 was 33 mm. As shown in Fig. 3, two kinds of KDs were used: R-KD with a disk-stagger angle to the transport direction of 45°, and N-KD with a disk-stagger angle of 90°. VCMT had the same length and outer diameter as those of R-KD, and the torsion angle was about 90° per 1L. As shown in Fig. 4, five-segment per one axis configuration was adopted, consisting of a forward screw, three mixing segments, and backward screw, in consideration of the minimum configuration for transporting and mixing.

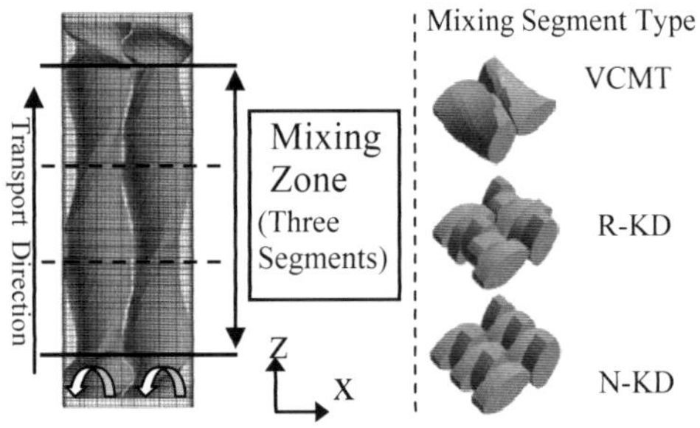

Fig.2 Computational Model (Mesh Arrangements and Mixing Segment Shapes)

A uniform velocity boundary was applied to the inlet, whereas a uniform pressure boundary was applied to the outlet. The barrel and screw surfaces were assumed to be non-slip planes. The mass flow rate Q was 70kg/h and the screw rotation N was 200rpm.

Commercial flow analysis software "POLYFLOW" version 3.12 using the finite element method was used for analysis. The number of elements was about 70000. It took about 130 hours to compute by one revolution with 9° increments in rotation phase, using a computer equipped with Xeon CPU from Intel with clock frequency of 3 GHz.

2.3 Particle tracking method

The dispersive mixing performance was evaluated using particle tracking method. Characteristics of the residence time distribution were also studied. While 1000 particles randomly arranged on the cross section at the upstream end of the mixing segments at the time t=0 reach the downstream end of the mixing segments, the trajectory and time history of each particle were calculated using the velocity field obtained by the flow analysis.

2.4 Mixing evaluation Indices

As mixing evaluation indices, we adopted the following evaluation indices, 1) residence time for the evaluation of uniformity, 2) maximum 1st principal stress for the evaluation of dispersive mixing, 3) probability distributions of each evaluation index on particles. With respect to 1) and 2), they were normalized by following way. The residence time was normalized using the theoretical residence time T_{th}, which is determined by dividing the effective volume by the volume flow. And The maximum stress was normalized using the apparent shear stress τ_{tip} on the assumption that the velocity distribution between the tip on the outer diameter and the barrel is linear.

Moreover, we proposed probability distributions of weighted indices (weighted probability distributions) as theoretically better evaluation. In the cases of conventional probability distributions drawn from indices just as they are obtained by particle tracking (non-weighted probability distribution), the probability is not corresponding with volume probability of polymer. In fact, as the residence time of a particle is shorter, the volume probability around the particle, which is obtaind at the outlet, becomes higher. On the other hand, volume probability is supposed to be important for more quantitative evaluation of the mechanical characteristics or quality of polymer. So, we drew the weighted probability distributions as a substitute for volume probability of the polymer. The reciprocal of the residence time of each particle was selected to weight indices.

When the number of particle that an index x is in the range of a to b is expressed $N(a<x<b)$, the probability that an index x is in the range of a to b is expresssed by eq.2.

$$P(a < x \leqq b) = \frac{N(a < x \leqq b)}{\Sigma N(a < x \leqq b)} \qquad \text{eq.2}$$

And probability of weighted index is expressed by eq.3.

$$wP(a < x \leqq b) = \frac{wN(a < x \leqq b)}{\Sigma wN(a < x \leqq b)} \qquad \text{eq.3}$$

Here, w means weighting. In this study, normalized reciprocal of the residence time T_{th}/t is applied as w.

3. Results and discussion

3.1 Characteristics of residence time distribution

3.1.1 Probability distribution of residence time

Fig.3 shows the non-weighted probability distributions of the residence time. And Fig. 4 shows the weighted probability distributions of the residence time.

For both probabilitye distribution, it can be mentioned that the probability of R-KD is higher than that of VCMT and N-KD around the non-dimensionless residence time t/T_{th} of 0.3-0.5. From these result, it is considered that R-KD has the lowest uniformity of the residence time among the three segments. It can be also mentioned that for N-KD and VCMT, both peak positions are around t/T_{th}=1. However, the probability at the peak of VCMT is larger than N-KD, and it is expected that VCMT has more uniform residence time.

On the other hand, the difference among segments in the weighted probability distributions is more emphasized than that in non-weighted ones because the volume probability that residense time is short becomes higher compared with the probability of particles. Especially, for R-KD, probability at t/T_{th} of around 0.3 increases to 18.07 from 7.1 by weighting.

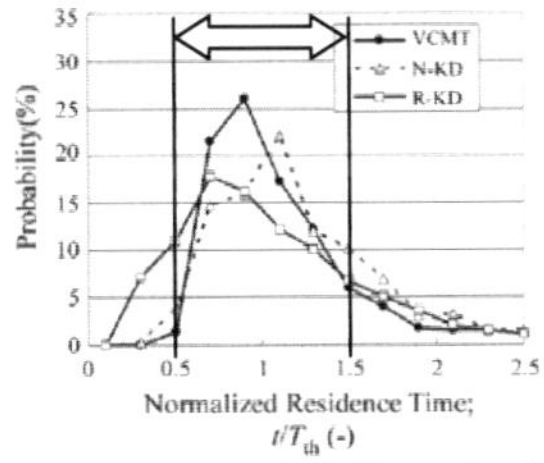

Fig. 3 Non-Weighted Probability Distributions of Normalized Residence time

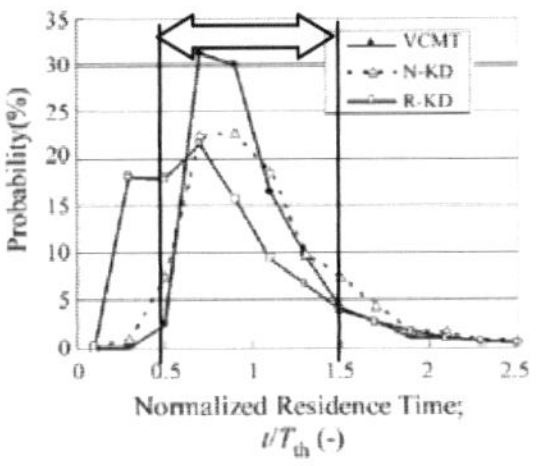

Fig. 4 Weighted Probability Distributions of Normalized Residence time

3.1.2 Quantitative evaluation of uniformity of residence time

It was expected that when the probability that the residence time of particles is close to the theoretical value is higher, the uniformity of residence time of fluid is higher. Accordingly, we evaluated quantitatively the uniformity of the residence time, by estimating the probability that t/T_{th} is in the range of 0.5 to 1.5 for each case. This range was determined so that the mixing characteristics of each segment could be distinguished. The results are summarized in Fig. 5.

Fig. 5 shows that the probability that the residence time of particles is close to the theoretical value increases by adopting weighting, for VCMT and N-KD. For R-KD on the other hand, it decreases because probability at t/T_{th} of around 0.3 increases very much by weighting. And among segments, it is highest for VCMT, regardless of the weighting. From such results, it would appear that VCMT is capable of mixing with highly uniform residence time compared with the two kinds of KDs, in the same way as discussed in 3.1.1.

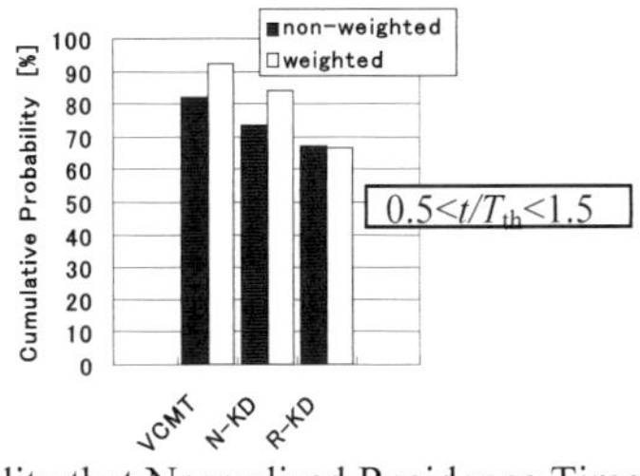

Fig. 5 Probability that Normalized Residence Time of Particles is between 0.5 and 1.5

3.2 Dispersive mixing performance

3.2.1 Probability distribution of maximum 1st principal stress

Fig. 6 shows the probability distributions of the maximum 1st principal stress $\sigma_{1,max}$. And Fig. 7 shows the weighted probability distributions of the maximum 1st principal stress $\sigma_{1,max}$.

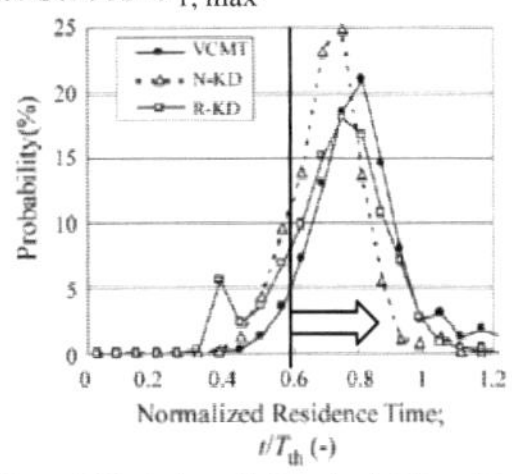

Fig. 6 Non-Weighted Probability Distribution of Normalized Maximum 1st Principal Stress

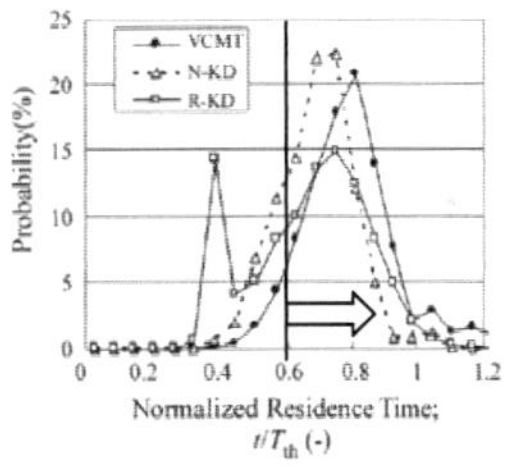

Fig. 7 Weighted Probability Distribution of Normalized Maximum 1st Principal Stress of Particles (R-KD)It was expected that the higher the probability that the residenc

Fig. 6 and 7 shows that, for R-KD, the particles are in the region where the dimensionless 1st principal stress $\sigma_{1,max}/\tau_{tip}$ is 0.3 to 0.4. For VCMT and N-KD, on the other hand, there are very few particles in the region where $\sigma_{1,max}/\tau_{tip}$ is 0.4 or less. The peak position of VCMT ($\sigma_{1,max}/\tau_{tip}$=0.81) is higher than that of N-KD ($\sigma_{1,max}/\tau_{tip}$=0.75). The probability of VCMT is higher than that of the two kinds of KDs in the region where $\sigma_{1,max}/\tau_{tip}$ is 0.8 or more. From such results, it is believed that VCMT is better in dispersive mixing performance compared with N-KD and R-KD.

As well as the probability distributions of residnce time, the difference among segments in the weighted probability distributions is more emphasized than that in non-weighted ones. For R-KD and N-KD, probability in small $\sigma_{1,\ max}/\tau_{tip}$ region in the weighted probability distribution is higher than that in non-weighted ones, which shows the correlation between the residence time t/T_{th} and the maximum 1st principal stress $\sigma_{1,\ max}/\tau_{tip}$. On the other hand, for VCMT, the profiles are almost same regardless of weighting, which shows less correlation between the residence time t/T_{th} and the maximum 1st principal stress $\sigma_{1,\ max}$.

Furthermore, we drew two-dimensional probability distributions diagrams with t/T_{th} on x coordinate and the $\sigma_{1,\ max}/\tau_{tip}$ on y coordinate, where the probability is expressed using contrasting shades. Fig. 8 is non-weighted one and fig. 9 is weighted one for R-KD.

From fig.8 it is found that the probability that t/T_{th} is in the range of 0.2 to 0.4 and $\sigma_{1,\ max}/\tau_{tip}$ is in the range of 0.36 to 0.42 is higher than other region. From this result, it can be mentioned that there is the correlation between the residence time t/T_{th} and the maximum 1st principal stress $\sigma_{1,\ max,}$ same as it is mentiond above. And from fig.9, it is found that probability in this range increase by adopting weighting.

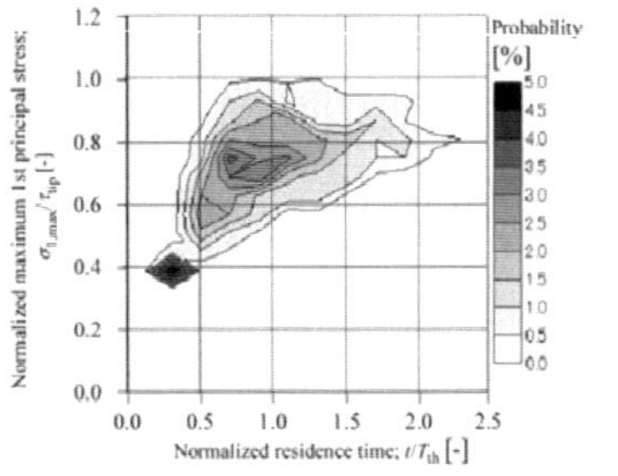

Fig. 8 Two-dimensional Weighted Probability Distribution of Residence Time and Maximum 1st Principal Stress (R-KD)

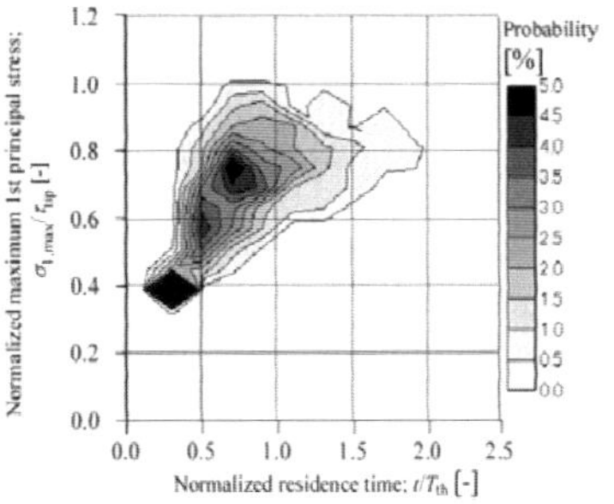

Fig. 9 Two-dimensional Non-Weighted Probability Distribution of Residence Time and Maximum 1st Principal Stress (R-KD)

3.2.2 Quantitative evaluation of dispersive mixing performance

It was expected that when the probability that the maximum 1st principal stress of particles $\sigma_{1,\ max}/\tau_{tip}$ is larger than the threshold is higher, the dispersive mixing performance is better. Accordingly, we evaluated quantitatively the dispersive mixing performance by estimating the probability that $\sigma_{1,\ max}/\tau_{tip}$ is 0.6 or more. The threshold was determined so that the mixing characteristics of the segments can be distinguished in the same way as discussed in 3.1.2, since this study did not target any particular filler. The results are summarized in Fig. 10.

Fig. 10 shows that the probability that the $\sigma_{1,\ max}/\tau_{tip}$ is larger than 0.6 decrease by adopting weighting, for N-KD and R-KD. For VCMT on the other hand, there is less difference by adopting weighting. And among segments, probability of VCMT is highest, regardless of the weighting. From such results, it would appear that VCMT is capable of highly dispersive mixing compared with the two kinds of KDs, in the same way as discussed in 3.2.1.

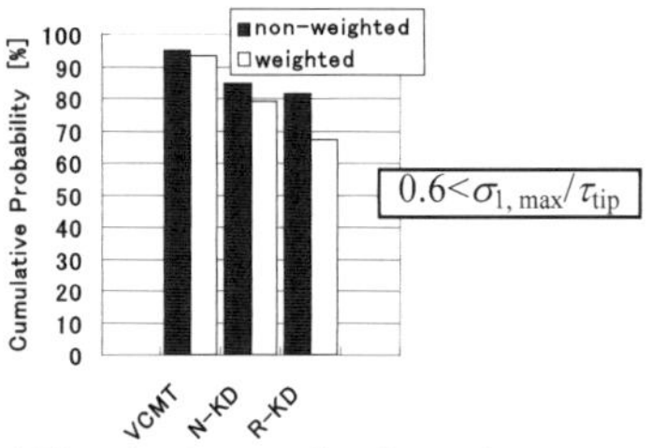

Fig. 10 Probability that Normalized Maximum 1st Principal Stress of particles is more than 0.6

4. Conclusion

By the flow analysis, the dispersive mixing performance of the special rotor segment (VCMT) was evaluated and the characteristics of the residence time distribution were studied. We suggested the weighted probability distributions as theoretically more quantitative evaluation. From the results, it was found that VCMT is excellent in dispersive mixing performance and uniformity of residence time compared with N-KD and R-KD . It is also found that the difference in probability distributions among segments is emphasized by adopting the weighted indices.

Reference

1. T. Kajiwara *Preprints of Seikei-Kakou Autumnal meeting 2008,* 2008, 63
2. Kohlgruber, K. *Co-Rotating Twin-Screw Extruders,* Hanser Publisher, 2008
3. K. Funatsu; S. Kihara; M. Miyazaki; S. Katsuki; T. Kajiwara *Polymer Engineering and Science,* 2002, *42,* 707
4. V. L. Bravo; A. N. Hrymak; J. D. Wright *Polymer Engineering and Science,* 2004, *44,* 779
5. H. F. Cheng; and I. Manas-Zloczower *Polymer Engineering and Science,* 1998, *38,* 926
6. X. M. Zhang; L. F. Fengm; W. X. Chen; G. H. Hu *Polymer Engineering and Science,* 2009, *49,* 1772
7. C. H. Yao; I. Manas-Zloczower *Polymer Engineering and Science,* 1998, *38,* 936
8. K. Inoue *Journal of the Japan Society of Polymer Processing,* 2004, *16,* 510
9. T. Ishikawa; S. Kihara, and K. Funatsu, *Polymer Engineering and Science,* 2000, *40,* 357

O-05-202

SILICON-BASED COMPOUNDS AS FLAME RETARDANT FOR POLYSTYRENE

S.A.C. Abarca[1,2], R.A.F. Machado[1], A.L.G. Prette[2], O. Flores[2] and G. Motz[2]

[1]*Dept. Chem Eng., Federal University of Santa Catarina, Brazil – silvia@enq.ufsc.br; ricardo@enq.ufsc.br*
[2] *Lehrstuhl Keramische Werkstoffe, Universität Bayreuth, Germany – andre.prette@uni-bayreuth.de; octavio.flores@uni-bayreuth.de; guenter.motz@uni-bayreuth.de*

Abstract – Fire resistance of polystyrene is very low when exposed to a heat source. As an alternative, halogen-based compounds used to be the best alternative to enhance fire resistance of polymers with high efficiency. Nevertheless, alternative molecules without halogen, therefore considered environmental friendly compounds, have been studied due to unsafe effects produced by halogen-based materials. Recent studies have shown that small amounts of silicon-based compounds can improve fire resistance of organic polymers. Those silicon-based compounds can be incorporated from many different ways or even used as main polymer matrix. Intending to improve fire resistance of polystyrene, this work purposed the incorporation of quantities of silazane compounds on polystyrene matrix by radical polymerization reaction. For the polymerization two different molecules were used, an inorganic (1,3,5-trivinyl-1,3,5-trimethylcyclotrisilazane) and an organic monomer (styrene). As radical initiator was used dicumyl peroxide. Subsequently, solvent was removed by vacuum distillation. Chemical and physical characteristics were analyzed by Thermogravimetric Analysis (TGA), Differential Scanning Calorimetry (DSC), Fourier Transformed Infrared (FTIR) and Flammability Analysis (in accordance with UL94). Through flammability analysis it was possible to observe an improvement of polystyrene flame retardancy containing 1,3,5-trivinyl-1,3,5-trimethylcyclotrisilazane compared to pure polystyrene. Additionally, some differences on thermal decomposition behavior between product synthesized by bulk and solvent polymerization were detected.

Keywords: Flame retardancy, Silicon-based compounds, Polystyrene

Introduction

Polymeric materials have been extremely necessary to industry, making possible creation of many products and also as a substitute for high cost raw material. Polymers also present a good combination of chemical and physical properties, can be easily processed and are recyclable, although some applications require characteristics sometimes not available. For polystyrene, one of the most used polymers nowadays, the flammability presented when this polymer is exposed to a heat source is extremely high.
The improvement of fire resistance for flammable polymeric material represents an alternative to allow some applications earlier not possible. Different alternatives have been used as halogen based compounds [1], metal hydroxides [2], nanocomposites [3], nitrogen based compounds [4], phosphorus based compounds [5] and silicon based compounds [6]. At the present time halogen based compounds have been an excellent alternative, with a high efficiency observed as FR. Nevertheless the combustion of these molecules generates the formation of smoke with a number of negative characteristics [2]. Nowadays, plastic industry is suffering hard pressures from health organizations which are against the use of halogen-based FR. The combustion of halogen-based compounds generates the formation of smoke with a number of negative characteristics [7]. Searching for an environmental safety FR option studies about different kind of halogen free flame retardant has started [8,9].

A great number of studies about silicon-based compounds, principally with siloxane [10], have been done. Some good characteristic can be achieved with inorganic compounds, e.g. these additives can form physical networks during polymer melt that may reduce dripping, act by dilution effect or even reduced heat feedback [11]. Meanwhile another group of silicon based compound, silazane molecules, also present interesting properties to flame retardancy field.
It is known that poly(vinyl)silazane, a preceramic polymer, have many physical and chemical properties, such as excellent thermal stability, good mechanical properties, good thermal conductivity and oxidation resistance [12].
Intending to reduce the flammability of polystyrene this work proposes the synthesis of a hybrid polymer, formed by an organic and an inorganic part, using radical polymerization.

Experimental

As flame retardant compound and inorganic monomer for hybrid polymer it was used 1,3,5-trivinyl-1,3,5-trimethylcyclotrisilazane (TVCS). Styrene was used as organic monomer. Both organic and inorganic compound were purchased from Sigma Aldrich and used as received, without any further purification.

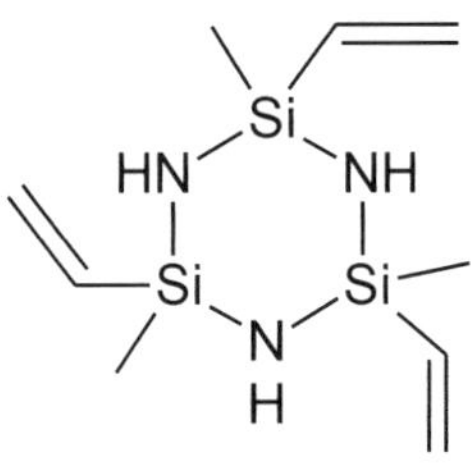

Figure 1 –1,3,5-trivinyl-1,3,5-trimethylcyclotrisilazane

The reaction was performed using a batch stirred glass reactor coupled with a condenser and an oil bath with controlled temperature. It was used a ratio of 3:1 of organic monomer and inorganic monomer, respectively. Both, solvent and bulk polymerization were used to obtain the hybrid polymer and dicumyl peroxide was used as radical initiator. When necessary, (1-methylethyl) benzene was applied as solvent phase. Firstly it was prepared the organic mixture in a glass reactor at room temperature. The reactor was heated and the inorganic compound were added and mixed under high rotation. After all compounds were mixed the temperature was increased until 135°C. After solvent polymerization, the solvent was removed with rotary evaporator and vacuum. All procedures were executed under Argon atmosphere.

Thermogravimetric analysis (TGA)

Thermogravimetric analysis (TGA) was performed by Lehrstuhl Keramische Werkstoffe using a thermogravimetric equipament from Linseis model Linseis L81 A1550. Samples (10-15 mg) for TGA analysis were heated at 10°C/min to 1000°C under Argon atmosphere. TGA analysis was important to observe the mechanism of action adopted by flame retardants.

Differential scanning calorimetry analysis (DSC)

Differential scanning calorimetry analysis (DSC) was performed by Lehrstuhl Polymer Werkstoffe – Universität Bayreuth using a differential scanning calorimetry equipament from TA Instruments model Q1000. Samples (5-20 mg) for DSC analysis were heated at 10°C/min to 300°C under Argon atmosphere. DSC analysis was used to monitor heat effects associated with transitions and chemical reaction as a function of temperature.

Fourier Transformed Infrared (FTIR)

Fourier Transformed Infrared (FTIR) was performed by Lehrstuhl Keramische Werkstoffe using FTIR equipment from Bruker Optics model Bruker Tensor 127. The FTIR was coupled with TGA to measure which substances were generated during TGA analyses.

Flammability analysis

Flammability test was performed with support of Frenzelit Werke GmbH in accord with UL94 – Test for Flammability of Plastic Materials for Parts in Devices and Appliances (horizontal samples). The analysis involves samples of hybrid polymer that was prepared as plaques with 150 mm x 50 mm X 13 mm. The apparatus is manually operated moving a burner ignition source under of the plaque sample to apply a flame from the source to the sample according to the UL 94 regulations.

Results and Discussion

Thermogravimetric analysis (TGA)

In Figure 2 are shown thermogravimetric results for pure monomers and hybrid compositions made by bulk and solvent polymerization.

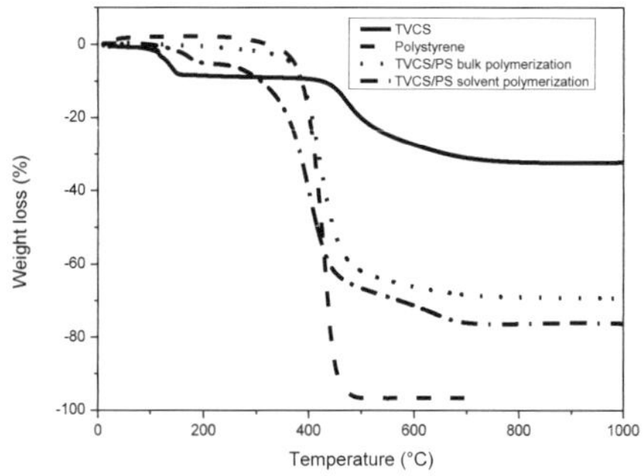

Figure 2 – Thermogravimetric analysis of pure 1,3,5-trivinyl-1,3,5-trimethylcyclotrisilazane, pure polystyrene, PS/TVCS bulk and PS/TVCS solvent polymerization.

It is possible to note that weight loss of TVCS was around 30%. This value is considered very low when compared to the polystyrene (almost 100% of weight loss) and was due to the capability of inorganic compound form, during the reaction, a cross-linked network. With thermic treatment, in this case the increase of temperature during TGA measurements, more cross-linking reactions take place. For TVCS, two stages of weight loss were observed, first at approximately 200°C (10% weight loss) and second at 430°C (20% weight loss). The second weight loss was probably due to a transformation to a dense ceramic. This behavior is well known and characterize TVCS as a highly crosslinkable precursor for SiCN ceramics [13].
Comparing with pure polystyrene was observed a decrease of weight loss when the inorganic monomer was reacted with polystyrene to make the hybrid compound. For TVCS/PS bulk polymerization was obtained 70% of weight loss and around 75% of weight loss for solvent polymerization. Another important point is the degradation temperature. For pure

polystyrene and TVCS/PS bulk polymerization, the initial degradation temperature was around 400°C. Although for TVCS/PS solvent polymerization was observed an initial degradation at 200° as a first step. This behavior is typically observed in condensed phase flame retardant systems where a protective layer is initially formed.

Differential scanning calorimetry analysis (DSC)

In Figure 3 are presented differential scanning calorimetry results for pure monomers and hybrid compositions made by bulk and solvent polymerization.

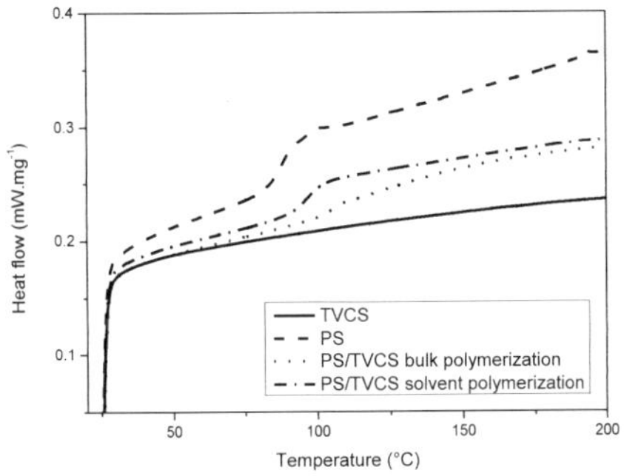

Figure 3 – Differential Scanning Calorimetry of pure 1,3,5-trivinyl-1,3,5-trimethylcyclotrisilazane, pure polystyrene, PS/TVCS bulk and PS/TVCS solvent polymerization.

Through DCS analysis was not possible to observe the degree of dispersion between both organic and inorganic compounds related to glass transitions values. The Tg signal intensity obtained for organic polymers (strong signal) when compared to inorganic precursor (weak signal) is significantly different and did not allow to observe a new Tg value for hybrid structure.

In table 1 are values of glass transition temperature for pure polystyrene and organic-inorganic compounds.

Table 1 – Glass transition temperatures values for pure polystyrene, PS/TVCS bulk and PS/TVCS solvent polymerization

Sample	Start (°C)	Onset (°C)	End (°C)
Polystyrene	83.51	87.50	92.10
PS/TVCS (solvent polymerization)	90.99	97.47	103.85
PS/TVCS (bulk polymerization)	101.74	108.96	116.18

A significant increase of Tg was observed when pure polystyrene is compared to both TVCS/PS bulk and solvent polymerization. The most significant increase was observed for TVCS/PS bulk polymerization. This behavior can be possibly due to the reaction between organic and inorganic part. During radical polymerization, double vinyl bond of styrene suffer a rupture caused by radical initiator. Those active monomers can react between themselves other can react also with double bonds from TVCS. Through this reaction is possible to obtain polystyrene with some quantity of inorganic monomer connected into polymeric matrix. Some additional analyses are necessary to confirm if this inorganic molecule is bonded as a pendant group which can react with other styrene monomers or with other TVCS molecules to execute a crosslinking reaction.

Thermalgravimetric analysis monitored by Fourier Transformed Infrared (TGA-FTIR)

The pyrolysis products formed during TGA analysis were monitored using TGA-FTIR. FTIR spectra of pyrolyzed products are shown in Figure 4.

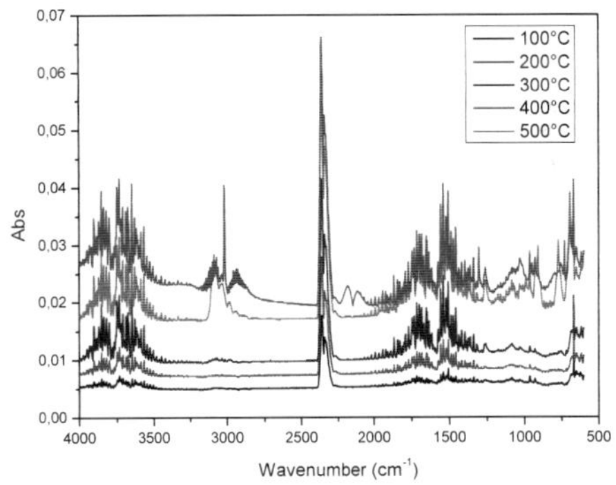

Figure 4 – FTIR-TGA PS/TVCS bulk and PS/TVCS solvent polymerization.

Pyrolysis products were monitored using TGA-FTIR. At 200°C and 300°C is possible to observe the formation of N-H groups (around 1500 cm^{-1}). The N-H groups are important volatiles for flame retardant systems for two reasons; at high temperatures ammonia is liberated and can act as diluents for oxygen and combustible gas mixtures, can also promote recombination with reactive radicals to form more thermally stable molecules. Some residues can be formed in the condensed phase resulting in endothermic process, another protection way of flame retardants. Some N-H groups are useful to help in intumescent process and formation of protective layers. At 400°C started the liberation of C-H groups (around 3000 cm^{-1}) and also another peak of N-H groups (around 1000 cm^{-1}). C-H liberated during pyrolysis is responsible to allow the formation of a protective layer with high thermal resistance. This protective layer is generally formed between the gaseous phase where occurs the combustion and the solid phase. The layer is important to limit the transfer of volatile gases and oxygen and can act as a physical barrier for solid products[14]. At 500°C another important group was released, Si-H group (around 2200 cm^{-1}). It is known that small quantities of silicon-based compound can improve flame resistance due to the capability of perform cross linking reactions with vinyl groups of styrene and also due to thermal resistance provided by silicon-based compounds.

Flammability analysis (UL94)

Flammability tests were executed with pure polystyrene and both samples, PS/TVCS bulk polymerization and PS/TVCS solvent polymerization. PS sample was totally burned with drops of polymer during the test. Samples of TVCS/PS presented some flame retardancy with a time to flame extinction below than 30s and no dropping was observed. According results the hybrid material was classified as a V-1. To achieve a higher degree of flame retardancy, another flame retardant can be applied to improve some deficiency of protection mechanism.

Conclusions

In this study was synthesized a hybrid organic inorganic compound adding some quantities of 1,3,5-trivinyl-1,3,5-trimethylcyclotrisilazane, an inorganic monomer with the aim to improve flame retardancy of polystyrene. TGA analyses show a decrease of weight loss of polystyrene when polysilazane was added. Bulk polymerization appears to be a better synthesis method to reduce the weight loss of polystyrene than solvent polymerization.

DSC analyses also shown a higher increase of glass transition temperature for PS/TVCS obtained by bulk polymerization. This behavior is probably due to cross linking reactions between organic and inorganic part, creating material with enhanced thermal resistance. Results from TGA-FTIR show that a quantity of important groups was liberated with the increase of temperature. In a flame retardant system those groups probably act in gas phase, reacting with combustible species and obtaining more stable compounds. Flammability analysis demonstrated that flame extinction was not the main mechanism of protection, once a big quantity of polysilazane was applied and extinction time was not efficient as expected. Although was observed no dropping during tests. This behavior is typically observed in flame retardants which act preferentially by condensed phase mechanism. An alternative to improve flame retardancy is the use of another flame retardant with high activity in gas phase mechanism. Synergic flame retardants systems with complementary protection mechanisms of two or more compound have been used constantly with good efficiency.

Acknowledgements

The author would like to thank CAPES for financial support, Federal University of Santa Catarina, Universität Bayreuth – Lehrstuhl Keramische Werkstoffe and Lehrstuhl Polymer Werkstoffe and Frenzelit Werke GmbH.

References

1. M.P.L. Cortemiglia, G. Camino, L. Costa, P. Roma, A. Rossi *Journal of Analytical and Applied Pyrolysis* 1987, 11, 511-526.
2. R.S. Bains, P.A. Cusack, A.W. Monk *European Polymer Journal* 1990, 11, 1221-1227.
3. M. Si, V. Zaitsev, M. Goldman, A. Frenkel, D.G. Peiffer, E. Weil, J.C. Sokolov, M.H. Rafailovich *Polymer Degradation and Stability* 2007, 92, 86-93.
4. S. Gaan, G. Sun, K. Hutches, M.H. Engelhard *Polymer Degradation and Stability* 2008, 93, 99-108.
5. M.W. Beach, N.G. Rondan, R.D. Froese, B.B. Gerhart, J.D. Green, B.G. Stobby, A.G. Shmakov, V.M. Shvartsberg, O.P. Korobeinichev *Polymer Degradation and Stability* 2008, 93, 1664–1673.
6. X. Chen, C. Jiao *Fire Safety Journal* 2009, 44, 1010-1014.
7. G. Camino, L. Costa, M.P. Luda Di Cortemiglia *Polymer Degradation and Stability* 1991, 33, 131-154.
8. X. Fu , S. Qutubuddin *Materials Letters* 2000, 42, 12–15.
9. X. Fu , S. Qutubuddin *Polymer* 2001, 42, 807–813.
10. Y. Tang, J. Zhuge, J. Lawrence, J. Mckee, J. Gou, C. Ibeh, Y. Hu *Polymer Degradation and Stability* 2011, 96, 760-770.
11. G. Marosi, A. Martón, P. Anna, G. Bertalan, B. Marosfői, A. Szép *Polymer Degradation and Stability* 2002, 77, 259-265.
12. Q.D. Nighiem, D. Kim, D.P. Kim *Advanced Materials* 2007, 19, 2351-2354.
13. Z.C. Wang, F. Aldinger, R. Riedel *Journal of American Ceramic Society* 2001, 84, 2179-2183.
14. F. Laoutid, L. Bonnaud, M. Alexandre, J.M. Lopez-Cuesta, Ph. Dubois *Material Science and Engineering R* 2008, 63, 100-125.

O-05-189

A MATRICIAL APPROACH OF FIBRE BREAKAGE IN TWIN-SCREW EXTRUSION OF GLASS FIBRES REINFORCED THERMOPLASTICS

A. Durin[1], P. De Micheli[1], J. Ville[2, 3], F. Inceoglu[4], R. Valette[1] and B. Vergnes[1]*

[1]CEMEF, MINES ParisTech, CS 10207, 06904 Sophia Antipolis Cedex, France – audrey.durin@mines-paristech.fr – pascal.demicheli@transvalor.com – rudy.valette@mines-paristech.fr – bruno.vergnes@mines-paristech.fr*

[2]LIMATB, Université de Bretagne Occidentale, France, [3]Polytechs SA, France – julien.ville@univ-brest.fr

[4]Cerdato, Arkema, France – fundainceoglu@kale.com.tr

Abstract - Limiting fibre breakage during composite processing is a crucial issue. The purpose of this study is to predict the evolution of the fibre-length distribution along a twin-screw extruder. This approach relies on using a fragmentation matrix to describe changes in the fibre-length distribution. The flow parameters in the screw elements are obtained using the simulation software Ludovic©. Evolution of an initial fibre-length distribution for several processing conditions was computed and the results were compared with experimental values. The computation gives satisfying results, even though more comparisons with experiments would be necessary.

Keywords: Glass fibres, Micro-mechanics, Computational modelling, Extrusion

Introduction

A classical use for glass fibres in industry is thermoplastic polymers reinforcement, mostly for injected parts. Long-fibre composites are known to offer better mechanical properties than short-fibre ones. Consequently, an important point is to preserve as much as possible long fibres during compounding, despite strong flow conditions, generally leading to severe break-up.

Fibre orientation for long-fibre thermoplastics in moulding process has already been studied and modelled [1-4] and a quantitative model predicting changes in fibre-length distribution during mould filling has been recently developed by Tucker *et al.* [5]. However, prior to injection moulding, other processes are used to compound glass fibres into polymer matrices. An important issue is thus to control fibre lengths in these processes to subsequently inject compounds exhibiting a suitable final length distribution. The most common of these processes is twin-screw extrusion, in which considerable fibre length degradation occurs [6-8]. Shon *et al.* [9] have been the firsts to develop an empirical model describing the average fibre length evolution in different continuous processes, including twin-screw extrusion. More recently, this approach was improved to calculate average fibre length evolution during twin screw extrusion and Buss kneader compounding [10-11]. However, these methods do not provide information on the whole fibre-length distribution. Therefore, the aim of the present study is to propose a computational method to predict changes in the fibre-length distribution along a twin-screw extruder.

Modelling

Theoretical model

Our model is based on the assumption that fibre breakage is only due to flow-induced buckling, as described by Forgacs and Mason [12]. According to this model, a rotating rigid fibre in a shear flow, which is prevailing in a twin-screw extruder, may break when oriented in the direction of compressive forces. Beyond a critical force, which depends on its mechanical properties and length, the fibre buckles and then breaks-up. Breakage occurs because of the severe tensile stress σ_s induced on the external surface of the fibre when it is bending. This stress depends on the fibre radius b, its Young modulus E and the local radius of curvature R:

$$\sigma_s(x) = -\frac{Eb}{R(x)} \qquad (1)$$

where x is the abscissa along the fibre principal axis. When the stress σ_s on the surface reaches the tensile strength value of the fibre σ_c, the fibre breaks-up. As the radius of curvature of the fibre is linked to the fibre deformation, the breakage phenomenon directly depends on this deformation.

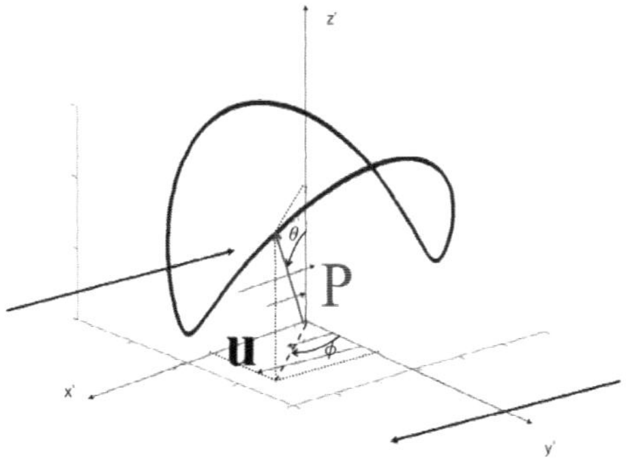

Figure 1 – Fibre end orbit in a simple shear flow. The shear plan frame (x', y', z') is translating with the fibre.

To determine when buckling (and then breakage) occurs, forces applied on the fibre as well as fibre orientation must be computed. Classically, the orientation **P** of a single ellipsoidal fibre of length $2a$ and radius b in a shear flow is obtained by solving Jeffery equation [13]:

$$\dot{\mathbf{P}} = \mathbf{\Omega}\cdot\mathbf{P} + \lambda\left[\dot{\boldsymbol{\varepsilon}}\cdot\mathbf{P} - \dot{\boldsymbol{\varepsilon}}:(\mathbf{P}\otimes\mathbf{P})\mathbf{P}\right] \qquad (2)$$

where **P** is the orientation vector of the fibre principal axis, $\mathbf{\Omega}$ the vorticity tensor, $\dot{\boldsymbol{\varepsilon}}$ the strain rate tensor. The aspect ratio is defined as $\beta = a/b$ and $\lambda = \dfrac{\beta^2 - 1}{\beta^2}$ is a parameter related to the aspect ratio.

The orientation vector **P** describes the fibre orientation in the reference frame. In the case of simple shear flow, this frame is defined as depicted in Fig. 1. This orientation can also be described by the angles θ and ϕ (which can be determined from **P**). As glass fibres are cylindrical, the ellipsoid aspect ratio β should be replaced in Eq. 2 with an equivalent aspect ratio for cylinders $\beta_e = 0.74\ a/b$, theoretically determined by Burgers [14].

The orientation vector was then used in the forces computation. The shear induced stress σ_B, integrated on a half-fibre, was given by Burgers [14], without further indication on the forces distribution along the fibre:

$$\sigma_B = \eta\dot{\gamma}\, M \frac{\beta^2}{\ln(2\beta) - 1.75} \qquad (3)$$

where η is the viscosity and $\dot{\gamma}$ the shear rate. M is defined as $M = \sin^2\theta \sin\phi\cos\phi$ where θ and ϕ were obtained from the orientation vector **P**.

In order to determine when buckling occurred, forces were assumed to be punctually applied at fibre ends and along its principal direction. Then, it was possible to obtain the buckling threshold by applying Euler buckling method. The bending momentum balance for a non-deformed configuration gives:

$$M_0(x) = F_B\, y(x) \qquad (4)$$

where $M_0(x)$ is the bending momentum and $y(x)$ the fibre deflection at point x. From this equation, the deformation below and beyond the buckling threshold can be obtained.

In this work, it was assumed that, when buckling occurs, the fibre systematically breaks-up because of the resulting huge deformation. In order to confirm that perfect (without any defect) rigid fibres cannot break-up when simply bending below the buckling threshold and also always break-up when buckling, the tensile stress σ_s applied on the external surface of the fibre when it bends and when it buckles was computed using the deformation expression obtained from Eq. 4 [15]. It was then compared to the tensile strength σ_c in order to check if the fibre does break. In the tested conditions, it was noticed that fibres were almost always breaking-up when buckling, justifying the proposed approximation that buckling is equivalent to break-up.

It was also observed [15] that fibres were breaking-up even when the angle θ between them and the z' axis is smaller than $\pi/2$ (fibre out of the shear plan, see Fig. 1). Moreover, fibres spend a long time along the flow direction during a Jeffery rotation period, and the angle θ at this moment is maximal. Consequently, the angle between most of the fibres and the z' axis is almost all the time close to $\pi/2$. Therefore, the parameter θ was set to $\pi/2$ in the computations, supposing that even if some of the fibres were slightly out of the plan, they would also break-up.

The buckling threshold was calculated under the assumption of "small" deformations (Euler buckling method), in which the bending momentum M_0 is approximated by:

$$M_0 = \frac{E\,I}{R} \approx E\,I\,y'' \qquad (5)$$

where I is the moment of inertia and R the radius of curvature. Combining Eq. 4 and Eq. 5 leads to a differential equation of y and then to the expression of the maximal stress σ_m^e before buckling:

$$\sigma_m^e = -E\left(\frac{\pi}{4\beta}\right)^2 \qquad (6)$$

Comparing the stress σ_B (Eq. 3) induced by the flow on the fibre with the maximal stress σ_m^e (Eq. 6) before buckling leads to a buckling parameter Bu:

$$Bu = \frac{\sigma_B}{\sigma_m^e} = -\left(\frac{4}{\pi}\right)^2 M \frac{\eta\dot{\gamma}}{E} \frac{\beta^4}{\ln(2\beta) - 1.75} \qquad (7)$$

which is greater than 1 for a buckling fibre.

As we had made the approximation that buckling is equivalent to break-up, it was possible to directly use the buckling parameter Bu to determine at which shear stress a perfect fibre would break for a given length.

Modelling approach

Because the minimal radius of curvature is located in the middle of a bending fibre (Eq. 1), a perfect fibre will always break-up at this point. However, a real fibre always presents defects which can be considered as weak points, leading to a possible break-up below the critical stress value. Therefore, introducing a statistical distribution of the breakage probability along the fibre was necessary in order to mimic the presence of defects. The usual breakage probability law for glass fibre reinforcement is the Weibull distribution [16]. However, this law is only valid for a homogenous stress along the fibre, i.e. for a fibre loaded in tension. Consequently, the breakage probability (Eq. 8) and the local breakage probability distribution along a fibre (Eq. 9) were assumed to be uncorrelated.

The fibre breakage probability is:

$$P(Bu) = \frac{1 - \exp(-Bu)}{1 - \exp(-1)} \qquad \text{if Bu} < 1$$

$$P(Bu) = 1 \qquad \text{if Bu} > 1 \qquad (8)$$

This probability law allows a fibre to break below the buckling threshold because of the defects. The fibre necessarily breaks-up once this value is reached, complying with the previous theoretical observations.

The local breakage probability distribution along the fibre was chosen as a Weibull-like distribution (Eq. 9), but it remains possible to choose another type of distribution, for example a normal distribution [5]. The breakage probability is maximal at the middle of the fibre (x = 0). In Eq. 9, m allows to change the distribution shape. In this first approach, m was arbitrarily set to 3.

$$P_l(x) = \frac{1-\exp\left[-\left(1-\left(\frac{x}{a}\right)^2\right)^m\right]}{\int_{-1}^{1}\left(1-\exp\left[-\left(1-\left(\frac{x}{a}\right)^2\right)^m\right]\right)d\left(\frac{x}{a}\right)} \quad (9)$$

In order to describe the entire distribution, a fragmentation matrix, based on the mass conservation [17] was used. Fibres were distributed into n classes according to their length. The minimal accessible length L_n was set to the minimal length below which a fibre cannot buckle for any applied stress (β = 2.88, Eq. 7). Then, the mass transfer from long fibre classes to shorter ones was expressed using the set of equations (Eq. 10), and solved with a Dormand-Prince method [18] using Matlab©.

The mass transfer set of equations is:

$$dm_i = \sum_{j=1}^{n}(\tau_j\, P_{ij}\, M_{ij}\, dt) - \tau_i\, m_i\, dt \quad (10)$$

where m_i is the mass ratio represented by the fibres of length L_i, P_{ij} the probability for breaking-up a L_j length fibre to generate a L_i length one, M_{ij} the mass transfer from the L_j length fibre class to the L_i length one (if a fibre breaks at the point required to generate a L_i length fibre), and τ_j the L_j length fibre breakage rate, that was considered to be equivalent with the L_j fibre length breakage probability per time unit. The quantities P_{ij}, M_{ij} and m_i are dimensionless and the breakage rate is expressed in s^{-1}. When a L_j length fibre breaks-up, generating a L_i length fibre, a part of its mass is transferred to the L_i length fibre class. The mass transfer between classes is then $M_{ij} = m_j\, L_i / L_j$. As the local breakage probability distribution on a fibre is symmetric relatively to its centre, the probability P_{ij} for a breaking-up L_j length fibre to generate a L_i length one is twice the probability P_l (Eq.9) to break at one of the two possible points, leading to $P_{ij} = 2P_l(x_i)$. The L_j length fibre breakage rate τ_j was considered as equivalent to the probability for a L_j length fibre to break-up during a rotation period t_r divided by this period. The breakage probability P_{tr} during a period was computed by adding the breakage probabilities P (Eq. 8) for each successive orientation **P** along the fibre orbit.

Input parameters for the simulation are the fibre characteristics, i.e. Young modulus (E = 72.4 GPa), initial length distribution and radius (b =5 μm). As explained previously, the angle θ was set to $\pi/2$. Estimates of the shear rate and residence time in the various sections of the extruder are also needed. To compute them, the Ludovic© software, developed in the laboratory and already largely used in different studies including fibre breakage [10], was used. The flow simulation with Ludovic© requires to define the screw profile, the polymer behaviour as well as processing conditions (feed rate, screw speed, barrel temperature). The matrix viscosity and the filled polymer viscosity were described by temperature-dependent Carreau-Yasuda laws:

$$\eta(T) = \eta_0(T)\left[1+(\lambda(T)\dot{\gamma})^a\right]^{\frac{n-1}{a}} \quad (11)$$

where η_0 is the zero-shear viscosity, λ a characteristic time, a the Yasuda parameter and n the power law index. η_0 and λ depend on temperature, according to Arrhenius laws. The parameters values correspond to a polyamide and to a polyamide filled with short glass fibres after mixing, and were experimentally measured. Computation starts from an experimental initial length distribution [10]. Then, flow conditions (shear rate, viscosity and residence time) along the twin-screw extruder were computed in a first step, using Ludovic© software. Then, from the obtained data, the rotation of a fibre of average length in a shear flow was computed using Jeffery's equation (Eq. 2). This "average" fibre was supposed to rotate in the shear plan (i.e. $\theta = \pi/2$). Afterwards, for each length class, the stress σ_B (Eq. 3) applied on a fibre and the limit stress before buckling σ_m^e (Eq. 6) were computed. The fibre breakage probability P (Eq. 8) for each length class and for each obtained orientation was then calculated. Subsequently, the probability P_{tr} to break during a rotation period was deduced for each length class. In parallel, for each length class, the local breakage probability distribution P_l along a fibre (Eq. 9) was computed and the probability P_{ij} for a L_j length fibre to break-up generating a L_i length fibre was deduced. Using all these values, the fragmentation matrix (Eq. 10) was generated. Finally, the set of equations was solved with Matlab© and the obtained mass distribution was injected as the starting point for the next extruder segment computation.

Results and Discussion

In order to validate the model, the evolution of fibre length distribution along a twin screw extruder was calculated and compared to experimental data previously obtained [10]. An example is shown in Fig. 2. The experiments were carried out on a laboratory scale twin-screw extruder (ThermoFisher PTW 24) for different processing parameters (screw speed N from 200 to 600 rpm, feed rate Q from 5 to 9 kg/h). Dried polyamide 12 (PA12) (Rilsan A, kindly supplied by Arkema, France) was fed through the main feeder and mixed with 30 wt% glass fibres, added through a second feeder after PA12 melting. The initial average fibre length (in mass) was 3.48 mm and the fibre Young modulus was 72.4 GPa. Several samples were taken along the screws, after stopping the extruder and opening the clamshell barrel. Fibre length measure-

ments were achieved by optical microscopy and image analysis, after burning the samples and dispersing the recovered fibres into glycerol [10, 11].
Generally speaking, the shapes of the simulated fibre length distributions corresponded quite well to the experimental ones. The effects of variation of the processing parameters were also correctly evaluated: a high rotation speed increased fibre breakage and a high feed rate limited it. Nevertheless, the experimental length distributions were very close from one set of parameters to another. Consequently, the accuracy of the model for different process conditions is not yet totally assessed.
The evolution of the average length in mass is shown on Fig. 3, as it is a common measure for fibre breakage study [9-11]. One again, the agreement between model and experiments is satisfactory. The same agreement was observed whatever the processing conditions. However, it is a poorer representation of the fibre length evolution, as the length distribution might be sometime bi-modal.

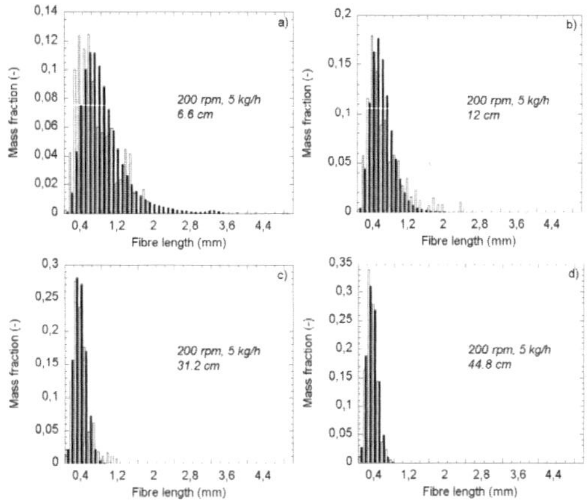

Figure 2 – Measured (white) and computed (black) length distributions. Screw rotation speed: 200 rpm, feed rate: 5 kg/h. Location from the fibres introduction: (a) 6.6 cm ± 0.5 cm, (b) 12 cm ± 0.5 cm, (c) 31.2 cm ± 0.5 cm, (d) 44.8 cm ± 0.5 cm

It appears that the beginning of the process was more difficult to predict as the breakage rate was very high and consequently the error was larger. As the experimental samples have a certain thickness (a few mm), the error on the sampling position (and on the sampling time) could be important and might explain a part of the difference between experimental and calculated values. Furthermore, the viscosity law parameters used in the computation were measured on polymer filled with already broken fibres, thus the viscosity is underestimated at the beginning of the process when the fibres are still long.

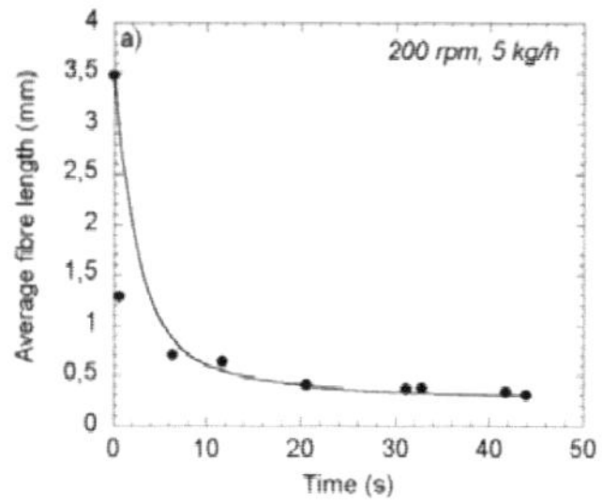

Figure 3 – Computed (full line) and measured (symbols) average lengths in mass. Screw rotation speed: 200 rpm, feed rate: 5 kg/h.

Conclusions

A new computational method to predict changes in the fibre-length distribution during compounding in a twin-screw extruder was proposed. This evolution was calculated for five different sets of processing conditions, without adjusting any parameter, and these results were compared to experimental data. A satisfactory agreement was found, except at the early beginning of the process. The experimental length distributions were very close from one set of parameters to another, so the model still needs to be validated more intensively.

Acknowledgements

This work was carried out in the context of the Fonlimics project, supported by FUI and labelled by the Plastipolis competitive cluster. We thank the DGE (France) for its financial support.

References

1. M. Vincent; E. Devilers; J.-F. Agassant *J. Non-Newt. Fluid. Mech.* 1997, *73*, 317.
2. S. G. Advani; C. L. Tucker III *J. Rheol.* 1987, *31*, 751.
3. F. Folgar; C. L. Tucker III *J. Reinf. Plast. Compos.* 1984, *3*, 98.
4. J. H. Phelps; C. L. Tucker III *J. Non-Newt. Fluid. Mech.* 2009, *156*, 165.
5. C. L. Tucker III; J. H. Phelps; A. I. Abd El-Rahman; V. Kunc in PPS 26 Proceedings, Banff, 2010, CD.
6. D. Wall *Polym. Comp.* 1989, *10*, 102.
7. K. Shon; J. L. White *Polym. Eng. Sci.* 1999, *39*, 1757.
8. U. Yilmazer; M. Cansever *Polym. Comp.* 2002, *23*, 61.
9. K. Shon; D. Liu; J. L. White *Int. Polym. Proc.* 2005, 20, 322.
10. F. Inceoglu; J. Ville; N. Ghamri; J. L. Pradel; A. Durin; R. Valette; B. Vergnes *Polym. Comp.* 2011, *32*, 1842.
11. J. Ville; F. Inceoglu; N. Ghamri; J. L. Pradel; A. Durin; R. Valette; B. Vergnes *Int. Polym. Proc.* 2012, *27*, 245.
12. O. L. Forgacs; S. G. Mason *J. Colloid. Sci.* 1959, *14*, 457.
13. G. B. Jeffery *Proc. R. Soc. A* 1922, *102*, 161-179.

14. J. M. Burgers in *On the Motion of Small Particles of Elongated Form, Suspended in a Viscous Liquid. Second Report on Viscosity and Plasticity*, Nordemann Publishing, New York, 1938.
15. A. Durin, PhD Thesis, Mines ParisTech, 2012.
16. W. Weibull *J. Appl. Mech.* 1951, 18, 293.
17. P. Blazy; J. Yvon; E. A. Jdid *Techniques de l'Ingénieur*, 2006, *J3050*.
18. J. R. Dormand; P. J. Prince *J. Comput. Appl. Math.* 1980, *6*, 19.
19. B. Vergnes; G. Della Valle; L. Delamare *Polym. Eng. Sci.* 1998, *38*, 1781.

O-05-184

ENHANCE PROPERTIES OF CROSSLINKED PP/UHMWPE BLEND

Anucha Khongrit[1], **Utai Meekum**[1,2]
[1] *School of Polymer Engineering, Suranaree University of Technology, Nakhon Rachasima, Thailand ; anuchaie18@gmail.com*
[2] *Center of Excellence on Petrochemical and Materials Technology, Chulalongkorn University, Bangkok, Thailand; umsut@g.sut.ac.th*

Abstract:The enhancement of HDT without scarifying the mechanical properties of the DCP/silane induced crosslinked PP in the presence of UHMWPE was explored. The effect of DCP in the crosslink system, blending with block PP and adding the talc filler were attempted. The mixing was performed in co-rotation twin screw extruder. The injection molded specimen was prepared. The sauna treatment was applied and compared. Both mechanical testing and morphological investigation using the standard procedures were conducted. It was found that increasing the DCP concentration on the silane/DPC crosslink system of PP/UHMWPE blending had vastly chain degraded via the free radical chain scission. Hence, the mechanical properties especially toughness of the blends were inferior. Using homo and block PP blended matrix mixed with UHMWPE to enhance the toughness of the tertiary blends was ineffective. The reverse effect was found. The toughness was deceased with copolymer loading. The chain degradation through the free radical induced chain scission was the major drawback. Adding talc filler into the PP/UHMWPE revealed that mechanical and thermal properties of the compound were increased with increasing the degree of fill. The well dispersion and decent adhesion of the talc particles in the polymer blend matrix is responsible for the superiority.

Keywords: PP/UHMWPE Blends; homo and block PP blend; talc filled PP compound; mechanical properties

Introduction

Crosslinking of polyolefins can be done to all polyolefin types. However, the vast majority of the publications are concerned with polyethylenes. Few studies concerning the crosslinking of PP have been reported. Crosslinking of PP by free radical reaction is difficult. This is most likely a consequence of the nature of the PP chain degradation[1]. During the mixing process, PP chain scission has accompanied as a main reaction. This would be negatively influence the properties of the final product. The macroradicals formed may subsequently follow two competing pathways. They can either initiate the grafting of monomers or undergo chain scission, which is the main side reaction, perhaps major reaction, during the process[2]. A polymeric alloy is one of the important routes for preparation alternative materials with synergistic and tailored properties. Developments of polymer blends have been mainly focused on two component systems; matrix and dispersed morphology[3]. Blending of different polymers has long been practiced in the manufacturing industry for various reasons including: (*i*) tailor made blends to meet specific processing and performance requirements that cannot be satisfied by a single component; (*ii*) blending of polymers can be used in the field of recycling post consumer wastes to upgrade their properties, (*iii*) scientific interests; and (*iv*) financial incentives[4]. The addition of fillers to polymers is a fast and cheap method to modify the properties of the polymeric materials. For this reason, filled polymers have subjected of increasing interest in both industry and research. Normally, strength, stiffness, thermal properties, among other properties, can be superior to the required values[5]. Talc is the most commonly used because it is the softest mineral, abundant and inexpensive filler. Filled products typically exhibit excellent properties such as high stiffness, surface aesthetics, excellent HDT, low coefficient of thermal expansion and shrinkage, improved scratch as well as resistance to some aggressive chemical agents[6]. In this research, attempting to enhance the HDT of PP without scarifying the mechanical properties of DCP/silane induced crosslinked polypropylene is the main objective.

Experimental

Materials

Main materials used in the study were PP, PP 700J, and PP, PP740J, which are homo and block copolymer, respectively. It was supplied from SCG Chemical Co., Ltd. The UHMWPE, UH900, was available from Asahi Chemical Industry. Dicumyl peroxide(DCP) and silane(VTMS), Silquest® A 171, were used as coupling/grafting system. They were also used as complementary crosslink agent. The chemicals were commercially available. Talc as filler with the average particle size of 1.4 μm was employed.

Specimen preparation

PPs was dry blended with DCP in warmed oven at 80°C for 5 min to allow DCP to liquidize and then was vigorously stirred in tight container. Then, 2.0 phr of VTMS was added into the polymer ingredients. After that, it was in corporately blended with UHMWPE and talc powder. The melt compounding was achieved in closely intermeshing co-rotation twin screw extruder, with L/D ratio and screw diameter of 25/1, 25 mm, respectively. The extruder barrel temperature profile from feed to die zone was at 190°C and screw speed at 10 rpm. Samples were injection molded on Tederic TRX60c injection molding machine. The barrel temperatures were profiled at 160, 165, 170,

170 and 170°C from feed to nozzle tip, gradually. The injection molded specimens were divided into 2 sets of samples. One was allowed to anneal at room temperature for a day before testing, assigned as "**original**". Latter was incubated in the water vapour saturated oven at 105°C for 12 hour to accelerate the competition of silane/moisture crosslink condensation reaction. Then, it was cooled down at room temperature for day priority to conduct the test. It was named as "**cured**" sample.

Materials testing
MFI test was performed at 170/2.16. HDT were measured at the 455 kPa. Unnotched and notched impact strengths were obtained. Flexural properties both strength and modulus sample were examined. The tests were conducted in accordance with ASTM standards. Phase morphology of the fractured surface of compound samples were examined by SEM.

Results and Discussion
Effect of DCP on the PP/UHMWPE blends
MFI and HDT of PP/UHMWPE blends with various DCP content is given in Fig. 1. It is seen that the MFI is exponentially increased with increasing the DCP content for both in the original and after sauna cured specimen. Sauna incubation does lower the index. Chain scission of PP by the free radical biting is the main suspect for this phenomenon. The higher in peroxide, hence free radical, concentration, the faster in chain degradation. HDT of the blends are more or less in depended on the DCP concentration, especially for the original samples. Within the critical chain length, the HDT of polymer does not depend much on the molar mass. Rising of HDT after sauna curing of the blends is two folds hypothesis explanation. The first one is the easily short chain crystallinity while annealing and the second is the chain recombination through the silane/moisture induced condensation. The later would cause the chain length to expand or perhaps forming network chain.

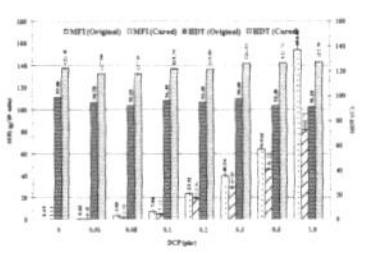

Figure 1 MFI and HDT of PP/UHMWPE blends with DCP contents

Fig. 2 reveals the impact strengths of the PP/UHMWPE. It is observed that both notched and unnotched impact strengths were lowered with increasing the DCP loading as expected. Because from the flow index, the short chain formation caused by the free radical scission would lower the material toughness. Chain recombination by sauna curing would slightly hurdle the impact values.

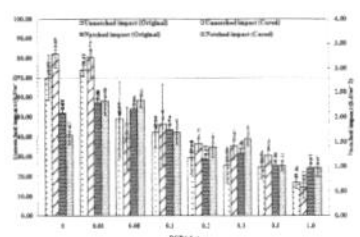

Figure 2 Impact strengths of PP/UHMWPE blends at given DCP contents

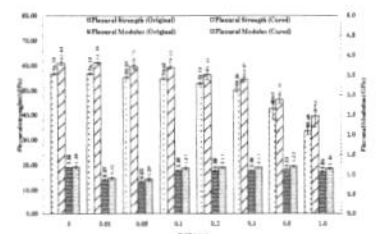

Figure 3 Flexural properties of the PP/UHMWPE blends and DCP contents

Similar to the impact outcome, the flexural strength and modulus of the PP/UHMWPE blends are presented in Fig. 3. The test results are illustrated that the strength is decreased but slightly increased in modulus with increasing the DCP. Results refine that the bending toughness of the materials is lost with adding the DCP. This observation is the common symptom of the brittle material caused by chain degradation.

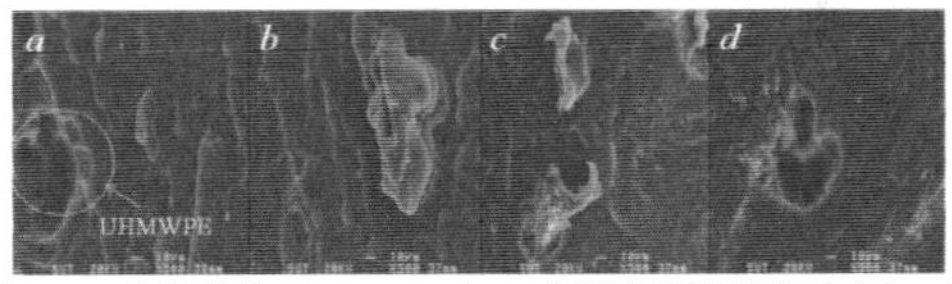

Figure 4 SEM photographs of PP/UHMWPE blends with; (*a*) 0.01 phr(original) (*b*) 0.01 phr(cured), (*c*) 1.0 phr(original) and (*d*) 1.0 phr(cured) of DCP, respectively.

Fig. 4 shows SEM pictures of PP/UHMWPE with 0.01 and 1.0 phr of DCP, before and after sauna incubation, respectively. It is seen that UHMWPE phase is obviously unmelted and dispersed in the PP matrix. It indicates that the UHMWPE powder could not be fused during the melt mixing. It becomes the foreign and non-adhere particles in the blend, creating more voids. Consequently, they accelerate the crake propagation. The brittle material fractured surface characteristic is observed on the PP phases. It is noticed that more rougher traces at higher DCP loading.

Effect of PP block on PP/UHMWPE Blends

The amount of the PP copolymer, 10 to 40 phr, was blended with homo PP in the PPs/UHMWPE blends. MFI and HDT of the blends obtained are presented in Fig. 5. It is found that MFI is gradually increased with increasing the PP block. Again, sauna incubation does lower the flow index values. According to the manufacturer data sheet, MFI of the block is much higher than homo PP. Attempting by using the rule of mixture between these two polymers, the calculated MFI of the blends suggest that the calculated MFI of blends are in same trend as the measured values. The higher in measured than the calculated ones is due to the greater chain scission via free radical effect.

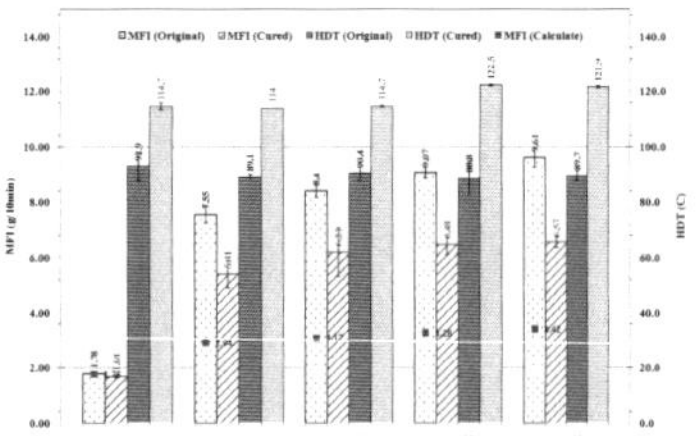

Figure 5 MFI and HDT of the PPs/UHMWPE blends and copolymer fraction

HDT of the original samples do not depend on the copolymer fraction but sauna treated specimen are fractionally increased with increasing the copolymer especially at the content above 30 phr. The increasing in HDT through the sauna curing process of the blends with high copolymer fraction might be due to the easily crystallization of ethylene segment in the PPs matrix after the given prolong annealing time at high temperature.

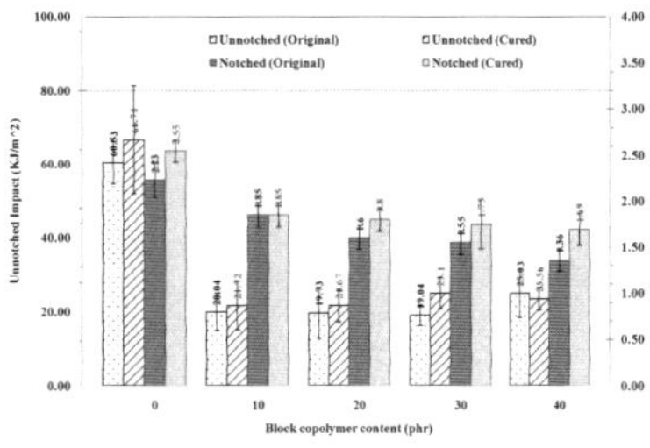

Figure 6 Impact strengths of the PPs/UHMWPE blends

Fig. 6 represents the notched and unnotched impact strengths of the PPs/UHMWPE blends. Unexpectedly, the notched strength is decreased with increasing the PP block loading but not much changed for the unnotched mode. The results indicate that the peroxide chain degradation would take the prime control of the notched impact strength of the blends over the characteristic of copolymer. Generally, the polymer degradation, especially during the melt processing, would lead to the decreasing in the material toughness.

Flexural strength and modulus of the PPs/UHMWPE blends is illustrated in Fig. 7. Similar to those impact strengths, the result reviews that flexural strength of both original and sauna cured samples are slightly decreased with increasing the block fraction. This is also contradicted to the fact that copolymer PP normally has the higher ductility than the homo. The modulus is slightly increased with increasing the block. In the flexural interpretation, if the strength decreases but increases in the modulus, it means inferior in the material toughness. Again, the free radical induced chain scission would be largely respond for the more brittle material found in this blends system.

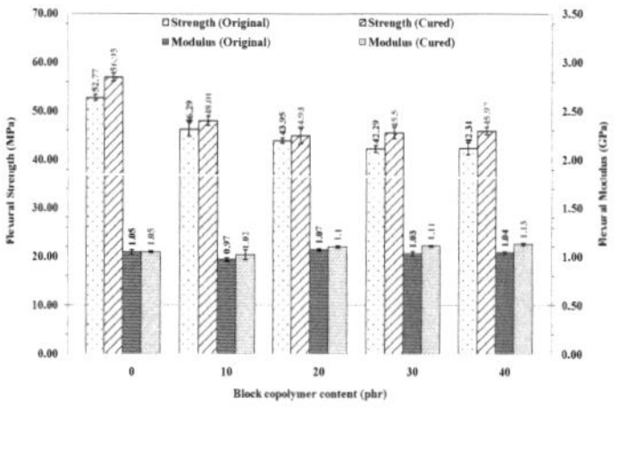

Figure 7 Flexural properties of the PPs/UHMWPE blends and block copolymer loading

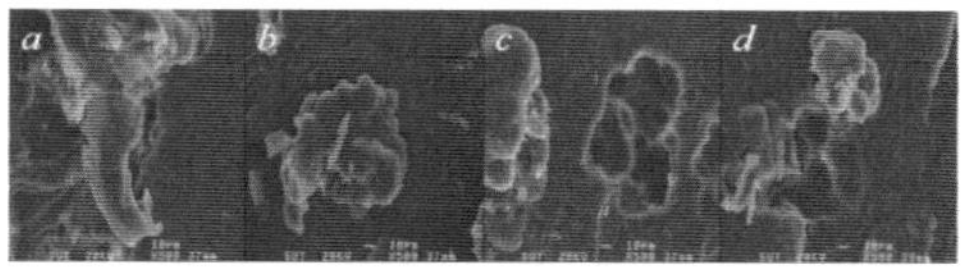

Figure 8 SEM of the PPs/UHMWPE blends with; (a) 10 phr and (b) 10 phr(cured), (c) 40 phr(original) and (d) 40 phr(cured) of PP copolymer, respectively.

According to the mechanical properties found for the PPs/UHMWPE blends. Phase morphology of the blends were examined and presented in Fig. 8. It is seen that the UHMWPE phase is still unmelted. Either by adding the copolymer or underwent sauna curing, the interfacial bonding between the matrix phase and UHMWPE particle does not obviously enhance. Thus, the toughening of the blends by UHMWPE cannot be achieved.

Talc filled PP/UHMWPE compound

Talc at 10 to 50 phr was added into the PP/UHMWPE blends to form the polymer compounds. MFI and HDT of the compound obtained are shown in Fig. 9. It is seen that the flow index is exponentially decreased with increasing the talc load. It indicates that the viscosity of the compound is increased by adding talc.

HDT is slightly increased with the talc loading. Both viscosity and HDT of the compound at the given talc loading are increased by sauna treatment. Commonly, adding particulate filler into the polymer compound can enhance the rheological and thermal properties of the material.

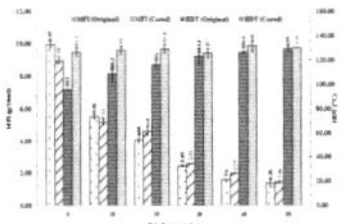

Figure 9 MFI and HDT of talc filled PP/UHMWPE compounds

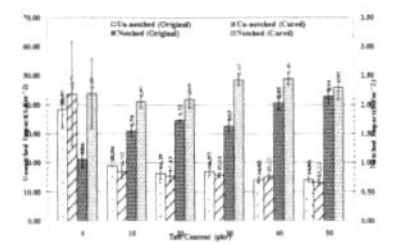

Figure 10 Impact strengths of the talc filled PP/UHMWPE compounds

Fig. 10 shows the impact strengths of the talc filled compound. As commonly found in the talc filled polymer compound, the notched impact strength is linearly increased, with increasing the talc loading. The contribution to the cracked propagation resisting by filler particles is respond for this positive effect. However, unnotched impact exhibits the contrarily responds. When poor adhesion occur between the polymer matrix and filler, the cracks tips would rapidly propagate and penetrate inside the sample. This phenomenon would easily fail the sample. The brittle material would be indentified.

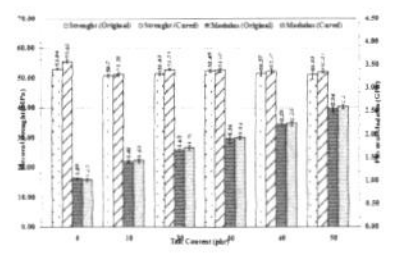

Figure 11 Flexural properties of the talc filled PP/UHMWPE compounds

Flexural properties of talc filled PP/UHMWPE compound are given in Fig 11. It is evidenced that the strengths of both original and cured samples is unchanged with the talc contents. On the other hand, the modulus is linearly increased with increasing the filler loading. Somehow, flat in the strength but rising in the modulus, it is interpreted that the materials indicate the superiority to the bending load.

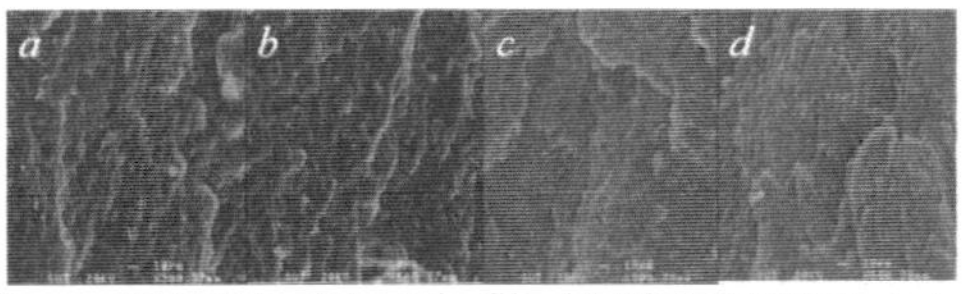

Figure 12 SEM of the talc filled PP/UHMWPE at; (a) 10 phr(original), (b) 10 phr(cured), (c) 50 phr(original), and (d) 50 phr(cured) loading, respectively

SEM investigation of 10 and 50 phr talc filled PP/UHMWPE compound before and after sauna incubation are presented in Fig. 12, respectively. From the morphological observation, with carefully observe at the traces, well distributed of fine particle of talc is seen on matrix phase. The particle has good adhesion with the matrix phase especially after the curing. Consequently, the fracture toughness of the samples would slightly increase with increasing the talc content.

Conclusions

In attempting to improve the HDT of the PP without scarifying the mechanical properties by blending and compounding techniques, it was found that increasing the DCP concentration on the silane/DPC crosslink system of PP/UHMWPE blending had vastly chain degraded via the free radical chain scission. Hence, the mechanical properties especially toughness of the blends were inferior. Using homo and block PPs blended matrix mixed with UHMWPE to enhance the toughness of the tertiary blends was ineffective. The reverse effect was found. The toughness was deceased with copolymer loading. The chain degradation through the free radical induced chain scission was the major drawback. Adding talc filler into the PP/UHMWPE revealed that mechanical and thermal properties of the compound were increased with increasing the degree of fill. The well dispersion and decent adhesion of the talc particles in the polymer blend matrix is responsible for the superiority.

References

1. N.C. Liu; G.P. Yao; H. Huang *Polymer.* 1996, *41,* 4537-4542.
2. M. Hu; Z. Wang; B. Qu; K. Hu *Reac & Func Polym.* 2006, *66,* 287-296.
3. M. Hemmati; H. Nazokdast; H. Shariatpanahi *J. Appl. Polym. Sci.* 2001, *82,* 1138-1146.
4. A-H.I. Mourad *Mater Des.* 2010, *31,* 918-929.
5. J. Gonzalez; C. Alban; M. Ichazo; B. Diaz *Eur. Polym.* 2002, *12,* 2465-2475.

M. Denac; V. Musil; I. Smit; F. Ranogajec *Polym deg. and Stab.* 2003, *82,* 263-270.

O-05-166

INVESTIGATION OF THE PLASTIFICATION BEHAVIOR OF POLYMERS IN HIGH-SPEED TWIN SCREW EXTRUDERS

G. Spohr[1], A. Knieper[2], C. Beinert[3]*

[1] *Mechanical and Process Engineering, Technical University Darmstadt, Germany – GuntramSpohr@gmail.com*
[2] *Dutch Polymer Institute (DPI), PO Box 902, 5600 AX Eindhoven, the Netherlands – Alexander.Knieper@lbf.fraunhofer.de*
[3] *Polymer Processing, Div. Plastics, Fraunhofer-Institute LBF, Germany – Christian.Beinert@lbf.fraunhofer.de;*

Abstract - A test stand to observe movement and deformation of thermoplastic polymers in a cross section of the plastification zone of a co-rotating twin screw extruder is presented. Multiple experiments to examine the initial melting processes in dependence of different process parameters such as rotational speed, temperature, kneading disk configuration, polymeric material, granule size and filling degree were conducted. The focus thereby was the visualization of underlying effects by high-speed imaging. This allowed gaining insights into the basic energy dissipation mechanisms and the most significant process parameters for the initial melting process.

Keywords: Twin-screw extruder, plastic energy dissipation, initial melting, visualization

Introduction

Research Objective

Since up to 80% of the energy input in a co-TSE extruder takes place in the plastification zone [1], the optimization of the plastification process offers the highest potential for an increase of the economic efficiency by energy savings, improvement of material properties and increase of throughput.

The objective of this work is therefore studying the process of initial melting by experimental investigations in the plastification zone of a high-speed co-TSE. Here the responsible mechanisms for energy input into the polymer are friction, heat transfer, plastic and viscous deformation [1]. The mechanisms are thereby coupled and change throughout the plastification zone. They are also dependent onto multiple parameters such as rotational speed, temperature, kneading disk configuration, material properties, granule geometry, filling degree, local pressure, residence time, etc. The differentiation of the four heating mechanisms is therefore difficult.

In order to have a basis for the development of new melting models which are suitable for high-speed co-TSE a visualization of the plastification zone was conducted. Because the effects especially in a cross section of the plastification zone have never been observed, a test-stand that allowed looking at the side faces of two kneading blocks was designed and built. High-speed imaging of deformation, movement and melting of the granules in the cross section of the plastification zone led to further insights into underlying effects.

State of the art

First models to describe plastification in extruders were based on models for single screw extruders [2], [3], [4]. The models are based on the assumption of continuous solid melting (CSM) which postulates a solid bed of granules surrounded by a pool of melt. These models are based on a continuous channel for material flow at low rpm. Thus they do not describe the melting processes adequately in twin screw extruders.

Therefore dispersed solid melting models (DSM) were developed [5], [6], [7]. They assume that granules are dispersed in melt. The melting of solids is thereby related to the dissipation of energy in the melt and conduction of energy from melt to solids. These models describe the melting process in twin screw extruders much better and are nowadays mainly used in commercial extruder software.

The main problem with the DSM models is that the initial melting is neglected in all these models. They are only valid for granules surrounded by a polymer melt matrix. Since at least 25% (for spherical geometries) of the material must be melted for granules to be embedded in melt, all energy dissipation mechanisms up to this point are neglected. But because plastic energy dissipation is the primary energy dissipation mechanism for initial melting it plays a significant role. It cannot be neglected for complete modeling of the melting process. Thought examination of plastic energy dissipation in polymers was already conducted by [8], [9], [10], [11], [12], [13], melting models still lack accuracy. In order to expand the knowledge of the underlying mechanisms and process parameters, experimental investigations on co-TSE extruders are necessary. This will allow generating melting models that integrate the initial melting.

Experimental

Experimental Setup

For the experiments a Coperion ZSK32 high-speed co-TSE was used. The overall length of the extruder was reduced to an L/D ratio of 21. The shafts protruded the extruder for approximately 70mm. The partial length of shaft inside the extruder was equipped with conveying elements; onto the screw tip kneading disks were mounted. This allowed positioning the first compression zone and thus the plastification zone in our test stand as seen in Fig. 1 and Fig. 3.

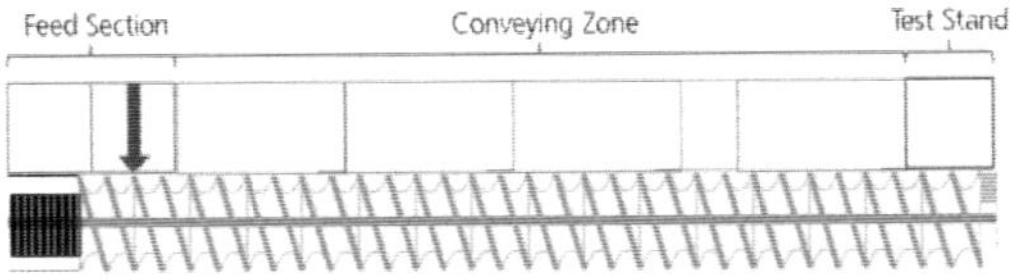

Figure 1 – Screw and extruder configuration

The test stand allowed positioning of a window 1/10th of a millimeter away from the side faces of the last kneading disks and thus allowed observation of the processes in the plastification zone.

Design of the Test Stand

The test stand (Fig. 2) consists out of three different parts:

•The first part is the "granule removal plate" (green). It is directly mounted onto the endplate (blue) of the extruder. It is necessary for running the extruder continuously, without filling of the "plastification zone". An opening allows exiting of granules out of the extruder. By inserting the slider (grey) into the test stand the experiment is started.

•The second part, the "plastification zone plate" (orange), allows positioning the last kneading disk pair in plane with the raise (in the middle). Also sliding rails (red) were welded onto the "plastification zone"

•The third part is the "sledge", in which a window is mounted. The sledge (and thus the window) slides on the rails (red). With screws (blue) the pressure of the window onto the raise on the "plastification zone plate" can be adjusted. This is done by tightening or loosening springs (not shown) which are mounted onto the screws (blue). At a predefined pressure inside the plastification zone the sledge thus moves away from the "plastification zone plate", allowing granules to exit the extruder at the end. This prevents brakeage of the window. For window material a 10mm polycarbonate plate was used. It was cut to fit into the sledge. Two holes were milled into the windows to allow the shafts to protrude the window. The windows were designed as wear and tear parts.

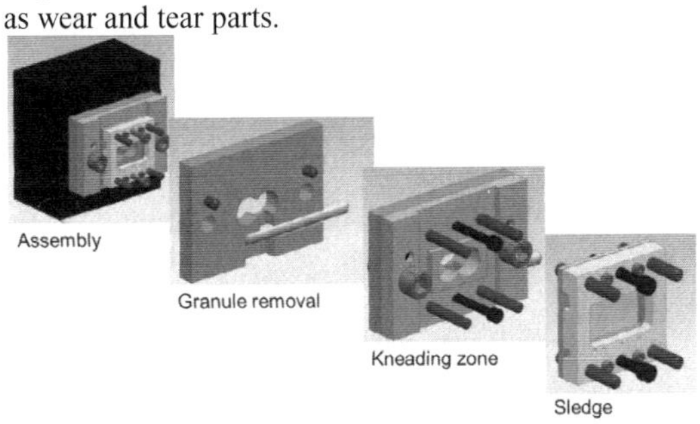

Figure 2 – Cad model of test stand

This setup made it possible to observe movement, deformation and melting of granules in a cross section of the plastification zone. The experiments were recorded by high-speed imaging with 3000 fps. This allowed recording of around 5 seconds. In this timespan the observation of the kneading disks from completely empty to completely filled and partly melted was possible. A window thereby inhibited any movement of granules out of the extruder until a predefined pressure was exceeded. It therefore acted as "back"-conveying elements.

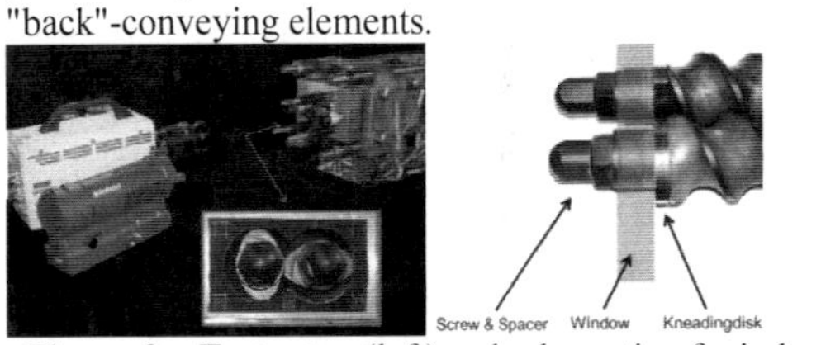

Figure 3 – Test setup (left) and schematic of window position at the screw tip (right)

A parametric study of the initial melting with variation of process parameters such as rotational speed, temperature, kneading disk configuration, polymeric material, granule size and filling degree inside the melting zone was conducted.

Results and Discussion

Plastic Energy Dissipation in the Intermeshing Zone

It was shown that the energy required for melting is primarily dissipated during one heavy, plastic deformation in the intermeshing zone. Granules get trapped between the active flanks and the upper or lower wedge, the intermeshing zone. The granules are thereby compressed into the void volume (~ 40% of bulk volume) and the compressed bulk is deformed (Fig. 4).

Both mechanisms lead to plastic deformation and thus plastic energy dissipation in approximately 0.005 seconds (@ 1200rpm). High strain rates of the material occur. Furthermore a hydrostatic stress is existent because granules are enclosed on all sides while compression and deformation take place.

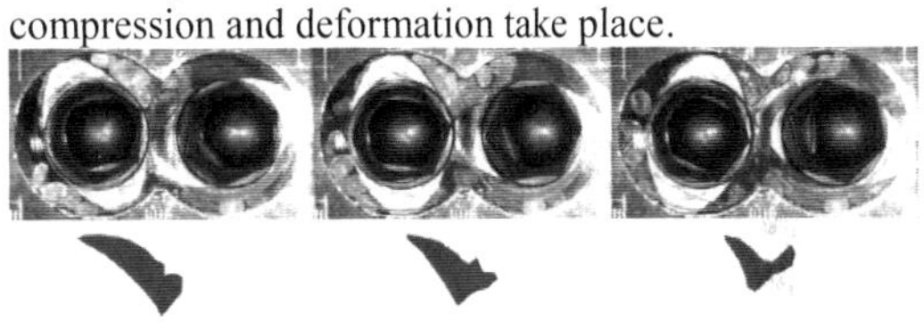

Figure 4 – Compression of polypropylene granules in the intermeshing zone

Compression in Front of Active Flank

Compression in front of the active flank (Fig. 5) can be observed during the experiments. Friction of granules on the cylinder wall compresses material in front of the active flank. Energy dissipation due to friction on the extruder wall was thereby not observed to heat material significantly. Heating only occurs due to "compression into the void volume".

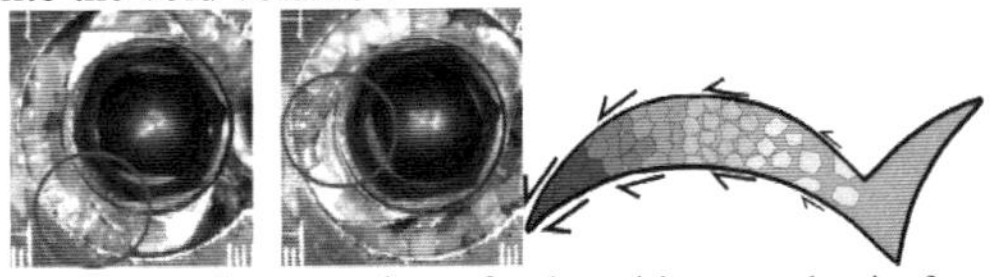

Figure 5 – Compression of polyamide granules in front of the active flank

Influence of Granule Size

Granule size has the biggest influence onto heating and melting of polymer material. Small granules thereby heat up slower than large granules. Reason for this is that small granules show an increased tendency to evade compression and deformation by flowing out of the intermeshing zone (Fig. 6).

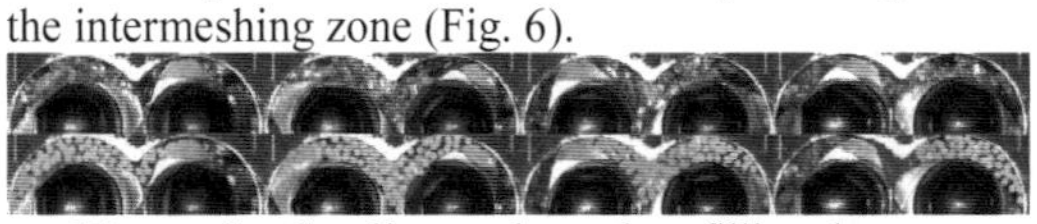

Figure 6 – Small granules at low filling degrees

Large granules by contrast tend to get caught between the kneading disks and the wedge even at low filling degrees (Fig. 7). Granules that get caught thereby stop any further movement of other granules through the intermeshing zone. The amount of material that is caught in the intermeshing zone is thus higher. Therefore the material has to undergo heavier deformation and heats up much faster.

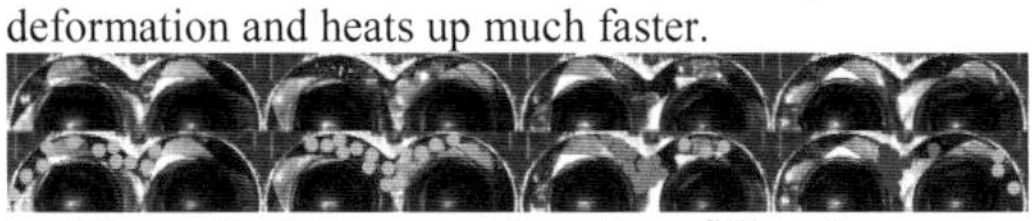

Figure 7 – Large granules at low filling degrees

This dependence of plastic energy dissipation onto granule geometry is a significant scale-up problem for initial melting. A melting process that is optimized for one extruder size, may not work for another.

Influence of Filling Degree / Pressure

The second important influence onto melting behavior of granules is the filling degree inside the plastification zone. It is observable that flow of (small) granules through the intermeshing zone decreases with increasing filling degree: Because all space is already filled by material, granules cannot evade compression and deformation. Additionally the melting degree increase with increasing amount of material used for experimentation (Fig. 8). While at low filling degrees and small granule sizes no plastic energy dissipation is present, it increases rapidly at highest filling degrees.

Figure 8 – Small granules at highest filling degrees

Only 10 to 14 compression and deformation cycles in a fully filled plastification zone are then needed to dissipate enough energy to partially melt all granules. This equals 5 to 7 rotations.

With increasing amount of material used during experimentation the pressure inside the plastification zone as well as the melting degree increased. It therefore seems reasonable that the hydrostatic stress, which increases the compressive strength, is relevant for increasing the amount of dissipated energy.

Influence of Rotational Speed on Granules

Experiments with two different rotational speeds (120 rpm & 1200rpm) were conducted. No dependence of plastic and frictional energy dissipation onto the rotational speed could be observed for granules. So although the compressive strength increases at high strain rates, no significant change of the amount of energy that was dissipated per rotation resulted.

The experiments were also conducted with white PA12 powder (diameter: 80-120μm) mixed with soot coagulates. Here a significant change of effects in dependence of the rotational speeds took place:

At lower rpm (120rpm) the soot coagulates were not destroyed. No significant amount of plastic or frictional energy was dissipated (Fig. 9).

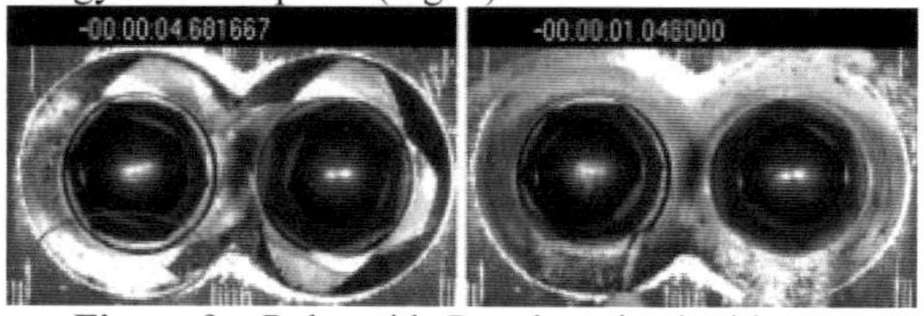

Figure 9 – Polyamide Powder mixed with soot coagulates at 120rpm

At 1200rpm however some coagulates were already ground in the conveying section (Fig. 10). High relative speeds of the screws lead to frictional destruction. In the plastification zone itself the powder turned black within only few rotations and the powder melted together. Plastic energy dissipation was not observed and is known to decrease with decreasing particle size. Therefore only friction remains as main energy dissipation mechanism.

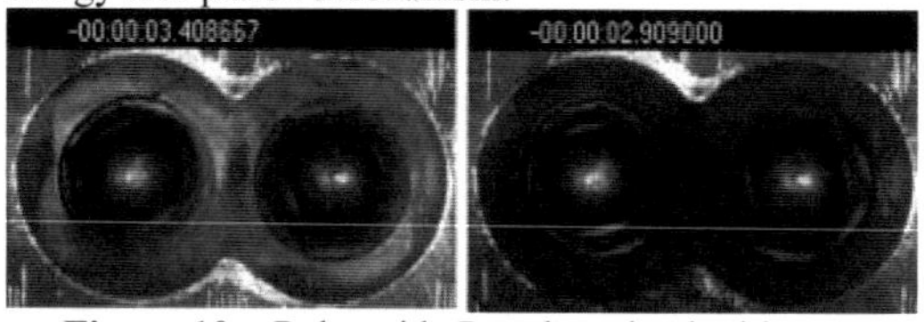

Figure 10 – Polyamide Powder mixed with soot coagulates at 1200rpm

Melting

It can be observed that the bulk of granules does not heat up evenly; but that a heated (transparent) partial quantity of granules is transported in front of the active flank of the kneading disk (similar to the continuous melting model of [4]). The energy is only dissipated during the compression and deformation cycles in front of the active flanks. The partial quantity of heated granules grows with each rotation until all material is transparent (Fig. 11).

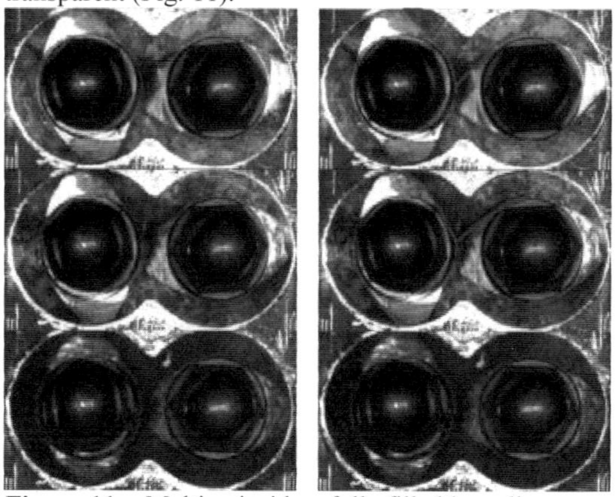

Figure 11 – Melting inside a fully filled kneading zone with small PP granules

Melting of the bulk of granules starts in front of the active flanks (highlighted in Fig. 11 on the right). During around 10 to 14 compression / deformation cycles (around 5-7 rotations) enough energy is dissipated for melting.

Furthermore it is observable that, the granules are embedded in a molten polymer matrix. In contrast to the side faces of the kneading disks one can distinctively see granules and melt (Fig. 12).

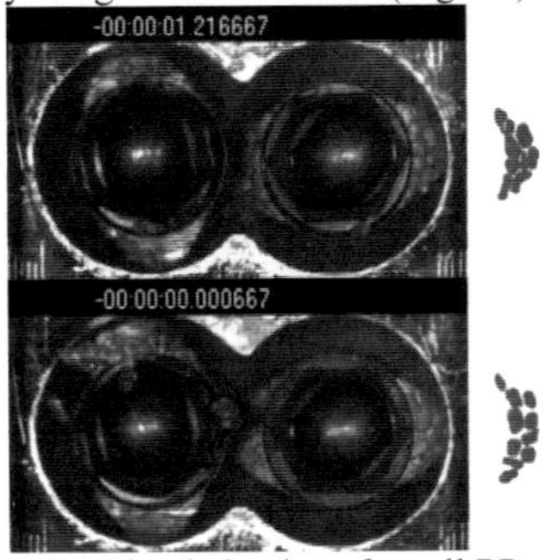

Figure 12 – Melting behavior of small PP granules at 1200rpm and heated extruder wall (120°C)

The partially melted mass thereby behaves at first like granular material. The bulk of granules is pushed in front of the active flank, it does not flow around the kneading disks.

After few further rotations the material starts to flow around the kneading disks.

Due to plastic and viscous energy dissipation the melting degree increases: The viscosity of the melt-granules mixture decreases and the mixture starts to behave like a liquid. The bulk (granular material) does not exist anymore. From this point on the melting behavior stands in very good comparison with the dispersed melting models (DSM).

Because material can flow around the kneading disks, material can also flow out of the intermeshing zone during the compression cycles. This results in receding plastic energy dissipation. Therefore all further energy can only be dissipated by viscous energy dissipation in the melt or by heat transfer from heated extruder wall.

Conclusions

It was shown that most energy is dissipated by plastic energy dissipation during one heavy deformation in the intermeshing zone. Here granules get trapped between the wedge and the two kneading disks and have to undergo compression as well as deformation. It was confirmed that plastic energy dissipation is the dominant mechanism for initial melting. The other energy dissipation mechanisms (friction, heat transfer) play only a secondary role for initial melting; viscous energy dissipation can only occur when melt is present. The rate of energy dissipation was thereby primarily dependent from the granule size. Large granules were compressed and deformed heavily and thus heated quickly, whereas small granules evaded compression and deformation by flowing out of the intermeshing zone. Significant heating of small granules occurred thereby only at highest filling degrees, which lead to higher hydrostatic stresses.

The rotational speed of the extruder did not have an observable effect onto the amount of plastic energy dissipation per rotation. Frictional energy dissipation increased due to higher relative speeds, but was thereby only significant for very small particle sizes (powder) and at high rpm.

The experiments allowed observation of filling of the plastification zone. The following plastic energy dissipation led to the initial melting. Observation of change from granular behavior to liquid behavior was possible. Therefore the complete initial melting process was visualized up to the point where the DSM models describe the further melting process.

Acknowledgements

This work is part of the Research Program of the Dutch Polymer Institute (DPI), Eindhoven, the Netherlands, project nr. #671.

The support from the Dutch Polymer Institute, Bayer Material Science, Bayer Technology Services and DSM is gratefully acknowledged.

References

1. B. Qian; C. Gogos, *Adv. Polym. Techn.* 2000, 4, 287.
2. H. Hermann; U. Burkhardt, *Kunststoffe* 1978, 68, 19.
3. H. Potente; J. Ansahl; B. Klarholz, *Intern. Polym. Proc.* 1994, 1, 11.
4. Z. Tadmor, *Polym. Eng. Sci.* 1966, 3, 185
5. H. Potente, *International Polymer Processing*, 2001, XVI.
6. P. Bleiman; M. Bulters; P. Elemans; H. Slot in SPE ANTEC 2002, 456
7. A. Thuemen, PhD Thesis, University Paderborn, 2008.
8. L. Zhu; K. A. Narh, *Simulation*, 2006, 8, 543.
9. D. Rittel, *Mechanics of Materials*, 1999, 31, 131.
10. D. Chenouf, AIF-Project, 2007, IGF 13751.
11. L. Zhu; X. Geng, *Adv. Polym. Techn.* 2002, 3, 188.
12. J. R. A. Pearson, *Polymer Engineering and Science*, 1978, 3, 222
13. S. Heiß; C. Beinert, Investigation of the plastification behaviour of polymers in high-speed co-rotating twin screw extruders, German Institute for Polymers, Dept. of Plastic Processing

O-05-164

OPTIMIZATIOM OF TESPT SILANE TREATMENT IN SILICA FILLED HNBR SYSTEMS CURED WITH PEROXIDE: EFFECTS OF TESPT LOADINGS AND MIXING CONDITIONS

M. Nillawong[1*], C. Sirisinha[1,2]

[1] *Department of Chemistry and Center of Excellence for Innovation in Chemistry, Faculty of Science, Mahidol University, Bangkok 10400, Thailand – manuchet.1@gmail.com*

[2] *Research and Development Centre for Thai Rubber Industry (RDCTRI), Faculty of Science, Mahidol University, Salaya Campus, Salaya, Nakhon Pathom 73170, Thailand – chakrit.sir@mahidol.ac.th*

Abstract

Hydrogenated nitrile butadiene rubber (HNBR) filled with precipitated silica (PSi) was prepared under various mixing conditions, i.e., normal conditions (discharge temperature = 80°C, mixing time = 15 minutes) and extreme conditions (discharge temperature = 140°C, mixing time = 30 minutes). Silane coupling agent, TESPT or the so-called "Si-69", was also incorporated at various loadings. It is found that increases in mixing time and temperature could significantly enhance bound rubber content (BRC) of HNBR compounds, indicating an improvement in rubber-filler interaction. The BRC is highest at mixing temperature and time of 140°C and 30 minutes, respectively. In this work, the peroxide was employed as crosslinking agent. Unlike conventional sulfur vulcanization, the peroxide curing reveals a cure retardation behavior with the addition of Si-69, and the retardation magnitude could be aleviated by increases in mixing temperature and time. Rheological properties as measured by Rubber Process Analyzer (RPA) exhibit elastic contribution of HNBR compounds and vulcanizates with increasing BRC. Moreover, tensile and dynamic properties are enhanced with increasing magnitude of rubber-filler interaction. Results obtained imply the importance of proper mixing conditions of PSi/Si-69 with rubber matrix for achiving good reinforcing and cure efficiencies.

Keywords: Hydrogenated nitrile butadiene rubber, Silica, Silane coupling agent, Bound rubber, Mixing conditions

Introduction

Nowadays, Hydrogenated nitrile rubber (HNBR) is used in high performance applications, such as engine seals, timing belts, gaskets, and paper mill rolls. Usually, reinforcing fillers are introduced in rubber in order to enhance properties of rubber products [1]. Precipitated silica (PSi) is well known as a non-black reinforcing filler. Compared with carbon black (CB), PSi gives higher tensile strength, abrasion resistance, and rolling resistance in cured rubbers [2]. However, due to strong interaction between PSi particles through silanol groups (Si-OH) on PSi surfaces, the difficulty in dispersion of PSi into rubber matrix is resulted. In order to tackle this problem, an introduction of silane coupling agent (SCA) is commonly used. Apart from the enahancement in mixing efficiency, the SCAs are known to improve filler dispersion and rubber-filler interaction [3]. Typically, a mercapto silane is usually used in sulfur curing system for the purpose of enhanced cure efficiency. Furthermore, it has been found that the in-situ coupling reaction under various mixing conditions leads to improvements in ruber-filler interaction and mechanical properties [4]. In order to quantify the magnitude of rubber-filler interaction, a measurement of bound rubber content (BRC) via solubility method is usually carried out [5]. However, not so many works focus mainly on a correlation of BRC, viscoelasic behavior, and mechanical properties of PSi filled elastomer. The main objective of this work is to study the influences of mixing conditions for maximizing the coupling efficiency of Si-69 in HNBR filled with PSi. As a highlight of this work, the peroxide was employed instead of conventional sulfur curing system to examine its influence on cure and silane coupling efficiencies. In addition, the correlations of BRC, viscoelastic behavior, and mechanical properties are established.

Experimental

Hydrogenated nitrile rubber (HNBR) with 85% hydrogenation and 36% acrylonitrile content was used in this work. Precipitated silica (PSi) was introduced as reinforcing filler. The Si-69 silane was used as coupling agent with varied concentrations of 2, 4, 6, 8, and 10 wt% of PSi. Other chemicals used in this work are shown in **Table 1**.

Table 1 Materials used in the present study

Materials	Amount (phr)
Hydrogenated nitrile rubber (HNBR)	100
Precipitated silica (PSi)	40
bis(triethoxysilylpropyl)tetrasulfide (TESPT or Si-69)	Variable
2, 2, 4-Trimethyl-1, 2-hydroquinoline (TMQ)	1
Stearic acid	1
Zinc oxide (ZnO)	5
Tri-2-ethylhexyl trimellitate (TOTM)	5
Trimethylolpropane trimethacrylate (TRIM)	10
Zinc diacrylate (ZDA)	6
Dicumyl peroxide (DCP)	2

HNBR compound was prepared under normal (discharge temperature = 80°C, mixing time = 15 minutes) and extreme conditions (discharge temperature = 140°C, mixing time = 30 minutes) using a laboratory-sized internal mixer (Haake Rheomix 90, Germany). Mixing was performed under rotor speed and fill factor of 40 rpm and 0.85, respectively. The compounding ingredients excluding TRIM, ZDA, and DCP were incorporated in the first mixing step. Thereafter, the rubber compound was mixed with co-agents and DCP via the use of Haake Rheomix 90 again at rotor speed and temperature of 40 rpm and 40°C, respectively. Cured HNBR specimens for mechanical tests were prepared by a hydraulic hot-press (Chai Suriya Rubber Factory, Thailand).

The BRC measurement was carried out by dissolving uncured compounds in acetone for 7 days, and then weighing the insoluble part. Rheological properties were monitored by Rubber Process Analyzer (RPA2000, USA) under strain sweep test at test frequency and temperature of 1 rad/s and 100°C, respectively. Swelling test was conducted to assess the magnitude of crosslink density in cured HNBR with the use of acetone as solvent. Tensile properties were measured according to ASTM D 412 using the universal tensile tester (Instron, USA). Heat build up of cured specimens was measured as per ASTM D 623 Test Method A by the use of BF Goodrich Flexometer (Model II, USA).

Results and Discussion

Bound rubber content

Fig.1 represents the BRC results of uncured HNBR compounds, in which the mixing HNBR with PSi/Si-69 under extreme conditions leads to a rise in BRC at any given Si-69 loadings. The results imply a development of rubber-interaction through high mixing temperature and long time. However, the HNBR compound with 2% Si-69 shows very low magnitude of BRC despite being prepared under extreme mixing conditions. This is due to insufficient amount of Si-69, leading to incomplete surface coverage on PSi particles.

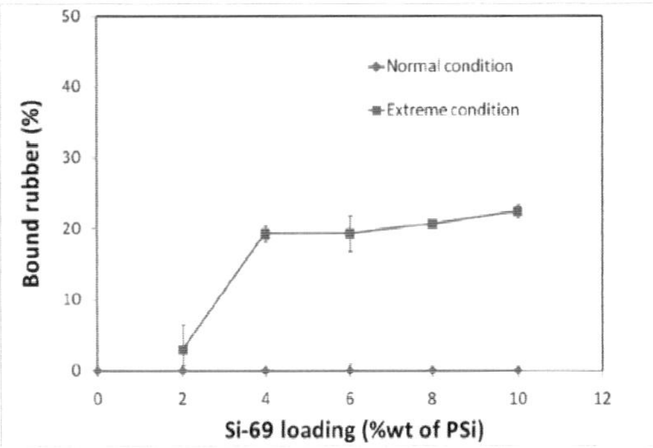

Figure 1 BRC of HNBR compounds prepared with different mixing conditions

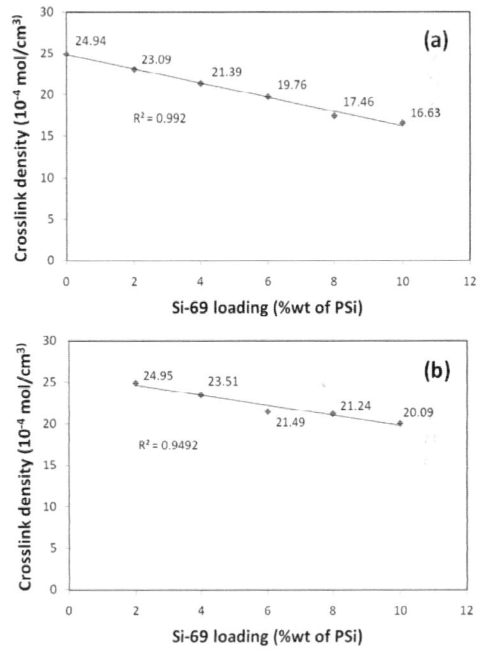

Figure 2 Crosslink density of cured HNBR specimens prepared from (a) normal (b) extreme conditions

Fig. 2 reveals crosslink density of cured HNBR specimens as measured by a swelling method. It is evident that the crosslink density decreases progressively with increasing amount of Si-69, due to a cure retardation behavior caused by released sulfur from Si-69 in peroxide curing system [6]. Nevertheless, the cured HNBR specimens prepared from the extreme conditions exhibit greater crosslink density. This implies a development in cure efficiency attributed to the smaller amount of released sulfur molecules available for retarding the crosslink reaction. Moreover, the magnitude of crosslink density measured by swelling test depends not only on chemical crosslink, but also on rubber-filler interaction in cured rubber [7]. The strong rubber-filler interaction leads to the tightly bound rubber and thus the low swelling ratio, i.e., high value of reported crosslink density.

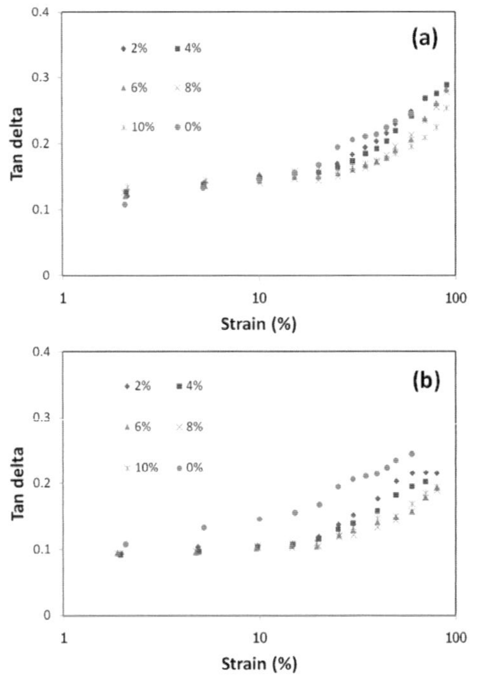

Figure 3 Tan delta as a function of strain of cured HNBR specimens prepared from different mixing conditions: normal condition (a); extreme condition (b)

Rheological behavior in terms of tan delta is demonstrated in **Fig. 3**. As evidenced in **Fig. 3 (a)**, tan delta of specimens with Si-69 is significantly lower than those without Si-69. This is due to the formation of rubber-filler through coupling reaction among silanol groups, silane and HNBR molecules leading to the increase in elastic response. Moreover, it is evident that the mixing performed under extreme conditions offer low tan delta indicating the enhancement in elastic contribution through improved cure efficiency in conjunction with strong rubber-filler interaction.

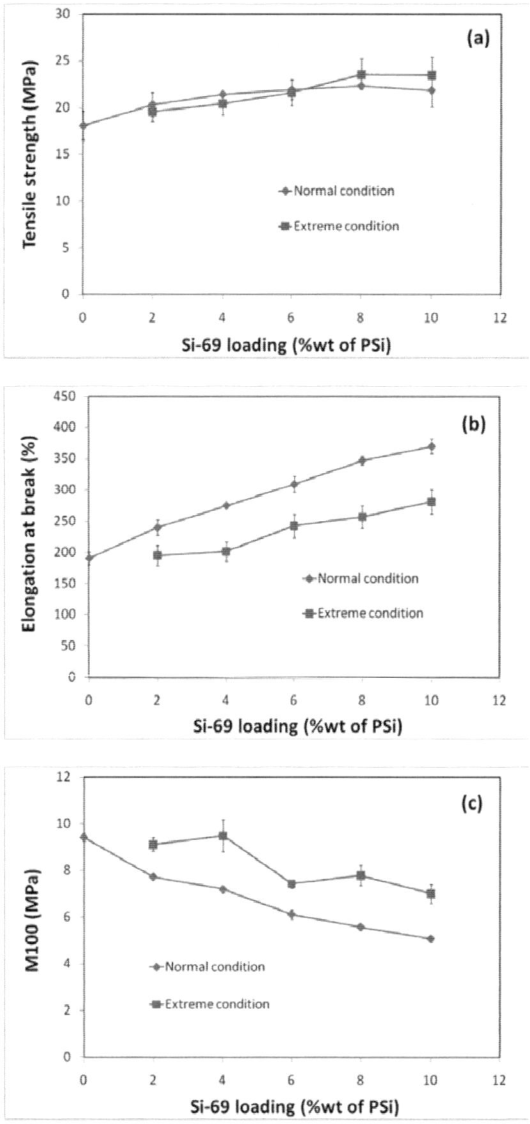

Figure 4 Tensile properties of cured HNBR specimens as a function of silane loading: (a) tensile strength; (b) elongation at break; (c) modulus at 100% elongation

Fig. 4 reveals tensile properties of cured HNBR specimens. With increasing Si-69, tensile strength and elongation at break increase, but modulus decreases. By the use of extreme mixing conditions, the occurrence of tightly bound rubber and enhanced cured efficiency contribute to the significant increase in modulus with the expense of elongation at break. Combined effects of enhanced rubber-filler interaction and cure retardation is probably responsible for the results.

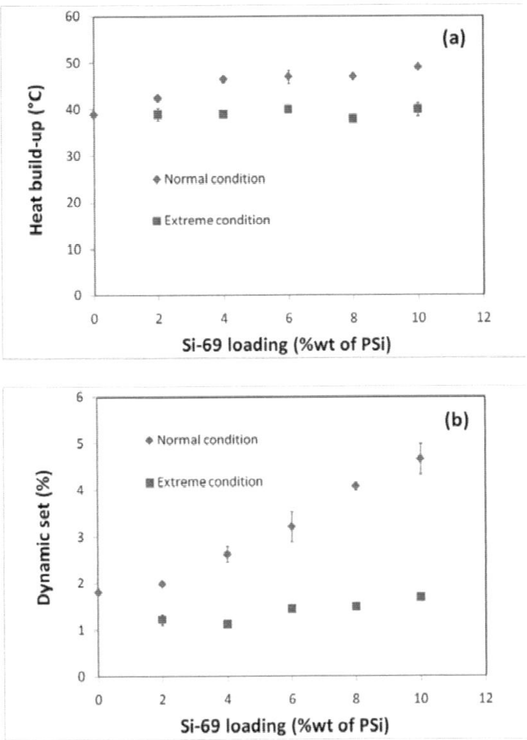

Figure 5 Dynamic mechanical properties of cured HNBR as a function of silane loading: (a) temperature rise; (b) dynamic set

Dynamic mechanical properties of cured HNBR specimens are shown in **Fig. 5**. The increase in Si-69 loading gives rise to the increase in temperature rise and dynamic set, implying a decrease in elasticity via the cure retardation behavior as discussed previously. However, the use of extreme mixing condition is capable of alleviating the undesirable cure retardation as a result of strong rubber-filler interaction together with improved cure efficiency. This leads to lower molecular mobility which is a major cause of energy dissipation via heat [8].

Conclusions

In this work, two different mixing conditions were used: normal and extreme conditions. As determined by BRC, the rubber-filler interaction is developed more efficiently under extreme mixing conditions, i.e., performing the mixing at high temperature and long time. Rheological properties of cured HNBR specimens are in good agreement with BRC results. The higher the BRC, the lower the tan delta, and thus the greater the elastic contribution. The crosslink density as measured by swelling method exhibits enhancements in cure efficiency and rubber-filler interaction caused by extreme mixing conditions as evidenced by results of tensile properties and dynamic mechanical behaviors.

Acknowledgement

Financial supports from Center of Excellent for Innovation in Chemistry (PERCH-CIC) and the Thailand Research Fund through the Royal Golden Jubilee Ph.D. Program are acknowledged.

References

1. D. C. Edwards *J. Mater. Sci.* 1990, *25*, 4175.
2. T. Mathew; R. Datta; et al. *Plasma. Chem. Plasma. P.* 2008, *28*, 273.
3. P. Vondráček; M. Hradec; et al. *Rubber. Chem. Technol.* 1984, *57*, 675.
4. S. Wolff *Rubber. Chem. Technol.* 1982, *55*, 967.
5. X. Liu; S. Zhao *J. Appl. Polym. Sci.* 2008, *108*, 3038.
6. M. M. A. Grima; A. G. Talma, et al. *Rubber. Chem. Technol.* 2006, *79*, 694.
7. J. L. Valentín; I. Mora-Barrantes; et al. *Macromolecules* 2009, *43*, 334.
8. M. J. Wang *Rubber. Chem. Technol.* 1998, *71*, 520.

KN-06-01

DYNAMIC EVOLUTION OF REACTION AND MORPHOLOGY FOR POLYMER REACTIVE PROCESSING IN A TWIN SCREW EXTRUDER

Guo-Hua Hu[1*], Sandrine Hoppe[1], Cailiang Zhang[2] and Lianfang Feng[2]

[1]*Laboratory of Reactions and Process Engineering, Université de Lorraine - CNRS, 1 rue Grandville, BP 20451, 54001 Nancy, France (guo-hua.hu@univ-lorraine.fr)*

[2]*State Key Laboratory of Chemical Engineering, Department of Chemical and Biochemical Engineering, Zhejiang University, Hangzhou 310027, China*

Abstract: This work develops a novel method to study the dynamic evolution of interfacial reaction and morphology of reactive polymer blends during a blending process in a twin screw extruder. The reactive blending system is composed of a chemically inert polystyrene (PS), an anhydride terminated polystyrene (PS-Anh) and amine terminated polyamide 6 (PA6-NH2). Fluorescent moieties such as anthracene are incorporated in the PS-Anh as a reactive compatibilizer-tracer Anh-PS-Ant An in-line fluorescence measuring device developed in the laboratory allows simultaneous measurement of the residence time distribution and the concentration of the in situ synthesized compatibilizer during the reactive blending process. The morphology of the resulting reactive blend at the die exit is measured offline. ppropriate compatibilizers under real industrial polymer blending conditions as well as scaling up and optimizing blending processes.

Keywords: Compatibilizer; reactive polymer blending; in-line measurement; interfacial reaction; morphology development

Introduction

Polymer blending is a very important polymer processing operation. It aims at preparing new polymer materials by blending existing polymers without the need of creating new molecules. The most commonly used polymer compatibilized method in the industry is reactive polymer blending, also called reactive compatibilization or in-situ compatibilization. In this method, a copolymer is formed in-situ at the interfaces between two functional polymers during blending. Reactive polymer blending is often carried out in a screw extruder. The latter is thus used both as a mixer and a reactor. It is a continuous process whose average residence is of the order of a few dozons of seconds to a few minutes. The biggest challenge for reactive polymer blending resides in the control of the kinetics of the interfacial reaction, the molecular structure of the resulting copolymer and the rate of the morpphology development under industrial or close-to-real blending conditions. Many studies use a batch mixer to study polymer reactive blending. The results obtained are not necessarily representative of what may take place in screw extruders especially in industrial-scale ones because of differences in thermo-mechanical histories.

Based on transient experiments for measuring residence time distribution (RTD), we developed a so-called compatibilizer-tracer concept for polymer blending processes which allows the use of very small amounts of compatibilizer-tracer to obtain the evolution of the dispersed phase domain size as a function of the graft copolymer concentration [1-3] in a twin screw extruder. This work attempts to extend this method to reactive polymer blending processes to probe both the degree of interfacial reaction and morphology development.

Experimental

Fig.1 shows the chemical structure of the reactive compatibilizer-tracer (Anh-PS-Ant).

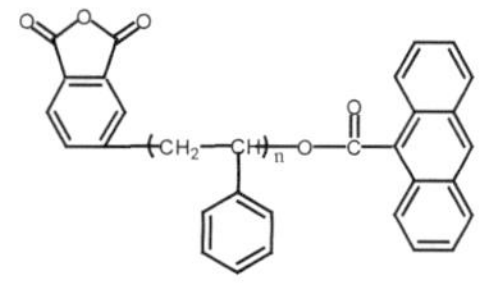

Figure 1 Schematic of the structure of the reactive compatibilizer-tracer.

In a transient experiment, when the extrusion of PS (containing Anh-PS-Ant or not)/PA6 blend in a twin-screw extruder reached a steady state, the Anh-PS-Ant was injected to the hopper as a pulse. At the same time, an in-line RTD measuring device started measuring the concentration of the Anh-PS-Ant and the corresponding extrudate was collected and then quenched immediately in liquid nitrogen to freeze-in its morphology. A scanning electron microscope of type FEI SIRION was used to measure the morphology of the extrudate in an offline matter. A semi-automatic image analysis method was used to determine the diameter of the dispersed phase domain size.

Results and Discussion

Fig.2 shows the RTD and the dispersed phase domain size distribution (DSD) measured using Anh-PS-Ant as a reactive compatibilizer-tracer for PS/PA6 blend. The DSD follows the overall trend of the RTD. A lower value of the DSD corresponds more or less to a higher value of the RTD. The emulsification curves can easily be deduced from both the RTD and DSD ones in Fig.2 and they are shown in Fig.3. For all the PS/PA6 blends, the dispersed phase domain size decreases with increasing reactive compatibilizer-tracer concentration, as expected. Moreover, the emulsification curve of the PS/PA6 (80/20 by mass) blend is located well above that of the PS/PA6 (90/10 by mass). More interestingly, each of the emulsification curves is not a single curve but a loop. It should be noted that these emulsification loops are obtained in the following manner. A RTD curve is divided into short and long time domains, which are demarcated by its maximum. In the short time domain, the reactive compatibilizer-tracer fraction increases with increasing residence time while in the long time domain it follows the opposite trend. When the dispersed phase domain size is plotted against the compatibilizer concentration, two curves are generated. It is seen that the curves corresponding to the long time domain are all systematically below those corresponding to the short time domain, indicating that a longer residence time and/or stronger mixing favors the reactive compatibilization. This is the opposite to what was observed for non-reactive compatibilization [1-3].

Fig.4 shows the effect of screw speed on the emulsification curve of the PS/PA6 (80/20 by mass) blend. Irrespective of the time domain, the emulsification curves obtained at 150 rpm are significantly below those obtained at 100 rpm, indicating that it is not the residence time but the mixing intensity that dictates the efficiency of the reactive compatbilization process.

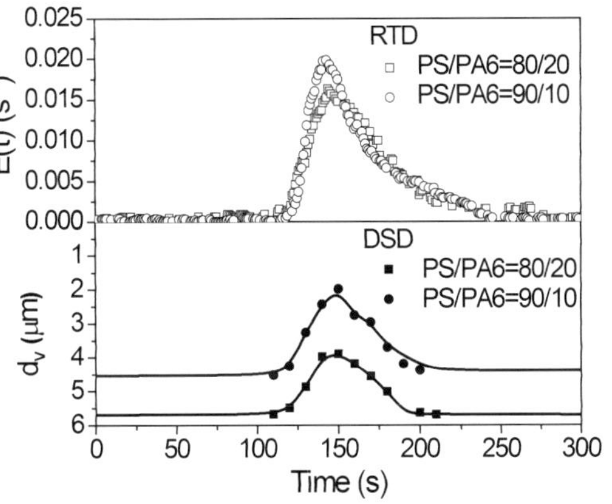

Figure 2 RTD and DSD curves of the PS/PA6 blend for two different compositions. Feed rate = 13 kg/h; screw speed = 100 rpm.

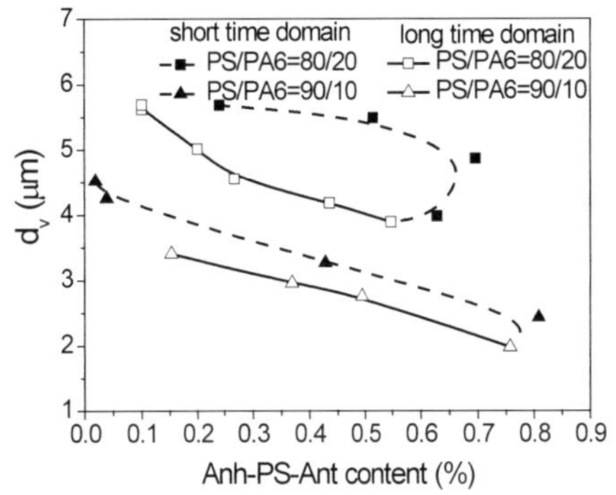

Figure 3 Emulsification curves of the PS/PA6 blend for two different compositions. Data are from Fig. 2.

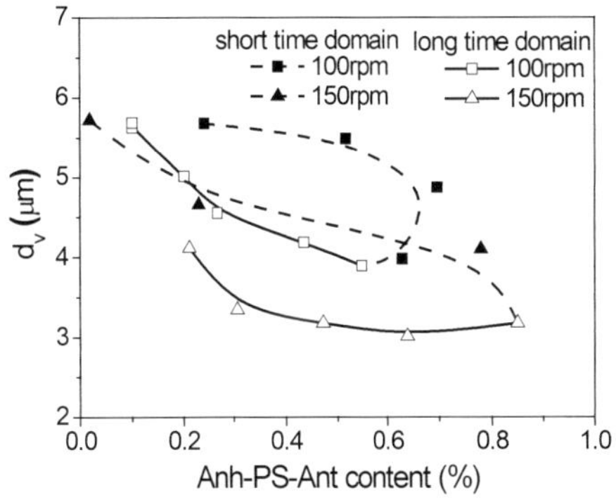

Figure 4 Effect of screw speed on the emulsification curve of the PS/PA6 (80/20 by mass) blend.

Conclusion

In summary, the above results show that the concept of reactive compatibilizer-tracer developed in this paper offers a very simple, convenient and reliable way to study reactive compatibilization processes in a continuous mixer such as a twin screw extruder. It also allows scaling up and optimizing such processes, which would otherwise be very difficult or impossible using classical approaches. It is expected that this concept will trigger more studies on or revision of some of the fundamental issues of reactive polymer blending processes.

Acknowledgements

The authors thank the Ministry of Science and Technology of China through an international cooperation program (grant number: 2001CB711203), the PHC CAI YUANPEI 2012 - 28021SM and the State Key Laboratory of Chemical Engineering (No. SKLChE-08D03) for their financial support.

References

1. C.L. Zhang; L.F Feng; S. HOPPE; G.H. Hu. *AIChE J.* 2012, *58,* 1931.
2. C.L. Zhang; L.F. Feng; X.P. GU; S. Hoppe; G.H. Hu. *Polym. Eng. Sci.*; 2012, *52*, 300.
3. C.L. Zhang; L.F. Feng; S. Hoppe; G.H. Hu. *AIChE J.* 2009, *55*, 279.

KN-06-02

MDIFICATION OF PP AND APPLICATION FOR PP/PLA ALLOY

Iwao Ito, Motonobu Furuta, Shotaro Nisituji and Takashi Inoue*
Yamagata University, Department of Polymer Science and Engineering,
4-3-16 Jonan, Yonezawa, 992-8510 Japan - tinoue@yz.yamagata-u.ac.jp

Abstract – A high-performance PP/PLA alloy was successfully developed by reactive extrusion with the aid of reactive compatibilizer, poly(ethylene-co-glycidylmethacrylate) (EGMA), and organic peroxide (dicumyl peroxide). It showed high Izod impact strength (12 kJ/m^2), large elongation at break (75%), and nice fluidity (melt index=50 g/10min @ 200°C). Three reactions to be involved in this reactive blending: (1) coupling between epoxide of EGMA and carboxyl acid chain end of PLA, (2) chain scission of PP, and (3) radical grafting of PP onto EGMA, may generate a desirable morphology of the fine particles of PLA core/EGMA shell, dispersed in the PP matrix of low viscosity.
Keywords: Polypropylene, Poly(lactic acid), Reactive blend, 40%-green materials

Poly(lactic acid)(PLA) is a green plastic material from renewable sources; however, it has a lot of drawbacks on the properties and processability; poor impact resistance (a few kJ/m^2), small elongation at break (a few %), low heat deflection temperature (HDT) (<60°C), and poor processability (high temperature mold is required for the crystallization to result in long cycle time for injection molding). Neat PLA is hardly applied for the injection molded plastics. Then, PLA is blended with commodity plastics (*e.g.*, polypropylene, PP). However, by increasing the amount of PLA, the blend properties seriously drop.

Recently, we modified PP by melt blending with small amount of phenol resin and organic peroxide (dicumyl peroxide, DCP) (both, less than 1 wt%) at 200 °C in an extruder to yield a super ductile PP (elongation at break, E_b >300%) and extremely high fluidity (melt index, MI>100 g/10min @ 200°C), as shown in Fig. 1. Thus, we succeeded to escape from the trade-off relationship between the fluidity and the ductility of plastics [1].

The phenol modification technique was successfully applied to render a high-performance PP80/PLA20 alloy (E_b=350%, MI=100) (Fig. 2). In this ductile matrix/brittle particle system, a "cold drawing" (ductile deformation) of brittle PLA particle may proceed by the compressive stress evolved at the equator [2,3]. By increasing PLA content to 40wt%, however, E_b decreased to 5% (Fig. 2).

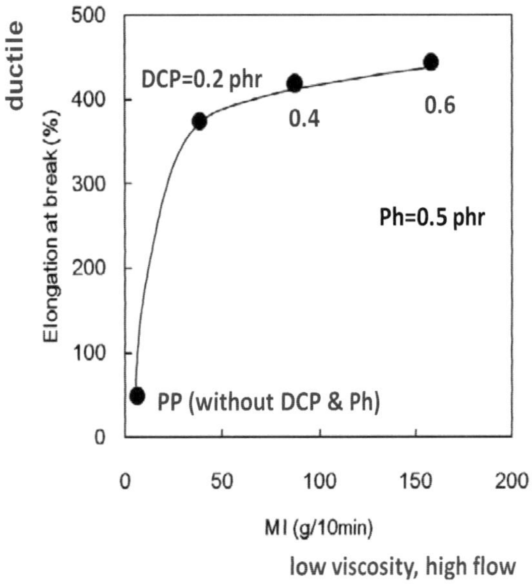

Fig. 1. Elongation of break *vs*. fluidity (MI) of phenol-modified PP (PP-Ph).

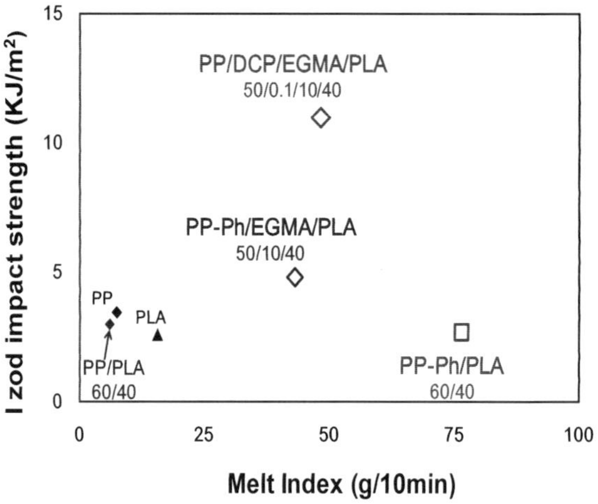

Fig. 2. Elongation of break *vs*. fluidity (MI) of PP/PLA blends.

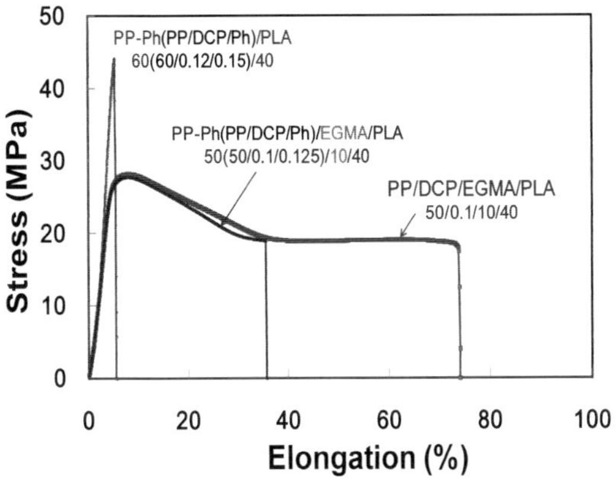

Fig. 3. Stress-strain curves of PP/PLA 60/40 blends.

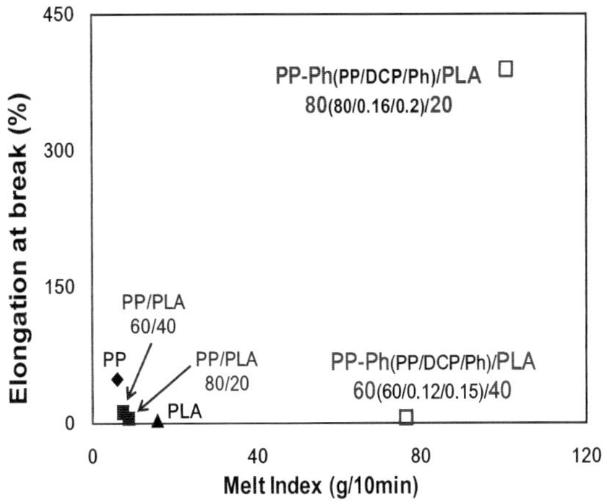

Fig. 4. Izod impact strength *vs*. fluidity (MI) of PP/PLA 60/40 blends

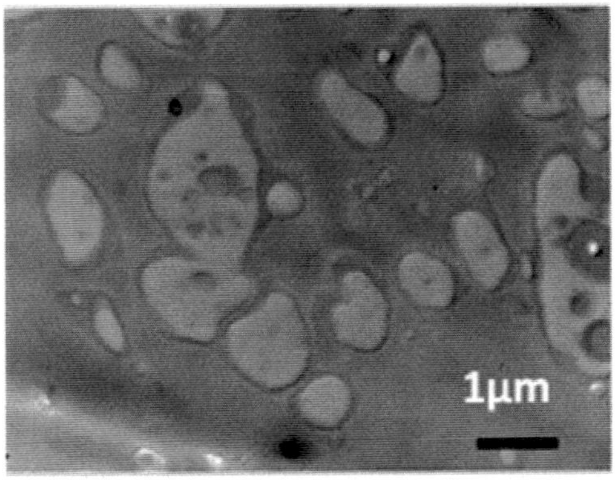

(a) PP-Ph/EGMA/PLA50/10/40

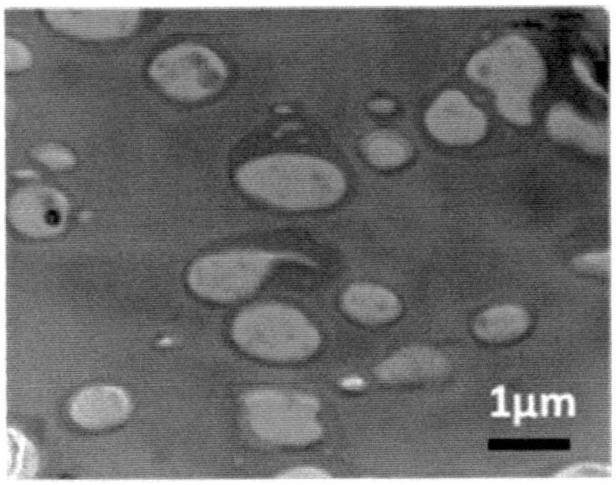

(b) PP/DCP/EGMA/PLA 50/0.1/10/40

Fig. 5. TEM micrographs (RuO_4)

By adding the third component, poly(ethylene-co-glycidylmethacrylate) (EGMA), as a reactive compatibilizer, E_b increased to 35% without drastic increase of melt viscosity (MI=50). Further, eliminating phenol resin (PP50/PLA40/EGMA10/DCP0.1), E_b increased to 75%, keeping the nice fluidity (Fig. 3). The Izod impact strength was also high (12 kJ/m^2) (Fig. 4).

TEM observation showed that EGMA encapsulated PLA particles and reduced the particle size. When phenol resin was incorporated, the encapsulated PLA particles partly aggregated (Fig. 5).

Plausible reactions in the reactive blending system are (1) coupling between the epoxide of EGMA and the carboxyl acid and/or hydroxyl chain ends of PLA, (2) coupling between the phenol group on PP and the epoxide of EGMA, (3) radical grafting of PP onto EGMA, and (4) chain scission of PP. The above results imply that a combination of the reactions (1), (3) and (4) seems to create the desirable morphology, which provides the high performance. The reaction (2) should not be involved, since it bridges the encapsulated particles to form aggregates, leading to harmful defects.

References

1. I. Itoh et al. *J. Polym. Sci. Technol.,Jpn.* 2012, 69, 55
2. K.K. Koo et al. *Polym. Eng. Sci.*, 1985, 25, 741
3.J.C. Angola et al. *J. Polym. Sci., Phys.* 1988, 26 807

O-06-523

EFFICIENCY OF GRAFT COPOLYMERS AS EMULSIFIERS FOR IMMISCIBLE POLYMER BLEND

Cailiang Zhang[1], Lianfang Feng[1*] and Guohua Hu[2*]

[1]*State Key Laboratory of Chemical Engineering, Department of Chemical and Biochemical Engineering, Zhejiang University, Hangzhou 310027, China*

[2]*Laboratory of Reactions and Process Engineering, Université de Lorraine-CNRS, 1 rue Grandville, BP 20451, 54001 Nancy, France (guo-hua.hu@univ-lorraine.fr)*

Abstract: A series of graft copolymers of polystyrene (PS) and polyamide 6 (PA6), denoted as PS-g-PA6, with different molecular structures were used as emulsifiers of immiscible PS/PA6 blends. The PS-g-PA6 was more efficient for the PS/PA6 (80/20) blend than for the PS/PA6 (20/80) one, indicating that a graft copolymer whose backbone and grafts match the matrix and the disperse phase polymers, respectively, has higher compatibilizing efficiency. Moreover, whatever the blend composition, for PS-g-PA6 graft copolymers with the same backbone and the same number of grafts per backbone, the longer the grafts, the higher their compatibilizing and stabilizing efficiency; for a given backbone/graft mass ratio, the longer the grafts and concomitantly the smaller the number of grafts per backbone, the higher the compatibilizing and stabilizing efficiency of the graft copolymer.

Keywords: Compatibilizing efficiency; Polymer blend; Graft copolymer; Emulsifier

Introduction

It is common practice to create new polymer materials with desirable properties by blending different polymers. However, most polymer pairs are immiscible and thermodynamically unstable, which leads to the phase separation and yields a material with poor mechanical properties. To address this challenge, block or graft copolymers are often used as emulsifiers to reduce the interfacial tension, promote the dispersion of one phase in another and stabilize resulting blends. Most experimental reports in this field are related to the effect of the molecular architecture of the block copolymers on the compatibilization of heterogeneous polymer blends because their molecular parameters are easy to control using anionic polymerization process. However, the prevailing method of in-situ compatibilization results mainly from the formation of graft copolymers. Therefore, it is very important to investigate the compatibilizing efficiency of graft copolymers with different molecular architectures and/or compositions.

Our previous work reported a novel method to synthesize a graft copolymer with a controlled structure by the activated anionic polymerization of ε-caprolactam (CL) using sodium ε-caprolactam (NaCL) as catalyst and 3-isopropenyl-α,α-dimethylbenzene isocyanate (TMI) bearing polypropylene (PP) or PS as activator (PP-g-TMI or PS-co-TMI) [1-6]. In this work, the compatibilizing efficiency of the graft copolymer (PS-g-PA6) with different structures and compositions as compatibilizer for PS and PA6 blends was investigated.

Experimental

Polymer blends were prepared by mixing the blend components in a Haake Rheomix equipped with a mixing chamber and two rotors inside the mixing chamber at 65 rpm and 230 °C for 8.5 min. Samples were taken from the mixing chamber and quenched in liquid nitrogen to freeze-in their morphologies. A scanning electron microscopy (SEM) of type JEOL JSM-T330A was used to characterize the blend morphologies.

Results and Discussion

A series of PS-g-PA6 with different molecular structures were used as the compatibilizers of immiscible PS/PA6 blends, as shown in Tab. 1.

Fig.1 compared emulsification curves of the PS/PA6 (20/80) and PS/PA6 (80/20) blend with PS-g-PA6a, PS-g-PA6b, PS-g-PA6c and PS-g-PA6d as compatibilizers. The result indicated whatever the blend composition, for graft copolymers with the same backbone and the same number of grafts per backbone, the longer the grafts, the higher their compatibilizing and stabilizing efficiency; for a given backbone/graft mass ratio, the longer the grafts and concomitantly the smaller the number of grafts per backbone, the higher the compatibilizing and stabilizing efficiency of the graft copolymer. This is in line with a conclusion drawn in the previous papers [7-10].

Moreover, a graft copolymer whose backbone and grafts match the matrix and the disperse phase polymers, respectively, has a higher compatibilizing efficiency. The first one lies in the way the PS-g-PA6 reaches the PS/PA6 interfaces. In the case of the PS/PA6 (80/20) blend, the compatibilizer can move toward the PA6 droplets and the latter can also move toward the former. In the case of the PS/PA6 (20/80) blend, the compatibilizer can reaches the PS/PA6 interfaces only by its own displacement through the PS dispersed phase. This indicates that it is easier for the compatibilizer to reach the interfaces of the PS/PA6 (80/20) blend than those of the PS/PA6 (20/80) one.

The second difference between the PS/PA6 (80/20) and PS/PA6 (20/80) blends resides in the way the compatibilizer is located at the interfaces. In the former blend, the backbone of the graft copolymer is located in the matrix phase and its grafts in the dispersed one while in the latter one it is the opposite. Before the graft copolymer reaches the final configurations at the interfaces, it is very likely that the configuration change it has experienced for PS/PA6 (20/80) blend. The above results seemly advise to avoid two things during the blending process: first, phase inversion between the matrix and the dispersed phases; second, significant configuration change for the compatibilizer to locate at the interfaces.

Table 1 Selected characteristics of the as-synthesized PS-g-PA6 graft copolymers

Copolymer designation	Composition (%)		Number of PA6 graft per PS backbone	Mn (kg/mol)	
	PS backbone	PA6 grafts		PS backbone	Each PA6 graft
PS-g-PA6-1	78.4	21.6	15.3	34.3	0.6
PS-g-PA6-2	76.5	23.5	7.1	33.9	1.5
PS-g-PA6-3	66.9	33.1	7.1	33.9	2.4
PS-g-PA6-4	71.2	28.8	3.8	32.0	3.4

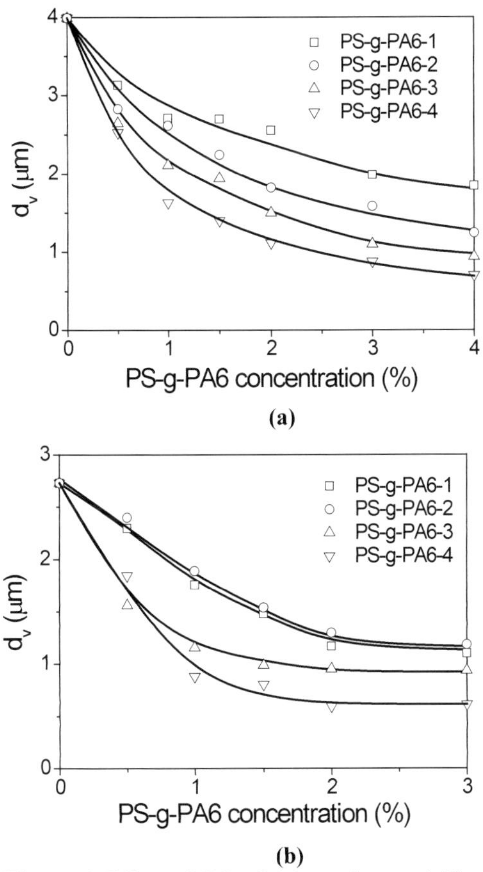

Figure 1 Effect of PS-g-PA6 on the emulsification curves of the PS/PA6 (20/80) (a) and PS/PA6 (80/20) (b) blends. Mixing temperature: 230 °C; mixing time: 8.5 min; rotation speed: 65 rpm.

Conclusions

This work focused on the effect of the structure and composition of a graft copolymer (PS-g-PA6) on its compatibilizing efficiency for PS/PA6 blends of different compositions. The results indicated that irrespective of the blend composition, for graft copolymers with the same backbone and the number of grafts per backbone, the longer the grafts, the higher their compatibilizing and stabilizing efficiency; for a given backbone/graft mass ratio, the longer the grafts and concomitantly the smaller the number of grafts per backbone, the higher the compatibilizing and stabilizing efficiency of the PS-g-PA6 graft copolymer. Moreover, a graft copolymer whose backbone and grafts match the matrix and the disperse phase polymers, respectively, has higher compatibilizing efficiency.

Acknowledgements

The authors thank the Ministry of Science and Technology of China through an international cooperation program (grant number: 2001CB711203), the PHC CAI YUANPEI 2012-28021SM and the State Key Laboratory of Chemical Engineering (No. SKLChE-08D03) for their financial support.

References

1. C.L. Zhang; L.F. Feng; X.P. GU; S. Hoppe; G.H. Hu. *Polym. Eng. Sci.*; 2011, *51*, 2261.
2. G.H. Hu; H. Li; L.F. Feng. *Macromolecules* 2002, *35*, 8247.
3. G.H. Hu; H. Li; and L.F. Feng, *Polymer*, 2005, *46*, 4562.
4. C.L. Zhang; L.F. Feng; X.P. Gu; S. Hoppe; G.H. Hu. *Polym. Test.* 2007, *26*, 793.
5. C.L. Zhang; L.F. Feng; G.H. Hu; Z.B. Xu. *J. Appl. Polym. Sci.* 2006, *101*, 1972.
6. C.L. Zhang; L.F. Feng; S. Hoppe; and G.H. Hu. *J. Polym. Sci. Part A: Polym. Chem.* 2008, *46*, 4766.
7. C.L. Zhang; L.F. Feng; J. Zhao; H. Huang; S. Hoppe; G.H. Hu. Polymer 2008, *49*, 3462.
8. C.L. Zhang; L.F. Feng; X.P. Gu; S. Hoppe; G.H. Hu. Polymer, 2007, *48*, 5940.
9. C.L. Zhang; L.F Feng; S. HOPPE; G.H. Hu. AIChE J. 2012, 58, 1931.
10. C.L. Zhang; L.F. Feng; X.P. GU; S. Hoppe; G.H. Hu. Polym. Eng. Sci.; 2012, 52, 300.

O-06-037

EFFECTS OF HMWPE ON MECHANICAL PROPERTIES OF BIMODAL PE

Yi Liu[1*], Juliane Braun[1], Sandeep Tyagi[1], Davide Tranchida[1] and Gottfried Kandioller[1]
[1]Borealis Polyolefine GmbH, St.-Peter-Strasse 25, A-4021 Linz, Austria
yi.liu@borealisgroup.com; juliane.braun@borealisgroup.com; sandeep.tyagi@borealisgroup.com; davide.tranchida@borealigroup.com and gottfried.kandioller@borealisgroup.com

Abstract In a bimodal PE system the high molecular weight (HMW) component provides mechanical performance while the low molecular weight (LMW) component provides the processability. It is desired to increase the MW of the polymer in order to improve mechanical properties. In the current work a HMWPE (MW of $1x10^6$ g/mol by intrinsic viscosity) was blended into a bimodal PE at various loadings with either a kneader or a twin screw co-rotating extruder. Light microscopy indicates that both approaches generated homogeneous blends particularly at a high loading rate of HMWPE (40%). Significant improvement in impact strength with a minor loss in tensile modulus of the bimodal PE is observed in the presence of HMWPE. It is found that at a HMWPE loading of 40%, twin screw extrusion retained the HMW fraction of the blend better than kneading, which in turn led to remarkably higher impact strength. Morphology studies including SEM and AFM was conducted on the extruded blends.

Keywords PE, bimodal, blending, impact strength

Introduction

A typical bimodal PE system comprises a high molecular weight (HMW) copolymer that provides mechanical performance and a low molecular weight (LMW) homopolymer that provides processability. In order for good mechanical properties increasing the molecular weight of the polymer is desired. This can be achieved by melt blending the bimodal PE with a polyethylene with even higher Mw than the original HMW copolymer in the bimodal PE.

From literature [1-3] polyethylene having molecular weight over 10^6 g/mol is classified as ultra-high molecular weight polyethylene (UHMWPE). However, according to ASTM standard (D 4020-05), UHMWPE decahydronaphthalene solution has a relative viscosity of greater than 1.44 at a concentration of 0.02 % and at 135°C. By calculation with this standard, UHMWPE should have a molecular weight more than $3.1x10^6$ g/mol. Nevertheless addition of polyethylene with molecular weight higher than 10^6 g/mol but lower than $3.1x10^6$ also improves the mechanical performance of HDPE [2, 4]. Such polymers are referred to as HMWPE in this paper.

Despite improvement in mechanical properties, homogeneously blending (U)HMWPE with other polyethylene is not easy due to a large viscosity difference. Typically, extended melt mixing[5], two-step blending[6] or solution blending[7] were implemented for better homogeneity. Addition of a third component as a compatibilizer also helps. For instance, Middle molecular weight polyethylene (MMWPE) was used for blending HDPE and HMWPE (with Mw of $1.5x 10^6$) in a two-step process, leading to improved homogeneity, processability and mechanical performance compared to direct processing[3]. This suggests that blending (U)HMWPE with a bimodal PE can be relatively easy since the HMW fraction would act as a compatibilizer. Indeed, mixing UHMWPE (Mw of $3.5x10^6$) with a bimodal pipe grade having low MFR was tried before[8], where UHMWPE was used as a nucleating agent at low loadings (up to 3%). In current work we blend up to 40% HMWPE (Mw of $1.1x10^6$) into a bimodal pipe grade using either a batch mixer or a co-rotating twin screw-extruder.

Experimental

1. Material
The base powder of a bimodal pipe grade from Borealis is used in this study. This pipe grade, in pellet form, has a Mw of $1.9x10^5$ g/10min and density of 959 kg/m^3. HMWPE is a commercial product from Jingchem that has Mw of $1.1x10^6$ g/mol and density of 934 kg/m^3. Both bimodal PE and HMWPE were in the form of powder. This facilitates dry mixing of them and hence favours the melt blending process.

2. Melt blending
2.1. Kneading
The mixture of bimodal PE base powder and HMWPE (loading from 10% to 40%), was kneaded using a Haake Rheomix 3000p at 230°C and 40 rpm for 8 minutes. A standard stabilization package was added; a carbon black masterbatch was added to provide contrast for microscopy.

2.2. Extrusion

The mixtures of bimodal PE base powder, HMWPE (loading from 10% to 40%) a standard stabilization package and a carbon black masterbatch, were extruded twice (i.e. 2-pass) using a Coperion ZSK 18 twin screw, co-rotating extruder at 230°C. The screw was designed in such a way to improve distributive mixing.

3. Characterisation

3.1. Rheology and GPC

Rheology (in a frequency sweep mode at 190°C) and GPC characterisation were conducted for structural information.

3.2. Mechanical testing

For tensile testing at room temperature and notched impact strength testing at 0°C, the materials were first compression moulded to plates and then cut into specimens of the required dimensions.

3.3. Morphology study

Light microscopy was implemented to study all samples that were microtomed to about 12 µm thick sections. For SEM study, pellets of selected samples were cut in the machine direction and $KMnO_4$ etched in H_2SO_4. Additionally, images were collected with AFM without sample treatment.

Results and Discussion

Average Mw (given by GPC) and complex viscosity at a shear rate of 0.05 s^{-1} (taken from rheology frequency sweep measurement) are plotted as functions of HMWPE loading in Figure 1. Addition of HMWPE to the bimodal PE was expected to increase the average Mw and viscosity. This was indeed seen for all blends prepared by extrusion but not for all those by kneading. Figure 1a indicates that the kneaded samples containing higher than 20% HMWPE do not follow the trend of increasing Mw and, particularly, the blend containing 40% HMWPE has a Mw even similar to the virgin bimodal PE. This suggests degradation of HMWPE in the kneading process. However, for both kneaded and extruded blends, the viscosity at 0.05 s^{-1} increases with increasing HMWPE loading (Figure 1b).

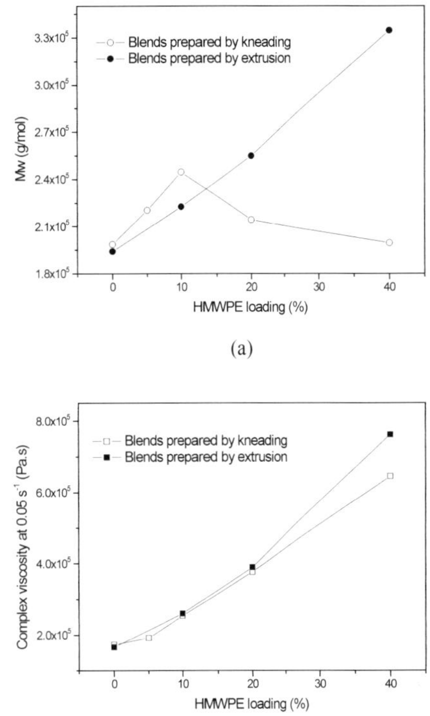

(b)

Figure 1 Mw (a) and complex viscosity at 0.05 s^{-1} (b) as functions of HMWPE loading

Viscosity at a low shear rate reflects the overall molecular weight of a material, whereas Mw given by GPC is more influenced by the high molecular weight chains. The inconsistency between Mw and viscosity of the blends with high loading of HMWPE (>20%) suggests that, during kneading, a number of high Mw chains from the HMWPE were cleaved but most of the cleaved chains still contribute to the viscosity; while during extrusion, a good number of high Mw chains were well retained. Indeed, comparison between the GPC curves shows, for kneaded blends with high HMWPE loading (40%), there is a cut-off of HMW tail while this was not seen for the reference bimodal material prepared by kneading (Figure 2).

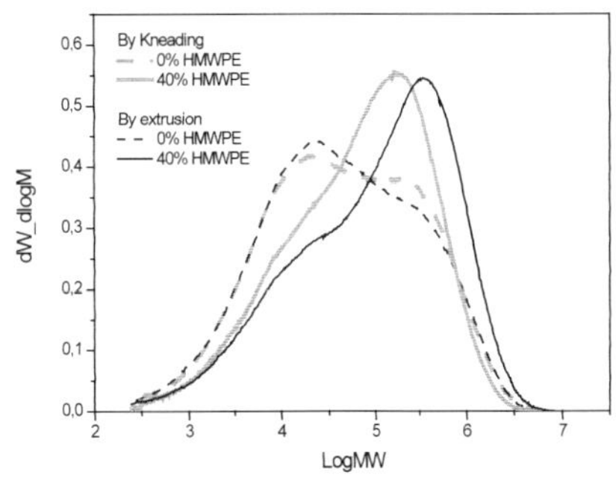

Figure 2 GPC curves of HMWPE blends.

The mechanical properties of all blends are summarized in Figure 3. A remarkable improvement in the impact strength is seen with addition of HMWPE into the bimodal PE. Up to a loading of 20% HMWPE, blends prepared both by kneading and extrusion have similar impact strengths. When the loading increases to 40% a dramatic deviation is seen in the impact strength where the extruded blend has more than double the impact strength in comparison with the kneaded counterpart. This coincides with the discrepancy in HMW tailing shown by GPC between the blends prepared in different ways (Figure 2). On the other hand, tensile moduli of the blends prepared in both ways are nearly identical and both decrease slightly with HMWPE loading. 40% HMWPE blend by extrusion shows a very good balance in stiffness and toughness: at a cost of about 10% tensile modulus, it has more than 4 folds' improvement in impact strength in comparison to the virgin bimodal PE.

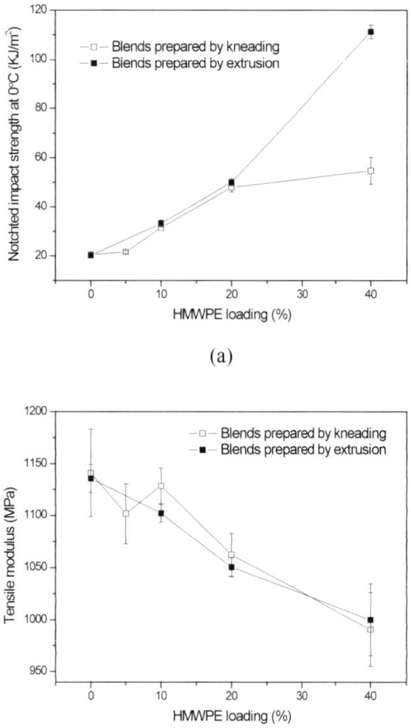

(a)

(b)

Figure 3 Notched impact strength (a) and tensile modulus (b) as a fucntion of HMWPE loading

In Figure 5, impact strength of all blends is plotted against viscosity at 0.05 s^{-1}. The kneaded blend having 40% HMWPE seems not to follow the linear increasing trend. As shown in Figure 4 and Figure 2, HMWPE in this material underwent severe degradation. The cleaved HMWPE chains keep its viscosity on a high level but its impact strength is much lower than it would be when the HMWPE chains are intact. This clearly shows that the long molecular chains are more effective in dissipating the impact energy than the shorter chains, even though the shorter chains, in a larger number, considerably increase the material's viscosity as the long chains do.

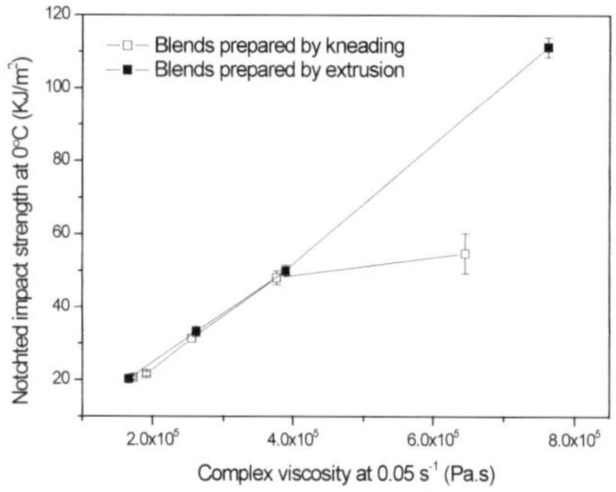

(a)

Figure 5 Notched impact strength v.s. complex viscosity at 0.05 s^{-1} of various HMWPE blends

Homogeneity of the blends were checked with light microscopy. Few white dots were discovered for nearly all blends. The blend with 40% HMWPE loading is particularly homogeneous (see Figure 6).

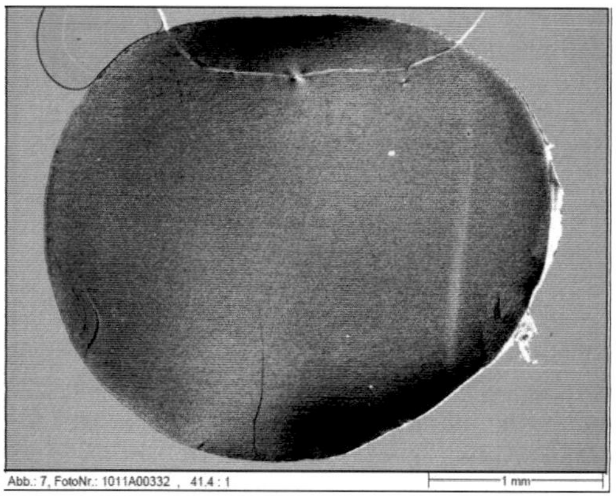

Figure 6 Micrograph of 40% HMWPE blend prepared by extrusion

Additional morphological characterization was carried out with SEM and AFM to image the lamellar arrangement of the three materials. SEM images of the extruded blends after standard etching are presented in Figure 7. Carbon black particles/agglomerates are visible in all samples, together with clearly defined edge-on and flat-on lamellae. In all cases considered, about 30 nm thick lamellae can be observed in the SEM images although this value might be affected by the etching procedure that could remove smaller ones. Figure 7c somewhat suggests that also smaller lamellae are formed in the case of 40% HMWPE loading.

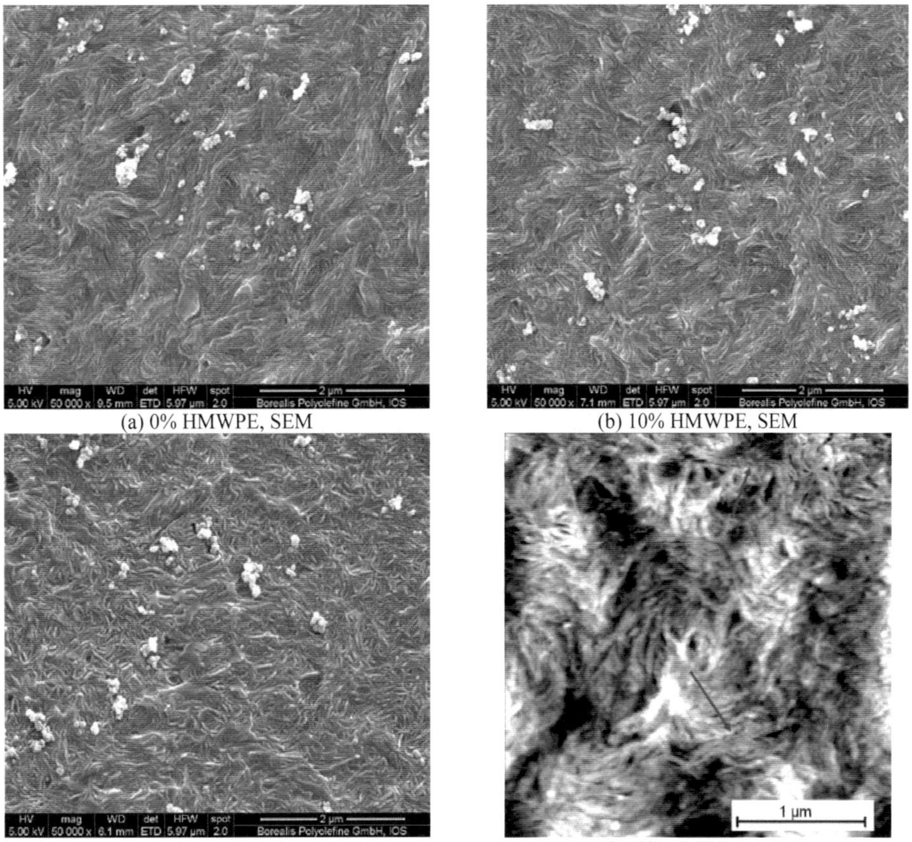

(a) 0% HMWPE, SEM (b) 10% HMWPE, SEM

(c) 40% HMWPE, SEM (d) 40% HMWPE, AFM

Figure 7 SEM of HMWPE blends prepared by extrusion: (a) 0% HMWPE (b) 10% HMWPE, (c) 40% HMWPE and (d) 40% HMWPE by AFM

Since the same experimental route has been followed for the three materials, though different susceptibility towards etching may be encountered, additional experiments were performed with AFM. In this case the sample was ultracryomicrotomed and imaged without further sample treatments. An AFM image of the material with 40% HMWPE is shown in Figure 7d. Smaller lamellae, in the order of 15 nm thickness, are indeed observed in this case as highlighted with the arrows.

Conclusions

Bimodal PE/HMWPE blends were homogeneously prepared by melt blending. Under the applied conditions, extrusion retained more HMW tail in the final blend than kneading particularly at high loadings of HMWPE. 40% loading of HMWPE, by extrusion, resulted in a homogenous blend with 4 fold's increase in impact strength at a cost of about 10% tensile modulus. AFM detects lamella populations of different thickness in this blend.

References

1. M. Mihailov and L. Minkova Colloid & Polymer Science 1987, 265, 681-685

2. Y. L. Huang and N. Brown Polymer 1992, 33, 2989-2997

3. J. D. Zuo, Y. M. Zhu, S. M. Liu, Z. J. Jiang and J. Q. Zhao Polym.Bull.(Heidelberg, Ger.) 2007, 58, 711-722

4. A. B. Boscoletto, R. Franco, M. Scapin and M. Tavan European Polymer Journal 1997, 33, 97-105

5. K. L. K. Lim, Z. A. Mohd Ishak, U. S. Ishiaku, A. M. Y. Fuad, A. H. Yusof, T. Czigany, B. Pukanszky and D. S. Ogunniyi Journal of Applied Polymer Science 2005, 97, 413-425

6. P. Vadhar and T. Kyu Polym.Eng.Sci. 1987, 27, 202-210

7. C. A. Avila-Orta, C. Burger, R. Somani, L. Yang, G. Marom, F. J. Medellin-Rodriguez and B. S. Hsiao Polymer 2005, 46, 8859-8871

8. S. Song, P. Wu, M. Ye, J. Feng and Y. Yang Polymer 2008, 49, 2964-2973

O-06-071

FLUORESCENCE QUENCHING OF FLUORANTHENE BY MALEIC ANHYDRIDE IN SOLUTION AND DURING NON-REACTIVE AND REACTIVE TWIN-SCREW EXTRUSION

F. Mighri[a,b,*], **A. Methenni**[a,b], **S. Elkoun**[a,c], **P. Cassagnau**[d]

[a] *Center for Applied Research on Polymers and Composites, CREPEC, QC, Canada*

[b] *Dep. of Chem. Eng., Laval University, QC, Canada; frej.mighri@gch.ulaval.ca*

[c] *Dep. of Mech. Eng., Université de Sherbrooke, QC, Canada*

[d] *Université de Lyon, CNRS, Ing. des Mat. Polym., Université Lyon 1, Villeurbanne, France*

Abstract - Fluorescence quenching of fluoranthene by maleic anhydride (MAH) in polypropylene (PP) during twin-screw extrusion (TSE) was investigated using an online fluorescence measurement system. The fluorescence quenching effect was clearly seen from the decrease of the peak intensity and integrated area of the residence time distribution (RTD) curves when MAH was introduced into PP. By using this quenching phenomenon, it was possible to measure online the grafting degree of PP with MAH in the TSE. The obtained results showed that MAH grafting degree and total conversion measured by both fluorescence technique and by Fourier transform infrared spectroscopy (FTIR) were in good agreement for MAH weight concentrations up to 0.8 wt.%.

Keywords: Fluorescence quenching, twin-screw extruder, residence time distribution (RTD), grafting, conversion.

Introduction

The fluorescence spectroscopy can provide insights into the environment, structure and dynamics of fluorophore probes that are either covalently bound or freely doped into the polymer system [1]. Fluorescence quenching of fluorophore molecules by electron deficient molecules were extensively studied in organic solutions [2]. Fluorescence quenching of phenanthrene (Ph) and anthracene (An) by MAH and n-Octadecenylsuccinic anhydride (ODSA) both in PP matrix and in chloroform solution were already studied [3]. It was observed that the fluorescence of Ph in solution was less efficiently quenched by ODSA than by MAH because the diffusion of ODSA was hindered by the long alkyl chain.

Real-time monitoring of the mixing performance in the TSE, such as the RTD, has been the subject of a number of studies because it is useful to understand the shear history and mixing quality during the extruding process [4, 5]. It is expected that the fluorescence quenching behaviour in the TSE should be similar to that in the organic solution. Bur et al. [6] and Hu and Kadri [7, 8] already used the fluorescence technique to measure the RTD behaviour of polymer melts during extrusion. They showed that the fluorescence intensity is proportional to the fluorescent tracer concentration. To the best of our knowledge, no study had been reported on the online characterization of polymer grafting level, such that of PP, using the fluorescence technique. The objective of this work is to characterize online the grafting degree of MAH on PP and the total conversion of MAH by measuring the fluorescence quenching of fluoranthene by MAH in the melted PP during the reactive extrusion process.

Experimental

Materials used: PP (PP1052 grade) was a commercial product by ExxonMobil Corp. Fluoranthene (99%) and MAH (99%) and PP-g-MAH (weight grafting degree: 0.29%, 0.5%, 0.7% and 0.9%) were purchased from Sigma-Aldrich Inc. 2,5-dimethyl-2,5-di(tert-butylperoxy) hexane (DHBP) (Trigonox101) was supplied by Akzo Nobel Chemical. A constant concentration of 0.5wt.% of DHBP was incorporated into the polymer. Chloroform was a HPLC grade purchased from Fisher Scientific.

Fluorescence measurement in solution: The fluorescence emission of fluoranthene solution in chloroform was carried out in a Cary Eclipse spectrofluorometer (from Varian Inc.). A fluoranthene solution (16.6 ppm) was firstly prepared, then the MAH was dissolved in the fluoranthene solution at various concentrations. The excitation wavelength was set at 365 nm and the fluorescence emission was scanned from 375 to 600 nm.

Fluorescence system and extruding conditions: The in-line fluorescence system was installed on the die of a co-rotating TSE, ZSE-27 (Leistritz, Germany). The barrel temperature was set at 180, 190, 200, 200, 200, 200, 200, 200, 190, 190°C from the feeding zone to the die. The screw RPM and feeding rate were respectively set constant at 120 rpm and 3 kg/hour. A small amount of tracer was injected to the hopper as a pulse after a stabilization period of 10 minutes. The tracer used here was a masterbatch mixture of PP/Fluoranthene (94wt.%/6wt.%) prepared in a batch mixer.

Grafting degree measured by FTIR: The grafting degree of PP-g-MAH blends was characterized by the FTIR technique [9] using a FTIR Specac FTLA 2000 at a resolution of 4 cm^{-1} with 16 scans. The grafted PP was purified by dissolution in xylene then precipitated in acetone and dried under vacuum at 90°C for 3 days. Dried PP-g-MAH was hot pressed to films of 100 μm in thickness, then heat treated at 130°C for 24 h to convert the acid groups to anhydrides. The calibration curve was constructed on blends of known concentrations of PP and n-Octadecenylsuccinic anhydride. The blends were prepared in the batch mixer and compressed to films, then tested in the FTIR. The calibration curve is shown in Figure 1.

Results and Discussion

Fluorescence quenching of fluoranthene by maleic anhydride in $HCCl_3$ solution: Figure 2 shows the fluorescence emission of fluoranthene in chloroform solution at constant fluoranthene concentration (16.6 ppm) but at various MAH concentrations (from 0.0 to 0.14mol/L). It is clearly seen that fluorescence intensity decreases with increasing the concentration of MAH. No new peaks corresponding to any new emitting species is observed. The fluorescence quenching effect, F_0 / F, which corresponds to the ratio between the fluorescence intensities of the unquenched (F_0) and quenched (F) emissions, is presented in Figure 3. The solid line corresponds to the following Stern-Volmer equation:

$$F_0 / F = 1 + K_{sv}[Q] = 1 + K_q \tau_0 [Q] \qquad (1)$$

where $Q\,(mol/L)$ is the quencher molar concentration, $K_{sv}\,(L/mol)$ is the Stern-Volmer constant, $K_q\,(L/mol.s)$ is the diffusion controlled bimolecular quenching constant and $\tau_0\,(ns)$ is the lifetime of the fluorophore in the absence of quencher. Optimized fitting was obtained with $K_{sv} = 211\,L/mol$. Compared to the fluorescence quenching of anthracene by MAH ($K_{sv} = 87\,L/mol$) [5], fluoranthene is more quenched by MAH due to its longer lifetime of around 50 ns [10] compared to 6 ns for anthracene.

Fluorescence quenching of fluoranthene by non-grafted and grafted MAH in twin-screw extruder: The characterization of fluorescence quenching in TSE was based on the RTD method. The reproducibility of the RTD measurements and the linearity between the amplitude of the fluorescence signal and tracer concentration incorporated at the main hopper of the extruder were already validated.

The case of non-grafted MAH: In Figure 4, the integrated areas of RTD curves are plotted as a function of fluoranthene tracer weight (g). As shown in the figure, linearity is observed between the fluorescence emission and the tracer weight (g). In Figure 5, the fluorescence intensity distribution is presented for a constant tracer amount (0.48 g) and various MAH weight concentrations (0.2-1.0 wt.%). It is clearly shown that fluorescence intensity decreases with increasing MAH wt.%. That is, the integrated area of the RTD curve consequently decreases with MAH concentration. It should be mentioned that the fluorescence signal was very small at MAH concentration of 1.0 wt.%, which was the highest concentration studied. If these fluorescence intensities are re-plotted in terms of the fluorescence quenching effect, F_0 / F, the latter shows a linear increase with MAH wt.%, as shown in Figure 6, and also follows the Stern-Volmer equation with a quenching constant $K_{sv} = 102.6\,L/mol$. This behaviour is identical to that observed in chloroform solution (Figure 3) but with a lower quenching constant, K_{sv}. According to Stokes-Einstein equation [11], the diffusion controlled quenching constant is inversely proportional to the solvent viscosity. Here, the viscosity of PP melt in the TSE is much higher than that of chloroform. However, the above quenching constant ($K_{sv} = 102.6L / mol$) in melted PP is only half of that in chloroform solution ($K_{sv} = 211\,L/mol$).

The case of MAH already grafted onto PP: In order to measure precisely the grafting degree of MAH onto PP during extrusion, we have to know the fluorescence quenching of fluoranthene by MAH already grafted onto PP, which is supposed to be lower than by non-grafted MAH. To do this, fluorescence quenching of fluoranthene by MAH was measured for four PP-g-MAH samples of known grafting degrees (0.29wt.%, 0.5wt.%, 0.7wt.% and 0.9wt.%). The corresponding RTD curves are presented in Figure 7, which shows that the fluorescence intensity decreases with increasing the grafting degree. So, even if MAH is grafted onto the PP backbone, it continues to quench the fluoranthene tracer but much less than free MAH. If the results of Figure 7 are re-plotted in terms of the fluorescence quenching effect, F_0 / F, a linear increase with MAH wt.% is also as in the case of free MAH but with much lower slope (i.e., lower quenching constant, $K_{sv} = 49.1L / mol$), as shown in Figure 8.

Fluorescence quenching of fluoranthene by MAH during reactive extrusion: The RTD curves of neat PP and PP with 0.5wt.% DHBP peroxide and various weight concentrations of MAH are shown in Figure 9. As a first observation, the fluorescence intensity is increased with the addition of 0.5 wt.% of peroxide to PP and shows a continuous decrease with increasing the amount of MAH. Figure 10 shows these fluorescence intensities in terms of the fluorescence quenching effect, F_0 / F, as a function of MAH wt.%, together with those already presented in Figure 6 (without peroxide). It can be seen that, for the same MAH wt.%, the quenching efficiency of MAH decreases with the incorporation of peroxide. This can be explained by the grafting of MAH onto PP and the decrease of free MAH concentration in PP melt. In our previous work, Fang et al. [3], fluorescence quenching of phenanthrene (Ph) and anthracene (An) by MAH and ODSA in chloroform solution was studied. It was found that the fluorescence of Ph and An in solution was less efficiently quenched by ODSA than by MAH. This is because the long alkyl chain on ODSA restricts its diffusion in solution. With the incorporation of peroxide, the grafting reaction is initiated and, as presented in Figure 8, grafted MAH is less efficient to quench the fluorescence emission of fluoranthene tracer. So, the fluorescence quenching is lower in the presence than that in the absence of peroxide.

Grafting degree measured by fluorescence quenching of fluoranthene tracer: As mentioned in the above section, the decrease in the quenching effect, F_0 / F, of MAH in the presence of peroxide is attributed to the grafting of MAH onto PP backbone, which leads to a decrease of the concentration of non-grafted MAH. If we take into account the small quenching effect of grafted MAH (presented in Figure 8), the grafting degree, $G_d\left(wt.\%\right)$, of PP-g-MAH can be estimated by using the following equation:

$$G_d(wt\%) = C_{MAH,0} - \frac{(F_0/F-1)_1}{(K_{sv})_1} - \frac{(F_0/F-1)_2}{(K_{sv})_2} \quad (2)$$

where $C_{MAH,0}(wt.\%)$ is the initial MAH weight concentration incorporated into PP, $(F_0/F-1)_1$ is the quenching efficiency of MAH in the presence of peroxide, and $(F_0/F-1)_2$ is the quenching efficiency of already grafted MAH. The corresponding Stern-Volmer quenching constants are $(K_{sv})_1 = 102.6\,L/mol$ and $(K_{sv})_2 = 49.1\,L/mol$. For the sake of comparison, MAH grafting degree and conversion (%) measured by FTIR and fluorescence techniques are presented in Figures 11 (a) and (b), respectively. For the MAH concentrations studied (up to 0.8wt.%), both techniques give close results for both grafting degree and conversion.

Conclusions

We studied in this work the fluorescence quenching of fluoranthene by MAH in chloroform solution and in PP melt flow during non-reactive and reactive TSE. It was observed that the quenching constant was lower in TSE melt flow than in solution due to the restricted collisional diffusion in melt flow. We also observed that studied the fluorescence quenching of MAH during reactive TSE process was largely decreased by the incorporation of peroxide. The characterization of the fluorescence quenching of already grafted MAH onto PP showed to us that grafted MAH had a small quenching effect. Consequently, the decrease of fluorescence quenching in the presence of peroxide was due to the grafting of MAH onto PP backbone and the decrease in the concentration of non-grafted MAH present in the melt flow. By taking into account the small quenching effect of grafted MAH, the grafting degree of PP-g-MAH and MAH conversion (%) were estimated and were compared to those characterized by FTIR technique. Good agreement between the two techniques was obtained for MAH weight concentrations up to 0.8 wt.%.

Acknowledgements

The authors are grateful to the Natural Sciences and Engineering Research Council of Canada (NSERC) for financial support of this work.

References

1. H. Morawetz, *J. Polym Sci part A: Polym Chem*, **37**, 1725 (1999).
2. B. Valeur, *Molecular Fluorescence: Principles and Application.* Chapter 4, Wiley-VCH, (2001).
3. H. Fang, F. Mighri, A. Ajji, *Polym. Eng. Sci.*, **47**, 192 (2007b).
4. A. V. Machado, M. Van Duin, J. A. Covas, *J. Polym. Sci., Part A: Polym. Chem.*, **38**, 3919 (2000).
5. D. Chang, J. L. White, *J. Appl. Polym. Sci.*, **90**, 1755 (2003).
6. A. J. Bur, F. M. Gallant, *Polym. Eng. Sci.*, **31**, 1365 (1991).
7. G. H. Hu, I. Kadri, *Polym. Eng. Sci.*, **39**, 930 (1999a).
8. G. H. Hu, I. Kadri, *Polym. Eng. Sci.*, **39**, 299 (1999b).
9. S. H. P. Bettini, J. A. M. Agnelli, *Polym. Testing,* **19**, 3 (2000).
10. H. D. Burrows, S. J. Formosinho, F. R. Paiva, E. J. Rsburn, *J. C. S. Faraday II*, **76**, 685 (1980).
11. N. Selvarajan, V. Ramakrishnan, *Indian J. Chem. A,* **18**, 340 (1979).

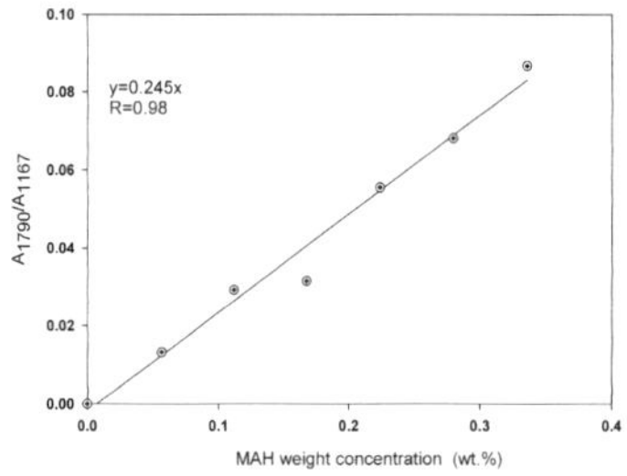

Fig.1. Calibration curve for the characterization of grafting degree of PP-g-MAH.

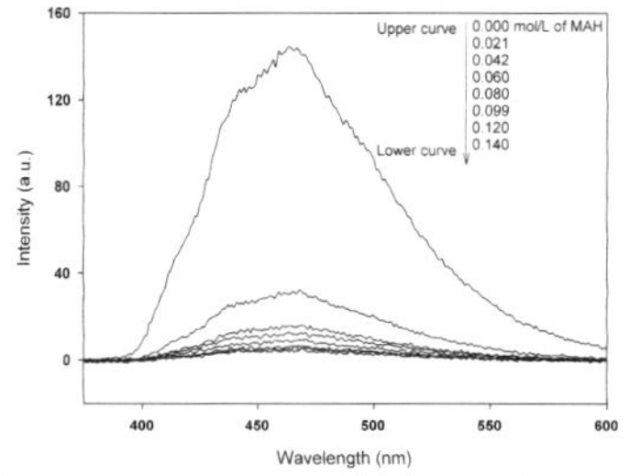

Fig. 2. Fluorescence spectra of fluoranthene in the absence and presence of maleic anhydride (0.0 to 0.14 mol/L) (a.u.: arbitrary unit).

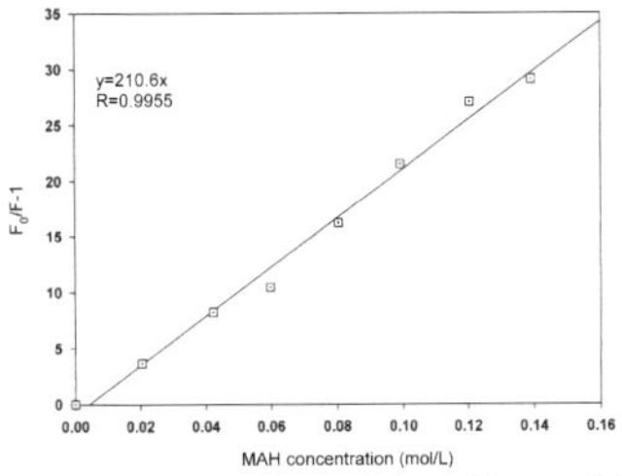

Fig. 3. Fluorescence quenching effect of fluoranthene by maleic anhydride in chloroform solution.

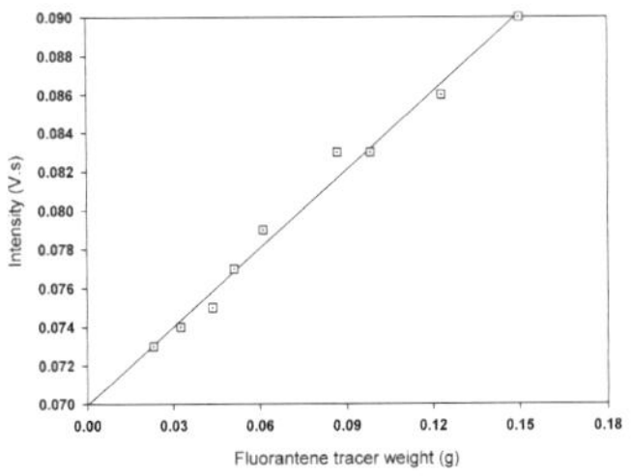

Fig. 4. Linearity between the fluorescence intensity and fluoranthene tracer weight in twin screw extruder.

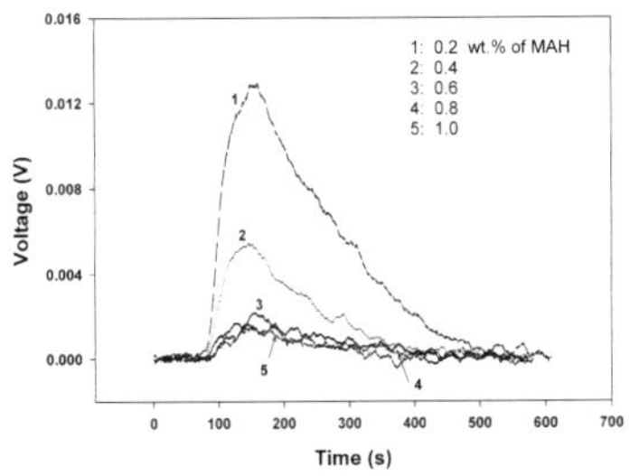

Fig. 5. RTD curves of neat PP and PP with different MAH concentrations.

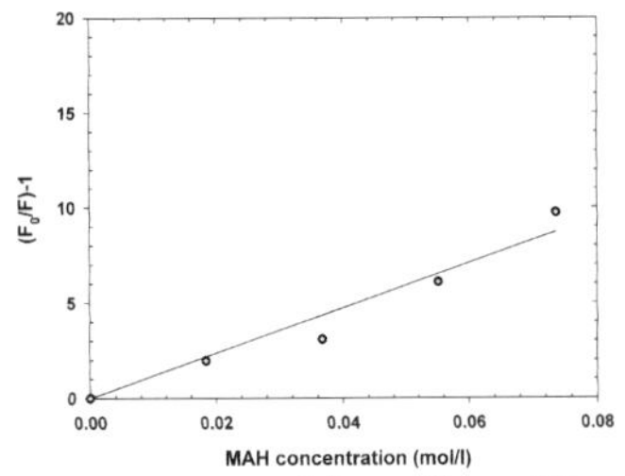

Fig. 6. Fluorescence quenching effect of fluoranthene by MAH in extruder.

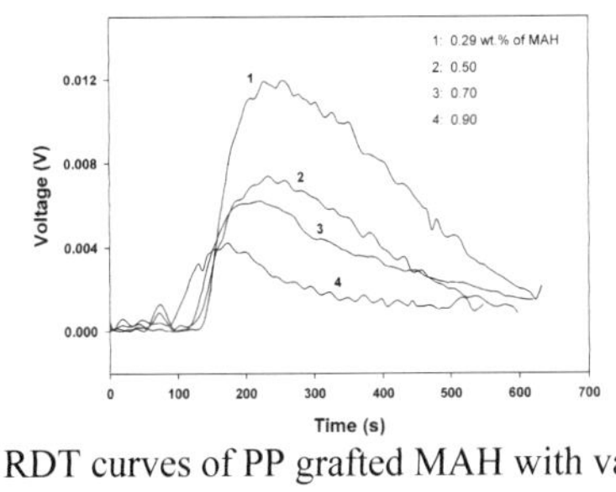

Fig. 7. RDT curves of PP grafted MAH with various grafting degrees.

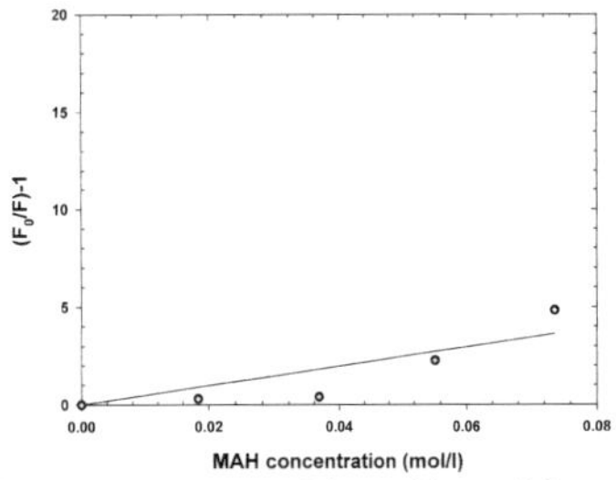

Fig. 8. Fluorescence quenching effect of fluoranthene by AMH grafted onto PP backbone in extruder.

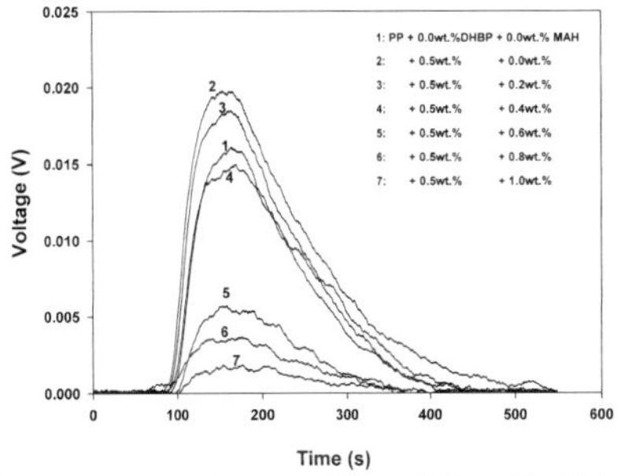

Fig. 9. RTD curves for neat PP and PP with 0.5wt.% of DHBP and various MAH concentrations.

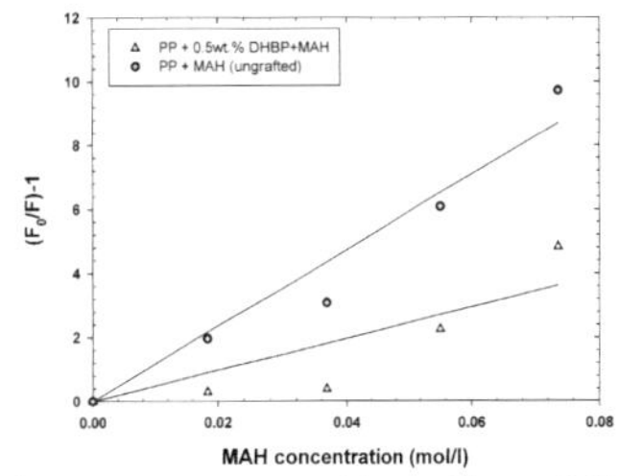

Fig. 10. Fluorescence quenching effect of fluoranthene by MAH in the absence and presence of peroxide.

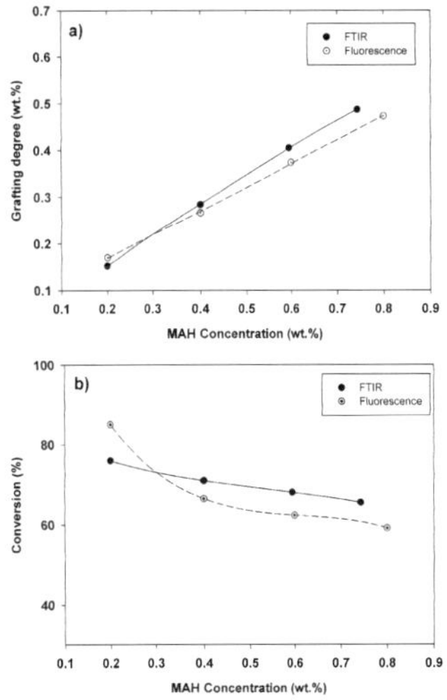

Fig. 11. Grafting degree (wt.%) **(a)** and conversion (%) **(b)** measured by FTIR and fluorescence techniques.

O-06-187

LAMINATED SANDWICH STRUCTURE BASED ON PC SKINS FOR IMPACT RESISTANT PANEL APPLICATIONS: INVESTIGATION OF CORE MATERIALS AND POLYMERIZATION PROCESSES

W. Angkura[1,2,*] and U. Meekum[1]

[1] *School of Polymer Engineering, Suranaree University of Technology, Thailand; umsut@g.sut.ac.th*

[2] *Center of Excellence on Petrochemical and Materials Technology, Chulalongkorn University, Thailand; wangkura@gmail.com*

Abstract: An impact resistance laminated sandwich panel consisted of PC skins and the blend between PMMA and epoxy based vinyl ester (EVE) as core had been attempted. In this report, 4 types of crosslinked PMMA/EVE core material produced by 2 blending methods; syrup and reactive, and then polymerized by either thermal or UV processed were explored and compared. The standard material testing procedures were adopted to evaluate the different between those core materials. AIBN and MEKP/CoNaph were used as free radical initiators for curing MMA and EVE resin, respectively. The crosslinked PMMA/EVE blends were verified by the gel test. The result showed that, using PMMA syrup blended with EVE under thermal curing, the blends were immiscible especially at EVE content above 15%. In contrast, reactive blending and underwent UV curing, the materials most likely to be miscible. The mechanical and thermal properties of the miscible blends were closely to obey the rule of mixture. SEM investigation confirmed that most of PMMA chain was only interpenetrated onto the network structure of cured EVE for the immiscible PMMA/EVE blends derived from syrup blending and thermally polymerization. The electron microscope also indicated that easily short chain PMMA was evidenced and explained when thermal induced polymerization was applied.

Keywords: PMMA/EVE blends; PMMA syrup blending; reactive blending; thermal and UV induced polymerizations

Introduction

Much attention has been paid to hybrid sandwich structure because it exhibits superior bending stiffness and strength compared with solid plates of the same mass [1]. Quite a lot of applications where the event of an impact resistance needs to be considered because the behavior of the core and the interface between the face sheet and the core play a crucial role in the way the impact energy is absorbed and dissipated [2]. Most of them are also required the optically clear to defeat incoming threats, withstand multiple impacts, and maintain optical clarity with minimal fragmentation and visual distortion for the user.

For decades, the component materials consisted of the glassy polymers poly(methyl methacrylate) (PMMA) and polycarbonate (PC) with respectable adhesives are interested by many applicators. The transparency and toughness of these materials are the main reasons. They are excellent choices for use in transparent applications to withstand impact [3]. Previous research, brittle resins or thermoplastic modification can provide effective toughening if correctly incorporated into the thermoplastic. Interactions between the dissolved polymers with one or both compounds of the epoxy formulation can be modified and led to variations in the phase separation process during curing. They carried out the ratio of PMMA/epoxy blends for improving impact strength, PMMA domain for modified epoxy resin containing up to 15% indicated completely transparent and high mechanical properties [4-5]. In additions, researchers studied vinyl ester based three component IPNs prepared from the VEN, DGEBA, and MMA that presented the good resistance to chemical reagents, good thermal stability, and good mechanical properties [6]. However, the effect of crosslink density was more significant for the improvement of mechanical properties than the effect of thermoplastic addition. The influence of curing conditions on mechanical properties of vinyl ester resins was explored. The properties of high temperature, UV and microwave curing processes were higher than at ambient condition [7-8].

However, there are no studies regarding the laminate structure made from three layered structures consisted of optically clear PMMA blends with epoxy vinyl ester resin (EVE) core and PC skins. The purpose of this publication is to compare the blending methods between PMMA and EVE and also the curing process for manufacturing the blends. Shorter processing time and superior in both performance properties and clarity is the main ambitious.

Experimental

A. Materials

MMA monomer was kindly supplied from Thai MMA Co., Ltd. The freshly recrystallized AIBN in methanol was used as free radical initiator for polymerization of PMMA. The epoxy based vinyl ester resin (EVE), Derakane 411-350, was commercially available from Ashland Inc. Methyl ethyl ketone peroxide (MEKP) and cobalt octoate (CoNaph) were supplied from AKPA Chemicals Co., Ltd. They were employed as co-catalyst for curing vinyl ester resin. Dimethyl aniline (DMA), activator, was obtained from Thai Mitsui Specialty Chemicals Co., Ltd.

B. Core Sheets Preparation

Core sheet was manufactured from PMMA/EVE blend. It was prepared by two mixing methods namely "**Reactive**" and "**Syrup**" blending respectively. Each of mixtures was polymerized by either thermal or UV induction. Totally, four differences PMMA/EVE cores were obtained. In reactive blending, solution of MMA with 0.2 phr of AIBN and the given % of EVE was polymerized in round bottom flask reactor at 70°C for 45 mins. The lightly viscous PMMA/EVE was obtained and cooled down. Then, 0.01 phr of DMA, 0.15 phr of CoNaph and 0.5 phr of MEKP with respect to EVE content was added into the pre-polymer liquid and vigorously stirred. The greenish mixture was thoroughly degassed. The flow ability mixture was casted in the Mylar^{TM} lined two glass plats mold. It was brought to complete polymerization by either thermal or UV induction processes. In the Syrup blending, MMA was pre polymerized into syrup by the same manner as previously described. Then, the greenish mixture of the pre calculated % of EVE with 0.01, 0.15 and 0.5 phr of DMA, CoNaph and MEKP, respectively was added into the PMMA syrup. The solution was thoroughly mixed and finally degassed. The bubble free liquid was casted on the glass mold as mentioned above.

Thermal induced polymerization was achieved by placing the mold in the oven at 40°C for 36 hrs, then increasing temperature to 50°C for 6 hrs and finally at 60°C for 6 hrs. The void free PMMA/EVE sheet with the thickness of 3 mm was demolded. UV oven equipped with two levels setting, low and high, of 800 watt UV lamp was in-house constructed. The UV induced polymerization was succeed by using low level of UV power, 400 watt. The cure schedule was followed. The molding was exposed to the light for 10 mins then resting by shutting the UV curtain for 10 mins, this process was repeated 4 times. After that, the semi hard sheet was thermally underwent annealing in the similar manner as explained earlier; 12 hrs at 40°C, 6 hrs at 50°C and 6 hrs at 60°C, consequently. The void free sheet was finally demolded. In total, approx. 25 hours was required for the UV induction process compare with 48 hours for the thermal one.

For the neat PMMA sheet, both thermal and UV cured sheets was manufactured. However, for the neat EVE sheet, the sample was promptly solidified at ambient temperature. For the sake of comparison, the readily hardened sheet was underwent in exact the same manner in manufacturing processes as for the rest of the samples.

C. Material Testing

Saw cut, well-polished, and post cured at 80°C for overnight specimen was obtained from the demolded sheet and used for mechanical and thermal properties testing. Mechanical tests by mean of flexural and impact strengths were conducted in according to ASTM methods. The Shore D surface hardness was also determined. Heat deflection temperature (HDT) was obtained at 455kPa. Morphological study was examined by SEM. Dichloromethane etching on the fractured surface of sample was performed and compared. The gel content, network density, was determined by ASTM D2765 in dichloromethane.

Results and Discussion

The physical appearance, degree of crosslinking, mechanical testing; flexural, impact and hardness and thermal properties were performed on the specimen obtained from the molded core sheets. The test results are summarized in Table 1.

A. Specimen Clarity

The physical appearance of the core sheets was visually inspected. Transparence, translucent and opaque were classified and reported in Table 1. From the observation was found that the clarity of the core sheet is incompetence when increasing the EVE content. When compare between reactive and syrup blending processes, it is seen that the loss in the transparency of PMMA/EVE core is very obvious for the syrup blending, especially when thermal induce polymerization is applied. In the reactive blending, the clarity of the obtained sheet is acceptable. It is indicated that phase separation between macho chain PMMA and networked EVE would be easily occurred. Vice versa, PMMA segments would be bonded within the network structure of EVE, therefore, good miscibility, better clarity, between the two components is obtained.

B. Gel Contents

EVE consists of two vinyl groups per molecule. Blending EVE with MMA and addition polymerization of the blend into polymers alloy would form the crosslink structure. According to the result for determining the crosslink density by means of gel fraction given in Table 1, it demonstrates that the PMMA/EVE blends had more than 95% of non-extractable solid compare with less than 1%, almost completely dissolved, for the neat PMMA. It could conclude that the PMMA/EVE core sheets manufactured by reactive and PMMA syrup blending and then thermally and UV curing are crosslinked in nature. It does not depend on the blending and polymerization procedures.

C. Mechanical and Thermal Properties

Mechanical and thermal properties of neat and PMMA/EVE blends are shown in Table 1. Commonly, cured EVE shows better flexural properties but lower in impact strengths than the PMMA. General observation, the mechanical properties of the miscible blends are likely to obey the rule of mixture. Meaning that increasing the superior fraction in the blends would increase the properties of the binary material. For example, strength of the blend is increased with increasing the higher flexural strength EVE fraction. However, the trend cannot take into account for the immiscible system, syrup blending. Because of the mechanical properties are not proportionally increased with increasing the EVE fraction. Phase separation would be taken into consideration for the results found. Vice versa, the impact strengths of the blends are in the reverse trend to the flexural values. But, exact explanation to the flexural testing would be considered. The better in miscibility is the better agreement on the combined properties of the blends. The shore D hardness is followed the same phenomenon. Due to the similarity in the HDT values between PMMA and cure EVE, the service temperature of the blends cannot be well defined by using the rule of mixture, namely Fox's equation. Nevertheless, HDT trend to be marginally decreased with increasing the EVE fraction. The observations found for the mechanical and thermal properties of the PMMA/EVE blends are regardless with the blending and polymerization methods as long as the blends are miscible.

D. Scanning Electron Microscopy (SEM)

Phase separation investigated on the fractured surface of the PMMA/EVE blends, both with and without dichloromethane etching, were determined using SEM technique. The SEM photographs of the blends are presented in Fig. 1. Taken the immiscible blend; 15% of syrup PMMA and 85% of EVE polymerized by thermal induction, Fig. 1a, into investigation, it is seen that no separated phased is observed on the fractured surface before etching. The rough surface is obviously evidenced on the etched specimen. it indicates that PMMA chain was dissolved by the solvent during the etching process. It reinforces the statement that PMMA was formed and interpenetration into the crosslinked chain of cured EVE resulted in the immiscible blend. On the other hand, when comparing the system with the UV induced polymerization shown in Fig. 1b, it is clearly seen that separated tiny particles are noticed on the surface before etching and they are removed after dichloromethane etching. By UV polymerization, the chain length of PMMA was further propagated and high molar mass polymer was formed.

Consequently, the phase separation between longer chain PMMA and networked EVE is more demonstrated than the shorter chain polymer by thermally polymerized.

In the reactive blending system, the blends are likely to be miscible. Most of the blends are transparence. At 15% PMMA and 85% EVE blends obtained by reactive blending and cured by thermal and UV processes shown in Fig. 1c and 1d, respectively. Both blends are reported in Table 1 as miscible. From

Table 1. Properties of the PMMA/EVE cores

Blend method	Blend process	Core composition MMA (%wt)	Core composition EVE (%wt)	Clarity	Gel content (%)	Mechanical: Flexural properties Strength (MPa)	Max. strain (%)	Modulus (GPa)	Izod impact strength Unnotched (kJ/m^2)	Notched (kJ/m^2)	Hardness Shore D	Thermal HDT (°C)
Thermal induced polymerization	Neat	0	100	○	94	125.13±2.12	6.06±0.13	3.17±0.11	9.47±0.27	1.70±0.51	88.03±0.58	102.3±0.6
		100	0	○	1	42.83±0.69	1.66±0.46	2.87±0.75	13.97±1.08	0.93±0.22	73.21±1.16	105.7±1.2
	Syrup blending	95	5	○	98	36.33±1.93	1.56±0.45	2.61±0.81	15.38±1.35	2.75±0.07	83.79±0.91	105.0±1.0
		90	10	**○**	95	44.03±0.98	3.16±1.09	2.17±0.21	13.32±0.50	1.36±0.16	82.55±0.79	104.3±0.6
		85	15	**○**	100	35.24±1.96	3.45±1.86	1.99±0.49	6.50±0.21	1.20±0.30	79.60±0.35	105.7±0.6
		15	85	●	95	50.10±6.49	1.64±0.15	3.18±0.12	4.26±0.87	1.17±0.21	86.06±0.43	101.4±0.6
		10	90	●	98	57.48±3.76	1.83±0.12	3.29±0.07	4.64±1.18	1.18±0.27	89.55±0.91	102.5±0.6
		5	95	**○**	97	65.91±6.31	2.16±0.24	3.20±0.04	9.05±0.67	1.03±0.10	88.73±0.54	103.1±0.6
	Reactive blending	95	5	○	97	66.06±1.99	2.09±0.11	3.39±0.10	16.50±2.00	0.85±0.10	84.36±1.45	108.5±0.7
		90	10	○	97	38.51±2.13	1.45±0.40	2.87±0.57	13.63±0.39	1.63±0.46	83.48±1.04	106.0±1.7
		85	15	**○**	95	51.66±3.87	1.69±0.16	3.21±0.22	6.43±1.63	1.63±0.24	81.82±0.86	107.5±0.7
		15	85	○	100	125.83±1.27	5.70±0.06	3.35±0.05	16.50±0.15	0.90±0.15	91.03±0.69	95.9±0.3
		10	90	○	100	125.84±1.42	5.59±0.31	3.34±0.04	18.51±1.27	1.13±0.16	91.21±0.71	97.8±0.3
		5	95	○	100	127.30±0.92	5.76±0.04	3.35±0.04	16.84±0.77	1.04±0.15	91.64±0.56	98.6±0.6
UV induced polymerization	Neat	0	100	○	96	136.68±4.22	6.30±0.63	3.18±0.10	23.00±0.95	5.65±0.26	86.06±1.21	102.7±1.0
		100	0	○	1	68.83±6.23	2.84±0.66	2.61±0.33	14.65±1.33	2.22±0.44	75.64±2.51	106.5±0.5
	Syrup blending	95	5	○	98	99.44±6.32	4.49±0.56	2.83±0.05	23.02±1.86	1.86±0.19	93.45±0.27	106.3±0.6
		90	10	**○**	99	61.22±6.21	2.24±0.24	2.83±0.10	9.26±1.71	1.39±0.18	91.88±0.69	105.0±1.0
		85	15	**○**	94	61.36±6.41	2.22±0.33	2.92±0.10	10.86±1.36	1.18±0.49	83.15±0.79	105.3±1.0
		15	85	**○**	100	119.53±0.64	5.57±0.29	3.30±0.01	24.69±3.51	0.75±0.16	91.27±0.31	95.7±0.6
		10	90	**○**	98	125.53±0.81	5.80±0.06	3.33±0.04	25.40±0.62	0.56±0.18	89.39±0.96	97.7±0.3
		5	95	**○**	99	124.91±1.89	5.84±0.07	3.30±0.03	26.86±4.59	0.60±0.14	91.39±0.38	97.3±0.9
	Reactive blending	95	5	○	100	81.08±2.47	3.33±0.26	2.67±0.16	19.82±1.53	1.44±0.38	93.33±0.43	105.7±0.8
		90	10	○	100	85.23±6.34	3.11±0.04	2.84±0.17	15.09±0.52	1.50±0.40	91.64±0.51	105.3±0.6
		85	15	○	97	47.37±4.71	1.63±0.14	3.06±0.05	6.36±0.59	0.00±0.00	89.09±1.36	105.0±1.0

		15	85	O	98	123.10±0.66	5.57±0.03	3.36±0.02	23.95±0.80	1.04±0.27	91.33±0.41	93.3±0.8
		10	90	O	100	122.56±0.58	5.60±0.05	3.35±0.02	34.81±1.79	0.96±0.01	91.21±0.52	93.1±0.1
		5	95	O	98	120.17±1.03	5.63±0.07	3.29±0.04	26.69±3.86	1.12±0.28	91.09±0.64	94.3±0.4

Note: O = transparent, **O** = translucent and ● opaque

a) 15%wt of PMMA syrup blended with 85%wt of EVE and thermally induced polymerization

b) 15%wt of PMMA syrup blended with 85%wt of EVE and UV induced polymerization

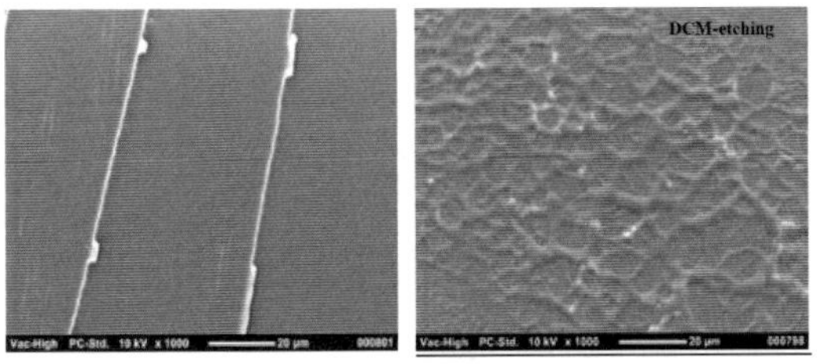

c) 15%wt of MMA with 85%wt of EVE reactive blended and thermally induced polymerization

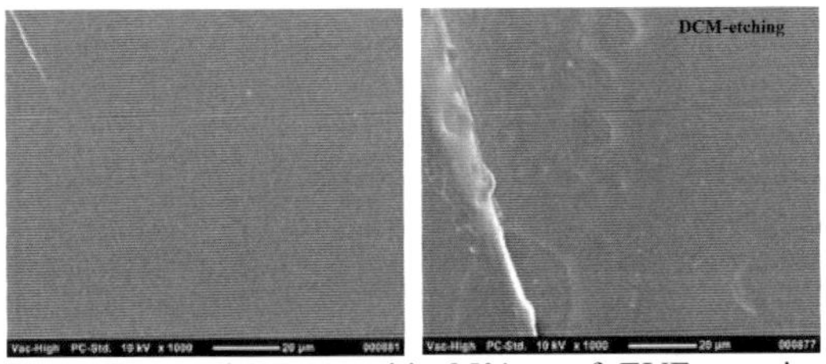

d) 15%wt of MMA with 85%wt of EVE reactive blended and UV induced polymerization

Figure 1. SEM micrographs of 15% of PMMA and 85% of EVE blends at X1,000.

the SEM of non-etched surfaces, they are illustrated that the blends are completely homogenous. Conversely, rough surfaces are noticed after the solvent etching. Thermal cured trace is rougher than the UV one. Incompletion polymerization of PMMA could be taken into consideration. As mentioned earlier, the conversion of MMA into PMMA by thermal driven reaction is slow compare with UV. Therefore, large amount of unreacted MMA or short chain PMMA, which is easily to be dissolved, is left behind. Hence, the toughness of the blend is normally inferior.

Conclusion

Four differences types of crosslinked PMMA/EVE core by mean of mixing methods, PMMA syrup and reactive, and polymerization processes, thermal and UV induction, were manufactured. The core obtained from PMMA syrup and cured by thermal process were immiscible, opaque, at EVE content above 10%. Reactive blending and UV curing of PMMA/EVE sheets were miscible and did not depend on the EVE loading. Mechanical and thermal properties of the miscible blends were followed the rule of mixture. Finally, SEM analysis revealed the miscibility and polymerizability of the blends obtained from 4 individual manufacturing methods.

Acknowledgements

This work is financially supported from Center of Excellence on Petrochemical and Materials Technology.

References

1. C. J. Yungwirth; J. O'Connor; A. Zakraysek; V. S. Deshpande; H. N. G. Wadley *J. Amer. Cera. Soc. 2011, 94,* s62.
2. R. S. Sharma; V. P. Raghupathy *ARPN J. Eng. App. Sci. 2011, 6,* 4.
3. J. S. Stenzler; N. C. Goulbourne *Int. J. Imp. Eng. 2011, 38,* 567.
4. I. Mondragon; P. M. Remiro; M. D. Martin; A. Valea; M. Franco; V. Bellenguer *Pol. Int. 1998, 47,* 152.
5. Siddaramaiah; F. L. Barcia; A. S. Sirqueira; C. M. Paranhos; B. G. Soares *J. App. Pol. Sci. 2007, 106,* 3808.
6. J. S. Parmar; G. R. Patel; A. K. Patel *Adv. Pol. Tech. 2004, 23,* 71.
7. M. Blanco; M. López; P. A. De Arcaya; J. A. Ramos; G. Kortaberria; C. C. Riccardi; I. Mondragon *J. App. Pol. Sci. 2009, 114,* 1753.
8. P. Li; X. Yang; Y. Yu; D. Yu *J. App. Pol. Sci. 2004, 92,* 1124.

O-06-204

HYBRID NANOCOMPOSITES PREPARATION: BATCH MIXER VS TWIN-SCREW EXTRUDER

M. Oliveira[1], A. V. Machado[1]*

[1] *Institute of Polymers and Composites (IPC) and I3N, University of Minho, Portugal – moliveira@dep.uminho.pt; avm@dep.uminho.pt*

Abstract - The present work aims to compare the preparation of hybrid polymer nanocomposites (HPN) in an internal mixer and in a twin-screw extruder. Nanocomposites, with the same composition, were prepared at the same temperature in both devices using PP grafted with maleic anhydride and aluminum isopropoxide, as organic and inorganic phases, respectively. Several analytical techniques were used to characterize the produced materials.
Chemical and morphological characterization showed that the sol-gel reaction in the extruder was more extensive, obtaining a nanocomposite with higher levels of crosslinking and smaller nanoparticles homogeneously dispersed in the polymer matrix.

Keywords: reactive extrusion, nanocomposites, internal mixer, extruder

Introduction

Hybrid nanocomposites are a recent class of materials whose developments started in 1940s in companies like, Dupont, Dow Corning, 3M and others. However, polymeric nanocomposites were only intensively studied in the eighties, when it was observed that a small amount of nano-size particles added to a polymeric matrix resulted in a significant performance improvement [1]. In fact a small amount of well dispersed inorganic filler can improve significantly the material properties, such as, gas permeability, thermal stability, flame retardance, mechanical performance or chemical resistance [2-4]. Nevertheless, the main difficulty of polymer nanocomposites (HPN) preparation is to obtain homogeneous dispersion of nanofillers in a polymeric matrix. In fact, particles agglomeration and voids formation at the interface of the particles and polymer matrix results in a material with worst properties than conventional polymers, which limits its effective application [5].
Regarding HPN synthesis there are three principal methods used to disperse nanoparticles in a polymeric matrix: solution dispersion, melting dispersion and in situ nanoparticles formation by sol-gel process [6,7]. In situ sol-gel reaction in the melt is the most suitable method to be applied at industrial scale, since it is solvent free and can be used in processes, like extrusion. In this case the hydrolysis-condensation reaction of metal alkoxide precursor occurs directly in the polymer melt [8].
The present study an extruder and an internal mixer were used as a continuous or batch chemical reactor, respectively, where the reaction between alkoxide precursor and polymer took place. Being reactive extrusion a complex process with many variables that can change during the process, the final material properties depend on the following parameters as residence time, pressure, temperature, mixing and diffusion [6]. Thus, the main objective of this work is to compare the preparation of hybrid nanocomposites in an internal batch mixer and in a twin-screw extruder. For that nanocomposites with the same composition were prepared in an internal mixer and twin-screw extruder at the same temperature. Characterization by techniques, such as, FT-IR, rheology, SEM and TEM was done to assess the main differences between the prepared materials.

Experimental

Materials and reagents

Polypropylene modified with maleic anhydride (PP-g-MA, Polybond 3200) with a melting temperature around 160 ºC and a MA content of 1 wt.%, was supplied by Crompton. Aluminium isopropoxide ($Al(Pr\text{-}i\text{-}O)_3$) was used as received and was supplied by Sigma Aldrich.

Nanocomposites preparation

Internal mixer

Nanocomposites were prepared in the melt, in a Haake batch mixer (Rheocord 90; volume 60 mL), equipped with two rotors running in a counter-rotating way. The rotor speed was 50 rpm and the set temperature was 180 ºC. The following procedure was adopted, first pellets of PP-g-MA were introduced into the hot mixer, after melting, the aluminium precursor was added (15 and 25 wt.%). The total sample was removed after 10 min mixing. The hybrid nanocomposites (HPN) synthesized in the internal mixer was called HPN-int.

Extruder

A modular co-rotating twin-screw extruder, Leistritz LSM 30.34, was used (Fig. 1). PP-g-MA and precursor were added separately. After melting of PP-g-MA, 15 wt.% of aluminium precursor was added using a side feeder. The barrel set temperature was 180 ºC, the screw speed 200 rpm and the throughput 3.8 kg/h. The average residence time was 160 seconds. The hybrid nanocomposite (HPN) synthesized in the extruder was called HPN-ext.

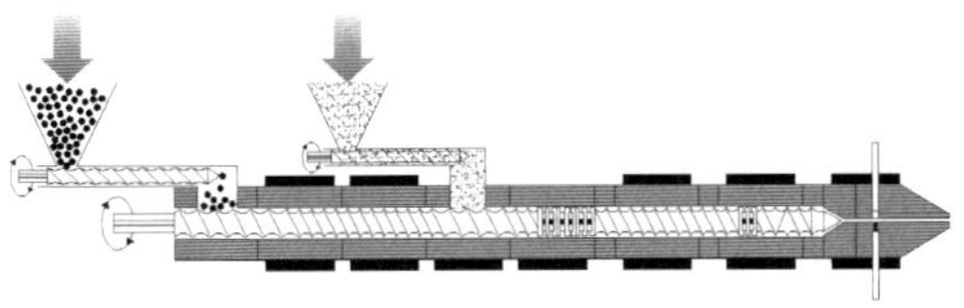

Figure 1 – Extruder layout.

Characterization

FT-IR

Infrared spectra of all HPNs were recorded in transmission mode, between 400 and 4000 cm^{-1}, using a Perkin Elmer 1610 spectrometer, with 32 scans and a resolution of 4 cm^{-1}. Thin films were previously prepared by compression molding in a hot press at 180 ºC and analysed directly using a solid film support.

Rheological properties

The rheological behaviour of the initial polymers and prepared HPNs were determined by oscillatory rheological measurements carried out in a Paar Physica MCR300 rheometer at 180 °C. The gap and diameter of the plates was 1 mm and 25 mm, respectively. Nitrogen atmosphere was used to prevent thermo-oxidative degradation. A frequency sweep from 0.1 to 100 Hz under constant strain in the linear viscoelastic region was performed for each sample.

Morphological characterization

Scanning electron microscopy (SEM) analysis was performed in a Leica Cambridge S360 microscope. The samples were previously fractured in liquid nitrogen and coated with a gold thin film.

Transmission electronic microscope (TEM) pictures were taken from samples of 70 nm thickness cut using a diamond knife, in a Leica ultramicrotome at -40 ºC. The cut sections were transferred onto copper grids and then analysed without staining in a Philips CM120, TEM.

Results and Discussion

The Fig. 2 depicts two zones of the FT-IR spectra of PP-g-MA, HPN-int and HPN-ext. For both nanocomposites two new bands between 1000 – 400 cm^{-1} and 3500 – 3000 cm^{-1}, corresponding to Al-O and OH stretching modes, respectively, are observed. Moreover, the OH band, with respective bending mode band at 1627 cm^{-1}, confirms that the precursor undergone hydrolysis and the presence of Al-O band indicates that sol-gel reaction was successfully carried out.

Regarding to OH band intensity, it is smaller for HPN-ext than for HPN-int, even though the reaction time was smaller. This can be related with reaction extension, which is higher in extruder associated to a more efficient by-product removal (2-propanol, C_3H_7OH) due to devolatilization applied on the extruder barrel.

The region 850 – 400 cm^{-1} also evidences differences between the two prepared materials. The aluminium oxide can exist in two different coordination states in the HPNs structure: tetrahedral, characterized by a broad band extending from 500 to 750 cm^{-1} and octahedral, bands at 500 and 640 cm^{-1} [9,10]. While for HPN-ext two bands indicating octahedral coordination can be seen, on the HPN-int spectrum only a broad band suggesting tetrahedral Al-O coordination is present. This difference is the result of different reaction extension, which affects the crosslinking density and consequently the distribution of the inorganic domains.

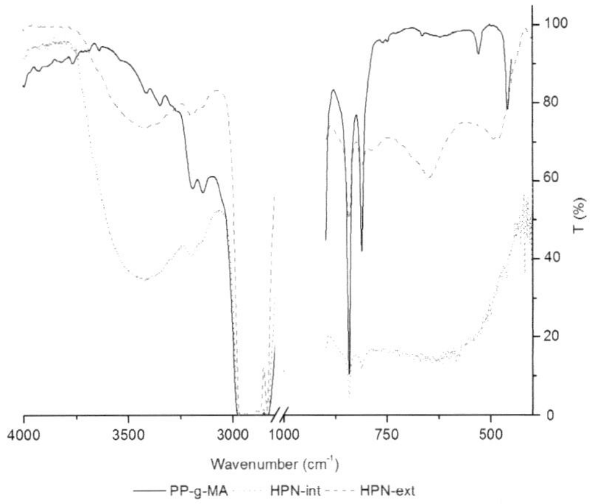

Figure 2 – FT-IR spectra of PP-g-MA, HPN-int and HPN-ext.

The rheological behaviour of the prepared HPNs gives important information about its molecular structure and consequently about crosslinking degrees. The results presented in Fig. 3 show a significant difference between the viscoelastic behaviour of HPN-ext, HPN-int and polymeric matrix. This significant improvement in viscosity and elasticity of the HPN-ext can be associated to the mixing efficiency in the extruder, which resulted in a more efficient sol-gel reaction and consequently more crosslinking.

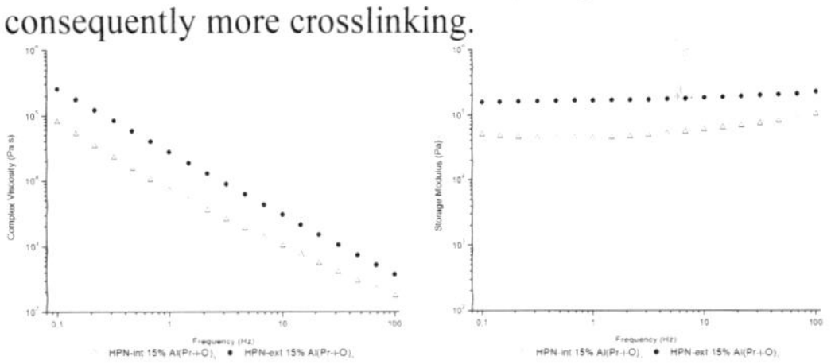

a) b)

Figure 3 – Rheological behaviour of PP-g-MA, HPN-int and HPN-ext; a) complex viscosity, b) storage modulus.

The microscopic technics were used to characterize the morphology of HPNs. SEM micrographs, in back-scattered electrons (BSE) mode, of HPNs fracture surfaces are presented in Fig. 4. HPN-int (Fig. 4a) has a heterogeneous surface, while HPN-ext has a homogeneous and smooth surface, more cohesive and regular (Fig. 4b). Inorganic particles agglomerations can not be observed for both materials, which indicates good and fine particles dispersion. To assess the size and dispersion of nanoparticles in the two preparation methods, TEM analysis was performed.

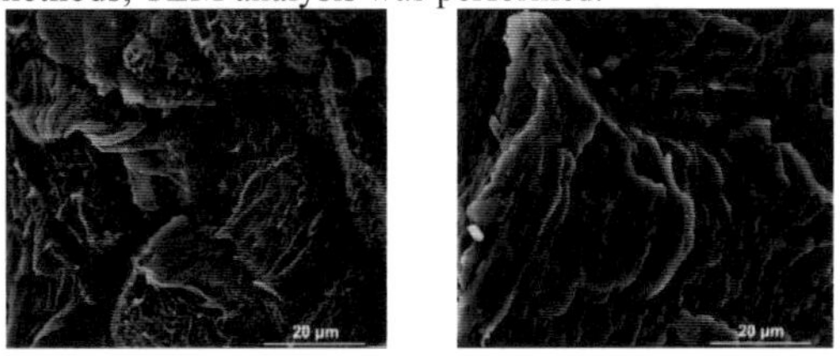

a) b)

Figure 4 – SEM micrographs of a) HPN-int and b) HPN-ext.

TEM micrographs presented in Fig. 5 confirms that in the extruder the reaction was more extensive than in the internal mixer resulting in better particle dispersion with smaller size, around 85 nm for HPN-ext compared to 155 nm of average particles size obtained for HPN-int.

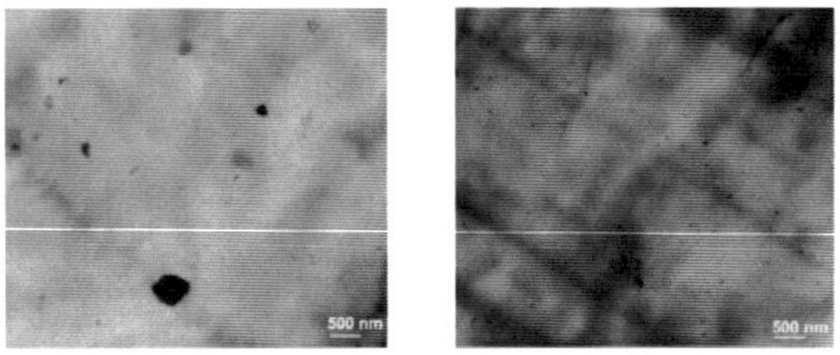

a) b)

Figure 5 – TEM micrographs of a) HPN-int and b) HPN-ext.

Conclusions

This study confirms that internal mixers are very useful to study the synthesis of hybrid polymer nanocomposites at the initial stage. However, in the extruder the sol-gel reaction will be more extensive, as observed by FT-IR rheological measurements. In the extruder, a material with smaller and well-dispersed nanoparticles and a more crosslinked structure was prepared.

Acknowledgements

The authors acknowledge the Foundation for Science and Technology (FCT) Project SFRH/BD/39085/2007 for the financial support.

References

1. Q. Dou; X. Zhu; K. Peter; D.E. Demco; M. Moller; C. Melin *J. Sol-Gel Sci. Tech.* 2008, *48(1-2)*, 51.
2. C. Sanchez; K. Shea; S. Kitagawa *Chem. Soc. Rev.* 2011, *40*, 696.
3. M. Sadeghi; G. Khanbabaei; A. Dehaghani; M. Sadeghi; M. Aravand; M. Akbarzade; S. Khatti *J. Membrane Sci.* 2008, *322(2)*, 423.
4. W. Bahloul; V. Bounor-Lagaré; L. David; P. Cassagnau *J. Poly. Sci. Part B: Poly. Phys.* 2010, *48(11)*, 1213.
5. H. Cong; M. Radosz; B. Towler; Y. Shen *Separation Purif. Tech.* 2008, *55(3)*, 281.
6. M. Oliveira; R. Nogueira; A.V. Machado *React. Funct. Polym.* 2012, *72*, 703.
7. M. Rong; M. Zhang; Y. Zheng; H. Zeng; R. Walter; K. Friedrich *Poly.* 2001, *42*, 167.
8. W. Bahloul; V. Bounor-legaré; L. David; P.J. Cassagnau *Poly. Sci. Part B: Poly. Phys.* 2010, *48*, 1213.
9. I.H. Joe; A.K. Vasudevan; G. Aruldhas; A.D. Damodaran; K.G.K. Warrier *J. Solid. State. Chem.* 1997, *131*, 181.
10. J.H. Wengrovius; M.F. Garbauskas; E.A. Williams; R.C. Going; P.E. Donahue; J.F. Smith; *J. Am. Chem. Soc.* 1986, *108*, 982.

O-06-210

HDPE/PA-6/EVOH BLEND PROPERTIES AND MORPHOLOGY

N. Jalalifar[1]*, Z. Ahmadi[1, 2]* and H.A. Khonakdar[3]

[1] *Department of Polymer Engineering, Mahshahr Branch, Islamic Azad University, Mahshahr, Iran - nadiyajalalifar@yahoo.com*

[2] *Color and Polymer Research Center, Amirkabir University of Technology, Tehran, Iran - zahmadi@aut.ac.ir*

[3]*Iran Polymer and Petrochemical Institute, 14965-115, Tehran, Iran - hakhonakdar@gmail.com*

Abstract -The mechanical properties of immiscible ternary blends of high density polyethylene/polyamide-6/poly ethylene co-vinyl alcohol has been investigated in the limits of a constant content of the major component while changing the weight ratio of the two minor constitutive polymers.

The particle size and morphology type of the all blends were then examined through the scanning electron microscopy (SEM) and image analysis techniques. Eventually, a core-shell structure was observed for HDPE/PA-6/EVOH blends and matrix disperse morphology in binary blends.

The relationship between mechanical properties and phase morphology of blends were explored. A SEM study has revealed that the encapsulation of EVOH around the PA-6 domains becomes gradually thin with increasing the ratio of PA-6 to EVOH. Thereupon, the low-strain tensile properties are decreased.

Keywords: Ternary polymer blend, Core-shell morphology, Mechanical properties.

Introduction

The ever growing demand of using polymers with special usage, and their unique changeable properties ability in the existing polymers in one vast vicinity with the different composition of blend components and various economical features are the main reasons for the growth of this industry.

In blending procedures, although the majority of polymers thermo-dynamically are not miscible, with respect to the better usage, the immiscible blends are far better than those polymers that are miscible. Additionally cause of availability of variable products and consequently more properties in comparison with the miscible blends they are preferred.

Also the new properties based on the unique morphological features will be produced. Therefore, the most important parameter in immiscible blends is the phase morphological control. [1-4]

It is well-known that the properties of polymer blends are greatly influenced by the morphology, which can be developed during the mixing process. [5]

In turn, the morphology is influenced by the following hierarchy of factors:

Interfacial tension > viscosity ratio > shear stress

It is obvious that, the ultimate property of polymer blends is dependent on their property components. Also, the composition of components and interfacial characteristics (e.g. interfacial adhesion) and morphology are influential. [6]

Most commercial blends have a morphology that consists of droplets dispersed in matrix phase, fibre-like and lamellar structures, and co-continuous morphology and also core-shell morphology is one of the other possibilities that can be taken into consideration in blends. [5]

Until present many different models for predicting the mechanical properties of polymer blends have been suggested. For instance the models based on the particulate composites can be mentioned. [7, 8]

So far various models predicting mechanical properties of particulate composites have also been frequently used for heterogeneous polymer blends. However, such models are not suitable for polymer blends because (i) they do not allow for an interval of the co-continuity of constituting phases and (ii) they cannot simultaneously predict several mechanical properties of a blend, thus relating them to a specific phase structure.

Therefore, individual models were developed to predict the mechanical properties of polymer blends. [9-12] These models were used to predict the properties of the binary blends, which are simpler than ternary blends.

Ternary blends behavior predication difficulty can be more complex structure due to the addition of the third component as variables.

This research considers the existing mechanical models for predicting mechanical properties in binary blends. The existing known binary models, employed for predicting the mechanical properties of HDPE/PA-6/EVOH ternary blends.

The objective of present study is, investigating morphology and mechanical properties of HDPE matrix and PA-6/EVOH dispersed phase ternary blends.

Experimental

Materials

The materials used in this study were: (i) High density polyethylene (HDPE), B4660 supplied by SABIC Petroleum Industries Corporation (King Saudi Arabic), (ii) polyamide (PA-6), MV14 supplied by Schulman Europe, (iii) poly ethylene co-vinyl alcohol (EVOH), Eval F101A supplied by Eval Company of America.

The physical and mechanical characteristics of the materials are listed in Table 1.

Table 1 - Specifications of the polymers

Polymer	Density (g/cm^3)	Tensile modulus (Mpa)	Stress at break (Mpa)	Strain at break (%)
HDPE	0.961	966.94	32.7	560.09
PA-6	1.13	1845.60	44.5	213.81
EVOH	1.19	3566.22	45.18	35.99

Processing

Binary and ternary blends were prepared by melt mixing in a Brabender co-rotating twin-screw extruder (model: TSE20, diameter of screw=2 and length / diameter ratio=40 cm), the extrudates were quenched in a cooling water bath and pelletized through a granulator in which; the screw speed was maintained at 100rpm.Before melting compounds, PA-6 and EVOH were dried in an oven for at least 16h ãC.8ÅAs shown in Table 2, in different blends, the matrix HDPE remains constant (80 wt %) whereas the amount of PA-6 and EVOH varies.

Table 2 - Composition of the binary and ternary blends

Sample code	Components	PA-6 (wt %)	EVOH (wt %)
1	HDPE/EVOH	0	20
2	HDPE/PA-6/EVOH	05	15
3	HDPE/PA-6/EVOH	10	10
4	HDPE/PA-6/EVOH	15	5
5	HDPE/PA-6	20	0

Morphology

The morphology of blend samples was characterized by Philips XL30 scanning electron microscope.

Prior to microscopic analysis, all samples were fractured in liquid nitrogen so that a brittle fracture could be obtained, thereby avoiding large deformations in the surface while breaking the samples.

Besides, the etched samples were coated with a thin layer of gold to make the samples conductive. In order to investigate the characteristic of the dispersed phases, approximately 100 domains were measured from each SEM micrograph using the MM-J software.

Mechanical testing

To characterize the mechanical properties of the blends and employed polymers, the tensile tests were carried out on the compression molded specimens using a universal testing machine Zwick 1456 according to the ISO 527, Section 2, Type 1BA, at cross head speed of 50 mm/min and modulus with speed of 1mm/min and the temperature of 240°c. The reported values are based on the average of at least five specimens.

Results and Discussion

Phase Morphology Analysis

Figs. 1 and 2 show that core-shell morphology in HDPE/PA-6/EVOH blends and matrix disperse morphology in binary blends, respectively.

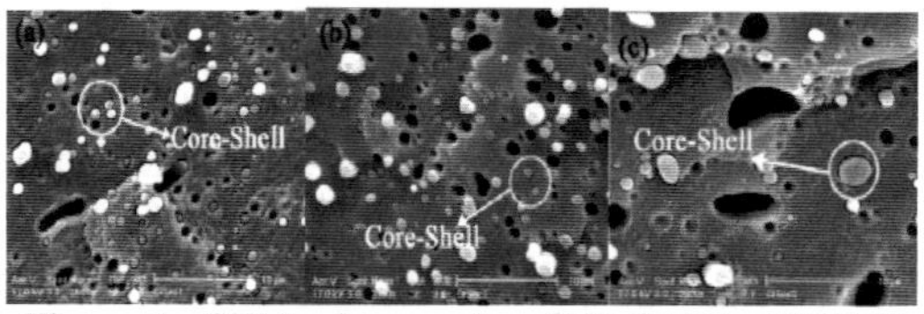

Figure 1 - SEM micrographs of HDPE/PA-6/EVOH blends after etching of EVOH (a) 80/5/15, (b) 80/10/10, (c) 80/15/5.

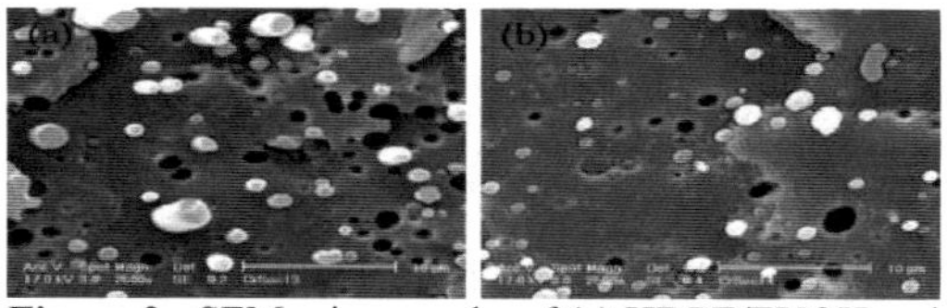

Figure 2 - SEM micrographs of (a) HDPE/EVOH and (b) HDPE/PA-6 binary blends.

It can be observed that the ternary blend compositions with higher content of PA-6 in dispersed phase show larger droplets and higher values of droplet average size. It should be noted that at higher contents of PA-6 and the average droplet size, as can be observed in the next section, this can substantially affect the mechanical properties of the blends.

Mechanical Properties Analysis

Mechanical properties are widely suited for analysis of multicomponent blends, through a comparison of experimental results and predictions based on various models. The application of various models gives insight into the properties of individual components. It also helps to check the assumptions regarding structure, mechanism and properties of the interface. Several theories have been proposed to predict the tensile properties in terms of various parameters. 'These theories can be classified into two categories: (i) based on composition and (ii) based on morphology. [13] Most of these theories assume perfect adhesion between the phases and a macroscopically homogenous and isotropic sample. The different models examined to predict the mechanical behavior of the blends include the Parallel and Series (Eqs.1, 2), Halpin-Tsai (Eq.4), Kerner-Nielsen (Eq.5), Takayanagi and Barentsen models (Eq.3). [8, 9, 14 -18]

$$(E = \phi_1 E_1 + \phi_2 E_2) \quad (1)$$

$$(E^{-1} = \phi_1 / E_1 + \phi_2 / E_2) \quad (2)$$

A combination of parallel and series as proposed by Takayanagi for two-dimensional and Afterward, Barentsen for three-dimensional geometries modified Takayanagi's model as follows:

$$E_a = E_m \frac{\lambda^2 E_d + (1-\lambda^2) E_m}{(1-\lambda)\lambda^2 E_d + (1-\lambda^2+\lambda^3) E_m} \quad (3)$$

$$E_b = (1-\lambda^2)E_m + \frac{\lambda^2 E_m E_d}{\lambda E_m + (1-\lambda)E_d}$$

Where:

$\phi_d = 1 - \phi_m = \lambda^3$, And E_m and E_d stand for modulus of matrix and dispersed phase, respectively. [9]

Halpin–Tsai equation is useful for modeling and determining the properties of polymer blends.

The model results allow a physical interpretation of the role of the dispersed phase in terms of the aspect ratio and of the stress concentration factors associated with the dispersed particles. [20]

$$\frac{E_1}{E} = \frac{(1 + A_i B_i \phi_2)}{(1 - B_i \phi_2)} \qquad (4)$$

In the Halpin–Tsai equation, subscripts 1 and 2 refer to the continuous and dispersed phases, respectively.

The constants A_i and B_i are defined by the morphology of the system. Mechanical properties of orthotropic lamina with uniaxially oriented fibers are usually calculated by the Halpin-Tsai equations (across the fibers).The latter equations are only formally analogous to those of Kerner and Nielsen for isotropic particulate composites. This model can be represented the mechanical behavior of the polymer blends. [8]

$$E_c = E_m\left(\frac{1 + A_f B_f \upsilon_f}{1 - P_f B_f \upsilon_f}\right) \qquad (5)$$

Where E_c is the modulus of the composite, E_m is the modulus of the matrix and υ f is the volume fraction of the filler. A_f, B_f, P_f are constant. [8]

The data used in modeling was sourced from our experimental measurements. The respective Poisson's ratios of the HDPE, PA6 and EVOH were 0.47, 0.45 and 0.38, respectively.

To calculate the modulus of the blends with core-shell morphology; first, the modulus of the isolated dispersed particles were calculated with EVOH as the matrix and PA6 as the dispersed phase, this value was then used as the dispersed phase modulus in a subsequent calculation with HDPE as the matrix.

Theoretical results for tensile modulus of blends by droplet-in-matrix models illustrated that, these models could not predicted the experimental result.

In other hand, the dominating particulate and fibrous composite models are not able to predict the tensile modulus of polymer blends. At the same time the amount calculated by Barentsen models which is a combination of Parallel and Series are not suitable.

Table 3 and Fig. 3 give comparisons between experimental results and the best theoretical predictions the modulus of binary and ternary blends (based on matrix-dispersed and core-shell morphology).

Should be noted, as seen in Table 3, only the binary blend of HDPE/EVOH with Parallel model is fitted .

In other blends, the experimental results are predicted with series model. But by increasing PA-6, the predictions were different from the values obtained by experiments.

Table 3 - comparison of experimental results and theoretical predictions the tensile modulus of binary and ternary blends

Sample Code	Model	**Theoretical Tensile Modulus (MPa)**	**Experimental Tensile Modulus (MPa)**
1	Parallel	**1408.82**	**1408.51**
2	Series	**1089.8**	**1084.29**
3	Series	**1076.76**	**1048.08**
4	Series	**1064.03**	**1020.67**
5	Series	**1052.1**	**1000.03**

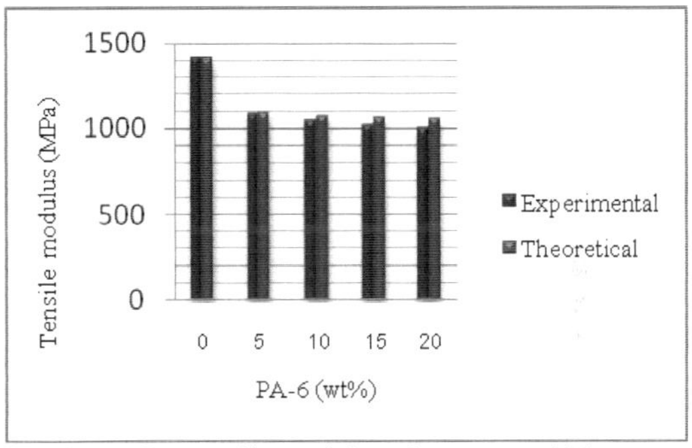

Figure 3 - Tensile modulus of ternary and binary blends

Further more, in order to have a better comprehension of assumption core-shell morphology it has to be seen that the experimental results are a suitable assumption or not. Therefore, it can be assumed that the dominant ternary morphological phase HDPE/PA-6/EVOH is not based on the core-shell morphology. Hence, with the usage of the previously reviewed models and assuming morphology two dispersed phase, the modulus of ternary blends can be predicted.

Whereas, the observed SEM and the result of predicted morphology was seen that the dominant ternary morphological system of HDPE/PA-6/EVOH are based on the core-shell morphology. But the predicted result derived from the mechanical models, on the assumption of two dispersed morphology, are extremely similar to the experimental results.

Therefore with respect to the above, relying only on the assumption of core-shell morphology and the usage of mechanical models would be insufficient.

Mechanical properties of the binary and ternary blends are summarized in Table 4.

As can be observed from Table 4 and fig. 1, by increasing the PA-6 to EVOH weight ratio, core forming droplets become bigger, whereas the shell thickness of EVOH diminishes slightly. Therefore, it is obvious that as the amount of PA-6 in all blends increased, the tensile modulus is decreased.

This is attributed to lower tensile modulus, tensile strength of PA-6 compared to that of EVOH.

The other factors affecting mechanical properties of blends include particle size and particle size distribution of dispersed phases, interfacial adhesion, phase separation and geometry of dispersed phases.

Table 4 - Mechanical properties of the binary and ternary blends

HDPE/PA-6/EVOH	Tensile Modulus (MPa)	Stress at break (MPa)	Strain at break (%)
80/0/20	1408.51	0.29	98.04
80/05/15	1084.29	0.63	141.24
80/10/10	1048.08	0.87	129.84
80/15/05	1020.67	8.97	64.60
80/20/0	1000.03	6.49	21.85

The effect of average particle size can be observed at high EVOH concentrations in which HDPE/PA-6/EVOH blends display higher elongation at break and lower values of strength at break (Table 4, Fig. 4).

Moreover, the results of image analysis technique regarding morphology of binary blends illustrates that number of measured particles EVOH lower than PA-6, and also measured values of radius for EVOH is higher than PA-6.

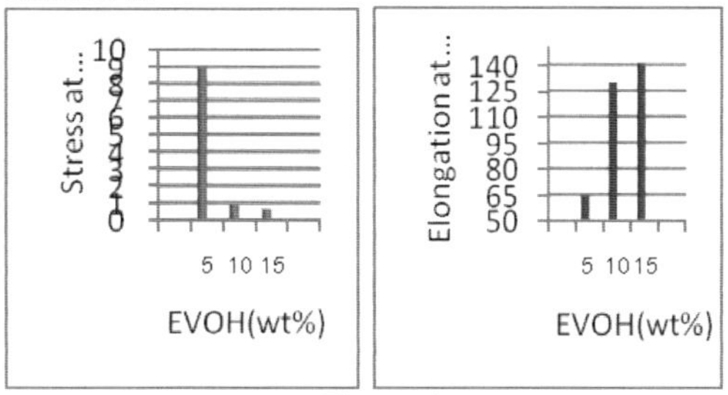

Figure 4

So, higher values of average particle size results in higher interfacial area and consequently higher modulus of elasticity in the system. This leads to better adhesion of EVOH to HDPE and therefore energy dissipation due to sufficient adhesion is increased.

Conclusions

Scanning electron microscopy showed core-shell type morphology, in which, PA-6 cores were engulfed by EVOH domains.

This work aims to predict the mechanical properties in ternary blends of HDPE/PA-6/EVOH by using modified mechanical models. The comparison of the models and experimental results shows that applied Series model for most blends were fitted to the experimental results. Whereas, the observed SEM indicates that the dominant ternary morphological system of HDPE/PA-6/EVOH are based on the core-shell morphology. But the predicted result derived from the mechanical models, on the assumption of two dispersed morphology, are extremely similar to the experimental results.

Therefore, relying only on the assumption of core-shell morphology and the usage of mechanical models would be insufficient.

Regarding the size of PA-6 core-forming constitutive as well as shell thickness of EVOH and interfacial adhesion between Components, it is found that the tensile modulus in all blends does not significantly change by composition of dispersed phase. Consequently, characterizes of matrix phase is of most importance parameter defining mechanical properties of blends.

References

1. H. F. Guo; S.Packirisamy *Polymer* 1997, *volume 38*, Issue 4,785.
2. T. Kojima; Y. Kikuchi; T. Inoue *Polym. Eng. Sci.* 1992, *32*, 1863.
3. A.K. Gupta; K.R. Srinivasan *J. Appl. Polym. Sci.* 1993, 167.
4. W. Zheng; Y. Leng; X. Zhu *Plast Rubber Compos Process Appl.* 1996, 490.
5. P. V. Van Puyvelde; P. Moldenaers, *Rheology Reviews* 2005, 101.
6. Chao Zhou et al. *Polym. Bull* 2009, *62*, 855.
7. S.Y. Fu et al. *Composites* 2008, Part B 39, 933.
8. Kolarik et al. *Polym. Adv. Technol.* 2000, *11,* 75.
9. H. Veenstra et al. *Polymer* 2000, *41*, 1817.
10. Colombini et al. *Macromolecules* 2001, *Volume 34*, 5916.
11. Colombini and Maurer *Macromolecules* 2002 *Volume 35*, No. 15, 5891.
12. Travasso et al. *J. Chem. Phys.* 2005, *122,*194906.
13. Z. Hashin *J. Appl. Mechan.* 1983, 481.
14. S. K et al. *Composites Science and Technology* 2007, *67,* 3071.
15. S.G. Lee; S.H. Kim *Polym Int.* 2003, *52*, 698.
16. R. A. Dickie *Appl. Polym. Sci* .1973, *17*, 45.
17. J.C. Halpin: J. L. Kardos *Polym. Eng. Sci.* 1976, 344.
18. R.M. Holsti; J.V. Seppala *Polym. Eng. Sci.* 1994, *34,*395.
19. S. George; N. R. Neelakantan; K. T. Varughese; S. Thomas *J. Polym. Sci.* 1997, *Part B 35*, 2309.
20. Miguel A. Lopez; J. Biagiotti; M. Arroyo; J.M. Kenny *Macromol.* 1999, *148,* 345

O-06-227

ELECTRON INDUCED REACTIVE PROCESSING: A PROMISING TECHNIQUE TO CREATE THERMOPLASTIC VULCANIZATES (TPVS)

U. Wagenknecht[1*], M. Mondal[1], U. Gohs[1], K. Naskar[1,2], R. Rajeshbabu[1,3], V. Thakur[1,4] and G. Heinrich[1]

[1] Leibniz-Institut für Polymerforschung Dresden e.V., Hohe Strasse 6, 01069 Dresden, Germany, wagenknt@ipfdd.de mailto:

[2] Rubber Technology Centre, IIT-Kharagpur-721302, West Bengal, India

[3] Apollo Tyres Limited, Limda, Vadodara-391760, Gujarat, India

[4] Lapp Engineering & Co., Hinterbergstrasse 15, CH-6330 Cham, Switzerland

Keywords: reactive processing, blend, high energy electron, thermoplastic vulcanizate

Introduction:

Electron irradiation technique is a well known technique for vulcanization in polymer industry. But this is only restricted to final product treatment. Here we take it to the next level by coupling a conventional internal mixer and a high energy electron accelerator. So the production of tailored material by using reactive processing is taken to one step ahead.

Experimental:

Figure 1 shows schematically the classical technology of reactive mixing based on addition of liquid or solid radical builders to the mixing partners. The radical generation is thermal induced. The control of this process is very limited caused by the inhomogeneous temperature field in the mixing chamber and the nonlinear building rate of radicals during the mixing period.

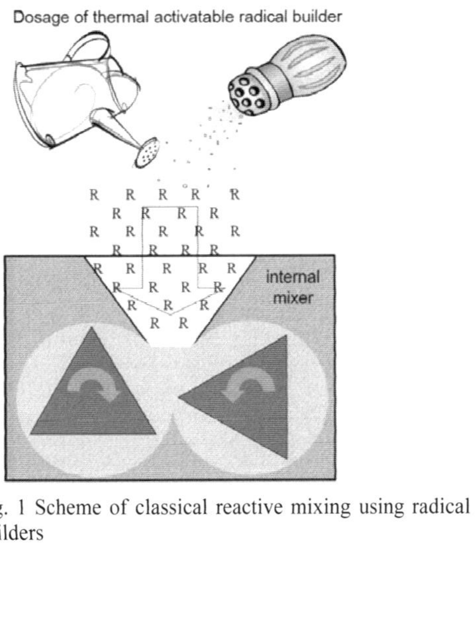

Fig. 1 Scheme of classical reactive mixing using radical builders

Irradiation with high energy electrons direct into the mixing chamber is an alternative way of radical generation during mixing. The main differences in relation to the classical process are:

- Additive free/environmental friendly
- Direct generation of reactive species at the polymer molecule
- Radical generation is independent from temperature and aggregate state
- Radical generation is exact controllable by time (Electron current)
- Radical generation is exact controllable by volume (Electron energy)

The process is schematically shown below.

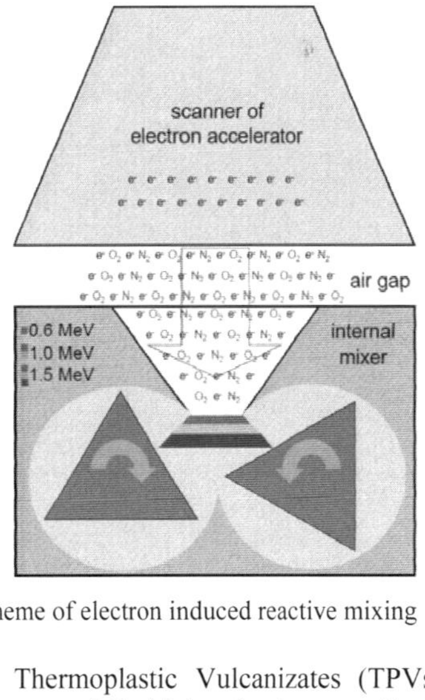

Fig. 2 Scheme of electron induced reactive mixing

Results: Thermoplastic Vulcanizates (TPVs) are itself a commercially high valued group of polymer blend. They render technological properties of conventional cross-linked rubber with the ease of thermoplastic melt processability.

The following TPVs

- polypropylene/ethylene propylene diene terpolymer (PP/EPDM) [1, 2]
- polypropylene/ethylene-octene copolymer (PP/EOC) [3,4]
- polypropylene/natural rubber (PP/NR) [5,6]

have been produced using this novel technology. Technological properties of these TPVs can be tailored according to the need of purpose by interplaying physical parameters of polymers and advanced high energy electron technology.
Stress-strain-curves (Fig. 3, 4, 5) of selected types of this three TPV's are shown as examples of the potential of the new technology.

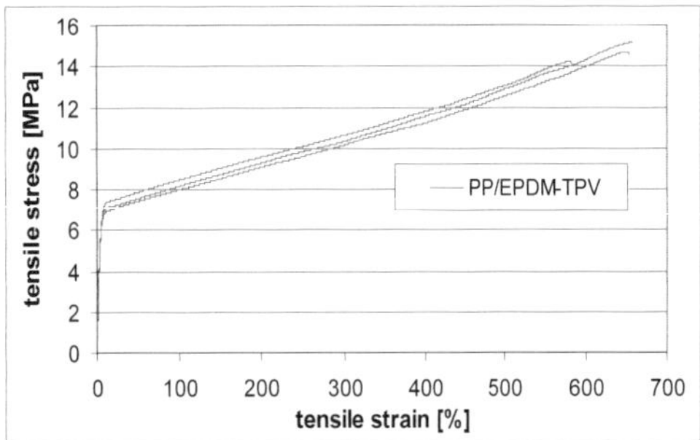

Fig. 3 Stress-strain-curve of a PP/EPDM-TPV

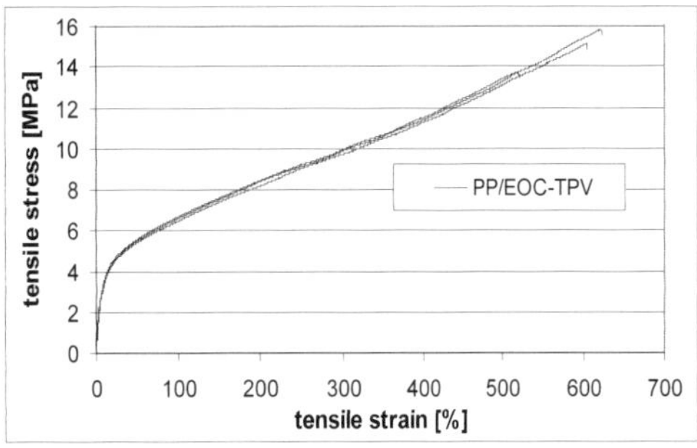

Fig. 4 Stress-strain-curve of a PP/EOC-TPV

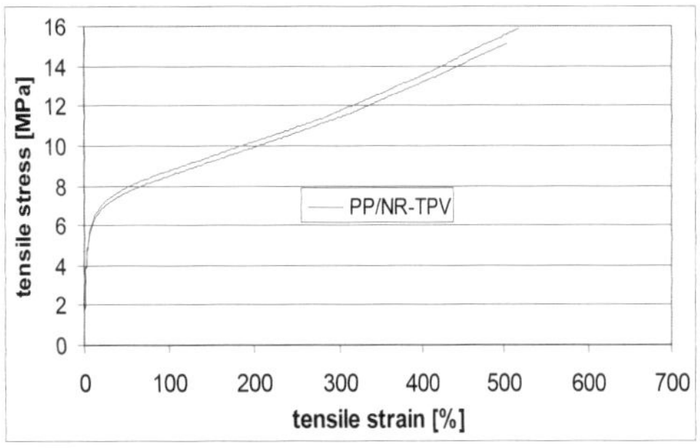

Fig. 5 Stress-strain-curve of a PP/NR-TPV

The size of the dispersed phase particles is one of the main factors determining the properties of a blend. In case of the PP/EPDM-TPV we found very polydisperse morphology. The smallest particles are about 30nm and the largest in µm range. Such small particles and the wide spread of particle size is unusual for polymer blends.

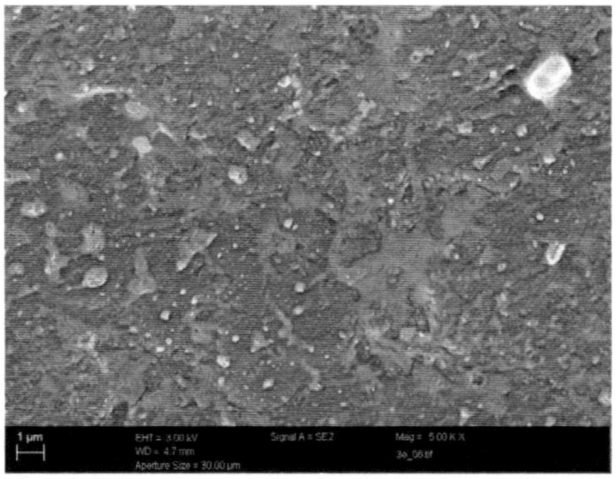

Fig. 6 SEM micrograph of a PP/EPDM-TPV made by the new technology

Conclusion:
Electron induced reactive processing enables

- radical generation independently from temperature as well as mixing conditions
- direct control of radical generation during melt mixing according to requirement of chemical reaction
- no use of reactive additives (→ food, medicine, biopolymers, life science)

The Electron Iinduced Reactive Processing is a promising way to high performance additive free thermoplastic vulcanizates

References:

1. K. K. Naskar et al. eXPRESS Polymer Letters 3 (2009) 677-683
2. V. Thakur et al. Polymer Journal 44 (2012) 439-448
3. R. Rajeshbabu et al., Rad. Phys. and Chem. 80 (2011) 1398-1405
4. R. Rajeshbabu et al., Macromol. Mater. Eng. 2011, DOI: 10.1002/mame.201100209
5. M. Mondal et al., Rad. Phys. and Chem. in press
6. M. Mondal et al., Polymer Journal in press

O-06-283

IMMOBILISATION OF POLYELECTROLYTE STABILIZED GOLD NANOPARTICLES ON A POLYMER SURFACE DURING MOULDING

F. Kroschwald[1]*, J. Nagel[2], C. Zimmerer[2], B. Voit[3] and G. Heinrich[3]

[1]*Dipl. chem., Leibniz Institut für Polymerforschung Dresden e.V., Germany Hohe Straße 6, 01069 Dresden– kroschwald@ipfdd.de*

[2]*Dr. rer. nat.; Leibniz Institut für Polymerforschung Dresden e.V., Germany – nagel@ipfdd.de; zimmerer@ipfdd.de*

[3]*Dr. rer. nat. habil.; Leibniz Institut für Polymerforschung Dresden e.V. Germany – voit@ipfdd.de; gheinrich@ipfdd.de*

Abstract – We synthesized gold nanoparticles in polyethylenimine with special properties like resistance against pH changing. We coated substrates via spin-coating with this solution and transferred the modifier layer to polycarbonate surface via injection moulding. The formed part had a red layer on top of the surface which indicated a complete transfer of the modifier. The created modifier layer was up to 800 nm thick.

Keywords: injection moulding, gold nanoparticles, surface modification, polyethylenimine.

Introduction

Gold nanoparticles are characterised by their high surface area and specific plasmon resonance. They are used for example as biosensors, in optoelectronics, etc[1] [2].

Experimental

Materials and Methods

Chloroauric acid ($HAuCl_4x3H_2O$; Fa. XYZ, USA), sodium citrate (Japan), Polyethylenimine (M_W=750.000; 50 Ma% H_2O; Germany), NaCl (Denmark) wurden were purchased from Aldrich and used as received. HCl (36%; Germany) were purchased from Merck. As solvent we used Milli-Q-Water with resistivities under 0,055 µS/cm. For injection moulding we used polycarbonate Makrolon LED 2025 (Bayer Material Science AG, Germany).

Synthesis of modifier

The synthesis of gold nanoparticles occurred in analogy to the method of *Turkevitch* and *Frens*[2]. In this special type, we reduced chlorauric acid with sodium citrate under reflux in an aqueous polyethylenimine (PEI) solution. The gold concentration in the solution ranged from 0.25 to 2 mM.

Results and Discussion

We made nanoparticles with diameters between 10 and 100 nm, depending on the synthesis conditions, stabilised with PEI. For this contribution we used nanoparticles with diameters of nearly 20 nm according to DLS and a maximum U.V.-Vis. absorbance of λ_{max}=525 nm.

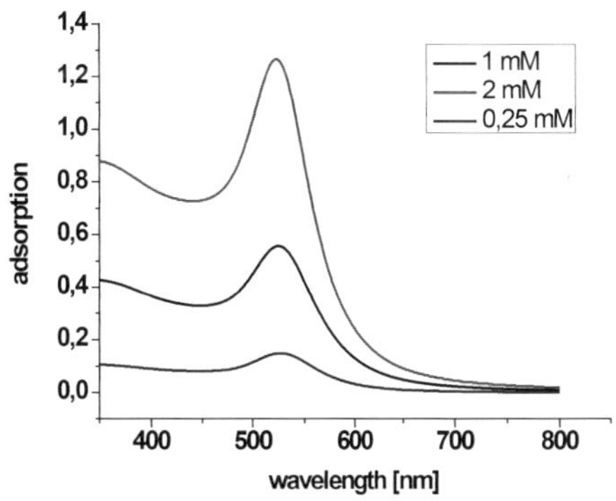

Abb. 1: UV-Vis spectra of AuNP/PEI solution with different concentrations. Maximum absorbance at wavelength of 525 nm was caused by dielectric environment of PEI.

The water was completely removed from the solution, and afterwards completely redispersed in water. Thereby the monodispers distribution was preserved. The particles were stable at pH values from 1 to 12. They had an isoelectric point at about 11. We also synthesized solutions with concentration of 4 mM, but these samples couldn't be dried and redispersed without changing properties. We found a shift of maximum absorbance in optical spectra from 525 nm to 650 nm, witch was an indicator for the formation of aggregates in the high concentrated solution. The concentration for the following experiments was 0,25 mM, 1 mM and 2 mM.

After the synthesis we prepared a modifier layer of PEI and gold nanoparticles on a mirror-polished steel substrate via spin-coating. Coating is also possible via dip-coating and spray-coating of the solution with different viscosities. The coated substrate was placed in the mould of an injection moulding machine and the modifier was transferred on a polycarbonate part by surface reactive injection moulding[4]. The melt temperature was 300°C, the mould temperature was 80°C. The formed part had a red layer on top of the surface which indicated a complete transfer of the modifier. After washing, the modifier layer thickness was reduced up to 25 nm of PEI, containing gold nanoparticles. No thickness reduction during washing was observed, if the layer was thermally treated at 110°C or chemically treated with alcoholic Diethylcarbonate solution prior to the washing step. The modifier layer was of up to 800 nm thick in those cases. It was permanently crosslinked and bonded to the polycarbonate surface. The characterisation occurred with scanning electron microscopy (SEM), atomic force microscopy (AFM) transmission electron microscopy (TEM), ultraviolet-visible spectroscopy (U.V.-Vis.) and attenuated total reflexion infrared spectroscopy (ATR-IR). SEM micrographs showed statistically distributed gold nanoparticles on the surface, measured with high accelerating voltage. With low accelerating voltage there were no nanoparticles visible in SEM micrograph. With U.V.-Vis. measurements we found no typical absorption band of agglomerated nanoparticles at λ=600 nm, so we proved the monodispersity of the particles. TEM micrographs of thin slice of modified polycarbonate parts confirmed the random and monodisperse distribution of the particles also in normal direction to the layer plane.

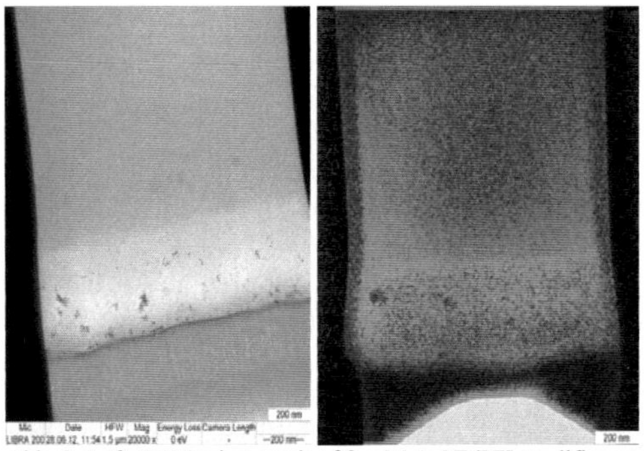
Abb. 2: Left: TEM micrograph of 2 mM AuNP/PEI modifier on PC. The lamella was prepared using focussed ion beam and was sputtered with carbon and coated with epoxy resin before cutting. Right: EFTEM micrograph of 2 mM AuNP/PEI modifier on PC. Red dots indicate oxygen, green dots stand for nitrogen.

Single particles in clusters within PEI environment had large distances to each other, so that plasmon waves around each particle did not interact. The main part by volume of the modifier layer was PEI. The Dark layer under the modifier was sputtered carbon and silica.
AFM micrographs showed that the nanoparticles are mostly embedded in the PEI layer underneath the surface. The ATR-IR spectra verified the successful modification and showed typical PEI bands next to PC bands.

Conclusions
With the presented method it is possible to immobilise nanoparticles permanently on a surface during the injection moulding process. The formed layer could be used directly for applications like as a laser adsorbent for selective heating, as seed crystals for chemical metallisation or as nanosized catalysts.

Acknowledgements
The funding of the work from the Deutsche Forschungsgemeinschaft under the reference number HE 4466/17-1 is greatly acknowledged. The authors thank Mr. Andreas Janke for taking the TEM and AFM micrographs and Mrs. Anja Caspari for the DLS measurements.

References
1. C. N. Ramachandra Rao, Giridhar U. Kulkarni, P. John Thomasa, Peter P. Edwards, *Metal nanoparticles and their assemblies*, Chem. Soc. Rev., **2000**, 29, 27-35
2. R. D. Averitt, D. Sarkar, N. J. Halas, "Plasmon Resonance Shifts of Au-Coated Au_2S Nanoshells: Insight into Multicomponent Nanoparticle Growth", Phys. Rev. Lett. **1997**, 78, 4217-4220
3. J. Turkevich, P. C. Stevenson, J. Hillier, "A study of the nucleation and growth processes in the synthesis of colloidal gold", Discuss. Faraday. Soc. **1951**, 11, 55-75
4. Norbert Herrlich, Johannes Kunz, Wolfgang Land, Walter Michaeli, *Kunststoffpraxis: Konstruktion, Band 2*, WEKA Media GmbH & Co. KG, Kissingen, **2004**

O-06-012

STRUCTURE-PROPERTY RELATIONSHIP OF DYNAMICALLY CROSS-LINKED PP/EOC AND PP/EPDM BLENDS (TPVS): A COMPARATIVE STUDY

R. R. Babu[1] **and K. Naskar**[2]*

[1] *Apollo Tyres Ltd, Vadodara, India – rajesh.babu@apollotyres.com*

[2] *Rubber Technology Centre, IIT Kharagpur, Kharagpur -721302, India – knaskar@rtc.iitkgp.ernet.in*

Abstract - Thermoplastic elastomers (TPEs) are one of the fastest growing polymeric materials which combine the elastic and mechanical properties of cross-linked rubbers with the melt processability of thermoplastics at elevated temperatures. TPE has its ability to reuse and recycle the production scrap and waste. Thermoplastic vulcanizates (TPVs) are a special class of TPEs, produced by simultaneously mixing and cross-linking a rubber with a thermoplastic at elevated temperature. As a result a typical morphology is formed, where the cross-linked rubber particles are finely dispersed in a continuous matrix of thermoplastic. The size of the dispersed rubber particles and their distribution in the matrix primarily govern the performance of TPVs. TPVs based on polypropylene (PP) with ethylene octene copolymer (EOC) and ethylene propylene diene rubber (EPDM) have been developed by coagent assisted dicumyl peroxide crosslinking system. The study was pursued to explore the influence of two dissimilar polyolefin polymers (EOC and EPDM) having different molecular architectures on the state and mode of dispersion of the blend components and their effects with special reference to morphological, thermal and mechanical characteristics. The effects of dynamic crosslinking of the PP/EOC and PP/EPDM have been compared by varying the concentration of crosslinking agent and ratio of blend components. The results suggested that the uncrosslinked and dynamically crosslinked blends of PP/EOC exhibit superior mechanical properties over PP/EPDM blends. From the hystersis experiments it was found that PP/EOC blends also perform better fatigue properties over PP/EPDM based blends.

Keywords: polymer blends, polypropylene, ethylene octene copolymer, ethylene propylene diene rubber, morphology

Introduction

Blending of polymers is often an important route for obtaining new and improved polymeric materials, which are difficult to obtain by direct polymerization process [1]. In addition to the family of synthesized thermoplastic elastomers (TPEs), advances in the rubber-plastic blends lead to the development of new avenues to prepare TPEs with tailor made properties by varying the blend ratio of the components. Thermoplastic vulcanizates (TPVs) are special class of TPEs prepared by dynamic vulcanization process which involves the crosslinking of rubber part while it is being mixed with the molten thermoplastic at high shear rate [2-5]. Amongst the various blends, polypropylene (PP)/ethylene propylene diene rubber (EPDM) blends have gained tremendous importance due to commercial interest because of the structural compatibility. A large number of reports have been published on the PP/EPDM uncrosslinked and dynamically crosslinked blends [5]. Recently, new families of ethylene-alpha olefin copolymers have gained significant attention due to its unique molecular architecture with narrow molecular weight distribution, uniform co-monomer distribution and long chain branching along the polymer backbone. Ethylene octene copolymer (EOC) in particular has been shown to provide better toughening characteristics to PP, when compared with the conventionally used EPDM rubber. Presently, EOC is used for the impact modification of PP for various applications especially in the automotive sectors. Moreover, these materials are available in the form of pellets that help in easy handling, faster mixing and compounding. Nevertheless, reports on dynamically vulcanized PP/EOC TPVs are scanty. In case of TPVs, in addition to the constituent polymers, type of curing agent plays a major role in determining the final performance of the TPVs. Among several crosslinking systems used in the preparation of PP/EPDM TPVs, phenolic resin gained more commercial success due to its crosslinking and compatiblizing tendency. However, in this particular PP/EOC blend system phenolic resin is ineffective, since the latter needs the presence of unsaturation to form a crosslinked network structure. Furthermore, use of phenolic resin in PP/EPDM blends causes black specks in the products, which motivates the development of other crosslinking systems. In that case, peroxides can be used which crosslink both saturated and unsaturated polymers without any reversion characteristics. However, the activity of peroxide depends on the type of polymer and presence of other ingredients in the system. It has been well established that PP exhibits β-chain scission (degradation) with the addition of peroxide. Hence the use of peroxide alone is limited in the preparation of PP based TPVs. In general, to minimize the unwanted side reactions, coagent assisted peroxide crosslinking system is recommended. Naskar et al. studied the influence of different peroxides and multifunctional peroxide (having peroxide and coagent functionality in a single molecule) in the conventional PP/EPDM blends [5]. Recently, Naskar and coworkers involved in the development of a novel PP/EPDM TPVs by electron beam induced reactive processing.

The present study primarily aims to evaluate the performance of the coagent assisted peroxide cured PP/EOC TPVs as a potential alternative to the conventional PP/EPDM TPVs. In this work, the morphological evolution and its influence on the mechanical and thermal properties have been investigated for both PP/EOC and PP/EPDM uncrosslinked and dynamically crosslinked blends. Detailed comparisons were pursued on morphological, thermal and mechanical properties. Investigation had been made to correlate the observed results with the microstructural architecture of the constituent polymers.

Experimental

Materials

Isotactic polypropylene (PP), ethylene-octene copolymers (EOC) and ethylene propylene diene rubber (EPDM) were used as blend components. Metallocene catalyzed ethylene octene copolymers (Exact® 5171) were employed in this study. The octene content of these Exact copolymers was determined from Proton Nuclear Magnetic Resonance Spectroscopy (^{1}H-NMR) measurement. The polymers taken for comparison are EOC and EPDM which were selected on the basis of number average molecular weight that lies relatively in the same range. Dicumyl peroxide (DCP) (Perkadox-BC-40B-PD) having active peroxide content of 40%; temperature at which half life time ($t_{1/2}$) is 1 hour at 138°C; specific gravity of 1.53 g/cm^3 at 23°C) was obtained from Akzo Nobel Chemical Company, The Netherlands. N-N'-m-phenylene dimaleimide (MPDM) (SR 525; specific gravity, 1.44 g/cm^3 at 23°C) was obtained from Sartomer Company, USA.

Preparation of TPVs

TPVs were prepared by melt mixing of PP with EOC/EPDM (rubbery material) in a Haake Rheomix 600s with a mixing chamber volume of 85 cm^3 at a temperature of 180°C. Dynamic vulcanization was pursued by adding co-agent assisted peroxide for 2 minutes. Mixing was continued for another 4 minutes to complete the vulcanization. After mixing, the blends were removed from the chamber at hot condition and sheeted out on a two roll mill in a single pass. Sheets were then cut and pressed in a compression molding machine (Moore Press, Birmingham, UK) at 190°C for 4 mins at 5 MPa pressure. Aluminum foils were placed between the mold plates. The molded sheets were then cooled down to room temperature under the same pressure.

Testing procedures

Morphology – Morphology studies were carried out using a scanning electron microscope (SEM) (JEOL JSM 5800, Japan). Molded samples of PP/EOC or PP/EPDM TPVs were cryofractured in liquid nitrogen to avoid any possibility of phase deformation during the fracture process. In case of uncrosslinked blends, the EOC or EPDM phase was preferentially extracted by treating with hot xylene at 50°C for 15 min, whereas in dynamically vulcanized blends the PP phase was preferentially and partially extracted by etching with hot xylene at 100°C for 45 min. The samples were then dried in a vacuum oven at 70°C for 5 hours to remove the traces of solvents present. Treated surfaces were then sputtered with gold before examination.

Differential Scanning Calorimetry (DSC) - DSC measurements were carried out in a TA instrument, USA (Model DSC Q100 V 8.1) to study the melting behavior. The samples of about 6 mg sealed in aluminum pans were heated from -80 °C to 200 °C at a scanning rate of 10 °C/min under N_2 atmosphere.

Mechanical Testing - The dumb-bell shaped specimens of the TPV used for testing were die cut from the compression molded sheet and the testing was done after 24 hours of maturation at room temperature. Tensile strength was measured according to ASTM D418-98A using a universal testing machine Hounsfield H10KS (UK) at a constant cross-head speed of 500 mm/min. Tension set was performed at room temperature with a stretching condition for 10 minutes at 100% elongation according to the ASTM D412-98 method.

Cyclic Tensile Testing - The tensile testing specimen was used to perform the hysteresis and fatigue life testing of the samples. The loading and unloading cross-head speed of 500 mm/min was maintained. All the values were measured at the room temperature (25 °C) in air without any pretreatment. The hysteresis measurements were done in two modes i.e, constant strain amplitude and step wise increase in strain amplitude.

Results and Discussion

Morphology

Fig. 1 shows the SEM photomicrographs of 50/50 blend ratio of PP/EOC and PP/EPDM uncrosslinked and dynamically crosslinked blends. In uncrosslinked blends, rubbery phase EOC or EPDM was preferentially extracted and in dynamically crosslinked blends PP phase was preferentially extracted. Both the uncrosslinked compound (YC and YD) results in the formation of co-continuous phase morphology. Depending on the viscosity ratio, various levels of co-continuity exist in the blend compounds. It is clear from the Fig. 1a and b that the PP/EOC (YC) contributes to have higher percentage of co-continuity than PP/EPDM (YD) blend. Because of higher viscosity ratio in YD, EPDM prefers to exist in dispersed phase. However by dynamic vulcanization, a complete phase transition (inversion) occurs i.e, co-continuous morphology is changed into dispersed phase morphology. The less viscous PP encapsulates the more viscous crosslinked rubber particles to minimize the mixing energy. At equal curative dosage of 2 phr DCP and compositional volume, PP/EPDM TPV shows coarser crosslinked rubber particle size than PP/EOC TPV (Figure 1c and d). By increasing the concentration of peroxide (4 phr) in order to increase the crosslink density of rubber phase, two opposite behaviors have been accounted in the particle size evolution for PP/EOC and PP/EPDM blend. Crosslinked rubber particles size increases in PP/EOC blends, whereas decreases in PP/EPDM blend as a function of peroxide concentration (Fig. 1e and f). It is well known that, PP undergoes chain-scission with the addition of peroxide. In case of PP/EOC blends when the concentration of peroxide increases from 2 phr to 4 phr, extent of degradation in the PP phase dominates over crosslinking in the EOC phase. Therefore, it becomes too difficult for a relatively very low viscosity PP matrix (low molecular weight PP) to break-up the crosslinked EOC rubber. But in case of PP/EPDM blends, in addition to the PP-matrix chain-scission, the inherent PP segments in the EPDM also undergo chain-scission which cause to form low molecular weight compound. These low molecular weight compounds have higher tendency towards fine dispersion by lowering the viscosity ratio.

Thermal properties

DSC measurements were performed to characterize the crystallization and melting behavior of uncrosslinked and dynamically vulcanized blends. Fig. 2 demonstrates the relevant thermal parameters as obtained from the DSC scans. It can be seen that neat PP exhibits a characteristic feature of semi-crystalline polymer with glass transition (T_g), melting temperature (T_m) and crystallization temperature (T_c) at around 1, 165 and 119 °C, respectively. Pristine EOC shows a strong β-transition (step in DSC heat capacity traces) due to its movement of octene branch points at -50 °C. On contrary, EPDM shows a transition at -45 °C due to the relaxation of short chain branch points. But, PP does not show a significant change in β-transition in the DSC traces due to its semi-crystalline characteristics. Both EOC and EPDM show the occurrence of T_m and T_c, suggesting the presence of crystalline phase. It is expected that, both EOC and EPDM seem to have long enough ethylene blocks, which are easy to crystallize in a usual manner. However, depending on its inherent molecular architecture T_c of EOC and EPDM are 45 °C and 20°C respectively.

At 50/50 rubber/plastic blend ratio (YC and YD), T_m, T_c and X_c found to vary significantly for both PP as well as for rubbery component. The possible inter-diffusion and inter-molecular mixing reduces the size and perfectness of crystals and consequently inhibits the crystallization process on either component as also reported by Bielinski. The results, therefore, suggest that PP blended with EOC or EPDM needs higher undercooling to crystallize (i.e., rubbery components may depress the crystallization of PP) and correspondingly decrease the T_c. It is worth noting that the addition of rubbery component broadens the melting endotherms of PP phase, which is probably due to the changes in the distribution of crystal lamella thickness.

Upon dynamic vulcanization, significant changes in the thermal properties were observed in both PP and rubbery phases. Apparently, as the peroxide dosage increases from 0 to 4 phr in PP/EOC blend, T_g marginally increases, T_m as well as X_c decrease continuously in the EOC phase due to low degree of chain alignment resulting from predominant crosslinking. As for PP matrix phase, variation in the T_g is marginal whereas T_m and X_c continuously decrease as the curative concentration increases. A similar trend is found for the PP/EPDM TPVs. It is very clear from the Table 4, the extent of decrease in X_c of PP is more pronounced for PP/EOC TPVs as a function of curative concentration i.e., X_c of PP decreased from 46.6% to 32.4% with addition of 4 phr of peroxide. Bulk crystallization temperatures of both the components were found to vary significantly by dynamic vulcanization. As for the rubbery phase, $T_{c\,(EOC)}$ was found to decrease continuously in PP/EOC TPVs. This may be associated with the prominent crosslinking reaction in the EOC phase which needs higher under-cooling to crystallize. Whereas in case of PP/EPDM TPVs, EPDM undergoes simultaneously crosslinking in the ethylene sequence and chain-scission in the propylene sequence. The combined effect of the above mentioned two competing reactions inconsistently affect the crystallization process of EPDM phase. As the peroxide dosage increases from 0 to 4 phr (with in the given concentration limit), the $T_{c\,(PP)}$ was found to increase and then decrease for PP/EOC based TPVs. At 4 phr concentration of peroxide, chain scission in the PP phase dominates over the crosslinking effect in the EOC phase. Therefore, the lower molecular weight PP chains have less under-cooling effect to crystallize. But in case of PP/EPDM TPVs, $T_{c(PP)}$ initially increases and then gets stabilized, which is mainly due to the competition between the chain coupling reaction in the polyethylene segments and chain scission reaction in the polypropylene segments in the EPDM phase.

Mechanical properties

When comparing the behavior of PP/EOC and PP/EPDM samples, in all the cases EOC based uncrosslinked and dynamically vulcanized blends exhibit better physical properties than PP/EPDM samples at identical composition. Since EPDM is an ethylene propylene diene random terpolymer, part of the propylene units undergoes chain-scission or degradation while using peroxide as curative. During dynamic vulcanization, as the concentration of curative increases from 0 to 4 phr, a higher level of degradation in the PP phase is coupled with the chain-scission in the PP segments in the EPDM polymer in PP/EPDM blends. Thereby PP/EPDM TPVs lose their mechanical strength despite containing the finest vulcanized rubber particles. Therefore, it is expected that dynamically vulcanized blends of PP/EPDM depict inferior physico-mechanical properties than PP/EOC TPVs. From the DSC and SEM analyses, it was found that EOC containing small crystals have good affinity and finely dispersed particles in the PP matrix phase. On the other hand, EPDM itself contains small crystals but its crystallinity might be too low (due to high degree of chain entanglement density) to form an effective reinforcing characteristics. Although mechanical properties are found to be superior for PP/EOC blends, set properties are better for PP/EPDM blends.

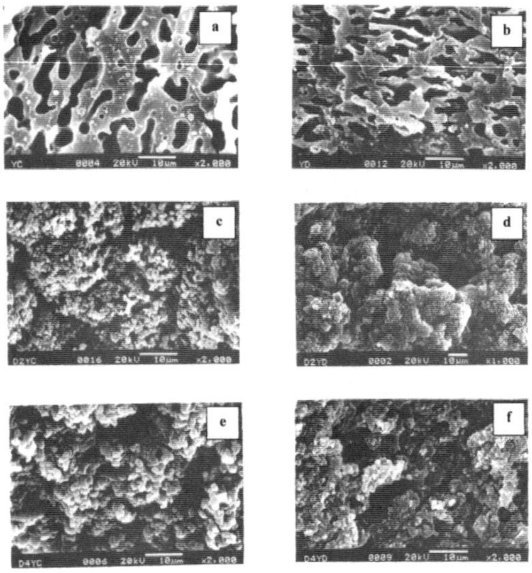

Fig. 1 – SEM photomicrographs of various blends

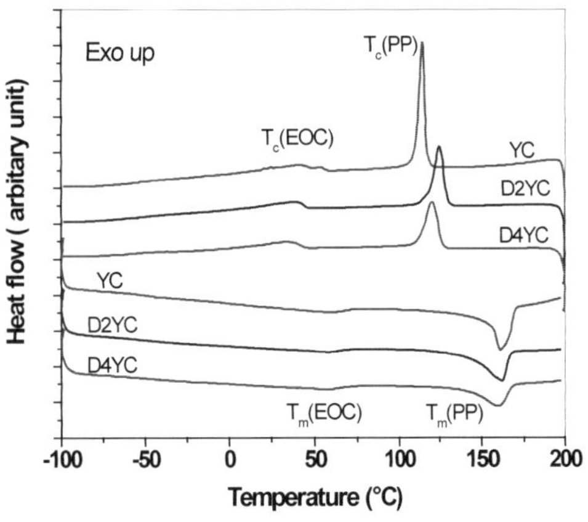

Fig. 2 – DSC thermograms of various uncrosslinked and crosslinked blends

Conclusions

The morphological, thermal and mechanical results are interpreted in terms of the balance between the crosslinking in the EOC and EPDM phases versus degradation in the PP phase. The morphology of PP/EOC and PP/EPDM blends indicate a two phase morphology in which the rubber phase is dispersed in continuous PP matrix at lower rubber/plastic blend ratio and followed by phase inversion with increase in rubber/plastic blend ratio. The 50/50 composition of rubber/plastic shows co-continuous phase morphology. However, after dynamic vulcanization all the samples exhibit a dispersed phase morphology in which crosslinked rubber particles are dispersed in continuous PP matrix. The dispersion characteristics of the dispersed crosslinked rubber phase is mainly influenced by the melt viscosity ratio, shear rate, curative dosage and the rubber content. When compared to the conventional PP/EPDM TPVs, PP/EOC TPVs show lower viscosity and elasticity in the melt state and better physico-mechanical properties in the solid state. Hystersis measurements have been found to be a useful tool to evaluate the structural changes during dynamic deformation and fatigue analysis. Dynamically vulcanized 50/50 rubber/ plastic blend shows lower energy loss and better set properties.

In short, PP/EOC shows better mechanical, and thermal properties which may be due to better interfacial interaction and unique molecular microstructure of the EOC phase. In addition to lower cost, EOC are supplied in free pellets which allow for both batch and continuous production of TPVs. So, peroxide cured PP/EOC TPVs combine better processing characteristics with superior performance in the solid state. Hence, peroxide cured PP/EOC can be considered as a potential alternative to the conventional PP/EPDM TPVs.

References

1. D. R. Paul; C. B. Bucknall, *Polymer Blends,* John Wiley & Sons, New York, 2000.
2. J. Karger Kocsis, *Polymer Blends and alloys,* Marcel Dekker, New York, 1999.
3. S. K. De; A. K. Bhowmick, *Thermoplastic Elastomers from Rubber Plastic Blends,* Horwood, London, 1990.
4. A. Y. Coran; R. Patel *Rub. Chem. Technol* 1982, *55,* 116.
5. K. Naskar *Rub. Chem. Technol* 2007, *80,* 504.

O-06-368

CO_2 ASSISTED BLENDING OF BIODEGRADABLE POLYESTERS

S.H. Murphy[1*], G.A. Leeke[2], C.A. Kelly[1] and M.J. Jenkins[1].

[1] *School of Metallurgy and Materials, University of Birmingham, Birmingham, B15 2TT – sxm517@bham.ac.uk, c.a.kelly@bham.ac,uk, m.j.jenkins@bham.ac.uk*

[2] *School of Chemical Engineering, University of Birmingham, Birmingham, B15 2TT – g.a.leeke@bham.ac.uk*

Abstract

The aim of this study is to encourage miscibility between two normally immiscible polymers, polycaprolactone (PCL) and polylactic acid (PLA), with the aid of CO_2. Initial investigations into the blending of numerous grades of polylactic acid (PLA 3051D, 2002D 4060D and 3801X) with polycaprolactone (PCL 6800) have been undertaken. Blends were prepared by a variety of different methods; solution casting with dichloromethane, melt blending with a single screw extruder and CO_2 assisted extrusion. Blends of 0/100, 25/75, 50/50, 75/25 and 100/0 of PLA/PCL were prepared using each method. Analysis of all blends were performed by environmental scanning electron microscope (ESEM) and Instron. Blends created by solvent casting with dichloromethane were found to be immiscible, however some transcrystallinity was present with blends containing PLA 3801X. Those produced by extrusion showed a dramatic improvement in mechanical properties over those produced by solvent casting. Initial ESEM analysis of CO_2 assisted extruder blends shows a significant reduction in the number and size of PCL droplets in a PLA matrix. However, a noticeable porous structure was observed. Initial experimentation of CO_2 assisted extrusion indicate that by injecting CO_2 into the polymer melt the polymer is plasticised, reducing the polymer viscosity and encouraging increased molecular interaction of the polymer chains. Consequently, it is able to improve the adhesion between phases at the microscale.

Keywords: Polylactic acid, Polycaprolactone, blends, carbon dioxide, extruder.

Introduction

Biodegradable polymers are becoming increasingly popular as their applications broaden into the biomedical and food packaging industries. Among these biopolymers, lactic acid based polymers (PLAs) are extensively used in various medical applications; such as for drug delivery and as scaffolds in tissue engineering. Additionally, there has been growing interest in the use of PLA as food packaging as PLA is derived from natural resources and can be composted as an alternative to landfill. With this in mind, PLA is set to play a large role in many future applications. It has been reported that PLA is relatively brittle in nature [1] and therefore has limited applications. In recent years, there has been research into blending PLA with other biodegradable polymers in order to improve its mechanical properties.

Blending enables a soft, elastomeric second polymer to be introduced into a brittle continuous phase as an approach to strengthen brittle polymers. The softer polymer acts as a stress concentrator, preventing brittle failure by enabling ductile yielding [2]. Polycaprolactone is a polymer possessing a high elongation at break and low tensile modulus, making it a good candidate to blend with PLA [3]. However, previous research with PLA and PCL has produced immiscible blends [4]. A solution to this problem is to use CO_2 which can act as a plasticiser, reducing polymer viscosity [5] and melting temperature [6]. This enhances mixing between the two polymers, encouraging miscibility. An additional advantage of using CO_2 is that it enables extrusion to be carried out at lower temperatures, preventing degradation.

Experimental

Materials

Polylactic acid (PLA) grades; 3051D, 2002D, 4060D and 3801X in the form of pellets were supplied by NatureWorks. Polycaprolcatone (PCL) grade CAPA 6800 was supplied by Perstorp carplocatones. CO2 (purity 99.9 %v/v) was obtained from BOC and used as received.

Method

Blends were prepared by a variety of different methods; solution casting with dichloromethane, melt blending with a single screw extruder and CO_2 assisted blending. Compositions of 0/100, 25/75, 50/50, 75/25 and 100/0 of PLA/PCL were prepared using each method.

Production of Solvent Cast Blends

Solution cast films were prepared and evaporated under a fume hood at constant air flow. Dicholoromethane was chosen as a common solvent to dissolve both polylactic acid and polycaprolactone. The polymer (4 g) was dissolved in dicholoromethane (40 ml) to give a 10 wt% solution. The mixture was covered and stirred for 2 hours to ensure complete dissolution of the polymer. Films were cast into petri dishes, 9cm in diameter, covered with glass lids and left to slowly evaporate for about 4 days.

Production of Melt Blends via Extrusion

Prior to extrusion, the PLA grades were subjected to drying to remove any residual moisture. Drying temperatures and times were obtained from NatureWork's processing data sheets. PLA 4060D and 3051D were subjected to 4 hours at 45 °C and 100 °C, respectively. PLA 2002D for 2 hours at 90 °C and 3801X for 3 hours at 80 °C.

Blends produced via extrusion were created in a Rondol Linear 25 mm rotating single-screw extruder with a L/D ratio of 25:1. Experiments were conducted with various screw speeds (10-40 rpm) and barrel and die temperatures set to 160 °C and 170 °C, respectively. Due to the higher melting temperature of PLA 3801X, the processing conditons for these blends were with the barrel temperature at 170 °C and the die at 180 °C. Samples were extruded through a 1 mm slit die and passed through cooled rollers onto a conveyor belt.

Production of CO_2 Assisted Blends

Blends were produced by CO_2 with the same extruder as discussed above. CO_2 was injected into the barrel via an unique design of inlets and ports in zones 2 and 3. Various CO_2 wts% (1, 5, 10 and 20%) were investigated at screw speeds of 20 and 30 rpm. The CO_2 flow rate in ml/min was calculated as a wt% of the measured polymer mass extruded per minute. CO_2 behaves as a plasticiser allowing the blends to be produced at lower temperatures, in this case, 150 °C in the barrel with the die temperature at 160 °C.

Analysis of Blends

Samples analysed by ESEM were prepared by submersing in liquid nitrogen in order to create a fracture surface. Observations of fracture surface morphology allowed analysis of the blends at a microscopic level. Tensile testing of the blends was performed using a 5566 Instron machine interfaced to a PC using merlin software. The solution cast films were cut into a standard 25 mm (L_o) dog bone shape. Width was measured as 4 mm and the thickness calculated as the average of three areas along the length. The strain rate of 10 mm/min was also used. Mechanical testing using an Instron allowed calculation and comparison of Young's modulus, elongation to break and ultimate tensile strength.

Results and Discussion

Blends created by solvent casting with dichloromethane were generally found to be immiscible with relatively large droplets (approximately 5 μm) observed on the polymer fracture surface (figure 1). However, some transcrystallinity occurred in the PLA3801X/PCL blends (figure 2). This is thought to be encouraged by the additive talc within this PLA grade. The semi-crystalline grades of PLA (3051D and 2002D) selected in this study exhibit extremely slow crystallisation kinetics and remain amorphous from the melt. PLA 3801X is a new grade produced by NatureWorks containing talc as a nucleating agent to facilitate crystallisation.

Figure 1. SEM image of solvent cast 75% PLA 4060D / 25% PCL 6800.

Figure 2. ESEM image of solvent cast with dichloromethane 25% PLA 3801X / 75% PCL 6800 illustrating transcrystallinity between droplet and matrix.

As expected, melt blending with a single screw extruder also produced immiscible blends indicated by analysis via ESEM (figure 3). However, a noticeable reduction in droplet size from solvent casting to melt blending was observed.

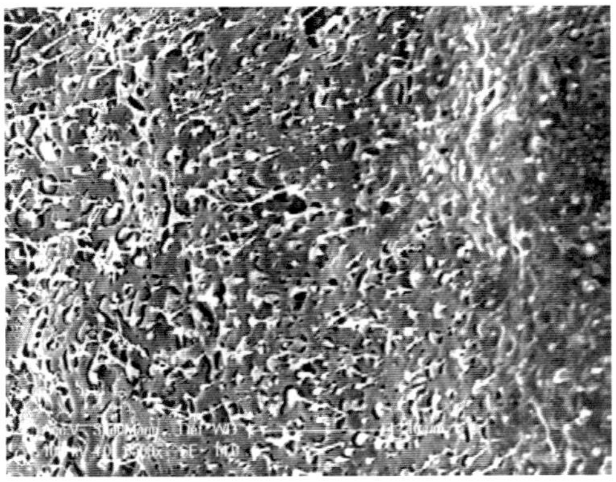

Figure 3. ESEM image of melt blend with a single screw extruder. 75% PLA 4060D / 25% PCL 6800. Extruder conditions of 160 °C and 20 rpm. Finely dispersed PCL droplets in a PLA matrix indicate immiscibility.

Mechanical testing of the blends showed a significant improvement in the elongation to break of polylactic acid with the addition of polcaprolactone (figures 4 and 5). This was observed for each PLA grade. PCL has a higher elongation than PLA. By blending these polymers together the properties combine resulting in a higher elongation for PLA. The blends produced by extrusion showed a dramatic improvement in mechanical properties compared to those produced by solvent casting. For example the solvent cast 25% 3801X blend had an elongation to break of 20% in comparison to the melt blend which produced samples with elongation around 920% (figure 4). This is because solvent casting produces extremely porous samples coupled with a higher crystallinity causing the sample to be more brittle.

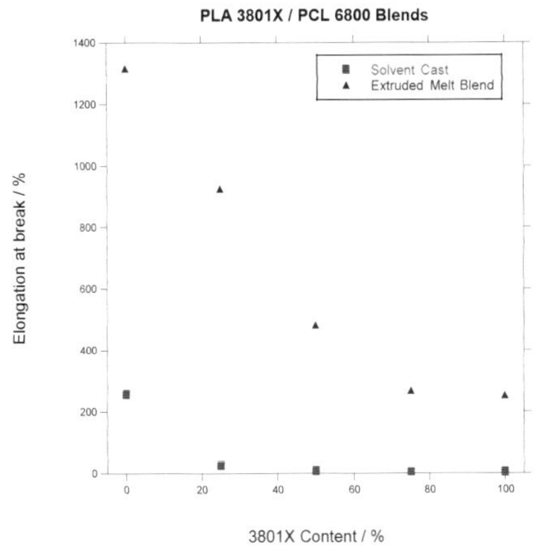

Figure 4. A comparison of elongation to break for solvent cast and melt blended PLA 3801X/PCL 6800 blend at 0, 25, 50, 75 and 100% composition.

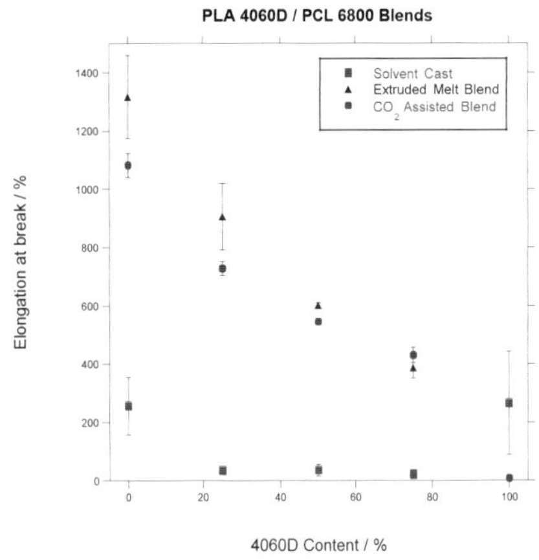

Figure 5. A comparison of elongation to break for solvent cast, melt blend and CO_2 assisted extruded PLA 4060D/PCL 6800 at 0, 25, 50, 75 and 100% composition.

Initial analysis of CO_2 assisted blends by ESEM show a significant reduction in number and size of PCL droplets in a PLA matrix (figure 6). However, a porous structure was noted in CO_2 samples when analysing by ESEM, which could attribute to the lack of improvement in elongation to break compared to the melt extruded samples. For the blend composition 75% PLA 4060D / 25% PCL 6800, CO_2 assisted extrusion provides the most effective method of blending. At this composition CO_2 assisted blending produces the greatest elongation to break and the fewest droplets seen by ESEM compared to the other blending techniques.

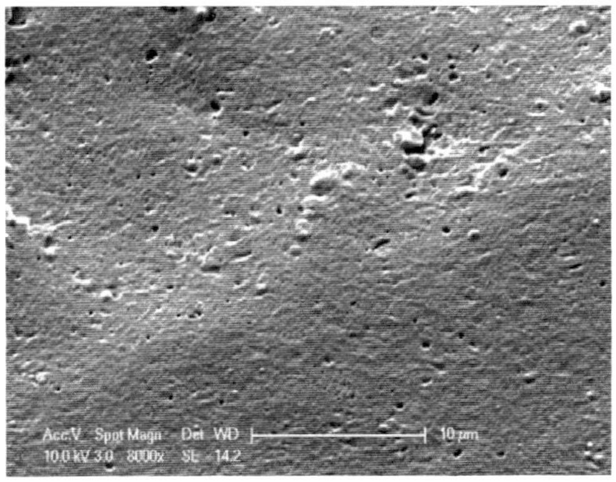

Figure 6. ESEM image of CO_2 assisted blend 75% PLA 4060D / 25% PCL 6800. Extruder conditions of 150 °C at 20 rpm with 10% CO_2.

Conclusions

Blends produced by solvent casting display immiscibility across all compositions. A significant improvement on mechanical properties and droplet size was observed through melt blending via a single screw extruder. Although the mechanical properties of the CO_2 assisted blends showed no improvement on those produced without CO_2, the ESEM images clearly indicate an improvement at the microscale. However, porous structures have been observed which potentially contribute to the lack of significant increase in mechanical properties.

The initial findings of CO_2 assisted extrusion indicate that by injecting CO_2 into the polymer melt the CO_2 is able to act as a molecular lubricant by reducing the polymer viscosity and encouraging increased molecular interaction of the polymer chains. Consequently it is able to improve the adhesion between phases at the microscale.

Analysis of further blend compositions and extruder conditions will also be presented with results from DSC, ESEM, Instron and FTIR. Studies will also include FTIR analysis of the influence of CO_2 pressure on the spectra of PCL and PLA.

Acknowledgements

I would like to thank Dr Mike Jenkins, Dr Gary Leeke and Dr Catherine Kelly for their guidance and support during my PhD. I would also like to give my appreciation to Frank Biddlestone who has provided me with invaluable technical assistance.

References

1. S.D. Park; M. Todo; K. Arakawa; M. Koganemaru *Polymer* 2006, *47*, 1357-1363.
2. N. Lopez-Rodriguez; A. Lopez-Arraiza; E. Meaurio; J.R. Sarasua *Polymer Engineering and Science* 2006, *46*, 1299-1308.
3. J.C. Meredith; E.J. Amis *Macromolecular Chemistry and Physics* 2000, *201*, 733-739.
4. D. Wu; Y. Zhang; M. Zhang; W. Zhou *European Polymer Journal* 2008, *44*, 2171-2183.
5. F. Kusmanto; M. Billham; H. Hornsby *J. Vinyl and Additive Technology* 2008, *14*, 163-166.
6. Z. Lian; S.A. Epstein; C.W. Blenk; A.D. Shine *Journal of Supercritical Fluids* 2006, *39*, 107-117.

O-06-538

MECHANICAL AND THERMAL PROPERTIES OF RECYCLED PET INJECTION MOLDINGS

K. Yamada[1]*, S. Tamada[2], N. Kunimune[2], and H. Hamada[1]

[1] *Department of Advanced Fibro-Science, Kyoto Institute of Technology, Japan - kazushi@kit.ac.jp, hhamada@kit.ac.jp*
[2] *Kunimune Co. Ltd., Japan - tamada@kunimune.co.jp, n.kunimune@kunimune.co.jp*

Abstract - Recycled poly(ethylene terephthalate) (RPET) is known to exhibit brittle behavior in the presence of notch and low thermal stability. Therefore, we tried to improve the toughness and thermal properties of RPET by incorporating with poly(ethylene-co-glycidyl-methacrylate) (E-GMA), talc filler and engineering thermoplastic, which aimed to use as impact modifier, for increasing the rigidity and heat distortion temperature of RPET. The blends and the composites were compounded in twin screw extruder and fabricated to dumbbell specimens by injection molding. The effect of talc filler and polymer blends contents on mechanical and thermal properties of the blends and the composites was investigated. From the result, these blends with E-GMA exhibited significantly higher stiffness and strength especially with increasing E-GMA content. In addition, the blends with talc filler indicated an enhancement of heat distortion temperature of the blends due to increase the crystallinity of RPET blends. Therefore, it was found that talc filler played an important role in enhancing the thermal resistance of RPET. Some injection molding parts, i.e. tray, chopstick, and so on, were successfully produced from these compound materials.

Keywords: Recycled PET, Polymer blends, talc, crystallization, thermal resistance.

Introduction

Poly(ethylene terephthalate) (PET) is widely used for synthetic fiber, beverage bottles, food and other liquid container and so on, which constitutes a large portion of post consumer wastes. In the present day, the volume of the PET bottle products greatly increased. For example, the annual volume of consumption of PET drinking bottle in Japan was 338,654 tons in 2000. However in 2008, a 50% increase in volume to 520,120 tons was recorded. Therefore recycling of PET bottles offers a very practical solution to reduce landfill waste and therefore preventing environmental problems.

PET is known to possess high elastic modulus, strength and toughness. However, recycled PET (RPET) tended to more susceptible to impact loadings and have lower heat distortion resistance as compare to virgin PET. Impact modifiers are often incorporated into RPET in order to enhance its toughness. In our previous work, the notched impact resistance of RPET was significantly enhanced when using poly(ethylene-co-glycidyl-methacrylate) (E-GMA) as impact modifier. However, E-GMA led to deterioration stiffness and yield strength of RPET/E-GMA material. Furthermore, this material would deform easily even when exposed to temperatures around 50 °C.

In this study, talc filler and poly(butylene terephthalate) (PBT) were used for improved the stiffness, heat distortion resistance and impact performance of RPET/E-GMA blends. The effects of talc loading, E-GMA and PBT contents on impact strength, static and dynamic mechanical performance, thermal properties and heat distortion temperature (HDT) of the composites were elucidated.

Experimental

Materials

RPET was supplied by Utsumi-Recycled- Systems Co., Ltd., Japan, which is a waste management company. E-GMA was used as an impact modifier, which was provided by Sumitomo Chemicals Co., Ltd., Japan. PBT (NOVADURAN®, 5010) was supplied by Mitsubishi Engineering Plastics Corporation, Japan. Fine talc fillers (Micro ACE series; the average particle size is 2.5 and 5.0 μm) were purchased from Nippon Talc Co., Ltd., Japan.

Sample Preparation

The sample designations and their compositions are shown in Table 1 and Table 2. The ratio of RPET/E-GMA was set constantly at 84/16 and 97/3. RPET/E-GMA was compounded with 0, 10, 15 and 20 wt% of talc or PBT in a twin screw extruder (TEX30 HSS, Japan Steel Works Co. Ltd., Japan). The barrel temperature of the extruder was set at 250-260 °C. The screw rotation speed was set at 250 rpm. After pelletizing, the blends were dried by using a dehumidifying drier at 80 °C for 5 hours prior to injection molded (UM50, Po Yuen Co. Ltd., China) into dumbbell specimens at a barrel temperature of 260-280 °C with injection speed of 100 mm/s. The dumbbell specimens were used for mechanical and thermal properties characterizations.

Characterization

(1) Static Mechaninal Properties

Tensile tests were performed by using an Instron 4206 universal testing machine in accordance to ASTM D638. The gauge length was 115 mm. The test was conducted at an extension rate of 50 mm/min. At least 5 specimens were used to ensure repeatability.

(2) Dynamic Mechanical Properties

Dynamic mechanical properties of the blend were studied with a dynamic mechanical analyzer (TA instruments, DMA2980) by using a dual cantilever method. The temperature range was set from 30 to 120

°C at the heating rate of 2 °C/min and the frequency of 1 Hz.

Table 1 Specimen designation for RPET/E-GMA/talc composites with 2.5 and 5.0 µm of talc fillers.

Specimen Designation	R-PET (wt%)	EGMA (wt%)	Talc (wt%)
E16-T0	84	16	0
E16-T5	79.8	15.2	5
E16-T10	75.6	14.4	10
E16-T15	71.4	13.6	15
E16-T20	67.2	12.8	20

Table 2 Specimen designation for RPET/E-GMA/talc composites with 2.5 µm talc filler and RPET/E-GMA/PBT blends.

Specimen Designation	R-PET (wt%)	EGMA (wt%)	Talc (wt%)	PBT (wt%)
E3-T0	97	3	0	0
E3-T5	92.1	2.9	5	0
E3-T10	87.3	2.7	10	0
E3-T15	82.5	2.5	15	0
E3-T20	77.6	2.4	20	0
PB-T5	92.1	2.9	0	5
PB-T10	87.3	2.7	0	10
PBT15	82.5	2.5	0	15
PBT20	77.6	2.4	0	20

(3) Izod Impact Performance

The specimens were obtained from the parallel regions of the dumbbell specimens. Notched Izod impact strength was determined for specimens at 2 mm notched depth (a/w=0.2). The tests were conducted by using a Toyo Seiki Izod impact tester with a 5.50 J pendulum at 23 °C in accordance to ASTM D256.

(4) Heat distortion temperature

Heat deflection temperature (HDT) of composites was investigated according to ASTMD648 by using the dual cantilever mode of dynamic mechanical analyzer (DMA2980). The standard stress of 0.45 MPa was constantly applied onto the specimens while temperature was increased at 2 °C/min from room temperature. The changes in specimen dimension as a function of temperature was determined. The heat deflection temperature is defined as the temperature at which the specimen deflects 0.25 mm or 0.2 % strain.

(5) Thermal Properties

Thermal properties were carried out by using a differential scanning calorimeter (TA Instruments, DSC2920) with temperature range from 30 to 300 °C under nitrogen. The heating rate was set at 10 °C/min.

Resuls And Discussion

Effect of Talc Filler on Properties of RPET/E-GMA/talc composites

Tensile Properties

Fig.1 and 2 show tensile properties of the RPET/E-GMA/talc composites at various talc contents. Tensile modulus of the composites steadily increased with incresing talc content as presented in Fig.1. It was due to the high rigidity of the filler. On the other hand, tensile strength of the composites was drastically reduced when incorpoarted 5 wt% of talc in the system as shown in Fig.2. However, with increasing talc content, the tensile strength of the composites was regained, especially when using a smaller talc particle size.

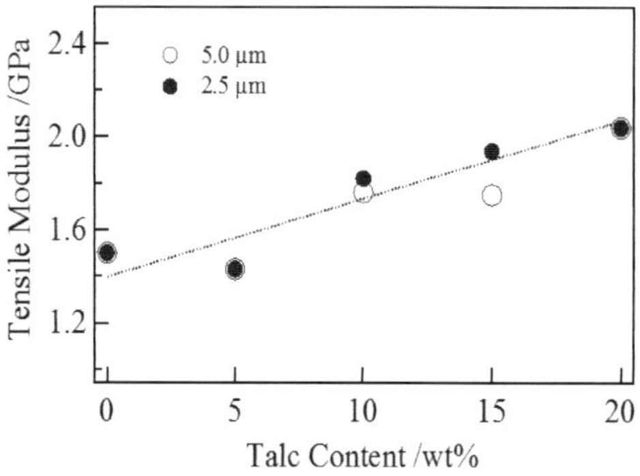

Fig.1 Effect of talc contents on tensile modulus of RPET/E-GMA/talc composites.

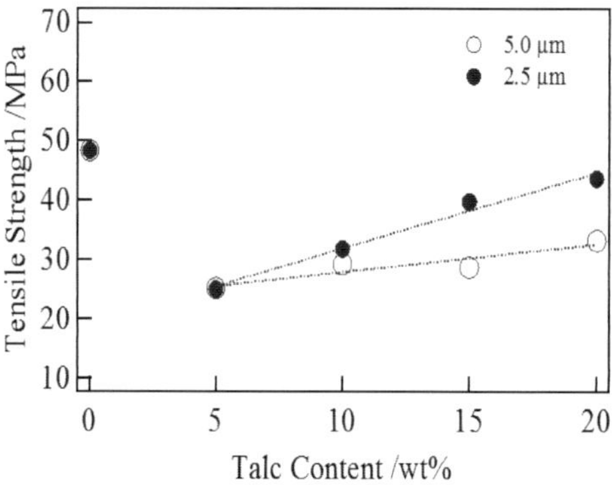

Fig.2 Effect of talc contents on tensile strength of RPET/E-GMA/talc composites.

Impact Properties

Fig.3 presents the impact strength of the RPET/E-GMA/talc composites at various talc contents. The incorporation of E-GMA could enhance notch sensitivity of RPET, which obtained 35 kJ/m^2 of notch Izod impact strength. On the contrary, with the presence of talc in the composites, the impact strength of the composites drastically decreased, while similar impact performance was recorded irrespective of subsequent increments in talc contents. The absence of a gradual deterioration in toughness during impact loading could be attributed to the notch sensitivity of RPET as well as the presence of micro-voids as a result of poor adhesion between the matrix and talc fillers. Furthermore, talc fillers acted as stress-concentration regions, which resulting in poor impact transfer loading. Nevertheless, the notched Izod impact strength of the composites were still two times higher than that of monotonic RPET, which is attributed to the more complex crack propagation path with the presence of talc. It should also be noted that the impact properties were not affected by talc particle size, which talc filler is an effective stress concentrator.

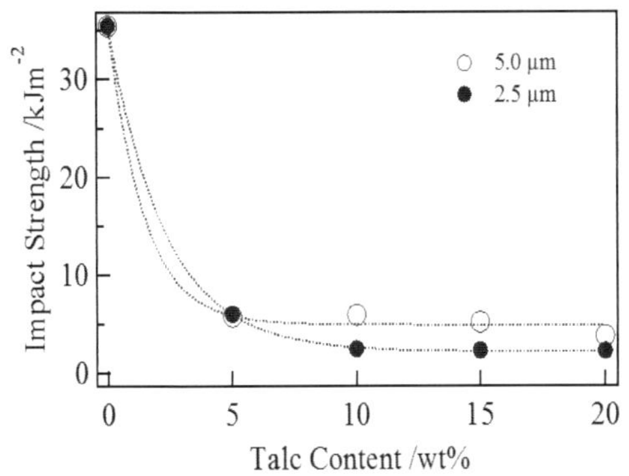

Fig.3 Izod impact strength of RPET/E-GMA/talc composites as a function of talc contents.

Dynamic Mechanical Properties

Fig.4 and 5 illustrate dynamic mechanical properties of the RPET/E-GMA/talc composites with different talc particle sizes. A gradual but significant increment in storage modulus could be observed with the increment of talc contents in RPET/E-GMA composites as presented in Fig.4. These results indicated that talc enhanced stiffness of the composites. Dynamic mechanical property results were good agreement with static tensile property. The storage modulus of the composites would gradually decrease when increasing temperature until around 70 °C. Then a sudden loss in storage modulus would occur, which indicated the onset of glass transition temperature of the composites. Fig.5 shows the tan δ peak of the composites. The decreasing of tan δ peak intensity corresponded to the reduction of molecular chain mobility in RPET due to the incorporation with talc. The usage of smaller talc particles would also result in lower tan δ peak intensities. It was attributed to the smaller talc particle size influenced on less molecular movement and exhibited higher rigidity of the composites.

Heat Deflection Temperature and Thermal Properties

The heat deflection temperature (HDT) test is a popular for industry application especially during the designing of products, which can be used as a simple comparison of the thermal flexural stability of the material. The incorporation of talc improved HDT of the composites as shown in Fig.6. For example, HDT of the composites was increased by 38% to 85 °C when incorporated 20 wt% of talc at 5.0 μm. This result suggested that the presence of talc would improve the dimensional stability of the composites by imparting resistance to molecular movement resulting from inducing crystallization of RPET in the composites when adding talc fillers. The improvement of HDT would be even more pronounced when smaller talc particle size was used in the composites. From the results, it is thought that these materials could be used for the production of kitchenware such as trays, bowls or plates, which were able to withstanding on temperatures up to 80 °C such as in dish washer or sterilization. It is interesting to note that the incorporation of talc can improve HDT value of the composites hence it is possible to use these RPET/E-GMA blend with talc filler in the application of high temperature usages.

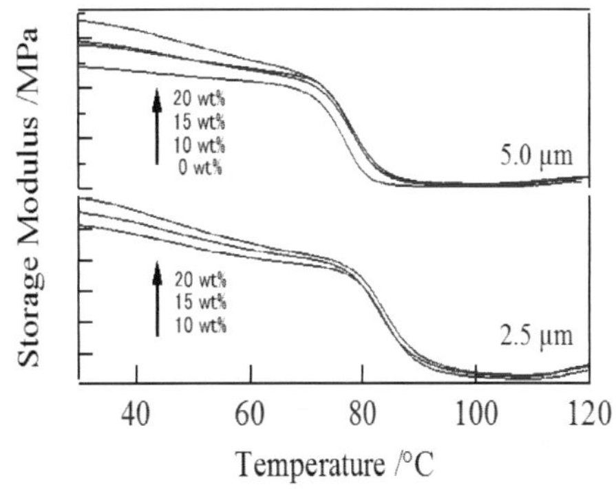

Fig.4 Storage modulus of RPET/E-GMA/talc composites.

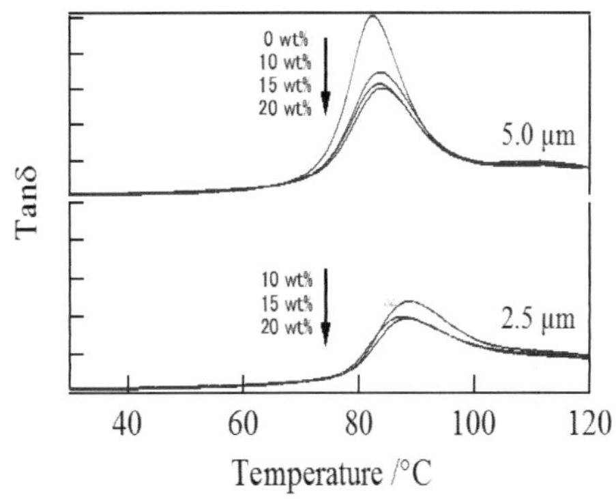

Fig.5 Tan δ of RPET/E-GMA/talc composites.

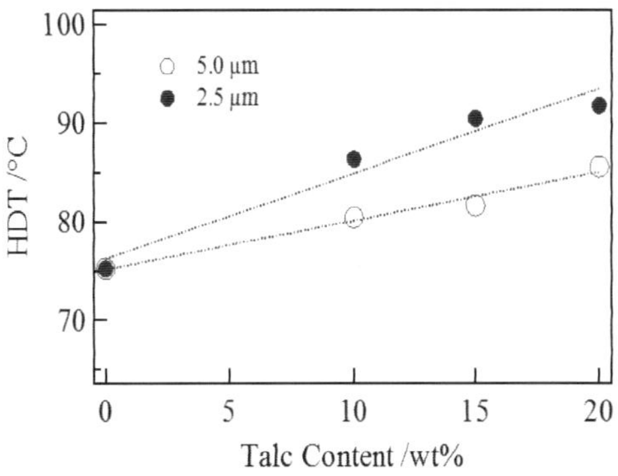

Fig.6 HDT of RPET/E-GMA/talc composites.

Fig.7 exhibits DSC thermograms of the composites with adding 2.5 μm talc filler. The exothermic peak of

RPET/E-GMA thermogram around 120 °C indicated the cold crystallization occurred. Nevertheless, cold-crystallization peak was diminished with increasing the contents of talc and the peak was almost never observed at 20 wt% talc content. From these results, it can be noted that the adding of talc to the system had improved the crystallization of RPET resulting in enhanced the crystallinity of the composites.

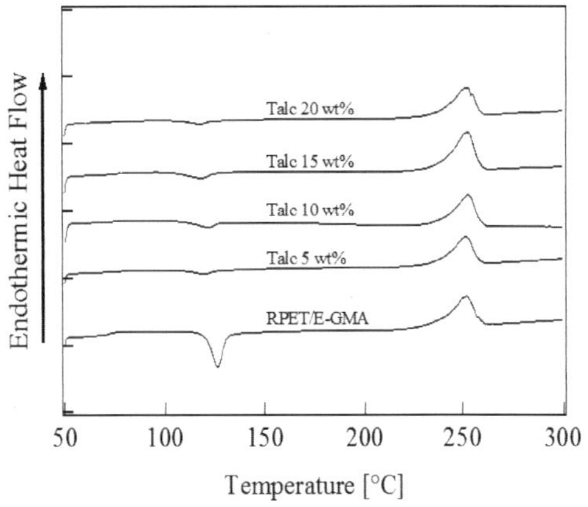

Fig.7 DSC thermograms of RPET/E-GMA/talc composites at various talc contents.

From above results, we indicated that talc filler played an important role in enhancing the crystallization of the composites, which influenced on improved heat resistance of the RPET/E-GMA/talc composites. However, the amount of E-GMA was fixed at 16 wt%, which was considered that this percentage of E-GMA is excess as general manufacturing in injection molding products. Therefore, we would like to reduce the E-GMA content but still remain its impact performance. Therefore, the content of E-GMA was studied at 3 wt%, while PBT was adding to this RPET/E-GMA system without compounding with talc filler. The effect of PBT contents on impact performance and thermal resistance of the blends was discussed.

Effect of PBT and Talc Filler contents on Properties of RPET/E-GMA blends

Fig.8 shows the comparison of impact performance between the adding of PBT and talc to the RPET/E-GMA blends. The results show that the impact strengths of RPET/E-GMA/PBT blends were almost unchanged when increasing PBT contents as presented in Fig.8. It was attributed to the compatibility between RPET and PBT as well as high performance of PBT, which would obtain the constantly of impact performance at different contents of PBT. On the contrary, as aforementioned, impact strength of the REPT/E-GMA blends decreased when compounding with talc as shown in Fig.8, which irrespective to E-GMA contents. It was due to the poor adhesion between filler and the matrix.

The effect of PBT contents on HDT of the blends was presented in Fig.9. HDT values of the blends were almost constant at around 75 °C. However, HDT values were slightly decreased with increasing the PBT contents. It is considered that PBT component prevent PET from crystallizing. Therefore, HDT of the blends was not improved due to unchanged in crystallinity of RPET/E-GME/PBT blends. On the other hand, HDT value for RPET/E-GMA/talc blends increased with increasing the talc contents. Moreover, the HDT value was obtained about 170 °C at 20 wt% talc content. It is interesting to note and can be concluded that talc plays an important role in enhancing the crystallization and heat resistance of RPET blends while E-GMA and PBT worked so well to improve the toughness of RPET blends by optimizing the additive amounts for RPET blends system.

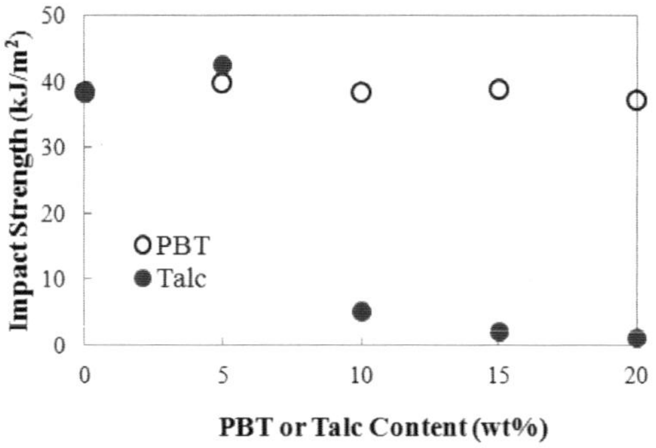

Fig.8 Izod impact strength for (a) RPET/E-GMA/PBT blends and (b) RPET/E-GMA/talc blends.

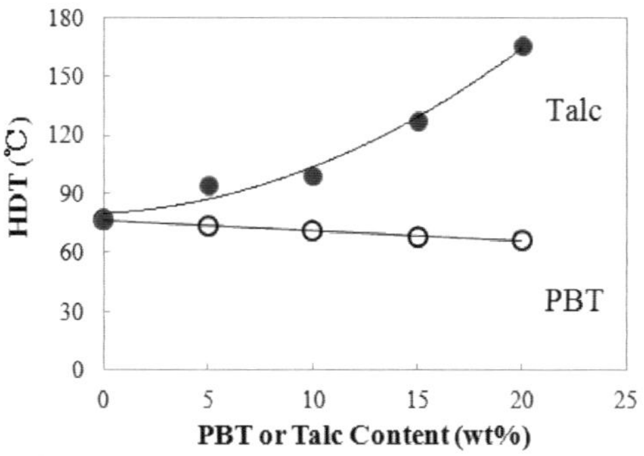

Fig.9 HDT properties for PBT or talc contents (wt%) on RPET/E-GMA(3 wt%) blends.

Conclusions

The effect of talc filler and polymer blends contents on properties of RPET injection molding was elucidated. E-GMA and PBT exhibited well defined for improved toughness of RPET. The heat deflection temperature and rigidity of RPET can be significantly enhanced with the incorporation of talc due to an improvement of crystallinity. These effects would be more pronounced if the talc particle size was small. However, the talc filler would cause significant reduction in notched impact resistance, thus low talc content is recommended for the commodity composites in order to attain a balance in terms of toughness, thermal and dimensional stability.

O-06-570

THE INFLUENCE OF SBS AS A COMPATIBILIZER ON THE RHEOLOGY AND MECHANICAL PROPERTIES OF HIPS/HDPE BLENDS

M. Ahmadi, A. Javadi*, H. Nazokdast and M. Bastani

Polymer and Color Engineering Department, AmirKabir University of Technology, P.O. Box 15875/4413, Tehran, Iran; ajavadi@aut.ac.ir; nazdast@aut.ac.ir; maryam.ahmadi.89132014@gmail.com

Abstract: Improvement of the mechanical properties and Environmental Stress Cracking Resistance (ESCR) of high impact polystyrene (HIPS) is important for various applications. A promising and economical solution to improve these properties is blending HIPS with polyolefins, especially polyethylene. Since HIPS and PE are incompatible, a suitable compatibilizer is necessary to achieve a blend with desirable properties. In this study, the effect of styrene-butadiene-styrene (SBS) as a compatibilizer for this system is investigated. Blend samples with different compositions were prepared using a twin-screw extruder and the rheology and mechanical properties of neat polymers, simple blends, as well as compatibilized blends were studied. It was found that the two polymers and their blends show shear thinning and display a pseudo-plastic behavior. As a result, their complex viscosity and storage modulus decrease by increasing the frequency. The viscosity and storage modulus of the compatibilized blends are higher than those of the simple blends, particularly in lower frequencies. This corresponds to more effective compatibilization at low frequencies. Results of the mechanical testing confirm that the impact strength and tensile properties of the simple blends of HIPS and HDPE are significantly lower than those of neat polymers. This could be attributed to the poor interfacial interactions between the two. In contrast, all compatibilized blends showed considerable enhancements in mechanical properties, especially in those blends with higher amounts of compatibilizer. It was demonstrated that greater degree of enhancement could be achieved for the samples with matrix-dispersed type morphology rather than co-continuous one.

Keywords: Blends, Compatibilizer, HDPE, HIPS, SBS.

Introduction

High impact polystyrene (HIPS) is one of the modified polymers that its toughness is improved by rubber modification, but when it has not high resistance to chemicals and when it exposes to oils, fats cracks over the time. Thus improvement of the mechanical properties and environmental stress cracking resistance (ESCR) of HIPS is important for various applications like using in refrigeration cabinets replacing with ABS. Methods such as rubber particle size change and changing crosslinking density of rubber particle can be used to solve this problem [1]. But such an approaches needs to design new lines of polymerization so these methods are very expensive. A promising and economical way to improve the impact strength and ESCR of HIPS, is blending it with polyolefins, especially polyethylene. Since these polymers are incompatible, many researchers have studied the effect of various types and also combination of different compatibilizers on these blends [2-10].

Chirawithayaboon et al. [2] showed that the mechanical properties of HDPE/HIPS blends increase with increasing the styrene-ethylene-butylene-tyrene copolymer (SEBS) contents, specially in blends cotaining more than 50 wt. % of HDPE. The morphology observation of these blends indicated a fine and homogeneous disperssion of HDPE particles in the HIPS matrix, in blends containing 5 to 20 phr of SEBS. Some researchers have investigated the effect of type of compatibilizer or a mixture of compatibilizers on the blend properties [3-6]. The effect of styrene–butadiene (SB) block copolymers and ethene–propene (EPM) random copolymers on the morphology and tensile impact strength of LDPE/PS (75/25) blends was studied. The results showed that the addition of an EPM/SB mixture containing EPM with a higher viscosity, leads to a higher improvement or at least the same improvement in the tensile impact strength of a compatibilized blend as the same amount of neat SB. Furthermore Perez et al. [3] showed that the addition of SEBS and SBS copolymers to HDPE/HIPS blends, resulting in enhancement of mechanical properties. Also the melting temperatures of compatibilized blends were reduced with increasing the amounts of compatibilizer.

Razavi et. al.[6] investigated the effect of compatibilization (SBS) and (SEBS) on blend properties. The results of linear viscoelastic properties showed negative deviation of viscosity and elasticity from the mixing rules in low HIPS content blends, while positive deviation was observed for high HIPS content blends. It was found that SEBS had better efficiency of compatibilization than SBS due to lower molecular weight and the presence of ethylene block in SEBS structure.

Effect of the Molecular Structure on Compatibilization Efficiency in compatibilized blends is one of the studies that have been conducted so far on blends. Fortelny et al.[5] was studied the effect of molecular structure SB block copolymers with various number of blocks and two lengths of styrene blocks on properties of polystyrene PS/HDPE blends. It was found that the localization of SB block copolymers in PS/PE/SB blends is strongly affected by molecular weight of

styrene blocks in the copolymer. The copolymers with short styrene blocks are preferentially located at the interface, the copolymers with long styrene blocks are mostly located in the PS phase.
Some studies investigated synergistic effect of simultaneous compatibilizer and stabilizer on the blends component from recycled material [7-9]. Pospisilet et al. [7] showed that the cooperative compatibilising system has a synergistic effect only in blends containing recycled or pre-aged LDPE and this system is not operative in the presence of virgin LDPE. Sometimes use of in situ compatibilized instead of employing the third component for compatibilizing. Gao et al. [10] used a Lewis acid catalyst, aluminum chloride ($AlCl_3$) to produce LLDPE-g-HIPS copolymers and used it as a compatibilizer for HIPS/LLDPE blends.
In this study we investigate the effect of SBS as a compatibilizer for the HDPE/HIPS blends to improve the mechanical and rheological properties of this blend.

Experimental

Materials

The blends constituents used in this study were of Industrial origin, HIPS 8350, MFI=3.76 gr/10min (200°C/5kg) and HDPE, BL3, MFI=1.2 gr/10min (190°C/5kg) were supplied by Tabriz petrochemical company. SBS-LG411 MFI<0.5 gr/10min (200°C, 5kg) from LG company, Korea were used as received.

Blend preparation

The melt compounding of neat HIPS/HDPE and HIPS/HDPE/SBS blends were carried out in a twin screw extruder at 60 rpm and a temperature range of 170-210°C .

Measurements of the Mechanical Properties

The standard samples for mechanical testings was prepared using an injection machine (Engle, Astria) at 210°C. Tensile tests were carried out with an universal tester (Galdabini, Sun 2500, Italy) according to ASTM D 638. The Izod impact strengths was measured with an impact tester (Ushima Seisakusho LTD, Tokyo, Japan) according to ASTM D 256. The maximum of the pendulum energy was 6 J for Izod impact tests for all samples.

Rheological studies

The flow behavior and melt properties of neat HIPS and HDPE, also HIPS/HDPE and HIPS/HDPE/SBS blends were investigated by using a Rheometric Mechanical Spectrometer (RMS; UDS 200, Anton Paar) equipped with parallel plate geometry (diameter=25mm, gap=1mm). The frequency sweep tests were performed in a range of 0.1–625 s^{-1} at 210°C and amplitude of 1% to maintain the response of materials in the linear viscoelastic regime.

Results and Discussion

Rheological observations

The rheological properties of neat polymers, binary blends and compatibilized blends were studied. It was shown that two polymers and their blends have shear thining behavior, so the complex viscosity and storage modulus of them decrease with increasing the frequency. The results show that at higher frequencies, the viscosity of binary blends is between neat polymers. Also it was found that viscosity of blends increase with increasing the amounts of polyethylene. In general, a difference between the rheological properties of binary blends is seen at higher frequencies, Fig. 1.

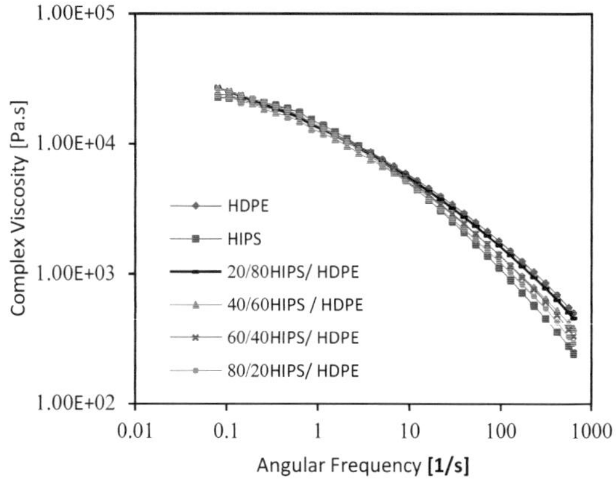

Figure1. Comparison between rheological Properties of neat polymers and the binary blends

It seems that at high frequencies, both phases are deformed and the response to shear is controlled by HIPS phase. So the viscosity decreases with increasing the HIPS content [5]. The viscoelastic response of the blends in low frequencies, where the effect of flow induced molecular orientation, on the viscosity and elasticity, can be used in evaluating of the interfacial interaction between phases .
The viscosity and storage modulus of compatibilized blends are higher than binary blends, especially in low frequencies. It shows that the compatibilization is more effective at low frequencies. Fig. 2 shows the variation of complex viscosity against angular frequency for neat polymers and the compatibilized blends. As can be seen, at low frequencies, the complex viscosity of compatibilized blends is increased by increasing the amount of HIPS. On the contrary, at higher frequencies, reverse trend is observed. It seems that because of compatibilization and improvement of adhesion between two phases, both of neat polymers are responsible to control the deformation, but in low frequencies, the effect of HDPE and in high frequencies the role of HIPS is more obvious.

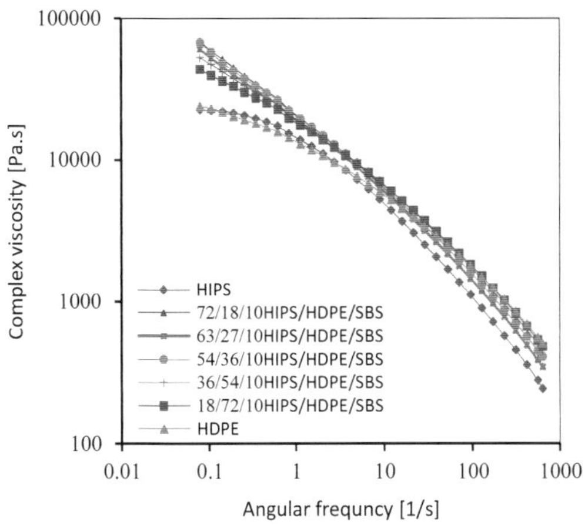

Figure2. Complex viscosity versus angular frequency in compatibilized blends

It is also shown that the storage modulus of blends increases with increasing the content of compatibilizer, with holding the constant ratio of polyethylene to polystyrene, Fig. 3.

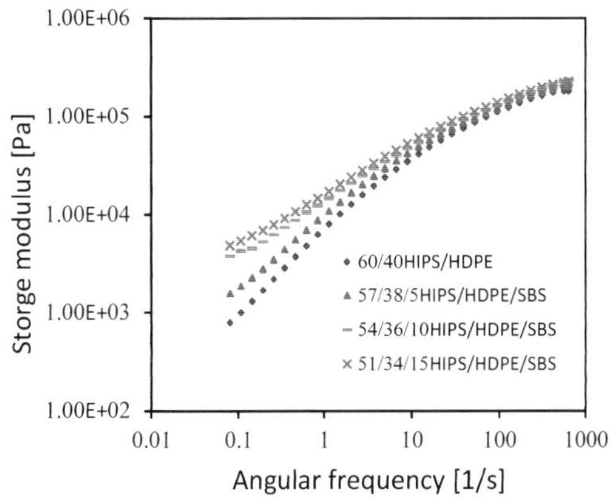

Figure3. Comparison HIPS/HDPE blends with different content of compatibilizer

Mechanical properties

The results of mechanical properties show that all binary blends (HIPS/HDPE) have lower impact strength in comparison to neat HIPS, which is due to poor interfacial interaction between HIPS and HDPE. The determination of tensile properties showed that the tensile strength of all binary blends is higher than neat HIPS, but the elongation at break of them is much lower than neat HIPS except the blends containing high content of HDPE (20/80, 30/70) which their elongation at break are about twice of neat HIPS. According to the results, all ternary blends (HIPS/HDPE/SBS) show a great improvement in mechanical properties. The impact strength of all blends increase by compatibilization and this improvement is more significant in HIPS/HDPE (20/80) and almost increased to eight times, Fig. 4.

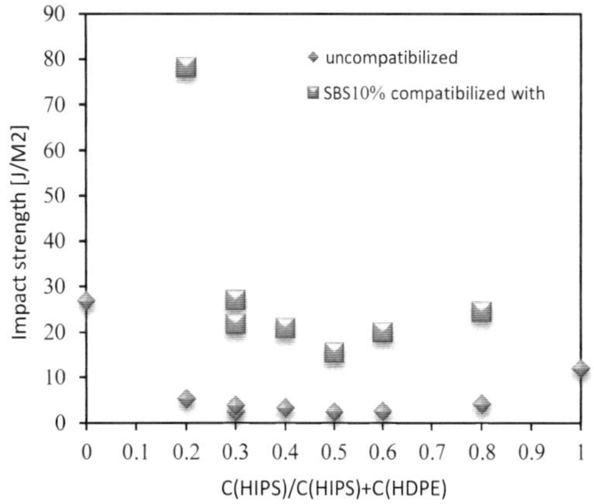

Figure4. Impact strength versus HDPE content (in binary and compatibilized blends)

Compatibilizing reduced the modulus and tensile strength of all blends, however the tensile strength is still higher than neat HIPS but modulus is reduced to less than neat HIPS.

To investigate the effect of the compatibilizer content, the mechanical tests was done on HIPS/HDPE (60/40) and (80/20) using 5 to 15 wt. % of SBS. The results showed that the mechanical properties of blends improve with increasing the amount of compatibilizer. The impact strength and elongation at break were increased with increasing the content of compatibilizer from 5 to 15 wt. % respectively, Fig 5. On the other hand, the tensile strength increased by using 5 wt. % of SBS but decreased with increasing of 10 and 15 wt. % of compatibilizer. Generally all ternary blends containing more content of HDPE (60/40) showed higher tensile strength than HIPS/HDPE (80/20) which it was predictable because of the higher tensile strength of neat HDPE in comparison to neat HIPS, Fig. 6.

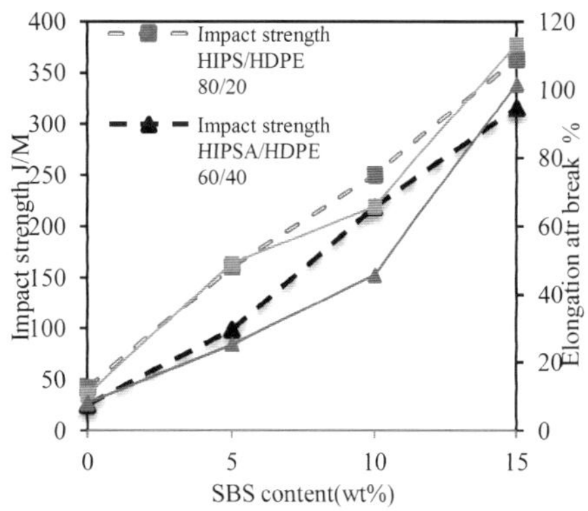

Figure5. The effect of various contents of compatibilizer on the Impact strength and Elongation at break of blends

The modulus of all blends is lower than the neat HIPS the reduction of modulus is very high in blends containing more HDPE. The HIPS/HDPE (80/20) blend modulus decreased with increasing 10 and 15 %wt SBS and increasing to maximum value with using 5 %wt compatibilizer. But modulus of the HIPS/HDPE (60/40) blend decreased with increasing content of compatibilizer.

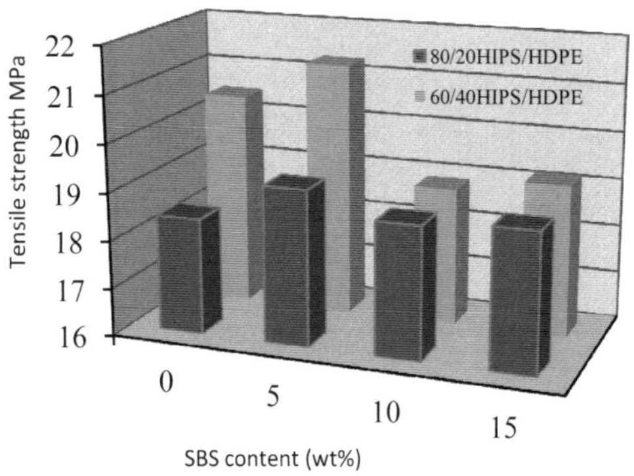

Figure6. The effect of various content of compatibilizer on the tensile strength of blends

Conclusion

The blend of immiscible HIPS/HDPE blends has poor mechanical properties in comparison to neat polymers. SBS is a suitable compatibilizer for this system which leads to decrease interfacial tension and increase interfacial adhesion between two polymers. As a result, the effective stress transfers between two phases and the great improvement in mechanical properties. specially in blends containing high content of HDPE is observed.

References

1. V .G .Grassi, M. F. Dal Pizzol, M. M. C. Forte , S. C. Amico, *J. of App. Polym. Sci., 121*, 2011.
2. A. chirawithyaboon , S. Kiatkamjomwong, *J. of App. Polym. Sci., 91* , 2004, *742.*
3. I. S. B. Perez , S. Manrich , S. Manrich , *Polym. Sci. and Tech.* , *18,* 2008, *207.*
4. J. Wu, B. Guo, C.Ming Chan, J. Li , H.S Tang, , *polymer, 42,* 2001, *8857.*
5. I. Fortelny , D. Michalkova, J. Hromadkova , F.Ledmcky, , *J. of App. Polym. Sci.*, 81, 2001, 570.
6. M . K . Razavi Aghjeh , M. Khodabandelou , M.K faridi, *J. of App. Polym. Sci ,114,* 2009, *2235.*
7. J. Pospisil, I. Fortelny, D. Michalkova, Z. Krulis, M. Slouf, *Polym. Deg. and stab., 90,* 2005, *244.*
8. D. Micha lkova , J. Pospisil , I. Fortelny, M. Slouf , Z. Krulis, *J. of vinyl & additive tech.,* DOI 10.1002/vnl, 2006.
9. D. Michalkova , J. Pospisil , I. Fortelny , J. Hormadkova , F. Lednicky , P. Schmidt , Z. Krulis, *J.of App. Sci., 94,* 2009, *1486.*
10. Y. Gao , H. Huang , Z. Yao ,D., Shi, Z. Ke, J. Yin, , *J. Appl. Polym. Sci*, 81, 2001, 570.

O-06-651

MOLECULAR INSTABILITY OF POLYAMIDE GRAFT COPOLYMERS UNDER POLYMER PROCESSING CONDITIONS

Guo-Hua Hu[1,2]*, Cailiang Zhang[3], Sandrine Hoppe[2] and Lianfang Feng[3]

[1]*School of Chemistry and Biological Engineering, University of Science and Technolgoy Beijing, Beijing 100083, China*

[2]*Laboratory of Reactions and Process Engineering, Université de Lorraine- CNRS, 1 rue Grandville, BP 20451, 54001 Nancy, France (guo-hua.hu@univ-lorraine.fr)*

[3]*State Key Laboratory of Chemical Engineering, Department of Chemical and Biochemical Engineering, Zhejiang University, Hangzhou 310027, China*

Abstract: One of the challenges polymer processing may face is macromolecular instability under thermo-mechanical conditions. Among the many polymer blends that have been studied in academia and industries, polyamide 6 (PA6) based polymer blends have probably received the most attention. This work shows that graft copolymers with PA6 as grafts are highly instable under melt polymer processing conditions. Once the graft copolymer is subjected to the high temperature and high shear strength, the degradation of the graft chain will take place, which will decrease the content of graft chain in copolymer.

Keywords: polymer processing; morphology; graft copolymer; instability; polyamide 6

Introduction

Nowadays, polymer blend receives much attention from the point of view of fundamental and practical interest. However, the majority of polymer blends is immiscible so that block or graft copolymer is often used as compatibilizer to minimize the interfacial intension and stabilize the phase morphology. The compatibilizing efficiency of a graft copolymer may depend very much on its structure variables such as backbone length, graft length, and graft density [1-3]. Polymer blend process often involved the thermo-mechanical conditions, which lead to the deterioration of physical and mechanical properties. Therefore, one of the challenges polymer processing may face is their molecular instability.

Among the broad variety of commercially available polymer blends, multi-phase systems consisting of polyamide 6 (PA6) represent an important class of engineering polymers. This work investigated the instability of graft copolymers with PA6 as grafts used as the compatibilizer of PA6 multi-phase systems under polymer processing conditions.

Experimental

The PS-g-PA6 graft copolymer was synthesized by the active anionic polymerization of ε-caprolactam (CL) from the copolymer (PS-co-TMI) of styrene and 3-isopropenyl-α,α-dimethylbenzene isocyanate in the presence of sodium ε-caprolactam as catalyst. Details on the synthesis and characterization procedures can be found in the literature [4-7].

The pure PS-g-PA6 was subjected to either static annealing in a hot press at 230 °C and 10 MPa or dynamic annealing in a mixer of type Haake torque rheometer at the same temperature and with a rotation speed of 65 rpm, corresponding to a nominal shear rate of 10 to 100 s^{-1}. The sample was quenched in liquid nitrogen to freeze-in its structure and morphology. Its morphology was characterized by transmission electron microscope (TEM) of type JEM-1200EX (Martinez-Salazar and Cannon, 1984). Prior to the TEM analysis, specimens were stained in the following manner. They were microtomed to films of less than 100 nm thick, and then placed on a drop of 2% phosphotungstic acid solution for 30 min at 50 °C using a specimen grid. Thereafter they were rinsed three times with distilled water.

Results and Discussion

It was well known that polymer is prone to subject to the degradation due to exposure to high processing temperature and high shear strength. For the researched copolymer system, PA6 occurs easily the degradation while the degradation of PS is very limited and often negligible below 250 °C. After static annealing and dynamic annealing, a fraction of copolymer PS-g-PA6 can soluble in formic acid, while there is hardly any material soluble in tetrahydrofuran (THF). Irrespective of static annealing or dynamic annealing, the PA6 graft's content in copolymer PS-g-PA6 decreases with annealing time, as shown in Fig.1. Moreover, the PA6 graft's content in PS-g-PA6 by static annealing is higher than that by dynamic annealing. This indicates that the mechanical shear can have a synergistic effect making the thermal degradation more easy and severe.

The phase morphologies of the copolymer PS-g-PA6 treated by the various methods are shown in Fig.2. In the micrographs, PA6 appears black and PS white. The as-synthesized pure PS-g-PA6 graft copolymer has a cylindrical morphology (A) looks like a co-continuous morphology. After extraction in formic acid, its morphology remains more or less unchanged (B). After the static or dynamic annealing, the morphology of the as-synthesized pure PS-g-PA6 had undergone much change (C and E). This mainly results from the aggregation of the homo-PA6 formed by the scission of the PA6 grafts of the PS-g-PA6 graft copolymer. The exaction of C and E in formic acid allows removing most of the homo-PA6 domains (see D and F) and the volume fractions of the PA6 grafts become 21.6 and 16.9%, respectively. After the static annealing, the morphology of the as-synthesized PS-g-PA6 becomes wormlike micelles. Upon dynamic annealing, its morphology exhibits ordered spheres.

The above results indicate that polymer processing conditions may induce significant changes in structure and composition of graft copolymers with PA6 as grafts and consequently their morphology. This may also explain the fact that polymer blends that exhibit desired morphology and properties under certain processing conditions may show undesired morphology and properties under other processing conditions.

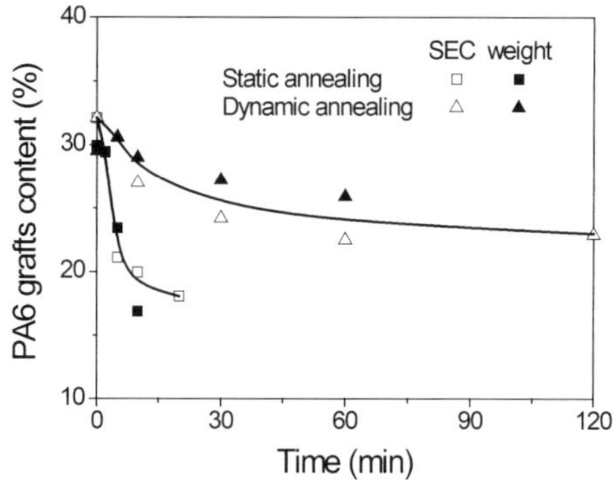

Figure 1 Content of PA6 grafts in PS-g-PA6 treated by static and dynamic annealing for different times at 230 °C and then extracted in formic acid for 12 h at 30 °C. The PA6 graft's content obtained by weight and SEC.

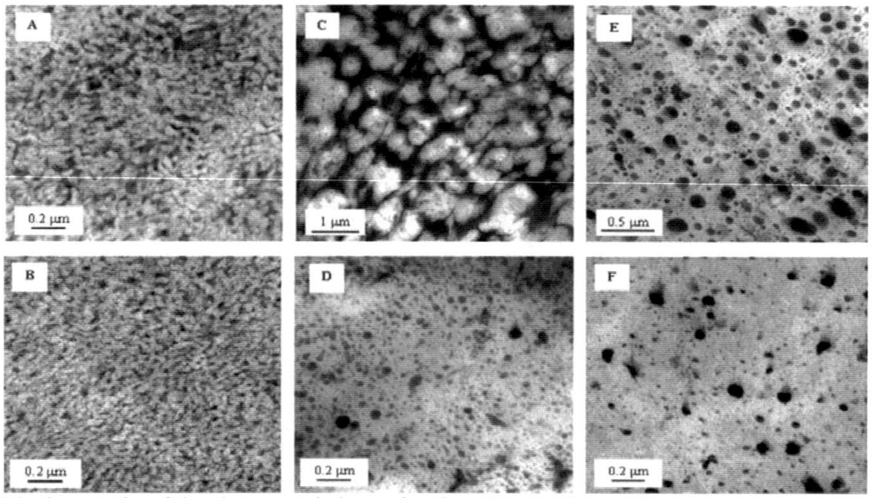

Figure 2 TEM micrographs of the phase morphology of various PS-g-PA6 by different treatments. A: Pure PS-g-PA6; B: Formic acid treatment 12 h at 30 °C; C: Static annealing for 2 h at 230 °C; D: Formic acid treatment 12 h at 30 °C after static annealing for 2 h at 230 °C; E: Haake mixing for 10 min at 230 °C; F: Formic acid treatment 12 at 30 °C after Haake mixing for 10 min at 230 °C.

Conclusions

This work reported on the instability of graft copolymers under polymer processing conditions. A graft copolymer with polystyrene as backbone and polyamide 6 as grafts (PS-g-PA6) is chosen as a model graft copolymer because of academic and industrial relevance. It is found that such a graft copolymer is highly unstable under polymer processing conditions. The instability of graft copolymers under processing conditions may be one of the reasons that polymer blends that exhibit desired morphology and properties under certain processing conditions may show undesired morphology and properties under other processing conditions.

Acknowledgements

The authors thank the Ministry of Science and Technology of China through an international cooperation program (grant number: 2001CB711203), the PHC CAI YUANPEI 2012-28021SM and the State Key Laboratory of Chemical Engineering (No. SKLChE-08D03) for their financial support.

References

1. C.L. Zhang; L.F. Feng, X.P. Gu, S. Hoppe, G.H. Hu. *Polym. Eng. Sci.* 2010, *50*, 2243.
2. C.L. Zhang; L.F. Feng; J. Zhao; H. Huang; S. Hoppe; G.H. Hu. *Polymer* 2008, *49*, 3462.
3. C.L. Zhang; L.F. Feng; X.P. Gu; S. Hoppe; G.H. Hu. Polymer, 2007, *48*, 5940.
4. C.L. Zhang; L.F. Feng; X.P. GU; S. Hoppe; G.H. Hu. *Polym. Eng. Sci.* 2011, *51*, 2261.
5. C.L. Zhang; L.F. Feng; X.P. Gu; S. Hoppe; G.H. Hu. *Polym. Test.* 2007, *26*, 793.
6. C.L. Zhang; L.F. Feng; G.H. Hu; Z.B. Xu. *J. Appl. Polym. Sci.* 2006, *101*, 1972.
7. C.L. Zhang; L.F. Feng; S. Hoppe; and G.H. Hu. *J. Polym. Sci. Part A: Polym. Chem.* 2008, *46*, 4766.

O-06-284

EFFECT OF SILANE AND UHMWPE ON THE CROSSLINKED HDPE/PTFE BLENDS

Nuntawut Chantanakorn[1,2], **Utai Meekum**[1]

[1] *School of Polymer Engineering, Suranaree University of Technology, Nakorn Ratchasima, Thailand; umsut@g.sut.ac.th*

[2] *Centre for Petroleum, Petrochemical, and Advanced Materials, Chulalongkorn University, Bangkok, Thailand; thang_kung@hotmail.com*

Abstract The peroxide initiated vinyl silane and moisture induced crosslinked HDPE/PTFE and HDPE/PTFE/UHMWPE were studied. The effects of silane contents at 2.0, 2.5, 3.0, 3.5 and 4 phr on the properties of the HDPE/PTFE blends were investigated. The contents of UHMWPE added on the HDPE/PTFE blends were also extendedly explored. Blending was manufactured on the co-rotation twin screw mixer. The standard test methods both mechanical and surface characteristics were employed. It was found that impact strength was increased with increasing the silane usage. Vice versa, MFI, flexural, HDT, surface properties of the blends were decreased with increasing the silane. Increasing in the network density via the silane/moisture condensation was the hypothesized explanation. At the given amount of silane, 4 phr, the HDPE/PTFE/UHMWPE having UHMWPE from 5 to 25 phr manifested that the MFI, tensile, HDT, flexural and the % wt loss were decreased with increasing the UHMWPE content. The notched impact and diameter of water drop ratio, however, slightly increased with the addition of the UHMWPE. The phase separation via adding the high molecular mass polymer into the blend was confirmed by SEM. The observation was used to explain properties incompetency of the UHMWPE for toughening the crosslinked HDPE/PTFE blends.

Keywords: Crosslinked HDPE/PTFE blend, HDPE/PTFE /UHMWPE blend, Mechanical properties and Surface characteristic.

Introduction

Contribution of the outstanding wear properties of PTFE in the polymer blends have been explored by the number of researchers. PEEK/PTFE blend indicated that the coefficient of friction in wear modes decreased with increasing the PTFE contents. At 30% of PTFE, the blend showed excellent combination of friction coefficient[1]. On the other hand, PEEK filled PTFE composite investigation showed that the lowest average friction coefficient($\overline{\mu}$) was obtained in the samples having PEEK at 50%(w/w). The lowest wear rate was obtained at 32% of PEEK filled sample[3]. Adding potassium titanate whiskers(PTW) into PEEK/PTFE composites seen that the PTW enhanced properties of the composite[4]. The blend between HDPE/UHMWPE was also published and found that the volumetric loss of the blends decreased with increasing the UHMWPE and it showed good mechanical properties[2]. Effects of untreated and pretreated carbon nanofibers(CNFs) on friction behavior and mechanical properties of UHMWPE/HDPE nanocomposites were also explored[5]. The enhancement in tensile strength of nanocomposites containing 0.5% of treated CNFs was four times higher than the untreated composite and pure sample. In this published work, the blend between HDPE/PTFE/UHMWPE was explored. The silane/peroxide crosslink system was added into the blends to enhance the compatibility and also mechanical inter lock via the chain networking formed. Main research objective was to obtain the blend material having the low surface adhesion characteristic.

Experimental

A. Materials

HDPE, H5814J, available from SCG-Chemical was used as the main polymer matrix. PTFE, Zonyl® MP 1300 and UHMWPE, UH 900, were purchased from DuPont and Asahi Kasei Sunfine and used as dispersed phase. Silane, Silquest® A-171, the vinyltrimethoxy silane(VTMS) with dicumyl peroxide(DCP) was used as crosslink system.

B. Sample preparation

For the HDPE/PTFE blends, PTFE was pretreated with 0.5 phr of DCP and 1 phr of silane in internal mixer at 100°C. Solution of 2.0, 2.5, 3.0, 3.5 and 4.0 phr of silane with 1 phr of DCP and dried HDPE pelltet were brought into cooperating and rigorously checked, respectively. Then, 20 phr of silane treated PTFE was added into the solid mixture and vigorously mixed. The solid ingredient was fed into twin screw extruder. The melt compounding was performed at screw speed of 10 rpm and constantly the barrel temperature at 190°C. The compound strand was cooled down and pelletized. The compound formula are shown in Table 1.

In order to study the effect of UHMWPE content in HDPE/PTFE blends, the master batch of UHMWPE at 60 phr in HDPE were manufactured at 200°C in twin screw mixer. The calculated amount of UHMWPE at 5,

10, 15, 20 and 25 phr, respectively, was added into the blend and compounded in the same manner as described in HDPE/PTFE blend. The designed compound ingredients of HDPE/PTFE/UHMWE are given in Table 2.

Table 1 Composition of the HDPE/PTFE blends with different silane content

Sample No.	HDPE (g)	PTFE (g) (phr)	DCP (g) (phr)	Silane (g) (phr)
HDPE/PTFE#1	400	80(20)	4(1)	8(2.0)
HDPE/PTFE#2	400	80(20)	4(1)	10(2.5)
HDPE/PTFE#3	400	80(20)	4(1)	12(3.0)
HDPE/PTFE#4	400	80(20)	4(1)	14(3.5)
HDPE/PTFE#5	400	80(20)	4(1)	16(4.0)

Table 2 Composition of the HDPE/PTFE/UHMWPE blends

Sample No.	HDPE(g)	PTFE (g) (phr)	DCP (g) (phr)	Silane (g) (phr)	UHMWPE (g) (phr)
HDPE/PTFE/UH#1	400	80(20)	4(1)	16(4)	20(5)
HDPE/PTFE/UH#2	400	80(20)	4(1)	16(4)	40(10)
HDPE/PTFE/UH#3	400	80(20)	4(1)	16(4)	60(15)
HDPE/PTFE/UH#4	400	80(20)	4(1)	16(4)	80(20)
HDPE/PTFE/UH#5	400	80(20)	4(1)	16(4)	100(25)

C. *Material testing*

The injection molding specimen was achieved at 190°C, 200°C, 210°C and 220°C from feed to nozzle, respectively and mold temperature of 30°C. Film samples with thickness of approx. 0.4 mm were fabricated by compression at 200°C and used for wear and liquid surface tension. Both samples were divided into 2 sets; annealed at room temperature and sauna cured at 65°C. Performance testing; tensile, flexural, impact testing, HDT and wear resistance were performed in accordance with the ASTM standards The liquid surface tension investigation was calculated from the micro drop, 1 μl, of distilled water(d_s) on the sample surface divided by the water diameter on referee material, PTFE sheet, (d_R).

Results and Discussion

A. Physical Properties

The effects of silane contents on the mechanical properties of HDPE/PTFE blends are summarized in table 3 to table 5, respectively. Generally, it is seen that the properties of the blends is fractionally increased via the sauna incubation. There might be two folds explanation, crosslinking via silane/moisture condensation and increasing in the crystallinity during the annealing process. From the results presented, the tensile properties seem to be independence on the silane contents. However, the flexural properties are lowered with increasing the contents. Implying that the materials had low in bending toughness. On the other hand, impact strength is marginally increased with increasing the silane content. Without the flow ability data, it might say that the increasing in crosslink density of the blends is suspected. Therefore, the material becomes more brittle. Nevertheless, from the MFI and HDT results shown in table 6, it is seen that the index does not obviously change with the silane addition. Consequently, the network density of the chain does not increase as expected. Another point of consideration could be taken into account is the interfacial bonding between polymer matrix and PTFE particle. This evidence will be presented by SEM data. For HDT outcome, it is observed that the HDT is decreased with increasing the silane for the original samples but it tendency to be increased after the sauna curing. Again, gaining the crystallinity via thermal prolonging would be the main reason.

Table 3 Tensile properties of HDPE/PTFE blends

Sample No.	Tensile strength(MPa)		Tensile modulus (GPa)	
	Original	Cured	Original	Cured
HDPE/PTFE#1	22.11±0.47	23.36±0.45	0.78±0.01	0.82±0.05
HDPE/PTFE#2	22.52±0.32	23.14±0.41	0.79±0.06	0.76±0.06
HDPE/PTFE#3	22.62±0.21	22.86±0.65	0.76±0.11	0.83±0.06
HDPE/PTFE#4	22.66±0.11	22.40±0.71	0.79±0.05	0.73±0.13
HDPE/PTFE#5	22.66±0.26	21.60±0.42	0.76±0.07	0.67±0.05

Table 4 Flexural properties of HDPE/PTFE blends

Sample No.	D_{sample}/D_{PTFE}		Weight loss (%)	
	Original	Cured	Original	Cured
HDPE/PTFE#1	1.03±0.09	1.01±0.08	0.56±0.02	0.25±0.03
HDPE/PTFE#2	1.03±0.11	1.01±0.02	0.48±0.02	0.25±0.02
HDPE/PTFE#3	1.04±0.05	1.00±0.12	0.48±0.05	0.23±0.05
HDPE/PTFE#4	1.04±0.05	0.98±0.24	0.41±0.06	0.18±0.03
HDPE/PTFE#5	1.04±0.04	0.96±0.07	0.40±0.08	0.11±0.03

Table 5 Impact strengths of HDPE/PTFE blends

Sample No.	Notched (kJ/m^2)		Unnotched (kJ/m^2)	
	Original	Cured	Original	Cured
HDPE/PTFE#1	4.01±0.25	4.18±0.18	NB	NB
HDPE/PTFE#2	4.02±0.21	4.49±0.21	NB	NB
HDPE/PTFE#3	4.35±0.13	4.77±0.19	NB	NB
HDPE/PTFE#4	4.43±0.14	4.99±0.25	NB	NB
HDPE/PTFE#5	4.46±0.11	5.04±0.07	NB	NB

Table 6 MFI and HDT of HDPE/PTFE blends

Sample No.	MFI@200/5.00	HDT (oC)	
	(g/10min)	Original	Cured
HDPE/PTFE#1	0.422±0.034	67.8±2.5	74.9±2.7
HDPE/PTFE#2	0.422±0.015	66.3±2.3	75.5±0.6
HDPE/PTFE#3	0.422±0.016	66.2±1.3	78.5±0.5
HDPE/PTFE#4	0.422±0.020	64.1±1.4	79.3±1.6
HDPE/PTFE#5	0.421±0.032	64.1±0.4	79.3±0.1

The mechanical properties of HDPE/PTFE/UHMWPE blends are shown in table 7 to table 9. Incorporating 5 to 25 phr of UHMWPE into the HDPE/PTFE blends. Again, it is seen that sauna curing has obvious positive effect to the properties. The tensile and flexural properties are slowly decreased with increasing the UHMWPE loading both original and cured samples. The phase separation between UHMWPE particle and the matrix due to the insufficient fusion energy of the high molecular weight polymer during the mixing process is explained. Recalling that, only 190°C mixing temperature was employed, compared with above 300°C of the UHMWPE processing temperature. Consequently, the ultra high mass polymer would only aggregate into the matrix phase. The toughening process could not be achieved. Vice versa, the impact strengths of the tertiary blends is raised by adding more of UHMWPE. It indicates that UHMWPE particles are acted as the crake energy absorber. Hence, they inhibit the crake growth. Flow ability and thermal properties of the blends are given in table 10. As expect, it is seen that the flow index of the blend is reversely affected by the ultra high polymer content. But the HDT is lowered by adding more UHMWPE for both original and cured samples. It is because UHMWPE has lower in HDT than the blend matrix.

Table 7 Tensile properties of HDPE/PTFE/UHMWPE blends

Sample No.	Tensile strength(MPa)		Tensile modulus (GPa)	
	Original	Cured	Original	Cured
HDPE/PTFE/UH#1	22.66±0.26	21.60±0.42	0.76±0.07	0.67±0.05
HDPE/PTFE/UH#2	19.44±0.91	20.12±0.16	1.33±0.11	0.71±0.02
HDPE/PTFE/UH#3	18.88±0.46	19.90±0.48	1.27±0.02	0.71±0.02
HDPE/PTFE/UH#4	18.71±0.57	19.49±0.23	0.98±0.38	0.68±0.03
HDPE/PTFE/UH#5	18.71±0.27	19.39±0.52	0.66±0.05	0.68±0.03
HDPE/PTFE/UH#6	18.66±0.46	19.34±0.33	0.63±0.02	0.68±0.04

Table 8 Flexural properties of HDPE/PTFE/UHMWPE blends

Sample No.	Flexural strength(MPa)		Fluxuralmodulus(GPa)	
	Original	Cured	Original	Cured
HDPE/PTFE/UH#1	36.68±0.99	38.16±1.30	1.91±0.16	2.03±0.09
HDPE/PTFE/UH#2	25.59±1.27	29.98±1.17	1.25±0.08	1.47±0.09
HDPE/PTFE/UH#3	25.02±1.03	29.54±0.67	1.23±0.08	1.47±0.05
HDPE/PTFE/UH#4	24.62±0.42	29.49±0.60	1.20±0.03	1.47±0.05
HDPE/PTFE/UH#5	25.96±0.79	29.31±0.17	1.24±0.04	1.46±0.01
HDPE/PTFE/UH#6	25.52±0.32	28.93±0.75	1.21±0.02	1.46±0.04

Table 9 Impact strength of HDPE/PTFE/UHMWPE blends

Sample No.	Notched (kJ/m^2)		Unnotched (kJ/m^2)	
	Original	Cured	Original	Cured
HDPE/PTFE/UH#1	4.46±0.11	5.04±0.07	NB	NB
HDPE/PTFE/UH#2	6.02±0.23	6.60±0.23	NB	NB
HDPE/PTFE/UH#3	6.41±0.11	6.80±0.15	NB	NB
HDPE/PTFE/UH#4	6.89±0.30	7.24±0.23	NB	NB
HDPE/PTFE/UH#5	7.22±0.04	7.64±0.22	NB	NB
HDPE/PTFE/UH#6	7.49±0.08	7.52±0.18	NB	NB

Table 10 MFI and HDT of HDPE/PTFE/UHMWPE blends

Sample No.	MFI@200/5.00 (g/10min)	HDT (°C)	
		Original	Cured
HDPE/PTFE/UH#1	0.421±0.032	64.1±0.4	79.3±0.1
HDPE/PTFE/UH#2	0.031±0.031	62.8±2.6	77.9±1.8
HDPE/PTFE/UH#3	0.031±0.021	58.9±0.6	76.9±0.1
HDPE/PTFE/UH#4	0.022±0.063	58.1±0.6	75.3±1.9
HDPE/PTFE/UH#5	0.021±0.012	57.3±0.3	75.3±0.9
HDPE/PTFE/UH#6	0.019±0.012	57.2±0.4	75.2±1.8

B. Surface Characteristics

The surface properties of HDPE/PTFE blends with against the silane loading by mean of water drop ratio and wear test are summarized in table 11. It is found that water diameter ratio and % wt loss are decreased when increasing the silane content. It also notices that the sauna annealing lower the tests values, especially the wear test. Deceasing in the diameter of water drop ratio means the lower in surface tension of the blends. Similarly, decreasing in % wt loss indicates the surface friction or good adhesion between PTFE particle and matrix. From the results, it suggest that the surface tension and wear properties of the blend is improved by adding more silane coupling agent. Table 12 presents the surface characteristic of HDPE/PTFE/UHMWPE blends. In contrast with the HDPE/PTFE blends, the water drop ratio and wear properties of the blends do not enhance by incorporating the UHMWPE. Moreover, the surface tension has tendency to be inferior by adding the polymer. It indicates that phase separation of UHMWPE due to the insufficient fusion energy would have the significant and negative effect to the surface properties of the blends. This evidence will be concluded by the SEM analysis later on.

Table 11 Surface characteristic of HDPE/PTFE blends

Sample No.	D_{sample}/D_{PTFE}		Weight loss (%)	
	Original	Cured	Original	Cured
HDPE/PTFE#1	1.03±0.09	1.01±0.08	0.56±0.02	0.25±0.03
HDPE/PTFE#2	1.03±0.11	1.01±0.02	0.48±0.02	0.25±0.02
HDPE/PTFE#3	1.04±0.05	1.00±0.12	0.48±0.05	0.23±0.05
HDPE/PTFE#4	1.04±0.05	0.98±0.24	0.41±0.06	0.18±0.03
HDPE/PTFE#5	1.04±0.04	0.96±0.07	0.40±0.08	0.11±0.03

Table 12 Surface characteristic of HDPE/PTFE/UHMWPE blends

Sample No.	D_{sample}/D_{PTFE}		Weight loss (%)	
	Original	Cured	Original	Cured
HDPE/PTFE/UH#1	1.03±0.42	0.96±0.07	0.49±0.07	0.40±0.03
HDPE/PTFE/UH#2	1.03±0.07	0.90±0.01	0.49±0.17	0.39±0.16
HDPE/PTFE/UH#3	1.03±0.06	0.90±0.05	0.49±0.16	0.38±0.17
HDPE/PTFE/UH#4	1.03±0.05	0.99±0.11	0.49±0.21	0.38±0.14
HDPE/PTFE/UH#5	1.05±0.09	1.04±0.04	0.49±0.03	0.37±0.24
HDPE/PTFE/UH#6	1.05±0.02	1.04±0.06	0.44±0.04	0.35±0.21

C. Morphology of Blends

SEM micrographs at X1000 of HDPE/PTFE blends with 2.0, 3.0 and 4.0 phr of silane are displayed in Figure 1**(a)** and 1**(b)** for original and cured, respectively. It is obviously seen that the PTFE particle is dispersed on the HDPE matrix. The PTFE pull out is the major crake phenomenon. According to the morphological observation, increasing in the silane contents do not manifest any vast improvement of the adhesion between the particle and the matrix. This indication could be used to explain why the tensile and flexural properties of the HDPE/PTFE blends do not superior by increasing the silane content as described earlier.

Closer investigation of the fractured surface, it is seen that the web like traces at the matrix, HDPE, is more obvious after sample undergo sauna incubation. Also, the number of webs do seemingly and directly depend on the amount of silane used.

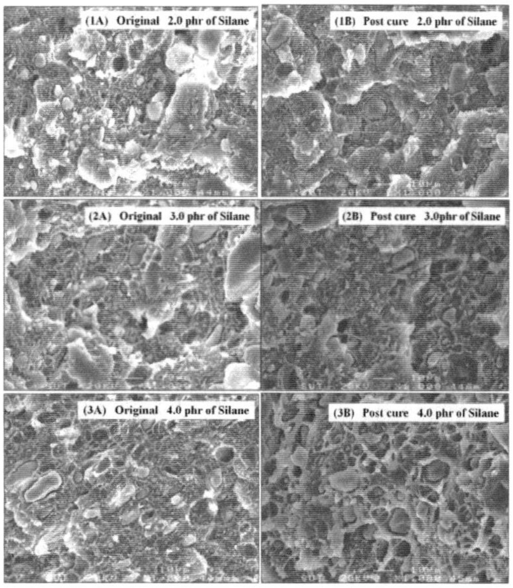

(a) (b)

Figure 1. SEM photographs of HDPE/PTFE blends with 2.0, 3.0 and 4.0 phr of silane both, (a) Original and (b) sauna cured samples, respectively.

The web like traces could be resulted from the silane/moisture crosslink reaction. The network density would increase with increasing the silane addition.

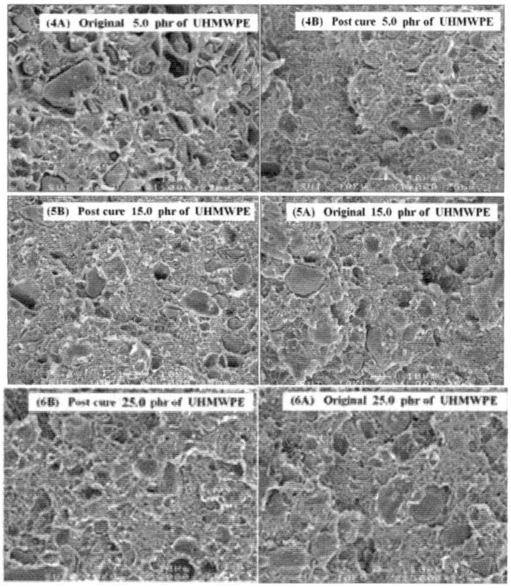

(a) (b)

Figure 2. SEM photographs of HDPE/PTFE/ UHMWPE blends; (a) Original and (b) cured sample.

Figure 2 is the SEM photographs of the HDPE/PTFE/UHMWPE blends at 5, 15 and 25 phr of UHMWPE with and without sauna curing, respectively. It does quit obvious to see that the UHMWPE particles are separated and dispersed on the HDPE matrix. This unmelted high molar mass chain would cause the inferior in the tensile and flexural properties of the blends. As mention before, the sauna incubation do not assist the adhesion between the dispersed particles but it promotes the crystallinity and the network formation via silane/moisture induced reaction at the matrix phase. Therefore, the cured process does not have much of the positive effect on the mechanical properties of the HDPE/PTFE/ UHMWPE blends as presented early on.

Conclusions

Increasing the silane in the HDPE/PTFE blend system found that the mechanical properties, tensile and flexural, are decreased. The increasing in the network formation via increasing the silane content were taken into explanation. The silane did not improve the adhesion between HDPE matrix and PTFE dispersed phase. In the tertiary blend system, HDPE/PTFE/ UHMWPE, it was found that the mechanical properties were also depleted at higher UHMWPE loading. The phase separation of unmelted UHMWPE particle was played the important role for the inferiority.

References

1. J. Bijwe, S. Sen and A. Ghosh: *Influence of PTFE content in PEEK–PTFE blends on mechanical properties and tribo-performance in various wear mode*. Wear 258 (2005) 1536-1542
2. A. de A. Lucas, J.D. Ambrósio, H. Otaguro, L. Costa and J.M. Agnelli: *Abrasive wear of HDPE/UHMWPE blends*. Wear. (2011).
3. D. L. Burris and W. G. Sawyer: *A low friction and ultra low wear rate PEEK/PTFE composite*. Wear 261 (2006) 410-418.
4. G. Sui, W.H. Zhong, X. Ren, X.Q.Wang and X.P. Yang: *Structure, mechanical properties and friction behavior of UHMWPE/HDPE/carbon nanofibers*. Materials Chemistry and Physics 115 (2009) 404-412.
5. G.Y. Xie, G.S. Zhuang, G.X. Sui and R. Yang: *Tribological behavior of PEEK/PTFE composites reinforced with potassium titanate whiskers*. Wear 268 (2010) 424-430.

KN-07-03

STRUCTURE AND PROPERTIES OF POLYMER-BASED COMPOSITE NANOFIBERS

Chan Kok Ho Kent, Yasuharu Ono, Nakagawa Ryouhei, Kotaki Masaya *

Department of Advanced Fibro Science, Kyoto Institute of Technology, Matsugasaki, Kyoto 606-8585, Japan m-kotaki@kit.ac.jp

Abstract - Coaxial electrospinning has been recognized as a technique to prepare composite nanofibers with core-sheath structure. This present work focuses on exploiting the coaxial electrospinning as a method to encapsulate functional nano-fillers into a continuous core channel. To investigate the optimum condition for effective confinement of the filler and development of core channel, processing parameters such as core and sheath fluid properties, along with spinneret sizes were examined in correlation to spinning behaviors and fiber morphologies. Microscopic camera was used to study the behavior of the fluids during the spinning process. Concealment of the fillers within the fiber interior was proposed to minimize development of geometric discontinuities and non-uniform fiber morphologies, associated with stress concentrations. To understand the unique mechanical properties associated with core-sheath structured composite fibers, tensile tests were used to evaluate the stress-strain behaviors of both the core-sheath structured fibers and composite fibers electrospun by single capillary spinneret. The nanotensile tester was used to characterize the mechanical properties of single nanofibers.

Keywords: Composite Nanofibers, Coaxial Electrospinning, Single Nanofiber Tensile Test

Introduction

Electrospinning is polymer processing technique commonly used to produce fine polymer fibers with diameters ranging from a few microns to tens of nanometers. A typical setup is composed of a high power voltage, fine capillary, and a counter electrode, which is used as collector for fibers. A syringe pump is used to transfer polymer solution to the fine capillary, used as a spinneret. Application of high voltage deforms the polymer solution suspended at the orifice of the spinneret into a conical shape, also known as Taylor cone. Fine electrically charged polymer solution jet can be generated from the tip of the cone when sufficiently high voltage is applied. The jet solidifies into fine polymer fibers as volatile organic used in the solution evaporates from the system.

For this study, coaxial electrospinning was employed as a technique to produce composite nanofibers with fillers confined within the fiber interior. It has been perceived that formation of filler aggregates along the fiber surface leads to geometric discontinuities and the development of stress concentrations. By intentionally confining the fillers to the fiber interior, it was anticipated that the discontinuities on the surface can be minimized and higher mechanical strength can be achieved. Presence of these geometric discontinuities can become prevalent in cases when high filler content is used for processing the material.

In this investigation, thermally conductive alumina nanoparticles were used as the fillers. Nanoparticles encapsulation was achieved by injecting fluid containing the fillers into core capillary during coaxial electrospinning. In an approach to promote high interfacial adhesion between the alumina nanoparticles and sheath polymer polyacrylonitrile, silane coupling agent was added to the core fluid. Mechanical properties of the fibers were evaluated via stress-strain behaviors based on single fiber tensile test. It should be highlighted that although mechanical characterization of coaxially electrospun fiber mats have been reported in a number of studies, limited works have been conducted to evaluate mechanical properties based on single coaxial fiber.

Experimental

Materials

Polyacrylonitrile (PAN) of Mw 150,000 was purchased from Sigma Aldrich. Organic solvents, dimethylformamide (DMF), dimethyl sulfoxide (DMSO) and methanol (MeOH) were purchased from Nacalai Tesque, Japan. Aluminum oxide nanoparticles (Aeroxide AluC) with average diameters of 15 to 30 nm were added to aminopropyethoxysilane (APES, Shinetsu Kagaku, Japan) and used for the preparation of the core fluid. Viscosity of core fluid was evaluated with Brookfield DV-II viscometer.

Electrospinning Process

Both single capillary and coaxial spinneret electrospinning were performed using commercial setup, Nanon (MECC, Japan). The distance between the tip of the spinneret and the electrically grounded collector was fixed at 15 cm.

Single Spinneret System

PAN dissolved at 10 wt% in DMF/DMSO of solvent ratio 9:1. DMSO was added to the polymer solution to suppress the formation of pores within the interior of the electrospun fibers. A stainless metallic needle with an inner diameter of 0.2 mm was used as the spinneret. A high voltage of 19-21 kV was applied by connecting a high-voltage supplier. Fluid flow rate was fixed at 0.5

ml h-1 by using a programmable pump. Flow rate was set at 1.0 ml/hr.

Coaxial Spinneret System

PAN dissolved at 10 wt% in DMF was used as the sheath solution PAN solutions with weight concentrations of 8 and 9 wt% were used for investigation related to the influence of sheath concentration on morphology. Materials used as the core fluid included aluminum oxide, DMF, MEOH and APES. The mixture of PAN and silane coupling were mixed with DMF and methanol in weight ratio of 9:1. Viscosities were measured with viscometer Brookfield DV-II. Coaxial spinneret produced by MECC (Japan) was used for the spinning process. Unless stated, inner diameters of the core and sheath capillaries were fixed at 2.0 mm and 0.37 mm respectively. Fig. 1 shows the schematic diagram of the coaxial electrospinning setup used in this work. To monitor the Taylor cone at the spinneret tip, microscopic camera (SKS-2000, Saito Kougaku) was placed close to the setup during the spinning process. Sheath and core fluid feed rate was set at 1.6 ml and 0.1 ml/hr respectively. Applied voltage was 18 to 21 kV.

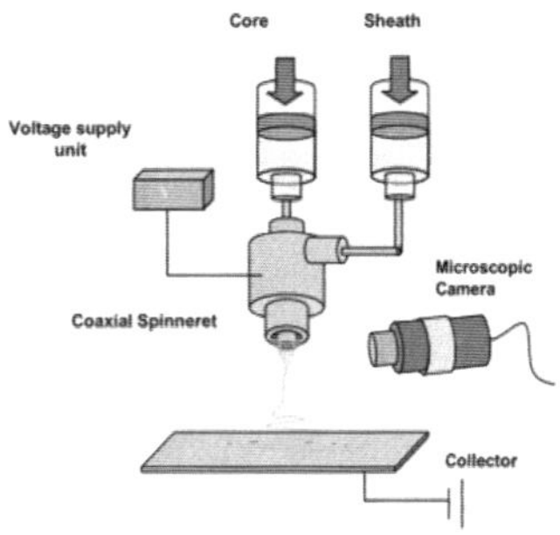

Figure 1. Schematic diagram of the coaxial electrospinning setup used in this study

Morphological Observation

Field emission microscope (FE-SEM) (Hitachi 3400, Japan) was used to observe the surface morphology of the electrospun fibers. Transmission electron microscope images were processed with the kind assistance of Nitto Denko

Mechanical Properties of Single Fiber

A modified disc collector with short gaps on the circumference was used to collect single nanofiber. The fibers deposited across each gap were transferred to a paper frame after short spinning time. To obtain a single fiber, the fibers were carefully removed under bright light till a single fiber can be observed on the frame. A segment of the fiber end was removed for FESEM observation. Gauge length of 10 mm was used for every test specimen. Tensile test were performed at a strain rate of 6 mm/min using a Nanomechanical Testing System (Nano UTM, MTS Systems Corporation) with 500 mN load range, and 50 nN load resolution. Ten samples were prepared under the same processing conditions for each tensile test. The fiber diameters were measured using image processing program, ImageJ based on FE-SEM micrographs.

Results and Discussion

Nanofiber *Morphologies*

Encapsulation process at the Taylor cone plays an important role in the development of core-sheath morphology of the fibers. To encourage effective encapsulation of the core material, two different approaches were taken for our study. One involved the use of solutions with high weight polymer weight concentration for the sheath fluid, while the other is related to the utilization of sheath capillary with increased orifice size.

One of the motivations for using polymer solutions with high polymer weight concentrations as the sheath fluid is associated with high solution viscosity which can be achieved with increased weight concentration. The high fluid viscosity was expected to minimize the diffusion of the core fluid at the core-sheath interface and encourage the formation of distinct core-sheath structure. Moreover, it has been highlighted that moderate sheath fluid viscosity is required to drive the core fluid into fine stream during spinning. In the case when non electrical conducting fluids such as olive oil are used for coaxial electrospinning, flow of core fluid during spinning was proposed to be largely induced by the viscous drag exerted by the sheath fluid.

Zussman et al had proposed that the precipitation at the core-sheath interface is crucial for the development of core-sheath structure. For our experiments, methanol, a non-solvent of PAN, was added to the core fluid. As solutions with high polymer concentrations show higher tendency to precipitate in presence of non-solvent, it was anticipated that the increase in polymer concentration in the sheath solution can assist in the formation of distinct core-sheath structure. Morphologies of coaxially electrospun PAN/alumina nanofibers prepared with sheath solution of different PAN weight concentrations can be seen in Fig. 2. Poor encapsulation of the core, as represented by the presence of filler particles on fiber surfaces, can be seen for fibers coaxially electrospun with PAN sheath solution of less than 10 wt% polymer concentration.

Our previous investigation demonstrated that effective encapsulation was achieved by maintaining large concentric gap between the sheath and core capillary. The concentric gap can be modified by using a sheath capillary with enlarged orifice or alternatively, a core capillary with a reduced orifice. Increasing the spacing between the sheath and core capillary ensures a more effective encapsulation, as the core fluid has to diffuse into a thicker sheath layer to break encapsulation. SEM observation indicated the presence of particle fillers on fiber surfaces spun from coaxial spinneret with sheath capillary inner diameter of 1.0 mm, suggesting that the use of sheath capillary with large orifice 2mm is more favorable for encapsulation of the core material.

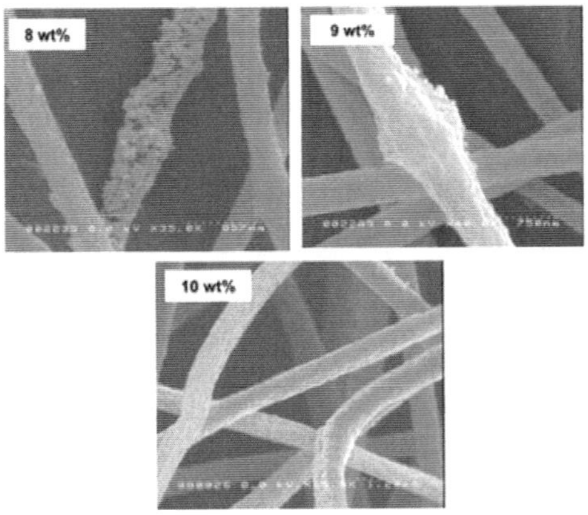

Figure 2. SEM micrographs showing electrospun PAN/alumina composite nanofibers produced by coaxial spinnerets with core and sheath capillary sizes of 0.37 and 2 mm respectively. PAN solutions with polymer concentrations of 8, 9 and 10 wt% were used to prepare the sheath fluids.

Filler morphologies of the composite fibers coaxial electrospun with different core fluids can be seen from the TEM micrographs found in Fig. 3. The results suggest that a correlation exists between filler morphology and spinning condition. Mechanical characterization of single core-sheath structured nanofibers was based on nanofibers prepared by using 20wt%/P-9 as core fluid.

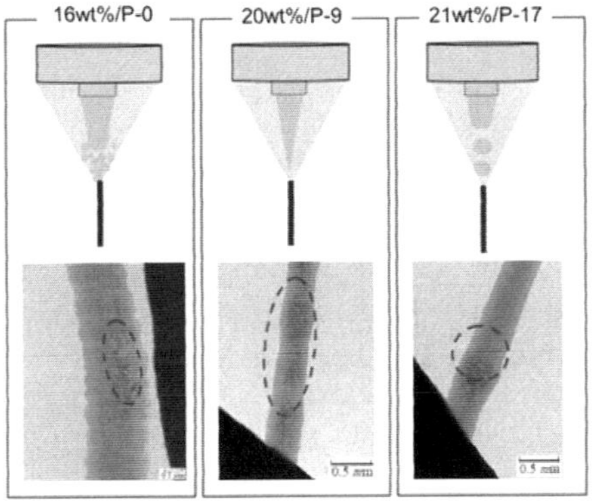

Figure 3. Schematic diagrams showing the behaviors at the core menisci for coaxial electrospinning processes prepared with different core fluids. TEM images of the coaxially electrospun alumina/PAN composite nanofibers can be found below the schematic diagrams.

Morphologies of PAN composite single fibers electrospun with different nanoparticle contents were observed. These fibers were electrospun with the conventional single capillary spinneret. Nanoparticle aggregates were embedded on surfaces of fibers electrospun at high filler contents. The non-uniform fiber morphologies and rough surface textures as observed in the micrographs are potential points of stress localization. Their presence is expected to accelerate fracture initiation and lower mechanical strength.

Mechanical Properties of Single Nanofiber

Fig. 4 shows the representative stress-strain curves for single electrospun composite fibers electrospun by both single capillary and coaxial electrospinning. The graphs comparing the tensile modulus, strain at break and ultimate tensile strength can be seen in Fig. 5 "Effective" filler content for coaxially electrospun fibers was calculated by determining the percentage area occupied by the fillers, as observed from the TEM micrograph.

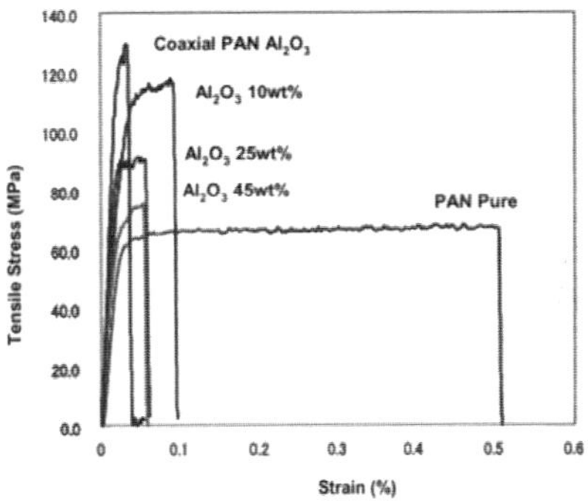

Figure 4. Representative stress strain curves compare the mechanical behavior of coaxially electrospun core-sheath structured nanofiber with PAN/alumina composite nanofibers electrospun with single capillary. The stress strain characteristics were evaluated by single fiber tensile test.

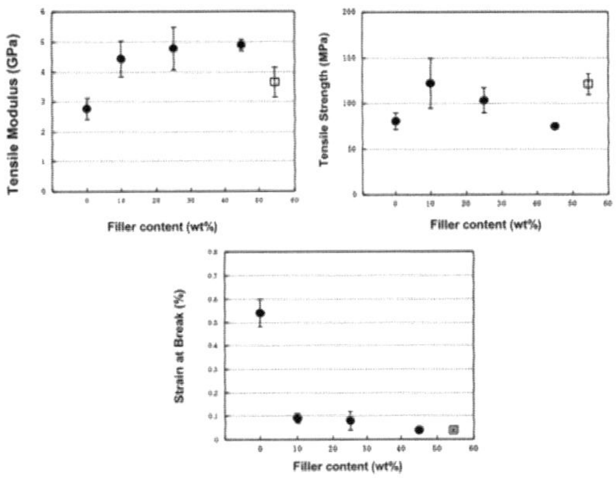

Figure 5. Graphs showing the tensile moduli, strains at break and tensile strengths of PAN/alumina composite fibers electrospun by single capillary and core-sheath structured coaxial electrospun fibers.

Single Capillary Electrospun Nanofibers

Increase in modulus for PAN/alumina composite fibers spun by single capillary can be attributed to the increase in weight concentration of the stiff alumina nanoparticles. However, the increment in modulus diminished with increasing filler content as seen in Fig. 10. This behavior may be related to the development of nanoparticle aggregates formed at higher fill content, resulting in poor dispersion of the fillers and deteriorating interfacial interaction between within filler and PAN matrix. Based on the proposed scenario, ineffective load transfer between polymer and filler network may cause the disproportionate increase in tensile modulus to with filler content.

The composite nanofibers exhibited relatively low strain at break compared to pure PAN nanofibers. Lowest strain at break was observed for the nanofibers prepared with the highest fill content of 45 wt%

Addressing the mechanical strength of the fibers, maximum tensile strength was observed for fibers with filler content of 10 wt%, at 122.15±27.35 MPa. The drop in strength at break for fibers with filler content of 25 wt% and 45 wt% can be interpreted as attributed to two factors. One is the poor polymer-filler interaction, arising from the onset of agglomeration, while the other is the development of localized stresses along the fiber, due to the presence of geometric irregularities on the surface.

Coaxially Electrospun Nanofibers

Core-sheath structured composite PAN/alumina nanofibers were produced using coaxial electrospinning, with the derived optimum processing condition mentioned in the earlier section. In contrast to single capillary electrospun composite fibers, coaxially electrospun fibers with an "effective fill" content of 55 wt% exhibited a relatively low tensile modulus of 3.55±0.46 GPa. Fig. 10 indicated that average tensile moduli for single capillary composite fibers electrospun with more than 10 wt% were observed to be more than 4 GPa. The low tensile modulus may be related to the unique core-sheath morphology of the fibers. Ohja et al reported an increase in modulus when tensile tests were performed for coaxially electrospun nanofibers with carbon nanotubes in the sheath layer.

It is interesting to note that the core-sheath structured alumina/PAN fibers exhibited low elongational strain at break of less than 0.5% with a relatively high mechanical strength of 121.05±11.50 MPa was observed. In contrast to non-structured composite fibers electrospun single capillary, highest mechanical strength was observed at filler composition of 10 wt%, at 122.15±27.35 MPa with a strain at break of about 0.1 wt%. The low strain at break for the core-sheath structured fibers may be due to the presence of a stiff "channel" packed with densely packed nanoparticles within the fiber core. Tensile data of the core-sheath structured fibers based on the effective filler concentration of 55 wt%, suggests coaxial electrospinning as a potential technique for the production of mechanically strong composite fibers with high filler concentration. Tensile strength for single capillary electrospun fibers with filler concentration of 25 wt% and 45 wt% were 103.42±13.53 and 75.04±0.43 MPa respectively.

One factor contributing to the high tensile strength may be related to the concealment of the alumina nanoparticles within the fiber interior. As mentioned in the earlier section, the effective encapsulation of the fillers within the interior of the fibers can minimize the development of non-uniform surfaces or stress concentrations responsible for initiation of mechanical failure. The authors, Ohja et al highlighted a decrease in the tensile strength for core-sheath structured nanofibers when they compared the mechanical properties of composite fibers produced with carbon nanotubes in the sheath layer to non-structured composite nanofibers spun by single capillary.

The outcomes in the tensile tests clearly highlighted the outstanding differences in mechanical characteristics between core-sheath structured PAN/alumina nanofibers and non-structured composite nanofibers produced by single capillary electrospinning.

Conclusions

In the present work, coaxial electrospinning has been used as a simple method to prepare core-sheath structured nanofibers with nanoparticle fillers packed into the core of the fibers. To effectively encapsulate and pack the fillers into a single continuous core channel within the fibers, electrospinning parameters such as spinneret size, sheath and core fluid properties were investigated. A microscopic camera was used to understand the meniscus behavior of the core fluid during the spinning process.

Mechanical tensile testing performed with single fibers showed that core-sheath structured nanofibers prepared with effective filler content of 55 wt% exhibited high tensile strength of 121.05±11.50 MPa. The results had demonstrated the potential of coaxial electrospinning as an effective technique to produce composite nanofibers with high filler content without compromising the mechanical strength. The concealment of the fillers within the interior of the fibers minimized the formation of stress concentrations, promoting higher mechanical strength.

As it has been anticipated that close packing of fillers within the core volume can encourage interconnectivity, further studies will be performed to understand conductivity related characteristics based on composite fibers produced by coaxially electrospinning.

KN-07-02

DESIGN OF NANOREINFORCEMENTS BY ENGINEERING AT NANOSCALE

Sadhan C. Jana[1,*], Sayantan Roy[1], Yannan Duan[1], and Rafael Benavides[1], Darrell Reneker[2]

[1]*Department of Polymer Engineering, University of Akron, Polymer Engineering Academic Center, 250 South Forge Street, Akron, OH 44325-0301*

[2] *Department of Polymer Science, University of Akron, Akron, OH 44325*

Extended Abstract - From almost twenty years of research experience in nanotechnology, we have finally learned how to put nanoparticles 'together' to achieve intended objectives in reference to development of polymer nanocomposites. This keynote address draws from several areas of research in our laboratory to show why simple mixing of nanoparticles with host polymers should fail to produce desired results. Instead, our research delves into engineering at nanoscale and develops fundamental understanding of how nanoscale reinforcement can be put to work effectively.

The first example describes how self-assembly of 1-3 nm size polyhedral oligomeric silsesquioxane (POSS) molecules can bring effective reinforcement and causes three times higher draw of isotactic polypropylene (iPP) fibers. In our work we considered a type of POSS that was not compatible with iPP but could easily form self-assembled aggregates due to hydrogen bonding interactions. A sorbitol nucleating agent, dibenzylidene sorbitol (DBS) was used to promote molecular dispersion of silanol-POSS molecules in non-polar iPP. It was found that DBS formed low viscosity amorphous complexes at iPP processing temperature of 180-200 °C with tetra-silanol phenyl-POSS (tetra-POSS) and tri-silanol phenyl-POSS (tri-POSS), although DBS, tri-POSS, and tetra-POSS were crystalline solids with melting temperature above ~220 °C. The low viscosity complex helped reduce the viscosity of the compounds during processing, formed separate domains during cooling, and formed elongated reinforcing entities in spun fibers, as see in **Figure 1** [1-3].

The second example considers selective localization of POSS nanoparticles at the neck regions of secondary silica particles in silica aerogels to bring large enhancements in compressive modulus and higher hydrophobicity. In this case, silanol-POSS molecules reacted with the residual Si-OH groups after silica networks were formed from a sol-gel process. The resultant materials were converted into modified silica aerogels by supercritical drying. It was found that POSS-silica reactions reinforced the silica networks to a factor of 3 increases in compressive modulus and factor of 5 increases in contact angle of water, **Figure 2**. The talk will present chemical aspects of POSS-silica reactions and present why some POSS molecules are more effective.

A third example describes how simple fluid mechanical manipulations can give birth to side by side, co-axial, and Janus nanofibers to open door for numerous applications. A new, simple, and effective method is reported for production of polymer fibers ranging from a few tens of nanometers to a few micrometers from polymer solutions. The method capitalizes on a high velocity expanding gas jet to turn polymer solutions streaming from nozzles into fibers with smooth or wrinkled fiber surface morphology and with core-shell and side-by-side arrangements. The polymer solution is brought in contact with the gas jet on a flat surface, at the tip of a circular needle, and at the surface a pendant drop (**Figure 3**). The fiber diameter bears relationship with capillary number of the liquid jet and polymer concentration in the solution. Several levels of fiber conglutination are observed as function of collection distance from the nozzle set up [4].

Keywords: nanoparticles; nanofibers; polypropylene, aerogels.

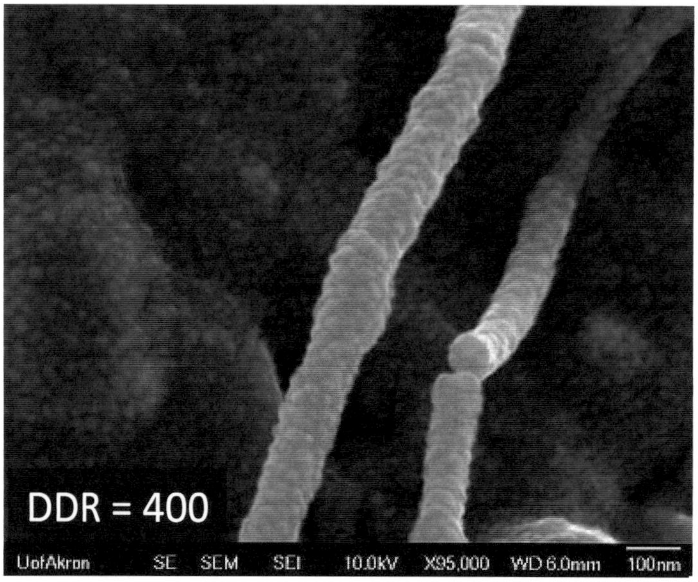

Figure 1. Cross-section of a melt-spun iPP fiber at draw-down ratio 400 showing ~100 nm fibrils of silanol-POSS/DBS amorphous complex.

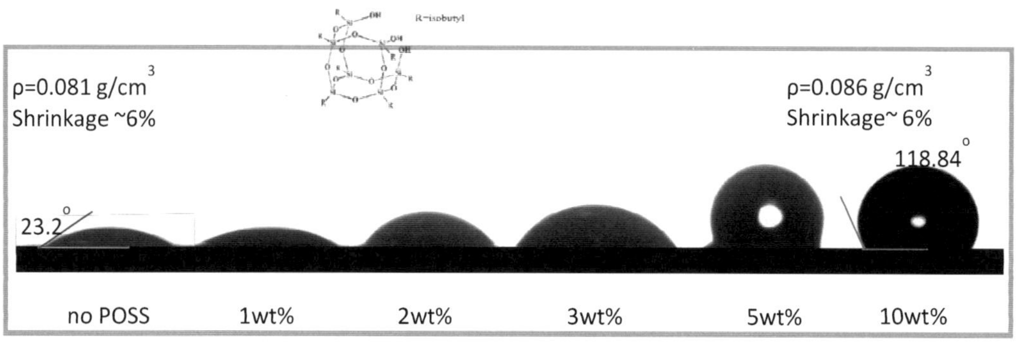

Figure 2. Dependence of water contact angle and density of aerogels on wt% of trisilanol isobutyl POSS in aerogel specimen. Contact angle was measured on compressed discs of aerogels.

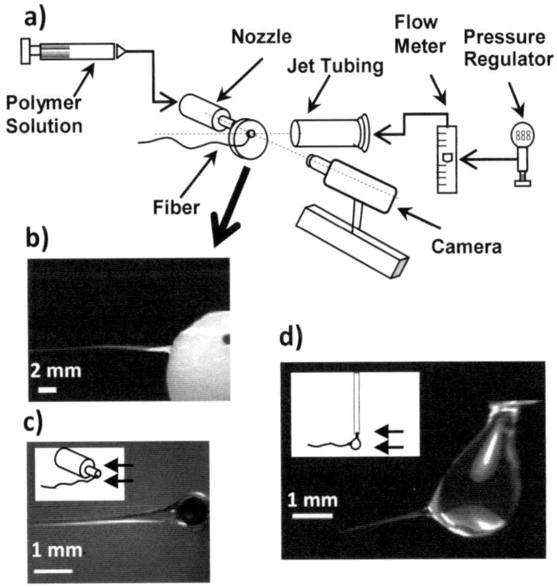

Figure 3. (a) Schematic setup of the GJF process with wall-anchored nozzle. (b–d) Images of polymeric liquid jet emanating from (b) wall-anchored nozzle, (c) needle-tip nozzle, and (d) pendant drop [4].

Acknowledgements
Financial assistance from U.S. National Science Foundation is gratefully acknowledged.

References

1. S. Roy, B.J. Lee, Z.M. Kakish, S.C. Jana, *Macromolecules*, 2012 45, 2420.
2. S. Roy, J. Feng, V. Scionti, S.C. Jana, C. Wesdemiotis, *Polymer*, 2012 53, 1711.
3. S. Roy, V. Scionti, S.C. Jana, C. Wesdemiotis, A.M. Pischera, M.P. Espe, *Macromolecules*, 2011, 44, 8064.
4. R. Benavides, S.C. Jana, D.H. Reneker, *ACS Macro Letters*, 2012, 1, 1032.

KN-07-01

ULTRAHIGH DENSITY ARRAY OF NANOPATTERNS ON BAROPLASTIC BLOCK COPOLYMER FILMS BY PHASE TRANSITION AT ROOM

Ara Jo and Jin Kon Kim*

National Creative Research Initiative Center for Block Copolymer Self-Assembly, Department of Chemical Engineering and Polymer Research Institute, Pohang University of Science and Technology, Pohang, Kyungbuk 790-784, Korea; jkkim@postech.ac.kr

Abstract We fabricated ultrahigh density array of nanopatterns on polystyrene-*block*-poly(n-pentylmethacrylate) copolymer (PS-*b*-PnPMA) film using an atomic force microscope (AFM) tip at room temperature. We also investigated the in-situ phase transition of the nanoscopic patterns during nanoindentation by using a transmission electron microscope (TEM) with a specially designed nanoindenter.

Keywords: Nanoscopic Pattern, Block Copolymer, Baroplasticity, in-situ TEM

Introduction

Various nanotechnologies such as electron-beam lithography, photolithography, microcontact printing, nanoimprinting, and scanning probe microscopy (SPM)-based lithography have been extensively employed for the fabrication of nano-sized patterns. For atomic force microscope (AFM)-based lithography, nanopatterns are generated by the mechanical contact between a nano-sized tip and a polymeric film surface at a predefined position.[1-3]

Here, we introduce a novel concept of "baro-lithography" for the fabrication of nanopattern on a polymeric substrate by AFM without using a heated tip. The nanopatterns are generated by the AFM tip at room temperature on a polymeric film and these patterns retain their original shapes at room temperature. This process is only possible by using a specific block copolymer polymeric film exhibiting baroplastic property that enables one to process at a relatively lower pressure and temperature by transforming microphase transition of the block copolymer. For this purpose, we used polystyrene-*block*-poly(n-pentyl methacrylate) copolymer (PS-*b*-PnPMA) thin film.[4-6] We clearly demonstrated via cross-sectional transmission electron microscopy (TEM) that the phase transition from nanodomain structures to disordered state indeed occurred during the indentation of AFM tip on the polymer film. We consider that the generated ultrahigh density array of nanopatterns could be used as next-generation storage media.

We also observed in-situ phase transition of the nanodomains on PS-*b*-PnPMA thin film during the indentation by using TEM with a specially designed nanoindentation holder. We found that the lamellar nanodomains were quickly transformed into the disordered region during the loading of the indenter tip on the film. Interestingly, the area corresponding to the disordered region is larger than the indented spots. Furthermore, the disordered region was slightly recovered during the unloading.

Experimental

Fabrication of nanopattern array on a baroplastic block copolymer thin film. A symmetric PS-*b*-PnPMA, having a styrene volume fraction of 0.5 and a number-average molecular weight (M_n) of 54,000, was prepared by anion polymerization. The 8 wt% of PS-*b*-PnPMA solution in toluene was spin-coated onto Pt-coated Si wafer at 3000 rpm for 90 seconds, and thermally annealed at 120 °C for two days which induced the film to have the lamellar microdomains oriented parallel to the substrate. Typical thickness of the film is ~270 nm. Nanopatterns are generated only by indentation of the AFM tip to the polymer film at room temperature. We used three kinds of AFM tips: point Probe Plus NCHR, AR5–NCHR, and AR5T–NCHR tip (purchased from Nanosensors™). The array of nanopatterns was fabricated in nanoindentation mode by using a probe microscope (SPM: XE-100, PSIA). Pattern depth was controlled by the maximum force of the indentation, which is changed by minimum value and maximum value in the indent control mode.

Characterization of Nanopatterns. Height images and depth profiles of the fabricated nanopatterns are obtained by using an atomic force microscope (AFM: XE100, PSIA). The cross-sectional image of PS-*b*-PnPMA film was observed by high resolution transmission electron microscope (HR-TEM: Jeol-1200 EX Ⅱ). For this purpose, the film was stained with ruthenium tetroxide (RuO_4) for 45 minutes; thus the dark region appearing in TEM images represents the PS phase. The patterned region was carefully cut by using the focused ion beam (FIB: SMI 3050, Seiko) and transferred on a TEM grid.

In-situ TEM Experiment: In-situ TEM Experiment was performed with TEM (JEOL JEM 2100F) working at 200 keV with a nanoindentation holder (Nanofactory™) having a diamond tip and operating in displacement-controlled mode with a nominal displacement rate of~0.5 nm/s. Videos of indentation process were recorded with a charge-coupled device (CCD) camera (Gatan Orius SC200D; Gatan, Pleasanton) at 40 frames per second.

Results and Discussion

Fig. 1a –d show AFM height images (each inset is the magnified image) and height profiles of nanopatterns prepared by the indentation of three different AFM tips at room temperature. Each pattern size in Fig. 1a is 4.5 nm in depth and the pitch distance is 62.5 nm; thus the pattern density is 170 Gb/in^2. Once a high aspect ratio tip with a tilted angle was used, the pattern density was increased to 260 Gb/in^2 (see Fig. 1b where each pattern is 4.0 nm in depth and the pitch between the patterns is 50 nm). The increased pattern density by using a tilted angle tip is attributed to the fact that this AFM tip can contact at a right angle (90°) to the substrate. If we carefully controlled the indentation force for a tilted angle tip, a pattern density up to 465 Gb/in^2 was generated (see Fig. 1c where each pattern is ~2 nm in depth and the pitch between the patterns is 37.5 nm). Furthermore, when another sharpened silicon tip was used, the pattern density was increased to 1.03 Tb/in^2, as shown in Fig. 1d, where each pattern is ~1.0 nm in depth and the pitch between the patterns is 25 nm.[7]

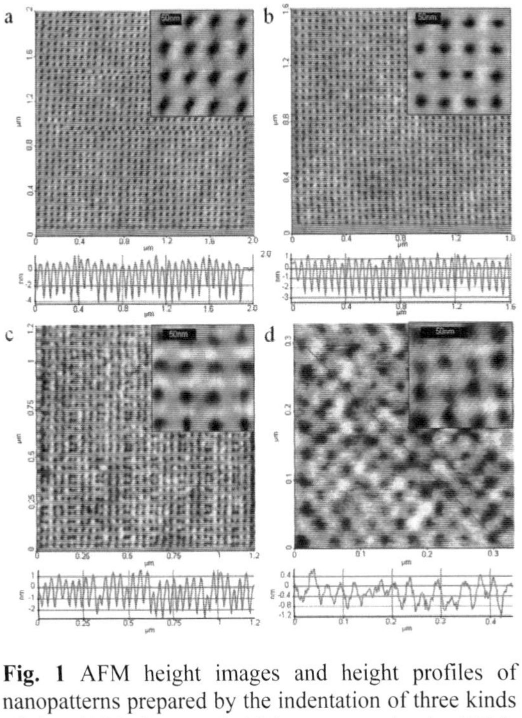

Fig. 1 AFM height images and height profiles of nanopatterns prepared by the indentation of three kinds of the AFM tips. **a.** A high aspect ratio tip (AR5-NCHR, Nanosensors^{TM}). **b, c.** A high aspect ratio tip with a tilted angle (AR5T-NCHR, Nanosensors^{TM}). **d.** A super sharpened silicon tip (SSS-NCHR, Nanosensors^{TM}).[7]

We found that the disordered state was clearly observed at the regions indented by the AFM tip, marked by the yellow dashed line. Namely, the original lamellar microdomains are completely destroyed and become the disordered state by the indentation of the AFM tip. This is because these regions have higher pressure than the critical pressure to induce the microphase transition into disordered state. Higher pressure could be achieved due to the smaller size of the AFM tip even though the indentation force is as small as the order of nN. Another evidence for the transformation of the lamellar microdomains into the disordered state during the indentation is the existence of the pile-up regions encircled by the blue dashed line near the nanopatterns. Also, the pile-up regions consist of the disordered state as well as substantial deformation of microdomains without transforming to the disordered state. If the microphase transition to the disordered state had not occurred by AFM indentation, the pile-up regions or flowing of the polymeric chains should not have been observed. This is the case of polystyrene-*block*-polyisoprene copolymer (PS-*b*-PI) film without having baroplastic properties. Namely, when the PS-*b*-PI thin film with parallel oriented lamellar microdomains was indented by the AFM tip at room temperature, the disordered state was not achieved, but the lamellar microdomains were just bent by larger forces.

From *in-situ* observation of TEM images of PS-*b*-PnPMA thin film during the indentation, we found that the disordered state is clearly observed at the regions indented by the diamond indenter tip. Very interestingly, the ordered lamellar microdomains are very quickly transformed to the disordered state, as soon as the tip contacts to the film surface. During the unloading of the indenter tip from the film, the generated disordered region is slightly (less than 10 % in area) reduced due to the strain recovery. These important results could not have been obtained without using the in-situ TEM experiment.

Conclusions

We have shown that a nanopatterned surface with an ultrahigh density array of nanoscopic indentations was fabricated on a baroplastic block copolymer thin film of PS-*b*-PnPMA at room temperature by AFM tip. The areal density was achieved up to 1.03 Tb/in^2. The nanopatterns with an ultrahigh density were achieved by the novel property of the baroplasticity of the block copolymer. Namely, the nanopatterns are generated by the microphase transition from lamellar microdomains to the disordered state by simple indentation of AFM tip. We consider that the generated a nanopatterned surface with an ultrahigh density array of nanoscopic indentations on baroplastic PS-*b*-PnPMA film could be used as next-generation storage media.

Acknowledgements

This work was supported by the National Creative Research Initiative Program

References

1. P. Vettiger; G. Cross; M. Depont; U. Drechsler; U. Durig; B. Gotsmann; W. Haberle; M. A. Lanntz; H. E. Rothuizen; R. Stutz; G. K. Binnig, *IEEE Trans. Nanotechnol.* 2002, *1*, 39.
2. B. Gotsmann; U. Duerig; J. Frommer; C. J. Hawker, *Adv. Mater.* 2006, *16*, 1499.
3. T. Altebaeumer; B. Gotsmann; H. Pozidis; A. Knoll; U. Duerig, *Nano Lett.* 2008, *8*, 4398.
4. D. Y. Ryu; U. Jeong; J. K. Kim; T. P. Russell, *Nat. Mater.* 2002, 1, 114.
5. D. H. Lee; H. J. Kim; J. K. Kim, *Macromol. Symp.* 2006, 240, 123.
6. D. Y. Ryu; D. J. Lee; J. K. Kim; K. A. Lavery; T. P. Russell, *Phys. Rev. Lett.* 2003, 90, 235501.
7. A. Jo; W. Joo; W. H. Jin; H. Nam; J. K. Kim, *Nature Nanotechnol.* 2009, 4, 727.

KN-07-05

MULTIFUNCTIONAL POLYMER COMPOSITES: HIERARCHICAL MICRO/NANO MULTI-SCALE HYBRID COMPOSITES FOR SENSING APPLICATIONS

E.T. Thostenson[1,2,3*], G. Pandey[1,2] and Q. An[1,3]

[1]*Center for Composite Materials,* [2]*Department of Mechanical Engineering,* [3]*Department of Materials Science and Engineering, University of Delaware, USA – thosten@udel.edu*

Abstract – Recent advances in the synthesis and characterization of nanostructured composite materials have enabled a broad range of opportunities for engineering the properties of polymer matrix materials. Building upon the science and technology of traditional fiber and particle-reinforced composites, the development of polymer composites reinforced with nanoscale materials has generated extensive research in industry, government and academia in studying their processing-structure-property relations. Carbon nanotubes are known to have exceptional mechanical, electrical and thermal properties. Because of their small size, carbon nanotubes can penetrate into regions between traditional micro-scale reinforcements and create a hierarchical micro/nano structure spanning several orders of magnitude in reinforcement scales. These hybrid composites have attracted broad interest because of their ability to tailor the matrix-dominated physical and mechanical properties of advanced fiber composites with minimal impact on the fiber-dominated mechanical properties. New approaches for processing these materials will be discussed and applications for *in situ* sensing of cracks and other changes presented.

Keywords: Carbon Nanotubes, Multifunctional Composites, Polymer Nanocomposites

Introduction

The field of nanotechnology has advanced rapidly over the past 10-20 years. These new nanoscale materials often possess unique material properties that are not observed in more macroscopic materials. Scientists and engineers have taken advantage of these novel properties by integrating the reinforcements in polymer systems to tailor their mechanical and physical properties. Polymer systems reinforced with nanomaterials, commonly referred to as nanocomposites, have opened-up a broad area of research related to their processing, characterization and modeling [1-3].

The change of the constituent reinforcing phase size from micrometers and larger of traditional composites to the nanoscale, where critical reinforcement dimensions are on the order of 100 nm and below, offers new opportunities in the tailoring of polymer mechanical and physical properties. Traditional composites based on glass or carbon fibers are typically aimed at improving structural (stiffness and strength) properties. Multifunctional materials are tailored to achieve both mechanical performance as well satisfy specific functional requirements. Carbon nanotubes have drawn specific interest in the development of multifunctional material systems due to their unique properties and fiber-like structure. Nanotubes can have extremely large aspect ratios (length/diameter) which is particularly desirable for both mechanical reinforcement and electrical property modification, where the nanotubes can form electrically conductive pathways in the insulating polymer matrix.

Taking advantage of their nanoscale size relative to traditional fibers along with their ability to form electrically conductive networks in the polymer matrix, carbon nanotubes can form nerve-like sensing networks surrounding the structural fiber (Fig. 1). The materials piezoresistive, in that the resistance measured during mechanical deformation is higher than expected based on a simple change in geometry, and have been introduced for sensor applications. The piezoresistivity is due to the nanotube/nanotube tunneling effect dominating the bulk resistance. As the nanotubes move away from each other under applied deformation the tunneling resistance changes and results in a large change in measured resistance.

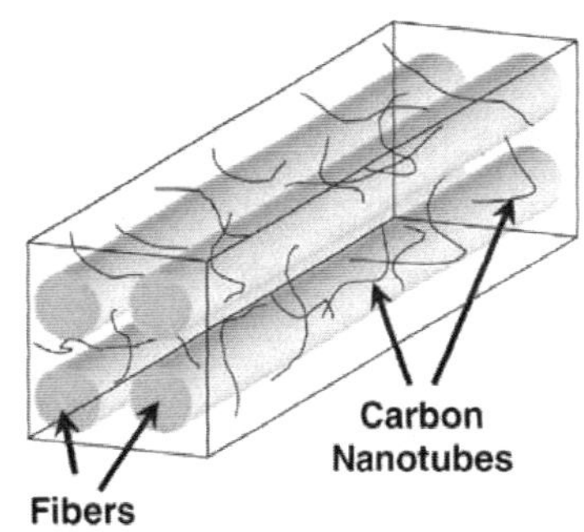

Figure 1 – 3-D model showing the formation of nerve-like conductive networks surrounding a micron-sized fiber reinforcement [2].

This growing interest in nanoscale hybridization with conventional reinforcement has highlighted the need to develop new processing techniques for integrating nanotubes with conventional fiber reinforcement and also modeling techniques which span the different reinforcement scales.

Processing of Multi-Scale Hybrid Composites
Recent progress in the development of techniques for producing multi-scale hybrid composites has been highlighted in an excellent recent review by Qian and co-workers [4], summarizing the processing and resulting properties of composites where there is continuous reinforcement of the polymer matrix as well as localized reinforcement due to growth or deposition of nanotubes directly onto fibers.

For the direct integration of nanomaterials with fiber reinforcing fibers two promising techniques have emerged and have been investigated in our laboratory. One approach involves the direct synthesis of carbon nanotubes onto fibers and the other approach involves the functionalization and dispersion of nanotubes in a solution followed by electrically-assisted deposition and integration into the fiber preform. After integration of nanotubes with the fibers a composite is then produced by infiltrating the polymer matrix.

Direct Chemical Vapor Deposition Synthesis
Thostenson *et al.* [5] first examined the growth of carbon nanotubes directly on the surfaces of carbon fiber and studied the influence of the local reinforcement on the load transfer at the fiber/matrix interface. For chemical vapor deposition (CVD) growth a catalyst was added to the carbon fibers and then the fibers were exposed to growth conditions in the presence of a hydrocarbon gas. The resulting structure (Fig. 2) shows carbon nanotubes grown on the fiber surface, forming a "fuzzy" fiber.

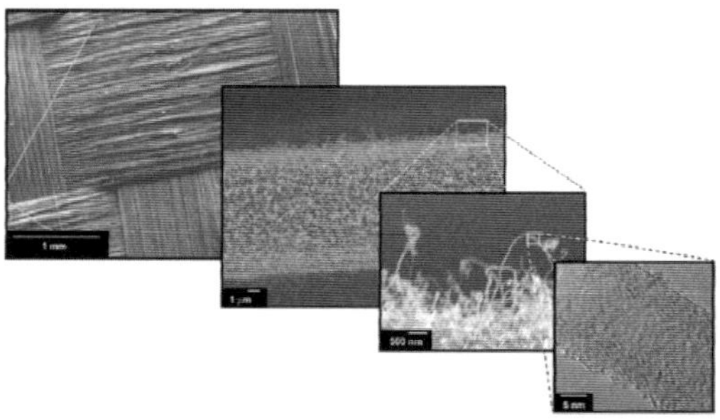

Figure 2 – Reinforcement scales from woven fabric to nanotubes – from left carbon fabric, single fiber with nanotubes grown via. CVD on the surface, tangled nanotubes, and nanotube wall structure [2].

One of the major disadvantages of this direct-growth process is that the exposure of the fibers to the CVD growth conditions can result in degradation of the fiber properties. In addition, the process, while scalable, is energy-intense, with growth temperatures typically in the range of 600-1000°C.

Electrophoretic Deposition
Electrophoretic deposition (EPD) is a widely used coating process utilized in applications ranging from automotive to electronics production. The EPD process deposits particles from solution and has the ability to homogenously coat complex shapes with well adhered films of controlled thickness and density and is a low-energy ambient temperature process. EPD has also been applied to produce multi-scale composite structures with carbon [6, 7] and glass [8, 9] fabrics using aqueous-nanotube solutions (Fig. 3).

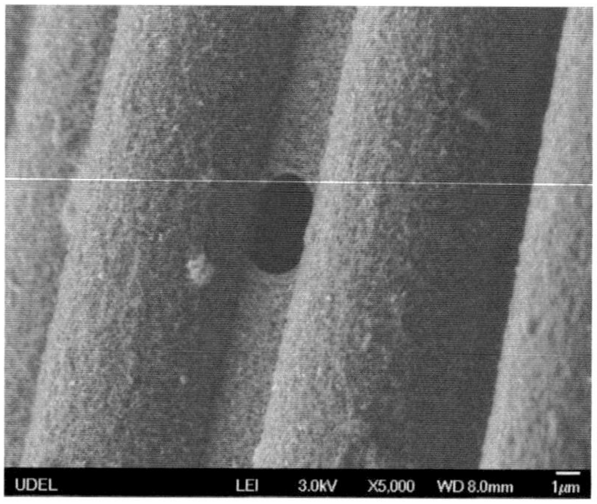

Figure 3 – Electrodeposited carbon nanotubes onto carbon fiber, showing uniform coverage and bridging of nanotubes between fibers [7].

Significant increases in the shear strengths and fracture toughness have been observed with the integration of nanotubes. Analysis of fracture surfaces reveals interlaminar regions with high levels of nanotubes and strong adhesion between the carbon nanotubes and sized carbon and glass fibers. The glass fibers are also electrically conductive, as conducting pathways have formed throughout the matrix phase.

Applications as Sensors
The ability of nanotubes to form conductive networks around structural fibers in polymer composites combined with the piezoresistive mechanical-electrical coupling enables nanotubes to act as a distributed sensor network for detecting both deformation and damage *in situ*. Thostenson [10] first reported the use of nanotubes as sensors for *in situ* sensing. Key experiments were conducted to evaluate the sensitivity of the nanotube network to damage accumulation in the form of transverse cracking and delamination. Results indicate high sensitivity to both the onset and

accumulation of damage. Experiments were later conducted to evaluate the electrical response under cyclic loading and a parameter was developed to quantitatively describe the state of damage in terms of the electrical behavior [11] (Fig. 4).

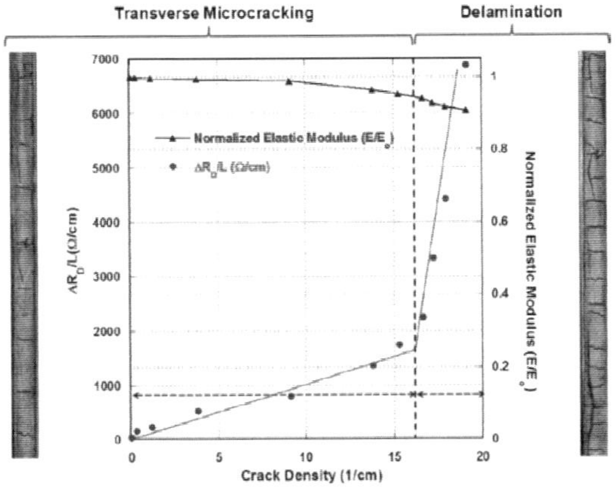

Figure 4 – Correlation of damage state and damaged resistance parameters in multi-scale hybrid composites with conductive networks of nanotubes [11].

As micro-scale damage propagates in the composite, the pathways of the conductive network of nanotubes become severed and result in permanent changes in resistance. The accumulation of micro-scale damage is difficult to detect using conventional non-destructive evaluation techniques whereas the changes in the electrical properties are very sensitive to the onset and accumulation of damage.

Pandey *et al.* [12] recently used electric time domain reflectometry (TDR) based measurement techniques for crack sensing in carbon nanotube modified composites. The technique has the potential to localize damage using only a single sensing line as opposed to the earlier approaches that yield a distributed measure of damage. The TDR technique was previously used to sense large-scale delaminations, but with the integration of carbon nanotubes it was possible to extend the approach to sense microscale cracks.

Conclusions

Advances in nanotechnology have facilitated structuring materials at nanometer-scales. Broadening of scales of reinforcements, from conventional fibers (microns) to nanoscale reinforcements (sub-100 nanometers), have presented profoundly new scientific and technical challenges and also new opportunities for designing materials for multifunctional applications. Before the multifunctional performance improvements can be fully realized new processing techniques and process-performance relations are needed. The application of these structured composites as self-sensing materials may result in improved reliability and safety of structural composites.

Acknowledgements

The authors would like to thank the support of the Air Force Office of Scientific Research (Young Investigator Grant FA9550-09-1-0218, Dr. Byung-Lip Lee, Program Director, and Phase II STTR Program FA9550-11-C-0003, Dr. David S. Stargel, Program Director), the US Army Program Executive Office: Soldier and the Army Research Laboratory (Cooperative Agreement W911NF-06-2-011), and the Office of Naval Research (N00014-06-1-1000) and the National Science Foundation National Science Foundation (Grant # 1138182, Dr. Mary Poats, Program Director).

References

1. E. T. Thostenson; Z. F. Ren; T. W. Chou *Compos. Sci. Technol.* 2001, *61,* 1899.
2. E. T. Thostenson; C. Y. Li; T. W. Chou *Compos. Sci. Technol.* 2005, *65,* 491.
3. G. Pandey; E. T. Thostenson, *Polymer Reviews* 2012, *52,* 355.
4. H. Qian; E. S. Greenhalgh; M. S. P. Schaffer; A. Bismarck *J. Mater. Chem.* 2010, *20,* 4751.
5. E. T. Thostenson; W. Z. Li; D. Z. Wang; *et al. J. Phys. D: Appl. Phys.*, 2002, *36,* 573.
6. E. Bekyarova; E. T. Thostenson; A. Yu; *et al. Langmuir,* 2007, *23,* 3970.
7. Q. An; A. N. Rider; E. T. Thostenson *Carbon,* 2012, *50,* 4130.
8. Q. An; A. N. Rider; E. T. Thostenson in 44th International SAMPE Technical Conference Proceedings, 2012, CD.
9. Q. An; A. N. Rider; E. T. Thostenson in American Society for Composites 27th Annual Technical Conference proceedings 2012, CD.
10. E. T. Thostenson; T. W. Chou *Adv. Mater.* 2006; *18*, 2837.
11. L. M. Gao; E. T. Thostenson; Z. G. Zhang *et al. Adv. Funct. Mater*, 2009, *19*, 123.
12. G. Pandey; M. Wolters; E. T. Thostenson; D. Heider, *Carbon*, 2012, *50,* 3816.

O-07-310

PARTIALLY RESOLVED CHARACTERIZATION OF THERMO-MECHANICALLY GRADED SELF-REINFORCED POLYPROPYLENE COMPOSITES

H.-P. Heim, A. Ries* and B. Rohde
Institute of Materials Engineering, Chair Polymer Technology, University of Kassel, Germany
heim@uni-kassel.de; angela.ries@uni-kassel.de; bjoern.rohde@uni-kassel.de

Abstract: Self-reinforced polypropylene composites possess enormous potential for the implementation of property gradation due to their basic structure. During the compression moulding process, varying processing temperatures and pressures are used to influence the self-reinforcement, and, ultimately, the later composite properties. In the course of this study, the employed basic materials will be introduced and the processing procedure for effective thermo-mechanical gradation will be explained. Subsequently, the material specific gradation potential will be presented and assessed using the summary of the localized strain measurements made using the GOM® method.

Keywords: Self-Reinforced Polypropylene Composites, Thermo-mechanical Gradation.

Introduction

Self-reinforced Polypropylene Composites (SR-PP) impress by means of their exceptional mechanical properties in combination with a low density of 0.91 g/cm³ [1]. In addition, they possess an interesting material concept. The composites are exclusively compacted using layered, thermoplastic, semi-finished textile products, for example fabrics or fleeces. They can be specifically tailored for almost any application by varying the textile layers and types. The fiber reinforcement is based solely on the self-reinforcement of the polymer, which was previously implemented by melt and solid phase deformation [2]. Here, the crystalline structures of the fibers and tapes are oriented on the macromolecular level, so-called shish-kebab structures are induced [3]. During the later manufacture of composites, the oriented sections are embedded in an identical polymer matrix and contribute to a significant improvement of the composite properties. Despite all named advantages, the self-reinforcement remains extremely temperature and pressure sensitive. Even small changes of the parameters can verifiably affect the later composite properties [4]. In turn, these changes also affect the aim to realize the component properties in accordance with the functional requirements [5].

Composite Manufacture

Basic Textile Materials

In this paper, three different self-reinforced polypropylene composites will be examined and compared in regards of their gradation potential. According to Table 1, three different semi-finished textile products on the basis of polypropylene, which have varying textile structures, will be used.

The plain weave with the name Pure is a tape fabric with optimised properties that is produced by the company Lankhorst, Netherlands [6]. Due to ist co-extruded, low-melting marginal layers, it is especially suitable for the hot compaction process. The internal, highly stretched core layer of the PP fabric tape and the exceptionally good mechanical properties are for the most part preserved and remain reproducible.

The mono-extruded semi-finished textile products react essentially more affinely for the thermo-mechanical gradation. Here, the tapes or fleece fabrics consist of an identical polymer phase. Therefore, the temperature and pressure exposure significantly influence the generation of the matrix phase and simultaneously also the embedding of the matrix identical reinforcement phases during the compression moulding process [7]. To be precise, these are the factors upon which the high gradation potential depends.

For this purpose, the mono-extruded tape fabric SG 30/30, produced by Bonar Technical Fabrics, Belgium, is employed [8]. This material is a 2/1 twill weave made from highly-oriented polypropylene tapes. Furthermore, the material system SNW 17, also produced by Bonar Technical Fabrics, Belgium, is used [9]. It consists of a needle-punched layered fleece that was manufactured using fine polypropylene filaments with a diameter of approximately 25µm.

Table 1. Semi-finished textile products employed for the compression moulding of SR-PP composites [6-9]

Product	Max. Elongation	Density	Area weight	Number of Layers in Composite
-	%	[g/cm³]	[g/m²]	-
Pure	6	0.760	105	16
SG30/30	17	0.909	124	16
SNW17	50	0.930	150	12

Compression Moulding Process

In order to be capable of achieving laminate thicknesses of approximately 2 mm, a total of 16 layers of fabric, or 12 layers of the fleece system are utilised. However, before the actual processing procedure is carried out, the highly-oriented semi-finished textile products must first be fixated in a tenter frame. This is necessary to prevent thermally induced relaxation processes, such as warpage. Moreover, it is meant to ensure the manageability of the otherwise flaccid basic products during processing.

The thermo-mechanical gradation of layered composites is realized in an ensuing compression moulding process with an upstream IR pre-heating sequence [10]. The exemplary processing parameters that are introduced in the course of the experiments can be found in Table 2.

During the pre-heating sequence (t_{IR}) partially thermal masking is induced by means of aluminium sheets that are manually inserted into one half of the IR pre-heating station for a predetermined masking time (t_{mask}). This relatively simple, yet effective implementation enables a thermal shading of the semi-finished textile products located beneath the sheets. This has been proven to verifiably affect the later composite properties [11].

Subsequently, the pre-heated, semi-finished textile products are transferred via the tenter frame into the moulding tool, which is already tempered to the processing temperature, and the compression moulding process begins. The moulding tool, which was especially designed for thermo-mechanical gradation, possesses a base area of 350 x 350 mm^2 and has a triangular geometry in its centre. The triangle has an aperture angle of 110° that leads to a reduction of pressure on the slants of the triangular geometry of consistent 18% (p_2) in contrast to tools with flat areas (p_1). Thus, it fulfils the requirements of mechanical gradation.

In addition, the compression mould is equipped with a total of two separate tempering circuits, which allow the tool halves (right and left) to be tempered to two individual and differing pressing temperatures (T_1 and T_2) for an effective thermal gradation during composite manufacture.

Here, it is important to note that a homogeneous temperature distribution is exclusively on hand on the tool halves. The desired gradual distribution of the temperature is achieved due to the construction of the triangular geometry slant, and is a consequence of the uneven heating of the modular tool application. Temperature creeping processes result on the peak of the triangular geometry of the tool. This is caused by the reciprocal influence of between the two, differently tempered tool halves.

Upon approaching the completion of processing, an isobaric cooling phase is initiated. Subsequently, the laminate can be removed from the pressing unit and demoulding completed.

All in all, four laminate zones, or sample positions, result from the performed thermo-mechanical gradation. They have been summarised with the according processing parameters (Table 2) in Figures 1 and 2.

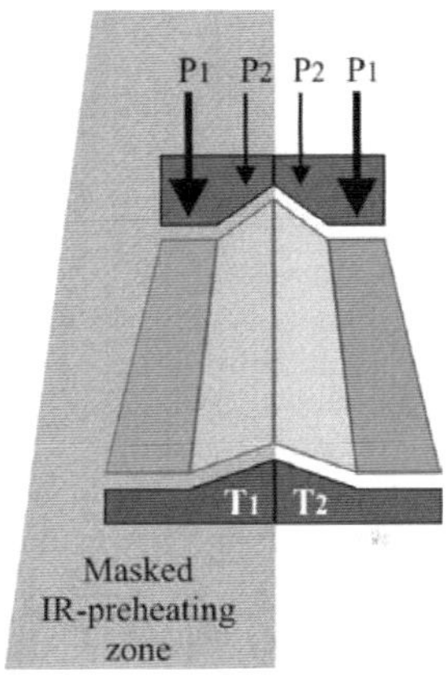

Figure 1: Process-induced gradation realised during processing; thermal gradation completed using a partially masked IR preheating station and thermo-mechanical gradation during the compression moulding process

Table 2. Processing parameter and designated sample positions of the thermo-mechanically graded composite

	IR Pre-heating			**Compression Moulding**				
	IR-Temp. T_{IR} [°C]	**IR-time** t_{IR} [sec]	**Masking Time** t_{mask} [sec]	**Pressing Time** t_{comp} [sec]	**Temperature / Pressure** T_1 [°C] / p_1 [MPa]	**Temperature / Pressure** T_1 [°C] / p_2 [MPa]	**Temperature / Pressure** T_2 [°C] / p_2 [MPa]	**Temperature / Pressure** T_2 [°C] / p_1 [MPa]
	160	165	90	89	185 / 1.2	185 / 0.8	200 / 0.8	200 / 1.2
Sample Position			a, b		a	b	c	d

Localized Strain Measurement

Sample Preparation

By means of utilizing the GOM® method during a standardised, quasi-statistical tensile test, the surface elongation of the SR-PP composites was measured and the completed gradation verified. Flat samples with an area of 2 x 15 x 160 mm³ were manufactured with packaging that required special handling. In order not to unnecessarily remove the layers of the consolidated layered composites during cutting, and risk a loss of samples, a two-step separation process was employed. The samples were roughly pre-cut on a circular saw with a thermoplastic saw blade, and, subsequently, fixated in a clamping device. Then, using a face cutter with a high rpm and medium feed, the samples were tailored to their finished size.

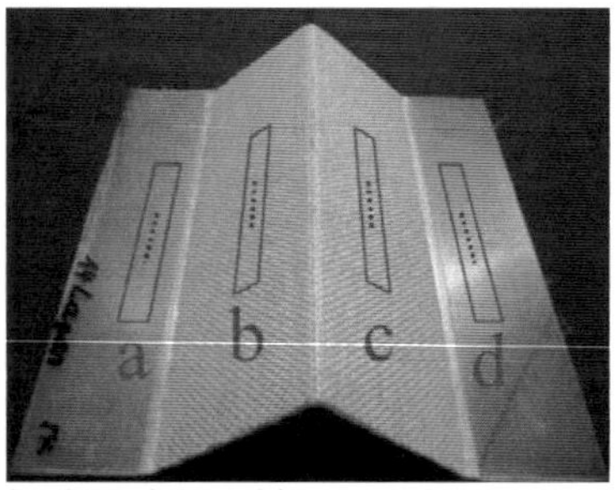

Figure 2. Sample position on the composite with according sample labelling and marking of the GOM® cutting line (red)

Furthermore, the sample surfaces had to be modified for the GOM® examinations, so that they would display a high-contrast and dense speckled pattern. Before testing, the samples were painted with a white prime coat, and, afterwards, the speckled pattern was applied with a black paint for contrast, see Figure 3.

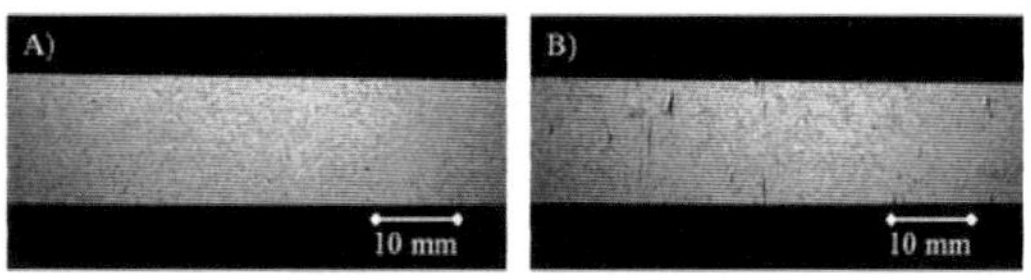

Figure 3. Coding on the sample SG30/30 b; A) intact sample preparation for GOM® measurements; B) Tearing of the sample coating caused by crack initiation in the case of higher strains

Performed Tensile Tests

In order for the material testing to be reproducible, the prepared flat samples were conditioned in accordance with DIN EN ISO 291 at room temperature (23°C) and 50% rel. humidity for 24 hours in a standard atmosphere. The strain properties were determined in a quasi-statistical tensile test related to DIN EN ISO 527-4 for isotropic and anisotropic fibre-reinforced plastic composite materials. The samples (2 x 15 x 160 mm³) were stretched along their main axis at a constant testing speed of v = 5 mm/min till failure occurred. The tensile tests were performed on a Schenck Hydropulser PC160M that is equipped with a GOM® camera system and a 16kN load cell. The clamping length of the samples was set at l_E = 100 mm, and the constant measuring length of the camera focus was set to approximately l_M = 50 mm. The image rates were consistently taken every 100 N. Therefore, it was possible to determine the exact occurrence of surface damage (tearing of the sample coating, Figure3, B) or even loss of the sample marking by means of chipping while controlling the strength. The evaluation was performed via ARAMIS® software.

Results and Discussion

Strain Distribution of Selected Samples Across Varying Strength Levels

In the following Figures 4 to 6 the strain distributions of the respective material systems Pure, SG 30/30 und SNW17 are illustrated across the strength levels using sample c as an example.

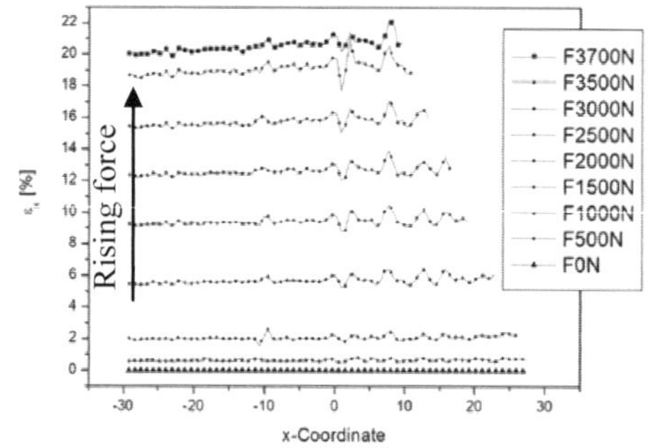

Figure 4. Strain distribution of Pure, Sample c in dependence of varying strength levels

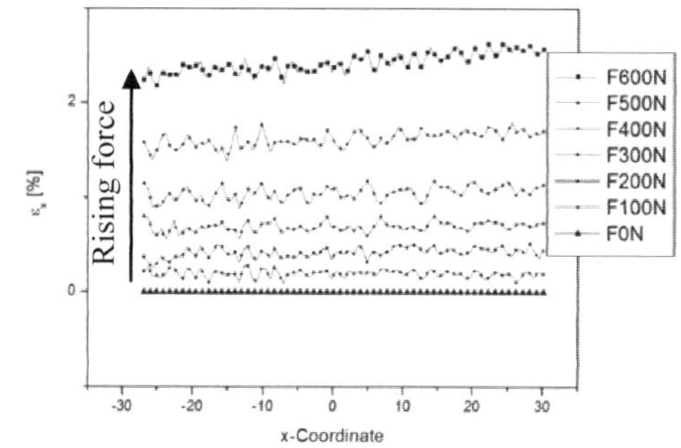

Figure 5. Strain distribution of SG 30/30, sample c in dependence of varying strength levels

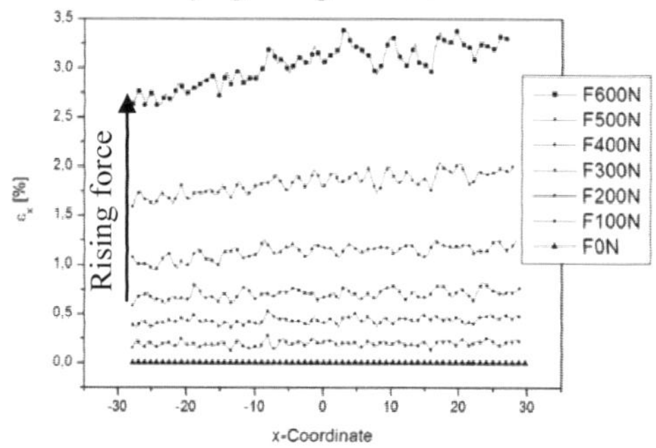

Figure 6. Strain distribution of SNW17, sample c in dependence of varying strength levels

This comparison reveals that the fleece-based composite system SNW 17 operates on an essentially higher strain level in than the fabric-based material systems Pure and SG 30/30 with comparable load. The above all highest strength level at F=3700 N, however, was achieved by the co-extruded Pure composite.

Surface Scans for the Observation of the Strain Distribution on the Laminate Surface

In order to obtain a comparative impression of the strain distribution across the laminate surface, the surface scans of three material classes across the sample positions a – d were compared in Table 3. Due to strongly differing material specific strain behaviour, individualized scaling was used here. Furthermore, the highest strength level of the according sample was selected, which is also shown in Table 3.

Table 3. Surface scans of the three material classes (Pure, SG 30/30 und SNW17) across the sample positions a – d

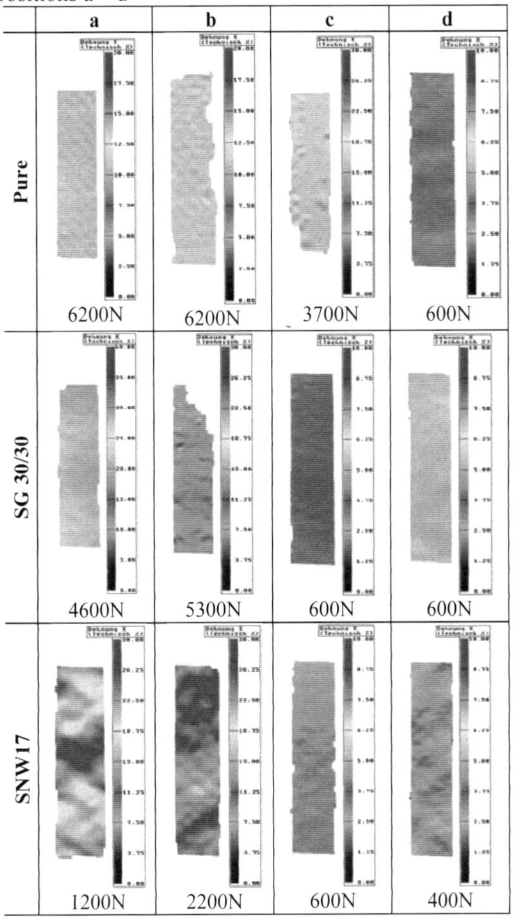

Conclusions

The comparatively high strength levels of the co-extruded Pure composite are a result of the material structure of the basic material. This enables an essential preservation of the inner self-reinforced material core throughout the processing procedure. Simultaneously, this is connected to a relatively low surface strain. Ultimately, a combination of a high processing temperature and pressure (sample d) shows that a significant strain loss is an indicator for a strongly consolidated compact composite.

The mono-extrudate based material systems SG 30/30 und SNW 17 display an inhomogeneous strain distribution in dependence of the processing parameters (samples a to d) and confirm their temperature and pressure sensitive material characteristics. In the case of the fleece system SNW17, this is manifested even stronger due to the fine filament cross-sections of approximately 25 µm. An effective thermo-mechanical gradation in regards of the strain behaviour can be verified for the mono-extrudates in this study, especially for the fleece system SNW 17.

Acknowledgements

This paper is based on investigations carried out in the collaborative research centre CRC/TRR 30, which is kindly supported by the German Research Foundation (DFG).

References

1. J. Klimke; *Kunststoffe.* 2002, *92,* 45.
2. A. Kmetty; T. Barany; J. Karger-Kocsis; *Prog. in Polymer Sci.* 2010, *35,* 1288.
3. G. W. Ehrenstein, *Polymer Werkstoffe*, Carl Hanser Verlag, München Wien, 1999.
4. H.-P. Heim; A. Ries; A.K. Bledzki in SEICO 11 Proceedings, Paris, CD.
5. A. K. Bledzki, D. Paßmann, A. Ries in *Functionally Graded Materials in Industrial Mass Production*, K. Steinhoff; H. J. Maier; D. Biermann, Ed.; Verlag wissenschaftliche Skripten, Auerbach, 2009; Vol. 1, 111-128.
6. N.N.: Technical data sheet, Pure, Lankhorst, NL-8600 AE Sneek – Netherlands
7. H.-P. Heim; A. Ries; B. Rohde in SEICO 12 Proceedings, Paris, CD.
8. N.N.: Technical data sheet, SG 30/30, Bonar TF, B-9240 Zele – Belgium
9. N.N.: Technical data sheet, SNW17, Bonar TF, B-9240 Zele – Belgium
10. A. K. Bledzki; A. Ries; D. Passmann; H.-P. Heim in ANTEC 2012 Proceedings, Orlando, CD.
11. A. K. Bledzki; D. Passmann; A. Ries; A. Chate; *MP Materials Testing*, 2008, *50,* 623.

O-07-175

POLY (VINYL ALCOHOL)/INORGANIC FILLERS COMPOSITES WITH HIGH PERFORMANCES PREPARED BY THERMAL PROCESSING

Li Li, Bin Wang, Ning Chen, Xiao Chen, Qi Wang*

State Key Laboratory of Polymer Materials Engineering, Polymer Research Institute of Sichuan University, Chengdu 610065, China – powerlily@scu.edu.cn; wangbin841124@126.com; ningchen@scu.edu.cn; chenxiao19871209@gmail.com; qiwang@scu.edu.cn

Abstract - This paper reported several novel highly filled PVA based composites with good comprehensive properties prepared by thermal processing based on molecular complexation and plasticization. The inorganic fillers used were talc, $CaCO_3$, $Mg(OH)_2$ and hydroxyapatite (HA), and the filler content could be up to 80wt.% without using any coupling agent or compatibilizer. The highly filled PVA based composites have wide thermal processing window and suitable melt viscosity for extrusion or injection molding. A special filler network could be formed in PVA matrix ascribing to the strong shearing force of screw, and apparently increased the heat deflection temperature (HDT) and the mechanical properties of PVA. By combining the thermal processing with solid state shear milling (S^3M) technologies, the well dispersion of nano-HA in PVA matrix was achieved and the micro-injection of PVA/nano-HA composite was realized.

Keywords: Thermal processing, highly filled PVA composites, high performances

Introduction

Poly (vinyl alcohol) (PVA), one of the few polymers that can be synthesized via non-petroleum routes, has excellent comprehensive properties[1, 2], such as superior mechanical properties and thermal properties, excellent barrier property, acid, alkali and organic solvents resistance, and good compatibility with organic and inorganic materials, etc., thus can be expected to be used as a novel plastic to prepare high performances PVA based materials without exter-addition of compatibilizers and coupling agents. The key is to realize the thermal processing of PVA.

As known, PVA is quite difficult to be thermal processed due to its close melting point and decomposition temperature, and thermal processing of PVA is still a big challenge in the world now. Based on molecular complexation and plasticization, our research group has successfully realized the thermal processing of PVA, and obtained high performance PVA fibers, films, sheets, bottles and foams via melt spinning, extrusion-blowing, injection molding and thermal foaming of PVA [3], which provides the possibility to prepare PVA based composites by thermal processing.

In this paper, we further develop the thermal processing of PVA by introducing inorganic fillers including talc, $CaCO_3$, $Mg(OH)_2$ and nano-hydroxyapatite (nano-HA) into PVA to prepare PVA/inorganic filler composites, especially highly filled PVA based composites, and studied the thermal processability, the structure as well as the properties of the composites. In order to obtain the ideal dispersion of inorganic fillers in PVA matrix, a novel solid state shear milling (S^3M) technology [4] developed in our laboratory was adopted, and the effect of S^3M on structure and properties of the composites were also studied.

Experimental

Materials

PVA used in this study was provided by SINOPEC Sichuan Vinylon Works with polymerization degree of 1750 and hydrolysis degree of 99%. Thermoplastic PVA (TPVA) was self-prepared in our laboratory.

Preparations of PVA based composites

Inorganic fillers were directly mixed with TPVA, or co-milled with PVA, and then mixed with TPVA in a high-speed mixer. The mixtures were then extruded and pelletized on a single or twin-screw extruder, and injected on different injection molding machines to prepare PVA based composites.

Measurements of modified PVA

Melt behavior of samples was analyzed on TA-Q20 (USA) under nitrogen atmosphere from 40°C to 230°C at a rate of 10°C/min. Thermogravimetric analysis of samples was conducted on TAQ-500 (USA) under nitrogen atmosphere from room temperature to 600°C at a heating rate of 10°C/min. The rheological behavior of samples was tested on Rheoraph2002 high-pressure capillary rheometer. The dispersion behavior of filler particles in the composites was observed by using an Inspect F (FEI) SEM instrument. Heat Distortion Temperature (HDT) of samples were tested on HDV2 (ATLAS, USA) at a heating rate of 2°C/min and a load of 0.455MPa. The mechanical properties of samples were tested on RG-L-10 (China) with the drawing speed of 50 mm/min.

Results and Discussion

Poly(vinyl alcohol) (PVA) is a special commodity polymer with multi-hydroxyl groups and strong polarity, thus are able to form hydrogen bonding or coordination interaction or electrostatic interaction with the polar groups and metal ion of inorganic filler particles. It is these special interactions that lead to the excellent compatibility of PVA with many inorganic particles. The interactions between inorganic particles and PVA show two different effects on the melt behavior of the composites. On the one hand, these interactions can further weaken the self-inter and intra-molecular interactions among the hydroxyl groups of PVA, and disturb the crystallization of PVA, which is likely to decrease the melting temperature of TPVA. On the other hand, these interactions can take the role of physical crosslinking points, thus inhibits the movement of PVA molecular chains to a certain extent and trends to increase the melting temperature of TPVA. In this two competing factors, the melting temperature of PVA based composites is almost unchanged or even decrease when compared with TPVA (Fig. 1). The new formed interactions between inorganic filler and PVA also reduce the dehydration reaction of the adjacent hydroxyl groups in PVA molecular chains, so apparently retards the first step degradation of PVA. That means PVA/inorganic filler composites have wider thermal processing windows than TPVA.

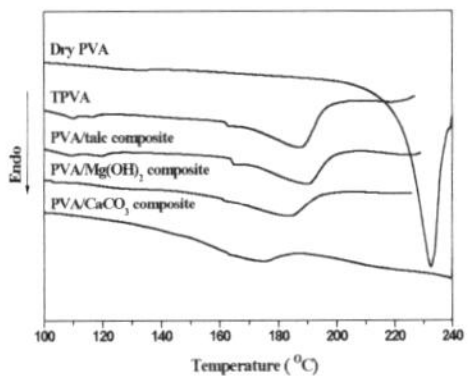

Figure 1 DSC heating curves of PVA and PVA based composites with 40wt% filler content.

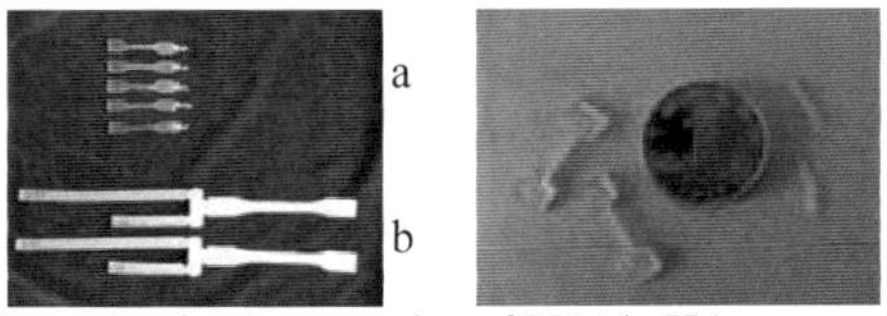

Figure 2 Injection samples of PVA/n-HA composites (a) small scale injection samples; (b) conventional injection samples; (c) micro-injection samples

However, the studies on the melt rheological behavior of PVA based composites show contrary results. The melts of TPVA and PVA based composites are typical shear-thinning fluid and show pseudoplastic characteristic. After adding inorganic filler, the stronger interactions formed between filler and PVA and the special network formed among filler particles would not only block the movement of PVA molecular chains, as mentioned above, but also act as a barrier to the diffusion of plasticizer in PVA by extending its diffusion path and suppress its plasticization on PVA during the melt process, thus increase the melt viscosities of the composites. Even though, the melt viscosity of the composites with up to 80wt% filler content still satisfies the requirement of thermal processing, and a lot of samples have been obtained through extrusion, conventional injection molding as well as small scale injection molding, and even micro-injection molding (Fig.2), which demands the better processabilities than conventional injection molding.

Generally speaking, the dispersion quality of fillers in polymer matrix has a direct effect on properties of the final composites. Different type of fillers may display different morphology in matrix. Compared with the particulate $CaCO_3$, $Mg(OH)_2$ and HA, the asymmetric lamellar structure of talc is liable to be dissociated and oriented along the melt flow direction, when suffered from the strong shearing and mixing force of the screw during the extrusion or injection processing (Fig. 3a). This special oriented filler network apparently increased the heat deflection temperature (HDT) and the mechanical properties of PVA. For example, when talc content was 50wt.%, the HDT, tensile strength and modulus of the PVA/talc composite were 115°C, 48MPa, 1.23GPa, respectively, increased by 92%, 16% and 150% as compared to the TPVA.

By further combining the thermal processing with solid state shear milling (S^3M) technologies, the well dispersion of filler particles, e.g. nano-HA, in PVA matrix was achieved (Fig.3b) , which could be used for micro-processing.

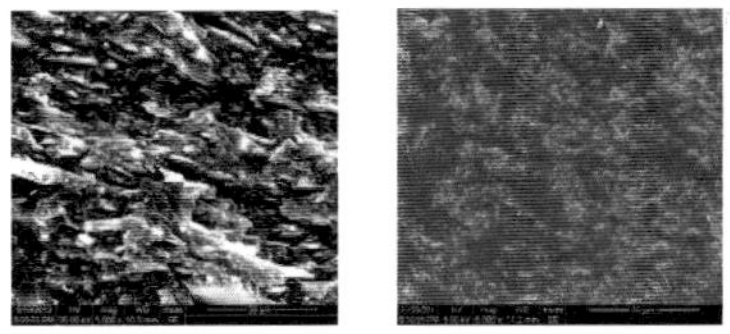

a b

Fig.3 SEM image of PVA based composites with 40wt% talc (a) and nano-HA (b).

Conclusions

Based on our innovated thermal processing of PVA, several high performance PVA/inorganic filler composites were successfully prepared by adopting molecular complexation and plasticization as well as S^3M technologies. The filler content was up to 80wt%. These highly inorganic filler filled PVA composites greatly expand the applications of PVA.

Acknowledgements

This work is supported by the National Natural Science Foundation of China (50833003, 51010004).

References

1. R. Yan: 'Water soluble polymers', 42; 1998, Beijing, Chemical Industry Publisher.
2. E. M. James. Polymer Data Handbook. New York & Oxford, Oxford University Press, 1999
3. N. Chen, L. Li, Q. Wang. Plastics, Rubber and Composites, 2007, 36(7-8): 283 ~ 290
4. X. Xu, Q. Wang. CN patent, ZL95111258.9, 2001

O-07-198

INJECTION MOLDING COMPOUNDING OF PP POLYMER NANOCOMPOSITES

M.G. Battisti[1]*, W. Friesenbichler[1]

[1] *Department Polymer Engineering and Science – Chair of Injection Molding of Polymers, University of Leoben (MUL), Austria – markus.battisti@unileoben.ac.at*

[1] *Department Polymer Engineering and Science – Chair of Injection Molding of Polymers, University of Leoben (MUL), Austria – walter.friesenbichler@unileoben.ac.at*

Abstract - The injection molding compounding process combines two processing steps; the material compounding and the injection molding process. The final part is processed directly by "only one heating process", therefore the material has not to be granulated and plasticized again after compounding. The molten material exits the extruder and is directly fed into a melt accumulator and further into the injection molding machine. The accumulator serves as a buffer element between the continuously-operating extruder and the discontinuously-operating injection molding machine. In contrast to many other industrial installations the worldwide unique MUL Polymer NanoComposite Injection Molding Compounder (PNC-IMC) features some special devices, which are necessary for the production of polymer nanocomposites.

One main objective of our research is to improve intercalation and exfoliation of layered silicates in polymer melts, thus the optimal potential of this mineral filler group can be realized. Furthermore, the differences in between the (single-heat) one-step process in the PNC-IMC and the conventional two-step process, using a compounder, has been monitored. Another experiment was the manufacturing of polypropylene nanocomposites in a masterbatch. The masterbatch pellets were fed into the twin-screw extruder again and then diluted with PP. Thermal conductivity measurements based on the "Line-Source" method as a function of pressure and temperature were carried out.

Keywords: Injection molding compounding, Polymer nanocomposites, Thermal conductivity

Introduction

Injection Molding Compounding

The injection molding compounder (IMC) concept was firstly presented by Krauss Maffei at the K1998 for long glass-fiber reinforced thermoplastics. Its characteristic feature is a twin-screw extruder, which is directly integrated into an injection molding system.

The injection molding compounding process combines two processing steps; the material compounding which normally takes place at the raw material manufacturer and the injection molding process which usually is done at the injection molder. The final part is processed directly "by only one heating process" and the material used has not to be granulated and plasticized again after the compounding step in the plasticizing unit of an injection molding machine. The molten material exits the extruder and is directly fed into a melt accumulator and further into the injection unit. The accumulator serves as a buffer element between the continuously-operating extruder and the discontinuously-operating injection unit. The outcome is greater flexibility for the processor when improving both the component structure and the quality of the molded part, as well as demonstrably lower production costs [1, 2].

The MUL polymer nanocomposite injection molding compounder (PNC-IMC) described below (Fig. 1) is the world's first special adapted IMC for manufacturing and injection molding of polymer nanocomposites. It consists of a ZSE 27 MAXX Leistritz-Compounder (Leistritz Extrusionstechnik GmbH, Nürnberg, Germany) and an 1800 kN fully-electric injection molding machine (Engel Austria GmbH, Schwertberg, Austria). A 3-way-valve and its special arrangement allows for using the injection molding compounder as "3-in-one" production line either for injection molding compounding, for injection molding or as a separately acting twin-screw compounder line. Furthermore, this PNC-IMC can also be used for the homogenization of fillers in high temperature polymers (HTP) in a temperature range up to 400 °C.

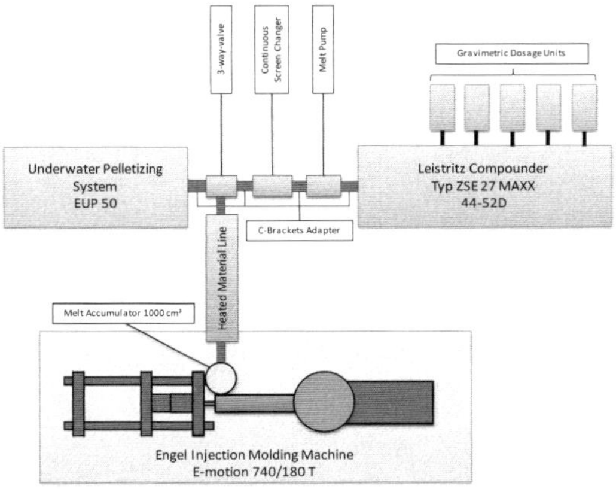

Figure 1. System schematic of the MUL Polymer NanoComposite Injection Molding Compounder (PNC-IMC)

Compounding of polymer nanocomposites
''Melt mixing'' (compounding) is industrially the most attractive method for polymer nanocomposites preparation due to its technological simplicity. The usage of common polymer processing machines is a much more pleasant way than using special equipment and procedures in chemical laboratories, which are required for ''in situ'' or ''solution'' methods [5].
From the economical point of view, the use of a co-rotating twin-screw extruder as a continuous processing technology is industrially preferred to melt mixing in a discontinuous kneader. It is obvious that for the preparation of polymer nanocomposites and the successful dispersion of silicate plates in a polymer melt high shear energy (usage of kneading blocks and high screw speed) and long residence time are necessary. Based on preliminary work at MUL it is now possible to realize high shear energy as well as long residence time (which are normally opposite in the extrusion process) by implementing a melt pump in front of the twin-screw extruder [6]. Further possibilities to diversify the residence time of the polymer melt are the screw length of the extruder, and the processing in an IMC, which both is possible with the PNC-IMC described.
There are many activities for the use of nanoclays in polymer systems ranging from the enhancement of structural and mechanical properties to functional materials with a tailored performance profile for thermostability, flame retardancy, barrier properties or enhancement of thermal conductivity [7, 8]. Our aim is on the one hand the improvement of the mechanical properties [9], and on the other hand, materials with higher thermal conductivity to reduce the cooling period of injection molded parts and blow-molded parts with medium to high wall thicknesses, respectively.

Thermal conductivity of polymers
The thermal conductivity is the most commonly used property that helps to quantify the transport of heat through a material. In polymers the thermal conductivity is based on two different mechanisms. On the one hand heat is transferred through Van-der-Waals forces, and on the other hand, phonons are stimulated through covalent bondings. By definition energy is transported proportionally to the speed of sound and therefore thermal conductivity follows the relationship

$$\lambda \approx c_p \cdot \rho \cdot v \cdot l \qquad (1)$$

where the thermal conductivity λ is directly proportional to the specific heat capacity c_p, the density ρ, the speed of sound v, and the molecular separation l of the elastic waves [10].
Due to the increase in density below the solidification of semi-crystalline thermoplastics, the thermal conductivity is higher in the solid state than in the melt. Furthermore, higher pressure increases the thermal conductivity too. Other factors such as temperature, morphology and molecular orientation are also affecting the thermal conductivity of polymers. Highly drawn semi-crystalline polymers like PP can have a much higher thermal conductivity in the direction of orientation of the polymer chains than perpendicular.
It is also important to know the thermal conductivity of filled compositions in order to model and analyze the heat transfer process during molding. Thermal conductivity can be influenced significantly by the addition of different fillers and additives since the thermal conductivity of most inorganic fillers is around 10 times higher than that of polymers. The increase of thermal conductivity depends on the type, on the amount as well as on the particle shape, the size and the distribution of the used fillers [11].

Experimental
Materials
An isotactic polypropylene (PP) homopolymer Bormed DM55pharm (MFR 2.8 g/10 min; 230°C/2.16 kg) used for the preparation of nanocomposites was supplied by Borealis (Linz, Austria). The used nanofiller was Nanofil 5 (montmorillonite intercalated with dimethyl distearyl ammonium chloride) supplied by Rockwood Clay Additives GmbH (Moosburg, Germany) and the compatibilizer Scona TPPP 2112 GA (PP grafted with 1.2 wt% of maleic acid anhydride, PP-g-MA, MFR 6 g/10 min; 190°C/2.16 kg) was supplied by Kometra (Schkopau, Germany).

Preparation of polypropylene nanocomposites
The preparation of the polymer nanocomposites had been carried out with the intermeshing, co-rotating twin-screw extruder of the PNC-IMC. The possibility to vary the length of the extruder opened a large area for scientific development and research and practical applications. In this case, we used 13 barrel segments for the compounding process where each barrel segment had a length of 4 times the screw diameter D. PP and compatibilizer were fed upstream through the main hopper, and the organoclay was fed downstream

at the 4th extruder barrel. All the components were fed by separately controlled gravimetric dosage units at an overall throughput rate of 6 kg/h.

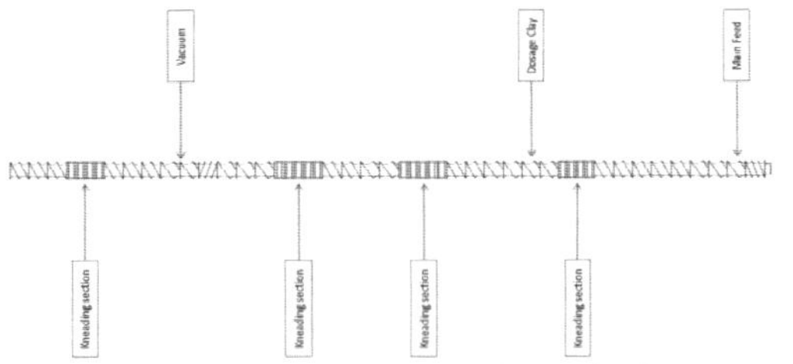

Figure 2. Schematical illustration of the screw geometry

The formulation of the compounded polymer nanocomposites was constant at 90 wt% polypropylene, 5 wt% organoclay and 5 wt% compatibilizer. The screw speed was set to 100 rpm, the counter-pressure of the melt pump to 75 bar and the screw geometry was held constant (Fig. 2). It is important to know, that the pressure gradient in the melt pump (Witte Pumps & Technology GmbH, Uetersen, Germany) is negative to increase the residence time and shear rate of the polymer compound leading to improved dispersion of the nanofiller. The minimal residence time was measured as the time between color masterbatch granulate insertion into the hopper and color change of the outgoing molten string at the 3-way-valve and the injected part, respectively. The extruder temperature profile was set at 170-210°C from hopper up to the underwater pelletizing system EUP 50, which is a water process & drying system and selfcleaning centrifuge purchased by Econ GmbH (Weißkirchen, Austria).
In order to monitor the differences to the processing with the PNC-IMC (processing in a single-heat), two compounding grades were manufactured both conventionally with the underwater pelletizing system and in the injection molding compounder. These experiments were made both with 5 wt% organoclay and 5 wt% compatibilizer as well as 10 wt% organoclay and 10 wt% compatibilizer additionally, to verify the influence of nanofillers on the thermal conductivity
For mixing PP, nanofiller and compatibilizer in one compounding process the abbreviation "3in1" is used. Another experiment was the manufacturing of polypropylene nanocomposites in a pre-stage master-batch process MB (50 wt% organoclay and 50 wt% compatibilizer). This masterbatch was then diluted in the PNC-IMC with PP to receive a compound with 5 wt% organoclay and 5 wt% compatibilizer.

Thermal conductivity
The thermal conductivity measurements based on the "Line-Source" method according to ASTM D5930-97 in the temperature range from 80°C to 240°C were carried out using a high pressure capillary rheometer (Rheograph 2002, Göttfert, Germany) with a single channel system having a diameter of 15mm. At the beginning of each measurement the high pressure capillary rheometer was heated to the specified temperature for 60 minutes. A specimen of defined weight was filled in the cylinder and compressed tightly while filling. Afterwards the material was compressed with a piston for 10 minutes and then the measurement started at the defined pressure stage. At least 3 measurements at each temperature were made, with a delay time of 12-13 minutes in between. The delay time between each temperature range was 20-30 minutes.

Results and Discussion

Effects of nanofillers at various pressures on thermal conductivity
This chapter deals with the increase in thermal conductivity influenced by pressure, conditions of the compounding process (3in1 and MB) and the one- and two-step processes, respectively. Table 1 displays the increase of thermal conductivity through the use of nanofillers for the one-step process and the 3in1 formulation. One can see an increase caused by 10 wt% nanofiller in the molten state of about 7.0 % compared to the molten virgin PP and about 10.2 % compared to the virgin solid PP. However, the best process (one-step process, 10 % MB) leads to an increase in the molten state of about 10.8 % and in the solid state of 19.4 %. It is striking to see, that the pre-stage masterbatch process leads to significant higher thermal conductivity values compared to the 3in1 formulations. Furthermore, the longer residence time of the one-step process slightly improved the thermal conductivity of the compounds.

Material	Melt state (%)	Solid State (%)
one-step_3in1_5 %	3.2	5.6
one-step_3in1_10 %	7.0	10.2
one-step_MB_5 %	7.6	12.7
one-step_MB_10 %	10.8	19.4
two-step_3in1_5 %	2.7	5.5
two-step_3in1_10 %	6.1	9.8
two-step_MB_5 %	7.4	13.4
two-step_MB_10 %	9.9	18.8

Table 1. Increase of thermal conductivity compared to the virgin unfilled PP DM55

The growth of thermal conductivity and the influence of pressure on the thermal conductivity are shown in Fig. 4. To simplify only 3 different materials, produced 3in1 with the one-step process, with 0 wt%, 5 wt% and 10 wt% at 0 bar and at 750 bar were plotted. The thermal conductivity strongly depends on the pressure applied. The increase in thermal conductivity at 0 bar from the melt state to the solid state was about 15 % and even 27.5 % at 750 bar measured for the virgin PP. The increase from 0 bar to 750 bar was about 10% in the melt state and about 25 % in the solid state for the three materials presented.

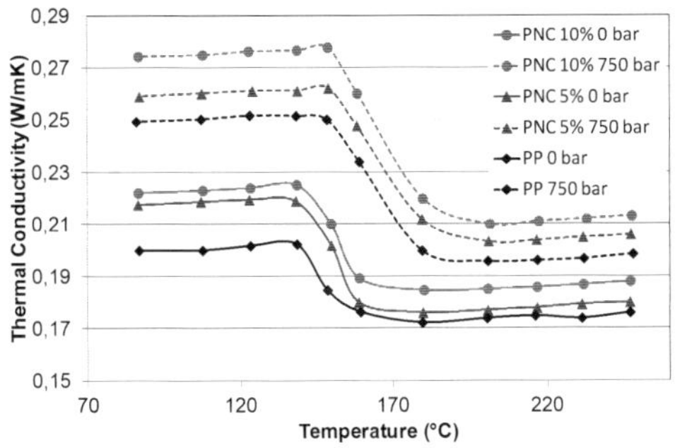

Figure 4. Effects of nanofillers and pressure on thermal conductivity

This is not really a new phenomenon, but should be considered in the production of injection molded parts regarding to cycle time reduction and cost savings. The cooling time equation (Eq. 2) shows, that the thermal conductivity λ is indirect proportional to the cooling time t_k so that a 10 % higher λ reduces t_k by 10 %.

$$t_k = \frac{s^2}{\left(\frac{\lambda}{\rho \cdot c_p}\right)_{eff,T_W} \cdot \pi^2} \cdot \ln\left[\left(\frac{4}{\pi}\right) \cdot \left(\frac{T_M - T_W}{T_E - T_W}\right)\right] \qquad (2)$$

with t_k = cooling time, s = wall thickness, λ = thermal conductivity, ρ = density, c_p = specific heat capacity, T_M = melt temperature, T_w = mean cavity wall temperature and T_E = demolding temperature.

Table 2 shows some calculated t_k values for plates of thickness 2, 4 and 8 mm, respectively. The thermal conductivity λ, specific heat capacity c_p and the density ρ were taken from the one-step 3in1 process, at 750 bar and 80 °C. It can be clearly seen, that the saving potential is enormous, especially for thick-walled parts.

Material	Thickness (mm)	Cooling time (s)	Δt_k (%)
Virgin PP DM55	2	7.3	-
one-step_3in1_5 %		6.2	15.1
one-step_3in1_10 %		5.5	24.7
Virgin PP DM55	4	29.2	-
one-step_3in1_5 %		24.9	14.7
one-step_3in1_10 %		22.1	24.3
Virgin PP DM55	8	116.8	-
one-step_3in1_5 %		99.5	14.8
one-step_3in1_10 %		88.3	24.4

Table 2. Calculated cooling times for 3 different materials at 750 bar and 80 °C

The increase of thermal conductivity has still potential to be improved towards 30 %. Therefore focused research in this area towards better intercalation and subsequent delamination of silicate platelets in the polymer matrix will be performed in near future.

Conclusions

The preparation of PP polymer nanocomposites has been carried out with the intermeshing, co-rotating twin-screw extruder of the polymer nanocomposite injection molding compounder (PNC-IMC) in order to improve the thermal conductivity. The PNC-IMC was used for processing of injected parts in a single-heat.
As shown in this paper, the thermal conductivity strongly depends on the pressure applied. The increase in thermal conductivity was about 27 % with the one-step pre-stage masterbatch process MB at 750 bar in melt state compared to the virgin PP. This leads to a calculated cooling time reduction of about 25% which motivates for further intensive investigations based on research and industrial interest.
Furthermore the use of the pre-stage masterbatch process, where a masterbatch of 50 wt% nanofiller and 50 wt% compatibilizer is produced and further diluted with PP, is more effective than the "3in1" process where PP, nanofiller and compatibilizer are mixed simultaneously.
Currently the chosen setup and process parameter set of the PNC-IMC production line was able to reach the material property improvements of the two-step masterbatch process. To bring the one-step process beyond the two-step process, further research will be done on:

- modification of the screw geometry,
- perfect adaptation of the mixing ratio between nanofiller and compatibilizer,
- and special treatment of the PNC melt between extruder and melt accumulator (flow induced improvement of intercalation and exfoliation).

Acknowledgements

We would like to thank Dr. G. R. Berger for his valuable comments on the manuscript.

References

1. J. Reitner, Kunststoffe Plast Europe, 2011, 9, 40-+.
2. M. Sieverding; E. Bürkle; R. Zimmet, Kunststoffe Plast Europe, 2005, 8, 34-37.
3. E. Bürkle; G. Scheel; L. Damedde, Kunststoffe Plast Europe, 2009, 2, 39-44.
4. H. Wobbe, Kunststoffe Plast Europe, 2003, 10, 60+-.
5. R.A. Vaia; K.D. Jandt; E.J. Kramer; E.P. Giannelis, Chemistry of Materials, 1996, 8, 2628-2635.
6. M. Kracalik; S. Laske; M. Gschweitl; W. Friesenbichler; G.R. Langecker, Journal of Applied Polymer Science, 2009, 133 (3), 1422-1428.

7. R. Sinha; M. Okamoto, Progress in Polymer Science, 2003, 28 (11), 1539-1641.
8. P. Kiliaris; C.D. Papaspyrides, Progress in Polymer Science, 2010, 35 (7), 902-958.
9. M.G. Battisti, W. Friesenbichler in 5th International Conference PMI 2012 Proceeding, Ghent, 2012.
10. T.A. Osswald; G. Menges in *Material Science of Polymers for Engineers,* Carl Hanser Verlag, Munich/Vienna/New York, 1996.
11. S. Laske, I. Duretek, A. Witschnigg, H. Mattausch, D. Tscharnuter, C. Holzer, Polymer Engineering and Science, 2012

O-07-213

STRAIN SENSING CAPABILITY OF CONDUCTIVE POLYMER COMPOSITES: THE EFFECT OF CONDUCTIVE NETWORK MORPHOLOGY AND INTERFACIAL INTERACTION

Hua Deng[a,*], Lin Lin[a], Emiliano Bilotti[b], Gao Xiang[a], Ton Peijs[b], Qiang Fu[a,*]
[a]*College of Polymer Science and Engineering, Sichuan University, State Key Laboratory of Polymer Materials Engineering, P.R. China*
[b]*Queen Mary University of London, School of Engineering and Materials Science, UK*
Corresponding authors: Huadeng@scu.edu.cn, and Qiangfu@scu.edu.cn

Abstract - Herein, we aim to investigate the effect of conductive network morphology and interfacial interaction on the strain sensing capability of CPCs based on thermoplastic elastomer. Using various post-processing methods (such as pre-straining and thermal annealing) and mixed conductive fillers (including carbon nanotube, carbon black and metallic particles) are shown to have significant effect on the conductive network morphology and strain sensitivity. It is noted that loosely entangled conductive networks formed with mixed filler shows higher strain sensitivity and post-processed spring-like conductive networks give rise to ultra-low strain sensitivity. To modify the interfacial interaction between conductive filler and polymer matrix, different thermoplastic elastomers are utilized. Polymer matrix with more polar group lead to higher orientation for conductive filler under strain, thus, higher strain sensitivity. The stress transfer between filler and matrix is thought to be responsible for this behavior.

Keywords: Conductive polymer composites, conductive network, morphological control, strain sensing

Introduction

Conductive polymer composites (CPCs) have attracted increasing amount of attention due to their wide range of applications and easy fabrication method [1-3]. In terms of strain sensing using CPCs, a wide range of applications have been demonstrated, including: smart textile, health monitoring, wearable electronics, movement sensor, flexible or stretchable conductor etc [4]. Different strain sensitivities are often required for various applications. In the effort to modify strain sensitivity, various methods have been proposed [5-9]. In above investigations, various issues have been considered for the control of strain sensitivity in CPCs, however, the range of tunable strain sensitivity is very small. Systematic and efficient method is still needed for the control of sensitivity. Therefore, fundamental aspect of strain-sensing CPCs needs to be re-considered for the preparation of high performance strain sensor with tunable strain sensitivity. The electrical conductivity in CPCs is provided by their conductive networks formed with conductive filler. Therefore, the morphology of conductive network might play crucial role on the strain sensitivity. Conductive network formed with mixed carbon fillers have been extensively investigated [10-12], it was concluded that different conductive network morphology can be constructed. However, the strain sensing capability of these networks still needs to be investigated.

Furthermore, the change of conductivity under strain originates from the deformation of conductive network, particularly, the change of local contacts or tunneling distance between conductive fillers [13], and the orientation of conductive fillers [14]. To achieve these changes in CPCs, stress needs to be transferred from polymer to conductive fillers. Therefore, the stress transfer efficiency should play an important role on strain sensitivity. Nevertheless, to the best of our knowledge, such an issue has yet to be investigated for strain-sensing CPCs despite of its vital importance.

Therefore, we aim to investigate the effect of conductive network morphology and interfacial interaction on the strain sensing capability of CPCs based on thermoplastic elastomer. Various methods such as: using pre-straining and thermal annealing, using mixed conductive fillers (including carbon nanotube, carbon black and metallic particles) and polymer matrixes, will be investigated for the preparation of high performance CPCs with tunable strain sensitivity.

Experimental

Materials
MWNTs (Nanocyl 7000, Nanocyl. S.A.) were used as conductive fillers in the composites. SBS with S/B ratio of 40/60 and trademark of SBS-792 was kindly provided by Baling Company of Sinopec Co. Ltd, China and used in this experiment as received. TPU (IROGRAN 455-200) was provided by Huntsman Company. OBC (INFUSE 9107) were produced by DOW Chemical Company. Eutectic alloy (mainly including bismuth and tin) with a melting temperature of 47 °C and density of 9.36 g cm^{-3} was bought from Jinyongan Industrial Co. (ShenZhen).

Sample Preparation

For the preparation of strain sensor based on different elastomer, firstly, the TPE-CNT composites were prepared by blending MWNTs with different elastomers (SBS, TPU, OBC) in Hakke internal mixer at 180 °C, 60 rpm for 10 min. Then, the composites containing different contents of CNT (4, 6 and 8 wt. %) were extruded as fibres with piston-mode Rosand RH70 (Malvern, Bohlin Instruments) capillary rheometer.

The stretchable conductive nanocomposites were fabricated with MWCNTs, ionic liquid and as-received TPU. MWCNTs (0.1g) were mixed with ionic liquid (1-butyl-3-methylimidazolium bisaimide, 1g) and subjected to grinding in an agate mortar for 1 hour, giving rise to a "bucky gel" state and enhancing the interface between MWCNTs and TPU. The pretreated MWCNTs were dispersed into N,N-dimethylformamide (50 mL) with a laboratory emulsion machine (WX500CY, Shanghai Weiyu Ltd.). Then, the dispersed MWCNTs were added into TPU (0.9 g) solution with N,N-dimethylformamide (50 mL) as solvent. The mixture was stirred at 40 °C for 10 h and sonicated for another 1 h. The resulting swollen gel was poured onto a polytetrafluoroethylene plate and dried at 80 °C for 2 h. ILs were subsequently recovered with Solxhlet extraction (ethanol as the solvent) and the resulting composites (without IL residue) were dried in the oven again at 80 °C for 4 h. Composites with other MWCNT contents (20, 30 wt.%) were fabricated in the same fashion.

The as-prepared composites were tailored into uniform samples with length of 70mm, width of 10mm and thickness of 0.1-0.2 mm. Samples were then stretched at indicated magnitude along one direction and then relaxed to the unstretched state. The samples were stretched in a SANS CMT4000 universal testing machine. To fabricate pre-strained samples, stretching and releasing were continuously repeated 10 times at room temperature with a rate of 5 mm/min. Afterwards samples went through thermal annealing in an oven under pre-strained to obtain the pre-strained & annealed samples. Annealing was conducted at a temperature of 100 °C for 1 hour.

Characterization

Composite films were clamped between a pair of aluminum electrodes, creating a gage length of 50 mm. Resistance was measured with a Keithley 6487 picoammeter under a constant voltage of 1 V to avoid strong electrical current within the sample. Silver paint was applied to both ends of the sample to ensure good contact. The resistance measurement set up and tensile test machine are both interfaced with a computer to record the in-situ resistivity-strain dependence of these CPC fibers.

Morphological studies were carried out using a SEM (JEOL JSM-5900LV) under an accelerating voltage of 20 KV. To investigate the conductive networks below the sample surface, un-coated specimens were used. The conductive networks in the polymer matrix are charged to emit enriched secondary electrons to make them visible.

Raman spectra were recorded on a micro-Raman spectrometer (JY HR800) equipped with a microscope and a 532nm excitation laser.

Two-dimensional wide-angle X-ray scattering (2D-WAXS) experiments were conducted on a Rigaku Denki RAD-B diffractometer. The wavelength of the monochromated X-ray from Cu KR radiation was 0.154 nm, and the sample-to-detector distance was 273 mm.

Results and Discussion

Figure 1 shows the strain-sensing behavior of all samples. The resistivity increases logarithmically with increasing strain. Plots for PU-5C and PU-3C are almost parallel to each other as the same resistivity–strain dependence is expected regardless of filler concentration. This is similar to the observed universal strain-sensing behavior in a TPU/CNT composite as reported previously [13]. When metallic particles are added in combination with CNTs into the fibers, an obvious increase in the slope can be observed for PU-1.5C-15A and PU-1C-10A composites. In spite of different starting points due to different filler concentration, the plots of these two samples are almost parallel to each other. A similar phenomenon is observed for mixed fillers containing a higher fraction of metallic particles, where a steeper slope is obtained.

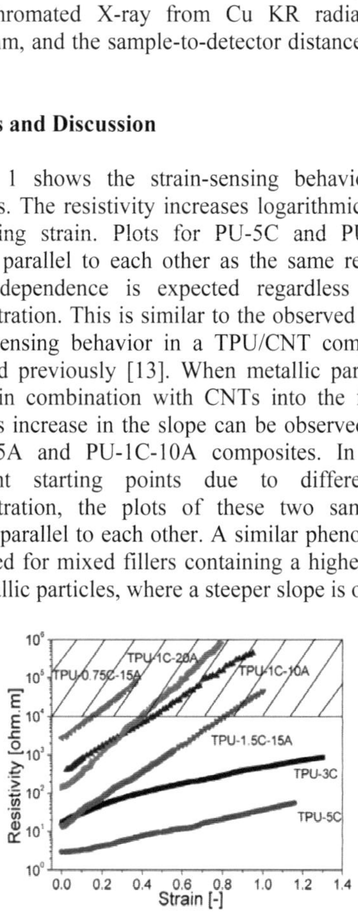

Figure 1-Resistivity versus strain and for various fillers and various concentrations in TPU composites.

To investigate the effect polymer matrix on the strain sensing behavior, CPCs containing 8 wt.% of MWNTs based different elastomer have been compared (see Figure 2). In spite of different starting points due to slight difference in the dispersion status of CNT in the composites, all samples demonstrated a definitive trend where the resistivity increases proportionally with increasing tensile strain. Moreover, it can be observed that these samples show different strain-sensing behavior (different slopes of the curves). Especially, the resistivity of SBS-8NT sample increases sharply with increasing strain and the slope of this curve was larger than that of the other two samples. This might originates from the higher interaction between SBS and CNT comparing with other matrices. The plateau appears after near 100% strain is quite unique and needs to be further investigated.

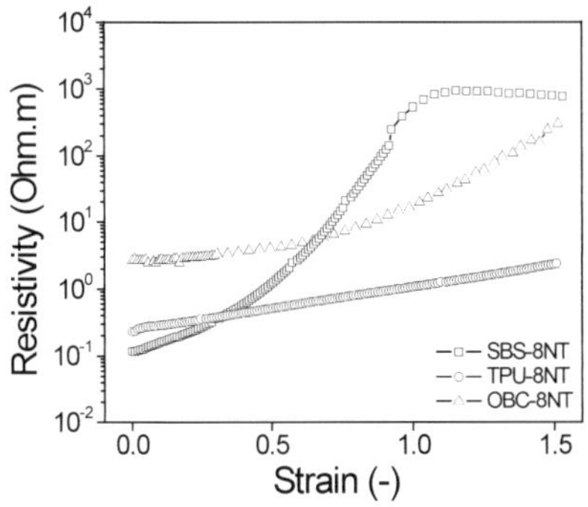

Figure 2-Resistivity versus strain curves for CPCs based on different elastomer (SBS, TPU and OBC), containing 8 wt.% of MWNTs.

To further investigate the resistivity–strain behavior of these CPCs, SEM was conducted to examine the morphology of the conductive networks under different strain. Different conductive networks are randomly distributed at zero strain (see Figure 3a. It has been reported that conductive networks in CPCs are often oriented during stretching [15-17]. In current study, MWNT bundles are oriented under 10% strain as indicated by the arrows in Figure 3b. Meanwhile, metallic particles are maintained as almost isotropic networks (Figure 3e). Nevertheless, both MWNT bundles and metal particle networks are oriented at 100% strain (Figure 3c and f). More importantly, the conductive networks are stretched apart at 100% strain, while only the network structure is deformed at 10% strain.

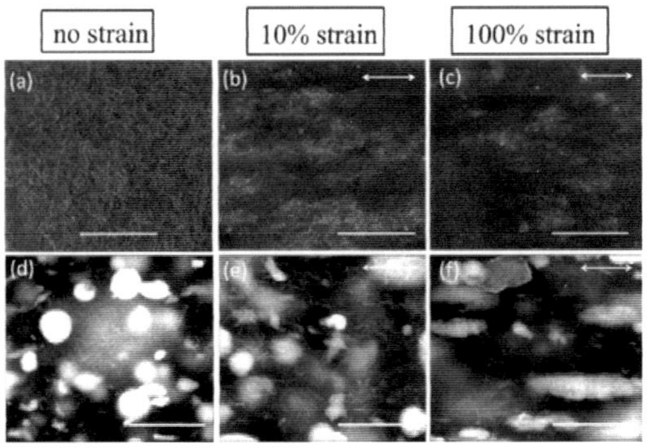

Figure 3-SEM images of (a–c) PU-5C and (d–f) PU-1.5C-15A, at (a, d) at zero strain, (b, e) 10% strain and (c, f) 100% strain. The samples were drawn along the arrow direction. The scale bars are 1 μm.

In terms of stretchable conductors, they can keep their excellent electrical conductivity while highly stretched. Their preparation is often complicated and expensive. Herein, we report an efficient method to prepare high performance stretchable conductor through morphological control of conductive network formed with carbon nanotubes (CNTs) in elastomer matrix. It is observed that an interface mediated method could be used to align randomly oriented filler during stretching and induce buckling of MWCNTs during relaxation. Further morphological studies indicate possible formation of CNT wavy structure induced by cyclic pre-straining. Subsequent thermal annealing is observed to collapse the oriented network and improve the local contacts between conductive networks. Through such a simple procedure, a conductivity of near 1000 S m-1 and stretch-ability of 200% can be achieved for composites containing 20 wt.% MWCNT (see Figure 4).

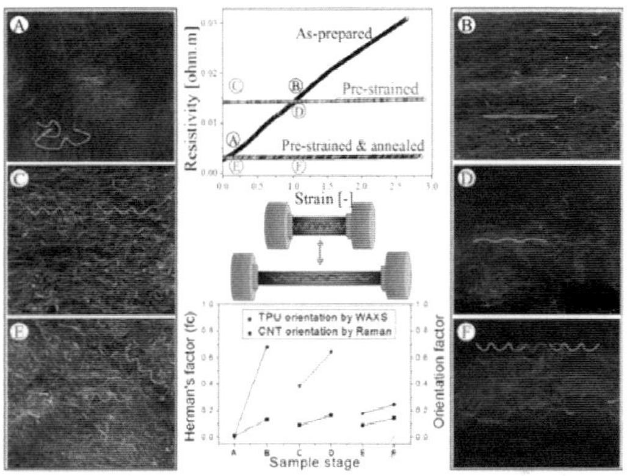

Figure 4-Control of conductive network morphology through pre-straining and annealing. Different conductive morphologies can be obtained (see A, C, E). They are able to demonstrate different resistivity-strain behavior under strain thanks to their special network structure (see B, D, F). Raman and WAXS have been used to characterize the orientation of TPU matrix and CNT, respectively.

Conclusions

The strain sensing capabilities of different CPCs have been investigated. Several methods are observed to be effective at tunneling the strain-resistivity behavior, including: using mixed filler, different polymer matrixes. These modifications in strain sensing behavior are thought to be related to conductive network morphology and interfacial interaction. Conductive network morphology in these CPCs can be modified through post-processing methods (pre-straining and thermal annealing). These different network morphologies give rise to different strain sensing behavior. An oriented and relaxed network is observed to be able to withstand more than 250% strain without any degradation in its conductivity. In terms of interfacial interaction, a higher interfacial interaction between filler and matrix is more likely to give rise to higher strain sensitivity. Nevertheless, more study is still needed to further investigate the mechanism responsible for these behaviors.

Acknowledgements

We express our sincere thanks to the National Natural Science Foundation of China for financial support (51273117, 51003063 and 51121001). This work was subsidized by the special funds for Major State Basic Research Projects of China (2011CB606006).

References

1. Byrne MT and Gun'ko YK. Advanced Materials 2010;22(15):1672-1688.
2. Bauhofer W and Kovacs JZ. Composites Science and Technology 2009;69(10):1486-1498.
3. Chou T-W, Gao L, Thostenson ET, Zhang Z, and Byun J-H. Composites Science and Technology 2010;70(1):1-19.
4. Bilotti E, Zhang R, Deng H, Baxendale M, and Peijs T. Journal of Materials Chemistry 2010;20:9449-9455.
5. Engel J, Chen J, and Liu C. Applied Physics Letters 2006;89(22).
6. Kollosche M, Stoyanov H, Laflamme S, and Kofod G. Journal of Materials Chemistry 2011;21(23):8292-8294.
7. Schwalb CH, Grimm C, Baranowski M, Sachser R, Porrati F, Reith H, Das P, Mueller J, Voelklein F, Kaya A, and Huth M. Sensors 2010;10(11):9847-9856.
8. Dang Z-M, Jiang M-J, Xie D, Yao S-H, Zhang L-Q, and Bai J. Journal of Applied Physics 2008;104(2).
9. Hwang J, Jang J, Hong K, Kim KN, Han JH, Shin K, and Park CE. Carbon 2011;49(1):106-110.
10. Etika KC, Liu L, Hess LA, and Grunlan JC. Carbon 2009;47(13):3128-3136.
11. Hermant M-C, van der Schoot P, Klumperman B, and Koning CE. ACS Nano 2010;4(4):2242-2248.
12. Sumfleth J, Adroher XC, and Schulte K. Journal of Materials Science 2009;44(12):3241-3247.
13. Zhang R, Baxendale M, and Peijs T. Physical Review B 2007;76(19).
14. Deng H, Skipa T, Bilotti E, Zhang R, Lellinger D, Mezzo L, Fu Q, Alig I, and Peijs T. Advanced Functional Materials 2010;20(9):1424-1432.
15. Deng H, Bilotti E, Zhang R, Loos J, and Peijs T. Synthetic Metals 2010;160(5-6):337-344.
16. Deng H, Zhang R, Reynolds CT, Bilotti E, and Peijs T. Macromolecular Materials and Engineering 2009;294(11):749-755.
17. Gao X, Zhang SM, Mai F, Lin L, Deng Y, Deng H, and Fu Q. Journal of Materials Chemistry 2011;21(17):6401-6408.

O-07-220

STUDY ON COMPOSITES OF GRAPHENE OXIDE (GO) AND CONDUCTIVE POLYMER

S. Konagaya[1*], M. Terada[1], M. Mutou[1] and T. Gotou[2]

[1] *Division of Applied Chemistry, Graduate School of Engineering, Nagoya University, Furo-cho, Chikusa-ku, Nagoya, 464-8603 JAPAN – konagaya@apchem.nagoya-u.ac.jp*

[2] *Mitsubishi Gas Chemical Company Inc., 1-1, Niijuku 6-chome, Katsushika-ku, Tokyo 125-8601, JAPAN*

Abstract - The authors have studied the composites (PEDOT:PSS/GO and PEDOT:PSS/GO/PEs), which were prepared from 3,4-ethylenedioxythiophene:polystyrene sulfonic acid (PEDOT:PSS), graphene oxide (GO) with an average diameter of 2.3×10^2 nm (and its reduced type :graphene (Gr)), a water dispersible polyester binder (PEs) and pyrogallol as a reducing agent in order to get thin film composites with higher conductivity. It was found that the addition of GO had no effect on the conductivity enhancement of the PEDOT:PSS/GO composites, but its reduced type Gr had an effect on the increase of the conductivity, which is attributable to π - π interaction between Gr and PEDOT:PSS. The Rs values of the heat treated PEDOT:PSS/GO/PEs/pyrogallol composites decreased with the increase of the GO/(GO+PEDOT:PSS) ratio up to 0.8. The Rs value of the reduced PEDOT:PSS/GO/PEs composites was lower than those of PEDOT:PSS/PEs and GO/PEs composites. Higher conductivity of the reduced PEDOT:PSS/GO/PEs is possibly attributed to the π - π interaction between PEDOT:PSS and Gr (reduced GO), and percolation of PEDOT:PSS/Gr particles around PEs nano-particles.

Keywords: graphene oxide, graphene, composite, conductive polymer, PEDOT

1. Introduction

Graphene (Gr) is a giant aromatic poly-molecule having high electrical conductivity, mechanical strength and optical absorption properties as shown in Fig.1. It can be incorporated into organic polymers to enhance electrical conductivity and mechanical strength. So Gr is a promising material taking the place of conventional transparent conductive material, Sn-doped In_2O_3 (ITO). Gr is provided from graphite with "Scotch tape" method[1]. However their Gr monolayers yield is so low that Gr is generally produced by the reduction of graphene oxide (GO) produced by the oxidation of graphite particles[2]. The conductivity of GO is lower than that of graphite since GO has lots of OH and COOH groups on its surface as shown in Fig.1, but GO is expected to have a good dispersibility in the aqueous solution and hydrophilic polymers.

The water soluble doped conductive polymer, Poly (3,4-ethyldioxythiophene):poly (styrene sulfonic acid) (PEDOT:PSS), possesses so high conductivity, transparency, and environmental stability. However the conductivity of PEDOT:PSS is required to be enhanced enough to take the place of ITO. So there are so many studies on additives and dopants such as ethylene glycol and sulfonated-graphene for the improvement of the conductivity of PEDOT:PSS [2,3,4,5].

Since GO (or its reduced type Gr) and PEDOT:PSS have benzene rings and thiophene rings, there is expected to be a strong π - π interaction between GO (or Gr) and PEDOT:PSS in the solid mixture (composite) of PEDOT:PSS and GO. So the composite of PEDOT:PSS/GO (or Gr) has a possibility to show higher conductivity than PEDOT:PSS and GO. Then the authors have been studying the composites of

GO, PEDOT:PSS and a polyester binder PEs to get thin film composites with higher conductivity.

In this paper the effect of GO and Gr on the conductivity of PEDOT:PSS and the PEDOT:PSS/polyester binder (PEs) composite will be discussed.

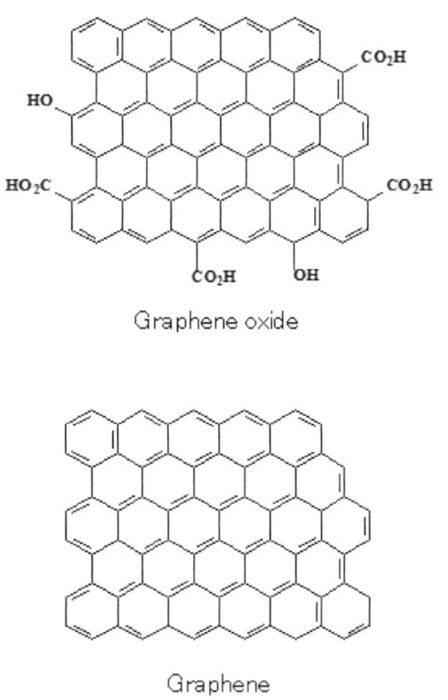

Fig.1 Chemical structure of Graphene oxide (GO) and its reduced type, Graphene (Gr)

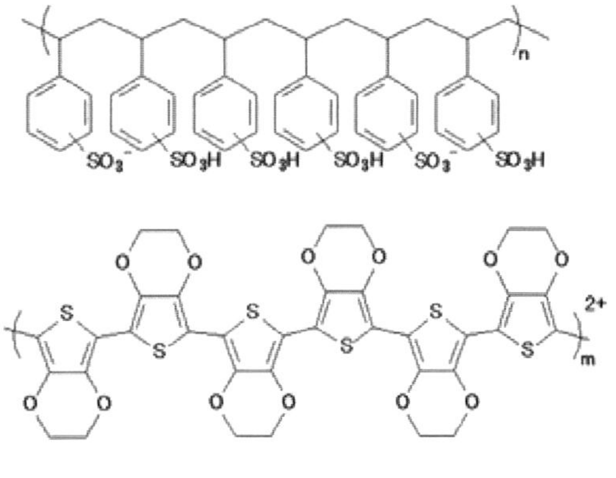

Fig2 Chemical structure of PEDOT:PSS (poly (3,4-ethyldioxythiophene):poly (styrene sulfonic acid))

2. Experimental

2.1 Materials

The GO /water slurry of 1 wt% was supplied by Mitsubishi Gas Chemical Company Inc.[6]. The aqueous PEDOT:PSS solution (conc: 1.3 wt %) was purchased by Aldrich Co., Ltd. Polyester nano-particles/water dispersion (PEs , conc: 34 wt. %) was supplied by TOYOBO Co. Ltd. Pure water was used for the preparation of PEDOT:PSS/GO and PEDOT:PSS/GO/PEs composites and zeta-potential measurement of GO nano-particles in water. Pre-cleaned micro slide glass plate with the size of 76 mm ×26 mm and the thickness of 0.9-1.2 mm was used as a substrate for the preparation of thin film composites.

2.2 Preparation of PEDOT:PSS/GO thin film composites on the glass plates.

The aqueous mixtures of GO/water slurry and PEDOT:PSS/water solution with the different weight ratio of GO/(GO+PEDOT:PSS) in the presence or absence of pyrogallol with the constant weight ratio (1.0) of pyrogallol/GO were dispersed well with ultrasonic waves for 20 min and coated with a bar coater on the glass plate. After that, above composites were dried and heated under the conditions of 70 ℃ for 0.5 h, 100℃ for 1 h and 170℃ for 1.5 h in air. The thickness of the PEDOT:PSS/GO composites were controlled to be around 2 μm.

2.3 Preparation of PEDOT:PSS/GO/PEs thin film composites on the glass plates.

PEs-water dispersion was added into the aqueous mixtures of aqueous PEDOT:PSS solution, GO/water with the different weight ratio of PEDOT:PSS/(GO+PEDOT:PSS) and pyrogallol with the constant weight ratio of pyrogallol/GO, where the weight ratio of PAS/PEs was kept constant (0.10, 0.15). The PEDOT:PSS/GO/PEs/pyrogallol/water mixtures were dispersed well with ultrasonic waves for 20 min. The mixtures were coated on the glass plate with a bar coater After that, above composites were dried and heated under the conditions of 70 ℃ for 0.5 h, 100℃ for 1 h and 170℃ for 1.5 h in air. The thickness of the PEDOT:PSS/GO/PEs/pyrogallol composites were controlled to be around 2 μm. The GO/PEs/pyrogallol (GO/PEs weight ratio: 0.10, 0.15) composites were prepared for the comparison with PEDOT:PSS/GO/PEs/pyrogallol according to the above method.

2.4 Measurement

2.4.1 Size and Zeta (ζ)-potential of GO dispersed in the water

The size and zeta-potential of GO nano-particle in the GO/water slurry with the concentration of 0.02 wt% were measured with the zeta-potential & size analyzer, ELS-Z (Photal Otsuka electronics Co. Ltd)

2.4.2 Surface resistivity (Rs) of thin film composites

Surface resistivity (Rs, Ω/□) of the thin film composites (PEDOT:PSS/GO, PEDOT:PSS/GO/PEs) laminated on the glass plate were measured with Lowresta (Mitsubishi Chemical Co. Ltd.) and Model 152 resistance/volume meter (Treck Japan Co. Ltd) under the following conditions; temperature: 20-25 ℃, humidity: 45-60 %, voltage: 100 V.

3. Results and Discussion

3.1 Characterization of GO/water slurry

The average diameter and zeta-potential of GO were estimated to be 2.3×10^2 nm and -38 mV. The absolute value of the zeta-potential (-38 mV) is large enough to give a good dispersion of GO/water slurry by the coulomb repulsive force among GO particles.

3.2 Effect of GO on the Rs value of PEDOT:PSS/GO composites

Fig.3 shows the conductivity of the PEDOT:PSS/GO composites with the different weight ratio of GO/(GO+PEDOT:PSS), which were prepared in the absence or presence of reducing agent, pyrogallol. The Rs values of PEDOT:PSS/GO composites increased with the weight ratio of GO/(GO+PEDOT:PSS) in the absence of pyrogallol, while their Rs values decreased in the presence of pyrogallol. In the absence of pyrogallol GO in the composite can be hardly reduced to Gr by the thermal treatment(170℃). No change of the conductivity of PEDOT:PSS/GO composites at the weight ratio of GO/(GO+PEDOT:PSS) ranging from 0 to 0.5 is possibly ascribed to the donor effect of GO to PEDOT:PSS. In the presence of pyrogallol, GO in the PEDOT:PSS/GO composite could be reduced to Gr with higher conductivity by the heat treatment. Then the Rs values of PEDOT:PSS/GO composites decreased with the increase of the GO/PEDOT:PSS weight ratio. The PEDOT:PSS/GO composites with GO/PEDOT:PSS ratio of more than 0.6 showed no change of the conductivity, of which GO was reduced to Gr by the heat treatment. This result showed that there was some interaction between PEDOT:PSS and Gr contributing to the increase of the conductivity. It can be assumed that there is π - π bonding between thiophene ring of PEDOT:PSS and benzene ring of Gr which promotes the electron hopping between them.

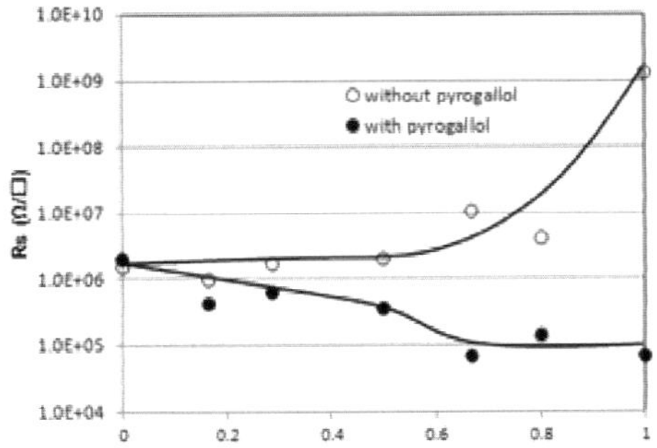

Fig.3 Effects of GO content and pyrogallol on Rs values of PEDOT:PSS/GO composites
Heat treatment: 170℃ × 1.5 h in the air

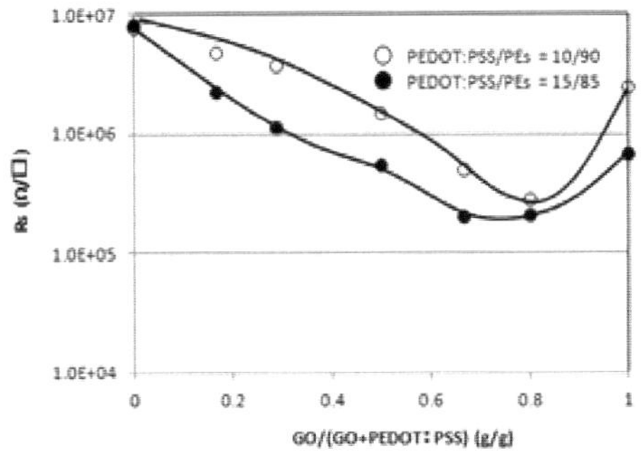

Fig.4 Effect of GO content on Rs values of PEDOT:PSS/GO (with pyrogallol)/PEs composites
Heat treatment:170℃ × 1.5 h in the air

3.3 Effect of GO on the Rs values of PEDOT:PSS/GO/PEs composites in the presence of pyrogallol

Fig.4 shows the conductivity of the PEDOT:PSS/GO/PEs composites with the different weight ratio of GO/(GO+PEDOT:PSS) prepared in the presence of pyrogallol, the weight ratio of PEDOT:PSS/PEs being kept constant (0.10 or 0.15). PEDOT:PSS/GO/PEs composites showed the minimum Rs value at the GO/(GO+PEDOT:PSS) weight ratio of about 0.8. Then GO in the PEDOT:PSS/GO/PEs was reduced to Gr by pyrogallol, and there could be the efficient formation of conductive networks based on the percolation of Gr particles.in addition to some interaction between PEDOT:PSS and Gr (reduced GO), which resulted in the superior conductivity of PEDOT:PSS/GO/PEs.

3.4 Action mechanism of GO on the conductivity of PEDOT:PSS/GO/PEs composites

The action mechanism of GO on the conductivity of PEDOT:PSS/GO/PEs composites is summarized as follows. When aqueous PEDOT solution is mixed with GO/water slurry, PEDOT:PSS/GO adducts are produced with higher dispersibility in water. PEDOT:PSS/GO adducts being mixed with PEs /water slurry, PEDOT:PSS/GO particles are distributed around the PEs nano-particles, and self-organized to make PEDOT:PSS/GO networks with a lower Rs value with the lapse of drying time. Moreover GO in the PEDOT:PSS/GO/PEs is reduced to Gr by pyrogallol to give PEDOT:PSS/Gr networks, which results in the enhancement of the conductivity.

4. Conclusions

The addition of graphene oxide (GO) and continued reduction of GO with pyrogallol at 170℃ was found to be effective for the improvement of the conductivity of PEDOT:PSS and PEDOT:PSS/PEs composites, which possibly resulted from the conductive network formation due to the strong π - π interaction and their percolation of PEDOT:PSS/Gr adducts.

Acknowledgements

The authors would like to thank Toyobo Company for the offer of a water dispersible polyester binder PEs.

References

1. Kian Ping Loh, Qiaoliang Bao, Priscilla Kailian Ang, Jiaxiang Yang, . *J. Mater. Chem.*, **20**, 2277-2289 (2010)
2. Jianyong Ouyang, Qianfei Xu, Chi-Wei Chu, Yang Yang, Gang Li, Joseph Shinar, *Polymer* **45** (2004) 8443-8450
3. Satoki Ahizawa, Masayoshi Ishihara, Hidenori Okuzaki, *Polymer Preprints Japan,* **51**, No12(2002) 3145
4. Yijie Xia, Jianyong Ouyang, *Applied Materials & Interfaces,* **2, No2** (2010) 474-485
5. Yanfei Xu, Yan Wang, Jiajie Liang, Yi Huang, Yanfeng Ma, Xiangjian Wan, Yongsheng Chen, *Nano Res,* **2009**, 2,343-348
6. Masukazu Hirata, Takuya Gotou, Shigeo Horiuchi, Masahiro Fujiwara, Michio Ohta, *Carbon* **42** (2004) 2929-2937, USP6596396

O-07-225

QUANTIFICATION OF CARBON NANOTUBE SHORTENING AFTER MELT COMPOUNDING

Beate Krause*, Tobias Villmow, Robert Socher and Petra Pötschke

Leibniz Institute of Polymer Research Dresden, Germany – krause-beate@ipfdd.de

Abstract - In this study, a relatively simple method is presented which allows determining the nanotube length distributions of as received MWCNTs as well of the same nanotubes recovered from melt processed composites using the same methodology. Different examples of composites containing polycarbonate or poly(caprolactone) and multiwalled carbon nanotubes were shown. It was found that during melt mixing always nanotubes shortening takes place. The degree of shortening varied in dependence on the initial nanotube length and the shear stresses applied on the as-produced MWCNT agglomerates. Furthermore, the relationship between electrical properties of the composites and the MWCNT aspect ratio could be shown.

Keywords: carbon nanotubes, aspect ratio, melt processing, polycarbonate.

Introduction

Due to the exceptional properties of carbon nanotubes (CNTs), such as high electrical and thermal conductivity and excellent mechanical properties, they are expected to have great potential as fillers for polymeric matrices. Carbon nanotubes are incorporated in electrically insulating polymer materials to achieve electrostatic dissipative behavior or electrical conductivity and improved mechanical properties. Current applications for such nanocomposites include electrostatically dissipative plastic housing or fuel lines, as well as lightweight and electrostatically paintable plastic components replacing metals in car panel applications. Incorporation of carbon nanotubes in polymers on an industrial scale is usually performed by melt processing. Industrial CNT materials are in most cases synthesized in an agglomerated and entangled state and have to be dispersed into single tubes during melt compounding in order to obtain the greatest benefit from intrinsic CNT properties. Electrically percolated networks are most effectively formed by secondary agglomeration of suitably dispersed nanotubes. Processes of wetting and diffusion of polymer chains into agglomerates as well as rupture and erosion of the CNT agglomerates occur before isolated CNTs can be distributed in the matrix. Both the properties of the matrix polymer (viscosity, polarity, surface energy, functional groups) and of the carbon nanotubes (wall number, purity, diameter, length, bulk density, waviness, dispersion, and functionalization) play a role in dispersion. To achieve good separation of the CNTs and a low electrical percolation threshold in polymers, the surface properties of the polymer and nanotubes should be in the same range and the melt processing parameters should be optimized according the desired nanocomposite property profile. For this purpose, the specific mechanical energy input (SME) during melt processing is of importance.

Due to the high heterogeneity within different types of carbon nanotubes, generalized conclusions about their properties (e.g. functionalization, dispersability, agglomerate strength, aspect ratio, degree of defects) can be only drawn to a limited extent. In the present study the length of CNTs was quantified using the same experimental approach before and after melt mixing with thermoplastics. As the nanotube's aspect ratio (length divided by diameter) significantly affects the achievable electrical percolation threshold in the composites, knowledge of CNT lengths and their reduction during processing is of interest. Furthermore, the relations between the specific mechanical energy input, CNT dispersion, CNT length, and electrical properties of melt mixed composites are discussed.

Experimental

The employed nanotubes represent industrial multiwalled CNTs (MWCNTs). They are Nanocyl™ NC7000 (Nanocyl SA, Sambreville, Belgium, >90% purity C, bulk density 66 kg/m^3) and C150P Baytubes® (Bayer MaterialScience AG, Leverkusen, Germany ; >95% purity C, bulk density 120-170 kg/m^3).

The effect of dry ball milling treatment of Nanocyl™ NC7000 on CNT length was studied and composites with PC (Lexan™ 141R, Sabic Innovative Plastics, The Netherlands) were prepared by masterbatch dilution as described in detail in Krause et al. [1].

Composites having different matrix melt viscosities were prepared through the incorporation of Baytubes® C150P in low viscous PC Makrolon® 2205 (PC1, Bayer MaterialScience AG) and high viscous PC Makrolon® 3108 (PC2, Bayer MaterialScience AG) using a DACA Microcompounder at 280°C, 250 rpm for 5 min [2].

The composites based on polycaprolactone (PCL, CAPA 6800, Perstorp) containing 0.5 wt% Nanocyl™ NC7000 were melt compounded using a Microcompounder Xplore DSM 5 (volume 5 ml) with different rotation speeds varying from 25 to 400 rpm at a mixing temperature of 90°C for 2 minutes [3].

Results and Discussion

The electrical percolation threshold of CNTs in composites is strongly dependent on the aspect ratio of the CNTs and thus on their lengths [4;5]. However, the determination of the length distribution of carbon nanotubes is more difficult than measuring their diameters and not often reported in literature. On the one hand, CNTs are entangled and have to be separated nondestructively before their length can be measured. For this purpose, quite often ultrasonic treatment is used [6-9], which was reported to lead to CNT shortening dependent on the applied energy[10-12]. On the other hand, the chosen method of length quantification has to be suitable to characterize the CNTs with their high aspect ratio, meaning that such magnifications have to be selected which show the full length of such individualized CNTs. For this e.g. scanning electron microscopy [6-8], atomic force [9;13], or transmissions electron microscopy (TEM) are applicable. To give a quantitative statement on the aspect ratio as a function of processing steps, it is necessary to determine the CNT length before and after processing using the same method, which was described by Krause et al. [14;15]. Fig. 1 illustrates the schema of the sample preparation used for the quantification of the length distribution of MWCNTs using TEM investigation [15].

As one example, the effect of treatment of MWCNTs using dry ball milling is shown [1;15]. It was concluded that the grinding of CNTs results in a significant reduction of the CNT length as shown on nanotube length distributions in Fig. 2 [15]. According to this figure, Nanocyl™ NC7000 material was shortened after 5 and 10 hours milling to 54% and 35% of its initial length, respectively (based on x_{50}-values). In the subsequent melt processing step for all three types of MWCNTs an additional length reduction could be observed, whereby the degree of reduction was the highest for the longest CNTs. When setting into relation the nanotube length and the achieved electrical percolation threshold in the composites based on the ball milled nanotubes [1], the lowest threshold was found for composites with untreated Nanocyl™ NC7000, for which the aspect ratio of 42 even after processing was the highest as compared to that of the CNTs already pre-shortened in the ball mill (37 and 31). In the composites containing the nanotubes with lower aspect ratios, higher percolation thresholds were found.

In previous literature, macrodispersion was improved with increasing specific mechanical energy input (SME) during the processing of melt mixed composites based on polyamide 6 [16] and polycarbonate [17;18]. At the same time the relation between electrical resistivity and specific mechanical energy input showed a minimum which was found in both the examples mentioned at about 1400-1800 J/cm^3 (0.4-0.5 kWh/kg). Further increase in SME resulted in increasing electrical resistivity values even if the macrodispersion improved with SME or leveled off at low values of A_A. Thus, a certain SME seems to be necessary in order to form a percolation network, but on the other hand too intense mixing resulted in an increase in the electrical resistivity. This effect is particularly pronounced for polyamide 6 composites.

On PC based composites the influence of matrix viscosity on the CNT shortening in a microcompounder was studied [2]. Due to the different matrix viscosities different levels of shear stresses and SME were introduced during melt mixing under constant mixing conditions. The higher viscosity of the PC2 matrix resulted in a twofold higher SME input during melt mixing compared to the low viscous PC1. This difference leads to a better macrodispersion indicated by a lower agglomerate area ratio A_A but also to more significant CNT shortening (Fig. 3). The electrical percolation threshold for the composites based on low viscous PC was between 0.5 – 0.75 wt.%, whereas it was at around 1 wt.% for the high viscosity grade. Based on the electrical volume resistivity values, for the composite based on the low viscous PC1 having longer nanotubes but worse CNT macrodispersion the formation of an electrical CNT network can be concluded at lower CNT content, whereas the composite based on PC2 has isolated nanotube structures. In particular, the x_{90} – values of the CNT length distributions show strong differences between both composites resulting in significant differences in the corresponding aspect ratios (Fig. 3). After processing, the CNTs in the composite with low matrix viscosity show an x_{90}-value of 1 µm, while in the composite with the high viscous matrix this value is only 570 nm. TEM studies showed that in the composite with high matrix viscosity a distinct individualization and separation of the CNTs occurred, while in the composite based on the low viscosity matrix a loosely packed network structure with secondary agglomerated CNTs was observed [2]. Thus, next to the effect of the matrix viscosity on CNT shortening and dispersion, also the secondary agglomeration seems to play an important role on the formation of the conductive network, which is pronounced at lower matrix viscosity.

The influence of the rotation speed during the melt mixing in small-scale and the related specific mechanical energy input on the CNT macrodispersion, the CNT length, and the electrical surface resistivity of the composites was studied on PCL composites filled 0.5 wt% Nanocyl™ NC7000 [3;15;19]. . This filling level was selected near the percolation region since changes in CNT aspect ratio due to CNT shortening have a very pronounced influence on the electrical resistivity espacially in this concentration range. Even during mixing at the lowest mixing speed of 25 rpm for 2 min a significant CNT shortening took place. The measured value after processing was 41% of the initial value (based on the x_{50}-value) (Fig. 4). The CNT length decreased up to 31% of the initial length at the highest SME of 0.7 kWh/kg achieved at 400 rpm. The lowest electrical surface resistivity of about $2 \cdot 10^3$ Ohm/square was found at a mixing speed of 75 rpm or a SME value of 0.5 kWh/kg. Above this minimum, with further increasing speed or SME, the electrical surface resistivity increases to values of about 10^{11} Ohm/square. The dependency of the agglomerate area ratio A_A on SME shows an exponential decrease up to about 100 rpm (0.7 kWh/kg) where it levels off. However, it can be also seen that only at very high screw speeds (400 rpm) dispersions nearly free of agglomerates can be achieved. The results indicate, that concerning the electrical resistivity, the effects of increasing macro-dispersion and length shortening are counteracting, leading to lowest resistivity values at SME values of about 0.5 kWh/kg, which interestingly was also the range of a minimum or start of the plateau in the examples discussed before for PA6 [16] and PC [17;18].

Figure

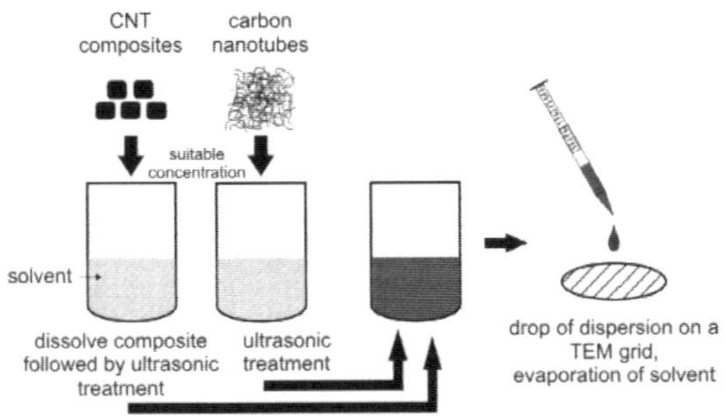

Fig. 1: Schema of the sample preparation used for the quantification of the length distribution of MWCNTs [15]

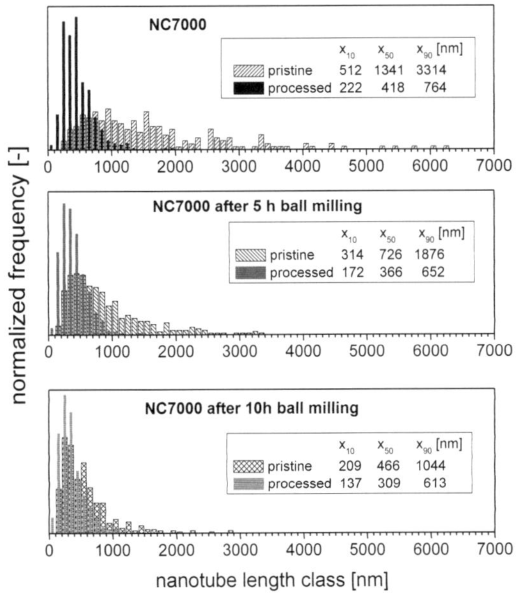

Fig. 2. Length distributions of as-grown, ball milled (5 or 10 hours), and processed Nanocyl™ NC7000 (dissolved from 2 wt% MWCNT/PC composites) [1;15]

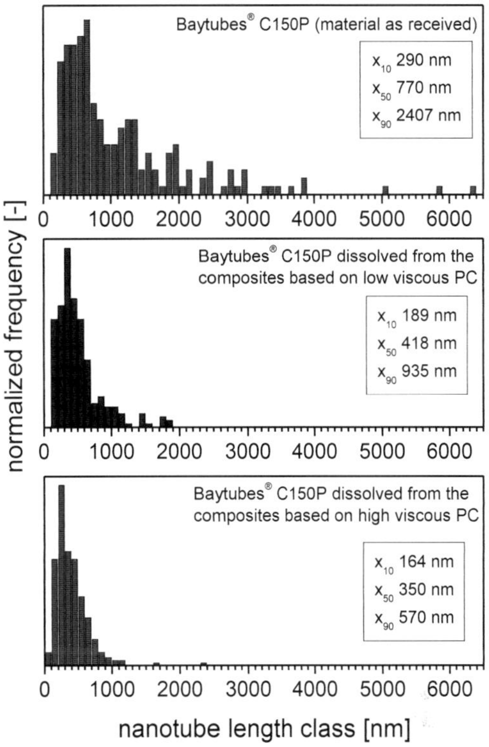

Fig. 3. Nanotube length distributions of as-grown Baytubes® C150P and MWCNTs dissolved from PC composites with 0.75 wt.% Baytubes® C150P [2]

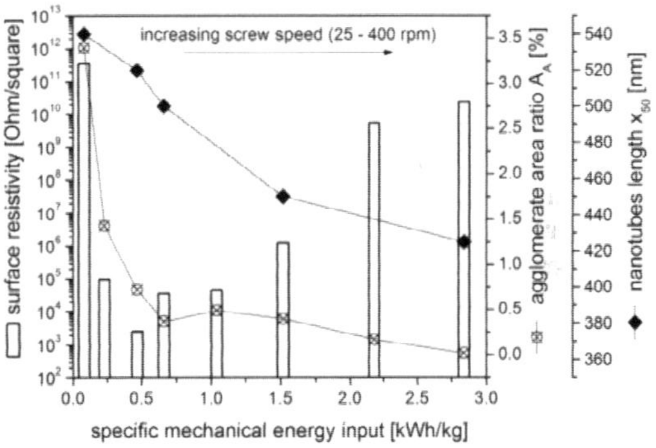

Fig. 4. Surface resistivity, agglomerate area ratio A_A, and nanotube length x_{50} in dependence on the specific mechanical energy input (SME) for PCL composites containing 0.5 wt% Nanocyl™ NC7000 [15]

Conclusions

The significant effect of shortening has been possibly underestimated so far in research on the properties of melt mixed CNT based composites and should be taken into consideration in further studies on that topic. This was certainly due in parts to the difficulties to measure the length after processing. However, the presented results clearly show a way to measure the length as well as the importance to consider this quantity.

In our study was found, that higher specific mechanical energy input resulted in all presented composites in a better CNT macrodispersion. At the same time, the CNTs were shortened significantly with increasing SME leading to higher resistivity values or even crossing the percolation concentration. In composites with shortened nanotubes, an increase in electrical percolation threshold was shown.

For practical applications, a balance of the counteracting processes of the better dispersion and more pronounced CNT length shortening with increasing SME has to be achieved. It has to be considered that for improved mechanical properties a good CNT dispersion is essential, since the remaining CNT agglomerates lead to defects in the composite. A good dispersion is achieved at high SME which on the other hand shortens the nanotubes more severely. However, for a high electrical conductivity of CNT composites the nanotubes should remain as long as possible which only can be achieved at low SME.

Thus, depending on the aspired property profile and in addition economic aspects suitable energy inputs and mixing conditions have to be selected. In addition, it has to be considered that final shaping in most cases requires a second processing step, like compression or injection molding. In this step especially the process of secondary agglomeration of dispersed nanotubes influences the development of the percolation network and the electrical properties of the parts, which again depends on the shaping conditions.

Acknowledgements

We thank the German Federal Ministry of Education and Research (project CarboDis, grant ID 03X0042M within the Innovation Alliance Inno.CNT,) and the German Federal Ministry of Economics and Technology (project grant ID AiF 336 ZBG) for financial support.

References

1. B. Krause; T. Villmow; R. Boldt; M. Mende; G. Petzold; P. Pötschke; *Composites Science and Technology* 2011, *71*, 1145.
2. R. Socher; B. Krause; M. T. Müller; R. Boldt; P. Pötschke; *Polymer* 2012, *53*, 495.
3. P. Pötschke; T. Villmow, in: ANTEC 2012 - Proceedings of the 70th Annual Technical Conference & Exhibition, Orlando, FL, USA, April 1-5, Society of Plastics Engineers
4. F. Y. Castillo; R. Socher; B. Krause; R. Headrick; B. P. Grady; R. Prada-Silvy; P. Pötschke; *Polymer* 2011, *52*, 3835.
5. D. Stauffer; A. Aharony, "*Introduction in percolation theory*", Taylor and Francis, London, 1994.
6. R. K. Duncan; X. G. Chen; J. B. Bult; L. C. Brinson; L. S. Schadler; *Composites Science and Technology* 2010, *70*, 599.
7. J. Albuerne; A. Boschetti-de-Fierro; V. Abetz; *Journal of Polymer Science Part B: Polymer Physics* 2010, *48*, 1035.
8. S.-Y. Fu; Z.-K. Chen; S. Hong; C. C. Han; *Carbon* 2009, *47*, 3192.
9. B. Lin; U. Sundararaj; P. Pötschke; *Macromolecular Materials and Engineering* 2006, *291*, 227.
10. S. Badaire; P. Poulin; M. Maugey; C. Zakri; *Langmuir* 2004, *20*, 10367.
11. K. L. Lu; R. M. Lago; Y. K. Chen; M. L. H. Green; P. J. F. Harris; S. C. Tsang; *Carbon* 1996, *34*, 814.
12. Y. Wang; J. Wu; F. Wei; *Carbon* 2003, *41*, 2939.
13. M. F. Islam; E. Rojas; D. M. Bergey; A. T. Johnson; A. G. Yodh; *Nano Letters* 2003, *3*, 269.
14. B. Krause; R. Boldt; P. Pötschke; *Carbon* 2011, *49*, 1243.
15. B. Krause; M. Mende; G. Petzold; R. Boldt; P. Pötschke, chapter 7, pp 212-233 in: Carbon Nanotube-Polymer Composites, D. Tasis, Ed.; RSC Publishing, London, 2013.
16. B. Krause; P. Pötschke; L. Häußler; *Composites Science and Technology* 2009, *69*, 1505.
17. G. R. Kasaliwal; S. Pegel; A. Göldel; P. Pötschke; G. Heinrich; *Polymer* 2010, *51*, 2708.
18. G. R. Kasaliwal; A. Göldel; P. Pötschke; *Journal of Applied Polymer Science* 2009, *112*, 3494.
19. P. Pötschke, T. Villmow, B. Krause, Melt mixed PCL/MWCNT composites prepared at different mixing speeds: characterization of rheological and electrical properties, molecular weight, CNT macrodispersion, and CNT length distribution, *Composites Science and Technology*, submitted 2012

O-07-234

THE EFFECT OF PROCESSING CONDITIONS ON THE EVOLUTION OF THE DISPERSION OF POLYMER-CLAY NANOCOMPOSITES ALONG THE EXTRUDER

J. M. Barbas[1], A. V. Machado[1], J. A. Covas[1]*

[1] Institute of Polymers and Composites (IPC) and I3N, University of Minho, Portugal – joana.barbas@dep.uminho.pt; avm@dep.uminho.pt; jcovas@dep.uminho.pt

Abstract – *Two polymer-clay nanocomposites, using polypropylene and polyamide 6 as matrices, are studied and the effect of the operating conditions on dispersion is investigated. In-line NIR and appropriate chemometrics are used to monitor dispersion in real time, thus avoiding subjecting samples to preparation/measurement routines that may modify their initial features. For each system, there is an optimal operating condition that maximizes dispersion.*

Keywords: On-line monitoring; rheology; near-infrared spectroscopy; polymer nanocomposites

Introduction

Melt mixing seems to be the practical method of choice for the preparation of polymer-clay nanocomposites due to its effectiveness in reaching good dispersion levels, continuous production and possibility of reaching high yields [1]. There is ample evidence that processing conditions such as screw speed, feed rate, temperature and screw geometry may significantly affect the dispersion levels achieved [1–3]. The conjugation of time and low viscosity (for diffusion) and shear (for delamination) seems necessary [4]. However, the role of process parameters on the onset and evolution of dispersion along the extruder axis is not well understood. Lertwimolnun and Vergnes [3] characterized post-mortem samples collected from various locations along a co-rotating twin screw machine by oscillatory rheometry and X-ray diffraction. They observed that both intercalation and exfoliation can reach relatively high levels immediately after melting. Also, depending on processing conditions, dispersion could evolve, remain constant or apparently revert along the screw. The latter was suggested to be caused by matrix degradation.

This work uses in-line NIR and appropriate chemometrics [5] to monitor the effects of screw speed (N) and throughput (Q) on the evolution of dispersion along a twin screw extruder. The aim is to contribute to a better understanding of dispersion in these machines using a fast response in-line technique and avoiding subjecting samples to preparation/measurement routines that may modify their initial features. Two matrices are studied, one based on polypropylene the other on polyamide, i.e., with distinct affinities with the organoclays.

Experimental

Processing

Nanocomposites based on PP (Moplen HP500N, Lyondell-Basell) compatibilized with Maleic Anhydride modified PP (PP-g-MA, Polybond 3200, with 1% wt. of MA, supplied by Crompton) and PA6 (Technyl C206, Rhodia) and organo-layered nanosilicates (Dellite 67G and 43B, Laviosa, added to PP and PA6, respectively), were prepared in a Leistritz LSM30.34 co-rotating twin-screw extruder. The compositions were kept constant at 90/5/5 in weight for the PP/PP-g-MA/D67G system, and at 95/5 in weight for the PA6/D43B system.

The nanocomposites were prepared under various screw speeds (50, 100, 200 and 300 rpm) at a constant throughput of 3 kg/h, and at various feed rates (1.5, 3, 6 and 9 kg/h) at a constant screw speed of 100 rpm. The barrel/die temperature was set at 200 ºC for PP and at 240 ºC for PA6.

In-line Monitoring

The extruder was fitted with modified barrel segments designed to allow for material sampling and connection of monitoring devices [6]. They also enable the measurement of the average melt temperature (by sticking a fast response thermocouple in the freshly collected melt) and an estimation of the minimum residence time (by counting the time elapsed between feeding a pigment tracer into the screws and identifying a local color change in the melt stream). A NIR Diffuse Reflectance probe (Hellma) with a sapphire window (diameter of 5.7 mm) was directly fixed to the extruder at L/D = 9, 10, 11, 19, 29 and die body (Figure 1) by a Dynisco-type thread and linked to a Matrix F Bruker Optics NIR spectrometer. Each measurement contained 50 spectra with a resolution of 8 cm^{-1} and accumulation of 4 scans (circa 2 seconds per spectrum). The method was presented and validated in [5].

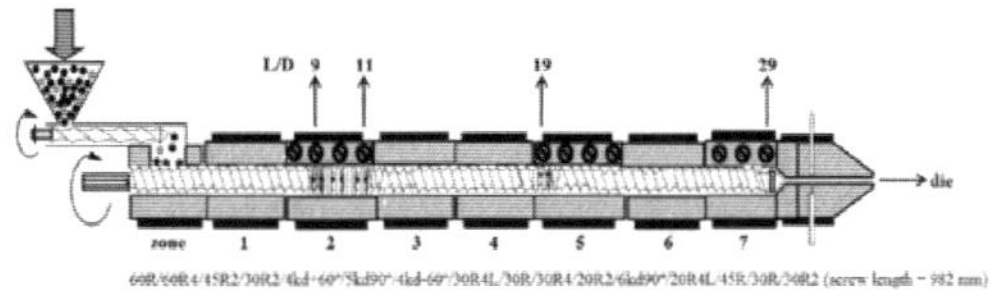

Figure 1 – Extruder layout and screw profile.

Off-line Characterization
Isothermal small amplitude oscillatory sweeps (ARG2, TA Instruments) from 0.1 to 100 rad/s were performed at 200 °C for the PP system and at 240 °C for the PA6 system, using parallel plates with a diameter of 25 mm and a 1 mm gap, under a constant strain (1% and 0.5% for the PP and PA6 systems). FT-IR spectra were obtained in transmission mode (Jasco FTIR4100) in the 4000 – 500 cm^{-1} wavenumber range with a resolution of 4 cm^{-1}, using 32 scans. A polarizer lens was used to eliminate the possible effects of clay orientation. The organoclays were analysed with a KBr mortar.
Thermal gravimetric analysis (TGA, TA Instruments Q500) of the organoclays (D67G and D43B) and polymer matrices was carried out under different conditions. Dynamic tests were made using a heating rate of 10 °C/min from 50 to 700 °C, and isothermal tests were performed during 10 min, both under a constant flow of nitrogen (60 ml/min).

Results and Discussion
Figure 2 shows the evolution of dispersion along the extruder for the two systems, as predicted by NIR. The results are based on a previously developed chemometric model built from data uncovering the effects of clay loading, compatibilizer content, screw speed and throughput and encompassing seven parameters derived from reference characterization techniques (oscillatory rheometry and FT-IR) and a thermo-mechanical index (specific mechanical energy) [5]. For PP nanocomposites, in the first part of the extruder (up to L/D=11) the dispersion develops and increases with screw speed (but is lower for 300 rpm). However, this behavior is not observed downstream. Nanocomposites prepared at 50 and 300 rpm show an axial increase and decrease of dispersion, respectively, while this remains approximately constant for the other two conditions. At the die, there is a general trend for a further decrease in dispersion.
The PA6 nanocomposites exhibit a quite distinct behavior, with an increase in dispersion until L/D=19, followed by a plateau and a decrease at the die exit. The effect of screw speed is not significant.
Reversion of clay dispersion at the die could be due to high shear rates and corresponding viscous dissipation, which would cause degradation of the clays surfactant. Previous studies have shown that degradation induces the collapse of the existing structure and/or re-agglomeration of the clay particles, regardless of the polymer matrix used [7]. Nevertheless, the composites based on PA6 attained higher dispersion levels at comparable operating conditions.

Figure 3 presents the evolution along the extruder of the storage modulus, G' (at 0.1 rad/s), for both PP and PA6 nanocomposites prepared with diverse screw speeds. In fact, the increase of elasticity at low frequencies is often attributed to the existence of fine suspended particles, i.e., higher dispersion levels [8 – 9]. For PP based materials (Fig. 3(a)), the evolution of the linear viscoelastic response is similar to that of the degree of dispersion as observed by NIR. The highest values of G' are achieved at 200 rpm. For the PA6 nanocomposites (Fig. 3(b)), a small increase of G' along the extruder for all screw speeds is seen. The PA6 nanocomposite prepared at 100 rpm has the highest G' values.

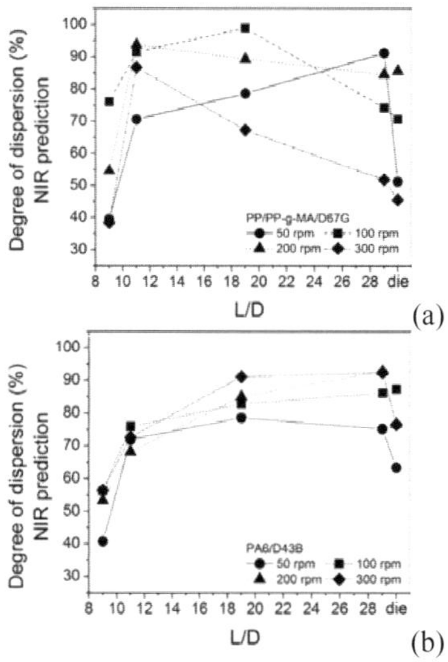

Figure 2 – Degree of dispersion along the extruder as predicted by NIR at various screw speeds: (a) PP matrix; (b) PA6 matrix.

Although not shown here, for the PP based nanocomposites the effect of feed rate is similar to that of screw speed. In the case of the PA6 system, varying the feed rate increases the dispersion levels along extruder, but the best outcome appears to occur at 6 kg/h. Again, for both systems and all conditions tested, the values decrease significantly at the die exit.
Therefore, there is an optimal operating condition that maximizes dispersion.

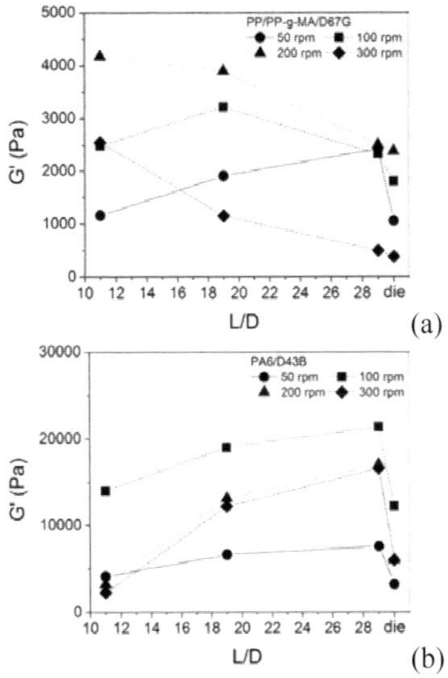

Figure 3 – Evolution of G' along the extruder at various screw speeds: (a) PP matrix; (b) PA6 matrix.

Conclusions

For the PP based nanocomposites, most of the conditions tested induced an increase of dispersion along the first mixing zone, followed by an apparent reversion downstream. The highest dispersion levels are achieved under intermediate shear/stress levels and high residence times. In the case of the PA6 system, reversion in dispersion could only be detected during flow along the die. This system benefits *a priori* from a better chemical interaction, dispersion appearing more related to stress levels.

References

1. J.W. Cho, D.R. Paul. *Polymer* 2001, 42, 1083
2. H.R. Dennis, D.L. Hunter, D. Chang, S. Kim, J.L. White, J.W. Cho, D.R. Paul. *Polymer* 2001, 42, 9513
3. W. Lertwimolnun, B. Vergnes. *Polymer Engineering and Science* 2007, 47, 2100
4. M. Bousmina. *Macromolecules* 2006, 39, 4259
5. J.M. Barbas, A.V. Machado, J.A: Covas. *Journal of Applied Polymer Science* 2012, DOI: 10.1002/app.38106
6. A.V. Machado, J.A. Covas, M. van Duin. *Journal of Applied Polymer Science* 1999, 71, 135
7. R.K. Shah, D.R. Paul. *Polymer* 2006, 47, 4075
8. L. Xu, H. Nakajima, E. Manias, R. Krishnamoorti. *Macromolecules*, 2009, 42, 3795

O-07-252

DEVELOPMENT ON DISPERSION OF CARBON NANOTUBES REINFORCED BIODEGRADABLE POLYMER NANOCOMPOSITES

S. Thumsorn[1]*, N. Srisawat[1], J. Wong On[2] and S. Pivsa-Art[1]

[1]*Faculty of Engineering, Rajamangala University of Technology Thanyaburi, Thailand – supaphorn.t@en.rmutt.ac.th; natee.s@en.rmutt.ac.th; sommai.p@en.rmutt.ac.th*

[2]*Faculty of Engineering, Pathumwan Institute of Technology, Thailand – jadzila@gmail.com*

Abstract - Carbon nanotubes (CNT) reinforced biodegradable polymer nanocomposites are prepared in this study. Poly(lactic acid) (PLA) was compounded with CNT through solutions cast film method. Dispersion of CNT is very important in order to obtain superior mechanical and thermal properties of PLA/CNT nanocomposites. In this study, dispersing agents was used for improved the dispersion of CNT on the PLA matrix. The effects of dispersing agent contents as well as CNT loading on morphological, thermal properties and electrical conductivity of PLA/CNT nanocomposites were investigated by using fourier transform infrared spectroscopy, scanning electron microscope, differential scanning calorimetry and resistivity meter. The electrical conductivity behavior of PLA/CNT nanocomposites was also elucidated. From the results, dispersion of CNT on PLA matrix enhanced with increasing the dispersing agent contents. SEM micrographs indicated good dispersion of CNT on the PLA matrix. Moreover, crystallization of PLA/nanocomposites could be clearly seen at higher contents of dispersing agents. Therefore, CNT would act as nucleating agents for PLA. Well dispersion of CNT would obtain higher crystallization process in the composites. However, the dispersing agents influenced on declination of electrical conductivity of PLA/CNT nanocomposites. It can noted that the results of this study would be valuable for applying systematic of PLA/CNT nanocomposites for wide area applications.

Keywords: carbon nanotubes, biodegradable polymer, poly(lactic acid), dispersion, nanocomposite

Introduction

Polymer nanocomposites are continuingly researched according to their attractive properties especially superior in mechanical properties for applied in various industrial applications.

There are various nano-sized additives for fabricating the polymer nanocomposite such as carbon nanotubes (CNT), nanosilica, nanoclay and nano-sized of inorganic fillers for example nano-$CaCO_3$ [1-6]. Carbon nanotubes are quite interesting for polymer nanocomposites especially for electrical and thermal conductivity applications i.e. sensor and conductivity polymer composites with high performance [7-8]. However, the different structure of CNT and organic polymer would obtain the composites with poor mechanical properties due to their incompatibles. There are various methods that used for modifying CNT in order to improve the compatibility with polymer matrix. Moreover, nano-sized structure is difficult to disperse on the polymer matrix. Chiu et al [1] studied on thermal, electrical amd mechanical properties of polyamide 46/muti-walled carbon nanotube (MWNT) nanocomposites. They modified MWNT by acid-treated with H_2SO_4 and HNO_3. Formic acid was used as solution for solution mixing technique. In this system MWNT was well dispersion and good dispersion on PA 46 matrix. The stiffness and electrical conductivity increased with increasing MWNT contents. Hu et al [2] used grafted multi-walled carbon nanotubes with polystyrene (PS). The various kinds of solution were used for preparing the nanocomposites. The grafting technique would improve dispersion of the MWNT in the nanocomposites. Furthermore, grafted MWNT improved thermal and electrical of PS/MWNT nanocomposites. Wu et al [3] researched on polylactide/muti-walled carbon nanotubes nano-composites. In this study acrylic acid grafted polylactide and multihydroxyl-functionalized MWNTs were prepared. The functionalized MWNT demonstrated significantly enhancement in thermal and mechanical properties of polylactide/MWNT nanocomposites. However, the dispersion of CNT is very important factor for preparing polymer/CNT nanocomposites.

Poly(lactic acid) (PLA) is a bio-based polymer derived from renewable resources. PLA has been popular due to its good recyclability and superior strength and clarity. PLA is promising materials for environmental friendly composites.

In this study, PLA/CNT nanocomposites were prepared. The dispersion of CNT in the PLA matrix was modified by using a dispersing agent. The effect of dispersing agent contents on the dispersion of CNT was investigated. The PLA/CNT nanocomposites properties were elucidated by using fourier transform infrared spectroscopy, differential scanning calorimetry, scanning electron microscope and electrical conductivity measurement.

Experimental

Materials and Composites Preparation

Poly(lactic acid) (PLA3052D was supplied from Nature Works LLC, USA. The MFR was 14 g/10 min. Multi-walled carbon nanotubes (MWCNT) (NANOCYL™ NC7000, Nanocyl S.A., Belgium) were purchased from Siam Extek Co., Ltd., Thailand. A dispersing agent, isotridecyl alcohol (isotrideceth-9, KOLB, Switzerland) was supplied by Connell Brother Co., Ltd., Thailand. The contents of dispersing agent used are varied from 0, 2, 4, 6, 8 and 10 wt%. Ethanol was supplied by Merck, Germany. Chloroform was supplied by Carlo Erba.

CNT was surface treated by ethanol in ultrasonic bath for 10 min then adding various contents of the dispersing agent, isotridecyl alcohol, in ultrasonic bath for 20 min. After that treated CNT was filtrated and dried at 80 °C in an oven for12 hours.

The specimens were fabricated by solution cast film process. PLA 5 wt% was dissolved in chloroform. After that 1 wt% of CNT with and without dispersing agent were added into PLA in chloroform. The solution was poured in Petri dishes and left at the room temperature. The cast film specimens were used for characterization.

Fourier Transform Infrared Spectroscopy (FTIR)
Fourier transform infrared spectroscopy (ATR-FTIR, PerkinElmer) was used for investigated PLA/CNT nanocomposites structure.

Thermal Properties
Thermal properties was analyzed using differential scanning calorimeter (DSC 200 F3, NETZSCH, Germany) at temperature range of 30-200 °C at a heating and cooling rate of 10 °C/min and 5 °C/min, respectively.

Morphology Observation
Morphology observation was investigated by scanning electron microscope (SEM) (JSMS410LV, JEOL, Japan). The specimens were vacuum dried before observation. Gold sputtering was performed on the specimens for electron conductivity.

Electrical Conductivity
Electrical conductivity of the specimens was carried out by Resistivity meter (KS-385B, KINGSOM, China). The surface resistivity values will be used for classified the conductivity of the nanocomposites.

Results and Discussion

Fourier Transform Infrared Spectroscopy Analysis
FTIR spectra of PLA and the nanocomposites are presented in Fig. 1 to Fig. 3. Fig. 1 shows FTIR spectrum of PLA. Fig. 2 and Fig. 3 show spectra of PLA/CNT nanocomposites without and with dispersing agent treated CNT. All spectra were similar, which presented the characteristic peaks of PLA structure at the (C-O) band at 1383 and 1181 cm^{-1}, the (C=O) band at 1750 and 1181 cm^{-1}, the (C-H) band at 2999 cm^{-1}.

However, it is not clearly seen on the absorption board band of (O-H). Therefore, there was not much alcohol remain in the nanocomposites as well as less adhesion between CNT and the dispersing agent.

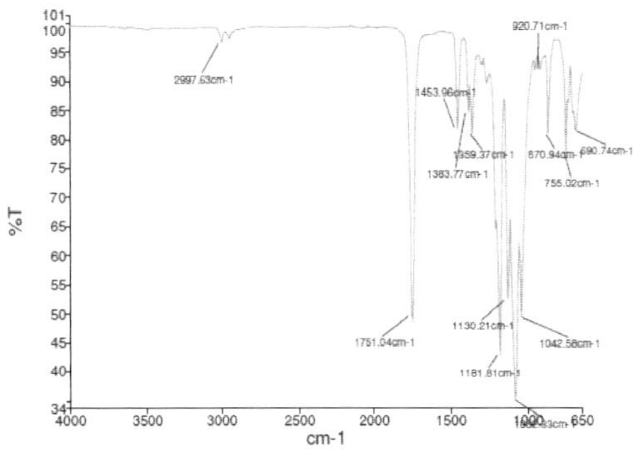

Fig. 1 FTIR spectrum of PLA.

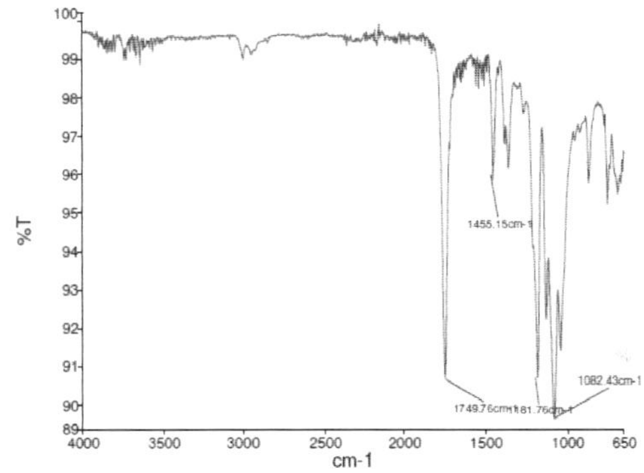

Fig. 2 FTIR spectrum of CNT-0 PLA nanocomposites.

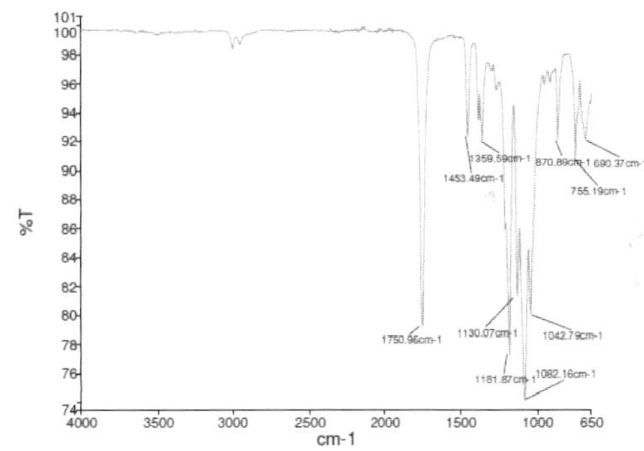

Fig. 3 FTIR spectrum of CNT-10 PLA nanocomposites.

Thermal properties of PLA/CNT nanocomposites
Differential scanning calorimetry was investigated thermal properties of PLA/CNT nanocomposites, which is tabulated in Table 1. From the results, PLA and all composites show melting endotherm at temperature of about 153-154 °C, which represent as melting temperature (T_m) of PLA. The effect of dispersing agent contents on crystallization temperature was also clearly seen from this table.

The crystallization temperature (T_c) of PLA increased with increasing the dispersing agent contents. Therefore, PLA/CNT nanocomposites could be easily crystallized than neat PLA. It was attributed to CNT acts as nucleating agent. Therefore, the dispersion of CNT in the nanocomposite was improved when adding higher contents of dispersing agent. Hence, these CNTs would be presented as nucleating site of PLA crystallization. However, crystallinity of the nanocomposites was lower than neat PLA. It might be due to the hindered of crystallization growth in the nanocomposites.

Table 1 Thermal properties of PLA/CNT nano composites.

Condition	Dispersing contents (wt%)	T_m (°C)	ΔH (J/g)	T_c (°C)	Crystallinity (%)
PLA	0	153	39.7	94	42.7
CNT-0	0	153	27.6	-	32.6
CNT-2	2	153	28.6	97	33.8
CNT-4	4	154	20.5	96	36.0
CNT-6	6	154	28.3	99	33.5
CNT-8	8	153	30.6	97	36.1
CNT-10	10	153	29.5	101	34.8

Morphology and dispersion of the nanocomposites

Fig. 4 presents photographs of dispersing agent treated CNT at various contents of the dispersing agent. CNTs dispersed in ethanol and then treated with dispersing agent of isotridecyl alcohol in ultrasonic bath. It can be seen that with adding higher contents of dispersing agent CNTs were well dispersed in the solution. Therefore, it could obtain the good dispersion of CNT on the PLA matrix.

Fig. 5 shows SEM micrographs of PLA/CNT nanocomposites at various contents of the dispersing agent. The micrographs exhibit good dispersion of CNT on the PLA matrix when adding more dispersing agent contents. Therefore, this dispersing agent could promote the dispersion of CNT on the PLA matrix. However, the dispersing agent could effect on electrical conductivity of the nanocomposites.

Fig. 4 Photographs of various dispersing agent contents treated CNT.

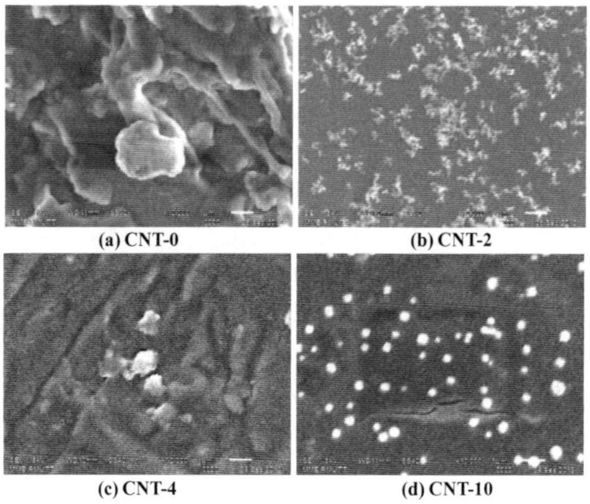

Fig. 5 SEM micrographs of PLA/CNT nanocomposites at various contents of dispersing agent. (a) 0 wt% (b) 2 wt% (c) 4 wt% and (d) 10 wt%.

Electrical Conductivity of the nanocomposites

The electrical conductivity of the nanocomposites was conducted with resistivity meter for comparison the conductivity of the nanocomposites before and after treated. The surface resistivity of the PLA/CNT nanocomposites is presented in Table 2. The value of surface resistivity can classified the electrical conductivity of the nanocomposites. From the results, surface resistivity of PLA indicated that PLA is polymer insulation. On the other hand, the incorporation of CNT exhibited enhanced surface resistivity of the nanocomposites, which presented the value of 10^5 Ω/square in CNT-0 and CNT-2. This value is represented for the conductivity of these CNT-0 and CNT-2 nanocomposites. However, the surface resistivity of the nanocomposites significantly decreased when adding higher contents of the dispersing agent. It can be noted that the dispersing agent might retard the conductivity of PLA/CNT nanocomposites at higher contents of the dispersing agent.

Table 2 Surface resistivity of PLA/CNT nano-composites.

Conditions	Surface resistivity (Ω/square)	Conductivity
PLA	> 10^{12}	Insulative
CNT-0	10^5	Conductive
CNT-2	10^5	Conductive
CNT-4	10^{12}	Dissipative
CNT-6	10^{12}	Dissipative
CNT-8	10^{12}	Dissipative
CNT-19	10^{12}	Dissipative

Conclusions

The dispersion of CNT in PLA/CNT nanocomposites was improved by using the dispersing agent, isotridecyl alcohol. The effect of dispersing agent contents on dispersion was elucidated. SEM micrographs indicated well dispersion of CNT on the PLA matrix when adding higher contents of the dispersing agent. The good dispersion of CNT in the nanocomposites would enhanced crystallization process of PLA/CNT nanocomposites, which indicated by the increment of crystallization temperature in the nanocomposites. The incorporation of CNT enhanced surface resistivity of the PLA/CNT nancomposites. However, the higher contents of the dispersing agent would effect on declination of electrical conductivity of PLA/CNT nanocomposites. It can noted that the results of this study would be valuable for applying systematic of PLA/CNT nanocomposites for wide area applications.

Acknowledgements

The authors would like to acknowledge Rajamangala University of Technology Thanyaburi (RMUTT), Thailand for supporting research fund. The authors would like to thank Mrs. Weraporn Pivsa-Art, Department of Chemical and Materials Engineering, Faculty of Engineering, RMUTT, Thailand for kindly provided multi-walled carbon nanotubes.

References

1. F.C. Chiu; G.F. Kao, *Composites: Part A*, 2012, *43*, 208.
2. H. Hu; K.N. Hui; K.S. Hui; S.K. Lee; W. Zhou, *Colloids and Surfaces A: Physicochemical and Engineering Aspects,* 2012, *396*, 177.
3. C.S. Wu; H.T. Liao, *Polymers*, 2007, *48*, 4449.
4. C.M.O. Muller; J.B. Laurindo; F. Yamashita, *Industrial Crops and Products*, 2011, *33*, 605.
5. M.Q. Zhang; M.Z. Rong; H.B. Zhang; K. Friedrich, *Polymer Engineering & Science*, 2003, *43*, 490.
6. W. Zhu; G. Zhang; J. Yu; G. Dai, *Journal of Applied Polymer Science*, 2004, *91*, 431.
7. H.Y. Chiu; P. Hung; H.W.C. Postma; M. Bockrath, *Nano Letters*, 2008, *8*, 4342.
8. L. Ci; J. Suhr; V. Pushparaj; X. Zhang; P.M. Ajayan, *Nano Letters*, 2008, *8*, 2762.

O-07-282

RHEOLOGICAL ANALYSIS OF DIFFERENT REACTIVITY OF POLYMER NANOCOMPOSITE TO THE ELECTRIC FIELD ACCORDING TO THE MOLECULAR STRUCTURES AND AFFINITIES BETWEEN POLYMER AND NANOPARTICLES

H.G. Ock, K.H. Ahn*

Department of Chemical and Biological Engineering, Seoul National University, Korea – ahnnet@snu.ac.kr

Abstract - Polymer nanocomposite is being continuously investigated because of many advantages of nanoparticles due to large surface area and high aspect ratio, compared to micro-sized functional fillers. In particular, organoclay has many benefits to make up physical properties, thermal properties, and gas permeability of polymers, if we can control exfoliated structures of clay particles in polymer matrix. The electric field has been used to facilitate the destruction of layer stacking and separation of silicate layers in polymer/clay nanocomposites. In this study, we investigated the response of nanocomposites to electric field, with various combinations of polymer matrices and clays. Three kinds of organically modified clays (Cloisite 10A, Cloisite 20A, Cloisite 30B) were compounded with two different polymers (poly(lactic acid) and polypropylene). The reactivity to electric field of polypropylene/clay was in the order of C20A, C10A and C30B, while the reactivity of poly(lactic acid)/clay was in the opposite order. Solubility parameter was introduced to explain this result based on the molecular structures of polymers and organic modifiers of clays. The microstructural change was observed during the dynamic time sweep test under electric field with time. The developed structure after electrical treatment was confirmed by the frequency sweep test. These results were also confirmed by the transmission electron microscopy morphology. The degree of rheological responses to the electric field was very different according to the combination of polymer matrices and clays. The most reactive combinations to electric field were poly(lactic acid)/Cloisite 30B and polypropylene/Cloisite 20A nanocomposites.

Keywords: Nanocomposites, Rheology, Poly(lactic acid), Clay, Electric field

Introduction

Polymer nanocomposite is being widely investigated to make high performances of polymers just using small amounts of nanoparticles. Because of large surface area of nanoparticle, it is expected to enhance various properties of base polymer more easily than micro-sized functional fillers. The nanoclays have many advantages to make up physical properties, thermal properties, and especially gas permeability of polymers.

To perform these features properly, it is essential to make exfoliated states of nanoclays. There are many chemical ways to get exfoliated structure of clay platelets, for example, clay modification, solution casting, in-situ polymerization, using compatibilizers. However, chemical method has limitation for industrial application because of very long processing time and high cost. Instead of that, physical methods to get exfoliation of clays, like applying high shear, ultrasonic, electric field, are being carried out. We are especially interested in using the electric field because it has many advantages like small amount of costs and possibility to be applied to the industrial field. The electric field was first found to facilitate the destruction of layer stacking and separation of silicate layers in polypropylene/clay nanocomposites [1]. To apply this concept for manufacturing of polymer/clay composites properly, we investigated the response of nanocomposites to electric field, with various combinations of polymer matrices and clays based on the structures of materials.

Experimental

Poly(lactic acid) and polypropylene were compounded with three different kinds of organically modified clays (Cloisite® 10A, 20A, 30B) using an Haake Rheocord 90 internal mixer with 100 rpm rotor speed, for 10minutes at 190℃ for PP and 210℃ for PLA. Each A.C. electric field strength (325, 550, 775, 1000V/mm) was applied using function generator and high voltage amplifier during dynamic time sweep test on RMS 800 with modified fixtures. The applied electric field was confirmed using a digital oscilloscope. The resulting structure was confirmed by the frequency sweep test at strain 10%, 170℃ for PP and strain 5%, 190℃ for PLA. Transmission Electric Microscopy morphology of each nanocomposites after electrical treatment was taken with Jeol JEM-2000EXII model.

Results and Discussion (Times New Roman 10, bold)

Fig. 1(a) shows the dynamic time sweep results of PP/C20A nanocomposites. For the first two minutes without electric field, the storage modulus was almost constant with time. However, the storage modulus after activating the electric field increased and then became saturated in a few minutes. Also, the saturated value of storage modulus under each electric field was increased with the magnitude of electric field strength. The dynamic frequency sweep test was conducted after electrical treatment to confirm structural evolution of each combination of nanocomposites in Fig.1(b). The storage modulus at low frequency region increased with electric field strength and the terminal slope approached to zero with increasing electric field strength.

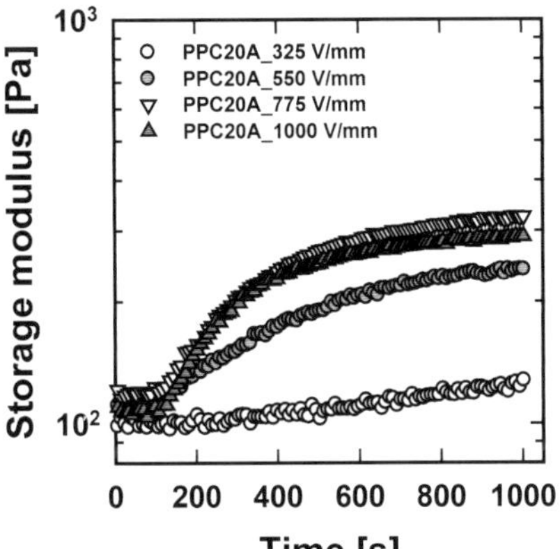

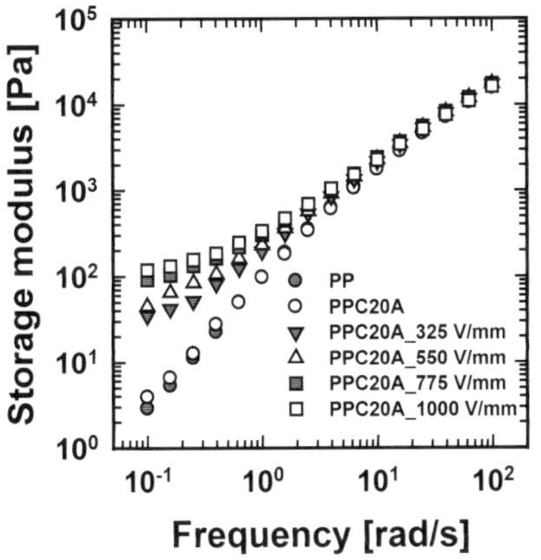

Fig.1. (a) Time sweep results of polypropylene/C20A composites before and after applying different electric field (325, 550, 775, 1000V/mm). (b) Frequency sweep results of polypropylene/C20A composites after time sweep tests.

The results showed that structure of polypropylene/clay nanocomposites became to form networked structure as electric field was increased. PLA/clay nanocomposites also showed similar behavior under frequency sweep test. In order to observe these micro-structural change with the magnitude of electric field strength, Transmission Electron Microscopy morphology from was taken in Fig.2. The images provided similar results with rheological properties. As an electric field increased, layer-stacking of clay tactoids destructed and layered silicate was separated. From these results – rheological data and TEM morphology, it can be inferred that clay is dispersed in polypropylene melt by activating electric field.

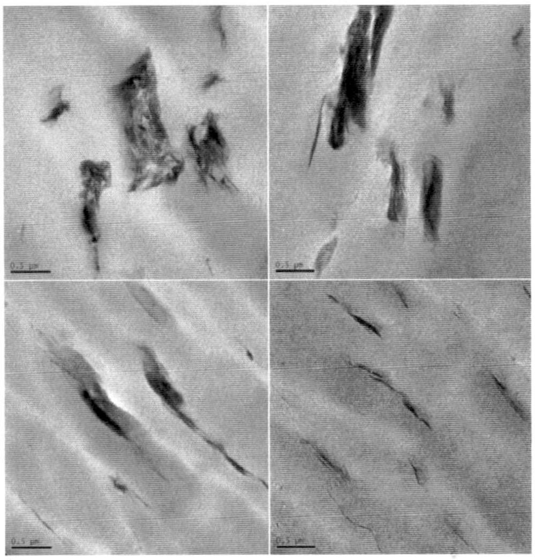

Fig. 2, Transmission Electron Microscopy morphology of PP/C20A after dynamic time sweep test under each electric strength (a) 325V/mm, (b) 550V/mm, (c) 775V/mm, (d) 1000V/mm.

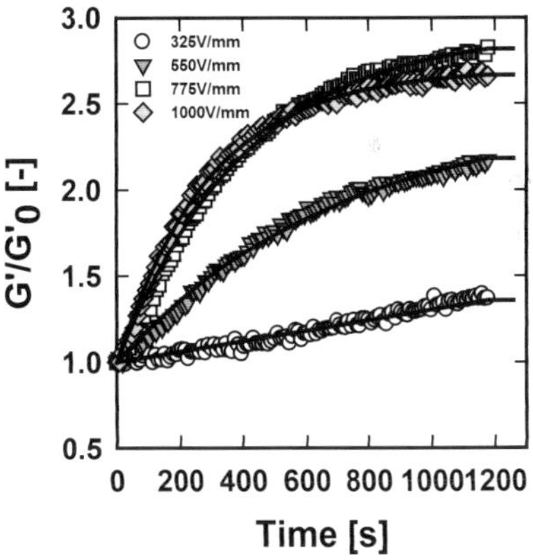

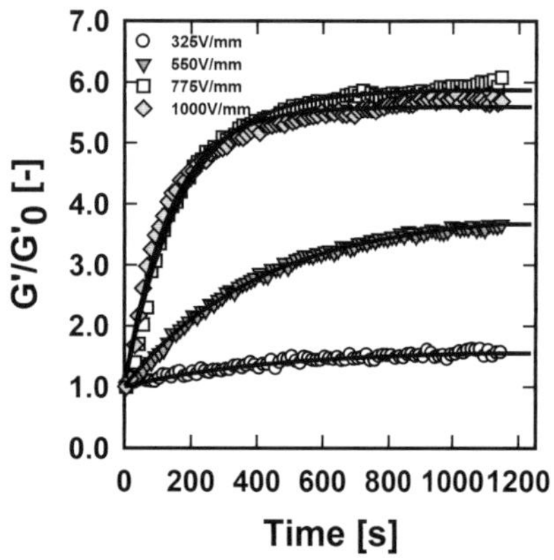

Fig.3. Reduced storage modulus with time for (a) PP/C20A (b) PLA/C30B

To specify the response of polymer nanocomposites to electric field, reduced storage modulus was plotted from nonlinear regression using exponential rising equation in Fig.3.

$$\frac{G'(t)}{G'_0} = 1 + a(1 - e^{-\frac{\lambda}{t}}) \quad (1)$$

From the equation, the initial storage modulus increasing rate (the initial rate, aλ) is suggested as a parameter which can represent the electrical response of nanocomposites. Fig.4 shows fitting results of initial rates. The initial rates of each material first increased with electric field strength and then became saturated after about 10 minutes. The reactivity of polypropylene/clay was in the order of C20A, C10A and C30B, while the reactivity of poly(lactic acid)/clay was in the opposite order.

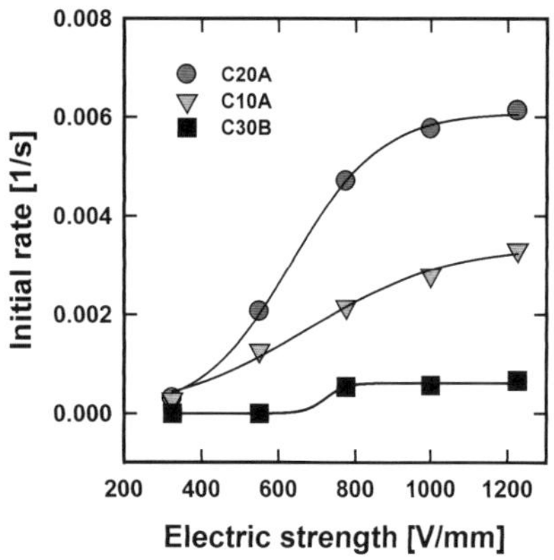

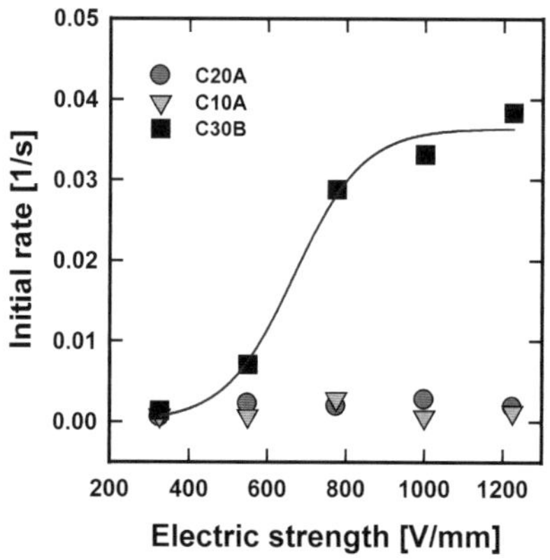

Fig.4. Initial storage modulus increasing rate (initial rate) with electric strength for (a) PP/clay (b) PLA/clay combinations

Based on the molecular structures of polymers and organic modifiers of clays, the structures of polypropylene and the organic modifier of C20A are more similar, While those of poly(lactic acid) and C30B are similar each other. The role of electric field for clay exfoliation in polymer melt was investigated in the previous research [2]. When the A.C. electric field was applied to the polypropylene/clay nanocomposites, polypropylene chain was slowly moving toward to galleries of clay platelet. Based on the result of this research, it might be suggested that moving rate of polymer chain is dependent upon affinities between base polymer and the functional group of clays. In order to explain the correlation between the structural similarity and initial rate, solubility parameter was introduced. The solubility parameter represents structural similarity and affinities between two materials and it could be applied to explain the affinities between base polymer and nanoparticle in polymer nanocomposites. It provides the affinities between two materials and it can be applied to explain the relation between base polymers and nanoparticles [3]. In this study, Hansen Solubility Parameter (HSP) was used to explain different response of polymer nanocomposites to electric field. The total solubility parameter (δ) is derived from dispersion (δd), polar interaction (δp) and hydrogen-bonding (δh) terms, where F is attraction constant, U in internal energy and V is Van der Waals volume.

$$\delta_d = (\textstyle\sum_z {}^zF_d)/V \quad (2)$$

$$\delta_p = (\textstyle\sum_z {}^zF_p^2)^{\frac{1}{2}}/V \quad (3)$$

$$\delta_h = (\textstyle\sum_z - {}^zU/V)^{\frac{1}{2}} \quad (4)$$

$$\delta = (\delta_d^2 + \delta_p^2 + \delta_h^2)^{\frac{1}{2}} \quad (5)$$

The Hansen parameters of each material were calculated in table 1 from references [4-5]. Polypropylene and C20A have nearly the same HSP value, while the value of PLA and C30B are similar. This result corresponds to the structural affinities of materials. Two materials which have similar solubility parameter mean that their structures are similar and they have affinities each other. Based on the functional groups of each material, the structures of PP and organic modifier of C20A are similar and those of PLA and modifier of C30B are also similar. The Hansen parameters can be matched to the results of different initial rates of each polymer/clay combinations.

Table 1 Calculated Hansen solubility parameters of each materials

	δ_d	δ_p	δ_h	δ
PP	15.84	0	0	15.84
PLA	15.71	18.42	10.30	26.31
C10A	19.14	2.28	5.47	20.02
C20A	16.9	1.53	5.23	17.66
C30B	21.60	3.16	10.89	24.39

Conclusions (Times New Roman 10, bold)

In this study, it is observed that the extent of rheological responses to the electric field was differently depending on the combination of polymer matrices and clays based on structural similarity. The best combinations were PP/C20A and PLA/C30B nanocomposites and the experimental result was explained by Hansen solubility parameter. In conclusion, compatibility of polymer and clay surfactant is predominant role of the efficiency of the electric field for clay exfoliation in polymer matrix.

References

1. D. H. Kim; K. H. Ahn; S. J. Lee *Macromol. Rapid Commun* 2003,*24*,388
2. J. U. Park; K. H. Ahn; S. J. Lee *Polymer* 2006,47,5145
3. B.N.Jang, D.Wang, Charles A. Wilkie, *Macromolucules* 2005,38,6533
4. Allan F.M. Barton, *Handbook of solubility parameters and other cohesion parameters,* CRC Press
5. Charles M. Hansen, *Introduction of Hansen solubility parameters : A user's handbook*; CRC Press

PREPARATION OF FLAME RETARDANT LDH/POLYOLEFIN NANOCOMPOSTIES

B. Kutlu[1,2*], N-J. Kang[3], D-Y. Wang[3] and A. Leuteritz[1]

[1]Leibniz-Institut für Polymerforschung Dresden e. V., Hohe Straße 6, D-01069 Dresden, Germany- kutlu@ipfdd.de

[2] Technische Universität Dresden, Professur für Polymerwerkstoffe und Elastomertechnik, Dresden, Germany

[3] Madrid Institute for Advanced Studies of Materials (IMDEA-Materials), Madrid, Spain

Abstract - Potentially flame retardant azo dye molecules were intercalated into layered double hydroxide (LDH) host following one-step synthesis route, which allows simultaneous substitution of interlayer ions with the desired organic species during co-precipitation of LDH layers. Organomodified LDH-maleic anhydride grafted polypropylene (PP-g-MA) mixture was refluxed in a common organic solvent (xylene) to prepare nanocomposites. Morphology, thermal stability and flame retardancy properties were investigated via X-ray scattering, transmission electron microscopy, thermogravimetric analysis, and microscale combustion calorimetry. In comparison to unmodified LDH (LDH-NO_3) or samples with a common modifier, sodium dodecylbenzenesulfonate, dye modified LDH reduces heat release rate (HRR), the total heat release (THR) and the heat release capacity (HRC) of the PP-g-MA nanocomposites up to 30% with a filler loading of 5 wt.-%, offering a promising way to prepare novel, non-halogenated and flame retardant composites.

Keywords: layered double hydroxide, flame retardancy, azo-dyes

Introduction

Layered double hydroxides consist of positively charged metal hydroxide layers, where the charge was balanced with interlayer anions. The charge-balancing anions, derived from inorganic acids [1], amino acids [2-4], or anionic polymers [5-6], are encapsulated in the interlayer space, and easily exchanged with various anions. LDHs can be represented by the chemical formula $[M^{2+}_{1-x}M^{3+}_x(OH)_2]^{x+}A^{n-}_{x/n}\cdot yH_2O$, where M^{2+}, M^{3+}, and A^{n-} are divalent metal cations, trivalent metal cations, and interlayer anions [7]. The species of M^{2+}, M^{3+}, and A^{n-} together with the value of x can be varied to give rise to a large class of isostructural LDHs.

LDH shows endothermic decomposition, when it is exposed to high temperature [8-10]. However, to be used as a nanofiller for polymers, it is necessary to modify the pristine LDH in order to enlarge the interlayer distance, to alter the surface properties and to facilitate the dispersion in the polymers [11-14]. LDH can be modified by a wide variety of anionic surfactants, such as fatty acid salts [15], sulfonates [16-17] and phosphates [18-19]. Among them, sodium dodecylbenzenesulfonate (SDBS) is most widely used [20-22]. However, SDBS is a flammable surfactant as well as most of the modifiers for LDH. Thus, there is a risk to deteriorate the flame retardancy of modified LDH and /or polymer-modified LDH nanocomposites by employing SDBS as modifier for LDH. In comparison to the structure of SDBS, most of the acid dyes contain nonfused aromatic, naphthalene or anthracene rings [23-24], possessing many rigid structures which may lead to form char-rich- residues during combustion or decomposition.

In the presented work, LDH was modified by two different char-forming modifiers and additionally by a a common modifier, SDBS. Nanocomposites were prepared by refluxing organomodified LDH and PP-g-MA in an organic solvent. Morphology, thermal stability and flame retardancy properties were investigated.

Experimental

The synthesis of dye-structure intercalated LDHs was carried out via a similar procedure as earlier reported [21]. Molar ratio between magnesium and aluminium was kept as 2:1 and corresponding organic modifier was introduced to the system with metal salts. pH value was kept to be 10±0.2 by adding 1 mol/L NaOH aqueous solution. The resulting slurry was stirred for another 30 min. and then thermally aged at 75°C for the following 18h. The product was filtered and washed with deionized water till pH value reaches 7.

Dye-structure intercalated LDH (d-LDH) was then used as a fire retardant nanofiller. Typical PP-g-MA/LDH composites were prepared by refluxing the mixture of desired amount of d-LDH and PP-g-MA in xylene for 24 h. The solution was then poured out into 100 ml ethanol. The precipitate was filtered and dried under vacuum at 80°C for 24h.

Results and Discussion

The FTIR spectra of LDH materials provide important information about the interlayer anions and hence are very useful to confirm the structure of these materials.

The FTIR of various reactants and products are shown in Figure 1. The broad band in the range 3000-3700 cm^{-1} originates from the O-H stretching of the metal hydroxide layer and interlayer water molecules. The vibration band of the interlayer H_2O is also reflected in the broad bands around 1624 cm^{-1}. In the spectrum of MgAl-NO_3-LDH, the characteristic band for interlayer nitrate (NO_3^-) was clearly observed (1381 cm^{-1}), which disappeares in cases of MgAl-Dye 1-LDH and MgAl-Dye 2-LDH, indicating that the nitrate is almost completely exchanged with dye anion. The band characteristics of the metal-oxygen bond stretching appear around 451 cm^{-1}. The main differences in the FTIR spectra between MgAl-NO_3-LDH and Dye-structure intercalated modified LDHs appeared due to the presence of dye ion in the host. It can be seen that all characteristic bands of Dye 1 and Dye 2 were also observed in the spectra of MgAl-Dye 1-LDH and MgAl-Dye 2-LDH, respectively. The results indicate the presence of dye molecules in all the modified LDHs.

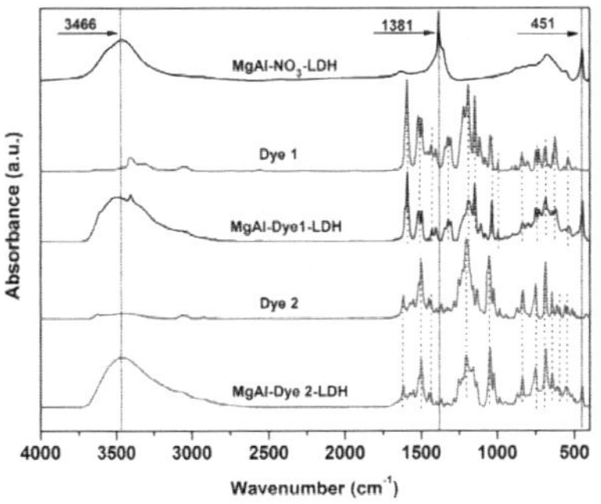

Figure 1 – FTIR spectra for samples of MgAl-NO_3-LDH, Dye 1, MgAl-Dye 1-LDH, Dye 2, and MgAl-Dye 2-LDH

Figure 2 shows the HRR plots for dye intercalated LDH, PP-g-MA, PP-g-MA/NO3-LDH, PP-g-MA/SDBS-LDH, and PP-g-MA/dye intercalated LDHs measured by microscale combustion calorimeter. It is clear that the pHRR value decreases significantly with addition of dye-intercalated- LDHs. While pHRR is 1043W/g in PP-g-MA, it decreases to 927W/g in PP-g-MA/SDBS-LDH, 812W/g in PP-g-MA/Dye 2-LDH, 702W/g in PP-g-MA/Dye 1-LDH, showing significantly improved flame resistance. The trend in THR values of PP-g-MA, PP-g-MA/Dye 2-LDH, and PP-g-MA/Dye 1-LDH composites is similar to the HRR and HRC. Additionally, the temperature at maximum pyrolysis rate (Tmax) shows no significant difference between the composites and the neat polymer.

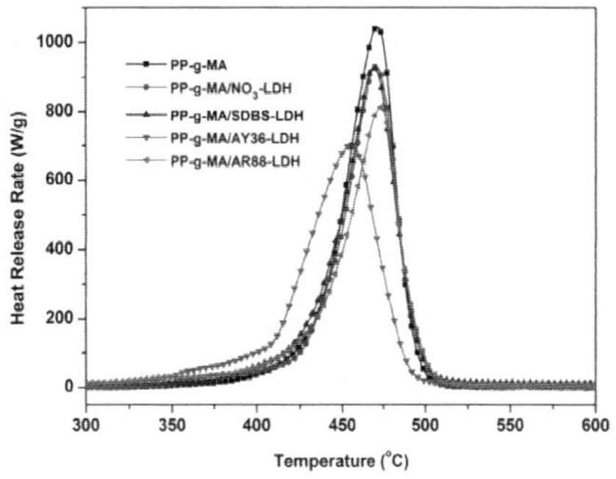

Figure 2 – HRR curves of PP-g-MA/LDH composites at 1K/s heating rate

Conclusions

Dye intercalated LDHs have been synthesized via one step method and used as additives for PP-g-MA. The PP-g-MA/dye intercalated LDH composites were prepared by solution compounding method. In comparison to PP-g-MA, PP-g-MA/dye intercalated LDH composites show lower pHRR, HRC, and THR values, indicating the flame retardancy of PP-g-MA was improved using the additive of dye intercalated LDHs.

References

1. F. R. Costa; U. Wagenknecht; G. Heinrich Polymer degradation and stability 2007, *92,*1813.
2. S. Aisawa; H. Kudo; T. Hoshi; S. Takahashi; H. Hirahara; Y. Umetsu; E. Narita Journal of solid state chemistry 2004, *177*, 3987.
3. T. Hibino Chem. Mater. 2004,*16*, 5482.
4. N. T. Whilton; P. J. Vickers, S. Mann J. Mater. Chem., 1997,*8*, 1623.
5. O. C. Wilson Jr; T. Olorunyolemi; A. Jaworski, L.Borum; D. Young; A. Siriwat; E. Dickens; C.Oriakhi; M. Lerner Applied clay science 1999, *15*, 265.
6. F. Leroux; J-P. Besse Chem. Mater. 2001, *13*, 3507.
7. C-X. Zhao; Y. Liu; D-Y. Wang; D-L. Wang; Y-Z. Wang Polymer Degradation and Stability 2008, *93*, 1323.
8. D.-Y. Wang; A. Das; A. Leuteritz; R. N. Mahaling; D. Jehnichen; U. Wagenknecht; G. Heinrich RSC Advances 2012, *2,* 3927.
9. C.M.C. Pereira; M. Herrero; F.M. Labajos; A.T. Marques; V. Polymer Degradation and Stability, 2009, *94*, 939.
10. C. Nyambo; E. Kandare; C. A. Wilkie Polymer Degradation and Stability 2009, *94*, 513.
11. V. Rives; M. A. Ulibarri Coordination chemistry reviews 1999, *181,* 61.

12. L. Qiu; W. Chen; B. Qu Polymer Degradation and stability 2005, *87*, 433.
13. C-S. Liao; W-B. Ye Journal of polymer research 2003; *10,* 241.
14. P. Ding; B. Qu Polymer Engineering and Science 2006, *46*, 1153.
15. N. Nhlapo; T. Motumi; E. Landman; S. M. C. Verryn; W. J Mater Sci 2008, *43*, 1033.
16. N. Iyi; Y. Ebina; T. Sasaki J. Mater. Chem. 2011, *21*, 8085.
17. M. Moujahid; J.-P. Besse; F. Leroux J. Mater. Chem., 2003, *13*, 258.
18. M. A. Woo; T. W. Kim; M.-J. Paek; H.-W. Ha; J.-H. Choy; S.-J. Hwang Journal of Solid State Chemistry 2011, *184*, 171.
19. J.-H. Choy; S.-Y. Kwak; J.-S. Park; Y.-J. Jeong; J. Portier J.Am.Chem.Soc. 1999, *121*, 1399.
20. F. R. Costa; M. Abdel-Goad; U. Wagenknecht; G. Heinrich Polymer 2005, *46*, 4447.
21. D.-Y. Wang; F. R. Costa; A. Vyalikh; A. Leuteritz; U. Scheler; D. Jehnichen; U. Wagenknecht; L. Haussler; G. Heinrich Chem. Mater. 2009, *21*, 4490.
22. A. Vyalikh; D.-Y. Wang; U. Wagenknecht; G. Heinrich; U. Scheler Chemical Physics Letters 2011, *509*, 138.
23. L. Gua; N. Zhua; L. Wang, X. Binga; X. Chena Journal of Hazardous Materials 2011; 198; 232.
24. T. Hihara; Y. Okada; Z. Morita Dyes and Pigments 2007, *75*, 585.

O-07-154

EFFECT OF NANOTUBE ASPECT RATIO ON PROPERTIES OF MELT-MIXED POLYCARBONATE NANOCOMPOSITES

Jiaxi Guo[1,2], Frank Yepez Castillo[1,2], Robert Socher[3], Beate Krause[3], Robert Headrick[4], Brian P. Grady[1,2] Ricardo Prada-Silvy[4], and Petra Pötschke[3]

[1]*Carbon Nanotube Technology Center (CaNTeC), University of Oklahoma, Norman, Oklahoma 73019, United States*

[2]*School of Chemical, Biological & Materials Engineering, University of Oklahoma, Norman, Oklahoma 73019, United States*

[3]*Leibniz Institute of Polymer Research Dresden (IPF), Hohe Straße 6, Dresden D-01069, Germany*

[4]*SouthWest NanoTechnologies, Inc. (SWeNT®), 2501 Technology Place, Norman, Oklahoma 73071, United States*

Abstract - The purpose of this work is to explore the question of how the aspect ratio affects various properties in carbon nanotube/polycarbonate composites. Aspect ratios (prior to mixing) of multi-walled carbon nanotubes varied between 50-500. Tubes were mixed with polycarbonate in a small conical twin-screw compounder and then compression molded into flat sheets. Electrical measurements, differential scanning calorimetry and dynamic mechanical analysis were performed on the as molded sheets. The aspect ratio correlated to both the electrical percolation threshold and the rubbery modulus in a simple linear fashion; however with a significant level of scatter. The glass transition behavior did not seem to be affected by the aspect ratio, although differences were seen which is likely due to the different sources of nanotubes and hence different levels of impurities, defects etc.

Keywords: Nanotube, polycarbonate, aspect ratio

Introduction

A great deal of research into polymers filled with carbon nanotubes (CNTs) has been performed because of the excellent mechanical and electrical properties of nanotubes. Due to their high aspect ratio resulting from length L up to the millimeter range and diameters D between 1 and 50 nm, CNTs can form an interconnected network at very low volume fractions. The lowest nanotube concentration at which such networks are first formed is known as the percolation threshold. Theoretically, the aspect ratio (=L/D) of fillers is directly related to their percolation concentration. According to continuum percolation theory, for randomly oriented ideal monodisperse penetrable rods with aspect ratios much larger than one the critical volume filler concentration for percolation ϕ_p can be well approximated by the following equation. [1]

$$\phi_p = D/(2L) \qquad (1)$$

The purpose of these experiments is to examine the effect of L/D on the properties of the nanocomposites, in particular the electrical conductivity.

Experimental

An injection molding grade polycarbonate (PC) with a medium viscosity, Makrolon® 2600 (Bayer MaterialScience AG, Leverkusen, Germany) was used in this study. The multi-walled nanotubes (MWCNT) used in this study were from a mix of commercial and experimentally available products, see Ref [2] for a list of the nanotubes and some specific properties; the other longer nanotubes were made from a process very similar to that used to make SMW-100 tubes and were provided by Southwest Nanotechnologies. Tubes with average aspect ratios 55, 69, 73, 94, 134, 250, 313 and 474 were used. In all cases, the average diameter was ~10 nm, except for the CNT materials with aspect ratios of 55(6.0 nm) and 73(9.8 nm).

Composites were made using the same melt mixing procedures, but on different equipment. Statements in parenthesis represent aspect ratios 50-150 and 250-500 respectively. After drying overnight at 80°C, PC was melt mixed with different MWCNT using a DSM twin-screw microcompounder (15 cm^3 or 5 cm^3). As found in previous investigations [3], high mixing temperatures and high mixing speeds result in a good CNT dispersion with low electrical percolation thresholds. Therefore, a mixing temperature of 280°C and a mixing speed of 200 rpm were applied. Considering the residence times in industrial melt mixing, a mixing time of 5 min was chosen. Granules obtained from the extruded strands were compression molded into plates of ~60 mm diameter and ~0.5 mm thickness using a pressing temperature of 280°C.

Details concerning electrical property measurements, differential scanning calorimetry (DSC) and dynamic mechanical analysis (DMA) were given in the previous reference [2].

Results and Discussion

Bulk electrical conductivities of the samples are shown in Figure 1. As indicated in the figure, composites made from the three longest nanotubes show statistically identical behavior in terms of the electrical conductivity as a function of nanotube concentration.

For these three composites, with 0.25 wt % nanotubes in the system, the bulk conductivity of the samples increases from 10^{-18} S/cm for the neat PC to approximately 10^{-3} S/cm and the percolation thresholds for these three composites are estimated to occur at nanotube concentration of about 0.15 wt. %. This percolation threshold is extremely low for a melt-mixed composite material. Part of our group (Pötschke et al.) has found an even lower percolation threshold 0.04 wt. % for polyamide/MWCNT composites [4]; however, for PC/MWCNT composites, to the best of our knowledge, the electric percolation threshold of the current work is the lowest ever found. This low threshold provides great opportunities to tailor the polymers having excellent electrical conductivity without undermining the flexibility of the polymers. Composites containing MWCNTs with lower aspect ratios show higher percolation concentration which nearly follows the order in the aspect ratio.

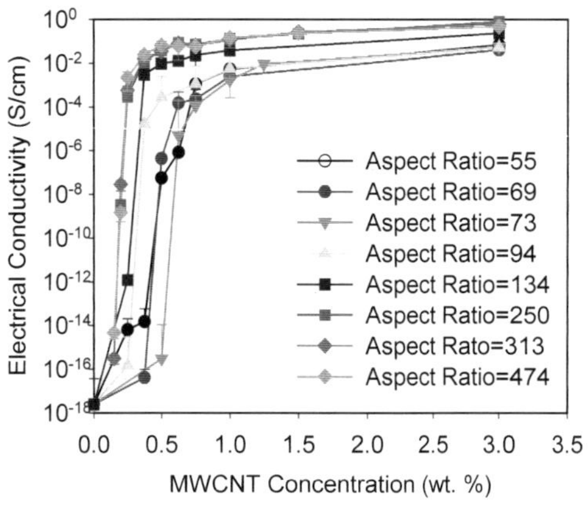

Figure 1: Graph showing bulk electrical conductivity of nanotube samples as a function of nanotube concentration. The initial CNT aspect ratios are indicated in the legend.

The percolation threshold was determined numerically by fitting the experimental data to Eqn. (2).

$$\delta(p) = B(p - p_c)^t \quad (2)$$

where δ(p) is the conductivity, B is a constant, p_c is the electrical percolation threshold, and t is the critical exponent.

The best-fit percolation thresholds from Equation 2 for the composites are shown in Table 1. Clearly Equation 2 overestimates the percolation threshold for the high aspect ratio samples; the number density of points in the percolation region is not large enough to make this fit more accurate and frankly precisely controlling the nanotube content to accurately reach the necessary number density is difficult because of the small volumes in the microcompounder. Figure 2 shows the data from Table 1, plotted against the theoretical percolation threshold calculated using Equation 1.

Table 1
Percolation thresholds and storage modulus of composites

Average aspect ratio	Percolation concentration	Storage modulus at 200°C for composites with 3 wt.% MWCNTs (MPa)
55	0.50 wt. %	0.52
69	0.49 wt. %	0.22
73	0.61 wt. %	0.35
94	0.33 wt. %	0.56
134	0.28 wt. %	1.01
250	0.23 wt. %	1.45
313	0.23 wt. %	1.22
474	0.23 wt. %	1.50

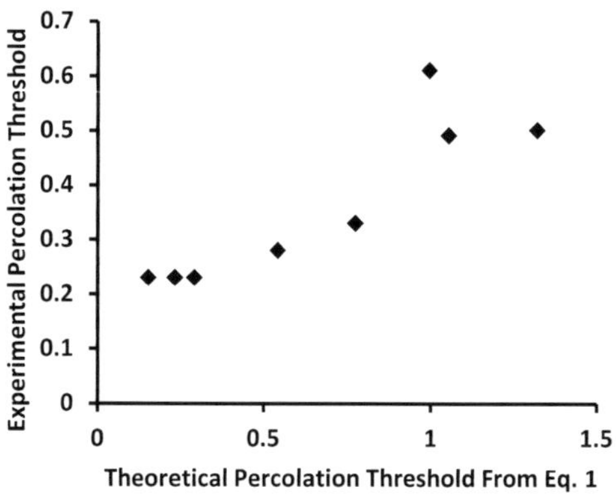

Figure 2: Experimental vs. theoretical percolation thresholds.

As can be seen in Table 1 and Figure 2, for the samples having aspect ratio lower than 250, the percolation threshold concentration increases with decreasing the aspect ratio. This behavior can be explained by the formation of a conductive network theory [1]: the higher the aspect ratio, the easier to form the network, leading to reducing the nanotube amount needed in the system to form such a network. On the other hand, for the samples having aspect ratio equal to or higher than 250, the percolation threshold seems to be independent of the aspect ratio, which can also be observed in Figure 1 for the experimental data. This behavior indicates that there could be a critical value above which the electrical percolation threshold concentration is independent of the aspect ratio, and such a critical aspect ratio for the present PC/MWCNT system is between 134-250 based on the experimental results. Our hypothesis is that processing in the conical twin-screw under the conditions used in this experiment fractionally reduces the CNT length of the material with the highest aspect ratio more than that of the other materials so that the actual aspect ratio in the composites are approximately the same for all of these high aspect ratio MWCNTs.

The same type of dependence on aspect ratio is found in the rubbery storage modulus, i.e. the storage modulus at temperature above the glass transition temperature (see Table 1). A sample graph showing some dynamic mechanical data for composites with aspect ratio [313] is presented in Figure 3 and clearly shows the increase in storage and loss moduli above the glass transition temperature.

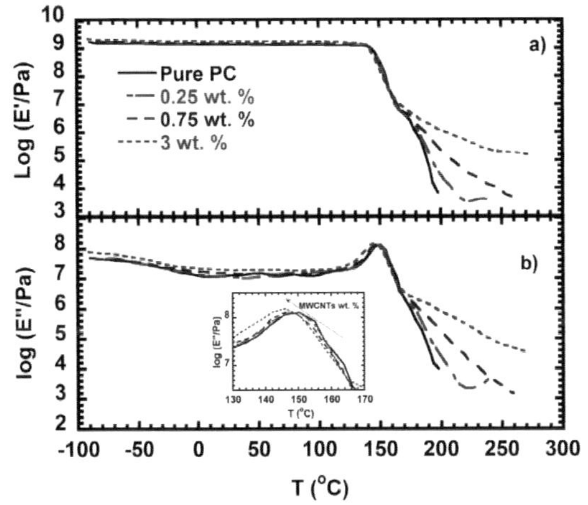

Figure 3: Storage a) and loss b) moduli measured as a function of temperature for the neat PC and its composites with MWCNTs having aspect ratio=313. The inset in (b) shows the shift in Tg as evidenced by the shift in the maximum of the loss modulus.

Data is given in Reference [2] that shows that the Tg decreased only very slightly for samples having aspect ratio up to and including 134. At higher aspect ratios, there is a clear drop in Tg with aspect ratio (rates of -0.83, -2.11, and -2.55 K/wt. % from DSC for the nanotubes having aspect ratio 250, 313, and 474, respectively) suggesting that the shear-induced degradation is higher for samples having higher aspect ratios, which is predictable given that the composites containing higher aspect ratio MWCNTs should have higher melt viscosities.

Changes in the heat capacity at the glass transition were small or negative; the latter indicating that a portion of the material is immobilized at the nanotube surface. There was no discernible pattern to the dependence of the change in heat capacity suggesting that the factors that influence this parameter (e.g. surface chemistry or perhaps degradation) were critical in determining this parameter.

Conclusions

Measurements expected to be sensitive to the aspect ratio of nanotubes, e.g. electrical percolation concentration and storage modulus above the glass transition, show that increasing aspect ratio effects those properties in the expected way, until a critical aspect ratio is reached above which the properties become insensitive to aspect ratio. This result suggests that nanotubes with larger aspect ratios are relatively more reduced in length than those with shorter aspect ratios. Such a result is not surprising giving the expected effect of MWCNT aspect ratio on the melt viscosity of the composite which in turn influences the breaking processes due to enhances shear stresses. Further, this result was shown previously in PC composites when comparing the shortening degree induced by melt mixing of different MWCNT materials in which different starting lengths were produced by a pretreatment using ball milling [5]. Another important result is that the reduction of Tg, believed to be due to degradation of the polycarbonate, scales with aspect ratio at higher aspect ratios, which is also likely due to the higher viscosity during processing.

Acknowledgements

We are grateful to the support given by the Department of Energy (Grant ER64239 0012293).

References

1. I. Balberg, *Philosophical Magazine B-Physics of Condensed Matter Statistical Mechanics Electronic Optical and Magnetic Properties*, 1987 *56*, 991.
2. F. Y. Castillo, R. Socher, B. Krause, R. Headrick, B. P. Grady, R. Prada-Silvy and P. Pötschke, *Polymer*, 2011 *52*, 3835.
3. G. Kasaliwal, A. Göldel and P. Pötschke, *J. Appl. Polym. Sci.*, 2009 *112*, 3494.
4. B. Krause, M. Ritschel, C. Täschner, S. Oswald, W. Gruner, A. Leonhardt and P. Pötschke, *Compos. Sci. Technol.*, 2010 *70*, 151.
5. B. Krause, T. Villmow, R. Boldt, M. Mende, G. Petzold, P. Pötschke, *Compos. Sci. Technol.*, 2011 71, 1145.

INFLUENCE OF SHORT GLASS FIBER ADDITION ON THE MORPHOLOGY AND PROPERTIES OF PC-MWCNT COMPOSITES

S. Sathyanarayana[1*,2], M. Wegrzyn[3], P. Weiss[1], E. Giménez[4], C. Hübner[1] and F. Henning[1,2]

[1]*Polymer Engineering Department, Fraunhofer Institute for Chemical Technology ICT, Germany – shyam.sathyanarayana; patrick.weiss; christof.huebner; frank.henning @ict.fraunhofer.de*
[2] *Institute for Vehicle System Technology (FAST), Karlsruhe Institute of Technology, Germany*
[3] *Instituto Tecnológico del Plástico (AIMPLAS), Spain – marcin@aimplas.es*
[4] *Instituto de Tecnología de Materiales. Universidad Politécnica de Valencia, Spain - enrique.gimenez@mcm.upv.es*

Abstract - Polycarbonate (PC) was compounded with 2 wt.% multiwalled carbon nanotubes (MWCNT) and reinforced with short glass fibers (10, 20 and 30 wt.%) in a twin-screw extruder and injection molded thereafter for testing and characterization. Two types of glass fibers were incorporated: "GA" and "GB" with a compatible and incompatible sizing for PC, respectively. SEM analysis showed very good dispersion of MWCNT in PC and a good interfacial adhesion between GA and PC. Incorporation of increasing content of glass fibers to PC-2 wt % MWCNT composite (PC-2C) significantly enhances the elastic modulus. Higher tensile strength values were observed in PC-2C reinforced with GA fibers due to the formation of a stronger interface than with GB fibers. However, the notched impact strength of PC decreased when reinforced with glass fibers, but 30 wt.% of GA's addition on PC-2C gave impact strength close to that of pure PC. The electrical volume resistivity of PC was reduced by 11 orders of magnitude on 2 wt.% MWCNT addition. With increasing glass fiber contents, the volume resistivity of PC-2C composite was not altered significantly. Thermal stability of PC-2C decreased with increasing glass content, but was better than pure PC's. The combined influence of short glass fibers and MWCNTs in PC presents a significant case for the development of multi-fucntional composite materials.

Keywords: polycarbonate, multiwalled carbon nanotubes, short glass fibers, composites, properties

Introduction

Commercially available carbon nanotubes (CNT) are highly entangled and hence their excellent dispersion in a polymer matrix is one of the principal objectives to translate their excellent electrical and mechanical characteristics to the composite. High strength of the CNT agglomerates and limitations to impart high shear with conventional twin-screw extruders and melt mixers to destruct these agglomerates due to a potential trade-off of CNT aspect ratio reduction [1], inert surface of the CNTs, and considerably high viscosity of the thermoplastic materials preventing a facile penetration into the agglomerates present a significant challenge in achieving a good multiwalled carbon nanotubes (MWCNT) in the host matrix.

Thermoplastics reinforced with CNTs have been widely investigated because of their potential applications in many fields like electronics, automobiles, aeronautics etc. The ability of the CNTs to form a conductive pathway on nanoscale re-agglomeration of initially dispersed fillers during compression molding results in excellent electrical properties at an extremely low filler concentration in a thermoplast [2]. Polycarbonate is among the widely investigated thermoplastic matrix owing to their lower interfacial energy difference with CNTs resulting in excellent filler dispersion. But, inspite of achieving an excellent primary dispersion, mechanical property reinforcements of PC-MWCNT composites are seldom reported and if reported the enhancements are not significant [3,4]. Injection molding is a widely employed processing and finishing step for compounded thermoplastic composite granulates. The shear associated with the process causing a potential reduction in the aspect ratio in addition to altering the primary dispersion quality and the formation of a skin-core structure [5] reduce the potential of CNTs to cause excellent mechanical and electrical property enhancements. Ci & Bai [6] report that it is very difficult to enhance the mechanical properties when CNTs are reinforced into an already stiff epoxy, this could also apply to PC.

The inabilities to achieve exceptional mechanical property reinforcements with MWCNTs in polymer matrices direct research attention towards incorporating a secondary filler carbon fibers or glass fibers in addition to MWCNTs. Carbon nanostructures grown insitu on the surface of glass and carbon fibers have been reported respectively in [7,8]. However, this approach is expensive. Glass fibers are economical and are known to create substantial mechanical reinforcements in polymers and hence, two different short glass fibers with compatible and incompatible sizings with PC are used in this work as additional fillers to MWCNTs for investigation on the morphology, mechanical, electrical and thermal properties of the composites.

Experimental

Materials

PC (Wonderlite PC-122) from CHIMEI-ASAHI Corporation, Taiwan was used as the matrix. Nanocyl NC7000 from Nanocyl S.A., Belgium was used as MWCNT. Chopped glass fiber OCV 995 (compatible sizing for PC) and OCV 968/968A (incompatible sizing for PC) from OCV reinforcements were used and named as GA and GB respectively. PC was dried for 4 hours at 120 °C before processing while the glass fibers and MWCNTs were used as-received.

Processing

Table 1 shows the nomenclature and the process conditions for pure PC and its composites processed on a Leistritz ZSE 27HP-52D twin-screw extruder with an L/D ratio of 52. The CNTs were feed at L/D = 16 and the glass fibers at L/D = 40 positions of the extruder barrel. The extruder barrels were set at 260 °C. PC and PC-2C were processed with a screw speed 500 rpm and throughput of 12.5 kg/h while the other composites were processed at 360 rpm and 20 kg/h. Extruded granules were injection molded with a mold and melt temperature of 70 °C and 265 °C respectively, injection speed in 10 steps of 3x25,2x20,5x15 cm^3/s, holding pressure of 70 bar and a holding time of 5 sec into specimens for tensile and impact testing on an Engel ES 200/60 HL ST injection molding machine.

Characterization

Field Emission Scanning Electron Microscopy (FESEM) was performed on broken charpy specimens (sputter coated with Gold for 120 s at 0.05 mbar and 40 mA) using Zeiss SUPRA 55 VP microscope. The notched impact strength (on atleast 10 samples) was evaluated using ISO 179-1/1eA norm on a CEAST Resil Impactor while tensile properties (on atleast 5 samples) were evaluated with ISO 527-1 standard on a Universal testing machine Inspekt Table 50KN from Hegewald & Peschke after conditioning the samples for 24 hours at 23 °C and 50 % RH. Electrical volume conductivity of the composites was measured according to the Van der Pauw's method with a FLUKE 8846A Digital Multimeter on atleast five injection molded bars (50 x 10 x 4 mm^3). Volume conductivity measurements ($<10^{-9}$ S/cm) were measured using Agilent Technologies' 4339B High Resistance Meter as 2-point measurements. Contact points were created with silver paint and samples were dried for 24 hours at 23 °C and 50% RH before measurements. Similar measurements were performed on extruded strands with elliptical geometry. Thermo Gravimetric Analysis (TGA) was performed in nitrogen on TGA Q5000. 10-12 mg samples were heated from 20 °C to 120 °C at 20 °C/min, held isothemally for 10 minutes and thereafter heated at 10 °C/min to 700 °C.

Results and Discussion

Morphology of the composites

Fig. 1 presents the SEM morphology of the composites of PC. Excellent dispersion of 2 wt.% MWCNT in PC due to the high polarity of PC and low surface energy difference between PC and MWCNTs is observed. The fracture morphology of 20 wt.% GA on PC shows excellent interfacial adhesion between PC and GA which is attributive to the PC compatible sizing on GA. On the other hand, the pull-out damage was dominant on the fractural surfaces of composites with GB fibers and produced holes or grooves with smooth walls, which indicates weak interfacial adhesion between the PC and the GB fibers. The addition of glass fibers to PC-2C does not affect the dispersion of MWCNTs in PC and the individual dispersive morphology of both MWCNTs and glass fibers in PC is maintained in the hybrid composite.

Mechanical properties of the composite

The elastic modulus (E) of PC (2.34 GPa) was enhanced by 10% when 2 wt.% MWCNT was incorporated. The addition of 20 wt.% short glass fiber enhanced the modulus of PC by 167% irrespective of its type. The strong interfacial adhesion of GA to PC as observed from the SEM images (Fig.1) has resulted in an 86% improvement in the tensile strength (σ_{max}) of PC compared to 33% enhancement when reinforced with GB. This is significantly higher than the 5% improvement in σ_{max} on 2 wt.% MWCNT additions. The E and σ_{max} of PC-2C increase significantly with increasing content of glass fiber incorporation, with the latter critically dependant on the type of sizing. The inability of 2 wt.% MWCNT to cause a significant reinforcing effect inspite of its excellent dispersion owing to reasons explained earlier is substantially compensated by the addition of low cost short glass fibers. By incorporating long glass fibers, further enhancement in the mechanical strength can be envisioned, however the complexities associated with processing long glass fibers needs to be considered.

The notched impact strength (I_{notch}) of PC was not affected with the addition of 2 wt.% MWCNT, while the addition of 20 wt.% GA lowered the crack resistive behavior of PC by 23 %. Similarly the I_{notch} of PC-2C decreased by 27% on incorporation of 20 wt.% GA, but I_{notch} increased with increasing GA additions. SEM micrographs and observations of σ_{max} indicate a comparatively stronger interface between PC and GA than PC and GB. In micro-composites, the Young's modulus and yield stress generally tend to increase with the filler content; toughness shows an opposite trend and usually decreases or remains at low levels. However, in the case of PC-2C reinforced with GA type fibers, we observe an increase in the toughness of the composite with increasing glass fiber loadings. The increase in the interfacial area with the increase in the glass fiber content could have resulted in increased energy absorption during the deflection of the cracks into the matrix resulting in higher composite toughness when PC-2C was reinforced with GA. On the other hand, the addition of 20 wt.% GB to PC has resulted in a 62% reduction in the crack resistive behavior of PC. The lack of a compatible interface between PC and GB has resulted in fiber pullout (Fig 1); the energy absorbed during this mechanism could have been lower than energy absorbed due to crack deflection and interfacial debonding when compared to PC reinforcements with GA. Similar effects could be attributed to lower toughness of PC-2C composites with GB reinforcements.

Electrical conductivity of the composites
The volume conductivity of injection molded PC-2C (9.52×10^{-3} S/m) was close to 11 orders of magnitude higher than that of pure PC. The electrical percolation threshold of these composites has already been achieved at 2 wt.% MWCNT loadings. The addition of short glass fibers to PC-2C slightly lowers the conductivity of the composites, but the conductivity increased with increasing content of glass fiber. It must be remembered that the addition of 30 wt.% of glass fibers to PC-2C had lowered the MWCNT concentration to 1.4 wt.% in the composite. The conductivity of the composite at the same glass fiber loading, irrespective of the type of fiber, indicates that the addition of glass fibers did not influence the electrical property created by the addition of MWCNTs. The extruded strands showed close to 3 orders of magnitude higher conductivity compared to the injection molded composites, but the trends observed in the composite conductivities remain the same. The orientation of the MWCNTs, the reduction in the aspect ratio of the MWCNTs due to injection molding shear and an outer isolative polymer skin [5], resulting from high temperature differences between the mold wall and the melt together, have an influence on the electrical conductivity of injection molded parts. The orientation morphology of the fillers in the extruded strands might vary depending on the filler concentrations, but this measurement is used as a reference to explain poor conductivities in an injection molded composite. Similar observations have also been reported earlier [4].

Thermal stability of the composites
The onset of decomposition (T_o) (temperature at 5 % weight loss) of pure PC increased from 465 °C to 484 °C on the addition of 2 wt.% MWCNT. Pötschke et al. had earlier reported reduction in the thermal stability of PC-MWCNT composites due to PC degradation caused by MWCNT addition. Assuming there is this possibility, the enhanced dispersion quality of MWCNTs which are known to be good free radical acceptors [9], presents enhanced surface area for accepting the free radicals resulting from the pyrolysis of PC resulting in enhanced thermal stability of the composite. The temperature of maximum weight loss (T_{max}) (temperature peak from dw/dT) of pure PC at 534 °C was not altered on MWCNT addition. PC-20GA and PC-20GB showed lower T_o at 442 °C and 456 °C respectively compared to pure PC. The addition of GA and GB to PC-2C results in thermal stabilities closer to those observed with PC-2C, but increasing glass fiber content (and consequently lower MWCNT wt.%) in the composite lowers the thermal stability of the composites. Hence, the thermal stability of the composite is dictated by the presence and the content of MWCNTs.

Conclusions
A bi-filler system of short glass fiber and MWCNT were successfully incorporated into PC to form a hybrid composite. Although CNT addition to PC does not significantly contribute to improvement in mechanical behavior, incorporating a glass fiber with a sizing compatible with that of the PC gives excellent mechanical properties to the composite. The electrical conductivity of PC on 2 wt.% MWCNT addition was enhanced by 11 orders of magnitude. Increasing content of either of the glass fiber types did not affect the conductive pathways created by MWCNT addition. The addition of MWCNT compensates for the loss in thermal stability of PC due to glass fiber addition. There is neither a strong synergy nor competing effects between glass and CNT in affecting the macroscopic properties, but the enhancement in electrical properties caused by CNT addition along with the improvement in mechanical properties due to glass fiber addition augurs well for a successful development of a multi-functional composite.

Acknowledgements
The research leading to these results has received funding from the European Community's Seventh Framework Programme (FP7-PEOPLE-ITN-2008) under grant agreement number 238363. The authors would like to thank OCV reinforcements for the supply of their glass fibers.

References
1. S. Sathyanarayana; G. Olowojoba; P. Weiss; B.Caglar; B. Pataki; I. Mikonsaari; C. Hübner; F. Henning *Macromol. Mat. Eng.* 2012, DOI = 10.1002/mame.201200018
2. G. Kasaliwal; A. Göldel, P. Pötschke *J Appl. Polym. Sci.* 2009, *112*, 3494.
3. F. Y. Castillo; R. Socher; B. Krause; R. Headrick; B. P. Grady; R. Prada-Silvy; P. Pötschke *Polymer* 2011, *52*, 3835.
4. C. Mack; S. Sathyanarayana; P. Weiss; I. Mikonsaari; C. Hübner; F. Henning; P. Elsner *IOP Conf. Ser.: Mater. Sci. Eng.* 2012, *40*, 012020.
5. T. Villmow; S. Pegel; P. Pötschke; U. Wagenknecht *Comp. Sci. Tech.* 2008, *68*, 777.
6. L. Ci; J. Bai *Comp. Sci. Tech.* 2006, *66*, 599.
7. E. J. Garcia; B. L. Wardle; A. J. Hart; N. Yamamoto *Comp. Sci. Tech.* 2008, 68, 2034.
8. M. Delmas; M. Pinault; S. Patel; D. Porterat; C. Reynaud; M Mayne-L'Hermite *Nanotechnology* 2012, *23*, 105604.
9. P. J. Krusic; E. Wassermann; P. N. Keizer; J. R. Morton; K. F. Preston *Science* 1991, *254*, 1183.

Table 1. Processed samples and their nomenclature

PC (wt.%)	MWCNT (wt.%)	PC-2C (wt.%)	Glass type and wt.%	Nomenclature
100				PC (Ref)
98	2			PC-2C
80			20GA	PC-20GA
80			20GB	PC-20GB
			10GA	PC-2C-10GA
			20GA	PC-2C-20GA
			30GA	PC-2C-30GA
			10GB	PC-2C-10GB
			20GB	PC-2C-20GB
			30GB	PC-2C-30GB

Table 2. Electrical conductivity of PC and its composites

	Volume Conductivity (Extruded strands) S/m	Volume Conductivity (Injected bars) S/m
PC (Ref)	E-14	E-14
PC-2C	2.48E+01	9.52E-03
PC-2C-10GA	7.58E+00	2.36E-04
PC-2C-20GA	7.81E+00	6.99E-04
PC-2C-30GA	1.11E+01	8.47E-04
PC-2C-10GB	8.62E+00	3.11E-04
PC-2C-20GB	8.85E+00	6.02E-04
PC-2C-30GB	8.13E+00	7.04E-04

Table 3. Thermal stability of PC and its composites

	PC (Ref)	PC-2C	PC-20GA	PC-20GB	PC-2C-10GA	PC-2C-20GA	PC-2C-30GA	PC-2C-10GB	PC-2C-20GB	PC-2C-30GB
T_o (°C)	465	484	442	456	484	479	471	483	479	476
T_{max} (°C)	534	533	508	506	525	522	512	512	518	515

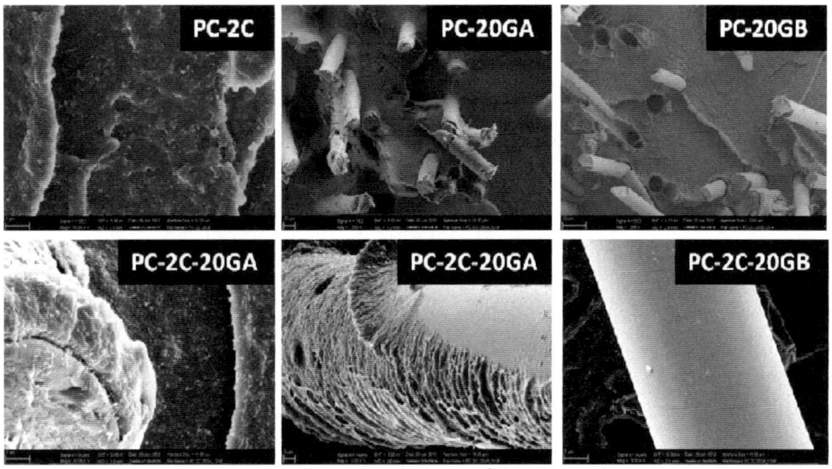

Figure 1. SEM morphology of the composites

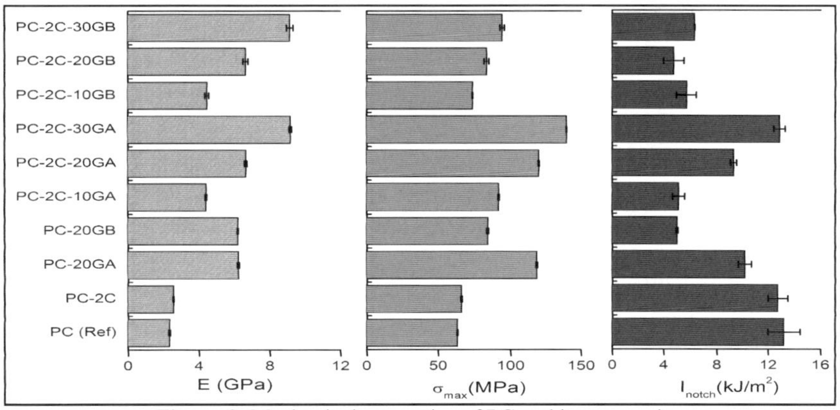

Figure 2. Mechanical properties of PC and its composites

O-07-351

MORPHOLOGY, ELECTRICAL AND THERMAL PROPERTIES OF POLYAMIDE6, EXPANDED GRAPHITE AND MULTIWALLED CARBON NANOTUBES COMPOSITES

M. S. Sreekanth, Arup R. Bhattacharyya* and Ajay S. Panwar

Department of Metallurgical Engineering & Materials Science, Indian Institute of Technology Bombay, Mumbai, India-400076, sreekanthms@iitb.ac.in; arupranjan@iitb.ac.in; panwar@iitb.ac.in

Abstract - Expanded graphite (EG) and multiwall carbon nanotubes (MWNTs) based hybrid nanocomposites were prepared using polyamide6 (PA6) matrix via melt-mixing strategy. EG+modified MWNTs based hybrid composites were prepared wherein EG:MWNTs ratio (wt/wt) was varied in 1:0.5, 1:1 and 1:4. MWNTs were modified with a novel modifier: Li-salt of 6-aminohexanoic acid (Li-AHA) in the ratio of 1:1 (wt/wt). EG+mMWNTs hybrid composites were characterized by transmission electron microscopy (TEM), scanning electron microscopy (SEM) and Raman spectroscopic analysis. The morphological studies of EG+mMWNTs composites showed the existence of MWNTs intercalated in the galleries of EG. The degree of exfoliation of EG was higher at higher MWNTs content as observed from the selected area electron diffraction (SAED) analysis. EG+mMWNTs based PA6 composites were characterized through AC electrical conductivity measurements to investigate the 3D 'network-like' structure of the hybrid filler. State of dispersion of the hybrid filler was assessed through SEM analysis. The nanocomposites exhibited the electrical percolation threshold between 3-4 wt% of EG+mMWNTs (1:4, wt/wt) due to the formation of interconnecting network of the hybrid filler in the PA6 matrix. A significant improvement in glass transition temperature of PA6 phase was recorded (up to 10 °C) with increased EG+mMWNTs concentration. The %crystallinity of PA6 phase of the composites increased (32% upto 40%) with increase in EG+mMWNTs concentration. A significant increase in the crystallization temperature of PA6 phase in the composite was noticed due to heterogeneous nucleating action of hybrid filler. Morphology and electrical conductivity relationship studies in these hybrid composites were addressed in this study.

Keywords: Polyamide6, Expanded graphite, Multiwall carbon nanotube, Nanocomposites

Introduction

The addition of carbon based nanofiller in to polymer matrix creates a novel class of material exhibiting superior mechanical, thermal and electrical properties at very low filler concentration. The outstanding mechanical, electrical and thermal properties associated with carbon nanotubes (CNTs) attracted them towards its use in polymer nanocomposites for various applications [1,2]. Expanded graphite (EG) exhibits excellent thermal and electrical properties. With appropriate surface treatment, dispersion of EG in a polymer matrix results in composites with excellent mechanical and electrical properties and high thermal conductivity [3]. EG is much easier to produce as compared to CNTs and has much lower prices.

The concept of hybrid filler has been introduced to make use of excellent properties of individual fillers. CNTs have superior mechanical properties, but must be dispersed uniformly and also form a 3D network to achieve electrical percolation. However, graphene or even graphite nanoplatelets have high surface area and remarkably high electron mobility at room temperature but poses a problem due to restacking. The hybrid composites showed synergistic properties than individual fillers. The hybrid composites prepared with graphene and CNTs showed excellent mechanical, electrical and thermal properties for various applications [4-6], such as sensors, solar cells, super capacitors, FET etc.

The objective of this work is to obtain electrically conducting high strength structural material by intercalating modified multiwall carbon nanotubes (MWNTs) in to EG gallery to form the 3D 'network-like' structure involving the hybrid

Experimental

Polyamide 6 (PA6) with a zero shear viscosity of 180 Pa s at 260 °C was obtained from GSFC, Gujarat, India (Gujlon M28RC, relative viscosity 2.8, M_v is 38642 in 85% formic acid). Multiwall carbon nanotubes (MWNTs) synthesized by CCVD were obtained from Nanocyl CA Belgium (NC-3100, average length: 1.5 μm, average diameter: 9.5 nm, purity: 95%). Expanded Graphite (EG) was supplied by Asbury carbon (3806, with carbon content 99.43%, average particle size <50 μm, surface area 23.39m^2/g). 6-aminohexanoic acid (AHA) (Merck, Germany, M_w = 131.17; purity: 99%) was neutralized using Lithium hydroxide (S D Fine Chem limited, India, purity: 99%) to obtain Lithium salt of 6-amino hexanoic acid (Li-AHA).

MWNTs were modified with Li-AHA in aqueous media in the ratio of 1:1 (wt/wt) (MWNTs:Li-AHA). EG and modified MWNTs hybrids were prepared in aqueous media in three compositions viz., 1:0.5, 1:1 and 1:4 (wt/wt) (EG:mMWNT). PA6 and EG+MWNT hybrid composites were melt-mixed in a conical twin-screw microcompounder (Micro 5, DSM Research, Netherlands) at 260 °C with a rotational speed of 150 rpm for 10 min. Table 1 shows the list of composition of hybrid filler and polymer composite.

Table 1: Composition of polymer hybrid composites

Sl. no	Sample Code	EG:MWNTs (wt/wt)	MWNTs: LiAHA (wt/wt)
1	EGmT(x:y)(a:b) Hybrid filler	x:y (1:0.5), (1:1) (1:4)	a:b (1:1)
2	P nEGmT(x:y)(a:b) Polymer composite*	x:y	a:b

*Polyamide6 with n wt% of hybrid filler (n = 1 to 7). P-PA6, EG-expanded graphite, mT-modified MWNTs

Characterization Techniques

The AC electrical conductivity measurements were performed on the injection molded samples (across the thickness) in the frequency range between 10^{-1} and 10^{6} Hz using Alpha high resolution analyzer coupled to a Novocontrol interface. The cryofractured surface was gold sputtered to avoid charging of the sample. These samples were then observed under scanning electron microscopy (SEM) using Hitachi S3400N. Transmission electron scanning (TEM) is used to study the morphology of filler at higher magnification. TEM was performed with JEOL JEM-2100 F at 200KV. The selected area electron diffraction (SAED) analysis was used to determine the crystalline morphology of graphitic structures. Differential scanning calorimetry (DSC) analysis was carried out using a DSC Q200 from (TA Instruments) under nitrogen atmosphere. The samples were heated till 260 °C and cooled to -40°C at a rate of 10 °C/min to record crystallization exotherm and again heated to 260 °C at a heating rate of 20 °C/min in order to calculate the glass transition temperature (T_g). The degree of crystallinity (X_c) of PA6 in multicomponent system was determined from the ratio of normalized heat of fusion ($\Delta H_{m,\ norm}$) to heat of fusion of 100% crystalline PA6, $(\Delta H_m)_0$, which was taken as 204.8 J/g. Raman spectroscopy was performed using Jobin Yovon (HR 800 micro Raman) on powder samples over a scanning range of 1000 to 3000 cm^{-1}. The laser light of wavelength 514nm was used with 10mW power. Zeta potential measurements were done on aqueous solution of hybrid filler using Beckman Coulter Delsa NanoC.

Results and Discussion

Characterization of Hybrid Filler

The EG+mMWNTs hybrid filler were characterized by Raman spectroscopy, TEM and SEM analysis. The incorporation of modified MWNTs to EG leads to the intercalation of MWNTs to the galleries of EG as shown in Figure 1. The degree of exfoliation of EG was higher at higher MWNTs concentration showed by SAED. The 1:4 (EG+mMWNTs) composition showed better exfoliation of EG sheets exhibiting the hexagonal structure in SAED similar to few layer graphene, whereas the compositions with less MWNTs showing multiple dots in SAED, indicating polycrystalline (stacked) nature of EG. SEM micrograph showed the existence of MWNTs in the galleries of EG. In all the compositions EG and MWNTs are observed to coexist.

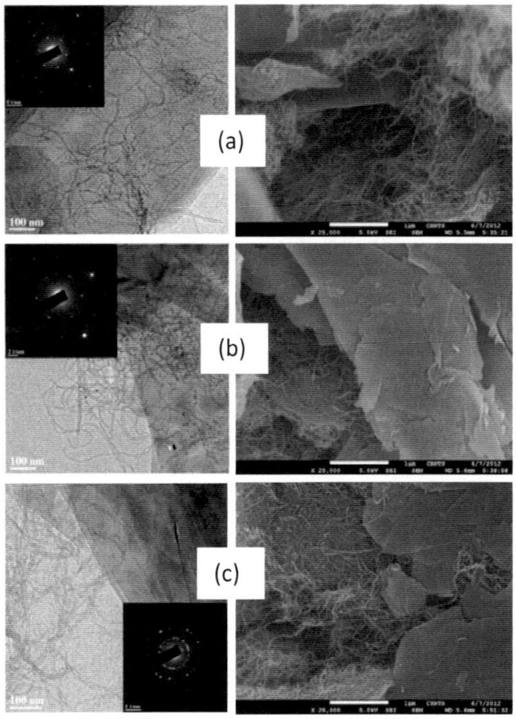

Figure 1: TEM and SEM of EG+mMWNTs hybrid filler with weight ratio (a) 1:4, (b) 1:1 and (c) 1:0.5

Raman spectroscopy is a powerful nondestructive technique to distinguish ordered and disordered crystal structures of carbon. The modification of MWNTs showed a shift in G band at higher wave number due to debundling of MWNTs in the presence of organic modifier [7]. The hybrid filler showed a shift in G band and 2D band in all the composition of EG and modified MWNTs as shown in Figure 2. The 2D band of hybrid filler shift to higher wave number with increasing EG concentration due to the presence of more ordered structure of EG. The I_D/I_G ratio is a measure of amount of defects present in the structure. EG with least defect showed a lesser I_D/I_G ratio of around 0.2, whereas MWNTs showed I_D/I_G value of around 1. The variation of I_D/I_G ratio in various compositions of hybrid filler is shown in Table 2. This observation is indicating the existence of MWNTs in the EG gallery as observed in SEM and TEM analysis. Aqueous stability of EG+mMWNTs hybrid filler was analyzed via zeta potential measurements. The EG+mMWNTs hybrids are found to be stable as shown in Table 3. Composition corresponding to 1:1 (EG+mMWNTs) is found to the most stable than other compositions.

Table 2: Intensity ratio of EG+mMWNTs hybrid filler

EG:MWNT	0:1	1:4	1:1	1:0.5	1:0
Intensity ratio (I_D/I_G)	0.90	0.76	0.67	0.59	0.19

Table 3: Zeta potential of EG+mMWNTs hybrid filler

EG:MWNT	1:4	1:1	1:0.5
Zeta potential (mV)	-44.3	-48.8	-42.9

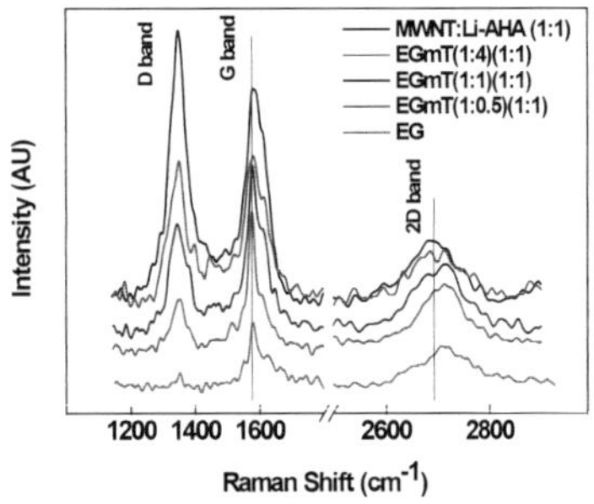

Figure 2: Raman spectra of EG+mMWNTs hybrid filler

Characterization of Polymer Hybrid Filler Composite

The state of dispersion of filler in polymer can be assessed through electrical conductivity measurements. The bulk electrical conductivity of pure PA6 increases with increase in frequency, as expected for an insulating material, with a value about 2E-13 Scm^{-1} at 0.01 Hz. The electrical conductivity of polymer composites arises due to the formation of 3D interconnecting network formation of filler in polymer matrix. The hybrid polymer composite of PA6 with 1:4 (wt/wt) EG+mMWNTs composition showed electrical percolation threshold of 3-4 wt% as shown in Figure 3, whereas composite with 1:1 (wt/wt) EG+mMWNTs showed the electrical percolation at around 5wt%. The composites of PA6 with 1:0.5 (wt/wt) EG+mMWNTs showed a percolation threshold around 6-7 wt% due to less amount of MWNTs. Lower percolation threshold in composites with higher MWNTs content is due to the presence of high aspect ratio (L/D) MWNTs, which helps to form a 3D 'network-like' structure at low MWNTs concentration.

The dispersion of hybrid filler in polymer matrix were studied through morphological investigation via scanning electron microscopy (SEM). The hybrid fillers has been uniformly dispersed in the polymer matrix as shown in Figure 4. The concentrations of MWNTs was more for PA6 with 1:4 (wt/wt) EG+mMWNTs and the presence of MWNTs was clearly visible as white dots. The concentration of EG was higher for PA6 with 1:0.5 (wt/wt) EG+mMWNTs samples as shown in Figure 4(c) and was clearly visible as plate like or flake structure. The EG+mMWNTs hybrid filler is showing a very good dispersion in PA6 matrix.

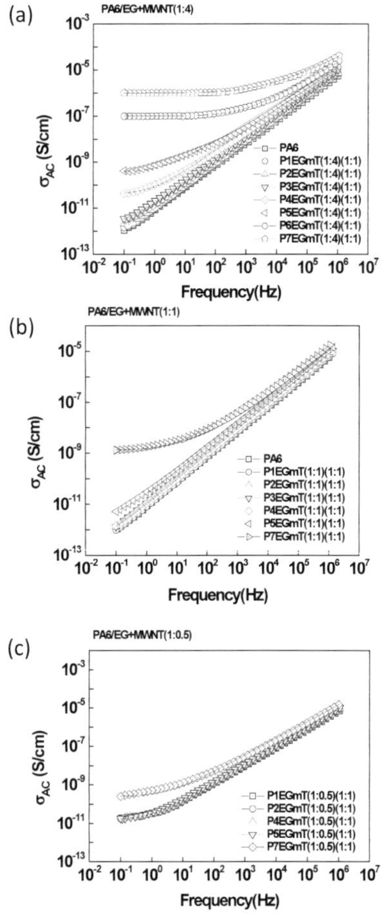

Figure 3: AC electrical conductivity Vs frequency of PA6 with increasing concentration of EG+mMWNTs hybrid filler (a) 1:4, (b) 1:1 & (c) 1:0.5

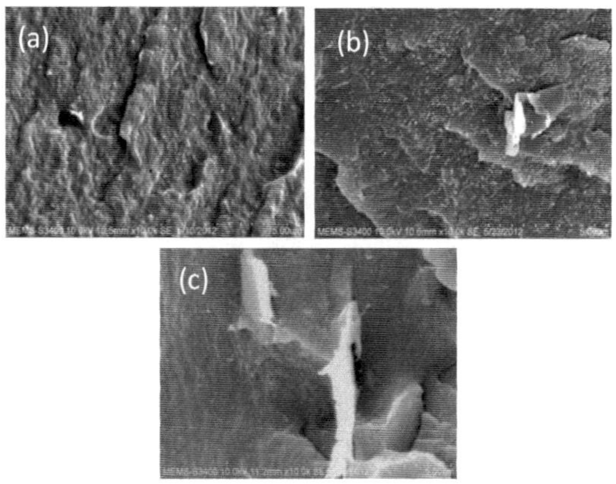

Figure 4: SEM of PA6 with 5 wt% EG+mMWNTs hybrid filler (a) 1:4, (b) 1:1 & (c) (1:0.5)

Differential scanning calorimetric analysis was carried out to study the nature of interaction and crystallization of PA6 in the presence of EG+mMWNTs hybrid filler. The variation of glass transition temperature (T_g) of composites with increasing concentration of EG+mMWNTs hybrid filler is shown in Figure 5(a). There was a significant improvement in T_g (up to 10 °C) with increasing EG+mMWNTs (1:4) (wt/wt) concentration due to improved interaction of hybrid filler with PA6 matrix. There is a significant improvement in crystallization temperature (T_c) of PA6 phase with filler addition due to heterogeneous nucleating action of hybrid filler in PA6 matrix as shown in Figure 5(b). The addition of MWNTs of hybrid filler leads to a second crystallization peak at 204 °C, which gets intensified and shifted to 207 °C with increasing hybrid filler concentration. The second crystallization peak was due to trans-lamellar crystallization of PA6 on MWNT surface [8]. There is no significant change in melting temperature (T_m) with increase in hybrid filler concentration. The degree of crystallinity of PA6 increased from 0.32 upto 0.40 with increase in the filler addition. The hybrid filler with all other compositions showed almost a similar trend in T_g and T_c with increasing hybrid filler concentration.

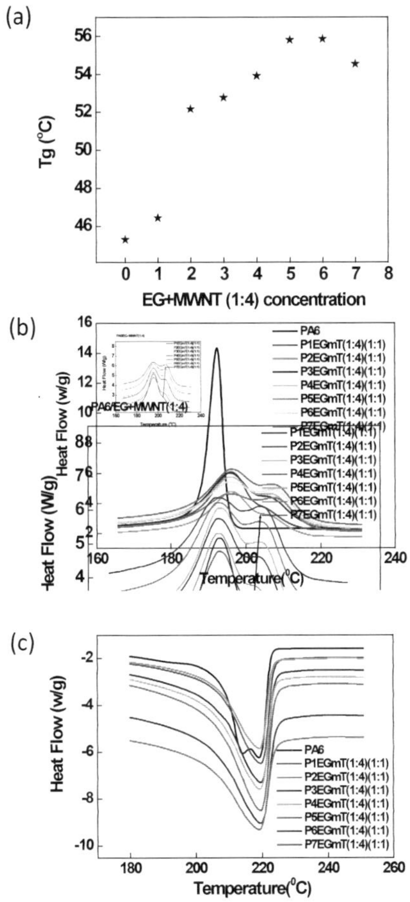

Figure 5: DSC heating and cooling scan (a) Variation of T_g with increasing hybrid filler concentration, (b) crystalization and (c) melting curves of PA6 with 1:4 (wt/wt) EG+mMWNTs composites.

Conclusions

PA6 with EG+mMWNTs hybrid filler composites were prepared via melt-mixing strategy. The morphological studies of EG+mMWNTs composites showed the existence of MWNTs intercalated in the galleries of EG. The degree of exfoliation of EG was higher at higher MWNTs content. Raman spectra of hybrid filler showed the existence of MWNTs within the EG gallery. The EG+mMWNTs hybrids were found to be stable in aqueous solution through zeta potential measurements The state of dispersion of hybrid filler was investigated through AC electrical conductivity measurements. The nanocomposites exhibited the electrical percolation threshold of around 3-4 wt% of 1:4 (wt/wt) EG+mMWNTs due to the formation of 3D network-like structure of hybrid filler in the PA6 matrix. Composites of hybrid filler with higher MWNTs content showed a better electrical conductivity due to high aspect ratio of MWNTs. State of dispersion of the hybrid filler in PA6 phase was also assessed through SEM which indicated a uniform dispersion of hybrid filler in the PA6 matrix. There was a significant increase in glass transition temperature (T_g) of PA6 (up to 10 °C) with the addition of EG+mMWNTs hybrid filler due to improved interaction of hybrid filler with PA6. There was a significant increase in crystallization temperature of PA6 with hybrid filler due to the heterogeneous nucleating action of hybrid filler. The degree of crystallinity of the composites increased with EG+MWNTs concentration (32% to 40%).

Acknowledgements

The authors would like to acknowledge Microcompounder Central Facility, IIT Bombay and SAIF, IIT Bombay for FEG-TEM, FEG-SEM and Raman spectroscopy.

References

1. R. H. Baughman, A. A. Zakhidov, W.A. de Heer, *Science*, 2002, *297*, 787.
2. S. B. Sinnott, R. Andrews, *Crit. Rev. Sold. Sta. & Mat. Sci.*, 2001, *26*, 145.
3. R. Senguptaa, M. Bhattacharya, S. Bandyopadhyay, A. K. Bhowmick, *Prog. Poly. Sci.*, 2011, *36*, 638.
4. M. K. Shin, B. Lee, S. H. Kim, S. Gambhir, *Nat. comm.*, 2012, *3*, 1.
5. Z. Antar, J.F. Feller, H. Noël, P. Glouannec, K. Elleuch, *Mat. Lett.*, 2012, *67*, 210.
6. S. Wooa, Y. R. Kima, T. D. Chunga, Y. Piaob, *Electrochimica Acta*, 2012, *59*, 509.
7. P. V. Kodgire, A. R. Bhattacharyya, A. Misra, *Chem. Phy. Lett.*, 2006, *432*, 480.
8. A. C. Brosse, S. T. Girault, P. M. Piccione, L. Leibler, *Polymer*, 2008, *49*, 4680.

O-07-356

HYBRID MIXTURES OF MULTIWALLED CARBON NANOTUBES AND CARBON BLACK IN HDPE-BASED CONDUCTIVE POLYMER COMPOSITES FOR INJECTION MOLDING

Isabell Kindgren[1], Martin Strååt[1]*

[1]*Dep. Polymeric materials and textiles, Swerea IVF, Sweden – isabell.kindgren@swerea.se; martin.straat@swerea.se*

Abstract –

Hybrid mixtures of carbon black (CB and carbon nanotubes (CNT) were formed by melt mixing in a twin-screw extruder with a rotational speed of 250 and 400 rpm using 3 material feed rates. The resistivity of compression molded and injection molded samples was compared as well as the notched Charpy impact toughness of the materials. The conclusion was that an increasing feed rate could give a material with lower resistivity but that the high shear forces involved had an influence on the impact toughness.

Keywords: carbon black, carbon nanotubes, high-density polyethylene, conductivity, injection molding

Introduction

Conductive polymer composites can be made with a variety of conductive fillers with carbon black (CB) being the most common and carbon nanotubes (CNT) increasingly gaining the attention from industry. It is well known that these two conductive fillers behave differently during processing where carbon nanotubes tend to align themselves in the flow direction which may cause a major change in resistivity in for example injection molded products [1]. It is also known that fillers will have an effect on impact toughness as well [2]. Hybrid mixtures of carbon black and carbon nanotubes could however offer an interesting combination of impact toughness and conductivity after injection molding.

This study shows the electrical and impact properties of injection molded samples. Hybrid mixtures with carbon black and carbon nanotubes based on HDPE were prepared using a twin-screw extruder. Samples were prepared by injection molding and compression molding and the electrical conductivity of the materials was determined. Charpy impact toughness (notched) was tested on injection molded samples.

Experimental

Materials

A commercial HDPE-based conductive masterbatch (CMB) was also used having density 1040 kg/m3, MFI 6 g/10min (190°C/21.6 kg) and Charpy impact toughness (notched) 40 kJ/m2.

A HDPE was used having density 964 kg/m^3, MFI 8 g/10min (190°C/2.16 kg) and Charpy impact toughness (notched) 12 kJ/m^2 according to the manufacturer.

The CNT used were Nanocyl 7000 from Nanocyl, supplied as a HDPE-masterbatch containing 15 wt% CNT. A reference material was also prepared containing 20wt% Vulcan XC-72 from CABOT inc, supplied as powder.

Compounding

HDPE was dry mixed with PL-CYL15 in a 2:1 ratio and then melt mixed in a twin screw extruder; Coperion ZSK 26K, Germany. The extruder screws had a configuration optimised for dispersion of particles. The temperature was kept at 190°C in all heating zones.

The compound produced was then melt mixed in a 1:1 ratio with the CMB. The result was a material containing 2.5 wt% CNT and 10 wt% CB. In order to check the influence of an extra compounding step, the CMB was also run through the process without adding anything. The resistivity of the materials was measured on compression molded samples, using silver paint to reduce contact resistance.

Table I Conductive materials prepared.

Material	*HDPE (wt%)*	*CB (wt%)*	*CNT (wt%)*	*Extr. speed (rev/min)*	*Feed rate (kg/h)*	*Comp. mold. res. (Ω.cm)*
CNT-1	96.25		3.75	250	8	9
CNT-2	95		5	250	8	5
CBCNT-1	87.5	10	2.5	250	8	11
CBCNT-2	87.5	10	2.5	250	5	18
CBCNT-3	87.5	10	2.5	400	5	16
CBCNT-4	87.5	10	2.5	400	8	14
CBCNT-5	87.5	10	2.5	400	15	11

CMB 1	80	20		-	-	8
CMB 1b	80	20		400	8	-
REF	80	20[a)]		250	8	7

[a)] Vulcan XC-72.

Injection molding

Plaques with dimensions 100 x 100 x 2 mm were made by injection molding using an ENGEL ES200/110HL Victory, Germany with melt temperature 190°C, mold temperature 30°C and injection time 0.9 s. Holding pressure was 80% of injection pressure, whilst holding time was 11 s.

Samples in the shape of 10 mm wide strips were cut in the flow direction from the edge and middle section. The resistivity and Charpy impact toughness was then measured. Triplet samples were used for each material.

Results and Discussion

The distribution of resistivity in injection molded samples can be seen in Figure 1. Resistivity of CNT-2 increases two orders of magnitude when injection molded which has been attributed to alignment of carbon nanotubes, which decreases the number of contact points. The mid section has a higher resistivity than the edges, probably due to backflow which could bring the CNT:s in contact with each other again.

The hybrid mixtures of CB and CNT all show a large increase in resistivity when injection molded, showed in Table II. By increasing the extruder speed from 250 rpm to 400 rpm, little was gained in achieving lower resistivities. Instead, increasing the feed rate (thereby increasing extruder fill ratio and residence time) gives lower resistivities in compression molded samples (Table I), but does not seem to have the same effect in samples that were injection molded. It is possible that the high shear forces involved may have caused a degradation of the material which is indicated by the increase in resistivity and decrease in impact toughness of CMB1b compared to CMB1.

Table II Resistivity and impact toughness of injection molded plaques, average values (n=3) are shown.

Material	*Comp mold Res (Ω.cm)*	*↔ Inj mold res, edge (Ω.cm)*	*↔ Inj mold res, mid section (Ω.cm)*	*Impact toughness (kJ/m^2)*
CNT-1	9	4284	12913	12
CNT-2	5	259	647	9
CBCNT-1	11	858	1858	16.5
CBCNT-2	18	673	1523	16
CBCNT-3	16	1644	4344	18
CBCNT-4	14	2242	5118	18
CBCNT-5	11	794	1817	13
CMB1	8	16	19	25
CMB1b	-	81	103	13
REF	7	6.7	10	4.5

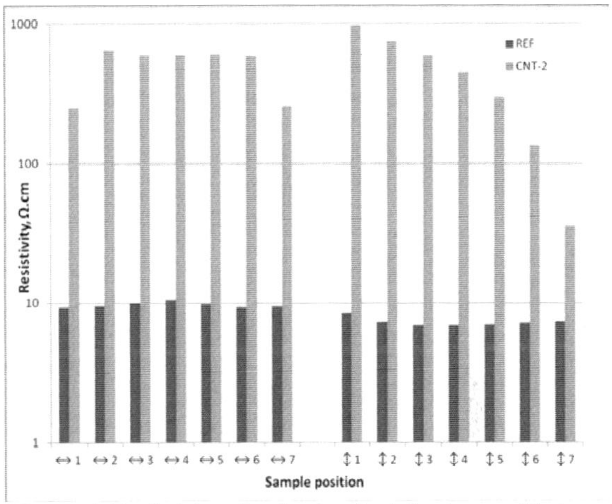

Figure 1 Resistivity distribution in injection molded plaques where ↔ denotes resistivity in the flow direction going from one edge (1) to the other (7) and ↕ denotes resistivity perpendicular to the flow direction going from the gate (1) to the opposing wall (7).

Conclusions

Hybrid mixtures of CB and CNT could potentially provide materials utilising the electrical conductivity of CNT:s also in injection molded products. However, careful mixing will be required to avoid degradation.

Acknowledgements

The author whish to acknowledge RISE Holding for funding via Swerea IVF.

References

1. T. Willmow; S. Pegel; P. Pötschke and U. Wagenknecht. *Compos. Sci. Tech.* 2008, *68*, 777.
2. D. Zhau; S. Wang; J. Wu; X. Bai. *Pigments and resin technology* 2009, *28*, 305.

O-07-358

INTERFACIAL LOCALIZATION OF MULTIWALL CARBON NANOTUBES IN POLYAMIDE6 AND ACRYLONITRILE BUTADIENE STYRENE CO-POLYMER BLENDS IN THE PRESENCE OF A REACTIVE COMPATIBILIZER

Amrita V. Poyekar[¥¶§], **Arup R. Bhattacharyya**[¥¶*], **Ajay S. Panwar**[¥¶], **George P. Simon**[¥§]

¥ IITB-Monash Research Academy, amritap@iitb.ac.in, amrita.poyekar@monash.edu
Indian Institute of Technology Bombay, Powai, Mumbai 400076, India
¶ Department of Metallurgical Engineering and Materials Science,
Indian Institute of Technology Bombay, Powai, Mumbai 400076, India, arupranjan@iitb.ac.in , panwar@iitb.ac.in
§ Department of Materials Engineering, Monash University, Clayton, VIC 3800, Australia, george.simon@monash.edu

Abstract - Blends of polyamide 6 (PA6) and acrylonitrile butadiene styrene co-polymer (ABS) with multiwall carbon nanotubes (MWNTs) were prepared via melt-mixing. In order to achieve improved dispersion of MWNTs in the blends, MWNTs were chemically modified through a non-covalent route using sodium salt of 6-aminohexanoic acid (Na-AHA) and polyaromatic molecule; pyrenecarboxaldehyde (PyCHO). The state of dispersion of MWNTs in the PA6/ABS blends with MWNTs was investigated through AC electrical conductivity measurements and morphological investigation. The modification of MWNTs has resulted in a decrease in the remaining aggregate size of MWNTs. The AC electrical conductivity of the blends has increased significantly after modification of MWNTs with Na-AHA and PyCHO. A classical approach of compatibilization was employed in order to localize a fraction of MWNTs at the interface. Modified MWNTs were encapsulated with styrene maleic anhydride co-polymer (SMA), which was found to be an efficient reactive compatibilizer for PA6/ABS blends. It was expected that during melt-blending, maleic anhydride moieties of SMA would react with amine functionality of Na-AHA in the case of Na-AHA modified-SMA encapsulated MWNTs. Further, there could be a possibility of π- π interactions between polyaromatic rings of PyCHO and styrene rings of SMA. This approach has resulted in the localization of a fraction of PyCHO modified MWNTs at the interface of PA6/ABS blends as observed through SEM and TEM analysis. This strategy demonstrates the possibility of localization of MWNTs at the interface of a co-continuous blend using a reactive compatibilizer.

Keywords: State of dispersion, MWNTs, chemical modifier, AC electrical conductivity, interface

Introduction

One of the most efficient strategies in order to develop new materials with multifunctional properties is to blend two or more polymers [1]. Incorporation of MWNTs in immiscible polymer blends offers multifunctional composites of greater strength, stiffness and electrical conductivity [2,3]. A lower electrical percolation threshold can likely be achieved through the restriction of MWNTs in one of the phases or more suitably at the interface [4,5]. Localization of the filler primarily depends on the differences in the interfacial energies between the filler and the constituent polymer phases. Smaller difference in interfacial energies facilitates the confinement of the nanotubes in the corresponding phase [6]. In addition, the melt viscosity and melt-mixing parameters have a significant role in governing the localization of fillers in the blends [5].

The objective of this study is to control the dispersion and localization of the multiwall carbon nanotubes (MWNTs) in the binary immiscible polymer blends of polyamide6 (PA6) and acrylonitrile butadiene styrene copolymer (ABS) by using a novel modifier, in combination with a suitable compatibilizer. In this context, two novel modifiers were used to control the dispersion of MWNTs through 'non-covalent' specific interactions and by interacting with the compatibilizer as well. By this mechanism, it is envisaged that MWNTs would be localized at the interface along with the compatibilizer.

Experimental

PA6/ABS blends with MWNTs in the presence of SMA copolymer were prepared by melt-mixing following simultaneous mixing protocol using (Micro 5, DSM Netherlands) a conical twin screw microcompounder at 260 °C at a screw speed of 100 rpm and mixing time of 15 minutes. The ingredients were dried under vacuum for 24 hours at 80 °C. The SMA-encapsulated MWNTs were prepared by sonication of MWNTs and SMA together in THF and subsequent removal of solvent. In the case of modified MWNTs, the nanotubes were initially modified with either Na-AHA or PyCHO by ultrasonicating the nanotubes in aqueous and THF solution respectively and further removal of the solvent. If encapsulation was required then the already modified MWNTs were further encapsulated with SMA as described above. The torque values were calculated by multiplying the force value (N) obtained during melt-mixing with the average screw radius (~ 0.5 cm). The FTIR measurements were carried out using the Vertex 80, Bruker FTIR equipment in the scanning range of 4000 cm^{-1} – 400 cm^{-1}, with the spectral resolution of 4 cm^{-1} (dried at 80 °C, 24 hours under vacuum, using KBr pellets). AC electrical conductivity measurements were performed on the injection molded samples in the frequency range between 10^{-1} to 10^{6} Hz, using Alpha A analyzer (3 μHz - 20 MHz) and Agilent E4991A RF analyzer (1 MHz - 3 GHz). The DC electrical conductivity was determined from the AC conductivity plots by fitting Jonscher's 'Universal Power law' equation ($\sigma_{AC} = \sigma_{DC} + A\omega^n$, $0<n<1$) [7]. The morphological features were analyzed using SEM, where one of the phases was selectively removed by using a solvent. PA6 phase was removed using formic acid and ABS phase was removed using THF. The blend samples were microtomed to obtain ultrathin sections at room temperature using 'Leica Utramicrotome Microsystems'. The butadiene phase of ABS was selectively stained with the help of OsO_4. TEM investigations were carried out on Philips JEOL JEM-2100 F. The crystallization behavior of the blends was investigated using TA Q200 differential scanning calorimeter. Table 1 enlists the sample codes and compositions.

Table 1. Sample codes

Sample	Code
N = PA6 , A = ABS, x = wt% of PA6, y = wt% of ABS, 'x' varies from 40 to 60 wt%	
PA6/ABS blends	NxAy
PA6/ABS blends + SMA (5 wt%)	NxAySm5
PA6/ABS blends + SMA encapsulated p-MWNTs (1:1 wt/wt, 5 wt%)	NxAySE5T5
PA6/ABS blends + SMA encapsulated Na-AHA modified MWNTs (1:1:1 wt/wt/wt, 5 wt%)	NxAySE5M5
PA6/ABS blends + SMA encapsulated PyCHO modified MWNTs (1:1:1 wt/wt/wt, 5 wt%)	NxAySE5Py5

Results and Discussion

Effect of compatibilizer: Melt-interfacial reaction and network-like structure of MWNTs

Fig. 1 shows the variation in the torque values for PA6/ABS blends with SMA encapsulated p-MWNTs (1:1 wt/wt), SMA encapsulated Na-AHA modified MWNTs and SMA encapsulated PyCHO modified MWNTs (1:1:1 wt/wt/wt). The torque values exhibit a decrease with increasing PA6 concentration in the case of neat PA6/ABS blends. However, in the presence of the SMA, the torque values are significantly higher as compared to the neat blends. Addition of SMA to neat PA6/ABS blends results in an increase in the torque value due to melt-interfacial reaction between amine end groups of PA6 and anhydride functionality of SMA copolymer. Interestingly, it is observed that blends with SMA encapsulated p-MWNTs show lower torque values as compared to blends with only SMA, which may be due to the plasticization effect or due to lower extent of interfacial reaction in the presence of p-MWNTs. Blends with SMA encapsulated and modified MWNTs results in further decrease in the torque values, wherein the plasticization of the PA6 phase due to moisture absorption of the organic modifier cannot be ruled out. It has also been reported earlier that the reactive blending in the presence of SMA alters the interphase of PA6/ABS blends and blends with MWNTs due to a varying extent of melt-interfacial reaction [8, 9].

Fig. 2 shows the FTIR spectra of neat PA6, PA6 and blends with SMA encapsulated p-MWNTs (1:1 wt/wt). The FTIR spectra indicate the presence of styrene aromatic ring through a characteristic peak at ~700 cm^{-1} [10]. Also, the presence of a characteristic peak at ~ 1710 and ~1778 cm^{-1} confirms the imide formation through melt-interfacial reaction between the amine group of PA6 and anhydride functional groups from SMA [11]. A signature of unreacted MA functionality of SMA copolymer is also noticed in the various compositions of SMA encapsulated MWNTs.

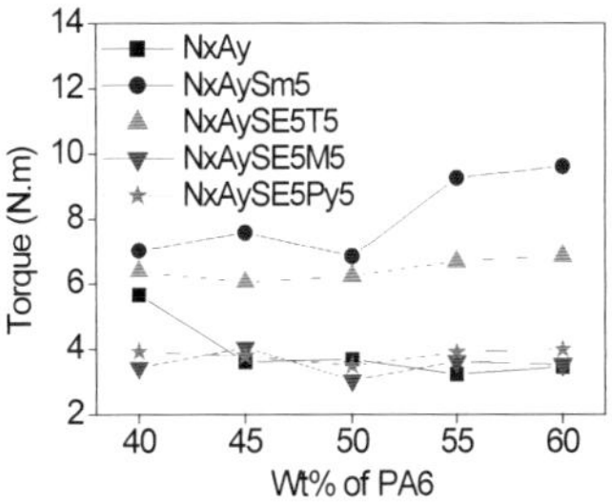

Figure 1. Torque values indicating the effect of SMA and SMA encapsulated MWNTs

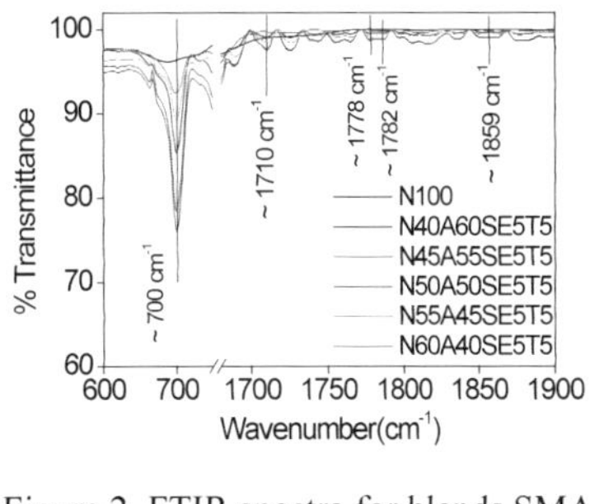

Figure 2. FTIR spectra for blends SMA encapsulated MWNTs (1:1 wt/wt) for varied blend composition

Fig. 3 (a) – (c) show the variation in the DC electrical conductivity with the increasing concentration of PA6 in the case of PA6/ABS blends with (a) SMA encapsulated p-MWNTs (1:1

wt/wt), (b) SMA encapsulated Na-AHA modified MWNTs and (c) SMA encapsulated PyCHO modified MWNTs (1:1:1 wt/wt/wt). All the blend compositions show a non-conducting behaviour in the presence of SMA encapsulated p-MWNTs. It is important to note here that PA6/ABS blends with p-MWNTs and in the absence of SMA showed blend composition dependant variation in DC electrical conductivity; with highest value of ~ 3.25E-6 S/cm for 60/40 PA6/ABS blends with p-MWNTs [12]. However, PA6/ABS blends with SMA encapsulated Na-AHA modified MWNTs exhibit a significantly higher electrical conductivity values for the complete composition range in accordance with the blends with only Na-AHA modified MWNTs.

Interestingly, the presence of SMA encapsulated PyCHO modified MWNTs (Fig. 3 (c)) indicate a blend composition dependence of DC electrical conductivity, which was also observed in the case of PA6/ABS blends with p-MWNTs [12]. Thus, it is likely that the presence of SMA might influence the dispersion of MWNTs, phase morphology and consequently the electrical conductivity values. In line with the findings, the encapsulation of MWNTs with SMA often leads to the lower electrical conductivity values [13]. The presence of aromatic rings in SMA makes it favorable for the compatibilizer to interact with the MWNTs surface and it leads to the wrapping of the MWNTs by the SMA co-polymer through "π-π" interaction [14]. In a consistent fashion, the SMA encapsulating layer also affects the nucleation ability of the nanotubes in the blends [13], which is discussed in the subsequent discussion.

The Na-AHA molecules are reported to improve the debundling of MWNTs through steric stabilization and electrostatic charge repulsion [15]. The state of dispersion exhibited in the solutions persists into the melt-blended state as well, thus, leading to a significantly higher electrical conductivity values [12, 16].

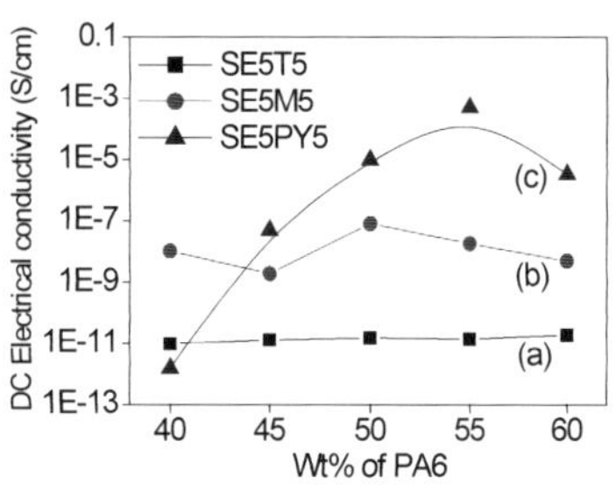

Figure 3. DC electrical conductivity versus blend composition for blends with SMA encapsulated MWNTs

Fig. 4 (a) shows the DSC crystallization exothermic peaks of the PA6 phase of PA6/ABS blends (60/40 wt/wt) with SMA encapsulated p-MWNTs, SMA encapsulated Na-AHA modified MWNTs and SMA encapsulated PyCHO modified MWNTs for varied blend compositions. PA6 in neat PA6/ABS blends shows a single crystallization exothermic peak at ~192 °C. In the presence of MWNTs, an additional crystallization exotherm is observed at higher temperature. The second crystallization peak appears due to the trans-crystalline organization of PA6 lamellae in the MWNTs surface due to the confinement. In general, the heterogeneous nucleation action of the MWNTs is also observed irrespective of blend compositions. The height of the second peak as compared to the first peak may indicate the higher extent of the trans-crystalline structure of PA6 lamellae in the surface of MWNTs. Further, the hetero-nucleating ability of MWNTs is dependent on the free surface area available for nucleation of PA6 chains even after encapsulation with SMA copolymer and also on the fraction of MWNTs localized in the PA6 phase. In the case of PA6, where Brosse et. al. [17] found that different PA6 crystallite organizations were present due to the matching crystallographic lattice parameters of PA6 with that of MWNTs.

In the presence of SMA, it is expected that there will be higher polymer chain entanglements due to the melt-interfacial reaction and the interface between the MWNTs and PA6 will be modified [9]. Also it is seen that for blends containing SMA encapsulated p-MWNTs, the height of the second crystallization peak has increased with the increased concentration of PA6 phase. The fact that the crystallization behavior of the PA6 phase in PA6/ABS blends with Na-AHA modified MWNTs has exhibited that the height of the crystallization peak at higher temperature is significant for all the blend compositions. In the case of blends with SMA encapsulated PyCHO modified MWNTs; there is no significant observation of the second peak in the case of 40/60 (wt/wt) PA6/ABS blend. Also, other blend compositions with SMA encapsulated PyCHO modified MWNTs showed a significant decrease in the height of second crystallization peak as compared to the corresponding compositions of SMA encapsulated Na-AHA modified MWNTs, which may be due to the lesser extent of free surface available in case of PyCHO modified MWNTs.

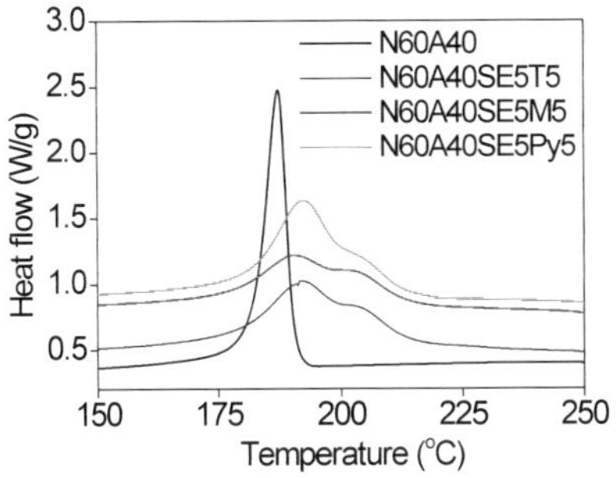

Figure 4. Crystallization behaviour of PA6 phase in PA6/ABS blends with SMA encapsulated MWNTs

Therefore, it can be inferred that the interaction between the MWNTs surface, the presence of SMA copolymer and the modifier molecule influence the crystallization behavior of PA6 phase in PA6/ABS blends. In the case of the p-MWNTs encapsulated with SMA, the SMA molecules may interact with the MWNTs surface, thus leading to an effective wrapping of the MWNTs. Consequently, the formation of an encapsulating layer on the MWNTs surface is likely to lead to a decreased surface area available for heterogeneous nucleating action and reduced chain mobility of the PA6 chains [9]. Further, a majority fraction MWNTs may eventually leave SMA phase and diffuse into PA6 phase during melt-mixing due to the lower surface energy difference between PA6 and MWNTs as compared to SMA copolymer and MWNTs. On the other hand, in the case of Na-AHA and PyCHO modified MWNTs, the modifier molecule which are present in the vicinity of the nanotubes may hinder the interaction between the MWNTs and SMA molecules.

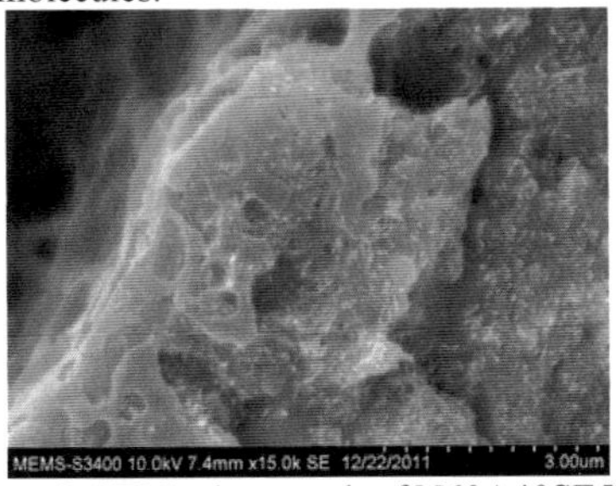

Figure 5. SEM micrograph of N60A40SE5Py5 (ABS phase removed by THF)

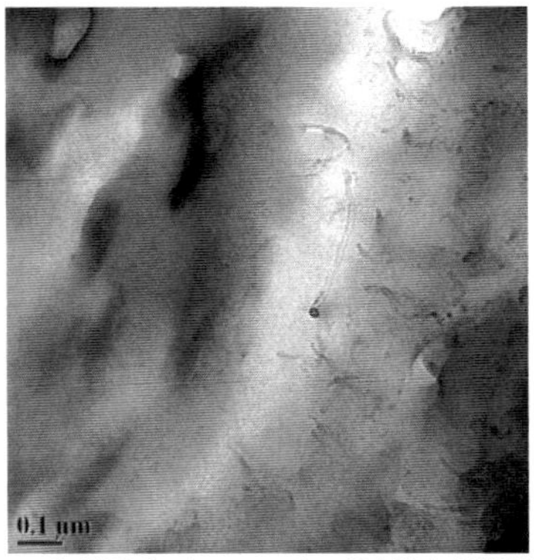

Figure 6. TEM micrograph of N60A40SE5Py5 (ABS stained using OsO_4)

Confinement, localization and state of dispersion of the MWNTs:

SEM micrograph (Figure 5) of 60/40 (wt/wt) PA6/ABS blend with SMA encapsulated PyCHO modified MWNTs, where the ABS phase has been removed using THF. It can be seen from the micrograph that the MWNTs are well-dispersed, which is also reflected in higher electrical conductivity values (Figure 3). Figure 6 shows the TEM micrograph of 60/40 (wt/wt) PA6/ABS blend with SMA encapsulated PyCHO modified MWNTs. A fraction of the MWNTs is found to be present at the interface of PA6/ABS blend as well.

Conclusions

The strategy of combining the reactive compatibilizer and the MWNTs through encapsulation in order to drive and confine them at the interface was utilized for MWNTs with and without modifiers. The assessment of the state of dispersion of MWNTs was carried out via electrical conductivity measurement. It showed varied behaviors for SMA encapsulated p-MWNTs, Na-AHA modified MWNTs and PyCHO modified MWNTs, which might be due to varied interactions between SMA, MWNTs and the modifier molecules. A fraction of the MWNTs localized at the interface as shown by the morphological studies for PA6/ABS blends with SMA encapsulated Na-AHA and PyCHO modified MWNTs.

Acknowledgements

The authors would like to thank Microcompounder central facility, SAIF and IIT Bombay.

References

[1] P. Potschke; D. R. Paul *J. Macromol. Sci. Part C: Poylm. Rev.* 2003, *C43*, 87.

[2] Z. Spitalsky; D. Tasis; K. Papagelis; C. Galeotis *Prog. Polym. Sci.* 2010, *35*, 357.

[3] M. Morcom; K. Atkinson; G. P. Simon *Polymer* 2010, *51*, 3540.

[4] M. A. Saleh; U. Sundararaj *Composites: Part A* 2008, *39*, 284.

[5] F. Fenouillot; P. Cassagnau; J.-C. Majesté *Polymer* 2009, *50*, 1333.

[6] S. Bose; A. R. Bhattacharyya; R.A. Khare; S. S. Kamath; A. R. Kulkarni *Polym. Engg. Sci.* 2011, *51*, 1987.
[7] A. K. Jonscher, *Nature* 1977, *267*, 673.
[8] S. Bose; A. R. Bhattacharyya; P. V. Kodgire; A. Misra *Polymer* 2007, *48*, 356.
[9] S. Bose; A. R. Bhattacharyya; P. V. Kodgire; A. Misra; P. Pötschke *J. Appl. Polym. Sci.* 2007, *106*, 3394.
[10] R. Schaefer; J. Kressler; R. Neuber; R. Muelhaupt *Macromolecules* 1995, *28,* 5037.
[11] B. Panda; A. R. Bhattacharyya; A. R. Kulkarni *J. Appl. Polym. Sci.* 2012, doi: 10.1002/app.38074.
[12] A. V. Poyekar ; A. R. Bhattacharyya; G. P. Simon; A. S. Panwar, Manuscript in preparation.
[13] S. Bose; A. R. Bhattacharyya; P. V. Kodgire; A. Misra, *Polymer* 2007, 48, 356.
[14] C.H. Xue; M.M. Shi; Q. X. Yan; Z. Shao; Y. Gao; G. Wu; X.B. Zhang; Y. Yang; H.Z. Chen; M. Wang *Nanotechnology* 2008, *19*, 115605.
[15] R. A. Khare; A. R Bhattacharyya; A. S. Panwar; S. Bose; A. R. Kulkarni *Polym. Eng. Sc*i. 2011, *51*, 1891.
[16] B. Krause; G. Petzold; S. Pegel; P. Pötschke; *Carbon* 2009, *47*, 602.
[17] A.C.Brosse; S.T.Girault; P.M. Piccione; L. Leibler *Polymer* 2008, *49*, 4680.

O-07-286

PREPARATION AND CHARACTERIZATION OF "PVC/TIO_2 NANORODS" NANOCOMPOSITES

A. Hasanzadeh[1], Soheil Pilehvar[2]*, Mehdi Shahrezaei[3], Mehdi Partovi[4], Abbas Moradi[5] , Farhad amanizadeh[2] Hamze jodat[2]

[1]*Department of Polymer engineering, Sahand University of Technology, Tabriz, Iran - a_hasanzadeh@sut.ac.ir*

[2*]*Polymer Engineering Group, School of Engineering, Science and Research Branch, Islamic Azad University - s.pilehvar1988@gmail.com,hamze.jodat@gmail.com,f.amanizadeh@gmail.com*

[3]*Department of Chemical engineering, Sahand University of Technology, Tabriz, Iran - mehdi_shahrezaei@yahoo.com*

[4]*Institute of Polymeric materials, Sahand University of Technology, Tabriz, Iran - partovi.mehdi@gmail.com*

[5]*Polymer Engineering Group, South Tehran Branch, Islamic Azad University, Tehran, Iran - amirabbas.moradi@yahoo.com*

Abstract

In this work, TiO_2 nanorods with different crystalline structure were synthesized from the commercial powder of TiO_2 (P25 Degussa company) by hydrothermal method and nanocomposites of PVC/TiO_2 nanorods, were prepared by melt-compounding. Effect of the synthesized TiO_2 nanorods on photochemical properties of PVC was investigated by surface morphology of PVC after photodegradation. The surface morphology of samples after 500 hr exposuring light showed that by increscent the radiation time, photostability of PVC matrix increases considerably which can be due to the presence of TiO_2 nanorods with a single crystalline structure (anatase type) of TiO_2. But TiO_2 nanorodes with combined of anatase and rutile crystalline structure (anatase as dominent structure) leading to high photochemical degradation of PVC matrix which can be attributed to their high photocatalytic properties.

Keywords: PVC, TiO_2, Nanorods, Photochemical, Photostability.

Introduction

Nanostructural materials have received much attention because of their novel properties which differ from those of bulk materials [1]. Titanium dioxide is an important inorganic functional material. It has possible uses in photocatalysis, photovoltaic cells, lithium ion batteries, gas sensors [2], hydrogen storage and hydrogen generation by water splitting [3] and etc.

The synthesis of TiO_2 nanomaterials, including nanoparticles, nanorods, nanowires and nanotubes are primarily categorized with the preparation method [3]. Hydrothermal method is one of the methods in synthesizing of nanorods.

Poly (vinyl chloride) (PVC) is one of the most popular thermoplastics used in the world, because of its low cost and the good performance [4]. These properties make it very suitable for outdoor applications, e.g. door and window profiles [5]. But these PVC products are degraded under sun's radiation and loss their mechanical properties [6].

Titanium dioxide (TiO_2) is one of the popular UV stabilizers and there is a large amount of literature regarding the photodegradation of PVC/TiO_2 nanocomposite [6, 7]. But, TiO_2 also can lead to the acceleration of oxidative ageing because of its photocatalytic properties [5]. Gesenhues [5] studied the SEM micrographs of irradiated PVC and illustrated that the photocatalytic properties of titanium dioxide can lead to the degradation of PVC. Several methods have been used to control the photocatalytic property of titanium dioxide [8, 9].

In this work, nanocomposites of PVC by synthesized TiO_2 nanorods with different crystalline structure were prepared by meltcompounding. Effect of the synthesized titanium dioxide nanorods on the photodegradation of PVC was investigated by morphological studies.

Experimental

Materials

Commercial titanium dioxide (P25 Degussa A6-Germany) nanopowder was added to sodium hydroxide (NaOH, %99 Merck) solution and then was poured into the Teflon holder of autoclave. The filtered material was thoroughly washed with distilled water until the washing solution achieved pH 7. Then, sample was washed with sulfuric acid (H_2SO_4, %98 Merck) and distilled water, respectively. The sample was dried in vacuum oven.

The synthesized TiO_2 nanorods were characterized by scanning electron microscopy and XRD to determine the crystalline structure and morphological studies.

The characteristics of the used materials for preparation PVC nanocomposites are given in Table 1.

Preparation of PVC nanocomposites

Two samples with different compositions according to Table 2 were premixed at 120 ^{0}C for 15 minute and then were melt-compound in the internal mixer type of Brabender at 170 ^{0}C for 8 minutes.

Artificial weathering

The artificially simulated weathering was carried out by using Osram 150 W fluorescent UV lamp. The samples were artificially aged in the chambers, for 500h of continuous irradiation. By using circulating air, the maximum temperature measured in the chamber was 40 ^{0}C.

Morphology analyses

The surface morphology of PVC nanocomposites were examined using VEGA\\TESCAN scanning electron microscopies.

Table 1 The characteristics of the used materials

Materials	Function	Supplier
Poly(vynilchloride) (PVC)	Polymer matrix	Bandar Imam petro. Co .
Titanium dioxide (TiO_2)	UV absorber	Degusa
Calcium carbonate (CaCO3)	Filler	Poodrkar Co.
Calcium stearate (CaSt2)	Thermal stabilizer	Poodrkar Co.
Zinc stearate (ZnSt2)	Thermal stabilizer	Poodrkar Co.
Paraffin wax	Internal lubricant	Poodrkar Co.
Stearic acid (StAc)	External lubricant	EPS IMPEX Co

Table 2 The compositions of the prepared PVC nanocomposites

Name	PVC (Phr)	TiO_2 Nanorod (Phr)	CaCO3 (Phr)	*CaSt2* (Phr)	*ZnSt2* (Phr)	Paraffin wax(Phr)	StAc (Phr)
NP1	100	A(3)	5	2.5	2.5	2.8	0.22
NP2	100	B(3)	5	2.5	2.5	2.8	0.22

St: Stearic, Ac: Acid, Phr: Part per houndered

Results and Discussion

Characterization of synthesized TiO_2 nanorods

SEM pictures of the TiO_2 nanorod, sample (A) and (B) have been shown in figure 1. The sample (B) with the lower concentration of the initial titanium dioxide nanoparticles has the suitable distribution of titanium dioxide nanorods with the unique diameters than sample (A) and the size of the nanorods diameter is also smaller and mainly situated in nanometric scale. High aspect ratio of these nanorods can provide a good direction for the using of this type of the nanostructure materials in the polymeric fields for producing nanocomposites with enhanced mechanical properties.

In Fig. 2. XRD patterns of the samples synthesized are shown. As it can be seen, sample (B) have a single phase of anatase type of structure of TiO_2 but in the XRD pattern of the sample (A) not only the anatase phase peaks and also rutile phase peaks at 2θ, 27.5° and 37 ° are significantly illustrated.

Morphology study of nanocomposites by SEM

Figure. 3. show the surface morphology of the irradiated "PVC/TiO_2" nanocomposites (NP1 and NP2) for 500 hours under air. The SEM images of NP1 reveal that the photocatalytic degradation of PVC leads to the formation of pits that might be induced by the escape of HCl during the elimination reactions from PVC matrix [6]. The pits are developed during the degradation and nanocomposites (NP1) consist of titanium dioxide nanorods (sample A), undergo the higher degradation, which can be due to the photocatalytic properties of TiO_2 nanorods. But, photo degredation of NP2 nanocomposite is lower than NP1 nanocomposite. This trend can be due to the presence of TiO_2 nanorods (sample B) with a single crystalline structure (anatase type) of TiO_2 nanorods. TiO_2 nanorodes with combined of anatase and rutile crystalline structure (anatase as dominent structure), sample (A) leading to high photochemical degradation of PVC matrix which can be attributed to their high photocatalytic properties than sample (B).

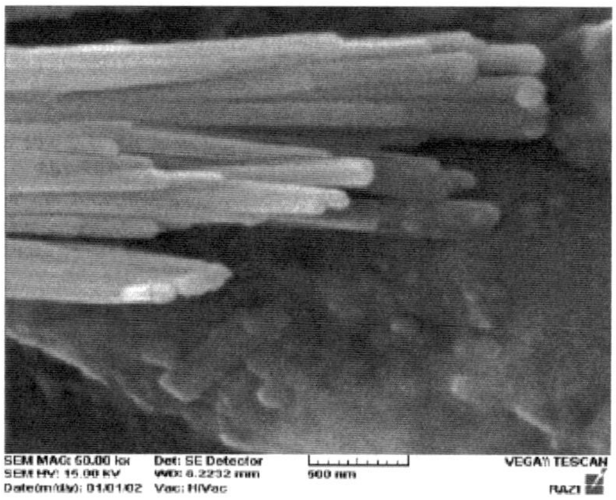

Sample (A)

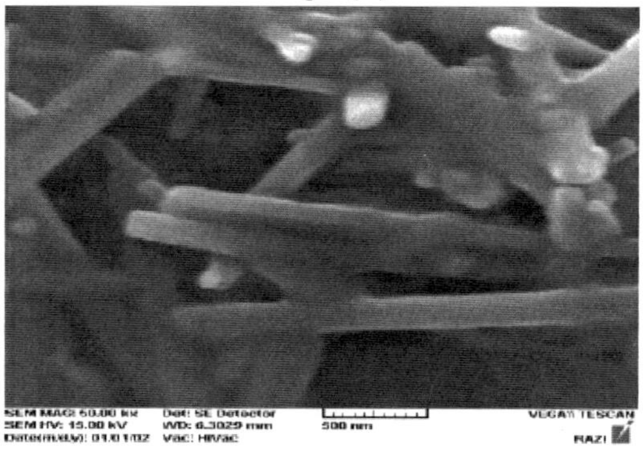

Sample (B)

Figure 1 - SEM images of TiO_2 nanorods, sample (A) and (B).

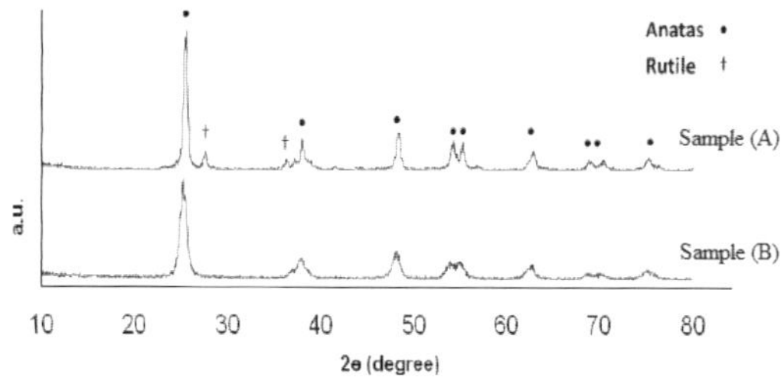

Figure 2 - XRD patterns of TiO_2 nanorods, sample (A) and (B).

Conclusions

TiO_2 nanorods with different crystalline structure were synthesized from the commercial powder of TiO_2 (P25 Degussa company) by hydrothermal method and nanocomposites of PVC/TiO_2 nanorods, were prepared by melt-compounding. The surface morphology of samples after 500 hr exposuring light showed that TiO_2 nanorodes with combined of anatase and rutile crystalline structure (anatase as dominent structure) leading to high photochemical degradation of PVC matrix which can be attributed to their high photocatalytic properties than TiO_2 nanorods with a single crystalline structure (anatase type).

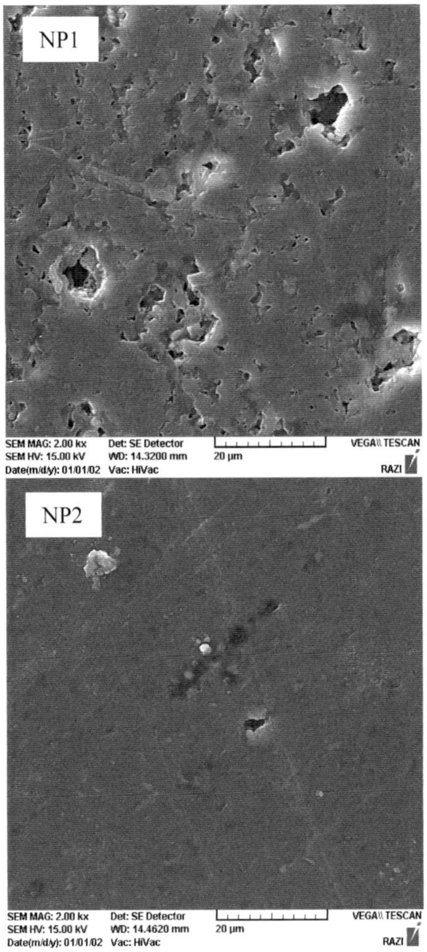

Figure 3 - SEM images of PVC nanocomposites after 500 hr of radation

Acknowledgements
The Author wish to thank Sahand University of Technology (SUT) for the financial support of this work. Also, thank coworkers and technical staff in the nanostructure material research center of SUT for their help during various stages of this work.

References

1. M. Wei, Y. Konishi, H. Zhou, Hideki Sugihara, H. Arakawa: *Solid State Communications*, 133 (2005) 493–497.
2. L. Gang, L. Zhongqing, Z. Zhao, Y. Xin: *Chin J Catal*, 30(1) (2009) 37–42.
3. X. Chen, S. S. Mao: *Chem. Rev*, 107 (2007), 2891-2959.
4. Luis Pimentel Reala, Ade lia Pereira Rochaa, Jean-Luc Gardetteb, *Polymer Degradation and Stability* 82 (2003) 235–243.
5. U. Gesenhues, *Polymer Degradation and Stability* 68 (2000) 185±196.
6. Fethi Kamisli, Cem Turan, *Journal of Materials Processing Technology* 159 (2005) 40–47.
7. C. Anton-Prinet, G. Mur, M. Gay, L. Audouin, J. Verduc, *Polymer Degradation and Stability* 61 (1998)211-216.
8. Daobao Chu, Ximei Yuan, Guoxu Qin, Mai Xu, Peng Zheng, Jia Lu, Longwu Zha, *J Nanopart Res* 10 (2008) 357–363.
9. Victor M.F. Evora, Arun Shukla, *Materials Science and Engineering* A361 (2003) 358–366.

PREPARATION AND PROPERTIES OF NANOCOMPOSITE FOAMS BASED ON NATURAL RUBBER (NR)/ETHYLENE-VINYL ACETATE (EVA) BLENDS

A. Masa[1], J. Julyanon[1], A. Kaesaman[1], C. Thongpin[2], T. Sakai[3] and N. Lopattananon[1*]
[1]*Department of Rubber Technology and Polymer Science, Faculty of Science and Technology, Prince of Songkla University, Pattani, Thailand - akim_psu@yahoo.com; juthapat_rubb@hotmail.com; kazizon@bunga.pn.psu.ac.th; lnatinee@bunga.pn.psu.ac.th*
[2]*Faculty of Engineering and Industrial Technology, Silpakorn University, Nakhon Pathom, Thailand - chanchai@su.ac.th*
[3] *Tokyo Office, Shizuoka University, 3-3-6 Shibaura, Minato, Tokyo 108-0023, Japan - mtsakai@ipc.shizuoka.ac.jp*

Abstract - Rubber nanocomposite foams based on ethylene vinyl acetate (EVA)/natural rubber (NR) blends with a ratio of 60/40 were prepared by melt mixing in an internal mixer and compounding on two-roll mill at 70°C, followed by compression molding. Effects of different nanofillers (Na-MMT and organoclay) and contents on filler dispersion, foam morphology and mechanical properties of the EVA/NR foams were studied. The obtained results were also compared with the rubber foams filled a commercial grade, china clay. The X-Ray diffraction (XRD) result showed that both nanoclay and organoclay were intercalated. The addition of filler into EVA/NR blends provided more cell density with smaller cell size. Increasing filler content increased cell density, specific stress at 100% and specific tensile strength. The compressive strength and elastic recovery of the rubber foams was found to decrease with increasing nanofiller loading, but increase with increasing china clay content. Based on the present work, it was suggested that the nanofiller content of 5 phr was recommended for the production of nanocomposite foams with optimum properties, and the use of nanofiller offered lighter weight and superior specific tensile strength when compared with the china clay.

Keywords: Ethylene-vinyl acetate; Natural rubber; Nanocomposite; Foam; Mechanical properties.

Introduction

Polymeric foams are widely used in various applications, due to their low density, superior relative mechanical properties, buoyancy, cushioning performance, impact damping, thermal and acoustic insulation properties, good energy absorption, and low cost if compared with unfoamed polymers. Foamed polymers are generally made up of solid phase and gas phase, which are practically produced by using either chemical or physical blowing agent.

Ethylene-vinyl acetate copolymer (EVA) is one of the most widely used polymers for manufacturing foamed products. EVA containing medium content of VA (30%) exhibits some properties of rubber but can be processed like ordinary thermoplastic. However, EVA has low melt strength and poor mechanical properties, i.e., high compression set, low flexibility etc. Therefore, it has been often blended with other polymers such as natural rubber[1] and EtBC (ethylene-1-butene copolymer)[2] to improve the mechanical properties of foamed materials.

Fillers are usually added in the polymer/rubber to improve foam morphology and certain properties. Recently, the incorporation of nano-size particle as a filler in polymer/rubber to enhance foaming process and mechanical properties was reported. [2, 3]

In this work, 60/40 EVA/NR blend were foamed using Azodicarbonamide (ACD) as a blowing agent. Nanoclay (Na-MMT) and modified nanoclay were used as fillers. The effect of different nanofillers and contents on foam structure and mechanical properties was studied. China clay was also used as a filler for preparation of the rubber foams.

Experimental

Preparation of EVA/NR blend foams

EVA pellet (POLENE Co., Ltd., Thailand) and NR (Khokpho-Maelarn, Pattani, Thailand) were firstly melt-mixed in an internal mixer at a temperature of 70°C with rotor speed of 60 rpm. Sodium montmorillonite (Na-MMT; grade Kunipia-F®) and alkyl dimethyl benzyl ammonium chloride modified clay (Grade Kunipia-B®) (Kunimine Industries CO., LTD., Japan) and other additives were added. The obtained EVA/NR blend masterbatchs were then compounded with ADC, dicumyl peroxide (DCP) and trimethylol propane trimethacrylate (TMPTMA) on a two-roll mill. The blends were kept at room temperature for 24 hours and then hot-pressed at 160°C to obtain EVA/NR foams. The foams were cut into test-specimens for density, mechanical property, XRD and SEM characterizations.

Characterizations

The X-ray diffraction (XRD) studies were performed by using a Rigaku (TTRAXIII) X-ray diffractometer. The X-ray beam was nickel-filtered CuKα (wave length = 0.154nm). Data were obtained for a 2θ range of 1-10°. The cell structure of the rubber foams was observed with FEI Quanta 400 scanning electron microscope (SEM). Density of foams was characterized by mean of mass to volume ratio of sample according to ASTM D1622-03. The tensile properties and compressive strength of the NR/EVA foams were performed using universal tensile testing machine (HOUNFIELD H10KS) at room temperature. The crosshead speed for the tensile measurement and compressive strength test were 500 and 12 mm/min, respectively. Five replicates of specimens were measured and the data were averaged. The compression set was carried out according to ASTM D395-03.

Results and Discussion

Clay dispersion

The XRD patterns of Na-MMT, organoclay and the nanocomposite foams are shown in Figure 1 and Figure 2.

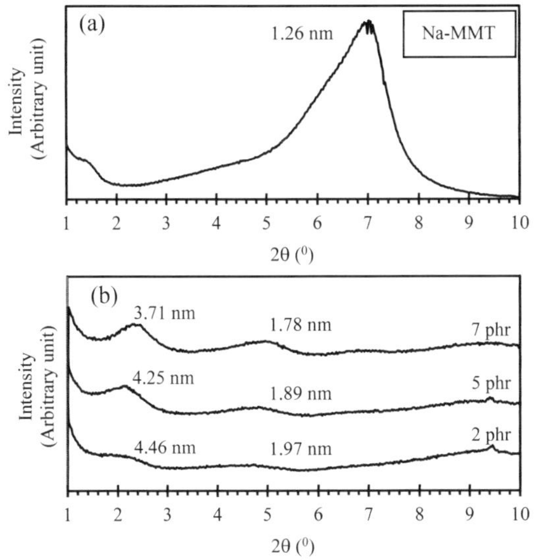

Figure 1. XRD patterns for Na-MMT and EVA/NR nanocomposite foams with various Na-MMT contents.

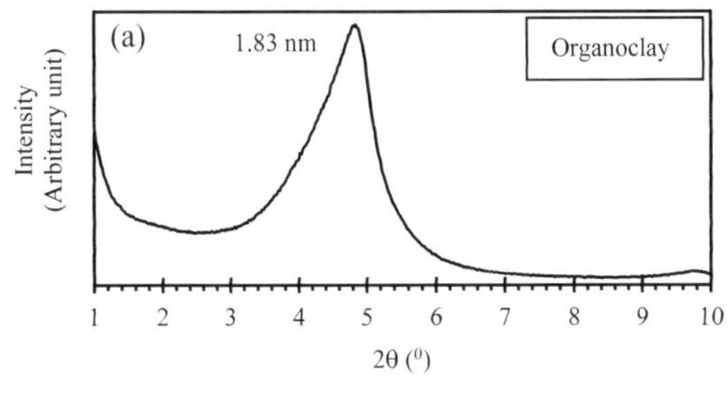

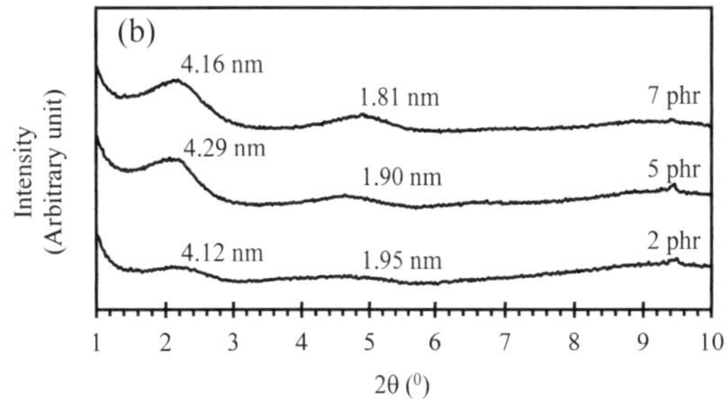

Figure 2. XRD patterns for organoclay and EVA/NR nanocomposite foams with various organoclay contents.

From Figure 1(a) and 2(a), Na-MMT and organoclay show a single diffraction peak at 7.04° and 4.84°, corresponding to a d-spacing of 1.26 and 1.83 nm, respectively. The XRD spectrum of the filled foams showed two diffraction peaks at lower angle than those of corresponding pristine nanofillers (Figure 1(b) and 2(b)). This indicated the formation of intercalated nanocomposites. It was also found that the expansion of clay interlayer was not significantly changed with filler grade and loading.

Foam morphology and properties

SEM photomicrographs of the unfilled and filled rubber foams are shown in Figure 3.

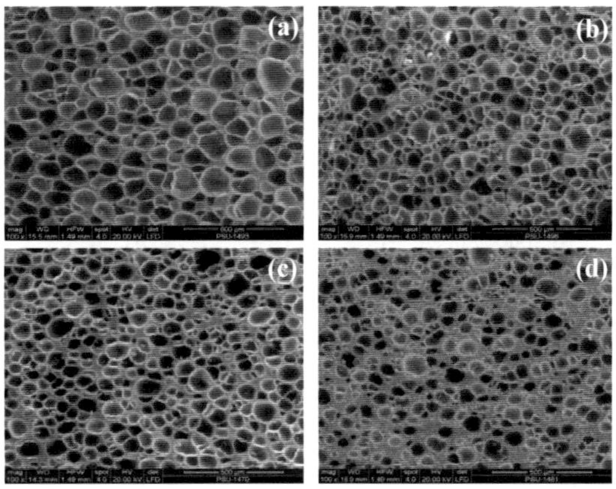

Figure 3. SEM images of (a) unfilled EVA/NR foam, (b) 5 phr Na-MMT, (c) 5 phr organoclay and (d) 5 phr china clay filled EVA/NR foams.

The SEM images revealed that the foamed rubber had closed cell structure. Figure 4 and 5 shows the numbers of cell per unit volume (cell density) and cell size as a function of filler content, respectively.

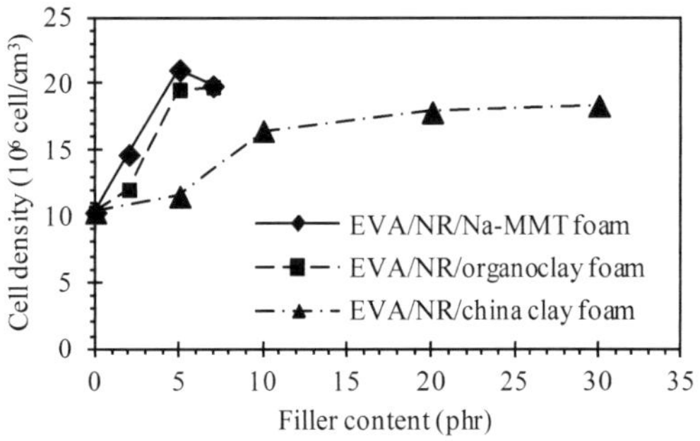

Figure 4. Cell density of different filled EVA/NR foams with various filler loadings.

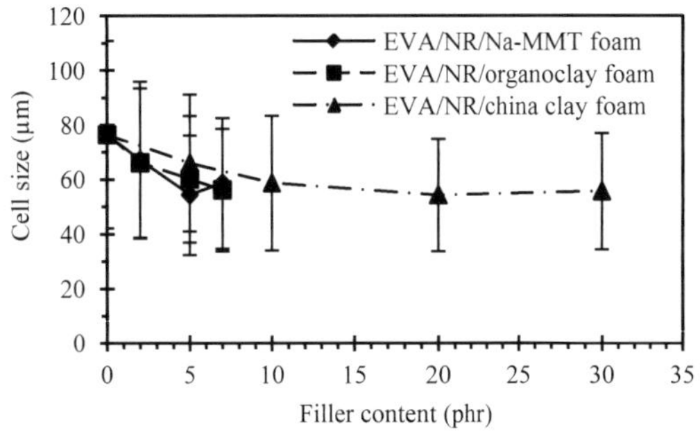

Figure 5. Cell size of different filled EVA/NR foams with various filler loadings.

It is seen that the cell density generally increased with increasing filler content, but the cell size decreased. The increase in cell density indicated that the filler served as nucleating agent in the foaming process. When the nucleation sites were more, the amount of gas available for a bubble growth would decrease accordingly, leading to smaller cell size. However, the cell density slightly decreased for the Na-MMT filled foam, while slowly increased for the organoclay and china clay filled ones at higher filler content. Both Na-MMT and organoclay gave much higher cell density, and smaller cell size than that of china clay because the nanofillers had much greater contacting surface area to nucleate the bubbles. The foaming of EVA/NR blends with Na-MMT showed a little more cell density when compared with that of the organoclay. For the Na-MMT filled foams, an aggregation of filler at high filler content possibly caused a slight decrease in cell density.

Density

Figure 6 shows the density of three kinds of filled foams with varying filler content.

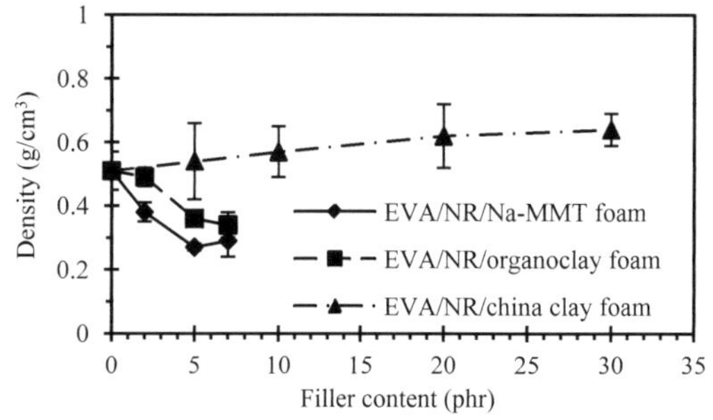

Figure 6. Density of different filled EVA/NR foams with various filler loadings.

Increasing nanofiller content decreased the density of rubber nanocomposite foams, but increased the density of china clay filled foams. The decrease was due to an increase of cell nucleation[4] and a reduction of gas loss by barrier effect of nanofiller. Foaming of rubber blends with Na-MMT gave lower density than that of organoclay. For the conventional composite foams, the increase of density with more filler loading was possibly due to higher filler density.

Mechanical properties

Figure 7-8 show the specific tensile stress at 100% elongation and specific tensile strength of different filled rubber foams.

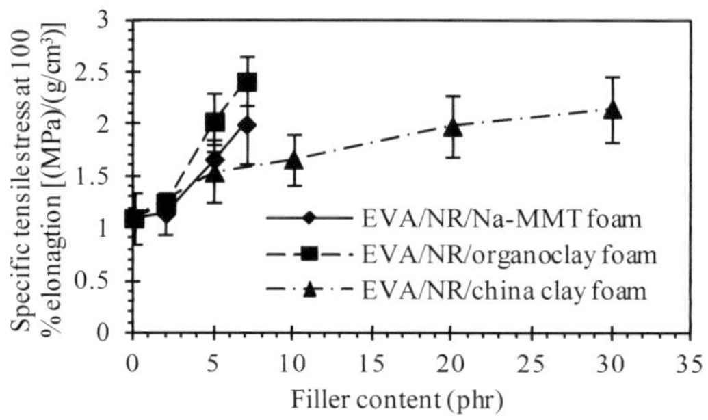

Figure 7 Specific tensile strength at 100%elongation of different EVA/NR foams with various filler loadings.

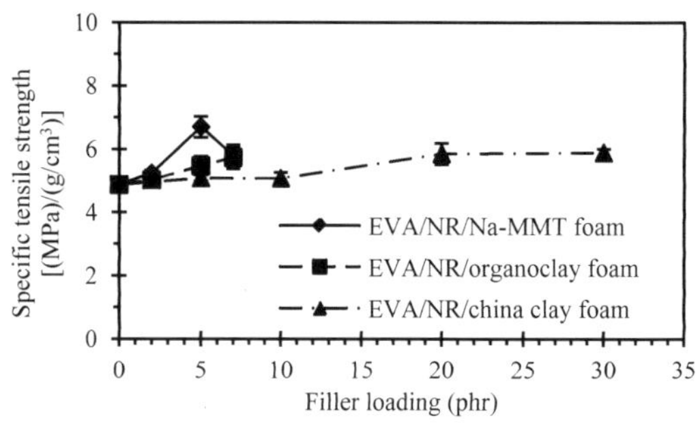

Figure 8 Specific tensile strength of different EVA/NR foams with various filler loadings.

It is seen that the specific tensile stress at 100% elongation (Figure 7) and specific tensile strength (Figure 8) increased with filler content. The specific tensile strength at 100% elongation increased because of reinforcing effect by the presence of filler. The best reinforcement was obtained when the organoclay was used as a filler. The addition of filler also increased specific tensile strength of EVA/NR foam, depending on grade and content of filler. The nanofiller provided higher specific tensile strength than china clay. The increase in specific tensile strength with more filler loading was due to smaller cell size[5] (Figure 5). However, an increase of Na-MMT filler over 5 phr decreased the specific tensile strength. The compression set of the filled foams is shown in Figure 9.

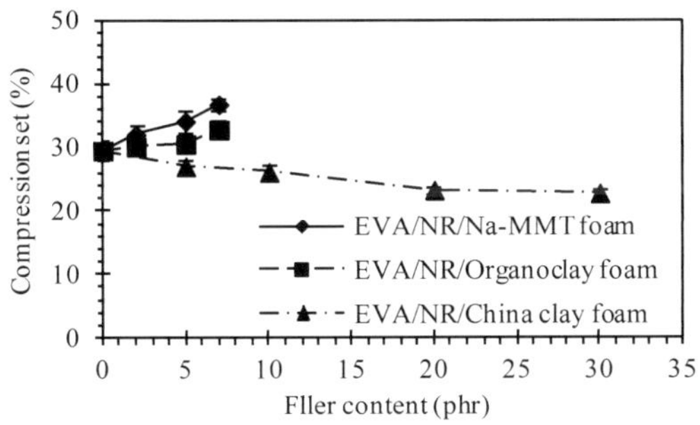

Figure 9 Compression set of different EVA/NR foams with various filler loadings.

Increasing nanofiller content increased %compression set, indicating lower elastic recovery of the nanofiller filled EVA/NR foams. However, the compression set was found to decrease with increasing china clay content. This observation can be explained by the density results shown in Figure 6. The rubber foam with higher density had more rubber material present in the cell wall, which could better resist the deformation of cell under compression during the test. Thus, the elastic recovery of the filled foams decreased with increasing nanofiller content, while it increased with increasing china clay content.

Compressive strength of different filled EVA/NR foam at various filler loading is shown in Figure 10.

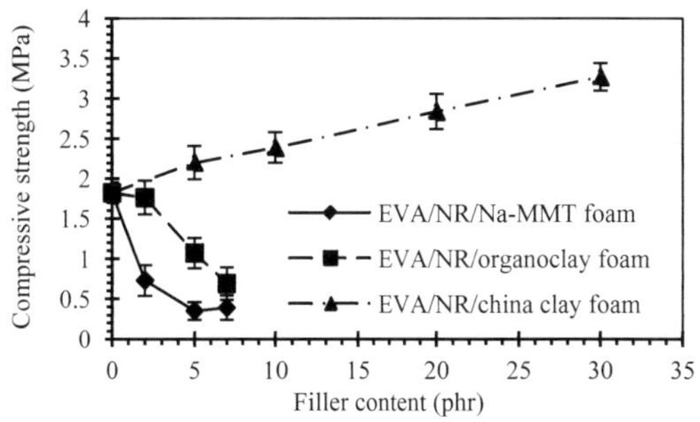

Figure 10 Compressive strength of different EVA/NR foams with various filler loadings.

The compressive strength of the nanocomposite foams decreased with increasing filler content, whereas the compressive strength of china clay containing foams increased with filler content. By increasing nanofiller content, the density decreased (Figure 6). According to this effect, the rubber in cell wall was less, and the cells easily bended and collapsed under compressive loading. On the other hand, the foams with china clay had thicker cell wall and more rubber in the cell wall as more clay was added. As a result, the china clay filled foam could withstand higher compressive loading.

Conclusions

Effects of different fillers (nanofiller: Na-MMT and organoclay and conventional filler: china clay) as well as filler content on the properties of EVA/NR foams were studied. It was found that the nanofiller was more effective in improving cell morphology and specific tensile strength of the rubber foams. The results suggested that the optimum nanofiller content to produce low density EVA/NR foam with acceptable strength was 5 phr.

Acknowledgements
The authors gratefully acknowledge the financial support of Prince of Songkla University (grant number SAT530106S).

References

1. M. S. Kim; C. C. Park; S. R. Chowdhury; G. H. Kim *J. Appl. Polym. Sci.* 2004, 94(5), 2212-2216.
2. W. K. Park; R. S. Chowdhury; C. C. Park; H. G. Kim *J. Appl. Polym. Sci.* 2007, 104(6), 3879-3885.
3. L. J. Lee; C. Zeng; Z. X. Cao; X. Han; J. Shen; G. Xu *Compos. Sci.* Technol. 2005, 65, 2344-2363.
4. P. H. Nam; P. Maiti; M. Okamoto; T. Kotaka *Polym. Eng. Sci.* 2002, 42, 1907-1918.
5. M. Shimbo; I. Higashitani; Y. J. Miyano; *J. Cell. Plastics.* 2007, 43, 157-167.

O-07-045

Compounding and Processing of Polymer / MWCNT Composites

X. Yang, H-J. Park, C. Schneider, C. Kujat, I. Hennig, H. Steininger, M. Weber
BASF SE, 67056 Ludwigshafen, Germany

Abstract
Carbon Nanotubes (CNTs) are in the focus of numerous studies due to their intrinsically extraordinary high thermal and electrical conductivity as well as their outstanding mechanical properties. These properties are of great interest for a wide range of applications, in particular if they could be transferred to polymer materials at macroscopic level. However, the composite properties achieved up to date are still far below the theoretically expected values related to pure CNTs [1].

In order to achieve the desired properties, appropriate technologies for processing of CNT composites have to be developed and transferred to industrial scale. For this work, specific know-how collected by BASF for spherical nano-particles could be adapted for the development of technologies for CNT compounds.

In this paper we will present the results from three different approaches for the development of polymer / Multi-Wall Carbon Nanotubes (MWCNTs) composites:

- the preparation of master batches and subsequent dilution
- the direct compounding of the MWCNTs
- the in-situ polymerization with MWCNTs

The properties of the obtained composites are measured and compared to different benchmark materials, e.g. commercial master batches and compounds containing conductive carbon blacks or carbon fibers.

Keywords: Carbon Nanotubes, composite, processing, industrial scale

Introduction

Carbon Nanotubes (CNTs) are in the focus of numerous studies due to their intrinsically extraordinary high thermal and electrical conductivity as well as their outstanding mechanical properties. These properties are of great interest for a wide range of applications, in particular if they could be transferred to polymer materials at macroscopic level. However, the composite properties achieved up to date are still far below the theoretically expected values related to pure CNTs [1].

In order to achieve the desired properties, appropriate technologies for processing of CNT composites have to be developed and transferred to industrial scale. For this work, specific know-how collected by BASF for spherical nano-particles could be adapted for the development of technologies for CNT compounds.

In this paper we will present the results from three different approaches for the development of polymer / Multi-Wall Carbon Nanotubes (MWCNTs) composites:

- the preparation of master batches via kneader and/or twin-screw co-rotating extruder and subsequent dilution
- the direct compounding of the MWCNTs in a polymer matrix via twin-screw co-rotating extruder
- the in-situ polymerization with MWCNTs.

CNTs form two different sources (Baytubes and Nanocyl) are applied to several BASF polymer matrixes: Polyamides (PA), Polybutylene terephthalate (PBT), and Polyethersulfone (PESU) for the composite preparation.

Optical properties, electrical conductivity as well as mechanical properties of the obtained composites are measured and compared to different benchmark materials, e.g. commercial master batches and compounds containing conductive carbon black or carbon fibers.

Moreover, a safe handling procedure for compounding of CNTs on industrial scale was developed and validated as crucial part of this study.

Experimental

Materials
Baytubes C150HP from the Bayer Material Science, Germany, and Nanocyl 7000 from the Nanocyl SA, Belgium are chosen as the investigated CNTs. Both

direct compounding of these CNTs as well as compounding with the commercially available master batches were performed in this study.

Printex XE2B from the Evonik Industries, Germany are used as reference carbon black. Fortafil 243 from the Toho Tenax Europe GmbH are applied as the reference carbon fibre.

Concerning the polymer matrix, PA6 (Ultramid® B27E), PA66 (Ultamid® A3W), PBT (Ultradur® B4500), PESU (Ultrason® E1010) from BASF SE are investigated.

Master batch preparation and their dilution
The master batches were prepared both with kneader and twin-screw extruder up to 15% CNT content. For the further investigations, the master batches were diluted to desired CNT concentration by using suitable extruder under proper processing conditions.

Direct compounding of CNT powders
The twin-screw co-rotating extruder was also used for the direct compounding of composites. For this purpose we developed a completely closed feeding system, as shown in Fig. 1. With this technology we are able to safely handle direct compounding of CNTs on industrial scale. This closed feeding system ensures zero contamination of CNT during the compounding work as proven by highly sensitive monitoring.

Figure 1 – The closed CNT feeding system

In the course of the composites compounding, the processing parameters, i.e. temperature, throughput, screw speed, and etc. were systematically varied in order to evaluate the influence of process parameter on the composite property. The composites were extruded directly to the final CNT content, e.g. 0.5wt% to 7.5wt%, which is relevant for the desired properties.

In-situ polymerization with MWNTs
A PA/CNT composite was also prepared by using in-situ polymerization. Carbon Nanotube / Caprolactam / water dispersions were prepared by pearl mill at room temperature. The dispersions were then polymerized in an autoclave at defined temperature and pressure and for specified time. Post-condensation of the reaction was carried out after while under defined time, temperature, and pressure. After the polymerization, the bulk sample was mechanically grinded roughly. Unreacted carprolactam from the grinded sample was extracted in water and the grinded sample was dried at 80 °C in vacuum condition.

Optical properties
Optical microscopy was mainly used to characterize the CNT dispersion in different composites. Thin cuts (2 to 5 μm) of the composite samples were prepared before observation. SEM was applied for determination of the CNT morphology.

Electrical conductivity
The specific volume resistance of the samples was determined by using a 4-point measurement as presented with Figure 2. To ensure a good contact between the electrodes and the sample, four silver-electrodes are directly painted on the sample using a conductive silver paste (Leitsilber 200 from Hans Wolbring GmbH). The measurement was performed either on pressed or on injection molded samples.

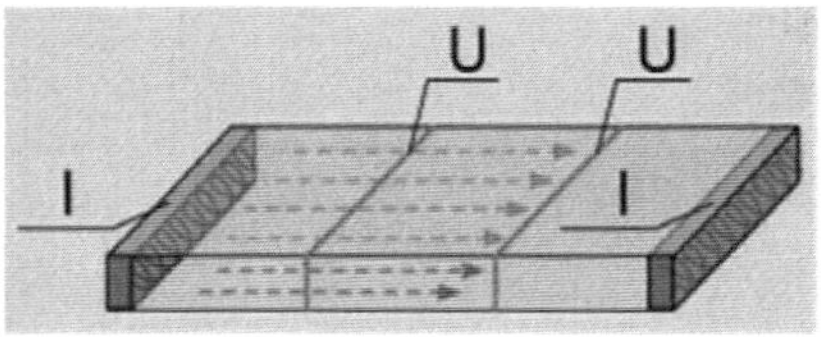

Figure 2 – Schematic of the 4-point conductivity measurement

Mechanical properties
Mechanical characterizations, e.g. tensile test, notched and un-notched Charpy of different composites samples were carried out under relavent test-standard conditions.

Results and Discussion

Figure 3 shows the results concerning the specific electrical volume resistance of different polymer matrixes containing varying CNT concentrations. All the samples are prepared with Nanocyl master batches diluted in a twin-screw extruder from 1wt% to 7.5wt% CNT content.

It is clear to see from the figure 3 that electrical percolation region of different polymer/CNT composites strongly depend on the polymer matrix. Polarity, melt-viscosity, crystallinity of the polymer, and the polymer-CNT interaction are all the important factors which influence the composites electrical property intensively.

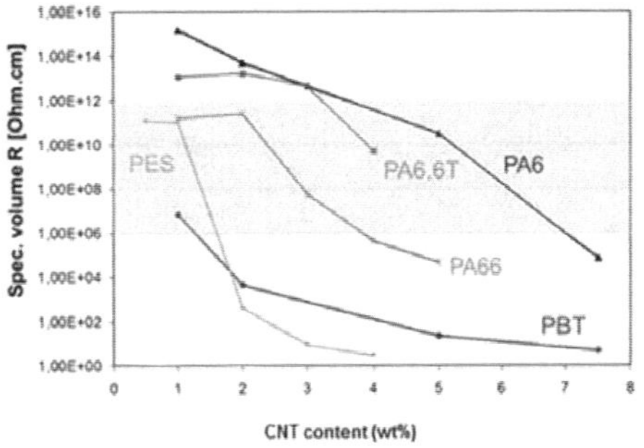

Figure 3 – The overview of specific volume resistance of different polymer matrixes containing varying CNT concentrations

Figure 4 illustrates the mechanical property, i.e., elongation at break (H_B) of different PA6 composites resulting from varying screw design. All the samples are produced with same extruder and different screw design. The results shown in blue are relating to screw A and red are corresponding to screw B. The screw B was optimized for better dispersion quality of C-particles. The first four groups of columns (dashed) are showing the H_B values of CNT - master batch dilution. The last four groups of columns are demonstrating the H_B values of direct compounding of carbon black and carbon fibre composites with same filler loading (2wt% and 5wt% respectively).

It can be concluded from figure 4 that the screw design doesn't change the H_B values of master batch diluted samples due to the pre-defined dispersion quality of the master batches. This is also confirmed by literature [2]. However, an optimized screw design can improves the dispersion quality in case of direct powder compounding. The higher H_B values of the red solid columns indicate this fact especially in the lower filler content region, where the composites are highly ductile.

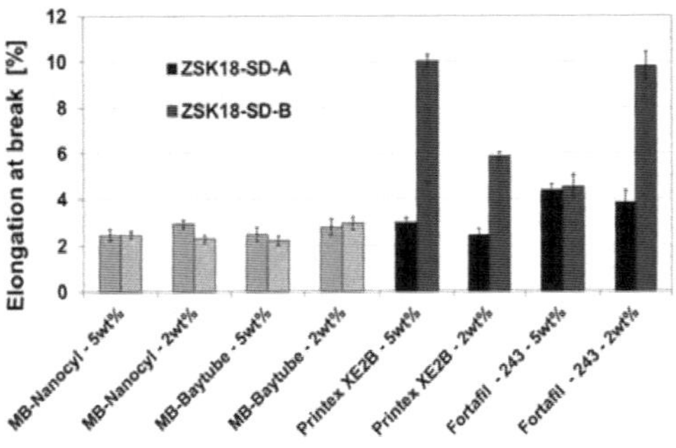

Figure 4 – The influence of screw design on the elongation at break of a PA6/CNT composite

Figure 5 displays the influence of processing parameters on the electrical property of PA6/CNT composites containing 4wt% either Nanocyl (blue columns) or Baytubes (red columns). The samples are compounded with same extruder and same screw design. The process parameters (screw speed and throughput) were varied systematically as shown in figure 5. During the compounding process, the shearing force within the extruder is mainly changed by changing screw speed and the residence time is mainly correlated to the throughput [3].

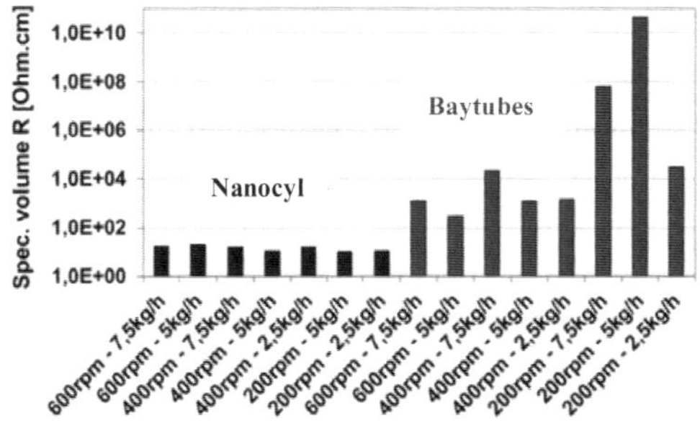

Figure 5 – The influence of processing parameters on specific volume resistance of PA6/CNT (4wt%) composites

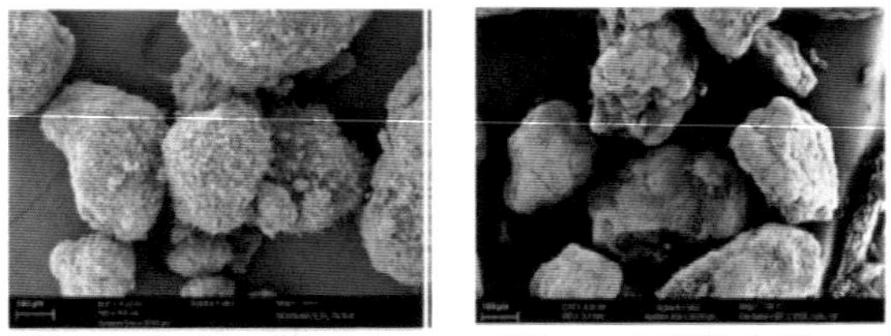

Figure 6 – The SEM pictures of Nanocyl 7000 (left) and Baytubes C150HP (right)

A clear trend can be observed from the figure 5: the specific resistance of Nanocyl composites is almost unaffected by process parameter variations, while Baytubes composites show significantly different values as a function of the process parameters. This phenomenon may be due to the difference in original CNT morphology. It can be seen from figure 6 that the Nanocyl CNT agglomerates are looking loose and Baytubes CNT existing in a dense state. These dense agglomerates require higher shearing force during extrusion for the disaggregating.

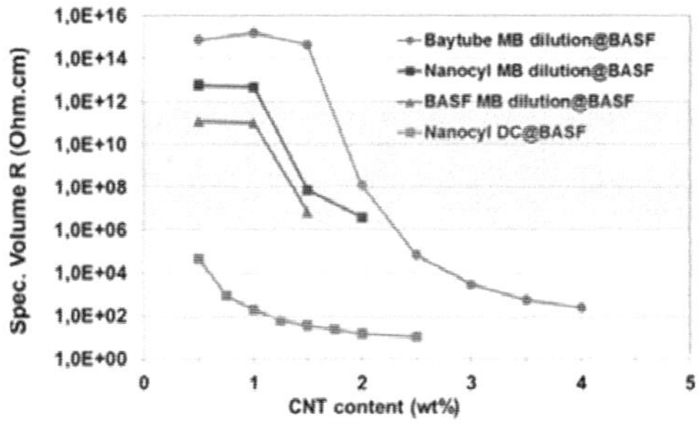

Figure 7 – The specific volume resistance of PESU/CNT composites prepared with different processing methods

The results regarding concerning electric properties of PSEU (Ultrason® E1010) / CNT composites are displayed in figure 7. The grey, red, and purple curves refer to the results for commercial Baytubes master batch, Nanocyl master batch and BASF prepared master batch (with Nanocyl CNT) diluted composites, respectively.

The blue curve indicats the results of CNT powder directly compounded into Ultrason® E1010 with various concentrations. The latter one shows obviously lower specific volume resistance which indicates the advantage of direct compounding process for improvement of electrical conductivity. A conductive Ultrason® E1010 / CNT composite can be already achieved at only 0.5wt% CNT loading by using this advantagious method.

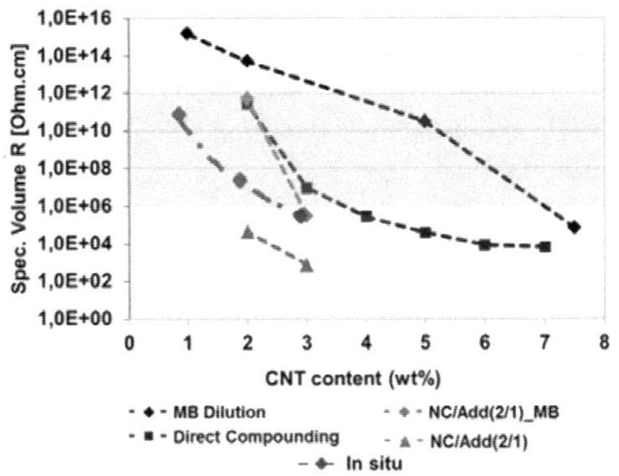

Figure 8 – The influence of processing methods on specific volume resistance of PA6/CNT composites

The specific volume resistances of PA6/CNT composites prepared with different methods with varying CNT content is presented in figure 8. All the composites are produced with Nanocyl 7000 CNT. The similar trend as for PESU/CNT composites can be observed for PA6/CNT composites, i.e. the direct compounding method (blue curve) results in better electrical conductivity in comparison to samples made by master batch dilution method (black curve) at same CNT loading.

A further improvement can be achieved through adding a processing additive. As shown in figure 8, the black curve can be shifted to the grey one and the blue curve can be shifted to the green one only by adding additive at half CNT content, i.e. CNT:Additive=2:1. In both cases the specific volume resistance values of the composites with same CNT-loading are reduced dramatically. By using the combination of direct compounding and adding processing additive (green curve), the conductive of PA6/CNT composite can be obtained by 2wt% CNT loading. This performance is even better than in-situ polymerization results (purple curve).

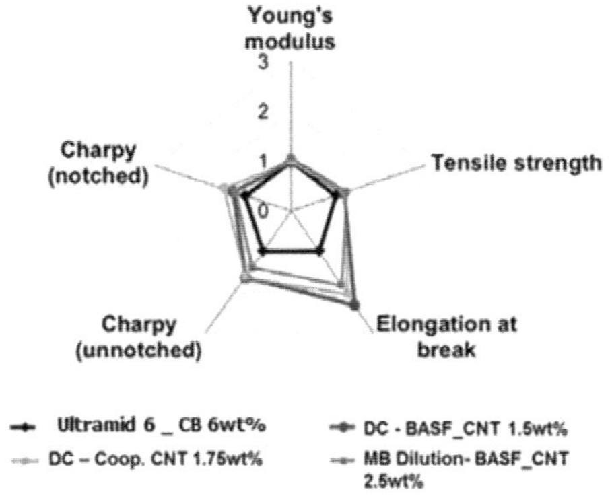

Figure 9 – The comparison of mechanical properties of PA6 /CNT and PA6/Carbon Black composites

Last but not least, the mechanical properties of different conductive PA composites were investigated. Five different mechanical values were measured for each sample as shown in Figure 9. All samples have similar electrical conductivity, i.e. specific volume resistance smaller than 10E06 Ohm*cm. A standard Ultramid® grade containing 6% of carbon black and certain amount of additives was chosen as reference (black circle). In order to achieve the similar electrical conductivity level, different CNT (only Nanocyl 7000 was used) loading was required by different processing methods, i.e. 2.5wt% by master batch dilution, 1.75wt% by direct compounding at external cooperation partner, only 1.5% by direct compounding at BASF. The same additives which used for carbon black composite were added for CNT composites. The largest elongation at break value can be obtained by direct compounding method using BASF own facilities (special designed screw combination and proper processing condition during extrusion). The overall mechanical properties of CNT composites are better than carbon black composite.

Conclusions

To sum up, this paper presents the improvement of electrical and mechanical properties of different polymer/CNT composites. Three approaches were applied for the composite preparation during this study. The direct compounding method in combination with optimized screw design and proper processing parameters turned out to be the best method. CNT composites show superior performance profile in comparison to reference material which containing carbon black and carbon fiber.

References

1. M. R. Loos; I. Manas-Zloczower, in PPS 27 Proceedings, KN-A-338, 2011, Marrakech, Morocco.
2. T. Villmow; P. Pötschke, and etc. *Polymer,* 2008, 49, 3500.
3. K. Kohlgrüber; *Co-Rotating Twin-Screw Extruders,* Carl Hanser Publisher, 2008.

O-07-048

PROPERTIES OF LAYERED DOUBLE HYDROXIDE MICRO- AND NANOCOMPOSITES

L. Moyo[1], W.W. Focke[1], D. Heidenreich[2], F.J.W.J. Labuschagne[1]* and H.-J. Radusch[2]

[1] *Institute of Applied Materials, Department of Chemical Engineering, University of Pretoria, Private Bag X20, Hatfield, 0028, Pretoria, South Africa – fj.labuschagne@up.ac.za*

[2] *Centre of Engineering Science, Martin Luther University Halle-Wittenberg, D-06099 Halle, Germany – hans-joachim.radusch@iw.uni-halle.de*

Abstract - Carbonate and stearate intercalated layered double hydroxides were used as fillers to prepare polymer micro- and nanocomposites respectively. The stearate modified starting material was bilayer-intercalated clay. During melt compounding excess stearates were released and the clay reverted to a monolayer-intercalated form. The exuded stearate acted as a lubricant lowering the melt viscosity of poly(ethylene-co-vinyl acetate) and linear low density polyethylene matrices. Strong hydrogen bond interactions between the chains of poly(ethylene-co-vinyl alcohol) and the clay platelet surfaces overwhelmed the lubrication effect and caused an increase in the melt viscosity of this matrix. The notched Charpy impact strength of this composite was almost double that of the neat polymer. It appears that this can be attributed to the ability of the highly dispersed and randomly oriented nano-sized clay platelets to promote extensive internal micro-cavitation during impact loading. The creation of a large internal surface area provided the requisite energy dissipation mechanism.

Keywords: Layered compounds, Composites, Intercalation reactions, Electron microscopy, Mechanical properties

Introduction

The dispersion of layered particles in polymeric matrices can enhance material properties and lead to improved gas barrier properties, better mechanical properties [1, 2], increased flame retardancy [3, 4, 5, 6, 7], UV stabilization [8] or enhanced photodegradation [10], etc. Layered double hydroxides (LDHs) is a group of layered materials with the general formula $[M^{2+}_{(1-x)}M^{3+}_{x}(OH)_2]_x(A^{n-}_{(x/n)})\cdot mH_2O$ where A represents a suitable anion, M^{2+} = Mg, Zn, Fe, Co, Ni, Cu and M^{3+} = Al, Fe, Cr, Mn, etc. [11]. They have a brucite-like structure of stacked sheets in which some trivalent cations isomorphously replaced some of the divalent ones. The net positive charge is counterbalanced by exchangeable anions located in the interlayer. Exchanging the interlayer anions with others modifies the interlayer properties-[12].

Three organo-clay nanocomposites microstructures are possible, i.e. phase separated; intercalated and exfoliated. For thorough clay sheet dispersion, highly favorable polymer-surface interactions are essential [13, 14]. Exfoliation of surfactant intercalated LDHs has been noted [14, 15, 16]. A greater degree of exfoliation and better dispersion of LDHs was obtained in more polar matrices.

The interlayer anions in surfactant intercalated LDHs showed thermotropic behavior [12, 17, 18]. LDH-fatty acids intercalated above the AEC levels apparently melt below polymer processing temperatures (120 °C) [12]. Droplet formation was observed in hot stage optical microscopy giving a façade of a completely molten LDH-stearate [12]. This can be attributed to the excess stearate acid exuding and forming a liquid droplet that surrounds the parent LDH-stearate platelets. The exudation of stearic acid has implications for the processing behavior and the dispersion of the clay in the polymer. It was therefore of interest to compare melt blending of LDH-CO3 and LDH-stearate in polymer matrices of varying polarity.

Experimental

Materials

The layered double hydroxide carbonate (LDH-CO3) Hydrotalcite HT-5 was supplied by Nkomazi Chemicals, South Africa. The 1-hexene random copolymer-based linear low density polyethylene (LLDPE) was Sasol Polymers rotomolding grade HR411. Poly(ethylene-co-vinyl acetate) (EVA) grade EV101 containing 18 mol% vinyl acetate was supplied by APC. Poly(ethylene-co-vinyl alcohol), (EVAL) grade T101B containing 68 mol% vinyl alcohol was obtained from Kuraray, Belgium. Technical grade stearic acid was supplied by Croda Chemicals, UK.

Preparation of LDH-stearate

The stearate intercalated layered double hydroxide (LDH-stearate) was prepared according to the method reported by Nhlapo et al. [12]. A typical intercalation procedure was carried out. The solids were separated by centrifugation, washed once with water and three times with ethanol and finally once with acetone and dried at room temperature.

Preparation of composites

Polymer composite containing 10 wt.% of either the pristine or the modified LDH were prepared by melt compounding. A TX28P 28 mm co-rotating twin screw extruder with an L/D ratio of 18 was used. The extruder was operated at 170 rpm and 100 °C at the feed to 220 °C at the die. The extruded strands were water cooled, granulated and dried overnight at 60 °C. Type IV ASTM dumbbells were injection molded using an Engel 3040 injection-molding machine with a clamping force 350 kN, with a temperature profiles of 200/210/220/220 °C for the LLDPE and EVAL composites and 140/150/160/170 °C for the EVA composites respectively.

Characterization

Phase identification was carried out using X-ray diffraction (XRD) analysis on a PANalytical X-pert Pro powder diffractometer. The X'Pert High Score Plus software was used for data manipulation.

Composite samples were cryogenically fractured in liquid nitrogen. They were coated times with gold using a Polaron Equipment Ltd. E5200 automatic sputter coater. Imaging of freeze-fractured surfaces was carried out in a JEOL 5400 scanning electron microscope (SEM). The side-view of impact test specimen was imaged on a Zeiss Stereo, Discovery V20 optical microscope. The polymer/clay composites were studied using a JEOL 2100F transmission electron microscopy (TEM) at an acceleration voltage of 200 kV. The samples were sliced with a diamond knife at -80 °C using a Leica-Reichert Ultracut R with EMFCS cryo-attachment to a nominal thickness of about 90 ± 10 nm. Each section was mounted on a 300 mesh copper/palladium grid.

Melt flow viscosity was determined with a Göttfert High Pressure Capillary Rheograph 2000 rheometer. The capillary die had a 180° entrance angle, a diameter of 1 mm and a length of 30 mm. Measurements were done at 190 °C with shear rates ranging from 1 s^{-1} to 5000 s^{-1}.

Tensile testing was carried out on a Lloyds Instruments LRX Plus machine according to ASTM D 638 using Type IV dumbbells. Five specimens were tested for each compound. Charpy impact testing was carried out on a Zwick Impact Tester using the 0.5 J hammer. Tensile impact tests were carried out according to ASTM D1822-06 on Type S and L test specimens.

Results and Discussion

The clay dispersion in the polymer composite and degree of clay layer separation was identified via X-ray diffraction (Fig. 1). The 2θ values of 13.5° and 27.2° typical for LDH-CO3 indicate a d-spacing (dL) of 0.76 nm. These reflections were also observed in the LDH-stearate indication their presence as an impurity phase. They were also noted in all the LDH-CO3 and LDH-stearate polymer composites indicating that this clay phase was either retained during melt compounding or more of it generated by decomposition of the LDH-stearate. The diffractogram for LDH-stearate showed three basal reflections positioned at 2.0°, 4.1° and 6.1° corresponding to a d-spacing of 5.1 nm and consistent with bilayer intercalation [12, 19, 20]. These reflections were absent in the diffractograms of the LDH-stearate composites. The first reflection observed at high 2θ values indicated a reduction in the d-spacing of the LDH-stearate in EVA and LLDPE composites (Fig. 1b). The interlayer spacing of 3.1 nm determined for the LLDPE composite is consistent with monolayer-intercalated stearic acid [20, 21]. This collapse in the d-spacing implies that the neutral stearic acid molecules initially present in the interlayer beyond the anionic exchange capacity (AEC) were removed during the melt compounding process. A collapse was also observed in the d-spacing value for the clay in the EVA composite. However, here the d-spacing was lower and this could be due to a less ordered arrangement of the intercalated chains and/or the loss of the interlayer water. Only vestiges of the LDH-stearate reflections were seen in the EVAL composite.

Figure 1

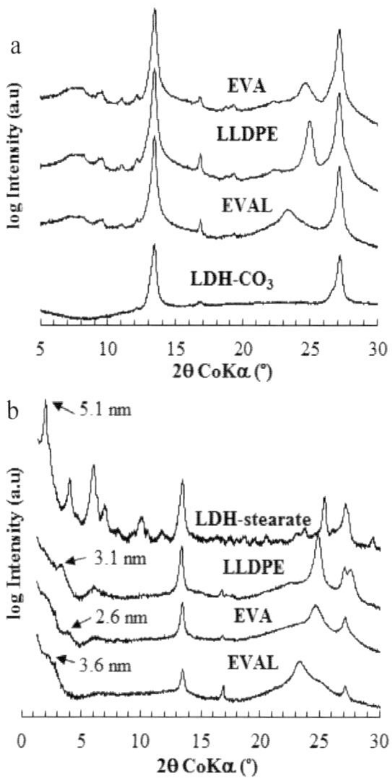

XRD diffractograms (WAXS) of (a) the LDH-CO3 filler and its polymer composites and (b) the LDH-stearate and its polymer composites

The LDH-CO3 composites featured distinct particles in the sub-micrometer range but also some particle agglomerates (Fig. 2). In the LDH-stearate composites the largest particles were much smaller. They appeared as planar tactoids with a length up to about 500 nm but with a thickness less than 100 nm (Fig. 2). This considerable reduction in dimensions has two possible origins. LDH platelets are weaker and less rigid than smectite clays and hence more prone to fracture under the high shear action [22]. Apart from such breakage, partial delamination may have occurred during the high-shear compounding process. The EVA and

LLDPE matrices of the LDH-stearate composites appeared to contain very few tactoids.

Fig. 3 shows SEM images that give a view of the general morphology and texture of freeze-fractured surfaces.

Figure 2

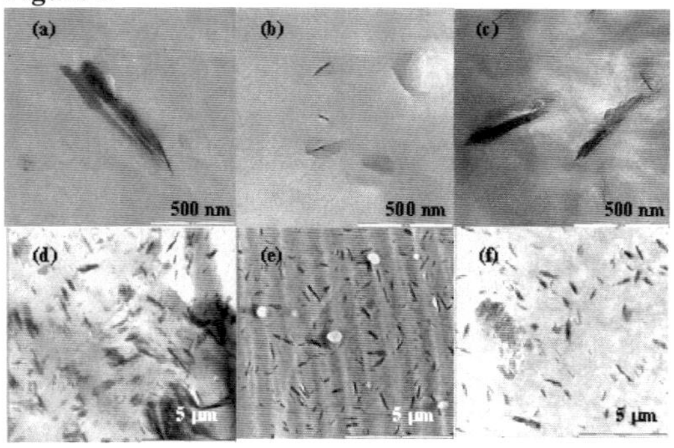

TEM images of the composites of (a) EVA/LDH-stearate; (b) EVAL/LDH-stearate; (c) LLDPE/LDH-stearate; (d) EVA/LDH-CO3; (e) EVAL/LDH-CO3, and (f) LLDPE/LDH-CO3

Figure 3

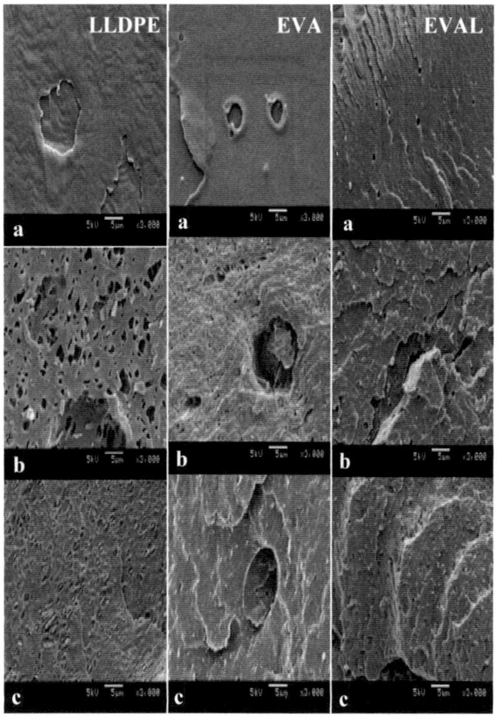

Freeze-fractured surfaces of (a) Neat polymer; (b) polymer/LDH-stearate and (c) polymer/LDH-CO3 composites (at 3000x magnification)

Poor interfacial adhesion is evident in both the LLDPE composites (Fig. 3). This is also the case for the EVA/LDH-stearate composite. In all the other samples there was relatively good adhesion between the matrix and the filler shown by the absence of cavities previously occupied by particles. Some spherical cavities were seen in all the EVAL samples including the neat polymer. However, they are attributed to volatilization of residual water during the molding process.

Fig. 4 reports the melt viscosity results for the composites obtained at 190 °C. All the samples showed strong shear-thinning behavior. The EVA and EVAL polymer/LDH-CO3 composites featured higher apparent viscosity values than the parent polymers.

Figure 4

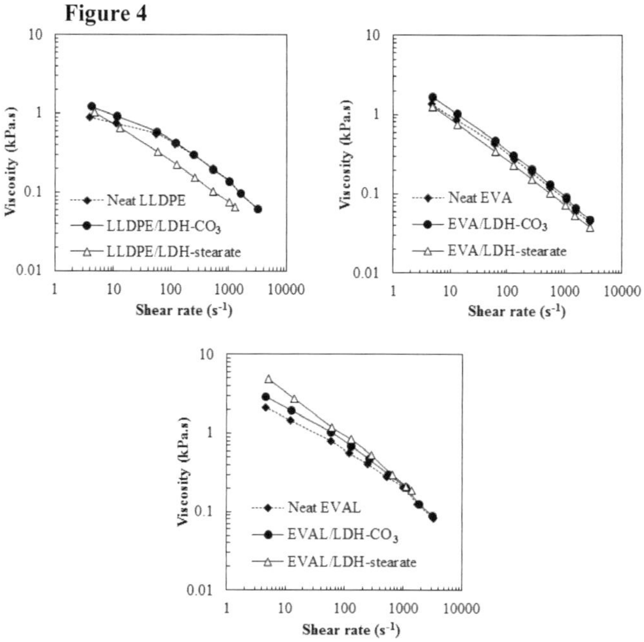

Effect of LDH incorporation on the melt viscosity of the polymer/LDH composites at 190 °C

The melt viscosity of the LLDPE/LDH-CO3 composite was marginally higher than that of the neat polymer. This is attributed to weak interactions between the polar filler surface and the nonpolar matrix. The apparent melt viscosities of the EVA and the LLDPE polymer/LDH-stearate composites were lower than that of the neat polymers. This is attributed to the lubricating effect of the exuded stearic acid.

In the EVAL composites, use of the LDH-stearate filler resulted in the highest increase in melt viscosity. This means that the interaction of the filler with the polymer chains must have overwhelmed the lubricating effect of the free stearic acid present. The surfaces of the LDH-stearate filler are at least partially covered with the chains of the electrostatically attached stearate anions. So it is expected that the interaction of these clay surface with the EVAL chains would weaker than the interaction with the uncoated surfaces of the LDH-CO3 particles. However, the melt viscosity of the LDH-CO3 composite was lower despite its higher surface polarity. Furthermore, according to the TG results, the inorganic content of the LDH-stearate is only 20 wt.%. In aggregate, these observations suggest that the clay surface area available for interaction with the polymer chains must have been much higher for LDH-stearate.

Mechanical properties are reported in Table 1. All the filled samples, except the EVAL/LDH-stearate composite featured higher tensile moduli than the neat polymers. Both LDHs fillers had a reinforcing effect on EVA and LLDPE as both the yield strength and the modulus increased. Better elongations were obtained in the EVA composites but a decrease was observed for the EVAL and LLDPE composites.

Table 1

Mechanical properties of LDH/polymer composites

Polymer Composite	Young's modulus	Yield strength	Strain to break	Charpy impact strength	Tensile impact strength
	MPa	MPa	%	kJ/m^2	kJ/m^2
EVA					
Neat	27 ±2	10 ±0.4	277 ±30	NB	3.8 ±0.14
+LDH-St	40 ±5	12 ±0.1	422 ±23	NB	4.3 ±0.19
+LDH-CO3	37 ±2	13 ±0.1	424 ±60	NB	5.2 ±0.06
LLDPE					
Neat	171 ±14	16 ±0.1	509 ±29	22 ±2	1.7 ±0.08
+LDH-St	196 ±23	17 ±0.4	495 ±43	12 ±1	1.2 ±0.07
+LDH-CO3	213 ±12	17 ±0.2	460 ±21	11 ±1	0.8 ±0.05
EVAL					
Neat	843 ±16	64 ±2	41 ±21	4.9 ±1	2 ±0.28
+LDH-St	718 ±40	59 ±1	18 ±5	7.9 ±3	0.8 ±0.33
+LDH-CO3	1025 ±18	76 ±0.4	37 ±2	4.6 ±1	0.7 ±0.29

NB = No break

Polymer toughness is governed by parameters such as the degree of particle dispersion within the polymer matrix, filler mobility, delamination and intrinsic changes to polymer properties promoted by the filler [26]. The EVA composites did not fracture in the notched Charpy tests because of the rubbery nature of matrix. The tests were basically carried out at room temperature well above the T_g of this polymer matrix. Tensile impact properties of the filled EVA composites were better than those of the neat polymer. However, the opposite was true for the other polymer matrices. In the tensile impact tests the LDH-CO3 filler gave better results than the LDH-stearate. See Table 1. Considering Fig. 3, this can be attributed to better interfacial adhesion between the polar filler and the polar EVA matrix. Strong interactions with the filler are equivalent to temporary crosslinks that dissipate energy when ruptured. Gersappe [24] proposed a theory that may explain the improved tensile impact observed in the EVA/LDH-stearate composite, relative to the neat polymer, despite minimal interaction between the polymer and filler. Since the clay particles are relatively free to move and rotate inside the cavities, they can align themselves in the direction of the applied stress. During the matrix deformation process the filler is squeezed tight by the cavity walls as it elongates and narrows. The resulting friction forces generate a region of enhanced strength that retards the growth of the cavity and thus delays polymer failure.

With the exception of the EVAL/LDH-stearate composite, all the other composite samples had worse notched Charpy impact properties than the parent polymers. Development of strong energy dissipation mechanisms are a prerequisite for good impact properties. The poor interface adhesion between both fillers and the polyethylene matrix is evident in Fig. 3. This means that the incorporation of the fillers led to the creation of internal flaws. These acted as stress concentrators that lead to premature mechanical failure. The LLDPE/LDH-stearate composite had slightly better impact strength properties than the LLDPE/LDH-CO3. This can be attributed to the slightly better compatibility with the polymer matrix contributed by the fatty acid coating on the filler particles (Fig. 3c). However, it is clear from the XRD diffractogram (Fig. 1) that the presence of this filler also affected the morphology of the parent polymer. This is confirmed by the broadening and shift in the position of main reflection attributable to the polymer matrix. This change in the morphology of the polymer matrix could also have affected the impact properties.

Unexpectedly, the notched Charpy impact strength of EVAL increased from 4.9 kJ m^{-2} to 7.9 kJ m^{-2} when 10 wt.% LDH-stearate was added (Table 1). Top-view SEM imaging of the fracture surfaces were inconclusive. They did not reveal a mechanism that could explain the improved impact behavior. Fig. 5 shows side-view images of the fracture surfaces of Charpy impact specimens obtained for the EVAL composites using an optical microscope. The neat EVAL sample and EVAL/LDH-CO3 composite sample showed clean fractures. In contrast, the EVAL/LDH-stearate sample showed an extensive stress-whitened damage region that extended deep into the sample. It seems that the presence of the LDH-stearate particles caused repeated branching of the crack fronts. This dissipates mechanical energy by creating numerous internal cracks with a very large total surface area. It seems that the clay particles transmitted incoming crack fronts as multiple crack fronts travelling in different directions. This ability of the LDH-stearate might derive from their sheet-like nature and random orientation in the glassy matrix. It is noteworthy that the stress whitened damage region was located in a narrow region towards the back of the sample. Extensive stress-whitening is usually also an indication of crazing and/or micro-fibril formation. The white appearance of the crazed region is due to light-scattering.

Conclusions

TEM results showed that microcomposites and nanocomposites were obtained using LDH-CO3 and LDH-stearate polymer melt compounding. The excess stearic acid in the clay was released and the clay reverted to a monolayer-intercalated form during melt compounding of the LDH-stearate composites as indicated by XRD data and the reduction in melt viscosity observed for the LLDPE and EVA polymer/LDH-stearate composites. The EVAL/LDH-stearate nanocomposite featured the highest melt viscosity. This suggests that the clay surface area available for interaction with the EVAL polymer chains must have been very high in the EVAL/LDH-stearate composite in order to overwhelm the lubrication effect.

Figure 5

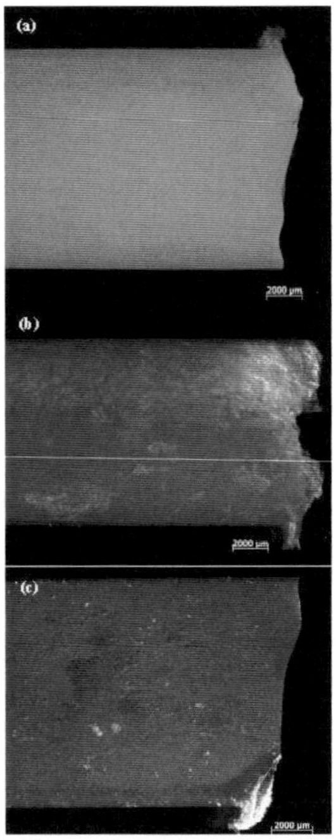

Side view of Charpy impact test specimens of EVAL and its LDH composites. (a) Neat polymer; (b) LDH-stearate, and (c) LDH-CO3

Both LDHs fillers had a reinforcing effect on EVA and LLDPE. Better elongations were obtained in the EVA composites but a decrease was observed for the EVAL and LLDPE composites. Both fillers improved the tensile impact strength of EVA. Unexpectedly, the notched Charpy impact strength of EVAL increased from 4.9 kJ m^{-2} to 7.9 kJ m^{-2} when 10 wt.% LDH-stearate was added. This is attributed to the ability of the highly and random dispersed clay nano-particles to promote extensive internal micro-cavitation during impact loading, creating large internal surface area which provides the requisite energy dissipation mechanism.

Acknowledgements
The National Research Foundation (NRF) through the Institutional Research Development Programme (IRDP), the .NRF and the Bundesministerium für Forschung (BMBF) for financial support.

References

1. H.-B. Hseuh, C.-Y. Chen, Polymer. 44 (2003) 5275-5283
2. G.-A. Wang, C.-C. Wang, C.-Y. Chen, J. Polym. Degrad. Stab. 91 (2006) 2443-2450
3. W. Chen, B. Qu, Chem. Mater. 15 (2003) 3208-3213
4. W. Chen, B. Qu, J. Mater. Chem. 14 (2004) 1705-1710
5. M. Zammarano, M. Franceschi, S. Bellayer, J.W. Gilman, S. Meriani, Polym. J. 46 (2005) 9314-9328
6. F. R. Costa, M. Abdel-Goad, U. Wagenknecht, G. Heinrich, Polymer, 46 (2005) 4447-4453
7. M. Zubitur, M.A. Gomez, M. Cortazar, J. Polym. Degrad. Stab. 94 (2009) 804- 809
8. S. Bocchini, S. Morlat-Therias, J. L. Gardette, G. Camino. Eur. Polym. J. 44 (2008) 3473-3481
9. B. Magagula, N. Nhlapo, W. W. Focke, J. Polym. Degrad. Stab. 94 (2009) 947-954
10. A. Manhique, W.W. Focke, Mol. Cryst. Liq. Cryst. 554 (2012)
11. W.T. Reichle, Chemtech, (1986) 58-63
12. N. Nhlapo, T. Motumi, E. Landman, S. M. C. Verryn, W. W. Focke, J. Mater. Sci. 43-3 (2008) 1033-1043
13. R.A. Vaia, E. P. Giannelis, Macromolecules. 30 (1997) 7990-7999
14. H. Fischer, Mat. Sci. Eng. C. 23 (2003) 763-772
15. F. Leroux, M. Adachi-Pagano, M. Intissar, S. Chauvière, C. Forano, J-P. Besse, J. Mater. Chem. 11 (2001) 105-112
16. A. Khan, D. O'Hare, J. Mater. Chem. 12 (2002) 3191-3198
17. W.W. Focke, N.S. Nhlapo, L. Moyo, S.M.C. Verryn, Mol. Cryst. Liq. Cryst. 521 (2010) 168-178
18. L. Moyo, W.W. Focke, F.J.W. Labuschagne, S.M.C. Verryn, Mol. Cryst. Liq. Cryst. (2012) Accepted
19. S. Carlino, Solid State Ionics, 98 (1997) 73-84
20. Z.P. Xu, P. Braterman, Appl. Clay Sci. 48 (2010) 235-242
21. P.S. Braterman, Z.P. Xu, F. Yarberry, in: S.M. Auerbach, K. A. Carrado, P.K. Dutta (Eds), Handbook of Layered Materials, CSC Press, Taylor & Francis Group, Boca Raton, 2004, pp. 373-449
22. S.A. Solin, D. Hines, S.K. Yun, T.J. Pinnavaia, M.F. Thorpe, J. Non-Cryst. Solids 182 (1995) 212-220
23. B. Chen, J.R.G. Evans, H.C. Greenwell, P. Boulet, P. V. Coveney, A. A. Bowden, A. Whiting, Chem. Soc. Rev. 37 (2008) 568-594
24. D. Gersappe, Phys. Rev. Lett. 89-5 (2002) 58301-5830

O-07-075

PERCOLATED NETWORK STRUCTURE FORMATION AND RHEOLOGICAL PROPERTIES IN NYLON 6/CLAY NANOCOMPOSITES

M. Okamoto[1]* and C. Mizuno[1]

[1]*Advanced Polymeric Nanostructured Materials Engineering, Graduate School of Engineering, Toyota Technological Institute, 2-12-1 Hisakata, Tempaku, Nagoya 468-8511, Japan – okamoto@toyota-ti.ac.jp*

Abstract - We have characterized the clay network structure in nylon 6-based nanocomposites (N6CNs) through TEM and FFT analyses. The volume fraction of montmorillonite (MMT) between 0.013 and 0.014 was the percolation threshold value for strong network formation. The volume spanning MMT network led to a very high value of the flow activation energy as compared with that of neat nylon 6, resulting in the pseudo-solid like response under molten state in N6CNs. We conducted the canonical NVT-MD simulation in the system made up by nylon 6 molecules/$Si(OH)_4$-molecules. The formation of the strong interfacial interaction between nylon 6 molecules and $Si(OH)_4$-molecules induced by OH groups was speculated.

Keywords: Molecular dynamics, , Percolated clay network, Rheology.

Introduction

Polymer/clay nanocomposites (PCNs) and their self-assembly behaviors have recently been approached to produce nanoscale polymeric materials. Additionally, these nanocomposites have been proposed as model systems to examine polymer structure and dynamics in confined environments [1]. However, the existence of percolated clay network structure, which is ignored in many literatures as always, is responsible for the linear viscoelastic response observed in PCN melts. The main objective of this study is to prove the effect of the percolated clay network structure formed by the randomly oriented exfoliated clay layers in PCNs on both solid and melt rheological properties. For this purpose, we first examined fast Fourier transform (FFT) analysis on digitally saved images of transmission electron microscope (TEM) micrographs using the image analysis, which allowed to provide us information equivalent to the scattering analyses. Using a molecular mechanics/molecular dynamics (MM/MD) program, we examined the dynamics of polymer chains in confined environment. We discussed the interfacial interaction between nylon 6 molecules and silicate particles, and calculated the potential and kinetic energies.

Experimental

Materials: The nylon 6-based nanocomposites (N6CNs) were kindly supplied by Dr. A. Usuki of Toyota Central R&D Labs., Inc. It was synthesized by in-situ polymerization of ε-caprolactam in lauryl ammonium intercalated montmorillonite (MMT) in the presence of small amount of 6-aminocaproic acid.

Characterization: For analyzing the features of the micrographs, we carried out fast Fourier transform (FFT) analysis on digitally saved images of TEM micrographs using the commercial image analysis software (Ultimage®, Graftek, France), which allowed to provide us information equivalent to the scattering analyses.

MD simulation: Using molecular mechanics program (MM2 Scigress, ver 2.2.0, Fujitsu Ltd.) we proposed the molecular structures of monomer, nylon 6, and silicate by taking their van der Waals radii into account. Optimization of the molecular structure was based on the minimization of the total energy of the molecular system. We conducted the canonical (constant particle number N, volume V, and temperature T)-molecular dynamics (MD) simulation at 298K using a MD program (Scigress, ver 2.2.0, Fujitsu Ltd.), and calculated the enthalpy (interfacial interaction), potential and kinetic energies. The product run of equilibrium MM/MD simulation was then executed for 100 ps with a time step of 0.2 fs. In all MD simulation, the atoms with non-bonding have the dispersive interaction described by Lennard-Jones (LJ) van der Waals potential. The most crucial and important aspect of these calculations is the method selected for sampling the relevant configurational phase space. Accordingly, the conformational search was carried out MM/MD protocol, in which the relaxed structures were subjected to the repeated all the simulation using NVT-MD conditions.

Results and Discussion

Flexibility of a single MMT layer

TEM images with two different magnifications are evident for the flexibility of the MMT particles (layers) (Fig.1). When viewed edge-on as in Fig. 1, single MMT sheet is apparent. The curvature ($1/R$) distribution of the dispersed MMT layers in N6CNs is characterized. The obtained value of $1/R$ ($\cong 8 \times 10^{-3}$ nm^{-1}) for N6C3.7 exhibits a large degree of flexibility of the mono-layered MMT sheet. It is well known that smectite clay sheet have a large degree of flexibility [2].

Dynamic mechanical properties

At the temperature range of –150 to 0 °C, there is a maximum of 170-240 % increment of *G'* for N6CNs in comparison to that of the matrix at well below T_g. Above T_g, when materials become soft the reinforcement effect of MMT particles becomes prominent, due to the restricted movement of the polymer chains, and hence strong enhancement (~ 400%) of modulus appeared (data not shown). This behavior is common for the nanocomposites reported so far.

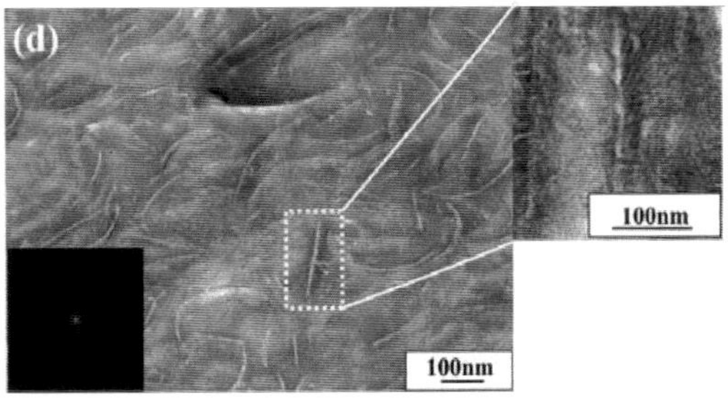

Fig. 1. Bright field TEM images of nanocomposite (N6C3.7): The enlarge part shown (Fig. 1 (d)) forms the indicated lamellae in the original image. The black strip inside the white part is an individual MMT particle. The inset is a computed FFT spectrum of the micrograph.

Fig. 2 shows the effect of MMT content on T_g of matrix polymer, represented by the tan δ peak location at the temperature range of 0 to 100 °C. Other two peaks in tan δ curves are observed at the range of temperature between –100 and –50 °C and at low temperature around –140 °C as a third peak, respectively. The former corresponds to the local motion of the amide groups possessing the formation of rather weak hydrogen bonding between amorphous chains, which is defined as β-relaxation. The latter corresponds to the local motion of the methylene (CH_2) segments, which is defined as γ-relaxation [3].

For T_g of matrix nylon 6, the presence of MMT particles leads to a significant peak shift to higher temperature accompanied with the depression of peak height of the tan δ curves for N6CNs compared to that of pure nylon 6. At the same time, a small remnant shoulder is observed around 20 °C (marked with the arrows). The magnitude of the shift estimated by tan δ is about 12 °C. The suppression of the glass transition (increase of T_g) may be ascribed to the restricted segmental motions at the nylon 6-MMT interfaces neighborhood of the exfoliated N6CNs. The β-relaxation exhibits same trend with loading of MMT. At the same time, the reduction of the thermal expansion coefficient (α) at the temperature range of –150 to 20 °C in N6CNs seems to stem from the mechanical constraint by the dispersed MMT particles. N6CNs show significant reduction of thermal expansion coefficient value and the value decreases up to ~ 20% for N6C3.7 as compared with that of neat nylon 6.

At the temperature range between 100 and 200 °C, the increasing of tanδ is due to the dampening effect of the MMT particles, as seen by the larger values of *G''* in this region.

MD simulation

For the generation of accurate model amorphous structures for nylon 6, the constitutive repeating unit (CRU) was built and its geometry optimized by energy minimization. The CRU was polymerized to a conventional degree of polymerization (DP_n) equal to 10. Although this chain may be too short to capture the genuine response of a long nylon 6 molecule, in the case of polystyrene (PS) it has been verified that a polymer with same DP_n is longer than average persistence length of PS in PS-based PCNs [4].

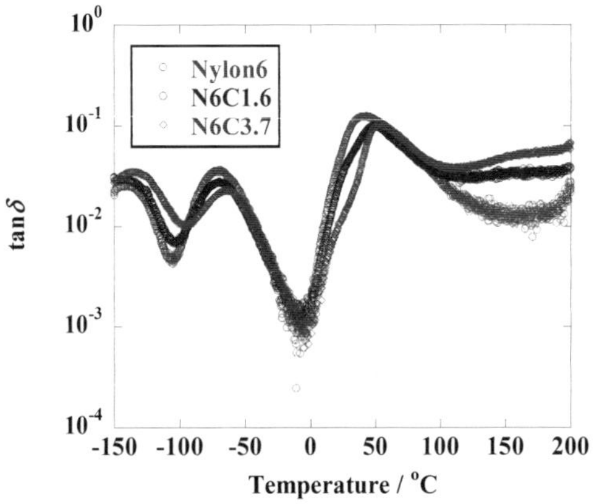

Fig. 2. Temperature dependence of tan δ for neat nylon 6 and nanocomposites.

The surface hydroxy concentration of MMT was estimated by titration with triethyl aluminum. The calculated value of a Si-OH density was 5(Si-OH)/nm^2 [5]. That is, 500-OH groups are localized near the edge surface of the individial silicate layers (~ 1x100 nm^2). These OH groups are particularly important because it improves the strength of the interface between nylon 6 matrix and MMT. For this reason, silica tetrahedral terminated with OH ($Si(OH)_4$) was modeled instead of layered silicate MMT, and overall system was built. For comparison, 50 ε-aminocaproic- acid monomers were used as a reference.

To illustrate the molecular conformations near $Si(OH)_4$ molecule, we reported two snapshots taken from a MD simulation for 10 ps at 298 K of the system made up by nylon 6 molecule (DP_n=10) of 5 chains/50 $Si(OH)_4$-molecules and 50 monomers/50 $Si(OH)_4$-molecules.

Fig. 3 shows the time variation of the energy components for the simulation. From the equilibrium conformation of the corresponding system we calculated the total energy of a system, which may be composed by the potential and kinetic energies. The results are presented in Table 1.

The large value of the total energy appeared in nylon 6 molecules as compared with that of the system in monomers. The potential energy in $Si(OH)_4$-molecules decreases in almost linear fashion with increasing molecules in the system. The interesting feature is the variation of the total energy generated in the systems when the system incorporates $Si(OH)_4$-molecule up to 50 units. The incorporation of $Si(OH)_4$-molecules induces the reduction of the total energy accompanied with decreasing of the enthalpy due to the negative potential energy of $Si(OH)_4$-molecules. The enthalpy having extremely large negative value (-5000kJ/mol) is evident in nylon 6/50 $Si(OH)_4$-molecules as compared with the system of monomers/$Si(OH)_4$-molecules (40000kJ/mol). Therefore it can be assumed that the OH groups induce the strong interfacial interaction between nylon 6 molecules and $Si(OH)_4$-molecules [6]. We can speculate the restricted segmental motions at the nylon 6-MMT interfaces neighborhood of the exfoliated N6CNs as discussed in above section.

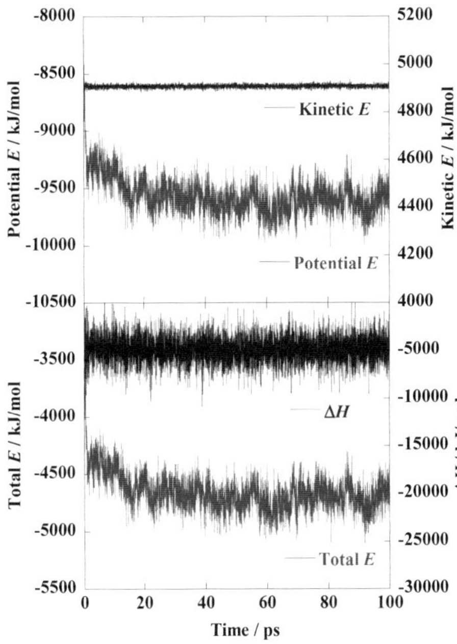

Fig. 3. Profiles of the energy components for the simulation of nylon 6 molecules/$Si(OH)_4$-molecules: (upper) potential and kinetic energies and (lower) total energy and enthalpy.

Conclusions

We have demonstrated the existence of percolated clay network structure in N6CNs through TEM and FFT analyses. The volume fraction of MMT between 0.013 and 0.014 was the percolation threshold value for strong network formation. For N6C3.7, the volume spanning MMT network led to a very high value of the flow activation energy as compared with that of neat nylon 6, resulting in the pseudo-solid like response under molten state. The dispersed mono-layered MMT sheets strongly contributed toward the enhancement of the dynamic storage and loss moduli, leading to the suppression of the glass transition. The magnitude of the shift estimated by tan δ was about 12 °C for N6C3.7 due to restricted segmental motions at the nylon 6-MMT interfaces neighborhood of the exfoliated nanocomposite. We have discussed the nylon 6 molecules in confined environment by NVT-MD simulation. The incorporation of $Si(OH)_4$-molecules into the system made up by nylon 6 molecules/$Si(OH)_4$-molecules induced the reduction of the total energy accompanied with decreasing of the enthalpy due to the negative potential energy of $Si(OH)_4$-molecules. The formation of the strong interfacial interaction between nylon 6 molecules and $Si(OH)_4$-molecules induced by OH groups was speculated.

Table 1. Energy components and enthalpy calculated from simulation of systems

Systems	$E_{potenrial}$/kJ mol^{-1}	$E_{kinetic}$/kJ mol^{-1}	E_{total}/kJ mol^{-1}	Enthalpy/ kJ mol^{-1}
Ny6/Si-5	2300	3350	5700	6000
Ny6/Si-10	1200	3500	4700	5000
Ny6/Si-50	-9500	4900	-4600	-5000
mono/Si-5	10000	3700	13500	95000
mono/Si-10	7300	3850	11000	83000
mono/Si-50	-9500	5200	-4300	38000

Acknowledgements

This work was supported by the Strategic Research Infrastructure Project of the Ministry of Education, Sports, Science and Technology, Japan (2010-2014).

References

1. RA.Vaia; EP. Giannelis *Macromolecules* 1997, 30 8000.
2. KA. Carrado; in Handbook of Layered Materials, SM. Auerbach KA. Carrado PK, Dutta Eds., Marcel-Dekker: NY, 2004; pp.1-38.
3. AE.Woodward; JM. Crissman; JA. Sauer J. Polym. Sci., 1960, 44, 23.
4. E. Manias; V. Kuppa, In Polymer Nanocomposite. ACS Symposium Series, RA.Vaia R. Krishnamoorti editors. vol. 804. Oxford: Oxford University Press, 2002 pp. 193-207.
5. MC. Hermosin; J. Cornejo Clays Clay Miner., 1986, 34, 591.
6. C. Mizuno; B. John; M. Okamoto Macrmole. Mater. Eng., DOI: 10.1002/ mame. 201200065.

O-07-086

NANOSTRUCTURE AND ADSORPTION BEHAVIOR OF NATURAL/SYNTHETIC ALLOPHANES

F. Iyoda[1], S. Arakawa[1] M. Okamoto[1]* and H. Hayashi[2]

[1]*Advanced Polymeric Nanostructured Materials Engineering, Graduate School of Engineering, Toyota Technological Institute, Japan – sd11404@toyota-ti.ac.jp; arakawa@toyota-ti.ac.jp; okamoto@toyota-ti.ac.jp*

[2]*New Product Development Center, Tsuchiya Co., Ltd., Japan – HidetomoHayashi@tsuchiya-group.co.jp*

Abstract – We synthesized three allophanes from precursors by a hydrothermal reaction at 100 °C for 48 h. The precursors were formed from the solution of Na_4SiO_4 and $AlCl_3{\cdot}6H_2O$ at different Si/Al molar ratios (0.5, 0.75, 1.0). The nanostructure of the synthetic allophanes was compared with that of natural allophanes from New Zealand and Japan by using X-ray diffractometry, energy dispersive X-ray spectroscopy, Fourier-transform infrared spectroscopy, thermogravimetry/differential thermal analysis, ^{29}Si and ^{27}Al nuclear magnetic resonance, field emission electron microscopy, and pore-size distribution based on the Cranston-Inkley method. The propensity of the allophanes to adsorb adenine and adenosine-5'-monophosphate (5'-AMP) was assessed by batch experiments. The adsorption data were fitted by the Freundlich equation and the adsorption parameters were discussed in relation to the properties of the natural and synthetic allophanes. The adsorption capacity (K_f) of the natural allophane for 5'-AMP was three times that of adenine. The average K_f value of the three synthetic allophanes for 5'-AMP was more than twice that of the natural allophane, possibly due to the higher purity of synthetic allophanes.

Keywords: Allophanes / Nanoparticles / Adenine / 5'-AMP / Adsorption

Introduction

In many parts of the world water bodies are becoming increasingly contaminated by chemicals from a wide range of human activities such as agriculture, manufacturing, energy consumption and waste disposal. A recent report from the World Health Organization (WHO) shows that at least 1.1 billion people in the world lack access to improved/safe water sources and more than 2.6 billion are without basic sanitation [1]. Thus, there is an increasing demand on research for cost-effective materials and technologies for water treatment.

Particle size and surface area are important features of adsorbent materials for use in water treatment to remove toxic and carcinogenic chemicals. A small particle size, a large surface area, and surface functional groups of materials are often desirable for their better performance in the adsorption of extraneous contamination. In this regard, allophane is the nanomaterial of choice for environmental applications because of its versatile ability of adsorbing cations (e.g., Cu^{2+}, Cd^{2+}), anions (e.g.. phosphate and arsenate), and organic molecules.

Allophane is a short-range-order clay mineral and occurs in some soils derived from volcanic ejecta. The primary particles of the allophane consist of hollow spherule with an outer diameter of 3.5-5.0 nm thick, which has perforations (~0.3 nm in diameter) with $(HO)Al(OH_2)$ groups, as shown in Figure 1. The surface area of allophane is as high as ~900 m^2/g, which is often larger than that of activated carbon. Further, the $(HO)Al(OH_2)$groups exposed on the wall perforations are at the source of the pH-dependent charge characteristics of allophane. They become $^+(H_2O)Al(OH_2)$ by acquiring protons on the acid side of the point of zero charge, and become $(HO)Al(OH)^-$ by losing protons of the alkaline side. Finally, the groups can form inner-sphere complexes with the carboxylic groups of organic compounds. The content of allophane in volcanic soils is variable, usually much lower than the 35-40% upper limit. In other words, natural allophane contains undesirable impurities like sand, silt, and clay-size particles that are difficult to be separated. Thus, pure natural allophane with consistent compositions and properties is not commercially available because natural allophane has a variable composition; its Si/Al atomic ratio changes with the environment where allophane occurs, but is typically between 0.5 and 1.0.

In this paper, three synthetic allophanes with different Si/Al ratios were prepared and characterized. The adsorption properties were assessed against a natural allophane, using adenine and adenosine 5'-monophosphate (5'-AMP) as adsorbates. 5'-AMP is a nucleotide and adenine plays an important role in biological systems.

Experimental

Allophanes

A New Zealand allophane (NP76) was obtained by dispersion of a subsoil sample taken from New Plymouth in the North Island. A Japanese allophane sample (AK70) produced from Mt. Akagi in Gunma was received from Shinagawa chemicals.

The precursor gels (Si/Al ratio = 0.5, 0.75 and 1.0) for the allophanes were prepared by mixing and stirring (for 1 h) of 100 mM of sodium silicate, ortho (Na_4SiO_4, Nacalai-Tesque) and aluminum chloride hexahydrate ($AlCl_3{\cdot}6H_2O$, Sigma-Aldrich). The sodium chloride formed was removed by centrifugation at a speed of 5000 rpm for 5 min. The precursors were then autoclaved at 100 °C for 48 h [2]. The synthesized samples were designated as 100Si/Al-x (x: Si/Al ratio;

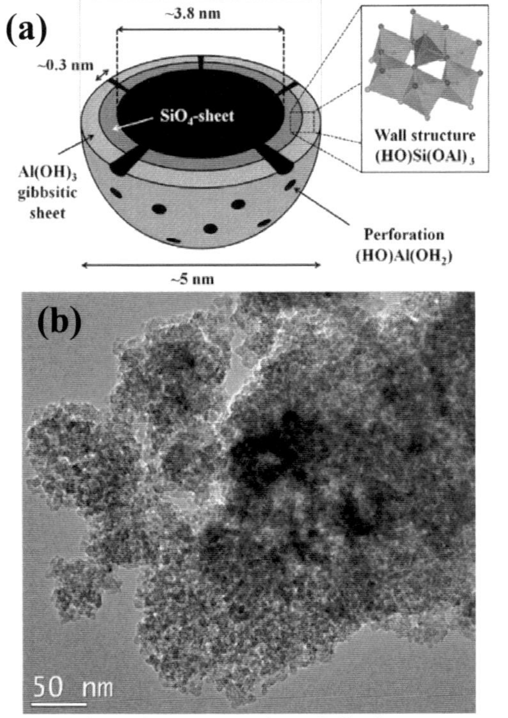

Fig. 1. Schematic representation of allophane structure (a) and high resolution TEM image of AK70 (b).

for e.g., 100Si/Al-1.0) and '*Pre-*' refers to the precursors.

Characterization

The crystal structure of allophanes was characterized by wide angle X-ray diffraction (WAXD), and structural changes were determined by Fourier-transform infrared (FTIR) spectra collected at a resolution of 2 cm^{-1} using a FTIR spectrometer. The morphological features of the allophanes were observed by field emission scanning electron microscopy (FE-SEM) equipped with elemental analysis by energy dispersive X-ray spectroscopy (EDX) after the samples were sputter-coated with platinum under an argon pressure of 10 Pa for 1 min at a current of 10 mA. The chemical and structural surroundings of both Si and Al atoms were measured using ^{27}Al and ^{29}Si magic angle spinning (MAS) solid-state nuclear magnetic resonance (NMR). Thermal behavior was determined by thermogravimetry/differential thermal analysis (TG/DTA) at a heating rate

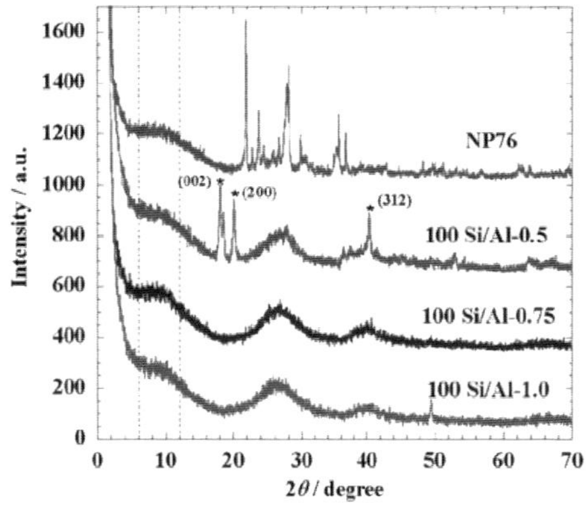

Fig. 2. WAXD patterns of natural and synthetic allophanes. The data on Y-axis were shifted to avoid overlapping. The dashed lines indicate the position of the remnant shoulder. The asterisks indicate the strong reflection of gibbsite.

of 5 °C/min. The pore volume and pore size distribution were measured by the Cranston-Inkley (CI) method [3] using a nitrogen adsorption-desorption approach. The specific surface area was estimated by the *t*-method.

Table 1. Properties of natural/synthetic allophanes

Samples	Si/Al ratio	Pore size / nm
NP76	0.53	3.82
AK70	0.58	3.82
100Si/Al-0.5	0.56	2.12, 3.82, 6.18
100Si/Al-0.75	0.82	3.64
100Si/Al-1.0	1.07	3.80
Pre-100Si/Al-0.5	0.63	nil

Adsorption experiments

Adenine was purchased from Kanto Chemicals, Japan, and 5'-AMP from Wako Pure Chemicals, Japan. 10 mg of a synthetic allophane, natural allophane or allophane precursor was dispersed in 10 ml of 0.1-2.0 mM aqueous solutions of adenine or 5'-AMP and shaken for 60 h at room temperature. The dispersions were then centrifuged at 5000 rpm for 10 min. The total organic carbon (TOC) and Total nitrogen (TN) in the supernatants were measured using the combustion method at 800 °C after four-point calibration.

Molecular Dynamics (MD) Simulation

Using a molecular dynamics program (Scigress, ver 2.2.0, Fujitsu Ltd.), we proposed the molecular dimensions of adenine, 5'-AMP and allophane by taking their van der Waals radii into account. Optimization of the molecular structure was based on the minimization of the total energy of the molecular system. We conducted the canonical (constant particle number N, volume V, and temperature T)-molecular dynamics (MD) simulation at 298 K using a MD program, and calculated the enthalpy (interfacial interaction), potential and kinetic energies. The product run of equilibrium MM/MD simulation was then executed for 100 ps with a time step of 0.1 fs. In all MD simulation, the atoms with non-bonding have the dispersive interaction described by Lennard-Jones (LJ) van der Waals potential. The most crucial and important aspect of these calculations is the method selected for sampling the relevant configurational phase space. Accordingly, the conformational search

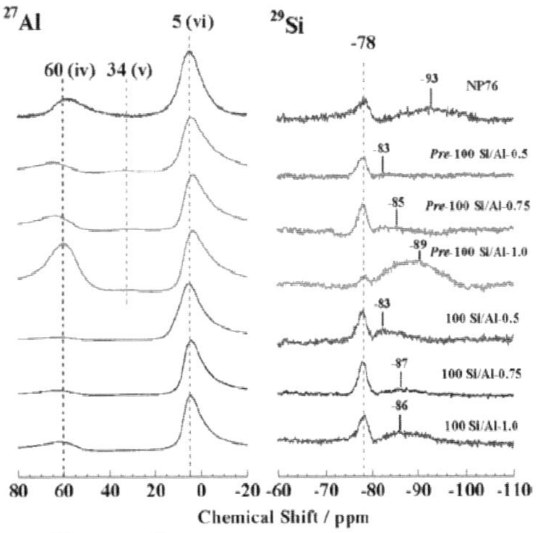

Fig. 3. ^{27}Al and ^{29} Si MAS solid-state NMR spectra of natural and synthetic allophanes.

was carried out MM/MD protocol, in which the relaxed structures were subjected to the repeated all the simulation using NVT-MD conditions.

Results and Discussion

The characteristic properties such as Si/Al ratio and pore size of allophanes are reported in Table 1. All allophanes analyzed in the present study had an average pore diameter of ~3.8 nm except the precursor which had none.

The WAXD patterns of Natural and synthetic allophanes showed a broad reflection at 2θ~26° (0.34 nm) and ~40° (0.23 nm), indicating the amorphous nature (Figure 2). However, a small shoulder at 2θ = 6-12° (0.74-1.5 nm) suggested the short-range order of allophanes [4]. Thus, allophanes are more appropriately defined as nanosized, short-range ordered aluminosilicates. The strong 002, 200 and 312 reflections at 2θ = 18.18°, 20.18° and 40.48° were assigned to gibbsite. The enhanced formation of a gibbsite phase at a high ratio of Al (Si/Al = 0.5) was accompanied by the appearance of three different pore size (2.12, 3.82 and 6.18 nm) (Table 1).

The sharp reflections observed in NP76 were due to the Fe-containing impurities, as also confirmed by the EDX data (data not shown). This natural allophane might have a significant structural variation due to irregularities in the pore structure.

Table 2. The Freundlich isotherm parameters of natural/synthetic allophanes

Adsorbates	Adsorbents	K_f mmol/g·mmol/L$^{-1/N}$	N -
Adenine	NP76	0.05	1.1
5'-AMP	NP76	0.18	1.3
	100Si/Al-0.5	0.39	1.2
	100Si/Al-0.75	0.31	1.2
	100Si/Al-1.0	0.48	1.5

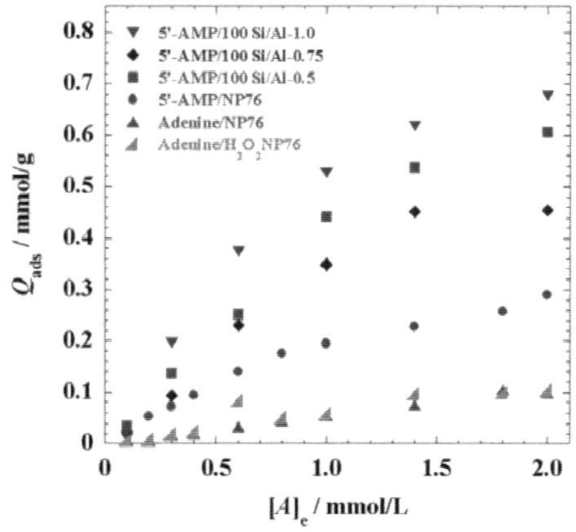

Fig. 4. Adsorption isotherms of adenine and 5'-AMP on natural and synthetic allophanes.

Fig. 3 shows ^{27}Al and ^{29}Si MAS NMR spectra of allophanes. The peak at -78 ppm of the ^{29}Si spectrum corresponds to the wall structure ((HO)Si(OAl)$_3$) consisting of Si tetrahedral attached to three aluminol groups (Al-OH) of the gibbsite sheet and one silanol group. No significant peak shift at -78 ppm was observed for all precursors and synthetic allophanes. There was a broad resonance between -83 and -89 ppm. This peak may suggest that the different chemical structure surroundings of the Si atoms shift the resonance to higher or lower values with the number of Al atoms bound to the central Si atom. With increasing Si/Al ratio, the broad peak around -85 ppm shifted towards a lower value. This trend was more obvious for the precursors than for the corresponding synthetic allophanes, indicating that allophanes were structurally more stable than precursors. In this study, a clear explanation could not be obtained.

In the case of NP76, the resonance at -93 ppm was related to the impurities. These impurities included volcanic glasses and silica gels.

In the ^{27}Al NMR spectra, the 5 ppm peak indicates Al atoms with octahedral coordination (^{vi}Al) while the peak at 60 ppm corresponds to tetrahedrally coordinated Al atoms (^{iv}Al) [5]. The ^{iv}Al atoms were substituting the Si atoms of the outer-sheet and represent the octahedral gibbsite layer. A weak peak at 34 ppm in the spectra of the precursors indicated fivefold coordinate Al atoms of the disordered structures. Due to the structural transformations of the precursor into the allophane formation, both iv and v Al atoms disappeared.

The adsorption isotherms of adenine and 5'-AMP on natural and synthetic allophanes (Fig. 4) were fitted by the Freundlich equation,

$$Q_{ads} = K_f [A]_e^{1/N} \quad (1)$$

where K_f is the relative adsorption capacity of the adsorbent and N is the adsorption intensity, which describes the shape of the isotherm (Table 2).

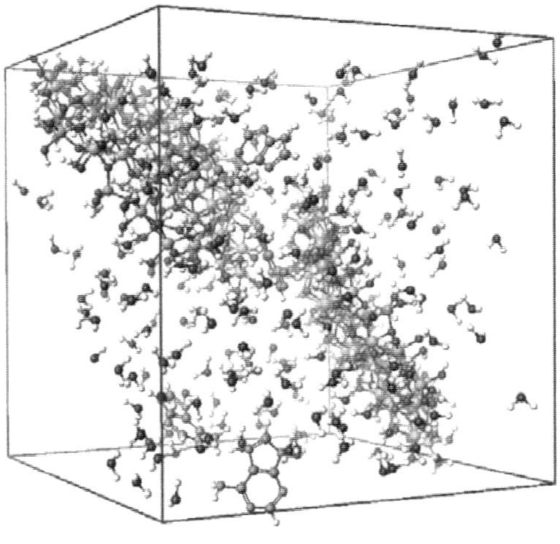

Fig. 5. Snapshot extracted from MD trajectory at 298 K of a system comprising allophane/adenine/H_2O molecules.

The adsorption of 5'-AMP by NP76 showed a much larger value of K_f (0.18) than that of adenine. The humic substances and the metallic impurities of NP76 significantly reduced the adsorption in comparison with synthetic allophanes. The absence of pores might have contributed to the low K_f value of the precursor.

Because of the incompatible dimensions of the wall perforations (~0.3 nm) of allophane and the size of adenine and 5'-AMP, the adsorbates could not enter the internal cavity of allophane. Thus, the Si-OH groups on the internal surface of allophane did not contribute to the adsorption of adenine and 5'-AMP. However, the allophane spherules existed as aggregates and not as single particles. The interstitial space between the spherules could accommodate adenine and 5'-AMP molecules. Electrostatic, hydrogen bonding and van der Waals interactions also contributed to the adsorption of these molecules.

Analysis of the adenine and 5'-AMP adsorption isotherms suggest that the adsorbed molecules likely form monolayers. The larger K_f value for 5'-AMP (Table 2) compared to adenine suggests a superior adsorption. The adsorption of 5'-AMP was facilitated by the interaction of the phosphate groups with the Al-OH groups on the wall perforations, involving the formation of inner-sphere complexes through a ligand-exchange reaction between Al-OH and $(HO)_2OP{=}O$ groups [6].

We constructed bilayer structure as a model of allophane single particle. The structure is consisted of Al octahedral (gibbstic) sheet and an inner Si tetrahedral sheet, where the wall structure ($(HO)Si(OAl)_3$) consisting of Si tetrahedra attached to three aluminol groups (Al-OH) of the gibbsite sheet and one silanol group. The active sites on allophane are associated with $(HO)Al(OH_2)$ groups at the surface defects, and overall system was built. The time variation of the energy components for the simulation was simulated. From the equilibrium conformation of the corresponding system we calculated the total energy of a system, which may be composed by the potential and kinetic energies. The snapshot taken from a MD simulation for 10 ps at 298 K of the system made up by 2 bilayer allophane sheets, 5 adenine molecules and 200 H_2O molecules (Fig. 5). The interesting feature is the variation of the total energy generated in the system. The system induces the reduction of the total energy accompanied with decreasing of the enthalpy due to the negative potential energy of the ternary components. The enthalpy having negative value (-859 kcal/mol) is evident as compared with the system consisting of adenine/H_2O (3186 kcal/mol). The formation of the strong interfacial interaction in the ternary system induced by perforations was speculated.

Conclusions

Three allophanes with Si/Al molar ratios of 0.5, 0.75 and 1.0 were synthesized and compared with a natural allophane sample. The natural and synthetic allophanes were poorly crystalline and consisted of hollow spherules of ~3.8 nm internal diameter and a characteristic wall structure composed of $(HO)Si(OAl)_3$ groups. The adsorption of 5'-AMP (K_f) on the natural allophane sample was three times higher compared to adenine. The strong adsorption of 5'-AMP by allophane may be due to the formation of inner-sphere complexes between the Al-OH and $(HO)_2OP{=}O$ groups. The average K_f value (0.39) for the 5'-AMP adsorption on the synthetic allophanes was distinctly higher than for the natural allophane (0.18). This may be explained by the higher purity of the synthetic allophanes. The influence of the Si/Al ratio on the adsorption of 5'-AMP by the synthetic allophanes was not consistent.

Acknowledgements

We are grateful to Dr. G. Yuan (Land care Research, New Plymouth, New Zealand) for providing natural allophane (NP76) samples. This work was supported by a Grant in TTI as a Special Research Project (2012).

References

1. WHO, http://www.who.int
2. F. Ohashi; S.I. Wada; M. Suzuki; M. Maeda; S. Tomura *Clay Minerals*, 2002, 37, 451.
3. R.W. Cranston; F.A. Inkley *Catalysis*, 1957, 9, 143.
4. S.J. Van der Gaast, K. Wada, S.I. Wada, Y.Kakuto *Clay and Clay Minerals*, 2002, 33, 237.
5. S. Hiradate; S.I. Wada *Clays and Clay Minerals*, 2005, 53, 401.
6. H. Hashizume; B.K.G. Theng Clays and Clay Minerals, 2007, 55, 599.

O-07-094

ELECTRICALLY CONDUCTIVE POLYETHYLENE NANOCOMPOSITES USING FERROMAGNETIC HIGH ASPECT RATIO NICKEL NANOWIRES AS FILLERS

Xiaoxiong Luo[1,2], **Uttandaraman Sundararaj**[1] *, **Jing-Li Luo**[2]

[1] *Department of Chemical and Petroleum Engineering, University of Calgary, Calgary, Alberta T2N 1N4, Canada – xluo@ualberta.ca; u.sundararaj@ucalgary.ca*

[2] *Department of Chemical and Materials Engineering, University of Alberta, Edmonton, Alberta T6G 2G6, Canada – xluo@ualberta.ca; jingli.luo@ualberta.ca*

Abstract - Ferromagnetic nickel nanowires (NiNWs) were synthesized by a scalable AC electrosynthesis method. After liberation and ultrasonication, the NiNWs exhibited a uniform diameter of ca. 38 nm and an average length of 2.68 μm, which translates to an average aspect ratio of 70. The high aspect ratio polycrystalline NiNW powder consisted of Ni metal, NiO and Ni2O3 with 92.7 wt. % Ni and 7.3 wt. % O elements. When dispersed in methanol, the NiNWs can be manipulated by a permanent magnet. The Curie temperature of the NiNW powder, determined by TGA, was 343.6 oC. Nanocomposites with NiNWs dispersed in linear low density polyethylene (LLDPE) matrix were synthesized. Electrical conductivity of the NiNW/LLDPE nanocomposite, with a NiNW volume fraction of 0.03, was 28.3 S/m. Backscattered SEM images of the freeze fractures and TEM images of the ultramicrotomed sections of NiNW/LLDPE nanocomposites revealed good dispersion and good network formation of the NiNWs.

Keywords: electrically conductive; nanocomposites; ferromagnetic; nickel nanowire

Introduction

Nickel nanowires (NiNWs) are one-dimensional nanostructures with ferromagnetic properties that attracted much attention in fundamental physical research [1]. Computer engineers are interested in applying NiNWs as ultra-high-density magnetic recording mediums [1, 2]. Another potential application for NiNWs is electromagnetic interference (EMI) shielding for electronic enclosures.

EMI shielding is the process of reducing the electromagnetic radiation in a space by blocking the radiation with barriers made of electrically conductive or magnetic materials. Three mechanisms are involved in EMI shielding, namely: reflection, absorption and multiple-reflection [3]. The absorption loss of the electromagnetic radiation increases with the product of electrical conductivity and relative magnetic permeability, and this product value is 20 times higher for Ni over Cu [4].

In this work, we used a scalable AC electrodeposition method to synthesize NiNWs in our custom designed porous aluminum oxide (PAO) templates. The NiNWs were liberated from the template; characterized using electron microscopy, X-ray diffraction, X-ray photoelectron spectroscopy and thermogravimetric analysis and then dispersed into LLDPE to create electrically conductive polymer nanocomposites.

Experimental

Synthesis of NiNWs

The PAO templates used for AC electrodeposition were prepared by a two-step anodization method [5–6] using 0.3 mol/L oxalic acid (≥ 97.0%, Sigma Aldrich) as the electrolyte under a DC anodization voltage of 40 V. The total time for the second anodization was 8 h. At the end of the second anodization, the DC voltage was decreased from 40.0 V to 10.0 V, at a rate of - 2 V/min, and held for 10 min at 10.0 V to decrease the thickness of the barrier layer.

The aqueous solution for AC electrodeposition was formulated with 300 g/L $NiSO_4 \cdot 6H_2O$ (98%, Alfa Aesar), 40 g/L H_3BO_3 (99.5% min., EMD Millipore) and 8 mg/L sodium dodecyl sulfate.

The electrodeposited NiNWs in the PAO templates were liberated by immersing the templates in 1 mol/L NaOH aqueous solution for 10 min at room temperature. The collected NiNWs were rinsed with methanol three times, then filtered and dispersed into methanol ultrasonication for 20 min.

Characterization of NiNWs

To prepare the NiNW sample for SEM imaging, a cover glass was fastened to the sample stage center of a Laurell WS-200-4T2 spin coater. Two drops of NiNW/methanol dispersion were dropped onto the cover glass, which was spinning at approximately 3000 rpm. After 1 minute spinning, the cover glass was removed and stuck onto an Al SEM stub with double-sided adhesive carbon tape. The cover glass/Al stub assembly was coated with gold and palladium to collect images using a Philips XL30 scanning electron microscope.

The length distribution of the NiNWs dispersed in methanol was determined from the SEM images. The lengths of over 1000 individual NiNWs were measured with ImageJ 1.44p software (National Institute of Health, USA).

TEM images were collected with a Tecnai F20 microscope. The sample for TEM was prepared by dropping very dilute NiNW/methanol dispersion onto a carbon coated Cu TEM grid.

X-ray diffraction (XRD) pattern of the NiNW powder was collected by a Rigaku Multiflex Diffraction System, with Cu Kα radiation.

Preparation of NiNW/LLDPE nanocomposites

LLDPE solutions were prepared by dissolving 2.0 g of polymer in 100 mL of xylene at 100 °C, under vigorous stirring. The NiNW/methanol dispersion of desired volume was added to the LLDPE/xylene hot solution drop by drop, at a rate of approximately 2 mL/min, in order to evaporate the methanol and leave the NiNWs dispersed in the solution. After dispersing NiNWs into the hot solution, 50 mL of methanol, at room temperature, was added into the mixture to precipitate the nanocomposite. The NiNW/LLDPE precipitates were ultrasonicated for 10 minutes, in a room temperature water bath, and then filtered. The filtered NiNW/LLDPE nanocomposites were then transferred to evaporation dishes for 24 h, in a fume hood, under ambient conditions; they were then moved into a vacuum oven at room temperature for 12 h to completely remove residual solvents.

The dry powder of NiNW/LLDPE nanocomposites was annealed under 138 MPa at 250 °C for 30 min, in a 10.5 mm diameter and 1.2 mm thick stainless steel mold.

Characterization of the NiNW/LLDPE nanocomposites

Backscattered SEM images of the NiNW/LLDPE nanocomposite fractured in liquid nitrogen and coated with gold and palladium were collected with the Philips XL30 scanning electron microscope.

To obtain desired cross sections of thickness about 100 nm, the NiNW/LLDPE nanocomposite sample for TEM observation was ultramicrotomed at -130 °C with an ultracut diamond knife. The ultramicrotomed slice was then transferred onto a Cu TEM grid and imaged with the Tecnai F20 TEM.

Electrical resistivity of the NiNW/LLDPE nanocomposites was measured by two different instruments. In order to minimize the impact of the contact resistance, a QPP four pin probe (MCP-TPQPP model, Mitsubishi Chemical Co., Japan) with an inter-pin distance of 1.5 mm was connected to the Loresta GP resistivity meter (MCP-T610 model, Mitsubishi Chemical Co., Japan) to measure the nanocomposites with resistivity lower than 10^4 Ω·m. For nanocomposites with higher resistivity, electrical resistance measurements were performed using a Keithley 6517A electrometer connected to an 8009 test fixture with the Keithley 6524 high resistance measurement software, conforming to ASTM D991-89 standards. The measured volume resistance R_v was converted to resistivity ρ_v.

Results and Discussion

The SEM, TEM images and the length distribution of the liberated NiNWs are shown in Fig. 1. The measured diameter of the NiNWs in the images was 38 ± 4 nm. The average length of the NiNWs after liberation in NaOH and ultrasonication in methanol was 2.68 μm. The average aspect ratio of the NiNWs, after liberation and ultrasonication, was approximately 70.

XRD pattern of the NiNW powder is shown in Fig. 2. The three sharp peaks of the fcc crystal structure of metal nickel can be clearly identified at two theta of 44.49, 51.85 and 76.38, corresponding to the planes of (111), (200) and (220), respectively (PDF# 65-2865). The peaks of NiO with low intensity can also be identified (PDF# 47-1049). Ni_2O_3 was not detected by XRD.

The NiNWs dispersed in methanol can be manipulated by a magnetic field. Fig. 3 shows the photographs of the NiNWs in methanol attracted to one side

Fig. 4 shows the electrical conductivity of the NiNW/LLDPE nanocomposites as a function of the volume fraction of NiNWs. Electrical conductivity of the NiNW/LLDPE nanocomposites, with NiNW volume fractions below 0.005, was close to that of the neat LLDPE matrix. However, a remarkable increase of over 12 orders of magnitude in the electrical conductivity can be observed when the volume fraction of NiNWs in the nanocomposites is increased from 0.005 to 0.0075, which indicates that the electrical percolation threshold of the NiNW/LLDPE nanocomposites was at a NiNW volume fraction of 0.005. An increase in the NiNW volume fraction from 0.0075 to 0.03 led to a gradual increase in the electrical conductivity of the NiNW/LLDPE nanocomposites from 6×10^{-2} S/m to 28.3 S/m.

The linear fitting of the electrical conductivity data of the nanocomposites with different volume fraction of NiNWs to the power law for electrical conductivity σ with percolation threshold ϕ_c = 0.005 is shown in the inset of Fig. 6, where the dot line represents the best fit. The fitting results were σ_0 = 4.34 × 10^5 S/m and critical exponent t = 2.59 ± 0.12. The critical exponent t = 2.59 ±0.12 (close to 3) indicated that the electrical conductivity in the NiNW/LLDPE nanocomposites mainly originated from contact between individual fibers.

Fig. 5 shows the backscattered SEM images of the freeze fractures and the TEM images of the ultramicrotomed section of the NiNW/LLDPE nanocomposites, containing 1 vol. % of NiNWs. All the images showed excellent dispersion of the NiNWs in the LLDPE matrix. Particularly, superior segregation and network formation of the NiNWs can be observed in both TEM images.

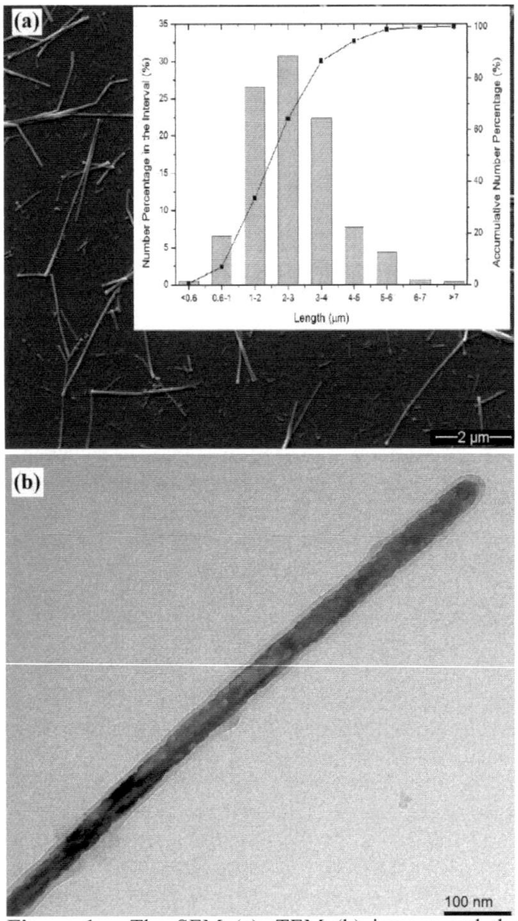

Figure 1 – The SEM (a), TEM (b) images and the length distribution (the inset in a) of the liberated NiNWs.

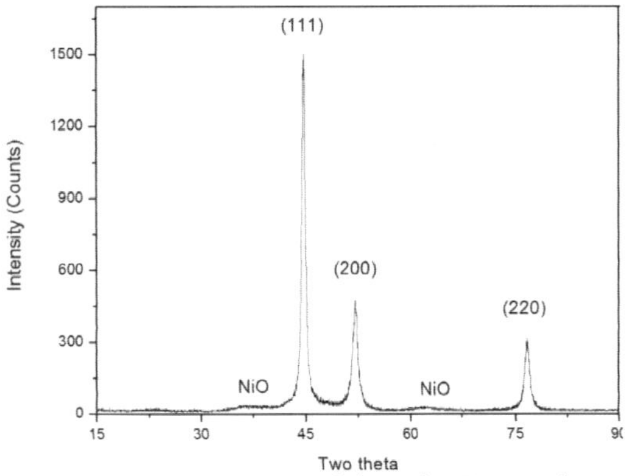

Figure 2 – XRD pattern of the NiNW powder.

Figure 3 – Photograph of the NiNWs in methanol attracted to one side of the vial by a permanent magnet.

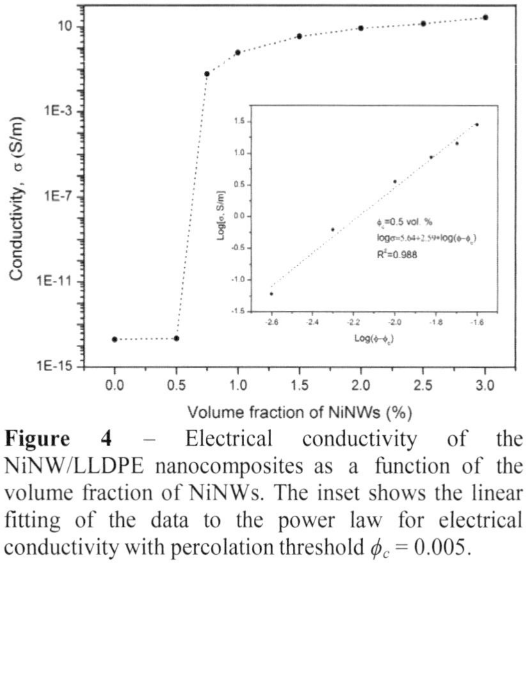

Figure 4 – Electrical conductivity of the NiNW/LLDPE nanocomposites as a function of the volume fraction of NiNWs. The inset shows the linear fitting of the data to the power law for electrical conductivity with percolation threshold $\phi_c = 0.005$.

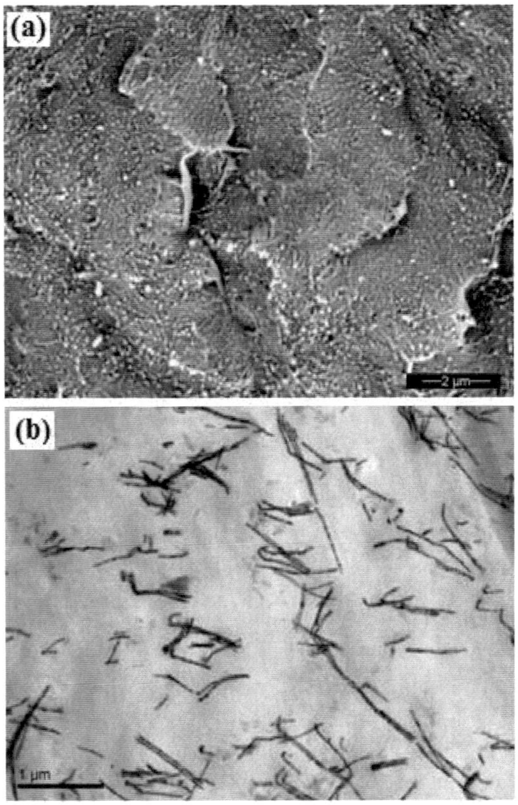

Figure 5 – The backscattered SEM images (a) of the freeze fractures and the TEM images (b) of the ultramicrotomed section of the NiNW/LLDPE nanocomposite, containing 1 vol. % of NiNWs.

Conclusions

NiNWs were synthesized by a scalable method of AC electrodeposition of nickel into the in-house prepared PAO templates. The synthesized NiNWs, after liberation and ultrasonication, exhibited a uniform diameter of ca. 38 nm and average length of 2.68 μm, which translates to an average aspect ratio of 70. XRD pattern indicated the fcc nickel crystal structure. The NiNWs dispersed in methanol can be manipulated by a permanent magnet.

NiNW/LLDPE nanocomposites were prepared by a modified MSMP method. The electrical percolation threshold is at a NiNW volume fraction of 0.005. The electrical conductivity of the NiNW/LLDPE nanocomposite with a NiNW volume fraction of 0.03 was 28.3 S/m. The power law fitting results implied that the electrical conductivity of the NiNW/LLDPE nanocomposites in this study mainly originated from contact between individual nanowires.

Acknowledgements

We thank Xerox Research Center of Canada and NSERC for the funding. We thank Nova chemicals for the polymer sample. The imaging facilities provided by Microscopy and Imaging Facility (MIF) of University of Calgary are acknowledged too.

References

1. P.M. Paulus; F. Luis; M. Kröll; G. Schmid; L.J. de Jongh *J. Magn. Magn. Mater.* 2001, 224, 180.
2. N. Tsuya; Y. Saito; H. Nakamura; S. Hayano; A. Furugohri; K. Ohta; Y. Wakui; T. Tokushima *J. Magn. Magn. Mater.* 1986, 54–57, 1681.
3. J.C. Huang *Adv. Polym. Tech.* 1995, 14(2), 137.
4. D.D.L. Chung *J. Mater. Eng. Perf.* 2000, 9(3), 350.
5. H. Masuda; F. Hasegwa; S. Ono *J. Electrochem. Soc.* 1997, 144(5), L127.
6. G.A. Gelves; Z.T.M. Murakami; M.J. Krantz; J.A. Haber *J. Mater. Chem.* 2006, 16, 3075.

O-07-171

TRIBOLOGICAL STUDIES OF PA66 NANOCOMPOSITES FILLED WITH LAYERED SILICATE CLAYS

S. Laske[1]*, R.K. Selvasankar[1], M. Kracalik[1], C. Holzer[1] and M. Winzer[2]

[1] *Department of Polymer Engineering and Science, Chair of Polymer Processing, Montanuniversitaet Leoben, Austria – Stephan.laske@unileoben.ac.at*

[2] *IWIS Motorsysteme GmbH & Co. KG, Munich, Germany*

Abstract. Incorporation of layered silicate nanoclays in polymer matrix can improve various physical and mechanical properties of the polymer. The influence of different structure modified organoclays on the tribological properties of Polyamide 66 (PA66) has been investigated in this paper. The PA66 nanocomposites with 5 wt.% organoclay were prepared by melt compounding on a co-rotating twin screw extruder. The tribological measurements were carried out on a pin-on-disc type tribometer by pressing a flat pin of steel at predefined loads against a polymer disc sample rotating at a predefined velocity. The disc samples were prepared by compression molding. The surface of both steel and polymer samples was grinded and polished to attain a defined surface roughness. The wear properties were studied at different loads under dry sliding conditions and the coefficient of friction (COF) was measured under dry sliding and oil lubricated conditions.
The results showed that most of the PA66 nanocomposites have superior tribological properties compared to the unfilled PA66. Tremendous improvements in the wear resistance of the nanocomposites were observed under dry sliding conditions. The reduction in COF for most of the nano composites under dry sliding conditions was obvious. With increase in the normal load the specific wear rate and COF reduces. The effect of nanoclays on reduction of COF in oil-lubricated conditions was more effective.

Keywords: Polyamide; Nanocomposites; Organoclay; Coefficient of friction (COF); Wear

Introduction

Intercalation of layered silicate nanoclays into polymer matrices is gaining more importance in the recent years because of their remarkable properties compared to pure polymers [1 - 4]. Due to this, polymer nanocomposites are being used as an alternative for metals in many engineering applications.
In applications where polymers come in contact with moving metal parts, friction and wear are important properties. A lot of research has been done on improving the tribological properties of polymers by adding different inorganic or organic nanosized fillers into the polymer matrix [5]. Studies have proved that polymer layered silicate nanocomposites can also improve the tribological properties of polymers to a great extend [3, 4, 6 - 8].
Polyamide (PA) is a semicrystalline thermoplastic material used for many engineering applications because of its high thermal and mechanical properties. Also in general they have good friction and wear properties. Recent investigations on the tribology of organoclay filled ,PA6 [3] show a low friction and high wear resistance. Also studies have revealed, that the tribological property of PA6 filled with clay depends on the filler size, normal load, sliding velocity and test medium [6]. It is also well documented that the wear and friction properties are influenced by the lubricating medium such as air, water or oil [9, 10]. Resent investigation on the tribological properties of PA66 filled with MMT (montmorillonite) modified by long alkyl ammonium bromide by Mu et al [8] has shown improvement in the coefficient of friction and wear rate of polymer nanocomposites.
This paper reports the tribological properties of PA66 filled with MMTs with different organic modifiers under dry sliding and oil-lubricated conditions.

Experimental

Materials

The matrix was a Polyamide 66 (PA66) type Ultramid PA66 A4H supplied by BASF. Commercially available layered silicates (Cloisite and Nanofil) with different structure modifications have been used in this study and were supplied by Rockwood Additives.
Cloisite and Nanofil fillers consist of organically modified, nanosized layered magnesium-aluminum silicate platelets [11]. The silicate platelets, where the additives are derived from are 1 nanometer thick and 70 – 150 nanometers across.
The Cloisite fillers belong to the same category with small modifications in their chemical structure, where the side chains are different.
Nanofil 116 is commercially available layered silica and was chemically modified at the Chair of Polymer Chemistry, Montanuniversitaet Leoben.

Compounding

The melt compounding of PA66 with 5 wt.% of filler was performed on a co-rotating twin screw extruder W&P ZSK 25D at the Chair of Polymer Processing, Montanuniversitaet Leoben. Before compounding, the polyamide material was dried by 80°C for at least 12 h. The melt mixed compound was extruded through a capillary die at a flow rate of 7 kg/h and 100 rpm screw speed. The extruded melt was passed through a water bath for cooling and granulated. The pellets were dried in a vacuum dryer prior to further usage.

Characterization of compounds

The thermal characteristics and the degree of crystallinity of the polymer nanocomposites were characterized using a differential scanning calorimetry DSC (DSC01, Mettler Toledo). The first heating of the polymer composites was carried out from room temperature to 300 °C at a heating rate of 10 °C/min and the first cooling was performed at 10 °C/min cooling rate until room temperature is reached. A second heating was done to study the thermal characteristics. The percentage of crystallinity K was determined, from the enthalpy curve measured during cooling, using the following eq.1:

$$K = \frac{\Delta Hm}{\Delta Hm_0} \cdot 100\ (\%) \qquad (1)$$

where ΔHm is the cooling enthalpy of sample in J/g and ΔHm_0 is the theoretical enthalpy for 100 % crystalline polymer. For PA66 this value is 255 J/g [12].

Sample Preparation

Samples of ~80 mm diameter and ~3 mm of thickness for the tribological testing were compression molded under vacuum on a compression molding machine (Polystat 300 S, Dr. Collin) at 290 °C.

Tribological Analysis

The tribological analysis were performed on a pin-on-disc type tribometer (UMT-2, CETR, USA), by pressing a flat pin of steel at predefined loads against a polymer disc sample rotating at a predefined velocity. The experiments in dry conditions were carried out at normal loads of 7 and 13 N and the sliding velocity was set to 1 m/s. The measurements were carried out for time duration of 30 minutes covering a total sliding distance of about 1.9 km.
The actual dynamic normal load and the friction force were monitored throughout the measurement time and the coefficient of friction (COF) was determined from the ration of these two values.
The sample´s weight was measured to an accuracy of four decimal points, before and after the measurements. From the loss in weight, the specific wear rate (*K*) was determined using eq. 2.

$$K = \frac{\Delta m}{L \cdot F \cdot \rho} \qquad (2)$$

where Δm is the weight loss in mg, L is the total sliding distance in m, F is the applied normal load in N and ρ is the density of the polymer sample in kg/m^3.

The COF measurements under lubricating conditions were carried out at normal loads of 5, 10 and 13 N at different sliding velocities. Since oil was used as lubricating medium, difficulties were encountered in measuring the wear loss. Due to this inaccuracy in the weight measurement, the wear rate calculation was neglected. Also measurements were carried out with varying velocities at constant load of 13 N. All the measurements were carried out in a temperature controlled chamber maintained at 23 °C and relative humidity of 45 %.

Results

DSC Analysis

The degree of crystallinity of PA66 and its composites, calculated from the first cooling enthalpy curves are given in table 1. Addition of layered silicate into the PA66 matrix does not change the crystallinity to a great extent, except in case of PA66/Cloisite 30B, which shows a 7% reduction in the degree of crystallinity compared to the neat PA66.

Table 1. Degree of crystallinity

Material	Degree of crystallinity in %
PA66	24
PA66+Cloisite 10A	23
PA66+Cloisite 93A	22
PA66+Cloisite 30B	17
PA66+Cloisite Na+	25
PA66+Nanofil 116	23
PA66+Nanofil 116m	21

Wear

The specific wear rate of unfilled PA66 and the PA66 nanocomposites at two different loads under dry sliding conditions are represented graphically in Figure 1.

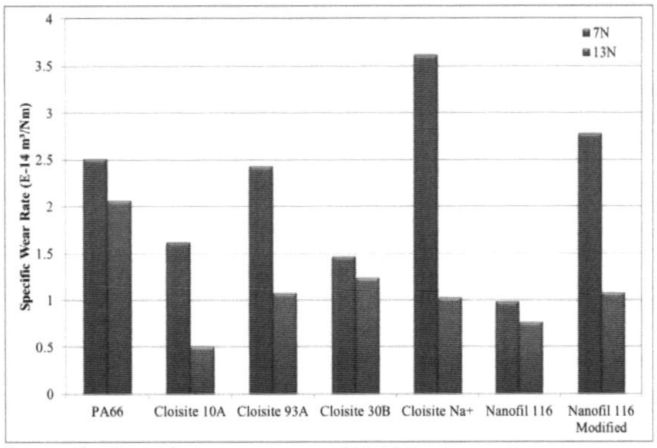

Figure 1: Specific wear rate of PA66 composites at different loads under dry sliding conditions

In contrary to the results obtained by Mu et al [8] higher wear loss is observed in some cases of 7 N. A tremendous reduction up to 76% in specific wear rate resulted in case of PA66 + 5 wt.% Cloisite 10A at a normal load of 13N

Coefficient of Friction

The COF measured under dry conditions for the polymer and its composites are shown in Figure 2. Measurements made at 13 N normal load showed lower COF. Cloisite 93A filler is the most effective in reducing the COF up to ~70%.

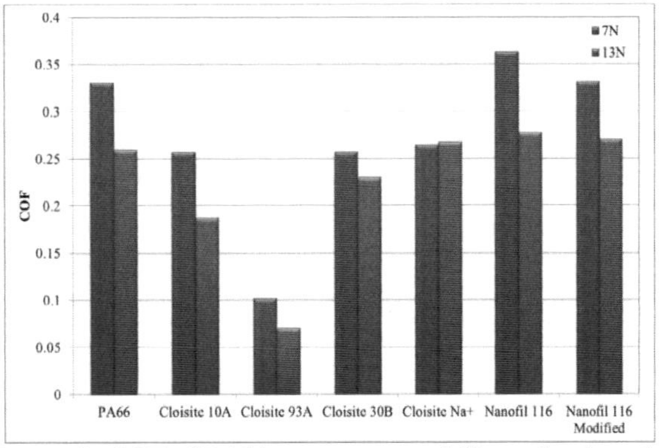

Figure 2: COF of PA66 composites at different loads under dry sliding conditions

In case of oil lubricated conditions, the measured COF decreased with increase in normal load except for Nanofil 116 and Colisite 93A.

Effect of load on COF under oil lubricating condition for a sliding velocity of 0.01 m/s is shown in Figure 3 below.

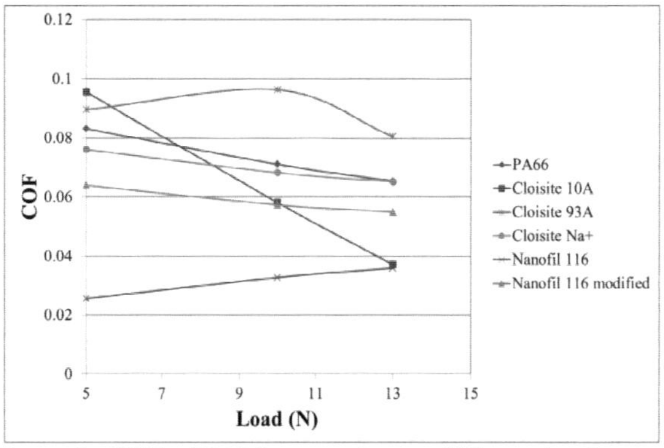

Figure 3: COF of PA66 composites at different loads at a sliding velocity of 0.01 m/s in oil lubricated condition

At higher sliding velocities, the influence of load on COF is less dominating (Figure 4, 5)

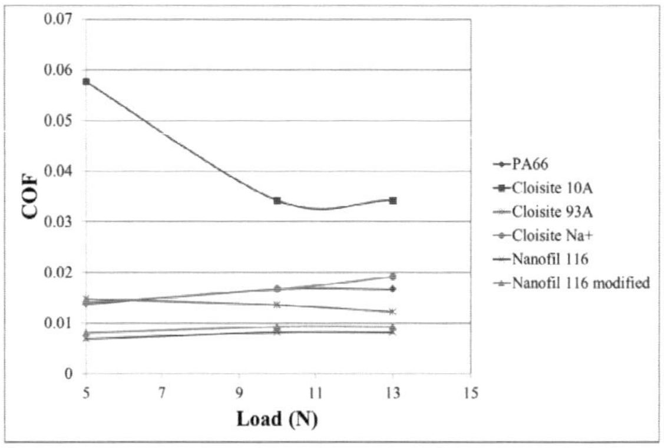

Figure 4: COF of PA66 composites at different loads at a sliding velocity of 0.2 m/s in oil lubricated condition

In case of PA66/Nanofil 116 composites the tendence was diffierent for different sliding velocities. At lower velocities the COF increases with increase in load and at a velocity of 0.6 m/s the tendence is inverse. Strong dependence of load on COF can been seen for PA66/Cloisite 10A compound for all sliding velocities.

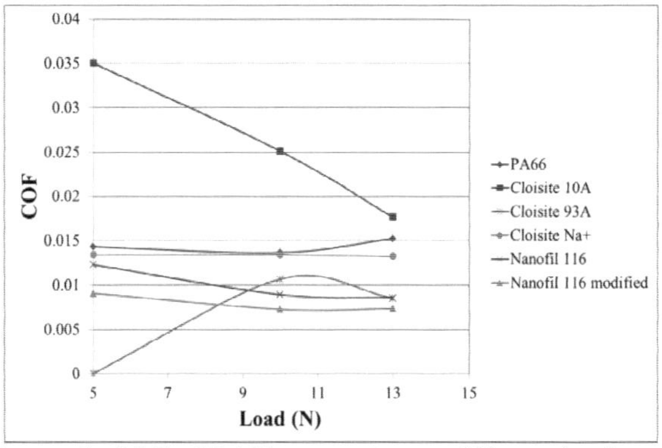

Figure 5: COF of PA66 composites at different loads at a sliding velocity of 0.6 m/s in oil lubricated condition

The Figure 4 shows the change in COF with respect to the sliding speed.

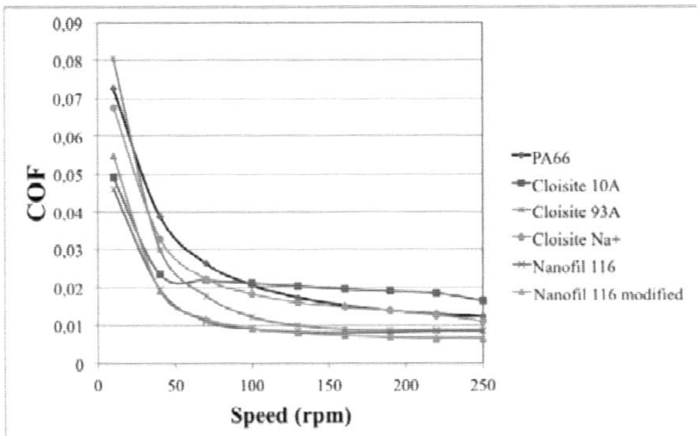

Figure 6: COF of PA66 composites at different sliding velocities by 13 N load in oil lubricated condition

The COF of most of the composites are clearly lower than the COF of unfilled PA66. Cloisite Na+ filler have a very small influence on changing the fricitonal properties of this PA66. With the addition on 5 wt.% Cloisite 10A, no decrease in the COF could be observed at higher sliding velocities. Tremendous reduction in friction value is achieved by the addition of Nanofil 116 natural and modified composites.

Conclusions

In this study, the incorporation of 5 wt. % of different organically modified MMT clays on the influence of wear and frictional properties of PA66 in dry and oil lubricated conditions were investigated.

Both COF and specific wear rate decrease with increase in load under dry sliding conditions and constant sliding velocity.

In case of oil-lubricated conditions, varying effects were observed on the influence of load on COF. Except Nanofil 116 and Cloisite 93A, the COF decreased for other composites with increasing load. However, the influence of normal load on the COF was dominating at low sliding velocities.

Addition of 5 wt.% of Nanofil 116 (both natural and modified) showed about 50% reduction in the COF of PA66.

References

1. Dasari, Aravind, Zhong-zhen Yu, and Yiu-wing Mai, "Clay exfoliation and organic modification on wear of nylon 6 nanocomposites processed by different routes." Composites Science and Technology 65, no. 2005 (2006): 2314-2328.
2. Jia, Q M, M Zheng, C Z Xu, and H X Chen, "The mechanical properties and tribological behavior of epoxy resin composites modified by different shape nanofillers." Polymers for Advanced Technologies (2006): 168-173.
3. Srinath, G, and R Gnanamoorthy, "Effect of nanoclay reinforcement on tensile and tribo behaviour of Nylon 6." Journal of Materials Science 0 (2005): 2897 - 2901.
4. Q. M. Jia, M. Zheng, C. Z. Xu and H. X. Chen," The mechanical properties and tribological behavior of epoxy resin composites modified by different shape nanofillers", Polymers for Advanced Technologies, 2006
5. Friedrich, K, Z Zhang, and a Schlarb, "Effects of various fillers on the sliding wear of polymer composites." Composites Science and Technology 65, no. 15-16, 2005
6. G. Srinath, R. Gnanamoorthy ,"Two-body abrasive wear characteristics of Nylon clay nanocomposites—effect of grit size, load, and sliding velocity", Materials Science and Engineering: A,2006
7. Mu, Bo, Qihua Wang, Tingmei Wang, Honggang Wang, Lingqi Jian, and Xianqiang Pei, "Preparation and Friction Properties of PBT / MMT Composites." Polymer (2009): 1-10.
8. Bo Mu, Qihua Wang, Tingmei Wang, Honggang Wang, Lingqi Jian ," The Friction and Wear Properties of Clay Filled PA66", POLYMER ENGINEERING AND SCIENCE, 2008
9. Srinath, G, and R Gnanamoorthy, "Sliding wear performance of polyamide 6 – clay nanocomposites in water." Composites Science and Technology 67 (2007): 399-405.
10. Jia, B, T Li, X Liu, and P Cong, "Tribological behaviors of several polymer–polymer sliding combinations under dry friction and oil-lubricated conditions." Wear 262, no. 11-12 (May 10, 2007):
11. http://www.nanoclay.com
12. G.W. Ehrenstein, G. Riedel, P. Trawiel, " Praxis der Thermischen Analyse der Kunststoffen", 2, Carl Hanser Verlag, Munich 2003.

O-07-107

THE RELATION BETWEEN COMPATIBILITY AND MAGNETIC PROPERTIES OF POLYMER BONDED MAGNETS FILLED Fe_3O_4 NANOPARTICLES

S. Mousavian[1], H. Ebadi-Dehaghani[2]*, D. Ashouri[1] and H. Sadeghipour[a]
[1] *Department of Chemical Engineering, Gachsaran Branch, Islamic Azad University, Gachsaran, Iran*
[2] *Department of Polymer Engineering, Shahreza Branch, Islamic Azad University, Shahreza, Iran-ebadi@iaush.ac.ir*

Abstract - The effect of binder on the magnetic properties of polymer bonded magnets (PBMs) was studied. Five PBMs containing 5 wt% of Fe_3O_4 nanoparticles were prepared by melt extrusion in a twin screw extruder. The selecting bonders were Polypropylene, Nylon 66, High impact polystyrene, Polycarbonate and Acrylonitrile butadiene styrene. Compression molded samples were prepared. The curve of saturation magnetization strength (Bs), versus surface energy was plotted for different nanocomposites. Bs had a good linear relationship with the surface energy. Regarding to the magnetic and rheological properties there was a correlation between homogeneity within the nanocomposites, which investigated via dynamic rheometry, and their magnetic properties. Magnetic properties of the nanocomposites at an equal loading content of nanoparticles depend on the interphase interactions and physical processes between particles.

Keywords: PBM, nanocomposite, Fe_3O_4, magnetic properties.

Introduction

The advantages of polymer bonded magnets (PBMs) over their metallic and ceramic counterparts include low weight and cost, resistance to corrosion, ease of machining and forming, and capability of high production rates [1]. Each materials has unique properties that make it more suitable for selected applications than other magnet and mechanical option such as, magnetic strength, resistance to demagnetization, density, physical strength, and flexibility [2]. The permanent magnets play an important role in the contemporary technology. The area of application of the hard magnetic materials grows along with the improvement of their magnetic, electrical, mechanical, and thermal properties. Permanent magnets are used, among others, in the instruments and devices for telecommunications, computer industry, as well as in the control and measurement technology. Their use gets lastly wider and wider in the electrical machines. The impregnation of magnetic fillers in the matrix imparts magnetic properties and modifies the physical properties of the matrix considerably. The mechanical properties of PBMs depend strongly on properties of polymer matrix, magnetic fillers, and interfacial condition between the components [3]. Also the surface area of the filler is the primary factor to control the mean distance between particles. The higher the surface area, the shorter the interaggregate distance would be [4]. Although there are many works done by researchers about PBMs [5-8], it is very little said about the effect of compatibility of matrix on the magnetic properties of PBMs in scientific literature. In the current work this effect was studied with different polymeric binders containing 5wt% Fe_3O_4 nanoparticles as a ferromagnetic material.

Experimental

Materials

Polypropylene (PP)was provided by Maroon Petrochemical Company (Iran) under a trade name of X30S, the density was 0.9 g/cm^3, melt flow index was 9 g/10 min. Nylon 66 was Tecnomide NA40 NLE grade (Eurotec Company, Turkey). High impact polystyrene (HIPS) was Tairirex HP 9450 grade (Taiwan). Polycarbonate (PC) was PC-110 Wonderlite grade (Chimeicorp, Taiwan). Acrylonitrile butadiene styrene (ABS) was SD- 0150 grade (Tabriz Petrochemical Co., Iran). Iron oxide (Fe_3O_4) nanoparticles was provided by Nano Pars Spadana Co. (Esfahan, Iran) and had a density of 0.65 g/cm^3, and a diameter mean of d=30-40nm.

Material Processing

All of the weighted materials were dried in an oven at 80°C for 12 hours before melt extrusion. Next the mentioned polymers' pellets and Fe_3O_4 nanoparticles (5wt% of the matrix) were dry blended and fed in a ZSK (ZSK 18 Megalab, Germany) co-rotating twin-screw extruder. The mixing was carried out at a rotor speed of 100 rpm. The extruded blends were pelletized at the die exit, dried and then compression molded for 2 min, to obtain circular discs for the rheological measurements. The samples were quenched in the cold press of the compression molding machine. For each type of nanocomposite, as well as for the neat polymer, the same procedure was applied so that the thermo-mechanical history of the nanocomposites and that of neat polymers remain similar.

Scanning Electron Microscopy (SEM)

A piece of the compression moulded samples containing 5 wt% of nanoparticles was cut for SEM studies. The dispersion quality of the nanoparticles within the matrix, and the nanostructures of the nanocomposites were investigated using a scanning electron microscope (SEM, Stereoscan-260, Cambridge, UK). The specimens were fractured in liquid nitrogen and coated with gold using a sputter coater, mounted and observed.

Melt dynamic rheometry

All rheological measurements were performed on a controlled strain rheometer (MCR 300 Rheometer, Anton Paar Physica, Otriš) with a torque transducer range of 1-15000 μNm using 25 mm diameter parallel plates. Testing sample disks with a thickness of 0.7 mm and a diameter of 2.5 mm were used. Oscillatory shear measurements were conducted within a 1% strain and frequency range of 0.03-600 rad/s. A sinusoidal strain of the form

$\gamma_{(t)}=\gamma_0 \sin(\omega t)$

where γ_0 is strain amplitude, ω is oscillatory frequency and t is time, was imposed on testing samples. Rotation mode measurements were conducted within a shear rate range of 0.01-100 s^{-1}. For ensuring the reliability of rheological data all measurements were conducted under nitrogen atmosphere to minimize oxidative degradation of the polymer and nanocomposites. Regarding to the melting temperature of used polymers different temperatures were used. The rheological measurements were carried out at 180, 230, 230, 260 and 280°C for PP, PC, ABS, HIPS and nylon 66 nanocomposites respectively.

Magnetic Properties

The magnetization crves of the samples were obtained in a Lakeshore 7300 Vibrating Sample Magnetometer (VSM) at room temperature.

Results and Discussion

Dispersion of nanoparticles in the matrices

The morphology of nanocomposites was evaluated by SEM to observe the distribution of nanoparticles within the extruded nanocomposites. Fig. 1 shows micrographs of a cryofractured surface of the nanocomposites. A partial fine dispersion was seen for any nanocomposite. However, partial coalescence was unavoidable in all nanocomposites. Several particles had undergone coalescence especially at low viscosity matrices. A good dispersion of nanofillers within the matrix could improve the magnetic properties of the nanocomposites even at lower nanofiller loadings owing to enhanced filler-matrix interaction.

Melt dynamic rheometry

Rheometry was used as a powerful tool for study the internal microstructure of polymer nanocomposites containing nanoparticles. Shear thinning behaviors of PBM nanocomposites are shown in Fig. 2. As seen the shear rate viscosities of PP and ABS nanocomposites are greater than that of Nylon 66, PC and HIPS nanocomposites. It is attributed to higher intristic viscosities of them and also good interactions with the Fe_3O_4 nanoparticles.

Shear stress-shear rate curves for the nanocomposites were shown in Fig. 3. As seen in the Figure there is a yielding point for any sample, but for HIPS nanocomposite yielding point happened at lower shear

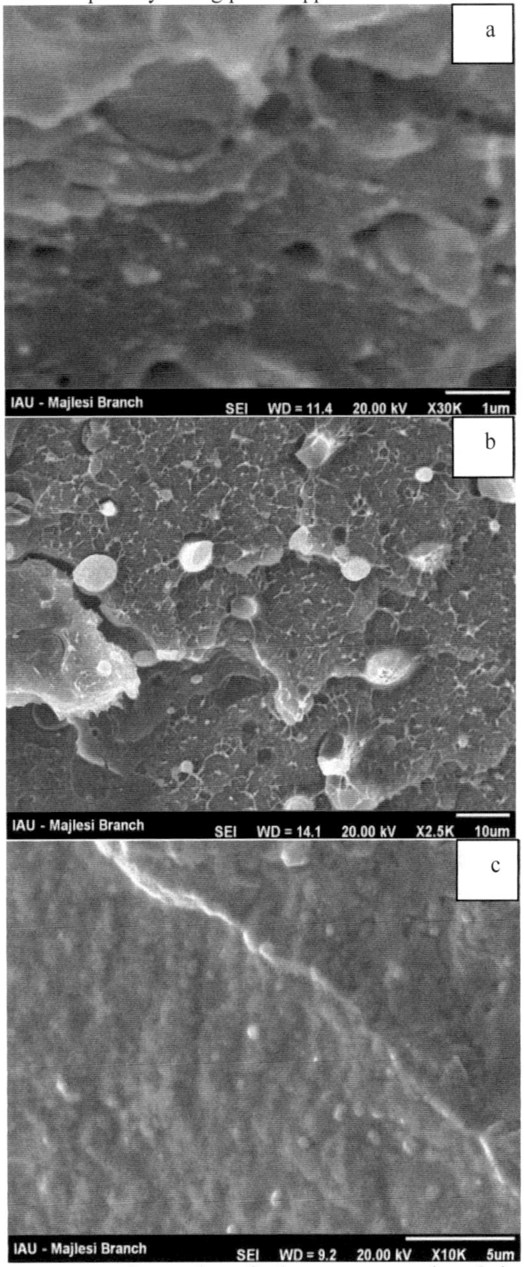

rates comparing to that of other nanocomposites. It is due to the weak interaction between nanoparticles and the HIPS matrix and also the coalescence of nanoparticles in this nanocomposite as seen in the SEM micrographs. The agglomerated particles could concentrate the imposed stress [9] so the yield stress reached at lower shear rates comparing to other nanocomposites. Similar results were obtained for

PP/$CaCO_3$ [10] and PP/ZnO [11] nanocomposites previously. In the nylon 66 nanocomposite the behavior was different and there wasn't any yielding point. It is attributed to a more compatibility between the matrix and nanoparticles comparing to that for other polymeric binders.

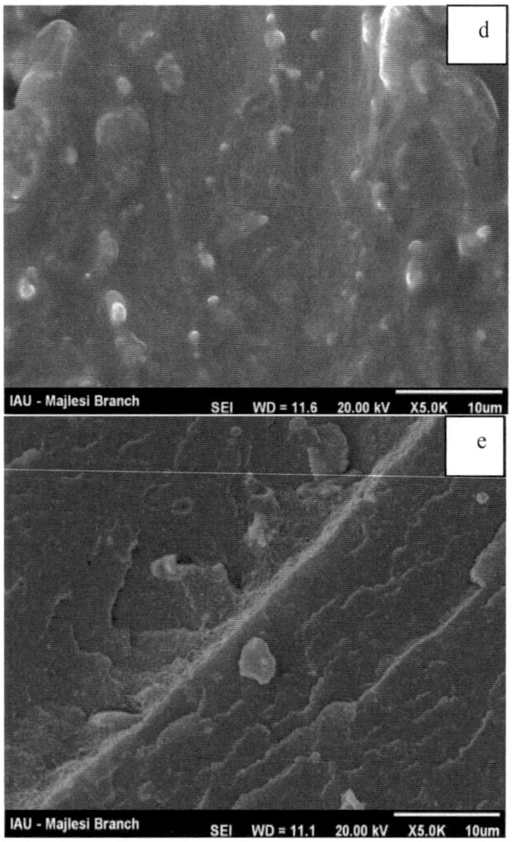

Figure 1 Scanning electron micrographs of a cryofractured surface of the PBM nanocomposites: a) PP/Fe_3O_4 b) HIPS/Fe_3O_4 c) Nylon 66/Fe_3O_4 d) PC/Fe_3O_4 e) ABS/Fe_3O_4

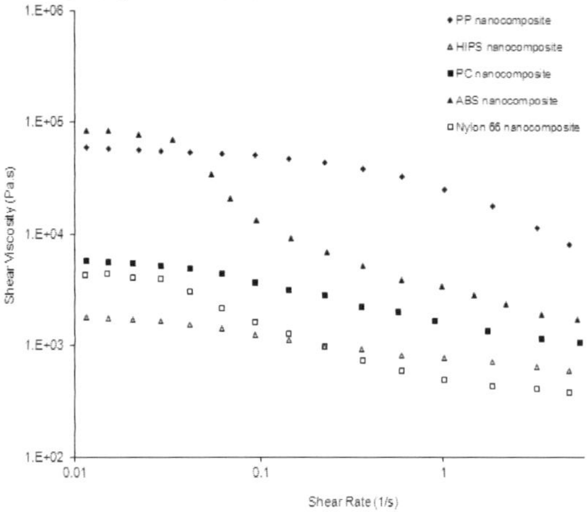

Figure 2 Shear viscosity vs. shear rate for PBM nanocomposites.

Magnetic Properties

Fig. 4 shows the magnetic hysteresis loops of polymer/nano-Fe_3O_4 composites. Clearly, by changing the matrix, magnetic properties of the polymer/nano-Fe_3O_4 composites changed. The composites show typical characteristics of paramagnetism. With the magnetic intensity increasing, saturation magnetization strength of the nanocomposites increased too. When the magnetic strength increased to 8,500 Oe, the value of saturation magnetization strength tends basically to steady.

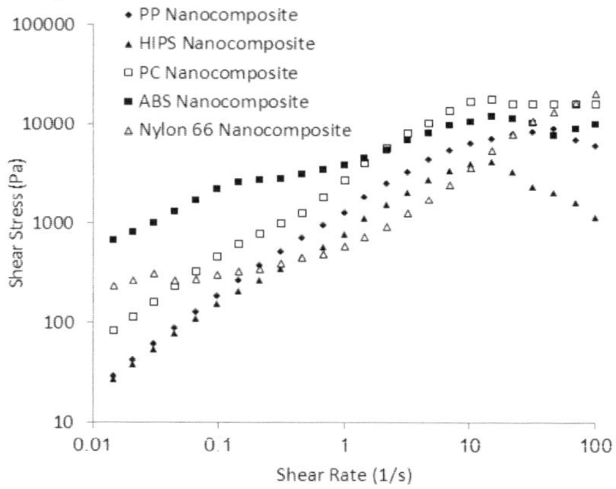

Figure 3 Shear stress-shear rate curves of PBM nanocomposites.

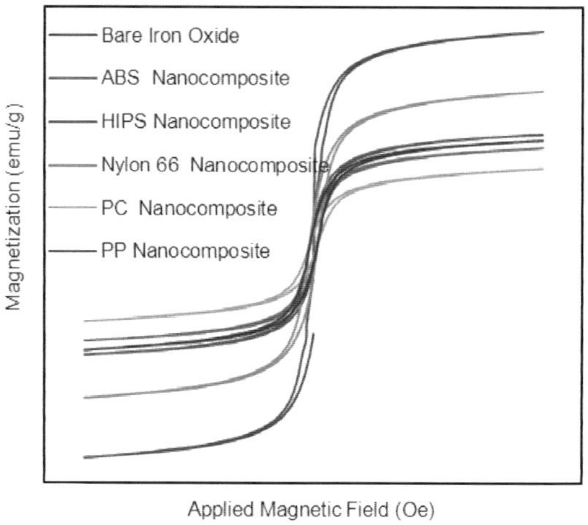

Figure 4 Magnetization curves of Fe_3O_4 and nanocomposites with different matrices and constant Fe_3O_4 contents at 300K.

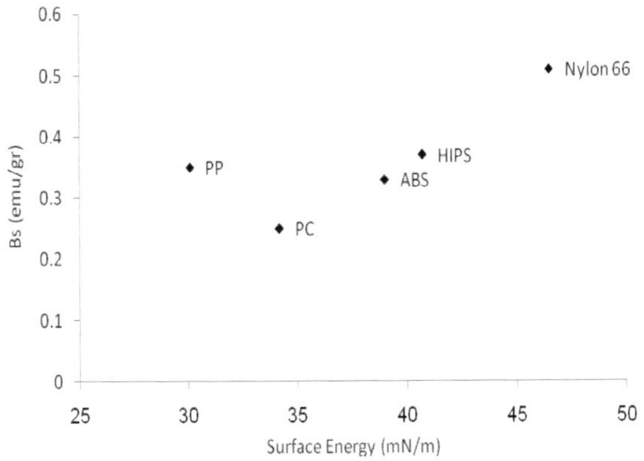

Figure 5 Saturation magnetization strength (Bs) versus surface energy for different nanocomposites.

Figure 5 shows the changing curve of saturation magnetization strength (Bs) versus surface energy of different nanocomposites' matrices [12]. As seen Bs has a good linear relationship with the surface energy. This is because magnetic properties of the nanocomposites at an equal loading content of nanoparticles depend on the interphase interactions and physical processes between particles. The OH groups in the Fe_3O_4 structure [13] could have a good interaction in the Nylon 66 matrix. So this polymer is a good binder for the magnetic nano-Fe_3O_4 particles. Meanwhile, nano-Fe_3O_4 particles are ferrimagnetics which have typical characteristic of high frequency; therefore, hysteresis losses are lesser. Regarding to the magnetic properties there is a correlation between melt viscosities of nanocomposites and their magnetic properties. The viscosity is corresponded to frictional interactions between the nanoparticles and also between the polymer chains and nanoparticles [14-17]. Therefore increasing the content of nanoparticles increases the number of frictional interactions between nanoparticles and between nanoparticles and matrix. It should be mentioned that any factor that confine the motions of the magnetic moments of nanoparticles within the matrix either in the melt or in the solid state can decrease the magnetic properties. So the higher the melt viscosity, the lower the magnetic property would be. Another point is the compatibility and homogeneity of nanocomposites that was seen in the SEM images and diagram of Bs versus surface energy of the nanocomposites.

Conclusions

Magnetic nanocomposites were prepared by melt extrusion in a twin screw extruder. Compression molded samples were prepared. Scanning electron microscopy (SEM) indicated a good dispersion of the modified clay nanofiller within the matrices, except for HIPS nanocomposite, that could enhance the rheological and magnetic properties of the nanocomposites even at low nanofiller loadings owing to enhanced filler-matrix interaction. Bs had a good linear relationship with the surface energy. Regarding to the magnetic properties there was a correlation between melt viscosities of nanocomposites and their magnetic properties. Regarding to the magnetic and rheological properties there was a correlation between homogeneity within the nanocomposites, which investigated via dynamic rheometry, and their magnetic properties. Magnetic properties of the nanocomposites at an equal loading content of nanoparticles depend on the interphase interactions and physical processes between particles and consequently compatibility with a matrix and particles.

Acknowledgements

The work described in this paper was supported by a grant from the Islamic Azad University. The authors gratefully acknowledge the Research Vice Chancellor of Islamic Azad University of Gachsaran, Shahreza and his co-workers. The valuable results were obtained by dynamic rheometry.

References

1. J. Otaigbe, H.S. Kim, J. Xiao *Polym. Comps.* 1999, *20*, 697
2. J. Ormerod, S. Constantinides *J. Appl. Phys.* 1997, *81*, 4816.
3. J. Xiao, J.U. Otaigbe *Polym. Comps.* 2000, *21*, 332
4. M.J. Wang *Rub. Chem. Tech.* 1998, *72*, 430.
5. K.A. Malini, M.R. Anantharaman, S. Sindhu *J. Mater. Sci.* 2001, *36*, 821.
6. R. Alizadeh, J. Beaudoin, L. Raki, V. Terskikh *J. Mater. Sci.* 2011, *46*, 460
7. H. Wang, W. Luo, J. Chen *J. Mater. Sci.* DOI 10.1007/s10853-012-6493-0, 2012
8. H. Wang, Y. Lan, M.A. Crimp, C.L. Lin, N. Shukla, T. Zeng, D. Wang, K. Li, Z. Ren, G. Chen *J Mater Sci* 2012DOI 10.1007/s10853-012-6498-8
9. L.E. Nielsen, F.L. Landel *Mechanical properties of polymers and composites* Marcel Dekker, New York, 2th edd, 1994
10. M. Mohagheghian, H. Ebadi-Dehaghani, D. Ashouri, S. Mousavian *Comps. B*, 2011, *42*, 2038.
11. S. Karamipour, H. Ebadi-Dehaghani, D. Ashouri, S. Mousavian *Polym. Test.* 2010, *30*, 110.
12. http://www.surface-tension.de/solid-surface-energy.htm
13. J. Sun, S. Zhou, P. Hou, Y. Yang, J. Weng, X. Li, M. Li *J. Biomedic. Mater. Res. A* DOI 10.1002/jbm.a
14. Y. Chen, H. Li *Polymer* 2005, 46, 7707.
15. V. Oprea, C. Simonescu *Plast. Kayt.* 1973, *20*, 179.
16. Utracki LA *Polymer Blends Handbook*, Kluwer Academic Publishers, Netherlands, 1973.
17. J.J. Aklonis, W.J. *Macknight Introduction to viscoelasticity*, Wiley, New York, chapter 2, 1983.

O-07-360

EFFECT OF PROCESSING STRATEGIES ON THE MORPHOLOGY AND THE ELECTRICAL CONDUCTIVITY OF THE CARBON NANOTUBE FILLED BLENDS OF POLYPROPYLENE AND POLYAMIDE 6

Nabaneeta Mukhopadhyay, Arup R. Bhattachryya* and Ajay S. Panwar
Department of Metallurgical Engineering and Materials Science, Indian Institute of Technology Bombay, Powai, Mumbai 400076, India– arupranjan@iitb.ac.in

Abstract - Blends of polypropylene (PP) and polyamide6 (PA6) (50/50, wt/wt) were melt-mixed with varying amount of multiwall carbon nanotubes (MWNTs) with the intention to develop conducting composites. Simultaneous, sequential and masterbatch mixing strategies were used to study the effect of various processing strategies on the electrical conductivity of the composites. The state of dispersion of MWNTs in these blends was assessed through scanning electron microscopy (SEM), AC electrical conductivity measurements and differential scanning calorimetric (DSC) analysis. Electrical conductivity measurements indicated that blends prepared with PP/MWNTs masterbatch showed lower electrical percolation threshold as compared to corresponding blends prepared by simultaneous mixing, due to the uneven distribution of the filler in the various phases. To understand the role of compatibilizer on the morphology and electrical conductivity of the blends, polypropylene grafted maleic anhydride (PP-g-MA) was utilized in the blend system. In order to improve the state of dispersion of MWNTs, an organic modifier: lithium salt of 6-aminohexanoic acid (Li-AHA) was used.

Keywords: Melt-mixing, Carbon nanotubes, Morphology, Electrical conductivity

Introduction

Polymer blending is a convenient route for the development of new polymeric materials, which combine the excellent properties of more than one existing polymer [1]. Most of the polymer blends are immiscible, where the constituent polymers exist as a phase-separated mixture in the blend. These incompatible blends are characterized by a two-phase morphology with sharp interface which exhibit poor physical and chemical interactions across the phase boundaries and poor mechanical properties [2]. Immiscible blends can develop different kinds of microstructures during processing ranging from matrix-droplet type and fibrillar to co-continuous morphologies. These microstructures determine various properties of the blends. Amongst the different microstructures, blends with co-continuous morphology have unique properties. To stabilize the blend morphology, incorporation of fillers is an efficient strategy. Selective localization of the conducting fillers in one of the phases or at the interface of co-continuous binary immiscible blends can lead to conducting polymer composites for several practical applications [3]. In this method, conducting composites can be developed using very low content of conducting fillers utilizing the concept of 'double percolation' [4], where the phase morphology of the blends depends on the concentration, shape, type, and dimension of the filler. Conducting polymer composites exhibit a sudden increase in the bulk electrical conductivity at a critical concentration of filler named as 'percolation threshold'. Recently, carbon nanotubes (CNTs) have attracted a great interest as conducting filler in polymer composites because of the possibility to achieve reasonably high electrical conductivity at low CNTs content due to its high aspect ratio (L/D).

Compatibilizers are known to stabilize the phase morphology in binary blends. These compatibilizers can improve the thermal, electrical and mechanical properties of the blends. They even alter the phase stability and morphology of the polymer blends. They restrict at the interface of the two polymers and stabilize the morphology by enhancing interfacial interaction.

Experimental

Materials

PP was obtained from Reliance Industries (H200MA) with melt flow index of 23 (2.16 kg load, 190°C). Polyamide 6 (PA6, zero shear viscosity = 180 Pa.s at 260°C) was obtained from GSFC, Gujarat, India (Gujlon M28RC, relative viscosity 2.8, M_v is 38,642 in 85% formic acid).MWNTs (NC 3100, average length: 1.5 lm, average diameter: 9.5 nm, carbon purity (95%) were obtained from Nanocyl CA. PP-g-MA (P-408) of density 0.91 g/ml, MAH content ranges from 1.6-2.5%, MFI=40 (P-408) was procured from Pluss Polymers, India.

Sample Preparation

Neat blends of 50/50 PP/PA6 (wt/wt) and blends with various amounts of MWNTs were prepared by melt-mixing in a conical twin-screw microcompounder

(Micro 5, DSM Research, Netherlands) at 260°C with a rotational speed of 150 rpm for 10 min. In the masterbatch dilution protocol, masterbatches of PP with 5 wt% MWNTs and PA6 with 5 wt% MWNTs were prepared (under the same extrusion conditions mentioned above). Required amount of the masterbatch was then subsequently diluted further in the next step of melt-mixing under the same extrusion condition mentioned above. Blends of 50/50 (wt/wt) PP/PA6 with masterbatch of PP-g-MA and MWNTs (unmodified and 1:1 Li-AHA modified MWNTs) were prepared using sequential mixing. Four mixing sequences were followed viz: i) Masterbatch was first mixed with PA6 for 10 minutes and then PP was added and mixed for 5 more minutes ii) Masterbatch was premixed with PP and then PA6 was added iii) one step mixing for 15 minutes iv) PP and PA6 were premixed and then masterbatch was added. All experiments were performed under nitrogen atmosphere to prevent oxidative degradation.

Table 1 - Codes and compositions of PP/PA6 blends with PP/MWNTs and PA6/MWNTs masterbatch

CODE	PP (P) wt%	MWNTs (C) wt%	PA6 (N) wt%
PP_5C	95	5	0
PA6_5C	0	5	95
xNmCyP PA6 masterbatch	x(30-70)	m(1.5-3.5)	y(70-30)
xPmCyN PP masterbatch	x(30-70)	m(1.5-3.5)	y(70-30)

Injection-molded samples (according to ASTM D 638, Type V) were prepared using mini injection-molding machine from DSM Research Netherlands. The injection-molding parameters maintained for all compositions were: injection pressure 3 bar, melt temperature 260°C, mold temperature 60°C, holding time 60 sec, and cooling time 2–3 min. The codes and composition details are listed in Table 1.

Characterization

The bulk electrical conductivity measurements were performed on injection-molded samples (across the thickness, 3 mm) in the frequency range between 10^{-1} and 10^{6} Hz using Alpha high-resolution analyzer coupled to a Novocontrol interface.

Morphology of PP/PA6 blends was studied through scanning electron microscopy (SEM), Hitachi S3400N instrument. Extrudate strands were cryofractured in liquid nitrogen and then selectively etched in respective solvents (xylene for PP phase and formic acid for PA6 phase). The etched surface was gold sputtered to avoid the charging of the sample.

DSC measurements were carried out using a DSC Q200 from (TA instruments) under nitrogen atmosphere. The extruded samples of about 5 mg were dried in a vacuum oven before experiment. The samples were heated till 260°C and were kept for 3 min to delete any previous thermal history followed by cooling to -50°C at a rate of 10°C/min and crystallization exotherms were recorded. Following this, samples were again heated to 260°C at a heating rate of 10°C/min to get the melting endotherms.

Transmission electron microscopic analysis was performed on a JEOL JEM-2100 F microscope at 200 KeV.The extruded samples were ultramicromed to get ~60 nm thick slice using Leica ultramicrotome at nitrogen atmosphere and finally the samples were put into 200 mesh copper TEM grids covered with thin amorphous carbon films. Further the PA6 phase was stained using phosphotungstic acid (1%).

Results and Discussion:

Part A: Effect of addition of MWNT and masterbatch approach

Morphological Analysis

Fig. 1 depicts the morphology of neat 50/50 (wt/wt) PP/PA6 blends. It shows a sea-island type of morphology when PA6 phase was etched out whereas in case of the blends where PP phase was removed selectively, shows continuous morphology.

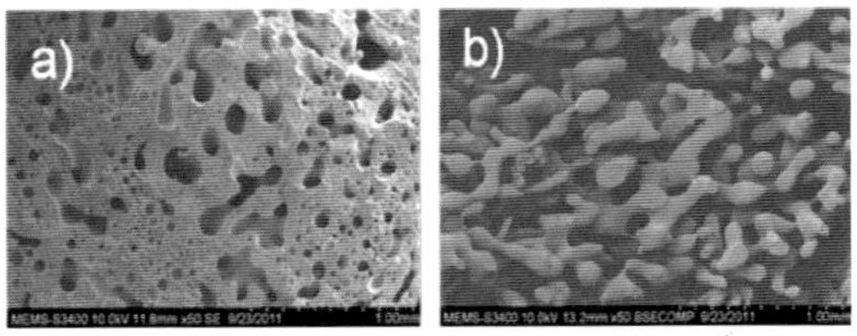

Figure 1 – SEM micrographs of neat blends of 50/50 (wt/wt) PP/PA6 a) PA6 phase etched with formic acid b) PP phase etched with hot xylene

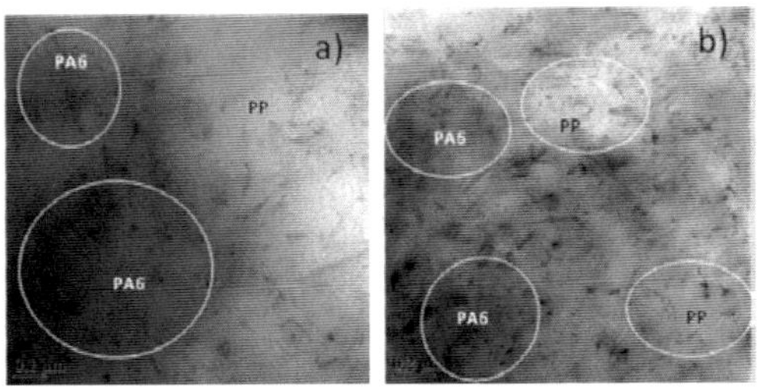

Figure 2 – HRTEM images of ultramicrotomed samples of 50/50 (wt/wt) PP/PA6 blends with PA6 masterbatch where PA6 phase was stained

50/50 PP/PA6 (wt/wt) blends with PP masterbatch showed a coarse co-continuous type of morphology whereas blends with PA6 masterbatch exhibited matrix-droplet type morphology [Fig. not shown]. At higher magnification, it was observed that in case of PP masterbatch based blends; nanotubes were mostly in the PP phase whereas in case of PA6 masterbatch based blends, large amount of nanotubes were present in the PP phase. So, for PA6 masterbatch based blends, significant amount of MWNTs has migrated to PP phase.

Incorporation of MWNTs in the PP phase was further confirmed by HRTEM images where PA6 phase was stained using phosphotungstic acid. Fig. 2 shows that in 50/50 (wt/wt) PP/PA6 blends with PA6 masterbatch (i.e. nanotubes were first mixed with PA6); MWNTs are present in the stained PA6 phase (darker appearance) as well as in the PP phase (lighter appearance).

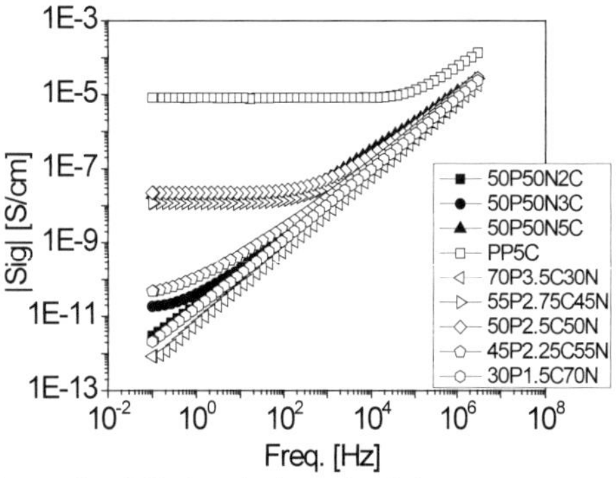

Figure 3 – AC electrical conductivity measurement of 50/50 (wt/wt) PP/PA6 blends with 2, 3 and 5 wt% MWNT (C) [one step mixing](filled symbols) and PP masterbatch based PP/PA6 blends (hollow symbols)

AC Electrical Conductivity Measurements

AC electrical conductivity measurements were performed to evaluate the state of dispersion of MWNTs in PP/PA6 blends. Fig. 4 shows AC electrical conductivity of PP/PA6 blends, where varying amount of MWNT is added to the blend using two different mixing protocol (one step mixing and of PP masterbatch based blends. For one step process electrical percolation threshold has been observed between 3-5 wt% for 50/50 PP/PA6 [Fig. 3 (filled symbols)], where as for PP masterbatch based blends , 50/50 PP/PA6 and 55/45 PP/PA6 showed conducting behaviour for 2.5 and 2.75 wt% of MWNTs respectively [Fig. 3 (hollow symbols)]. So, the percolation thershold is less for PP masterbatch based blends as compared to one step processed blends for 50/50 PP/PA6. In blends with PA6 masterbatch, the blends showed insulating behaviour [Fig. not shown].

Crystallization and Melting Behaviour

DSC analysis showed that the crystallization temperature (T_c) for PP has shifted to 131°C from 114°C with incorporation of 5 wt% MWNT (PP masterbatch) [Fig. 4]. This is because surface of MWNTs provides nucleation sites for PP to crystallize. But, in case of PP/PA6 blends with PP based masterbach, there is not much change in T_c of PP (around 123°C). This is due to the fact that PA6 itself acts as a heterogeneous nucleation site for crystallization of PP. So, when MWNTs have been incorporated along with PA6, the effect of various amounts of MWNTs is not significant as PA6 phase has already provided significant amount of heteronucleating effect to PP. Similar trend was observed for PA6 based masterbatch [Fig. not shown]

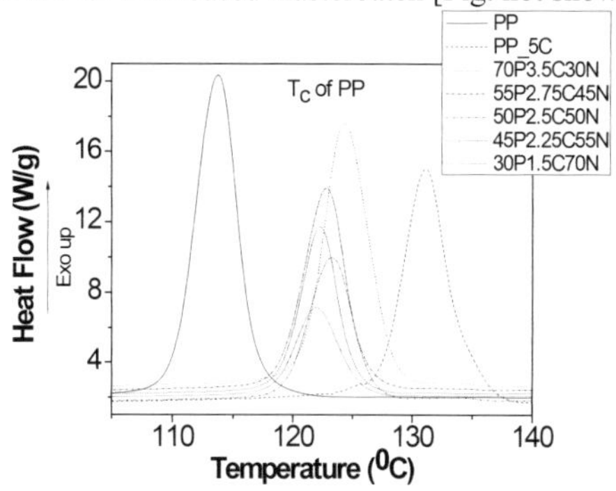

Figure 4 – Crystallization exotherm of PP in PP masterbatch based blends

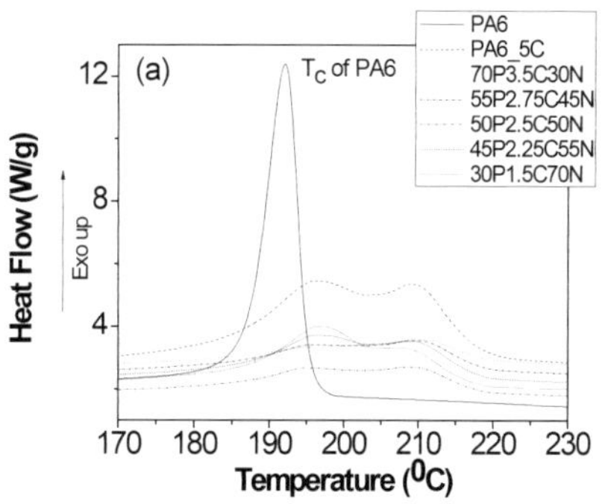

Figure 5 – Crystallization exotherm of PA6 in PP masterbatch based blends

Fig. 5 shows that addition of MWNTs results in an additional crystallization peak of PA6 at 209°C. This second crystallization peac occurs due to formation of transcrystalline lamella of PA6 near the interface of MWNT nucleated directly by the surface of MWNTs [5]. No significant change was observed in melting behaviour for PP masterbatch and PA6 masterbatch based blends. [Fig. not shown]

Part B: Effect of addition of compatibilizer

Masterbatch of PP-g-MA with 5 wt% MWNTs (unmodified and 1:1 Li-AHA modified) was incorporated into 50/50 (wt/wt) PP/PA6 blends via sequential mixing.

The torque values obtained during melt-mixing are indicative of melt viscosity of the system. It is observed that the mixing sequence affects the interaction between the components. Blend where PP-g-MA masterbatch is first mixed with PA6 shows highest torque values indicating good interaction between the PA6 chain and MA [Fig. not shown]. This trend was observed in both blends with PP-g-MA masterbatch of unmodified and Li-AHA modified MWNTs.

Morphological Analysis

50/50 (wt/wt) PP/PA6 blends with PP-g-MA masterbatch showed droplet-dispersed kind of morphology in all the samples with different mixing sequence. Droplet size is significantly reduced as compared to the neat 50/50 PP/PA6 (wt/wt) blends due to the compatibilization reaction of PP-g-MA. [Fig. not shown.]

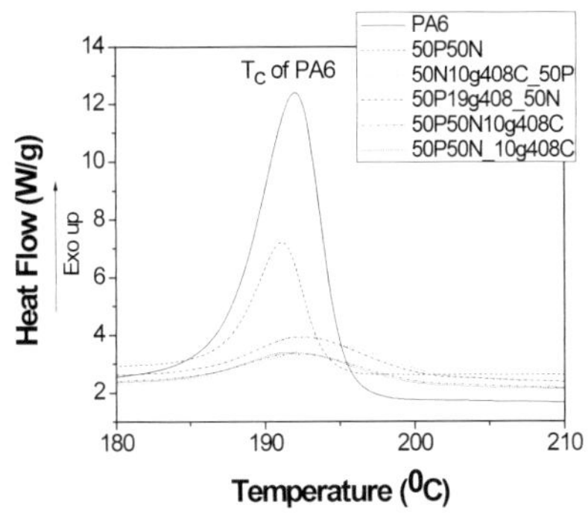

Figure 6 – Crystallization exotherm of PA6 in 50/50 (wt/wt) PP/PA6 blends with PP-g-MA/MWNTs masterbatch

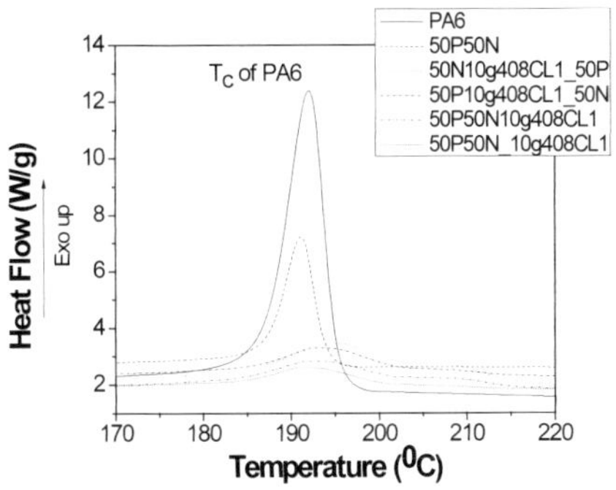

Figure 7 – Crystallization exotherm of PA6 in 50/50 (wt/wt) PP/PA6 blends with PP-g-MA/Li-AHA modified MWNTs masterbatch

Crystallization and Melting Behavior

It was observed that T_c of PP increases in the presence of PP-g-MA due to nucleating action of PP-g-MA [Fig. not shown]. Highest T_c of PP is observed in case of the blends where PP-g-MA masterbatch was premixed with PA6 phase. T_c of PA6 is delayed due to melt-interfacial reaction indicating difficulty in chain folding. No significant changes were observed in the melting behaviour.

In case of blends with PP-g-MA masterbatch, double peak of PA6 does not occur [Fig. 6]. In PP-g-MA/Li-AHA modified MWNTs masterbatch based blends, double peak of PA6 was observed for blends prepared by one step mixing and blends where PP-g-MA masterbatch is premixed with PA6 [Fig. 7]. This indicates that interaction of Li-AHA with PA6 phase leads to availability of more bare surfaces of MWNTs for PA6 to form crystilline lamella.

AC Electrical Conductivity Measurements

Masterbatch of PP-g-MA with 5 wt% MWNTs shows very high electrical conductivity. However, 50/50 (wt/wt) PP/PA6 blends with 10% of PP-g-MA/MWNTs masterbatch do not exhibit conducting behaviour due to inadequate amount of MWNTs (0.5wt%) which was unable to form a percolating network in the blends [Fig. not shown].

Conclusions

PP/PA6 (wt/wt) 50/50 blends with MWNTs were prepared using one step mixing and masterbatch approach. Blends with PP masterbatch showed better dispersion of MWNTs in the blends in terms of electrical percolation, whereas blends with PA6 based masterbatch showed insulating behaviour. Masterbatch of PP-g-MA with 5 wt% MWNTs shows very high electrical conductivity. However, 50/50 (wt/wt) PP/PA6 blends with 10% of that masterbatch do not exhibit conducting behaviour beacuse the effective amount of MWNTs present in those blends was 0.5 wt% which was inadequate to form a percolating network.

Shift in T_c in the crystallization studies of PP is observed in case of 50/50 (wt/wt) PP/PA6 since PA6 acts as heterogenious nucleating site for PP. Further shift in T_c of PP occurs in case of blends with PP-g-MA for the same reason. Also droplet size has decreased significantly due to compatibilizing action of PP-g-MA. In presence of compatibilizer, T_c of PA6 has shifted to lower temperature indicating delayed crystallization because of melt-interfacial reaction, which may be due to difficulty in chain folding.

Acknowledgements

The authors thank Microcompounder Central facility, SAIF and CRNTS, IIT Bombay.

References

1. L.A. Utracki, *Polymer Alloys and Blends*, Hanser, Munich 1989.
2. S. George, R. Joseph, S. Thomas, K. T. Varughese., *Polymer*, 1995, 36, 4405.
3. N.F. Colaneri and L.W. Shacklette, *IEEE Transact. Instrumentation Measurement*, 1992, 49, 292.

4. M. Sumita, K. Sakata, Y. Hayakawa, S. Asai, K. Miyasaka and M. Tanemura, *Colloid Polym. Sci.*, 1992, 270, 134.
5. A C Brosse, S Tencé-Girault, P. M. Piccione, L. Leibler *Polymer.* 2008, 49, 4680.

EFFECTS OF BAMBOO CHARCOAL ON THE ELECTRICAL AND RHEOLOGICAL PROPERTIES OF EVA COPOLYMER

Sofiane Belaïd[1, 2], Philippe Cassagnau[1, 2] and Gisèle Boiteux[1, 2]
[1] Université de Lyon, 69003 Lyon, France
[2] Université de Lyon 1, CNRS UMR 5223, Ingénierie des Matériaux Polymères, 69100 Villeurbanne, France

Abstract - The electrical and rheological properties of ethylene vinyl acetate copolymer (EVA) filled with bamboo charcoal were investigated. The composites were prepared by melt process in an internal batch mixer. Size distribution analysis showed that $d_{(50)}$ and $d_{(90)}$ values of bamboo charcoal particles are 12.7 μm and 40 μm respectively with a mean diameter of 22 μm. Scanning electron microscopy proved that the particles of bamboo charcoal present rectangular shape. The electrical percolation threshold was observed at 0.18 volume fraction (35 %wt) of bamboo. Beyond the percolation threshold, a considerable increase in electrical properties was observed up to limit value of 10^{-2} S/m. The rheological percolation was studied from different rheological models. As a result, the rheological percolation threshold was observed at 0.3 volume fraction (50 %wt) of bamboo charcoal contents. So, the electrical percolation occurs before the rheological percolation. This is, principally due to the filler's characteristics such as the specific surface area, the aspect ratio and the surface properties. Finally, the bamboo charcoal confers high electrical properties to EVA composite without inducing strong changes in its viscoelastic properties.

Keywords: Bamboo Charcoal; EVA; percolation; electrical properties; rheological properties; biocomposites.

Introduction

Electrically conductive polymers have a great interest in a lot of applications like pressure-sensitive sensor, transducer and EMI-shielding material. To improve the electrical conductivity of polymer materials, it is common practice to load polymer matrices with conducting fillers like metals fillers or more recently carbon fillers: carbon black, graphite, carbon filler and carbon nanotubes. At a critical volume fraction of the conductive filler, an abrupt increase of the composite conductivity is observed, indicating the percolation threshold corresponding to the formation of a conductive filler network in the polymer. For example, Sohi et al. [1] show that electrical percolation threshold of Ethylene Vinyl Acetate-Carbon Black composite was 0.14 volume fraction of CB.

Generally, carbon fillers are produced by the incomplete combustion of heavy petroleum products. So, it will be interesting to found an environmentally functional material that could have the same properties. Actually, bamboo fiber is the most promising filler. It is one of the world's best known natural engineering materials. Bamboo charcoal obtained by pyrolysis of bamboo gives good properties like prevention from electromagnetic wave, adsorbent, highly caloric fuel and antibacterial [2]. In 1878, a filament of bamboo charcoal has been used already by Thomas Edison for the first incandescent lamp. However, its utilization has not been fully explored and electrical conductive properties of bamboo charcoal based composites are expected. Consequently, our work aims to study the electrical and rheological behavior of composites filled with bamboo charcoal. The matrix used in the present study is an ethylene vinyl acetate (EVA) copolymer in order to compare our results with the works of Sohi et al.

Experimental

The ethylene vinyl acetate (EVA) used in this study was supplied by Arkema Evatane ®. The EVA 2803 containing 28%wt of vinyl acetate was used in the present study. The bamboo charcoal (BC) was kindly supplied by Bamboo Fibers Technology. The mean diameter measured using Mastersizer 2000 was 22 μm ($d_{(50)}$ = 12.7 μm). The different composites were prepared by dry mix process at 120°C in an internal batch mixer (Haake Rheomix 600) to obtain 20, 30, 35, 40, 50, 60 and 70%wt of bamboo fillers containing composites. The composition of composites was precisely determined by Thermogravimetric analysis from TA Instrument (TGA Q500). The ac-conductivity of different composites was performed using a TA Instruments DEA 2970 Dielectric Analyzer. The complex shear modulus (storage modulus G' and loss modulus G") of different composites were measured using an AR2000 Rheometer from TA Instruments. The composites were also studied by Scanning Electron Microscopy (SEM), HITACHI S800 (HITACHI Corp.).

Results and Discussion

Electrical properties

At low filler concentration, the electrical response of the composite is like the matrix response, it is insulator because there is no possible contact between particles. At a critical volume fraction of filler around 0.2, the conductivity increases abruptly due to the conductive filler network formation (conductive path). The composite becomes electrically conductor. This critical volume fraction is defined like the electrical percolation threshold. This can be clearly evidenced from the variation of dc-conductivity versus the filler's concentration as shown in Fig. 1.

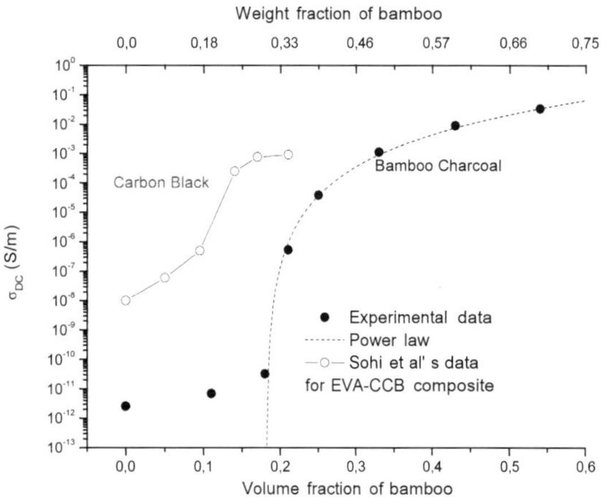

Figure 1 – Variation of the Dc-conductivity of the composite versus the volume fraction of bamboo charcoal. The dashed curve is the power law. For comparison with carbon black, the result of Sohi et al [1] have been plotted.

By using an power law, we derive an electrical percolation threshold at 0.18 volume fraction of bamboo. Sohi et al. [1] showed that the percolation threshold of Carbon black-EVA 2803 composite was observed at $\Phi c=0.14$. Compared with the present results, this slight difference is also probably due to low specific area of bamboo charcoal (0.82 $m^2.g^{-1}$) compared with CB one used in their work (≈240 $m^2.g^1$). However, above the percolation, the conductivity of EVA-Bamboo charcoal composite was higher than EVA-CB composite. So, bamboo charcoal is as efficient as carbon black to be used as conductive filler.

Rheological properties

Increasing the bamboo charcoal loading, a filler-filler and filler-polymer network are formed and the appearance of a transition from a liquid-like behaviour to a solid-like behavior is expected from the apparition of a secondary plateau of the storage modulus at low frequency which is obviously higher than the loss modulus [3]. This secondary plateau on storage modulus G' is clearly observable for 50, 60 and 70%wt (33, 43 and 54 %v respectively) of filler contents. Thus, we can say that the sol-gel transition, which defines the rheological percolation threshold, appears between 0.25 and 0.33 volume fraction (respectively 40 %wt and 50 %wt) of bamboo charcoal in the EVA matrix. On the other hand, the change from liquid-like to solid-like behavior due to the formation of the filler-filler and filler-polymer network is also observable in Fig. 2 from the variation of the absolute complex viscosity versus frequency. Actually, the solid-like behavior response is subsequently an apparent yield stress and a shear-thinning dependence on viscosity. According to Larson [4], it was suggested that when a yield stress is observed, that means when the volume fraction of particle is just enough to observe filler-filler contacts.

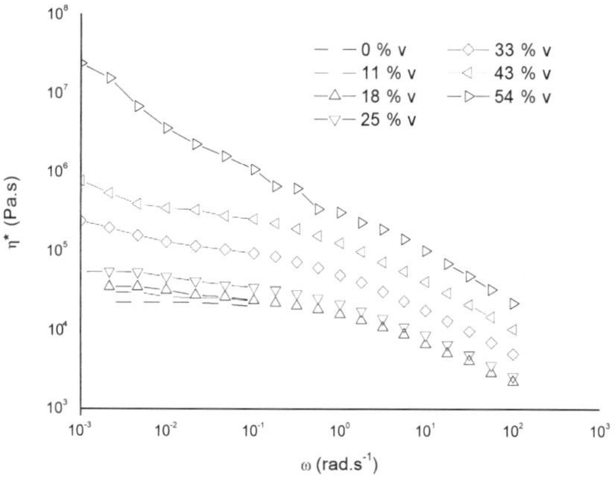

Figure 2 – Variation of the absolute complex viscosity |η*| versus frequency for different bamboo charcoal loadings.

The Carreau-Yasuda model with yield stress describes the frequency dependence of the absolute complex viscosity of nanocomposites. Fig. 3a shows that below the percolation threshold the hydrodynamic interactions are the dominant ones as a simple Carreau-Yasuda model ($\sigma_0=0$) fits well the variation of the absolute complex viscosity. Fig. 3b shows that Carreau-Yasuda model with yield stress fits perfectly the rheological behavior of the EVA composites. Actually, this model means that the hydrodynamic interactions dominate in the high frequency range whereas the particle-particle interactions are dominant in lower frequency due to the formation of the particle network.

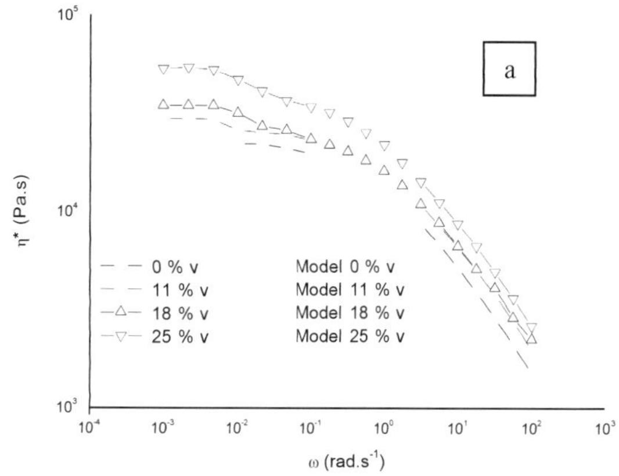

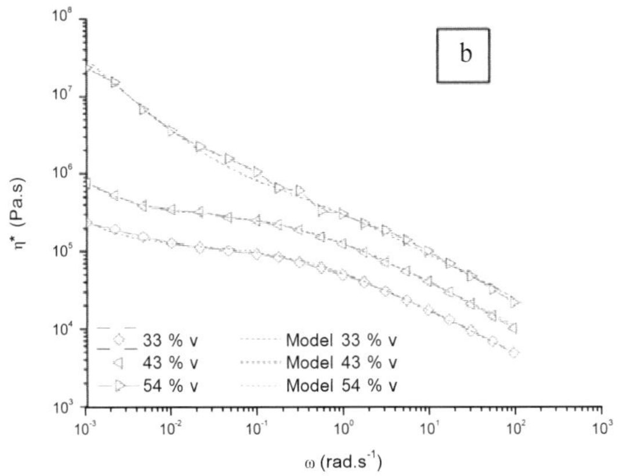

Figure 3 – Modeling of the variation of the absolute complex viscosity using the Carreau-Yassuda's model (dashed line) at different bamboo charcoal concentrations.

a) Concentrations lower than the percolation threshold, $\sigma_0 = 0$

b) Concentration higher than the percolation threshold.

Finally, it can be concluded that the formation of the three-dimensional filler network, defined as rheological percolation, appears at nearly $\Phi_R \approx 0.3$ volume fraction of bamboo charcoal. This rheological percolation threshold must be compared to the electrical percolation threshold $\Phi_C \approx 0.18$ as determined in the first part of the present work. This strong difference could be explained by the filler-filler distance required to induce electron transfer in the first case and stress transfer in the latter [5].

Conclusions

The electrical and rheological properties of EVA copolymer filled with bamboo charcoal particles were investigated. SEM and size distribution analysis showed that bamboo charcoal particle presents a rectangular shape with low specific surface area (0.82 m^2.g-1).

The electrical percolation threshold was observed at $\Phi_C \approx 0.18$ volume fraction of bamboo charcoal composite contents. Furthermore, the maximum of the electrical conductivity at higher concentration was observed around 10^{-2} S/m. This level of electrical conductivity proves that bamboo charcoal is as efficient as carbon black and carbon nanotubes for electrical applications of polymer composites.

The rheological percolation threshold is more difficult to define from a relevant criteria point of view. The yield stress behavior of the absolute complex viscosity was used as the rheological criteria to determine the percolation threshold. Furthermore, the rheological behavior was successfully modeled from a modified Carreau-Yasuda equation. The rheological percolation was found at $\Phi_R \approx 0.3$ volume fraction of bamboo charcoal composite contents.

Finally, this work shows that electrical percolation occurs before rheological percolation. This difference can be explained from the microscopic scale of the bamboo charcoal particles. As a result, bamboo charcoal confers high electrical properties to polymer matrix without inducing strong changes in its viscolelasticity properties. The final conclusion of the present could be the following question: Why to try desperately to disperse nanoscale particles while micrometric particles and cheaper particles such bamboo charcoal are easier to disperse and give better electrical conductivity properties, at least for bulk samples?

Acknowledgements

The authors thank Bamboo Fibers Technology for their bamboo charcoal providing.

References

1. N. J. S. Sohi; S. Bahdra; D. Khastgir *Carbon* 2011, *49,* 1349-1361.
2. G. Mingjie, *Manual for Bamboo Charcoal Production and Utilization,* Bamboo Engineering Research Center, 2004.
3. P. Cassagnau *Polymer* 2008, 49, 2183-2196.
4. R.G. Larson in *The structure and rheology of complex fluids*, Oxford University Press, New York, 1999.
5. S. Huang; Z. Liu; C. Yin; Y. Wang; Y. Gao; C. Chen; M. Yang *Macromolecular Materials and Engineering 2012*, 297, 51-59.

O-07-144

POLYAMIDE AS COMPATIBILIZER FOR POLYOLEFINE NANOCOMPOSITES

A. Witschnigg[1*], S. Laske[1], I. Duretek[1], C. Holzer[1]
[1]*Department of Polymer Engineering and Science, Chair of Polymer Processing, Montanuniversitaet Leoben, Austria – andreas.witschnigg@unileoben.ac.at*

Abstract - Layered silicates used as nanofillers are one of the most emphasized research topics in polymer engineering and science in the last decades. Nevertheless to achieve their great potential of enhancement a homogenous dispersion of the nanofiller in the polymer matrix is essential. The achievable material improvement is in great extend depending on the particle distribution (agglomerated, intercalated or exfoliated) of the filler occurring in the nanocomposite. It is well known that layered silicates are very well and easily dispersed in polyamides. For polyolefines, specific compatibilizers are needed to achieve a good particle distribution of the nanofiller. In most of the cases the polyolefine is grafted with maleic acid anhydride to adapt the polarity between the polar layered silicate and the nonpolar polyolefine. Nevertheless the enhancement level of polyamide nanocomposites could not be reached with this commonly known combatibilizers.
In this study the effect of polyamide as compatibilizer in polyolefine nanocomposites was investigated. Therefore different types of Polyethylene and Polypropylene nanocomposites were produced. The difference between polyamide as compatibilizer and a maleic acid anhydride polyolefin as compatibilizer was investigated by diverse methods such as tensile test, rheological measurements as well as thermal conductivity. It could be seen that polyamide as a compatibilizer for polyolefine has quite a potential of increasing material properties.

Keywords: nanocomposites, layered silicates, polyolefine, polyamide, material properties

Introduction

Nanoclays (especially layered silicates) are the most common nanofiller used in polymer processing due to their great potential of enhancing material and the fact that they are less expensive compared to other nanofillers. The proper use of layered silicate can enhance mechanical properties. Especially in Polyamids the enhancing effect of layered silicates is very high [1-4].
Due to the polarity of the layered silicates a successful use in non-polar polyolefines is exceedingly linked with the use of compatibilizers to adjust the polarity differences for a good particle distribution. Commonly polyolefines grafted with maleic acid anhydride are used as compatibilizer [5, 6].
In this study the effect of Polyamide as compatibilizer in polyolefines was investigated.

Experimental

Materials

The Polypropylene BA212E from Borealis Inc. Austria, with a melt flow rate (MFR) of 0.3 g/10 min; 230 °C/ 2.16 kg has been used. The second material used for the investigations was a high density Polyethylene BorPure MB6561 with a MFR of 1.5 g/10 min; 190 °C/2.16 kg. Two layered silicate types have been used. The first was a modificated montmorillonite type intercalated with dimethyl distearyl ammonium chloride (Nanofil5, Rockwood Clay Additives, United Kingdom). The second filler was an untreated natural montmorillonite with the brand name Cloisite Na^+ supplied by Rockwood Clay Additives. The polyamide type used for the production of the masterbatches was a PA 6 Schulamid HM natural. For the second series of masterbatches the compatibilizer Scona TPPP 2112 GA (PP grafted with maleic acid anhydride) respectively Scona TSPE 2002 GAHD (PE grafted with maleic acid anhydride) was used.

Preparation of the compounds

The nanocomposites were prepared via melt intercalation over a masterbatch step. For the preparation of this masterbatch an intermeshing, co-rotating twin screw extruder ZSK25 (Werner&Pfleiderer, Stuttgart, Germany) was used. The feed rate was set at 4 kg/h with a screw speed of 200 rpm. The masterbatch formulation was 50 wt.% compatibilizer or polyamide and 50 wt.% layered silicate. The melt temperature was set to 220 °C for the polyamide masterbatches and 180 °C for the common compatibilizer. The exact temperature profile can be seen in table 1 and table 2.

Table 1 – Temperature profile for the polyamide masterbatches

Cylinderzones	Temperature (°C)
1 (hopper)	cooling
2	180
3	200
4	220
5	220
6	220
die	220

Table 2 – Temperature profile for compatibilizer masterbatches

Cylinderzones	Temperature (°C)
1 (hopper)	cooling
2	160
3	180
4	200
5	200
6	200
die	200

Kneading

The finally produced nanocomposites were prepared via melt intercalation in a Haake Polylab System 3000P laboratory kneader from Thermo Fisher Scientific, Waltham, USA. The kneading temperature was set at 175 °C for the compatibilizer materials and 220 °C for the Polyamide composites. The kneading time was set to 5 min with a rotation speed of 60 rpm. The kneading times were measured from the time of adding the masterbatch. The formulation of the finally produced nanocomposites was 90 wt% polyolefine, 5 wt% compatibilizer or polyamide and 5 wt% layered silicate.

Production of plates

For the mechanical investigations plates with a thickness of 4 mm were prepared using a hydraulic vacuum press (Collin 200 PV, Dr. Collin, Ebersberg, Germany). The specimens for the different mechanical investigations were then cut out of these plates.

Mechanical properties – tensile test

A universal tensile testing device UTS EuroTest 250 (Hegewald & Peschke, Nossen, Germany) was used to carry out tensile tests according to ISO 527-1. All specimens were tested under standardized conditions (23 °C/ 50 %r.H.).

Measuring of the thermal conductivity

The thermal conductivity was measured with the K-System II based on the line-source method in melt and solid state according to ASTM D5930-97 in the temperature range from 80 °C to 200 °C.

Rheological investigations

The rheological measurements were carried out using a Physica MCR 501 (Anton Paar, Graz, Austria) rotational rheometer. All tests were performed using a 25 mm plate-and-plate geometry and gap height of 1 mm. To avoid oxidative degradation of the polymers nitrogen was used as a protective gas. The measurements were conducted in the dynamic oscillatory mode in the frequency range from 0.1 rad/s to 500 rad/s at a temperature of 200 °C.

In all measurements, the linearity of the sample response with respect to stress amplitude was verified. The data has been evaluated from at least 3 measurements for each sample.

Results and Discussion

Influence on Young's Modulus

Table 3 – Young's Modulus of the investigated materials and its change compared to the unfilled material

Material	Young's Modulus (MPa)	Change (%)
PE	1152	
PE-PA\Nanofil5	1200	4.1
PE-SCONA\Nanofil5	1189	3.2
PE-PA\Cl. Na+	1190	3.3
PE-SCONA\Cl. Na+	1203	4.5
PP	1474	
PP-PA\Nanofil5	1660	12.6
PP-SCONA\Nanofil5	1742	18.2
PP-PA\Cl. Na+	1648	11.8
PP-SCONA\Cl. Na+	1846	25.2

It can be seen (table 3) that the Young's Modulus is only increasing very slightly (around 4%) for the polyethylene nanocomposites. Due to the fact that this increase is lying in the range of the standard deviation no differences between the different mixtures could be observed.

Regarding the polypropylene nanocomposites the Young's Modulus is increasing around 12 % for the polyamide based nanocomposites for both filler types. Compared to the results for the polyethylene the Scona compatibilizer achieves higher values (increase of 18 and 25 %) compared to the polyamide counterpart. A possible explanation may be side reactions during processing leading to chain scission and therefore changes in molecular mass distribution. It can also be seen that the untreated natural layered silicate is leading to better values for the Young's modulus.

Table 4 – Tensile strength of the investigated materials and its change compared to the unfilled material

Material	Tensile strength (MPa)	Change (%)
PE	16.9	
PE-PA\Nanofil5	21.8	28.5
PE-SCONA\Nanofil5	23.6	39.2
PE-PA\Cl. Na+	25.4	49.7
PE-SCONA\Cl. Na+	18.4	8.7
PP	24.7	
PP-PA\Nanofil5	22.6	-8.4
PP-SCONA\Nanofil5	23.5	-4.7
PP-PA\Cl. Na+	22.6	-8.4
PP-SCONA\Cl. Na+	25.3	2.5

Table 4 is showing the results for the tensile strength. Regarding polyethylene very good results could be achieved. It can be seen that with polyamide and the untreated layered silicate Cloisite Na+ the tensile strength could be increased by nearly 50 %. The treated layered silicate Nanofil5 led to an increase of circa 29 %. For the Scona compatibilizer the best results could be achieved with the use of the treated layered silicate (39 %).

The polypropylene results are showing that both polyamide types are decreasing the tensile strength around 8 %. A reason for that, as mentioned before, might the processing at 220 °C leading to accelerated chain scission. The Scona based nanocomposites are showing different behavior. For this material class the untreated silicate could slightly increase the tensile strength.

Influence on thermal conductivity

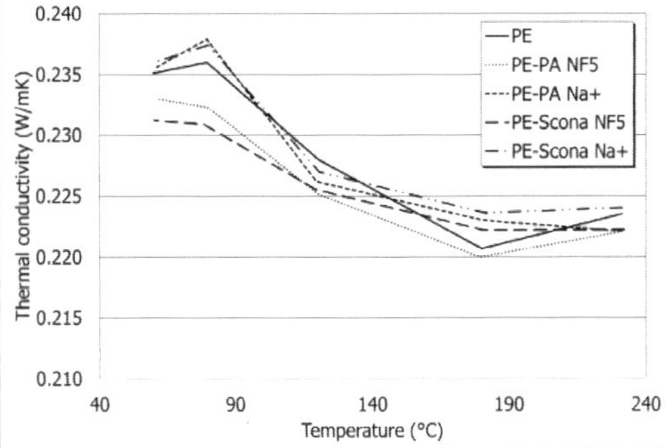

Figure 1 – Thermal conductivity for polyethylene nanocomposites

Fig. 1 is showing the thermal conductivity for the polyethylene nanocomposites. The results are very close to the unfilled polymer and a slight increase could only be observed for the materials with the untreated montmorillonite Cloisite Na^{+}. The organomodified Nanofil5 is tending to decrease the thermal conductivity for this polyethylene type.

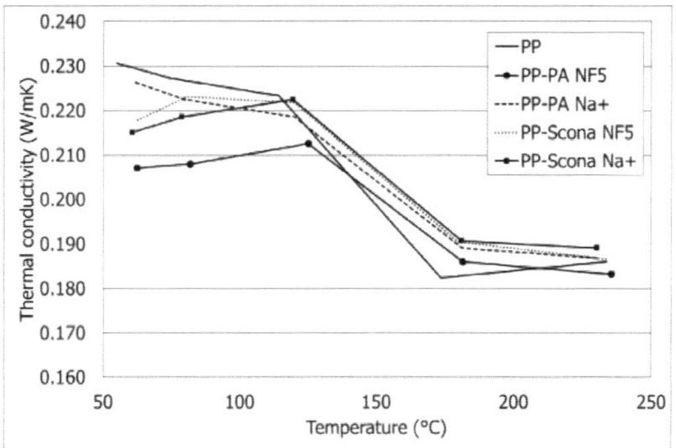

Figure 2 – Thermal conductivity for polypropylene nanocomposites

Fig. 2 is showing the thermal conductivity of polypropylene nanocomposites. It can be seen that the thermal conductivity of the nanocomposites compared to the unfilled PP is slightly higher in the melt state and slightly lower in the solid state. A difference between the nanofiller type or compatibilizers type could not be observed. The exception here is the PP-PA Nanofil5 curve. The reasons for that will be highlighted in the rheological part of the paper.

Influence on rheological behavior

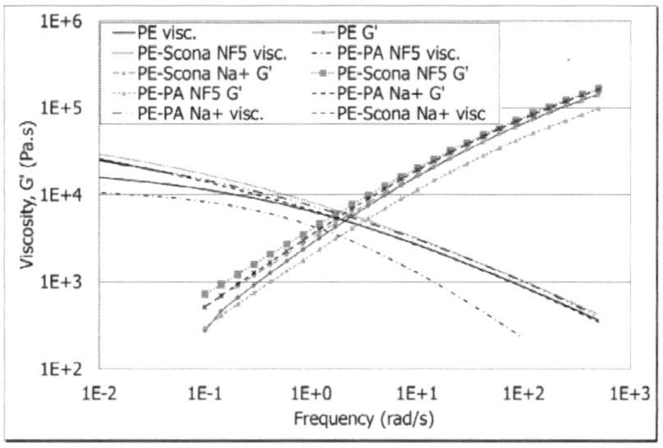

Figure 3 – G' and viscosity curve for polyethylene nanocomposites

Table 5 – Cross-Over points of the different polyethylene nanocomposites

Material	Frequency (rad/s)	G'=G'' (Pa.s)
PE	49.3	44600
PE-PA\Nanofil5	65.0	36200
PE-SCONA\Nanofil5	38.6	46500
PE-PA\Cl. Na+	44.0	48000
PE-SCONA\Cl. Na+	39.3	39300

Fig. 3 is showing G' and the viscosity curves for the polyethylene nanocomposites. The zero shear viscosity is slightly increasing for the PE-nanocomposites whereas the shape of the curves is nearly the same compared to the unfilled PE. The viscosity curve of PE-PA Nanofil5 is different compared to the others.

Due to the fact that a lower zero shear viscosity and lower values of viscosity when reaching higher values of shearing is normally based on shorter polymer chains due to chain scission during processing, this is likely to be the reason for the shape of this viscosity curve. The Cross-Over point (table 5) for these materials is not changing significantly compared to the unfilled material. A slight reduction in the frequency can be observed, which can be interpreted as an increase in chain length. Due to the fact that the polymer chains are normally shortened during processing the "virtual" chain extension measured can be caused by the intercalated layered silicates, which are clamping the polymer chains, when intercalated, between them leading to virtual longer chains. One exception is the PE-PA with Nanofil5 composite which has as slight shift to the right indicating a small reduction in polymer chain length.

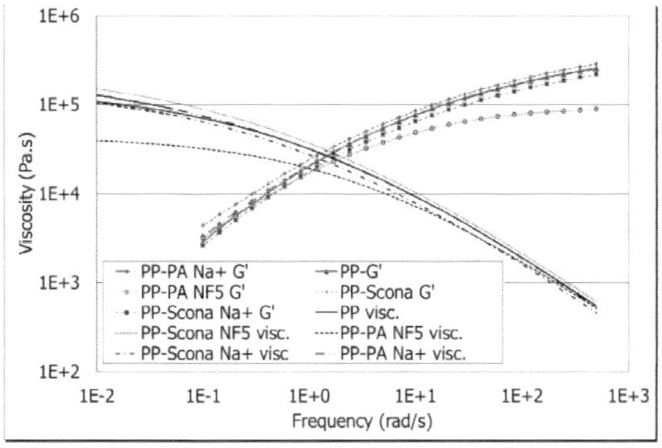

Figure 4 – G' and viscosity curve for polypropylene nanocomposites

Table 6 – Cross-Over points of the different polyethylene nanocomposites

Material	Frequency (rad/s)	G'=G'' (Pa.s)
PP	1.5	28500
PP-PA\Nanofil5	7.4	43500
PP-SCONA\Nanofil5	1.9	35600
PP-PA\Cl. Na+	2.0	32600
PP-SCONA\Cl. Na+	2.2	29100

Fig. 4 is showing G' and the viscosity curves for the polypropylene nanocomposites. The zero shear viscosity is slightly increasing for the PP-nanocomposites. The viscosity curve of PP-PA Nanofil5 is different compared to the others. As discussed above the lower values of zero shear viscosity is presumably justified in shorter polymer chains. The Cross-Over points (table 6) for these materials are changing slightly. PP-PA with Nanofil5 is the exception because of the fact that the cross over point is moving to the upper right, which is an indicator for a massive reduction in chain length and average molar mass. It can also be seen that the untreated filler is changing the rheological behavior of the material less than the modified silicate layer.

Conclusions

The investigations are showing that the use of Polyamide as compatibilizer is leading to diverging results depending on the type of layered silicate used and on the type of polyolefines investigated. It could be seen that Polyamide as a compatibilizer for polyolefine has quite a potential of increasing material properties. The investigations on Polyethylene showed that the mechanical properties achieved with polyamide can be even surpassing those of the commonly used compatibilizers. It could also be seen that if a "virtual chain extension" can be achieved the material properties are increasing. Regarding polypropylene the mechanical properties of the Polyamide type nanocomposites could not match those of the common compatibilizer. To achieve better mechanical properties the processing should be adjusted to get a better, gentle incorporation of the filler.

All the results showed that the untreated layered silicate with the brand name Cloisite Na+ is able to enhance the mechanical properties better than organomodified montmorillonite Nanofil5.

The thermal investigations showed that the nanofiller type is reacting different for Polyethylene compared to Polypropylene. Nevertheless a difference between the compatibilizers influencing thermal conductivity could not be observed.

Further investigations should tie up on this results and lay emphasis on the processing as well as on the rate of compatibilizer to nanofiller. If this rate as well as the processing is optimized the mechanical properties should be increased in even bigger extend than shown in this paper.

Acknowledgements

We want to thank Selina Moellnitz as well as Alexander Eder for their support during production and investigations of the nanocomposites.

References

1. T.X Liu; Z.H Liu; K.X Ma; L. Shen; K.Y Zeng; C.B He *Comp. Sci. Techn.*, 2003, *63*, 331.
2. M. Debowska; J. Dolega; J. Rudzinska-Girulska; J. Piglowski *Acta phys. Pol.* 2008, *113*, 1321.
3. G.M. Kim; S. Goerlitz; G. H. Michler *J. App. Poly. Sci.*, 2007, *105*, 38.
4. E.P. Giannelis *Appl. Organometal. Chem.*, 1998, *12*, 675.
5. M. Kracalik; S. Laske; M. Gschweitl; W. Friesenbichler; G.R. Langecker *J. App. Poly. Sci.* 2009, *113*, 1422
6. A. Witschnigg; S. Laske; M. Kracalik; C. Holzer *Poly. Eng. & Sci.* 2012, *52*, 1155.

O-07-147

SMART MULTILAYER FREQUENCY-SELECTIVE ABSORBER FOR MICROWAVE BASED ON HYBRID STRUCTURE OF POLYMER AND CARBON NANOTUBES

Y. Danlée[1,2*], I. Huynen[1], and C. Bailly[2]

[1] *Information and Communications Technologies, Electronics and Applied Mathematics, université catholique de Louvain, Belgium – yann.danlee@uclouvain.be; isabelle.huynen@uclouvain.be*

[2] *Institute of Condensed Matter and Nanosciences, université catholique de Louvain, Belgium – christian.bailly@uclouvain.be*

Abstract - Telecommunication systems use advanced signal processing techniques based on multiple band operation/access, implying that frequency-selective EMI shielding solutions are preferred to classical broadband absorption. Mathematical considerations show that an arrangement of successive dielectric and conductive layers stands out as an attractive solution. A smart multilayer arrangement of polymer and carbon nanotubes produces an absorber where transmitted and blocked wavelengths can be controlled. The conductive layers are stacked in gradient of conductivity in order to limit the reflection of incident microwave and so increase absorption index. Thickness and permittivity of the dielectric polymer open up narrow transmission windows in the spectrum. It is therefore possible to isolate a room or building from a wide range of telecommunication systems except for narrow bands using a material with a total thickness lower than λ.

Keywords: Carbon nanotubes, multilayer, electromagnetic bandgap filter.

Introduction

The current telecommunication tools often require advanced technology beyond simple shielding or complete absorption [1]. The material should ideally filter out unwanted frequencies while being transparent to useful waves. The periodic structure of metamaterials is conducive to the creation of electromagnetic bandgap filters. We present here a thin microwave absorber based on carbon nanotubes (CNT) and a dielectric polymer. The polymer confers interesting mechanical properties to the device because of its strength, flexibility and easy thermoformability. The CNT network is obtained by depositing a waterborne CNT-based ink (AquaCyl™) on the substrate. These components are used to form submillimetric conductive and dielectric layers which are successively stacked. A smart periodic arrangement with increasing nanofiller concentration is obtained to absorb selected frequencies in the GHz range despite its overall millimetric thickness. This gives an advantage over recently described filter systems [2]. The system presented in this paper is producible on a large scale and relatively cheap. It is designed to cover wide surface such as an entire room or even a building.

Experimental

Electromagnetic absorption is defined as $A = \frac{P_{abs}}{P_{in}} = \frac{P_{in}-P_{ref}-P_{tr}}{P_{in}}$ where P_{tr} and P_{ref} are respectively transmitted and reflected power. It imposes the choice of a material with a low electromagnetic transmission but also a low reflection. In other words, this amounts to producing a conductive material (for low P_{tr}) with low permittivity (for low P_{ref}). Nonetheless no simple material gets those properties. An arrangement of successive dielectric and conductive layers stands out as an attractive solution to reach the aimed properties. Absorption increases further by reducing reflection if the stacking is organized in gradient of conductivity [3]. Making a gradient system requires producing a series of thin films having conductivity between ~1 and ~10^3 S/m. This range of conductivity is achievable thanks to carbon nanotubes composites or inks.

Most polymers have an intrinsic low permittivity. In this work, the dielectric layers are composed of pure polycarbonate (PC) chosen for its attractive mechanical properties. It is easily molded and thermoformed. Conductive layers are made from a dried carbon nanotubes (CNT) based ink deposited on PC films. For different reasons the nanotubes are probably the most suitable carbonated elements for this type of electromagnetic applications [4][5]. The wire-shape of CNT reduces the electrical percolation threshold. In addition, CNT have an extremely high intrinsic conductivity. The company NanoCyl provides the multi-walled nanotubes used in this paper. AquaCyl™ is a CNT water-based ink from the same company. Its composition consists of 98% of pure water, 1% of multi-walled CNT and 1% of ionic surfactant. AquaCyl™ is applicable on wide surfaces as a conventional ink. .

Actually, polycarbonate is hydrophobic; its wetting angle must be modified to allow a significant amount of CNT to be deposited as a homogeneous film on the polymer. A surface treatment is hence used to increase its wettability. A PC sheet is first activated by plasma. After several tests, it appears that a H_2O plasma is more efficient than O_3 or Ar plasma. Exposure for 5 minutes at a power of 150W alters the surface chemistry and roughness, so that the contact angle goes from 88° to 56°. The modified surface film is afterwards dipped in a solution of 3-aminopropyltriethoxysilane coupling agent during at least 1 hour at room temperature [6]. Thanks to the coupling agent, the prepared PC sheet can be dip coated in pure AquaCyl™ and enough CNT adhere to the surface to reach the upper limit of desired conductivity. A series of dip coatings is then performed with different dilutions of AquaCyl™. The amount of CNT attached to the polycarbonate surface is a function of ink dilution. As explained above, the conductivity of the films must cover a range of σ-values. In this manner, we produce coated films covering a range of conductivities for the gradient stacking. The resulting structure has the advantage of being especially thin since the conductive coat has a thickness of 2 to 6 µm. Figure 1 shows an example of a carbon nanotubes layer. The CNT density in the ink layer is also very high which implies a large σ.

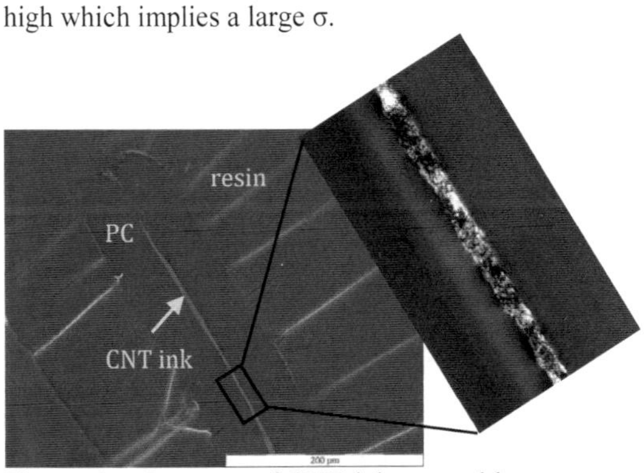

Figure 1 – Structure of CNT ink trapped between two polycarbonate sheets. The optical microscope image is zoomed 20x and the TEM one is zoomed 5000x.

Figure 2 confirms this observation since measurements of conductivity as function of concentration after dilution show that σ rises up to ~1500 S/m at GHz-frequency.

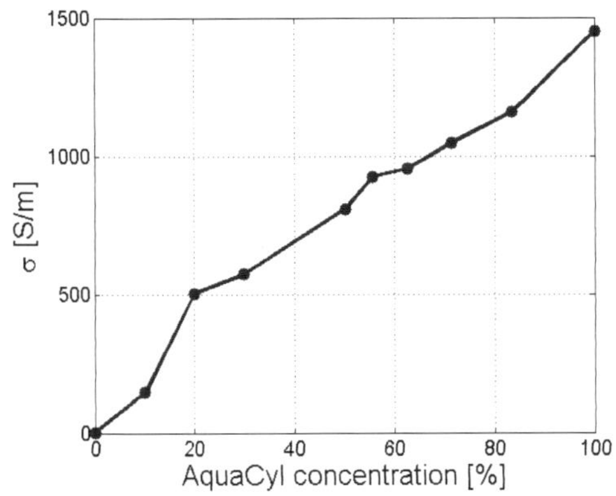

Figure 2 – Graph of conductivity at 33 GHz of carbon nanotubes layer dip coated on polycarbonate sheet according to dilution of AquaCyl™

Results and Discussion

The nanotubes deposits sandwiched between the polycarbonate sheets form the conductive films of the multilayer. To create the bandgap filter, conductive films are separated by insulating dielectric sheets made of a RF commercial substrate from Rogers Co. (RT/duroid®6010.2LM). The choice of this polymer is dictated by its high dielectric constant (ε = 10.2) in order to reduce the overall thickness of the bandgap filter.

Figure 3 illustrates the absorption index as a function of frequency and thickness of the dielectric. The simulation is based on a multilayer with 10 insulating dielectric sheets separated by 9 conductive layers showing linear progressive increase of conductivity from 1 to 10^3 S/m. For this example, we fix the thickness at 1.65 mm and we then observe two peaks of non-absorption.

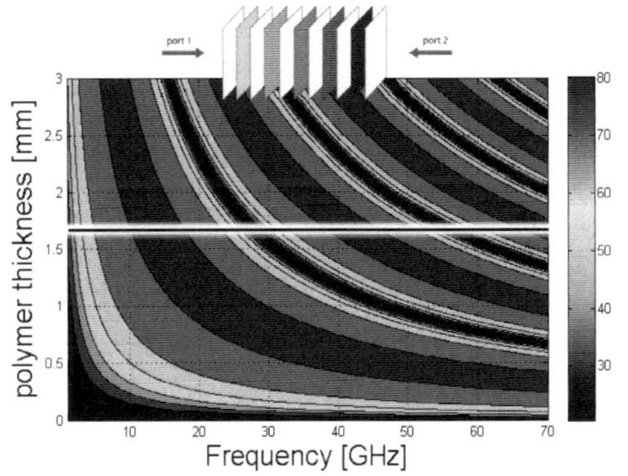

Figure 3 – simulation of absorption according to frequency and thickness of dielectric layer for whose permittivity is 10.2. The black line illustrates absorption corresponding to a dielectric thickness of 1.65 mm (as an example).

Electromagnetic characterization is performed with the help of a Vector Network Analyzer Model PNAX-70 from Agilent, on a multilayer having the same geometry and parameters than those considered for the analytical simulation above. Figure 4 shows measurement compared with theory.

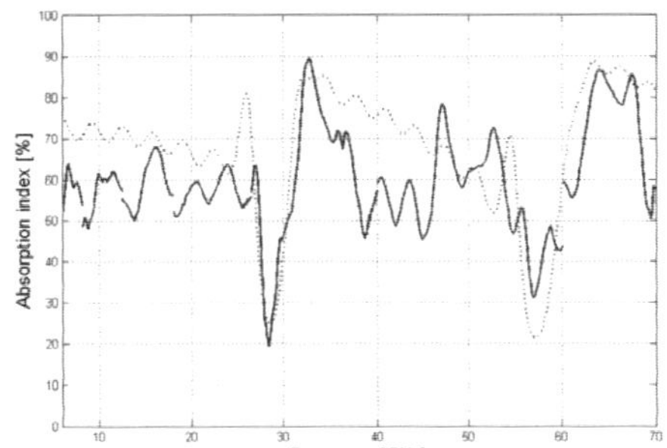

Figure 4 – Measured absorption of a AquaCyl™ based composite gradient multilayer, the thickness of the dielectric (ε = 10.2) is 1.65 mm compared with theory (dotted line).

The thickness of the dielectric plays an important role in defining the frequency bands where absorption occurs. It sets the distance between two conductive layers. But when the wave passes through the material, it is "trapped" between the conductive layers by multiple reflections. If this reflection region extends over n.λ/2 where n ∈ ℤ, the incident and reflected waves are in phase. At frequencies matching this condition, the absorption is minimal. The phenomenon appears at 28 GHz and its harmonics corresponding to a half wavelength in the dielectric medium.

Theoretical analysis of metamaterial behavior proves that structural modification shifts the absorption peaks. The analytical simulation in Figure 5 shows a high absorption of Wi-Fi and almost transparency at lower frequencies like the ones used for mobile phones networks. This multilayer is produced from only two conducting layers separated by a dielectric layer of 9 mm and a permittivity of 10.2. The filter maintains a thickness of 27 mm while it works at frequencies at the entrance of GHz-range. The selected dielectric corresponds to Rogers substrate because the filter requires a high permittivity material to minimize the overall thickness. Ceramics such as granite or concrete would also reach such high values of permittivity. It is therefore possible to produce a filter from granite plates alternated with films of carbon nanotubes. They form quite reasonable candidates for the construction of isolated rooms. This theoretical study of multilayers is thus very encouraging because it demonstrates the versatility of our proposed system.

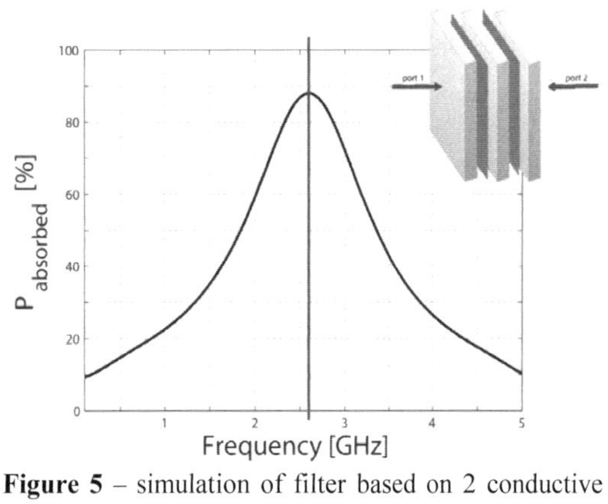

Figure 5 – simulation of filter based on 2 conductive layers separated by 3 dielectric substances 9 mm thick with ε=10.2

Two CNT coats are surrounded by the high-ε substrate that have a thickness of $\lambda/4 = 9$mm. This increases the "anti-resonance" effect in the 3 substrates and it reduces drastically incident reflection. The result is directly seen on the graph of absorption index dramatically increasing from 10 to 90% in the Wi-Fi frequency band.

Conclusions

The concept of lossy metamaterials opens up new avenues for microwave absorption. In this paper, we have developed an effective thin and flexible bandgap filter solution. This way, it should be possible to cover electronic, medical devices or others to isolate them from external and internal electromagnetic interferences. On a larger scale, companies could prevent data leakage from wireless LANs by isolating dedicated rooms or entire buildings. An analytical model has been developed to simulate the structures, using the simple but accurate chain matrix formalism. Its prediction of the metamaterial frequency response is in good agreement with measurements.

Acknowledgements

The authors are grateful to the National Fund for Scientific Research (F.R.S.-FNRS) and FRIA for supporting this research. We thank T. Pardoen and P. Bollen for their collaboration as well as N. Quiévy, I. Molenberg, F. Tao and L. V. Mohite, A. Bahrami, P. van Velthem and P. Lipnik for fruitful discussions. The support of P. Simon, P. Gerard and D. Spote for electromagnetic measurements is also sincerely acknowledged.

References

1. G. Matthaei, E.M.T. Jones, L. Young, "Microwave Filters, Impedance-Matching Networks, and Coupling Structures"
2. S. Şimşeka, E. Topuzb, E. Niverc, "A novel design method for electromagnetic bandgap based waveguide filters with periodic dielectric loading", International J. of Elec. & Com., 2011
3. Y. Danlée, C. Bailly, I. Huynen, "Thin smart multilayer microwave absorber based on hybrid structure of polymer and carbon nanotubes", Appl. Phys. Lett. 100, 213105 (2012).
4. A. Saib, L. Bednarz, R. Daussin, C. Bailly, X. Lou, J-M. Thomassin, C. Pagnoulle, C. Detrembleur, R. Jerome, I. Huynen, "Carbon nanotube composites for broadband microwave absorbing materials", Microwave Conference, 2005
5. J-M.Thomassin, X. Lou, C. Pagnoulle, A. Saib, L. Bednarz, I. Huynen, R. Jerome, C. Detrembleur, "Multiwalled Carbon Nanotube/Poly(ε-caprolactone) Nanocomposites with Exceptional Electromagnetic Interference Shielding Properties", Journal of Physical Chemistry B, vol. 111, n° 30, pp. 11186-11192, 2007
6. M. R. S. Castro, H. K. Schmidt, "Preparation and characterization of low- and high-adherent transparent multi-walled carbon nanotube thin films", Mater. Chem. Phys., 111 (2008), pp. 317–321

O-07-389

A NOVEL METHOD TO IMPROVE DIELECTRIC PROPERTIES IN CONDUCTIVE FILLER/POLYMER COMPOSITES

Mohammad Arjmand[a], Mehdi Mahmoodi[b], Simon Park[b], Uttandaraman Sundararaj[a]*

[1]Department of Chemical and Petroleum Engineering, University of Calgary, Canada – marjmand@ucalgary.ca; u.sundararaj@ucalgary.ca

[2] Department of Mechanical and Manufacturing Engineering, University of Calgary, Canada – mahmoom@ucalgary.ca; simon.park@ucalgary.ca

Abstract - In this study, we present conductive filler alignment as a novel approach to reduce the dissipation factor of conductive filler/polymer composites and to widen the typically narrow concentration window near the percolation threshold, which is used to tune the dielectric properties, i.e., real permittivity and imaginary permittivity. The effects of multi-walled carbon nanotube (MWCNT) alignment on the dielectric properties for MWCNT/polystyrene composites in the X-band (8.2 to 12.4 GHz) were investigated by comparing the dielectric properties of injection molded samples, where MWCNTs were aligned, versus compression molded samples, where MWCNTs were randomly distributed. Raman spectroscopy technique was employed to verify partial alignment of MWCNTs in the injection molded samples. It was observed that the injection molded samples showed much lower imaginary permittivity than the compression molded samples. Low imaginary permittivity of the injection molded samples over the wide MWCNT concentration range confirmed that MWCNT alignment can provide a broad concentration window to regulate the dielectric properties. Moreover, the results showed that MWCNT alignment improved the dielectric properties by reducing the dissipation factor. For instance, at MWCNT concentrations of 0.50, 1.00 and 2.00 wt%, the compression molded samples showed dissipation factors of 0.06, 0.23 and 0.59, respectively, while the injection molded samples presented the dissipation factors considerably lower and equal to 0.01, 0.03 and 0.07, respectively. This study shows that injection molding process, as an industrial technique, can be employed to improve significantly the dielectric properties of conductive filler/polymer composites for charge storage applications.

Keywords: Carbon nanotubes; Nano composites; Electrical properties; Injection molding; Dielectric properties

Introduction

Consumers are demanding lighter weight and smaller electronic devices in today's marketplace; therefore, printed circuit board (PCB) space is becoming a scarce resource. As capacitors are the main component in PCBs; accordingly, industry is now moving toward replacing large surface mounted capacitors with miniature capacitors embedded into PCBs [1]. The material requirements for embedded capacitors include high real permittivity, low leakage current (imaginary permittivity) and process compatibility with PCBs.

Recently, conductive filler/polymer composites (CPCs) have been proposed as candidates for embedded capacitors as they can satisfy well the requirements of charge storage materials at low cost and weight [2]. According to the percolation theory, a high real permittivity with a low leakage current in CPCs can only be achieved at filler loadings very close to the percolation threshold [3]. This poses a challenge to use CPCs as charge storage materials, because of the typically narrow insulator-conductor transition window around the percolation threshold.

Herein we present conductive filler alignment as a novel approach to reduce imaginary permittivity and to hinder the sharp insulator-conductor transition in CPCs. In this study, a multi-walled carbon nanotube/polystyrene (MWCNT/PS) composite, as a typical CPC, was employed to investigate the effects of conductive filler alignment on dielectric properties. The alignment of MWCNTs was induced by applying a high shear/drag force using an injection molding machine. The results showed that the MWCNT alignment led to a tremendous decrease in the dissipation factor (imaginary permittivity/real permittivity) of the molded samples arising from lower probability of the MWCNTs neighboring or contacting each other. The alignment of the MWCNTs also impeded the sharp increase in the imaginary permittivity near the percolation threshold. This feature of the MWCNT-aligned samples broadens the narrow filler concentration window near the percolation threshold in CPCs, which is used to adjust the dielectric properties.

Experimental

Materials and composite molding

A masterbatch of 20.0 wt% MWCNT in PS was obtained from Hyperion Catalysis International, Cambridge, MA, USA. The masterbatch was diluted with a pure PS (Styron® 610) kindly provided by

Americas Styrenics LLC, in order to prepare nanocomposite samples with different MWCNT concentrations. Prior to mixing, all the materials were dried at 50 °C for at least 4 h under vacuum. The composites with different concentrations of MWCNT, i.e., 0.10, 0.30, 0.50, 1.00, 2.00, 3.50, 5.00, 10.00 and 20.00 wt% were prepared employing a 25 mm Coperion ZSK co-rotating intermeshing twin-screw extruder operated at a barrel temperature, extruder speed and residence time of 200 °C, 150 rpm and 2 min, respectively.

Our previous investigations showed that there is a direct relationship between MWCNT alignment and volume resistivity [4,5]. This relationship and also the inverse relationship between volume resistivity and imaginary permittivity were the inspirations for the investigation of the effects of MWCNT alignment on the dielectric properties. Our previous study showed that the melt temperature followed by the injection velocity had the greatest impact on the MWCNT alignment of the injection molded MWCNT/PS nanocomposites [6]. On the other hand, the mold temperature and injection/holding pressure did not significantly affect the MWCNT alignment. Given the remarkable effects of the melt temperature and injection velocity on the MWCNT alignment in the injection molded MWCNT/PS nanocomposites, the injection molding process was carried out at the lowest possible melt temperature, i.e., 215 °C and the highest possible injection velocity, i.e., 240 mm.sec^{-1} to obtain the MWCNT/PS nanocomposites with the greatest MWCNT alignment. The mold temperature and injection/holding pressure employed were 60 °C and 100 bar, respectively. The injection molded samples were used to investigate the effect of MWCNT alignment on the dielectric properties.

An injection molding machine (Boy 12A) was used to inject the MWCNT/PS nanocomposite melt into a rectangular cavity. The cavity was fed with an edge gate and had dimensions of 22.86 × 10.16 × 2.0 mm. To achieve a more comprehensive picture of the effects of MWCNT alignment on the dielectric properties, the dielectric properties of the aligned injection molded samples were compared with those of the compression molded samples, where MWCNTs were randomly distributed. A Carver compression molder (Carver Inc., Wabash, IN) was employed to fabricate the compression molded samples with the same dimensions as the injection molded samples. The compression molding process was carried out at 210 °C for 10 min under 38 MPa pressure.

Raman spectroscopy

Raman spectroscopy was employed to verify the alignment of MWCNTs in the injection molded samples. A Renishaw spectrometer equipped with an inVia Raman microscope was used to obtain the Raman spectra from the molded samples. The samples were excited by a near-infrared (NIR) laser beam in regular mode. The Raman intensity measurements were performed at two normal orientations of the laser beam with respect to the flow direction, i.e., parallel and perpendicular, to obtain information about the MWCNT alignment.

Dielectric properties measurements

To evaluate the potential of CPCs as charge storage materials for high-frequency range applications, it is essential to characterize the high-frequency dielectric properties of CPCs. Accordingly, the dielectric properties in the X-band (8.2 to 12.4 GHz) frequency range were investigated in this study. The complex permittivity measurements in the X-band were carried out in a WR-90 rectangular waveguide using an Agilent programmable network analyzer (Model E8364B).

Results and Discussion

Raman spectroscopy

The Raman spectra of MWCNT/polymer composites provide two important features: the *D* band (disorder band) and *G* band (graphite band). The *D* band, correlating to disorder in the sp^2 hybridized graphitic structure, is more responsive to the alignment of MWCNTs than the *G* band, which corresponds to the in-plane vibration of the graphitic wall [7]. Higher intensity ratios of $D_{\parallel}/D_{\perp}$ and $G_{\parallel}/G_{\perp}$ (parallel/perpendicular to the flow direction) correspond to higher MWCNT alignment. The injection molded samples showed $D_{\parallel}/D_{\perp}$ and $G_{\parallel}/G_{\perp}$ ratios equal to 1.66 and 1.51, respectively, whereas the compression molded samples exhibited $D_{\parallel}/D_{\perp}$ and $G_{\parallel}/G_{\perp}$ of 1.01 and 1.01, respectively. From the Raman spectroscopy results, it can be claimed that the MWCNTs were randomly distributed in the compression molded samples, while they were partially aligned in the injection molded samples.

The effects of MWCNT alignment on the dielectric properties

As mentioned previously, a high real permittivity with a low imaginary permittivity in CPCs can only be achieved at filler loadings very close to the percolation threshold. The increased real permittivity observed in CPCs near the percolation threshold results from the formation of a large number of nanocapacitors, i.e., conducting clusters isolated by thin layers of polymer [1]. These nanocapacitors enable CPCs to store large amount of charges. The insulator-conductor transition that occurs in CPCs at the percolation threshold leads to a drastic variation in the volume resistivity and imaginary permittivity; thereby it prohibits using CPCs as charge storage materials above the percolation threshold. Accordingly, there is a very narrow concentration window near the percolation threshold for high aspect ratio fillers, such as MWCNTs, to adjust the dielectric properties.

Figs. 1(a) and 1(b) show the imaginary and real permittivities, respectively, of the compression molded and injection molded samples as a function of MWCNT concentration in the frequency range of the X-band. As can be observed in Fig. 1(a), the imaginary permittivity increased with increase in the MWCNT concentration for both types of samples. In general, the imaginary permittivity of CPCs can result from the polarization loss, e.g., distortional and interfacial, and/or Ohmic loss. Increase in the MWCNT concentration is equivalent to an increase in the amount of mobile charge carriers (Ohmic loss) and the number of nanocapacitors (polarization loss), both of which can account for an enhancement in the imaginary permittivity.

Fig. 1(a), however, shows that the imaginary permittivities of the injection molded samples were significantly lower than those of the compression molded samples. For instance, the imaginary permittivities for the compression molded samples at the MWCNT concentrations of 0.50, 1.00 and 2.00 wt% were 0.21, 1.16 and 6.14, respectively; whereas, the imaginary permittivities for the injection molded samples at these concentrations were significantly lower and equal to 0.02, 0.12 and 0.35, respectively. Lower imaginary permittivity of the injection molded samples can be ascribed to MWCNT alignment which led to inferior conductive network formation. In another words, the movement scales of the electrons in each half cycle of the alternating field in the injection molded samples, because of the poorer network formation, was lower than those of the compression molded samples. This is what led to lower energy dissipation in the injection molded samples. Low imaginary permittivity of the injection molded samples over the wide MWCNT concentration range verified that MWCNT alignment can provide a broad concentration window to regulate the dielectric properties.

Fig. 1(b) shows that the real permittivity increased with increases in the MWCNT concentration. The enhancement of the real permittivity with an increased MWCNT concentration is well-established, and is attributed to an increase in number of nanocapacitors and a decrease in the thickness of insulative polymer gaps (nanodielectrics), both of which contributed to greater charge polarization. Moreover, it is believed that the real permittivity of MWCNT/polymer composites is influenced by the polarization within the MWCNTs, and this also contributes to the greater real permittivity at higher MWCNT concentrations [8].

As presented in Fig. 1(b), it was surprising to observe that the MWCNT alignment had an adverse influence on the real permittivity. At MWCNT concentrations of 0.50, 1.00 and 2.00 wt%, the compression molded samples showed the real permittivities equal to 3.58 and 5.06 and 10.37, respectively, while the real permittivities of the injection molded samples were 3.21, 3.70 and 5.24, respectively. The difference between the real permittivities of the compression molded and injection molded samples can be ascribed to the lower probability that MWCNTs are in close proximity to each other in the injection molded samples. In the narrow insulative gaps between the conductive fillers, there may be a buildup of very high field strength, which is higher than the macroscopic field strength by a factor of M (i.e., the ratio of the average size of the conducting MWCNT aggregates to the average gaps width) [9]. This high field strength significantly contributed to the electronic polarization of the PS matrix. In the injection molded samples, the insulative gaps of the polymer were thicker leading to a lower applied field and poorer electronic polarization of the PS matrix. Therefore, the injection molded samples showed lower real permittivity than the compression molded samples.

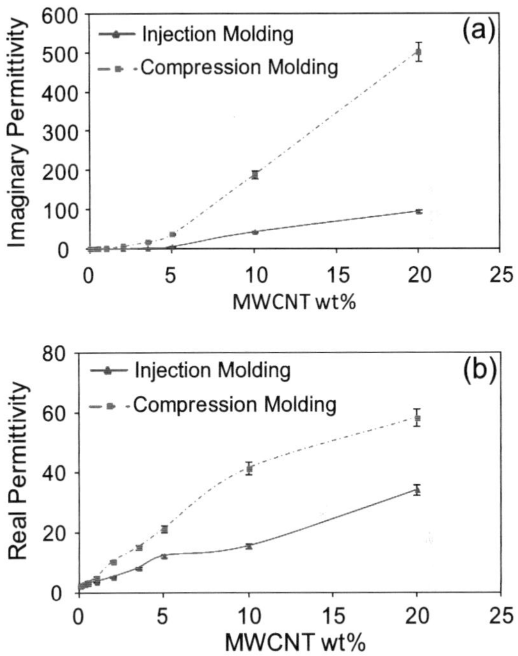

Figure 1 – (a) Imaginary permittivity and (b) real permittivity, as a function of MWCNT concentration, for the compression molded and injection molded samples of the MWCNT/PS composites in the X-band.

MWCNT alignment reduced both the real permittivity and imaginary permittivity; therefore, it is necessary to evaluate the overall impact of MWCNT alignment on the dielectric properties. Hence, the dissipation factors (imaginary permittivity/real permittivity) of the compression molded and injection molded samples at different MWCNT concentrations were compared with each other. As can be observed in Fig. 2, the MWCNT alignment decreased the dissipation factors at all the MWCNT concentrations, demonstrating its positive effect on the dielectric properties. For instance, at the MWCNT concentrations of 0.50, 1.00 and 2.00 wt%, the dissipation factors of the compression molded samples were 0.06, 0.23 and 0.59, respectively, while the injection molded samples presented the dissipation factors significantly lower and equal to 0.01, 0.03 and 0.07, respectively. These results prove that the positive effect of the MWCNT alignment on reducing the dissipative energy dominated its adverse effect on decreasing the capacitive energy.

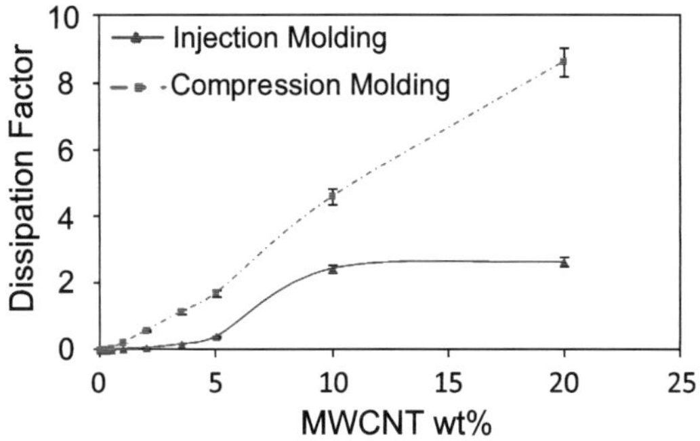

Figure 2 – Dissipation factors for the compression molded and injection molded samples of the MWCNT/PS composites as a function of MWCNT concentration in the X-band.

Conclusions

In conclusion, it was shown that the MWCNT alignment, induced by an injection molding machine, in the MWCNT/PS composites positively influenced the dielectric properties. MWCNT alignment widened the typically narrow concentration window near the percolation threshold, which is used to tune the dielectric properties, thereby reducing challenges and risks in manipulating CPCs as charge storage materials. It was also shown that the MWCNT alignment reduced both the real permittivity and imaginary permittivity; nonetheless, the positive effect of the MWCNT alignment on reducing the imaginary permittivity overshadowed its negative effect of reducing the real permittivity. The positive impact of MWCNT alignment on the dielectric properties presented in this article is industrially significant because injection molding is one of the most common fabrication methods for polymer nanocomposites and it can be used to control and tune dielectric properties.

Acknowledgements

The authors acknowledge support for this project from Natural Sciences and Engineering Research Council of Canada (NSERC). We would like to gratefully acknowledge Thomas Apperley and Dr. Michal Okoniewski for assistance with the dielectric properties measurements. We are grateful to Americas Styrenics LLC, for providing the neat polystyrene. We would like to thank Dr. Samaneh Abbasi of Ecole Polytechnique (Montreal, Canada) for assistance with Raman spectroscopy.

References

1. Z. M. Dang; L. Wang; Y. Yin, Q. Zhang; Q. Q. Lei *Adv. Mater.* 2007, *19*, 852.
2. J. B. Kim; J. W. Yi; J. E. Nam Thin Solid Films 2011, *519*, 5050.
3. Z. M. Dang; S. H. Yao; J. K. Yuan; J. B. Bai. *J. Phys. Chem. C.* 2010, *114*, 13204.
4. M. Arjmand; M. Mahmoodi; G. A. Gelves; S. Park; U. Sundararaj. *Carbon.* 2011, *49*, 3430.
5. M. Arjmand; T. Apperley; M. Okoniewski; U. Sundararaj *Carbon.* 2012, *50*, 5126.
6. M. Mahmoodi; M. Arjmand; U. Sundararaj; S. Park *Carbon.* 2012, *50*, 1455.
7. S. Abbasi; P. J. Carreau; A. Derdouri. *Polymer.* 2010, *51*, 922.
8. M. J. Jiang; Z. M. Dang; M. Bozlar; F. Miomandre; J. B. Bai. *J. Appl. Phys.* 2009, *106*, 084902.
9. E. K. Sichel; J. I. Gittleman; P. Sheng *Phys. Rev. B.* 1978, *18*, 5712.

O-07-159

EFFECT OF WOOD POWDER DIAMETER AND COMPOUNDING SCREW GEOMETRY ON THE MECHANICAL PROPERTY OF WOOD POWDER/PP COMPOSITE

Y YU[1*], Y YANG[2], M MURAKAMI[1], T OTA[3], T UMEMURA[4], K HARA[5], M NOMURA[5], and H HAMADA[1]

1: Kyoto Institute of Technology, Matsugasaki, Sakyo-ku, Kyoto, 606-8585, JAPAN, yuying_yoyo@hotmail.com; mm60murakami@gmail.com; hhamada@kit.ac.jp*

2: Donghua University, amy_yuqiu_yang@hotmail.com

3. Chuou Business Group, promotl@gold.ocn.ne.jp

4. Ryoko chemical co.jp, umemura@ryoko.co.jp

5: Green plaisir, manabu.nomura01@ilcc.co.jp

Abstract - Filler reinforced thermoplastics especially for natural filler reinforced plastics have been frequently used to improve the physical and thermal properties of polymer materials in plastic industry due to their low density, low cost and environmental friendliness. At current study, a preliminary investigation on the mechanical properties and morphologies of polypropylene (PP) reinforced by wood powder were carried out. Two different compounding screws with different mixing sections were used to evaluate the effect of compounding screw geometry on the appearance and mechanical properties of the wood/PP composites since the screw section geometry has an effect on the final mixing condition of filler/matrix, filler damage and scorch. Additionally, the effect of particle size on the mechanical properties of wood/PP composites was also evaluated based on the tensile and Izod impact tests. The reflection-type optical microscope and scanning electron microscope (SEM) observation results on the specimen surfaces were used to discuss the interface properties and powder distribution degrees of the composites.

Keywords: wood powder, particle size, screw geometry, distribution, injection molding

Introduction

Owing to low prices, ecological and economical advantage, less health hazards, and the steadily rising performance of technical and standard plastics, the demand for natural fiber composites has been increased in various application areas such as automotive components, building materials, and the aerospace industry. More recently, the critical discussion about the preservation of natural resources and recycling has led to a renewed interest concerning natural materials with the focus on renewable raw materials. Among all the natural fibers, wood is considered one of the attractive alternatives for reinforcing thermoplastics. The use of wood for the reinforcement of the composites has received increasing attentions recent years both by the academic sector and the industry as inexpensive filler to increase the mechanical properties and thermal stabilities of thermoplastic or to reduce raw material costs since the wood flour can be derived from sawdust, pulp mill wood residue, bark, nutshell, and straw. Therefore, as potential applications of the composites which are furniture components, door moldings, floor systems for light frame construction, and packaging pallets, wood reinforced thermoplastic composites are commercially attractive for high volume applications. As a filler form, it is well known that the reinforcing effect of the fillers depends on a lot of factors and the reinforcing mechanism is quite complicated. Several factors have already been found could affect the mechanical behavior of filler reinforced thermoplastics such as the characteristics of matrix and filler, interfacial bonding, filler content, filler size, aspect ratio or filler shape, filler concentration and distribution. Therefore, investigation on the effect of the inherent in filler on the mechanical property of filler reinforced composites is of great importance [1-3].

In the other hand, regards molding method, extrusion compounding and injection molding are conventional techniques for fast manufacturing of natural fiber reinforced thermoplastics. Before injection molding, extrusion compounding can be made using an extruder for the corresponding compound containing a few components. Extrusion compounding employs a screw as a transport mechanism due to the ability of the screw geometry to perform all the elementary steps of polymer processing: feeding, pumping, melting, and mixing. Thus, the geometry of the screws is one of the important factors in achieve proper mixing for injection products. Especially for natural fiber, it was found that it would be difficult to get a constant fiber volume and uniform fiber distribution for natural fiber by normal injection process. Not only because the surface of the natural fiber is rough and they are easy to bother each other during injecting compared to synthetic fiber, but also, natural fiber will orient during feeding process, in details fiber orientation will vary from one location of the injected part to another. Moreover, unlike glass fiber, natural fiber are easily to get burnt and breakage due to the high temperature and pressure during compounding. Therefore, getting a much better dispersive mixing and less scorch in compounding process is benefit for improving the properties of the products [4-6].

Therefore, at current study, two different compounding screws with different mixing sections were used to evaluate the effect of compounding screw geometry on the appearance and mechanical properties of the wood/PP composites, additionally, the effect of particle size on the mechanical properties of wood/PP composites were also evaluated based on the tensile and Izod impact tests.

Experimental

Materials and fabrication

Polypropylene pellet was used as matrix because it is a widely used plastic in the automotive industry that can be easily injection molded, and cellulose powder (NIHON SEISHI Chemical CO., Ltd) and wood powder were used as reinforcements to fabricate the cellulose/PP and wood/PP dumbbell shape specimens by injection molding method. Two kinds of the wood powders were used. One is the normal powder which can be obtained from the Japanese market, the other one is wood powder made from trees in Hiroshima prefecture, Japan, which called riguno in this study. Before injection molding, reinforcement pellets was compound with PP through extrusion compounding machine. In particularly, for cellulose/PP, during compounding, two different compounding screws with different mixing sections were used. Figure.1 shows the different geometry of two screw extruders, three different mixing sections were used. Screw design (a) has a right-handed element (RHE) in section I and II, and one gear mixing elements (GME) in section III. While, screw design (b) has a left-handed element (LHE) in section I, II and III. Those different combinations gave different intensities of mixing and residence times in the mixing sections. Screw (b) provides a higher intensity of mixing than screw (a) because of the backup effect of the LHE.

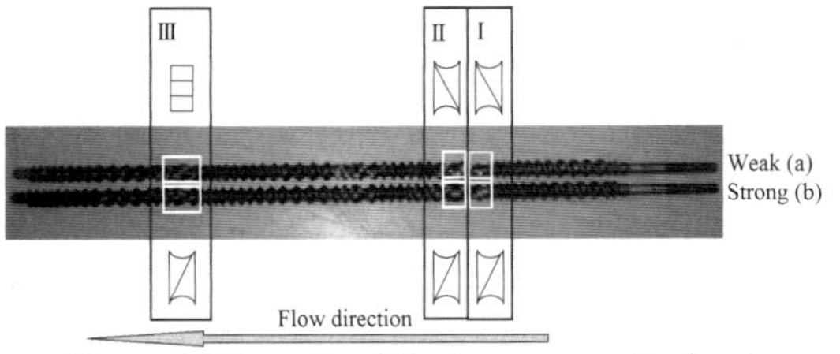

Figure.1 Schematic of the two screw extruders in compounding process with three different mixing sections. (a) weak mixing; (b) strong mixing

6 kinds of cellulose/PP and wood/PP dumbbell shaped specimens with a filler content of 30 wt% were prepared as listed in Table.1. i.e., the composition of C1 and C2 is the same, but C1 is compounded by using strong mixing screw extruder, while C2 is compounded by weak mixing one. In the other hand, for W specimens, W1 and W2 are fabricated by using normal wood powder, and W3 and W4 are fabricated by using riguno powder. Both of the wood types had two different particle sizes. The particle size distribution of wood powder was measured using an ultrasonic vibration particle size distribution sieving machine and the results were displayed in figure.2. Results show that the particle length of 50% of the total mass of W1 and W2 were smaller than 150 and 87.2 μm, respectively. While for W3 and W4 were smaller than 58.2 and 22.8 μm respectively. Additionally, the SEM photos of the wood powder (W3 & W4) were shown as examples in figure.3. It is noted that W3 has a higher aspect ratio, and much like fiber shape compared to W4. The aspect ration of W1, W2, W3 and W4 were 9.57, 8.36, 6.30 and 4.37, respectively.

All of the dumbbell test pieces were fabricated by using a 30-ton TOYO PSS TT-30F6 injection molding machine at a barrel temperature of between 150-210℃.

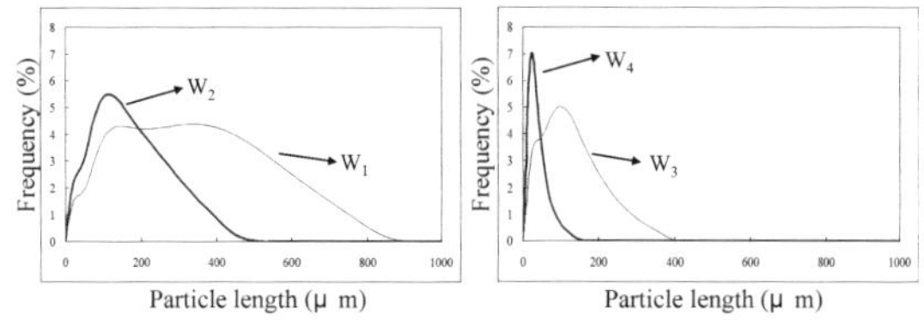

Figure.2 Particle size distribution of wood powder

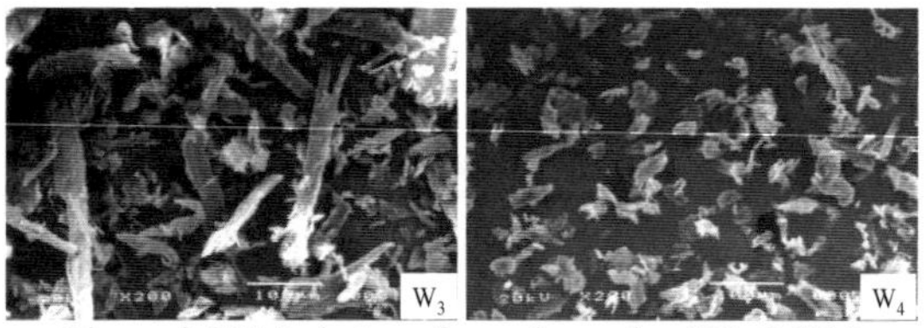

Figure.3 SEM photos of wood powder (W_3 & W_4)

Table.1 Material composition

Specimen ID	Polypropylene (wt%)	Wood powder (wt%)	Cellulose (wt%)	Others (wt%)	Screw description
C_1	68	/	30	2	Strong mixing
C_2	68	/	30	2	Weak mixing
W_1	68	# 40: 30	/	2	Strong mixing
W_2	68	# 80: 30	/	2	Strong mixing
W_3	68	Riguno: 30	/	2	Strong mixing
W_4	68	Riguno (Grade): 30	/	2	Strong mixing

Tensile test

For tensile test, the gauge length is 115mm, and strain was measured using extension measurement, which were fixed onto the central surface of the specimens by a rubber band. Tensile test was conducted on an Instron Universal Testing Machine (Type 4206) under a nominal test speed of 1.0mm/min. 3 pieces were repeated.

Izod impact test

Izod impact test was performed on the Impact tester (Toyoseiki) pendulum 5.5J in accordance with ADTM D 256-05. The size of the specimen was 10×60mm, V-notch was used. Five specimens were repeated.

Attenuated Total Reflectance Fourier transform infrared spectrum (ATR-FTIR)

ATR-FTIR was conducted on Spectrum GX (Perkin Elmer) to analyze the scorch condition of the wood powder during molding in this study. The surface of the specimens were polished and dried in order to let the infrared light wave go across the samples with sufficient reflection.

Results and Discussion

Effect of the screw geometry on the mechanical properties of cellulose/PP composite

The photo image of the cellulose/PP specimens was showed in figure.4. It is obviously that compared to C1 specimen, the color of the C2 is much lighter which is considered due to the lower intensity of mixing and smaller residence time in the mixing section of screw (a). Moreover, FTIR spectroscopy measurements were performed and the spectrum of cellulose/PP composites was given in figure.5. It is known that the possibility of the cellulose to oxidize during compounding extruding and injection molding is relatively high due to the high temperature. Since cellulose is hydrophilic and contains lots of hydrogen groups, under certain condition, it could be oxidized and transformed to carboxylic acid, aldehyde, and/or ketone groups which would depend on the nature of the oxidant and reaction conditions. From figure.5, although the differences between the two composites are not obviously, the absorption peak of the carbonyl group near 1720 cm-1 of C1 specimen is slight intense than that of C2. That means the scorch of C1 specimen is a bit serious than C2.

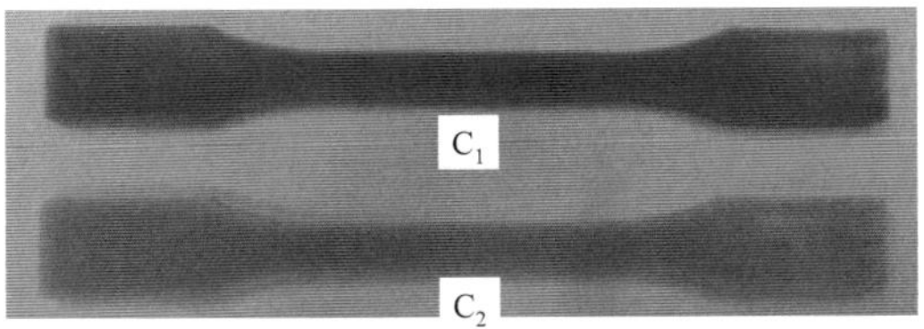

Figure.4 Photo image of the cellulose/PP specimens

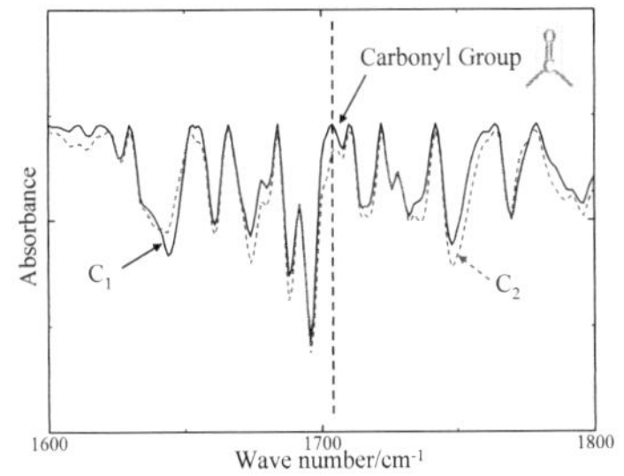

Figure.5 FTIR spectrum of cellulose/PP composites

In order to understand the distribution of the cellulose powder in the PP matrix, a thin cross-section of cellulose/PP sample was observed by reflection-type optical microscope. Sample was cut along the transverse directions to the flow direction and mounted with epoxy resin. Then it was polished to a thickness of less than 0.5mm and observed. Optical images were shown in figure.6. It is clearly that the distribution of cellulose powder was improved compounded by using the strong mixing screw. Additionally, the size of the cellulose powder was also much better uniform in C1 composite.

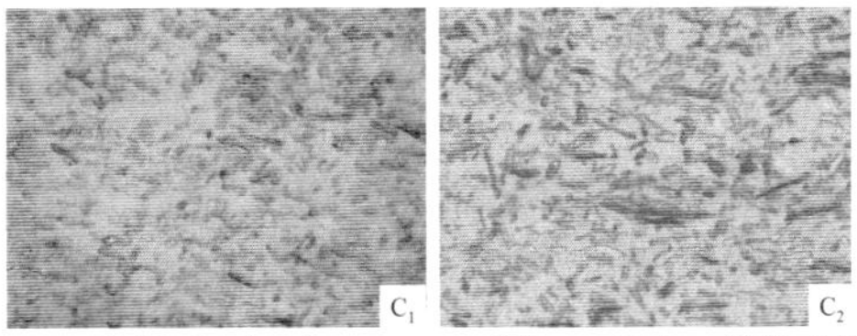

Figure.6. Optical microscope of the cellulose/PP at the transverse direction

In the other hand, the tensile properties of cellulose/PP compounded by using different extruder screws were compared in figure.7. The detail values in the table are the average values. Results showed that there are not significant differences between two specimens. In details, C2 owned a bit higher tensile strength, while the elongation of C1 specimen is larger than C2 specimen.

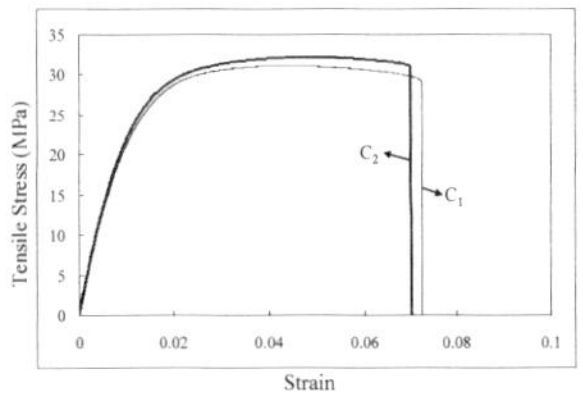

Specimen ID	Modulus (GPa)	Strength (MPa)
C_1	3.25	31.20
C_2	3.22	32.06

Figure.7 Tensile properties of cellulose/PP compounded by using different extruder screws

Effect of the wood particle size on the mechanical properties of the wood/PP composites

The effect of the wood particle size on the tensile properties of the composites is shown in figure.8. Although there are two types of wood in this study, it is seen that for both kinds, the effect of the wood particle size on the tensile properties of the composites is not significant. The big particle size reinforced wood/PP showed a little higher tensile property than the small ones. It suggested that in general the aspect ratio or filler shape along with the particle size can significantly affect the mechanical properties of the composites. Therefore, for both types of wood, the higher aspect ratio of the wood powder is considered resulted in the higher tensile property of the composites. In the other hand, the Izod impact properties, which presented in figure.9, showed that the big particle size reinforced wood/PP had a higher Izod strength. In details, it could be seen that W1 can provide a higher impact resistance, almost 10% higher Iozd strength than that of the W2 composite. While, W3 provide 8% higher value than that of the W4 composites.

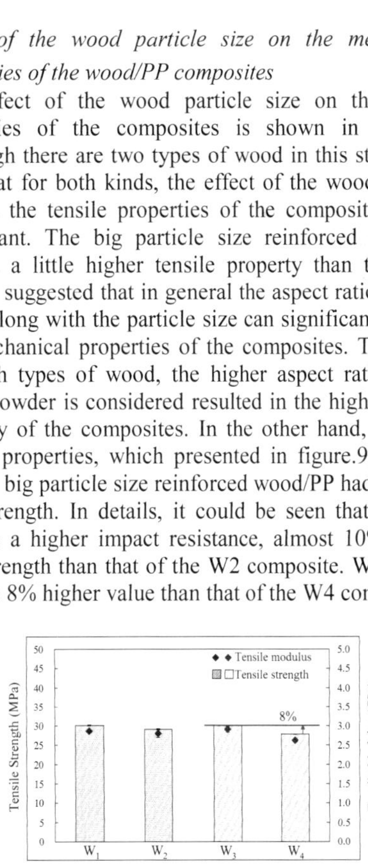

Figure.8 Tensile properties of wood/PP reinforced by wood with different particle size

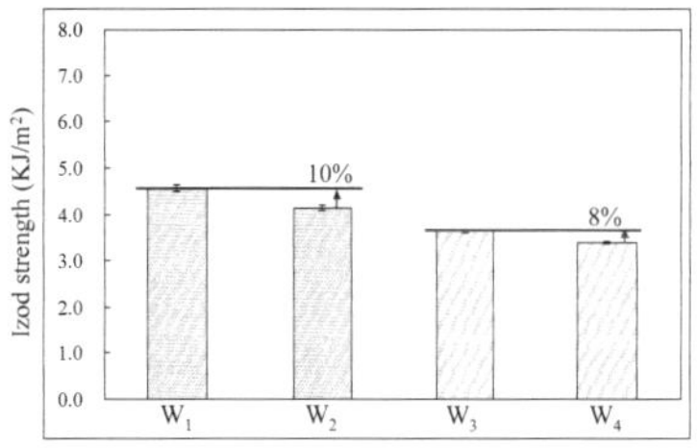

Figure.9 Izod impact properties of the wood/PP.

The fracture surface observations after etching process of wood/PP (W3 & W4) at the transverse direction showed in figure 10, it was obviously that after etching, the wood powder was etched and left behind lots of voids on the fracture surface. Those voids distribution of the specimen after etching was considered equal to wood distribution. It can be seen that the wood powder in both composites distributed well. However, the size of the wood powder in W4 composite is much uniform. It is considered in Izod impact test, when the pendulum attacked the composite, the big size of the wood powder could absorb comparable high impact energy and may result in better impact resistance and damage tolerance compared to the small size wood powder reinforced composites.

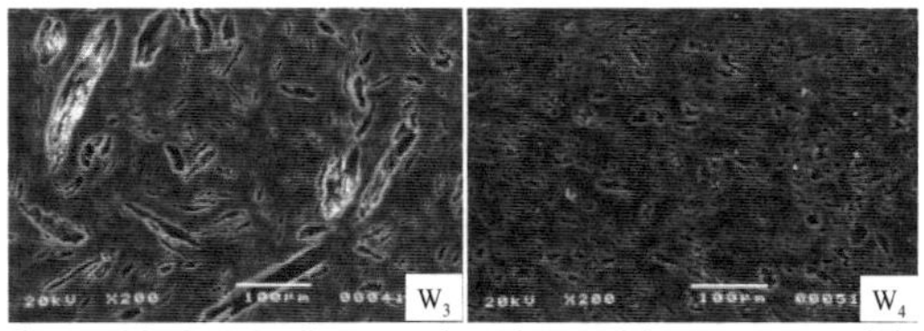

Figure.10. SEM micrographs after etching process of the wood/PP (W3 & W4) at the transverse direction

From SEM micrographs of wood/PP reinforced by different particle size after Izod impact test (figure.11), it was clearly that compared to W4, the wood powder in W3 composite is much like fiber form. The wood powder after Izod test was still bended to the matrix and especially it is found that in W4 composite, the matrix was found surrounded with the wood powder.

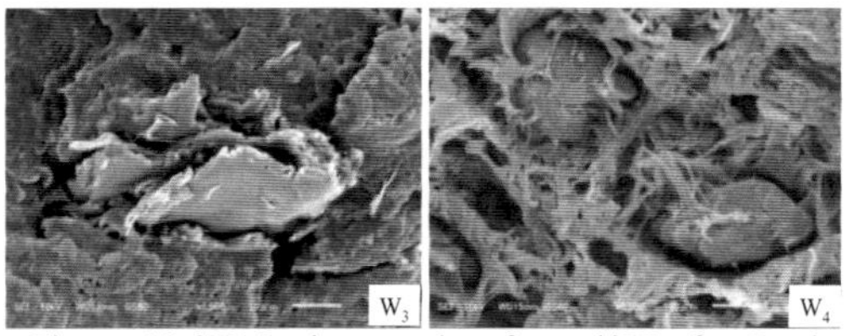

Figure.11 SEM micrographs of wood/PP after Izod impact test (W3 & W4)

Conclusion

Filler reinforced thermoplastics have been frequently used to improve the physical and thermal properties of polymer materials in plastic industry. At current study, two different compounding screws with different mixing sections were used to evaluate the effect of compounding screw geometry on the appearance and mechanical properties of the wood/PP composites. Additionally, the effect of particle size on the mechanical properties of wood/PP composites was also evaluated based on the tensile and Izod impact tests.

It was found that the scorch of the cellulose/PP specimen molded by the strong extruder screw is a bit serious than that molded by the weak extruder screw. However, the distribution of cellulose powder was improved a little compounded by using the strong mixing screw. Additionally, there are not significant differences of the tensile properties between two specimens.

In the other hand, the aspect ratio of the filler can affect the mechanical properties of the composites. For wood powder, the higher aspect ratio of the wood powder resulted in the higher tensile property. It is also considered the big size of the wood powder could absorbed comparable high impact energy in the Izod impact test. In particularly, although the wood powder distributed well in both big and small powder reinforced composites, the size of the wood powder in small powder reinforced composite is much uniform.

References

[1] W. Dziadur, A. Tabor. *The Effect of wood filler behavior on structure and fracture of polyethylene.* Journal of Achievements in Materails and Manufacturing Engineering (2006), Vol.17, 53-56

[2] A.K. Bledzki, J. Gassan, and S. Theis. *Wood-filled thermoplastic composites*. Mechanics of Composite Materials (1998), Vol.34, 563-568

[3] Velichko Hristov, John Vlachopoulos. *Effect of polymer molecular weight and filler particle size on flow behavior of wood polymer composites*. Polymer Composites (2008), Vol. 29, 831-829

[4] C.P.J.M Verbraak, H.E.H.Meijer. *Screw design in injection molding*. Polymer Engineering and Science(1989), Vol.29, 479-487

[5] Luke M. Miller, Walter S. Smith, Timothy W. Womer, Robert A. Sickles Jr. *Evaluation of different injection molding screw design for better color mixing*. ANTEC Conference Proceedings (2007), Vol.2, 708-711

[6] Shao-Yun Fu, Bernd Lauke and Yiu-Wing Mai. Science and engineering of short fiber reinforced polymer composites

AN INVESTIGATION ON ELECTROMAGNETIC INTERFERENCE SHIELDING MECHANISMS OF INJECTION MOLDED MULTI-WALLED CARBON NANOTUBE/POLYSTYRENE COMPOSITES

Mohammad Arjmand[1], Thomas Apperley[2], Michal Okoniewski[2], Uttandaraman Sundararaj[1]*

[1]*Department of Chemical and Petroleum Engineering, University of Calgary, Canada – marjmand@ucalgary.ca; u.sundararaj@ucalgary.ca*
[2]*Department of Electrical and Computer Engineering, University of Calgary, Canada – tjmapper@ucalgary.ca; okoniews@ucalgary.ca*

Abstract - Due to inevitable flow-induced alignment of multi-walled carbon nanotubes (MWCNTs) in injection molding process, the study of electromagnetic interference (EMI) shielding of injection molded MWCNT/polystyrene composites was based on MWCNT alignment. To obtain better understanding from the effects of MWCNT alignment, the EMI shielding properties of the injection molded samples were compared with those of compression molded samples, where MWCNTs were randomly distributed. Raman spectroscopy technique verified the alignment of the MWCNTs in the injection molded samples. It was observed that the EMI shielding had an inverse relationship with MWCNT alignment. The injection molded samples showed lower EMI shielding than the compression molded samples. The EMI shielding results showed that the shielding by reflection was almost the same at different MWCNT alignments, i.e., various MWCNT connectivities, which was related to the same amount of interacting surface charges for different MWCNT alignments. Contrary to reflection, absorption decreased tremendously with an increase in MWCNT alignment. Lower absorption at greater MWCNT alignments was ascribed to inferior MWCNT network formation.

Keywords: Carbon nanotubes; Nanocomposites; Electromagnetic interference shielding; Injection molding.

Introduction

Electromagnetic interference (EMI) occurs when undesirable signals superimpose upon a signal of interest. EMI effects can range from interruption of operation to degradation of electronics or electrical equipments; accordingly, EMI has become a significant technical challenge given the rapid development of electronic devices, such as laptops, cell phones, weather radars, TV picture transmitters, and the like [1].

Recently, conductive filler/polymer composites (CPCs) have attracted great interest to be used for EMI shielding applications due to their light weight, low cost and improved design options. Different conductive fillers have been embedded into polymer matrices in order to get CPCs with desired EMI shielding properties [2-4]. Among different conductive fillers used in CPCs, carbon nanotubes (CNTs) have fascinated researchers due to their unique electronic structure and extraordinary properties.

Structural perfection, aspect ratio, dispersion, distribution and alignment of CNTs are among the most important properties that have great influence on the EMI shielding properties of CNT/polymer composites [5]. Among these parameters, the effect of alignment of CNTs on EMI shielding properties is a concept that has not been comprehended well and requires further investigation.

The inevitable flow-induced alignment of CNTs in injection molding process was the inspiration to investigate the effects of CNT alignment on the EMI shielding properties of CNT/polymer composites. Most of investigations on the EMI shielding properties of CPCs have been devoted to compression molded samples which have randomly distributed conductive filler; this has led to results that are not applicable to injection molded samples, where conductive filler is aligned [5,6]. Therefore, this study compares the EMI shielding properties of injection molded versus compression molded multi-walled carbon nanotube/polystyrene (MWCNT/PS) composites in terms of MWCNT alignment.

Experimental

Materials and composite molding

A masterbatch of 20.0 wt% MWCNT in PS was obtained from Hyperion Catalysis International, Cambridge, MA, USA. The masterbatch was diluted with a pure PS (Styron® 610) kindly provided by Americas Styrenics LLC, in order to prepare nanocomposite samples with different MWCNT concentrations. Prior to mixing, all the materials were dried at 50 °C for at least 4 h under vacuum. The composites with different concentrations of MWCNT, i.e., 0.10, 0.30, 0.50, 1.00, 2.00, 3.50, 5.00, 10.00 and 20.00 wt% were prepared using a 25 mm Coperion ZSK co-rotating intermeshing twin-screw extruder operated at a barrel temperature of 200 °C, extruder speed of 150 rpm and residence time of 2 min.

Our previous investigation [7] showed that the melt temperature followed by the injection velocity had the greatest impact on the MWCNT alignment of the injection molded MWCNT/PS nanocomposites; however, the mold temperature and injection/holding pressure did not significantly affect the MWCNT alignment. Given the significant effects of the melt temperature and injection velocity on the MWCNT alignment in the injection molded MWCNT/PS nanocomposites, the injection molding process was carried out at the lowest possible melt temperature, i.e., 215 °C and the highest possible injection velocity, i.e., 240 $mm \cdot sec^{-1}$ to obtain the MWCNT/PS nanocomposites with the greatest MWCNT alignment. The mold temperature and injection/holding pressure employed were 60 °C and 100 bar, respectively.

An injection molding machine (Boy 12A) was used to inject the MWCNT/PS nanocomposite melt into a rectangular cavity. Fig. 1 shows a schematic of the mold design employed in the injection molding process and the mold dimensions are listed in Table 1. The thickness of cavity was 2.0 mm. To achieve a more comprehensive picture of the effects of MWCNT alignment on the EMI shielding properties, the EMI shielding properties of the aligned injection molded samples were compared with those of the compression molded samples, where MWCNTs were randomly distributed. A Carver compression molder (Carver Inc., Wabash, IN) was employed to fabricate the compression molded samples with the same dimensions as the injection molded samples. The compression molding process was carried out at 210 °C for 10 min under 38 MPa pressure.

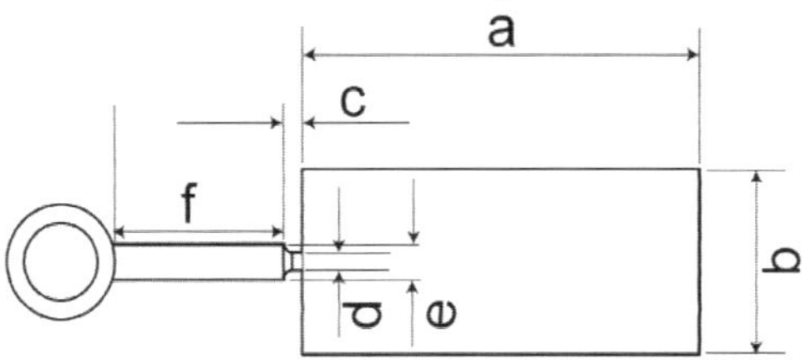

Figure 1 – A schematic view of the designed mold.

Table 1 – Dimensions of the designed mold.

Parameter	Value (mm)
a	22.86
b	10.16
c, d	1
e	2
f	10

In order to obtain detailed information about the alignment of MWCNTs, Raman spectra were collected from the compression molded and injection molded samples. A Renishaw spectrometer equipped with an inVia Raman microscope was employed to obtain the Raman spectra. Excitation was provided by a near-infrared (NIR) laser beam (785 nm) in regular mode. Measurements were performed at two normal orientations of the laser beam with respect to the flow direction in the samples. The MWCNT alignment was determined by comparing the Raman spectra obtained from parallel and perpendicular directions.

The in-flow electrical conductivity measurements were performed using two different setups. For the samples with a conductivity of more than 10^{-2} $S \cdot m^{-1}$, the conductivity measurements were conducted according to the ASTM 257-75 standards, using a Loresta GP resistivity meter (MCP-T610 model, Mitsubishi Chemical Co., Japan) connected with a four-pin probe. For a conductivity of less than 10^{-2} $S \cdot m^{-1}$, the measurements were performed using a Keithley 6517A electrometer connected to a Keithley 8009 test fixture (Keithley Instruments, USA).

EMI shielding effectiveness (SE) is expressed in dB and is the logarithm of the ratio of the incident power to the transmitted power. EMI SE was found in the X-band (8.2-12.4 GHz) for the samples used in the conductivity experiments. EMI SE measurements in the X-band frequency range were carried out in a WR-90 rectangular waveguide using an Agilent programmable network analyzer (PNA) (Model E8364B). To measure the EMI SE, the MWCNT/PS samples were placed inside the cross section of the rectangular waveguide. The S-parameters of each sample were recorded and used to calculate the EMI SE using the Nicolson-Ross-Weir method [8]. Contributions of reflection and absorption to overall EMI SE were also calculated using relevant incident and transmitted powers for each mechanism.

Results and Discussion

Raman spectroscopy

The Raman spectra of MWCNT/polymer composites provide two important features: the *D* band (disorder band) and *G* band (graphite band). The *D* band, correlating to disorder in the sp^2 hybridized graphitic structure, is more responsive to the alignment of MWCNTs than the *G* band, which corresponds to the in-plane vibration of the graphitic wall [9]. Higher intensity ratios of $D_{\|}/D_{\perp}$ and $G_{\|}/G_{\perp}$ (parallel/perpendicular to the flow direction) correspond to higher MWCNT alignment. The injection molded samples showed $D_{\|}/D_{\perp}$ and $G_{\|}/G_{\perp}$ ratios equal to 1.66 and 1.51, respectively, whereas the compression molded samples exhibited $D_{\|}/D_{\perp}$ and $G_{\|}/G_{\perp}$ of 1.01 and 1.01, respectively. From the Raman spectroscopy results, it can be claimed that the MWCNTs were randomly distributed in the compression molded samples, while they were partially aligned in the injection molded samples.

Electrical Conductivity and EMI SE

The formation of a conductive network in CPCs can be clarified with the concept of the electrical percolation threshold [10]. Electrical percolation is the concentration at which the filler particles come into contact with each other and form a continuous conductive pathway in the composite, which allows the electrical current to pass through the sample. As shown in Fig. 2, the compression molded samples showed higher electrical conductivity and lower percolation threshold than the injection molded samples. The percolation threshold in the compression molded samples obtained from the percolation theory was 0.7 wt%; however, it was interesting to observe that the percolation threshold in the injection molded samples was about six times greater than the percolation threshold in the compression molded samples. The lower electrical conductivity and higher percolation threshold in the injection molded samples can be attributed to the lower probability of MWCNTs neighboring or contacting each other due to MWCNT alignment [11].

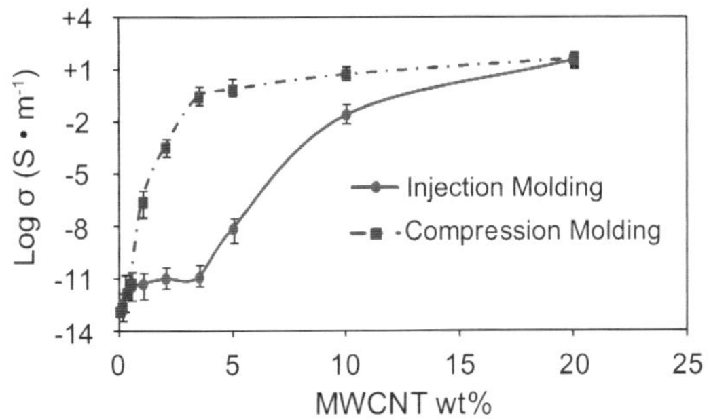

Figure 2 – Electrical conductivity for the compression molded and injection molded samples of the MWCNT/PS composites as a function of MWCNT concentration.

Fig. 3 shows the EMI SE of the compression molded and injection molded samples as a function of MWCNT concentration. The EMI SE results are the average values over the X-band frequency range. The EMI SE increased with increase in MWCNT concentration for both the compression molded and injection molded samples. For instance, the EMI SE for the compression molded samples at 5.00 wt% was 25.4 dB while it increased to 63.6 dB at 20.0 wt%. This increase was mostly due to greater amount of interacting mobile charge carriers at higher MWCNT concentrations [12].

It was interesting to observe that the EMI SE in the injection molded samples was significantly lower than the compression molded samples, particularly at high MWCNT concentrations. For instance, the injection molded samples at 5.00 and 20.0 wt% MWCNT exhibited EMI SE around 15 and 30 dB, respectively, lower than compression molded samples. Even more interesting was the fact that both types of samples had the same electrical conductivity at 20.0 wt% but exhibited significantly different EMI SE. For example, although both the compression molded and injection molded samples had the electrical conductivity around 35 $S \cdot m^{-1}$ at 20.0 wt% MWCNT, the EMI SE was double for the compression molded sample versus the injection molded sample. This discrepancy between the compression molded and injection molded samples suggests that the compression molded samples had a more extensively connected MWCNT network than the injection molded samples.

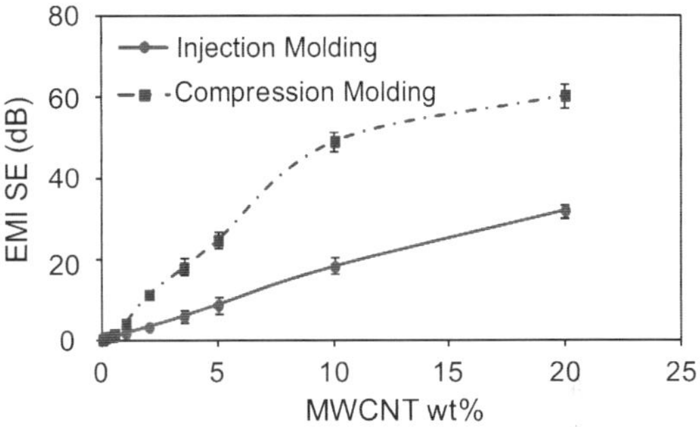

Figure 3 – EMI SE for the compression molded and injection molded samples of the MWCNT/PS composites as a function of MWCNT concentration. The thickness of all the samples was 2.0 mm.

In view of Figs. 2 and 3 and Raman spectroscopy results, it can be claimed that the CPCs with greater MWCNT alignment showed lower electrical conductivity and EMI SE. In addition, it can be interpreted that greater MWCNT connectivity in the compression molded samples resulted in higher EMI SE in the compression molded samples than the injection molded samples. These results are in agreement with researchers who believe that EMI shielding does not *require* filler connectivity; however, it *increases* with filler connectivity [13,14].

Effects of MWCNT alignment on shielding mechanisms in MWCNT/PS composites

To investigate the effects of MWCMT alignment on EMI shielding more precisely, the contributions of the reflection and absorption to the overall EMI SE were examined as a function of MWCNT concentration and alignment (Figs. 4(a) and (b)). The difference in the connectivity of the MWCNTs in the compression molded and injection molded samples is the key to interpret the reflection and absorption mechanisms.

To shield by reflection, the material must have mobile charge carriers (electrons or holes) to interact with incoming electromagnetic wave [14,15]. As shown in Fig. 4(a), the shielding by reflection increased with increase in MWCNT concentration which can be related to higher amount of mobile charge carriers at greater MWCNT concentrations. The shielding by reflection was 2.7 dB at 2.00 wt.% of MWCNT, which increased to 7.8 dB at 20.0 wt.% of MWCNT.

As there is a direct relationship between electrical conductivity and shielding by reflection in conductive monolithic materials [15], a higher shielding by reflection was expected in the compression molded CPCs than the injection molded ones. However, as shown in Fig. 4(a), it is surprising to observe that the shieldings by reflection of the compression molded and injection molded samples were almost the same at different MWCNT concentrations. The similar shieldings by reflection in the compression molded and injection molded samples may be explained by comparable area of the MWCNT surface projection normal to the incoming electromagnetic wave whether the MWCNTs were aligned or not.

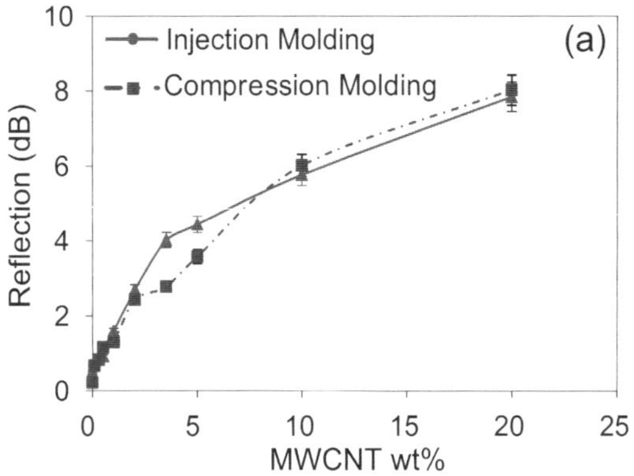

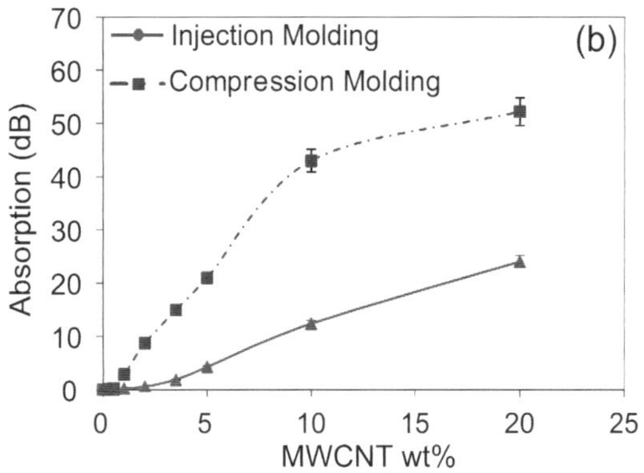

Figure 4 – Contributions of (a) reflection and (b) absorption to the overall EMI SE for the compression molded and injection molded samples of the MWCNT/PS composites as a function of MWCNT concentration. The thickness of all the samples was 2.0 mm.

Fig. 4(b) shows that the shielding by absorption increased with increase in MWCNT concentration and decrease in MWCNT alignment. Increase in absorption by increasing MWCNT concentration is a well-established concept which has been ascribed to higher amount of mobile charge carriers at higher MWCNT concentrations [13,14]. According to Fig. 4(b), unlike the shielding by reflection, the shielding by absorption was highly sensitive to MWCNT alignment and the injection molded samples showed much lower absorption than the compression molded samples. These results suggest that shielding by absorption is related to MWCNT connectivity.

In the compression molded samples, the electrons had a greater mean free path in which to move according to the direction of electric field in each half cycle and, consequently, could dissipate more electrical energy [12]. Greater electrons' mean free path at lower MWCNT alignments can be related to lower thickness of insulative gaps between MWCNTs, i.e., greater MWCNT network formation, leading to larger conduction current via the mechanisms of conduction, hopping and tunneling. The greater electrical energy loss by free electrons in the compression molded samples, due to greater electrons' mean free path, contributed to higher shielding by absorption in these samples.

Conclusions

Comparing the EMI shielding properties of the injection molded samples, where the conductive filler was aligned, versus compression molded samples, which had randomly distributed conductive filler, of the MWCNT/PS composites showed that MWCNT alignment had an adverse effect on EMI shielding properties. Inferior EMI shielding properties, i.e., properties such as EMI SE and electrical conductivity, of the injection molded samples relative to the compression molded samples were related to MWCNT alignment which led to poorer MWCNT network formation.

Since the electrical conductivity and EMI SE of the compression molded samples were greater than those of the injection molded samples, it can be concluded that designing a mold for injection molding that achieves a random MWCNT distribution is crucial in order to attain high EMI shielding properties.

Acknowledgements

The authors acknowledge support for this project from Natural Sciences and Engineering Research Council of Canada (NSERC). We would like to thank Mr. Mehdi Mahmoodi and Dr. Simon Park in the Department of Mechanical and Manufacturing Engineering at the University of Calgary for assistance with the mold design and manufacturing and also performing injection molding of the nanocomposites. We would like to thank Dr. Samaneh Abbasi of Ecole Polytechnique (Montreal, Canada) for assistance with Raman spectroscopy. We are grateful to Americas Styrenics LLC, for providing the neat polystyrene.

References

1. S. Y. Yang; K. Lozano; A. Lomeli; H. D. Foltz; R. Jones. *Compos Part A*. 2005, *36*, 691.

2. G. A. Gelves; B. Lin B; U. Sundararaj, J. A. Haber. *Adv. Funct. Mater*. 2006, *16*, 2423.
3. M. H. Al-Saleh; G. A. Gelves; U. Sundararaj. *Compos Part A*. 2011, *42*, 92.
4. Y. Chekanov; R. Ohnogi; S. Asai; M. Sumita. *J. Mater. Sci*. 1999, *34*, 5589.
5. N. Li; Y. Huang; F. Du; X. B. He; X. Lin; H. J. Gao. *Nano Lett*. 2006, *6*, 1141.
6. W. S. Jou; H. Z. Cheng; C. F. Hsu. *J Alloys Compd*. 2007, *434–5*, 641.
7. M. Mahmoodi; M. Arjmand; U. Sundararaj; S. Park. *Carbon*, 2012, *50*, 1455.
8. W. B. Weir. *Proc. I. E. E. E*. 1974, *62*, 33.
9. S. Abbasi; P. J. Carreau; A. Derdouri. *Polymer*, 2010, *51*, 922.
10. M. Weber, M. R. Kamal, *Polym. Compos*. 1997, *18*, 711.
11. M. Arjmand; M. Mahmoodi; G. A. Gelves; S. Park; U. Sundararaj. *Carbon*, 2011, *49*, 3430.
12. M. Arjmand; T. Apperley; M. Okoniewski; U. Sundararaj. *Carbon*, 2012, *50*, 5126.
13. D. D. L. Chung. *J. Mater. Eng. Peform*. 2000, *9*, 350.
14. M. H. Al-Saleh; U. Sundararaj. *Carbon*, 2009, *47*, 1738.
15. K. L. Kaiser, *Electromagnetic shielding*, CRC Press, Boca Raton, 2006.

O-07- 552

NANOCOMPOSITE FIBERS AND NANOFIBERS FOR BIOMEDICAL APPLICATIONS

Mangala Joshi*, Kasturi Saha and BS Butola
Department of Textile Technology, Indian Institute of Technology, Hauz Khas, New Delhi 110016, India
Email: mangala@textile.iitd.ac.in; poddar.kasturi@gmail.com;bsbutola@gmail.com

Abstract: Polymer nanocomposites are a unique new class of materials with an ultrafine dispersion of nanomaterials in a polymeric matrix. They exhibit much superior properties in terms of increased strength and modulus, improved heat resistance, decreased gas permeability and flame retardance at very low loadings of <5 wt% of nanofillers as compared to neat polymers or conventional composites. We have developed nanocomposite fibers based on modified layered silicate nanoclays with antimicrobial properties. These polyamide and polypropylene nanocomposite fibers show 100% antimicrobial activity against both gram positive and gram negative bacteria and show durability to washing. The fibers retain their mechanical strength and flexibility and can be converted to multifilament yarn and knitted structure for various applications.

These bioactive fibers can also be used for drug delivery applications, when loaded with suitable drugs such as sulphanilamide. These fibers show desired bioactivity and controlled release of drug over a period of time. When the bioactive nanocomposites are converted to nanofibrous forms such as webs or membranes, they show different kinetics of drug release as compared to microfibers. Nanofibers webs are more suitable for wound dressing because of high surface area to volume ratio, flexibility and porosity. In this study, we are developing novel polymeric nanocomposites in microfibers and nanofibrous forms, which can have diverse applications in biomedical fields such as sutures, wound dressings, bandage and health and hygiene fabrics.

Keywords: Nanocomposites Fibers, Nanofibres, Biomedical applications, Drug Delivery

Introduction

Nanoscience and nanotechnology combined, have revitalized material science and led to the development and evolution of a new range of improved materials including polymers and textiles through nanostructuring and engineering. Organized nano structures as exhibited by either nanocomposite fibers, nanocoatings, nanofinishing and nanofibers seem to have immense potential to revolutionize the textile industry with new functionality such as self cleaning surfaces, conducting textiles, antimicrobial properties, controlled hydrophilicity or hydrophobicity, protection against fire, UV radiation etc. without affecting the bulk properties of fibers and fabrics.

Advances in medical textiles and polymer science have led to the development of a variety of novel drug delivery systems [1]. Delivery of pharmaceutical agent to the systemic circulation and consequently to the site of action to produce a desired pharmacological effect, is the ultimate goal of drug delivery. At present, 95% of all new therapeutics has poor pharmacokinetics [2] and biopharmaceutical properties. Therefore, there is a need to develop suitable drug delivery systems that distribute the therapeutically active drug molecule only to the site of action, without affecting healthy organs and tissues. Recent developments in the field of nano-biotechnology [3] are actively focused towards in vivo drug delivery technique through nanofillers to protect the drug entity in the systemic circulation, restrict access of the drug to the chosen sites and to deliver the drug at a controlled and sustained rate to the site of action.

Clays can be used both as excipients [4] and active agents in drug delivery. They have high retention capacities as well as swelling and colloidal properties so they can be effectively used to delay or target drug release or even improve drug solubility. Clay is inherently having some of the antibacterial properties [5] as it was used for curing wounds, soothe irritation, inhibition of hemorrhages and as a cure of snakebites. The extensive use of the clay in medicine and its properties motivated us to use clay as a carrier of drug such as smectite (montmorillonite) clays can be loaded with drugs by the cation exchange mechanism [6]. The positive ions in the clay galleries can be exchanged with the cationic drugs and these clays can then be used with the polymer to form nanocomposites fibers or filaments.

Our present paper focuses on the intercalation of sulfanilamide (SA) drug into the intergallery of montmorillonite (MMT) clay via cation exchange route. The drug loaded clay- SA-MMT was characterized by X-ray diffraction (XRD) and Thermogravimetric (TGA) technique. The antimicrobial activity of SA-MMT hybrid composite was determined against both gram positive and gram negative bacteria. Also in comparison with pure drug loaded polymeric fiber, Nylon 6 and polypropylene nanocomposite fiber shows a controlled release of the drug in the phosphate buffer solution. The produced nanocomposite fibers with controlled drug release and bioactivity against microbes can have wide application in the area of medical textiles.

Materials

Sodium Montmorillonite (Na-MMT) clay with cation exchange capacity (CEC) of 120 meq/100 gm was procured from Kunimine Industries Co. Ltd. (Japan).

Sulfanilamide drug (SA; white powder; M.W. 172.20; M.P. 165°C) was kindly supplied by LOBA Chemie Pvt. Ltd. (Mumbai, India) and used as received. The chemical structure ($C_6H_8N_2O_2S$) of the model drug is shown in Fig. 1.

Nylon 6 (Density: Amorphous 1.084 g/cm^3, Crystalline 1.23 g/cm^3; M. P. 220°C) is a semicrystalline polyamide was kindly supplied by M/s SRF Limited, Chennai, India.

Polypropylene (Density : Amorphous 0.855g/cm^3, Crystalline 0.946 g/cm^3; M.P. 160°C) is a thermoplastic polymer was kindly supplied by Sigma Aldrich Pvt. Limited, Bangalore, India.

All other chemicals and solvents were of reagent grade and used without further purification.

Experimental

Intercalation of Sulfanilamide into Na-MMT

The modified clay (SA-MMT) was prepared through cation exchange reaction of Sulfanilamide drug (SA) with Sodium Montmorillonite (Na-MMT). In the preparation of the SA-MMT, 1g of Na-MMT was dispersed in 30 ml of distilled water with vigorous stirring for 0.5 h at room temperature. Then 2g of SA drug was dissolved in desired amount of distilled water. These two solutions were mixed together (pH value adjusted to 4.1 with acid) using a magnetic stirrer for about three hours at 80°C. After cooling the mixture was centrifuged at 5000 rpm. All products were washed with water, dried at 80°C and ground in an agate mortar for further characterization.

Characterization of drug loaded clay

X–ray diffraction studies of pure clay and modified clay were carried out on a PANalytical X–Ray Powder Diffractometer using Ni–filtered CuK_α radiation of wavelength 1.5418Å, working voltage 40 kV and working current 30 mA .Scanning was carried out in the range 2θ values from 2° – 50° at 2°/min for all the samples.

Perkin Elmer, TGA-7 system was used to determine amount of drug loading and thermal stability of pure Sulphanaliamide and drug loaded clay. Testing was carried out using a heating rate of 20°C per minute up to 600 ° C in Nitrogen atmosphere.

Antibacterial activity assay

Antibacterial activity of the modified clay was tested against both gram positive bacteria, *S. aureus* and gram negative bacteria, *E. coli* by the Disc Diffusion test (AATCC 90). 0.1 grams of each sample was taken and a paste was made using DI water. Paste was uniformly spread over onto paper disk (10 mm diameter). Nutrient agar solution was made by suspending 28 grams of Luria broth and 0.5 % of agar-agar in 1000 ml of DI water. About 20 ml of nutrient agar solution was uniformly spread in the Petri dishes. 20 μl of bacterial solution was evenly applied over the nutrient agar solution. Concentration of *S. aureus* around (1.5×10^6 CFU)/ml and *E. coli* around (1.8×10^6 CFU)/ml were used for conducting this experiment. The clay discs made earlier were then placed onto the agar plate. The agar plates were kept for incubation at 37°C for 24 hrs. The zone of inhibition was measured after 24 hrs incubation period.

Preparation of Nylon 6 & PP nanocomposite filaments

Nylon 6 and Nylon 6/clay nanocomposite filaments were prepared by melt intercalation route using Twin Screw Micro Compounder (Make: DSM Research, Netherlands). About 10% of pure SA & SA-MMT clay was used for loading. The SA & SA-MMT were dispersed in acetone and ultrasonicated for 30 min. The nylon chips were added in the clay dispersion and ultrasonicated for again 30 min. The physically mixed nylon clay chips were oven dried at 80°C for 3 hr and vacuum dried at 110°C for 12hr.

PP and PP/clay nanocomposite filaments were prepared by melt intercalation route using Twin Screw Micro Compounder. About 10% of pure SA & SA-MMT clay was used for loading. The loading of pure drug (SA) & modified clay (SA-MMT) onto PP was same as above.

The Nylon and PP clay nanocompsoite filaments were spun on laboratory scale DSM 5 Micro Twin Screw Compounder. Table 1 indicates process conditions used during melt intercalation route for both Nylon & PP.

SEM analysis of filament

Morphological appearance of the neat and the drug-loaded Nylon & PP filament was observed by a Zeiss EVO 50 scanning electron microscope (SEM). Each of the filaments was sputtered with a thin layer of gold using a EMITECH K550X sputtering device prior to SEM observation.

Release Profile Study of filaments

Drug release behavior of PP & Nylon 6 nanocomposite filaments were tested in phosphate buffer solution (pH 7.4). About 0.1g of each filament was taken and put in different conical flasks each containing 10 ml of buffer solution and kept in an incubation chamber at 37°C with stirring at 200rpm. About 2 ml of the solution was taken at regular time interval and the corresponding absorbance value was measured in UV spectrophotometer at 259 nm, which is the characteristic peak of sulphanilamide. The drugs concentration was back-calculated from the predetermined calibration curve of the model drug.

Results & Discussion

XRD analysis

The XRD patterns of MMT and SA-MMT are shown in Fig. 2. The X-ray diffraction pattern of the clay was identified with interlayer spacing of 12.8 Å, a characteristic value for MMT. After ion exchange, the diffractogram results of the modified clay showed that the spacing of SA-MMT increased from 12.8 Å to 14.36 Å, an increase of 15.46 % .This larger d value indicated that SA was successfully intercalated in the montmorillonite interlayer spacing.

Thermogravimetric analysis

TGA shown in Fig. 3 was done with an objective to determine the thermal stability of drug before and after loading onto clay. The TGA of pure sulphanilamide reveals that the drug is stable upto a temperature of 220°C, while the drug loaded clays are stable upto temperature of 340°C without any significant weight loss. This shows that loading of drug onto clay has a very positive effect in terms of thermal stability.

The pure clay contains 88% residue at 700°C temperature while only 32% residue is remaining in pure drug. But after drug loading, amount of residue leftover is 76 % in modified clay. This result concluded that amount of drug loading percentage is around 12% by weight in case of modified clay. This also indicates that the drug entity becomes stabilized after intercalation into clay interlayer spacing.

Antibacterial analysis

Disc Diffusion results of MMT and SA-MMT are shown in Fig. 4 and Fig. 5.

The inhibition zone of SA-MMT on *S. aureus* and *E. coli* were 32 and 28 mm. The results indicated that SA-MMT has strong activity against both Gram-positive and Gram-negative bacteria. This is also an indication of the diffusion of the antibacterial agents from treated MMT into the agar.

SEM analysis

Fig. 6 shows the SEM images of neat nylon, pure SA drug and 10% SA-MMT loaded nylon & PP nanocomposite filaments. For higher percentage of clay loading, the clay agglomerates are found to move towards the surface and helping in easier release of drug.

Release studies of nanocomposite filaments

The release profile of drug from 10% SA-MMT/Nylon6, 10% SA-MMT/PP & 10% pure SA/PP nanocomposite fibers in saline solution are shown in Fig. 7.

It has been observed that there is very minimal release of drug from pure SA/PP filaments while there is a controlled and good release of drug from both the PP & Nylon nanocomposite fibers. This study was done with an objective to know the role of clay in controlled release and it is evident from the result that the drug loaded clay nanocomposite fibers show a controlled release of drug compared to drug/polymer combination where the release of the drug is very slow. This may be because of drug entities getting embedded inside the fiber and not able to diffuse out. It has been also observed that the drug slowly releases from the nanocomposite fiber surfaces over a period of 7 days. The concentration of drug release from the nanocomposite fibers increases with the increase in loading percentage, as seen from the release profile.

Table 1. Process Parameters of Melt Intercalation route

Sample	Nylon 6	Polypropylene
Extruder rpm	10	10
Mixing rpm	150	150
Temperature	230° C	200° C
Residence Time	3 min	3 min

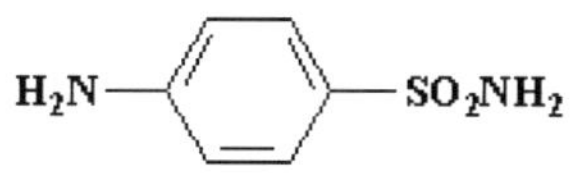

Figure 1. Chemical structure of Sulphanilamide (SA)

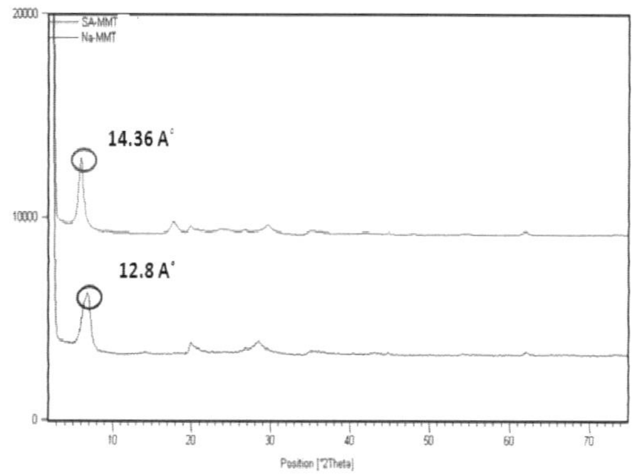

Figure 2. X-ray diffraction patterns of Na-MMT and SA-MMT

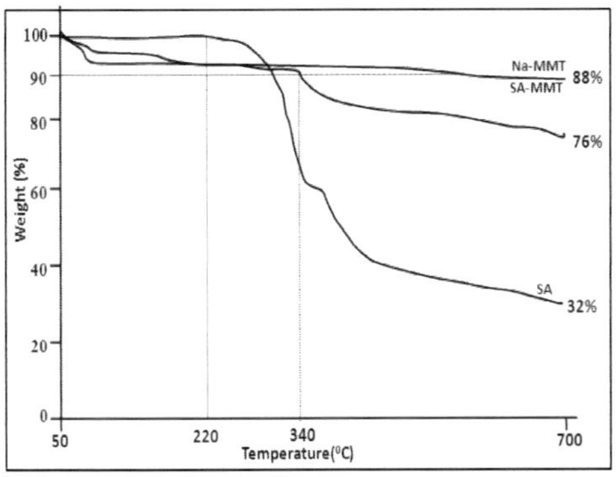

Figure 3. TGA curves of Na-MMT, SA-MMT and SA drug

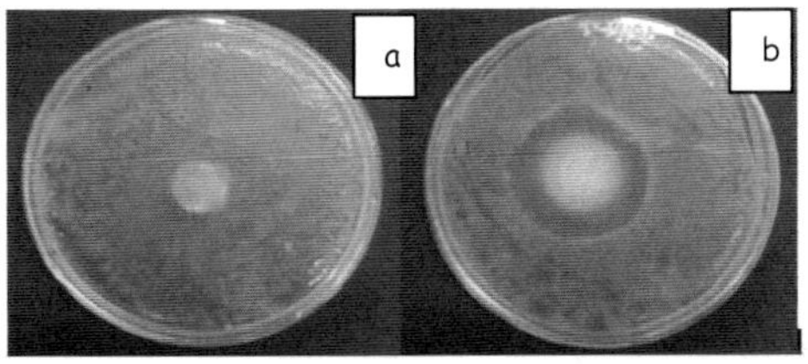

Figure 4. Antibacterial test of (a) MMT and (b) SA-MMT against *S. aureus*

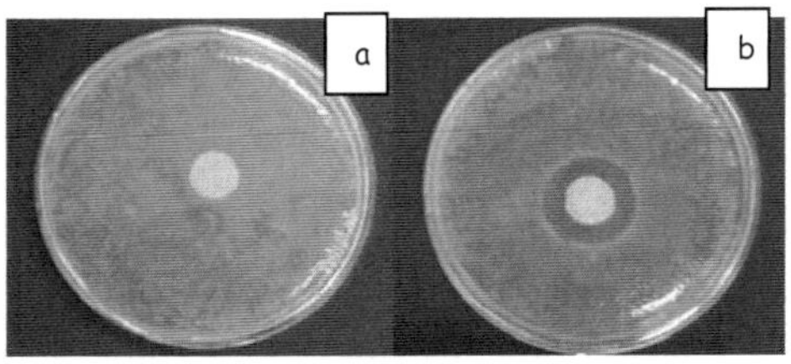

Figure 5. Antbacterial test of (a) MMT and (b) SA-MMT against *E. coli*

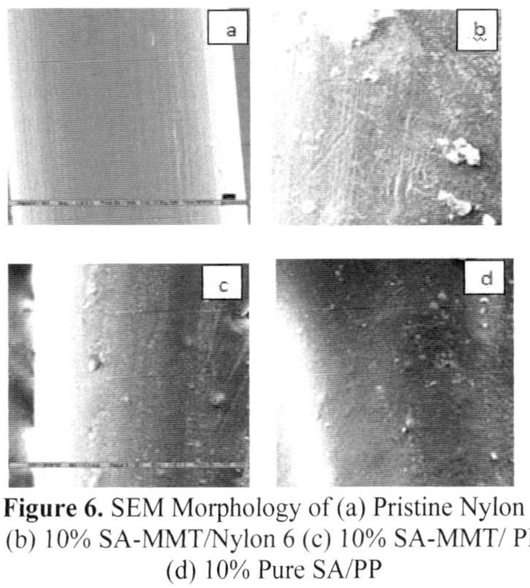

Figure 6. SEM Morphology of (a) Pristine Nylon 6 (b) 10% SA-MMT/Nylon 6 (c) 10% SA-MMT/ PP (d) 10% Pure SA/PP

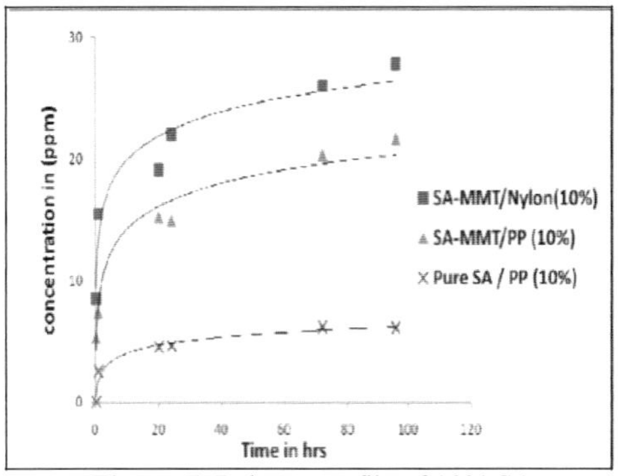

Figure 7. Release profile of 10% SA-MMT/Nylon6, 10% SA-MMT/PP & 10% pure SA/PP nanocomposite filament

Conclusions

Hybrid nanoparticle shows 100% antibacterial activity both against Gram positive *Staphylococcus aureus* and Gram negative *Escherichia coli* bacteria. In comparison with pure drug loaded polymeric fiber, Nylon 6 & PP nanocomposite fiber shows a controlled release of the drug in the phosphate buffer solution (pH 7.4) for a period from 1 hour to 7 days. The produced nanocomposite fibers with controlled drug release and bioactivity against microbes can have wide application in the area of medical textiles. These nanocomposites can be further converted to nanofibrous forms which are expected to show improved kinetics of drug release because of very large surface area to volume ratio of Nanofibers in comparison to microfibers and the work is in progress.

References

1. D.J. Brayden *Drug Discov. Today*, 2003, *8,* 976.
2. G.A. Hughes *Nanomedicine,* 2005, *1,* 22.
3. Z. Xinhua; K. Kwangsok; F. Dufei; R. Shaofeng; S.H. Benjamin; C. Benjamin *Polymer,* 2002, *43,* 4403.
4. J.W. McGinity; J.L. Lach *J. Pharm. Sci.,* 1976, *65,* 896.
5. N. Castela-Papin; S. Cai; J. Vatier; F. Keller; C.H. Souleau; R. Farinotti *Int. J. Pharm.,* 1999, *182,* 111.
6. F.H. Lin; Y.H. Lee; C.H. Jian *Biomaterials,* 2002, *23,* 1981.

O-07-568

CRYSTALLIZATION BEHAVIOR OF POLYAMIDE 6/MICA NANOCOMPOSITES USING DYNAMIC RHEOMETRY AND DIFFERENTIAL SCANNING CALORIMETER

Tayebeh Sadat Kermanian[a], **Hamid Garmabi** [*a]

[a] *Department of Polymer Engineering and Color Technology, AmirKabir University of Technology, P.O. Box 15875-413, Tehran, Iran - kermanian@aut.ac.ir; garmabi@aut.ac.ir .*

Abstract

This study considers the crystallization behavior of Polyamide-6(PA6)/Mica nanocomposites. The crystallization of nanocomposites has been investigated using dynamic rheometry and differential scanning calorimetery (DSC). The nanocomposites (with 1, 3 and 5 wt% mica) were prepared by melt mixing the layered silicate with PA6 matrix in a co-rotating twin-screw extruder. The obtained granules were injection molded using an injection-molding machine. Rheological measurements were carried out to study the isothermal crystallization of PA6/mica nanocomposites. The isothermal crystallization kinetics of the samples was studied on the basis of the rheological data and applying Avrami model. The non-isothermal analysis was carried out using a differential scanning calorimeter (DSC). The non-isothermal crystallization behavior was investigated applying identical cooling rate (10 °C/min) for neat PA6 and its nanocomposites. The DSC thermograms of non-isothermal crystallization for PA6/mica (3%wt) at different cooling rates (5, 10 and 15 °C/min) were obtained.

Keywords: Polyamide 6, Mica, Crystallization, Isothermal, Non-Isothermal.

1. Introduction

In recent years, organic–inorganic nanoscale composites have attracted great interest since they frequently exhibit unexpected hybrid properties synergistically derived from two components [1]. One of the most promising composite systems is the hybrid based on organic polymers and inorganic clay minerals consisting of layered silicates [1]. In this article, the Polyamide-6(PA6)/mica nanocomposites were prepared and the main objective was to focus on the crystallization behaviors of the prepared nanocomposites. It is well known that the physical and mechanical property of crystalline polymers depends on the morphology, the crystalline structure and the degree of crystallization [1]. The overall crystallization process in semi crystalline polymers can be divided into two main processes: primary crystallization and secondary crystallization. The primary crystallization process is the macroscopic development of crystallinity as a result of two consecutive microscopic mechanisms: primary nucleation and secondary nucleation (i.e., subsequent crystal growth). The secondary crystallization process is mainly concerned with the crystallization of interfibrillar melt, which was rejected and trapped between the fibrillar structure formed during the growth of crystalline aggregates (e.g., axialites, spherulites, etc.). It should be noted that, if the crystallization time becomes very long, other types of secondary crystallization (i.e., crystal perfection and crystal thickening) may become significant, enough to increase the ultimate absolute crystallinity [2].

In the present paper, using RMS and DSC, the isothermal and non isothermal crystallization kinetics of polyamide 6 and its nanocomposites were investigated.

2. Experimental

2.1. Materials

High molecular weight PA6 (HMW PA6), Ultramid B 50L01 used in this study was manufactured by BASF, Germany. Organically modified layered silicate, Somasif MEE (Synthetic Mica) was provided by Co-op Chemicals, Japan. Somasif MEE is synthetic fluorine mica, covered with a methyl bis-2-hydroxyethyl coco quaternary ammonium surfactant (28.5 wt%, determined with a Perkin–Elmer TGA-7).

2.2. Sample preparation

The nanocomposites including PA6 and Somasif MEE were prepared by mixing the layered silicate with PA6 in a Brabender co-rotating twin-screw extruder (d= 20 mm, L/D=40). The extruder was operated at a screw speed of 150 RPM. The temperature profile was set between 240 °C, and 260 °C. Then, using an injection molding machine, extruded composite pellets were injection molded.

2.3. Testing

2.3.1. DSC experiments

Differential scanning calorimetry (DSC) measurements were performed on a Mettler DSC protected by nitrogen atmosphere. About 10 mg of the polymer sample was weighted in the aluminium DSC pan and placed in the DSC cell. It was heated from 25 to 260°C at a rate of 10°C/min, under nitrogen atmosphere. The sample was kept for 10 min at 260°C to eliminate the heat history before cooling at a specified cooling rate. Constant cooling rates of 5, 10 and 15°C/min were applied. The thermograms corresponding to the heating and cooling cycles were recorded and analyzed.

2.3.2. Rheological analysis

The rheological behavior of the nanocomposites was examined using dynamic rheometery. Rheological tests were carried on a strain-controlled rheometer (Anton paar-Physica MCR 501) at a temperature of 210 °C. Small-amplitude oscillatory shear measurements were performed using parallel plate geometry .The linear viscoelastic material functions of PA6 and PA6/Mica nanocomposites were characterized by employing small-amplitude oscillatory shear. The isothermal crystallization experiments were performed in oscillatory mode, time sweep test at frequencies of 1, 3 and 5 rad/s and initial strain of 1%.

3. Results and Discussion

The isothermal crystallization measurements using rheological tests was performed at 210 °C, where the induction time was high enough to avoid undesired premature crystallization during cooling. For isothermal melt crystallization, the samples were quickly heated from 25 to 260 °C at a heating rate of 70 °C/min and maintained at 260 °C for 10 min to eliminate any nuclei that might act as seed crystals. The samples were then cooled to 210 °C at constant rates of 10 °C/min. The isothermal crystallization experiments were performed in oscillatory mode, time sweep test at frequency of 1 rad/s and initial strain of 1%. The Avrami equation can well describe the isothermal crystallization behavior. We analyzed crystallization kinetics of samples based on the Avrami equation [3]:

$$X_c (t) = 1 - \exp(-kt^n) \quad (1)$$

Where X_c is the weight fraction crystallized at time *t*, *k* is the Avrami rate constant and *n* is the Avrami exponent. The Avrami exponent 'n' provides information on the nature of nucleation (nucleation type) and growth processes during crystallization. On the other hand, 'K' is dependent upon the shape of the growing crystalline entities (for e.g., whether they are spheres, discs or rods), as well as the type and amount of nucleation (sporadic or predetermined) and the linear growth rate 'G' of the growing crystalline moieties [3]. Note that in Equation 1, the parameters *k* and *n* are coupled. Thus a modified Avrami equation was used to remove this coupling [3]:

$$X_C = 1- \exp(-Kt)^n \quad (2)$$

Where *K* is the composite Avrami rate constant, such that $k = K^n$. Using the modified Avrami equation the values of *K* (with units of inverse time) were obtained, using Equation 2 which is independent of *n*. Flory theory suggests that the storage modulus would be proportional to the degree of cross-linking, and hence crystallization from the melt can be considered to be similar to the development of a network structure with the crystallites acting as cross-links/junction points. Thus the change in the storage modulus during the shear-induced crystallization process can be taken to be indicative of the rate of crystallization on the basis of a normalized modulus, $G_n(t)$ [3]:

$$G_n (t) = \frac{G'(t) - G'(0)}{G'(\infty) - G'(0)} \quad (3)$$

Where G′(t) is the storage modulus at time t, G′(0) is the storage modulus at t=0 and the $G'(\infty)$ is the storage modulus value at the conclusion of the shearing process. Using this normalized modulus, an Avrami-type equation for the normalized storage modulus can be written as [3]:

$$G_n (t) = 1 - \exp(-Kt)^n \quad (4)$$

Figure 1 shows the typical shear-induced crystallization results obtained for the storage modulus versus time for PA6 and PA6-mica nanocomposites. Samples collected at a constant frequency of 1 rps and 1% strain amplitude at 210 °C. As shown in Figure 1, an increase in G′ with time was observed. The required values were calculated by plotting $\ln[-\ln(1-G_n(t)]$ vs. ln(t) and evaluating the slope, the Avrami exponent, n, and the intercept, the constant ln(K). The "n" exponent represents the nucleation mechanism and growth dimension. Generally, for most of crystallization polymers, the value of Avrami exponent "*n*" was found to vary between 1 and 4 [4]. It's observed that the values of the Avrami exponent "*n*" for nanocomposites (higher than 4) are significantly higher than PA6. The increase of "*n*" suggests that the crystallization mechanism has changed from inhomogeneous nucleation into homogeneous nucleation, i.e. the nucleation rate is an increasing function against time [5]. It has been suggested that the collapse of spherulite induced by flow occurs and generates small nuclei resultantly [5]. A higher value of the Avrami exponent is indicative of the state of preferred orientation of the polymer chains during the crystallization process and the resulting increase in the nucleation and growth rates [3].

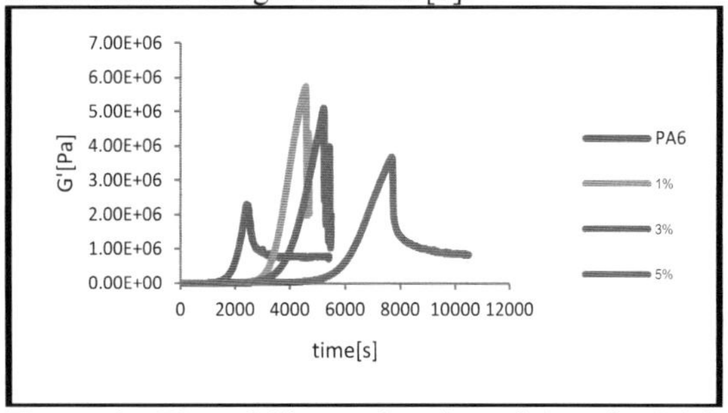

Figure 1: Plot of G' as a function of time for PA6 and PA6CN.

Figure 2 shows the time-dependent evolution of G' for the 3% mica nanocomposite samples subjected to small amplitude oscillatory shear at 210 °C and 1% strain amplitude at frequencies of 1, 3 and 5 rps. For the nanocomposite samples it is clearly observed that the crystallization rate decreased and the induction time for crystallization increased with increasing frequency. A plot of ln [-ln (1-G_n (t)] versus ln (t) gives the slope *n*, the Avrami exponent, and the intercept ln (K). It's observed that "*n*" is higher than 4 for all frequencies.

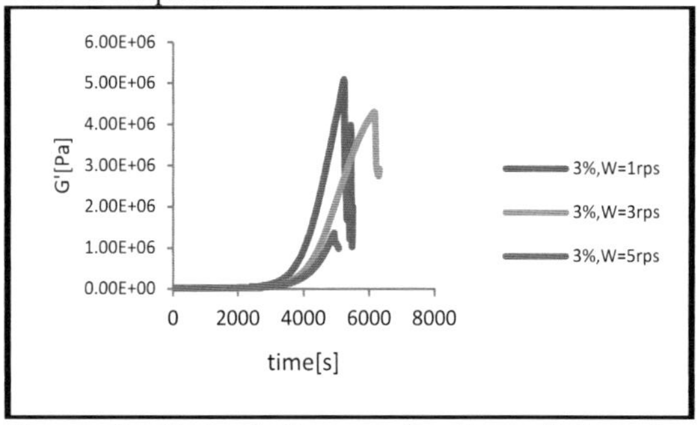

Figure 2: Plot of G' as a function of time for PA6/3%mica nanocomposite at different frequencies.

The DSC thermograms of non-isothermal crystallization for PA6 and PA6CN at 10°C/min cooling rate are presented in Figure 3 and thermal characteristics were listed in Table 1. In the cooling trace of DSC measurements (Figure 3(a)), the nanocomposites shows an exothermic peak (crystallization) in a narrower temperature range in comparison with neat PA6, which indicates a high crystallization rate for the nanocomposites [6]. In comparison with the neat PA6 sample in Figure 3(b), in the heating trace, the nanocomposite samples possessed a lower melting peak. It has been widely reported that the γ crystalline form is responsible for this phenomenon (Li et al., 2005) [6]. It seems as the mica content increases, the rate of formation of the γ-form becomes faster than that of the α-form [7]. The addition of silicate layers favored the formation of the γ crystalline form (Khanna, 1992; Lincoln et al., 2001; Fornes et al., 2002) [6, 7, 8, and 9]. The incorporation of mica had little effect on the glass transition and melting temperatures and increased the crystallization temperature of the PA6/mica nanocomposites. This result confirms that the incorporation of a very small quantity of mica enhances the nucleation of PA6 crystallization. The mica promotes the formation of heterogeneous nuclei, with lower energy consumption required to reach critical stability for crystal growth ,resulting in them functioning as effective nucleating agents in the PA6 matrix[10].

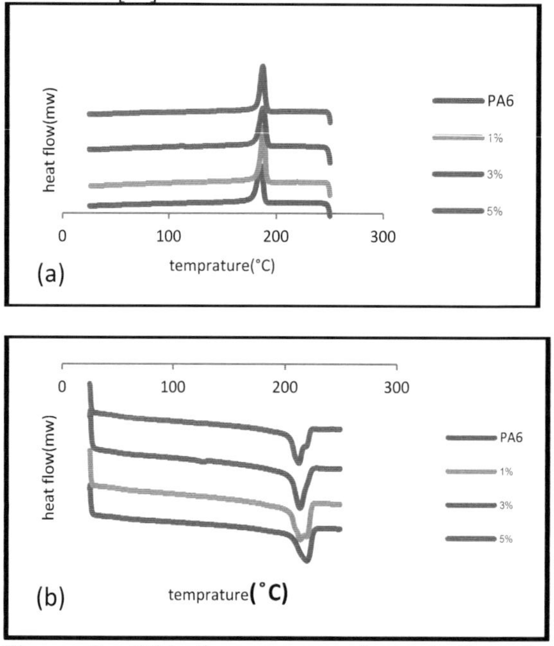

Figure 3: DSC thermograms of non-isothermal crystallization for PA6 and PA6CN.

Sample	Tc(°C)	Tm(°C)-α	Tm(°C)-γ	X_c%	T_g(°C)
PA-6	185.64	219.90	-	29.12	52.44
PA-6(1%)	188.09	219.65	214.14	28.20	51.41
PA-6(3%)	187.52	-	213.61	24.20	~ 51
PA-6(5%)	187.26	219.96	212.69	24.69	50.76

Table 1: Melting Temperatures (T_m), Degree of Crystallinity, (X_c), Crystallization Temperature (T_c) and glass transition temperature (T_g) of PA6 and PA6 Nanocomposites from DSC Analysis.

The DSC thermograms of non-isothermal crystallization for PA6/3% mica nanocomposite at different cooling rates are presented in Figure 4.

From these curves, the crystallization peak temperature (T_c), the melt temperature (T_m), the Crystallinity percent (X_c%) and the glass temperature (T_g) are obtained and listed in Table 2. Figure 4(a) reveals The crystallization peak (T_c) for PA6 and its nanocomposites with mica cooled at 5,10 and 15 °C/min. T_c peak becomes wider, more intense and shifts to lower temperatures as the cooling rate increases. This phenomenon can be explained as follows: at the lower cooling rate, the PA6 molecular chains may have enough time to pack up in a unit cell and then their nuclei grows up at higher temperatures; on the contrary, at higher cooling rate, before the PA6 molecular chains get ready to arrange regularly or to form crystal nuclei, the temperature may have gone down to a lower point, resulting in a lower T_c [11]. The increasing of frequency had little effect on the glass transition temperature of the PA6/3%mica nanocomposites. The PA6-3%Mica nanocomposite at 5 °C/min cooling rate exhibits an endothermic approximately at 50.93°C corresponding to the T_g and the enhancement of T_g transition of the nanocomposites are found slightly.

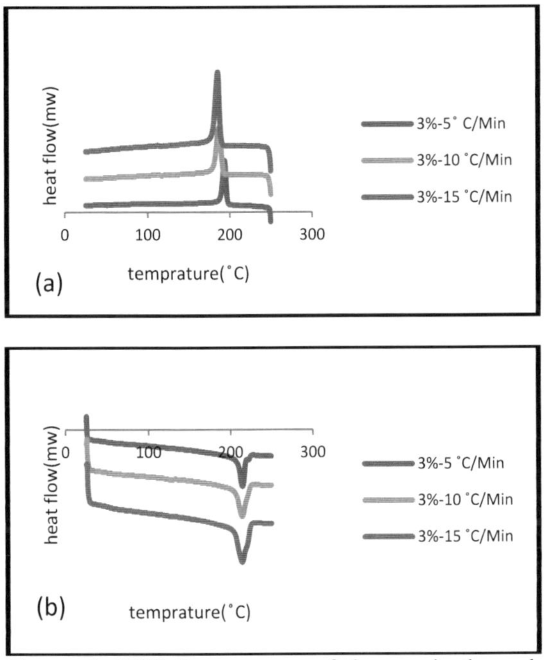

Figure 4: DSC thermograms of the non-isothermal crystallization for PA6 (a) and its nanocomposite (b) at various cooling rates.

Table 2: Thermal properties of PA6/3%mica nanocomposite at different frequencies.

sample	Tc(°C)	Tm(°C)-α	Tm(°C)-γ	X_c%	T_g(°C)
PA-6(3%)-5rps	193.66	221.46	214.30	26.49	50.93
PA-6(3%)-10rps	187.52	-	213.61	24.20	~ 51
PA-6(3%)-15rps	185.56	222.21	-	21.28	51.63

Conclusions

In this study, the nanocomposites with Somasif MEE were prepared by mixing the layered silicate with PA6. The following conclusions can be drawn from the investigation into the effects of inclusion of Mica nanoparticles on thermal properties, isothermal and non isothermal melt crystallization kinetics of PA6:

1. The isothermal crystallization kinetics of the samples based on the Avrami equation was analyzed. The values of the Avrami exponent "*n*" for nanocomposites are significantly higher than PA6. Also it's observed that "*n*" is higher than 4 for PA6-3%Mica nanocomposite and all frequencies. The increase of "n" suggests that the crystallization mechanism is changed from inhomogeneous nucleation into homogeneous nucleation.
2. The non-isothermal crystallization of the PA6/mica nanocomposites depends on the mica content and cooling rate. The crystallization temperature for the PA6/mica nanocomposites decreased with increasing cooling rate for given mica content and increased with mica content for a given cooling rate.
3. The addition of mica decreased the glass transition temperature of PA6 and slightly decreased the melting temperature of PA6.

References

1. X. Liu; Q. Wu; European Polymer Journal, 2002,38,1383.
2. P. Supaphol; J. E. Spruiell; MACROMOL. SCI.—PHYS., 2000, B39 (2), 257.
3. G. Mago; F. T. Fisher; D. M. Kalyon; Macromolecules, 2008, 41, 8103.
4. X. L. Jiang; S. J. Luo; K. Sun; X. D. Chen;eXPRESS Polymer Letters;2007; 1; No.4 247.
5. Y. Masubuchi; K. Watanabe; W. Nagatake; J. I. Takimoto; K. Koyama; Polymer, 2001, 42, 5023.
6. H. Uno; K. Tamura; H. Yamada; K. Umeyama; T. Hatta; Y. Moriyoshi; Applied Clay Science, 2009, 46, 81.
7. K. Tamura; H. Uno; H. Yamada; K. Umeyama; Journal of Polymer Science: Part B: Polymer Physics, 2009, 47, 583.
8. F. M. Uhl; Q. Yao; H. Nakajima; E. Manias; C. A. Wilkie; Polymer Degradation and Stability, 2005, 89, 70e84.
9. S. Pavlidou; C. D. Papaspyrides; Progress in Polymer Science, 2008, 33, 1119.

10. J. Y. Kim; H. S. Park; S. H. Kim; Polymer, 2006, 47, 1379.

11. H. Meng; G. Suiy; G. Xie; R. Yang, J. Mater. Sci. Technol., 2009, 25, No.2.

O-07-594

INVESTIGATION OF SORBITOL/POLYHEDRAL OLIGOMERIC SILSESQUIOXANE/CARBON NANOTUBE / POLYPROPYLENE NANOCOMPOSITE SPUN FIBERS

H. Mattausch[1] and S.C. Jana[2]*

[1]*Department of Polymer Engineering and Science, Chair of Polymer Processing, Montanuniversitaet Leoben, Austria – hannelore.mattausch@unileoben.ac.at*

[2]*Department of Polymer Engineering, The University of Akron, Ohio, United States – janas@uakron.edu*

Abstract - The use of isotactic polypropylene (iPP) nanocomposites filled with polyhedral oligomeric silsequioxane (POSS) or carbon nanotubes (CNT) have attracted research interest over the last decade. Nanofillers can enhance the material's properties only with a small amount of filler compared to conventional micro-scaled fillers. Sorbitol acts as heterogeneous nucleating agent for iPP due to the formation of a nanofibrillar network. Synergistic combinations of sorbitol and POSS in iPP form a low viscosity liquid complex above the crystalline melting temperature due to the noncovalent interactions such as hydrogen bonding and π-π interactions between sorbitol and POSS. This formed complex decreases the complex viscosity of the composite dramatically [1]. iPP filled with CNT, on the other hand, increases the complex viscosity significantly. In this research, synergistic combinations of sorbitol/POSS/CNT in iPP are produced to investigate the interactions between these fillers. CNT can form several structures from completely ordered to agglomerated arrangements in the low viscosity complex. To investigate the role of CNT, the crystallinity was characterized with differential scanning calorimeter (DSC), wide angle X-ray diffraction (WAXD) and Scanning electron microscopy (SEM) pictures were taken of the fibers to reveal CNT dispersion. The draw down ratio of the fibers during spinning was a very important aspect in the estimation of interactions between the single fillers. The mechanical properties of the fibers were measured via tensile testing. The complex viscosity was seen enhanced with the addition of CNT in the sorbitol/POSS composite.

Keywords: complex viscosity, carbon nanotubes, sorbitol, polyhedral oligomeric silsesquioxane, fiber spinning

Introduction

Polypropylene nanocomposites (PNC) are in the focus of interest in many fields of possible applications. The reason is that polypropylene (PP) offers good performance in many different applications (automotive, biomedical components, home appliances) and a good cost-performance ratio [2]. But this engineering material presents practical limitations such as low toughness, low service temperature and high flammability for many different applications [3]. To overcome these disadvantages nanofillers can be used. But there are still many challenges such as homogenous dispersion that must be analyzed to reach the full potential of the PNC.

Polyhedral oligomeric silsesquioxane (POSS) are hybrid organic-inorganic fillers and leads to greater temperature and oxidation resistance, surface hardening and reduction in flammability [4]. Nucleating agent di-benzylidene-sorbitol (DBS) forms a nanofibrillar network and a low viscosity liquid complex in combination with POSS.

Carbon based nanofillers such as carbon nanotubes (CNT) have remarkable mechanical properties [5] and enhances the electrical and thermal properties [6]. In this research work the influence of CNTs on the low viscosity liquid complex, formed by POSS and sorbitol, is investigated. Due to the strong correlation between morphology and property enhancement, different characterization methods such as rheology, differential scanning calorimeter or tensile tests were used.

Experimental

Materials

The isotactic polypropylene P4G2Z-159 (iPP) was supplied by Flint Hill Resources (Wichita, Kansas, US) with MFR 1.95 g/ 10 min and crystalline melting temperature (TM) of 165°C. This iPP does not contain any nucleating agent.

The used di-benzyliden-sorbitol DBS Millad 3905 (M_w=358.4 g/mol, T_M=225 °C) was obtained from Milliken Chemicals (Spartanburg, SC, US) and it has two free –OH groups with phenyl side rings.

The tri-silanol phenyl-polyhedral oligomeric silsesquioxance (POSS; M_w= 930.07 g/mol and T_M=230°C) were supplied by Hybrid Plastics (Hattiesburg, MS, US). POSS has three silanol groups.

The multiwall carbon nanotubes (MWCNT) Graphitstrength® C100, which were supplied by Arkema (King of Prussia, PA, US) was used with a diameter of 10 to 15 nm and a length of 0.1 to 10 µm.

Preparation of Polymer Nanocomposites

The materials were produced in an internal mixer Brabender Plasticorder Typ PL2000 (Duisburg, Germany). The processing temperature was 185°C and the screw rotation speed was 80 rpm. After melting the iPP, the POSS and DBS were added in the mixing chamber and mixed for 10 min. At the end the CNTs were added and mixed for 5 min. Before mixing all additives were dried at 80°C for 24 h. The compounds were cooled down to room temperature and ground into granulate for further investigations.

Fiber spinning
The fibers were spun with the help of capillary rheometer ROSAND RH7 (Malvern Instruments Ltd., Malvern, UK) and a home-made take up device. The length to diameter ratio was 32:2 and the plunger speed was varied between 0.2 and 2 mm/min. The take-up velocity was between 10 rpm and 120 rpm at a melt temperature of 200°C. The draw down ratio (DDR) of the melt spun fibers was calculated as in eq. 1 with the velocity of fiber take-up (v_{fiber}) and the velocity of melt (v_{melt}).

$$DDR = \frac{v_{fiber}}{v_{melt}} = \left(\frac{D_{die}}{D_{fiber}}\right)^2 \quad (1)$$

Crystallinity
Using differential scanning calorimeter (DSC), DSC 2920 from TA Instruments (New Castle, DE, US), the crystallinity was calculated as in eq. 2 with the enthalpy of nanocomposite (ΔH_m) to the enthalpy of fusion of 100% crystalline iPP (ΔH_0=177 J/g) [7].

$$\chi_c = \frac{\Delta H_m}{\Delta H_C} * 100\ \% \quad (2)$$

The materials were heated up to 200°C (heat rate 10°C/min), kept for 5 min at 200°C and cooled down to 30°C (cool rate 10° C/min) under nitrogen atmosphere. At least 2 samples of each material were characterized.

Wide angle X-ray diffraction
The orientation factors of the fibers were analyzed using a Bruker wide-angle X-ray diffraction equipment with 40 kV and 40 mA with a beam monochromatized to CuKα radiation wavelength (λ) of 1.54 Angstroms. The fiber samples were mounted around a metal frame and measured for 5 min. For comparison the total diameter of the fiber bundle in each measurement was kept constant. The orientation factors were calculated using Herman-Stein orientation factor (see eq. 3) [8-12] and Wilchinsky equation (see eq.4 and eq.5) [13, 14]. The (110) and (040) have the strongest diffraction intensities. Wilchinsky developed the eq. 5 for iPP and this relationship was obtained for the average cosine squared of the angle between the fiber axis and the c-axis [11, 14].

$$f_{b,z} = \frac{3\cos^2\chi_{b,z} - 1}{2} \quad (3)$$

$$\cos^2\chi_{b,z} = \frac{\sum_{\chi=0}^{90} I(\chi)_{040} \sin\chi \cos^2\chi}{\sum_{\chi=0}^{90} I(\chi)_{040} \sin\chi} \quad (4)$$

$$\langle \cos^2\chi_{c,z}\rangle = 1 - 1.099\,\langle \cos^2\chi_{110,z}\rangle - 0.901\,\langle \cos^2\chi_{040,z}\rangle \quad (5)$$

Scanning electron microscopy
The scanning electron microscopy (SEM) images were taken with JEOL JSM 5310 (JEOL, Tokyo, Japan) at an operating voltage of 10 kV. For sample preparation the fibers were fixed epoxy resin and fractured after soaking in liquid nitrogen. The surface of the sample was sputter-coated for 1.5 min using Sputter coater ISI 5400 under argon gas atmosphere.

Mechanical properties
To investigate the mechanical properties of the iPP nanocomposites tensile tests were carried out on an instron tensile tester Model 5577 (Instron, Norwood, Ma, USA) with a 5 N load cell and 26 mm gauge length. The chosen speed was 50 mm/min according to ASTM D3833 method. The fibers were fixed on a preformed paper and the paper was attached to the grips of the tensile testing machine. At least 10 specimens were characterized for each material and the average values were calculated.

Complex viscosity
The complex viscosity of the nanocomposites was investigated with rotational shear rheometer ARES (TA Instruments, New Castle, DE, US). The disk-shaped specimens of 25 mm diameter und 2mm thickness were kept at 200°C for 5 min. The strain rate was fixed at 2%. The frequency sweep was measured from 100 to 10 rad/s. At least 2 samples were characterized.
The complex viscosities were also measured in the dependency of the temperature. The data were taken during cooling from 220°C to 120°C at 10°C/min and an angular frequency of 10 rad/s.

Results and Discussion

Influence of CNT on the low viscosity liquid complex
Rheological properties are sensitive to the morphology, particle size, particle shape and surface properties and to the dispersion state of filler in the polymer matrix. The complex viscosities of the iPP compounds in dependency of the temperature are depicted in Fig. 1. The sudden increase of the complex viscosities at lower temperature is because of the onset of iPP crystallization [16]. Roy et al. [17] found out that the complex viscosity of the PP/DBS/POSS compound decreases due to the formation of low viscosity liquid complex. The reduction of the complex viscosities at about 220°C, which can be only seen in the compounds with DBS, is due the dispersion of the low viscosity liquid DBS in iPP [17, 18]. With the addition of CNT in the iPP compounds the complex viscosity increases. This indicates that the formation of an effective network between the nanofiller and the polymer matrix [19]. The complex viscosity of PP/DBS/POSS compounds increases with the addition of CNT. Adding 2 wt% CNT the complex viscosity increases to the level of unfilled iPP. At the curves of complex viscosities for PP+1 wt%DBS+5 wt% POSS+1 wt% CNT and PP+1 wt%DBS+5 wt% POSS+2 wt% CNT a shoulder at about 150°C can be seen. This shoulder can be attributed to the development of a gel state by the fibrillar networks of DBS, which can be normally

observed in PP/DBS compounds [20-22]. The same trend of the complex viscosities can be observed with the frequency sweep method (2% strain). This can be attributed to the insufficient formation of the low viscosity liquid complex. So the whole amount of DBS cannot be found by POSS under presence of CNT.

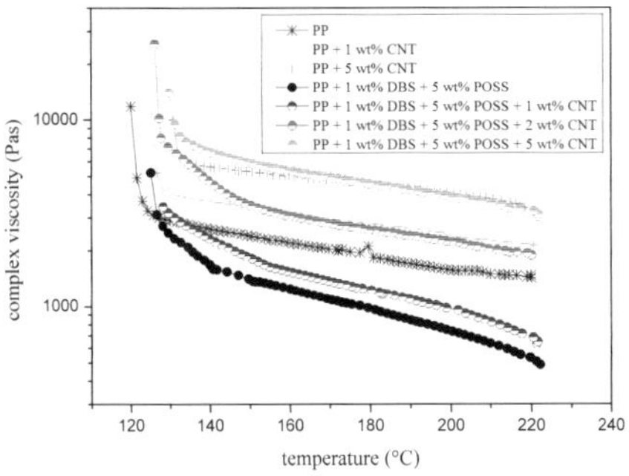

Figure 1 – complex viscosities of the compounds in dependency of the temperature (cooling rate 10°C/min, angular frequency of 10 rad/min)

Draw Down Ratio

In Fig. 2 the maximum DDR and the corresponding diameter of the fibers are depicted. It can be seen that with increasing DDR the diameter of the fibers decrease. The highest DDR can be achieved with PP/DBS/POSS compound, which corresponds with the results of the complex viscosity, where PP/DBS/POSS showed the lowest complex viscosity.

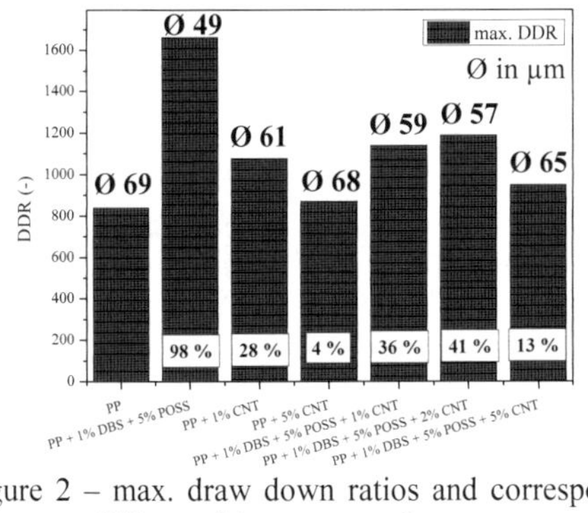

Figure 2 – max. draw down ratios and corresponding diameter of fibers of the compounds

Morphology

The Herman orientation was calculated for the c-axis of the monoclinic iPP crystals. Unity describes perfect parallel orientation, zero specifies random orientation and perpendicular orientation is identified by -0.5. In Fig. 3 the orientation factor fc for all fibers at three different DDR are displayed. With the addition of the various filler the orientation factor changes to a more random morphology. With higher DDR a more random orientation can be seen. In Fig. 4 the morphology of PP/DBS/POSS compound with 2 wt% CNT is depicted. It can be seen that the CNTs form agglomerates with a random alignment and that they are not connected with the DBS/POSS particles. This confirms the results of the different other characterization methods.

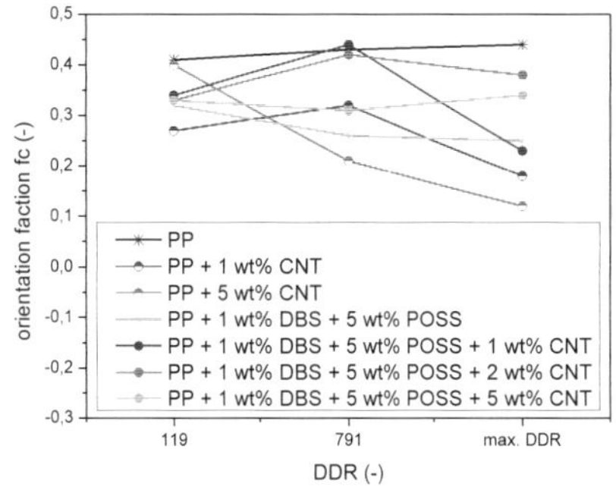

Figure 3 – orientation factor fc of the fibers at different DDRs (119, 791 and max. DDR)

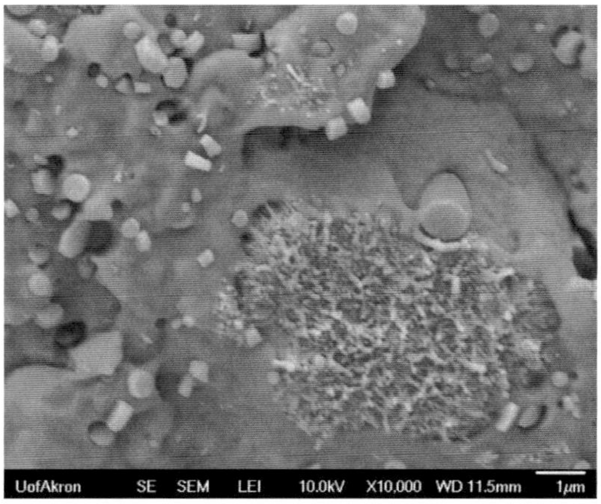

Figure 4 – PP+1 wt% DBS+5 wt% POSS+2 wt% CNT (DDR of 119)

Crystallinity

In Fig. 5 the crystallinity of the melt spun fibers of the different compounds can be observed. It can be seen that there is no significant change in the crystallinity of the melt spun fibers. So there is no significant change in formation of crystals during cooling and DBS act not as nucleating agent. However CNTs can accelerate the nucleation and crystal growth of the polymer. The CNTs would get coated with a layer of polymer of uniform because the crystal nucleation occurs on the surface of the CNTs [23, 24].

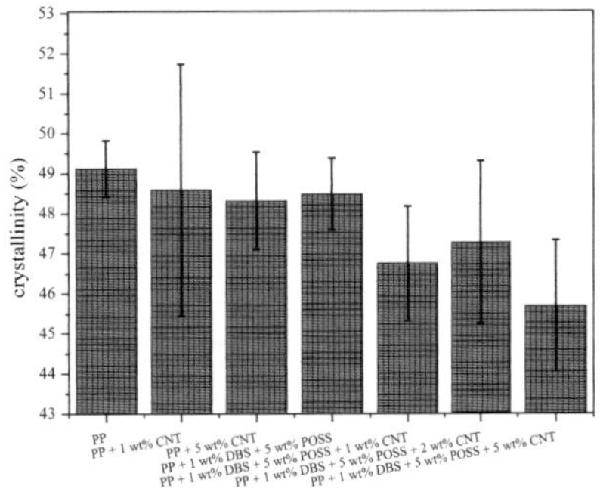

Figure 5 – crystallinity of melt spun fibers (DDR 791)

Young's modulus

The Young's moduli of the fibers at two different DDR are depicted in Fig. 6. The changes in the Young's modulus are only marginal, also due to the huge error bar. The differences between two different DDR are also not significant enough. So the CNT has no additional reinforcement factor. Wang et al. [25] discovered that CNT only improves the mechanical properties of polymers by parallel alignment of the CNTs to the applied stress, which correlates with the orientation factor.

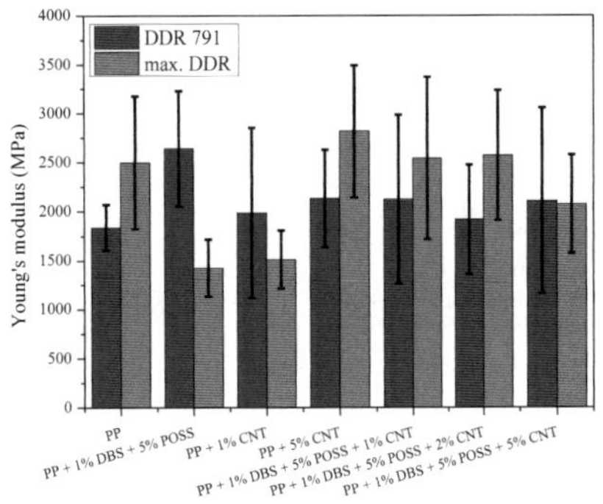

Figure 6 – Young's modulus of melt spun fiber with two different DDR (791, max. DDR)

Conclusions

This study investigated the influenced of CNTs on the low viscosity liquid complex, which is formed by DBS and POSS. The various results show that due to the addition of CNT the formation of the low viscosity liquid complex is disturbed. The whole amount of POSS and DBS does not interact with each other the proper way and hinders the CNT to get dispersed well.

Acknowledgements

We acknowledge Dr. Rong Bai from the Department of Polymer Engineering for her help with several characterization methods and Rafael Benavides Gonsalez from the Department of Polymer Engineering for his assistance in taking SEM pictures. We also want to thank Dr. Sayantan Roy for several discussions and his feedback regarding PP/DBS/POSS compounds. Special thanks to the Marshall Plan Scholarships for the realization of the research at The University of Akron.

References

1. S. Roy; V. Scionti; S.C. Jana; C. Wesdemiotis; A.M. Pischera; M.P. Espe *Macromolecules* 2011, *44*, 8064.
2. A.S. Luyt; M.D. Dramicanin; Z Antic; V. Djokovic *Polym. Test.* 2009, *28*, 348.
3. E. Onder; N. Sarier; M.S. Ersoy *Thermochim. Acta* 2012, *543*, 37.
4. S.W. Kuo; F.C. Chang *Prog. Polym. Sci.* 2011, *36*, 1649.
5. S. Iijima *Nature* 1991, *354*, 56.
6. K. Prashantha; J. Soulestin; M.F. Lacrampe; P. Krawczak; G. Dupin; M. Claes *Compos. Sci. Technol.* 2009, *69*, 1756.
7. J. Brandrup; *I. E. Polymer Handbook*, John Wiley & Sons, New York, 1989.
8. R.S.J. Stein *J. Polym. Sci.* 1958, *31*, 327.
9. L.E. Alexander, *X-ray diffraction methods in polymer science*, John Wiley & Sons, New York, 1969.
10. Z.W. Wilchinsky *J. Appl. Phys.* 1960, *31*, 1969.
11. H.P. Nadella; H.M. Henson; J.E. Spruiell; J.L. White *J. Appl. Polym. Sci.* 1977, *21*, 3003.
12. M. Afshari; R. Kotek; B.S. Gupta; M.H. Krish; H.N.J. Dast *J. Appl. Polym. Sci.* 2005, *97*, 532.
13. Z.W. Wilchinsky *J. Appl. Phys.* 1959, *30*, 792.
14. S.A. Arvidson; S.A. Khan; R.E. Gorga *Macromolecules* 2010, *43*, 2916.
15. Z.W. Wilchinsky, *Advances in X-ray Analysis*, Plenum Press, New York, 1963.
16. J.E. Perilla; B.J. Lee; S.C. Jana *J. Rheol.* 2010, *54*, 761.
17. S. Roy; B.J. Lee; Z.M. Kakish; S.C. Jana *Macromolecules* 2012, *45*, 2420.
18. L. Balzano; G. Portale; G.W.M. Peters; S. Rastogi *Macromolecules* 2008, *41*, 5350.
19. M. Ganß; B.K. Satapathy; M. Thunga; R. Weidisch; P. Pötschke; D. Jehnichen *Acta. Mater.* 2008, *56*, 2247.
20. M. Kristiansen; T. Tervoort; P. Smith; H. Goossens *Macromolecules* 2005, *38*, 10461.
21. M. Kristiansen; M. Werner, T. Tervoort; P. Smith; M. Blomenhofer; H.W Schmidt *Macromolecules* 2003, *36*, 5150.
22. Y. Zhao; D.A. Schiraldi *Polymer* 2005, *46*, 11640.
23. Y.A. Kim; T. Hayashi; M. Endo; Y. Gotoh; N. Wada; J. Seivama *Mater.* 2006, *54*, 31.
24. L. Valentini; J. Biagiotti; J.M. Kenny; S. Santucci *J. Appl. Polym. Sci.* 2003, *87*, 708.
25. Q. Wang, J. Dai, W. Li, Z. Wie. J. Jiang *Compos. Sci. Technol.* 2008, *68*, 1644.

O-07-601

THE SHEAR FLOW INDUCED NANOCLAY ORIENTATIONOF PE NANOCLAY NANOCOMPOSITES IN FILM BLOWING PROCESS

Nima Mostofi[1], Hossein Nazockdast[2*] and Ali Asghar Katbab[3]

[1]*Polymer Engineering and Color Technology Department, Amirkabir University of Technology, No. 424, Hafez Ave, Tehran-Iran, n_mostofi@aut.ac.ir*

[2] *Polymer Engineering and Color Technology Department, Amirkabir University of Technology, No. 424, Hafez Ave, Tehran-Iran, Tel: +98 21 64 54 24 32, Fax: +98 21 66 46 96 23, Email: nazdast@aut.ac.ir*

[3] *Polymer Engineering and Color Technology Department, Amirkabir University of Technology, No. 424, Hafez Ave, Tehran-Iran, katbab@aut.ac.ir*

Abstract - Effect of interfacial interaction and shear rate on the shear flow induced nanoclay orientation in Polyethylene-Nanoclay nanocomposites during film blowing process were studied by means of linear and non-linear Rheology. The PE-clay nanocomposite samples compatibilized with Polyethylene grafted maleic anhydride, (PE-g-MA) and nanocomposite samples using PE-g-MA as matrix with the same nanoclay loading, all prepared by melt compounding were considered. All the samples were prepared by melt intercalation using modular twin screw extruder. The samples were converted into films using different blow up ratios. The results of x-ray diffraction (XRD) showed a great extent of melt intercalation and good nanoclay dispersion in the samples.

The results of dynamic oscillating and transient stress tests performed on the samples, prepared at different blow up ratios exhibited different linear and non-linear viscoelastic behavior. While the un-drawn samples exhibited a low frequency solid like response due to 3d network structure, the melt linear viscoelastic behavior of drawn samples was found to be varied with blow-up ratio. These results indicated that in the drawn samples there is no sign of 3D network but flow induced oriented nanoclay remain intact during the frequency sweep test period. This was supported by the results of time sweep tests which showed that the time required for the nanoclay restructuring is about 3600S which was much longer than the frequency sweep test time 3600S. These results were found to be consistent with non-linear viscoelastic results obtained from startup flow and flow-reversal experiments.

The PE-g-MA based nanocomposite samples in spite of greater extent of melt intercalation exhibited a low frequency non terminal storage modulus. This could be attributed to much greater interfacial interaction between nanoclay and the matrix; the parameter which could accelerate the reorientation in the favor of 3D network restructuring. Also the greater number of individual platelets compared to tactoids in the PE based samples can also play a roll.

Keywords: *Nanoclay Orientation, Film Blowing, Nanocomposite, Rheology*

Introduction

Understanding the orientation change of a concentrated suspension, has been the focus of scientific research for several decades. It has been demonstrated [1–3] that the material properties of a product are strongly dependent on the fibre orientation within the polymer matrix, therefore it is essential to understand the orientation throughout the fully processed product.

The standard fiber orientation model is based on Jeffery's equation for the motion of a single fiber in an infinite Newtonian matrix[4] with an isotropic rotary diffusion term to account for fiber-fiber interactions.[5]

Nanoclay filled polyethylene represent a major class of growing importance since conventional equipment (extruder and injection machines) can be used to produce barrier films. The mechanical properties along with barrierity properties of the produced films depend strongly on the process itself, on the rheological properties of the filled polymer and on the orientation of nanoclays during processing.

Many researchers have studied the orientation of the fibers in suspensions and their rheological models in shear flows. However, the researches done on the orientation of nanoclay in film processing process are much lower compared to fiber orientation process. Since the mechanical properties as well as barrier properties of the films are highly dependent on the nanoclay orientation, in this study we tried to better understand the parameters affecting the orientation in the film blowing process.

Experimental

Materials:

Film grade linear low density polyethylene (LLDPE), with MFI (190°C/2.16 Kg) = 2.2gr/10min, (LL0220 AA) supplied by the Tabriz Petrochemical Co. Iran, was used as major component. Maleic Anhydride grafted low density polyethylene (PE-g-MA) (E-100 supplied by Fusabond) was utilized as the compatibilizer.

Natural montmorillonite organically modified with a quaternary ammonium salt (OMMT) with commercial grade name of Cloisite 20A, supplied by Southern Clay Company, USA, was used to prepare nanocomposites.

Blend Preparation:

All of the materials were dried in an oven at temperature of 80°C for at least 8 hours before melt blending. All the blend samples were melt blended using a Brabender twin scew extruder (DSE25) with the screw speed of 150 RPM. The composition of the sample prepared was LLDPE/E100/20A=80/15/5 wt%

Rheological Measurements:
The linear and nonlinear melt viscoealstic properties of the blend components and blend samples were carried out using a Rheometeric Mechanical Spectrometer (RMS) equipped with a parallel plate with a gap of 1 mm and a plate diameter of 25 mm. The linear measurements were performed at constant amplitude of 1%, in the frequency range of 0.01-625 s^{-1} at 180°C under dry nitrogen atmosphere. Nonlinear tests were performed in rotational mode with the shear rate of 0.1 s^{-1}.

WAXD Experiments
The WAXD experiments were conducted using a Philips X' pert by Cu-Kα ray with the wave length of 0.154nm.

Film Blowing Process
Films with different blow up and draw up ratios were prepared using a Brabender single screw extruder equipped with a film blowing die head and a Brabender film drawing equipment. The film sample names are shown in

Table 1

Results and Discussion

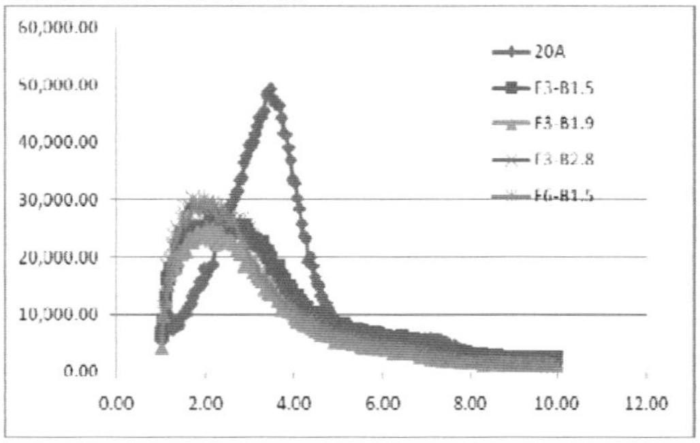

Fig. 1 shows the XRD patterns for the neat Cloisite 20A, nanocomposite sample and the blown films with different draw-up and blow-up ratios. The results of the XRD patterns are summarized in Table 2. As it can be seen after melt blending the organoclays are intercalated and the d-spacing has been increased. Moreover, the film blowing process increases the d-spacing of the organoclay platelets; but as we will discuss later, this process will also orient the organoclays parallel to the films surface.

Fig. 2 show the storage modulus vs. angular frequency for nanocomposite sample as well as the blown films. As it was expected, the nanocomposite sample showed a non-terminal plateau in low frequency regions, due to network formation. As it was shown earlier, by film blowing process, the nanoclay platelets intercalate more, and the d-spacing increases, so we expect this non terminal plateau increases. But as it can be seen, not only the amount of non terminal plateau did not increase, but also it decreases. This is due to orientation of the organoclay in the LLDPE matrix and hence breaking the network formed between the nanoclay particles and formation of a new network between polymer chains and nanoparticles.

The results of time sweep tests performed on the films are presented in Fig. 3. As it can be seen the extent of storage modulus increases by increasing the blow up ratio, due to more network formation. Moreover the extent of change in the storage modulus vs. time for these samples is summarized in Table 3. As it can be seen, by increasing the blow up ratio, this amount increases. These results indicate that the extent of organoclay orientation increases by increasing blow up ratio.

The above mentioned results are supported by the results of non-linear startup flow tests as shown in Fig. 4. The stress overshoot seen for the undrawn nanocomposite samples, almost vanishes in the film sample with blow up ratio 1.5; by increasing the blow up ratio, the extent of stress overshoot peak increases. As mentioned before this increase can be attributed to the better intercalation of organoclay, but the amount of stress overshoot peak as well as steady state shear rate is always lower than the undrawn nanocomposite due to the orientation of the organoclay in the elongation flow field of film blowing process.

The mentioned results were also confirmed by the results of the flow reversal tests. The stress overshoot peak increases by increasing the shear rate in the flow reversal tests, but the amount of peak in the reversal test almost shows no change (Fig. 5). The mentioned results show the network formed between particles is broke and the nanoclay particles start to orient in the direction of the flow field. As it can be seen in the same figure, by increasing the shear rate extent of the decrease in the stress after overshoot increases indicating more orientation of the organoclay in the flow field.

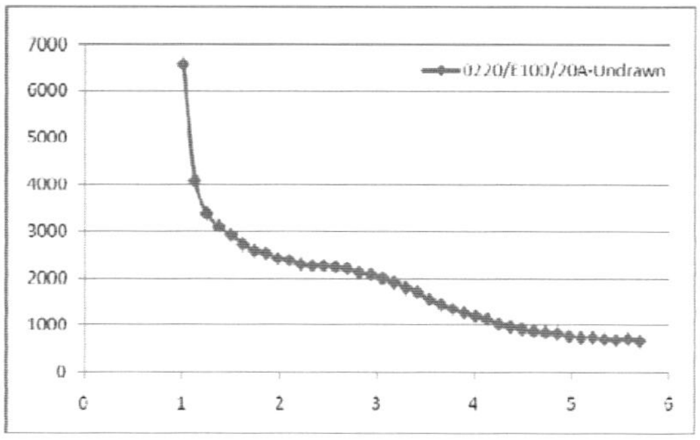

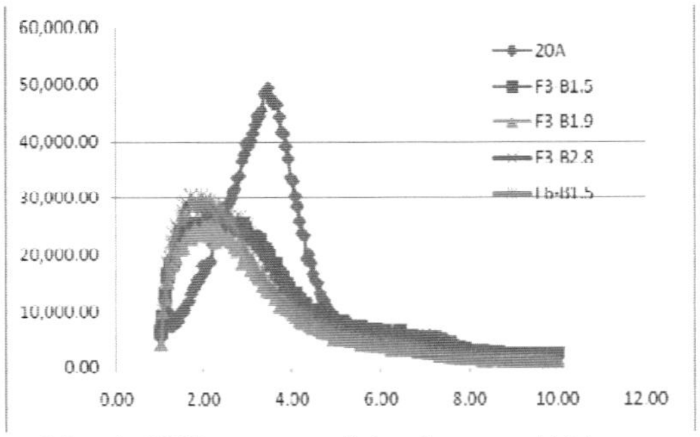

Fig. 1: XRD pattrens of a)undrawn and b)drawn nanocomposites.

Conclusions

The results of x-ray diffraction (XRD) showed a great extent of melt intercalation and good nanoclay dispersion in the samples. Also these results showed ehnanced melt intercalations for films with high blow up ratios.

The results of dynamic oscillating and transient stress tests performed on the samples, showed that the un-drawn samples exhibited a low frequency solid like response due to 3d network structure, however the melt linear viscoelastic behavior of drawn samples was found to be varied with blow-up ratio. These results indicated that in the drawn samples there is no sign of 3D network but flow induced oriented nanoclay remain intact during the frequency sweep test period. This was supported by the results of time sweep tests which showed that the time required for the nanoclay restructuring is about 3600S which was much longer than the frequency sweep test time 3600S. These results were found to be consistent with non-linear viscoelastic results obtained from startup flow and flow-reversal experiments.

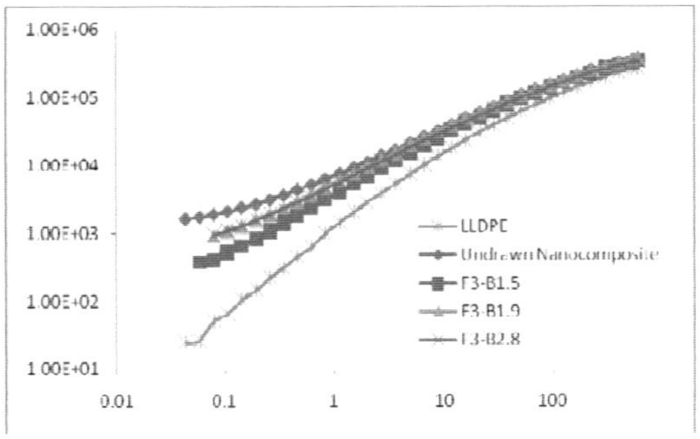

Fig. 2: Storage Modulus vs. Angular Frequency for different samples in 175°C

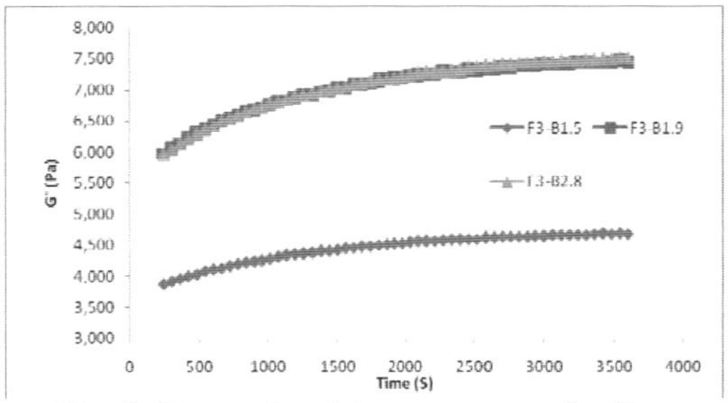

Fig. 3: The results of time sweep tests for films performed in 175°C, with angular frequency of 1 S^{-1}

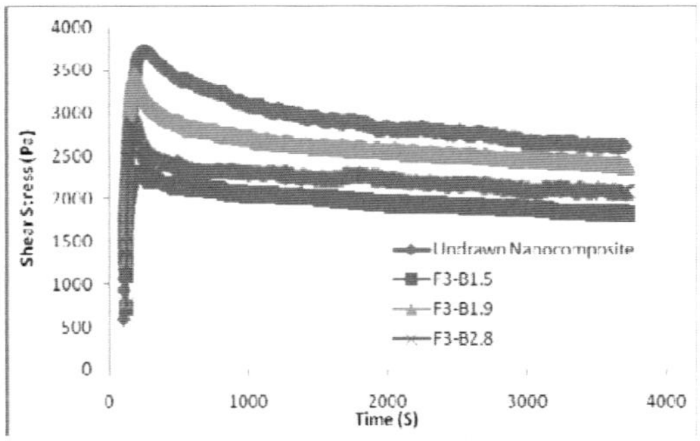

Fig. 4: Results of startup flow tests performed on film samples in 175°C and shear rate of 0.1 S^{-1}

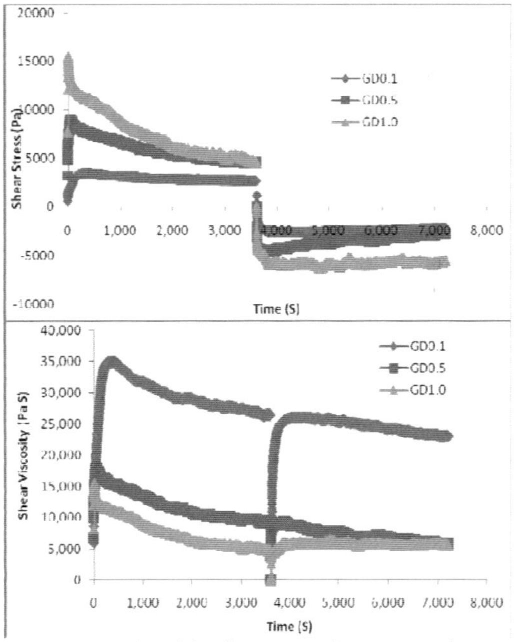

Fig. 5: Results of the flow reversal tests on undrawn nanocomposite samples, a) stress and b) visocosity vs. time in 175°C

Table 1: Sample name for blown films

Sample Name	Draw-Up speed (m/min)	Blow Up Ratio
F3-B1.5	3	1.5
F3-B1.9	3	1.9

F3-B2.8	3	2.8
F6-B1.5	6	1.5

Table 2: d-spacing for different samples

Sample	d-Spacing (Å)
20A	24
Nanocomposite	29.5
F3-B1.5	39
F3-B1.9	43
F3-B2.8	48
F6-B1.5	48

Table 3: The exent of increase in storage modulus during time sweep test

Sample	Extent of G' increase during the time sweep test
F3-B1.5	20%
F3-B1.9	26%
F3-B2.8	28%

Acknowledgements

The authors would like to appreciate Ms. S. Mohammadi and L. Zare for performing the RMS tests.

References

[1] M. J. Solomo.; A. S. Almusalla; K. F. Seefeldt; A. Somwangthanaroj; P. Varadan; Macromolecules 2001, 34, pp. 1864-1872.

[2] G. L. Hand, Arch. Rat. Mech. Anal. 1961, 7, pp. 81–86.

[3] H. Giesekus; Rheol. Acta 1962, 2, pp. 50–62.

[4]Jeffery G. B. Proc. R. Soc. Lond. 1922, A102, pp. 161-179.

[5] F. P. Folgar F. P.; C. L. Tucker; J. Reinf. Plast. Compos. 1984, 3, pp. 98–119.

[6] S. G. Advani; C. L. Tucker ; J. Rheol. 1990, 34, pp. 367–386.

[7] S. G. Advani; C. L. Tucker; J. Rheol. 1987, 31, pp. 751–784.

[8] J. Wang; J. F. O'Gara; C. L. Tucker; J. Rheol. 2008, 52, pp. 1179-1200

O-07-629

CURING BEHAVIOUR OF PMMA-GRAFTED NATURAL RUBBER TOUGHENED EPOXY/CLAY NANOCOMPOSITES

N. Y. Yuhana[1,*], S. Ahmad[2], M. R. Kamal[3], S.C. Jana[4] and Shamsul Bahri A. R.[5]

[1] *Dept. of Chemical and Process Engineering, Universiti Kebangsaan Malaysia, Malaysia – yuliana@eng.ukm.my*
[2] *Dept. of Material Science, Universiti Kebangsaan Malaysia – sahrim@ukm.my*
[3] *Dept. of Chemical Engineering, McGill University, Canada - musa.kamal@mcgill.ca*
[4] *Dept. of Polymer Engineering, University of Akron, United States - janas@uakron.edu*
[5] *Rubber Research Institute Experimental Station, Selangor, Malaysia – shamsul@lgm.gov.my*

Abstract - Curing behavior of ternary systems containing epoxy, PMMA grafted natural rubber and nanoclay were studied. Natural montmorillonite (Cloisite Na) and chemically modified montmorillonite (Cloisite 30B) were used at 5 phr, while the amount of rubber was fixed at 5 phr. Poly(etheramine) was used as the curing agent at the equivalent and 1.05 amine-epoxy molar ratios. Mechanical stirring and ultrasonic mixing were used to achieve dispersion. Both unmodified epoxy/clay nanocomposites and cured toughened epoxy/clay nanocomposites were evaluated using differential scanning calorimetry (DSC. It was found that organoclay accelerates the curing process. The presence of rubber in the excess of curing agent, however resulted in a decrease in exothermic heat value, that can be due to the restricted heat release in high viscosity mixture. Also, it was found that the organic modifier can accelerate the thermal decomposition of epoxy resin in binary and ternary systems.

Keywords: Epoxy; Nanocomposite; Organoclay

Introduction

The incorporation of dispersed rubbery particles in a layered silicate / epoxy nanocomposite has received significant attention [1-6]. Therefore, this study explores the effect of the incorporation of rubber in epoxy systems. In polymer nanocomposites, a small amount of silicate nanoparticles (less than 5 wt. %) is commonly used. Some researchers observed a catalytic effect of organically modified clay [7-11]. A study on the influence of the stoichiometric epoxy/curing agent ratio on clay dispersion was done by Chin et al. [12]. They observed that addition of high amount of curing agent leads to intercalated structure of clay in epoxy.

This report is part of a general study dealing with the curing behaviour, morphology and properties of a diglycidyl ether of bisphenol A (DGEBA) type of epoxy cured with stoichiometric and above-stoichiometric amounts of hardener, in the presence of PMMA grafted natural rubber and nanoclay. In this paper, we report the results of a DSC curing study of the ternary systems. The parameters considered are the onset of reaction temperature (T_o), peak temperature (T_p) and heat of reaction (ΔH_r). The presence of exotherm and endotherm peaks are also analysed.

Experimental

Materials

Diglycidyl ether bisphenol A (DGEBA) epoxy resin, Epikote 828, was obtained from Hexion Specialty Chemicals, Korea. The epoxide equivalent weight was 187 g/mol. The liquid curing agent, Baxxodur TM EC301 was supplied by BASF. PMMA-grafted natural rubber Megatex 30 was obtained from Green HPSP (M) Sdn. Bhd. Organically modified montmorillonite Cloisite 30B and natural montmorillonite Cloisite Na were purchased from Southern Clay Products, USA.

Sample Preparation

The compositions of the four prepared systems are given in Table 1. Sample preparation methods are given elsewhere [13]. The curing behavior study was performed using differential scanning calorimetry (DSC). Liquid mixtures containing epoxy, with and without 5 phr clay, 5 phr rubber and curing agent, either stoichiometric or above-stoichiometric concentration, were heated from 25 to 300 °C, at rate of 20 °C/min. In this study, we reported the cure properties of the epoxy, epoxy-rubber, and epoxy-clay-rubber, in the absence and presence of curing agent mixtures.

Table .1 Nomenclature and composition of prepared samples

Nomenclature	Weight of component (phr gram of epoxy resin)			EW Epoxy : Amine
	Rubber	Cloisite Na	Cloisite 30B	
Epoxy 1	0	0	0	1:1
Ep5MG301	5	0	0	1:1
Epoxy 2	0	0	0	1:1.05
Ep5MG302	5	0	0	1:1.05
Ep5Na1	0	5	0	1:1

Ep5Na2	0	5	0	1:1.05
Ep530B	0	0	5	1:1
Ep730B	0	0	7	1:1
Ep5MG305Na1	5	5	0	1:1
Ep5MG305Na2	5	5	0	1:1.05
Ep5MG30230B	5	0	2	1:1
Ep5MG30530B	5	0	5	1:1

Results and Discussion

The effect of nanoclay and rubber on heating process of epoxy

Both melting and evaporation of low molecular weight compounds occurred during the heating process. This is indicated by the presence of several endothermic peaks, as shown in Figure 1(a).

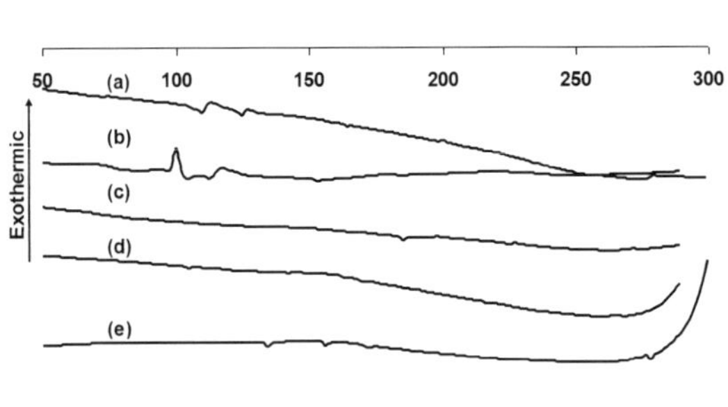

Figure 1 DSC analysis on (a) Epoxy (b) Epoxy-5Na (c) Epoxy-5Na after 1 hour sonication (d)Epoxy-530B (e) Epoxy-530B after 1 hour sonication

In the presence of 5 phr Cloisite Na, a small exothermic peak around 100 °C was detected, with the heat of reaction about 0.6 J/g, as demonstrated in Figure 1(b). The exothermic heat also could be due to swelling of clay in epoxy. This was not reported by others. However, the small exothermic peak disappeared for liquid mixtures that had undergone ultrasonic treatment, as the swelling process took place during sonification during the treatment. This is illustrated in Figure 1 (c).

Homopolymerization is the curing of the epoxy resins, based on the reactions between the epoxide molecules. The addition reaction is between the epoxide group and other kinds of reactive molecules, in the presence or absence of catalyst. A weak exothermic reaction was detected in the presence of Cloisite 30B, with the ΔH_r before and after sonication treatment, which are 7.28 and 382 J/g respectively, as given in Figure 1 (d-e). These results suggest that the Cloisite 30B organoclay plays a catalytic role in the homopolymerization of DGEBA. The larger value of exothermic heat can be attributed to improved clay dispersion. The mixture of Epoxy-Cloisite 30B started to decompose above 250 °C. This indicated the catalytic activity of ammonium ion salt on decomposition of epoxy.

An endothermic peak is observed in Figure 2, upon heating a liquid mixture containing epoxy, silicate, and rubber. The endothermic peak could refer the melting point of PMMA.

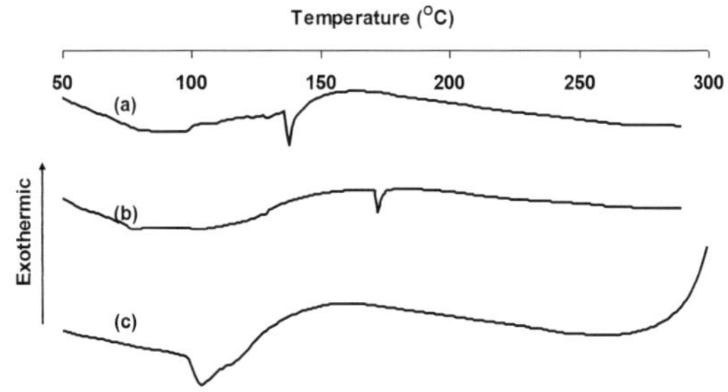

Figure 2 DSC analysis on (a) Epoxy-5MG30 (b) Epoxy-5MG30-5 Cloisite Na (c) Epoxy-5MG30 - 5Cloisite30B

In the absence of silicate, the endothermic peak is at 135 °C, as shown in Figure 2 (a). But the peak is shifted to higher temperature, 172 °C, in the presence of Cloisite Na, as illustrated in Figure 2 (b). The peak shifting may be attributed to the ability of dispersed Cloisite Na to resist heat transfer in the mixture. In contrast, it is obvious that Cloisite 30B addition shifts the endothermic peak to a lower temperature, 101 °C, as shown in Figure 1(c). The explanation for this is the possibility of ammonium ion salt to interact with epoxy, and catalyze the decomposition process. This will ease the melting process of PMMA, and subsequently the decomposition process starts at 250 °C. These findings indicate that the Cloisite 30B plays a catalytic role while the Cloisite Na plays an incatalytic role.

The effect of concentration of curing agent, rubber and montmorillonite on curing behavior of epoxy

The reaction between the epoxide group and polyetheramine as curing agent is exothermic, as can be observed in DSC thermograms in Figure 3. The results of DSC cure analysis on epoxy and epoxy-rubber blend are summarized in Table 2.

Table 2 Cure properties epoxy and epoxy-rubber blend

Graph	Liquid Sample	ΔH_r (J/g)	T_o (°C)	T_p (°C)
(a)	Epoxy 1	432	37	140
(b)	Epoxy 2	421	32	143
(c)	Ep5MG301	460	32	137
(d)	Ep5MG302	392	32	132

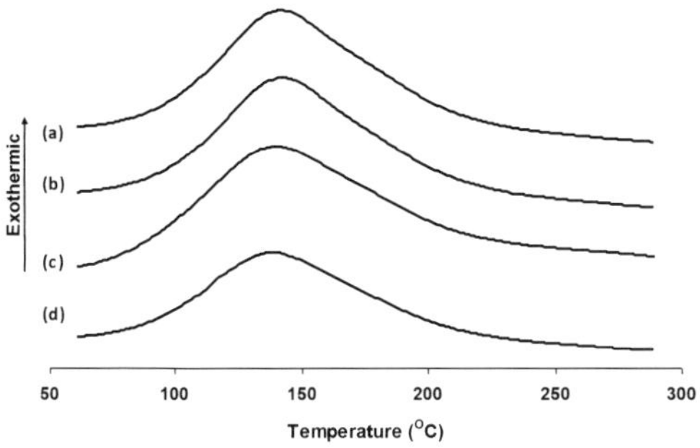

Figure 3 DSC analysis on (a) Epoxy 1 (b) Epoxy 2 (c) Ep5MG301 (d) Ep5MG30 2

The addition of rubber and above-stoichiometric concentration of curing agent contributed to lowering T_o, by about 5 °C. The addition of rubber, with stoichiometric concentration of curing agent, increased ΔH_r by 30 J/g. The ester group of PMMA may react with amine to form amide. This is an exothermic reaction.

The presence of rubber with excess of curing agent, however resulted in a decrease in ΔH_r by about 70 J/g. This can be due to the restricted heat release in high viscosity mixture. Also, there is a possibility that some amount of heat is used to heat the rubber component.

The T_p may indicate the temperature at the maximum of polymerization rate. It was observed that the presence of MG30 rubber and excess curing agent could shift the T_p to lower temperature. This may be attributed to the higher reaction rate in the presence of excess curing agent, and rubber.

Figure 4 shows the DSC thermograms for the exothermic reaction between the epoxide group and polyetheramine as curing agent, in the presence of nanoclay and rubber. The characteristics of the cure reaction of involving the epoxy ternary systems are summarized in Table 3.

Table 3 Cure properties of binary and ternary systems

Graph	Liquid Sample	ΔH_r (J/g)	T_o (°C)	T_p (°C)
(a)	Ep5Na 1	420	57	139
(b)	Ep5Na 2	434	43	139
(c)	Ep530B	382	40	135
(d)	Ep5MG305Na1	364	65	135
(e)	Ep5MG305Na2	368	40	135
(f)	Ep5MG30530B	380	52	135

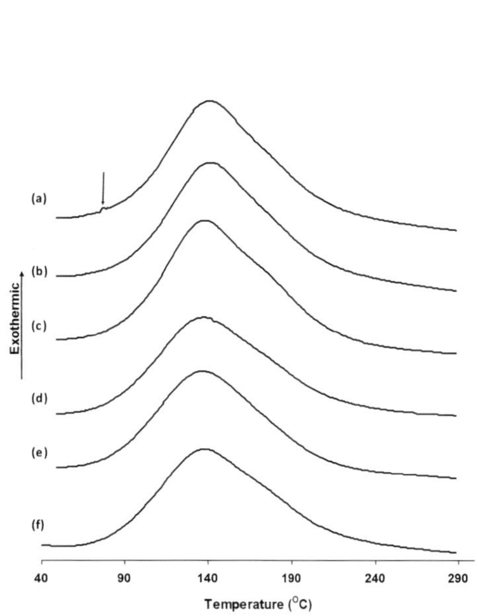

Figure 4.4 DSC analysis on (a) Ep5Na1 (b) Ep5Na2 (c) Ep530B (d) Ep5MG305Na1 (e) Ep5MG305Na2 (f) Ep5MG30530B.

The results show that the ΔH_r for epoxy containing 5 phr Cloisite Na 2, was effectively unaffected. On the other hand, ΔH_r was greatly reduced in the presence of Cloisite 30B. Possibly, some heat of reaction is absorbed to activate the reaction between the organic modifier of clay and the epoxy. Similar results were reported by Harsch et al. [14]. They suggested the possibility that fillers are act as inhibitors for the reaction between the epoxy and curing agent, by adsorption of the reactants on the surface of the filler and thus changing the stoichiometry. Another possible factor for the reduction of the measured exothermic ΔH_r may be due to solvent evaporation [15].

Our previous finding showed that the elasticity modulus for the binary system containing 5 phr Cloisite Na was around 5.5 GPa, as compared to the one contaning Cloisite 30B system which is about 3.8 GPa [13]. The effective strengthening of the composite material containing Cloisite Na, can be due to higher crosslinking density, that can be observed from the greater value of ΔH_r.

The higher T_o for all binary and ternary nanocomposite systems, compared to the neat epoxy and epoxy-rubber blend, suggest that the fillers acted as inhibitors for the curing process. However, the T_p is lower, for binary nanocomposite containing organoclay. This may show the capability of the organoclay to act as catalyst for the curing reaction. The author proposed that the hydroxyl group presence may catalyze the curing reaction between epoxide group and amine, as shown in Figure 5.

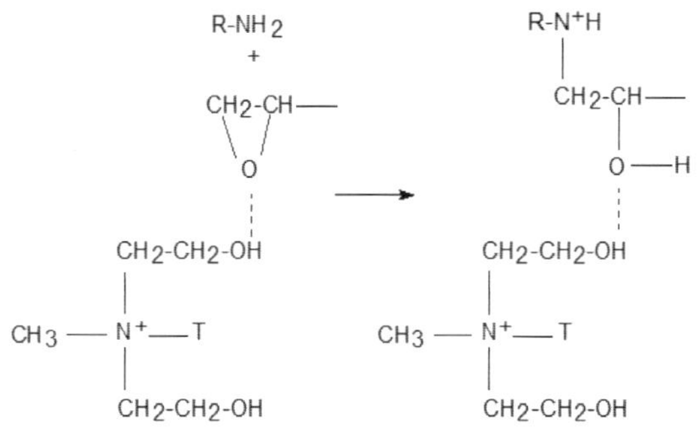

Figure 5. Chemical interaction between hydroxyl groups of organoclay and epoxide.

Conclusions

The organoclay acts as catalyst in the homopolymerization of epoxy and epoxy-amine reaction. Also, the organic modifier accelerates the thermal decomposition of epoxy resin in binary and ternary systems. However, Cloisite Na plays an incatalytic role. The melting point of PMMA in the epoxy-rubber blend mixture was lowered, in the presence of organoclay. The presence of rubber in the excess of curing agent, however resulted in a decrease in ΔH_r, that can be due to the restricted heat release in high viscosity mixture, and some amount of heat is used to heat the rubber component. The ΔH_r was greatly reduced in the presence of Cloisite 30B. This is possibly because some of heat of reaction is consumed for the reaction between the organic modifier of clay and the epoxy.

Acknowledgements

The authors would like to acknowledge help from Mr. Wan Mohd Nazir Bin Wan Taha and Mr. Mohd. Razif Maafol for their kind assistance. Also, the authors wish to acknowledge Universiti Kebangsaan Malaysia, Malaysia Public Service Department (PSD) and the Ministry of Science, Technology and Innovation (MOSTI) for SLAI scholarship and Science Fund Grant (03-01-02-SF0059) to conduct the research.

References

1. Becker, O. and G.P. Simon, *Epoxy Layered Silicate Nanocomposite.* Advanced Polymer Science, 2005. **179**: p. 29-82.
2. Liu, W., S.V. Hoa, and M. Pugh, *Morphology and performance of epoxy nanocomposites modified with organoclay and rubber.* Polymer Engineering and Science 2004. **44**(6): p. 1178-1186.
3. Froehlich, J., et al., *High-performance epoxy hybrid nanocomposites containing organophilic layered silicates and compatibilized liquid rubber.* Journal of Applied Polymer Science, 2004. **92**(5): p. 3088-3096.
4. Balakrishnan, S. and D. Raghavan, *Acrylic, elastomeric, particle-dispersed epoxy-clay hybrid nanocomposites: Mechanical properties.* Macromolecular Rapid Communications 2004. **25**(3): p. 481-485.
5. Balakrishnan, S., et al., *The influence of clay and elastomer concentration on the morphology and fracture energy of preformed acrylic rubber dispersed clay filled epoxy nanocomposites.* Polymer, 2005. **46**(25): p. 11255-11262.
6. Ratna, D., et al., *Nanocomposites based on a combination of epoxy resin, hyperbranched epoxy and a layered silicate.* Polymer, 2003. **44**(24): p. 7449-7457
7. Lan, T., P.D. Kaviratna, and T.J. Pinnavaia, *Epoxy self-polymerization in smectite clays.* Journal of Physics and Chemistry of Solids, 1996. **57**(6-8): p. 1005-1010.
8. Juwono, A. and G. Edward, *A study of Clay-Epoxy Nanocomposites consisting of Unmodified Clay and Organo Clay*, in *Makara, Sains*2006. p. 6-12.
9. Chen, D. and P. He, *Monitoring the curing process of epoxy resin nanocomposites based on organo-montmorillonite – a new application of resin curemeter.* Composites Science and Technology, 2004. **64**(16): p. 2501-2507.
10. Weibing, X., H. Pingsheng, and C. Dazhu, *Cure behavior of epoxy resin/montmorillonite/imidazole nanocomposite by dynamic torsional vibration method.* European Polymer Journal, Volume 39, Issue 3, March 2003, Pages 2003. **39**(3): p. 617-625.
11. Chen, D.Z., P.S. He, and L.J. Pan, *Cure kinetics of epoxy-based nanocomposites analyzed by Avrami theory of phase change.* Polymer Testing, 2003. **22**(6): p. 689-697.
12. Chin, I.-J., et al., *On exfoliation of montmorillonite in epoxy.* Polymer, 2001. **42**: p. 5947-5952.
13. Yuhana, N.Y., *Curing Behaviour, Morphology and Properties of PMMA grafted Natural Rubber Toughened Epoxy/Clay Nanocomposites*, in *Material Science*2012, Universiti Kebangsaan Malaysia: Bangi.
14. Harsch, M., J. Karger-Kocsis, and M. Holst, *Influence of fillers and additives on the cure kinetics of an epoxy/anhydride resin.* European Polymer Journal 2007. **43**(4): p. 1168-1178.
15. Hong, S.-G. and C.-S. Wu, *DSC and FTIR analysis of the curing behaviors of epoxy/DICY/solvent open systems.* Thermochimica Acta, 1998. **316**(2): p. 167-175.

Conundrum of Clay Nanofillers and Durability of Polyolefin Nanocomposites

S. Al-Malaika* and H. H. Sheena

Aston University
Polymer Processing & Performance Research Unit
Birmingham, B4 7ET, England, UK

*S.Al-Malaika@aston.ac.uk; Corresponding Author

Abstract

The effects of the environment on the performance of polymer products are well documented but much less is available in the open literature on the durability and performance characteristics of products based on polymer nanocomposites (PNCs). It is generally accepted that the presence of layered silicate nanofillers in PNCs offer significant advantages in terms of barrier properties, fire retardancy, mechanical properties. However, in polyolefin-based clay (e.g. montmorillonite, MMT) nanocomposites these nanofillers are themselves the source of inherent instability when compared to their unfilled polymer counterparts in both stabilised and unstabilised formulations.

The effects of accelerated weathering and thermal ageing on the durability (long-term oxidative stability) and morphology of melt extruded polypropylene and polyethylene nanocomposites (PONCs) are examined and compared with that of their unfilled analogues. The challenges of stabilization of PONCs and the effect of different stabilisation systems are discussed. Both thermal and photoxidative stabilities of compatibilised organically-modified monmorillonite (OMMT)-containing polyolefin nanocomposites were found to be lower than the uncompatibilised analogues and these in turn exhibited lower stability than the neat polymer or the polyolefin-compatibiliser polymer blend. The major factors contributing to the higher extent of oxidation of the nanocomposites are discussed.

Keywords: Polyolefin nanocomposites, Processing, Oxidative stability, Long term durability

Introduction

Clay-reinforced polymer nanocomposites (PNC), particularly the plate-like montmorillonite (MMT), have attracted great interest in the recent past due to the low nanofiller concentration typically required (1-5% compared to 30-40% for conventional fillers and fibres) and the abundance and low cost of the nanoclay fillers combined with significant benefits in terms of the performance properties of the resultant nanocomposites, and especially fire retardancy, thermal and barrier properties [1]. To achieve PNCs with good performance properties, the clay platelets must be sufficiently dispersed with exfoliated, intercalated, or a mixture of both exfoliated and intercalated (hybrid) morphologies [2]. However, the nanoscale dimension, the high aspect ratio and the hydrophilic character of the MMT clays make their dispersion during processing in hydrophobic non-polar polymers such as polyolefins quite difficult. In view of this, the surface chemistry of MMT is normally modified through ion exchange reactions with organic cation surfactants, e.g. quaternary ammonium salts having typically long organic chains (referred to as quats), in order to obtain a more organophilic MMT (referred to as OMMT) clays. The quat surfactants (intercalants) promote nanoclay-polymer interactions because of increased compatibility and their tendency to bring about intercalation of the clay particles. In polyolefins, the process of dispersion of OMMT is further improved by the addition of a compatibilising agent, typically a polyolefin graft copolymer containing maleic anhydride or ethylene-(meth)acrylic acid comonomers (high or low molar mass depending on the level of the anhydride or acid functions) [3].

The advantages of the compatibilisation of PONCs and that of the organic modification of their clay nanofillers by quat molecules are counterbalanced by challenges in terms of the adverse effect they pause on the long-term thermal- and photo-oxidative stability and durability of these nanocomposites [4, 5]. Much of the literature on processing of polyolefin nanocomposites tends to focus mainly on the clay morphology and final properties of the nanocomposites and not on their long term durability [6, 7]. For outdoor applications, e.g. interior and exterior automotive parts, it is very important that clay nanocomposite materials possess both thermal and photoxidative stability. Clay-polymer nanocomposites based on polyolefins, however, have been shown not to have sufficient photoxidative stability to fulfil this role [5, 8]. To address this, issues relating to compatibilisers, organic modifiers (surfactants such as quats), clay reactivity and clay-polymer interfacial chemistry, and their roles on the long-term thermal and photo stability of polyolefins need to be better understood and controlled.

Effect of Processing on Thermal Stability of Polyolefin-Clay Nanocomposites

A good clay dispersion and good thermal stability during processing are vital for achieving good performance properties for polyolefin-based nanocomposite products. Different analytical methodologies_have been used to determine the extent of dispersion of nanoclays in polymers and these have their advantages and drawbacks. X-ray (**XRD**) results are used widely for this purpose but this method may not detect agglomerated clay particles thus, could give a false impression of delamination. Scanning and Transition electron microscopic (**SEM** and **TEM**) analyses, which give visual images, are therefore often used in conjunction with **XRD** to give a more accurate, albeit qualitative, picture of the 'real' microstructure of the nanocomposites [9]. For example, in the case of PE, **TEM** shows that the presence of both quat-modified clay (OMMT) and MA-g-PE compatibiliser give rise to a well dispersed clay structure in the matrix having a hybrid morphology containing both exfoliated (some single clay platelets are observed) and intercalated clay structures with increased contact surfaces between the delaminated clay platelets and the matrix, **Fig. 1** [5].

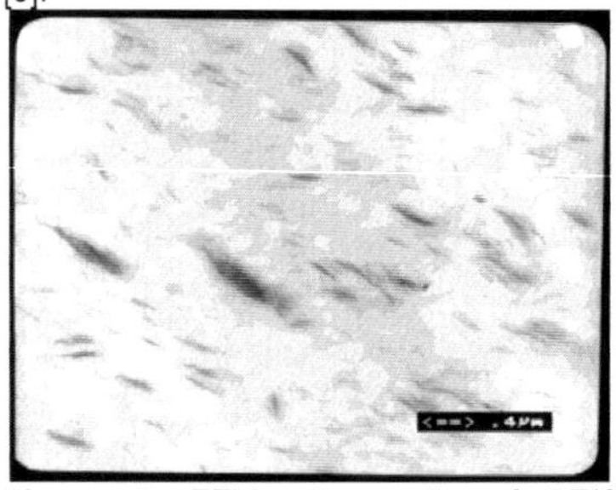

Figure 1: TEM micrographs of PE/OMMT/PEgMA melt extruded nanocomposites [5]

More recently, the use of attenuated total reflectance – Fourier transform infrared spectroscopy (**ATR-FTIR**) has been applied to examine the extent of clay delamination in polymer-organoclay nanocomposites [5, 10]. The methodology used here focuses on measuring, in reflectance mode, infrared absorption of the clay silicon-oxygen (Si-O) at the surface of a polymer-clay nanocomposite. ATR-FTIR of a sample of pristine montmorillonite powder shows only one broad band in the region 950-1150 cm^{-1}. By contrast, aqueous clay dispersions show several distinct absorption bands with somewhat narrow bandwidth indicating a higher state of dispersion and delamination of the clay platelets [10]. A similar methodology was applied to examining the state of delamination of the nanoclay platelets in PE-clay nanocomposites. **Fig. 2a** shows a narrow band that appears around 1007 cm^{-1} corresponding to the Si-O absorption frequency of the different OMMT clays. However, when the OMMT is incorporated in PE in the presence of a compatibiliser (PE-g-MA), the ATR-FTIR spectrum of the extruded film of the PE-nanocomposite shows a sharply contrasting spectral profile of the Si-O absorption region with strong signatures marking both the in-plane and out-of-plane vibrational frequencies of Si-O bond around 1075 and 1045 cm^{-1}, respectively, both of which appear to be distinctly resolved and separated, in contrast to an Si-O absorption envelop normally observed in the neat clays, **Fig. 2b** [5].

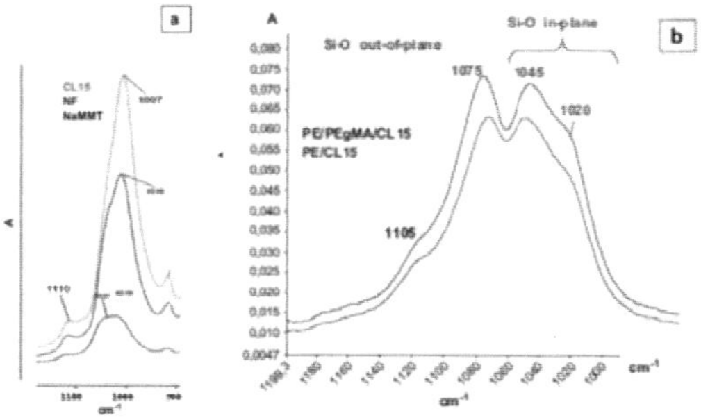

Figure 2: ATR-FTIR showing the Si-O absorption region of (a) neat OMMT clays and (b) the surface of thin films of the compatibilised PE-OMMT nanocomposite

The resolved Si-O absorption bands at around 1045 and 1075 cm^{-1} can then be used to monitor the state of dispersion of the layered silicate clays, **Fig. 2b**. The importance of the ATR-FTIR technique is that, in addition to being a relatively simple complementary off-line analytical method for assessing the state of dispersion of clays in nanocomposites, it can also be adapted for in-line monitoring of the extrusion process of clay-polymer nanocomposites thus offering the capability of fine-tuning the processing parameters to obtain the optimum dispersion, quality and consistency of clay nanocomposites during production [11].

Rheological properties of polymers provide a fundamental understanding of their melt processing behaviour. The rheological characteristics of nanocomposites are known to be sensitive to their state of dispersion and the surface characteristics of the dispersed phase [12]. Melt rheology therefore provides a further methodology to assessing the state of dispersion of the nanofillers during processing of polymer melts.

Plots of different rheological properties, e.g. storage modulus, G', complex viscosity, η^*, and loss modulus, G", as function of frequency for polyolefin nanocomposites having different compositions (e.g. compatibilised, non-compatibilised samples containing different montmorillonite clays) show that at high frequencies there are minimal differences between the changes in all the rheological properties for the different samples and this is a consequence of their response under these conditions being dominated by the polymer matrix only. However, at low frequencies (lower than 1 rad/s) differences appear to be more significant indicating a solid-like rheological behaviour. This is illustrated for the storage modulus in **Figure 3** [5] which shows significant differences at low frequencies with compatibilised PE-nanocomposites containing OMMT (e.g. PE-PEgMA-CL15) showing significantly higher storage modulus than their counterparts that have their clays unmodified (containing NaMMT) which is attributed to percolation of the randomly distributed silicate layers due to their anisotropy [5, 12]. Rheological measurements, therefore, confirm and support results obtained from the other methodologies discussed above to show that the state of exfoliation is higher in compatibilised nanocomposites containing organically modified clays than is the case either in the absence of a clay surfactant or a compatibiliser.

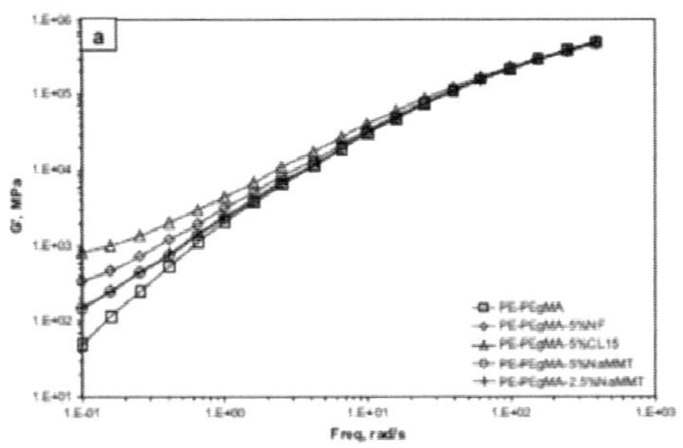

Figure 3: Storage Modulus, G' vs frequency of compatibilised PE-nanocomposites with organically modified nanofillers (OMMT; CL15 or NF) and non-modified Na-MMT [5]

The melt processing has a direct effect on the thermal stability of polyolefin nanocomposites and on the extent of the dispersion of the clay. The thermal stability of the clay organic modifier (quat) is known to be low under typical polymer melt processing conditions of high temperature and shear thus, this would affect both the extent of exfoliation (which influences the mechanical properties) and the long-term durability and performance of the nanocomposite product [13]. It has been suggested, based on thermogravimetic studies, that the thermal decomposition of quaternary alkyl ammonium ions starts at about 180 °C and that the main decomposition occurs between 200 and 500 °C [13]. This issue is further complicated by the additional use of compatibilisers in these nanocomposites which introduce a further problem with respect to the stability of the nanocomposite system in spite of their apparent beneficial effects in improving nanoclay dispersion and distribution.

The thermal degradation of surfactant (quat) on the organoclay is known to occur through a Hoffman elimination mechanism, **reaction 1** and once the □-olefin (a vinyl moiety) is formed, it can contribute to polymer degradation by a number of known degradative reactions [14]. We have shown [5] that the vinyl concentrations in PE-OMMT are much higher than that developed in PE extruded under the same conditions alone or with the non-organically modified NaMMT clay and the difference is therefore entirely due to the formation of vinyl unsaturation from the organic surfactant on the clay through the Hoffman elimination reaction above.

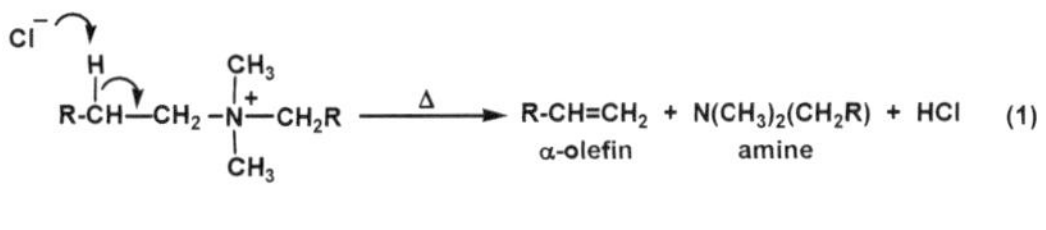

Effect of Accelerated Ageing and Weathering on the Oxidative Stability of Polyolefin-Clay Nanocomposites

In spite of the improvement in properties that can be achieved from polyolefin-clay nanocomposites, these materials do have one main drawback which is their poor long term thermal and photo-oxidative stability in outdoor applications which is lower than that of their unfilled counterparts [5, 15, 16]. It has been shown that the OMMT and the MA-based compatibiliser do not affect the mechanism of photo- or thermal oxidation as the oxidation products formed in all cases are very similar, however, they have a major influence on the rate of their oxidation.

Different analytical methodologies are used to assess the oxidative stability, state of clay dispersion and changes in the mechanical characteristics upon ageing of polyolefin nanocomposites. Amongst the most important are infrared methods including surface ATR measurements, bulk transmission FTIR and IR-microscopy, as well as DMA, tensile measurements, embrittlement time, TEM, SEM.

The rate of photoxidation during exposure to weathering or accelerated photo-ageing conditions is normally monitored by following the changes in oxidation products, mainly carbonyl-containing compounds and hydroxyl group, by infrared spectroscopy. **Figure 4** illustrates this and shows a significant increase in the carbonyl indicies of compatibilised PE- clay nanocomposites reflecting a much higher rate of photoxidation, and also lower thermoxidative embrittlement times (EMT at 125 °C) in the case of PPNC compared to those of the corresponding pristine polymers. Furthermore, in the case of photoxidation, the presence of the compatibiliser in PE-OMMT nanocomposite (e.g. PE-PEgMA-CL15) causes a more significant increase in the rate of photoxidation compared to the absence of the compatibiliser (e.g. PE-CL15), **Fig. 4a**. The negative effect of PE-g-MA on the rate of photoxidation of nanocomposites is due to a large extent, to the photoxidative instability of maleic anhydride, see **reaction 2,** and to unfavourable interactions between the anhydride group and the clay under the UV-exposure conditions used [5]. The much higher rate of photooxidation of the polyolefine nanocomposites is further supported by a similar trend, i.e. decrease, in mechanical properties, under the same UV exposure conditions.

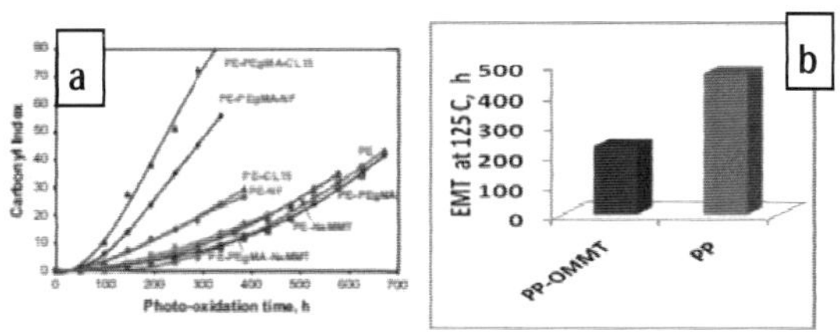

Figure 4: Effect of accelerated photoageing of PENCs in a QUV device (a) and thermal ageing (at 125 °C) of PPNC (b) and their pristine counterparts on their photoxidative and thermoxidative stabilities.

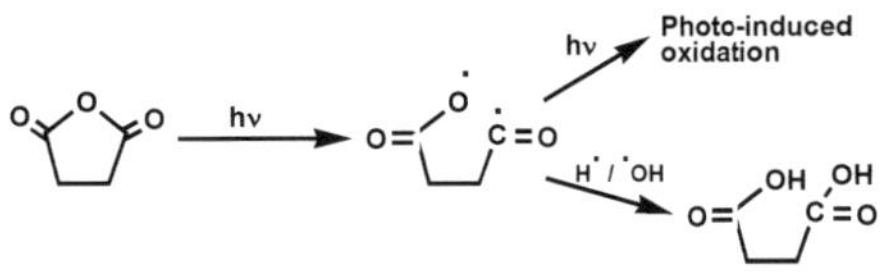

(2)

Transmission FTIR-microscopy of aged PP-clay nanocomposites shows a significant narrowing and separation of the Si-O absorption band [17] which indicates the formation of a much more exfoliated morphology following thermal and photoxidative ageing; this is attributed to increased contribution from the small hydrophilic products formed during photoxidation which would favour increased interaction with clay platelets leading to enhanced dispersion and greater degree of exfoliation. This was confirmed also to be the case in PE-nanocomposites as observed from TEM examination of photo-aged samples [5]. The much higher extent of exfoliation observed in photo-aged samples gives rise to exposure of larger areas of interfacial regions to UV light; this could additionally contribute to greater extent of release of photoxidation products from the quat into the matrix and resulting in a higher overall level of photoxidation.

Overall, the poor photoxidative stability of polyolyolefine-clay nanocomposites is due to a complex combination of factors including the instability and degradation of the clay-organic modifier, the photo-instability of the MA-containing compatibiliser, the presence of catalytic metal ion, e.g. iron, impurities and possible adverse interactions between the various components and their thermal- and photo- oxidative products.

It is important to point out, however, that the thermal and photo degradation of clay/polyolefin nanocomposites can be reduced in the presence of appropriate stabilising system. However, the performance of conventional antioxidants is typically compromised in the presence of the nanoclay and compatibiliser, hence stabilised polyolefin nanocomposites generally have lower overall stability than their unfilled polyolefin polymer counterparts. The use of UV-absorbers in the presence of typical melt stabiliser antioxidants gives a much improved stability compared to the unstabilised counterpart, **Figure 5.** However, it is clear that the addition of a metal deactivator to the same UV-stabilising system shows an even more significant increase in the photoxidative stability of the nanocomposite. This indicates that the presence of metal ion impurities in the clay play an important role in accelerating the photo-degradation of the polymer. The inferior performance of antioxidants when used in polyolefin nanocomposites compared to their performance in the unfilled analogues is due to several reasons, in addition to those contributing to the degradation of nanocomposites mentioned earlier, including possible adsorption of the antioxidants on the very large clay surfaces and interactions with the clay organic modifier and its degradation products.

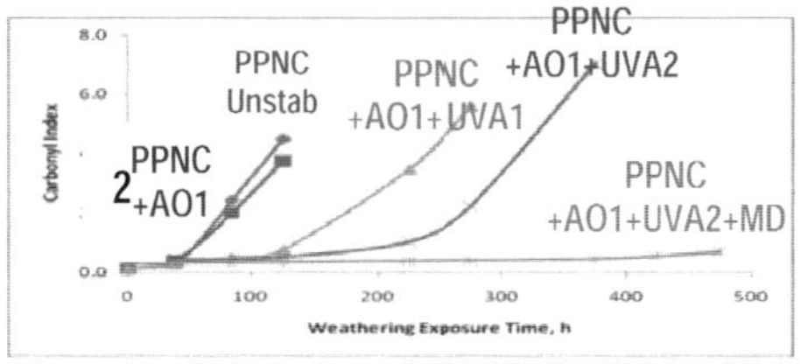

Figure 5: Stabilisation of PPNC (PP/PPgMA/OMMT) under accelerated weathering conditions using different stabiliser systems

REFERENCES

1. Leszczynska, J. Njuguna, KP. Pielichowski, JR. Banerjee, Thermochim Acta, **454**, 1 (2007)
2. S. S Ray, M Okamoto, Progress in Polym. Sci., **28**, 1539 (2003)
3. M. Zhang, U. Sundararaj, Macromol Mater Eng, **291**, 697 (2006).
4. R. K. Shah, DR. Paul, , Polymer, **47**, 4075-4084 (2006).
5. N Tz Dentcheva, S. Al-Malaika, F LaMantia, Polym Degrad Stab, 1571, **94** (2009)
6. P. Peltola, E. Valipakka, J. Vuorinen, S. Syrjala, K. Hanhi, *Polym. Eng. Sci.*, **46,** 995 (2006).
7. W. Wenyi, Z. Xiaofei, W. Guoquan, Ch. Jianfeng, J Appl Polym Sci, **100**, 2875 (2006).
8. S Bocchini, S Morlat-Therias, JL Gardette, G Camino, Polym Degrad Stab, **92**, 1847 (2007).
9. A B Morgan, J W Gilman, J Appl Polym Sci, **87**, 1329 (2003)
10. I Wouter, S Kemnetz, D Benderly, Polym Eng. Sci, **46**, 1031 (2006)
11. D. Fischer, G. Camino, D. Tabuani, E, Masarati, P. Cassagnau, S. Al-Malaika, EU-FP6 Project Internal Report, Contract No. IP-026685-2 (2009)
12. S S Ray, M Okamoto, Progress in Polym. Sci., **28**, 1539 (2003)
13. R K. Shah, DR. Paul, Polymer, **47**, 4075 (2006).
14. S. Al-Malaika, X. Peng and H. Watson, Polym Deg and Stab, **91,** 3131 (2006)
15. H. Qin, Z. Zhang, M. Feng, F. Gong, S. Zhang, M. Yang, J Polym Sci Part B: Polym Phys, **42**, 3006 (2004).
16. S. Morlat-Therias, B. Mailhot, J L. Gardette, C. Da Silva, B. Haidar, A. Vidal, Polym Degrad Stab, **90**, 78 (2005)
17. S.Al-Malaika and H H Sheena, Unpublished work

O-07-572

AN INVESTIGATION ON THE MORPHOLOGY AND RHEOLOGY OF NANOCOMPOSITE OBTAINED BY MELT MIXING

A. Abbasi, A. Javadi *, H. Nazockdast
Polymer Engineering Department, Amirkabir University of Technology, P.O. Box 5875/4413, Tehran, Iran;ajavadi@aut.ac.ir; nazdast@aut.ac.ir; abbasimoudaref@gmail.com

Abstract
Nanocomposite of PA6/PP/MWNT with different compositions were made via melt mixing. A neat blend of PA6/PP with a volume ratio of (60:40) was chosen as a baseline to which 1 and 2 Vol. % MWNTs were added. The rheological response of nanocomposite was measured using Rheomechanical Spectroscopy (RMS) with the aim of investigating the influence of MWNTs concentration on the chain mobility of nanocomposite. The curve of storage modulus versus frequency reaches a plateau at fairly low frequencies, which is an indication of pseudo-solid like behavior. All samples exhibited shear-thinning behavior at higher frequencies. The morphology of the blend matrix as well as dispersion and orientation of MWNTs were investigated through transmission Electron microscopy (TEM). TEM images reveal excellent dispersion and random orientation of MWNTs throughout the matrix. The dispersion of nanotubes as indicated by TEM micrographs suggests the formation of a nanotube network in the PA6/PP matrix. In addition, observation of the SEM images confirm a co-continuous morphology of PA6/PP nanocomposites.

Keywords: Nanocomposite, blend morphology , rheology, PA6/PP

Introduction

Polymer nanocomposites containing carbon nanotubes (MWNTs) have attained great interest in material development due to a unique property profile of the nanotubes, like excellent mechanical, electrical, magnetic, optical and thermal properties. Adding small amount of carbon nanotubes can turn an electrically insulating polymer to a conductive composite which is not possible with the same amount of conventional (lower aspect ratio or spherical) conductive fillers like carbon black (CB) [1]. Nanocomposites containing MWNTs are promising materials with potential applications as electromagnetic shielding coatings, electrostatic dissipative materials, aerospace structural materials and active elements in sensors [2]. The combination of exceptionally high interfacial areas of MWNTs and small inter particle distance, obtained at low filler loadings (5 wt% maximum), yields hybrids possessing unique properties, typically not shared by their more conventional microscopic counterparts so many researchers are intrested to prepare conductive nanocomposites at very low volume percent of MWNTs [1-8]. Recent studies has showed that through double percolation phenomena, the conductive nanocomposites can be obtained even at lower volume percent which is more desireble [4-6].

2. Experimental

2.1 Materials

CCVD (catalytic carbon vapor deposition) synthesized thin purified MWNT were obtained from Nanocyl CA Belgium (NC 3100, purity 90%, Average diameter of 9.5 nm and the length of 1.5 μm). PA6, akulon F136-C1, was purchased from akzo. PP with MFI of 7.5 g/10 min (230°c, 2.16 kg) was supplied from Jam petrochemical company.

2.2. Composite preparation

Neat blends of PA6/PP (60/40) and it's nanocomposites containing 1-3 Vol% with MWNTs were melt-mixed in an internal mixer at 230 °c with a rotational speed of 90 rpm for 8 min. First PA6 and MWNTs were mixed and then PP was added to the mixture.

2.3. Rheology

The rheological measurement were performed using Rheometric mechanical spectrometer (Paar Physica USD 200, Austria) at 245 °c with parallel plate geometry (plate diameter of 25 mm, gap of 1 mm). The melt linear viscoelastic properties of samples were measured under controlled shear deformation oscillatory mode at frequencies varying from 0.0588 to 1000 rad/s. The strains were chosen in order to be within the linear viscoelastic range.

2.4. Scanning electron microscopy (SEM)

The state of dispersion was studied by scanning electron microscopy (SEM; Seron, AIS-2100, Korea). Samples were cry fractured using liquid nitrogen and etched in formic acid for 24 h at room temperature to remove PA6 phase of the blends and nanocomposites. The etched samples were also observed under SEM.

2.5. Transmission electron microscopy (TEM)

Ultra thin sections were cut out of the samples using an ultra microtome. For all samples, bright field transmission electron microscopy at an acceleration voltage of 80kV was carried out using a Zeiss 902 TEM.

3. Results and Discussion

3.1. Rheology

In order to have better understanding about the microstructure of nanocomposites, oscillatory rheological properties of samples were investigated. As it was expected, adding MWNTs had its effects on rheological properties. Fig. 1 shows the variation of the complex viscosity of samples with different volume ratios of nanotubes. All samples exhibited shear thinning behavior with increasing frequencies. While neat blend showed the plateau in complex viscosity at low frequencies, in nanocomposites nearly plateau region disappeared. This type of behavior was seen by Kota and his coworkers [3] for PS/MWNT. Zhang et. al. [4] and Syryasarathi and his coworkers [5] found the same behavior for the PA6/PP/MWNT nanocomposites too. A non-terminal behavior was observed in all samples at low frequencies, even when the content of nanotubes was as low as 1 vol%. This confirmed the existence of a percolated network in polymer matrices. It has been reported that interconnected structures of an isotropic fillers can result in an apparent yield stress which is visible in dynamic measurement by a plateau of G' versus frequency at low frequencies [4].

Fig. 2 shows the variation of the storage modulus of neat PA6/PP blend and PA6/PP/MWNTs nanocomposites with different volume ratios of nanotubes. The non-terminal behavior which is seen here, confirms the existance of network again.

To investigate the affinity of CNTs to PP and PA matrices, PA6/MWNTs and PP/MWNTs nanocomposites with 97/3 volume % for both was prepared and the rheological behavior of them was studied. It was found that the dispersability and blending conditions was effective enough to disentangle MWNTs and disperse them throughout polymeric matrices but rheological measurements showed that the PA matrix were more successful in this way. This can be concluded from greater increase of storage modulus at low frequencies in case of PA6. Fig. 3 shows the variation of storage modulus of both mentioned nocomposites. As it can be seen, the storage modulus has increased in both cases, but this increase is much more greater in case of PA6 matrix. At frequency of 2 rad/s, the ratio of storage modulus of Nanocomposite to storage modulus of polymeric matrix for PA6 is 2000 while it is 27.8 for PP matrix. This greater increase can be attributed to greater thermodynamic tendency of MWNTs toward PA6 matrix.

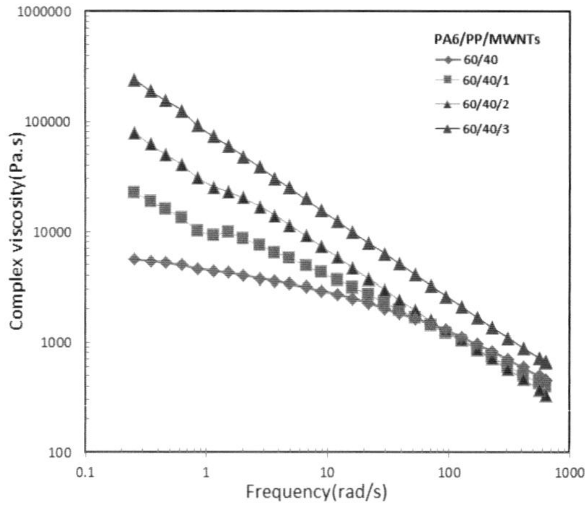

Fig. 1. Complex viscosity versus frequency for nanocomposites with different amounts of MWNTs

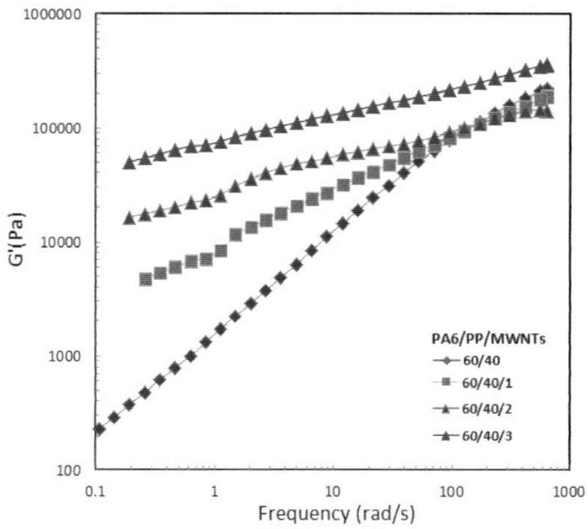

Fig. 2. Storage modulus versus frequency for neat PA/PP blend and its nanocomposites with different amounts of MWNTs

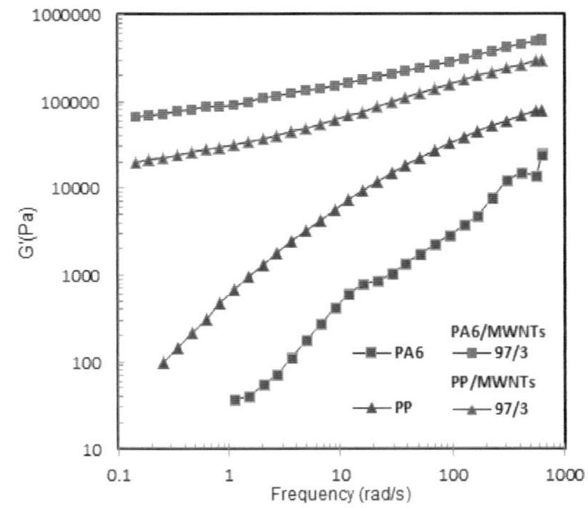

Fig. 3. The storage modulus of PA6/MWNTs and PP/MWNTs nanocomposites versus frequency

3.2 Morphology

3.2.1 SEM

To investigate the morphology of PA6/PP/MWNT Nanocomposites, the scanning electron microscopy (SEM) was used to observe the fractured surface of samples. Fig. 4 shows the SEM micrograph of formic acid etched surfaces of 60/40/3 PA6/PP/MWNT nanocomposite. It is evident that Nanocomposite shows co-continuous structure.

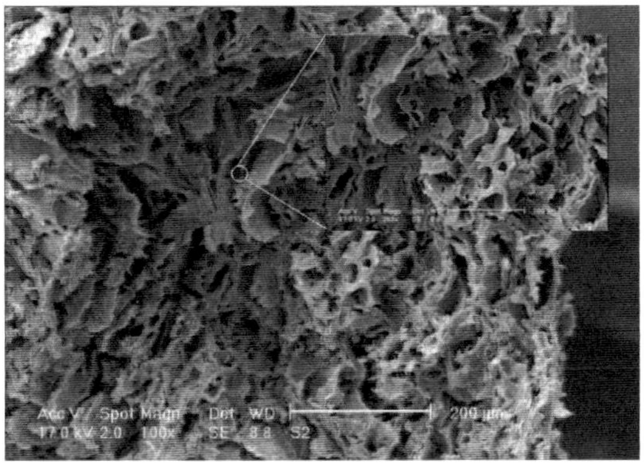

Fig. 4. Etched surface of 60/40/3 PA6/PP/MWNT Nanocomposite

3.2.2 TEM

Fig. 5 and 6 show the TEM images of 60/40/1 and 60/40/2 of PA6/PP/MWNT nanocomposites. As it is obsevbed, the dispersion of MWNTs in both nanocomposites is excellent.

TEM images also showed random orientation of MWNTs inside polymeric matrix which allows better interconnectivity between MWNTs nanoparticles. As a result, the formation of electrical path happens at lower volume percent of MWNTs which is favorable. Also lack of orientation of MWNTs toward specific direction cause the nanocomposite to become anisotropic.

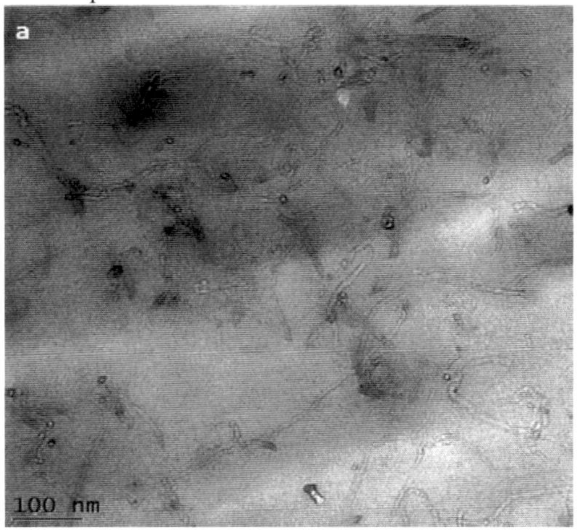

Fig.5 . TEM images of PA6/PP/MWNTs Composites: 60/40/1

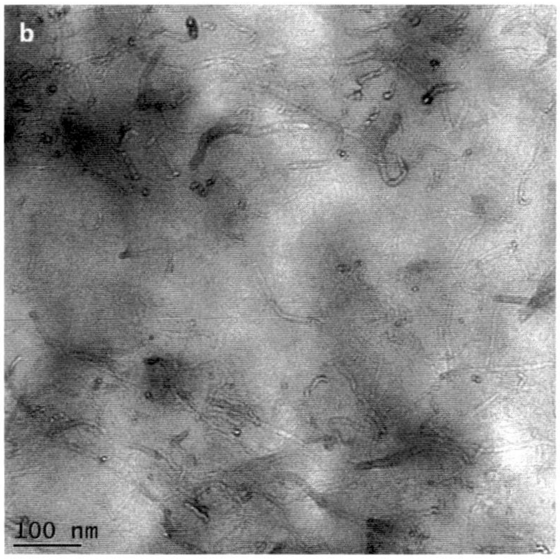

Fig.6 . TEM images of PA6/PP/MWNTs Composites: 60/40/2

4. Conclusions

The rheology and morphology of PA6/PP/MWNTs nanocomposite were investigated. The rheological results showed a viscosity upturn and a non-terminal behavior for all samples at low frequencies in complex viscosity and storage modulus curves respectively. This confirmed the existence of a percolated network in polymer matrices. From result of SEM images, it was observed that morphology of nanocomposite at PA6/PP/MWNT 60/40/3 is co-continuous which is of intrest for electrical conductivity at lower amount of MWNT. TEM images revealed homogenous dispersion of MWNTs inside polymeric matrix and also random orientation of MWNTs that facilitates the interconnectivity between MWNTs nanoparticles and leads to higher electrical conductivity.

5. References

[1] I. Alig, P. Pötschke, D. Lellinger , T. Skipa, S. Pegel, G. Kasaliwal, T. Villmow, ", *Polym.,* in press 2011.

[2] E. Logakis, C. Pandis, V. Peoglos, P. Pissis, C. Stergiou, J. Pionteck, P. Pötschke, M. Micusik, M. Omastová, , *J. Of. Polym. Sci. Part B: Polym. Phys. 47,* 2009, *764.*

[3] A. K. Kota, B. Cipriano, M. Duesterberg, A. Gershon , D. powell, S. Raghavan, H. Bruck, *Macromolecules, 40*, 2007, *7400.*

[4] L. Zhang, C. Wan, Y. Zhang, , *Compos. Sci. and Tech., 69,* 2009, *221.*

[5] S. Bose, A. R. Bhattacharyya, A. R. Kulkarni, P. Pötschke, , Composites Sci. and Tech. , *69,* 2009*365.*

[6] L. Zhang, C. Wan, Y. Zhang, *Polym. Eng. Sci., 49,* 2009, *1919.*

[7] W.Bauhofer, J. Kovacs, , Progress in *polym. Sci. 35,* 2010, *357.*

O-07-468

PREDICTIVE MODELING OF ELECTRICAL PERCOLATION IN COMPOSITES OF POLYMERS AND CARBON NANOTUBES USING 2D AND 3D NUMERICAL MODELS

Ciara Tighe, Savko Malinov* and Tony McNally*
School of Mechanical and Aerospace Engineering, Queens University Belfast, Belfast, BT9 5AH, UK
ctighe03@qub.ac.uk; s.malinov@qub.ac.uk; t.mcnally@qub.ac.uk

Abstract - Two-dimensional (2D) and three-dimensional (3D) computer models and programs were developed to simulate the formation of the nanostructure of composites of polymers and carbon nanotubes (CNTs). The formation of this nanostructure is modeled and simulated numerically, using MATLAB as a platform, by nucleation and then growth of CNTs in the polymer matrix. Further developments incorporate options for simulation of composite nanostructure and CNT distribution as a function of varying length, aspect ratio and degree of entanglement. This allows for simulation of complex nanostructures for different CNT volume fraction, and investigate cases with different degrees of alignment and orientation in 2D and 3D space. The models predict electrical percolation with the ultimate aim to predict electrical conductivity/resistivity in these materials taking into consideration CNT volume loading, dispersion and distribution, orientation and deformation. The results show correlation with a host of published experimental data.

Keywords: Carbon Nanotube, Polymer, Modeling, Percolation, Electrical conductivity

Introduction

In recent years (CNTs) have attracted much interest due to their outstanding electrical properties; harnessing this capability will be crucial to their implementation in future technologies. As a material they can be more conductive than cooper wire [1], and introducing them into insulating materials such as ceramics, glass and polymer can significantly enhance the electrical properties of these matrices. In order to create effective conducting pathways in such matrices it is essential for there to be a percolating interconnected network of CNTs. Percolation itself is a well studied and documented field of statistical mathematics that has been used extensively in engineering and science; however, successfully linking percolation and optimum conductivity requires further development.

With varying CNT aspect ratios, percolation thresholds as low as 0.002 wt% in an epoxy resin have been reported [2]; however percolation is only the point at which the sample will start to conduct and not the point of maximum conductivity. There are other factors at play, including the number of CNTs involved in obtaining this threshold, matrix properties, method of composite preparation, CNT defects, as well as the contact resistance between CNT and matrix.

In this paper, we report our initial findings on the investigation of CNT percolation and connectivity using numerical 2D and 3D models and computer programs that have been developed on a modular fashion using MATLAB as a platform.

Experimental

There have been recent studies on the modeling of CNTs using MATLAB [3]. 2D schematic diagrams currently offer a 'sticklike' representation of CNTs, as means of studying percolation and the various properties that are governed by it, for example tunneling resistance. The work described here has developed this further by taking into consideration the curvature and aspect ratio of a model CNT, based on NANOCYL™ NC7000.

2D Model

The model and the computer programs are developed using MATLAB. The composite nanostructure is modeled and simulated numerically, where each CNT has an initial randomly assigned location in both the x and y axis using the *rand* function in MATLAB which acts like a nucleation point. It is from this location that the direction of orientation and "growth" of the CNT is initiated. There are eight possible directions in which the CNT can grow with each direction having a different probability of growth. The likelihood of growth in these directions can be controlled with a random or an assigned probability. This level of control allows for CNTs to be aligned or simply orientated, ultimately increasing the complexity of calculating the overall conductivity of the sample. Orientated CNTs are controlled so that growth occurs at 0°, 45° and 90° and aligned CNTs are controlled so that growth can only be at 0° and 45° from the original site of nucleation. Entangled CNTs can move in any direction, as can be seen in Fig.1. The simulation allows for this change of direction to occur after every pixel. The models incorporate options for simulation of composite nanostructure as a function of varying CNT length, aspect ratio and degree of entanglement. It can be seen from Fig.1 that the model illustrates the natural formation of individual CNTs and CNT agglomerates similar to that observed experimentally.

Since the earliest observation of percolation in CNT composites much work has been done attempting to estimate percolation thresholds. The unit space in the model is divided into CNT and polymer matrix rich areas, the edges of the network can be open or closed with a probability, p, and the point at which a network is statistically probable to percolate is p_c, the critical percolation for which there has been a suggested correlation between this threshold and the maximum conductivity [4].

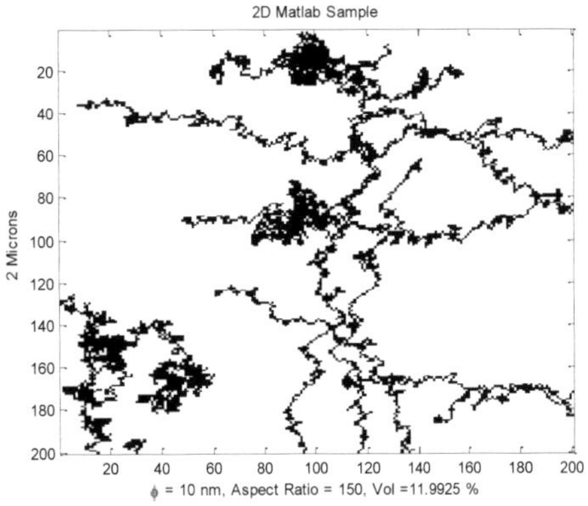

Figure 1 – A 2D Simulation illustrating CNT growth (localization) and composite structure. Φ is the CNT diameter.

The percolation of the network in the model is only considered after an interconnected CNT network has formed through the sample. The network is mapped from the top of the model and a color change is associated with the connectivity of tubes, from black to grey, i.e. the point(s) at which 2 CNTs touch can be identified and thus the connectivity of CNTs through the sample. The point where two CNTs meet is not taken into consideration when the path is being mapped, it will however be looked at closer when the conductivity is being calculated. The number of electrically percolating pathways is determined by the number of tubes that are grey in color at the bottom of the 2D space modeled. This however does not refer to the overall conductivity of the network and the number of nodes in the critical path will need to be considered before estimation of conductivity can be determined.

3D Model

The 3D model and associated computer programs have also been developed. The CNTs are modeled and simulated numerically, with each of them randomly assigned a nucleation point in the x, y and z axis. From the point of nucleation the CNT is then grown is all 3 dimensions which gives them 26 directions in which to grow. As with the 2D model the *switch* function is used, initially giving equal and random probability that growth will occur in each direction, it is this that is used to control the curvature of the CNTs. Similar to the 2D model there are inputs for the length of the CNTs allowing for variable aspect ratio. However, in reality, a single batch of CNTs will contain CNTs of varying lengths, and has been taken into consideration within a range between 1μm-9μm. The alignment and general orientation of the CNTs is controlled in a similar manner to the 2D model; the orientated CNTs can grow in 0°, 45° and 90° and the aligned CNTs are only allowed to grow at 0° and 45° from the original growth site. Orientated CNTs therefore have 17 directions in which they can grown in either the x, y or z direction and the aligned CNTs are restricted to 9 directions, meaning the tubes are less likely to touch one another. Fig.2 illustrates regions of entangled CNTs, which are allowed to move randomly in any given direction allowing for the formation of both individual CNTs and CNT agglomerates. To ensure that there has been no preference made in the random location or the growth of the CNTs in the model, the 3D model or the 2D slices of the 3D images generated; each slice has a thickness equal to a CNT diameter, 10nm were examined, thus evaluating CNT rich zones.

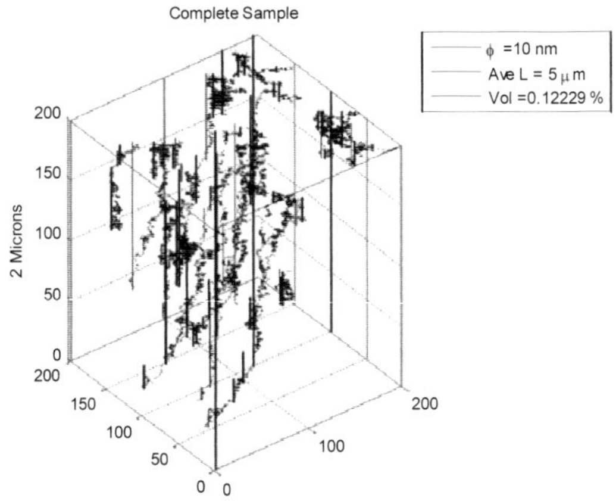

Figure 2 – A 3D Simulation, illustrating CNT growth (localization) and composite structure. Ave L is the average length of the CNT.

Percolation has been studied through the 3D model. The network is evaluated from the top of the unit volume modeled and in each 2D slice the CNTs that are connected (touching) are changed from black to grey, a slice at a time. The number of pathways through the model has been calculated as the area of the bottom slice that is connected to the first layer of the sample. The electrical tunneling effect and contact resistance play a significant role in interfacial behavior of CNTs. Electron tunneling requires a CNT-CNT distance to be less than 5nm to ensure conductivity [5], this is a scale that is currently not represented in the model but these effects will still need to be taken into consideration at each node when calculating conductivity. The increase in resistance is a function of the distance between the two tubes, as can be seen in Eq.1 [3] and even if there is contact at the circular edge of the tube means, the point of touch is relatively small and the surrounding tubes need to be considered.

(1)

where R_t is the tunneling resistance, m is the mass of an electron, e is the elementary charge, h is Planck's constant, d is the distance between CNTs, A the cross-sectional area at point of contact and λ is the square tunneling barrier height. The model can also be adopted to examine the effect of mechanical deformation on CNT localization and ultimately predict mechanical properties of composites of polymers and CNTs. The stresses that may be applied will cause a change in both the shape of the unit volume of matrix and the nucleation points of the CNTs within host material. The localization of CNTs will change (alignment and orientation) as they are deformed in a given direction.

Results and Discussion

From the models developed to date, there have been several studies completed in both 2D and 3D format.

2D Model

The primary study within the 2D model allows for a network of percolating CNTs to be identified. Fig.3 shows the network for a sample 2D model, where the one site of nucleation along the top of the sample is translated into three paths along the bottom (as highlighted by the rows in Fig.3), through the use of connecting CNTs and nodes.

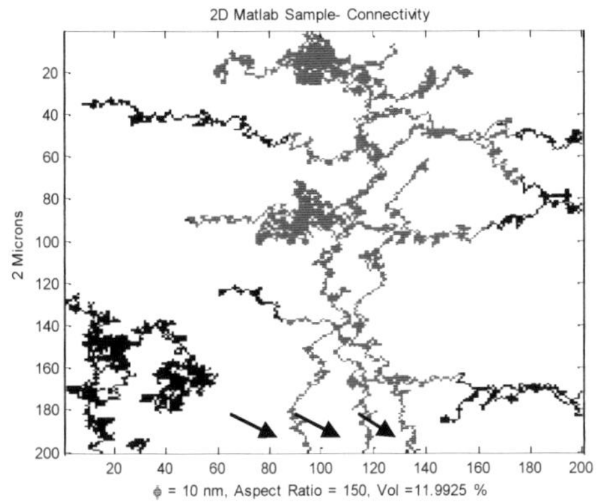

Figure 3– The 2D simulation shown in Fig.1 but with a percolating network visable in grey.

The image in Fig.3 shows CNTs with a100% connectivity through the sample, however CNTs have inherent defects as a consequence of their synthesis which cause them to lose conductivity along their length. From the literature it has shown that CNT defects can reduce electrical conductivity by as much as 50% [6]. This is represented in the model with a color change to each CNT. A CNT that is 100% non-conductive is black in color and a decrease in the shade of the CNT indicates a reduction in conductivity, hence grey CNTs will have conductivity as low at 50% of the stated range. The application of these color changes is random with an equal probability; this is done with the aim to predict the conductivity more accurately as shown in Fig.4.

Furthermore, the 2D simulation takes into consideration a visualization of the interfacial area around CNTs at the polymer-CNT interface. Lower than predicted mechanical properties of polymer-CNT composites have been linked to poor interfacial bonding and stress transfer at the polymer-CNT interface.

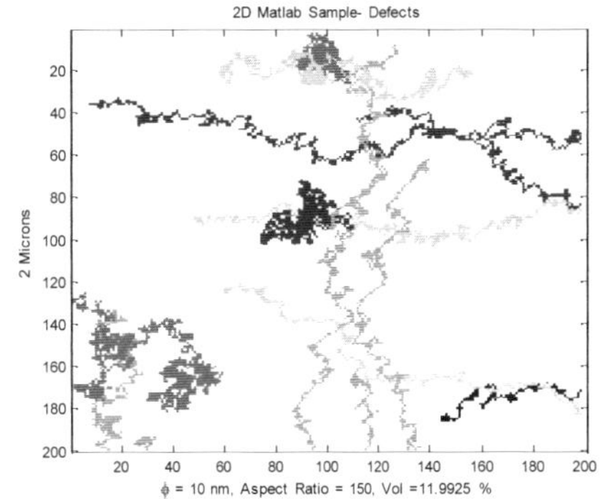

Figure 4- The 2D simulation shown in Fig.1 but illustrating the variation in electrical resistance as a function of defect density.

The interfacial bond area between polymer and CNT can also be illustrated, see Fig.5, with the CNT being pictured in black and the interface a dark grey, where some of the CNT may not be bonded to the polymer matrix.

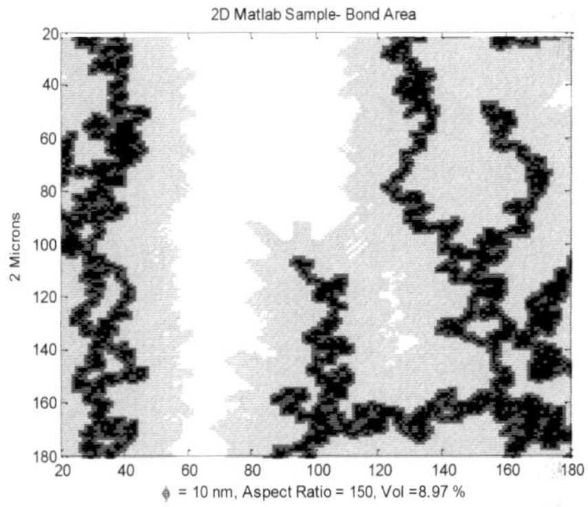

Figure 5- A 2D simulation showing the interfacial region (dark grey) between polymer (white) and CNT (black) rich zones.

3D Model

The 3D model was also applied to the same primary percolation study; however, there are a number of steps that have been taken to ensure the sample is representative of an isotropic CNT-polymer composite. A study showing the distribution of CNTs in a unit volume of matrix (3D) and took slices examining layer by layer the percentage fill showing the homogenous distribution of CNTs. This distribution study was completed in the x, y, and z direction to ensure that no grouping of CNTs where found in the 3D sample, as can be seen in Fig.6. It was confirmed there was no preference in the positioning of CNTs. The layers of the 3D sample are the same thickness as diameter of the CNT itself and as can be seen in Fig.7 are completely different to the previous 2D model that can be seen in Fig.1. Fig.7 shows the initial start point of several CNTs showing their small cross-sectional area.

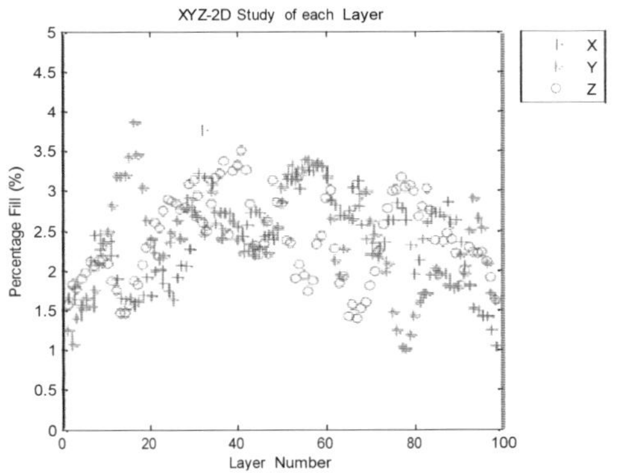

Figure 6- The distribution of CNTs in each layer and in all 3D's - for a CNT loading of 3 vol%.

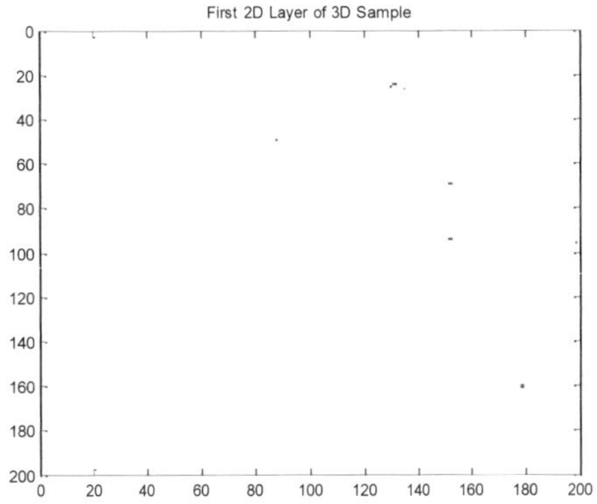

Figure 7- The first layer in the 3D sample shown in Fig.2 in the z-direction, for a CNT loading of 0.12 vol%.

Further to Fig.7, Fig.8 shows the final layer of the sample; it can be seen when the CNT is forced against the top layer of the 3D sample, it continues to grow only in the x and y axis producing the pattern as can be seen, it is from this that number of pathways through the sample, can be determined, ensuring percolation has been attained.

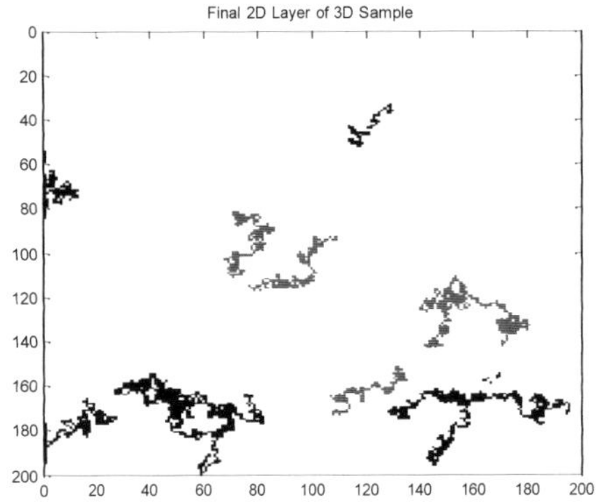

Figure 8- The final layer of the 3D sample, for a CNT loading of 0.12 vol%

Composites of polymers and CNTs will find wide application in strain sensing [7], and will undergo deformation during secondary processing. The alignment and orientation, re-localization, of CNTs as a consequence of uniaxial and biaxial deformation was also investigated. Fig.9 shows the effect of deformation on composite structure, and the effect of both uniaxial and biaxial stresses on the orientation of the CNTs.

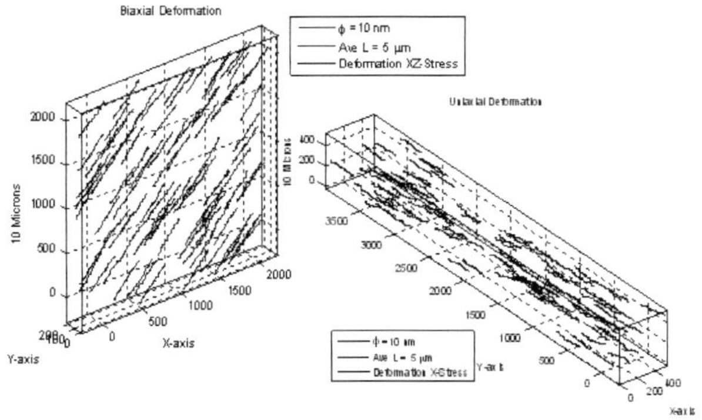

Figure 9- Biaxial and uniaxial stresses applied to a homogeneously dispersed sample of CNTs.

Conclusions

We have reported initial findings for the use of MATLAB, numerical modeling for the study of the structure of composites of CNTs and polymers, in both 2D and 3D. An understanding of the key features controlling CNT connectivity (electrical conductivity) and prediction of percolation have been gained. The potential of MATLAB as a tool to predict electrical and mechanical properties of these composites has been demonstrated. The outputs from detailed studies using this approach will be published shortly.

Acknowledgements

The authors thank Nanocyl S.A. Belgium for helpful discussions.

References

1. Y Bin, M Kitanaka, D Zhu, and M Matsuo. *Macromolecules 36*, 2003, 6213-6219.
2. C.A. Martin, J.K.W. Sandler, M.S.P. Shaffer, M.-K. Schwarz, W. Bauhofer,K. Schulte, A.H. Windle. *Composites Science and Technology 64*, 2004, 2309–2316.
3. Servati, Rubaiya Rahman and Peyman. *Nanotechnology 23,*2012, 1-9.
4. W. Bauhofer, J.Z. Kovacs. Composites Science and Technology 69, 2009, 1486–1498.
5. Q.Zhang, S. Rastogi, D. Chen, D. Lippits, P.J. Lemstra. *Carbon 44,* 2006, 778–785.
6. Paul C. P. Watts, Wen-Kuang Hsu, Harold W. Kroto, and David R. M. Walton. *Nano Letters 3*, 2003, 549-553.
7. I. Kang, M.J Schulz, J.H Kim, V. Shanov and D. Shi. *Smart Mater. Struct. 15,* 2006, 737–748.

AUTHOR INDEX

AUTHOR INDEX

AUTHOR INDEX

AUTHOR INDEX

AUTHOR INDEX

AUTHOR INDEX

AUTHOR INDEX

AUTHOR INDEX

28th Polymer Processing Society Annual Meeting 2012 (PPS-28)

Pattaya, Thailand
11 – 15 December 2012

Volume 2 of 2

ISBN: 978-1-5108-4261-8

Printed from e-media with permission by:

Curran Associates, Inc.
57 Morehouse Lane
Red Hook, NY 12571

Some format issues inherent in the e-media version may also appear in this print version.

Printed by Curran Associates, Inc. (2016)

For permission requests, please contact Polymer Processing Society
at the address below.

Polymer Processing Society
810 E. 10th St.
Lawrence, KS 66044
USA

Phone: +1-800-627-0326
+1-785-865-9403
Fax: +1-785-843-6153

pps@allenpress.com

Additional copies of this publication are available from:

Curran Associates, Inc.
57 Morehouse Lane
Red Hook, NY 12571 USA
Phone: 845-758-0400
Fax: 845-758-2633
Email: curran@proceedings.com
Web: www.proceedings.com

TABLE OF CONTENTS

VOLUME 1

EXTRUSION AND EXTRUSION PROCESSES

BLOW MOLDING, ROTOMOLDING AND THERMOFORMING

FIBERS, FILMS AND FOAMS

INJECTION AND MICROINJECTION MOLDING AND MOLDS

MIXING AND COMPOUNDING

REACTIVE PROCESSING, POLYMER BLENDS AND ALLOYS

POLYMER COMPOSITES AND NANOCOMPOSITES

VOLUME 2

RHEOLOGY AND RHEOMETRY

MORPHOLOGY AND STRUCTUR DEVELOPMENT

PROCESS MODELING AND SIMULATION

BIO-BASED & BIODEGRADABLE POLYMERS AND POLYMER RECYCLING

FRONTIER PROCESSING TECHNOLOGY

SYMPOSIUM HONORING PROF. MUSA KAMAL

MEMORIAL SESSION OF PROF. TOSHIRO MASUDA

PLENARY SESSION

O-07-031

COMPARISON OF REACTIVE AND NON-REACTIVE POSS TYPES ON THE MECHANICAL, THERMAL AND MORPHOLOGICAL PROPERTIES OF PLA AND PLASTICIZED PLA

Mehmet Kodal[1], **Hümeyra Şirin**[1] **and Güralp Özkoç**[1,*]

[1]Department of Chemical Engineering, Kocaeli University, 41380, Turkey

*mehmet.kodal@kocaeli.edu.tr; humeyra.sirin@gmail.com; *guralp.ozkoc@kocaeli.edu.tr*

Abstract: Two types of polyhedral oligomeric silsesquioxanes (POSS), aminopropylisobutyl-POSS (A-POSS) as the reactive one and the octaisobutyl-POSS (O-POSS) as the non-reactive one (1-10 wt% filler loadings) were compounded with poly(lactic acid) (PLA) and poly(ethylene glycol) (PEG) plasticized PLA (PLA/PEG) composites with a twin a screw extruder. PEG amount was kept constant at 10 wt% in PLA matrix. The effects of POSS type and loading on the mechanical, thermal and morphological properties for the POSS filled PLA and PLA/PEG composites were investigated. The mechanical test results showed that in most cases, the e-modulus and yield strength of the PLA/PEG/POSS and PLA/POSS composites were lower than pure PLA. However, 1 wt% A-POSS filled PLA composites have higher yield strength values compared to neat PLA. Moreover, a significant improvement in Izod impact strength of plasticized and unplasticized of PLA composites was obtained by incorporation POSSs particles (regardless of POSS type) into PLA. SEM analysis showed that POSSs dispersed homogeneously in polymer matrix at all loadings. Melt viscosity of neat PLA decreased with using POSSs. Differential scanning calorimetry (DSC) analysis revealed that POSS particles acted as a nucleating agent for PLA/PEG composites.

Keywords: PLA, POSS, plasticization, composite, mechanical properties

Introduction

Plastic consumption is expected to be approximately 365 million tons worldwide in 2015 [1]. This increasing rate of plastic usage leads to a serious solid plastic waste problem. Poly(lactic acid) (PLA), obtained from renewable resources, is a linear, semi-crystalline and biodegradable aliphatic polyester that offers a potential solution for the management of plastic waste pollution. PLA is a brittle material and generally needs to be plasticized to be used in apllication where toughness is a must, for example in flexible packaging. Poly(ethylene glycol) (PEG) and citrate esters are widely used in plasticization of PLA as the environmental plasticizers [2, 3]. However, a drawback of plasticization is the sharp reduction of stiffness of the material. Reinforcement with nanoparticles is one of the newly used approach to enhance the stiffnes/toughness ratio. Polyhedral oligomeric silsesquioxanes (POSS) nanocages are alternative candidates for reinforcement of polymers [4].

In the current study, for the first time in the literature, it was aimed to investigate and compare the mechanical, thermal and morphological properties of PEG plasticized and unplasticized PLA compounded with reactive and non-reactive POSS types, i.e. aminopropylisobutyl-POSS (A-POSS) as the reactive one and the octaisobutyl-POSS (O-POSS) as the non-reactive one. The POSS content was varied as 1, 3, 7 and 10 wt %. The composites were processed in molten state to mimic the industrial methods. The mechanical, thermal, morphological properties and flow properties of composites were investigated.

Experimental

Materials and Sample Preparation

Poly(lactic acid) (PLA) in pellet form with a melt flow index of 10-30 g/10 minute at 190 °C, 2.16 kg was used for the matrix material . It was purchased from the NaturePlast Company. Poly (ethylene glycol) (PEG) with the molecular weight of 8000 g/mol used as plasticizer was supplied from Sigma-Aldrich. PEG content was 10 wt% in all composites. PLA and PLA/PEG composites were filled with two types of POSS; aminopropylisobutyl-POSS (A-POSS) as the reactive one and the octaisobutyl-POSS (O-POSS) as the non-reactive one. POSSs were supplied from Hybrid Plastic. The POSS content was varied as 1, 3, 7 and 10 wt %. Table 1 shows the chemical structure of POSSs used in this study.

A twin screw microcompounder (15 ml, DSM Xplore model) was used to prepare PLA based composites; PLA/PEG, PLA/PEG/A-POSS, PLA/PEG/O-POSS,

PLA/A-POSS and PLA/O-POSS. The barrel temperature of microcompounder was set to 180 °C. The compounds were blended at a screw speed of 100 rpm and mixing time of 3 min. At the end of this period, the extrudate was taken by changing the position of the valve to guide the polymer through the die. Extruded composites were injection molded on a DSM Xplore 10 mL injection molding machine at 180 °C and holding pressure of 10 bar. The samples were abbreviated with respect to the type and loading level of POSS, for instance A-POSS1 represents the composite including 1 wt % aminopropylisobutyl-POSS.

Characterization

Melt viscosity of the composites was determined from vertical force (VF) data recorded in every minute after feeding. The tensile properties were determined using an Instron Universal Testing machine following the procedure ISO 527. The crosshead speed of 25 mm/min was used. The Izod impact strength was measured with a Zwick/Roell test instrument according to ISO 180. Morphological studies of fracture surfaces of impact samples were carried out using a scanning electron microscope (SEM, JEOL JSM 6060). Differential scanning calorimetry analysis (DSC) were run using a Mettler Toledo DSC-1 Star Model instrument at the temperature range of 25°C to 190°C with the heating rate of 20 °C/min under nitrogen flow. The percent crystallinity (%X_C) and the nucleation rate of the samples were calculated according to Eq. 1 and Eq. 2, respectively:

)x100 (1)

where ΔH_m is melting enthalpy (J/g), ΔH_c is cold crystallization enthalpy (J/g), ΔH_m^* is heat of fusion for PLA (93 J/g) [5] and φ is the fraction of the PLA in the composite,

$$Nucleation\ rate = T_{cc} - T_{cc,onset}$$
(2)

where $T_{cc,onset}$ is the crystallization onset temperature and T_{cc} is the crystallization peak temperature [6].

Table1. Chemical structure of POSSs used in this study

POSS Type	Chemical Structure
Aminopropyl isobutyl-POSS (A-POSS)	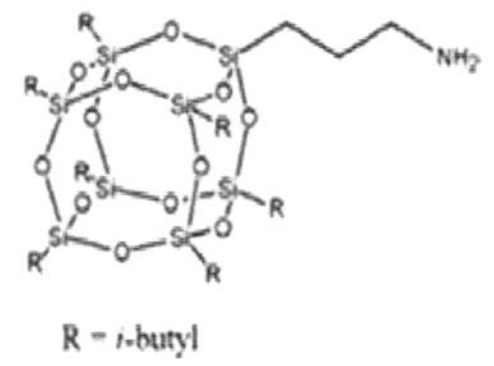
Octaisobutyl-POSS (O-POSS)	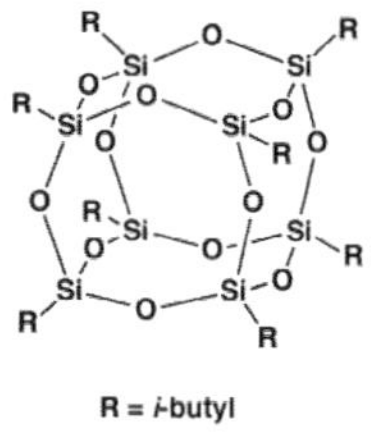

Result and discussions

Vertical force measurements

The incorporation of PEG into PLA resulted in decrease in vertical force (VF), in other words melt viscosity of pure PLA. Further reduction in VF values was observed by adding POSS to PLA/PEG composites. It was obvious that the VF of PLA/PEG/A-POSS and PLA/PEG/O-POSS composites decreased with increasing POSS loading (regardless of POSS type) from 1 to 10 wt% as seen in Fig. 1 (a) and (b). This behavior can be explained by the lubricating effect of POSS particles in PLA matrix. Reduction in melt viscosity in the presence of POSS plays an important role in power consumption during melt processing. However, the VF of PLA/A-POSS and PLA/O-POSS composites increased with particle loading to 1 wt% and 7 and 10 wt%, respectively. This can be attributed to the possible covalent bonding between PLA and POSSs [Figure 1(c) and (d)].

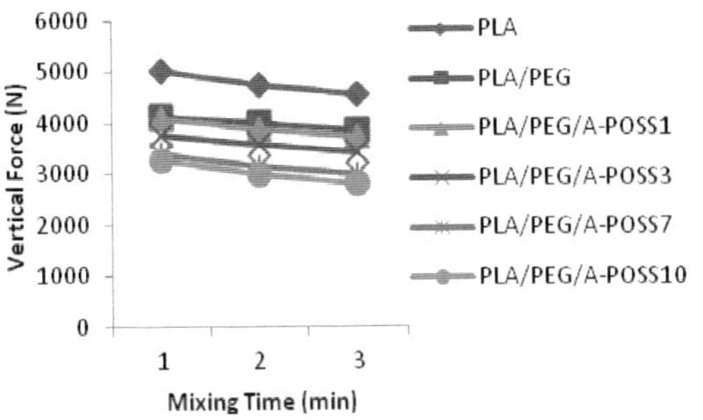

Figure 1(a): The variation of vertical force (N) of neat PLA, PLA/PEG and PLA/PEG/A-POSS composites.

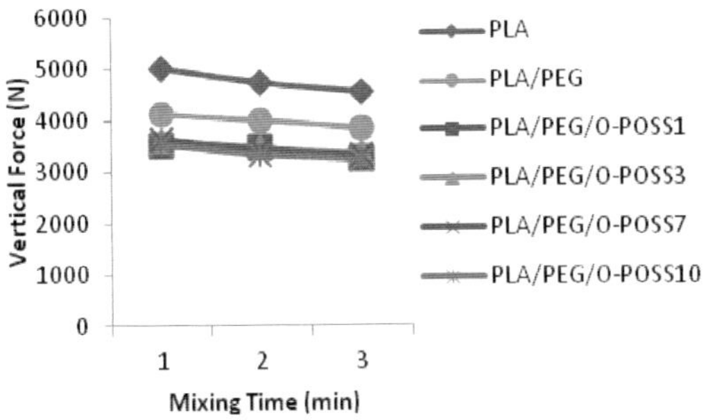

Figure 1(b): The variation of vertical force (N) of neat PLA, PLA/PEG and PLA/PEG/O-POSS composites.

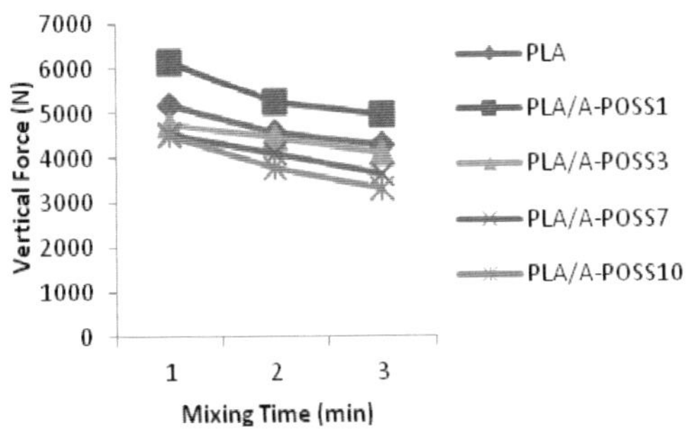

Figure 1(c): The variation of vertical force (N) of neat PLA and PLA/A-POSS composites.

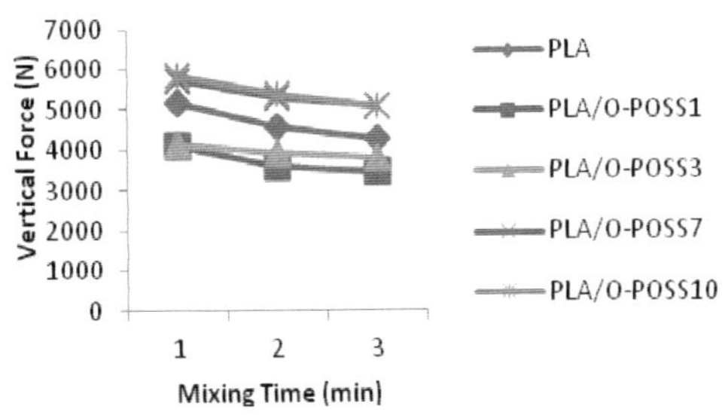

Figure 1(d): The variation of vertical force (N) of neat PLA and PLA/O-POSS composites.

Tensile and impact test results

As seen in Table 2, tensile test results showed a decrease in yield strength and E-modulus values of neat PLA by the addition of PEG. For PLA/PEG/POSS composites, there is a slight decrease in yield and E-modulus with increasing POSS loadings. However, when 1wt% A-POSS and 1 wt% O-POSS were added, this reduction in E-modulus of PLA/PEG composite was recovered app. 10% and 5%, respectively. Besides, POSS filled PLA/PEG composites have higher elongation at break (EAB) values compared to neat PLA and PLA/PEG composite. Moreover, significant improvements in Izod impact resistance of plasticized and unplasticized PLA composites were obtained by adding POSS particles. The similar trends in mechanical properties were obtained for PLA/POSS composites (Table 2). The mechanical analysis showed that enhancement in stiffness/toughness ratio of PLA was achieved by incorporating POSS types into PLA.

Table 2: Mechanical test results of PLA and PLA/PEG and PLA/PEG/POSS and PLA/POSS composites

Materials	Yield Strength (MPa)	Modulus (MPa)	Elongation at break (%)	Izod impact strength (kJ/m^2)
PLA	73.7	1203	9.08	2.53
PLA/PEG	67.5	1105	7.68	3.87
PLA/PEG/A-POSS1	61.2	1191	18.45	3.95
PLA/PEG/A-POSS3	56.6	1161	21.81	4.17
PLA/PEG/O-POSS1	66.5	1159	18.70	3.77
PLA/PEG/O-POSS3	64.5	1114	19.67	4.16
PLA/A-POSS1	77.5	1154	10.00	2.74
PLA/A-POSS3	71.2	1147	15.10	2.43
PLA/O-POSS1	67.3	1116	13.14	3.23
PLA/O-POSS3	62.8	1082	16.87	3.66

Morphological analysis

SEM analysis showed that there is not any phase separation between PLA and PEG (Fig. 2). Moreover, A-POSS and O-POSS particles were sub-micron size at 3 wt% loading and homogeneously dispersed in PLA matrix (Fig. 3 and Fig. 4).

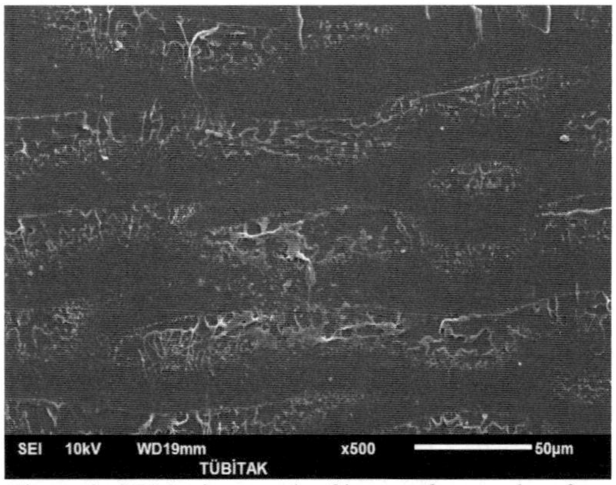

Figure 2: SEM micrograph of impact fractured surface of PLA/PEG composites at X 500 magnification.

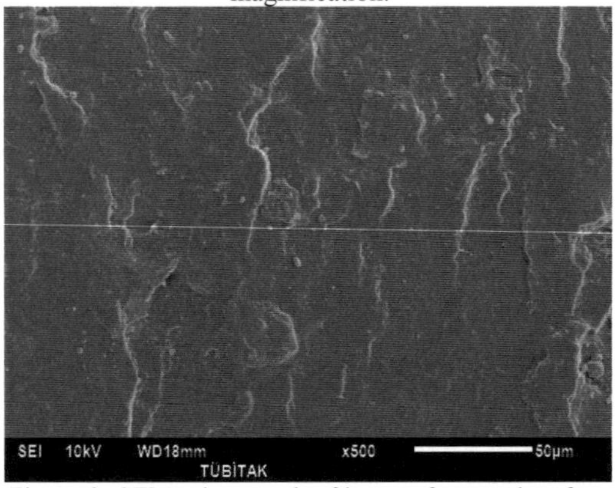

Figure 3: SEM micrograph of impact fractured surface of PLA/PEG/A-POSS3 composites at X 500 magnification

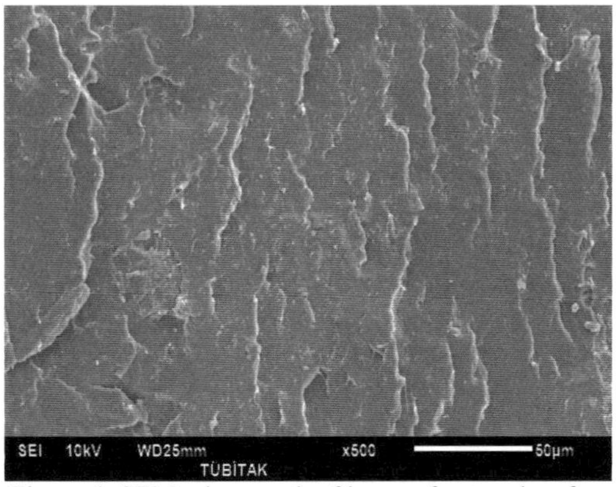

Figure 4: SEM micrograph of impact fractured surface of PLA/O-POSS3 composites at X 500 magnification

Thermal analysis

DSC test results revealed that glass transition temperature (T_g) of PLA and PLA/PEG composites decreased with adding POSS particles. Beside, POSS particles accelerated the nucleation rate of the pure PLA. Percent crysallinity of POSSs filled PLA and PLA/PEG composites were higher than neat one as seen in Table 3.

Table 3: DSC Results of PLA, PLA/PEG, PLA/PEG/POSS and PLA/POSS composites

Materials	T_g	T_{cc}	$T_{cc,onset}$	T_{cc}-$T_{cc,onset}$	T_m	X_c (%)
PLA	52.9	114.7	103.3	11.3	151.1	2.3
PLA/PEG	50.1	101.0	92.5	8.5	151.6	6.8
PLA/PEG /A-POSS3	48.4	100.8	92.7	8.1	152.6	5.2
PLA/PEG /A-POSS10	44.8	97.67	89.47	8.20	152.6	14.3
PLA/PEG /O-POSS3	41.6	96.51	89.97	6.54	151.5	4.75
PLA/PEG /O-POSS10	38.4	95.7	87.7	8.0	151.8	5.44

Conclusions

The SEM analysis showed that both types of POSS were well dispersed in the PLA. Addition of both POSS types decreased the melt viscosity of PLA. Mechanical tests indicated an increase in yield strength of 1 wt% of A-POSS/PLA. The use of both types of POSS remarkably improved the impact strength of PLA. Besides, elongation at break and energy at break values of PLA increased by using both types of POSS. DSC analyses showed that glass transition temperature and cold crystallization temperature of PLA decreased by addition of POSS. In addition, both POSS types acted as a nucleating agent by accelerating the nucleating rate of PLA.

Acknowledgement

The authors thank to the Scientific and Technological Research Council of Turkey for supporting this study with contract grant number of 111M514.

References

1. Pardos Marketing in Rapra International Conference, Brussels Belgium, 2005.
2. Y. Hu; Y.S. Hu; V. Topolkaraev; A. Hiltner; E. Baer; *Polymer*. 2003, *44*, 5681-89.
3. N. Ljungberg; B. Wesslen; *Polymer*, 2003, *44*, 7679-88.
4. J.H. Lee; G.Y. Jeong; *J Appl Poly Sci*, 2010,*115*, 1039.
5. C.D. Migliaresi; D. Cohn; A. De Lollis; L. Fambri; *J Appl Poly Sci*, 1991, *43*, 83.

6. J. Zeng; S. Kumar; S. Iyer; D.A. Schiraldi; R.I. Gonzales; *High Performance Polymers*, 2005, *17*, 403-24.

O-07-388

PMMA NANOCOMPOSITES BASED ON 2nd GENERATION LAYERED SILICATES

B. Fischer[1*], M. Ziadeh[2], J. Breu[2] and V. Altstädt[1]

[1] Department of Polymer Engineering, University of Bayreuth, Universitätsstraße 30, 95447 Bayreuth, Germany – bianca.fischer@uni-bayreuth.de

[2] Department of Inorganic Chemistry I, University of Bayreuth, Germany – josef.breu@uni-bayreuth.de

Abstract - Poly(methyl methacrylate) (PMMA) nanocomposites based on a synthetic shear-stiff, mica-like, large aspect-ratio K-fluorohectorite clay (K-hect) were prepared by melt-compounding. This new synthetic type of clay offers very huge lateral dimensions and therefore high aspect ratios above 600, whereas a tailored surface modification of the clay yields a good compatibility to the PMMA matrix. The influence of different clay loadings (0 – 4 wt.-%) on the mechanical behaviour of PMMA/clay nanocomposites was studied and compared to natural montmorillonite (MMT) clay, which has an aspect ratio of ≈50. It was found, that the use of these novel nanoplatelets leads to a high dispersion quality in PMMA and to a significant increase in fracture toughness of about 70 % in comparison to neat PMMA, without sacrificing tensile strength. The incorporation of natural MMT increases the stiffness but has an dentrimental effect on the tensile strength and the fracture toughness. The presence of the novel nanofiller varies the local stress state in the matrix and promotes additional energy dissipating mechanisms like crack deflection, crack pinning as well as debonding effects with platelets pull-out leading to enhanced fracture toughness.

Keywords: PMMA/clay nanocomposites, large aspect ratio filler, mechanical properties

Introduction

Since the first studies performed by researchers from Toyota in 1985, the rapid development of polymer/clay nanocomposites brings continuous improvement of the overall properties of such materials along [1;2]. Besides the mechanical properties of the filler itself, the key issues that significantly influence the resulting properties are the aspect ratio and the quality of the surface modification of the nanofiller. Commercially available organically modified clays have the drawback of limited aspect ratios < 100 and a high heterogeneity of surface charge. Furthermore, during melt compounding the delamination of tactoids is hardly ever complete, which reduces the maximal possible aspect ratios and therefore the desired properties. So it is an important step to use synthetic clay, which allows high dispersion quality in the polymer matrix together with much higher aspect ratios as compared to natural clay.

In this study, a novel synthetic modified, mica-like, shear-stiff K-hect with an aspect ratio of ~ 600 is used and melt-compounded into PMMA. In order to improve the dispersion of the nanofiller a strategy was applied with no need for exfoliation and intercalation. In addition to the better dispersion, by achieving a high aspect ratio beside an increased E-modulus also an increased fracture toughness of PMMA is expected. The tailored surface modification is based on anchoring of polymer chains bearing multiple quaternized amino groups on the external basal planes of the synthetic K-hect, which yielded a synthetic organo-clay and assured a high quality of dispersion. A series of PMMA/clay nanocomposites was prepared by melt compounding in a discontinuous micro-compounder with various clay concentrations. For a comparison and highlighting the advantages of the novel synthetic clay, also PMMA nanocomposites based on natural clay, i. e. Bentone, were prepared. The microstructure and mechanical properties of the nanocomposites were investigated and the micromechanics of the observed deformation behaviour was analysed in detail.

Experimental

Materials

A commercial PMMA (Plexiglas POQ62 supplied by Evonik Industries) has been used as matrix material with a molecular weight (M_w) of 73 kg/mol and a polydispersity index of 1.63 (measured by gel permeation chromatography). The commercial clay Bentone 38 modified with alkyl-ammonium chains was supplied by Elementis Specialties, which has a CEC of 70 mequiv. / 100g. The starting material for the preparation of the novel nanofiller is a synthetic Na-fluorohectorite (Na-hect) with the chemical formula $Na_{0.5}[Mg_{2.5}Li_{0.5}]Si_4O_{10}F_2$ and a CEC of 110 mequiv. / 100g [3]. The Na-hect was suspended in deionized water and then processed in a stirred media mill in order to exfoliate the tactoid stacks by applying shear forces [4]. This was done using a highly hydrated 'shear-labile' state, represented by the presence of Mg^{2+} cations (Mg-hect) in the interlayer space. Subsequent to the exfoliation process the clay samples were exchanged with K^+ cations which transformed the clay to a non-swollen 'shear-stiff', mica-like state with no intracrystalline reactivity [5]. Due to the inertness of the interlayer space, this mica-like material allows for a selective cation exchange limited to the external basal planes of the clay. Consequently, the hydrophilic nanoplatelets can be converted to hydrophobic fillers compatible with the organic matrix by applying polycations in the cation exchange. The surface modifier chosen by us was a polycationic copolymer, poly(2-(2-bromoisobutyryloxy)ethyl meth-acrylate)-stat-(2-dimethyl-aminoethyl methacrylate) (PBD). Moreover, the surface modification concomitantly triggered flocculation of K-hect. PBD ($1 gL^{-1}$) was quaternized and dissolved in water by adding equimolar amounts of concentrated acetic acid and this solution was added to the aqueous dispersions of K-hect obtained after milling until the clay was fully flocculated. Subsequently, the surface modified shear-stiff, mica-like O-hect was washed several times with deionized water and finally freeze-dried.

Sample preparation

The freeze-dried synthetic clay as well as the commercial clay were melt-compounded at various concentrations with PMMA in a discontinuous counter-rotating twin-screw micro-compounder (DSM Xplore, 15 mL microcompounder) at a temperature of 190 °C, a mixing speed of 210 rpm and a mixing time of 3 min. The material was added stepwise to the running microcompounder and during each cycle a batch of 7.3 g was processed. After extrusion, the extruded flakes were compression-moulded with a Hot-Plate-Press (190 °C, under vacuum) into CT (compact-tension) specimens to determine the critical stress intensity factor K_{Ic} or injection-moulded with a microinjector (DSM Xplore 12 ml injection moulding machine; melt temperature: 190 °C; mould temperature: 40 °C; injection pressure: 8 bar) into dumbbell specimens (75 mm x 5 mm x 2 mm) for tensile testing.

Characterisation

Monitoring of agglomerates was carried out using a Skyscan X-Ray Microtomograph 1072 with a maximum spot resolution of 5 μm per pixel. The measurements were done at a voltage of 80 kV and a current of 94 μA. The cylindrical specimens were 4 mm in diameter and height. The 3D-data is obtained by acquisitions every 0.23° with an exposure time of 1.5 s up to a complete rotation of 360°. 3-D volumetric images were digitally reconstructed by using the software "NRecon" from Skyscan.

A scanning electron microscope Leo 1530 Gemini (Zeiss) was used to image the fracture surfaces of the PMMA/clay nanocomposites to analyse the micromechanical reason for fracturing.

Tensile modulus, tensile strength and elongation at break were measured with a Universal Tensile Tester according to ISO 527 with a strain rate of 1 mm/min. For each material at least 5 samples were tested. The elongation was determined with a macro-displacement-transducer.

Fracture toughness tests were performed according to ISO 13586 to obtain the mode-I critical stress intensity factor (K_{Ic}) of the PMMA/clay nanocomposites. At least 5 notched, compact tension specimens were tested with a strain rate of 10 mm/min. The thickness of the specimens was 4 mm.

Results and Discussion

For investigation of the nanocomposite morphology, μ-CT measurements were performed in order to generate 3-dimensional representation of filler dispersion (Fig. 1). In case of the high aspect ratio clay, almost no agglomerates could be detected indicating good dispersion of clay platelets. However, in case of natural clay, a huge amount of agglomerates was observed.

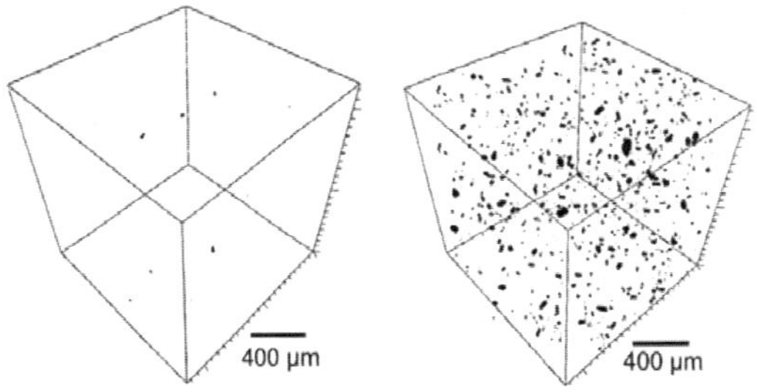

Figure 1. 3-dimensional reconstruction exhibiting dispersion quality of different nanofillers; left: synthetic nanofiller; right: natural nanofiller (μ-computer tomography).

By the incorporation of the novel clay, the tensile modulus as well as the fracture toughness of PMMA/clay nanocomposites shows a significant enhancement compared to neat PMMA. The tensile strength is not negatively influenced by the addition of synthetic clay. The incorporation of natural clay leads to an increase in tensile modulus but the values for tensile strength and fracture toughness are significantly reduced. This can be explained by the presence of large agglomerates due to inefficient compatibility of the filler, acting as stress concentrators.

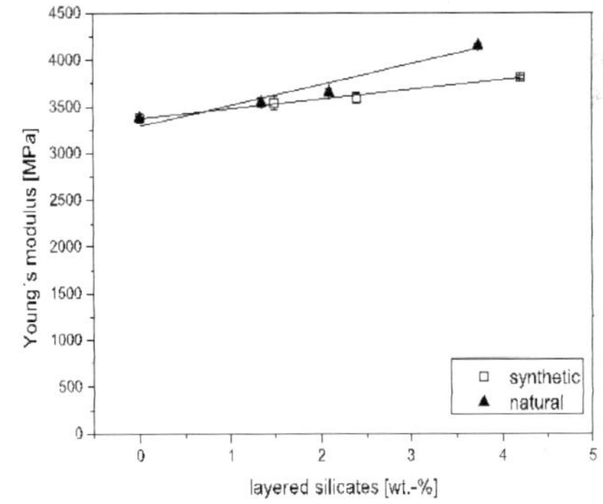

Figure 2. Effect of clay content on Young's Modulus of PMMA/clay nanocomposites for natural and for synthetic nanofiller.

In general, the addition of rigid nanofillers to a brittle matrix leads to an increase in modulus at the expense of strength, strain and toughness as embrittlement takes place [6;7]. This phenomenon is also observed in this study for PMMA nanocomposites based on natural clay. For synthetic clay nanocomposites an increase of

Young's modulus by 11 % could be achieved at 4 wt. % clay loading.

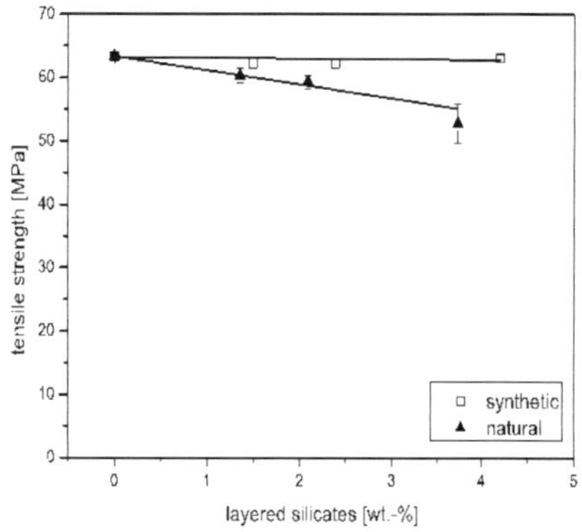

Figure 3. Effect of clay content on tensile strength of PMMA/clay nanocomposites for natural and for synthetic nanofiller.

Concerning the tensile strength, the addition of the novel clay platelets does not show a negative influence (Fig. 3). This phenomenon can be explained by the high dispersion quality of the nanofiller. In general, large agglomerates lead to additional local stress concentrations and consequently to lower tensile strengths in the case of brittle polymer matrices [8;9]. The use of natural clay shows this typical behaviour and leads to a decrease of tensile strength by 16 % at a clay amount of 4 wt.-% due to the presence of large agglomerates caused by inefficient dispersion.

An increase of the tensile strength is expected to take place in the case of strong interfacial adhesion between the nanofiller and the matrix in order to generate effective stress transfer from the matrix to the filler. In this study it can be assumed that the adhesion between the novel clay platelets and PMMA chains is relatively weak as no change in T_g could be detected by DSC measurements and therefore also no increase in tensile strength could be observed.

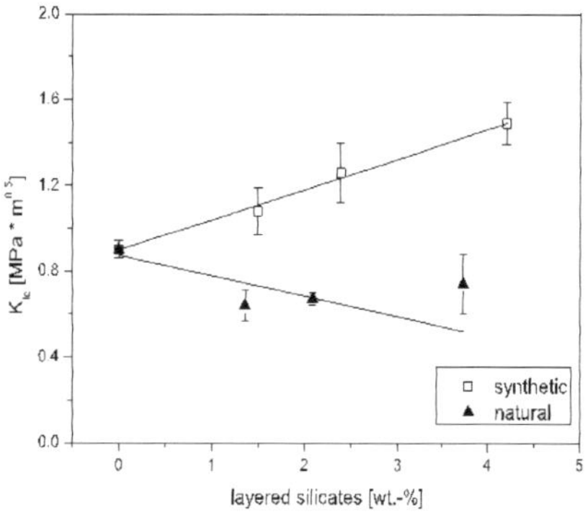

Figure 4. Effect of clay amount on stress intensity factor of PMMA/clay nanocomposites for natural and for synthetic clay.

According to literature, a decrease in fracture toughness as expressed by K_{Ic} could be expected due to the increasing modulus of PMMA with increasing clay content [6;10]. This behaviour can be clearly observed in case of the nanocomposite based on natural clay (Fig. 4). Surprisingly, a significant improvement in K_{Ic} fracture toughness was observed by the incorporation of synthetic clay into PMMA whereas an improvement in fracture toughness of about 66 % at a clay loading of 4 wt.-% was observed (Fig. 4).

These results demand consideration of the dominating toughening mechanisms. To analyse the occurring toughening mechanisms, the micro-morphology of the fractured surfaces from the K_{Ic} tests of neat PMMA and nanocomposites were investigated via scanning electron microscopy. The crack-propagation direction in all images is from left to right. Fig. 5 shows the micrographs of neat PMMA and PMMA nanocomposites taken from the centre of the fracture surface in the vicinity of the notched region.

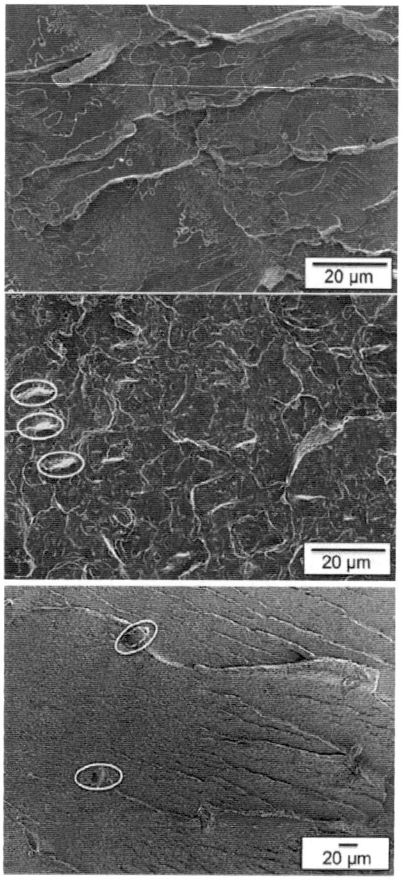

Figure 5. SEM micrographs of fracture surfaces of PMMA and PMMA/clay nanocomposites at a clay loading of 4 wt.-%. top: neat PMMA; middle: synthetic nanofiller; bottom: natural nanofiller; white cycles highlight selected agglomerates for natural clay and platelets for synthetic clay.

As can be seen neat PMMA possesses the typical smooth failure surface morphology formed by craze formation and growth. One can clearly observe that the fracture surfaces of the nanocomposites based on the novel kind of clay are rougher than those of neat PMMA, indicating the creation of additional surface area. In case of natural clay large agglomerates are present.

As the clay platelets are sticking out of the matrix, it reveals that during the fracture process, the layered silicates were debonded and pulled out of the PMMA matrix. The surface of the platelets is smooth, which indicates a weak adhesion between the nanofiller and the PMMA matrix which confirms the DSC results.

In the case of PMMA nanocomposites based on natural clay, the incorporation of the filler leads to a reduction in fracture toughness due to the large agglomerates caused by unsatisfying dispersion of the clay in the matrix. As a consequence of the extreme stress concentrations at these agglomerates, only a few but large crazes are initiated that break down into a microcrack without much energy dissipating.

For improving the fracture toughness of PMMA, it must be ensured that massive crazes are initiated over a huge volume of material ahead of the main crack tip and hence more energy can be dissipated before fracture occurs. In the novel PMMA/clay nanocomposites, the highly dispersed rigid layered silicates act as stress concentrators that initiate a large number of crazes as indicated by the rough surface structure (Fig. 5). This results in a high amount of localized energy dissipation. Other important fracture events, which were observed and contribute to energy dissipation in the nanocomposites are crack deflection and crack pinning. During crack deflection, the crack is forced to alter the initial propagation plane by striking platelets leading to an increased total fracture surface. Pinning the surfaces of the cracks by the platelets slows down the development of crazes into microcracks and a higher stress level is needed for further craze growing. Additional stress loading leads to debonding along the PMMA/clay interface due to weak adhesion and subsequent to the formation of microvoids and finally to platelets pull-out. The initiation of microvoids consumes additional energy as well as the extension of the microvoids and the pull-out of platelets. Debonding occurs, as the interfacial strength between PMMA and layered silicates is lower than the fracture strength of PMMA.

Conclusions

This study focused on the melt incorporation of tailored large aspect ratio, shear-stiff, mica-like nanoplatelets in PMMA and its impact on the mechanical behaviour. The prepared novel PMMA/clay nanocomposites offer improved toughness and stiffness over the neat PMMA without reduced strength. However, the incorporation of commercial natural clay leads to detrimental effects due to inefficient dispersion.

To the best of our knowledge, this is the first report on the preparation of PMMA/clay nanocomposites conducted by the industrial relevant melt compounding process that yielded a significantly improved toughness. Due to additional energy dissipating mechanisms an increase in fracture toughness of 66 % could be detected in case of the large aspect ratio, mica-like clay.

Acknowledgements

The authors thank Hussein Kalo from the Department of Inorganic Chemistry I for producing the synthetic clay and Stephan Weiß form the Department of Macromolecular Chemistry II for synthesising the macroinitiator.

Special thank goes to Florian Billing for his experimental contribution and Annelise Lang for the SEM pictures. The financial support from the German Research Foundation, in the frame of the Collaborative Research Centre (SFB) 840: "From particulate nanosystems to mesotechnology" is highly acknowledged.

References

1. K. Hedicke; H. Wittich; C. Mehler; F. Gruber; V. Altstädt *Composites Science and Technology* 2006, *66*, 571.
2. S. Lietz; J-L.Yang; E. Bosch; J.K.W. Sandler; Z. Zhang; V. Altstädt *Macromolecular Materials and Engineering* 2007, *292*, 23.
3. H. Kalo; M.W. Möller; M. Ziadeh; D. Dolejš; J. Breu *Applied Clay Science* 2010, *48*, 39.
4. M. Ziadeh; B. Chwalka; H. Kalo; R.M. Schütz; J. Breu J. *doi:10.1180/claymin.2012.047.3.05*
5. M.W. Möller; D. Hirsemann; F. Haarmann; J. Senker; J. Breu *Chemistry of Materials* 2009, *22*, 186.
6. J.H. Park; S.C. Jana *Polymer* 2003, *44*, 2091.
7. J. H. Liaw; T.Y. Hsueh; T-S. Tan; Y. Wang; S-M. Chiao *Polymer International* 2007, *56*, 1045.
8. S-Y. Fu; X-Q. Feng; B. Lauke B; Y-W. Mai *Composites Part B: Engineering* 2008, *39*, 933.
9. S. Tanoue; L.A. Utracki; A. Garcia-Rejon; J. Tatibouët; M.R. Kamal *Polymer Engineering & Science* 2005, 45, 827.
10. R.R. Tiwari; U. Natarajan *Polymer International* 2008, 57, 738.

O-07-392

POLY(METHACRYLIC ACID)/EGGSHELL HYDROGEL COMPOSITES: A NEW ABSORBENT FOR REMOVING OF WATER POLLUTANTS

Atefeh Darvishi[1], Hadi Bakhshi[2]*

[1] *Young Researchers Club, Karaj Branch, Islamic Azad University, Karaj, Iran – darvishi.chem@yahoo.com*
[2] *Iran Polymer and Petrochemical Institute, P.O. Box: 14965/115, Tehran, Iran – hadibakhshi@yahoo.com; h.bakhshi@ippi.a.ir*

Abstract - In the present study, hydrogel composites based on poly(methacrylic acid) and eggshell (ES) particles as a low-cost and waste material were prepared and their potential to remove dyes as water pollutants was evaluated. For this propose, hydrogel composites were prepared through in situ free radical polymerization of methacrylic acid monomer in the presence of different content of ES particles using methylene bisacrylamide as crosslinker and ammonium persulphate as initiator at 75 °C. The obtained hydrogel composites were characterized via FTIR spectroscopy and SEM. FTIR spectroscopy confirmed fully consumption of methacrylic acid monomer during radical polymerization (gel content > 99%) as well as combination of ES particles in the hydrogel composites. SEM images showed good distribution of ES particles in the poly(methacrylic acid) matrix. Moreover, incorporation of ES particles resulted in more porosity of the hydrogel matrix in composition to unfilled poly(methacrylic acid). The absorption of methylene blue and crystal violet as dyes onto hydrogel composites were studied. Although incorporation of ES particles into the hydrogels decreased the absorption capacity for MB, increasing ES particles content in the composites lead to higher absorption capacity. Nevertheless, addition of ES particles into the hydrogels increased the absorption capacity for CV, while lower absorption capacity was observed for composites with higher ES particles content.

Keywords: Poly(methacrylic acid), Eggshell, Composites, Water pollutant, Waste materials

Introduction

Waste water is one of prevalent environmental predicaments from the viewpoint of pollutant such as dyes and toxic ions. A wide variety of industries such as textile, paper, food and agricultural manufactories discharge dyes into sewage [1]. Therefore, different treatments including coagulation, adsorption, ozone treatment and so on have been employed to eliminate such hazardous materials. Polymeric adsorbents, on account of fundamental properties such as high adsorption capacity and mechanical stability, have been concidered by many researchers for treatment of waste water. However, utilization of these synthetic absorbents is limited due to their high cost.

Natural-based adsorbents due to its low cost, biodegradability and accessibility in extended amounts were indicated benefits in comparison to synthetic ones [2]. Among natural adsorbents, application of eggshell (ES) has been kept at the forefront in recent years. The porous structure of ES makes it an attractive material to employ as an adsorbent for removal of dyes [3].

Hence, present study focused on preparation of an effective sorbent to remove of dyes from aqueous solution. In this regard, some hydrogels based on poly(methacrylic acid)/eggshell composites were prepared and their potential for removing of model dyes from aqueous solution was investigated.

Experimental

Eggshells (ES) were collected from restaurant of Iran Polymer and Petrochemical Institute, washed with distilled water, air dried, ground into powder and sieved into different sizes. Hydrogel composites were prepared through in situ free radical polymerization of methacrylic acid (MAA, 6 g) monomer in the presence of different content of ES particles using methylene bisacrylamide (BMA, 1 g) as crosslinker and ammonium persulphate (APS, 0.06 g) as initiator at 75 °C. The formulation of prepared hydrogel composites is tabulated in Table 1.

The equilibrium water absorption (EWA, g/g) was calculated using the following equation:

$$EWA\ (g/g) = [(W_1 - W_0)/W_0] \quad (1)$$

, where W_0 and W_1 are the weights of the dried and swollen hydrogel composites, respectively.

The adsorption capacity of the prepared hydrogel composites for methylene blue (MB) and crystal violet (CV) dyes as model water pollutants was determined using a UV-visible spectrometry assay. The dye absorbency (Q_e, mg/g) was calculated using following equation:

$$Q_e(mg/g) = (C_0 - C_e) \times V/m \quad (2)$$

, where, C_0 and C_e are the concentrations of the dye in the solution (mg/L) before and after equilibrium absorption, respectively, V is the volume of the solution used for the adsorption (L) and m is the mass of the dried hydrogel composite (g).

Results and Discussion

The ES particles besides obtained hydrogel composites were characterized via FTIR spectroscopy and SEM.

FTIR spectroscopy confirmed fully consumption of MAA monomer during radical polymerization (gel content > 99%) as well as combination of ES particles in the hydrogel composites. According to SEM images (Fig. 1), ES particles had non-uniform and porous microstructure. furthermore, SEM images showed good distribution of ES particles in the PMAA matrix. Moreover, incorporation of ES particles in the PMMA matrix resulted in more porous morphology in composition to unfilled one (sample A1). It is supposed that these pores increase the regions of water contact and therefore enhance the absorption rate and capacity of water pollutants.

The water absorption values of the hydrogel composites were given in Table 2. Although hydrogels containing ES particles showed higher EWA values in comparison to pure PMMA hydrogel (Sample A1), increasing the ES particles content in the composites decreased their water absorbency. Also, samples based on smaller ES particles showed higher EWA values and expectedly.

The absorption rate and capacity of MB and CV dyes onto the hydrogel composites were studied (Table 2). Although incorporation of ES particles into the hydrogels decreased the absorption capacity for MB, increasing ES particles content in the composites lead to higher absorption capacity. On the contrary, addition of ES particles into the hydrogels increased the absorption capacity for CV, while lower absorption capacity was observed for composites with higher ES particles content.

Table 1. Formulation of the hydrogel composites.

Sample Code	MAA (g)	MBA (g)	APS (g)	ES (g)	ES particle size (μm)
A1	6.0	1.0	0.6	0	–
A2	6.0	1.0	0.6	2.8	<75
A3	6.0	1.0	0.6	2.8	75-150
A4	6.0	1.0	0.6	2.8	150-212
A5	6.0	1.0	0.6	1.4	<75
A6	6.0	1.0	0.6	4.2	<75

Table 2. Swelling and dye absorbency of the hydrogel composites.

Sample Code	EWA (g/g)	Q_e (mg/g)	
		MB	CV
A1	7.97	1.922	1.815
A2	13.62	1.586	1.904
A3	11.43	1.746	1.912
A4	12.60	1.873	1.883
A5	14.15	1.572	1.928
A6	10.13	1.639	1.865

Figure 1. SEM images of ES particles (mesh >200) (a, b), and samples A1 (c) and A2 (d).

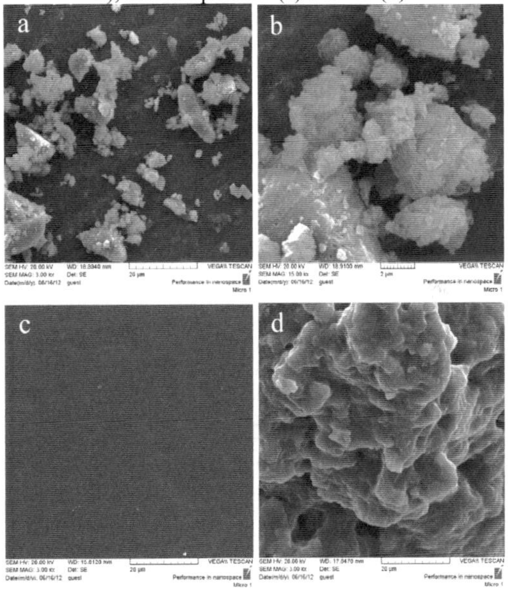

Conclusions

These new hydrogels based on poly(methacrylic acid)/eggshell composites showed good potency for removing of water pollutants from aqueous solution. Appropriate absorption rate and capacity, low cost and using waste raw material are some benefits of prepared hydrogel composites.

References (Times New Roman 10, bold)

1. B.L. Rivas and S. Villegas, *J. Appl. Polym. Sci.* 2004, *91*, 3679.
2. A. Gomez-Valdemoro; V. Calderon; N. San-Jose; F.C. Garcia; D.L.P. J and J. Garcia, *J. Polym. Sci. Part A: Polym. Chem.* 2009, *47*, 670.
3. M.F. Elkady; A.M. Ibrahim and M.M.A. El-Latif, *Desalination* 2011, *278*, 412.

Novel TPO nanocomposites based on PP/EPDM/Nanodiamond with well-balanced mechanical properties and excellent thermal resistivity

F. Hejazi[a], A. A. Katbab[a,*]

[a] *Department of Polymer Engineering and Color Technology, No. 424, Hafez Ave, Amirkabir University of Technology, Tehran, Iran, fhejazi@aut.ac.ir, Katbab@aut.ac.ir*

Abstract:

Nanoscale diamond is a valuable additive to be used in multifunctional polymer composites especially in applications where a combination of enhanced mechanical, thermal and tribological properties is required. The drastic improvement in stiffness and strength even with the addition of small percentage of nanodiamond to a polypropylene matrix can be obtained but it is traded-off with significant reductions in fracture toughness. It is believed that the presence of this stiff cage-like nano-filler will restrict the mobility of the surrounding matrix chains, and thus limit its ability to undergo plastic deformation, thereby decreasing their fracture toughness. Conversely, the incorporation of a rubber phase increased the toughness at the expense of strength and stiffness. Therefore, this has led to the idea of preparing ternary nanocomposites by adding a soft elastomeric dispersed phase to polymer/diamond systems to obtain well-balanced mechanical properties. In this work graphitized nanodiamond with the cage-like structure employed to prepare thermoplastic olefin (TPO) nanocomposites based on polypropylene (PP) and ethylene-propylene-diene rubber (EPDM) and maleated polypropylene as compatibilizer. Incorporation of diamond nanoparticles, in addition to peculiarly enhancing the strength and stiffness, affected the dispersion and distribution of the rubber droplets in PP continuous phase which leads to significant reduction in size of EPDM droplets, leading to the increase in impact strength. The tensile properties and impact strength of these toughened nanocomposites are discussed in terms of the ND and rubber contents and morphology. In addition to enormous improvement in mechanical properties, effect of the presence on nanodiamond particles on Thermal properties of these TPO composites were also studied and results revealed that all interfacially compatibilized nanodiamond composites showed enhanced heat distortion temperature (HDT) and heat resistivity. Thermogravimetric analysis showed that incorporation of only 3% nanodiamond powder into the TPO blend, increase its thermal decomposition temperature by about 50°C.

Keywords: Nanodiamond, TPO, Morphology, Impact resistivity, Thermal stability

INTRODUCTION

Polymer-based filled with particles composite materials are widely used in the modern industry and technologies, including nanotechnologies, thanks to their light processing into final products. Thermoplastic elastomer (TPE) blends based on polypropylene (PP) and EPDM rubber, have been used in a wide range of applications such as automotive industry, where enhanced mechanical properties, thermal stability, scratch and wear resistance are essential. In TPE materials, the thermoplastic phase forms the main matrix, and the elastomer is dispersed in the form of aggregated droplets. The dispersed elastomer phase serves to improve and control the toughness and low-temperature impact resistance. Among other species, the promising components for the polymer nanocomposites are nanodiamond prepared by the detonation synthesis (ND). ND, also known as ultradispersed diamond, combines unique properties of diamond core such as superior hardness and thermal conductivity with large readily modifiable surface. ND can be used as effective modifying agents providing improved mechanical, thermal and Tribological characteristics of the filled polymer systems.

The aim of this work is to apply diamond nanoparticles as an additive for the PP/EPDM thermoplastic elastomer and study the effect of the presence of this nanopowder on morphological, mechanical and thermal properties of the polymer matrix.

Experimental

The polypropylene used in this work was isotactic and injection grade with commercial name of EP2X 83CI (Arak, Iran). The EPDM rubber with trade name of KEP 330 having 7.9 wt % ENB as diene monomer and mooney viscosity of ML(1-4)100 =24 produced by KELTAN Company (Korea) was used as the dispersed rubber phase. Maleic anhydride grafted polypropylene with melt flow index (MFI) of 40 and maleation percent of 0.5-0.8 manufactured by Pluss Polymer Company (India) with the commercial name of OPTIM P406 was employed as compatibilizer. Ultra dispersed graphitized diamond supplied by ALIT Inc. (Ukraine) was employed as nanofiller.

Polypropylene and EPDM with the ratio of 70 to 30 (w/w %) respectively, were melt blended with different percentage of nanodiamond (3 and 5% wt) in a Brabender Plasticator PL 2200 model internal mixer. Blending process was carried out at 175 and mixing speed of 60 and 120 rpm for 10 min. Samples were then injection moulded into proper shapes for mechanical testing using an ES 15060 model Injection machine (Australia). Thin sheets with the thickness of 3 mm were also prepared via compression moulding method by means of a hydraulic press at 200 °C.[1]

Results and Discussion

Morphology

The morphologies of the TPO composite and ND/TPO nanocomposite are shown in figure 1. A comparison between these SEM images is interesting. As can be seen, incorporation of 5% wt nanodiamond significantly reduced the size of EPDM droplets from 1.7 μm to less than 0.5 μm. With increase in EPDM content of nanocomposite from 30% wt to 35% wt, the interparticle distance of this dispersed rubber phase also reduced.

Impact strength

Results of impact test are summarized in table 1. In samples containing PP:EPDM with the ratio of 70:30, despite very small size of EPDM droplets (according to SEM images), because of the big interparticle distance of the dispersed rubber phase, toughening mechanism didn't occur. In nanocomposite with the ratio of 65:35 (for PP:EPDM), despite the fact that interparticle distance of EPDM droplets is also in the required range (0.3 μm) for obtaining a super tough sample[2] impact strength didn't changed. This result is attributed to antiplastisization mechanism of nanodiamond physical networks that leads to stress concentration around filler clusters and so, restriction the occurrence of toughening mechanism of EPDM phase. To strengthen the polymer-nanoparticle interface, higher content of compatibilizer was used (PP-g-MAH to Nanodiamond ratio of 3:1). In this case higher impact strength was observed. It can be conclude that by increasing the interactions in the system, higher impact strength can be achieved.

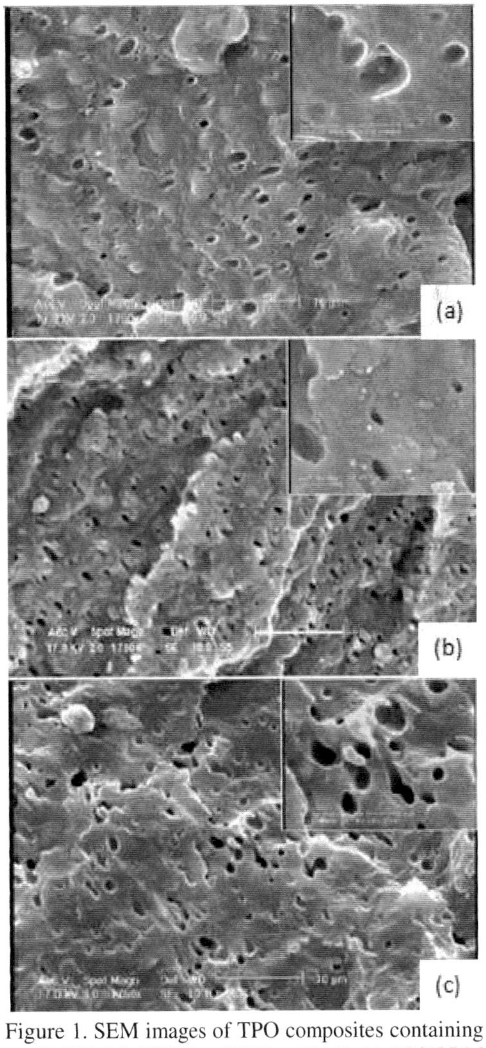

Figure 1. SEM images of TPO composites containing a) PP/EPDM (70/30) +5%PP-g-MAH, b) PP/EPDM (70/30)+ 5% PP-g-MAH+5%ND and c) PP/EPDM(65/35)+5% PP-g-MAH+5%ND

Table 1. Summary of the impact strength of TPO/Nanodiamond composites

sample	Energy (Kgf/cm)
PP/EPDM(70/30)+5% PP-g-MAH	21.7
PP/EPDM(70/30)+5% ND+5% PP-g-MAH	20.25
PP/EPDM(65/35)+5% PP-g-MAH	27.9
PP/EPDM(65/35)+5% ND+5% PP-g-MAH	20.75
PP/EPDM(65/35)+15% PP-g-MAH	28.2
PP/EPDM(65/35)+15% ND+5% PP-g-MAH	29.85

Stress–Strain Properties

The representative stress–strain plots are shown in Figure 2. As can be seen in the this plot, the addition of 5% nanodiamond to TPO composite, results in the increase in Tensile Modulus, Tensile Strength and Fracture stress.

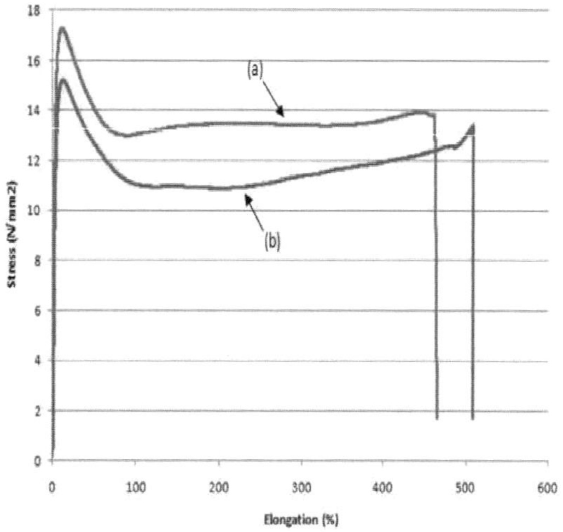

Figure 2. Representative stress–strain plots of (a) PP/EPDM(70/30)+5%ND+5% PP-g-MAH, (b) PP/EPDM+5% PP-g-MAH

HDT Results

As thermal stability of polymer composites is one of the important factors in their applications, Heat Distortion Temperature test applied on TPO composites and the effect of the presence of ND in this matrix verified. As can be seen in table 2, Heat Distortion Temperature of TPO composite increases with incorporation of nanodiamond particles. This observation is also because of the cage-like structure of nanodiamond particles that create physical network in the system and inhibit the mobility of entrant chains.

Thermal Stability of the composites

The TGA thermograms of the various diamond nanocomposites are compared with those of without nano. As summarized in table 3, thermal decomposition temperatures and the residual yields of TPO/ND nanocomposites increased compare to bare TPO. In addition, these factors increased with increasing ND content, implying that thermal decomposition of the TPO/ND was retarded by incorporating ND into the TPO matrix with higher residual yield. This incident can be explaned according to the thermal shielding role of nanodiamond in the polymer matrix. In the other word, nanodiamond, because of its high thermal stability, bear a large portion of the heat, so, increase the degradation temperature of the whole sample.

Table 2. Table 3. Results of HDT analysis

sample	Heat Distortion Temperature (°C)
PP/EPDM+5% PP-g-MAH	54.7
PP/EPDM+5% ND+5% PP-g-MAH	61

Table 3. Thermal gravimetric analysis Results

sample	Thermal decomposition temperature (°C)	Residual yield (%)
PP/EPDM+5%PP-g-MAH	343.53	0.37
PP/EPDM +15%PP-g-MAH	358.73	0.40
PP/EPDM+3%ND+3%PP-g-MAH	383.99	3.2
PP/EPDM+5%ND+5%PP-g-MAH	388.61	4.68
PP/EPDM+10%ND+10%PP-g-MAH	388.71	8.6
PP/EPDM+5% ND+15%PP-g-MAH	389	4.7

Conclusion

Presence of diamond nanoparticles in the PP/EPDM TPE matrix has a significant effect on dispersion and distribution of rubber droplets in thermoplastic phase. An acceptable improvement in impact strength can be obtained by the incorporation of nanodiamond powder into thermoplastic elastomer phase that can be the idea of achieving a supper though material by strengthening the interfacial surface of the nanoparticles and polymer matrix. Additionally, this nanocomposite shows good mechanical properties as well as peculiar thermal resistance. Higher heat distortion temperature and thermal decomposition temperature are other souvenirs of nanodiamond for PP/EPDM blend.

References

1. Fateme Hejazi, Ali Asghar Katbab, "Nanodiamond based PP/EPDM thermoplastic Elastomer (TPE) composites: Microstructure, Tribo-dynamic and thermal properties", Journal of Applied Polymer Science, Vol. 125, 1942–1950 (2012)
2. Wu S. Polymer 1985;26:1855.

O-07-418

CHARACTERIZATION OF SOY PROTEIN ISOLATE/CARBOXYMETHYL CELLULOSE/ORGANOCLAY BIOCOMPOSITE FILM

R. Suriyatem[1], A. Somwangthanaroj[2] and P. Rachtanapunc[3,4*]
[1]*Div. Food Science and Techology, Chiang Mai University, Thailand – suriyatem.k@gmail.com*
[2]*Dep. Chemical Engineering, Chulalongkorn University, Thailand – Anongnat.S@chula.ac.th*
[3]*Div. Packaging Technology, Chiang Mai University, Thailand – p.rachta@chiangmai.ac.th*
[4]*Materials Science Research Center, Chiang Mai University, Thailand*

Abstract - Soy protein isolate/carboxymethyl cellulose/organoclay (SPI/CMC/clay) composite films were prepared. The effect of clay contents were used in a film forming solution on chemical structure, crystallinity, sorption characteristic, and water vapor permeability (WVP) of SPI/CMC/clay composite films was evaluated. Chemical structure was confirmed by Fourier Transform Infrared spectra. The XRD pattern of each SPI/CMC/clay showed a shift to lower angles of the diffraction peak. Sorption isotherm of SPI/CMC/clay films provided the characteristic of sigmoid-shaped type II. The WVP and a percent elongation at break (EB) of composite film decreased with an increase of organoclay content (0-6g clay/100g total solid). However, they came to increase when organoclay was introduced over 6g/100g total solid. The trend of tensile strength of the composite film was contrast to EB value.

Keywords: Biopolymer, carboxymethyl cellulose, organoclay, biocomposite film, soy protein

Introduction

Most synthetic polymers are produced from petrochemicals and harmful to environment. The growth of nature concerns over non-biodegradable petrochemical based plastic packaging materials has raised interest. Soy protein is one of the most promising materials in renewable source-based biopolymer because it has low cost and is normally biodegradable [1-3]. Protein film has better gas and oil barrier than some synthetic films. However, brittleness and low water vapor resistant of the pure protein films are normally insufficient for food packaging applications. One option to improve the properties and also save the costs is an incorporation of other materials into the polymer matrix.

Carboxymethyl cellulose (CMC) is cellulose ether which is water-soluble, biodegradable and non-toxic. The CMC has been used in polymer blending for improving the properties of bio-based film [1, 4]. Inorganic clay, which is commonly used, is montmorillonite. Montmorillonite which has alumino-silicate structure consists of two tetrahedral layers sandwiching an octahedral layer and Na^+ or Ca^{2+} ions residing in the interlayer. Polymer/clay composites reveal better mechanical, thermal, barrier and other physicochemical properties than neat polymer [5-7].

There are many publications on bio-based composites [1, 4-7] but no work studies on properties of SPI/CMC incorporated with organoclay. This research investigated the effect of clay content (0-15g/100g of SPI/CMC) on water vapor permeability (WVP), sorption isotherm, crystallinity, chemical structure, morphology and mechanical properties of SPI/CMC/clay biocomposite films.

Experimental

Film preparation: Film forming solutions (5% w/v SPI and CMC) with different organoclay contents were prepared to yield different SPI/CMC/clay composite films (Table 1). A mixture of SPI and CMC solution was prepared according to Rachtanapun and Suriyatem [1] by using glycerol as a plasticizer. The different amounts of organoclay were added into the mixture and then homogenized at 6500 rpm for 30 min. The mixture was degasified, cast on silicone plate, and dried at 25°C for 36 h.

Fourier transform infrared (FTIR) spectroscopy: Infrared (IR) spectra of SPI/CMC composite films were run on Fourier-Transformed Infrared Spectrophotometer (Bruker Equinox 55) with DTGS detector and Omnic software. Transmission was measured at the wave number range of 4000–400 cm^{-1}.

Crystallinity: X-ray diffraction patterns of SPI/CMC composite films were recorded with an X-ray diffractometer (X'Pert MPD, Philips, Germany). The samples were respectively scanned at the angle (2θ) of 5 to 35° with a scanning rate of 2° min^{-1}.

Water vapor permeability (WVP): The WVP of the blended films was investigated according to ASTM E96 as previously described by Rachtanapun and Suriyatem [1]. The test was carried out at 25°C for 7 days. All measurements were performed in triplicate.

Moisture sorption isotherm. A film specimen (size 2mm x 2mm) was dried and placed in desiccators containing saturated salt solutions (LiCl, $MgCl_2$, $Mg(NO_3)_2$, $NaNO_3$, NaCl and KCl) to create the specific RH (11, 34, 55, 65, 75 and 86%) at 25°C. The film was weighed every 24 h until an equilibrium was achieved. The equilibrium moisture content (EMC) was calculated [1].

Mechanical properties: The examined mechanical properties (Tensile strength (TS) and elongation at break (EB)) were determined using an Instron

Universal Testing Machine Model 1000 (H1K-S, UK) with the procedure according to ASTM D882-80a. The initial gauge length and cross-head speed was set at 50 mm and 20 mm/min, respectively. Ten specimens were tested for each sample.

All data was analyzed by one-way analysis of variance (ANOVA) and Tukey's multiple range test ($p \leq 0.05$) using statistical program SPSS version 11.

Results and Discussion

The SPI/CMC/clay-0 composite film (without clay) was translucent and very light yellow (Fig. 1). The other SPI/CMC/clay composite films showed similar appearance with that of SPI/CMC/clay-0 film (data not shown).

FTIR spectroscopy analysis: The FTIR spectroscopy analysis was performed to better understand the influences of organoclay filler on the structure of the SPI/CMC/clay composites film. Each spectra of SPI/CMC/clay composotes gave the same characteristic FT-IR spectrum, the spectrum had characteristic of both SPI and CMC film (Fig. 2). They displayed the characteristic two strong bands of the amide groups around 1625 and 1596 cm^{-1} ascribed respectively to the C=O stretching and –NH bending. The band at 1045 cm^{-1} is due to –O– stretching, respectively. In the presence of the organoclay as filler, the bending modes of –O– bonds of the SPI/CMC/clay composite film are shifted towards higher wavelengths; from 1045 to 1052, 1049, 1048, 1051, and 1049, respectively, for SPI/CMC/clay composite film containing 3, 6, 9, 12, and 15 g/100g solid content. This shift are due to organoclay–SPI/CMC interactions through hydrogen bonds. The wavelength shift in each SPI/CMC/clay composite films is larger than that of SPI/CMC based film due to the presence of the raw organoclay. This phenomenon indicated an increase in the organoclay–SPI/CMC interaction, due to a better dispersion of the intercalated organoclay. This observation is in accordance with literature for cassava starch–kaolinite composite film [8].

Fig. 1. Characteristic of SPI/CMC/clay-0 composite films

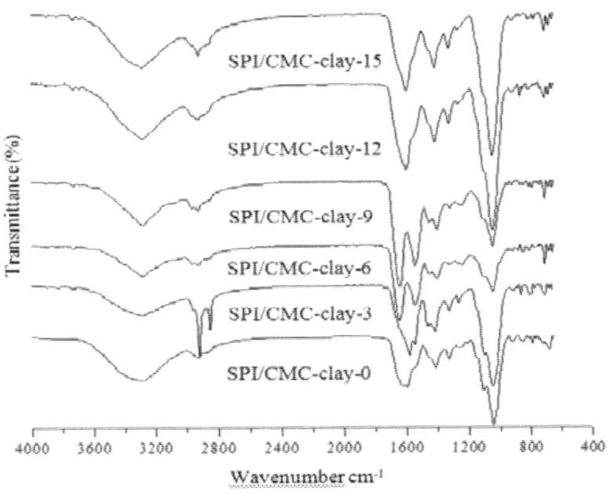

Fig. 2. IR Spectrum of SPI/CMC/clay composite films.

XRD pattern: The XRD pattern of SPI/CMC/clay-0 (without clay) revealed the diffraction peak at ϴ=20° (Fig. 3). All SPI/CMC/clay composite films showed similar pattern of XRD (Fig. 3) because major portion of the composite films composes of the biopolymer matrix, whereas only the decrease of intensity of diffraction peak was occurred after blended with clay. Powders of organoclay showed characteristic crystallinity peaks at ϴ=20°, 22°, and 27° (Fig. 3). These peaks were shifted toward lower angle at 14°, 16.5°, and 21.6°, respectively, after formation of SPI/CMC/clay composite film with clay concentration from 6 to 15 g/100g solid content. The angle of diffraction peak was not noticeably affected by the clay content but only the intensity of the diffraction peak increased when clay concentration in SPI/CMC/clay composite film increased [7, 9]. Similar result of increase in peak intensity with increase in clay concentration was observed with agar/clay (Cloisite Na+) nanocomposite films [7] and starch/CMC/Cloisite Na+ nanocomposite films [10].

Water vapor permeability (WVP): Poor water vapor barrier ability is one of the major limitations in employing protein-based films as food packaging materials, thus reduction in WVP is desirable for potential applications in food packaging.. The WVTR of the SPI/CMC/clay composite films ranged between 37.8 to 51.0 g/day.m^2 (Table 1). The SPI/CMC/clay-0 composite film had a WVP of 3.87 g.mm/m^2.mHg.day. The WVP of composite films decreased with increase of the clay concent (0-6g/100g SPI and CMC) (Table 1), and then it gradually increased when the clay was used over 6g/100 g solid content. The values of WVP for films with 3, 6, and 9g/100g solid content reduced by 31.5%, 44.7%, and 26.4%, respectively, compared with that of the SPI/CMC/clay-0 composite film. The SPI/CMC/clay-12 showed not differ WVP with that of SPI/CMC/clay-0 film. Kumar et al. [3] reported a reduction of 22.1% and 30.8%, respectively, for bio-nanocomposite films based on SPI and 5% Cloisite

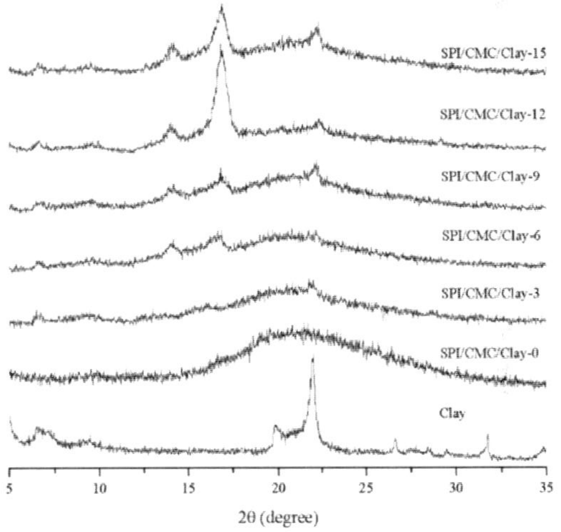

Fig. 3. XRD patterns of SPI/CMC/clay composite films.

Na+, and SPI and 5% Cloisite 20A. The reported that the reduction in WVP by modified montmorillonite has been ascribe to the creation of a tortuous pathway for water vapor to diffuse out of the bio-nanocomposite matrix. This increases the sufficient path length for diffusion of water vapor molecules therefore WVP reduced [6]. At 3-12g organoclay/100g solid content, the reduction in WVP is attributed to the hydrophobic nature of organoclay, but for SPI/CMC/clay-15 composite film, the WVP was 1.2x higher than that of SPI/CMC/clay-0 composite film. Such phenomenon of decrease in WVP with incorporation of clays have frequently observed with various biopolymers [2, 3, 7, 11]. Yano et al. [6] reported that the WVP of polymer/clay nanocomposite films generally decreases exponentially when clay content or aspect ratio of the clay increase.

Moisture sorption isotherm: Changes in the environmental humidity influence the moisture sorption isotherm for the SPI/CMC/clay composite films. For each film different clay concentrations, an increasing percent relative humidity (RH) also increased the percent of EMC of the films (Fig. 4). The sorption isotherm of all composites displayed the characteristic sigmoid-shaped type II isotherm (Fig. 4) [4]. When compared at an environmental RH of 86%, the EMC of the composite films decreased with increasing clay concentrations between 0 and 6g/100g SPI and CMC with the minimum EMC obtained in SPI/CMC/clay-6 film (i.e. the lowest WVP value). The increase of clay content beyond 6 g/100g SPI and CMC effected the moisture content increased. The SPI/CMC/clay-6 composite film had the greatest water resist nature among each other composites. The trend of EMC of SPI/CMC/clay composites was related to WVP reported above. These results indicated that the addition of clay improve the water resistance of the SPI and CMC matrix. Similar trend has been reported by the other researchers worked with starch/CMC/clay nanocomposite films [10] and agar/clay nanocomposite films [7]. Almasi et al. [10] tested water sorption of starch-CMC- clay films with varying amount of MMT (0–7%) and they found that the water vapor absorption rate of nanocomposites films was slower than the film without clay. They gave the reason of the observations that the starch and CMC are able to form hydrogen bonds with the hydroxyl groups of the MMT layers and this strong structure could reduce the diffusion of water molecules in the material. Cyras et al. [9] also noted that the trend of decreasing the water vapor uptake ratio of the agar film with increasing the clay content is coincided with that of the WVP. The decrease in water vapor uptake ratio of the nanocomposite films is probably partly due to the tortuous structure formed by the nanoclay dispersed through the polymer matrix [9].

Mechanical properties: Mechanical properties (TS and EB) of SPI/CMC/clay composite films with various organoclay content were showed in Fig. 5, respectively. The mechanical properties were greatly influenced with the amount of clay addition [7]. SPI/CMC/clay-0 (without clay) composite film had a TS of 12.9±0.8 MPa and EB of 13.9±1.1%. Films containing organoclay had significantly higher TS as compared to that of SPI/CMC/clay-0 composite film, except SPI/CMC/clay-15 composite film. The value of TS increased from 12.9±0.8 MPa to 34.5±2.1 MPa (167%) as the clay content increased from 0 to 6 g/100g solid content. A similar behavior of increase in TS with increasing in the amount of clay has been observed in a various biopolymer-based nanocomposite films [2, 3, 7, 10]. Rhim et al. [7] and Almasi et al. [10] explained that this behavior may be attributed to the resistance exerted by the clay itself and the stronger interfacial interaction through hydrogen or ionic bonds between polymer matrix and intercalated layered silicate with vast interfacial area. However, such reinforcing effect of clay was reduced above certain level of clay addition (6 g/100g solid content). This is owing to the formation of stacked clays without complete dispersion through polymer matrix at such high level of clay concentration [7]. On the contrary, flexibility of the SPI/CMC/clay composite films, as determined in the EB, displayed reversed pattern of the TS. The EB decreased with increase in clay content up to 6 g/100g solid content, and increased above 6 g/100g solid content of clay

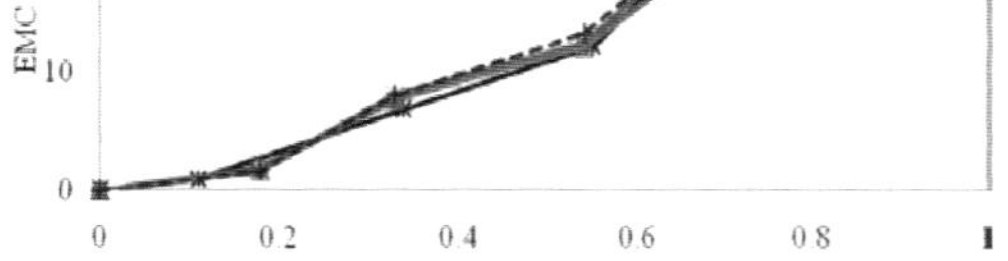

Fig. 4. Moisture adsorption of SPI/CMC/clay composite films.

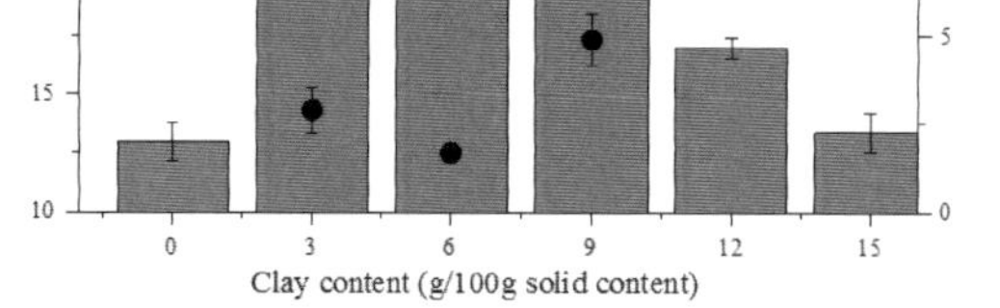

Fig. 5. Mechanical properties of SPI/CMC/clay biocomposite films

incorporation.

Conclusions

The SPI/CMC/clay composite films were prepared to study the effect of clay content on the composites' properties. The presence of clay caused better mechanical properties and moisture barrier for dipersion of clay in SPI/CMC biocomposites. The highest TS and the lowest EB, WVP, and EMC values were obtained from the SPI/CMC/clay composite incorporated with 6g clay/100g solid content. The authors believe that this new biopolymer is useful for industrial utilization possibly as a film-forming agent and fruits or vegetables coating solution to substitute the synthetics and extend a shelf life of food products.

Acknowledgements

The authors gratefully acknowledge the financial support (budget of 2011) from National Nanotechnology Center (NANOTEC), Bangkok, Thailand. We also wish to thank the National Research University Project under Thailand's Office of the Higher Education Commission for financial support.

References

1. P. Rachtanapun; R. Suriyatem in SLIM 2012 Proceedings, Changwon, South Korea, 2012, Book.
2. P. Kumar; K.P. Sandeep; S. Alavi; V.D. Truong; R.E. Gorga, *J. Food Sci.* 2010, *75(5)*, N46-N56.
3. P. Kumar; K.P. Sandeep; S. Alavi; V.D. Truong; R.E. Gorga, *J. Food Eng.* 2010, *100(3)*, 480-489.
4. W. Tongdeesoontorn; L.J. Mauer; S. Wongruong; P. Rachtanapun, *As. J. Food Ag-Ind.* 2009, *2(4)*, 501-514.
5. S. Arunvisut; S. Phummanee; A. Somwangthanaroj, *J. Appl. Polym. Sci.* 2007, *106*, 2210-2217.
6. K. Yano; A. Usuki; A. Okada, *J. Polym. Sci. Pol. Chem.* 1997, *35(11)*, 2289-2294.
7. J.-W. Rhim, *Carbohyd. Polym.* 2011, *86*, 691–699.
8. J.A. Mbey; S. Hoppe; F. Thomas, *Carbohyd. Polym.* 2012, *88(1)*, 213-222.
9. V.P. Cyras; L.B. Manfredi; M.-T. Ton-That; A. Vázquez, *Carbohyd. Polym.* 2008, *73(1)*, 55-63.
10. H. Almasi; B. Ghanbarzadeh; A.A. Entezami, *Int. J. Biol. Macromol.* 2010, *46(1)*, 1-5.
11. R. Sothornvit; S.-I. Hong; D.J. An; J.-W. Rhim, *LWT-Food Sci. Technol.* 2010, *43(2)*, 279-284.

Table 1. Water vapor permeability of SPI/CMC/clay biocomposite films

Type of films	Clay content (g/100g solid content)	Thickness (mm)	(WVTR) (g/day.m^2)	(WVP) (g.mm/m^2.mHg.day)
SPI/CMC/clay-0	0	0.246 ± 0.010	37.849 ± 0.005 a	3.872 ± 0.219 a
SPI/CMC/clay-3	3	0.097 ± 0.006	42.071 ± 1.825 b	2.651 ± 0.109 b
SPI/CMC/clay-6	6	0.084 ± 0.006	39.394 ± 2.247 ab	2.140 ± 0.056 c
SPI/CMC/clay-9	9	0.111 ± 0.007	39.712 ± 0.769 ab	2.851 ± 0.224 b
SPI/CMC/clay-12	12	0.139 ± 0.009	40.880 ± 0.573 ab	3.674 ± 0.222 a
SPI/CMC/clay-15	15	0.147 ± 0.005	51.032 ± 0.698 c	4.874 ± 0.140 d

Values followed with different letters in the same column are significant differences at the 5% level, according to Tukey ($P < 0.05$).

O-07-434

SULFONATED MESOPOROUS BENZENE-SILICA EMBEDDED SULFONATED PEEK MEMBRANES FOR ENHANCED PROTON CONDUCTION AND ANTI-DEHYDRATION

Jung Hwa Park, Seong Yeon Han and Dukjoon Kim*
School of Chemical Engineering, Sungkyunkwan University, Suwon, Kyunggi 440-746, Republic of Korea – djkim@skku.edu

Abstract - The performance of proton exchange membranes in the fuel cell system plays an important role both as an electrolyte and a separator. Nafion® membranes have several drawbacks such as the reduced conductivity at high-temperature operation, high cost, and the high methanol fuel crossover in direct methanol fuel cell (DMFC) system. In this study, sPEEK based composite membranes containing sulfonated mesoporous benzene-silica (sMBS) were prepared and evaluated as an electrolyte for DMFC applications. Sulfonated cubic mesoporous benzene-silica (sMBS) were prepared by a post-sulfonation of mesoporous benzene-silica synthesized from 1,4-bis(triethoxysilyl) benzene and in the presence of F127 PEO-PPO-PEO block copolymer template in acidic condition. The well-ordered cubic mesostructure of sMBS and the characteristic length (i.e. the average distance between ionic clusters) of sPEEK matrix were confirmed by small angle X-ray scattering. All sPEEK-based sMBS composite membranes showed lower methanol permeability than that of Nafion®117. All sMBS composite membranes showed better proton conductivity than that of pristine sPEEK membrane at a high relative humidity of 95%. Moreover, the proton conductivity of composite membrane was maintained at low relative humidity and high temperatures up to 70 – 80 °C.

Keywords: fuel cell, membrane, mesoporous material, PEEK

Introduction

Polymer Electrolyte Fuel Cell (PEFC) has been extensively studied as a convenient and efficient electrochemical power source. In the PEFC system, proton exchange membrane plays an important role [1-3] as a proton transport layer but fuel barrier. It is well-known that the commercial Nafion membrane has excellent properties such as high conductivity and high mechanical and chemical stability so that they are widely used. However, Nafion also has some drawbacks in association with high fuel crossover, high cost, and poor properties at high temperature above 100 C. Its proton conductivity is reported to be decreased quite a lot at such high temperatures due to dehydration of the membrane. Because of this, researchers have been trying to discover new alternative polymers [4-14] to replace Nafion. Poly(ether sulfone), polybenziimidazole, polyimide, poly(vinyl alcohol), poly(ether ether ketone) are among them.

In this study, highly sulfonated mesoporous benzene silica was prepared. As the sulfonation position was not only phenyl but Si-end group, it will allow higher degree of sulfonation than the previous one. sPEEK based composite membranes were prepared by incorporating the sulfonated 2-D hexagonal mesoporous benzene silica. Mesoporous materials were synthesized using a commercial cetyltrimethylammonium chloride (CTAC) as a template, and 1.4-bis (triethoxysilyl)beneze (BTEB) and (3-Mercaptopropyl)- trimethoxysilane (MPTMS) as precursor via former reported method. Mesoporeswere expected to retain water at high temperature due to the capillary force. A variety of bulk and transport properties were investigated including water uptake, methanol permeability, ion conductivity, thermal stability, and mechanical stability, etc.

Experimental

Materials. PEEK (Vitrex 450PF) was purchased from ICI Company (Rotherham, UK). Its molecular weight was 100,000 g/mol. Cetyltrimethylammonium chloride solution (CTAC) (25wt%, Aldrich), 1.4-bis(triethoxysilyl)benzene (BTEB) (Aldrich), (3-Mercaptopropyl)trimethoxysilane (MPTMS) (95wt%, Aldrich), sodium hydroxide (99wt%, Duksan Reagents & Chemicals, Korea), hydrogen peroxide (30wt%, Duksan), hydrochloric acid (37wt%, Aldrich), sulfuric acid (Duksan), methylsulfonic acid (MSA) (Acros Organics, USA), N,N-dimethyl acetamide (DMAC) (Fluka Chemie AG) were purchased and used without further treatment.

Preparation of mesoporous benzene-silica material (MBS). MBS powder was synthesized using cetyltrimethylammonium chloride solution (CTAC) (25wt%) as template via the method reported by Inagaki et al. Typically, 7.63g CTAC was dissolved in a mixture of 56.75g DI water and 0.66g sodium hydroxide at room temperature. After stirring for 2 hours, 1.00g 1.4-bis (triethoxysilyl) benzene (BTEB) and 0.73g (3-Mercaptopropyl)trimethoxysilane (MPTMS) were added. The mixture was kept stirring for another 12 hours. Next, the products were aged for 24 h at 90℃ in a oil bath. Then a mixture of 200 mL ethanol and 3.00g HCl was employed to remove the surfactant and CTAC template at 6℃ under stirring for 6 h. The next step is to extract the solids using a suction flask and wash them by DI water and then acetone several times. The product was dried at 120℃ for 24 h. Mesoporous benzene-silica (MBS), 1g, was added in 10 g of hydrogen peroxide (30wt%) and kept stirring for 12 h at room temperature. After stirring, a suction flask was used to filter the mixture, followed by washing it with DI water and acetone several times. The product was then added into 500 mL sulfuric acid fuming SO_3 reagent. After stirring for 12 h, it was filtered and washed repeatedly with DI water and acetone. The final solid product was dried at 120℃ for 24 h.

Preparation of composite membrane. The commercial PEEK powder was sulfonated as reported elsewhere. Typically, 10g PEEK powder was dissolved in 50 mL methylsulfonic acid (MSA) under stirring. After the powder was completely dissolved, the solution was mixed with 400 mL of sulfuric acid (98%). The mixture was kept in a three-neck flask with nitrogen gas flow under room temperature for 80 h. Then, the solution was dropped into DI water slowly, followed by washing with a suction flask. The polymer obtained was washed by DI Water several times and dried at 100℃ for 24 h.

Composite membranes were prepared by a cast method. Typically, 0.3g sulfonated PEEK (sPEEK) powder was dissolved in 9.7 g N,N-dimethyl acetamide (DMAC) with stirring under room temperature for 2 h. Then, sMBS powder was added into the solution in the weight ratios of 5, 10, 15, and 20 wt% to sPEEK. After stirring for 2 h at room temperature, the mixtures are sonicated for 30 minutes and then cast on the glass dishes, followed by drying at 100℃ for 12 h. After drying for another 2 h at 120℃ under vacumm, the final product, composite membrane was obtained.

Characterization of SMBS and sPEEK-based composite membranes. The transmission electron microscopy (TEM) images were obtained using an FE-TEM (JEOL JEM2100F) operated at an accelerating voltage of 200 kV.

The scanning electron microscopy (SEM) images were obtained with a field emission SEM (JEOL JSM6700F) operating at an accelerating voltage at 30kV. Vertical section was observed for SMBS dispersion in polymer.

The nitrogen adsorption–desorption isotherm was obtained for the sulfonated mesoporous benzene-silica on a Micromeritics 2010 analyzer. The sulfonated benzene-silica mesoporous sample was degassed at 150℃ to a vacuum less than 30_mHg. The BET (Brunauer–Emmet–Teller) specific surface area was calculated from the adsorption data in a relative pressure range from 0.05 to 0.2. The total pore volume was evaluated from the amount of gas adsorbed at a relative pressure of 0.99. The pore size distribution (PSD) curves were calculated from the adsorption branches of the isotherms using the BJH method, and the pore diameter was estimated from the maximum of the PSD curve.

The solid-state NMR spectra were obtained with a Bruker DSX400 spectrometer using a 4 mm magic angle (MAS) spinning probe at the Korea Basic Science Institute. The solid-state ^{29}Si MAS NMR spectra were obtained with a 400 MHz solid state NMR spectrometer using a 4 mm magic angle(MAS) spinning. All of the samples were spun at rates of 4.5–7 kHz and the chemical shifts were obtained with respect to the mercapto propyl and phenyl group reference peak. The ^{13}C CP-MAS NMR spectra were measured under the following experimental conditions: a contact time of 2 ms, a recycle delay of 3 s, and 3000 scans; and the ^{29}Si MAS NMR spectra were obtained with a recycle delay of 50 s and 1500 scans.

The mechanical properties of the fully hydrated membranes were measured using a universal tensile machine (UTM-model 5565, Lloyd,Fareham, UK). Samples with a dimension of 20 mm×50 mm were tested with a 250 N load cell pulled at 50 m/min, within a 21 cm gauge length. To ensure measurement accuracy, all membranes were tested three times and the average values taken.

The methanol permeability was measured at room temperature using a glass diffusion cell where the membrane was located in the middle. The left compartment was filled with 50 ml of 2 M methanol aqueous solution, and the right one with 50 ml of distilled water. The methanol permeability (P) was calculated from the slope of the linear plot of methanol concentration versus permeation time from the following equation:

$$C_B(t) = (A/V_B)(P/L)\,C_A(t - t_0)$$

C_A and C_B are the methanol concentrationin the left (MeOH) and the right (water) compartments. A is the surface area of membranes, L is the thickness of membranes, and VB is the volume of the right compartment.

A BEKKTECH cell was used to measure the membrane resistance so that the ion conductivity was determined. The gas resource used was hydrogen. A Nitrogen flow was employed with 1000 cc per minute to avoid the precipitation of water on the membrane. The resistance was measured at 90% relative humidity condition. The following equation was used to calculate the ion conductivity.

$$\sigma\ (S/cm) = L/(R \times W \times T)$$

L is the distance between the electrodes which is fixed as 0.425cm in the BEKKTECH cell, R is the resistance of membrane, W is the width of the membrane, and T is the thickness of the membrane.

A thermo gravimetric analyzer (TGA, TGA7, Perkin-Elmer) was used to analyze the dehydration rate of the composite membranes. The weight of the samples was scanned at100 ℃ for 30 min under a nitrogen flow.

Results and Discussion

Characteristics of SulfonatedMesoporous Benzene-silica (SMBS). The nitrogen adsorption-desorption isotherms for SMBS is shown in Figure 1. The isotherm graph is a typical IV type. The BET surface area of SMBS is around 481.3m^2/g, and the total pore volume is around 0.5046cm^3/g. The pore size diameter determined is around 3.338nm.

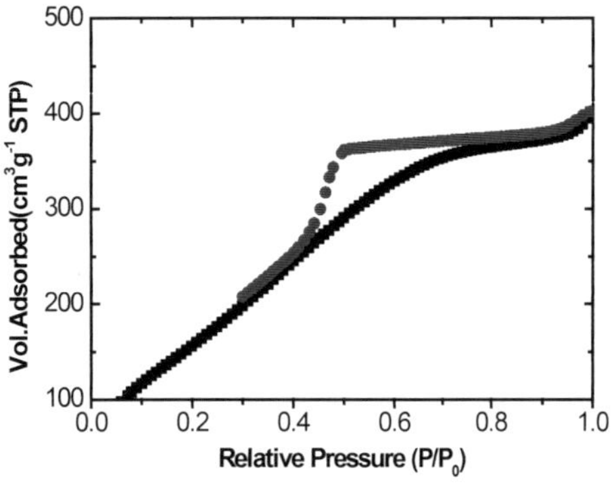

Figure 1. Nitrogen adsorption-desorption isotherm of SMBS

Characterization of composite membranes. Composite membranes are indexed as SMBS5, SMBS10, SMBS15, and SMBS20 which represent the weight percentage of SMBS in sPEEK65 matrix. The FE-SEM images of pure and composite membrane showed the well dispersed spherical shaped SMBS materials. If incorporated particles have disordered distribution, the transportation ability will be reduced in a great scale due to the blocking effect. In this sPEEK based composite membranes, the similarity of chemical structures allows SMBS powder to disperse homogenously inside the membranes.

Mechanical strength of the composite membranes is shown in Figure 2. The tensile strength decreased with increasing SMBS contents. When more SMBS was added, polymer density in membrane is decreased. But, this decreased mechanical strength was no problems with fuel cell running.

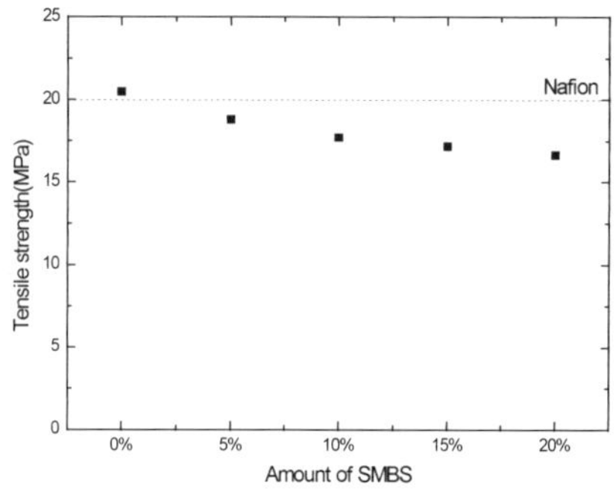

Figure 2. Tensile strength of composite membranes

All composite membranes were weighted (W_0) and then immersed in DI water for 12 hours at room temperature followed by weighting (W). So the water uptake can be calculated by the equation below:

$$wt\% = \frac{W - W_0}{W} \times 100\%$$

All composite membranes showed higher water uptakes than that of pure sPEEK membrane because of the sulfonated mesoporous benzene-silica inside. Pristine membrane showed from 19.75 to 59.56 % and SMBS20 showed 25 to 104.69 % of water uptakes. The water uptake value increases with the SMBS contents and temperature. It showed that SMBS could keep water in composite membrane. SMBS had high sulfonation degree and it influenced water uptake of composite membrane. It showed possibility that composite membrane could hold water in membrane at high temperature operation.

The composite membranes showed relative higher conductivity value than that of pure sPEEK membrane.sPEEK based membranes had narrow, less-separated channels along with more dead-end channels than Nafion as reported earlier. There is no clear relationship between the conductivity and content of SMBS. As mentioned earlier, SMBS powder had high sulfonation degree and water storage ability. Both of factors influenced ion conductivity of membrane and ion conductivity was increased. Especially, SMBS20 at 80℃ showed 0.08233S cm^{-2} which value was 99.2% of nafion at 80℃. This value increase was similar with water uptake increase. All membranes were immersed in DI water for 24 hours and their weight loss were measured for 30 minutes at 100℃. The weight of all membranes was reduced because of free water evaporation. Although pure sPEEK membrane showed lower water uptake than composite membranes, pristine membrane showed higher weight loss. Moreover, the value of water uptake increases with the content of SMBS. It showed that the capillary of PMO had positive effect to keep water in composite membranes.

Conclusions

sPEEK based sulfonatedmesoporous benzene-silica composite membranes were prepared by casting method using N,N-dimethyl acetamide. Sulfonatedmesoporous benzene-silica (SMBS) was prepared with two position sulfonation. The ordered mesostructure and properties of SMBS were confirmed using transmission electron microscopy and nitrogen adsorption-desorption isotherm experiment. As the contents of SMBS increased, more SMBS particles were observed in polymer matrix without aggregation.SMBS worked like methanol barrier and lowest methanol permeability value was about 1.4% of Nafion 117. In addition, composite membranes had higher water uptake and lower dehydration rate than pristine sPEEK.Water uptake value was increased with SMBS contents in membrane. Increased water contents in membrane work like medium of ion transport and it made ion conductivity increased. Highest ion conductivity was 0.08233Scm^{-2} and this value was 99.2% of Nafion. SMBS in composite membrane held water at high temperature. The results of ater uptake and dehydration rate showed that SMBS worksfor water storaging in polymer. This composite membrane could support high temperature operation or low humidity system for PEFC application.

Acknowledgements (Times New Roman 10, bold)

This work was sponsored by the National Research Foundation of Korea Grant funded by the Korean Government (MEST) (NRF-2009-0093033 and NRF-2011-0029478).

References (Times New Roman 10, bold)

[1] R.F. Service, *Science*. 2002, *296*, 1222–1224.
[2] R. Dillon, S. Srinvasan, A.S. Aricom, V. Antonucci, *J. Power Sources*.2004, *127*112–126.
[3] V.Neburchilov, J. Martin, H.J.Wang, J.J. Zhang, *J. Power Sources*. 2007, *169*, 221–238.
[4] Y. Yin, J.H. Fang, Y.F. Cui, K. Tanaka, H. Kita, K. Okamoto, *Polymer*. 2003, *44*, 4509–4518.
[5] M. Gil, X. Ji, X. Li, H. Na, J.E. Hampsey, Y. Lu, *J. Membr. Sci*.2005,*234*, 75–81.
[6] B.R. Einsla, Y.S. Kim,M.A. Hickner, Y.T. Hong, M.L. Hill, B.S. Pivovar, J.E. McGrath,*J. Membr. Sci*.2005, *255*, 141–148.
[7] K. Okamoto, Y. Yin, O. Yamada, M.N. Islam, T. Honda, T. Mishima, Y. Suto, K.Tanaka, H. Kita, *J. Membr. Sci*.2005, *258*, 115–122.
[8] S. Vetter, B. Ruffmann, I. Buder, S.P. Nunes, *J. Membr. Sci*.2005, 260, 181–186.
[9] A. Carbone, R. Pedicini, G. Portale, A. Longo, L. D'Ilario, E. Passalacqua, *J. PowerSources*. 2006*163*, 18–26.
[10] F.X. Zhai, X.X. Guo, J.H. Fang, H.J. Xu, *J. Membr. Sci*.2007, *296*, 102–109.
[11] L. Li, J. Zhang, Y.X. Wang, *J. Membr. Sci*. 2003, *226*, 159–167.
[12] P. Mustarelli, E. Quartarone, S. Grandi, A. Carollo, A. Magistris, *Adv. Mater*. 2008, *20*, 1339–1343.
[13] K.N.T. Do, D. Kim, *J. Power Sources*.2008, *185*, 63–69.
[14] D.X. Luu, E.-B. Cho, O.H. Han, D. Kim, *J. Phys. Chem. B*.2009,*113*, 10072–10076.

O-07-437

POLYURETHANE/GRAPHENE OXIDE NANO-COMPOSITES WITH GAS BARRIER PROPERTIES

Poria Kaveh[1], Mehrzad Mortezaei[1]*, Behrouz Esmaeili[1], Soheil Pilehvar[2]

[1]Polymer Engineering Center, Islamic Azad University, South Tehran Branch, Tehran, Iran – poriakaveh@gmail.com; mhrmorteza@gmail.com; behrouz.esmaeili@yahoo.com

[2]Polymer Engineering Center, Islamic Azad University, Science and Research Branch, Tehran, Iran – s.pilehvar1988@gmail.com

Abstract - In the present study, the effect of graphene oxide on gas barrier performance of thermoplastic polyurethane/graphene oxide (TPU/GO) nano-composites was investigated. For this propose, TPU was reinforced with various contents of GO (0-1 wt.%) through solvent blending procedure. Transmission electron microscopy (TEM) and wide angle X-ray reflectance diffraction (WAXD) showed that graphene was fully exfoliated in the nano-composites. The physical and mechanical properties of the prepared nano-composites were investigated through dynamic mechanical analysis (DMA) and differential scanning colorimetery (DSC), as well as, tensile strength and gas permeability measurements. The addition of GO into TPU was found to enhance the storage modulus, tensile strength and tensile modulus, while, decreased the strain at break, gas permeability of TPU. Enhanced gas barrier properties of nano-composited in comparison to pure TPU were attributed to the formed torturous paths for the gas molecules. Fourier transform infrared spectroscopy (FTIR) confirmed the interaction between the TPU matrix and GO plates. Finally, Results showed that this new materials are suitable for high gas barrier applications.

Keywords: Thermoplastic Polyurethane, Graphene oxide, Nano-Composite, Gas Barrier

Introduction

Polymeric materials have long been utilized for packaging applications. However, overwhelming demands in the marketplace have put significant pressure for material developers to pursue novel routes to manufacture low cost packaging films with greatly improved barrier properties [1].

The emergence of polymer nano-composites introduces a new opportunity to improve barrier properties of polymers. When the nano-platelets are well exfoliated, in addition to the improved barrier property, polymer nano-composites usually exhibit enhanced mechanical properties and better thermal stability [1, 2].

Graphene has a high basal plane elastic modulus, $E \approx 1$ TPa; ultimate strength, $\sigma \sim 130$ GPa; and room temperature charge-carrier mobility, $\mu \approx 10\ 000$ cm^2/V s. Adding highly exfoliated carbon layers can significantly alter mechanical and electrical properties of polymers at extremely small loading [3]. Unlike carbon nanotubes that can provide similar mechanical and electrical benefits, these impermeable two-dimensional sheets can reduce gas permeability of host membranes [4].

In the present study, the effect of graphene oxide (GO) on gas barrier performance of thermoplastic polyurethane/grapheme oxide (TPU/GO) nano-composites was investigated. For this propose, TPU was reinforced with various contents of GO (0-1 wt.%) through solvent blending procedure. The physical and mechanical properties of the prepared nano-composites were investigated through transmission electron microscopy (TEM), wide angle X-ray reflectance diffraction (WAXD), dynamic mechanical analysis (DMA) and differential scanning colorimetery (DSC), as well as, tensile strength and gas permeability measurements.

Experimental

Dispersion of GO into TPU was conducted via solvent blending procedure. In solvent mixing, GO were dispersed in DMF solvent (50 mL) and stirred with a magnetic bar for 30 min and sonicated through a probe ultrasonicator for 4 min. Dry TPU powders (5 g) were separately added to DMF solvent (50 ml) and stirred with a magnetic bar for 30 min. Then, GO dispersion and TPU solution were mixed and stirred with a magnetic bar for another 30 min. Finally, DMF solution was cast on a heated silicone plate (60 ˚C). DMF removal from the cast films continued for 24 h in an oven at 50 ˚C before characterization. Temperatures were kept low because GO decomposes even below 100 ˚C. The formulation of prepared TPU/ GO nano-composites is given in Table 1.

Results and Discussion

The physical and mechanical properties of the prepared nano-composites were investigated through transmission electron microscopy (TEM), wide angle X-ray reflectance diffraction (WAXD), dynamic mechanical analysis (DMA) and differential scanning colorimetery (DSC), as well as, tensile strength and gas permeability measurements.

Transmission electron microscopy (TEM) and wide angle X-ray reflectance diffraction (WAXD) showed that GO was fully exfoliated in the nano-composites. The addition of the GO into TPU was found to enhance the storage modulus, tensile strength and tensile modulus, while, decreased the strain at break, gas permeability of TPU. Enhanced gas barrier properties of nano-composited were attributed to the formed torturous paths for the gas molecules, after incorporation of GO. Fourier transform infrared (FTIR) spectroscopy confirmed the interaction between the TPU matrix and GO plates.

Table 1 - The formulation of prepared TPU/GO nano-composites.

Sample code	TPU (g)	GO (g)
S1	5.0	0.00
S2	5.0	0.015
S3	5.0	0.025
S4	5.0	0.050

Conclusions

Results showed that this new materials are suitable for high gas barrier applications.

References

1. L. Sun; W.-J. Boo; A. Clearfield; H.-J. Sue and H.Q. Phamc, *J. Membr. Sci.* 2008, *318*, 129.
2. H. Kim; Y. Miura and C.W. Macosko, *Chem. Mater.* 2010, *22*, 3441.
3. S. Stankovich; D. A. Dikin; G.H.B. Dommett; K.M. Kohlhaas; E.J. Zimney; E.A. Stach; R.D. Piner; S.T. Nguyen and R.S. Ruoff, *Nature* 2006, *442*, 282.
4. H. Kim and C.W. Macosko, *Macromolecules* 2008, *41*, 3317.

O-07-529

MECHANICAL PROPERTIES OF POLYPROPYLENE COMPOSITES CONTAINING RECYCLED PET FIBERS

R. C. Nonato; B. C. Bonse*

prebbonse@fei.edu.br; renato_nonato@hotmail.com

Centro Universitário da FEI, São Bernardo do Campo, SP, Brazil

Abstract - The effect of recycled PET fiber (PETr) and compatibilizer (maleic anhydride grafted PP) content on the mechanical properties of PP/PETr composites was analyzed using multiple regression statistical analysis 2^2 with central point. The investigated fiber and compatibilizer levels were 10 and 18wt % and 1 and 4wt%, respectively. A composition containing 5wt% fibers and 4wt% compatibilizer was also prepared for comparison. Composites were obtained from a master blend containing 20wt% PETr and 80wt% PP, which was prepared in a HAAKE mixer. After grinding the master, compositions were adjusted during extrusion and composites were injection molded. Effects of the variables on mechanical properties were assessed by means of tensile strength, elongation at break, flexural strength, flexural modulus and impact strength. Neat PP, subjected to the same composite processing conditions, was used as control. All investigated composite properties showed to be affected by both fiber and compatibilizer content.

Keywords: Polypropylene, recycled PET fibers, composites.

Introduction

Thermoplastic composites reinforced with different types of fibers and/or recycled materials are increasingly gaining attention, especially in applications where strength to density ratio is important, such as the automobile industry. These composites tend to have low density, high specific strength and modulus and their importance, in addition to the obviously environmental reasons, include economic benefits, ease of processing, tailor-made properties etc. [1]. However, lack of adequate bonding between the polymeric matrix and fibers may result in fiber pull out without breaking, when the composite is subjected to mechanical load, with no enhancement in composite properties. To solve this problem, surface treatments, coupling agents and compatibilizers may be used.

Polypropylene composites containing short recycled PET (PETr) fibers may be a potential material for the aforementioned markets; however there are few works in this area and usually fiber content do not exceed 10wt%.

In this context, this research was performed to analyze the effect of PETr fibers and compatibilizer on the mechanical properties of polypropylene composites containing PETr fibers, using multiple regression statistical analysis 2^2 with central point and PETr fiber content higher than commonly encountered in literature.

Experimental

Materials

Isotactic polypropylene pellets were donated by Braskem under code H503, with density of 0.905 g/cm³ and Melt Flow Rate (MFR) of 3.5 g/10min (230°C / 2.16 kg). PETr fibers with length of 10 mm, titer of 7 dtex and density 1.38 g/cm³ were purchased from São Marino Comércio de Fibras Têxteis Ltda. Compatibilizer maleic anhydride grafted PP (PP-g-MA), Polybond 3200, density 0.92 g/cm³ and MFI = 110 g/10min (190°C and 2.16 kg), was purchased from Crompton-Uniroyal Chemical (São Paulo, Brazil).

Methods

PETr fibers were oven dried for 4 hours and 150°C to prepare a 20/80 PETr/PP master blend. The master was prepared in a HAAKE PolyLab OS RheoDrive model PTW 16 mixer at 200ºC and 80 rpm. The master was then ground and dried for 4 hours at 120°C for subsequent extrusion to prepare the compositions shown in Table 1. The extrudate was cooled with a fan and pelletized. Specimens were injection molded in a H60/350 Battenfeld machine.

Tensile and flexural tests were performed in an Instron 2267, at a rate of 5 mm/min and 1.3 mm/min, respectively, according to ASTM D638 and D790. Impact testing was performed in a Charpy VEB Werkstoff Prüfmaschinen Leipzig impact tester, according to ASTM D6110. For tensile fatigue testing an MTS 810 was used; a minimum load of 50 N was used to avoid specimen buckling and a maximum of 1000 N. HDT testing was carried out in CEAST HDT Vicat 6 apparatus.

Tensile fractured surfaces were covered with gold and analyzed in a CamScan CS3200 LV Scanning Electronic Microscope (SEM).

Table 1 – Codified and decodified variables of the assessed compositions

Assay	codified variable		decodified variable	
	x_1	x_2	PETr (wt%)	PP-g-MA (wt%)
PP control	-	-	0	0
5-4	-	-	5	4
10-1	-1	-1	10	1
10-4	-1	1	10	4
14-2.5	0	0	14	2.5
18-1	1	-1	18	1
18-4	1	1	18	4

Results and Discussion

Tensile tests results are shown in Figure 1.

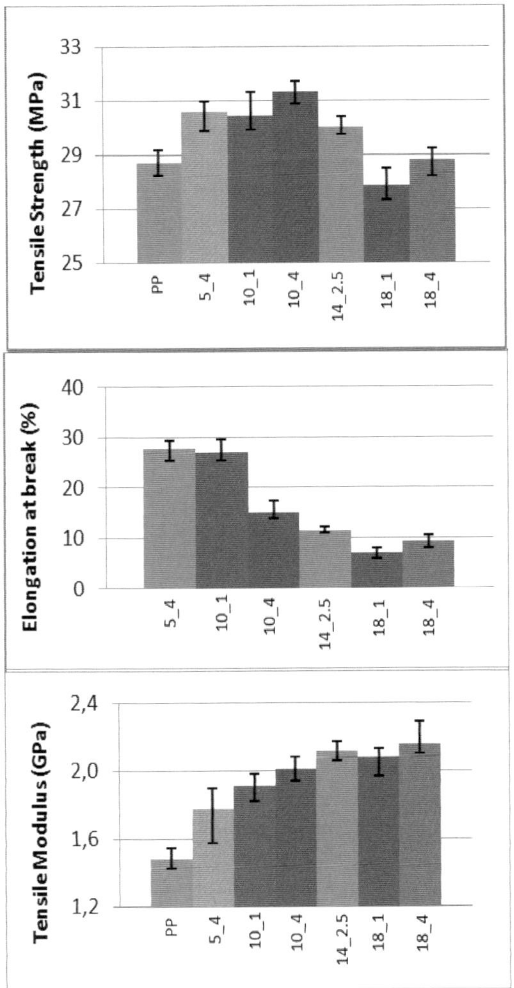

Figure 1 – Results of tensile strength, elongation at break and tensile modulus for PP control and PP/PETr composites. (x_y: x wt% fiber _ y wt% PPgMA)

In relation to PP control tensile strength increased more than 10% for compositions with 10 and 14wt% PETr fibers. For compositions containing 18wt% PETr fibers no significant change was observed and for composition 18-1 the value was lower than that of PP control. In compositions with higher fiber content, compatibilizer content might not have been sufficient for proper adhesion. The non-compatibilized fibers might have acted as stress concentrators, resulting in decrease in tensile strength. Similar behavior was observed by Santos and Pezzin [3]. Elongation at break values of PP control were omitted in Fig. 1 because values exceeded 100% deformation without break. For the composites, the elongation decreased as more fibers were added; this phenomenon was expected because PETr fibers have a higher modulus than PP. Tensile modulus increased for all compositions, in the order of 30% when compared with PP control. The proximity of modulus for compositions 18-4 and 14-2.5 indicates that both fiber and compatibilizer content improve modulus. Similar behavior was observed by Lopez-Manchado and Arroyo [4].

Flexural test results are shown in Figure 2.

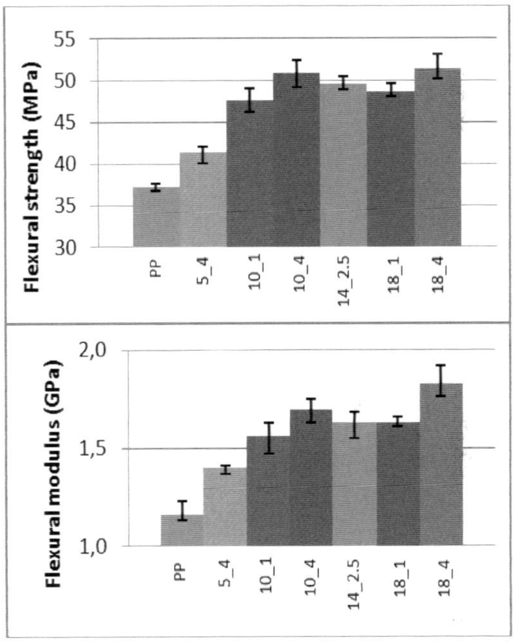

Figure 2 –Flexural test results for PP control and PP/PETr composites. (x_y: x wt% fiber _ y wt% PPgMA)

Fig. 2 shows that flexural strength and modulus also increased in relation to PP control, in the order of 30 and 35% respectively. These results also indicate that the ratio between fiber and compatibilizer content is important.

Impact energy results in Figure 3 show that for composition 5-4 in relation to PP control impact energy increased around 20%. This behavior at low fiber content was also observed by Saujanya and Radhakrishnan [5] and they attributed this to PETr fiber induced nucleation in PP, resulting in smaller spherulite size and hence higher flexibility. For compositions containing more than 10wt% PETr fibers, opposite behavior was observed and impact energy decreased almost 40%.

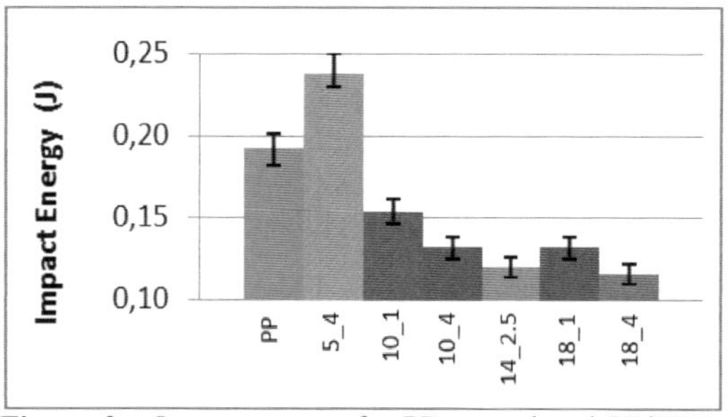

Figure 3 – Impact energy for PP control and PP/PETr composites. (x_y: x wt% fiber _ y wt% PPgMA)

The results of tensile fatigue tests are show in Figure 4:

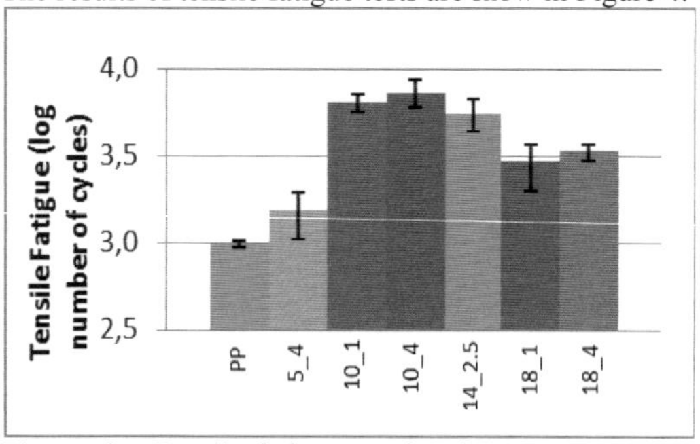

Figure 4 – Tensile fatigue results for PP control and PP/PETr composites. (x_y: x wt% fiber _ y wt% PPgMA)

For tensile fatigue adequate fiber/matrix bonding is necessary, this can be seen in Fig. 4 as higher PETr fiber content resulted in a decrease in this property. Tensile fatigue depends on dispersion of micro fissures, internal failures and residual tension; an excess of non-compatibilized PETr fibers will lead to composite failure.

Results of HDT tests are shown in Figure 5.

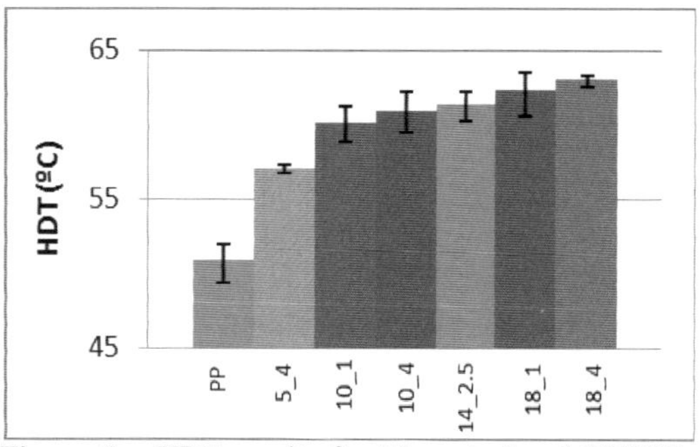

Figure 5 – HDT results for PP control and PP/PETr composites. (x_y: x wt% fiber _ y wt% PPgMA)

All compositions had higher HDT values than PP control, as shown in Fig. 5, representing increased thermal stability of the composites promoted by the PETr fibers.

Addition of compatibilizer improves adhesion between fiber and matrix; this is evidenced not only by the mechanical and thermal results, but also by SEM. Fig. 6 shows improvement in adhesion when comparing (a) composition 10-1 with (b) composition 10-4. Improved adhesion can be seen in (b).

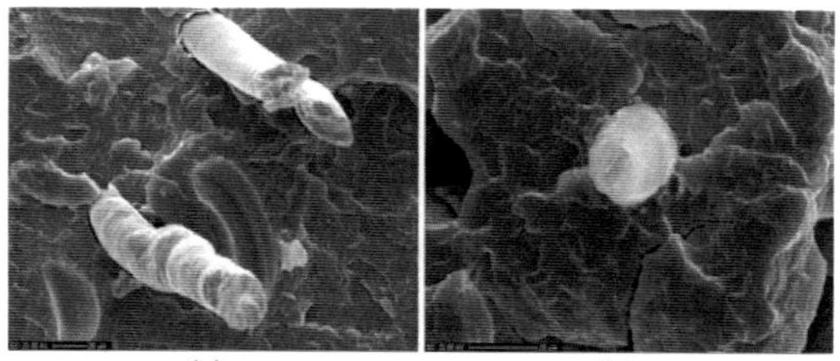

(a) (b)

Figure 6 – SEM micrograph of tensile fractured specimens: (a) composite with 10wt% PETr fibers and 1wt% compatibilizer and (b) composite with 10wt% PETr fibers and 4wt% compatibilizer.

The assessed properties could be fitted by a second order model, at significance level of 5%. Tab. 2 contains the coefficients of the fitted polynomial equations from the multiple regression analysis performed, where x_1 and x_2 are the codified variables referent to PETr fiber and PPgMA content, respectively.

Table 2 - Coefficients of the regression model $\mathbf{a_0+a_1x_1+a_2x_2+a_3x_1x_2}$ used for the different properties

	a_0	a_1	a_2	a_3	R^2
TS	29.72	– 1.26	0.47	--	0.902
εbreak	14.49	– 6.04	– 2.18	3.75	0.937
E	2063.1	79.4	41.9	--	0.640
IE	0.130	– 0.009	– 0.009	--	0.572
FM	1670.5	– 53.8	87.18	--	0.670
σflex	49.78	--	1.72	--	0.593
Fatigue (Log_{10} cycles to failure)	3.68	– 0.19	--	--	0.848
HDT	61.41	1.02	--	--	0.527

TS: tensile strength; ε_{break}: strain at break; E: tensile modulus; IE: impact energy; FM: flexural modulus at 0.3% strain; σflex: flexural stress at 5% strain

The a_1 values in Table 2 indicate PETr fiber content has a positive effect on tensile modulus and HDT, and negative effect on tensile strength, strain at break, impact energy, flexural modulus and fatigue. The a_2 values indicate that PP-g-MA content has a positive effect on tensile strength, tensile modulus, flexural modulus and flexural stress and negative effect on strain at break and impact energy. The a_3 values show that the interaction between fiber and compatibilizer content is significant only for strain at break, with a positive effect.

Conclusions

Incorporation of PETr fibers and PP-g-MA compatibilizer in PP/PETr composites resulted in increase in tensile strength, tensile modulus, flexural strength, flexural modulus, tensile fatigue and HDT, except for compositions 18-1 and 18-4, where excess of non-compatibilized PETr fibers likely resulted in a decrease in these values, probably because the PETr fibers acted as stress concentrators.

Increasing PETr fiber and PP-g-MA content resulted in decrease in elongation at break and impact strength, except for composition 5-4 which showed an increase in impact strength, likely because of PETr fiber induced nucleation in PP.

SEM micrograph showed that a relatively high compatibilizer content fiber/matrix adhesion improved. The results showed that recycled PET fibers can be used as reinforcement in polypropylene when adequate amount of compatibilizer is used.

Acknowledgements

To Braskem for having donated the PP used in this work.

References

1. Bonse, B. C.; Mamede, M. C. S.; da Costa, R. A.; Bettini S. H. P. In: PPS-26 Proceedings, Banff (Canada) 2010, pen-drive.

2. Arzondo, L.M.; Perez, C.J.; Carella, J.M. Polymer Engineering and Science, April, 2005.

3. Santos, P.; Pezzin, S.H. Journal of Materials Processing Technology 143–144 (2003) 517–520.

4. López-Manchado, M.A.; Arroyo, M. Polymer 41 (2000) 7761-7767.

5. Saujanya, C.; Radhakrishnan, S. Polymer 42 (2001) 4537 – 4548.

O-07-461

ANALYSIS OF THE PRODUCTION OF COMPOSITE PARTS WITH FUNCTIONAL ELEMENTS USING THE GAP IMPREGNATION PROCESS

Ch. Hopmann, R. Bastian*, K. Fischer, R. Schoeldgen

Institute of Plastics Processing at RWTH Aachen University (IKV), Aachen, Germany – bastian@ikv.rwth-aachen.de

Abstract - Fiber reinforced plastics offer a high potential for lightweight structures. By using Compression Resin Transfer Molding Processes (C-RTM), a reduction of cycle times can be achieved.
One derivate of the C-RTM is the Gap Impregnation Process, which is currently developed at the Institute of Plastics Processing (IKV) at RWTH Aachen University from basic research up to a series production status.
By using a special mold technology and an automated process control, it facilitates a fast and stable impregnation of net-shape preforms. Combined with fast-curing epoxy resins, this leads to a significantly reduced cycle time compared to established RTM-Processes.
After placing the net-shape preform into the mold, the resin system is injected into a flow gap generated between the preform and the upper mold. A special feature of the process is, that the orientation of the mold can be adjusted, so that the flow front is inclined relative to the force of gravity. By this, a homogeneous flow front of the resin can be achieved inside the cavity.
Starting with planar parts (500 mm x 500 mm), the complexity of the parts has been increased during the development of the Gap Impregnation Process. Next to curved composite parts with varying thicknesses, components with integrated functional elements like cut-outs, inserts and stiffening profiles are investigated.
Within this paper, results of a process analysis of parts with cut-outs and inserts are presented. The optimization of the process provides a total cycle time of less than 5 minutes. Only 14 seconds of the process time are necessary for the impregnation of the preform including injection and compression to a fiber volume content of 50 %. The fast impregnation and compression facilitates the processing of fast-curing resins in an isothermally heated mold without heating and cooling cycles. By this, only 3 minutes are needed for the curing of the resin.

Keywords: automation , composites, functional integration, high volume production

Introduction

Liquid composite molding processes have a high potential for the economical manufacturing of planar continuous fiber reinforced plastics (FRP) parts with a high fiber volume content [1,2,3,4]. Common for all liquid composite molding processes is the use of a preform, a near net-shape dry semifinished textile, impregnated with a liquid resin system and cured under heat. In comparison with the prepreg technology, wherein the semifinished textile is pre-impregnated and cured under heat, more beneficial semifinished products can be used since air-conditioned storage is not required. Conventional liquid composite molding processes are the Resin Infusion Process (RI) and the Resin Transfer Molding Process (RTM).
A fully automated process chain is mandatory for a mass production with a high production volume, starting with the composition of the textiles and ending with the finished composite parts [5,6]. Figure 1 shows the production chain of a RTM Process [7].

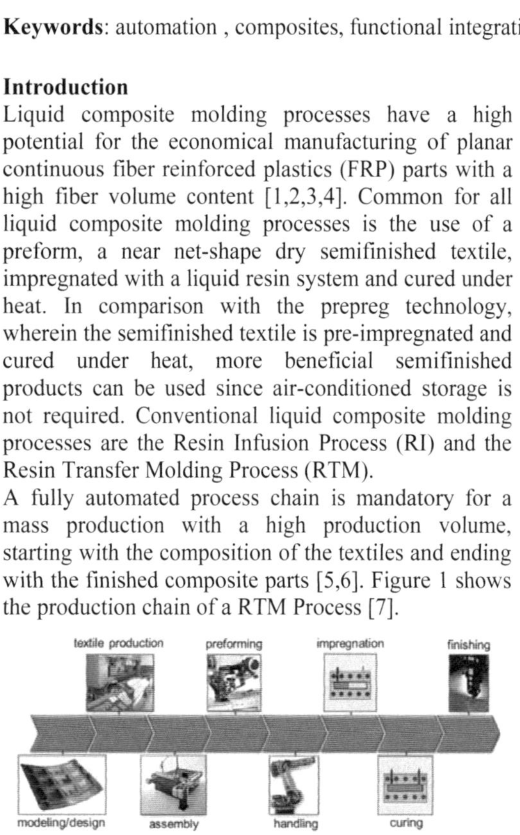

Figure 1 – Production chain of a RTM process

In order to produce parts with a low level of post-processing steps, automated manufacturing and an appropriate mold technology are indispensable for a stable and reproducible process.
The research unit 860 with the title: "New Process Chains for Production of Continuous Fiber Reinforced Plastic Parts: Integration of Preforming, Impregnation, Molding and Curing" has the production of FRP composite parts as one of the research objectives. The entire process is examined holistically from the production of the semifinished textiles to the production of the composite parts for the purpose of automated manufacturing. The research is carried out in collaboration with the Institute of Plastics Processing at RWTH Aachen University, the Institut fuer Textiltechnik (ITA) of the RWTH Aachen University and the Fraunhofer Institute for Production Technology (IPT). The research unit 860 pursues different concepts in three different process chains to achieve a mass production of composite parts with fiber volume contents over 50 % in a cycle of less than 10 min. The different processes are the Gap Impregnation Process, the Resin Spray Prepregging Process and the Resin Transfer Prepregging Process.

Introduction of the Gap Impregnation Process

The Gap Impregnation Process is a RTM-like process which combines the advantages of the RTM Process (the option of automation, high surface qualities on all

sides, and narrow production tolerances) and the RI Process (fast planar distribution of the resin, impregnation of the preform perpendicular to the surface) by employing a special mold technology and process control.
For the production of monolithic components, the mold consists of a lower part with a cavity (female mold) and a movable upper part (male mold). Using a silicone o-ring in a flute in the male mold, the mold can be sealed to evacuate the cavity by applying a vacuum. The schematic procedure for producing composite parts in the Gap Impregnation Process is shown in figure 2.

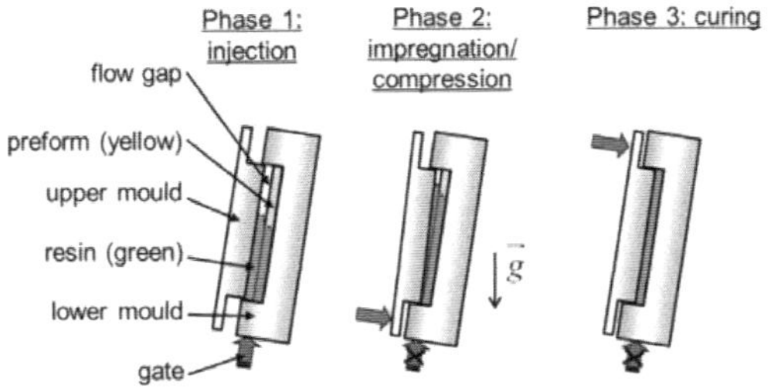

Figure 2 – Schematic procedure of the Gap Impregnation Process

At first a net-shape preform is placed into the cavity of the female mold, the cavity is closed and a defined flow gap (order of magnitude ~1-3 mm) above the preform is generated by lifting the male mold into a predefined position. The gap facilitates a fast and low-resistance resin injection. In addition, the preform is not compressed in contrast to the RTM Process, thus the porosity and the permeability are increased. This allows a more efficient and faster impregnation. After the evacuation of the cavity, a defined amount of resin is injected into the flow gap which distributes smoothly over the preform. Transferring the resin system into the flow gap facilitates a fast, low-resistance injection into the cavity. The actual impregnation of the preform takes place in the thickness-direction of the preform, resulting in short flow paths within the preform. In order to realize a homogeneous and fast injection of the resin system and an optimum evacuation of air bubbles, the mold is moved into an inclined position in which the flow direction of the resin system is oriented against gravity. In addition to the flow gap the uncompressed preform supports a very quick distribution of the resin system throughout the cavity. After the injection of a defined amount of resin into the cavity, the gate is shut and the flow gap is closed in a wedge-shape manner starting from the lowest point of the mold. The resin system is further distributed by closing the flow gap in the direction of extraction and impregnates the remaining dry areas of the preform. In the last step the suction is closed. At this point, plane pressure can be applied to the part and the desired fiber volume content is set. After curing of the resin system, the mold is opened and the part ejected.
Figure 3 shows the Gap Impregnation unit at the laboratory of IKV.

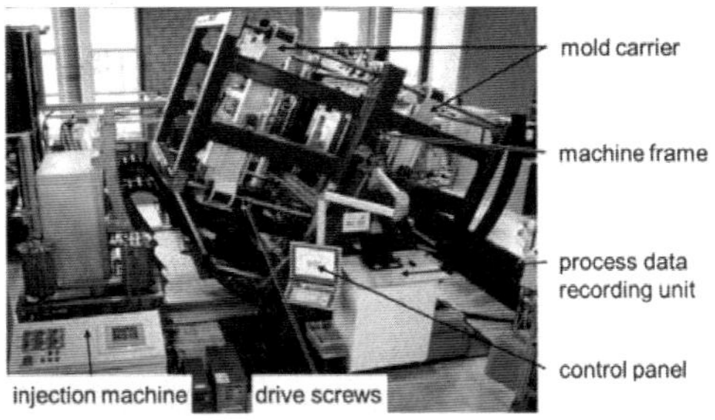

Figure 3 – Gap Impregnation unit at the laboratory of IKV

Experimental and Results

Monolithic and curved composite parts

In the beginning of the research unit 860, general investigations have been conducted to ensure a correct motion sequence, temperature control of the mold halves and the resin system injection sequence for different shaped parts. The produced part was a planar composite part of 2 mm thickness. The outer dimensions were 500 mm x 500 mm. Afterwards the acquired knowledge was transferred onto parts with a varying thickness and subsequently curved parts with a radius of 500 mm. In order to shorten production cycles and increase the production energy efficiency the whole manufacturing process is isotherm. Hence parts are produced in less than five minutes. Only 14 seconds of the manufacturing process are needed for the injection of the resin system and the complete and homogeneous impregnation of the preform. By using a fast curing resin system, the curing of the parts has been reduced to three minutes. The whole cycle time of the process takes only 280 seconds. With this process configuration, composite parts with a fiber volume content of 50 % could be manufactured in high quality.

Functional elements

The traditional manufacture of complex metallic parts usually requires post processing. Surfaces and load bearing elements as well as functional elements are manufactured in separate steps and combined in a subsequent process to form the finished product.
Another advantage of the Gap Impregnation Process lies in the reduction of these necessary steps. The design of FRP parts allows a more direct approach for the integration of numerous functional elements into the finished product in one step. Most required structures can already be placed in the preform and are completed during the impregnation and curing phase of the manufacturing process. This ensures a higher integrity of the finished parts and a reduction of process steps and production time.
In the Gap Impregnation Process, different substructures were integrated successfully. Those were cut-outs, connecting elements like inserts and stiffening profiles, which are all integrated into the preform prior impregnation. For the research on these

functional elements, a modular mold was constructed, that allows the integration of these elements.

Mold technology

The mold technology plays a vital part in the realization of any FRP mass production process. Not only surface-quality of the cavity, but also mold-dimensions, tolerances, heating and sealing-concepts, and even the motion-sequence of the mold parts have a high impact on the part quality and the possibility of automation of any FRP part production process.
The modular mold for FRP composite parts with functional elements is shown in figure 4.

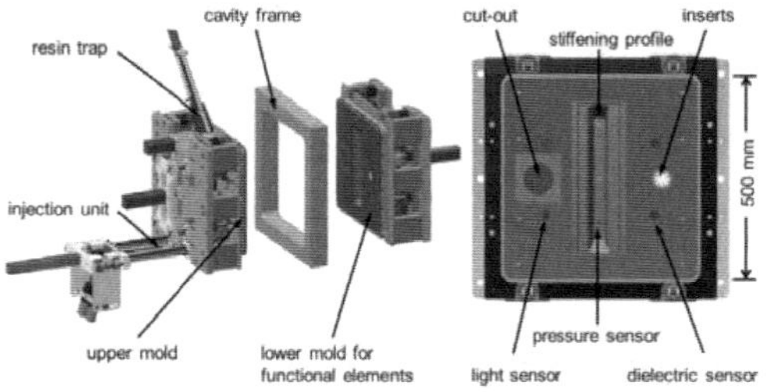

Figure 4 – Modular mold for the Gap Impregnation Process

For the process control and the process analysis, three different types of sensors were used in the process and integrated in the mold. Those include pressure sensors, light sensors to record the impregnation process and dielectric sensors, which make it possible to record the curing of the resin system in real time.

Cut-outs

The influence of three different cut-out geometries has been investigated based upon the research on monolithic and curved composite parts with the installed mold technology. Those include rectangular cut-outs in flow direction, rectangular cut-outs perpendicular to flow direction and circular cut-outs. To achieve a good part quality combined with a fast process, different process variations, like variation of the mold temperature, the flow gap height and the closing speed were investigated. With the appropriate combination of parameters, composite parts for all different cut-out geometries have been produced. The quality of the composite parts was ensured by ultrasonics, determination of the interlaminar shear strength and microscopy. All of the functional elements were integrated without any negative interference for the manufacturing process. Even in the direct proximity of the cut-outs, the impregnation was good and no defects were detected. Figure 5 shows the different cut-out geometries, which were installed in the modular mold, and the produced composite parts.

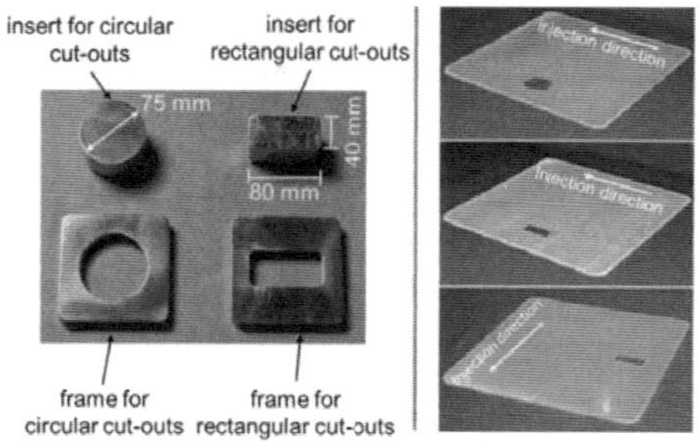

Figure 5 – Different cut-out geometries and produced composite parts

Integrated inserts

To introduce a force into composite parts, two different ways are possible by principle. The use of onserts, which are fixed onto the cured composite part by using a glue system. Or the integration of a load transfer element (insert) into the dry preform, which increases the level of integration and reduces the manufacturing steps. In the research unit 860, an aluminum insert was placed into the preform during the preforming process. The insert has a screwed bolt with an internal screw thread (M8) and 8 holes in the base plate which lead to a better impregnation of the parts on the rear side of the insert.
Based on the previous research, different process variations were executed. Next to the enhancement of the mold temperature to reduce the curing time, the resin injection speed and the plane stress were under examination. With optimized process parameters, inserts were embedded firmly and without pores in the cured composite part. The inserts are connected cohesively by the resin. Figure 6 shows a composite part with an integrated insert and two detail images of the front and back side of the composite part.

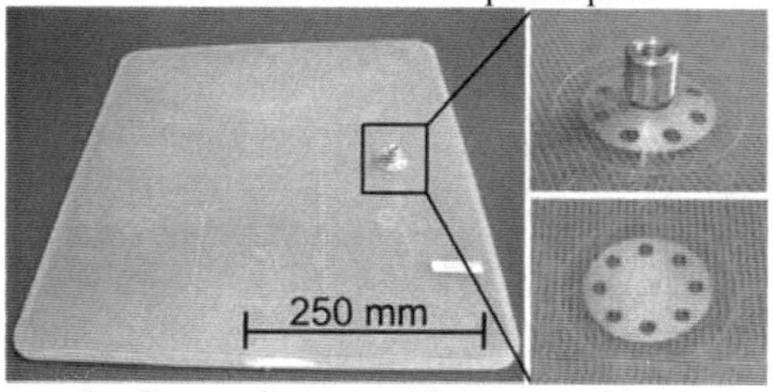

Figure 6 – Composite part with integrated insert

The quality of the composite parts was investigated by ultrasonic, microscopy and pull out strength tests.
A microscopic image of the load transfer element (perpendicular view) is shown in figure 7.

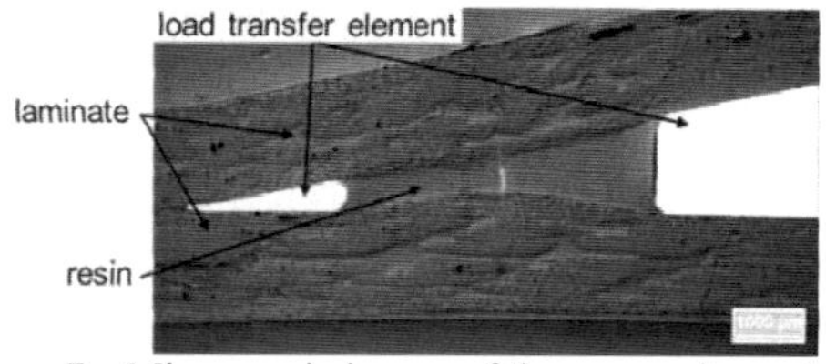

Figure 7 – Microscopic image of the composite part with integrated load transfer element

With this process, a very good impregnation of the fibers and a positive connection due to an ondulation of the fibers into the impregnation holes are achieved. These results are further supported in the pull out strength tests.

Stiffening profiles

In the last step of the research unit 860 composite parts with stiffening profile are studied. For the stiffening profiles two different flank angles of 45° and 70° are considered.

The integration of an impermeable foam core reinforcement structure facilitates the production of sandwich panels with a high bending stiffness. The challenge for the manufacturing process and mold technology is to achieve a mutual impregnation of the preform on both sides. Thus, a movement of both mold halves is required to generate a homogeneous flow in the flow gaps. Additionally, the preform is fixed centrally in the cavity to allow a homogeneous flow front on both sides of the preforms. Figure 8 shows a preform with an integrated stiffening profile manufactured at IKV.

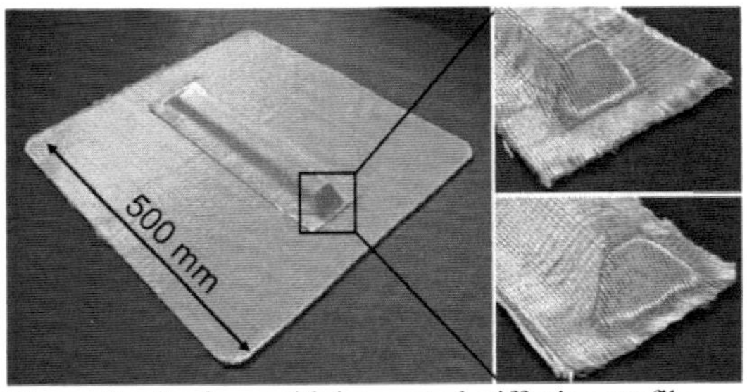

Figure 8 – Preform with integrated stiffening profile with two different flank angle (45° and 70°)

Actually the research of different process parameters, like mold temperature, injection speed, flow gap height and the closing speed are carried out. The quality of the composite parts with stiffening profiles will be analyzed afterwards by using ultrasonics and microscopy analysis.

Conclusions and Outlooks

The machine and mold technologies developed within the research unit 860 for the Gap Impregnation Process allows the production of composite parts with high fiber volume content in large quantities with desirable qualities and short production cycles. An isothermal process ensures high energy efficiency and almost eradicates any reoccurring heating and cooling of the mold.

The integration of functional elements (cut-out, insert and stiffening profile) into the preform before the impregnation leads to an economic manufacturing process of composite parts. After the research on stiffening profiles, the results of the Gap Impregnation Process will be transferred to more complex composite parts.

In those next steps to an industrial realization of the Gap Impregnation Process the manufacturing of a fully integrated carbon fiber reinforced engine hood is pursued. This engine hood will feature reinforcements and a curved shape. In addition inserts for the mounting of the hood to a car will be integrated in the finished product. The mold technology and production unit constitutes new challenges for an implementation into a mass production process, due to the dimensions of the hood (102 cm x 158 cm).

Acknowledgements

The presented investigations have been funded by the Deutsche Forschungsgemeinschaft (German Research Foundation, DFG) within the research unit 860 "New Process Chains for Production of Continuous Fiber Reinforced Plastic Parts: Integration of Preforming, Impregnation, Molding and Curing". This support is gratefully acknowledged by the authors. We also thank all companies that supported this research by providing materials and other equipment.

References

1. W. Michaeli; M. Wegener in *Einführung in die Technologie der Faserverbundwerkstoffe,* Carl Hanser Verlag, München, Wien, 1990
2. M. Flemming; G. Ziegmann; S. Roth in *Faserverbundbauweisen - Fertigungsverfahren mit duroplastischer Matrix,* Springer Verlag, Berlin, Heidelberg, 1999
3. M. Neitzel; P. Mitschang in *Handbuch Verbundwerkstoffe - Werkstoffe, Verarbeitung, Anwendung,* Carl Hanser Verlag, München, Wien, 2004
4. U. Eberth; A.S. Herrmann in *Technische Textilien für Luftfahrt- und Fahrzeuganwendungen, Technische Textilien, chapter 6: Mobiltextilien,* Deutscher Fachverlag GmbH, Frankfurt am Main, 2006
5. G.W. Ehrenstein in *Faserverbund-Kunststoffe - Werkstoffe, Verarbeitung, Eigenschaften,* Carl Hanser Verlag, München, Wien, 2006
6. P. Mitschang in 8. Internationale AVK-Tagung, Baden-Baden, 2005
7. Ch. Hopmann; K. Fischer; M. Emonts in 14. Internationale AVK-Tagung, Stuttgart, 2011

O-07-526

HIGH PRESSURE RTM – A MANUFACTURING PROCESS FOR LARGE SCALE PART PRODUCTION

F. Henning[1,2]*, R. Chaudhari[1,2], M. Karcher[1,2], P. Elsner[1,3]

[1]*Fraunhofer Institute for Chemical Technology ICT, Pfinztal, Germany – frank.henning@ict.fraunhofer.de; raman.chaudhari@ict.fraunhofer.de; michael.karcher@ict.fraunhofer.de; peter.elsner@ict.fraunhofer.de;*

[2]*Institute for Vehicle System Technology, Karlsruhe Institute of Technology, Germany*

[3]*Institute for Applied Materials, Karlsruhe Institute of Technology, Germany*

Abstract - The paper addresses new variants of the resin transfer molding (RTM) process namely high-pressure injection resin transfer molding (HP-IRTM) and high-pressure compression resin transfer molding (HP-CRTM) for manufacturing of continuous fiber reinforced composites with high fiber volume content. Both these processes utilize high pressure RTM equipment for precise dosing and mixing of highly reactive epoxy resin and amine hardener with relatively high throughput rates. The paper includes an introduction to the high-pressure RTM (HP-RTM) processes for reduction of the cycle time significantly and to improve the performance of the composite by increasing the achievable fiber volume content. The results present the influence of the process variables on the quality of the laminates and the mechanical properties. Finally the applicability of the HP-RTM process variants for high volume manufacturing of composites is discussed. The paper discusses the influence of different injection geometries such as film gate compared to pin gate as well as a direct comparison between the HP-IRTM process and the HP-CRTM process.

Keywords: RTM, HP-IRTM, HP-CRTM, infiltration, preforming, fiber volume content

Introduction

The resin transfer molding (RTM) process is well established for low volume manufacturing applications and has recently attracted significant interest in the manufacturing of higher volumes. This interest is strongly motivated by the need for "light-weighting" in the automotive and aerospace industries. The classical RTM process in its current status, however, is limited by the low volume production capacity of the preforming processes, the long impregnation time and the lack of robust injection equipment. These issues hinder the use of the RTM process for the high volume manufacturing of automotive components. In the classical RTM process the three-dimensional preform is placed into the cavity and the mold is completely closed. The mold closure leads to compaction of the preform to the final part thickness, thereby reducing the preform permeability. The reactive resin (premixed thermoset resin and hardener) is then injected into the cavity under a pressure of typically 1 to 20 bar [1]. The resin flows along the length of the preform and impregnates the fibers. Ideally, once the impregnation process is completed the resin undergoes a curing reaction, and after a defined time interval the part can be demolded. The complete closure of the mold prior to the resin injection leads to a reduction in the permeability of the preform, therefore resulting in longer resin injection and preform impregnation time. The longer impregnation time in the classical RTM process does not allow the use of fast-curing resins for high volume manufacturing of the RTM components. An alternative approach to achieve fast resin injection into the mold cavity is to use high-pressure RTM equipment which allows injection of the resin into the cavity with a high throughput rate. Using this processing equipment two different variants of the RTM process, namely high-pressure injection resin transfer molding (HP-IRTM) and high-pressure compression resin transfer molding (HP-CRTM) can be implemented for the manufacturing of continuous-fiber reinforced composites.

The HP-IRTM process differs from the classical RTM process only in terms of the processing equipment for the resin system. In the classical RTM process, low pressure mixing and dosing equipment is used for processing resins, whereas in the HP-IRTM process resin mixing (up to 180bar) and dosing (up to 230bar) is carried out using high-pressure RTM equipment. Similar to the classical RTM process, in the HP-IRTM process the preform is placed into the mold cavity and then the mold is completely closed to compact the preform to final part thickness. Once the mold is closed, the resin and hardener mixture is injected into the cavity under high injection speed. The high throughput rate allows the cavity to fill quickly, significantly reducing the resin injection time. As the resin can be injected into the cavity in a shorter time, this process variant allows the use of resins with relatively high reactivity. After the resin curing reaction is completed the part can be demolded [2, 3].

The HP-CRTM process is a combination of resin transfer molding (RTM) and compression molding in which, similar to the HP-IRTM process, high-pressure RTM equipment is used for processing of resins. If conducted without the use of high-pressure RTM equipment, the process is referred to in the literature as the Compression RTM process. In this process, the preform is placed into the mold cavity and then the mold is closed partially, leaving a small gap between the mold surface and the fiber preform. The resin is introduced into the gap, flows easily over the preform and may partially impregnate it. Once the required amount of resin has been injected into the gap and the injection point is closed, the mold closes further and applies high compression pressure to squeeze the resin into the preform. In this step, the preform is compacted to achieve the desired part thickness and fiber volume fraction. The part can be demolded after the resin has cured [4].

Various studies have been conducted by different researchers to investigate the effect of certain process parameters on the quality of manufactured laminates in the Compression RTM (CRTM) process. Kang et al. have attempted to analyze the CRTM process by studying the experimental and numerical data. Experiments were performed using a vinylester resin and complicated three dimensional preform based on continuous randomly-oriented glass fiber mats. The resin was injected into the cavity in 30 s and had a gel time of approx. 9 min. The fiber volume content achieved in the CRTM parts was approximately 17%. The varied process parameters were compression force and compression speed. A close agreement was found between the experimental and numerical data by Kang et al [5]. The effect of mold gap and gap closure speed on the void content and flexural properties of the CRTM laminates was investigated by Ikegawa et al. However the compression pressure and the cavity pressure used were extremely low and varied between 1 to 13 bar. The fiber volume content in the laminates produced was 33% [6]. Chang et al. investigated the effect of injection pressure, mold gap, resin temperature, compression pressure, mold temperature and curing temperature on the CRTM process. Experiments were performed using epoxy resin and bidirectional woven glass fabric and the fiber volume content obtained in the laminates was approximately 40%. However the compression pressure used was once again very low, varying only between 1 to 2 bar [7]. The reactivity of the chosen resins was, however, very low in most of these studies. Current author have investigated HP-CRTM processing using epoxy resins with different viscosities, different mold gaps and different laminate layups. For the bidirectional laminates manufactured by the HP-CRTM process, selection of high-viscosity epoxy resin led to roving displacement in the 90° layers (layers oriented perpendicularly to the mold length) due to the resin flow under compression pressure. This also caused poor quality impregnation in the areas where rovings were displaced. The poor quality impregnation and roving displacement resulted in measurement of higher fiber volume content (63 vol.%) compared to the fiber volume content measured in the well impregnated sections (59 vol.%). Reduction of the resin viscosity did not lead to any roving displacement. For the unidirectional laminates, the selection of different mold gaps and different resin viscosities did not influence the fiber volume content and mechanical properties of the laminates [8]. In the case of the HP-IRTM process only limited literature is available. Kacaras et al. have investigated the effect of two different epoxy resins and resin injection time on the HP-IRTM and HP-CRTM processes. Mainly the resin injection time was varied between 30 s to 11 s for these investigations. It was found that the faster injection of resin into the cavity led to poor impregnation quality of the HP-RTM laminates for one resin system, whereas equivalent impregnation quality was obtained for second resin system at all resin injection times [9]. The goal of the current study was to evaluate the effect of resin injection time on the properties of HP-RTM laminates manufactured using carbon fibers, as well as to demonstrate the high volume production capability of the HP-RTM process variants.

Experimental

For the current HP RTM process studies, an epoxy resin with the trade name Biresin® CR 170 and an amine-based hardener with the trade name Biresin® CH150-3 were used. A carbon-fiber-based biaxial non-crimp fabric from Hexcel Corporation was used as a reinforcement material for studying the HP-RTM process variants. This biaxial non-crimp fabric is manufactured using Toray carbon fiber type T620S-24K-50C. The non-crimp fabric used had a surface area weight of 330 g/m² and was available in two different orientations (0°/90° as well as 45°/-45°). A plate mold specially designed and constructed for studying HP-RTM process variants was used to compare the HP-IRTM process with HP-CRTM process. The cavity of this mold measured 910 mm x 560 mm and the cavity height varied between 1.55 mm to 1.75 mm near and away from the injection point respectively if the mold was completely closed. The injection point of this mold was designed in the middle of the lower mold half. This mold facilitated vacuum application through vacuum ports which were located on either side of the lower mold half. However, these vacuum ports were not used for studying the HP-RTM process variants. Four p-T sensors were situated in the lower mold half. Two of these sensors were located symmetrically to each other at a distance of 180 mm from the injection point (hereafter referred as sensors near injection point) and other two sensors were located symmetrically to each other at a distance of 410 mm from the injection point (hereafter referred as sensors away from the injection point). A hydraulic compression press from the company Dieffenbacher GmbH (press type - DYL630/500) was used to mount the RTM mold as well as to precisely apply defined compression force and to control the mold gap and gap closure speed during the HP-CRTM process study. For the mixing of the chosen epoxy resin and curing agent, and for injecting the reactive resin into the mold cavity, HP-RTM equipment from the company KraussMaffei Technologies GmbH was used.

Table 1 – Process parameter matrix used for evaluation of the HP-RTM process variants

Number of textile layers	Gate geometry	HP-CRTM		HP-IRTM	Resin amount [g]	Resin flow rate [g/s]
		Mold gap	Inj. time [s]	Inj. time [s]		
5 layers	Point gate	0.3	30	30	570	19
		0.3	15	15	570	38
		0.3	7.5	7.5**	570	76
8 layers	Point gate	1	30	30**	670	22
5 layers	Film gate	-	-	30	570	19
8 layers	Film gate	1	30	30**	670	22

Table 1 summarizes the process parameters used for studying the HP-RTM process variants. Two different laminate layups based on 5 layers of reinforcement [0/90#45/-45#0/90#-45/45#90/0] and 8 layers $[0/90\ 45/-45\ -45/45\ 90/0]_s$ of reinforcement were used for manufacturing the laminates using chosen process parameters of HP-RTM process variants. The table summarizes the used resin injection time and resin amount, mold gap and selected gate geometries for manufacturing of the laminates using both the process variants. The pressure on resin and hardener in the mixing was set to approximately 120 bar. The experimental series marked with "**" were not feasible due to an increase in the mixing head pressure to more than 230 bar which led to equipment safety shut down. A press force of 4200 kN was used for manufacturing HP-IRTM laminates, and a press force of 3100 kN was used for manufacturing HP-CRTM laminates. All the laminates were cured for 300 s after resin injection in the mold at a mold temperature of 120°C.

The manufactured laminates were characterized to evaluate fiber volume content (DIN EN ISO 7822), flexural properties (DIN EN ISO 14125) and inter-laminar shear strength (DIN EN ISO 14130) to evaluate the effect of selected process variants on the laminate mechanical properties.

Results and Discussion

Process analysis for HP-IRTM and HP-CRTM processes

Fig. 1 (a) shows the summary of the process data from the injection equipment for a laminate manufactured using a HP-IRTM process with 30 s resin injection time and point gate geometry. This figure is divided into three important sections, namely press closure, resin injection and curing. In the "press closure" step the mold was closed as the upper half of the press was moved down from its maximum opening height, and a defined press force of 4200 kN was applied on the mold, which contained 5 layers of textile reinforcement with defined laminate layup. In this step the high-pressure mixing head remained closed and the equipment re-circulated the resin and hardener through the high-pressure mixing head under a defined mixing pressure which was 117 bar for resin and 124 bar for hardener.

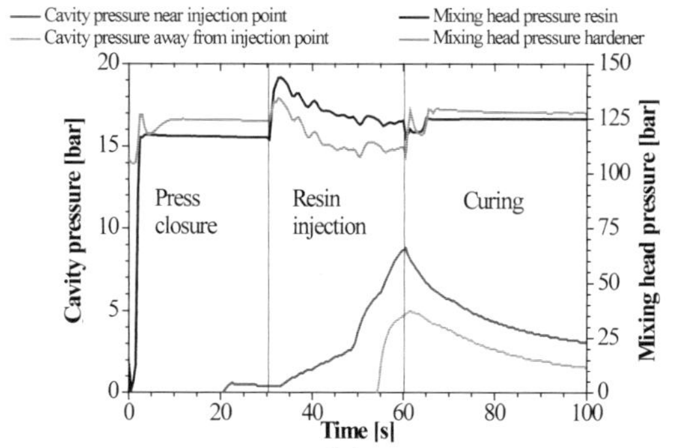

(a)

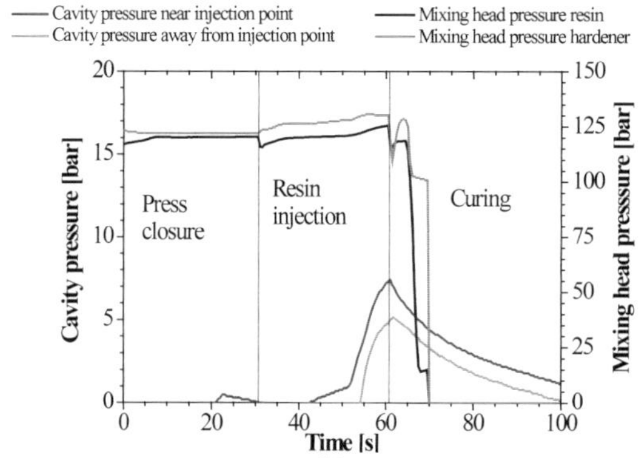

(b)

Figure 1 - HP-IRTM process analysis for manufacturing lamiantes with 5 layers of reinforcement and 30 s resin injection time; (a) point gate, (b) film gate

During the "resin injection" step, the high-pressure mixing head opened, resulting in resin and hardener mixing under high pressure, and the reactive resin mixture was injected into the cavity at a defined flow rate. As can be observed in fig. 1 (a), as the resin injection step started the resin and hardener pressure increased suddenly to 142 bar from an initial value of 117 bar, and to 132 bar from an initial value of 124 bar respectively, and during the whole resin injection step the resin and hardener pressure fluctuated. In addition, during this step the cavity pressure increased continuously. As the pressure sensors were placed at a distance of 180 mm and 410 mm from the injection point, an increase in the cavity pressure was realized with a time delay of 3 s near the injection point and 24 s away from the injection point after the resin injection step was started. Once the required amount of resin was injected into the cavity the mixing head closed again and the resin and hardener were shifted back to re-circulation mode. After the resin injection step was completed, the "curing" step followed during which the laminate was cured for 300 s and then demolded. During the curing step a continuous drop in the cavity pressure was observed which is attributed to a combination of resin shrinkage and minor resin flow out of the cavity through the gaskets. As only a drop in the cavity pressure was observed in the curing step, the complete time scale for process analysis in fig. 1 (a) is limited to 100 s only from the process start. A strong displacement of the rovings was observed directly at the injection point for the manufactured laminate after demolding. This roving displacement can be correlated with a sudden increase of resin and hardener pressure at the start of the resin injection step, and with subsequent fluctuation in the resin and hardener pressure throughout the whole injection step. As mentioned earlier the mold, when closed completely, exhibited a cavity height from 1.55 mm to 1.75 mm near and away from the injection point respectively. Hence a complete closure of the mold with a press force of 4200 kN led to strong compaction of the textile layers in the area near the injection point. This strong compaction of the textile can also be observed in the intensity increase of the cavity pressure near the injection point during the press closure step, which does not occur away from the injection point during the press closure step. This compaction of the textile layers near the injection point reduced the permeability of the textile layers in this area drastically, and as a result at the start of the resin injection step itself the resin and hardener pressure increased immediately. The later drop in the resin and hardener pressure and pressure variations can be explained by the displacement of rovings in the textile layers, leading to continuous changes in the local permeability of the textiles. The variations in the pressure can also be correlated with differential macroscopic and microscopic resin flow through textiles. As the resin flows through free spaces between rovings (macroscopic flow through regions with high permeability) the resistance to resin flow decreases, leading to pressure drop, and as the resin flows through a roving and between fiber filaments (microscopic flow through regions with low permeability) the resistance to resin flow increases, leading to pressure increase. Further experiments were conducted using point gate geometry and 15 s resin injection time. The reduction of resin injection time led to further significant increase of pressure on resin and hardener in the mixing head and thus led to higher fiber displacement in the laminates at the injection point.

Fig.1 (b) shows a summary of the process data for a laminate manufactured using HP-IRTM process and film gate geometry. As can be seen, if the gate geometry was changed from point to film geometry, a drastic change in pressure profile of resin and hardener was observed. The pressure on resin and hardener did not show any sudden increase and fluctuations during resin injection step. In contrast to the laminates manufactured using point gate geometry, no displacement of the rovings was observed at the injection point for the laminates manufactured using film gate geometry.

Fig. 2 shows a summary of the process data for a laminate manufactured using the HP-CRTM process with 30 s resin injection time. The process data for the HP-CRTM is divided into four important sections, namely press closure to defined mold gap, resin injection, compression and curing.

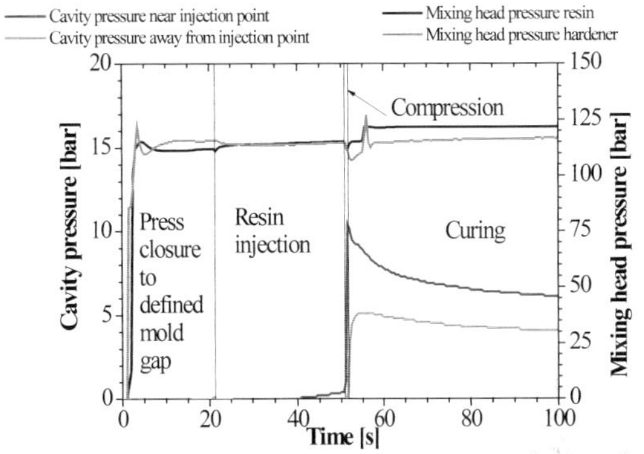

Figure 2 - HP-CRTM process analysis for manufacturing lamiantes with 5 layers of reinforcement and 30 s resin injection time; point

In the HP-CRTM process the first step is "press closure to defined mold gap". In this step the press was closed from its maximum opening height to leave a mold gap of 0.3 mm between upper and lower mold halves. Similar to the HP-IRTM process, in this step of press closure to the defined mold gap, the high-pressure mixing head remained closed and the equipment re-circulated the resin and hardener through the high-pressure mixing head under a defined mixing pressure which was 111 bar for the resin and 115 bar for the hardener. As it can be seen, 21 s were required for the press closure step to a mold gap of 0.3 mm. During the "resin injection" step, the high-pressure mixing head opened, resulting in resin and hardener mixing under high pressure. The reactive resin mixture was injected into the cavity at a defined flow rate of 19 g/s. Unlike the HP-IRTM process, as the resin injection step started, the resin and hardener pressure did not increase during this HP-CRTM experiment with 30 s resin injection time, and the resin and hardener pressure during the injection step remained constant at 115 bar. Almost no strong increase of cavity pressure near or away from the injection point was observed during the resin injection step. Once the required amount of resin was injected into the cavity the mixing head closed again and the resin and hardener were shifted back to re-circulation mode. Immediately after completion of the resin injection step, the "compression" step was carried out. The press closed the partially open mold completely, applying a compression force of 3100 kN.

During the compression step, which lasted for 2-3 s, there was an immediate increase in the cavity pressure. After the compression step, the "curing" step was carried out, in which the laminate was cured for 300 s and then demolded. In comparison to laminates manufactured by the HP-IRTM process with 30 s and 15 s resin injection time, very little roving displacement was observed in the area near the injection point. Also, as in the HP-IRTM process, during the HP-CRTM process no increase of cavity pressure intensity during press closure to the defined mold gap was observed. This indicates that there was no compaction of the textile layers as the press closed to a mold gap of 0.3 mm, and enough permeability of the textiles must have been retained homogeneously over the entire laminate area. Also, as no variations in the resin and hardener pressure during the resin injection step were observed and no cavity pressure built up during the resin injection step, it appears likely that the macroscopic resin flow was dominant during the resin injection step and that only a limited microscopic impregnation of the laminates was obtained during this step. As the cavity pressure rose significantly during the compression step, the major part of the microscopic impregnation of the roving filaments must have taken place during the compression step. Further experiments were conducted using point gate geometry and 7.5 s resin injection time. The reduction of resin injection time led to no significant increase of pressure on resin and hardener in the mixing head. However higher fiber displacement was observed at the injection point than in the laminates manufactured using HP-CRTM process and 30 s resin injection time. One probable explanation for the displacement of the rovings at 7.5 s resin injection time is a high flow rate of the reactive resin mixture through the injection gate, causing rovings to displace in the area near the injection point to accommodate the higher volume of resin entering the mold cavity. It should also be noted that although no strong cavity pressure built up was observed during the resin injection step. The sensor near the injection point was located 180 mm from the injection point, and a local increase of cavity pressure in the area near the injection point due to a high flow rate of the reactive resin is likely to be high, possibly causing displacement of the rovings.

Fiber volume content in the HP-RTM laminates

Fig. 3 (a) shows part thickness and fiber volume content in the laminates manufactured using 5 layers of reinforcement. As the laminates showed slight thickness variations over the entire surface, the presented fiber volume content is normalized to a thickness of 1.7 mm. The properties were measured sections near the injection point (NIP) and away from the injection point (AIP) which represents the area. The section near the injection point is defined as an area having 180 mm flow length from either side of injection point (180 mm is also the position of nearest pressure sensor in the mold to the injection point) and mold width of 560 mm. The section away from the injection point is defined as an area having between 180 mm to 400 mm flow length from either side of injection point (400 mm is the position of farthest pressure sensor in the mold from the injection point) and mold width of 560 mm. As it can be seen, all the laminates exhibited nearly equivalent fiber volume content in the sections NIP and AIP of the laminates independent of selected HP-RTM process variant and gate geometry. Fig. 3 (b) shows part thickness and fiber volume content in the laminates manufactured using 8 layers of reinforcement. In this case as well all the laminates exhibite nearly equivalent fiber volume content in both the sections of the laminates independent of selected HP-RTM process variant and gate geometry.

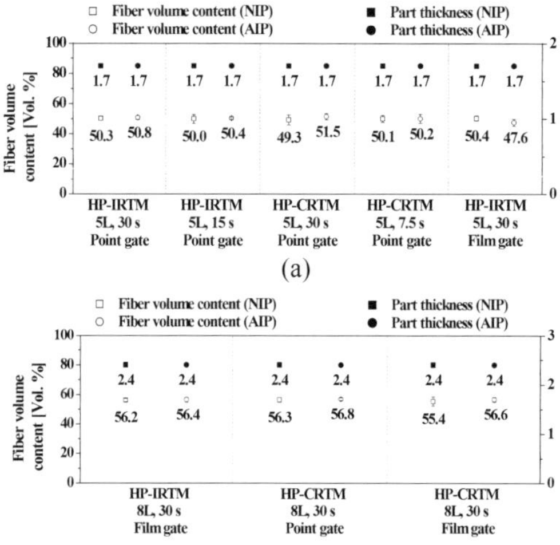

(b)

Figure 3 - Part thickness and fiber volume content in laminates manufactured by the HP-IRTM and HP-CRTM process variants; (a) laminates with 5 layers of reinforcement, (b) laminates with 8 layers of reinforcement

ILSS properties of the HP-RTM laminates

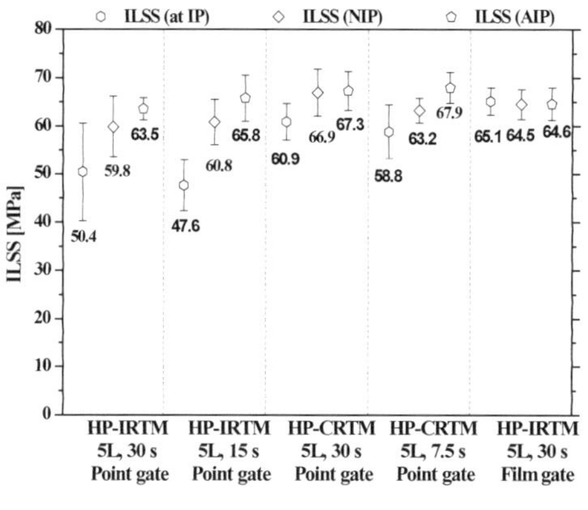

Figure 4 - Inter-laminar shear strength (ILSS) of laminates manufactured by the HP-IRTM and HP-CRTM process variants

Fig. 4 shows the ILSS properties of the HP-RTM laminates. The ILSS properties of the laminates were measured directly at the injection point (IP), near the injection point (NIP) and away from the injection point (AIP). It can be observed for the HP-IRTM process variant, the ILSS properties at the injection point tend to drop with reduced resin injection time (30 s resin injection time compared to 15 s resin injection time) if a point gate geometry is used. Also the ILSS values at the injection point show relatively high standard deviation as compared to ILSS values in other regions (NIP and AIP) of the laminate. This drop in ILSS values and high standard deviation for HP-IRTM laminates can be attributed to the higher roving displacement observed in the laminates when the point gate is used. The roving displacement resulted in local variations of resin content (and hence fiber volume content) in the test specimen, thereby leading to a larger deviation in the ILSS properties. In comparison to the HP-IRTM process, the HP-CRTM process at 30 s resin injection time and point injection gate geometry leads to nearly equivalent ILSS properties in all regions of a laminate. Also, the average ILSS values of the HP-CRTM laminates are higher than HP-IRTM laminates at 30 s resin injection time. However with 7.5 s resin injection time, the ILSS properties at the injection point show again a high standard deviation, attributed to increased roving displacement. In case of HP-IRTM process, if the point gate geometry was changed to film gate geometry, constant ILSS properties were observed in all the three sections of the laminates. This indicates that homogeneous quality was observed in the laminates manufactured using film gate geometry. The presentation will include an overview of further process data for the manufactured laminates, impregnation quality of laminates manufactured using 5 layers and 8 layers of reinforcement and flexural properties of the laminates.

Conclusions

The HP-RTM process variants were studied using a fast curing resin system to evaluate their applicabilty for high volme manufacturing of high-performance composites. Using both the process variants, laminates having fiber volume content of approximatley 50 to 56% were manufactured using 5 layers and 8 layers of textile reinforcement respectively. In case of the HP-IRTM process a reduction of the resin injection time from 30 s to 15 s was achieved whereas for the HP-CRTM process a reduction of the resin injection time from 30 s to 7.5 s was realized if a point gate geometry was used for laminate manufacturing using 5 layers of textile reinforcement. A minor decrease in the cavity height of the mold near the injection point led to a significant problem of roving displacement in the HP-IRTM process, if point gate was used, due to the reduced permeability of the textile layers. Due to the displacement of the rovings, the properties of the laminates dropped significantly at the injection point. As the point gate was replaced by a film gate geometry, then no displacement of rovings was observed and nearly constant properities were obtained in all the sections of the laminate. In comparison to the HP-IRTM process, the problem of roving displacement at 30 s resin injection time was not as significant as in the HP-CRTM process. However with a strong reduction in the resin injection time, problem of roving displacement was also observed in the HP-CRTM process. The use of film gate led to homogeneous quality in the laminates manufactured using 8 layers of reinforcment in both the HP-RTM process variants.

Acknowledgements

These investigations are carried out as part of the R&D activities of the "Technologie-Cluster Composites Baden-Württemberg" (TC²) project. The title of the project is "RTM-InnoHoch – RTM-Fertigungsprozesskette – Innovativ, hochintegriert". This project is funded by the European Union through the program "European Funds for Regional Development" as well as the state government of Baden-Wuerttemberg in Germany. The administrative agency of this program is the Ministry of Rural Development, Food and Consumer Protection. For more information visit www.rwb-efre.baden-wuerttemberg.de. We are also grateful to Daimler AG for providing the textile reinforcements and Sika AG for providing the resin system for conducting these investigations.

References

1. W. Young; C. Tseng *Journal of Reinforced Composites*, 1994, *13*, 467-482
2. M. Graf in 10th SPE ACCE Proceedings, Troy (MI), 2010, CD
3. S. Schmidhuber in 18th National Symposium SAMPE Deutschland Proceeding, Karlsruhe, 2012, CD
4. P. Simacek; S. Advani; S. Iobst *Journal of Composite Materials*, 2008, *42*, 2523-2548
5. M. Kang; W. Lee *Polymer Composites*, 1999, *20*, 293-304
6. N. Ikegawa; H. Hamada; Z. Maekawa *Polymer Engineering and Science*, 1996, *36*, 953-962
7. C. Chang; L. Hourng *Journal of Reinforced Plastics and Composites*, 2006, *25*, 1027-1037
8. R. Chaudhari, in CFK Valley Stade Convention Proceeding, Stade, 2011, CD
9. A. Kacaras, Diploma thesis, Karlsruhe Institute of Technology, 2012

O-07-483

ELECTROMAGNETIC INTERFERENCE SHIELDING EFFECTIVENESS OF MWCNT-FILLED POLY (ETHER SULFONE) AND POLY(ETHER IMIDE) BASED NANOCOMPOSITES

Susanta Banerjee*[a], Amit Das[b], Aruna Kumar Mohanty[a], Pravin Sawai[a], Anindita Ghosh[a], Kapil Pareekh[a], Brigitte Voit[b]

[a]*Materials Science Centre, Indian Institute of Technology, Kharagpur 721302, India - *susanta_b20012001@yahoo.com*

[b]*Leibniz-Institut für Polymerforschung Dresden e.V., Hohe Strasse 6, 01069, Dresden, Germany*

Abstract - Multiwall carbon nanotubes (MWCNTs) filled poly(ether sulfone) (PES) and poly(ether imide) (PEI) composites were prepared with different MWCNT weight fractions. Their electrical conductivities, electromagnetic interference (EMI) shielding effectiveness (SE), return loss (RL), and absorption loss (AL) were investigated. The electrical conductivity of these composites was measured at room temperature and the electrical percolation threshold values achieved in these systems were the lowest so far. The measurements of shielding effectiveness (SE) were carried out in the frequency range of 8-12 GHz (X-band range). It is observed that SE of the composites is frequency dependent and it increases with increasing frequency. The SE also increases with the increase in filler loading. The nanocomposites having 3 wt% of MWCNT exhibit shielding levels between 30-37 dB. The return loss and loss due to absorption were also measured as a function of frequency in the X-band range. Morphologies of the fracture surfaces of nanocomposites were studied by scanning electron and transmission electron microscopy.

Keywords: polymer nanocomposite, MWCNT, electromagnetic interference (EMI), shielding effectiveness (SE), electrical properties.

Introduction

In recent years, a revolutionary advancement of science and technology in the field of data and communication devices has been noticed. For equipment and devices applied in commerce, health care, defence and aerospace it is indispensable to use electromagnetic waves ranging from the microwave to the millimeter wave. Often, this equipment comes across the common phenomenon of electromagnetic wave interference (EMI) which obstructs them to function normally [1]. For example, when a cell phone is brought near to a television (TV) set, an audio-visual disturbance (noise) happens in the TV. The effect of EMI can range from degradation to interception and obstruction of the performance of electronic or electrical equipment. Therefore, in order to alleviate these troubles the development of EMI shielding materials for microwave and millimeter waves are receiving increasing attention.

EMI shielding refers to the reflection and/or adsorption of electromagnetic radiation by a material, which thereby acts as a shield against the penetration of the radiation through it. Depending on the shielding mechanism, a shielding material should have at least one the following properties such as moderate conductivity and/or electric or magnetic permeability (high dielectric constant) [1]. Since, the polymer matrix is usually electrically insulating and does not contribute directly to the resultant shielding effect, extensive research in the CNT composites has been taken place with different polymer matrices in recent years for their end-use application as effective light weight EMI shielding material. For any conducting filler, the EMI shielding effectiveness increases with increasing filler concentration in the composites, but the maximum filler loading is limited by the poor mechanical properties of the composite at high filler loadings resulting from the poor filler-matrix bonding. For materials and process cost saving and good mechanical properties, the attainment of a high shielding effectiveness at a low filler loading is desirable. The EMI shielding efficiency (SE) of a composite material depends on many factors, including the filler's intrinsic conductivity, dielectric constant, and aspect ratio [1]. In this type composites, the main contribution to overall EMI shielding is due to reflection.

Polyimides/ poly(ether imides) (PEI) and poly(ether sulfone) (PES) are known to be high performance polymers due to their outstanding set of properties [2,3]. In this paper PEI and PES are chosen as polymer matrices because of their good thermo-oxidative stability, good chemical resistance, high mechanical strength and most importantly very good processability for preparation of polymer/MWCNT composites for EMI shielding application. Though, extensive articles have been published for PEI/MWCNT composites, to the best of our knowledge, the EMI-shielding application based discussion is not done yet. Similarly, metal particles based PES composites [4] have been studied, but no studies are found on PES/MWCNT for EMI shielding application. So, in the preset investigation the influence of different weight percentages of MWCNT and different polymer matrices on the conductivity and EMI shielding has been studied for a frequency (8-12 GHz) in the X-band region. The electrical, thermo-mechanical and the morphological properties of the resulting polymer composites have also been investigated.

Experimental

Materials

The dihalide 4,4′-dichlorodiphenyl sulfone (DCDPS) and 1,3-phenylenediamine (mPD) (99 %) were purchased from Aldrich (USA) and were used as received. 4,′4 -Isopropylidenediphenoxy)bis(phthalic anhydride) (BPADA) (97 %) was purchased from Aldrich, USA and oven dried overnight at 120 °C before use. Bisphenol namely′ 4,4 - isopropylidenediphenol (BPA) was purchased from Loba Chemical Company (India) and was recrystallized from toluene. K_2CO_3 and dichloromethane (DCM) (E. Merck, India) were used as received. N-methyl-2-pyrrolidone (NMP) (E. Merck, India) was purified from NaOH and distilled from P_2O_5 before use. Toluene (Merck, India) was refluxed over Na metal to remove water and freshly distilled before use. Ortho dichloro benzene and methanol were from (E. Marck, India) and used as received. Multiwalled carbon nanotubes were purchased from Nanocyl S.A., Sambreville, Belgium. According to the provided data sheet from the manufacturer, the average diameter, average length, carbon purity, metal oxide impurity, and surface area are 9.5 nm, 1.5 μ, 90%, 10%, and 250-300 m^2/gm respectively.

Synthesis of PEI and PES

PEI and PES was prepared following the standard polymerization protocol as reported elsewhere [3,5]. GPC results indicate the formation of high molar mass (PEI, M_w ~ 75,000 g/mol and PES, M_w ~ 135,000 g/mol). The DSC analysis of PEI and PES showed T_g as 215 °C and 196 °C. The structure of the repeat unit of the synthesized PEI and PES is represented in Figure 1(a) and (b).

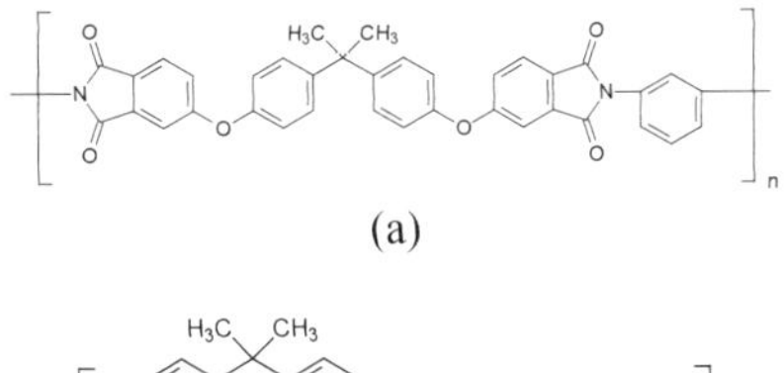

(a)

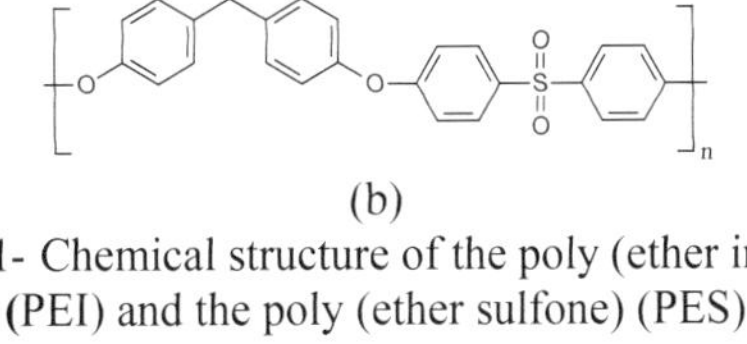

(b)

Fig. 1- Chemical structure of the poly (ether imide) (PEI) and the poly (ether sulfone) (PES)

Preparation of nanocomposite films and EMI specimens

Nanocomposite films having different concentration loading (0.5, 0.75, 1.0, 1.5, 2.0 wt%) of CNT have been prepared from ultrasonication induced solution dispersion technique using DCM as solvent for both PEI and PES. The uniform suspension of MWCNT-polymer was poured on flat bottomed Petri dishes and the solvent was allowed to evaporate at a controlled rate at 30 °C overnight to obtain composite films. Similarly, EMI waveguide samples having different loading (3, 4, 5 wt%) were prepared by filling the adequate amount corresponding composite films into the specimen and then, by hot compression molding technique. Samples with compact and smooth surface were obtained in all the cases.

Testing procedures

Electron microscopy – Scanning electron microscopic analysis was done using JEOL JSM-5800 SEM. This was done to observe the morphology of fracture surfaces of the composites and dispersion of CNT in the composites. Samples were fractured in liquid nitrogen and the fractured surfaces were gold plated, and then mounted over aluminum stub using double sided electric adhesive tape. The vacuum was in the order of 10^{-4} – 10^{-6} mm Hg during scanning of the composite samples.

The dispersion of the MWCNT in the polymer matrix was further studied by a transmission electron microscope (HRTEM, JEM-2100, JEOL, Japan) operated at an accelerating voltage of 200 kV. The composite sample was ultramicrotomed under cryogenic condition with a thickness of 100 nm. Since the MWCNT has much higher electron density than the polymers, it appeared dark in the TEM images.

Dynamic mechanical analysis (DMA) - Dynamic mechanical analysis was performed with an Eplexor 2000N) Gabo Qualimeter, Ahlden, Germany) using a constant frequency of 10 Hz in a temperature range from 20 to 230 °C. The heating rate was 2 °C/min. This experiment was done in tension mode with samples of a thickness of ~0.5 mm. For the measurement of complex modulus, E*, a static load at 1% pre-strain was applied and then the sample was oscillated to a dynamic load at 0.5% strain.

AC electrical conductivity - The dielectric properties were measured with a broadband dielectric spectrometer, BDS 40 system (Novocontrol GmbH, Germany). The experiments were done with samples of a thickness of 0.5 mm and a diameter of 20 mm. To ensure good contact, a gold plated brass electrode was used. The samples were sputtered with a thin layer of gold on both sides. The frequency range was 0.1 Hz to 10 MHz and the real part of the conductivity was directly obtained from the instrument.

EMI shielding measurement - The EMI SE was measured with a set-up consisting of a sweep oscillator (model HP 8350B, Hewlett Packard), a power splitter, a detector and a scalar network analyzer (model HP 8757 C/E, Hewlett Packard). The SE was measured using the co-axial cable line method. The SE in the frequency range 8–12 GHz was measured using X-band wave-guide as sample holder. Samples having a thickness of 4.96 mm were used in the measurement. EMI shielding measurement was carried out for each sample continuously sweeping the frequency over ranges of 8–12 GHz. When SE was found to be greater than 40 dB, some noise was encountered during the measurement; otherwise, the result obtained is a smooth curve of SE against frequency.

Results and Discussion

Morphology

Figure 2(a) and (b) show FESEM images of the cryo-fractured surface of the compression molded PEI/MWCNT and PES/MWCNT composites loaded with 5 wt. % MWCNT, respectively. The images give the impression of a percolative morphology due to overlapping of the tubes. Though it is difficult to comment on the dispersion of MWCNT in the PEI matrix from these images with high percentage of loading, clearly a random distribution and dispersion of the tubes could be found (Figure 2b) for the PES matrix. A lower viscosity of the PES-MWCNT solution (during the solution processing) could be the reason behind this fact. As a whole, it seems that strong van der Waals forces between the polymer and the nanotubes during exceed the existing inter-tube forces, thus allowing a good dispersion during sonication and preventing tube aggregation which results in a controlled morphology. Thicker diameter of MWCNT (25-40 nm) in micrograph compared to the as received MWCNT (~ 10 nm) confirms wrapping of plymer. The wrapping with polymer (PEI or PES) is due to π-stacking and/or non-covalent interactions between the nanotubes and the polymer matrix.

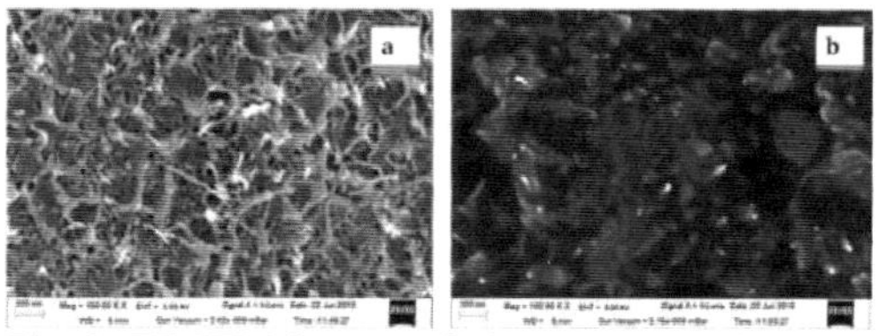

Fig. 2- FE-SEM images of 5% nanocomposite (a) PEI (b) PES

The bulk morphology of the composites was further investigated by transmission electron microscopy (TEM). TEM micrographs revealed that MWCNTs were randomly well dispersed in both PEI and PES matrix. At low loading, the MWCNTs are dispersed at an individual level. As the nanotube loading increased, these randomly dispersed carbon nanotubes come closer (in other words, decrease in path length between nanotubes) and form a percolative network structure. In contrast, in the TEM micrographs for PEI/MWCNT composites the MWCNTs appear to be straight in the PEI matrix. This may be due to differences in the physical properties like viscosity, structural rigidity of the two polymers. The relative variation in the length of MWCNTs in both PEI (longer) and PES (shorter), even as the same MWCNT may also be indication of the entangled and curved MWCNT structure in the PES matrix which could be cut several times and appeared as one or more individual CNTs in TEM images.

Thermo-mechanical properties

DMA measurements indicated single glass transition curve for individual composites of the both the polymer matrix due to the amorphous glassy to rubbery state of the polymer matrix. In the course of heating after reaching the glass transition regime, the polymers flowed out from the sample holder resulting in a sharp drop of the storage modulus values. It is interesting to note that PEI/MWCNT show improvement of ~10-15°C in the Tg as compared with pure PEI, where as PES/MWCNT show an identical glass transition behavior irrespective of their MWCNT content. Despite the fact that in the glassy region the reinforcement effect of MWCNTs did not show, in the rubbery region the storage modulus increased with the increase of the MWCNT loading in the PES polymer. This fact indicates that the higher chain length of the PES macromolecules led to intensive entanglements.

AC Electrical conductivity

PEI/MWCNT composite behaved like a semiconductor even at a loading as low as 0.5 wt%. Owing to the presence of MWCNT, the formed three dimensional CNT network is carrying charge by hopping and/or tunneling mechanisms. It is also observed that with the increase of MWCNT content, the conductivity of the PEI matrix is increasing and the value reaches to ~ 0.003 S/m for the composites filled with 2 wt. % MWCNT. A value of 0.03 S/m is shown by PES which is 10 times higher than PEI composites at the same loading of MWCNT. There may be two reasons for the higher conductivity of PES/MWCNT composite compared to PEI/MWCNT composite. First, it may be due to an orientation effect [31,35,36] caused by the significant alignment of nanotubes in the PEI matrix resulting a partial anisotropic structure Second, the absolute conductivity values achieved after reaching the percolation threshold in MWCNT composites depend on the path difference between the tubes. Most probably the MWCNTs are well wrapped by PEI polymer chains and the direct contact area between the tubes is reduced resulting in inferior conductivity of PEI composites compared to the PES composites. Whereas, in PES/MWCNT composites, MWCNT are not so tightly covered by the polymer chains and more direct contacts between the tubes exist which enhances the charge carrying properties of the composites.

EMI shielding effectiveness

An electromagnetic shielding material is a material that attenuates radiated electromagnetic energy. The SE of CNT nanocomposites was measured over the X-band region (8-12 GHz). It was observed that SE increased with MWCNT loading at all frequencies of the incident radiation for both PEI and PES. The MWCNT composites from PEI and PES exhibited mostly a frequency independent EMI SE performance. An appreciable increase in SE was found when the frequency was changed from the lower to higher frequency regime; for example, the nanocomposites of PEI at 3 wt% of MWCNT exhibited a shielding level between 30-37 dB over 8-12 GHz, while for the nanocomposites of PES at 4% of MWCNT, the shielding value measured was between 35-43 dB over 8-12 GHz. This increment of the EMI shielding effectiveness is ascribed mainly due to increase in the formation of conductive interconnected nanotube networks in the polymer matrix with increase in MWCNT loading. The conductive network formed due to the dispersion of MWCNT behaves like a conductive mesh which intercepts electromagnetic radiation. The size of the mesh generally determines the effectiveness of shielding. A small mesh size generally increases the shielding effectiveness. However, the size of the mesh is also related to frequency of the incident radiation which will be intercepted.

EMI SE of the material has three contributions: reflectivity (R), Absorptivity (A) and the multiple internal reflections (B) [6]. For, approximate calculation of contribution behavior, the internal reflection (B) part could be neglected (particularly when $A \geq 15$ dB) and approximate A value can be calculated from experimental reflectance (R) and transmittance (T) value using the equation $A = 1 - R - T$. It is seen that for high frequency (12 GHz), at low wt% loading of PEI/MWCNT composite, both A and R increase with increasing loading. But interestingly, when the loading is high (5 wt%), an increase in A and decrease in R is observed. For example, for 3 wt% MWCNT loading, the R value is as high as 93% and A is 6%. When the loading is increased to 4 wt%, the R and A values changed slightly to 94% and 5%, respectively. But when the loading is further increased to 5 wt%, a shift in contribution behavior is noticed. : the R value decreased from 94% to 83% and the A value increased to 13% which is almost double to what has been observed at 3 wt% loading. So, overall, for the entire PEI/MWCNT composite, the major contribution to total SE comes from reflection rather than absorption quite similar to other polymer matrix CNT composites. However, the results of R and A measured for PES/MWCNT composite at high frequency (14 GHz) are quite interesting: for low loadings there is a gradual increase in the R value and a small change in the A value. But when the loading is increased (5%), reversal contribution of R and A values to the total SE takes place, i.e. at 5 wt% loading, the R value is as low as 13 % and the A value is 86%. The cause for the reversal of the contribution is still unknown and hence, requires further research.

Conclusions

The EMI SE of PEI and PES based MWCNT composites have been studied as a function of both concentration loading of filler and frequency over X-band region. EMI SE has been found to increase with with increasing the weight fraction of MWCNT at room temperature. The MWCNT composites of PEI and PES showed very high electrical conductivity at very low loading of MWCNT. It is found that the percolation network of MWCNT in the composites is achieved around 0.5 wt% due to the high aspect ratio of MWCNT. Furthermore, differences in the morphological microstructure features and network structures played a major role in deciding the conductivity and EMI SE. The nanocomposites comprised of 5 wt% of MWCNT exhibit EMI SE in between 42-45 dB at 8GHz. In almost all nanocomposites, the main contribution to total SE comes from reflection except in the PES/MWCNT composite with 5 wt% loading. However, particularly at high frequency, composites showed gradual shift from the factors contributing to EMI from reflection to absorption with increasing MWCNT loading. This can be attributed to an increase in dielectric loss values with increasing MWCNT fraction. Because of high EMI SE (in the range 29-47 dB) along with excellent physical, chemical properties and good molding property of PEI and PES, these composites are thus promising light weight shielding material for use in both civil and military applications.

References

1. D. D. L Chung, *Carbon* 2001, 39, 279.
2. N. G. Sahoo; S. Rana; J. W. Cho; L. Li; S. W. Chan, *Prog Polym Sci* 2010, 35,837.
3. R. Srivastava; S. Banerjee; D. Jehnichen; B. Voit; F Bohme, *Macromol Mater Eng* 2009, 294, 96.
4. L. Li; D. D. L Chung, *Composites* 1994, 25, 215.
5. A. Ghosh; S. Maji; S. Banerjee; B. Voit, *Polym Adv Technol* 2011, 22, 794.
6. P. Sawai; S. Banerjee, *J Appl Polym Sci* 2008, 109, 2054.

O-07-497

IMPROVED DISPERSION OF CELLULOSE MICROCRYSTALS BY SURFACE ACETYLATION IN POLYLACTIC ACID (PLA) BASED COMPOSITES

T. Mukherjee*, N. Kao, Rahul K. Gupta, N. Quazi, and S.N. Bhattacharya
Rheology and Materials Processing Centre,
School of Civil, Environmental and Chemical Engineering, RMIT University, Melbourne, Victoria 3001, Australia
tapashi.mukherjee@rmit.edu.au; nhol.kao@rmit.edu.au; rahul.gupta@rmit.edu.au; nurul.quazi@rmit.edu.au; sati.bhattacharya@rmit.edu.au

Abstract – Design of sustainable bioplastics can be achieved by preparing composites from renewable materials like microcrystalline cellulose (MCC) fibre and biopolymer such as Polylactic acid (PLA). The key driving factor that affects their performance is the quality of dispersion of MCC and the interfacial interaction between the matrix and the reinforcement phase. In the previous study of surface modification of MCC, one way to facilitate improved dispersion was attempted by esterification reaction using acetyl chloride. In this study, the degree of improved surface acetylation is carried out by dimethyl formamide and sodium hydride. Confirmation of acetylated group is accompanied by FTIR study. Morphological study is conducted by ESEM. The change in crystalline property is observed by XRD study. The results are compared with composites prepared by surface acetylation of MCC with acetyl chloride. Improved dispersion is observed at an optimal loading of 2.5 wt%, with the new acetylation reaction. This study is aimed to enhance the surface acetylation technique for improved dispersion.

Keywords: Dispersion; Cellulose Microcrystals; Acetylation; Poly (lactic acid); Biopolymers

Introduction

Cellulose as available from natural fibres has been much explored as a reinforcing material in biopolymer based matrix like polylactic acid or polylactide (PLA) [1] to produce a completely biodegradable composite. Microcrystalline cellulose (MCC) fibre, as the name suggests, is commercially available in 20-50 micron level [2]. Due to the presence of hydroxyl groups on their surface, MCC has the tendency to stick together or agglomerate resulting in crack or failure of the composite [3]. In previous study, surface modification was carried out on microcrystalline cellulose (MCC) by substituting hydroxyl groups with acetyl chloride and fatty acids [4], [5]. In this study, improved srface acetylation is accompanied by reaction with sodium anhydride and dimethyl formamide. Confirmation of acetylated group is accompanied by FTIR study. Morphological study is conducted by ESEM. The change in crystalline property is observed by XRD study. The results are compared with composites prepared by surface acetylation of MCC with acetyl chloride. Improved dispersion is observed at an optimal loading of 2.5 wt%, with the new acetylation reaction. This study is aimed to enhance the surface acetylation technique for improved dispersion.

Experimental

Material:

PLA biopolymer (Naturework PLA Polymer 4032D) with a density of 1.24 gcm^{-3} and a melting temperature of 160 °C was chosen as a matrix. Microcrystalline cellulose (MCC) with a mean particle size of 20 µm, supplied in powder form by Sigma Aldrich was used as a raw material. Acetic anhydride, sodium hydride, and dimethyl formamide were purchased from Sigma Aldrich.

Sample Preparation:

In a one litre round bottom flask, 15 g of MCC and 100 ml of dimethyl formamide were stirred at 60 °C for half an hour. Sodium Hydride (7 g, 60 % dispersion in mineral oil) was added portion wise in the reaction mixture and stirring was continued overnight (16 h) in a nitrogen atmosphere. Next day, acetic anhydride (10 ml) was added and stirred for further 3 hours.
The above reaction mixture was cooled to room temperature with continuous stirring. Finally ethanol (50 ml) was added carefully to deactivate excess sodium hydride. The reaction mixture was poured into 400 ml of water and stirred; the solid was collected by filtration and washed with water and acetone, dried at 80 °C in a vacuum oven, to give the acetylated MCC (~15g). The new surface acetylated MCC were coded as AC-MCC-NaH to separate them from previous acetylation where, they were coded as AC-MCC.

Preparation of Composites:
Composites were prepared by solvent casting technique, using dichloromethane (DCM) as the medium. The desired amount of AC-MCC was added along with 20.0 g of PLA into DCM to produce a mixture. The mixture was then conditioned overnight to eliminate bubbles and was then cast into a petri dish. The DCM was allowed to evaporate at ambient

temperature (25 °C) for 24 h. Finally the solidified films, with a thickness of around 1 mm, were vacuum dried overnight, and then kept in a desiccator containing silica gel. Composites with pure cellulose are coded as PLA-MCC, while composites with acetylated cellulose with acetyl chloride and sodium hydride are coded as PLA-AC-MCC and PLA-AC-MCC-NaH, respectively.

Characterization

FT-IR Spectroscopy:
FTIR spectra of the powdered MCC, AC-MCC and AC-MCC-NaH were recorded using a PerkinElmer FTIR spectrophotometer (TA 8000). The powders were measured using a KBr-drift method, in the range of 4000-450 cm^{-1}. The spectrum was analysed with a Perkin Elmer Spectrum software.

Environmental Scanning Electron Microscopy:
Morphology of MCC and the fractured surface of the MCC based PLA composites at different loading of weight percentage, were examined using an FEI Quanta 200 Environmental Scanning Electron Microscope (ESEM), at an acceleration voltage of 30 kV and pressure of 0.50 Torr. MCC powder was sprayed in a carbon tape to study its morphology. The fractured surface was prepared by immersing the specimen in liquid nitrogen and manually fracturing the specimen. Samples were then mounted onto a sample holder with fractured surface facing upwards. The sample holder was then placed in the ESEM sample chamber. Various sample surfaces were scanned to obtain a visual impression of fibre fracture, distribution and the appearance of the fibre/polymer interface.

X-Ray Diffraction Pattern Study (XRD):
XRD patterns were obtained using a Bruker D4 Endeavor X-ray diffractometer in the angular range of 6 ° to 90° (2θ) at a voltage of 40 KV and current of 100 mA. Peak intensities were every 0.02° at sweep rates of 1.0° 2θ/min, to see the change in crystallinity after acetylation, if any.

Shear Rheology:
Dynamic Frequency Sweep tests were conducted using ARES Rheometer (TA Instrument), based on constant strain rate. The tests were conducted using 25 mm parallel plate at temperature of 170 °C. All measurements were performed using a force transducer with a range of 0.2 to 200 g-cm torque. Prior to any tests, the zero gaps between the parallel plates were calibrated at the required temperature to take care of the thermal expansion. Dynamic Strain Sweeps were conducted to check the linear viscoelastic region (LVR) of these composites. All tests were conducted only in the LVR. Dynamic frequency sweep tests were conducted between the frequency of 0.1 and 100 rad/s at 170 °C.

Results and Discussion

Acetylation of MCC (FT-IR Analysis):
The reaction scheme of acetylation acetyl chloride along with FTIR results is represented in Fig.1. The spectral curve shows the appearance of a new peak in the carbonyl area around 1752 cm^{-1}, as associated with the formation of the ester group. Intensity of the peaks located around 3350 cm^{-1} was assigned to –OH stretching, which subsequently decreased after acetylation.

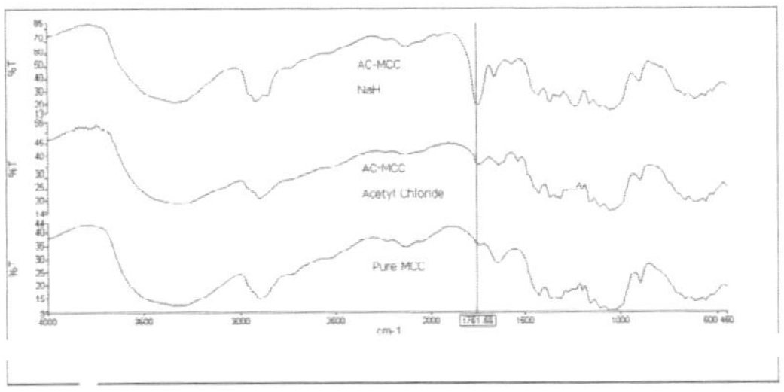

Fig.1. FTIR Study of the AC-MCC with acetyl chloride and sodium hydride as compared to pure MCC.

From the FTIR study, it was further observed that the carbonyl peak for the new reaction is more intense than that of the previous reaction by acetyl chloride, indicating enhanced degree of substitution of the ester group replacing the hydroxyl group.

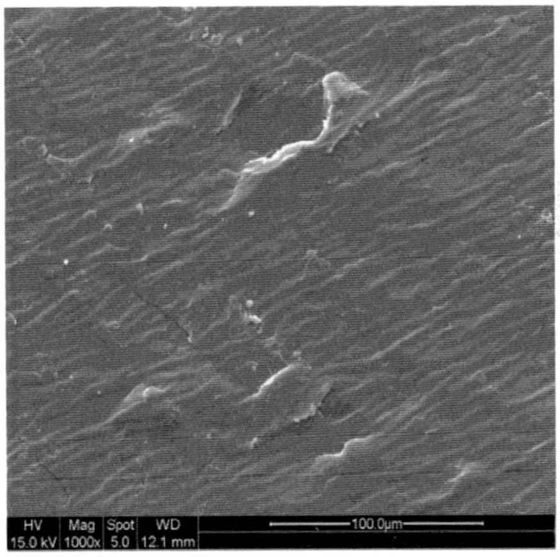

(a)

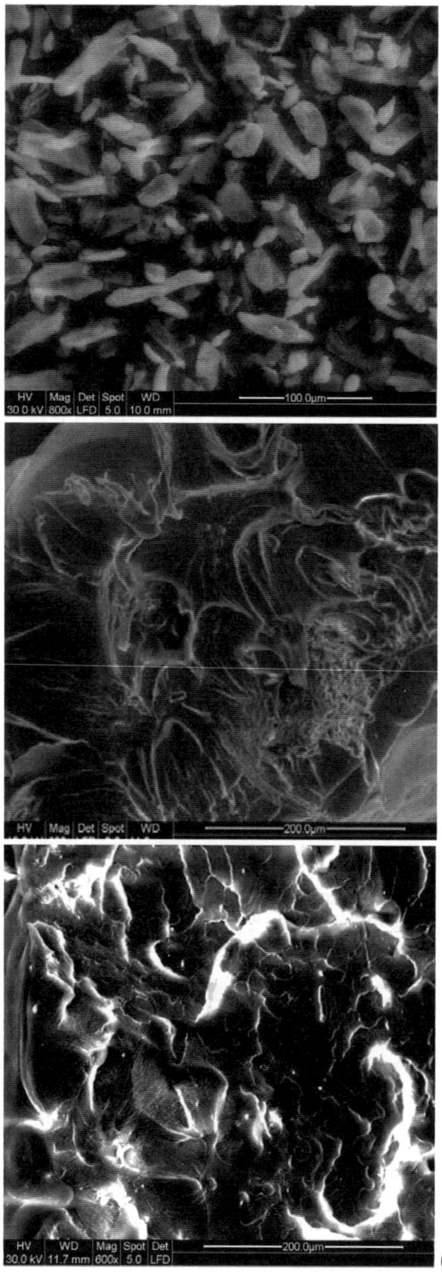

(b) (c) (d)

Fig.2.(a) Neat PLA Surface, (b) Morphology of MCC (c) PLA-AC-MCC (2.5 wt %)(with acetyl chloride) (d) PLA-AC-MCC (2.5 wt %)(with sodium hydride and Dimethyl formamide)

Morphology and Structural Changes:

Fig. 2 (a) shows neat fractured surface of PLA, while Fig. 2 (b) represents the morphology of MCC. From Fig. 2 (b) aspect ratio as calculated appears in the range of 4.45 -7.5µm. Fig 2 (c) and Fig. 2 (d) represent PLA-AC-MCC composites at 2.5 wt% with the previous reaction with acetyl chloride and the new reaction with dimethyl formamide and sodium hydride, respectively. From the ESEM picture, it is evident that the new reaction enhanced the dispersion of the MCC as compared to the reaction with acetyl chloride.

In XRD analysis, neat PLA was mainly comprised of the amorphous polymer structure with intermittent crystalline structure characteristics. It exhibited three diffraction peaks located at about 16.4°, 19.1° and 22.6° and there is an appearance of another peak at 44.2° of 2θ as shown in Fig 3(a). Acetylated MCC, when introduced at a different loading (1.5 - 5 wt%), improved the crystallinity of neat PLA in overall. With increase in level of loading, the acetylated cellulose microcrystals exhibited two distinct diffraction peaks located at around 64° and 82.6° of 2θ, particularly at a loading above 1 wt%. The peak around 44.2° became more significant above that loading as well. Appearance of these peaks with increased loading of acetylated MCC is indicative of the enhanced crystalline structure of PLA-AC-MCC-NaH composites.

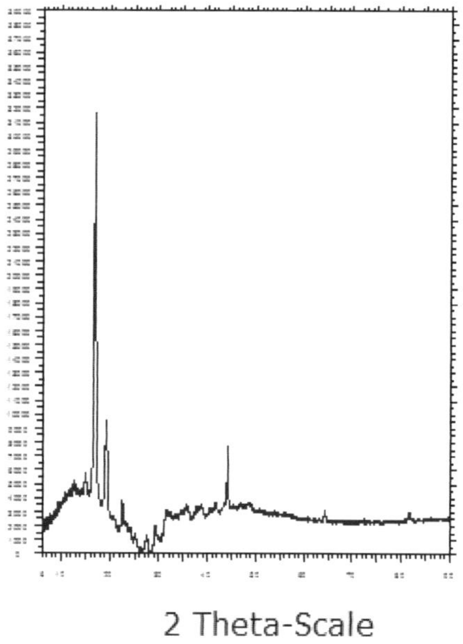

Fig.3.(a) Neat PLA XRD Pattern

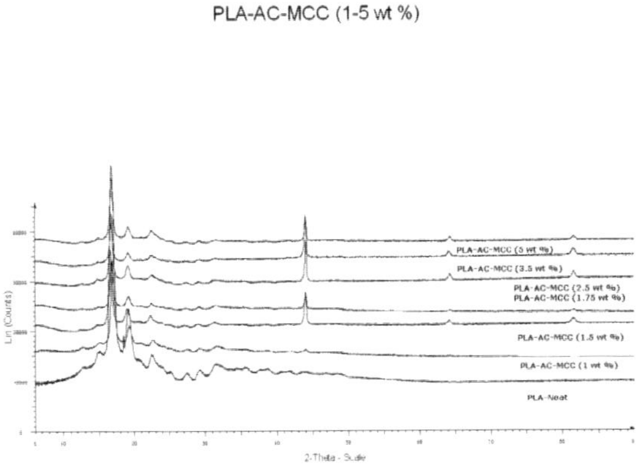

Fig.3. (b) PLA-AC-MCC XRD Pattern

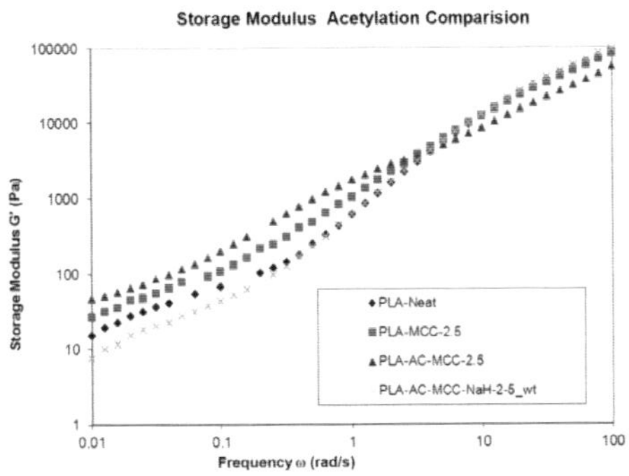

Fig.4.(a) Comparison of Storage Modulus (G')

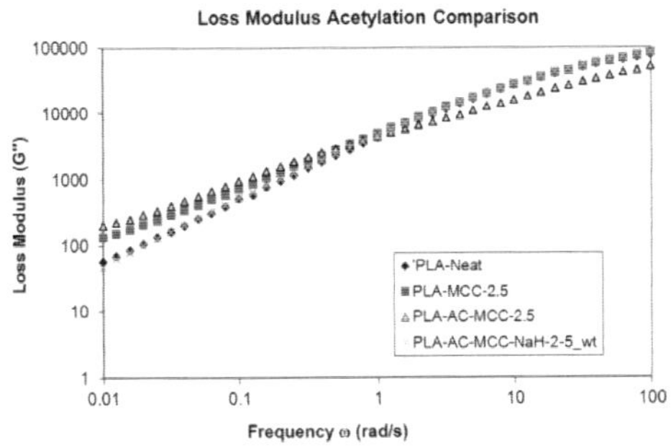

Fig. 4 (b) Comparison of Loss Modulus (G'')

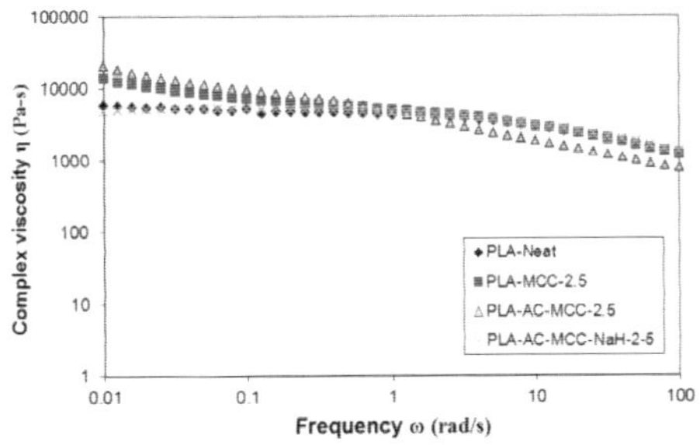

Fig. 4(c) Comparison of Complex Viscosity (η *)

Shear Rheology:

Fig.4 (a - c) compares the storage (G'), loss modulus (G'') and complex viscosity (η*) of the acetylated MCC reinforced PLA based composites with that of pure MCC based PLA composites at a loading of 2.5 wt%, which was found to be the optimal loading from the previous study [2].

As evident from this figure, both G', G'' and η* values are higher in the composite containing new acetylation of MCC with dimethyl formamide and sodium hydride, in comparison to the composite containing acetylation with acetyl chloride at a higher frequency region (1 – 100 rad/s). At a higher frequency, the hydrodynamic force dominates over the fibre/matrix adhesion force. From the XRD data, it is also evident that AC-MCC-NaH has got its highest crystallinity from 2.5 wt %. High crystallinity contributes to the high storage modulus. There could also be a possibility that shear induced crystallization is contributing to the high storage modulus (G') at a higher frequency.

Conclusion

The new acetylation reaction of MCC with dimethyl formamide and sodium hydride has its both advantages and disadvantages. Surface acetylation is more intense as can be revealed from FTIR and morphological analysis. There is a significant improvement in crystalline structure and storage modulus of the polymer matrix as observed in XRD and shear rheology.

References

1. A.N. Frone; S. Berlioz; J.F. Chailan; D.M. Panaitescu; D. Donescu, *Polym. Comp.* 2011, *32,* 976
2. L. Peterson; I. Kvien; K. Oksman; *Comp. Sci. & Tech.* 2007, 67, 2535
3. G. Siqueira; J. B. Dufresne; *Polymers*, 2010, 2, 728
4. T. Mukherjee in Condensed Matter and Materials Meeting, Wagga Wagga, Australian Institute of Physics, http://www.aip.org.au/wagga2012/
5. T. Mukherjee in PPS America Proceedings, Niagra Falls, Ontario, 2012, Canada.

O-07-510

EFFECT OF REPROCESSING ON THE PROPERTIES OF POLYETHYLENE/CLOISITE 15A NANOCOMPOSITES

F.P. La Mantia[1]*, M.C. Mistretta[1], M. Morreale[2]

[1] *Department of Civil, Environmental, Aerospace, Materials Engineering, University of Palermo, Italy – francescopaolo.lamantia@unipa.it, mariachiara.mistretta@unipa.it*

[2] *Faculty of Engineering, Architecture and Motor Science, Kore University of Enna, Italy - marco.morreale@unikore.it*

Abstract - Recycling of polymer based materials is, nowadays, a consolidated technique and can lead to significant environmental and economical advantages. On the other hand, the increasing interest for polymer-based nanocomposites can lead to concerns regarding their behavior upon recycling, especially when multiple reprocessing cycles are carried out.

In this work, the effects of reprocessing on the rheological and mechanical properties of LDPE/Cloisite 15A nanocomposites have been investigated. The components have been processed using a batch mixer at different temperatures. Rheological and mechanical tests showed a significant influence of the processing temperature on the properties of the obtained materials.

Keywords: nanocomposites, reprocessing, rheological properties

Introduction

Polymer nanocomposites can lead to some important improvements if compared to the neat polymer matrices. For this reason, polymer nanocomposites are starting to gain interest in the field of industrial applications [1].

Since the use of polymer based nanocomposites these materials in the industry is gradually increasing, the recycling of these materials is beginning to gain importance.

The scientific literature reports only few studies on the effects of recycling and reprocessing on polymer nanocomposites [2-4].

Furthermore, there are no systematic studies on the behavior of low-density polyethylene (LDPE)/ modified clay nanocomposites. In this work, LDPE/Cloisite 15A systems were prepared in a batch mixer, operating at different temperatures and processing speed. The behavior was investigated by rheological, mechanical, and morphological analysis.

Experimental

Materials

The polymer matrix used was a low density polyethylene (melt index 0.26 dg/min, produced by Versalis, Italy), while the nanofiller was an organo-modified clay (Cloisite® 15A, produced by Southern Clay Products, average diameter of 8 μm, organo - modifier concentration 125 meq/100 g clay).

Processing

All the systems were prepared by processing in a Brabender batch mixer operating at different temperatures (160, 210 and 280 °C), with processing time of one hour and speed equal to 60 rpm. Some samples were prepared, for comparison purposes, at a lower speed (10 rpm), in order to assess the effects upon reducing the mechanical stress.

Characterization

Rheological properties included MFI (melt flow index), detected by using a Ceast equipment, and rheological curves in shear obtained using a ARES (TA Instruments) in a plate-plate mode at a test temperature of 190 °C, in the range 0.1-500 rad/sec.

SEM micrographs were obtained by using a FEI Quanta200 ESEM.

Mechanical tests were performed with an Instron 3365 machine on samples prepared by compression molding.

Results and Discussion

The reprocessing was simulated by using a batch mixer operating for long time periods, rather than performing multiple extrusions, since it is known that these lead only to small changes when LDPE is used.

The filled materials show a slightly intercalated morphology, due to the limited affinity between the non-polar matrix and the polar filler. Moreover, the sample processed at lower temperature had a very poor dispersion of the particles in the matrix (micrographs not shown here for sake of conciseness). After one hour processing, the filled materials show a better dispersion and a small increase of the intercalation.

In order to understand the effect of processing temperature and processing speed (and thus, mechanical stresses), Figure 1 reports the melt flow rate for the samples processed at 160, 210 and 280 °C, at lower (10 rpm) or higher (60 rpm) processing speed. The increase of temperature (from 160 °C to 210 °C) leads to a decrease of MFI and then an increase of viscosity. This means that the effect of the lower temperature is to significantly raise the mechanical stresses during processing, leading to chain scission phenomena, resulting in a reduced viscosity. Furthermore, the effect of processing speed is in the same direction: the increase of speed causes higher mechanical stresses and thus increased chain scission phenomena, resulting in lower viscosities-higher MFIs.

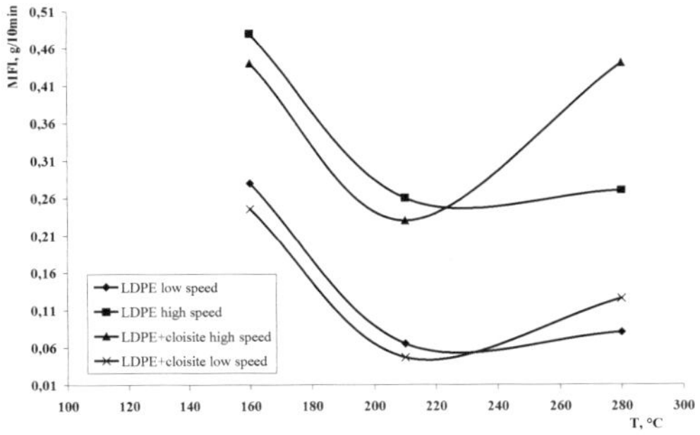

Fig. 1 – Melt Flow Index of the investigated materials

In addition, all of the curves have the same shape and trend, this means that the presence of the nanosized clay does not affect the degradation path during processing, which is therefore controlled by the degradation of the matrix; this, in turn, depends mainly on the decrease of the molecular weight (due to chain scission) or to the formation of chain branching. These two mechanism are basically in competition, and the final result depends on the prevalence of one or another. Both of them rely on the formation of radicals following shear and/or thermal degradation [4-6]; here, the obtained results suggest that macromolecular chain scission prevails at lower temperature. On the other hand, MFI at the higher temperatures (210 and 280 °C) is significantly lower, and may be explained with an increase of branching (and/or crosslinking) phenomena., with no significant differences between 210 °C and 280 °C.

However, the situation is different for the nanocomposites. Actually, the behavior is quite similar to the unfilled samples up to 210 °C while, upon increasing the temperature, the trend is completely different: the melt flow index increases, and thus the viscosity decreases. This is particularly striking at higher processing speed.

In this case, an explanation could be found by taking into account the role of the filler, which was negligible at the lower temperatures (as discussed above). In particular, it is known form the literature that the organically modified clay can undergo degradation phenomena involving the organic modifier, known as "Hoffmann's degradation". The presence of these degradation products (for instance, α-olefins) can increase the rate of formation of radicals and, therefore, the rate of chain scission reactions. The following reduction of molecular weight leads, in turn, to the increase of MFI reported.

Table 1 shows the main results of the mechanical tests for the nanocomposites processed at the three investigated temperature and higher processing speed. It can be stated that the aforementioned degradation phenomena can explain also these results, as they actually reduce the tensile properties, in particular the elongation at break.

	TS [MPa]	EB [%]
LDPE + Cloisite (160 °C)	8.1	110
LDPE + Cloisite (210 °C)	10	86
LDPE + Cloisite (280 °C)	8	60

Table 1 - Mechanical properties of the nanocomposites

Conclusions

The behavior of polymer nanocomposites after reprocessing/recycling (here simulated by conducting long-time batch mixing operations at low and high processing speed) depends on several factors, such as the processing conditions and the degradation behavior of every component. For instance, in the case of clay-filled nanocomposites investigated in this work, the degradation of the organo-modified clay can give rise to the production of radicals, and thus the degradation depends also on the interactions between the degradation products of the two components (polymer and filler). This effect increases upon increasing the processing temperature, while at lower temperatures the degradation of the polymer matrix (including different mechanisms) prevails.

The results show also that the level of degradation and of change of properties increase significantly upon increasing the processing time.

References

1. R. Pfaendner, *Polym. Degrad. Stab.* 2010, *95*, 369.
2. G.M. Russo, V. Nicolais, L. Di Maio, S. Montesano, L. Incarnato, *Polym. Degrad. Stab.* 2007, *92*, 1925.
3. E.-K. Karahaliou, P. A. Tarantili, *J. Appl. Polym. Sci.* 2009, *113*, 2271.
4. N. Touati , M. Kaci , S. Bruzaud, Y. Grohens, *Polym. Degrad. Stab.* 2011, *96*, 1064.
5. G.R. Rideal, J.C. Padget, *J. Polym. Sci.* 1976, *57*, 1.
6. A.V.Machado, J.A. Covas, M.Van Duin, *J. Appl. Polym. Sci.* 2001, *81*, 51.

O-07-511

COUPLED EFFECT OF HUMIDITY EXPOSURE AND CURE PRESSURE ON THE FORMATION OF MICROVOIDS IN COMPOSITE LAMINATES

J.P. Anderson and M.C. Altan*

School of Aerospace and Mechanical Engineering, University of Oklahoma, USA- jande@ou.edu; altan@ou.edu

Abstract - Microvoids are common manufacturing defects found in thermoset composite components and are known to adversely affect mechanical properties. The moisture level in composite prepregs before curing and the pressure applied during cure are the two primary factors that influence the final void content of a cured composite. It has been well-established that the formation of microvoids can be reduced if the prepregs are not exposed to a humid environment and the pressure applied during cure is increased. In this work, the coupled effect of humidity environment and cure pressure on microvoid formation during cure was investigated. This was achieved by fabricating eight-ply laminates in a hot press at cure pressures ranging from 69 kPa to 345 kPa. For each cure pressure, a laminate was fabricated from prepreg which was unconditioned (direct from storage) and conditioned at 25%, and 99% relative humidity for 24hrs at 25°C to induce different initial levels of moisture in the lay-up. The void content of these laminates were found to vary from 1.6-5.0% depending on humidity environment and the applied cure pressure. It was observed that with increased cure pressure, the void faction in all laminates approach an asymptotic value of approximately 1.6% regardless of the humidity environment. Existing predictive models for the process-induced microvoids needed to be modified to account for this coupled effect of humidity and pressure. Using an asymptotic void fraction value at high cure pressures and a pressure reduction coefficient to distinguish actual resin pressure from the applied cure pressure resulted in an accurate void model that showed excellent agreement with experimental data.

Keywords: micro-voids, composite laminate, humidity exposure

Introduction

Voids are one of the most common defects in composite materials and are known to degrade their mechanical performance [1-5]. Many studies have addressed the causes of void formation in components manufactured by autoclave processing of prepregs [6-8]. Due to a growing need to find lower cost alternatives to autoclave processing, other manufacturing processes such as hot-press molding or vacuum-bag curing have been utilized [9]. However, the current void formation models developed for autoclave curing may not be applicable or accurate for other composite laminate processing techniques.

The models currently available to predict void content often rely on a diffusion-based void growth model proposed by Kardos et al. [7]. This model takes into consideration resin moisture content, temperature, pressure and time. In order to determine void content, Boey and Lye [6] suggested the use of a unit matrix volume along with the Kardos model. This approach was recently confirmed experimentally by Grunenfelder and Nutt [10]. However, in its present form the model makes or implies assumptions which are not necessarily applicable to all manufacturing processes for composite laminates. The following three assumptions are the most commonly used: (i) the resin pressure is equal to that of the applied pressure, (ii) the fiber reinforcement bed does not carry any of consolidation load, and (iii) the resin moisture content is not time dependent, i.e., it is based solely on the relative humidity at the time of manufacturing. In this study, the void content prediction model developed by Boey and Lye [6] for the autoclave process is generalized to account for difference in the applied pressure and internal resin pressure as well as the effect of the moisture absorbed by the prepregs before fabrication. The generalized model can be used to predict the process-induced void content in laminates fabricated by press curing, vacuum bag curing and autoclave curing.

Proposed Model Framework

The void prediction model as proposed by Boey and Lye [6] and can be expressed as

$$V_v\% = \left(\frac{A*RH^2+B}{P}\right)^3 \frac{4}{3}\pi \qquad \text{Eq. (1)}$$

where $V_v\%$ is the final void content, $A*RH^2$ is the moisture content in the resin, B is a constant and P is the applied pressure.

In this study, it is of interest to determine if Eq. (1) could be modified such as

$$V_v\% = \left(\frac{C*M_c+D}{P(1-P_R)}\right)^3 \frac{4}{3}\pi + V_{min}\% \qquad \text{Eq. (2)}$$

where C and D are experimental constants, M_c is moisture content, P_R is the pressure reduction coefficient and $V_{min}\%$ is the minimum void content possible for all conditions and is dominated by the load carried by the fiber bed.

Experimental

Materials
TenCate® BT250/7781, a glass/epoxy prepreg, was used to manufacture all samples in this study. This material is composed of a balanced eight-harness woven E-glass fabric and an epoxy matrix which cures at 121°C.

Fabrication Procedure
Nine 8-ply laminates (L=152mm W=152mm) were manufactured by the hot-press process using curing pressures of 69 kPa, 207 kPa, and 345 kPa. For each of the cure pressures, a laminate was fabricated from prepregs which were: (i) unconditioned (direct from storage), (ii) conditioned at 25% and, (iii) conditioned at 99% relative humidity for 24hrs at 25°C before curing to induce different initial levels of moisture in the lay-ups. A cure cycle of a 1°C/min ramp from 25°C to 121°C, a two-hour isothermal hold at 121°C, and a 10°C/min ramp down to 60°C was used for all laminates.

Moisture Content
The time-dependent moisture content of the prepregs was determined by exposing prepreg strips (L=31.8mm W=25.4mm) to humidity environments of 25%, 99%, and 45% and recording their moisture contents with time until moisture saturation. The moisture contents of the laminate lay-ups were also determined at their time of manufacturing by using the 1-D Fickian moisture diffusion model [11].

$$G = 1 - exp\left[-7.3\left(\frac{D_x t}{2h^2}\right)^{0.75}\right] \qquad \text{Eq. (3)}$$

$$M_c = G(M_m - M_i) + M_i \qquad \text{Eq. (4)}$$

where D_x is the diffusivity through the thickness, h is the prepreg thickness, M_m and M_i are the saturation and initial moisture content respectively.

All the moisture measurements were determined using a Computrac Vapor Pro® moisture analyzer. The analyzer functions by circulating a dry carrier gas across a heated sample and measuring the relative humidity of the outflowing gas. Combining the humidity data, gas flow rate, and sensor temperature, allows for the moisture transmission rate from the sample to be determined. When this data is integrated with time the total moisture mass of the sample is found. For the experiments in this study, the gas flow rate was set to 75 mL/min, the samples were heated to 121°C, and the moisture content measured until the moisture transmission rate decreased below 0.5 µg/s.

Void Content Measurement
From each laminate, eight specimen (L=31.8mm W=12.7mm) were cut to allow for the void contents of the manufactured laminates to be determined. To accomplish this, the density of each composite specimen was determined by suspending each in a solution of Cargill Labs 2.49g/cc Heavy Liquid and water. Similarly, the matrix density was determined by suspending void free matrix specimens in a water/glycerol solution. The subsequent measurement of the solution density allowed for the accurate measurement of the bulk density of each specimen. The density of the fiber reinforcement was determined by helium pycnometer.

The matrix (ρ_m) and fiber (ρ_f) reinforcement densities were found to be 1.224 g/cc and 2.635 g/cc respectively. The void contents in the specimen were then found by determining the matrix (W_m) and fiber (W_f) weight contents of each specimen. This was done by using the matrix burn-off technique according to ASTM D2584 and using Eq. (5) as recommended by ASTM D2734.

$$V_v\% = 100 - \rho_c\left(\frac{W_f}{\rho_f} + \frac{W_m}{\rho_m}\right) \qquad \text{Eq. (5)}$$

Results and Discussion

Moisture Diffusion in Uncured Prepregs
The moisture content of TenCate® BT250/7781 for different humidity enviroments with the progression of time and the 1-D Fickan model fit is shown in Fig. 1.

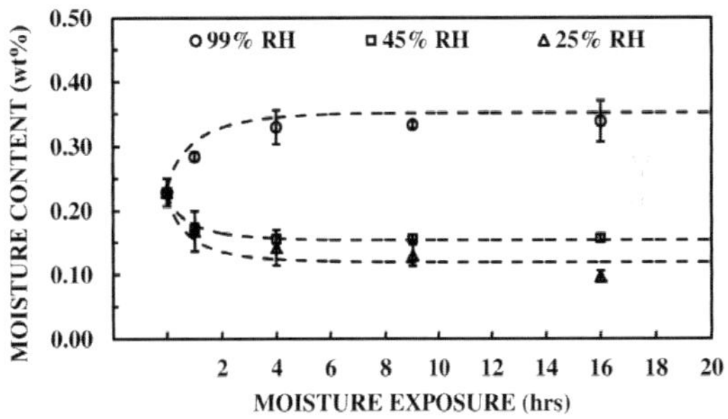

Figure 1 – Moisture content of TenCate® BT250/7781 exposed to different humidity environments until saturation is reached. The model prediction obtained using Eq. (4) is also shown.

In Fig. 1, it is found that the diffusion of moisture into and out of the prepreg is Fickian and has a diffusivity of 4.01E-2 mm²/hr. This finding indicates that a relatively short exposure time to a humid environment can potentially have a drastic impact on final part quality. For example, the prepreg in this study was found to become saturated with moisture after just 10 hours of exposure for all three humidity environments. In addition, knowing prepreg diffusivity allows for the moisture content of a prepreg to be determined provided its humidity exposure is monitored over time.

Humidity and Pressure Effect on Void Formation

The relationship between void content and processing pressure is shown in Fig. 2 along with the fit obtained for Eq. 2 to the experimental data using MatLabs' optimization tool.

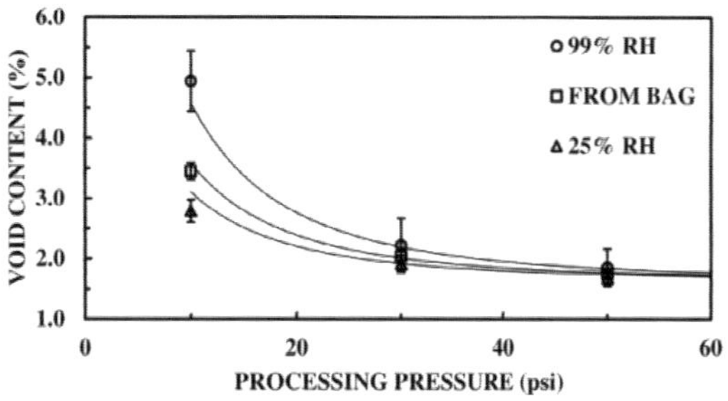

Figure 2 – Effect of humidity environment and processing pressure on the formation of microvoids.

In Fig. 2, the proposed improvements to the current predictive model, as presented in Eq. 2, are found to very accurately predict the final void contents of the fabricated laminates. The parameters obtained from the fit are displayed in Table 1.

Table 1 – Fitting parameters of Eq. 2 for the experimental data.

P_R	V_{min}%	C (atm)	D (atm)
0.055	1.644	1.958	0.816

From Table 1, the importance of the pressure reduction coefficient becomes obvious. This term shows that the pressure in the resin is less than the applied pressure. This is logical as the edges of the laminate are unconstrained in hot-press manufacturing. The magnitude of the reduction coefficient is, therefore, expected to be related to the permeability of fabric during curing as discussed by Gutowski et al. and Dave et al. [12, 13].

Also, Table 1 shows that the minimum theoretically possible void content is 1.644%. This finding indicates that as processing pressure is increased, and the fiber volume content with it, the fiber bed begins to carry an increasing amount of the applied pressure [14]. This shows that the effectiveness of applied pressure in reducing void content is diminished as the fiber volume fraction of a composite component manufactured by hot-press is increased.

Model Validation

The proposed model, shown in Eq. 2, was found to be capable of accurately predicting the void contents of the laminates used in its development. However, in order to further validate the model, two additional laminates were manufactured, after having been cycled through one of the two humidity cycles shown in Fig. 3. The moisture content predicted by the the 1-D Fickian model for each of these laminates as time progressed is shown in Fig 4. After conditioning these laminates were processed at a pressure of 69 kPa using the the same cure cycle as the previous nine laminates.

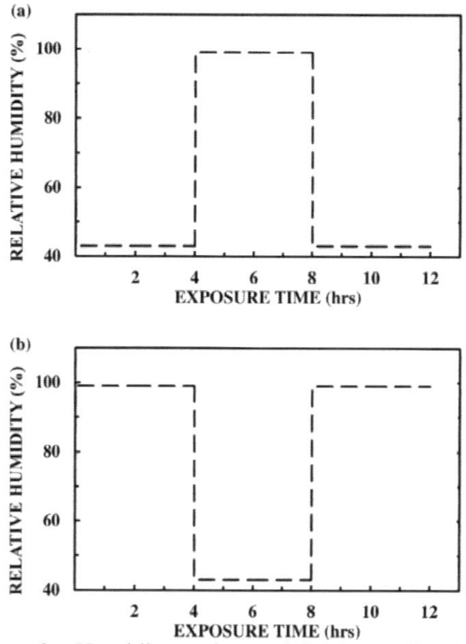

Figure 3 – Humidity cycles used for the verification of the proposed model.

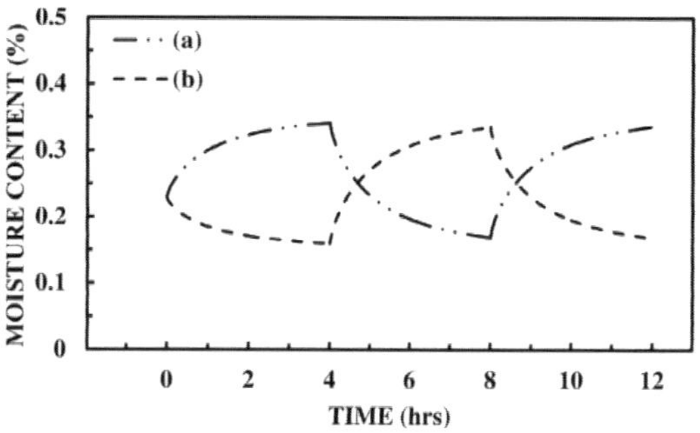

Figure 4 – Predicted moisture content of the verification laminates during their humidity cycles using Eq. (4).

Using the predicted moisture contents, the two laminate void contents were predicted and compared to the measured void contents below in Table 2.

Table 2 – Predicted void contents compared to the actual void content of the laminates exposed to separate humidity cycles.

Humidity Cycle	Predicted Void% Using M_c from Fickian Model	Actual Void%
(a)	3.23	3.27±0.42
(b)	4.99	4.65±0.69

Conclusions

The coupled effect of humidity exposure and cure pressure on the formation of voids in glass/epoxy laminates was explored. It was found that a prepreg ply subjected to a humidity environment will rapidly approach its saturation moisture for that humidity level. The void content of the manufactured laminates were found to reduce to an asymptotic value regardless of initial moisture content. This phenomenon was explained by the fiber bed carrying an increasing amount of the applied pressure as the fiber volume content was increased. A modified void content prediction model was proposed which accounts for the coupled effect of humidity and cure pressure on the formation of voids. This model was found to be capable of predicting final laminate void content to within 0.70%.

Acknowledgments

The authors would like to acknowledge the donation of the prepreg material by TenCate Composite. This work is partially funded by a grant to SEAM Center at the University of Oklahoma by the Oklahoma EDGE program.

References

[1] P.O. Hagstrand, F. Bonjour, J.A.E. Månson, Composites Part A: Applied Science and Manufacturing, 36 (2005) 705-714.

[2] Ghiorse, Sr., SAMPE QUARTERLY, 24 (1993) 54-59.

[3] N.C.W. Judd, W.W. Wright, SAMPE Quarterly, 14 (1978) 10-14.

[4] J.P. Anderson, M.C. Altan, Journal of Engineering Materials and Technology, 134 (2012) 7.

[5] K.A. Olivero, H.J. Barraza, E.A. O'Rear, M.C. Altan, Journal of Composite Materials, 36 (2002) 2011-2028.

[6] F.Y.C. Boey, S.W. Lye, Composites, 23 (1992) 261-265.

[7] J.L. Kardos, R. Dave, M.P. Dudukovic, in: T.G. Gutowski (Ed.) Manufacturing Science of Composites, ASME, Atlanta, GA USA, 1988, pp. 41-48.

[8] L. Liu, B.-M. Zhang, D.-F. Wang, Z.-J. Wu, Composite Structures, 73 (2006) 303-309.

[9] M.G. Bader, Composites Part A: Applied Science and Manufacturing, 33 (2002) 913-934.

[10] L.K. Grunenfelder, S.R. Nutt, Composites Science and Technology, 70 (2010) 2304-2309.

[11] C.-H. Shen, G.S. Springer, Journal of Composite Materials, 10 (1976) 2-20.

[12] T.G. Gutowski, T. Morigaki, Zhong Cai, Journal of Composite Materials, 21 (1987) 172-188.

[13] R. Dave, J.L. Kardos, M.P. Duduković, Polymer Composites, 8 (1987) 29-38.

[14] T.G. Gutowski, G. Dillon, Journal of Composite Materials, 26 (1992) 2330-2347.

O-07-513

CARBON NANOTUBE CONDUCTIVE NETWORKS IN PP AND PP/CBT BLENDS

Samaneh Abbasi[1], Abdessalem Derdouri[2], Pierre J. Carreau[1]

[1] *CREPEC, Department of Chemical Engineering, Ecole Polytechnique Montreal, Montreal, Canada*
pierre.carreau@polymtl.ca;samaneh.abbasi@polymtl.ca

[2] *Industrial Materials Institute, National Research Council Canada, Boucherville, Quebec, Canada*
salim.derdouri@imi.cnrc-nrc.gc.ca

Abstract - We investigated the electrical conductivity and percolation behavior of nanocomposites based on multiwalled carbon nanotubes (MWCNTs) mixed in a polypropylene (PP) matrix and in a blend of PP with cyclic butylene terephthalate (CBT). Scanning electronic microscopy (SEM) was used to examine the quality of the dispersion and the formation of nanotube networks within the polymer blend. Transmission electronic microscopy (TEM) micrographs showed that for the diluted PP/MWCNT composites the nanotubes were well aligned in sheared samples. The electrical conductivity was also significantly altered by the nanotube alignment in samples prepared by micro-injection molding under high shear rate conditions; the conductivity decreased and the percolation threshold rose in highly sheared samples when compared to samples prepared by compression molding where orientation and alignment were minimal. The electrical conductivity of PP/CBT blends was improved through a double percolation that is the basic requirement for the conductivity of the ternary nanocomposites. It was also observed that the nanotubes affected the morphology of the blends resulting in a smaller and more elongated dispersed phase. It was found that the carbon nanotubes would preferentially be distributed within the CBT phase as predicted by Young's equation for the wetting coefficient.

Keywords: micro-injection molding, carbone nanotubes, polypropylene, electrical conductivity

Introduction

Carbon nanotubes (CNTs) are very strong materials which can be either metals or semi-conductors depending on their electronic structure. These features give them a very high potential to be used in a variety of nano-electronic applications [1,2]. During the last decade, they also generated a huge interest from polymer scientists in order to develop high performance polymer nanocomposites due to their size scale and aspect ratio combined with their outstanding mechanical, electrical and thermal properties. For example, the incorporation of carbon nanoparticles into non-miscible polymer blends at fairly low concentrations can make them electrically conductive as a result of a double percolation phenomenon [3-5]. In a blend of two polymers the difference in affinity of the added conductive particles to each polymer component results in a heterogeneous distribution of the conductive filler. The resulting morphology of the nanocomposite is directly controlled by the respective interfacial energies of its components.

In this work, the electrical properties of MWCNT-filled PP/CBT blends are studied in relation to the morphology developed. As nanotubes intertwine into agglomerates that are difficult to disperse due to strong van der Waals forces, we used premixed PP/carbon nanotubes masterbatches and diluted them to the required concentrations by adding the base polymer. This insured consistent and reproducible results. The prepared samples were then micro injection molded and compression molded. The electrical conductivity of the PP/MWCNT was measured in terms of nanotubes loading for both processes for comparison purposes. The choice of a cyclic oligomer such as CBT was motivated by the very low viscosity it displays in contrast to the much more viscous PP. It was expected that by adding and mixing CBT to MWCNT reinforced PP a much better dispersion and distribution of the nanotubes would result because of a more efficient wetting of the nanoparticles by oligomers. The selective location of carbon nanotubes and the double percolation threshold of the nanocomposite blend have a direct influence on the blend morphology and its electrical conductivity as is shown in the results section.

Experimental

Masterbatches of 20 wt% MWCNT in PP purchased from Hyperion Catalysis International, Cambridge, MA. The masterbatches were diluted with an isotactic polypropylene (PP HD120) supplied by Borealis Co. to prepare nanocomposite samples of various loadings. Cyclic butylene terephthalate (CBT) was graciously supplied by Cyclics Co. Ternary nanocomposites of PP/CBT/MWCNT were produced by diluting the PP/MWCNT masterbatch with proper amounts of PP and CBT using a laboratory twin-screw extruder, Leistritz ZSE 18HP, operating at 185°C and 250 rpm. Microinjection molding was then used to prepare dog-bone shaped tensile bars under high shear conditions. Compression molded disks were also prepared to assess the effects of orientation on the microstructure and electrical conductivity of the prepared samples. The morphology of nanocomposites samples was studied through scanning electron microscopy (SEM) using ahigh resolution Hitachi S-4700 microscope. The surfaces of the PP and PP/CBT nanocomposite samples were etched with a potassium permanganate solution to eliminate the amorphous phase from the surface. Transmission electron microscopy (TEM) was also done on ultrathin PP/MWCNT cross sections for a better observation of the morphology at higher resolution. The volume resistivity of nanocomposite samples was determined by measuring the DC resistance across the thickness of the samples. When the MWCNT concentration was low (0 to 1%) a voltage within a range of 1000V was required and a Keithley electrometer model 6517 equipped with a two probe test fixture was used. Samples with more than 2wt% MWCNT were tested using the Keithley electrometer model 6220, at a voltage of 100V.

Results and Discussion

The aggregation of carbon nanotubes in the polypropylene matrix can be seen in the scanning electron micrograph of Fig. 1 for the PP/3%MWCNT compression molded sample. The crystals of which the lamella edges are shown by the arrows, reject the nanotubes during their growth. Thus the nanotubes form a continuous path in the matrix amorphous phase where they are confined. The concentration of fewer nanotubes in a conductive network in the amorphous phase will determine the percolation threshold for electrical conductivity.

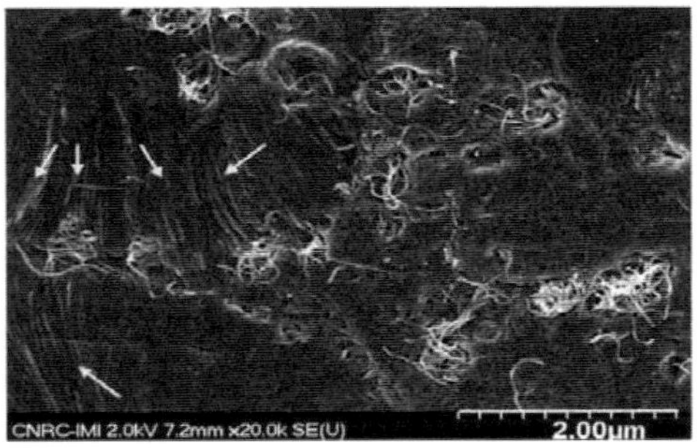

Fig. 1 SEM micrographs of etched surface of PP/ 3wt% MWCNT nanocomposite

Transmission electron microscopy photographs of compression molded and micro injection molded PP/3%MWCNT samples are shown in Fig. 2. For the compression molded case (Fig. 2 (a)), the nanotubes are well dispersed as aggregates with no discernable orientation. On the other hand, for the microinjected sample of Fig.2 (b), the nanotubes, while still aggregated, are overall clearly oriented in the flow direction indicated by the black arrow. The effect of alignment of the nanotubes has been shown to be detrimental to the formation of an effective conductive network [6].

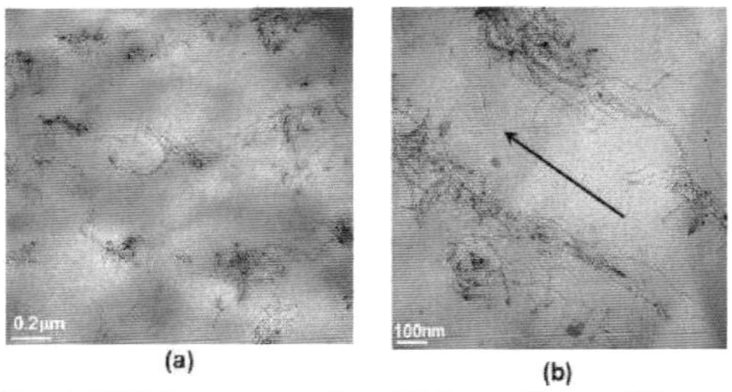

Fig. 2 TEM micrographs of thin sections of the PP/ 3wt% MWCNT nanocomposite (a) compression molded (b) microinjection molded sample

This effect can be seen in Fig. 3 where the electrical conductivity of PP/MWCNT is plotted versus nanotubes loading for the compression molded and micro injection molded samples. It is obvious that the compression molded case displays a percolation threshold of about 1% while the micro injection case does not, but rather the electrical conductivity increases continuously over a range of concentration.

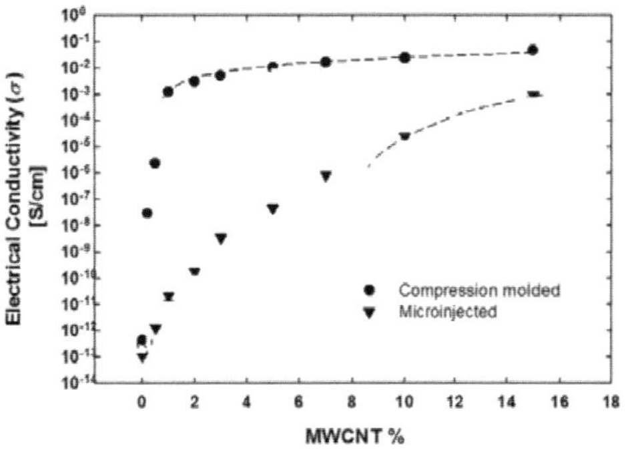

Fig. 3 Electrical conductivity of compression molded (●) and micro injection molded (▼) PP/MWCNT nanocomposites.

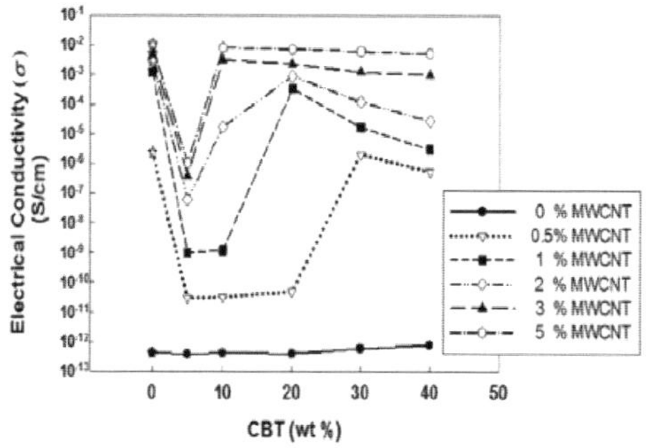

Fig. 4 Electrical conductivity of MWCNT-filled PP/CBT blends versus CBT concentration at various nanotube loadings

The electrical conductivity for the PP/MWCNT/CBT nanocomposites is shown in Fig. 4 as a function of the wt% nanotubes loading. We observe an initial drop in conductivity when CBT is added indicating a disruption of the initial conductive networks. Electrical conductivity is restored to the previous levels as CBT is added up to about 20wt% CBT where the largest jump in electrical conductivity is observed between 0.5 and 1wt% MWCNTs. This increase over about 8 decades signals the presence of a renewed nanotube conductive network and is due to the double percolation phenomenon [4]. The decrease in conductivity beyond the 20wt% CBT threshold can be explained by a dilution effect particularly at low level of nanotubes loading. The data of Fig. 4 also show that the conductivity of the composite blend is always lower than that of the initial PP nanocomposite. Following Bueche [7] we may write an equation for the conductivity of the system as:

$$\sigma \cong \sigma_c \omega (1 - V_m) \quad (1)$$

where σ_c is conductivity of the initial conducting phase, ω is a fraction of the initial conducting particles and V_m the volume fraction of the matrix. Eq. 1 implies that the conductivity of the ternary nanocomposite system, formed by the initial conducting system PP/MWCNT and the non-conduction component CBT, is always less than that of the initial conducting system.

Fig. 5 shows SEM micrographs of a PP/CBT blend without nanotubes (Fig. 5a) and with 1 wt% MWCNTs (Fig. 5b) when the CBT concentration is at the percolation threshold of 20%. The dark zones constitute the polypropylene matrix while the white domains form the minor CBT phase. As it can be observed in Fig. 5a, the CBT dispersed phase is in the shape of about 10μm droplets indicating that at this composition the blend is still far from that of a co-continuous morphology. However, the incorporation of nanotubes into the blend results in a completely different morphology (Fig. 5b). The CBT domains are no longer spherical in shape but rather elongated domains on the verge of merging into a more continuous one within the PP matrix. The CNTs are preferentially located within or at the edges of the CBT domains. This means that the nanotubes migrated from the PP phase where they were originally located to the much less viscous CBT phase suggesting that the carbon nanotubes have a greater affinity with CBT (higher wettability) than with PP. As a result of this process the CNT strings we see in the right micrograph of Fig. 5b form bridges between CBT domains across the PP phase. Rejection of the CNTs by PP, possibly enhanced during crystallization as can be seen in Fig. 1, helps in establishing this network of bridges between CBT domains. The smaller size of the CBT domains is also an enhancing factor as is the lower viscosity of CBT, which allows for more mobility. It seems that the nanotubes act as compatibilizers connecting the two phases or increasing compatibility between them.

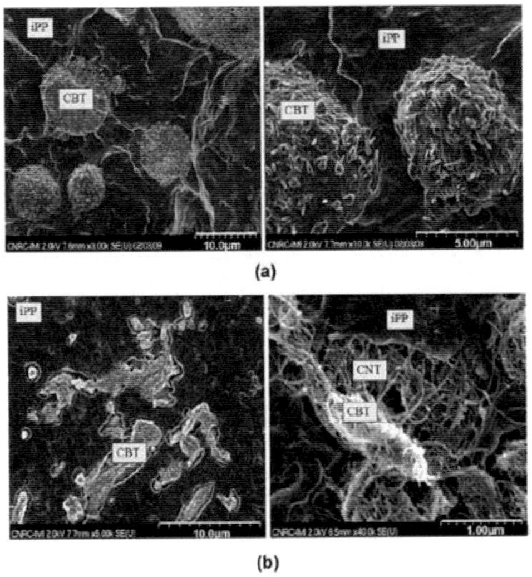

Fig.5. SEM micrograph of iPP/CBT (80/20) without nanotubes (a) and filled with 1 wt% MWCNT (b).

For the particular case of the PP/MWCNT/CBT composite the distribution of the nanotubes at equilibrium can be predicted from the interfacial free energy by calculating the wetting coefficient at equilibrium using Young's equation:

$$\omega_{12} = \frac{\gamma_{C2}-\gamma_{C1}}{\gamma_{12}} \quad (2)$$

Indices 1 and 2 stand for CBT and PP respectively and C for carbon nanotubes. The interfacial tension between polymer-polymer or polymer-solid phases can be estimated from the harmonic mean equation given by Wu [9]:

$$\gamma_{ij} = \gamma_i + \gamma_j - 4\left[\frac{\gamma_i^d\gamma_j^d}{\gamma_i^d+\gamma_j^d} + \frac{\gamma_i^p\gamma_j^p}{\gamma_i^p+\gamma_j^p}\right] \quad (3)$$

If $\omega_{12} > 1$ the carbon nanotubes would preferentially be located in CBT, while if $\omega_{12} < -1$ they would distribute within the PP phase.When the wetting coefficient is such that $-1 < \omega_{12} < 1$ they distribute at the interface between CBT and PP. The interfacial energies for each component were estimated from the literature and are summarized in Table I. Substituting the data into eq. (2) and (3) it is found that $\omega_{12} = 1.23$, predicting that the carbon nanotubes would preferentially be distributed within the CBT phase.

Table 1 Estimated surface tensions of PP, CBT and MWCNT

	γ^p, mJ/m^2	γ^d, mJ/m^2	$\gamma = \gamma^p + \gamma^d$, mJ/m^2	Refs
PP (190°C)	0.0	20.5	20.5	[9], [10]
MWCNT	26.9	18.4	45.3	[11]
CBT (185°C)	9.8	31.9	41.7	[12]

Conclusion

The electrical conductivity of PP and PP/CBT nanocomposites based on multi-wall carbon nanotubes was investigated. The presence of a crystalline phase and nanotube alignment both affected the electrical conductivity of the PP nanocomposites. Micro injection molded samples did not display a sharp percolation threshold as did compression molded samples and their thermal conductivity was lower. This result was attributed to the nanotubes alignment due to the high shear rate in the injection molding process. The effect of nanotube loading on the electrical conductivity of PP/CBT blends was also analyzed and it was found that a double percolation is the basic requirement for the conductivity of these nanocomposites. Furthermore, the calculated wetting coefficient of the carbon nanotubes with respect to PP and CBT indicates that they are more likely to be located in the CBT phase.

Acknowledgements

Financial support from NSERC (Natural Science and Engineering Research Council of Canada) is gratefully acknowledged. We are also thankful to Mrs. Weawkamol Leelapornpisit for her great help in the morphological studies.

References

1. Meyyappan M. Carbon nanotubes : science and applications. Boca Raton: CRC Press 2005.
2. Dresselhaus MS, Dresselhaus G, Saito R. Carbon. 1995; 33(7):883-891.
3. Sumita M, Sakata K, Asai S, Miyasaka K, Nakagawa H. Polymer Bulletin. 1991; 25(2):265-271.
4. Sumita M, Sakata K, Hayakawa Y, Asai S, Miyasaka K, Tanemura M. Colloid Polym. Sci. 1992; 270(2):134-139.
5. Potschke P, Bhattacharyya AR, Janke A. Carbon 2004; 42:965-969.

6. Abbasi S, Carreau PJ, Derdouri A. Polymer. 2010; 51(4):922-935.
7. Bueche F., J. Appl. Phys. 1972; 43(11):4837-4838
8. Fenouillot F, Cassagnau P, Majesté JC. Polymer. 2009;50(6):1333-1350.
9. Wu, S. ``Polymer Interface and Adhesion``, Marcel Dekker, New-York, 1982; Ch.3
10. Oda, Y., Hata, T. Preprints 267, 17th Annual Meeting of the High Polymer Society of Japan, May 1968
11. Nuriel S., Liu L., Barber A.H., Wagner H.D., Chem. Phys. Lett. 2005; 404:263-266
12. Hobbs S.Y., Dekkers M.E.J., Watkins V.H., Polymer. 1988; 29(9):1598-1602

O-07-367

MULTIWALL CARBON NANOTUBES BASED POLYPROPYLENE COMPOSITE FIBER: INFLUENCE OF A NOVEL MODIFIER AND A REACTIVE COMPATIBILIZER

Arup R. Bhattacharyya[¶*], Pankaj B. Tambe[¶], Indradev Samajdar[¶], Ajit R. Kulkarni[¶], Sreekumar T. Veedu[§], Kingsuk Mukhopadhyay[§], Mukesh K. Sinha[§], Anurag Srivastav[§]

¶ Department of Metallurgical Engineering and Materials Science, Indian Institute of Technology Bombay, Powai, Mumbai 400076, India, arupranjan@iitb.ac.in , pbtambe2010@gmail.com, indra@iitb.ac.in, ajit.kulkarni@iitb.ac.in

§ Defence Materials & Stores Research & Development Establishment (DMSRDE), G T Road, Kanpur 208013, India, kingsuk.mukhopadhyay@gmail.com, sreekumar.veedu@gmail.com, muk120in@yahoo.co.in, anurag_dmsrde@rediffmail.com

Abstract – Polypropylene (PP)/multiwall carbon nanotubes (MWNTs) composite fibers of 0.5 wt% MWNTs content were made by melt-mixing followed by melt-spinning and post-drawing. In order to improve the dispersion of MWNTs and also to improve interfacial interaction between PP and MWNTs, sodium salt of 6-amino heaxanoic acid (Na-AHA) was utilized along with polypropylene-g-maleic anhydride (PP-g-MA), which acts as a compatibilizer. It has been observed that Na-AHA modified MWNTs increases the mechanical properties of melt-spun and post-drawn PP/MWNTs composite fiber in presence of PP-g-MA at 0.5 wt% MWNTs. The superior mechanical properties achieved in this fiber might be due to better dispersion of MWNTs in presence of Na-AHA, which also engages in the melt-interfacial reaction with PP-g-MA, thus enhances the interfacial interaction between PP and MWNTs. The improved interfacial reaction led to a higher drawability of the melt-spun composite fiber, was has manifested in higher orientation of PP chain & MWNTs along the fiber axis. The higher degree of crystallinity (X_c) of PP phase might also contribute towards higher mechanical properties in PP/Na-AHA/MWNTs/PP-g-MA composite fiber. A combination of all these factors led to the development of PP/MWNTs composite fiber of tensile strength and tensile modulus of 973 MPa and 10.4 GPa respectively at 0.5 wt% MWNTs content in PP/Na-AHA/MWNTs/PP-g-MA composite fiber of draw ratio of 20.

Keywords: PP/MWNTs composite fiber, melt-spinning, mechanical properties, crystallization

Introduction

In recent years, enormous research initiatives have been documented in carbon nanotubes based (CNTs) polypropylene (PP) composite to develop high strength, high modulus PP/CNTs composite fiber [1-2]. It has been realized that as-spun PP/CNTs melt-spun composite fiber needs significantly higher orientation of the PP chain [3-5] and that of CNTs in the fiber axis direction in order to translate the outstanding properties of CNTs in the composite fiber. However, mechanical properties of the melt-spun and post-drawn PP/single wall carbon nanotube (SWNTs) composite fiber did not show any appreciable improvement at 0.8 wt% SWNTs content despite the presence of highly oriented SWNTs along the fiber axis [6]. Kearns and Shambaugh [7] have reported a significant improvement in mechanical properties of PP/SWNTs composite fiber, which has been prepared via two-step processing (solution processing followed by melt-spinning) to achieve a better dispersion of SWNTs in PP matrix. Lee et al. [8] have observed a significant increase (25%) in tensile strength and modulus in PP/SWNTs composite fiber containing 0.1 wt% of SWNTs (nitric acid treated) in the presence of polypropylene-g-maleic anhydride (PP-g-MA). Further, PP/amine functionalized multiwall carbon nanotube (NH_2-MWNTs; f-MWNTs) composite fiber showed a significant improvement in tensile strength and tensile modulus at 0.5 wt% f-MWNTs content with an increase in draw ratio during post-drawing operation after melt-spinning [9]. It has also been understood that interfacial interaction, state of dispersion of CNTs along with higher orientation factors of both PP as well as CNTs are the key issues, which primarily control the improvement in mechanical properties in the composite fiber.

The state of dispersion of MWNTs has been found to be significantly improved while modifying MWNTs with an organic modifier (Na-salt of 6-amino heaxanoic acid, Na-AHA), which has manifested in lower electrical percolation threshold in PP matrix [10]. The objective of this work is to investigate the role of Na-AHA modified MWNTs along with a reactive compatibilizer (PP-g-MA) in investigating the mechanical properties of melt-spun and post-drawn PP/MWNTs composite fiber.

Experimental

PP was obtained from Reliance Industries Ltd., India (REPOL H200FG, melt flow index (MFI) of 20 g/10 min at 230 °C with a load of 2.16 kg). PP-g-MA was obtained from Pluss Polymer Ltd., India (OPTIM:P-405, MFI of 14g/10 min at 230 °C with a load of 2.16 kg, the degree of MA grafting = 0.9%). Purified multiwall carbon nanotubes (MWNTs, purity > 84%, diameter ~ 8.5 nm, length ~ 1.5 micron) were synthesized at DMSRDE, Kanpur, India Representative compositions of PP/MWNTs composite fibers are shown in Table 1.

PP/MWNTs composite fibers were prepared in two stages: melt-mixing was carried out in a conical twin-screw microcompounder (Micro 5, DSM Research, Netherlands) at 260 °C with a rotational speed of 200 rpm for 15 min followed by melt-spinning. Melt-mixed PP/MWNTs samples were dried at 80 °C under vacuum oven for 24 h. PP/MWNTs composite (0.5 wt% MWNTs) were melt-spun using a small-scale spinning machine manufactured by the Bradford University Research Ltd. UK, using a single hole spinneret of 1 mm diameter. Melt-spinning using MWNTs was carried out at 200 °C with a ram speed of 1.4 mm/min. The take-up speed was kept at 20 m/min. Melt spun PP/MWNTs composite fibers were subsequently post-drawn at 120 °C and at a draw ratio (DR) of 8 (feed roller speed = 1 m/min, collector speed = 8 m/min). Control PP sample was melt-spun at the identical melt-spinning conditions used to prepare PP/MWNTs composite fiber. PP/MWNTs/Na-AHA/PP-g-MA (0.5 wt% of MWNTs) composite fiber was melt-spun at 220 °C at a ram speed of 4.6 mm/min. The take-up speed was kept at 40 m/min. Due to the difference in melt-rheology of PP/MWNTs/Na-AHA/PP-g-MA composite (as compared to PP/MWNTs composite) the corresponding composition was melt-spun at a different melt-spinning conditions used for both pure PP and PP/MWNTs composite. PP/MWNTs/Na-AHA/PP-g-MA composite fibers were subsequently post-drawn at 120 °C and at DR of 12, 16, and 20 respectively.

Mechanical properties of post-drawn fibers were tested on a Favimat fiber tester (Textechno, Germany) with a single filament fiber with gauge length of 20 mm and at a cross-head speed of 20 mm/min. The data reported here are averages of 25 tests for each composition; the standard deviation is less than 5%. Raman spectroscopy was performed using Jobin Yovon (HR 800 micro-Raman) in the scanning range of 200–2000 cm^{-1} with incident laser excitation wavelength of 514 nm. Scanning electron microscopic (SEM) analysis was performed with Hitachi S3400, operated at 15 kV. Orientation studies were carried out using PANalytical X-ray Texture instrument. Differential scanning calorimetric (DSC) measurements were carried out using a DSC Q10 from TA instruments. The extruded samples of about 5 mg were dried in a vacuum oven prior to experiment. In the 1st heating run all samples were recorded in the temperature range from 25°C to 260 °C at a scan rate of 10 K/min under nitrogen atmosphere and annealed at the final temperature (260 °C) for 3 min to delete the previous thermal history. Subsequent cooling scan was recorded in the temperature range from 260 °C to -20 °C at a scan rate of 10 K/min under nitrogen atmosphere. Second heating scan was recorded from -20 °C to 260 °C at a scan rate of 10 K/min under nitrogen atmosphere. The degree of crystallinity of PP phase was calculated from the heat of fusion from the second heating run. The heat of fusion (ΔH_m) of PP phase was normalized to the fraction of PP present in the composite. The degree of crystallinity (X_c) of PP phase was determined from the ratio of normalized heat of fusion ($\Delta H_{m,norm}$) to the heat of fusion of 100% crystalline PP, (ΔH_m)100%, which was taken as 207 J/g [11].

Table 1. Representative compositions along with the orientation factor of PP

Sample	PP orientation factor along (110) crystalline plane
MWNTs = 0.5 wt%, in all the composite fiber	
PP	0.80
PP/MWNTs (DR = 8)	0.77
PP/MWNTs/Na-AHA (1:15) (DR = 8)	0.72
PP/MWNTs/Na-AHA (1:15)/PP-g-MA (DR = 16)	0.71
PP/MWNTs/Na-AHA (1:10)/PP-g-MA (DR = 20)	0.86

Results and Discussion

Effect of Na-AHA modified MWNTs on the mechanical properties and crystallization behaviour of PP/Na-AHA/MWNTs/PP-g-MA composite fiber

Fig. 1 shows the transmission electron micrograph of MWNTs, which shows entangled network of MWNTs, which also exhibits an average diameter of ~ 8.5 nm. The ratio of intensity of G-band as that of D-band (I_G/I_D) in the Raman spectroscopy is a measure to estimate the crystalline fraction of MWNTs; which is found to be 1.31 for MWNTs studied here.

It has been earlier observed that Na-AHA modified MWNTs are dispersed well in PP matrix, wherein PP-g-MA acts as a compatibilizer. The finer dispersion of MWNTs has been manifested in lower electrical percolation threshold in PP/MWNTs composite [10]. Based on our earlier observation, PP/MWNTs composite fibers were prepared through melt-mixing followed by melt-spinning utilizing Na-AHA modified MWNTs and PP-g-MA. A series of composite fibers were made by varying MWNTs:Na-AHA ratio. The draw ratio was also varied during post-drawing operation after melt-spinning of the composite fiber.

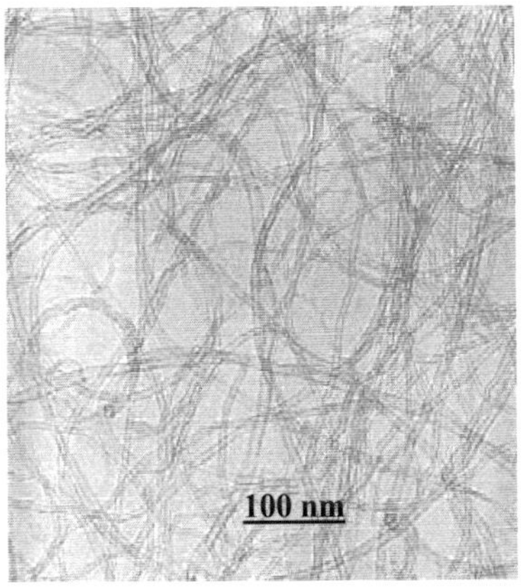

Figure 1. Transmission electron micrograph of MWNTs

The Herrman's orientation factor of PP chain for the (110) crystalline plane along the fiber axis for PP/MWNTs/Na-AHA composite fiber is found to be lower as compared to both pure PP and PP/MWNTs composite fibers (Table 1). However, the PP orientation factor is increased with increase in Na-AHA content. It is also observed that the Herman's orientation factor of the PP chain is increased in PP/MWNTs/Na-AHA/PP-g-MA composite fiber with increase in draw ratio (DR) during post-drawing.

Mechanical properties of the various compositions of melt-spun and post-drwan PP/MWNTs composite fiber have been conducted in presence of Na-AHA and PP-g-MA (the corresponding stress-strain plots and the data are not shown). Tensile modulus and tensile strength of PP (DR = 8) are found to be 4.5 GPa and 389 MPa respectively. However, tensile modulus and tensile strength of PP/MWNTs composite fiber of 0.5 wt% MWNTs are found to be 6.2 GPa and 482 MPa respectively. In contrast, PP/MWNTs/Na-AHA composite fibers of 1:1 and 1:4 MWNTs:Na-AHA exhibit a decrease in tensile strength and tensile modulus as compared to the corresponding PP/MWNTs composite fiber. It is observed that PP/MWNTs/Na-AHA/PP- g-MA composite fibers of (1:4), (1:10) and (1:15) ratios of MWNTs:Na-AHA exhibit a significant increase in tensile strength and tensile modulus with an increase in draw-ratio.

Tensile strength and tensile modulus of the PP/MWNTs/Na-AHA/PP-g-MA composite fibers are increased with increasing content of Na-AHA in the MWNTs/Na-AHA mixture. However, the tensile strength and the tensile modulus of the PP/D-MWNTs/Na-AHA/PP-g-MA composite fibers are found to be maximum for fibers of the ratio (1:4), (1:10) and (1:15) of MWNTs:Na-AHA, post drawn at DR of 20. The highest tensile strength and the tensile modulus are observed in the composite fiber of (1:15) ratio of MWNTs:Na-AHA, post-drawn at DR of 20. Thus, the highest tensile modulus and tensile strength have been achieved for PP/D-MWNT/Na-AHA/PP-g-MA composite fiber of 1:15 D-MWNT:Na-AHA, which exhibit 10.4 GPa of tensile modulus and 973 MPa tensile strength.

These observations can be explained on the basis of higher orientation factor of PP chains (and also MWNTs) along the fiber axis in the composite fiber processed at higher DR. Additionally, MWNTs are found to be debundled more with increase in Na-AHA concentration, which presumably reacts with PP-g-MA during melt-interfacial reaction, thus, improves the interfacial adhesion between PP and MWNTs.

Figure 2. Tensile fractured surface of PP/MWNTs/Na-AHA/PP-g-MA composite fiber

Highly oriented fibrils of PP/MWNTs are observed (Figure 2) in the tensile fractured surface of PP/MWNTs/Na-AHA/PP-g-MA composite fiber. MWNTs wrapping by PP is also observed, which is due to the melt-interfacial interaction between PP-g-MA and Na-AHA.

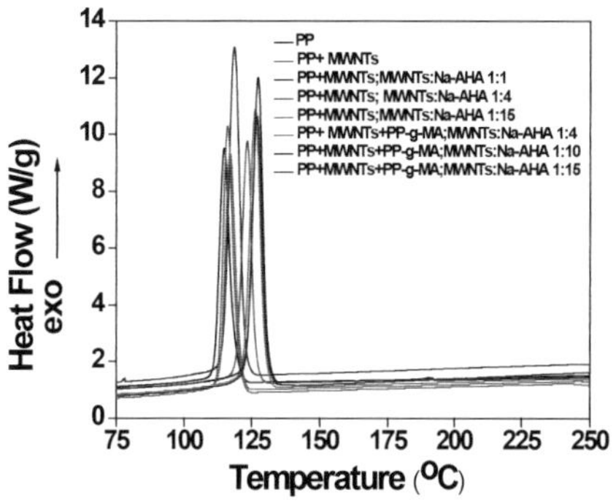

Figure 3. Crystallization behaviour of PP phase in PP/MWNTs, PP/MWNTs/Na-AHA and PP/MWNTs/Na-AHA/PP-g-MA composite fibers

The crystallization exothermic peak of PP phase for PP, PP/MWNTs and PP/MWNTs composite fibers in presence of either Na-AHA or in presence of Na-AHA and PP-g-MA are shown in Figure 3. It is observed that PP crystallizes at ~ 115 °C, whereas T_c of PP is significantly higher (~123-125 °C) in presence of MWNTs. This indicates heterogeneous nucleating action of PP in presence of MWNTs.

It is interesting to note that the T_c of PP is decreased in PP/MWNTs/Na-AHA composite as compared to corresponding PP/MWNTs composite. However, T_c of PP is increased in PP/MWNTs/Na-AHA/PP-g-MA composites. This can be explained on the basis that the heteronucleating action of MWNTs is decreased due to the adsorbed Na-AHA on MWNTs surface. The value of T_c (~ 116-119 °C, which is still higher than pure PP) in this composition also indicates that MWNTs surface is not fully covered by Na-AHA, and the heteronucleating action of MWNTs is observed in this case as well. However, one can observe a better nucleating action of MWNTs in presence of both Na-AHA and PP-g-MA. Since Na-AHA may partially desorbs from the MWNTs surface during melt-interfacial reaction with PP-g-MA, the heteronucleating action of MWNTs is increased which is manifested in higher T_c (~ 126-128 °C) in the corresponding composite fibers of PP/MWNTs/Na-AHA/PP-g-MA. PP-g-MA alone could also act as a heteronucleating agent for PP which is indicated in the increase in T_c (~126-128 °C).

The degree of crystallinity (X_c, the corresponding data are not shown) of PP is increased marginally in PP/MWNTs composite fiber as compared to neat PP with any appreciable change in the crystalline melting point (T_m). T_m of PP is decreased in presence of Na-AHA. However, T_m of PP is increased marginally in Na-AHA and PP-g-MA based composite fiber as compared to pure PP. X_c is increased in PP/MWNTs/Na-AHA composite as compared to corresponding PP/MWNTs composite. However, X_c is also increased in presence of Na-AHA and PP-g-MA. These observations indicate a possible plasticization effect of Na-AHA in PP/MWNT/Na-AHA composite. However, Na-AHA modified MWNTs act as a better nucleating agent due to higher surface area available in debundled MWNTs.

Conclusions

This study clearly showed the role of Na-AHA modified MWNTs in improving the mechanical properties of melt-spun and post-drawn PP/MWNTs composite fiber in presence of PP-g-MA, which acts as a compatibilizer. The superior mechanical properties achieved in this fiber might be due to better dispersion of MWNTs in presence of Na-AHA, which also engages in the melt-interfacial reaction with PP-g-MA, thus, enhances the interfacial interaction between PP and MWNTs. The improved interfacial reaction led to a higher drawability of the melt-spun composite fiber, which has manifested in higher orientation of PP chain & MWNTs along the fiber axis. The higher X_c of PP could also contribute towards higher mechanical properties in PP/Na-AHA/MWNTs/PP-g-MA composite fiber. All these factors led to the development of PP/MWNTs composite fiber of tensile strength and tensile modulus of 973 MPa and 10.4 GPa respectively at 0.5 wt% MWNTs content in PP/Na-AHA/MWNTs/PP-g-MA composite fiber of DR of 20.

Acknowledgements

The authors would like to thank Microcompounder central facility, SAIF and CRNTS IIT Bombay. One of the authors (ARB) would like to acknowledge Director, DMSRDE Kanpur for the financial assistance towards the project 10DRDO10 and 10DRDO14.

References

[1] O. Breuer; U. Sundararaj *Polym. Compos.* 2004, 25, 630.
[2] J.K.W. Sandler; S. Pegel; M. Cadek; F. Gojny; M. van Es; J. Lohmar; W.J. Blau; K. Schulte; A.H. Windle; M.S.P. Shaffer *Polymer* 2004, 45, 2001.
[3] Z. Jin; K.P. Pramoda; G.S. Xu; H. Goh *Chem. Phys. Lett.* 2001, 337, 43.
[4] R. Blake; Y.K. Gun'ko; J. Coleman; M. Cadek; A. Fonseca; J.B. Nagy; W.J. Blau *J. Am. Chem. Soc.* 2004, 126, 10226.
[5] S. Tilo; P. Ton; J. Piet; L. Joachim *Macromolecules* 2004, 37, 1810.
[6] A.R. Bhattacharyya; T.V. Sreekumar; T. Liu; S. Kumar; L.M. Ericson; R. Hauge; R.E. Smalley *Polymer* 2003, 44, 2373.
[7] J.C. Kearns; R.L. Shambaugh *J. Appl. Polym. Sci.* 2002, 86, 2079.

[8] G. Lee; S. Jagannathan; H.G. Chae; M.L. Minus; S. Kumar *Polymer* 2008, 49, 1831.
[9] P.B.Tambe; A.R. Bhattacharyya; S. S. Kamath,; A.R. Kulkarni; T.V. Sreekumar; A. Srivastav; K.U. Bhasker Rao; Y. Liu; S. Kumar *Polym. Eng. Sci.* 2012, 52, 1183.
[10] P.B.Tambe; A.R. Bhattacharyya; S. S. Kamath; A.R. Kulkarni; T.V. Sreekumar; K. Mukhopadhyay; M. Sinha; A. Srivastav; K.U. Bhasker Rao in *Carbon nanotubes based nanocomposites: recent development* ed by K. K. Kar; A. Hodzic 2011, pp. 297-314, Research Publishing Services, Singapore.
[11] B. Wunderlich *Thermal Analysis* 1990, Academic Press, New York.

KN-08-02

RHEOLOGY AND PHASE BEHAVIOR OF POLYMER BLEND

Wei Yu*, Chongwen Huang, Nan Wang and Chixing Zhou
Advanced Rheology Institute, Department of Polymer Science and Engineering
Shanghai Jiao Tong University, Shanghai 200240, P. R. China

Abstract - Morphology of polymer blends will have a significant change during the phase transitions, which makes it possible to be studied by rheology from the appearance of characteristic long relaxation time. Different rheological methods used for the phase transition will be studied here and its validity will be discussed. It is found that the accuracy of rheological methods is affected by the viscoelasticity and dynamics of polymers. For dynamic symmetric blends, concentration fluctuation and interfacial tension can be determined from the variation of dynamic moduli both at the high frequency and low frequency. For dynamic asymmetric blends, the abnormal evolution of storage moduli is ascribed to the change of components' rheology due to the variation of effective glass transition temperatures as well as the morphology development.

Keywords: phase separation, polymer blend, rheology

Introduction

The phase behavior of partially miscible polymer blends is crucial in their application. The phase separation of polymer blends can be inferred from different techniques including optical microscopy, light scattering and rheology. Optical methods rely on the difference in the refractive index while the rheological methods depend on the viscoelastic asymmetry and the relative contribution of concentration fluctuation or interface to the moduli or viscosity [1-3]. However, one should be cautious about the application of rheological method in phase separation due to the complicated relationship between phase separation kinetics, the polymer dynamics and rheological measurements. Two examples on the complex rheological behavior in phase separation are studied in this paper.

Experimental

Poly(methyl methacrylate) (PMMA, T_g=96°C) and poly(styrene-co-acrylonitrile) (SAN, T_g=105°C, containing 28.4wt.% acrylonitrile) were both supplied by LG Chemical Ltd. Poly(styrene-co-maleic anhydride) (SMA12, T_g=125°C), containing 12 wt% of maleic anhydride, was donated by SINOPEC Shanghai Research Institute of Petrochemical Technology. Poly(styrene-co-maleic anhydride) (SMA28, T_g=165°C), containing 28wt% of maleic anhydride, was supplied by Polyscope Polymers. PMMA/SAN, PMMA/SMA12 and PMMA/SMA28 blends with different compositions are melt mixed in a batch mixer under suitable temperature. The samples for rheological testing were prepared by compression molding into a disc-like form with diameter 25 mm and 1mm thick at 150°C under 10MPa. All rheological test were performed using parallel plate geometry on a rotational rheometer (Bohlin Gemini 200HR). Optical observation was conducted on a Leica optical microscopy (DM2500P) with the temperature controlled by a Cambridge Linkam shear cell (CSS450).

Results and Discussion

Usually, it is believed that the phase diagram determined from different methods could be different due to the difference in the spatial resolution. Optical method is restricted by the wavelength of light, while the sensitivity of rheological method depends on the frequency. However, both methods are also affected by the kinetics of phase separation. Figure 1 shows the phase diagram of PMMA/SAN blend. The rheological transition temperatures are determined using power-law scaling analysis for near-critical compositions and shifted Cole-Cole plot for off-critical compositions [3]. The optical transition temperatures are determined from the change in transmittance of light. Apparently, there is evident discrepancy between two transition lines, and the optical transition temperatures are about ten degrees below the rheological transition temperatures. The test time scale of optical microscopy is around 4000s, while it is about 2400s for a typical frequency sweep in rheological experiment. It means that the spatial resolution of rheological method is larger than that of the optical method in this blend. It is ascribed to the dynamic symmetry and high viscosity of PMMA and SAN. High symmetry of the blend implies that the concentration fluctuation (CF) to the dynamic moduli is weak, while high viscosity of components means the contribution from CF is relatively small, both of which delays the detection of phase separation in rheology [2]. Such result implies that the sensitivity of the rheological method depends on the phase separation kinetics, dynamic symmetry and the components viscoelasticity.

Figure 2 shows the time dependency of storage modulus in PMMA/SMA12 and PMMA/SMA28 blends in a phase separated temperature, where the former has weakly dynamic asymmetry while the later has strong dynamic asymmetry. The evolution of storage modulus is quite different in two blends. Storage modulus decreases all the time both in PMMA/SMA12 80/20 and 20/80 blends, while it increases in PMMA/SMA28 20/80 blend but decreases in PMMA/SMA28 80/20 blend. The difference is also ascribed to the dynamic asymmetry. The increase in the storage modulus in PMMA/SMA28 20/80 blend can be attributed to the increase of the effective glass transition temperature of the major component SMA28 during phase separation.

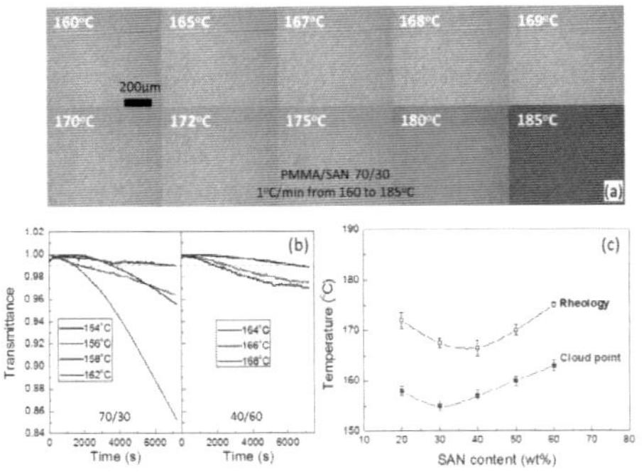

Figure 1 Optical microscopy image of PMMA/SAN 70/30 blends during heating (a), transmittance of light with time for blends with different compositions (b) and phase transition temperature determined from rheology and optical microscopy (c).

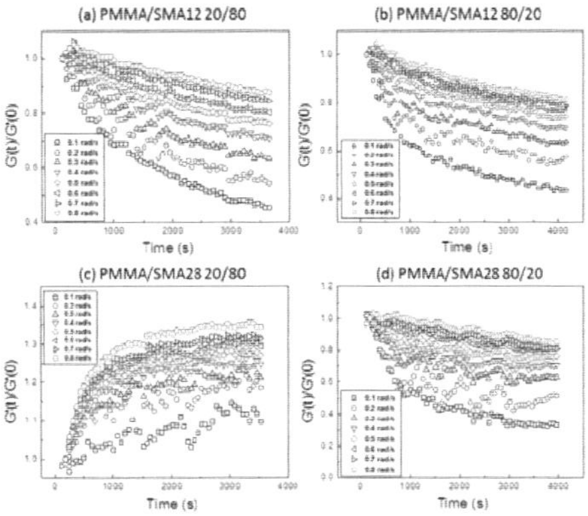

Figure 2 Evolution of dynamic modulus with time at different frequencies for two polymer blends.

Conclusions

It is illustrated in this paper that the rheological behaviors of phase separating fluid are very complex. The evolution of dynamic moduli during phase separation is affected by multiple factors including the dynamic asymmetry of components, the phase separation kinetics and the rheology of components.

Acknowledgements

The authors would like to thank the support from the NSFC (No. 50930002), 973 Program (No. 2011CB606005). W. Yu is supported by the Program for New Century Excellent Talents in University and the SMC project of Shanghai Jiao Tong University.

References

1. W. Yu; R.M. Li; C.X. Zhou, *Polymer* 2011, *52*, 2693.
2. W. Yu; C.X. Zhou, *Polymer*, 2012, *53*, 881.
3. J.P. Gao; C.W. Huang; N. Wang; W. Yu; C.X. Zhou, *Polymer*, 2012, *53*, 1772

O-08-499

ANOMALOUS NORMAL STRESS BEHAVIOUR FOR POLYMER NANOCOMPOSITES AND LIQUID CRYSTALLINE POLYMER COMPOSITES

Esmaeil Narimissa, Ahmed Rahman, Rahul K.Gupta*, Sati N Bhattacharya
Rheology and Materials Processing (RMPC) Centre, School of Civil, Environmental and Chemical Engineering, RMIT University, Melbourne, Vic, Australia
**Corresponding author: E-mail: rahul.gupta@rmit.edu.au*

Abstract - First normal stress difference (N_1) behaviour of polymer nanocomposites and liquid crystalline polymers is a measure of elasticity and is affected by shear stress due to the change in the morphology at the molecular and nanostructure level. Generally for macromolecules the first normal stress difference increases with increase in stress level. However, in polymeric nanocomposites, N_1 is affected by the level of loading and the dispersion of nanofillers. The relationship between N_1 and filler contents could be used as a scale to verify the level of dispersion as N_1 behaves independent of filler contents in intercalated systems but is dependent in exfoliated systems.

In this study, the steady shear rheological behaviours of polylactide (PLA) and nanographite platelets (NGP) bio-nanocomposites with 1, 2, 3, and 5 wt% filler contents have been investigated. In addition the shear rheological properties of glass filled liquid crystalline polymers (LCPs) have also been examined. The objective has been to obtain a correlation between N_1, filler contents, and shear rate of the measurements. The results suggest that the N_1 in PLA/NGP bionanocomposites is dependent on the level of loading and on shear rate beyond a critical value. However, a similar correlation between N_1 and shear rate for liquid crystalline polymer composites could not be obtained in both high and low rate regions. The anomalous behaviour of N_1 for liquid crystalline polymers will be discussed.

Keywords: First normal stress difference, PLA, LCPs, shear rheology.

Introduction

The shearing of viscoelastic materials between two parallel surfaces at a considerable shear rate results into the formation of viscous shear stress (τ_{12}) and normal stress differences ($N_1= \tau_{11}- \tau_{22}$, $N_2= \tau_{22}- \tau_{33}$). The flow direction and the perpendicular surface at which the fluid is sheared are indicated by 1 and 2, and 3 is the neutral direction [1]. N_1 is the largest normal stress differences and is responsible for rod climbing behaviour of viscoelastic materials. In the case of isotropic materials, the magnitude of N_1 is greater or equal to zero and in parallel plate and cone and plate rheometers, it pushes apart the contact surfaces. The magnitude of the second normal stress difference (N_2) is usually much smaller than N_1 and it is always negative. According to Keentok et al. (1980) and Ramachandran et al (1985) [2, 3] the relationship between N_1 and N_2 becomes $0.05 < \frac{-N_2}{N_1} < 0.3$. In the zero shear viscosity region of steady shear tests where $\eta = \frac{\tau_{12}}{\dot{\gamma}}$ becomes independent of shear rate, the first and second normal stress differences approach the limit $N_1, N_2 \propto \dot{\gamma}$ at small shear rates, consequently, the normal stress coefficients (Eq. 1) approach constant values:

$$\psi_1 \equiv \frac{\tau_{11} - \tau_{22}}{\dot{\gamma}^2},\ \psi_2 \equiv \frac{\tau_{22} - \tau_{33}}{\dot{\gamma}^2} \qquad (1)$$

Although at low shear rates η, ψ_1 approach constant magnitudes (η_0, $\psi_{1,0}$), they exhibit a continuous and dramatic decrease in their size at higher shear rates (nonlinear shear thinning behaviour). The deviation of zero-shear behaviour occurs only when both shear rate $\dot{\gamma}$ and strain $\gamma = \dot{\gamma} t$ are not low any more. At low shear rate region, the steady state (η_0) shear viscosity and $\psi_{1,0}$ can be evaluated from low frequency linear viscoelastic measurements:

$$\eta(\dot{\gamma}) = \eta'(\omega);\quad as\ \ \omega = \dot{\gamma} \to 0 \qquad (2)$$
$$\psi_1(\dot{\gamma}) = \frac{2G'(\omega)}{\omega^2};\quad as\ \ \omega = \dot{\gamma} \to 0$$

After a step shear strain of polymer melts and solutions, the shear stress and the first normal stress difference (N_1) can be correlated to each other thru Lodge-Meissner relationship [4]:

$$N_1 = \tau_{12}\gamma = \gamma^2 G(t) \qquad (3)$$

Where, γ is the shear strain and G(t) is the relaxation modulus. Lodge-Meissner rule suggests that at sufficiently low strains, the strain ratio tends to become equal to the strain [4]. Thus, for isotropic perfectly elastic solids, the stress ratio (N_1/τ) is always equal to strain and the Doi-Edwards theory predicts that for all strains, the stress ratio is equal to strain [5]:

$$\frac{N_1(t,\gamma)}{\tau(t,\gamma)} = \gamma \qquad (4)$$

Consequently, if the time-strain superposability is valid, the transient first normal stress difference in terms of its shear damping function (h (γ)) is [5]:

$$N_1 = \gamma^2 h(\gamma) G(t),$$

$$\psi_1(t,\gamma) \equiv \frac{N_1}{\gamma^2} = h(\gamma) G(t) \qquad (5)$$

Indeed, the above relationship has successfully described the experimental data for LDPE for strain up to 30, polybutadiene solution for strain up to 3.3 and polystyrene solution for strain up to 8 [5-8]. It must be taken into consideration that the measurements for transient normal stress difference is very challenging owing to the errors caused by the temperature variations and compliance of the instrument [9]. Furthermore, the "wall slip" also causes a large deal of error in step-strain experiments and the true strain could be even lower than the nominal strain caused by the displacement of a rheometer's surface. Therefore, when the Lodge-Meissner rule is not obeyed, it can be assumed that the "wall slip" has occurred; hence, for reporting normal stress difference and shear stress data nominal strain would be replaced by stress ratio.

Due to experimental difficulties of measurements, the elasticity of melt polymeric systems has not been studied the same way as their shear measurements [10, 11]. Some of the common ways of evaluating the elasticity are normal force measurements, capillary flow entrance pressure drop and extrudate swell. One approach to measure the elasticity of melt polymers in steady shear experiment is to measure N_1 of the samples.

In this study, the steady shear rheological behaviours of polylactide (PLA) and nanographite platelets (NGP) bio-nanocomposites in addition to glass filled liquid crystalline polymers (LCPs) have also been examined to explore the likely existence of a correlation between N_1, filler contents, and shear rate of the measurements.

Experimental

.

<u>Materials</u>

PLA/NGP

Poly (L,L-lactide)- (PLA) was supplied by NatureWorks LLC and the grade of PLA used was 3051D with melt index and specific gravity of 10-30g/10min and 1.24 respectively. Nanographite platelets (NGPs) were supplied by XG Sciences, Inc US Michigan and the grade of NGP used for this study was "M" with characteristics: average thickness of approximately 6 - 8 nanometers and a typical surface area of 120 to 150 m^2/g. According to the material data sheet, Grade M (xGnp-M) is available with average particle diameters of 5, 15 or 25 microns.

LCP

Most liquid crystalline molecules are highly anisotropic, and to a good approximation can be regarded as rigid rods. Four aromatic nematic thermotropic LCPs in the form of pellets were chosen for this study. Due to propriety rights, only the major constituents of TLCPs are mentioned. The following nomenclatures are used for different filled and unfilled LCPs in our study.

1. LCP Polymer A unfilled (copolymer of p-hydroxy benzoic acid (HBA), 6-hydroxy-2 naphthoic acid (HNA), terephthalic acid (TA) and hydroxyl quinone,HQ)
2. LCP Polymer A filled with 35% glass fibre.
3. LCP Polymer B unfilled (copolymer of p-hydroxy benzoic acid (HBA), 6-hydroxy-2 naphthoic acid (HNA), terephthalic acid (TA) and hydroxyl quinone, HQ)
4. LCP Polymer B filled with 30% glass fibre.

Processing

PLA/NGP

PLA pellets and nanographite platelets (NGP) were dry mixed in the desired composition before melt blending in 700g batches. Table 1 shows the compositions (nominal wt%) and the codes of neat PLA and PLA/NGP samples (hereafter, samples are referred to according to their sample codes). Samples were melt-blended in a Brabender Twin Screw extruder. The speed and temperature of the extruder were set at 180 o C and 40 RPM. Dried pellets were compression moulded (temp.=180 oC , compression = 80 kN for 5 minute) into 2 mm thick circular plaque with 20 mm diameter specimens.

LCP

The LCP materials were dried in a vacuum dryer at 120 °C for eight hours to achieve target moisture content of less than 0.02 % prior to any use. The 2 mm and 3mm thick compression moulded samples were cut into 25 mm and 50 mm diameter discs. Shear rheological properties were studied using the Advanced Rheometrics Expansion System (ARES). The tests were conducted using 25 mm diameter plates as well as 50 mm diameter plates at temperatures between 330 °C and 380 °C for the various LCP materials. For each test, the sample was placed between preheated fixtures for 10 minutes to reach thermal equilibrium, and then the gap was set. For shear rheology experiments, the shear rates ranged from 0.01 to 10 s^{-1} because in the parallel plate geometry, slippage occurred at the shear rate above 10 s^{-1}.

<u>Rheological measurements</u>

Steady shear measurements were performed through an Advanced Rheometrics Expansion System (ARES) rheometer (TA Instruments) with parallel plate geometry using 25 mm diameter plates. A force transducer with a torque range of 0.2 to 200 gm-cm was applied to all measurements. All rheological tests were performed at the working temperature of PLA (180 °C). The tests were executed within the strain rate of 0.01-100 s^{-1} measuring 20 points per decade. Delay before measurement and measuring time of each point were set at 15 and 45 seconds respectively.

Results and Discussion

PLA/NGP

Figure 1 demonstrates the steady shear viscosity of neat PLA versus its NGP composites. At low shear rate region (< 2 s-1), neat PLA and PLA/NGP composites exhibited Newtonian behaviour. However, the zero shear viscosity of samples increased according to the amount of their filler loadings. In a study conducted by Ramsay (1986), the shear viscosity of intercalated and delaminated poly (dimethyl siloxane)/2M2T-MMT composites with varying silicate contents were measured at 28 °C [12, 13]. The absence of Newtonian behaviour at low shear rate region (shear thinning behaviour) of intercalated poly (dimethyl siloxane) composites was associated with the alteration of the relaxation dynamics of the polymer matrix as a result of highly anisotropic intercalated layered silicate particles causing shear thinning behaviour of composites at all shear rates [13]. The alteration of relaxation dynamics occurs at long relaxation times (corresponding to low shear rate region) where the behaviour of relaxation modulus determines the likelihood of the occurrence of zero shear viscosity behaviour. Nonetheless, fully delaminated poly (dimethyl siloxane)/2M2T-MMT nanocomposites exhibited a systematic increase in their steady-shear viscosity behaviour, yet; they obeyed Newtonian behaviour in low shear rate region. Sinha Ray (2006) mentioned that such behaviour indicates that delaminated filler layers do not affect the relaxation dynamics of polymer matrix; thus, the rheological behaviour of these nanocomposites shows resemblance to the conventional filler filled systems [13]. Therefore, the observation of Newtonian like behaviour of samples in figure 1 may indicates that NGP fillers of this research could have become delaminated within the polymer matrix to some extent.

Figures 2 and 3 illustrate the first normal force difference of the samples versus steady shear rate and shear stress. In the case of polymer composites, the contributing factors to the elasticity of the filled systems are fillers' size, concentration and shape. Isotropic fillers such as glass beads, carbon black, CaCO3 and TiO2 have caused decrease of elasticity of their polymer composites compared to neat polymeric systems [14]. On the other hand, due to the strong orientation in the direction of flow- resulting in hydrodynamic particle effects of fibre-like fillers- the N_1 value of composites become higher than unfiled systems [15]. However, the lower elasticity of isometric-filler polymer composites is because of their normal component of stress being zero while the shear component has a finite value, hence lower elasticity [16]. However, in case of anisometric fillers (e.g., fibres and platelets), the elasticity is escalated due to the increase in their N_1. Although fillers' morphology plays an important role in elasticity of the composites, excessive (higher than optimum) level of filler concentration has an adverse effect on the elasticity of such systems [17, 18]. This rise in rigidity as a result of excessive filler loading is due to the reduced mobility of polymer chains in the presence of the fillers [10, 17]. The dispersion of fillers within polymeric matrix can also play an important part in the elasticity of polymer composites. Furthermore, the near independence of N_1 from filler content at all shear rates in intercalated polymer composite systems has been associated with the intercalation of fillers in the composites as a result of ability of two dimensional silicate layers to preferentially orient in the direction of flow [19, 20]. Therefore, the effect of filler-filler interactions and their influence to elasticity is diminished. Nonetheless, according to Prasad et al. (2005), in exfoliated systems the magnitude of N1 has an inverse relationship with filler contents; i.e., the value of N_1 declines with increasing level of silicate loading. They associated this to the reduced mobility of polymer chains as a result of enhanced interaction between silicate fillers and polymer chains [21].

Anomalous Behaviour of PLA/NGP Composites

Consequently, the dependence of N1 behaviour on the NGP loading at low shear rate region (< 2 s^{-1}) could indicate that the dispersion of NGP within PLA matrix has reached exfoliation to some extent. Nevertheless, the fall in N_1 of composites did not reveal a direct relationship with nanofiller loading. The N1 magnitude of PLA01 did not exhibit a noticeable difference from neat PLA while PLA02 showed a considerable increase from neat PLA; yet, it was reduced significantly for PLA03 and PLA05 whilst the lowest N1 value of samples was recorded for PLA03. Nevertheless, at higher shear rates (> 2 s^{-1}) and shear stress (> 2E03 Pa) the value of N_1 of all samples reached a near equivalence. This near equivalence and independence of filler content of N_1 at higher shear rates were associated with preferential orientation ability of the two-dimensional nanofiller layers to the direction of shear flow [13, 22].

Figure 4 shows the steady shear viscosity and first normal stress difference of PLA02 (as a typical sample) versus shear rate. This figure demonstrates that as the shear rate rises, the shear thinning behaviour of shear viscosity and stress became evident. However, N_1 exhibited an utterly different behaviour upon which it increased steadily and reached a steady state plateau at very high rates (10-100 s^{-1}) where no steady viscosity data was available. Therefore, the observation of steady state behaviour in addition to N_1's near equivalence (at high shear rates) is another indication of the independence of N_1 behaviour of polymer matrix from both shear rate and the incorporated nanofillers, regardless of their loading and level of dispersion.

LCP

Figures 5 and 6 represent the experimental first normal stress difference values for two filled and two unfilled thermotropic LCPs. It was observed that for filled system, in the low shear rate region LCP composites exhibit negative normal stress difference. For both LCP composites the N_1 values become positive after 10 s^{-1}. Whereas, for both unfilled LCPs positive N_1 values are observed at low shear rate region.
Negative value of N_1 is usual for lyotropic LCPs. On the other hand the negative N_1 for thermotropic LCPs is rather abnormal [23]. The value of N_1 also depends on the concentration as well as shear rate. Beak and Magda (1993) showed how N_1 of lyotropic LCPs change from negative to positive with shear rate as well as concentration.
Usually Newtonian fluids, do not exhibit any N_1 [24]. On the other hand, filled Newtonian fluids exhibit significant values of N_1. The magnitude of first normal stress difference is also dependent on the concentration and aspect ratio of fibre as well as on the elastic behaviour of the matrix [25].

Table 1 – Compositions of PLA/NGP composites and their corresponding sample codes.

	Sample Code				
Sample Compositions	neat PLA	PLA01	PLA02	PLA03	PLA05
PLA content (wt%)	100	99	98	97	95
NGP content (wt%)	0	1	2	3	5

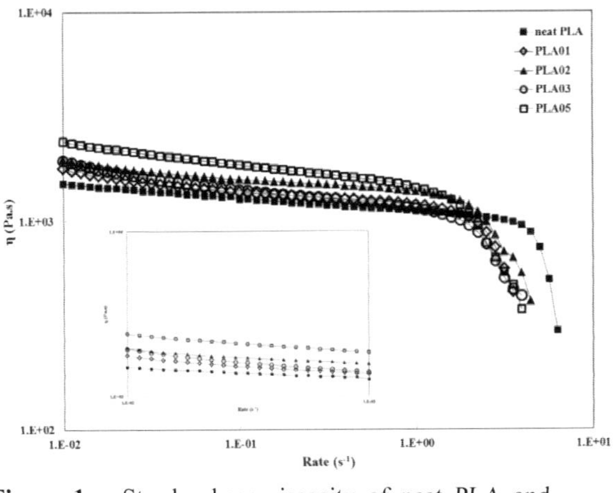

Figure 1 – Steady shear viscosity of neat PLA and PLA-NGP composites at T= 180 °C.

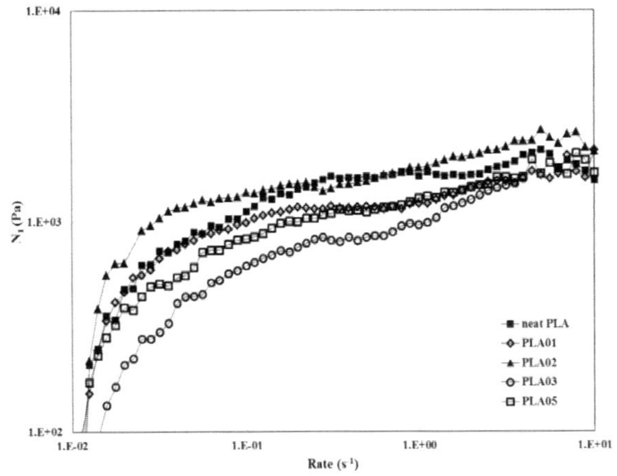

Figure 2 – Illustration of first normal stress difference of neat PLA and PLA/NGP composites versus shear rate at T= 180 °C.

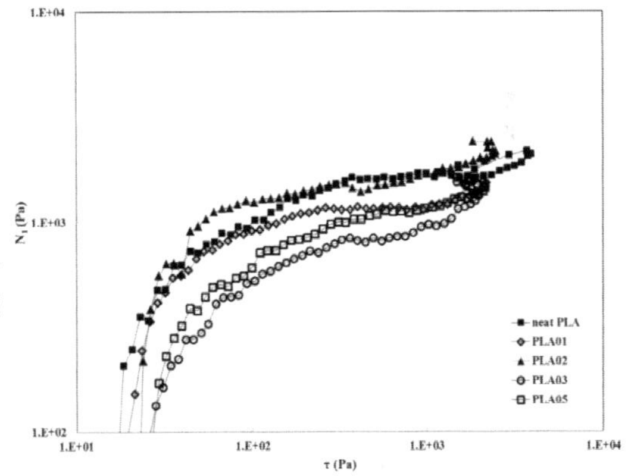

Figure 3 – Illustration of first normal stress difference of neat PLA and PLA/NGP composites versus shear stress at T= 180 °C.

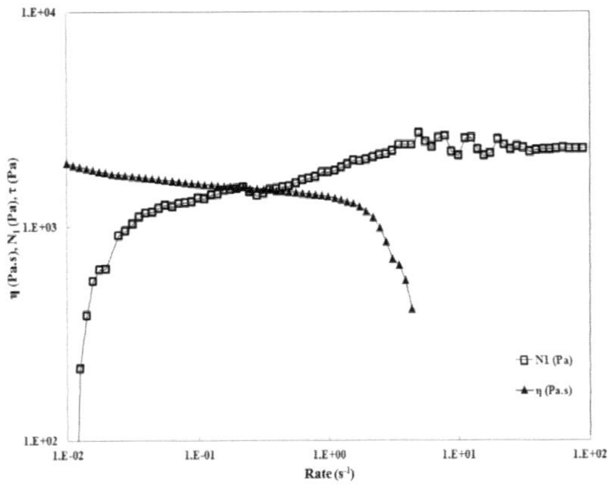

Figure 4 – Illustration of steady shear rheological behaviour (N_1 and shear viscosity) of PLA02 versus shear rate at T= 180 °C.

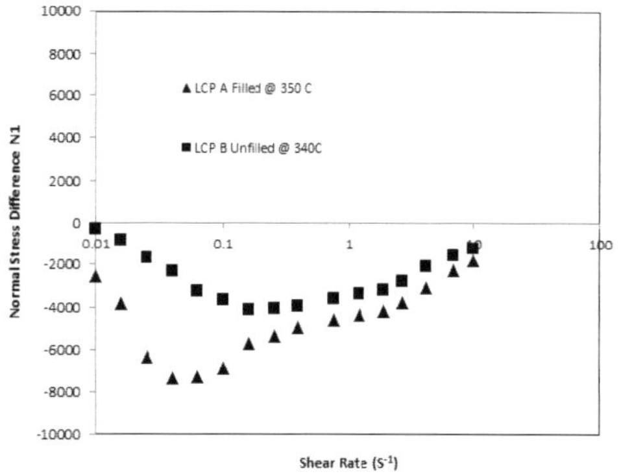

Figure 5 – Normal stress data for LCP A and LCP B composites @ 350 °C.

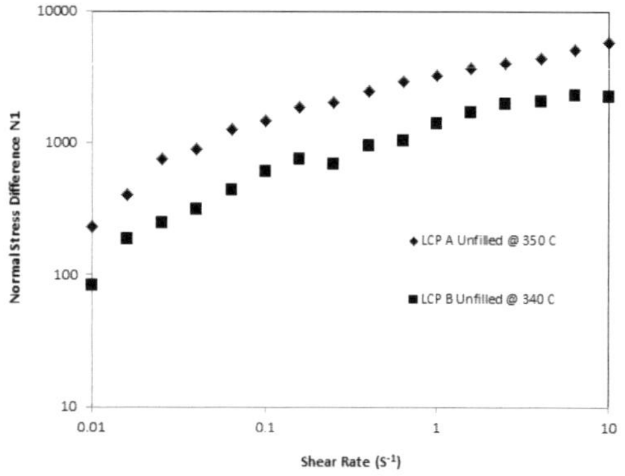

Figure 6 – Normal stress data for LCP A and LCP B Unfilled @ 350 °C.

Conclusions

The results suggest that the N_1 of PLA/NGP bionanocomposites is dependent on the level of loading and on shear rate beyond a critical value. This observation as well as the existence of Newtonian behaviour of PLA/NGP composites at low shear rate could indicate the occurrence of delamination of NGP nanofillers within PLA matrix. However, this study could not obtain a similar correlation between N_1 and shear rate for liquid crystalline polymers in both high and low rate regions.

In the case of LCP, Fibre- fibre interactions were competing with the elasticity of rigid rod like molecule at low shear rate region. Overall, this interaction seems to suppress the normal stress and leads to negative values of N_1.

References

1. C.W. Macosko, Rheology: Principles, Measurements, and Applications, Wiley-VCH, New York, 1994.
2. M. Keentok; A.G. Georgescu; A.A. Sherwood; R.I. Tanner Journal of Non-Newtonian Fluid Mechanics 1980, 6, 303-324.
3. S. Ramachandran; H.W. Gao; E.B. Christiansen Macromolecules 1985, 18, 695-699.
4. A.S. Lodge; J. Meissner Rheologica Acta 1972, 11, 351-352.
5. J.M. Dealy; R.G. Larson, Structure and Rheology of Molten Polymers: From Structure to Flow Behaviour and Back Again, Hanser, Munich, 2006.
6. H.M. Laun Rheologica Acta 1978, 17, 1-15.
7. K. Osaki; N. Bessho; T. Kojimoto; M. Kurata Journal of Rheology 1979, 23, 617-624.
8. C.M. Vrentas; W.W. Graessley Journal of Non-Newtonian Fluid Mechanics 1981, 9, 339-355.
9. L.J. Kasehagen; C.W. Macosko Journal of Rheology 1998, 42, 1303-1327.
10. A.V. Shenoy, Rheology of filled polymer systems, Kluwer Academic Publisher, London, 1999.
11. J.M. Dealy; K.F. Wissburn, Melt Rheology and Its Roles in Plastics Processing, Chapman and Hall, New York, 1990.
12. J.D.F. Ramsay Journal of Colloid and Interface Science 1986, 109, 441-447.
13. S. Sinha Ray Journal of Industrial and Engineering Chemistry 2006, 12, 811-842.
14. S.N. Bhattacharya; R.K. Gupta; M.R. Kamal, Polymeric Nanocomposites: Theory and Practice, Hanser Gardner Pubns, Munich, 2007.
15. J.L. White; L. Czarnecki; H. Tanaka Rubber Chemistry and Technology 1980, 53, 823-835.
16. S.A. Khan; R.K. Prud'Homme Reviews in Chemical Engineering 1987, 4, 205-270.
17. C. Dae Han Journal of Applied Polymer Science 1974, 18, 821-829.
18. H. Tanaka; J.L. White Polymer Engineering and Science 1980, 20, 949-956.
19. R. Krishnamoorti; J. Ren; A.S. Silva Journal of Chemical Physics 2001, 114, 4968-4973.
20. J. Ren; R. Krishnamoorti Macromolecules 2003, 36, 4443-4451.
21. R. Prasad; R.K. Gupta; F. Cser; S.N. Bhattacharya Journal of Polymer Engineering 2005, 25, 305-330.
22. E.P. Giannelis; R. Krishnamoorti; E. Manias Adv. Polym. Sci. 1999, 138, 107-147.
23. A.M. Donald; A.H. Windle; S. Hanna, Liquid Crystalline Polymers, Cambridge University Press, 2006.
24. P.R.A. Eberle; D.G. Baird; P. Wapperom Ind. Eng. Chem. Res. 2008, 47, 3470-3488.
25. S.A.A. Ramazani; A. Ait-Kadi; M. Grmela Journal of Rheology 2001, 45, 945-962.

O-08-070

FUNDAMENTAL UNDERSTANDING OF ENCAPSULATION IN MULTIPHASE POLYMERS

Khalid Lamnawar[2], Mosto Bousmina[3], Abderrahim Maazouz[1,3]

[1]Université de Lyon, F-69361, Lyon, France; CNRS, UMR 5223, Ingénierie des Matériaux Polymères, INSA Lyon, F-69621, Villeurbanne, France;

[2]Laboratoire de Mécanique des Contacts et des Structures-Université de Lyon, CNRS, INSA-Lyon, LaMCoS UMR CNRS#5259, Groupe de Recherche Pluridisciplinaire en Plasturgie, F69621, France

[3] Hassan II Academy of Science and Technology, Rabat, Morocco.

Abstract: The aim of the present work has been to gain a fundamental understanding of the mechanisms governing encapsulation in the multiphase systems as a blend or multilayer structures. The model systems chosen for this study are based on (i) Newtonian poly (dimethylsiloxane) polymers of varying molar masses, (ii) high molecular weights, viscoelastic and compatible pair polymers of PVDF and PMMA. The same approach was applied to functionalized polymers to investigate the effect of physic-chemical affinity on two pairs of asymmetrical reactive polymers based on PE-GMA (Glycidyl methacrylate)/PVDF-g-MA (Maleic anhydride) and a PE/PVDF as a reference. The linear viscoelastic and surface properties of the neat and bilayer model systems structures have been investigated. The optical obeservations of the encapsulation kinetic of two drops were recorded using a home-made device. Specific experiments were carried out to follow up the kinetic of encapsulation and the results were rationalized as function of the effect of the viscosity, elasticity ratios, drop geometry, the interfacial tension and the physico-chemical affinity. Throughout all the experiment, the mechanisms were purposed for each system and discussed based on the theories of molecular forces or Brownian motion governing diffusion and Ostwald ripening in contrast to the theories of coalescence. The viscosity ratio coupled to the drop geometry of the material were found to be a key parameter and that It has to be linked to the interfacial tension and spreading parameters. Furthermore, the encapsulation appeared to be hindered by the interdiffusion process in the case of compatible pair system despite their elasticity and surface tension contrast. Finally, the encapsulation kinetic could be reduced or eliminated by the creation of a copolymer at the interface for a reactive system. The results obtained by the optical investigation of two drops corroborated the rheological data of the bilayer systems. Hence, the obtained results rendered it possible to decouple the influence of the viscoelastic parameters to flow, interfacial tension, thereby highlighting a number of macroscopic effects that were governed by the interdiffusion or reaction of macromolecular chains at the interface to give a better understanding of encapsulation phenomenon in multiphase systems during blend extrusion and especially coextrusion.

Keywords: Encapsulation; interfacial Rheology; multilayer; optical observation; interdiffusion

Introduction

Encapsulation phenomena has been observed in experiments involving a wide range of materials from oil and water to polymer melts for both high and low Reynolds numbers. It can also used as a technique in the food, biomedical, and pharmaceutical industry where small particles, cells, and cell agglomerates are produced and dispersed in an ambient medium. Encapsulation efficiency is also of primary importance in the biomedical industry, as reviewed by Orive et al [1]. The phenomenon has been also observed in other systems as in bubbles and cells [2,3].

The encapsulation subject is old in principle; the pioneering works show that lower viscosity material will eventually encapsulate the higher viscosity fluid [4](Tomotika et al. (1935)). In the case of multiphase polymer system, we shall quickly list some investigations of close relevance to the present study:

* Encapsulation in multilayer structure:

Previous work on layer thickness uniformity has primarily been devoted to the effects of differing polymer viscosities in the individual layers under shear [5-9]. These differences in viscosities give rise to a phenomenon known as "viscous encapsulation" in producing nonuniform layer thicknesses. there has only been performed a very limited number of experimental studies dealing with encapsulation phenomena [10-12], and most of these research efforts have suggested that, irrespective of the stability of the interface, the more viscous fluid tends to push into its less viscous counterpart, eventually leading to the less viscous fluid encapsulating the more viscous components. In other words, encapsulation is caused by the tendency of the less viscous polymer to enter the region of high shear (i.e., the wall). The above-mentioned phenomenon is mainly driven by a jump in viscosity but also by a difference in elasticity between the two fluids [13]. In an early study, White et al.[14] argued that the shape of the interface can also be influenced by the jump in secondary normal stress difference. These experimental observations are consistent with numerical studies with Newtonian and purely inelastic constitutive equations [15,16]. However, examples of coextruded structures with nonuniform layer thicknesses have been observed even when the viscoselastic properties of the materials are very similar [17]. This implies that another factor affects the flow of the materials during process. An investigation of the effect of polymer viscoelasticity and physicochemical affinity on the layer thickness uniformity of multilayer coextruded structures would thus be of great interest.

* Encapsulation in blend:

In studies of multicomponent polymer blends of more than two phases, papers dedicated to this subject have shown that the morphology resulting from the competition between drops collision under flow followed by coalescence /encapsulation or drop breakup, depend mainly on the interfacial and rheological properties. The final droplet size distribution, under mixing, results from the balance between the present phenomena [18]. Recently, the effect of compatibilisation on the deformation and breakup of drops in step-wise increasing shear flow has been investigated [19].

Hobbs et al.[20] first reported the encapsulation concept on the spontaneous development of the composite droplet morphology in immiscible as. They suggested that, under equilibrium mixing conditions, interfacial forces might play an important role in establishing the phase morphology of multiphase polymer blends. By using the concept of a spreading coefficient, they rewrote the Harkin equation (equation 1) in which two dissimilar phases are dispersed within a third matrix material by substituting the appropriate interfacial tensions for the surface tension values

$$\lambda_{31} = \sigma_{12} - \sigma_{31} - \sigma_{13} \quad (1)$$

Here, σ_{12}, σ_{32} and σ_{13} are the interfacial tensions for each component pair, and λ_{31} is defined as the spreading coefficient for the case of component 3 encapsulating component 1. The index 2 refers to the matrix. λ_{31} must be positive for component 1 to be encapsulated by component 3. The authors successfully demonstrated the usefulness of the spreading coefficient concept for predicting the various morphologies that were formed, and in particular the encapsulation effect.

Recently, a numerical simulation of particle encapsulation due to liquid thread breakup has been published to investigate the effect of the particle spacing, viscosity ratio, Reynolds number, and waveform of interfacial perturbations [21].

Recent papers on ternary systems have suggested that composite droplet formation may be related to the viscosity of the dispersed components. Nemirovski et al.[22] studied the phase morphology of a number of three-component (thermoplastic/thermotropic) systems and suggested that the dispersed component A would encapsulate the dispersed component B (in a matrix C) when both the thermodynamics, expressed by means of a positive λ A/B spreading coefficient, and the kinetic effects, expressed by means of a dispersed phase viscosity ratio smaller than 1, acted cooperatively. However, in some cases, a dispersed phase viscosity ratio greater than 1 was found to hinder the development of the core–shell structure (A encapsulating B) even though encapsulation had been predicted by a positive λ A/B spreading coefficient.

The reverse effect was observed by Gupta and Srinivasan [23] for ternary blends of styrene–ethylene–butylene terpolymer (SEBS) and polycarbonate (PC) dispersed within a PP matrix. The SEBS component formed a boundary layer at the surface of the PC droplets when PC was less viscous than SEBS, whereas the minor components were separately dispersed when PC was more viscous than SEBS. However, most authors have failed to find any influence of the dispersed phase core–shell viscosity ratio on the composite droplet structure [24].

Recently, Reignier et al. [25] investigated the influence of the molecular weight of the dispersed phase components on encapsulation effects in the composite droplet phase for a ternary blend composed of PS and PMMA dispersed in HDPE. Nevertheless, arguments based on the effect of the viscosity ratio or the absolute viscosities of the phases could not explain the above results.

Finally, as briefly summarized above, significant progress in understanding the encapsulation phenomenon in multiphase systems has been obtained over the past decades. Few papers have been dedicated to polymer systems. To the best of our knowledge, few research efforts have, with regard to fundamental and experimental aspects, been dedicated to this subject. Despite the interesting nature of this kind of research, it is of no help when attempting to comprehend either the generation of the encapsulation decoupled to shear or their connection with the physicochemical affinity and viscoelasticity. A thorough understanding of physicochemical aspects as well as of the full rheological parameters of the encapsulation phenomena have yet to be established. In order to attain this, the attention of the present study has been focused on performing original experiments with well-characterized model polymer fluids with different molar masses and functionalities. The aim of this work thus involved decoupling the influence of viscoelastic parameters and flow; thereby highlighting a number of macroscopic effects coupled to the physico-chemical affinity which can affect the encapsulation phenomenon. The purpose of the investigation has been to present simples experimental studies in an attempt to gain an understanding of the mechanisms governing encapsulation decoupled to these parameters. The obtained optical results will be analyzed based on the surface properties, and rheological properties of mono and bilayer systems. The details of the experimental procedures are presented in the next section, followed by the main results and a discussion.

Experimental section and main results

Base Materials

1. The model fluids

Three types of PDMS with varying viscosities were chosen. Table 1 lists their designations and viscosities. A complete investigation of their rheological properties was carried out.

Table 1. The designations of the 3 PDMS materials and their respective viscosities.

Fluid	Newtonian viscosity (Pa.s)
PDMS1	10,6
PDMS2	28,9
PDMS3	80,8

2. High molecular model polymers :

Two industrial systems were kindly supplied by Arkema (Serquigny, France):

(i) The first pair was based on PVDF/PMMA and was completely miscible.

(ii)The second system was based on two pairs of reactive polymers at interfaces of PE-GMA (glycidyl methacrylate)/PVDF-g-M A(maleic anhydride). The PE/PVDF was also used as a nonreactive system.

The rheological properties (in both dynamic and steady modes) of these materials were investigated and reported.

Rheological investigations

Rheological properties of the pure polymers:

All rheological measurements were carried out in the linear viscoelastic region in small amplitude oscillatory shear mode as was verified by preliminary strain sweep measurements. Polymer degradation was avoided by continuously purging the oven with nitrogen. The measured rheological functions included the complex viscosity η^*, its real and imaginary components η' and η'', as well as the elastic modulus G' and loss modulus G":

1) The rheological behavior of model fluids was studied. Flow tests were performed with a stress-controlled rotational rheometer AR100 from TA Instruments, with a cone and plate geometry (ϕ=60 mm, $\alpha=2°\approx0.035$ rad, gap=66 µm). To obtain the largest range of frequencies, the tests were carried out from 25°C to 65°C. A master curve was plotted for a reference temperature of 25°C using a time-temperature superposition procedure. Newtonian viscosities and relaxation times at 25°C for each fluid were then deduced from this curve. All three fluids presented a large Newtonian plateau. As could be expected, the higher the viscosity, the smaller was the plateau. For PDMS1, the upper limit of the plateau was about 250 rad/s, whereas this value decreased to 6 rad/s for PDMS3. In the following section, only the results of the PDMS1/PDMS3 pair are presented.

2) Rheological measurements for High molecular model polymers were performed using a strain-controlled rotational rheometer ARES from TA Instruments with a plate-plate geometry. Rheological properties were explored within a range of temperatures spanning from 180°C to 240°C. temperature. Newtonian viscosities and relaxations times were calculated at several temperatures. The corresponding activation energies of viscous flow calculated from the master curve at such a reference temperature are respectively, 51,5; 70 Kj/moles for PVDF and PMMA. 53 and 56 Kj/mole for PE and PEGMA.

Rheological studies of bilayer systems:

The rheological experiments of the bilayer systems were performed with a strain-controlled rheometer: ARES (Advanced Rheometrics Expansion System, Rheometrics Co.) using a parallel plate geometry (Φ=25mm). Sample discs were placed between the plates and melted. For the sake of clarity, additional details of the experimental procedure are given in previous articles [26-29]

Optical observations

The optical experiments were conducted on a regulated hot stage on which the drops or the particles were deposited. A Zeiss binocular and a Pantera CCD camera linked to a computer allow to record images at regular intervals. More details of this device are reported in paper of Muller et al. [30].

Drops are deposited on the substrate. The entire process of coalescence is recorded. At short times, i.e. when neck radius grows quickly, all images are analysed. When the sintering rate decreases, data are exploited every 5 or 10 images. All tests were repeated at least 3 times to ensure reproducibility. The presented results corresponded to the mean values of the 3 tests since this decreased the experimental error (4%).

Model Fluids:

The experiments were carried out at 25°C for the model fluids. Because of their relatively low viscosities, the drops adopted - more or less instantaneously - a spherical shape. This facilitated the sintering kinetics measurements at short times. Only one substrate based on PTFE with varying surface tension values was used [31].

High molecular model polymers :

The optical observations of the high molecular model polymers were performed at a variety of temperatures ranging from 180°C to 240°C. It was thus possible to examine the effects of viscosities, relaxation times, and interfacial reaction in the case of the reactive system. The same substrates were employed.

Main results and discussions:

Highlight of the encapsulation kinetics of the model fluids:

The encapsulation phenomenon was recorded by optical observations as a function of the viscosity ratio and the polymer/substrate interfacial tension. Figure 1 displays an illustration of the coalescence evolution of the PDMS1/PDMS3 pair, for which the viscosity ratio reached a value of 8.

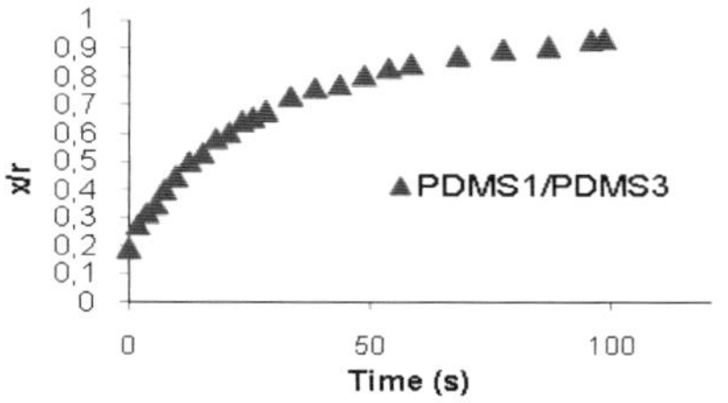

Fig. 1.The encapsulation kinetic curve of PDMS1/PDMS3 drops at 25°C.

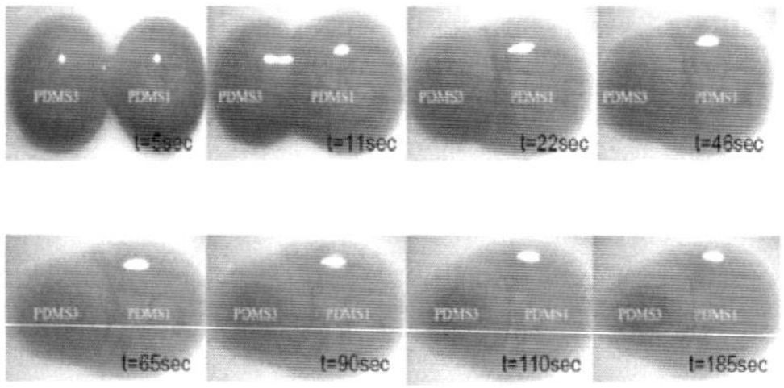

Fig. 2. An illustration of the encapsulation at 25°C in the case of the PDMS1/PDMS3 pair [32].

Figure 2 provides an experimental illustration to the fact that the less viscous polymer tended to encapsulate its more viscous counterpart but at zero shear rate. These results also confirmed that the viscous dissipation mechanism given in the literature was not the only one present. Finally, the coupling between rheological and optical studies showed that the viscosity of the material was a key parameter and that it had a considerable influence on the encapsulation kinetics.

Experimental investigation of High molecular model polymer :

In order to investigate the effect of the physical compatibilization on the encapsulation phenomenon, studies were performed on the industrial materials based on PVDF/PMMA, which were completely miscible. The kinetics of coalescence was assessed at several temperatures (i.e., viscosity ratios) and was found to be less pronounced as compared to those of PVDF/PVDF and PMMA/PMMA. This result can be explained on the basis of the diffusion mechanism which is significant in the case of symmetrical polymer/polymer interfaces as compared to those of asymmetrical polymer pairs.

An example of typical curves is given in Figure 2. We have shown that the viscosity ratio of PVDF/PMMA was higher at 190°C than at 240°C. Indeed, the kinetic of coalescence was slower than at 240°C where there was a high mobility of chains at the polymer/polymer interface. Furthermore, no encapsulation effect could be seen as confirmed by Figure 3, despite the difference in viscosity at 190°C. The observed development represented the signature of the kinetics of diffusion where the macromolecular chains intertwined and gradually became replaced by a more robust interphase.

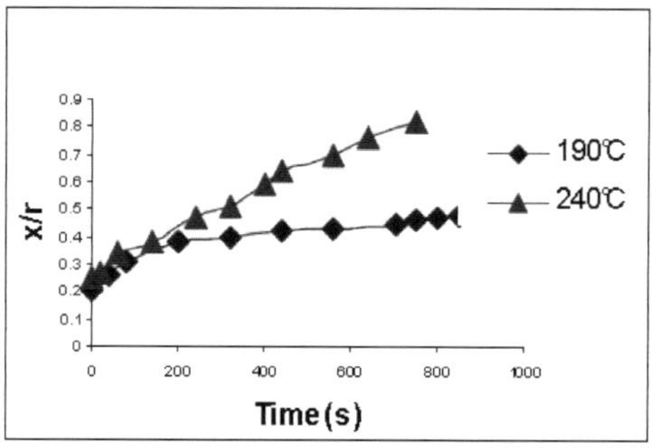

Fig. 2. Coalescence/encapsulation curves of PVDF/PMMA drops at 190°C and 240°C.

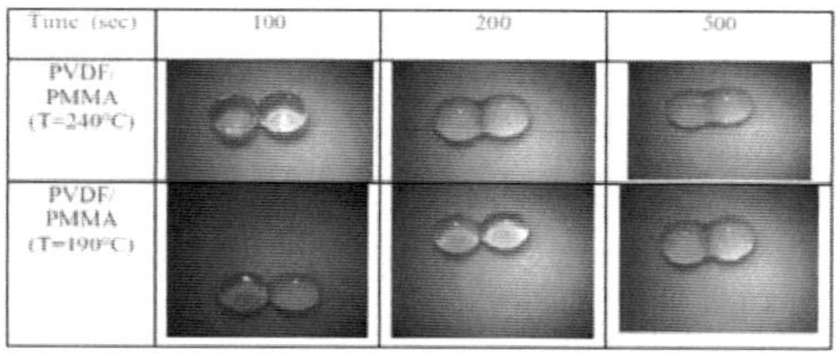

Fig. 3. An illustration of the coalescence kinetics of PVDF/PMMA drops at 190°C and 240°C.

On the other hand, the same approach was applied to functionalized polymers in order to investigate the effect of reactive compatibilization on the encapsulation phenomenon. Two pairs of reactive polymers at interfaces based on PE-GMA (glycidyl methacrylate)/PVDF-g-AM (maleic anhydride) as well as a nonreactive material at an interface based on PVDF/PE were selected. The kinetics of coalescence was evaluated at different temperatures. For example, Figure 4 shows a comparison of the coalescence kinetics of these systems at 240°C. The interfacial reaction between MA and GMA functions certainly attenuated the coalescence kinetic.

Furthermore, results from an optical investigation of the coalescence kinetics of two drops corroborated those obtained through rheology of the bilayer systems according to the procedure given in our previous work [27-29]. Indeed, the observed phenomena and the results can be analyzed based on the physico-chemical mechanisms involved in the diffusion/reaction interfaces. As can be seen from Figure 5, the elastic modulus of PE-GMA/PVDF-g-AM increased with time as compared to the nonreactive system based on PE/PVDF. In fact, no significant increase in viscosity was found in the case of the latter, thus confirming that the observed phenomenon was caused by the reactivity at the interface at a low oscillation frequency. Since a reactive polymer bilayer generates graft or block copolymers near the interface, each block (or graft) component can easily entangle with (or stick to) the corresponding homopolymer located near the interface. Thus, such an interface should be strengthened as compared to an interface without graft or block copolymers. This could increase the viscosity of the reactive polymer bilayer as opposed to that of a nonreactive polymer bilayer. Consequently, such a behavior was evidence of the formation of a copolymer from the reaction of PE-GMA and the carboxylic and maleic anhydride functions of functionalized PVDF.

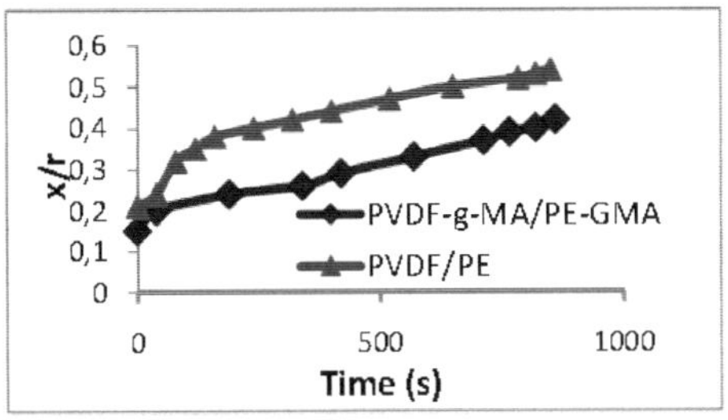

Fig. 4. A comparison of coalescence curves at 240°C featuring PVDF-g-AM /PE-GMA and PVDF/PE drop systems.

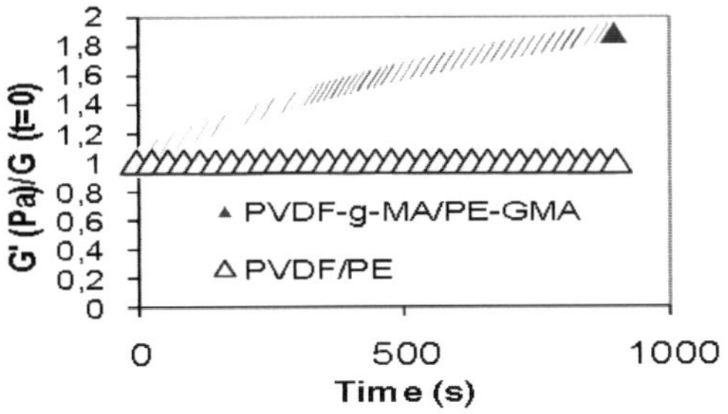

Fig. 5. A comparison of the evolution of G'(Pa)/G' (t=0) *vs* time at 240°C for a reactive bilayer PE-GMA/PVDF-g- system and a nonreactive one based on PE/PVDF.

Finally, no encapsulation was noted for PVDF-g MA/PE-GMA (contrary to PE/PVDF) as was also confirmed by Figure 6, despite that its zero shear viscosity ratio was close to 1.2.

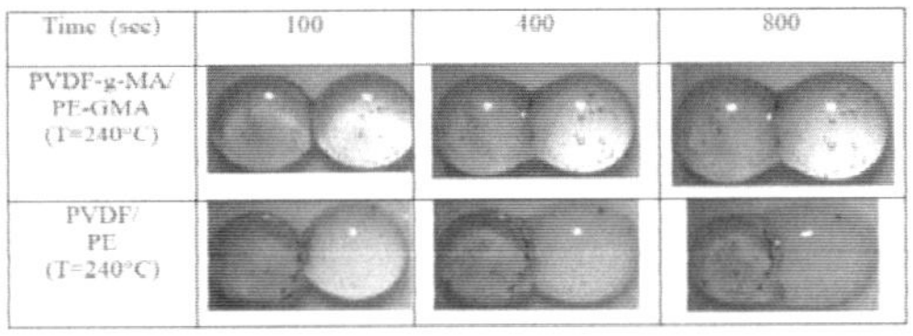

Fig. 6. An illustration of the coalescence kinetics of PVDF-g-MA/PE-GMA and PVDF/PE drops at 240°C.

This seemed to confirm that it was essential to couple the viscosity ratio to the physico-chemical affinity and viscoelastic parameters in order to gain a better understanding of the encapsulation defects in multiphase systems. In addition, the role of the viscosity ratio, elasticity ratio, and layer ratio should also be investigated, thereby coupling to the reaction rate/compatibilization phenomena of the polymer/polymer interface. Indeed, for a reactive system, the encapsulation defects could be reduced or eliminated by the creation of a copolymer at the interface [32].

References:

1. Orive G, Hernandez RM, Gascon AR, Callafiore R, Chang TMS, Hortelano G ., Nat Med, 9,104 (2003).
2. Goosen MFA, O'Shea GM, Gharapetian HM, Chou S, Sun AM. Biotechnol Bioeng 27(2),146 (1985)
3. Deng Q, Anilkumar AV, Wang TG, J Fluid Mech., 578, 119 (2007)
4. Tomotika S. Proc Roy Soc A, 150,322 (1935).
5. Southern J. H. and Ballman R. L. J. App. Polyrn Sci, 20,175(1973)
6. Southern J. H. and Ballman R. L. . J. Polym Sci, 13,863(1975).
7. Everage A. E., Trans. Soc. Rheo. 17, 629 (1973)
8. Lee B. L. and White J. L., 'Dm. Soc. Rheo., 18, 3, 467 (1974).
9. Dooley J., Hyun K.S., and Hughes K., Polym. Eng. Sci., 38, 1060 (1998).
10. Wilson, G.M., Khomami, B., J. Non-Newt. Fluid Mech., 45, 355 (1992)
11. Theofanous G., Nourgaliev R., Khomami B. J. Non-Newtonian Fluid Mech. 143, 131 (2007).
12. Yih, C.S., J. Fluid Mech., 27 (2), 337 (1967).
13. Khomami, B.. J. Non-Newt. Fluid Mech.,, 37, 19(1990).
14. White J.L., Ufford R.C., Dharod K.R., Price R.L., J. Appl. Poly. Sci., 16, 1313 (1972).
15. Joseph D.D, K. Nguyen, G.S. Beaves, J.Fluid Mech, 141, 319 (1984).
16. Karagianis, Hrymak A.N., Vlachpoulos J., Rheol. Acta 29, 71 (1990).
17. Lamnawar, K.; Maazouz A. Polym. Eng. Sci.; 49, 727 (2009).
18. Jeelani S.A.K., Windhab E.J. Chem. Eng. Sci. 64,2718 (2009)
19. Sourki Abbassi F., Michel A. Huneault, Bousmina M.., Polymer, 50,645 (2009).
20. Hobbs SY, Dekkers MEJ, Watkins WH. J Mater Sci, 23,1219 (1988)
21. Haoxiang L., Pozrikidis C. Computers & Fluids, 38, 564 (2009)
22. Nemirovski N, Siegmann A, Narkis M. J Macromol Sci, Phys B, 34,459 (1995).
23. Gupta AK, Srinivasan KR . J Appl Polym Sci. 47, 167 (1993)
24. Hemmati M, Nazokdast H, Panahi HS., J Appl Polym Sci.,82, 1129 (2001)
25. Reignier J., Basil D. Favis, Marie-Claude Heuzey, Polymer, 44, 49 (2003).
26. Qiu, H.; Bousmina, M. Macromolecules; 33, 6588 (2000)
27. Lamnawar K, Maazouz A, Rheol. Acta, 45, 411 (2006).
28. Lamnawar K, Maazouz A, Rheol. Acta, 47, 383 (2008).
29. Lamnawar K., Baudoin A., Maazouz A. European Polymer Journal, 46, 1604 (2010).
30. Muller J-D., Bousmina M., Maazouz A. Macromolecules, 41 (6), 2096 (2008).
31. Wu, S. Polymer Interface and Adhesion, Marcel Dekker INC., New-York (1982)
32. K. Lamnawar, M. Bousmina, A. Maazouz. 2D Encapsulation in Multiphase Polymers: Role of Viscoelastic, Geometrical and Interfacial

Properties. *Macromolecules*, 2012, *45 (1)*, 441–454.

VISCOELASTIC MELT RHEOLOGY AND TIME-TEMPERATURE SUPERPOSITION OF POLYCARBONATE – MULTI-WALLED CARBON NANOTUBE NANOCOMPOSITES

G.Y.H. Choong[1]*, D.S.A. De Focatiis[1] and D.G. Hassell[2]
[1]*Division of Materials, Mechanics, and Structures, University of Nottingham, UK – eaxyhgc@nottingham.ac.uk; davide.de.focatiis@nottingham.ac.uk*
[2]*Petroleum and Chemical Engineering, Institut Teknologi Brunei, Gadong, Brunei Darussalam – zdavid.hassell@itb.edu.bn*

Abstract – This work investigates the linear and non-linear viscoelastic melt rheology of four grades of polycarbonate-multi-walled carbon nanotube nanocomposites with varying molar mass but constant filler content (3 wt%). Small amplitude oscillatory shear rheology was employed to study the time and temperature dependence of the viscoelastic response. Mastercurves are constructed from isothermal frequency sweeps using both vertical and horizontal shifting. Although all nanocomposites exhibit a second plateau at ~100kPa, the relaxation times estimated from the peaks in loss tangents are not statistically different from those estimated from cross-over frequencies on pure melts: all relaxation times scale with molar mass in the same way, evidence that relaxation of the polymer network is the dominant mechanism in these nanocomposites.

Keywords: Melt rheology, time-temperature superposition, polycarbonate, nanocomposite

Introduction

Carbon nanotubes (CNTs) can render thermoplastics conductive at much lower loading content than conventional fillers. The attraction for the industrial community arises from the potential to generate new, melt processable materials with enhanced properties relative to commodity plastics.

Time-temperature superposition (TTS) is used to produce a mastercurve that extends the viscoelastic response of polymers over a wider range of frequencies that can be measured experimentally. It relies on the principle of thermorheological simplicity [1]. If the relaxation times have the same temperature dependence, frequency or time-dependent viscoelastic data can be shifted to any reference temperature T_{ref} with an appropriate horizontal shift factor. A vertical shift factor may be required because of the effect of temperature on the modulus.

TTS is well established on pure polymer melts, but there have been mixed conclusions in its application to nanocomposites. For example, the viscoelastic moduli of polycarbonate + 2 wt% multi-walled carbon nanotubes (PC-MWCNTs) measured by Handge and coworkers at 190 °C and 210 °C did not superpose satisfactorily [2], whereas Wu and coworkers found that TTS was possible in a poly(butylene terephthalate)-MWCNT between 230 °C to 240 °C only [3]. None of the many studies attempting TTS in the literature discuss the application of vertical shifting.

This work aims to explore TTS, and in particular the need for vertical shifting, by carrying out new isothermal frequency sweeps at a range of amplitudes measuring the viscoelastic properties of a range of PC-MWCNT nanocomposites with varying matrix molar mass, and constructing mastercurves.

Experimental

Materials

The materials used in this study are four grades of PC supplied by Bayer MaterialScience AG. NC7000 MWCNTs were manufactured by Nanocyl S.A., Belgium, via a catalytic chemical vapour deposition process. The molar mass of each of the supplied PC grades was measured in THF using GPC, and is shown in Table 1.

Table 1 – Molar mass of PC grades measured by GPC.

Materials	M_n (g mol^{-1})	M_w (g mol^{-1})	PDI
Makrolon 2205	13900	33600	2.46
Makrolon 2405	14200	38700	2.73
Makrolon 2805	18000	45700	2.59
Makrolon 3105	21400	50500	2.40

Nanocomposites were melt-compounded by Nanocyl with an Industrial Leistritz ZSK-MAXX co-rotating twin screw extruder with a screw length to diameter ratio of 48:1. A constant mass fraction of 3 % MWCNTs was added to all PC grades. The screw speed was fixed at 300 rpm and the barrel temperature was 280 °C.

Methods

After drying at 80 °C for at least 8 hours, circular discs 30 mm in diameter and 0.5mm in thickness were compression moulded at 250 °C. Dynamic rheometry was performed using a Bohlin C-VOR Instruments rheometer in oscillatory shear mode using a 25 mm parallel plate geometry with a 0.5 mm gap size in an air atmosphere. Each disc was trimmed to size with a sharp blade prior to commencing the test after acclimatising at the highest temperature for 5 minutes.

Isothermal frequency sweeps were carried out at a fixed 0.5% strain amplitude (within the linear viscoelastic limit identified from amplitude sweeps in [4]); 5% and 50% strain amplitude (probing the non-linear region). All grades of pure PC and nanocomposite were measured in decreasing 20 °C steps across 160 – 260 °C, and 160 – 300 °C respectively. The frequency range logarithmically increased from 0.2π to 80π rad s^{-1}. A 5.5 minutes delay was imposed for thermal equilibration between each test temperature.

Results and Discussion

Vertical shifting

When a van Gurp-Palmen (VGP) plot of loss angle, δ vs complex modulus, G^*, produces overlapping curves without any shifting, thermorheological simplicity can be inferred [5]. Fig. 1 illustrates VGP plots of unfilled and filled PC. The unfilled PC exhibits a reasonably good overlap across the temperature range, although this fit does improve very slightly with a density correction (not shown here, details described in [4]). The vertical shift factors b_T used in the density correction are shown in Fig. 2.

The VGP plot for the PC-MWCNT shows a distinct lack of overlap that cannot be resolved by density alone. Instead, the vertical shift factors are obtained by horizontally shifting G^* in the VGP plot to superpose the measurements [1], as shown in Fig. 3 for filled Makrolon 2205. b_T values obtained for measurements on all filled grades from this manual shifting are shown in Fig. 2.

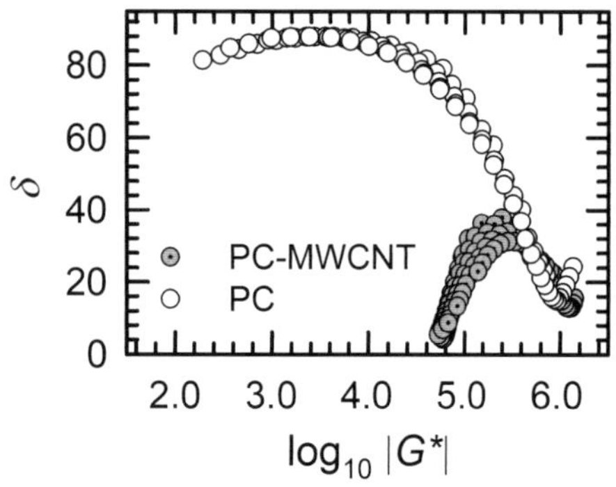

Figure 1 – Van Gurp-Palmen plot of unshifted data.

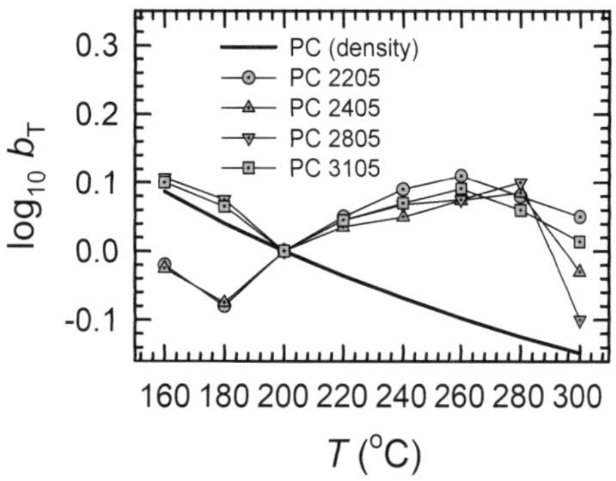

Figure 2 – b_T determined from density for PC and from manual VGP shifts for filled PC.

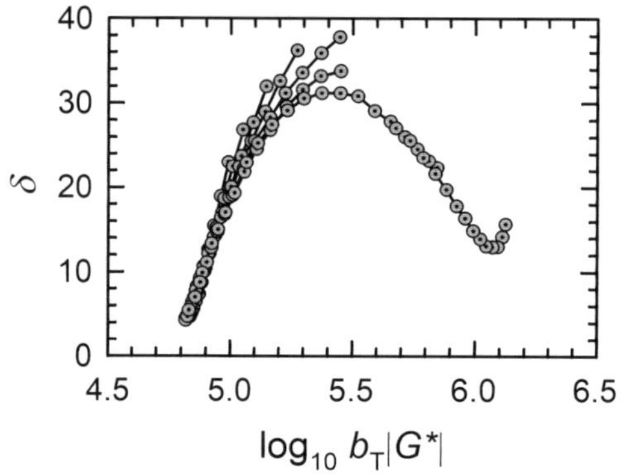

Figure 3 – Manually shifted van Gurp-Palmen plot for PC-MWCNT to determine b_T at T_{ref} = 200 °C.

Although it is possible to achieve overlap in the low frequency data, it is challenging to achieve overlap in the high frequency range in all temperatures except for 160 °C. There are no significant differences in either b_T or the quality of fit between matrix grades at temperatures above 200 °C; at temperatures below 200 °C there appears to be some effect of molar mass.

Horizontal shifting

A numerical minimisation procedure based on that suggested by [6] was developed and employed to aid the production of mastercurves for all materials, and is described in detail elsewhere [4]. Horizontal shift

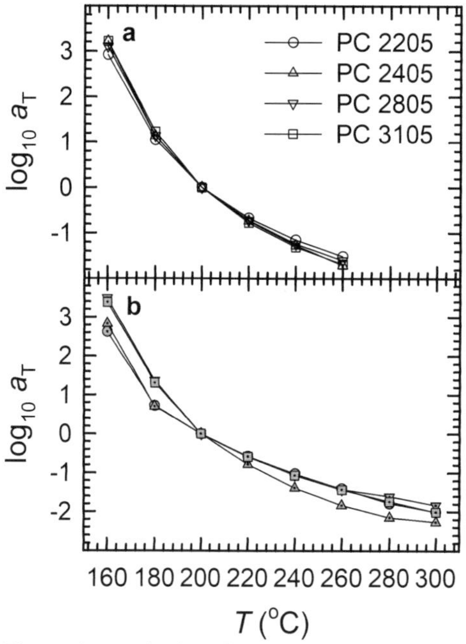

Figure 4 – a_T obtained for (a) unfilled PCs and (b) nanocomposites at T_{ref} = 200 °C.

factors, a_T shown in Fig. 4 for all materials, are found to be relatively independent of both matrix polymer and of the presence of nanotubes.

The similarity of shift factors irrespective of the presence of CNTs is a signature that the relaxation processes probed in this temperature range have the same temperature dependence (even though the moduli do not, see Fig. 6).

Mastercurves

The mastercurves obtained on pure PC grades exhibit properties typical of polymer melts, with characteristic gradients in the terminal zone of $G' \propto \omega^2$ and $G'' \propto \omega$. The nanocomposite mastercurves, shown in Fig. 5, differ from the unfilled materials in two important aspects: (1) the presence of a second plateau instead of a zone of terminal flow; (2) a loss tangent that is less than unity throughout the extended frequency range. The differences between matrix grades are not large, but it is possible to discern relaxation occurring at lower reduced frequencies with increasing molar mass.

At low reduced frequencies the second plateau has a modulus of ~10^5 Pa, indicating the formation a percolated network within the matrix system [7]. This second plateau has a modulus approximately 1.5 orders of magnitude smaller than the rubbery plateau arising from the entanglement network, and its stiffness is probably related to nanotube bending modes [8].

Fig. 6 displays the viscoelastic moduli of filled PCs expressed as a ratio to the moduli of their respective unfilled matrices. The influence of the CNT

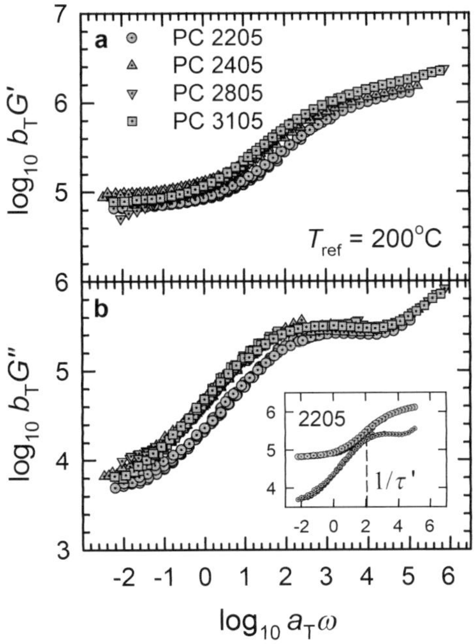

Figure 5 – Mastercurves of (a) G' and (b) G'' of PC-MWCNTs of varying molar mass at γ = 0.5%. Inset shows the absence of a cross-over frequency (example is for Makrolon 2205, but absent in all grades).

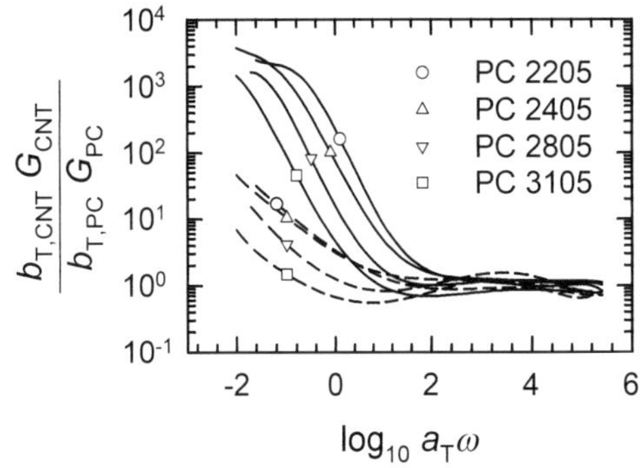

Figure 6 – G' (solid line) and G'' (dashed line) expressed as ratios of filled PC nanocomposites to unfilled PC, as a function of reduced frequency.

on the viscoelastic response is greatest at the lowest reduced frequencies. The onset of the deviation from unity occurs at a lower reduced frequency (longer timescale) with increasing molar mass of PC and the influence of the nanotubes on the low frequency viscoelastic response is apparent.

Relaxation times

In PC melts, a characteristic relaxation time τ was estimated from the inverse of the cross-over frequency for each grade. For filled PC melts, a similar but not identical timescale τ' was determined corresponding to the inverse of the frequency at which the peak in tan δ occurs (since there is no crossover). Linear regression was then employed to identify the coefficient n of a power-law relationship of the forms $\tau \propto M_w^n$ and $\tau' \propto M_w^n$, as shown in Fig. 7. n was identified as 5.66 ± 0.22 (where ± refers to the standard deviation) for PC and 5.25 ± 0.62 for PC-MWCNT. The pure PC value is somewhat higher than reported literature values of 3.4-3.75 [9], possibly due to the polydispersity of the commercial resins.

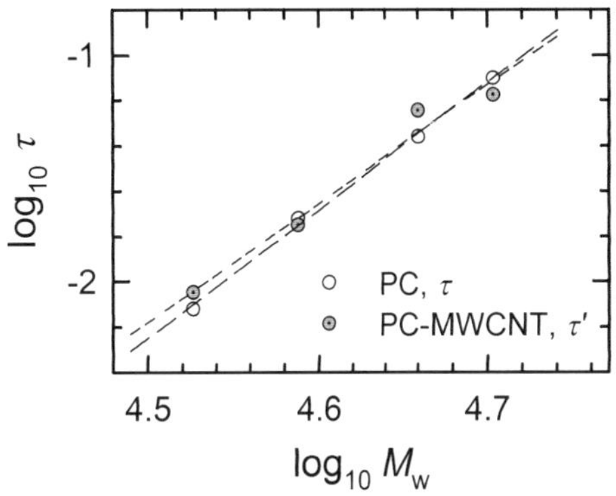

Figure 7 – Characteristic relaxation times of unfilled PC and PC-MWCNT as a function of molar mass. Dashed lines represent linear regressions through the data sets.

Nevertheless, the two power-law coefficients can be shown to be not statistically different (two-tailed Student's t-test at 90% confidence level, t = 0.48, n = 4), suggesting that the same dependence of molar mass on the relaxation mechanism applies to both unfilled and filled PC melts. This fact highlights the key role played by polymer-polymer interactions even in nanocomposite systems whose viscoelastic response differs strongly to that of the matrix polymer.

Non-linear response

Mastercurves were also constructed from isothermal frequency sweeps at strains outside the linear viscoelastic limit (5% and 50%, mastercurves not shown). The vertical shift factors, shown in Fig. 8, are most affected by strain amplitude at the highest test temperatures. At larger strain amplitudes there is increased disruption of the nanotube network and the moduli are reduced. The kinetics of nanotube network formation are highly temperature sensitive, as has been repeatedly observed in flocculation experiments [10]. The horizontal shift factors (not shown) were relatively unaffected by the increased strain amplitudes. This is consistent with experimental evidence of the much more limited effect of strain amplitude on moduli of pure polymer melts, and reinforces the view that relaxation of the polymer matrix dominates the relaxation processes in the nanocomposite. Thus it is a reasonable assumption that polymer mobility is the main driver for filler network rearrangement due to the substantial differences in size between polymer chains and CNTs.

The strain and temperature history experienced by a nanocomposite undoubtedly plays an important role in defining the extent of development of the nanotube network, and hence the degree to which it contributes to the response. This effect was clearly observable by comparing moduli obtained from amplitude and frequency sweeps performed at the same reduced frequency and strain amplitude, but with a different strain and temperature history [4].

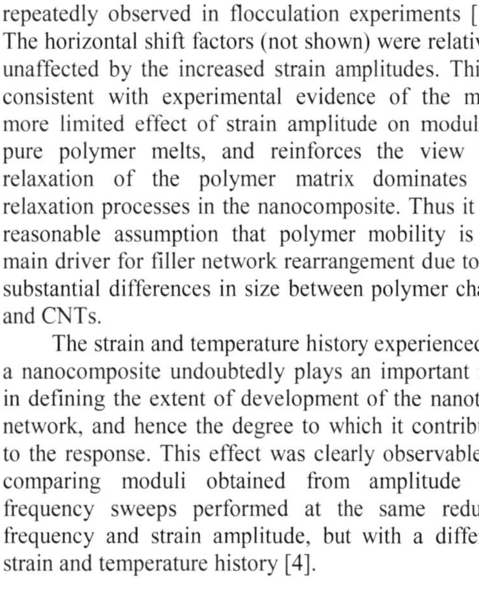

Figure 8 – b_T obtained for Makrolon 2205 filled with MWCNTs (3 wt%) for varying strain amplitudes.

Conclusions

Isothermal frequency sweeps were performed on both MWCNT-filled and unfilled PC and mastercurves were constructed for both materials. The effect of temperature on modulus was accounted for by employing a density correction to the modulus in pure materials, and a manually obtained shift obtained from van Gurp-Palmen plots in the filled materials. The similarity of horizontal shift factors irrespective of the polymer molar mass and the presence of MWCNTs suggests that the relaxation processes have the same temperature dependence. Processes in filled and unfilled grades also exhibit a very similar molar mass dependence on relaxation times. This is evidence that the dominant relaxation mechanism in PC-MWCNT is associated with the polymer network. PC-MWCNT mastercurves have a low frequency plateau at ~100kPa, and no cross-over is present.

Large amplitude isothermal frequency sweeps were also carried out. There was a significant effect of amplitude on the vertical shift factors, particularly at higher temperatures. This is attributed to the different levels of nanotube network build-up and disruption with larger amplitude oscillations.

Acknowledgements

The authors gratefully acknowledge the contribution of Nanocyl S.A. (C.Y. Lew) for the supply and compounding of PC-MWCNTs; and of N. Birkin from the University of Nottingham for assistance with the GPC measurements. GYHC acknowledges the financial contributions of the British Society of Rheology's Student Travel Bursary, of the Polymer Composites Research Group and of the Faculty of Engineering at the University of Nottingham.

References

1. Dealy, J. and Plazek, D. (2009) *Rheol. Bull.* 78(2):16-32.
2. Handge, U. and Pötschke, P. (2007) *Rheol. Acta* 46(6):889-898.
3. Wu, D., Wu, L. and Zhang, M. (2007) *J. Polym. Sci., Part B: Polym. Phys.* 45(16):2239-2251.
4. Choong, Y.H.G., De Focatiis, D.S.A. and Hassell, D.G. (2012), submitted to *Rheol. Acta.*
5. van Gurp, M. and Palmen, J. (1998) *Rheol. Bull.* 67:1-5.
6. Gergesova, M., Zupančič, B., Saprunov, I. and Emri, I. (2010) *J. Rheol.* 55(1):1-16.
7. Pötschke, P., Fornes, T.D. and Paul, D.R. (2002) *Polymer* 43(11):3247-3255.
8. Rubinstein, M. and Colby, R.H., *Polymer Physics*. Vol. 1. 2003: Oxford University Press.
9. Jordan, T.C. and Richards, W.D., in *Handbook of Polycarbonate Science and Technology*, Legrand, D.G. and Bendler, J.T., Editors. 2000, Marcel Dekker: New York. p. 179 -224.
10. Richter, S., Saphiannikova, M., Jehnichen, D., Bierdel, M. and Heinrich, G. (2009) *Express Polym. Lett.* 3(12):753 - 768.

O-08-133

THE FLOW BEHAVIOR OF ELECTRORHEOLOGICAL (ER) SUSPENSIONS

Youngwook P. Seo and Yongsok Seo
[1]*Intellectual Textile System Research Center (ITRC) and School of Materials Science and Engineering, College of Engineering, Seoul National University, Shillim-9-dong 56-1, Kwanakgu, Seoul, Republic of Korea 151-744*

Abstract - A new rheological model was applied to the analysis of the electrorheological behavior of a fluid containing silica nanoparticle-decorated polyaniline nanofibers. A model's predictions were compared with the experimental data, revealing that the proposed model correctly predicted the shear stress behavior both quantitatively and qualitatively. The shear stress data of the electrorheological fluid showing aligned fibers' structural reformation as a function of the shear rate agreed well with the new model which required fewer parameters than the CCJ (Cho–Choi–Jhon) model. The static yield stress was found to be quadratically dependent on the field strength, in agreement with the predictions of the polarization model. A scaling function was used to model the yield stress behavior of the electrorheological fluid over a range of electric fields, and it correctly predicted the static yield stress behavior both quantitatively and qualitatively.

Keywords: ER fluids, Universal scaling, Constitutive equation

Electrorheological (ER) fluids are suspensions of very fine particles in an electrically insulating fluid.[1-2] An electric field induces attractive interactions between the suspended particles that enable them to rapidly form a solid-like network of fibers aligned along the direction of the electric field. This reversible structural transition varies the apparent viscosity of ER fluids by three to four orders of magnitude in response to an electric field.[4,5] Typical ER fluids transform from a liquid to a gel state upon application of an electric field. The process is reversed when the electrical field is removed, with a response time on the order of milliseconds. Choi et al. recently described the ER characteristics of silica nanoparticle-decorated polyaniline nanofibers (PASi2). They reported that nanosilica modified PASi2 exhibits higher ER effects than pure PASi2 suspensions because silica and PASi2 are each electro-responsive ER materials in their own right.We reanalyze the experimental data presented in Choi et al. to characterize the ER characteristics of PASi2-based ER fluids using a simple model that describes the ER fluids flow (in particular the appearance of minima in the shear stress versus shear rate curves) with fewer parameters than are present in the CCJ model. To describe an ER fluid over a wide range of shear rates, Choi et al. proposed an empirical equation containing six parameters as follows (the CCJ model):[1]

$$\tau = \frac{\tau_y}{(1+(t_1\dot{\gamma})^{\alpha})} + \eta_{\infty}\left(1 + \frac{1}{(t_2\dot{\gamma})^{\beta}}\right)\dot{\gamma} \quad (1)$$

where t_1 and t_2 are time constants, $\dot{\gamma}$ is the shear rate, τ_y is the yield stress, and η_{∞} is the viscosity at high shear rates. The first term on the right hand side represents the shear stress behavior at low shear rates. Under these conditions, the shear stress decreases with the shear rate after exhibiting yield behavior. The second term describes the shear thinning behavior at high shear rates.[1] The yield stress values calculated using the CCJ model were occasionally lower than the shear stresses in the low shear rate region because the stress arose from a dynamic yield stress.[2] Dynamic yield stress is not equivalent to the yield stress of an ER fluid with a shear rate dependent stress. The dynamic yield stress is usually obtained by extrapolating the shear stress versus shear rate curve back to the shear stress intercept at a shear rate of zero, that is, in the absence of flow. As revealed previously, once the shear stress exceeds the yield stress, the fluid should be in a flow regime of Newtonian or shear-thinning fluid. If fluid deformation occurs, the first term on the right hand side of Eq.(1) is inapplicable because it is used to calculate the dynamic yield stress in a region without flow (the Bingham yield stress). This inconsistency is due to the inappropriate use of the Bingham model, which predicts a discontinuity in the stress at the critical shear rate. Structural deformation and reformation proceed, and the flow commences when the stress reaches the static yield stress.[2]

Structural reformation processes begin in a yielding regime, wherein portions of the material may not be in deformed state when the aligned structures reform. This situation is similar to the yield stress singularity problem. Papanastasiou explained that the difficulties associated with the Bingham model arise from the need to track randomly distributed yield material surfaces.[17] We proposed a new model to avoid the discontinuities in the flow curves and to incorporate the static yield stress, the the yield criterion, and stress changes associated with the structural realignments:

$$\tau = \tau_{sy}\left(1 - \frac{(1-\exp(-a\dot{\gamma}))}{(1+(a\dot{\gamma})^{\alpha})}\right) + \eta\dot{\gamma} \quad (2)$$

where τ_{sy} is the static yield shear stress, η is the shear viscosity, a is the time constant (i.e., the reciprocal of critical shear rate for aligned particle structure deformation), and α is the power-law index used to decide the degree of shear thinning.[2] This model was devised to accommodate the static yield stress and stress variation as a function of the shear rate described by the Papanastasiou model. The power-law form of the first term in the denominator was used to account for the breakup and reformation of the aligned particle structures, similar to the von-Mises criterion for the constitutive stress-strain behavior relation. Depending on the parameter values, Eq.(2) can describe either a Newtonian fluid ($\tau_{sy} = 0$) or a Bingham fluid ($a = 0$). The model can also simulate the De Kee-Turcotte model ($\alpha = 0$, $a \neq 0$) and the Herschel-Bulkley model ($a = 0$, $\eta = m\dot{\gamma}^{n-1}$ where m and n are the consistency parameter and the power-law index, respectively).

The model was applied first to our previous study for comparison.[5] We synthesized conducting polyaniline (PANI)/Barium titanate ($BaTiO_3$) composites via an in situ oxidative polymerization. Conducting PANI/ inorganic composites are excellent candidate ER fluid materials because of their physical properties, unique structure, and the combined properties of the two phases.[5] The experimental data were reanalyzed using Eq.(2). Fig. 2 shows the fitting results. The proposed model provides a good fit to the flow curves over range of shear rates and electric field strengths. The CCJ model also provided a good fit. The parameters used in the CCJ and the proposed model are summarized in Table 1. Variations in the CCJ model parameters arose because the six parameters were optimized simultaneously to minimize the total error. The values reflected local minima in the relative error. The parameter □ in the CCJ model should fall between 0 and 1, because $d\tau/d\dot{\gamma} \geq 0$; however, the values determined by fitting exceeded 1. The other power index, α, increases steadily with the electric field. The fitted yield stress (τ_y) values at high electric field strength were lower than the measured shear stress at low shear rates. In contrast, the new model predicted a yield stress that was higher than the measured shear stress at low shear rates. The yield stress increases steadily with the electric field strength. The critical shear rate required to induce deformation can be expressed as the inverse of the time constant a, i.e., $\dot{\gamma}_y^* = 1/a$, the point at which the rheological behavior undergoes a transition. The critical shear rate also increased (a decreased) with the electric field strength, indicating that the yielding behavior occurred at higher shear rates for stronger electric fields, in agreement with the Bingham fluid behavior. The viscosity, η, decreased initially because aligned structures became disrupted. The viscosity later increased due to the rapid reformation of chains in the presence of a high electric field. In the absence of an applied electric field, the viscosity was 0.183 Pa·s. The power-law index decreased with the electric field strength and this decrease was even more pronounced in the presence of structural deformations.[1] The proposed model, which included fewer parameters than the CCJ model, afforded good agreement with the experimental data and consistent parameter values over the range of shear rates and applied electric field strengths.

The model was applied to the flow curves of the ER fluids reported by Choi et al. (silica nanoparticle-decorated polyaniline nanofiber (PASi2) suspensions)[2] Polyaniline nanofibers were synthesized through an interfacial polymerization method. The nanofibers had diameters of 300–400 nm and lengths of 1–5 mm. The fibers were fully wrapped by smaller silica nanoparticles at least 10 nm in size. The ER properties of the fluid prepared by dispersing the product in a silicone oil were measured using a rotational rheometer (MCR 300, Anton Paar, Germany) equipped with a high direct current voltage generator. The flow curves were obtained by measuring the shear stress as a function of the shear rate.

The experimental data were then reanalyzed using Eq.(2). Fig. 3 shows the fitting results. The proposed and the CCJ models provided comparable fits to the flow curves over the range of shear rates and electric field strengths tested. The parameters used in the CCJ and the proposed models are summarized in Table 2. The proposed model agreed well with the experimental data using fewer parameters than the CCJ model. The yield stress, τ_y, predicted by the CCJ model increased with the electric field strength, although some values were lower than the shear stress at high electric field strengths. In contrast, the new model predicted a yield stress higher than the shear stress at high electric field strengths. The stress decreased at the starts of the flow. The static yield stress also increased consistently with the electric field strength. The critical shear rate that marked the onset of deformation could be expressed as the inverse of the time constant a ($\dot{\gamma}_y^* = 1/a$, the point at which the rheological behavior changed), although uncertainties in the experimental data prevented elucidation of the dependence on the electric field strength. At the low electric field strength (0.5 kV/mm), few structures formed from aligned fibers or gelled structure were observed, and the minimum of the shear stress was more difficult to determine than at high electric field strength. The viscosity, η, increased with the electric field strengths due to the rapid reformation of chains in the presence of an electric field. Without an applied electric field, the viscosity was 0.115 Pa·s. The power-law index, α, decreased with the electric field such that the stress was less sensitive to the structural deformation.

As a conclusion, a new rheological model which was proposed to describe the deformation (break-up and reformation) of aligned structures at low shear rates and the yield behavior at high shear rates was applied to analyze the behavior of the flow curves of ER fluids (silica nanoparticle-decorated polyaniline nanofiber (PASi2) suspensions) described by Choi et al.[12] The four parameter model (Eq.(2)) accurately described the

ER flow structural deformation and reformation. The model could describe the experimental data and provided phenomenological explanations. The static yield stress depended quadratically on the electrical field strength, as predicted by the polarization model. This result illustrated the assertion that deviations from the theoretical predictions depends on the material and are not generally observed. Flow curve analysis using the proposed model combined with dimensional analysis enabled qualitative and quantitative predictions of the ER fluid behavior with relatively few experimental measurements. The yield stress behavior was correctly fit using a single-parameter nonlinear model described by Seo.

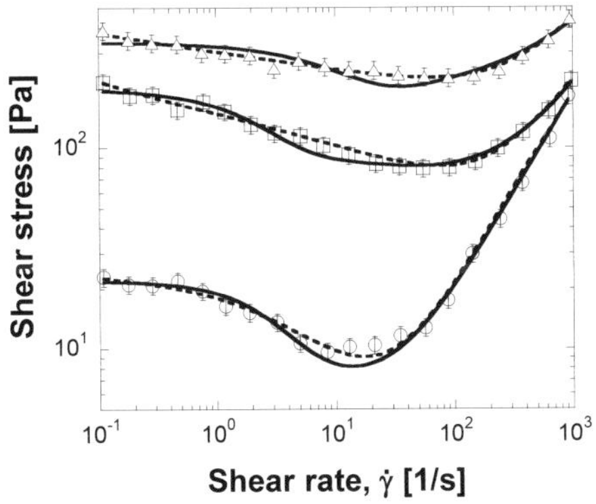

Figure 1. Shear stress vs. shear rate for PANI/$BaTiO_3$ (25wt%) based ER fluids: ((○) 0.5 kV/mm, (□) 1.5 kV/mm, (△) 2.5 kV/mm. The experimental data are from our previous study.[5] (They are average values of three times measurements. The error is mostly within the size of the symbol.). The lines are fits by the current model (—) and the CCJ model(- - -). The optimal parameter values are in Table1.

Table 1. The optimal parameters obtained from the flow curves of PANI/$BaTiO_3$ (25wt%) based ER fluids.

Model	Parameter	0.5 kV/mm	1.5 kV/mm	2.5 kV/mm
CCJ[a)]	τ_{dy}	22.6	192	247
(Eq.1)	t_1	0.225	0.096	0.0027
	α	0.832	0.347	0.1666
	η_∞	0.189	0.192	0.298
	t_2	0.915	0.0618	0.0036
	β	1.659	1.575	1.16
Current model (Eq.2)	τ_{sy}	21.8	198	338.5
	a	0.465	0.504	0.0834
	α	-0.582	-0.164	0.208
	η	0.188	0.165	0.192

a)The CCJ model parameter values are different from ref.[5]. These are obtained from our corrected simulation.

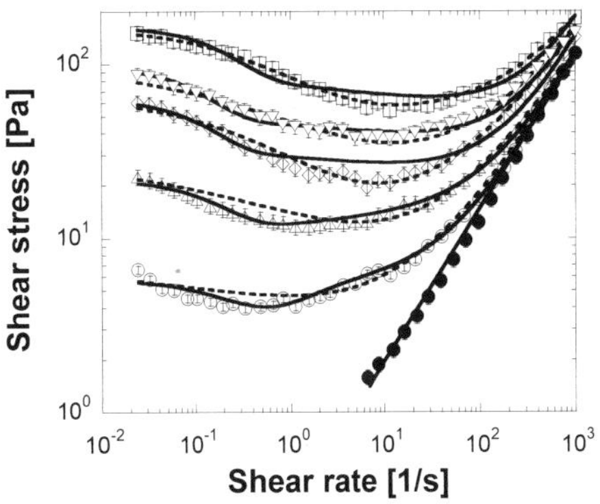

Figure 2. Shear stress vs. shear rate for PASi2 (10 vol%) based ER fluids: ((●) 0 kV/mm, (○) 0.5 kV/mm, (△) 1 kV/mm, (◇) 1.5 kV/mm, (▽) 2.5 kV/mm, (□) 3.5 kV/mm). The experimental data are from Choi et al.[12] (They are average values of three times measurements. The error is mostly within the size of the symbol.). The lines are fits by the current model (—) and the CCJ model(- - -). The optimal parameter values are in Table 2.

Table 2. The optimal parameters obtained from the flow curve of PASi2 (10 volume %) based ER fluid at various electric field strengths.[a)]

Bingham	τ_y	6	23	55	87	150
	η_{pl}	0.11	0.10	0.10	0.10	0.19
CCJ	τ_y	8	28	65	98	150
(Eq.1)	t_1	1.5	2.3	2.5	2.6	0.23
	α	0.25	0.40	0.60	0.45	0.35
	η_∞	0.08	0.08	0.10	0.10	0.12
	t_2	0.01	0.004	0.0025	0.002	0.001
	β	0.5	0.6	0.65	0.70	0.7
Current model(Eq.2)	τ_{sy}	5.9	22	64	98.1	165
	a	1.98	5.3	7.8	8.4	4.5
	α	1.	0.10	-0.09	-0.092	-0.109
	η	0.11	0.10	0.126	0.133	0.14

a) The Bingham and the CCJ model parameter values are from ref.[12] except 2.5 kV/mm case which was obtained from our simulation.

Acknowledgements

This work was supported by NRF (RIAM NR03-09 (0417-20090027) , the SRC/ERC program (R11 - 2005-065)) and by MIKE (Original Material Technology Program (RIAMI-AC14-10 (0417-20100043)).

References

1) Y.Seo, H.J. Choi, Y.Seo *Soft Matter*, **8**, 4659 (2012)
2) Y.Seo, Y.Seo, *Langmuir*, ***28***, 3077 (2012)

O-08-224

EFFECT OF BENZYL TRIPHENYL PHOSPHONIUM CHLORIDE ON MELT RHEOLOGY OF POLYVINYLIDENE FLUORIDE (PVDF)

S. K. Tiwari[1], A. Misra[2] and D. V. Khakhar[1]

[1]Department of Chemical Engineering, Indian Institute of Technology Bombay, Powai, Mumbai, 400076, India – khakhar@iitb.ac.in, saurabh.tiwari@iitb.ac.in

[2]Intellectual Ventures, No. 701 Raheja Paramount, 138 Residency road, Bangalore 560 025, India - amisra@iitb.ac.in

Abstract - The effect of onium salt, benzyl triphenyl phosphonium chloride (BTPC), on the melt rheology of polyvinylidene fluoride (PVDF) has been discussed. Different amounts of BTPC were melt mixed into PVDF using a micro twin screw extruder. The storage and loss modulus of the blends, in the melt phase, both increase with increase in frequency at different temperatures. At low frequencies, the storage modulus becomes nearly independent of frequency, which indicates solid like behaviour resulting from network formation due to interaction between PVDF and BTPC. At high frequencies, the loss tangent values are nearly the same and independent of BTPC concentration. At low frequencies each composition shows a different peak and that relaxation peaks can be attributed to the glass transition temperature (Tg) of the PVDF/BTPC blends. The frequency corresponding to the peak increases and the peak value decreases with increasing BTPC concentration, starting with pure PVDF. The complex viscosity and the shear viscosity both show a well defined power law region in the higher frequency range and both decrease with increase in weight fraction of BTPC. However, both the viscosities are larger than that for pure PVDF even at the highest concentration. Shift factors derived from time temperature superposition show a good fit to the Arrhenius equation. The flow activation energy and coefficient of thermal expansion of PVDF/BTPC blends are not significantly affected by addition of BTPC.

Keywords: Melt rheology, Flow activation energy, coefficient of thermal expansion, PVDF/BTPC blends.

Introduction

The polymer polyvinylidene fluoride (PVDF) is chemically inert, melt processable and has good piezoelectric properties [1, 2]. The applications of PVDF films thus include sensors, actuators and finger pressure switches. PVDF is also used as a piping material in ultra pure environments (food and pharmaceutical industries), cable insulation, sports and leisure goods [3]. PVDF exhibits a complex crystalline polymorphism as α, β, γ and δ phases of which α and β are most common [2]. The conformation of the α phase is a slightly distorted TG^+TG^- with a unit cell that is nonpolar due to the antiparallel packing of the two chains contained in the cell [4].

The measurement of rheological properties of any polymeric material in the molten state is important in order to get the basic understanding of processability of that material. In order to find the rheology of PVDF, Shen et al. [5] studied the rheological properties of two different PVDF resins and they observed that the viscosity flow curve shows a Newtonian behavior in low shear rate range and non Newtonian behavior in higher shear rate range. They also found that the temperature dependence of shift factor could be well understood by the Arrhenius relation from which the flow activation energy of both PVDF resins was calculated. A study of rheological behavior of clay based PVDF/PMMA blend showed at lower frequencies, the storage modulus increased with PMMA concentration composite and at 10 % PMMA the frequency dependency of storage modulus showed solid-like behavior [6]. The melt viscosity and dynamic modulus of PVDF/MWCNT nanocomposites are proportional to the MWCNT loading, and loss tangent values decrease with increasing MWCNT content especially at low frequencies [7]. Judovits et al. [8] studied the effect of alkyl and aryl onium salts on PVDF melt rheology that shows the low shear effect as compare to PVDF. We analyze the effects of addition of BTPC on the rheology of PVDF melt. The role of BTPC as an additive has not been previously examined and would be useful for practical applications.

Experimental

Materials

Polyvinylidene fluoride (Kynar 720, Arkema Group) having melt flow index 12 g/10 min (230 °C/5 kg load) and the onium salt, benzyl triphenyl phosphonium chloride (BTPC) obtained from Sigma-Aldrich with melting point of 323-332 °C were used for the study.

Preparation of PVDF/BTPC blends

Melt blending of PVDF/BTPC was performed in a microcompounder equipped with two counter rotating screws (DSM, Holland) at 210 °C and 125 rpm for 10 min. The weights of PVDF and BTPC used in the different samples along with resulting weight percentage of BTPC are given in Table 1. These blends were used only for the melt rheology study.

Table 1 – Composition of PVDF/BTPC blends.

Blend	PVDF (g)	BTPC (g)	BTPC (wt%)
PVDF/BTPC	5	0	0
PVDF/BTPC	4.975	0.025	0.5
PVDF/BTPC	4.95	0.05	1
PVDF/BTPC	4.85	0.15	3

Rheological measurements

The rheological measurements were performed on a rheometer (Physica MCR 301) using cone and plate and parallel-plate geometries with a diameter of 25 mm for steady shear and dynamic oscillatory shear measurements, respectively. The gap between parallel plates and cone angle of cone and plate geometry were 0.8 mm and 2.005° respectively. Steady shear measurements were performed at strain rates ranging from 0.01 to 10 s^{-1} using the cone and plate attachment. Dynamic oscillatory shear measurements were performed using the parallel plate attachment in a frequency range of 0.01 to 100 s^{-1} with a constant strain of 5 %, which is within the range of strain exhibiting linear viscoelastic behavior for the materials.

Results and Discussion

Fig. 1 shows the plots of complex viscosity versus frequency at different temperatures for different BTPC concentrations. Each curve exhibits a Newtonian plateau at low frequencies and a power law decay at high frequencies. Further, the viscosity decreases with increase in temperature in all cases. Addition of 0.5% BTPC results in a significant increase in the complex viscosity (Fig. 1b) but further addition does not result in much overall change (Fig. 1c, d).

Assuming time-temperature superposition [9, 10] the data of Fig. 1 are shifted to collapse them to a master curve for each BTPC concentration. The master curves are calculated for a reference temperature of 200 °C using the individual shift factors a_T (horizontal direction) and b_T (vertical direction). Horizontal shifting, along the axis of the angular frequency, is dominant. The considerably smaller shift in vertical direction is caused mainly by the temperature-dependent density change of the polymer [10].

Fig. 2 depicts the master curves of complex viscosity as a function of frequency for the PVDF/BTPC blends. The shifted data merge well into a single curve for each BTPC concentration indicating the validity of time-temperature superposition in this case. The viscosity curve clearly shows a well defined power law region at higher frequencies along with a low shear rate plateau. At low frequencies, there is a significant increase in viscosity with reducing frequency for the samples with BTPC in contrast to the case of pure PVDF. Such an increase is typically seen in fluids with a yield stress, indicating solid - like behavior. In the present case the viscosity increase indicates that BTPC promotes the creation of physical crosslinks leading to network formation.

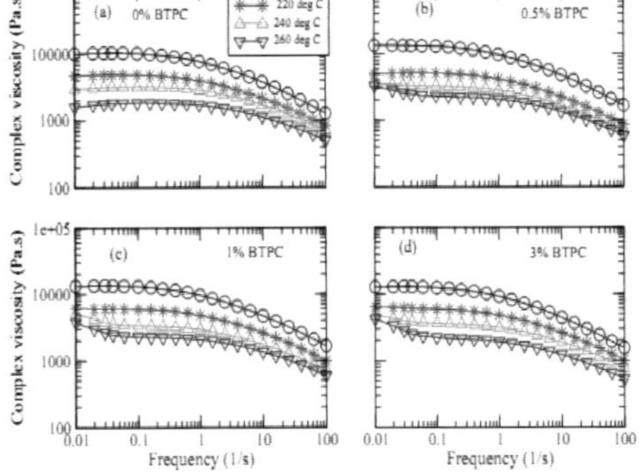

Figure 1 –Complex viscosity at different temperatures, (a) for pure PVDF (b) for PVDF with 0.5 % BTPC by weight (c) PVDF with 1 % BTPC by weight (d) PVDF with 3 % BTPC by weight.

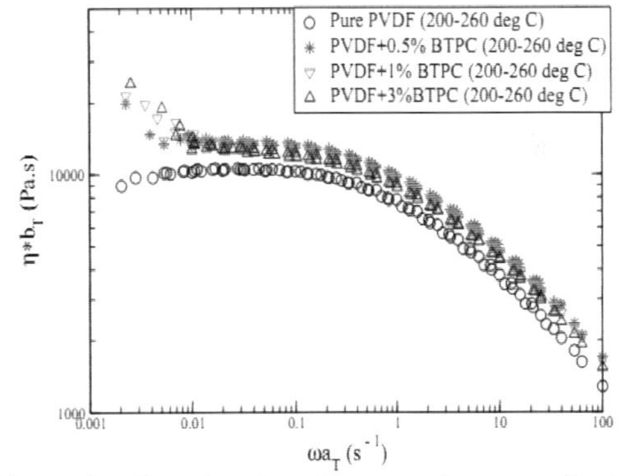

Figure 2 – Complex viscosity versus frequency for the different weight fraction of PVDF/BTPC blend at different temperatures.

As noted above, there is a significant increase in the complex viscosity upon addition of 0.5 % BTPC. Further increase in the BTPC concentration results in a small decease in the viscosity, but the viscosity even at the highest concentration is larger than that for pure PVDF [8].

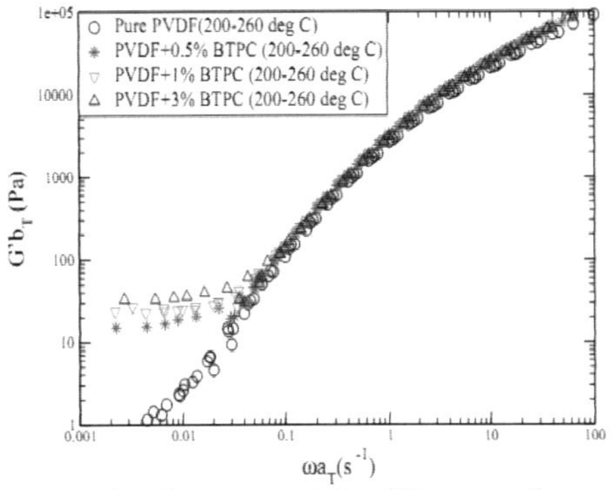

Figure 3 – Storage modulus (G') versus frequency for the PVDF/BTPC blend at different temperatures.

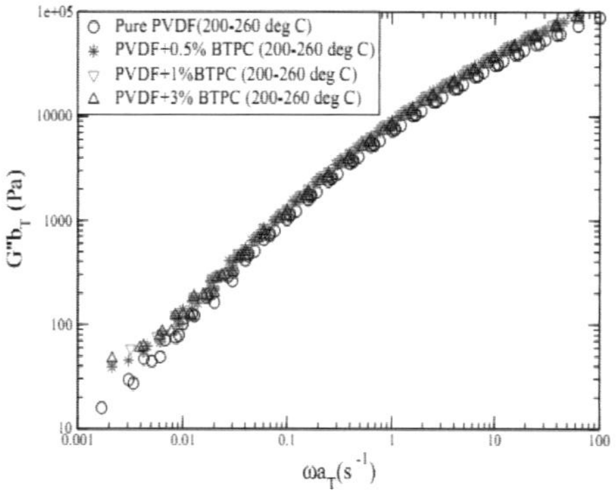

Figure 4 – Loss modulus (G") versus frequency for the PVDF/BTPC blend at different temperatures.

Fig. 3 and Fig. 4 show a master curves of storage and loss modulus as a function of frequency for PVDF and its blends at different temperatures using the same shift factors (a_T, b_T) as in Fig. 2 for the complex viscosity. There is good overlap of the shifted curves in this case as well. The storage and loss moduli of pure PVDF and PVDF/BTPC blend increase with increasing frequency. At low frequencies the storage modulus of samples with BTPC becomes independent of frequency. This is indicative of network formation and is consistent with the rise in viscosity seen at low frequencies. Similar behavior has been reported for reinforcing fillers like fumed silica with polymethylsiloxynes (PDMS) and PMMA/rubber, at low volume fraction shows a weak physical network and at high concentration of silica, storage modulus is independent of angular frequency (secondary plateau, G'p) due to its cross-linked network [11, 12]. The trends in the variation of the loss and storage moduli with BTPC concentration are the same as seen for the complex viscosity.

Fig. 5 shows the master curves for the loss tangent as a function of frequency for PVDF and the PVDF/BTPC blends. The curves for the different temperatures superimpose well upon shifting, as above. At high frequencies, the loss tangent values are nearly independent of BTPC concentration and decrease with increasing frequency.

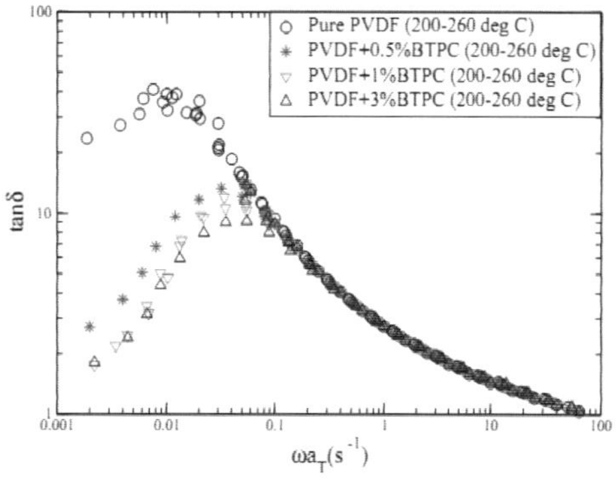

Figure 5 – Loss tangent versus frequency for the PVDF/BTPC blend at different temperatures.

At low frequencies each composition shows a different peak. The frequency corresponding to the peak increases and the peak value decreases with increasing BTPC concentration, starting with pure PVDF. The peak trend for PVDF and PVDF/BTPC blend is close to the previously reported trends of other polymer blends [13]. The data indicate a lower dissipation for BTPC added samples implying a larger elastic component and thus solid-like behavior, resulting from formation of a network structure.

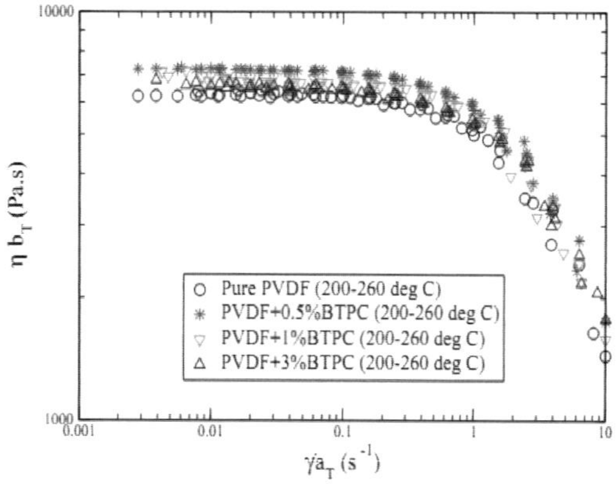

Figure 6 – Shear viscosity as a function of shear rate for melt of PVDF/BTPC blends at various temperatures.

Fig. 6 shows the master curves of the shear viscosity versus shear rate from data obtained for steady shear flow in a cone and plate rheometer for different temperatures and BTPC concentrations. The master curves show a constant viscosity at low shear rates, in contrast to the increase in complex viscosity at low frequencies for BTPC blends from small amplitude oscillations (Fig. 2). This is because the BTPC mediated network does not form in the presence of flow. Shear thinning behavior is obtained at higher rates as in the case of the complex viscosity. The variation of shear viscosity with BTPC at a given shear rate is the same as that for the complex viscosity.

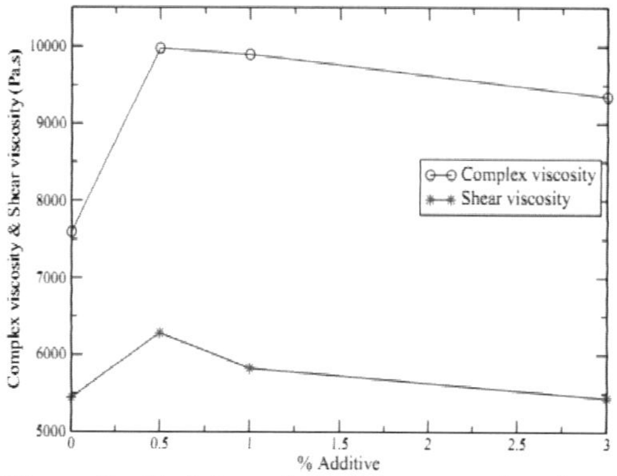

Figure 7 – Variation of the plateau complex viscosity and shear viscosity of PVDF/BTPC blends with weight percent of BTPC.

The effect of BTPC concentration on the complex viscosity and the shear viscosity of PVDF/BTPC blends at low frequencies (ωa_T =0.01-100 s^{-1}) and low shear rates (γa_T =0.01-10 s^{-1}) are shown in Fig. 7. In both cases, the viscosity increases sharply on addition of BTPC and then decreases with increase in weight fraction of BTPC. The highest complex viscosity and shear viscosity are obtained for 0.5 % fraction of BTPC, as noted above.

The horizontal, a_T and vertical, b_T, shift factors were calculated for the different temperatures are shown in Fig. 8. As expected the vertical shift factor is very close to unity and is a weak function of the temperature, while the horizontal shift factor shows Arrhenius-like behavior, given by Eq. 1

$$\ln a_T = \frac{E}{R}\left(\frac{1}{T} - \frac{1}{T_R}\right) \qquad \text{......... 1}$$

where R is the universal gas constant, T_R is a reference temperature and E is the flow activation energy. The log a_T and log b_T versus 1/T data fall on straight lines (Fig.8) and linear least squares fits yield the following flow activation energy values E for PVDF and its blends: 28.7 kJ/mol (PVDF), 28.4 kJ/mol (0.5 % BTPC), 27.6 kJ/mol (1% BTPC), 27.8 kJ/mol (3% BTPC). Thus, the flow activation energy is not significantly affected by addition of BTPC.

The coefficient of thermal expansion is related to the vertical shift factor (b_T) and calculated by Eq. 2 [10]

$$\ln\left(b_T T_R / T\right) = -\alpha\left(T - T_R\right) \qquad \text{........2}$$

where α is coefficient of thermal expansion and T_R is a reference temperature. The experimental values of α obtained by fitting Eq. 2 to the data in Fig. 8 in temperature range of 200-260 °C are negative and are as follows: $-4.0*10^{-3}$ K^{-1} (PVDF), $-4.1*10^{-3}$ K^{-1} (0.5 % BTPC), $-4.5*10^{-3}$ K^{-1} (1% BTPC), $-4.4*10^{-3}$ K^{-1} (3% BTPC). Such negative coefficients of expansion (contraction upon heating) have been reported previously for different polymer melts [14]. The coefficient of thermal expansion is not significantly affected by addition of BTPC.

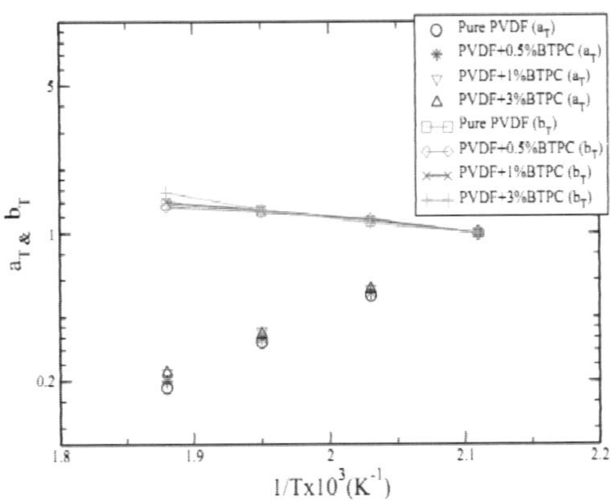

Figure 8 – a_T and b_T versus 1/T for the PVDF/BTPC blends.

Conclusions

The dynamic viscoelastic properties of PVDF and BTPC filled blends were studied. In PVDF/BTPC melt blend samples, at low frequencies the storage modulus of samples with BTPC becomes independent of frequency. This is indicative of network formation and is consistent with the rise in viscosity seen at low frequencies. At higher frequency very less change has been observed in loss and storage modulus with increase in weight fraction of BTPC. Complex and shear viscosities both shows a well defined power law region at higher frequency range and both decreases with increase in weight fraction of BTPC but both the viscosities even at the highest concentration are larger than that for pure PVDF. No significant change has been found in loss tangent with increasing weight fraction of BTPC at higher frequencies. At lower frequency the peak value decreases with increase in BTPC concentration. The flow activation energy and coefficient of thermal expansion of PVDF/BTPC blend is not significantly affected by addition of BTPC.

Acknowledgements

The authors acknowledge the use of the Micro-compounding Central Facility and Rheometer Facility, Department of Chemical Engineering, IIT Bombay.

References

1. H. Kawai; *Jpn. J. Appl. Phys.* 1969, *8,* 975.
2. A. J. Lovinger; *Poly (vinylidene fluoride) In Developments in crystalline Polymers-I*, D. C. Bassett, Ed.; Applied science publisher, London, 1982; 195-273.
3. G. Gerliczy; R. Betz Sens. *Actuators* 1987, *12,* 207.
4. R. Hasegawa; Y. Takahashi; Y. Chatani; H. Tadokaro *Polym. J.* 1972, *3,* 600.
5. L. Shen; Y. Song; Y. Qian; F. Qiao; J. Zhang; Q. Zheng *Chinese J. Polym. sci.* 2008, *26,* 639.
6. N. Mossaif; G. Groeninckx *Polym.* 2003, *44,* 7899.
7. J. N. Martins; T. S. Bassani; G. M. Barra; R. V. Oliveira *J. Polym. Int.* 2010, *60,* 430.
8. L. H. Judovits, N. J. Rodak, S. Zerafati, R. R. Gupta, U.S. Patent WO/2007/145668, 2008.
9. Y. M. Alexander, I. I. Avraam, *Rheology: Concepts, Methods and Applications,* Chemtec Publishing, Canada, 2006.
10. J. M. Dealy, R. G. Larson, *Structure and rheology of molten polymer: From structure to flow behavior and back again,* Hanser garden publications, Inc. USA, 2006.
11. I. A. Mirta; M. Elsi; V. D. Jon; W. M. Christopher *J. Rheol.* 1992, *6,* 1165.
12. B. Mosto; R. Mullar Rheol. *Acta.* 1996, *35,* 369.
13. G. Wu; Y. Song; Q. Zheng; M. Du; P. Zhang *J. Appl. Polym. sci.* 2003, *88,* 2160.
14. S. Singamaneni; M. C. Lemiuex; H. Jiang; T. J. Bunning; V. V. Tsukruk *Chem. Mater.* 2007, *19,* 129.

O-08-264

MULTISCALE SIMULATION OF FLOW IN ENTANGLED POLYMER MELTS AROUND A CYLINDRICAL OBSTACLE

Takahiro Murashima[1,3] and Takashi Taniguchi[2,3]*

[1] *Graduate School of Science, Tohoku University, Sendai 980-8578, JAPAN – murasima@cmpt.phys.tohoku.ac.jp*

[2] *Graduate School of Engineering, Kyoto University, Kyoto 615-8510, JAPAN – taniguch@cheme.kyoto-u.ac.jp*

[3] *Core Research for Evolutional Science and Technology, JST - Kawaguchi, Saitama, 332-0012, Japan*

Abstract - We have developed a new multiscale simulation technique to investigate the history-dependent flow behavior of polymer melt using a smoothed particle hydrodynamics simulation with microscopic simulators that account for the dynamics of entangled polymer chains acting on each fluid element. The multiscale simulation technique is applied to polymer melt flow passing a circular object in a two-dimensional periodic system. It is found that the strain-rate history-dependent stress of the polymer melt affects its flow behavior, and this stress memory causes nonlinear behavior even in the region $\mathbf{Wi} \leq 1$. The multiscale simulation method enables us to evalute the macroscopic spatial distribution of microscopic quantities, such as the spatial distribution of the number of entanglement.

Keywords: multiscale simulation, micro-macro simulation, polymer chain entanglements, Lagrangian method

Introduction

Products made of polymeric materials have been indispensable for our daily life. One of the important characteristics of thermo-plastic polymeric material is that it can be easily molded and processed by controlling its state, which is beneficial in a variety of practical applications. The melt of polymeric material can exhibit a wide range of flow behaviors depending on the dynamic response of the polymer's microscopic internal states under an imposed strain or strain-rate. It is difficult to predict a flow of polymeric fluid because the macroscopic flow behavior depends highly on the dynamic behavior of the microscopic internal states, and the states of polymers with very high molecular weight are complex at the microscopic level. When a polymer melt consists of polymers with a molecular weight M higher than the entanglement molecular weight Me, the polymer molecules are entangled, and the relaxation time of the polymer conformation is long compared to that of dilute polymer solutions because of the entanglement. Therefore, even for a homogeneous bulk polymer melt, it is very difficult to predict the rheological behavior by molecular dynamics without any coarse-graining procedures accounting for the microscopic internal degrees of freedom. The original reptation theory and the extended ones have succeeded in explaining the complex behavior observed in many experiment of polymer melts. However, it is difficult to apply the theory to arbitrary polymer melts because it is too mathematically difficult to incorporate molecular pictures of polymer with arbitrary architectures into the Fokker-Planck equation. On the other hand, Langevin stochastic models have been developed and reproduced the rheological properties of various polymeric materials with branches and/or molecular weight distributions [1,2]. These models use efficient numerical computation to predict the bulk rheology of polymer melts; however, they are not applicable to macroscopically inhomogeneous flows. The macroscopic flow behavior of polymer melt is usually predicted using a fluid dynamics simulation with a constitutive equation that describes a nonlinear relation between the stress and strain-rate fields without microscopic details. No general constitutive equation that is applicable to arbitrary polymer melts exists because a general polymer melt has extremely complex molecular states, which makes the stress response unpredictable even in a simple flow pattern. As described above, each of the microscopic and macroscopic approaches has limited applications. As an alternative to using a constitutive equation, we propose a new simulation technique that incorporates Langevin course-grained simulators into a fluid dynamics simulation. This approach is a type of micro-macro simulation [3] that was developed to consider history-dependent flow behaviors of polymer solutions. Our approach uses the Lagrangian fluid particle method to solve the macroscopic fluid dynamics, and therefore the dependence of the stress on the history of the strain-rate is directly involved [4].

The aim of the present study is to demonstrate how our proposed multiscale simulation works and to show physical information that can only be obtained in this multiscale simulation method.

Simulation Results and Discussion

In our multiscale simulation method, to maintain the microscopic information such as entanglements and deformation in polymers, we use a Lagrangian fluid particle simulation technique in which each fluid particle has a microscopic level simulator that accounts for the internal states of the fluid particle. The local stress tensor that is needed to solve the macroscopic flow can be evaluated from microscopic variables describing polymer dynamics under an instantaneous local strain-rate. As shown in Fig. 1, the fluid system is composed of a large number of fluid particles with the radius a. The flow behavior is described by the motions of the set of fluid particles. The main procedures of the simulation are summarized briefly as follows:

(1) Calculate the local density and the strain rate tensor at the position of each Lagrangian particle in SPM manner at the macroscopic level.
(2) Obtain the pressure from the density distribution at the macroscopic level.
(3) Update the local slip-link model of the i-th fluid particle under the local strain-rate at the position and then obtain the local stress tensor from the resulting configuration of the slip-link model. This procedure is executed on each fluid element in turn.
(4) Calculate the force coming form stress tensor and the pressure at the macroscopic level.
(5) Update the velocity and position of all the fluid particles by using the flowing equation :

$$\rho_i \frac{d\boldsymbol{v}_i}{dt} = -\nabla p_i + \nabla\cdot\boldsymbol{\sigma}_i + \boldsymbol{F}_{bi} \qquad \frac{d\boldsymbol{r}_i}{dt} = \boldsymbol{v}_i \qquad (1)$$

(6) Return to the item (1).

By using the above algorithm, we consider a polymer melt flowing in the x-direction and passing an infinitely long cylinder oriented in the z-direction with a radius r_c. Because of the symmetry of the system, we can treat the system as two-dimensional, and the flow can be described as two-dimensional flow in the xy-plane. We assume a non-slip boundary condition for the velocity on the surface of the cylinder and periodic boundary conditions at the boundaries of the system. As a microscopic simulator we used the slip-link model that can accurately describe the dynamics of entangled polymers.

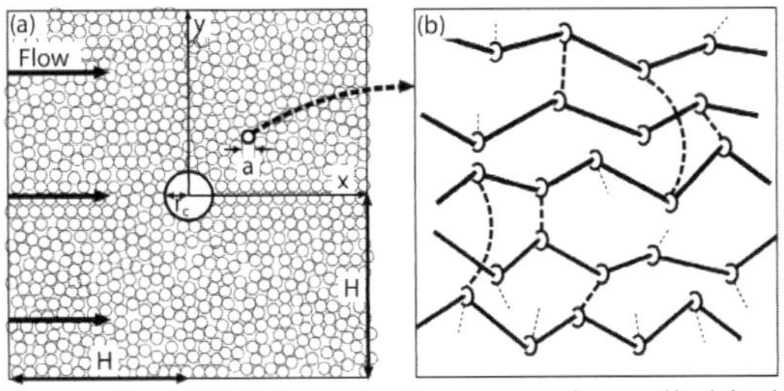

Figure 1 – (a) Polymeric flow around a cylindrical obstacle with a radius r_c. The small circles represent Lagrangian fluid elements with radius a, each of which have a microscopic simulator schematically shown in (b). (b) Schematic view of our used Langevin type model describing a well-entangled polymer melt. A ring denotes a slip-link describing entanglement point and the dotted line in (b) means the two chains are entangled at the point.

The dimensionless parameters governing the present problem are the Reynolds number **Re**=$\rho U r_c/\eta^0$, the Weissenberg number **Wi**=$\tau_d D_{xy}$, and the shear viscosity ratio η_d/η^0, where U is the average flow velocity. The zero shear viscosity η^0 of the polymer melt is given by $\eta_p^0+\eta_d$, where η_p^0 is the zero shear viscosity of a polymer melt described only by the slip-link model. From a rheological data derived from the slip-link simulation with average Z=7 entanglements in the bulk, we obtain the longest relaxation time $\tau_d \simeq 200t_e$ and the zero shear viscosity $\eta_p^0 \simeq 17.5\sigma_e t_e$ where t_e is the time unit of the slip-link model. The cylinder radius r_c was set to $3a$, where a is the unit length in the fluid particle simulation, and we assign the unit mass m to all fluid particles. The wall of the cylinder consists of fixed fluid particles evenly spaced on the perimeter. Each fluid particle consists of 10,000 polymers, enough to describe the bulk rheological properties of the polymer melt under an imposed shear and/or extensional deformation. About 1000 fluid particles with 10,000 polymers each are evenly placed in the system; therefore, we need to simultaneously solve the dynamics of ca. 10,000,000 polymers. Each slip-link simulation can be performed independently of the others, making parallel computing effective in this multiscale simulation. In order to induce a flow around the obstacle, the fluid is subject to the external field described by the body force $\boldsymbol{F}_b$=$(5\times10^{-4},0)$ $(\eta_0/t_e a)$, the flow becomes steady-state in about $1000t_e$. The average flow velocity in steady-state in this system is nearly equal to $(0.04,\ 0)a/t_e$ for a fully Newtonian flow $\eta^0 = \eta^d$ (entanglement stress=0), and $(0.055,\ 0)a/t_e$ for a polymer melt flow with $Z = 7$ and $\eta_d/\eta^0 = 0.1$. In both cases, **Re** is less than 0.2, and the flow is laminar. In the polymer melt flow case, the average flow velocity is higher than that of Newtonian flow, i.e., the flow exhibits shear thinning behavior. Because the strain-rate is larger than $1/\tau_d$ in the vicinity of the cylinder, **Wi** is larger than 1. From simulations it is found that the shear stress σ_{xy} of the polymer melt exhibits an apparent asymmetry between the upstream and downstream regions. The nonlinear relationship between σ_{xy} and D_{xy} is observed near the cylinder where **Wi**>1. If the asymmetry in the stress distribution is caused only by the nonlinear relationship, the stress distribution should be symmetric in the regions where **Wi**≤1. However, the asymmetry is also observed in the regions where **Wi**≤1. Judging from the analyses of memory effect (that is not shown here) we conclude that the asymmetry in the stress distribution is caused by the history-dependent stress in the polymer melt.

In order to extact a relationship between microscopic state and macroscopic flow from the present multiscale simulation, we investigate ⟨Z⟩ the spatial distribution of entanglements. We find that in the vicinity of the cylinder, the number of entanglements per chain slightly decreases. When $\tau_R(d\gamma/dt)$ □ 1 (τ_R : Rouse relaxation time, and $d\gamma/dt$: strength of a local shear rate), the contour length of the polymer is significantly longer than its equilibrium value, $a\langle Z\rangle_{eq}$. The chain retraction process which attempts to return the chain contour length to its equilibrium value causes disentanglement. In the upstream and downstream regions around the cylinder, the reduction of ⟨Z⟩ is also observed because of nonzero $|D_{yy}|$. Advection due to the flow broadens the region where $\langle Z \rangle < \langle Z \rangle_{eq}$ in the downstream from the cylinder. The tail length in the downstream region can be estimated to be $\tau_R U \square r_c$. The spatial distribution of $\langle Z \rangle$ is quite similar to that of $Tr(\sigma)$, because the number of entanglements depends on the elongation of polymers in this multiscale simulation. We hope that the innovative advances in the micro-rheology techniques will enable the observation of the spatial distribution of entanglements in the future.

Conclusions

We have developed a new multiscale simulation method to investigate history-dependent flow behavior in polymer melt. This multiscale simulation consists of a smoothed particle hydrodynamics simulation and a slip-link model describing microscopic dynamics of polymer chains in each fluid element. The multiscale simulation technique has been applied to polymer melt flow passing a circular object in a two dimensional periodic system to demonstrate the effects of history-dependent stress. We have found that the strainrate history-dependent stress in the polymer melt affects its flow behavior, and the stress field is retarded from the strain-rate field. The presented multiscale simulation is applicable to various polymer melts, because the slip-link model or the alternative course-grained models can address a variety of polymer architectures, e.g., linear and/or branched polymers, polymer blends, and polydispersed polymers. This multiscale simulation is advantageous because it employs a fully Lagrangian method at the macroscopic level, while conventional Eulerian-based techniques which have difficulties accounting for the macroscopic advection of microscopic internal states.

Acknowledgements

References

1. M. Doi; J. Takimoto, *Philos. Tras. R. Soc. London, Ser. A*, **361** (2003) 641.
2. Y. Masubuchi et al., *J. Chem. Phys.*, **115** (2001) 4387.
3. M. Laso; H. Ottinger, *J. Non-Newtonian Fluid Mech.*, **47** (1993) 1, P. Halin *et al.*, *ibid.*, **79** (1998) 387. M. A. Hulsen *et al.*, *ibid.*, **70** (1997) 79.
4. T. Murashima; T. Taniguchi, *Europhysics Letters*, 2011, **96**, 18002. T. Murashima; T. Taniguchi, *J. Polym. Sci. B*, 48 (2010) 886. T. Murashima; T. Taniguchi, *J. Phys. Soc. Jpn.* **81** (2012) SA013.

O-08-375

THE ROLE OF NANO CLAY ON KINETIC OF POLYETHYLENE CRYSTALLIZATION BY MEAN OF RHEOLOGICAL APPROACH

S. Masoumi[a], H. Nazockdast[a*], A. Gooneie[a] and N. Mostofi[a]
[a] *Polymer Engineering and Color Technology Department, Amirkabir University of Technology, Tehran-Iran, nazdast@aut.ac.ir, Tel: +98 21 64 54 24 32, Fax: +98 21 66 46 96 23*

This study represents the rheological characteristics of a liquid-solid transition (LST), of polyethylene/organoclay nanocomposites. To achieve this goal, X-ray diffraction (XRD) and scanning electron microscopy (SEM) are used along with rheological experiments to study the microstructure of high density polyethylene nanocomposites over a range of organoclay content of 3-7 wt%. Crystallization, a LS transition of polymers, has important implications both on rheology and mechanical properties. Crystallization behavior of polyethylene and its nanocomposites was studied by means of rheology, wide angle X-ray diffraction (WAXD) and differential scanning calorimetry (DSC). Near the liquid-solid transition, polymeric materials exhibit a very distinct relaxation pattern which reflects the material structure; the motions of the entire molecules (or of large chain segments) and of small scale details of the molecules contribute to the longer and shorter modes respectively.

The linear viscoelastic response of the nanocomposites exhibits an increase in the values of the complex viscosity, storage and loss moduli as the filler content increases. Non-Newtonian behavior of the complex viscosity at low frequencies first appears at 5 wt% which is consistent with the dominance of a percolated nano-polymer network. The same trends are also reflected in the loss tangent vs. frequency data. To better understand the microstructure evolution of nanocomposites in processing, nonlinear rheological response of the samples under simple shear flows were examined. Furthermore, the effect of the nanoclay content on the transient shear stress response is discussed. By increasing nanoclay content the strength of the percolated network increases and thus higher shear stresses are required to break the structure. Temperature sweep at a fixed frequency was applied to capture the rheological behavior of the nanocomposites in near liquid-solid transition. Complex viscosity vs. temperature shows a maximum value at LST which is attributed to the different elastic modulus of the solid and liquid near the transition region. It was found that the filler content, shear flow and rate of crystallization do not additively affect the LST behavior, but synergistically enhance it. This non-additive mechanism is believed to dominate the formation of oriented morphological structures. Indeed, DSC analysis reveals the combined effect of nanoclay and shear flow on the isothermal crystallization process of the matrix supporting the rheological data.

1. Introduction

During the last two decades, polymer/clay nanocomposites have gained great interest due to attractive technological applications, superior properties and improved performance of these materials compared with virgin polymers or conventional micro and macro composites [1-3].

High density polyethylene (HDPE) is one of the most popular polymers with applications in flexible packaging films, regarding to its good balance of optical and mechanical properties as well as low permeation compared with other types of polyolefins. Incorporation of a soft elastomeric material and rigid clay particles into the HDPE could induce a good balance of toughness and stiffness [4].

The aim of this work is to evaluate the crystallization behavior and Rheological properties of HDPE/ α-Olefin /organoclay nanocomposites. Morphology of nanocomposites has also been studied.

2. Experimental

2.1. Materials

The high density polyethylene was HD7255 (MFI=4.0 g/10 min; 190°C, 5 Kg) obtained from Jam Petrochemical Company, Iran. The α-Olefin was Tafmer DF640, obtained from Mitsui Chemical Grope, Singapore. Maleic anhydride grafted polyethylene, Fusabond E100, from Dupont company, USA, was used as a compatibilizer. Sodium montmorillonite clay modified with methyl tallow bis-2-hydroxy ethyl quaternary ammonium (Cloisite 20A) was provided by Southern Clay Products, Inc. (Gonzales, Texas).

2.2. Preparation of nanocomposites and reference samples

All the samples were prepared in an internal mixer (Brabender W50EHT) with a rotor speed of 70 rpm at temperature of 170 °C for 15 min. All the polymeric components were first dry mixed and fed into the chamber. Formulation and notation of samples are reported in Table 1.

2.3. Charactrization

A Rheometric Mechanical Spectrometer (US200: Anton Paar) was used to measure the rheological characteristics of nanocomposites and reference samples using parallel plate geometry. The diameter of both upper and lower plates was 25 mm, and the gap between two parallel plates was 1mm. Wide angle X-ray scattering (WAXS) analysis was carried out using an X-ray diffractometer (Simens D5000- Germany) operating at 30 kV and 30 mA for Cu Kα radiation (λ = 0.154 nm) under the diffraction angle, 2θ, in the range of 1–11°. The crystallization and melting behaviors of the samples were measured using a METTLER DSC, Germany. During the crystallization experiments, the specimens were first melted at 150 °C at the rate of 10 °C/min, and then cooled to room temperature at the rate of 10 °C/min. The crystallized specimens were subsequently heated at the same rate for the corresponding melting behavior investigations and the degree of crystallinity (Xc) of the specimens were calculated. The fracture surface of the blend nanocomposites and reference blends were inspected in a scanning electron microscope (SEM; XL30, Philips, Holland).

3. Results and Discussion

XRD patterns for the pure clay and its nanocomposites are presented in **Figure 1** The results show that the virgin clay exhibits a single diffraction peak at 2θ of 3.42°, which corresponds to a d-spacing of 2.43 nm. The diffraction pattern of HDPE/ α-Olefin /Clay nanocomposite has shifted to lower angles and it displays lower intensity compared to pure clay, representative of intercalated and/or a mixture of intercalated and exfoliated morphology. The shape of the diffractogram obtained for the HDPE/ α-Olefin /Clay nanocomposite, indicates a high degree of the organoclay dispersion. So it can be concluded that the most of clay tactoids have been exfoliated. **Figure 2-a** shows the cooling section of the DSC thermograms of neat HDPE, α-Olefin, their blend and the nanocomposites as a function of clay concentration. This indicated that the added organoclay acted as effective nucleation sites. However, it significantly deteriorated the perfection of the crystalline structure. On the heating curves shown in **Figure 2-b** for neat HDPE only one endothermic peak (T_m), was observed at 133°C. It is found that the crystallization peak (T_m) shift toward lower temperature, and the areas of exothermic curve decrease with increasing Nanoclay content in the blends. This indicates that the crystallization of blends is influenced by Clay.

As the results of **Figure 3** show, adding of 3 wt % clay remarkably increased the storage modulus and complex viscosity of HDPE/ α-Olefin blend and causes to a more pronounced shear thinning and solid-like behavior at low frequencies. The results showed that, the extent of increscent in elasticity of HDPE/ α-Olefin blend with clay loading is much more than those of neat HDPE and HDPE/ α-Olefin. Figure 3 depicts the Storage viscosity (G') of HDPE, α-Olefin, their blends with different clay content. It is seen that the Storage viscosity increasing with increasing of frequency. This is due to the strong shear thinning behavior of the polymers at the molten state. Entanglements between branches and chain segments of α-Olefin and HDPE may occur randomly, modifying the entanglement density of these blends under shear flow. Consequently, a nonlinear rheological behavior is observed. It is also seen that α-Olefin is more viscose than HDPE and Storage viscosity of the blends are intermediate to those of the pure components. **Figure 4** shows SEM micrographs of the nanocomposites for HDPE/ α-Olefin blend with 3 phr of organoclays. In these images the light areas refer to nanoclay layers phase and dark areas are related to Matrix. The nanocomposite containing Cloisite 20A shown in images 1 with different magnifications reveals the complete exfoliation degree of this organoclay.

4. Conclusion

HDPE/ α-Olefin /Clay nanocomposites could be prepared via a simple direct melt compounding. PE-g-MA acted as an effective compatibilizer for clay dispersion. XRD patterns confirmed presence of different levels of intercalated structures for all the Nanocomposites. Rheological results revealed that α-Olefin was more viscose than HDPE and all samples showed shear thinning behavior. Clear correlation between the shear thinning behavior and extent of exfoliation was observed. The rheological behavior of the nanocomposites changed from a liquidlike to a solid-like when the structure changed from a phase separated state to an intercalated and/or an exfoliated structure possibly due to percolation.

Table 1 Formulation and notation of the samples

sampels	HDPE	α-Olefin	PE-g-MA	Closite 20A
1	100	-	-	-
2	-	100	-	-
3	-	-	100	-
4	-	-	-	100
5	80	20	-	-
6	72/8	18/2	6	3
7	68	17	10	5
8	75/05	18/76	6/19	-
9	71/58	17/89	10/53	-

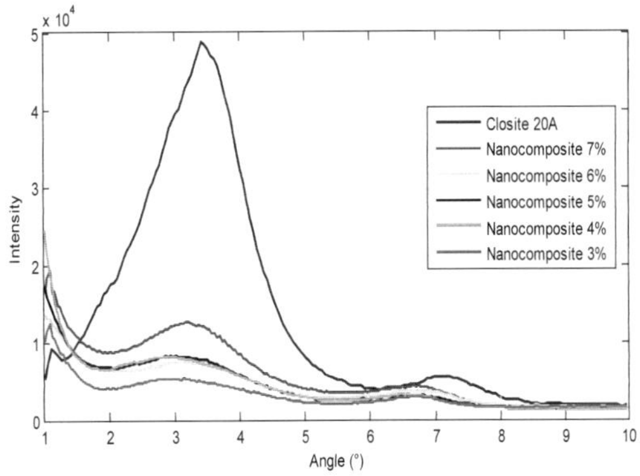

Figure 1 XRD patterns for the pure clay and its nanocomposites

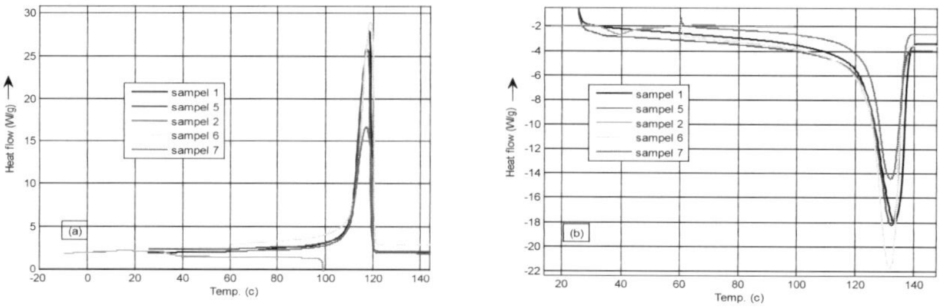

Figure 2 DSC thermo grams

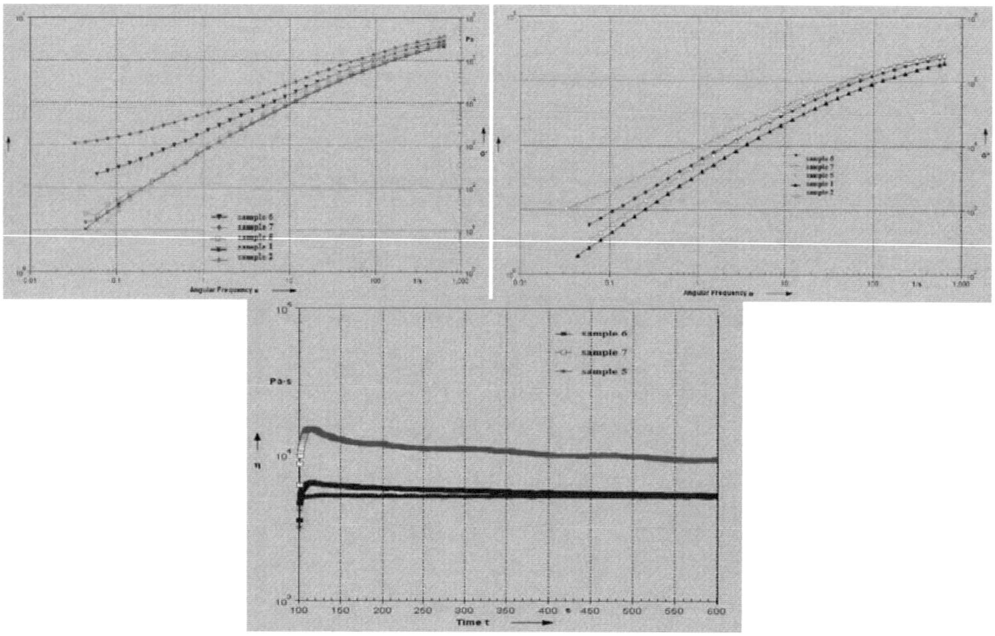

Figure 3 the results of rheological studies.

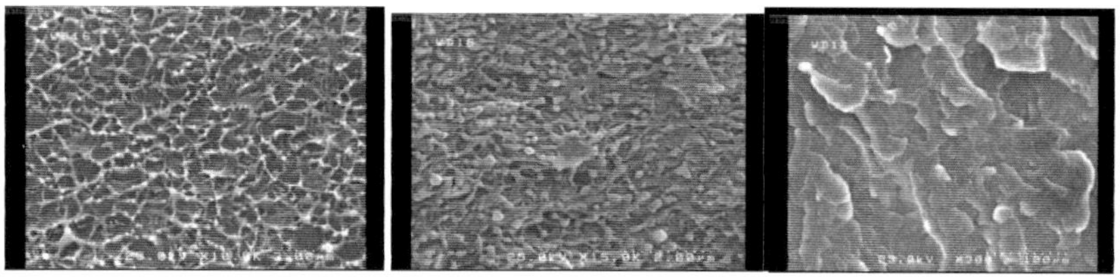

5. References

1. Mittal, V. European Polymer Journal 2007, 43, 3727.

2. Mittal, V. Journal of Applied Polymer Science 2007, 107, 1350.

3. Maged, A.; Osman, A.; Mittal, V.; Ulrich, W. S. Macromolecular Chemistry and Physics 2007, 208, 68.

4. Y. C. Ahn, D. R. Paul, Polymer 2006, 47, 2830–2838.

IMPACT OF SALINITY AND TEMPERATURE ON THE RHEOLOGY OF PARTIALLY HYDROLYZED POLYACRYLAMIDE SOLUTIONS

M. Shahzad Kamal[1], Abdullah S. Sultan[2], Usamah A. Al-Mubaiyedh[1], Ming Han[3] and Ibnelwaleed A. Hussein[1*]
[1]Department of Chemical Engineering, King Fahd University of Petroleum & Minerals, 31261 Dhahran, Saudi Arabia, shahzadmalik@kfupm.edu.sa; usamah@kfupm.edu.sa; ihussein@kfupm.edu.sa
[2] Department of Petroleum Engineering and Center of Petroleum & Minerals, King Fahd University of Petroleum & Minerals, 31261 Dhahran, Saudi Arabia, sultanas@kfupm.edu.sa
[3] EXPEC Advanced Research Center, Saudi Aramco, 31311 Dhahran, Saudi Arabia, ming.han@aramco.com

Abstract - Polymer flooding is one of the promising enhanced oil recovery (EOR) techniques used to recover residual oil after water flooding from both sandstone and carbonate reservoirs. Depending upon the objectives of applications in specific reservoir conditions, surfactants and/or alkalis may be combined with polymers to improve the oil recovery. Mobility of displacing flood should be lower than the displaced fluid to avoid viscous fingering. So, it is important to study the rheological and thermal behavior of polymer solutions prior to core flooding. Partially hydrolyzed polyacrylamide (HPAM) is widely used in chemical EOR applications. Experimental studies were conducted to evaluate the rheological properties of three different polyacrylamide-types of polymers with different molecular weight and structure. Polymer concentration ranged from 0.05 to 0.25 % (w/v). Effects of temperature, salinity, molecular weight, polymer concentration and polymer structure on rheological properties of three different polymers were investigated by using both dynamic and steady shear measurements. About 60% drop in viscosity of partially hydrolyzed polyacrylamide was observed when temperature was increased from 50°C to 90°C. However, reduction in viscosity for a copolymer of acrylamide and acrylamido tertiary butyl sulfonic acid (ATBS) was only 15% for the same temperature range. Time sweep measurements showed instability of partially hydrolyzed polyacrylamide at 90°C while ATBS based copolymer was stable at the same temperature. In salt-free water, viscosity and storage modulus of all polymers in the temperature range 50°-90°C were very high as compared to viscosity in saline water.

Keywords: rheology; polyacrylamide; salinity; enhanced oil recovery; polymer flooding

Introduction

In the past, contribution of chemical enhanced oil recovery (cEOR) was very small as compared to other EOR techniques like thermal and gas flooding due to high chemical cost, unavailability of huge amount of chemicals, low oil prices and insignificant understanding of process mechanism [1-4]. Due to depletion of oil reservoir, advancement in technologies high oil prices and the fact that billion barrels of oil are unlocked and could be recovered by EOR, it is expected that cEOR will contribute significantly in meeting the world's oil demand. China is leading in cEOR with implementation of many successful cEOR projects [5]. Polymer flooding is one of the most promising cEOR techniques which are successfully applied in some oil fields. In the Middle East, only Oman has started recovering oil by polymer flooding and many other countries in the region have started lab scale research on polymer flooding [6].

For almost all polymer flooding applications water soluble acrylamide based polymer, copolymer and terpolymer are used due to their high viscosity at low concentration [7-8]. These polymers are available in different grades containing different molecular weight, degree of hydrolysis and structure. The main objective of using these polymers is to increase the viscosity of displacing fluid and increase the macroscopic displacement efficiency. It has been proved in some recent researches that these polymers are not only improving macroscopic displacement efficiency but also microscopic efficiency due to elasticity [9-12].

Rheological properties are very important in screening the polymers for cEOR applications. The objective of this work is to study the influence of salts and temperature on steady state and dynamic rheological properties of different polyacrylamide solutions having different structure and molecular weight.

Experimental

Materials

Water soluble polymers Flopaam 3630S and Flopaam An125SH manufactured by SNF FLOERGER were used and received as powder. Salt solution was prepared using laboratory grade sodium bicarbonate, sodium sulphate, sodium chloride, calcium chloride and magnesium chloride with a total salinity of 57,638 ppm.

Preparation of polymer samples

Polymer solution was prepared by using magnetically driven stirrer. Polymer was added slowly on the shoulder of vortex formed by deionized water or salt solution to avoid formation of slubs. As soon as the entire polymer was added the rotor speed was reduced from 300 to 80 rpm to avoid mechanical degradation. Stirring was turned off after 3 hours and solution was kept at room temperature for the next 48 hours for complete hydration. Finally, the solution was filtered and diluted to desired concentration.

Rheological measurements

Rheological measurements were carried out using TA instruments discovery hybrid rheometer (DHR-3). Concentric cylinder geometry was used to measure both steady state and dynamic measurements. The range of shear rate for steady state viscosity measurements was from 10^{-3} to 10^{3} s^{-1}. Frequency from 0.01 to 100 rad/s was applied. All reported data points are within the torque limit of machine. Oscillation time experiment was also performed to check the thermal stability for two hours.

Results and Discussion

Flopaam 3630S is partially hydrolyzed polyacrylamide (HPAM) and it is a copolymer of acrylamide and acrylate while Flopaam AN125 SH is a copolymer of acrylamide and acrylamido tertiary butyl sulfonic acid (ATBS). Performance of these polymers was evaluated at two different temperatures 50°C and 90°C. Both steady state and dynamic rheological properties were evaluated to see the effect of temperature. It is evident from Fig.1 that increasing temperature from 50°C to 90°C causes a major drop in steady shear viscosity of HPAM at low shear rate. At low shear rate, viscosity at 90°C is about 60 % less than corresponding viscosity at 50°C. However, for AN125 SH, very small change in viscosity was observed when temperature was increased from 50°C to 90°C. These steady state measurements show that temperature effect is more significant on HPAM as compared to copolymer of acrylamide and ATBS. Effect of temperature on dynamic properties was also investigated. Frequency sweep tests were preceded by strain sweep tests to determine the linear viscoelastic range. A typical strain sweep curve is shown in Fig. 2. Figure 3 demonstrates the effect of temperature on storage and loss moduli of HPAM. At low frequency, about 50% reduction in storage modulus of HPAM was observed. Elastic behavior was dominating at 50°C for all frequency range. At 90°C the crossover was observed at about 0.06 rad/s while at 50°C it is somewhere less than 0.02 rad/s. This smaller crossover frequency at 50°C correspond to higher relaxation time which shows polymer is more structured at 50°C as compared to polymer at 90°C.

Time sweep experiments were run for two hours for both polymers at 50°C and 90°C. Figure 4 presents results obtained at 50°C while Fig. 5 shows time sweep experiments at 90°C. Both polymers were stable within the experimental time at 50°C as shown in Fig. 4. Storage moduli for both polymers were higher than loss moduli. At 90°C completely different results were found. A continuous decrease in storage modulus of HPAM was observed as experiment proceeds. Within the experimental time storage modulus of HPAM was higher than loss modulus but the cross over is expected after some time as indicated from the trend. From the above results it can be concluded that at 50°C both polymers showed good stability and can be used at this temperature. At 90°C HPAM cannot be used as it shows a continuous decrease in rheological properties. Copolymer of acrylamide and ATBS was more stable at 90°C.

Polymers were evaluated for their salt tolerance at different temperatures. For this purpose solution of 57,638 ppm typical to sea water was prepared. This solution contains 650 ppm Ca^{+} ions and 2083 ppm Mg^{+} ions. As evident from Fig. 6 both polymers were not tolerant to this salinity and complete collapse of rheological properties was observed. Most probably it is due to high amounts of divalent cations. Relative effect of individual ions on rheological properties of these water soluble polymers is a focus of our ongoing research. There is strong repulsion among the molecules of chain due to the presence of negative charges in salt free polymer solution. Salts bring counterions in the solution that decrease the repulsion and molecules attain more compact conformation resulting in lowering of viscosity. From Fig. 7 it is apparent that addition of salt causes a major drop in both storage and loss modulus even at 50°C. It can also be observed from Fig. 7 that in deionized water polymer elastic behavior is more dominant while in salt solution the reverse was detected. So, addition of salt has adverse effect on rheological properties in two ways. Firstly, addition of salt causes decrease in viscosity which may cause viscous fingering and low macroscopic displacement efficiency. Secondly, the decrease in elastic properties also causes lowering of microscopic displacement efficiency.

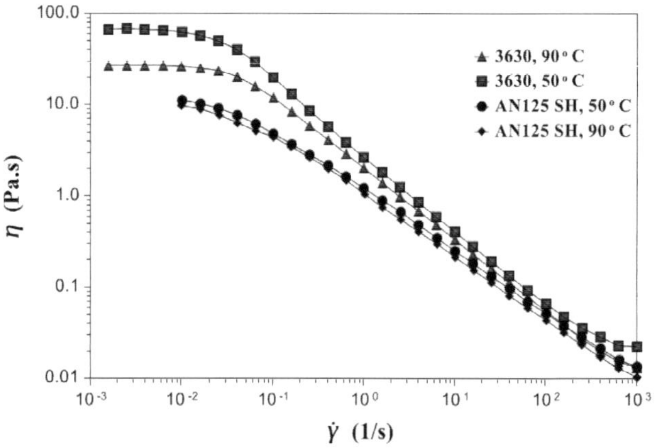

Figure 1- Effect of temperature on steady state viscosity of polymer solutions (0.25%) in deionized water

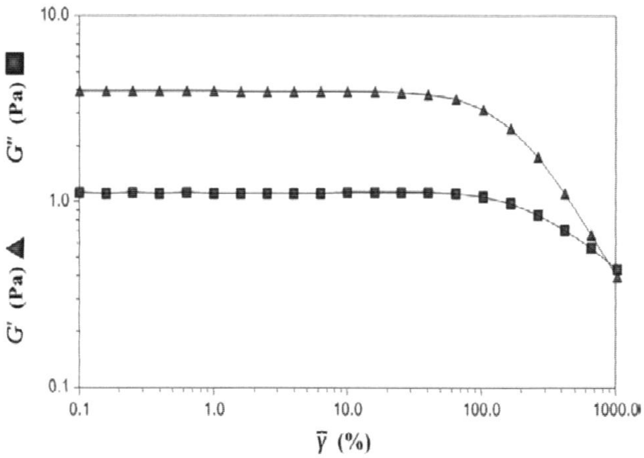

Figure 2- Strain sweep curve of 0.25% Floppam3630S at 50°C and angular frequency of 1 rad/s

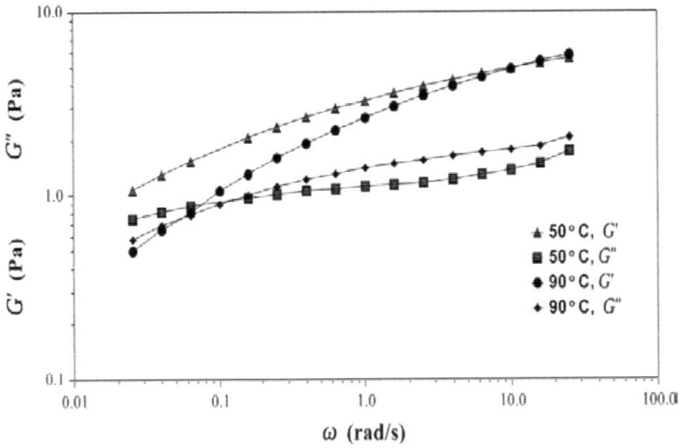

Figure 3- Effect of temperature on storage and loss modulus of Flopaam3630S (0.25%) in deionized water

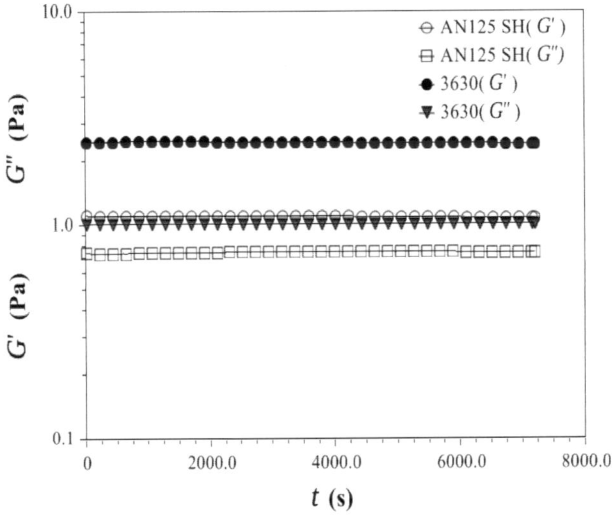

Figure 4- Comparison of storage modulus of two polymers (0.25%) by time sweep measurements for 2 hours in deionized water at 50°C

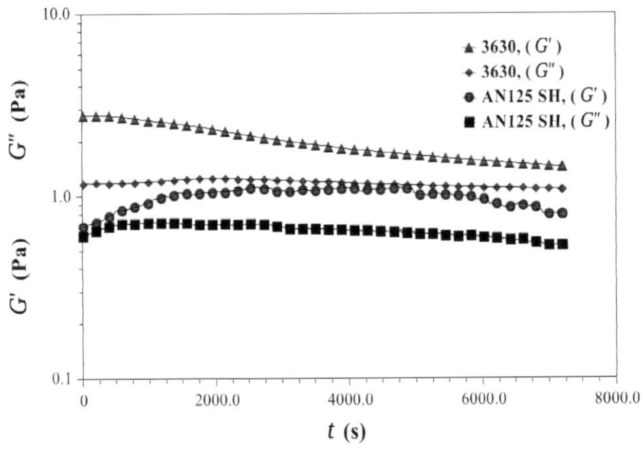

Figure 5- Comparison of storage modulus of two polymers (0.25%) by time sweep measurements for 2 hours in deionized water at 90°C

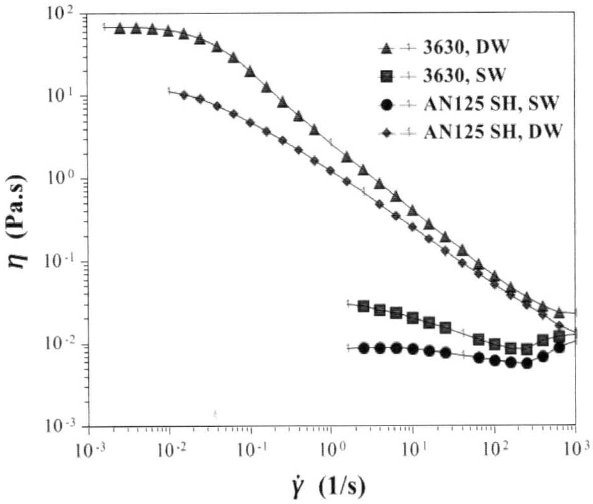

Figure 6- Effect of salinity (57638 ppm) on steady state viscosity profile of two different polymer solutions (0.25%) at 50°C

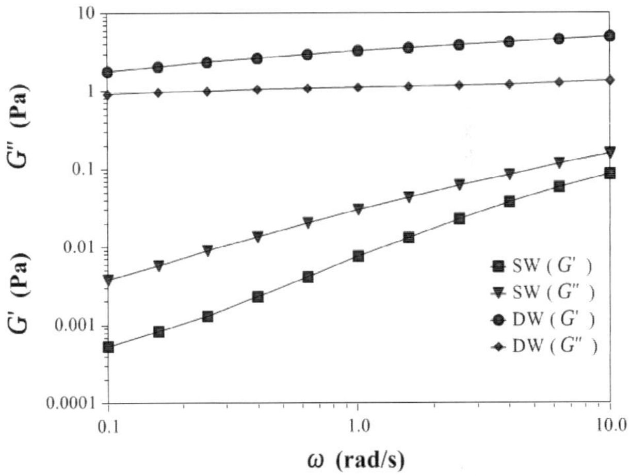

Figure 7- Effect of salinity (57638 ppm) on storage and loss modulus of Flopaam3630 (0.25%) at 50°C

Conclusions

Impact of salinity and temperature on rheological properties of polyacrylamide solutions was investigated. For HPAM increase in temperature from 50°C to 90°C causes a notable reduction in viscosity. For copolymer of acrylamide and ATBS this increase in temperature causes a minute drop in zero shear viscosity. From time sweep experiments it can be concluded that at 90°C HPAM was not stable while ATBS based copolymer showed stability at the same temperature. Addition of 57638 ppm salts causes a complete loss in rheological properties. A major decrease in steady shear viscosity storage modulus and loss modulus was noted on addition of salts for both polymers. The Above results are very important for application of these polymers in EOR. These results provide us information on the temperature and salinity limit of these polymers. For example we cannot use HPAM (3630S) at 90°C and both polymers cannot be used where water has very high salinity.

Acknowledgements
Authors would like to thanks Center of Petroleum and Minerals, King Fahd University of Petroleum & Minerals and Saudi Aramco for financial support. Authors would like to thank SNF for providing polymer samples.

References

1. A. A. Adasani; B. Bai, Journal of Petroleum Science and Engineering, 2011, 79, 10.
2. J.J.Taber; F.D.Martin; R.S.Seright, SPE Reservoir Engineering, 1997, 189.
3. V. Alvarado; E. Manrique, *Energies,* 2010, 3, 1529.
4. E. Manrique; C. Thomas; R. Ravikiran; M. Izadi, M. Lantz; J. Romero; V. Alvarado, SPE Improved Oil Recovery Symposium proceedings, Tulsa, 2010.
5. G. Degré; M. Morvan; J. Beaumont; A. Colin; G. Dupuis; A. Zaitoun; R. S. Al-Maamari; A. R. Al-Hashmi; and H. H. Al-Sharji, SPE Asia Pacific Oil and Gas Conference and Exhibition proceedings, Jakarta, 2001.
6. S.M. Al-Mutairi; S. L. Kokal; SPE EUROPEC/EAGE Annual Conference and Exhibition proceedings, Vienna, Austria, 2011
7. S. Lee; D.H. Kim; C. Huh; G.A. Pope, SPE Annual Technical Conference proceedings, Louisiana, 2009.
8. S.K. Veerabhadrappa; T.S. Urbissinova; J.J. Trivedi; E. Kuru; SPE Western North American Regional Meeting proceedings, Alaska,2011.
9. W. Demin; C. Jiecheng; X. Huifen; L. Qun; S. Jingping, SPE Asia Pacific Improved Oil Recovery Conference proceedings, Kuala Lumpur, 2001
10. T.S. Urbissinova; J. Trivedi; E. Kuru, SPE Western Regional Meeting proceeding, California, 2010.
11. W. Demin; X. Huifen; L. Zhongchun; Y. Qingyan, SPE Asia Pacific Oil and Gas Conference and Exhibition proceedings, Jakarta, Indonesia, 2001.
12. X. Huifen, W. Demin; W. Junzheng; K. Fanshun, SPE Asia Pacific Oil and Gas Conference and Exhibition proceedings, Perth, 2004

O-08-413

STUDY ON MIGRATION OF NANO SILICA IN ANNEALING TIME BY RHEOLOGICAL PROPERTIES CHARACTERIZATION

P. Kiani, H.Nazockdast[1]*, B.Ranjbar

[1] *No. 424, Hafez Ave., P.O. Box: 15875-4413*
Department of Polymer Engineering and Color Technology, Amirkabir University of Technology
Tehran, Iran

Abstract
An attempt was made to study the effect of the surface tension of nano silica on morphology development in poly butylene terephthalate /polypropylene (PBT/PP) blends. For this purpose the sample of blend's with blend ratio of 20/80 (w/w) with two type of nano silica that are differing in surface modification (hydrophilic and hydrophobic) and concentration (1/3/6%w/w) were considered. All samples were prepared by melt compounding using an internal mixer. Firstly the masterbaches of PBT or PP with nano silica were prepared which then were used in preparation of blends. blend containing nano particles in PBT droplets, are characterized by a 3-time frequency sweep test without any time interval between the tests. about system containing 1% and 3% nano particles in PBT droplet, the result of first r ound of frequency sweep showed all nano particles located in droplets and second and third rounds of frequency sweep test showed the location of nano particles in pp matrix however base on thermodynamic rules it was expected at last hydrophilic nano particles be remain in PBT phase. it is attribute this to self aggregation of nano particles in PBT droplet which cause higher hydrophobicity, as a consequence the nano particle migrate to pp matrix so the hydrophilicity of pp increase then other nano particles migrate to pp matrix and then aggregate, but for blend which pp contains nano particle as matrix, the results show no migration from matrix to PBT droplet, however it was expected that hydrophilic silica particles migrate to PBT droplets it can be concluded that self aggregation of silica particles suppress the migration. T he SEM result confirm above rheological observation.

Keywords: nano silica, migration, polymer blend, linear viscoelasticity

Introduction

Blending of immiscible polymer is one of the ways to obtain new material with special properties. However the biggest their problem is weakness of interface due to their immiscibility. In recent years a new method to compatibles immiscible polymer blend is using from inorganic nano particles. In 2004 it was showed the compatibility of PP/PS improved with presence of nano silica particles the time of mixing and silica concentration play significant role in this compatibility. In 2007 it was showed that addition of nano silica in PIB/PDMS blend can cause the clustering of droplets so the blend shows gel-like behavior. In the same time effect of silica nano particles on morphology and rheological properties of PP/PS blend was investigated. Two type of nano silica (hydrophilic and hydrophobic) were used. A migration of hydrophilic silica from PP phase to PS droplets was observed, due to chemical affinity between PS and silanol group of hydrophilic nano silica. In 2008 this groups observed convective force migration of nano silica from matrix to droplets is higher than migration from droplets to matrix and they discussed mechanism of migration. It has been shown the rheological properties of multiphase system are strongly influenced by the phase morphology, which depends on thermodynamic interaction among the constituent component and Flow history. it has been investigated the location of nano particle controlled by thermodynamic and kinetic, through the mixing. it has proved that nano particles can migrate from matrix to inter face or to dispersed phase, when the layer interaction be stronger, that the wetting parameter can predict the thermodynamically favorable place of nano particles, but the mechanisms of migration discussed rarely. The aim of this study is to discussed this migration.[1,2]

Materials and methods

A commercial polypropylene grade V30S, from Maron petrochemical and poly butylene terfetalate (PBT) from Lanxess Deutschland with melting point 225^{0}c, and fumed nano-silica, Aerosil 200 and Aerosil R816 by Degussa GmbH in this study be used. The Aerosol R816, m-silica, is a hexadecylsilane surface treated hydrophobic nano silica based on Aerosil 200. The nano silica particles have a spherical surface and are free of pores. All samples with same concentration of PP/PBT (35/65v/v%) but difference in type and concentration of nano particles were prepared by melt blending in Barabender internal mixer at 240 0c and speed of rotors was 60 rpm for approximately 5 and 10 minute and direct mixing was used. the morphology of the virgin and filled samples examined by scanning electron microscopy using Hitachi F4160 model in Tehran university that operated at 15 kv and under high vacuum. For rheological measurement from a stress/strain controlled Rheometer, Paar Physica model uds 200,

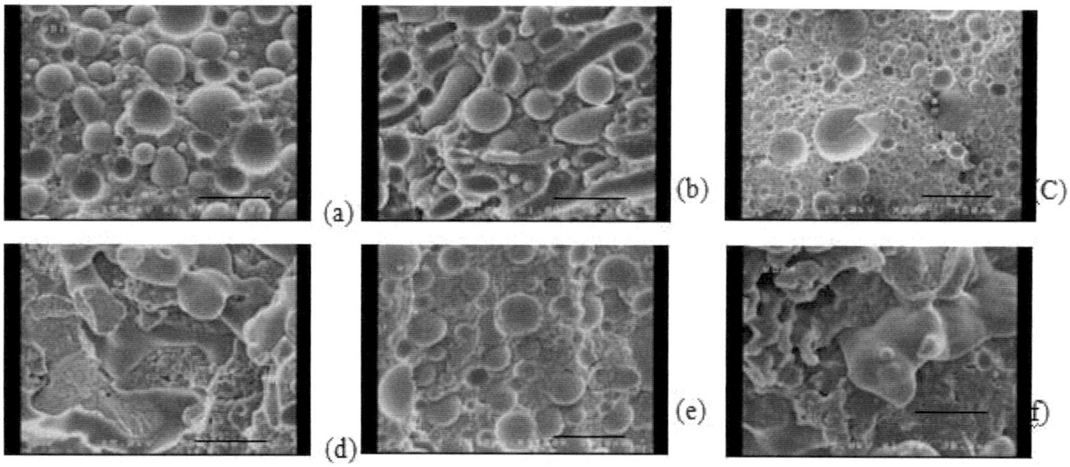

Fig.1) morphology of PBT/PP (35/65/x (v %v%/w%) blend after mixing, (a) x=0,(b) x= 1% R816,(c)x=3% R816,(d) x=6%R816,(e) x=1% A200,(f) x=6%A200. After 5min mixing, All scale bar =30μm, c =150μm,d =60μm

Equipped with parallel plate geometry with 25 mm diameter and 1mm gap at 260 0C was used.

Result & discussion

Wetting parameter for system filled with hydrophilic nano silica and hydrophobic nano silica is equal to -2.39 and 0.52 respectively. This means hydrophilic nano silica thermodynamically tends to locate in PBT droplets and hydrophobic particles locate in interface. The SEM results show that, (fig1), first decreasing and then increasing size of droplet as increasing A200 particles concentration, about 1%A200 location of nano particle in interface by suppressing the coalescence decreases droplet's size. About 6% A200, nano particles partially aggregate in pp matrix duo to low time of mixing (5 min) their concentration, then viscosity ratio increased this cause bigger size of droplets and location of nano silica in PBT droplets causes the droplet breakup be harder. About system filled with R816 it can be observed that, first (1%) droplet without spherical shape due to non homogeneously distribution of R816 in interface and then (with increasing nano particles concentration) some particles locate in matrix and then bigger size of droplets (3%, 6%), on the other hand could said co-continuity increased. it supports the results of [3, 4]. at 3% R816 domain phase have spherical shape due to3% R816 homogenously locate in interface. About 6% R816 the bigger domain size be observed, it is suggested partially of particles aggregate and then migrate from interface to pp matrix then the coalescence increased this means kinetic controls morphology higher than thermodynamic. For study on migration of nano particles after 5 and 10 min mixing, the rheological properties of sample under strain 1% and frequency between 0.05-1000 at 260 0C for 3-time was measured, while before every starting 1min relaxation time be spotted. For samples with 5 min mixing time and concentration higher than 6%fig (2,a), the results show increasing storage modulus at low frequency. it means the self agglomeration in matrix be it is due to low time of mixing and location of particles in pp matrix and interface this shows migration of nano particles from interface to matrix and migration of particles through the matrix. The result of this test for samples after 10 minute mixing to be observed, (fig.2.b). The non terminal at middle frequency region shows the location of nano particles in PBT droplets and its reduction at second run shows migration of nano particles to pp matrix. The increasing of storage modulus at third run shows aggregation of nano particles that were migrated to matrix. The results for 6% R816 filled system after 10 minute mixing is shown in fig.3. In second run the lower storage modulus at middle frequency region is observed. It is suggested this means migration of nano particles from interface and droplets to matrix due to disappear the terminal region through 1-10 frequency that appeared due to bigger droplets and then increasing of modulus. And A200 filled system, about 1% is not observed significant change in storage modulus. About 3% increasing is attributed to migration of nano particles from interface and droplet to matrix.

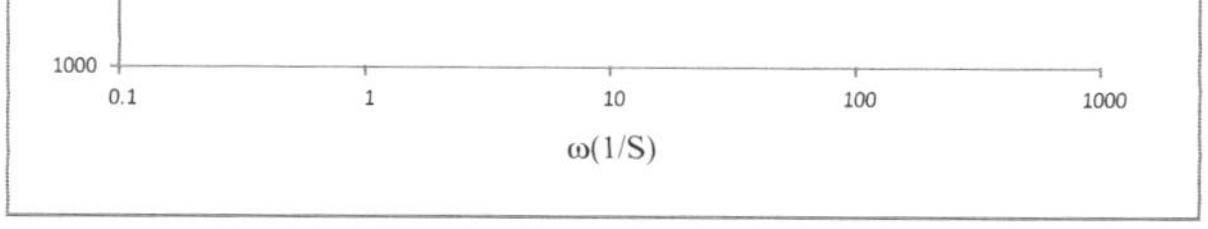

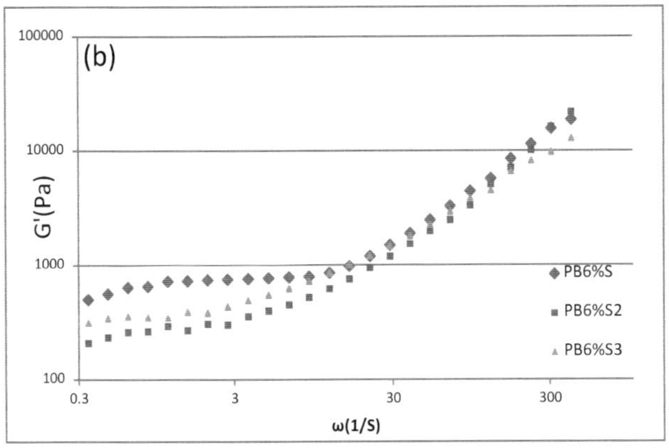

Fig2, shows the storage modulus vs frequency (a) system filled with 6% A200 after 5 min mixing (b) system filled with 6% A200 after 10 min mixing

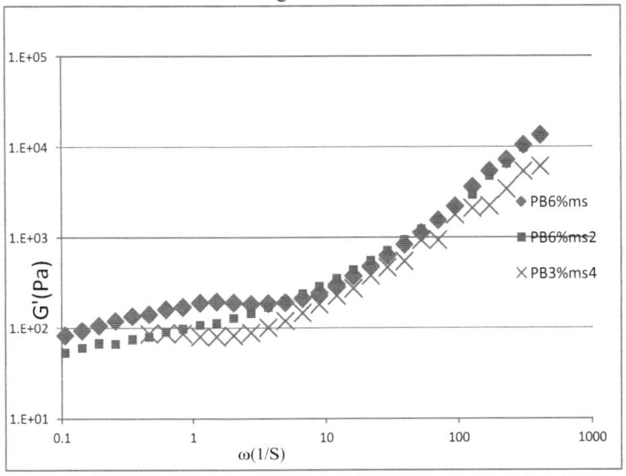

Fig3, shows the storage modulus vs frequency . system filled with 6% R816 after 10 min mixing

For study on migration of nano silica particles in annealing time, first master batches of poly propylene and poly butylene terfetalate with nano silica particles by internal mixer with 60 rpm rotor's speed at 240 ^{0}C prepared and then mixed with second polymer for 2.5 minute. A3-time frequency sweep test through annealing time for checking of nano silica particle's migration for every sample was used. Fig 4 shows results for sample consist of PP/ (PBT/1%silica). Increasing of non terminality at low frequency region means nano silica particles migrate from PBT droplets to interface and PP matrix but higher in interface. Fig 5 shows effect of nano silica's concentration on migration time. However it has been expected the migration of nano silica increase with enhancement nano silica concentration but increasing nano silica from 1 to 3percente didn't obey this trend. It is attributed this to self aggregation of nano silica in PBT droplets so migration of nano silica's aggregates was harder, the lower ramp of storage modulus's increasing about 3% nano silica confirm this. About presence of 6% nano silica the storage modulus start from higher point, this means migration started sooner than other samples due to high nano silica's concentration.

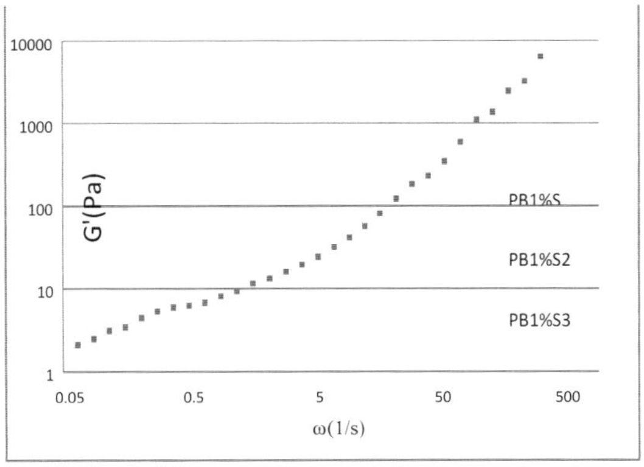

Fig. 4) a 3-time Frequency sweep experiment results of PP /PBT/ 1% hydrophilic silica samples at 260°C.

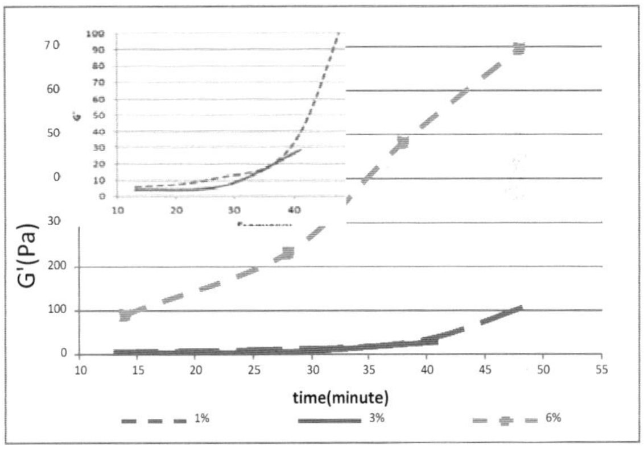

fig.5) variation of storage modulus verses time of three concentrations of hydrophilic nano silica at 0.192 (s^{-1}) angular frequency and 1% of stain.

Fig 6) shows the variation of modulus verses time of three concentration (1,3,5%) of hydrophobic nano silica that first locate in PBT droplets at 0.192 (s^{-1}) and strain 1%. The effect of concentration here is clearer than presence of hydrophilic nano silica. this figure shows enhancement of initial point of storage modulus with increasing percent of nano silica also shows clearer effect of affinity between nano silica and pp matrix, then this verify aggregation of hydrophilic nano silica in PBT matrix when it's concentration was 3% w/w.

However up on surface tension it has been expected didn't any migration to interface and matrix from droplet, this migration verifies firstly location of nano silica in interface regardless surface treatments and this confirm the concentration of nano silica plays significant role in location of nano particles therefore properties of product.

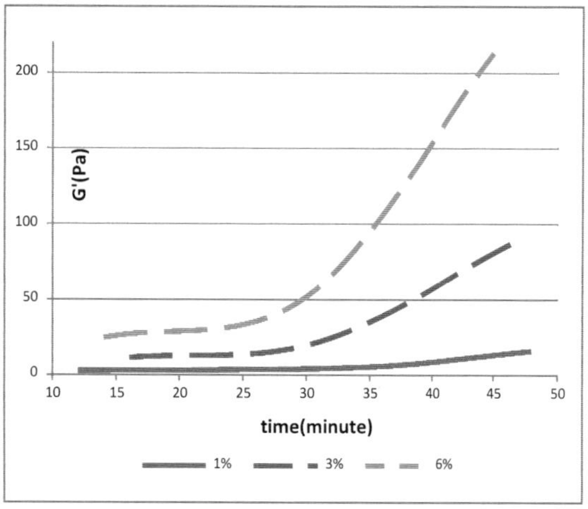

Fig. 6) variation of storage modulus verses time of three concentrations of hydrophobic nano silica at 0.192 (s^{-1}) angular frequency and 1% of stain.

Conclusion

It has be shown the location of nano silica depend of nano silica concentration and processing condition, that control the morphology of product. It has be shown the rheology can be used to show the migration of nano particles.

References

1. Wei Tong.Yajiang Huang.Chenglei Liu. Xiaolian Chen. Qi Yang. Guangxian Li ,Colloid Polym Sic (2010) 288:753-760
2. Elias, F. Fenouillot, J.-C. Majeste´, G. Martin,1 P. Cassagnau Journal of Polymer Science: Part B: Polymer Physics, Vol. 46, 1976–1983 (2008)
3. Liu w, Huang Y, Wu D, Gao L, Yang Q, Li G, Acta Polymerica Sinica 2009;6:553-9.
4. Hui S, Chaki T, Chattopadhyay S, J Appl Polym Sci 2008;110: 825-36

O-08-068

MODELING OF THE INTERDIFFUSION AND THE INTERPHASE DEVELOPMENT BY RHEOLOGY OF A SYMMETRICAL AND AN ASYMMETRICAL BILAYER BASED ON PMMA AND PVDF MODEL POLYMERS

Huagui Zhang[1], Khalid Lamnawar[2]* and Abderrahim Maazouz[1,3]*

[1]*Université de Lyon, F-69361, Lyon, France; CNRS, UMR 5223, Ingénierie des Matériaux Polymères, INSA Lyon, F-69621, Villeurbanne, France;*

[2] *Université de Lyon, F-69361, Lyon, France; CNRS, UMR 5259, INSA-Lyon, LaMCoS, Laboratoire de Mécanique des Contacts et des Structures, Groupe de Recherche Pluridisciplinaire en Plasturgie, F69621, Villeurbanne, France*

[3] *Hassan II Academy of Science and Technology, Rabat, Morocco.*

Abstract -Small amplitude oscillatory shear measurement (SAOS) was employed to probe the interdiffusion process and the development of the triggered interphase taking place at a molten polymer/polymer interface between a symmetrical bilayer and an asymmetrical bilayer consisting of two compatible model polymers, poly (vinylidene fluoride) (PVDF) and poly (methyl methacrylate) (PMMA) of varying molecular weights, respectively. For the symmetrical bilayer, rheological modeling has been realized by correlating the complex modulus(G*) of the bilayer to the diffusion coefficient; the kinetics of the interdiffusion and the effect of annealing conditions such as healing time, angular frequency, temperature and molecular weight have been highlighted and interpreted under the framework of Doi-Edwards theory. For the asymmetrical bilayer, rheology method, Lodge-McLeish model and test of the time-temperature superpositon (tTS) principle were employed to probe the thermorheological complexity of this polymer couple and calculate the monomeric friction coefficient of each species in mixed state, which have been examined to be close in the present experimental conditions. Hence, a quantitative rheological model modified from a primitive Qiu–Bousmina's model was proposed to determine the mutual diffusion coefficient (D_m) by using a presumably apparent friction resistance factor and taking into account the rheological behavior of the interphase. The obtained results of D_m confirm the prediction of the fast-mode theory and are consistent with the literature. Meanwhile, rheology and properties of the generated interphase are able to be revealed. Scanning electron microscopy coupled with energy dispersive X-ray analysis (SEM-EDX) characterizations were confronted to the rheological results in terms of concentration profile and interphase thickness, where a good corroboration between rheology and SEM-EDX analysis was obtained.

Keywords: Interdiffusion process; Rheological modeling; Diffusion coefficient; Interphase; Viscoelasticity

Introduction

The interdiffusion process at a molten polymer/polymer interface is quite of academic and industrial significance in understanding the chains' motions and in polymer processing like co-extrusion, co-injection molding, etc. [1-2]. This is especially true in the case of reactive processing since it may affect the interfacial reaction and the generation of the interphase [3-4]. Hence, to understand the interfacial phenomena occurred during polymer processing, it is essential to firstly probe and quantify the kinetics of the interdiffusion process between polymers.

Besides the sophisticated spectral techniques [5-6], rheometry has also been demonstrated to be a reliable tool to quantitatively investigate the interdiffusion process between polymer melts [7-8]. In this study, fundamental investigations and rheological modeling of the interdiffusion process and the interphase development are conducted based on both symmetrical bilayer and asymmetrical bilayer structures using PMMA and PVDF as model compatible polymers.

Experimental

Rheological measurements were carried out by a strain-controlled rheometer: ARES (Advanced Rheometrics Expansion System) with a parallel-plate geometry (Φ=25 mm) at varying temperatures from 180 °C to 240 °C under a nitrogen atmosphere. The rheological characterizations of neat polymers and PMMA/PVDF blends were performed within the linear viscoelastic regime.

Symmetrical PMMA/PMMA or PVDF/PVDF bilayer assembly and asymmetrical PMMA/PVDF bilayer assembly for the interdiffusion process were prepared by bringing the given round disks into intimate contact at room temperature with a desirable configuration before loading them between the plates and annealing them in the oven of the rheometer. Evolution of the interdiffusion process was tracked by performing dynamic time sweep tests at a given temperature and a given angular frequency in the linear viscoelastic regime for a certain time.

The PMMA/PVDF assembly specimens quenched after the annealing process were microtomed normal to the interfacial plane before being exposed in a Hitachi S800 FEG scanning electron microscopy (SEM) with an EDX microanalysis system. The electron-beam induced x-ray fluorescence of the characteristic atoms present in the sample permits direct measurement of the concentration profile of components in the interfacial region.

Results and Discussion

Interdiffusion process at symmetrical bilayer structures

The kinetics of the interdiffusion was experimentally obtained by following the evolution of the complex modulus (G*) of the bilayer assembly versus the healing time. From the G*, an apparent diffusion coefficient (D_a) was obtained using following expressions [8]:

$$D_R = \left(\frac{Nb^2}{24}\right)\left(\frac{\rho RT}{M}\right)^2 \frac{\omega}{\left[G^*(\omega)\right]^2} \quad (1)$$

for the Rouse model, and

$$D_{rep} = \frac{Nb^2\omega}{3\pi^2}\left[\left(\frac{8G_N^0}{\pi^2 G^*(\omega)}\right)^2 - 1\right]^{1/2} \quad (2)$$

for the reptation model. N, b, ρ, RT, M, ω and G_N^0 are the number of segments per chain, the effective bond length, the density, the temperature defined in energy units, the molecular weight, angular frequency and the plateau modulus. It is worth mentioning that the Rouse model applies to the motions of short chains in the melt, corresponding to the case of chains with relaxation times shorter than τ_e (Rouse relaxation time between entanglements) and the reptation model is applicable for $t > \tau_e$.

From a typical example of PMMA/PMMA symmetric bilayer in Fig.1, we can see that the evolution of G∗ (or D_a) with time exhibited three clear stages: at the beginning, G∗ increased (or D_a decreased) monotonically with time at a rapid rate, then the increase (or decrease) slowed down and gradually transformed into the third stage, where a plateau occurred. The G∗ (or D_a) approached a saturation value, within experimental errors, expected to be close to the G∗ value (or the self-diffusion coefficient D_s) of the neat polymer.

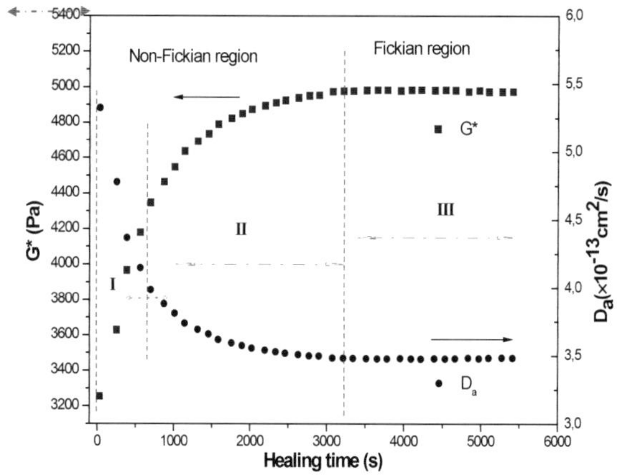

Figure 1-An example of complex modulus G∗ (square) and apparent diffusion coefficient D_a (circle) versus healing time for a PMMA/PMMA symmetric bilayer at ω= 0.1 rad/s and T = 200°C

Fig.2a portrays the variations of the apparent diffusion coefficient versus healing time at different angular frequencies ranging from 0.05 rad/s to 100 rad/s at 200 °C for PMMA/PMMA assembly. Strikingly, the diffusion coefficients remained nearly independent of frequency at the terminal zone ($\omega < 1/\tau_{rep}$, i.e. reptation relaxation time), whereas it increased pronouncedly with frequency at $\omega > 1/\tau_{rep}$. The diffusion coefficients detected at different frequencies reveal the diffusion processes of different chains: short chains at high frequencies and longer chains at lower frequencies. Especially, when $\omega < 1/\tau_{rep}$, all chain types are involved.

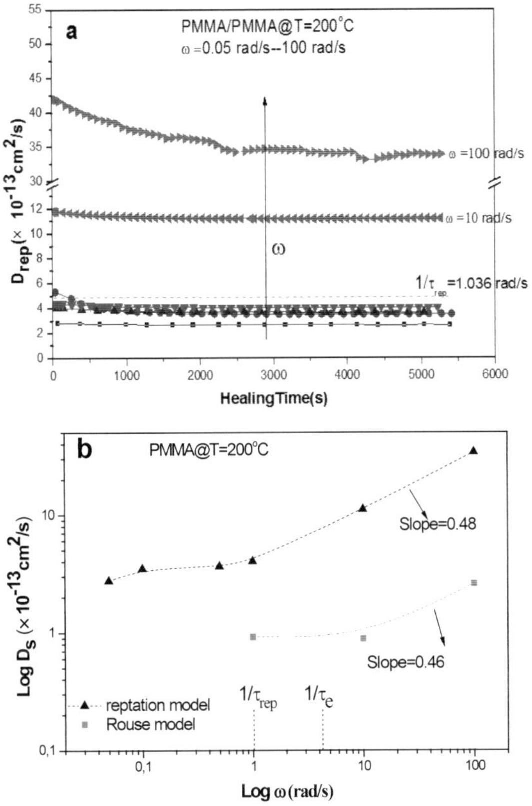

Figure 2–(a) D_a versus healing time at various frequencies (from 0.05 rad/s to 100 rad/s) (The dotted line is a virtual line used to delimit the frequency zone at $1/\tau_{rep}$) and (b) D_s versus angular frequency in log-log scale ranging from 0.05 rad/s to 100 rad/s at 200 °C for PMMA/PMMA bilayer structure

In addition, when we plot the self diffusion coefficients(D_s) (i.e. the plateau value of D_a) versus frequency in a log-log scale, we can see that it agrees well with Doi-Edwards theory[9], that is, D_s is proportional to the square root of ω when ω is higher than $1/\tau_e$ and independent of ω when ω is higher than $1/\tau_{rep}$ (Fig.2b). Apart from the difference in absolute values, a similar relation $D_s \sim \omega^{1/2}$ was also observed for Rouse model at $\omega > 1/\tau_e$, which was in perfect accord with the Doi-Edwards theory. The above discussion have made it clear that only the diffusion coefficient determined at the terminal zone represents the self-diffusion coefficient of the entire material.

Moreover, our experimental data indicated that the temperature dependence of D_s can be described by Arrhenius law. The data were also found to be consistent with those determined by Graessley's model [10] from zero-shear viscosity as well as those reported in literatures [8].

Interdiffusion process at asymmetrical bilayer structures

Firstly, the thermorheological complexity of the PMMA/PVDF couple was probed before rheological modeling. The monomeric friction coefficient of each species in the blend, $\zeta_i(\varphi,T)$, was extracted from the rheological data of blends via "double reptation" model and theoretically predicted using Lodge-Mcleish model[11-12] (Figure 3). The $\zeta(\varphi,T)$of PMMA and PVDF components in the blend have been examined to vary with composition and temperature and to be close in the present experimental conditions, at least, lying in the same order of magnitude.

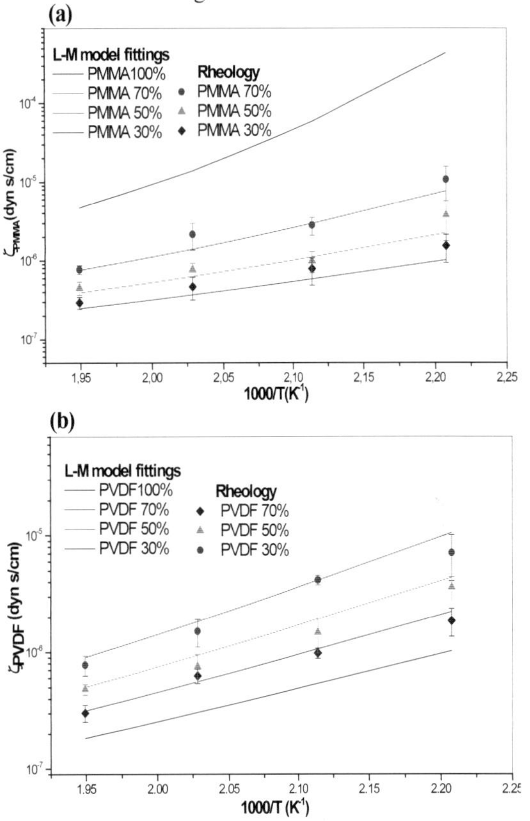

Figure 3-Monomeric friction coefficient $\zeta(\varphi,T)$ in mixed state extracted from rheology experiments of blends and those predicted using Lodge-McLeish model plotted versus temperature for (a) PMMA and (b) PVDF.

Provided the similar $\zeta(\varphi,T)$ of each component in the PMMA/PVDF mixtures, a quantitative model that connected the mobility in the mixed state to the properties of the matrix was used for the mutual diffusion coefficient [7]:

$$D_m = \frac{N_b^e N_b^3 b_b^4 K^{1/2}}{\pi^2 e_b^2}\left[\frac{(1-\varphi)}{N_A}+\frac{\varphi}{N_B}\right]\left[\frac{(1-\varphi)}{N_A}+\frac{\varphi}{N_B}-2\chi\varphi(1-\varphi)\right] \quad (3)$$

with
$$K=\omega^2\left[\left(\frac{8G_N^0}{\pi^2 G^*(\omega)}\right)^2-1\right] \quad (4)$$

where N_b^e, N_b, b_b, e_b and $G_{N,b}^0$ are the average number of repeat units between entanglements, the repeat unit numbers, the effective bond length, the step length of the virtual tube and the plateau modulus of the interphase (or equivalent blend), which is also a function of the composition.

For a multilayered polymer system, the rheology can be described as follow[3]:

$$\frac{1}{G_{s,t}^*}=\frac{\varphi_{A,t}}{G_{A,t}^*}+\frac{\varphi_{B,t}}{G_{B,t}^*}+\frac{\varphi_{I,t}}{G_{I,t}^*}=\frac{h_A^{'}/H}{G_{A,t}^*}+\frac{h_B^{'}/H}{G_{B,t}^*}+\frac{h_I^{'}/H}{G_{I,t}^*} \quad (5)$$

where $G_{i,t}^*$ .and $\varphi_{i,t}$ represent respectively the complex modulus and volume fractions of layer i at a diffusion time t>0 with an initial complex modulus of layer i as $G_{i,0}^*$ at t=0; $h_i^{'}$ represents the thickness of layer i developing with the elapsed time, and H is the total thickness of the sandwich (1.2 mm for the PMMA/PVDF assembly used in their work). The expression relating the interphase thickness to the mutual diffusion coefficient is:

$$h_I^{'}=2(D_m t)^{1/2} \quad (6)$$

which reduces eq.(5) to

$$\frac{1}{G_I^*(t)}=\frac{H}{2(D_m t)^{1/2}}\left(\frac{1}{G_{s,t}^*}-\frac{1}{G_{s,0}^*}\right)+\left(\frac{\varphi_A}{G_{A,0}^*}+\frac{\varphi_B}{G_{B,0}^*}\right) \quad (7)$$

Since the interdiffusion process occurs in the interfacial region (the interphase). In this study, the D_m was calculated on the basis of the rheology of the interphase by substituting $G^*(\omega)$ in eq.(4) with the complex modulus of the interphase $G_I^*(t)$ from eq.(7) and rearranging eq.(3), a new expression for the mutual diffusion coefficient was obtained:

$$D_m=\left[\frac{(2/3)^{1/3}p}{\left(9q+\sqrt{3}\sqrt{-4p^3+27q^2}\right)^{1/3}}+\frac{\left(9q+\sqrt{3}\sqrt{-4p^3+27q^2}\right)^{1/3}}{2^{1/3}3^{2/3}}\right]^2 \quad (8)$$

with

$$p=\frac{8\delta\omega G_N^0}{\pi^2}\left(\frac{\varphi_A}{G_{A,0}^*}+\frac{\varphi_B}{G_{B,0}^*}\right) \quad (9)$$

$$q=\frac{8\delta\omega G_N^0}{\pi^2}\frac{H}{2t^{1/2}}\left(\frac{1}{G_{s,t}^*}-\frac{1}{G_{s,0}^*}\right) \quad (10)$$

where $\delta=\frac{N_b^e N_b^3 b_b^4}{\pi^2 e_b^2}\left(\frac{\varphi_B}{N_A}+\frac{\varphi_A}{N_B}\right)\left(\frac{\varphi_B}{N_A}+\frac{\varphi_A}{N_B}-2\chi\varphi_A\varphi_B\right) \quad (11)$

Fig.4 depicts the evolution of the determined mutual diffusion coefficient (D_m) versus annealing time for the PMMA/PVDF bilayer system in this study. A D_m value of 7.08×10^{-9} cm^2/s was obtained at the equilibrium state at 220°C and w=1.0rad/s. This value was somewhat greater than that determined from Qiu-Bousmina's model and was reported to be comparable to the results presented in the literature as obtained by sophisticated methods [2].

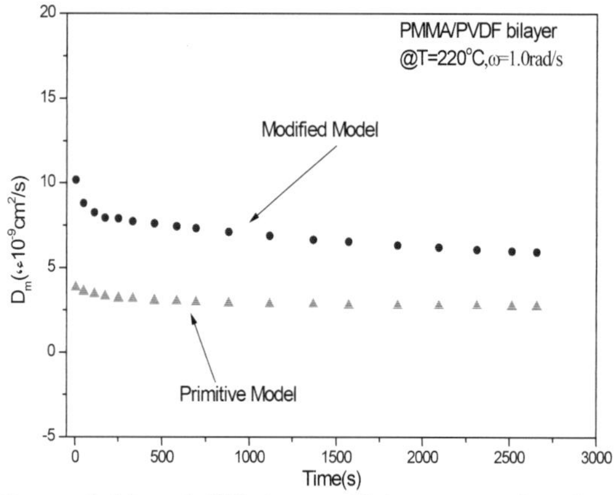

Figure 4- Mutual diffusion coefficient versus time for the PMMA/PVDF bilayer at 220 °C,ω=1.0 rad/s.

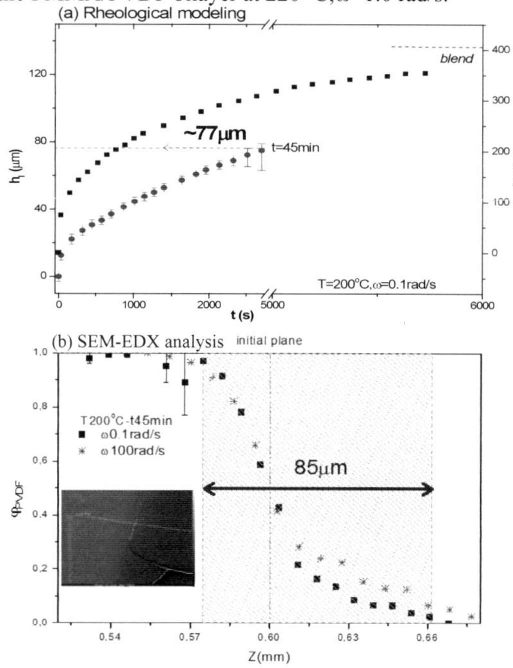

Figure 5- (a) The G*$_I$(t) and the h_I(t) of the interphase with time for a PMMA/PVDF assembly at T=200 °C, ω=0.1 rad/s. (b) Concentration profile of PVDF vs. position of quenched specimens at ω=0.1 rad/s and ω=100.0 rad/s, respectively(SEM-EDX result); the shadow zone designates the interphase (~85um) of the specimen at ω=0.1 rad/s.

Moreover, viscoelastic properties and thicknesses of the interphase could be quantified on the basis of the modified model. Fig. 5a shows a typical sample of the resulting thicknesses and the resulting complex modulus of the interphase triggered in a PMMA/PVDF assembly at 200 °C. w=0.1rad/s. After mutual diffusion for 45min, one can observe that the thickness of the interphase could reach several dozens of microns (e.g., ~77 μm at 200 °C), a similar order of amplitude as that found in the literature. The G*$_I$(t) of the interphase experienced a rapid initial increase before rising at a moderate rate and then approaching those of its equivalent blend. This implies that a short-time interdiffusion along with the establishment of entanglements by interchains within the interphase gave rise to greatly enhanced properties of the interphase. Moreover, the experimental concentration profile of the interphase by SEM-EDX was compared to the mathematical one calculated from the D_m of rheology (Fig.5b). The experimental result of the interphase thickness (~85um) at same condition indicated a good correlation between theory and experimental results.

Conclusions

In this study, rheometry has been demonstrated to be a reliable tool for monitoring the interdiffusion process at symmetrical and asymmetrical polymer/polymer interface and exploring the interfacial rheology of the generated interphase. Quantitative rheological models based on the viscoelastic variations of the bilayer assemlbly were developed for self diffusion coefficient and mutual diffusion coefficient, respectively. Effects of the annealing factors and structural properties on the diffusion kinetics and interphase development have been focused within the framework of Doi-Edwards theory. For asymmetrical bilayer system, a new and original modification of the model consists on the calculation and the modeling of the friction coeifficients of each component in the multiphase systems which are correlated to the diffusion quantifications. The D_m corroborates with the prediction of the fast-mode theory as well as the literatures. Rheological modeling was validated by the SEM-EDX analysis in terms of interphase thickness and concentration profile.

References

1. J.K. Kim; C.D. Han Polym. Eng. Sci.1991, 31, 258
2. H. Zhang; K. Lamnawar; A. Maazouz Macromolecules (Accepted) Doi: http://dx.doi.org/10.1021/ma301620a
3. K. Lamnawar; A. Baudoin; A. Maazouz Eur. Polym. J. 2010, 46, 1604
4. K. Lamnawar, M. Bousmina, A. Maazouz. Macromolecules, 2012, 45 (1), 441–454
5. E.J. Kramer; P. Green; C.J. Palmstrom Polymer 1984, 25, 473
6. R. Schnell; M. Stamm; C. Creton Macromolecules 1998, 31, 2284
7. H. Qiu; M. Bousmina Macromolecules 2000, 33, 6588
8. H. Zhang; K. Lamnawar; A. Maazouz Rheol Acta. 2012, 51,691
9. M. Doi; S.F. Edwards The theory of polymer dynamics. Clarendon, Oxford,1986
10. W.W. Graessley J Polym Sci Polym Phys Ed, 1980, 18, 27
11. Lodge, T. P.; McLeish, T. C. B.; *Macromolecules* 2000, *33*, 5278-5284
12. Watanabe, H.; Chen,Q.; Kawasaki,Y.; Matsumiya Y. *Macromolecules* 2011, *44*, 1570–1584.

EXTENSIONAL RHEOLOGICAL ANALYSIS OF BIODEGRADABLE POLYLACTIDE-NANOGRAPHITE PLATELET NANOCOMPOSITES

Esmaeil Narimissa, Rahul K.Gupta*, Nhol Kao and Sati N. Bhattacharya
Rheology and Materials Processing (RMPC) Centre, School of Civil, Environmental and Chemical Engineering, RMIT University, Melbourne, Vic, Australia
**Corresponding author: E-mail: rahul.gupta@rmit.edu.au*

Abstract – This paper discusses the production and rheological investigation of polylactide (PLA) and nanographite platelets (NGP) based bionanocomposites. Melt intercalation mixing process was used to produce PLA/NGP composites at different filler contents (1, 3, 5, 7 and 10 wt%). This method demonstrated a mechanical and rheological percolation threshold around 3 wt% NGP content. However, the morphological and thermal investigations of the samples exhibited poor dispersion of nanofillers, through the presence of tactoids and agglomerates, as well as considerable decreases in the crystallinity of the composites. Uniaxial extensional experiments have been carried out to analyse the impacts of the nanofillers on linear viscoelastic envelope (LVE) and non-linear viscoelastic behaviour (strain hardening region) of the nanocomposites. Finally, the modelling of strain-hardening behaviour of neat PLA and its nanocomposites have been investigated using steady shear viscosity, relaxation spectrum and damping function based on Papanastasiou-Scriven- Macosko (PSM) version of Kaye-Bernstein-Kearsley-Zapas (K-BKZ) model.

Keywords: PLA, Nanographite platelets, Extensional rheology, Strain hardening

Introduction

Homogenous dispersion and alignment of reinforced fillers have significant contributing roles in determining the flow direction in polymeric composites [1]. The presence of extensional deformation as well as shear flow in polymeric composites makes it indispensable to investigate both shear and extensional properties of such systems in order to design processing methods for fibre spinning, film and sheet production, etc. Due to enormous difficulties involved with the measurement of extensional viscosity of polymer melt, the presence of a substantial knowledge gap in the study of extensional flow properties of polymers has remained broad [2]. Polymer melt extensional flow occurs when the material is not in contact with solid boundaries i.e., drawing of sheets, filaments of fibre spinning and film blowing. There are two distinct regions in extensional viscosity measurement of the polymeric samples. The first region is called Linear Viscoelastic Envelope (LVE) where a linear and steady increase of elongation viscosity (η_E) occurs at elongation time (t_E). This region is independent of extensional Hencky strain rate ($\dot{\varepsilon}$) and follows Trouton's ratio (η_E at LVE $\approx 3\ \eta_0$) and is followed by an up-rising time ($t_{\eta E}$) at which a sharp and rapid upward deviation of η_E occurs. The rapid increase of the extensional viscosity in this region follows a nonlinear pattern and is called the strain hardening region [3]. Trouton's ratio can be satisfied at lower extensional rates; however, this could deviate for polymeric melts at higher extensional rates [2]. The occurrence of strain hardening behaviour during extensional measurements is a function of Hencky rate as well as the material properties of the polymeric matrix such as branch content, width of relaxation time distribution and molecular weight distribution (MWD). As a rule of thumb, the higher the Hencky rate, the smaller the uprising time of strain hardening region. Hence, the extensional analysis of polymeric materials would lead to thorough understanding of the flow behaviour of the systems and it is essential for the developments and applications of polymer nanocomposites.

The K-BKZ rheological constitutive integral equation has been broadly used to investigate the linear and nonlinear rheological characteristics of several polymer melts. In this study, the applicability of constitutive equations for modelling of extensional viscosity behaviour of PLA/NGP composites is examined. The samples undergo controlled and highly accurate uniaxial extensional measurements by means of Meissner-type extensional rheometer (Rheometrics Extensional Rheometer (RME)) [4].

Theoretical Background

K-BKZ constitutive equation modelling

The following equation presents the K-BKZ model used in this research:

$$\sigma(t) = \int_{-\infty}^{t} \mu(t,t')C^{-1}(t,t')h(I,II)dt' \qquad \text{(Eq. 1)}$$

Where $\sigma(t)$ is a stress tensor, $\mu(t,t')$ is a memory function, h(I, II) is a damping (strain) function , $C^{-1}(t,t')$ is the fingers strain tensor which is the transpose of Cauchy strain tensor and I, II are the first and second invariants of the Finger strain tensor. The memory function in terms of generalized Maxwell model is usually expressed as an exponentially fading term, i.e.

$$\mu(t,t') = \sum \frac{G_i}{\lambda_i} \exp(-(t-t')/\lambda_i) \qquad \text{(Eq. 2)}$$

Where, G_i and λ_i are relaxation strength and time components from relaxation spectrum of the material. The damping function can be experimentally measured from the stress-relaxation test; hence, the damping function in shear deformation is [5]:

$$h(I,II) = h(\gamma) \qquad \text{(Eq. 3)}$$

Where $h(\gamma)=\frac{G(t,\gamma)}{G^0(t)}=\frac{1}{(1+\alpha\gamma^2)}$ and G(t, γ) and $G^0(t)$ are the relaxation modulus of the samples in nonlinear and linear shear flow, respectively. Finally, the extensional viscosity of the samples is determined from the following equation:

$$\eta_E^+(t)=\frac{\sigma(t)_1-\sigma(t)_2}{\dot{\varepsilon}}=\frac{1}{\dot{\varepsilon}}\int_{-\infty}^{t}\{\sum\frac{G_i}{\lambda_i}\exp(-(t-t')/\lambda_i)\}\frac{\alpha}{(\alpha-3)+\beta I+(1-\beta)II}(e^{2(\dot{\varepsilon}(t-t')}-e^{(-\dot{\varepsilon}(t-t'))})dt'$$

(Eq. 4)

Experimental

Materials

Poly (L,L-lactide) was supplied by NatureWorks LLC and the grade of PLA used was 3051D with melt index and specific gravity of 10-30g/10min and 1.24, respectively. Nanographite platelets were supplied by XG Sciences, Inc. US Michigan and the grade of NGP used for this study was "M" with characteristics: average thickness of approximately 6 - 8 nanometers and a typical surface area of 120 to 150 m^2/g.

Processing

PLA pellets and nanographite platelets were dry mixed in the desired composition before melt blending in 700g batches. Table 1 shows the compositions (nominal wt%) and the codes of neat PLA and PLA/NGP samples (hereafter, samples are referred to according to their sample codes). Samples were melt-blended in a Brabender twin screw extruder at 40 RPM and 180 °C. Dried pellets were compression moulded (at 180 °C, 80 kN for 5 minutes) into 2 mm thick circular plaque with 20 mm diameter specimens. In addition, the pellets were moulded into 57 x 7 x 1.5 mm^3 rectangular cavities to be tested in RME.

Table 1: Compositions of PLA/NGP composites and their corresponding sample codes.

Sample Compositions	Neat PLA	PLA01	PLA02	PLA03
PLA (wt%)	100	99	98	97
NGP (wt%)	0	1	2	3

Extensional Rheology

The extensional uniaxial viscosity of the samples were measured at constant strain rates of 0.1, 0.5, 1 S^{-1}. The measuring temperature was the same as the shear measurements (150°C). The measurements were carried out via Meissner-type extensional rheometer (Rheometrics Extensional Rheometer (RME)) [4].

Results and Discussion

Stress Relaxation test

The shear damping function of samples, h(γ), was evaluated through computing the vertical shift required to superpose the relaxation curves on the log-log plot of relaxation modulus versus time for both linear and nonlinear strain regions (Eq.3). The value of damping function starts from unity in linear viscoelastic strain region and decreases gradually in nonlinear region. Fig. 1 demonstrates the time-strain superposition of relaxation modulus of neat PLA both inside and outside the linear viscoelastic region. It shows a plot of relaxation modulus versus time at strains from 0.1-40. The application of time-strain separability leads to the presence of two distinct regions of strain-dependent (t = 0.01-0.2 s) and time-dependent (t = 0.2-1000 s) regions. Although the time-strain separability occurred at t = 0.2 s, the results of this study is in agreement with previous studies of similar systems where the time-dependency occurred over a broad and complex time range [2, 6].

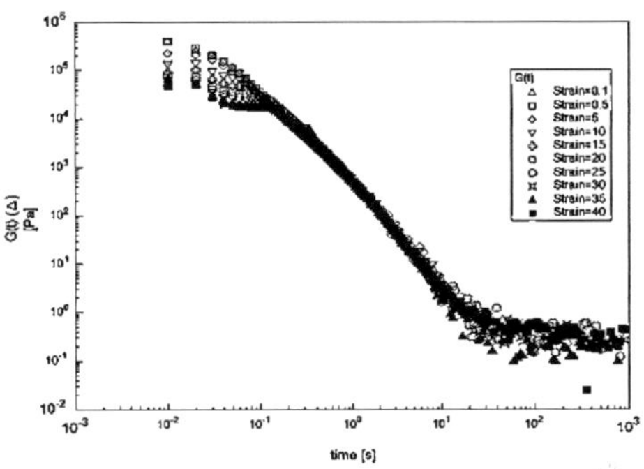

Fig. 1: Evaluation of linear and non-linear relaxation moduli of neat PLA via stress-relaxation test at different step shear strain (γ).

Fig. 2 shows the shear damping functions of neat PLA and composite samples evaluated from Eq. 3. The composites displayed a moderate drop in their material parameter α from neat PLA; it was not a significant decrease (Table 2). However, the large magnitude of parameter α indicates the strength of stain hardening behaviour of PLA matrix.

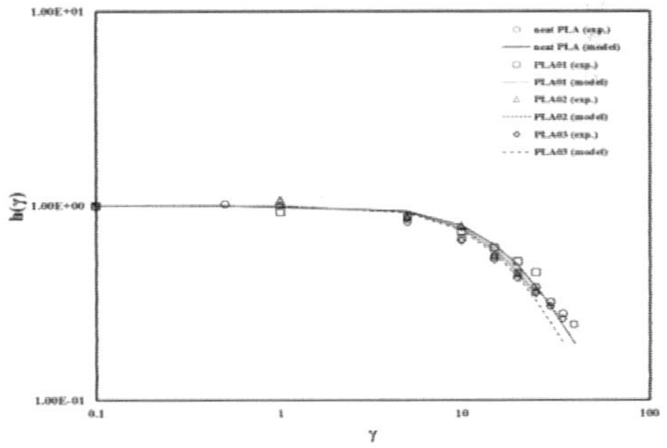

Fig. 2: Determination of shear damping function of the samples and computational evaluation of parameter α.

Table 2: Estimated values of "α" of the samples.

Sample	Neat PLA	PLA01	PLA02	PLA03
α	385	349	323	294

Extensional Viscosity

Figs. 3-5 show the extensional viscosity of the samples versus their elongation time at Hencky rates 0.1, 0.5, 1 s^{-1}. The results did not express a considerable variation between neat PLA and its composites at different Hencky rates. The samples demonstrated a pronounced presence of strain hardening behaviour at all Hencky rates. However, the occurrence of strain hardening behaviour was highly dependent on the magnitude of Hencky rate. As the magnitude of Hencky rate rises, the $t_{\eta E}$ falls and consequently the Hencky strain remains constant (ε~3). The absence of considerable deviation in extensional behaviour of neat PLA and its composites is consistent with shear rheological results at the same temperature (150 ± 2 °C).

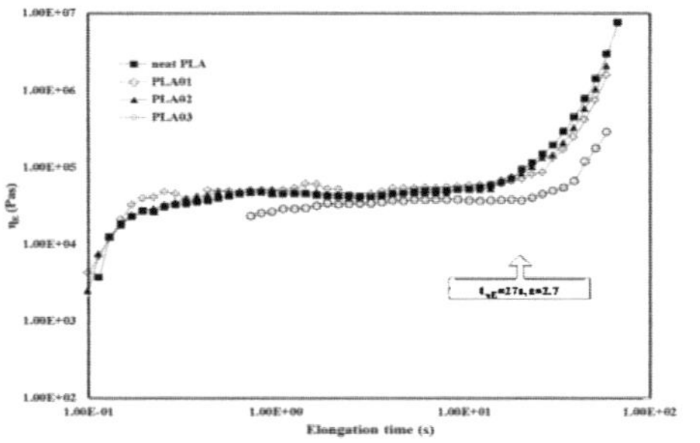

Fig. 3: Extensional viscosity growth of samples under Hencky rate=0.1 s^{-1} in transient extensional viscosity test at 150 °C.

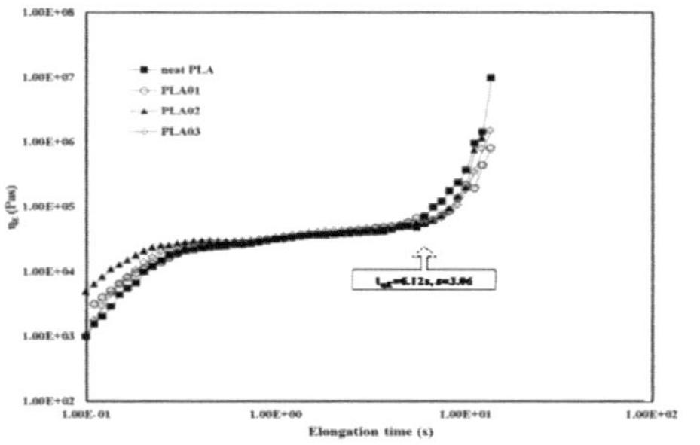

Fig. 4: Extensional viscosity growth of samples under Hencky rate=0.5 s^{-1} in transient extensional viscosity test at 150 °C.

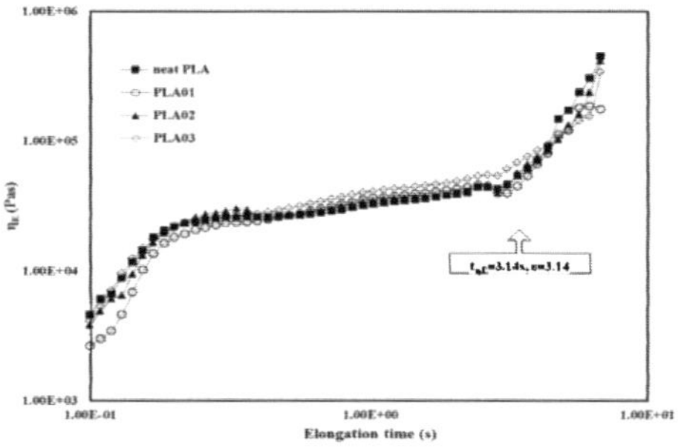

Fig. 5: Extensional viscosity growth of samples under Hencky rate=1 s^{-1} in transient extensional viscosity test at 150 °C.

Nevertheless, at an elevated temperature of 180 °C shear rheological analysis of the samples revealed a noteworthy difference between neat PLA and PLA composites at which the existence of a percolation threshold became evident at around 3 wt% filler content. This lack of difference in extensional behaviour could be due to the dispersion of nanofillers within polymeric matrix and its direct relationship with the testing temperature. Although the extensional tests were performed at the melting temperature of the polymeric matrix, the dispersion of NGP nanofiller was not improved at this temperature and the fillers behaved as isolated islands of agglomerates within the neat PLA matrix.

K-BKZ Modelling

Figs. 6-9 demonstrate the application of K-BKZ extensional modelling on the samples' experimental extensional results. The model was evaluated through PSMLT damping function of neat PLA and PLA01-03. The material parameter β signifies the existence of the steady state extensional behaviour at longer elongation time. Even though the steady state behaviour was not achieved in this research, the model suggests that this behaviour is expected to occur at longer elongation times. One of the constraining factors regarding the absence of such behaviour (before sample rupture) could be the small size of the samples in RME. Neat PLA, PLA01 and PLA02 samples showed satisfactory agreement with K-BKZ model; yet, the PLA03 did not conform well into this model. The poor behaviour of PLA03 composites at all strain rates (0.1, 0.5, 1) could suggest that the incorporation of NGP platelets into the PLA matrix has an extensional rheological percolation threshold of 2 wt%.

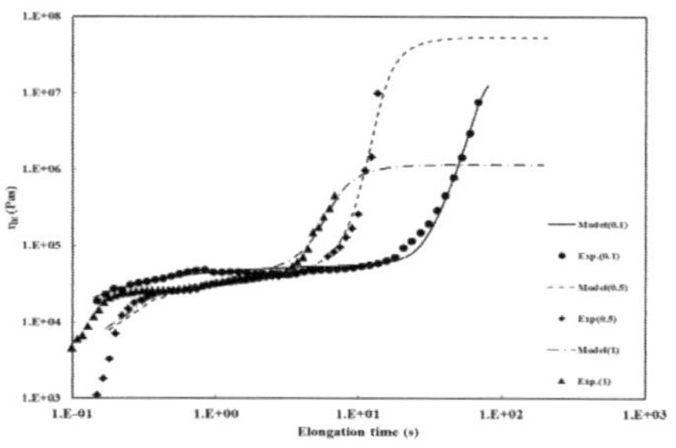

Fig. 6: Extensional viscosity growth predicted by PSMLT damping function of K-BKZ model at different Hencky rates for neat PLA at 150 °C.

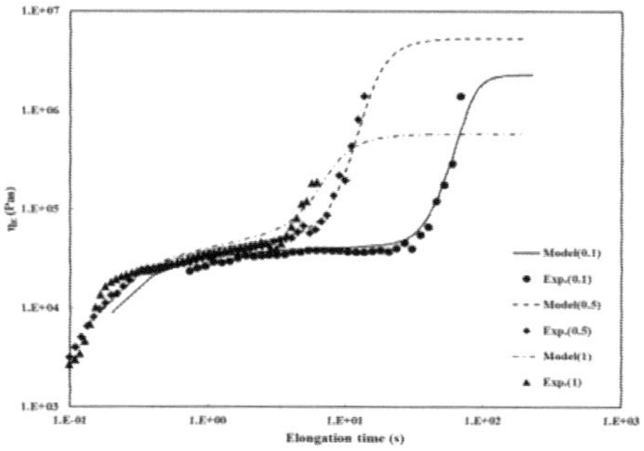

Fig. 7: Extensional viscosity growth predicted by PSMLT damping function of K-BKZ model at different Hencky rates for PLA01 at 150 °C.

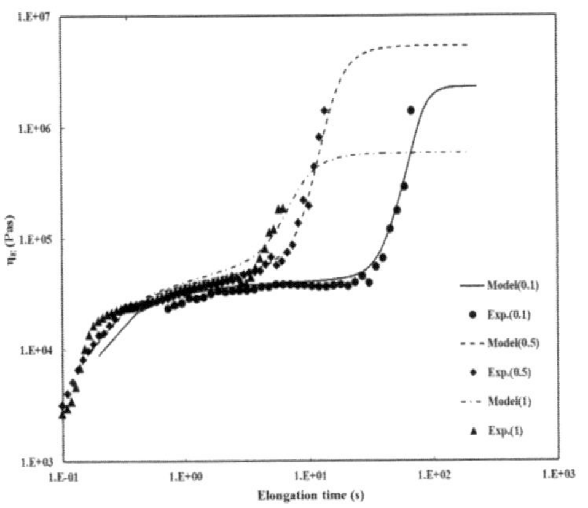

Fig. 8: Extensional viscosity growth predicted by PSMLT damping function of K-BKZ model at different Hencky rates for PLA02 at 150 °C.

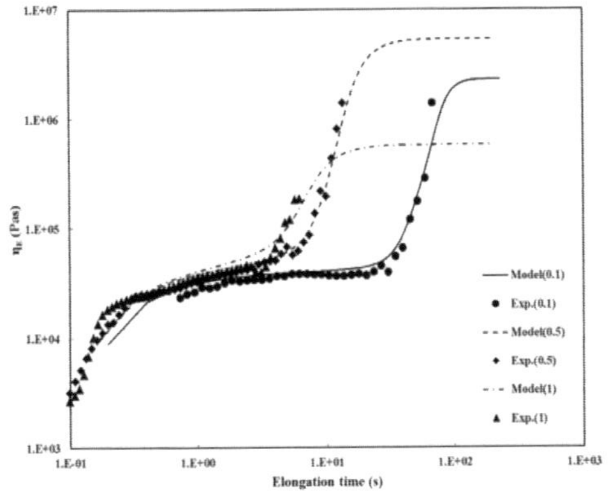

Fig. 9: Extensional viscosity growth predicted by PSMLT damping function of K-BKZ model at different Hencky rates for PLA03 at 150 °C.

According to Papanastasiou, Kasehagan, Macosko and Isaki, the values of parameter α increases with increasing the branch content and it is related to the width of relaxation time distribution branching and molecular weight distribution (MWD) [7-9]. Therefore, the strength of the strain hardening characteristic of the polymeric system has a direct relationship with the value of α. On the other hand, the value of material parameter β has an inverse correlation with the same phenomena; i.e., the closer it is to zero, the more accentuated the strain hardening behaviour of the material [8]. As the results of this study suggest, the diminishing values of α indicate that the addiction of NGP platelets to the PLA matrix could result to a gradual decrease in the strength of the strain hardening behaviour of PLA/NGP composites.

Moreover, the noteworthy deviation of extensional viscosity of the composites at elevated level of filler contents (> 2 wt%) from K-BKZ model also suggest that the incorporation of plate-like nanofillers such as NGP could have detrimental effect on the extensional behaviour of nanocomposites. Thus, the enhanced shear rheological, mechanical, electrical, thermal properties [10-12] as well as augmented barrier properties of the PLA/NGP composites must be compromised with their reduced extensional properties should the extensional needs rise.

Conclusions

In order to determine the uniaxial extensional behaviour of neat PLA and PLA/NGP nanocomposites, the samples underwent shear and extensional analysis at 150 °C. The extensional viscosity of the samples was determined at Hencky strain rates of 0.1, 0.5 and 1 s^{-1}. Dynamic frequency sweep tests in linear viscoelastic region as well as stress-relaxation tests in both linear and non-linear regions led to the determination of the shear damping function and consequently parameter α of the samples. By means of PSMLT version of K-BKZ constitutive extensional model, the validity of the experimental tests in addition to the evaluation of parameters α and β of the samples was accomplished. Results suggest that the addition of nanofillers to the composites may lead to a reduction in strain hardening behaviour of the composites. Furthermore, the extensional rheological percolation threshold of the composites, the limit of NGP doping where the validation of KBK-Z model was satisfied, was determined to be at around 2 wt%.

References

1. V. Pasanovic-Zujo; R.K. Gupta; S.N. Bhattacharya International Polymer Processing 2004, 19, 388-394.
2. S.H. Lee; S.Y. Kim; J.R. Youn Polymer Composites 2009, 30, 1426-1436.
3. S. Sinha Ray Journal of Industrial and Engineering Chemistry 2006, 12, 811-842.
4. J. Meissner; J. Hostettler Rheologica Acta 1994, 33, 1-21.
5. A. Nishioka; T. Takahashi; Y. Masubuchi; J.I. Takimoto; K. Koyama Journal of Non-Newtonian Fluid Mechanics 2000, 89, 287-301.
6. M.T. Islam; J. Sanchez-Reyes; L.A. Archer Journal of Rheology 2001, 45, 61-82.
7. A.C. Papanastasiou; L.E. Scriven; C.W. Macosko Journal of Rheology 1983, 27, 387-410.
8. T. Isaki; T. Takahashi Journal of the Society of Rheology, Japan 2003, 31, 185-194.
9. L.J. Kasehagen; C.W. Macosko Journal of Rheology 1998, 42, 1303-1327.
10. E. Narimissa; M. Jollands; N. Kao; R. Gupta, PPS27, Morocco p. 1-8.
11. E. Narimissa; R.K. Gupta; H.J. Choi; N. Kao; M. Jollands, Polymer Composites 2012, 33, 1505-1515.
12. E. Narimissa; R. Gupta; M. Bhaskaran; S. Sriram Polymer Degradation and Stability 2012, 97, 829-832.

O-08-500

RHEOLOGICAL SIMULATION OF LIQUID CRYSTAL POLYMERS (LCPs) AT HIGH SHEAR RATES

Ahmed Rahman, Rahul K Gupta, Sati N. Bhattacharya, Shishir Ray[1] and Franco Costa[1]
Rheology and Materials Processing Centre,
School of Civil, Environmental and Chemical Engineering,
RMIT University, Melbourne 3001, Vic, Australia
[1] Autodesk Pty Ltd.
259-261 Colchester Rd Kilsyth 3137, Vic, Australia
Corresponding Author's E-mail: rahul.gupta@rmit.edu.au

Abstract - The rigid and rod like molecules of Liquid Crystal Polymers (LCPs) orientate in different ways at different shear rates. Under low shear rate, the molecules tend to align in the direction of the flow as well as tumble around their own axis. The abnormal orientation of molecules also depends upon temperature. These behaviours lead to unusual rheological properties of LCPs. But with the increase of shear rate, the molecules are oriented in the direction of flow. The molecular orientation at high shear rate is considered as anisotropic flow. Recently developed Leonov's constitutive theory of weakly viscoelastic nematodynamics has been considered as the standard local approach of thermodynamics to describe the rheological properties of LCPs. The complicated rheological properties and characteristic anisotropic properties of LCPs are modeled by a set of quasi–linear viscoselastic constitutive equation. Simulation has been carried out in Mathematica software and the characteristic anisotropic properties of LCPs have been identified at low and high shear rates. The experimentally measured viscosities at high shear rate have been validated with model predictions.

Keywords: Anisotropic flow, high shear rate, Leonov Model

Introduction

In the last two decades, liquid crystalline polymers (LCPs) have received a great deal of attention from both industry and academiaTAI-S [1]. LCPs whose phase transition to the liquid crystalline phase occurs under change in temperature, are called thermotropic while a phase transition by changing the polymer concentration in a solvent (as well as temperature) is called lyotropic liquid crystal [2]. Three major mesophases have been identified to describe the liquid crystalline phase: nematic, cholestoric and smectic [3]. In nematic phase, molecules rotate in long range orientational order since it has no positional order. But in cholestoric phase, molecules possess different positional order and different positions exhibit different orientational order. In smectic phase molecules rotate with positional and orientational order[4].

For Lyotropic LCPs, by increasing the shear rate from very small values to very large ones, the viscosity first decreases (region I), then levels off (region II), and again decreases (region III)[5]. In region I, LCPs exhibit shear thinning behaviour at low shear rate, i.e., the viscosity decreases with increasing shear rate. The rigid rod like molecules can easily tumble around their own axis with low shear. On the other hand, the conventional polymer show plateau viscosity at the low shear rate region. Then the viscosity is almost constant with shear rate (Newtonian behavior), which is referred to the region II. With the increase of shear rate, coalescence of the domains of molecules also increased. For this reason, the number of small domains also reduced. At region III (very high shear rate zone), the molecules of LCPS start to move with flow aligning directions, which leads to strong shear thinning behaviour[6].

The orientation and direction of LCP molecules depend upon the shear rate. In the low shear rate region the molecules exhibit pronounced oscillations or negative first normal stress difference. All the abnormal orientation of rigid rod like molecules ceases with increase of shear rate. The anisotropic structure (different types of orientation) of molecules can be represented by a dimensionless unit vector ***n*** called the *director* i.e. average direction of rods. The unit vector ***n*** represents the direction of preferred orientation of molecules in the neighborhood of any point. Because there is no physical polarity along the director axis, ***n*** and ***-n*** are fully equivalent [4]. The action of external field or flow causes the orientation of mesogens, which induces uniaxial anisotropy, with an additional degree of freedom, the internal rotation [7]. The directors' exhibit tumbling, kayaking and wagging, flow-aligning, or log-rolling types of motion depending on the applied shear rates.

The above unusual characteristic of LCP molecules makes the rheological properties of LCPs exceptional compared to conventional polymers. The viscosities of LCPs also exhibit abnormal temperature dependency indicating that LCPs are not thermorheologically simple[8]. In the last two decades several rheological models have been developed to take these factors into account. The oldest is the Leslie Ericson (LE) phenomenological theory. The LE theory was developed based on the viscous stress (Erickson theory), which was later combined with Frank theory for the elastic stress by Leslie. LE theory does not consider the non-linear viscoelastic region for LCPs but it is very useful for explaining the role of polydomain textures on rheological properties of LCPs. The LE theory is limited to only slow flows (low shear rates) and weak molecular distortions [9].The other abnormal rheological properties of LCPs cannot be explained by the LE theory.

The Doi theory takes into account the effects of flow, Brownian motion, and intermolecular forces on the molecular orientation distribution. Thus, it gives a good representation of the molecular viscoelasticity[10]. However, it does not include the so-called distortional elasticity, which represents the crystalline characteristic of these materials. Since there is no elastic penalty for spatial distortion, the Doi theory is valid only in the limit of spatial homogeneity[9]. It is especially noteworthy that spatial distortion of nematics often takes the form of orientational defects known as disclination, which is perhaps the most spectacular and significant feature of polymeric and small-molecule nematics [11]. The Doi rigid rod theory has been extensively tested[7] and has been successfully applied for lyotropic LCPs of small and moderate concentrations. On the other hand, the modelling of thermotropic LCPs cannot be done well using the Doi theory due to the complex nature of the equations [12-14].

Leonov (2008) developed a continuum theory of non-symmetric weak viscoelastic nematodynamics of Maxwell type fluid, where the viscous and elastic stresses coincide with the total stress. In the newly developed Leonov's viscoelastic nematic continuum theory, the assumptions of the smallness of transient (elastic) strains and relative rotations were taken into consideration. On the other hand, this theory avoided the well-known general balance equations of angular momentum for nematics as well as related rotational inertia effects [15-17]. These assumptions seem to be appropriate for LCPs with rigid enough macromolecules, and for slow flows of viscoelastic suspensions with uniaxially shaped particles. This theory utilized a specific viscoelastic and nematic kinematics and ignored inertia effects and in cases of LCPs, the effects of Frank elasticity[18].

The aim of this study is to find out the values of anisotropic properties of LCPs using the recently developed Leonov's viscoelastic nematodynamics model. This model, developed for flows of LCPs, has been numerically analysed for high shear steady flows of polymer nematics. Comparison between model predictions and experimental data has also been carried out.

Simulation of Model Parameters

The simulations of rheological data in this and following sections are based on the continuum theory of weak viscoelastic nematodynamics [13, 18, 19] in absence of any magnetic field. In absence of magnetic field, the theory is reduced to the symmetric stress and director. The de Gennes elastic potential (Helmoltz free energy density) can be expressed as [13].

$$f = \frac{1}{2} G_0 |\underline{\underline{e}}|^2 + G_1 \underline{nn}\,\underline{\underline{e}}^2 + G_2(\underline{nn} : \underline{\underline{e}})^2 - 2G_3 \underline{nn}(\underline{\underline{e}} . \underline{\underline{\Omega}}_e) - G_5 \underline{nn}\, \underline{\underline{\Omega}}_e^2 \qquad (1)$$

Here, G_K is moduli (K = 0, 1..5), $\underline{n}$ is director, $\underline{\underline{e}}$ is elastic strain and $\underline{\underline{\Omega}}_e$ is elastic rotation. The equation of symmetric stress can be expressed as

$$\underline{\underline{\boldsymbol{\sigma}}} = \delta f / \delta \underline{\underline{e}} = G_0 \underline{\underline{e}} + G_1 [\underline{nn} : \underline{\underline{e}} + \underline{\underline{e}} : \underline{nn} - 2\,\underline{nn} :(\underline{\underline{e}} . \underline{nn})] + 2(G_1 + G_2) \times (\underline{\underline{e}} : \underline{nn})(\underline{nn} - 1/3\underline{\underline{\delta}}) \qquad (2)$$

In simple shearing flow, equation (2) can be re written in component form as

$$\theta_0 \frac{d\sigma_{12}}{dt} + \theta_0 \dot{\gamma} \frac{\sigma_{11} - \sigma_{12}}{2} + \sigma_{12} + (r_1 - 1)[\sigma_{12} + \sigma_{11})] + (\sigma_{11} + n_1{}^2 + \sigma_{22} n_2{}^2 + 2\sigma_{12} n_1 n_2)$$

$$n_1 n_2 [\frac{3}{2}(r_2 - 1) - 2(r_1 - 1)] = \frac{\dot{\gamma}\eta_0}{2}(1 + \alpha + 2n_1{}^2 n_2{}^2 (\beta - 2\alpha - \frac{3}{2}))$$

3(a)

$$\theta_0 \frac{d\sigma_{11}}{dt} + \theta_0 \dot{\gamma}\sigma_{12} + \sigma_{11} + 2(r_1 - 1)(\sigma_{11} n_1^2 + \sigma_{12} n_1 n_2) + [\frac{3}{2}(r_2 - 1)(n_1^2 - 1/3) - 2(r_1 - 1)n_1^2)]$$

$$\times(\sigma_{11} n_1^2 + \sigma_{22} n_2^2 + 2\sigma_{12} n_1 n_2) = \dot{\gamma}\eta_0 n_1 n_2 (\alpha(1 - 2n_1^2) + (\beta - \frac{3}{2}) \times (n_1^2 - \frac{1}{3})$$

3(b)

$$\theta_0 \frac{d\sigma_{22}}{dt} + \theta_0 \dot{\gamma}\sigma_{12} + \sigma_{22} + 2(r_1 - 1)(\sigma_{22} n_2^2 + \sigma_{12} n_1 n_2) + [\frac{3}{2}(r_2 - 1)(n_2^2 - 1/3) - 2(r_1 - 1)n_2^2)] =$$

$$\times(\sigma_{11} n_1^2 + \sigma_{22} n_2^2 + 2\sigma_{12} n_1 n_2) = \dot{\gamma}\eta_0 n_1 n_2 (\alpha(1 - 2n_2^2) + (\beta - \frac{3}{2}) \times (n_2^2 - \frac{1}{3})$$

3(c)

Equations 11(a) to 11(c) were analytically solved in simple steady state shear flow ($\delta\sigma / \delta t = 0$) using "Mathematica" software [19]. The stress components σ_{11}, σ_{12} *and* σ_{22} are expressed as function of shear rate $\dot{\gamma}$ with the values of constitutive parameters $\theta_0, \alpha, \beta, r_1, r_2, \lambda_e, \lambda_v$ *and* η_0. Here θ_0 *and* η_0 represent relaxation time and zero shear viscosity respectively. The other parameters

λ_e *and* λ_v represent the tumbling for elasticity and viscosity. Rest of the characteristic parameters α, β, r_1, r_2 represent anisotropic properties. Rod like molecules tumble and oscillate under shear flow and r_1 and r_2 represent the oscillation during tumbling. The last six parameters cannot be measured by phenomenological theory. Among the eight parameters, only relaxation time and zero shear viscosity are determined from experimental data. The other six parameters can be obtained from curve fitting data using the software.

Materials

Most liquid crystalline molecules are highly anisotropic, and to a good approximation can be regarded as rigid rods. In general, the units of liquid crystal compounds consist of two or more aromatic rings [1]. Four aromatic nematic thermotropic LCPs in the form of pellets were chosen for this study. Due to propriety rights, only the major constituents of TLCPs are mentioned. The following nomenclatures are used for different filled and unfilled LCPs in our study.

1. LCP Polymer A Unfilled (mainly copolymer of p-hydroxy benzoic acid (HBA), 6-hydroxy-2 naphthoic acid (HNA), terephthalic acid (TA), hydroxyl quinone,HQ) and other chemicals
2. LCP Polymer A with 35% glass fibre.
3. LCP Polymer B Unfilled (copolymer of p-hydroxy benzoic acid (HBA), 6-hydroxy-2 naphthoic acid (HNA), terephthalic acid (TA) and hydroxyl quinone, HQ)
4. LCP Polymer B with 30% glass fibre.

Experiment:

The high shear rate rheology tests were conducted at the Autodesk research facilities using a CEAST double piston-driven capillary rheometer (with circular dies). The high shear rate ranges were approximately from 10 s^{-1} to 6000 s^{-1}. Two different temperature ranges were used for the two groups of polymers (A and B). The tests for LCP polymer A was conducted from 340^{0}C to 360^{0}C with 10^{0}C increments, and the tests for LCP polymer B was conducted from 330°C to 350°C with 10°C increments using different length to diameter (L/D) ratios. In this study, L/D ratios of 5, 20 and 40 were used. Bagley correction was applied to get the true shear stress and shear rate data.

Comparison of experimental data and model prediction

Figures 1 to 4 represent the experimental viscosity values and the viscosities from model predictions at the high shear rate region. It is observed that in the high shear rate region (10 -2000 s^{-1}) the model predictions show a good fit with the experimental data except at a very high shear rate (above 2000 s^{-1}) where this model predicted more shear thinning behaviour compared to the experimental data.

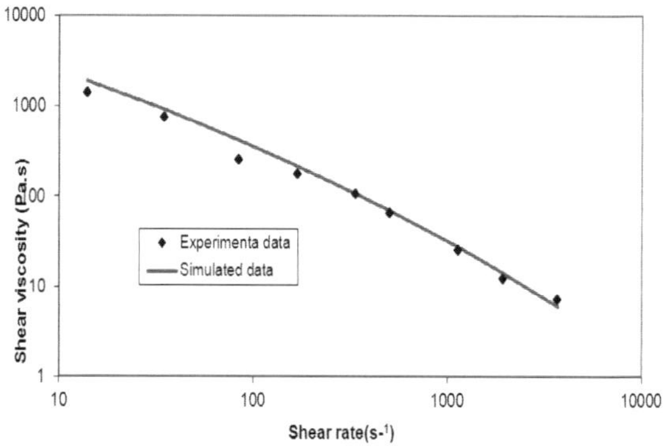

Figure 1: Shear viscosity of LCP A Filled at high shear rate @ 350°C.

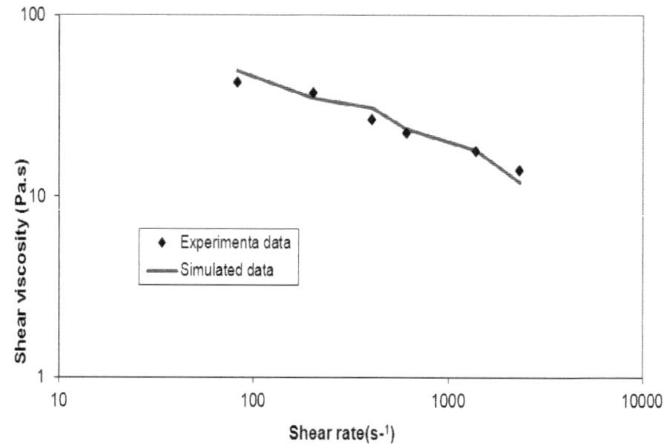

Figure 2: Shear viscosity of LCP A Unfilled at high shear rate @ 350°C.

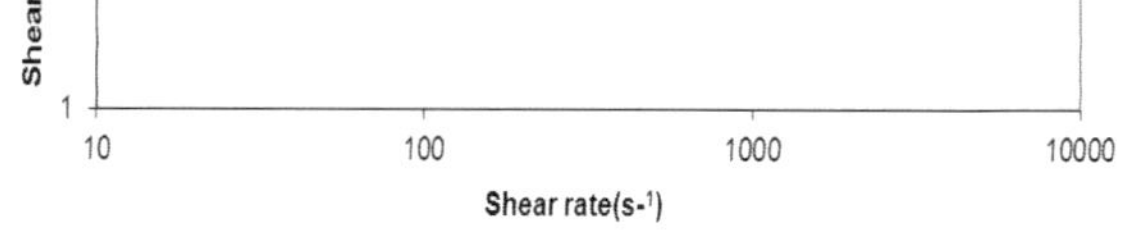

Figure 3: Shear viscosity of LCP B Filled at high shear rate @ 330°C.

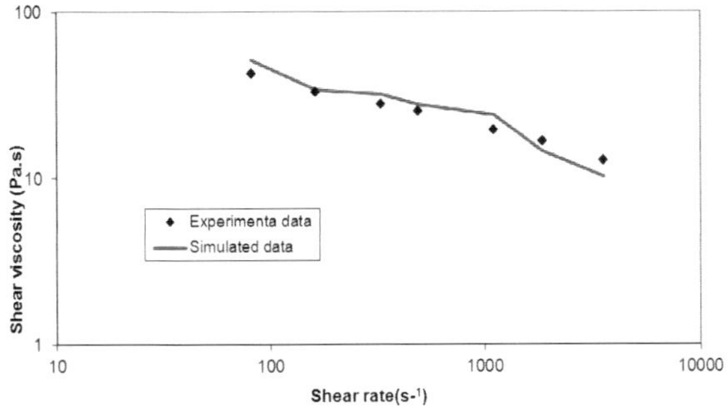

Figure 4: Shear viscosity of LCP B Unfilled at high shear rate @ 330°C.

Table 1: Model parameters for LCPs

	α	β	r_1	r_2	λ_e	λ_v
LCP A Filled	1.53	3.06	1.72	3.38	1.58	1.97
LCP A Unfilled	2.51	2.42	3.90	2.15	2.32	1.82
LCP B Filled	2.19	3.85	2.78	1.26	1.62	2.08
LCP B Unfilled	1.70	2.60	1.93	1.18	1.13	2.15

Conclusion:

Currently available theories such as Leslie-Ericksen theory and Doi theory for LCPs are not suitable to model the rheological behaviour of thermotropic LCPs. The Leonov model of weak viscoelastic nematodynamics combines the de Gennes-type elastic potential and the LEP viscous nematodynamic approach in one constitutive equation. Particularly this theory is modelled by a set of quasi-linear anisotropic viscoelastic constitutive equations (CE) with anisotropy described by a director, whose viscoelastic evolution is coupled with the CE. The numerical simulation for the flow of four LCPs has been carried out and the predicted viscosities have been compared with experimentally measured viscosity data, indicating a good fit in the shear rate range of 10 to 2000 s^{-1}

References

1. TAI-S, C. POLYMER ENGINEERING AND SCIENCE, 1986. **26**(13).
2. Brostow, W., *"An Introduction to Liquid Crystallinity" in Liquid Crystalline Polymers: From Structures to Applications,* , ed. A.A. Collyer. 1992, New York: Elsevier Applied Science.
3. Singh, S., *Phase Transitions in Liquid Crystals*, in *Physics Reports* 2000. p. 107-269.
4. Chandrasekhar, S., *Liquid Crystals*. 2nd ed. 1994, Cambridge: Cambridge University Press.
5. Onogi, S. and T. Asada, *Rheology*, ed. G. Astanta, G. Marrucci, and L. Nicolais. Vol. 1. 1980, New York: Plenum Press.
6. Muir , M.C. and R.S. Porter Mol. Cryst. Liq. Cryst., 1989. **169**: p. 83-95.
7. Larson, R.G., *The Structure and Rheology of Complex Fluids*. 1999, New York: Oxford University Press.
8. Brostow, W., et al. Journal of Chemical Physics, 1999. **110**(19): p. 9706-9712.
9. A. V. Bhave, et al. The Society of Rheology, 1993. **37**: p. 413-432.
10. Feng, J. and L.G. Leal. J. Rheol., 1997. **41**(6): p. 1317-1336.
11. Feng, J., C.V. Chaubal, and L.G. Leal. Journal of Rheology, 1998. **42**(5): p. 1095-1119.
12. Einaga, Y., G.C. Berry, and S.-G. Chu. Polymer (Japan), 1985. **17**: p. 239.
13. Leonov , A.I. Math Phys Anal Geom, 2008. **11**: p. 87-116.
14. Sgalari, G., G.L. Leal, and J. Feng. J. Non Newtonian Fluid Mech, 2002. **102**: p. 361-382.
15. de Gennes, P.G. and J. Prost, *The Physics of Liquid Crystal*. 2nd ed. 1993, Oxford: Clarendon Press.
16. Gennes, P.G. and J. Prost, *The Physics of Liquid Crystals*. 1993: Oxford: Clarendon Press.
17. Kle'man, M. Rep. Prog. Phys, 1989. **52**: p. 555-654.
18. Leonov, A.I. Z. angew. Math. Phys., 2008. **59**: p. 333-359.
19. Leonov, A.I. and H. Chen, *Modeling and Simulation in Polymers*, ed. P.D. Gujrati and A.I. Leonov. 2010, Weinheim: WILEY-VCH Verlag GmbH & Co. KGa, .

O-08-509

VISCOELASTIC FLOW SIMULATION OF POLYMER MELTS IN A ROTATIONAL-TYPE RHEOMETER

S. Tanoue[1*], H. Uematsu[1] and Y. Iemoto[1]

[1]*Graduate School of Engineering, University of Fukui, Japan – tanoue@matse.u-fukui.ac.jp*

Abstract - Polymer samples in the jigs begin to protrude when the heat is turned up for measuring the rheological characteristics of polymer melts using rotational type rheometers. To clarify the effect of this protruding part size on the estimated rheological characteristics, we tried to estimate the measurement error of rheological data when a polymer sample in the rotational type rheometer protrudes from the jig by using the viscoelastic flow simulation. The Phan-Thien Tanner (PTT) model was employed as the constitutive equation. The deviation of shear viscosity between the estimated value from the simulation results and that of the PTT model increased with the size of the protruding part. On the contrary, the existence of the protruding part has a little influence on the primary normal stress difference although the deviation of the primary normal stress difference also increased with the size of protruding part. Next, we discussed the case of frequency sweep testing using the parallel type rheometer. The storage modulus *G'* and loss modulus *G''* increases with the size of the protruding part of the polymer melt specimen at the same frequency. The absolute deviations of the estimated *G'* and *G''* using the simulation results were up to 20% at the protruding part size of 1 mm when the plate radius is 12.5 mm and the distance between the plates is 2 mm. In addition, we discussed the utility of the cup type jig for measurement of the flow characteristics using the viscoelastic flow simulation.

Keywords: Rotational-type rheometer, Viscoelastic flow simulation, Polymer melt

Introduction

It is convenient to use a disc sample made with injection molding and compression molding when the rheological properties of the polymer melt are measured with rotational-type rheometers, e.g., the parallel plate type and the cone-and-plate type. The placed sample begins to protrude when the heat is turned up after a disc sample is set from the jig. When the rheological data are measured concretely with rotational-type rheometers, it is necessary to remove this protruded part from the jig. However, it is difficult to remove the protruded part. Then, we need to handle the effect of protruded part on the measurement rheological data. One of the methods for solving this problem is to use the numerical simulation of polymer melt flow in a rotational-type rheometer.

In this study, we carried out a viscoelastic flow simulation of polymer melts in a rotational-type rheometer and attempted to quantitatively evaluate the effects of protruding part size on the measurement rheological data. In addition, we also discussed the utility of the cup type jig by using viscoelastic flow simulation. The advantage point of using the cup type jig is to reduce the setting error of specimens.

Governing equations and calculation method

This flow problem was considered as a non-isothermal or isothermal flow simulation problem of viscoelastic and incompressible fluid. The Phan-Thien Tanner (PTT) model proposed by Sugeng et al. [1] was employed as the constitutive equation.

The Galerkin finite element method was employed as the calculation methods in this study. We used the under-relaxation method for rate-of-deformation tensor proposed by Tanoue et al. [2] for the purpose of stable calculation.

Effect of protruding part size on the measurement rheological data

Test fluid and calculation model

The HDPE melt was employed as the test fluid. The experimental data of flow characteristics for the test fluid were referred from reference [3]. These experimental data were used for the determination of the material constants for the PTT model. Other material constants, e.g. density and thermal conductivity, were referred from reference [4].

Figure 1 shows the whole geometry of the calculation region for the parallel plate rheometer. The flow is assumed to be axisymmetric with respect to the rotation axis of the plates. Here, *R* is the plate radius, *H* is the distance between the plates, and is the angular velocity of the plate. The protruding part size was defined by the *E*. In this study, some assumptions are also imposed as follows; (i) The interface shape between a liquid and a gas is represented by the quadratic function and the shape is not changed. (ii) The full slip condition is imposed on the interface

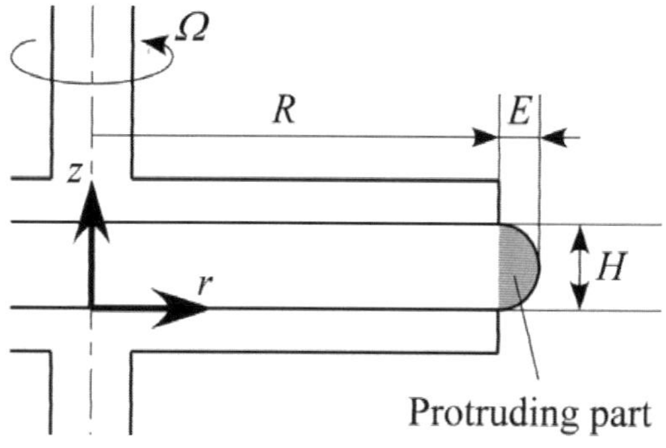

Figure 1 The whole calculation region for the polymer melt flow simulation in a parallel plate rheometer. The protruding part size is also defined in this figure.

between a liquid and a gas. (iii) Non-slip conditions are imposed on the wall. (iv) The forced-convection heat transfer conditions are imposed on this liquid-gas interface.

Evaluation of steady-state rheological data estimated from the viscoelastic flow simulation results

First, we discuss the effect of protruding part size on the steady-state rheological data by using the steady-state viscoelastic flow simulation. Here, we chiefly discuss the polymer melt flow in the parallel plate type rheometer with a plate diameter of 25 mm (R = 12.5 mm) and the distance H = 2 mm between the plates. From results of the temperature distributions, the rise in temperature was a maximum of 0.5 °C at a shear rate of 10 s^{-1}. Therefore, viscous dissipation hardly influences the polymer melt flow in the parallel plate rheometer. The same tendencies can be seen with a cone-and-plate type rheometer with a cone angle of 0.1 rad. In addition, the secondary flow has little influence on the measurement of rheological data.

The shear viscosity and the first normal stress difference at each shearing speed are estimated from the simulation results, and we discuss the influence of the protruding part size on the estimate value. The estimation method is shown in the reference [5]. Figure 2 shows the comparison of the shear viscosity and the primary normal stress difference between the estimated values from the viscoelastic flow simulation results and the flow characteristics of the PTT model at E = 0.0 mm and 1.0 mm. The estimated shear viscosity from the simulation results in the case of E = 1.0 mm is on the whole larger than that of the flow characteristics of the PTT model at the same shear rate. The estimated primary normal stress difference at E = 1.0 mm seems to be almost the same as that at E = 0.0 mm on the whole shear rate region. The same tendencies can be seen in the case of the cone-and-plate rheometer.

To quantitatively evaluate difference in the rheological data between the estimated values by the

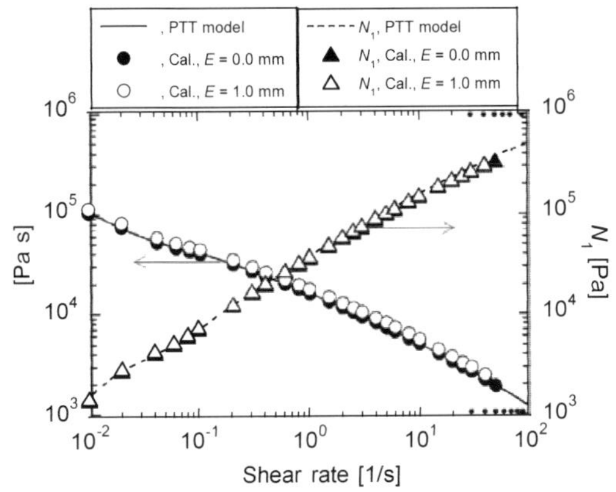

Figure 2 Comparison between the shear flow characteristics (the shear viscosity and the primary normal stress difference N_1) estimated from the simulation results of polymer melt flow in the parallel plate rheometer and those of the PTT model [5].

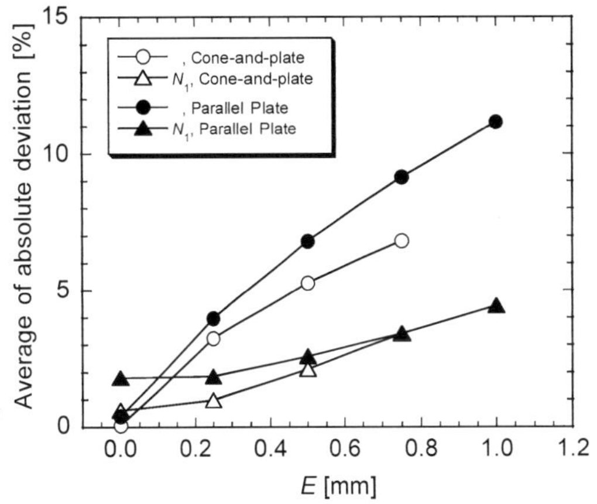

Figure 3 The average values of absolute deviation of shear viscosity and primary normal stress difference N_1 as a function of the protruding part size E for parallel plate and cone-and-plate rheometer [5].

simulation results and the flow characteristics of the PTT model, the deviations of the shear viscosity and the primary normal stress difference are calculated by the following equation.

$$\text{Deviation of } A \quad \frac{A_{\text{Cal.}} \quad A_{\text{PTT}}}{A_{\text{PTT}}} \quad 100 \ [\%] \tag{1}$$

where A indicates the shear viscosity or the primary normal stress difference, the subscript cal. of A is denoted as the estimated value A using the viscoelastic flow simulation result and the subscript PTT of A is denoted as A of the PTT model at the same shear rate. To comprehensively evaluate the effect of protruding part size E on the estimated values of steady-state shear flow characteristics, the average of the absolute deviations of the shear viscosity and the primary normal stress difference in the shear rate region of 0.01~50 s^{-1} are calculated at each E by using the simulation results. Figure 3 shows the average of absolute deviation of shear viscosity and primary normal stress difference N_1 as a function of the protruding part size E for the parallel plate and the cone-and-plate rheometer. The average of absolute deviation of N_1 increases a little with E for both rheometer types. The effect of the protruding part size on the estimated N_1 with a parallel plate rheometer is slightly larger than that with a cone-and-plate one. In addition, the averages of absolute deviation of and N_1 of estimated rheological data with the cone-and-plate rheometer are smaller than those with the parallel plate one. This is the obvious results as we consider the shear flow state in the rheometer.

Evaluation of dynamic rheological data estimated from the oscillatory viscoelastic flow simulation results

Next, we discuss the effect of protruding part size on the dynamic rheological data by using the oscillatory viscoelastic flow simulation results. Here, we focused the case using a parallel plate rheometer. The calculation was conducted for three sizes of protruding part, E = 0.0, 0.5 and 1.0 mm, and 5 different frequencies = 0.01, 0.1, 1, 10 and 100 rad/s. The amplitude of strain $_0$ was set up at 10 %. Other geometry conditions are the plate radii R = 12.5 mm and the distance between the plates H = 2.0 mm.

For calculating storage modulus G' and loss modulus G'' from the simulation results in the parallel plate type rheometer, we applied the theory of SAOS (Small Amplitude Oscillatory Shear). The detail information is shown in the reference [6]. Figure 4 shows the comparison of G' and G'' as a function of frequency between the data calculated from the simulation results (named by "Cal.") and the data of the test fluid (named by "Fluid") at E = 0.0 and 1.0 mm. Both simulations give good conformity with experimental data for the test fluid although there is a difference between protruding part size.

For the concrete consideration of the effect of the protruding part size on the measurement data, we estimated the deviations of storage and loss modulus by the formula:

$$\text{Deviation of } G' \quad \frac{G'_{Cal.}\quad G'_{Fluid}}{G'_{Fluid}}\ 100\ [\%] \qquad (2)$$

$$\text{Deviation of } G'' \quad \frac{G''_{Cal.}\quad G''_{Fluid}}{G''_{Fluid}}\ 100\ [\%] \qquad (3)$$

where $G'_{Cal.}$ and $G''_{Cal.}$ are storage and loss modulus estimated from simulation and G'_{Fluid} and G''_{Fluid} are reference storage and loss modulus, respectively. Figure 5 shows the deviations of storage and loss modulus from the characteristics of the test fluid at various protruding part sizes. We can see the deviations of G' and G'' predicted by simulation increase with the protruding part size and the frequency. The absolute deviation of G'' is larger than that of G' at each frequency except for E = 0.0 mm, i.e. no protruding part case. The deviation of G' is less than about 12 % at E = 1.0 mm and that of G'' is less than about 20 % at E = 1.0 mm. The average values of absolute deviations increase with the protruding part size. These average values are ca. 10 % for G' and ca. 15 % for G'' in this study.

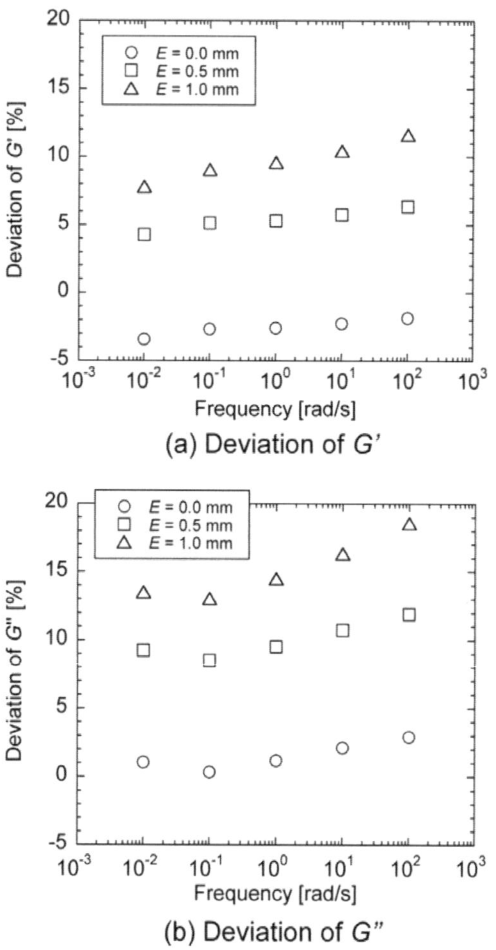

Figure 5 Deviations of G' and G'' as a function of frequency at various protruding part sizes E [6].

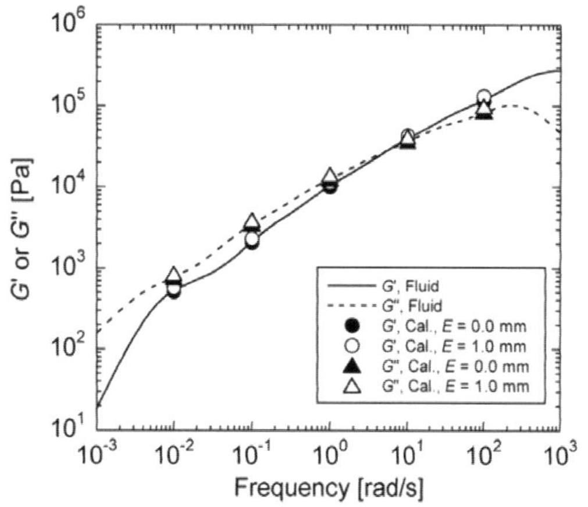

Figure 4 Comparison of storage modulus G' and loss modulus G'' estimated from the viscoelastic flow simulation result (Cal.) and the flow characterisrics of the test fluid (Fluid) [6].

Evaluation of cup-type jig for parallel plate rheometer by viscoelastic flow simulation

Next we discuss the evaluation of cup-type jig for parallel plate rheometer by viscoelastic flow simulation. The PET (Polyethylene Telephtarate) melt was focused as the test fluid. The experimental data of flow characteristics for the test fluid were referred from the reference [7]. Figure 6 shows the whole geometry of the calculation region for the parallel plate rheometer with cup-type jig. The flow is assumed to be axisymmetric with respect to the rotation axis of the plates. Here, R is the plate radius, h is the distance between the plate (disk) and the bottom of cup-type jig, and E_c is the different length between the cup-type jig radius and the disc radius shown in Fig.6. Boundary conditions are as follows; 1) Non-silp conditions on the wall. 2) The full slip condition on the interface between a liquid and a gas, 3) The shape of this interface is assumed to be unchangeable.

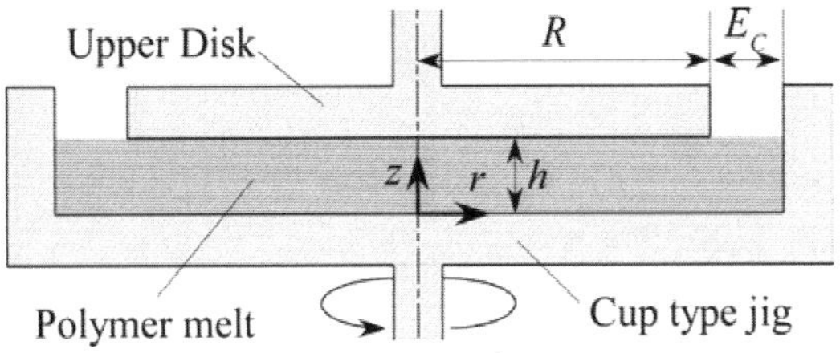

Figure 6 The whole geometry of the calculation region for the parallel plate rheometer with cup-type jig

Figure 7 shows the comparison between the steady-state shear flow characteristics estimated from the simulation results in case using the cup-type jig and that in case using the disk with radius R, i.e. in case of parallel plate rheometer with no-protruding part. The equations for estimating these flow characteristics in case of a parallel plate type rheometer were used in both estimations. The estimated shear viscosity in case using the cup-type jig is larger than that in case using the disk with radius R at same shear rate. On the whole, the estimated shear flow characteristics in case using the cup-type jig almost agree with those in case using the disk with radius R. These results imply the utility of a cup-type jig for measurement of shear flow characteristics.

In order to discuss the utility of the cup-type jig quantitatively, we estimated the deviations of shear flow characteristics at same shear rate by the formula:

$$\text{Deviation of } A \quad \frac{A_{\text{Cup}} \quad A_{\text{Rheo.}}}{A_{\text{Rheo.}}} \quad 100\ [\%] \qquad (4)$$

where A indicates the shear viscosity or the primary normal stress difference N_1, the subscript Cup of A is denoted as the estimated value A using the simulation result in case using the cup-type jig, and the subscript Rheo. of A is denoted as the estimated value A using the simulation result in case using the disk with radius

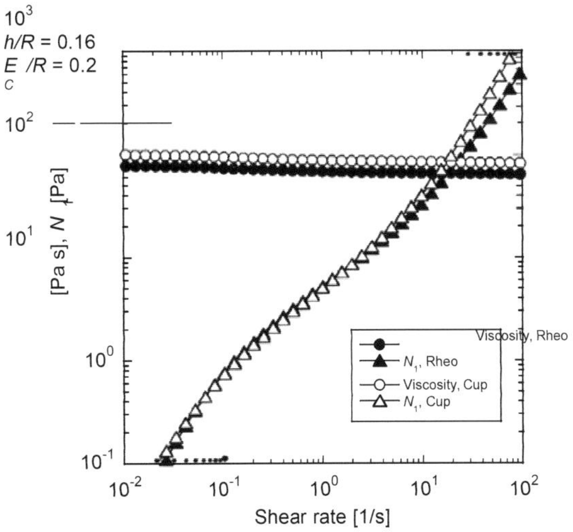

Figure 7 Comparison of steady-state shear flow characteristics between the values estimated from the simulation results in case using the cup-type jig (Cup) and those in case using the disk with radius R (Rheo).

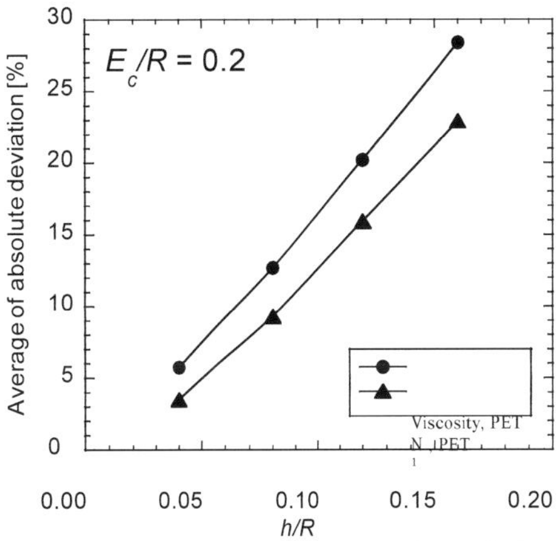

Figure 8 Average of absolute deviations of shear flow characteristics for parallel plate rheometer with the cup-type jig as a function of h/R at $E_c/R = 0.2$

R at the same shear rate. To comprehensively evaluate the effect of the cup-type jig size on the estimated steady-state shear flow characteristics, the average of absolute deviations of and N_1 in the shear rate region of 0.01~100 s^{-1} were calculated at each cup-type jig size. Figure 8 shows the average of absolute deviations as a function of the dimensionless distance h/R between the plate and the cup-type jig at the dimensionless length $E_c/R = 0.2$ between the cup-type jig radius and the disc radius. The deviation of increases linearly with h/R. In addition, the average of absolute deviation of N_1 also increases with h/R.
In addition, the deviation becomes constant in the region of $E_c/R > 0.2$ at $h/R = 0.8$. This tendency can be seen in and N_1. And the deviation of N_1 is larger than that of .

Conclusions

On the whole, we could evaluate the effect of protruding part of the sample on the estimated data of flow characteristics, and the utility of the cup-type jig for measuring the shear flow characteristics of polymer melt by using the viscoelastic flow simulation.

References

1. F. Sugeng; N. Phan-Thien; R. I. Tanner *J. Rheol.* 1987, *31*, 37.
2. S. Tanoue; T. Kajiwara; Y. Iemoto; K. Funatsu *Polym. Eng. Sci.* 1998, *38*, 409.
3. R. J. Koopmans *Polym. Eng. Sci.* 1992, *32*, 1741.
4. A. Goublomme; M. J. Crochet *J. Non-Newton. Fluid Mech.* 1993, *47*, 281
5. S. Tanoue; M. K. Zielinski; Y. Iemoto *Polym. Eng. Sci.* 2009, *49*, 1573.
6. S. Tanoue; M. K. Zielinski; N. Washida; Y. Iemoto *Nihon Reoroji Gakkaishi* 2009, *37*, 113.
7. G. Hu; C. Zhao; S. Zhang; M. Yang; Z. Wang *Polymer* 2006, *47*, 480.

O-08-617

SHEAR AND ELONGATIONAL FLOW BEHAVIORS OF LONG CHAIN BRANCHED ETFE MELTS

Masataka Sugimoto[a*], Eiichi Nishi[b], Takashi Satou[b], Satoru Imai[a], Kiyohito Koyama[a]
a Yamagata University, Yonezawa 992-8510, Japan – sugimoto@yz.yamagata-u.ac.jp
b Asahi Glass Co., Ltd, Research and Development Div., Ichihara 290-8566, Japan

Abstract Rheological behavior of alternating ethylene-tetrafluoroethylene copolymer (ETFE) under shear and elongational flow was studied. Linear ETFE showed typical rheological behaviors as well as other linear polymer such as polyethylen and polypropyrene. ETFE with long chain branch, however, showed outstanding behaviors. The branched ETFE indicated high G' at low frequencies, slow relaxation of relaxation modulus G(t), weak strain dependency of damping function. Under elongational flow the branched ETFE showed strong deviation of elongational viscosity from linear viscoelastic regime. This suggests enhanced processability of ETFE in film casting, film blowing, foaming, blow molding.

Keywords: ETFE, fluoropolymer, rheology, dynamic viscoelasticity, elongational viscosity

Introduction

Ethylene-tetrafluoroethylene copolymer (ETFE) is widely used in many fields such as automobile, modern construction, airplane, solar battery panel areas due to the high chemical and thermal resistant, flame retardancy, weathering, an electric insulation properties. Therefore ETFE is intensively explored on the molecular parameters, microstructures, crystallization behaviors, flow behaviors under steady shear, flow instabilities. However, ETFE is one of typical linear polymers and the improvement of the rheological behavior by introducing the long chain branch has not been reported.

Now more advanced control of rheological behavior of ETFE is required for the wider applications and precise processings in which ETFE has not be utilized owing to the limited rheological properties so far. In this paper we try to introduce long chain branch into ETFE and the shear and elongational flow behaviors are reported.

Experimental

We used conventional ETFE (C88AX, Asahi Glass Co., LTD, Japan) and ETFE copolymerized with divinyl compound, which is indicated as ETFE-B. ETFE-B of weight percentages of 5, 10 and 30 was blended with ETFE by twin-screw extruder, indicated by notations of ETFE5, 10, 30, respectively.

Dynamic strain, frequency and temperature sweep measurements were performed by Physica MCR-301 (Anton Paar). Dynamic frequency tests were conducted from 0.1 to 100 rad/s at constant strain less than 0.05 and temperatures of 270 °C using parallel fixture. Elongational viscosity was measured by rotational rheometer (Ares, TA instruments) at constant rates and temperature of 270 °C.

Mono-layer T-die cast film extrusion was carried out by twin-screw extruder (φ=15mm, ULT-nano, Technovel Co., Japan). External gear pump was used for constant volumetric flow rate.

Results and Discussion

Fig.1 shows storage (G') and loss modulus (G") of ETFE and ETFE-B at 270 °C. ETFE indicated terminal flow behavior at low frequency range. On the other hand, ETFE-B synthesized with divinyl compound showed much elastic viscoelastic behavior. Within the measured frequency G' was always higher than G". G' is parallel to G" at high frequency and the slope of G' leveled off with decreasing with frequency. This means an existence of very long time relaxation component.

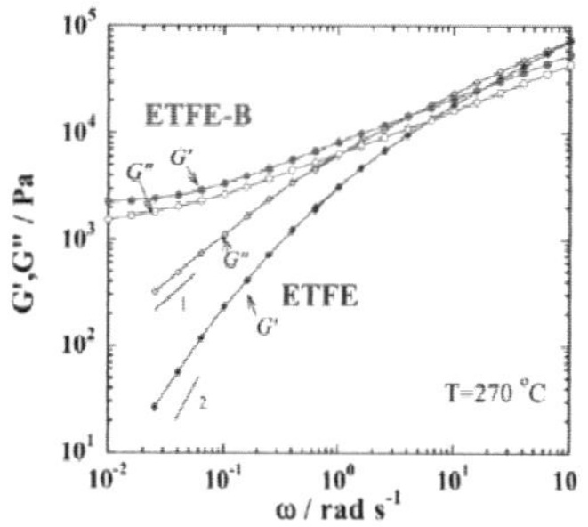

Fig.1 Storage and loss modulus of ETFE and ETFE-B at 270 °C

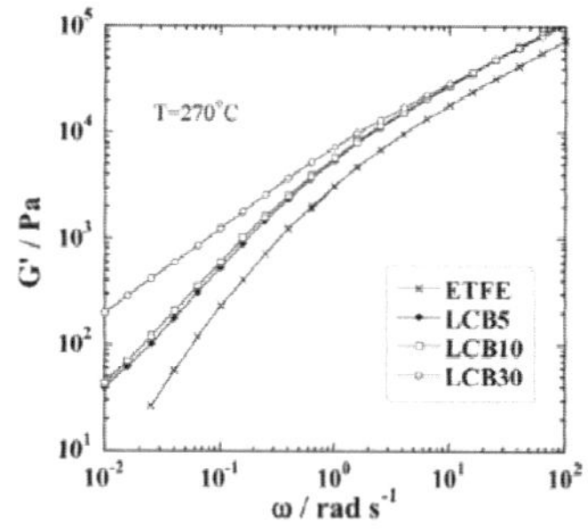

Fig.2 Storage and loss modulus of ETFE blended with ETFE-B (5, 10, 30 wt%) at 270 °C

Fig.2 shows linear viscoelasticity of ETFE blended with ETFE-B (5~30 wt%) at 270 °C. At high frequency G's of the blend samples were slightly higher than ETFE. G' of the blends at low frequency increased with increasing the weight ratio of ETFE-B. This is due to introduction of long relaxation time component of ETFE-B which dispersed in ETFE.

Transient elongational viscosity of ETFE and ETFE-B blends are illustrated in Fig.3. Linear ETFE mere showed smooth increase of elongational viscosity. The curves were independent of strain rate and merged into one curve. The elongational viscosity of ETFE blended with EtFE-B exhibited steep increase at a certain time, depended on the strain rate. ETFE30 showed the strong increase of elongational viscosity in comparison with ETFE10. This strong strain hardening is nonlinear behavior and never expected to linear ETFE. A small amount of the divinyl compound was used to copolymerize with ETFE. The branch structure can be expected for copolymerization of ethylene, tetrafluoroethylene and divinyl compound. If the star-type branching structure was formed for ETFE-B, the nonlinear viscoelastic behavior should not changed the way graft, pom-pom and random types branced polymer do under elongational flow, while the linear viscoelastic response was changed. Therefore we consider that ETFE-B includes long chain branched molecules with two branching point at least.

We carried out T-die film casting extrusion of ETFE and ETFE-B blends. It was found that the neck-in length was reduded according to the incorporated ratio of ETFE-B, comparing with that of linear ETFE. In the film casting uniaxial flow is dominant at edge part of the film, while the rest is mainly subjected by planar flow. We consider that the strain hardening behavior of ETFE-B suppressed the neck-in length of casted film.

Conclusions

ETFE was modified by the incorporation of divinyl compound in the copolymerization process. ETFE-B copolymerized with divinyl compound and the blends with ETFE showed very long relaxation time and pronounced strain hardening behavior under elongational flow which can never be expected for conventional fluorinated polymer. We confirmed that the enhanced rheological behavior provide the beneficial effect on the T-die film casting extrusion process.

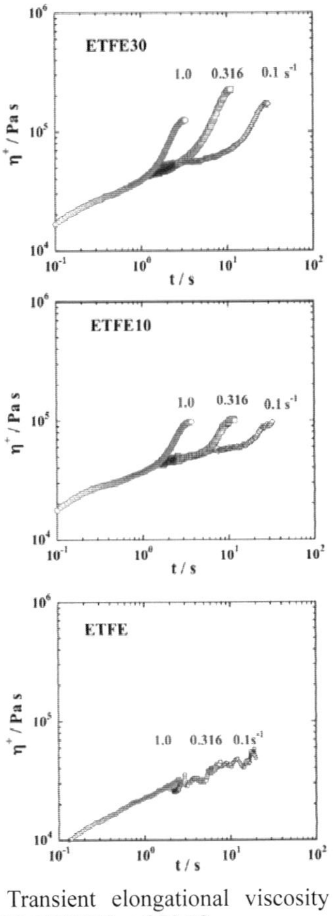

Fig.3 Transient elongational viscosity of ETFE, ETFE10, ETFE30 at 270 °C

O-08-665

COMPARISON IN SLIP CHARACTERISTIC DETERMINATION OF HDPE BY FOUR DIFFERENT TECHNIQUES

D. Tabtippawon[1*], W. Tanthapanichakoon[1], W. Cheevasrirungruang[1], B. Thitisuk[1], A. Fuongfuchat[2], P. Muangcharoen[2] and T. Makmoon[2]

[1]*Polymer Innovation, ThaipolyethyleneCo.,Ltd.(TPE), Thailand – dumronta@scg.co.th; Boonyakt@scg.co.th; watchche@scg.co.th; winyut@scg.co.th*

[2]*National Metal and Materials Technology Center (MTEC), Thailand - asiraf@mtec.or.th; pongphm@mtec.or.th; thidaram@mtec.or.th*

Abstract

Slip characteristic can be observed by different methods including 1) wall slip measurement by capillary rheometer 2) Δn comparison using Carreau-Yasuda curve fitting of low/high shear rate region 3) Calculation of friction coefficient using experimental extrusion data of an annular die and 4)Computer simulation for determination of friction coefficient based on capillary result. Each technique has its own difficulties, advantages and disadvantages to find coefficient of friction for study of slip characteristics.

Keywords: wall slip, coefficient of friction, dynamic coefficient of friction, capillary simulation

Introduction

In the extrusion of thermoplastics, there is a region of low shear rate which plastics melt adheres to wall surface while flowing through the die. Above a critical shear stress, [1] described that the adjacent layer begins to lose adhesion with the wall and slip flow occurs. In this study, slip characteristic of HDPE has been observed by four rheological methods.

First method is to find slip velocity from the result of capillary rheometer. A set of flat dies with various diameters is used and dynamic friction coefficient is eventually calculated according to Mooney equation.

Slip characteristic is also determined by difference of power law exponent, n, of LVE data and non-LVE. LVE data of a HDPE represents non-slip behavior while it is assumed that non-LVE range shows slip flow. Hence in same sample, the difference in slope of the graph indicates how strong of its slip characteristic.

Another technique to observe slip characteristic is to calculate directly from extrusion experiment. Calculation of wall slip is described [2]. Pressure drop in extrusion die and output can be measured, and with simplified die geometry, coefficient of friction can be calculated.

Computer simulation also poses itself as a new remarkable option since it requires much less resource. Only a set of interested shear rate range data from a common capillary rheometeris enough to be an input to calculate for optimized coefficient of friction.

Experimental

Wall Slip Measurement by Capillary Rheometer

Wall shear stress at apparent shear rate range of 10 – 10 000 1/s was measured at 190°C using Rosand RH 2200 capillary rheometer(Bohlin Instruments). Wall slip analysis based on Mooney equation [3] was carried out on the data from series of flat dies having 0.5, 1.0 and 2.0 mm diameter and aspect ratio (L/D) of 16. Preheat sequence were compressed sample, 2 min preheat, compressed samples again and 4 min preheat. For capillary flow of polymer melts, apparent shear rate can be defined as follows:

$$\dot{\gamma}_A = \dot{\gamma}_{A,S} + \frac{8u_s}{D} \qquad \text{Eq.1}$$

Slip velocity, u_s, can be thus determined from the plot of $\dot{\gamma}_A$ versus 1/D, where D is die diameter and $\dot{\gamma}_A$ is apparent shear rate.

Dynamic friction coefficient in shear thinning region was then calculated using the following equation [4]:

$$\tau_w = k_D u_s \qquad \text{Eq.2}$$

where, τw is wall shear stress and kD is dynamic friction coefficient per unit area between melts and wall.

Delta n comparison using Carreau-Yasuda curve fitting of low/high shear rate region

The flow curve at 190°C was measured from Rotational rheometer (ARES Rheometric, TA Instrument)with 25 mm parallel plate.The steady shear experiment was conducted at shear rate 0.01-10 1/s to determine low shear region. While Oscillation mode was also tested at angular frequency of 0.1-100 rad/s with 5% strain within LVE strain deformation and convert into shear rate via Cox-merz rule.

For high shear region, a capillary rheometer (Rosand RH2200 - Bohlin Instruments) with double bore of 16-mm/ 1mm flat die and 1 mm orifice diewas used to determine flow curve at 10-10000 1/s. Preheat sequence were compressed sample, 2 min preheat, compressed samples again and 4 min preheatAll corresponding curves were fit using Carreau-Yasuda model. LVE data represents non-slip behavior and non-LVE one for slip flow.

$$\eta = \eta_\infty + (\eta_0 - \eta_\infty)[1 + (\lambda\dot{\gamma})^a]^{\frac{n-1}{a}} \quad \text{Eq.3}$$

The difference of nbetween slip and non-slip curve was analyzed to determine degree of wall slip.

Calculation of friction coefficient using experimental extrusion data of an annular die

The analytical mathematical model considering wall slip is initially from Coulomb's law of friction of a solid body on a wall forming the equilibrium between viscosity and friction forces for a flow through a tube. This study adopted this concept and created a mathematical model for a constant cross section annular die.

$$P_0 = \frac{1}{H}\left(\frac{2(m+2)\dot{V}}{\emptyset\pi R H^2}\right)^{\frac{1}{m}} \cdot \left[2L + \frac{(R+r)}{\mu_C}\left[1 - \ln\left[\frac{(R+r)}{P_{atm} H \mu_C}\left(\frac{2(m+2)\dot{V}}{\emptyset\pi R H^2}\right)^{\frac{1}{m}}\right]\right]\right] \quad \text{Eq.4}$$

$$m = \frac{1}{n}$$

where $\emptyset = k^{-m}$ and

Slip coefficient, μ_G, can be calculated at a specific volumetric flow rate, $\dot{V}$, and initial die pressure, P_0.

Computer simulation for determination of friction coefficient based on capillary result

The required data for this technique are the pressure drops at various flow rates of polymer melt flowing through the capillary die. These data are readily available when the capillary rheometer is used to determine the shear viscosity of polymer melt.

The slip characteristic of polymer melt can be explained by Navier's law:

$$F(V) = -k * V^e \quad \text{Eq.5}$$

where $F(V)$ is friction force; a function of fluid velocityat the wall, V;k is the first slip coefficient, and e is the second slip coefficient.

To obtain the slip information, the optimization routine in Polyflow (ANSYS Inc.) was used to simulate the flow of polymer melt in a capillary die at different flow rates. The input needed by Polyflow simulation was the rheology properties, consisting of shear viscosity as a function of shear rate, and storage modulus and loss modulus as functions of force frequencies.

In the simulation, Navier's law, Eq.5 was applied at the wall boundary condition. The optimization routine then worked by optimizing k and e values to match the calculated pressure to the capillary pressures.

Results and Discussion

Two extrusion grades of HDPE which have dissimilar flow behavior were observed with each technique and finally compared.

Wall Slip Measurement by Capillary Rheometer

Fig. 1 shows Flow curve of Sample A and Sample B from 3 different dies. In the studied shear rate range, i.e., 10 – 10 000 1/s, all the melts showed power law behavior. Pressure oscillations were observed at shear rates > 100 1/s. The determination of slip velocity, u_s, was thus carried out using the data from shear rates ≤100 1/s, i.e., in stable flow region. In this range of shear rates, wall shear stress was lower when using a narrower die, which implied that slip velocity was higher when the melts flowed in narrower dies.

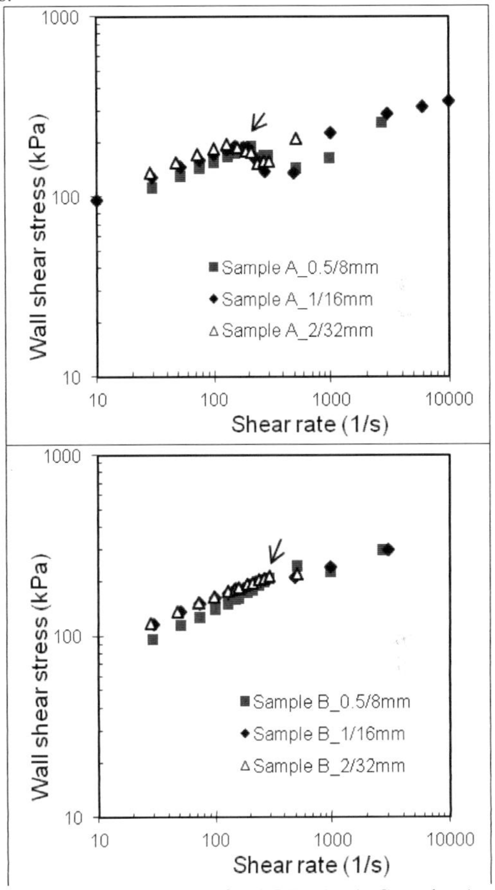

Figure 1- Flow curve of HDPE, (top) Sample A and (bottom) Sample B at 190°C from 0.5/8 mm, 1/16mm and 2/32 mm dies (Arrows indicate onset of pressure oscillations; The number marked in legend refers to the length to diameter ratio of capillary die. For example, L/D = 0.5/8mm means die length = 0.5 mm and die hole diameter = 8 mm.).

Values of slip velocity of sample A and Sample B at different wall shear stress (see Table 1) were calculated from Eq.1. Dynamic friction coefficients per unit area between melts and wall, k_D, were then calculated using

Eq.2. The k_D of Sample A and Sample B are 10.6 x 10^9 and 62.9 x 10^9 g/m^2.s, respectively.

Table 1 – slip velocity at different wall shear stress

τ_w (kPa)	u_s (mm/s)	
	Sample A	Sample B
10	0.96	0.16
50	4.77	0.80
70	6.68	1.11
100	9.55	1.59
150	14.32	2.39
200	19.09	3.18

Delta n comparison using Carreau-Yasuda curve fitting of low/high shear rate region

LVE curve and non-LVE curve at 190°C were fit separately as shown in Fig. 2. Difference between curves implies to degree of slip of a material. Table 2 shows that Sample A, with Δn of 0.22, exhibits more slip flow than Sample B at 190°C.

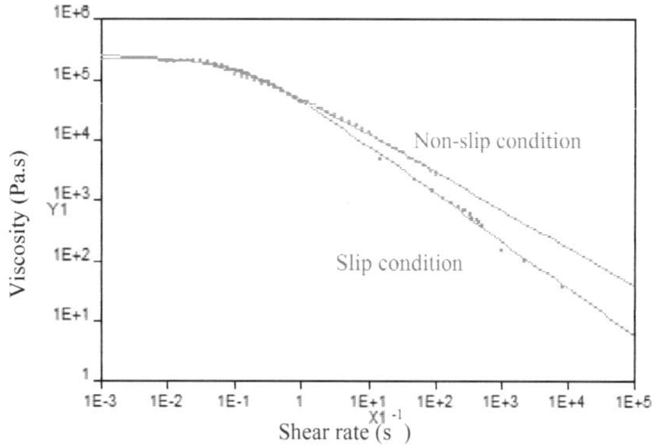

Figure 2– Carreau-Yasuda curve fitting on Sample B at 190°C

Table 2– Carreau-Yasuda curve fitting result

Parameter	Sample A		Sample B	
	LVE	Non-LVE	LVE	Non-LVE
η_0 (Pa.s)	194050	188898	174607	165473
a	0.79	0.76	0.85	0.91
n	0.34	0.11	0.35	0.24
Δn	0.22		0.11	

Calculation of friction coefficient using experimental extrusion data of an annular die

Varying output provides range of friction coefficient as in Fig.3. The lower coefficient of friction of Sample A means that it shows more slip behavior than Sample B in all given flow condition in the extruder.

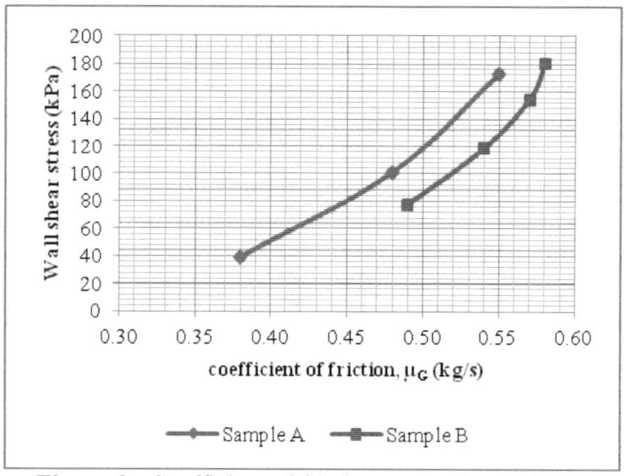

Figure 3– Coefficient of friction at 190°C at different processing condition

Wall Slip Measurement by Capillary Rheometer

The optimized *k* and *e* were determined at given flow rates. When the *k* values were plotted as a function of the *e* value, as shown in Fig. 4, the cross points of these curves (each curve represented each flow rate) are the optimized slip coefficients where are valid for the range of these given flow rates.

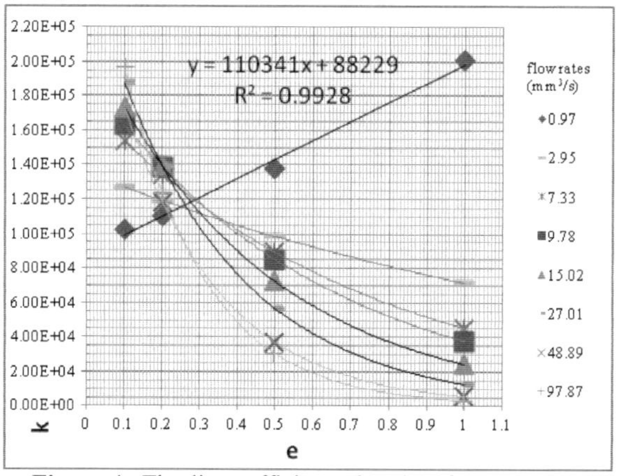

Figure 4– Theslip coefficients, *k* and *e* of Sample B at 190°C

Fig.4 shows the curves of the slip coefficients of Sample B at 190C. For the range of flow rates of 2.95-9.78 mm^3/s, the optimized slip coefficients are k=1.055x10^5 and e=0.36. This range is corresponding to the range of shear rate of 20-80 s^{-1}, where is the shear rate range of the customer's machine.

Table 5 –Resultcomparison for all techniques

Technique	Parameter	Unit	Sample A	Sample B
Capillary	Dynamic friction coefficients, k_D	g/m^2.s	10.6 x 10^9	62.9 x 10^9
Curve fitting	Δn	-	0.22	0.11

Extrusion	Coefficient of friction, μ_G	kg/s	0.47	0.52
Simulation	Coefficient of friction, *k*	g/s	1.00x10^5	1.055x10^5
	Power factor,e	-	0.225	0.36

Conclusions

These four techniques point to the same trend that Sample A shows more slip behavior than Sample B. However, for implementation of coefficient of friction in further study, only the ones calculated from extrusion experiment and simulation can be used. The other estimations of slip characteristic techniques require a calibration or correlation curve to convert k_D or Δn to coefficient of friction. Each technique has different advantages and disadvantages upon using as described in Table 6.

Table 6 – Comparison of slip characteristic techniques

Technique	Advantages	Disadvantages
Capillary	· Less sample	· Dynamic slip · Comparison only · Empirical model · Special precaution for die cleanliness
Curve fitting	· Highest accuracy · Capable of applying to all range of shear rates · Closest to real customer die · Fastest technique	· Much more required sample · Heater limitation of extruder
Extrusion	· Less sample · Lowest cost	· Comparison only · Semi-empirical model
Simulation	· Less sample · High accuracy (viscous heating included)	· Require computer simulation program

References

1. S. G. Hatzikiriakos; John M. Dealy J. Rheol. 1992, 36, 703.
2. E. Uhland RheolActa. 1976, 15, 30
3. S. G. HatzikiriakosPolm. Sci. 2012, 37, 624-643
4. L.A. Archer in *Polymer processing instabilities, control and understanding*, S. G.Hatzikiriakos and KB Migler, Ed.; Marcel Dekker, New York. 2005; 73 – 120.

O-08-470

A TOOLBOX FOR PARAMETER-FREE PREDICTIONS OF SOLID-STATE PROPERTIES OF ORIENTED GLASSY POLYMERS

D.S.A. De Focatiis[1*] and C.P. Buckley[2]

[1]*Division of Materials, Mechanics, and Structures, University of Nottingham, UK – davide.defocatiis@nottingham.ac.uk*

[2]*Department of Engineering Science, University of Oxford, UK – paul.buckley@eng.ox.ac.uk*

Abstract – The focus of this study is to develop genuinely parameter-free predictive capability for solid-state properties of melt-processed glassy polymers. A toolbox is assembled from a previously proposed melt-solid constitutive model that provides the necessary residual stress components required for predictions of birefringence and craze initiation stress. The Likhtman-McLeish theory for linear rheology of entangled polymers is employed to generate the low reduced frequency part of the linear viscoelastic spectrum, the only molar mass dependent input parameter. All other parameters are obtained by experiment or from literature, and can be considered material constants. Toolbox predictions are compared to new experimental data on two grades of linear monodisperse polystyrene of known molar mass but unknown rheology and to literature data, and can account for the role of molar mass on birefringence and craze initiation stress of polystyrenes subjected to supra-entanglement orientation processes.

Keywords: polystyrene, orientation, constitutive model, birefringence, crazing.

Introduction

Polymer processing engineers are well versed with both intentional and unintentional frozen-in molecular orientation arising in polymeric products as a direct consequence of flow processes during melt-processing. Orientation leads to anisotropy in Young's modulus [1], yield stress [2, 3], crazing stress [4, 5], optical birefringence [2, 6, 7], strain-hardening modulus [8, 9] and fracture toughness [10], and these properties are sensitive to both the grade of polymer employed and to the flow history encountered [5, 7, 8].

The present authors recently proposed a fully 3-D constitutive model for predicting the non-linear melt-state and solid-state constitutive response of glassy monodisperse polymers [8]. By fitting this model to the linear viscoelastic spectrum of two monodisperse linear polystyrenes (PS), the authors demonstrated that it is possible to predict the optical birefringence [7], the craze initiation stress [5], and the solid-state mechanical response [8] of a polymer of given molar mass subjected to an arbitrary process history, with a small number of molar mass independent material constants.

For the special case of linear monodisperse polymers, the melt region of the linear viscoelastic spectrum can be predicted accurately with the Likhtman-Mcleish (L-M) theory [11]. Here we tested whether it is possible to employ the L-M theory to replace the rheological experimental data that was previously necessary to provide input to the constitutive model as far as solid-state property predictions are concerned. This critical step enables connectivity between theories whose input parameters are *all* material constants associated with a particular polymer species. The toolbox predictions were compared with experimental data consisting of birefringence and craze initiation stress measurements on two monodisperse PS grades of unknown rheology but significantly different molar mass, subjected to various uniaxial melt orientation processes.

Experimental

The materials studied were two samples of mono-disperse linear atactic PS with M_w = 121 kDa (BE) and M_w = 966 kDa (BB) and narrow polydispersities (PDI = M_w/M_n < 1.1). An in-house near net-shape moulding technique was employed for the production of parallelepipedic 80 × 6 × 0.5 mm^3 bars [12], and isotropic specimens were produced according to ref. [13].

Oriented specimens were produced by uniaxial melt-drawing in an Instron 5985 testing machine fitted with an environmental chamber, at a range of temperatures T above T_g between 105 °C and 135 °C. All stretching was carried out at a constant crosshead velocity corresponding to a nominal strain rate of 0.02 s^{-1}, to a fixed stretch ratio of λ = 3. After stretching, specimens were immediately quenched using a cold spray (cooling rate ~15 °C s^{-1}).

Optical birefringence was measured on the central region of the specimens using an Olympus BX51-P transmission optical polarizing microscope fitted with polarizer and compensator under white light [7], and values are reported in Figure 1. Additional experimental data from ref. [7] is also shown.

A previously described technique was employed for the measurement of craze initiation stress of miniature specimens subjected to isochronal 3-point bending creep [14]. Miniature beams approximately 6 × 2 × 0.5 mm^3 were cut from the oriented specimens using a custom jig, soaked in reagent grade diethylene glycol (DEG) for 24 hours, and loaded in 3-point bending creep for 300 s with the beam axis aligned with the orientation direction. After unloading, the length of each crazed region was measured using optical microscopy, allowing determination of the craze initiation stress σ_c. Measurements of $\sigma_c - \sigma_c^{iso}$ (where σ_c^{iso} is the isotropic craze initiation stress) are reported in Figure 2 as averages from a minimum of four beams per experimental condition.

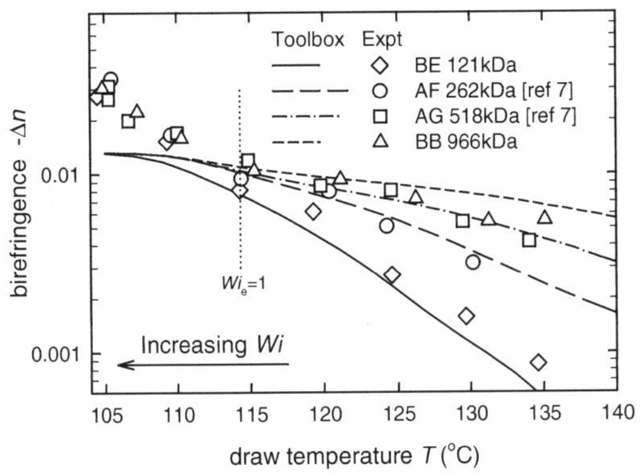

Figure 1 Frozen-in optical birefringence measured on PS specimens melt-drawn at temperature T.

Constitutive model

A constitutive model for the prediction of multiple solid-state properties of glassy polymers with process-induced molecular orientation was developed by the present authors [8], and is shown schematically in Figure 3. It describes the material response to a deformation gradient tensor $\mathbf{F}$ in terms of the Cauchy stress tensor $\boldsymbol{\sigma}$, separated into volumetric $\sigma_m = \frac{1}{3}\mathrm{trace}(\boldsymbol{\sigma}) = K \ln J$ (K is the bulk modulus and $J = \det \mathbf{F}$) and deviatoric $\mathbf{S} = \mathbf{S}_b + \mathbf{S}_c$ parts ($\mathbf{S}_b$ and $\mathbf{S}_c$ are the bond-stretching and conformational contributions respectively). The deviatoric rate of deformation $\bar{\mathbf{D}}$ is given by

$$\bar{\mathbf{D}} = \frac{\hat{\mathbf{S}}_{b,j}}{2G_b} + \frac{\mathbf{S}_{b,j}}{2G_b \tau_j} \tag{1}$$

for each mode j of a spectrum of Maxwell modes. Here $\hat{\mathbf{S}}_j^b$ is the Jaumann rate of stress, G_b is the bond-stretching shear modulus, and τ_j is the jth glassy relaxation time. The deviatoric stress is computed as the sum over all M bond-stretching modes, as $\mathbf{S}_b = \sum_{j=1}^{M} \upsilon_j \mathbf{S}_{b,j}$ where υ_j is the relative volume

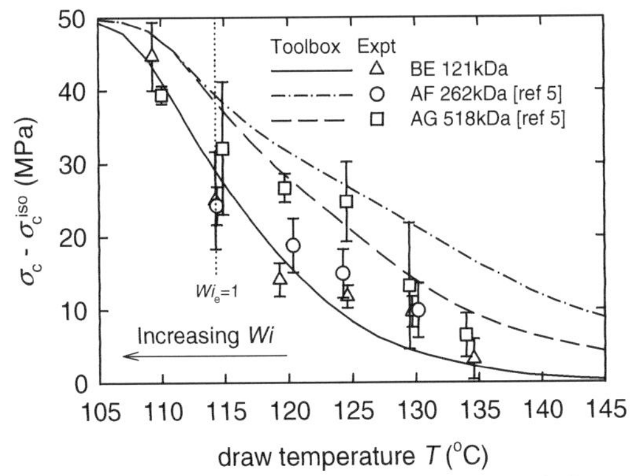

Figure 2 Increase in craze initiation stress relative to isotropic of PS specimens melt-drawn at temperature T. Additional experimental data from ref. [5].

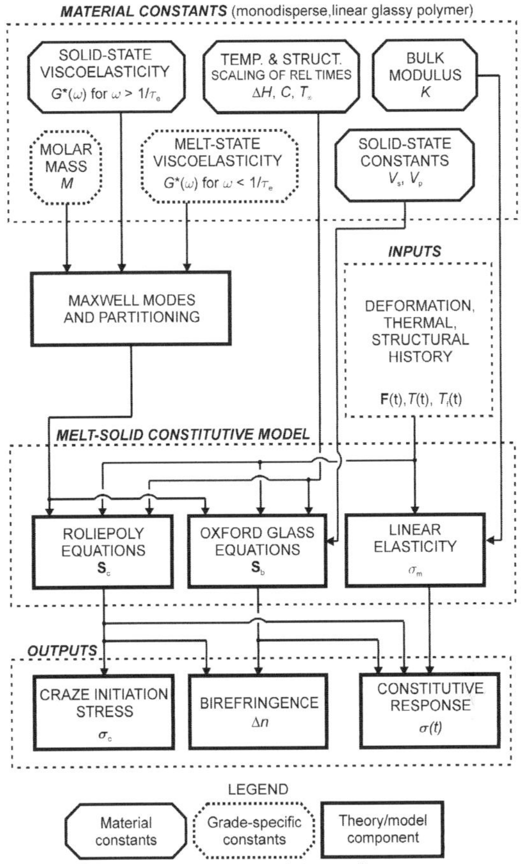

Figure 3 Schematic outline of the toolbox.

fraction of the jth glassy mode. The reader is referred to refs [15-17] for a more detailed presentation.

The conformational stress is computed from a modified Roliepoly constitutive model that accounts for finite extensibility of chains [7, 18]. A conformational stress is computed for each mode k across a spectrum of N discrete modes by numerical integration of the governing differential equation

$$\dot{\mathbf{T}}_k = \bar{\mathbf{L}} \cdot \mathbf{T}_k + \mathbf{T}_k \cdot \bar{\mathbf{L}}^{\mathrm{T}} - \left(\mathbf{T}_k - \lambda_k^2 \mathbf{I}\right) / \tau_{d,k} - 2\mathbf{T}_k \left(F(\lambda_k) - 1\right) / \lambda_k \tau_{R,k} \tag{2}$$

where $\tau_{R,k}$ and $\tau_{d,k}$ are the Rouse and reptation times associated with the kth mode, $\lambda_k = \sqrt{\frac{1}{3}\mathrm{tr}(\mathbf{T}_k)}$ and $F(\lambda_k)$ is a finite chain extensibility correction factor. The conformational stress $\mathbf{S}_{c,k}$ is the deviatoric part of $G_e\left[\mathbf{T}_k F(\lambda_k) / \lambda_k - \mathbf{I}\right]$, where G_e is the entanglement modulus. The conformational contribution to the total deviatoric stress is computed as the sum over all N conformational modes, as $\mathbf{S}_c = \sum_{k=1}^{N} \upsilon_k \mathbf{S}_{c,k}$, where υ_k is the relative volume fraction of the kth Rolie-Poly mode, $\sum_{k=1}^{N} \upsilon_k = 1$.

Rouse and reptation times are referred back to reference values τ^* at a temperature T^* and structural state T_f^* through shift factors a_T and a_{struct} through $\tau_{R,k} = a_T a_{struct} \tau_{R,k}^*$ and $\tau_{d,k} = a_T a_{struct} \tau_{d,k}^*$. Glassy relaxation times are referred back to reference temperature, structural state and (unstressed) stress state $\tau_{j,0}^*$ through $\tau_j = a_T a_{struct} a_{\sigma,j} \tau_{j,0}^*$ where $a_T = \exp\left[(\Delta H / R)(1/T - 1/T^*)\right]$ (ΔH is the activation enthalpy), $a_{struct} = \exp\left[C\left((T_f - T_\infty) - (T_f^* - T_\infty)\right)\right]$ (C is the Cohen Turnbull constant, T_f is the fictive temperature and T_∞ is the Vogel temperature), and $a_{\sigma,j} = \left(V_s \tau_{oct,j}^b / 2RT\right)\exp\left(-V_p \sigma_m / RT\right) / \sinh\left(V_s \tau_{oct,j}^b / 2RT\right)$ ($\tau_{oct,j}^b$ is the octahedral shear stress acting on the j th mode, and V_s and V_p are the shear and pressure activation volumes). In the simulations, we used constants from the literature [8, 19], and the PS was taken to be in equilibrium for temperatures $T \geq 98.9$ °C, and at $T_f = 98.9$ °C for $T < 98.9$ °C. G_e (entanglement modulus), M_e and τ_e (molar mass and Rouse time of one entanglement length), were taken as $G_e = 317.9$ kPa, $M_e = 13.14$ kDa and $\tau_e = 10.57$ s at $T^* = 120$ °C. λ_{max}, the finite chain extensibility, was 4.09 [8]. Whole molecule Rouse and reptation times were computed using $Z = M / M_e$ as $\tau_{R,1}^* = Z^2 \tau^e$ and $\tau_{d,1}^* = 3Z^3\left(1 - 2.38Z^{-0.5} + 4.17Z^{-1} - 1.55Z^{-1.5}\right)\tau_e$.

Moduli and relaxation times have previously been obtained by fitting a discrete spectrum of Maxwell modes to a viscoelastic spectrum obtained from experimental data from DMA and shear rheometry [8]. The high frequency part of the spectrum is insensitive to molar mass, but the low frequency linear melt rheology was required to be measured experimentally for each grade, since it is strongly dependent on molar mass. In this toolbox, the linear viscoelastic frequency response was *predicted* directly by the L-M theory for a monodisperse polymer of given molar mass, and hence Z, using the material constants G_e, M_e and τ_e [20]. A discrete relaxation spectrum was then fitted to a combination of the experimental linear viscoelastic data for high frequencies and the theory predictions for low frequencies. Starting at the longest relaxation time, $\tau_d = \tau_{d,1}$, one mode per decade of time was assigned to cover the frequency range of interest using an in-house optimiser. The condition $\sum_{k=1}^{N} G_k = G_e$ was imposed, restricting Rolie-Poly modes to the conformational response. The viscoelastic spectra of materials BE and BB reconstructed from the Maxwell modes are reproduced in Figure 4.

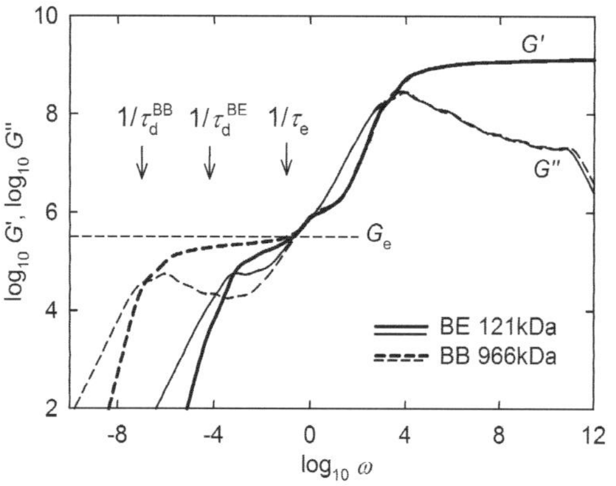

Figure 4 The full linear viscoelastic spectrum for polymers BE and BB used in the simulations.

Results and Discussion

The toolbox is employed to simulate the three stages of the process by which orientation is frozen into the specimens: an isothermal uniaxial orientation through to $\lambda = 3$ at a rate of $\dot{\varepsilon} = 0.02$ s^{-1} for a range of temperatures between $105 - 145$ °C; a fast quench through to 20 °C at fixed deformation, at a rate of 15 °C s^{-1}; and a fast unloading through to zero net stress at a deformation rate of $\dot{\varepsilon} = -1$ s^{-1}. The simulation predicts a state of self-stress for unloaded oriented specimens in the glassy state, with $\mathbf{S}_b = -\mathbf{S}_c$.

Birefringence

Optical birefringence arises from both bond stretching distortions and from conformational alignment of the molecules, and is computed using stress-optical coefficients C_b and C_c (8.3x10^{-12} Pa^{-1} and -4.5x10^{-9} Pa^{-1} respectively) as the sum of the two contributions from the in-plane principal bond-stretching and conformational stress differences, ΔS_b and ΔS_c respectively, as

$$\Delta n = C_b \sum_{j=1}^{M} \upsilon_j \Delta S_{b,j} + C_c \sum_{k=1}^{N} \upsilon_k \Delta S_{c,k} H(\lambda_k / \lambda_{max}) / H(1/\lambda_{max})$$

where $H(x) \simeq 1 + x^2/3 + x^4/3$ is a correction factor accounting for finite chain extensibility [7]. Computations of birefringence from toolbox simulations are shown in Figure 1.

The toolbox successfully accounts for the role of molar mass on birefringence, and correctly predicts the increase of Δn with molar mass at a given draw temperature. The numerical values are close to the experimental data for draw temperatures $T \geq 115$ °C (the drawing regime for which $Wi_e < 1$). The region on the left ($Wi_e > 1$) represents drawing with sub-entanglement molecular stretch, and the toolbox underpredicts. This discrepancy highlights the current limitations in the theoretical understanding of orientation processes at very short length scales. Where $Wi_e < 1$ and supra-entanglement stretch dominates, the average difference between toolbox prediction and experiment is 17%, and is largest at the smallest Δn

values, where there is greatest experimental uncertainty due to the measurement accuracy of the compensator.

Craze initiation stress

In oriented polymers, a back-stress acts to increase the stress required for craze formation σ_c relative to the isotropic σ_c^{iso}, and this increase is proportional to birefringence for a wide range of orientations [4]. For moderate levels of birefringence the conformational birefringence is directly proportional to the conformational deviatoric back stress, and the increase in crazing stress is given by $\sigma_c - \sigma_c^{iso} = \Delta S_c / \beta$ where β is a constant, previously determined to be 0.059 ± 0.002 for PS [5]. Thus, knowledge of the frozen-in conformational stress tensor $\mathbf{S}_c$ and of the isotropic craze initiation stress enables predictions of craze initiation stress in uniaxially melt-oriented specimens. This is computed using the toolbox for the full range of experimental orientation temperatures, and is shown in Figure 2.

The toolbox accounts for the role of molar mass on craze initiation stress, and correctly predicts increasing crazing stress with molar mass for a given draw temperature. The predictions also match the experimental observation that the effect of molar mass on craze initiation stress is greatest at intermediate draw temperatures, and smallest at the lowest and highest draw temperatures. At the lowest draw temperatures, relaxation times during melt-drawing are very long, and molecular orientation occurs with little to no relaxation (and hence independently of molar mass). At very high temperatures, there is almost complete relaxation during melt-drawing, and the oriented polymers craze at a similar stress to isotropic polymers of the same molar mass. There is an average difference between toolbox and experiment of 5.8 MPa, or 34% of the matching measured experimental value, across the full range of conditions.

Conclusions

This study has presented a toolbox for the prediction of birefringence and craze initiation stress in oriented monodisperse linear polymers of known molar mass but unknown rheology, using only polymer-dependent material constants and theory. The toolbox was applied to simulations of melt-drawing at a range of temperatures for grades of PS of unknown rheology. Outputs were combined with existing theories to produce predictions of birefringence and of craze initiation stress, and compared to new experimental measurements on two grades of linear monodisperse polystyrene of substantially different molar mass, and to the literature.

The toolbox predictions accounted for all observed experimental trends, and in particular for the role of molar mass on both birefringence and craze initiation. Birefringence predictions were on average within 17% of experimentally measured quantities for all supra-entanglement orientation states and molar masses probed. Crazing predictions were on average within 34% of experimentally measured quantities for the full range of orientations and polymers explored.

Acknowledgements

The authors gratefully acknowledge the contributions of Prof. Mackley of the University of Cambridge, and of Dr Hutchings of the University of Durham for the supply and characterisation of the polystyrenes, and of the Microscale Polymer Processing consortium for helpful discussions. This work was supported by Engineering and Physical Sciences Research Council grant EP/J007978/1.

References

1. H. Wright; C.S.N. Faraday; E.F.T. White; L.R.G. Treloar *J. Phys. D Appl. Phys.*, 1971, 4(12), 2002-2014.
2. A. De Francesco; R.A. Duckett *Polymer*, 2004, 45(23), 8005-8011.
3. E.M. Arruda; M.C. Boyce *Int. J. Plasticity*, 1993, 9(6), 697-720.
4. C. Maestrini; E.J. Kramer *Polymer*, 1991, 32(4), 609-618.
5. D.S.A. De Focatiis; C.P. Buckley *Polymer*, 2011, 52(18), 4045-4053.
6. A. De Francesco; R.A. Duckett *Polymer*, 2004, 45(12), 4297-4306.
7. D.S.A. De Focatiis; C.P. Buckley *Macromolecules*, 2011, 44(8), 3085-3095.
8. D.S.A. De Focatiis; J. Embery; C.P. Buckley *J. Polym. Sci. Pol. Phys.*, 2010, 48(13), 1449-1463.
9. P.J. Hine; A. Duckett; D.J. Read *Macromolecules*, 2007, 40(8), 2782-2790.
10. J. Embery; R.S. Graham; R.A. Duckett; D. Groves; M. Collis; M.R. Mackley; T.C.B. McLeish *J. Polym. Sci. Pol. Phys.*, 2007, 45(4), 377-394.
11. A.E. Likhtman; T.C.B. McLeish *Macromolecules*, 2002, 35(6332-6343.
12. D.S.A. De Focatiis *Polym. Test.*, 2012, 31(4), 550-556.
13. D.S.A. De Focatiis; C.P. Buckley *J. Polym. Sci. Pol. Phys.*, 2012, in press.
14. D.S.A. De Focatiis; C.P. Buckley *Polym. Test.*, 2008, 27(2), 136-145.
15. C.P. Buckley; D.C. Jones *Polymer*, 1995, 36(17), 3301-3312.
16. C.P. Buckley; P.J. Dooling; J. Harding; C. Ruiz *J. Mech. Phys. Solids*, 2004, 52(10), 2355-2377.
17. J.J. Wu; C.P. Buckley *J. Polym. Sci. Pol. Phys.*, 2004, 42(11), 2027-2040.
18. A.E. Likhtman; R.S. Graham *J. Non-Newtonian Fluid Mech.*, 2003, 114(1-12.
19. J.J. Wu; C.P. Buckley *J. Polym. Sci. Pol. Phys.*, 2004, 42(11), 2027-2040.
20. A.E. Likhtman; S.T. Milner; T.C.B. McLeish *Phys. Rev. Lett.*, 2000, 85(4550-4553.

KN-09-02

MICROSCOPICALLY-VIEWED STRUCTURAL EVOLUTION PROCESSES IN THE CRYSTALLIZATION PHENOMENA OF GREEN POLYMERS INCLUDING POLY(L-LACTIC ACID) AND *trans*-1,4-POLYISOPRENE

Kohji Tashiro*, Kaewkan Wasanasuk, and Paramita Jaya Ratri Department of Future Industry-oriented Basic Science and Materials, Toyota Technological Institute, Nagoya 468-8511, Japan

Abstract – The structural evolution processes in the crystallization phenomena were investigated for poly(L-lactic acid) (PLLA) and *trans*-1,4-polyisoprene (TPI) by performing the time-resolved wide-angle X-ray diffraction and FTIR spectral measurements. In the case of PLLA, the mesophase, prepared by melt-quenching into ice water, transformed to the δ form and then to the α form in the heating process. In the isothermal crystallization, the melt was found to change to the intermediate phase once, which transformed to the α form and δ form depending on the degree of supercooling. In the case of TPI, the similar regularization was detected in the isothermal crystallization process from the melt, that is, the melt changed to the intermediate phase and then to the α form or to the mixture of α and β forms in the isothermal crystallizations at 40°C and 30°C, respectively. The β crystalline form of TPI was found to transform to the α form *via* the oriented amorphous (or intermediate) phase in the heating process.

Keywords: Green Polymers; Poly(L-lactic Acid), *trans*-1,4-Polyisoprene, Crystallization, Structural Evolution

Introduction

Crystal structure, phase transition and crystallization behaviors of some green polymers have been investigated in detail by using an organized combination of wide-angle X-ray diffraction (WAXD), small-angle X-ray scattering (SAXS) and Fourier-transform infrared spectral data. The polymers studied were poly(L-lactic acid) (PLLA) and *trans*-1,4-polyisoprene (TPI).

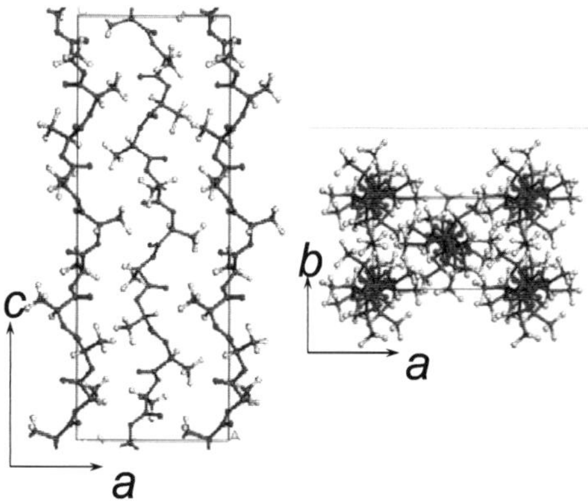

Figure 1. Crystal structure of PLLA α form [1].

Results and Discussion

(1) PLLA

Crystal Structures The crystal structures of the α and α' (δ) forms and mesophase were analyzed in detail using the 2-dimensional WAXD data [1, 2]. As shown in Figure 1, the α form takes a (10/3) helical conformation with some degree of disorder along the chain axis. These chains are packed together in the pseudo-orthorhombic unit cell. In the crystal lattice of the δ form, the molecular chains take more highly disordered conformation although the basic (10/3) helical feature is mainteined. The

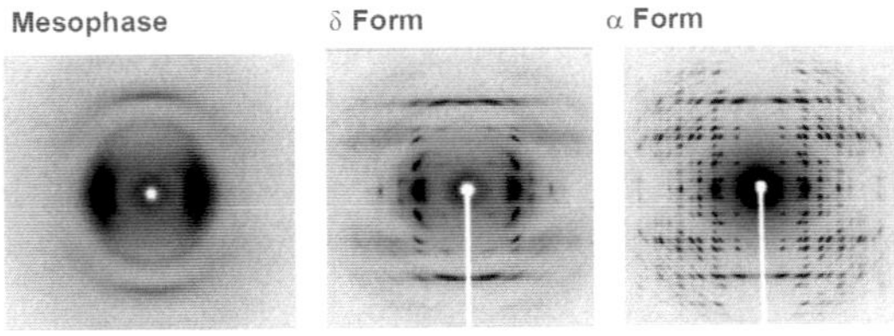

Figure 2. X-ray diffraction patterns of PLLA α form, δ forms and mesophase [2, 3].

diffuse streak lines observed in the wide-angle X-ray diffraction pattern of the δ form can be interpreted reasonably by introducing the relative height disorder among the crystalline domains of a finite size [2]. The mesophase, which can be obtained by quenching the melt into ice-water temperature, consists of the conformationally-disordered helical chains without any strong correlation between the neighboring chains as known from its characteristic X-ray diffraction pattern (see Figure 2) [3].

Structural Regularization and Mechanical Property Change in the Heating Process

When this mesophase was heated stepwisely, the X-ray diffraction peaks showed the drastic changes in the particular regions [3]. The crystallite size along the chain axis as well as that in the lateral direction was evaluated by using the half widths of the corresponding reflections. The crystallite size increased remarkably in the temperature regions where the regularization occurred from the mesophase to the δ form and then to the α form (see Figure 3). This regularization behavior was found to reflect on the change in the mechanical property of the bulk samples as shown in Figure 4, where the Young's moduli of the bulk samples consisting of these crystalline phases were measured in the heating process using a dynamic viscoelastometer [4]. In parallel the crystallite moduli of these phases were estimated using the X-ray diffraction technique under the tensile force: the crystallite moduli of the α and δ forms and mesophase were 13.76 ± 0.17 GPa,

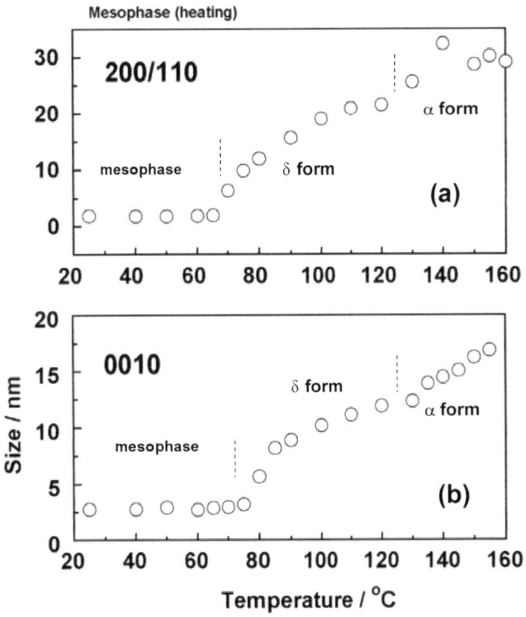

Figure 3. Temperature dependence of the crystallite sizes estimated for the oriented PLLA crystalline forms along the *a*- and *c*-axes in the heating process starting from the mesophase [3].

12.58 ± 0.15 GPa, and 7.47 ± 0.20 GPa, respectively. As seen in Figure 4, the Young's modulus of the α form decreased at *Tg* (70°C) but did not show a remarkable change up to the melting temperature region. On the other hand, the modulus of the δ form increased slightly at the transition point of

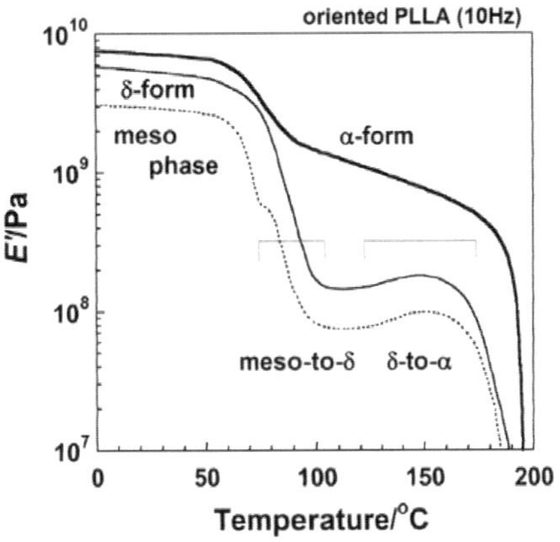

Figure 4. Temperature dependence of the Young's modulus measured for the PLLA α form, δ form and mesophase in the heating process [4].

δ-to-α form and α-to-melt. This transient increase of Young's modulus was detected since the crystallite modulus of the α form is higher than that of the δ form. Similar change was observed also for the mesophase, where the 3-step changes were detected in the temperature regions of the mesophase-to-δ form, the δ form-to-α form and α form to the melt.

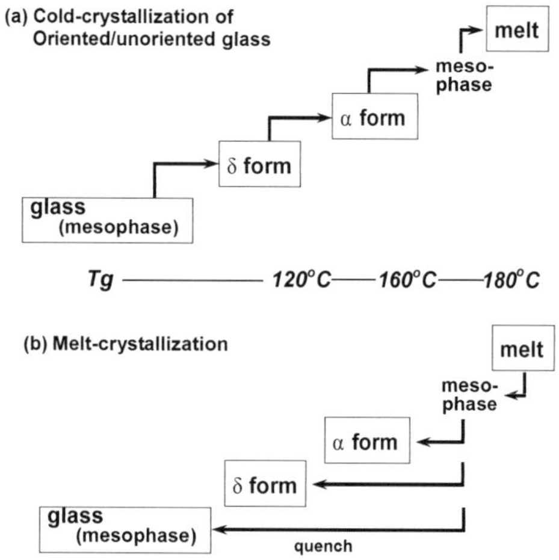

Figure 5. Crystallization processes of PLLA [3].

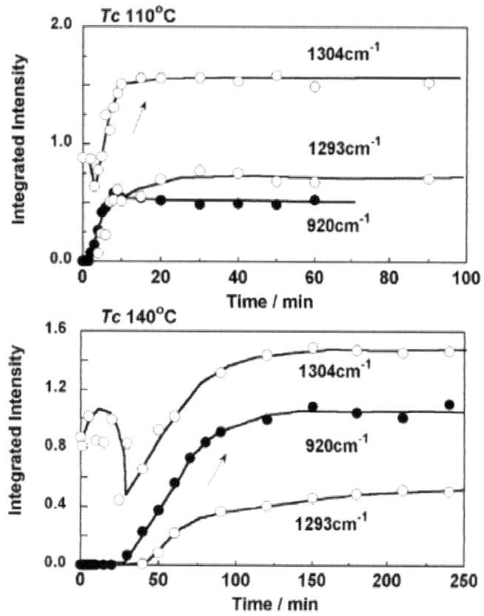

Figure 6. Melt-isothermal crystallization of PLLA at Tc = 110 and 140°C.

Isothermal Crystallizations from the Melt

The structural changes in the isothermal crystallization from the melt were investigated as a

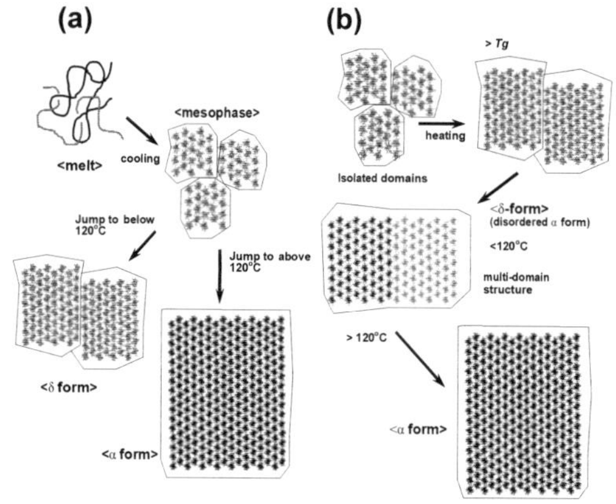

Figure 7. Schematic illustrations of structural evolution of PLLA in the (a) melt-crystallization and (b) cold-crystallization processes. The small domains are gathered together to form the domains of larger size, where the structural regularization also occurs simultaneously about the chain conformation and chain packing mode in the crystal lattice. It is noticed here the domain size increases not only in the lateral directions but also along the chain axial direction.

function of time. Figure 5 summarizes the structural evolution processes in the crystallization from the melt (and from the glass). The molten sample changed at first to the mesophase, which changed to the crystalline α form when the isothermal temperature (Tc) was higher than 120°C. When the Tc was about 110°C, the mesophase transformed to the δ form. As shown in Figure 6, the transition from the melt to the mesophase occurred so quickly, while the regularization process from the mesophase to the α form occurred slowly, which is the main reason why PLLA shows the slow crystallization. Figure 7 summarizes the structural changes in the crystallization of PLLA from the melt and glass [3].

(2) Structural Changes in Phase Transition and Isothermal Crystallization of TPI

As shown in Figure 8, the details of the crystal

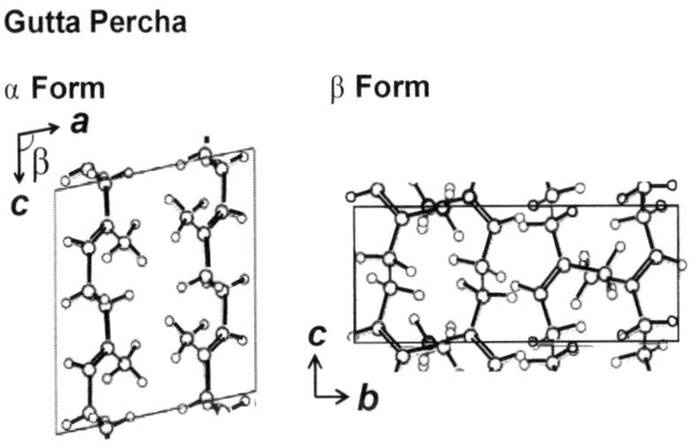

Figure 8. Crystal structures of α and β forms of TPI [5].

structures of the α and β forms were analyzed using the WAXD data with more numbers of highly-qualified X-ray diffraction spots [5]. The β form was found to transform to the α form *via* the oriented amorphous phase.

The structural evolution process was investigated in detail in the isothermal crystallization phenomenon starting from the molten state. Similarly to the above-mentioned PLLA case, the melt changed to the mesophase at first, which transformed to the α form when the crystallization temperature was about 40°C, while it changed to the mixture of the α and β form at 30°C. The detailed structural evolution was investigated by analyzing the time-resolved FTIR, WAXD and SAXS data collected in the isothermal crystallization process at *Tc* = 40°C. The quantitative analysis of the SAXS data was performed in the different crystallization stages of domain formation, domain approach and lamellar formation [6]. The structural evolution was almost the same as the case of *isotactic* polypropylene [6]. In the melt, some domains with higher density were formed and they approached each other with the correlation length ξ. The correlation length ξ became shorter with the passage of time, and finally the stacked lamellar structure was formed from these domains. Figure 9 illustrates the regularization process in the isothermal crystallization of TPI from the melt [7].

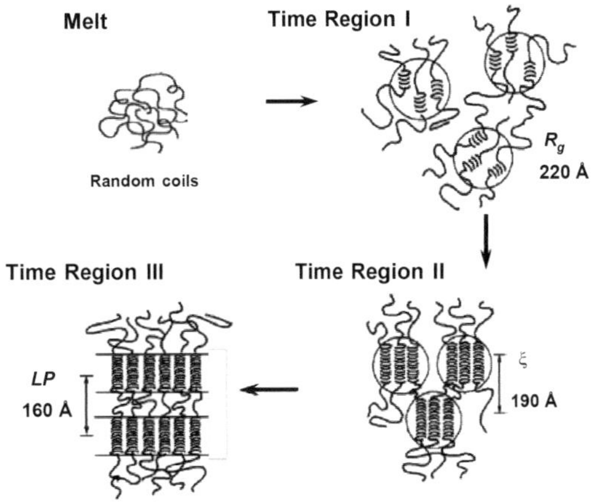

Figure 9. Schematic illustration of structural evolution of TPI in the isothermal crystallization from the melt.

Acknowledgements

This study was supported financially by MEXT "Strategic Project to Support the Formation of Research Bases at Private Universities (2010-2014)".

References

1. K. Wasanasuk; K. Tashiro, M. Hanesaka; T. Ohhara; K. Kurihara, R. Kuroki; T. Tamada; T. Ozeki, T. Kanamoto, *Macromolecules*, 2011, *44*, 6441.
2. K. Wasanasuk; K. Tashiro, *Polymer*, 2011, *52*, 6097.
3. K. Wasanasuk; K. Tashiro, *Macromolecules*, 2011, *44*, 9650.
4. K. Wasanasuk; K. Tashiro, *Macromolecules*, 2012, *45*, in press.
5. P. Jaya Ratri; K. Tashiro, M.Iguchi, *Polymer*,2012, *53*, 354.
6. R. K. Reddy; K. Tashiro; T. Sakurai; N. Yamaguchi; S. Sasaki; H. Masunaga; M. Takata, *Macromolecules*, 2009, *42*, 4191.
7. P. Jaya Ratri; K. Tashiro, *Polym. Prepr. Jpn*, 2012, *61*, 3584.

KN-09-01

IN SITU SYNTHESIS OF INORGANIC FILLERS IN MOLTEN POLYMERS: PROCESSING AND MORPHOLOGY DEVELOPMENT

V. Bounor-Legaré, Ph Cassagnau*
Université de Lyon, Univ Lyon 1, CNRS, Ingénierie des Matériaux Polymères (IMP - UMR 5223), 15 Boulevard Latarjet, 69622 Villeurbanne Cedex, France, philippe.cassagnau@univ-lyon1.fr

Abstract - This lecture aims to present some relevant results on the nanocomposite and hybrid preparation from sol-gel chemistry under polymer processing. Different methods and strategies of in situ synthesis of inorganic particles are discussed from experimental results. Finally, in situ preparation of nanocomposites and/or hybrids materials from reactive extrusion processes is original and elegant ways to prepare new polymer materials with some functionnalities such as viscoelastic, biocide and fire retardancy properties.

Keywords: in situ nanocomposites, reactive extrusion, properties

Introduction

Dispersion of inorganic nanoparticles in molten polymer is usually not a trivial process and more particularly with apolar polymers such as polyolefins. Because of the poor compatibility of hydrophilic nanoparticles with conventional hydrophobic polymer matrices, the control of the dispersion scale is generally difficult to access under usual processing conditions. An original way to overcome these problems is the in-situ synthesis of the inorganic filler in the molten polymer matrix by sol-gel chemistry. Compared with a top-down usual dispersion, this bottom-up approach by in situ synthesis of the inorganic phase from organic precursors is then an elegant way to process nanocomposites or organic/inorganic structures.

Unfortunately, one of the major drawbacks of the sol-gel process is the use of an organic solvent, which limits the application to coating of glass, metal and polymer substrata. However, news strategies have been recently developed by many authors to extent the sol-gel chemistry to bulk materials without any solvents. We report in the present work, some strategies that we have developed in our laboratory, including sol-gel chemistry in twin screw extruder.

Sol-gel reaction

The sol-gel chemistry is based on the hydrolysis and condensation reactions of metal alkoxides, M(OR): where M is a metal ion (Al, Ti, Si, ...) and R an alkyl group (R = Me, Et . .). Let us take titanium alkoxide for example.

First,the hydrolysis reaction leads to the formation of Ti-OH groups:

$$\text{Ti-OR} + H_2O \rightarrow \text{Ti-OH} + \text{ROH}. \quad (1)$$

The condensation reactions lead to the formation of oxo bridges, Ti-O-Ti represented by the following reactions:

Oxalation or water elimination

$$2\ \text{Ti-OH} \rightarrow \text{Ti-O-Ti} + H_2O \quad (2)$$

And alcoxalation or alcohol elimination

$$\text{TiOH} + \text{TiOR} \rightarrow \text{Ti-O-Ti} + \text{ROH} \quad (3)$$

Finally, the overall generalized reaction can be written

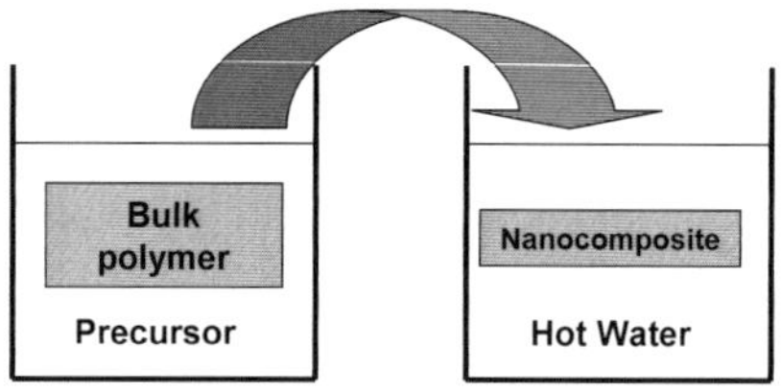

as:

$$Ti(OR)_4 + 2\ H_2O \rightarrow TiO_2 + 4\ ROH \quad (4)$$

Bulk sol-gel methods

Diffusion/Swelling: Different methods have been developed to perform sol-gel reaction in bulk conditions. For example, the bulk sample is placed in the organic precursor solution (Fig 1). Consequently, the organic precursor diffuses and swells the polymer sample. Following this, the swell sample is placed in an aqueous solution at hot temperature (T~80°C).

Fig 1: Nanocomposite preparation from polymer swelling followed by the hydrolysis/condensation reaction in hot water

The formation of the inorganic network from the organic precursor is then a two-step process: swelling to equilibrium followed by the hydrolysis/condensation reactions. Actually, this method is suitable for crosslinked elastomers and has already been reported for the preparation of nanocomposites based silicone rubber, polyisobutylene and several diene rubbers. However such a method cannot be developed in a continuous process.

Melt Mixing: The synthesis of the polymer nanocomposite can be achieved by mixing the inorganic precursor with the molten polymers under processing conditions followed by the sol-gel synthesis of the inorganic particle in solid polymer placed in hot water (Fig 2). The hydrolysis/condensation reactions are then carried out in hot water. Actually, this process is nearly the same that the first one (swelling, Fig 1), the only difference is the diffusion/mixing of the precursor in the polymer under usual processing conditions of thermoplastic polymers.

Fig 2: Nanocomposite preparation from mixing polymer/Organic precursor under molten processing conditions followed by the hydrolysis/condensation reaction in hot water

Reactive extrusion: The reactive extrusion process consists in using an extruder (single or twin-screw) as a continuous chemical reactor [1]. Besides the classical functions of a screw extruder (solid conveying, melting, mixing, pumping), additional functions are involved, such as the development and the control of a chemical reaction. Nevertheless the residence time in such continuous process is short and in magnitude order of few minutes depending on the processing conditions (flow rate and screw speed). Therefore, the hydrolysis/condensation reaction of alkoxysilane cannot be processed directly in extruder as these reactions last few hours even at high temperatures. Fortunately, the hydrolysis/condensation of titanium n-butoxide precursor occurs in few minutes at temperatures higher than 200°C as proved by Balhoul et al [2]. Therefore, these kinetics showed that the hydrolysis condensation of titanium butoxide can be carried out in a twin extruder to prepare in situ and directly a polymer/TiO_2 nanocomposite. Nevertheless, the hydrolysis-condensation reaction is not completed (<90%) at the die exit of the extruder. In order to improve the reaction conversion, water was directly introduced in the extruder as shown in Fig 3.

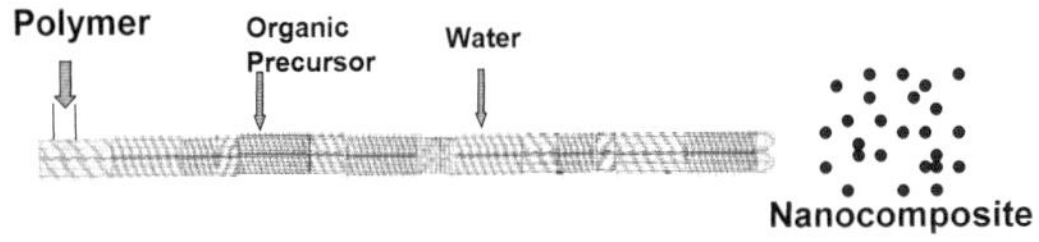

Fig 3: PP/TiO_2 synthesis by one step of reactive extrusion

Properties of in situ nanocomposites

It is well known that polymer nanocomposites can attain substantially greater properties such as stiffness, strength, thermal stability (including fire retardancy), barrier properties, etc….at very low nanofiller content compared to plastics filled with traditional micrometer-sized particles. A large volume of literature on the relationships between the nano-scale structural variables and macroscale physical and mechanical properties of polymer nanocomposites has been published over the last 15 years. Furthermore, the nature of the effects caused by adding solid nanoparticles into amorphous polymer matrix has been recently reviewed by Jancar et al [3].

Viscoelastic behavior: Fig 4a shows a clear modification of the viscoelastic behavior of in situ PP/TiO_2 nanocomposites (I-2.9% TiO_2,I-6%TiO_2 and I-9.3%TiO_2,). Actually, the terminal relaxation zone (Variation of the storage modulus G') of the nanocomposites has been strongly altered. For the highest concentration of in situ generated TiO_2 particles (6%TiO_2 and 9.3%TiO_2 samples) the viscoelastic behaviour does not show any terminal flow zone, the elastic character of this suspension becomes predominant at low frequencies with the appearance of a secondary plateau ($G_0 \approx 1 \times 10^4$ Pa and 3×10^4 Pa respectively). Actually, these nanocomposites show a solid-like behavior response which includes a non–terminal zone of relaxation leading to apparent yield stress and a shear-thinning dependence on viscosity. Furthermore, these nanocomposites exhibit a strong non-linear behavior. We then demonstrated [4] that the particle-particle interaction is the dominant mechanism through the TiO_2 fractal nanostructuration and finally explained this typical viscoelastic behaviour.

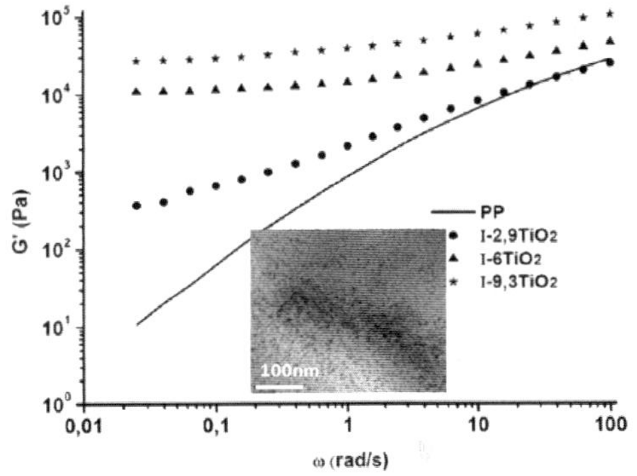

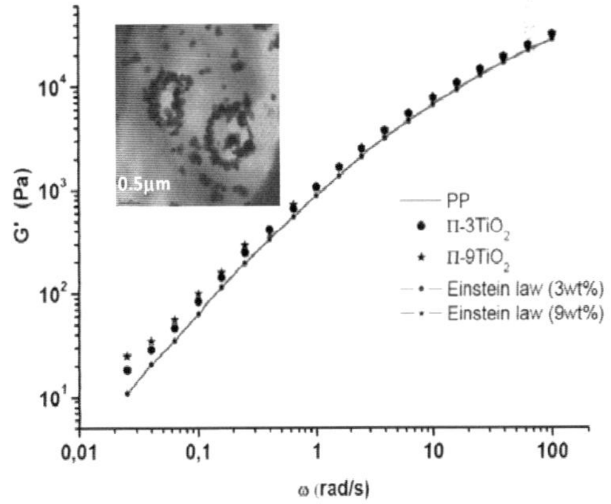

Fig 4: Viscoelastic behavior of PP filled with TiO_2 particles a): Nanocomposite from TiO_2 in situ synthesis b): Micro-composites from melt dispersion of commercial TiO_2 particles. From Bahloul et al [4]

On the contrary, Fig 4b shows that the simple dispersion of titanium dioxide (anatase) with 3 and

9%wt respectively does not considerably affect the linear viscoelastic behaviour of the PP matrix. The reinforcement of the TiO_2 micro-particles (average size~0.3µm) is weak, in agreement with the Einstein law: $G^*(\omega) = G_m^*(\omega)\left(\frac{1+3/2\phi}{1-\phi}\right)$. This result means that there are no interactions between the micro-size TiO_2 particles and no significant interactions between the particles and the polymer matrix.

To conclude, the appearance of a second plateau modulus at low frequency is mainly correlated to the formation of an aggregate-particle network from primary TiO_2 nanoparticles.

Antibacterial properties

Numerous studies have demonstrated the antibacterial power of TiO_2 particles. Actually, the antibacterial properties are very dependent on the test characteristic (irradiation or not, nature of the bacteria..) and on the inorganic domains specificity such as the size the crystalline structure, the surface area and the particle concentration. In our previous work [5], we studied the antibacterial properties of the PP, PP/TiO_2 (anatase) and *in situ* PP/TiO_2 samples submitted to E. coli and S. Aureus bacteria. As a result, it can be observed in Fig 5 that the antibacterial effect after 24 hr for PP and PP/TiO_2 (anatase) is relatively weak whereas this antibacterial property increases remarkably for the *in situ* PP/TiO_2 nanocomposite. For example, in the case of E.Coli bacteria, only 30 CFU from 33.10^6 CFU survived after 24 hr of incubation.

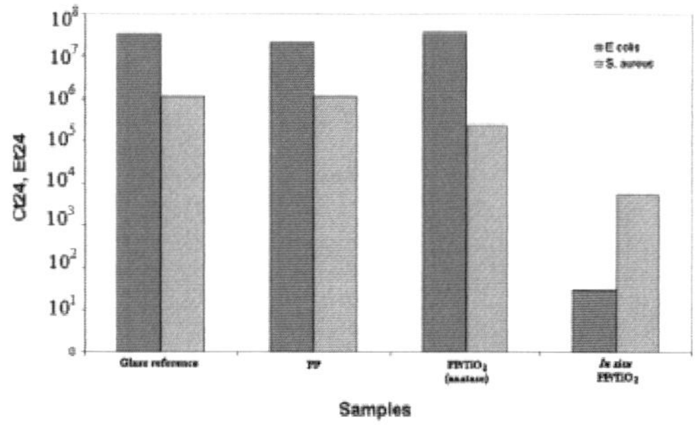

Fig 5: Antibacterial activity of PP, PP/TiO2 (anatase) and in situ PP/TiO2 (Ct24: number of survived bacteria after 24 h of incubation in class reference, Et24: number of survived bacteria in different samples after 24 h of incubation). From Bahloul [5]

Fire retardancy properties

The fire retardancy properties have been investigated on a PA6 Nanocomposites. This nanocomposite was prepared by extrusion via in-situ generation of silica functionalised with phosphorous groups. Fig 5 (cone-calorimeter measurements) shows a strong improvement of the fire retardancy properties of the PA6 nanocomposites, even at low concentration of phosphorous content (<1%).

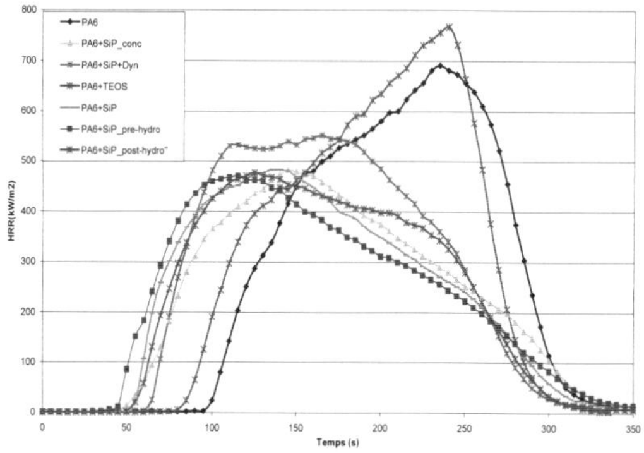

Fig 5: Fire retardancy properties of PA/nanocomposite by in-situ generation of silica functionalised with phosphorous groups. With the cooperation of the Prof G. Camino (Alessandria, Italy)

Conclusions

This survey aimed to present some relevant results on the nanocomposite and hybrid preparation from sol-gel chemistry under polymer processing. Different methods and strategies for in situ synthesis of inorganic particles have been discussed from experimental results. Actually the sol-gel chemistry is based on the hydrolysis and condensation reactions of metal alkoxides and the respective inorganic oxide metallic particles can be then synthesized in soft conditions. The present survey was focused on the processing methods for the synthesis of TiO_2 particles. For example, it was shown than a reactive process with the addition of water in extruder allowed us to complete (conversion >0.97) the hydrolysis-condensation reactions inside the extruder.

Finally, these new methods for the in situ preparation of nanocomposites and/or hybrids materials are original ways to prepare new polymer materials with some functionalities such as fire retardancy and biocide properties. This last point will be developed in details during the lecture

Acknowledgements

This lecture is dedicated in memory of Dr Daniela Tabuani (Alessandria, Italy)

References

1. Cassagnau P, Bounor-Legaré V, Fenouillot F., International. Polymer Processing, 2007, 22, 218
2. Bahloul W, Bounor-Legaré V, Seytre G, Cassagnau P, J Sol-Gel Sci Tech, 2011, 57, 86
3. Jancar J, Douglas JF, Starr FW, Kumar SK, Cassagnau P, Lesser AJ, Sternstein SS, Buehler MJ, Polymer, 2010, 51, 332
4. Balhoul W, Bounor-Legaré V, David L, Cassagnau P, J. Polym Sci: Polym Phys, 2010, 48, 1213
5. Bahloul W, Mélis F, Bounor-Legaré V, Cassagnau P, Materials Chemistry and Physics, 2012, 134, 399

KN-09-03

MORPHOLOGY OF POLYMER BLENDS TAILORED BY NANOPARTICLES IN QUIESCENT AND SHEAR FLOW

Guangxian Li[1, 2*], Yajiang Huang[1], Miqiu Kong[1], Yuan Mei[1], Wei Tong[1], Tian Xia[1], Yusong He[1] and Qi Yang[1,2]
[1] *College of polymer science and engineering, Sichuan University, China – guangxianli@scu.edu.cn;*
[2] *State key laboratory of polymer materials engineering of China, Sichuan University, China – guangxianli@scu.edu.cn*

Abstract - Utilizing inorganic nanoparticles such as silica nanoparticles and organoclays to control the microstructures and properties of multicomponent polymeric systems has aroused extensive interest from both scientific and industrial points of view in the past decades. One of the main challenges in this field is to reveal the underlying mechanism and dynamics of morphology evolution for polymer blends tailored by nanoparticles. We have investigated the morphology-rheology correlation of several immiscible polymer blends (PA6/PS, PIB/PDMS and PBT/PS) modified with silica nanoparticles in quiescent and shear flow via in situ optical-shear technique. It has been demonstrated that the changes in the interfacial tension, viscoelasticity and crystallization of polymer blends induced by silica nanoparticles play important roles in determining the structure dynamics of blends. For partial miscible polymer blends, it is suggests that changing the dynamical asymmetry of components by nanoparticles may be a promising approach in controlling their morphology-structure relationship.

Keywords: blends; nanoparticles; flow; morphology; rheology.

Introduction

The promising application of nanoparticles in controlling the microstructures and properties of multicomponent polymeric systems has aroused extensive interest from both scientific and industrial points of view in the past decade[1-3]. One of the main challenges in this field is to reveal dynamics of morphology evolution and the underlying mechanism for polymer blends tailored by nanoparticles during melt processing. On-line morphology characterization under flow is expected to provide deep insight into the role of nanoparticles in determining the morphological evolution of blends. We have investigated the morphology-rheology correlation of several polymer blends in the presence of nanosilica, including polyamide 6/polystyrene (PA6/PS), polyisobutylene/ poly(dimethylsiloxame) (PIB/PDMS) and polybutylene terephthalate/PS (PBT/PS), in quiescent and flow via *in-situ* optical-shear technique. The changes in interfacial tension, viscoelasticity and crystallization of these blends induced by nanosilica and their relative contribution on the morphology have been examined. Additionally, the nanosilica-induced morphology-rheology relationship in phase separating PS/poly(vinyl methyl ether) (PVME) blends is also exploited.

Experimental

Materials. PIB (PIB1300) from Daelim Industrial Co. Ltd and PDMS from Aldrich were used as liquid systems. PS (GP5250) was supplied by Taihua Plastic (Ningbo) co., Ltd, while PA6 (YH800) was provided by Yueyang petrochemical co., Ltd. PBT (B4500) was supplied by BASF. PS and PVME for phase separation experiments was provided by Aldrich and Tokyo Chemical Industry Co., Ltd, respectively. Fumed nanosilica (Aerosil R972, hydrophobic; Aerosil A200, hydrophilic silica) were supplied by Degussa Corp.

Sample preparation. PIB/PDMS blends were prepared by stirring the blends fiercely with a spatula by hand and then vacuumed at room temperature to remove all air bubbles. Solid PA6/PS and PBT/PS blends with different nanosilica loadings were melt mixed by extruder or internal mixer. Disk samples for rheological test were prepared by compression molding.

Rheological analysis. A strain-controlled rheometer (ARES) was used to characterize the viscoelasticity of samples. Dynamic sweeps including strain, frequency and time were performed.

Interfacial tension. The interfacial tension of blends in the presence of nanosilica were determined by the deformed droplet retraction method (DDRM).

Morphological analysis. The *in-situ* visualization of microstructural evolution of nanosilica-filled blends was accomplished on a microscope (Olympus BX51) equipped with a CSS-450 shearing system (Linkam Scientific). For droplet dynamics tests in quiescent, solid fibers or powders of dispersed phase were first sandwiched between two disks and then melted in the shear cell. Then the droplet stability of PA6/PS and PBT/PS blends were investigated. Coalescence, breakup, deformation and relaxation process of PIB/PDMS blends were carried out under flow

Results and Discussion

Relaxation and stability of nanosilica-filled droplets

Model experiments with nanosilica-filled dispersed phases allow a reliable evaluation about the relative contribution of viscoelasticity and interfacial tension on the morphology of nanosilica-filled blends. Figure 2 showed that the relaxation dynamics of slender PA6 droplets was dramatically slowed down when the nanosilica content exceeded 3 vol.%. This retarded retraction dynamics indicated that the shape stability of PA6 droplets was enhanced after the addition of nanosilica [4]. Figure 2 indicated that the relaxation mechanism of PA6 droplet was changed even if 1 vol.% nanosilica was added. Table 1 showed that the critical break aspect ratio (AR_{crit}) of filled PA6 droplets increased as a function of nanosilica concentration especially for 3~5 vol.% nanosilica filled ones.

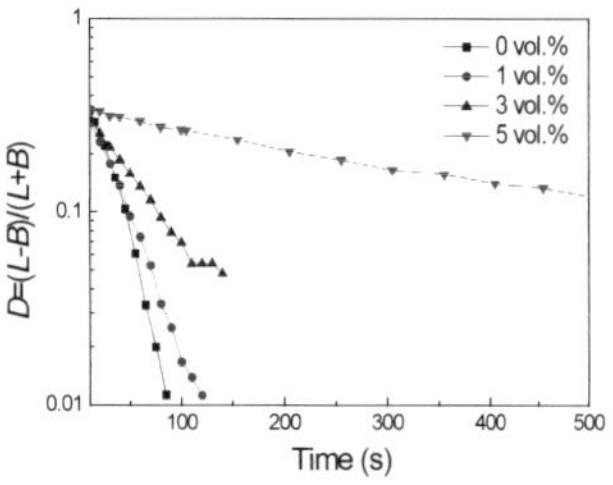

Figure 1. Deformability D of deformed PA6 droplets with different nanosilica contents as a function of time at 230 °C. The equilibrium radius of droplets is 80 μm.

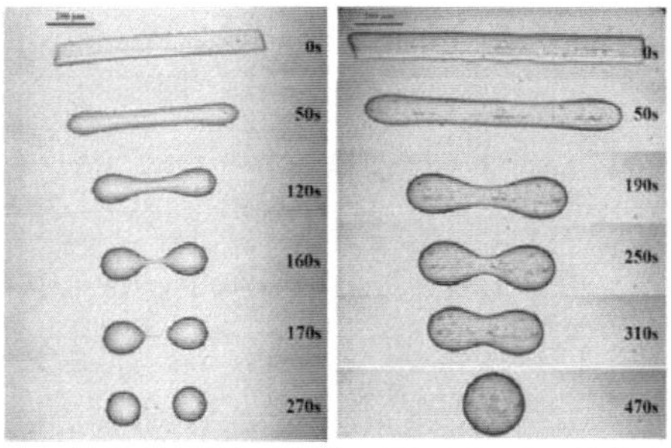

Figure 2. Relaxation of (a) a PA6 droplet (AR=10.8) and (b) a PA6 droplet (AR=11.5) filled with 1 vol.% nanosilica at 230 °C. The scale bar denotes 200 μm.

The characteristic relaxation time τ_d of a droplet is defined as

$$\tau_d = \frac{(2p+3)(19p+16)\eta_m}{40(p+1)} \frac{R_0}{\sigma} \quad (1)$$

and the AR_{crit} could be predicted qualitatively by Willemse's theory [5] in the light of viscosity ratio,

$$AR_{cr} = \left(\frac{L}{B}\right)_{cr} = \left(\frac{\lambda_m}{B}\right) = \frac{\pi}{X_m(p)} \quad (2)$$

Table 1 showed that the addition of nanosilica significant improved the viscoelasticity of PA6 droplets while the interfacial tension decreased only slightly. According to Eq. (1) and (2), the more pronounced change in viscoelasticity PA6 droplets in the presence of nanosilica should be responsible for the enhanced shape stability observed while the contribution from the reduction in interfacial tension may be the secondary. The experimental AR_{crit} was qualitatively agreed with the prediction of Willemse's theory and was closer to the experimental result reported by Stone *et al.* [6].

Table 1. Zero shear viscosities, viscosity ratios and interfacial tension and AR_{crit} at 230 °C.

SiO_2 (vol.%)	η_{PA} (Pa·s)	η_{PS} (Pa·s)	p	σ (mN/m)	AR_{crit}
0	667	1095	0.61	7.4	9
1	685	-	0.63	5.8	12
3	2046	-	1.87	5.6	15
5	18638	-	17.0	5.4	22

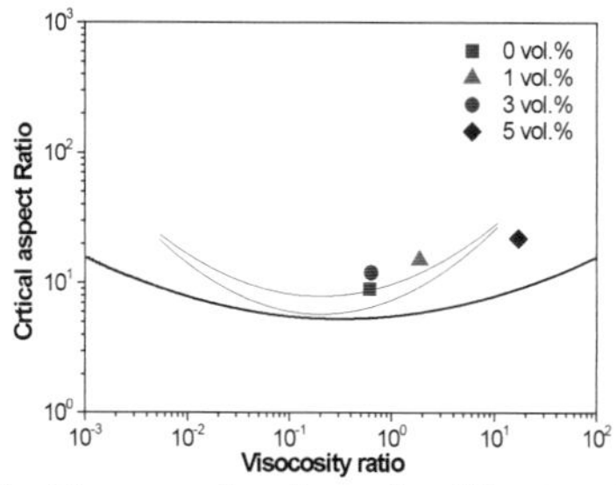

Figure 3. AR_{crit} as a function of p. The two thin lines corresponds to the experimental results of Stone *et al.*[6]. The scatter points indicates our experimental values at different nanosilica contents.

For semicrystalline polymers such as PBT, the morphology stability of their fibrils filled with nanosilica particles is also studied[7]. The addition of inorganic nanoparticles usually affects the viscoelastic properties and crystallization behavior simultaneously. Therefore, the addition of inorganic particles into crystallizable fibrils provides a possibility for improving their shape stability in a more comprehensive way [8]. Figure 4 showed that upon the addition of only 0.05 wt.% nanosilica, the shape stability of PBT fibrils during quiescent annealing at 211°C increased noticeably. As shown in Figure 5, upon the addition of nanosilica the smallest diameter for PBT fibril without breakup after annealing at 211°C for 944 s decreased while the retention of fibrils after annealing increased noticeably. It is suggested that the improved stability of fibrils is due to the accelerated crystallization in these filled PBT droplets.

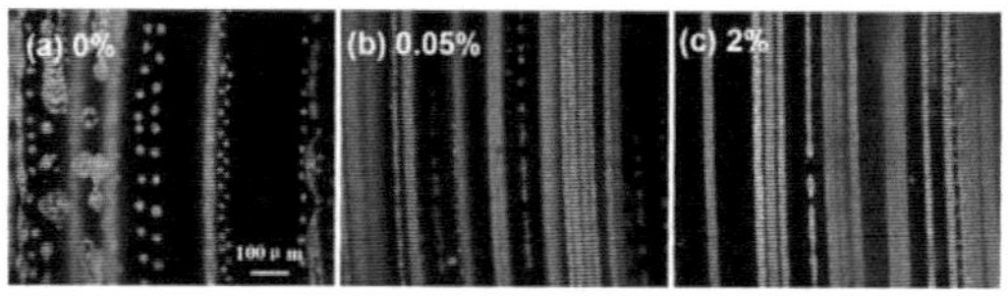

Figure 4. Polarized micrographs of annealed PBT fibrils. Nanosilica content: (a) 0 wt.%; (b) 0.05 wt.%; (c) 2 wt.%.

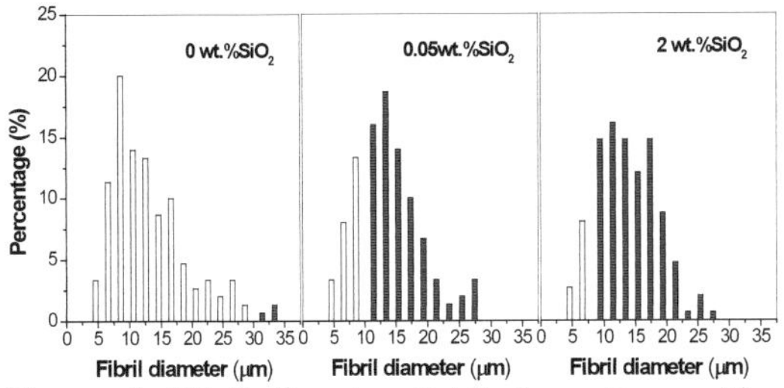

Figure 5. Fibril diameter distribution of nanosilica filled PBT fibrils after preshearing. After annealing at 211 °C for 944 s, fibrils with diameters denoted by white bars will breakup while those fibrils with diameters denoted by red bars will survive.

Figure 6 illustrated that the nanosilica decreased the half-time of crystallization of PBT at three different temperatures (216, 211 and 205°C). At the same time, the nanosilica only induced a slight increase in viscoelasticity. Figure 7 showed that the addition of silica nanoparticles suppressed the development of nodular morphology on PBT fibrils under shear flow caused by heterogeneous crystallization. It has been suggested that the loading of a relative high content of silica nanoparticles, the application of a rapid quenching and low rate shear flow are in favor of the shape stability of PBT fibrils under shear flow.

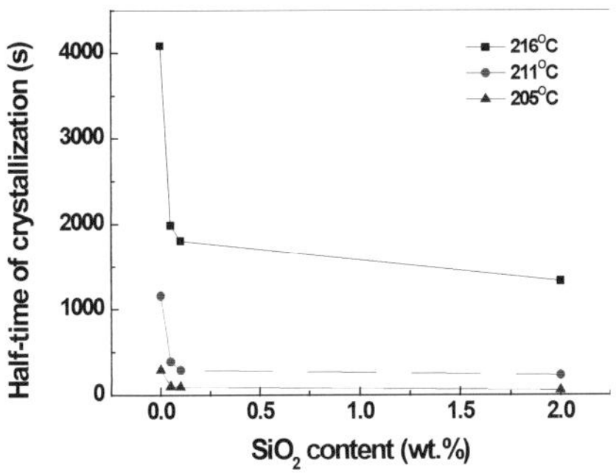

Figure 6. Half-time of crystallization for PBT.

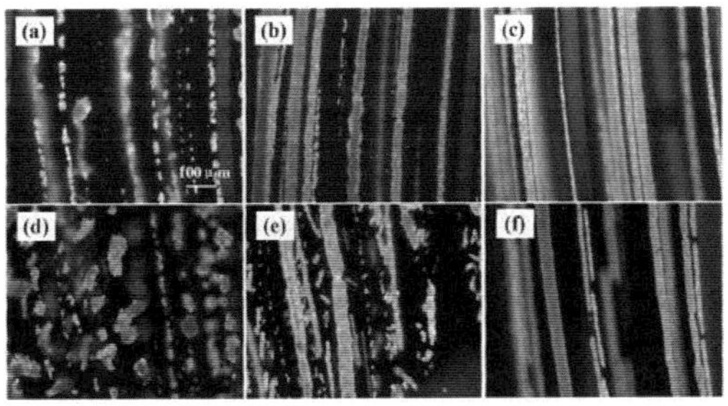

Figure 7. Polarized micrographs of PBT fibrils during crystallization at 205°C and 0.1 s^{-1} for 100 s: (a) 0 wt.%; (b) 0.05 wt.%; (c) 2 wt.%. (d)~(f) were corresponding morphology after shearing at 0.1 s^{-1} for 300 s.

Coalescence and breakup behavior of nanosilica-filled blends under flow

The morphology evolution of immiscible polymer blends under flow is closely related with the dynamic competition between the coalescence and breakup behavior of dispersed phase. Figure 8 showed that the coalescence of PIB droplets in PIB/PDMS(10/90) blends under flow was suppressed efficiently by incorporating a very small amount of hydrophobic or hydrophilic nanosilica (less than 0.4 wt%) [9]. At a higher nanosilica content, the PIB phase in blends filled with hydrophobic nanosilica obtained a equilibrium droplets size while the blends filled with more than 0.8 wt% hydrophobic nanosilica formed clusters of PIB droplets and then larger elongated PIB threads along with the flow direction. The later case suggested that the coalescence of PIB droplets was promoted. It indicated that the discrepancy in the morphology evolution of blends upon the addition of nanosilica is controlled not only by the surface chemistry of nanosilica, but also by their concentration in the blends. Similar phenomenon was also reported independently by Velankar *et al.* [10] at the same time, that depending on the contents and surface nature of nanosilica the size of dispersed phase can increase or decrease under flow. We proposed that the hydrophobic nanosilica stabilized the blends by forming a particle layer at the interface (Figure 9). By contrast, the hydrophilic nanosilica prohibited the coalescence of collided PIB droplets via the bridging effect of particle aggregates and the formation of droplet clusters at low nanosilica contents. The promoted coalescence of PIB droplets at higher nanosilica contents might be attributed to the easy migration of hydrophilic nanosilica aggregates towards the PDMS matrix at higher contents which leads to the rapid droplet coalescence. The morphological hysteresis behavior of immiscible polymer blends reflects the dependence of their steady-state morphology on the shear protocol applied. Compared with particle-free blend, the morphology hysteresis zone of filled blends was found to be expanded by the addition of hydrophobic or hydrophilic nanosilica [11]. It was suggested that the expansion of the morphology hysteresis zone in hydrophobic nanosilica-filled blend stemmed from the suppression of droplet coalescence. However, a less expansion in the morphological hysteresis zone for hydrophilic nanosilica-filled blend was due to that the breakup of PIB droplets became more difficult than that in particle-free blend.

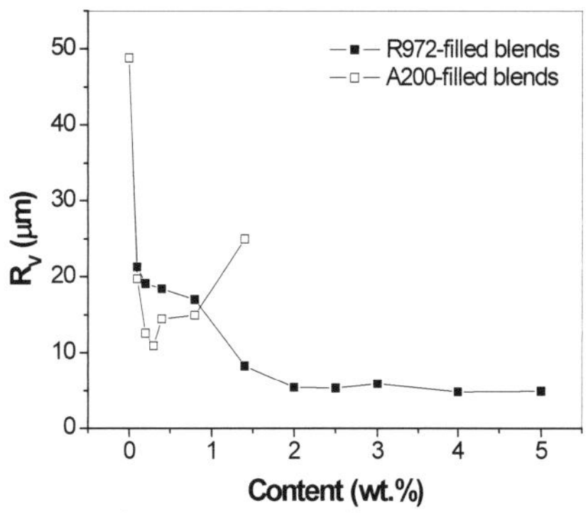

Figure 8. Volume average radius (R_V) vs. nanosilica loading in PIB/PDMS(10/90) blends after shearing at 0.05 s^{-1} for 40000 s.

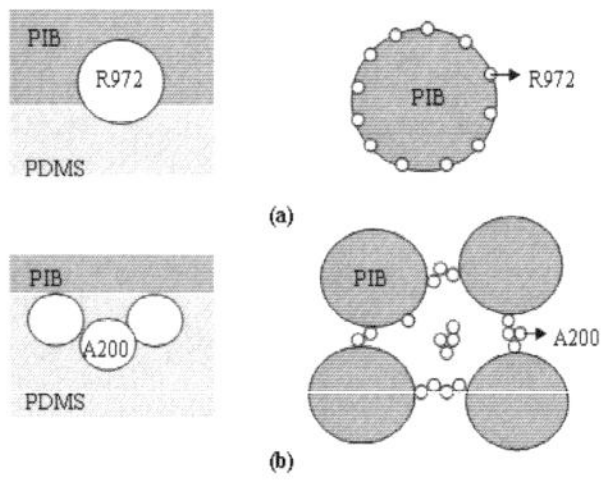

Figure 9. Schematics of the location of nanosilica in PIB/PDMS blends.

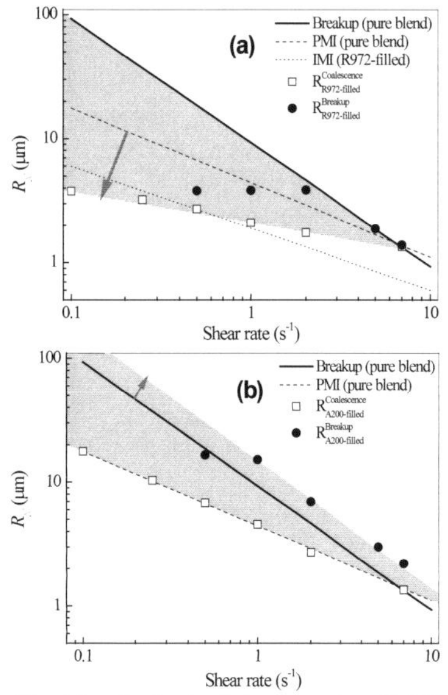

Figure 10. Morphological hysteresis of PIB/PDMS (10/90) blends filled with (a) R972 and (b) A200 nanosilica. Gray areas and those with shade line denote the hysteresis zone of pure and nanosilica-filled blends.

Morphology transition induced by nanosilica in phase separating blends

The preferential affinity between nanosilica and polymer chains is expected to effect the structure formation in polymer blends undergoing phase separation. Figure 11 showed that the addition of 4 wt.% hydrophilic nanosilica into a dynamically asymmetric PS/PVME blends changed the morphology from typical network structure formed by viscoelastic phase separation to more common island-sea structure [12].

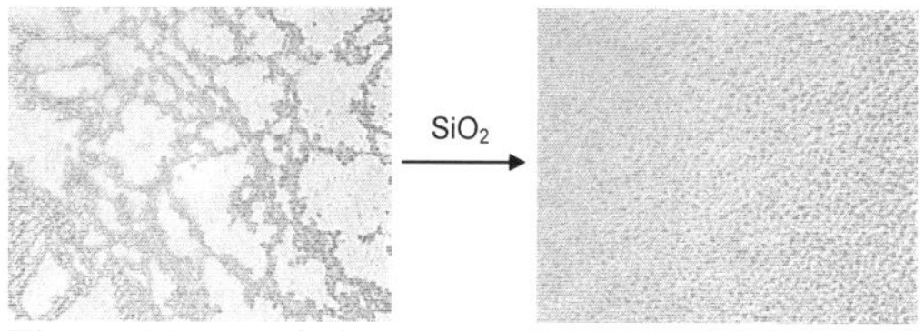

Figure 11. Morphology transition in PS/PVME(10/90) blend during phase separation induced by the addition of 4 wt.% hydrophilic nanosilica.

Table 2. Relaxation times of PS-rich phases and PVME-rich phases determined by rheology at 120°C.

Blends	Relation time (s)		ξ
	PS-rich	PVME-rich	
Pure blend	0.15	<0.002	>75
Filled blend	0.25	12.92	52

The significant nanoparticle-induced decrease of dynamic asymmetry characterized by rheological relaxation time (Table 2) was proposed to be the primary origin of the disappearance of viscoelastic phase separation in blends. Moreover, it was found that the original fast PVME component with a higher weight fraction became the slow component, which may also be responsible for this morphology transition. This result suggested that besides the dynamic asymmetry of components, the content of fast component could also play an important role in the pattern formation of these kinds of polymer blends.

Conclusions

The changes in the interfacial tension, viscoelasticity and crystallization of polymer blends induced by silica nanoparticles play important roles in determining the structure dynamics of blends. For blends filled with a small amount of nanosilica, diverse flow-induced morphologies can be observed depending on the surface nature and concentration of nanosilica added. For blends filled with a large amount of nanosilica, the enhanced viscoelasticity induced by nanosilica rather than the interfacial tension determines the morphology stability of droplets. The accelerated crystallization induced by nanosilica is also found to be in favor of the shape stability of crystallizable fibrils at quiescent and under shear flow. Changing the dynamical asymmetry of blends by nanosilica may be a promising approach in controlling the morphology and structure of similar materials.

Acknowledgements

This work was supported by the National Science Foundation of China (51003062, 51121001).

References

1. L. Elias; F. Fenouillot; J. C. Majeste; P. Alcouffe; P. Cassagnau. *Polymer* 2008, *49*, 4378.
2. P. Thareja; S. Velankar. *Rheol. Acta* 2008, *47*, 189.
3. S. Vandebril; J. Vermant; P. Moldenaers. *Soft Matter* 2010, *6*, 3353.
4. M. Q. Kong; Y. J. Huang; G. L. Chen; Q. Yang; G. X. Li. *Polymer* 2011, *52*, 5231.
5. R. Willemse. *Polymer* 1998, *40*, 2175.
6. H. A. Stone. *Annu. Rev. Fluid. Mech.* 1994, *26*, 65.
7. Y. Deyrail; R. Fulchiron; P. Cassagnau. *Polymer* 2002, *43*, 3311.
8. Y. Mei; Y. J. Huang; Q. Yang; G. X. Li. *Polymer* 2012, *doi: 10.1016/j.polymer.2012.09.028.*
9. W. Tong; Y. J. Huang; C. L. Liu; X. L. Chen; Q. Yang; G. X. Li. *Colloid Polym. Sci.* 2010, *288*, 753.
10. P. Thareja; K. Moritz; S. Velankar. *Rheol. Acta* 2010, *49*, 285.
11. Y. S. He; Y. J. Huang; Q. Li; Y. Mei; M. Q. Kong; Q. Yang. *Colloid Polym. Sci.* 2012, *290*, 997.
12. T. Xia; Y. J. Huang; X. J. Peng; G. X. Li. *Macromol. Chem. Phys.* 2010, *211*, 2240.

O-09-378

SINTERING OF UHMWPE: INFLUENCE OF COCRYSTALLIZATION AND RE-ENTANGLEMENT ON THE MECHANICAL PROPERTIES

Tiana Deplancke[1*], Olivier Lame[1], François Rousset[2], Roland Seguela[1] and Gérard Vigier[1]

[a] *Université de Lyon, CNRS, INSA-Lyon, MATEIS, UMR5510, F-69621, Villeurbanne, France tiana.deplancke@insa-lyon.fr, olivier.lame@insa-lyon.fr, roland.seguela@insa-lyon.fr, gerard.vigier@insa-lyon.fr .*

[b] *Université de Lyon, CNRS, INSA-Lyon, CETHIL, UMR5008, F-69621, Villeurbanne, France Université Lyon 1, F-69622, France françois.rousset@insa-lyon.fr .*

Abstract

One of the main issues of processing ultra-high-molecular-weight polyethylene (UHMWPE) is to overcome its very high viscosity that precludes injection or extrusion. For that reason, sintering is often required for processing UHMWPE parts. However, sintering mechanisms remain incompletely understood. Indeed, the two main mechanisms generally mentioned for interparticle welding, namely re-entanglement and cocrystallization can hardly be observed separately. In order to evaluate the influence of sintering parameters on the mechanical properties, a specific device has been designed. It enables to set different temperature, time and pressure conditions. Fortunately, due to its very high molecular weight, UHMWPE exhibits an exceptionally broad rubbery plateau so that mechanical tensile tests can be easily performed above the melting point. As a consequence, the influence of cocrystallites binding the initial powder particles can be removed or not, depending on the temperature of the test.

In the present paper, two methods are proposed to evaluate the influence of sintering parameters (temperature and time) on the mechanical properties of the material. On the one hand, monotonic tensile tests at a temperature above and below the melting point are carried out and on the other hand dynamic mechanical analysis at very small shear strain is performed. New features of the sintering mechanisms are discussed.

Keywords: sintering, UHMWPE, cocrystallization, re-entanglement.

Introduction

Processing UHMWPE is problematic due to its very high viscosity that prevents processing by conventional means such as extrusion or injection molding. Therefore processes inspired from powder metallurgy such as sintering have been developed. However, the sintering mechanisms remain incompletely understood.

Sintering is a diffusive process analogous to the diffusion of long polymer chains across interfaces (Fig. 1) in the case of welding. The sintering phenomenon is generally modeled by the reptation theory [1,2].

The mechanical characteristics are generally supposed to mainly depend on the interpenetration distance in the interfacial region [3]. It is even often assumed that the bulk mechanical properties only depend on the interpenetration distance [2,3].

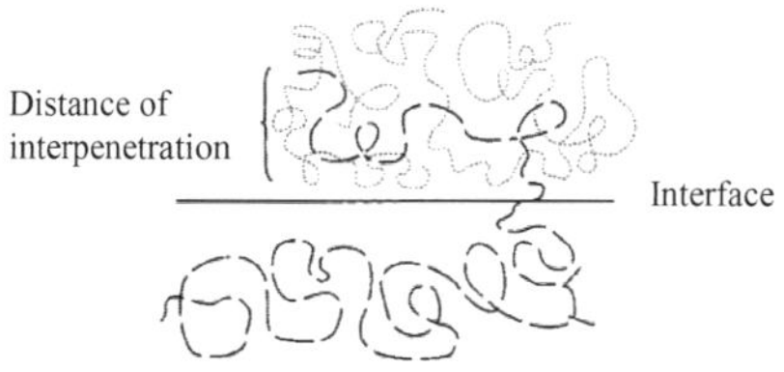

***Figure 1**: Sketch of chain diffusion across an interface.[2]*

Moreover, in the case of semi-crystalline polymers, the interfacial strength also depends on the occurrence of cocrystallization [4].

In the present work a sintering procedure has been developed. Specific mechanical tests that allow to characterize the interface strength have been carried out. Particle welding is probed and the kinetics of re-entanglement is characterized.

Method of sintering

A nascent GUR 4113 powder produced by Ticona (Oberhausen, Germany) was used to produce sintered parts. The molar mass of GUR 4113 deduced from viscosimetric measurements is $3.9.10^6$ g/mol. The average grain size is 50μm.

The sintering process is often described by two steps [3]. In a first time the densification stage corresponds to the wetting of the surfaces between the powder grains. In a second time the interface consolidation takes place via chain diffusion. This second step is the most influential in the completion of sintering.

In order to discriminate these two main stages the forming has been done in two steps. First, densification (Fig. 2) was carried out under vacuum. The temperature was kept below the melting temperature so as to neglect diffusion of chains across the interface. This procedure was the same for all samples.

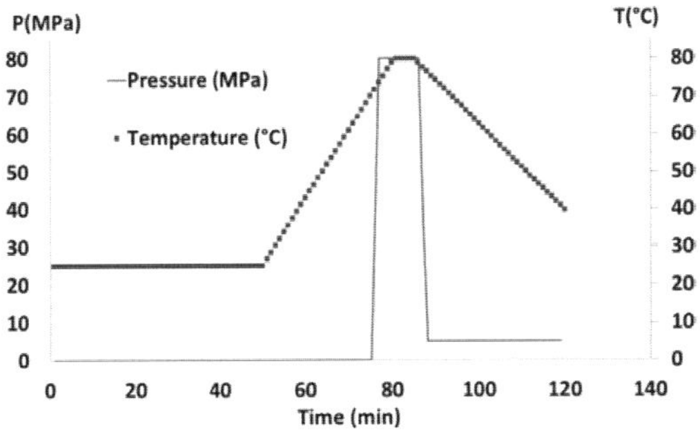

Figure 2: Procedure of densification (common to all samples).

The density after this first step is approximately $0.95 g.cm^{-3}$. Considering that crystallinity is about 60%, according to DSC measurement, this result confirms the hypothesis that densification and wetting have been correctly achieved.

During the sintering step two variable parameters have been studied: time and temperature (Fig. 3). The pressure cycle is the same for all samples: 5MPa during sintering and 20MPa during subsequent cooling in order to prevent shrinkage cavities.

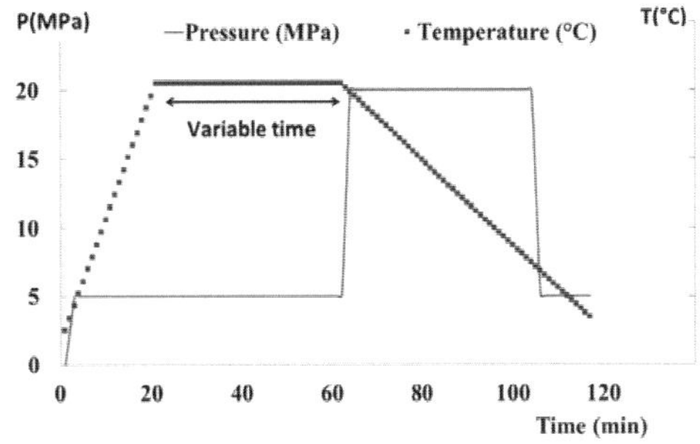

Figure 3: Procedure of sintering. The variable parameters are the time (15mn to 20h) and the temperature (150°C to 250°C).

Cylinders having 2 mm in length and 25 mm of diameter have been prepared.

Mechanical characterization

The two mechanisms generally mentioned for interparticle welding are re-entanglement and cocrystallization. Due to its very high viscosity, UHMWPE may be mechanically tested in the rubbery region, i.e. above the melting temperature, as well as in the solid state. The main interest of this dual mechanical approach is to discriminate the two types of interparticle reinforcement, i.e. cocrystallization and re-entanglement.

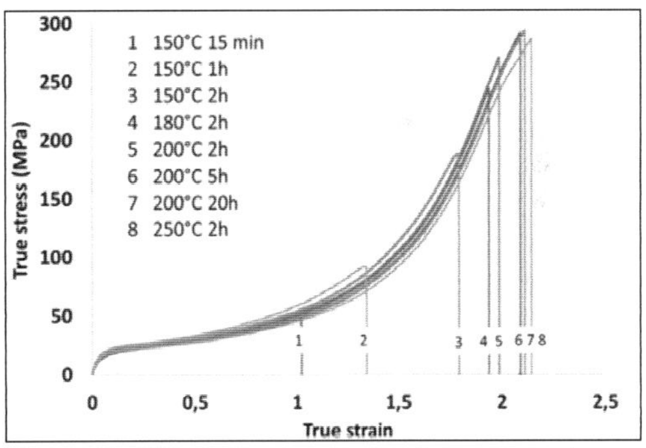

Figure 4: True stress-strain tensile curves at room temperature for different sintering conditions. The sintering characteristics are given in the legend. The strain rate is $0.35\% s^{-1}$.

The stress-strain curves at room temperature (Fig. 4) show that cocrystallization ensures a good interparticle cohesion. It is worth noticing that the tensile modulus and the yield stress are very similar for all samples regardless of the type of sintering. In contrast, the elongation at break seems to depend on sintering conditions.

The yield stress is not well defined then the flow stress at strain level 30% is considered for comparison of different samples. These stresses are close to the value (25 2 MPa). This flow stress does not significantly change with sintering condition (time and temperature).
This fact can be explained by cocrystallization. Indeed cocrystallization can only take place after chain interdiffusion over distances shorter than that required for re-entanglement. Hence, cocrystallization can hide the mechanical contribution of re-entanglement.

Tensile test experiments carried out at 150°C for allow focusing on the re-entanglement phenomenon only owing to the melting of the crystallites.

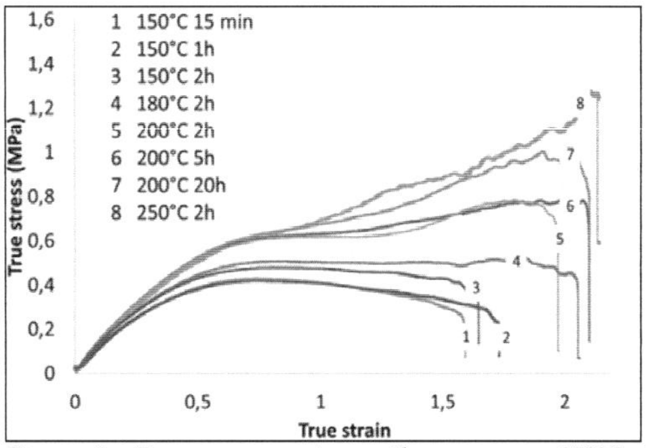

Figure 5: True stress-strain tensile curves at 150°C for different sintering conditions. The sintering characteristics are given in the legend. The strain rate is 1% s^{-1}.

The tensile curves display two different behaviors:

- Without strain hardening for sintering performed at low temperature and/or short times (curves 1, 2, and 3 for example).
- With strain hardening for sintering performed at higher temperature and/or at longer time.

This is a piece of evidence that re-entanglement increases with temperature and duration of sintering.

The elastic modulus is a relevant parameters but its determination is not an easy task.
So, the stress at strain 0.75 have been chosen (pseudo-plateau can be observed around this value). This stress is called stress at pseudo-plateau (for further.

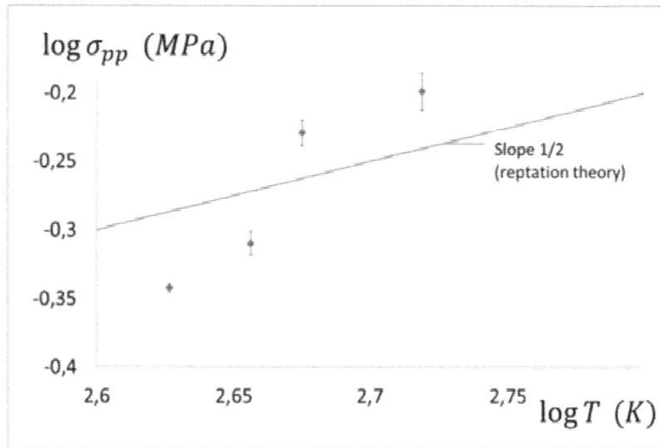

(a)

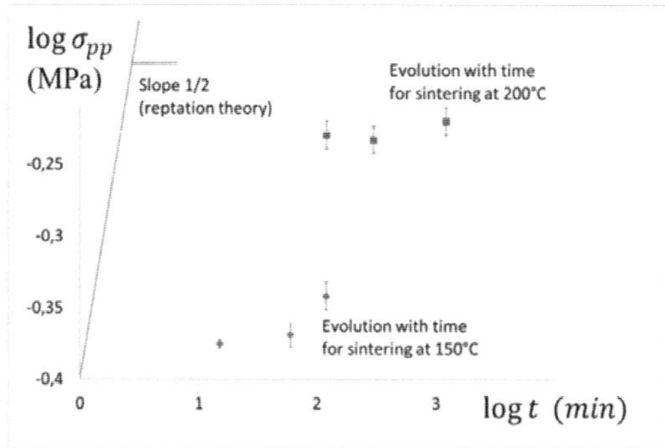

(b)

Figure 6: The stress at pseudo-plateau as a function of temperature of sintering (a) and as a function of time duration of sintering. This is obtained by the strain-stress curves at 150°C. Each point is a mean value computed from four tensile tests.

By considering that chains diffusion by reptation obeys; a 0.5 power dependence of time and temperature is obtained [5]. So a straight line with a slope 0.5 is drawn in figure 6a and 6b for the sake of comparison.

The evolution of this stress with temperature and time of sintering are different: the time dependence is very weak whereas the temperature dependence.

The elongation at break is known to be strongly dependent of re-entanglement in the rubbery state and some authors assume that this also applies to the elongation at break in the semi-crystalline state. This is corroborating by the observation in Figure 7 that elongation at break is roughly the same for all experiment at room temperature and 150°C. The evolution of elongation at break is similar to that observed on figure 6 a and b. This increase is more sensitive to temperature than time.

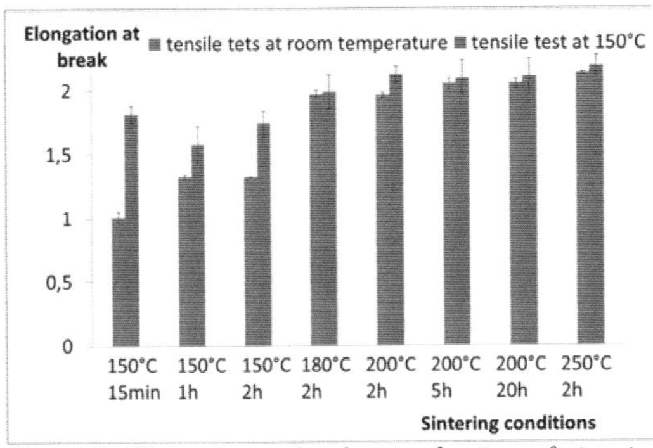

Figure 7: Elongation at break as a function of sintering conditions. Each point is a mean value computed from four tensile tests.

Conclusions

A sintering protocol was developed that envolves as variable parameters time and temperature of sintering. The samples were then tested at room temperature and in the rubbery region, which allows following the evolution of re-entanglement and cocrystallization.

The cocrystallization allows reaching good mechanical properties quickly with low temperature and short time of sintering. This can be explain by cocrystallization which takes place after chain interdiffusion over distances shorter than that required for re-entanglement.

In the rubbery state re-entanglement play a major role and it has been given evidence that temperature is much more efficient than time in the mechanism of sintering.

References

1. P.G. De Gennes ; Reptation of a Polymer Chain in the presence of Fixed Obstacles, *The journal of chemical physics,* 1971; 55, 572
2. M. Bousmina; H. Qiu; M. Gmela; Klemberg-Saphieha; Diffusion at Polymer/polymer Interfaces Probed by Rheological tools, *J. E. Macromolecules,* 1998;31, 8273-8280
3. Y.H. Kim, R. Wool; A Theory of Healing at Polymer-Polymer Interface, *Macromolecules* 1983; 16, 1115-1120.
4. Y.Q. Xue, T.A. Tervoort, S. Rastogi and P.J.Lemstra; Welding Behavior of semicrystalline Polymer. 2. Effect of Cocrystallization on Autoadhesion; *Macromolecules,* 2000; 33, 7084-7087.
5. M. Doi and S.F. Edwards; The Theory of Polymer Dynamics; *Oxford science publications,* 1986; 188-217.

O-09-065

PROCESSING OF POLYPROPYLENE/RAPESEED STRAW COMPOSITES BY EXTRUSION AND MOULDING METHODS AND ITS INFLUENCE ON THE STRUCTURE OF POLYPROPYLENE MATRIX

D. Paukszta[1], M. Szostak[2]*

[1]*Institute of Chemical Technology and Engineeringy, Poznan University of Technology,60-965 Poznan, Sklodowskiej-Curie 1sq, Poland – dominik.paukszta@put.poznan.pl*

[2]*Institute of Materials Technology, Poznan University of Technology,60-965 Poznan, Piotrowo 3 street, Poland – marek.szostak@put.poznan.pl*

Abstract - Biodegradbility of fillers, interesting physical and mechanical properties of polypropylene/lignocellulosic materials are essential motives for undertaking studies of various properties of such materials. Up to now, there have been no complex investigations defining processing of polypropylene/rapeseed straw composites. This study presents the results of research into finding optimal technological parameters of extrusion and injection moulding methods.
The structural investigations were done using the Wide Angle X-ray Scattering method. Crystallization of polypropylene in composites filled with rapeseed straw is connected with interfacial interactions, such as shearing forces at the boundary of phases: polypropylene matrix – mould wall – lignocellulosic filler. The value of the MFR for polypropylene influences the crystallization structure of the crystallization in the injection moulding process. Moreover, various parts of dumbbells have different structure of the matrix, which is caused by shearing forces and cooling conditions during the polypropylene processing.

Keywords: composites, polypropylene, rapeseed straw, extrusion, injection moulding, WAXS investigations

Introduction

In recent years, thermoplastic polymer composites filled with natural lignocellulosic materials have been increasingly applied and have become the object of scientific research. The advantages of using natural lignocellulosic materials include, among others, reduction of energy consumption, lower costs and biodegradability. Various kinds of wood, jute, coconuts, curaua, hemp, flax, sisal and kenaf are most commonly used [1,2].
Large areas of oilseed rape cultivation in countries of temperate climate (China, Canada, Australia, Germany, France, Poland and Great Britain) make it possible to obtain large quantities of straw as well as seeds. One of the directions of rapeseed straw application could be production of composites with the addition of this filler.
The first aim of this study was to determine the optimum conditions for obtaining composites by extrusion, as well as finding the right processing parameters of injection moulding process. The second objective was to observe the effect of the injection moulding processing parameters on the crystalline structure of the polypropylene matrix.

Experimental

Materials

Commercial polypropylenes with various MFR index were used as a matrix: Malen F-401, Malen S-702, Malen HP-548R and Malen HP 648T (PKN Orlen Poland). They were characterized by the following MFR indices: 2,4-3,2 g/10 min for Malen F-401, 11-16 g/10 min for Malen P S-702, 25 g/10 min for Moplen HP-548R and 50 g/10 min for Moplen HP-648T.
Lignocellulosic materials derived from various winter oilseed rape varieties were used as fillers.

Experimental techniques

The composites containing 25 wt.% of rapeseed straw and 75 wt.% of polypropylene were prepared in a single-screw extruder Fairex (Mc Nell Akron Repiquetn), D = 25 mm, L/D = 25. The samples for structural investigations were made by the injection moulding method, using Engel ES 80/25 HL S machine.
The X-ray analysis of composites were performed with a horizontal diffractometer TUR M-62, equipped in a computer control. The measurements were taken using the Wide Angle X-ray Scattering method (WAXS) at a CuKα radiation with use of nickel filtering. The following parameters were applied: voltage 30 kV, anodic current 25 mA, angle range 2Θ: 10-30°, counting step (2Θ): 0.04 counting time: 3s, with remaining standard parameters. Deconvolution of multipeaks was performed with a method proposed by

Hindeleh and Johnson [3], improved and programmed by Rabiej [4].
The proportions of the β polypropylene form in composite matrices as well as the extent of propylene matrix crystallinity were analysed. X-ray measurements were carried out in various places of mouldings, at different distances from the place of injection as well as for external and internal layers of mouldings.

Results and Discussion

Extrusion

The composite materials containing 25% wt of rapeseed straw were obtained during extrusion process from a mixture of commercial granulate of polypropylene and wooden parts of rapeseed straw.
It was known from literature data that the filler temperature should not exceed 200°C due to possibilities of occurrence of degradation processes.
In Table 1, processing parameters considered to be the best for obtaining the PP/rapeseed straw composites are shown.

Table 1. Temperatures on extruder zones during extrusion of composites

Extruder zone temperatures [°C]			
zone I	zone II	zone III	Head
140	180	195	200

It is worth remembering that when the extrusion method is employed, β form of isotactic polypropylene develops in the extrudate composite which is in agreement with earlier experiments of other researchers [5]. This, however, has no practical significance from the industrial viewpoint because the granulate is only a semi-finished product and the polypropylene matrix requires re-melting in the course of application of various processing techniques leading to manufacture of finished articles, including, for example: injection and pressing or extrusion resulting in obtaining continuous profiles.

Injection moulding

Table 2, presents the injection moulding parameters of obtaining PP composites filled with comminuted rapeseed straw. These composited were recognized as optimal after many tests.
It is evident from the data presented in the above two Tables that the temperature of 200°C was not exceeded neither in the extrusion nor injection process. In addition, the obtained mouldings were characterised by repeatability of shapes and aesthetic appearance.

Table 2. Injection moulding parameters

Injection pressure	110 MPa
Injection speed	80 mm/s
Holding pressure	7.5 MPa
Holding time	3 s
Cooling time	35 s
Screw rotation speed	100 rpm
Plasticizing pressure	1 MPa
Screw movement	65 mm
Temperatures on plasticizing unit zone I / zone II / zone III /injection nozzle	185/ 200/ 200/ 190

Structural investigations

Examinations of the matrix crystalline structure carried out in the obtained mouldings demonstrated that the structure of the crystalline phase depended on the applied polymer as well as on the place of the sample which was taken for investigations. The following two crystalline forms were found present in the examined mouldings: monoclinic α form and hexagonal β form. It was found that in the case of the composite characterised by the lowest MFR index, the highest quantity of the β form ranging from 17 – 21% was present depending on the place of measurement in the moulding. In addition, it w as observed that the moulding structure was different inside samples and in the near-wall layer. The observed phenomenon should be attributed to different viscosity of polymers characterised by a different MFR index as this caused the occurrence of shearing forces of differing values resulting in polypropylene crystallisation in the β form. Moreover, different parts of dumbbells had a different matrix structure which was caused by shearing forces and cooling conditions during the polypropylene processing. For example, in the case of the polypropylene composites characterised by MFR of 25 g/min and 50 g/min, this form was absent in internal layers. It should be emphasised that the obtained results were in keeping with thermal investigations carried out with the assistance of the differential scanning calorimetric technique [6].
In addition, it was found that the filling of the examined polypropylenes using lignocellulosic material decreased the extent of polypropylene crystallinity. It turned out that the higher was the value of the MFR index, the greater was the difference between the crystallinity degree for the non-filled polypropylene and for the polypropylene used as the composite matrix.

Acknowledgements

This research project was supported by the grant of Polish National Science Centre no. 32-6190/2011-2013.

References

1. A.K. Bledzki, J. Gassan; Prog Polym Sci 1999, 24, 221.
2. H. S. Yang, M. P. Wolcott, H. S. Kim, S. Kim, H. J. Kim; Polym Test 006, 25, 668.
3. A.M. Hindeleh, D. Johnson; Polymer 1974, 15, 697.
4. S. Rabiej: Eur. Polym. J. 1991, 27, 947.
5. J. Garbarczyk, S. Borysiak; Polimery, 2004, 49, 541.
6. D. Paukszta, J. Zielińska-Mackowiak; J Therm Anal Calorim DOI 10.1007/s1073-012-2371-x.

O-09-152

SECONDARY CRYSTALLIZATION IN ZINC-NEUTRALIZED ETHYLENE-CO-METHACRYLIC ACID IONOMERS

Yuan Rui, Brian P. Grady

[2]*School of Chemical, Biological & Materials Engineering, University of Oklahoma, Norman, Oklahoma 73019, United States*

Abstract - The purpose of this work is to explore the question of whether the fractional crystallization in a approximately random copolymer is not a function of chemical identity of the comonomer, assuming that everything else (amount of comonomer, location of comonomer along the chain) is equal. Ethylene-methacrylic acid ionomers, ethylene copolymers with a small mole fraction of ionic groups pendant to the main chain, can be used to answer that question by neutralizing the ionic groups with metal cations to a substantial degree. Neutralization has two molecular-level effects relavant to crystallization: (1) significantly increasing the diffusion constant and (2) causing microphase separation of the ionic groups into ionic-rich domains. As the neutralization level increases, the primary crystallization peak becomes smaller and the the secondary crystallization peak (which in these materials occurs at a temperature ~50 C lower than the primary crystallization peak) becomes larger. The rate of crystallization of the latter feature is extremely low; noticeable changes in crystallinity occur after 100 days of storage at room temperature while the primary crystallization peak does not change significantly with time. Is the total amount of crystallization independent of the chemical identity of the comonomer with the only shift being between the amount of primary and secondary crystallites? Our work shows that the answer is no, which disagrees with the famous paper by Flory. However, because of phase separation, this is really not a fair test of Flory's theory.

Keywords: Ionomer, crystallization, Avrami

Introduction

Ionomers are polymers that have a small mole fraction (typically less than 10 mole %) of ionic groups covalently bonded to the polymer backbone as pendant moieties. Because of energetic incompatibility between the polar ionic groups and the nonpolar polymer backbone, in most ionomers the ionic groups microphase separate into ionic-rich domains. The subjects of this study are copolymers of ethylene and methacrylic acid (E-MAA); these ionomers contain pendant carboxylic acid groups which in this study are partially neutralized with zinc. The acid form of the copolymer, i.e. hydrogen neutralizes all acid groups in the sample, is typically not classified as an ionomer. One reason for not classifying acid copolymers as ionomers is morphological, namely a separate ionic aggregate phase does not appear to occur in the acid form.

The fraction of material that is contained in crystallites changes with the introduction of acid groups; specifically a reduction in the fractional crystallinity relative to the unionized homopolymer occurs due to the copolymer effect as discussed at length in the classical paper by Flory.[1] In fact, the Flory paper makes the themodynamic-based argument that in a random copolymer where the non-crystallizable component cannot crystallize, the fractional crystallinity will be controlled exclusively by the crystallizable sequence length distribution at temperatures far below melting. Over the last almost 60 years, other authors have made observations that seem to support this statement; for example Crist and Howard[2] noted that kinetic stimuli such as rate of cooling have very little ability to shift the fractional crystallinity of random copolymers where the same is not true for homopolymers, while Mirabella[3] noted that processing in a stress field also does not change the fractional crystallinity of copolymers unlike with homopolymers. Although the fractional crystallinity does not reach the level predicted by Flory at the comonomer levels relevant for these ionomers (~5 mol %) which has been attributed to the inability of the longest crystallizable units to form extended chain crystals,[2,4] it is reasonable to state that the majority of evidence seems to support the view of Flory in his original paper.

The ionomers in this study represent a proper way to test Flory's hypothesis in the sense that the chains are identical and the identity of the comonomer is the only thing that varied. Although no claim is made that the ionomers are perfectly random; there is no reason to believe that these ionomers are any more or less random than any other "random" ethylene copolymer. However, Flory certainly did not anticipate his theory to apply to the case where another thermodynamic phase exists in the system, i.e. the ionomer aggregate phase. In other words, this paper is an extremely tough test of the Flory theory; does the copolymer sequence distribution control the fractional crystallinity even in the case where a separate phase (e.g. not crystalline or amorphous) is present.

The effect of ionic groups on crystallization is very important commercially; ionomers are well-known for forming optically clear materials because of the small crystalline regions. The crystallization of E-MAA ionomers is quite complicated since ionomers will form two distinct crystallite populations; one that melts at around 90-100°C and another that melts at a temperature around 50°C. In general these two peaks are characterized as the primary and secondary peaks, which is reasonable since in isothermal crystallization experiments the primary crystals form on a timescale of minutes to tens of minutes while secondary crystals form on a timescale of days to tens of days

In this paper, detailed crystallization kinetic expressions are fit to the primary and secondary peaks and compared. Tour knowledge, such a task has not been performed previously. The former experiments were performed by measuring isothermal crystallization exotherms at temperatures between 74 and 86°C, well above the melting temperature of the low temperature isotherm. The latter were performed at room temperature only. The overall crystallinity was determined not by summing the primary and secondary crystallization, rather it was possible to draw a baseline that encompassed both peaks as will be described in more detail in the next section.

Experimental

Zinc-neutralized ionomers were supplied by the DuPont Chemical Company at levels of 31%, 58%, 77% and 86%. The molar degree of substitution of the acid comonomer was reported by DuPont as 3.9 mol%. The zinc ionomer with 31% neutralization level converted to the acid polymer using the ion exchange resin Amberlyst 15 ion exchange resin. Samples were compression molded at ~150°C at 10000 pounds pressure in a 6" x 6" mold and then slowly cooled to room temperature.

Differential Scanning Calorimetry (DSC) Experiments

Normal crimped pans were used for the samples, and approximately 15 mg (measured to the closest 0.01 mg) samples were used and each sample was about 0.3 mm thick. The sample was heated to 160°C, held for 5 minutes then cooled at -60°C/minute to -50°C. The sample was then heated to 35°C and removed from the cell and stored at room temperature for a defined period of time. At the appropriate time, the sample was cooled to -70°C and then heated at 10°C/min and the heat capacity determined as a function of temperature. One sample was heated immediately after cooling and is designated the 0 hour sample. The overall crystallinity was determined by integrating the area between the DSC trace and the baseline heat capacity as shown schematically in the bottom three plots of Figure 1. Polyethylene enthalpy of 293 J/g was assumed[5] and was adjusted where appropriate by the weight fraction of crystallizable segments. Determination of the exotherms for the secondary and primary peak were not done using the areas under the curve, due to the difficulties of accurately determining these values as noted elsewhere.[6] Instead, the peak value of the heat capacity was used, and was subtracted from the baseline value at that temperature and is given the symbol ΔC_p. This procedure is shown in the top plot of Figure 1

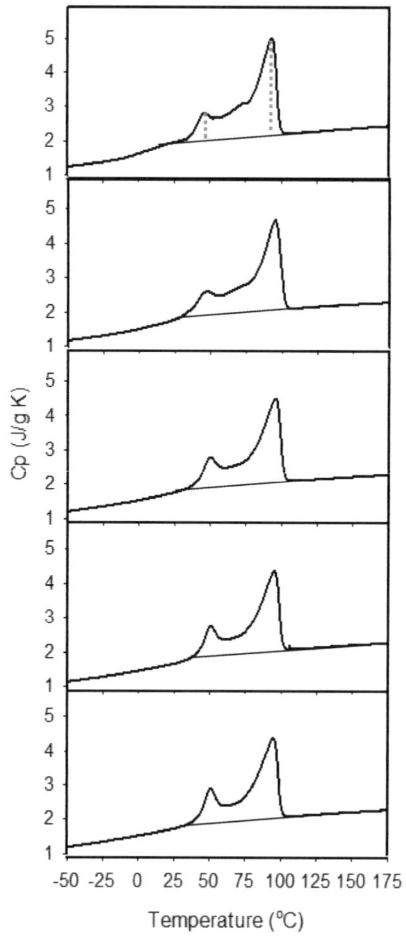

Figure 1 – Data after 1 day of room temperature annealing showing the drawing of the baseline for determination of fractional crystallization. Black vertical lines on top graph represent determination of ΔCp for determination of relative amount of primary and secondary crystals.

Results and Discussion

Figure 2 shows important parameters relevant for the primary (higher temperature) crystallization peak, i.e. the temperature corresponding to the maximum in heat capacity (heat flow) and the relative amount of primary crystals as measured by the heat capacity. The 0 hour sample has been designated as 0.4 hour in this graph and in all subsequent graphs where such a sample appears. The 0.4 hour designation comes from examination of the enthalpy of melting and determining the annealing time.

Neither the fraction of primary crystals nor the thickness of the crystals as measured by heat capacity or temperature respectively changes a large amount with annealing time. With the higher neutralized samples, there does seem to be a slight upward trend in both however. Not unsurprisingly, as the neutralization level decreases, the fraction of primary crystals increases. For the zinc-neutralized samples, at short times samples with lower amounts of neutralization have slightly lower melting temperatures but this difference disappears at long annealing times. More surprising is the clearly lower melting temperature of the primary crystal endotherm which indicates more poorly formed crystals for the acid. In other words, the absence of phase separation allows more poorly formed segments to enter the primary crystalline lamellae or chain folding is somehow more facile in the non phase-separated acid form.

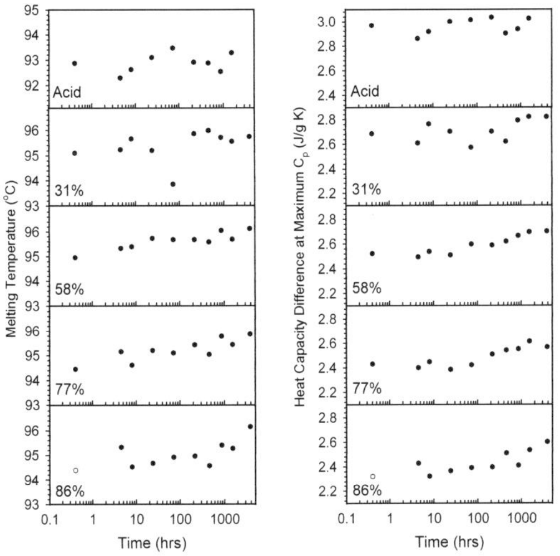

Figure 2 – Left: Primary crystallization (high temperature) melting temperature, i.e. where the heat capacity is a maximum. Right: Heat capacity at melting temperature – baseline heat capacity.

Figure 3 shows the behavior of the melting temperature and fractional crystallinity. Regarding the former, previous papers by Marand et al.[7,8] on ethylene copolymers have shown that a plot of the peak temperature of the low temperature secondary crystallization- crystallization temperature vs. time on a logarithmic scale yields a straight line as shown here. The data point at 160 days (8640 hours) was not used in the calculation of the slope for the ionomers since the data suggested that there was no change in any of the crystallinity (secondary, primary or overall) at this time, for the acid sample the 64 day (1536 hours) point was not included.

The fractional crystallization is shown on the right side of Figure 3. The Avrami equation is typically used to fit such data:

$$\varphi_c = 1 - \exp\left(-kt^n\right) \qquad \textbf{Eq. 1}$$

where φ_c is the relative crystallinity (defined to be 1 when crystallization stops at a given isothermal temperature), k is the Avrami isothermal crystallization rate constant, t is time and n is the Avrami index.

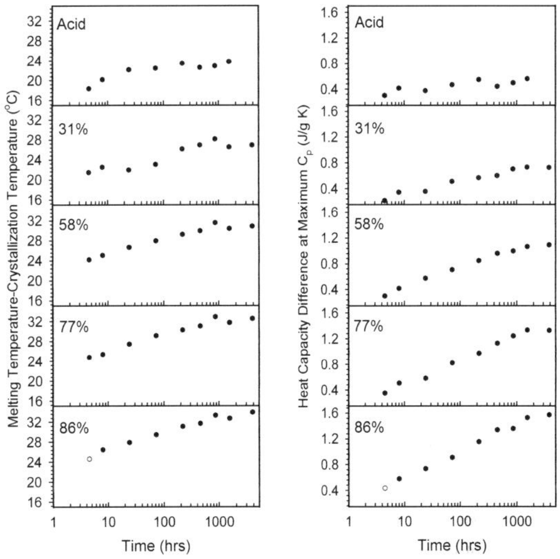

Figure 3 – Left: Secondary crystallization (low temperature) melting temperature minus crystallization temperature (room temperature). Right: Heat capacity at melting temperature minus baseline heat capacity.

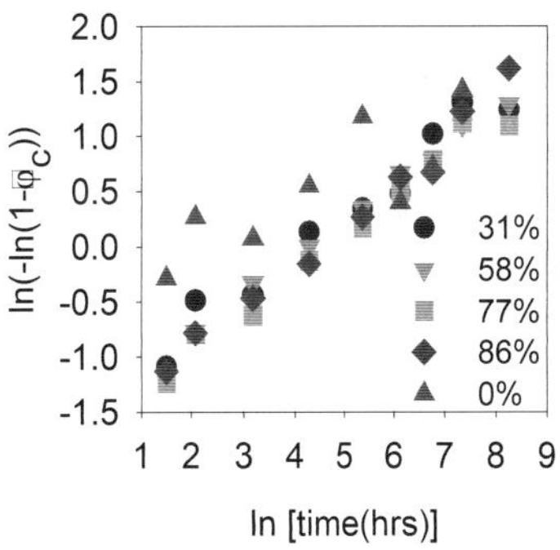

Figure 4 – Avrami plots for the various ionomers for the secondary crystallization peak.

The result of this Avrami analysis is shown in Figure 4. For the ionomers, the best-fit lines for the different samples are almost on top of one another even though the absolute rates are very different (see Figure 3). We were very surprised that the relative rates of crystallization were identical for all ionomers. The data in the top plot of Figure 4 which represents the acid form is much more scattered; it should be noted that the error bars in this graph would be larger than in the other graphs since the amount of secondary crystals is much less in this sample. Even so, the slope and intercept are significantly different. Earlier, it was stated that the last datapoint was not used because crystallization was complete in all aspects. The data presented here does not clearly confirm this assertion; however it was felt that the majority of the evidence supported the appropriateness of ignoring the highest time point. The Avrami expression by its very nature will never actually reach its ultimate value except at infinite time. The extremely low value of the Avrami rate constant (~0.4) is roughly the same value as that found previously for the growth of a low-temperature secondary crystallization endotherm in ethylene-octene copolymers.[7]

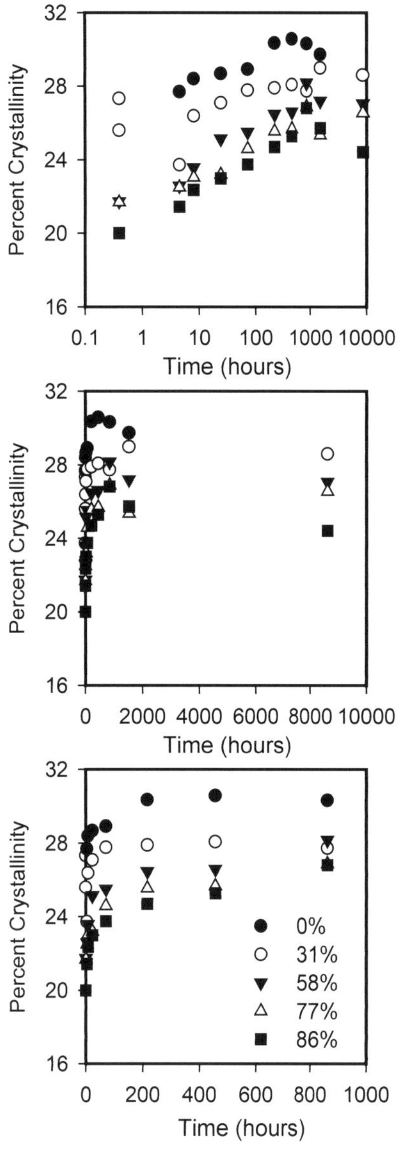

Figure 5 – Overall crystallinity vs. time plotted in different ways

The final plots from room temperature annealing experiments are shown in Figure 5 in both linear and logarithmic form. Clearly, these plots show that the final value of the fractional crystallinity depends on neutralization level. There is a significant difference between the acid form and the ionomers, while the ionomers differ slightly with a higher level of neutralization leading to a lower degree of crystallinity. Hence, the identity of the comonomer does play a role in determining the final level of crystallinity, at least in a case where the comonomer can form a separate phase. Flory almost certainly did not consider this possibility in his analysis; in fact implicit in Flory's analysis is that the system starts from a state which has a spatially random distribution of monomer segments, which is not the case with ionomers since the ionic aggregates persist in the melt. In other words, one could argue that this experiment is not really a fair test of the efficacy of Flory's theory.

Conclusions

Detailed calorimetric investigations of zinc-neutralized ionomers made from an ethylene-methacrylic acid copolymer with 3.9 mol% acid groups have showed that the fraction of repeat units located in crystalline regions changes as a function of neutralization level, which is in disagreement with Flory's theory of crystallization in copolymers. However, these ionomers contains a phase other than crystalline or amorphous phases, which is not considered in the treatment. In other words, a qualifier "as long as another phase does not form" should be added to Flory's original hypothesis about crystallization only being dependent on comonomer content. Other features include the surprising result that the Avrami parameters were independent of neutraliztion level for the ionomer for the low temperature secondary crystals, while being much different for the acid precursor.

References

1. P. J. Flory, *Transactions of the Faraday Society*, 1955, *51*, 848.
2. B. Crist and P. R. Howard, *Macromolecules*, 1999, *32*, 3057.
3. F. M. Mirabella, *J Polym Sci Pol Phys*, 2001, *39*, 2800.
4. B. Crist and D. N. Williams, *J Macromol Sci Phys*, 2000, *B39*, 1.
5. B. Wunderlich and C. M. Cormier, *Journal of Polymer Science Part a-2-Polymer Physics*, 1967, *5*, 987.
6. M. W. Spencer, M. D. Wetzel, C. Troeltzsch and D. R. Paul, *Polymer*, 2012, *53*, 569.
7. A. Alizadeh, L. Richardson, J. Xu, S. McCartney, H. Marand, Y. W. Cheung and S. Chum, *Macromolecules*, 1999, *32*, 6221.
8. H. Marand and Z. Y. Huang, *Macromolecules*, 2004, *37*, 6492.

O-09-167

EFFECT OF NANOCLAY ON THE MICROSTRUCTURE AND RHEOLOGICAL PROPERTIES OF PBT/EPOXY IN-SITU HYBRID NANOCOMPOSITES

M.M. Salehi[1,2,*], **H. Nazockdast**[2], **G.R. Pircheraghi**[2]

[1] *Polymer and Science Technology Division, Research Institute of Petroleum Industry (RIPI), Tehran, Iran – salehimm@ripi.ir; mehdisalehi@aut.ac.ir*

[2] *Department of Polymer Engineering and Color Technology, Amirkabir University of Technology, Tehran, Iran – nazdast@aut.ac.ir; greza@aut.ac.ir*

Abstract - The role of organoclay and epoxy curing agent on the microfibrillation process of epoxy phase in Poly(butylene terephtalate), PBT, matrix was studied. The PBT/epoxy blend samples with constant blend ratio (60/40) were prepared by melt mixing in a twin screw extruder. The elongational flow imposed by drawing the extrudate at the die exit permitted controlled generation of fibril morphology of the dispersed epoxy phase, with a fiber diameter of 0.8 µm and an aspect ratio greater than 100. The SEM results showed smaller epoxy droplet size for samples containing nanoclay compared to simple blends. This could be explained in terms of hindrance effect of nanoclay in reducing the coalescence as a dominating parameter due to presence of nanoclay in both phases. The linear viscoelastic results obtained for uncured samples, prepared by the PBT/epoxy blend with different draw ratio, exhibited the same rheological behavior and it was reason to destroy the fibrils of the epoxy when they did not stabilized with curing agent. However, when the organoclay was used in the polymer blend system, the rheological behavior of the systems showed different trend and the storage modulus and complex viscosity of nanocomposites were enhanced as draw ratio increase. This mainly attributed to special 3-D network formed between microfibrils whose shape was stabilized by either presence of nanoclay or accelerated curing process of fibrillated epoxy. From SEM micrographs and linear viscoelastic measurements it can be deduced that curing process has small changes on the extent of fibrillation epoxy droplet in stretched samples. But it leads to improvement of interfacial interaction between two phases in the nanocomposites.

Keywords: Microfibril, Morphology, Epoxy, In-situ, Nanocomposite.

Introduction

Combining materials with different characteristics to enlarge the spectrum of properties of the final material has been used for years. Blending and alloying of polymer blends, as well as incorporating solid substrates in polymeric matrices to make composite materials, are among the possibilities made available to scientists and engineers working in the area of polymeric materials [1].

As most polymers are immiscible, their blends form multi-phase systems with various morphologies. The nature of these structures created during processing depends on several factors: the structural parameters of the neat polymers (interfacial tension viscosity ratio, elasticity ratio), the processing conditions (shear rate and mixing time) and the volume fraction. The different morphologies that are built play a key role in inducing properties. Co-continuous morphology exhibits properties for conductivity or permeability. Lamellar morphology yields barrier properties. Since the droplets/matrix morphology improves the impact properties of blends as in the case of high-impact polystyrene, this structure has been widely studied. The situation is quite different for fibrillar morphologies although they exhibit interesting properties for conductivity or permeability, the microfibrils increase the channel length of gas or vapor passing through the materials, and good tensile properties [2]. These particular blends are sometimes called in-situ microfibrillar composites because the lamellae or the fibers are not added to the polymer, but are directly formed into the thermoplastic matrix during the processing step that is usually extrusion [3]. In-situ microfibrillar composites have attracted a great deal of interest because they can solve some problems that arise during the processing of conventional fiber-reinforced composites [4]. However, the stability of the microfibrils in these composites, when they used in the second process such as injection molding, are necessary. In this regards, the high melting point polymers (engineering thermoplastics such as nylons (PA), poly(ethylene terephthalate) (PET), poly(butylene terephthalate (PBT), polycarbonate (PC), liquid crystalline polymer (LCP), etc.) [5] are used as the fibril component in these composites. The idea of using thermoset polymers in in-situ microfibrillar composites, in order to stabilize the shape of fibrils during further processing, is very interesting and was proposed by Fenouillot for polystyrene/epoxy systems [3].

Nano-size particulates filled polymers (nanocomposites) are of great interest due to their lower density, excellent performances and attractive prospect, compared to conventionally filled polymers

[6]. It has been reported that nano-fillers such as nanoclay can also act as a compatibilizer in immiscible blends during mixing process [6].
In this regards, we investigated the effects of curing of epoxy and presence of nanoclay on the fibril formation of epoxy in the PBT/epoxy blends and rheological behavior of these hybrid nanocomposites

Experimental

1 Materials

The basic materials used in this work were: PBT (Pocan® B 1505), supplied by Bayer, epoxy based on a diglycidyl ether of bisphenol A (DGEBA) kindly supplied by Nan YA Plastic Corporation, an organophilic methyl, tallow, bis-2-hydroxyethyl, quaternary ammonium exchanged montmorillonite (MT2EtOH Cloisite 30B supplied by southern products) was used as the organoclay and the diamine was 4, 4'-methylenebis [2, 6-diethylaniline] (MDEA) supplied by Aldrich.

2 Sample Preparation

The PBT/epoxy/organoclay (60/40/5) hybrid nanocomposite samples were prepared by melt compounding in a Brabender (Germany) DSE25 intermeshing, corotating twin- screw extruder with a length/diameter ratio of 32. The throughput of the extruder was kept constant by means of an independent dozing system. The temperature profile along the barrel of the extruder was: 130, 230, 240, 250 and 240 ºC. The extruder is equipped with 4 mm diameter capillary die. The extrudate is pulled out through a drawer roller and the pulling speed is adjusted to impose a draw ratio to the extrudate. All samples were prepared at the same processing condition (screw speed = 100 rpm and throughout rate = 13.3 g/min). The draw ratio (1, 3 and 5) was defined as $\lambda = V_r/V_e$, where V_r is the linear velocity of the rolls and V_e is the linear velocity of the extrudate. The length between the die and the rolls was 0.2 m. All materials are fed simultaneously to the extruder. The curing agent was added to the nanocopmosites based on stoichiometric ratio. The hybrid nanocomposite samples were post-cured in an oven. The time and temperature of the post-curing process were 120 minutes and 150°C, respectively.

3 Sample Characterization

WAXD Measurements

WAXD experiments were conducted on a Phillips X'pert diffractometer with Cu Ka radiation of wavelength (λ) 0.154 nm generated at 40 kV and 40 mA. The diffractograms were scanned in the 2θ range from 1° to 10° at the rate of 1°/min at ambient temperature and the measurements were recorded every 0.04°.

Morphology Characterization

The morphology of the blends was investigated using scanning electron microscope, SEM, (PHILIPS XL30, Netherlands) with 20kV accelerating voltage. The samples were kept in liquid nitrogen and the brittle fractured. The cryogenically fractured surfaces were coated with gold for enhanced conductivity using SPI sputter coater.

Rheological measurements

The melt viscoelastic properties of the samples were studied with a rheometric mechanical spectrometer (Paar Physica USD 200, Austria) equipped with parallel plate geometry with diameter of 25 mm. All samples were vacuum dried under 80 °C in an oven, in order to prevent moisture induced degradation. The rheological measurements were carried out in linear and nonlinear viscoelastic regions. The melt linear viscoelastic behavior of the samples was executed in a frequency range of 0.1 – 1000 sec^{-1} using strain amplitude of 1%, proven to be in the linear viscoelasticity range by means of strain sweep measurements at temperatures of 260 °C.

Results and Discussion

1 Residence time distribution in extruder and kinetics of curing of epoxy

The mean residence time distribution of materials in the twin extruder and compatibility of this time with the time for curing of epoxy is the crucial processing parameter to attain desired morphology in the thermoplastic/thermoset polymer blends. We determined the rate of curing of epoxy in our previous work, [7]. The gel time of epoxy, epoxy/nanoclay and PBT/epoxy samples were about 780 s, 480 s and 1200 s, respectively. The mean residence time distribution for the feeding rate of 13.3 g/min and screw speed of 100 rpm was 440 s, [8]. From comparing these results one can realize that the curing time is compatible with the use of the curing agent in the twin screw extruder and these samples could hardly be cured during the melt blending and extrusion process. In other word, in order to achieve complete curing of epoxy, the post curing process is indispensable.

2 PBT/epoxy blends

2.1 Observation of the morphology

Fig. 1 shows the SEM images of cryogenically fractured surface of PBT/Epoxy blends with different draw ratios.

Draw ratio = 1 *Draw ratio = 3* *Draw ratio = 5*

Figure 1 - SEM images of cryogenically fractured surface for PBT/Epoxy blend samples at different draw ratios.

As it can be seen, the non-stretched blend sample (λ=1) shows a matrix-dispersed type morphology in which the epoxy particle with average diameter (3.5 μm) is finely dispersed in PBT matrix. The stretched sample (λ=3) shows the fibrillar morphology with a mean diameter of 1 μm. The aspect ratio is estimated to be higher than 50. It was impossible to determine it precisely from the SEM micrographs because of the length of the fibers. With increasing the draw ratio from 3 to 5, the morphology of the blend did not changed drastically but the average diameter of the fibrils decreased to 0.8 μm.

2.2 Rheological properties of the PBT/epoxy blend samples

The storage modulus, G', and complex viscosity, η*, of the PBT/epoxy blends with different draw ratios are shown in Fig. 2. As it can be seen, these three samples, in spite of the different morphology, showed the same rheological behaviour. This may be attributed to the un-stabilizing the fibrillar morphology in the stretched samples and recovery of the fibrils after melting the blend. In other words, after melting the stretched samples, the shape of fibrils are changed and the morphology of the samples changed from fibrillar type morphology to the matrix-dispersed type morphology. As a result, stabilizing the morphology with the curing of the epoxy phase is necessary.

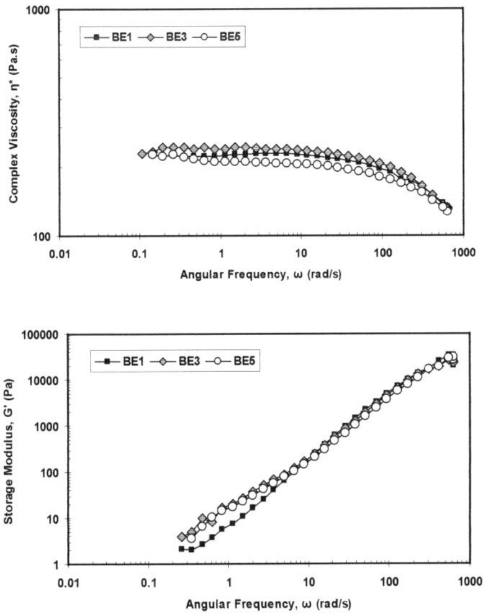

Figure 2 - Plots of storage modulus, G', and complex viscosity, η*, vs. angular frequency for the uncured PBT/epoxy blends at different draw ratios.

By addition of the curing agent to the PBT/epoxy blends was expected the morphology of the stretched samples are stabilized and with remelting the blends the shape of the fibrils of epoxy did not changed. This phenomenon could be clearly seen in the linear viscoelastic properties of cured samples were shown in Fig.3.

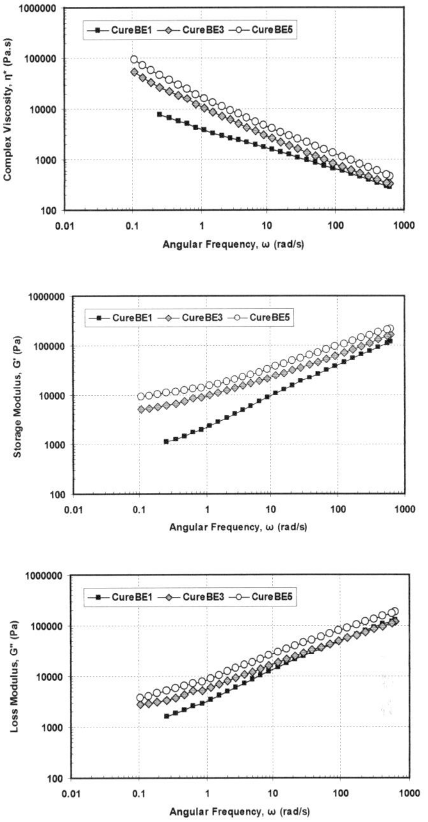

Figure 3 - Plots of storage modulus, G', loss modulus, G", and complex viscosity, η*, vs. angular frequency for the cured PBT/epoxy blends at different draw ratios.

Both stretched samples (CureBE3 and CureBE5) exhibit much greater viscosity and melt elasticity with pronounced solid body behaviour at low frequency range. The storage modulus (G') values of sample with draw ratio of 3 (CureBE3) was higher than that of for non-stretched sample (CureBE1). This can be attributed to the contribution of epoxy fibrils, with a high aspect ratio, towards the elastic properties of the samples. CureBE1 contains only spherical domains of epoxy which are not capable of contributing to the storage modulus like the fibrils especially at lower and higher frequencies.

Interestingly the difference in loss modulus (G") values for the samples is negligible. This indicates the viscous behavior at higher frequencies is identical for the stretched samples irrespective of the draw ratio. The difference in G" values of CureBE3 and CureBE5 with CureBE1 at low frequencies is lower than the corresponding difference in the G' values. This indicates that the epoxy fibrils have a more significant effect on the elastic behavior than the viscous behavior of the blend. The complex viscosity (η^*) of the stretched samples are higher than that of the non-stretched sample. This may be attributed to the reinforcing effect of epoxy fibrils on the PBT matrix.

3 PBT/epoxy/nanoclay nanocomposites

3.1 WAXD Results

Fig. 4 shows the WAXD pattern of organically modified montmorillonite (Cloisite 30B) and PBT/epoxy/nanoclay hybrid samples. As it can be seen the organoclay exhibits one broad characteristic peak at $2\theta = 4.83°$. By considering these results, one may notice that the characteristic peak of the organoclay is changed into two peaks for the prepared samples. The presence of higher order peaks (peak 2) confirm the presence of ordered tactoid in the nanocomposites. Extent of intercalation is almost equal for BN and EN samples and this indicates that PBT and epoxy have the same ability for intercalation process. This intercalated microstructure for BN and EN samples obtained via two different mechanisms. On the other hand, in EN sample, the combination of mass diffusion and more thermodynamic affinity between epoxy and organoclay are the affecting parameters in formation the intercalated microstructure and increasing the d-spacing of nanoclay. In contrast, in BN sample, the convective forces originated from the flow field have a key role in increasing the d-spacing of nanoclay.

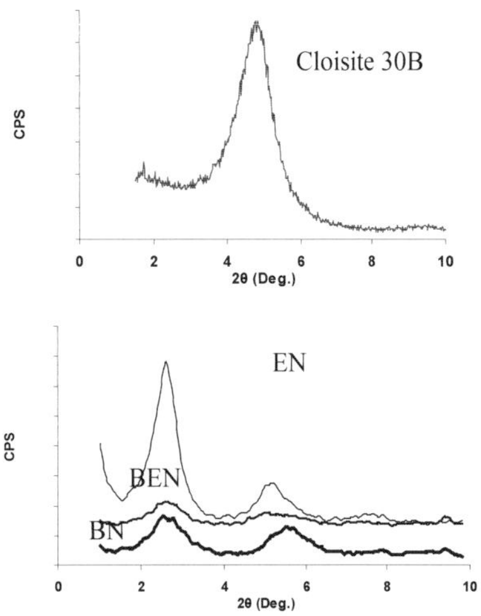

Figure 4 - XRD patterns for the organoclay and nanocomposite samples.

3.2 Observation of the morphology

Fig. 5 shows the SEM images of cryogenically fractured surface of PBT/epoxy nanocomposites with different draw ratios.

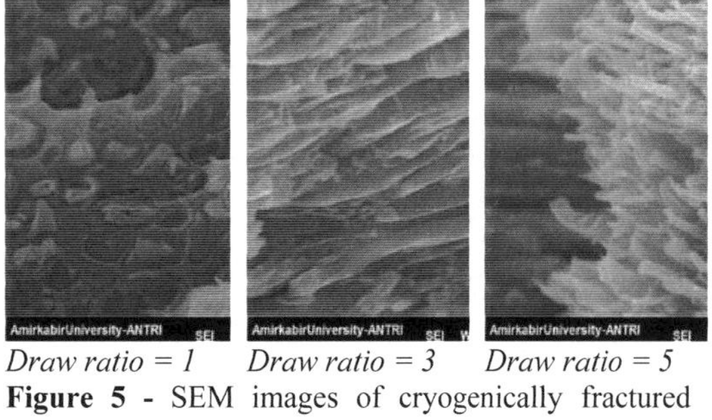

Draw ratio = 1 *Draw ratio = 3* *Draw ratio = 5*

Figure 5 - SEM images of cryogenically fractured surface for PBT/epoxy nanocomposite samples at different draw ratios.

As it can be seen, the non-stretched nanocomposite sample (BEN1) shows a matrix-dispersed type morphology similar to that observed for BE1 blend samples, in which the epoxy particle with average diameter (4 μm) is finely dispersed in PBT matrix.

These results indicate that, in the uncured nanocomposite samples, presence of organoclay play different role in determining the epoxy droplet size. This effect arises from the changing in viscosity and nanoclay concentration in each phase. On the other hand, in BEN1 sample, compared to BE1 sample, presence of nanoclay in the epoxy suppress the droplet breakup process. The stretched sample (λ=3) shows the fibrillar morphology with a mean diameter of 1.3 μm. The aspect ratio is estimated to be higher than 50. With increasing the draw ratio from 3 to 5, the morphology of the blend did not changed drastically but the average diameter of the fibrils decreased to 1 μm.

3.3 Rheological properties of the PBT/epoxy hybrid nanocomposite samples

The storage modulus, G', and complex viscosity, η^*, of the PBT/epoxy nanocomposites with different draw ratios are shown in Fig. 6. As it can be seen, these three samples showed the different rheological behaviours. This in contrast with the results obtained for PBT/epoxy blend samples. For BE samples the epoxy fibril did not stabilize hence when they melting for the rheological analysis the morphology of the stretched samples (BE3 and BE5) changed to the non-stretched sample (BE1).

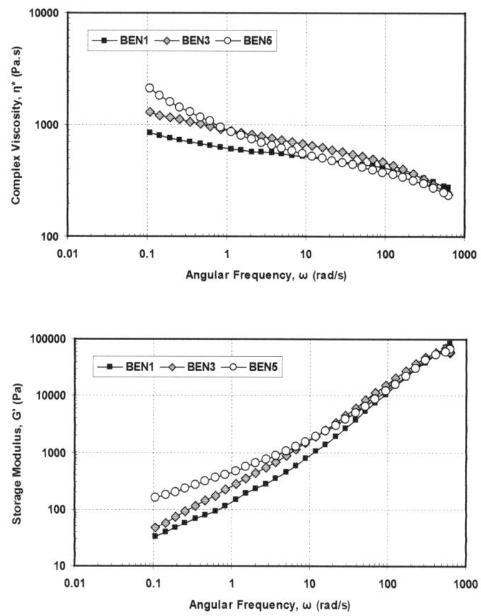

Figure 6 - Plots of storage modulus, G', and complex viscosity, η^*, vs. angular frequency for the uncured PBT/epoxy nanocomposites at different draw ratios.

But interestingly the epoxy fibrils in stretched nanocomposite samples (BEN3 and BEN5) seem stabilize in the uncured samples. In other words, the nanoclay led to stabilize the epoxy fibrils in the nanocomposite samples. Nanoclay can perform this stabilization via two different mechanisms; 1) stabilizing the epoxy fibril with nanoclay layers and platelets, 2) silicate layers catalyzed the curing process of epoxy, [7], then the presence of nanoclay lead to occur partial curing of epoxy fibrils.

By addition of the curing agent to the PBT/epoxy nanocomposites was expected the morphology of the stretched samples are completely stabilized. The same results can be seen for the cured nanocomposites (Fig. 7).

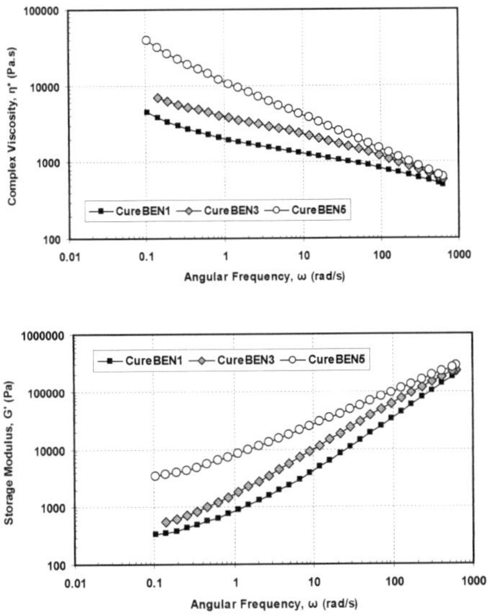

Figure 7 - Plots of storage modulus, G', and complex viscosity, η*, vs. angular frequency for the cured PBT/epoxy nanocomposites at different draw ratios.

It can be deduced that the curing process only led to stabilize the epoxy fibrils. However, increasing the complex viscosity and storage modulus for cured samples compared to those of uncured samples mainly due to curing of epoxy droplets and fibrils and increasing the interfacial adhesion between two phases.

Conclusions

The stretched PBT/epoxy blend (60/40) showed a microfibrillar type morphology in which the epoxy microfibrils were finely dispersed in the PBT matrix that was consistent with their melt linear viscoelastic behavior. It was shown that when the curing agent was injected into the extruder, its reactivity with epoxy at high temperature was high enough and selectively affected on the epoxy phase. The linear viscoelastic results obtained for uncured samples, prepared by the PBT/epoxy blend with different draw ratio, exhibited the same rheological behavior and it was reason to destroy the fibrils of the epoxy when they did not stabilized with curing agent. However, when the organoclay was used in the polymer blend system, the rheological behavior of the systems showed different trend and the storage modulus and complex viscosity of nanocomposites were enhanced as draw ratio increase. On the other hand, the organoclay led to stabilize the fibril morphology of epoxy without any recovery in the fibrillar shape of the epoxy phase. Finally, when the curing agent was injected to the extruder, the morphology of epoxy fibrils stabilized and the linear viscoelastic results exhibits a behaviour similar to those reported for microfibrillar composites

References

1. M. F. Boyaud, A. Ait-Kadi, M. Bousmina, A. Michel, Ph. Cassagnau, Polymer 2001, *42*, 6515-6526.
2. S. Chaput, C. Carrot, M. Castro, F. Prochazka, Rheol. Acta 2004, *43*, 417-426.
3. F. Fenouillot, H. P. Camby, Polym. Eng. Sci. 2004, *44*, 625-637.
4. Y. Seo, H. J. Kim, H. Kim, H. W. Rhee, Polym. Eng. Sci. 2002, *42*, 951-960.
5. L. Zhang, K.C. Tam, L.H. Gan, C.Y. Yue, Y.C. Lam, X. Hu, J. Appl. Polym. Sci. 2003, *87*, 1484-1492.
6. J. Chen, P. Chen, L. Wu, J. Zhang, J. He, Polymer 2007, *48*, 4242-4251.
7. M. M. Salehi, H. Nazockdast, Gh. R. Pircheraghi, J. Macromolecular Sci. Part B: Physics 2012, *51*, 906-925.
8. M.K. RazaviAghjeh, H. Nazockdast, H. Assempour, Int. Polym. Process. 2004, *4*, 335-341.

O-09-215

CONTROL OF HIERARCHICAL STRUCTURE OF POLYMER ARTICLES VIA NOVEL PROCESSING: FROM TRADITIONAL PROCESSING TOWARD "STRUCTURING" PROCESSING

K. Wang, Z.M. Li, M.B. Yang and Q. Fu*
College of Polymer Science & Engineering, Sichuan University, State Key Laboratory of Polymer Materials Engineering, Chengdu, 610065, P.R. China – qiangfu@scu.edu.cn

Abstract - New demands of ultrahigh performance, versatility of property, functionalization and facile manufacturing, have been proposed for the modern polymer processing area. To meet these high-quality demands, the novel "structuring" processing methodology, has recently been developed, in which besides the common processing variables (temperature, pressure, strain, shear velocity, duration time and so on), some unique external fields are applied by using ingenuity of processing instruments and/or particular processing procedure. In this talk, three kinds of "structuring" processing methods, viz. in-process morphology manipulation during injection molding, layer multiplying coextrusion (LMCOE) and in-situ microfibrillation processing, have been discussed. The in-process morphology manipulation is a powerful method to induce highly oriented microstructure in the injection molded bars, eventually resulting in prominent mechanical reinforcement. The LMCOE technique is benefited for multilayered film with alternating microlayers of two or three different polymer species or compositions. For the in-situ microfibrillation processing, the microfibrils-reinforced composites are fabricated by using conventional processing equipments but adopting special stretching procedure. Since the "structuring" processing methodology enables the processing-structure-property relationship more variable and controllable, it represents an advanced idea of polymer processing, and should be regarded as a promising tendency in future.

Keywords: molding processing, direct structuring assembly, hierarchical structure, externally applied field

1. Introduction: state-of-the-art in polymer processing

For all of polymer-based materials including neat polymers, blends, composites and hybrids, the subject of processing is crucial importance to link physical/chemical properties of polymer materials and their practical applications. By adopting different processing techniques, such as extrusion, heat compression, film stretching/blowing, injection and fiber spinning, distinctly different modes of temperature and/or stress fields have been imposed on the melts of polymer materials, leading to the extreme changes in microscopic structure and macroscopic properties of the prepared products. A good example is given as Fig. 1 for a semicrystalline polymer, polyethylene (PE). With the unchanged chemical composition and molecular chain architecture, PE can possess a broad range of Young's modulus from 0.01 GPa to 200 GPa because the PE crystalline structure in the processed products is varied from spherulite to shish-kebab and to extended-chain cylinders obtained via different processing methods of gas-assisted foaming, extrusion-co-extension and fast fiber spinning, respectively. Therefore, the same PE grade can be suitable for versatile applications as elastic foam, extrusion tube and ultrahigh strengthening fiber, etc. After a long time research and application of polymer processing, it is extensively aware of that the internal morphology/structure and ultimate performances of polymer material articles are strongly sensitive to the molding processing procedures/techniques utilized. According to the processing-structure-property relation of polymer material as schematically summarized in Fig. 2, the coupling of various externally applied fields will significantly affect the formation and evolution of structure/morphology during the proceeding of melt processing, and finally resulting in the specific, thermodynamically nonequilibrium morphology/structures which are frozen in the processed products.

In order to well realize the controls of hierarchical structure in processed polymer products, the most of works on polymer processing are correlated to the influences of externally applied fields. By in-situ monitoring the evolution of morphology/structure during processing or precisely revealing the microstructure change along the special directions, such as thickness and shear flow direction, the roles of processing variables like temperature gradient, shear velocity, stress gradient, strain rate, etc., on hierarchical structure can be ascertained and demonstrated clearly, and the correlations between processing variables and hierarchical structure feature can be established at least partially. This methodology is referred to the relatively traditional processing techniques, and has been developed for several decades [1,2], and some new results are still reported in recent years [3-5]. However, all most no effort is made to purposely control the microstructure/morphology in traditional processing techniques, this methodology is characteristic as "externally applied field impacted hierarchical structure". While in another methodology, the control of hierarchical structure is achieved through a different pathway. Due to continuously improved processing technology and profoundly understanding of processing theory, the fine mechanical designs of processing apparatus can introduce many specific externally applied fields, such as oscillation/vibration shearing [6], microlayer extrusion/extension [7], microfibrillation [8], on the melts of polymer materials, leading to the significantly improved ability of tailoring microstructure/morphology. It is characteristic as "externally applied field determined hierarchical structure", i.e. "structuring" processing. In this paper, we would like to review the most recent progress in "structuring" processing area, which includes three special processing strategies: (1) in-process morphology control of injection molded bar; (2) multilayer coextrusion; (3) in-situ microfibrillation.

2. An overview of "structuring" processing

Some new demands of ultrahigh performance, versatility of property, functionalization and facile manufacturing, have been proposed for the modern polymer processing area. To meet these high-quality demands, the morphology/structure of processing articles should be extremely controllable. However, due to the technological limitations in the traditional processing methods, the morphology/ structure control is often difficult to attain in the conventional processing articles. Alternatively, for the processing methodology of "structuring" process, besides the common processing variables some unique external fields were applied by using ingenuity of processing instruments and/or particular processing procedure. Thus the as-received morphology/ structure significantly differed to that in the conventional processed articles. In recent years, three kinds of "structuring" processing methods, viz. in-process morphology manipulation during injection molding, layer multiplying coextrusion and in-situ microfibrillation during molding process, have attracted intensive attention.

The in-process morphology manipulation is a powerful method to induce highly oriented microstructure in the injection molded bars, eventually resulting in prominent mechanical reinforcement. In particular, the shear controlled orientation in injection molding (SCORIM) technique, one of the in-process morphology manipulation methods, favors to the formation of fine shish-kebab superstructure and/or epitaxial interlocking crystalline morphology. Furthermore, the oscillation shear field during SCORIM process was employed to manipulate the biphase separation behavior of miscible polymer blends, the dispersed phase morphology of rubber/elastomer in thermoplastics, the dispersion and orientation of nanoparticles in polymer matrix, and the hybrid crystal superstructure consisted of semicrystalline polymer and fibrous filler. An interesting bamboo-like bionic structure was generated by the SCORIM technique in the injection molded bars of iPP/β-NA blend. The synergetic effect of strengthened-shell/toughened-core was responsible for the comprehensive mechanics of excellent rigidity-toughness balance.

The layer multiplying coextrusion (LMCOE) technique is benefited of multilayered film with alternating microlayers of two or three different polymer species or compositions, whose thickness varied within a wide range from tens microns to ten nanometers. Because the microlayer thickness could be modulated exactly, the multilayered films prepared by LMCOE were used as the ideal model systems to explore the micron/nanoscale confinement crystalline behavior of semicrystalline polymers. In general, the lamellae orientation mode varied from isotropic to on-edge and to in-plane with decreasing the microlayer thickness. A dramatic improvement of two or three orders of magnitude in gas barrier was achieved by the appearance of large aspect ratio in-plane lamellae. Interestingly, a transition of mechanical failure fashion from crazing to shear yielding occurred at a critical layer thickness, 200 nm, which was responsible to significant increase of ductility.

In the in-situ microfibrillation processing, the in-situ microfibrils-reinforced composites (MFCs) were fabricated by using conventional processing equipments but adopting special stretching procedure. Good strength, modulus, rigidity and thermo-size stability were available, due to the in-situ formation of microfibrils inside polymer matrix. For the in-situ MFCs of semicrystalline polymer blend, the interfacial crystallization event likes the generation of interface TCL and/or shish-kebab structure offered new understanding in the interfacial enhancement between microfibrils and polymer matrix. The electronically conductive fillers, CB and CNT, have been premixed with the microfibrillated component to prepare electronic conductivity microfibrils. The assembling 3-D conductive network by electronic conductivity microfibrils brought extremely low volume resistance in the in-situ MFCs of CB/PET/PE. It is notable that the selective distribution of CB particles at the surface of micrfibrils induced a remarkable decrease in the threshold of conductive percolation, as compared to the cases of the distribution inside microfibrils and/or in both of matrix and fibrillated components. Meanwhile, the unique temperature-resistivity behaviors including PTC and NTC effect were detected in the anisotropically conductive polymer composites.

In summary, the "structuring" processing methodology affords plenty of means to generate particular hierarchical structure in processing articles, and enables the processing-structure-property relationship more variable and controllable. It represents an advanced idea of polymer processing, and should be regarded as a promising tendency in future.

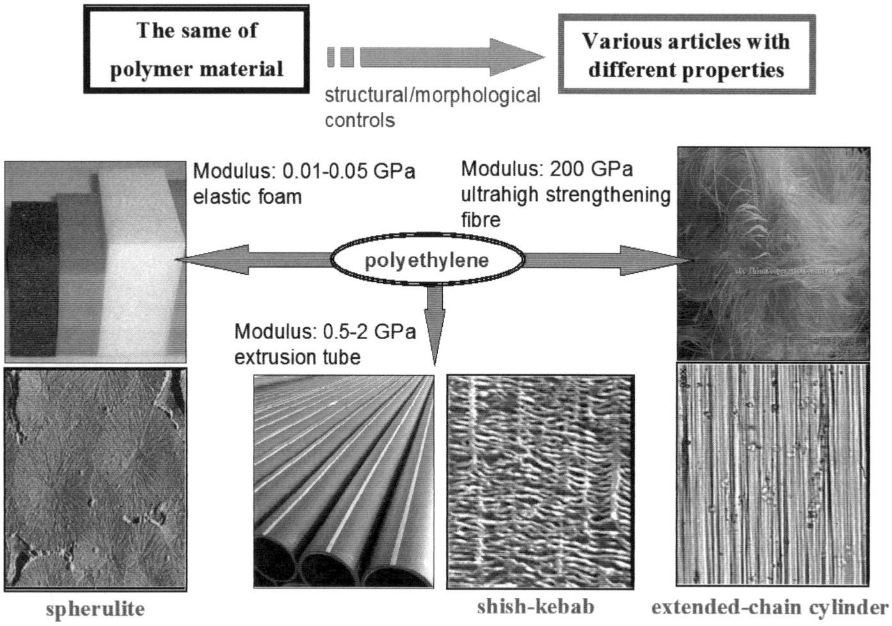

Figure 1 – Microscopic structure and macroscopic properties of polymer articles are strongly dependent on their molding processing methods: an example of polyethylene

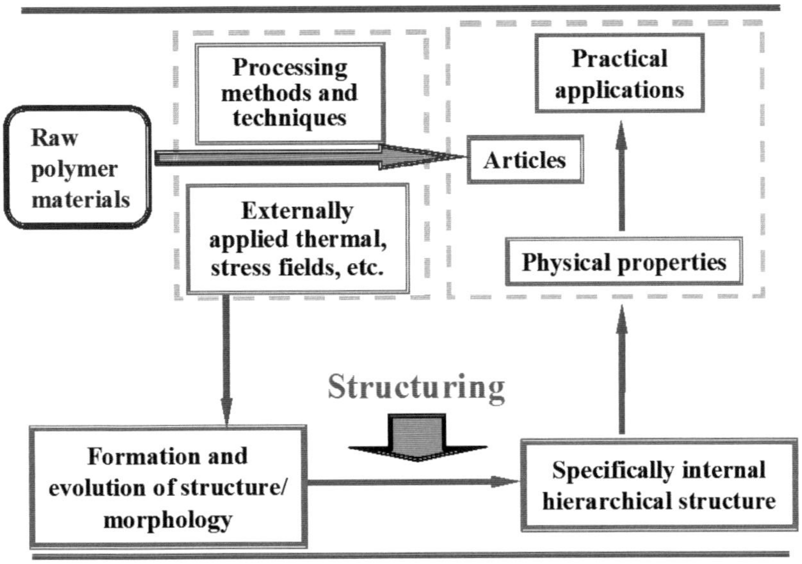

Figure 2 –The molding processing-structure-property relation of polymer materials: a key issue of "structuring"

Acknowledgements
This work was supported by the Special Funds for Major State Basic Research Projects of China (2011CB606006), and National Natural Science Foundation of China (51121001).

References

1. J. Kolnaar; A. Keller; S. Seifert; C. Zschanke; H. Zachmann *Polymer* 1995, 36, 3969.
2. Z. Wang; B. Hsiao; N. Stribeck; R. Gehrke *Macromolecules* 2002, 35, 2200.
3. R. Cermak; M. Obadal; P. Pon zil; M. Polaskova; K. Stoklasa; A. Lengalova *Eur Polym J* 2005, 41, 1838
4. Y. Wang; B. Steinhoff; C. Brinkmann; I. Alig *Polymer* 2008, 49, 1257.
5. R. Pantani; I. Coccorullo; V. Speranza; G. Titomanlio *Prog Polym Sci* 2005, 30, 1185.
6. D. Angstadt; J. Coulter *Polym Eng Sci* 2006, 46, 1691.
7. M. Gupta; Y. Lin; T. Deans; E. Baer; A. Hiltner; D. Schiraldi *Macromolecules* 2010, 43, 4230.
8. K. Friedrich; M. Evstatiev; S. Fakirov; O. Evstatiev; M. Ishii; M. Harrass *Compos Sci Technol* 2005, 65, 107.

O-09-232

MICROSTRUCTURE SIMULATION OF SEMI-CRYSTALLINE THERMOPLASTICS

Ch. Hopmann[1], M. Spekowius[1]* and K. Küsters[1]
[1]Institute of Plastics Processing (IKV) at RWTH Aachen University, Germany – spekowius@ikv.rwth-aachen.de

Abstract - The injection molding process is economically the most important method for the production of plastic components. To allow a precise component dimensioning and to exploit the full potential of the used material it is necessary to simulate the part's behavior. The mechanical behavior of semi-crystalline parts is strongly influenced by inner properties e.g. degree of crystallization and morphology. The inner properties themselves depend on the processing conditions of the injection molding process. In order to improve the prediction of the part's behavior it is necessary to simulate the local microstructure taking into account boundary conditions from the injection molding process. As part of the current cluster of excellence in Germany „Integrative Production Technology for High-Wage Countries" a simulation software for the precise prediction of the local microstructure is developed. Using the mesh and calculation results from an injection molding simulation the software allows the time-resolved 3D calculation of the microstructure in quiescent melt. A cellular automaton with a nucleation and growth model is used to calculate the crystallization process. Comparison with polarization microscopy pictures shows that the simulation results correspond qualitatively well.

Keywords: microstructure, crystallization, integrative simulation, injection molding, cellular automaton

Introduction

Today, plastics belong to the most important materials available. From all the processing procedures the injection molding is particularly important because it has a huge economic impact. An important cost factor of this procedure is the development of the mold. After the fabrication of the mold a lot of cost intensive adjustments are needed to create a distortion-free part. Thus the simulation of the injection molding process is an important tool to predict the later part properties in the virtual design process and avoid cost intensive adjustments. Today's simulation tools allow the simulation of the filling and cooling process and also a qualitative prediction of the part's shrinkage and warpage. Unfortunately the prediction is quite poor for semi-crystalline thermoplastics [2]. One of the reasons for this is the inhomogeneous microstructure in the part that strongly influences the mechanical properties but is not considered in the simulation. To improve the simulation of the part's properties it is necessary to simulate the microstructure and offer these results to a structural analysis so that a calculation with a local, microstructure depending distribution of mechanical properties is possible.

In the last years the simulation of the solidification process of semi-crystalline thermoplastics has been the research focus of multiple groups. Depending on the research goal the developed tools simulate the solidification process on various scales. Molecular simulations at a nanometer scale are used to simulate the chain-folding of single molecules to get information about the resulting crystal structure. By prealigning the molecules a change from a spherulitical crystal structure to a Shish-Kebab structure can be simulated [3]. Others try to describe the crystallization in a macroscopic way by using characteristic values like the crystallization degree or the avrami exponent. By reducing the entire morphology with a volume of several cubic millimeters to a few values it is possible to predict the mechanical behavior with a rough integration of the crystalline structure for entire plastic components [4]. A third group of scientists tries to close the gap between the molecular and macroscopic simulation. The goal is to predict the morphology at the micrometer scale of a millimeter sized simulation area based on a small set of boundary conditions. A common simulation technique for these kinds of problems is the usage of a cellular automaton. Cellular automata are a technique that is often used to simulate crystallization processes of melts. In general, cellular automata are systems of cells with defined states whose temporal development is depending on the states of the neighboring cells. The state changes are applied by a global transformation rule that is executed for every timestep of the simulation process. In the material science, cellular automata have their main application in the simulation of recrystallization and grain size of metals [5]. In polymer science they are used to calculate the spherulitical microstructure that develops during the solidification of semi-crystalline thermoplastics. An example for such a simulation tool is the software SphäroSim that has been developed at the Institute of Plastics Processing for many years. The software is part of the AixViPMaP platform and is used in an integrative multiscale simulation chain to offer a high resolution microstructure that can be used for a structural analysis [6].

Crystallization of semi-crystalline thermoplastics

There are two major processes involved in the crystallization of polymers: nucleation and crystal growth. During the nucleation process small nuclei are created that start growing in the crystal growth process as soon as their size exceeds a critical value. It is necessary to distinguish between nuclei made of aligned and unaligned molecules. Nuclei consisting of aligned molecules, e.g. because of shear forces, have a thread like shape. During the crystal growth process it is possible that Shish-Kebab structures grow out of them (Fig 1a) [3]. Unaligned nuclei result in a spherulitical microstructure (Fig 1b).

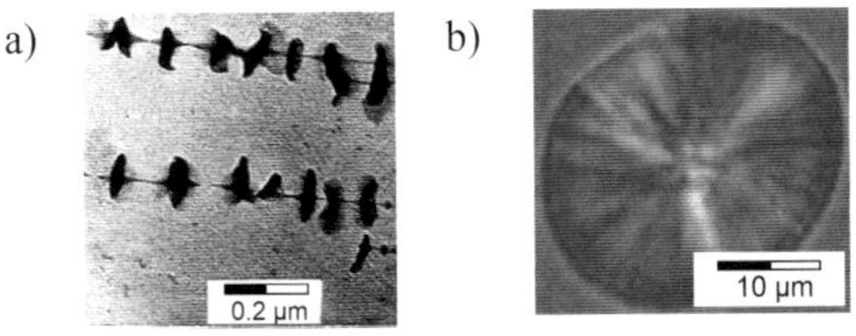

Figure 1: a) Electron micrograph of polyethylene showing Shish-Kebab structures after shear treatment of a moderately concentrated solution [12]. b) Microscopy taking of a single spherulite of Polypropylene.

Normally, in injection molded parts the spherulitical structure is dominant, with oriented structures occurring in areas of high shear forces. Unfortunately the process how the nuclei are created is not fully understood yet. Hoffman, Davis and Lauritzen presented a nucleation model based on the rate theory [7]. However, the rate theory can only be applied for temperatures near the melting temperature of the material [8]. Thus most research groups use empirical models to describe the nucleation rate $\alpha(T)$ and nucleus density ρ_N [1, 9]. Here we use an empirical approach for the nucleation rate that is suggested by Moneke [10]:

$$\alpha(T) = \frac{A}{\sqrt{2\pi \cdot B^2}} \cdot e^{\frac{-(T-T_c)^2}{2 \cdot B^2}} \quad (1)$$

T is the temperature, A, B and T_C are model parameter that have to be adjusted for each material. Using the nucleation rate it is possible to calculate the nucleus density.

$$\rho_N(t) = \int_{-\infty}^{t} dz\, \alpha(T(z)) + N_A \quad (2)$$

N_A is a constant value used to describe the nuclei that are induced by impurities in the melt. The crystal growth process is described by the growth speed of a single nucleus using the Hoffman, Davis and Lauritzen theory [7].

$$G(T) = G_0 \cdot e^{\frac{U^*}{R \cdot (T - T_G - T_\infty)}} \cdot e^{\frac{\kappa_G \cdot T_m^2 \cdot (T_m + T)}{2 \cdot T^2 \cdot (T_m - T)}} \quad (3)$$

The four parameters U*, κ_G, G_0 and T_∞ have to be estimated for each material. Furthermore it is necessary to know the glass transition temperature T_G and the crystal melting temperature T_m. R equals the gas constant.

Microstructure simulation with SphäroSim

Starting from an injection molding simulation several elements of the enmeshed geometry are chosen. These elements define the simulation area. The result data from the injection molding simulation (like for example the temperature in the corner points of each element) are transferred to the microstructure simulation software SphäroSim. Here, each element of the simulation area is subdivided into a high resolution grid of equally sized cells (Fig. 2a).

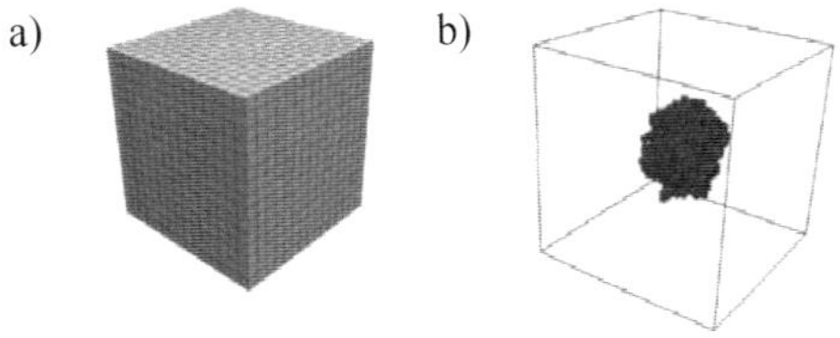

Figure 2: a) Each element is subdivided into a high resolution grid of uniform cells. b) Growth of a single spherulite using a cell size of $\lambda = 2.5$ µm.

The volume of a cell V_{cell} is equal in the entire simulation area and defined by the length of the cell edge λ.

$$V_{cell} = \lambda^3$$

Typical values for λ range from 2 µm to 10 µm, depending on the desired resolution and size of the simulation area. Because of the regular arrangement each cell within an element can be identified by three indices (a|b|c), one for each direction of the Cartesian coordinate system. The number of cells in each direction $N_{x,y,z}$ is depending on the element size $d_{x,y,z}$.

$$N_{x,y,z} = \frac{d_{x,y,z}}{\lambda}$$

It should be noted, that the edge size λ has to be chosen in a way, that $N_{x,y,z}$ is a whole number. Otherwise parts of the simulation area are not considered in the microstructure simulation. Assigned to each cell is a cell state $\sigma_{a,b,c}$ which describes the phase in the cell volume.

$$\sigma_{a,b,c} = \begin{cases} 0 \; for \; liquid \\ 1 \; for \; solid \end{cases}$$

All the cell states within a single element define a set of states that describes the entire phase of the simulation area at a given point in time.

$$\Omega = \{\sigma_{a,b,c} | a = 1..N_x; b = 1..N_y; c = 1..N_z\}$$

The temporal development of the cell states is done with a Monte Carlo transformation rule $P(\Omega, t_i)$.

$$\Omega'(t_i) = P(\Omega(t_i), t_i) = \{\theta(p_{a,b,c}(t_i) - r_{a,b,c}(t_i)) | \forall (a, b, c)\}$$

θ is the Heavi-Side step function using $\theta(0) = 1$, $p_{a,b,c}(t_i)$ the switching probability for a state change which is 1 if the cell is solid and ranging from 0 to 1 if liquid, $r_{a,b,c}(t_i)$ is a random number ranging from 0 to 1 that changes every time the transformation rule is applied. The switching probability $p_{a,b,c}(t_i)$ significantly depends on the amount of solid neighboring cells. To determine this number the van Neumann neighboring Ψ of the cell is considered.

$$\Psi_{a,b,c} = \{\sigma_{k,l,m} | (k|l|m) = (a \pm 1|b|c); (a|b \pm 1|c); (a|b|c \pm 1)\}$$

This allows to calculate the number of solid neighbors $Z_{a,b,c}$.

$$Z_{a,b,c} = \sum_{\sigma \in \Psi_{a,b,c}} \sigma$$

In case of $Z_{a,b,c} > 0$ the cell is part of a crystal front of a spherulite. Thus the switching probability depends on the crystal growth speed $G_{a,b,c}(t_i)$ (Eq. 3).

$$p_{a,b,c}(t_i) = \frac{Z_{a,b,c}}{\lambda} \cdot G_{a,b,c}(t_i) \cdot (t_i - t_{i-1}); Z_{a,b,c} > 0$$

In a more visual way of saying, the switching probability equals the amount of volume covered by the growing crystal front. In addition to the growing process there is also the nucleation process described by the nucleation rate (Eq. 1). This is simulated by applying the transformation rule using the nucleation switching probability.

$$p_{a,b,c}(t_i) = \alpha_{a,b,c}(t_i) \cdot (t_i - t_{i-1}) \cdot V_{cell}$$

Thus the transformation rule is applied two times for a point in time t_i to allow the simultaneous occurrence of crystal growth and nucleation. Iterating t_i in discrete timesteps from 0 s to the moment all the cells of the simulation area are solid allows a calculation of the final microstructure. Unfortunately how the timestepsize Δt_i is chosen is important for the quality of the simulation results.

$$t_{i+1} = t_i + \Delta t_i$$

If the value of Δt_i is too big it is possible to calculate switching possibilities bigger than one. This means the crystal front either grows over a length of more than one cell during the timestep or there has been more than one nucleus appeared in the cell volume according to the nucleation rate. Both effects lead to inaccurate results because they are not explicitly treated in the simulation. If Δt_i is chosen too small the transformation rule is applied a lot of times leading to a lot of Monte-Carlo decisions on a limited number of cells in the simulation area. To get accurate results it would be necessary to use a high spatial resolution of cells which would dramatically increase the calculation time. To avoid too small and too big timesteps the timestep is calculated for each simulation step.

$$\Delta t_{i,G} = min\left(\frac{\lambda}{Z_{a,b,c} \cdot G_{a,b,c}(t_i)}\right); \forall(a,b,c)$$

$$\Delta t_{i,\alpha} = min\left(1/\left(V_{cell} \cdot N_{Liquid} \cdot \alpha_{a,b,c}(t_i)\right)\right); \forall(a,b,c)$$

$$\Delta t_i = min(\Delta t_{i,G}; \Delta t_{i,\alpha})$$

The parameter N_{Liquid} equals the number of liquid cells in the element. Using the growth speed and nucleation rate for the current timestep the fastest growing crystal front and the fastest nucleation rate are identified. The time needed for 100% switching probability because of the crystal growth process and the average time needed to switch a single cell of the element because of nucleation are calculated. From these two values the smaller one is chosen as timestep allowing the simulation of the growing process of spherulites (Fig. 2 b).

Results and Discussion

For testing the cellular automaton a simulation chain made of the injection molding simulation software SigmaSoft v4.8 from the company Sigma Engineering GmbH, Aachen, Germany and SphäroSim is setup. A plate geometry with at thickness of 2 mm and the dimensions 11.5 x 11.5 cm² is chosen as the part. Due to the symmetry only half of the part must be simulated. The material for the simulation is isotactic Polypropylene. For validating the simulation results the part geometry is manufactured using a KM 160-1000 CX injection molding machine from the company Krauss Maffei Technologies GmbH, Munich, Germany. The mold temperature is 40 °C and the melt temperature 240 °C using a holding pressure of 400 bar and a holding time of 10 s. As material Polypropylene PP 505 P from the company Sabic, Bergen Op Zoom, Netherlands, is chosen.

In the first step of the simulation the filling and cooling simulation is executed and the resulting temperatures are transfered to SphäroSim with a temporal resolution of 0.1 s. Before this data can be used for the microstructure simulation it is necessary to calibrate the nucleation model α(T) and growth model G(T). The nucleation model is calibrated via the nucleus density which requires the growth model for the calculation. Thus the growth model is calibrated first, using a glass transition temperature of T_G = -10 °C and an equilibrium melting temperature of T_m = 194 °C [11]. Luckily the growth speed is nearly the same for all isotactic Polypropylenes of the different manufacturers when the solidification is mainly taking place in the α-Phase. Thus it is possible to take literature values for the growth model (Eq. 3) [11].

$$T_\infty = 51.6\,K \qquad \frac{U^*}{R} = 2068.8\,K$$

$$G_0 = 4.36 \cdot 10^8 \frac{m}{s} \qquad \kappa_G = 2.7979$$

Using this configuration it is possible to calibrate the nucleation model α(T) by executing simulation with isothermal boundary conditions for different temperatures. A cubic simulation area of 250 µm edge size is created and a cell size of λ = 2.5 µm is used. For each temperature the nucleus density is calculated by counting the total number of spherulites in the element.

These values are compared to literature values taken from Janeschitz-Kriegl [1]. The nucleation parameters are adjusted until the simulated nucleus density is in good agreement to the measured one (Eq. 1 and Eq. 2).

$$A = 5 \cdot 10^{19} \frac{K}{s \cdot m^3} \qquad B = 19\,K$$

$$T_c = 280\,K \qquad N_A = 5 \cdot 10^{10} \frac{1}{m^3}$$

These adjusted values result in a good agreement between simulation and measurement (Fig. 3).

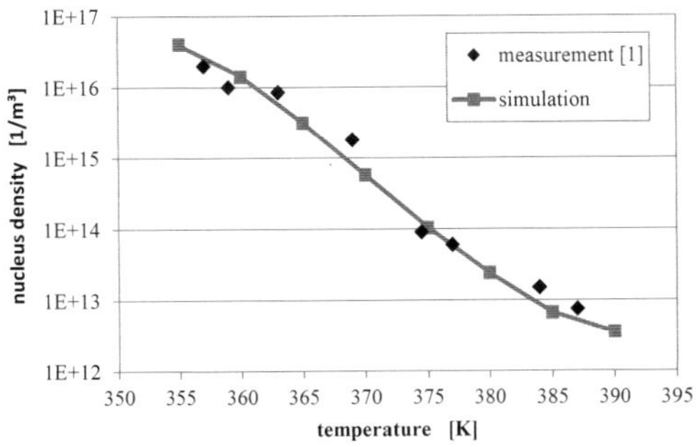

Figure 3: Simulated nucleus density for several isothermal scenarios.

After the calibration the microstructure can be simulated for the part geometry using an edge size of $\lambda = 2.5$ µm. For validating the simulation method five elements are taken which cover the entire plate thickness (Fig. 4).

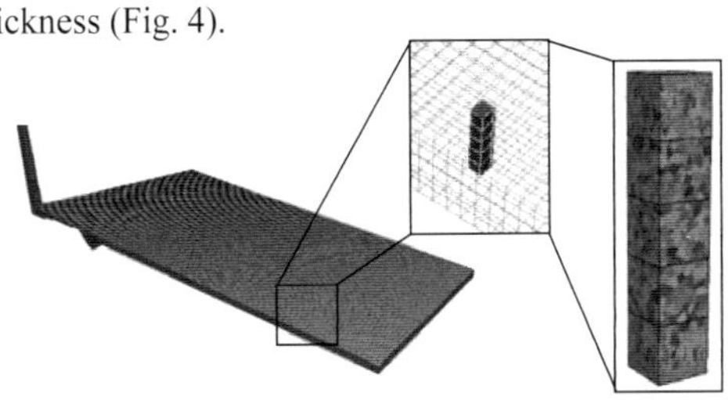

Figure 4: The plate geometry is subdivided in 268,715 cubic elements with an edge size of 400 µm. Five elements are taken for the high resolution microstructure simulation.

The simulation results show a smaller spherulite size at the mold surface and a bigger size in the interior area of the part because of the lower cooling rate. The comparison to the microscopy shows a qualitatively good agreement by predicting a finer spherulite diameter for the first 100 µm at the edge layer and a larger diameter up to 49.7 µm in the middle layer of the part (Fig. 5).

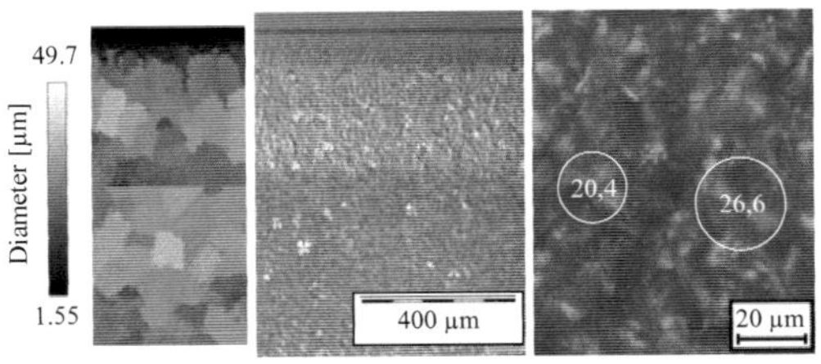

Figure 5: Comparison between the simulated microstructure (left) and polarized light microscopy images of the plate geometry from the edge layer to the middle layer (middle) and only the middle layer with two representative spherulites (right).

However, the calculated spherulites in the middle layer of the plate are too large (Fig. 5, right). This is because of the parameter choice of the nucleation model that was using data from a different industrial i-PP [1]. Furthermore the data was acquired by isothermal experiments in quiescent melt which misses boundary conditions occurring during the injection molding, like shear forces and impurities in the melt.

Conclusion

The presented software SphäroSim allows the simulation of the microstructure of semi-crystalline thermoplastics. A cellular automaton with a Monte-Carlo switching rule is used for simulating the growth of spherulites. As part of the AixViPMap platform it can be used in integrative multiscale simulation chains. The calculated microstructure is in qualitatively good agreement to microscopy takings of injection molded plates. Next steps will be a parallelization of the software to allow the high computing requiring calculation of entire parts and an adjustment of the material models, especially of the nucleation model, to calculate the effects of shear forces.

Acknowledgements

The depicted research is funded by the Deutsche Forschungsgemeinschaft (DFG) as part of the Cluster of Excellence "Integrative Production Technology for High-wage Countries". We would like to extend our thanks to the DFG.

References

1. Janeschitz-Kriegl, *Crystallization Modalities in Polymer Melt Processing,* Springer-Verlag, Wien, 2010
2. T. Baranowski, PhD Thesis, RWTH Aachen University, 2011
3. W. Hu; D. Frenkel; V.B.F. Mathot *Macromolecules* 2002, *35*, 7172-7174
4. S. Hoffmann, PhD Thesis, Thesis, RWTH Aachen University, 2002
5. D. Raabe *Ann. Rev. Mater. Res.* 2002, 32-53
6. C. Hopmann; W. Michaeli in *Integrative Computational Materials Engineering*, G. J. Schmitz; U. Prahl, Ed.; Wiley, Weinheim, 2012
7. J. D. Hoffman; G. T. Davis; J. I. J. Lauritzen in *Treatise on Solid State Chemistry*, New York: Plenum Press, 1976, 497-614
8. K. Kaji; K. Nishida; T. Kanaya; G. Matsuba; T. Konishi; M. Imai *Adv. Polym. Sci.*, 2005, *191*, 187-240
9. J. W. Housmans, PhD Thesis, Technische Universiteit Eindhoven, 2008

10. M. Moneke, PhD Thesis, Technische Universität Darmstadt, 2001
11. G. Lamberti *Polym. Bulletin,* 2004, *52*, 443-449
12. J. D. Hoffman; R. L. Miller in *Advancing Materials Research,* P. A. Psaras; H. D. Langford, Ed.; National Academy Press, Washington D.C., 1987

O-09-235

MODELING FLOW-INDUCED CRYSTALLIZATION OF POLY(1-BUTENE) IN SHORT-TERM AND CONTINUOUS SHEAR AND EXTENSIONAL FLOW

P.C. Roozemond[1] and G.W.M. Peters[1]*
[1] *Materials Technology, Eindhoven University of Technology, Eindhoven, The Netherlands – g.w.m.peters@tue.nl*

Abstract - We apply a modeling framework for flow-enhanced nucleation of polymers to a broad set of data from literature. Creation of flow-induced point-like nuclei is coupled in a phenomenological way to backbone stretch of the high-molecular weight tail of the material, calculated with a rheological constitutive model. As the flow-induced nuclei grow, the crystalline volume fraction increases and with it the viscosity of the material. This is accounted for by describing the material as a suspension of soft spheres in a viscoelastic matrix. Simulations are compared with experimental data from literature on three grades of poly(1-butene). First, a parameter set is determined by fitting to flow-induced nucleation densities from short-term shear experiments. Next, this parameter set is used to validate the framework in continuous flow experiments, in which viscosity is monitored during a constant flow rate. The model is able to quantitatively predict experimental results in both shear and extensional flow. The results illustrate the importance of a thorough rheological characterization of the material in interpreting results from flow-induced crystallization experiments. We show that the observed viscosity upturn in extensional flow at constant strain, is caused by strain hardening because the extension rate exceeds the inverse of the Rouse time of the longest chains.

Keywords: Flow-induced crystallization, poly(1-butene), extensional flow

Introduction (Times New Roman 10, bold)

The mechanical and optical properties of semi-crystalline polymer products strongly depend on crystalline structure [1], which is largely determined by processing conditions [2]. The development of a quantitative model for flow-induced crystalline structure development is therefore a vital step towards predicting properties of polymer products.

The majority of experimental work on flow-induced crystallization (FIC) has followed the short-term shear protocol proposed by the group of Janeschitz-Kriegl [3] which spawned from the idea that if the crystallization time is much longer than flow time, it can be assumed that the only effect of flow is creation of structure; crystalline growth rate is unaffected. Because of the short flow times, this method is suitable for investigating FIC up to flow rates that are relevant for industrial processing conditions. Alternatively, FIC can be investigated in a continuous flow experiment. In these types of experiments the material is subjected to a constant flow rate in a rheometer and the viscosity is monitored. At a certain point in time the viscosity will show an upturn due to crystallization. The time at which this happens is a measure for the crystallization-enhancing effect of flow. These types of experiments have been performed in rotational [4,5] as well as extensional [6,7,8] flow. Flow times are typically higher than for short-term flow experiments, which at high shear rates might result in destruction of the sample. Therefore this approach is mainly suited for low shear or extension rates.

Zuidema derived a phenomenological model for flow-induced crystallization where the creation rate of flow-induced nuclei is coupled to the stretch of high-molecular weight chains on a continuum level [9]. In this paper we apply this modeling framework and combine it with a suspension model that predicts viscoelastic properties from crystalline volume fraction. In this way we are able to capture experimental data in both short-term and continuous flow.

Theory

The present model is a simplified version of the nucleation model first developed by Zuidema [9] and improved by Steenbakkers [10]. The polymer chains orient and stretch as the material is subjected to deformation. Regions with highly stretched molecules can act as precursors to nucleation. In a polydisperse melt, this effect is most pronounced for the high molecular weight tail (hmw) of the material, because long chains with high relaxation times are more susceptible to deformation than shorter chains. In this paper we do not consider the intermediate state of nucleation precursors, which may either relax or become nuclei. Instead, we only consider active precursors that nucleate instantly. The creation rate of flow-induced nuclei is coupled to the backbone stretch in the high-molecular weight tail of the material. This phenomenon is captured by a simple expression,

$$\frac{dN_n}{dt} = g_n\left(\Lambda_{hmw}^4 - 1\right) \quad (1)$$

where N_n is the number density of nuclei and g_n is a scaling parameter which depends on temperature, pressure, and the chemical properties of the material, such as isotacticity and the presence of nucleating agents. In this work we implemented an Arrhenius-like temperature dependence of g_n,

$$g_n(T) = g_{n,ref} \exp\left(\frac{E_{a,g}}{R}\left(\frac{1}{T} - \frac{1}{T_{ref}}\right)\right). \quad (2)$$

Λ_{hmw} is the backbone stretch of the long chains in the material, which can be calculated on a constitutive scale from flow rate using a rheological model. In this work we use the Rolie-Poly model [11]. This approach can be used in any type of flow, be it shear, extension or any combination of the two.

Physical cross-link

The mobility of the long chains which are part of a flow-induced nucleation site is restricted. Consequently, the nucleation sites act as physical cross-links, decreasing the mobility of the long chains and hence increasing the relaxation times,

$$\frac{\lambda_{k,hmw}}{\lambda_{k,hmw,0}} = \left(1 + \alpha N_n\right) \quad (3)$$

With $k \in \{rep, R\}$. α is a scaling parameter. This was introduced on empirical basis by Zuidema to be able to capture the experimental data. Although the exact mechanism remains unproven, there are clear indications for a mechanism that alters the rheology of the crystallizing melt during the early stages of crystallization when the spherulites are still too small to have a noticeable effect on the viscosity if the behavior of the material is purely suspension-like [12,13].

We now require 3 parameters for each material to calculate flow-induced nucleation rate; $g_{n,ref}$, $E_{a,g}$ and α.

Continuous flow

The viscosity during continuous flow is calculated from crystalline volume fraction using a model for a suspension of spheres, the Generalized Self-Consisted Method [14,15]. The GSCM was shown to describe well the evolution of viscosity during spherulitic crystallization in iPP as well as HDPE [12,16]. The crystalline volume fraction is calculated from flow-induced nucleation rate using the Schneider rate equations under the assumption that crystalline growth rate remains constant.

Experimental

Materials

Two commercial grades of isotactic poly(1-butene), produced by LyondellBasell, were used in the experiments to which simulations are compared in this study, PB800 (M_w=85kg/mol, M_w/M_n=5.1, $\lambda_{rep,hmw@98°C}$=0.25s) and BR200 (M_w=762kg/mol, M_w/M_n=19.6, $\lambda_{rep,hmw@98°C}$=1248s. A multi-mode Maxwell spectrum was fitted to rheological data from literature. The reptation time of the slowest Maxwell mode, corresponding to the inverse of the oscillatory frequency at the onset of the terminal regime, was taken as the relaxation time representing the high molecular weight tail for our nucleation model. The corresponding Rouse time was calculated using [17]

$$\lambda_R = \frac{\lambda_{rep}}{3Z} \quad (4)$$

where Z is the number of entanglements per chain.

Short-term shear flow experiment

Short-term shear experiments were performed at the University of Genova on grades PB800 and BR200 in a Linkam CSS 450 rotational shearing device [18]. Thermo-mechanical history was erased at 180 °C before cooling down to a temperature where a shear pulse was applied. The shear rate, shear time and shear temperature were varied. For PB800 $T_{shear} \in \{83,97\}$ °C, for BR200 $T_{shear} \in \{100,160\}$ °C. After shear, the material was monitored using optical microscopy. The number of nuclei in the observation window was counted from which nucleation density was estimated. PB800 was held at the shear temperature to crystallize, the BR200 samples were cooled down to 96 °C because of the low growth rate at the shear temperatures.

Continuous flow experiments

Simulations are compared with results from continuous flow experiments from various sources [5,6]. In these types of experiments the polymer melt is subjected to a constant shear or extension rate in a rheometer while the viscosity is monitored. For shear a plate-plate set-up is used, uniaxial extensional flow experiments are done using an extensional viscosity fixture. We will compare predictions from the model and experimental data for the transient viscosity signal for rotational and extensional rheometry.

Results and Discussion

First, parameter sets for PB800 and BR200 are determined by fitting to data from short-term shear experiments which give nucleation density for various shear rates, shear times and temperatures. Next, simulations for continuous flow experiments are presented and compared to experimental data. The same parameter sets for the nucleation model as for the short-term shear experiments are used.

Short-term flow

The parameters in the nucleation model were determined by least squares fitting to the short-term shear experiments for PB800 and BR200. Fig. 1 shows experimental data and simulations for both materials at one temperature. The experimental results and simulations show good agreement. For the majority of shear rates and temperatures the model is able to capture the experimental data quantitatively, however at the highest shear rates at some temperatures the simulations consistently under or over predict the nucleation densities.

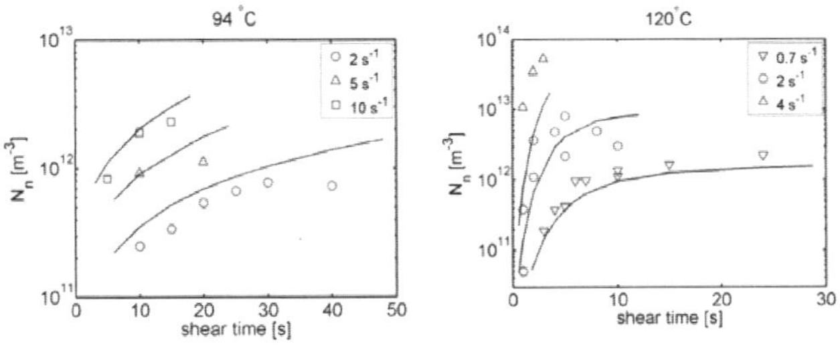

(a) (b)

Figure 1 – Flow-induced nucleation densities for (a) PB800 and (b) BR200. Symbols indicate measurements, lines are simulations.

A striking qualitative difference between the two materials is that PB800 shows a nucleation density which can within reasonable accuracy be described with a linear relation of nucleation density versus shear time on a log-linear scale. On the other hand, the nucleation density for BR200 clearly exhibits a high slope for short shear times which levels off at longer shear times. This phenomenon could be interpreted as saturation, which has been observed in a number of studies [19,20,21]. Saturation has been hypothesized to be caused by depletion of polymer chains above a certain critical molecular weight. However, in the current experiments the saturation-like behavior can be explained purely from a rheological standpoint. As the longest Rouse time for BR200 is much higher than for PB800 it takes much longer for the backbone stretch to reach a steady state. Consequently, the time-dependency of nucleation rate for BR200 is much more pronounced than for PB800.

Continuous shear flow

We validate our modeling framework for continuous flow with continuous shear flow experiments on PB800 performed by Acierno et al. [5]. The parameter set that was determined from short term shear flow was used, the quiescent nucleation density was fit to the experiment with the lowest shear rate. The results are presented in Fig. 2. The model performs well over a range of shear rates of nearly three decades, especially considering the fact that the parameter set was fit to experiments done in a different group with a different experimental technique.

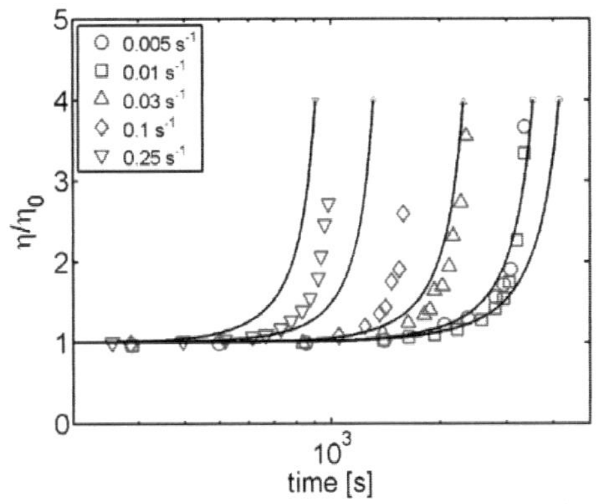

Figure 2 – Viscosity normalized with melt viscosity for PB800 during continuous shear flow. Symbols indicate measurements, lines are simulations.

Continuous extensional flow

Using an extensional viscosity fixture fitted to a conventional rheometer, Hadinata et al. performed continuous uniaxial extensional flow experiments on grade BR200 [6]. Fig. 3 shows the experimental data and results from our simulations with the same flow conditions. The parameters in the nucleation model are the ones determined from the short-term shear flow experiment. The simulations capture the experimental data very well, especially considering the parameter set was determined in shear flow.

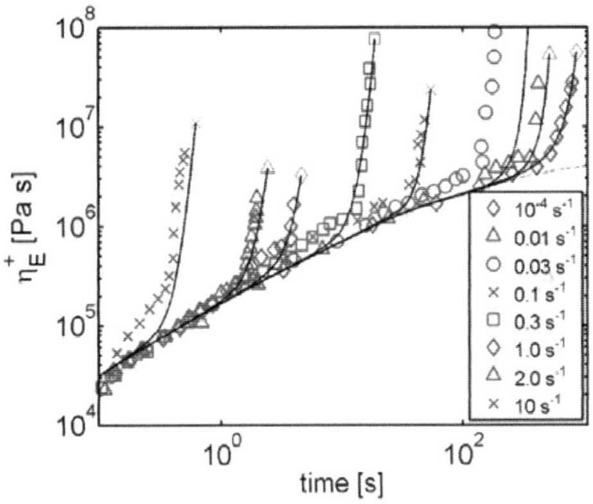

Figure 3 – Viscosity during continuous uniaxial extension for BR200 at 97 °C. Symbols indicate measurements, lines are simulations.

However, the viscosity increase in the simulations is not solely caused by nucleation and growth. At higher strain rates the viscosity increase in the simulations is caused by strain hardening, because the strain rate exceeds the inverse of the longest Rouse relaxation time. The longest Rouse time at 97 °C is 10.3 s. Hence we can expect strain hardening to occur at strain rates of 0.1 s^{-1} and above. Hadinata et al. observed that the viscosity upturn appears a constant strain for these strain rates. We believe that the explanation lies in the fact that it is caused by long chains becoming fully stretched. In our simulations this directly causes the viscosity upturn, in experiments the viscosity upturn might be caused by a combination of strain hardening and very fast nucleation of oriented crystals.

Conclusions

A modeling framework was presented to calculate flow-induced nucleation density and the evolution of viscosity under the influence of crystallization-enhancing flow. The model couples nucleation rate to molecular deformation, which is calculated from flow conditions via rheological characteristics, i.e. relaxation times. This intermediate step of calculating molecular deformation from a rheological constitutive model has proven to be vital in understanding flow-induced crystallization experiments.

Simulations show good agreement with various types of experiments from different research groups. Parameters for two materials were determined by fitting to nucleation densities from short-term shear experiments. With these parameters, the model predicts a viscosity signal that is in quantitative agreement with experiments in continuous shear and extensional flow for a range of deformation rates and temperatures.

The model provides a tool to interpret some experimental observations. First, the saturation effect in flow-induced number density, at least in the present experiments, can be explained solely from the stress overshoot after flow start-up. Because the polymer chains experience large deformations during this overshoot, the nucleation rate also shows an overshoot before reaching steady state, thus causing a saturation-like effect in flow-enhanced nucleation density. This phenomenon is predicted to be more pronounced in materials with higher relaxation times, which is confirmed by experiments. The saturation level, i.e. the value at which nucleation density seemingly reaches a plateau, is highly dependent on shear rate. Second, we believe the upturn in extensional viscosity that occurs at constant strain during continuous flow to be caused by the long chains becoming fully stretched.

Acknowledgements

We would like to thank Prof. G.C. Alfonso for kindly sharing the data for the short-term shear experiments. This work is part of a collaboration with the group of Prof. Julia Kornfield at the California Institute of Technology. The project is supported by the Dutch Technology Foundation (STW), grant no. 08083 and the National Science Foundation (NSF), grant no. DMR-0710662.

References

1. B. Schrauwen; L. van Breemen; A. Spoelstra; L. Govaert; G. Peters; H. Meijer. *Macromolecules.* 2004, *37,* 8618.
2. J. van Meerveld; G. Peters; M. Huetter. *Rheol. Acta.* 2004, *44,* 119.
3. S. Liedauer; G. Eder; H. Janeschitz-Kriegl; P. Jerschow; W. Geymayer; E. Ingolic. *Int. Polym. Proc.* 1993, *8*, 236.
4. C. Hadinata; C. Gabriel; M. Ruellman; H. Laun. *J. Rheol.* 2005, *49*, 327.
5. S. Acierno; S. Coppola; N. Grizzuti. *J. Rheol.* 2008, *52,* 551.
6. C. Hadinata; D. Boos; C. Gabriel; E. Wassner; M. Ruellmann; N. Kao; M. Laun. *J. Rheol.* 2007, *51,* 195.
7. M. Derakhshandeh; S. Hatzikiriakos. *Rheol. Acta.* 2012, *51,* 315.
8. E. White; H. Winter; J. Rothstein. *Rheol. Acta.* 2012, *51,* 303.
9. H. Zuidema; G. Peters; H. Meijer. *Macromol. Theor. Simul.* 2001, *10,* 447.
10. R. Steenbakkers; G. Peters. *J. Rheol.* 2011, *55,* 401.
11. A. Likhtman; R. Graham. *J. Non-Newton. Fluid.* 2003, *114,* 1.
12. P. Roozemond; V. Janssens; P. van Puyvelde; G. Peters. *Rheol. Acta.* 2012, *51*, 97.
13. C. Carrot; J. Guillet; K. Boutahar. *Rheol. Acta.* 1993, *32,* 566.
14. R. Christensen; K. Lo. *J. Mech. Phys. Solids.* 1979, *27,* 315.
15. R. Christensen. *J. Mech. Phys. Solids.* 1990, *38,* 379.
16. R. Steenbakkers; G. Peters. *Rheol. Acta.* 2008, *47,* 643.
17. M. Doi; S. Edwards, *The theory of polymer dynamics,* Clarendon Press, Oxford, 1986.
18. P. Ottonello, MSc Thesis, University of Genova, Italy, 2003.
19. J. Housmans; R. Steenbakkers; P. Roozemond; G. Peters; H. Meijer. *Macromolecules.* 2009, *42,* 5728.
20. J. Baert; P. van Puyvelde. *Macromolecules*. 2006, *39,* 9215.
21. G. Kumaraswamy; A. Issaian; J. Kornfield. *Macromolecules*. 1999, *32,* 7537.

O-09-240

ORIENTATION ANALYSIS OF METALLIC FLAKES IN INJECTION MOLDED PRODUCTS USING X-RAY CT TECHNOLOGY

K. Taira[1]* and H. Yokoi[1]

[1] *Institute of Industrial Science, the University of Tokyo, Japan – ktaira@iis.u-tokyo.ac.jp; hiyokoi@iis.u-tokyo.ac.jp*

Abstract - In plastic injection molding, various external defects are known to occur due to the flow of resins. In metallic injection molding which mixes metallic flakes with resins to enhance the appearance of molded articles, as the metallic flakes are oriented along the flow direction of resins inside molded products, these external defects tend to be emphasized even more according to the reflection characteristics of light.

In this study, the authors developed a method of measuring and quantifying the three-dimensional orientation distribution of metal flakes inside metallic molded products. Using this method, we paid attention to several external defects and analyzed their correlation with the orientation of metallic flakes inside molded products to clarify the usefulness of the developed method. To measure the three-dimensional distribution of metallic flakes, imege of the flakes were taken using an X-ray CT device and the orientation directions were calculated by analyzing these images. In addition, the correlation between the transition of the orientation vector in each area and external defects was also analyzed.

Keywords: Orientation analysis, X-ray CT, Metallic flake, Injection molding

Introduction

In recent years, the method of compounding pigmented flakes such as metal flakes (hereafter called MF) in resins to realize metallic appearances to give plastic products luxurious appearances or replace plating is being widely applied. In the injection molding of MF molding materials, as the flat surface of the MF is oriented along the flow direction of resins inside molded products, external defects tend to be emphasized even more according to the reflection characteristics of light. To date, studies have been conducted on the conditions and mechanisms by which external defects occur through observations of the external views and cross-sections of molded products using a microscope, etc[1]. However, regarding MF orientation inside molded products, no studies have been conducted on details of the orientation and behavior and accurate distribution of MF.

In this study, the authors developed a method of measuring and quantifying the three-dimensional orientation distribution of MF inside metallic molded products using X-ray CT. Using this method, we imaged and analyzed the orientation distribution of MF at the weldlines and flow marks of metallic molded products containing aluminum flakes. This report discusses the results of external view observations and correlation with MF orientation in the products.

Experimental Method

Experimental device

We used the Shimadzu SMT-130CT for the microfocus X-ray CT device, and VolumeGraphics' VGStudio MAX 1.2.1 for processing and displaying the images taken. To measure the reflection intensity at the flow marks in external view observations, we used the Hamamatsu Photonics photoelectric detector C10475.

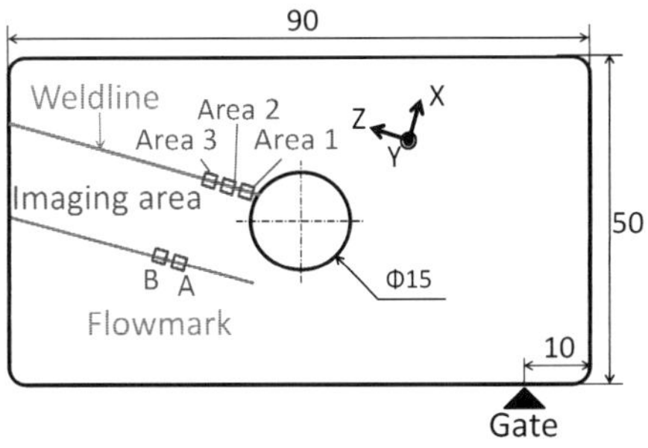

Fig.1 Outline of molded sample [unit: mm]

Calculation of orientation

We calculated MF orientation from the three-dimensional digital data obtained with the X-ray CT device by the following procedure; First, the images of the flakes and resin were differentiated and represented in binary, then each flake was assigned a number and separated, and three-axe moment calculation was performed for each flake. Next, the molded product was divided into square or cubic areas, the orientation direction (plane-normal vector) of each area was averaged, and the orientation vectors in three-dimensional spaces were displayed.

Test samples and analyze areas

In the experiments, we use the flat metallic injection molded plate 2 mm in thickness, back of which had embossing. PP mixed with 0.48wt% of aluminum flakes with an average particle size of 40μm (PYS-XX-0304, OK-KASEI Co., Ltd) was used as the molding material. Figure 1 shows the external view of the whole molded product and Figure 2 shows a picture of the external view of one part of the product. The external view of the weld lines and flow marks of this molded product was observed, and X-ray CT imaging

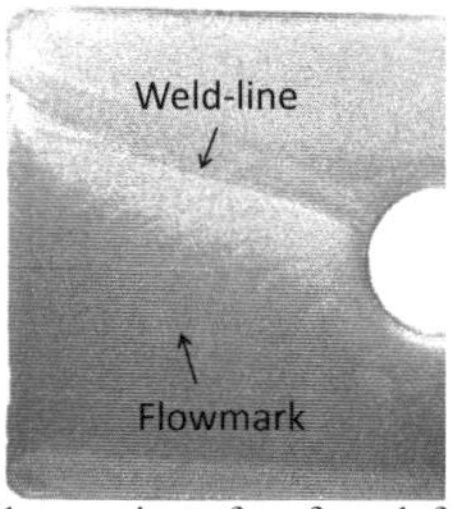

Fig.2 Observation of surface defects in metallic flake molded sample

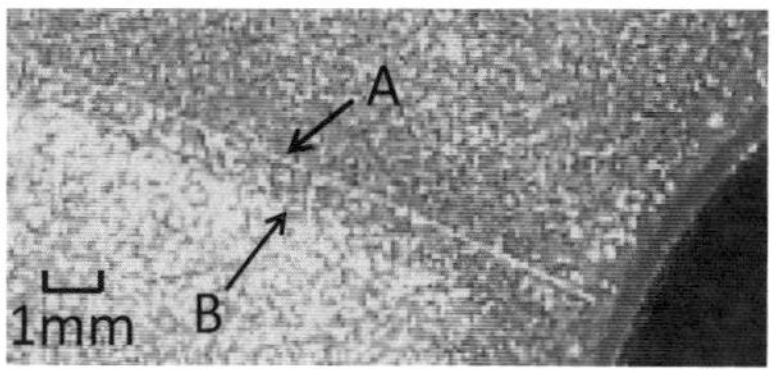

Fig.3 Enlarged view of weldline

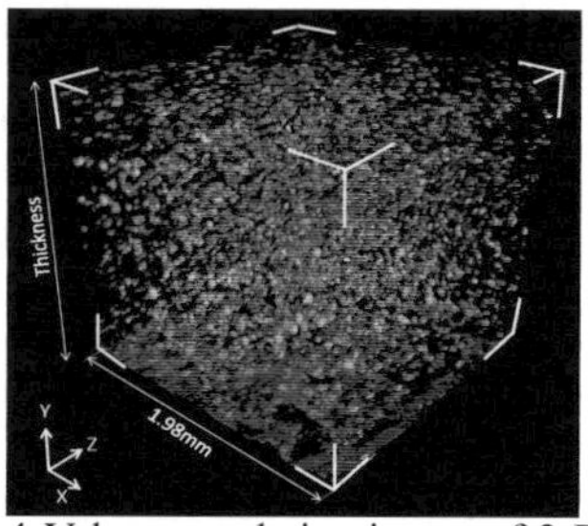

Fig.4 Volume rendering image of 3-D distribution aluminum flakes

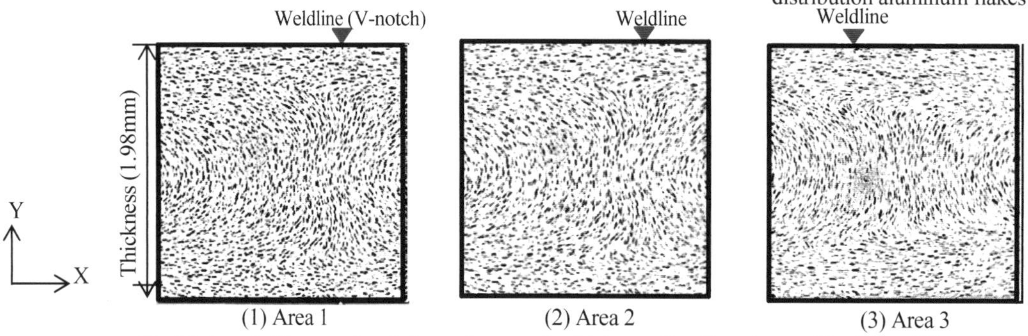

Fig.5 Integrated image of each layer along cavity thickness direction observed by X-ray CT

and orientation analysis were carried out for the part shown in Figure 1.

Results and Discussion

Weld lines

Figure 3 shows an enlarged photo of the area observed with weld lines. A clear weld line A and line B which runs parallel to A and has different reflection characteristics can be seen to form double lines. These lines are thought to be caused by the orientation of the flat surface of MF in along the thickness of the product from the outer to inner layers of the product. The three areas 1 to 3 (1.98×1.98×1.98mm) near the obstacle pin in this weld line area were imaged so that their right sides in Figure 1 were 5, 10, and 15mm away from the edge of the obstacle pin respectively.

Figure 4 shows the three-dimensional volume rendering image of only the MF based on the imaging data of Area 2. Black-and-white reversed processing images (X-Y projection plane) projecting all flakes and integrating along the weldline direction (Z-direction) were compared with Figure 5 for Areas 1 to 3 respectively. The V-notch-shaped weld on the surface is indicated by an arrow. In the inner layers of the molded product, MF can be seen to orient vertically from the S-shaped swelling, resulting in the deviation from the V-notch position. This deviation is thought to be due to the re-flow process of the core layer during the pressure holding stage (i.e., compensation flow process) after the formation of the V-notch during filling, because the gate is positioned asymmetrically to the obstacle pin. In the skin layer at the outermost surface layer, MF is oriented more or less parallel to the surface, while the position at which vertical orientation starts to occur is closer towards the center

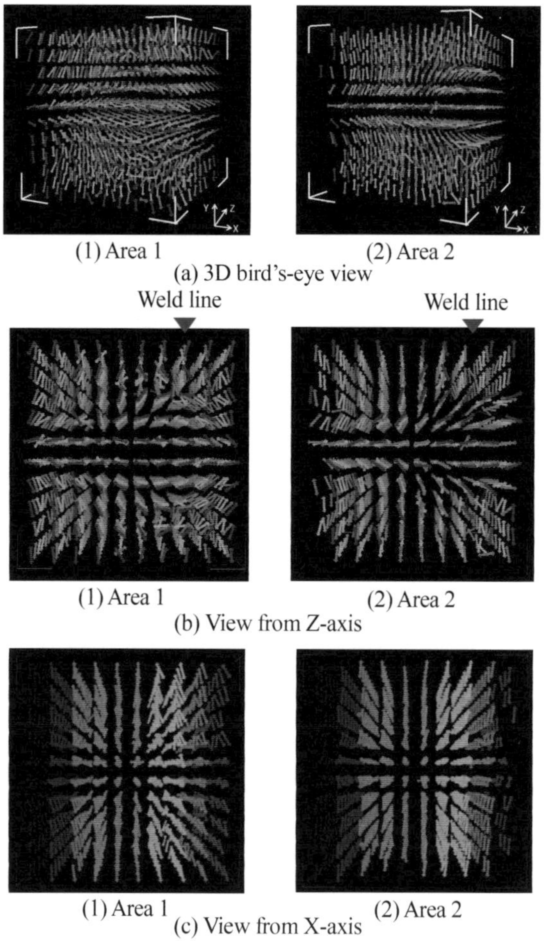

Fig.6 Three-dimensional orientation vectors image

of the thickness of the molded product from Area 1 (which is close to the obstacle pin) towards Area 3.

Orientation was analyzed in these three areas, and Figure 6 shows the results of Area 1 and Area 2 which showed marked differences. Areas1 and 2 are cut into 10× 10×10 areas (one side of each area is about 0.2mm), and the average orientation direction of MF in these areas is indicated as surface normal vectors. Each layer in the Z direction is colored with a different color. (a) is a bird's eye view and (b) is the view in the Z direction. (1) and (2) of Figure 5 are therefore from the same visual field, and shows the orientation in more detail. Figure 5 shows the changes in the depth at which vertical orientation occurs, but in Figure 6 (b), it can be seen that the vector starts inclining from several layers under the surface layer. In Area 1, orientation vectors become unstable at the first layer from the top (the layer nearest the surface), and vertical orientation starts moderately from the second layer. In respect to this, in Area 2, no vertical orientation can be seen until the second layer, but from the third layer. However, in the second layer from the top in Area 2, the vector can be seen to be inclining slightly at about two layers (0.4mm) left from the V-notch of the surface. This deviation in the vertical orientation position and V-notch position are thought to appear on the external view of the molded product as the weld line and line running parallel to it. Fig.6 (c) shows the view from the X direction. The flow is smooth in the Z-direction in Area 2 but not in Area 1 which is close to the obstacle pin. The average value of the components of the orientation vector (unit vector) in the Z-direction in each area is 0.17 in Area 1 and 0.06 in Area 2. It can be seen that orientation is more disorderly in Area 1. This may be due to the effects of disorderly flow caused by the obstacle pin in Area 1.

Based on these results, the external defects A and B in Figure 3 are thought to be caused by reflection anisotropy resulting from the penetration of the reflective light through the skin layer in areas with strong orientation in the vertical direction in the core layer.

Flow marks

When vertical light is irradiated on the sheet surface and the test sample is observed from directly above, no flow marks can be seen. However, when observed from the inclined direction, a band of flow marks can be clearly seen. Of the flow mark areas, Figure 7 shows the results of measuring the reflected light intensity along the line shown in Figure 1 at the projection angle of 0° and detecting angle of 70°. The vertical axis is the reflected light intensity and the horizontal axis is the measured position. It can be seen from this graph that the reflected light intensity is inconsistent. X-ray CT imaging and analysis were conducted on each area along the line where the reflection intensity was measured, and show the results of areas A and B in Figure 1. Figure 8 shows the results of black and white reversal processing of cross-sectional images projected and integrated in the X-

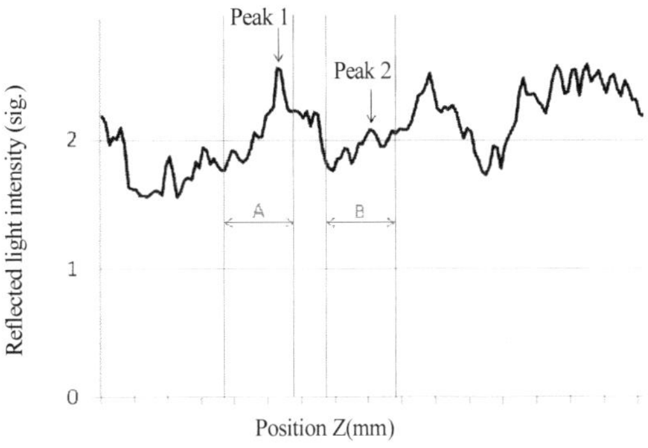

Fig.7 Measurement results of reflected light intensity distribution along flow mark

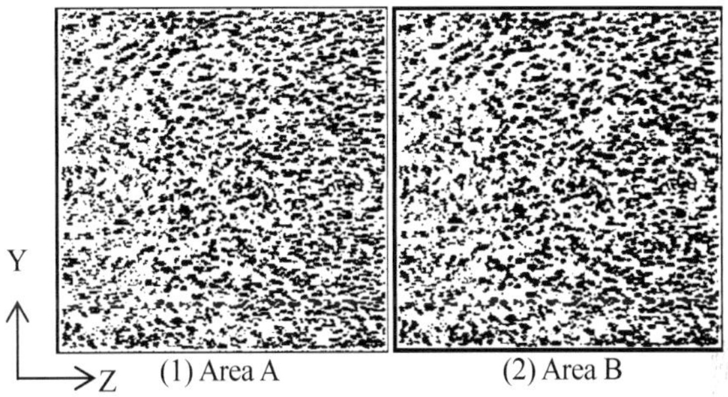

Fig.8 Integrated image of each layer along cavity thickness direction observed by X-ray CT

direction (Y-Z projection plane). In the areas where flow marks have generated, the orientation of MF is more moderate compared to the weldline area. Particularly in area B, no MF inclination can be seen in the skin layer, suggesting that the imbalance of reflection intensity of the surface may be affected by the orientation of MF in the inner layers of the molded products.

Next, both areas were divided into 10×10×10 parts, and the average orientation of each part was calculated. For the five layers from the surface in the Z-X plane (each layer is about 0.2mm thick), the orientation results were colored according to the size of the inclination in respect to the flow direction (Z-direction). Figure 9 shows the comparison of the results. Layew 1 is the layer closest to the surface, and they are the extraction of the middle area at which reflection intencity showed in Figure 7 was measured. The gray area is the area where orientation was difficult to measure due to noise, lack of resolution, etc. The areas closer to the normal vector towards the Y-axis are shown in white, the colors of areas are shown changing from green to blue as the more they incline in the flow direction (Z-direction) , and those in the opposite direction are displayed in yellow to red. It is thought that the reflection intensity measured in Figure 7 is stronger in the darker areas. In the red or blue areas, MF is inclined by more than 70°, and this is thought to have more or less no effect on the reflection in the measured direction. In Figure 7, in area A, the

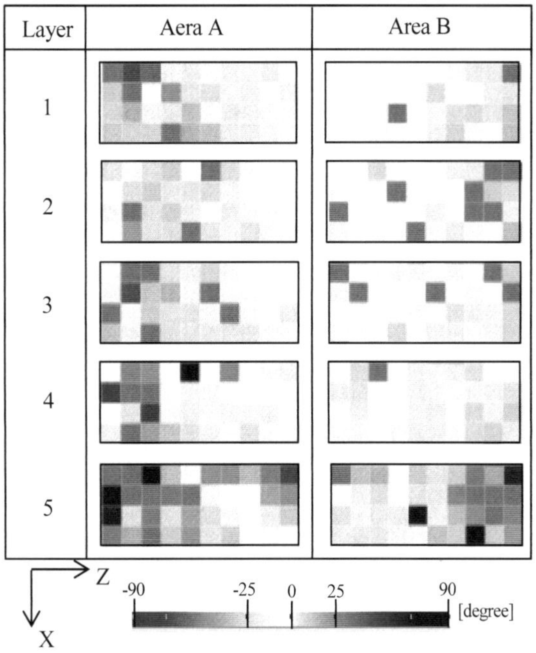

Fig.9 2D distribution patterns of orientation angle at each layer

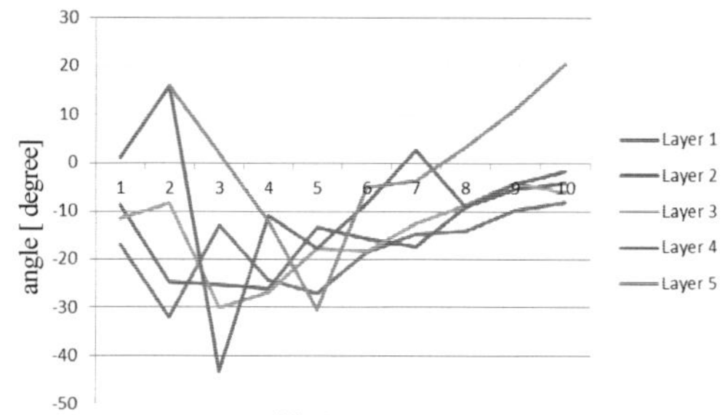

(1) Area A

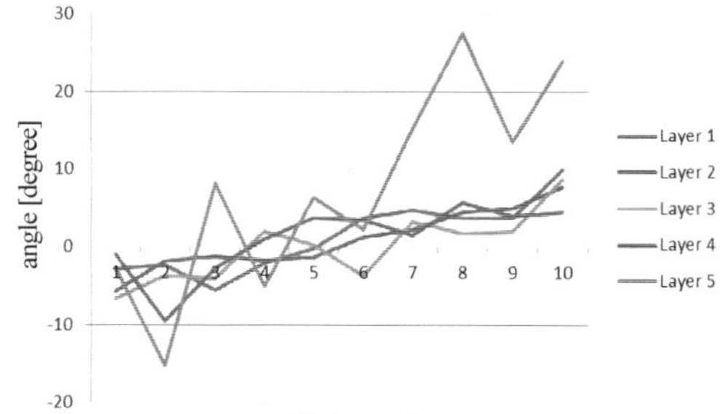

(2) Area B

Fig.10 Distribution curves of average angle of orientation vectors at each layer

reflection intensity peaks 1 at the one-fifth position from the right of the area. However in Figure 9, the area around the second row from the right corresponds to this position. Looking at each layer, layers 1 to 3 are white or yellow on the whole, with the left side particularly yellow. In layer 4, a green area appears for the first time from rows 4 and 9. In layer 5, the green areas increase but there exist a white and yellow layer between them, which correspond to peak 1 and the small peak on the left side in Figure 7. In addition, the reflection intensity is rising upwards to the right probably due to the effects of the distribution of yellow and white in layers 1 to 3. On the other hand, area B has weaker reflection intensity than area A and shows a moderate peak 2. In layers 1 to 4 in Figure 9, the overall color changes from light yellow to light green , and the reflection intensity increases moderately towards the right. In layer 5, a dark green color appears in the fourth row from the right corresponding to peak 2.

Figure 10 shows the results of averaging the orientation angle per row at each layer in Figure 9 except for the red and blue areas. Even from Figure 10, it can be seen that orientation occurs in respect to peaks 1 and 2 in layer 5.

In both areas, the MF orientation 0.8 mm deep from the surface is thought to have created the peak reflection intensity causing the flow mark defect. The fact that orientation in the inner layers can cause such an effect on the external view is due to the penetration of light to the inner layers of the product. Because of the low MF content in the resin compared to the MF size, there are many gaps between MF. However, there remain some areas where MF cannnot exactly be measured due to inadequate resolution of the X-ray CT device. If imaging can be carried out more accurately, moreprecise analysis about correlation between the reflection anisotropy and the MF orientation should also be possible.

Conclusions

In this study, the authors proposed a method of quantifying the three-dimensional orientation of metal flakes inside metallic molded products using X-ray CT and three-dimensional image processing. Using the proposed method, the correlation between the area of weldlines in molded products, internal orientation of flow mark areas, and generation of reflective anisotropy was analyzed to demonstrate the usefulness of the method. Owing to the high universality of this method and the increasing improvement of X-ray CT resolution, this method is expected to be widely applied to the analysis of filler orientation such as metal flakes and analysis of external defects in the future.

Acknowledgements

The authors would like to thank OK-KASEI Co., Ltd. for providing the various metallic molded products, and Nihon Visual Science Inc. for cooperating in developing the image analysis software. This research was conducted as a subproject of the U'11 project of the Institute of Industrial Science, The University of Tokyo.

References

1. F. Pisciotti; A. Boldizar; M. Rigdahl; I. Arino *Polymer Engineering and Science*, 2005, *45*, 1557.

O-09-245

CRYSTALLIZATION OF IPP IN PROCESSING CONDITIONS. EXPERIMENTS AND MODELLING

F. De Santis, V. Speranza', R. Pantani and G. Titomanlio*
Dipartimento di Ingegneria Industriale, Università di Salerno, Italy

Abstract- The effect of shear flow on nucleation density, spherulitic growth rate was analysed in rotational configuration. The experiments were carried out with an iPP, T30G, deeply caractherized as far as effect of cooling rate and pressure on crystallization kinetic functions.
Final morphology of samples isotermally crystallised after step shear was detected. Spherulite dimension was detected as function of the radius.
Results are interpreted on the basis of a model which describes the effect of flow upon crystallization through an increase of the thermodynamic crystallization temperature by effect of the molecular stretch determined by the flow. Model predictions reproduce most of the features of final morphology after solification.

Keywords: structure development, crystallization, flow induced crystallization, isotactic polypropylene, injection moulding.

Introduction

Many characteristics of polymeric objects are determined by morphology distribution induced by their manufacturing process.
The evolution of morphology is determined by the thermomechanical history up to solidification, which in polymer processing operations may take place at very high cooling rates and pressures and under the effect of flow. Indeed, in the last decades large interest was devoted by researchers to clarify main aspects of the effect of cooling rate, pressure and flow on morphology evolution of thermoplastic polymers.
The study of the effect of flow on crystallization may be schematized as follows: i) the identification of the threshold below which morphology evolution keeps quiescent features; ii) above such a threshold, how are spherulites nucleation and growth rates affected by the flow; iii) the identification of a second threshold above which crystallization characteristics change into fibrillar morphology and related fibrillar cristallization kinetics.
An experimental analysis of the effect of flow on spherulitic nucleation and growth rate of an iPP is presented in this paper, and a model to describe the effect of flow on spherulitic morphology evolution is underlined; predictions of a Flow Induced Crystallization Model are compared with morphology distribution in discs isothermally crystallized after rotational shear steps.

Experimental

All tests considered in this work were carried out with an iPP, T30G, supplied by Basell. The material was characterized for crystallization kinetics under very high cooling rates and pressure [1].
The material was subjected to isothermal (at 140°C) step shear experiments in a Linkam rotational cell; it was kept in the cell at constant temperature up to complete solidification. Two samples were produced, both with a gap of 400micron and a shear rate (at R=7.5mm) equal to 5/s. The first sample was kept under shear for 20s (sample_30); the second one for a time of 30s (sample_20). The morphology of both samples was characterized by means of polarized optical microscopy (POM). The dimensions of the spherulites along the midplane of the samples was measured as a function of the distance from the centre (R=0).

Experimental results

Some POM pictures of Sample_20 are reported in fig. 1 and clearly show that the dimensions of the spherulites decrease on increasing the distance from the centre.
Fig. 2 repors the measured diameters of the spherulites in the two samples as a function of the distance from the centre (namely as a function of the shear rate) It can be evinced that for both samples the dimensions decrease on increasing the distance from the centre (that is on increasing the shear rate). At each shear rate, the effect of a longer shearing time is toward a reduction of the dimensions of the spherulites.

The Flow Induced Crystallization Model (FICM)

The main assumption of the model is that the increase of both the growth and the nucleation rates are related to molecular strain.
For a given strain rate spherulitic growth rates were measured under steady shear conditions at several temperatures [2-3]. The effect of flow upon spherulitic growth rate was analised on the basis of a flow dependent crystallization temperature, Tm, which enters into the Hoffman-Lauritzen equation of growth rate at each temperature

$$G[T] = G_0 \exp\left[-\frac{U}{R\cdot(T-T_\infty)}\right]\exp\left[-\frac{K_g\cdot(T+T_m)}{2T^2\cdot(T_m-T)}\right] \quad (1)$$

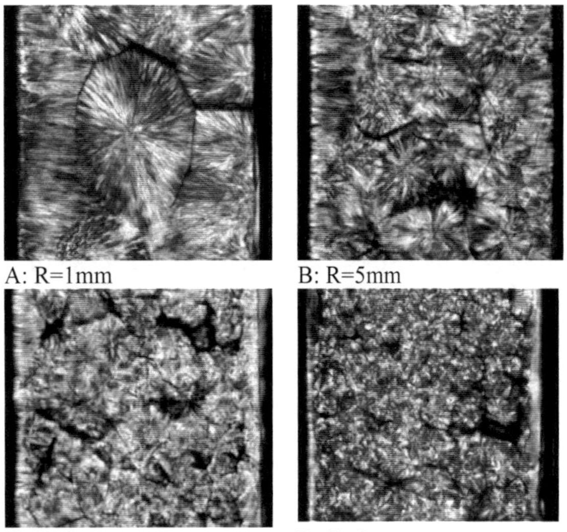

A: R=1mm B: R=5mm

C: R=7mm D: R=8mm

Fig. 1 POM micrographs of Sample_20 at several distances from the centre. Pictures refer to the whole thickness of the sample, which is 400micron.

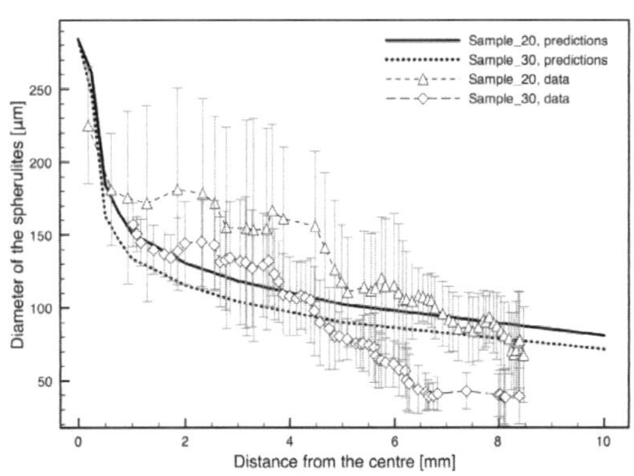

Fig. 2 Dimensions of the spherulites in function of the distance from the centre. Thick lines refer to the results of the simulations

In other words Eq. 1 was assumed to be valid also under flow conditions, limiting the effect of flow just on the melting temperature Tm. Such a flow induced melting temperature Tm, was then associated to a molecular strain parameter Δ □□chosen as the difference between the two largest eigenvalues of the molecular conformation tensor, where the molecular conformation tensor was described by a Maxwell equation:

$$\frac{D}{Dt}\bar{\bar{A}} - (\nabla\bar{v})^T \times \bar{\bar{A}} - \bar{\bar{A}} \times (\nabla\bar{v}) = -\frac{1}{\lambda}\bar{\bar{A}} + (\nabla\bar{v}) + (\nabla\bar{v})^T \quad (2)$$

with the relaxation time λ □function, in its turn, of the strain parameter Δ according to the equation:

$$\lambda = \frac{\lambda_0}{\left[1 + (a\,\Delta)^b\right]^c} \quad (3)$$

Under simple shear the parameter Δ is:

$$\Delta = \sqrt{A_{11}^2 + 4A_{12}^2} \quad (4)$$

As under steady shear $\bar{\bar{A}}_{11} = 2\lambda^2\dot{\gamma}^2$ and $\bar{\bar{A}}_{12} = \lambda\dot{\gamma}$; Δ becomes

$$\Delta = 2\dot{\gamma}\lambda\sqrt{1 + (\dot{\gamma}\lambda)^2} \quad (5)$$

The correlation between Tm and Δ is reported in fig.2:

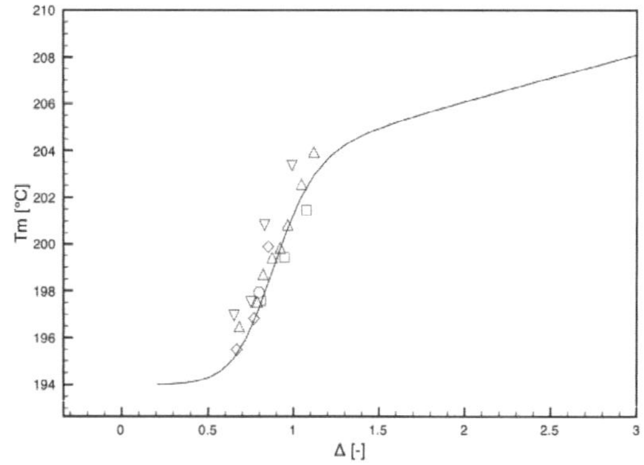

Fig.3 Correlation between Tm and Δ

Heterogeneous nucleation was predominant in the material adopted for the experiments under quiescent conditions; namely, nucleation density under quiescent conditions was found constant at constant temperature. Vice versa, nucleation rates were observed and measured under steady shear conditions In the same experiments carried out for the measurement of growth rate. A correlation among the excess of Growth rate (with respect to quiescent conditions) and the nucleation rate under the same conditions (of shear rate and temperature) was identified as reported in fig.4.

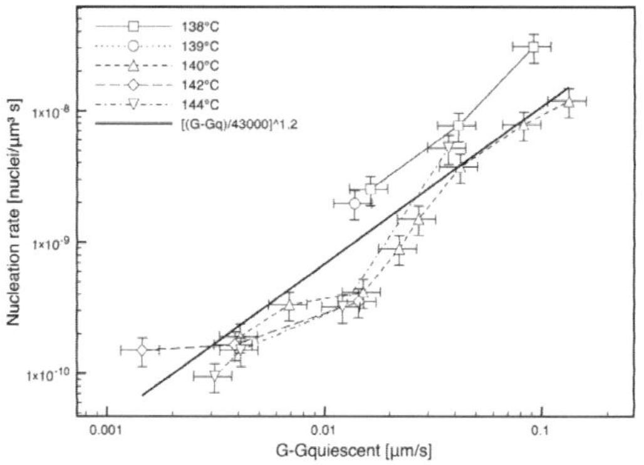

Fig.4 Correlation between Nucleation rate and excess Growth rate

The existence of such a correlation is consistent with with the fact that also nucleation rate is usually described by H/L equation.

The model allows to describe the average dimensions of the spherilites according to the following equation

$$\bar{R} = \sqrt[3]{\frac{3\xi_{\alpha,\mathrm{final}}}{4\pi N_{\mathrm{a,final}}}} \qquad (6)$$

where $\xi_{\alpha,\mathrm{final}}$ is the final value of the relative crystallinity degree of the α phase and $N_{\mathrm{a,final}}$ is the final number of active nuclei calculated as

$$N_{\mathrm{a,final}} = \int_0^{t_{\mathrm{final}}} \frac{\mathrm{d}N[T(s)]}{\mathrm{d}s}(1 - \xi(s))\mathrm{d}s \qquad (7)$$

Comparison and discussion

The moel described above was adopted to describe the final average dimension of the spherulites for both the samples analyzed in this work. The results of the simulations are compared in fig. 2 with the experimental results.

It can be noticed that the order of magnitude of the dimaeters is correctly predicted by the model and also the reduction of the dimensions with the radius (namely with the shear rate) is correctly predicted. Furthermore, the model also describes the effect of a longer shearing time.

Conclusions

In this work, an experimental analysis of the effect of flow on spherulitic nucleation and growth rate of an iPP is presented. A model to describe the effect of flow on spherulitic morphology evolution is described;and the predictions of a Flow Induced Crystallization Model are compared with the experimental data of morphology distribution. The main features of experimental data are correctly described by the model.

References (Times New Roman 10, bold)

1. R Pantani, I Coccorullo, V Speranza, G Titomanlio, Progress in Polymer Science (2005) vol. 30 (12) pp. 1185-1222.
2. R Pantani, I Coccorullo, V Volpe, G Titomanlio, Macromolecules (2010) vol. 43 (21) pp. 9030-9038.
3. I Coccorullo, R Pantani, G Titomanlio, Macromolecules (2008) vol. 41 (23) pp. 9214-9223

O-09-288

A NEW PARADIGM FOR THE TOUGHENING OF POLYLACTIDE-BASED MATERIALS

J. Odent, Y. Habibi, J-M. Raquez, Ph. Dubois

Laboratory of Polymeric and Composite Materials (LPCM) ; Center of Innovation and Research in Materials and Polymers (CIRMAP), University of Mons, Place du Parc 23, B-7000 Mons, Belgium – jeremy.odent@umons.ac.be

Abstract - In this work, we report immiscible polymer blends made of polylactide (PLA) toughened with impact rubber-like modifiers in order to elaborate ultratough PLA-based blends mediated with nanoparticles. Accordingly, a commercially available impact modifier specifically designed for PLA was selected as a model. In this study, 10 wt% of Biomax Strong 100 (BS) and silica nanoparticles (SiO_2) were melt-blended within PLA using microcompounder in order to investigate the microstructure and related mechanical performances for the resulting blends upon the processing methods and nanoparticles content. A synergistic toughening effect of impact modifier and silica nanoparticles is highlighted, together with a clear alteration of the blend morphology. This led to the exclusive conversion of rubbery spherical cavities into elongated structures in the presence of SiO_2. Related with the microstructure of rubbery microdomains, the use of PLA/BS-based systems containing 10% weight silica nanoparticles resulted in toughness improvement, actually reaching a eleven-fold increase in impact strength. The improvement in mechanical properties could be attributed to the morphological modification of the PLA-based blends mediated by the presence of the dispersed nanoparticles, and the concomitant fine tuning of the dispersed rubbery microdomains in terms of shape, average size and related size-distribution.

Keywords: impact modifier, morphology, polylactide, nanoparticles, ultratoughness.

Introduction

Due to its renewability and biodegradability, polylactide (PLA) is the most extensively investigated biopolymer as a biosourced substitute to conventional petroleum-based polymers in different types of applications such as in automotive and electronic industries [1, 2]. However, the applications of PLA are currently limited by its brittle nature [3, 4]. Among the strategies developed to overcome this limitation and to greatly enhance impact strength of PLA, the addition of impact modifiers is widely used [5, 6] and dispersed in the polyester matrix as rubbery nodules [7-9]. When the toughness efficiency is considered, it is recognized that there is an optimum size of these rubbery nodules at which the impact strength of the blend is maximized [10, 11]. Accordingly, the rubber phase morphology (domain size and related size-distribution) and polymeric matrix–rubbery domain interfacial properties must be properly controlled [12-15]. Recently, compatibilization methods have been reported via the use of nanoparticles into immiscible polymer blends [16, 17]. This represents a universal platform to tailoring the phase morphology of polymer blends, and therefore to control the material performances.

In this work, we investigated polymer blends made of PLA and impact rubber-like modifiers designed in the presence of nanoparticles in order to elaborate ultra-tough PLA-based blends. The role of nanoparticles is to specifically localize at the interface of both polymeric partners, to strengthen the interfacial adhesion between both partners and therefore the performance materials [18, 19]. Hence, an appropriate control about the nature of nanoparticles and their processing on blending provides a straight strategy to tailoring the phase morphology of polymer blends, and therefore to controling the material performances. We have recently demonstrated that the use of as-synthesized impact modifiers made of hydrolytically degradable poly(ε-caprolactone-*co*-D,L-lactide) (P[CL-*co*-LA]) copolyesters can enhance the toughness of PLA materials and reveal peculiar morphologies for the resulting blends (from round-like to oblong structures for rubbery phase) [20]. The most interesting feature was that the highest enhancement in impact strength achieved in the case of oblong structures within PLA matrix, indicating the importance for controlling the affinity and the interfacial compatibilization between the blend components. Hence, the present work aims at extending our approach to design ultratough PLA-based materials through an appropriate incorporation of nanoparticles into PLA-based materials toughened with Biomax Strong 100, which is a commercially available impact modifier specifically designed for the polyester matrix. In this study, we will report some efforts to control the phase structure of PLA-based materials containing 10 wt% of Biomax Strong 100 (BS) and silica nanoparticles (SiO_2) by adjusting the interfacial interaction within the resulting blend upon the processing methods and nanoparticles content. The relationship between morphological features and the synergistic toughening effect of rubbery-like system and nanoparticles are investigated.

Experimental

Materials: A commercially available extrusion-grade PLA (NatureWorks 4032D) was used as received ($\overline{M_n}$=133,500±5,000 g/mol, Đ=1.94±0.06 as determined by size-exclusion chromatography, 1.4±0,2% D-isomer content as determined by the supplier). Biomax Strong 100 (BS), a commercially available impact modifier specifically designed for PLA was provided by Dupont. CAB-O-SIL TS530 (Silica nanoparticles, SiO_2) is a high surface area fumed silica (225 m²/g) which has been surface modified with hexamethyldisilazane and was supplied by Cabot. Cloisite 20A (CL20A), a dimethyl-dihydrogenated tallow-ammonium treated MMT (modifier concentration: 95 meq/100 g), Cloisite 25A (CL25A), a dimethyl-dihydrogenated tallow-(2-ethylhexyl)-ammonium treated MMT (modifier concentration: 95 meq/100 g) and Cloisite 30B (CL30B), a methyl-tallow-bis(hydroxyethyl)-ammonium treated MMT (modifier concentration: 90 meq/100 g) were supplied by Southern Clays Products INC.

Melt-blending of PLA-based materials: Prior to extrusion, PLA pellets were dried for at least 12h at 80°C in an oven. Both binary and ternary systems based on PLA were prepared in a DSM twin-screw micro-compounder (15 cc) at 200°C and 60 rpm for 3 min. Two processing methods were employed to prepare ternary PLA/BS/nanoparticles composites upon how these nanoparticles were dispersed, i.e., either directly into the blend (so-called one-step process) or first into one both partners before blending (so-called two-step process) into PLA/BS-based systems. In the one-step process, both impact modifier (BS) and nanoparticles were directly compound with PLA, while in the two-step process, nanoparticles were mixed first in one of both partners by dissolving the compounds in a minimal of chloroform followed by a slow evaporation of solvent under stirring and then blended following the same procedure as the one-step process. Standard samples of resulting PLA-based materials were then prepared by compression molding at 200°C for 10 min.

Characterization techniques: Differential scanning calorimetry (DSC) was performed using a DSC Q2000 from TA Instruments at heating and cooling rates of 10°C/min under nitrogen flow (2nd scan). Notched Izod impact tests were performed according to ASTM D256 using a Ray-Ran 2500 pendulum impact tester (E=4J, mass=0.668kg and speed=0.46m/s). Tensile tests were performed according to ASTM D638 using a Zwick universal tensile testing machine (speed=1m/s and preload = 5N). Room-temperature impact-fractured surfaces of specimens were examined for morphological structure through transmission electron microscopy (TEM). For recording TEM images, the samples were cryomicrotomed at -100 °C by a Leica UCT microtome. Transmission electron microscopy was carried out using a Philips CM20 microscope operated at 200 kV. Analysis software ImageJ is used for the analysis of TEM images to estimate the particles size and their distribution within the matrix.

Results and Discussion

In this work, 10 wt% of Biomax Strong 100 (BS), a commercially available impact modifier specifically designed for PLA, was used and melt-blended at 200°C within PLA using a twin-screw microcompounder (15 cc). The resulting blend was recovered and compression molded at 200°C. Toughness of the PLA-based materials was determined by notched Izod impact testing using a pendulum impact tester, while tensile properties were determined at a tensile rate of 1 mm/min (**Fig. 1**). Unfilled PLA itself behaves as a stiff and glassy materials characterized by a high Young's Modulus of ca. 1860 MPa, a tensile strength of ca. 58 MPa, an elongation at break of ca. 4% and an impact strength of 2.7 kJ/m². When loaded at 10 wt% BS as impact modifier, a significant improvement in toughness are highlighted (impact strength passing from 2.7 for PLA matrix to 11.3 kJ/m²), together with an increase in elongation at break (at ca. 24 % in elongation) and a decrease of the Young's modulus (at ca. 1780 MPa in modulus). When nanoparticles were loaded into the binary blend, the tensile properties of resulting ternary PLA-based nanocomposites are not altered and the Young's modulus is even preserved. For instance, Young's modulus passes from ca. 1780 to ca. 1700 MPa for PLA/BS blend and PLA-based ternary nanocomposites containing 5 wt% silica nanoparticles, respectively. The most striking feature concerns the significant increase, up to seven-fold, in the impact strength recorded in the case of PLA-based ternary nanocomposites (**Fig. 1**). For instance, in the presence of only 5 wt% of Silica nanoparticles, a very significant increase in the impact strength was recorded, i.e., 18.5 kJ/m² vs. 2.7 kJ/m² for the unfilled PLA. An eleven-fold increase in impact strength was even recorded at 10 wt% silica nanoparticles (30.2 kJ/m²), together with Young's modulus at ca. 1720 MPa.

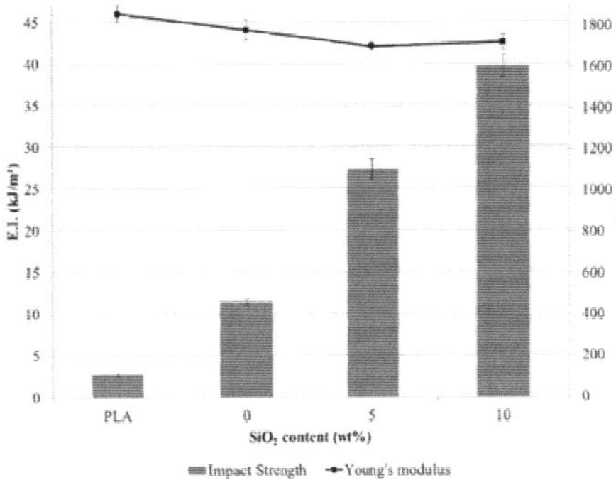

Figure 1 - Influence of the Silica nanoparticles content on the notched Izod impact strength and Young's modulus of PLA-based materials containing 10 wt% of BS impact modifier.

Importantly, the ultra-toughness behavior is highly dependent on how silica nanoparticles are incorporated into PLA-based melt-blends (**Fig. 2**). When impact modifier and nanoparticles were used together, both simultaneous increase of the toughness and stiffness in ternary composites were only achieved by using an appropriate processing method, i.e. by adding first silica nanoparticles into PLA, followed by a consecutive melt-blending with BS (two-stage melt-process). Changing the mode of silica dispersion in the melt-blend allows tuning the toughness. For instance, the direct dispersion of 5 wt% silica nanoparticles into the blend of PLA and BS (90/10, wt/wt) or the other two-stage melt-process, i.e., first melt-mixing of Silica into BS, followed by processing within PLA resulted in a drastic decrease in toughness. The close relationship between mechanical performance and microstructure of these PLA-based blends may explain why processing methods appear to be suitable techniques to reach acceptable toughness. Accordingly, an appropriate processing method in two-step, in which silica nanoparticles were first added into PLA partner, is required to greatly enhance the mechanical performances by adjusting the morphological features and promoting optimum mechanical performances.

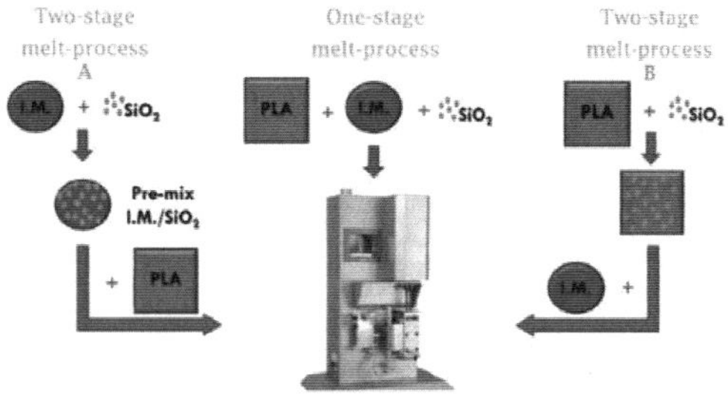

Figure 2 – Investigated processing method of PLA-based materials containing an impact modifier (I.M.) and silica nanoparticles (SiO_2). Samples produced via direct dispersion of Silica into both PLA and I.M. partners (middle), via pre-mixing of I.M. and SiO_2 followed by blending within PLA (left) and via pre-mixing of PLA and SiO_2 followed by blending within I.M. (right).

As attested by differential scanning calorimetry (DSC), silica nanoparticles loading has no real effect on the thermal properties of the PLA/BS blends (**Tab. 1**). The melting (T_m) and glass transition temperatures (T_g) of PLA matrix remain unchanged at 169 and 61°C, respectively. Regarding the cristallinity degree for PLA-based blends, a slight increase is recorded by adding a rubber-like impact modifier into PLA (from 3% to 13%). However, the crystallinity degree for ternary PLA-based composites is not modified by the incorporation of silica nanoparticles. This indicates that the enhancement in toughness is directly related to the formation of peculiar morphologies for the rubbery phase in the presence of silica nanoparticles, without any modification of crystallinity degree for PLA/BS blends in the presence of silica nanoparticles.

Table 1. Thermal properties of unfilled PLA, PLA/BS blends (10 wt%) and the corresponding ternary blends.

Entry	SiO_2 content (wt%)	T_g [a] (°C)	T_c [a] (°C)	ΔH_c [a] (J/g)	T_m [a] (°C)	ΔH_m [a] (J/g)	χ [b] (%)
1	PLA	62	113	29	170	32	3
2	PLA/BS	62	114	28	169	32	11
3	+ 5 wt%	62	107	18	168	27	11
4	+ 10 wt%	62	103	8	168	27	13

a) As determined by DSC (under N_2 flow, heating rate of 10°C/min, second scan).

b) As determined on the first scan and calculated taking into account the contribution of both impact modifier and silica nanoparticles with $\chi = (\Delta H_m - \Delta H_c)/\Delta H^{\circ}_m$ and $\Delta H^{\circ}_m = 93.6$ J/g.

Transmission electron microscopy (TEM) analyses of PLA/BS blend shows the presence of round-like nodules regularly dispersed within the PLA matrix that are ascribed to the dispersed rubbery phase. When silica nanoparticles were dispersed by a two-stage melt-process (by adding first Silica into PLA partner), morphological features (domain shape, average size and related size distribution) are clearly affected, mediated by the presence of nanoparticles. TEM analyses show that silica nanoparticles are mostly located at the interface PLA/BS, promoting the formation of peculiar morphologies. According to our previous work, the improvement in PLA toughness was namely explained by blend-morphology of the dispersed phase in PLA matrix, where silica nanoparticles were previously highlighted as a interesting strategy to tailoring the phase morphology of resulting PLA/impact modifiers blends. Recently, we have already demonstrated that the co-addition of a P[CL-*co*-LA] copolyester (72/28 mol% CL/LA, 10 wt%) and silica nanoparticles in PLA through a two-step process increased significantly the material toughness by promoting spherical nodules commonly observed in immiscible blends into nanosized ribbon-like structures [21]. Surprisingly, such positive effect of silica nanoparticles can be achieved when commercially available impact modifiers like Biomax Strong was used to toughen PLA-based materials instead of P[CL-*co*-LA] copolyester. It is moreover highlighted that the impact properties can be directly correlated to morphological alterations of the PLA-based blends, not only upon the average size and related size-distribution, but also upon the shape of the dispersed rubbery microdomains. Accordingly, a fine tuning of the morphological issues (control in terms of domain shape, average size and size-distribution of the dispersed rubbery microdomains) is governed by the presence of nanoparticles and endows PLA-based materials with new impact performances.

Another interesting feature about the co-addition of BS and Silica into PLA is to check whether such positive effect of nanoparticles can be achieved with other nanoparticles within PLA-based materials. Remarkably, the addition of organomodified layered aluminosilicate (clays) has also shown a significant improvement in the impact strength and leads to the formation of ultra-tough materials (**Tab. 2**). For instance, a value of 37.6 ± 2.1 kJ/m² has been achieved with only 1 wt% of CL25A; this value is in the same range of the toughness obtained with 10 wt% of silica nanoparticles. From these results, it becomes evident that silica and other nanoparticles should be combined with conventional impact modifiers in order to develop synergistic systems with high efficiency and acceptable impact strength.

Table 2. Notched Izod impact strength of unfilled PLA, PLA/BS blends (10 wt%) and the corresponding ternary blends containing 1 wt% of clays.

Entry	Type of Clay	E.I.[a] (kJ/m²)
1	PLA	2,7 ± 0,1
2	PLA/BS	11,3 ± 0,8
3	+ 1 wt% CL20A	32,6 ± 1,0
4	+ 1 wt% CL25A	37,6 ± 2,1
5	+ 1 wt% CL30B	21,9 ± 1,2

a) As determined by Notched Izod impact tests.

Conclusions

In this paper, we report the toughness and phase morphology of PLA/impact modifier/nanoparticles ternary nanocomposites. In this respect, 10 wt% of BS and silica nanoparticles were melt-blended within PLA using microcompounder in order to investigate the microstructure and related mechanical performances of the resulting blend upon the processing methods (one-step or two-step) and nanoparticles content. From this investigation, only a two-step process with an addition of Silica into PLA first, followed by this into BS leads to high impact performances. In this way, the co-addition of a commercially available impact modifier (BS) and Silica within PLA significantly increase the material toughness as attested by the recorded eleven-fold increase in impact strength, while maintaining high level of rigidity. This synergistic toughening effect of BS and Silica are observed together with a clear alteration of the blend morphology, promoting new peculiar morphologies in the presence of silica nanoparticles. Directly related to the microstructure of rubbery microdomains and the morphological modification of the PLA-based blends mediated by the presence of nanoparticles (fine tuning of the dispersed rubbery microdomains in terms of shape, average size and related size-distribution), an unprecedented toughness enhancement was achieved. Moreover, the material performances show similar behavior when an as-syntesized hydrolytically degradable P[CL-*co*-LA] statistical copolyester are employed unstead of BS; highlighting the use of nanoparticles as a simple and universal strategy to endowing PLA-based materials with new mechanical properties. In this regards, the use of other nanoparticles like nanoclays can be also beneficial for applications requiring high performance polymeric materials for environmentally friendly polymers like PLA.

Acknowledgements

The authors are grateful to the ''Région Wallonne'' and European Community (FEDER, FSE) in the frame of ''Pôle d'Excellence Materia Nova'' and in the excellence program OPTI²MAT for their financial support. CIRMAP thanks the ''Belgian Federal Governement Office Policy of Science (SSTC)'' for general support in the frame of the PAI-6/27. J. Odent thanks F.R.I.A. for his financial support thesis grant. J.-M. Raquez is ''chercheur qualifié'' by the F.R.S.-FNRS (Belgium).

References

1. K. Madhavan Nampoothiri; N. R. Nair; R. P. John, *Bioresour. Technol.* 2010, *101*, 8493.
2. R. Auras; B. Harte; S. Selke, *Macromol. Biosci.* 2004, *4*, 835.
3. R. M. Rasal; A. V. Janorkar; D. E. Hirt, *Prog. Polym. Sci.* 2010, *35*, 338.
4. A. Södergård; M. Stolt, *Prog. Polym. Sci.* 2002, *27*, 1123.
5. K. S. Anderson; K. M. Schreck; M. A. Hillmyer, *Polym. Rev.* 2008, *48*, 85
6. H. Liu; J. Zhang, *J. Polym. Sci., Part B: Polym. Phys.* 2011, *49*, 1051.
7. S. Ishida; R. Nagasaki; K. Chino; T. Dong; Y. Inoue, *J. Appl. Polym. Sci.* 2009, *113*, 558.
8. J. Z. Liang; R. K. Y. Li, *J. Appl. Polym. Sci.* 2000, *77*, 409.
9. A. Alfarraj; E. Bruce Nauman, *Polymer* 2004, *45*, 8435.
10. S. Wu, *Polymer* 1985, *26*, 1855.
11. C. B. Bucknall; D. R. Paul, *Polymer* 2009, *50*, 5539.
12. Z. H. Liu; X. D. Zhang; X. G. Zhu; Z. N. Qi; F. S. Wang; R. K. Y. Li; C. L. Choy, *Polymer* 1998, *39*, 5047.
13. S. Wu, *Polymer Engineering & Science* 1990, *30*, 753.
14. K. Cho; J. Yang; S. Yoon; M. Hwang; S. V. Nair, *J. Appl. Polym. Sci.* 2005, *95*, 748.
15. Y. Okamoto; H. Miyagi; M. Kakugo; K. Takahashi, *Macromolecules* 1991, *24*, 5639.
16. L. Ashabi; S. H. Jafari; B. Baghaei; H. A. Khonakdar; P. Pötschke; F. Böhme, *Polymer* 2008, *49*, 2119.
17. X.-Q. Liu; W. Yang; B.-H. Xie; M.-B. Yang, *Materials & Design* 2012, *34*, 355.
18. H. Yang; X. Zhang; C. Qu; B. Li; L. Zhang; Q. Zhang; Q. Fu, *Polymer* 2007, *48*, 860.
19. I. Kelnar; J. Rotrekl; J. Kotek; L. Kaprálková; J. Hromádková, *Eur. Polym. J.* 2009, *45*, 2760.
20. J. Odent; P. Leclère; J.-M. Raquez; P. Dubois, *European Polymer Journal (submitted)* 2012.
21. J. Odent; Y. Habibi; J.-M. Raquez; P. Dubois, *Applied Materials and Interfaces (submitted)* 2012.

O-09-291

INTEGRATION OF NANOFILLERS IN POLYURETHANE FOAMS FOR DIFFUSION CONTROL#

Regine Boldt[a,*], **Andreas Leuteritz**[a], **Jörg Finnberg**[b], **Thomas Thieme**[c] **and Gert Heinrich**[a,d]
[a] *Leibniz-Institut für Polymerforschung Dresden e.V., Hohe Str. 6, 01069 Dresden, Germany*
[b] *IMA Dresden GmbH, Wihelmine-Reichard Ring 4, 01109 Dresden, Germany*
[c] *GEF-Ingenieurgesellschaft für Energietechnik und Fernwärme, Dresdner Str. 113, 09131 Chemnitz, Germany*
[d] *TU Dresden, Institut für Werkstoffwissenschaft, Helmholtzstr. 7, 01062 Dresden, Germany*

Polyurethanes (PURs) are unique polymer materials with a wide range of applications due to their varied physical and chemical properties. Polyurethane foams are used for building insulation, appliances, transportations, furniture, etc. The properties relevant for applications like barrier effects, thermal resistance and fire behavior can be improved by using anisotropic nanofillers in PUR-foams. In our ongoing project the control of gas permeation is the key to develop a new kind of preinsulated plastic jacket pipe for district heating. Due to the loss of cyclopentane and carbon dioxide as well as intrusion of air the insulation action is reduced.

In this work the effect of several nanofillers in combination with various kind of mixing procedures on dispersion of platelets for the retardation of diffusion was analyzed. Two different nanofillers were used with different chemical composition – pristine and organically modified. The fillers were dispersed either in the polyol component or in isocyanate to investigate the influence of mixing parameters on the degree of dispersion of the fillers with the aim to place the flake-like particles in the "windows" of the PUR-foam. For a detailed description of such composites the following techniques are used at different states of PUR-foam preparation. SAXS/WAXS investigations are used after mixing the platelets either with isocyanate or polyol in order to determine the intercalation/exfoliation behavior of the nanoparticles. In addition, rheological measurements contribute to evaluate the state of dispersion and the inter-particle relation before foaming.

Electron microscopy is used to localize the particles within foamed samples. Energy dispersive X-ray analyses make the visualization of the platelets possible by representing the distribution of fillers in terms of elemental maps (see Figure 1).

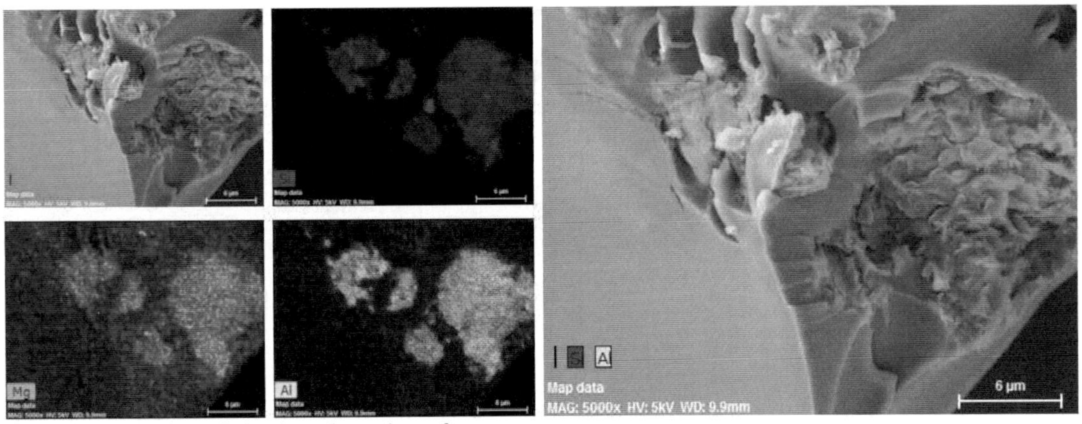

Figure 1: distribution of clay in polyurethane foams

This project is funded by the Federal Ministry of Economics and Technology under grant 03ET1080A-C

Keywords: polyurethane foam, nanofiller, district heating, plastic jacked pipes

***Corresponding Author**: boldt@ipfdd.de

O-09-019

THE INFLUENCE OF TALC AS A NUCLEATION AGENT ON THE NON-ISOTHERMAL CRYSTALLIZATION AND MORPHOLOGY OF ISOTACTIC POLYPROPYLENE: THE APPLICATION OF THE LAURITZEN-HOFFMANN, AVRAMI AND OZAWA THEORIES

N. Kocic[1]*, K. Kretschmer[1], M. Bastian[1] and P. Heidemeyer[1]

[1]*SKZ - Das Kunststoff-Zentrum, Würzburg, Deutschland – n.kocic@skz.de; k.kretschmer@ skz.de; m.bastian@skz.de; p.heidemeyer@skz.de*

Abstract - In this article we propose a new method, which can be used to analyze the influence of different additives and fillers on the non-isothermal crystallization of semi-crystalline polymers. The composites of talc in isotactic polypropylene (i-PP) were compounded and subsequently injection molded. PP morphology and talc dispersion were visualized using optical microscopy and computed tomography. WAXS and SAXS measurements provided an insight into the crystal structure of PP. The data obtained from non-isothermal DSC measurements were fitted to the Avrami model for the non-isothermal case. The calculated Avrami's exponent (n) was used in the combination of Lauritzen-Hoffman and Ozawa models to calculate the nucleation parameter (K_g). As expected, a reduction of the K_g with increased talc concentration was found, as a consequence of the talc activity toward nucleation of i-PP. Using the proposed model extension, the comparison between the nucleation activities of three talcs with different particle sizes was made. As expected, the K_g values decrease with the size of talc particles. Furthermore, it is shown that delaminated talc with a higher specific surface and the same particle size is more efficient in nucleation of i-PP compared to the conventionally micronized grade. Thus, the introduced model extension could be a useful tool for comparing of nucleation ability of different additives in the non-isothermal crystallization of semi-crystalline polymers and it proves to be superior to previous approaches regarding filled i-PP.

Keywords: crystallization, morphology, modeling, composites, thermoplastics

Introduction

In order to form a stable nucleus during the polymer crystallisation at temperature T_c, the energy barrier of nucleation (ΔG^*) has to be overcome: [1]:

$$\Delta G^* \propto \frac{K_g}{T_m^0 - T_c} \qquad \text{Eq.1}$$

, where T^0_m is the equilibrium melting temperature and K_g is the nucleation parameter, which denotes the energy (in K^2) requiered for the formation of a nucleus of a critical size. Nucleation agents presented in the melt reduce K_g and therefore ΔG^* [2]. This raises the crystallization temperature [3-5], increases the number of nuclei [6] and causes the growth of smaller, evenly sized and thicker crystals [7]. As a consequence, the physical properties of the polymers are improved, like stiffness, impact strength and transparency [8-11]. Because of its obvious practical importance, crystallization of polymers has been widely examined. The Avrami macrokinetic model has been used to analyze the overall kinetics of polymer crystallization under isothermal conditions [12-13]:

$$\log[-\ln(1 - X(t))] = \log K + n \cdot \log t \qquad \text{Eq.2}$$

where X(t) is the relative degree of crystallinity as a function of time. Avrami´s exponent (n) and Avrami´s rate constant (K) are constants specific to a given type of nucleation and growth of crystalls [14]. If the model adequately represents the process, the plot of the left-hand side of the Eq.2 versus log t should be a straight line, with the Avrami´s exponent n as the slope and log K as the intercept. Some authors used the Avrami approach for the non-isothermal crystallization [15-16]. However, in this case Avrami parameters K and n don't have the same physical meaning, as when speaking about isothermal crystallization [17].

Monasse and Haudin [18] proposed an extension of Ozawa´s theory [19] to predict the nucleation parameter (K_g) under non-isothermal conditions:

$$\frac{2}{3}\ln[-\ln(1-X(T))]-\ln\left(-\frac{d(\ln(1-X(T)))}{dT}\right)- \\ -\frac{U}{R(T-T_\infty)}=C+\frac{K_g}{T(T_m^0-T)} \qquad \text{Eq.3}$$

, where X(T) denotes the relative degree of crystallinity as a function of temperature, T_∞ is defined as $T_\infty = T_g - 30$, T_g is the glass transition temperature, $T^0_m - T$ denotes the supercooling whereby T^0_m is the equilibrium melting temperature, U is the activation energy for the transport of polymer segments to the site of crystallization, R is the universal gas constant and C is a constant. As regards pure PP, Monasse and Haudin plotted experimental values on the left-hand side of the Eq.3 against $1/(T^0_m - T)$ and obtained K_g as the slope of the straight line.

In this article, we modified the Monasse-Haudin approach for the case of filled PP. The influence of talc particle size as well as talc specific surface on the crystallization of PP was examined using using the proposed model extension.

Experimental

Materials

Isotactic polypropylene HD120MO (M_w = 365 kg/mol, M_w/M_n = 5,4 [20]) was used in this study. The main properties of the talc types used are summarized in Table 1:

Table 1: The main properties of the talc used (manufacturer data): a: d_{98}; b: d_{95}

Type	d_{50} (μm)	Top c. (μm)	S. surface (m^2/g)
Fintalc M05SLC	2,3	10[a]	9,5
Fintalc M10E	2,8	13[a]	8
Fintalc M15E	4,2	14[a]	6,5
Fintalc M30SL	10	27[a]	3,5
Jetfine 8CF	2	5,7[b]	8
HAR T84	2	11,3[b]	16

Composite preparation

All composites were prepared by melt mixing on a ZSE 18 Hpe (Leistritz) co-rotating twin screw extruder. After compounding, the pellets were air dried and injection molded using the Unilog 4000 (Battenfeld) machine. To isolate the influence of talc, all the process parameters were held constant.

Characterization

An analysis of the crystallization of the PP composites was performed in a Netzsch 200F3 Maia differential scanning calorimeter. Changes in the PP morphology and quality of talc dispersion were visualized using Axioskop 2 MAT Zeiss optical light microscopy equipped with AxioCam MRc camera. The agglomeration of talc particles was visualized with exaCT s50HR computer tomography and analysed with the software VG Studio MAX 2.0. For the characterization of the crystal structure of PP and the determination of overall degree of crystallinity, we used a Siemens D 5005 WAXS. The crystalline long period was measured using a Bruker Nano Star small angle X ray scattering (SAXS).

Results and Discussion

Mathematical development

Using the Eq.3 we attempted to find the influence of talc on the nucleation of i-PP. Unfortunately, fitting the DSC-data into this model didn't give the satisfactory linear correlation necessary to find a nucleation parameter K_g. During the development of Eq.3, a instantaneous nucleation and 3-dimensional crystal growth were assumed. Nevertheless, the way of crystal growth could be changed, due to the influence of the filler. Thus, we tried to adapt the Eq.3 to our conditions. Combining the Ozawa [19] and the Lauritzen-Hoffman model [21], we obtained [22]:

$$\ln\left(\frac{d(\ln(1-X(T)))}{dT}\right)-\frac{m}{m+1}\ln(-\ln(1-X(T)))+ \\ +\frac{U}{R(T-T_\infty)}=[\ln(G_0)+\phi]-\frac{Kg}{T(T_m^0-T)} \qquad \text{Eq.4}$$

We further assume that crystal growth under the influence of talc deviates from the pure three-dimensional growth. Concerning this assumption, factor m in Eq.4 cannot be taken as a constant, but must be calculated experimentally. The best linear fit of Eq.4 was obtained calculating factor m as n - 2, where the parameter n was obtained from experimental data using Eq.2. In this study the T_g value of PP used was 253 K, T^0_m = 481 K, U = 6280 J/mol, and X(T) and T were obtained from DSC non-isothermal experiments.

Experimental results

For all talc concentrations, only α - form of i-PP

was detected using WAXS, with characteristic peaks at 2θ = 14, 17 18,5, 21,5°. The crystallization peak temperature (DSC), long period (SAXS), overall degree of crystallinity (WAXS) and lamellae thickness (calculated as a product of long period and degree of crystallinity) in pure PP and in the composite containing 10 wt.% of Fintalc M05 SLC are presented in Table 2:

Table 2: Crystallization peak temperature (T_c), long period (D), overall degree of crystallinity (X_c) and lamellae thickness (l_c) in pure PP and in the composite containing 10 wt.% of Fintalc M05 SLC

Talc wt.%	T_c (°C)	D (Å)	X_c (-)	l_c (Å)
0	111	123	0,49	60,3
10	123	133	0,54	71,8

As one can see, the crystallization peak temperature is shifted to higher temperatures as a consequence of the nucleation activity of the talc. This was further reflected on the structure and size of crystalline units, whereby the higher amount of thicker crystals was formed.
The time and the relative degrees of crystallinity (t and X(t)) obtained from DSC-cooling cycle were fit to Avrami´s equation (Eq.2) to find the parameter n as the slope, which was further used to fit Eq.4 with experimental results. The result of fitting is presented in Fig. 1 for PP filled with 5 wt.% of Fintalc M05 SLC.

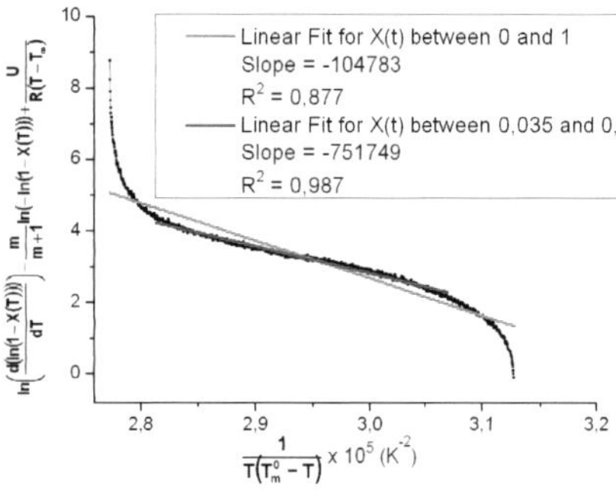

Fig. 1: The fitting of the experimental data with equation Eq.4 for PP fillled with 5 wt.% of Fintalc M05 SLC

As can be seen, Eq.4 gives satisfactory agreemet with experiments, except at the beginning and at the end of the process. The correlation coefficient R^2 was not less than 0,87 in all cases. However, for our further calculations, we used the linear part in Fig. 1, which corresponds to the relative degree of conversion between 0,035 and 0,98. K_g values were determined from the slopes of the straight part and presented in Fig. 2 for all talc concentrations:

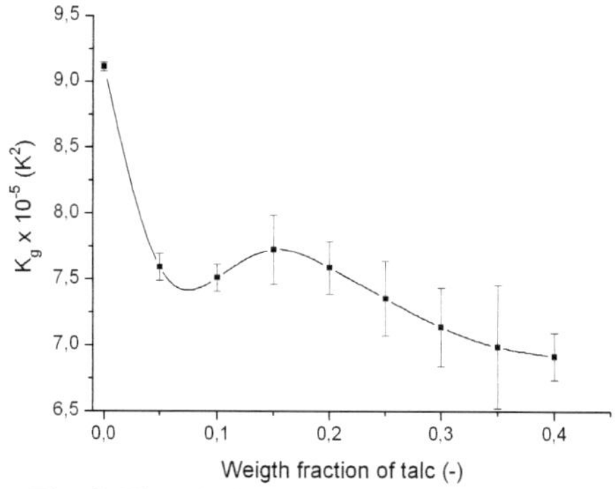

Fig. 2: The obtained K_g values in respect to concentration of Fintalc M05 SLC

As can be seen in Fig. 2, there is a clear tendency for K_g to decrease as the amount of talc is increased, as a consequence of nucleation activity of talc. After the sharp decrease of K_g between 0 and 5 % of talc, there is a plateau up to approximately 20 % of talc. For the concentrations higher than 20 wt. % of talc, K_g further decreases, probably due to the deagglomeration of talc particles, which occurs as a consequence of increased shear stress between the numerous particles. To prove this, light microscopy pictures for the composites of 5 and 40 wt. % were made and presented in Fig. 3.

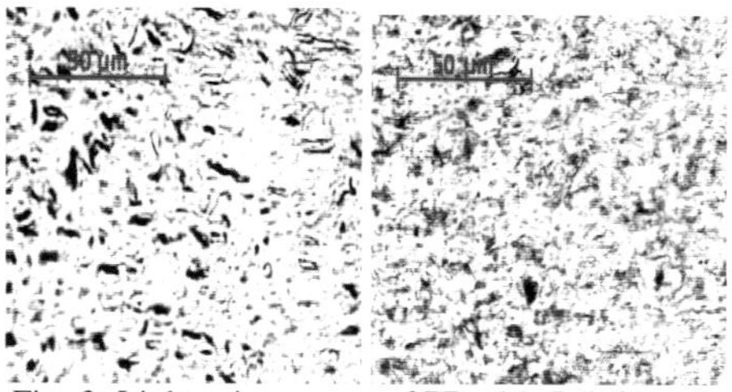

Fig. 3: Light microscopy of PP with 5 (left) and; 40 wt.% (right) of Fintalc M05 SLC.

The comparison between microscopy pictures shows that talc particles were much smaller in the composite with 40 wt.% than in the composite with 5 wt.% of talc.. This deagglomeration could explain further decrease of K_g for the concentrations higher than 20 wt.%, because finer particles are known to be more efficient as nucleation agent [23].

Furthermore we investigated whether it is possible with the new model to see the difference in nucleation ability between three talc grades with different particle sizes. The results are presented in Fig. 4.

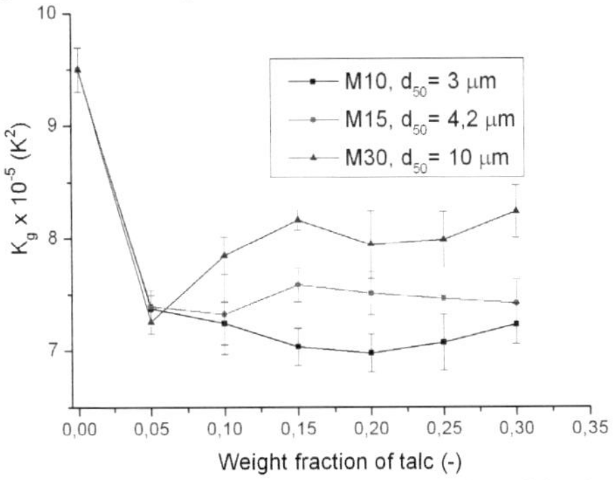

Fig. 4: Comparison of nucleation ability of 3 talc grades with a different d_{50}

As expected, the reduction of the energy required for the formation of a nucleus of a critical size is more pronounced for the talc with smaller particles (lower d_{50}), so that the nucleation parameter decreases in the order K_g (M30) > K_g (M15) > K_g(M10). Contrary to other talc grades, a strong increase of K_g occurs between 5 and 10 wt.% of M30 talc. Thus, pronounced agglomeration could be expected in the case of 10 wt.% of this talc grade. To prove this, the agglomeration of talc particles was analyzed using CT. The results are presented in Fig. 5. As it can be clearly seen, the agglomeration of filler particles as regards 10 wt.% occurred to a grater extent compared to the sample with 5 wt.% of M30 talc. This was reflected by a higher value of K_g of 10 wt.% composite.

Fig. 5: CT images of PP with 5 (A) and 10 (B) wt.% of Fintalc M30. The color of the particles is determined by particle size (red for the coarse and blue for the fine particles).

We further wanted to examine the difference in nucleation ability of two Luzenac talcs with the same d_{50}, but different specific surface. Because of the larger specific surface, the nucleation of the polymer molecules on the HAR T84 should be enhanced, which means that energy required for the formation of a nucleus of a critical size (K_g) should decrease. The results obtained are presented in Fig. 6.

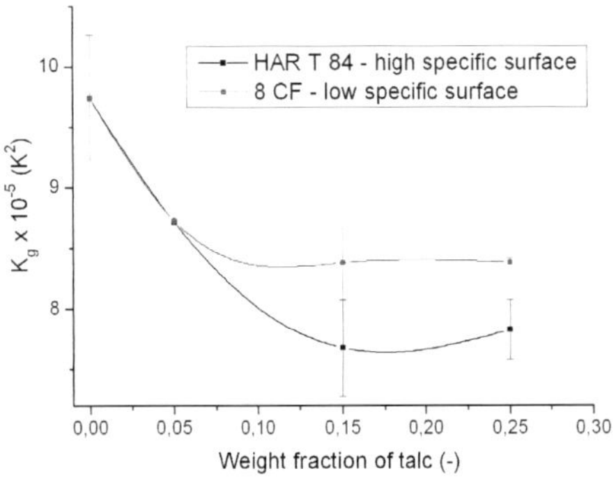

Fig. 6: A comparison of the nucleation ability of Luzenac talcs with the same d_{50} but different specific surface

Indeed, it can be seen that K_g from conventionally produced talc 8 CF and delaminated HAR T84 shows the same trend, but at the same time Har T 84 possess lower K_g values, which means that this grade shows a higher nucleation ability as regards i-PP.

Conclusions

The proposed model extension proves to be applicable for filled or additive modified semi crystalline polymers and non-isothermal crystallization. It can be used to quantify and compare the nucleation effect of different additives. Furthermore, it could be one new way to establish the mathematical correlations between the crystallization and final morphology of semi crystalline polymers.

References

1. D. C. Basset, *Principles of Polymer Morphology,* University Press, Cambridge, 1981
2. G. C. Papageorgiou; D. S. Achilias; D. N. Bikiaris; G. P. Karayannidis *Thermochimica Acta* 2005, *427*, 117.
3. M. Mucha; J. Marszalek; A. Fidrych *Polymer* 2000, *41*, 4137.
4. B. Pukánszky; J. Móczó *Macrom Symp* 2004, *214*, 115.

5. B. Pukánszky; I. Mudra; P. Staniek *Journal of Vinyl and Additive Technology* 1997, *3*, 53.
6. B. Fillon; B. Lotz; A. Thierry; J. C. Wittmann. *Journal of Polymer Science: Part B: Polymer Physics* 1993, 1395.
7. K. Nagarajan; K. Levon; A. S. Myerson *Journal of Thermal Analysis and Calorimetry* 2000, *59*, 497.
8. J. Karger-Kozsis, *Nano-and Micro-mechanics of Polymer Blends and Composites*; Carl Hanser Verlag, München, 2009.
9. H. Botkin; N. Dunski; D. Maeder,. *Improving molding productivity and enhancing mechanical propertiesof polypropylene with nucleating agents*, Ciba Speciality Chemicals Inc., Basel, 2002.
10. J. Kotek; I. Kelnar; J. Baldrian; M. Raab *European Polymer Journal* 2004, *40*, 679.
11. M. Raab; J. Ščudla; J. Kolařík. *European Polymer Journal* 2004, *40*, 1317.
12. L. H. Sperling, *Introduction to Physical Polymer Science 3rd Edition*, John Wiley & Sons Inc, New York, 2001.
13. Y. Feng; X. Jin; N. Hay *J Appl Polym Sci* 1998, *69*, 2089.
14. B. Wunderlich, *Macromolecular Physics,* Academic Press, New York, 1976; Vol. 2.
15. P. Cebe *Polym Compos* 1988, *9*, 271.
16. H. Yiping; C. Guangme; Y. Zhen; L. Hongwu; W. Yong *European Polymer Journal* 2005, *41*, 2753.
17. M. L. Di Lorenzo; C. Silvestre *Prog Polym Sci* 1999, *24*, 917.
18. B. Monasse; J. M. Haudin, J. *Colloid and Polymer Science* 1986, *264*, 117.
19. T. Ozawa *Polymer* 1971, *12*, 150.
20. J. W. Housmans; M. Gahleitner; G. W. M. Peters; H. E. H. Meijer *Polymer* 2009, *50*, 2304.
21. J. D. Hoffman; G. D. Davis; J. I. Lauritzen in *Treatise on Solid State Chemistry*, N. B. Hannay, Ed.; Plenum Press, New York, 1976, Vol. 3, Chap. 7.
22. N. Kocic, N; K. Kretschmer; M. Bastian; P. Heidemeyer *J. Appl. Polym. Sci.* 2012, doi: 10.1002/app.36880.
23. E. Ferrage; F. Martin; A. Boudet; S. Petit; G. Fourty; F. Jouffret; P. Micoud; P. De Parseval; S. Salvi; C. Bourgerette; J. Ferret; Y. Saint-Gerard; S. Buratto; J. P. Fortune *J Mater Sci* 2002, *37*, 1561.

O-09-370

INFLUENCE OF DIFFERENT NUCLEATING AGENTS ON THE BARRIER PROPERTIES OF SEMICRYSTALLINE POLYMERS

M. Kersch[1], F. Richter[2], H.-W. Schmidt[2] and V. Altstädt[1]*

[1] Department of Polymer Engineering, University of Bayreuth, Universitätsstraße 30, 95447 Bayreuth, Germany – altstaedt@uni-bayreuth.de

[2] Department of Macromolecular Chemistry I, University of Bayreuth – hans-werner.schmidt@uni-bayreuth.de

Abstract - Different additives have been investigated with regard to their influence on the barrier properties of polyamides. Talc and sodium phenyl phosphinate are known to nucleate polyamides and have therefore been chosen for investigations. Samples were prepared by injection molding and films were obtained by subsequent compression molding. It could be shown that even though talc did not increase the degree of crystallinity, a remarkable improvement of barrier properties towards carbon dioxide of up to 37 % was achieved. Sodium phenyl phosphinate did not have an influence on the permeability of polyamide films, but resulted in reduced permeation of the test gas through injection-molded samples as a result of a modified crystalline morphology in the outer sample layer.

Keywords: Nucleation, semicrystalline polymers, barrier properties, sodium phenyl phosphinate, micro-talc.

Introduction

Due to their outstanding properties polyamides (PA) are widely used for numerous engineering applications. High chemical and temperature resistance coupled with a high tensile strength and toughness justifies their role as one of the most important classes of engineering plastics [1]. One tool to modify optical or mechanical properties of semi-crystalline polymers in general is heterogeneous nucleation, which has been the subject of many studies over the last decades. A nucleating agent (NA) shortens the induction time and increases the rate of conversion from the molten to the solid state. Polyamides (PAs) show a moderate crystallization rate comparable to polypropylene (PP), of which the nucleation behavior has already been extensively studied.

In contrast to PP there are only a few reports on the nucleation of polyamides. The crystallization behavior of nucleated PA6 was investigated for a dozen of different nucleating agents by three different working groups.

In addition to the aforementioned excellent mechanical properties, PAs exhibit good barrier properties and are frequently used as packaging materials. The permeation properties of polymers in general are strongly dependent on the chemical nature [2], free volume [3], crystallinity, chain orientation, size of the migrant and migrant clustering [2], additives such as antiplasticizers [4] and blend morphology [5]. Regarding PA, the barrier properties had been improved inter alia by the incorporation of clay in the polymer sample [6]. For example, Alexandre et al. investigated the transport of nitrogen through polyamide 12 (PA12)/clay nanocomposites with a clay fraction of up to 5 wt.-% and found a decrease of nitrogen permeability with increasing clay content, due to the increase of tortuosity of the diffusion path [7]. Another possibility to alter the barrier properties lies in the surface modification of the material via plasma treatment. By this, Dreux et al. modified the surface of PA12 samples by CF4 plasma treatment to improve water and toluene barrier properties. The resulting hydrophobic surface revealed less attraction to water and toluene molecules and yielded a material with lower permeability [8].

Talc is a prevalently used additive and is known to act as a nucleating agent for polyamides [9] and was therefore used in this study. Quite surprisingly, to our knowledge no work can be found in the literature regarding the influence of talc on the permeation behavior of polyamides. Sodium phenyl phosphinate (SPP) is a commonly used additive for PA6 in the chemical industry, but to the best of our knowledge, no work can be found in the literature regarding the permeation behavior of PA6 nucleated with SPP. Since the use of additives influences the crystalline structure when they act as NAs, we investigated the relationship between morphology and barrier properties of neat PAs and PAs containing nucleating agents. To determine the influence of the shape of the additive, a plate-like (talc) and a non plate-like (SPP) nucleating agent were chosen.

Experimental

Materials

Polyamide 12 (PA12) (Vestamid L1600) was supplied as pellets by Evonik Degussa GmbH. Polyamide 6 (PA6) (Ultramid B40) was supplied as pellets by BASF SE. Both polyamides did not contain any additives. The talc used in this study (Micro-Talc IT Extra, Mondo Minerals OY) had a plate-like shape, a median particle size of 1.7 µm (d50%) and a specific surface area of 9.2 m^2/g. Phenylphosphinic acid was supplied by abcr GmbH & Co. KG. The nucleating agent used in this study, sodium phenyl phosphinate (SPP), was obtained through reaction with sodium hydroxide

(Pellets, Merck KGaA). The polymer was dried at low pressure (1 mbar) at 80 °C for 24 h prior to processing.

Sample Preparation
The concentrations of 0.5, 1.0 and 2.0 wt.-% talc in PA12 and 1.0 wt.-% SPP in PA6 were obtained through preparation of a masterbatch (5 wt.-% for talc and 3 wt % for SPP) and subsequent dilution with neat polymer by melt compounding. Circular plates with 30 mm diameter and a thickness of 1 mm were injection molded using a Krauss Maffei KM 65/ 180/ 55 CX V injection molding machine with a mold temperature of 100 °C. Films were obtained by subsequent compression molding. The films had a thickness of 100 μm.

Characterization
The transition temperatures and the degree of crystallinity of the samples were determined using a Mettler Toledo DSC/SDTA 821e DSC. The samples were subjected to a two step drying procedure prior to performing the actual measurement according to the method described by Khanna et al [10].
Thin films with a thickness of 7 μm (27 mm circular plates) were prepared from the injection molded samples with a Leica RM2255 rotary microtome. The morphology of the samples was characterized using an optical microscope with crossed polarizers (Nikon, DIAPHOT 300) equipped with a digital camera (Nikon, DMX1200).
The permeation properties of the injection molded samples and films for carbon dioxide were carried out at 23 °C and ambient pressure and 0 % r.h. using equipment from MOCON: MOCON PERMATRAN-C® 4/41 (according to ASTM F-2476-05). The sample thickness was 1 mm for the injection molded samples and 100 μm for the films and the test area was 5 cm^2. The measurements were carried out for at least 24 hours on a minimum of four samples for each material.

Results and Discussion

Influence of talc on the permeability of carbon dioxide through polyamide films
DSC measurements were carried out to determine the crystallization temperature and degree of crystallinity of the samples.

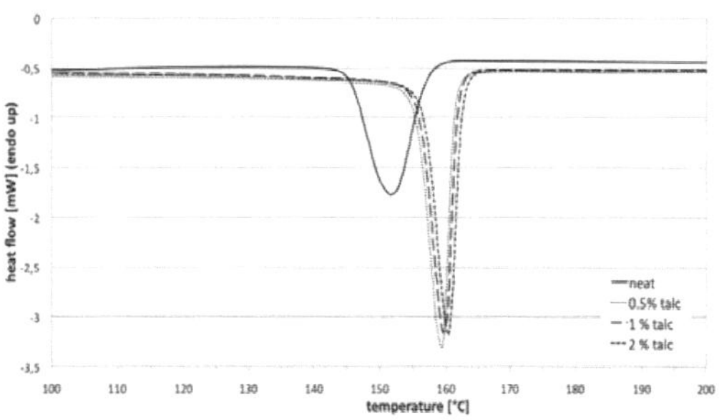

Figure 1 – DSC thermograms of neat PA12 and PA12 with different concentrations of talc

The DSC measurements revealed a clear nucleating effect of talc. The crystallization temperature of the samples containing talc (159 °C for 0.5 and 1 % talc, 160 °C for 2 % talc) was significantly higher than of the neat PA12 (152 °C), figure 1. The incorporation of talc had no significant influence on the degree of crystallinity of the samples.

With respect to the permeability of carbon dioxide, a significant decrease for all three permeates was found when talc was added, figures 6 – 8. The permeability of carbon dioxide was reduced from 3.5 x 10^{-3} mol mm/(m^2 day atm) for the neat material to 2.2 x 10^{-3} mol mm/(m^2 day atm) for the samples containing 0.5, 1.0 and 2.0 wt.-% of talc. Higher concentrations of talc did not improve the barrier properties any further. This could be explained by an insufficient dispersion of the additive at higher concentrations.
Similar effects have been observed by Gill et al. for HDPE and talc [11]. The change in permeation was explained by the relatively high aspect ratio of the additional impermeable sites and the resulting increase in tortuosity of the diffusion path, figure 9. This effect has also been explained in theory by Nielsen for polymers filled with plate-like particles [12].

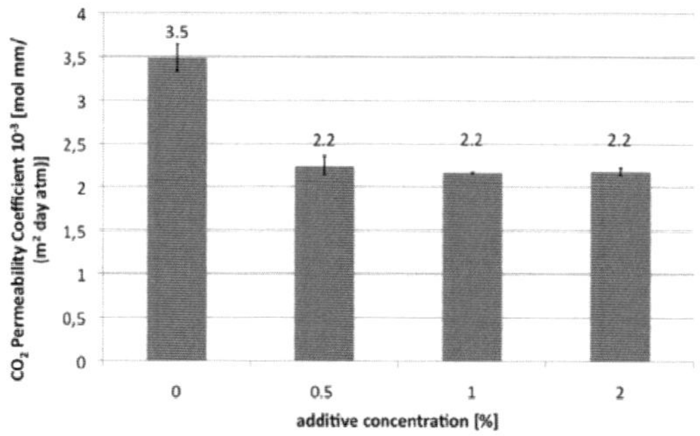

Figure 2 – Permeability coefficients of carbon dioxide through neat PA12 films and samples with different amounts of talc

To distinguish between effects caused by changes in the crystalline structure of the polymer and tortuosity effects resulting from the additive, diffusion and solubility of carbon dioxide were determined, table 1. Changes in the crystalline structure would mainly affect the solubility of the permeates, whereas changes in the diffusion coefficient can be mainly related to the presence of additional impermeable sites that increase the tortuosity. The solubility of carbon dioxide in the samples did not change from the neat samples to those containing different amounts of talc, but a clear decrease in diffusion could be observed when talc was added. No further decrease could be achieved at higher additive concentrations compared to 0.5 wt.-%. Taking these results into account, it can be concluded, that the increased length of the diffusion path resulting from the addition of the high-aspect ratio talc particles is the main reason for the reduced permeability.

Table 1 – Diffusion coefficients and solubility of carbon dioxide in PA12

	Diffusion Coefficient 10^{-12} [m^2/sec]	Solubilty [cm^3/cm^3]
PA12 neat	2.18 (± 0.68)	0.19 (± 0.05)
0.5 wt.-% talc	1.36 (± 0.04)	0.19 (± 0.01)
2 wt.-% talc	1.47 (± 0.29)	0.18 (± 0.03)

Influence of sodium phenyl phosphinate on the permeability of carbon dioxide through polyamide films and injection molded samples

DSC measurements only showed a slight increase in crystallization temperature when SPP was incorporated in the samples. The crystallization temperature was increased from 187 °C for the neat PA6 to 188 °C for the polymer containing 1 % of the additive.

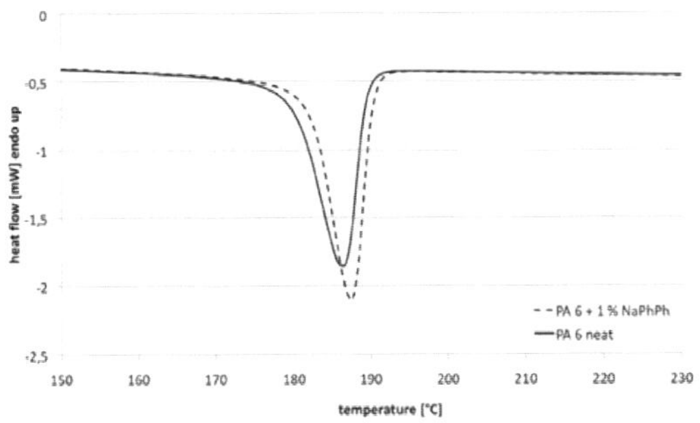

Figure 3 – DSC thermograms of neat PA6 and PA6 with 1 wt.-% of SPP

In this study, the influence of sodium phenyl phosphinate as a nucleating agent on the permeability of PA6 has been investigated for the first time. For a consideration of the influence of sample thickness and preparation method on the permeability, injection molded samples with a thickness of 1 mm and compression molded films with a thickness of 100 μm were analyzed. Figure 4 shows the carbon dioxide permeability coefficients for the films.

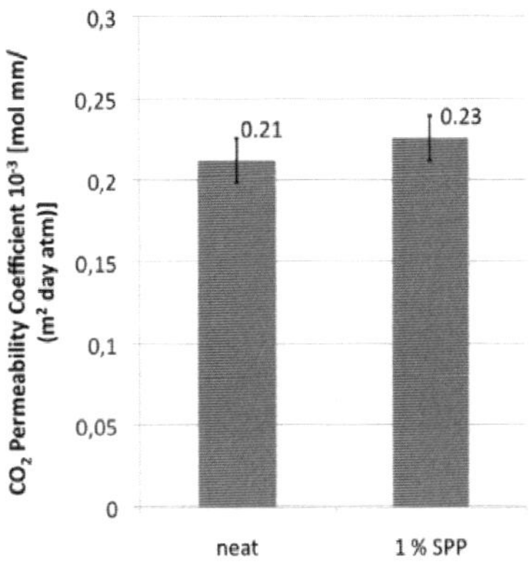

Figure 4 – Permeability coefficients of carbon dioxide through films of neat PA6 and PA6 containing 1 wt.-% of SPP

Within the standard deviation, no difference can be observed between the neat samples and those containing 1 wt.-% of SPP. The main factor influencing the permeability of semi-crystalline polymers is the degree of crystallinity, since permeation only occurs in the amorphous regions. The crystallinity of the samples however was not affected by the presence of the nucleating agent. The insoluble additive increases the amount of impermeable sites, but at low concentrations of 1 wt.-% this effect is negligible for low aspect ratios. Therefore no change in permeability could be observed between the neat and nucleated samples.

Compared to the films, the injection molded samples exhibited a different tendency. Figure 5 shows the permeability coefficients of carbon dioxide through the neat and nucleated samples. A clear decrease in permeation was achieved for both permeates by the addition of 1 wt.-% of sodium phenyl phoshinate. The permeability coefficient of carbon dioxide was reduced from 0.94 10^{-3} mol mm/(m^2 day atm) for the neat material to 0.75 10^{-3} mol mm/(m^2 day atm) for the samples containing 1.0 wt.-% of SPP.

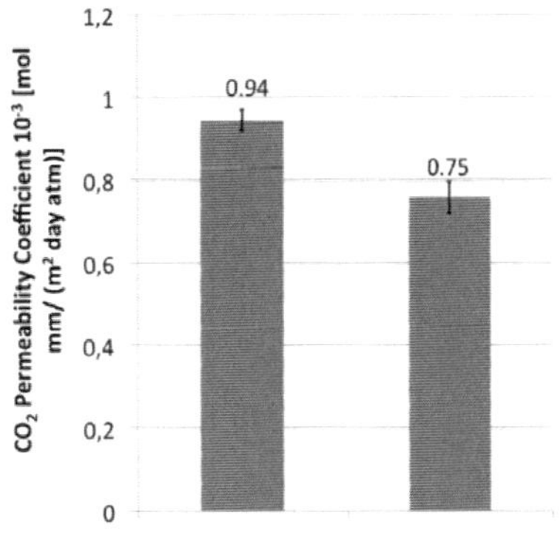

Figure 5 – Permeability coefficients of carbon dioxide through injection molded samples of neat PA6 and PA6 containing 1 wt.-% of SPP

As verified by the permeability measurements on films, the NA itself or a possible influence of the additive on the crystal size cannot be responsible for the reduced permeability. The difference between the injection molded samples and the compression molded films can be explained by the heterogeneity in the crystal structure across the sample cross section in the thicker samples due to different cooling rates at the surface and in the middle of the samples. It is known that as a result of the high cooling rates at the sample surface in the injection molding process, a surface layer consisting of oriented crystals is formed whereas in the middle of the sample no orientation of the crystal structure can be observed. Eby et al. showed that permeability increases with increasing thickness of this oriented surface layer [13]. The addition of SPP as a nucleating agent promotes heterogeneous nucleation leading to a thinner surface layer. Optical microscopy images of neat and nucleated samples give evidence for this (figure 6). A reduction of about 20 % in thickness of the surface layer of the injection molded samples can be observed as a consequence of the addition of 1 wt.-% SPP. This was considered as the main reason for the reduced permeability of the nucleated samples.

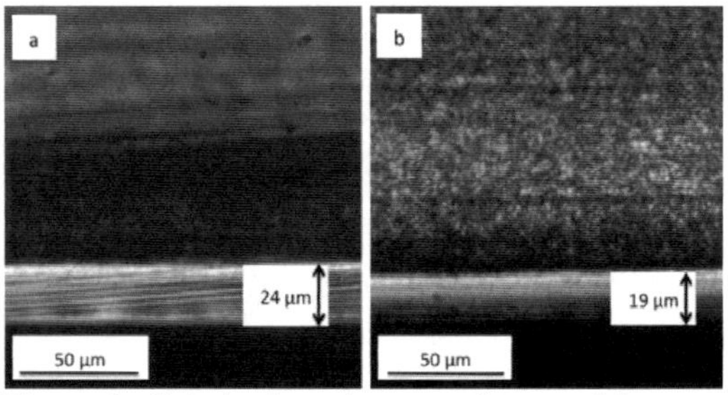

Figure 6 – Polarization microscopy images of the cross section of injection molded samples (a) neat PA6, (b) PA6 1 wt.-% SPP.

Conclusions

The influence of two different nucleating agents on the permeability of carbon dioxide through polyamide samples was investigated.

Even though talc had no influence on the degree of crystallinity of the samples, a reduced permeability of carbon dioxide through PA12 films could be observed. It could be shown that an increase in the length of the diffusion path through the addition of talc was the main reason for the reduction in the permeability.

For the first time, the influence of SPP on the permeation behavior was investigated. It could be shown that the additive had no influence on the barrier properties of PA6 films but a clear reduction in permeability was achieved in injection molded samples due to a reduction of the thickness of the oriented surface layer by 20 %.

Acknowledgements

The authors are grateful to the German Research Foundation (DFG) for financial support within the Collaborative Research Center 840 (SFB 840), project B4.

References

1. D. K. Platt, *Engineering and High Performance Plastics Market Report,* Rapra Technology, Shawbury, 2003.
2. W.L.H. Moll *Kolloid-Zeitschrift und Zeitschrift für Polymere* 1963, 195, 43.
3. K.-S. Chang, Y.-H. Huang, K.-R. Lee, K.-L. Tung *Journal of Membrane Science* 2010, 354 (1–2), 93.
4. M. H. Cohen, D. Turnbull *Journal of Chemical Physics* 1959, 31, 1164.
5. D. Jarus , A. Hiltner, E. Baer *Polymer* 2002, 43(8), 2401.
6. B. Alexandre, S. Marais, D. Langevin, P. Médéric, T. Aubry *Desalination* 2006, 199(1–3), 164.
7. B. Alexandre, L. Colasse, D. Langevin, P. Médéric, T. Aubry, C. Chappey, S. Marais *The Journal of Physical Chemistry B* 2010, 114(27), 8827.
8. F. Dreux, S. Marais, F. Poncin-Epaillard, M. Métayer, M. Labbé *Langmuir* 2002, 18(26), 10411.
9. G. Gurato, D. Gaidano, R. Zannetti *Die Makromolekulare Chemie* 1978, 179(1), 231.
10. Y. P. Khanna, W. P. Kuhn *Journal of Polymer Science Part B: Polymer Physics* 1997, 35, 2219.
11. T. S. Gill, M. Xanthos *Journal of Vinyl & Additive Technology 2* 1996, 248.
12. L. E. Nielsen *Journal of Macromolecular Science, Part A* 1967, 1, 929.
13. R. K. Eby *Journal of Applied Physics* 1964, 35 (9), 2720.

O-09-406

EFFECT OF SLIPPING AGENT ON SURFACE FRACTURE BEHAVIOR OF INJECTION MOLDED POLYPROPYLENE (PP) UNDER SCRATCH LOADING

V. Chivatanasoontorn[1], S. Kato[1], S. Chiba[2], H. Nakajima[2] and M. Kotaki[1]*

[1]*Department of Advanced Fibro-Science, Kyoto Institute of Technology, Kyoto – vadee@kit.ac.jp; m0651009@edu.kit.ac.jp; m-kotaki@kit.ac.jp*

[2] *Petrochemicals Research Laboratory, Sumitomo Chemical Co.,Ltd., Chiba – chibas@sc.sumitomo-chem.co.jp; nakajimah4@sc.sumitomo-chem.co.jp*

Abstract - A progressive load scratch test according to ISO 19252 was performed in this study for investigating surface fracture behavior of injection molded PP incorporated with slipping agents, i.e. oleamide (OA), erucamide (EA) and behenamide (BA). The results showed that scratch properties were improved by slipping agents. PP incorporated with EA led to higher scratch properties than those with OA and BA. Surface mechanical properties, i.e. indentation modulus and shear strength near the surface, were measured by a surface/interface mechanical testing machine. To understand the influence of slipping agents on structure-property relationship at the surface, Raman spectroscopy and wide angle X-ray diffraction (WAXD) were utilized to evaluate molecular orientation and β–phase content near the surface. It was observed that the molecular orientation near the surface was not influenced by slipping agents, whereas the β–phase content near the surface of PP incorporated with EA was found to be higher than those with OA and BA.

Keywords: Injection molding, Slipping agent, Crystal structure, Surface properties and Scratch.

Introduction

Recently, a quantitative evaluation on scratch properties can be obtained by a progressive load scratch test according to ISO 19252, which has been developed by Sue and co-workers [1-3]. It was reported that various kinds of scratch damage transitions can be observed through this test. Moreover, it provides a quantitative evaluation of critical load for various scratch damage modes especially onset of scratch visibility [3]. The scratch behavior of polymers is very complex and depends on various parameters, which can be classified into scratch conditions [4], surface characteristics [5], material properties [6] and type and amounts of fillers and additives incorporated into polymers [7].

Polypropylene (PP) is one of the most widely used commodity polymers for manufacture of plastic products because of its good mechanical properties, processability and low cost [8]. However, scratches on plastic surface and visible scratch damages are easily formed in PP when compared with other materials like metal and ceramics. An improvement of scratch resistance for injection molded PP has become an important issue for various applications that demand high surface quality [9].

A way to improve the surface characteristics of PP is to introduce slipping agents into the polymer systems. They are added to the polymer matrix in molten stage and later migrate to the surface, which is known as "blooming" [10]. The unique thermochemical properties and the lack of compatibility between them and the polymer matrix are the reason for their migration to the surface [11]. After the slipping agents migrate to the polymer surface, they form a thin film layer which leads to decrease of coefficient of friction (CoF).

In this study, the surface fracture behavior of injection molded PP and PP incorporated with slipping agents under progressive load scratch test has been investigated, in order to understand the influence of slipping agents on the scratch behavior. In addition, the effect of slipping agents on structure-property relationship near the surface is discussed.

Experimental

Material

PP and PP incorporated with three different slipping agents, i.e. oleamide (OA), erucamide (EA) and behenamide (BA), were prepared by injection molding. The amount of slipping agents used in this study was set at 0.2 wt%. Table 1 shows all the sample systems and their composition.

Table 1 Material Compositions

	Slipping agent	
Sample	**Type**	**Melting temperature (T_m, °C)**
PP	-	-
PP-OA	OA ($C_{18}H_{35}NO$)	66 – 72
PP-EA	EA ($C_{22}H_{43}NO$)	79 – 85
PP-BA	BA ($C_{22}H_{45}NO$)	110 – 113

Scratch Test

A scratch test machine (scratch machine model KK-01, Kato-tech co., Ltd.) was used to perform progressive load scratch tests according to the ISO 19252 in this study. All samples were tested under the load range from 1 N to 50 N at room temperature. The scratch direction was parallel to the flow direction (FD). A stainless-steel ball-bearing scratch tip with 1 mm in diameter was used. The scratch length and scratch velocity were set at 100 mm and 100 mm/sec, respectively.

To evaluate the scratched damage transitions, a commercial flatbed scanner (GT-X970 Photo PC scanner, EPSON) was used to scan the scratched surfaces at a resolution of 3200×3200 dpi in 8-bit grayscale level. A threshold function in ImageJ software was adapted to scanned images for measuring onset of scratch visibility. In addition, field emission scanning electron microscopy (FESEM; S–4200, Hitachi) was operated at an accelerating voltage of 8 kV to observe the surface deformation behavior at a sufficiently high resolution.

Coefficient of Friction Test

The scratch test machine was adapted to perform constant load tests for investigating coefficient of friction (CoF). A constant normal load at 2 N and scratch velocity of 10 mm/s were utilized. The scratch length was set at 100 mm and a test was conducted by using a stainless-steel ball-bearing tip with 10 mm in diameter. The scratch length from 40–60 mm was selected to evaluate CoF. CoF was obtained from the ratio of the tangential load to applied normal load in a sliding process.

Surface Mechanical Properties

To study surface mechanical properties of polymeric materials, a surface/interface mechanical testing machine (Surface and interface cutting analysis system (SAICAS) model DN-20S; DAIPLA Wintes Co., Ltd.) was adapted to perform micro-indentation test and micro-cutting test. All samples are indented under a constant speed of 0.1 μm/sec into 10 μm in depth with a stainless-steel spherical indenter of 1 mm in diameter. The indentation modulus near the surface in this study is calculated from the following equation;

$$Indentation\ modulus = \frac{\sqrt{\pi}}{2} \times \frac{S}{\sqrt{A}} \qquad \text{(Eq. 1)}$$

Where, S is the slope of the unload curve at the maximum displacement point and A is the projected area of contact.

In this test, the projected area is measured from the maximum depth of penetration of the spherical indenter into the specimen surface by using an equation as follow:

$$Projected\ area\ of\ contact\ (A) = \pi r^2 \qquad \text{(Eq. 2)}$$

$$r = \sqrt{R^2 - (R - indentation\ depth)^2} \qquad \text{(Eq. 3)}$$

Where, R is a radius of indenter.

The micro-cutting test was conducted with the vertical speed and horizontal speed of 1 μm/sec and 0.1 μm/sec, respectively. The observation area was cut along the FD. The equation for calculating the shear strength distribution near the surface is described as follow;

$$Shear\ strength\ (\tau) = \frac{F_H}{2 \times W \times d \times \cot\theta} \qquad \text{(Eq. 4)}$$

Where, F_H is a horizontal force, W is a width of the blade, d is the depth of the cut, and θ is a shear angle.

Raman Spectroscopy

Raman spectra were measured at room temperature by Laser Raman Spectrometer (NRS-3100, JASCO). The light source was green laser (= 532 nm). The diameter of the focused laser beam was 2 μm. All spectra were recorded by 1-3 time integrations after 3-second irradiation to eliminate laser noise. Molecular orientation near the surface can be observed from intensity ratio of 844 cm^{-1} to 813 cm^{-1} (I_{844}/I_{813}).

Wide angle X-ray Diffraction (WAXD)

To observe crystal structure near the surface, wide angle X-ray diffraction (WAXD) patterns were measured by an X-ray diffractometer (RINT-2100-FSL, Rikagu) under ambient condition. A nickel-filtered CuKα radiation with a wavelength of 0.1542 nm, from a Rigaku RAD 2C sealed beam X-ray generator operating at 40 kV and 20 mA. The relative amount of the β–phase (K) was evaluated by the method of Turner Jones and co-workers;

$$K = \frac{I_{\beta\,(300)}}{I_{\alpha\,(110)} + I_{\alpha\,(040)} + I_{\alpha\,(130)} + I_{\beta\,(300)}} \qquad \text{(Eq. 5)}$$

Where, $I_{\alpha(110)}$, $I_{\alpha(040)}$, and $I_{\alpha(130)}$ are the intensity of the reflection peaks at (110), (040) and (130) of α–phase, respectively. $I_{\beta(300)}$ is the intensity of the reflection peak at (300) of β–phase.

Results and Discussion

Scratch Property

Three typical scratch damage modes in PP under progressive load, i.e. onset of fish–scale pattern, onset of scratch visibility and transition to cutting mode, were observed and selected as criteria to evaluate scratch property in this study. Fig.1 shows a quantitative evaluation of the critical normal load for these scratch damage modes. It was observed that PP incorporated with slipping agents exhibited an increase in scratch resistance. In addition, comparison of the scratch property between slipping agents shows that in the case of PP–EA, the scratch resistance to the surface deformation was better than those of PP–BA and PP–OA, respectively. Thus, the incorporation of EA into PP has a positive effect on enhancing scratch resistance. In order to understand the influence of slipping agent on the scratch behavior, the critical scratch load at onset of scratch visibility for PP and PP incorporated with slipping agents is plotted as a function of CoF (Fig.2). The CoF decreased from 0.27

for the PP to 0.11 when EA or BA incorporated into PP, whereas PP–OA has similar CoF as PP. The results show that there are the ranges of critical normal load, which independent of the CoF, i.e. the critical normal load increased without the decrease of CoF.

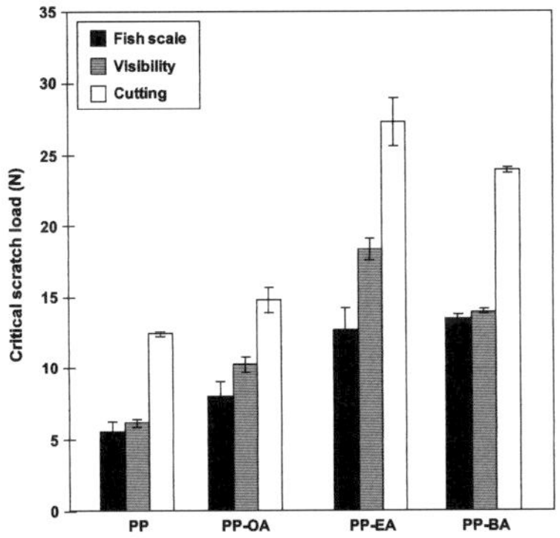

Fig. 1 Influence of slipping agent on the critical scratch load of three scratch damage modes; onset of fish-scale pattern, onset of scratch visibility and transition to cutting mode.

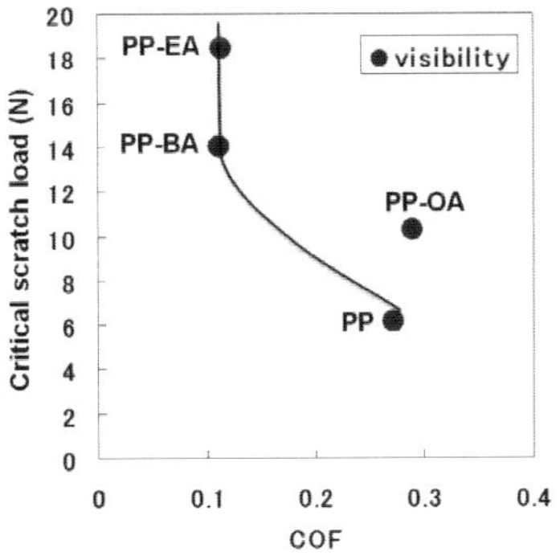

Fig. 2 Correlation between critical scratch load at onset of visibility and CoF.

Surface Fracture Behavior

Fig.3 shows the difference in surface deformation behavior of PP and PP incorporated with slipping agents scratched under progressive load, which observed at 12 N by SEM.

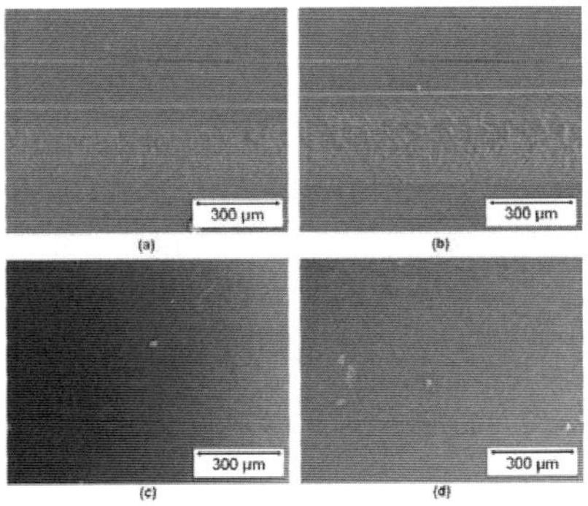

Fig. 3 Influence of slipping agents on the surface deformation behavior observed at 12 N (a) PP, (b) PP-OA, (c) PP-EA and (d) PP-BA.

PP and PP–OA clearly showed a parabolic pattern (fish–scale pattern) and increase of surface roughness inside the scratch groove. However, the scratched surface of PP–EA and PP–BA exhibited smooth surface after scratch tip passed and compressed, which term as mar deformation. Photographs of the deformation behavior under scratched surface for PP–EA and PP–BA at the region before the transition to cutting mode are shown in Fig.4. The fish–scale pattern was clearly observed in both cases. However, these photographs were taken at different scratch load, i.e. PP–EA was observed at 26.8 N and 20.2 N in the case of PP–BA. This result indicated that the ductility near the surface can be enhanced when EA is incorporated into PP, which results in the increase of scratch resistance.

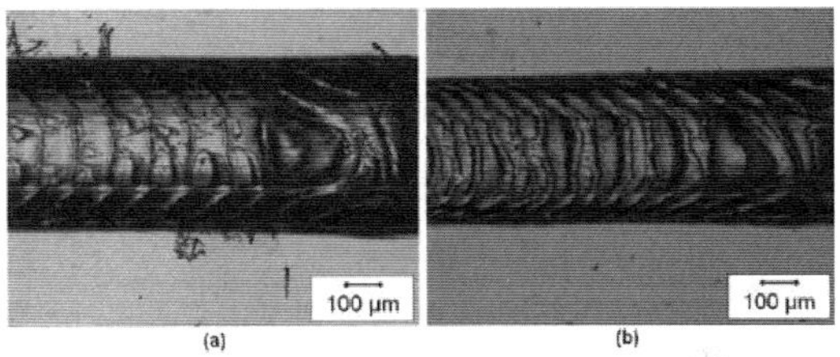

Fig. 4 Influence of slipping agents on the surface deformation behavior observed at the transition area of fish-scale pattern to cutting mode (a) PP-EA at 26.8 N and (b) PP-BA at 20.2 N .

Surface Mechanical Properties

The influence of slipping agents on the surface mechanical properties, i.e. indentation modulus and shear strength near the surface, was investigated as shown in Fig.5. The results in Fig.5(a) showed that there was no significant difference in the indentation modulus of PP and PP incorporated with slipping agents. This means that the slipping agent migrated to the surface did not influence the modulus near the surface. Furthermore, it was found that the shear strength has been shown to be decreased as slipping agents incorporated into PP, especially in the case of PP–EA and PP–BA(Figure 5(b)). However, there was not a great difference between PP and PP–OA.

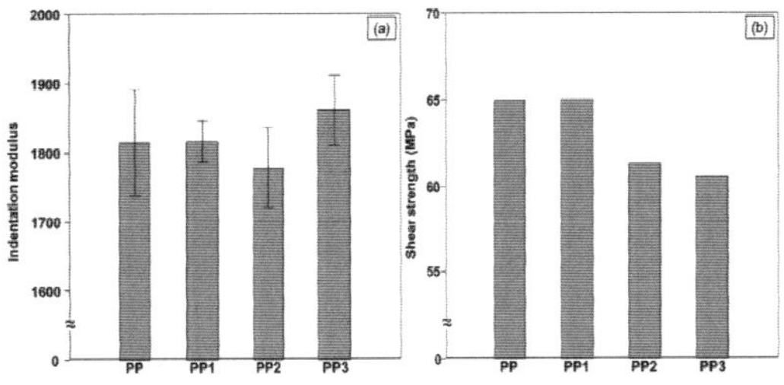

Fig. 5 (a) Surface indentation modulus and (b) Surface shear strength of PP and PP incorporated with slipping agents.

Molecular orientation and crystal structure near the surface

To study the effect of slipping agents on the molecular orientation near the surface, the depth profiles of the intensity ratio (I_{844}/I_{813}) at different position from the surface to ~500 μm in depth were investigated by Raman spectroscopy (Fig.6). It can be seen that all of the samples in this study had similar profile of the intensity ratio, i.e. the molecular orientation near the surface was not influenced by slipping agents. The crystalline orientation near the surface was observed by WAXD.

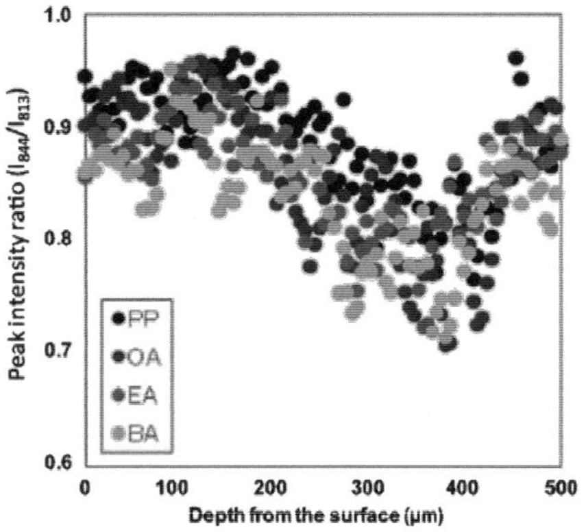

Fig. 6 Relative intensity ratio of I_{844}/I_{813} of PP and PP incorporated with slipping agents.

The typical 2DWAXD patterns of the PP and the PP–EA near the surface (~30 μm in depth) are shown in Fig.7. The diffraction arcs observed at (110) and (040) in PP (Fig.7(a)) corresponds to the α–phase. From Fig.7(b), it was observed that the diffraction arcs of β–phase (300) were clearly detected when EA was incorporated into PP. These results indicated that the slipping agent induced the formation of β–phase near the surface.

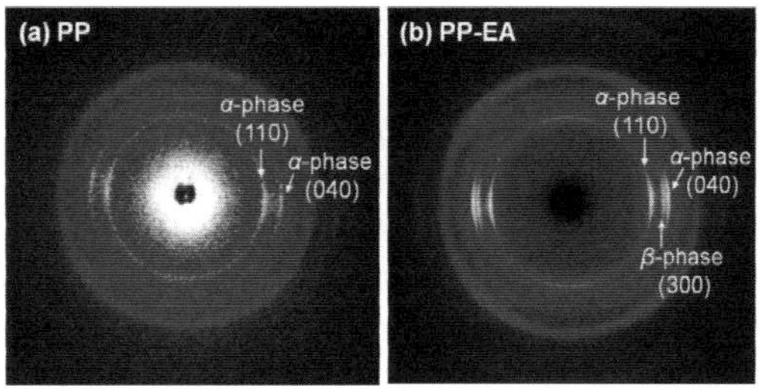

Fig. 7 Two-dimensional WAXD patterns of PP.

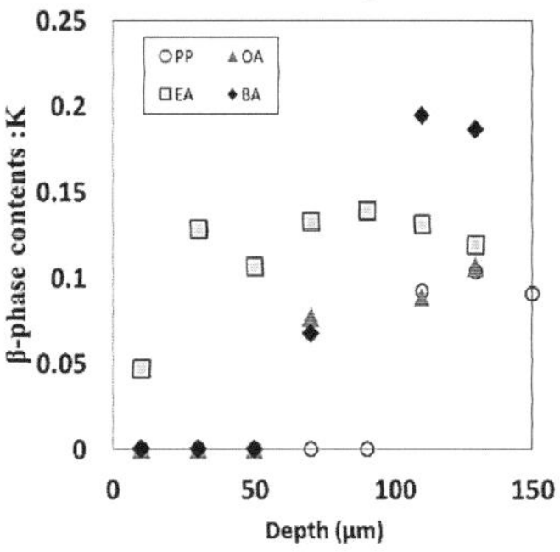

Fig. 8 β-phase content as a function of distance from the surface of PP and PP incorporated with slipping agents.

Figure 8 shows the influence of slipping agents on β–phase content near the surface in depth of 150 μm in PP and PP incorporated with slipping agents. It was observed that the β–phase content of PP incorporated with slipping agents was higher than that of PP in depth of 100 μm. The results indicated that the relative amount of β–phase content near the surface in depth of 100 μm was affected by slipping agents. In addition, the region in depth of 70 μm, the β–phase content of PP–EA was much higher than that of PP–BA. These results showed that EA was an effective slipping agent, which induced the formation of β–phase near the surface in depth of 70 μm and then resulted in the improvement in resistance to the surface deformation under scratch loading.

Conclusions

The influence of slipping agents on the scratch behavior of injection molded PP was investigated according to ISO 19252. The incorporation of slipping agents into PP had an effect on enhancing scratch resistance due to the decrease of CoF on the surface. The investigation on micro-structure near the surface, i.e. molecular orientation and crystal structure, of injection molded PP showed that no obvious effect was seen on molecular orientation near the surface but the formation of β–phase was enhanced as slipping agents was incorporated. The results showed that the increase of β–phase content near the surface in depth of 70 μm was achieved when EA was incorporated into PP.

References

1. Wong, M.; Lim, G.T.; Moyse, A.; Reddy, J.N.; Sue, H.-J. *Wear* **2004**, 256 (11-12), 1214-1227.
2. ISO 19252, "*Plastics — Determination of scratch properties*", International Organization for Standardization.
3. Wong, M.; Moyse, A.; Lee, F.; Sue, H.-J. *J. Mater. Sci.* **2004**, 39, 3293-3308.
4. Briscoe, B.J.; Pelillo, E.; Sinha, S.K. *Polym. Eng. Sci.* **1996**, 36 (24), 2996-3005.
5. Harlin, P.; Carlsson, P.; Bexell, U.; Olsson, M. *Surf. Coat. Tech.* **2006**, 201, 4253-4259.
6. Chu, J.; Rumao, L.; Coleman, B. *Polym. Eng. Sci.* **1998**, 38 (11), 1906-1914.
7. Briscoe, B.J.; Pelillo, E.; Ragazzi, F.; Sinha, S.K. *Polymer* **1998**, 39, 2161-2168.
8. Pasquini, N. Polypropylene Handbook, 2nd edition, Hanser Gardner Pubns, Germany **2005**.
9. Jiang, H.; Browning, R.; Sue, H.-J. *Polymer* **2009**, 50 (16), 4056-4065.
10. Edenbaum, J. *Plastic Additives and Modifiers Handbook*, Chapman & Hall, London, **1996**, 107, 865.

11. Llop, C.; Manrique, A.; Navarro, R.; Mijangos, C.; Reinecke, H. *Polym. Eng. Sci.* **2011**, 51, 1763-1769.

O-09-422

INFLUENCES ON MORPHOLOGY IN LASER SINTERED PEEK MATERIALS

M. Kroh[1]*, C. Bonten[1] and P. Eyerer[2]

:[1]*Institute für Kunststofftechnik (IKT), University of Stuttgart, Germany – michael.kroh@ikt.uni-stuttgart.de;*

[2]*Fraunhofer Institute for Chemical Technology (ICT), Germany – peter.eyerer@ict.fraunhofer.de*

Abstract - The additive assisted laser sintering was recently developed at IKT: A carbon black (CB) additive is used to adjust the polymer´s laser absorption behaviour with the aim to improve the interconnection of sintered powder layers. In a parameter study, Polyetheretherketone (PEEK) samples were prepared with different contents of carbon black and were laser sintered. The samples were investigated under different thermal treatments on a hot stage device and differential scanning calorimetry.

On the one hand, contrary to an expected nucleating effect, a reduced degree of crystallinity was observed for PEEK with CB additive mixtures. On the other hand an influence on the morphology at the border areas of particles and intersections of laser sintered layers was noticed by means of light microscopy. Depending on viscosity of the raw material, CB content and thermal treatment, different shapes of lamellae were observed. These (trans-) crystalline or polymorph structures, respectively, influence the thermal and mechanical behaviour of the virgin PEEK.

Keywords: Selective laser sintering, additive assisted laser sintering, polyetheretherketone; morphological alterations

Introduction

Additive manufacturing (AM) is the designation for a group of primary shaping technologies without the need for building costly moulds known e.g. from injection moulding, so called solid free-form manufacturing. The basic principle of all AM-techniques is a layer wise building process. 3D-CAD-models are transformed into the surface tessellation language (STL). In a next step these STL-models are sliced into thin layers of typical 0.1 to 0.2 mm. In a laser sintering machine these layer informations are scanned by a laser onto a prepared powder layer. So the powder layer starts to melt at the scanned geometry. After that a new powder layer is wiped on the powder bed. That process is sequentially repeated until the initial designed model is built.

The presented study deals with the additive assisted laser sintering (ASLS) of polyetheretherketone (PEEK) using a Nd:YAG-laser source and carbon black (CB) as NIR absorber to enable an adjustable optical absorption behavior of the polymer powder [7 - 9]. Due to its specific wavelength of λ = 1.064 µm, the Nd:YAG-laser penetrates into the polymer powder and heats the incorporated absorber particles, which leads to a volume like heat source within the powder layer.

To avoid the so called 'curl' effect [3], the powder bed is held at a temperature between the melting point and above the onset of crystallization during the sintering process. Usually, the building process runs for several hours depending on part size and resolution. To avoid oxidation and thermal degradation, the building chamber is flooded with nitrogen.

During that building process, the powder and the molten polymer is more or less annealed. This iso-thermal treatment leads to different changes in material properties. As reported by [4 - 6] PEEK shows multiple minor endothermic peaks in addition to the main melting peak, if annealed above T_c or slowly cooled (< 3 K/min). Due to that a lamellar polymorphism of PEEK was observed and correlated to mechanical properties [4, 10].

Experimental

For all experiments commercially available standard viscous PEEK™450PF (SV-PEEK) and low viscous PEEK™150PF (LV-PEEK) powder from Victrex Plc. were used. An overview of the properties is given in table 1. As NIR-absorbing additive Cabot Vulcan XC72R, Worlee Chemie, (CB) was mixed with varying mass-% into the polymer powder. The powder was sieved to an average particle size of 60 µm and a maximum diameter of 90 µm.

Table 1 - Properties of PEEK

melting point	343 °C
continuous operating temperature	up to 260 °C
tensile strength (23 °C)	97 MPa
Young's modulus	3600 MPa
Melt viscosity SV-PEEK	450 Pa s
Melt viscosity LV-PEEK	130 Pa s

Transmission electron microscopy (TEM)

To investigate a nucleating effect of CB on lamellar structures of the PEEK samples TEM micrographs were prepared with Zeiss EM-10C/CR at an accelerating voltage of 80 kV. The specimens have been contrasted with RuO_4. Samples were cut by an ultra microtome (Reichert Jung, Ultracut E, FC 4 D) with a diamond knife into thin slices of 80 nm.

Differential scanning calorimetry (DSC)

For DSC measurements, a Mettler Toledo DSC 821 (STAR software V9.10) was used. The sample masses were about 8 mg. Samples were heated from RT up to 380 °C at 10 K/min. The degree of crystallinity K was measured and calculated according to equation (1) with $\Delta H^0_{m,PEEK} = 130\,J/g$, [2]

$$K = \frac{\Delta H_m}{\Delta H^0_m} \qquad (1)$$

Determination of heat conductivity

The temperature conductivity was determined by a light flash device (NETZSCH LFA 447/2-4/InSb NanoFlash™) in the range from RT up to 300 °C. Heat conductivity was calculated by equation (2), with heat conductivity λ, bulk density of the powder mixtures ρ, temperature conductivity a and specific heat capacity c_p:

$$\lambda(T) = \rho \cdot c_p(T) \cdot a(T) \qquad (2)$$

Hot stage microscopy

To determine the sintering behaviour of PEEK particles a hot stage device (Linkam TMS94) was used. By measuring the relation of particle diameter, sintering time and viscosity the surface tension was calculated according to Frenkel [1].

Reference Samples

As described in [10] to compare the thermal, mechanical and morphological characteristics of the investigated materials, reference samples have been sintered in heating press without any pressure with respect to the pressure less sintering characteristic of the SLS process. The powder was heated up to 380 °C with 80 K/min, held for 5 minutes and finally sintered for 30 min at 330 °C. From these plates tensile bars have been machined.

Laser Sintering of Test Specimen

A specially designed laser sintering system was used that allowed an independent variation of parameters [3,4]. Single layer tensile bars have been sintered, according to ISO 527. The sintering strategy was based on a bidirectional scanning in and across tensile direction. For every parameter setting four tensile bars have been sintered with a constant laser power of $P_L = 30$ W and a hatch distance of $h_s = 0.02$ mm (laser spot diameter 30 µm). With respect to equation (3), the mean energy density e_m was calculated by the scan speed v_s.

$$e_m = \frac{P_L}{h_s \cdot v_s} \qquad (3)$$

Based on preliminary tests [4], the process parameters have been varied as listed in table 2. All experiments have been carried out in nitrogen atmosphere to avoid thermal oxidation.

Table 2- Range of testing parameters

Parameter	Symbol	Range
Powder bed temperature	T_{bed}, °C	23 - 330
Scan speed	v_s, mm/s	2,500 – 6,000
Carbon black content	C, mass-%	0.05 – 0.15
Melt viscosity	η, kPa s	0.13 – 0.44

To improve mechanical properties like tensile strength and elongation at break, the annealing time ΔT_H after laser treatment was varied as discussed in [12]. Mechanical testing was performed according to ISO 527 at 1 mm/min testing speed on a universal testing machine (Zwick, type 1455).

Results and Discussion

The measured values of tensile strength, Young's-Modulus and sintering depth have been analyzed by a multiple non linear regression [11]. Due to that the main influence of CB-content and mean energy density on mechanical behavior and sintering depth can be described by a regression plane as illustrated in Figure 1 to Figure 3.

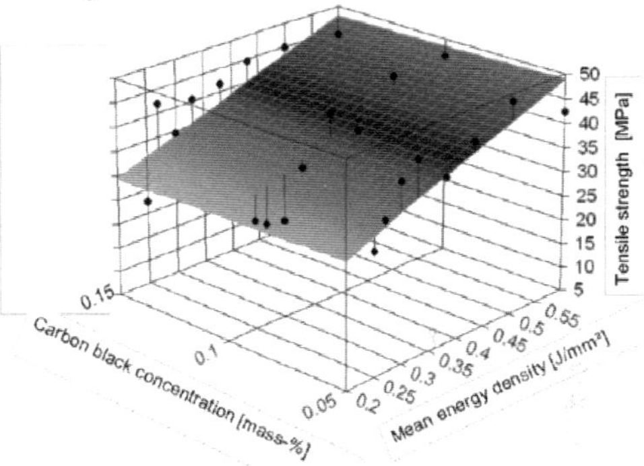

Figure 1 – tensile strength for SV-PEEK at T_{bed} = 330 °C depending on CB-content and e_m

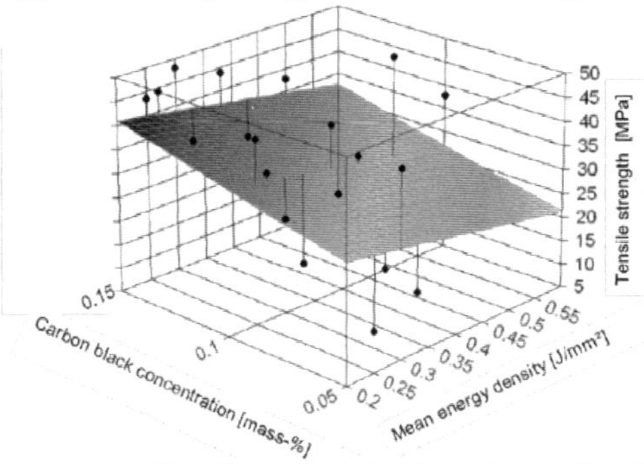

Figure 2 – Tensile strength for low viscous PEEK at T_{bed}=330 °C depending on CB-content and mean energy

Regarding the tensile strength in detail it can be shown that the standard viscous material has a rather reduced variance for the measured values compared to the low viscous specimens. As illustrated in Figure 1, an increasing of the CB content results in higher tensile strength values, but shows a rather alternating behaviour at a low energy density of 0.2 J/mm²; a quite good and constant strength was achieved at medium energy density of 0.35 to 0.45 J/mm². An increasing the mean energy density up to 0.55 J/mm² did not show a further increase. In contrast, lower strength and slightly burned specimens were obtained. The LV-PEEK showed a very high variance all in all (Figure 2). But up to a mean energy density of 0.45 J/mm², an increased tensile strength was observed with increased CB-content, similar to SV-PEEK. Higher energy values caused some weakening of the specimen by a localized burning.

Similar results were observed for the Young's modulus. The SV-PEEK showed a rather linear increase with increased mean energy density and slightly improved values with increased CB for low and medium energy values. In contrast to that, the LV-PEEK showed a rather high variance of the measured values which is ascribed to its brittle behaviour. Nevertheless, best results were obtained for an average energy value of 0.35 J/mm² which is assumed to be the best processing window for both materials.

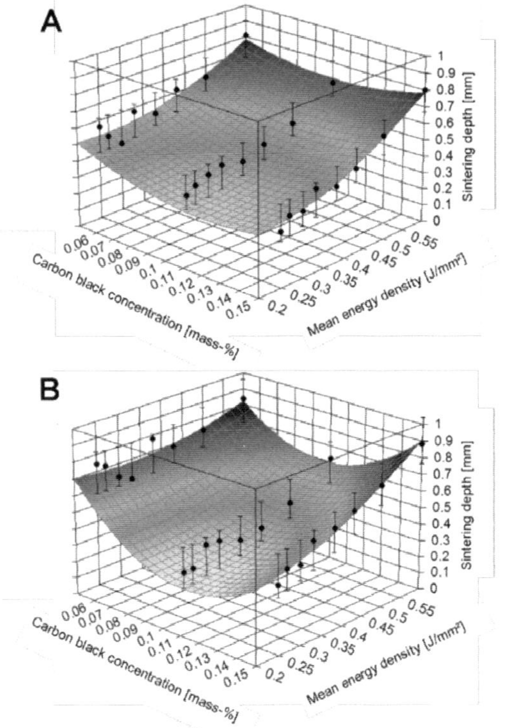

Figure 3 - Influence of CB content on sintering depth of LV (A) and SV-PEEK (B) at T_{bed} = 330 °C

As described in [11] by a non linear multiple regression, significant differences regarding the effective sintering depth at a powder bed temperature of 330 °C were observed (Figure 3). According to the fundamentals of ASLS [10], thinner specimens with increased CB-concentrations, were expected. But both materials showed a re-increased thickness at CB concentrations of 0.15 mass-%. It was pointed out in [11] that this behaviour is a possible phenomenon of increased heat conduction. But here it showed only a neglectable decrease with increased CB-content.

With an increased CB-content, an increase of the surface tension was observed. At a content of 0.15 mass-%, a step change in the sintering time of two particles was measured. This caused a significantly increased surface tension according to Frenkles correlation [1] as indicated in Figure 4. It is assumed that the higher surface tension of the slightly molten particles leads locally to a faster sintering process; so more surrounding particles and deeper layers are sintered.

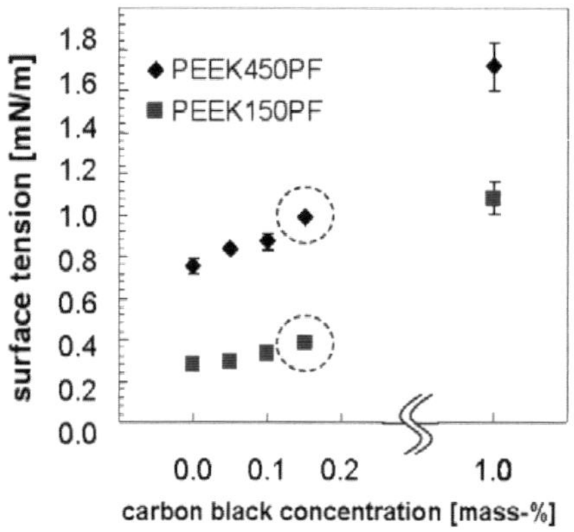

Figure 4 – Surface tension of PEEK particles depending on the CB concentration

The annealing of every slice of the test specimen after laser treatment leads to an increased tensile strength until an optimum was reached at an interval of Δt_H = 60 sec. A longer annealing interval leads to a decrease of tensile strength again, as reported in [12]. This effect was correlated to the degree of crystallinity as shown in Figure 5. It is supposed that the annealing leads to a higher degree of molten phase of the polymer particles, so the initial *K* of the raw material is reduced.

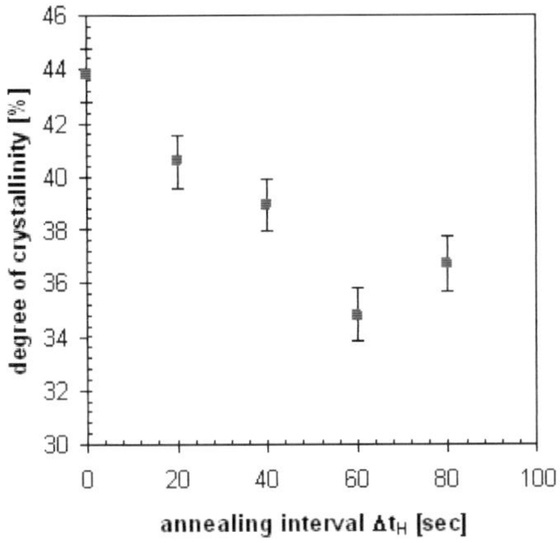

Figure 5 - Degree of crystallinity K of laser sintered multilayer tensile bars of SV-PEEK over Δt_H

Due to the wiping process of new powder onto the upper layer of molten area, the melting process is shortly interrupted. This effect becomes less dominant by increasing the annealing interval up to 80 sec. In that case the particles were molten in the particle centre and could partly re-crystallize even after the wiping process. On the other hand, it is assumed that the reduced crystalline phase enables a better diffusion of the macromolecules between the polymer particles and finally leads to a better mechanical performance.

Morphological characterization

As indicated by DSC experiments [10], no nucleating effect was observed by adding CB. E.g. a lamellar growth through and around CB-particles were found by TEM investigations as indicated in Figure 7,B.

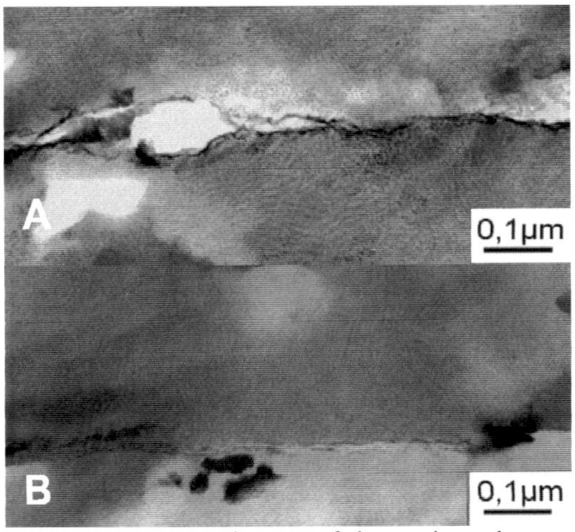

Figure 6 – TEM-pictures of laser sintered cross sections of tensile bars of SV-PEEK with (A) 0.05 and (B) 0.10 mass-% CB – particle boundaries

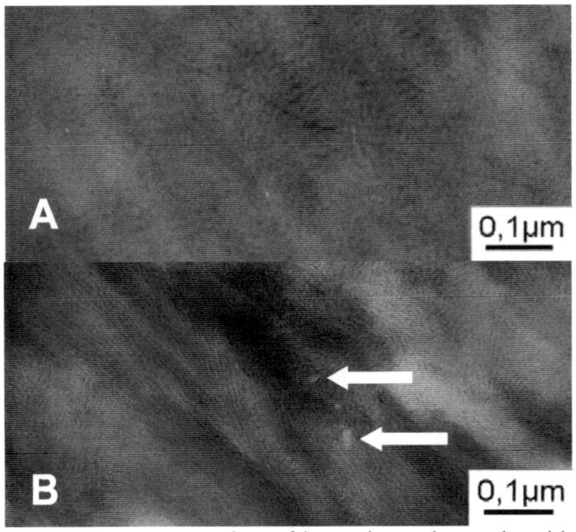

Figure 7 – cross section of laser sintered sample with 0.10 mass-% CB of (A) LV- and (B) SV-PEEK

In contrast to reference samples the laser sintered specimens indicated a more oriented lamellar structure. Moreover the degree of orientation was increased by the content of CB, as illustrated at the interconnection of layers in Figure 6. With an increased content of 0.10 mass-% CB there were significant orientations in an orthogonal direction to the welding line of the particles. Compared to that, the LV-PEEK is indicated by a less oriented lamellar structure as shown by Figure 7.

Conclusions

The additive assisted laser sintering offers a new way to improve the tensile strength and Young's- modulus of laser sintered specimens. The basic effect was shown to be altered morphological structures, in particularly the oriented (trans-) crystalline lamellar structures in the interconnection area of laser sintered specimen which can be influenced by the processing. Increasing these structures by increasing the CB-content from 0.05 to 0.15 mass-% leads to an optimized mechanical behaviour with respect to an energy value of 0.35 to 0.45 J/mm². So, further investigations by e.g. varying the polymer, CB-type and –concentrations should be done to improve the ASLS-process and quality of laser sintered products, respectively.

Acknowledgements

S. Osterloh for preparing and discussing the transmission electron micrographs.

References

1. J. Frenkel *J.Physics.* 1945, *9, 385.*
2. D. Blundell, B. Osborn *Polymer*, 1983, 24, 953
3. G. Alscher, PhD Thesis, University of Essen, 2000
4. T. Ko, E. Woo *Polymer,* 1996, 37, 1167
5. A. Jonas, T. Russel, D. Yoon *Macromolecules*, 1995, 28, 8491

6. H. Marand, A. Prasad *Macromolecules*, 1992, 25, 1731
7. N. Woicke, M. Kroh, P. Eyerer in PPS 22, Proceedings, Yamagata, Japan, 2006
8. M. Kroh, N. Woicke, P. Eyerer, Entwicklung und Erprobung innovativer Produkte – Rapid Prototyping, Springer, Berlin, 2007
9. M. Kroh in PPS 24 Proceedings, Salerno, Italy, 2008
10. M. Kroh; P. Eyerer in PPS 26 Proceedings, Banff, Canada, 2010.
11. M. Kroh; C. Bonten; P. Eyerer in ANTEC 2011 Proceeding, Boston, 2011, 1806-1811
12. M. Kroh, P. Eyerer in 21. Stuttgarter Kunststoffkolloquium, Proceedings, Stuttgart, 2009, 2/V6

O-09-438

MORHOLOGY CONTROLLED TRIPLE AND QUADRUPLE SHAPE-MEMORY POLYMER MATERIALS BASED ON ELECTRON IRRADIATION CROSS-LINKED HETEROGENEOUS POLYOLEFIN BLENDS

H.-J. Radusch[1]*, I. Kolesov[1], U. Gohs[2], G. Heinrich[2]

[1]*Department of Engineering Sciences, University of Halle-Wittenberg, D-06099 Halle, Germany*
hans-joachim.radusch@iw.uni-halle.de
[2]*Leibniz Institute of Polymer Research Dresden, D-01069 Dresden, Germany*
gheinrich@ipfdd.de

Abstract - Polyolefine blends on the basis of high density polyethylene (HDPE), ethylene octene copolymer (EOC) and polycyclooctene (trans-polyocteneamer) (TOR) were produced by melt mixing and subsequent cross-linking by electron irradiation (EBR) for generation of shape-memory materials. Electron microscopy investigations, DSC and DMTA showed that the blends contained separated phases with different crystalline species or crystal populations, respectively, embedded in cross-linked amorphous phase. Thus, superpositioned amorphous network and crystal network have been generated as the basis for the entropy driven recovery after material programming and subsequent heating. The blends evinced pronounced steps of modulus-temperature characteristics as well as distinct DSC melting peaks. The melting temperatures of the crystalline phases served as shape recovery switching temperatures. The thermo-induced unconstrained shape-memory (SM) behavior of the blends was characterized by the parameters shape fixity ratio, shape recovery ratio and shape recovery rate. Material programming and determination of the shape-memory parameters was performed in a dynamic-mechanical spectrometer. The materials exhibited well developed triple- and quadruple SM behavior even after one-step programming (OSP) process both for binary HDPE/TOR and EOC/TOR as well as ternary HDPE/EOC/TOR blends. All investigated blends showed high values of strain fixity and strain recovery ratios of 95 to 99 %. Comparative investigations of the same blend systems cross-linked by means of chemical cross-linking agents (peroxide) showed that cross-linking via high energy electrons results in a more pronounced and stronger SM effect.

Keywords: shape-memory, polyolefin blends, cross-linking, electron irradiation.

Introduction

The performance of the thermally triggered recovery – or shape memory effect (SME) - in cross-linked semi-crystalline polymers strongly depends on the properties of crystalline structure and of covalent molecular network [1-3]. In the first instance, these are melting temperature and crystallinity as function of temperature, which are dependent on crystal size and perfection and their distribution as well as cross-linking density and connected mechanical properties of network such as modulus, strain at break and residual strain [1,2,4]. Recently, multiple switchable SM polymer materials were developed, which evince triple-shape [5] and even quadruple-shape behavior [6]. The main disadvantage of such SM polymers is the need for a multi-step programming (MSP) process for the appearance of appreciable steps of SM recovery at the different switching temperatures (T_{sw}) of the system [5,6]. Along with the technical complexity of the MSP and in many cases very diffuse indistinct steps and peaks of SM recovery strain and rate, respectively, the main disadvantages of MSP compared to OSP process are the relatively low values of strain fixing (R_f) and strain recovery (R_r) ratios. This undesirable effect is resulting from the drastic increase of irreversible deformation parts caused by plastic deformation of already crystallized phase/phases in the second and third programming step. In complex multiphase AB polymer networks created lately on the basis of poly(ε-caprolactone) and poly(cyclohexyl methacrylate) the shortcomings mentioned above were eliminated. As a result, these networks demonstrate the pronounced triple-shape behavior with good SM characteristics after simple one-step programming [7]. A scheme of a triple-shape memory behavior after multi-step as well as one-step programming is displayed in Fig. 1.

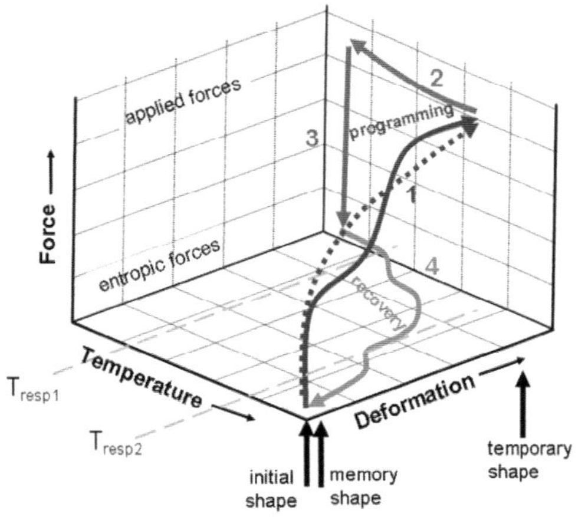

Figure 1 – Scheme of a triple-shape memory behavior after multi-step (solid red line) and one-step programming (dotted red line), respectively.

Multiple SM polymer materials are highly innovative smart materials, which permit some novel applications in comparison to single SME materials [5,7]. SM polymer materials with relatively low switching temperature (T_{sw}) are of particular interest for a number of engineering, medical and pharmaceutical applications. For blends of linear and short-chain branched polyethylenes cross-linked by peroxide a distinct multiple SM behavior could be not achieved by OSP up to now [6]. In the frame of the present work SM polymer materials were generated on the basis of covalently cross-linked polyolefin blends on the basis of linear polyethylene (HDPE) and/or short-chain branched polyethylenes (EOC), which show distinctly multiple SM behavior with high performance, as well as polycyclooctene (PCO) containing polyoctenamer (TOR), which is thermodynamically incompatible with polyethylenes and has a low melt viscosity. Electron radiation cross-linking of the blends was applied subsequently after blend preparation and morphology fixation after targeted cooling.

Experimental

Materials

Polymers used HDPE KS 10100 UE (Dow Chemical) with the melting temperature T_m=132°C, short-chain branched EOC AFFINITY™ PL 1280G (Dow Chemical) with approx. 30 hexyl branches per 1000 C and T_m=96°C as well as PCO containing TOR VESTENAMER® 8012 (Evonik Degussa) with T_m=54°C.

Processing

Binary and ternary blends of TOR, EOC and/or HDPE with different mass ratios of the components (Table 1) were prepared by melt mixing in laboratory kneader PL 2000 (Brabender) for 5 min at 140°C. Films of TOR, EOC and HDPE as well as their blends with thickness of 1 mm were compression molded and subsequently quenched (q) or slowly cooled with 1 Kmin^{-1} (sc) from the melt. Thus, two series of samples with two different thermal histories and accordingly with different phase morphologies were prepared.

Table 1 – Composition of TOR/EOC30/HDPE blends

Blends	Content, [%]		
	HDPE	EOC	TOR
50 EOC/50 TOR	-	50	50
30 HDPE /70 TOR	30	-	70
50 HDPE/50 TOR	50	-	50
50 HDPE/50 EOC	50	50	-
20 HDPE/40 EOC/40 TOR	20	40	40

The samples were cross-linked by EBR by means of high energy electrons with a radiation dose of 200 kGy. The irradiation for all materials was carried out at room temperature using an electron accelerator ELV-2 (Budger INP, Novosibirsk). During irradiation the sheets were placed in nitrogen gas filled plastic bags to avoid molecular degradation by air oxygen.

Characterization

Melting and crystallization of the blends were characterized by differential scanning calorimetry (DSC) using a DSC Q1000 (TA Instruments) between -80 and 180 °C and 10 Kmin^{-1} heating and cooling rate, respectively. The temperature dependences of storage modulus in 1st cooling and subsequent 2nd heating runs were measured in tensile mode using a mechanical spectrometer MARK III (Rheometric Scientific) at 1 Hz and 2 Kmin^{-1} heating and cooling rate, respectively. The morphology of the blends was characterized by scanning (SEM) and transmission electron microscopy (TEM). SEM images were obtained using an ULTRA plus microscope (Zeiss). The surface of SEM specimens was cryo-cutted at –160 °C and stained with RuO_4. TEM investigations were performed by a Libra 200MS (Zeiss). The TEM samples were also stained with RuO_4 before cutting and sliced with a diamond knife to ultra-thin sections of 60 nm thickness at -160 °C by the cryo-microtome Ultracut S (Leica). Shape-memory investigations were carried out in tensile mode using a mechanical spectrometer MARK III (Rheometric Scientific). The samples with a cross-section area 2.0x1.0 mm^2 were tested during the specific cyclic thermo-mechanical experiment at an initial distance of the clamps of 12 mm. The samples were programmed by a tensile strain of ε_{pr} = 100% during a standard dwell period of 120 s at the programming temperature T_{pr} of 125 °C, and cooled down to the lowest temperature T_{low} of 0 °C in the constant deformed state at an average cooling rate of 11 Kmin^{-1}. The values of programming stress σ_{pr} were measured at T_{pr} for 120 s after loading of the sample. The stretched specimens were thermally equilibrated for 10 min at T_{low}, unloaded and heated to 25 °C. After a dwell time of 10 min at 25 °C the remaining strain ε_v for the estimation of strain fixing ratio R_f was determined, and then the unconstrained thermally-induced recovery of the completely programmed sample in a heating run at a rate of 2 Kmin^{-1} and 'zero' stress of 70 Pa was measured. The final value of ε_r was used for the determination of strain recovery ratio R_r according Lendlein et al. [1].

Results and Discussion

Melting and crystallization behavior

Melting and crystallization behavior of the irradiated binary and ternary HDPE/EOC/TOR blends, slowly cooled or quenched after compression molding, have been investigated. Fig. 2 shows the melting behavior of the blends in 2nd heating run. Melting peaks corresponding to the components of both binary and ternary blends are slightly more discrete for sc than for q samples. The irradiated TOR containing blends, especially HDPE/TOR and EOC/TOR binary blends, show significant discrete melting and crystallization peaks for all compositions (50HDPE/50TOR, 30HDPE/ 70TOR and 50EOC/50TOR) in comparison to the 50HDPE/50EOC blend. Discrete melting peaks point to the separation between the crystalline TOR phase and crystalline HDPE/EOC phases due to thermodynamic incompatibility between TOR and polyethylene.

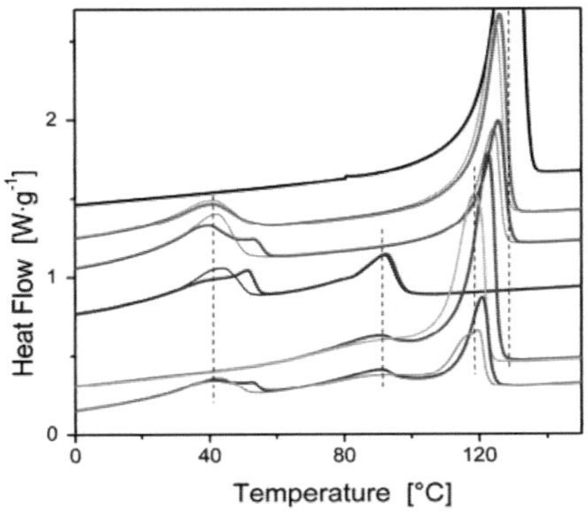

Figure 2 – DSC thermograms of EBR cross-linked binary and ternary blends on the basis of HDPE, EOC and TOR, slowly cooled (sc - thick lines) or quenched (q - thin lines), 2nd heating runs.

Dynamic mechanical behavior.

The thermal loops of storage modulus of TOR, EOC and HDPE as well as of their blends show pronounced steps nearby melting and crystallization temperatures of the components. As an example for all compositions, Fig. 3 shows two distinctly separated melting and crystallization transition steps for blends of HDPE/TOR (ca. 140/60°C and 110/30°C).

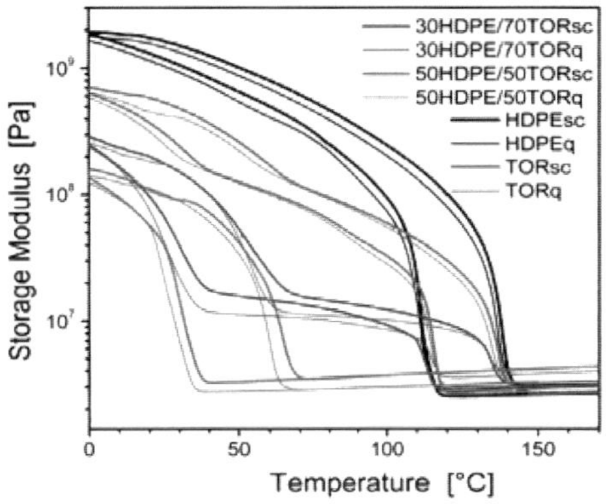

Figure 3 – Storage modulus of EBR cross-linked blends of HDPE and TOR and their components in dependence on temperature. Heating and cooling runs. sc - slowly cooled, q – quenched.

Discrete melting transitions for ternary blend (20HDPE/40EOC/40TOR) are not so pronounced although weak steps of decrease in storage modulus were observed. All materials showed rubber-like plateaus of the storage modulus at temperatures above the melting or crystallization temperature of the blend component with the highest T_c and T_m value, respectively, that is typical for covalently cross-linked polymers. The blends possess different moduli depending on their composition and thermal history. As expected, with increasing content of HDPE the modulus of all blends increased. Furthermore, a slight influence of thermal history on modulus is observed in terms of an increase of modulus at slowly cooling.

Phase morphology

SEM images of binary 30HDPE/70TOR and 50HDPE/50TOR blends, which showed discrete melting and crystallization peaks for both components in DSC and a stepwise decrease in storage modulus curves, reveal the incompatibility between HDPE and TOR (Fig. 4). Droplet-like or extended spots (Fig. 4 a,b) with a size of 1 to 50 μm and identified as HDPE indicate the dispersion of HDPE in the TOR matrix.

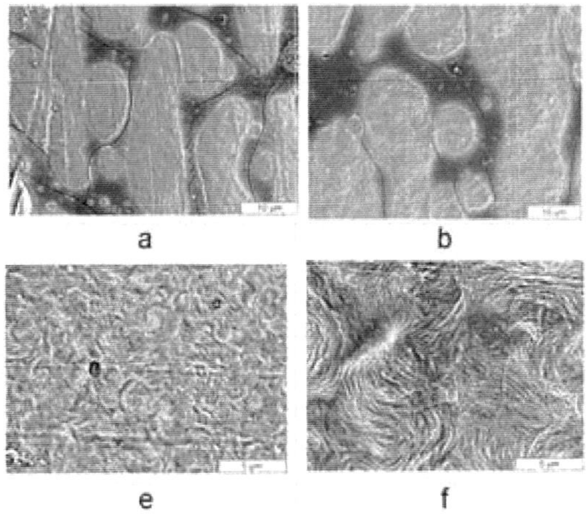

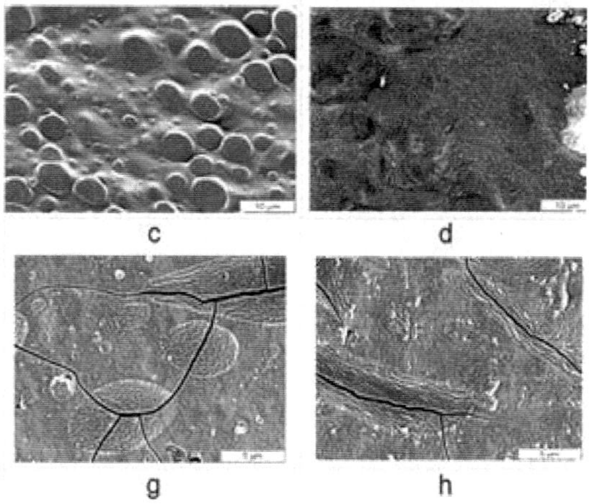

Figure 4 – SEM images of 50HDPE/50TOR (a, b), 30HDPE/70TOR (c, d), 50HDPE/50EOC (e, f) and 20HDPE/40EOC/40TOR (g, h) blends, quenched (a, c, e, g) or slowly cooled (b, d, f, h).

The disperse HDPE phase is segregated from the continuous TOR phase as a result of thermodynamic incompatibility, and this segregation results in a multiple melting behavior, expressed in the DSC and DMA runs. In the 50HDPE/50EOC blend the separate phase was not able to reveal by SEM. The SEM images of the 50HDPE/50EOC blend (Fig. 4e,f) demonstrate a distorted, chaotically oriented lamellae-like supermolecular structure, as produced by co-crystallization of HDPE and EOC. The SEM image of the ternary 20HDPE/40EOC/40TOR blend (Fig. 4g,h) supports the assumption of thermodynamic incompatibility between TOR and HDPE/EOC. The partial compatibility of HDPE and EOC is seen in the same contrast on the 20HDPE/40EOC/40TOR micrograph (Fig. 4g,h).

Shape memory behavior

The results of the thermo-mechanical tests for the characterization of the multiple SM behavior in blends are presented in Fig. 5 and Table 2. Because TOR forms the continuous phase and has the lowest melting temperature of all components, it does not constrain the SM recovery by melting of the disperse phase/phases, and the blends evince a pronounced multiple SM behavior (see Fig. 5).

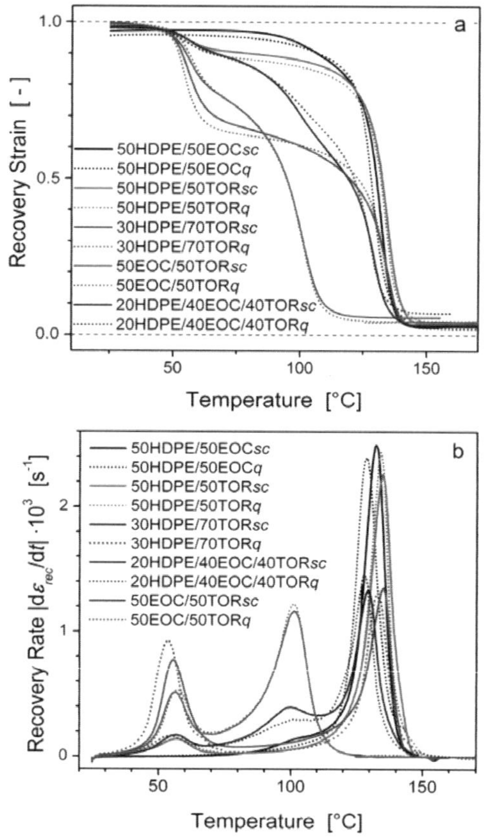

Figure 5 – SM recovery strain (a) and SM recovery rate (b) of binary and ternary blends on the basis of HDPE, EOC and TOR as function of temperature cross-linked by EBR with 200 kGy.

The SM blends cross-linked by EBR show high values of strain fixing and strain recovery ratios of 95 to 99 %. The TOR/HDPE and TOR/EOC30 blends of q preparation demonstrate more pronounced triple-shape behavior in comparison to blends of sc preparation (see Fig. 5) that is also reflected in the higher peak values of SM recovery rate in the particular TOR phase (see Table 2). Such a triple-shape behavior correlates with the stronger phase separation in quenched blends.

Table 2 – Shape memory characteristics of EBR cross-linked blends of HDPE, EOC and TOR and their components.

Blends	TH*	R_f %	R_f %	T_{sw} [°C]			$\mid d\varepsilon_{rec}/dt \mid_{max} \cdot 10^3$ [s^{-1}]		
				TOR	EOC	HDPE	TOR	EOC	HDPE
100HDPE	sc	95.4	96.2	-	-	134.5	-	-	3.65
	q	95.8	96.0	-	-	135.1	-	-	4.04
100EOC30	sc	96.5	94.4	-	100.9	-	-	2.07	-
	q	97.1	95.9	-	100.3	-	-	2.11	-
100TOR	sc	98.8	98.1	53.0	-	-	3.38	-	-
	q	99.5	98.2	54.5	-	-	4.02	-	-
30HDPE/70TOR	sc	99.5	96.8	55.5	-	135.2	0.77	-	1.35
	q	99.9	97.8	53.1	-	132.8	0.93	-	1.31
50HDPE/50TOR	sc	97.8	96.2	56.8	-	134.7	0.14	-	2.26
	q	97.9	96.0	56.5	-	133.9	0.17	-	2.43
50EOC30/ 50TOR	sc	98.8	94.5	56.1	101.1	-	0.51	1.16	-
	q	99.3	95.8	57.6	100.7	-	0.53	1.12	-
20 HDPE/ 40EOC30/ 40TOR	sc	98.3	97.1	56.6	99.8	129.5	0.17	0.39	1.33
	q	98.1	93.0	54.7	101.4	128.3	0.17	0.29	1.44
50HDPE/ 50EOC30	sc	96.9	97.5	-	104.8	132.1	-	0.16	2.49
	q	95.5	98.0	-	**	128.3	-	**	2.39

*- TH = thermal history; ** - not measurable

Conclusions

A pronounced multiple shape memory effect can be observed in polyolefin blends cross-linked by electron irradiation after one-step programming procedure, under the condition that different, well segregated crystalline phases having different melting temperatures can be created in such blends. For this purpose, binary and ternary blends of HDPE, EOC and TOR, which are thermodynamically incompatible, were prepared by melt mixing and processed by compression molding and subsequent quenching or slowly cooling. After morphology formation the specimens were cross-linked by means of electron beam radiation at room temperature.

EBR cross-linked HDPE/EOC/TOR blends show a pronounced multiple SM behavior after one-step programming, and very good SM performances with high values of strain fixing and strain recovery ratio of 95 to 99 %. Present research has shown that multiple SM behavior is enforced by blending of thermodynamically incompatible materials as well as an optimal thermal history at cooling after mixing.

References

1. A. Lendlein; S. Kelch *Angew. Chem. Intern. Ed.* 2002, *41,* 2034.
2. C. Liu; H. Qin; P.T. Mather *J. Mater. Chem.* 2007, *17,* 1543.
3. H.-Y. Jiang; A.M. Schmidt in *Shape-Memory Polymers and Multifunctional Composites*, J. Leng, Y. Liu (Eds.), CRC Press, Boca Raton, 2010; 21-63.
4. I.S. Kolesov; K. Kratz; A. Lendlein; H.-J. Radusch *Polymer* 2009, *50*, 5490.
5. I. Bellin; S. Kelch; R. Langer; A. Lendlein *Proc. Natl. Acad. Sci. USA (PNAS)* 2007, *103*, 18043.
6. I.S. Kolesov; H.-J. Radusch *eXPRESS Polym. Lett.* 2008, 461.
7. M. Behl; I. Bellin; S. Kelch; W. Wagermaier; A. Lendlein *Adv. Funct. Mater.* 2009, *19,* 102.

O-09-471

A MODEL EXPERIMENT FOR HIGH-PRESSURE POLYMER CRYSTALLIZATION

S.A.E. Boyer[1,2]* and J.-M. Haudin[2]*

[1 present address] *P PRIME Institute, ISAE-ENSMA, UPR CNRS 3346, Department of Physics and Mechanics of Materials, 1 Avenue Clément Ader, 86961 Futuroscope, France – severine.boyer@ensma.fr*

[2] *MINES ParisTech, Centre for Material Forming, UMR CNRS 7635, 1 Rue Claude Daunesse, 06904 Sophia Antipolis, France – jean-marc.haudin@mines-paristech.fr*

Abstract –

An important aspect of 'extreme' processing conditions that can be reproduced thanks to the development of model experiments is the pressure influence. Mainly, the experimental setups can be classified into four families: "simple" cells, dilatometric set-ups, differential thermal analysis and diamond anvil plus *in-situ* measurement.

As such *in-situ* measurement cell, an original optical high-pressure cell, named CRISTAPRESS, has been designed and constructed. High pressure up to 200 MPa with a fine temperature control up to 300 °C can be applied. Concomitantly, *PVT* control and *in-situ* light depolarizing microscopic observations are possible.

A first illustration of the ability of the CRISTAPRESS cell consists in following high-pressure isothermal crystallization. Time-resolved measurements of isotactic polypropylene crystallization at the spherulitic scale have been carried out.

The preliminary experimental results show the role of the barometric and thermal histories on the key kinetic parameters of polymer crystallization, that are the number of activated nuclei $N_a(t)$, the spherulitic growth rate $G(T,P)$, and in a more global manner the overall kinetics $\alpha(t)$, as well as on the subsequent morphologies. Discussion dealing with the Avrami equation and the Hoffman & Lauritzen theory is introduced.

Keywords: Crystallization, Polymer Morphology, High Pressure, Model Experiment.

Introduction

As temperature, pressure is a major thermodynamic parameter to control equilibrium between phases in all types of materials. The first papers on the behavior of polymers under pressure were published in the 1930s, but active research in the field actually began in 1949. Reviews on early work were published by Wunderlich & Davidson [1], and Bhateja & Pae [2]. The more general paper of Bhateja & Pae concerns compressibility, crystallization and melting under pressure of various polymers, while Wunderlich & Davidson's review was mainly devoted to pressure-crystallization of polyethylene (PE).

Since, by studying the relevant literature documents dealing with the investigation of crystallization and melting of polymers under pressure, a concise classification of experimental setups is possible as following:

1. "Simple" cells or bombs [3-6], with a *post-mortem* analysis of the polymers. But some peculiarities are observed as the crystallization bomb of Edwards & Phillips [4] with subsequent observation by transmission electron microscopy (TEM) and *in-situ* nuclear magnetic resonance (NMR) measurements by Brown & Jonas [5].
2. Dilatometric setups [7-10]. In addition, Eckel et al. [8] performed combined Raman and dilatometric measurements.
3. Differential thermal analysis (DTA) [11] and differential scanning calorimetry (DSC) [12].
4. Diamond anvil and *in situ* measurements using different facilities: optical microscopy [13,14], depolarized light intensity measurements (DLI) and small-angle light scattering (SALS) [15], X-ray diffraction [16], etc.

Concerning in more details polymer crystallization mechanisms, overall kinetics as well as the elementary steps of nucleation and growth are of great importance. Overall crystallization kinetics under pressure were mainly determined by dilatometry [10,17-18], but also by DTA [19] and even DLI measurements, X-ray diffraction and RMN [5]. In contrast to overall kinetics, which has been considered in a significant number of papers, little work has been devoted to the elementary steps of nucleation and growth. The total number of activated nuclei has been determined *post mortem* by counting the number of spherulites in an optical microscope [10,18].

In order to explore the fundamental stages of polymer crystallization under the influence of pressure, the present work aims at presenting a model experiment for high pressure polymer crystallization. The model experiment consists in an original optical high-pressure cell, named CRISTAPRESS. It is classified in the fourth family 'diamond anvil and *in situ* measurements' of high-pressure experimental setups described above. The cell has been especially constructed to observe phase transitions under simultaneously controlled pressure, volume and temperature. It has been applied to the crystallization of isotactic polypropylene (iPP). The spherulite growth rate has been measured in various temperature and pressure conditions. The new data collection has been used in well-known polymer crystallization modeling, allowing thus a critical analysis of the latter.

Experimental

Materials and preparation

An isotactic polypropylene (iPP) grade for injection molding is selected as a model polymer. This polymer was selected because it produces spherulites easily observable by optical microscopy. Its commercial reference is 3250 MR1 (ATOFINA Company, France). *In-situ* polymer crystallization from direct optical observation requires that the unoriented polymer film is thin enough, 50-μm-thick, to allow two dimensional (2D) observations.

High pressure constraint

The actual design process is a synthesis of criteria which concern four main areas: (i) materials morphology and optical properties; (ii) constraints imposed by optical equipment and optical requirements of the chamber windows; (iii) mechanical strength and methods for generating pressure and volume change of the system; and (iv) temperature requirements.

The high-pressure optical cell is used on the stage of a commercial optical microscope (Leica-DMRX 300, Leica-Microsystèmes, France) with an attached DFC 420C color digital microscope camera. The birefringence resulting from the development of semi-crystalline entities is optically observed in the transmission mode between crossed polarizers.

Temperature-pressure procedure

Runs are made with samples crystallized at a uniform temperature and constant pressure. The thermo-barometric protocol followed in crystallization experiments is given in **Figure 1**.

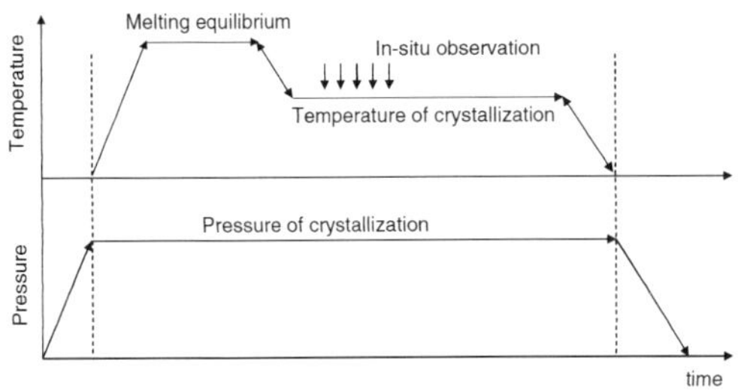

Figure 1 – Temperature-pressure evolutions at constant programmed temperature and pressure rates.

Results and Discussion

Results

A series of kinetics measurements of different films of i-PP in a selected range of T and in a selected range of moderately high pressures, i.e., up to 90 MPa was experimented.

The application of pressure results in moving the growth kinetics to higher temperatures and in increasing the maximum growth rate. The rates are sensitive to temperature and the extent to which crystallization proceeds is controlled by the supercooling. Calculation on growth kinetics was done by selecting isolated circular entities as described in our previous work dealing with high-cooling rate induced crystallization under atmospheric pressure [20]. Selected growth rates are plotted versus the pressure in **Figure 2**.

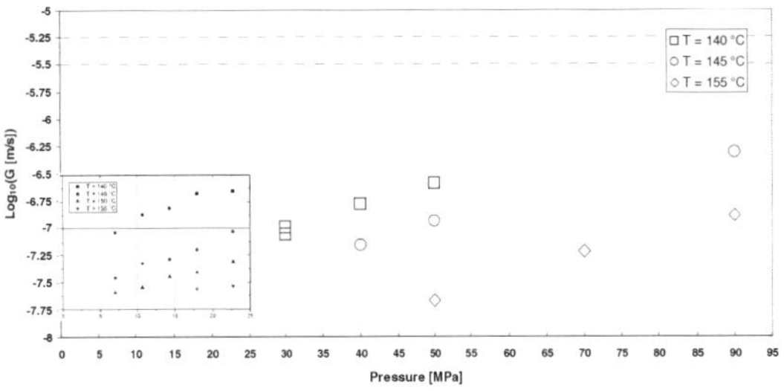

Figure 2 – Growth rates versus pressure at constant programmed temperature. Insert: data from Eder et al. [10].

The collected data are compared to some possibly comparable data from literature. For instance, Eder et al. [10] analyzed the growth rate *versus* pressure by applying a modified bar press to study the influence of isotropic pressure on the crystallization kinetics of an industrial polypropylene under isothermal condition. They calculated the growth rate from the crystallization half-time and from the *post-mortem* measurement of the number of nuclei using an Avrami analysis. Results from the present work are in agreement with the estimates done by Eder et al. However, a small shift between the two sets of curves is observed. This could be due to the different methods used or to intrinsic differences between the two polymers.

Moreover, the melting point under pressure has been estimated as the temperature where the spherulitic structure completely disappears. From these optical observations it is shown that pressure shifts the melting temperature T_m to higher values and the rate of change of the melting temperature dT_m/dp decreases with increasing pressure, as illustrated in **Figure 3**. A comparison is done with both the alpha and gamma phase melting temperatures, provided by Nakafuku's work [21]. As a matter of fact, it is known that, by applying pressure, iPP can crystallize into two phases namely the well-known monoclinic α- and the less frequent orthorhombic γ-phase [22]. The melting temperature tendency is in agreement with the previous high-pressure works of Nakafuku [21], Mezghani & Phillips [23] and of Fulchiron et al. [9] for equilibrium melting temperature with a $T^{0Patm}{}_m$ of 208 °C [24]. Nakafuku estimated, from DTA investigation, a quadratic dependency of T_m with the pressure as: T_m= $164.1 + 0.0366P - 2.548.10^{-6}P^2$ and of $150.6 + 0.0395P - 2.980.10^{-6}P^2$, with P in bar, for alpha and gamma forms, respectively [21]. Fulchiron et al. estimated the equilibrium melting temperature T^0_m as T^{0Patm}_m + $0.283P - 2.08.10^{-4}P^2$, with P in MPa, from conventional piston-type dilatometry (PVT100) [9].

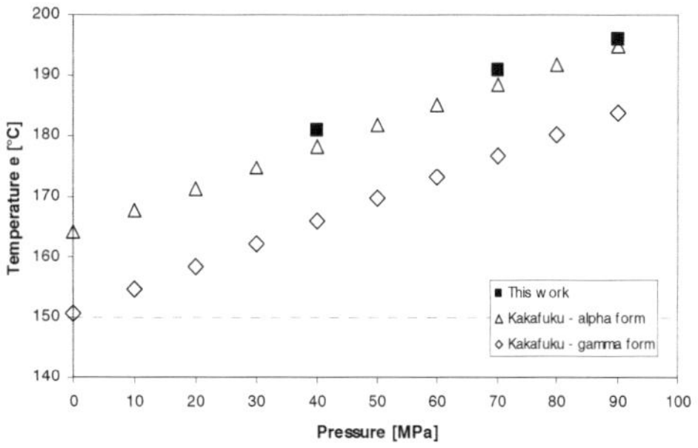

Figure 3 – Pressure dependence of melting temperature of iPP spherulites from *in-situ* observation with the CRISTAPRESS cell. Comparison with Nakafuku's work [21].

Discussion: mechanisms of nucleation and growth

Thanks to the new analysis extended to pressures higher than 90 MPa, it would be possible to propose a new reliable crystallization kinetics law by simultaneously considering the thermodynamics, the kinetics and the polymer structure. This would permit to illustrate the trend of growth rates to merge into a single curve looking like a part of a bell-shaped curve, as expected from Hoffman–Lauritzen's theory at atmospheric pressure [24,25].

By revisiting Hoffman–Lauritzen's theory, two modifications are proposed [18]:

- modification 1: since the equilibrium temperature is pressure dependent, the growth parameter (K_g) is pressure dependent as well, as described in Eq.1:

$$\ln G + U^* / R(T_c - T_\infty)\ vs.\ T^0_m(P)/T_c(\Delta T) \quad \text{(Eq.1)}$$

Under this assumption, Hoffman & Lauritzen plot leads to a single curve but not exactly, meaning that additional parameters are pressure dependent.

- modification 2: the effect of pressure on the transport term is then taken into account, as described in Eq.2:

$$\ln G + \frac{U^* + \alpha P}{R(T_c - T_\infty)} + \beta P\ vs.\ T^0_m(P)/T_c(\Delta T) \quad \text{(Eq.2)}$$

Under the second assumption, the main difficulty is to define the right expression of the mobility term dependence with pressure, namely the two α and β coefficients.

Conclusions

This paper describes the development of a miniature, pressure-volume-temperature-controlled (300°C, 200 MPa) apparatus with a new data collection on polymer crystallization. Experiments were conducted to study *in-situ* and *in real-time* the development of semi-crystalline entities, i.e., spherulites of polymers, from the melt under pressure. The applied procedure provides growth kinetics for each selected T and P run. Gathered data allow us to distinguish the pressure effect from the temperature effect on the growth rate $G(T,P)$.

Future Works

Beyond the scope of the present paper, the emphasis being laid on growth rate to see the aptitude of the equipment for following kinetics of crystallization, quantification of the nucleation event would show the versatility of the new equipment. The crystals formed at low temperature under pressure could act as nuclei for a higher temperature under atmospheric pressure, thus resulting in a controlled nucleation. Density of nuclei would depend more on supercooling that varies with the shift of the melting point imposed by the pressure than with the pressure itself.

With the fact to couple high-pressure technology with *in-situ* morphology and size distribution of the growing entities, next step is to implement the original physical analysis into the kinetic modeling of polymer crystallization based on Avrami-Evans equations [26].

Acknowledgements

CNRS, MINES ParisTech, Conseil Régional de Provence-Alpes-Côte d'Azur, Conseil Général des Alpes-Maritimes.

References

1. B. Wunderlich, T. Davidson *J. Polym. Sci., Part A-2*, 1969, *7*, 2043.
2. S. K. Bhateja, K. D. Pae *J. Macromol. Sci. - Revs. Macromol. Chem.*, 1975, *C13(1)*, 77.
3. B. Wunderlich *Rev. Sci. Instr.*, 1961, *32*, 1424.
4. B. C. Edwards, P. J. Phillips *Polymer*, 1974, *15*, 491.
5. D. R. Brown, J. Jonas *J. Polym. Sci., Polym. Phys. Ed.*, 1984, *22*, 655.
6. T. Kazmierczak, A. Galeski *J. Appl. Polym. Sci.*, 2002, *86*, 1337.
7. S. Matsuoka J. *Polym. Sci.*, 1962, *57*, 569.
8. R. Eckel, M. Buback, G. R. Strobl *Colloid Polym. Sci.*, 1981, *259*, 326.
9. R. Fulchiron, E. Koscher, G. Poutot, D. Delaunay, G. Régnier *J. Macromol. Sci. Phys.*, 2001, *B40 (3&4)*, 297.
10. G. Eder, P. Hierzenberger, P. Amorim, PPS-27, 2011, 27th World Congress of the Polymer Processing Society, May 10-14, 2011, Marrakech, Morocco
11. T. Davidson, B. Wunderlich *J. Polym. Sci., Part A-2*, 1969, *7*, 377.
12. G. W. H. Höhne, S. Rastogi, B. Wunderlich *Polymer*, 2000, *41*, 8869.
13. D. C. Bassett, S. Block, G. J. Piermarini *J. Appl. Phys.*, 1974, *45*, 4146.

14. K. Mezghani, P. J. Phillips *Polymer*, 1997, *38*, 5725.
15. H. T. Tseng, P. J. Phillips *Macromolecules*, 1985, *18*, 1565.
16. M. Hikosaka , S. Tamaki *J. Phys. Soc. Jpn*., 1981, *50*, 638.
17. J. He, P. Zoller *J. Polym. Sci., B Polym. Phys.*, 1994, *32*, 1049.
18. C. Angelloz, R. Fulchiron, A. Douillard, B. Chabert, R. Fillit, A. Vautrin, L. David *Macromolecules*, 2000, *33*, 4138.
19. S. Sawada, T. Nose *Polym. J.*, 1979, *11*, 477.
20. S. A. E. Boyer, J.-M. Haudin *Polym. Testing*, 2010, *29*, 445.
21. C. Nakafuku *Polymer*, 1981, *22*, 1673.
22. E. Piorkowska, S. A. E. Boyer, P. Sowinski, J.-M. Haudin, K. Zapala, PPS-28, 2012, 28th World Congress of the Polymer Processing Society, December 11-15, 2012, Pattaya, Thailand
23. K. Mezghani, P. J. Phillips, *Polymer*, 1998, *39*, 3735.
24. S. A. E. Boyer, P. Ganet, P. Robinson, J.-P. Melis, J.-M. Haudin *J. Appl. Polym. Sci.*, 2012, *125*, 4219.
25. J. I. Lauritzen Jr., J. D. Hoffman *J Res Natl Bur Stand A*, 1960, *64*, 73.
26. J.-M. Haudin, J.-L. Chenot *Int. Polym. Process*, 2004, *19*, 267.

O-09-472

A MODEL EXPERIMENT FOR NUCLEATION OF THE HIGH-PRESSURE GAMMA FORM IN ISOTACTIC POLYPROPYLENE

E. Piorkowska[1]*, S.A.E. Boyer[2,3]*, P. Sowinski[1], J.-M. Haudin[3]* and K. Zapala[1]

[1] *Polish Academy of Science, Centre of Molecular and Macromolecular Studies, Sienkiewicza 112, 90 363 Lodz, Poland – epiorkow@bilbo.cbmm.lodz.pl*

[2 present address] *P PRIME Institute, ISAE-ENSMA, UPR CNRS 3346, Department of Physics and Mechanics of Materials, 1 Avenue Clément Ader, 86961 Futuroscope, France – severine.boyer@ensma.fr*

[3] *MINES ParisTech, Centre for Material Forming, UMR CNRS 7635, 1 Rue Claude Daunesse, 06904 Sophia Antipolis, France – jean-marc.haudin@mines-paristech.fr*

Abstract –

The present study aims at exploring the fundamental stage of polymer nucleation during crystallization. Concern is on the determination of nucleating ability of α- and β- nucleating agents on high pressure crystallization of iPP in the γ form.

Three nucleating agents are used, *i-ii)* two which are known to nucleate efficiently crystallization of iPP in the monoclinic α form under atmospheric pressure, namely Hyperform HPN-20E (produced by Milliken Chemicals) and poly(tetrafluoroethylene) submicron particles, Dispersez–200W2 (purchased from Polysciences Inc), and *iii)* one which is known to nucleate efficiently crystallization of iPP in the hexagonal β form under atmospheric pressure, namely calcium pimelate (synthesized at CMMS, Lodz). The high pressure crystallization of iPP is carried in the crystallization cell that consists of a barrel and a piston, and a loading frame of the tensile testing machine (Instron), under various pressure and temperature conditions.

The first investigations demonstrate that both α- nucleating agents accelerated crystallization of iPP in the high pressure orthorhombic γ form, possibly by nucleating the α 'seeds" facilitating further crystallization of the γ crystals. Especially, poly(tetrafluoroethylene) is known to nucleate the iPP α phase through epitaxy involving (110)α plane, which does not apply for the γ form.

Keywords: High Pressure, Gamma Form, Polypropylene, Nucleating Agents.

Introduction

In polymer liquid to solid transition, nucleation is a fundamental stage. This stage dictates solidification temperature and time, size of polycrystalline aggregates, and it can influence crystallographic forms of crystals.

Relevant literatures dealing with advanced nucleating agents for polypropylene (PP) have been recently reviewed [1]. The mechanism and efficiency of nucleating compounds can be classified according to their tendency to generate the most widespread monoclinic α-, the less stable hexagonal β- and the rare orthorhombic γ-PP phases. The α-nucleators are sorbitol-based compounds. N,N(-dicyclohexyl-2,6-naphthalene dicer-boxamide, series of calcium carboxylates [2], Ca-malonate (Ca-mt) [3], Ca-salt of suberic (Ca-sub) and pimelic acid (Ca-pim) [4-9] were found to be β-nucleators. The γ-modification is the most unexplored modification. More especially, this phase can be obtained with specific nuleators as bicyclo[2.2.1]heptane dicarboxylate salt, commercially known as HPN-68 [10,11], also when iPP is crystallized at elevated pressures [12] and even with a decrease in PP isotacticity [13].

In the aim to explore this fundamental stage of polymer nucleation during crystallization, the present work aspires at determining the effect of nucleating ability of α- and β- nucleating agents on high pressure crystallization of iPP in the γ form.

Experimental

Materials

The considered polymer matrix was an isotactic polypropylene (3250 MR1, Atofina, France). Three nucleating agents were selected. Two are known to nucleate efficiently crystallization of iPP in the monoclinic α form under atmospheric pressure, namely *i)* an organic carboxylic acid salt (Hyperform HPN-20E, produced by Milliken Chemicals) and *ii)* an aqueous dispersion of poly(tetrafluoroethylene) (PTFE) submicron particles (Dispersez–200W2, purchased from Polysciences Inc) [14]. One is known to nucleate efficiently crystallization of iPP in the hexagonal β form under atmospheric pressure, *iii)* namely calcium pimelate (CaPim) that we have synthesized at CMMS, Lodz.

Sample preparation

iPP was mixed with nucleating agents with of amount of 0.2 wt% according to a one-step melt-blending in a Brabender batch mixer at the speed of 60 rpm for 5 min under N_2 flow. The temperature was set at 195 °C. Three samples were prepared: iPP/hyperform-0.2 wt% (noted as PP/H), iPP/PTFE-0.2 wt% (noted as PP/T) and iPP/Ca-pim-0.2 wt% (noted as PP/C). Additionally, iPP without nucleator was processed in a similar way to produce a blank control sample.

High pressure constraint

The high pressure crystallization of iPP was carried in a custom-built crystallization cell. The high pressure consists of a barrel and a piston, and a loading frame of the tensile testing machine (Instron) under various pressure and temperature conditions [15].

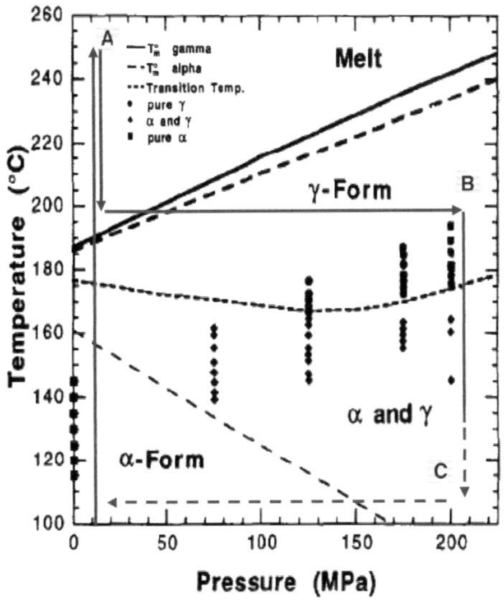

Figure 1 – Phase diagram for iPP [Mezghani & Phillips, *Polymer*, 1998, *39*, 3735].

A thermo-mechanical treatment was strictly followed to preferentially favor the γ-PP phase formation, as illustrated in the phase diagram proposed by Mezghani & Phillips in **Figure 1**. The pressure-temperature (PT) way consisted in A. melting iPP at 250 °C for 5 min under atmospheric pressure, B. followed by a rapid cooling down to 200 °C with a pressurization at a selected pressures in the range of 100 to 300 MPa for a selected time of 2 min, 15 min and 4 hrs. Then, C. the polymer was cooled down to 50 °C under the selected and constant high-pressure, and then depressurized.

Wide-angle X-ray diffraction (WAXD) characterization

WAXD diffraction patterns of the non- and nucleated iPP samples were recorded. A wide-angle goniometer coupled to a Phillips PW3830 sealed tube X-ray generator (Eindhoven, The Netherlands) was used. It was operated in a reflection mode at a voltage of 40 kV and a filament current of 35 mA. The X-ray beam consisted of CuKα radiation (0.154 nm) filtered by a Ni filter and electronically. Radial scans of intensity versus diffraction angle (2θ) were recorded in the range of 10-30°. The relative contents of the crystalline forms were expressed in terms of the K value proposed by Turner-Jones *et al* [16] and Obadal *et al* [17].

Results and Discussion

Results

The nature of nucleating agents affects the WAXD patterns of iPP. As an illustration, WAXS diffractograms of iPP, PP/T, PP/H and PP/C annealed under 100 MPa at 200 °C for 15 min are displayed in **Figure 2**.

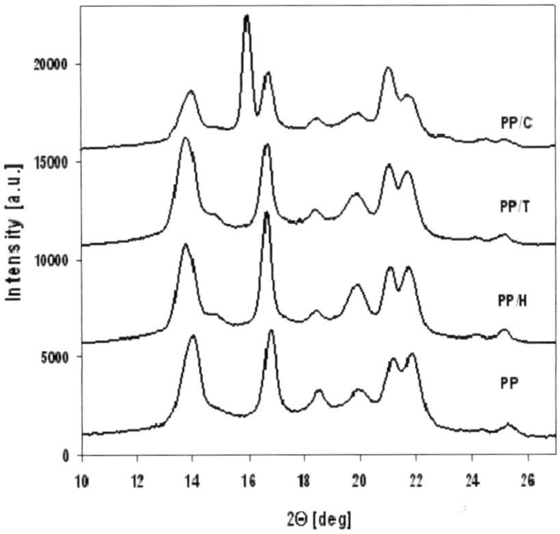

Figure 2 – WAXS diffractograms of PP, PP/T, PP/H and PP/C annealed under 100 MPa, 200 °C, 15 min.

On can see that the three monoclinic α-, hexagonal *β*- and orthorhombic γ-PP phases are observed. The calculated contents of each phase are plotted in **Figures 3** and in **Figure 4**.

Figure 3 shows the calculated phase contents in PP, PP/H, PP/T, PP/C *versus* crystallization pressure (200 °C, 15 min); X_c is the crystallinity degree.

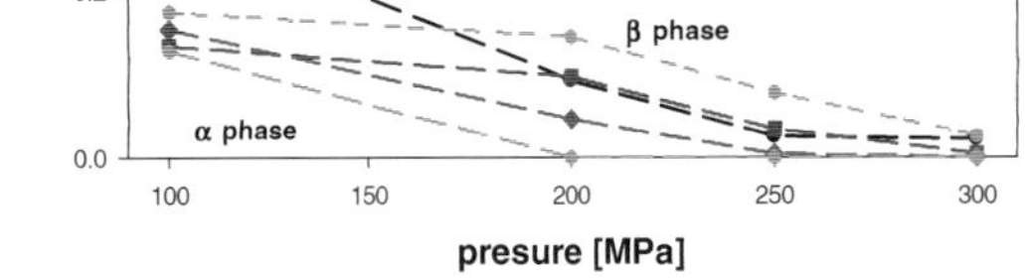

Figure 3 – Phase contents in PP, PP/H, PP/T, PP/C *versus* crystallization pressure (200 °C, 15 min).

Figure 4 demonstrates the calculated phase contents in PP, PP/H, PP/T, PP/C *versus* crystallization time (200 °C, 200 MPa); X_c is the crystallinity degree.

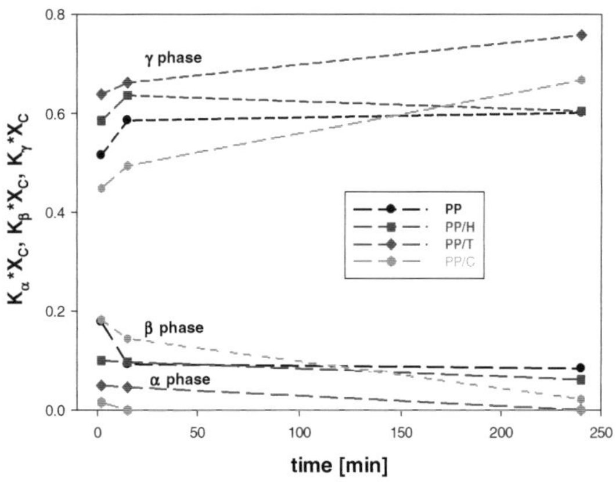

Figure 3 – Phase contents in PP, PP/H, PP/T, PP/C *versus* crystallization time (200 °C, 200 MPa).

Discussions: Analysis of the nucleation mechanism under pressure

The first investigations show that:

1. After crystallization under atmospheric pressure only α formed in neat PP and PP/T whereas in PP/C β formed together with a small fraction of α;
2. After crystallization under elevated pressure α and γ were found in neat PP, PP/H and PP/T; most probably α formed during cooling;
   - after annealing at 200 MPa and 200°C the content of γ phase weakly depended on time
   - in samples annealed at 200 °C for 15 min the γ content increased with elevation of pressure, which increased undercooling and shortened time required for crystallization
   - the γ content in PP/H and PP/T was higher than in neat PP; especially for short crystallization time (200 MPa and 200 °C) and small pressure (200 °C, 15 min);
3. α, β and γ were found after crystallization under elevated pressure in PP/C
   - the γ content increased with increasing time (200 °C, 200 MPa) and with increasing pressure (200 °C, 15min)
   - the γ content in PP/C was smaller than in neat PP.

Both α- nucleating agents accelerated crystallization of iPP in the high pressure orthorhombic γ form, possibly by nucleating the α 'seeds" facilitating further crystallization of the γ crystals [18]. Especially, poly(tetrafluoroethylene) is known to nucleate the iPP α phase through epitaxy involving (110)α plane, which does not apply for the γ form [19].

Conclusions

The presence of HPN-20E and PTFE particles enhanced crystallization of iPP in the γ form under high pressure *via* nucleation of α seeds, on which the γ form grows, or *via* direct nucleation of the γ form. PTFE nucleates the α form through epitaxy involving (110)α crystallographic plane; this epitaxy does not apply for the γ phase [20]. Hence, in the case of PTFE the first mechanism seems to be more probable.

The presence of CaPim did not enhance crystallization of iPP in the γ form under high pressure.

Future Works

Future works will deal on studies of morphology for better understanding of the nucleation process.

Acknowledgements

PICS Project n°4702: Polish Academy of Sciences (PAS, Poland) & Centre National de la Recherche Scientifique (CNRS, France)..

References

1. D. Libster, A. Aserin, N. Garti, *Polym. Adv. Technol.* 2007, *18*, 685.
2. X. Li, K. Hu, M. Ji, Y. Huang, G. Zhou, *J. Appl. Polym. Sci.* 2002, *86*, 633.
3. Q. Dou, Q.-L. Lu, *Polym. Adv. Technol.* 2008, *19*, 1522
4. 4 A. Menyhard, J. Varga, G. Molnar, *J. Therm. Anal. Cal.* 2006, *83*, 625.
5. J. Karger-Kocsis, J. Varga J., *Appl. Polym. Sci.* 1996, *62*, 291
6. T. A. Huy, R. Adhikari, T. Lupke, S. Henning, G.H.Michler *J. Polym. Sci. Part B Polym. Phys.* 2004, *42*, 4478
7. E. Lezak, Z. Bartczak, *Text. East. Eur.* 2005, *13*, 51
8. E. Lezak, Z. Bartczak, *J. Polym. Sci. Part B Polym. Phys.* 2008, *46*, 92
9. Z. Cai, Y. Zhang, J. Li, Y. Shang, H. Huo, J. Feng, S. S. Funari, S. Jiang, *J. Appl. Polym. Sci.* 2012, (accepted)
10. D. Libster, A. Aserin, N. Garti, *J. Colloid Interface Sci.* 2006, *299*, 172.
11. D. Libster, A. Aserin, N. Garti, *J. Colloid Interface Sci.* 2006, *302*, 322.
12. K. Mezghani, P.J. Phillips, *Polymer* 1997, *38*, 5725.
13. R.G. Alamo, M.H. Kim, M.J. Galante, J.R. Isasi, L. Mandelkern, *Macromolecules*, 1999, *32*, 4050.
14. R. Masirek, E. Piorkowska, *Eur. Polym. J.* 2010, *46*, 1436.
15. M. Psarski, A. Galeski, E. Piorkowska, *Macromolecules* 2000, *33*, 916.
16. A. Turner-Jones, J.M. Aizlewood, D.R. Beckett. *Makromol. Chem.* 1964, 75, 134
17. M. Obadal, R. Čermák, K. Stoklasa, *Macromol. Rapid Commun.*, 2000, *526*, 1253.
18. E. Lezak, Z. Bartczak, A. Galeski, *Macromolecules* 2006, *39*, 4811.
19. S. Yan, F. Katzenberg, J. Petermann, D. Yang, Y. Shen, C. Straupe, J.C. Wittman, B. Lotz, *Polymer* 2000, *41*, 2613.
20. C. Mathieu, A. Thierry, J.C. Wittmann, B. Lotz, *Polymer*, 2000, *41*, 7241.

O-09-519

UNUSUAL TUNING OF MECHANICAL PROPERTIES OF ISOTACTIC POLYPROPYLENE UTILIZING COMPETITIVE GROWTH OF ALPHA- AND BETA-CRYSTALS IN BETA-NUCLEATED ISOTACTIC POLYPROPYLENE UNDER SHEAR FLOW

Zhongming Li [1]*, Yanhui Chen [1] and Ganji Zhong [1]

[1] *College of Polymer Science and Engineering and State Key Laboratory of Polymer Materials Engineering, Sichuan University, Chengdu, 610065, China - zmli@scu.edu.cn; sweety520025@163.com; ganji.zhong@gmail.com*

Abstract - Compared with the common α-crystals of isotatic polypropylene (iPP), another crystal form, i.e. β-crystals, exhibits improved elongation at break and excellent impact strength, which attracts lots of research interests. β-crystals can be induced by shear flow or β-nucleating agent. However, coexistence of shear flow and β-nucleating agent will decrease the content of β-crystals. Through in-situ synchrotron X-ray diffraction study on the crystallization process of β-nucleated iPP after shear flow, we figure out in the nucleation stage, the amounts of α-and β-nuclei are in the same order of magnitude. As a result, α- and β-crystals grow competitively in the initial crystallizations. In the growth stage, the growth rate of β-crystals is faster than that of α-crystals where the epitaxial growth of β-crystals on α-crystals also occurs. Consequently, the content of β-crystals becomes dominant in the limited growth space; however, it is still less than that formed in the quiescent isothermal crystallization of β-nucleated iPP. Utilizing the mechanism of the depression of β-crystals under the coexistence of shear flow and β-nucleating agent, we adopted a modified injection molding machine called oscillation shear injection molding (OSIM) to fabricate β-nucleated iPP sample. This kind of samples has a large amount of shish-kebabs in the surface and intermediate layer while numerous β-crystals form in the core layer, thus forming a 'stiff surface and soft core' structure. This bionic bamboo-like structure leads to prominent reinforcement and toughening for iPP samples.

Keywords: Polypropylene; Flow-induced crystallization; Reinforcement; Toughening.

Introduction

Isotactic polypropylene (iPP) is a polymorphic material, having at least four modifications: α, β, γ and smectic.[1] The α-crystals with inferior toughness are usually present in industrial products prepared by conventional processing techniques. Toughening of iPP has thus become an important research subject.[2] Compared with α-crystals, β-crystals exhibit improved elongation at break and excellent impact strength due to the β-α polymorphous transition and the loose structure. Therefore, β-crystals show a great potential to toughen iPP. Addition of β-nucleating agent is efficient to induce a great amount of β-crystals while shear flow can also induce the formation of β-crystals. During processing, it is expected that the existent shear flow can give rise to more β-crystals when β-nucleating agent is present in iPP. However, it is first reported by Varga et al. that the high shear rate in fact restrained the formation of β-crystals.[3] Huo et al. performed a systematic study on crystallization of β-nucleated iPP, and observed that as the shear rate increased, the content of β crystals consistently decreased, showing a counteraction effect between the shear flow and β-nucleating agent for the formation of β-crystals.[4] The underlying mechanism of this counteraction is not clear until now.

Moreover, this counteraction does not necessarily mean any setback to simultaneously enhance strength and toughness of iPP. We attempt to fabricate iPP with considerably increased strength and toughness, for the first time, utilizing the counteraction of shear induced orientation crystallization and β-nucleating agent on β-form nucleation. The crystallization process of iPP is manipulated by two stages, one is flow-induced crystallization for high molecular orientation, and the other is β-nucleating agent induced nucleation for high β-crystal content. More specifically, the oscillation shear supplied by an oscillation shear injection molding (OSIM) (its detailed definition is included in the experimental section) is imposed on β nucleated iPP melt in the mold during injection molding. Initially, due to occurrence of shear flow and β nucleating agent, β-form formation is restrained; thereby, the oriented α-crystals first form in the skin and intermediate layers of the sample. When the gate of the mold freezes, the shear ceases, and hereby, β-nucleating agent revives. The remainder of iPP melt in the core layer undergoes β nucleation induced crystallization. The results demonstrate that a large amount of shish-kebab structure appears in the surface and intermediate layers of the sample, at the same time, numerous β-form crystals form in the core layer, thus leading to prominent reinforcement and toughening for iPP.

Experimental

Materials. iPP, model T30S, was purchased from Dushanzi Petroleum Chemical Co., China, with a melt flow rate (MFR) of 3 g/10min (230 °C, 21.6 N), Mw = 39.9×10^4 g mol^{-1}, and Mw/Mn =4.6. The β-nucleating agent is aryl amide compounds (TMB-5), and was supplied by Fine Chemical Institute, Shanxi, China.

Preparation of nucleated iPP pellets. TMB-5 was first melt mixed with iPP in a twin-screw extruder to form a 1.0 wt % β-nucleating agent masterbatch. The masterbatch was further melt compounded with fresh iPP to produce iPP (pellets) with 0.2 wt % of TMB-5. The screw speed was fixed at 82 rpm and the processing temperature profile was 170-180 °C from hopper to die. The pure iPP pellets without nucleating agents were also prepared under the same processing conditions for comparison purpose.

Preparation of injection molded samples. The iPP pellets were injection molded into dumb-bell bars in an injection temperature profile of 170-200 °C from hopper to nozzle. At packing stage of an injection molding cycle, the oscillation shear provided by the OSIM machine whose picture is shown in Fig. 1 is continuously imposed on the melt. The primary feature of OSIM is that the hot melt is subjected to high pulse shear stress in the mold, which is given by two pistons moved reversibly at the same frequency. Generally, the OSIM can provide a peak shear rate in a single cycle from several s-1 up to hundreds of s-1. In this work, the shear rate is about 220 s-1. Only when the gate of the mold solidifies, the shear ceases. The conventional injection molding (CIM) was also carried out under the same processing conditions (only without oscillation shear) for comparison.

Figure 1. 3D schematic illustration for the mold with an oscillation shear supplier.

X-ray measurement. In order to study the crystal growth, WAXD and SAXS measurements were carried out at the Advanced Polymers Beamline (X27C, λ= 0.1371 nm) in the National Synchrotron Light Source (NSLS), Brookhaven National Laboratory (BNL). A MAR CCD X-ray detector (MARUSA) was employed for detection of 2D-WXAD and 2D-SAXS images, having a resolution of 1024 × 1024 pixels (pixel size = 158.44 μm). The data acquisition time was 20 s for each scattering pattern (image). The sample-to-detector distance was 112.6 mm and 2330 mm for WAXD (calibrated by an aluminum oxide (Al2O3) standard) and SAXS (calibrated by a silver behenate (AgBe) standard), respectively. The scattered intensities were in the range of scattering angles from 0 to 30°.

Linear WAXD and SAXS profiles were obtained from circular integration of intensities from 2D-WAXD and 2D-SAXS imagess, respectively. The intensity was plotted as a function of the scattering vector, q, where $|q|=4\pi\sin\theta/\lambda$ with λ being the wavelength of the incident beam, and 2θ being the scattering angle. Subsequently, through deconvoluting the peaks in linear WAXD profiles, the overall crystallinity Xc was calculated by

$$X_c = \frac{\sum A_{cryst}}{\sum A_{cryst} + \sum A_{amorp}} \quad (1)$$

where A_{cryst} and A_{amorp} are the fitted areas of crystal and amorphous peaks, respectively. The relative amount of the β-crystals K_β was evaluated by the method of Turner Jones et al. and modified by Hsiao et al.[5,6]

$$K_\beta = \frac{A_\beta(300)}{A_\beta(300) + A_\alpha(110) + A_\alpha(040) + A_\alpha(130)} \quad (2)$$

$A_\beta(300)$ represents the area of the (300) reflection peak; $A_\alpha(110)$, $A_\alpha(040)$, and $A_\alpha(130)$ representthe areas of the (110), (040), (130) reflection peaks, respectively. Meanwhile, the crystallinity of β-crystals (X_β) and the crystallinity of α-crystals (X_α) were given by

$$X_\beta = K_\beta X_c \quad (3)$$

The injection molded samples was also ex-situ characterized by X-ray scattering layer by layer, i.e. from the edge to the center of the samples, in order to show their hierarchic structures.

Experimental Procedures. A Linkam CSS-450 high-temperature shearing stage modified for in situ X-ray scattering studies. The experimental temperature profiles were set as follows: (1) heating at a rate of 30 °C/ min from room temperature to 210 °C; (2) holding the temperature at 210 °C for 300 s to eliminate residual structure; (3) cooling at a rate of 30 °C/ min down to 134 °C (4) holding the temperature at 134 °C for at least 900 s for X-ray measurements. We adopted two shear rates: $15s^{-1}$ and $30s^{-1}$. Duration is 5s.

Results and Discussion

The depression of β-crystal content

A representative series of 2D-WAXD images of isothermal crystallization of iPP0 and iPP02 samples at 134 °C with and without shear flow (0, 15, 30 s-1) are shown in Fig. 2. Fig. 2a1-4 show that iPP0 crystallized under quiescent condition. The characteristic lattice planes of α-crystals, (110), (040), (130), (111), and (-131), from inner to outer, developed homogeneously (Fig. 2a1-3), until the completion of crystallization (Fig. 2a4), without the appearance of β-crystals. When a shear flow was applied to iPP0, despite different shear rates (15, 30 s-1), the samples shared the same growth process (Fig. 2b and c). The initial images (Fig. 2b1 and c1) showed diffuse scattering. Although the melt might be oriented, it was not detected due to the limitation of the technique. After 40s, sharp α-crystal lattice planes (040) and (110) were clearly observed, which exhibited the arc-like diffractions of oriented α-crystals in Fig. 2b2 and c2. The diffraction intensity of both α- and β-crystals became stronger with time due to the growth of the crystallites. After a certain time, the diffraction intensity of all crystal reflections remained constant, indicative of completion of crystallization (Fig. 2b4 and c4). For the iPP02 sample, even under quiescent isothermal condition (Fig. 2d), the (300) and (131) lattice planes of β-crystals still grew fast and the diffraction intensity was visible enough at t = 20 s (Fig. 2d2) due to the strong heterogeneous nucleation of β-nucleating agent. The growth of β-crystals was always overwhelming until the crystallization was finished (Fig. 2d3, 4). The ultimate crystallites were mainly in the form of unoriented β-crystals. Upon imposition of shear, an interesting phenomenon was observed in which the initial crystals were unoriented β-crystals (Fig. 2f1) instead of α-crystals, subsequently the intensity of (300) lattice plane got stronger and the oriented (040) lattice plane of α-crystals appeared (Fig. 2e2 and f2). The diffraction intensity of both α- and β-crystals became stronger with time due to the growth of the crystallites (Fig. 2e3 and f3).

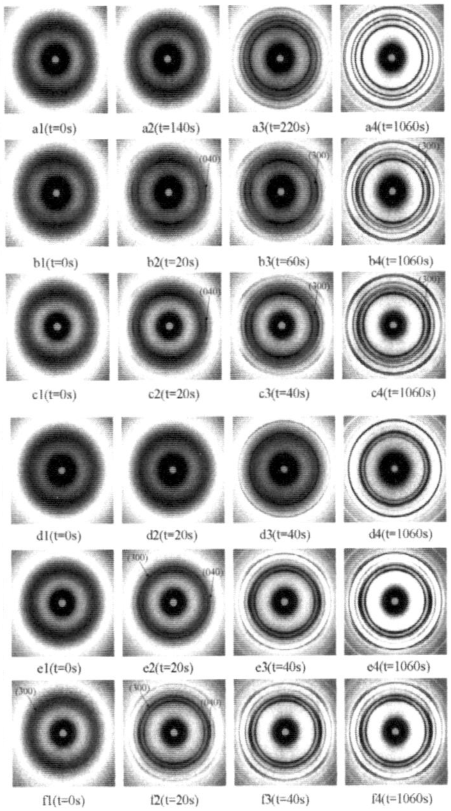

Figure 2. Selected 2D-WXAD patterns showing the crystal growth process of iPP isothermally crystallized at 134 °C. (a) iPP0 without shear, (b) iPP0 at a shear rate of 15 s^{-1}, (c) iPP0 at a shear rate of 30 s^{-1}, (d) iPP02 without shear, (e) iPP02 at a shear rate of 15 s^{-1}, and (f) iPP02 at a shear rate of 30 s^{-1}.

To determine the evolution of β-crystals, β-phase crystallinity (X_β)as a function of crystallization time are present in Fig. 3, calculated by Eq.1-3. The quiescently, isothermally crystallized iPP0 showed no β-crystals as observed from Fig. 3a. When application of shear, the β-crystals appeared, and their content increased with the crystallization time (Fig. 3b), until it reached a plateau at about t = 140 s, with X_β= 0.082. As the shear rate increased to 30 s^{-1} (Fig. 3c), the content of β-crystals increased up to X_β= 0.096, meanwhile the saturation time was shortened to around 120 s. It showed the content of β-crystals of iPP0 increased as the shear rate. In the case of iPP02 (Fig. 3d), β-nucleating agent greatly speeded up the growth of β-crystals, as a result, the content of β-crystals already achieved the saturation at about 100 s. The final crystallites were mainly β-crystals with X_β= 0.48 Unexpectedly, when shear (shear rate = 15 s^{-1}) was applied (Fig. 3e), the content of β-crystals decreased to 0.40 (X_β). As the shear rate further increased to 30 s^{-1} (Fig. 3f), greater reduction of the β-crystal content was observed (X_β= 0.30), and the saturation time was shortened to 60 s. It is indicative of the depression of β-crystals in β-nucleated iPP sample under shear flow field.

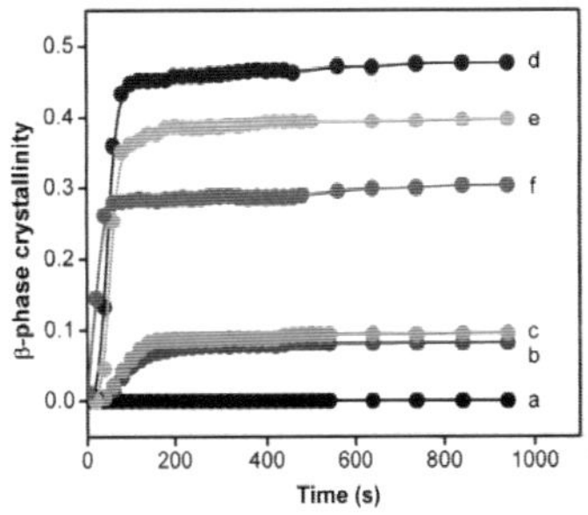

Figure 3. β-phase crystallinity in the iPP melt as a function of crystallization time, obtained from their 1D-WAXD obtained from Fig.2.

The competitive growth of α and β-crystals

From the traditional point of view, crystallization is a two-step dynamic process: 1) formation of stable nuclei (named nucleation step) followed by 2) crystal growth characterized by the crystal growth rate (G).[5] In our case, the sources of α- and β-crystals are briefly described in Fig. 4. Based on the spherulite growth model of α- and β-crystals, The nucleation density N can be extracted from the progress of crystallinity, adopting Kolmogoroff-Avrami-Evans equation: [6]

$$N = -\frac{3}{4\pi G^3}\frac{\ln(1-X)}{t^3} \quad (4)$$

In which X = 0.1 and t = $\tau_{1/10}$ in order to reduce the interference between growing crystallites and the effect of $N_{hom}(T_c,t)$.

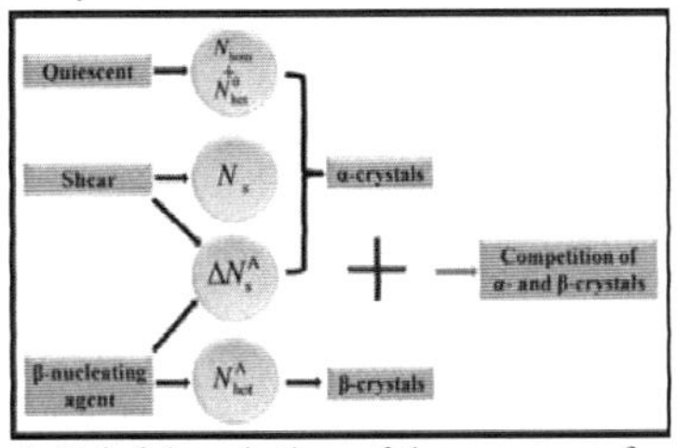

Figure 4. Brief description of the sources of α- and β-crystals and the competition of α- and β-crystals.

As we can see from Tab.1, when the shear flow was applied to iPP02, shear-induced nuclei and ΔN_s^A (9.5 × 10^{16} nuclei/m^3 i.e. the last column in Table 2 minus

N^{A}_{het}) for α-crystal growth and β-nucleator-induced nuclei for β-crystal growth (assumed to be 5.5 × 10^{16} nuclei/m^3) coexisted in iPP melt and were at the same order of magnitude, thus competitive growth of α- and β-crystals happened.

Table 1. Experimentally determined number of nuclei per cubic meter at X= 0.1.

	$\tau_{1/10}$(s)	$N_{hom}(T_c,t)$ + $N^{0}_{het}(T_c)$	N_s	N^{A}_{het}	N^{A}_{het} + N_s + ΔN^{A}_{s}
a	150	1.2 × 10^{14}			
b	21.6		4.0 × 10^{16}		
c	19.1		5.7 × 10^{16}		
d	17.4			5.5 × 10^{16}	
e	13.2				15 × 10^{16}
f	10.5				29 × 10^{16}

Skin-core structure

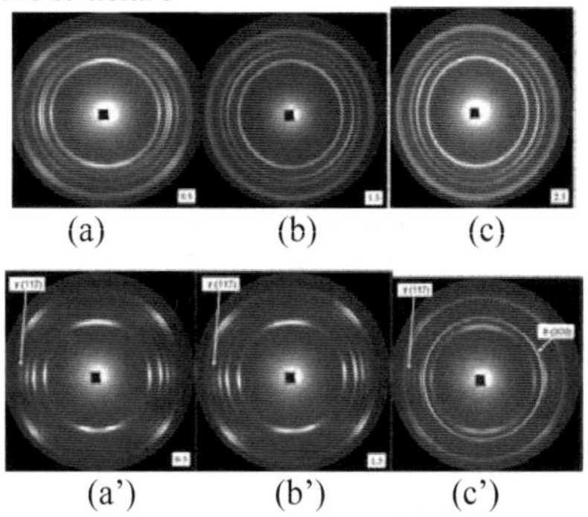

Figure 5. 2D-WAXD patterns of skin layer (a, a'), intermediate layer (b, b') and core layer (c, c') of conventional injection molding (CIM) iPP (a, b, c) and OSIM β-iPP (a', b', c'). The number of each picture represents the distance to the edge of samples (mm).

Compared with CIM iPP samples, there is obvious azimuthal diffraction intensity in every layer of OSIM β-iPP sample (Fig. 5a'-c'), which indicates the strong orientation of molecular chains throughout the sample. The orientation should undoubtedly be ascribed to the oscillation shear on the melt before the gate of the mold freezes. Strong reflections of β crystals emerge in the core layer. This phenomenon inidcates that combination of shear field and β-nucleating agent suppresses the formation of β crystals in the skin and intermediate layers. In the core layer, the molecular chains relaxed so that the β-nucleating agent takes effect, leading to β-phase.

Mechanical properties

Compared with CIM iPP samples, OSIM β-iPP samples exhibit superior mechanical properties as shown in Figure 2. The tensile strength of OSIM β-iPP samples reaches 55.6 MPa, which is 19.3 MPa higher than that of CIM iPP samples (36.3 MPa). The impact strength of OSIM β-iPP samples (29.2 kJ/m^2) is over seven times of CIM iPP samples (4.1 kJ/m^2), showing super tough and brittle transition between these two samples. The tensile modulus rises from 1.42 GPa to 2.10 GPa. Only the elongation at break decreases from 87.3 % to 51.6 %, but it still fractures in a ductile manner. These data indicate that the mechanical properties of OSIM β-iPP fall into the range of engineering plastics. Hence, conclusions can be safely obtained that, by means of the counteraction of shear induced molecular orientation crystallization and β-nucleating agent induced β-form crystals, the simultaneous reinforcement and toughening of iPP have been successfully achieved. The schematic diagram for skin-core bionic bamboolike structure is diplayed in Fig. 7.

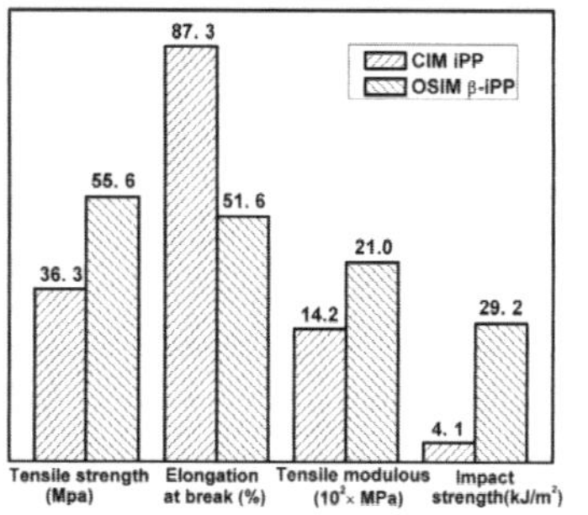

Figure 6. the mechanical properties including tensile strength, elongation at break, tensile modulus, and impact strength for CIM iPP and OSIM β-iPP.

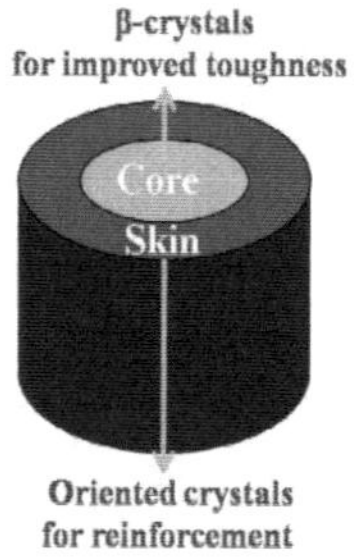

Figure 7. Schematic diagram for skin-core structure.

Conclusions

Utilizing the counteraction mechanism of β-nucleating agent and shear flow inducing the formation of β-crystals, we fabricated simultaneous reinforcement and toughening iPP samples, through adopting an modified injection molding machine.

Acknowledgements

This work is financially supported by State Key Program of National Natural Science of China (Grant No. 51033004) and Outstanding Youth Foundation of Natural Science Foundation of China (Grant No. 50925311).

References

1. S. Bruckner; S.V. Meille; V. Petraccone; B. Pirozzi. *Prog. Polym. Sci.* 1991, *16*, 361.
2. C. Grein. *Adv. Polym. Sci.* 2005, *188*, 43.
3. J. Varga. *J. Therm. Anal.* 1989, *35*, 1891.
4. H. Huo; S.J. Jiang; L.J. An; J.C. Feng. *Macromolecules* 2004, *37*, 2478.
5. J.I. Lauritzen; J.D. Hoffman, *J. Res. Natl. Bur. Stand.* 1960, *64A*, 73.

6. A.N. Kolmogoroff. *Isv. Akad. Nauk SSSR, Ser. Math.* 1937, *1*, 355.

O-09-550

PRODUCTION OF KETOPROFEN-Β CYCLODEXTRIN MICROPARTICLE USING PARTICLES FROM GAS SATURATED SOLUTION (PGSS) TECHNIQUE

Sumarno*, P.R Pradnyandari, R. Tetrisyanda, P.N Trisanti, and F. Kurniawansyah.
Chemical Engineering Department, Institut Teknologi Sepuluh Nopember
(ITS) Sukolilo, Surabaya, Indonesia –onramus@chem-eng.its.ac.id

Abstract –Particle formation of drug composite processed under supercritical fluids carbon dioxide applications was studied. The technique is has been recognized as safe and suitable for mostly application in pharmaceutical and food chemical industries. In this study, particles from gas saturated solution (PGSS) micronization technique were used to produce ketoprofen-β Cyclodextrin micro-particles. Both materials were blended using supercritical carbon dioxide at 50 ^{O}C. The effect of drug/polymer compositions, and saturation pressures (80-200 bars) on particle morphology and drug release were investigated. Product was characterized by particle size analyzer (PSA) and Scanning Electron Microscopy (SEM) to determine size and morphology of micro-composites. Drug release was determined by using Spectrophotometric UV-Vis. The results showed that morphology of the micro-particle drug was irregular shape, with a mean particle size between 1,09 and 6,855 μm. The highest dissolution rate was achieved by sample processed under 200 bars and 1:3 % (w/w).

Keywords: Micro-particles, composite, PGSS, Ketoprofen, β-Cyclodextrin

Introduction

In the pharmaceutical industry, drug processing technology using non-organic solvents is needed. It is triggered by regualtion and safety demand to apply the methode which is better and more benign along the process also the resulting product. One of the technologies is supercritical fluids for drug particles formation. This fluid can be used as solvent, anti-solvent and solute. It is possible to produce micron or nano size particles by using supercritical fluids, especially with supercritical carbon dioxide. Supercritical CO_2 has been proved already as drug processing media in pharmaceutical because of it is inexpensive and inert in nature.

Particle from gas saturated solution (PGSS) method was used in this work to produced drug particle of ketoprofen-β_Cyclodextrin composite system. The development of this technique based on two reasons; first is free organic solvent in processing. The second is based on the fact that carbon dioxide was provenly able and powerful to promote the inclusion process of organic particle inside β_Cyclodextrin .

Production of particle use PGSS method has been published widely. Shekunov (2006) used PGSS method for making composite particles to improve controlled release of acetaminophen, ketoprofen, and trypsinogen. PEG, Eudragit RS, and tripalmitin as excipients. The results showed that the particle size is smaller when it produced with the smaller of nozzle size diameter.

Additionally, Bounaceur-Rodier et al. conducted research for ketoprofen and β-cyclodextrin system on the pressure 150-200 bar, temperature 65-75 ° C and the water is added at the end of the process. This study concluded that an increasing temperature will increase the rate of ketoprofen-cyclodextrin composite formation.

The aim of this study is to investigate the potential of PGSS in the production of ketoprofen-β-cyclodextrin micro particles.

Experimental

Materials

Ketoprofen having molecular weight 254.28 gr/mol and melting point 92-96 °C, were donated from P.T. Kimia Farma Indonesia. β-Cyclodextrin was (Mw =10000-18000 Da) was got from Roquette (France). CO_2 and N_2 with purities 99% were purchased from PT. Tri Gases (Indonesia).

Micronization with PGSS technique

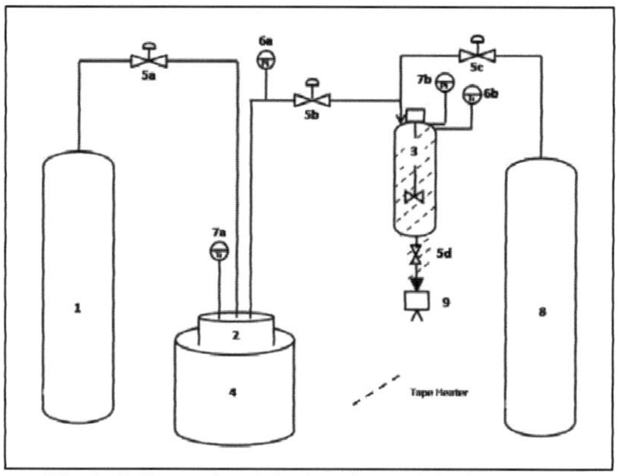

Fig.1. Schematic of the batch Particle from Gas Saturated Solution (PGSS) process; (1) CO_2 tank,

(2) Gas supply, (3) Saturation vessel, (4) Water bath, (5) Valve, (6) Pressure Indicator, (7) Temperature Indicator, (8) N_2 storage, (9) Nozzle

Experiments were started by condensing CO_2 gas in a 1 liter of supply tank. After one hour, the supply tank was heated with hot water. The increasing of vessel temperature lead to an increasing of CO_2 vapor pressure and can be delivered or injected into the saturation vessel (volume of 0.5 liter), where sample of ketoprofen:β-cyclodextrin in 1:3 and 1:5 (w/w) ratio was placed. At the temperature of 50°C and pressures of 80 – 200 bars, sample was in contact for mixing purpose with (supercritical) carbon dioxide for 2 hour time in saturation vessel. The vessel was equipped with 3 glass ball to promote intensive mixing and perfect solubilization of carbon dioxide. After the mixing time was achieved, mixture of drug-polymers was expanded from high pressure in saturation vessel into atmosphere. To ensure no drastic pressure decrease along decompression, the nitrogen gas was by injecting nitrogen gas. Nitrogen may facilitate rapid drying of droplets during formation. Samples were captured at the outlet of the 0.5 mm ID nozzle by plastic bag, and then being stored in a dry box before characterization process.

Particle size and particle size distribution

The particle size and size distribution of the micro particles were measured by Zetasizer ZS90 (Malvern, United Kingdom).

Scanning electron microscopy (SEM)

Sample of drug-polymer composites were observed by *Scanning Electron Microscopy* (EVO MA 10, Carl Zeiss, Jerman). The SEM Samples were fixed onto a metallic stub with double-sided conductive tape and covered with gold using a sputter coater (Emitech, Dubai). This analysis was conducted to determine the morphology of the resulting micro particles and the dispersion pattern of particles produced.

Dissolution studies

Dissolution studies were carried out according to the paddle method using phosphate buffer solution at pH 7.4. 600 ml of the buffer solution and the temperature was adjusted at 37±1°C. The stirring speed was controlled at 100 rpm. At appropriate time intervals (5, 10, 15, 30, 45, 60, 90, and 120 minute), 10 ml of each sample was taken and replaced by fresh dissolution media. The cumulative release rate of ketoprofen composites was determined with UV–visible spectrophotometer (Genesys 10uV scanning, Thermo Scientific UV Spectrophotometer) at 262 nm.

Results and Discussion

Ketoprofen-*β*-cyclodextrin micro particles have been produced using PGSS technique. The experiments were carried out in a saturation pressure range from 80-200 bar and at temperature 50°C.

From figure 2, show the morphology of ketoprofen and β-cyclodextrin before PGSS process. It can be seen that the morphology of both processed material were irregular respectively.

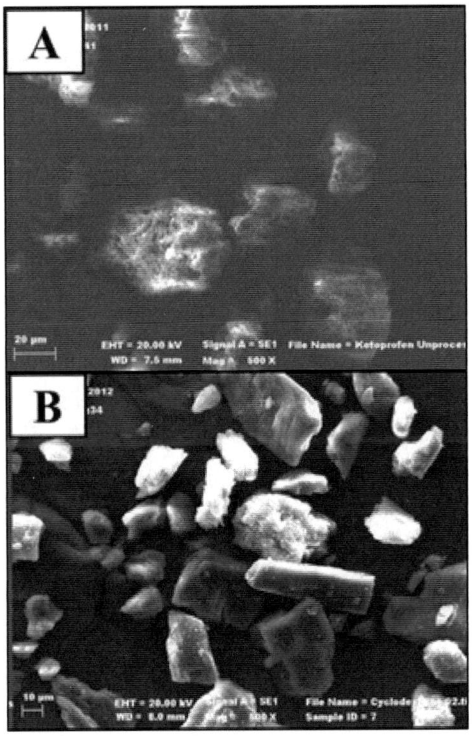

Fig. 2. SEM pictures of (A) ketoprofen, and (B) β-cyclodextrin unprocess.

The effect of saturation pressure on particle morphology was observed on ketoprofen-β-cyclodextrin micro particles at saturation pressures ranging between 80 and 200 bar, at 50°C. The results were as shown in SEM images in Fig. 3. All experiments were performed at high pressure and under mixing with 3 glass ball to ensure inclusion process homogeneously occurred and can be expanded to be smaller micro particles. Fig. 3 show that size of particles after PGSS process was changed. But the morphology still irregular. It can be seen that for composites with composition of 1:3 (w/w) having a size which is larger than composition of 1:5 (w/w). This is because β-cyclodextrin have great interaction with supercritical carbon dioxide than ketoprofen. Since the amount of β-cyclodextrin is more, the inclusion probability is higher, and affect to the smaller particle produced along the expansion process. Furthermore, when the saturation pressure increases the particle size obtained decrease. This is due to increasing of saturation pressure, the solubility of carbon dioxide in the mixture will raise as well. The amount of CO_2 molecules in the fluidization of CO_2 and mixture of ketoprofen-β-cyclodextrin along decompression increase also. Hydrodynamic effect resulted in nucleation of particles become smaller.

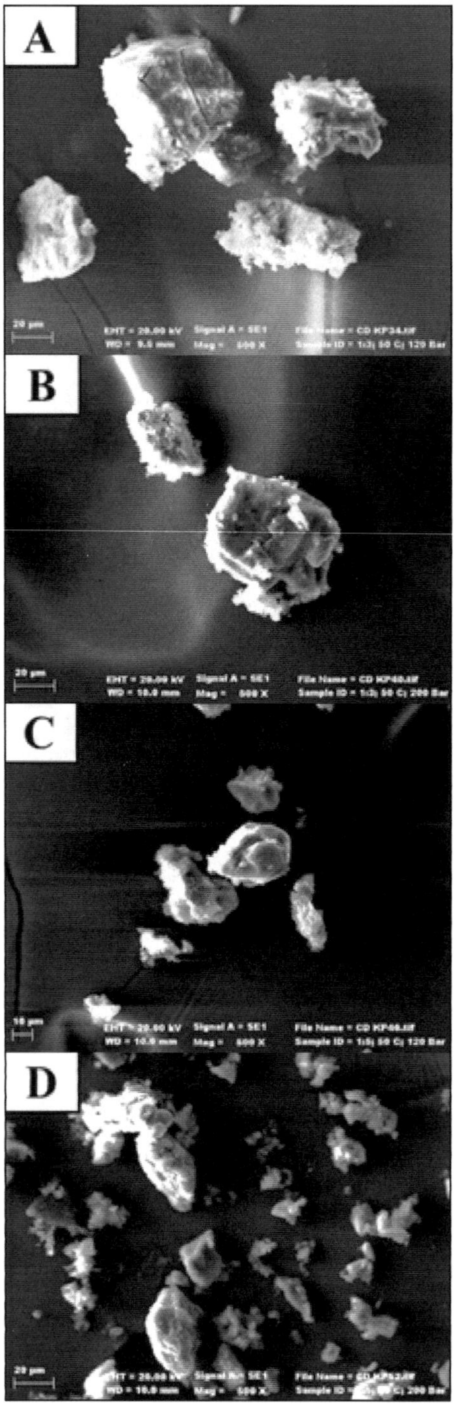

Fig. 3. SEM pictures of ketoprofen:β-cyclodextrin at 50°C (A) 120 bar (1:3 (w/w)), (B) 200 bar (1:3(w/w)), (C) 120 bar (1:5(w/w)), (D) 200 bar (1:5(w/w)).

Tabel 1.Average particle size diameter ketoprofen:β-cyclodextrin micro composite in various composition and various saturation pressures.

Composition	Pressure (bar)			
	80	120	160	200
1:3	6,855	4,808	2,002	1,572
1:5	1,853	1,922	1,09	1,357

*in micrometer(μm)

Average particle size is shown in Table 1. Average diameter for composition 1:3 (w/w) is 1.572-6.855 μm and 1:8 (w/w) are 1.09-1.922 μm. It can be seen that particle size decrease with an increase in saturation pressure.

Fig.4 and 5 show the particle size distribution of ketoprofen:β-cyclodextrin. For composition of 1:3 (w/w) (Fig.4), shows that particle size distribution is bimodal and narrow. But the particle size isn't uniform. In the lower pressure, the size of particle was the largest. For composition 1:5 (w/w) (Fig. 5) produce bimodal and uniform particle size. But in the higher pressure has widely particle size distribution.

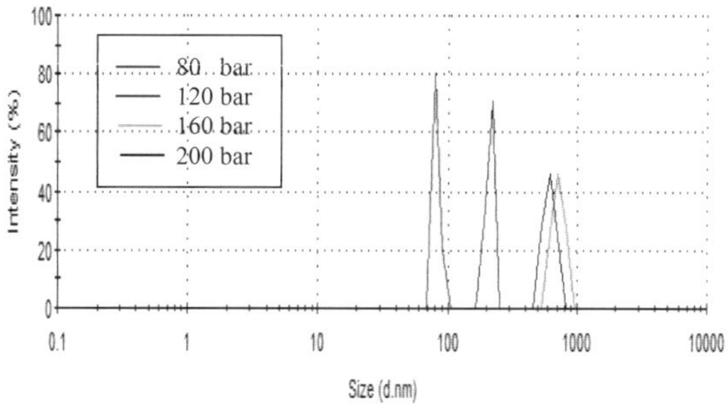

Fig. 4. Particle size distribution of ketoprofen:β-cyclodextrin (1:3) at 50°C and various pressure.

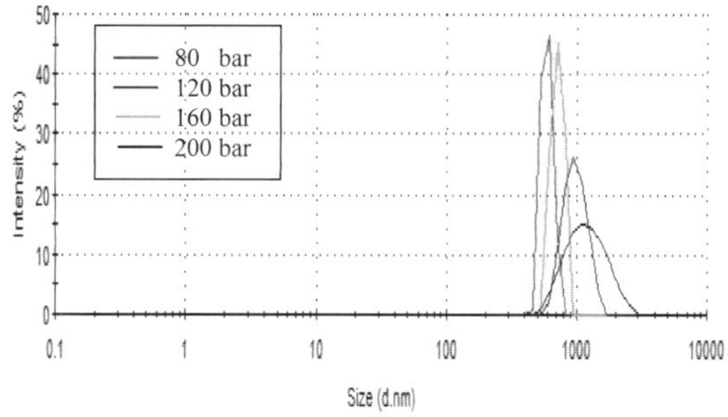

Fig. 5. Particle size distribution of ketoprofen:β-cyclodextrin (1:5) at 50°C and various pressure

Fig 6 and fig. 7 presents the result of the dissolution rate of unprocessed ketoprofen and the micro composite of ketoprofen-β-cyclodextrin produced by particle from gas saturated solution (PGSS) technique.

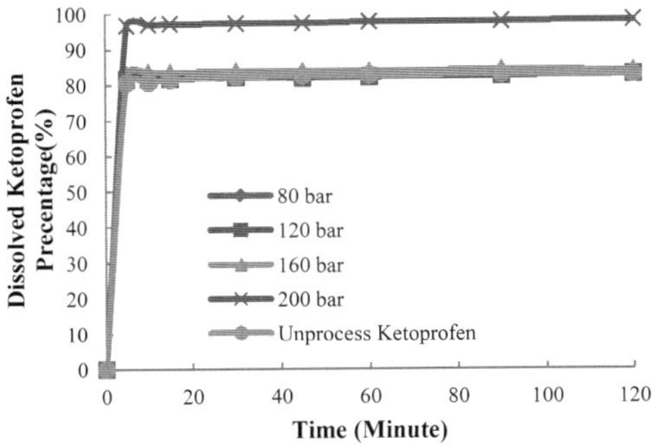

Fig. 6. Dissolution rate of dissolved ketoprofen in ketoprofen-β-cyclodextrin (1:3 (w/w)) composite.

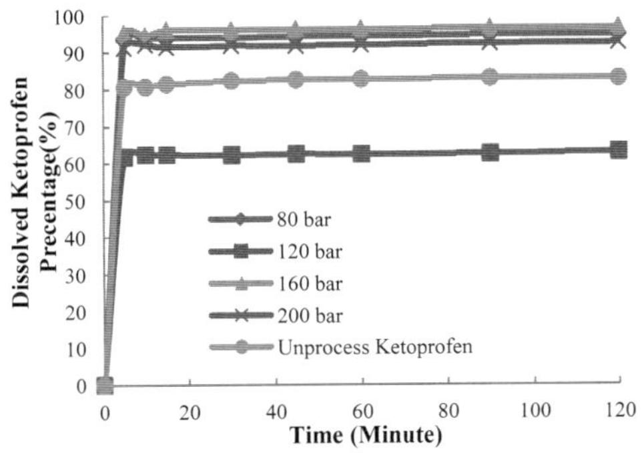

Fig. 7. Dissolution rate of dissolved ketoprofen in ketoprofen-β-cyclodextrin (1:5 (w/w)) composite.

From Figure 6, for composition 1:3 (w/w) seen that most composites have the same dissolution rate with unprocessed ketoprofen. However, the highest saturation pressure conditions have a higher rate of solubility than unprocess ketoprofen. This doesn't occur in the composite with the composition 1:5 (w/w) (Fig. 7). Some composites have a higher dissolution rate than unprocessed ketoprofen. However this is not the case in the composite saturation pressure of 120 bar. Dissolution rate is influenced by several factors, such size of particles product, composition of polymer and interaction between the drugs and β-cyclodextrin.

Conclusions

PGSS technique, a supercritical process, was successfully used to produce ketoprofen:β-cyclodextrin microparticles. Experimental parameter such as saturation pressure and ratio of component was studied. The results showed that morphology of the micro-particle drug was irregular shape, with a mean particle size between 1.09 and 6.855 μm. The highest dissolution rate was achieved by sample processed under 200 bars and 1:3 % (w/w).

References

1. Bounaceur, A., Rodier, E., Fages, J., "Maturation of Ketoprofen/β-Cyclodextrin Mixture With Supercritical Carbon Dioxide", *J. of Supercritical Fluids*, **41**, 429-439 (2007).
2. Fages, J., H .Lochard, J.J. Letourneau, M. Sauceau, E. Rodier, "Particle Generation for Pharmaceutical Application Using Supercritical Fluid Technology", *Powder Technology*, **141**, *Issue 33*: 219 (2004).
3. Hirayama, F. dan Uekama, K., "Cyclodextrin-based Controlled Drug Release System", Advanced Drug Delivery Reviews, **36**, 125-141 (1999).
4. Jung, J. dan M. Perrut, "Particle Design Using Supercritical Fluids: Literature and Patent Survey", *J. of Supercritical Fluids*, **20**, 179-219 (2001).
5. Kerc, J., S. Srcic, Z. Knez., P. Sencar Bozic., "Micronization of Drugs Using Supercritical Carbon Dioxide", *International Journal of Pharmaceutics*, **182**, (1):33-39 (1999).
6. Sauceau, M., Rodier, E. And Fages, J. "Preparation of Inclusion Complex of Piroxicam withCyclodextrin by Using Supercritical Carbon Dioxide", Journal of Supercritical Fluids, 47, 2, 326-332 (2008).

O-09-555

SELECTIVE LOCALIZATION AND MIGRATION OF NANOCLAY IN HNBR/ENR BLENDS: AN ONLINE EVALUATION

Zulfiqar Ali[1*], Hai Hong Le[2], , Sybill Ilisch[2], Hans-Joachim Radusch[2]
[1]*Department of Chemical Engineering COMSATS, Institute of Information Technology, 54000 Lahore, Pakistan*
[2]*Martin Luther University Halle-Wittenberg, D-06099 Halle (Saale), Germany*

Abstract
During the preparation of nanoclay filled blends based on hydrogenated acrylonitrile butadiene rubber (HNBR)/epoxidized natural rubber (ENR) electrical conductance (OMEC) of the blends was recorded online during mixing by means of a sensor system installed inside the mixing chamber. A characteristic chart of the online measured conductance has been observed. Taking into consideration the results of different structural investigations like atomic force microscopy (AFM), transmission electron microscopy (TEM), and small angle X-ray scattering (SAXS) the physical background of the online measured electrical conductance of organoclay filled rubber compounds and blends is discussed. The observed online conductance is ionic in nature that has arisen due to the release of surfactants from the clay galleries. It was found that the online electrical conductance depends on the phase specific localization of nanoclay and development of blend morphology. The favorable localization of nanoclay in the HNBR phase can be monitored by the chart of OMEC and visualized by microscopic methods. It is corresponding well with the filler localization predicted by means of surface tension based model proposed and applied in our previous works. Thus, the method of the online conductance promises to be a powerful tool to study the nanoclay dispersion and localization in rubber nanocomposites during the melt mixing process.

Keywords: *Online conductance, blend nanocomposites, dispersion kinetics*

1. Introduction

Exorbitant market demands for new polymeric materials that can be met either by synthesizing new materials or blending and filling the existing ones. The latter is easy to process and finds commercial applications. Recently, the incorporation of nanoclay into rubber blends is finding increasing interest in industrial and scientific research [1,2]. In literature a large number of works reported about improvement of mechanical and tribological properties [3,4] as well as thermal and chemical stability [56] can be found. For nanocomposites based on rubber blends, it has been extensively reported that nanoclay preferentially resides in the blend phase that has better affinity to clay or if clay shows the same affinity to both blend phases it concentrates dominantly at the interphase [7,8]. Organoclay at the interface causes an effective size reduction of the dispersed phase and narrowed size distribution [9,] as well as in improvement of the compatibility in the blend by reducing the surface energy gap between the blend components [,10]. Generally, the dispersion and distribution of organoclay in polymer blends as well as the development of blend morphology is characterized by post processing techniques e.g. TEM, AFM and SEM etc. Owing to increasing demands with respect to productivity, quality and commercialization of the products, it is highly required to have an online technique that provides real time information about development of morphological changes at different stages of mixing process. In our previous works [11,] we introduced a method of online measured electrical conductance (OMEC) for monitoring the mixing process of carbon black in nonpolar/nonpolar EPDM/SBR blends and nanoclay in polar/nonpolar HNBR/NR blends. In the present work the phase specific nanoclay localization as well as the development of blend morphology in polar/polar HNBR/ENR blends will be characterized by the OMEC and other methods.

2. Experimental

2.1. Materials

* Corresponding author. Tel: +92 (42) 111 001 007 (886). Fax: +92 42 353 210 48.
E-mail address: zulfiqar,ali@ciitlahore.edu.pk

Hydrogenated acrylonitrile butadiene rubber Zetpol 2030L (Zeon Europe GmbH, Germany) with acrylonitrile content of 36 wt% and epoxidized natural rubber (Weber & Schaer GmbH & Co. KG, Germany), with epoxidation grade of 50 wt% were used as rubber matrix. The epoxy groups are statistically distributed along the main chain of ENR [12]. Organoclay Nanofil 9 (Süd-Chemie Germany) modified by stearyl benzyl dimethyl ammonium chloride as surfactant was used as filler. The data of the clay provider as well as our investigations show that the organoclay has a basal spacing of 2.0 nm before compounding. Peroxide Luperox 101 (Atofina Chemicals) was used as curing agent for investigated rubber compounds.

2.2. Sample preparation

To characterize the clay dispersion in single rubber compounds, pure HNBR and ENR were mixed with 1.8 wt% peroxide in an internal mixer PolyLab System Rheocord (Thermo Electron/Haake) at an initial chamber temperature of 50 °C and a rotor speed of 70 rpm. Clay was added to the rubber after 7 min. A concentration of 4.7 wt% clay was kept constant for all nanocomposites.

For preparation of HNBR/ENR/clay blends a ENR-clay masterbatch containing 9.1 wt% clay was mixed with pure HNB. In all these blends the ratio of HNBR/ENR is kept at 50/50 having a total clay concentration of 4.7 wt%. In order to investigate the kinetics of organoclay dispersion and the blend morphology, the samples were taken out at different mixing times. The mixing time denotes the time after mixing of the ENR-clay masterbatch and unfilled HNBR. The mixtures were compression-molded and vulcanized at 170 °C.

2.3. Testing methods

2.3.1 Online measurement of electrical conductance (OMEC)

During the mixing processes, the electrical conductance signal of the mixture has been measured.

2.3.2 Atomic force microscopy

Morphological investigations were carried out by an atomic force microscope Q-Scope 250 (Quesant), operated in intermittent mode with a scan-head of 40 µm. Samples were produced by cutting in a cryo-chamber CN 30 of a rotary microtom HM 360 (Microm) with a diamond knife at –100 °C.

2.3.3 Transmission electron microscopy

The bulk morphology of the sample was analyzed using a transmission electron microscope JEM 2010 (Jeol). Thin sections of 100 nm were prepared at -100 °C using an ultracut microtome (Leica), equipped with a 45 ° diamond knife.

2.3.4 X-ray diffraction

Small-angle X-ray scattering measurements were performed at room temperature using a rotating anode X-ray source RU-3HR (Rigaku) equipped with a nickel-filtered CuKα tube (λ = 0.154 nm) for detection of the state of exfoliation. The generator voltage was 40 kV and generator current was 60 mA. The scattering vector q is defined by *q=4□/□sin □*The first order Bragg peak in the Lorentz corrected intensity curves were fitted by using a Gaussian function with linear background subtraction, giving the interlayer distance and the relative peak strength. All samples had a uniform thickness of 1.0 mm, i.e. the obtained peak area corresponds to the amount of ordered structures.

2.3.5 Tensile properties

The tensile modulus was determined by stress-strain measurements according to ISO 37 using a tensile tester Z005 (Zwick/Roell) with a cross-head speed of 200 mm min^{-1} at room temperature. The test specimens had a thickness of 1 mm and an initial length of 50 mm. All data presented are the average of five measured specimens for each sample.

3. Results and discussion

3.1. Clay dispersion kinetics in HNBR and ENR

The OMEC curve of the HNBR- and ENR-clay nanocomposites versus the mixing time is shown in Fig. 1. Considering the result of the previous work [13] the rapid increase of the conductance signal just after adding organoclay is attributed to the ionic conductance

Fig. 1b shows the development of online conductance of ENR-clay nanocomposites as a function of mixing time. The lower level of OMEC of ENR-clay nanocomposites is due to the lower conductivity of ENR compared to HNBR. It can be noted that the increase of OMEC in ENR-clay nanocomposites is due to the release of surfactants, but at the same time the surfactants are consumed due to the reaction as well as tendency to form H-bonding between epoxy groups of ENR and surfactants [14]. In the start of mixing process the surfactant release process is faster as compared to the consumption process resulting in an increase of conductance. After reaching a local maximum, the release process of surfactant slows down but the consumption process still goes on. Thus, in this mixing period more surfactants are consumed than released, and as a result, the OMEC decreases.

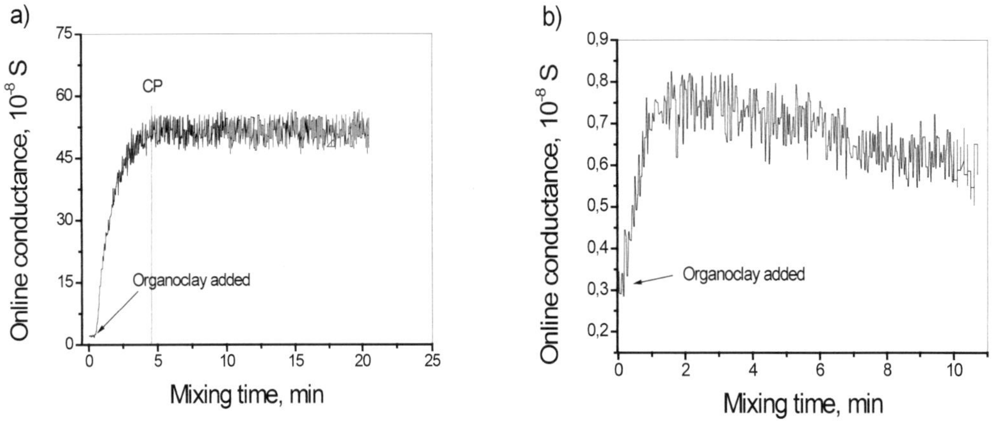

Fig. 1: OMEC of (a) HNBR-clay and (b) ENR-clay nanocomposites as function of mixing time.

In Fig. 2a the SAXS analysis shows a broad peak of the HNBR-clay compounds representing the interlayer spacing of the organoclay in the nanocomposites.

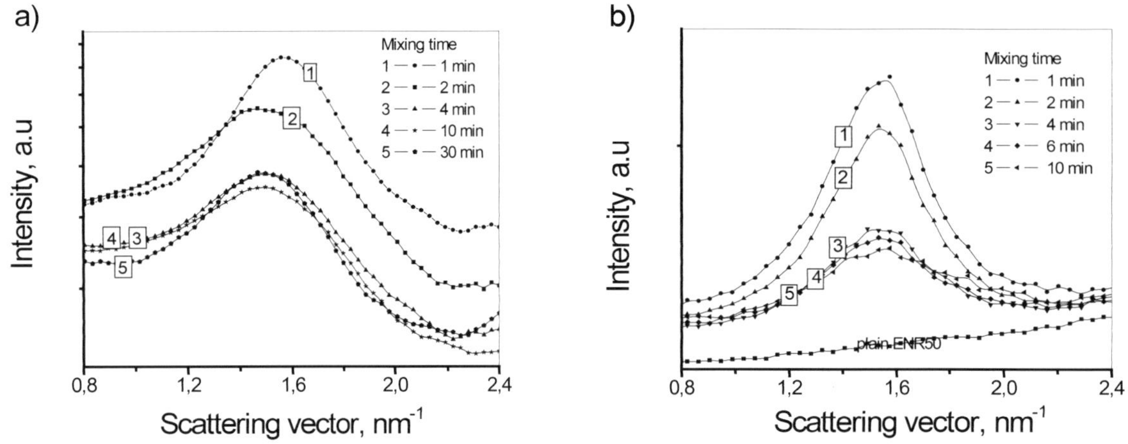

Fig. 2: SAXS analysis of (a) HNBR-clay and (b) ENR-clay nanocomposites as function of mixing time

The SAXS analysis of ENR-clay nanocomposites depending on mixing time is shown in the Fig. 2b. The peak maximum for all blends is found at scattering vector 0.15 A^{-1} as well and its height is decreasing with increasing mixing time. The decrease of peak height in the period up to 4 min implies that the clay layers are exfoliated in the polymer matrix. However, it can be observed that the peak position and shape do not change after 4 min of mixing. This is related to the fact that a final morphology has been reached after 4 min of mixing time.

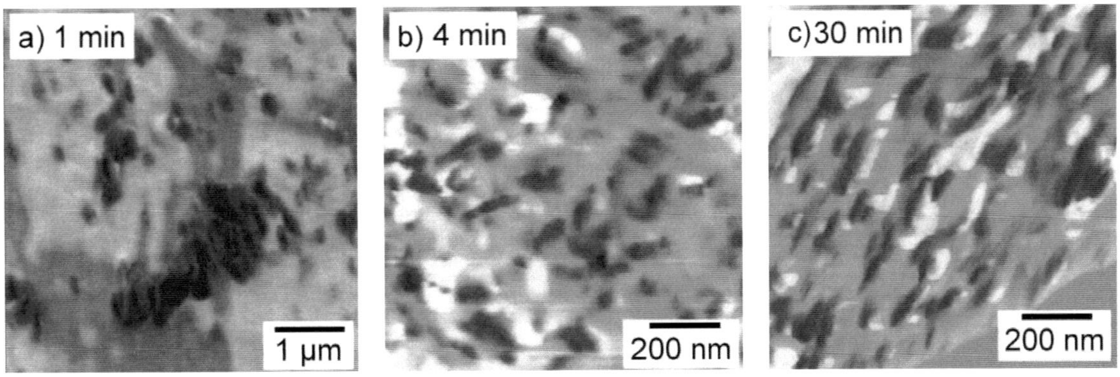

Fig. 3: AFM-images of HNBR-clay nanocomposites in dependence on the mixing time

The microdispersion of clay in HNBR was characterized by AFM as presented in Fig. 3. It can be seen that after 4 minutes of mixing time, morphology seems to remain unchanged with mixing time. That could explain the plateau of the online conductance in this period (Fig. 1a).

Microdispersion of organoclay in ENR matrix has been characterized by transmission electron microscopy.

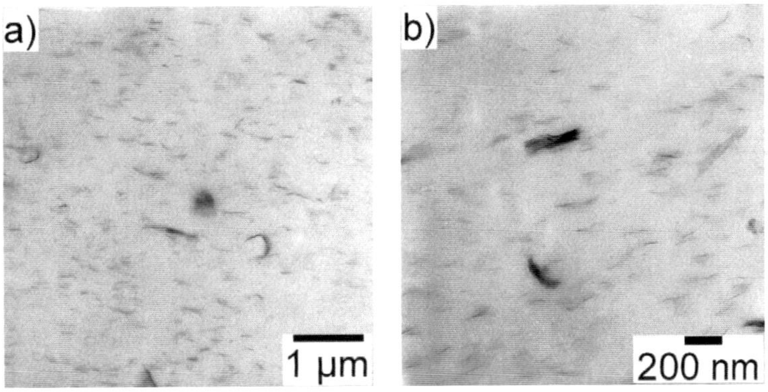

Fig. 4: TEM images of ENR-clay nanocomposites after 4 min of mixing time at (a) low magnification and (b) high magnification

3.2. Effect of organoclay on morphology development in HNBR/ENR blends

3.2.1. Online conductance of unfilled and filled HNBR/ENR-clay blend

Online measured electrical conductance of unfilled HNBR/ENR and HNBR/(ENR-clay masterbatch) blends as a function of mixing time is shown in the Fig. 5. The OMEC signal of unfilled HNBR/ENR (50/50) blends (curve 1) is quite low, which is mainly originated by the HNBR phase, because the online conductance of ENR is very small compared to that of HNBR. When ENR-clay masterbatch was mixed with plain HNBR, the online conductance increases and reaches a plateau value after a short time (curve 2).

In order to understand the background of the online conductance of the HNBR/(ENR-clay masterbatch), the curve 2 is divided into two parts, i.e. part I and part II, respectively.

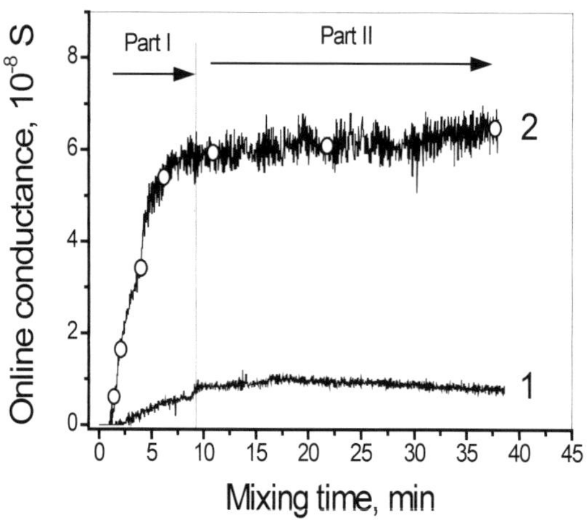

Fig. 5: OMEC of the unfilled HNBR/ENR blend (curve 1) and HNBR/(NR-clay masterbatch) blend (curve 2) in dependence on mixing time

In order to investigate the localization of the clay experimentally, AFM micrographs of HNBR/(ENR-clay masterbatch) at the beginning of mixing process (after 1.5 min) and at the end of mixing process (40 min) were obtained and are shown in Fig. 6a and 6b, respectively.

Fig. 7a shows the SAXS analysis of unfilled HNBR/ENR and HNBR/(ENR-clay masterbatch) blends at different mixing time. The SAXS pattern shows that the peak position and its high do not change with the mixing time that means the intercalated structures nearly remain constant, when clay migrates from the ENR to HNBR phase.

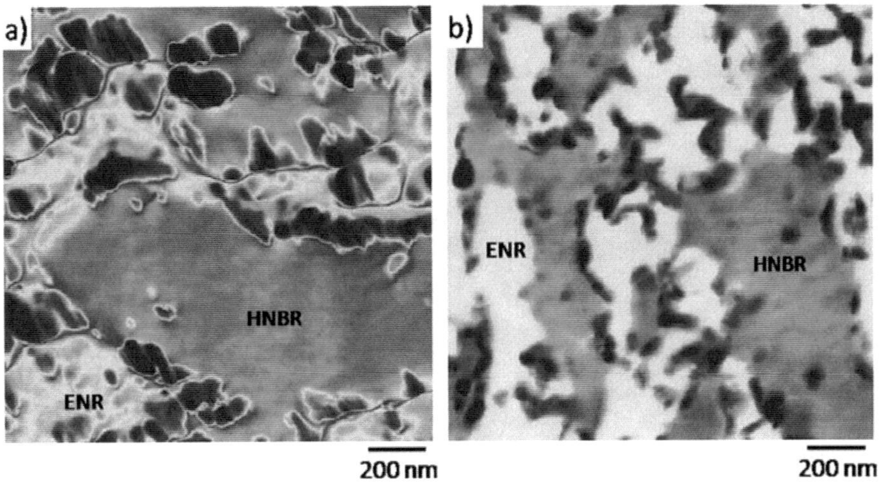

Fig. 6: AFM micrographs of HNBR/(ENR-clay masterbatch) blends after (a) 1.5 min (b) 40 min (light area: ENR, grey area: HNBR, black area: nanoclay)

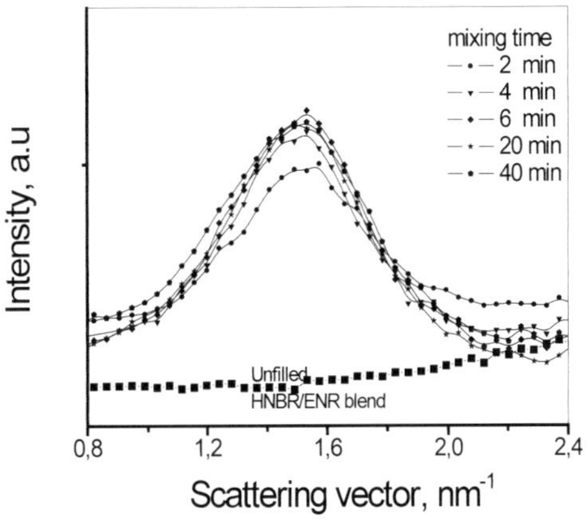

Fig. 7: SAXS analysis of unfilled HNBR/ENR and filled blends of HNBR/(ENR-clay masterbatch) at different mixing time

3.2.3. Morphology development in organoclay modified HNBR/ENR blends

The development of morphology of the HNBR/(ENR-clay masterbatch) blend as a function of mixing time is shown in the Fig. 9. Analyzing the kinetics of morphological changes of the blends not only leads to understand the part II of OMEC curve (Fig. 6) but also guides us to realize the mechanism of clay transfer from ENR- to HNBR-phase as well as the phase inversion of blend morphology.

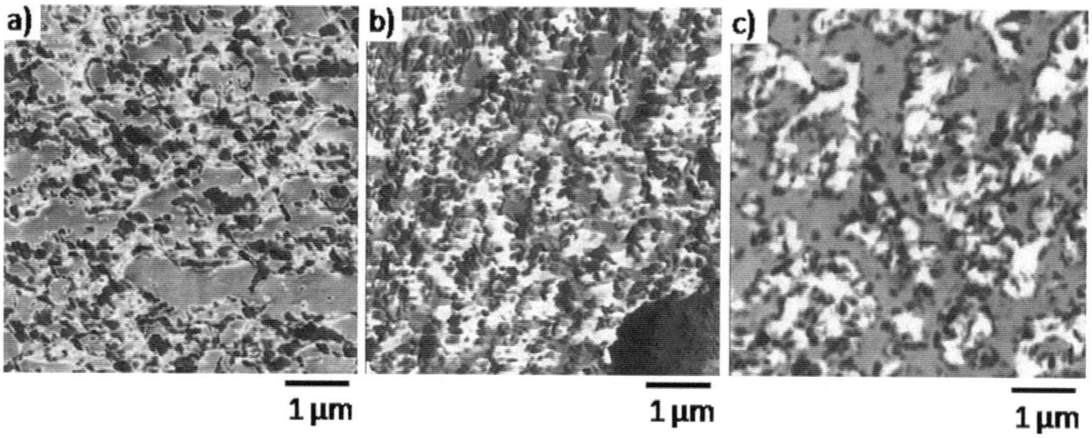

Fig. 8: AFM micrographs taken at (a) 1.5 min, (b) 10 min and (c) 40 min of HNBR/NR-clay masterbatch) blends (light area: ENR, grey area: HNBR, black area: nanoclay)

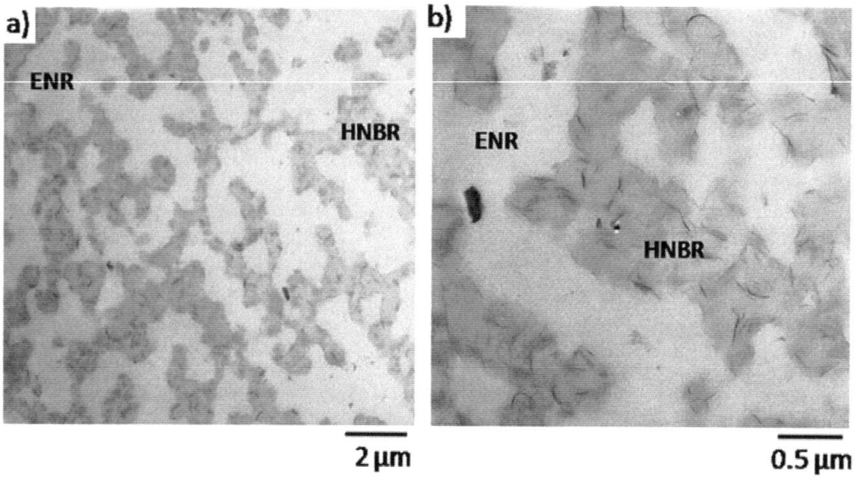

Fig. 9: TEM micrographs of HNBR/ENR-clay masterbatch) blend after 40 min mixing time at (a) low and (b) high magnification.

It can be clearly seen in Fig. 9a and 9b, that after 40 min mixing time most of clay platelets have transferred to the continuous HNBR phase. This also explains the plateau of online conductance shown in Fig. 5.

3.2.4. Investigation of mechanical behavior of the blend nanocomposites

The tensile properties of the virgin and HNBR/(ENR-clay) blends were investigated and reported in Fig. 11a and 11b. Fig. 10a shows that the modulus of HNBR/(ENR-clay masterbatch) increases by loading of 4.7 wt% of organoclay than that of virgin blends. With increasing mixing time, the modulus goes through a local minimum value, which may be related to the co-continuous morphology.

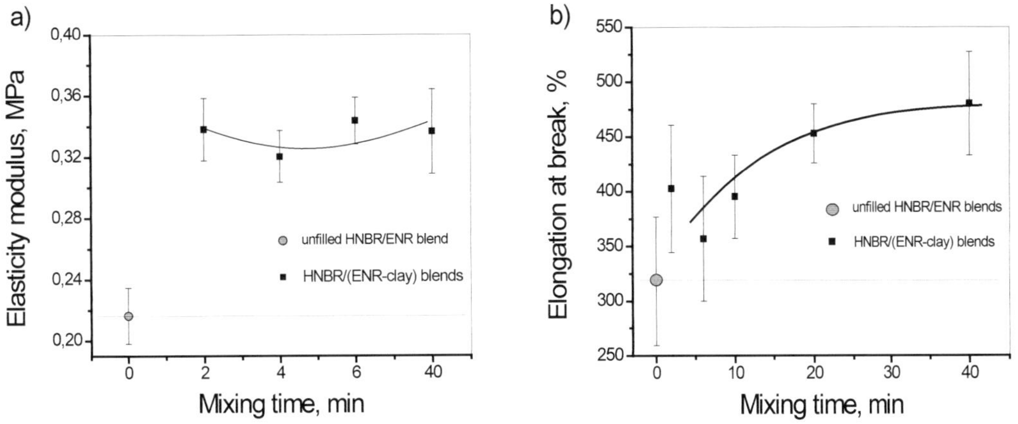

Fig. 11: Tensile test results of HNBR/(ENR-clay masterbatch) as function of mixing time and corresponding unfilled blend; (a) tensile modulus, (b) elongation at break

Fig. 11b shows the elongation at break as function of mixing time. The localization of organoclay at the interface strengthens the phase adhesion and increases the resistance to failure. The increasing of strain at break along with the mixing time could be due to better dispersion and distribution of organoclay in the HNBR-phase and improved interfacial adhesion between the phases as it has been reported by Ganter et al. [15] in the case of nanoclay filled SBR.

4. Conclusion

The kinetics of migration and selective localization of nanoclay in HNBR/ENR blends has been characterized during the mixing process by the online measured electrical conductance method. The observed online conductance is ionic in nature that has arisen due to the release of surfactants from the clay galleries. It was found that the online electrical conductance depends on the phase specific localization of nanoclay and development of blend morphology. The favorable localization of nanoclay in the HNBR phase can be monitored by the chart of OMEC and visualized by microscopic methods. The results correspond well with the filler localization predicted by means of the surface tension based model proposed in our previous works. Thus, the method of the online conductance together with the surface tension based prediction model promise to be powerful tools for the characterization of nanoclay dispersion and localization in rubber nanocomposites during the melt mixing process.

5. References:

1 Usuki A, Kawasum M, Kojima Y, Fukushima Y, Okada A, Kurauchi A, Kamigaito O. Swelling behavior of montmorillonite cation exchanged for ω-amino acids by ε-caprolactam. J Mater Res 1993;8(5):1174–1178.

2 Paul DR, Robeson LM. Polymer Nanotechnology: Nanocomposites. Polymer 2008;49(15):,3187–3204.

3 Herrmann W, Uhl C, Heinrich G, Jehnichen D. Analysis of HNBR-Montmorillonite Nanocomposites, Morphology, Orientation and Macroscopic Properties. Polym Bull 2006;57(3):395-405.

4 Tsuruk VV. Nanocomposite polymer layers for molecular tribology. Tribology Letters 2001;10(1-2):127-132.

5 Wang S, Zhang Y, Peng Z, Zhang Y. New method for preparing polybutadiene rubber/clay composites. J Appl Polym Sci 2005;98(1):227-237.

6 Karger-Kocsis J, Wu CM. Thermoset rubber/layered silicate nanocomposites. Status and futute trends. Polym Eng Sci 2004;44(6):1083-1093.

7 Chow WS, Mohd Ishak ZA, Ishiaku US, Karger-Kocsis J, Apostolov AA. The effect of organoclay on the mechanical properties and morphology of injection-molded polyamide 6/polypropylene nanocomposites. J Appl Polym Sci 2004;91(1):175-189.

8 Si M, Araki T, Ade H, Kilcoyne ALD, Fisher R, Sokolov JC, Rafailovich MH. Compatibilizing bulk polymer blends by using organoclays. Macromolecules 2006;39(14):4793-4801.

9 Wu S. Formation of dispersed phase in incompatible polymer blends: Interfacial and rheological effects. Polym Eng Sci 1987;27(5):335-343.

10 Wang Y, Zhang Q, Fu Q. Compatibilization of immiscible poly(propylene)/polystyrene blends using clay. Macromol Rapid Commun 2003;24(3):231-235.

11 Le HH, Qamer Z, Ilisch S, Radusch H-J. Carbon black distribution in the components of rubber blends monitored by online measured electrical conductance. Rubber Chem Technol 2006;79(4):621-630.

12 Epoxidized natural rubber, Weber & Schaer GmbH & Co., http://www.weber-schaer.com.

13 Ali Z, Le HH, Ilisch S, Radusch H-J. Morphology development in nanoclay filled rubber compounds and rubber blends detected by online measured electrical conductance. J Mater Sci 2009;44(23):6427-6436.

14. Mohd Ishak ZA, Wan PY, Wong PL, Ahmad Z, Ishiaku US, Karger-Kocsis J. Effects of hygrothermally decomposed polyurethane on the curing and mechanical properties of carbon black-filled epoxidized natural rubber vulcanizates. J Appl Polym Sci 2002;84(12):2265-2276.

15. Ganter M, Gronski W, Semke H, Zilg T, Thomann C, Mülhlhaupt R. Surface-compatibilized layered silicates: A novel class of nanofillers for rubbers with improved mechanical properties. Kautsch Gummi Kunstst 2001;54(4):166-171.

O-09-299

DEVELOPMENT OF PHASE STRUCTURE IN THERMOPLASTIC ELASTOMER BASED ON POLY(VINYLIDENE FLUORIDE)/EPOXIDIZED NATURAL RUBBER BLENDS AND THEIR PROPERTIES

S. Salaeh[1,2,*], C. Nakason[1], G. Boiteux[2], T. Tamet[2] and P. Cassagnau[2]

[1]*Center of Excellence in Natural Rubber Technology (CoE-NR), Faculty of Science and Technology, Prince of Songkla University, Pattani Campus, 94000, Thailand – subhan.salaeh@gmail.com; ncharoen@bunga.pn.psu.ac.th*

[2]*University of Lyon, Université Claude Bernard Lyon1, Ingénierie des Matériaux Polymères (IMP), UMR CNRS 5223, 43 Boulevard du 11 Novembre 1918, 69622 Villeurbanne, France – gisele.boiteux@univ-lyon1.fr; thierry.tamet@univ-lyon1.fr; philippe.cassagnau@univ-lyon1.fr*

Abstract - Thermoplastic elastomer (TPE) based on blending of poly(vinylidene fluoride) (PVDF) and epoxidized natural rubber (ENR) were prepared. The ENR with 50 mol% epoxide (i.e., ENR-50) was used to blend with PVDF using blend ratios of PVDF/ENR-50 = 80/20, 70/30, 60/40, 50/50, 40/60, 30/70 and 20/80 wt%. Morphological, dynamical mechanical properties of TPE were investigated. The observation phase structure of PVDF/ENR-50 blends by SEM show phase structure development and phase inversion of the bends. The evaluation of phase inversion by storage modulus and tan δ at room temperature are precisely detection and correlated well with phase morphologies by SEM. Temperature dependent of tan δ plot showed two relaxation peaks which relate to characteristic of PVDF and ENR-50. This indicated phase separation phenomenon in the PVDF/ENR-50 blends. Furthermore, the glass transition temperatures (T_g) of PVDF and ENR-50 phases were moved slightly toward higher and lower temperature. The shift of T_g different from the individual polymer in the blends indicated the formation of partially compatible between two components. When barium titanate ($BaTiO_3$) particle are added to the blends, $BaTiO_3$ particle selectively locates in ENR-50 phase, which leads to change phase morphology and properties of the blends.

Keywords: poly(vinylidene fluoride), epoxidized natural rubber, thermoplastic elastomer.

1. Introduction

Polymer blending is a common way to combine the excellent properties of two polymer may result in new polymer blends. The blending of thermoplastic and elastomer has been widely investigated which is called thermoplastic elastomer (TPE). These blending are immiscible blends and containing multi-phase separated structure. The phase structure of TPEs play importance role to control their properties which is depend on several parameters, such as interfacial adhesion, blend ratio, viscosity and processing condition [1]. Therefore, it is very important to control the morphology of TPEs. Separated phase structure exhibit rubber-like characteristic and can be processed by thermoplastic equipment. Several structure of TPE can be obtained over range of blends composition. A dispersed-matrix structure the properties are maily controlled by matrix. However, The condition for an elastomeric material is a three-dimensional, co-continuous structure with the continuous hard phase providing the strength and the continuous soft phase providing the flexibility [2].

Co-continuous phase region in polymer blends , so called phase invesion concentratio and can be developed at initial, intermidiate and final stage of the blending depending on mixing condition, interfacial emusifier and type of blender [3]. A wide rang techniques have been describe to predict and detect the cotinuous phase morphologies in immiscible polymer blends. The most commonly method is phase extraction, which is use solvent to dissolve one phase and unable to dissolve on the other phase.. This method is widely used for study co-continuity on three dimension scale is observed by scanning electron micrograph (SEM). Dynamic mechanical properties are also useful method to observe phase inversion by study relationship storage modulus/loss tangent (tan δ) and blends compositions [3]. The transition region of storage modulus or loss tangent is formation of co-continuous phase structure. In addition, incorporation of filler particle in immiscible polymer blends is also interesting to understanding selectively locates of filler particles in polymer blends which impact to morphology and mechanical properties.

The aim of this work is to study phase development and phase inversion in immiscible poly(vinylidene fluoride)/epoxidized natural rubber blends. Morphological and dynamic mechanical properties of the blends were investigated. The incorporation of $BaTiO_3$ particles on phase structure and blends properties were also investigated.

2. Experimental

2.1 Materials

The poly(vinylidene fluoride) (PVDF) and epoxidized natural rubber (ENRs) rubber used were commercially available Kynar 740 (Arkemar, France) and Epoxyprene 50 (Muang Mai Guthrie Co., Ltd., Thailand), respectively. Barium titanate ($BaTiO_3$) was purchased from Alfa Aesar GmbH, Germany.

2.2 Blends preparation

The blends were prepared by melt mixing at 180°C in an internal mixer (Haake Rheomix R600) with a mixing capacity of 50 cm^3 at a rotor speed 50 rpm. Different blends ratio of PVDF/ENR-50 were studied at 80/20, 70/30, 60/40, 50/50, 40/60, 30/70 and 20/80 (wt%). PVDF at was first melted for 3 minutes and then ENR was added and the mixing was continued for 5 minutes. The effect of $BaTiO_3$, 80/20 and 50/50 of PVDF/ENR-50 were selected to study phase structure and blends properties by using 20wt% of $BaTiO_3$ in the blends. The blends were fabricated to be a thin sheet with 1 mm thick by electrical compression molding at 180°C.

2.3 Morphological properties.

The cryogenic-fractured surface was prepared in liquid nitrogen. The fractured surface of sample were extracted by using toluene (ENR-50 lower than 60wt% in the blends) and DMF (ENR-50 at and higher than 60wt% in the blends) to remove ENR phase and were then dried in an air oven at 70°C for 48 hr to eliminate the solvent. Scanning electron micrograph observations were determined by using a scanning electron microscope (SEM) FEI Quanta 400 instrument.

2.4 Dynamic mechanical properties

Dynamic mechanical properties of parent polymers and the PVDF/ENR-50 blends were charecterized using a TA instrument dynamic analyzer (DMA; model Q800) in a tension mode under the test temperature ranges of -90 to 100°C at a frequency of 1 Hz under stream of liquid nitrogen. The heating rate was 4 °C/min for -90 to -50 °C and 10 to 100 °C and was in 2 oC/min for -50 -10 °C. Lower heating rate in range of glass transition temperature were used for good respond in the relaxation region of the blends.

3. Results and Discussion

3.1 Effect of PVDF/ENR-50 blends ratio.

3.1.1 Morphological properties

SEM micrograph of PVDF/ENR-50 blends are shown in Fig. 1. All the micrograph distinctively show a very clear interface between PVDF and ENR-50 due to immiscible blends. At ENR-50 content lower than 50wt%, It can be also seen that small particle of ENR-50 were dispersed in PVDF matrix. The size of rubber particle increased as rubber content increase for SEM micrograh of blend ratio at 20/80, 30/70, 40/60 and the shape of rubber phase significantly deviating from the spherical. Co-continuous phase structure can be observed at blend ratio 60/40 to 50/50 wt% of PVDF/ENR-50 due to co-continuity rather than at single point. Rubber content is higher than 50wt% rubber phase become a matrix and PVDF exhibit dispersed phase. The size of PVDF phase decreased with rubber content increased.

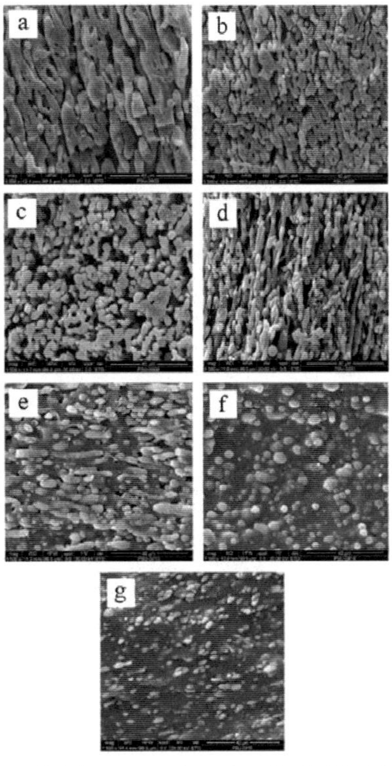

Figure 1 SEM micrograhp of PVDF/ENR-50 blends with differenct blends ratio (a) 80/20, (b) 70/30, (c) 60/40, (d) 50/50, (e) 40/60, (f) 30/70 and (g) 20/80

3.1.2 Dynamic Mechanical properties.

The storage modulus and tan δ as function of temperature are shown in Fig. 2. It can be seen that storage modulus decrease with increasing temperature due to mobility of molecular segment increased. For storage modulus of PVDF small fall off at 40 °C due to glass transition temperature (T_g) while ENR-50 exhibited drastically drop in storage modulus at -10 °C. It can be seen that blends exhibited double drop of storage modulus around -40 and -10 °C which is correspond to T_g of PVDF and ENR-50 respectively. The decreasing sharply of storage modulus correlated well with two relaxation peaks around -35 and -10 °C in tan δ plot. The presence of two relaxation peaks correspond to α-relaxation (T_g) of PVDF and ENR-50 which is contribution from segmental motion in amorphous region of PVDF and ENR-50 and confirm that blends are immiscible blends. Furthermore, It can be also seen the amplitude of tan δ increase as rubber content increase. This is due to increasing of amorphous phase in the blends which is correlated well with the reduction of storage modulus in transition region. The maximum position of tan δ peak is commonly used to indicate glass transition temperature of materials. It was found that T_g of PVDF and ENR-50 shifted to higher and lower temperature, respectively (as shown in Fig. 3). This indicates PVDF and ENR-50 are partially compatible between two phases

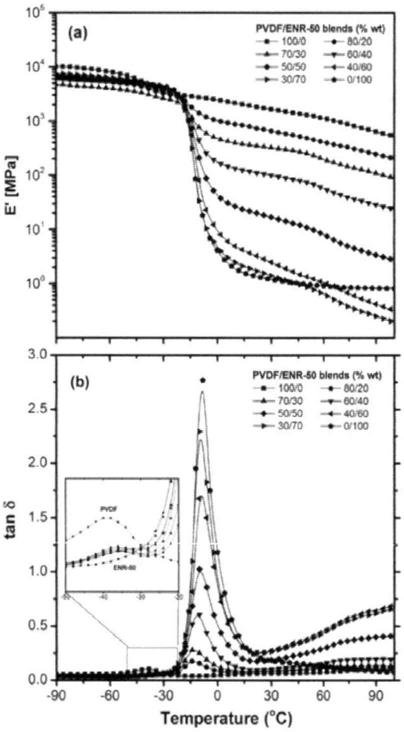

Figure 2 (a) Storage modulus and (b) tan δ of PVDF/ENR-50 blends as function of temperature with various blends ratio.

For comparison, it was found that at room temperature higher than glass transition temperature, the storage modulus decrease as rubber content increase while ten δ increased. The storage modulus and tan δ which is determined at room temperature as function of PVDF content are shown in Fig. 4. At the beginning, storage modulus and tan δ slightly decrease and increase, respectively due to PVDF is matrix phase and the properties the blends are controlled from matrix phase. The both value exhibited radically decreasing and increasing around 60 to 45 wt% of PVDF which is correspond to co-continuous phase structure are formed at those PVDF content. This transition confirms the dual co-continuous phase structure as shown in SEM micrograph [3].

3.2 *Effect of $BaTiO_3$ in PVDF/ENR-50 blends*

3.2.1 *Morphological properties*

SEM micrographs of PVDF/ENR-50 blends filled with 20 wt% of $BaTiO_3$ are shown in Fig 5. It can be seen that 80/20 of PVDF blends show ENR-50 particles dispersed in PVDF matrix with presence $BaTiO_3$ particle in ENR-50 phase as well as at interface. The presence of $BaTiO_3$ particle in ENR-50 and interface leads to increase physical compatibilization between both phases. However, 50/50 PVDF/ENR-50 in Fig. 5c and 5d, $BaTiO_3$ particles dispersed in ENR-50 phase and ENR-50 exhibit matrix phase due to incorporation of $BaTiO_3$ caused to increase viscosity of ENR-50.

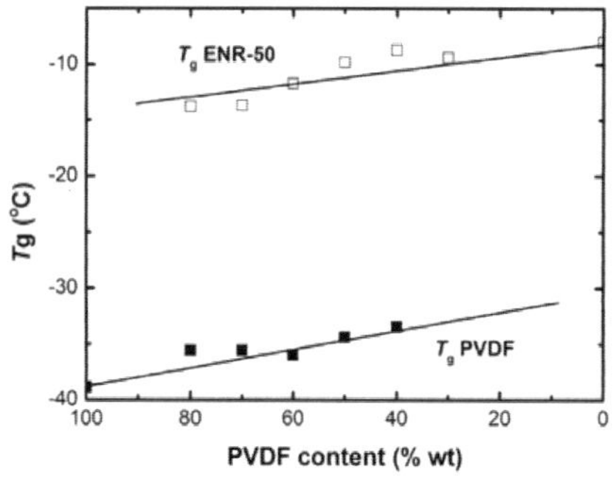

Figure 3 Glass transition temperatures of PVDF phase and ENR-50 as function PVDF content in the blends.

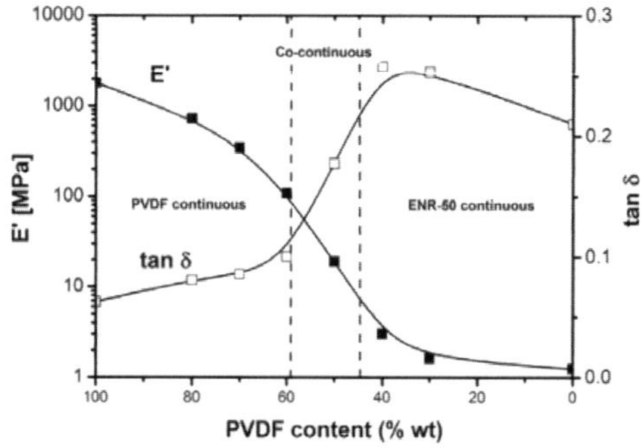

Figure 4 Storage modulus and tan δ as function of PVDF content in the blends

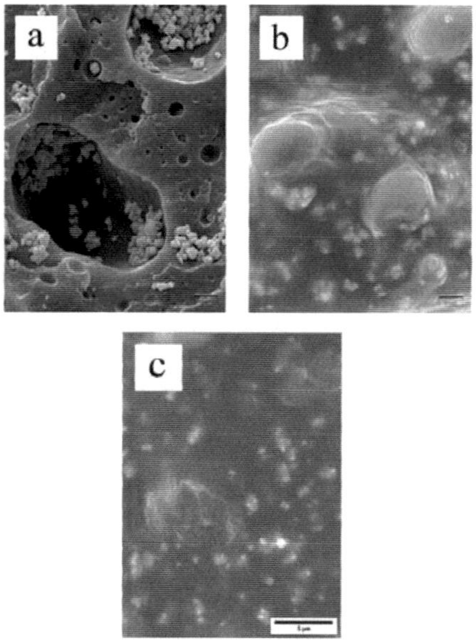

Figure 5 SEM micrograph of PVDF/ENR-50 blends filled with 20 wt% $BaTiO_3$ at different blends ratio

80/20 (a), 50/50 with extraction (b) and 50/50 without extraction (c)

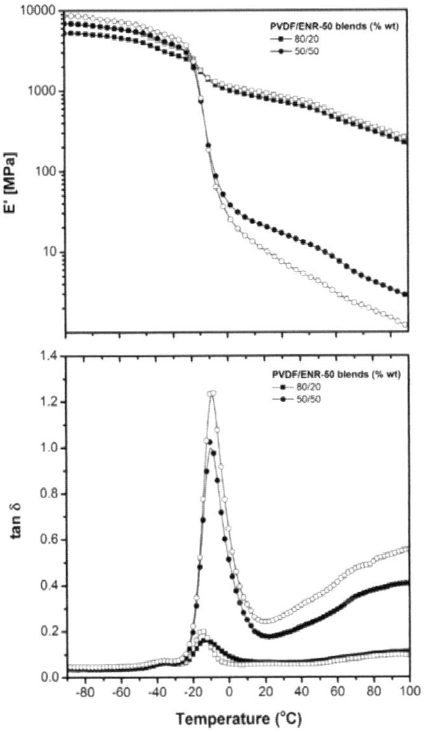

Figure 6 Storage modulus and tan δ of 80/20 and 50/50 PVDF/ENR-50 blends without (full symbol) and filled with $BaTiO_3$ (open symbol).

3.2.2 Dynamic mechanical properties

Temperature dependence of storage modulus and tan δ of PVDF/WNE-50 blends with and without $BaTiO_3$ are show in Fig. 6. At rubbery stage, it was found that 80/20 PVDF/ENR-50 blends with filled $BaTiO_3$ show higher storage modulus than that of the blends without filler. This due to presence of $BaTiO_3$ at interface leads to increase physical compatibilization between two phases as result in SEM micrograph [4, 5]. However, incorporation of $BaTiO_3$ in 50/50 PVDF/ENR-50 cause to decrease storage modulus due to ENR-50 exhibit as matrix phase as shown in Fig. 5c and 5d, the storage modulus of the blends is dominated by ENR-50. It was also found that the blends show two transition zones around -40 and -10 which is correspond to glass transition temperature of PVDF and ENR-50. The glass transition temperatures of the blend are summarized in Table 1. It was found that T_g of ENR-50 phase in 80/20 PVDF/ENR-50 shifted to lower temperature due to physical compatibilization and increase interfacial tension in the presence of $BaTiO_3$ while T_g of ENR-50 in 50/50 shifted to higher temperature. This is due to $BaTiO_3$ was localized in ENR-50 leads to increase phase separation of the blends.

Table 1 Storage modulus at 25 °C of the blends and glass transition temperature (T_g) of PVDF and ENR-50 phases.

PVDF/ENR-50	E' [MPa]	T_g [°C]	
		PVDF phase	**ENR-50 phase**
80/20	756	-37.0	-12.0
80/20+$BaTiO_3$	887	-37.0	-15.3
50/50	18	-34.5	-9.6
50/50+ $BaTiO_3$	9	-36.0	-8.5

4. Conclusions

PVDF/ENR-50 blends with different blends ratio were prepared. Morphological observation using SEM clearly show phase structure of PVDF/ENR-50 blends, which exhibit co-continuous phase within range 60/40 to 45/55 wt% of PVDF/ENR-50. Above T_g, The storage modulus and tan δ of the blends decreased and increased, respectively with increasing rubber content. Storage modulus and tan δ exhibit sharp decrease and increase when reach to co-continuous structure. The glass transition of PVDF and ENR-50 shifted to higher and lower temperature, respectively due to partial compatible between two phases. The incorporation of $BaTiO_3$ particle in the blends was studied by SEM and DMA. The results show that $BaTiO_3$ particle were localized in ENR-50 phase and correlated well with dynamic properties of filled blends.

Acknowledgements

Financial support from the Thailand Research Fund through the Royal Golden Jubilee Ph.D. Program (Grant No. PHD/0245/2549) and French embassy in Thailand to Mr.Subhan Salaeh and Assoc.Prof.Chareon Nakason is acknowledged. This work was also supported by the Higher Education Research Promotion and National Research University Project of Thailand, Office of the Higher Education Commission by Centre National de la Recherche Scientifique (CNRS) UMR 5223 "IMP@Lyon1". We also thank to the Center of Electronic Microscopy of UCBL for performing electronic microscopy analysis.

References

[1] S. Chaput, C. Carrot, M. Castro, F, Prochazka *Rheol. Acta*. 2004, *43*, 417.

[2] J. G. Drobny, *Handbook of Thermoplastic Elastomers* (William Andrew Pub., USA, 2007).

[3] T.S. Omonov, C. Harrats, P. Moldenaers, G. Groeninckx *Polymer* 2007, *48*, 5917.

[4] G.matin, C. Barres, P. Sonntag, N. Garois, P. Cassagnau *Mater. Chem. Phys*. 2009, *113*, 889.

[5] S. S. Ray, S. Pouliot, M. Bousmina, L. A. Utracki *Polymer* 2004, *45*, 8403

O-10-484

3D FINITE ELEMENT SIMULATION OF THE ELECTROFUSION PROCESS OF POLYETHYLENE FITTINGS

Z. Chebbo[1], Y. Tillier[1], M. Vincent[1]*, A. Boujlal[2], D. Gueugnaut[2]

[1]Cemef, Mines ParisTech, 06904 Sophia Antipolis, France – michel.vincent@mines-paristech.fr, yannick.tillier@mines-paristech.fr, ziad.chebbo@mines-paristech.fr

[2] GDF-SUEZ, R&I Division, France – dominique.gueugnaut@gdfsuez.com, adil.boujlal@gdfsuez.com

Abstract – In the electrofusion technique, the energy necessary to melt the polymers is provided by an electrical current in a wire which is embedded in a polymer fitting. After a certain time, the electrical current is cut, and the material cools down. The main objective of the work is to develop a 3D finite element simulation of melting and solidification, and to compute a criterion based on the macromolecular inter diffusion theory, that is able to determine whether a good welding occurred or not at the end of the welding cycle. A specific version of the 3D simulation software Forge® was developed in order to take into account the fitting parameters (wire characteristics and spacing, fittings dimensions…), the polyethylene parameters (melting and crystallization kinetics, thermal conductivity, thermal expansion) and the electrical and geometrical parameters. The thermal equation including enthalpy of crystallization, and the crystallization kinetics equation are solved and the welding criterion is computed. A well refined anisotropic mesh adaptation is applied to capture the contact condition between the wire and the polyethylene and inside the heating polyethylene layer.

Computation is applied to different geometries, including real 3D ones, and to a more simple academic planar geometry, which has been experimentally realized.

Keywords: Welding – Electrofusion – Polyethylene – Numerical simulation

Introduction

Electrofusion is widely used in the gas distribution networks in order to weld polyethylene components. For example, the fitting for joining two pipes is a polyethylene cylinder with an inner diameter a little bit larger than the outer tube diameter. The cylindrical fitting covers both pipe ends. An electrical wire is inside the fitting, close to its inner surface, and more or less parallel to the hoop direction. The energy necessary to melt the polymers is provided by an electrical current in the wire. After a certain time, the electrical current is cut, and the material freely cools down.

The main objective of the work is to develop a 3D finite element simulation of melting and solidification, and to compute a criterion based on the macromolecular inter diffusion theory, that is able to determine whether a good welding occurred or not at the end of the welding cycle.

Equations

Heat transfer equation

The energy equation writes as follows:

$$\rho C_p \frac{dT}{dt} = k\Delta T + \Delta H \qquad (1)$$

ρ is the density, Cp the heat capacity, k the thermal conductivity, ΔH is the crystallization enthalpy. It is calculated according to the Ozawa law. For the contact with air, a boundary condition with a heat transfer coefficient is used (the heat flux density is proportional to the temperature difference between the air and the polymer surface). A thermal contact resistance is taken into account between the electrical wire and the polymer, and between the two polymer parts. The initial temperature is the room temperature. A given heat flux is applied inside the electrical wire. This heat flux is related to the electrical tension and intensity which are imposed.

Inter diffusion

Interdiffusion of macromolecules is the main phenomena responsible for welding quality in a range of temperature where molecular motions are allowed. Temperature and time are the main macroscopic parameters which can influence the welding quality. Following Saint-Royre et al. [1], we calculate the interdiffusion parameter p as:

$$p = \frac{\langle l^2 \rangle}{KD_0} = \int_{t_1}^{t_2} \exp\left(-\frac{E}{RT(t)}\right) dt \qquad (2)$$

where $\langle l^2 \rangle$ is the average depth of interdiffusion, t_1 and t_2 are the initial and final interdiffusion times, K is a coefficient depending on the material, D_0 is the diffusion coefficient of the material, E is the activation energy of diffusion, R is the gas constant. A good welding occurs when the parameter p exceeds a threshold value which depends on the material.

Numerical resolution

A specific version of the 3D simulation software Forge® was developed in order to take into account the fitting parameters (wire characteristics and spacing, fittings dimensions…), the polyethylene material parameters (melting and crystallization kinetics, thermal conductivity, thermal expansion) (after [2]) and the electrical and geometrical parameters (initial gap between the pipes and the fitting, tension and intensity of the electrical current).

The computational domain is discretized using enhanced 4-nodes tetrahedron elements (P1+/P1). The discretized domain includes the wire itself (Fig. 1).

The mesh with adequate element sizes is generated with Gmsh software and an anisotropic mesh adaptation improves the precision with an acceptable number of elements. Three domains are considered: the fitting (polyethylene with embedded electrical wire), the heating wire and the tube. The electrical power is specified inside the wire (grey area on Fig. 1). Conformal meshes at each interface leads to coincident nodes (Fig. 1).

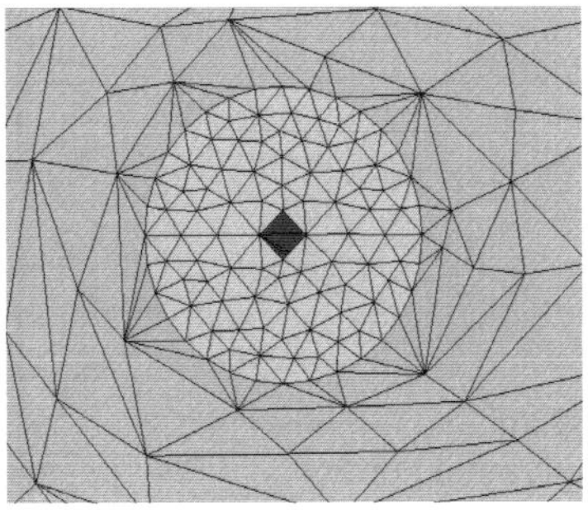

Figure 1 – Conformed mesh at wire/polymer interface.

Numerical results

Fig. 2 shows an example of result in an annular geometry (2.3 millions elements). Colors show the temperature field in a scale from 20 °C (in blue) to 132 °C (in red) at a certain time during the heating phase. This example shows that a real industrial geometry can be accounted for.

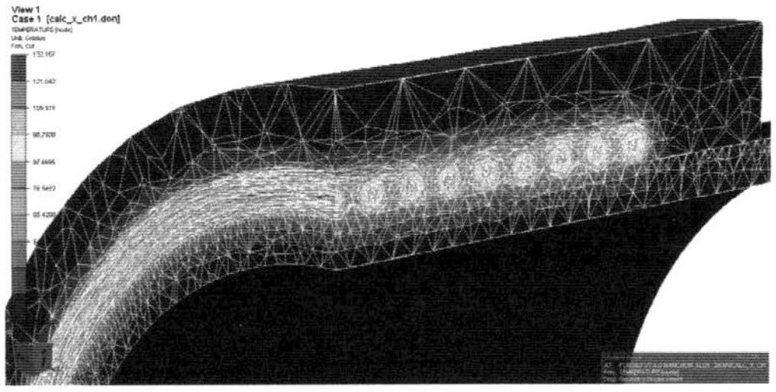

Figure 2 – Temperature field during heating.

Experimental device

In order to validate the computation, we molded polyethylene plaques 1 mm thick. We fastened the electric wire at the surface. Then we overmolded this plaque in an 8 mm thick rectangular cavity. We obtain therefore a plaque which includes a wire about 1 mm from the surface, comparable to what can be found in real fittings. This side of the plaque is put in contact with an 8 mm plaque (without wire) (Fig. 3). The electrical current leads to the fusion of the polyethylene, and after cooling, the two 8 mm plaques are welded.

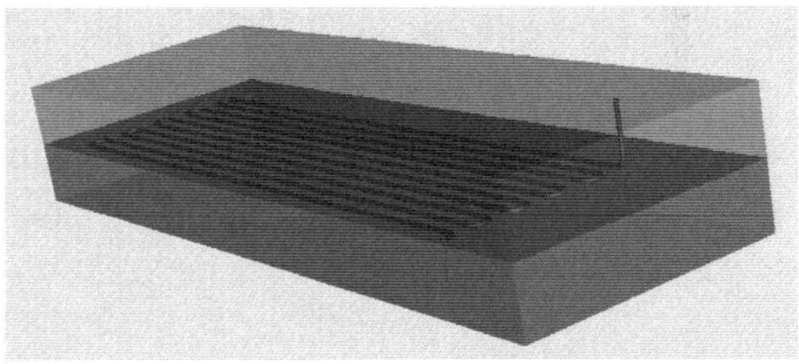

Figure 3 – Half of experimental geometry composed by two plaques (the top one has embedded wire).

Fig. 4 shows that for a low electrical power of 165 W, and a long heating time (110 s), the molten area seems to be correct (left), but the welding criteria defined in Eq. 2 is not satisfied over the whole welding surface (right). The computation allows optimizing the couple electrical power and heating time.

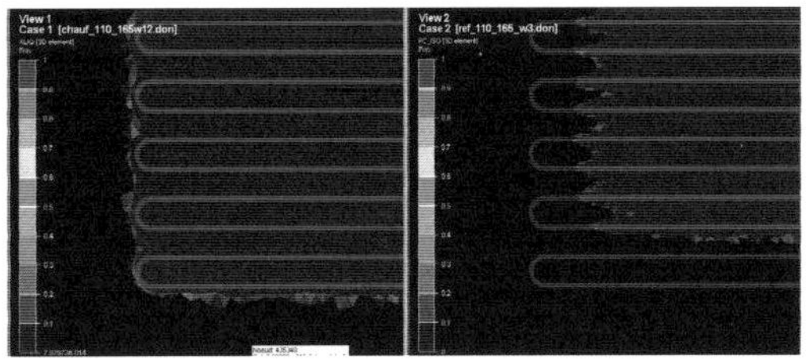

Figure 4 – Molten area (left) and welding criteria (right)

Conclusions

A 3D finite element resolution of the electrofusion welding process has been presented. It is applied to different geometries, including real 3D industrial parts. A more simple planar experimental device will allow the validation of the computation.

References

1. D. Saint-Royre; D. Gueugnaut; D. Reveret *J. App. Polym. Sci.* 1989, *38*, 147.
2. O. Hehn, PhD Thesis, Ecole des Mines de Paris, 2006.

O-10-355

SIMULATION OF ASSEMBLY INJECTION MOLDING PROCESS (AIM) OF POLYMER-METAL-HYBRID PARTS

Ines Kühnert[a,*], **Torsten Maenz**[b,c] **and Tim A. Osswald**[c]
[a] *Leibniz-Institut für Polymerforschung Dresden e. V. (IPF), Hohe Straße 6, 01069 Dresden, Germany*
[b] *Technische Universität Dresden, 01062 Dresden, Germany*
[c] *University of Wisconsin-Madison, Polymer Engineering Center (PEC), Madison, WI 53706, United States of America*
**Corresponding Author: Dr. Ines Kühnert, Kuehnert@ipfdd.de*

Abstract - In Polymer-Metal-Hybrids advantages of properties and functionalities shall be combined to a symbiosis. The challenge of this unlike bond is to get god adhesion between the materials. At the IPF Dresden (Germany) an adhesive coat was developed to improve the bonding strength between metal and polymer. Furthermore, a special multi material specimen (see Fig. 1) was designed that consists of a material combination from a thin metal sheet with the special adhesive coat, a thermoplastic polyurethane as intermediate soft layer to absorb interfacial stress caused by different thermal expansion and a thermoplastic functional layer (unreinforced or reinforced Nylon 6) on top.

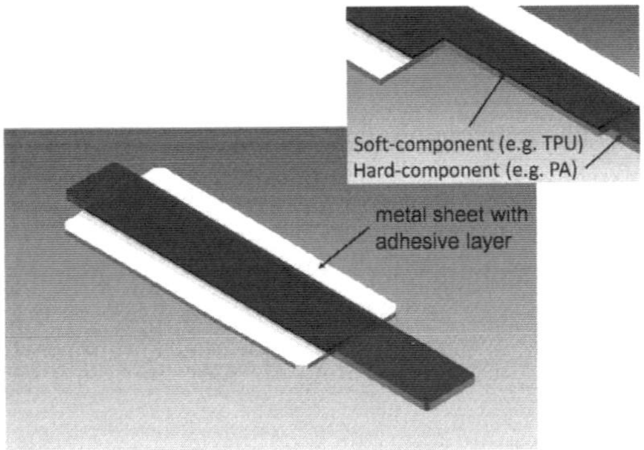

Fig. 1: Multi material specimen of the IPF

The complexity of the multi-step process, which is shown in Fig. 2, is not feasible without a systematic analysis with numerical simulations. Each processing step must be included with the appropriate boundary conditions. Especially, it is required that the contact interferences change during the processing (metal to polymer, polymer to polymer or metal to metal). During mold filling the heat transfer conditions are constantly changing.

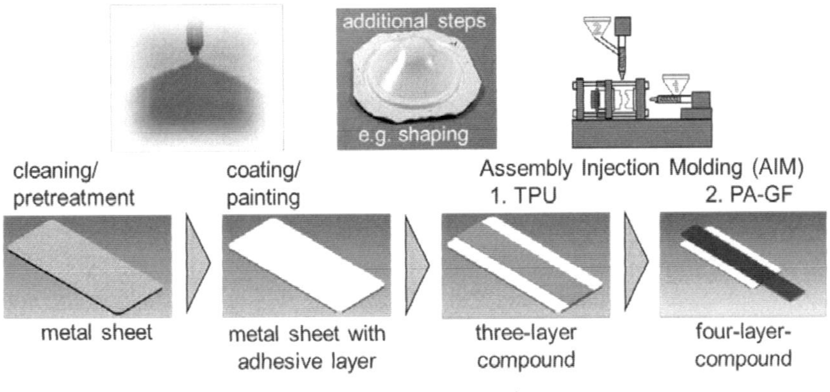

Fig. 2: Multi-step process for producing of the IPF specimen

In this work and under cooperation with the PEC Madison (USA) two different tools were used to simulate the assembly injection molding process for a better understanding especially of the influence of different material data on simulation results. So, the pvT behavior is of particular importance for the prediction of shrinkage and warpage. Fig. 3 shows the different in the pvT-data in different software databases (Cadmould 3D-F and Moldex3D). Furthermore, the contact temperatures between different components during assembly injection molding depend strongly from the heat transfer settings.

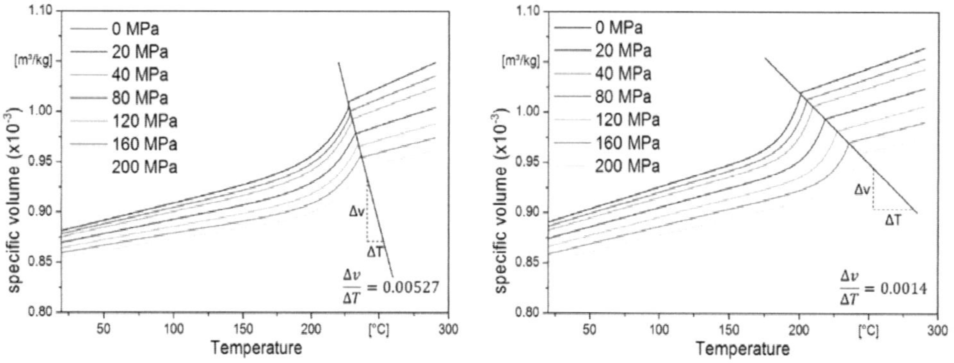

Fig. 3: Comparison of the pvT-data (unreinforced Nylon 6) of two different sources (Cadmould 3D-F and Moldex3D)

In addition to the challenges of the process simulation also experimental results will be shown which were improved with the use of the process simulation.

Keywords: process simulation, assembly injection molding, pvT behavior, polymer metal hybrid, heat transfer

O-10-239

COMPARISON OF DYNAMIC FOAMING BEHAVIOR IN MICROCELLULAR INJECTION MOLDING PROCESS FROM SIMULATION AND EXPERIMENT

Tai-Yi Shiu [1]*, Yuan-Jung Chang[1], Chao-Tsai Huang[1], Chia-Hsiang Hsu[1], Rong-Yu Chang[2], Shyh-Shin Hwang[3]

*[1]CoreTech System (Moldex3D) Co., Ltd., Hsinchu, Taiwan – *lloydshiu@moldex3d.com;danchang@moldex3d.com;cthuang@moldex3d.com;davidhsu@moldex3d.com*

[2]Department of Chemical Engineering, National Tsing Hua University, Hsinchu, Taiwan – rychang@moldex3d.com

[3]Department of Mechanical Engineering, Ching-Yun University, Chung-Li, Taiwan – hwang002@ms10.hinet.net

Abstract

Microcellular injection molding process is a promising solution for products with special requirements such as weight reduction and extra thin wall. Despite microcellular foaming application used in reciprocating screw injection molding machine was built more than a decade, simulation tool for microcellular injection molding was not successful in earlier CAE application due to insufficient physical can computational considerations, limited by complicated bubble growth mechanism. In this paper, we present the recent development of three-dimensional prediction of foaming process in injection molding process. Predictions of microcellular injection molding process are carried out with polyolefins with nitrogen supercritical fluids (SCFs) according to current experimental results for validating the capability of 3D simulation. Comparisons of simulations between microcellular injection molding with and without counter pressure are discussed to provide insights into the correlation of surface quality and cell size distribution near the surface of product. Furthermore, comparisons between predictions of foaming properties and experimental results of real products are in good agreement.

Keywords: Microcellular injection molding, Foaming, Simulation

Introduction

Microcellular foaming technology has been enormously developing since the microcellular batch processing technology has brought out by Dr. Nam Suh and co-workers at MIT in the early 1980s [1]. The technology applied in injection molding process is now widely used in many special required products such as products with both weight reduction and extra thin wall products. Despite microcellular foaming application used in reciprocating screw injection molding machine was built by Trexel and Engel in 1998 [2], the reliable computer aided engineering application is not well developed due to limited understanding of complex foaming mechanism. Venerus [3] reviewed numerous diffusion-controlled modeling studies of polymer foaming and showed the diffusion-induced bubble growth in viscoelastic liquids numerically having good agreement with experimental data. Taki [4] studied the effects of pressure release rate on bubble density and sizes. This study applies the well developed models of cell foaming to couple with 3D flow motion technology for microcellular injection molding. The validation of 3D modeling is to compare the experimental data done by Turng and co-workers [5].

Numerical Modeling

The details of several model developments can be found in the papers by Taki [4]. We consider the single phase fluid is of binary constitution, SCFs dissolved in polymer melt, before onset of bubble nucleation. The polymer melt is assumed as General Newtonian Fluid (GNF). Hence the non-isothermal 3D flow motion can be mathematically described by the following governing equations:

$$\frac{\partial \rho}{\partial t} + \nabla \cdot \rho \mathbf{u} = 0 \qquad \text{(Eq.1)}$$

$$\frac{\partial}{\partial t}(\rho \mathbf{u}) + \nabla \cdot (\rho \mathbf{uu} - \boldsymbol{\sigma}) = \rho \mathbf{g} \qquad \text{(Eq.2)}$$

$$\boldsymbol{\sigma} = -p\mathbf{I} + \eta\left(\nabla \mathbf{u} + \nabla \mathbf{u}^{\mathrm{T}}\right) \qquad \text{(Eq.3)}$$

$$\rho C_P\left(\frac{\partial T}{\partial t} + \mathbf{u} \cdot \nabla T\right) = \nabla \cdot (\mathbf{k}\nabla T) + \eta\dot{\gamma}^2 \qquad \text{(Eq.4)}$$

where **u** is the velocity vector, T is the temperature, t is the time, p is the pressure, σ is the total stress tensor, ρ is the density, η is the viscosity, **k** is the thermal conductivity, Cp is the specific heat and $\dot{\gamma}$ is the shear rate. The Finite Volume Method (FVM) due to its robustness and efficiency is employed in this study to solve the transient flow field in the complex three-dimensional geometry.

The microcellular foaming process happens after melt is injected into mold cavity. The 3D numerical simulation is applied for describing dynamic behavior of bubble growth which is coupled with polymer melt flow. Radius of bubble growth is given as:

$$\frac{dR}{dt} = \frac{R}{4\eta}\left(P_D - P_C - \frac{2\gamma}{R}\right) \qquad \text{(Eq.5)}$$

where bubble radius is R, viscosity is η, bubble pressure is P_D, ambient pressure is P_C, and surface tension is γ. Thin boundary layer condition is assumed and dissolved gas concentration profile along the radial direction of thin shell is described by a diffusion equation as described below.

$$\frac{\partial c}{\partial t} = D\left[\frac{1}{r^2}\frac{\partial}{\partial r}\left(r^2\frac{\partial c}{\partial r}\right)\right] \qquad \text{(Eq.6)}$$

where c is dissolved gas concentration and D is the diffusion coefficient.

The dynamic bubble growth behavior is described by the mass transfer at the interface of gas bubble, and is proposed by Han and Yoo [6].

$$\frac{d}{dt}\left(P_D R^3\right) = \frac{6D(R_g T)(c_\infty - c_R)R}{-1 + \left\{1 + \frac{2/R^3}{R_g T}\left(\frac{P_D R^3 - P_{D0}R_0^{\ 3}}{c_\infty - c_R}\right)\right\}^{1/2}} \qquad \text{(Eq.7)}$$

The concentration has the following relation,

$$\frac{c_\infty - c}{c_\infty - c_R} = \left(1 - \frac{r - R}{\delta}\right)^2 \qquad \text{(Eq.8)}$$

and δ is the concentration boundary thickness.

Bubble nucleation, describe by equation (9), appears when pressure of molten polymer decrease from sprue to mold cavity during filling process. The cell nucleation and bubble growing is a competition mechanism and can be proposed as a classical exponential function which is coupled with a mass conservation of dissolved gas as given in equation (10).

$$J(t) = f_0\left(\frac{2\gamma}{\pi M_w / N_A}\right)^{1/2} \exp\left(-\frac{16\pi\gamma^3 F}{3k_B T\left(\bar{c}(t)/k_H - P_C(t)\right)^2}\right) N_A \bar{c}(t) \qquad \text{(Eq.9)}$$

where f_0 and F are fitting parameters, $\bar{c}$ is average dissolved gas concentration, N_A is Avogadro number, k_B is Boltzmann constant, and M_w is gas molecular weight.

$$\bar{c}(t)V_{L0} = c_0 V_{L0} - \int_0^t \frac{4\pi}{3} R^3(t - t', t')\frac{P_D(t - t', t')}{R_g T} J(t')V_{L0}dt' \qquad \text{(Eq.10)}$$

where V_{L0} is the volume of polymer matrix.

Simulations are done by a FVM based algorithm developed by Chang and Yang [7]. The validation of 3D modeling of microcellular injection molding simulation is performed on a CAE software Moldex3D theoretically and the simulation results are compared to the experimental results done by Turng and co-workers [5]. In addition, a comparison of simulation of microcellular and conventional injection molding is for illustrating the benefits of using microcellular injection molding process.

Results and Discussion

Calculated cell density and cell size versus dissolved gas amount are compared with experimental data of microcellular injection molded dog-bone samples reported by Turng and co-workers [5] as shown in Fig. 1. For PP/N_2 system, the calculated cell size well agrees with experimental data while the calculated density well quantitatively agrees experiment for lower dissolved gas amount and qualitatively agree with experiment for higher gas concentration.

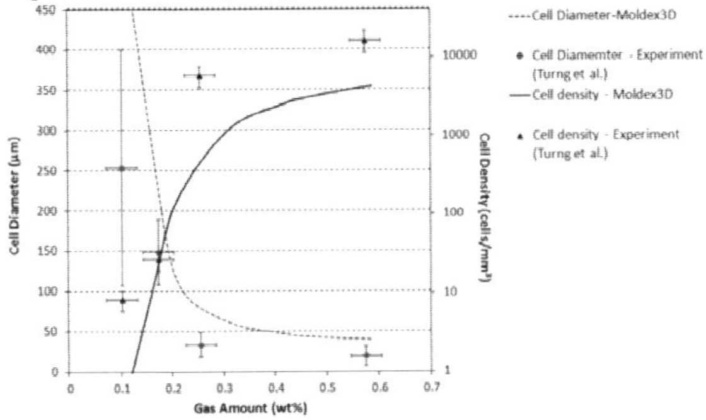

Fig. 1 Comparison of calculated and experimental data of cell sizes and densities for PP/N2 system. (Y-axis error bar is 0.03wt%)

Simulation results of dissolved gas concentration 0.57wt% for PP/N_2 system of microcellular injection molded dog-bone samples are analyzed for the cell nucleation and bubble growing behaviors. In Fig. 2, cell density distribution of molded part is from 2.28×10^6 to 5.36×10^6 cells/cm^3, and average value is about 3.71×10^6 cells/cm^3, which is mean value calculated from nodes of three dimensional mesh. The maximum value appears at the position where pressure driven flow stops flowing and molten polymer starts to fulfill the cavity by cell expansion; also, the minimum value happens at the end of cavity where is last filled by melt. The density distribution of core section of molded part, as shown in Fig. 3, shows broader density range than that of surface of molded part. The cell size distribution results are given in Fig. 4 and Fig. 5 for surface and core section respectively. The radius of molded part is from 3 μm to 32 μm and the average radius is 20 μm; so that, one might observe that cells having size smaller 10 μm almost appears near the surface (frozen layer).

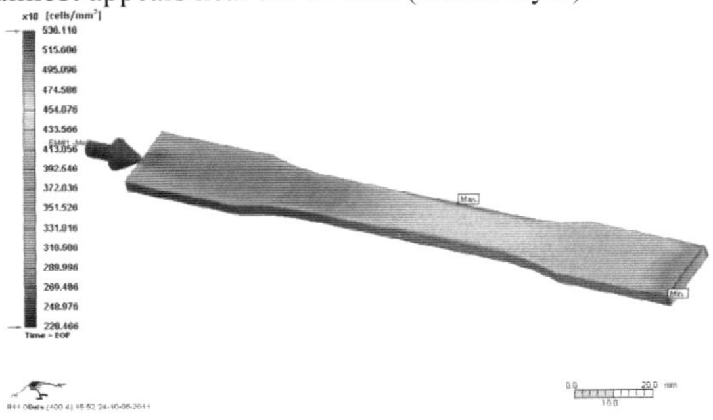

Fig. 2 Cell density distribution of sample surface for 0.57 wt% N2 SCF dissolved in PP. (Legend scale is 2.28x103 ~ 5.36x103 cells/mm3)

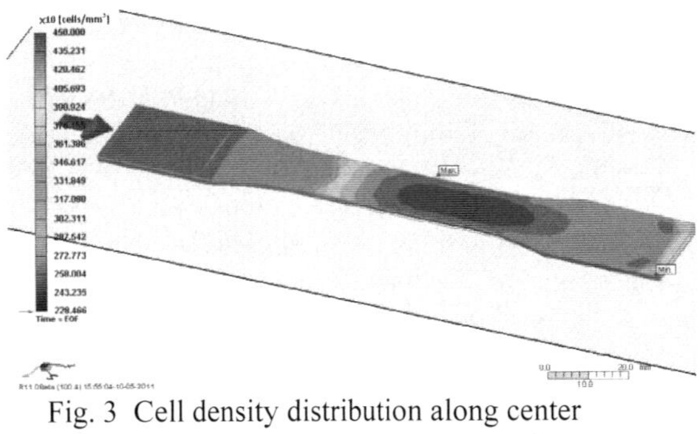

Fig. 3 Cell density distribution along center section of sample for 0.57 wt% N2 SCF dissolved in PP. (Legend scale is 2.28x103 ~ 4.50 x103 cells/mm3)

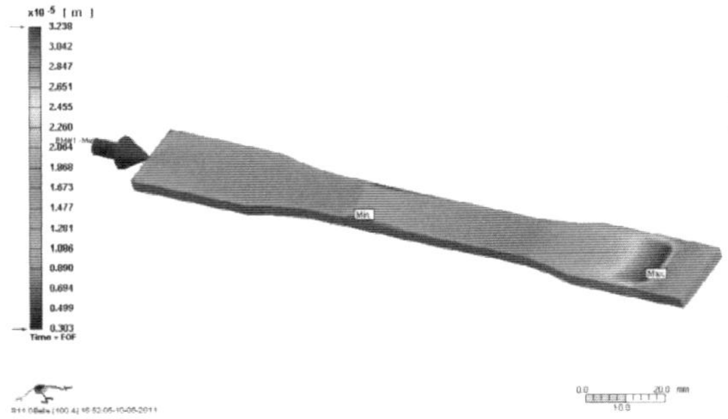

Fig. 4 Cell radius distribution of sample surface for 0.57 wt% N2 SCF dissolved in PP. (Legend scale is 3.03 ~ 32.38 μm)

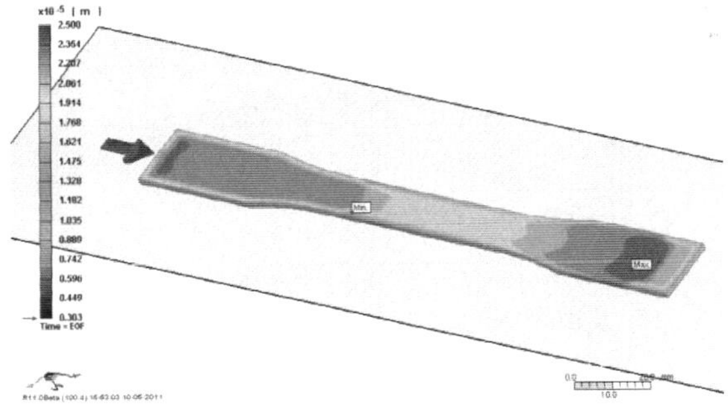

Fig. 5 Cell radius distribution along center section of sample for 0.57 wt% N2 SCF dissolved in PP. (Legend scale is 3.03 ~ 32.38 μ m)

In the average sense, higher dissolved SCF concentration gives higher cell density and smaller cell size than that of lower concentration.

The phenomena can be interpreted by the competition behavior between cell nucleation and bubble growing. In addition, lower dissolved SCF gives more uniformed cell size distribution. Comparison of Fig. 4 and Fig. 5 also tells the cells near surface, i.e. frozen layer, are much smaller than average cell size, which is the phenomena observed in a real microcellular injection molded part.

The comparisons of microcellular injection simulations with and without considering counter pressure are illustrated in Fig. 6 and Fig. 7. In Fig. 7, simulation result is of dissolved gas concentration 0.57wt% for LDPE/N_2 with counter pressure 2MPa. The average radius of part surface with counter pressure 2MPa, as shown in Fig. 6, is obviously smaller than that without counter pressure as shown in Fig. 7. This comparison agree that counter pressure application improve the surface quality of microcellular foaming product.

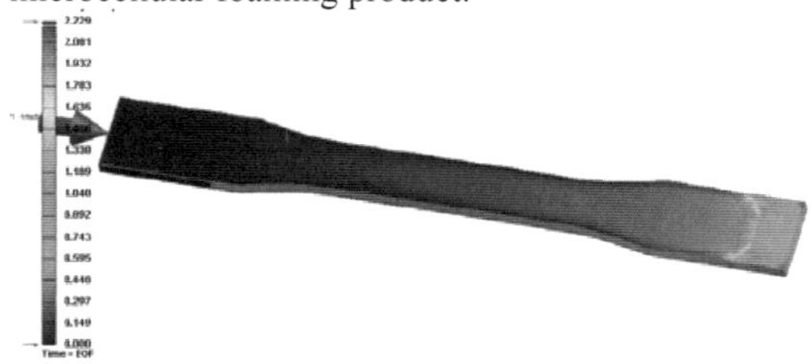

Fig. 6 Cell radius distribution of sample surface for 0.57 wt% N2 SCF dissolved in LDPE with counter pressure 2 MPa. (Legend scale is 0 ~ 22.3 μm)

Fig. 7 Cell radius distribution of sample surface for 0.57 wt% N2 SCF dissolved in LDPE. (Legend scale is 0 ~ 22.4 μm)

From Fig. 8 and Fig. 9 are SEM photos of a microcellular molded part and were taken from the position as indicated in Fig. 10; furthermore, simulation results as shown in Fig. 11 and Fig. 12 are given for comparisons. For all the locations, it's clear to see the difference of the cell size distribution from core zone to skin layer. No surprisingly, the bigger cells appear in the core zone and cell size is very small along the skin layer due to molten polymer are solidified in a short time. The comparisons between experimental results, Fig. 9 and Fig.10, and simulation results, Fig. 13 and Fig.14, provide a consistent conclusion that cell size in the core zone are bigger for the flow end position than that of middle way of flow. Also, the core zone is narrower in the flow end region, which are both shown in Fig. 10 of SEM and in Fig. 14 of simulation.

Fig. 8 SEM of middle part (position A as indicated in Fig. 10) for 0.2 wt% N2 SCF dissolved in PP

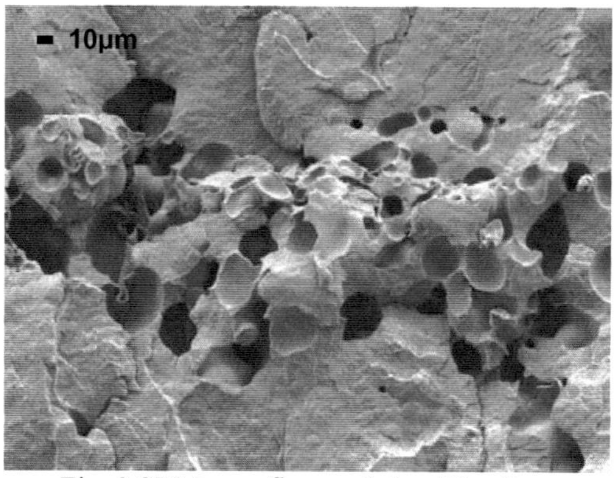

Fig. 9 SEM near flow end (position B as indicated in Fig. 10) for 0.2 wt% N2 SCF dissolved in PP

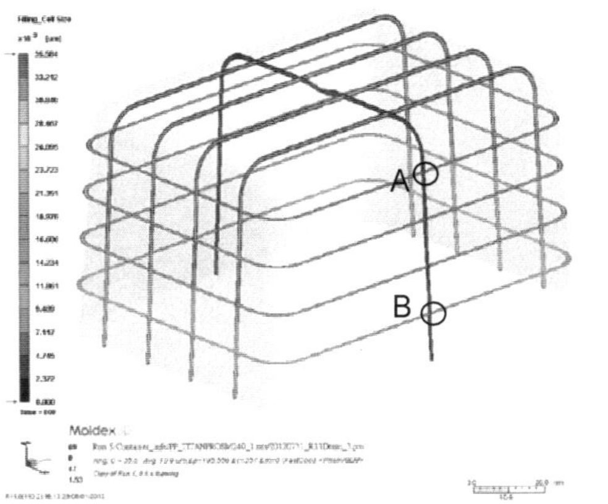

Fig. 10 SEM of middle part for 0.2 wt% N2 SCF dissolved in PP

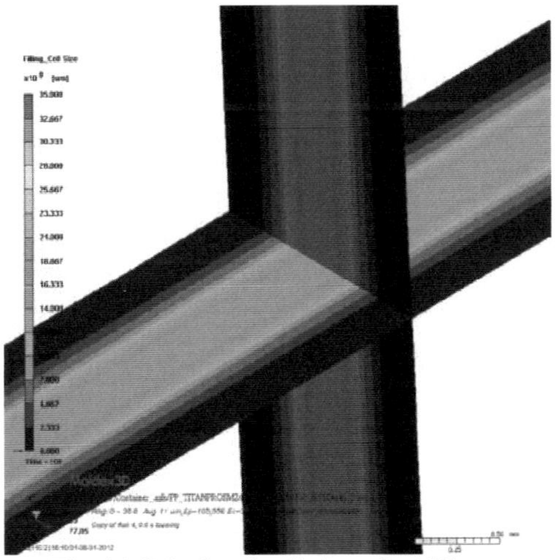
Fig. 11 Calculated cell size of middle part (position A as indicated in Fig. 10) for 0.2 wt% N2 SCF dissolved in PP

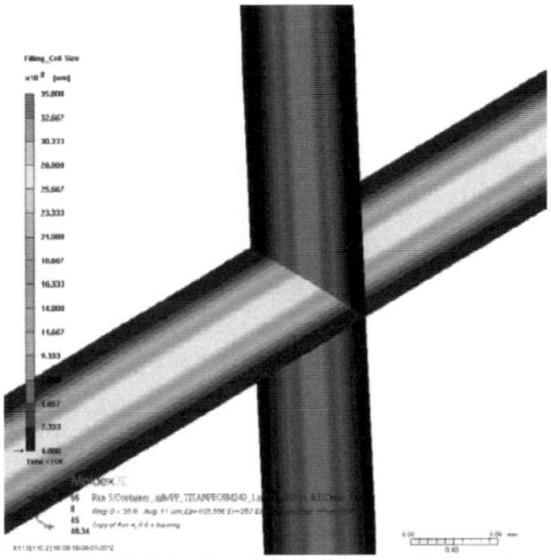
Fig. 12 Calculated cell size near flow end (position B as indicated in Fig. 10) for 0.2 wt% N2 SCF dissolved in PP

Conclusions

This study presents the 3D simulation capability of predicting dynamic behavior for microcellular injection molding process. Simulation results of microcellular injection molding for nitrogen SCF dissolved in PP show good agreement with experimental data. In the industrial application, the validation of simulation results with experimental data meet the criteria of prediction capability for further application of microcellular injection molding simulation for a complex geometrical product.

We also carried out the comparisons of calculated results for microcellular injection molding with and without considering counter pressure. The comparison provides an correlation of surface quality and surface cell distribution of microcellular injection molding process.

Furthermore, a microcellular molded part are studied for its cell size distribution by SEM and simulation. The results are consistent and 3D simulation can provide the cell size variation along the direction perpendicular to the flow.

References

1. N. P. Suh, Hanser/Garden Publications, Cincinnati, Chpter 3 (1996).
2. J. Xu, John Wiley and Sons, New Jersey (2010).
3. Venerus, *Cellular Polymers*, **22**, 9 (2003).
4. K. Taki, *Chemical Engineering Science,* **63,** 3643 (2008).

5. J. Lee and L.-S. Turng et al, *Polymer*, **52**, 1436 (2011).
6. C. D. Han and H. J. Yoo, *Polymer Engineering and Science*, **21**, 518 (1981).
7. R.-Y. Chang and W.-H. Yang, *Int. J. Numer. Meth. Fluids*, **37**, 125 (2001).

O-10-238

COOLING SYSTEM EFFECTS IN A REAL HOT RUNNER SYSTEM USING A 3D NUMERICAL METHOD

Fu-Hung Hsu[1*], Hung-Chou Wang[1], Chao-Tsai Huang[1], and Rong-Yeu Chang[2]

[1] *CoreTech System (Moldex3D) Co., Ltd., ChuPei City, Hsinchu 302, Taiwan – andrewhsu@moldex3d.com; kentwang@moldex3d.com; cthuang@moldex3d.com*
[2] *Department of Chemical Engineering, National Tsing-Hua University, Hsinchu 30043, Taiwan – rychang@moldex3d.com*

Abstract

Hot runner system is widely used nowadays in injection molding. However, the system is quite complex including many components. Thus, it is usually not easy to simulate its mechanism in a full scheme. There are several critical issues in a hot runner system such as temperature control, flow balance, and material degradation. Heating coils and cooling system play an important role in temperature control. If not controlled properly, product defects may occur. In the current study, the concept of conformal cooling system was introduced to optimize cooling effects close to nozzle tips. A 3-dimensional numerical method was applied. The results were first validated with a real hot runner system. Then few different cooling systems were designed to investigate their temperature and product quality effects. The results provide a helpful reference for a better hot runner system design.

Keywords: Conformal cooling, hot runner, nozzle tip, 3D numerical method

Introduction

There are two types of runner system in injection molding – cold runner and hot runner. A hot runner system has many benefits over a cold runner system [1-3]. However, it is very complex which can be constructed over a thousand components. In addition to heating components such as heating coils, cooling channels may exist in the system in order to provide a more efficient cooling. For example, gate temperature control is very critical since it can cause stringing if the temperature is too high and difficulty in injecting if the melt temperature is too low. It is not always neccessary for a cooling channel around nozzle tips. For amorphous materials, the operating temperature range is larger than crystalline material. Thus, a nozzle tip cooling channel is needed to control the gate temperature precisely. In the case of smaller parts, the nozzle tip cooling system design can also play an important role in shortening cycle time and avoiding part defects such as sink mark and warpage.Here the concept of conformal cooling is introduced in order to have the best result. With laser sintering method, it is feasible to manufacture cooling channels in arbitrary geometry [4-6]. In this research, we modified the cooling system of an existing hot runner system. A three dimensioinal numerical method was used to show the improvement of cooling efficiency and product quality [7].

Numerical Simulation

A three dimensional numerical tool (Moldex3D) was used for this analysis. Each major component of this single drop hot runner system can be molded in the preprocessor including heating coil power, cooling system, material properties, and contact boundary condition. Figure 1 shows the details of the settings. Notice that the bottom heating coil has a higher power (610W) than other two since it is closer to the cooling channel. In the current study, the heating coil power is controlled at a temperature range from 278℃ to 282℃.

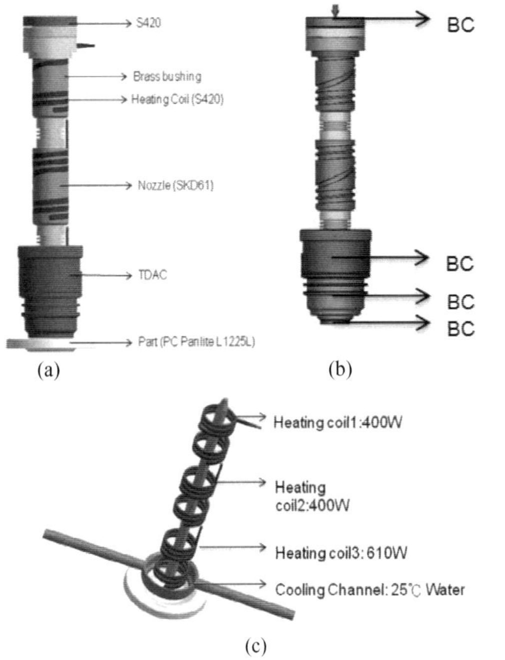

Fig 1. (a) Component material setting (b) Contact BC setting (c) heating coils and cooling system settings

The heat transfer phenomenon is governed by a three-dimensional Poisson equation [8] as followed:

$$\rho C_P \frac{\partial T}{\partial t} = k\left(\frac{\partial^2 T}{\partial x^2} + \frac{\partial^2 T}{\partial y^2} + \frac{\partial^2 T}{\partial z^2}\right) \quad \text{(Eq.1)}$$

where T is the temperature, t is the time, x, y, and z are the Cartesian coordinates, ρ is the density, C_P is the specific heat, k is the is the thermal conductivity. The initial part temperature distribution is obtained from the analysis results at the end of filling and packing stages.

$$T(0,\vec{r}) = \begin{cases} T_c, & for\ \vec{r} \in \Omega_m \\ T_p(\vec{r}), & for\ \vec{r} \in \Omega_p \end{cases} \quad \text{(Eq.2)}$$

In this study, the effect of thermal radiation is ignored. The conditions defined over the boundary surfaces and interfaces of the mold are specified as,

$$for\ t \geq 0, \quad -k_m \frac{\partial T}{\partial n} = h(T - T_o) \quad \text{(Eq.3)}$$

Where n is the normal direction of mold boundary, h is heat transfer coefficient. On the exterior surfaces of the moldbase,

$$h = h_{air}, T_0 = T_{air} \quad for\ \vec{r} \in \Gamma_m \quad \text{(Eq.4)}$$

For the fluid part, it is considered to be incompressible, Newtonian (for water) or generalized Newtonian (for polymer melt). The governing equations for 3D transient non-isothermal motion are:

$$\frac{\partial \rho}{\partial t} + \nabla \cdot \rho \mathbf{u} = 0 \quad \text{(Eq.5)}$$

$$\frac{\partial}{\partial t}(\rho \mathbf{u}) + \nabla \cdot (\rho \mathbf{u}\mathbf{u} + \tau) = -\nabla p + \rho \mathbf{g} \quad \text{(Eq.6)}$$

$$\rho C_P\left(\frac{\partial T}{\partial t} + \mathbf{u} \cdot \nabla T\right) = \nabla(\mathbf{k}\nabla T) + \eta\dot{\gamma}^2 \quad \text{(Eq.7)}$$

where $\mathbf{u}$ is velocity vector, T is temperature, t is time, p is pressure, $\boldsymbol{\tau}$ is stress tensor, ρ is density, η is viscosity, k is thermal conductivity, C_p is specific heat and $\dot{\gamma}$ is shear rate. For the polymer melt, the stress tensor can be expressed as:

$$\tau = -\eta(\nabla \mathbf{u} + \nabla \mathbf{u}^T) \quad \text{(Eq.8)}$$

The modified-Cross model with Arrhenius temperature dependence is employed to describe the viscosity of polymer melt:

$$\eta(T, \dot{\gamma}) = \frac{\eta_o(T)}{1 + (\eta_o \dot{\gamma}/\tau^*)^{1-n}} \quad (9)$$

with

$$\eta_o(T) = BExp\left(\frac{T_b}{T}\right) \quad (10)$$

where n is the power law index, η_o is the zero shear viscosity, τ^* is the parameter that describes the transition region between zero shear rate and the power law region of the viscosity curve.

Results and Discussion

In the previous study [7], this hot runner system had been validated with good temperature agreement with experimental results. In the current study, we focus on the cooling system effect on part temperature distribution, warpage, and gate temperature control.

Figure 2 shows the five measure points on the part. Point 1 is at the center and point five is at the border. Figure 3 shows the original cooling system design while figure 4 and 5 show two alternative designs.

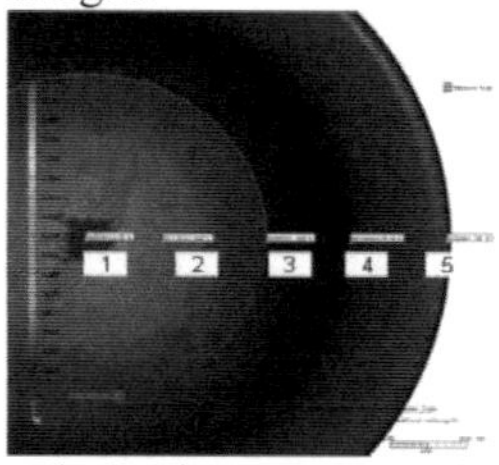

Fig 2. Five measure points

Fig 3. Original design

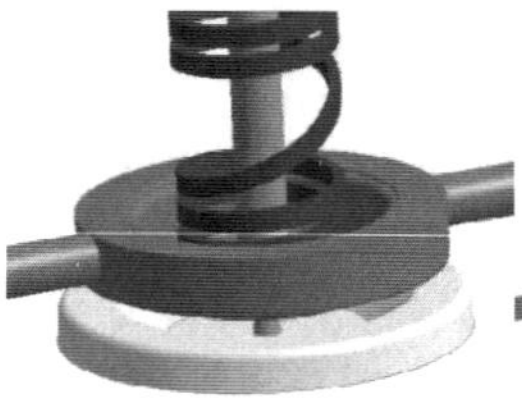
Fig 4. Design 1

Fig 5. Design 2

Design 1 has wider contact surface with the part than the original design; however, it does not keep the same distance to the part. Design 2 is so-called "conformal" cooling design, the distance between cooling channel and part surface is essentially the same throughout the part. This assures the most uniform cooling effect.

Figure 6 to Figure 8 show the result comparisons of different cooling system designs. In addition to the three curves under three different designs, the fourth curve is also added based on design 2 with shorter cooling time (50% and 10 seconds shorter).

In Figure 6, it can be observed that Design 2 has a much more uniform surface temperature distribution than other two. Also, with shorter cooling time, there is also no effect on surface temperature except the center.

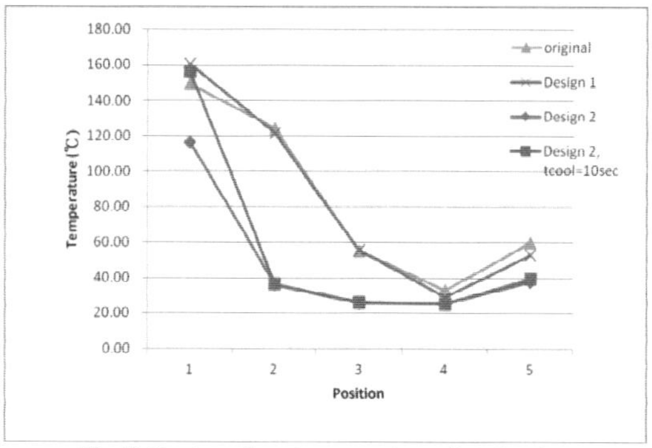

Fig 6. Part surface temperature distribution

Figure 7 shows the comparisons of volumetric shrinkage rate. Generally, lower temperature areas have lower volumetric shrinkage. Design 2 has lower shrinkage rate than other two throughout the part. Also, except for the center point, all other measure points have fairly close values. Figure 8 shows the total displacement results. Obviously, design 2 has the lowest deformation among all designs. Even with a shorter cooling time, the difference is not significant.

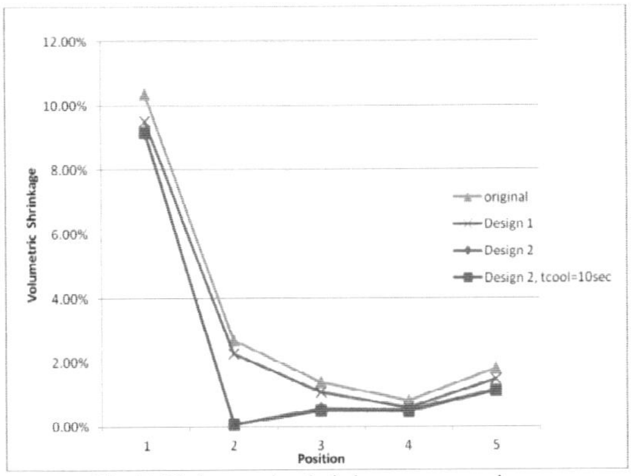

Fig 7. Volumetric shrinkage comparison

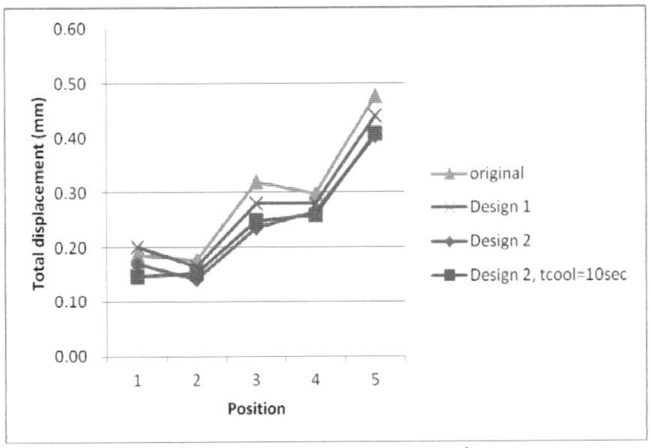

Fig 8. Total displacement comparison

All of the cases above have the same cooling process conditions including flow rate, coolant temperature, and cooling time except for the last case (cooling time is 10 seconds). This is because that the gate temperature is too low for design 2 if same flow rate is kept. Figure 9 shows the gate temperature history with the original cooling system design. After ten cycles, the temperature at the end of cooling is 203 ℃. This temperature is a little higher than the glass transition temperature.

However, for design 2, the gate temperature will be to low if the same cooling time (20 seconds) is applied. Hot runner gate temperature need to be precisely control. If the temperature is too high, then a drooling or stringing problem can occur. On the other side, if the temperature is too low, then it will have difficulty in filling the cavity. Thus, in order to raise the temperature, a ten seconds shorter cooling time is applied. As shown in Figure 10, it turned out a fairly close temperature (201℃) to the original design at the end of cooling. Also, temperature reaches a steady cycle even faster with a 10-second cooling time.

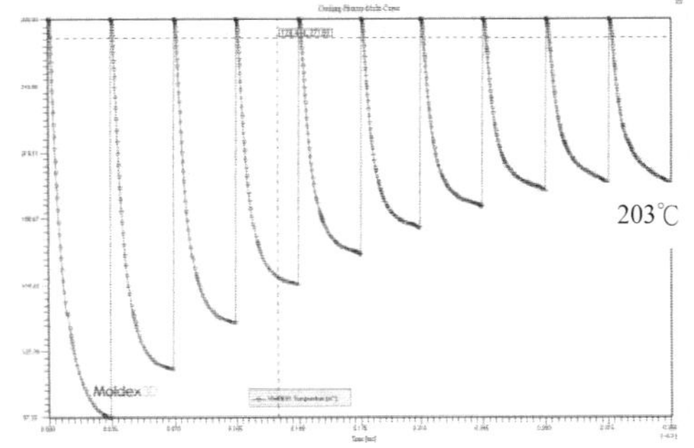

Fig 9. Temperature history at gate (original design)

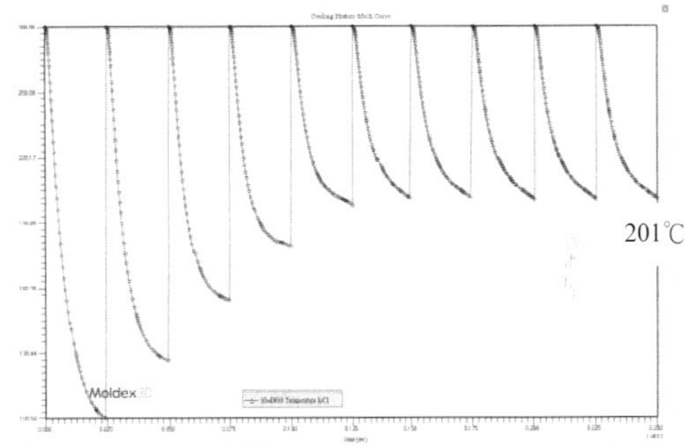

Fig 10. Temperature history at gate (Design 2)

Conclusions

For amorphous plastic materials, a hot runner tip cooling channel is usually applied in order to precisely control gate temperature. In the current study, the concept of conformal cooling was applied on a new hot runner tip cooling system design. The three dimensional simulation results show that this design can make the part temperature more uniform and reduce part warpage. In addition, cooling time can be even shorter (50% reduction in this case). Although the cost of conformal cooling manufacturing can be higher than traditional one, the benefits can compensate the cost and increase product competitive strength.

References

1. D. Frenkler, H. Zawistowski, "Hot Runners in Injection Moulds", Rapra publishers, Shawbury 1998, p.267.
2. Peter Unger, "Hot Runner Technology", Hanser Gardener Publication, 2006.
3. John P. Beaumont, "Runner and Gating Design Handbook", Hanser Gardener Publication, 2007

4. K.W. Dalgarno and T.D. Stewart, "Manufacture of Production Injection Mould Tooling Incorporating Conformal Cooling Channels via Indirect Selective Laser Sintering", proceeding of the institution of mechanical engineers, 2001, Vol. 215, part B, pp. 1323-1332
5. Emanuel Sachs, Edward Wylonis, Samuel Allen, Michael Cima, and Honglin Guo, "Production of Injection Molding Tooling with Conformal Cooling Channels Using the Three Dimensional Printing Process", Polymer Engineering and Science, 2000, Vol. 40, No.5, pp.1232-1247
6. Siegfried Mayer, "Optimized Mould Temperature Control Procedure Using DMLS", EOS whitepaper
7. Tzu-Chau Chen, "TEMPERATURE INVESTIGATION IN REAL HOT RUNNER SYSTEM USING A TRUE 3D NUMERICAL METHOD "ANTEC Tech. Papers 2012
8. F. P. Incropera and D. P. DeWitt, "Introduction to Heat Transfer", John Wiley & Sons Inc, 1996.

O-10-237

NEW POSSIBILITIES OF TRANSIENT SIMULATION IN DESCRIBING COLOR AND MATERIAL CHANGE PROCESSES IN EXTRUSION TECHNOLOGY

L. Szöke*, O. te Heesen, C. Ketteler and J. Wortberg

Institute of Product Engineering (IPE), University of Duisburg - Essen, Germany – levente.szoeke@uni-due.de

Abstract Although the assumption of stationary conditions is common to describe extrusion processes theoretically, it is shown that transient modeling leads to more accurate explanations of the occurring phenomena and can even lead to important information to develop new die designs. A practical example is presented for color and material changes due to the highly dynamical process behavior. Product changes have significantly won in interest over the last decades due to decreasing lot sizes and higher number of changes. A selective optimization of the die geometry is a promising way to improve the change characteristics of the system. Therefore the prediction of the process is highly important. This possibility is now given by modern computational calculation software which is commercially available. By expanding the common stationary three dimensional flow simulation to a transient multiphase description of the flow the interaction between two different melt phases can be represented. It is possible to identify geometrical regions with negative influence to the change processes. This leads to the opportunity to optimize the geometry and improve purge characteristics. In addition the multiphase approach enables the user to predict the chronological progress of the change process itself. Qualitative statements can easily be done to compare different die geometries. To make quantitative predictions of the purge time it will be necessary to calibrate the computation by experiments to include the influence of pigments and additives which are not yet implemented in the model describing the material properties.

Keywords: color and material change, die geometry optimization, transient CFD simulation, multiphase flow

Introduction

The manufacturing of packaging is one of the most important product fields of the plastics industry. In the recent years well over a third of all plastic products were produced for this segment. In this field an increasing function-oriented specialization of material formulations can be observed regarding the application of the products. So beside aspects of marketing this causes decreasing lot sizes and thereby increasing numbers of color and material changes. In terms of economic producing this means a great challenge to the plastics industry. This is for example best seen at the production of very disturbance sensitive products such as thin films with multiple functional layers with a vast quantity of possible recipes. For every product change the decision is to be made whether the change is executed in the ongoing operation and the waste of material is accepted or whether a disassembly and cleaning of the whole system is necessary. A disassembly and manual cleaning results in considerable economic losses mainly through downtimes and personnel costs depending on the size of the plant. However a product change under operation conditions also requires the loss of valuable production time and material resources until the process returns to manufacturing with reasonable quality. Therefore the knowledge of the system behavior during a change is a highly important designing criterion. In [1] this is already described for flow channels with a simple basic geometry and a simplified analytical calculation approach is proposed. The potential of commercial CFD tools to examine complex geometries with integrated multiphase models offers the opportunity to analyze highly dynamical flow situations in various applications. This is already common for injection molding to describe the filling of the mold for example [2]. In extrusion technology transient simulations are used to predict the melting behavior in extruders itself [3]. Therefore in this paper complex three-dimensional flows in transient change processes are observed to evaluate the possible gain in knowledge concerning the optimization of die geometries.

Experimental and computational methods

Transient modeling of color and material change processes in extrusion dies

The transient simulation of an extrusion die presented in this work includes a multiphase approach, which leads to a representation of the displacement of one melt phase by another. Therefore it is possible to predict the purging behavior of different combinations of materials in specific die variants. The calculation of the multiphase flow is based on the Volume of Fluid method which is used to track the interface between the two melt phases [4]. Both phases are considered as immiscible continua what leads to a strict boundary inside the die. For each cell of the computational mesh there are variables called volume fraction for each of the phases. A value of 1 means that the cell is completely filled with the specific phase while a value of 0 stands for an empty cell. Because of the conservation of mass the sum of volume fractions for each cell has to be 1. Near the interface the boundary inside the cells is approximated by a linear function leading to a volume fraction between 0 and 1.

Therefore it is necessary to define at which point the cells are assumed to be completely purged. Depending on the fineness of the computational mesh a global boundary value for the volume fraction has to be specified above which the cell is declared as completely filled with the secondary phase.
To validate the approach explained above computational and experimental results considering a simple geometry have to be compared. This was done by a simple straight rod. The simulation was executed in 2D with a high resolution of the near-wall layers. The cell size next to the wall starts with 10 μm in radial dimension. It increases with a constant factor while closing in on the axis of symmetry to reduce the overall number of cells. The results of the computed wall layer thickness are compared to experimental results in the following section.

Experimental methods to validate the approach

Due to the differences of the melt flow velocities (very low velocities at the wall regions) the melt from the channel center actually arrives outside the die at first and forms the exterior surface of the probes. So the usage of data from outside the channel may not be directly comparable to the calculation inside the extrusion die. It is therefore necessary to somehow measure the flow conditions in the channel in order to validate the calculation results. To visualize the flow conditions it proves to be helpful, that there are multiple materials involved in the changing process, as the material extruded before the change (further referred to as starting material) is colored with a masterbatch as a tracer. The starting and the target material used in this paper are both low density polyethylene (PE-LD). The changing process is interrupted at certain time steps and the filled die is cooled down rapidly. Through optical measurement of the frozen layers at the wall it is possible to compare calculation to experiment. Before looking into complex three dimensional geometries at first a straight rod die of 20 mm diameter and 200 mm length is analyzed. The design allows to open up the die into two halves. The retrieved probe is milled down to one half and the surface showing two melt phases frozen in a particular time-step is scanned with high resolution. The thin layer of the starting material containing the tracer is measured by digital image processing. To quantitatively validate the approach a color change process is interrupted in six different time stages. The layers of starting material colored by the tracer masterbatch are measured from the scanned images. At the same time not only the layer thickness is of interest but also the position in the probe rod. Due to the knowledge of the exact location a comparison with the simulated results can be drawn. For complex three dimensional geometries such as spiral mandrel dies only qualitative comparison is possible due to the high complexity of the experimental procedure. On one hand the contribution of both material phases is compared at the outflow region over a broad period of time during a change. On the other hand the extrusion process through the die is stopped during a change and the cooled down melt is observed from the dismantled die. The presented method is limited by the shrinkage, but nevertheless capable of supplying valid data due to the fast cooling of the near-wall layers.

Results and Discussion

Potential to predict the changing behavior

The progress of color and material changes is mainly driven through the behavior of the melt flow near the channel walls. The measured data of the layer thickness at the channel wall therefore characterizes the change process. As shown in Fig. 1 the experimental data for the layer thickness at the channel wall is very similar to the computed values. The maximal relative deviation lies at 7.7 %. A more general and in regard of the methods of measurement more independent view can be achieved through the analysis of the wall layer reduction rate. This parameter characterizes the velocity of the reduction of the starting material layer thickness over the flow time. As the interface approaches the zone near the wall, the reduction rate reaches a very low value. So the transient behavior of the wall layer thickness can be explained. The similarity between experiment and computation, as shown in this paper, is high at the very beginning of the changing process. But it is nevertheless very likely that this similarity decreases for very small layer thicknesses. This is mainly because of the influential effects of the topological properties of the surface and the tracer material attributes. In the literature such influences were quantitatively proven in [1].

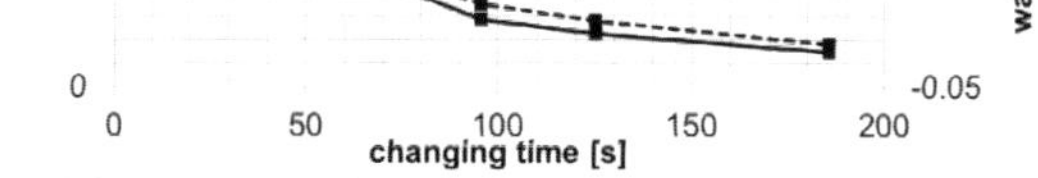

Fig. 1: comparison between experimental and calculated data

To virtualy observe and be able to analyse the terminal phases of changing processes, these effects have to be taken into consideration through appropriate calibration of the calculation method. Furthermore it is necessary to change the method of validation towards the end of the change, because of the increasing inaccuracy of the thickness measurement in very thin layers.

Qualitative validation of the results

To adapt the presented approach to a complex geometry a flat die was used. The analyzed die is shown in Fig. 2. The characteristic property of the geometry is the overlapping flow channel which leads to a particular development of the flow behavior at the outlet of the die. At point (1) in Fig. 2 the inlet of the die is shown. The plastic melt is distributed in two channels in circumferential direction (2). Parallel to that the melt flows in a small gap in radial direction (3). At point (4) the two channels overlap. Point (5) is the outlet of the die.

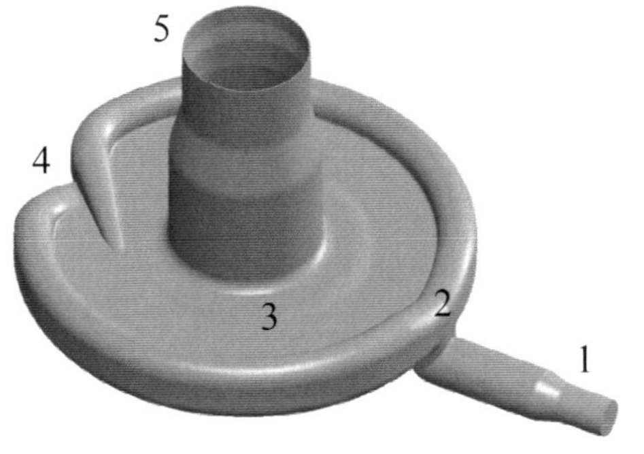

Fig. 2: analyzed geometrie of a flat die

Fig. 3 to 5 show the displacement of the natural starting material with black colored target material at the outlet of the die. The upper pictures represent the results of the experimental changing process at different time steps while the lower image is the corresponding simulation result. These simulation results show the volume fraction of the target material where a volume fraction above 0.95 is plotted in black.

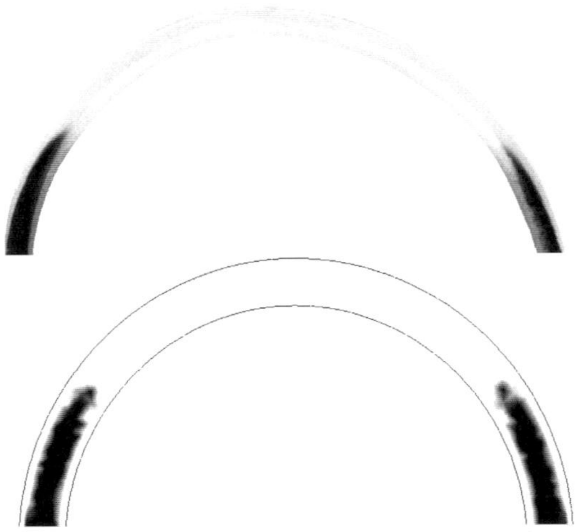

Fig. 3: changing process in experiment and simulation after 180 s

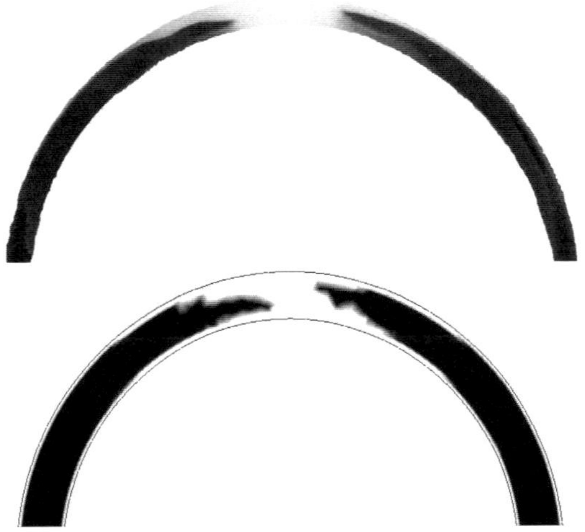

Fig. 4: changing process in experiment and simulation after 430 s

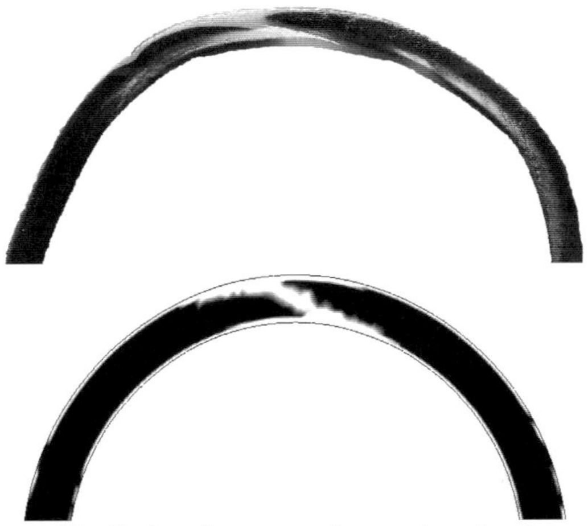

Fig. 5: changing process in experiment and simulation after 650 s

As seen in the figures above the transient calculation shows a good correlation to the experimental results. It is shown that the first target material reaches the outlet at the side directed to the inlet of the die. The further the process proceeds the two melt fronts continue to extend towards each other.

The beginning of the characteristic formation of the melt front can be seen in Fig. 4 while indentations develop at the inner wall on the right side of the melt front and on the outer wall on the left side. These phenomena occur because of the overlapping flow channel geometrie in the back area of the die. In the further changing process the two melt fronts flow past each other while a thin layer of the starting material remains, seperating the two fronts as shown in Fig. 5. Theses effects can be declared as the major factors increasing the changing time.

With the knowledge of the development of the occuring phenomena it is possible to detect the regions inside the geometry which cause theses problems. So the transient modelling of the changing process inside the die gives the opportunity to optimize the geometry at the crucial locations.

Conclusions

By describing the continuous extrusion process with transient flow simulations it is possible to describe the complex interaction of different melt phases in the various components of an extrusion line. Considering the fact that the residence time behavior has great importance on color and material change processes as well as on the product quality, this leads to the possibility of specific optimizations of these components.

The presented approach is already sufficient to compare different analyzed variants in a qualitative manner. In order to achieve quantitative results such as an exact purging time, a calibration of the applied simulation becomes necessary. This calibration has to be able to take into account the influence of certain additives in the compounds as well as the characteristics of the flow channels surface.

Further investigations may also lead to a transient description of the material data. By calculating the transient development of the flow, thermo-rheological changes in material properties could be implemented by a suitable model. These changes will for example have a great influence on the flow development if one of the involved materials shows a time-dependent increase of viscosity while the other materials viscosity decreases over time.

Acknowledgements

The flat die used for the practical tests in this paper was designed and manufactured in the following project:

NRW Ziel-2 Program

CheK.NRW – Chemie und Kunststofftechnik

ERDF – European Regional Development Fund

Title: Computer-Aided Optimization of Extrusion-Dies in Consideration of Manufacturing Restrictions

Project-No. 280809302

Special Thanks:

- Sabic Europe (PELD 2100TN00)
- Basell Polyolefine GmbH (HDPE Hostalen 9255 Plus)
- A. Schulmann GmbH (Polybatch blue)

for supporting this research project with material.

References

1. J. Wortberg, C. Groß, *Optimizing purging time at product change in blown film extrusion by selective modification of process parameters*, Annual Technical Conference ANTEC of the Plastics Engineers SPE, Chicago USA, 2004
2. M. R. Kamal, E. Chu, P. G. Lafleur, M. E. Ryan, *Computer Simulation of Injection Mold Filling for Viscoelastic Melts with Fountain Flow,* Polymer Engineering and Science, vol. 26 no. 3, p. 190-196, 1986
3. F. Ilinca, J.-F. Hétu, *Three-dimensional Finite Element Solution of the Flow in Single and Twin-Screw Extruders*, International Polymer Processing, vol. 25 no. 4, p. 275-286, 2010
4. C. W. Hirth, B. D. Nichols, *Volume of Fluid (VOF) Method for the Dynamics of Free Boundaries*, Journal of Computational Physics, no. 39, p. 201-225, 1981

O-10-236

CFD-BASED OPTIMIZATION OF EXTRUSION DIES BY ANALYZING THE DWELL TIME DISTRIBUTION AND FLOW HISTORY

O. te Heesen*, J. Wortberg

Institute of Product Engineering (IPE), University of Duisburg-Essen, Germany – oliver.teheesen@uni-due.de

Abstract – Nowadays three-dimensional simulations of the complex flow in extrusion dies by usage of computational fluid dynamics (CFD) are used in the design process. The results of such CFD-simulations have to be analyzed for every design proposal by experienced engineers. The geometry optimization only based on the basic post processing tools (view of path lines, contour plots, etc.) is very time consuming and depends on the background knowledge of the CFD-engineer. Therefore methods are needed to directly analyze the resulting product quality by scalar values without the need of human interpretation. The content of this publication is the optimization of extrusion dies by scalar values for the dwell time distribution. Furthermore a possibility for increasing the die quality by detail optimization of critical areas is given. Up to now the volume flow distribution at the outlet of the die is used to predict the wall thickness of the extruded product. In this paper it is shown that for some plastics it is necessary to additionally analyze the dwell time distribution at the outlet of a die. Based on that it is possible to analyze the flow history along specific path lines which start in zones of high dwell times. By a reverse analysis of the flow history along these path lines, critical areas in the die-geometry can be located. It is also shown that critical zones for the purge process of the die could be located the same way. A validation of these methods is given by a comparison between experimental tests and CFD-simulation results of the same die geometry.

Keywords: extrusion, die optimization, flow history analysis, dwell time dependency, CFD-based die design

Introduction

One of the main components in an extrusion line is the die, which has significant influence on the product quality. The wall thickness distribution of an extruded tube for example is directly influenced by the melt distribution in the die, so in the process of die design the resulting product quality is set. Nowadays CFD simulations of the melt flow in the die are used in the design process. These available simulation tools helps to understand the complex flow behavior of the plastic melt. To predict the product quality of the resulting die directly in the design process analysis methods for the simulation results are needed which correlate to properties of the extrudes product. In the past such values were presented by the IPE (for example the angle of weld-line overlapping). [1]

The content of this publication is the optimization of extrusion dies by scalar values for the dwell time distribution. To predict the wall thickness distribution of the product usually the CFD result of the volume flow distribution at the outlet of the die is analyzed. In this paper it is shown that for some plastics it is necessary to additionally analyze the dwell time distribution at the outlet of a die. Besides the dependency of temperature, some plastics show a time dependent behavior of the shear viscosity and modulus. [2] In a steady CFD-simulation this phenomena is neglected, because the viscosity is calculated depending on shear rate and temperature only. A method is shown that allows to calculate the dwell time distribution at the die-outlet. By calculating the root mean square deviation of this distribution different design proposals could be compared. In this paper the methods and experimental tests are discussed at a geometrie of spiral mandrel die. Because of the generality of the analysis methods it is possible to use the same methods for other die geometries like dies for profile extrusion.

Conventional die design

The workflow of extrusion die design is based on an analytical (pre-)design process and often followed by a process of three-dimensional simulation and optimization. [3, 4, 5] As mentioned above, the CFD results of the volume flow distribution at the outlet of the die are analyzed to predict the wall thickness distribution of the product. In the following picture the velocity distribution at the outlet of a spiral mandrel die which is simulated in an 3D-CFD-simulation is shown.

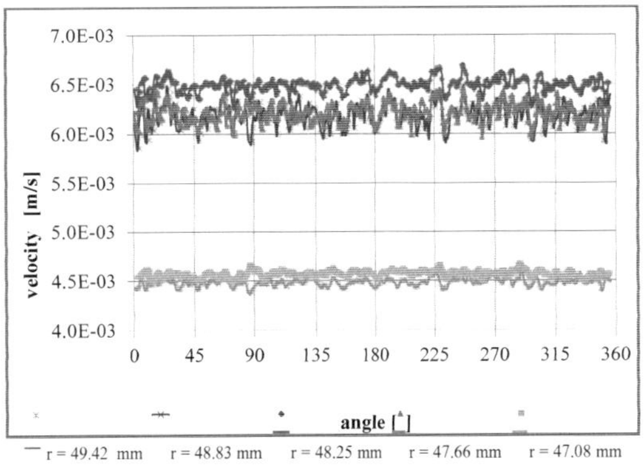

Figure 1: velocity distribution at the outlet of a spiral mandrel die

For the experiments and simulations in this paper a spiral mandrel die with four mandrels is used. The outer radius of the die is 50 mm, the inner radius is 46,5 mm. Unless otherwise mentioned a LD-PE (polyethylene with low density, sabic 2100 TN00) with a mass flow rate of 15 kg/h is used. The fvm-mesh used for the simulations consists of 8.1 m elements with a boundary layer of five prisms.

For a detail analysis the velocity distribution is calculated along five circles at the outlet of the die with an result for every 0.5 deg. The circles are equal spaced along the gap, so the velocity of each point is proportional to the volume flow. As you can see in Fig. 1 the distribution is uniform considering numerical inaccuracies because of the discretization.

In Fig. 2 the wall thickness distribution of a LD-PE blow film extruded with the die is shown. The blow up ratio of the bubble is two and the nominal thickness is 100 μm.

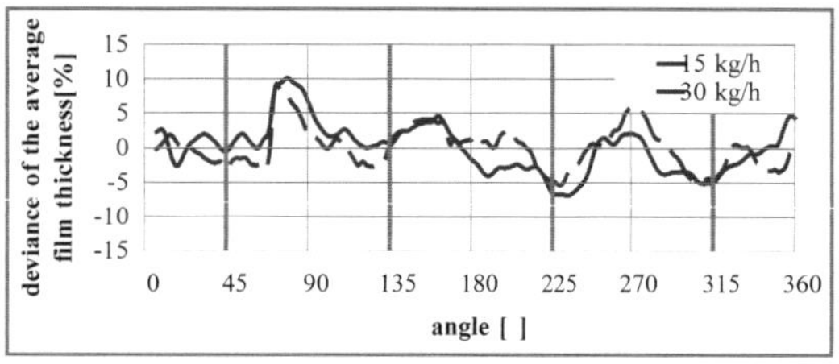

Figure 2: wall thickness distribution of a LD-PE film

With a frequency of 90 deg. there are four local minima and four local maxima of the distribution detected. The distribution is not as equal as predicted by the analysis of the velocity distribution of the CFD-simulation. The reason for this unequal distribution must be a geometric element which has also a frequency of four. The influence of the cooling air can be neglected because there are six flow channels for the air. On this account it is obvious that the mandrel structure of the die causes the thicker and thinner sections in the extruded film. Up to now there is no possibility to predict this behavior, because no design criterion is able to calculate the reason for that.

Advanced analysis methods

The results of the CFD-simulation are based upon the carreau model. The calculated viscosity and so the whole flow field is calculated by dependency of shear rate and temperature only. The influence of the dwell time in the die on the material parameter is neglected. Up to now there is no way to involve this dependencies into a calculation with on the market available CFD-software.

Time dependent measurements of the shear viscosity of the used LD-PE in the working range (shear rate between 1 and 100 1/s, dwell time up to 1 h) show a rising viscosity up to 2 %. In addition to that time dependent measurements of the storage modulus are shown in Fig. 3 to get a general impression of the thermo stability of the LD-PE. Changes of the storage modulus indicate changes in biaxial stretching properties within the blow film extrusion process. [6]

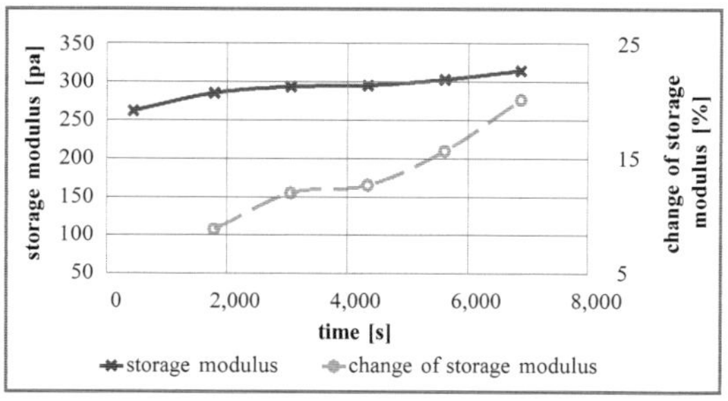

Figure 3: time dependence of storage modulus

The measured values were taken at a temperature of 190 °C and a shear rate of 0.01 rad/s. To determine the maximal dwell time of the plastic melt in the die experimental purge tests have been conducted. Therefore the LD-PE was colored with a blue non-organic master batch which serves as tracer for the purge process. During the film extrusion process the material feed was changed from blue colored LD-PE to natural LD-PE. The resulting film was analyzed with a spectral photometer to identify the start- and endpoint of the change. So the purge time is calculated. This method is discussed in detail in [7].

For the LD-PE the maximal experimentally determined dwell time is 3,900 s for a mass-throughput of 15 kg/h and 3,500 s for 30 kg/h. The minimal dwell time based on the first completely purged sector of the film probe is 1,800 s for a mass-throughput of 15 kg/h and 1000 s for 30 kg/h. Compared to the time dependence of the storage modulus does it rise up 12.7 % in the spectrum of dwell time (see Fig. 3).

To include these time dependencies into the process of die design it is necessary to calculate the dwell time distribution at the outlet of the die. Although the material models used for the CFD-simulation are time independent, it is goal-orientated to calculate the dwell time distribution with the aim to homogenize the distribution.

In Fig. 4 the average dwell time distribution at the outlet of the die is shown. It is calculated by the usage of the same CFD-simulation-results shown in Fig. 2.

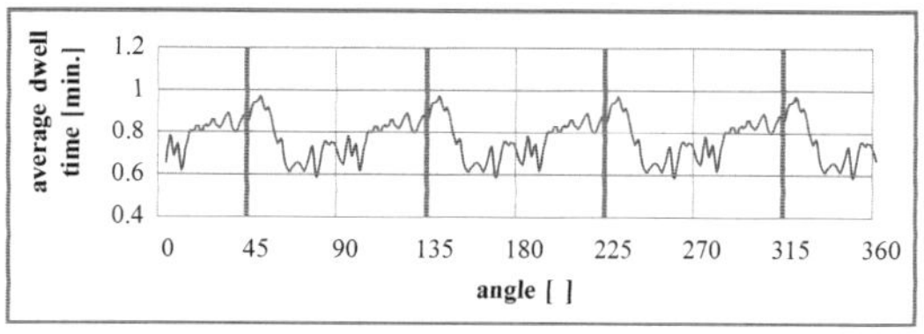

Figure 4: average dwell time distribution at the outlet of a spiral mandrel die

Differently to the velocity distribution the dwell time distribution shows periodic inequalities. Analog to the thin segments of the film which shows a periodicity of

90 deg. (see Fig. 2) the dwell time has local maxima at these points with the same periodicity.
The average dwell time shown in Fig. 4 is calculated for 360 sectors of 1 deg. at the ring gap at the outlet of the die. For the calculation the dwell time along a constant number of path lines for each sector is integrated. The average time is calculated for each sector. The scale of the results is directly dependent of the used number of path linens and their distance to the wall of the geometry. For the calculation of a exact dwell time this method has to be calibrated with the results of the experimental tests with the die. To just compare design proposals to get a "better" or "worse" information it is only necessary to use a constant number of path lines and a constant wall distance for each design proposal. Also statistic methods like calculating the standard deviation of the distribution could be used for a comparison of different geometries.

Detail optimization by analyzing the flow history
To analyze details of the geometry which are responsible for an unequal dwell time distribution it is necessary to analyze the whole flow history. Therefore the local dwell times along a path line are plotted as function of path line length. By analyzing the slope of the plot for each path line it is possible to detect differences and connect these with features or regions of the die geometry.
In Fig. 5 a sector of path lines is shown. They represent the angle from 40 to 50 deg. (see Fig. 4).

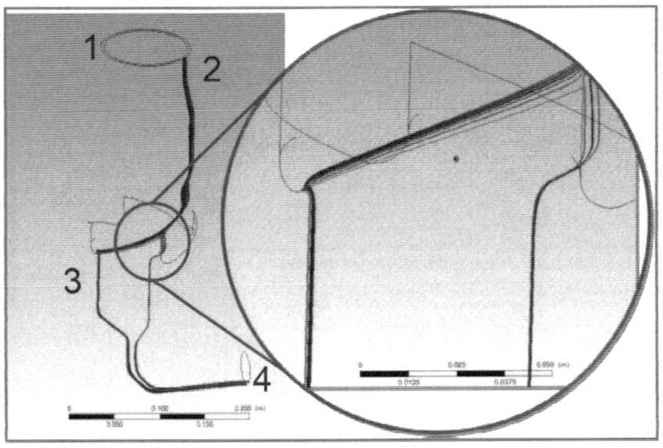

Figure: 5: sector of path lines for flow history analyze

At point (1) the outlet of the die is shown. The path lines start at point (2) in reverse direction. At point (3) the spiral mandrel die ends and the pre-distribution begins. Point (4) is the inlet of the die. The red marked path lines are that ones with the highest dwell time.
In Fig. 6 the integrated dwell time along each path line is plotted over the length of each path line. It is noticeable that for two of the curves the slope changes abruptly. Along those two paths the highest dwell time is reached (see Fig. 5, red path lines). At the changing points of the slope the geometry has a significant influence on the flow field. By comparing the length information of the changing points of the slope with the 3D-CAD geometry of the die it is possible to identify those areas or features of the die.

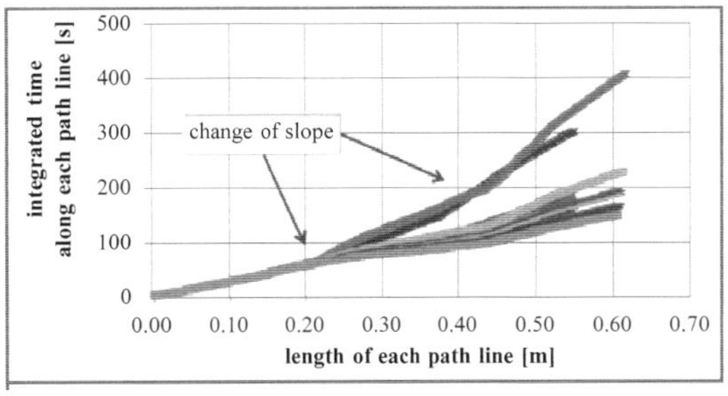

Figure 6: dwell time along path lines

A first experimental validation of this optimization method was carried out by a purge experiment. The result is shown in the following picture.

Figure 7: frozen plastic melt out of the die

A purging test was conducted in which a black colored LD-PE was purged out by a natural LD-PE. The experiment is stopped shortly before the change is complete, so that the areas with the highest dwell dime in the die are still filled with black colored LD-PE. After the experiment the cooled die has been unassembled. On Fig. 7 a rayed photo of the frozen plastic melt out of the die is shown. The red marked sections of the highest dwell time correspond to the calculated paths from the CFD-simulation (compare between Fig. 4, 5 and 7). This comparison shows a very good qualitative fit between the experimental test and the analysis method of the CFD-simulation.

Conclusions
Caused by increasing computer power nowadays three-dimensional simulations of the complex flow in

extrusion dies are used more and more often in the design process. Besides existing post-processing tools of the software systems methods are needed which are able to predict the quality of the resulting product (e.g. wall thickness distribution of a pIPE or blow film).
It was shown that a connection of rheological background knowledge with worked out analysis methods are able to allow an knowledge based interpretation of simulation results. By analyzing the results of the dwell time distribution of a steady CFD-simulation in which time dependent changes of material parameters are neglected it is possible to fit the calculation results with experimental tests. Based on that it is possible to track back inequalities in the wall thickness distribution of an extruded film to differences in the dwell time distribution and flow history.
For future work it is planned to expand the experimental validation with different materials (e.g. PE-HD) and different or optimized dies.

Acknowledgements
The spiral mandrel die used for the practical tests in this paper was designed and manufactured in the following project:
Project Sponsor: German National Science Foundation (Deutsche Forschungsgemeinschaft, DFG)
Title: Integrative coupling of design, calculation and simulation for the design of flow channels in the plastics processing
Project-No. WO-302-37-2

Special thanks goes to:

- Sabic Europe (LD-PE 2100TN00)
- Basell Polyolefine GmbH (HD-PE Hostalen 9255)
- A. Schulmann GmbH (Polybatch blue, Polyblak)

for supporting this research project with material.

References

1. J. Wortberg, K. Saul, *Improved Performance and Product Quality by Automated Geometry Optimization of Extrusion Dies*, Proceedings of Polymer Processing Society 26th, Banff Canada, 2010

2. A. Ya. Malkin, S. G. Kulichikhin, *Rheokinetics*, Hüthig & Wwpf Publishign house, Oxford USA, 1996

3. K. Saul, PhD Thesis, *Automatic design of extrusion dies*, Universiry of Duisburg-Essen Germany, 2011

4. W. Michaeli, P. Blömer, *Flat spiral dies – rheological design with network theory*, Journal of Polymer Engineering, Vol. 24, Nos. 1-3, 2004

5. J. Vlachopoulos, *Romancing the Rheological Design of Extrusion Dies*, 26th annual meeting of Polymer Processing Society, Banff Canada, 2010

6. H. H. Winter, *A collaborative study on the relation between film blowing performance and rheological properties of two low-density and two high-density polyethylene sampels*, Pure & Appl. Chem., Vol.55, No.6, 943-976, 1983

7. J. Wortberg, C. Groß, *Optimizing purging time at product change in blown film extrusion by selective modification of process parameters*, Annual Technical Conference ANTEC of the Plastics Engineers SPE, Chicago USA, 2004

O-10-229

FILLING BEHAVIOR OF LONG FIBER-REINFORCED THERMOPLASTICS IN THE INJECTION MOLDING PROCESS

Ch. Hopmann[1], E. Juntasaro[2], K. Küsters[1], J. van Haag[1]* and T. Thienthong[2]

[1]*Institute of Plastics Processing (IKV), RWTH Aachen University, Pontstraße 55, 52062 Aachen, Germany – vanhaag@ikv.rwth-aachen.de*

[2]*Mechanical Engineering (Simulation and Design), The Sirindhorn International Thai-German Graduate School of Engineering (TGGS), King Mongkut's University of Technology North Bangkok, Bangkok, Thailand*

Abstract - In recent years long fiber-reinforced thermoplastics (LFT) play an increasingly important role in lightweight design. For those kinds of reinforced materials the flow-induced properties, especially the fiber orientations (FO) and the local fiber length distribution (FLD), show a high impact on the mechanical properties. To predict the mechanical behavior of LFT numerical models are needed which take the local FO as well as the local FLD into account. However, the resulting FLD of LFT after the injection molding process cannot be precisely predicted yet by any commercial software. The aim of this paper is to i.) investigate the flow-induced behavior of LFT by determining the FO and FLD in injection molded parts and ii.) compare the FO results from available orientation models to those obtained from experimental data. Three different geometries are used to analyze the flow behavior of LFT. A center-gated plate is used to analyze the FO while investigations on a lateral-gated plate focus on both, FO and FLD. As a third test geometry the lateral-gated plate is extended by fitted ribs from different dimensions. This plate is used to investigate the filling behavior of those hard-shaped deflections resulting e.g. from ribs. Elected sections of the described molded parts are examined under Computed Tomographic (CT) and microscopic analyzers for FO and FLD results, respectively. Further the FO will be predicted by using an anisotropic rotary diffusion - reduced strain closure model (ARD-RSC) and an improved anisotropic rotary diffusion – retarding principal rate model (iARD-RPR) implemented in the commercial filling software Moldex3D, from CoreTech System Co., Ltd., Taiwan. In future research the results from these simulations will be compared to those from the experimental investigations.

Keywords: long fiber-reinforced thermoplastics (LFT); filling behavior; fiber orientation; fiber length distribution; injection molding process.

Introduction

In recent years short- and long fiber-reinforced thermoplastics (SFTs and LFT) have been widely used in automotive, consumer and industrial application for a lightweight design of injection molded parts. The main reason for this lays in the favorable mechanical properties - especially strength and durability - when using long fibers instead of short fibers. Especially in European countries the market for LFT materials has been continuously increased during the last years [1]. The mechanical behavior of LFT is strongly influenced by the flow-induced properties like local fiber orientations (FO) and fiber length distributions (FLD) resulting from fiber damage during the molding process [2]. Thus and also shown in the study of Nguyen el al., obtaining precise information about the FO and FLD is crucial to predict the mechanical properties of LFT accurately [3].

To predict the FO in injection molded parts different numerical models have been continuously developed [4]. However, caused by the much slower orientation kinetics of long fibers compared to short fibers, the widely used Folgar-Tucker orientation model does not show proper results for the prediction of FO in LFT parts [5]. Thus, there are two models available which have been recently developed to predict the FO especially of LFT and have also been implemented into commercially available software: an anisotropic rotary diffusion - reduced strain closure model (ARD-RSC) [5] and an improved anisotropic rotary diffusion - retarding principal rate model (iARD-RPR) [6]. While the prediction of FO recently shows advancement, the calculation of FLD has not been established in any commercial software yet. Amongst others these circumstances can lead to an insufficient prediction of the mechanical properties of LFT currently. Although investigations have been conducted to obtain the FLD in LFT parts and also to predict the mechanical properties based on these FLD, not every aspects of fiber damage in injection molding of LFT materials are pointed out so far. The most investigations have been conducted on conventional geometries e.g. disks and plates [7-9]. Thus, this paper will examine the flow-induced behavior of LFT (FO and FLD) in injection molded parts which have a more complex geometry by adding ribs onto a conventional plate. Especially these areas are examined regarding FLD and FO. Further the predictions of the ARD-RSC and the iARD-RPR orientation model are compared and evaluated.

Methodology

This paper deals with the flow-induced behavior of LFT in the injection molding process and focuses basically on the local FO and FLD. The experiments are conducted with a long glass fiber-reinforced polypropylene (PP-LGF) and observe two discontinuous points of injection flow rates (low and high) and a variation of three different cavity designs.

After producing injection molded parts, selected sections were cut from the part and examined under Computed Tomographic (CT) analyzer for FO. To conduct the analyses regarding FLD, parts have been prepared by the procedure of Kunc et al. [5] which will be described below. Moreover, the FO within the tested geometries are predicted by the ARD-RSC and the iARD-RPR model implemented in the filling simulation software Moldex3D (Release11), CoreTech System Co., Ltd., Taiwan and are compared to the experimental results. A detailed explanation of the conducted investigations is given in the following sections.

Test geometries

Three different test geometries have been produced in an injection molding process: a center-gated plate and a lateral-gated plate (fan gate) with and without ribs (Fig. 1).

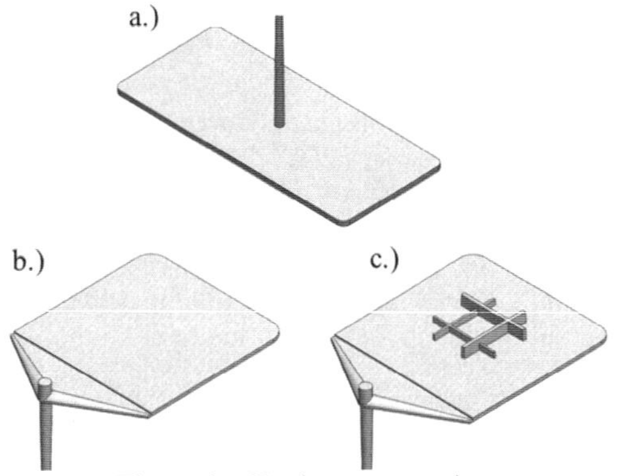

Figure 1 – Used test geometries:
a – center-gated plate (C), b – fan-gated plate without ribs (F), and c – fan-gated plate with ribs (R)

To produce the test specimens shown above, polypropylene pellets with 20 wt% of long glass fibers (type: Celstran PP-GF40-0414 P10 blended to 20 wt% with Tipplen K199, Ticona GmbH, Germany) have been produced on a standard injection molding machine.

The used pellets were cylindrically shaped with a nominal length of 10 mm. However, the length of the pellets was statistically measured by using a vernier caliper (accuracy ± 0.02 mm). These measurements resulted in an average pellet-length of 10.4 mm (standard deviation: 0.017 mm).

While the nozzle temperature was kept constant at 285 °C and the mold surface temperature was held at 50 °C in every case, other process conditions were consistently maintained according to Table 1. The injection flow rate was varied at two levels: low rate at 15.0 cc/sec and high rate at 120.0 cc/sec.

The center-gated plated has the diameters of 80 mm in width and 300 mm in length and a thickness of 3.5 mm. The fan-gated plate has a quadratic shape of 115 mm length and a thickness of 2.6 mm. The thickness of the major ribs at the test geometry c) was designed by using 95 % of the base thickness (= 2.47 mm) with a draft angle of 1.5 ° while the minor rib using 60 % of the base thickness (= 1.56 mm) and the same draft angle. A certain kind of case will be referred by 4 letters. The first letter indicates the type of geometry: C – center-gated plate, F – fan-gated plate without ribs and R – fan-gated plate with ribs. The second letter stands for the injection speed: H – high injection speed and L – low injection speed. The last two numbers indicate the weight percentage of fibers.

Parameters	CH20	CL20	FH20	FL20	RH20	RL20
Injection time [sec]	0.52	3.75	0.45	3.16	0.47	3.28
Injection speed [cc/sec]	120	15	120	15	120	15
Holding time [sec]	6		10		10	
Holding pres. [bar]	250		300		350	
Cooling time [sec]	35					
Screw rotation speed [mm/sec]	101					

Table 1 – Process parameters of conducted investigations

In order to conduct any FO and/or FLD measurements, from each geometry test specimens were cut out at four or five locations, respectively: A – at the gate or nearest to the gate, B – at a defined location in the center of the investigated cavity, C – at the end of the flow path and R – at the runner. The locations for these cuttings are shown in Fig. 2. The cuttings' dimensions were set to 12 x 12 mm.

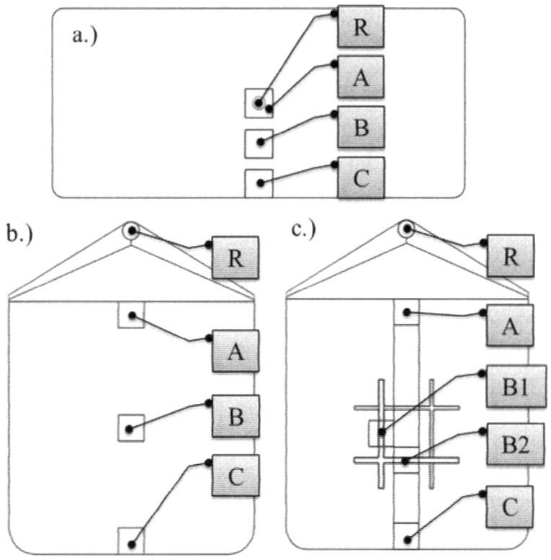

Figure 2 – Cutting layout for:
a – center-gated plate, b – fan-gated plate without ribs, and c – fan-gated plate with ribs

FO and FLD were measured with two different methods which will be explained in the following.

Fiber Orientation Measurement

The CT-Scans for visualizing the fibers inside the polymeric matrix have been conducted by using a desktop X-Ray microtomograph (type: SkyScan1172, Bruker-microCT N.V., Belgium). The raw data pictures have been reconstructed by the use of the software Nrecon (SkyScan). An example of 3D-rendering after reconstruction is shown in Fig. 3. There are also orthogonal 2D-views of the same test specimen depicted in Fig. 4, which give a better view to the local orientations of the fibers in each spatial direction.

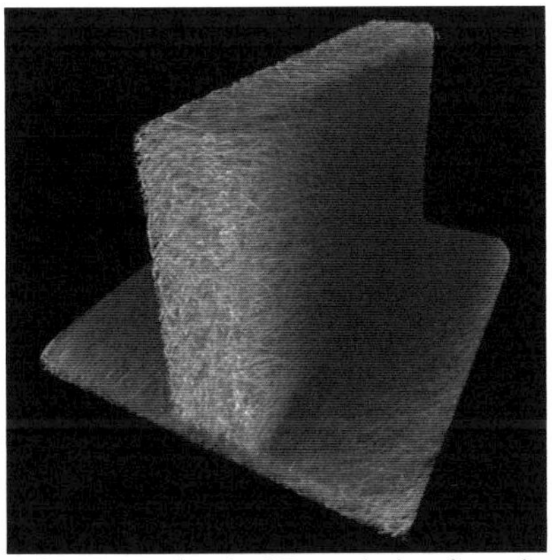

Figure 3 – Example of a 3D-rendering from the case RH20B1

Fiber Length Distribution Measurement

To measure the FLD, the polymer matrix has to be removed from the fibers. Amongst others there are two established methods to achieve this aim: a chemical solving of the polymer and a thermal treatment by pyrolysis. In the presented experiments the test specimens have been ashed at 500 °C using a high temperature oven. After the ashing process the fibers were carefully dispersed to allow further microscopic analyses. For these a digital microscope (type: Keyence VHX-600, Keyence corp., Japan) has been used at a magnification of 30 to measure the fiber length. The determination of fiber length has been provided by certain analyzing software which allows to measure the length of an object by marking at least the start and the end point of it. The accuracy of this method is about ± 0.01 mm. At least 1000 fibers have been measured manually for each sub-case to decrease the overall statistical error.

Simulation Method

In the presented research, the injection molding simulation software, Moldex3D (Release 11) was used to predict the local FO. The boundary conditions were referred from the injection molding conditions. In term of discretized models, the hex-dominant mesh was applied to all simulated cases in the cavity section while the runner was meshed with tetrahedral elements. The ARD-RSC and iARD-RPR orientation models have been used to predict the FO.

Results and Discussion

Fiber Orientation Results

After reconstructing the raw data from the CT-analyses to those orthogonal projections shown in Fig. 4, the overall orientation of the fibers in each spatial direction can be visualized easily. The shown image represents the middle plane (all directions) of the RH20B1 specimen.

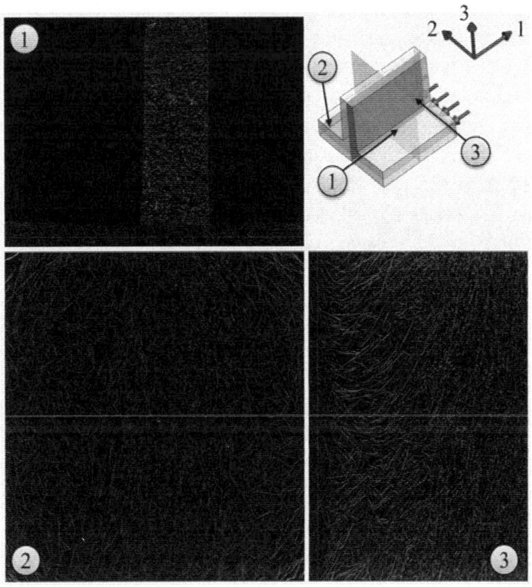

Figure 4 –Reconstructed image of fibers from the CT-analyses, case: RH20B1,
1: flow direction, 2: in plane, and 3: along thickness

From Fig. 4 the spatial FO can be seen qualitatively. A clearly visible borderline can be determined at the transition point from the plate into the rib region. This aspect is discussed later on in comparison to the simulation results.

Fiber Length Distribution Results

In Fig. 5 the measured FLD for the location B1 of the fan-gated plate with ribs is depicted. It can be seen that the initial fiber length of approx. 10.4 mm is decreased excessively.

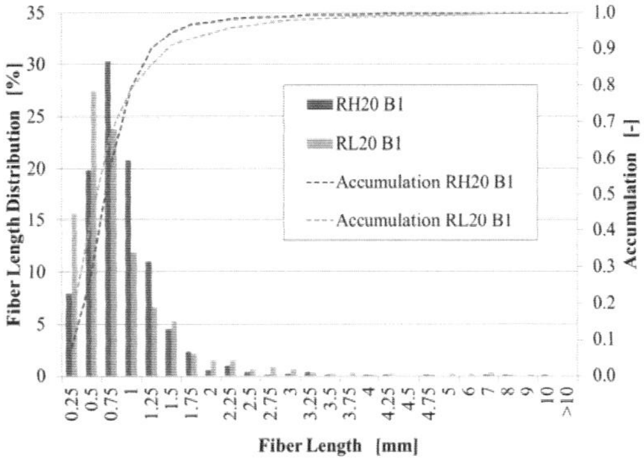

Figure 5 – Fiber length distribution at the location B1 for the cases RH20 and RL20

It is also obvious that with higher injection speed the amount of lower fibers increases while the number of longer fibers decreases. The weight averaged fiber length is 1.25 mm for the case RH20 and 1.59 mm for the case RL20.

Compared to the other locations of the test geometry with ribs the weight averaged fiber length at the location B1 represents the lowest measured fiber length average for the high injection speed (Fig 6). Overall a slightly decreasing fiber length over the flow path from the inlet (location: R) to the end of the cavity (location: C) can be seen, while the fiber weight fraction increases over the flow paths. From this can be drawn the conclusion that there is occurring a progressive fiber damage during melt flow due to that short and broken fibers are transported to the end of the cavity. For the location B1 the fiber damage seems to be additionally increased because of the influence of the rib.

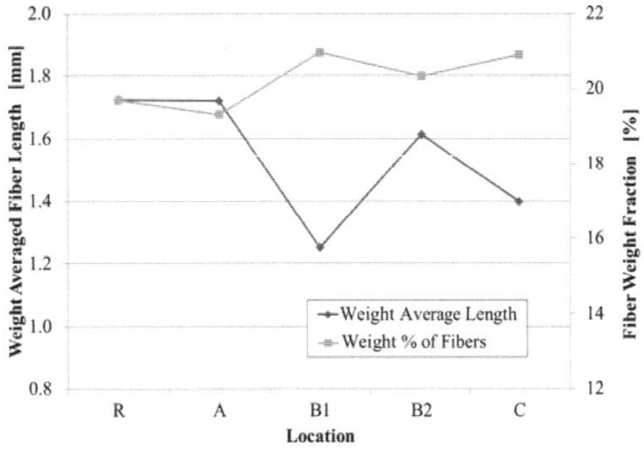

Figure 6 – Weight average length and fiber weight fraction, case: RH20

Simulation Results

The prediction of the FO at the location B1 with the iARD-RPR and the ARD-RSC orientation model is shown in Fig. 7, respectively. The shown parameter A_{ii} represents the orientation eigenvalues for each direction in space which indicate the degree of orientation (1 := fully oriented). Both models show the highest orientation of fibers in the middle of the cavity. Overall both models show very similar FO results

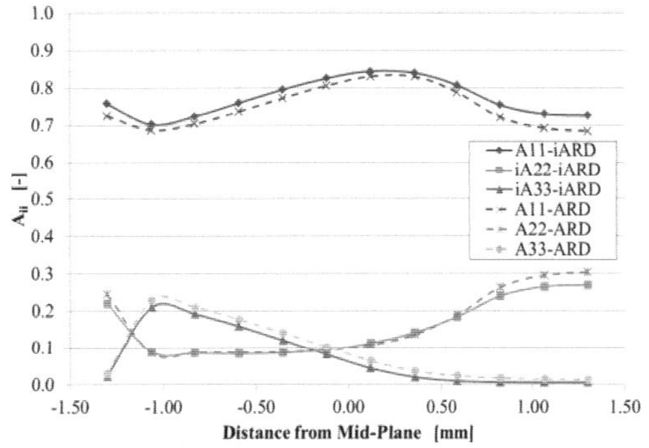

Figure 7 – Spatial fiber orientations predicted from the iARD-RPR and ARD-RSC model, case: RH20B1

while the iARD-RPR model predicts a slightly higher orientation in flow direction.

A qualitative comparison to the visualized fibers in Fig. 4 shows an agreement in the disappearing orientation in 3-direction at the transition point from the plate into the rib. This can be seen from the clearly visible borderline to the rib region within the CT-scans. However, a quantitative comparison cannot be conducted yet.

Conclusions and Outline

Next to a strong reduction of the average fiber length during the injection molding process, an increase of the fiber weight fraction can be observed with longer flow paths. This is likely caused due to a progressive damage of fibers during the injection phase and an afterwards transportation along the melt flow. From these investigations it can be shown that ribs provoke significant additional fiber damage.

The FO predictions of both investigated orientation models show similar results. A qualitative comparison of the predicted FO with CT analyses shows reasonable results. However, a final statement to the accuracy of these models can only be given if the FO are measured quantitatively.

In the next step the three dimensional FO will be obtained from the 2D CT-Images e.g. depicted in Fig. 4 with the use of image analyzing software.

Acknowledgements

The authors are very grateful to SimpaTec GmbH, Germany for providing the Moldex3D license used in this research.

References

1. M. Schemme *Reinforced Plastics*, 2008
2. B. N. Nguyen; J. D. Holbery; K. I. Johnson; M. T. Smith *SPE Automotive Comp.*, 2005
3. B. N. Nguyen; S. K. Bapanapalli; J. D. Holbery; et al. *J Comp. Mat.*, 2008, 42, 1003–1029
4. J. M. Park; S. J. Park, Math. Problems in Eng., 2011, 1-14
5. J. H. Phelps; C. L. TuckerIII *J of Non-Newtonian Fluid Mech.*, 2009, 156, 165-176
6. Sun, T. *Moldex3D R11 European Webinar Series*, 2012
7. V. Kunc; B. Frame; B. N. Nguyenet al. *SPE Automotive Comp.*, 2007
8. B. N. Nguyen; S. K. Bapanapalli; J. D. Holbery et al. *J Comp. Mat.*, 2008, 42, 1003–1029
9. B. N. Nguyen, S. K. Bapanapalli; V. Kuncet al. *J Comp. Mat.*, 2009, 43, 217–246

O-10-163

DIMENSIONLESS MODEL OF THE SOLIDS THROUGHPUT OF SINGLE-SCREW SMOOTH BARREL EXTRUDERS REGARDING THE FEED SECTION PARAMETERS

V. Schöppner[1], J.-S. Leßmann[1]*, R. Weddige[2] and A. Porsch[3]

[1] *Kunststofftechnik Paderborn, University of Paderborn, Warburger Str. 100, 33098 Paderborn, Germany – volker.schoeppner@ktp.upb.de; johann.sebastian.lessmann@ktp.upb.de*

[2] *KraussMaffei Technologies GmbH – robert.weddige@kraussmaffei.com*

[3] *Konstruktion und Kunststoffmaschinen, University of Duisburg-Essen, Germany – alexander.porsch@uni-due.de*

Abstract - In a smooth barrel extruder, the throughput is generally dominated by the metering section's pumping ability. Within the feed zone, more pellets can always be conveyed than can melt in the compression zone and be discharged in the metering zone. At high screw speeds, this effect will only occur if the pellets flow quickly enough out of the hopper into the screw channel and this is fully filled. In this case, the solids throughput up to maximum screw speed shows an approximately linear relationship to the speed. These linear solids conveying characteristics necessitate constructional changes to the hopper opening. In this article, solids conveying processes up to speeds of 2,000 min^{-1} (peripheral velocities up to 10m/s) are simulated with various feed hopper geometries. The simulations are carried out by means of DEM (Discrete Element Method) in which the pellets are approximated by single spherical particles. From the simulation results, a dimensionless model is derived for describing the throughput as a function of the geometry and process parameters of the solids feed zone.

Keywords: Extrusion, simulation, DEM, solids conveying throughput, modeling

Introduction

Several models exist to mathematically describe the solids conveying throughput in extruders. The first model from Darnell and Mol [1] assumes, for the solids throughput, that the pellets are conveyed as a solid block. This assumption is derived from the fact that the internal coefficients of friction (plastic-plastic) are much higher than the external coefficients of friction (plastic-steel). Relative movements within the pellets are therefore not taken into account.

Darnell's and Mol's model has been optimized over the years, e.g. in [2-7], but all the models assume a compacted solids bed. However, a solids bed is only formed when the screw channel beneath the hopper is completely full. At high speeds, however, the channel is not always full resulting in the conveying behavior shown in Fig. 2.

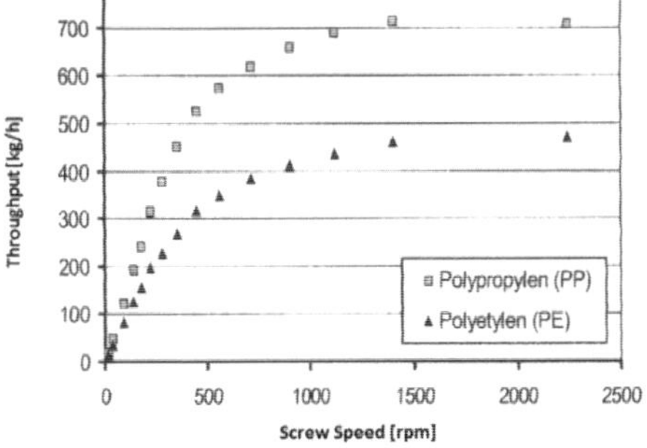

Figure 1 - Solids conveying for higher screw speeds

Obviously, the throughput increases only degressively at higher screw speeds. There are two reasons for this effect: Firstly, at high screw speeds, the pellets' acceleration by gravity does not suffice to fill the channel within its time under the hopper opening. Secondly, through the rising flight, the pellets receive an impulse back towards the hopper. Such effects require constructional modifications to the feed opening such as lengthening the hopper or providing pockets (marked in Fig. 3).

In this article, based on results from [8] a dimensional analysis model is presented that describes the throughput as a function of various geometric values of the opening and the feed section up to high screw speeds. The data base for this model is achieved by DEM (Discrete Element Method) extrusion throughput simulations which allow much more data to be generated than in experimental tests with a smaller effort. Using the DEM, in contrast to conventional models, particle-particle interactions are also considered. The solid bed assumption is therefore no longer necessary.

The starting point for modifications to the geometry is the hopper opening shown in Fig. 3, where L_p is the length of the opening and H_p the width of the pocket. The other values are explained in Table 1.

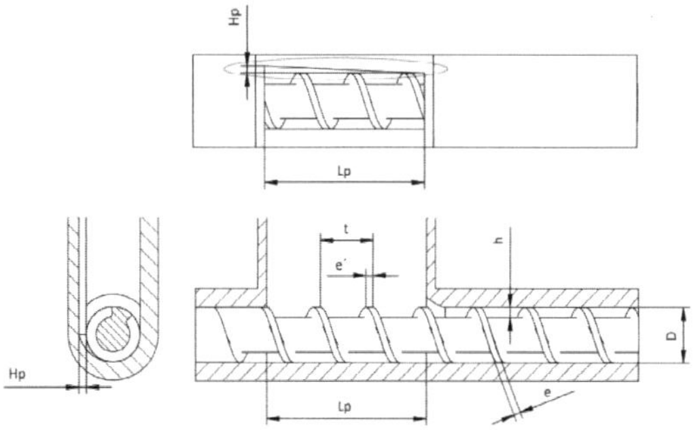

Figure 2 – Geometry of the simulation extruders

The pocket width H_p decreases along the length of the hopper in the conveying direction so that the plan view shows a trapezoidal cross section. The values of D, L_p and h (flight depth) and screw speed are varied in extensive simulation studies in order to detect their influence on the throughput. In these throughput simulations, the authors make use of the DEM, utilizing the software EDEM (DEM Solutions Ltd.). Finally, the simulation results for the solids conveying throughput are used to develop a mathematical model describing the throughput as a function of D, L_p and h and N (screw speed) using dimensional analysis.

Simulations and Experimental

In the authors' simulations (using the DEM software EDEM), each pellet is approximated by a rigid sphere. Screw and barrel are seen as rigid bodies with rigid walls. Interactions between two or more particles or between particle and wall are modeled by impulse and frictional forces resulting from collisions of these particles. In contrast to FEM, a collision is not modeled by elastic or plastic deformation but by a virtual overlap of the rigid spheres. Resulting contact forces (normal and tangential frictional, elastic and damping forces) are calculated based on this virtual overlap δ (force-displacement relation) using a physical model of those contact forces. This model regards coefficients of restitution (COR) and friction.

For all particles, the Newtonian motion equations are solved in every iteration step taking into account the present position and velocity as well as the presented contact forces. This gives the position and velocity for the next iteration step. Details on the method of DEM in general can be found in [9-12]. Details on the used force-displacement models and their integration in the Newtonian motion equations are published in [13-14]. Fig. 4 shows as a practical example pellets approximated by spherical particles in a single-screw extruder. This model was used in the authors' DEM simulations of the solids conveying processes.

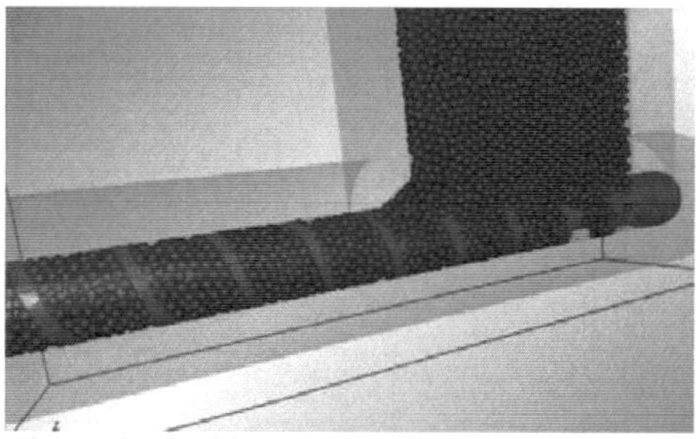

Figure 3 - Solids conveying simulation extruder

Many *simulations* and *experimental* throughput measurements have been conducted to prove their conformity. The chart in Fig. 4 shows as an example the screw speed-throughput behavior of an LDPE determined in solids conveying tests and simulations.

The experimental throughput has been determined by weighing the mass of pellets that was conveyed by a solids conveying screw in a defined time. All relevant simulation parameters are shown in Table 1. To calculate the simulated throughput, the conveyed spheres were counted over a defined period. The throughput results from the conveyed mass (resulting from the number of spheres, their volume and density) during the definite period of time.

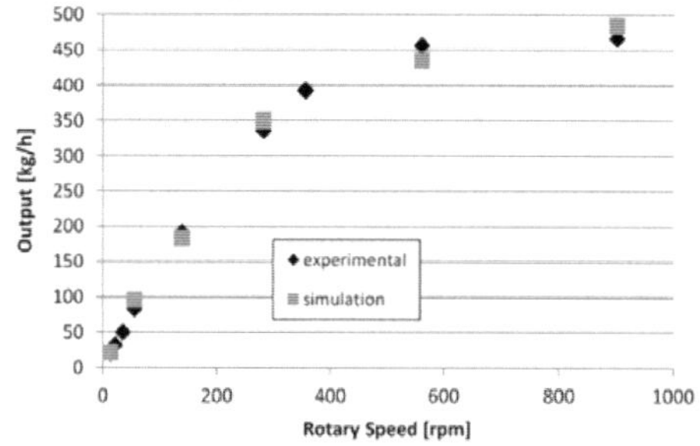

Figure 4 – Measured and simulated throughput

Obviously simulation and experiment show good conformity while using realistic material data, given in Table 1. DEM is thus suitable for studying extrusion processes. Through the simulations, the number of complex experimental tests can be reduced significantly, enabling a relatively close-meshed variation of the simulation parameters. Consequently, the information provided by a subsequent dimensional analysis description of the extrusion throughput as a function of various parameters of the solids conveying can be considered very reliable.

Table 1 – Simulation parameters LDPE

parameter	unit	value(s)
screw diameter D	mm	50
flight depth h	mm	14.4
pitch t	mm	47
flight width e	mm	6
pocket length L_p	mm	ø 50
pocket width H_p	mm	0
screw speed N	min^{-1}	14 ... 900
granules diameter d	mm	4
plastic density ρ	kg / m³	918
plastic shear modulus G	MPa	100
internal friction coef. μ_i	-	0.5
friction coef. barrel μ_b	-	0.195
friction coef. screw μ_s	-	0.18
Poisson's ratio	-	0.5
COR	-	0.5
gravity g	m / s²	9.81

The following simulations have been conducted with the range of parameters given in Table 2 in order to develop the solids conveying model. For all simulations, the throughput has been evaluated and is presented in the next section.

Table 2 – Varied ranges of the simulation parameters

parameter	unit	value(s)
screw diameter D	mm	19.8...250
flight depth h	mm	4.3...16.9
pitch t	mm	30...250
flight width e	D	≈ 0.1
pocket length L_p	mm	17.7...625
pocket width H_p	mm	0
screw speed N	min^{-1}	50...2,000
granules diameter d	mm	4
plastic density ρ	kg / m³	1,500
plastic shear modulus G	MPa	100
coefficients of friction $\mu_{i,e}$	-	0.5
Poisson's ratio	-	0.5
COR	-	0.5
gravity g	m / s²	9.81

Results and Discussion

Detection and increase of transition screw speeds

Simulations were initially carried out for a high-speed 30mm extruder, the geometry of which is shown in Fig. 3. The first simulations were performed with a particularly simple opening geometry, in other words without any pocket (H_p=0mm). The pitch is initially 30mm, the flight depth 5.6mm and the land width 4mm. With this extruder, simulations were carried out at screw speed intervals of 50 min^{-1}. The length of the hopper opening is initially 1D=30mm. Up to a speed of 450 min^{-1}, the throughput follows a linear pattern [4]. The gradient of the curve then decreases significantly. The simulated throughput at 500 min^{-1} is only around 85% of the throughput calculated according to [4]. This is where the transition screw speed is defined.

Simulations with higher speeds are carried out with a hopper opening extended in each case by 0.5D. As a result, the transition speed is shifted to higher ranges. Fig. 5 shows the simulated throughput/speed profiles up to a hopper opening length of L_p=3D. The transition screw speeds increase in size with increasing hopper opening length. At the same time, the threshold values for the respective throughputs also increase. These two effects show that the previously held opinion that hopper opening lengths larger than 1.5D do not make sense is incorrect.

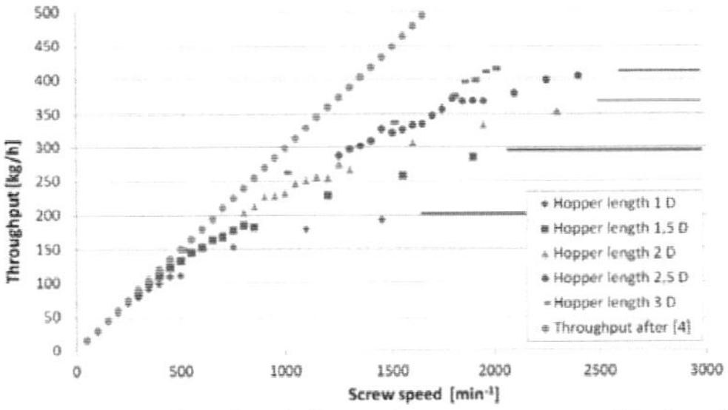

Figure 5 – Simulated throughput-screw speed behavior for D=30mm and L_p=1D to L_p=3D

Deriving a dimensionless throughput model

To be able to find a dimensionless model describing the solids conveying throughput, further simulated throughputs have been evaluated regarding the varied parameters D, h, L_p an N (Table 2). For all operating points, so called dimensionless π-values are determined. The dimensionless π-values for throughput and screw speed take into account the throughput influencing parameters and are constructed as follows:

For the dimension-related throughput, the following formula applies if the influence of the screw flight is neglected:

$$\dot{m} \sim \rho_S \cdot V_{channel} \cdot N \sim \rho_S \cdot h \cdot \pi \cdot D \cdot t \cdot N \sim \rho_S \cdot h \cdot D^2 \cdot N \quad with\, t = D \tag{1}$$

ρ_s is the bulk density and N the screw speed. The geometric values are explained in Fig. 2. The conveyed mass flow is thus proportional to the product of bulk density, screw speed and channel volume conveyed per revolution. For this reason, the dimensionless throughput can be described, in line with Schneider [2], as

$$\pi_{\dot{m}} = \frac{\dot{m}}{\rho_S \cdot h \cdot D^2 \cdot N} \tag{2}$$

The dimension-analytical formulation of the screw speed is based on the theoretical backgrounds to Fig. 5. There, the two borderline cases of the solids conveying are outlined in each case:

1. Linear increase in throughput with screw speed
2. Reaching of a threshold throughput at very high speeds.

A theoretical transition screw speed results from the points of intersection of the two straight lines, in other words from the ratio of the amount of material falling into the hopper to the amount of material conveyed per revolution. In order to determine this ratio, the describing mathematical formulations of the two cases are made equal. For the speed-dominated mass flow, the following relationship again applies.

$$\dot{m} \sim \rho_S \cdot h \cdot D^2 \cdot N \tag{3}$$

To describe the threshold mass flow, the mathematical modeling of hopper exit flows is used, similar to those that occur in silos. In line with [15], the following equation applies for the maximum mass flow from longitudinal hopper openings

$$\dot{m}_{max} \sim \rho_S \cdot g^{0,5} \cdot L_p \cdot D^{1,5} \tag{4}$$

where L_p is the length of the hopper opening, and D its width, which corresponds to the screw diameter. By putting Eq. 4 equal to Eq. 5 and by rearranging the equation, we get

$$N_{tr,theor} \sim \frac{L_T}{D} \cdot \frac{D}{h} \cdot \left(\frac{g}{D}\right)^{0,5} \quad \textbf{(5)}$$

This theoretical transition speed is used for the dedimensionalising the simulated process speed. The formula is thus

$$\pi_N = \frac{N}{N_{tr,theor}} = \frac{N}{\frac{L_p}{D} \cdot \frac{D}{h} \cdot \left(\frac{g}{D}\right)^{0,5}} \quad \textbf{(6)}$$

In this modeling, the pellet size has so far not been taken into account. Its influence over the factor $(h/d)^{0.25}$ is estimated.

Two other dimensionless factors improve the modeling of the theoretical transition speed:

- $(h/D)^2$ to take account of the "available area for the material to fall into the channel below the screw". In the plan view (Fig. 2, above), this is the area between the barrel liner and the screw root, which is available for the pellets to fall into. The larger this area, the faster the channel can be filled and the higher its transition speed.
- $(D/D_{max})^1$ to take account of the curvature of the trajectory, which the pellets must travel to fill the lower half of the barrel. The larger the diameter, the more the pellets must travel along curved – and thus longer – trajectories. This factor takes into account the size of the reverse pulse experienced by the pellets due to the peripheral velocity of the rising screw flight in the direction of the hopper opening. D_{max} is the maximum diameter of the validity range of the model, here 250mm.

Overall, the dimensionless screw speed can be summarized as

$$\pi_N = \frac{N}{N_{tr,theor}} = \frac{N}{\frac{L_p}{D} \cdot \frac{D}{h} \cdot \left(\frac{g}{D}\right)^{0,5} \cdot \left(\frac{h}{d}\right)^{0,25} \cdot \left(\frac{h}{D}\right)^{2} \cdot \frac{D}{D_{max}}} \quad \textbf{(7)}$$

For all simulations carried out with the value ranges given in Table 2, the graphic relationship shown in Fig. 6 results from plotting $\pi_{\dot{m}}$ against π_N. It can be simply described mathematically by a regression function of the type of an exponential function. With a coefficient of determination of R^2=0.89, this gives, for the dimensionless solids conveying model:

$$\pi_{\dot{m}} = 1{,}18 \cdot \exp\left(-0{,}015 \cdot \frac{N}{\frac{L_T}{D} \cdot \frac{D}{h} \cdot \left(\frac{g}{D}\right)^{0,5} \cdot \left(\frac{h}{d}\right)^{0,25} \cdot \left(\frac{h}{D}\right)^{2} \cdot \frac{D}{D_{max}}}\right) \quad \textbf{(8)}$$

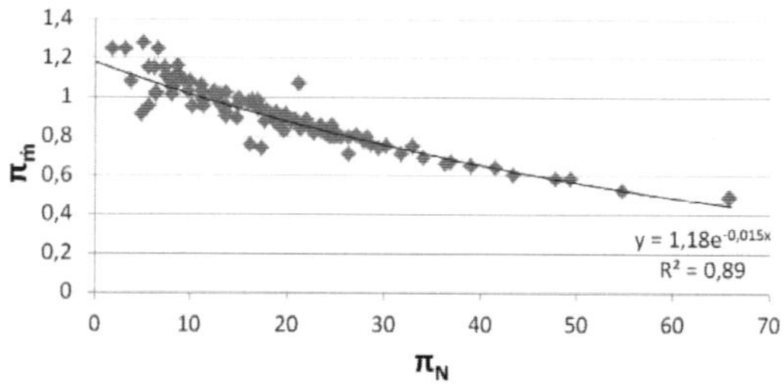

Figure 6 – Relationship between dimensionless throughput and dimensionless speed

The model has so far applied to single-flight screws with diameters of 30mm to 250mm, constant flight depths in the usual dimensions and peripheral velocities of up to 10m/s.

Conclusions

The throughput simulations in comparison to measurements have shown that, DEM is suitable as a method for simulating solids conveying processes. Through the extensive simulations with various hopper opening lengths, it is demonstrated that the widespread belief that hopper opening lengths greater than 1.5D make no sense, is incorrect, at least for small screw diameters. Finally, with the help of the throughput simulations, a dimensionless model was determined for describing the pressureless solids conveying throughput. This model should be expanded in subsequent projects concerned with pitch and pressure dependence.

Acknowledgements

We wish to thank the Deutsche Forschungsgemeinschaft, which has made these studies possible by financing the SCHO 551/13-1 project.

References

1. W. H. Darnell; E. A. J. Mol *SPE Journal* 1956, *4*, 20.
2. K. Schneider, PhD Thesis, University of Aachen, 1968.
3. Z. Tadmor; E. Broyer *Polym. Eng. Sci.* 1972, *12*, 378.
4. V. Schöppner, PhD Thesis, University of Paderborn, 1995.
5. K. S. Huyn; M. A. Spalding in SPE ANTEC Proceedings, 1997, CD
6. J. Jungemann, PhD Thesis, University of Paderborn, 1999.
7. H. Rahal, PhD Thesis, University of Duisburg-Essen, 2008.
8. J.-S. Leßmann et al., *Int. Polym. Proc.* 2012, *27*, 469
9. P. A. Moysey; M. R. Thompson *Polym. Eng. Sci.* 2004, *44*, 2203.
10. P. A. Moysey; M. R. Thompson *Powder Technol.* 2005, *153*, 95.
11. P. A. Moysey; M. R. Thompson *Chem. Eng. Sci.* 2004, *62*, 3699.

12. P. A. Moysey; M. R. Thompson *Polym. Eng. Sci.* 2008, *48*, 62.
13. A. Di Renzo; F. P. Di Maio *Chem. Eng. Sci.* 2004, *59*, 525.
14. Y. Tsuji; T. Tanaka; T. Ishida *Powder Technol.* 1992, *71*, 239.
15. D. Schulze, *Pulver und Schüttgüter,* Springer Verlag, Berlin, 2009.

O-10-148

POLYMERE SELECTIVE LASER MELTING SIMULATION USING THE C-NEM METHOD

D. Defauchy[1]*, G. Regnier[1], P. Peyre[1], I. Amran[1] and A. Ammar[2]

[1] *Arts et Métiers ParisTech, PIMM, France – denis.defauchy@ensam.eu – gilles.regnier@ensam.eu*
patrice.peyre@ensam.eu – amran.illoul@ensam.eu
[2] *Arts et Métiers ParisTech, LAMPA, France – amine.ammar@ensam.eu*

Abstract - Direct manufacturing technology using Selective Laser Melting (SLM) on polymer powders allows obtaining final parts in a short time, with a high degree of flexibility concerning shapes, evolution of parts and with classical polymer density. The physical base of this process is the coalescence of grains, which initiates the densification of powder during SLM. This study will present a bi-dimensional simulation of the selective laser melting process. The C-NEM method is implemented for both hydrodynamic and thermal calculations. A particular work is done to adapt the mesh to well describe surface tension action in the vicinity of the interface. The code is validated simulating a pendant drop and basic thermal problems. The process simulation is finally presented. The degree of densification is predicted with a curvature analysis, the material thermal history is studied and the welding depth is evaluated. The characteristic times of coalescence, densification and welding are obtained. As the material has to stay melted until adding the new powder layer, good combination of parameters is determined.

Keywords: Laser, Melting, Polymer, C-NEM, Simulation.

Introduction

Polymer selective laser sintering process enables to create very light parts in a short time with complex geometries. A recent review [1] presents a good overview of the process and its difficulties. The powder densification is driven by capillary flows, whereas viscosity dissipates the energy and slows down the densification phenomenon. The coalescence phenomenon is at the origin of the densification and its kinetics has been presented in a former study [2].

Polymer powder from 20µm up to 200µm particles diameters is used. Thin layers of 100µm to 300µm are usually deposited and preheated some degrees under the material melting temperature. A laser pass melts the selected powder area to create a new slice of the final part. Coalescence, densification, welding and air diffusion [3] occurs as the material cools down and may crystallize. The material has to stay melted the longer as possible in order to obtain a good final density.

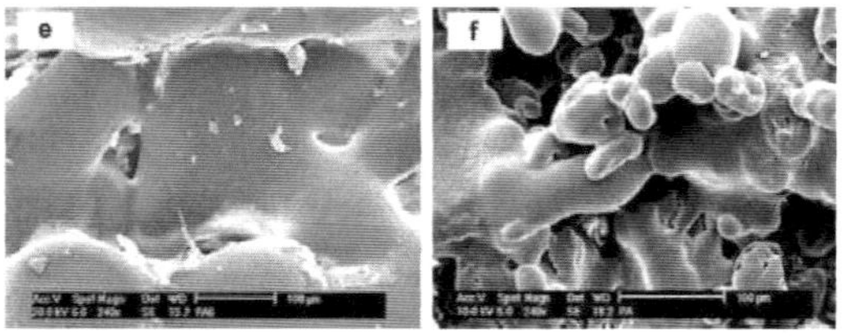

Figure 1 [4] Micrograph of PA6 (e) and PA12 (f) melted powder for same process parameters

Fig.1 presents the micrograph of PA6 and PA12 powders melted by laser sintering for the same process parameters. It is shown that the densification rate is extremely different. The quality of parts is driven by a lot of process parameters (laser power, pre-heating temperatures, laser speed, hatch distance, laser beam diameter, layer thickness, layer spreading velocity, time between layer spreading) and material parameters (viscosity, surface tension, powder geometry, thermal conductivity, heat capacity, laser absorptivity, density). Even with fixed parameters, the repeatability of parts characteristics is not easy to obtain [3] & [5].

Experimental studies are also made in order to understand the effect of each parameter and to create parts with good density and mechanical characteristics.

Simulations of the process integrating the whole local physics don't exist yet. Our first objective is to understand the role of each phenomenon and determine their kinetics. The second one is to study the influence of the process and material parameters on the densification rate, welding depth and material solid/fluid state evolution.

Numerical development

Numerical method

The C-NEM method (Constrained Natural Element Method) [6,7] is based on Voronoï cells around nodes. It looks like a classical finite element method (FEM) but the shape functions (Sibson) are different. This difference permit to better take into account the participation of all nearest neighbors nodes in streams interpolations. The main advantage of this fact is that calculations are not as dependent as FEM considering distorted elements. Indeed, the mesh nodes can move with high deformation rates without remeshing properly.

The development of the method is presented in a former paper [8]. The guide lines are described here.

Hydrodynamic development

The polymer is considered as an incompressible Newtonian fluid. The basic equations which drive the fluid movements (eq.1) are integrated on the domain and the system to solve (eq.2) is introduced.

$$\begin{cases} \nabla . \underline{\underline{\sigma}} + \underline{f_d} = 0 \\ \underline{\underline{\sigma}} = -p\underline{\underline{I}} + 2\mu \underline{\underline{D}} \\ div\underline{V} = 0 \end{cases} \quad (1)$$

where $\underline{\underline{\sigma}}$ is the strain tensor, $\underline{f_d}$ the volume force in the domain, $\underline{\underline{D}}$ the deformation tensor and μ the viscosity.

$$\underline{\underline{K}}\,\underline{U} = \underline{F} \quad (2)$$

where $\underline{\underline{K}}$ is the hydrodynamic matrix, $\underline{U}$ the nodal velocity and pressures and $\underline{F}$ the external surface actions vector.

The surface tension is integrated in the vector $\underline{F}$ as a pressure $\overrightarrow{P_T}$ which value is calculated (eq.3) using Laplace equation and interface equilibrium.

$$\overrightarrow{P_T} = \frac{\Gamma}{R}\vec{n} \quad (3)$$

where Γ is the material surface tension, R the local curvature ray of the interface and $\vec{n}$ the normal vector of the interface.

As the problem is discretized, R and $\vec{n}$ are evaluated with a circle adjusted to the node of the evalutaion and its two neighbors (Fig.2).

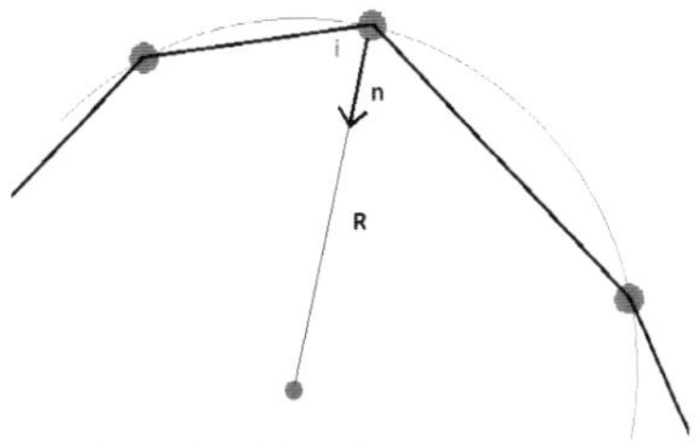

Figure 2 Discretized interface curvature ray R and normal $\vec{n}$ calculation

The velocity determination of all nodes enables to displace the nodes with a Taylor development (eq.4).

$$x(t+dt) = x(t) + v(t)dt + \gamma(t)\frac{dt^2}{2} \quad (4)$$

where $x(t)$ is the coordinates of each node at time t, v and γ its velocity and acceleration, and dt the time step.

dt is calculated at each time step using the interface nodes velocities. The criterion is an angle value (<30°) and the local rotation of interface segments aroud each node is limited to this value.

The viscosity is related to the temperature with an arrhenius law (eq.5).

$$\eta(T) = \eta_0(T)e^{\frac{E}{R}\left(\frac{1}{T}-\frac{1}{Tm}\right)} \quad (5)$$

where η_0 is the viscosity at melting temperature, E the activation energy, R the perfect gaz constant, T the temperature and Tm the melting temperature

Lagrangian conditions are added to take into account cavities volumes assumed as constant for the moment and symmetry conditions on the domain.

Thermal development

The heat transfer equation (eq.6) is integrated on the domain and an implicit method is used to solve the system obtained. The new nodal temperature at each time step is obtained (eq.7).

$$\lambda\Delta T + P = \rho c\frac{\partial T}{\partial t} \quad (6)$$

where λ is the thermal conductivity, P the heat source, ρ the volume mass and c the heat capacity

$$\overline{T}^{n+1} = \left[\frac{\underline{\underline{M}}}{dt} + \underline{\underline{K}}\right]^{-1}\left(\frac{\underline{\underline{M}}T^n}{dt} + \underline{B}\right) \quad (7)$$

where T^n is the nodal temperature at time step n, $\underline{\underline{M}}$ the thermal matrix, $\underline{\underline{K}}$ the conductivity integration matrix and $\underline{B}$ the surface heat flux and volume heat flow integration vector.

The laser action is modelized by a volume heat flow $I(z)$ with a Beer-Lambert law (eq.8).

$$I(z) = (1 - R_e)I_0e^{-kz} \quad (8)$$

where R_e is the material laser reflectivity, I_0 a constant depending on the laser power, k the obsorption coefficient and z the depth in the material under laser action.

The welding quality S (eq.9) of the material is estimated regarding former work on PEEK welding study [9].

$$S = \int_{T_f}^{T_c} \frac{1}{t_r(T)}dt \quad (9)$$

where $t_r(T) = Ae^{\frac{E}{RT}}$, $A = 7{,}3.10^{-6}$s and $E = 59000\ J.mol^{-1}$

Meshing specifications

The C-NEM method has no meshing specifications. However, the surface tension has to be well evaluated on the nodes at the interface. Thus, it is necessary to remesh the interface and the volume in its vicinity.

An interface remeshing is performed with a local criterion in degrees per node in order to well describe the local curvature. A volume remeshing is performed, based on local mesh elements dimensions, in order to well take into account the effect of surface tension in calculations.

Simulation validation

The code is implemented in Matlab and a validation is performed before analyzing the process.

The hydrodynamic calculation is validated simulating the bi-dimensional pendant drop (Fig.3). Analyzing the shape evolution of the drop, it is possible to calculate the surface tension initially input in the model and to find the initial value imposed.

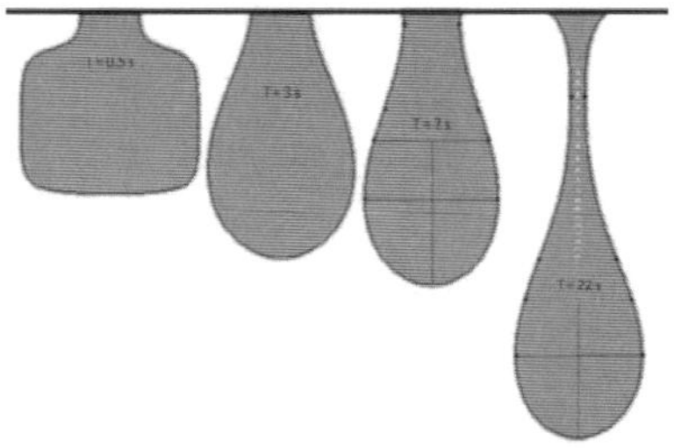

Figure 3 Pendant drop simulation

The thermal calculation is validated simulating three simple tests. Simulations are compared with analytic solutions for a plane at initial temperature, setting firstly a boundary temperature condition on a side, secondly an air temperature condition. Finally, we simulate a laser pass and compare the temperature evolution with a commercial finite element code. The interaction between laser and polymer was determined by laser absorption measurements through powder and is modeled with a Beer-Lambert law.

Results and Discussion

Introduction
A single Selective Laser Melting process simulation is presented. It allows understanding the phenomenon and their time scale acting in the process.

Material and process parameters
The material and process parameters are presented in table 1 and 2. The parameters used are the one found in the literature for the melting of polymers like PEEK (Polyetheretherketone). The initial mesh has been obtained using images realized by X-ray micro tomography on a PEEK powder. The studied area is a square which side is 500µm.

Table 1 – Material parameters.

Surface tension	30 mN.m^{-1}
Conductivity	0.25 W.m^{-1}.K^{-1}
Heat capacity	1320 J.Kg^{-1}K^{-1}
Air/Polymer convection coefficient	15 W.m^{-2}.K^{-1}
Melting Temp.	Tm = 340°C
Crystallization Temp.	300°C
Viscosity	$\eta(T) = 810e^{\frac{82200}{R}\left(\frac{1}{T}-\frac{1}{Tm}\right)}$
Polymer density	1300 Kg.m^{-3}
Laser reflectivity	16%
Laser absorption coefficient	20000 m^{-1}

Table 2 – Process parameters.

Pre-heating temperature	310 °C
Air temperature	50 °C
Laser power	5W
Laser speed	0.5 m.s^{-1}
Laser beam diameter	300 µm

Simulation images
The initial domain is at the pre-heating temperature (Fig.4).

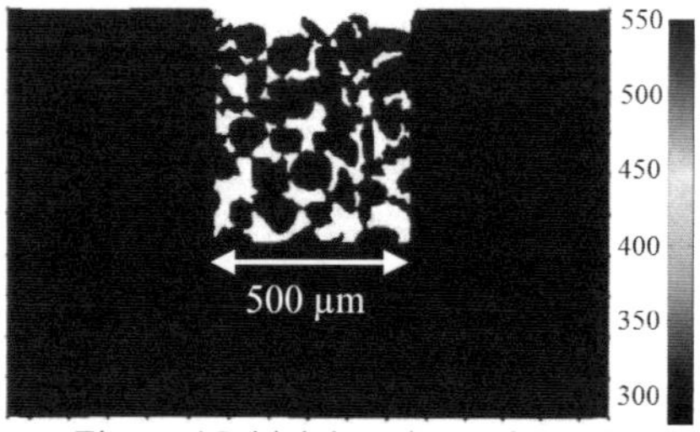

Figure 4 Initial domain pre-heated

The laser pass occurs during about 10 ms from left to right (Fig.5)

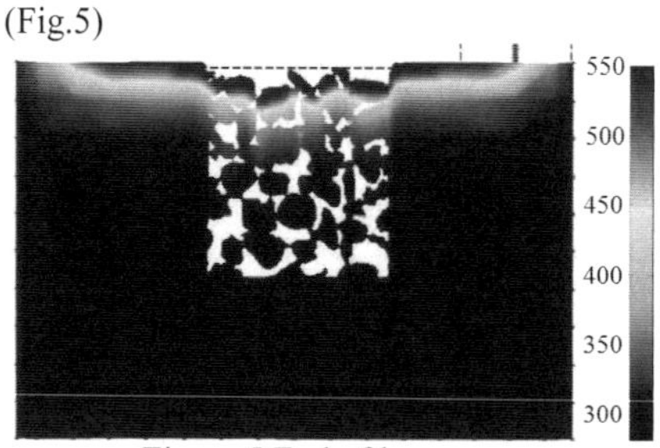

Figure 5 End of laser pass

The maximal temperature obtained after laser pass is 665°C. The material cools down extremely rapidly (Fig.6)

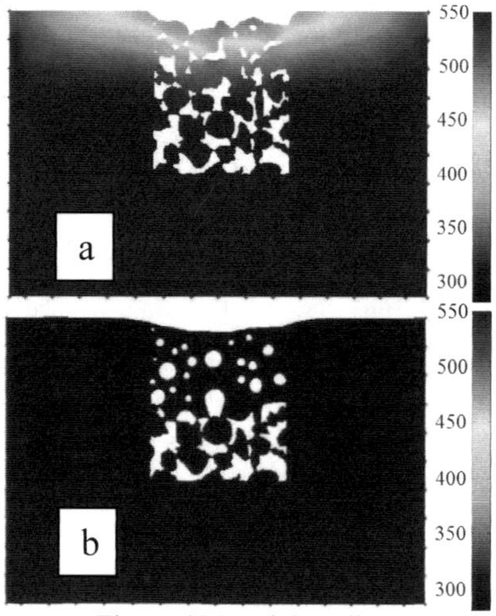

Figure 6 Material cooling
a) 21 ms after laser pass – b) 3 s after laser pass

Simulation results
This simulation enables to understand the characteristic times of each phenomenon during the process and evaluate the welding of the material.
The interface curvature adimensioned by its initial curvature evolution is presented (Fig.7). The evolution is very fast. In less than a second, coalescence of grains is much advanced.

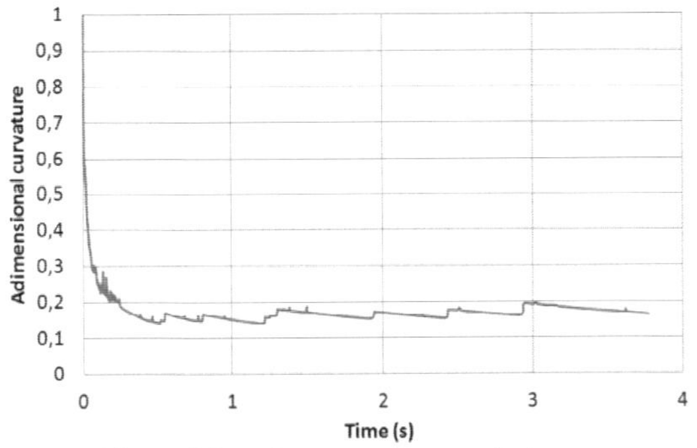

Figure 7 Interface curvature evolution

The welding of the material is also extremely rapid. Quite all the material melted has had enough time to get a good welding state. The good welded area is presented (Fig.8) 3 seconds after laser pass. A value of 1 represents a good welding.

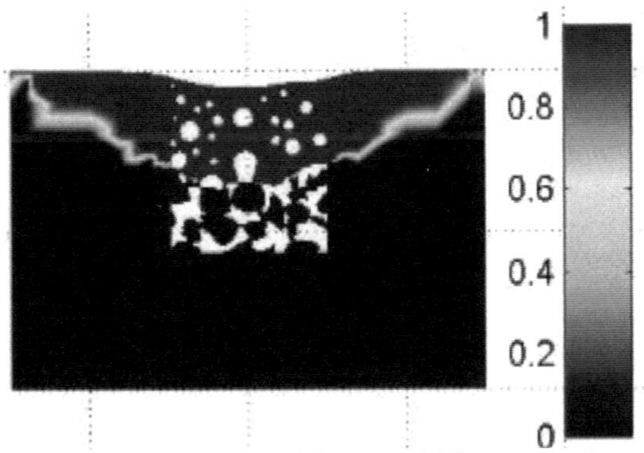

Figure 8 Welding state of the material 3 seconds after laser pass

A good welding depth is obtained until 370 μm. It corresponds quite well to experimental results. A good welding state has been obtained 84ms after the laser pass. Polymer has been melted until a depth of 385μm. The evolution of the welding depth with time is presented (Fig.9). Thi curve is dependent on the nodes position.

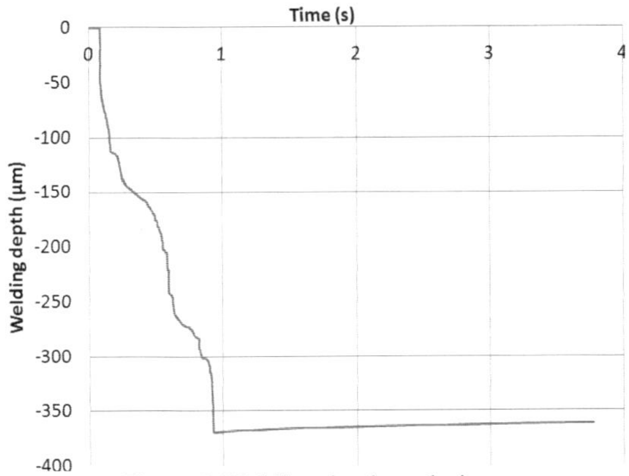

Figure 9 Welding depth evolution

After one second, the maximal welding depth is obtained.

In this situation, the melted polymer stays in a liquid state after the laser pass when the temperature has reached the initial temperature of the powder bed.

Conclusions

The process simulation has been performed in two dimensions. A lot of numerical difficulties (time step calculation and interface remeshing particularly) have been solved to be able to enable the calculation. These difficulties would be much more complicated to solve in 3D. Results in 2D are a first approach of the process. The coalescence of grains is a rapid phenomenon taking place in some seconds, as well as the welding of the material. As coalescence phenomenon is fast, porosities should disappear rapidly. The presence of porosities in real parts can be explained by the apparition of closed ones. Then, their volume decreases with time through an air diffusion phenomenon. This phenomenon cannot be prompt in a bi-dimensional simulation.

Further work has to be done to compare experimental data and simulations, in order to confirm the correlation of the simulation with reality.

Future work will be done to compare different material and process parameters effects on the process, their effect on interface curvature and welding evolution particularly.

It will be also possible to determine the conditions inducing the solidification of the melted polymer, situation not suitable as the polymer has to stay melted as long as possible in order to increase density (air diffusion) and to reduce parts distortions.

Acknowledgements

This study takes place in the FUI Project FADIPLAST. We thank Francisco Chinesta as he participated and supported this study.

References

1. R. Goodridge; C. Tuck; R. Hague *J. Progress in Materials Science* 2012**,** *57,* 229.
2. D. Defauchy; G. Regnier; P. Peyre; I. Amran; A. Ammar in PPS 27 proceedings, Marrakech, 2011.
3. S. Dupin, PhD Thesis, INSA de Lyon, 2012.
4. G. Salmoria; J. Leite; R. Paggi *Polymer Testing* 2009**,** *28-7,* 746.
5. D. Leigh in Additive Manufacturing International Conference, 2010.
6. J. Yvonnet, D. Ryckelynck, P. Lorong, F. Chinesta *International Journal for Numerical Methods in Engineering 2004, 60, 1451.*
7. Alfaro I., Yvonnet J., Chinesta F., Cueto E. *International Journal for Numerical Methods in Engineering 2007, 71, 1436.*
8. D. Defauchy; G. Regnier; I. Amran; P. Peyre; A. Ammar; F.Chinesta in COMPLAS XI proceedings, Barcelone, 2011.
9. C. Nicodeau, PhD Thesis, PIMM, 2005.

O-10-143

FINITE ELEMENT SIMULATIONS VS. SIMPLIFIED MODELS FOR THE EVALUATION OF FORCES INVOLVED IN THE NANOIMPRINT PROCESS

H. Teyssèdre[1,2]*, P. Gilormini[1], G. Régnier[1]

[1]PIMM, Arts et Métiers ParisTech, CNRS, Paris, France – pierre.gilormini@ensam.eu; gilles.regnier@ensam

[2]Silsef, Archamps, and CEA/LETI, Grenoble, France – hubert.teyssedre@gmail.com

Abstract - It is shown that assuming simplified flows may lead to quite incorrect evaluations of nanoimprint forces, even for simple periodic patterns and a Newtonian behavior. The classical result of the lubrication theory does not account for the range of thickness-to-width ratios of interest, especially at the beginning of the process, and finite element simulations demonstrate limitations of two more elaborate models of the literature. These simulations also show that two flow modes can be defined, according to whether or not the polymer touches the mold sidewalls.

Keywords: nanoimprint, finite elements, simplified models.

Introduction

Nanoimprint lithography is a process introduced by Chou et al. [1], where nanometric patterns are engraved into a very thin polymer film. In the variant considered here, thermal nanoimprint, the polymer is spin coated on a substrate before impression at high temperature. The surfaces involved are of the order of hundreds of square centimeters to allow for the simultaneous impression of a very large number of patterns, and defects may arise from mold deflection over relatively long distances. This motivates the study of the forces that the flowing polymer applies on the mold.

Two approaches can be used to evaluate these forces. The first one relies on analytical expressions which allow putting more computational resource at the upper level of mold deflection. The risk is a too crude evaluation of the efforts involved in the many configurations that may arise in nanoimpression. Basically, results taken from the lubrication theory are employed in this context. For instance, Schultz et al. [2] adapted this theory to model the flow below an elementary periodic pattern, and Leveder et al. [3] used it in a reverse manner to deduce the polymer viscosity from the applied force. This type of flow was also combined along two perpendicular directions by Hsin and Young [4] to account for the filling of pattern cavities by power-law fluids. The other approach, which allows especially to overcome the limitation of the lubrication theory to small thickness-to-width aspect ratios below patterns, performs a full numerical simulation of the polymer flow like in Jeong et al. [5], Rowland et al. [6], or Young [7], for example.

Problem Statement

Nanoimprint lithography can be used to reproduce a variety of patterns, like periodic arrays of pillars or holes with circular or square cross-sections, among others (see [8] for examples), but simple lines are very frequently used to demonstrate the technique, as in the pioneering papers by Chou et al. [1, 9], and to test models ([5-7], for instance). As shown in Fig. 1, the pattern is characterized by its pitch W, width L and depth D, and the initial thickness of the polymer h_0 defines the last geometric parameter of the problem.

The polymer is considered as a linear viscous fluid in this study, with its viscosity η being the only material parameter, and, surface tension is neglected, like in many other studies. Inertial effects and gravity can also be neglected in microflows and nanoflows of viscous liquids ([10], for instance), where extremely low Reynolds numbers are involved, and a Stokes flow without body forces is considered here, consequently.

A very approximate model for the two-dimensional flow between mold and substrate can be defined by a mere squeeze process, like in [3] for instance. As illustrated in Fig. 1b, this amounts to continuously trimming off, during the mold descent, the polymer volume that is not located beneath the indenter. Therefore, the efforts are likely to be underestimated, the mold depth D has no effect, the force will depend on the current thickness h of the film, without any history effect related to its initial value h_0, and it will be given by the classical result of the lubrication theory:

$$F_L = \eta V \left(\frac{L}{h}\right)^3 \quad (1)$$

This will be used as a reference force below.

Finite element simulations

In this study, finite element simulations have been conducted with the Abaqus [11] code to simulate nanoimpression. This code offers the possibility, among numerous features, to simulate the flow of quasi-incompressible viscous fluids, including the Newtonian special case. In the volume of fluid method used, two parameters must be prescribed that are not listed above: a density and a velocity for wave propagation. The time increment is excessively small if realistic values are taken for the density and wave speed, because of the length scale involved. Therefore, dummy values were defined, which did not alter significantly the initial problem to solve and allow reasonable computation times. First, the density was chosen in order to have a Reynolds number of 0.1, since this was checked to be a good approximation to a Stokes flow (zero Reynolds number) in our case. Then, the wave speed was adjusted to get both a good volume preservation and time increments that lead to reasonable computation times.

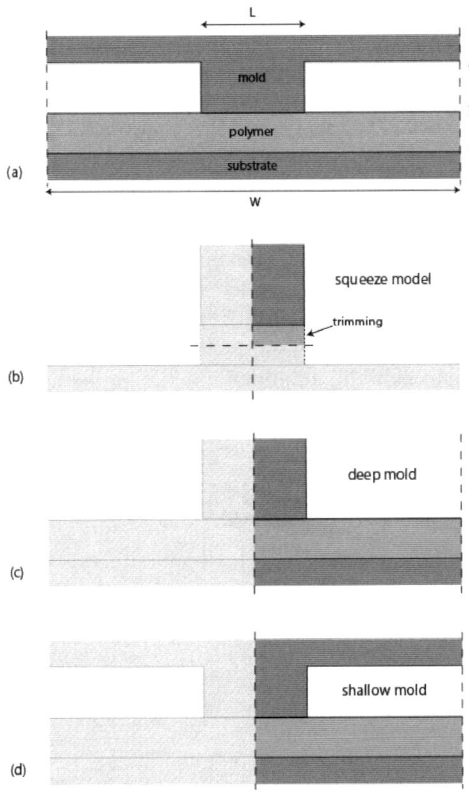

Figure 1 - Cross section of a single pattern period. Dashed lines define symmetry axes and darker shades of gray show the minimal parts to be considered under various assumptions.

Initially, the amount of polymer that opposes the lateral flow occurring below the indenter depends on W and on the initial polymer thickness h_0 only. Later in the impression process, for a polymer thickness $h < h_0$ below the indenter, a fraction of the polymer has moved above the indenter level, and this may increase the force further. In these conditions, the current force may not be a function of h only, but also of h_0. Fortunately, the finite element simulations that we performed with deep patterns (see Fig. 1c) demonstrate that the influence of h_0 can be neglected below a value that depends on the W/L ratio. For a ratio of W/L=2, for instance, Fig. 2 (open symbols) shows that the evolution of the applied force is a function of h/L only provided that h_0 is below a value of about 0.6, since the forces obtained are close to a single curve. This means that the polymer located above the indenter level does not contribute significantly to the force required for the indentation to carry on.

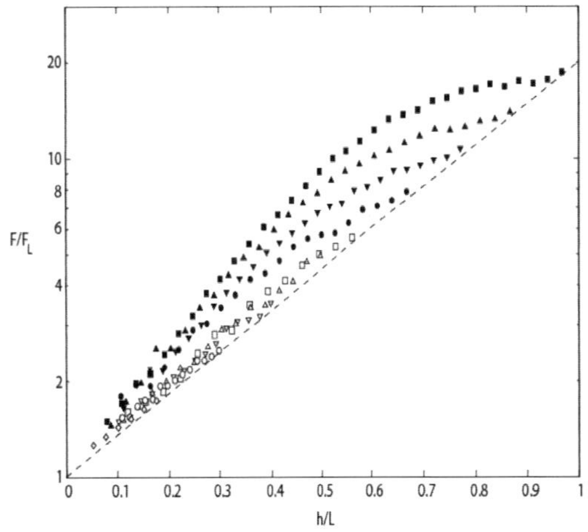

Figure 2 - Force computed by finite elements and normalized by the lubrication theory result for a deep pattern with W/L=2, starting from various h_0/L values. The straight dashed line is just a guide for the eye.

For an initial thickness beyond the critical value, the nanoimpression force still starts from an extension of the result for mode A (shown as a dashed line in Fig. 2), but then the force keeps above the curve obtained for mode A and is higher for larger initial thicknesses (solid symbols in Fig. 2). This corresponds to a change in the flow mode: below the critical h_0/L value (open symbols in Fig. 2), in mode A, the flow does not contact the mold sidewalls, whereas this contact occurs for larger h_0/L values (solid symbols in Fig. 2) and defines mode B. The shear induced by this additional contact leads to a higher impression force. Moreover, volume preservation implies, for a given h/L value, a longer contact length if the initial thickness is larger, which induces a larger force. When h/L becomes extremely small, the force required to squeeze the very thin film below the indenter increases without limit, since it varies as $(L/h)^3$ as recalled in Eq. 1, and the additional force induced by the polymer located beyond the sidewalls, with or without contact with the latter, becomes negligible comparatively.

A (dashed) straight line has been obtained for mode A when W/L=2 with the axes used in Fig. 2, at least in the h/L range considered, but this was incidental, since Fig. 3 shows a curved line when W/L=4, i.e., for a larger distance between successive indenters. The other trends are qualitatively similar to Fig. 2, but with the transition between the two flow modes appearing for a larger h_0/L value, about 1.8. Simulations were also carried out with a shallow pattern (see Fig. 1d) for W/L=4, with a pattern depth of D/L=2/3. This corresponds to a filled cavity for a mold descent of $L/2$, since a complete filling is obtained for

$$h = h_0 - D\left(1 - \frac{L}{W}\right) \quad (2)$$

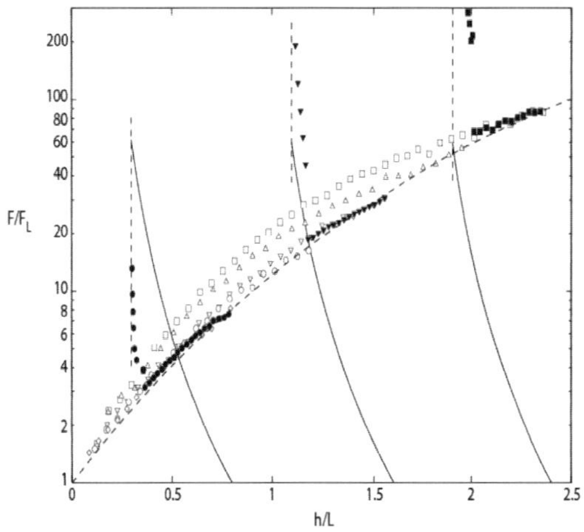

Figure 3 - Same as Fig. 2 for *W/L*=4, using either a deep mold (open symbols) or a shallow mold with *D/L*=2/3 (solid symbols). Comparison with the model of Schultz et al. [2] for a shallow mold (solid lines).

Fig. 3 compares the evolutions of the force when using a shallow pattern or a deep pattern (solid vs. open symbols), for three different initial thicknesses h_0/L. Of course, the evolutions are identical before the polymer touches the top of the cavity. Then, the normalized force increases sharply for a shallow pattern, whereas it keeps decreasing for a deep pattern.

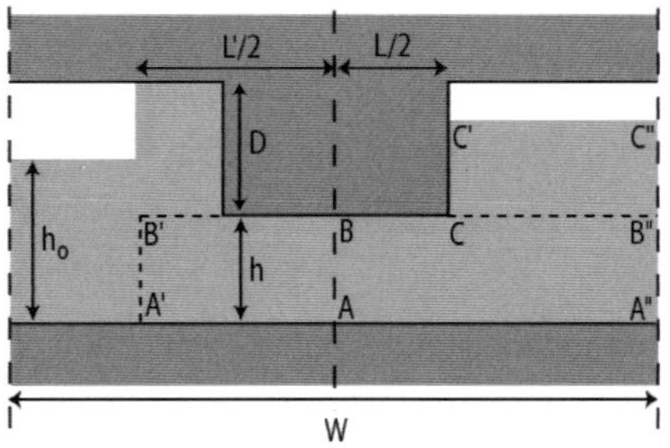

Figure 4 - Notations used for the models of Schultz et al. [2] (left) and of Hsin and Young [4] (right).

Comparison with previous models

The above discussion of two flow modes for indentation with a deep periodic mold can be compared with the results obtained by Rowland et al. [6]. These authors also defined two flow modes, but with the free surface within one pattern period showing either one or two peaks. This definition puts more emphasis on the free surface shape than on the indentation force, and Rowland et al. [6] found that the mode change is obtained for $h_0/L = (W/L-1)/2.4$ with our notations. This gives $h_0/L \approx 0.4$ when $W/L = 2$, and we did obtain a dual-peak surface for $h_0/L = 0.2$ and 0.3, and a single peak for $h_0/L = 0.4$ and beyond, but contact with the indenter sidewalls nevertheless occurred for $h_0/L = 0.7$ at least. This indicates that mode transition for the force occurs for h_0/L values that are larger than for the shape transition observed in [6], since single-peak free surfaces may be associated with mode A flow. This is confirmed in the $W/L = 4$ case of Fig. 3, where [6] would predict a change in peak numbers for $h_0/L = 1.25$, while we obtained a transition for $h_0/L \approx 1.8$ for the force evolution.

A first model that uses a simplified flow and which can be compared with the finite element results is that of Schultz et al. [2]. As indicated in Fig. 4, it assumes that the volume of polymer that is occupied by the new position of the mold is moved near the corner of the mold. Volume preservation provides the polymer-mold contact width L' readily, and the model considers a squeeze of the polymer in the ABB'A' region, using the lubrication theory result, which leads to the total force per unit length

$$F_S = \eta V \left(\frac{L'}{h}\right)^3 = \eta V \left(\frac{L}{h}\right)^3 \left(\frac{D}{D-h_0+h}\right)^3 \qquad (3)$$

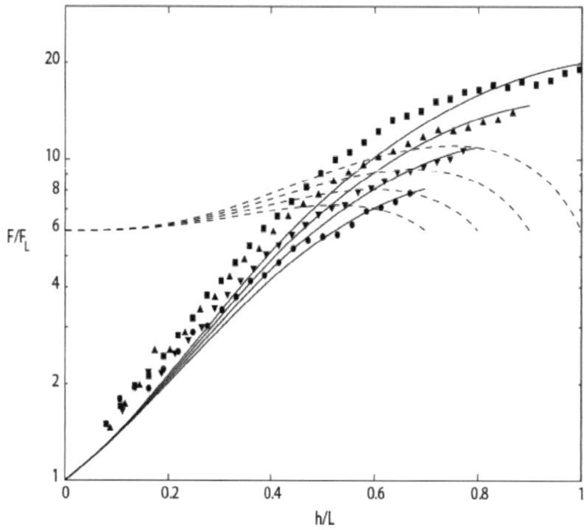

Figure 5 - Comparison of the finite element results of Fig. 2 (symbols) with the model of Hsin and Young [4] (dashed lines) and with the proposed model (solid lines), for a deep pattern.

When normalized by F_L, the force on shallow molds is equal to 1 when $h=h_0$, i.e., the force evolution starts from the lower axis in Fig. 3, which is far too low compared to the finite element results, and it equals $(W/L)^3$ for any h_0 when the cavity is filled, according to Eq. 2. Moreover, the F_S/F_L ratio keeps increasing when h decreases, whereas Fig. 3 (solid symbols) shows a decrease before the ratio increases up to complete cavity filling. It may thus be concluded from the comparison performed in Fig. 3 in three cases that the very simple flow assumed in [2] does not allow a good evaluation of the nanoimpression force involved with shallow molds for the h_0/L ratios considered here.

A different simplified polymer flow was considered by Hsin and Young [4] for power-law fluids, where the free surface of the polymer keeps parallel to the substrate, as shown in Fig. 4. In the simple case of a Newtonian behavior, horizontal and vertical plane Poiseuille flows are assumed in the AA"B"B region and in twice the CB"C"C' region (since B"C" is a symmetry axis), respectively, leading to pressure drops

$$p_0 - p_1 = 12\eta \frac{Q}{h^3}\frac{W}{2} \quad (4)$$

and

$$p_2 = 12\eta \frac{Q'}{(W-L)^3} W \frac{h_0-h}{W-L} \quad (5)$$

where p_0, p_1 and p_2 denote pressure along AB, A"B" and B"C, respectively. The above computation of p_2 assumed a zero pressure on the free surface C'C" and applied the preservation of volume to compute the CC' length. In these expressions, Q and Q' denote planar flow rates (per unit mold length normal to the cross section considered). Assuming now that $p_2 = p_1$ and $Q'=2Q=VL$ gives

$$p_0 = 3\eta VWL\left[\frac{1}{h^3} + 4\frac{h_0-h}{(W-L)^4}\right] \quad (6)$$

and

$$p_1 = 12\eta VWL\frac{h_0-h}{(W-L)^4} \quad (7)$$

Pressure p_1 applied along CB" equilibrates the shear force induced by the vertical plane Poiseuille flow on sidewall CC' and induces a vertical force F_v. In addition, Hsin and Young [4] compute the vertical force F_h due to the horizontal plane Poiseuille flow on BC by applying p_0 uniformly (although pressure decreases linearly, actually). These two contributions lead to the following expression for the total force per unit length, for one period of the mold pattern:

$$F_H = F_h + F_v = 3\eta VWL^2\left[\frac{1}{h^3} + 4\frac{W}{L}\frac{h_0-h}{(W-L)^4}\right] \quad (8)$$

which does not tend to F_L when h tends to 0, but rather to $3F_LW/L$. A comparison is made in Fig. 5 with finite element results in four cases where sidewall contact was established, and a strong disagreement appears clearly.

Proposed model

The latter model, with its contributions from two distinct flows, can be modified as follows. It may be noted first that assuming a plane Poiseuille flow in the AA"B"B region is not compatible with the symmetry conditions that apply along AB and A"B". Actually, the force due to the flow in the AA"B"B region corresponds to the A mode flow described above, and therefore it can be evaluated from the dashed lines shown in Figs. 2 and 3. Since a straight line was obtained for a W/L ratio of 2, the following fit yields readily:

$$F_h = 10^{1.3h/L} F_L \quad (9)$$

which provides F_h for this special W/L value only.

In contrast with the horizontal flow, the vertical flow assumed in [4] is in reasonable agreement with the velocity profiles found in the finite element simulations. This part of the Hsin and Young [4] model should be corrected, though, because the flow rate Q' must be computed with respect to the sidewall, which is moving vertically with velocity V, for a relevant computation of the pressure drop. Therefore, $Q'=VW$ applies, rather than $Q'=VL$. Introducing this correction in Eq. 5 leads to the following vertical component for two sidewalls:

$$F_v = 12\eta VW^2\frac{h_0-h}{(W-L)^3} \quad (10)$$

with an amplifying factor of $W/L-1$ with respect to Eq. 8. The total force F_h+F_v that results from Eqs. 9 and 10 is shown in Fig. 5 as solid lines for four h_0 values. Although the agreement is still imperfect, which may be due to an interaction between the two flows, one notes that the trends are correct.

Unfortunately, this model is not fully predictive, since the contribution F_h of mode A has to be obtained from finite element simulations. This should stimulate further effort to develop appropriate simple models for mode A, or to fit F_h for a large set of W/L ratios.

Conclusions

It has been shown that assuming simplified flows may lead to quite incorrect evaluations of the force involved in the nanoimprint of simple periodic patterns on Newtonian polymers for given mold velocity. It is recalled that an underestimation of the effort for a fixed mold velocity implies an overestimation of the mold velocity for a constant applied force, which is the usual experimental condition. The mere use of the classical result of the lubrication theory significantly underestimates the efforts in most cases. Finite element simulations have demonstrated limitations of two more elaborate models of the literature. These simulations have also shown that two flow modes could be defined, according to whether or not the polymer touches the mold sidewalls. A deeper analysis of these two modes may help the definition of a more appropriate simplified model in the future.

Acknowledgments

Support from the Agence Nationale de la Recherche through project SINCRONE is acknowledged.

References

1. S.Y. Chou; P.R. Krauss; P.J. Renstrom *Appl. Phys. Lett.* 1995, *67*, 3114.
2. H. Schulz; M. Wissen; N. Bogdanski; H.-C. Scheer; K. Mattes; Ch. Friedrich *Microelectronic Eng.* 2006, *83*, 259.
3. T. Leveder; S. Landis; L. Davoust; N. Chaix *Microelectronic Eng.* 2007, *84*, 928.
4. I.-C. Hsin; W.-B. Young *Int. Polym. Processing* 2008, *23*, 24.
5. J.-H. Jeong; Y.-S. Choi; Y.-J. Shin; J.-J. Lee; K.-T. Park; E.-S. Lee; S.-R. Lee *Fibers Polym.* 2002, *3*, 113.
6. H.D. Rowland; A.C. Sun; P.R. Schunk; W.P. King *J. Micromech. Microeng.* 2005, *15*, 2414.
7. W.-B. Young *Microelectronic Eng.* 2005, *77*, 405.
8. Guo, L.J. *J. Phys. D: Appl. Phys.* 2004, *37*, R123.
9. S.Y. Chou; P.R. Krauss; P.J. Renstrom *J. Vac. Sci. Technol. B* 1996, *14*, 4129.
10. G. Karniadakis; A. Beskok; N. Aluru, *Microflows and Nanoflows. Fundamentals and Simulation*, Springer, 2005.
11. Abaqus Software, version 6.9, Dassault Systèmes Simulia Corp., Providence RI, USA, 2009.

O-10-067

PIPE COOLING SIMULATION FOR ENERGY SAVINGS AND ENHANCED PRODUCT QUALITY

G. Hiesgen[1]*, K. Saul[2#], M. Spitz[3#], P. Weiß[4#]
**University of Duisburg-Essen, Institute for Mechatronic, Lotharstr. 1, 47057 Duisburg, Germany*
#SHS Energieeffizienz, Neukölner Str. 118, 46147 Oberhausen, Germany
[1]gregor.hiesgen@uni-due.de, [2]kenny.saul@shs-energieeffizienz.de, [3]martin.spitz@shs-energieeffizienz.de,
[3]patrick.weiss@shs-energieeffizienz.de

Abstract
Within the polymer pipe extrusion process the thermoplastic material is heated up to a temperature-level of oftentimes more than 200 °C. After shaping the final product geometry in the extrusion die, for example a pipe geometry with a special pipe diameter and wall thickness (SDR), the product is cooled within up to six different cooling tanks in a cooling line. The product quality (geometric shape, surface quality, stiffness, degree of cristallinity, density) is influenced significantly within this cooling line. In general the cooling fluid (water) is prepared by electrical driven compression chillers, whereby the operation-costs can reach more than 100.000€/year. SHS Efficiency develops strategies for increasing the energy-efficiency. Therefore a finite difference method based simulation was implemented to determine the pipe temperature in predefined radial layers within the whole cooling facility. The simulation tool is able to calculate temperature-profiles for different materials, cooling facility set-ups or mass-throughputs, taking into account temperature dependent material models. The main challenge is located in the determination of the heat transmission and conduction coefficients for the different segments of the cooling facility. In order to identify these coefficients, the pipe temperature profile is measured during the production process. These measurements represent the basis for the parameter optimization with a hybrid optimizer. Finally, the simulation results from the calibrated simulation software can reach an impressive accuracy; a simulation of the overall cooling process is possible. This enables the operator to improve the process and to save energy.

Keywords: energy efficiency, pipe cooling, simulation, finite difference method

Introduction

Energy costs increase continuously. Particularly with regard to the electric energy used to realize product cooling in the polymer processing industry. 20-25% of the electrical energy consumption is used by chillers in an average polymer extrusion plant [12]. For this reason, industrial companies are forced to reduce their energy consumption and to raise the degree of independence from the energy costs escalation. The manufacturers of cooling solutions for the polymer industry provide the market with more and more energy efficient systems (see [9], [10] and [11]). For the companies it is a complicated decision, whether an investment in new technologies is profitable in an adequate time period or not.

In this contribution, the application of a finite difference method simulation to obtain the potentials of cooling energy savings in the polymer pipe production process with the application of conventional cooling equipment is realized. The software delivers results for the 3D-temperature distribution, the material quality and the heat flow in the polymer pipe during the production process and the demand of cooling power respectively.

The intuitive graphical user interface (GUI) enables the operator to define an arbitrary process for the simulation. Hence, the creation of a virtual production line is possible. Furthermore, different optimization objectives can be defined, for example the final inner surface temperature of the pipe, the cooling medium temperature for which maximal mass throughput by taking into account lowest energy costs can be realized. The underlying material characteristics are summarizes in material models and adequate look up tables.

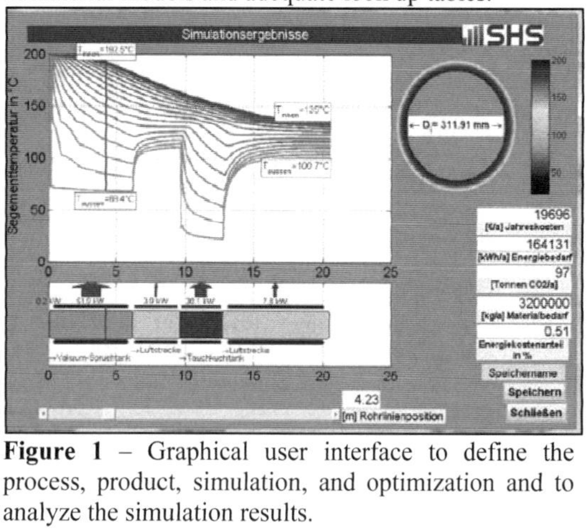

Figure 1 – Graphical user interface to define the process, product, simulation, and optimization and to analyze the simulation results.

Process and Simulation Software Basics

The simulation software is customized for the polymer pipe production process. The implementation of the finite differences method to the cylindrical pipe geometry delivers the temperature distribution in discrete radial layers, and the number of layers or discretization width respectively.

Further simulation results are the cooling temperature and cooling gradient for each layer at every position in the production line. The demand of cooling power and the thermal losses can be obtained in detail from the 3D-temperature distribution and the underlying heat flow. The heat conductivity and transmission coefficients are essential model parameters. They are optimized with respect to the polymer material characterization and temperature measurements during the real production.

An exemplary polymer pipe production facility with an overall length of approximately 21m is illustrated in Fig. 1. The software provides the possibility to create a virtual facility with all relevant parameters, for instance the material, pipe dimensions, output rate, facility geometry, cooling media, temperatures, ambient conditions, etc. The most common polymer material characteristics are already provided in a data base.

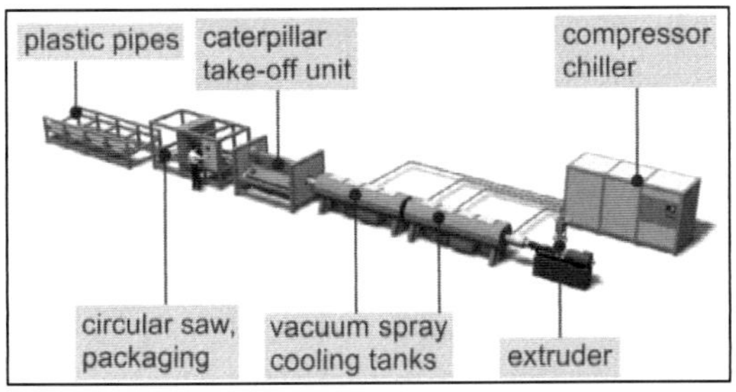

Figure 2 – Example for the polymer pipe (PE/PP/PVC) production process. The overall length of the production line exceeds 20 m in most application cases.

The discretization of the pipe can be performed by the user, however to guarantee stability in the simulation algorithm, the software provides suitable simulation parameters to guarantee convergence. Fig. 3 shows a schematic example for the finite difference discretization into five layers for the radial pipe direction.

The geometry in radial direction is defined by the outer and inner diameter ($D_{a,i}$) and the discretization, where symmetry of the shape is assumed. The heat conductivities in each layer $\lambda_i = \lambda_i(T, \dot{T})$ depend on the material, temperatures and temperature gradients. Furthermore, the system boundaries have to be defined. For example the initial inner and outer cooling media temperatures $T_{\infty,cool,i,a}$ have to be settled.

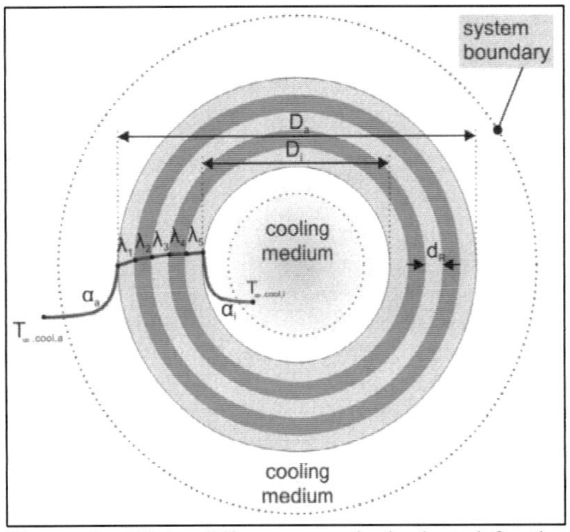

Figure 3 – Finite differences method adapted for the pipe extrusion process in radial direction and based on [1].

The stationary temperature distribution can be calculated with respect to [1] and [2], the radial discretization index m, the total number of radial elements M, the extrusion direction discretization index j, the heat conduction coefficient λ, the absolute radial discretization Δr, the layer radius r_i, the density ρ, the thermal capacity c_P, the conductibility of temperature

$$a = \frac{\lambda}{\rho \cdot c_P}$$

and the term

$$W = \frac{\Delta\lambda}{\Delta T}\frac{1}{\lambda}$$

to

$$T_{m,j} = T_{m,j-1} + \frac{a\cdot\Delta t}{\Delta r^2}\Big[T_{m+1,j-1} - 2T_{m,j-1} + T_{m-1,j-1} + \frac{1}{2\left(\frac{r_i}{\Delta r}+m\right)}\cdot\left(T_{m+1,j-1} - T_{m-1,j-1}\right) + \frac{W}{4}\cdot\left(T^2_{m+1,j-1} - 2T_{m+1,j-1}\cdot T_{m-1,j-1} + T^2_{m-1,j-1}\right)\Big].$$

The boundary condition for $r = r_A = \frac{D_A}{2}$ $(m = M)$ with s being the pipe wall thickness, the heat transmission coefficient α_a and the BIOT-number $Bi = \frac{\alpha_a \cdot s}{\lambda}$ reads

$$T_{M,j} = T_{\infty,cool,a} + \frac{T_{M-1,j} - T_{\infty,cool,a}}{1+\frac{1}{M}\cdot Bi}.$$

The corresponding boundary condition for the inner pipe surface can be calculated with

$$T_{1,j} = T_{\infty,cool,i} + \frac{T_{2,j} - T_{\infty,cool,i}}{1+\frac{1}{M}\cdot Bi}.$$

The heat transmission coefficients $\alpha_{i,a}$ in the different facility sections are obtained from the optimized calibration based on suitable measurements with thermal imaging (Fig. 5), innovative infrared sensors (Fig. 6) and the minimization of the objective function

$$J = \sum_{i=1}^{n}\left(T_{measure,i} - T_{simulation,i}\right)^2,$$

with a hybrid optimization, consisting of a GENETIC [5] and a NELDER-MEAD simplex [6] algorithm.

Simulation Results and Discussion

For an exemplary PE/HD-pipe production facility,

- containing two vacuum spray cooling tanks (6m and 4m length),
- pipe dimension of 400x3.6mm,
- underlying electricity costs of 0.11 €/kWh,
- a production time of 8000 hours per year and an
- output rate of 150 kg/h,

the simulation is progressed.

The temperature profiles along the facility for each radial layer are shown in Fig. 4. The process does not comprise an inner pipe cooling, so that the inmost layer (coded with the darkest blue color) is the heat center along the whole facility. Another essential phenomenon, which is covered by the simulation, can be found in the crystallization plateau of the used polymer at approx. 138°C. Due to the low temperature conductivity of the processed PE, the temperature of the outside layer and the outer surface respectively increases in the uncooled air sections between the primary and secondary cooling tank. The results from the reference thermal imaging analysis of the outer pipe surface are drawn with green rhombs (for example see Fig. 5), the inner pipe measurements with an infrared sensor are marked with dark blue rhombs. The innovative measurement device to determine the inner pipe temperatures during the production process is illustrated in Fig. 6.

The associated simulation results are highlighted with small red and pale blue dots, in order to realize a local visual comparability between measurement and simulation results. The simulation parameters are optimized as described before.

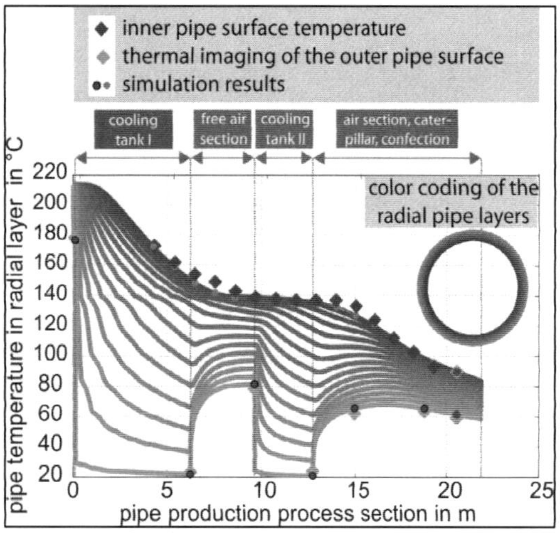

Figure 4 – Simulation results with optimized heat conduction and transmission coefficients.

Figure 5 – Thermal image of the PE-pipe cross section after confection with a circular saw.

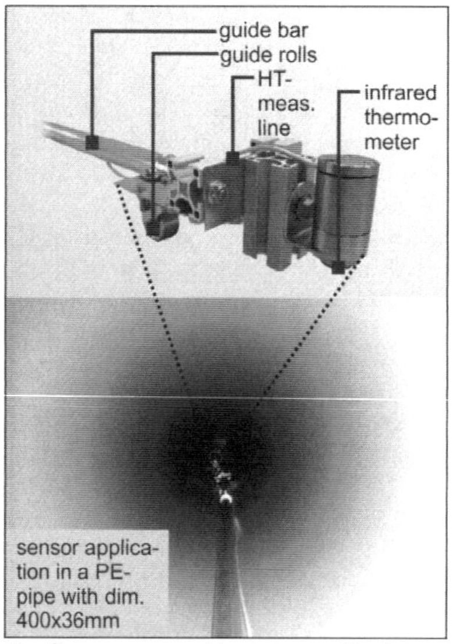

Figure 6 – Sensor concept to measure the inner surface pipe temperature during the production process.

Further simulation results are presented in Fig. 7. The demand of cooling power and thus the consumption of electric energy basically correlate with the temperature of the cooling medium. The electric energy consumption for conditioning the coolant in the introduced facility is analyzed for different cooling medium temperatures in the primary vacuum spray tank. The temperature is virtually varied from 10 to 65°C, while the coolant temperature in the secondary tank is kept constant at 15°C. As shown in Fig. 7 the overall cooling power demand decreases continuously from 54.4 down to 44.2kW, this equates to energy cost savings of 18.75%. The cooling power in the primary tank decreases with the rising coolant temperature, this effect is based on the reduction of the temperature gradient for the heat transmission from the outer pipe surface to the coolant. The raise of the cooling power demand in the secondary tank depends on the higher outer surface temperature of the pipe, when entering the tank. As a result, the distribution of the cooling power demands in both tanks is more balanced. The overall cooling energy demand reduction can lead to economies of approximately 9,000 € per year for the assumptions made prior and an electric compressor chiller with an energy efficient ratio of three.

At this point has to be mentioned, that the increase of the coolant temperature is not possible for every process. It has to be guaranteed, that the pipe characteristics still comply with the norms and the confection is possible without a pipe-deformation. Nevertheless, plenty of production facilities are driven with highly varying products and mass throughputs, so that an individual optimization of the processes implies high potentials for energy savings.

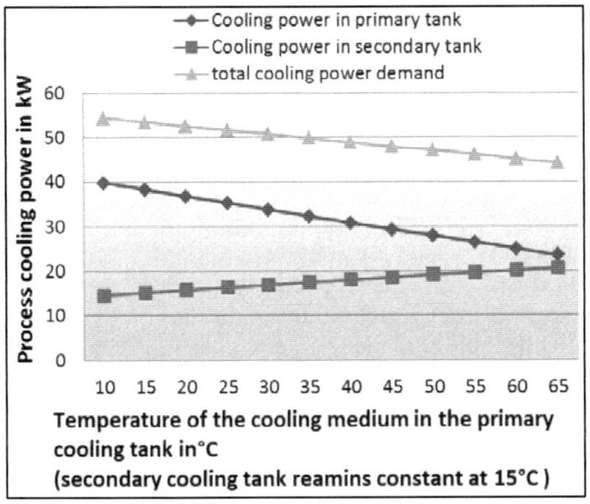

Figure 7 – Demand of cooling power with varying coolant temperature in the primary cooling tank.

A further advantage of the simulation is the prediction of the cooling gradients in each layer. In principle does the degree of crystallization increase with decreasing cooling gradients. This leads to an enhanced stability of the final product, which can improve the results from the internal pressure creep rupture test (DIN EN ISO 9080 [8]). Fundamentally can be proposed, that the residual stresses increase with a faster cooling of the outer pipe surface as presented in Fig. 8 and based on [7].

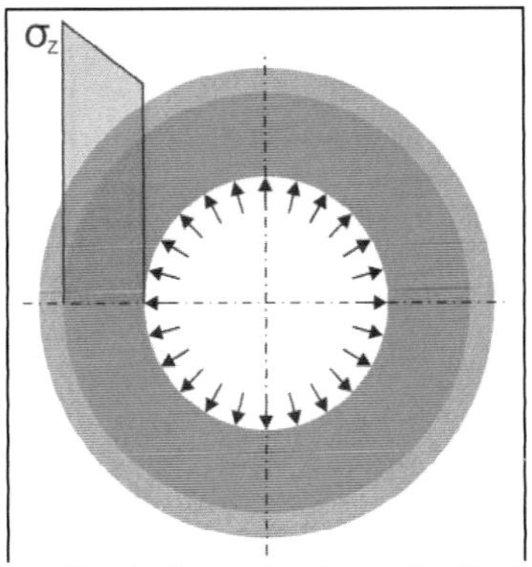

Figure 8 – Residual tensile stress distribution for a polymer pipe with outer calibration and internal pressure based on [7].

Furthermore, for products with a demand of low residual stresses tempering section are needed. The adaption of the coolant temperatures and the cooling gradients respectively can face this challenge inline.

Conclusions

The energy costs escalate continuously, especially with regard to the electric energy costs. The proposed simulation software is able to determine the cooling process for polymer pipes. A finite difference method based simulation is realized in combination with detailed material models, measurement techniques and a multi objective optimizer to realize optimal simulation results. The software can be accessed user friendly by a graphical user interface. The simulation results (temperature profiles, demand of cooling power) can be easily evaluated by the user. Furthermore, an optimized process for polymer pipes with respect to predefined objectives can be conducted. The simulation results indicate a high quality as proofed by the comparison to the measurement data in an exemplary process. In particular, the cooling power and energy consumption for the conditioning of the cooling fluids can be obtained. This directly enables the user to estimate the potentials for energy savings under the guarantee of a stable process with low residual stresses.

References

1. U. Kleindienst; *Untersuchung des Abkühlvorganges und dessen Einfluss auf das Eigenspannungsfeld in der Wand extrudierter Kunststoffrohre*; PhD Thesis; 1976.
2. E. Haberstroh; *Analysis of Cooling Sections in Extrusion Lines*; PhD Thesis; Aachen; 1981.
3. VDI-Gesellschaft Kunststofftechnik; *Kühlen von Extrudaten*; VDI-Verlag; 1978.
4. VDI-Gesellschaft Verfahrenstechnik; *VDI-Wärmeatlas*; Springer Verlag; 2006.
5. A. J. Chipperfield et. al.; *A Genetic Algorithm Toolbox for MATLAB;* in International Conference on Systems Engineering proceedings, Coventry, UK; 1994.
6. J. C. Lagarias; J. A. Reeds; M. H. Wright, and P. E. Wright; *Convergence Properties of the Nelder-Mead Simplex Method in Low Dimensions*; SIAM Journal of Optimization; Vol. 9; 1998.
7. P. Eyerer, T, Hirth, P. Elsner; *Polymer Engineering*; Springer; 2008
8. DIN EN ISO 9080:2003-10; *Plastics piping and ducting systems - Determination of the long-term hydrostatic strength of thermoplastics materials in pipe form by extrapolation*; 2003.
9. V. Mergl; *Effizienz von Kälteanlagen*; Kälte Klima Aktuell, Ausgabe 03/210-Technik; 2010.
10. R. Dzuban; *Abgeschaltete Kältemaschinen sind bares Geld für Kunden;* Kunststoffverarbeitung Deutschland; ONI Wärmetrafo GmbH; 2008
11. Rehsler Kühlsysteme; *Moderne Energiesparkonzepte zur Reduzierung des Treibhauseffektes und optimaler Nutzung unserer Primärressourcen;* URL:www.rehsler.com; 2012
12. J. Wortberg, K. Saul, S. Kruppa: *Im Zusammenhang betrachtet, Energieeffizienz in der Extrusion,*

O-10-054

COMPUTATIONAL DEVELOPMENT OF A COUNTER FLOW COOLING SYSTEM FOR THE BLOWN FILM PROCESS

M. Janas[1]*, L. Fehlberg[2] and J. Wortberg[1]

[1]*Institute of Product Engineering (IPE), University Duisburg-Essen, Germany – marius.janas@uni-due.de*
[2]*Institute for Mechatronics and System Dynamics, University Duisburg-Essen, Germany – lisa.fehlberg@uni-due.de*

Abstract – The cooling capacity limits the production throughput of extrusion lines. Hence, extruders can provide more polymer melt than it is possible to cool the extrudate. By optimizing the cooling system, the mass flow rate can be increased considering all quality issues of the product. Conventional cooling systems for the blown film extrusion guide the cooling airflow only in draw up direction. In contrast, a counter flow system also provides the air in direction towards the die. The airflow is separated into two individual flows. Therefore, it is possible to cool the film intensively over a longer film surface. In addition, an interaction of the jet flow and the ambient air can be reduced so that thickness variations are decreased. A process model for the blown film extrusion, being able to predict the bubble shape in the tube formation zone, is used to develop such a cooling system numerically. The simulation model is based on two submodules. A CFD-analysis computes the temperature in the film as well as the pressure on the film surface over the height. These results set the input data for the contour calculation module which uses the framework of Pearson and Petrie, as well as a modified rheological description from Phan-Thien and Tanner. Both calculation modules interact in a loop until a final quasi-stationary blown film contour is found. For any given cooling geometry, the developed simulation model can predict a realistic blown film behavior. Hence, it is possible to optimize the guiding lip geometry to achieve the best possible cooling condition. This approach is able to reduce the development time and can afford novel designs of cooling systems for the blown film extrusion.

Keywords: blown film extrusion, process model, counter flow cooling system, Phan-Thien and Tanner, simulation of the blown film contour

Introduction

The blown film extrusion is the major production technique to manufacture continuously thin polymeric film at high processing rate. Molten polymer is pushed through an annular die to form a tube. Nip rolls draw the film in vertical direction with a higher velocity than the polymer leave the die. Simultaneously, the tube is inflated by an internal pressure and cooled by external air rings.

In the 70s Pearson and Petrie [1, 2] developed a model to simulate the blown film process. This model sets the theoretical framework for most subsequent studies. Using the thin shell theory, the kinematic and dynamic behavior within the tube formation zone is described. For the rheological description, Pearson and Petrie used an isothermal Newtonian fluid. Their simulation results show a blown film contour with a long neck and sudden blown-up. Further authors enhanced their process model, changing the simple rheological description by non-isothermal differential material models like the upper convected Maxwell [3], the Phan-Thien and Tanner [4, 5] or the Pom-Pom [6, 7] model, to obtain a realistic film contour prediction. One important issue for the good correlation of these models between simulation and experiment is the temperature dependent material behavior. In the majority of the studies, the necessary temperature profile of the melt reaching from the die exit up to the frost line is calculated by an equation for the energy balance [5, 7 and 8], using constant convection coefficients and air jet velocities. Another approach that is used in this paper is the coupling of a CFD-analysis, which calculates the cooling conditions during the film processing and the rheological description [4]. Wolf et al. [9] developed a two dimensional CFD model to simulate the rotational, symmetric cooling jet flow acting on the film for a stationary consideration. Moreover, the CFD-analysis offers a realistic pressure profile on the film surface.

In addition, to correct thermodynamic values and realistic acting forces, the essential aspect to gain a good numerical prediction of a blown film contour, accurate material parameters are required. In the literature standard measurement methods are mostly used to characterize the polymer properties in steady shearing, oscillatory shearing and uniaxial elongation [10, 11]. Hence, these material data are not measured at realistic process conditions. E.g. the viscosity of the polymer is a function of uniaxial elongation. However, during the tubular blown film process, the polymer chains are not only oriented in one direction, they are biaxially stretched [4]. Therefore, it is important to record these material properties for different process conditions and stretching ratios that are close to reality respectively.

In this paper a simulation model for the blown film process is shown, which is able to predict a realistic bubble behavior dependent on the cooling situation. Therefore, the simulation model is qualified for the virtual optimization and benchmarking of new cooling geometries using the example of a counter flow system.

Process Model for the Blown Film Extrusion

The process model for the blown film extrusion is based on two sub-modules. Both modules interact in a loop until a quasi-stationary final blown film contour is found. In Figure 1, the simulation schema is shown to compute an unknown process state.

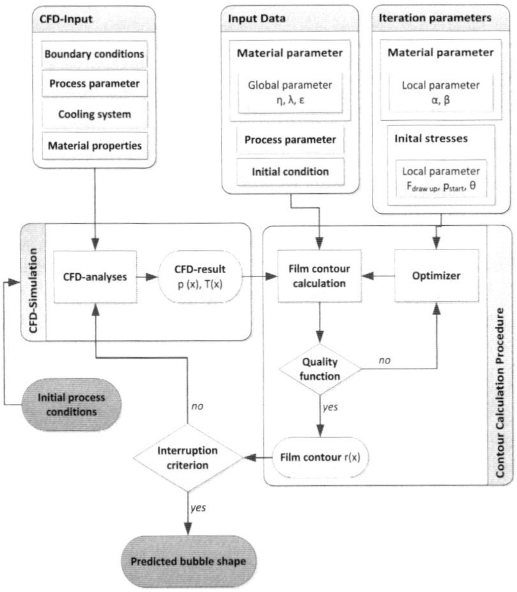

Figure 1 – Blown film simulation schema

Starting from initial conditions (start contour and a cooling geometry) a CFD-analysis computes the thermodynamic behavior of the film and the fluid dynamic properties of the cooling jet flow. The temperature ($T(x)$) in the film over the height and the pressure profile ($p(x)$) on the film surface set the initial conditions for the film contour calculation procedure. Next to the CFD-results, global material and process parameters are required. Computing a realistic film contour an optimizer varies anisotropic local material parameters (α, β), factors for the initial stresses ($F_{draw\ up}$, p_{start}) and the start blowing angle (θ). These parameters depend on the calculated process state and each iteration step [3, 6 and 8]. A quality function controls the procedure until a film contour is calculated. The quality criteria are the blow-up ratio (BUR) and the final film thickness. So, the simulation model does not know a complete target contour. In the last step, an interruption criterion checks whether the computed film contour does not change anymore compared to the iteration before. If a final quasi-stationary contour is determined, the simulation is finished. Otherwise, the new film curve replaces the old one in the CFD-simulation and the procedure is repeated.

Film Contour Model

As mention above, the contour calculation model uses the framework from Pearson and Petrie. The membrane theory (Eq.1) allows the description of the interaction between the rheological stresses in the film (σ_{11}, σ_{33}), the geometrical properties of the bubble (R_{11}, R_{33}), the film thickness (h) and the pressure load on the surface of the film (Δp).

$$\frac{\Delta p}{h} = \frac{\sigma_{11}}{R_{11}} + \frac{\sigma_{33}}{R_{33}} \qquad (1)$$

Converting all necessary boundary conditions (cf. [8]) in Eq.1, the following first order nonlinear differential equation for the blowing angle of the film contour (θ) is achieved.

$$\Theta' = \frac{1}{\sigma_{11}}\left(\frac{\sigma_{33}}{r} - \frac{\Delta p}{h \cdot \cos\theta}\right) \qquad (2)$$

Thereby, the blowing angle is defined as the arc tangent of the bubble radius' derivative (6).

$$r' = \tan\theta \qquad (3)$$

Using the mass conservation in differential form, the film thickness can be calculated as shown in Eq.4.

$$h' = -\left(\frac{v'}{v} + \frac{r'}{r} + \frac{\rho'}{\rho}\right) h \qquad (4)$$

Compared to former papers, the force balance in extrusion direction is neglected. Consequently, the velocity of the film cannot be calculated in a coupled manner. Here, another approach is used. A master curve of a measured film velocity for a known process condition is taken [12]. In a first approximation, this curve can be adjusted to other process condition for the same material, different melt throughputs and various film end thicknesses. Each polymer has a characteristic bubble shape and a typical film velocity profile [13]. Such an approach achieves a realistic computation of the bubble shape and the correct film thickness profile. Moreover, the simulation is much more robust.

The rheological stresses are computed using a modified material model from Phan-Thien and Tanner (PTT) (Eq. 5 to 7).

$$\tau_{11}' = 2\,\eta_{11}\frac{v'}{v}\frac{1}{\lambda_{11}} - \frac{\tau_{11}\, e^{\left(\frac{\varepsilon\,\lambda_{11}}{\eta_{11}}(tr\,\tau_{ii})\right)}}{\lambda_{11}\, v\, \cos\theta} + 2\,(1-\xi)\,\tau_{11}\frac{v'}{v} \qquad (5)$$

$$\tau_{22}' = 2\,\eta_{22}\frac{h'}{h}\frac{1}{\lambda_{22}} - \frac{\tau_{22}\, e^{\left(\frac{\varepsilon\,\lambda_{22}}{\eta_{22}}(tr\,\tau_{ii})\right)}}{\lambda_{22}\, v\, \cos\theta} + 2\,(1-\xi)\,\tau_{22}\frac{h'}{h} \qquad (6)$$

$$\tau_{33}' = 2\,\eta_{33}\frac{r'}{r}\frac{1}{\lambda_{33}} - \frac{\tau_{33}\, e^{\left(\frac{\varepsilon\,\lambda_{33}}{\eta_{33}}(tr\,\tau_{ii})\right)}}{\lambda_{33}\, v\, \cos\theta} + 2\,(1-\xi)\,\tau_{33}\frac{r'}{r} \qquad (7)$$

The standard PTT-model describes the elongation viscosity for uniaxial stretching correctly [14] but not for a biaxial blown film process. Thus, the rheological model is enhanced by an approach to improve the description of the anisotropic material behavior (Eq. 8 to 11).

$$\eta_{22} = \alpha \cdot \eta_{11} \quad (8) \qquad \eta_{33} = \beta \cdot \eta_{11} \quad (10)$$

$$\lambda_{22} = \alpha \cdot \lambda_{11} \quad (9) \qquad \lambda_{33} = \beta \cdot \lambda_{11} \quad (11)$$

Further information about the simulation model and the used equations are presented in [4, 15 and 16].

Material Calibration

Simulating a realistic blown film contour, accurate material parameters are needed. Due to the process condition during the manufacturing of thin tubular film, polymer chains are biaxially stretched. Standardized measurement methods cannot gain correct data at process conditions (see above). Therefore, another approach is developed to calibrate the material model [4, 15]. The film contour is the result of all acting influences on the bubble, i.e. the impact of the air jet flow, the take-up force and of course the material behavior. Thus, a set of experimentally recorded blown film contours with various process conditions (calibration space) have to be recomputed using only the contour calculation model without the CFD-module. Therefore, all parameters have to be optimized to achieve the best fit of the whole calibration space at once. In this case thirteen states are recomputed simultaneously. The global material parameters (η, λ, ε) are for every process state the same compare to the local iteration ones (α, β, $F_{draw\ up}$, p_{start}, θ) (cf. Fig.1).

Numerical Development of a Counter Flow Cooling System

The idea of a counter flow cooling system is to cool the blown film intensively over a longer film surface, without interactions of the jet flow and the ambient air. So, the active film cooling surface can be increased and film thickness variations can be decreased. Hence, the cooling jet flow is separated into two independent ones. One jet flow is directed in draw up direction as known in conventional systems and another one is oriented towards the die.

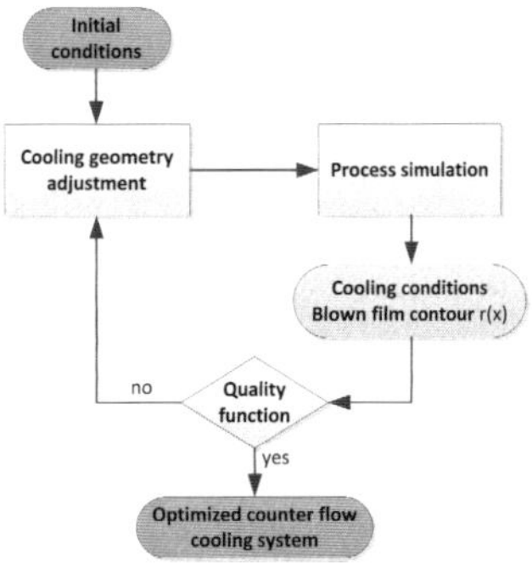

Figure 2 – Optimization schema for a new cooling geometry

Figure 2 shows the manual optimization for a counter flow cooling systems. Starting with a given cooling design, a complete process simulation of an unknown state has to be done as report above and in [16]. If the quasi-stationary film contour is achieved, the results for the frost line height and the temperature behavior in the film are checked. The idea of the quality function is to reduce the frost line height in changing the cooling geometry at constant cooling boundaries like cooling volume and cooling air temperature. Geometrical parameters to modify the cooling design are e.g. the opening angel of the cooling guiding geometry or the length of the cooling lips (Fig.3). After a geometrical adjustment of the cooling lips, the simulation is started again. This optimization loop is repeat until no reduction of the frost line height can be detected.

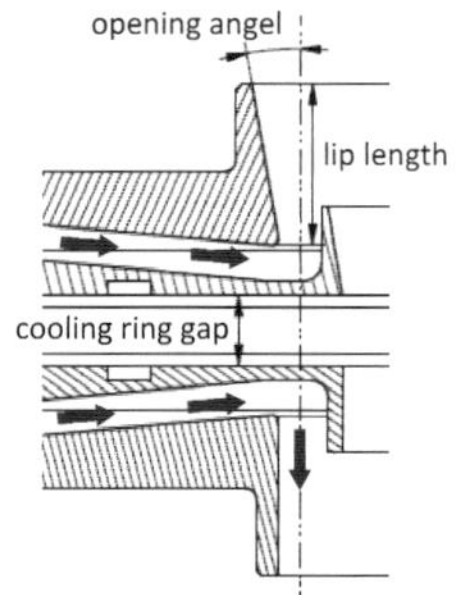

Figure 3 – Design of guiding lips for a **c**ounter **f**low **c**ooling system (CFC-1)

Figure 3 shows the initial cooling lip geometry for the optimization. It is obvious that the jet flow is guided on the film in 80° in extrusion direction and 90° towards the die. In addition, there is a gap between the two air rings to avoid a low pressure between cooling system and flexible film. A low pressure could cause that the tubular film could tear. Because of the gap, ambient air can stream through the gap and eliminates the low pressure.

Table 1 – Simulation settings

cooling air volume	upwards	700	[m³/h]
	downwards	400	[m³/h]
cooling air temperature	upwards	300	[K]
	downwards	300	[K]
polymer		PELD	
BUR		2	[-]
film end thickness		100	[µm]
melt mass flow rate		35	[kg/h]

All important simulation settings for the CFD-analysis and the film contour calculation are listed in Table 1. The material parameters for the modified PTT-model are used that are calibrated for a PE-LD calibration space and a close guiding cooling system [16].

Simulation and Experiment Results

Starting from the initial cooling systems (Fig.3), several different geometry variations of a counter flow cooling device are simulated. Each process simulation needs an average of ten iteration steps to gain a quasi-stationary film contour. In Figure 4, the frost line heights for several cooling geometries are shown.

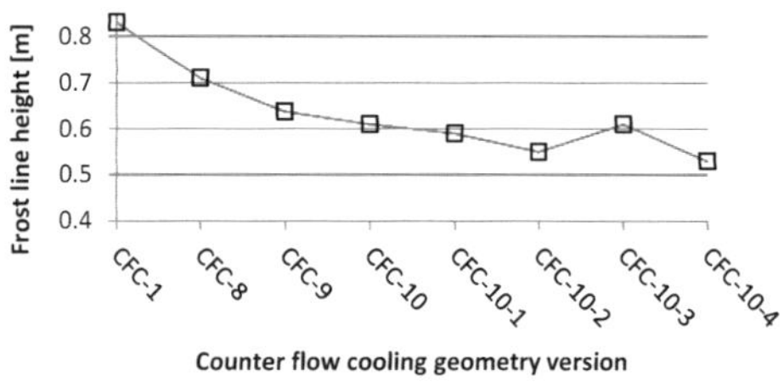

Figure 4 – Frost line heights for different cooling geometries

By changing the lip geometry of the cooling system a frost line height reduction of 30 cm is achieved, using the same cooling configuration (cf. Tab.1).

Figure 5 shows a CFD velocity plot for the initial cooling system (CFC-1) and the optimized version (CFC-10-4). It is obvious that the jet flow for the optimized system has got a higher velocity for the same boundary conditions at the output section of the lip as for the CFC-1 system. In addition, the jet can be provided on the bubble surface over a longer distance. So, an intensive cooling of the film can be achieved, especially for the downwards directed jet flow.

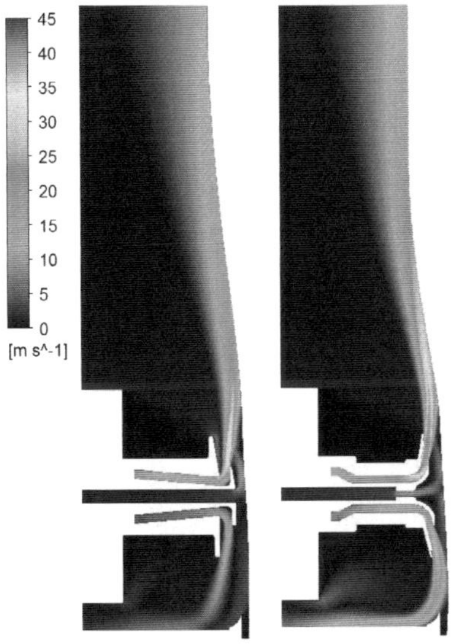

Figure 5 – CFD velocity plot, left) initial cooling system CFC-1, right) optimized system CFC-10-4

After the optimization is done, the system "CFC-10-4" is manufactured and tested on the laboratory blown film line using same process conditions as in the simulation. The blown film contour is recorded with the aid of a digital photo camera. Five images are taken for one process condition. An edge tracking software is used to measure the experimental film curve. Figure 6 shows the results of the simulation and experiment. It is obvious that the simulation can predict the experimentally recorded film contour in a very good manner.

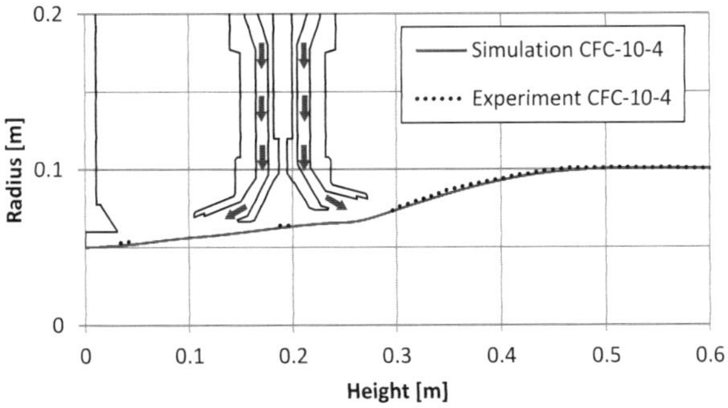

Figure 6 – Simulated and experimentally recorded film contour

The correct frost line height is almost hit.

Conclusion

In this paper, a simulation procedure for a numerical development of a counter flow cooling system is achieved. Therefore, a process model for the blown film extrusion is used. The model is able to predict a realistic bubble shape behavior according to the current cooling configuration. Using a manual optimization different cooling lip geometries are simulated and benchmarked according to the bubble shape and the frost line height. The optimized geometry is tested on the blown film line and the results of experiment are compared with the simulation. Summarized, the computed film contour is in good agreement with the experiment. Hence, a numerical development of a new cooling system is successfully achieved.

Acknowledgements

This project is supported by the German National Science Foundation (Deutsche Forschungsgemeinschaft, DFG), project-nr. WO 302/39-1. We are also thankful for the material support from A. Schulman, BASF, DOW, LyondellBasell and SABIC.

References

1. J. Pearson, C. Petrie, *J. Fluid Mech.* 1970, *40(1)*, 1
2. J. Pearson, C. Petrie, *J. Fluid Mech.* 1970, *42(3)*, 609
3. J. C. Bennett, PhD Thesis, RMIT University, Melbourne, Australia 2008
4. M. Bussmann, PhD Thesis, University of Duisburg-Essen, Germany 2010
5. J. Sung Lee, H. Wook Jung, J. Chun Hyun, *AIChE Journal* 2011, *57(12)*, 3565
6. L. K. Henrichsen, PhD Thesis, University of Illinois at Urbana-Champaign, Illinois, USA 2006
7. F. Sharif, S. Sarafrazi, *Intern. Polym. Process.* 2008, *XXIII(1)*, 30
8. I. A. Muslet, M. R. Kamal, *J. Rheol.* 2004, *48(3)*, 525
9. D. Wolf, B. Feron, J. Wortberg, *Intern. Polym. Process.* 1997, *XII(1)*, 38
10. T. Kanai, J. L. White, *Polym. Eng. Sci.* 1984, *24(15)*, 1185
11. P. P. Tas, PhD Thesis, Eindhoven University of Technology, Netherland 1994
12. J. Spirgatis, PhD Thesis, University of Duisburg-Essen, Germany 2004
13. A. Ghaneh-Fard, P. J. Carreau, P. G. Lafleur, *Polym. Eng. Sci.* 1997, *37(7)*, 1148
14. S. A. Khan, R. G. Larson, *J. Rheol.* 1987, *31(3)*, 207
15. M. Bussmann, M. Janas, J. Wortberg, in PPS-26, Banff, Canada 2010
16. M. Janas, M. Bussmann, L. Fehlberg, L. Mikelsons, J. Wortberg, in PPS Americas Conference, Niagara Falls, Canada 2012

KN-11-02

INJECTABLE PH/TEMPERATURE-SENSITIVE HYDROGELS FOR BIOMEDICAL APPLICATIONS

Cong Truc Huynh[a] and Doo Sung Lee*

Theranostic Macromolecules Research Center, Department of Polymer Science and Engineering, Sungkyunkwan University, Suwon, Gyeonggi-do 440-746, South Korea. C.T. Huynh: huynhcong_truc@yahoo.com; D.S. Lee: dslee@skku.edu

[a] *current address: Australian Institute for Bioengineering and Nanotechnology, The University of Queensland, St Lucia, Brisbane, QLD 4072, Australia.*

Abstract - Over the last decade, injectable stimuli-sensitive polymeric hydrogels have attracted considerable attention, because of their potential biomedical and pharmaceutical applications, such as in drug/protein delivery and tissue engineering. These polymeric aqueous solutions exhibit a sol-gel phase transition in response to external stimuli, such as pH, temperature, glucose, electric fields, magnetic fields, ionic strength or a combination of them. Hydrogel precursors can be formulated with bioactive molecules or cells, and then simply injected into the body at specific sites for the application purposes. In this paper, we report recent progress in injectable block copolymer hydrogels responding to pH and temperature, which were developed by our group, and their potential pharmaceutical and biomedical applications. The copolymers contain pH-sensitive moieties and many different chemical groups, such as ester, amide, urethane, urea…, were designed to control the hydrogel properties, including biodegradable, mechanical, in vitro and in vivo stability, cytotoxicity and release behavior. These copolymer aqueous solutions existed in the sol state at acidic (cationic hydrogels) or basic (anionic hydrogels) pH and low temperatures with low viscosity, which are suitable for formulation with bioactive molecules, but turned into the gel state under physiological conditions (37 °C, pH 7.4) or after being injected into the body, which could let them serve as proteins/bioactive molecules depots for long term sustained release.

Keywords: biodegradable, biomaterials, delivery system, injectable hydrogels, pH/temperature-sensitive.

Introduction

Hydrogels are three-dimensional hydrophilic polymeric networks that can absorb a considerable amount of water or biological fluids while maintaining their semi-solid morphology [1-3]. Injectable polymeric hydrogels are the most attractive materials for biomedical and pharmaceutical applications, especially for drug/protein delivery and tissue engineering [1-6]. Such injectable hydrogel systems are of particular interest, because of the advantages they offer: i) they require no organic solvent in processing; ii) they require no surgical or implantation procedures; iii) their precursor solutions can be easily mixed with bioactive molecules such as drug, protein, DNA, or cells prior to gelation or injection; iv) the formed gels become drug carriers for localized delivery or act as scaffolds for tissue regeneration; v) the formed gels have the ability to release both hydrophobic and hydrophilic drug and ionic bioactive molecules, with sustained release behaviour. The aqueous solutions of these polymeric hydrogels exhibit a sol-gel phase transition in response to external stimuli, such as pH, temperature, glucose, electric fields, magnetic fields, ionic strength or a combination of them. Bioactive molecules, such as drug, protein, DNA or cells, can be easily mixed with the hydrogel precursor solutions, prior to in vitro gelation or injection into the body, as shown in Fig. 1. The formed gels become carriers of therapeutic agents for localized delivery, or scaffolds for tissue regeneration [1-6].

Injectable hydrogels can be classified into two basic types, depending on their crosslinking method: chemical or physical. Chemically crosslinked hydrogels can be generated via a Schiff base [7], enzymes [8], photopolymerization [9], and Michael-addition or chemical reactions [10,11], and usually exhibit a noticeable volume change. Although chemically crosslinked hydrogels possess strong mechanical properties, the requirement of crosslinking agents, enzymes, photoinitiators and/or organic solvents, which may damage cells as well as denature incorporated bioactive molecules, may limit their applications. In contrast, physically crosslinked hydrogels, formed via the self-assembly of amphiphilic block/graft copolymers in response to external stimuli, such as temperature or pH/temperature, offer a mild method of preparing hydrogels. Moreover, such preparations exhibit a sol-gel transition without any significant volume change [1-5].

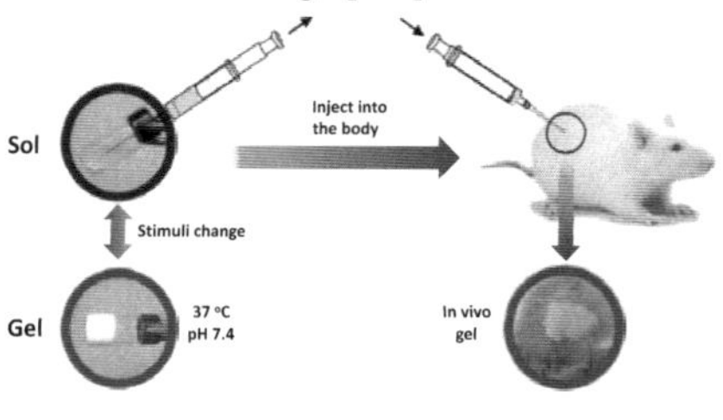

Figure 1: General sol-gel phase transitions of injectable stimuli-sensitive hydrogels. Adapted from Huynh et al. [1] by permission of American Chemical Society.

In this paper, we report recent progress in injectable block copolymer hydrogels responding to pH and temperature, which were developed by our group, and their potential pharmaceutical and biomedical applications. The copolymers contain pH-sensitive moieties and many different chemical groups, such as ester, amide, urethane, urea…, were designed to control the hydrogel properties, including sol-gel phase transition behaviour, biodegradation, mechanical properteis, in vitro and in vivo stability, cytotoxicity and drugs/proteins release behavior.

Experimental

Synthesis of block copolymers

In the past few years, many different kinds of block copolymers have been synthesized and examined the potential application by our group. They are block copolymers, consisting of hydrophilic poly(ethylene glycol) (PEG) and ionizable blocks, such as poly(amido amine) (PAA), poly(amino urethane) (PAU), poly(amino urea urethane) (PAUU), poly(β-amino ester) (PAE), poly(amino ester urethane) (PAEU), oligosulfamethazine (OSM), oligopicolyamine (OPCA), or some low molecular weight gelators, such as oligo(β-amino ester urethane) (OAEU), oligo(amido amine)s (OAAs) and oligo(amido amine amino ester) (OAAAE) [12-40]. Specially, the amphoteric copolymer with dually cationic and anionic properties was also reported [41].The detailed synthesis procedures can be found in the appropriate references.

Hydrogel properties

The procedures for checking the properties of the injectable hydrogel systems, including sol-gel phase transition, rheological properties, degradation, in vitro cytotoxicity, in vivo injectability, in vitro and in vivo release of biomolecules, have been mentioned in the previous reported [26-29,35-41]; thus, they would be kept the details in this paper.

Results and Discussion

Developed hydrogel systems by our group

"Smart" polymers that are capable of responding to multiple stimuli, especially pH and temperature, have drawn considerable research interest, because of their advantages in terms of chemical structure, and ability to form an ionic complex with therapeutic agents. This dual response can prevent gelation within the injection needle. These copolymer aqueous solutions existed in the sol state at acidic (cationic hydrogels) or basic (anionic hydrogels) pH and low temperatures with low viscosity, which are suitable for formulation with bioactive molecules, but turned into the gel state under physiological conditions (37 °C, pH 7.4) or after being injected into the body, which could let them serve as proteins/bioactive molecules depots for long-term sustained release. The sol-gel phase transition diagram of these hydrogel systems can easily be controlled by many different factors, such as molecular weight of PEG, hydrophobic/hydrophilic ratios, monomer structure, presented chemical groups and copolymer concentration.

Cationic pH/temperature-sensitive hydrogels, using block copolymer composed of PEG and PAA (Fig. 2A) or PAU, were developed in our group [12-15]. These copolymer aqueous solutions exist in the sol state at low pH (e.g. pH 6.5) in the whole range of experimental temperature (0-70 °C), but exhibited a sol-to-gel-to-sol phase transition with increasing temperature at pH≥7.0, due to hydrophobic interactions and hydrogen bonding between the PAA or PAU blocks. However, by adding urea groups into PAU to form PAUU (Fig. 2B), the aqueous copolymer solution existed in the sol state at acidic pH and low temperatures, but exhibited a sol-to-gel and a gel-to-sol phase transition with increasing pH and temperature, respectively [16]. The disappearance of the sol regions at newtral pH and low temperatures in PAUU hydrogel was attributed to the more hydrophobicity and less solubility of PAUU segments. These hydrogels may have some limitation in the applicaiton because of their non-biodegradability.

Figure 2. Chemical structure of several developed ionizable pH/temperature-sensitive copolymers.

Biodegradable cationic poly(β-amino ester) (PAE) is non-cytotoxic and forms a good ionic complex with negatively charged proteins or pDNA at physiological pH. Aqueous solutions of PAE-PCL-PEG-PCL-PAE (Fig. 2C), or PAE-PCLA-PEG-PCLA-PAE pentablock copolymers, or graft copolymer of (PAE-g-PCL)-PEG-(PAE-g-PCL) [PCL: poly(ε-caprolactone); PCLA: poly(ε-caprolactone-co-lactide)] undergo a sol-to-gel-to-sol phase transition as a function of both pH and temperature (Fig. 3A) [17-21]. However, aqueous solutions of PAE-PEG-PAE or PEG(-PAE)$_4$ exhibit a sol-to-gel and gel-to-sol phase transition with increasing pH and temperature, respectively [22,23]. These two polymers are easily dissolved in water at a relatively low pH, because of free hydrophobic blocks. Multiblock copolymer [PCL-PEG-PCL-PAU]$_x$ was reported to show a sol-to-gel-to-sol phase transition in aqueous solution [24]. However, copolymers containing PCL(A)-PEG-PCL(A) are difficult to dissolve in water. Efforts to prepare biodegradable PAU-based copolymers without hydrophobic blocks led to the recent development of poly(β-amino ester urethane)-based copolymers (PAEU) (Fig. 2D) [25-29]. The [PEG-PAEU]$_x$ multiblock and PAEU-PEG-PAEU triblock copolymers are easily dissolved in water at acidic pH, and exhibit a sol-to-gel phase transition with changing pH and temperature to physiological conditions.

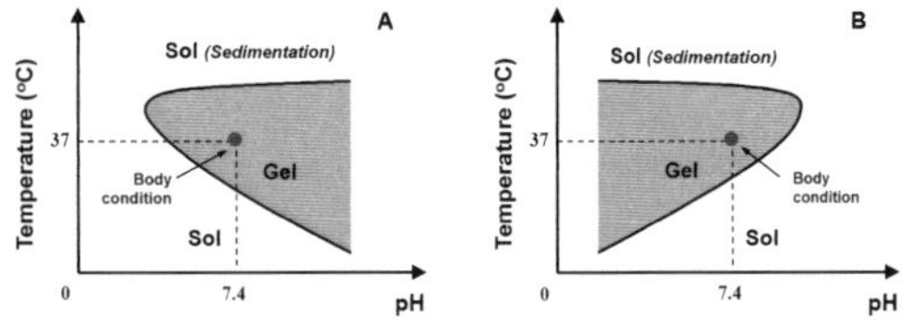

Figure 3: Sol-gel phase diagran of typical cationic hydrogel (A) and anionic hydrogel (B).

In contrast with cationic pH-sensitive polymers, anionic polymers can dissolve in water at basic pH (e.g. pH 8.0) and form ionic complex with growth factors, mostly positively charged, but become insoluble at physiological pH. The typical example of anionic biodegradable pH/temperature-sensitive copolymer hydrogel is oligomer sulfamethazine-based (OSM) copolymers. Aqueous solutions of pentablock copolymers OSM-PCLA-PEG-PCLA-OSM (Fig. 2F) or OSM-PLGA-PEG-PLGA-OSM exhibit a reversible sol-to-gel-to-sol phase transition as a function of both pH and temperature (Fig. 3B) [30-36]. The existence of a gel at physiological conditions lets it serve as a drug carrier, especially as a growth factor depot for cells culture.

pH/temperature-sensitive block copolymer hydrogels possess complicated structures, and their preparation requires several steps. In contrast, low molecular weight gelators, such as cationic oligo(β-amino ester urethane) (OAEU), oligo(amido amine)s (OAAs) and oligo(amido amine amino ester) (OAAAE) (Fig. 2E), have simple structures, and can be prepared via a simple one-step synthesis [38-40]. Their aqueous solutions exist in sol state at low pH (e.g. pH 6.5), but turn into gel state in physiological conditions, and thus serve as depot carriers for bioactive molecules.

Very recently, a new generation of hydrogel, dually cationic and anionic hydrogel, has been reported [41]. This copolymer, called poly(urethane amino sulfamethazine) (PUASM) (Fig. 2G), bears both acidic and basic pH-sensitive moieties. Aqueous copolymer solutions existed in the sol states at mildly acidic (e.g., pH 6.5) or basic pH (e.g., pH 8.5), which allow for homogeneously formulation with bioactive molecules and injection into the body, but exhibited specific cloosed-loop sol-gel-sol phase transitions with increasing temperature at the pH range of 6.8-8.2 (Fig. 4A). The ionization of basic portions containing amine groups in the PASM at slightly acidic pH (Fig. 4B1) and of acidic portions composed of sulfonamide groups in the PASM at slightly basic pH (Fig. 4B2) renders the polymer hydrophilic, resulting in a sol state in water. In contrast, relative deionization of both basic and acidic portions of the PASM and increases in hydrophobicity of the PCLA at physiological conditions led to formation and interconnection of nanostructured polymeric micelles, resulting in gelation (Fig. 4B3). This novel hydrogel system is expected to form ionic complex for sustained release wiht both anionic and cationic bioactive molecules.

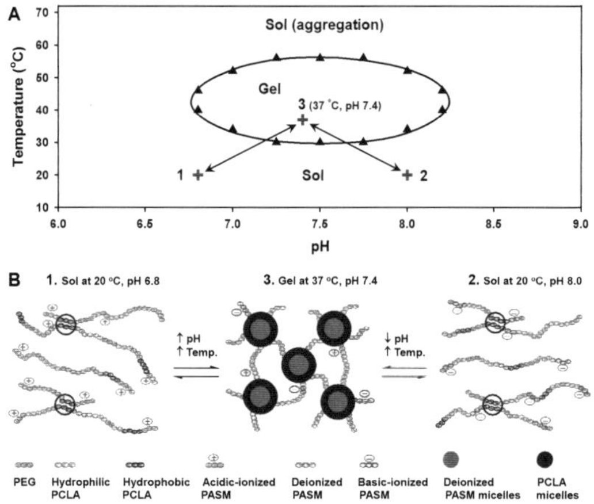

Figure 4. A) Specific closed-loop sol-gel-sol phase diagram of PUASM hydrogel. The bioactive molecules and copolymer solution can be formulated and injected into the body at low pH (e.g., 1, pH 6.8) or high pH (e.g., 2, pH 8.0), and a gel forms rapidly at physiological conditions (3, 37 °C and pH 7.4) for delivery purposes. B) Schematic showing reversible sol-gel-sol phase transition of the PUASM hydrogel with changing temperature and pH. Adapted from Huynh et al. [41] by permission of Royal Society Chemistry.

Potential applications of developed hydrogel systems as drug delivery systems

The developed injectable pH/temperature-sensitive hydrogel systems have been examined the potential application as drug delivery systems. Non-degradable

$[PEG\text{-}PAU]_x$ hydrogels released chlorambucil, an anticancer drug, for 2 weeks in vitro [14]. PAA-PEG-PAA and PAE-PEG-PAE hydrogels exhibited bioadhesive properties in mucosal surfaces and released flubiprofen and lidocaine for 1 day in vitro, respectively [13,22]. $[PEG\text{-}PAUU]_x$ hydrogels showed the ability to sustained release FITC-BSA in the course of 6 weeks in vitro [16]. With a single injection of insulin-loaded PAE-PCL-PEG-PCL-PAE hydrogels into Sprague-Dawley (SD) rats, the insulin concentration in serum was maintained at a high level for 2 weeks without an initial burst, whereas insulin-loaded PCL-PEG-PCL hydrogels showed a remarkable initial burst with complete insulin release within 1 day. This sustained release of insulin was attributable to the formation of an ionic complex between insulin and degradable PAE segments [17]. $[PCL\text{-}PEG\text{-}PCL\text{-}PAU]_x$ copolymer hydrogels showed the sustained in vitro release of PTX over 1 month [24]. $[PEG\text{-}PAEU]_x$ hydrogels were found to be non-cytotoxic, and released DOX for more than 5 weeks in vitro while PAEU-PEG-PAEU hydrogels could controlled release human growth hormone (hGH) for 5 day in vitro and for over 3 days in SD rat [26-28]. PTX-loaded OSM-PCLA-PEG-PCLA-OSM hydrogels showed antitumor efficacy for up to 2 weeks after one injection into tumor-bearing mice [33]. Injection of OSM-PCLA-PEG-PCLA-OSM hydrogels incorporating human MSCs and recombinant human bone morphogenetic protein-2 (rhBMP-2) into mice resulted in high levels of alkaline phosphate up to 7 weeks after injection [35] Biodegradable OAEU hydrogel released DOX in vitro for 1 week [38] and OAAAE hydrogel was reported to release insulin for up to 1 week in vitro. With a single injection of the insulin-loaded OAAAE hydrogel into SD rats, serum insulin concentrations were maintained for 4 days without an initial burst [40]. In additional, with a single injection of hGH-loaded dually cationic and anionic PUASM hydrogel, the hGH concentration in the serum of the SD rats was maintained at higher effective concentration (≥1 ng/ mL) for more than 3 days with a minimal initial burst [41]. The controlled release of hGH from the complex hydrogel was governed by ionic complexation between the anionic hGH and cationic moieties in the PUASM copolymer.

Conclusions

Many different injectable pH/temperature-sensitive hydrogel systems have been developed. These copolymer aqueous solutions existed in the sol state at acidic (cationic hydrogels) or basic (anionic hydrogels) pH and low temperatures with low viscosity, which are suitable for formulation with bioactive molecules, but turned into the gel state under physiological conditions (37 °C, pH 7.4) or after being injected into the body, which could let them serve as bioactive molecules depots for long-term sustained release. Their ability in the release of drugs/proteins may offer potential biomedical and pharmaceutical applicability.

Acknowledgements

This study was supported by the Basic Science Research Program through a National Research Foundation of Korea (NRF) grant funded by the Korean Government (MEST) (2010-0027955) and the Converging Research Center Program funded by the Ministry of Education, Science and Technology (2011K000817).

References

1. C.T. Huynh; M.K. Nguyen; D.S. Lee, *Macromolecules*, 2011, 44, 6629.
2. C. He; S.W. Kim; D.S. Lee, *J. Controlled Release*, 2008, 127, 189.
3. M.K. Nguyen; D.S. Lee, *Macromol. Biosci.*, 2010, 10, 563.
4. M.H. Park, M.K. Joo; B.G. Choi; B. Jeong, *Acc. Chem. Res.*, 2012, 45, 424.
5. L. Yu; J. Ding, *Chem. Soc. Rev.*, 2008, 37, 1473.
6. S. Van Vlierberghe; P. Dubruel; E. Schacht, *Biomacromolecules*, 2011, 12, 1387.
7. Y. Zhang; L. Tao; S. Li; Y. Wei, *Biomacromolecules*, 2011, 12, 2894.
8. N.Q. Tran; Y.K. Joung; E. Lih; K. D. Park, *Biomacromolecules*, 2011, 12, 2872.
9. B.L. Dargaville; C. Vaquette; H. Peng; F. Rasoul; Y.Q. Chau; J.J. Cooper-White; et al., *Biomacromolecules*, 2011, 12, 3856.
10. R. Jin; L.S.M. Teixeira; A. Krouwels; P.J. Dijkstra; C.A. van Blitterswijk; M. Karperien; J. Feijen, *Acta Biomater.*, 2010, 6, 1968.
11. D.K. Wang; F. Rasoul; D.J.T. Hill; G.R. Hanson; C.J. Noble; A.K. Whittaker, *Soft Matter*, 2012, 8, 435.
12. M.K. Nguyen; D.K. Park; D.S. Lee, *Biomacromolecules*, 2009, 10, 728.
13. M.K. Nguyen; D.S. Lee, *Macromol. Res.*, 2010, 18, 284.
14. K. Dayananda; C. He; D.K. Park; T.G. Park; D.S. Lee, *Polymer*, 2008, 49, 4968.
15. C.T. Huynh; M.K. Nguyen; D.P. Huynh; S.W. Kim; D.S. Lee, *Polymer*, 2010, 51, 3843.
16. C.T. Huynh; Q.V. Nguyen; S.W. Kang; D.S. Lee, *Polymer*, 2012, 53, 4069.
17. D.P. Huynh; M.K. Nguyen; B.S. Pi; M.S. Kim; S.Y. Chae; K.C. Lee; et al., *Biomaterials*, 2008, 29, 2527.
18. D.P. Huynh; G.J. Im; S.Y. Chae; K.C. Lee; D.S. Lee, *J. Controlled Release*, 2009, 137, 20.
19. D.P. Huynh; M.K. Nguyen; B.S. Kim; D.S. Lee, *Polymer*, 2009, 50, 2565.
20. D.P. Huynh; M.K. Nguyen; D.S. Lee, *Macromol. Res.*, 2010, 18, 192.
21. Y. Zheng; M.K. Nguyen; C. He; C.T. Huynh; D.S. Lee, *Macromol. Res.*, 2010, 18, 1096.
22. M.K. Nguyen; C.T. Huynh; D.S. Lee, *Polymer*, 2009, 50, 5205.
23. C.T. Huynh; M.K. Nguyen; D.P. Huynh; D.S. Lee, *Colloid Polym. Sci.*, 2011, 289, 301.
24. K. Dayananda; C. He; D.S. Lee, *Polymer*, 2008, 49, 4620.
25. Y. Zheng; C. He; C.T. Huynh; D.S. Lee, *Macromol. Res.*, 2010, 18, 974.
26. C.T. Huynh; M.K. Nguyen; I.K. Jeong; S.W. Kim; D.S. Lee, *J. Biomater. Sci. Polym. Edn.*, 2012, 23, 1091.
27. C.T. Huynh; M.K. Nguyen; J.H .Kim; S.W. Kang; B.S. Kim; D.S. Lee, *Soft Matter*, 2011, 7, 4974.
28. C.T. Huynh; S.W. Kang; Y. Li; B.S. Kim; D.S. Lee, *Soft Matter*, 2011, 7, 8984.
29. C.T. Huynh; D.S. Lee, *Colloid Polym. Sci.*, 2012, 290, 1077.
30. W.S. Shim; J.S. Yoo; Y.H. Bae; D.S. Lee, *Biomacromolecules*, 2005, 6, 2930.
31. W.S. Shim; J.-H. Kim; H. Park; K. Kim; I.C. Kwon; D.S. Lee, *Biomaterials*, 2006, 27, 5178.
32. W.S. Shim; S.W. Kim; D.S. Lee, *Biomacromolecules*, 2006, 7, 1935.
33. W.S. Shim; J.-H. Kim; K. Kim; Y.-S. Kim; R.-W. Park; I.-S. Kim; et al., *Inter. J. Pharm.*, 2007, 331, 11.
34. K. Dayananda, B.S. Pi; B.S. Kim; T.G. Park; D.S. Lee, *Polymer*, 2007, 48, 758.
35. H.K. Kim; W.S. Shim; S.E. Kim; K.-H. Lee; E. Kang; J.-H. Kim; et al., *Tissue Eng. Part A*, 2009, 15, 923.
36. D.P. Huynh; W.S. Shim; J.H. Kim; D.S. Lee, *Polymer*, 2006, 47, 7918.

37. D.P. Huynh; C.T. Huynh; D.S. Lee, *Macromol. Res.*, 2010, 18, 589.
38. C.T. Huynh; M.K. Nguyen; D.S. Lee, *Acta Biomater.*, 2011, 7, 3123.
39. M.K. Nguyen; D.S. Lee, *Chem. Commun.*, 2010, 46, 3583.
40. M.K. Nguyen; C.T. Huynh; G.H. Gao; J.H. Kim; D.P. Huynh; S.Y. Chae; et al., *Soft Matter*, 2011, 7, 2994.
41. C.T. Huynh; M.K. Nguyen; D.S. Lee, *Chem. Commun.*, 2012, 00, 000-000.

KN-11-01

INJECTION MOLDING OF BIODEGRADABLE STENTS FOR AIRWAY REMODELING

Kuo-Sheng Liu[1], Fu-Jun Chiang[2], Shih-Jung Liu[2]

[1]Department of Thoracic and Cardiovascular Surgery, Chang Gung Memorial Hospital

[2]Department of Mechanical Engineering, Chang Gung University

Tao-Yuan 333, Taiwan

Email address: shihjung@mail.cgu.edu.tw

Abstract

The purpose of this report was to develop novel balloon-expandable self-lock poly(□-caprolactone) stents for airway remodeling. To fabricate the biodegradable stents, polycaprolactone components were first fabricated by a lab-scale micro-injection molded machine. They were then assembled and hot-spot welded into mesh-like stents of 3 and 5 mm in diameters. A special geometry of the components was designed to self-lock the assembled stents and to resist the external pressure of the blood vessels after being expanded by balloons. Characterization of the biodegradable PCL stents was carried out. PCL stents exhibited comparable mechanical property to that of metallic stents. No significant collapse pressure reduction and weight loss of the stents were observed after being submerged in PBS for 12 weeks. By adopting the novel techniques, we will be able to fabricate biodegradable PCL stents of different sizes for various clinical applications.

Keyword: biodegradable stents, polycaprolactone, mechanical properties, characterization

Introduction

Despite the metallic stents preserve luminal patency and provide a larger and more regular lumen, they may also induce various long-term side effects such as the prevention of the lumen expansion associated with late favorable remodeling and the preclusion of surgical revascularization which may become necessary at a later time [1,2]. Since the development of the first biodegradable stent in the 1980s, various efforts have been done to develop biodegradable stents for cardiovascular and trachea applications. In this study, a novel balloon-expandable, self-lock stent made of biodegradable poly(□-caprolactone) was developed and fabricated using micro-injection molding technology. To manufacture the bioabsorbable stents, a lab-scale micro-injection molding machine was designed and built in our lab especially for this study.

Materials and Method

The biodegradable polymers used were poly(□-caprolactone) with a molecular weight (M_n) of 80,000 (Sigma-Aldrich, U.S.A.). To fabricate a biodegradable stent, two types of stent element designs (Type A and Type B that consisted of an ellipse connected with two rings at the sides, as shown in Figures 1a and 1b respectively) were first developed.

Result and Discussion

Figure 2 shows photographically the expansion process of Type B stent, while Figure 3 shows the variation of stents' external diameter with the balloon pressure [3,4]. Scanning electron micrograph of the surface of the virgin PCL stents is shown in Figure 4. The external surface of the specimen is relatively smooth, except for the debris that apparently resulted from the injection molding of the polymer during manufacture. It also shows micrographs of the PCL stents after 12 weeks elution in PBS at 37°C. No obvious sign of material degradation was observed in the specimens. As shown in Figure 5, while there was no significant strength variation for Type B stent, the compression strengths of Type A stent reduced by approximately 50% after 12 weeks elution.

Conclusions

A special geometry of the components was designed to self-lock the stents and to resist the external pressure of the blood vessels after being expanded by balloons. Characterization of the biodegradable PCL stents was carried out. No significant collapse pressure reduction and weight loss of the stents were observed after being submerged in PBS for 12 weeks.

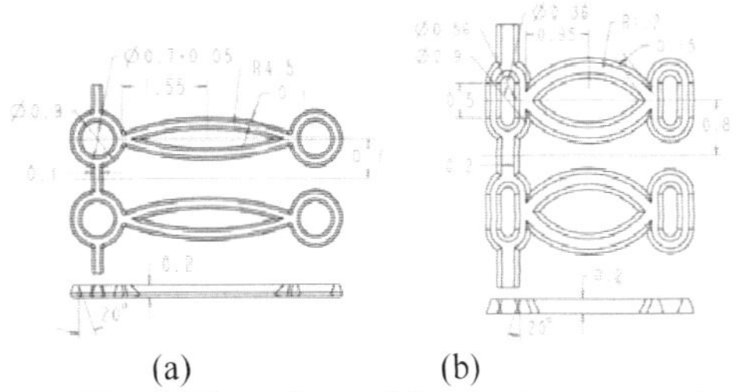

(a) (b)

Fig. 1 Dimensions of the stent components.

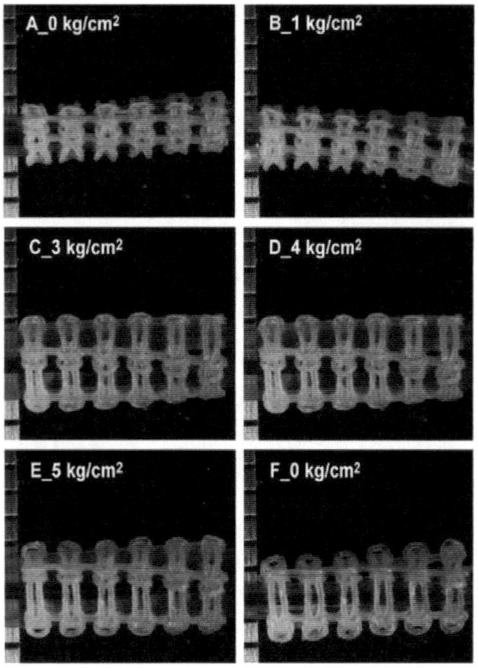

Fig. 2 Photos show the balloon expansion process of type A stent.

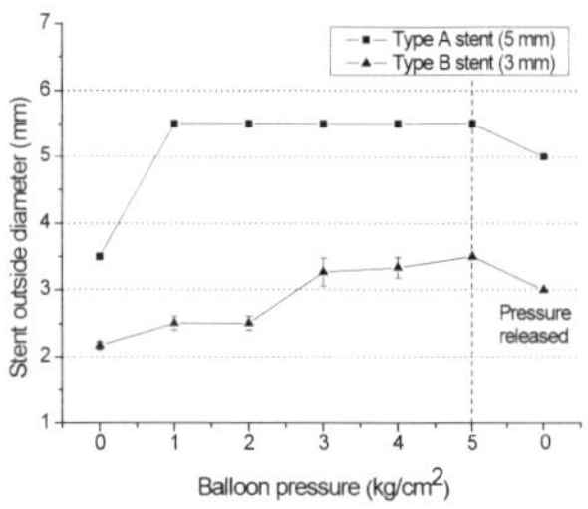

Fig. 3 Variation of stent diameter with the inflation pressures.

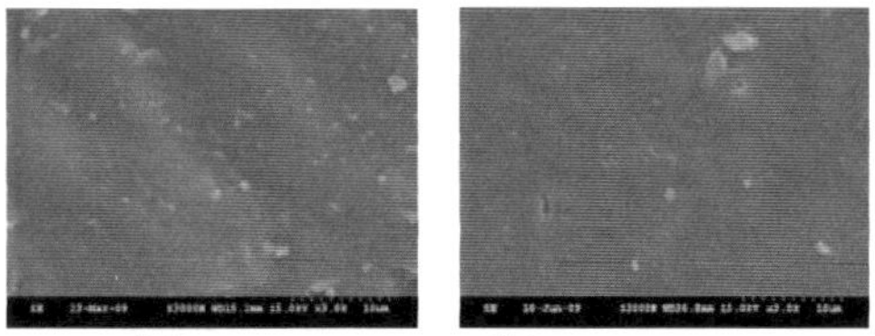

(a) (b)

Fig. 4 SEM photo of the stent surfaces, (a) after micro-injection molding, and (b) 12 weeks after the elution process.

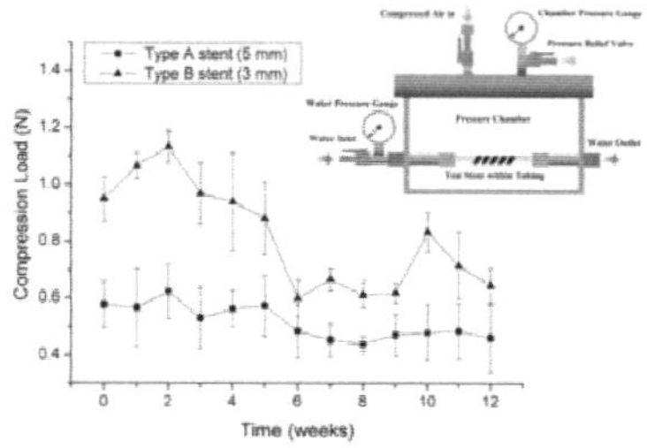

Fig. 5 Variation of stent diameter with the inflation pressures.

Reference

1- A. Colombo, E. Karvouni, in *Circulation* **2000,** Vol. 102, Ch. 4, pp. 371-3.

2- F. Alixis, S.S. Venkatraman, S.K. Rath, F. Boey, in *J. Contr. Release* **2004**, Vol. 98, pp. 67-74.

3- S.J. Liu, F.J. Chiang, C.Y. Hsiao, Y.C. Kau, K.S. Liu. *Annal. Biomed. Eng.* **2010**, 38, 3185-3194.

4- K.S. Liu, Y.H. Liu, Y.J. Peng, S.J. Liu, *J. Thora. Cardiovas. Surg.* **2011**, 141, 463-468.

O-11-090

Melt Strength Modification of Poly(Lactic Acid) Blends for Packaging Film Production

K. Sungsanit[1,*], N. Kao[2], S.N. Bhattacharya[2] and S. Pivsaart[1]

[1] *Faculty of Engineering, Rajamangala University of Technology Thanyaburi (RMUTT) Rangsit-Klong6, Thanyaburi, Pathumthani, Thailand*
[2] *Rheology and Materials Processing Centre, School of Civil, Environmental and Chemical Engineering, RMIT University, Melbourne, Australia*

**Corresponding author:kullawadee.s@en.rmutt.ac.th*

Abstract - This research explores the possibility of improving PLA's properties to satisfy and extend its processing applications such as in blown film production. Branched PLA (B-PLA) and linear/branched PLA (L,B-PLA) samples were blended with polyethylene glycol (PEG) with Mw of 1,000 g/mol, as a plasticizer, in various PEG concentrations (0, 5 and 10 wt%). Melt strength measurement by Rheotens experiment was used to verify the possibility of using some of those blends for packaging film production by blown film extrusion. The results showed that the neat B-PLA and L,B-PLA exhibited similar level of melt strength properties. The increase of PEG content decreased the melt strength values, but increased the drawability. The L,B-PLA system showed greater strain hardening behaviour than B-PLA system. Hence, the most beneficial composition (L,B-PLA with 10 wt% of PEG) was used to produce film at varying blow-up ratios of the bubble at 2.5:1 and 3.2:1. It was found that blending of linear and branched PLA, with the addition of PEG as a plasticizer, can be tailored to meet the requirements for film manufacturing as well as the achievement in film extensibility.

Keywords - Branched Poly(lactic acid), Plasticization, Melt Strength, Strain Hardening, Blown Film Extrusion.

Introduction

Poly(lactic acid) (PLA) is a biodegradable polymer belonging to the family of aliphatic polyesters, with some excellent mechanical properties and transparency which is now being produced on a large scale from fermentation of corn to lactic acid and subsequent chemical polymerization [1,2]. Unfortunately, PLA film is difficult to produce from a blown film extrusion because of its low melt strength which leads to bubble breakage during the blown film extrusion process and the film produced is usually brittle as well [3-5]. In order to improve the melt strength as well as some other rheological properties of PLA, many researchers have investigated the method of polymer blending, modification of molecular structure and addition of some additives [4-17]. Nevertheless, the biodegradable characteristic of PLA must be considered.

The melt strength is very important to polymer processing operation in which extensional deformation occurs such as in film blowing, blow moulding, or thermoforming. On the other hand, extensional behaviour is a sensitive indicator of properties associated to the molecular structure such as high molecular weight tails or long-chain branching. The role of molecular structure of the melt strength (obtained via Rheotens tests) and uniaxial extensional viscosity were described and studied by many authors [18-21]. Recently, Sungsanit et al. (2010) [22], introduced branching to PLA via peroxide melt-modification. Branched PLA exhibited higher impact resistant, melt viscosity and dynamic moduli than linear PLA. These were the results of chain entanglement among themselves in branched structure. Blends of linear and branched PLA were plasticized with PEG to increase their ductility, hence they can be used in blown film extrusion.

It is well known that extensional melt flow properties can be assessed with the Rheotens test, a non-isothermal fibre-spinning experiment. As a result from this test, melt strength (Ms) and the melt extensibility, as break velocity (Br.Vel), are obtained. The melt strength of a polymer is a measure of its resistance to extensional deformation. The test is useful in obtaining meaningful comparisons of the drawing behaviour of different polymers [23]. The melt strength is considered as the drawdown force required to break the melt strand [18, 21].

Therefore, this research investigates the effect of introducing branch structure into PLA by plasticization with PEG on the response of a melt to a stretching flow that is not revealed in shear flow. In addition, the film produced from the most beneficial PLA blend composition has been investigated.

Experimental

Materials:

Branched PLA (B-PLA) was produced directly from linear PLA (4032D-grade, biaxially oriented films from NatureWorks produced by Cargill Dow LLC) by reactive extrusion [22]. A poly (ethylene-glycol), PEG 1000 used was supplied by Sigma-Aldrich.

Procedures:

Two types of PLA and PEG were melt-blended using a Brabender twin-screw extruder in the ratio of 100/0 (neat PLA), 95/5 and 90/10 where the first and second numbers represent PLA and PEG by weight percentage, respectively. The extruder had three controlled temperature zones which were set from 180 °C (next to the feeding segment), 190 °C (compression zone) and 200 °C (metering zone). The screw speed was maintained at 30 rpm for all runs. Subsequently, plasticized PLA pellets were dried again under vacuum at 50 °C overnight prior to melt strength measurement.

The melt strength was measured using a four-wheeled Gottfert Rheotens 71.97 tester in combination with a high pressure capillary rheometer as seen in **Fig.1.** The Rheotens 71.97 was located close to the exit of the capillary. The additional pair of pull-off wheels was integrated to prevent sticking of the elongated polymer strand, once it had passed the first pair of pulleys during the test. **Table 1** lists the test conditions for the extensional experiments used to measure the melt strength and velocity at break. The polymer samples (pellets) were melted in the heated test cylinder and pressed with a test piston at constant speed through a capillary. The melt strand was continuously drawn down at a linear exponentially accelerating velocity between the two counter-rotating wheels of the device, which were mounted on a balanced beam. The force and velocity resulting from the melt strand breakage are known as ''Melt strength (Ms)'' and ''velocity at break (Br.Vel)'', respectively, for the tested condition.

Fig. 1 A four-wheeled Gottfert Rheotens 71.97 tester in combination with a high pressure capillary rheometer

The blown film experiments were performed using a lab-scale blown film unit manufactured by Strand Plast Maskiner, Sweden. A mandrel die was mounted between the 90° crosshead and the annular die to ensure a homogeneous throughput through the annular die. The latter has a gap of 1.0 mm. A spiral flow blown-film die with a 40.75 mm diameter was used to produce the film. The blow-up ratios (BUR) of 2.5:1 and 3.2:1 were used to produce a bubble with an average film thickness of 60 μm and 50 μm, respectively.
The tensile properties were determined for blown film samples, which were tested under the temperature of 25 ±2 °C and followed the ASTM standard specification (ASTM D882), with Instron tensile tester and cross-head speed of 500 mm/min.

Table 1 Set up of the extensional experiments.

Capillary Rheometer	
Diameter of Piston, D_p	20 mm
Piston velocity, v_p	0.33 mm/s
Diameter of die, D_d	2 mm
Length of die, L_d	30 mm
Rheotens	
Length from die exit to pull wheels, L_s	120 mm
Speed of the strand, v_0	40 mm/s
Speed of the wheels, v_S	50 mm/s
Acceleration,	60 m/s^2

Results and Discussion

Melt Strength Measurement:

Fig. 2 shows the melt strength of all PLA systems at the temperature of 175 °C. As take-up speed of strand increases, the melt strength (or tension) increases until the strand breaks. **Fig. (2a)** shows that the unprocessed B-PLA possesses higher melt tension (tensile force) than the other samples. On the other hand the highest drawability was found for the B-PLA sample containing 5 wt% PEG. The melt strengths of each sample (unprocessed B-PLA, 0 wt%, 5 wt% and 10 wt% of PEG) are approximately 3.0, 2.5, 2.2 and 0.7 cN, respectively.

Therefore, it should be pointed out that the observed difference of melt strength between those four PLA blends is due to the entanglement origin and plasticizer content. Similarly, the increased in melt strength as the take-up speed of strand increased was also observed in the L,B-PLA system (50:50 wt% of linear and branched PLA blend with various PEG contents) as shown in **Fig. (2b)**. In particular, the unprocessed L,B-PLA showed higher melt tension than the others. On the other hand the highest drawability was found on the L,B-PLA with 10 wt% PEG sample. The melt strengths of each sample (unprocessed L,B-PLA, 0 wt%, 5 wt% and 10 wt% of PEG) are approximately 2.8, 2.5, 2.3 and 1.3 cN, respectively.

In addition, the evidence of draw resonances (flow instability) were experienced at lower concentrations of PEG and/or without PEG sample as shown in **Fig.2.** Draw resonance is defined by periodic fluctuations in the diameter of the stretched filament just before its breakage point in stretched condition. This means that the addition of PEG has reduced the draw resonance of the blends due to the decreasing of die swell ratio of extruded melt strand. It has been reported that the polymer melts with low molecular weight, low branched chains and the addition of plasticizers will

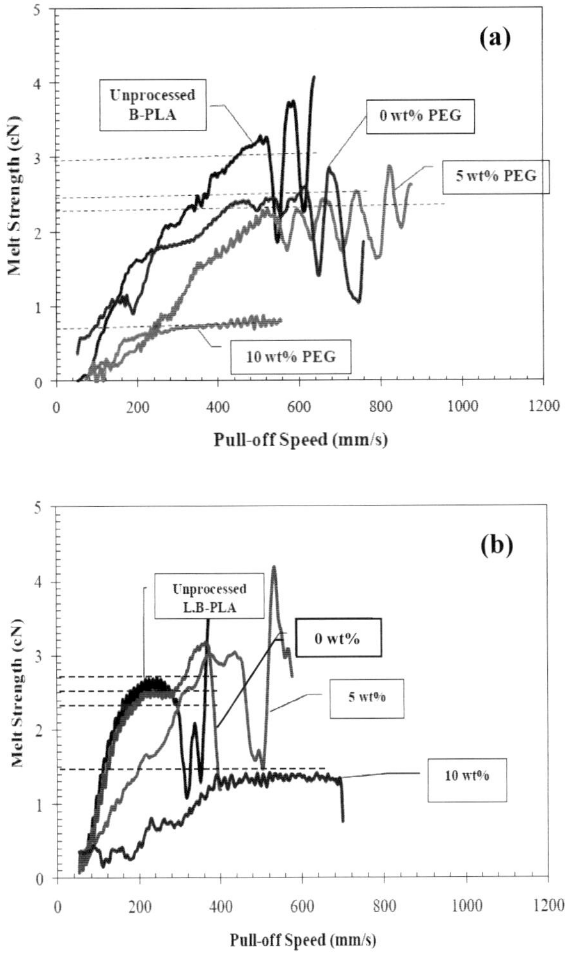

Fig. 2 Melt strength curves of the (a) B-PLA system and (b) L,B-PLA system, as a function of pull-off speed at the temperature of 175 ºC.

have lower melt elasticity and therefore the die swell ratio would be decreased [24, 25].

The melt extensibility depends on the existence of a high level of molecular entanglement as in the presence of long chain branching in B-PLA system. Therefore, in case of plasticizied PLA, this has certainly resulted in decreasing of melt strength due to enhanced molecular chain mobility of the blend. Consequently, the increase of drawability was observed with the addition of more PEG contents.

As a result of the Rheotens experiments, the modification of PLA with branching and addition of a plasticer has enabled the high draw down ratio in blown film extrusion to be carried out due to its strain hardening behaviour. Therefore, the L,B-PLA with 10 wt% of PEG films were examined to determine the impact of BUR on the tensile properties.

Apparent Extensional Viscosity from the Rheotens Measurement:

An "apparent extensional viscosity" can be defined as follows:

$$\eta_e = \frac{\sigma_z}{\dot{\varepsilon}} \qquad \textbf{(Eq.1)}$$

Where the σ_z is the tensile stress of melt strand and $\dot{\varepsilon}$ is the apparent extensional rate, both calculated at a position near the take-up wheels. The typical range of strain rates for all materials, at a position near the take-up wheels achieved with the rheotens conditions explored in this work, are 0.0002 – 30 1/s. From this Rheotens experiment, a strain hardening behaviour was observed in both PLA systems as seen in **Fig.3 (a)** and **Fig.3 (b)**. The strain hardening behaviour increases with an increase of strain rate, reaching a maximum. Moreover, the maximum point of strain hardening character was different for the various PEG content and the extensional viscosity of these materials.

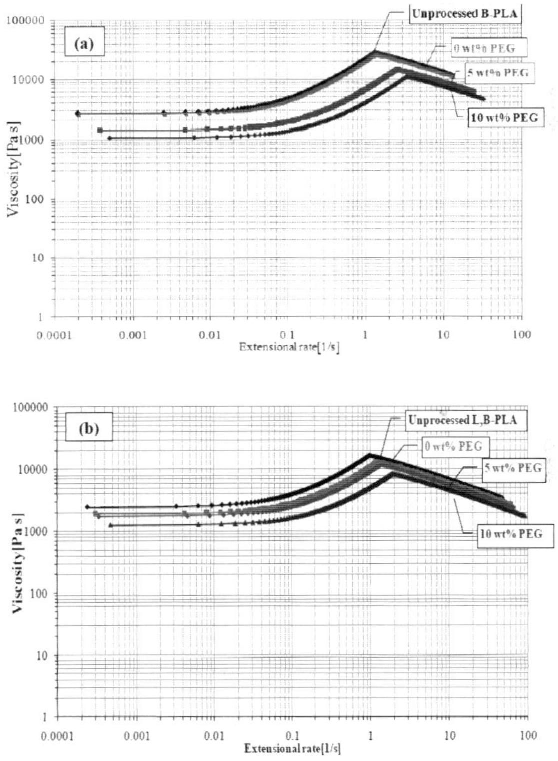

Fig.3 The calculated extensional viscosity from Rheotens experiment, (a) B-PLA system and (b) L,B-PLA system.

Tensile Properties:
The two different blow-up ratios were evaluated at 2.5:1 and 3.2:1. Frequently, the BUR ranging in 2 – 4 are almost used to produce PLA films [26]. The properties, evaluated only in machine direction, are shown in **Table 2**.

Table 2. A comparison of the tensile properties of PLA films

Tensile Properties	L,B-PLA/10 wt% PEG		Neat PLA
	45-60μm (BUR:3.2)	65-75 μm (BUR:2.5)	(Auras, 2004) (23 μm)
Tensile Strength (MPa)	35 ± 4.5	25 ± 2.6	72
Young's Modulus(GPa)	1.2 ± 0.24	0.5 ± 0.13	2.1
% Elongation	45 ± 12	200 ± 21	10.7

Upon comparison of tensile properties (Table 2), the L,B-PLA with 10 wt% PEG displayed a lower tensile strength than the neat PLA, as reported in Auras (2004) [27] of different film thickness, but the significant improvement of % elongation at break was observed. The improvement in elongation and decrease in tensile strength were the influence of addition of PEG into the PLA matrix and also the introduction of branched structures into PLA molecules as well. In addition, the production of L,B-PLA/PEG film illustrated the effects of two different BUR. The higher BUR indicated the increase in tensile strength. This could be explained with the differences of the degree of molecular deformation and the original stresses in the film direction.

Eventaully, the properties of the films produced in this study were able to be manipulated by varying the conditions (BUR) at which the films were stretched. Therefore, the introduction of branched structures into PLA molecules and the addition of PEG as a plasticizer can be tailored to meet the requirements for film manufacturing.

Conclusion

The B-PLA and L,B-PLA exhibits similar level of melt strength properties. This is due to a high level of molecular entanglements due to the presence of long chain branching. However, the increase of PEG content decreased the melt strength values, but increased the drawability. The modified PLA film, which was produced from the L,B-PLA sample containing 10 wt% of PEG, exhibited improvement in % elongation at break. The tensile properties of modified PLA film depend on the BUR. The film production with higher BUR (3.2:1) exhibited better tensile strength and modulus, because the lower of residual stresses in machine direction was induced and the higher of molecular chain network or the orientation of the film was generated.

Acknowledgements

The authors express thanks to Mr. Mike Allan and Dr. Rahul Gupta from RMIT University for their support during the melt strength and tensile properties investigations.

References

1. D. E. Henton ; P. Gruber ; J. Lunt, *Natural Fibers, Biopolymers and Biocomposites*. CRC Press, Boca Raton, FL., 2005.
2. S. J. Huang, *Poly(Lactic Acid) and Copolyesters*. Smithers Rapra Technology Crewe, UK., 2005.
3. P. R. Gruber ;M. O'Brien, *Polylactides. "NatureWorks^{TM} PLA"*. Wiley-VCH Verlag GmbH, Weinheim, 2002.
4. D. Carlson ; P. Dubois ; L. Nie, *J. Appl. Polym. Sci.* 1999, *72*, 477-485.
5. N. Ljungberg ;B. Wesslen; *Polymer* 2003, *44*, 7679–7688.
6. L. V. LABRECQUE ; R. A. KUMAR ; V. DAVE', *Journal of Applied Polymer Science* 1997, *66*, 1507-1513.
7. G. V. Larverde, Master of Science. University of Massachusetts Lowell, Massachusetts, 1997.
8. M. H. Hartmann, in *Biopolymers from renewable resources.*, D. L. Kaplan,Ed.; Springer-Verlag, Berlin/Heidelberg, 1998, 367-411.
9. S. Jacobsen ;H. G. Fritz; *Polym. Eng. Sci.* 1999, *39*, 1303-1310.
10. N. Ljungberg ;B. Wesslen; *J. Appl. Polym. Sci.* 2002, *86*, 1227-1234.
11. M. Baiardo ; G. Frisoni ; M. Scandola, *J. Appl. Polym. Sci.* 2003, *90*, 1731-1738.
12. I. Pillin ; N. Montrelay ;Y. Grohens; *Polymer* 2006, *47*, 4676-4682.
13. E. Piorkowska ; Z. Kulinski ; A. Galeski, et al.; *Polymer* 2006, *47*, 7178-7188.
14. S. L. Yang ; Z. H. Wu ; W. Yang, *Polymer Testing* 2008, *27*, 957-963.
15. J. Liu ; L. Lou ; W. Yu, *Polymer* 2010, *51*, 5186-5197.
16. D. Wu ; Y. Zhang ; M. Zhang, *European Polymer Journal* 2008, *48*, 2171–2183.
17. S. Y. Gu ; K. Zhang ; J. Ren, *Carbohydrate Polymers* 2008 *74*, 79-85.
18. S. Muke ; I. Ivanov ; N. Kao, *J. Non-Newtonian Fluid Mech.* 2001, *101*, 77-93.
19. A. Malmberg ; C. Gabriel ; T. Steffl, *Macromol.* 2002, *35*, 1038-1048.
20. W. L. Zhu ; J. Wang ; C. B. Park, *SPE-ANTEC Tech. Papers* 2009, 41-45.
21. F. H. Su ;H. X. Huang; *Polym. Eng. & Sci.* 2010, *50*, 342-351.
22. K. Sungsanit ; N. Kao ; S. Bhattacharya, *Korea-Australia Rheology Journal* 2010, *22*, 187-195.
23. J. M. Dealy ;K. F. Wissbrun, *Melt Rheology and Its Role in Plastics Processing*. Van Nostrand Reinhold, New York, 1990.
24. J. K. Sears ;J. R. Darby, *Technology of Plasticizers*. John Wiley & Sons New York, 1982.
25. Y. J. Lee ; H. S. Sohn ;S. H. Park; *Korea-Australia Rheology Journal* 2000, *12*, 181-186.
26. A. Sodergard ;M. Stolt; *Prog Polym Sci* 2002, *27*, 1123-1163
27. R. A. Auras, Doctor of Philosophy. Michigan State University, Michigan, 2004.

O-11-233

SUSTAINABLE SEALANT FILM PROCESSING BY RICE FLOUR COMPOUNDING

Ken Miyata*, Motohiro Noguchi, Akihiro Nishioka and Tomonori Koda

Department of Polymer Science and Engineering, Yamagata University, Japan – ken@yz.yamagata-u.ac.jp

Abstract - The packaging materials are spoiled in a short cycle. An environment packaging materials are demanded. The economic cost is one of problems to use bio-resource sustainable plastics. In this study the potential of rice flour as compounding filler for sealant polymer was examined. Poly(butylene succinate)(PBS) / rice flour composite biodegradable film was processed directly by twin-screw extruder equipped with T-die. Heat seal property of the biodegradable film was investigated. As a composition filler material, the effect of amorphous and semi-crystal rice flour on heat seal property was studied. At 100 °C. of heat seal temperature, the PBS / rice flour composite film showed high heat seal strength comparing with neat PBS film. By scanning electron microscopy observation, we summarized bled out rice flour at film surface affected heat seal property. However, compounded rice flour dropped heat seal strength for HDPE / rice flour composite film. The exposed rice flour particles worked effectively for PBS / rice flour composite film.

Keywords: Sealant film, Heat seal, Poly(butylene succinate), Filler, Rice flour

Introduction

The packaging materials are spoiled in a short cycle. The dumping of packaging plastics is an existing social problem. An environment packaging materials are demanded. The economic cost is one of problems to use bio-resource sustainable plastics. In this study the potential of rice flour as compounding filler for sealant polymer was examined. We focused on poly(butylene succinate) as a base biodegradable material and rice flour as a bio-resource filler. PBS / rice flour composite biodegradable film was processed directly by twin-screw extruder equipped with T-die. Heat seal property of the biodegradable film was investigated. As a composition filler material, amorphous and semi-crystal rice flour was studied.

Materials and Experimental

Materials

As a sealant base material PBS supplied by Showa Denko K.K.; Bionolle #1001; was chosen. PBS is developed as biodegradable polymer for film, sheet, foaming and filament for many plastic products. As compounding filler rice flour ground from grains was tested in this study. Rice flour is also biodegraded. Accordingly PBS / rice flour composite film is biodegraded. High density polyethylene supplied by Prime Polymer Co., Ltd.; Hizex 3300F; was also compared as general oil resource polymer.

Film processing

PBS / rice flour composite film was processed directly by twin-screw extruder equipped with T-die (KZW-30; Technovel Co., Ltd.). The compounding of rice flour component was varied at 0, 10 and 25 wt %. The thickness of extruded film was controlled to 50 μm by a cooling roll. The casting speed was 0.5 m/min. The processed composite film was almost isotropic molecular orientation. The screw rotating speed was controlled at 140 rpm to keep a constant throughput rate. The ratio of length and diameter, L/D, of the screw was 45. Temperature inside the screw cylinder was varied from 140-170 °C. The T-die temperature was also controlled at 170 °C. HDPE / rice flour composite film was also processed in the same conditions.

Experimental

PBS / rice flour composite film and HDPE / rice flour composite film were heat sealed by using a heat seal tester (TP-701-B; Tester Sangyo Co., Ltd.). Two pieces of films were heat sealed near their melting temperature. The heat bar temperature was controlled at 95, 97 and 100 °C for PBS / rice flour composite film. The seal pressure was kept at 0.2 MPa constantly. The heat seal time for applying heat and pressure was varied 0.1 to 2.0 seconds.

To evaluate heat seal properties, "T-peel" strength was measured using a tensile tester (3342; Instron Co., Ltd.). The heat sealed film was peeled at the rate of 300 mm/min. Ten tests were performed at each condition and the average strength was adopted. The rice flour particles condition at film surface was observed by polarized optical microscopy and scanning electron micrscopy.

Results and Discussion

Fig.1 showed effect of rice flour components on heat seal strength in peeling tests for PBS / rice flour composite film by extrusion. Heat seal strength increased at 100 °C of seal temperature comparing with 95 and 97 °C. Failure mode in peeling tests was changed cohesive failure at sealed interface to tearing of film at 100 °C. Heat seal strength did not depend on rice flour components ratio. Rice flour content does not prohibit heat seal strength.

Fig. 2 showed effect of rice flour components on heat seal strength in peeling tests for PBS / rice flour composite film by hot press manufacturing. At 100 °C of seal temperature the PBS / rice flour composite film was not heat sealed. Heat seal strength did not increase at 100 °C of seal temperature. To compare film surface condition, Fig. 3 showed scanning electron microscopy images of PBS / rice flour composite film surface processed by extrusion (top) and hot press (bottom). Compounded rice flour particles were bled out at film surface for the extruded film in the bottom image. No rice flour particles were observed at hot pressed film surface in the top image. The bled out rice flour particles affected seal strength.

Fig.4 showed effect of rice flour components on heat seal strength for HDPE / rice flour composite film by extrusion at 100 °C of seal temperature. Heat seal strength did not increase for 1.0 s at 100 °C. When heat seal time exceeded 2.0 s, seal strength increased. The exposed rice flour particles on film surface worked on sealing for PBS / rice flour composite film. Heat seal property for PBS / rice flour composite film was affected by interaction between PBS and rice flour. For HDPE / rice flour composite film, the rice flour particles did not work effectively.

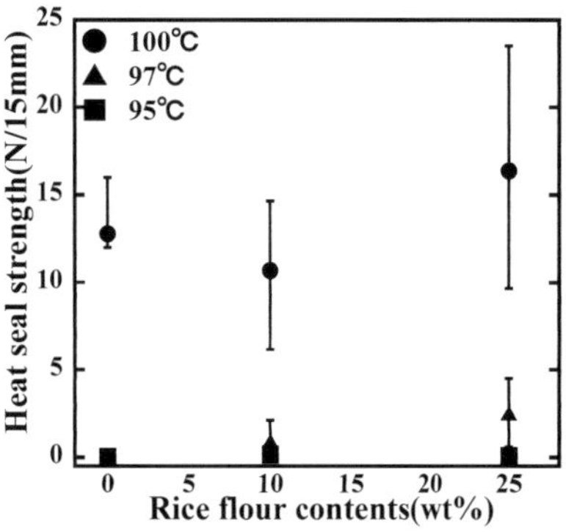

Figure 1 – Heat seal strength of PBS / rice flour composite film by extrusion for 1.0s of seal time.

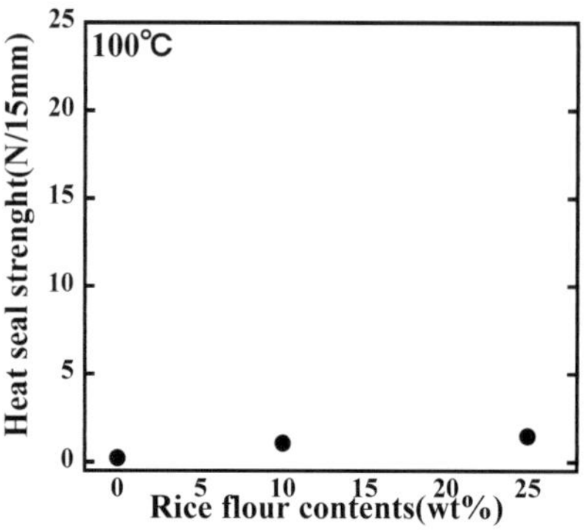

Figure 2 – Heat seal strength of PBS / rice flour composite film by hot press.

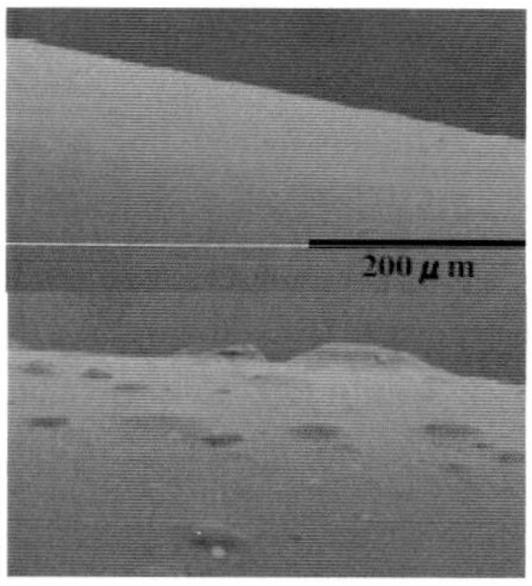

Figure 3 – Scanning electron microscopy images of PBS / rice flour composite film by extrusion (top) and hot press (bottom).

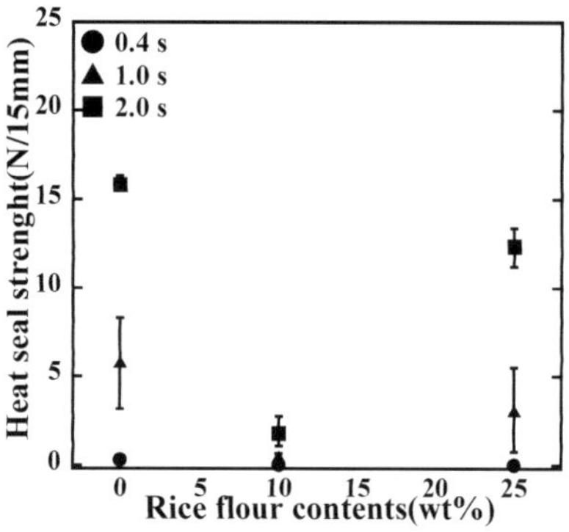

Figure 4 – Heat seal strength of HDPE / rice flour composite film by extrusion at 100°C of seal temperature..

Conclusions

PBS / rice flour composite film processed by extrusion showed high heat seal strength. However, compounded rice flour dropped heat seal strength for HDPE / rice flour composite film. The exposed rice flour worked effectively for PBS / rice flour composite film.

References

1. C. Mueller, G. Capaccio, A. Hiltner, E. Baer, J. Appl. Polym. Sci., **70**, 2021-2030(1998)
2. F. C. Stehling, P. Meka, J. Appl. Polym. Sci., **51**, 105-119(1994)
3. K. Miyata, T. Ozama, K. Katsuno, A. Nishioka, T. Koda, G. Murasawa, ANTEC Tech. paper, 741-744(2009)
4. K. Miyata, T. Hiyama, K. Yamada, Yew W. Leong
   ANTEC 2011, Tech. paper, 1472-1478(2011)
5. K. Miyata, T. Ozama, K. Katsuno, A. Nishioka, T. Koda, G. Murasawa, ANTEC Tech. paper, 406-411(2010)
6. K. Miyata, T. Abe, H. Tokanai, A. Nishioka, G. Murasawa, T. Koda
   ANTEC 2011, Tech. paper, 1467-1471 (2011)
7. M. Kakuda, T. Takahashi, K. Miyata, M. Sugimoto, T. Taniguchi, K. Koyama, J. of The Adhesion Society of Japan, **42**, 506(2006)

O-11-116

MICROCELLULAR FOAMS OF POLY(LACTIC ACID) BASED COMPOSITES FOR BIOMEDICAL APPLICATIONS

Jutika Goswami[1,*], **A.K.Ghosh**[1], **N. Bhatnagar**[2] **and S. Mohanty**[3]
[1]*Centre for Polymer Science and Engineering, IIT Delhi, New Delhi-110016, India, anupkghosh@gmail.com*
[2]*Mechanical Engineering Department,IIT Delhi, New Delhi-110016, India, narbhat@hotmail.com*
[3]*Stem Cell Facility, AIIMS, New Delhi-110029, India, drmohantysujata@gmail.com*
[*]*jutikagoswami@gmail.com*

Abstract
Poly(lacic acid)(PLA) has been melt blended with poly(ε-caprolactone)(PCL) and addition of wollastonite as filler into the blend in a compounding process using twin-screw extruder resulted in development of bioactive composites. PLA/PCL has been blended in 85/15 proportion and the filler added was in 1, 4 and 8 phr designated as PLCL85(blend), PLCLW1, PLCLW4 and PLCLW8 respectively. Moderate improvement in load bearing capability in the composites was found in tensile test with the incorporation of filler. The composites were batch-foamed using compressed CO_2 at specific saturation conditions to prepare the materials as 3D-porous structures. The foamed constructs were used for *in vitro* study by cell seeding method so as to develop these as biocompatible scaffolds.

Keywords: Composites, Biomedical, Filler

Introduction

Poly(lacic acid)(PLA), the most researched "green polymer", offers great potential for biomedical scaffolding application due to its biocompatible and biodegradable nature. Being of bio-origin (from corn starch etc.), it provides with a continuous supply from renewable resources that has been rightfully exploited for different application avenues like packaging and agriculture as well as medical devices [1-2].

Studies have been done reporting PLA scaffold preparation by solvent based methods like solution casting, freeze drying etc. where complete removal of the organic solvent is a tedious job and the presence of it may be harmful towards body tissues as well as the performance of the scaffold.

Melt blending, like extrusion, is a continuous process devoid of use of any solvent making the resultant product solvent-free. Poly(ε-caprolactone) is another biodegradable elastomeric polymer that can be blended with PLA to improve the toughness property of the pristine PLA as well as to prepare scaffolds for long-term applications.

Organic polymers have very low affinity to be strongly bonded to body tissues due to its very nature. Performance of a body implant gets reduced without proper adherence to the host tissue and the presence of a bioactive filler has a role at this point. The filler acts as bone-bonding material between the polymer and the body tissues. CaO and SiO_2 are reported as two main bone-bonding materials and wollastonite as $CaSiO_3$ is a potential additive for this purpose. [3-7].

A scaffold is a 3D-networked structure that acts as support to damaged tissues and the tissues rebuild the destroyed body part by regeneration of tissues with the support of the scaffold [8].

The PLA/PCL/Wollastonite composites have been prepared in the current study using twin-screw extruder in different compositions and the developed composites have been studied as foams and as scaffold by seeding cells on it so as to study their cell viability.

Experimental

Poly(lactic acid)(PLA) (ESUN™ A-1001) and poly(ε-caprolactone)(PCL)(CAPA-6800) were supplied by Shenzhen BrightChina Industrial Co. Ltd, China and Perstorp Chemicals, UK, respectively. Wollastonite (Kemolit K-1010) was supplied by Wolkem India Limited, Udaipur, India as white powder of size 9.07micron (D-97).

The composites were prepared in a twin-screw extruder at 190°C temperature and 200 rpm and the compositions prepared were in PLA(%)/PCL(wt%)/W(phr) ratio of 85/15/1, 85/15/4 and 85/15/8 designated respectively as PLCLW1, PLCLW4 and PLCLW8. The blend composition PLCL85 (PLA/PCL=85/15) was also compounded as controlled composition. Injection moulding was done to prepare samples for mechanical testing and batch foaming of the composites.

Batch foamed samples were prepared using 0.75mm thickness sheet of samples using compressed CO_2 as physical blowing agent. The samples were saturated with the blowing agent for specific time at room temperature and these were then foamed in hot glycerin bath at 160°C foaming temperature. The foams were used as scaffold for *in vitro* analysis to study their cell viability.

Results and Discussion

Tensile test experiment was performed as per ASTM standard D638 type1and the result is reported in Fig. 1 as load-displacement plot. The test revealed gradual improvement in carried load of the composites with increased incorporation of the acicular filler, wollastonite. The composites showed higher displacement as compared with the blend composition. This enhaced strain is be due to the presence of filler that helps in flowability of the polymers on load application.

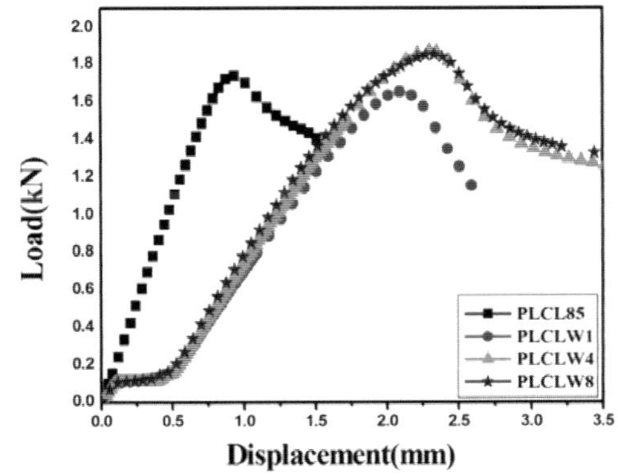

Figure 1. Load-displacement plots of the polymers from tensile test.

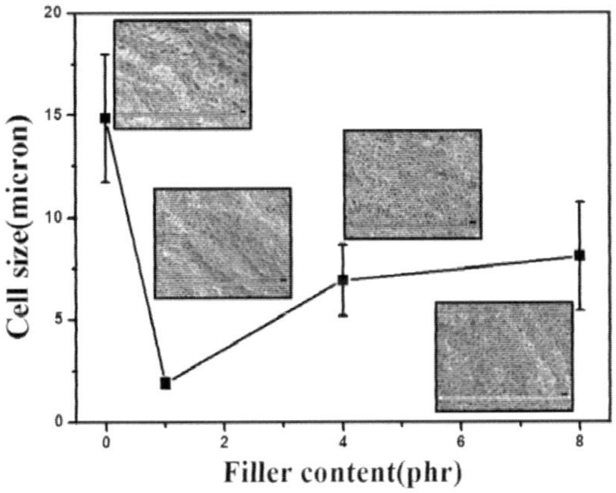

Figure 2. SEM images of composite foams a)PLCL85, b)PLCLW1, c)PLCLW4, and d)PLCLW8.

Scanning electron microscopy of cryofractured surfaces of the foam samples was performed using Carl ZEISS EVO 50 microscope at 20kV and the samples were sputtered with gold before scanning so as to disperse the charge produced. The microscopic images are shown in Fig. 2 and as can be seen, addition of filler created smaller cells with higher cell density as compared with that of the blend. Being inorganic in nature, the filler acts as nucleating agent for CO_2 absorption by introducing inhomogeneity in the composites. But the nucleating effect of the filler is more pronounced at lower phr only and with further addition, cell size starts increasing even though these sizes are still smaller than the blend foam.

Mesenchymal stem cells (MSCs) were seeded onto the prepared porous scaffolds and two representative SEM images are shown in Fig. 3. The cells were attached on the surfaces and the number of attached cells increased with higher filler content as can be seen in PLCLW8.

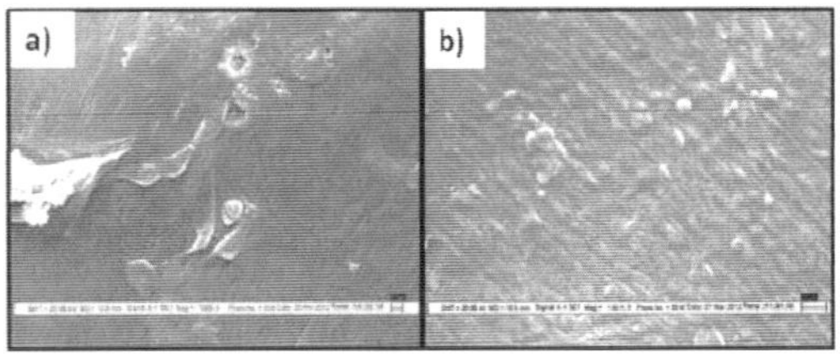

Figure 3. Representative SEM images of composites clearly showing cell growth on their surfaces a)PLCLW1 and b)PLCLW8.

Conclusion

Better mechanical behavior was found in the composites due to the presence of stiffer inorganic filler. The filler also helped in batch foaming of the polymers by creating nucleation sites. Uniform cell morphology could be found in the composites as compared to the blend. The *in vitro* experiment justified the use of the composites for scaffold applications that proved biocompatibity on the surfaces.

References

1. F. Carrasco; P. Pagès; J. G. Pérez; O.O. Santana; M.L. Maspoch *Polymer Degradation and Stability* 2010, 95, 116.
2. K. M. Nampoothiri; N. R. Nair; R. P. John *Bioresource Technology* 2010, 101, 8493.
3. L.T. Lim; R. Auras; M. Rubino *Progress in Polymer Science* 2008, 33, 820.
4. H. T. Liao; C. S. Wu *Materials Science and Engineering A* 2009, 515, 207.
5. H. J. Sung; C. Meredith; C. J. Zorin; S. Galis *Biomaterials* 2004, 25, 5735.
6. M. Wang *Biomaterials* 2003, 24, 2133.
7. C. R. Kothapalli; M.T. Shaw; M. Wei *Acta Biomaterialia* 2005, 1,653.
8. E. Reverchon; S. Cardea; C. Rapuano *J. of Supercritical Fluids* 2008, 45, 365.

O-11-250

PREPARATION AND HYDROLYTIC DEGRADATION OF POLY-LACTIC ACID BASED SCAFFOLD FOR SOFT TISSUE ENGINEERING

V. Brucato[1], F. Carfì Pavia[1] and V. La Carrubba[2]

[1]*Dip. di Ingegneria Chimica Gestionale Informatica Meccanica, University of Palermo, Italy – valerio.brucato@unipa.it; francesco.carfipavia@unipa.it*

[2]*Dip. di Ingegneria Civile Ambientale, Aerospaziale, dei Materiali, University of Palermo, Italy – vincenzo.lacarrubba@unipa.it*

Abstract - Blends of Poly-L-Lactic Acid (PLLA) with Poly-Lactic Acid (PLA) in different proportions (90/10 and 60/40) were utilized in order to produce porous biodegradable scaffolds for soft tissue engineering applications. The scaffolds were produced via thermally induced phase separation (TIPS) starting from ternary systems where dioxane and water were the solvent and non-solvent respectively. Morphology (average pore size and interconnection) was evaluated by Scanning Electron Microscopy. DSC analysis was carried out on the as-obtained scaffold in order to obtain information about theirs thermal properties (enthalpy of melt and crystallization). The results showed that is possible to prepare scaffolds of PLLA/PLA via TIPS. Moreover, the PLA modifies the demixing temperatures and consequently the TIPS process.

An in-vitro hydrolytic degradation in Phosphate Buffer Saline (PBS, pH 7,4) of as-obtained porous scaffolds was carried out and studied by measurements of weight loss and DSC. Through these measurements, it was possible to highlight the different degradation kinetics showed by the various samples examined. Specifically, the degradation rate increases when the PLA (amorphous polymer) content is increased.

These data confirm that the degradation rate as well as the morphology and the crystallinity of the scaffold can be tuned using Poly-lactic acid blends.

Keywords: Tissue Engineering, Poly-Lactic Acid, Phase Separation, Hydrolytic Degradation.

Introduction

Tissue engineering is an interdisciplinary field that applies the principles of engineering and life science toward the development of biological substitutes that restore, maintain and or improve tissue function [1]. The production of an engineered tissue starts by the design and the formation of a structure able to support the migration and growth of cells that will originate the new tissue [2]. An interconnected pore network able to guide, after the degradation of the scaffold, the implanted cells to form a new tissue showing a well-integrated structure, characterizes those structures. A crucial aspect of the tissue engineering regards the scaffold biodegradability. After implantation, the scaffold has to be gradually replaced by cells and extra cellular matrix and it is crucial that this replacement takes place with the appropriate dynamics. As a matter of fact, a premature degradation could lead to a collapse of the structure, as the newly generated tissue could not have reached yet suitable mechanical properties. Conversely, a long degradation time could depress or completely interrupt the development of the new tissue. For these reasons, it could be interesting to design an effective method to tune the degradation rate of the scaffold. Poly-Lactic-Acid (PLA) is a biodegradable thermoplastic polyester that can be produced through ring-opening polymerization of lactic acid. Since lactic acid is a chiral molecule, it exists in two forms, Poly-L-Lactic Acid (PLLA) and Poly-D-Lactic-Acid (PDLA). It is known that the properties of PLA are highly affected by the stereo-isomeric L/D ratio of the lactate units. In fact, PLLA and PDLA, consisting only of L- and D-lactate units, respectively, are highly crystalline with identical chemical and physical properties, while poly (DL-lactide) (PDLLA or simply PLA), consisting of racemic sequences of D and L lactate units, is rather amorphous. The involvement of D- and L-units in the sequences of PLLA and PDLA, respectively, exerts a profound effect on their thermal and mechanical properties [3]. Generally, the increased stereo-isomeric ratio decreases the crystallinity. For the aforementioned reasons, blending PLLA with PDLLA is an effective method for controlling the polymer crystallization, morphology, and hydrolysis nature [4,5].

In the present work Poly-L-lactic acid (PLLA) and Poly lactic acid (PDLLA or PLA) based blends (100/0, 90/10 and 60/40) were utilized to prepare porous foams for tissue engineering applications via Thermally Induced Phase Separation (TIPS). The foams obtained were characterized via Scanning Electronic Microscopy (SEM) and Differential Scanning Calorimetry (DSC) in order to verify if PLA influences theirs final morphology and the thermal properties. Moreover, in order to understand the different degradation kinetics between the samples prepared at different PLLA/PLA weight ratios, an *in-vitro* hydrolytic degradation in Phosphate Buffer Saline was carried out.

Experimental

Materials

Poly-L-lactic-acid (PLLA, ResomerTN L 209 S), Poly Lactic Acid (RESOMER® MD Type R 208) (both kindly supplied by Boehringer Ingelheim Pharma KG) were used for preparing PLLA/PLA blends. 1,4 dioxane (Sigma) and double distilled water were utilized as solvents to prepare the ternary solution.

Scaffold preparation and characterization

A homogeneous ternary solution composed by PLLA/PLA, dioxane and water was prepared, with a constant dioxane to water volumetric ratio of 84,5/15,5. The concentration of PLLA was 6% wt/wt. The solution was initially kept at 84 °C. The temperature was then suddenly lowered to a value within the unstable region for 30 minutes. Later a quench by pool immersion of the mixture in an ethyl alcohol bath at a temperature of -25 °C was performed. The foams obtained were rinsed in deionised water and dried at 20 °C under vacuum to completely remove the solvents from the samples.

Pore size and morphology of each foam were analyzed by scanning electron microscopy (SEM Philips 505 Microscope) on a sample cross section obtained by fracturing it in liquid nitrogen and gold sputtering the surface. Enthalpies of melting and crystallization were obtained via DSC analysis, utilizing a JADE Perkin Elmer instrument applying a rate of 10 °C/min both to heating and cooling.

Hydrolytic degradation

Disc-shaped scaffolds of (diameter 12 mm, thickness 1 mm) were obtained and every sample was weighed in an analytical balance. The samples were placed into a well containing Phosphate Buffer Saline (PBS, pH 7,4). The multi-well containing the foams was placed in an incubator set at 37 °C. The medium inside the well was renewed each week. After 1, 2, 4, 7 and 10 weeks, at least 3 samples were extracted from the medium, rinsed in deionized water for 48 h, dried at 30 °C under vacuum for 48 hours and weighed. Afterward, a DSC analysis was carried out on each sample at the same conditions described before.

Results and Discussion

Figure 1 shows the DSC heating scans of PLLA and PLA films obtained by solvent evaporation from a homogeneous solution. PLLA exhibits a melting peak located around 180 °C, whereas pure PLA exhibits the typical behavior of an amorphous polymer with a glass transition temperature located at about 55 °C.

In figure 2 SEM micrographs of the cross sections of the foams prepared at PLLA/PLA weight ratios (90/10 and 70/30) are shown. The open porosity with interconnected pores detected in the foams indicates that the phase separation process was correctly addressed. Two fundamental process parameters have had to be modified with respect to pure PLLA [6] for obtaining these structures. In order to prevent the collapse of the foams, the polymer concentration in the starting ternary solution was increased to 6 %. As a matter of fact lower polymer concentrations led to not homogeneous structures with an irregular and not controllable porosity. Moreover, in order to obtain the aforementioned morphology a different range of demixing temperatures has been detected for each blend at the different PLLA/PLA ratios.

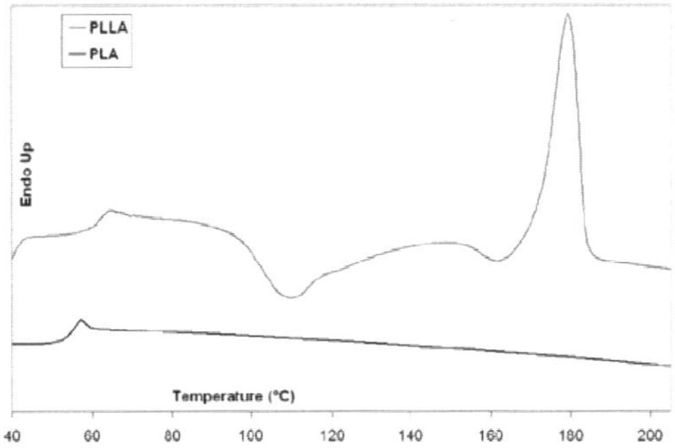

Figure 1 – DSC heating scan of PLLA and PLA.

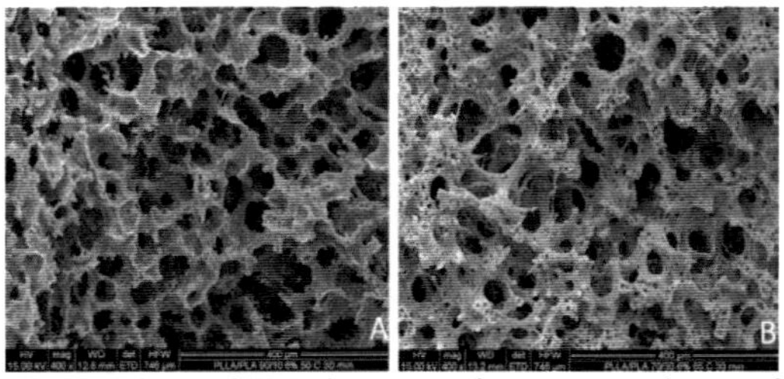

Figure 2 – SEM images of cross sections of PLLA/PLA foams at different polymer ratios: 90/10 (a) and 60/40 (b).

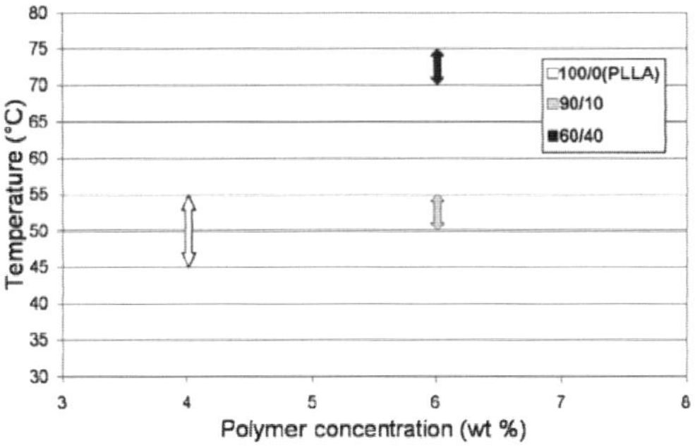

Figure 3 – Demixing temperatures at which phase separation carries out.

In figure 3 the temperature ranges where a satisfactory foaming behavior was observed are reported and compared with the range for pure PLLA. It is easy to notice that when PLLA/PLA 90/10 blends are used, the demixing temperature at which phase separation takes place is the almost the same of pure PLLA with some reduction of the span. Conversely, when increasing

PLA content to 40 %, the temperature has to be increased to 65-70 °C. This behavior is probably linked to the characteristic of the materials employed (Mwd, crystallization kinetics in solution, final levels of crystallinity) and deserves deeper analysis.

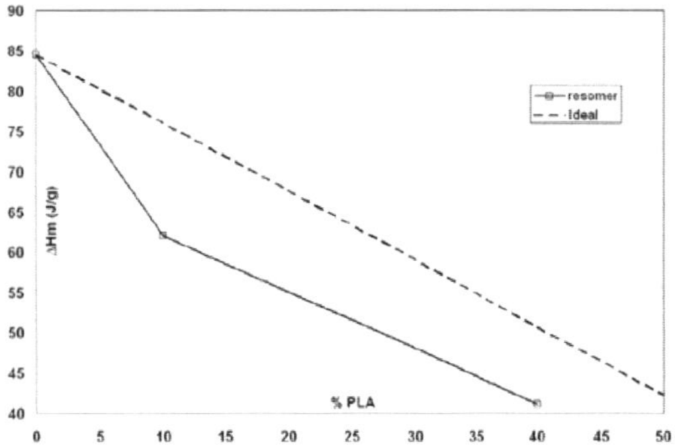

Figure 4 – Enthalpies of melting for PLLA/PLA scaffolds prepared at different PLLA/PLA ratios.

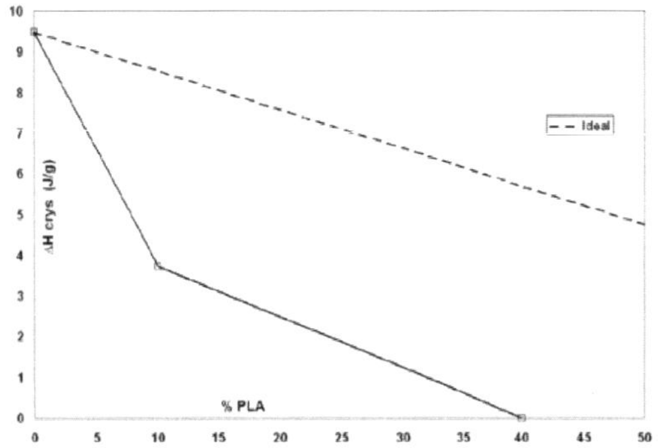

Figure 5 – Enthalpies of crystallization for PLLA/PLA scaffolds prepared at different PLLA/PLA ratios.

In figure 4 the latent heats of melting (ΔHm) and crystallization of the PLLA/PLA foams at the different PLLA/PLA ratios are shown. It is worth noticing, that the ΔHm decreases by increasing the PLA/PLLA ratio. This behavior was also observed for the enthalpy of crystallization (ΔHcrys) (see figure 5). In both cases, however, it has been noticed a crystallinity level lower than the level predicted by ideal mixing rules (dotted line). It is reasonable to suppose that blending prevents the crystallization kinetics of the scaffolds.

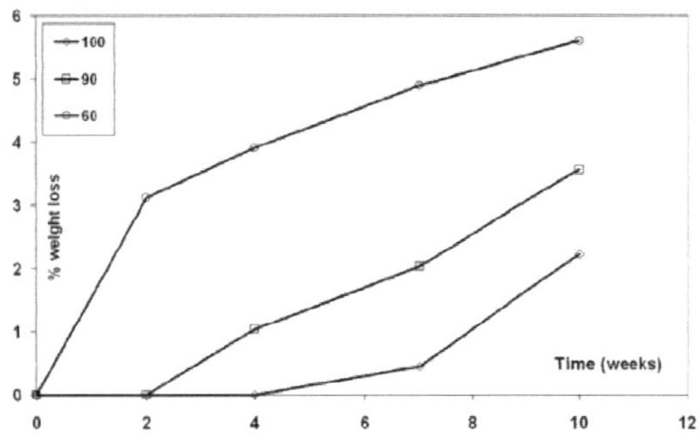

Figure 6 – Weight loss of PLLA/PLA scaffold after hydrolytic degradation at different times.

Figure 6 shows the weight loss of PLLA/PLA scaffold after hydrolytic degradation at different times. It is easy to notice the different behavior among the three composition tested. Specifically, for pure PLLA no weight loss occurs in the first four weekss. An increasing weight loss was observed from week 7 to week 10 (almost 2%). The weight of the 90/10 samples is maintained still the second week while an increasing weight loss was observed starting from the fourth week. The 60/40 samples showed an early degradation starting from the first observation forward.

No significant differences were detected in the enthalpies of melting of the samples during the degradation period. Moreover, it was observed that the temperatures of the melting peaks of the different foams do not differ significantly by changing the PLA content, an hint of the immiscibility of the two components in the solid state. In addition, the peak temperatures of the different foams, do not show significant differences along the degradation process (data not shown).

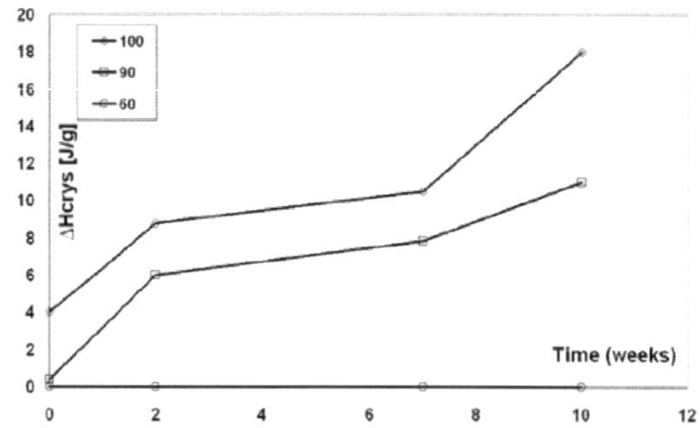

Figure 7 – Enthalpies of crystallization for PLLA/PLA scaffolds after hydrolytic degradation.

Figure 7 shows the enthalpies of crystallization of the foams. Also in this case different pathways were highlighted. An increase of enthalpy was observed from the first weeks for pure PLLA and 90/10 foams, whereas no crystallization peaks were detected in the case of 60/40 foams.

The early weight loss of the 60/40 foams can be attributed to the fact that the amorphous regions, mainly due to the presence of the PLA, start to hydrolyze quickly. This behavior was not observed in the 90/10 mixtures and pure PLLA foams where the semi-crystalline structure predominates. Moreover, for the latter the increase in the enthalpy of crystallization indicates a better ability of the foams to crystallize from the melt. With the progress of the degradation time, an hydrolysis of the crystalline portion of PLLA present in the mixture, which is prevalent in these two types of foams, occurs with random breaking of the polymer chains. Therefore the increased ability toward crystallization observed is probably due to the chain shortening and lower dynamic hindering. The 60/40 foams does not show peaks of crystallization. For these foams one can reasonably suppose that the amorphous portion of PLA, present in larger quantity, is degraded before that of PLLA protecting this last in the early process weeks.

From these results, it can be assumed that, during the processing of polymeric foams of PLLA / PLA, a segregation of the two polymers occurs. Furthermore, it is reasonable to state that, since the PLLA is less soluble of the PLA, in the dioxane/water solution, PLLA precipitate first orienting the segregation of the PLA towards the outside of the solid phase. The so-obtained foams could show a skin-core structure with a core mainly consisting of PLLA and a PLA predominant skin.

Conclusions

Porous PLLA/PLA foams were produced via Thermally Induced Phase Separation. For each blend the experimental conditions at which phase separation process gives rise to porous and interconnected structures were assessed. The data have shown that when blends are used the polymer concentration has to be increased to 6%, with respect to pure PLLA (4%), in order to prevent the structure from collapsing. Moreover, the demixing temperature has to be increased, when increasing the content of PLA in the blend from 10% to 40%. The differential scanning calorimetry analysis revealed a decrease of enthalpy of melting and crystallization on increasing the PLA content. Crystallinity level lower than predicted by ideal mixing rule were found, showing that blending prevents the crystallization kinetics of the scaffolds. Hydrolytic degradation tests highlighted different behaviors among the foams featuring different PLLA / PLA ratios. As a matter of fact, a faster degradation was observed on raising the PLA content. These information confirm that blending PLLA with different amounts of PLA is a way to control the degradation kinetics of the scaffolds.

References

1. R. Langer Pharm. Res. 1997, 14, 840.
2. J.J. Marler; J. Upton; R. Langer; J.P. Vacanti Adv. Drug. Deliv. Rev. 1998, 33, 165.
3. H. Yamane; K Sasai Polymer 2003, 44, 2569.
4. H. Tsuji; Y. Ikada, Polymer 1996, 37, 595.
5. H. Tsuji; K. Ikarashi, Biomaterials 2004, 25, 5449.
6. F. Carfì Pavia; V. La Carrubba; V. Brucato; S. Piccarolo, J. Biomed. Mat. Res. A 2008, 86, 459.

O-11-097

Mechanical properties of modified banana fiber/ polymer composites

Takeshi Yasuda[1*] and Tetsuto Kajiyama[2]

[1]*Tama Techno Plaza, Tokyo Metropolitan Industrial Technology Research Institute, 3-6-1, Azumacho, Akishima-city, Tokyo, 196-0033, Japan – yasuda.takeshi@iri-tokyo.jp*

[2]*Research and Development Department, Tokyo Metropolitan Industrial Technology Research Institute, 2-4-10, Aomi, Koto-ku, Tokyo, 135-0064, Japan – kajiyama.tetsuto@iri-tokyo.jp*

Abstract More than one billion tons of banana leaves are discarded per year. Banana fibers can be made from banana leaves. In the present study, banana fibers without being processed (BF), banana fibers with an alkali treatment (BF-A), banana fibers treated in an alkali aqueous solution with tension (BF-AT), and polybutylene succinate (PBS) were used. BF/PBS composites, BF-A/PBS and BF-AT/PBS composites were made by applying a twin extruder, respectively. The bending test and the charpy impact test were performed. The results of bending test show that when banana fiber were increased in composites, both bending strength and bending modules increased. Bending strength of composites contained BF was higher than that one composites contained BF-A. Bending strength of composites contained BF-A was higher than composites contained BF-AT. The results of charpy impact test show that when banana fibers were increased in composites, impact strength increased. Impact strength of composites contained BF-A was higher than that of composites contained BF. Impact strength of composites contained BF-AT was higher than that of composites contained BF-A.

Keywords: Banana fiber, Polybutylene succinate, Alkali treatment, Mechanical properties

Introduction

Polymer products are necessary in our life. It is a matter of great urgency to reduce emissions of carbon dioxide. Biodegradable resin are expected as the material which can decrease some environmental loads.

On the other hand, more than one billion tons of banana leaves are discarded per year. Banana fibers[1] can be made from banana leaves(Fig. 1(a)). It is expected to make use of banana fibers in making new products.

In the present study, two kinds of banana fibers and polybutylene succinate composites were made. Three point bending test and charpy impact test to the obtained composites were performed.

Materials and Experiments

Fig. 1(b) shows banana fibers. Three kinds of banana fibers were used. One of them was not being processed (BF), other one was treated with 10 wt. % NaOH aqueous solution (BF-A), the rest one was treated with 10 wt. % NaOH aqueous solution with tension(BF-AT). BF, BF-A and BF-AT were cut to 5 mm. Polybutylene succinate (PBS, Bionolle #1020, Showa Denko K.K.) were also used. BF/PBS composites (BF: PBS = 10: 90 (10 wt. %))and 20:80 (20 wt. %)) , BF-A/PBS (BF-A: PBS = 10: 90 (10 wt. %) and 20:80 (20 wt. %)) and BF-AT/PBS (BF-AT: PBS = 10: 90 (10 wt. %) and 20:80 (20 wt. %)) composites were made by applying a twin extruder (PCM-30, L/D = 31.5, D = 30 mm, L: length of screw, D: diameter of screw, Ikegai, Co., Ltd) at 140 °C. Specimens in size of 80×10×4 mm were prepared using a injection molding machine (Netstal 150/75, Maximum clamping force 75 ton, Sumitomo Heavy Industries, Ltd.) at 140 °C.

Surface morphology of BF, BF-A and BF-AT were observed using a scanning electron microscope (SEM, VE-9800, KEYENCE Corporation).

Specimens of PBS and the composites were assessed by three points bending test (Autograph AG-10TD, Shimadzu corporation) and notched charpy impact test (DG-UB, Toyo seiki seisaku-sho, Ltd.).

Results and Discussion

Fig. 2 shows SEM photographs of three kinds of banana fibers (BF, BF-A and BF-AT). Surface of BF(Fig. 2 (a)) was smooth. Surface of BF-A(Fig. 2 (b)) was bumpy. Lignins[2] of surface of banana fiber were removed by alkaline treatment. Surface of BF-AT(Fig. 2 (c)) were more smoother than that of BF-A. Anatomy of cellulose of banana fiber were oriented by alkaline treatment with tension.

Fig. 3 shows stress-strain curves of BF/PBS composites on three points bending test. When the content of BF increased, bending modules and bending strength increased. BF/PBS composites of 20 wt. % were broken under strain 0.1.

Fig. 4 shows stress-strain curves of BF-A/PBS composites on three points bending test. When content of BF-A increased, the bending modules and the bending strength increased. The bending modules and the bending strength of BF/PBS composites were higher than that of BF-A/PBS composites with the same contents of banana fiber.

Fig. 5 shows stress-strain curves of BF-AT/PBS composites on three points bending test. When content of BF-AT increased, the bending modules and the bending strength increased. The bending modules and the bending strength of BF-AT/PBS composites were higher than that of BF-A/PBS composites with the same contents of banana fiber. Broken strain of BF-

AT/PBS composites of 20 wt. % were less than that BF-A/PBS composites of 20 wt. %.

Fig. 6 shows results of charpy impact strength of BF/PBS composites, BF-A/PBS composites and BF-AT/PBS composites. Charpy impact strength of PBS was higher than that of BF/PBS composites, BF-A/PBS composites and BF-AT/PBS composites. When banana fibers are increased in composites, charpy impact strength increased. The charpy impact strength of composites contained BF-A was higher than that of composites contained BF. The charpy impact strength of composites contained BF-AT was higher than that of composites contained BF-A.

(a) Banana leaves (b) Banana fibers

Figure 1 Photographs of banana leaves and banana fibers used in the present study

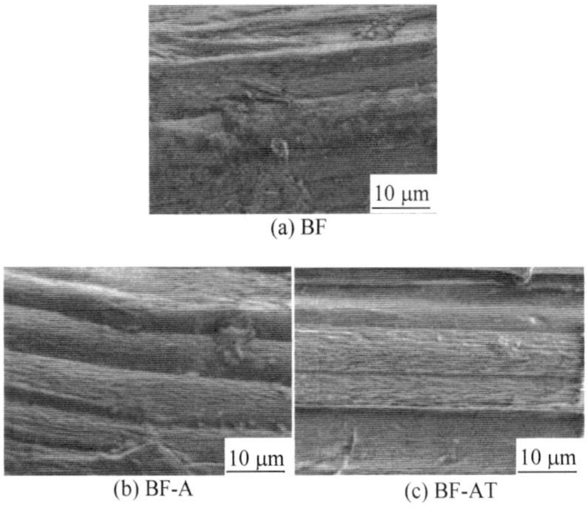

Figure 2 SEM photographs of surface of BF, BF-A and BF-AT

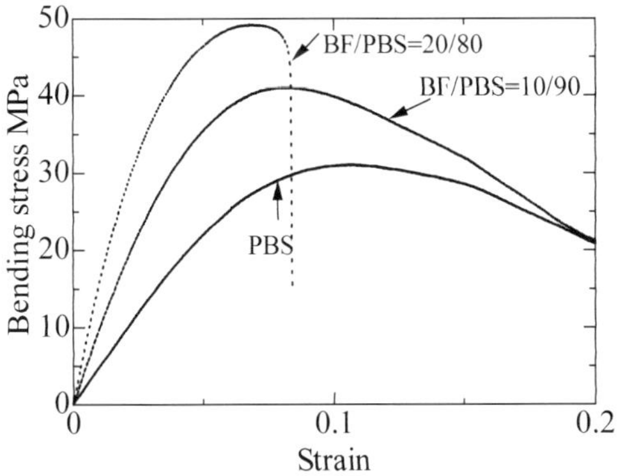

Figure 3 Stress-strain curves on three point bending test of BF/PBS composites

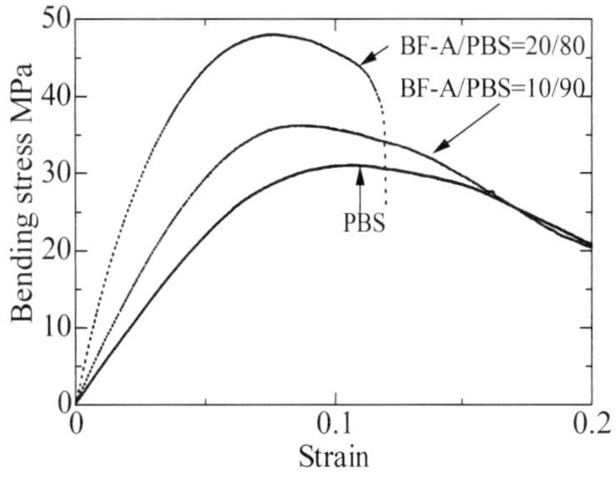

Figure 4 Stress-strain curves on three point bending test of BF-A/PBS composites

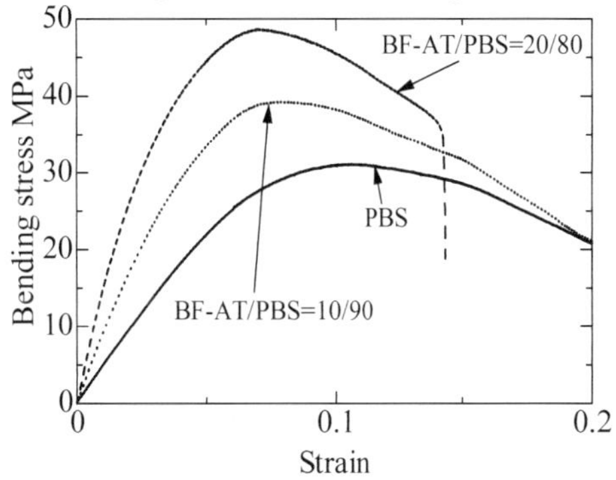

Figure 5 Stress-strain curves on three point bending test of BF-AT/PBS composites

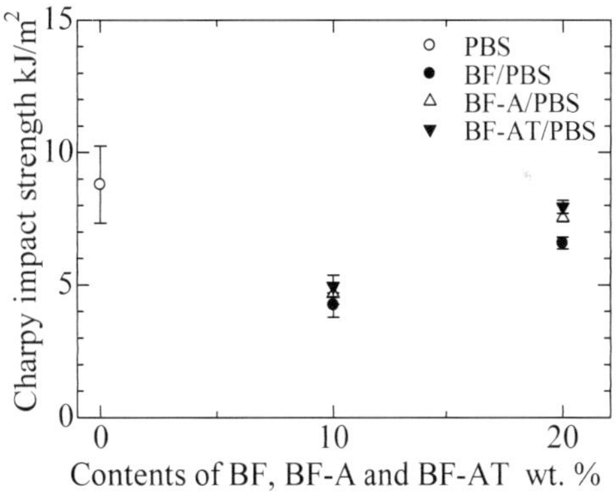

Figure 6 Results of charpy impact test

Conclusions

Banana fiber is a promising material to be used as fortification to resin. In order to obtain high mechanical strength composites, it is necessary to treat of the surface of banana fibers.

References

1. A. G. Kulkarni, K. G. Satyanarayana, P. K. Rohatgi and K. Vijayan: J. Mat. Sci. **18** (1983) 2290
2. K. Shimaji and S. Sudo, Mokuzai no soshiki, Morikita shuppan Ltd., (1976); 47 (in Japanese)

O-11-088

FABRICATION OF POLYLACTIDE BASED BIODEGRADABLE THERMOSET SCAFFOLDS FOR TISSUE ENGINEERING APPLICATIONS

Reika Sakai[1], Baiju John[1], Masami Okamoto[1]*, Jukka V. Seppälä[2], Jayasheelan Vaithilingam[3], Husnah Hussein[3], Ruth Goodridge[3]

[1] *Advanced Polymeric Nanostructured Materials Engineering, Graduate School of Engineering, Toyota Technological Institute, 2-12-1 Hisakata, Tempaku, Nagoya 468 8511, Japan – okamoto@toyota-ti.ac.jp*

[2] *Department of Biotechnology and Chemical Technology, School of Chemical Technology, Aalto University, Finland – jukka.seppala@jvs-polymers.fi*

[3] *The Wolfson School of Mechanical and Manufacturing Engineering, Loughborough University, UK – R.D.Goodridge@lboro.ac.uk*

Abstract - The fabrication of tissue engineering scaffolds based on the polymerization of crosslinked polylactide using leaching and batch foaming to generate well-controlled and interconnected biodegradable polymer scaffolds was reported. The scaffolds fabrication parameters were studied in relation to the pore morphology, interpore connectivity, and structural stability. In vitro cell culture and in vitro degradation were used to analyze the biocompatibility and biodegradability of the thermoset PLA scaffolds. The scaffolds prepared with 30 μm size porogen particles possessed a stable internal architecture and well-developed pore interconnectivities. The scaffolds with porosity of nearly 50% were favorable for bone scaffolds. The in vitro cell culture test was conducted on the scaffolds with human mesenchymal stem cells. The scaffolds possessed nontoxic nature and promoting the cells ability to attach and propagate in the scaffolds structure. The biodegradability of the scaffolds was investigated in phosphate buffered saline solution at 37.5 °C. Both the scaffolds showed significant degradation after two weeks of immersion in PBS due to the matching of the glass transition temperature of PLA thermoset scaffolds with the in vitro set temperature. This is an added advantage as it gives relevant flexibility when used for tissue repairs and regeneration. This new crosslinked PLA thermoset scaffolds with porosity of nearly 50% were favorable for bone tissue engineering applications due to the complex internal architecture, interpore connectivity, thermal stability and biocompatibility.

Keywords: tissue engineering, scaffold, crosslinked polylactide

Introduction

Tissue engineering plays a significant role modern medical advancement with biological scaffolds for repair or replaces the failing organs or tissues [1]. Pore size, porosity, interconnectivity between pores and surface area of the scaffolds shall help in cell seeding, migration, growth, mass transport and tissue formation. Thermoplastic polylactide nanocomposite foams and scaffolds are becoming popular these days with its biodegradable and biocompatible nature [2]. However, the research on PLA based thermoset scaffolds are at its infancy, which may find more suitable with predictable degradability, thermal stability and mechanically viable as compared with the thermoplastic ones.

In this research, the fabrication of highly porous and interconnected biodegradable polymer scaffolds based on the crosslinked PLA fabricated through salt leaching method and the one prepared by a combination of batch foaming followed by leaching [3]. The micro architectural properties of the crosslinked PLA scaffolds, in particular its morphology, porosity and structural stability, were analyzed and correlated with the results of the in vitro cell culture using the colonization of human mesenchymal stem cells (hMSCs) and degradation using phosphate buffered saline (PBS) solution.

Experimental

Fabrication of PLA thermoset scaffolds

The crosslinked PLA thermoset scaffolds were prepared by curing the Lait-X resin [4] using initiator with NaCl (the particle size ~30 μm) as the porogen. The ratio of resin to NaCl ratio was 1:1.2. The salt porogen was leached through water leaving the porous scaffolds. One of the cured samples with the salt particulate was taken for foaming using batch process under supercritical CO_2. The prepared disks were soaked in ultrapure water to leach the salt particulate, and subsequently dried and stored. The samples were designated as Lait-X/a and Lait-X/b, which were prepared by salt leaching and batch foaming followed by leaching respectively.

Characterization

The morphological features of the Lait-X scaffolds were analyzed with scanning electron microscope (SEM). The thermal properties were analyzed using the differential scanning calorimetry (DSC).

Pore size distribution was measured by mercury porosimetry and calculated using Washburn's equation [5]

$$\pi r^2 p = -2\pi r \gamma \cos\theta \qquad \text{(Eq. 1)}$$

This equation assumes that the mercury intrude a cylindrical pore. The terms r, γ, θ, p represent the radius of the cylindrical pore, surface tension (γ =

485 × 10^{-5} mN·m^{-1}), contact angle (θ = 130°), and pressure of mercury, respectively.

In vitro tests

hMSCs were used for in vitro cell culture to analyze the biocompatibility. The cell activity was checked on the 6th day, using alamarBlue® assay and a plate reader in the wavelength range of 570-600 nm. PBS (pH 7.4), with similar ionic concentrations of the human body was used to assess the biodegradability. All containers with scaffolds were kept in an incubator at 37.5 ± 0.5 °C. In order to estimate the weight losses and to observe the scaffolds with field emission SEM (FE-SEM), the scaffolds were taken out and air dried with the duration of two and four weeks.

Results and Discussion

Characterization

Fig. 1 shows the magnified views on the internal morphology of Lait-X/a and Lait-X/b scaffolds. The Lait-X/a and Lait-X/b have the same percent of salt particulate with similar particle size; the former was turned into scaffold through simple leaching while the latter was through batch foaming followed by leaching. Both showed a well-developed porosity and interconnectivity with varying pore size spanning over a very wide range, from few μms to hundreds of μms. The interconnectivity of pores is a very significant parameter for porous scaffolds which plays a significant role in tissue development through cell seeding, adsorption of H_2O, nourishments, and desorption of wastes.

Table 1 presents the median pore diameter, the porosity, and the glass transition temperature (T_g) of Lait-X/a and Lait-X/b. The median pore diameter and the porosity were estimated by the results of the pore size distribution (Fig. 2). Fig. 2 shows the relation between the pore size diameter to the cumulative and differential intrusions of mercury in Lait-X/a and Lait-X/b scaffolds. The maximum intrusion and interconnectivities of Lait-X/a was from 0.1 to 1 μm pore diameter regions and Lait-X/b it was from 0.1 to 10 μm. The median pore diameter and the porosity for Lait-X/a calculated by mercury porosimetry were 0.520 μm and 43.0% whereas for Lait-X/b it was 1.26 μm and 49.0%, respectively (Table 1). The shift in the values of Lait-X/a and Lait-X/b was due to the effect of batch foaming which lead to the movement of salt particulate during foaming resulting in the increased porosity and total intrusion volume. The value of porosity that is best suited for the bone tissue scaffold lies in the range of 50% [6] which matches with both the porosity values of Lait-X/a and Lait-X/b.

The thermal behavior and the T_g of the scaffolds were analyzed using DSC thermograms (Table 1). There were no melting points, which demonstrates that both scaffolds were amorphous and thermoset. The T_g of Lait-X/a and Lait-X/b were 37.6 °C and 37.4 °C with no significant variation. It was clear that there was no structural change in the methods

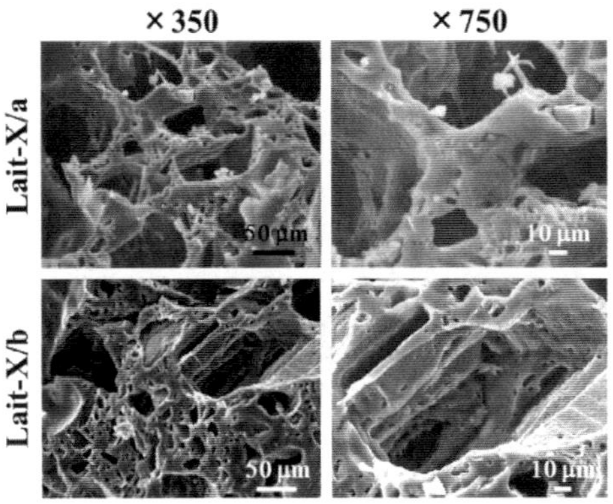

Figure 1 SEM images of crosslinked PLA scaffolds.

Table 1 Physical properties of Lait-X/a and Lait-X/b.

Sample	median pore diameter (volume) / μm	porosity / %	T_g / °C
Lait-X/a	0.520	43.0	37.6
Lait-X/b	1.26	49.0	37.4

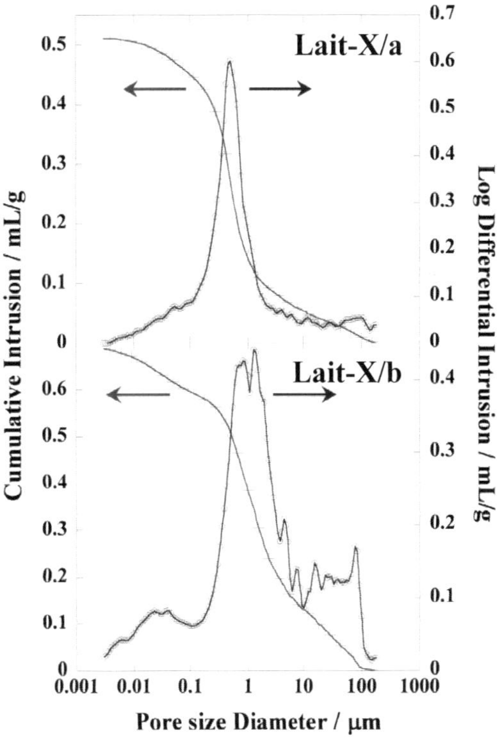

Figure 2 Pore size distribution in Lait-X/a and Lait-X/b scaffolds.

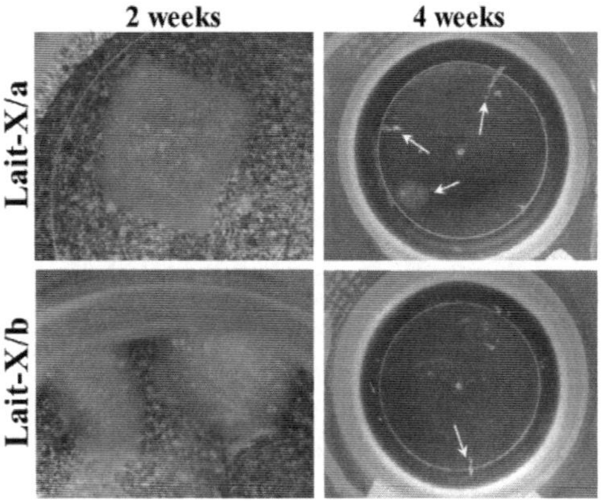

Figure 3 – Optical images of scaffolds after two and four weeks of immersion in PBS.

of scaffolds preparation. The Lait-X thermoset scaffolds were thermally stable, possess a T_g that matches with the human body temperature, was an added advantage towards the potential application as a tissue engineering scaffold.

In vitro tests

The cell activity of Lait-X/a and Lait-X/b was checked with alamarBlue® assay using a plate reader. The control with the scaffolds in mesenchymal stem cell basal medium (MSCBM) without the hMSCs showed no cell activity while the ones with cells showed a cell activity. This confirms that the Lait-X/a and Lait-X/b were nontoxic and the cells seeded into the scaffolds exhibited the ability to attach and propagate in the scaffold structure. The absorbencies of Lait-X/a and Lait-X/b evaluated by a plate reader were 0.601 and 0.419, respectively. It confirms that Lait-X/a showed more cellular activity than the Lait-X/b even though both had same in composition and varies only in the preparation method. This may be due to the effect of pore wall regions in Lait-X/a prepared by only salt leaching that can promote the cellular adhesion and growth.

The biodegradability of porous thermoset scaffolds were investigated in PBS solution at 37.5 °C by analyzing the loss of weight and pore morphology throughout degradation. Fig. 3 shows the optical images of both scaffolds after two and four weeks immersion in PBS media. Both the scaffolds showed significant degradation after two weeks of immersion in PBS and total disintegration in another two weeks. The weight ratios of Lait-X/a and Lait-X/b after in vitro degradation were nearly 50% with two weeks of exposure. Fig. 3 also shows the increased disintegration of Lait-X/b over Lait-X/a, which was due to the increased surface area and porosity attained through foaming. This measurement reveals that there were no harmful or non-disintegrative byproducts that harm or remain in the biological system with its use. Moreover, the rate of biodegradation of thermoset PLA scaffolds is best suited for bone tissue regenerating application in children where the ones with faster regeneration are preferred. This was caused by the matching of the T_g of PLA thermoset scaffolds with the in vitro set temperature (human body temperature). This is an added advantage as it gives relevant flexibility when used for tissue repairs and regeneration.

Conclusions

The highly porous crosslinked PLA scaffolds were successfully prepared through particulate leaching (Lait-X/a) and foaming followed by leaching methods (Lait-X/b). The scaffolds were porous with good interconnectivity and thermal stability. The SEM images confirmed the pore connectivity and structural stability of the crosslinked PLA scaffold. The in vitro cell culture demonstrated the ability of the scaffolds to support hMSCs adhesion confirming the biocompatibility through the cell–scaffold interaction. The in vitro degradation test revealed biodegradable nature of the PLA thermoset scaffolds. The matching of the T_g of PLA thermoset scaffold with the human body temperature added an advantage as it gives relevant flexibility and the fast degradation rate suited for treatment in children. The results suggests that novel crosslinked PLA thermoset macroporous scaffolds with a porosity of 50% discussed in this study were best suited for the bone tissue engineering applications.

Acknowledgements

This work was supported by KAKENHI (22656148) and the Strategic Research Infrastructure Project of the Ministry of Education, Sports, Science and Technology, Japan (2010-2014).

References

1. D. W. Hutmacher *Biomaterials* 2000, *21*, 2529.
2. Y. Fujimoto; S. Sinha Ray; M. Okamoto; A. Ogami; K. Yamada; K. Ueda *Macromol. Rapid Commun.* 2003, *24*, 457.
3. A. Salerno; S. Iannace; P. A. Netti *Macromol. Biosci.* 2008, *8*, 655.
4. A. O. Helminen; H. Korhonen; J.V. Seppälä *J. Appl. Polymer. Sci.* 2002, *86*, 3616.
5. E. W. Washburn *Phys. Rev.* 1921, *17*, 273.
6. K. Whang; K.E. Healy; D.R. Elenz; E.K. Nam; D.C. Tsai; C.H. Thomas; G.W. Nuber; F.H. Glorieux; R. Travers; S.M. Sprague *Tissue Eng.* 1999, *5*, 35.

O-11-074

COMPOUNDING AND MELT STRENGTHENING OF POLY (LACTIC ACID): SHEAR AND ELONGATION RHEOLOGICAL INVESTIGATIONS FOR FORMING PROCESS.

Benoit Mallet[1], Khalid Lamnawar[2], Abderrahim Maazouz[1,3]*

1Université de Lyon, Ingénierie des Matériaux Polymères, IMP UMR CNRS #5223, INSA Lyon, 20 Avenue Albert Einstein, Groupe de Recherche pluridisciplinaire en Plasturgie/GRPP-INSA de LYON, France

2 Université de Lyon, Laboratoire de Mécanique des Contacts et des Structures-Université de Lyon, CNRS, INSA-Lyon, LaMCoS UMR CNRS#5259, Groupe de Recherche Pluridisciplinaire en Plasturgie/GRPP-INSA de LYON, France

3Hassan II Academy of Science and Technology Rabat, Morocco

Introduction

Poly (lactic acid), PLA, is a polymer that possesses many physical characteristics rendering it suitable for replacing commodity polymers [1]. Beside it being biocompatible and available from renewable resources, PLA also demonstrates a high strength, thermoplastic processability, good crease-retention, grease and oil resistance and an excellent aroma barrier [2-5]. The large amount of crystallinity in PLA is the reason for its high modulus and strength, but is also responsible for its brittle nature and lack in toughness. These drawbacks are especially disadvantageous in the film extrusion industry. However, there are means of improving the flexibility and toughness in PLA, for instance by blending with a plasticizer.

Previous laboratory studies have mainly been focused on modifying the resin through branching reactions, polymer blending and charge filling6-8. There are also numerous reports on blending PLA with plasticizers for biomedical [9-11] as well as for film applications [12-16]. When synthesizing PLA from corn starch, its molecular weight distribution is generally not well controlled and the polydispersity of common commercial grades is often quite broad. This specificity, in addition to degradation problems [17] arising from PLA's polyester nature, puts the studies of this material apart from those of classical resins.

A prerequisite in order to allow a larger industrial use of PLA is melt strengthening, so as to enlarge its processing window. The aim of the present study was thus to investigate a commercial branching agent presented in masterbatch form. The selected pre-mixed compound is dedicated to industrial usage where the dosing process and the subsequent reactive extrusion need to be as simple as possible. PLA was modified through reactive extrusion with this chain brancher and the resultant material was characterized with regard to its molecular weight distribution (MWD), as well as by means of rheological analysis.

To this end, the forming process as well blown extrusion and foaming of PLA was investigated in this study. Through this work, we present some promising routes to enhance its processing ability which presents several challenges mainly due to the poor shear and elongation properties of this biopolymer. To our knowledge, there is few papers dedicated to the investigation of foaming an/or blown extrusion of PLA that involves structural, rheological and thermomechanical properties.

Experimental Part

Materials and reactive extrusion

The PLA used in this study was supplied by Natureworks® (PLA 3051D®) and is referenced as an injection grade. This chain extender called a Joncryl was provided by BASF. It is a multifunctional (f>9) epoxide. To prevent the hydrolytic degradation, PLA and joncryl were both dried for a minimum of 4 h before processing. Nucleants agents were also used as referenced in the INSA-patent. TALC (powder <10 um), N,N'-Ethylenbis(stearamide) and PEG ($\overline{M_W}$=1450 g/mol) were purchased from Sigma-Aldrich. TALC and N,N'-Ethylenbis(stearamide) were used as nucleant, PEG as plasticizer. N,N'-Ethylenbis(stearamide) will be referred to as EBSA in this work.

Forming process

(a) **<u>Foaming:</u>** a commercial-grade of CO2 provided by Air Liquide® was used as the blowing agent during batch foaming. Batch foaming process. The foaming was performed in an autoclave (Model

4565C, Paar Instrument) with a volume of 300 cm^3. The maximum working pressure was 200 bars and the limit in temperature was 350°C. The reactor was thermally regulated with the help of double-wall socket in which oil circulates and that is which was connected to a cryo-thermostat (Phoenix C35P, Haake). In these conditions, cycles with a heating ramp of 1.3°C.min-1 and a cooling ramp of 1°C.min-1 could be set. The CO2 pressure was set in the autoclave from a gaseous CO2 cylinder connected to a pneumatic gas booster (GR CRA 4AG25, Haskel Milton Roy). The pressure release was computer-controlled via a pneumatic back pressure regulator (Tescom), allowing a maximum pressure drop rate of 2 MPa.s-1. The autoclave was equipped with a thermocouple and a pressure transducer connected to the computer in order to monitor the temperature and pressure during all the experiments.

In order to obtain similar foaming conditions, a basket holding four samples was designed to simultaneously foam the neat and modified PLAs. The specimens were cut from the hot pressed plates into discs with 6 mm diameter and placed into the reactor. A CO2 purge was performed prior to all experiments. The first step was the saturation of polymer with CO2 during 2 hours at the temperature of 165°C and pressure ranging from 9.6 MPa up to 14.2 MPa depending on the targeted pressure during the foaming step. This saturation period was assumed to be long enough to insure sorption equilibrium in the entire sample. The second step was a thermal stabilization of the CO2/PLA solution at the foaming temperature TF. An isothermal of 20 minutes was respected before starting the polymer expansion step. Foaming was thus initiated by a pressure quench down to atmospheric pressure at a rate of about 2 MPa.s-1. Due to this pressure drop, PLA matrix became supersaturated with CO2, which led to cell nucleation and growth. The foamed samples were then removed from the reactor and allowed to cool to ambient conditions.

(b) **Blown extrusion:** The blowing extrusion using a tubular 25mm diameter blow die coupled to a melt pump and twin screw extruder to process the neat and the prepared formulations of PLA and chain extender (Joncryl®). Molten neat and modified PLA with varying formulations was extruded using an annular die to form a tube. The temperature profile was kept constant as the step (i) and the temperatures of the tubular die was changed from 150° to 180°C. By blowing air through the die head, the tube is inflated into a thin tubular bubble and cooled. The blown film is then flattened between nip rolls and pulled up by the winder. Different forms of instability are defined and the effect of film blowing parameters is illustrated. These parameters are the Freeze line height (FLH), the blow-up ratio (BUR) which is defined as the ratio of the final bubble diameter to the die diameter, and the take-up ratio (TUR) or Draw ratio (DR) which is the ratio of the take-up velocity to the extruded material velocity at the die exit. BUR and TUR are obviously two key parameters of the film blowing process. Increasing the values of these parameters is explicitly desirable in a commercial film production. Typical BUR values in industry for HDPE fall between 1.5 and 3.5, while TUR values range from 8 to 50. By varying the BUR, screw speed, air pressure and winder speed, films of different thicknesses between 10 to 150μm and degree of orientation can be achieved.

Main results and discussion:
Various formulations of PLA with multifunctionalized epoxy, nucleants and plasticizer were prepared and characterized on the basis of their linear viscoelasticity and extensional properties. The balance of chain extension and branching has been also investigated using solution viscosimetry, Steric exclusion chromatography (SEC) and rheology (relaxation spectrum, Van Gurp Palmen curves….). We pushed further by characterizing both the structure and thermomechanical properties of PLA formulations. All these studies brought informations about the presence of branching in modified polymers depending on the Joncryl amount and the reaction time. These plots revealed curves those were characteristic of long-chain branched polymers. Indeed, they exhibited a typical feature of a mixture of linear and randomly branched polymers.
On one hand, a batch foaming assisted with supercritical CO2 was achieved following a full characterization in physicochemical, rheological and thermal domain, The influence of the foaming parameters, the extent of chain modification as well as the contribution of crystallization on cell morphology was evaluated. Based on these parameters, structures ranging from micro to macro-cellular-cell were obtained (Figure 1).

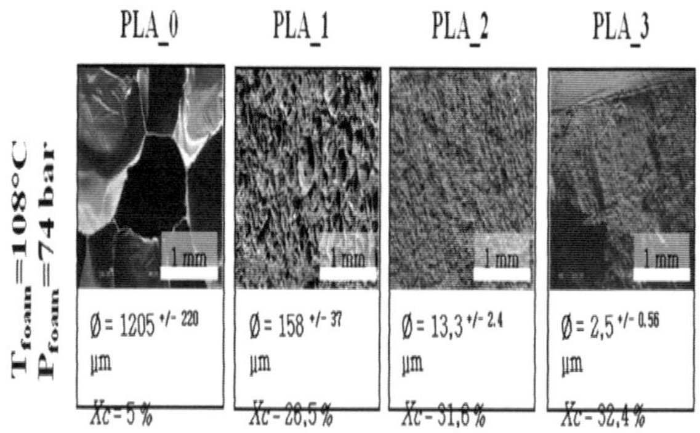

Figure 1- example of cellular morphology of neat and modified PLA obtained by supercritical CO2 fluid

On the other hand, the stability maps of blown processing for neat and modified PLA were established at different die temperatures. We have achieved a great enhancement of the blown processing windows of PLA with high BUR (Blow Up Ratio) and TUR (Take Up Ratio) attained (Figure 2).

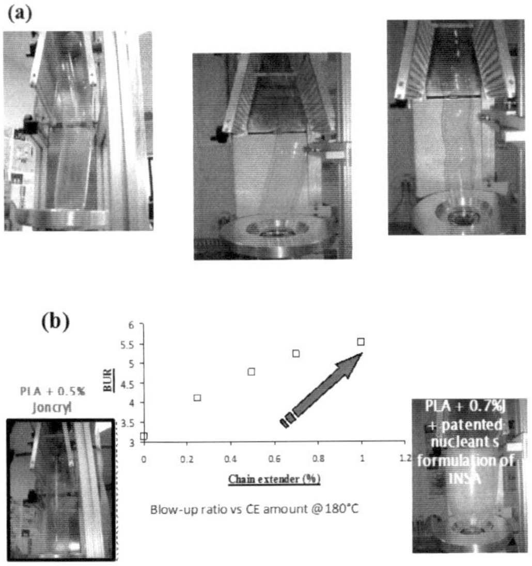

Figure 2- Improvement of the blown extrusion properties with the modified PLA: (a) neat PLA with defects (b) stable blown films of modified PLA.

Furthermore, we were able to demonstrate that a higher kinetic of crystallization can also be reached for chain-extended and branched PLA formulated with adequate amounts of nucleants and plasticizers. Induced crystallization during process was also demonstrated. Through this work, blown films with interesting thermomechanical and mechanical properties have been produced using an optimal formulation for PLA [21-23].

Conclusion

Polylactic acid (PLA) was modified through reactive extrusion and chain branching was achieved with the help of an epoxide additive. An optimized processing temperature was identified as a key parameter to obtain the desired efficiency of the reaction, since this would reduce the competition between chain extension and thermal chain scission.

Chain branching drastically modified the PLA structure by shifting the molecular weight distribution from a monomodal to a bimodal population, and this new chain distribution had an enlarging effect on the relaxation time spectrum. Moreover, the melt rheology of the materials was influenced by these molecular couplings. On the one hand, the low shear viscosity could be drastically improved by increasing the content of the branching agent. On the other hand, converging values were obtained for high shear rates. The melt thus obviously became stronger with increasing epoxide content. The thermal properties of the modified PLA materials were found to diverge from those of neat PLA. For instance, the crystallization nucleation was strongly increased when branching reactions were performed. This phenomenon induced a kinetic improvement but influenced the final crystallinity, which was found to become lowered when the branching level was increased. To finish, the branched materials featured a broader crystallization temperature window as compared to that of the neat PLA.

On one hand, the foaming ability of neat PLA has been enhanced by chain extension through reactive extrusion. The resulting rheological improvements were found to be key parameters to enlarge the processing window over foaming temperatures. Elsewhere, the crystallization behaviour under sc-CO2 was found to drastically differ from neat to highly chain extended PLAs. This is of a particular importance for the micro scaled foams obtained at low foaming temperatures for the chain extended PLAs. In this case, the foaming seems to be governed by nucleation

On the other hand, the improvement of both processing stability and microstuctural properties of the blown film based on the PLA. As we have firstly demonstrated in this paper, the extrusion blowing of various grades of the neat PLA is limited, a proper correlation between melt shear and extensional rheological properties is very important to get a better handle of this process. Our challenge was to make this process stable at higher melting temperature with the highest BUR and TUR ratios sustainable, using blending of PLA with appropriate additives. It was clearly observed that the addition of multifunctionalized epoxide (Joncryl®) has a direct influence on the rheological behavior of PLA. Furthermore, the stability maps of processed blends were established. When compared to a PLA/Joncryl® blend, the ternary system leads to a clear enlargement of the processing window as bubbles were obtained at higher BUR ratios. Of most interest is the significant increase in the maximal value for stable BUR ratios, going from 7 to 12 at high TUR. The maximal BUR ratio which can possibly be attained is higher as we reached the limitations of our equipments. This surprising behavior is in agreement with the effects of chain extension branching and the role of nucleants and plasticizers. The blown bubbles also showed a translucent aspect which can be associated with a semi-crystalline structure which was characterized using DSC, WAXS and FTIR. By using a ternary system with chain extension/branching reaction, a higher crystallization kinetic for PLA was achieved. This can induce crystallization during the blowing process at higher BUR and TUR ratios (e.g. streching and orientation of chains). These results are of high enough interest to warrant a thorough analysis of the microstructural properties of such blown-processed films, which will be the subject of the work in progress. Thermo-mechanical properties of PLA and the PLA-based optimized blends have also been investigated confirming the presence of a crystalline structure induced during processing. Mechanical properties of the PLA-based optimized blend showed a more flexible material with a ductile behavior compared to the brittle neat PLA

References

1. Vert, M.; Schwarch, G.; Coudane, J. J. Macromol. Sci., Part A. Pure Appl. Chem. 1995, 32, 787-796.
2. Lunt, J., Large-scale production, properties and commercial applications of polylactic acid polymers. Polymer Degradation and Stability 1998, 59, (1-3), 145-152.
3. Wehrenberg, R. H., Lactic acid polymers: strong, degradable thermoplastics. Mater. Eng. 1981, 94 (3), 63-66.
4. R. E. Drumright, P. R. G., D. E. Henton, Polylactic Acid Technology. Advanced Materials 2000, 12, (23), 1841 - 1846.
5. E.S. Lipinsky, R. G. S., Is lactic acid a commodity chemical? Chem. Eng. Prog. 1986, 82, 26–32.
6. John R. Dorgan, J. S. W., Melt rheology of poly(lactic acid): Entanglement and chain architecture effects. J. Rheol. 1999, 43, (5), 1141-1155.
7. Z. F. Zhou, G. Q. H., W. B. Xu, F. M. Ren, Chain extension and branching of poly(L-lactic acid) produced by reaction with a DGEBA-based epoxy resin. eXPRESS Polymer Letters 2007, 1, (11), 734–739.
17. A.Lopez Arraiza, J. R. S., J. Verdu, X. Colin, Rheological behavior and modeling of thermal degradation of poly(e-caprolactone) and poly(L-Lactide). Polymer processing XXII 2007, 5, 389-394.
18. Campbell, J. D. D., J. A, Deyoung D. J. , Giannakitsas, I. Dean R. S, D D. Teymour, F Villalobos, M. A. Process for producing polymers by free radical polymerization and condensation reaction, and apparatus and products related thereto. 2002.
19. M. Villalobos, A. A., T. Greeley, G. Turco, G. Deeter, Oligomeric chain extenders for economic reprocessing and recycling of condensation plastics. Energy 2006, 31, 3227–3234.
20. Y.-M. Corre; A. Maazouz; J. Duchet; J. Reignier. Batch foaming of chain extended PLA with supercritical CO2: Influence of the rheological properties and the process parameters on the cellular structure. *J. of Supercritical Fluids* 2011,*58,* 177 - 188
21. A. Maazouz, K. Lamnawar, B. Mallet Patent: C08L67/00; C08J5/10. FR2941702 (A1); 2010.
22. B. Mallet; K. Lamnawar; A. Maazouz. Compounding and processing of biodegradable materials based on PLA for packaging applications: In greening the 21st century material's world, *Frontiers in Science and Engineering,* 2011,*1-2*, 1-44
23. Mallet; K. Lamnawar; A. Maazouz. Improvement of blown extrusion processing of PLA: structure-processing-properties relashionships. Polymer engineering and Science (To appear 2013).

O-11-060

POLY(LACTIC ACID)/SILICONE BLEND FOR SUB ZERO TEMPERATURE APPLICATIONS: DESIGN OF EXPERIMEN

U. Meekum[1], A. Khiansanoi[1,2]

[1]*School of Polymer Engineering, Suranaree University of Technology, Nakorn Ratchasima, Thailand*

[2]*Centre for Petroleum, Petrochemical, and Advanced Materials, Bangkok, Thailand*

Abstract: The toughness improvement of PLA/silicone rubbers blends for sub zero temperature applications was attempted. The statistical approach namely 2^k factorial design of experiment was used to optimize and quantify the PLA/silicone blend ingredients. In this study, three parameter, k = 3, consisted of silicone(A), silane(B) and polyester polyols(C) as chain extender were assigned. Then, eight PLA/silicone formulas were compounded into pellets by co-rotation twin screw extruder at 190°C and the test specimens were injection molded. The original and sauna cured samples were obtained. The sauna cured was achieved by incubation in moisture saturated oven at 65°C for 12 hrs. The tensile properties, HDT and impact strengths both at room and at sub zero temperatures were conducted and utilized as the statistical responds for the analysis. The ANOVA conclusions showed that silicone and polyols were the positive effect on the impact properties both at room and sub zero temperatures. Moreover, the mechanical properties were improved via the sauna curing. The PLA/silicone blends were translucent. Scanning electron microscopy(SEM) verified that the silicone rubber was dispersed onto PLA matrix as micro droplets.

Keywords: PLA/Silicone Blend, Design of Experiment, Sub Zero Temperature Application and Sauna Curing

Introduction

Poly(lactic acid)(PLA) is a biodegradable plastic and renewable from natural materials. It has an incompetency due to its brittleness in nature. Therefore, it has not been commercially and wildly used as packaging applications especially at low temperature. Nevertheless, the flexibility of PLA can be improved by modifying its properties through several approaches including copolymerization and blending or compounding[1, 2]. Blending or compounding with other polymers or filler normally offers a cost effective way to enhance the hardness and ductility of PLA. Silicone rubber blended polymeric material normally offers the advantages in thermal characteristics that are contributed from the rubber[3-5]. In this studied, PLA/silicone rubbers blend were explored. Silicone is broader thermal stability and retain it toughness at low temperatures. The mechanical properties by mean of tensile and impact strength both at room and sub zero temperatures was investigated. Also, the effect of silicone on morphology, melt rheology, and thermal properties of PLA/silicone blend will be discussed. The statistical approach by mean of design of experimental(DOE) on the parameters that effect the properties of PLA/silicone was the prime interests. The experimental design on PLA/silicone blends used 2^k factorial experimental design was obtained and verified the significant effect on the mechanical properties of PLA/silicone blends.

Experimental

Material

Polylactic acid(PLA), 2002D was available from Nature Works®. Two packs silicone, SILASTIC® 3483, was kindly provided from Dow Corning Corporation. The amino silane, Silquest A-1100, was available from Crompton Corporation and used as crosslink agent. Aliphatic polyester polyols, Rayelast A 8770, was kindly supplied from IRPC Plc. The commercial glycerol as plasticizer was added. The blended of Irganox 1076 and Irgafos 168, supplied from Ciba specialty chemicals, at weight ratio of 1:1 were added as thermal/processing stabilizer.

Design of experiment

The 2^k factorial design of experiment(DOE) was employed to optimize and quantify the amount of the compound compositions and the blend properties. Three parameters, k = 3, explicitly as silicone(A), silane(B) and polyols(C) contents were assigned. Each parameter was divided into two levels; high(+1) and low(-1), respectively. In each level, it was also subdivided into two levels as shown in table 1. Total eight formula matrices for the design was constructed and shown in table 2. The statistical analysis was assisted and performed by using commercial software, Design Expert™.

Sample preparation

PLA pellets was underwent dehumidification in vacuum oven at 80°C for 2 hrs. The assigned amounts of silane and high viscous polyester polyols were manually and vigorously premixed. The dried PLA pellet and silicone were mixed by mechanical stirrer. The coated silicone PLA was then blended with the silane liquid. Finally, thermal/processing stabilizer powder was added and again vigorously stirred for a few minutes. The well premixed ingredient was melt compounded in co-rotating intermeshing twin screw extruders consisted of three kneader zones. The mixing was achieved at 10 rpm and barrel temperature profile at 150, 160, 170, 180, 190°C from feed to die zones, respectively. The blend granulate was obtained. The dehumidified pellet was molded into test specimen by injection molding. The molded specimens were equally divided into two sets of sample; (*i*) allowed to annealing at room temperature for overnight and (*ii*) incubated at 65°C in moisture saturated oven for 12 hrs. The former was called original and the later called sauna cured sample, respectively. For the sub zero temperature test, the sample was immersed in liquid N_2/methanol mixture for at least ten minutes. Then, the deep frozen sample was taken out and immediately tested within a fraction of second.

Properties measurement: Melt flow index(MFI), tensile, impact strength both notched and unnotched at room and sub zero temperatures, HDT at 455 kPa were tested in accordance with ASTM standards. Morphology of the fractured surface was examined by SEM.

Parameters	**Low level(-1)**		**High level(+1)**	
Silicone(phr)	0.5	1.0	2.0	3.0
Silane(phr)	0.2	0.5	1.0	1.5
Polyol(phr)	1.0	2.0	4.0	5.0

Table 1.Parameters and levels of DOE

Run	**Silicone(A)**	**Silane(B)**	**Polyol(C)**	**Stabilizer**
1	0.5(-1)	0.5(-1)	2.0(-1)	1
2	1.0(-1)	0.2(-1)	5.0(+1)	1
3	1.0(-1)	1.5(+1)	1.0(-1)	1
4	0.5(-1)	1.0(+1)	4.0(+1)	1
5	2.0(+1)	0.2(-1)	1.0(-1)	1
6	3.0(+1)	0.5(-1)	5.0(+1)	1
7	3.0(+1)	1.0(+1)	2.0(-1)	1
8	2.0(+1)	1.5(+1)	4.0(+1)	1

Table 2 DOE matrix

Results and Discussion

DOE analysis

Table 3 summarizes the MFI, tensile, impact strengths and HDT for both original and sauna cured samples. Only notched impact strength was obtained at both room and low zero temperatures testing. Quick observation, the test values indicate that sauna incubation obviously enhance the mechanical and thermal properties of the blends. It is emphasised that crosslink reaction through silane/water condensation reaction had been taken place during the brewing process and hence superior in the properties are achieved. In comparison with the film grade LDPE, most of the compound formula manifests the better tensile strength, modulus, and impact properties, except the % elongation at break. Moreover, the PLA/Silicone blends also reveal the superior in toughness than neat PLA. For the sake of our prime interest for this research work, the notched impacts strength, obtained at room temperature is taken as the statistically responds and then analyzed by the Design Expert^{TM}. The normal probability against the standardized effects is plotted and shown in Figure 1(a) and 1(b), for original and cured sample, respectively. Also, Figure 2(a) and 2(b) are the statistically results of the notched impact respond tested at sub zero temperature, respectively. At the ambient temperature respond, the plots show that silicon is obviously a positive effect(**A**+) on the impacts strength of the PLA/silicone blends both before and after sauna cured. According to ANOVA conclusion result at 95% of confidential(p-value $<$ 0.05), it confirms that the designed model is significant and also the parameter A, silicone content, is the significant effect on the notched impacts strength of the blend. The rests of the parameters are lined on the linear trend. It indicates that there is no significantly effect on the impact of the sample. At sub zero temperature testing results, the impact effect plot for the original sample is given in Figure 2(a). The statistical analysis reveals that all of the calculated effects for all parameters are approximately closely fitted into the linear trend line. It indicates that the designed model is not significant or perhaps the assigned levels of parameters have no significant effect on the impact of both samples at low temperature testing. This statement is reinforced with the calculated ANOVA results as listed in the graph. However, Figure 2(b) demonstrates that parameters A+, C- and BC- are probably and noticeably suspected to be significant effects on the impact values of the sample after sauna curing. Again, the ANOVA figures conceal that the designed model is significant and those parameters are also significant effects on the notched impact strength of the sauna cured PLA blends . Moreover, whenever the model is significant, the statistical approach can also be utilized to construct the regressed model for prediction the impact strength of the PLA/silicone blends as a function of the designed parameters and their contents. The prediction model is listed in Table 4. By applying the exact approach described above, the regressed models for the other responds are also included in Table 4.

Scanning electron microscopy (SEM)

Visual observation of the molding, the translucent piece of specimen was obtained from the PLA/silicone blends. The SEM photographs of the fractured surface of the samples; neat PLA and Run#6 both original and cured, are presented in Figure 3(a) -3(c), respectively. Figure 3(a) illustrates the surface of glassy clear neat PLA. The trace fracture surface is smooth indicating brittle glassy materials. However, the surfaces of PLA/silicone blend at high silicone and polyols but low silane contents, Run#6, exhibited better impact strength both at room and low temperatures, are illustrate on Figure 3(b) and 3(c) for original and cured, respectively. They signify rougher fracture traces than the neat PLA. The tiny separated particles are also observed. The dispersed droplets are probably the silicone rubber. It would dissipate the impact energy and hence improve the impact resistance of the sample as evidence in the testing results. The size of the dispersed phase is too tiny that still allow the light to pass through. Therefore, the compound is apparently translucent. Accordingly, it would possibly to be applied for frozen food packaging without limiting its clarity.

Run No.	MFI (g/10m.)	Tensile Strength (MPa)		Elongation @Break (%)		Tensile Modulus (GPa)		Notched Impact (kJ/m^2)		Unnotched Impact (kJ/m^2)		Notched Impact @low temp (kJ/m^2)		HDT (^{o}C)	
	original	original	**cured**	original	**cured**	original	**cured**	original	**cured**	original	**cured**	original	**cured**	original	**cured**
LDPE	2.86	8.38	8.38	83.0	82.2	0.17	0.17	Nb*	Nb*	Nb*	Nb*	3.92	4.43	32.1	41.2
PLA	1.41	52.3	59.4	4.9	4.3	2.67	2.58	2.49	3.45	15.9	21.1	2.60	5.73	52.0	52.4
1	2.07	43.9	47.6	4.4	6.5	2.49	2.33	2.54	3.23	20.3	22.2	5.40	5.36	49.2	59.3
2	2.43	38.2	41.0	4.3	8.7	2.48	2.21	3.07	3.05	21.5	24.2	3.16	4.99	48.0	57.8
3	2.64	38.9	45.2	6.0	9.3	2.52	2.43	2.31	2.96	22.0	23.2	5.40	6.70	49.3	57.6
4	2.43	37.4	43.1	6.9	8.1	2.50	2.23	2.37	3.45	26.0	28.7	4.75	3.86	47.9	60.0
5	2.45	34.2	40.9	7.5	16.8	2.52	2.48	3.66	6.34	29.7	32.7	6.05	6.40	48.5	57.9
6	3.22	29.5	38.6	9.1	12.4	2.45	2.31	2.81	5.69	26.0	36.2	6.45	6.60	47.3	58.0
7	3.28	31.6	39.8	12.9	12.8	2.53	2.39	3.56	5.93	25.3	28.8	5.24	7.38	48.5	60.3
8	2.48	34.4	42.2	5.1	9.9	2.39	2.38	3.67	3.93	20.6	26.1	6.16	5.67	47.8	58.6

*Nb = specimen not broken

Table 3 Test results for original and cured samples

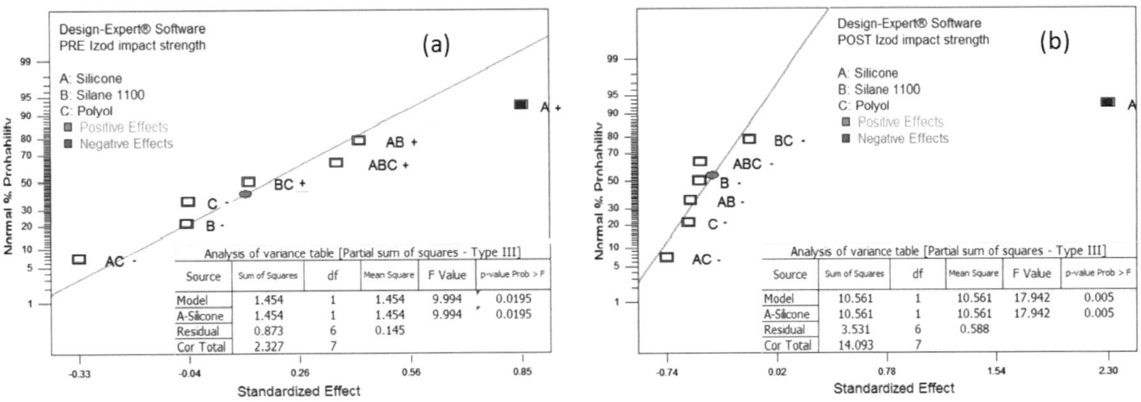

Figure 1 Normal probability plots of notched impacts tested at room temperature for; (*a*) original and (*b*) cured samples of PLA/Silicone compound, respectively.

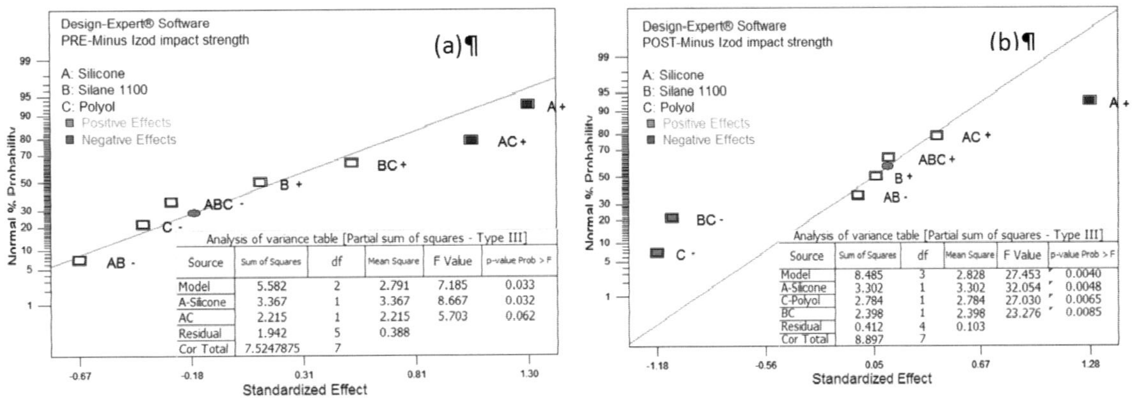

Figure 2 Normal probability plots of notched impacts tested at sub zero temperature for; (*a*) original and (*b*) cured samples of PLA/Silicone compound, respectively.

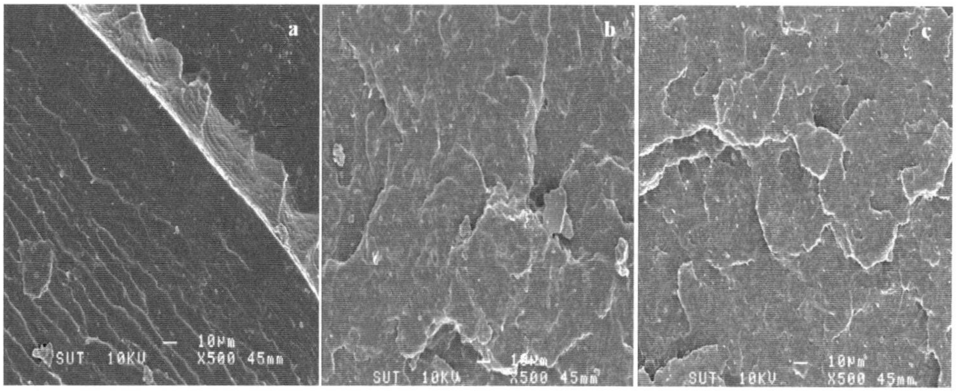

Figure 3 SEM micrograph of; (a) neat PLA (b) Run#6(original) (c) Run#6(cured) at X500, respectively.

Properties	Regressed models	
Room Temperature:	**Original**	**Cured**
MFI	2.63+0.23(A)-0.27(BC)	na
Tensile Strength	36.06-3.60(A)	42.34-1.92(A)-1.09(C)+1.10(AC)+1.16(BC)
Elongation@Break	No significant	70.62+2.41(A)
Tensile Modulus	2490.12-30.69(C)-22.99(AC)	2350.69-63.82(C)
Notched Impact	3.00+0.43(A)	4.32+1.15(A)
Unnotched Impact	23.97+1.47(A)-2.00(AB)-1.69(AC)	27.82+3.19(A)-2.44(AB)
HDT	48.35-0.30(A)-0.56(C)	58.71+0.45(B)-0.69(ABC)
Sub Zero Temperature:		
Notched Impact	5.33+0.65(A)+0.53(AC)	5.87+0.64(A)-0.59(C)-0.55(BC)

Table 4 Regression models for prediction properties of PLA/silicone blends

Conclusions

Blending PLA with silicone rubber in the present of polyester polyols plasticizer and silane coupling/crosslink agent gave rise to the blend having better properties than the conventional LDPE and virgin PLA. Translucent and flexible appearance of the blend was obtained. The silane/water crosslink reaction through the sauna treatment also proved to enhance the blend properties. The statistical approach by mean of DOE concluded that the incorporated ingredients; silicone, polyester polyols and silane, played the important role in the PLA/Silicone blends with the desired properties. The formula having silicone and polyols content at high levels but low level of silane content showed the promising properties. Further exploring to assess the best possible final formula is needed.

References

1. Rasal. et al. *Pro in Pol Sci*, (n.d.), 35(3), 338.
2. I. S. Miles, and S. Rostami, *Multicomponent polymer systems,* Longman Scientific & Technical, Singapore, 1992.
3. M. Miwa, *et al. Composites*, 1995, 26(5) 371.
4. T. Takahashi, *et al. J. of Euro. Cer. Soc.* 2001, 21(16), 2821
5. S. Kole, S. Roy, A. K. Bhowmick *Polymer.* 1994, 35(16), 3423

O-11-055

EXTRUSION OF BLENDS OF POLYOLEFINS REINFORCED WITH NANOCELLULOSE FROM SISAL PLANTS

Alcides Leao[a,*], Bibin M. Cherian[a], Sivoney F. de Souza[b], Mohini Sain[c], Suresh Narine[d]
[a] *UNESP – Sao Paulo State University, Botucatu, Brazil*
alcidesleao@fca.unesp.br
[b] *UFABC – Federal University of ABC, Santo Andre, SP, Brazil*
[c]*University of Toronto, Toronto, Canada*
[d]*Trent University, Peterborough, Canada*

Cellulose nanocrystals have been evaluated as reinforcement material in polymeric matrixes due to their potential to improve the mechanical, optical, and dielectric properties of these matrixes as well as its environmental positive footprint. These characteristics are due to its renewability source; can be produced from residues, such as primary sludge from pulp and paper industries and other agricultural by-products, including sugar cane bagasse, pineapple leaves, banana pseudo stem, sisal and many others agro-industrial raw-materials. Finally, its process production is not harmful to the environment. This work describes how some of the nanocellulose from different sources can be used to replace others not so friendly materials in many applications including, biomaterials, automotive industries and packaging. The process used is very mild to the environment and consists of a high pressure defibrillation followed by a chemical purification which affects the fiber morphology, going from macro and micro to nanoscale. Many fibers characterization processes were used including microscopy techniques and X-ray diffraction to study the structure and properties of the prepared nanofibers and composites. Microscopy studies showed that the used individualization processes lead to a unique morphology of interconnected web-like structure of the fibers. The resulted nanofibers are bundles of widths (5 to 30 nm) and estimated lengths of several micrometers. Further research are going on testing new applications such as dental, biomedical (implants, membranes), automotive, etc. Blends of polypropylene and sisal nanofibers were prepared in a twin-screw extruder, model ZSK-25, Coperion, L/D ratio of 25, at 180 rpm and the temperature profile of 190, 190, 180, 180 and 190 ºC. Those were the constant parameters achieved initially for all the extrusion zones. The isoprene modified nanofibers were incorporated into the PP in dried form with minimum agglomeration.

It is possible to produce composites based on vegetable fibers and uniform all the characteristics and the production of a new material in the technical sense of the word. The final properties differ strongly of processing and distribution, but also the previous history of the materials, harvest, part of the fiber and use.

Keywords: Nanocellulose, Nanocomposites, Natural Fibers, Sisal, Polypropylene

***Corresponding Author**: Alcides Lopes Leao – alcidesleao@fca.unesp.br

Introduction

The promising performance of cellulose nanofibres and their abundance encourages the utilization of agricultural waste residue, which acts as the main source of cellulose. In nature, a large number of plants and animals synthesize extra-cellular high-performance skeletal biocomposites consisting of a matrix reinforced by fibrous biopolymers [1]. Cellulose is a classical example where the reinforcing elements exist as whisker-like microfibrils that are biosynthesized and deposited in a continuous manner [2]. Sisal fibre obtained from Sisal plant is one of the major underutilized raw materials which is composed of cellulose nanofibre which constitutes 40% of its dry weight. Total world sisal fibre production is 250,000 tonnes per year and out of it, 80% of the fibre is contributed by the coastal region of Asia [3]. The raw fibres have been reported to be used in the field of micro composite application [4]. Efforts are going on for exploring wider export markets for sisal and sisal products but still most of the raw sisal fibre remain underutilized. Judged from the increase in production and employment, the progress has been rather slow and exports in physical terms have remained more or lets static.

In an earlier work, nanocellulose was prepared from banana fibres by steam explosion. Sisal is another important source of cellulose. The sisal fibre comprise mainly of cellulose, lignin and hemicellulose. The cellulose part is crystalline and the lignin part is completely amorphous in nature. Due to its heterogeneity, the direct utilization of the biomass is extremely low. Sisal fibre industry mainly uses the fibre as it is obtained from the sisal plant and no other extraction procedure has been reported yet. However, such separation is mandatory for its effective utilization in the current nanotechnological areas. The extracted lignin is a suitable source for renewable energy production. In this paper we report on a novel technique, steam explosion, for the separation of nanocellulose from sisal fibre there by expanding its application fields. In the various pre-treatment technologies for the isolation of nanocellulose fibres, steam explosion is an attractive choice. We used mild acids with low concentration which avoids toxicity. In the present work we intend to obtain nanocellulose1 by a green pathway which makes it important.

Experimental

Materials

Sisal fibers were collected from locally available suppliers.

Chemical analysis of the fibres

Chemical constituents of fibres were determined according to ASTM standards. α-cellulose (ASTM D 1103-55T), hemicellulose (ASTM D 1104-56), lignin (ASTM D1106-56), Moisture content (ASTM D 4442-92). The cellulose, hemicelluloses, lignin and moisture content of the fibre in the untreated raw, steam exploded and acid hydrolysed stages of the sisal fibre were determined.

Methods for the isolation of nano cellulose from raw fibres

Alkali treatment of the fibre

The optimum alkali treatment is both a very effective surface modification and a low cost surface treatment for sisal fibres. Locally available sisal fibres were cut it into pieces and subjected for treatment. Sisal fibres were soaked in 2% caustic soda and placed for six hours at a temperature of 25°C.

Steam explosion of the alkali treated fibre

A laboratory autoclave which can work with 137 Pa (20 lbs) pressure was used for steam treatment. Steam explosion technique was applied on the alkali treated fibre for one hour. Steam pre-treatment was performed by loading the lignocellulosic material directly into the steam gun and treating it with high pressure steam at temperatures within 200 to 250°C. The paper refers 'steam exploded fibre' which means the alkali treatment followed by steam exploded sisal fibre.

Bleaching of the steam exploded fibre

Alkali treated and steam exploded samples are then subjected to bleaching. After the successive chemical treatments, the bleaching treatment with a chlorine dioxide ($NaClO_2$) solution (pH 2.3) for 1 h at 50°C was performed to remove the remaining lignin. In the bleaching step, the absence of elemental chlorine is accomplished by using $NaClO_2$.

Acid treatment followed by steam explosion

The bleached sample is then subjected to mild acid treatment. 5% oxalic acid was used for the acid hydrolysis followed by second step of steam explosion for one hour. A pressure of 137 Pa (20 lbs) was used, followed by sudden release of pressure. The fibres were then washed thoroughly by water and then subjected to mechanical stirring followed by sonication.

Preparation of Nonwoven Nanofibril Mats

The nonwoven mats of cellulose fibers and nanofibrils were formed by filtering the water-nanocellulose slurry. For the nanocellulose, special membrane filter (Polyethersulfone, 0.1 μm) was used for mat formation of fibrils was used for drying and forming the mat. The slurry was agitated during the filtering process to distribute the nanocellulose evenly on the mesh. The nanocellulose formed mat could easily be peeled off from the mesh after drying due to strong hydrogen bonds. The thickness of the fiber mat was adjusted by altering the cellulose concentrations on the water–nanocellulose slurry. The thickness of the mats varied between 0.05–0.2 μm.

Fourier Transform- Infra Red Spectroscopy (FTIR)

Fourier Transform- Infra Red spectroscopy (FTIR) spectra of the fibers were recorded by a Shimadzu IR-470 IR spectrophotometer. The fibre particles were then mixed with KBr and pressed into a small disc about 1 mm thick.

X-Ray diffraction technique (XRD)

X-ray equatorial diffraction profiles of the fibres were collected by a JEOL diffractometer, Model JDX 8P, using CuKα radiation at the operating voltage and current of 30 kV and 20 mA, respectively. The diffraction intensities were recorded between 2 and 80° (2θangle range).

Scanning Electron Microscopy (SEM)

Morphological analysis of untreated and steam exploded fibres were done by scanning electron microscopy. SEM micrographs of fiber surface were taken using a scanning electron microscope model JEOL JSM-35 C and Cambridge 250 MK3 stereo scan operated at 20 kV. Prior to SEM evaluation, the samples were coated with gold by means of a plasma sputtering apparatus.

Scanning Probe Microscopy (SPM)

The morphological study of the final nano cellulose fibrils were done by scanning probe microscopy. The scanning probe microscopy images of the nano fibres were made with a Multimode SPM (Veeco Inc. Santa Barbara, USA) with a Nanoscope IV controller in tapping mode. A dilute solution of nano dispersion which was sonicated just before the experiment was conducted.

Dynamic light scattering (DLS)

The particle sizing of nanocellulose dispersion in water was recorded by dynamic light scattering (DLS) system Inc. Santa Barbara, Calif., USA. The apparent hydrodynamic diameter was calculated according to the Stokes-Einstein equation:

$dh = kT/3\pi\eta\, Dapp$

with *D*app being the apparent diffusion coefficient and Z the viscosity of the solution. Run time of the experiment was 5-7 min. The measurements were performed with an angle of 160° by using laser (4mW) and the wavelength of the scattering light used was 639 nm and the count rate was 284 kHz. The temperature of the system was 20°C. The channel width was 200.0 μ sec and the index of refractive was 1.427.

Thermo Gravimetric Analysis (TGA)

TGA analyses were carried out using a Mettler TG 50 module attached to a Mettler TC 11 4000 thermal analyzer (Switzerland). The thermal analyses were done in a nitrogen atmosphere under a flow rate of 100 mL/min using an alumina crucible with a pinhole. A constant heating rate of 10°C/min was maintained from room temperature to 700°C.

Differential Scanning Calorimetric (DSC) Analysis

DSC measurements were carried out on a DSC Q100 (TA Instruments, DE, U.S.A.) equipped with a refrigerated cooling system. All the DSC measurements were performed under a dry nitrogen gas atmosphere. The samples were heated at a rate of 10 °C/min from 25 to 400 °C/min. DSC measurements were performed with a modulation amplitude of 1 °C/min and a modulation period of 60 s at a rate of 3 °C/min.

Results and Discussions

Chemical analysis of the fibres

Table 1 describes the chemical composition and moisture content of the sisal fibres at different processing stages. Chemical constituents of fibres were determined according to ASTM standards. The raw sisal fibre is mainly composed of cellulose (44.7%), lignin (9.2%) and other components. Upon alkali treatment, the lignin starts to dissolve out and increases the relative percentage of crystalline cellulose components. It is clearly demonstrated in the data presented in the table 1. The fine structure of cellulose materials is composed of crystalline and amorphous regions.

Table 1 Constituents of the fibres in different stages

Fibre Stage	Cellulose (%)	Lignin (%)	Hemicellulose (%)	Moisture Content (%)
Raw sisal fibre	44.7 (±2)	9.2 (±6)	23 (±0.5)	9.8 (±1)
Alkali treated fibre	62.8 (±3)	7.1 (±4)	18.7	10.3(±1)
Steam expl. fibre	84.4 (±3)	3.6 (±3)	-	8.8 (±1)
Bleached fibre	93.7 (±3)	0.2 (±0.1)	-	8.5 (±1)

The amorphous regions easily absorb chemicals such as dyes and resins, whereas the compactness of the crystalline regions makes it difficult for chemical penetration [5]. The mercerization of the sisal fibers may result in the removal of the surface impurities, the swelling of the crystalline region, and alkalisation of the peripheral hydroxyl groups. The common trend from the observation is the gradual decrease of amorphous components like lignin and hemicellulose from raw fibre to bleached fibre.

The lignin will react with $NaClO_2$ and an oxidative fragmentation of lignin takes place and some part of lignin will dissolve out as lignin chloride. The percentage increase of the pure cellulose component (44.7 to 93.7%) and decrease of the lignin components (9.2 to 0.2%) with each processing steps are the main observation.

Moisture content shows an increase from raw to alkali treated fibre followed by a gradual decrease. During alkali treatment, the swelling of the fibre takes place, which promote the absorption of moisture by capillary action. The alkali treatment of the fibre will lead to the swelling which facilitate the breakdown during acid hydrolysis.

During every processing step there is an increase in the percentage of crystalline cellulose content. Since most hydroxyl groups in the final stage are bonded by intra and inter molecular hydrogen bonding, moisture absorption decrease.

Acid hydrolysis of cellulose leads to hydrolytic cleavage of glycosidic bond between two anhydroglucose units. Thus the amorphous portion gets dissolved by acid hydrolysis, leaving behind the crystalline regions. Acid hydrolysis followed by mechanical treatment (sonication) results in disintegration of the cellulose structure into nanocrystalline form [5]. From the calculations for the yield of the product formed, it was found that 364 gm of dry nanocellulose is obtained from 1 kg of dry raw sisal fibre.

FT-IR Analysis

Figure 3 gives the FT-IR spectrum of sisal fibre at different processing stages. As discussed in chemical analysis, the main components in the sisal fibre are cellulose, hemicellulose and lignin. These three components are mainly composed of alkanes, esters, aromatics, ketones and alcohols, with different oxygen-containing functional groups. Erdtman, Klemm and Oh have reported on the infrared spectra of cellulose, hemicellulose and lignin [6-8]. Lignin present in the sisal fibre gives characteristic peaks at 1240, 1650 and 1740 cm^{-1} corresponding to the aromatic skeletal vibration and carbonyl group. It is clear from the spectra that these three peaks are completely absent in the bleached fibre. The peak present at 1650 cm^{-1} in the spectrum corresponding to the raw fibres is due to the presence of C=O linkage, which is a characteristic group of lignin and at 1740 cm^{-1} is due to hemicellulose. Another possibility is that carboxyl or aldehyde absorption could be arising from the opened terminal glycopyranose rings or oxidation of the C–OH groups. Alkali treatment reduces hydrogen bonding due to removal of the hydroxyl groups by reacting with sodium hydroxide. This results in the decrease of the -OH concentration, evident from the decreased intensity of the peak between 3300 and 3320 cm^{-1} bands compared to the untreated fibre [10-11]. Since the bleached fibre is purely crystalline cellulose, its spectra gives the transmittance of the three hydroxyl groups. The exposed hydroxyl group of the cellulose is clearly seen from the increased intensity of the 3315 cm^{-1} peak.

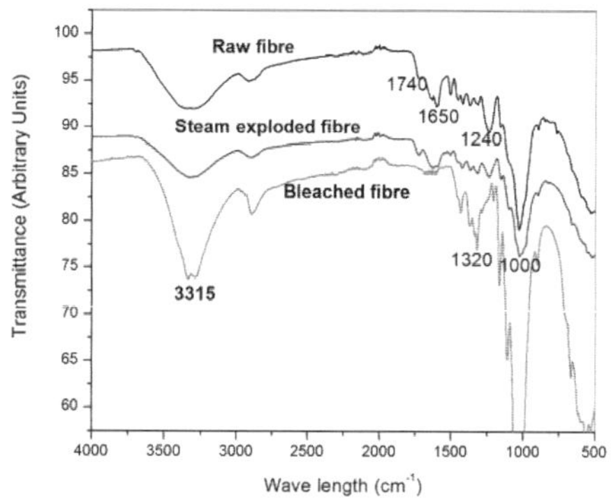

Figure 3 FTIR spectra of sisal fibre

The three hydroxyl functional group present in the cellulose structure is primarily responsible for this peak in addition to aromatic hydroxyl groups present in the lignin and the water absorbed in the case of raw and steam exploded fibres. The successive treatments leads to the exposure of the three hydroxyl groups which is gradually increased from raw fibre to bleached fibre. These hydroxyl groups are tightly bound by intermolecular hydrogen bonding which gives a reduced moisture absorption in the final crystalline cellulose fibre. Acid hydrolysed fibre gives almost the same spectra as the bleached fibre. In addition, due to the presence of functional groups such as methoxyl $-O-CH_3$, C–O–C and aromatic C=C; also peaks in the region between 1250 $cm^{-1}$and 1600 cm^{-1} was observed in the raw and steam exploded fibre [9,10]. But the destruction of the lignin-cellulose complex during alkali treatment and bleaching steps facilitate the hydrogen bonding by enhancing the cellulose cellulose interaction. From the FTIR analysis it has been concluded that there is a reduction in the quantum of binding components present in the fibres due to the process of steam and chemical treatment. The raw fibres have a characteristic peak in between 1240-1740 $cm^{-1}$and at 1240 cm^{-1}.

These peaks are mainly due to the hemicellulose and lignin components. The increased intensity of the peak at 1000 cm- 1 and the appearance of the new peak at 1320 cm^{-1} are due to the isolated cellulose components.

X-ray Diffraction analysis

The effect of alkali and acid treatments on the fibre surface is clearly envisaged by XRD. Figure 4 shows the X-ray diffraction analysis of the sisal fibre at various processing stages. In raw fibre, crystalline cellulose components are oriented in the matrix of lignin, hemicellulose, pectin, etc. It is completely an amorphous material. During chemical treatment, the cementing materials (matrix) will be dissolved, and the remaining pure crystalline particles isolated. These particles show increasing orientation along a particular axis, due to its similarity in shape. This pure crystalline particles show increasing orientation along a particular axis as the fibres are treated under different processing conditions. Alkali treatment of the sisal fibre will lead to the swelling of the fibre and subsequent increase in the absorption of moisture.

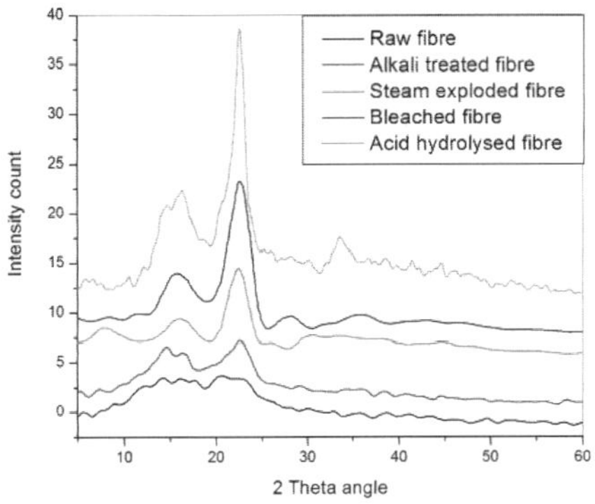

Figure 4 XRD analysis of sisal fibre

Treatment of the fibres with 2% alkali and successive bleaching steps removed the lignin completely. Alkalization of plant fibres changes the surface topography of the fibres and their crystallographic structure. Apart from truly crystalline and truly amorphous, there are some regions of intermediate order where the molecular configuration is liable to change by the chemical treatment. Degree of crystallinity of the fibre goes on increasing in both the processing stages. This can be believed to be due to the removal of lignin which acts as a cementing material and on delignification, an ordered arrangement of the crystalline cellulose in the structure takes place. From the figure it is clear that the raw sisal fibre has almost zero crystalline nature and the crystallinity keeps on increasing gradually with processing stages. As can be seen, the overall shape of all diffraction patterns is quite similar except for untreated raw sisal fibre. The baseline of the raw sisal fibre diffraction patterns almost flat, suggesting a higher content of amorphous material in the sample. On removing the non-cellulosic constituents (lignin and hemicellulose) from the fibres by chemical treatment, the degree of crystallinity and crystallinity index will change. The degree of crystallinity, i.e., the amount of crystalline cellulose present in a cellulosic fibre cannot be exactly defined, as neither the crystalline portions are perfect crystals nor the noncrystalline portion completely disordered. XRD analysis of the alkali treated fibres also reveals an increase in the crystallinity index of the sisal fibre. Other researchers have noted an improvement in the order of the crystallites as the cell wall thickens upon alkali treatment [12]. In the present case, an optimum concentration of 10% oxalic acid was found to give the best dispersion in the acid hydrolysis step. The effect of various treatments on the crystallinity of the fibres was also calculated. Table 2 shows the values of the crystallinity index obtained for the fibre at different processing stages.

The crystallinity index of the fibre has been calculated using the formula:

Crystallinity index = {(I crystalline – I amorphous)/ I crystalline} 100

The crystallinity was found to vary depending on the conditions applied. The maximum crystallinity was obtained when acid hydrolysis was carried out on the bleached fibre. In natural cellulose fibres, the regions of intermediate order in the structure plays an important role in the determination of the degree of crystallinity. The crystallographic transformation of the cellulose is negligible since the percentage of the alkali treatment is very low (2%).

Table 2 The crystallinity index of the fibres

Material	**Crystallinity index (*Ic*)**
Raw sisal fibre	3.2 (±1)
Alkali treated sisal fibre	12.6 (±2)
Steam exploded sisal fibre	28.2 (±3)
Bleached sisal fibre	42.1 (±3)
Acid hydrolysed fibre	60.3 (±2)

From the peak intensity of the variously treated fibres that were observed, it has been found that acid hydrolysis changes the fibre diameter as well as the crystallinity. It has been concluded from the X-ray diffractograms that as raw sisal fibres undergo steam explosion, bleaching and further acid hydrolysis there is a decrease in the fibre diameter as well as an increase in the crystallinity.

Morphological analysis by SEM and SPM

Figure 5 shows the SEM micrograph of the sisal fibre at different processing stages. Figure 5 (A) shows the SEM micrograph of the untreated raw sisal fibre. The fibre surface appears to be smooth due to the presence of waxes and oil. However, the presence of pores could be observed on the surface. Though sisal fibre has high lignin content (49.2), the moisture absorption capability is comparable (9.2%) with the other fibres. The hemicellulose and cellulose are the main moisture absorbing components in the lignocellulosic fibres [13]. The presence of these pores is responsible for the high value of moisture content in the raw sisal fibre. Figure 5 (B) is the alkali treated fibre. The arrangement of cellulose fibres within the matrix of lignin is clearly seen in this picture. The removal of cementing materials, primarily lignin, from the surface of the raw fibre occurs during this step. Figure 5 (C) represents the SEM of fibres subjected to steam treatment after alkali treatment. During the steam explosion with alkali at a high temperature, the hemicellulose is hydrolyzed and become water soluble. The steam explosion facilitates the ease of access of alkali to the interior part of the fibre. The clear demonstration of the defibrillation and depolymerisation by steam explosion is seen in this figure. Figure 5 (D) shows the SEM micrograph of the bleached sisal fibre. The lignin gets depolymerized at this step. As a result, defibrillation of the fibre occurs because of the removal of the cementing materials which can be seen from the SEM micro photograph.
Bleached fibre is composed of several microfibrils with diameters in the range of 3–12μm. Each elementary fibre shows a compact structure; exhibiting an alignment in the fibre axis direction with some non-fibrous components in the fibre surface [14]. It was previously shown that during the chemical treatment (alkalization) most of the lignin and hemicellulose were removed. Mechanical treatment (Steam explosion) further removed the amorphous materials (lignin, hemicellulose, tannin, pectin, etc.) from the inner part of the fibre via depolymerisation and defibrillation. Lignin is rapidly oxidized by bleaching agent. Lignin oxidation leads to lignin degradation and thereby to the formation of hydroxyl, carbonyl and carboxylic groups, which facilitate the lignin solubilisation in an alkaline medium [15].

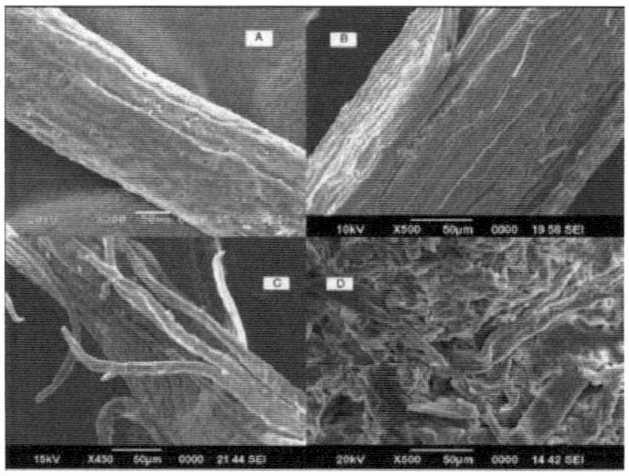

Figure 5 SEM pictures: **(A)**. Raw sisal fibre, **(B)**. Alkali treated sisal fibre, **(C)**.Steam exploded sisal fibre and **(D)**. Bleached sisal fibre

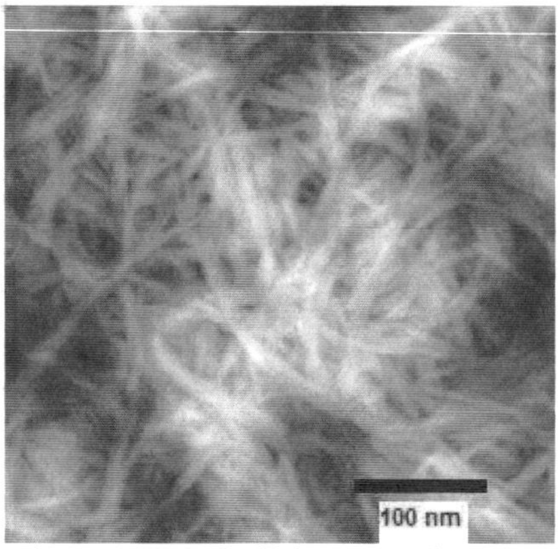

Figure 6 SPM of cellulose nano dispersion

Fibre diameter was again reduced in the acid hydrolysis followed by steam explosion and pure cellulose fibre with a diameter of less than 100 nm was obtained. Thus cellulose microfibrils of the original fibres were separated from each other to produce fibrils with diameters around 5–50 nm. The scanning probe microscopy (SPM) image of the final nanocellulose is shown in Figure 6. The nanofibres are separated since the sample was subjected to sonication just before the experiment was conducted to avoid the agglomeration. The fibres are dispersed randomly and are white in colour. SPM pictures show the final cellulose nano fibrils with diameters of 5–50 nm which supports the evidence for the isolation of individual nanofibres from raw sisal fibre.

Dynamic Light Scattering (DLS) and Zeta potential

For particle size distribution, the water suspended nanocellulose particles were diluted appropriately and analyzed in DLS particle size analyzer. The intensity of light scattered in a particular direction by dispersed nanocellulose tends to periodically change with time. These fluctuations in the intensity vs. time profile are caused by the constant changing of nanocellulose positions brought on by Brownian motion. DLS instruments obtain, from the intensity counts vs. time profile, a correlation function.

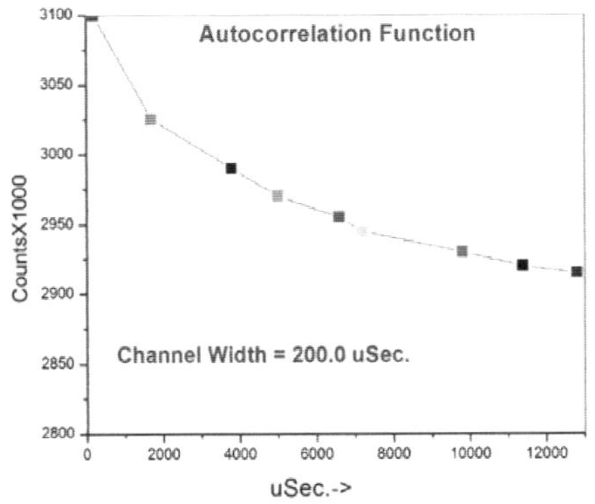

Figure 7 Autocorrelation function of DLS analysis

This exponentially decaying correlation function is analyzed for characteristic decay times, which are related to diffusion coefficients and then by the Stokes–Einstein equation, to a particle radius. The physical mechanism that is used to stabilize most aqueous nanocellulose particles systems is electrostatic repulsion. The correlation function of water suspended nanocellulose particles is given in Figure 7 while the Gaussian size distribution is given in Figure 8. Both the results confirmed that the dimensions of the nanocellulose particles are in the range of 4-100 nm with an average value of 37.8 nm. The standard deviation for the particle sizing is 2.1 nm (5.5%) with a *Chi Squared* of 2.114 for Gaussian summary. The linearity in particle size reduction is observed for dilute solution. The increase in the concentration of the suspension resulted in partial agglomeration. The colloidal nanocellulose particles of interest are charged, resulting in their mutual repulsion at extended distances. Ideally, the repulsive forces are sufficiently strong to prevent the nanocelluloses from diffusing close to each other, where short-range van der Waals attractive forces dominate and lead to agglomeration.

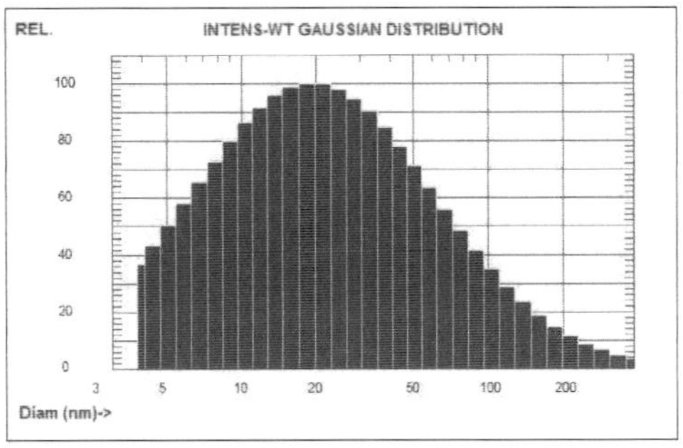

Figure 8 Gaussian distribution curve of nanocellulose fibre obtained from sisal fibre

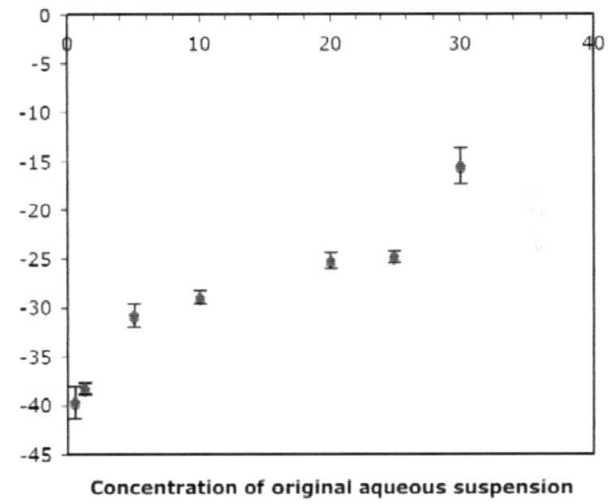

Figure 9 Zeta potential as a function of concentration (%) of sisal nanocellulose suspension in 0.1 mM KCl electrolyte

Zeta potential is derived from measuring the mobility distribution of the dispersion of charged nanocellulose particles as they are subjected to an electric field. In the present study, the average zeta potential of nanocellulose prepared by acid hydrolysis is −18.3mV. Zeta potential measured as function of concentration of original aqueous sisal nanocellulose suspension in 0.1 mM KCl electrolyte is shown in Figure 9. From the experiments it has been found that using a concentration of between 10 and 20% appears most promising for measuring zeta potential *f* (pH) tests. The negative charge on the surface of nanocellulose prepared by mild oxalic acid hydrolysis indicates the attachment of oxalate groups on its surface. The cellulosic surfaces generally show bipolar character with prevalent acidic contribution due to the proton of the hydroxyl functional group as well as of present carboxyl groups from oxalic acid.

Thermal studies of the fibres

Thermal decomposition parameters were determined from the TG, DTG curves as described below at a heating rate of 10°C/min. The TG and DTG curves of the raw sisal fibre and the nanocellulose extracted from it are shown in Figures 10 and 11 respectively. In the present study, decomposition of the untreated sisal fibre shows several stages, indicating the presence of different components that decompose at different temperatures. In both cases, a small weight loss was found in the range of 40–120°C due to the evaporation of water or low molecular weight compounds. In the case of cellulose nanofibers, the moisture-loss peak shifted to a lower temperature.

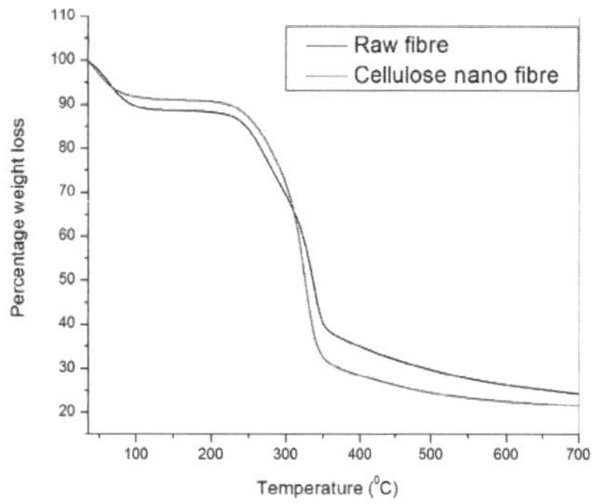

Figure 10 TGA of raw sisal and nanocellulose fibre

This tendency towards releasing moisture at a lower temperature might be caused by an increase in the surface area of the split fibers, facilitating easier evaporation of moisture at a lower temperature. The decrease in percentage moisture loss in the nanocellulose could be the effect of increased crystallinity, as observed with X-ray diffraction. Due to the differences in the chemical structures between lignin, cellulose and hemicellulose they usually decompose at different temperatures. The TGA shows the sisal fibre with the degradation of lignin, its major cementing material, at a temperature of 270°C and the α-cellulose component degrades at 335°C.

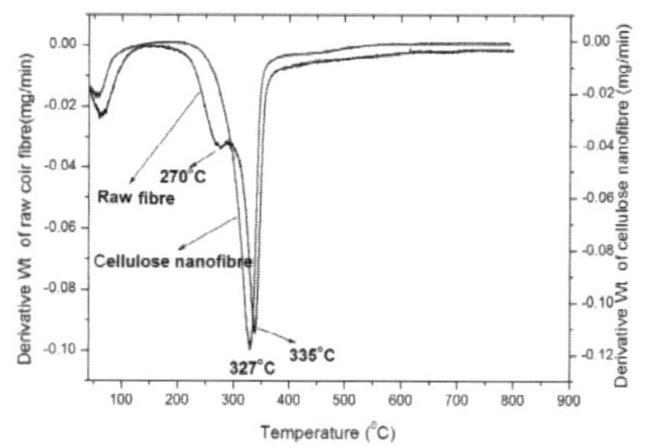

Figure 11 DTG of raw sisal and nanocellulose fibre

Hemicellulose and lignin are thermally less stable than cellulose whose decomposition start at 310°C and persist until 400°C [10]. It has been observed that the percentage of the solid residue left after the degradation of the raw sisal is higher than he percentage weight of the solid residue left after the pyrolysis of the nanocellulose which is extracted from sisal fibre. Degree of polymerisation of hemicellulose and lignin are very low when compared with cellulose. The higher amount of residual char or raw sisal is due to the strong cellulose-lignin complex and the aromatic nature of the lignin complexes. Lignin starts its decomposition at 220°C and continues up to 300°C. The decomposition peak reach the maximum mass loss (30 wt. %) at 270°C, showing a 65 wt. % of solid residuals at 300°C. Actually, the lignin decomposition extend to the whole temperature range, starting well below 200°C and persisting above 500°C. The wide temperature range observed during lignin decomposition is due to the different activities of the chemical bonds present on its structure [8]. Here the degradation of extracted nanocellulose has a lower value, 327°C, than the α-cellulose embedded in the raw sisal fibre (335°C). This reduction in the degradation temperature is due to the presence of strong lignin-cellulose complex which exist in raw sisal fibre. Raw sisal fibre has a lignin content of 45-50% of weight. Lignin has an aromatic skeletal structure and the cellulose and lignin are packed in the raw sisal as a stable lignin-cellulose complex which is more thermally stable than pure α-cellulose.

This is reflected in the increased amount of residual char. This lignin-cellulose complex can be believed to protect the α-cellulose from degradation to a considerable extent. The degree of polymerization of the extracted nanocellulose has a lower value due to the low cellulose content of the fibre which also results in decreased degradation temperature of the extracted fibre. An increase in residual char formation and lowering of the degradation temperature during the pyrolysis of treated cotton has also been reported [14-17]. It might be caused by an increased rate of formation of free radicals that are stabilized by condensed carbon ring formations in the char [17]. The percentage degradation remain almost the same for the α-cellulose but decrease for lignin, confirming the observation that >45% lignin was removed by the treatment. At 400°C almost all cellulose was pyrolyzed, and the solid residual char for nanocellulose is 22.53 wt. % at 600°C. This value of residual char is little lower than the total mass of cellulose and lignin in the raw sisal (26.29 wt. %). The residual char left at 600°C decreased considerably from 26.29% to 22.53% in the case of nanofibres. A similar observation was reported by Saha et al. [18] who explained that alkali treatment followed by steam explosion removed the lignin to a considerable extent, reduced the strength of lignin–cellulose complex, thereby making the product less stable than the raw sample, and this was reflected in the decreased amount of residual char. Thus generally nanocellulose obtained from sisal fibre has lower thermal stability than the α-cellulose present in the untreated sisal fibres. In addition, they showed lower amounts of residual solids.

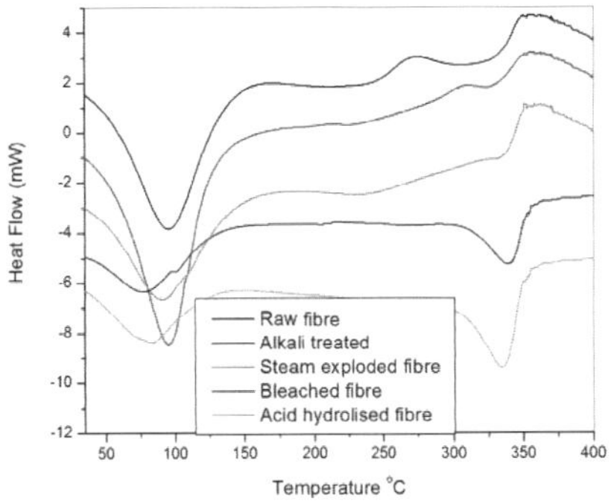

Figure 12 DSC analysis of sisal fibre at various processing stages

In the first half of the degradation process (30-300°C) nanocellulose shows higher thermal stability than raw fibre and in the second half (300-600°C) later has higher thermal stability. An increase in residual char formation and the degradation temperature for raw sisal during the pyrolysis may be due to the increased rate of formation of free radicals at higher temperature that are stabilized by condensed carbon ring formations in the char [19]. Nanocellulose extracted from raw fibre shows reduced pyrolysis temperature, increased weight loss during pyrolysis, and reduced residual char formation with a lowering in the formation of the flammable volatiles, and thus it can be easily concluded that the thermal stability of α-cellulose embedded in raw sisal shows increased thermal stability than the α-cellulose extracted from it.

The DSC analyses of the raw sisal fibre at various processing stages are shown in Figure 12. An endothermic peak below 100°C was observed for all the fibres, resulting from moisture loss. The raw fibre shows endothermic peaks at 254.9°C and at 366.4°C caused by the lignin and cellulose decomposition, respectively [18, 20, 21] In case of the alkali treated fibre, the lignin decomposition peak is diminishing and the cellulose decomposition peak remains unchanged. The alkali treated followed by steam exploded fibre shows a decreased value for the cellulose decomposition peak at 356.4°C and the lignin decomposition peak again shifted to lower temperature. The intensity of the lignin decomposition peak shows a gradual decrease. For the bleached and acid hydrolysed fibre, the cellulose decomposition peak is gradually shifted to lower temperature. The intensity of the endothermic peak of the cellulose decomposition is also increased because of the absence of the lignin-cellulose complex in the bleached and acid hydrolysed samples. Another interesting observation is the presence of an exothermic peak at 370°C in the raw, alkali treated and steam exploded fibres where the lignin-cellulose complex is existed. The destruction of the lignin-cellulose leads to the liberation of heat which is responsible for this peak. The corresponding exothermic peak is completely absent in bleached and acid hydrolysed fibres. It is, therefore, apparent that the removal of the non-cellulosic constituents, such as lignin and hemicellulose during alkali and bleaching treatment, leads to the destruction of the chemical linkages between the constituents. This might have some influence in the endothermic α-cellulose degradation peak followed by an exothermic degradation. The presence of lignin-cellulose complex is the key player of this peak. In the case of acetylation or cyanoethylation, where only the hydroxyl groups took part in the chemical reaction, no such peak was observed [22, 23]. The enthalpy associated with moisture absorption, α-cellulose, and lignin decomposition showed an increasing trend up to alkali treatment but decreased for the further steps of the process. The enthalpy associated with raw fibre is responsible for the strong lignin-cellulose complex and this enhanced enthalpy retained in the fibre structure until the complex structure degraded by alkali treatment and bleaching steps. In addition, increase in the enthalpy of the alkali treated and steam exploded fibres are again facilitated by the closer packing of the cellulose chains, resulting in increased crystallinity of the fibres, as observed from X-ray diffraction [24]. The increased H-bonding between the closely packed cellulose chains tends to stabilize the structure and increase the enthalpy values. Basch and Lewin [25] and Leao et al. [26], reported a similar observation in the case of cotton and coir fibre respectivelly. They also concluded that a correlation might exist between the DSC peak areas and the percentage crystallinity of the fibers. They observed that the fine structure of cellulose had considerable influence on its pyrolytic behaviour and that the stabilizing effect of crystallinity was clearly evident in the DSC behaviour of the fibers. Even though nanocellulose has highest crystallinity and closely packed arrangements, the absence of lignin-cellulose complex dominate over the other properties which facilitate the enhancement in enthalpy change.

Conclusions

The objective of this work was to study the effect of steam explosion on the structure and the thermal properties of sisal fibre and thereby enhance its effective utilization and to avoid the retting process of the sisal fibre. Steam explosion has been found to be successful in obtaining fibres in the nano dimension from raw sisal fibres.

Morphological and dynamic light scattering analyses of the fibres at different processing stages revealed that the gradual removal of the cementing materials and isolation of cellulose nano fibres occurs in the final step of the process as an aqueous suspension. FT-IR and XRD analysis revealed that the pre-treatments led to the removal of lignin and gradual increase in the crystallinity of the fibre at each processing stages. The presence of lignin-cellulose complex in the raw sisal fibre plays a major role in its structural and thermal properties. The thermal stability of the extracted nanocellulose has a lower value than the nanocellulose present in the raw sisal fibre because of the presence of this strong lignin-cellulose complex in raw fibre. Homogenous cellulose nanofibrils with a diameter of 5-50 nm are obtained from sisal fibre by this process.

Since the pure cellulose has a variety of properties, its nano form can find applications in various industrial and biomedical fields by exploring the nanotechnological possibilities.

The produced nanocellulose and its composites confirmed to be a very versatile material having the wide range of medical applications, including scaffolds for tissue engineering, cardiovascular implants and vascular grafts. These implants were produced from bioresorbable and/or biodegradable materials. Progressive degradation of the implant material may then be accompanied by the formation of the new tissues. The developed material can also be utilized for construction of non latex condoms, breathable wound dressing, surgical gloves, surgical gowns or drapes, medical bags, organ retrieval bags and medical disposables.

Acknowledgements

The authors are grateful to the National Council of Technological and Scientific Development (CNPq), Foundation for Research Support of São Paulo (FAPESP) Government of Brazil and International Science and Technology Partnerships Canada Inc (ISTPCanada) for the financial funding of this research work.

References

(1) Mohanty, A. K.; Misra, M.; Hinrichsen, G. *Macro. Mat. & Eng.* **2000**, *276/277*, 1–24
(2) Itoh, T.; Brown, R. M. Jr.; *J. of Cell. Biol.* **1984,** *160*, 372–81
(3) Hon, D. N.-S.; D.N.-S.; *Cellulose* **1994,** *1*, 1.
(4) Geethamma, V. G.; Thomas Mathew, K.; Lakshminarayanan, R.; Thomas, S. *Polymer* **1998**, *39*, 1483-1491
(5) Dieter Klemm, D.; Schumann, D.; Kramer, F.; Hebler, N.; Hornung, M.; Schmauder, H. P.; Marsch, S. *Adv. Polym. Sci.* **2006,** *205*, 49–96
(6) Erdtman, H. *J. of Poly. Scie. Part B: Polymer Letters*, **2003,** *10*, 228 - 230
(7) Klemm, D.; Philipp, B.; Heinze, T.; Heinze, U.; Wagenknecht, W.; *Comprehensive Cellulose Chemistry, Volume 1,* Wiley-VCH Verlag GmbH, 2004; Section 2.4–2.4.3.
(8) Oh, S. Y.; Yoo, D. I.; Shin, Y.; Seo, G. *Carbohyd. Res.* **2005,** *340*, 417–428
(9) Reddy, N.; Yang, Y. *Polymer* **2005**, *46*, 5494–5500
(10) Lojewska, J.; Miskowiec, P.; Lojewski, T.; Pronienwicz, L. M. *Poly. Degra. & Stability* **2005**, *88*, 512–520
(11) Mwaikambo, Y.; Ansell, M. P. *Angew. Makromol. Chem.* **1999**, *272*, 108
(12) Nishiyama, Y.; Langan, P.; Chanzy, H. *Journal of American Chemical Society* **2002**, *124*, 9074–9082
(13) Kostic, M. M.; Pejic, B. M.; Asanovic, K. A.; Aleksic, V. M.; Skundric, P.D. *Industrial Crops and Products* **2010,** *32*, 169-174
(14) Garcıa-Jaldon, G.; Dupeyre, D.; Vignon, M.R *Bioma. Bioenerg.* **1998**, *14*, 251–260

(15) Lawther, J. M.; Sun, R. *Industrial Crops and Products* **1996,** *5,* 87-95
(16) Sherely Annie Paul.; Doreen Piasta.; Stefan Spange.; Laly A. Pothan.; Sabu Thomas.; Cornelia Bellmann. *Biomacromolecules* **2008,** *9*, 1802–1810.
(17) Hinojosa, O.; Arthur, J. C.; Mares. T. *Tex Res J.* **1973**, *43*, 609
(18) Saha, S. C.; Ray, P. K.; Pandey, S. N.; Goswamy, K. *J Polym Sci.***1991**, *42*, 2767
(19) Jawaid, M.; Abdul Khalil, H. P. S. *Carbohydrate Polymers* **2011,** *88,* 1-18
(20) Basch, A.; Lewin, M. *J Polym Sci Polym Chem.* **1973**, *11*, 3095
(21) Ray, D.; Sarkar, B. K. *J Appl Polym Sci.* **2001**, *80*, 1013
(22) Mitra, B. C.; Basak, R. K.; Sarkar, M. *J Appl Polym Sci.* **1998**, *67*, 1093
(23) Sikdar, B.; Basak, R. K.; Mitra, B. C. *J Appl Polym Sci.* **1995**, *55*, 1673
(24) Ray, D.; Sarkar, B. K. *J Appl Polym Sci.* **2001**, *80*, 1013.
(25) Basch, A.; Lewin, M. *J Polym Sci Polym Chem.* **1973**, *11*, 3095.
(26) Leao, A. L.; Cherian, B. M.; Souza, S. F.; Narine, S. 20120. Coir Nanocellulose for High-End Applications. In: International Coir Technology Expo 2012, Kochi, India, 2012.

O-11-111

MODIFICATION OF ADHESION BETWEEN BANANA FIBER AND POLY(BUTYLENE SUCCINATE)

Tetsuto Kajiyama[1]* and Takeshi Yasuda[2]

[1] *Research and Development Department, Tokyo Metropolitan Industrial Technology Research Institute, 2-4-10, Aomi, Koto-ku, Tokyo, 135-0064, Japan – kajiyama.tetsuto@iri-tokyo.jp*

[2] *Tama Techno Plaza, Tokyo Metropolitan Industrial Technology Research Institute, 3-6-1, Azumacho, Akishima-city, Tokyo, 196-0033, Japan – yasuda.takeshi@iri-tokyo.jp*

Abstract - We prepared banana fiber (BF)/poly(butylene succinate) (PBS) composites using hexamethylene diisocyanate (HDI) as a compatibilizer. Effect of concentration of NaOH aq on tensile strength and elongation at break, effect of amount of HDI on tensile strength and elongation at break of BF/PBS/HDI and BF-A/PBS/HDI composites were studied. When contents of HDI in composites were increased, tensile strength increased. Although the mechanical properties of PBS were first degraded with the addition of BF, alkaline treatment to BF and the addition of HDI improved the mechanical properties. This probably due to the promotion of the adhesion between BF and PBS. The surface morphology of the fracture surfaces of the composites were observed with a scanning electron microscope (SEM).

Keywords: Banana fiber, Poly(butylene succinate), Compatibilizer

Introduction

Recently, many kinds of biodegradable polymers have been developed. Moreover, a number of studies on the composites using these polymers with natural fibers, including wood, kenaf, jute fibers and starch have been reported[1]. On the other hand, composite materials usually lack interfacial adhesion. Therefore, it is important to incease the interfacial adhesion of the composites, and consequently improve mechanical properties of the composites.

In the present study, banana fiber (BF)/PBS composites using hexamethylene diisocyanate (HDI) as the reactive compatibilizer were investigated.

Materials and Experiments

Banana fibers were from the Republic of the Philippines. Two kinds of banana fibers were used, unprocessed banana fiber (BF) and alkaline treated banana fiber (BF-A). PBS (Bionolle #1020, Showa Denko, Co., Ltd., Tokyo, Japan), HDI (Kanto Chemical, Co., Ltd., Tokyo, Japan) were purchased and used without further purification.

100 g of Banana fiber was treated with 3000 ml of 10, 20 wt% NaOH aqueous solution (aq), respectively at room temperature for 2 hour. The resulting alkaline treated banana fiber (BF-A-10, 20) was washed with water and then dried overnight in a vacuum oven at 70 °C. The BF and the BF-As were cut into the length of 5 mm. BF/PBS composites were made by applying a batch type twin-kneader (IMC-188B, Imoto seisaku-sho, Co., Ltd.) at 130 °C. In BFs of composites (20 wt% BF/80 wt% PBS/1 per hundred resin (phr) HDI) of Run 1, BF, 10 wt% and 20 wt% alkaline treated banana fiber were used. And blend time of Run 1 was 10 min. In Run 2, 1, 2.5, and 5 phr HDI were added to composites (20 wt% BF/80 wt% PBS). 20 wt% alkaline treated banana fiber was used, and the blend time of Run 2 was 10 min. Specimens of pure PBS and the composites for mechanical test were prepared by a injection molding machine (BIRMINGHAM, Mining and Chemical Co., Ltd.) at 130 °C. Specimens of PBS and the composites were assessed by a tensile tester (Tensilon RTC-1325, Orientec, Co., Ltd.). The surface morphology of the banana fibers and the fracture surfaces of the composites were observed with a scanning electron microscope (SEM, VE-9800, Keyence, Co., Ltd.).

Results and Discussion

Fig. 1 and Fig. 2 show the effect of concentration of NaOH aq on tensile strength and elongation at break, respectively. Tensile strength and elongation at break of PBS and BF-A-20/PBS/No-HDI composite are shown in Fig. 1 and Fig. 2, respectively. It is found that tensile strength of the composites and PBS ranged from 32.2 MPa to 42.9 Mpa and elongation at break of composites and PBS ranged from 22.1 % to 61.9 %. Tensile strengths of composites containing BF-A-10, 20 were higher than those of PBS and composite containing BF. And, elongation at break of BF/PBS composites were lower than PBS.

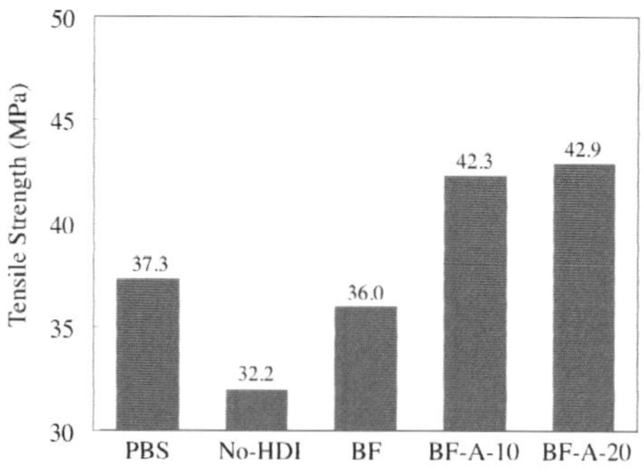

Fig. 1 Effect of concentration of NaOH aq on tensile strength.

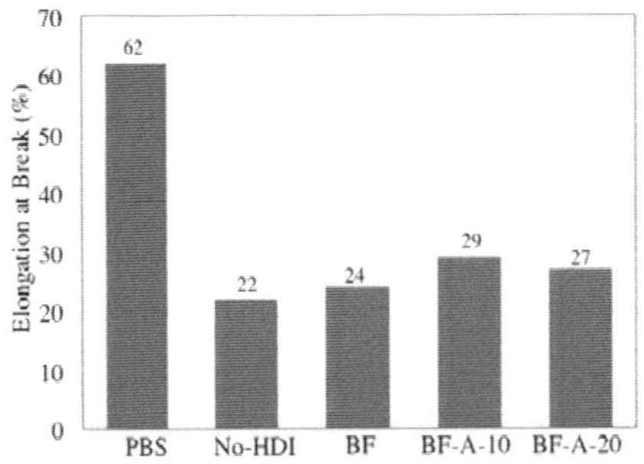

Fig. 2 Effect of concentration of NaOH aq on elongation at break.

As shown in Fig. 1 an addition of BF to PBS decreased the tensile strength probably due to insufficient adhesion between these materials. An addition of HDI improved the tensile strength of the composite, but it was still lower than that of PBS. In case of using BF after alkaline treatment, however, the mechanical properties of the PBS were improved, showing that BF-A could surely enforce the strength of PBS.
Fig. 3 and Fig. 4 show the effect of amount of HDI on tensile strength and elongation at break, respectively. BF-A-20 was added to the composites. Tensile strength and elongation at break of PBS and BF-A-20/PBS/No-HDI composite are shown in Fig. 3 and Fig. 4, respectively.

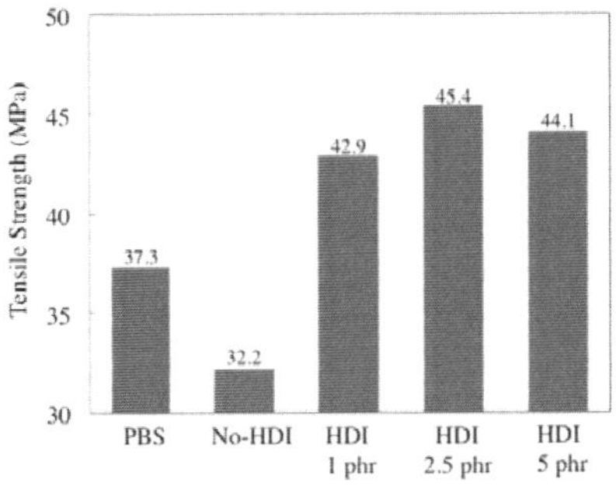

Fig. 3 Effect of addition of HDI on the tensile strength of BF-A-20/PBS/HDI composites.

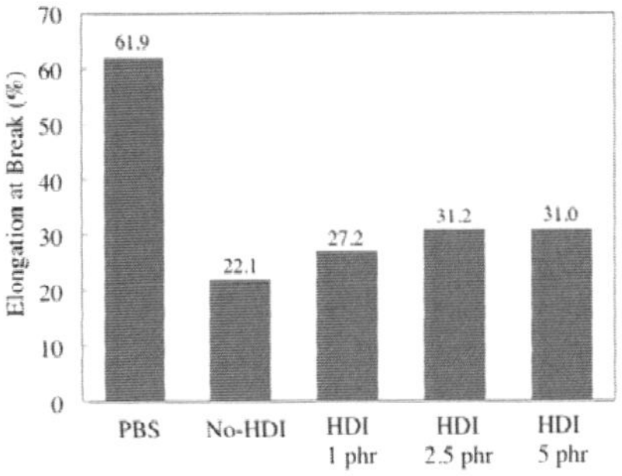

Fig. 4 Effect of addition of HDI on the elongation at break of BF-A-20/PBS/HDI composites.

It is found that tensile strength of the composites and PBS ranged from 32.2 MPa to 45.4 MPa, and elongation at break of composites and PBS ranged from 22.1 % to 61.9 %. Tensile strength of 1.0 phr HDI composite was higher than that of PBS and No-HDI. Tensile strength increased when the amount of HDI was raised to 2.5 phr, but decreased when the amount of HDI reached 5 phr. In addition, elongation at break of BF/PBS composites were lower than PBS. It can be considered that the two isocyanate groups (-NCO) of HDI reacted with the end hydroxy group (-OH) of PBS and/or hydroxy groups of banana fiber (Fig. 5).

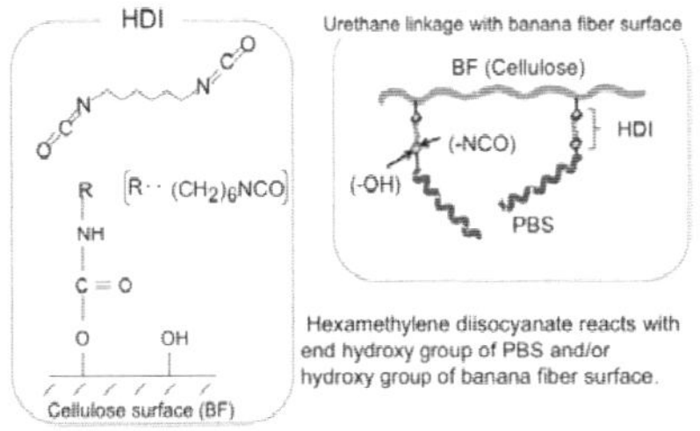

Fig. 5 Reaction scheme of BF and PBS with HDI

Fig. 6 shows SEM images of broken-out sections of banana fiber in BF-A-20/PBS/non-HDI composite and BF-A-20/PBS/1phr HDI composite after the izod impact test.

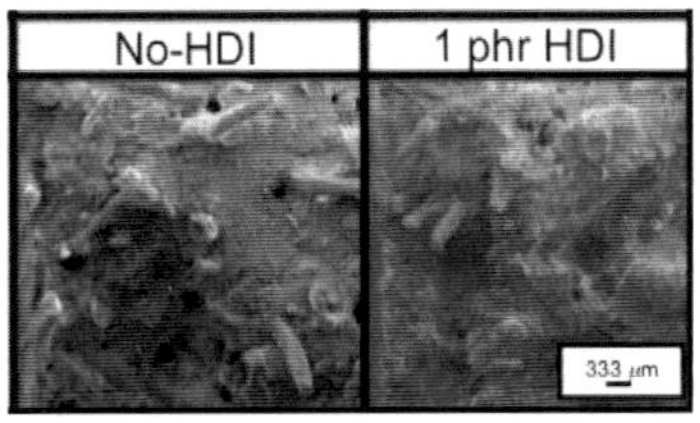

Fig. 6 SEM images of the fracture surfaces for BF-A-20/PBS composites with and without HDI

The images show that HDI improved adhesion between PBS and BF-A-20. From these SEM observations, higher mechanical properties can be expected from compatibilized composites based on BF-As with aliphatic polyesters. Banana fibers were exfoliated from the matrix in the sample without HDI, while such a exfoliation was not observed in the HDI-added sample, showing that HDI improved the adhesion between PBS and BFA-20.

Conclusions

Although the mechanical properties of PBS were first degraded with the addition of BF, alkaline treatment and the addition of HDI improved the mechanical properties. This probably due to the enhanced adhesion between BF and PBS. These results suggest that BF can be used as a reinforcement agent for plastics like PBS.

References

1. S. H. Lee; S. Wang *Composites Part A* 2006, *37*, 80.

O-11-393

THE EFFECT OF SONICATION PROCESS ON HIDROTHERMAL REACTION OF CHITOSAN

Emma Savitri*, Azra Yuliana, Linggar Septy, Anitarakhmi, Sumarno, and Achmad Roesyadi
Chemical Engineering Department, Sepuluh Nopember Institute of Technology, Indonesia – savitri_ma@yahoo.com

Abstract - Chitosan is a linier copolymer (1→4)–linked 2-acetamido-2-deoxy-β-D-glucan (GlcNAc) and 2-amino-2-deoxy-β-D-glucan (GlcN) units in varying proportions. It is considered to be the most widely distributed biopolymers, non toxic, reactive and biodegradable material. Because of the advantages of chitosan, it has received much attention as a functional biopolymer for many diverse applications. Considering that original chitosan has high molecular weight and strong intra and intermolecular hydrogen bonding, sonication becomes one of alternatives to modify its molecular weight and structure to achieve high conversion of hydrothermal reaction. In this study, chitosan was pre-treated by sonication before subjected to hydrothermal process. Chitosan was dispersed in water and sonicated for several periods (30 - 120 min) at temperature 60 °C. Chitosan solid residue was characterized by viscosimetry, X-ray diffraction, Fourier Transform Infra Red, and Scanning Electron Microscopy. After sonication, hydrothermal reaction was carried out in a batch reactor at temperature 200 °C and 250 °C, pressure of 25 MPa, and reaction time of 10 – 30 s. The effects of temperature and reaction time were observed. The degradation of chitosan was determined by total organic carbon and mass of residual solid. The experimental results suggested that there were alterations on molecular weight, degree of crystallinity, and morphology of degraded chitosan. The changes of chitosan molecular weight and structure after sonication give some influences on hydrothermal reaction.

Keywords: chitosan, sonication, hydrothermal

Introduction

Chitosan, (1→4)-2-acetamido-2-deoxy-β-D-glucan (GlcNAc) and 2-amino-2-deoxy-β-D-glucan (GlcN), is a natural copolymer generally produced by deacetylation of chitin isolated from crustacean shells. Chitosan and its derivatives have interesting properties such as non toxicity, biocompatibility, controllable biodegradability, and non antigenicity. These properties cause chitosan as an attractive and potensial biopolymer for wide area of applications such as biotechnology, pharmaceutical, wastewater treatment, cosmetics, agriculture, food science, and textiles [1].

Generally, chitosan obtained from the deacetylation of chitin has a high molecular weight and low solubility in water or neutral pH solution, which restricts its applications, especially in medicine, pharmacy and in the food industry. There are some requirements to be fulfilled to apply chitosan in medicine or pharmacy such as degree of deacetylation > 80 % and low degree of polymerization, since low molecular weight chitosan (LMWC) and chitooligomers (COS) are responsible for chitosan biological activity. Muzzarelli *et al* demonstrated that chitosan with an average molecular weight in the range of 1,000 – 10,000 Da is most interesting for a number of medical and biotechnological applications [2]. Moreover, low molecular weight of chitosan and chitosan oligomers might possess high solubility in water. Thus the development of efficient processes for reducing the molecular size of chitosan, without altering its chemical structure, are of great interest.

Like other polysaccharides, chitosan is susceptible to a variety of degradation techniques, including acid hydrolysis [3], oxidative degradation [4] and enzymatic methods [5]. Hydrothermal reaction has been attracting worldwide attention because of the fascinating characteristics of water as reaction medium at elevated temperatures and pressures [6]. The degradation process by hydrothermal reaction proceeds through a variety of mechanisms including hydrolysis, oxidation and gasification [7]. Hydrolysis refers to breaking up of the organic substances into smaller organic fragments in water. Hydrothermal decomposition also acts on the large organic molecules reducing them into smaller fragments, some of which dissolve in water. Hydrothermal reaction is an effective method for the degrading of organic subtances. Sonication is also particularly effective in splitting up aggregates, and reducing the size and polydispersity of nanoparticles. It is generally agreed that sonication causes main chain scissions at the 1,4-glycosidic bond [8] without affecting the DD of chitosan. The process has, therefore, been conveniently applied to produce low molecular weight of chitosan of on the same DD [9].

Recently, hydrothermal was used to degraded biomass such as cellulose, starch, protein, chitin, etc to produce many valuable products. Generally, the polysaccharida has strong intra and intermolecular hydrogen bonding and it might affect on hydrothermal process. Besides degrading, sonication also give alterations on physical and structural properties of chitosan. The changes on physical and structural properties of chitosan might make it easy to be hydrolized on hydrothermal process.

The main aim of this study is to explore the effect of sonication process on physical properties of chitosan and it might affect on the performance of hydrothermal process.

Experimental

Material used was medical grade chitosan and purchased from Biotech Surindo (Cirebon, Indonesia). Chitosan as initial material was used directly without any treatments. The degree of deacetylation (DD) and viscosity-average molecular weight (Mv) of chitosan are 80 % and 3,783 kDa.

All sonication experiments were conducted in a 400 mL pyrex glass as reactor at 60 °C and different time periods (30 – 120 min). Sonication was performed with a high-intensity ultrasonic processor VCX 500 Sonics and Materials Inc, USA (500 W, 20 kHz, 50 % Amplitude) equipped with Titanium Alloy probe transducer. The converter was made of piezoelectric lead zirconate titanate crystals. The reactor was immersed into a water bath equipped with circulating water thermostat model 9102 (PolyScience, USA). Treated samples were filtered and dried before subjected to FT-IR, X-RD, SEM and viscosimetry. The reactor for hydrothermal was a tube reactor (SAF 2507, 4.6-mm i.d. 35-mm length, 5.8-mL capacity) equipped with a thermocouple, valves, and a pressure gauge. 4 mL of 1 % sonicated slurry was charged into the reactor, and the atmosphere in the reactor was replaced with nitrogen gas. Then, the reactor was sealed after nitrogen gas was pressurized to the prescribed initial pressure. The reactor was heated in two steps using electrical heater which were heated at 180 °C and the prescribed reaction temperature 200 °C and 250 °C, respectively. After the prescribed reaction time (10, 20, 30 s), the reactor was immediately cooled in a water bath to quench the reaction.

FTIR 8400S Shimadzu, Japan was used to examine the signal variations of the degraded chitosan with the KBr (at ratio 1:100) disc technique. FTIR spectroscopic was conducted in range 4000 – 400 cm^{-1}.

X-ray diffraction (XRD) measurements were carried out on PANalytical PW 3373/00 X'Pert X-ray diffractometer, Netherland and used a CuKα = 1.54 at 40 kV-30 mA. The relative intensity was recorded in the scattering range (2θ) of 0 – 90°. The degree of crystallinity was calculated as the ratio of the integrals of the intensities of the crystalline component and that of the sample studied.

The morphology and topography of chitosan and degraded residue was observed under a SEM FEI INSPECT S-50, Netherland. The powder was dispersed onto double sided carbon tape and then placed on a sample holder.

The intrinsic viscosities were determined using an Ubbelohde viscosimeter by capillary viscometry Schott Gerate type 53210 (Germany) at 25 °C. Prior to measurement of flow times, dried chitosan residues were dissolved in 1 % v/v acetic acid solution at various concentration. The molecular weight, M_v was calculated according to Eq. 1:

$$[\eta] = KM_v^{\alpha} \quad (1)$$

The content of the total organic carbon (TOC) in the liquid phase was utilized to determine the amount of organic carbon dissolved after sonication. TOC was analyzed by a SHIMADZU-TOC-VCPN. TOC liquid were analyzed from the slurry after sonication. Before subjected to be analyzed, slurry was centrifuged and filtered to separate the solid residue. Mass of solid residues was recorded as unreacted chitosan

Results and Discussion

The FT-IR spectra of the original and degraded chitosan at 60 °C for an hour of sonication displays in Fig.1. The spectra was recorded to investigate any structural changes during the sonication process.

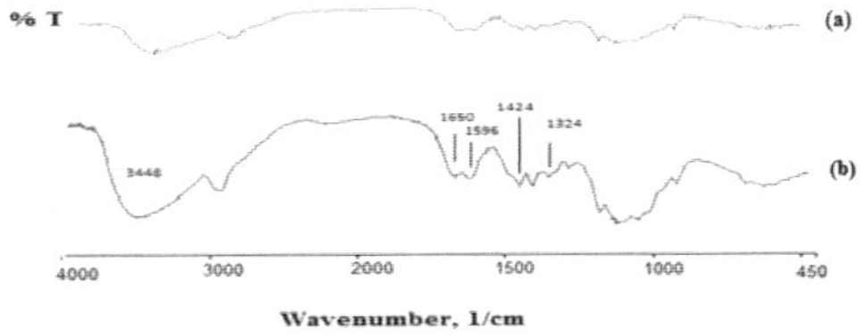

Figure 1 - FT-IR spectra of chitosan : (a) degraded chitosan by means of sonication for a hour at temperature 60 °C (b) original chitosan

As shown fig. 1, there were no significant differences in functional group between the residues of chitosan before and after sonication, unless there were reducing in absorption peaks in the range of spectra. The peak at 3340 – 3570 and 3230 - 3310 cm^{-1} was $-NH_2$ bond stretching overlapping with valence vibration of H-bonded O-H, indicating intermolecular and intramolecular hydrogen bonding of chitosan molecule. Those bands of parent chitosan had a strong intramolecular and intermolecular O-H band, but degraded chitosan at 60°C was weaker. This indicated that sonication gave significant alteration to chitosan structure by decreasing inter and intra molecular hydrogen bonding. The results showed that an increase in temperature gave a significant degradation effect by lowering hydrogen bonding of chitosan. Hydrogen bonding in chitosan is one of factors that is responsible on the strong chitosan structure. The results is suitable with Price *et al* (1992) that maximum temperature reached from cavitation process in sonication system is proportional to ambient temperature of sonication system. The higher temperature produced from cavitation give increasing degradation rate because of increasing radical formation that responsible on breaking glycosidic bond in chitosan [10].

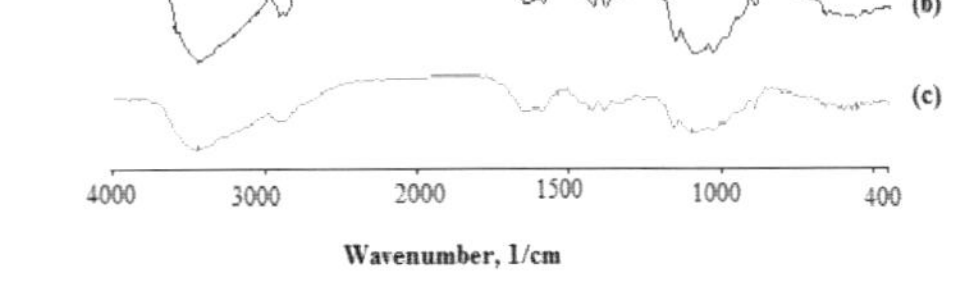

Figure 2 - FT-IR spectra of chitosan : Degraded chitosan by means of sonication at 60 °C for sonication time period (b) 30 min, (c) 60 min, and (d) 120 min

Fig. 2 shows the development of spectra during sonication time. The significant changes on the development of absorption peaks indicated that inter and intramolecular hydrogen bond and no new band bands in the spectra. Reducing inter and intramolecular hydrogen bonding because chitosan molecular chain degrade into small parts and it decreases hydrogen bond of chitosan. When chitosan is subjected to high frequency of sonication, it undergoes degradation (main chain scission). This manifests itself as a decrease in molecular weight. We followed degradation by direct determination of changes in molecular weight by viscosimetry. The molecular weight of initial chitosan was ca. 3,700 kDa.

The degradation of chitosan by sonication at 60°C is shown in Fig. 3. Within the investigated time range of sonication, the average viscosity molecular weight decreases with sonication time. When the initial chitosan was exposed to sonication, the molecular weight of chitosan was decreased to ca. 2,800 kDa for 120 min sonication.

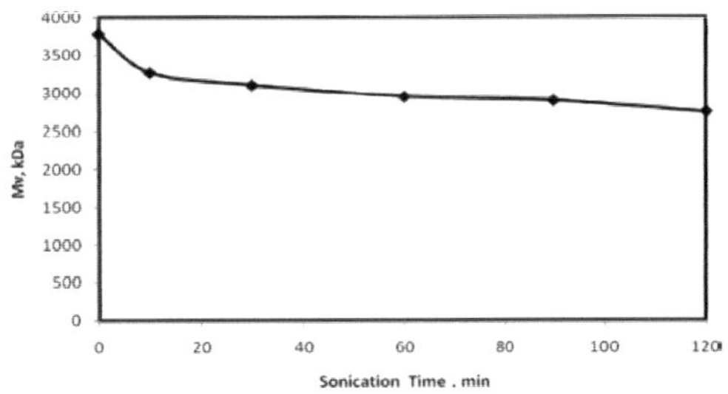

Figure 3 - The average viscosities molecular weight versus sonication time at 60 °C

This was fully understandable, since more of OH radicals were formed locally in the cavitation bubbles and present in the immediate vicinity of these bubbles at high concentrations contact with chitosan for longer time of sonication, and it would yield more scissions in chitosan molecules. But the reaction becomes gradually less efficient with sonication time, what is expected for mechanochemical effects. Only long chains are vulnerable to breakage by the action of hydrodynamic forces, while chains below some limiting critical size cannot be degraded in this way [11], this limiting size, presumably under pure mechanochemical conditions.

Fig. 4 illustrates the X-ray diffraction patterns of original and degraded chitosan prepared by ultrasonication at different sonication time. The X-ray diffraction pattern of parent chitosan showed its characteristic peaks at 2θ = 11.44°, 19.71°, and 29.51° which coincided with the pattern of the "L-2 polymorph" of chitosan reported previously [12]. Otherwise, degraded chitosans treated at 60 °C for 30 – 120 min sonication displayed the same peaks as those of original chitosan and also a new characteristic peak at 2θ = 21.5°. All the patterns of degraded chitosan are characterized a chitosan polymorph, which referred to as the "tendon hydrate polymorph" [13]. The intensity of their characteristic peaks at 2θ = 10.29° increased more than that of parent chitosan and the peaks around 2θ = 10 ° of degraded chitosans were shifted to the left that indicated the increase in d-spacing because of bounded water.

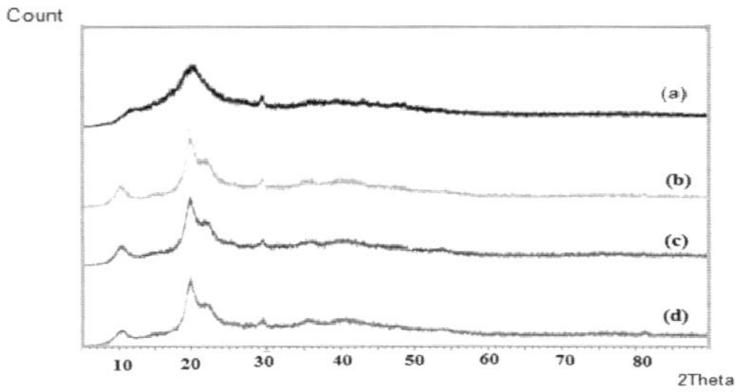

Figure 4 - X-ray diffraction pattern of (a) Original and Degraded Chitosan at 60 °C for : (b) 30 min, (c) 60 min, (d) 120 min

The reduction of crystallinity can be shown by fig. 5. The degree of crystallinity decrease during sonication process. Mechanochemical mechanism of sonication might break the regularity of chitosan structure and change the structure of chitosan into amorphous form. It can be validated by decreasing hydrogen bonding on FTIR spectra of degraded chitosan residue. The alteration of chitosan structure would be expected to facilitate the hydrothermal process run effectively.

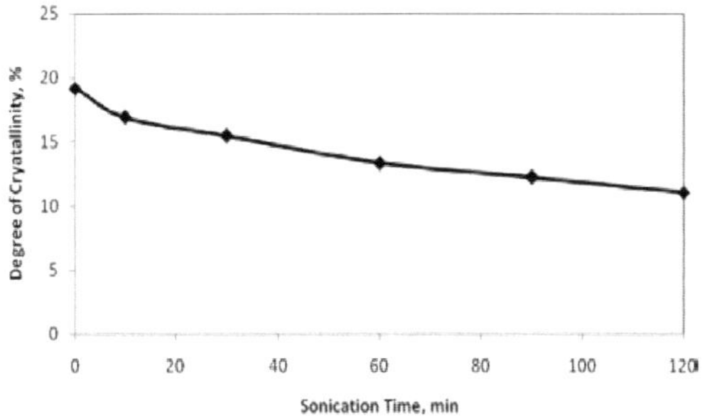

Figure 5 - The degree of crystallinity during sonication at 60 °C

In order to describe the phenomena by which the sonic waves affect the morphology and structure of chitosan, the degraded chitosan treated by sonication at 60 °C was displayed by fig.6. Figures were taken after sonication treatment for 30 min up to 120 min. Degraded chitosan up to 60 min sonication still showed compact structure with littele cleavege in the edge of particles. But fig.6c gave significant effect on the chitosan morphology. The image of particle suggests that depolymerisation of chitosan occurred via peeling off the surface layer by layer. It occurred because intermolecular hydrogen bondings of degraded the cleavages and chitosan structure. The phenomena assigned that sonication is combination mechanochemical processes. Chemical effect of sonication will break down the weakest linkage in the molecule, in this case the end glycosidic linkages is degraded faster than other glycosidic linkage.

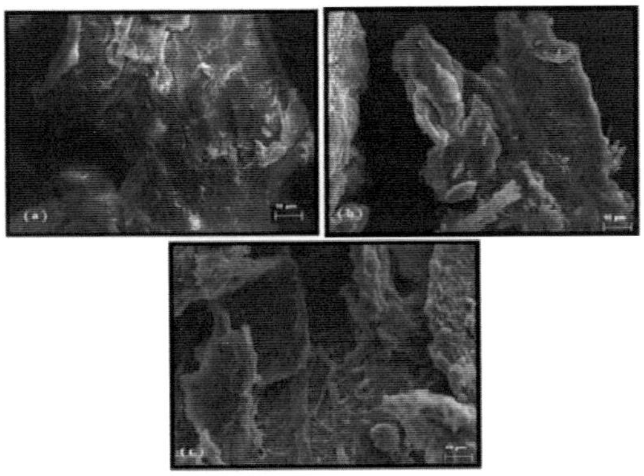

Figure 6 - SEM micrographs of degraded chitosan at 60 °C using magnification of 5,000 (a) after 30 min, (b) after 60 min, and (c) after 120 min sonication. Scale bars : 10 µm

On mechanical aspect, cavity implosion generates a powerful liquid jet when the cavity is formed near a solid surface and create some erosion on the surface especially on the edge of solid. Stephanis, C.G., et.al (1997) had reported that the cause of the erosion is due to the jets created by the assymmetric implosion of cavities, especially those adjacent to the plate and with a radius smaller than 4 µm [14].

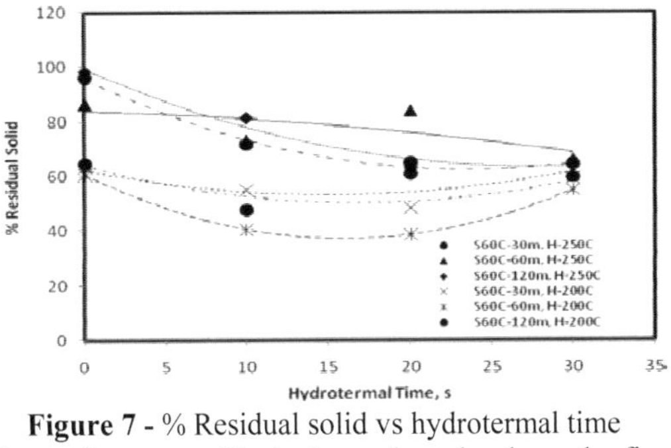

Figure 7 - % Residual solid vs hydrotermal time

The performance of hydrothermal can be shown by fig. 7 and 8. From the figure 7, it can be described that along hydrothermal time, solid residue of chitosan decrease and it meant that chitosan had been converted into dissolved substance. Figure 7 also shows that % residual solids at 250 °C were higher than those of 200 °C. It might describe that at higher temperature, depolymerization of chitosan proceed slower than that of 200 °C. It was appropriate with Arora, et al (2011) that decomposition of chitosan is exothermic event [15].

Figure 8 ilustrates the dissolved substances as a result of hydrolysis process. According to analysis, it could be seen that dissolved substances of hydrothermal was in the range 30 – 80 ppm and the experiment of S60C-120m,H-200C had produced relatively good product compared to others. Unfortunately, the effect of the alteration structure of degraded chitosan to hydrothermal product has been random and still not clearly seen. We plan to conduct further analysis of dissolved substance to explore deeper in the hydrothermal depolymerization of chitosan.

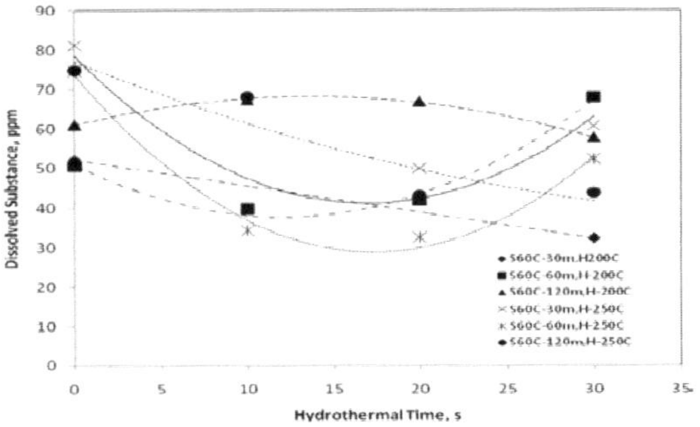

Figure 8 – Dissolved substance vs hydrothermal time

Conclusion

Sonication gave some effects on the alteration of physical and structural of chitosan such as the changes in hydrogen bonding, viscosimetric average molecular weight, degree of crystallynity and morphology of chitosan. Sonication at 60 °C for 120 min gave significant alteration on degraded chitosan and this condition at hydrothermal 200 °C gave relatively higher dissolved substance

Acknowledgements

Financial support from University of Surabaya is gratefully acknowledged.

References

1. Q. Li; E.T. Dunn; E.W. Grandmaison; M.F.A. Goosen in *Applications of Chitin and Chitosan*, M.F.A. Goosen Ed.; Technomic Publishing, Lancaster, 1997, 3–29.
2. R.A. Muzzarelli; M. Tomasetti; P. Iiari *Enzyme Microb. Tech.*1994, *16*, 110
3. K.M., Varum; M.H.Ottoy; O. Smidsrod *Carbohydr. Polym.*,2001, *46*, 89
4. G.G. Allan; M. Peyron *Carbohydr. Res.*1995, *277*, 257
5. W. Xia, P.Liu, J. Liu. *Bioresource Tech., 2008, 99*, 6751
6. P.E. Savage in *Chem Res.* 1999, *9*, 603
7. S. Jomaa, A. Shanableh, W. Khalil, B. Trebilco Adv. Environ. Res. 2003, *7*, 647.
8. R.H. Chen; J.R. Chang; J.S. Shyur *Carbohydr Res.*1997, *299*, 287
9. M.L.Tsaih; R.H.Chen *J.Applied Polym.Sci.* 2003, *90*, 3526
10. V.K. Mourya; N.N. Inamdar; Y.M. Choudhari *Nat.Polym.*, 2011, *53*, 583
11. G.J. Price; P.F. Smith *Polymer* 1993, *34*, 4111
12. R.H. Chen; J.S. Chen *Adv. Chitin Sci.* 2000, *4*, 361
13. E. Belamie; A. Domard; M. Giraud-Guille, *J. Polym. Sci. Part A : Polym. Chem.* 1997, *35*, 3181

14. F.M. Goycoole; W. ArgOelles-Mona; C. Peniche I. Higuera-Ciapara, *Novel Macromol. Food Systems*, 2000, 265-308
15. S. Aroraa; S. Lal; S. Kumar;M. Kumar; M. Kumar Scholars Research Library, 2011, *3*, 188

O-11-450

IMPACT MODIFICATION OF POLY (LACTIC ACID) BY USING MODIFIED NATURAL RUBBER

W. Chareonsalung[1], N. Euaphantasate[2] and C. Pattamaprom[1*]

[1]*Deparment of Chemical Engineering, Faculty of Engineering, Thammasat University, Thailand - chawilaiporn@gmail.com; cattalee@engr.tu.ac.th*

[2]*National Metal and Materials Technology Center, Pathum Thani 12120, Thailand - nukule@mtec.or.th.*

Abstract - Poly (lactic acid) (PLA) is one of the biodegradable thermoplastic polyester which is derived from renewable plant sources. PLA exhibits many excellent physical and mechanical properties such as good transparency and high strength. Nevertheless, the major disadvantages of PLA include low toughness and slow crystallization. This study focused on enhancement of impact resistance by blending with 10wt% natural rubber(NR). Two types of modified natural rubber including maleic anhydride grafted natural rubber (NR-g-MA) and poly (methy methacrylate) grafted natural rubber (NR-g-PMMA) were used in this study. The effects of rubber modifications was investigated for morphology and impact resistance. It was found that all the blends showed higher impact resistance than neat PLA and that unmodified NR possessed the highest efficiency in toughening PLA than other modified rubbers.

Keywords: poly (lactic acid), natural rubber, modified natural rubber, impact resistance

Introduction

Polylactide or poly (lactic acid) (PLA) is one of the biodegradable thermoplastic polyesters, which is derived from renewable plant sources. PLA has been studied because of its biocompatibility and high mechanical properties. In addition, PLA can be processed the same way as petroleum-based polymers, such as extrusion, film blowing and injection molding. Therefore, PLA is widely used for packaging and medical applications. However, the significant disadvantages of PLA include high brittleness and slow crystallization. The high brittleness of PLA at room temperature is because its glass transition temperature is above room temperature (55 - 65 °C). There are many methods for improving toughness of PLA such as blending with plasticizers[1,2] or other polymers[3,4]. Numerous researches have investigated methods for increasing impact resistance of PLA by blending with biodegradable polymers such as poly(ethylene glycol) (PEG)[1,2], polycaprolactone[3] and poly(butylene succinate)[4]. In addition, natural rubber (NR) have also been used for toughening PLA [5,6,7] due to its high molecular weight, low glass transition temperature, high flexibility and biodegradability. Suksut et al.[5] studied effect of nucleating agents (talc, calcium carbonate ($CaCO_3$) and cyclodextrin) on crystallization, mechanical properties and morphology of PLA and 10 wt% NR blend. These study reported the enhanced toughness of PLA by blending with NR and the increase in degree of crystallinity of PLA in the blend by adding talc and $CaCO_3$. Bitinis et al.[6] used NR to improve the physical and mechanical properties of PLA. Their results showed that the optimal NR content to improve the brittleness of PLA was 10 wt%. This concentration of NR could improve the elongation at break from 5% for pure PLA to 200% for the blend. Moreover, they also reported that presence of NR could increase crystallisability of PLA. Jaratrotkamjorn et al.[7] used NR and modified rubber including epoxidized natural rubber (ENR25 and ENR50) and natural rubber grafted with poly(methyl methacrylate) (NR-g-PMMA) in their study to improve toughness of PLA. Their study reported that all blends showed higher impact strength than PLA and that unmodified NR was found to be the best toughening agent. The optimal NR diameter should be 0.5-1.0 μm. The most important properties of impact modifier include the rubber glass transition temperature, the rubber particle size in polymer matrix, the quality of dispersion and compatibility to the polymer matrix [8]. Therefore, decreasing the size of rubber particles dispersed in PLA matrix led to high flexibility, high crystallinity and better compatibility. Our study focuses on using melt compounding for preparation of PLA/NR blend to enhance impact resistance. Considering that PLA and NR have different polarity, the surface-modified NR or compatibilizer may be required for higher compatibility. One of the chemicals commonly used as coupling agent is maleic anhydride (MA), which is a polar monomer and can react with double bonds or free radicals, to improve hydrophilicity and interfacial adhesion between binary (or more) phases. Therefore, grafting of MA onto various polymers have been used in reactive blending as a blend compatibilizer [9]. MA grafted onto NR molecules were called maleated natural rubber (NR-g-MA). It has been used to prepare blends with polar polymers such as nylon 12 [10], polypropylene (PP) [11], ethylene-vinyl acetate (EVA) [12]. The another chemical modification of NR, the grafting of methyl methacrylate (MMA) onto NR (NR-g-PMMA), exhibited core-shell structure, where NR was the core and poly(methyl methacrylate) (PMMA) was the shell. This is believed to increase miscibility of PLA-rubber blends because PMMA could be missible with PLA [13,14]. Furthermore, MA was considered grafting onto PLA (PLA-g-MA). PLA-g-MA was used as compatibilizer to improve compatibility and adhesion between PLA and other polymers or fillers such as starch [15,16], talc [17], and layered-silicate (filler) [18]. Although PLA-g-MA was used to compatibilize PLA with various polymers or fillers, based on our knowledge, there was no report using PLA-g-MA to compatibilize between PLA and NR.

The objective of this study was to investigate modified NRs (NR-g-MA and NR-g-PMMA) as the second-phase polymer in PLA and PLA-g-MA to improve interfacial adhesion in PLA and rubber blends. These blends were studied on the basis of morphological properties and impact resistance.

Experimental

Materials

Poly (lactic acid) (PLA 3051D, Nature Work) with the density of 1.24 g/ml was used in this study. Standard Thai Rubber 5L (STR 5L) and high ammonium concentrated latex were supplied by Rubber Estate Organization, Thailand. Methyl methacrylate (MMA, 99%, stabilized) was manufactured by Acros organics. Maleic anhydride (MA) was purchased from Merck. Dicumyl peroxide (DCP) and Tetraethylenepentamine (TEPA) purchased from Acros organics were used as initiator. Dichloromethane, acetone and methanol (Carlo erba reagents) were used as solvents. PLA was dried at 80 °C overnight in an oven before use.

Modification of Natural rubber

The grafting of MA onto NR molecules (maleated natural rubber, NR-g-MA) was prepared by blending MA and NR in an internal mixer (Haake Rheomix 3000 OS). The NR was masticated at 135 °C with a rotor speed of 60 rpm for 2 min. The MA (6 phr) was added into internal mixer and mixing was continued for 8 min [9]. After that the blend product was cut into small pieces before blending with PLA.

The grafting of MMA onto NR latex (NR-g-PMMA) was synthesized through emulsion polymerization by using DCP and TEPA (ratio 1:1) as redox initiator. The quantity of MMA and DCP were 3.0 and 3.3×10^{-2} mmol/g of rubber, respectively. The time to polymerization is 5 hours. This method was reported on W. Arayapranee et al. [19]. After that the grafted product was dried in vacuum oven at 40 °C for 8 hours.

PLA-g-MA Preparation

MA was grafted onto PLA backbone (PLA-g-MA) according to an adapted experiment from other studies [16, 18]. The grafting was prepared in an internal mixer (Haake Rheomix 3000 OS) at 160 °C and rotor speed of 80 rpm. First, the PLA was molten inside the chamber, after that the chemical initiator DCP (0.35 %wt) was added and mixed for 5 minutes. Next, MA (10 %wt) was added and the reaction was continued for 5 minutes. The excess moisture was removed from PLA-g-MA at 80 °C overnight before use.

Blend preparation

All of the blends were prepared by melt blending in Haake Rheomix 3000 OS internal mixer at the optimal processing conditions which were 160 °C with rotor speed of 60 rpm and blending time of 15 minutes [6]. The concentration of the rubber in the blends was fixed at 10 %wt with and without 5 %wt MPLA. The samples were prepared for impact resistance test by compression molding at 190 °C for 5 minutes and were then cooled to room temperature.

Table 1 - Sample codes and compositions of PLA blends

Sample code	Weight Ratio				
	PLA	NR	NR-g-MA	NR-g-PMMA	PLA-g-MA
A	90	10	-	-	-
B	90	-	10	-	-
C	90	-	-	10	-
D	85	10	-	-	5
E	85	-	10	-	5

Blend Characterization

The interfacial adhesion and the dispersion of rubber in the PLA matrix were observed using scanning electron microscope (SEM) at room temperature. The samples were broken in liquid nitrogen and the fractured surfaces were coated with gold before analysis.

The impact resistances of samples were measured according to ASTM D256 using pendulum energy of 2.7 J at 25 °C. The samples were kept in desiccators overnight before testing.

Results and Discussion

Fracture surface analysis

The fractured surfaces of the blends were shown in Figure. 1. The black domains represent the position of extracted rubber phase. In Figure 1 (a), PLA and unmodified NR blend (sample A) showed phase separated morphology where small rubber droplets were dispersed in PLA matrix. This proves poor interfacial adhesion due to the hydrophobicity of NR while PLA is a little hydrophilic and has very different viscosity and molecular weight. Figure 1 (b) of PLA and NR-g-MA blend (sample B) showed large droplets and flat fracture surfaces. This may be attributed to high gel content occurred from the reaction of MA onto NR molecules. This gel led to higher viscosity and lower dispersability of the grafted NR in PLA [9]. The fractured surfaces of the PLA blend containing NR-g-PMMA (sample C) was shown in Figure 1 (c). This blend showed very flat fracture surface with multiple small holes which seem to be more compatible to PLA than unmodified NR. The smaller droplets were because PMMA could be miscible in PLA.

(a)

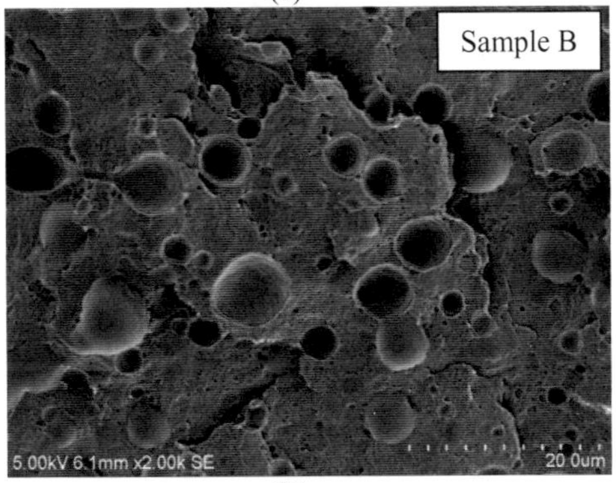

(b)

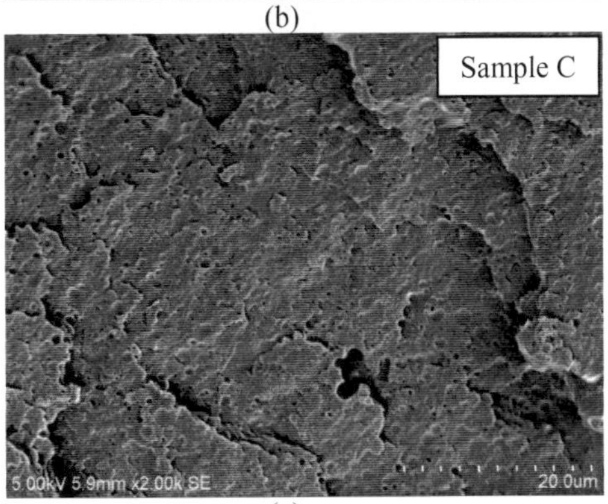

(c)

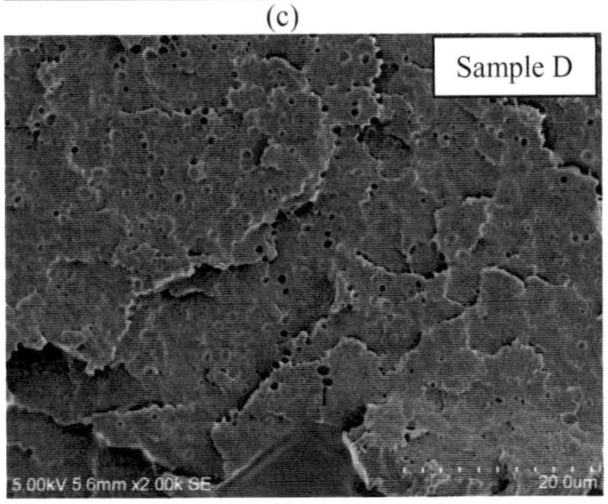

(d)

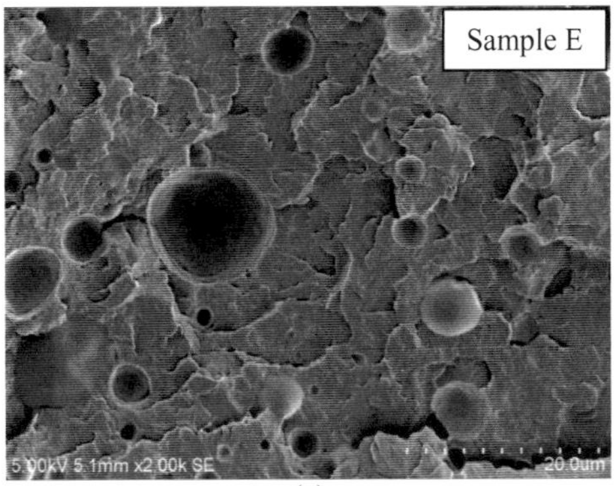

(e)

Figure 1 – SEM micrographs of freeze fractured surfaces of the blends: (a) sample A (b) sample B (c) sample C (d) sample D and (e) sample E

The figures 1 (d) and (e) are for PLA containing NR and NR-g-MA when using PLA-g-MA as compatibilizer (samples D and E). The blends showed flatter fractured surfaces than the blends without compatibilizer indicating higher brittleness. The reason is that PLA-g-MA had caused even higher polarity difference between the two phases. The average rubber particle sizes obtained by image analysis are listed in Table 2.

The impact resistance properties

The impact resistances of the blends are summarized in Table 2. All of the blends showed higher impact resistance than neat PLA. The improvement in impact resistance could mostly be explained by the micrographs of the fracture surfaces (Figure 1). The samples B and E, which contained NR-g-MA, presented low efficiency in toughening PLA because the large rubber sizes led to lower interfacial adhesion. The samples C and D possessed better compatibility than unmodified NR blend, but they exhibited higher brittleness consistent with their flat fractured surfaces, corresponding to brittle fracture with low dissipated energy [20]. This is due probably to the miscible PMMA portion. Therefore, NR blend exhibited the highest impact resistance consistent with the work by R. Jaratrotkamjorn et al. [5], who also found that unmodified NR is the best toughening agent for PLA.

Table 2 - Impact resistance and particle diameter of the blends containing different rubbers

Sample	Impact resistance (J/m)	Average particle diameter (μm)
PLA	25.96 ± 1.91	-
A	84.80 ± 4.33	0.63 ± 0.11
B	48.85 ± 1.15	3.89 ± 1.13
C	39.04 ± 5.70	0.65 ± 0.18
D	53.89 ± 6.03	0.56 ± 0.19
E	26.15 ± 1.42	3.73 ± 1.51

Conclusion

In this study NR was found to have the highest efficiency in toughening PLA than other modified rubbers. Even though the NR-g-MA could provide better adhesion of NR to PLA, the large particle size of NR posed problems in reducing impact property of NR/PLA blend. On the other hand, the NR-g-PMMA and PLA-g-MA contributed to more brittle blends than simple NR/PLA blend. Future improvement in NR/PLA blend should be focus on improving compatibility and lowering brittleness simultaneously.

References

1. M. Baiardo; G. Frisoni; M. Scandola; M. Rimelen; D. Lips; K. Ruffieux; E. Wintermantel *J Appl Polym Sci.* 2003, *90*, 1731.
2. Y. Hu; Y. S. Hu; V. Topolkaraev; A. Hiltner; E. Baer Polymer. 2003, *44*, 5681.
3. Y. H. Na; Y. He; X. Shuai; Y. Kikkawa; Y. Doi; Y. Inoue *Biomacromolecules*.2002, *3*, 1179.
4. J. W. Park; S. S. Im *J Appl Polym Sci* 2002, *86*, 647.
5. B. Suksut; C. Deeprasertkul *J polym Environ* 2011, *19*, 288-296.
6. N. Bitinis; R. Verdejo; P. Cassagnau; M. Lopez-Manchado. *Mater Chem Phys*, 2011, *129*, 823-831.
7. R. Jaratrotkamjorn; C. Khaokong, V. Tanrattanakul *J Appl Polym Sci* 2012, *124*, 5027-5036.
8. T. Li; L. S. Tumg; S. Gong; K. Erlacher *Polym Eng Sci*, 2006, 1419-1427.
9. C. Nakason; S. Saiwaree; S. Tatun; A. Kaesaman *Polym Testing* 2006, *25*, 656-667.
10. R. Magaraphan; C. Totanapoka; Alexander; M. Jamieson *Des Monomer Polym*, 2004, *7*, 165-180.
11. C. Nakason; S. Saiwari; A. Kaeseman *Polym Eng Sci* 2006, 594-600.
12. C. Nakason; W. Kaewsakul; A. Kaesaman *J Elastomv Plast* 2011, *00*, 1-25.
13. G. Zhang; J. Ahang; S. Wang; D. Shen *J Polym Sci Part B:Polym Phys* 2003, *41*, 23.
14. S. H. Li; E. M. Woo *J Polym Sci Part B: Polym Phys* 2008, *46*, 2355.
15. J.F. Zhang; X. Sun *Biomacromolecules* 2004, *5*, 1446-1451.
16. V.H. Orozco; W. Brostow; W. Chonkaew; B.L. Lo'pez *Macromol. Symp* 2009, *277*, 69-80.
17. A.C. Fowlks; R. Narayan *J Appl Polym Sci* 2010, *118*, 2810-2820.
18. L. Petersson; K. Oksman; A.P. Mathew *J Appl Polym Sci* 2006, *102*, 1852-1862.
19. W. Arayapranee; G. L. Rempel *J Appl Polym Sci.* 2008, 110, 2475–2482.
20. M. Todo; A. Harada; H. Tsuji in 16 th international conference on composite material proceeding, Kyoto 2007.

1. [1] M. Baiardo; G. Frisoni; M. Scandola; M. Rimelen; D. Lips; K. Ruffieux; E. Wintermantel *J Appl Polym Sci.* 2003, *90*, 1731.

2. [2] Y. Hu; Y. S. Hu; V. Topolkaraev; A. Hiltner; E. Baer Polymer. 2003, *44*, 5681.

3. [3] Y. H. Na; Y. He; X. Shuai; Y. Kikkawa; Y. Doi; Y. Inoue *Biomacromolecules*.2002, *3*, 1179.

4. [4] J. W. Park; S. S. Im *J Appl Polym Sci* 2002, *86*, 647.

[5] B. Suksut, C. Deeprasertkul, "Effect of nucleating agents on physical properties of Poly(lactic acid) and Its blend with Natural rubber", J polym Environ (2011), 19:288-296.

[6] (N. Bitinis, 2011), "Structure and properties of polylactide/natural rubber blends, Materials chemistry and Physics, 129(2011),823-831.

[7] Rudee Jaratrotkamjorn, Chuanpit Khaokong, Varaporn Tanrattanakul, "Toughness Enhancement of poly(lactic acid) by melt blending with natural rubber", Journal of Applied Polymer Science, Vol. 124, 5027-5036 (2012)

[8] Tongnian Li, Lih-Sheng Tumg, Shaoqin Gong, Kurt Erlacher, "Polylactide, Nanoclay, and core-shell rubber composites", Polymer Engineering and Science, 2006(1419-1427)

[9] C. Nakason, S. Saiwaree, S. Tatun, A. Kaesaman, "Rheological, thermal and morphological properties of maleated natural rubber and its reactive blending with poly(methyl methacrylate), Polymer testing, 25(2006):656-667

[10] Rathanawan Magaraphan, Chatchawan Totanapoka, Alexander, M. Jamieson, "Morphology of nylon12/natural rubber blends compatibilized by reactive processing, Designed monomers and polymer, Vol.7, No.1-2,165-180(2004)

[11] Charoen Nakason, Sitisaiyidah Saiwari, Azizon Kaeseman, "Thermoplastic Vulcaanizates based on maleated natural rubber/polypropylene blends : Effect of blend ratios on rheological , mechanical and morphological properties", Polymer engineering and science 2006,594-600

[12] Charoen Nakason, W. Kaewsakul, and A. Kaesaman, "Thermoplastic natural rubber based on blending of ethylene-vinyl acetate copolymer with different type of natural rubber", Journal of elastomers and plastics Vol.00-month 2011 pp. 1-25

[13] Zhang, G.; Ahang, J.; Wang, S.; Shen, D. J Polym Sci Part B:Polym Phys 2003, 41, 23.

[14] Li, S. H.; Woo, E. M. J Polym Sci Part B: Polym Phys 2008, 46, 2355.

[15] J.F.Zhang, X. Sun, "Mechanical properties of poly(lactic acid)/starch composites compatibilized by maleic anhydride", Biomacromolecules, 5(2004),1446-1451

[16] V.H. Orozco, W. Brostow, W. Chonkaew, B.L. Lo'pez, "Preparation and characterization of poly(lactic acid)-g-maleic anhydride starch blends", Macromol. Symp, 277(2009),69-80

[17] A.C. Fowlks, R. Narayan, "The effect of maleated polylactic acid (PLA) as an interfacial modifier in PLA-talc composites", Journal of Applied Polymer Science, 118(2010),2810-2820

[18] L. Petersson, K. Oksman, A.P. Mathew, "Using maleic anhydride grafted poly(lactic acid) as a compatibilizer in poly(lactic acid)/layered-silicate nanocomposites", Journal of Applied Polymer Science, 102(2006),1852-1862

[19] Wanvimon Arayapranee,1 Garry L. Rempel, "Preparation of a Natural Rubber Core/Polymer Shell in a Nanomatrix by Graft Copolymerization", Journal of Applied Polymer Science,Vol. 110, 2475–2482 (2008

[20] Mitsugu Todo, Ayako Harada, Hideto Tsuji, "Fracture characterization of biodegradable PLLA

polymer blends, 16 th international conference on composite material.2007 ,Kyoto Japan

O-11-501

UV IRRADIATION FORMATION OF POLY(VINYL ALCOHOL)/ CHITOSAN HYDROGEL THIN FILMS FOR BIOMEDICAL APPLICATION

Ngoc-Thang Nguyen, Jui-Hsiang Liu*

Department of Chemical Engineering, National Cheng Kung University, Taiwan

thangnn-tex@mail.hut.edu.vn; jhliu@mail.ncku.edu.tw

Abstract - The work presented herein aimed to investigate and fabricate a series of poly (vinyl alcohol) (PVA)/Chitosan (CTS) hydrogel thin films by ultraviolet (UV) irradiation method using acrylic acid (AAc) monomer as a crosslinker, without adding any photoinitiator. The effect of reaction conditions such as the concentration of AAc and the irradiation time were investigated by Fourier Transform Infrared (FTIR), X-Ray Diffraction (XRD), and Scanning Electron Microscope (SEM). FTIR spectra confirmed that poly(acrylic acid) (PAAc) was polymerized successfully and the chemical crosslink was formed in hydrogel film. XRD analyses indicated that the crystalline degree of the hydrogel film decreased with the increase of the PAAc content. SEM micrographs showed the uniform interconnected pore formation through the entire hydrogel structure and the gradient in crosslink density through the film thickness was observed by extent irradiation time. With the increase of UV irradiation time, hydrogel thin films remained the interlinked pores in the interior and showed the larger inter-pores on the surface. The prepared transparent and thin films constituted from CTS and PVA without using a specific photo-initiator can be a promising method to make PVA/CTS hydrogel thin film for biomedical applications.

Keywords: Poly(vinyl alcohol); Chitosan; Ultraviolet irradiation; Hydrogel thin films; Biomedical applications.

Introduction

Hydrogels known as cross-linked polymeric networks that are able to adsorb water without dissolving, have been extensively used in a variety of biomedical applications due to their unique biocompatibility, flexible methods of synthesis, range of constituents, and desirable physical characteristics [1-2]. They can be fabricated by many methods such as irradiation, thermal annealing, freeze-thawing, as well as by using crosslinking chemical agents to generate the physical or chemical crosslinked hydrophilic polymer networks. In which irradiation (e.g. γ-rays, electron beam, or UV irradiation) is considered as an effective tool [3-7]. By using of UV irradiation to crosslink or chemically modify polymer systems, a photoinitiator and a crosslinker are usually used to induce free radicals for covalent crosslinking performance. Most of the crosslinkers and chemical initiators are toxic and difficult to remove from polymer networks; and only a free trace of them may be a harmful factor for biomedical-related applications [8]. To overcome this shortcoming, a desired route is choosing proper monomers which directly form free radicals by adsorption of UV irradiation. In this work, acrylic acid monomer was employed as a crosslinker to fabricate poly(vinyl alcohol)/chitosan hydrogel thin films via ultraviolet (UV) irradiation without using photo-initiator or additional crosslinking agents.

Experimental

1. Material

PVA (Mw=145 000, fully hydrolysis) was obtained from Merck. Chitosan (Mw = 100 000 ÷ 300 000, deacetylation ≥ 85 %) and acrylic acid were purchased from Acros Organics. Deionized (DI) water from a Millipore-Q system was used as the solvent in all experiments. All chemicals were used as received except that AAc was vacuum distilled to remove the polymerization inhibitor before use.

2. Preparation of the hydrogel thin films

PVA/CTS hydrogel thin films were prepared by UV irradiation by which crosslinked PAAc chains were formed in the hydrogel. The feed composition and designation of the films are listed in Table 1. The homogeneous solutions were casted onto Petri dishes and exposed to a 1000 W UV lamp (1000 UV curing, Taiwan) with a wavelength range of 250 ÷ 420 nm, at a distance of 15 cm, under nitrogen atmosphere. The result films of 45 ± 5 μm in thickness were characterized by FTIR (Jasco VALOR III), XRD (RigakuUltimaIV XRD) and SEM (JEOL HRFESEM JSM-6700F).

Results and Discussion

1. FTIR

Fig. 1 shows FTIR spectra of PVA, CTS, PAA, S20-30 and S23-30. The absorption bands of C=O in PAAc are located at 1766 cm^{-1}. This band for S23-30 hydrogel is shifted to a lower frequency of 1720 cm^{-1} with high intensity indicating that a strong hydrogen bonding and

Table 1. The feed composition and designation of the PVA/CTS thin films

Code[a]	Molar ratio [OH]+[NH_2]:[COOH]	Weight (gram)		
		PVA	CTS	AAc
S20-xx[b]	2: 0[c]	0.5	0.2	0.0
S21-xx	2 : 1	0.5	0.2	0.44
S22-xx	2 : 2	0.5	0.2	0.88
S23-xx	2 : 3	0.5	0.2	1.32

[a] 5 wt % of PVA and 2 wt% of CTS was prepared for all solutions.
[b] xx - the irradiation times were 15, 30 or 45 min, respectively.
[c] 3 wt% of acetic acid was used to dissolve CTS.

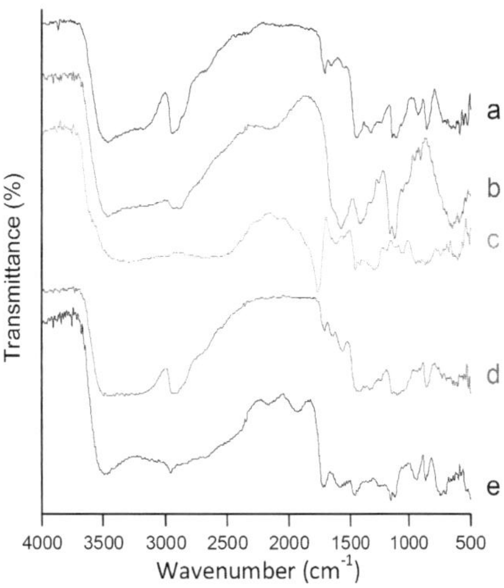

Figure 1. FTIR spectra of (a) PVA, (b) CTS, (c) PAA, (d) S20-30 and (e) S23-30.

also electrostatic interaction should occur among PVA, CTS and PAAc in this hydrogel. The ionization of primary amino group of CTS in the presence of carboxylic group of PAAc can be observed around 1544 cm^{-1} due to symmetric deformation of $-NH_3^+$ [9]. In contrast, the C=O stretching band of S20-30 hydrogel is located at 1710 cm^{-1} with low intensity even an equivalent amount of acetic acid compared with AAc monomer to dissolve CTS has been used. On the other hand, after extracting S23-30 hydrogel for 72 hrs at 50°C in DI water, a high intensity of the C=O band located around 1720 cm^{-1} conformed that the AAc has been polymerized and crosslinked under UV irradiation [10].

2. XRD

The X-ray diffraction patterns of PVA, CTS, PAAc and various PVA/CTS hydrogel thin films after 30 min UV irradiation are showed in Fig. 2. The characteristic peaks of PVA are located at $2\theta = 11.4^o$, 19.7^o, 22.9^o and 41° while those of CTS are exhibited at $2\theta = 10.6^o$ and 19.6^o. On the other hand, diffraction pattern of PAAc does not show any peaks indicating that it is totally amorphous [6]. As showed in Fig. 2d-2g, the effect of the PAAc content in the crystallinity of the result films are obtained obviously. Two peaks for the various hydrogel films are found: One peak with high intensity appears at $2\theta = 19.7^o$ and another peak with very low intensity appears at $2\theta = 41^o$. With increasing the content of PAAc in the hydrogel, the relative intensity of the peaks decreases significantly. It can be explained that the decrease in crystallinity of these hydrogel films are attributed to the deformation of the strong hydrogen bonding between hydroxyl and amino groups in both CTS and PVA due to interaction of these groups with carboxyl groups of PAAc, as well as the formation of PAAc network in the hydrogel films. Similar discussions for the deformation of crystal

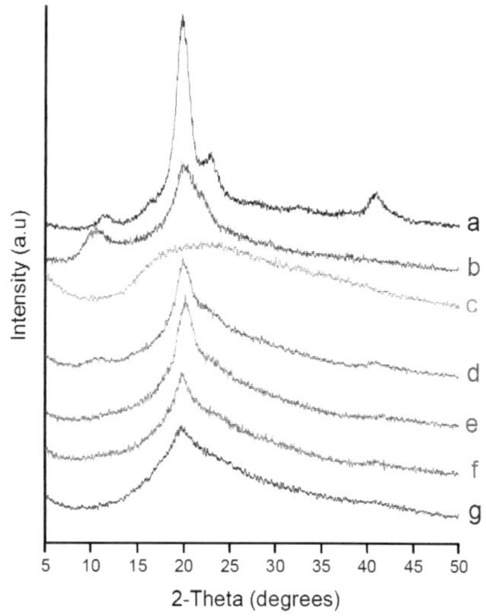

Figure 2. XRD patterns of (a) PVA, (b) CTS, (c) PAA, (d) S20-30, (e) S21-30, (e) S22-30 and (e) S23-30

structure had been reported in several references [11-13].

3. SEM

The effect of irradiation time on the surface and cross-section morphologies of PVA/CTS hydrogel thin films are obtained obviously in Fig. 3. All the hydrogels exhibit an interlinked pore structure, however, the pore size seems to be increases with extending of irradiation time. As can be seen in Fig. 3, after 15 and 30 min UV irradiation, the interconnecting pores are observed through the entire hydrogel structure; however, the surface image shows lager inter-pores of the 30 min irradiated hydrogel film. After 45 min irradiation, the top surface of the thin film shows a dense, noporous structure while the interlinked pores are still remained inside hydrogel film. It indicates that the microstructure of hydrogels film is strong effect by the irradiation time. These changes in morphology of hydrogel can be explained by the formation of PAAc and its crosslinked network in hydrogel film. It is known that, PAAc has the excellent water swelling property, thus, the more PAAc content in the film, the higher water retaining in the hydrated hydrogel and as a result, the larger pore size will be after lyophilization [14].

It is reported that PAAc, like most polymers, was degraded when exposed to the 254 nm UV irradiation in the presence of oxygen containing atmosphere [15-17]. In this case, the hydrogel thin film that was obtained by UV irradiation for 45 min in a nitrogen atmosphere is dense in the surface and loose in the interior because chemical crosslinking at surface is more dominant than that inside the hydrogel film. The results clearly exhibited that a gradient in crosslink density exists through the film thickness even though it is in microscale. The porous structure and biocompati-

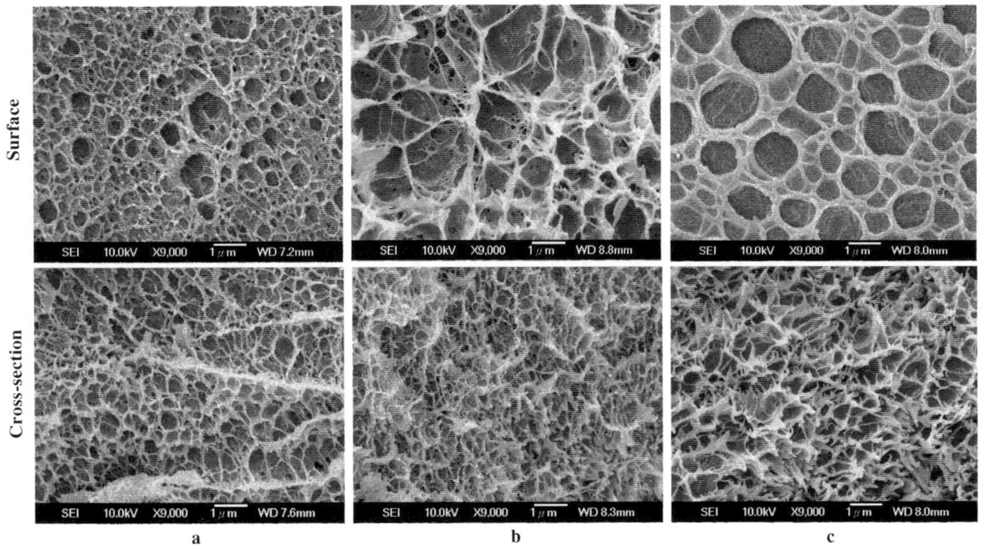

Figure 3. SEM micrographs of the surface and cross-section of various PVA/CTS hydrogel films containing same composition and made by different UV irradiation times: (a) S23-15, (b) S23-30 and (c) S23-45.

ble property of PVA/CTS hydrogel thin film can bring it to be used in biomedical application such as wound dressing.

Conclusions

A series of PVA/CTS hydrogel thin films were synthesized by using AAc monomer as a crosslinker under UV irradiation without using photoinitiator or addition of crosslinking agents. The properties of these films including chemical bonds, molecular structure, crystallinity degree, the morphology of the surface and internal structure, as well as their relationship to AAc content were studied by FTIR, XRD and SEM analysis. It is found that PAAc had been polymerized and chemical crosslink was formed in the hydrogel film. By this approach, PVA/CTS hydrogel thin film can be used as biomaterial for medical applications because of its inherent biocompatible and purity.

Acknowledgements

The authors would like to thank the National Science Council (NSC) of the Republic of China (Taiwan) for financially supporting this research under Contract No. 98-2221-E006-007 and 99-2923-E006-004.

References

1. J. J. Grodzinski, *Polym. Adv. Technol.*, 2010, *21*, 27-47.
2. B. V. Slaughter; S. S. Khurshid; O. Z. Fisher; A. Khademhosseini; N. A. Peppas, *Adv. Mater.*, 2009, *21*, 3307-3029
3. X. Yang; Q. Liu; X. Chen; F. Yu; Z. Zhu, *Carbohyd. Polym.*, 2008, *73*, 401-408.
4. N. Sheikh; L. Jalili; F. Anvari, *Radiat. Phys. Chem.*, 2010, *79*(6), 735-739.
5. Y.M. Lee; S. H. Kim; C. S. Cho, *J. Appl. Polym. Sci.*, 1996, *62*, 301-311.
6. J. W. Lee; S. Y. Kim; S. S. Kim; Y. M. Lee; K. H. Lee; S. J. Kim, *J. Appl. Polym. Sci.*, 1999, *73*, 113-120.
7. A. Gestos; P. G. Whitten; G. M. Spinks; G. G. Wallace, *Soft Matter*, 2010, *6*, 1045-1052.
8. J. Berger; M. Reist; J. M. Mayer; O. Felt; N. A. Peppas; R. Gurny, *Eur. J. Pharm. Biopharm.*, 2004, *57* (1), 19-34.
9. L. J. Bellamy, *Infrared Spectra of Complex Molecules*. Chapman & Hall, New York, 1980.
10. J. Liu; W. Wang; A. Wang, *Polym. Adv. Technol.*, 2011, *22*, 627-634.
11. M. Y. Jen; Z. W. Hung; C. Y. Chun, *J. Memb. Sci.*, 2008, *322*(1), 74-80.
12. J. H. Kim; J. Y. Kim; Y. M. Lee; K. Y. Kim, J. *Appl. Polym. Sci.*, 1992, *45*, 1711-1717.
13. J. A. Ratto; C. C. Chen; R. B. Blumstein, *J. Appl. Polym. Sc.*, 1996, *59*, 1451-1461.
14. H. A. Abd El-Rehim; E. A. Hegazy; F. H. Khalil; N. A. Hamed, *Nucl. Instr. Meth. Phys. Res. Sect. B*, 2007, *254* (1), 105-112.
15. D. O. H. Teare; C. Ton-That; R. H. Bradley, *Surf. Interface Anal.*, 2000, *29*, 276-283.
16. H. Y. Nie; M. J. Walzak; B. Berno; N. McIntyre, *App. Surf. Sci.*, 1999, *144-145*, 627-632.
17. R. Srinivasan; B. Baren; D. Seeger, et. al., *Macromolecules*, 1986, *19*(9), 16-921.

O-11-528

EFFECT OF REPROCESSING ON PROPERTIES OF SUGARCANE BAGASSE FIBER REINFORCED POLYPROPYLENE COMPOSITES

B. C. Bonse*; G. Capecchi; W. Y. Mardegam
prebbonse@fei.edu.br
Centro Universitário da FEI, São Bernardo do Campo, SP, Brazil

Abstract - Recyclability of sugarcane bagasse fiber filled polypropylene composites were assessed by submitting these to six reprocessing cycles. Polypropylene composite specimens containing 40wt% sugarcane bagasse fibers and 4wt% compatibilizer (maleic anhydride grafted polypropylene) were manufactured by extruding a blend of the composite components in a co-rotating twin-screw extruder, with subsequent pelletizing and injection molding. Composite specimens were then ground in a cutting mill and re-injected. For control neat PP was also submitted to the same processing and reprocessing conditions. Heat deflection temperature (HDT), tensile, TGA, bending, impact and fatigue (F_{max}=1200N, f=6Hz) testing were performed. SEM analyses revealed fiber breakage with reprocessing, however no significant differences have been observed between virgin and reprocessed composites regarding fiber-matrix adhesion. For some properties a pronounced drop was seen after the first reprocessing cycle, likely because of more intense fiber breakage in this cycle with fiber size stabilizing afterwards. Composites containing sugarcane bagasse fiber showed inferior performance compared to bamboo fiber filled ones. It has been shown that natural fiber composites can be reprocessed without great losses in property for applications not subject to fatigue and/or impact loading.

Keywords: Composite, Polypropylene, Sugarcane bagasse fiber, Bamboo fiber, Recycling.

Introduction

Plant fibers are increasingly becoming attractive in relation to synthetic fibers, giving rise to bio composites [1]. The main advantages of plant fibers are: low production costs, low density, low wear of processing equipment, they are renewable, non-toxic, and have excellent strength to density ratio, to name a few [2]. Negative aspects are the low degradation temperature, processing should not exceed 200 ° C, inhomogeneity in physical and mechanical properties which may vary with climate and planting site, and high moisture absorption [3]. In addition to being easily obtained, inexpensive and having good mechanical properties, PP can be easily processed below 200 °C and is thus a suitable candidate as matrix for these composites. However, hydrophobic PP and hydrophilic bamboo are incompatible and composites require compatibilizers such as maleic anhydride grafted PP (PPgMA) which reduces interfacial tension and improves the adhesion between polymer and fiber. To assess the extent to which these composites can be considered recyclable, the current study aimed to prepare and study the reprocessibility of polypropylene composites containing 40 wt% bamboo fiber and 4 wt% compatibilizer. The physical and mechanical properties of the composites were accompanied over six reprocessing cycles. Results were also compared with those of bamboo fiber reinforced polypropylene [4].

Experimental

Materials

Isotactic polypropylene pellets were donated by Braskem under code H503, with density of 0.905 g/cm³ and Melt Flow Rate (MFR) of 3.5 g/10min (230°C / 2.16 kg). Sugar cane fibers were obtained from waste of sugarcane juice producing microenterprises. Compatibilizer maleic anhydride grafted PP (PP-g-MA), Polybond 3200, density 0.92 g/cm³ and MFI = 110 g/10min (190°C and 2.16 kg), was purchased from Crompton-Uniroyal Chemical (São Paulo, Brazil).

Methods

Previously one-week sun-dried sugarcane waste was ground in a MAK-250 grinder from Kie Máquinas e Plásticos, using a 3-mm screen. Grinding required pre-drying the fibers for 4 hours at 105°C. PP pellets (56 wt%), compatibilizer PPgMA (4wt%) and ground sugar cane bagasse fibers (40 wt%), dried for 12 hours at 105°C, were pre-mixed in a PowderMix Y-mixer for five minutes and subsequently extruded in a Thermo Scientific HAAKE PolyLab OS RheoDrive co-rotating twin-screw extruder. Temperature profile used was: 170°C, 175°C, 180°C, 180°C, 190°C and 185°C. Rotor speed was 250 rpm.

The extrudate was cooled with a fan and pelletized. Specimens were injection molded in a H60/350 Battenfeld machine. A sufficient amount of specimens was set aside for testing and the remaining part was ground, dried at 105 °C for 12 hours and re-injected at the same conditions of the first injection molding. This procedure was repeated until completion of six reprocessing cycles.

Tensile and flexural tests were performed in an Instron 2267, at a rate of 5 mm/min and 1.3 mm/min, respectively, according to ASTM D638 and D790. Impact testing was performed in a Charpy VEB Werkstoff Prüfmaschinen Leipzig impact tester, according to ASTM D6110. Tensile fatigue testing was carried out in an MTS 810; a minimum load of 50 N was used to avoid specimen buckling and a maximum of 1200 N. HDT testing was carried out in CEAST HDT Vicat 6 apparatus.

Tensile fractured surfaces were covered with gold and analyzed in a CamScan CS3200 LV Scanning Electronic Microscope (SEM).

Thermogravimetric analyses (TGA) were performed in a TA Instruments SDT 600 equipment at a heating rate of 10°C/min. and nitrogen purge of 100mL/min.

Results and Discussion

Tensile and flexural tests results of sugarcane bagasse fiber reinforced PP (PP+SF) are shown in Fig. 1 and 2, along with those of bamboo fiber reinforced PP (PP+BF) and PP control. These results (Fig.1 and 2) clearly show the expected reinforcing effect of the fibers when compatibilized, i.e., increase in strength and modulus in relation to PP matrix.

Apparent tensile strength (TS) of neat PP is seen to increase slightly after the first reprocessing cycles, which may be attributed to increase in crystallinity due to chain length reduction caused by thermo-mechanical degradation. From cycle 4 on TS decreases slightly, whereas the decrease in modulus (TM) is more pronounced. This drop in mechanical properties is likely due to more significant molecular weight reduction, resulting in loss in TS and TM.

With regard to the sugercane bagasse fiber composites a different trend from the neat PP is observed. TS decreases slightly up to cycle 4, after which a steeper drop is observed

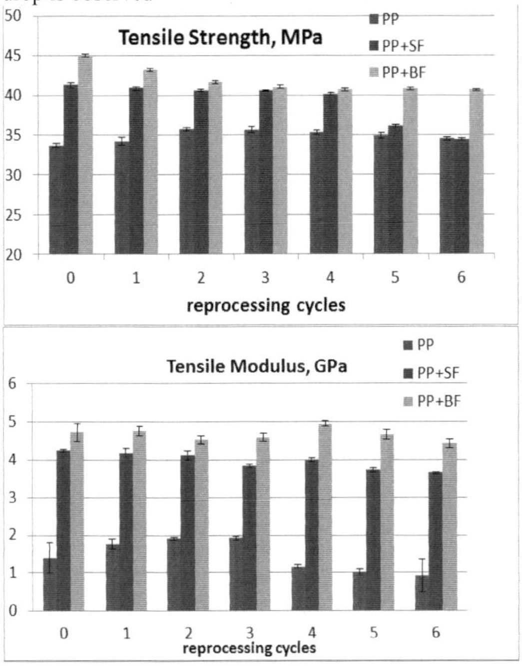

Figure 1 –Tensile strength and modulus for PP control PP/SF, and PP/BF composites as a function of reprocessing.

In contrast to tensile and flexural results, impact and fatigue behavior are strongly affected by reprocessing, as can be seen in Fig. 3 and 4, respectively, and Table 1. After 6 reprocessing cycles fatigue life reduces almost by 90% for both bagasse and bamboo fiber reinforced composites. Loss in properties can mainly be attributed to fiber breakage and thermo-mechanical degradation of the matrix. No evidence was found of adhesion deterioration after reprocessing, as shown in Fig. 6c and d. Similar results were obtained for PP/BF composites [4].

Overall properties of PP/SF composites showed to be more strongly affected than PP/BF by reprocessing (Table 1). Fiber size does not seem to be a significant variable, as can be observed by the SEM micrographs in Fig. 6a and d. Sugarcane bagasse fiber size is not that much different from bamboo fiber size. HDT results (Fig. 5) indicate that thermal stability of the fiber might be an important factor. This is evidenced by TGA results. Fig. 7 shows that onset of degradation for sugarcane bagasse fibers is at lower temperature compared to bamboo fibers. Hence lower processing temperatures and perhaps matrix with higher MFR should be used for sugarcane bagasse fiber.

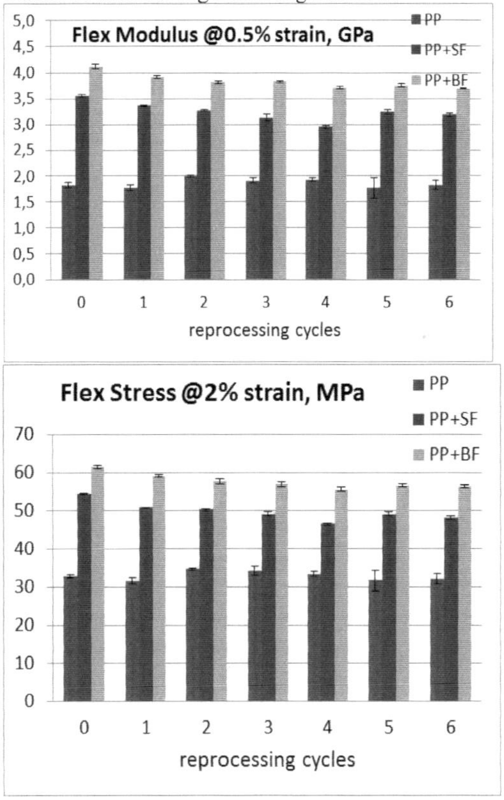

Figure 2 – Flexural strength and modulus for PP control, PP/SF and PP/BF composites as a function of reprocessing.

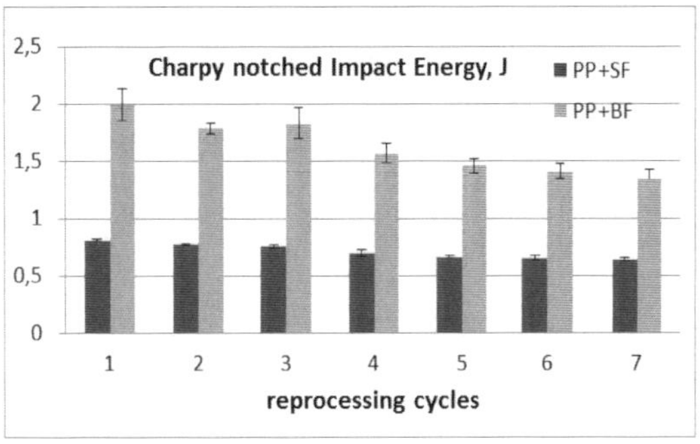

Figure 3 – Charpy notched impact energy for PP/SF, and PP/BF composites as a function of reprocessing.

The results of tensile fatigue tests are shown in Figure 4:

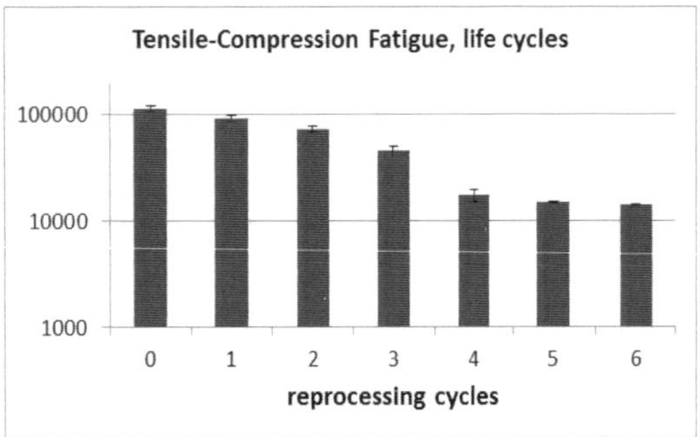

Figure 4 – Tensile fatigue results for PP sugarcane reinforced composites.

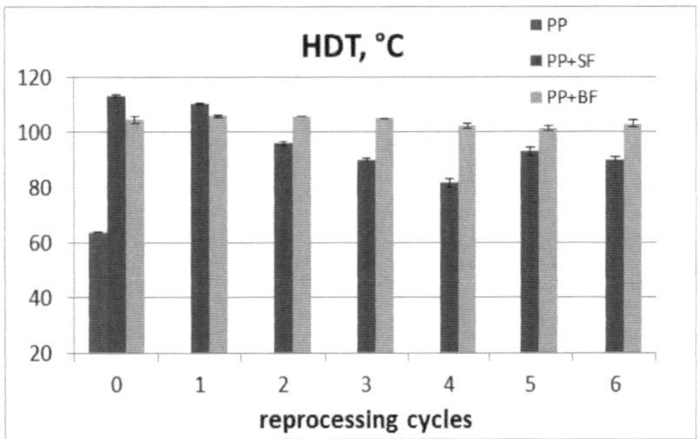

Figure 5 – HDT results for PP control PP/SF, and PP/BF composites as a function of reprocessing

Table 1 – Comparison between sugarcane bagasse (SF) and bamboo fiber (BF) reinforced PP composites, regarding drop in properties after 6 reprocessing cycles

property	PP/SF	PP/BF
Tensile Strength	17%	10%
Tensile Modulus	14%	9%
Flex Stress @2% strain	13%	8%
Flex Modulus	11%	11%
Impact Energy	22%	33%
HDT	26%	1,3%
Fatigue life	88%	86%

a)

b)

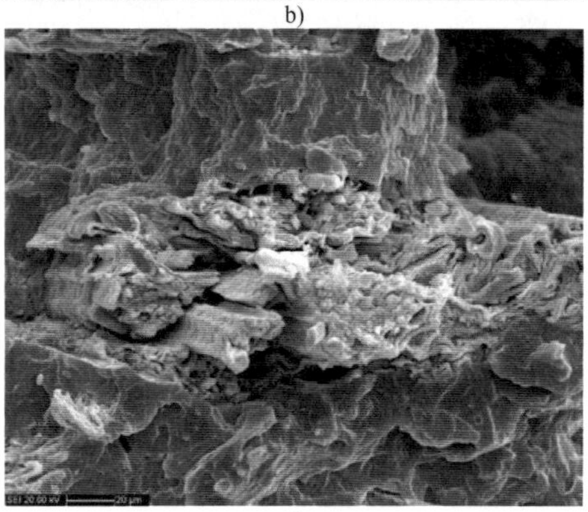

c)

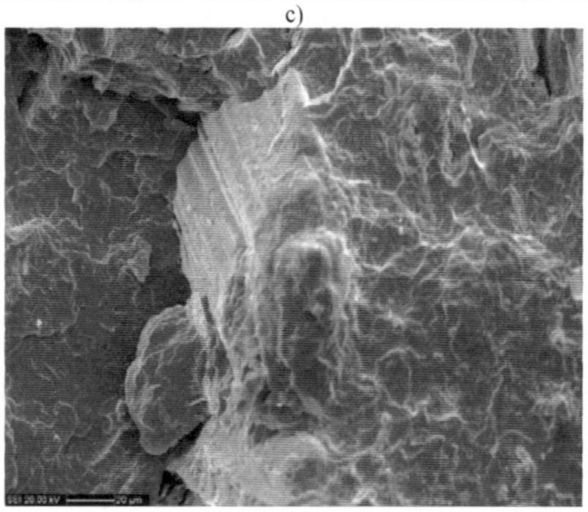

d)

Figure 6 – SEM micrographs of fibers: (a) bagasse and (b) bamboo; and impact fractured PP/SF composites, reprocessed: c) 0 and d) 4 times.

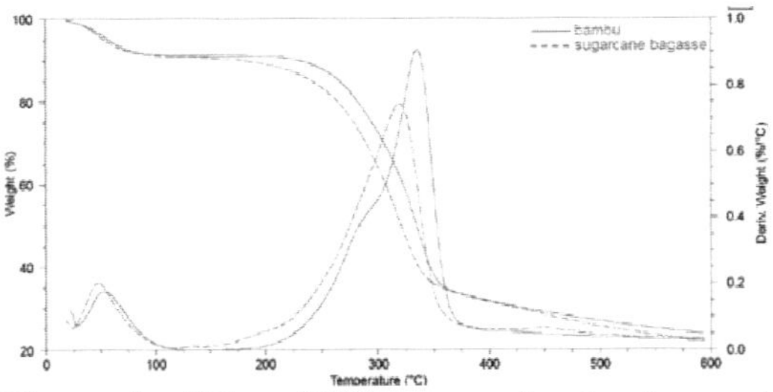

Figure 7 –TG e DTG curves for bamboo and sugarcane bagasse fiber

Conclusions

Natural fiber reinforced composites can be reprocessed without great losses in properties for applications not subject to fatigue and/or impact loading. Loss in mechanical properties is mainly due to fiber breakage and thermo-mechanical degradation of the matrix. Fiber/matrix adhesion does not seem to be affected by reprocessing. Sugarcane bagasse fiber does not perform as well as bamboo fiber after several reprocessing, likely due to lower thermal stability.

Acknowledgments

To Braskem for having donated the PP used in this work.

References

1. Bonse, B. C.; Mamede, M. C. S.; da Costa, R. A.; Bettini S. H. P. In: PPS-26 Proceedings, Banff (Canada) 2010.
2. Lucas, A. A.; Ambrosio, J. D.; Bonse, B. C.; Bettini, S. H. P. Natural Fibers Polymer Composites Technology applied to the Recovery and Protection of Tropical Forests allied to the Recycling of Industrial and Urban Residues. In: Dr. Pavla Tesinova (Vozkova). (Org.). Advances in Composite Materials - Analysis of Natural and Man-Made Materials. 1ed.: InTech, 2011, p. 163-194
3. Mohanty A. K., Natural Fibers, Biopolymers and Biocomposites, CRC Publishers, U.S.A. (2005).
4. Bonse, B. C.; Capecchi, G In: 7th Modest Proceedings, Prague 2012.

O-11-551

COMPARISON OF BLEACHED BAGASSE CARBOXYMETHYL CELLULOSE PROPERTIES FROM SPRAY DRYING AND NON-SPRAY DRYING PROCESSES

S. Kamthai[1] and R. Magaraphan[1*]
[1]*Petroleum and Petrochemical College, Chulalongkorn University, Bangkok, Thailand*
Suphat_61@hotmail.com ; *Rathanawan.K@chula.ac.th

Abstract -In this present work, bleached bagasse carboxymethyl cellulose (CMC_B) is extracted from conventional process and spray drying process for evaluate CMC_B film properties. To investigate the effectiveness of spray drying process on the CMC_b properties, the bleached bagasse cellulose powder was synthesized using different alkaline concentration; 20 - 40% (NaOH, w/v) in sequential reactions with monochloroacetate (MCA) at desired condition and then CMC_B solution was sprayed at feed flow rate of 20 ml/min and inlet temperature of 150 ± 5 °C. The results revealed the CMC_B productions gave CMC_B with high degree of substitution (DS), CMC content, but the spray drying CMC_B yield and viscosity had less than non-spray drying CMC_B. The experiment presented the spray drying CMC_B film had a high water solubility and good tensile strength.
Keywords: Bleached bagasse pulp, Carboxymethyl cellulose, Spray drying

Introduction

Thailand is agricultural country that has many kinds of lignocellulosic agricultural by products. They are copious and cheap sources for cellulose fibers, for example by products from the cultivation of corn, wheat, rice, sorghum, sugarcane and so on. These are the major sources of agro-based bio-fibers and their chemical compositions and physical properties are suitable for further uses such as pulp and papermaking, composites, textile, food additive and etc. One of those agricultural residues in Thailand is bagasse, the sugarcane waste from sugar production. In 2011 the sugarcane plantation area was estimated about 1.3 million hectares and its production was 11.75 tons/hectare [1].Bagasse is common non-woods or hardwood substitutes known as lignocellulosic (LC) fiber containing high cellulose content and it consisted about 32 – 48% of cellulose [2]. Therefore bagasse cellulose is abundant for cellulose derivative production. Generally, cellulose is linear, high molecular weight and it has a high crystallinity. Cellulose is insoluble because the hydrogen bonds associated with structure of molecular cellulose; it neithermelts nor dissolves readily in common solvents [3, 4] but its derivative can be soluble in water. In order to increase cellulose application, cellulose modification by chemicals that are reactive with the hydroxyl groups of the glycosidic units to produce cellulose derivatives is one preferred method.

However, the most utilized cellulose derivative is the carboxymethyl cellulose (CMC), which is commonly converted from cellulose by etherification in aqueous NaOH to produce swollen alkali cellulose followed by reaction with monochloracetic acid (MCA) or its sodium salt in a surplus of organic solvent (e.g. isopropanol, or ethanol) [5]. Generally the using of NaOH has widely range such as 10 - 50% (w/v). Some experiment is observed that the highest DS of CMC at 25%NaOH (w/v) is 0.82 [6]. Above 40% NaOH (w/v) the DS is reached a steady value, about 0.7 [7]. However, additional NaOH may cause reduction in DS because polymer degradation is occurred at high NaOH concentration [8,9].

The CMC is extensively used in several activities according to its DS and it is soluble in water when the DS is higher than 0.5, giving high viscosity in dilute solutions. The purified CMC is a white- to cream-colored,tasteless, odorless, free-flowing powder [10]. CMC has a thickening effect, film-forming ability, and excellent behavior as protecting colloid and adhesive [11]. The application of CMC is found in the pharmaceutical, medical, food and non-food industries. For example it uses to suspend solids in aqueous medium, stabilizeemulsions, absorb moisture, help tosolubilize proteins and stabilize their solutions [12], and coat tablet to control drug-release pills [13] etc.In this paper, the research is focusing on the production of CMC from bleached bagasse pulp (CMC_B) at appropriate condition by using spray drying technique and to characterize the properties of the spray-dried CMC_B.

Experimental

1. Bagasse pulp preparation

The bleached bagasse pulp was cut to a size 1 x 1 cm and then soaked for 24 hours After a specified time for pulp soaking, it was disintegrated by disintegrator with a speed around 5,000 rpm. Next bleached bagasse pulp was dewatered, torn into small pieces and dried in the hot air oven at 80 °C for 24 hours, respectively. The dried bleached bagasse was ground with wood grinder and screened the bleached bagasse particles to be less than

300 mesh. The moisture content of the bleached bagasse powder was determined by TAPPI T236 om-99 and then it was stored in a sealed container for carboxymethyl cellulose extraction.

2. Synthesis of CMC

About 10 g (OD. weight) of bleached bagasse powder was weighed and 40 ml of aqueous sodium hydroxide was then added drop wise. The suspended solution was stirred for 5 minutes. Next 300 ml of isopropanol was added. The carboxymethylation was started by adding 12 g of monochloroacetic acid (MCA) into the solution. The mixture temperature was adjusted in an oven to 55 °C. Then, the obtained solid phase was suspended in 300 ml of methanol, and neutralized with glacial acetic acid. The CMC was purified by washing with 70% ethanol to remove undesired by-product. Next, the CMC was sprayed at feed flow rate of 20 ml/min and inlet temperature of 150 ± 5 °C.

3. Preparation of Bleached Bagasse CMC film

About 2 g of CMC_B was weighed and dissolved in 100 of distilled water at 70 ± 5 ° by using a magnetic stirrer. To remove the entrapped air bubbles during mixing, the solution was cooled down to 40 °C and mixed gently for 20 min. Then, about 31 ± 0.1 ml of the CMC solution was poured into a glass plate to cast CMC film with 0.04 ± 0.01 mm thick by drying at 50 ± 5 °C in a hot air oven for 8 hours. After that, the resulting CMC_B films were peeled off from the glass plate.

4. Characterization of CMC properties

The degree of substitution of carboxyl group in CMC_B was determined according to the USP XXIII method. The methods divided into two steps which were (1) titration and (2) residue on ignition. The calculation of DS was triplicated and based on the following equation; DS = A + S; where A is the degree of acid carboxymethyl substitution which can be calculated using equation, A = 1150M/(7120 - 412M - 80C) and S is the degree of sodium carboxymethyl substitution which can be calculated using equation, S = (162 + 58A)C/(7120 - 80C); where M is the net milliequivalent of base required for neutralization of 1 g CMC_B as determined by titration test and C is the percentage of residue on ignition of CMC_B determined as residue on ignition testing.

For titration testing, about 1 g of CMC_B, accurately weighed, was filled in a 500 ml/Erlenmeyer flask and then 300 ml of NaCl solution (10% w/v) was poured and stirred for 5 min. Five drops of m-cresol purple and 15 ml of HCl solution (0.1N) were then added in the stirred solution. If the solution was still violet, HCl solution (0.1N) was filled until the solution became yellow. Next, the solution was titrated with NaOH (0.1N) until the solution returned to violet at the endpoint. The net amount of milliequivalent base required for neutralization of 1 g CMC_B (M) was calculated by eq.; M(mEq) = m.mole x Valence where m is 10^{-3}; mole is mass in grams per molecular weight of NaOH and the valence of NaOH is 1.

The step (2) was ignition residue testing and it started by adding about 1 g of OD. weight of CMC_B in a clean crucible (burning in the furnace at 800 °C for 1 hr and leaving it cool down in a desiccator). To obtain the absolutely black residue, the bleached bagasse crucible was maintained at 400 °C about 1-1.5 hr and cooled down in the desiccator. The volatization of CMC_B was completed by white fume occurred after heating and together by using the H_2SO_4 to moisten the entire residue. The crucible with residue was ignited at 800 ± 25 °C to obtain the white residue and placed in the desiccator. All treatments were triplicated and calculated the percentage of residue on ignition by using the eq.;

C = (Weight of residue)/(Weight of CMC) x 100

To determine CMC_B yield (%), it was calculated and triplicated base on oven dry weight [14]. The moisture content of CMC_B was determined by applying the method of ASTM D4442-97 and based on the following eq.; CMC yield (%) = (W_1/W_2) x 100;where W_1 is weight of dry bleached bagasse CMC and W_2 is weight of dry bleached bagasse pulp.

To measure CMC and NaCl content, about 1.5 g of CMC_B was added to 100 ml of 80% methanol, stirred, kept for 10 min and filtered. Cake of CMC_B was washed by 100 ml of 80% methanol at the finally stage and dried. The CMC percentag calculated by using the eq.; CMC% = (m_2/m_1) x 100 where m_1(g) is the weight of dried sample and m_2 (g) is the weight of washed sample. Next, to determine NaCl amount, 2g of CMC_B was added to 250 ml of 65%methanol and kept for 5 hrs and then 100 ml of liquid phase was neutralized by diluted HNO_3 and tritrated by 0.1 $AgNO_3$solution. The NaCl percentage calculated by using the eq.; NaCl% = 1.46V/m where V (ml) is the amount of $AgNO_3$ and m (g) is the weight of dried sample [15].

The viscosity of CMC_B was determined and about2 g of CMC_B was dissolved in 100 ml of distilled water. Brookfield Rotational Viscometer was used to measure its viscosity using the test conditions which were 20.7%, torque, 25 ± 1 °C, 02; spindle and 250 rpm, [16].

5. CMC_B Film Properties Test

All CMC_B films were conditioned according to ASTM D618-05 (ASTM, 2005) for film permeability and mechanical testing. The sample was conditioned at 20 ± 1 °C and 55 ± 1% RH prior to further testing.

To test for water solubility of CMC_B film, the pieces of CMC_B were cut into 20 mm x 20 mm and dried at 70 °C in an oven for 2 hours, and then weighed to the nearest 0.0001 g for the initial dry mass of CMC film. Film was submersed into 20 ml of distilled water in a 50 ml/Erlenmeyer flask. The flask was shaken gently with a shaking water bath set at 25± 2 °C for 24 hours. Then, the solution was filtered through a filter paper to get an insoluble part. After that the filter paper was dried at 103± 2 °C overnight [17]. The film water solubility (%FS) was calculated using the following eq.; %FS = $(W_i - W_f)/ W_i$ x 100 where W_i is the initial CMC film weight (g) and W_f (g) is the final weight of the undissolved CMC portion after drying.

Mechanical properties of the CMC_B film such as tensile strength (TS) and percent elongation (%E) at break were determined using an Instron Universal Testing Machine Model 1000 (H1K-S, UK) at the room temperature following ASTM D882-97. The initial grip was 100 mm and the crosshead speed was set at 5 mm/min for film stretching.

Result and Discussion

1. Evaluation on CMC_B Properties

To compare between the influences of non-spray and spray drying technique on CMC_B properties, Table 1 shows the characterization of CMC_B at 20% NaOH (w/v) from both processes. The results indicate that the spray dried CMC_Bhas higher DS and %CMC than those of non-spray dried CMC_B. They were 0.68 and 92.88% respectively. On the other hand, the yield, and viscosity of spray dried CMC_B has lower values than those of non-spray dried CMC_B. This is attributed tothe spray drying can generate the small particle size of CMC_B as a spherical shape that has homogeneous surface and it is easily to replace by other functional groups in carboxymethylation reaction.Generally, the hydroxyl groups in cellulose are usually replaced by carboxymethyl groups in the order of C2 > C6 > C3 [18]. In work accomplished by Vehring reported the spray drying process was a dedicated technology to prepare a trypically spherical particles and its particle morphology depended on spray drying conditions [19].

Table 1.The characterization of CMC_B at 20% NaOH (w/v).

Properties	CMC_B processing	
	Non-spray drying	Spray drying
Degree of Substitution (DS)	0.60 ± 0.01	0.68 ± 0.001
CMC (%)	83.12 ± 0.02	92.88 ± 0.70
NaCl (%)	1.11 ± 0.03	1.02 ± 0.12
Viscosity (cps)	635 ± 0.5	347.5 ± 0.5

In the experiment, the DS of spray dried CMC_B was 0.68 and it means that CMC_B can be disslove in water. When the DS of CMC was below 0.4, the CMC was swellable but insoluble and it was above this value CMC was fully water soluble [20]. The high DS of CMC refers to high replacement of carboxymethyl groups in cellulose structures or destroying of cellulose molecules resulting in the viscosity reduced. Basically, the decrease of CMC viscosity can attribute to the breakage of macromolercular aggregation and it is occurred in polyelctrolytes derived from polysaccharide in aqueous solution [21].

2.Charaterization of spray dried CMC_B

The previous data obviously demonstrated that not only the spray drying process had more influence on the CMC_B properties such as DS and %CMC but also the alkaline pretreatment process at different NaOH concentration (20 – 40% w/v) affected the properties of spray-dried CMC_B. For Instance, Bagasse pulp was treated with 20 to 50% NaOH (w/v) in carboxymethylation reaction resulting in its declined DS from 0.69 to 0.59. [22]. The table 2 reveal that the alkaline pretreatment has no effect on the DS of spray dried CMC_B and it was around 0.69. In contrast to the DS of sago waste CMC, it is continued to rise with the increasing of NaOH concentration and it reached a maximum DS of 0.82 at 25% NaOH (w/v) [6]. Similar observation had been reported that the DS of cotton fiber CMC increased with NaOH concentration from 5 to 40% NaOH (w/v). They found that the DS values were around 0.2 – 0.7 [7]. From the results, the obtained DS values are near to the DS of CMC from palm kernel cake which was about 0.67 when it was synthesized by 17.5%NaOH (w/v) [16]. This research informs that the results of the CMC_B synthesis having low chemicals requirement, high DS and %CMC implying, that synthesis offers lower chemical cost and better green process or more environmental friendly.

Table 2. The characterization of spray drying CMC_B at different NaOH% (w/v).

Properties	NaOH% (w/v)		
	20	30	40
Degree of Substitution (DS)	0.68 ± 0.001	0.68 ± 0.001	0.69 ± 0.001
CMC (%)	92.88 ± 0.7	91.07 ± 0.7	88.82 ± 0.3
NaCl (%)	1.02 ± 0.12	1.18 ± 0.07	1.39 ± 0.04
Viscosity (cps)	347.5 ± 0.5	320 ± 0.5	300 ± 0.5

However, the CMC% tended to be reduced but the %NaCl continued to rise with increasing NaOH concentration. Table 2 reports the %CMC and %NaCl at the highest NaOH concentration (40%,w/v) which are 88.82% and 1.39%, respectively. It is clear that increasing NaOH concentration leads to an increase in by-product content e.g. %NaCl and sodium glycolate [23] while the DS is not significantly changed.

At high concentration of NaOH, formation of sodium glycolate was increased because more NaMCA molecules tended to react with NaOH [8]. The reaction can be explained by 2 equations (1) a cellulose reacts with sodium monochloroacetate (NaMCA) to obtain CMC_B illustrated in the equation:

$$\text{Cellulose} + ClCH_2COONa \rightarrow \text{Cell-O-}CH_2COONa + NaCl + H_2O$$

and (2) the NaMCA reacts with NaOH to form sodium glycolate as a by-productshown in the equation:

$$NaOH + NaMCA \rightarrow \underset{\text{sodiumglycolate}}{HO\text{-}CH_2COONa} + NaCl$$

Moreover, it was found that viscosity of CMC_B decreased with increasing NaOH concentration and it was reduced from 348 to 300 cps. This reduction was caused by the degradation of cellulose moleculues resulting in its declined molar mass. The high concentration of NaOH could result in more NaOH residue left in the produced CMC_B and thus more chain degradation was occurred.

3. CMC_B Film Properties

The preliminary properties of 0.04 mm thick CMC_B film prepared by solution casting are listed in Table 3. The CMC_B chosen to make this film was the one with the highest DS (produced by 40 %w/v NaOH). It shows very good water solubilty about 85.8%. Its mechanical strength is quite outstanding 45.62 MPa. However, the elongation at break of this CMC_B film is rather small 5.5%. This is attibuted to the nature of cellulose from bagasse which is a high crystallinity fiber, approximately 65.5 CI% [24]. As such the CMC_B film is rather brittle and not flexible.

Table 3. The propertisof spray drying CMCB film at 40% NaOH (w/v).

Film properties	40 %NaOH (w/v)
Thickness (mm)	0.04± 0.001
Film water solubility (%)	85.84 ± 0.37
Tensile strength (MPa)	45.62 ± 1.95
Elongation (%)	5.50 ± 0.58

Conclusion

The results reveal that the spray drying and alkaline pretreatment processes in carboxymethylaton have significantly influence on the CMC_B properties, particularly its CMC yield, DS, CMC% and viscosity. Furthermore, the CMCb film of DS 0.69 is fully soluble in water. It shows good tensile strength but not flexible

Acknowledgment

The author are grateful for laboratory and the equipment support from Division of Packaging Technology, School of Agro-Industry, Faculty of Agro-Industry, Chiang Mai University, Petroleum and Petrochemical College, Chulalongkorn University and thank the Biodegradable Packaging for Environment Co., Ltd. that supplied raw materials.

References

[1] The office of cane and sugar board, Ministry of Industry, Thailand. The annual report of sugar canefarming2011/2012.http://www.ocsb.go.th/upload/journal/fileupload/923-5826.pdf. Accessed 30 January 2012

[2] Rowell, R.M. et al. *In* Paper and Composites from Agro-Based Resources, CRC Press, Boca Raton, Florida, 1997, pp. 7 – 427.

[3] Hattori, K., E. Abe, T. Yoshida and J. Cuculo. *Polymer J.,* 2004, 36 (2), 123-130.

[4] Adinugraha, M.P. and D.W. Marseno. Haryadi.*CarbohydrPolym.*, 2005, 62: 164-169.

[5] Heinze T, Pfeiffer K. *Die AngewandteMakromolekulareChemie*, 1999, 266(4638), 37–45.

[6] Pushpamalar V, Langford SJ, Ahmad M, Lim YY. *CarbohydrPolym.*, 2006, 64, 312–318.

[7] Heydarzadeh HD, Najafpour, Nazari-Moghaddam AA (2009. *World ApplSci J.*, 2009, 6(4), 564-569

[8] Baria BK, Singhal RS, Kulkarni PR. *CarbohydrPolym.,* 1997, 32, 299-231.

[9] Adinugraha MP, Marseno DW, Haryadi. *CarbohydrPolym.,* 2005, 62, 164–169.

[10] Keller, J. D. *In* M. Gliksman.*Food hydrocolloids* (Vol. 3) (pp. 45–104). CRC Press, Boca Raton, Florida, 1986.

[11] Rinaudo M. and Reguant J. *In*: Frollini E., Leao A. and Mattoso L.H.C. (eds). EmbrapaInstrumentacaoAgropecuaria, 2000, Sao Carlos, Brazil, pp. 15–39.

[12] Feddersen, R.L., Thorp, S.N. *In*: Whistler, R.L., Bemiller, J.N. (Eds.), Industrial Gums, Polysaccharides and their Derivatives. Academic Press, New York, 1993, pp. 537–578.

[13] Dapia S, Santos V, Parajo JC. Bioresour Technol., 2003, 89, 289-296.

[14] Bono A, Ying PH, Yan FY, Muie CL, Sarbatly R, Krishnaiah D. *Adv in Natural and Applied Science.,* 2009, 3(1), 5-1.

[15] Far, B. M. Master Thesis. GaziUnivrsity, Ankara, 1992.

[16] Bono A, Ying PH, Yan FY, Muie CL, Sarbatly R, Krishnaiah D. *Adv in Natural and Applied Science.,* 2009, 3(1), 5-11

[17] Jangchud, A., Chinnan, M.S. *J. Food Sci.,* 1999, 64, 153-157.

[18] Tezuka Y, Tsuchiya Y, Shiomi T. *Carbohydr Researc.,*1999, 291, 99–108.

[19] Vehring R. *Pharmaceutical Research.,* 2008, 25, 999–1022.

[20] Varshney VK, Gupta PK, Sanjar N, Ritu K, Amit B, Soni PL. *CarbohydrPolym.*, 2006, 63, 40–45.

[21] Almedia, A. R., Dias, M. L. *Polymer Degradation and Stability*, 1997, 56, 331–337.

[22] Kamthai, S. and Magarapha R. in 6th PACCON e-Abstract, Chiang Mai, Thailand, 2012, CD.

[23] Hasan, T., Nurhan A. *CarbohydrPolym.,* 2003, 54, 73-82.

[24] Roberta C.C., Fabiana M. M., Clarissa P., Celso A., Wanderley S., Yuri A., Elba P. B., Viridiana F. L.*Biotechnology for Biofuels.*, 2012,5:36, 1-8

O-11-564

PNEUMATIC TRANSFER MOLDING TO FABRICATE THREE-DIMENSIONAL BIODEGRADABLE POROUS SCAFFOLDS

M.T. Lin, R.H. Hong, P.L. Chen, S.Y. Yang*

Department of Mechanical Engineering, National Taiwan University, Taiwan – syyang@ntu.edu.tw

Abstract - Porous 3D biomedical Scaffolds are essential and critical for tissue engineering. The scaffold can be used to culture autologous cell in vivo. Stimulated by growth factors and signals, cells can grow into the desired tissue morphology so that they can be transplanted back into the body. Conventional scaffolds are mostly prepared by organic solvents or rapid prototyping forming. There are problems such as residual solvent, complicated process and high cost. In order to overcome these problems, a pneumatic transfer molding is designed and implemented to fabricate scaffolds. The biomaterial polymers poly(DL-lactide-co-glycolide(PLGA)) were used in this research. It was blended with porogenic agent sodium chloride with size ranges from 200μm to 500μm. The experiment included two parts: In the first part, we used different mixing ratio and forming parameters to fabricate scaffolds. Secondly, cell viability study was performed. The weight ratio of PLGA:NaCl=1:11 yielded the best result. The scaffolds have porosity over 90 % with pore size 200 ~ 500μm, which satisfies the demanded of medical chondrocytes. The investigation of cell growth shows that the scaffold successfully facilitates cell growth and extracellular matrix content.

Keywords: Pneumatic Transfer Molding, Porosity, Moisture, Scaffold, Cell.

Introduction

Scaffold with pores of high porosity and interconnectivity are essential for facilitating cell seeding, adhesion, growth, differentiation and eventual tissue regeneration. Tissue engineering is playing critical roles in the cartilage reconstruction and tissue repair. Fabrication of 3D porous scaffolds requires the integration of material science, engineering, and medical science. There are many reported methods for the fabrication of scaffolds [1] such as solvent-casting particulate-leaching [2], membrane lamination, chemical foaming and solution casing. Cytotoxic organic solvent such as acetone and chloroform were used in most of these methods which often resulted in residual solvent [3]. Other methods like rapid prototyping require expensive machine, complicated and time-consuming procedures, and the scaffolds are often weak in mechanical strength.

The goal of the present study is to combine this existing salt particulate leaching technique with polymer molding technology without using organic solvents. A pneumatic transfer molding machine is designed and implemented for fabrication of scaffolds. This device has the following characteristics: no solvent involved, no screw/ piston, user-friendly and low cost.

Experimental

2.1 Pneumatic transfer molding machine

In this study, porous scaffold were fabricated via a pneumatic transfer molding machine at appropriate temperature followed by particulate leaching. The setup of the pneumatic transfer molding machine, which is manufactured from 6061 Aluminum Alloy, as shown in Fig. 1., is described as follows:

(a) The functions of feeding tube are material feeding and gas injection. The feeding tube is coated by Diamond-like Carbon which provides anti-corrosion ability and anti-sticking property.

(b) The outer feeding tube is heated by a spiral heating tube (E-heater, Taiwan) to melt scaffold materials and to maintain appropriate temperature.

(c) The mold is implemented mica heater (E-heater, Taiwan) to maintain mold temperature for forming. The mold insert is designed for rapid mold exchange.

The schematic and the photo of the pneumatic transfer molding machine are shown in Fig. 1(a) and Fig. 1(b), respectively.

2.2 Molds

The mold is composed of mold bases, made of aluminum 6061, and the mold insert. There are two types of mold insert. One is for molding the disk with 5mm in diameter and 2mm in thickness. The other is for molding the ear-shaped made of PDMS (Sylgard184, Dow Corning, U.S.A). The upper mold is tapered to fit the feeding tube with higher accuracy.

2.3 Materials

The Biomaterial polymers used were poly(DL-lactide-co-glycolide(PLGA) with a DL-lactide acid to glycolic acid ratio of 85:15 with molecular weight of 206,000 (Green Square Materials Inc.,Taiwan). The biomaterial polymers were available in a granular form. Porogenic agent was sodium chloride (Chang Gung Medical Foundation). The salt were sieved by standard sieves of 40 mesh. The size ranges from 200 μm to 425 μm.

2.4 Mixing, molding, and leaching procedures

The PLGA copolymers were premixed with Porogenic agent NaCl using a Teflon beaker by DLC coated mixing bar in oven at 140℃. After melted, the PLGA/NaCl mixture was transferred into mold by gas with pressure of 1~6 kgf/cm^2. In the end, the scaffold specimen was leached in the deionized water temperature of 35 ℃ (setting is far below Tg point). No sodium chloride residue discovered after 18 hours of leaching, determined by silver nitrate-sodium chloride residue test.

First, we investigated scaffolds with different mixing ratio of PLGA/salt mass ratio from 1:9~1:12. The process parameters were: The heating tube temperature of 200 ℃, pneumatic injection pressure 4kgf/cm^2 (holding pressure), mold temperature 90℃.

After finding the best weight ratio (PLGA /NaCl), with the same mixing and leaching before and after, a series of transfer molding experiments was conducted with different forming parameters shown in Table 1.

2.5 Characterization of scaffolds

Since the characteristic of scaffold influence the rate and quality of new tissue formation, the following facilities were used to characterize the scaffold:

(1) Scanning electron microscope (JEOL JSM-6500F, Japan) was used to examine the surface and cross section of scaffold. The scaffolds were prepared by a surgical blade before being observed by with SEM.

(2)Precision electronic scales was used to measure porosity of scaffold by soaking alcohol and deionized method. Porosity and moisture(water uptake ratio) affect the mass transportation of oxygen and nutrients into the interior of scaffold which supports cellular grow in pore region. The porosity(P) and moisture(M) of scaffold were determined after processing precision weighting using the following equations:

$$P = (V - W / \rho) / V \times 100\% \qquad \text{-Eq.1}$$

Where V is the volume of the scaffolds, W is the weight of the scaffold, and ρ is the density of PLGA, 1.3 g/ml.

$$M = (W - W_0 / W_0) \times 100\% \qquad \text{- Eq.2}$$

Where W is the absorption DI-water of scaffold, and W_0 is the dry weight of the scaffolds.

2.6 Cell Culture planning

The porcine chondrocytes were isolated from the freshly harvested articular cartilage of baby pigs, weight 7-9 kilogram. Chondrocytes cell were plated at a density of 1x10^6 cells/mL and cultured until passage 2 for further experiments. Three aspects were examined:

(1) 3-(4,5-Dimethylthiazol-2-yl)-2,5-dipheny-ltetrazolium bromide (MTT, Sigma-Aldrich) reagent was used to test cells on the Scaffold at 1, 3 ,7 ,14 and 28 days. The purpose is to investigate the affixed cell attachment and grow rate situations. At each time point, the supernatant was removed from the plate. Scaffolds were washed with PBS solution and transferred into a new 24-well plate. Each scaffold was then immersed in 0.1 ml of MTT solution and 0.4 ml of cell medium (5:1 of PBS solution and MTT stock mixture), and incubated for 4hrs at 37℃. Scaffolds were then transferred to an adjustable pipette(Eppendorf). 0.5 ml of Dimethyl Sulfoxide (DMSO, Sigma-Aldrich) was added to dissolve intracellular formazan crystals. The supernatant was then measured by spectrophotometer at 540nm.

(2) In Live & Dead / SEM section, after seeding 1x10^6 cell/ml onto each scaffold. Cells and ECM secretion status was observed during 2, 4 and 8 weeks. Scaffolds were washed with PBS solution and transferred into a new 24-well plate. Wrapped the plate by thin foil and stored at incubator for 15 minutes after added 1ml of Calcein and Ethidium (Calcein 4 mM and Ethidium 2 mM in 10 ml PBS). Before viewed under a fluorescent confocal microscope, specimens were washed with PBS.

Results and Discussion

3.1 The characteristics of scaffolds with different mixing ratio

The porosity and moisture were measured and the results were shown in Table 2. The disk with a mixing ratio of 1:11 yield a porosity up to 85.59%, with pore size 200-500 μm. Note that matrices with the mixing ratio 1:12 yielded low porosity. Little open porous were found on the external surface.

The scaffolds morphology were observed in surface and cross sections under SEM. The surface pore structures of scaffold with mixing ratio 1:11 showed smaller wall thickness between pore and pore and better interpore connections, which facilitated flow transport of nutrients and metabolic waste. The decrease of wall thickness was followed by increasing the content of particulates NaCl.

From the porosity, moisture and SEM image result , we know the mixing ratio 1:11 matrices exhibited the best porosity and moisture compared to mixing ratio 1:9, 1:10, 1:12, so we choose mixing ratio 1:11 to carry out forming parameter testing.

3.2 The characteristic of different forming parameter

When fixing the mold temperature at 90℃, and changing packing pressure from 1 kgf/cm^2 to 6 kgf/cm^2, we found out similar results, therefore, we changed the mold temperature from 60℃ to 140℃ with intervals of 20℃. The porosity and moisture improved when mold temperature increased, as mold temperature arrived at 120℃ with the best effect porosity 97.13% and moisture 395.34%, as shown in Table 3. However, when the mold temperature reaches 140℃, the PLGA polymer startes to degrade and produced the phenomenon of yellowing, therefore, the porosity and moisture value decreased to79.76% and 218.79%.

After measuring the porosity and moisture, we observed the scaffolds morphology in surface section and cross section. Fig. 2. shows SEM image in forming parameter: Mold temperature 120℃, packing pressure 4kgf/cm^2, at this parameter the scaffold produced 200-500μm abundant interconnected macropores were distributed present in the surface and cross section region.

3.3 The result of ear-shaped of scaffolds forming

The ear-shaped scaffolds were fabricated by the best forming parameter (mold temperature 120℃, packing pressure 4 kgf/cm^2) and observed the morphology by SEM. Fig. 3 shows the ear-shaped scaffold fabricated by pneumatic transfer molding machine and the SEM image of the scaffold. The pore size is 200-500 μm and interconnected.

3.4 The results of scaffolds with cartilage cell seeded

(1)MTT assay of cartilage cell seeding

To quantitatively compare the viability, the MTT agent was normalized by the cell numbers at each point and is presented in Fig. 4. It shows that the viability per 10^6 cells (cell density 6.5×10^7 cells/ml)in the scaffolds at all the time points is always larger. Only the cells number decreased in the scaffold at 7 days in order to adjust the environment impact.

(2)Live & Death of cartilage cell seeding

To evaluate the cartilage cells and ECM distribution in the PLGA scaffold after cell seeding 2, 4, and 8 weeks, the cartilage cells were labeled with fluorescence and imaged by confocal laser scanning microscopy (Wavelength 540nm). Labeled cartilage cells showed uniformly green fluorescence, indicating that the cells secreted numerous(red color represents the dead cells), as shown in Fig. 5. From the confocal image results, the surface section on scaffold have abundant cartilage cells after 2 weeks. After cartilage cell seeding for 2 weeks, sporadic cartilage-like tissues were observed in the cross section, indicating that the cells starts to grow from the outer of the scaffolds to the innerpart of the scaffolds. After 4 weeks, the cells keep growing deeper into the scaffold and the cartilage-like tissue grew further inside. The chondrocyte cells proliferation and viability increased along with the culture time 2 and 4 weeks. After 8 weeks of in vitro culture, homogenous cartilage-like tissue with abundant cartilage specific extracellular matrices (ECM) observed in the PLGA scaffold. In the meantime, the cells proliferation gradually grew in accordance with the cartilage properties.

(3)SEM of cartilage cell seeding

After 2, 4, 8 weeks incubation of cartilage cell scaffold, SEM examination showed that the seeded cells adhered to scaffolds and covered the pores compare to scaffold without cell seeding as shown in Fig. 6. The cartilage cells in the cross section of PLGA scaffolds after 2 weeks of seeding only covered a few pores, whereas the cartilage cell in the cross section of PLGA scaffolds after seeding for 4 and 8 weeks covered most of the pores and produced more ECM matrixes. Higher magnification showed that cartilage cells and ECM adhered even better after 8 weeks than 2 and 4 weeks. The cell and ECM amount increased along with increasing of time. These indicted that PLGA scaffold possessed exellent biocompatibility for attachment of cartilage cell in vitro.

Table 1 –forming parameters planning

	Fixed Parameter		Variation of Parameter	
	Feed Tube Temperature (℃)	Packing Time(min)	Mold Temperature (℃)	Packing Pressure(kgf/cm^2)
Parameter 1	200	5	60℃	1~6
Parameter 2			80℃	1~6
Parameter 3			100℃	1~6
Parameter 4			120℃	1~6
Parameter 5			140℃	1~6

Table 2 –Measured and calculated characteristic of different mixing ratio scaffold

Mixing Ratio	Porosity (%)	Moisture (%)	PoreSize (μm)
1:9	64.91	249.63	200~500
1:10	75.89	301.59	200~500
1:11	85.59	344.07	200~500
1:12	55.54	150.61	200~500

Table 3 –Measured and calculated characteristic of scaffold by fabrication of variation mold temperature

Mold Temp.(℃)	Porosity (%)	Moisture (%)	PoreSize (μm)
60	84.59	236.95	200~500
80	90.16	298.23	200~500
100	95.90	380.32	200~500
120	97.13	395.34	200~500
140	79.76	218.49	200~500

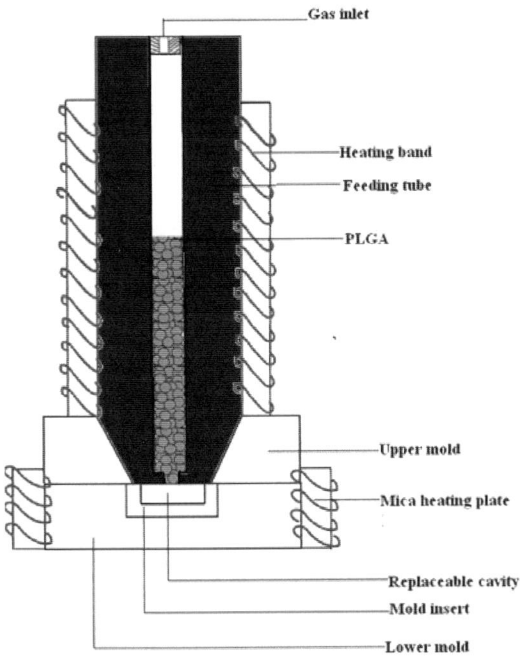

Figure 1(a) –Schematic of pneumatic transfer molding machine

Figure 1(b) –Photo of pneumatic transfer molding machine

Figure 2 –The SEM of the mixing ratio 1:11 of PLGA scaffolds

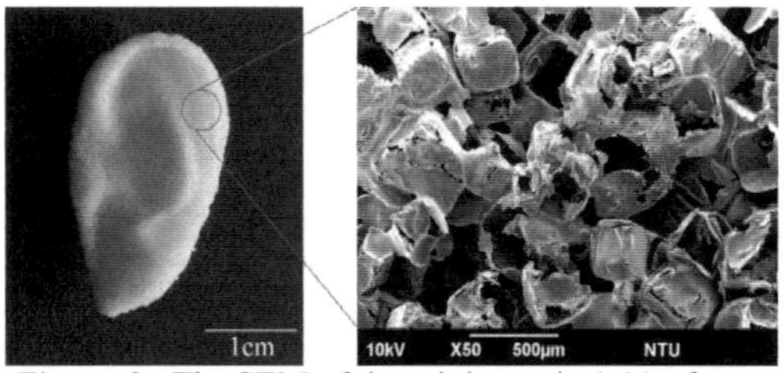

Figure 3 –The SEM of the mixing ratio 1:11 of ear-shaped PLGA scaffolds

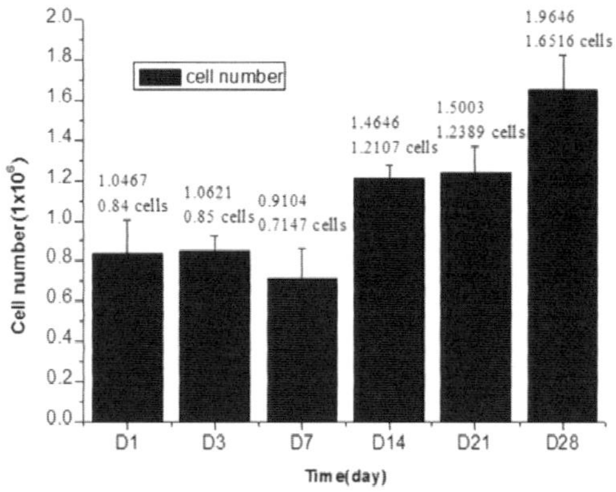

Figure 4 –MTT assay of scaffolds at different days

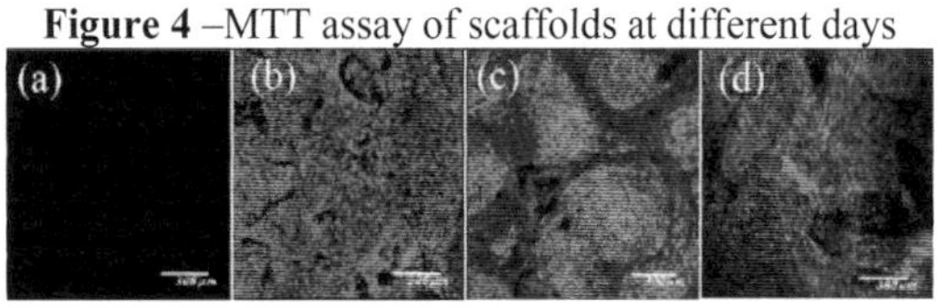

Figure 5 –Live & Death image of cartilage cell seeding at (a) without cell (b)2 weeks (c) 4 weeks (d) 8 weeks

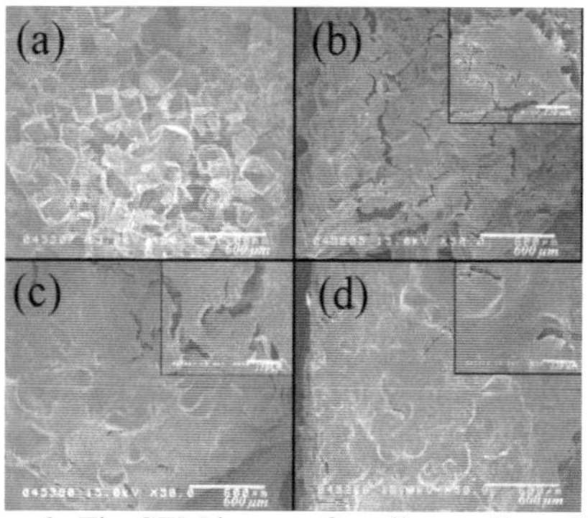

Figure 6 –The SEM image of cartilage cell seeding at (a) without cell (b)2 weeks (c) 4 weeks (d) 8 weeks

Conclusions

This article has adopted a solvent-free method of processing biodegradable PLGA polymers as chondrocyte tissue reconstruction. The experiment successfully fabricates the three-dimensional porous scaffold by using pneumatic transfer molding. The scaffold has appropriate porosity, moisture and morphology. This biodegradable PLGA scaffold is a suitable scaffold providing adequate support for the cartilage attachment and proliferation after cells seeding testing for 2, 4, 8 weeks in vitro.

In the future, a polymeric scaffold material can permit application of a solid-free form fabrication technology, so that a porous scaffold with any desired 3D geometry can be design and fabrication by using pneumatic transfer molding machine and changing the rapid mold.

Acknowledgements

We would like to thank the National Science Council of the Republic of China (Taiwan) for financial support of this research.

Assistance provided by Dr. Chih-Hao Chen was greatly appreciated. Special thanks are extended to the staff of Chang Gung Medical Foundation.

References

1. H.Y. Cheung; K.T. Lau; T.P. Lu; D. Hui *Compos. Part B-Eng.* 2007, *38*, 291
2. S.S Kim; M.S. Park; O. Jeon; C.Y. Choi; B.S. Kim *Biomaterials,* 2006, *27*, 1399
3. J. Manson; D. Dixson *J Biomater. Appl.* 2012, *26*, 623

O-11-567

STUDY OF MICROSTRUCTURE AND PHYSICAL/MECHANICAL PROPERTIES OF PLA/TPS/NANOCLAY NANOCOMPOSITES

F. Afsharnajafi, H. Garmabi*

Polymer and Color Engineering Department, AmirKabir University of Technology, P.O. Box 15875/4413, Tehran, Iran, garmabi@aut.ac.ir

Abstract - Urgent need for developing biodegradable polymers to replace petrochemical-based polymers has led to an increasing interest in using polylactic acid (PLA) and thermoplastic starch as packaging material. The addition of nanoclay to form a polymer nanocomposite is recently developed due to significant improvement in mechanical, barrier and various other properties as compared with those of virgin polymers. In this investigation, two extrusion stages were carried out to compounding polylactic acid (PLA), sorbitol-plasticized thermoplastic starch (TPS) and natural montmorillonite (MMT). Blend morphology, tensile properties as well as dynamic rheological properties were measured. XRD patterns showed intercalated microstructure for the prepared nanocomposites. Dynamic rheological measurements showed higher viscosity and more pronounced elastic properties for nanocomposites compared to pure PLA. PLA/TPS showed lower modulus and strength than neat PLA, as expected. The incorporation of MMT in TPS leads to an increase of the Young's modulus of the nanocomposites.

Keywords: polylactic acid, thermoplastic starch, nanoclay, mechanical properties, microstructure.

Introduction

Polylactic acid is the most extensively researched and utilized biodegradable and renewable thermoplastic polyester, with potential to replace conventional petrochemical-based polymers. PLA is biodegradable, biocompatible and compostable and it has better thermal processibility compared to other biopolymers. However, some of its properties, like high gas permeability, poor toughness, slow degradation rate and high price limit its use in certain applications [1,2].
Starch is a low-cost and abundant biopolymer which is attracting interest as an effective packing material. Various plasticizers have been used with starch to convert it into thermoplastic starch (TPS), like water, glycerol, sorbitol, urea [3-5]. TPS offers better material processibility and better dispersion than neat starch in PLA. The main disadvantages of TPS are its moisture sensivity, limited temperature stability, plasticizer lost and starch recrystallization with time leading to embrittlement [6,7].
PLA blends with biodegradable polymers like TPS have been extensively investigated because they offer property improvements without compromising biodegradability [1]. However, hydrophobic PLA and hydrophilic TPS are thermodynamically immiscible, which generally leads to poor adhesion between the two components and thus poor performance. Various compatibilizers and additives to improve the interfaces between PLA and TPS have been evaluated, such as polyvinyl alcohol, acetyl triethyl citrate and maleic anhydride [8-11]. But it was found that the sorbitol-plasticized TPS can uniformly distribute in the PLA even in absence of any compatibilization [12].
Organic-inorganic hybrids have attached great interests since they exhibit unexpected hybrid properties synergistically derived from the two components [13]. With only a few percent of clay, polymer-layered silicate exhibit improved mechanical, thermal and barrier properties compared with the pristine polymers [14].
Park et al. investigate the influence of the natural and organically-modified MMT on the properties of TPS and showed natural MMT had the best dispersion in the TPS that lead to higher tensile properties and lower water vapor permeability than the TPS [15].
It was generally observed that in the case of PLA organically-modified MMT had better dispersion than natural MMT [16-18].
Arroyo et al. investigated the properties and structure of PLA/TPS/MMT nanocomposites using a twin-screw extrusion process. It was found that the clay was preferentially located in the TPS phase or at the blend interface [19].
This research work, considers the preparation of PLA/TPS/MMT nanocomposites in two extrusion stages. The objective of this study is to investigate the microstructure, physical and mechanical properties of the prepared nanocomposites.

Experimental

Materials

The polylactic acid (PLA 6350D) supplied by NatureWorks. Corn starch was provided by Mojallali Chemical Lab. D-Sorbitol was obtained from Roquette Company with a purity of 98.5%. Natural MMT, Cloisite Na^+, supplied by Southern clay products.

Blend preparation

PLA/TPS/MMT nanocomposites were prepared in two stages. First corn starch and sorbitol (36wt %) were mixed at 3000 rpm for 4 min. The powder mixture were then manually fed into the lab scale twin-screw extruder (L/D=40) at temperature profile of 130-150^{0}C and screw speed of 100 rpm. Then PLA, TPS (30, 45wt%) and MMT (1, 3, 5wt%) were totally mixed inside the same twin-screw extruder at temperature profile of 130-180^{0}C and screw speed of 100 rpm. PLA and blends of PLA/TPS were prepared using the same condition as reference compounds. All components were dried perior to compounding in an oven at 70^{0}C for 24 h.

Morphological Characterization

The morphology of the PLA/TPS blends and nanocomposites examined by scanning electron microscopy using Hitachi F4160 model operated at 15 kV and under high vacuum. The samples surfaces were microtomed at -100^{0}C and then the TPS phase was extracted with hydrochloric acid of 1M to improve contrast.

X-ray diffraction measurements (XRD) were performed to investigate the intercalation and exfoliation of MMT. XRD patterns were obtained by using a Poland Philips Xpert X-ray diffractometer with Cu Kα generator, scanning from 1 to 10^0.

Rheological studied

The rheological properties of neat PLA, PLA/TPS blends and PLA/TPS/MMT nanocomposites were investigated using a rheometric mechanical spectrometer (RMS; UDS 200, Anton Paar) equipped with parallel plate geometry (diameter= 25mm, gap=1mm). The frequency sweep tests were performed in the range of 0.1–625 s^{-1} at temperature of 190°C and with amplitude of 1% to maintain the response of materials in the linear viscoelastic regime.

Tensile Characterization

Tensile tests were performed according to the ASTM D638 test method with a crosshead speed of 5 $mmmin^{-1}$. The samples were injected molded into dumbbell-shaped test specimens at temperature of 180^{0}C. Results from 3 specimens were averaged.

Results and Discussion

Morphology

SEM photomicrographs of PLA/TPS and PLA/TPS/MMT nanocomposites (3 wt% MMT) are shown in Figure 1. The TPS phase presented a uniform size distribution with particle size between 0.2 and 0.4 µm for 30wt% TPS blend and between 0.7 and 0.9 µm for 45wt% TPS blend. The particles generally had regular spherical shapes. Lack of Sorbitol migration from the TPS to the PLA leads to a low viscosity ratio of the dispersed TPS phase and the PLA matrix which in turn lead to uniform and homogenous blend morphologies that is unexpected for uncompatibilized PLA/TPS blends [12]. The slight decrease in TPS phase size was observed upon MMT addition.

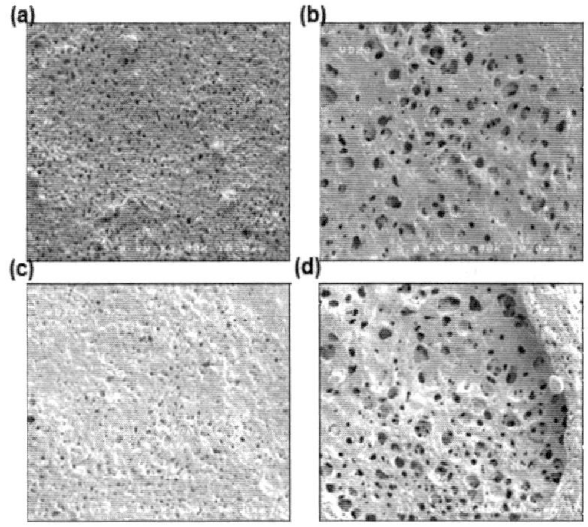

Figure 1. Morphology of PLA/TPS blends and nanocomposites, (a) PLA/30TPS (b) PLA/45TPS (c) PLA/30TPS/3MMT (d) PLA/45TPS/3MMT.

X-ray analysis

Figure 2 shows the XRD patterns of MMT and PLA/TPS/MMT nanocomposites. MMT showed the principal peak on 2θ=9.06^0 corresponding to an interlayer spacing of 0.98 nm. This peak was shifted to lower angles in all formulations indicating the intercalation of polymer chains inside the clay layers.

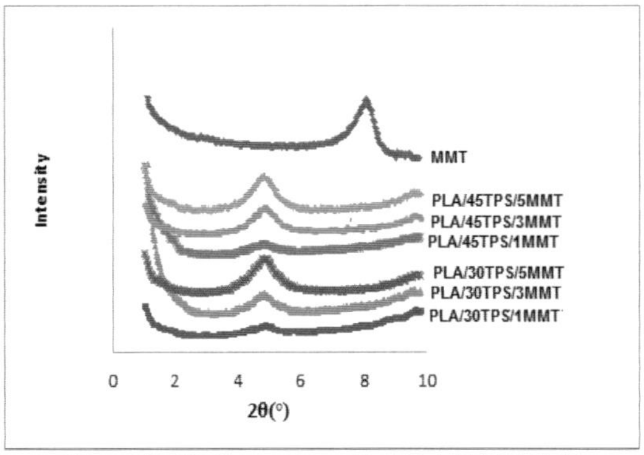

Figure 2. XRD spectrum for PLA/TPS/MMT nanocomposites.

Rheological properties

Figure 3(a,b) showed the storage modulus G' and complex viscosity η* of PLA, PLA/TPS blends and PLA/TPS/MMT nanocomposites as a function of frequency at 190 ^{0}C. It was clear that complex viscosity of neat PLA exhibited Newtonian liquid behavior at low frequency region. The complex viscosity of the PLA/TPS blends were higher than that of PLA. It could be seen that storage modulus of PLA/TPS blends increased with the increase of TPS content. The increase in viscosity and elasticity of the PLA/TPS blends compared with PLA can be attributed to the increase of the interfacial adhesion between PLA and TPS, caused by the formation of entanglements between TPS and PLA chains [20].

The complex viscosity of the nanocomposites with different amount of MMT loading apparently showed non-Newtonian shear-thinning behavior. The slope of G' in the low frequency region decreased to values close to zero for all nanocomposites that indicated solid-like behavior of nanocomposites on long time scales [21].

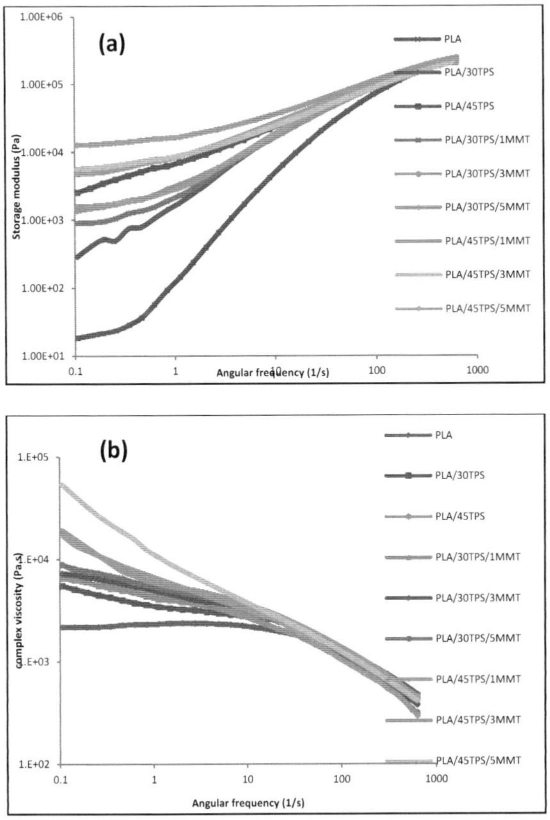

Figure 3. rheological properties of the specimens.

Mechanical properties

The tensile properties of PLA/TPS blends and PLA/TPS/MMT nanocomposites were given in Table 1. PLA showed high modulus and strength and low elongation at break. PLA/TPS blends exhibited lower modulus and strength and higher elongation at break duo to soft nature of TPS.

The MMT incorporation into PLA/30TPS increased the modulus and the tensile strength . The addition of MMT led to reduction in elongation at break. The MMT incorporation into PLA/45TPS showed the similar effect as into PLA/30TPS.

Table 1. mechanical properties of investigated materials.

material	Modulus (GPa)	Strength (MPa)	Elongation at break (%)
PLA	3.1 (0.5)	64.5 (0.3)	4.6 (1.7)
PLA/30TPS	2.2 (0.8)	51.2 (0.6)	4.7 (0.8)
PLA/45TPS	2.0 (1.5)	43.3 (0.9)	5.9 (0.7)
PLA/30TPS/1 MMT	2.5 (0.2)	49.4 (1.2)	3.6 (0.4)
PLA/30TPS/3 MMT	2.6 (1.4)	60.7 (1.2)	3.5 (0.4)
PLA/30TPS/5 MMT	2.9 (0.7)	49.2 (0.8)	3.0 (1.4)
PLA/45TPS/1 MMT	2.3 (0.5)	44.6 (0.2)	4.2 (1.7)
PLA/45TPS/3 MMT	2.4 (0.9)	51.9 (1.1)	3.9 (1.2)
PLA/45TPS/5 MMT	2.8 (0.9)	43.5 (1.8)	3.8 (0.5)

Conclusions

Two extrusion stages were carried out to prepare PLA/TPS/MMT nanocomposites. Sorbitol-plasticized TPS showed good distribution in the PLA. Intercalated structure observed for all prepared nanocomposites that lead to higher viscosity and more pronounced elastic properties than pure PLA and also tensile properties improvement.

References

1. Rahul M. Rasala,b, Amol V. Janorkarc, Douglas E. Hirta. Progress in Polymer Science ,2010, 35, 338.
2. K. Madhavan Nampoothiri, Nimisha Rajendran Nair, Rojan Pappy John. Bioresource Technology ,2010, 101, 8493.
3. Farayde Matta Fakhoury , Silvia Maria Martelli, LWT - Food Science and Technology, 2012, 49 149.
4. K. Krogars, J. Heinamaki, International Journal of Pharmaceutics, 2003, 251, 205-508.
5. X.F.Ma, J.G.Yu, Carbohydrate Polymers, 2005, 60, 111.
6. O. Martin, L. Averous, Polymer, 2001, 42, 6209.
7. N. Wang, J. Yu, Carbohydrate Polymer, 2008, 71, 109.
8. T. Ke, X. Sun, Journal of Polymer Environ, 2003, 11, 7.
9. J. Zhang, X. Sun, Macromol Biosci, 2004, 4, 1053.
10. C. S. Wu, Macromol Biosci, 2005, 5, 352.
11. M. A. Huneault, H. Li, Polymer, 2007, 48, 270.
12. H. Li, M. A. Huneault, Journal of Applied Polymer Science, 2011, 119, 2439.
13. A. Ova, Y. Kurokawa, Journal of material science, 2000, 41, 1365.
14. G. Jimenez, N. Ogata, Journal of Applied Polymer Science, 1997, 64, 2211.

15. H. Park, X. Li, Macromol. Mater. Eng, 2002, 287, 553.
16. L. Zaidi, Journal of Applied Polymer Science, 2010, 116, 1357.
17. K. Das, Journal of Applied Polymer Science, 2010, 118, 143.
18. J. Rhim, LWT - Food Science and Technology, 2009, 42, 612.
19. O.H. Arroyo, M.A. Huneault, Polymer Composites, 2010, 31, 114.
20. H. Zhang, N. Liu, Journal of Applied Polymer Science, 2012, 125, 550.
21. L. Zaidi, S. Bruzaud, Journal of Applied Polymer Science, 2010, 116, 1357.

O-11-627

EFFECT OF MATERIAL PARAMETERS ON MECHANICAL PROPERTIES OF BIODEGRADABLE POLYMERS/NANOFIBRILLATED CELLULOSE (NFC) NANOCOMPOSITES

Yottha Srithep[1], Ronald Sabo[2], Craig Clemons[2], Lih-Sheng Turng[1,*], Srikanth Pilla[1], and Jun Peng[1]
[1]*Polymer Engineering Center, Dept. of Mechanical Eng., University of Wisconsin–Madison, Madison, WI 53706, USA*
srithep@wisc.edu; turng@engr.wisc.edu (corresponding author); pilla@wid.wisc.edu; jpeng4@wisc.edu
[2]*USDA Forest Service. Forest Products Laboratory, Madison, WI 53726, USA; cclemons@fs.fed.us; rsabo@fs.fed.us*

Abstract - Using natural cellulosic fibers as fillers for biodegradable polymers can result in fully biodegradable composites. Biodegradable composites were prepared using nanofibrillated cellulose (NFC) as the reinforcement and poly (3-hydroxybutyrate-co-3-hydroxyvalerate, PHBV) as the polymer matrix. The objective of this study was to determine how various additives (i.e., blending PHBV with chain extender (CE) and/or Ecovio—a commercially available PLA/PBAT blend) affect the mechanical properties of injection molded PHBV/NFC nanocomposites. Tensile test bars of PHBV/NFC nanocomposites were produced via a novel process involving freeze drying and high-shear followed by injection molding. A three-variable, two-level, full-factorial design was used to investigate the effects and interactions of CE, Ecovio, and NFC content on the mechanical properties of the nanocomposites. At the level investigated, Ecovio had the largest influence of all of the additives studied on the mechanical properties of PHBV. Two significant interactions between variables were also found. The tensile modulus and strength increased slightly with 3% NFC content in neat PHBV but more considerably in PHBV with Ecovio. Moreover, 1 wt% CE increased the tensile modulus of neat PHBV slightly but had no effect on PHBV with NFC or Ecovio.

Keywords: Nanofibrillated cellulose (NFC), Biobased and biodegradable polymers, nanocomposites, poly (3-hydroxybutyrate-co-3-hydroxyvalerate), PHBV.

Introduction

The development of commercially viable biodegradable plastics is an important effort in preserving and revitalizing our global environment and economy. Polylactide (PLA) and polyhydroxy-alkanoates (PHAs) have attracted much attention over the last two decades mainly due to increasing environmental concerns [1, 2]. Generally, PHBV is less crystalline and more flexible than highly crystalline, extremely brittle, and relatively hydrophobic poly(3-hydroxybutyrate) PHB; however, higher toughness is required for PHBV to broaden its applications.

Poly(butylene adipate-co-terephthalate) (PBAT) is an aliphatic–aromatic copolyester that is bio-degradable [3]. Due to its high toughness and biodegradability, PBAT is an excellent candidate for toughening biodegradable and biobased polymers such as PLA and PHBV. The PHBV/PBAT blends were shown to have decreased tensile strength and modulus; however, their elongation and toughness dramatically increased [4].

Generally, the thermal degradation of PHBV is considered as an inevitable side effect of normal melt processing conditions. Inspired by the work of using chain extenders (CE) to control the degradation of neat PLA and recycled poly(ethylene terephthalate) (RPET) [5, 6], this study employed CE in the material formulation to reduce the effects of thermal degradation on PHBV.

Reinforcing fibers of various sizes and forms have been effectively used in polymer composites as reinforcing agents [7]. Nanofillers, however, are found to be preferable in many applications due to their high surface area-to-volume ratio. Nanofibrillated cellulose (NFC) is a biologically derived nanofiber reinforcer suitable for polymer materials. NFC is an interconnected web with fibrils having diameters in the range of 10 to 50 nm [8]. NFC film has a high Young's modulus (ca. 3.7 GPa as reported in previous study [8],) which makes it a suitable candidate for polymer reinforcement.

This study investigates the effects of adding CE, Ecovio, and NFC, on the mechanical properties of PHBV. Characterization techniques including tensile testing, Gel permeation chromatography (GPC) and scanning electron microscopy (SEM), were employed.

Experimental

Materials

The materials used in this study were PHBV, NFC, Ecovio, and CE. PHBV under the trade name Y1000P was purchased in both pellet and powder forms from Ningbo Tianan Biologic Material, Co. Ltd. NFC was prepared at the U.S. Forest Service Forest Products Laboratory (Madison, WI) according to a procedure described by Saito and Isogai [9]. An aqueous suspension of NFC fibers at 0.4 wt% was used in this study. Ecovio (a compatibilized blend of PLA/PBAT at a weight ratio of 45/55) was obtained from BASF and had a melt viscosity of 2.5 to 4.5 ml/10 min at 190 °C. The chain extender (Joncryl ADR 4368C) supplied by BASF was a multi-functional reactive polymer with Tg = 55°C, obtained in flake form. Joncryl ADR-4368 used to increase the melt strength of polyesters. The chain extender was used as received.

Experimental Design

A two-level, full-factorial statistical design (DOE) was used to establish the main effects and interactions of three additives on the mechanical properties of PHBV composites (cf. Table 1). Additional PHBV specimens filled with different Ecovio contents, (i.e. 25%, 50%, and 100%) were prepared for comparison.

Table 1: Material compositions (by weight).

Exp. No.	Sample	PHBV (%)	NFC (%)	Ecovio (%)	CE (%)
1	PHBV	100	0	0	0
2	PHBV+3%NFC	97	3	0	0
3	PHBV+75%Ecovio	25	0	75	0
4	PHBV+75%Ecovio+3%NFC	22	3	75	0
5	PHBV+1%CE	99	0	0	1
6	PHBV+1%CE+3%NFC	96	3	0	1
7	PHBV+1%CE+75%Ecovio	24	0	75	1
8	PHBV+1%CE+75%Ecovio+3%NFC	21	3	75	1

Sample Preparation

Composites were prepared using a two-step process. Master batch preparations using a freeze drier were followed by melt compounding, as described below.

Preparation of the master batch

To prepare the composites, the water medium had to be removed from the corresponding aqueous suspension of NFC. PHBV powder was first dispersed in distilled water, stirred for 1 hour, and then mixed with an aqueous suspension of NFC to reach a dry weight ratio of 85% PHBV and 15% NFC. The mixture was then stirred overnight using a magnetic stirrer. Afterwards, this suspension was quickly frozen by liquid nitrogen to prevent the PHBV powder from settling. In addition, rapid freezing might avoid NFC aggregation during freeze drying, as reported by Pääkkö et al. [10].

The frozen mixture was then freeze dried using a 4.5L Labconco FreeZone freeze drier to remove water thoroughly. Dried PHBV/NFC masterbatch was later diluted to 3 wt% NFC content by melt compounding.

Melt compounding and preparation of specimens

The PHBV pellets were combined with the NFC masterbatch, CE, and/or Ecovio in a variety of formulations. PHBV was dried in an oven for 2 hours at 90 °C before processing. Prior to injection molding, materials were melt compounded using a thermokinetic mixer. They were compounded in 50 g batches and discharged when the temperature reached 180 °C. fter discharge, the molten blend was subsequently granulated. Tensile bars (ASTM D638 Type V) were injection molded using a micro injection molding machine (DSM Xplore). The molding was done at 180 °C with a mold temperature of 25 °C, a cooling time of 15 seconds, and a holding pressure of 7 bar.

Characterization

Tensile testing

Tensile tests were performed on the injection molded tensile specimens following the ASTM D638 standard. The static tensile modulus, strength, and strain-at-break were measured at atmospheric conditions on an Instron 5865 mechanical testing instrument. The tensile tests were performed using an initial load of 0.5 N and a constant crosshead speed of 1 mm/min. Five specimens of each sample group were tested and the average results were reported.

The significant main effects and interactions of the material variables were determined for the tensile properties. The DesignExpert 6.0.101 software was used in the data analysis.

Gel permeation chromatography (GPC)

The number- and weight-average molecular weights (Mn and Mw, respectively) and the polydispersity index (PDI) for molded PHBV, and molded PHBV + 1 wt% CE were determined. Samples were dissolved chloroform via continuous stirring in a constant temperature sand bath (60 °C) for 1 hour. The sample solution was filtered through a 0.2 mm PTFE membrane filter. With an eluent flow rate of 1.0 mL/min, 100 mL samples were injected into a Viscotek model VE2001 with Model 302-050 tetra detector array (differential refractive index (RI)).

Scanning electron microscopy (SEM)

SEM images were examined using an SEM (JEOL Neoscope Benchtop) operated at 10 kV. All specimens were sputter-coated with a thin layer of gold prior to examination.

Results and Discussion

Freeze Dried Aqueous Suspension of NFC

The SEM images of PHBV powder, freeze-dried NFC, and freeze-dried PHBV + 15% NFC masterbatch are shown in Figures 1(a), (b), and (c), respectively. As shown in Figure 1 (a), the individual PHBV powder is smaller than 1 μm. Freeze-dried NFC (Figure 1 (b)) shows an interconnected fibrillar skeleton structure with diameters on the order of 1 μm although some nano-scale fibers were still present and others form 2-dimensional extended sheet-like structures.

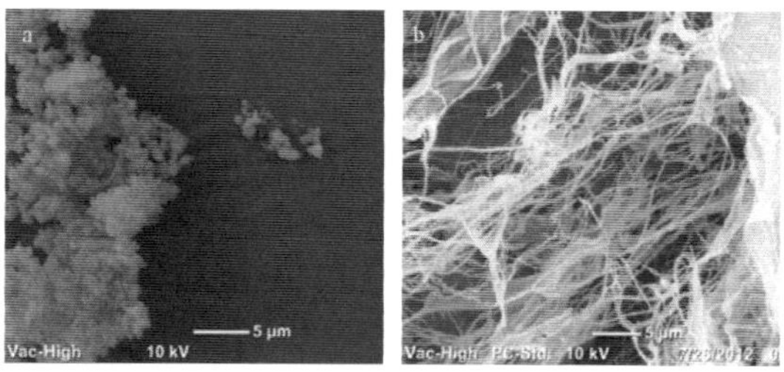

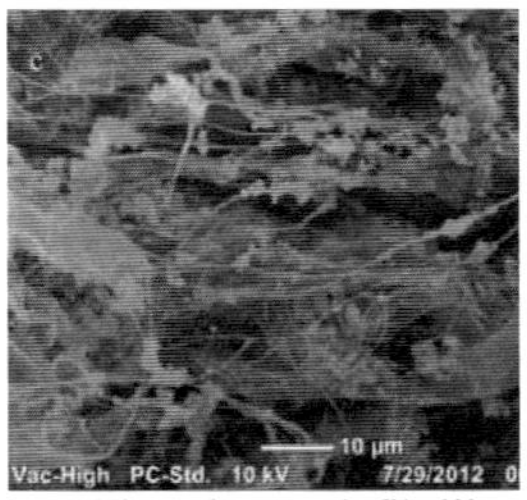

Figure 1: SEM images of: (a) PHBV powder, (b) freeze-dried NFC, and (c) freeze-dried PHBV + 15% NFC.

The observed fibrillar diameter after the freeze-dried process was much larger than what was observed (e.g., 5–10 nm) in the aqueous gels using transmission electron microscopy (TEM) in a previous study [8]. This indicates that some aggregation of nanofibers took place in the process of freeze drying. Similar behavior in freeze drying results has been reported in another study [10]. For the freeze-dried PHBV + 15% NFC, Figure 1 (c) shows that some of the PHBV powder attached to the fibrillar NFC network, while some of it aggregated. This PHBV + 15% NFC was used as the masterbatch in the subsequent melt compounding process.

Tensile Properties

As shown in Figure 2, Ecovio content (at the 75% level) had the largest influence among main effects and interactions. The influence of Ecovio content on modulus, strength and strain, can be seen in more detail in Figure 3, which shows tensile curves for a range of PHBV:Ecovio weight ratios. Strains increased from 3-21% but modulus and strength decreased with an increase in Ecovio content, demonstrating the toughening effect of PBAT in the Ecovio [4]. However, since Ecovio interacted with other variables, its effect was not consistent at different levels of the other variables. Hence, these interactions must be considered to adequately describe the behavior. For example, Figure 4 shows the interaction between Ecovio and NFC content for the modulus measurements. The addition of NFC into PHBV did not significantly change the tensile properties of PHBV. This might be due to the agglomeration of fibers which was confirmed by SEM.

However, NFC did increase the tensile modulus and strength of PHBV +75% Ecovio by 60% and 26%, respectively, suggesting good fiber–matrix adhesion [11]. Nevertheless, the strain at break for PHBV+75% Ecovio reinforced with 3%NFC decreased. Similar effects by other wood fiber on the tensile properties of PBAT, which is the matrix component in Ecovio, have been reported previously [11, 12]. Moreover, as shown in Figure 2 (Exp. Nos. 1 and 5), adding 1 % CE did not cause any difference in the tensile strength but induced a slight increase in the tensile modulus of PHBV by 12% (e.g., Figure 5) though not in the blends of CE with PHBV+75% Ecovio.

GPC results revealed that CE increased the average molecular weight of PHBV slightly. PHBV exhibited a number average molecular weight of 132,000±1500 before processing. After processing, the number average molecular weight decreased by 33%. As discussed, PHBV is susceptible to thermal degradation during melt blending and injection molding. However, the addition of 1% CE increased the number average molecular weight by 10%. The polydispersities of unprocessed PHBV pellets, molded PHBV, and molded PHBV + 1% CE were 2.4, 2.3, and 2.1, respectively.

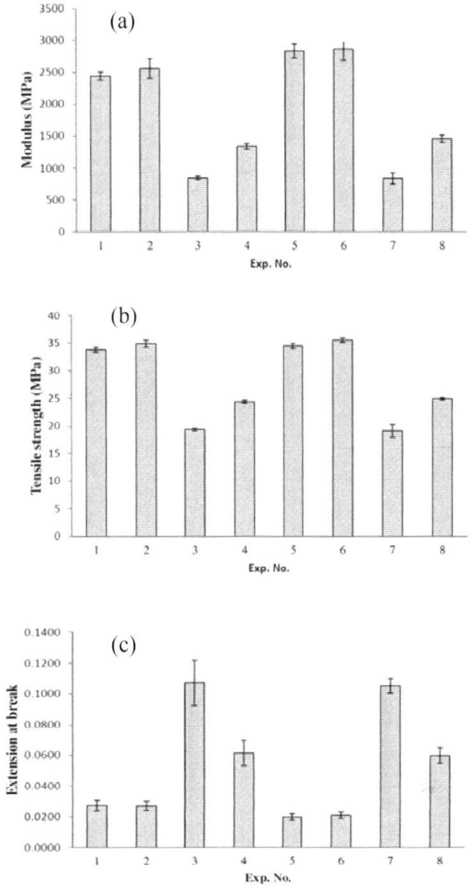

Figure 2: Mechanical properties in terms of (a) tensile modulus (MPa), (b) tensile strength (MPa), and (c) strain-at-break for (1) PHBV, (2) PHBV + 3% NFC, (3) PHBV + 75% Ecovio, (4) PHBV + 75% Ecovio + 3% NFC, (5) PHBV + 1% CE, (6) PHBV + 1% CE + 3% NFC, (7) PHBV + 1% CE + 75% Ecovio, and (8) PHBV + 1% CE + 75% Ecovio + 3% NFC (cf. Table 1).

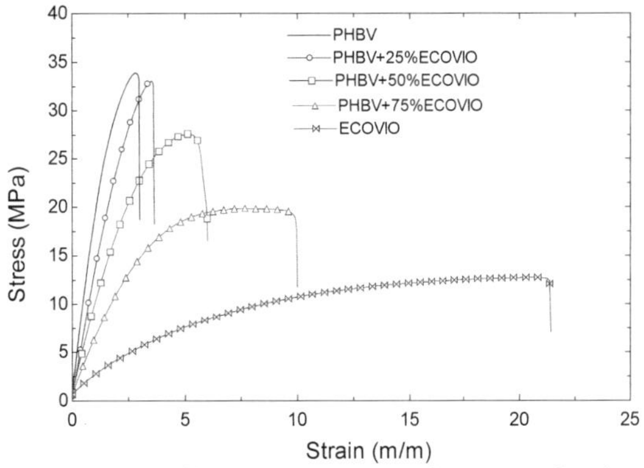

Figure 3: Tensile stress versus strain curves for the PHBV and Ecovio blends.

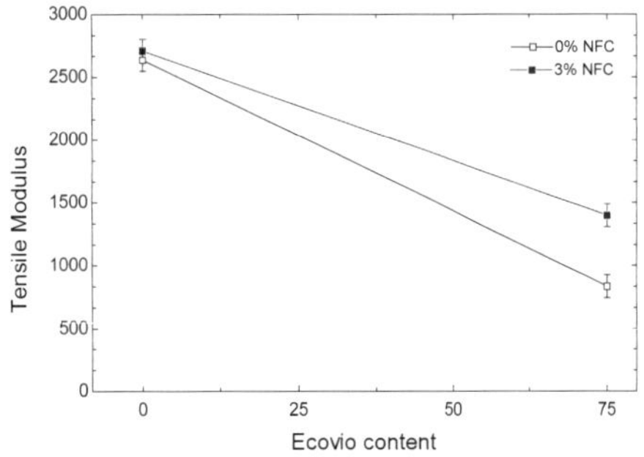

Figure 4: Ecovio content × NFC content interaction for the tensile modulus. Symbols represent the averages of the different levels of Ecovio and NFC content. Error bars show plus and minus one standard deviation.

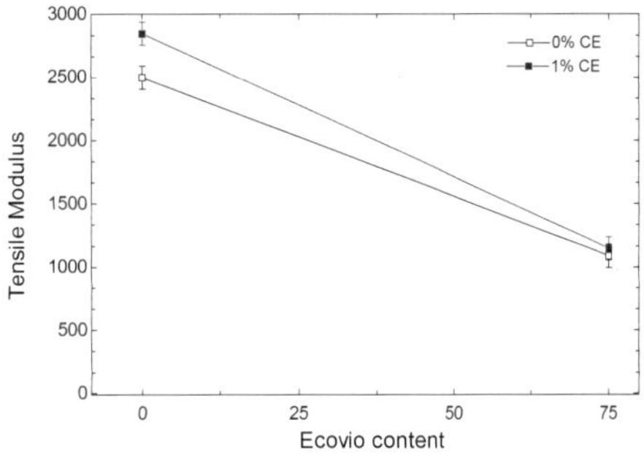

Figure 5: Ecovio content × CE content interaction for the tensile modulus. Symbols represent the averages of the different levels of Ecovio and CE content. Error bars show plus and minus one standard deviation.

Morphology of the Fracture Surfaces

Figure 6 shows the SEM images of the tensile fracture surfaces of the PHBV blends. The tensile fracture surfaces of PHBV + 75% Ecovio (Fig 6 (c)) and PHBV + 75% Ecovio + 3% NFC (Fig. 6 (d)) exhibited ductile fractures. This observation was supported by the values of strain-at-break obtained from the tensile tests. Furthermore, Figure 6 (b) shows that NFC aggregates were visible on the fracture surface and a few fibril aggregates were pulled-out.

Conclusions

A two-level, full-factorial DOE was performed to determine how NFC, Ecovio, CE, and their interactions affected the mechanical properties of PHBV blends and composites. The average tensile modulus of PHBV was 12% higher when 1% CE was added, which was due to the increase of PHBV's molecular weight. The Ecovio, because of its large content level, had the largest influence on the properties studied. Increasing Ecovio content increased the strain-at-break and reduced the tensile modulus and strength. These results were not surprising since Ecovio contains 55% PBAT, which is commonly used as a toughening agent. NFC, which tended to aggregate in PHBV composites, was found to reinforce PHBV + 75% Ecovio. To summarize, adding NFC, Ecovio, and/or CE to PHBV was found to be an effective way of modifying the mechanical properties of biobased PHBV.

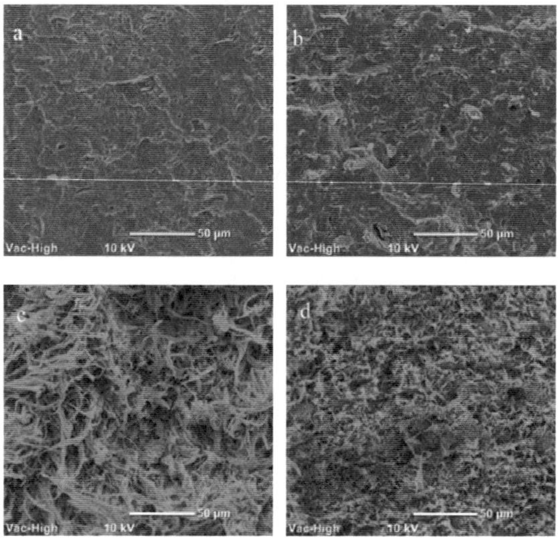

Figure 6: Tensile fractured surfaces of (a) PHBV, (b) PHBV + 3% NFC, (c) PHBV + 75% Ecovio, and (d) PHBV + 75% Ecovio + 3% NFC.

Acknowledgements

This research is supported by the United States Department of Agriculture National Institute of Food and Agriculture Award (No. 2011-67009-20056).

References

1. A. Mohanty, M. Misra, and L. Drzal, *J. Polym. Environ.* **10**, 1 (2002)
2. A. Javadi, Y. Srithep, J. Lee, S. Pilla, C. Clemons, S. Gong, and L.S. Turng, *Compos. Part. A Appl. Sci. Manuf.* **41**, 8 (2010)
3. S.Y. Gu, K. Zhang, J. Ren, and H. Zhan, *Carbohydr. Polym.* **74**, 1 (2008)
4. A. Javadi, A.J. Kramschuster, S. Pilla, J. Lee, S. Gong, and L.S. Turng, *Polym. Eng. Sci.* **50**, 7 (2010)
5. N. Najafi, M. Heuzey, P. Carreau, and P.M. Wood-Adams, *Polym. Degrad. Stabil.* **97**, 4 (2012)

6. Y. Srithep, A. Javadi, S. Pilla, L.S. Turng, S. Gong, C. Clemons, and J. Peng, *Polym. Eng. Sci.* **51**, 6 (2011)
7. E. Ten, J. Turtle, D. Bahr, L. Jiang, and M. Wolcott, *Polym.* **51**, 12 (2010)
8. Y. Srithep, L.S. Turng, R. Sabo, and C. Clemons, *Cellulose*, **19**, 4 (2012)
9. T. Saito, Y. Okita, T. Nge, J. Sugiyama, and A. Isogai, *Carbohydr. Polym.* **65**, 4 (2006)
10. M. Pääkkö, J. Vapaavuori, R. Silvennoinen, H. Kosonen, M. Ankerfors, T. Lindström, L.A. Berglund, and O. Ikkala, *Soft Matter*. **4**, 12 (2008)
11. E. Sykacek, W. Schlager, and N. Mundigler, *Polym. Compos.* **31**, 3 (2010)
12. E. Bodros, I. Pillin, N. Montrelay, and C. Baley, *Compos. Sci. Technol.* **67**, 3 (2007)

PREPARATION OF COMPOSITES OF POLY(LACTIC ACID) AND POLY(BUTYLENE SUCCINATE) REINFORCED WITH JUTE FIBRE

S. Pivsa-Art[1]*, N. Srisawat[1], W. Worrasettapong[2], S. Mathurosemontri[1], R. Wongpajan[1] and S. Thumsorn[1]

[1]*Faculty of Engineering, Rajamangala University of Technology Thanyaburi, Thailand – sommai.p@en.rmutt.ac.th, natee.s@en.rmutt.ac.th, cussu_think@hotmail.com, supaphorn.t@rmutt.ac.th*

[2]*Faculty of Science and Technology, Rajamangala University of Technology Thanyaburi, Thailand*

Abstract - Natural fiber such as jute is currently being developed as a possible reinforcement for polymeric materials due to their low cost, low density, and biodegradability. Natural fiber reinforced thermoplastic composites have gained wide interest in recent years. However, the problem with producing natural fiber reinforced thermoplastic composite is the difficulty to impregnate the fibers due to the high melt viscosity of the thermoplastic resin. In this circumstance, intermediate materials such as commingled yarns, have been developed to improve impregnation state in the composites. In this study, a continuous intermediate material was developed by aligning resin fibers alongside the reinforcing fiber with the braiding technique. This intermediate material is known as the "micro-braided-yarn" (MBY). The biodegradable polymers used in this study were poly(lactic acid) (PLA) and poly(butylene succinate-co-adipate) (PBSA). The content of PBSA in the hybrid composites were used at 0, 10 and 15 wt%. The composites prepared from alternate braiding and neat PLA or PLA/PBSA resin fibers MBYs, tensile modulus was increased**.** The ratios of PLA:PBSA studied were 90:10 and 85:15. The polymer blends were subjected to analysis of tensile strength, thermal property and morphology study. SEM photographs exhibited ductile behavior in PLA/PBSA blend, which there was no phase separation of the blend with PBSA content up to 15 wt%. It can be note that the mechanical performance of the composites is significantly enhaced by jute yarns as the natural fiber reinforcement through micro-braid yarn technique. The environmental friendly composites show satisfied high mechanical properties.

Keywords: Poly(lactic acid), poly(butylene succinate), jute, composite, micro-braided yarn

Introduction

The composite materials have been continuously used because of their excellent mechanical strength, high thermal resistance and light weight. Composites made from glass fiber or carbon fiber reinforced non-degradable polymer such as epoxy resin, polypropylene are not degradable and are the cause of environmental problems from their post-consumers.

Biodegradable polymers are synthesized from renewable resources or petroleum-based chemical such as poly(lactic acid) (PLA), polycaprolactone (PCL), poly(hydroxybutyrate-co-hydroxyvalerate) (PHBV), polyesteramide (PEA) and poly(butylene succinate-*co*-adipate) (PBSA). The products from these polymers are biodegradability and environmental friendly [1-5].

Natural fiber is also environmental friendly and generally used as a reinforcement for polymer composites [6-8]. Therefore, natural fiber reinforced biodegradable composites are quite promising as the novel composites for today.

Micro-braided yarn is a technique that polymer fiber is braided around the reinforcement fiber [9-10]. Therefore, the reinforcement at the center would perform a superior properties of the micro-braided yarn composite.

In this study, PLA/PBSA blend fiber was used as the polymer matrix, which reinforced by jute yarn using micro-braided yarn technique. The effects of blend ratio on morphology and mechanical properties of the micro-braided yarn composites were investigated. The performance of the micro-braided yarn on properties of the composites was elucidated.

Experimental

Materials

PLA (grade 2002D) was supplied by NatureWorks LLC, USA. PBSA (Binolle® 3000) was supplied by Showa HighPolymer Co., Ltd., Japan. Jute yarn (170 tex) was provided by Janata Jute Mill Ltd., Bangladesh.

Specimens Preparation

PLA and PBSA were dried in an oven at 80 °C for at least 4 hours. PBSA contents were varied at 0, 10 and 15 wt%. PLA and PBSA were melt spun in a single screw extruder equipped with multi-filament spinneret (24 holes). The barrel temperature was set at 150-180 °C with screw speed of 9 rpm. Take up speed was set at 25 m/min.

PLA/PBSA blends at various ratio were braided around the jute yarn by braiding machine to perform a micro-braided yarn as shown in Figure 1. The braiding speed was 30 rpm.

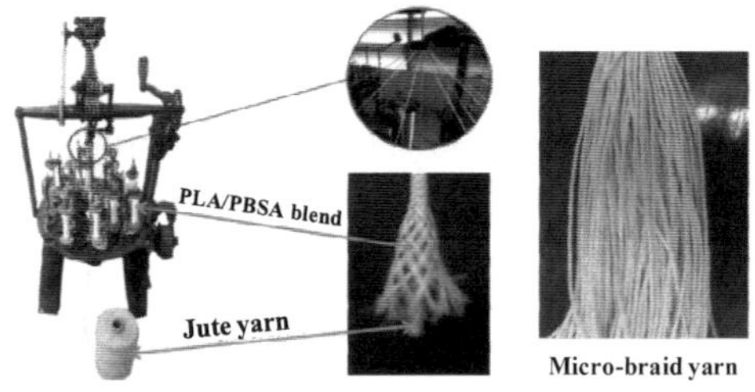

Figure 1 SEM photographs of PLA/PBSA blend.

The micro-braided yarn were wound around a metal frame 40 times then put in the metal mold. The mold was hot pressed at 150 °C. The mold was preheat for 3 minutes before full pressing at 20 MPa for another 3 minutes. The cooling time was 3 minutes. PLA/PBSA micro-braided yarn composite is shown in Figure 2.

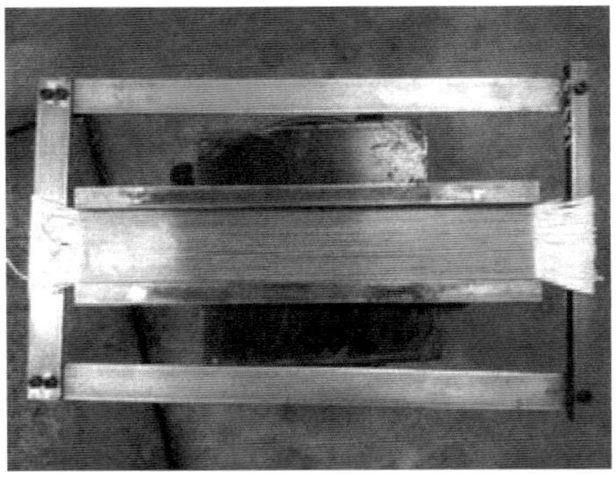

Figure 2 Compression moled of PLA/PBSA micro-braided yarn composites.

Characterization

Thermal properties of the blend were evaluated by using a differential scanning calorimeter (NETZSCH DSC 200). The specimens were heated from 30 to 200 °C under nitrogen. The heating rate was set at 10 °C/min.

Morphology of the blends was characterized by observing the cryogenic fractured surface of the specimens. A scanning electron microscope (JEOL, JSM6400) was used for surface observation. Gold was sputtered onto the specimens for electron conductivity.

Tensile properties of the blends fiber, micro-braid yarn and the composites were carried out using LLOYD tensile testing machine.

Results and Discussion

Figure 3 shows SEM photographs of the fractured surface of PLA/PBSA blends. Figure 3 (a) reveals smooth surface of PLA, which indicated a morphology of brittle fractured surface. On the contrary, Figure 3 (b) exhibited ductile behavior of PBSA fractured surface. In the blend, it can be seen that there is no phase separation in PLA/PBSA blends as shown in Figure 3 (c) and (d). The morphology of the blends are ductile characteristic, which the fractured surface is more ductile at higher content of PBSA in Figure 3 (d).

Table 1 presents the weight fraction of the blend fiber and jute yarns in micro-braided yarn. It can be seen that the micro-braid yarn are including with the blend fiber fraction of about 73% and the jute yarn fraction of about 27% of all different blend fibers.

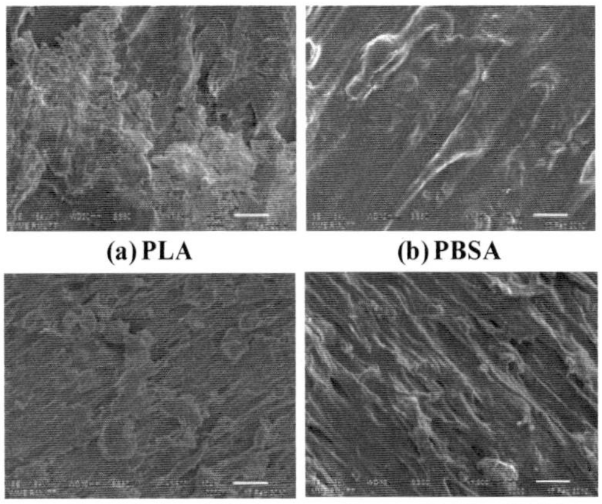

(a) PLA **(b) PBSA**

(c) PLA/PBSA (90/10) **(d) PLA/PBSA (85/15)**

Figure 3 SEM photographs of PLA/PBSA blend.

Table 1 Weight fraction of blend fiber and jute yarn in micro-braid yarn.

PLA/PBSA	Blend fiber (wt%)	Jute yarn (wt%)
100/0	72.4	27.6
90/10	72.0	28.0
85/15	73.9	26.1

Figure 4 to Figure 6 show the comparison of tensile properties between PLA/PBSA blend fibers and micro-braided yarn at various contents of PBSA. Tensile modulus and tensile strength of micro-braided yarns were higher than PLA/PBSA blend fibers as shown in Figure 4 and Figure 5, respectively. It was due to stiffness and toughness of the jute yarn enhanced both tensile modulus and tensile strength of micro-braided yarns. However, the tensile modulus and tensile strength of both specimens decreased when increasing PBSA content, which was due to high elasticity of PBSA.

Figure 6 shows elongation at break of micro-braided yarns and PLA/PBSA blend fibers. Elongation at break of the micro-braided yarns was lower than the blends fibers. It was attribute to jute yarn is tough but less elasticity, which exhibited less deformation in micro-braided yarns. In addition, the elasticity of PBSA could enhance elongation in both specimens.

Table 2 tabulates tensile properties of micro-braided yarn composites. It is interesting to note that both tensile modulus and tensile strength of the composites was significantly higher than the tensile properties of the individual fiber and yarn. It was due to the jute yarn was successfully effect in order to reinforce the polymer blend matrix. However, the tensile performance of the composites decresed when increasing the PBSA content in the blend fiber.

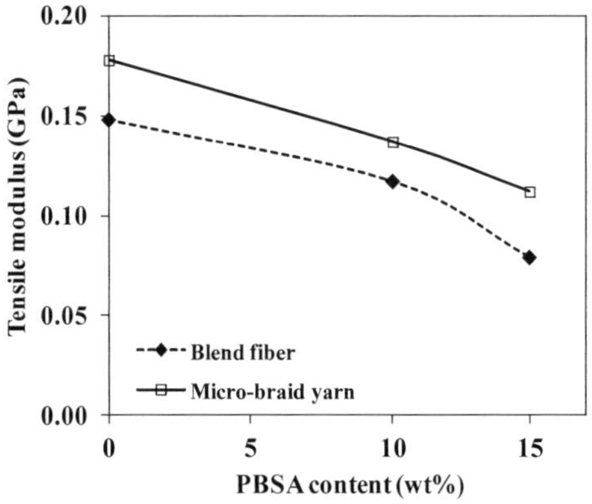

Figure 4 Tensile modulus of blend fiber and micro-braid yarn.

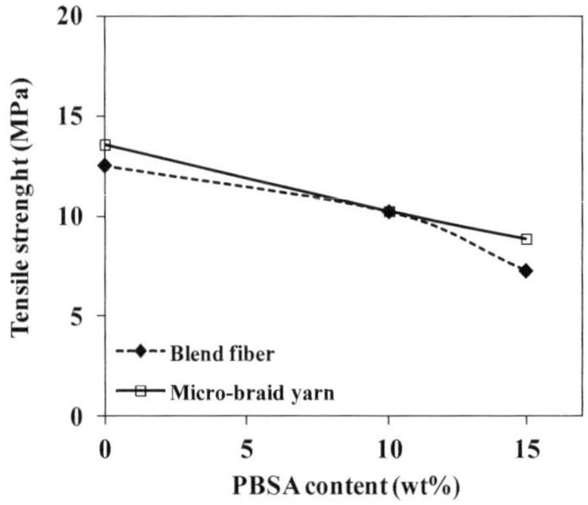

Figure 5 Tensile strength of blend fiber and micro-braid yarn.

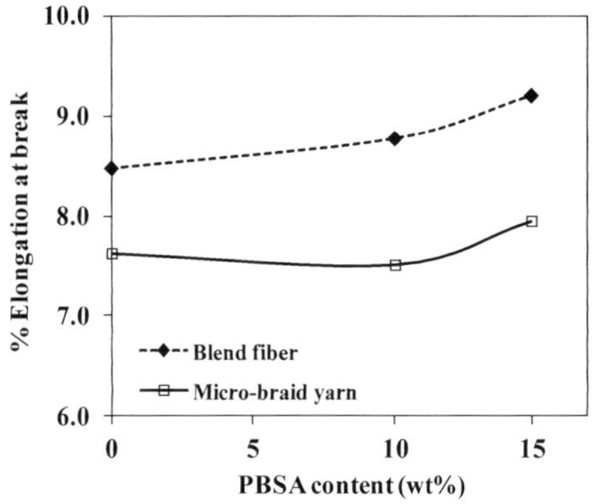

Figure 6 %Elongation at break of blend fiber and micro-braid yarn.

Table 2 Tensile properties of micro-braid yarn composites.

PLA/PBSA	Tensile modulus (GPa)	Tensile strength (MPa)
100/0	3.84	89.2
90/10	3.73	82.0
85/15	3.66	73.3

Conclusions

Natural fiber reinforce PLA/PBSA blend are successfully prepared through micro-braided yarn composites. The jute yarn presented as a reinforcement in this composite, which the stiffness and the toughness of the jute yarn enhanced tensile modulus and tensile strength of the composites. The tensile properties of the PLA/PBSA blend micro-braid yarn composites shows the highest values at PBSA 10 wt%, which exhibited tensile modulus of 3.73 GPa and tensile strength of 82.0 MPa. From this study, the environmental friendly composites of PLA/PBSA blend micro-braid yarn show satisfied high mechanical properties.

Acknowledgements

The authors would like to thank National Metal and Materials Technology Center (MTEC), Thailand for supporting the research fund.

References

1. W. Amass; A. Amass; B. Tighe, *Polym. Int.* 1998, *47*, 89.
2. R.E. Drumright; P. R. Gruber; D.E. Henton, *Advan. Mater.* 2000 *12*, 1841.
3. M. Avella; M.E. Errico, *J. Appl. Polym. Sci.*, 2000, *77*, 232-236.
4. L. Averous; N. Fauconnier; L. Moro, *J. Appl. Polym. Sci.*, 2000, *76*, 1117.
5. V. Tserki; P. Matzinos; E. Pavlidou; D. Vachliotis; C. Panayiotou, *Polm. Degrad. Stab.*, 2006, *91*, 367.
6. P. Wambua; J. Ivenes; I. Verpoest, *Comp. Sci. Tech.*, 2003, *63*, 1259.
7. L. Liu; J. Yu; L. Cheng; X. Yang, *Polm. Degrad. Stab.*, 2009, *94*, 90.
8. T. Nishino; K. Hirao; M. Kotera; K. Nakamae; H. Inagaki, *Comp. Sci. Tech.*, 2003, *63*, 1281.
9. O.A. Khondjer; U.S. Ishiaku; A. Nakai; H. Hamada; *Composites Part A.*, 2006, *37*, 2274.
10. K. Fujihara; Z.-M. Huang; S. Ramakrishna; K. Ramakrishna; H. Hamada, *Biomaterials*, 2003, 24, 2661.

O-11-638

DEVELOPMENT OF POLYMER BLENDS OF POLY(LACTIC ACID) AND POLY(BUTYLENE SUCCINATE) USING TALCUM AS ADDITIVE FOR INJECTION MOLDING APPLICATIONS

W. Pivsa-Art[1*], S. Thumsorn[2], J. Ratanapisit[1], H. Yamane[3] and H. Ohara[3]

[1]*Department of Chemical and Materials Engineering, Faculty of Engineering, Rajamangala University of Technology Thanyaburi, Thailand - weraporn.p@en.rmutt.ac.th, juraivan.r@en.rmutt.ac.th*

[2]*Department of Textile Engineering, Faculty of Engineering, Rajamangala University of Technology Thanyaburi, Thailand - supaphorn.t@rmutt.ac.th*

[3]*Department of Biobased Materials Science, Kyoto Institute of Technology, Kyoto, Japan - hyamane@kit.ac.jp, ohara@kit.ac.jp*

Abstract - Polymer blends of poly(lactic acid) (PLA)/poly(butylene succinate) (PBS) were prepared via the twin screw extrusion process with ratios of PLA:PBS of 90:10, 80:20 and 70:30. Talc has been added to improve mechanical properties and reduce production cost. The amount of talc was varied from 5 to 20 wt% of the polymer matrix. The blends with talc additives were compressed at 220°C and samples were subjected to physical and mechanical property analysis. Addition of talc had no significant effect to melt flow of PLA/PBS blend systems. Morphology investigation revealed well distribution of talc in the polymer matrix. Modulus of the blends increased while increasing the amount of talc additives. The impact strength of the blends was not significantly reduced with addition of talk but slightly increased with 5wt% of the mineral additive. The results of physical and mechanical investigation confirmed adequate properties for injection molding applications.

Keywords: Poly(lactic acid), poly(butylene succinate), talcum, additive, injection molding

Introduction

Recently, poly(lactic acid) (PLA) has become a polymer of large interest not only in the traditional areas of biomedical and pharmaceutical applications but also in the commodity area, i.e. mainly packaging applications. Indeed PLA has biodegradability, annually renewable availability, and process ability following standard melt processes such as injection moulding, film blowing or melt spinning [1,2]. However, application of PLA is limited due to its brittleness and melt viscosity for processing. Therefore, blending of PLA with higher toughness biodegradable polymer will overcome the problems and may enhance process ability of the polymer blends. Poly(butylene succinate) (PBS) is one of the aliphatic polyesters having good potential to blend with PLA.

Poly(butylene succinate) (PBS) is known under the trade name ''Bionolle'', synthesized through the polycondensation reaction of 1,4-butanediol, and succinic acid [3-5]. This white crystalline thermoplastic polymer exhibits not only a melting point similar to that of low-density polyethylene (LDPE), but also glass transition temperature (Tg) and tensile strength between those of polyethylene (PE) and poly-(propylene) (PP), and stiffness between that of LDPE and high density polyethylene (HDPE). In addition, this biodegradable polymer possesses satisfactory strength and toughness, close to those of LDPE, which is considered highly promising as a commercial commodity polymer [6-7].

However, the polymer blends of PLA and PBS still show insufficient mechanical and thermal properties as well as high production cost of biodegradable polymer components. Therefore, addition of inorganic additive is considered to be one of the most effective methods.

Talc or talcum is an organophilic inorganic filler, water repellent and chemically inert mineral, which is widely used as polymer additive including biodegradable polymer. Lamellar talc with high aspect ratio improves the deflection temperature of polyolefins to a greater extent than talc with alower aspect ratio. Moreover, talc is water repellent and chemically inert which is very important for the direct contact of mineral filled packaging material with food-stuffs [8]. We have studied the addition of talcum to the polymer blends of PLA and PBS to improve various properties and lower the production cost. Talcum will not only enhance the polymer properties but also acting as the nucleaing agent.

Experimental

Materials

Poly(lactic acid) (3052D) and Poly(butylene succinate) (1020MD, Bionolle™) were commercial grades supplied by Nature Works LLC, and Showa Denko K.K., respectively. Fine talc powder (Micro ACE series P-3) with an average particle size of 5 μm was purchased from Nippon Talc CO., Ltd.

Methods

Prior to mixing, PLA and PBS were dried at 80 °C under vacuum for 12 h and talcum was dried at 100 °C under vacuum for 6 h. The polymer blend according to Table 1 were compounded by simultaneous addition of all components to twin-screw extruder. PLA and PBS blends were prepared with weight percentage ratios and talcum was added to the blends. The temperature profile during compounding of all blends was 170-190 °C. The screw rotation speed was fixed at 70 rpm. The extruded materials were preparated into standard tensile, hardness and impact specimens by compression moulding using compression mould of 20 x 20 x 0.03 cm and a hot-plate hydraulic press at 220°C and allowed to cool to room temperature under pressure.

Table 1 Compositions of PLA/PBS blends

PLA	100	90	80	70	0
PBS	0	10	20	30	100
Talcum	5, 10, 20				

Analysis

1. *Rheology*

The PLA, PBS and polymer blend samples with specified amount are preheated for 5 min at 210°C before analysis. The standard compress weight was 2.16 kg.

2. *Morphology*

The morphology of samples from impact tests were analyzed by a scanning electron microscopy (SEM) (JEOL JSM-6510).

3. *Mechanical property*

Tensile Testing

The tensile test was carried out at a rate of 10 mm/min according to ASTM D638. The samples were compression molded Standard type I dumbbell-shaped samples with a thickness of 3.0 mm.

Izod Impact Strength Testing

The samples were compression molded Standard shaped samples (ASTM D256) with a thickness of 3.2 mm. A pendulum swings on its track and strikes at a notched, cantilevered plastic sample. The energy lost (required to break the sample) as the pedulum continues on its path is measured from the distance of its follow through.

Results and Discussion

Melt flow Index

Melt flow index (ASTM D256) of PLA, PBS, polymer blends and polymer blends with talcum additives are shown in Fig 1.

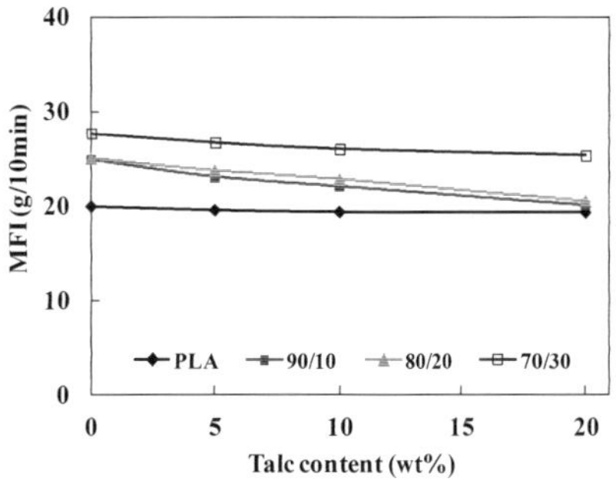

Figure 1 Melt Flow Index

The addition of talc in PLA, and PLA/PBS polymer blends did not change the melt flow of polymer significantly. Although PLA is easily degraded at high temperature but the results confirmed that talc stabilized the thermal stability of polymers. The melt flow of PBS was too high compared with PLA and polymer blends.

Morphology

The morphology of the polymer blends were investigated using scanning electron microscopy (SEM) [7, 9]. Figure 2 shows the SEM photo of PLA/PBS blends without talc. PLA and PBS pure is clear. As the PBS wt% increases, the phase separation is quite evident in these blends.

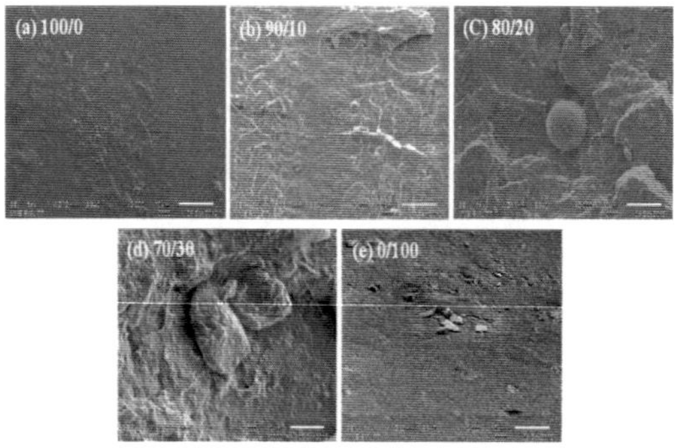

Figure 2 SEM photographs of PLA/PBS blends (a) PLA, (b) 90/10, (c) 80/20), (d) 70/30, (d) PBS (scale bar is 10 μm).

Figure 3, 4 and 5 show morphology and dispersion of talc with 5, 10 and 20%wt in PLA/PBS polymer blend matrix, respectively.

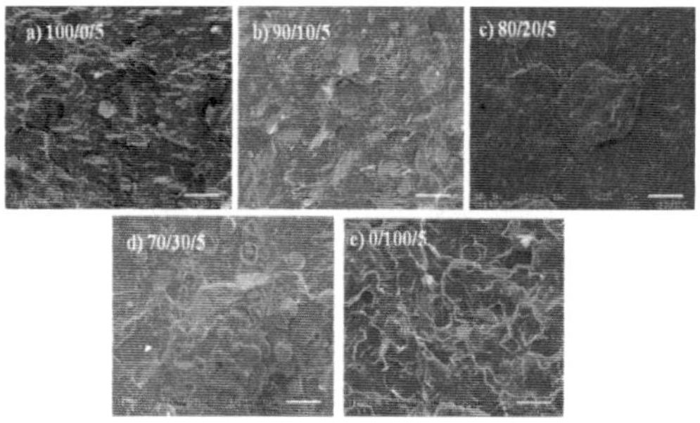

Figure 3 SEM photographs of PLA/PBS blends (a) PLA, (b) 90/10, (c) 80/20), (d) 70/30, (d) PBS with talc 5 wt% (scale bar is 10 μm).

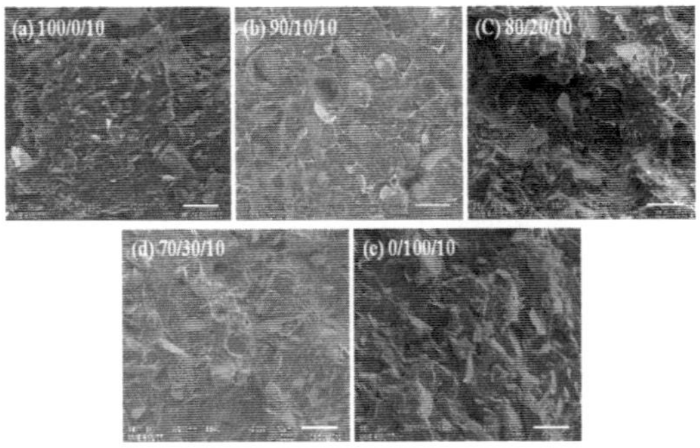

Figure 4 SEM photographs of PLA/PBS blends (a) PLA, (b) 90/10, (c) 80/20), (d) 70/30, (d) PBS with talc 10 wt% (scale bar is 10 μm).

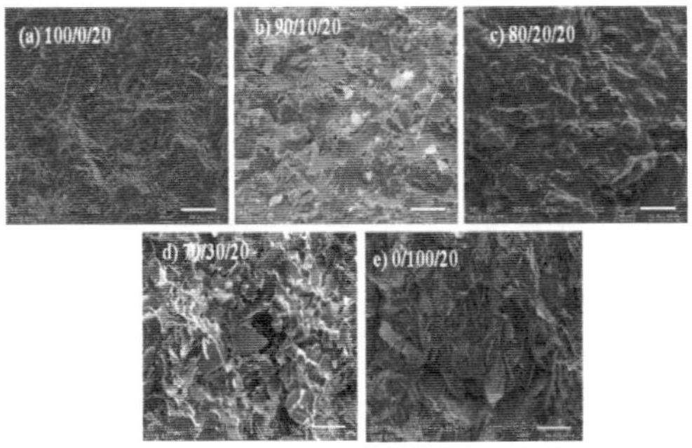

Figure 5 SEM photographs of PLA/PBS blends (a) PLA, (b) 90/10, (c) 80/20), (d) 70/30, (d) PBS with talc 20 wt% (scale bar is 10 μm).

The SEM photographs show similar morphology of distribution of talc 5 to 20 wt%. The structure of talc was well dispersed in the polymer blend matrix in all samples. Talc consists of a layer or sheet of brucite ($Mg(OH)_2$) sandwiched between two sheets of silica (SiO_2) [5]. The lamellar structure of talc improves the strength of polymer more than the particle fliiers [10].

Tensile tests

Tensile strength analysis were performed at 25°C (room temperature) to study tensile strength and tensile modulus of the polymer blends with talc additives. All tests were done according to ASTM D638 standard. Tests of neat PLA, neat PBS and polymer blends were carried out with a crosshead speed of 10 mm/min. Property values reported here represent an average of the results for tests run on ten specimens.

The addition of PBS in PLA decreases the tensile modulus of the blends because PBS is a soft polymer having lower modulus. It was found that addition of talc increased the tensile modulus of the polymer blends clearly.

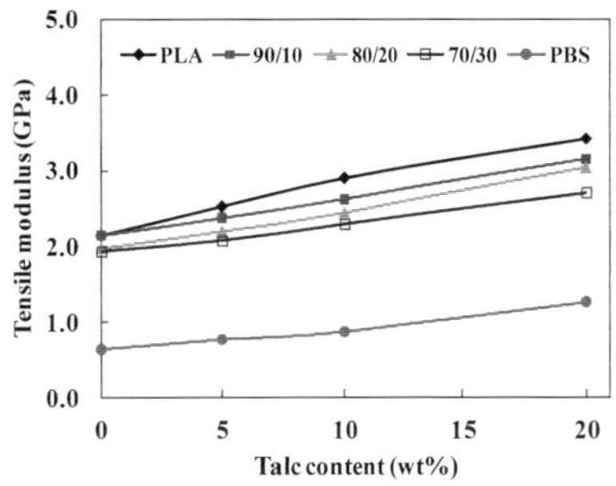

Figure 6 Tensile modulus of PLA/PBS blends with addition of talc 5, 10, 15 and 20 wt%.

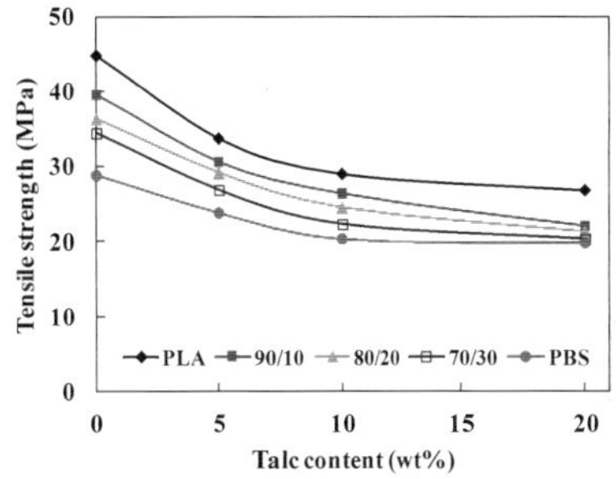

Figure 7 Tensile Strength of PLA/PBS blends with addition of talc 5, 10, 15 and 20 wt%.

Figure 7 shows tensile strength of PLA/PBS blends and PLA/PBS blends with talc 5, 10, 15 and 20 wt%. It was found that the strength decreased with increasing amounts of talc. The phenomena may due to weak adhesion of talc to the polymer matrix as can be seen from the SEM pictures. The addition of PBS to PLA formed the phase separation of two polymers.

Izod Impact Strength

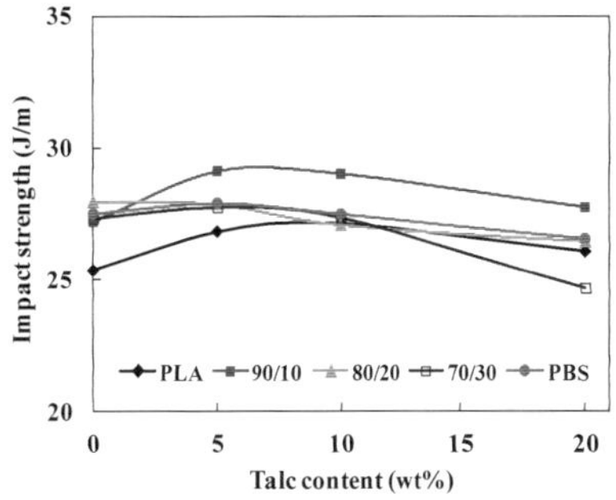

Figure 8 Izod impact strength tests of PLA/PBS blends with talc additive.

Figure 8 shows impact strength of PLA, PBS and PLA/PBS blends with talc. Higher PBS contents polymer show increase of impact strength due to flexibility of PBS. The addition of did not reduced the impact property of the blend but slightly increased with 5 and 10 wt% of talc. The results may due to the lamella structure of talc distributed in the polymer matrix absorbed and dispersed the impact force.

Conclusions

The effect of talc additive to the physical and mechanical properties of polymer blends of poly(lactic acid) (PLA)/poly(butylene succinate) (PBS) were investigated. The amount of talc studied was 5 to 20 wt% respected to the polymer weight. Melt flow of polymer blend was unchanged significantly with the amount of talc additions which can be implied with the stabilization of polymer of talc. Morphology investigation showed well distribution of talc in the polymer matrix. Modulus of the blends increased while increasing the amount of talc additives. The impact strength of the blends was not significantly reduced with increasing amounts of talk. The physical and mechanical investigation confirmed adequate properties for injection molding applications.

Acknowledgements

Department of Chemical and Materials Engineering, Faculty of Engineering, Rajamangala University of Technology Thanyaburi, THAILAND.

Department of Materials Science and Engineering, Faculty of Engineering and Industrial Technology, Silpakorn University, THAILAND.

Department of Biobase Materials Science, Kyoto Institute of Technology, JAPAN.

References

1. R. Auras; B. Harte, S; Selke S. *J. Mac. Bio.* 2004, 9, 835–64.
2. F. Signori; M. B. Coltelli; S. Bronco. *J. Polymer Degradation and Stability*. 2009, 94, 74–82.
3. Bhari, K; H. Mitomo; T. Enjoji; F. Yoshii and K. Makuuchi. *J. Polymer Degradation Stability*. 1998, 62, 551-557.
4. Doi, Y; K. Kasuya; H. Abe; N. Koyama; S. Ishiwatara; K. Takagi and Y. Yoshida. *J. Polymer Degradation Stability*. 1996, 51, 281-286.
5. A. Bhatia; R. K. Gupta; Sati. N. Bhattacharya and H. J. Choi. *J. Korea-Australia Rheology*. 2007, 19, 125-131.
6. T. Ujimaki. *J. Polym Degrad Stab*. 1998, 59, 209.
7. Y.F. Shih; L.S. Chen; R.J. Jeng. *J. Polymer*. 2008, 49, 4602–4611
8. http://www.mondominerals.com. Technical Bulletin 1301. *Talc in Plastics*. Mondo Minerals B.V.
9. L. Bokobza. *J. Polymer*. 2007, 48, 4907.
10. B. Sornil. *Plastic Technology*. Technology Promotion Association (Thailand-Japan), 2003.

O-11-273

MECHANICAL PROPERTIES AND STRUCTURAL CHARACTERIZATION OF THERMOPLASTIC PROTEIN DERIVED FROM BLOODMEAL WITH TRI-ETHYLENE GLYCOL AS A PLASTICISER

J. M. Bier*, C. J. R. Verbeek and M. C. Lay[1]

[1]*University of Waikato, Private Bag 3105, Hamilton 3240, New* Zealand– Jmb101@waikato.ac.nz; jverbeek@waikato.ac.nz; mclay@waikato.ac.nz

Abstract –The aim of this research was to replace water with a less volatile plasticizer, tri-ethylene glycol (TEG) in thermoplastics derived from bloodmeal, which is a by-product of the meat industry.. Mechanical properties, glass transition temperature and the chemical structure of injection moulded specimens were used to determine the effect of replacing water with TEG. Thermoplastic protein with 0, 10, 20 or 30 g TEG per hundred grams bloodmeal (pph_{BM}) was prepared using extrusion. The total plasticizer (water plus TEG) content was kept constant at 60 pph_{BM}. Glass transition temperature was determined from peaks in tanδ using dynamic mechanical analysis (after 7 days conditioning at 23°C and 50% relative humidity) to be 84°C, 76°C, 64°C and 57°C at 0, 10, 20 and 30 pph_{BM} TEG respectively. Ultimate tensile strength without TEG was 16 MPa, however this material was very brittle and had a strain at break of only 0.01. TEG increased strain at break while reducing strength and 20 pph_{BM} was chosen as an appropriate compromise between strength and ductility resulting in a ultimate tensile strength of 6 MPa, Young's modulus of 250 MPa, toughness of 2.8 MPa and strain at break of 0.53. Wide angle x-ray scattering (WAXS) and Fourier transform infrared spectroscopy revealed that some protein secondary structures were present in the processed plastic. WAXS revealed that inter-structure packing of α-helices and β-sheets were reduced, as evident from a reduction in peak intensity, corresponding to 10 Å.

Keywords: Protein, Bioplastic, Biodegradable plastic

Introduction

As with energy research, concerns around finite fossil fuel supply and projected price increases along with widespread awareness of sustainability motivates research into bio-based alternatives to petroleum derived polymers [1]. Commercial viability of these alternatives hinges on being able to process them with conventional thermoplastic processing techniques such as extrusion and injection moulding [2].

Thermoplastic materials can be produced from biomass via one of the following three methods: Firstly, materials such as polylactide (PLA) and bio-based polyethylene require fermentation to produce small molecules which can then be purified, modified and polymerized. Secondly, fermentation can also be used to produce intra-cellular polymers in bacteria, such as polyhydroxyalkanoates (PHA), which can be extracted and purified. The third approach is to take biological macromolecules, such as starch or proteins, and modify them in such a way that they will behave as thermoplastics [3].

Bloodmeal, a low value by-product of the meat processing industry, is a readily available source of inedible protein. Treatment with a protein denaturant, surfactant and reducing agent modifies bloodmeal into a thermoplastic protein based material which can be extruded and injection moulded with standard plastics processing equipment [4-5]. Due to the hydrophilicity of proteins, this material is moisture sensitive and it becomes brittle during storage when water is used as plasticizer [6]. The aim of this research was to investigate the effect of replacing water with a less volatile plasticizer, tri-ethylene glycol (TEG), on mechanical properties, glass transition temperature and the structure of conditioned injection moulded thermoplastic protein derived from bloodmeal.

Experimental

Sample preparation

Different formulations of thermoplastic protein were prepared to assess the effect of varying TEG content, at constant total plasticizer content (TEG+water) (Tab. 1). The non-TEG additives were mixed with dry bloodmeal for five minutes in a high speed mixer, then TEG was added, followed by an additional five minutes mixing. The resulting mixture was extruded (Thermoprism TSE-16-TC twin screw extruder) and injection moulded (BOY - 35A injection moulder) into Type 1 tensile test specimens [7] as previously described [5, 8]. Test pieces were conditioned at 50% RH and 23°C for seven days. Samples with 0 and 20 parts per hundred bloodmeal (pph_{BM}) TEG were ground and freeze dried for structural analysis.

Table 1 – Design of formulations used to assess effect of variation in TEG content.

Sample Name	*Sample formulation (pph_{BM})*					
	Blood meal	Water	Urea	SS	SDS	TEG
V0	100	60	10	3	3	0
V1	100	50	10	3	3	10
V2	100	40	10	3	3	20
V3	100	30	10	3	3	30

Analysis

Mechanical properties were measured in tension at 5 mm/minute strain rate in an Instron model 33R4204 fitted with a 50 mm gauge length extensometer. A secant modulus was calculated between 0.0005 and 0.0025 strain.

Dynamic mechanical analysis (DMA) was performed in a Perkin Elmer DMA8000 fitted with a high temperature furnace and cooled with liquid nitrogen. Samples of approximately 3.5x6.5x30 mm were cut from larger injection moulded pieces and tested between -100 and 120°C at a programmed heating rate of 2 °C/min in single cantilever bending mode (1 Hz, 12.5 mm free length, 0.03 mm dynamic displacement).

Fourier Transform Infrared (FTIR) spectra of freeze dried samples in KBr pellets were collected in a Perkin Elmer spectrophotometer using 16 scans from 4000 to 400 cm^{-1} at 4 cm^{-1} resolution and automatic background subtraction. Spectra were baseline corrected and normalised to the maximum absorbance between 900 and 1900 cm^{-1} in Microsoft excel.

Powder wide angle x-ray scattering (WAXS) was collected using a Phillips X'Pert system with a wavelength of 1.54 Å (PW3373/00 Cu LFF DK233995 x-ray tube at 40 kV and 40mA). Samples were scanned from 2θ = 2° to 2θ = 60° at 0.020° steps.

Results and Discussion

Mechanical Properties

A consistent total plasticizer (water + TEG) content of 60 pph_{BM} was used for processing in all samples. However, moisture evaporates during conditioning at 23°C and 50% relative humidity, reducing the amount of plasticizer present in conditioned tensile specimens. TEG is less volatile, so samples with a higher ratio of TEG to water retain a higher level of plasticization after conditioning (Tab. 2), partly due to TEG also being hydrophilic, thereby having higher equilibrium moisture content. This results in the mechanical properties seen in Fig. 1.

Although material prepared without TEG is the stiffest and strongest, it is also very brittle, as evidenced by low strain at break and has very low toughness, implying little energy is required to cause fracture. Increased plasticization after conditioning increases chain mobility which reduces strength, but also allows for some chain rearrangements in response to the applied force resulting in a tougher material with a higher strain at break. 20 pphBM was chosen as good compromise between strength and toughness with the best strain at break of the four formulations. 10 pphBM was stronger and also resulted in good toughness, but showed greater variation between samples. At 30 pph_{BM} plasticization reduces chain entanglements too extensively, reducing overall toughness.

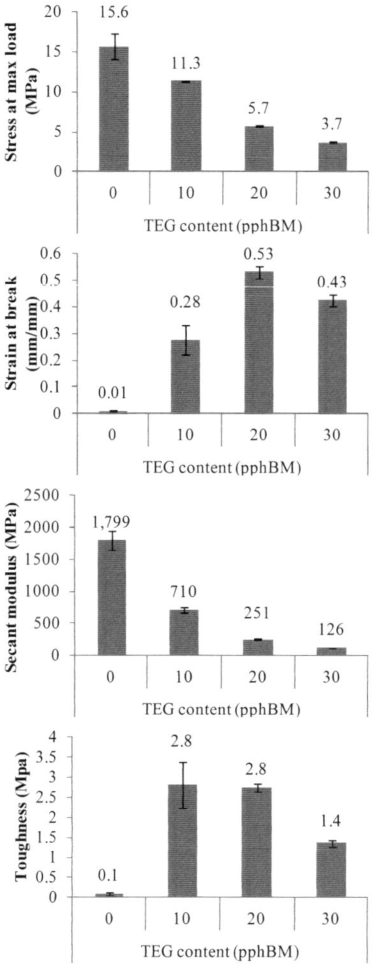

Figure 1– Mechanical properties of conditioned test pieces. Error bars denote standard error of the mean.

Dynamic Mechanical Analysis

Dynamic mechanical analysis of conditioned samples shows two transitions in the temperature range of 200-400 K (Fig. 2).The upper transition is characterised by a peak in tanδ and a drop of two orders of magnitude in E', both features consistent with the glass transition (T_g). Increased TEG broadens this transition and shifts it to lower temperatures, despite only a small difference in moisture content (Tab. 2), demonstrating the plasticizing effect of TEG. The second lower transition was thought to be a β-transition, associated with short range relaxations and therefore mostly unaffected by TEG.

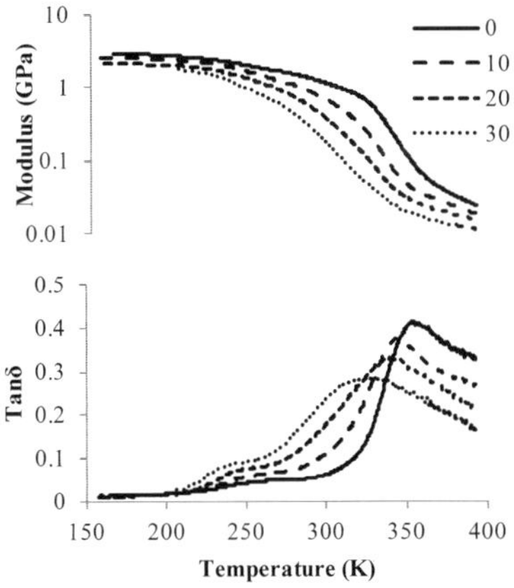

Figure 2 – Dynamic mechanical analysis of conditioned bar samples. Differently dashed lines denote formulation in pph_{BM} TEG.

Table 2 – Transition temperatures, moisture content and TEG content of conditioned samples.

TEG (pphBM)		*0*	*10*	*20*	*30*
Moisture content		6%	6.50%	8%	8.50%
TEG (%)		0%	7%	14%	19%
T_g	E'	324	313	290	280
	tanδ	354	344	339	329

WAXS

X-ray scattering enables estimation of inter-chain distances (d) in glassy polymers. This is done by identifying peak positions in the inter-chain correlations region (1–2.5 $Å^{-1}$) of the WAXS scattering vector, s, where $s = 4\pi \sin \theta/\lambda$ [9]. For the wavelength used in this study, this corresponds to a 2θ range of approximately 5–35°. Using Braggs law, $n\lambda=2d\sin\theta$, the scattering angle of these peaks can be related to an inter-chain distances.

Two such peaks are typical in proteins, one at $s \approx 0.6 - 0.7$ $Å^{-1}$ ($d \approx 10$ Å), and one $s \approx 1.5$ $Å^{-1}$ ($d \approx 4$–5 Å). The first is attributed to inter-helix packing in helix rich proteins and inter-sheet separation in beta sheet rich proteins. The second peak is similarly attributed to either to hydrogen bonding along the backbone in helices, or to bonding between strands in beta sheets [10]. With the wavelength used in this study, these correspond to the peaks seen at 2θ = 9° and 2θ = 19° for bloodmeal and thermoplastic derived from bloodmeal (Fig. 3). The former peak reduced in size when TEG was included, suggesting some disruption to the packing of ordered secondary structures when this TEG is included. To further clarify, a Gaussian peak representing scattering by amorphous regions was fitted under the curve and used to estimate the degree of crystallinity giving 34% for bloodmeal, 27% for thermoplastic without TEG and 17% for material with 20 pph_{BM} TEG. After fitting the halo, it is also apparent the peak corresponding to hydrogen bonding in ordered structures is decreased in the plastic, and decreased further with 20pph_{BM} TEG.

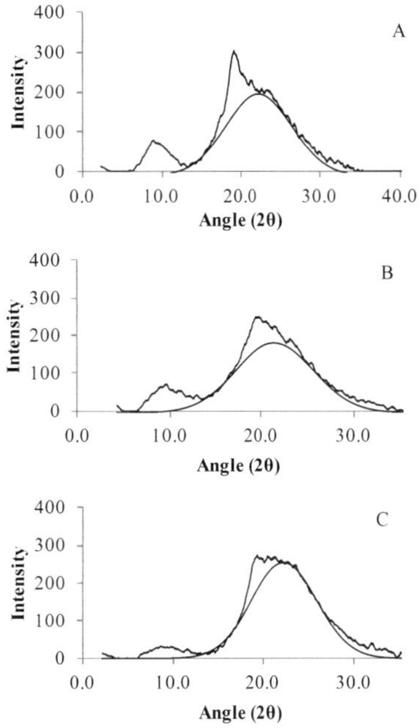

Figure 3 – WAXS scattering by A) bloodmeal, B) thermoplastic protein without TEG, C) Thermoplastic protein with 20 pphBM TEG.

FTIR

Both bloodmeal and the processed thermoplastic revealed peaks within the amide III region (1200-1350 cm^{-1}) which corresponded to secondary structures usually found in proteins (Fig. 4). The peak between 1295 and 1330 cm^{-1} was assigned to α-helices and the peak between 1220 and 1245 cm^{-1} was assigned to β-sheets [11]. The ratio of these peaks was approximately 0.9 in bloodmeal, but reduced to 0.7 in the thermoplastic, whether or not TEG was included. This suggests a reduction in alpha helices, relative to beta sheets due to the combined effect of all processing additives and processing itself. Alpha helices are stabilized by hydrogen bonds along the backbone, whereas beta sheets are stabilized by many parallel bonds between adjacent sheets, so it makes sense that as hydrogen bonding is decreased, alpha helices are the first to be destabilized. A further observation on the FTIR spectra is the peak broadening, or emergence of a shoulder at 1250 cm^{-1} with TEG (Fig. 4). This may correspond to an increase in random coils [11]. Together with the WAXS results, this shows that ordered structures (alpha helices and beta sheets) has decreased compared to bloodmeal, and that TEG has contributed to a reduction in interactions between chains, due to competition for hydrogen bonding sites.

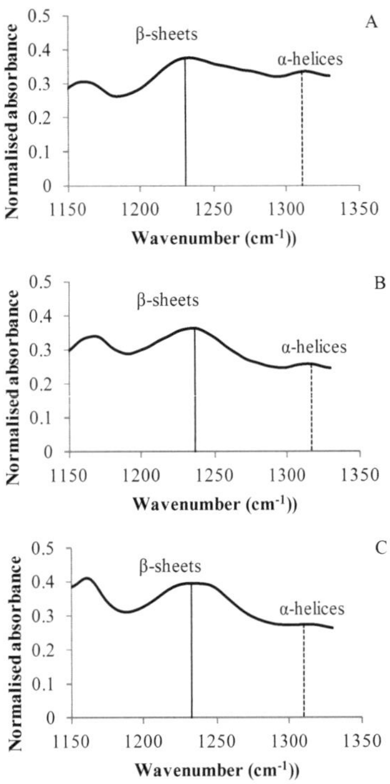

Figure 4 – Amide three regions for A) bloodmeal, B) thermoplastic protein without TEG, C) Thermoplastic protein with 20 pphBM TEG.

Conclusions

TEG was used successfully as a plasticizer, reducing glass transition temperature and producing a more ductile and tougher conditioned plastic than water and the other additives alone.

FTIR revealed a decrease in α-helices relative to β-sheets in the processed plastic, compared with bloodmeal, both with and without TEG. WAXS revealed that inter structure packing of alpha helices and beta sheets were reduced when TEG was included, as a result of the plasticizing effect of TEG. This shows that the combination of additives, extrusion and injection molding have changed interactions between protein chains from those in bloodmeal. Inclusion of TEG further reduces such interactions in the processed plastic. This allows increased chain mobility, which contributes to changing mechanical properties from brittle (without TEG) to ductile (with TEG)

A TEG content of 20 pphBM was found to produce a material with a reasonable compromise between strength and toughness. This is recommended as the standard formulation for ongoing work.

Acknowledgements

The Authors wish to thank New Zealand Universities and the Claude McCarthy Fellowship for travel funding assistance to attend PPS-28

References

1. A. Gandini, *Macromolecules*, 2008, 41, 9491-9504.
2. C. J. R. Verbeek; L. E. van den Berg, *Macromol. Mater. Eng.*, 2010, 295, 10-21.
3. E. Rudnik, *Compostable polymer materials*, Elsevier, Amsterdam ;London, 2008.
4. C. J. R. Verbeek; C. Viljoen; K. L. Pickering; L. E. van den Berg, New Zealand Patent 551531, 2007.
5. C. J. R. Verbeek; L. E. van den Berg, *J. Polym. Environ.*, 2010, 1-10.
6. C. J. R. Verbeek; L. E. van den Berg, *Macromol. Mater. Eng.*, 2011, 296, 524-234.
7. ASTM International, ASTM International,, West Conshohocken, PA 19428-2959, United States., 2004.
8. J. M. Bier; C. J. R. Verbeek; M. C. Lay, *J. Therm. Anal. Calorim.*, 2012, 1-13.
9. A. Elicegui; J. J. Del Val; J. L. Millan; C. Mijangos, *J. Non-Cryst. Solids*, 1998, 235, 623-627.
10. W. M. Elshemey; A. A. Elfiky; W. A. Gawad, *Protein J.*, 2010, 29, 545-550.
11. S. W. Cai; B. R. Singh, *Biophysical Chemistry*, 1999, 80, 7-20.

O-12-647

EXTRACTING THE STRESS-STRAIN PROPERTIES OF NON-LINEAR VISCOELASTIC MATERIALS USING THE BUBBLE INFLATION TEST

S.M. Goh*, B.S. Cheng
Curtin University Sarawak, Miri, Malaysia – aaron.smgoh@gmail.com

Abstract – The characterization of stress-strain properties from the bubble inflation test typically requires the measurements of pressure in the bubble and the geometry of the bubble. In this work, a novel inverse method to circumvent the need to measure the geometry of the bubble has been developed and tested. The inverse method is based on the Levenberg-Marquardt algorithm and the finite element method. A numerical experiment was performed in which the strain dependent behaviour was characterised by a two-parameter Mooney-Rivlin hyperelastic model while the time dependent behaviour was characterised by a three-parameter power law equation. It was found that good estimates of the viscoelastic properties could be obtained using only one set of bubble pressure–piston displacement data.

Keywords: biaxial – bubble inflation – membrane – viscoelastic – inverse identification.

Introduction

The bubble inflation test, also called the bulge test, the membrane inflation test, the biaxial inflation test or the inflation test in short, is a widely used method to characterise the mechanical properties of soft, thin materials including rubbers, polymer melts, foods and tissues. The test is attractive because it approximates the biaxial tensile conditions attained in practice for these materials. For example, the test is used for characterising the behaviour of dough as it is perceived to simulate the expansion of the air cells in the dough during the baking process [1].

In the inflation test, a sample in the shape of a thin disc or a membrane is held by a clamp at its circumference and then inflated to a bubble using pressurised air [1]. The pressurised air can be achieved via for example a piston mechanism. Typically, the pressure inside the bubble and the shape of the bubble are measured experimentally. Whereas the former measurement can be done quite easily, the latter measurement is more tedious and new methods are still being developed [e.g. 2,3]. Current research is also being done on the theoretical models of the tests, in particular the extension of the elasticity solutions to incorporate non-linear viscoelastic behavior [e.g. 4].

The objective of this work was to develop a novel method to extract the stress-strain properties from the inflation test. The advantage of such a method is that only the bubble pressure is required and the need to perform tedious experimental setup to measure the bubble is circumvented. This paper presents the method and some preliminary data that have been obtained to test the method.

Methodology

Finite Element Model

The finite element analyses were performed using the commercial package ABAQUS. The material was modelled to be in contact with a gas which was enclosed in a cylinder and driven with a piston at a predetermined speed. The material was modelled using axisymmetric, eight-noded elements while the gas was modelled using bi-nodal hydrostatic elements. Both the cylinder and piston were modelled as rigid bodies. Unless otherwise stated, the simulations were performed assuming a piston speed of 25 mm/min. The total piston displacement was 300 mm and the time duration was 12s.

From the simulations, values of bubble pressure and piston displacement were obtained from the ABAQUS code variables PCAV and U2 respectively. These variables relate to experimental measurements of pressure in the bubble and piston displacement that can normally be obtained without significant experimental set-up.

The inverse methodology pursued in this work was based on matching predicted bubble pressure–piston displacement using guesses for the material properties to 'experimental' data. The 'experimental' data here refer to the exact finite element computations of PCAV and U2 using known values of material properties.

Material Model

Simple models to characterise viscoelastic behaviour typically assume that the strain and time dependent behaviour is separable such that under a step-strain relaxation test, the stress is described by:

$$\sigma(\varepsilon,t)=\sigma_0(\varepsilon)g(t)$$

where σ is the stress at true strain ε and time t. The strain dependent function, $\sigma_0(\varepsilon)$, has dimensions of stress and the time dependent function, $g(t)$, is dimensionless.

The strain dependent behaviour of the material was assumed to be described by the Mooney-Rivlin hyperelastic potential, where the true stress as a function of the stretch λ in the equibiaxial deformation state is given as:

$$\sigma_0(\lambda) = 2C_{10}\left(\lambda^2 - \frac{1}{\lambda^4}\right) + 2C_{01}\left(\lambda^4 - \frac{1}{\lambda^2}\right)$$

where C_{10} and C_{01} are constants. Incompressibility of the material was assumed.

The time dependent behavior was assumed to be described by a power law equation:

$$g(t) = g_\infty + (1 - g_\infty)\left(1 + \frac{t}{t_r}\right)^{-n}$$

where n is the power law exponent, t_r is a reference time and g_∞ defines the equilibrium component.

Inverse Method

The inverse method was used to determine the two strain dependent parameters, C_{10} and C_{01}, and the three time dependent parameters, n, t_r and g_∞. The Levenberg-Marquardt algorithm [5] was utilised to minimise the objective function, Φ, which is given by:

$$\Phi = \frac{1}{2}\sum_{i=1}^{100}[r_i(\mathbf{p})]^2 = \frac{1}{2}\mathbf{r}^T\mathbf{r}$$

where $\mathbf{p}^T = \{C_{10}, C_{01,}, n, t_r, g_\infty\}$ and $\mathbf{r} = \mathbf{P}^* - \hat{\mathbf{P}}$, where $\mathbf{P}^*$ and $\hat{\mathbf{P}}$ are respectively the predicted and experimental values of bubble pressure corresponding to the same piston displacement. The total piston displacement was divided into 100 points and the objective function was summed over all the points.

A general outline of the inverse method is presented in Fig. 1. Guess values were initially assigned to the material parameters C_{10}, C_{01}, n, t_r and g_∞. In ABAQUS [6], viscoelasticity is modeled using the Prony series. In order to convert the power law model to the Prony series, a least squares error fitting was performed in a Microsoft Excel spreadsheet using the Solver function. When not stated, seven exponential terms in the Prony series were used corresponding to the time constants of 0.01, 0.1, 1, 10, 100, 1000 and 10000s respectively. The constants in the Prony series and the strain dependent parameters were then input into the finite element models of the bubble inflation to obtain the predicted bubble pressure-piston displacement relationship. This was repeated in subsequent iterations as the material parameters were changed in the algorithm to match the predicted relationship to the experimental measurements. If no solution was found after twenty iterations, the inverse method was stopped and the last computed material parameters are reported.

The inverse method was implemented in a Python script which interfaced with the ABAQUS software to perform the finite element simulations and with Microsoft Excel to fit the Prony series to the power law equation using the Solver function. In order to maintain stability of the material models and obtain physically meaningful values, the following constraints were applied: $C_{10} \geq 0$, $C_{01} \geq 0$, $t_r > 0$ and $g_\infty \geq 0$.

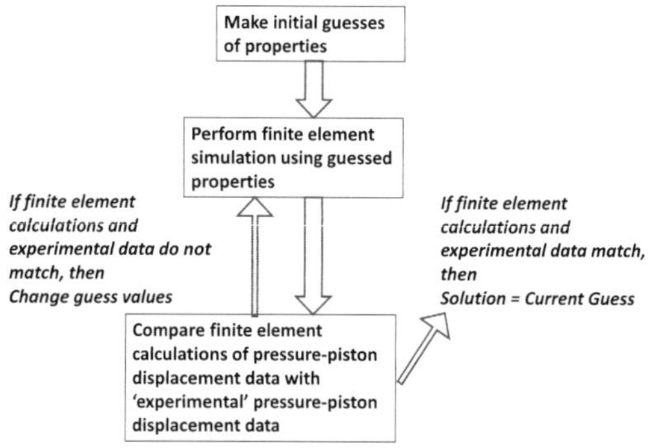

Fig. 1 Outline of inverse method.

Numerical Experiments

To evaluate the feasibility of applying the inverse method to the bubble inflation technique, a numerical experiment was conducted, i.e. finite element computations of the bubble pressure-piston displacement based on known parameters were used as the 'experimental' data. The parameters used to generate the experimental data were C_{10}=100kPa, C_{01}=1kPa, n=-0.5, t_r=0.5s and g_∞=0.2.

Twelve different sets of initial guesses were attempted to test the inverse method (see Table 1). Initial values were mostly either multiples by a factor of 2 or 10, or fractions by a factor of 0.1 or 0.5. In most sets, only one of the parameters was different from the exact solution. In the two sets, all of the initial guesses were higher than the solution or lower. The different sets of combination are referred to as 'Cases' from this point onwards.

Results and Discussion

The results of the inverse method are shown in Table 1. Cases 4 and 8 gave an exact match to the solution while some discrepancies between the solution and the results are found for the other Cases. However, these discrepancies are not reflected in significant deviation in the pressure-piston displacement relationship as shown in Fig. 2. Cases 2 and 12 did not complete.

Table 1 Summary of values for the different cases in the numerical experiments. The units are kPa for C_{10} and C_{01} and seconds for t_r. Both n and g_∞ have no units. Where applicable, the values are expressed up to two decimal places.

Case	Initial guesses					Final values					Iter-ation
	C10	C01	n	tr	g	C10	C01	n	tr	g	
1	1000	1	-0.5	0.5	0.2	106	0.74	-0.39	0.24	0.19	21
2	10	1	-0.5	0.5	0.2	91	14.92	0.49	1.94	1.65	-
3	100	10	-0.5	0.5	0.2	98	1.56	-0.49	0.62	0.12	21
4	100	0.1	-0.5	0.5	0.2	100	1.00	-0.50	0.50	0.20	16
5	100	1	-1	0.5	0.2	99	0.97	-0.55	0.57	0.22	21
6	100	1	-0.3	0.5	0.2	101	1.04	-0.46	0.43	0.18	21
7	100	1	-0.5	5	0.2	97	1.32	-0.86	1.24	0.24	21
8	100	1	-0.5	0.1	0.2	100	1.00	-0.50	0.50	0.20	13
9	100	1	-0.5	0.5	0.4	101	0.61	-0.58	0.48	0.29	21
10	100	1	-0.5	0.5	0.1	100	1.09	-0.49	0.50	0.18	21
11	1000	10	-1	5	0.4	102	0.28	-0.79	0.56	0.38	21
12	10	0.1	-0.3	0.1	0.1	88	0.62	0.54	4.04	2.80	-

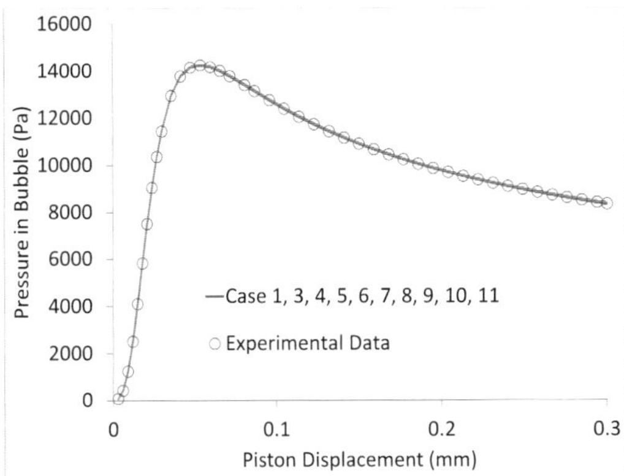

Fig. 2 Comparison of bubble pressure-piston displacement relationships between the 'experimental' data and the results from the inverse method.

The biaxial stress-strain relationships corresponding to three different strain rates of 0.01, 0.1 and 1/s and calculated based on the results in Table 1 are shown in Figs. 3-5. For the strain rate of 0.1/s, a strain of one is achieved in 10s, which is of a similar order of magnitude to the duration of the inflation test. It can be seen that the results from the inverse method agree closely with the solution at this particular strain rate. However, at lower strain rates in particular, i.e. when the deformation is slower, the results from the inverse method showed a larger discrepancy compared to the solution. This observation is in line with previous work [7] which suggested that the time dependent parameters are sensitive only within the relevant times encountered in the experiments. Since the duration of the inflation test was only 12s, the time dependent behaviour at larger times would not be expected to be characterised accurately.

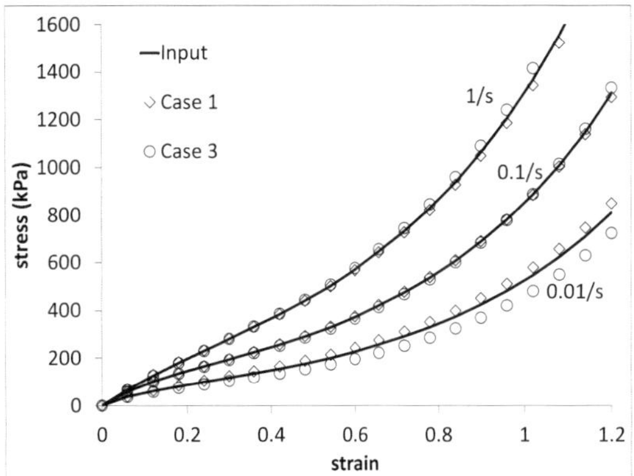

Fig. 3 Comparison of biaxial stress-strain relationships between the exact solution and the results from the inverse method for Cases 1 and 3.

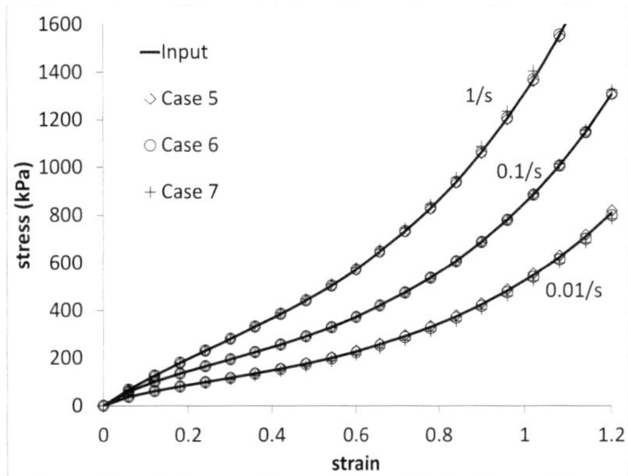

Fig. 4 Comparison of biaxial stress-strain relationships between the exact solution and the results from the inverse method for Cases 5, 6 and 7.

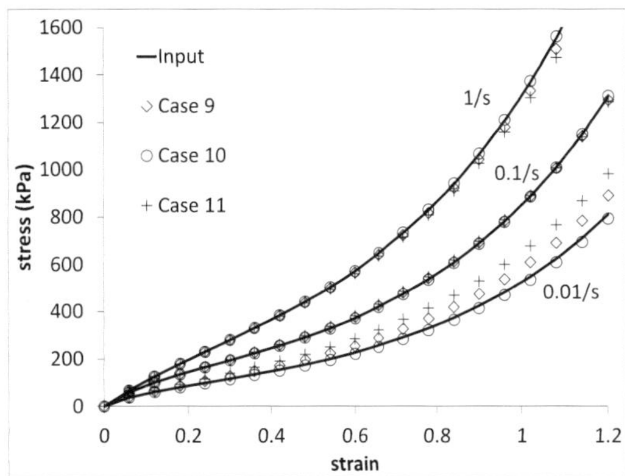

Fig. 5 Comparison of biaxial stress-strain relationships between the exact solution and the results from the inverse method for Cases 9, 10 and 11.

Although the numerical experiment was relatively simple, the results suggest that the inverse method was particularly sensitive to the constant C_{10}, i.e. when the initial guess for C_{10} was low, there was a tendency for the inverse method to overcompensate for the other parameters, leading to unreasonable values. The lack of the constraint $g_{\infty} \leq 1$ might have also contributed to this problem. The inverse method is currently being tested with other sets of initial guesses and the results will be reported elsewhere.

The use of the power law equation facilitates the application of inverse methods to viscoelastic materials as it requires only three independent variables to characterise the time dependent behavior. Although the Prony series is generally more popular, the fewer and fixed number of variables make the power law equation a more convenient choice for use in inverse methods. The disadvantage of the power law equation is that it provides a smooth characterisation of the time dependent behaviour and is not as flexible as the Prony series. A potential error in the current fitting procedure is the lack-of-goodness of fit of the Prony series to the power law equation. This error could be overcome by increasing the number of terms in the Prony series. However, this would result in a longer computational time to perform the fitting procedure.

Inverse identification methods which combine optimisation algorithms and finite element analysis have previously been found to be useful for estimating material parameters from the bubble inflation test [8]. However, previous inverse methods were limited to characterising only the strain dependent behaviour. For highly viscoelastic materials, excluding the time dependent behaviour can lead to inaccurate material characterisation. The results obtained here suggest that it is possible to apply the inverse method to obtain both the strain and the time dependent material parameters. In particular, the parameters could be predicted without the need for information regarding the bubble geometry. Only data describing the pressure in the bubble and the piston displacement are necessary. The use of the piston displacement instead of the bubble volume or height implies that the pressurised air is incompressible. In the case of the material being soft, the volume of the air in the bubble could be assumed to be equal to the volume of air displaced by the piston [1]. For stiffer materials, the bubble pressure-piston displacement relationship would either need to be corrected for the compressibility of the air or the air can be replaced by an incompressible fluid like oil [e.g. 9].

A somewhat surprising finding in the current work is that very reasonable identification of material parameters could be achieved despite using data from one piston speed. This would most probably be due to the range of strains and strain rates that were encountered during the bubble inflation. The availability of a larger set of experimental data would be expected to improve the accuracy of the inverse method and reduce the 'uniqueness' problem. Thus additional experiments may be needed to fully characterise the material parameters. However, additional data sets will lead to an increase in computational efforts as each parameter has to be perturbed in each iteration, requiring a new finite element simulation each time. An alternative would be to judiciously vary the experimental conditions in a single test. For example, the inflation could be stopped at a predetermined stage and the stress (pressure) relaxation of the bubble could be measured in the same experiment.

Conclusions

This study has applied an inverse method to extract both the strain and the time dependent behaviour from the bubble inflation test. The strain dependent behaviour was characterised by a hyperelastic potential while the time dependent behaviour was characterised by a power law equation. The numerical experiments showed that good estimates of the material parameters could be obtained from the inverse method.

References

1. M.N. Charalambides; L. Wanigasooriya; J.G. Williams *Rheol. Acta*, 2001, *41*, 532.
2. P. A. O'Connell; G. B. Mckenna *J. Rheol.* 2008, *31*, 113.
3. G. Machado; D. Favier; G. Chagnon *Exp. Mech.* 2012, *52*, 865.
4. A. Wineman *Comput. Math. Appl.* 2007, *53*, 168
5. D.S. Schnur; N. Zabaras *Int. J. Numer. Meth. Eng.* 1992, *33*, 2039.
6. ABAQUS's user manual ver 6.4. Hibbitt, Karlsson and Sorensen (UK), Cheshire 2004.
7. S.M. Goh *J.O.S.T.T.* 2008, *4*, 135.
8. M. Rachik; F. Schmidt; N. Reuge; et al. *Polym. Eng. Sci.* 2001, *41*, 532.
9. C.D. Denson; D.C. Hylton *Polym. Eng. Sci* 1980, *20,* 53

O-12-633

CONTROLLING OF VISCOSITY AND GELATION IN PASTEURISED HIGH-CALORIE OIL-IN-WATER EMULSION: USE OF OKARA AS DIETARY FIBER

P. Hongsprabhas[1]* and P. Hongsprabhas[2]

[1]*Dept. Food Science and Technology, Faculty of Agro-Industry, Kasetsart University, Thailand - parichat.h@ku.ac.th

[2]Dept. Internal Medicine, Faculty of Medicine, Khon Kaen University, Thailand - pranithi@kku.ac.th

Abstract – Phase separation characteristics of water-in-water dispersion (w/w) in the bulk aqueous phase of oil-in-water (o/w) high-calorie emulsion were investigated. Okara, the soy residue from soymilk production, was used as dietary fibre source in pasteurised multi-phase liquid emulsions containing calorie density of 0.9-1.5 kcal/mL from carbohydrate to protein to lipid at the ratio of 37-55:15-22:30-45, respectively. Gelation of such high-calorie emulsion, which contained high molecular weight polymers of starch, maltodextrin, soy and milk proteins, was inhibited by modifying the characteristics of okara through acidification, enzymatic hydrolysis and dehydration. It was found that phase separation between proteins and polysaccharides in the w/w bulk aqueous phase, generated by charged soluble polysaccharide and hydrolysed cellulose in okara, was the main factor determining the distribution of protein in the bulk aqueous phase, protein at the oil-water interface, and subsequent flow behaviours of the high-calorie liquid emulsions.

Keywords: emulsion; flow; okara; phase separation, protein

Introduction

Okara is a by-product of soymilk and tofu preparation collected after filtration. Dried okara contains 25.4-28.4% protein, 9.3-10.9% oil, 40.2-43.6% insoluble fiber, 12.6-14.6% soluble fiber, and 3.8-5.3% soluble carbohydrates [1, 2, 3]. Therefore, okara is a rich source of dietary fiber. Okara also contains high-quality protein. It is used as an ingredient in many food products such as pickles, tempeh, salads, sauces, baked goods and desserts [4]. Extractions of proteins or soluble polysaccharides from okara have been reported [3, 5, 6]. However, the highly structured fiber and high water-retention capacity of okara promotes a highly viscous product. As a result, the application of okara in liquid foods is quite limited.

We have investigated the potential use of okara powder (OKP) and pectinase-hydrolysed okara powder (PHO) – on the physicochemical characteristics of a high-calorie emulsion (1 kcal/mL) stabilised by soy and milk proteins, with caloric distribution from carbohydrate to protein to lipid of 55:15:30. It was found that during *in vitro* protein digestion by pepsin and trypsin, the use of OKP induced the formation of a thick paste under acidic conditions of stomach, and oil could be released after 15 min digestion by pepsin. On the other hand, the presence of PHO increased the viscosity of the emulsion under acidic pH and delayed the release of oil for up to 105 min during peptic digestion. The oil could be released from PHO-containing digesta after tryptic digestion, both in the absence and presence of bile acid. Our previous study indicated that the rate of nutrient release could be controlled by the presence of dietary fiber source. This was due to the pectic-like structure of soy soluble polysaccharides (SSP), which enhanced depletion flocculation and bridging flocculation of the oil droplets under acidic condition, particularly when the charged SSP were shortened by pectinase hydrolysis [1, 5].

It is apparent that the presence of highly structured fiber in okara affected not only the stability of the high-calorie emulsion, but also post-prandial responses of nutrients due to the controlling of protein digestibility and oil release under gastro-intestinal conditions. The objectives of this study were to further explore modification methods employed to okara to be used as the fiber source in high-calorie emulsions. The insights obtained would reveal the mechanisms involved in the phase separation and flow properties of the high-calorie o/w liquid emulsion.

Experimental

Materials

Fresh okara was obtained from 2 soymilk factories in Thailand, designated as VTM okara (V) and LTS okara (L). The samples were frozen at -18 °C. Protein sources used in the high-calorie emulsion were commercial sodium caseinate (High Viscous, Gansu Hualing Milk Products Group; Gansu, China). Cassava maltodextrin with dextrose equivalent (D.E.) of 10 (Neo-Maldex®, Neotech Food Co.; Bangkok, Thailand) was used as a carbohydrate source; while refined rice bran oil (King, Thai Edible Oil; Bangkok, Thailand) was used as lipid source in the high-calorie emulsion. Lactic acid used for acidification of okara prior to enzymatic hydrolysis was of food grade (L-LAC FG 88, Shanxi Leda Biochemical; Taiyuan, China). Commercial pectinase from *Aspergillus aculeatus* (Pectinex® Ultra SP-L with 26,000 PG/mL activity) and cellulase from *Trichoderma reesei* (Celluclast® with 700 endo-glucanase units/g) were donated from Novozymes A/S; Dittingen, Switzerland.

Bromelain (E.C.3.4.22.32) was purchased from Sigma-Aldrich, Inc., Saint Louis, Missouri, USA.

Effect of fresh okara on the characteristics of high-calorie emulsion

Liquid emulsions were prepared using okara from de-hull soybean in the laboratory (LAB), VTM and LTS. They were used as the protein source in the pasteurised liquid emulsions using different formulation as described in **Table 1.** Sodium caseinate was used to standardise the protein content. The o/w emulsions were prepared using a two-stage heating process as described earlier [7]. The apparent viscosity of the emulsion was recorded using a Brookfield RVDV-III Ultra Rheometer (Brookfield Engineering Laboratories Inc.; Stoughton MA, USA) equipped with a UL-adapter.

Table 1 Formulations for high-calorie liquid emulsion using fresh okara as protein source.

Specifications for macronutrients	Standard formula (STD)	Diabetes forbula (DM)	Hemodialysis formula (HD)
Calories (kcal/mL)	1.5	0.9	1.0
Carbohydrate (kcal)	50	37	48
Protein (kcal)	15	18	22
Lipid (kcal)	35	45	30

Modification and dehydration of okara from industrial production of soymilk

Room temperature-thawed okara was modified at high water content originally present in fresh okara at 30 °C for 20 h using methods shown in **Table 2** and that described earlier when pectinase was used [2]. The acidification was performed to lower the pH and to optimise the pH to the required pH for enzymatic hydrolysis according to the recommendations for each enzyme. After modification, the okara was dried in a tray-dryer at 80 °C for 3-4 h to the moisture content below 10 % (w.b.), sifted through 100 mesh seive and analysed for the chemical constituents [8]. The dried modified okara was used as the fiber source in the pasteurised o/w liquid emulsions prepared using method described earlier [7].

Table 2 Detailed modification methods investigated

Source of commercial okara	Treatment no.	Treatment
V or L	1	Control : no pH adjustment
V or L	2	Acidified by lactic acid to pH 4
V or L	3	Acidified by lactic acid to pH 4 and added with celluclast (1% by weight)
V or L	5	Acidified by lactic acid to pH 4 and added with 0.1 g bromelain in 1 mL distilled water

Effect of cellulase hydrolysis of okara on multi-phase structure and viscosity of pasteurised liquid emulsions

Okara obtained from commercial production of soymilk from LTS factory was modified by cellulase hydrolysis. After hydrolysis, the modified okara was mixed with mungbean starch at a 1:1 ratio of starch to total solid in okara using a kitchen aid mixer to separate okara particle, sorb water and enhance fluidising during drying. The starch granules acted as a sorbent and spouting material during fluidised-bed drying [2]. The sample (300 g) was dried at 60 °C in a 5 L stainless steel chamber of laboratory fluidized-bed dryer (Sherwood Scientific Ltd., Cambridge, UK) using air velocity of 3.50 m/s for 25-45 min until the moisture content was around 10 %. The dried okara was ground in a hammer mill, sifted through 100 mesh screen and kept at -20 °C before used.

The o/w emulsions were prepared using formulations described in **Table 1** and prepared using method previously described [7]. The microstructure of steamed soy grain and multi-phase pasteurised liquid emulsions was observed under confocal laser scanning microscope (LSM 5 PASCAL, Carl Zeiss Pte. Ltd.; Jena, Thüringen, Germany).

Statistical analysis

The experiments were carried out in two separate trials. Each trial was run in duplicates. The data were analysed by analysis of variance (ANOVA) with significance at $p<0.05$. All statistical analyses were performed using SPSS Software Version 12.

Results and Discussion

Effect of fresh okara on the characteristics of high-calorie emulsion

The chemical composition of the oven-dried laboratory prepared okara (LAB) and okara obtained from commercial-scale production of soymilk (VTM) was shown in **Table 3**. The dried LAB okara had protein content high enough to be the sole protein source in the high-calorie emulsion. The dried VTM, however, had a much lower protein content and a high content of crude fiber and carbohydrate.

Table 3 Chemical composition of the oven-dried okara

Constituent (% w.b.)	Laboratory okara (LAB)	Industrial okara (VTM)
Moisture	7.04±0.61	7.55±0.09
Fat	20.23±1.40	1.85±0.19
Crude fiber	3.29±0.06	16.94±0.29
Ash	3.96±0.00	3.05±0.01
Protein	55.95±0.28	22.57±1.91
Carbohydrate by difference	9.53±1.79	48.04±1.57

Means ± std. dev. (n = 2/trial; total of 2 trials)

Due to the use of okara as sole protein source, the emulsions of all formulae prepared using LAB okara had crude fiber around 0.32 %, which was quite low. However, all formulae had viscosity below 100 mPa·s as shown in **Fig. 1**. The viscosity range was within the drinkable range and the emulsion could flow under gravity. The flow behaviour of all formulae indicated Bingham flow ($p<0.01$) with low yield stress as shown in **Fig. 2**.

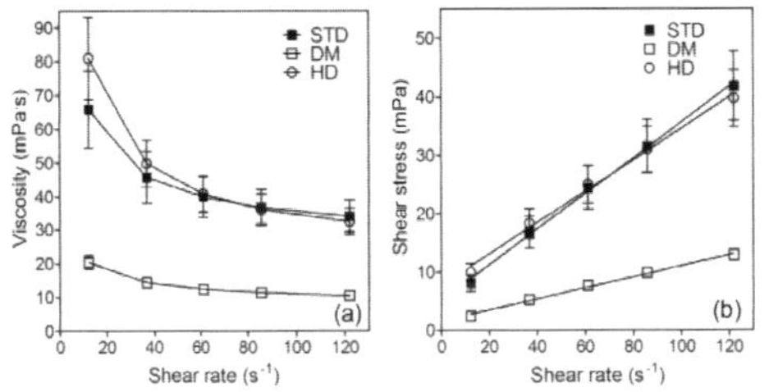

Figure 1 Flow behaviour of high-calorie liquid emulsions containing different calorie density and distribution; STD = standard formula, DM = diabetes formula, and HD = hemodialysis formula.

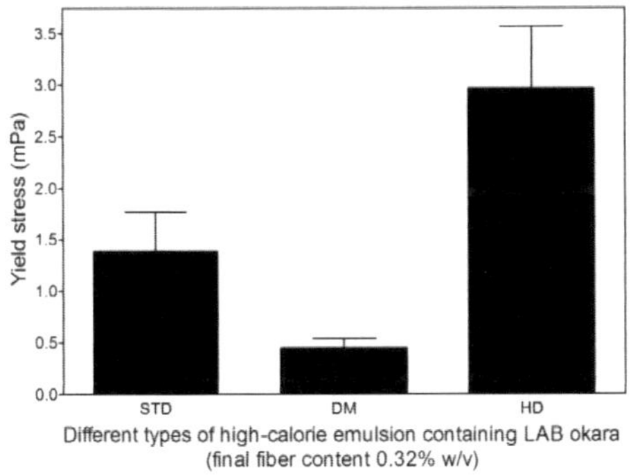

Figure 2 Yield stress of high-calorie liquid emulsions containing different calorie density and distribution; STD = standard formula, DM = diabetes formula, and HD = hemodialysis formula.

When the VTM okara was used as the protein and fiber source in the high-calorie formula, the high content of fiber in VTM okara induced gelation of all formulae shown in **Fig.3**. The high water absorption of okara was likely due to the soy soluble fraction [1, 3, 5, 6], as well as the alterations in the semicrystalline structure of cellulose in the husk after high temperature treatment (**Fig.4**). The birefringence under polarised light of the husk were less intensed after steaming.

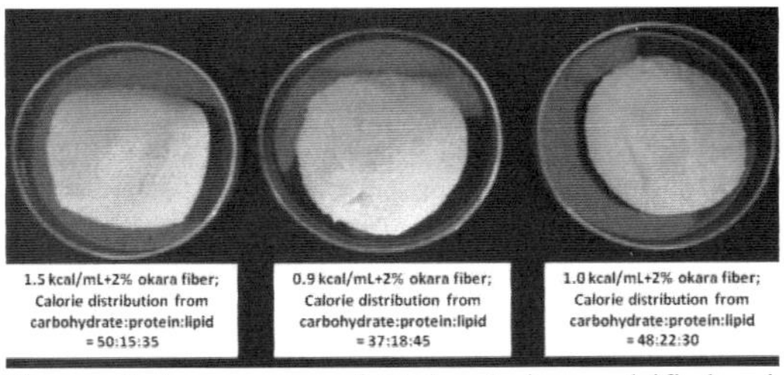

Figure 3 Appearance of high-calorie emulsified gel prior to pasteurisation. Final fiber content from VTM okara was 2 %(w/v).

Figure 4 Light micrographs of (a, b) soy endosperm and (c, d) soy husk.

Modification and dehydration of okara from industrial production of soymilk

Okara from commercial-scale production had different constituents (**Table 3**), particularly protein, crude fiber and carbohydrate contents. All of which was responsible for water absorption of the emulsified gel (**Fig.1**). Nonetheless, modification of VTM and LTS okara by acidification, enzymatic hydrolysis and dehydration resulted in the loss of water-holding capacity of okara (results not shown). Consequently, the high-calorie emulsions could flow after pasteurisation and storage at 5 °C (**Fig.5**). Using cellulase was the most effective process in modifying industrial okara to be used for high-calorie emulsions regardless of the history of how okara/soymilk was produced.

Table 4 Chemical composition of fresh okara obtained from commercial scale production

Constituents (%)	VTM	LTS
Moisture content	84.64 ± 0.19	84.54 ± 0.20
Protein	3.20 ± 0.03	4.15 ± 0.06
Lipid	0.49 ± 0.02	1.45 ± 0.03
Ash	0.53 ± 0.01	0.55 ± 0.00
Crude fiber	2.54 ± 0.05	3.52 ± 0.09
Carbohydrate	8.59 ± 0.21	5.78 ± 0.35

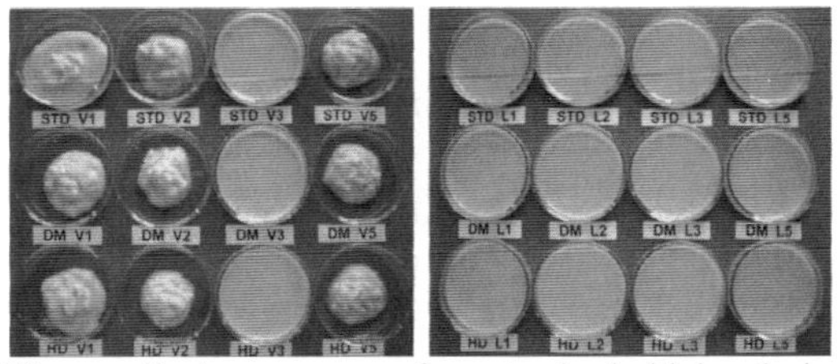

Figure 5 Appearance of pasteurised high-calorie emulsions containing different calorie distribution. Each formula contained 2% (w/v) fiber from unmodified and modified VTM and LTS okara powder; STD = standard formula; DM = diabetes formula; and HD = hemodialysis formula. V1, L1 = unmodified okara; V2, L2 = acidified okara; V3, L3 = acidified okara + cellulase treated; and V5, L5 = acidified okara + bromelain treated.

Effect of cellulase hydrolysis of okara on multi-phase structure and viscosity of pasteurised liquid emulsions
Due to the presence of mungbean starch as sorbent and spouting materials during fluidised-bed dehydration step, the high-calorie emulsions had apparent viscosity around 140 – 400 mPa·s, depending on the formulation (**Table 5**). Although the modification did not alter the apparent viscosity of STD and DM formulae, the hydrolysis of cellulose resulted in the dramatic increase in the viscosity of HD formula, which had the highest protein to oil ratio.

Table 5 Effect of cellulase hydrolysis (CH) of LTS okara on apparent viscosity of high-calorie emulsion containing 1.3% fiber and mungbean starch as part of calorie-bearing carbohydrate source. STD = standard formula, DM = diabetes formula, HD = hemodialysis formula.

For mula	Calorie density (kcal/ mL)	Calorie distribution from carbohydrate :protein:lipid	App. viscosity (mPa·s)	
			Un-modified okara	CH-okara
STD	1.5	50:15:35	142^d	141^d
DM	0.9	37:18:45	286^b	294^b
HD	1.0	48:22:30	170^c	409^a

Means followed by different superscript are significantly different ($p<0.05$).

Both starch and protein phase in the emulsions co-existed as separated phases in the high-calorie emulsions (**Fig. 6** and **Fig. 7**). CLSM revealed the alterations of oil droplets (dark circle) and proteins (light colour), both in bulk aqueous phase and at the oil-water interface (**Fig. 7**). The formulae added with unmodified okara powder had oil droplet size within the range of 2-10 µm, depending on the ratio of protein to oil ratio. However, hydrolysis of cellulose in the okara resulted in the coalescence of the oil droplets in the emulsions. The oil droplets coalesced to the larger size of almost 25 µm with a wide range of size distribution in STD and HD formulae.

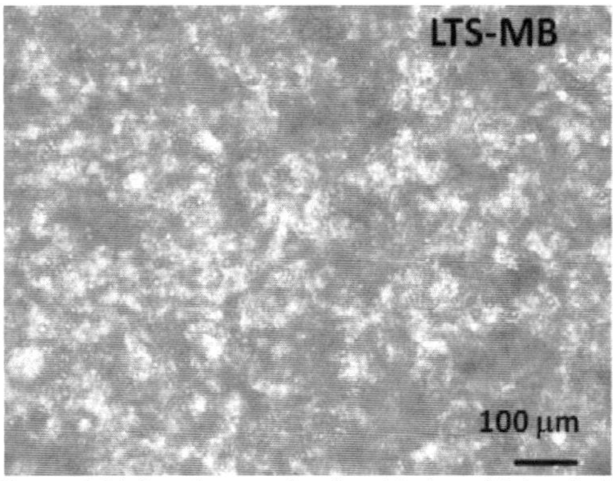

Figure 6 Light micrograph of multi-phase high-calorie liquid emulsion containing fiber from LTS okara and mungbean starch as sorbent and spouting material during fluidised-bed drying. Dark colour represents Lugol's iodine dyed starch fraction.

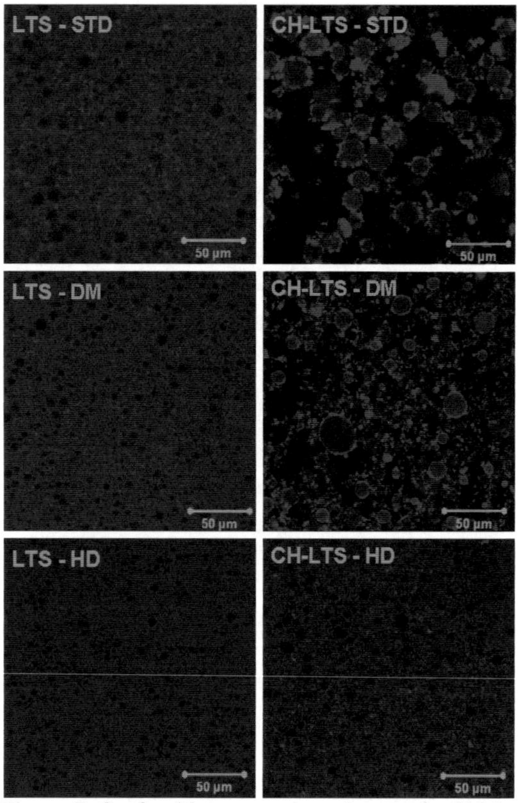

Figure 7 Confocal laser scanning micrograph of high-calorie emulsion containing fiber from industrial okara (LTS) or cellulase-modified okara (CH-LTS). Proteins are fluoresced in red by rhodamine B.

It is likely that the addition of cellulose-modified okara caused phase separation in the bulk aqueous phase. The protein was excluded from w/w dispersion and deposited at the oil-water interface; while the maltodextrin, starch and short-chain glucan and oligosaccharide were rich in the bulk aqueous phase. In HD formula, the residual protein in bulk aqueous phase could interact with the modified okara, causing an increase in viscosity due to the *thermodynamic incompatibility* between charged proteins and non-ionic short chain glucan or oligosaccharide of hydrolysed cellulose.

Conclusions
Dietary fiber, both pectic-like structure of soy soluble polysaccharides and short chain glucan obtained from the hydrolysis of cellulose played significant roles in the distribution of proteins in bulk aqueous phase, at the oil-water interface and subsequent viscosity of pasteurised high-calorie liquid emulsions.

Acknowledgements
The authors gratefully acknowledge financial support from the National Center for Genetic Engineering and Biotechnology, Thailand, under Project no. BT-B-01-FG-11-4802 and assistance from Ms.Kamolwan Israkarn.

References (Times New Roman 10, bold)

1. W. Tudthong; P. Hongsprabhas; P. Hongsprabhas, P. In *Proceeding of the 45th Kasetsart University Annual Conference*, 2007, http://www.cababstractplus.org/abstracts/Abstract.aspx?AcNo=20073176534.
2. N. Na Nakornpanom; P. Hongsprabhas; P. Hongsprabhas. *Food Res. Intern.*, 2010, *43*, 26.
3. C.-Y. Ma; W.S. Liu; K.C. Kwok; F. Kwok. *Food Res. Intern.*, 1997, *29*, 799.
4. D.K. O'Tool. *J. Agricul. Food Chem.*, 1999, *47*, 363.
5. AOAC. Association of Official Analytical Chemists, Arlingtion, 2000.
6. N. Fafaungwithayakul; P. Hongsprabhas; P. Hongsprabhas P. *Food Biophys*, 2011, *6*, 407.
7. N. Na Nakornpanom; P. Hongsprabhas; P. Hongsprabhas. In *Chemistry, Texture and Flavor of Soy* K.R. Cadwallader and S.K.C. Chang (Eds.), ACS Symposium Series Vol.1059, 2010, p. 45-60.
8. F. Yamaguchi; Y. Ota; C. Hatanaka. *Carb. Polym.*, 1996, *30*, 265.

O-12-622

EFFECT OF GELLING CONDITIONS ON PHYSICAL PROPERTY AND NETWORK STRUCTURE IN POLYSACCHARIDE GELS

Shingo Matsukawa*, Ryota Shimada, Kanae Matsui
Tokyo University of Marine Science and Technology,4-5-7 Kohnan,Minato-ku Tokyo 108-8477,Japan; matsukaw@kaiyodai.ac.jp

Abstract - Physical property and network structure in polysaccharide gels with different gelling conditions were studied by rheological measurements and gradient NMR measurements. Gellan gum was used as a sample polysaccharide which is a linear heteropolysaccharide composed of repeating units (glucose, glucuronic acid, glucose, rhamnose) with a carboxyl group from glucuronic acid. The effect of cooling rates on the properties and network structure of gellan gum were investigated by preparing the samples in three different cooling rates (Quench(*55°C/min),1°C/min,0.1°C/min)*, and we measured their gel strength and the network structures were discussed based on the NMR measurements.

Keywords: gelling condition/ polysaccharide gel/ viscoelasticity/ network structure/ diffusion coefficient

Introduction

Polysaccharides in solution take a random coil conformation at high temperature, and with decreasing temperature take high ordered structure such as double-helix structure. The aggregation of this double-helix is thought to lead the formation of network structure to induce the gelation. The Polysaccharide chains are involved into the aggregates accompanied with the decrease of molecular mobility, and the network structure by the aggregates is formed to induce the gelation before reaching the thermodynamic stable state. Therefore, the network structure in the gel changes to the stable state slowly and the gelation conditions such as cooling rate should affect the network structure and the physical property of gel.

Gellan gum is a microbial polysaccharide and a linear anionic polymer consisting of tetrasaccharide repeating units (D-glucose, D-glucuronic acid, D-glucose and L-rhamnose) [1,2]. Cooling hot gellan gum solutions in the presence of cations yields transparent, irreversible gels which are not easily digested by acid. As such, gellan gum finds extensive use in the food industry as a dietary fiber [3]. During cooling, gellan gum undergoes a random-coil to helix transition, forming in the initial aggregation step double helices [4]. In the presence of cations, the double-helices are further associated in junction zones, a process which can yield a gel network [5] above a critical concentration. Structural changes in the gel on a larger length-scale continue to develop in time after the on-set of gelation, so that the gel structure depends on the holding time and temperature [3].

In this work, the relationship between the physical property, network structure and molecular mobility was investigated for gellan gum gels made by three different cooling rates by measuring gel strength, the diffusion coefficient of pullulan added in gellan gum solutions as a probe polymer and the echo intensity of gellan were obtained by PFG(Pulsed field gradient)-Ste ^{1}H NMR measurements.

Experimental

Sample Preparation

Gellan gum was purchased from San-ei Gen FFI Co. (Osaka, Japan). Pullulan was purchased from Showa Denko Co. (Tokyo, Japan). Gellan solution preparation was as previously reported [6,7]. Gellan gum powder was added to D2O to a concentration of 2 % (w/w) and stirred at 40 °C overnight. Additional stirring for 2 h at 70 °C and 30 min at 90 °C effected complete dissolution. The gellan solution was subsequently mixed and stirred for 30 min at 90 °C with 0.2 % pullulan (Mw = 112,000) containing10 mM $CaCl_2$ to give a solution with 1% gellan and 0.1 % pullulan with 5 mM $CaCl_2$. The solution was cool down at 55(Quench), 1 and 0.1 °C/min to give gels with different cooling rates.

Measurements

Gel strength was measured at 25 °C by Fudho rheometer RTC-2002DD (Rheotech Co., Japan) with parallel plates at compressing rate 50mm/min on syndical gels with 10 mm height and 11mm diameter. The pulsed-fi eld-gradient 1H NMR spectra were recorded on a Bruker Avance 400 WB spectrometer equipped with a field gradient system operating at 400.13 MHz. The diffusion coefficient measurements were performed at 25 °C as reported previously [8]. The following experimental set-up was used for the PFG-Ste: pulse duration time 1 ms, gradient pulse interval 10 ms, repetition time 0.48 s, while the field gradient strength was varied from 200 to 2000 G/cm.

Results and Discussion

Fig. 1 shows the Young's modulus for gels made at cooling ratio of 55(Quench), 1 and 0.1 °C/min obtained from the gel strength measured on the day of preparation and after a week. The Young's modulus measured on the day of preparation became larger with decreasing the cooling rate. In these measurements, the braking strain became smaller with increasing the Young's modulus resulting the breaking energies were almost same for all samples (data not shown). In Fig.2, it can be also seen that the Young's modulus measured after a week became larger than that measured on the day of preparation for each sample. From these results, it can be said that sample preparation by slow cooling make the gellan gels hard and brittle and the storage also make the gels hard and brittle.

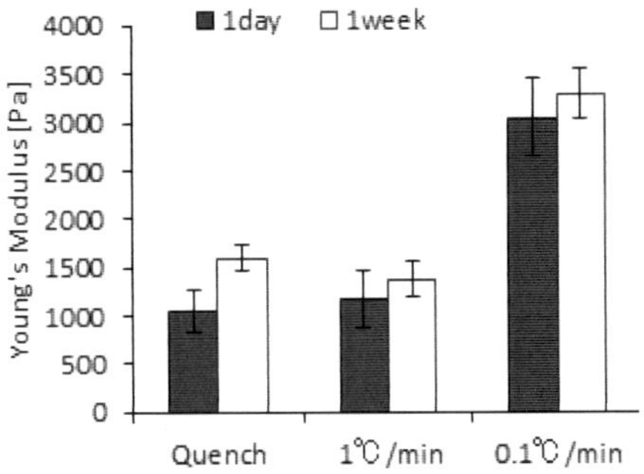

Fig.1 Young's Modulus for 1% gellan gels made at various cooling rates after storage times of 1 day and 1 week.

In order to clarify the relationship between the change of physical property by conditions in the preparation and storage time and the change of network structure in the gels, PGF-Ste ^{1}H NMR measurements were carried out for gel samples containing pullulan as a probe polymer. In Fig. 2, the echo intensity of gellan $I_{0,\mathrm{gell}}$ and the diffusion coefficient of pullulan added in gellan solutions as a probe polymer D_{pull} are plotted by closed circle on the left hand axis and open circle on the right axis, respectively.

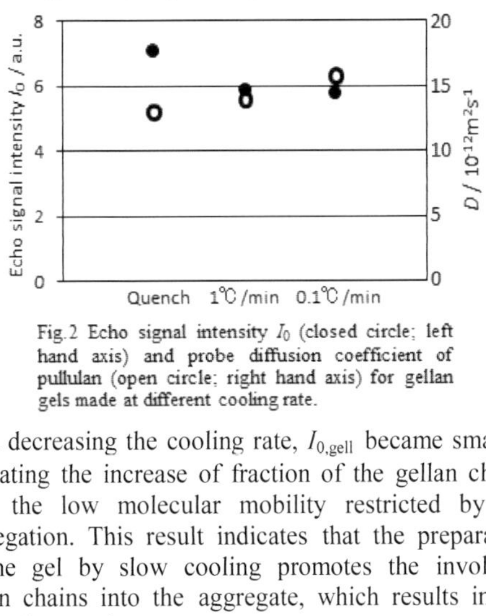

Fig.2 Echo signal intensity I_0 (closed circle; left hand axis) and probe diffusion coefficient of pullulan (open circle; right hand axis) for gellan gels made at different cooling rate.

With decreasing the cooling rate, $I_{0,\mathrm{gell}}$ became smaller, indicating the increase of fraction of the gellan chains with the low molecular mobility restricted by the aggregation. This result indicates that the preparation of the gel by slow cooling promotes the involving gellan chains into the aggregate, which results in the high gel strength.

On the other hand, D_{pull} increased with decreasing the cooling rate, indicating the decrease of restriction by gellan chain. This means that the involving gellan chains in higher level results in the coarse network structure where the restriction on pullulan diffusion becomes smaller [8,9,10].

Strength measurements for gels made by the cooling with holding times at various temperatures indicated that the holding at 45 °C is most effective to make the coarse structure. Fig. 3 shows stress-strain curves for the gel prepared by the cooling with a holding time of 3 h at 45 °C. The curve for continuous compression shows a typical projection at around 0.1 in strain, which contributes the high Young's modulus. In the second compression, which applied after the first compression up to 0.15 in strain and setting time of 1h, the projection disappeared but showed an almost same curve with the continuous compression under the strain of more than 0.15, suggesting that the structure which contributes the high Young's modulus was irreversibly destroyed by the first compression.

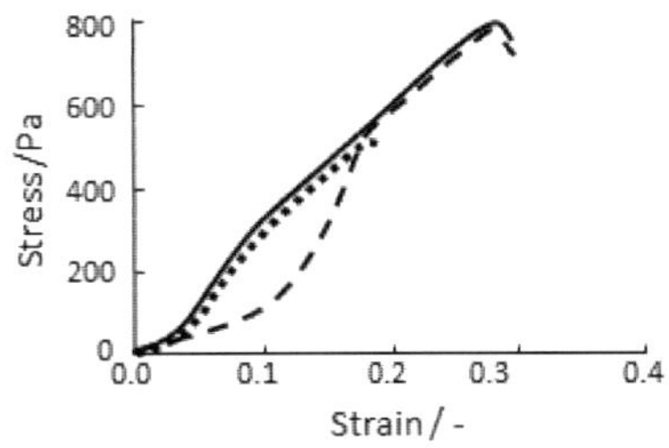

Fig.3 Stress-strain curves for 1% gellan gels made with a holding time of 3 h at 45 °C. Doted, dashed and solid lines are for the first, second and continuous compression, respectively.

Conclusions

In this study, the change of gel strength with the cooling rate has been attributed to the formation of thick aggregates in the network structure, where NMR measurements gave a useful information about the molecular mobilities.

Acknowledgements

The authors are thankful for the financial support of a Grant-in-Aid for Scientific Research (No.22240076) from the Ministry of Education, Culture, Sports, Science and Technology, Japan (MECSST).

References

1. Jansson, P-E., Lindberg, B., Sandford, P. A. *Carbohydrate Res.*,1983, *124*,135.
2. O'Neil, MA., Selvendran, RR., Morris, VJ., *Carbohydrate Res.*, 1983,*124*:123.
3. Sanderson, G. R.. *GELLAN GUM,* In: P. Harris (ed) Food gels. Elsevier, London, 1990,201.
4. Carroll, V., Miles, M. J., Morris, V. J.., *Macromolecules*, 1982,*4*:432.
5. Grasdalen, H., Smidsrod, O.., *Carbohydr. Polym.*, 1987,*7*: 371.
6. Miyoshi, E., Takaya, T., Nishinari, K. , *Food Hydrocolloids*, 1994,*8*:505.
7. Matsukawa, S., Tang, Z., Watanabe,T. , *Progr Colloid Polym Sci*, 1999, *114*:15.
8. Dai, B., Matsukawa, S.. *Food Hydrocolloids*, 2012,*26*, 181.
9. Shimizu, M., Brenner, T., Liao, R., Matsukawa, S. *Food Hydrocolloids*, 2012,*26*, 28.
10. Zhao,Q.,Matsukawa,S. *Polym. Jour.*, 2012,*44*, 901.

O-12-485

FLOW PROPERTIES OF COCOA BUTTER / SILICA DISPERSION AS A MODEL CHOCOLATE

I. Kaneda*, H. Miyazawa, and M. Ito
Department of Food Science, Rakuno Gakuen University, Japan – kaneda-i@rakuno.ac.jp

Abstract: To study the rheological properties of chocolate, we constructed a model chocolate comprised of fine spherical silica particles and cocoa butter. By using spherical silica dispersants to represent the sugar and cocoa particles, we would analyse the flow properties of the model chocolate by changing the intrinsic viscosity parameters and comparing with a theoretical model. We studied the effects of a sugar ester (specifically, sucrose erucate) on the flow properties of the model chocolate. We found that the surfactant reduced the viscosity at high shear rate as reported in a previous study. Furthermore, we estimated the intrinsic viscosity of the silica particles by using the Krieger–Dougherty equation and found the value to be around 3.5 in agreement with previous report. We concluded that the model chocolate system is a useful tool for studying chocolate rheology for application in the chocolate manufacturing process.
Keywords: chocolate, rheology, sucrose fatty acid ester

Introduction (Times New Roman 10, bold)

Chocolate consists of the dispersions of fine solid particles of sugar and cacao-mass, about 70% in total, in a continuous cocoa butter phase. As the chocolate melts in the mouth, lipids and fine solid particles coat the oral epithelial surfaces and affect the perception of taste sensation. Therefore, the rheological properties of chocolate have a strong influence on the sensation of smoothness that chocolate leaves in the mouth of taster. On the other hand, the rheological properties of molten chocolate are also important to the manufacture in the processing of chocolate, particularly during the enrobing, moulding and panning procedures. Because chocolate is a relatively high cost product, it is necessary to reduce the processing loss during manufacturing.

The rheology of molten chocolate has been studied by food scientists. In particular, the effect of emulsifiers on the rheological properties of molten chocolate is of interest [1]. Although the rheological properties of molten chocolate can be adjusted by changing the contents of the solid particles (sugar or cacao–mass) or fat, such changes may affect the flavour and taste of the chocolate. However, adding small amounts of certain emulsifiers to chocolate can also change its rheological properties. For example, lecithin is widely used in modern chocolate confectionery. Adding lecithin dramatically changes the rheological properties of the chocolate by reducing chocolate viscosity [2 - 4]. Another ingredient used is polyglycerol polyricinoleate (PGPR); It does not have a large effect on the viscosity but can reduce the yield stress [3 - 4].

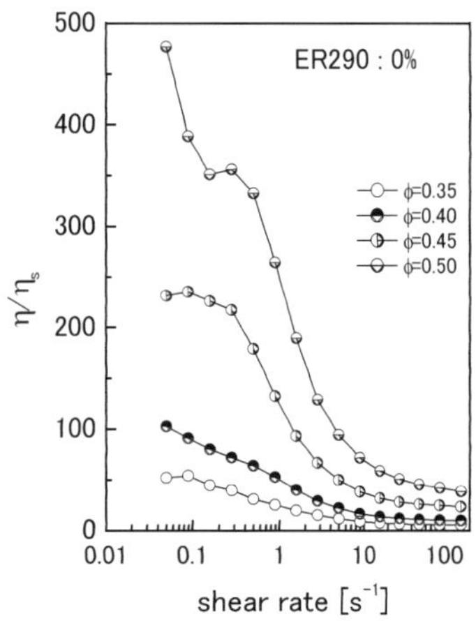

Fig.1 The flow curves of the model chocolate containing silica particles in various volume fractions.

Previous studies have shown that adding a surfactant to chocolate affects the rheological properties of the chocolate [1]. However, further studies to examine the fundamental rheological aspects of chocolate are required. A serious problem that arises when studying chocolate rheology is the complexity of chocolate. Because the dispersants of chocolate have quite complicated shapes and physicochemical surface properties, it is difficult to study the precise effects of the surfactants that are added in chocolate manufacturing. In this study, we constructed a model chocolate to study the effects of added surfactant in chocolate. The main component of the solid dispersant (fine sugar particles) was replaced with spherical silica particles. The silica particles were chosen to represent the sugar particles because sphere shaped silica particles are easy to obtain and have similar surface properties as those of sugar (both surfaces have many OH groups). In this paper, we report the effects of adding a sucrose fatty acid ester, which has a similar effect of lecithin [5] to our chocolate.

Experimental (Times New Roman 10, bold)

Materials

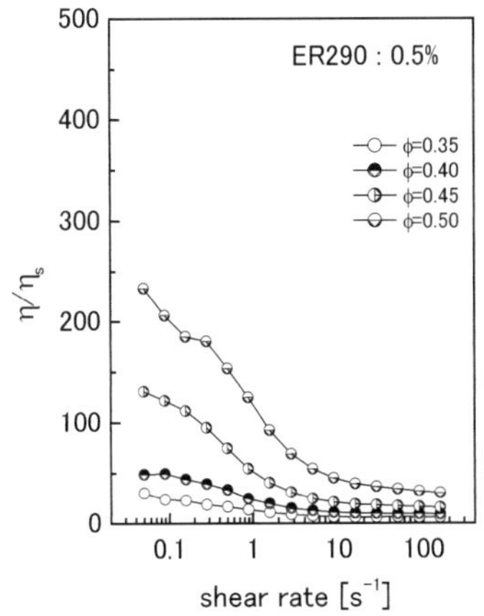

Fig.2 The flow curves of the model chocolate containing ER-290 (0.5wt%) and silica particles in various volume fractions.

Cocoa butter (Nisshin Kako Co. Ltd., Osaka, Japan), and sucrose erucate ; Ryoto Sugar Ester ER-290 (Mitsubishi Kagaku Foods, Tokyo, Japan) were used without further purification. For the fine silica particles, we used EXCELICA (Tokuyama Co.Ltd., Yamaguchi, Japan). The average diameter of the silica particles was 10.6 μm and the specific surface area was 1.9 m^2/g.

Model chocolate

The model chocolate was prepared as follows; cocoa butter and silica particles were mixed with a hand mixer at 50°C. The volume fraction of the silica particles in the model chocolate was calculated according to Eq.1.

$$\phi = \frac{w_s(1/\rho_s)}{w_{cb}(1/\rho_{cb}) + w_s(1/\rho_s)} \qquad (1)$$

where w_s, and ρ_s are the weight and density of the silica particles, and , w_{cb}, and ρ_{cb} are the weight of the weight and density of the cacao butter, respectively. The density of the samples was measured with a pycnometer at 50 °C. The surfactant was added the molten cocoa butter and then, the silica particles were added to the cocoa butter-surfactant mixture.

Rheology

The steady state viscosity was measured with an ARES rheometer (TA instruments, USA). A parallel plate fixture was used (gap: 0.5mm). The measurement was performed at 50°C. The intrinsic viscosities of the samples were obtained using the Krieger–Dougherty equation (Eq.2) [6]

$$\frac{\eta}{\eta_s} = \left(1 - \frac{\phi}{\phi_m}\right)^{-[\eta]\phi_m} \qquad (2)$$

where η, η_s, ϕ, ϕ_m and [η] are the viscosity of the model chocolate at the high-shear-rate plateau, viscosity of the continuous phase (with or without surfactant) at the high-shear-rate plateau, volume fraction of the silica particles in the model chocolate, maximum packing fraction of sphere, and intrinsic viscosity of the dispersant, respectively. In this study, ϕ_m was fixed at 0.63 and [η] was estimated using the experimental data by the non-linear least squares method.

Results and Discussion

Flow curves

The flow curves of the model chocolates without surfactant are shown in Fig.1 and the result of the sample containing ER-290 at 0.5 wt% is shown in Fig. 2. Clearly, the viscosity at low shear rate decreased when ER-290 was added. A sugar ester shows a viscosity reduction effect similar to that of lecithin [5]. The model chocolate that contained only cocoa butter and silica particles showed results that agreed with those of a previous study [5] that was conducted using real chocolate. As shown in Fig. 1, it is difficult to estimate the zero shear viscosity from the experimental results. Therefore, we focused on the plateau region of the flow curves at high shear rates, namely η_∞ ($\dot{\gamma} \rightarrow \infty$). The viscosity reduction effect of ER-290 for the samples containing the silica particles in various volume fractions is shown in Fig. 3. The viscosity

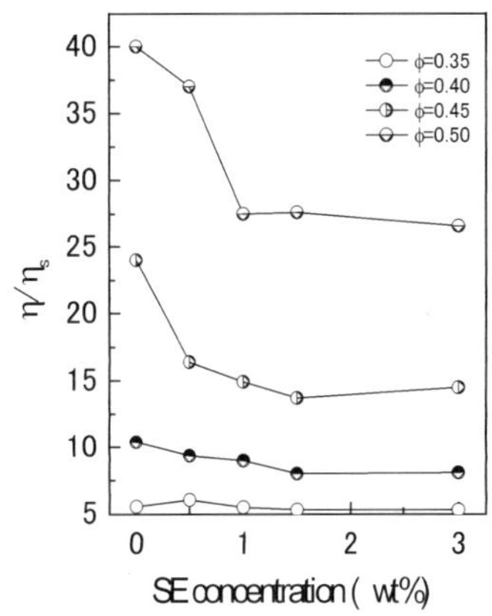

Fig.3 The viscosity reduction effect of ER-290 in various concentrations

reduction effect was more significant in the high volume fraction sample. These results suggest that the viscosity reduction effect of ER-290 is due to the change in the surface property of the silica particles, because the mechanical friction between the particles may influence the bulk viscosity of the high volume fraction samples. On the basis of these results, we would expect ER-290 to be absorbed on the surfaces of the silica particles. Although it is reported that adding lecithin or a sugar ester to chocolate improves its yield value [2-4], no such phenomenon was observed in this study. This could be due to our use of relatively low volume fraction samples (ϕ = 0.2 - 0.5). To investigate

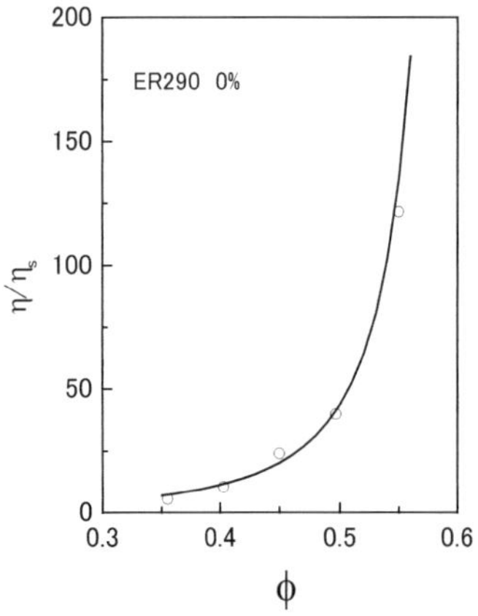

Fig.4 The volume fraction dependency of the relative viscosity for the model chocolate without surfactant. Circles and line denote the experimental results and fitting result with K-D equation, respectively

the yield value, higher volume fraction samples close to the maximum packing fraction should be studied. Such a study could easily be conducted our model chocolate system.

Intrinsic viscosity

By using spherical silica particles in our model chocolate, we were able to analyse the chocolate's flow properties by changing the parameter in our model chocolate and comparing with a theoretical model. To check the dispersibility of the silica particles in the model chocolate, we estimated the intrinsic viscosity of the silica particles. According to the Einstein's formula [7], the intrinsic viscosity of a spherical particle is 2.5. A deviation from the theoretical value is considered as a parameter of dispersibility for the silica particles in the model chocolate. Fig. 4 and 5 show the results for

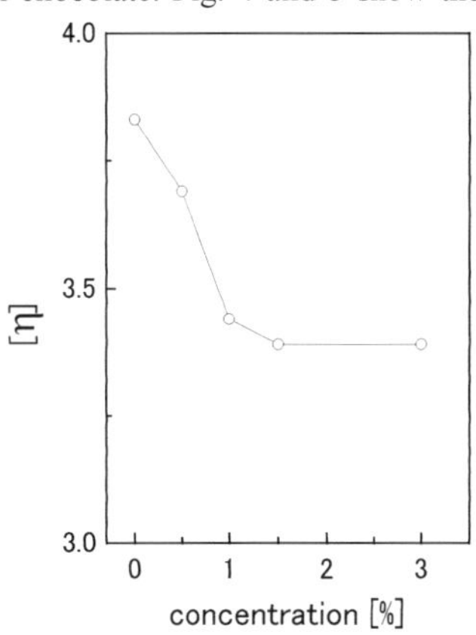

Fig.6 The surfactant concentration dependency of the intrinsic viscosity for silica particle in the model chocolate.

the model chocolate with and without surfactant, respectively, along with the theoretical curves. The calculated values were in good agree with the

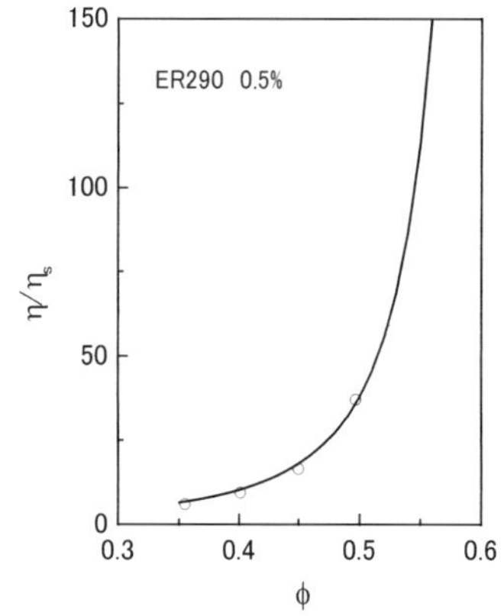

Fig.5 The volume fraction dependency of the relative viscosity for the model chocolate with ER-290 (0.5wt%). Circles and line denote the experimental results and fitting result with K-D equation, respectively

experimental results. Changing the concentration of ER-290 in the model chocolate (Fig. 6) clearly showed that intrinsic viscosity reduced with increasing surfactant concentration. The intrinsic viscosity of the sample without surfactant was about 3.7 indicating that the aspect ratio of the dispersant was elatively large. When ER-290 was added at 1 wt%, the intrinsic viscosity reduced to about 3.4. The intrinsic viscosity depends on the aspect ratio of the particles. The intrinsic viscosity of a relatively large (40 μm in diameter) spheres is 3.28 [8]. This value is much closer to our experimental result. We focused on the viscosity at high shear rate in this study. This rheological property strongly influences the enrobing and panning process in chocolate processing.

Conclusions

We developed a novel model chocolate system to study its rheological properties. To simplify the problem of chocolate rheology, we employed spherical silica particles to simulate the fine sugar particles. One of the biggest advantages of using this model is that we can quantitatively evaluate the dispersibility of the solid dispersant in chocolate by changing the parameter of intrinsic viscosity.

The results of our steady state viscosity experiment using model chocolate containing sugar ester confirmed the usefulness of the model system that can be valuable improving the chocolate manufacturing process.

Acknowledgements

Materials used in this study were kindly provided by Mitsubishi Kagaku Foods Co. Ltd.

References (Times New Roman 10, bold)

1. E. Ohene Afoakwa; A. Paterson; M. Fowler *Trens in Food Sci & Tech* 2007, *18*, 290
2. J. Chevalley *Industrial Chocolate manufacture and use*, S.T. Beckett, Ed.; Blackwell Science, Oxford, 1999: pp.182-200
3. D. Rector *The manufacturing Confectioner* 2000, *80 (5)*, 63
4. B. Scantz; H. Rohm *Lebebsmilltel Wissenschaft und Technologie* 2005, *38*, 41
5. J.O. Swer; Y. Reddy S *J Food Sci Technol* 2008, *45 (5)*, 447.
6. I.M. Krieger; T.J. Dogherty *Trans Soc Rheol* 1959, *3*, 137
7. A. Einstein *Ann Physik* 1906, *19*, 289
8. H. Giesekus *Physical properties of foods*, R. Jowitt et al, Ed.; Allied Science Publishers, 1983: Chap. 13

KN-12-01

POLYMER SCIENCE APPROACH TO FOOD RHEOLOGY AND PROCESSING

K. Nishinari

[1]*Dept Food and Health Science, Graduate School of Human Life Science, Osaka City University,* 3-3-138Sugimoto, Sumiyoshi, Osaka, 558-8585 Japan - katsuyoshi.nishinari@gmail.com

Abstract – Recent progress in rheology of gels and gelling processes of proteins and polysaccharides, gelatinization and retrogradation of starch, and mixed food polymers is overviewed and the endeavour of creating foods for patients with difficulty in mastication and deglutition is briefly described.

Keywords: gel, polysaccharide, protein, mixture, aspiration.

Introduction

Since the palatability of foods is determined mainly by texture and flavour, rheology plays an important role in food processing. Texture is thought to be a sensory property, and the relation between the sensory evaluation and instrumental measurement has been studied. Recently, the muscle activity, movement of organs in the oral cavity during mastication and swallowing have been studied by electric physiological measurements. Hydrocolloids with thickening and gelling function have been used to control the texture of foods. Flavour intensity was found to be changed by the change in texture.

In the sensory evaluation, the ambiguity of texture terms sometimes causes the difficulty to get reliable results, and the relation between rheological terms and daily used terms should be checked.

Results and Discussion

Gelling process of foods

Kinetic studies on gelling process of soybean protein and casein micelles have been useful for improving the manufacturing of tofu (soybean curd) and cheese. Gelling processes of other food gels have also been studied to improve the manufacturing. Dynamic viscoelastic measurements with small amplitude have been employed to understand the gelling process. The information obtained is useful to select the better raw material, processing conditions such as temperature, pH, ionic strength and so on.

Fig.1 shows the gelation process of soybean milk in the presence of glucono-delta-lactone (GDL).

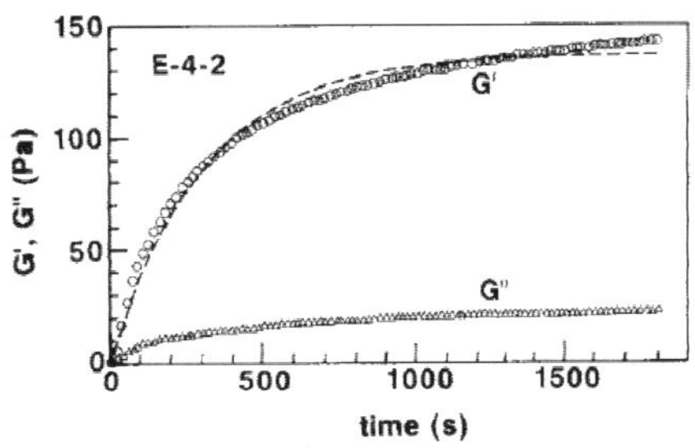

Fig,1 Storage and loss shear moduli, G' (▲),the loss modulus G'', of 4% soymilk in the presence of 0.4% GDL at 80°C. The time zero was taken as the time at which moduli began to deviate from the base line.

The storage modulus increased with the lapse of time and could be approximated well by a first order kinetic equation (Nishinari et al,1991).

Gelation of gellan

Gellan is an anionic extracellular bacterial poly-saccharide discovered in 1978. Deacylated gellan (most commercially available gellan) converts on cooling from disordered coils to 3-fold double helices. The coil–helix transition temperature (T_m) is raised by salt: divalent cations are more effective to raise T_m than monovalent cations. Gelation occurs by aggregation of double helices. Reduction in pH promotes aggregation and gelation by decreasing the negative charge on the polymer and thus decreasing electrostatic repulsion between the helices.

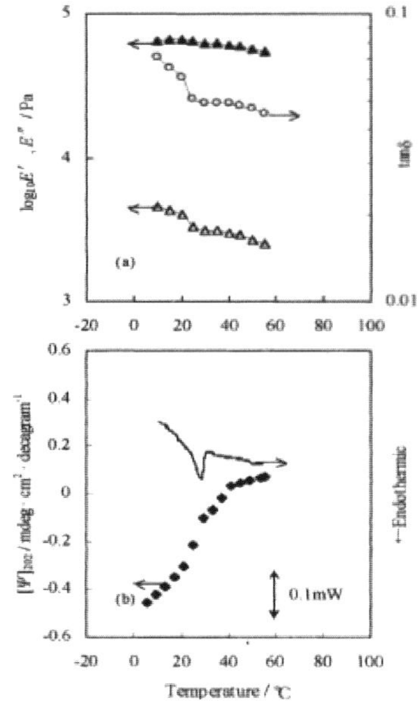

Fig.2 (a)Temperature dependence of the storage modulus E' (▲), the loss modulus E'' (△) and mechanical loss tanδ (○) on heating and (b) the heating DSC curve (solid curve) and temperature dependence of the specific ellipticity at 202nm, $[\Psi]_{202}$, (◆) on heating for 1.6%(w/w) gellan gum gels. E' and E'' were measured after 15min equilibration at each temperature Heating rate (DSC&CD): 0.5°C /min (Nitta et al, 2001).

Fig.2 shows the temperature dependence of storage and loss Young's moduli, E' and E'', specific ellipticity, and heating DSC curve for a deacylated gellan gum (1.6%). E' and E'' were measured by the observation of longitudinal vibration of a cylinfdrically moulded gel. Since the gel keeps the cylindrical shape below 60°C, the endothermic peak in a heating DSC curve and the steep change in the specific ellipticity must be caused by helix-to-coil transition in a solid gel., indicating that there is a sufficient space for conformational change from stiff chains to flexible chains.

As menioned above, salt raises T_m and increases the elastic modulus. Fig.3 shows that the addition of salt increases E', but the excessive addition decreases E'.

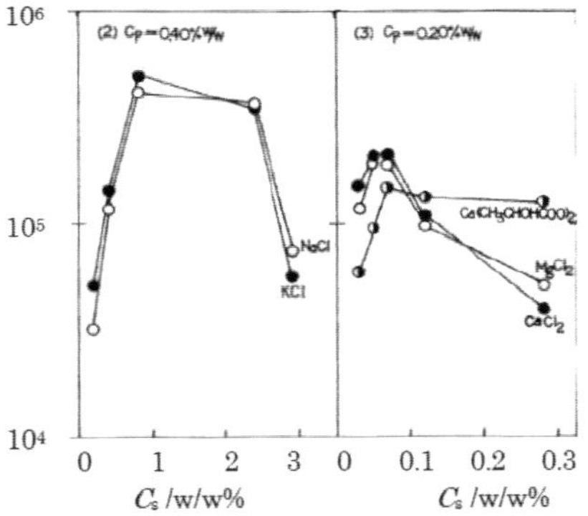

Fig.3. Storage Young's modulus of gellan gels as a function of salt concentration (Moritaka et al, 1991)

V.J.Morris et al (1995) suggested that the structure of gellan gels became heterogeneous at higher salt concentration. This phenomenon has also been observed on progressive addition of alkali metal salts to κ-carrageenan (Watase & Nishinari, 1982), and maxima in gel strength on varying salt concentration and/or pH are commonly observed in thermogelation of globular proteins (Foegeding, Bowland, & Hardin, 1995). Indeed, such maxima seem to be universal, and are perhaps to be expected. Biopolymer gels form part of a continuum from solutions of individual molecules at one extreme to close-packed solids at the other as the extent of intermolecular association is increased (by varying conditions such as pH and ionic environment). Optimum crosslinking will occur somewhere along the continuum; less association will give a weaker network; greater association will give larger aggregates, with consequent reduction in the effective number of individual junctions, until ultimately the network collapses into a solid precipitate. This interpretation is consistent with the apparent "phase separation" observed for gellan gels at very low pH or with a large excess of salt, and with the observation (Ohtsuka & Watanabe, 1996) that the gels become turbid at the point where their strength begins to decrease.

High acyl gellan was also studied (Noda et al, 2008). Nitta et al (2010) recently reported the anisotropic structure formation in gellan. Gelation of gellan was recently reviewed (E.R.Morris et al, 2012).

Rheology of Starch

Starch is a major carbohydrate reserved in plant tubers and seed endosperm and used as thickener, water binder, emulsion stabilizer and gelling agent. Main sources of starch are maize, wheat, potato, tapioca and rice. In food processing, many hydrocolloids are added to improve the organoleptic and physiological properties of final products. Among many hydrocolloids, starch is the most important polysaccharides used in food systems. Maize starch is most widely used .

Hirashima et al (2005) studied the effects of sugar, salt, acids on rheological properties of maize starch pastes. They found that when sucrose was added before heating of starch dispersions, the viscosity of the pastes increased with increasing sucrose concentration up to 20 wt% and decreased with increasing sucrose concentration above 20 wt%, although the viscosity of samples in the presence of sucrose was greater than that of the control at higher shear rate. No decrease in the viscoelasticity of the pastes above 20 wt% sucrose concentration was observed by adding sucrose after heating of starch dispersions. The viscoelasticity of maize starch paste to which sucrose was added after heating increased with increasing sucrose concentration (Fig.4).

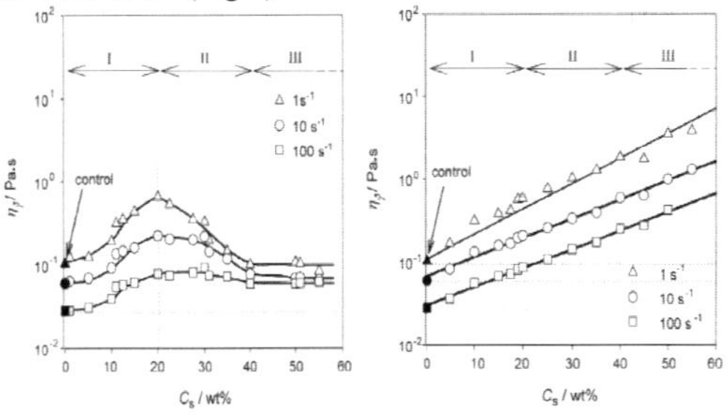

Fig.4. (Left)Sucrose concentration (C_s) dependence of shear viscosity of 3.0 wt% maize starch pastes at the shear rates of 1, 10, and 100 s^{-1}, respectively. Sucrose was added before heating. (Right) Sucrose was added after heating. Measurements were made at 25.0°C.

Singh et al (2006) studied the structural, thermal and viscoelastic characteristics of starches separated from

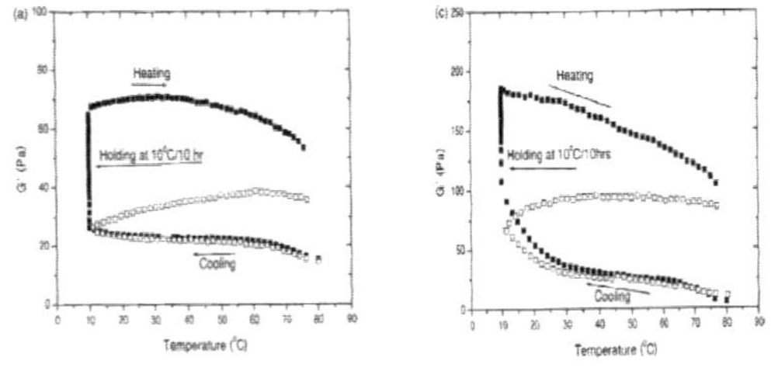

Fig.5 Changes in G' during cooling and reheating process with holding at and reheating of 5% paste of normal and maize starch (closed squares). Open squares represent the curves without holding at 10°C for 10h (Singh et al, 2006).

normal, sugary and waxy maize, and concluded that normal and sugary maize starches showed higher storage and loss moduli than waxy type starches. The change in the moduli during cooling and reheating of pastes cooked at different temperatures revealed low disintegration of granular structure in starch with higher amylose and amylose–lipid complex as well as low crystallinity.

Tsutsui et al (2012) examined the effects of annealing on rice starch suspensions. When the annealing time t_a increased, for the sample which was annealed at almost the same as the gelatinization temperature (= T_{gel}; 62°C), 58 or 62°C, the onset temperature at which storage modulus, G', started to increase shifted to higher temperatures, and G' in the temperature range from 65 to about 75°C decreased.

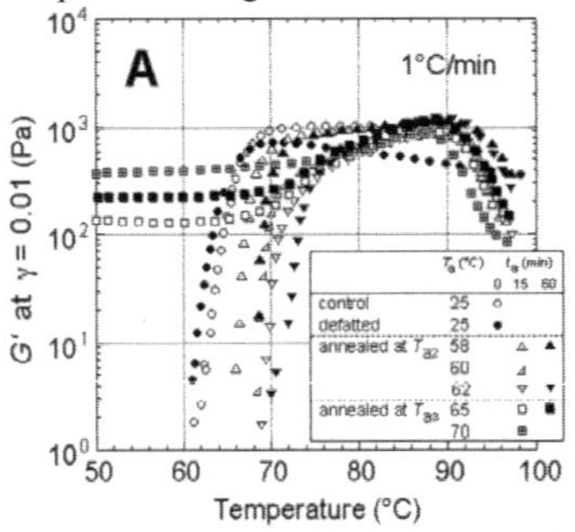

Fig.6 Twmperature dependence of G′ of 10wt% rice starch suspension annealed at various temperatures.

Further, G' of the sample annealed at higher than T_{gel}, 65°C, increased in the temperature range from 50 to 75°C (Fig.6). The amorphous and thermally unstable portion in the starch decreased with increasing t_a. Effects of annealing on the temperature dependence of steady shear viscosity were found almost the same as those of G' reported previously. Shear thinning hardly occurred at the shear rate range from 1 to10 s^{-1} for the annealed sample.

Chantaro et al (2012) studied the gelatinization and retrogradation of tapioca starch (TS) with and without xanthan. They suggested that an RVA (rapid visco analyser) failed to show the retrogradation, network structure formation behaviour in the mixture of TS and xanthan, which is triggered by the gelation of amylose.

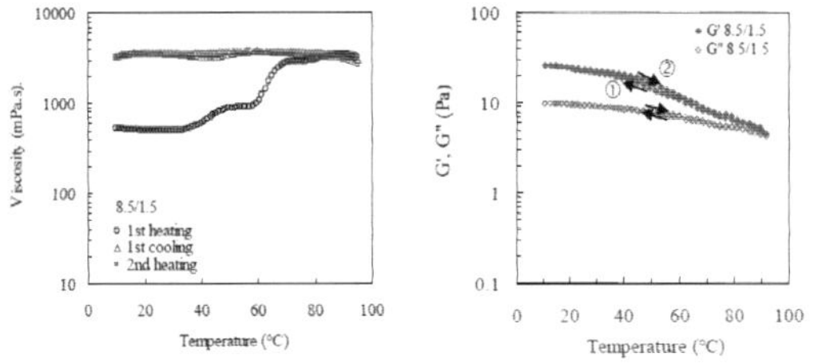

Fig.7 (Left)Temperature dependence of steady shear viscosity of 5% w/w TS/Xan mixtures: 8.5/1.5 Shear rate: 10 s^{-1}. ① heating from 10 to 95°C ② cooling from 95 to 10°C, ③ and finally reheating from 10 to 95°C at the rate of 1 °C/min. (Right) Temperature dependence of G' and G'' of 5% w/w TS/Xan pastes 8.5/1.5. ① cooling from 90 to 10°C and ② subsequently heating from 10 to 90°C. (frequency: 1 rad/s, scan rate at 1 °C/min, 5% strain) (Chantaro et al, to be published)

Fig.7 shows the structural formation of TS in the presence of xanthan on cooling is inhibited in the steady shear viscosity measurements whilst it is clearly seen in the small deformation oscillatory measurement.

Mixed systems

When a single biopolymer cannot achieve the function, a mixture with another biopolymer can realize the required property. In other words, the mixing of two biopolymers can create a new gelling or a thickening agent which can control the texture of foods. Kim et al (2006) examined the mixture of xanthan/xyloglucan and found that the mixture of xyloglucan and pyruvate-free xanthan showed no synergistic interaction. It was found that this synergistic interaction is caused by the intermolecular binding between xyloglucan and xanthan, and, in the heterotypic junction zones, the xanthan side chain becomes a new state that is different from both the coil and helix states. Annable et al (1994) examined the mixture of konjac glucomannan and xanthan. The results indicate that xanthan/KM interaction occurs at temperatures much lower than the conformational transition. The rheological properties (i.e., storage and loss moduli) and gelation temperature are greatly influenced by the presence of electrolyte(Fig.8). In all instances, both storage and loss moduli, gelation temperature are reduced when electrolyte is present, with divalent cations having a greater effect than monovalent cations. These observations are explained by the fact that electrolyte promotes xanthan self-association at the expense of xanthan/KM interaction.

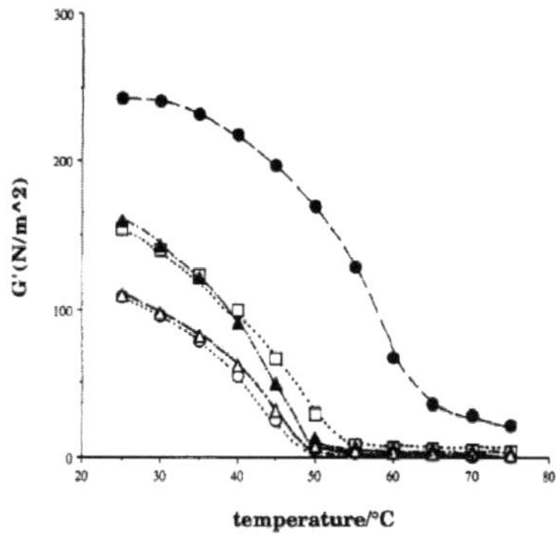

Fig.8 G' as a function of temperature for xanthan-KGM mixtures (1:1, 0.6 wt %) in the presence and absence of monovalent cations (0.04 mol dm^{-3}: H_2O (●); NaCl (○); KC1(□); CsCl (▲); NH_4Cl (Δ) (Annable et al, 1994)

Brenner et al (2012) performed small and large deformation tests on polysaccharide- based gels containing konjac glucomannan, locust bean gum and κ-carrageenan. Low pH degradation of κ-carrageenan during heating led to lower gel strength. When κ-carrageenan degradation was avoided, the linear-regime rheology was dominated by the κ-carrageenan content, and the storage Young's modulus was found to decrease with decreasing κ-carrageenan concentration. The large deformation behaviour, however, was dominated by the konjac glucomannan concentration, and the rupture strain increased strongly with increasing konjac concentration. These two opposing trends led to a maximum in rupture stress in extension tests.

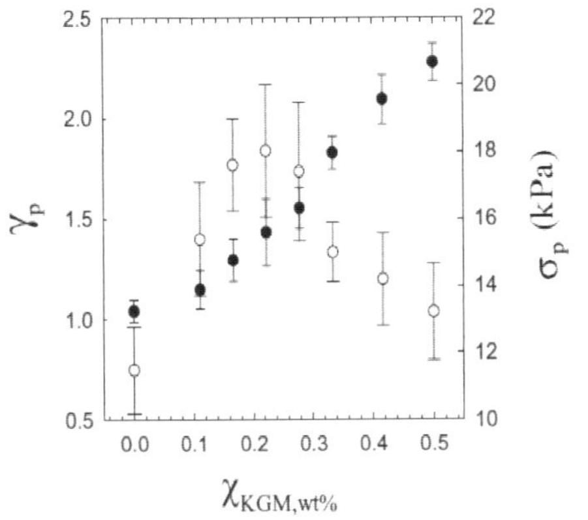

Fig.8 Rupture strain γ_p (•) and stress σ_p (○) in ring extension tests, plotted as a function of KGM concentration χ_{KGM} for constant total polysaccharide content = 1.2%. (Brenner et al, 2012)

Foods for patients with difficulty in mastication and deglutition

With the advent of aged society, the aspiration is becoming the most important cause of death for elderly. The use of thickening agents to alter the rheologic properties of food is a common practice in dysphagia treatment although the mechanism of dysphagia is still not well understood. We have proposed a method to judge the risk of aspiration by videofluorography using three solutions of polysaccharides widely used in food industry with different shear thinning behaviours (Nishinari et al, 2011). This study showed the possibility to reduce the risk of aspiration in dysphagic patients by controlling the food rheology. Sonomura, Mizunuma et al (2011) showed the possibility of numerical simulation of dysphagia.

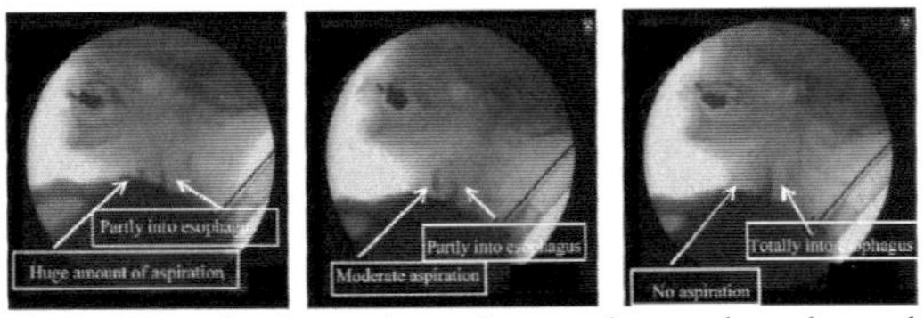

Fig.9 Videofluolography of a patient who showed aspiration for two samples a, left (G1); b, middle (G2), and without aspiration for c, right (X2) (Nishinari et al, 2011)

A research group of Tohoku University uses menthol, capsaicin to shorten the latent time of swallowing. The combination of rheological and chemical methods should be effective to prevent the aspiration. It is first necessary to understand the mechanism of aspiration.

Acknowledgements

The author wishes to thank coworkers. This study was supported by JSPS Grants-in-Aid for Scientific Research A19200051, Improvement of QOL in meals of persons with difficulty in deglutition by preparation of solutions for judging aspiration symptoms (2007-2010), A 22240076 The relation between the flavour release and the structure/physical properties of food gels in the mastication (2009-2012), and by MAFF Research and Development Projects for Application in Promoting New Policy of Agriculture Forestry and Fisheries 22026 (2009-2012).

References

Annable, P., Williams, P.A. & Nishinari, K. *Macromolecules*,1994, *27,* 4204.

Brenner, T., Achayuthakan, P. & Nishinari, K. *J Texture Studies*, 2012, in press

Chantaro, P., Pongsawatmanit, R. & Nishinari, K. *Food Hydrocolloids,* 2012, to be published

Ebihara,T., Ebihara,S., Watando,A., Okazaki, T., Asada,M., Ohrui, T., Yamaya, M. & Arai,H. *British J. Clinical Pharmacology*, 2006, *62,* 369.

Foegeding, E.A., Bowland, E.L., & Hardin, C.C. *Food Hydrocolloids*,1995, *9*, 237.

Hirashima,M.,Takahashi,R.&Nishinari,K.*Food Hydrocolloids*, 2005, *19*, 777.

Kim,B.S., Takemasa,M. & Nishinari, K. *Biomacromolecules*, 2006, *7*, 1223.

Moritaka,H., Fukuba,H., Kumeno,K., Nakahama, N. and Nishinari,K. *Food Hydrocolloids,* 1991, *4*, 495.

Morris, E., Nishinari, K. & Rinaudo, M. *Food Hydrocolloids,* 2012, *28*, 373.

Morris,V.J., Tsiami, A & Brownsey, G.J. *J Carbohydr. Chem.* 1995. *14*, 667.

Nishinari, K., Hayakawa, F., Xia, C-F., Huang, L., Meullenet, J-F., Sieffermann, J-M. *J.Texture Studies* 2008, *39*,530.

Nishinari, K. , Takemasa, M., Su, L., Michiwaki, Y., et al. *Food Hydrocolloids*, 2011, *25*, 1737

Nitta,Y., Ikeda,S., Takaya, T. & Nishinari,K. *Transactions of MRS-J*, 2001, *26*, 621.

Nitta, Y., Takahashi, R. & Nishinari, K. *Biomacromolecules*, 2010, *11,* 187.

Noda, S., Funami, T., Nakauma, M., Asai, I.et al *Food Hydrocolloids*, 2008, *22*, 1148.

Ohtsuka, A. & Watanabe, T. *Carbohydr. Polym.*1996, *30*, 135.

Singh, N., Inouchi, N. & Nishinari, K. *Food Hydrocolloids*, 2006, *20*, 923.

Sonomura ,M., Mizunuma, H., Numamori, T., Michiwaki, Y. et al *J. Texture Studies*, 2011, *42*, 203.

Watase,M. & Nishinari,K. *Rheol. Acta*, 1982, *21*, 318.

Tsutsui,K. et al *J.Texture Studies,* 2012, *In press*

KN-13-01

LIQUID MIXING: A LOW COST, ENERGY SAVING AND EFFECTIVE METHODOLOGY FOR PREPRATION OF ELASTOMERIC POLYMER NANOCOMPOSITES

Liqun Zhang[1,2*],Yiqing Wang[1], Youping Wu[1], Yonglai Lu[1] and Ming Tian[1]
[1]*The Key laboratory of Beijing City on Preparation and Processing of Novel Polymer Materials, Beijing University of Chemical Technology, Beijing, 100029, China*
[2]*State Key Laboratory of Organic/Inorganic Composites,Beijing University of Chemical Technology, Beijing, 100029, China. zhanglq@mail.buct.edu.cn*

Abstract - Reinforcement, especially nano-reinforcement, is very important for rubber matrixes [1-3]. In such a case, the dispersion of nano-filler in rubber matrix and interfacial interaction between filler and rubber are two critical factors to affect the final performance of composites. It is always an interesting field to develop novel techniques to achieve fine dispersion of nano-filler and strong interfacial interaction. Recently, an universal method of preparing rubber nanocomposites has been developed in our laboratory, i.e. co-coagulating rubber latex and target filler aqueous suspension (namely LCM). In this paper, rubber-clay nanocomposites, rubber-attapulgite (AT) nanocomposites, rubber-starch nanocomposites, rubber-graphite nanocomposites, rubber-graphene nanocomposites and rubber-silica nanocomposites were prepared by LCM and their structures and properties are investigated. Compared with direct blend method, LCM can remarkably improve the filler dispersion. In addition, the designed interfacial interaction can be obtained through the in-situ modification during latex compounding. As a result, the composites prepared by LCM exhibit excellent mechanical properties. LCM is proved to be a very effective way to prepare rubber nanocomposites with many advantages, such as low cost and low pollution, etc.
Key word: latex rubber dispersion nanocomposites

Introduction

As far as the mechanical properties (tensile strength, stiffness, abrasion resistance and fatigue resistance, etc.) are concerned, most elastomers would be of no practical value if they were not reinforced. For many years, fine nano-sized particles of carbon black, and more recently, silica, have been incorporated into rubber compounds at about 20 – 25 % volume concentration to give major improvements in the physical properties. Generally, fine-particle carbon black and silica can enhance the tensile strength of non-self-reinforcing elastomers by up to ten times[1-3].

The commonly-employed processing method of direct mechanical mixing the filler particles into rubber may be termed a mechanical blending compounding technique. Since most elastomers have high viscosity, the strong shearing forces break up the particle agglomerates and thus the nano-particles are easy to be dispersed in the rubber matrix in comparison with mechanical mixing of powders in molten plastics. In addition, the high melt viscosity impedes re-agglomeration of dispersed nano-particles.

However, for nano-particles with much smaller size, the lower apparent density of agglomerates and the higher viscosity of rubber result in obvious shortcomings by using this compounding technique, for example: flying dust, long mixing times (1/2～1/3 of the total mixing time of the compound), high energy consumption in processing, and generation of some aggregates or agglomerates of nano-particles. Especially at a relatively high volume concentration, a network of nano-particles is created within the rubber matrix.

As we know, most rubbers have their latex forms, such as natural rubber, styrene-butadiene rubber, nitrile rubber, chloroprene rubber, etc. In rubber latex, rubber particles with 0.05 micrometer to 0.5 micrometer are dispersed in water medium. Mixing target nano-filler with rubber latex to form a pre-dispersed filler/rubber masterbatch possibly avoids high energy consumption and dust pollution in comparison with traditional blending method. In fact, the carbon black/rubber masterbatch has been applied in rubber industry for a long time. A typical example is Carbot Elastomer Composites (CEC), which exhibits better dispersion and good processing ability.

In this paper, rubber nanocomposites with different kind of nanofiller were prepared via latex compounding method, [4-16]. Their dispersion in matrix and properties are introduced. Compared with direct blend method, co-coagulation method can remarkably improve the filler dispersion. In addition, the designed interfacial interaction can be obtained by using the in-situ modification during latex compounding. The composites prepared by LCM exhibit excellent mechanical properties and satisfactory processing performance.

Experimental

The preparation and subsequent processing, structure and morphology characterization, performance and application of different nanocomposites with different nanofillers can be found in detail in the listed reference [4-16]. Here, SBR-clay nanocomposite was shown as an example to illustrate the preparation process [4-9].

Na^{+}-montmorillonite (Na^{+}-MMT) was dispersed in water at 25°C under continuous mechanical stirring for 5 hours to obtain a clay suspension, in which Na^{+}-MMT content is of about 2%. Then the clay aqueous suspension and the rubber latex were mixed and vigorously stirred for 30minutes. After that, the mixture was co-coagulated in the electrolyte solution (2% dilute sulfuric acid solution), washed with water and dried in an oven at 50℃ for 20h, and then the rubber-clay nanocompound (uncured nanocomposite) was obtained.

The vulcanizing ingredients and other additives were mixed into the nanocompound on a 6-inch two-roll mill, according to the relative recipe. Then the compound was vulcanized in a standard mold at 150℃. The vulcanizates are referred to as SBR-clay nanocomposites.

Results and Discussion

Rubber-clay nanocomposites

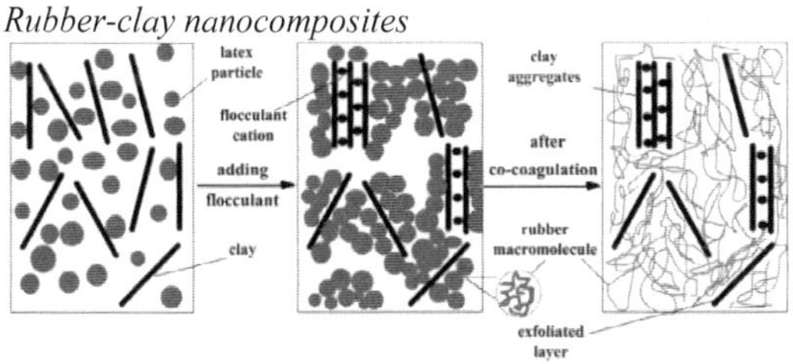

Figure 1 Schematic illustration of the mixing and co-coagulating process[4]

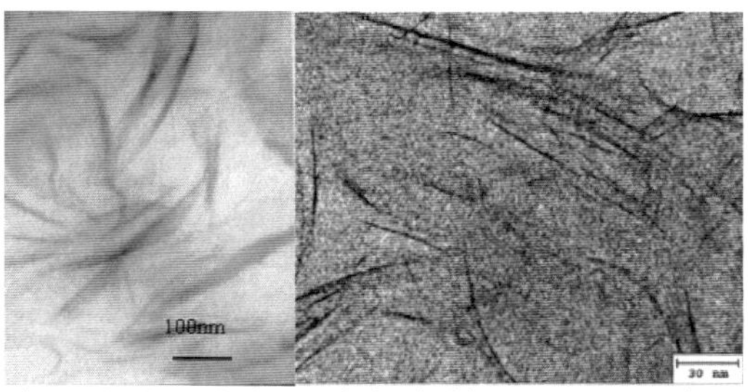

Figure 2. Morphology of clay/rubber nanocomposites: 2a nanocomposite with low loading clay (4vol%) by LCM; 2b exfoliated structural nanocomposite by LCM.[4,9]

Table 1. Mechanical properties of rubber-clay

sample	SBR-Clay		NR-Clay	
	MC	NC	MC	NC
Stress at 300% /MPa	2.1	7.4	2.7	12.3
Tensile strength /MPa	2.4	14.5	11.6	26.8
Elongation at break/%	400	548	568	644
Shore A hardness	52	60	41	54
Tear strength/(KN/m)	16.5	45.3	22.8	44.1

nanocomposites (NC samples) and conventional rubber-clay composites (MC samples) with 20 phr clay

Figure 3. Application of nanocomposites in road-off tire and transport belts

Rubber-attapulgite (AT) nanocomposite

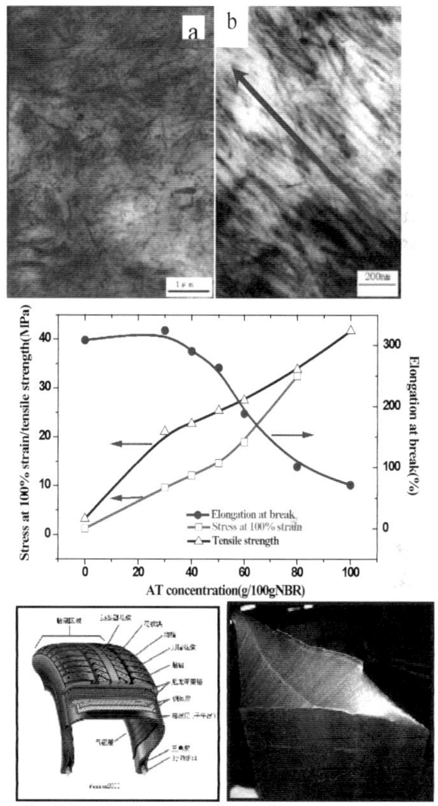

Figure 4. Morphology and performance of fibril silicate/rubber nanocomposite: (a) and (b) morphology; (c) mechanical properties with loading.[10-11] (d) application of AT/rubber nanocomposites

Rubber-starch nanocomposites

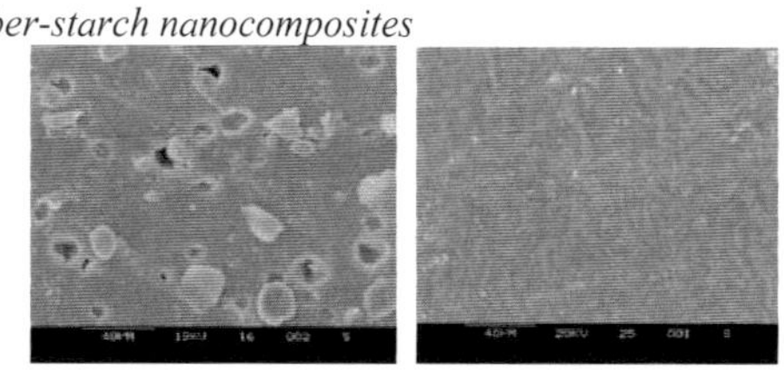

Figure 5. SEM micrographs of the tensile fracture surface of starch/rubber composites [12]. (a) starch/SBR composite by direct blending (b) starch/SBR composite by LCM

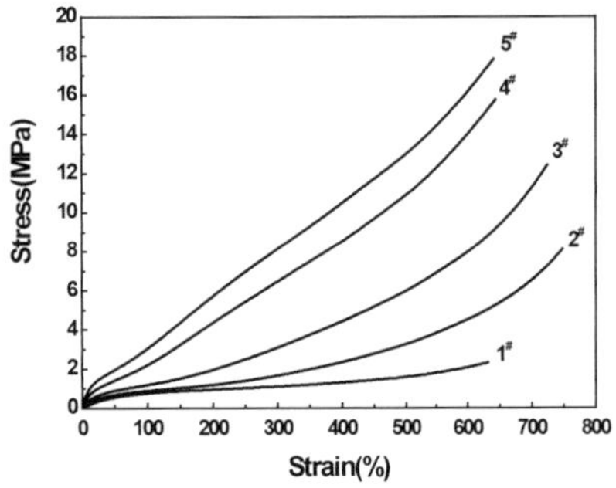

Figure 6 Strain-Stress curves of SBR and starch/SBR composites[13,14]

starch/RF/KH-792/SBR : 1#—0/0/0/100; 2#—5/0.6/3/100; 3#—10/1.2/3/100; 4#—20/2.4/3/100; 5#—30/3.6/3/100;

Rubber-graphite nanocomposites

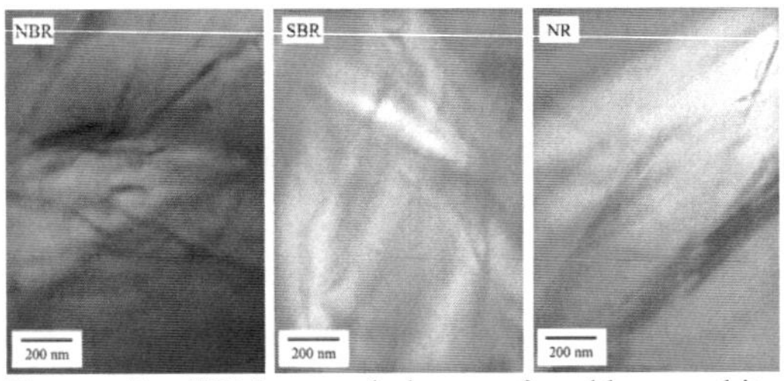

Figure 7. TEM morphology of rubber-graphite nanocomposites[15]

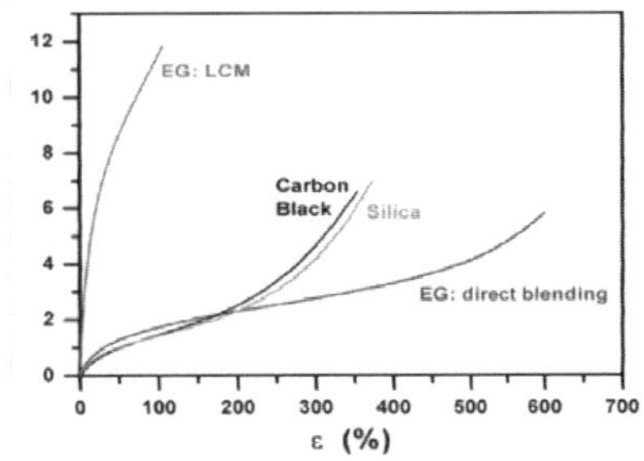

Figure 8 Comparisons of reinforcement efficiency of different fillers at 10phr loading [15]

Table 2 Performance of rubber-graphite nanocomposites

EG content (phr)	0	3	5	10	10[a]
N_2 Permeability ($10^{-17}m^2\cdot Pa^{-1}\cdot s^{-1}$)	3.19	—	1.98	1.21	1.83
Friction coefficient [b]	1.72	—	—	1.36	1.38
Wear rate ($mm^3/N\cdot m$)	5×10^{-2}	—	—	8×10^{-4}	3.5×10^{-2}
Electrical conductivity (S/cm) [c]	×	×	2×10^{-3}	3×10^{-1}	×
Thermal conductivity ($W\cdot m^{-1}\cdot K^{-1}$)	0.19	—	—	0.30	0.23

Rubber-graphene nanocomposites

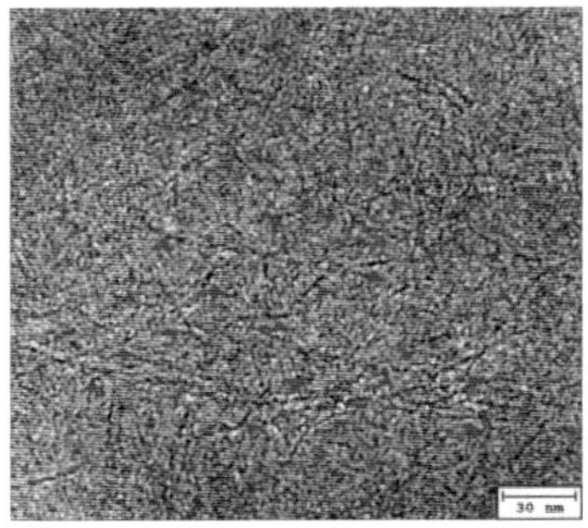

Figure 9. HRTEM morphology of rubber-oxided graphite nanocomposites[16]

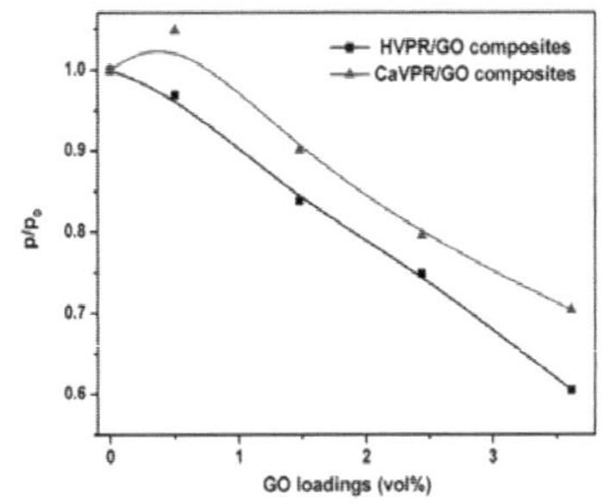

Figure 10. Improvement of gas barrier property of rubber by oxided graphene[16]

Rubber-silica nanocomposites

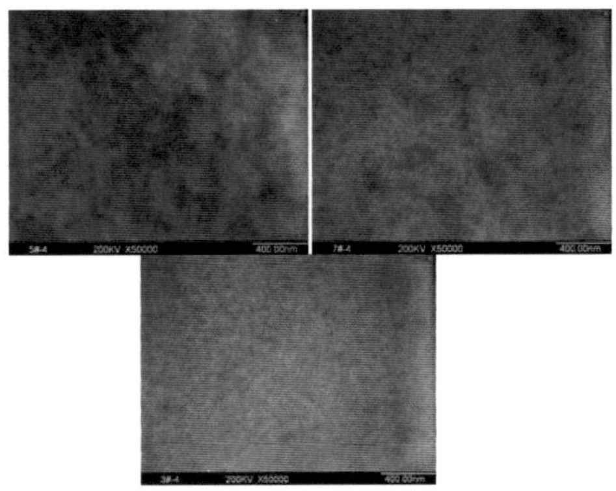

Figure 11 TEM morphology of rubber-silica nanocomposites by LCM with different siliane coupling agent (KH550, KH570, KH590, separately)

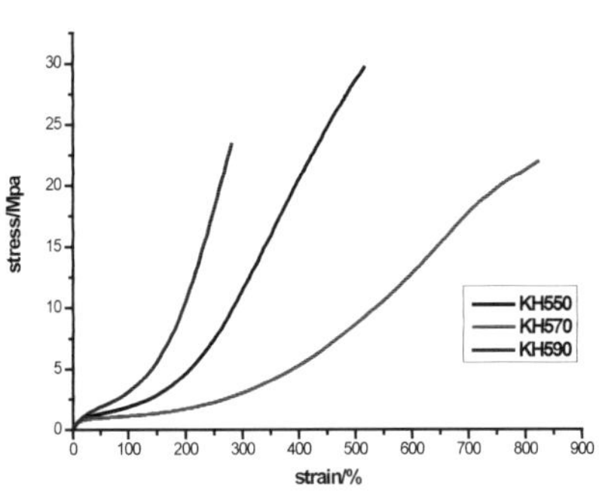

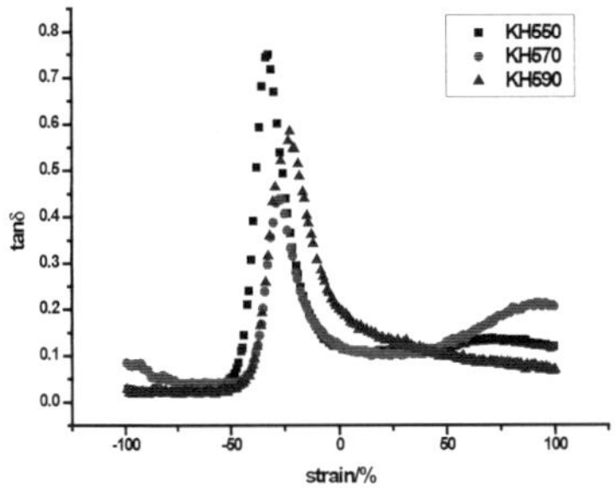

Figure 12 Static and dynamic performance of rubber-silica nanocomposites by LCM

Rubber-carbon nanotube nanocomposites

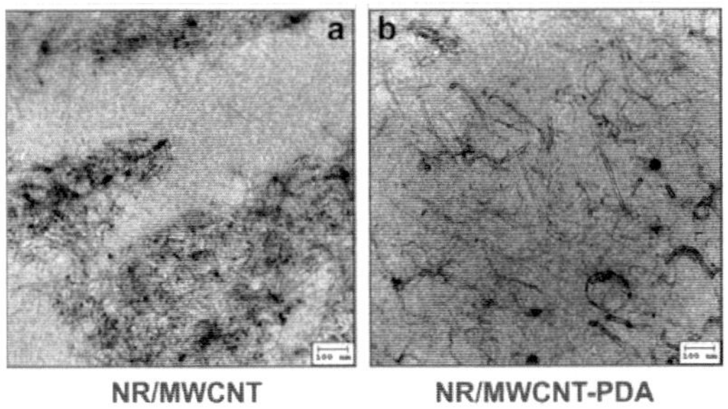

Figure 13 TEM morphology of rubber-MWNT nanocomposites by LCM:(a) NR/raw MWCNT,(b)NR/ MWCNT-PDA

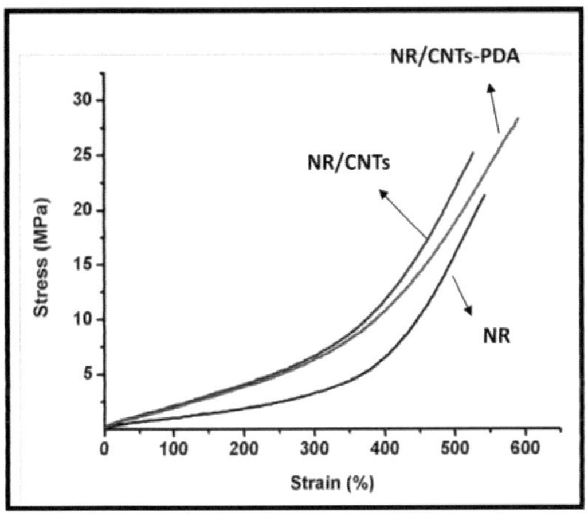

Figure 14 Stress-strain curves of NR, NR/CNTs and NR/CNTs-PDA

Conclusions:

Through LCM technique, many kinds of rubber nanocomposites with outstanding mechanical properties are prepared. The key in LCM is to prevent the nano-fillers from re-aggregation or re-assembly by controlling the mixing parameters, mixture parameters and recipe, and to improve the interfacial interaction between nano-fillers and rubber. LCM is a method to prepare rubber based nanocomposites with low energy consumption and low pollution, compared with traditional mechanical blending method. It is also a method that can make nanocomposites from micro-scaled target filler provided this filler could be dispersed into nano-dispersion in water medium.

Acknowledgements The authors gratefully acknowledge the financial support of the Distinguished Youth Research Foundationof the NSF (50725310) and the National High Technology Research and Development Program of China (863 Program, 2009AA03Z338).

References (Times New Roman 10, bold)

1. Z.H. Wang; J. Liu; L.Q. Zhang; et al Phys. Chem. Chem. Phys. 2010; 12; 3014.
2. J. Liu; D.P. Cao; L.Q. Zhang; et al Macromolecules. 2009; 42; 2831.
3. J. Liu; D.P. Cao; L.Q. Zhang; et al Phys. Chem. Chem. Phys. 2011; 13; 13058.
4. Y.P. Wu; Y.Q. Wang; H.F. Zhang; et al Compos. Sci. Technol. 2005; 65; 1195.
5. L.Q. Zhang; Y.Z. Wang; Y.Q. Wang; et al J. Appl. Polym. Sci. 2000; 78; 1873.
6. Y.Z. Wang; L.Q. Zhang; C.H. Tang; et al J. Appl. Polym. Sci. 2000; 78; 1879.
7. J. Ma; P. Xiang; Y.W. Mai; et al Macromol. Rapid. Comm. 2004; 25; 1692.
8. Q.X Jia; Y.P. Wu; Y.L Xu; et al Macromol. Mater. Eng. 2006; 291; 218.
9. S.Q. He; Y.Q. Wang; Y.P. Feng; et al Nanotechnology. 2010; 21; 115601.
10. M. Tian; C.D. Qu; Y.X. Feng; et al J. Mater. Sci. 2003; 38; 4917.
11. M. Tian; W.L. Liang; G.Y. Rao; et al Comp. Sci. Technol. 2005; 65; 1129.
12.Y.P. Wu; M.Q. Ji; L.Q. Zhang; et al. Macromol. Rapid. Comm. 2004; 25; 565.
13. Q. Qi; Y.P. Wu; J. Ma; et al Polymer. 2006; 47; 3896.
14. Y.P. Wu; Q. Qi; L.Q. Zhang; et al Carbohyd. Polym. 2006; 65; 109.
15. J. Yang; Z.Z. Yu; L.Q. Zhang; et al Acta. Mater. 2007; 55; 6372.
16. Z.H. Tang; X.H. Wu; B.C. Guo; et al J. Mater. Chem. 2012; 22; 7492.

O-13-408

RUBBER/GRAPHENE COMPOSITES BY LATEX MIXING

H.S. Xia*, Y.H. Zhan and N. Yan
State Key Laboratory of Polymer Materials Engineering, Polymer Research Institute of Sichuan University, Chengdu 610065, China–xiahs@scu.edu.cn

Abstract - The vulcanized graphene (GE)/natural rubber (NR) composites with the conductive segregated network exhibiting good electrical conductivity, water vapor permeability and high mechanical strength are prepared by self-assembly in latex and static hot pressing. The fatigue crack propagation of unfilled NR and GE filled NR composite are examined under uniaxial cyclic tension. For the first time, we observe that GE has an opposite effect on crack growth resistance of NR with regard to different fatigue strains. The micro-focus hard X-ray diffraction beamline was utilized for the study of strain induced crystallization around the crack tip at various macroscopic extensions corresponding to the fatigue strains. Fatigue crack resistance is correlated to strain induced crystallization in a novel way.

Keywords: Graphene; Rubber; Electrical conductivity; Fatigue crack growth mechanism.

Introduction

Graphene (GE) combines the extrodinanary electrical conductivity of carbon nanotubes and excellent gas barrier properties of layered structure of clays, together with its intrinsic high mechanical strength, and it is hopefully to be the best nanofiller for polymer. GE/polymer composites can have excellent electrical conductive properties.[1] Despite that a range of thermoplastic polymer/GE composites were prepared, surprisingly, GE/rubber composites were rarely reported. The big challenge to make GE/rubber composites is the difficulty in the dispersion of GE and the formation of GE conductive network or path in the crosslinked rubber matrix.[2]

2D or 3D graphene network construction in the thermoplastic polymer matrix is a key parameter to reduce the conduction threshold.[3] The construction of GE network in polymer matrix is still a big challenge, especially in crosslinked thermoset rubber matrix. On the other hand, the incorporation of two-dimensional layered fillers can significantly reduce gas permeation of polymer materials, which is especially important for the rubber used for seal or airproof. In addition, the crystallization results in a crack branching allowing the dissipation of much more energy which is believed to be the main reason for the reinforcement of the crack growth resistance of NR. Strain induced crystallization of NR has been widely investigated using a number of techniques. However, these studies mainly concerned the crystallization in the bulk rubber, the strain induced crystallization occurred at the crack tip is rarely reported, possibly due to the experimental difficulty.

We report a newly developed approach, i.e. self assembly in latex, to obtain high stretchable, good electrical conductive and water vapor barrier rubber composites with conductive segregated GE network. The electrical conductivity of the composite at a 1.78 vol% content of GE prepared by this approach is ~5 orders of magnitude higher than that of the composites prepared by conventional methods.[4] The microfocus hard-X-ray diffraction beamline (MF-XRD) was utilized for the investigation of strain induced crystallization at the crack tip. Our study provides a correlation of strain induced crystallization at the crack tip and fatigue crack growth resistance, and a new insight into the mechanism of fatigue crack growth was proposed.

Experimental

Preparation of the crosslinked GE/NR composites

The preparation of GO and GE was described in our recent work and the yield of GE is about 40% from GO.[2] For the vulcanized GE/NR composites, the preparation method is following: the dried GO powder was added into the aqueous solution by sonication. Then NR latex was dispersed into the GO solution by sonication to obtain the light brown GO/NR latex. Hydrazine hydrate was added into the GO/NR latex and the mixture was subjected to an in-situ reduction reaction under ultrasonic irradiation to obtain the black GE/NR latex. And the aqueous suspension containing crosslinked agent sulfur and other additives were dispersed into the GE/NR latex and then the mixed latex was coagulated. After filtering, the solid mixture was dried. These composites were compressively molded and vulcanized.

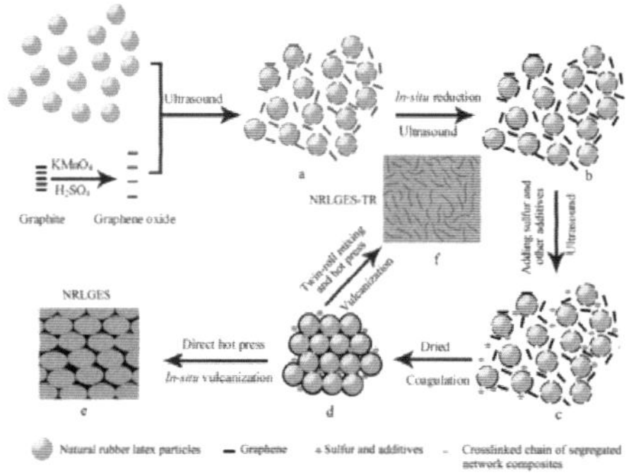

Scheme 1. The preparation of GE/rubber composites with the conductive segregated network of GE by self-assembly in latex and static hot press.

Characterization

TEM was performed using a FEI Tecnai G2 F20 S-TWIN transmission electron microscope. The measurement of mechanical properties was conducted on a universal testing machine (Instron 5567, US) at room temperature. The fatigue test was performed in MTS810 material test system with a displacement-control mode under the uniaxial tensile cyclic loading condition at room temperature. Strain-induced crystallization around crack tip of NR and GE/NR has been measured by MF-XRD at the Shanghai Synchrotron Radiation Facility.

Results and Discussion

Microstructure of GO and NR/GE Composites

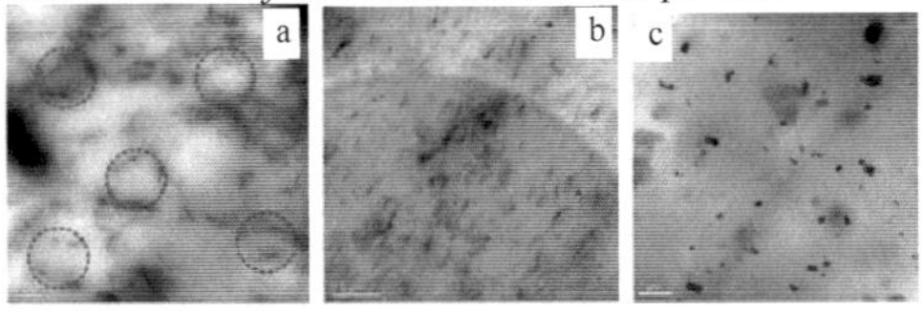

Figure 1 TEM images of the GE/NR composites with 3.50 vol% GE prepared by self-assembly in latex and static hot pressing (a), self-assembly in latex & twin-roll mixing (b), conventional two-roll mixing (c).

Figure 1a confirms that the segregated network structure of GE is formed in the GE/NR composites prepared by static hot pressing, as shown in Scheme 1e. The GE shells coated on the surface of the rubber particles, which are evenly distributed in the rubber matrix, are connected to form a conductive network. Figure 1b show that after the twin-roll mixing, the segregated networks are destroyed under the dynamic shear force and the thinner GE layers (~3 nm) are uniformly distributed in the rubber matrix as illustrated in Scheme 1f. To illustrate the advantage of the ULMR process, a control sample (NR/GE-T) prepared by the conventional twin-roll method was also studied by TEM. As shown in Figure 1c, the GE agglomerates with a size range of ~300-800 nm in the rubber matrix can be observed, which suggests that the GE cannot be dispersed in the rubber matrix in an exfoliation state by conventional twin-roll mixing.

Electrical conductivity of NR/GE Composites

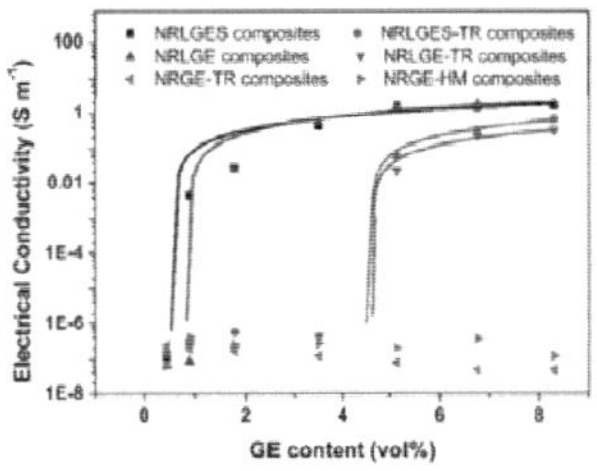

Figure 2 The electrical conductivity as a function of the GE content for the GE/NR composites prepared by the different methods.

The sample NRLGES with segregated network, corresponding to Scheme 1e and Fig. 1a, obtained by self-assembly in latex and static hot pressing, has a much higher conductivity and a lower conductivity threshold than that of the sample NRLGES-TR with well dispersed GE platelets obtained by self-assembly in latex and twin-roll dynamic mixing, corresponding to Scheme 1f and Figure 1b. For example, the electrical conductivity of the NRLGES-TR-2 composites is ~5.7×10^{-7} S/m, while for NRLGES-2 it reaches ~0.03 S/m, showing an significant enhancement of ~5 orders of magnitude. The electrical conductivity of the NRLGES-2 composites with a GE content of 1.78 vol% was also ~5 orders of magnitude higher than that of the NR/GE composites at the same loading fraction reported by Robert Prud'Homme.8 the percolation threshold of NRLGES with the GE network is only ~0.62 vol%, much lower than that of NRLGES-TR without a GE network (~4.62 vol%). This suggests that the GE segregated network plays an important role in the electrical conductivity of the NR/GE composites.

Mechanical properties of NR/GE Composites

Table 1. Mechanical properties for NR composites with 2 wt% different nanofillers and two preparation methods.

Sample	Tensile strength (MPa)	Elongation (%)
Neat NR	17.1	579
NR/CB (2 phr)	18.2	586
NR/CNT (2 phr)	18.6	534
NR/GE (2 phr)	25.2	564
NR/GE-T(2 phr)*	18.8	600

*: The NR/GE-T composites prepared by a direct conventional twin-roll mixing.

The mechanical properties of NR composites with 2 wt% different nanofillers prepared by an ULMR process are listed in Table 1. The pure NR is a control sample. Compared with pure NR, the tensile strength of NR/GE composites with 2 wt% GE content is increased by ~47%. In contrast, the tensile strength of NR/CB and NR/MWCNT composites was increased by only ~6 and 9%, respectively.

Crack growth of NR/GE Composites

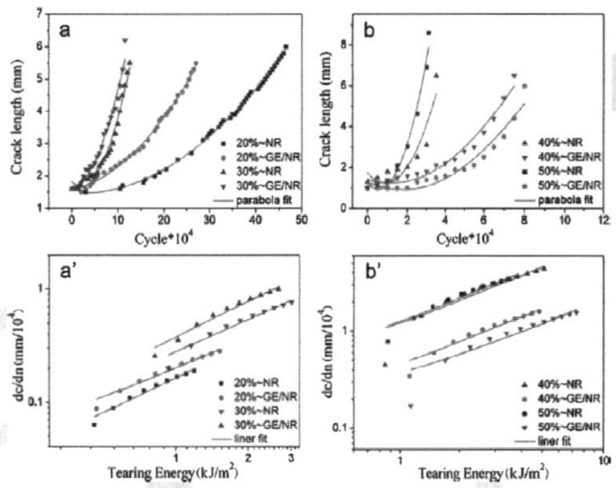

Figure 3 Crack length versus fatigue cycles (a, b) and crack growth rate versus peak tearing energy (a', b') for unfilled NR and GE/NR composite at various fatigue strains, (a, a') 20, 30, (b, b') 40, 50%.

Crack growth of the specimens was investigated under uniaxial cyclic stretching with a positive loading ratio. Crack length (c) versus loading cycles (n) for unfilled NR and GE/NR under different fatigue strains are presented in Figure 6a and b. They were all parabola-fitted preferentially and differentiated to obtain the crack propagation rate (dc/dn, shown in Figure 3a' and b'). Figure 6a shows that the crack of GE/NR grows more quickly with the fatigue cycles than unfilled NR at the lower fatigue strains. The situation becomes quite different at higher fatigue strains 40 and 50%. Figure 6b shows that the crack of GE/NR grows slower with the fatigue cycles than unfilled NR at the higher fatigue strains. It is noteworthy that the crack growth rate of GE/NR at a given tearing energy at the fatigue strain of 50% is even lower than that of 40% as shown in Figure3b'.

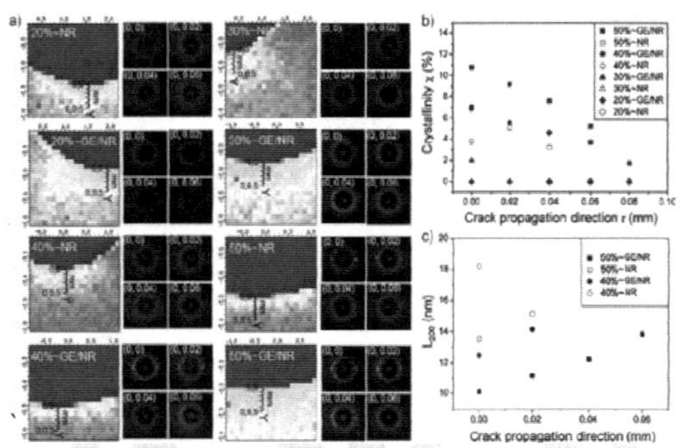

Figure 4 (a) Contour maps of the crack tip and two dimensional XRD patterns for unfilled NR and GE/NR composite. (b) Crystallinity x of NR and GE/NR composite near the crack tip. (c) Lateral crystallite size (L200) versus r along Y-axis for the samples NR and GE/NR at various stretch strains.

MF-XRD, which has high intensity, high luminance, and high temporal and spatial resolution, was used to characterize the crystallization at the crack tip of samples under various strains. In this study the sample is statically exposed in the X-ray at a given elongation corresponding to the maximum fatigue strains.

To characterize the strain induced crystallization around the crack tip of unfilled NR and GE/NR composite, the samples with a crack length of 2 mm during fatigue test were used. The contour maps of crack tip at various stretch strains are presented in Figure 4a. The crack tip is taken as the origin of coordinates, the diffraction patterns were recorded by scanning the sample in Y-axis perpendicular to the loading direction. Each displacement is 20 mm.

For more clarity, the crystallinity x was calculated according to the diffraction patterns. Figure 4b shows the crystallinity x near the crack tip of the samples at different stretch strains along Y-axis of the crack. It was found that almost no crystallization occurs at a low strain (a<30%) for the NR and GE/NR samples. This is consistent with the report of Trabelsi et al.,[6] which demonstrated that a crystallized zone appears at stretch ratios greater than 1.3 for crack tips of initial cut lengths 0.5 and 1 mm. When the strain increases to 40%, the crystallinity at the crack tip of the samples NR and GE/NR were calculated as 3.8 and 6.9%, and increased to 7.0 and 10.8% at the strain of 50%, respectively. It is noteworthy that strain induced crystallization of NR does not happen at a strain of 30%, while it starts to occur at the same strain for the GE/NR sample, suggesting that the addition of GE significantly promotes the strain induced crystallization. This critical strain of crystallization occur-rence is also the transition point at which the effect of GE on the crack growth resistance of NR changes.

Moreover, Figure 4b shows that the crystallinity decreases continuously with increasing r value, i.e., near the crack tip the crystallization is strong, and far from the crack tip it becomes weak. At the strain of 50%, the r values at which crystallization disappears for NR and GE/NR are 0.04 and 0.08mm, respectively. This result shows that GE/NR has a larger crystalline zone around the crack tip, approximately twice than that for NR at the same strain.

Conclusions

We have developed a novel approach, i.e. self-assembly in latex and static vulcanization, to prepare the electrical conductive, water vapor barrier and highly stretchable crosslinked rubber composites with segregated graphene network. GE platelets are assembled on the surface of the rubber latex particles with a size of several hundred nanometers to form the prototype of the GE network. The fatigue crack growth of NR and GE/NR was studied based on the tearing energy concept. For the first time, we observed the two opposite effects of GE on the crack growth resistance of NR: at lower strains, the inclusion of GE accelerates the crack growth; at higher strains, the crack growth is retarded with the addition of GE. The strain induced crystallization at crack tip was used to explain the two opposite effects of GE. Our findings will be useful for the design of the advanced anti-fatigue materials.

Acknowledgements
H.S. Xia acknowledges financial support from the National Natural Science Foundation of China (NNSFC 51010004) and the International Science & Technology Cooperation Program of China (2010DFA92860).

References

1. S. Stankovich; A.D. Dikin; G.H.B. Dommett; et al *Nature*, 2006, *442*, 282.
2. Y.H. Zhan; J.K. Wu, H.S. Xia; et al *Macromol. Mater. Eng.*, 2011, *296*, 590.
3. Y.H. Zhan; M. Lavorgna; G. Buonocore; H.S. Xia *J. Mater. Chem.*, 2012, *22*, 10464.
4. N. Yan; H.S. Xia; Y.H. Zhan; *Macromol. Mater. Eng.*, 2012, DOI: 10.1002/mame. 201200044
5. J.B. Le Cam; B. Huneau; E. Verron; L. Gornet *Macromolecules*, 2004, *37*, 5011.
6. S. Trabelsi; P.A. Albouy; J. Rault, *Macromolecules*, 2002, *35*,10054.

O-13-118

UTILIZATION OF PETROLEUM-BASED SULFUR AS BY-PRODUCT FROM REFINERY PROCESS IN RUBBER VULCANIZATION

P. Pangamol_[a,*] and C. Sirisinha [a,b]

[a] *Faculty of Science, Mahidol University, Thailand – p.pangamol@gmail.com; chakrit.sir@mahidol.ac.th*

[b] *Research and Development Center of Thai Rubber Industry, Mahidol University, Thailand*

Abstract - Petroleum-based sulfur treated as by-product from refinery process is shown for its potential utilization as vulcanizing agent in rubber industry by substituting the elemental sulfur (rhombic sulfur) usually used in rubber vulcanization. According to XRD, FT-IR, and TGA results, the chemical structures of the petroleum-based and rhombic sulfurs are similar. As vulcanizing agent in natural rubber (NR), the petroleum-based ground sulfur and rhombic sulfur from 3 manufacturers are compared in the aspects of cure efficiency and its influences on mechanical properties of NR vulcanizates. Results revealed that cure characteristics (scorch time and cure time) and crosslink density of the NR vulcanized by petroleum-based sulfur were similar to those vulcanized by rhombic sulfur. Moreover, viscoelastic properties as well as mechanical properties of NR vulcanizates are comparable with the commercial sulfur. Therefore, it could be summarized that the petroleum-based sulfur was capable of functioning as vulcanizing agent in rubber products in similar manner to the conventionally used rhombic sulfur.

Keywords: Sulfur; Vulcanization; Cure characteristics; Mechanical properties

Introduction

Generally, raw rubber cannot be used as-received because of its unstable shape, low elasticity, and poor mechanical properties. Thus, the raw rubber needs to be transformed into vulcanized rubber through the vulcanization process [1-4], resulting in crosslink between rubber molecules. Enhancement in elasticity and mechanical properties of cured specimens are resulted. Generally, there are 3 major curing agents used, i.e., sulfur, peroxide, and metal oxide [1-4]. However, sulfur is the most widely used for curing diene rubbers because of its superiority in mechanical properties and ease of curing behavior adjustment [5-9].

Actually, there are 2 main sources of sulfur which are natural source (sulfur flow) and petroleum-based refinery [10]. The sulfur originated from the former is a major source of sulfur widely used in rubber industry while that from the latter is generally treated as by-product from refinery process, and sold in low cost for preparing sulfuric acid as reagent [11-14].

Therefore, this work aims to add the value of petroleum-based sulfur by investigating the potential utilization of such sulfur as vulcanizing agent in rubber industry to substitute the elemental sulfur (rhombic sulfur) usually used in rubber vulcanization.

Experimental

Materials

Natural rubber (STR 5L) was purchased from Union Rubber Product Co., Ltd. Conventional rhombic sulfurs used in this research consisting of RS, SGS, and SP400 were supplied by Chemmin Corporation Ltd., Cosan (Thailand) Co.,Ltd., and Miwon Commercial Co., Ltd., respectively while a petroleum-based sulfur (PS) was supported by IRPC Plc. Co., Ltd. Other chemical ingredients, namely, stearic acid and zinc oxide (ZnO) were supplied by Chemmin Corporation Ltd., and N-tret-butyl-2-benzothiazole sulfonamide (Santocure-TBBS) was by Reliance Technochem Co., Ltd. Except for the PS, all ingredients were used as-received. In the case of PS, the as-received PS was ground with ball-milling process to achieve desirable particle size.

Compounds preparation and testing

Compounding recipes were shown in **Table 1**. Two-roll mill (LabTech, model LRM 150 electric mill, Thailand) was used as a mixer with the mixing time of 10 min at 50°C. Viscoelastic properties were investigated using Rubber Process Analyzer (RPA) (Alpha Technologies model RPA 2000). Storage modulus (G') of the rubber compounds was measured as a function of strain (from 0.56 to 1200%) at test frequency and temperature of 1.0 Hz and 100°C, respectively. Cure characteristics were examined by Moving Die Rheometer (MDR) (TechPro MD+) at 150°C.

Characterization of sulfur characteristics

The chemical structure of PS was characterized, and compared with RS by the uses of x-ray diffractometer (XRD) (Bruker AXS Model D8 Discover) at 25°C, Fourier transform infrared spectrometer (FT-IR) under a diffuse reflectance infrared Fourier Transform (DRIFT) mode (Bruker Equinox 55) at 25°C, and Thermogravimeter (TGA) (Mettler Toledo model TGA/SDTA851[e]) at heating rate of 20°C/min. Temperatures were scanned from ambient temperature to 600°C under nitrogen atmosphere. After 600°C, the atmosphere was switched to oxygen and the test was carried out until 800°C.

Testing of rubber vulcanizates

Hardness was measured using Shore A Durometer (Wallace model COGENIX) as per ASTM D2240. Tensile properties of cured specimens were determined using Universal Testing Machine (Instron model 5566) according to ASTM D412 Die C at crosshead speed and load cell of 500 mm/min and 1 kN, respectively. Dynamic mechanical properties of the vulcanizates were examined using dynamic mechanical analyzer (GABO EPLEXOR 25N) under strain sweep test mode at 30°C and frequency of 5 Hz. Heat build-up test was conducted using BF Goodrich Flexometer (Model II) as per ASTM D623-07.

Table 1 Formula of rubber compounds

Chemicals	Loading (phr*)			
	RS	**PS**	**SGS**	**SP400**
NR	100	100	100	100
ZnO	3	3	3	3
SA	1	1	1	1
TBBS	1	1	1	1
Sulfur (RS)	1.75	-	-	-
Sulfur (PS)	-	1.75	-	-
Sulfur (SGS)	-	-	1.75	-
Sulfur (SP400)	-	-	-	1.75

*phr = part per hundred of rubber

Determination of crosslink density

Cured specimens (aka. vulcanizates) were cut into rectangles with dimensions of 10x10x2 mm^3, and weighed before and after soaked in toluene for 7 days. Volume fraction of rubber in the swollen network (υ_r) was calculated according to **Eq. 1** [15]. The crosslink density was calculated using the Flory-Rehner equation as shown in **Eq. 2** [15].

$$v_r = \frac{\left(\frac{w_1}{\rho_d}\right)}{\left(\frac{w_1}{\rho_d} + \frac{(w_2 - w_1)}{\rho_s}\right)} \quad (1)$$

where w_1 is the mass of rubber before swelling, w_2 is the mass of the swollen rubber, ρ_d is the rubber density before swelling, and ρ_s is the toluene density.

$$v_e = \frac{-\left[\ln(1-v_r) + v_r + \chi_1 v_r^2\right]}{\left[v_1\left(v_r^{1/3} - \frac{v_r}{2}\right)\right]} \quad (2)$$

where υ_e is the network chain density, υ_1 is molar volume of toluene, and χ_1 is the Flory–Huggins interaction parameter between rubber and toluene.

The swelling ratio (S) could be calculated following **Eq. 3** [15].

$$S = \frac{(w_2 - w_1)}{w_1} \quad (3)$$

Results and Discussion

Chemical structure of PS

Fig. 1 shows the XRD pattern comparison of RS and PS. Evidently, both RS and PS give similar XRD pattern, implying similar in chemical structure.

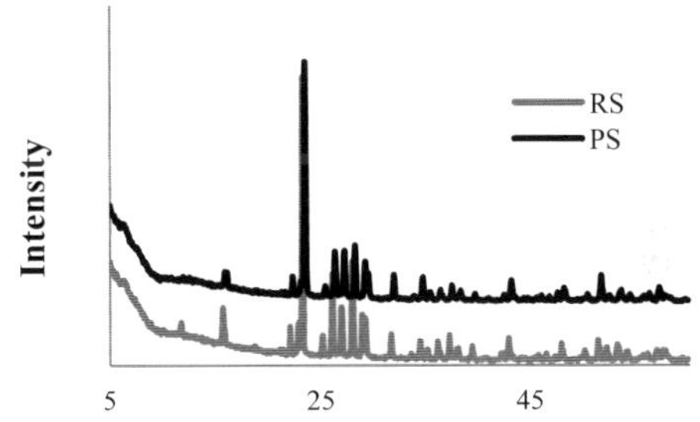

Figure 1 XRD patterns of RS and PS

Fig. 2 exhibits the FT-IR spectra of RS and PS. The spectrum of the RS show characteristic peak of sulfur at 1101, 845, and 468 cm^{-1}, which are also observed in that of PS. The results suggest similarity in functional groups on RS and PS.

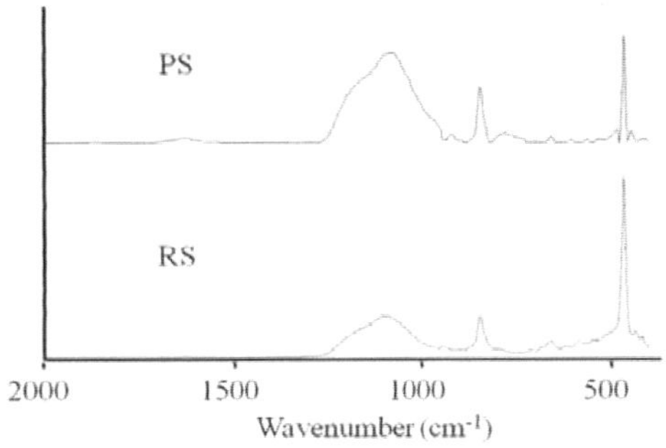

Figure 2 FT-IR spectra of RS and PS

Fig. 3 displays similarity in TGA thermograms of RS and PS, supporting FTIR and XRD results that the chemical structure of both RS and PS are in common with the decomposition temperature of approximately 370°C.

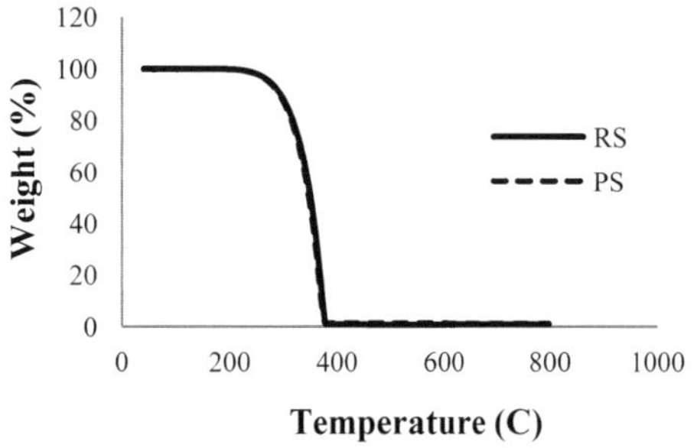

Figure 3 TGA curves of RS and PS

Properties of compounds and vulcanizates

Table 2 illustrates a comparison of cure characteristics of rubber compounds incorporated with different types of sulfurs. Obviously, there are no significant differences between scorch time and cure time in all compounds. Similarly, there is no disparity of crosslink density observed, indicating the comparable cure reactivity of all sulfurs.

Table 2 Cure characteristics, swelling ratio (S), and calculated crosslink density (υ_e) of vulcanizates

Formula	**Cure characteristics**		**S**	υ_e **(10^{-4} mol.cm^{-3})**
	Scorch time (min)	Cure time (min)		
RS	11.59	17.26	3.80	1.28
PS	12.43	17.10	3.94	1.23
SGS	12.01	17.10	3.64	1.33
SP400	12.01	16.39	3.75	1.11

Fig. 4 shows viscoelastic properties as a function of shear strain of rubber compounds incorporated with different sulfurs. It is evident that, regardless of sulfur types used, the storage modulus (G') of all compounds is superimposable suggesting no profound effect of sulfur on rheological behavior of rubber compounds. In other words, the processability of rubber compounds incorporated with all sulfurs is similar.

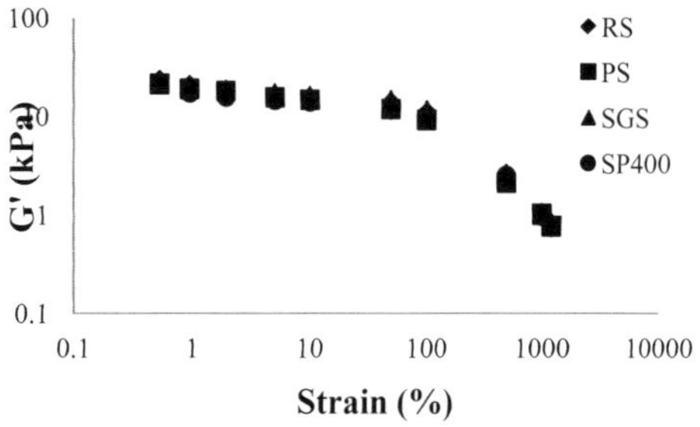

Figure 4 Storage modulus as a function of strain in rubber compounds incorporated with different types of sulfurs

Fig. 5 exhibits dynamic mechanical properties as a function of shear strain of vulcanizates incorporated with different sulfurs. Obviously, the storage modulus (E') of vulcanizates shows similar trend which has no change while increasing strain. The results imply that the PS can be used as vulcanizing agent by replacing the commercial rhombic sulfurs while still offers similar dynamic mechanical properties to conventional rhombic sulfurs.

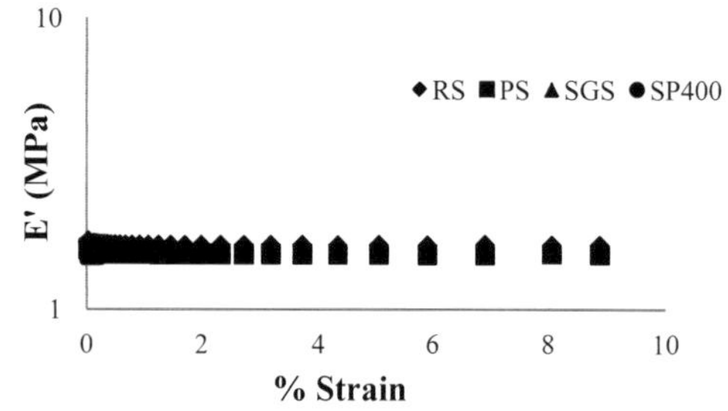

Figure 5 Storage modulus as a function of strain in vulcanizates incorporated with different types of sulfurs

Fig. 6 shows hardness of vulcanizates prepared with different sulfurs. Compared with the other rhombic sulfurs, the PS gives the vulcanizates with similar hardness range which is in good agreement with the results of crosslink density as discussed previously in **Table 2**. Moreover, tensile strength, elongation at break and tear strength of vulcanizates with PS and RS are comparable as depicted in **Figs. 7-9**.

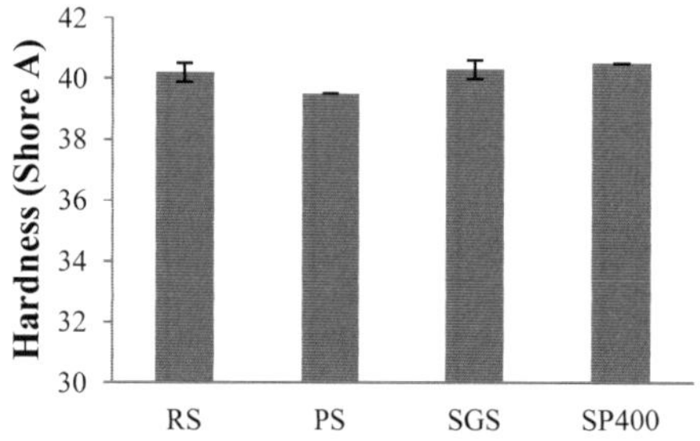

Figure 6 Hardness values of vulcanizates

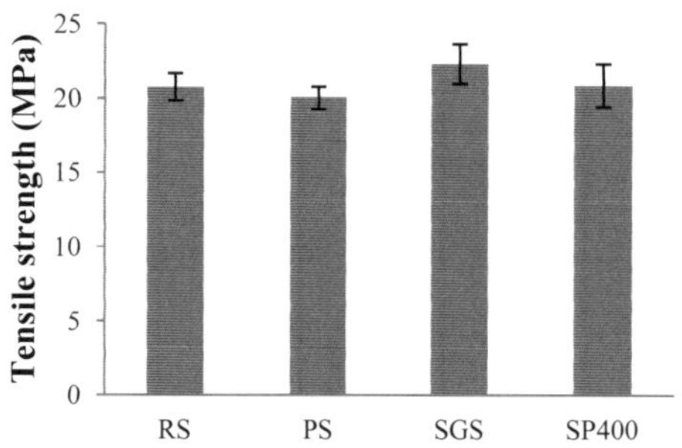

Figure 7 Tensile strength of vulcanizates

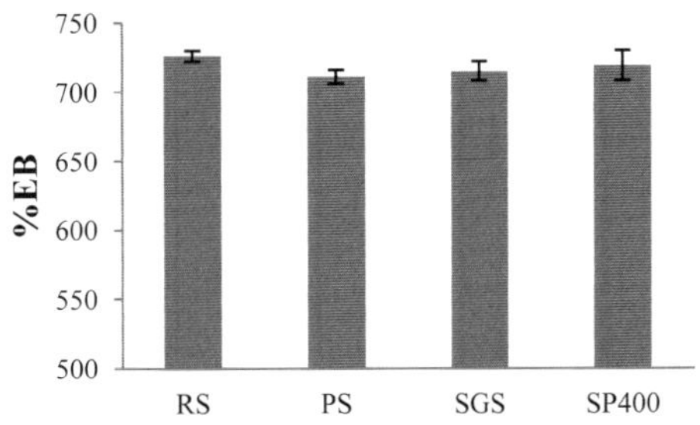

Figure 8 Elongation at break of vulcanizates

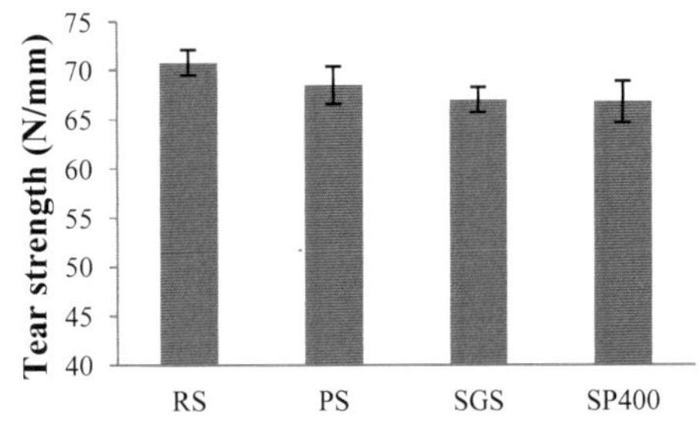

Figure 9 Tear strength of vulcanizates

Fig. 10 exhibits heat build-up (HBU) test results demonstrating similar in PS gives similar trend of HBU performance which is in agreement with the results of viscoelastic and mechanical properties.

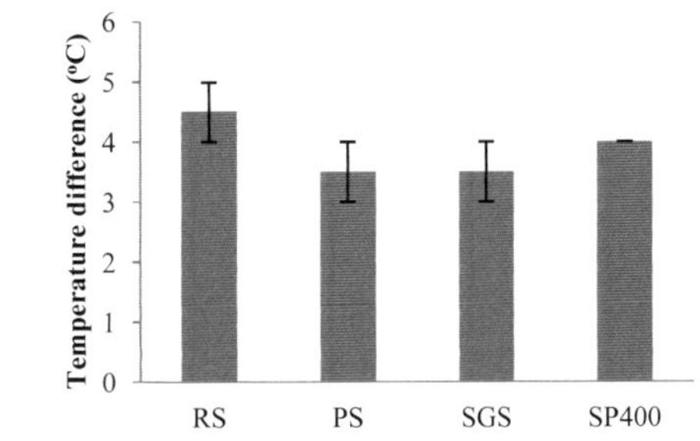

Figure 10 Heat build-up (HBU) of vulcanizates

Conclusions

Petroleum-based sulfur (PS) as a by-product from refinery process was used as vulcanizing agent in rubber vulcanization process. According to XRD, FTIR-DRIFT and TGA results, the chemical structure of PS and RS are similar. Thus, both offer comparable cure behavior, viscoelastic and mechanical properties. Thus, it can be summarized that the PS possesses the potential utilization to replace rhombic sulfur in rubber industry.

Acknowledgements

The authors would like to express our profound gratitude toward IRPC Plc. Co., Ltd. for supporting petroleum-based sulfur. Financial support from Thailand Research Fund through the Royal Golden Jubilee PhD Program (Grant No PHD/0116/2553) is also acknowledged.

References

1. W. Hofmann, *Rubber Technology Handbook*, Hanser Publishers, Munich, 1980.
2. K. Nagdi, *Rubber as an Engineering Material: Guildline for Users*, Hanser Publishers, Munich, 1993.
3. A. Y. Coran, in *The Science and Technology of Rubber 3rd*, J. E. Mark, B. Erman, F. R. Eirich Ed.; Elsevier Academic Press, London, 2005; 321 - 366.
4. W. Hofmann, *Vulcanization and Vulcanizing Agents*, Palmerton Publisher Co., Inc., New York, 1967.
5. A. P. Susamma; V. T. Elizabeth Mini.; A. P. Kuriakose, *J. Appl. Polym. Sci.* 2001, 79, 1.
6. S. C. Debnath; S. K. Mandal; D. K. Basu, *J. Appl. Polym. Sci.* 1995, 57, 555.
7. A. S. Aprem; K. Joseph; T. Mathew; V. Altstaedt; S. Thomas, *Eur. Polym. J.* 2003, 39, 1451.
8. J. M. Vergnaud; L. D. Rosca, *Rubber Curing and Properties*, CRC Press, New York, 2009.
9. A. Y. Coran, *J. Appl. Polym. Sci.* 2003, 87, 24.
10. W. N. Karel, *Ullmann's Encyclopedia of Industrial Chemistry*, Wiley-VCH Verlag, 2006.
11. V. Simanzhenkov; R. Idem, *Crude Oil Chemistry*, Marcel Dekker Inc., New York, 2003.
12. J. G. Speight, *Handbook of Petroleum Analysis*, John Wiley & Sons Inc., New York, 2001.
13. J. G. Speight, *Handbook of Petroleum Product Analysis*, John Wiley & Sons Inc., New York, 2002.
14. J. G. Speight, *The Chemistry and Technology of Petroleum*, CRC Press, New York, 2006.
15. Y. Zhan; J. Wu; H. Xia; N. Yan; G. Fei; G. Yuan, *Macromol. Mater. Eng.* 2011, 296, 590.

O-13-131

EFFECT OF CARBON BLACK TYPE ON THE THERMAL PROPERTIES OF FLUOROELASTOMER

Talat Khalkhali*, Ali Samadi,Mohammad Mahdi Salehi, Hossein Pious, Mahmoud Hemmati
Polymer and Science Technology Division, Research Institute Petroleum Industry, Tehran, Iran
**Correspounding authour:khalkhalit@ripi.ir*

Abstract
The main objective of the present work is to study the effect of carbon black (CB) microstructure on the thermal properties of Fluoroelastomer (Viton A). In order to study the microstructure effect of CB, three types of CB (N330, N660 and N990) were selected. Melt mixing method was employed for compound preparation. Vulcanization, mechanical properties before and after oxidative aging in oil and oven conditions, and thermal stability of the compounds were evaluated. Vulcanization characteristics of compounds revealed that CB with higher structure has higher minimum and maximum torque and lower vulcanization rate. Mechanical stability in oxidative aging and resistance to thermal degradation were all increased in CB containing composites. It can be attributed to CB characteristics and CB-polymer interaction which is different in various CB containing samples.

Keywords: fluoroelastomer, Viton, mechanical stability, thermal stability

Introduction
Fluoroelastomers (FKM or F-polymers) are used in many applications such as gaskets, seals, O-rings and fuel hoses in petroleum industry. This is due to their unique combination of useful mechanical properties coupled with the high resistance to fuel/oil and heat [1]. F-polymers are generally prepared by high-pressure, free radical emulsion polymerization [2]. One of these fluoroelastomers, Viton A, is a copolymer of vinylidene fluoride (VDF) and Hexafluoropropylene (HFP) developed by DuPont and was made available commercially in 1955. Many reports have been published on thermal degradation properties of FKMs [3-6], where attempts have been made to establish degradation mechanisms. Many other reports have focused on chemical degradation of FKM rubbers in specific aggressive aqueous chemical environments [7-12].
In the rubber industry, fillers are used to reinforce the polymer matrix to achieve the required mechanical properties. Viton A is normally reinforced by carbon black (CB) as the active filler. In the lifetime of FKM products, they are exposed to sever conditions such as high temperature and chemical environments.
Hence, in the present work, different CBs having various surface energies, particle sizes and structures were taken and their effect on the thermal properties of fluoroelastomer (Viton A), vulcanization characteristics and required performances such as mechanical properties, stability in oxidative aging in oil and oven conditions and stability in thermal degradation were investigated. Vulcanization characteristics revealed that CB with lower particle size and higher structure has higher minimum and maximum torque and lower vulcanization rate in the compounds. Mechanical stability in oxidative aging and resistance to thermal degradation were all increased in CB containing composites. It is may be attributed to CB characteristics and CB-polymer interactions that are different at various CB containing samples.

Experimental
Materials and Methods
Fluoroelastomer (Viton A, DuPont, USA), Carbon black (CB N-330, N660, Ahvaz CB Co., Iran; N990, Cancarb, Canada), calcium hydroxide (Merck, Germany), magnesium oxide (Merck, Germany), Bisphenol AF (Sigma-Aldrich, Germany) and TPBPC (Merck, Germany) were mixed according to the recipe given in Table 1.Surface energy component of carbon blacks are shown in Table2.

Table1.Formulation of different composites and their designation

Component	Green (phr)	CB990 (phr)	CB660 (phr)	CB330 (phr)
Viton	100	100	100	100
CB N990	0	30	0	0
CB N660	0	0	30	0
CB N330	0	0	0	30
MgO	3	3	3	3
Ca(OH)2	6	6	6	6
P Salt	0.5	0.5	0.5	0.5
BFAF	1.8	1.8	1.8	1.8

Table2. Surface energy component of carbon blacks

	N330	N660	N990
Specific surface area N_2 (BET) (m^2/g)	76.5	39.4	7.9
Dispersive component of surface energy at 150 'C (mJ/m^2)	196.9	124.7	78.7
Polar component of surface energy at 150 'C (mJ/m^2)	85.9	71.1	58.8

Melt intercalation mixing method was employed for compound preparation using Brabender laboratory two-roll mixing mill with a friction ratio of 1:1.4 at room temperature. Vulcanization characteristics of compounds were determined by Monsanto Oscillating Disk Rheometer 100 at 177 C. Rubber samples were vulcanized in an electrically heated hydraulic press at 177 °C for 10 minute. Uniaxial tension properties were studied by a Zwick/Roell 2030 tensile testing instrument according to ASTM D412, (5 specimens from each sample). IRHD hardness was measured with a Zwick hardness tester according to the standard ASTM D2240. Oven aging condition was applied in a hot-air oven at 120 °C for 24 hours at atmospheric pressure. Oil aging condition was applied in a hot turbine oil at 120 °C for 24 hours at atmospheric pressure. Heat resistance was studied by a polymer laboratory series 1500 thermo gravimetric analysis (TGA) in a temperature sweep from room temperature to +900 °C.

Results and Discussion

Effect of CB Type on Vulcanization Characteristics

Vulcanization characteristics of the FKM and FKM/carbon black composites were summarized in Table 3. Minimum and maximum torques and the differences between them (MH–ML) for all CB containing samples are higher than unfilled FKM compound. This shows the reinforcing effect of this filler in the rubber compounds. Also, scorch and optimum cure times are longer when CB is present in the compound compared to the compounds without CB. During melt mixing process of uncured FKM with CB, FKM chains have contact with CB particles, and entangle or trap between carbon black aggregates. The rubber chains become highly immobilized and localized in the form of occluded or rubber shell [13]. The physical crosslinking between CB and FKM hinders the mobility of rubber chains and restrains the deformation of rubber. Lower deformation capability of FKM chains at certain torques, increase minimum requirement torques of rheometric vulcanization characterization. Hence, it is understandable that fillers with higher surface area have higher reinforcing effect on rubber. Therefore, the surface area of carbon black is a major reinforcing property of CB. Most properties, such as Mooney viscosity and 300% modulus of the filled rubber depend on the specific surface area of carbon black and ML is the one of them [6]. As the surface area of carbon black increases, the number of rubber chains entangling with carbon black aggregates as well as that of the cross-links climbs up[14]. ML increases correspondingly with increasing surface area of carbon black, as the number of rubber chains immobilized on the carbon black surfaces grows up. It can be seen in Table 2 that both the BET specific surface area and external surface area of N330 are higher than N660 and these properties of N660 are higher than those of N990. ΔM, the difference between MH and ML, is considered as the parameter to demonstrate the degree of chemical cross-linking. Chemical bonding of rubbers with the functional groups on carbon blacks is occurring during vulcanization [15-17]. The chemical bonding of rubbers with carbon blacks increases the degree of chemical crosslinking that is higher at N660 compound. Cure Rate Index (VC1) is based on the difference between optimum vulcanization and incipient scorch time, VC1 = 100/(cure time – scorch time). Higher structure CB (N330) filled FKM has lower vulcanization rate and the Green FKM has the highest vulcanization rate. CB decreases the vulcanization rate of FKM that may be attributed to diffusion limitation and lower chain mobility of FKM introduced by CB- FKM interactions. Diffusion limitations introduced by filler structure reduce the accessibility of vulcanizing agents to already limited polymer chains.

Table3. Vulcanization characteristics of FKM and FKM/carbon black composites at 177 °C

	Green	CB990	CB660	CB330
Minimum Torque, ML (N.m)	1.97	2.40	2.87	3.25
Maximum Torque, MH (N.m)	5.96	8.37	9.29	9.56
ΔM =MH-ML (N.m)	3.99	5.97	6.42	6.31
Scorch Time, T1 (Min)	2.89	2.55	2.42	2.30
Optimum Cure,T90 (Min)	4.4	4.6	5.7	7.8
Cure Rate Index, V_{C1}	66.2	48.8	30.5	18.3

Mechanical Properties and Stability in Oxidative Aging

Many FKM products are exposed to oxidative aging at oil or ozone conditions. Failure can happen due to oxidative aging which results in loss of mechanical properties. Therefore, strong mechanical properties and aging stability are necessary in FKM compound design.

Table4. Mechanical properties at various aging condition

	Green	CB990	CB660	CB330
Hardness before Aging	64	80	90	94
Hardness after Hot Air Aging	64	80	90	94
Hardness after Hot Oil Aging	61	80	90	94
Maximum Stress before Aging (MPa)	3.26	13.24	15.66	14.97
Maximum Stress after Hot Air Aging (MPa)	4.99	12.84	16.11	13.35
Maximum Stress after Hot Oil Aging (MPa)	5.22	13.77	16.57	14.79
Maximum Stress Change after Hot Air Aging (%)	53.18	-2.96	2.93	-10.80
Maximum Stress Change after Hot Oil Aging (%)	60.14	4.07	5.82	-1.23

Mechanical properties of vulcanized compounds before and after aging in oxidative aging at oil and oven conditions are shown in table4. Hardness increases correspondingly with increasing the surface area of carbon black, as the number of rubber chains immobilized on the carbon black surfaces grows up. Thus, the largest area of rigid rubber shell is obtained on the N330 surface since it has the largest external surface area, which is accessible for rubber molecules. This suggests that the external surface area of carbon blacks is indeed a major factor determining the hardness of FKM/carbon black composites. In fact, this generally agrees with the reports in the literature [16]. Oxidative aging has decreased the hardness of unfilled FKM (Green) at oil aging condition. Composites containing CB show no change in hardness as a result of aging compared to the one without CB. Hardness stability of CB containing composites can be correlated to the heat stability resulted from presence of CB particles.

CB has increased the maximum stress of FKM. N660 containing FKM has the highest maximum stress before and after aging among all. Effects of oxidative aging on stability of maximum stresses are shown in Figure3. It can be noticed that aging has increased the modulus of Green compound. However, aging effects in composites containing CBs are less than their corresponding composites with no CB. Less change in mechanical properties of composites containing CBs can be correlated to heat stability resulted from presence of CB particles.

Effect of CB Type Thermal Properties of Composites

Result of thermo-gravimetric analysis (TGA) of all used CBs and vulcanized composites are plotted in Figure 1. Carbon Black N330 loses weight at a lower temperature than N660 and N660 loses weight at a lower temperature than N990. However, the lower structure CB has higher thermal stability, but not any differences between thermal stability of CB containing rubbers. It can be seen that Green fluoroelastomer loses weight at a lower temperature than other composites. Diffusion limitations introduced by inclusion of CB structure, reduced the exchange rate of oxygen and produced degradation gases which in turn limit further rubber degradation.

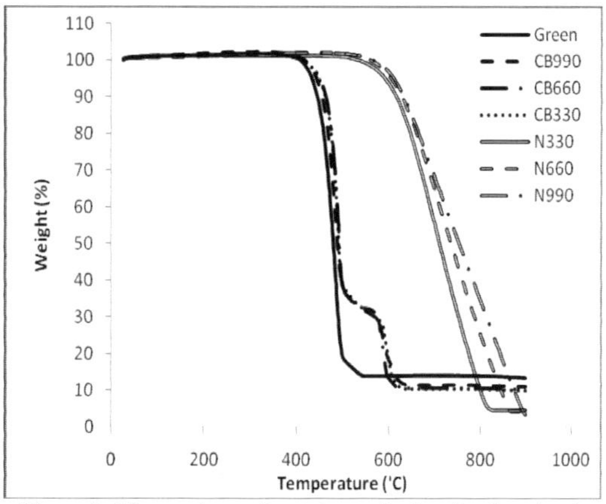

Figure1. Thermal gravimetric analysis of vulcanized rubber compounds

Conclusions

To study the microstructure effect of CB, three types of CB (N330, N660 and N990) were selected and mixed with melt mixing method. Vulcanization characteristics results revealed that CB with higher structure has higher minimum and maximum torque and lower vulcanization rate in the compounds. Aging effects in composites containing CBs are less than their corresponding composites without CB. It can be attributed to CB characteristics and CB-polymer interaction that is different at various CB containing samples. Less change in mechanical properties of composites containing CBs can be correlated to heat stability resulted from presence of CB particles. Thermo-gravimetric analysis results showed ,the lower structure CB has higher thermal stability, but not any differences between thermal stability of CB containing rubbers.

References

1. B. Améduri; B. Boutevin; G. Kostov, *Progress in Polymer Science* 2001, *26(1)*, 105-187.
2. J. G. Drobny, Technology of Fluoropolymers. CRC Press, 2001.
3. I. Banik,; A. K. Bhowmick; S. V. Raghavan; A. B. Majali; V. K. Tikku, *Polymer Degradation and Stability* 1999, *63(3)*, 413-421.
4. A. Cuccuru,;F. Sodi, *ThermochimicaActa* 1975, *12(3)*, 281-286.
5. A. Cuccuru; F. Sodi, *ThermochimicaActa* 1976, *15 (2)*, 253-256.
6. M. A. Kader; A. K. Bhowmick, *Polymer Degradation and Stability* 2003, *79(2)*, 283-295.
7. B.N. Dinzburg; K.T. Gillen *Society of Tribologists and Lubrication Engineers*, 2002.
8. H.J. Harwood,*American Society for Testing and Materials*, 1983.
9. S. Mitra; A. Ghanbari-Siahkali; P. Kingshott; K. Almdal; H. KemRehmeier; A. G. Christensen, *Polymer Degradation and Stability* 2004, *83(2)*, 195-206.
10. T. Sugama, *Materials Letters* 2001, *50(2–3)*, 66-72.
11. T. Sugama; B. Sullivan, *Journal of Materials Science Letters* 2001, *20(18)*, 1737-1740.
12. G. C. Smith; D. Park; K. J. Titchener; R. E. Davies; R. H. West, *Applied Surface Science* 1995, *90(3)*, 357-371.
13. M.-J.Wang, *Rubber Chemistry and Technology* 1998, *71(3)*, 520-589.
14. V. M. Litvinov; P. A. M. Steeman, *Macromolecules* 1999, *32(25)*, 8476-8490.
15. A. Y. Coran, Vulcanization. Burlington, Academic Press, 2005.
16. J. Donnet; A. Vidal, Carbon black 1986.
17. D. C. Edwards, *Journal of Materials Science* 1990, *25(10)*, 4175-4185.

O-13-146

THE CHALLENGES OF SILICA-SILANE REINFORCEMENT OF NATURAL RUBBER

S.S. Sarkawi[1,2], W.K. Dierkes[1]* and J.W.M. Noordermeer[1]

[1] *Elastomer Technology and Engineering, University of Twente, P.O.Box 217, 7500AE Enschede, the Netherlands – W.K.Dierkes@utwente.nl; S.S.Sarkawi@utwente.nl; J.W.M.Noordermeer@utwente.nl*

[2] *Malaysian Rubber Board, RRIM Research Station, Sg. Buloh, 47000 Selangor, Malaysia – ssalina@lgm.gov.my*

Abstract - In recent years, highly-dispersible silica has become the preferred alternative to carbon-black as reinforcing filler for low rolling-resistance tires. However, the application of this filler system is so far limited to passenger car tires, as their treads contain styrene butadiene rubber (SBR). In contrast to this, truck tires are mainly made from natural rubber (NR); this is the main application of the currently used 11 million tons of natural rubber. Unfortunately, the combination of NR with silica and a coupling agent remains a challenge. Natural rubber is a durable, natural resource, but has the disadvantage of containing a variety of non-rubber components such as proteins. An in-rubber study of the interaction of silica with proteins present in natural rubber shows that the latter compete with the coupling agent during the silanization reaction; the presence of proteins makes the silane less efficient for improving dispersion and filler-polymer coupling, and thus negatively influences the final properties of the rubber material. Furthermore, the protein content influences the rheological properties as well as filler-filler and filler-polymer interactions. Stress strain properties also vary with protein content, as do dynamic properties. With high amounts of proteins present in NR, the interactions between proteins and silica are able to disrupt the silica-silica network and improve silica dispersion. High amounts of proteins reduce the thermal sensitivity of the filler-polymer network formation. The effect of proteins is most pronounced when no silane is used; however, proteins are not able to replace a coupling agent. In order to achieve a good balance of properties, the presence of a coupling agent is essential.

Keywords: silica, silane, natural rubber, protein, rolling resistance.

Introduction

Compared to carbon black filled materials, mixing of silica compounds involves many difficulties due to the large polarity difference between silica and rubber. A bifunctional organosilane such as bis-(triethoxysilyl propyl)tetrasulfide (TESPT) is commonly used as coupling agent for enhancing the compatibility of silica and rubber, by chemically modifying silica surfaces and eventually creating a chemical link between silica aggregates and the rubber chains. Complications arise during mixing silica compounds as several chemical reactions need to take place, all at their appropriate time slots during rubber processing, namely the silica and silane reaction or silanization, silane-rubber coupling and crosslinking between the rubber chains [1]. The highly-dispersible silica technology, as it is used today, employs mainly solution-polymerized synthetic rubber, and is currently not feasible with Natural Rubber (NR). It was postulated that non-rubber constituents contained in NR such as proteins compete with the coupling agent for reaction with the silica during mixing, so disturbing its reinforcement action.

NR derived from *Hevea Brasiliensis* latex contains about 3-5% of non-rubber constituents [2], essentially proteins and phospholipids. The structure of a linear NR chain consists of a long sequence of 1000 - 3000 cis-1,4 isoprene units, two trans-1,4 isoprene units, with α- and ω-chain ends [3]. The α- and ω-terminals are associated with phospholipids and proteins respectively [4-5], and are presumed to play a part in the branching and gel formation in NR [6].

In the present investigation, the influence of non-rubber constituents in NR, particularly proteins, on silica reinforced compounds in the presence and absence of coupling agent is illustrated. NR is compared to purified NR from deproteinization (DPNR), as well as skim rubber with high protein content. The effect of mixing dump temperature on silica reinforcement is highlighted.

Experimental

Materials

The compound was based on a truck tire tread compound recipe consisting of 100 phr of NR (Malaysian Rubber Board), 55 phr of silica (Ultrasil 7005, Evonik), and 5 phr of bis (triethoxysilylpropyl) tetrasulfide or TESPT silane coupling agent (Evonik). NR's with different protein contents were compared as shown in Table 1. For skim rubber, the formulation is adjusted to 112 phr to take into account the high protein content.

Table 1 – Protein Content of NR's used

NR type	Protein content,%
NR (SMR 20)	1.3
DPNR (Deproteinized NR)	0.4
Skim Rubber	12.9

The compounds were mixed in 2 stages. In the first stage mixing, all ingredients except the curatives, were mixed in a Brabender Plasticorder 350S internal mixer with 60 rpm rotor speed and 0.7 fill factor. The starting temperature was varied from 70 to 120°C to obtain variable temperature histories and dump temperatures. The curatives were added during the second stage mixing on a two-roll mill.

Testing and Characterization

Mooney viscosity was measured at 100°C with a Mooney viscometer 2000E (Alpha Technologies) using a large rotor for compounds and small rotor for masterbatches. Vulcanisation curves were measured using a Rubber Process Analyser, RPA 2000 (Alpha Technologies) at 150°C, under condition of 0.833 Hz and 2.79% strain. The Payne effect was measured prior and after curing in the same equipment. Before curing, the sample was heated to 100°C in the RPA and subsequently subjected to a strain sweep at 0.5 Hz. The Payne effect was calculated as the difference between the storage modulus, G', at 0.56% and at 100.04% strain. The Payne effect after cure was measured after vulcanization in the RPA at 150°C for 10 minutes and subsequent cooling to 100°C, making use of the same strain sweep conditions.

The Bound Rubber Content (BRC) measurements were performed on unvulcanized samples by extracting the unbound rubber with toluene at room temperature for seven days in both normal and ammonia environment. The ammonia treatment of BRC was done in order to obtain the chemically BRC as ammonia cleaves the physical linkages between rubber and silica [7-8]. The amount of BRC (g/g filler) was calculated by:

$$BRC\ (g/g\ filler) = \frac{w_{dry} - w_{insolubles}}{w_o \times \frac{w_{filler,phr}}{w_{total,phr}}} \quad \text{Eq.1}$$

Where w_o is the initial weight of the sample, w_{dry} is the dry weight of the extracted sample, $w_{insolubles}$ is the weight of insolubles (mainly filler) in the sample, $w_{filler,phr}$ is the total filler weight in phr and $w_{total,\ phr}$ is the total compound weight in phr. The physically BRC was taken as the difference between untreated BRC and ammonia treated BRC.

The vulcanizates were prepared by curing the compounds for their respective t_{95} at 150°C using a Wickert laboratory press WLP 1600/5*4/3 at 100 bar. Tensile properties of the vulcanizates were measured using a Zwick Z020 tensile tester according to ISO-37. The hardness of the cured samples was determined according to DIN-53505. The tan delta value calculated as the ratio of the loss modulus G" to the storage modulus G' at 60°C was measured using the RPA by applying a frequency sweep at 3.49% strain after first curing in the RPA at 150°C.

Results and Discussion

Processability

In terms of processability of the masterbatches, NR and DPNR are comparable, but skim rubber has a lower viscosity. In Fig. 1, the increase in viscosity of the masterbatches with rising mixer dump temperature up to a temperature of 150ºC is a combination of the hydrodynamic effect and silanization rate of the silica. More silica is hydrophobized by TESPT when the dump temperature is raised, and this results in a higher compatibility between silica and rubber and consequently increment of the viscosity. However, the viscosity of the masterbatches of NR and DPNR start to decrease above the optimum dump temperature, but in the case of skim rubber it levels off. One explanation is the degradation of the NR chains at higher temperatures, which seems to be inhibited by a high protein content.Once the curatives are added to the compounds, the viscosities drop to processable levels, mainly due to the remilling step. In spite of the overall lower Mooney viscosities of the skim rubber masterbatches after the first mixing step, the Mooney viscosities of the compounds with curatives after mill mixing are almost comparable with those of the NR and DPNR compounds.

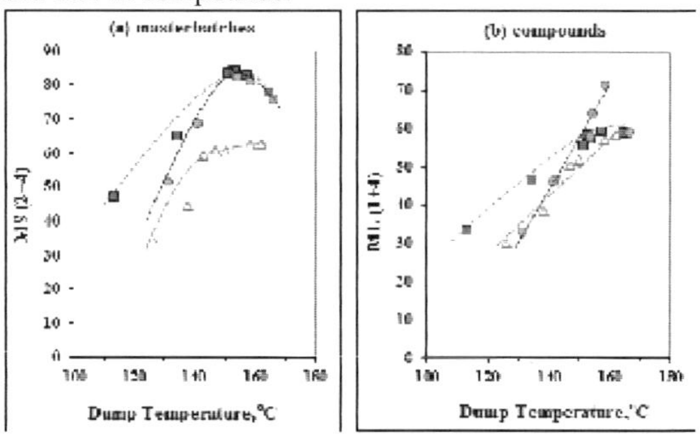

Figure 1 – Mooney viscosity of:(a) masterbatches after 1st mixing step, and (b) compounds after 2nd mill mixing of silica-filled NR at varying protein contents: (•): 0.4% (DPNR); (■): 1% (NR); (Δ): 12% (SkimRubber).

Vulcanization properties

The influence of proteins in NR on the silica-silica interaction can be clearly observed from the vulcanization curve as depicted in Fig.2. The clear two-step curve for NR-silica and DPNR-silica compounds without silane is due to the silica flocculation or re-agglomeration [9-10] and strong silica networking. With high amounts of proteins present in the compound, the silica-silica interaction is disrupted and this is shown with no sign of flocculation at the beginning of vulcanization for the skim rubber-silica compound without silane.

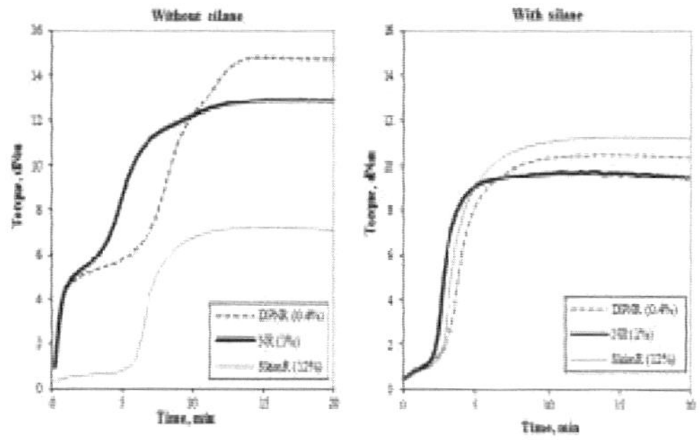

Figure 2 – Comparison of vulcanization curves of silica compounds with and without silane.

The use of a silane, TESPT in this case, in the NR-silica compound results in less pronounced silica flocculation and this is demonstrated by a small initial torque rise at the beginning of vulcanization (Fig. 3). As compared to the silica compounds without silane, the flocculation of silica in the compounds with TESPT is small due to hydrophobation of the silica surface by TESPT. The effect of protein on the cure behavior of the silica compounds fades with presence of TESPT.

Filler-filler interaction

Filler-filler interaction is normally indicated by a decrease in storage modulus of filled rubber upon an increase in strain, or the so-called Payne effect, due to breakage of physical bonds between filler particles.

The use of silica without silane modification in rubber results in a high Payne effect due to strong inter-aggregate interaction of silica. With TESPT modification, the Payne effect of the silica-filled compounds is greatly reduced as the silica surface is hydrophobized by TESPT, and the silica-silica network is disrupted as schematically shown in Fig.3. What is interesting is that the same effect can be seen in presence of a high amount of proteins: the Payne effect of the silica compound without silane is lowered. There is a relation between the amount of protein and the decrease of silica-silica interaction. This indicates a strong interaction of proteins and silica, as well as the role of proteins in hydrophobizing the silica surface.

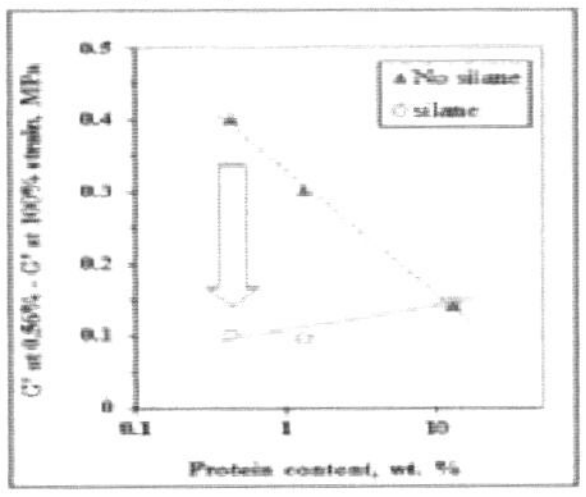

Figure 3 –Influence of proteins on the Payne effect of NR-silica compounds

For silica compounds with silane, the Payne effect decreases sharply with increasing dump temperature, as is also seen in synthetic rubber / silica compounds and taken as a sign of reaction and consequent hydrophobation of the silica by the silane coupling agent [11-12]. No effect of mixing temperature is perceived on filler-filler interaction for the skim rubber compound (Fig.4), this is observed for the unvulcanized as well as for the vulcanized compounds. This again indicates a strong interference of the proteins in skim rubber with the filler-filler network. For skim rubber, the silica-silica network is not influenced by dump temperature even in the presence of silane because silanization is hindered. It can be seen in Fig. 4 that the Payne effect of vulcanized skim rubber compound is higher than those of NR and DPNR compounds. The proteins in the skim rubber prevent the modification of the silica surface by the silane coupling agent. The logical explanation is that the interaction between the two overrules the coupling agent and that the protein is shielding the silica surface.

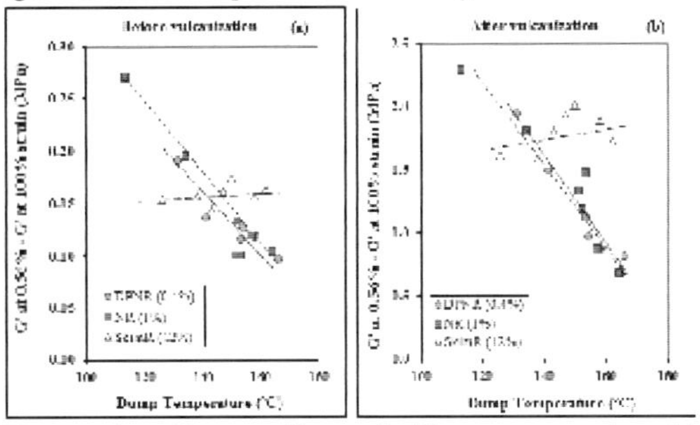

Figure 4 – Payne effect of silica compounds with TESPT at varying protein contents in natural rubber: (a) unvulcanized samples; (b) vulcanized samples.

Rubber-to-filler interaction

The filler to rubber interaction of silica-filled NR with varying protein content can also be judged on basis of the chemically and physically bound rubber content (BRC). Most of the bound rubber formed in a NR-silica-TESPT compound is chemically attached as shown in Table 2. This is obviously due to the hydrophobation of the silica surface as a result of silanization with TESPT. The increase in silica-TESPT coupling consequently results in more filler-to-rubber interaction. This corresponds well with the lower Payne effect of the silica compounds with TESPT.

Without silane in the compounds, there is still silica-rubber interaction, as indicated by the physically bound rubber in Table 2. It demonstrates that the proteins contained in NR do interact with the silica, make it less hydrophilic and thus increase the rubber-silica interaction. However, no chemically bound rubber was obtained for the silica compound without silane after ammonia treatment. This means that without silane in the compound, only loosely or physically bound rubber is formed. This again indicates that the interaction of silica with NR in the absence of coupling agent is weaker than in a compound with silane.

Table 2 – Bound Rubber Content (BRC) of silica compounds

	NR	DPNR	Skim Rubber
Without silane			
Physically BRC (%)	57	45	51
Chemically BRC (%)	0	0	0
With silane			
Physically BRC (%)	11	11	13
Chemically BRC (%)	68	76	80

Physical Properties

The use of TESPT as a coupling agent improves the vulcanizate properties of silica-filled compounds. Vulcanizates without silane exhibit inferior tensile strength than those with silane as depicted in Fig. 5. NR-silica-TESPT vulcanizates mixed at higher dump temperatures exhibit slightly lower tensile strength and modulus values. Surprisingly, DPNR-silica-TESPT vulcanizates show less influence of dump temperature and more constant physical properties. As expected, the skim rubber vulcanizates perform overall much worse compared to NR and DPNR vulcanizates in terms of physical properties, due to lack of silica-rubber coupling and lower molecular weight of the polymers to start with.

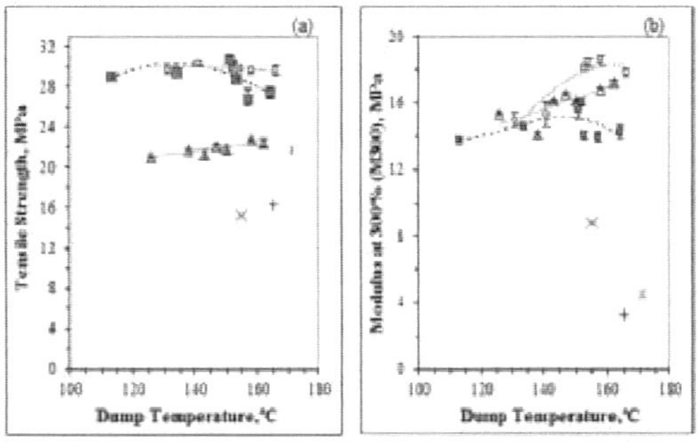

Figure 5 –Comparison of (a) tensile strength, (b) modulus at 300% for silica-TESPT-NR vulcanizates with different amounts of protein contents:
(●): 0.4% (DPNR); (■): 1% (NR); (Δ): 12% (Skim Rubber); and silica-NR vulcanizate without silane:
(+): NR ; (☓) DPNR; (x): SkimRubber.

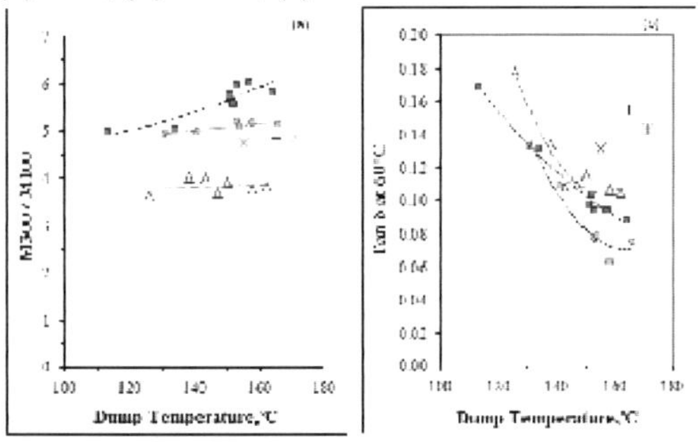

Figure 6 –Comparison of : (a) reinforcement index (M300/M100) and (b) tan δ at 60°C for NR vulcanizates with varying protein content; symbols as in Fig.5.

Commonly, the dynamic mechanical loss angle tan δ at 60°C of cured compounds is employed as indication for the rolling resistance of tires: the lower tan δ at 60°C, the lower the rolling resistance expected in real tire performance. NR vulcanizates show a strong decrease in tan δ at 60°C with increasing dump temperature regardless of the amount of protein in the rubber (Fig. 6b). Improvement in tan δ at 60°C can still be achieved with higher mixing temperature, like with synthetic rubber. This must obviously be the result of more coupling of silica to the rubber with greater silanization efficiency at high mixing temperatures. With low protein content, the DPNR vulcanizates exhibit the lowest tan δ at 60°C at high dump temperature. This actually relates well with the higher chemically bound rubber content of DPNR than of the NR compound. Still with all the protein contained in skim rubber, the tan δ at 60°C is significantly lowered by mixing temperature history and only marginally worse than for NR and DPNR.

Conclusions

Coupling agent and proteins show an antagonistic effect in silica reinforcement of rubber. When high amounts of protein are present in NR, the interactions between proteins and silica are able to disrupt the silica-silica networking. The effect of proteins is most pronounced when no silane is used in NR-silica compounds. The temperature development is an important parameter in mixing NR-silica with the aid of TESPT as coupling agent, as silica-silica interaction is reduced through silanization at sufficiently high mixing temperatures. This is clearly the case for NR and low protein content rubber DPNR. However, mixing temperature has little influence on the properties of a high protein-content skim rubber compound. Consequently, the hydrophobation of the silica surface by silane may be partially hindered due to silica-protein interactions.

Acknowledgements

The authors would like to express gratitude for financial support from the Malaysian Rubber Board for this project. The authors also thank Evonik for supplying the silica fillers.

References

1. J.W. Noordermeer; W.K. Dierkes in *Rubber Technologist's Handbook,* J. White, S.K. De; K. Naskar, Ed.; Smithers Rapra Technology, Shawbury, UK, 2008;Vol. 2, Chap.3.
2. A.B. Othman; C. Hepburn *Plast. Rub. Compos. Pro.* 1993, *19*, 185.
3. Y. Tanaka; L. Tarachiwin, *Rubber Chem. Technol.* 2009, *82*, 283.
4. J.Sakdapipanich *J. Biosci. Bioeng,* .2007, *103*, 287.
5. S. Amnuaypornsri; J. Sakdapipanich; S. Toki; B.S. Hsiao; N. Ichikawa; Y. Tanaka *Rubber Chem. Technol.* 2008, *81*, 753.
6. L. Tarachiwin; J. Sakdapipanich; Y. Tanaka *Kautsch. Gummi Kunstst.,* 2005, *58*, 115.
7. K.E. Polmanteer; C.W. Lentz *Rubber Chem. Technol*.1975, *48*, 795.
8. S.Wolff; M.-J. Wang; E.-H. Tan *Rubber Chem. Technol*.1992,*66*, 163.
9. C.J. Lin; W.L. Hergenrother; E. Alexanian, G.G.A. Böhm, *Rubber Chem. Technol.* 2002, *75*, 865.
10. S.Mihara; R.N.Datta; J.W.M.Noordermeer *Rubber Chem. Technol.* 2009, *82,* 524.
11. L.A.E.M. Reuvekamp; J.W. ten Brinke; P.J. van Swaaij; J.W.M.Noordermeer *Rubber Chem. Technol.* 2002, *75,* 187.
12. W. Kaewsakul; K. Sahakaro; W.K. Dierkes; J.W.M. Noordermeer *Rubber Chem. Technol.* 2012, *85,* 277.

O-13-168

COMPRESSION INDUCED HEATING OF RUBBER COMPOUNDS UNDER PROCESSING CONDITIONS

L. Perko[1*], W. Friesenbichler[1] and C. Fellner[1]
[1]*Chair of Injection Molding of Polymers, Department Polymer Engineering and Science, Montanuniversität Leoben, Austria – leonhard.perko@unileoben.ac.at*

Abstract - The bulk temperature is one of the most important processing parameters for rubber compounds. A variation of 10 °C can change the curing time by a factor of two. The bulk temperature in a process like injection molding is not only influenced by heat conduction but also by dissipative and compression induced heating. Little scientific efforts have been made to quantify and explore the nature of the last. In this work measurements of the compression induced heating of an SBR and an EPDM compound were carried out in both a high pressure capillary rheometer (HPCR) and a rubber injection moulding machine. The temperature rise was measured by a thermocouple which was welded into a HPCR die. This design enabled the temperature measurement as well as the sealing of the die chamber. The pressure was applied to the material by the HPCR plunger. To scale up the experiment, analogue tests were carried out on a rubber injection molding machine using a specially designed mould. For the calculation of the compression induced heating, a formula for the adiabatic case was used. The correlation between both experimental set ups was very good. The measured compression induced heating was around 6°C per 1000 bar pressure rise for both compounds. A good match of these values to the results of the calculation was achieved. Furthermore, the experiments show that a significant temperature rise can already occur at typical injection moulding pressure conditions, thus the compression induced heating should be taken into account in the simulation of the injection molding process of rubber compounds.

Keywords: compression induced heating, compression heat, rubber compound, thermodynamic property, injection molding simulation.

Introduction

It is common knowledge that temperature is an essential parameter in polymer processing, especially if crosslinking systems like rubber compounds are used. According to a wide spread rule of the thumb the rate of crosslinking nearly doubles when the temperature increases about 10°C. This law is sometimes referred as van´t Hoff´s law.
According to this the exact knowledge of the actual bulk temperature of the rubber compound is crucial for the simulation and design of processes like injection molding and extrusion.
During the processing of rubber compounds high pressures above 1000 bar can occur, thus compression induced heating may be significant.
Only few scientific efforts have been made for the characterization of this phenomenon in polymers and rubber compounds so far. Yue & Wood [5] found a linear increase of the melt temperature for LDPE, HDPE and PS. Jegadeesan [3] developped a method for the measurement of compression induced heating of thermoplastics on the HPCR.
Rudolph et al [4] presented some recent research on the compression induced heating of amorphous and semicrystalline thermoplastics where temperature rises of 12°C at a pressure level of 2000 bar were reported.

Theory

Eq. 1 shows the law of conservation of energy [1] for the cylindrical coordinate system

$$\underbrace{\rho \cdot c_V \cdot \frac{DT}{Dt}}_{\text{witnessed change of energy over time}} = \underbrace{\sigma_{rz} \cdot \frac{\partial v_Z}{\partial r} + \sigma_{zz} \cdot \frac{\partial v_Z}{\partial z}}_{\text{dissipative heating}} - \underbrace{T \cdot \left(\frac{\partial p}{\partial T}\right)_V \cdot \frac{\partial v_Z}{\partial z}}_{\text{(de)compression}} - \underbrace{\frac{1}{r} \cdot \frac{\partial (r \cdot q_r)}{\partial r}}_{\text{heat conduction}} + \underbrace{\dot{S}}_{\text{inner sources of energy}} \quad (1)$$

where z = flow direction; r = radius in m ; ρ = material density in kg/m^3; c_v = heat capacity in J/(kg*K); T = Temperature in K; t = time in s; σ_{rz} = stress in the r-plane into the z-direction in N/m^2; v_z = flow rate into the z-direction in m/s ; p = pressure in Pa*s; q_r = specific heat flow in W/m^2 and $\dot{S}$ = inner heat sources in W/m^3.

This equation describes the thermodynamic changes in a fluid that flows through a circular conduit, like the flow of the polymer melt or a rubber compound through a runner system during injection molding. The term on the left side describes the temperature change in the material whereas the terms on the right side describe the different mechanisms of heat creation. The dissipative heating is created by the deformation of the fluid, like for example in a shear flow of a runner system. Heat conduction can occur between the surface of the mold and the rubber compound or within the fluid. Inner sources of energy can be any chemical reaction like the exothermic crosslinking. There is also a term for the heat generated by compression and decompression. The quantity and the influence of this mechanism are characterized and measured in this work for rubber compounds.

The adiabatic temperature rise due to compression and the temperature decrease due to expansion are called the Joule-Thompson-Effect [2]. The formula for the calculation of the temperature change is given below (Eq.2)

$$\Delta T = \frac{\alpha \cdot v}{c_V} \cdot \overline{T} \cdot \Delta p \qquad (2)$$

With ΔT = temperature rise in K; α = thermal expansion coefficient in K^{-1}; v = specific volume in m^3/kg; $\overline{T}$ = average bulk temperature in K and Δp pressure rise in Pa*s.

The specific volume at different temperature and pressure levels can be measured in a pvT-device. The thermal expansion coefficient is also ascertained from pvT curves by using Eq.3.

$$\alpha = \left| \frac{1}{v} \cdot \left(\frac{\partial v}{\partial T} \right) \right|_p \qquad (3)$$

Experimental

Material

The experiments were carried out using two industrial rubber compounds. The first one is an SBR compound with 52 phr of carbon black and 62 phr of non-carbon filler. The second is an EPDM compound with 60 phr carbon black. Both compounds contain additional components of unknown quantity. Crosslinking agents were not added to eliminate the influence of crosslinking on the measurement.

The compounds were produced in an industrial scale with a GK 190E internal mixer (Harburg Freudenberg Maschinenbau GmbH, Freudenberg, Germany).

Table 1 – Mixing conditions for the rubber compounds

	SBR	EPDM
Mixing time in s	300	240
Max. bulk temperature in °C	150	130

pvT measurements

The pvT diagrams for the calculation of v and α were measured on a PVT100 (SWO Polymertechnik GmbH, Krefeld, Germany) device and a temperature range from 50°C to 180°C and pressures from 200-2000 bar.

Experiments on the HPRC

The measurements of the compression induced heating were carried out on a Rheograph 2002 (Goettfert, Buchen, Germany) using a type J thermocouple which was welded into a HPCR die (Fig.1). It also sealed the die completely. The pressure was applied by the plunger of the HPCR and measured by the transducer shown in Fig.1. The experiments were carried out at start temperatures of 80, 100 and 120°C.

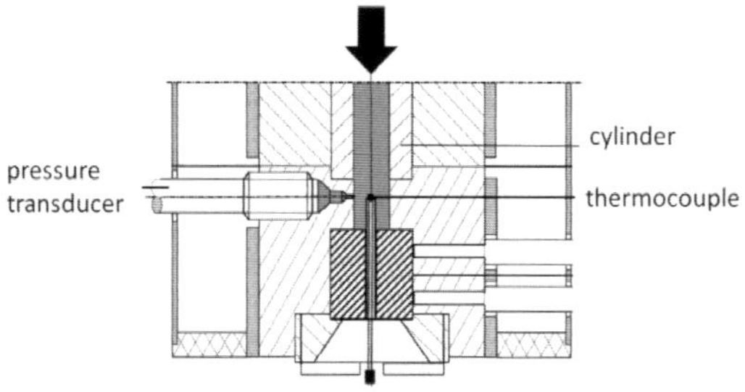

Figure 1 – Cross-section of the experimental set up on the HPCR [3]

Experiments on the injection molding machine

The upscaled experiments were carried out on a MTF 750/160 edition (Maplan GmbH, Ternitz, Austria) using a special testing mold (Fig.2). For both, the measurement of the bulk temperature and the sealing of the cavity, the same thermocouples as in the HPCR were used. The pressure was applied by the injection unit of the machine onto a shot volume of 200 cm^3. Again, the experiments were carried out at start temperatures of 80, 100 and 120°C, where the mold was temperated by the heating platens and a heater band.

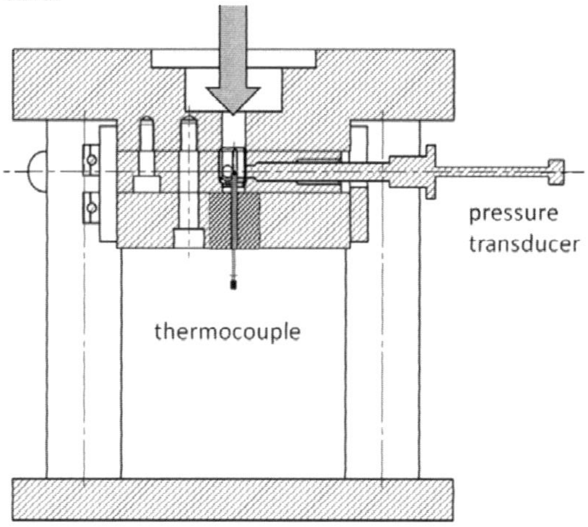

Figure 2 – Cross-section of the test mold for the injection molding machine

Results and Discussion

Both, the measured temperature and the pressure profiles at one single pressure level is shown in Fig.3

for the HPCR and in Fig.4 for the injection molding machine. The pressure induced temperature rise can be clearly seen. The slope of the temperature curves differs due to the different ramp speeds. On the HPCR the slower pressure build-up is dominant for the temperature rise. In the injection molding machine the pressure rise is very fast due to the high compression speed of the injection unit whereas the temperature rise is a bit slower because of the inertia of the thermocouple. The temperature level in the HPCR falls below the start level due to the heat flow into the die during the compression phase. However, in the injection mold heat conduction is negligible because of the larger dimensions of the pressure chamber. The diameter of the HPCR cylinder was 9 mm and the area where the thermocouples were mounted into the mold was 15 mm in diameter.

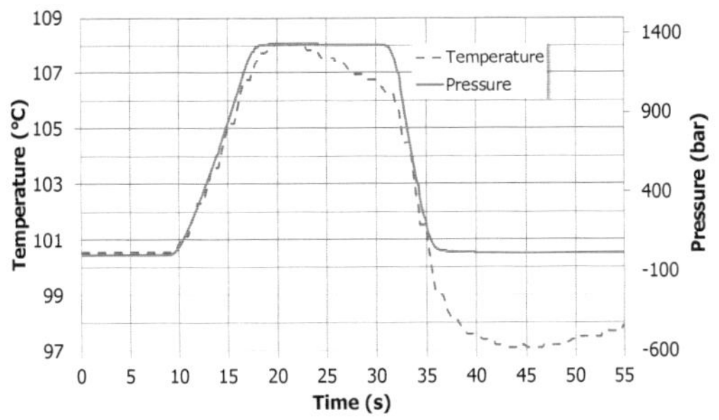

Figure 3 – Compression induced heating measurement. Transient pressure and temperature development on the HPCR for the SBR compound at a start temperature of 100°C

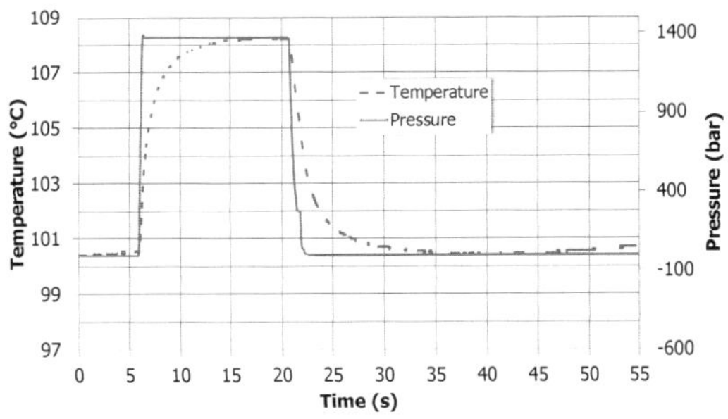

Figure 4 – Compression induced heating measurement. Transient pressure and temperature development in the injection molding machine for the SBR compound at a start temperature of 100°C

The temperature increase at different pressure levels is plotted for the SBR compound in Fig.5 and for the EPDM-compound in Fig. 6.

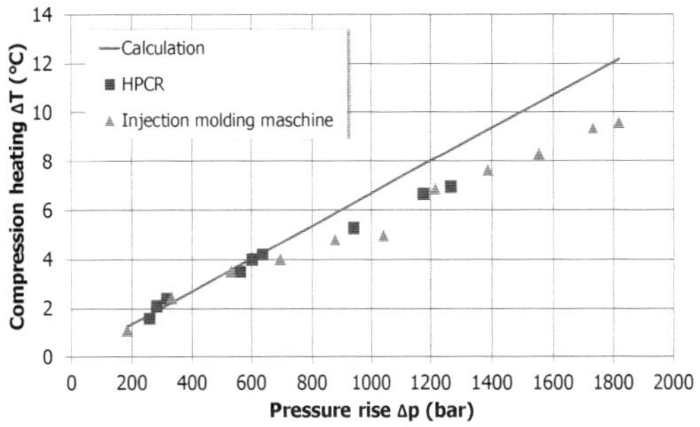

Figure 5 – Compression induced heat vs. pressure for the SBR compound at a start temperature of 80°C

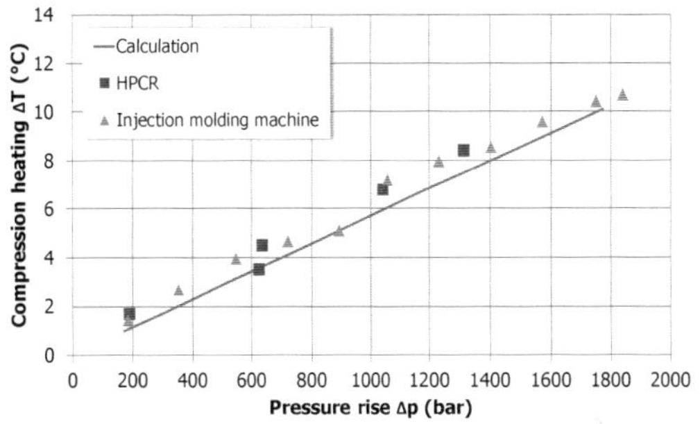

Figure 6 – Compression induced heat vs. pressure for the EPDM compound at a start temperature of 80°C

A good correlation between the measurement methods was detected, moreover, the results for the EPDM were well predicted by Eq.2. The deviations between calculated and measured compression heat of the SBR can be caused by inaccurate material properties. The average temperature rise is 0.58 K/100bar for the SBR and 0.59 K/100bar for the EPDM both at a start Temperature of 80°C.

According to Eq.2 the quantity of the compression induced heating should be proportional to the average bulk temperature which could not be verified clearly as Fig. 7 demonstrates.

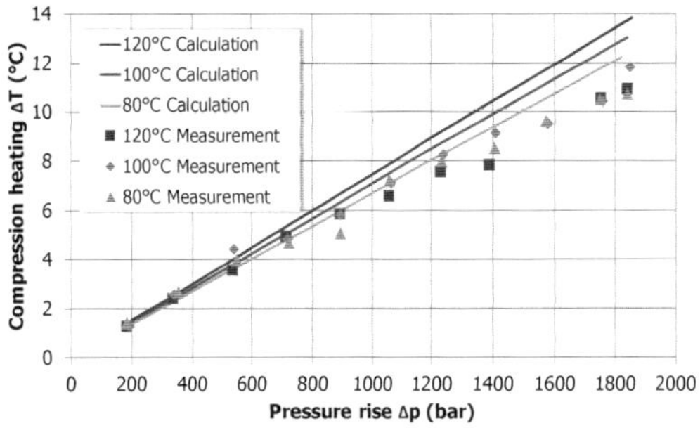

Figure 7 – Compression induced heating vs. pressure for the EPDM compound at three different start temperatures; measurement in the injection molding machine and calculation

The temperature dependence of the compression heat is weak according to the calculations. This dependence cannot be seen in the measurement values due to their scattering. The reason for this can either be the inaccuracy of the measurement or the nature of the material. Rubber compounds are highly filled and the influence of the fillers on the compression heat has not been characterised yet.

Conclusions

The compression induced heat of rubber compounds was measured for the first time in technical literature. Similar results were obtained in both, a HPCR and in a larger scale on an injection molding machine. The correlation to the temperature rise estimated by an analytical approach was quite satisfying thus the pressure induced heating can be predicted very well by the formula given above.
The highest temperature rise was about 10°C for SBR and EPDM compounds as a pressure of 1800 bar was applied which may double the crosslinking rate.
This concludes that the effect of compression induced heating has to be taken into account whenever high pressures are reached in the processing of rubber compounds.

Acknowledgements

This work was done with the support of the Austrian Research Promotion Agency (FFG) and an industrial consortium of Maplan GmbH, Semperit technische Produkte GmbH and Erwin Mach Gummitechnik.

References

1. R.B. Bird, *Transport Phenomena,* Jon Wiley & Sons, New York, 2007
2. G.W. Ehrenstein, G. Riedel, P. Trawiel, *Praxis der thermischen Analyse von Kunststoffen,*Munich, 2003
3. R. Jegadeesan, Master Thesis, Montanuniversität Leoben, 2005
4. N. Rudolph, K. Vetter, I. Kühnert *Zeitschrift Kunststofftechnik* 2009, *5,* 4
5. M. Yue, A.K. Wood, *Polymer Bulletin*, 1994, *33*, 127-132.

EFFECT OF CROSSLINKING JUNCTIONS AS A MOLECULAR SINGULARITY POINT ON MECHANICAL PROPERTIES

O. Chaikumpollert[1,2], Y. Yamamoto[3], K. Suchiva[1] and S. Kawahara[2*]

[1]*National Metal and Materials Technology Center, 114, Thailand Science Park, Paholyothin Rd., Klong 1, Klong Luang, Pathumthani, 12120 Thailand*

[2]*Department of Materials Science and Technology, Faculty of Engineering, Nagaoka University of Technology, Nagaoka, Niigata 940-2188, Japan*

[3]*Department of Chemical Science and Engineering, Tokyo National College of Technology 1220-2 Kunugida-machi, Hachioji-shi, Tokyo 193-0997, Japan*

Abstract - Molecular singularity of crosslinked natural rubber was proposed to be a significant crosslinking junction of the rubber. The crosslinking junctions of the vulcanized natural rubber were analyzed by solid-state NMR spectroscopy equipped with a new field gradient-fast-magic angle spinning (FG-FMAS) probe. The signals at 40 and 44 ppm were assigned to C4 of *trans*-1,4-isoprene units and secondary carbons adjacent to carbons linking to S atom, respectively, and the signals at 58 ppm were to the tertiary and quaternary carbons linking to sulfur atoms. Furthermore, the ^{13}C-signals at 58 ppm were distinguishably correlated to the ^{1}H-signals at 3.4 and 4.2 ppm, which were assigned to the –C-**CH**-Sx group and =C-**CH**-Sx group, respectively. Consequently, the crosslinking junctions of the vulcanized natural rubber were found to be not only tertiary carbons but also secondary and quaternary carbons. It was found to play an important role in the most outstanding mechanical properties of the rubber, when the relationships between structure of the crosslinking junctions and mechanical properties for various crosslinked rubbers were investigated. The various crosslinked natural rubbers were prepared by conventional vulcanization (CV), efficient vulcanization (EV) and semi-efficient vulcanization (SemiEV) to be almost similar crosslink density. The mechanical properties of the CV rubber were the most excellent; hence, the quaternary carbons linking to sulfur as crosslinking junction was assigned to the molecular singularity.

Keywords: Natural rubber, Crosslinking junctions, NMR, Mechanical properties

Introduction

Crosslinking junction of natural rubber vulcanizates may be a principal constitutive unit that governs mechanical properties of the rubber, i.e. tensile strength, tear strength, permanent set and so forth; hence, it has been analyzed to elucidate relationship between the crosslinking junctions and the mechanical properties. For instance, nuclear magnetic resonance (NMR) spectroscopy was mainly applied to analyze the crosslinking junctions, since NMR spectrum was found to depend on chemical environments of individual atoms. Recently, we developed a novel technique to analyze the crosslinking junctions of the vulcanizates by solid state NMR spectroscopy with a field gradient magic angle spinning (FG-HS-MAS) probe[1].

In the present work, the relationship between the crosslinking junctions and the mechanical properties was investigated for various natural rubber vulcanizates, which were prepared with conventional vulcanization (CV), efficient vulcanization (EV) and semi-efficient vulcanization (Semi-EV) systems.

Experimental

NR sample used was high ammonia natural rubber (HANR). The dried rubber sample was vulcanized by accelerated sulfur vulcanization system. The compound formulation is given in Table 1.

The vulcanization characterizations of the rubber compound were measured by using a rotorless rheometer (RLR-4) at 150°C. Optimum curing time was taken as the time at which the rheometer torque increases to 90% of total torque change following curing of rubber.

Table 1 Compound formulations of NR samples studied

Ingredient	CV (phr)	EV (phr)	Semi-EV (phr)
NR	100	100	100
ZnO	3	3	3
Stearic acid	1	1	1
TBBS	1	1	1
TMTD	0.3	2	0.8
Sulfur	2.2	0.2	1

The tensile properties of the specimens were measured according to ASTM D 412-89a, using an Instron universal tester. The crosshead rate was 200 mm/min.

The crosslink densities of NR vulcanizates were determined from swelling method. The sample was immersed in 100 mL of toluene in the dark for 1 week at room temperature. The crosslink density was calculated using the Flory-Rehner equation[2].

Solid-state ^{1}H and ^{13}C-NMR measurements were carried out with a JEOL ECA-400 NMR spectrometer with FG-HS-MAS probe at 50°C, scanning times of 10,000 scans for ^{1}H measurement and 90,000 scans for ^{13}C-NMR measurement and repetition time of 7 and 5 sec, respectively. A spinning rate of the sample tube for the solid –state measurement was about 20 kHz. To prepare the solid state NMR samples, the samples were crushed by JFC-300 cryogenic sample crusher before measurement.

Results and Discussion

Solid-state ^{13}C-NMR spectra for the vulcanized natural rubbers; HA-CV, HA-EV and HA-SemiEV, measured with FG-HS-MAS probe, are shown in Figure 1. Five major signals in the spectrum were assigned to C atoms of *cis*-1, 4-isoprene units. Values of half width of the signals and signal to noise (S/N) ratio of the solid-state ^{13}C-NMR spectrum for the natural rubber vulcanizates were quite similar. In the figure, small signals at 40, 44 and 58 ppm appeared in the solid-state ^{13}C-NMR spectra for the vulcanized CV and Semi-EV whereas only signal at 40 ppm appeared in the solid-state ^{13}C-NMR spectrum for the EV vulcanizate. In spite of the crosslink density of the CV, EV and Semi-EV vulcanizates (1.42, 1.31 and 1.37x 10^{-4} mol/cm^{3}, respectively) was almost similar, the intensity of the signal at 40 ppm showed almost similar in every vulcanization systems whereas the intensity of signals at 44 and 58 ppm in the solid-state ^{13}C-NMR spectrum decreased in Semi-EV vulcanizate spectrum. These signals disappeared for EV vulcanizate spectrum. The signals at 40 and 44 ppm were assigned to the secondary carbons and the signal at 58 ppm to the tertiary and quaternary carbons. This may be implied that the crosslinking junction of CV and Semi-EV vulcanizates were not only secondary carbon but also tertiary and quaternary carbons whereas that of EV vulcanizate was only secondary carbon.

Mechanical properties of the compounds, the stress-strain curve of HA vulcanizates cured with CV, EV and Semi-EV vulcanization systems are presented in Figure 3. The results showed that tensile strength at break, elongation at break and modulus at 100% strain of the CV vulcanizate are higher than those of EV and Semi-EV vulcanizates even though every vulcanizates contained almost similar crosslink density. Relationship between crosslinking structure and mechanical properties were also studied. Three vulcanizates have the same crosslink density but the mechanical properties were different. HA-CV showed the highest stress at break and elongation. The better mechanism properties of the conventional compound may be corresponded to tertiary and quaternary carbon of crosslinking junction showed at 58 ppm. Moreover, in the case of Semi-EV vulcanizate, it also showed the same signals with CV vulcanizate but the intensity of the signals was different. High intensity of the signals of the tertiary and quaternary carbon may be related to higher mechanical properties of CV vulcanizate than Semi-EV vulcanizate. However, the crosslinking junctions of the efficient compounds presented only the

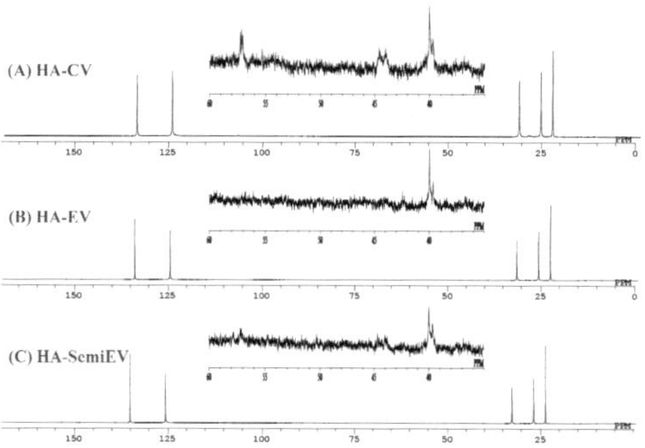

Figure 2 Solid-state ^{13}C-NMR spectra for the vulcanized natural rubber: (A) HA-CV, (B) HA-EV and (C) HA-SemiEV.

secondary carbon that may be showed lower mechanical properties in the vulcanizate.

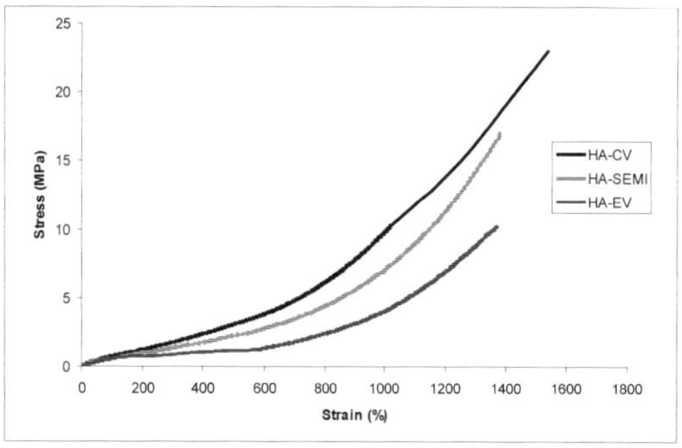

Conclusions

The crosslinking junctions of CV and Semi-EV vulcanizates were not only secondary carbon but also tertiary and quaternary carbons whereas those of EV vulcanizate were only secondary carbon. Relationship between crosslinking junctions and mechanical properties showed that in sprit of almost the same crosslink density, high intensity of the signals of the tertiary and quaternary carbon may be related to higher mechanical properties of CV vulcanizate than Semi-EV vulcanizate showed at 58 ppm. However, the crosslinking junctions of the efficient compounds presented only the secondary carbon that may be showed lower mechanical properties in the vulcanizate.

References

1. S. Kawahara, O. Chaikumpollert, S. Sakurai, Y. Yamamoto and K. Akabori, Polymer, 50, 1626 (2009).
2. G. Kraus, Rubb Chem Technol, 30, 928 (1957).

PROCESSABILITY AND PROPERTIES OF PEROXIDE CURED EPDM RUBBER: EFFECT OF MIXING TEMPERATURE AND PEROXIDE TYPE

T. Saleesung[1*], C. Sirisinha[1,2]

[1] *Department of Chemistry, Faculty of Science, Mahidol University, Rama 6 Rd., Bangkok 10400, Thailand- Thanisararat.sal@student.mahidol.ac.th*

[2] *Research and Development Centre for Thai Rubber Industry (RDCTRI), Faculty of Science, Mahidol University, Salaya Campus, Phutthamonthon 4 Rd., Salaya, NakhonPathom, 73170, Thailand- chakrit.sir@mahidol.ac.th*

Abstract - Time and temperature selected for a crosslinking reaction initiated by peroxide are dependent strongly on half-life of peroxide. The mixing conditions, and particularly the mixing temperature, are crucial parameters to be chosen in order to avoid a pre-crosslinking reaction (scorch) of rubber compound in the processing equipment, including internal mixer. In this work the effect of mixing temperature (i.e., 50, 70, 90, and 120°C) on processability of EPDM compounds (i.e., Mooney viscosity, rheological properties, scorch and cure times), and on their vulcanisate properties (i.e., tensile properties, compression set, and crosslink density) were investigated. Two types of peroxide, namely, mono- and bi-functional peroxides were used as crosslinking agents. Results obtained reveal that the mixing temperatures play a dominant role in compound processability and vulcanisate properties. With increasing mixing temperature, the scorch time significantly decreases whereas the Mooney viscosity and storage modulus increase because of a partial crosslinking reaction occurring in the mixer. Also, thermal degradation of EPDM prepared at a high mixing temperature of 120°C leads to a decreased crosslink density, and thus to a reduced tensile properties in association with increased compression set. The mono-functional peroxide offers poorer compound processability and vulcanisate properties than does the bi-functional peroxide.

Keywords: Mixing conditions, EPDM, Scorch, Peroxide, Processability, Crosslink.

Introduction

Ethylene propylene diene terpolymer (EPDM) is widely used in automotive applications such as hoses and seals. This is due to its excellent ozone and thermal resistances in comparison with diene elastomers. A peroxide crosslinking system is widely used for curing EPDM rubber to achieve good thermal ageing resistance in conjunction with low magnitude of compression set. Peroxide crosslinking involves a free radical process initiated by homolytic peroxide decomposition, and is capable of curing both unsaturated and saturated elastomers via addition reaction and hydrogen abstraction reaction, respectively. The efficiency of peroxide crosslinking depends strongly on the half-life profile of peroxide. Basically, the peroxide initiated crosslinking cannot be hastened by the addition of cure accelerator to rubber compounds, as is the case with sulphur vulcanisation. The only practical way to achieve faster crosslinking is by increasing the cure temperature, which lowers the half-life of the peroxide. However, this has to be performed carefully, since the peroxide with too-rapid crosslinking can also generate a pre-crosslinking reaction (aka. scorch). Thus, the mixing conditions are crucial factors to be considered in order to avoid a scorch reaction in the mixer. Some studies have focused on the effects of mixing conditions on the properties of the resulting elastomer. Gatos and his coworkers [1] studied the effect of mixer type and mixing temperature on the properties of EPDM nanocomposites. They found that increases in temperature and shear mixing could improve the mechanical performance of the rubber nanocomposites. The effect of processing conditions on the properties of styrene butadiene rubber (SBR)/organoclay nanocomposites has been studied by Ghasemi and his coworkers [2]. They reported that the mixing parameters (time and temperature) have a predominant role on properties and morphology of the nanocomposites. Unfortunately, there are only a few published works dealing with the effect of mixing conditions, and especially the mixing temperture, on the rheological and mechancal properties of peroxide-cured EPDM sytems. Therefore, this becomes parts of objectives in this work. Two peroxides were used in this work which are mono- and bi- functional peroxides.

Experimental

Materials

EPDM rubber (Mitsui EPT 3090EM; ML(1+4)–59; ethylene content=48%; diene (ENB) content=4.7 %; oil content=10 phr) was supplied by MDR International Co., Ltd., Thailand. Carbon black (N550) was purchased from Kitpiboon Chemical Part Ltd., Thailand. Paraffinic oil was bought from Lucky Union Co., Ltd., Thailand. Antioxidant (Naugard 445) and Coagent (SR533) were obtained from Chemical Innovation Co., Ltd., Thailand. Luperox DCP (Dicumyl peroxide), Luperox F (a,a' Di(t-butylperoxy) diisopropylbenzenes), and Luperox 101 (2,5-dimethyl-2,5-Di-(t-butylperoxy)hexane) were kindly supplied by Arkema Pte. Ltd., Thailand. Their half-life times of 1 hour were achieved at 137, 139, and 140 °C, respectively.

Formulation

The compounding recipe as shown in **Table 1** was used for preparing EPDM compounds.

Table 1 Compounding recipe

Material	Amount (phr)
EPDM	100
N550 carbon black	115
Paraffinic oil	60
Antioxidant (Naugard 445)	1
Co-agent (SR 533)	4
Peroxide	10

Sample preparation

Compounding ingredients were mixed in internal mixer (Brabender Plasticorder R2000, Germany) at various mixing temperatures of 50, 70, 90, and 120 °C. The rotor speed, mixing time and fill factor used were kept constant at 50 rpm, 10 minutes, and 0.75, respectively. Vulcanisates were prepared by compression moulding the rubber compounds using hydraulic hot-press at 160 °C and 15 MPa clamping pressure. Cure time used (t_{c99}) was pre-determined from the Moving Die Rheometer (MDR; TechPro, Ohio, USA).

Property measurement

The properties of rubber compounds were measured using the following methods:
1) *Cure characteristics*: A Moving Die Rheometer (MDR; TechPro, Ohio, USA) was at a test temperature of 160°C according to ASTM D5289.
2) *Mooney viscosity*: A Mooney viscometer equipped with small rotor (TechPro, Ohio, USA) was used at 100°C, and torque measured at 4 minutes was reported as MS 1+4, referred to ASTM D 1646.
3) *Rheological behaviour*: A Rubber Process Analyser (RPA 2000; Alpha Technology, Ohio, USA) was utilised using a frequency sweep test from 0.5 to 200 rad/s at a strain of 0.5% (within the linear viscoelastic region) at test temperature of 100°C.

The properties of rubber vulcanisates were measured as follows:
1) *Tensile properties*: a universal tensile tester (Instron model 5569, USA) was used as per ASTM D412 with the crosshead speed of 500 mm/min.
2) *Compression set*: a compression set test was performed in accordance with ASTM D395(method B) and specimen types 2.
3) *Equilibrium solvent swelling*: The swelling test was conducted for determining the crosslink densities of rubber vulcanisates. A rectangular piece of vulcanisates with 2 mm thickness was allowed to swell in toluene for 7 days at room temperature. Initial weight, swollen weight and de-swollen or dried weight were determined, and used for calculating the crosslink density via the modified Flory-Rehner equation [3-4].

Results and Discussion

The variation of Mooney viscosity with various mixing temperatures and peroxide types is shown in **Figure1**. Compared with mixing temperature of 120°C, the mixing temperatures of 50, 70, and 90°C demonstrate less influence on Mooney viscosity.
At mixing temperature of 120°C, the relatively high Mooney viscosity could be explained by: (i) the increase in magnitude of peroxide decomposition; (ii) the rise in molecular mobility of the rubber chains, and consequently the rate of reactant diffusion. These could facilitate the diffusion controlled crosslinking process [2]. Among the 3 different types of peroxides, the compound prepared with DCP exhibits greatest Mooney viscosity followed by that with Luperox 101 and Luperox F, respectively. This is attributed probably to the relatively low molecular weight of DCP (Mw=270), compared with Luperox 101 (Mw=290) and Luperox F (Mw=338). So DCP is capable of diffusing throughout the rubber matrix more easily, and thus the scorch magnitude.

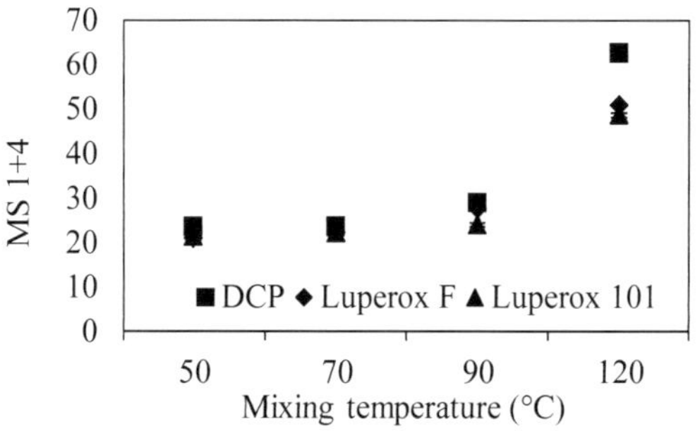

Figure 1 Mooney viscosity (MS1+4) against mixing temperature of EPDM compounds with 3 types of peroxides

Plots of storage modulus (G′) versus angular frequency for EPDM compounds prepared with DCP, Luperox F, and Luperox 101 at various mixing temperatures are shown in **Figures 2-4**. An increase in G′ with increasing frequency could be observed in all samples. This is not unusual, due to the insufficient time available for completing the molecular relaxation at high frequencies. It is noticeable that differences in G′ of the compounds are pronounced in the low frequency region. Among various mixing temperatures, the mixing temperature of 120 °C shows the highest G′ with smallest magnitude of frequency dependence. This implies an increase in the elastic response arising from the crosslinks in the compound during the mixing process (scorch phenomenon). Additionally, the DCP gives higher shear moduli than Luperox F and Luperox

101, respectively. These trends correspond with those discussed previously in **Figure 1**. The dependence of scorch time and cure time on mixing temperature for the 3 peroxides is demonstrated in **Table 2**. Luperox 101 and Luperox F give longer scorch and cure times than does DCP, suggesting superiority in process safety. This is because each bi-functional peroxide molecule could decompose twice to generate more primary radicals [5] leading to the longer time to reach an equilibrium state than with the mono-functional peroxide. At the mixing temperature of 120°C, for all compounds, the cure and the scorch times are lowest corresponding to the results of Mooney viscosity. In other words, at high mixing temperature, the peroxide decomposes and diffuses faster, yielding more crosslinking and thus shorter scorch and cure times.

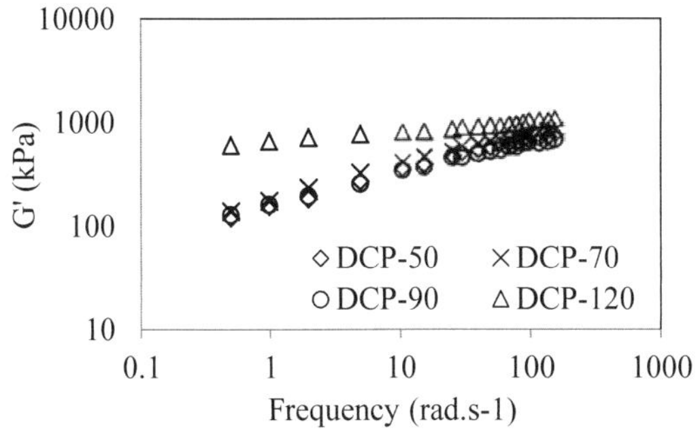

Figure 2 Storage modulus (G′) versus test frequency in EPDM compounds initiated with DCP at various mixing temperatures

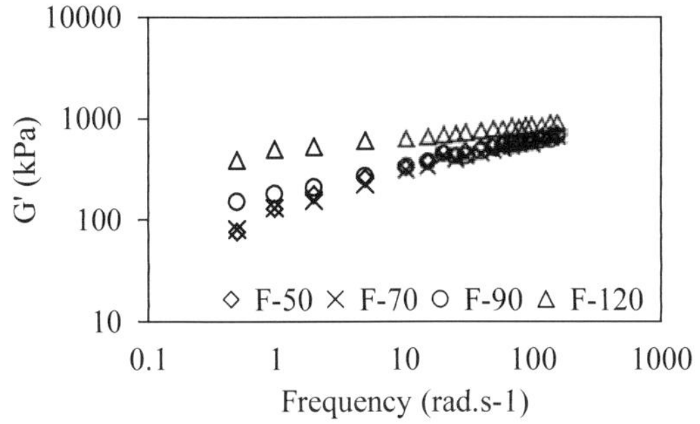

Figure 3 Storage modulus (G′) versus test frequency in EPDM compounds initiated with Luperox F at various mixing temperatures

Table 2 Scorch time (t_{s2}) and Cure time (t_{c99}) of EPDM compounds

Mixing temperature (°C)	Scorch time (t_{s2}) (min.)			Cure time (t_{c99}) (min.)		
	DCP	F	101	DCP	F	101
50	3.3	4.1	5.0	33.9	62.8	91.5
70	3.2	4.1	4.8	34.3	63.6	91.8
90	2.8	3.6	4.8	32.6	63.9	91.6
120	2.2	2.7	3.3	32.3	60.0	86.5

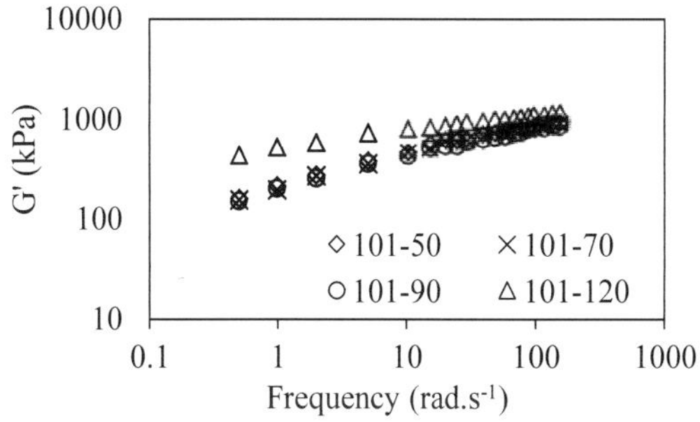

Figure 4 Storage modulus (G′) versus test frequency in EPDM compounds initiated with Luperox 101 at various mixing temperatures

The influences of peroxide type and mixing temperature on mechanical properties of EPDM vulcanisates are shown in **Figures 5-6**. It is generally acknowledged that magnitude of such effects depends on nature of the rubber, as well as on types and concentrations of peroxides [6]. It is observed from **Figure 5** that tensile strength of all test specimens increases initially, reaches a maximum value at mixing temperature of 90°C, and then dramatically decreases at temperature of 120°C. It is believed that the decrease in tensile strength at high mixing temperature is the result of thermal degradation of rubber molecules. Also, the chemical structure of rubber can be altered by thermal degradation giving the decrease in measured tensile strength. The tensile strength of the system with DCP initiation exhibits the lowest strength because the bi-functional peroxides (i.e., Luperox F and Luperox 101) offer larger amounts of crosslinking radicals over time meaning the higher cure efficiency. The modulus results are shown in **Figure 6**, revealing the initial increase in modulus with mixing temperature up to 90°C, followed by the slight decrease in modulus at mixing temperature of 120°C. Bi-functional peroxides yield vulcanisates with higher modulus than the mono-functional peroxide, which corresponds to the result trend of tensile strength.

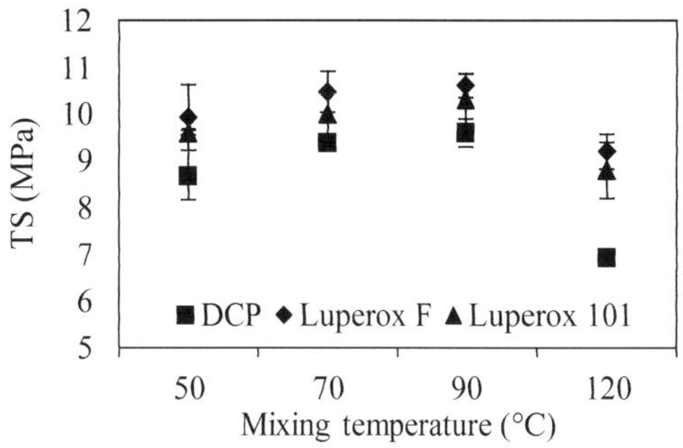

Figure 5 Tensile strength (TS) against mixing temperature of EPDM compounds with 3 types of peroxides

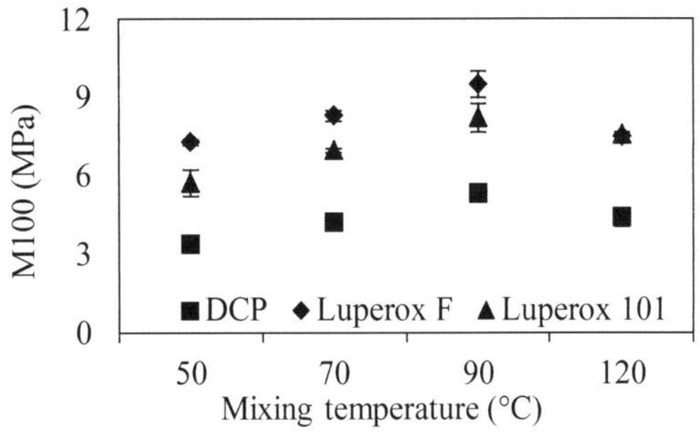

Figure 6 Modulus at 100% strain (M100) against mixing temperature of EPDM compounds initiated with 3 types of peroxides

Compressions set results are illustrated in **Figure 7**. As mixing temperature increases to 70°C, the reduction in compression set is observed in all peroxide types. Beyond 70°C, the compression set becomes to increase with mixing temperature, which is probably be caused by thermal degradation of the rubber chains. Noticeably, at any given mixing temperature, the DCP initiation leads to greatest compression set, followed by the Luperox 101 and Luperox F. This is because the relatively large amount of reactive radicals provided by Luperox 101 and Luperox F, giving the vulcanisates with relatively high crosslink densities as evidenced by **Figure 8**.

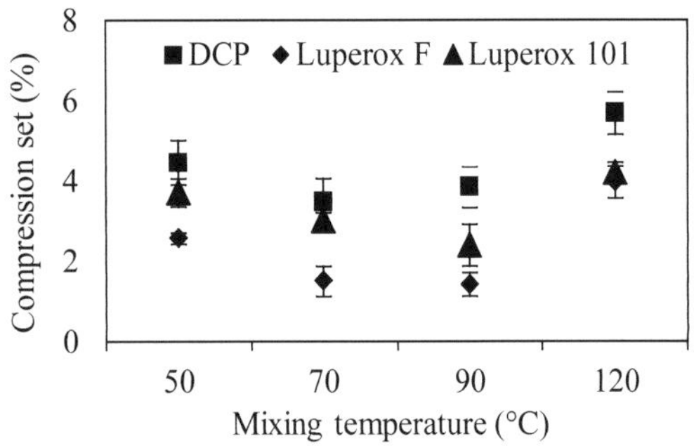

Figure 7 Compression set against mixing temperature of EPDM compounds with 3 types of peroxides

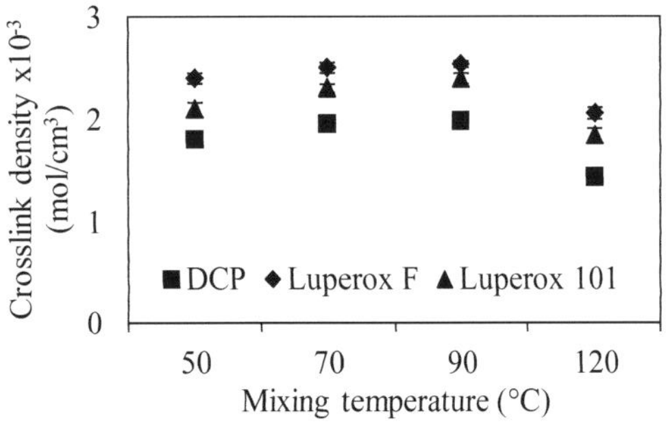

Figure 8 Crosslink density against mixing temperature of EPDM compounds with 3 types of peroxides

Conclusions

Effects of mixing temperature and type of peroxide on processability (Mooney viscosity, storage modulus, cure and scorch time) and vulcanisates properties (mechanical properties, and crosslink density) of EPDM rubber were investigated. Mono- and bi-functional peroxides as the crosslinking agents were compared for their effect on properties of EPDM compounds and vulcanisates. The following conclusions could be drawn:

1. The mixing temperatures influence the properties of EPDM compounds and vulcanisates. For the compounds, the high mixing temperature leads to a high rate of peroxide decomposition, leading to an increase of bulk viscosity and storage modulus with concomitant reduction of scorch and cure times. For vulcanisates, the high mixing temperature of 120°C leads to the thermal degradation of rubber chains, resulting in a decrease of tensile properties in conjunction with an increase in compression set.

2. Bi-functional peroxides (Luperox 101 and Luperox F) provide more radicals capable of crosslinking the rubber, so lead to higher crosslink density than does the mono-functional peroxide (DCP).

Acknowledgements

The authors would like to express their gratitude to The Royal Golden Jubilee Ph.d Program (RGJ-Ph.D.), Thailand for the financial support, and the Arkema Pte Ltd., Thailand for supplying the organic peroxides used in this research.

References

1. K.G.Gatos; R. Thomann; J. Karger-Kocsis*Polym.Int.* 2004, 53, 1191.
2.I. Ghasemi; M. Karrabi; M. Mohammadi; H. Azizi *Express Polym. Lett.* 2010, 4, 62.
3.S.B. Kwak; N.S. Choi *Int. J. Mod. Phys. B.* 2010, 24, 2597.
4. H. Peng; L. Liu; Y. Luo; X. Wang; D. Jia *Polym. Compos.* 2009, 30, 955.
5. L. Cavin; A. Rouge; Th. Meyer; A. Renken *Polymer* 2000, 40, 3925.
6. R.R. Babu; N.K. Singha; K. Naskar *Express Polym. Lett.* 2008, 2, 226.

O-13-105

THE EFFECT OF NON-ISOTHERMAL VULCANIZATION ON STATE OF CURE AND MECHANICAL PROPERTIES OF VULCANIZED RUBBER VIA FINITE ELEMENT SIMULATION

Weerawuth Nabpetch, Tulyapong Tulyapitak and Nattapong Nithi-Uthai*

Department of rubber technology and polymer science, Faculty of science and technology, Prince of Songkla University, Pattani campus, Thailand - nnattapo@bunga.pn.psu.ac.th

Abstract - This study utilized finite element method to simulate state of cure and predict mechanical properties of solid rubber vulcanizied under time dependent non-isothermal condition. The simulation used 2D axisymetric model in commercial finite element software, Ansys PolyFlow. The software simulated heat transfer then used resulted temperature to calculate state of cure. In this research state of cure equations were modified to take into account of over cure via subroutine alteration. The resulted state of cure simulation was coupled with a database of relationship between mechanical properties versus state of cure which were done by experiments earlier. The simulation can then show results of state of cure together with mechanical properties. Simulation results were well within 20% difference compared to experimental.

Keywords: State of cure, Non-isothermal vulcanization, Finite element simulation.

Introduction

Curing process a significant and directly affect on the properties and quality of products. The curing process of solid rubber such as solid tire occurs under non-isothermal condition. This lead to the difference in the state of cure for each position in the product hence difficult to evaluate the optimum cure. Many research, have tried to use mathematical calculations to predict the state of cure in each position of the rubber specimen via temperature history in order to calculate the state of cure from the relationship between temperature and time [1, 2, 3]. However, monitoring temperature at each position in rubber specimen was difficult. Therefore, it is necessary to use finite element simulation to assist in estimating the temperature and the state of cure in the rubber specimen. This experiment used Ansys PolyFlow software for simulating the state of cure in rubber specimen by using 2D axisymmetric mesh. Mechanical properties such as tensile strength, 300% modulus and hardness were then incorporated into the model in order to demonstrate final properties of vulcanized rubber.

Experimental

Materials

Ribbed Smoked Sheet No. 3 (RSS#3) grade natural rubber was obtained from Tavorn Rubber Industry (1982) Co., Ltd. Industrial grade zinc oxide, stearic acid, N-tert-Butyl-2-benzothiazolesulfenamide (TBBS), and sulfur were obtained from Flexsys Industry. N330 grade carbon black was obtained from Thai Carbon Black Public Co.,Ltd.

Rubber compounding

The rubber compounds were prepared on a two-roll mill at room temperature. The nip of the rolls was kept at 2 mm. at a speed of 40 rpm. One minute was required to soften the rubber. After softening, zinc oxide, stearic acid, TBBS, carbon black, and sulfur were added sequentially. The total mixing time was 20 minutes. The two-roll mill was cooled by cold water circulation during mixing. The compounds formula shown in Table 1. The preparation of compounds based on ASTM D3184.

Cure simulation process

Cure simulation of rubber sample can be divided into two main steps; heating and cooling. Firstly, the rubber was heated to the required temperature (140, 160, and 180 °C) in the heated mould. After the heating step, rubber sample was demoulded and allowed to cool down at room temperature. Fig. 1 shows a cross section of the rubber sample with the dimension 5 x 5 cm. By taking advantage of symmetry, only a half of the cross-section was modeled and the 2-D axisymmetrical elements were used to represent the model. The boundary conditions were applied as shown in Table 2. Material properties used in the simulation were listed in the Table 3.

The state of cure is obtained from finite element software shown in Eq. 1[4].

$$x(t) = \begin{cases} 0, \; for\; t < t_i \\ \left(\dfrac{1 - Exp\left[-\frac{\int_{t_i}^{t} k_0 Exp\,[-E_a}{RT\,(t)}]dt\right]}{1 + Exp\left[-\frac{\int_{t_i}^{t} k_0 Exp\,[-E_a}{RT\,(t)}]dt\right]} \right)^2, \; for\; t \geq t_i \end{cases} \qquad (1)$$

However, state of cure is obtained from Eq. 1 can not predict state of cure after reaching 100% cure hence no information. Therefore, to incorporate an over cure, state of cure is given by

$$x(t)_{tot} = [x(t)^{\infty} \times x(t)_{over}] + x(t) \quad (2)$$

Where $x(t)_{over}$ was such that

$$x(t)_{over} = 1 - \left(\frac{\tau(t)-\tau(0)}{\tau(\alpha)-\tau(0)}\right) \quad (3)$$

Where			
	$x(t)$	is	state of cure
	$x(t)_{tot}$	is	total state of over
	$x(t)_{over}$	is	state of over cure
	τ	is	torque (N-m)

Table 1 Compounds formula

Chemical	Non-filled	filled
RSS#3	100	100
Stearic acid	0.5	0.5
Zinc oxide	6	6
TBBS	0.7	0.7
Sulfur	2.25	2.25
Carbon black N330	-	35

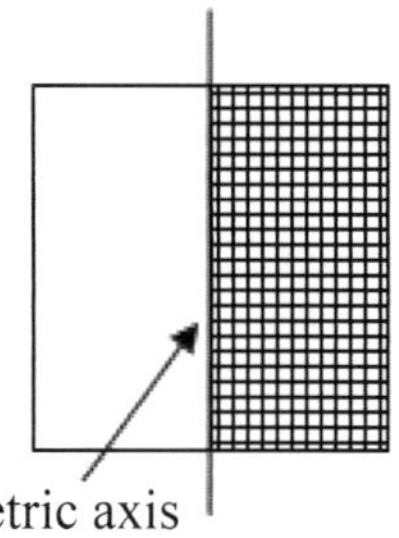

Fig. 1 Cross section of the rubber sample.

Table 2 The boundary conditions applied in the simulation.

Simulation step	Boundary conditions
Heating	1. The initial temperatures of all internal nodes were set at room temperature (33 °C), except the nodes located at the boundary surface in which their initial temperatures were set at the required temperature (140, 160, and 180 °C). Intimate contact between rubber and hot metal surface was assumed so that the interface thermal resistance could be neglected [5]. 2. No heat was allowed to transfer across the surfaces of symmetry. 3. Material properties such as density, thermal conductivity, heat capacity, frequency factor for rate constant of cure reaction, activation energy for rate constant of cure reaction, frequency factor for induction period of cure reaction, and activation energy for induction period of cure reaction were then applied to the model.
Cooling	1. The initial temperatures of all nodes were the same as the final nodal temperatures extracted from the heating step. 2. The film temperature along the exposed edges was defined by room temperature (33 °C). 3. No heat was allowed to transfer across the surfaces of symmetry. 4. Material properties such as density, thermal conductivity, heat capacity, frequency factor for rate constant of cure reaction, activation energy for rate constant of cure reaction, frequency factor for induction period of cure reaction, and activation energy for induction period of cure reaction were then applied to the model.

Table 3 The material properties required in the simulation.

Properties	Non-filled	Filled
density (ρ, kg/m3)	950	1050
thermal conductivity (k, J/min m °C)	20 + 0.25T	40 + 0.25T
heat capacity (Cp, g/°C)	2000	1800
frequency factor for rate constant of cure reaction (K_0, 1/min)	8.6399×10^{13}	1.0608×10^{12}
activation energy for rate constant of cure reaction (E_a, J/mol)	115664	100607
frequency factor for induction period of cure reaction (t_{io}, 1/min)	2.4009×10^{-10}	3.8500×10^{-7}
activation energy for induction period of cure reaction (E_i, J/mol)	81385	54489

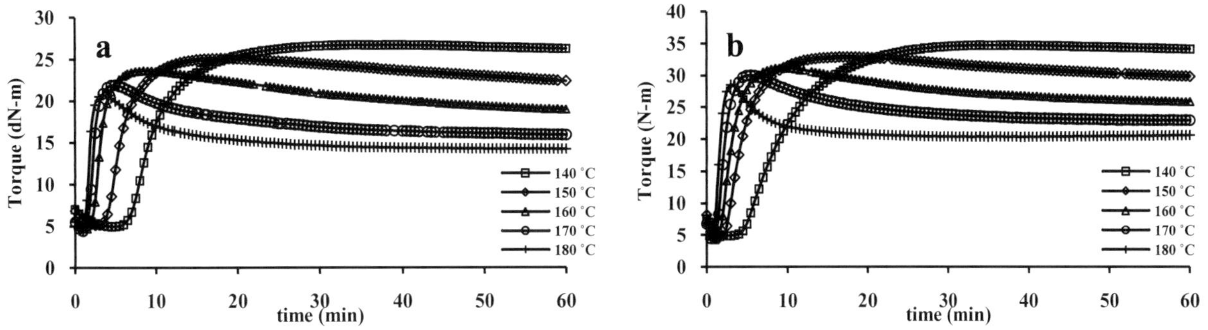

Fig. 2 Cure characteristic of non-filled (a) and carbon black filled (b) rubber compounds.

Results and Discussion

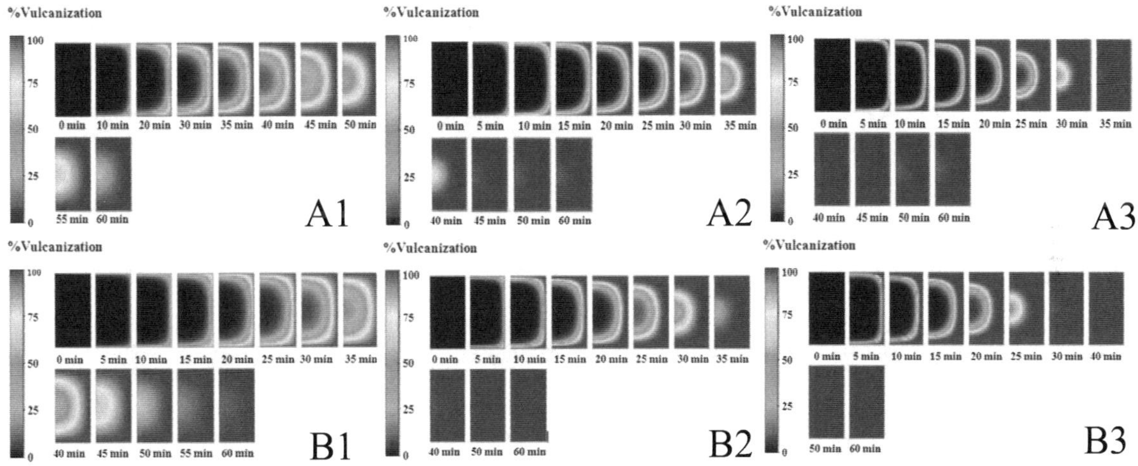

Fig. 3 The contour of cure distribution of rubber sample via Ansys PolyFlow software; at 140 °C of unfilled rubber compounds (A1), 160 °C of unfilled rubber compounds (A2), 180 °C of unfilled rubber compounds (A3), 140 °C of filled rubber compounds (B1), 160 °C of filled rubber compounds (B2), and 180 °C of filled rubber compounds (B3).

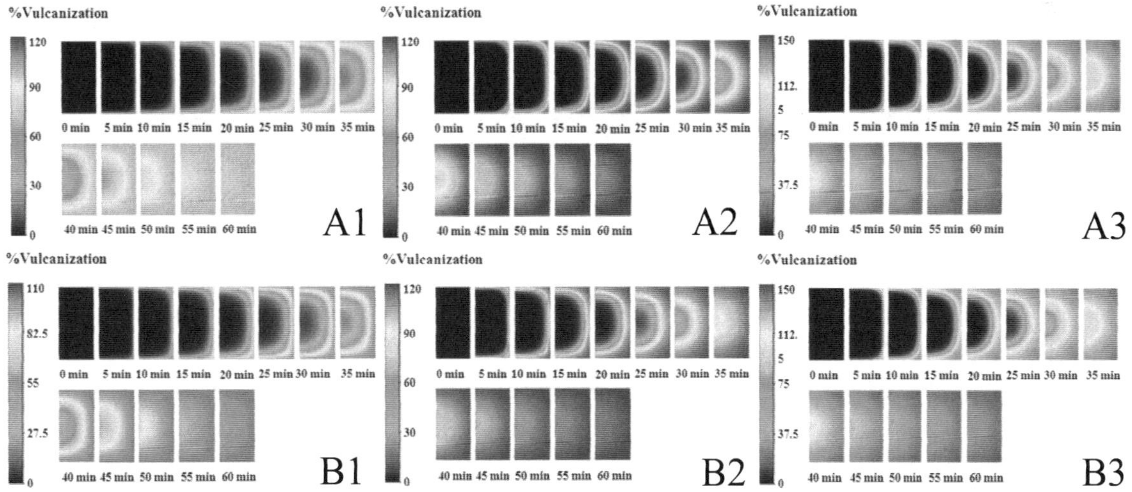

Fig. 4 The contour of cure distribution of rubber sample after modified the equation; at 140 °C of unfilled rubber compounds (A1), 160 °C of unfilled rubber compounds (A2), 180 °C of unfilled rubber compounds (A3), 140 °C of filled rubber compounds (B1), 160 °C of filled rubber compounds (B2), and 180 °C of filled rubber compounds (B3).

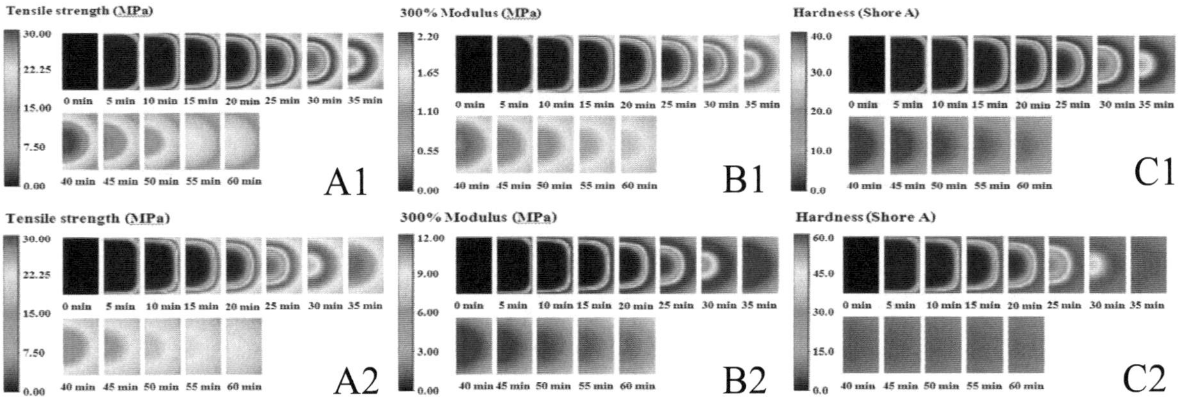

Fig. 5 The contour of mechanical properties distribution of rubber sample; tensile strength (A1), 300% modulus (B1), and hardness (C1) of non-filled rubber compounds and tensile strength (A2), 300% modulus (B2), and hardness (C2) of filled rubber compounds used cure temperature at 160 °C

Fig. 3 show the contour of cure distribution of rubber samples via Ansys PolyFlow software. The increasing of moulding temperature lead to the increasing in rate of vulcanization and decreasing in cure time. However, the state of cure was obtained from Ansys PolyFlow software has maximum at 1 (100% vulcanization) and could not predict the over cure state. As a better of fact, curing occurred at all time even after maximum cure hence over cure. Therefore, the state of cure in rubber sample during curing process was allowed by Eq. 2. The contour of cure distribution of rubber sample after modified the equation for predict the state of cure shown in Fig. 4. From the results, the %vulcanization beyond 100% represent the over cure condition of rubber sample which is advantage of this simulation technique. The result shows clearly the over cure condition of samples when increasing moulding temperature due to the increase in decomposition rate. Moreover, the state of cure and temperature change in the center of rubber sample shows good correlation between Ansys PolyFlow software and experimental were well within 20%.

Fig. 5 shows the contour of mechanical properties such as tensile strength, 300% modulus, and hardness of rubber sample obtained from simulations with the Ansys PolyFlow software. From the results, increasing of times of cure reaction lead to the increasing in tensile strength, 300% modulus, and hardness of rubber sample. However, when times of cure reaction were over than optimum cure time of rubber sample, decreasing of tensile strength, 300% modulus, and hardness at surface of rubber sample due to the decomposition of rubber sample can be easily detected.

Conclusions

The experiments show the good prediction of Ansys PolyFlow software for predicting changes in state of cure. The advantages of this technique was to predict the over cure by modified the equation in the software and the different of state of cure between the simulations and experimental were well within 20%. The increasing of cure reaction time lead to the increasing in tensile strength, 300% modulus, and hardness of rubber sample. However, cure reaction time were over than optimum cure time of rubber sample, decreasing of tensile strength, 300% modulus, and hardness at surface of rubber sample. Finite element simulation was suitable to be used in the thick moulding application such as on solid tire industry for evaluating the state of cure of each position in the solid tire and improve the product quality.

Acknowledgements

The authors wish to thank Department of Rubber technology and polymer science, Faculty of Science and technology, Prince of Songkla University, RDO Prince of Songkla University, Thailand Research Fund (TRF) and The Royal Bangkok Sports Club for permitting and supplying the outlay for use of Ansys PolyFlow Software and scholarships.

References

1. W. E. Claxton; J. W. Liska. *Journal of Rubber Age*. 1965, 95(2), 237.
2. Sh. Nozu, H. Tsuji, M. Itadani, W. Fujiwara, K. Ohnishi. *Journal of Materials Processing Technology*. 2008, 201, 270.
3. P. M. Abhilash, K. Kannan, B. Varkey. *Materials Science and Engineering B*. 2010, 168, 237.
4. W. J. Toth; J. P. Chang; C. Zanichelli. *Tire Science and Technology*. 1991, 19(4), 178.
5. S. D. Gehman. *Rubber Chemistry Technology*. 1976, 40, 36.

O-13-253

THE PREDICTION OF RHEOLOGICAL PROPERTIES OF MASTICATED NATURAL RUBBER BY DUAL CONSTRAINT MODEL

C. Pattamaprom*, **N. Saengklin**
Department of Chemical Engineering, Faculty of Engineering, Thammasat University, Klong Luang, Pathumtani, Thailand – cattalee@engr.tu.ac.th

Abstract - Mastication of natural rubber before compounding with other chemicals and fillers is a common process to reduce viscosity and molecular weight of rubber. This research quantified the molecular weights of masticated natural rubbers by using GPC technique. The rheological properties of masticated natural rubbers were measured at a wide range of molecular weight and the rheological parameters of natural rubber were obtained. Moreover, the rheological properties of natural rubber were predicted by using the dual constraint model. The rheological properties of natural rubber predicted by using dual constraint model indicated that the dual constraint model could be used to predict the rheological properties of natural rubber, especially at low molecular weight (after mastication) and it could be use as an indicator for the presence of gels.

Keywords: Natural rubber, rheological parameter, molecular weight, dual constraint model

1. Introduction

Natural rubber has been used as a raw material for many products such as tires, thread, seals. In rubber industries, mastication of natural rubber before compounding with other chemicals and fillers is a common process as the molecular weight of natural rubber is usually very high. The structure of natural rubber after mastication is deformed by three mechanisms, which are the mechanical breakdown by shearing action, thermal degradation and thermal oxidation. The mechanical breakdown by shearing and the thermal degradation lead to shorter rubber chains, while the high-temperature thermal oxidation produces longer chains. At low to moderate temperatures, which are the general conditions for rubber compounding, the mechanical breakdown dominates the phenomena. The shorter chains are desirable for incorporating fillers and chemicals into the rubber compounds; however, the chains that are too short are undesirable for good mechanical strength.

In the past, rheological models have been investigated for explaining flow behavior of polymers, elastromers, synthesis of polymers, and polymer blends. Because the rheological predictions for random branched and random gelled polymers depends on several factors including the homogeneity of molecular architecture, quantitative predictions of commercial random branched and gelled polymers are not quantitatively possible. Therefore, the highest potential for quantitative predictions is in linear polymers, where several models have shown to provide good predictions by using the least amount of fitting parameters. In this paper, we choose to investigate the prediction of rheological properties of natural rubbernusing the rheological model called “the dual constraint model” [1,2,3]. If the model prediction is valid for natural rubber, we can use this model to predict viscosities and other rheological properties of natural rubber from their molecular weights at all temperatures and, in the other way around, we can predict the molecular weight of natural rubber by measuring on the rheological properties without needs for the complicated GPC.

2. Theoretical Description: The Dual Constraint Model

The dual constraint model was originally developed by T.J. Van Dyke, R.G. Larson, D.W. Mead and M. Doi and refined by Pattamaprom et al.[1,2]. This model provides good predictions for monodisperse, bidisperse, and polydisperse linear polymers and star-branched polymers for a broad

**corresponding author: cattalee@engr.tu.ac.th*

range of molecular weights. The required parameters for this model are the plateau modulus (G_N^0) and the relaxation time for an entanglement segment (τ_e), where these two parameters are independent of molecular architecture, molecular weight or molecular weight distribution. The plateau modulus could be looked up from the polymer handbooks or obtained experimentally, whereas τ_e could be calculated from the friction coefficient ζ of Rouse theory or could be allowed as the only one fitting parameter for a single type of polymer at one temperature. The governing equations for this model are shown below, where the most-updated full description of the model could be found in Pattamaprom et al. (2008) [3].

The dual constraint model combines two forms of constraint release. The first is "double diffusion" which augments "double reptation" by inclusion of primitive path fluctuations as a mechanism of constraint release. The second form of constraint release is "dynamic dilution," which was introduced by Ball and McLeish [4] to describe the time-dependent loosening of the effective entanglement network, or widening of the tube. The upper limit of the combined constraint release rate from both mechanisms is controlled by Rouse-like motions of the tube containing the test chain; these motions are called "constraint-release Rouse" processes [5,6]

The diffusion equation for the tube survival probability p_i of chain type i including constraint release is rewritten as

$$\frac{\partial p_i(s_i,t)}{\partial t} = \frac{D_i}{L_i^2}\frac{\partial^2 p_i(s_i,t)}{\partial s_i^{\,2}} - \frac{p_i(s_i,t)}{\tau_{\xi,i}(s_i)} \qquad (1)$$

where $p_i(s_i,t)$ is the survival probability of a tube segment occupied by a chain of type i as a function of time t and contour distance s_i, where s_i ranges from 0 at the center of a linear polymer chain to ½ at the end , or from 0 at the branch point to 1 at the end of a star branch. Note that the variable ξ_i in Table 1 is given by ξ_i = 1-s_i for a star arm and is ξ_i = 1-2s_i for a linear molecule. D_i, the curvilinear diffusion coefficient, is given by $\frac{L_i^2}{\tau_{d,i}\pi^2}$, where L_i is the average contour length of the tube and $\tau_{d,i}$ is the reptation time constant [7]. $\tau_{\xi,i}$ is the time constant for contour-length fluctuations in the presence of constraint release consisting of the early-time (τ_{early}) and late-time (τ_{late}) fluctuations. $\tau_{\xi,i}$ is obtained from τ_{early} and τ_{late} using a crossover formula reported in our earlier work. (Note that in Pattamaprom et al. (2000), due to a typological error, a square root is missing from the second line of the crossover formula in eq. 4, where $\tau^*_{\xi,i}$ should equal $\sqrt{\tau_{early,i}\cdot\tau^*_{late,i}}$). $\tau_{d,i}$, where τ_{early} and τ_{late} are related to the relaxation time of one entanglement segment, τ_e, by the equations described in Table 1 of Pattamaprom et al. (2008).[3]

The overall survival probability $\phi_i(t)$ of a tube occupied by chain i can then be calculated from $p_i(s_i,t)$ by

$$\phi_i(t) = \int_0^1 p_i(s_i,t)ds_i \qquad (2)$$

and the average survival probability $\phi(t)$ of all tubes is

$$\phi(t) = \Sigma w_i\,\phi_i(t) \qquad (3)$$

where w_i is the weight fraction of chains of type i.

The relaxation modulus G(t) from the contributions of reptation, contour-length fluctuations, and constraint release can be obtained from:

$$G(t) = G_N^0\,\phi(t)\phi'(t) \qquad (4)$$

where $\phi'(t)$ is the average survival probability modified by accounting for the constraint-release Rouse mechanism proposed by Viovy et al. [5]. A full description of $\phi_i(t)$, as well as a complete explanation of the model, can be found in Pattamaprom et al. (2008).[3] The model parameters are the plateau modulus G_N^0 and the Rouse relaxation time of a tube segment τ_e, which are both independent of the polymer molecular weight.

3. Experimental

3.1 Materials

Natural rubber used in this research was Standard Thai Rubber 5L (STR 5L) grade supplied by the Rubber Estate Organization Thailand. No chemicals or fillers were added to natural rubber during mastication.

3.2 Sample preparation

Natural rubber (STR5L) was masticated in the Brabender Plasti-corder Lab-station internal mixer, constituted of two banbury-type counter-rotors in a 55 cm^3 chamber. A fill factor of 0.8 was chosen, which permitted good homogenization in the mixer. The rubber was fed into the internal mixer at the starting temperature of 50°C. Then, the rubber was masticated at a constant speed (40, 70, 100 rpm) for a fixed time of 20 min. Simultaneously, the torque and temperature profiles were recorded. Subsequently, rubber was kneaded in the two-roll mill to make thin sheet and cut in to disks of 25 mm in diameter.

3.3 Characterization

The characterization includes the determination of gel content, molecular weight and molecular weight distribution of natural rubber before and after mastication. The gel content experiment was carried out by dissolving small pieces of rubber in toluene (0.1% w/v) and keeping this mixture in the dark at room temperature without shaking for one week. The gel fraction was separated from the solution and dried at 40°C for 24 hours. The percentage of gel in the original rubber piece was estimated as the gel content. The molecular weight (MW) and molecular weight distribution (MWD) of natural rubber were determined by gel-permeation chromatography (GPC) using the Breeze water, having exclusion limits of $2x10^3$ to $4x10^6$. Measurements were made using tetrahydrofuran as an eluent with a flow rate of 3.33 µL/min. Commercially obtained standard polystyrenes were used for calibration.

3.4 Rheological measurements

Rheological properties were measured by a stress-controlled rheometer, Malvern-Kinexus Pro, equipped with parallel-plate configuration. The plate diameter is 25 mm and the gap is 1 mm. The dynamic oscillatory measurements were done over the frequency range of 0.001 to 150 Hz in controlled stress mode. Experiments were carried out in dynamic conditions over a temperature range of 5-125°C. A strain of 1% which was found to be within the linear viscoelastic region is imposed on the sample.

4. Results and Discussion

This research investigated the chain reduction of natural rubber when masticated at different speeds. The analysis was carried out in two parts, which were the changes in natural rubber properties after mastication and the investigation of the dual constraint model for predicting natural rubber rheological properties from their molecular weights.

4.1 The changes in natural rubber properties after mastication

The gel content of natural rubber before mastication was found to be as high as 40 %w/w. Nevertheless, after mastication for 20 minutes, no gels have been found at any of the rotor speeds in this study. The decrease in the gel content during mastication could be attributed to random polymer chain scission [8].

The molecular weight (MW) and molecular weight distribution (MWD) of natural rubber were determined by gel-permeation chromatography (GPC). We found that the molecular weight of natural rubber masticated at the rotor speed of 40 rpm was higher than that masticated at the rotor speed of 70 and 100 rpm, respectively. At the lowest rotor speed (40 rpm), where degradation was supposed to be minimal, the molecular weight of natural rubber was still high as the long chain structures of natural rubber survived under shear. On the other hand, at 100 rpm, the molecular weight of natural rubber was reduced by more than half of its initial value with narrower molecular weight distribution. The molecular weight and molecular weight distribution of natural rubber at different speeds were shown in Table 1. As can be seen, mastication at higher rotor speeds leads to lower molecular weight of natural rubber due to the effects of mechanical breakdown and thermal degradation.

Table 1 Molecular weight and molecular weight distribution of natural rubber masticated at different speeds by GPC

Rotor speed (rpm)	M_w (g/mol)	M_n (g/mol)	MWD
0	886229	410712	2.16
40	644288	272291	2.37
70	621635	261434	2.38
100	368776	194658	1.89

4.2 The prediction of natural rubber's rheological properties from their molecular weights by the dual constraint model

The storage (G') and loss (G") modulii of natural rubber masticated at difference speeds were carried out by dynamic frequency sweep over the temperature range of 5, 15, 25, 50, 75, 100 and 125°C and shifted experimentally to the master curves at the reference temperature of 25°C. The rheological properties of natural rubber after mastication were dependent on the rotor speed. We found that the storage and loss modulii decreased when the rotor speed increased, consistent with the reduction in molecular weight reported by the GPC result.

By using the experimental value of plateau modulus (G_N°) (2.75x10^5 Pa.), the molecular weight between entanglements (M_e) was calculated by Ferry (1980)'s definition [9] to be 8382.9 g/mol. Figure 1 shows the prediction of the dual constraint model using only a single fitting parameter, which is the relaxation time between entanglement segment (τ_e). For natural rubber at 25 °C, the best fit value of τ_e is xxxxx sec. As can be seen, the prediction of the dual constraint model for linear polymer could predict the rheological properties of natural rubber masticated at 70 and 100 rpm well, while the predictions at lower mastication speed and for the unmasticated rubber deviated more and more in the terminal regime. This result indicated that mastication at high speeds could make natural rubber behave more like linear polymer, implying the reduction in the detectable and undetectable gel content in masticated natural rubber.

5. Conclusion

Mastication of natural rubber expectedly reduced the molecular weight of the rubber. We found that the mastication at 100 rpm, which produced high shear and thus high internal temperatures could reduce the molecular weight of green natural rubber by half. We also found that, by mastication, the gel content could reduce from 40% to and undetectable level even at the lowest experimental speed of 40 rpm. However, considering the dual constraint model prediction for linear polymer, the fits for the rubber sample at low mastication speed (40 rpm) and for the unmasticated rubber deviated more from the experimental data indicating that the rubber structures before mastication and masticated at low speed still contain some detectable and/or undetectable gels.

References:

1.C. Pattamaprom; R.G. Larson; T.J. van Dykes *Rheologica Acta.* 2000, 39, 517-531

2.C. Pattamaprom; R.G. Larson *Rheologica Acta.* 2001, 40, 516-532

3.C. Pattamaprom; R.G. Larson; A. Sirivat *Rheologica Acta.* 2008, 47, 689-700

4.R.C. Ball; T.C.B. McLeish *Macromolecules.* 1989, 22, 1911-1913

5.J.L. Viovy; M. Rubinstein, R.H. Colby *Macromolecules* 1991, 24, 3587-3596

6.S.T. Milner, T.C.B. McLeish, R.N. Young, A. Hakiki, J.M. Johnson *Macromolecules* 1998, 31, 9345-9353

7.M. Doi, S.F. Edwards (1986) The Theory of Polymer Dynamics: Clarendon, Oxford.

8.J. Tangpakdee, Y. Tanaka, *J. Rubber Chem. Tech.*1993, 70, 707-713.

9.Ferry JD (1980) Viscoelastic Properties of Polymers. Wiley, New York.

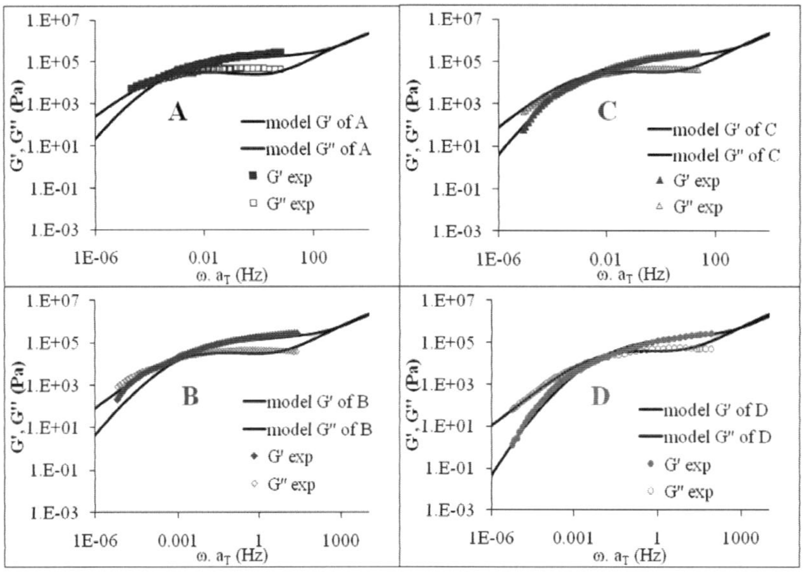

Figure 1: The comparison between the prediction of the dual constraint model and the experimental storage (G') and loss (G") modulii of (a) as-received STR 5L grade natural rubber, and the natural rubber masticated at (b) 40 rpm, (c) 40 rpm, and (d) 100 rpm. The solid lines are the model predictions and the symbols are experimental values.

O-13-433

VISCOELASTIC PROPERTIES OF RUBBER BLENDS FILLED WITH SILICA DEPENDING ON THEIR MORPHOLOGIES

Pauline Grau[1]*, Olivier Sanséau[1] and Didier R. Long [1]

[1]Laboratoire des Polymères et Matériaux Avancés (LPMA), CNRS/Rhodia, member of the Solvay Group
85 avenue des Frères Perret, 69192 Saint Fons, France - pauline.grau-exterieur@eu.rhodia.com

Abstract -

Blending of rubber is often used in tire industry to enhance gum properties by combining characteristics of each component. The addition of reinforcing filler in this kind of blends could lead to an uneven distribution of particles between phases. It has been show that filler distribution, as well as rubber phase morphology, could influence blends properties. In the present work, the structuration of NR/SBR blends filled with silica is investigated. First, we focus on how to characterize rubber blends morphologies, and especially silica distribution. Then, we study the impact of process on blend morphologies, and the relation between these morphologies and viscoelastic properties.

Keywords: Rubber, Blends, Silica, Morphologies, Viscoelasticity, Process

Introduction

The tradeoff in tire tread properties is notably controlled by the dependence in frequency of the rubber dissipation: a high dissipation of energy improves wet grip at high frequencies while it may increase rolling resistance at low frequencies. Using silica instead of carbon black has allowed an improvement of this tradeoff, but the physical mechanisms responsible of these properties are still under discussion (polymer-filler networks [1], glassy layer [2-3] ...). Furthermore, the use of rubber blends is a well-known technique to obtain materials with tailored properties, allowing for example tuning of the dissipation range. These unique properties depend not only on the miscibility state of rubbers, but also on particles distribution between phases [4-6].
In this context, the purpose of this study is to generate various blend morphologies where silica location differs between phases, and to study their impact on blend properties, especially damping ones. By varying interactions parameters between the three components, we could obtain different morphologies. Moreover, it had been shown that final morphologies depend not only on these interaction parameters, but also on systems rheology [7]. Thus, we studied the impact of process and rubber viscosities in order to decorrelate the impact of rheology and interactions on blend structuration.

Experimental

Materials

Various systems based on Natural Rubber (NR) and different Styrene Butadiene Rubber (SBR) are presented. The NR used is a standard one from the Malaysian production, with a controlled Mooney Viscosity of 60MU, that we have to peptize to ease filler incorporation. The SBR, provided by Lanxess, is composed of 25% of styrene and 50% of 1,2 vinyls units (Table 1). We reinforced our blends with 40phr a silica with a small specific surface area, the Z1085 from Rhodia. Because of the low compatibility with diene rubbers and polar silica, interfacial agents are added to the formulations.

Table 1 – Neat rubbers properties and δ calculated from the group contribution method

Materials	**Tg** (°C)	**Mooney Viscosity** (MU 100 1+4)	**δ** (mJ/m²)
NR CV60	-65	40.86	17,52
SBR 5025-0	-21	67	17,42

Process

Two different procedures are used to obtain various morphologies and orientate silica location. The first one is the classic "One Pot" technique, which consists in introducing silica simultaneously in both elastomers. This technique could allow us to study the competition between viscosity of phases and interactions between rubbers and filler on the blend structuration. The second technique, known as "Masterbatch" technique, consists in introducing all the filler into only one rubber before blending it with the second one. Thanks to this process, it could be possible to orientate silica location inside one specific rubber. Blends are elaborated in an internal blender before adding the curing system on two rolls mill, and then vulcanized under press.

Morphologies characterization

Morphologies are first characterized by Electronic Microscopy. The study of pure gum blends shows a distribution of different rubber domains, sign of the immiscibility of these NR/SBR systems Figure 1.a). When adding silica, rubber phases are not distinguishable any longer with the same kind of classic MET observations (Figure 1.b). Thus, we developed a technique based on selective staining to observe silica distribution between phases.

Figure 1 – Microscopic analysis of 50:50 NR/SBR blends : a) pure gum blend, b) filled with 40phr of Silica

Results and Discussion

Impact of process on blends morphologies

Microscopic analysis presented on Figure 2 show that different morphologies are obtained depending on the process. In case of the masterbatch technique, where silica is premixed in one specific rubber, it could be seen that particles are preferentially distributed into this specific phase. Here, rubber high viscosities prevent from silica migration, and allow controlling its location between phases.

Figure 2 – Morphologies analysis of 50/50 NR/SBR blends filled with 40phr of Z1085 Silica & OCTEO, depending on process

Impact of blend structuration on viscoelastic properties

Then we focus on the relation between blends structuration and their viscoelastic behavior with temperature at small deformation (linear domain). These properties are measured by DMA between -80°C and 80°C at 3°C/min. The analyses were performed at a deformation rate of 0.089% and a frequency of 10Hz.

Figure 3 presents the behavior of 50-50 NR/SBR blends where silica location differs. The distinction of two peaks confirms the immiscibility of the blends. The lowest transition peak corresponds to the NR (~ -50°C) while the highest one is the SBR (~ -15°C). This figure also reveals that the morphologies characterized by microscopy for 50/50 NR/SBR blends are in good agreement with viscoelastic properties. Indeed, when silica is distributed in one specific rubber (for example in the NR for the blend based on a NR masterbatch), an increase of its dissipation peak is observed.

Figure 3 – Viscoelastic properties of NR/SBR 50/50 blends filled with 40phr of Z1085&OCTEO, depending on their morphologies

Conclusions

To conclude, this study presents a new technique that allows a fine characterization of NR/SBR rubber blends filled with silica. With this technique, we can clearly see the impact of process on silica distribution. We also observed a direct relation between blend morphologies and the evolution of their loss modulus E''.

Acknowledgements

Ludovic Odoni, Rémi Sellier and Laetitia Tribolet from Rhodia CRTL

Laurent Guy, Aurélie Papon, Benoît Moreaux and Alain Monton from Rhodia Silcea

References

1. Wang M.J.; Rubber Chemistry and Technology ; i 3, V71, p520 (1999)

2. Berriot, J, Montes H, Lequeux F, Long D, and Sotta P ; Macromolecules ; i 23, V35, p9756 (2003)

3. Lequeux, F., Long, D., Sotta, P., and Saalwachter, K.; Angewandte Chemie-International Edition ; i 37, V50, pA63 (2011)

4. Hess, W. M. and Chirico, V. E.; Rubber Chemistry and Technology ; i 2, V50, p301 (1977)

5. Klüppel M., Schuster RH, and Schaper J. Carbon; Rubber Chemistry and Technology ; 72, p91 (1999)

6. Nordsiek, K. H.; Kautschuk Gummi Kunststoffe ; i 3, V38, p178 (1985)

7. Hess, W. M., Herd, C. R., and Vegvari, P. C.; Rubber Chemistry & Technology ; i 3, V66, p329 (1993)

8. Le, H. H., Ilisch, S., Heidenreich, D., Wutzler, A., and Radusch, H. J.; Polymer Composites ; i 10, V31, p1701 (2010)

O-13-440

HIGH-THROUGHPUT SEC ANALYSIS USING SEMI-MICRO COLUMN PACKED WITH MONO-DISPERSE POLYSTYRENE PARTICLE WHICH HAS A BROAD PORE SIZE DISTRIBUTION

H. Tomizawa[1]*, K. Tokunaga[2], T. Tsuda[2] and H. Moriyama[2]
[1]*Bioscience division, Tosoh Corporation, Japan –hiroshi-tomizawa-jp@tosoh.co.jp*
[2] *Separation Center, Tosoh Corporation, Japan*

Abstract - Size exclusion chromatography (SEC) is a convenient and reproducible method to analyze and determine relative molecular mass and molecular mass distribution of polymers, then widely used for R&D and QC in polymer industry. We, Tosoh Corporation have a history to develop SEC columns and equipment over 40 years and have been perusing high-throughput, reproducible and easy-to-use SEC system. One of the key factor for especially high-throughput is column technology and we developed polystyrene/divinylbenzene packing materials with 3 micron as particle size and packed in semi-micro columns (4.6 mm I.D. x 15 cm) few years ago. The columns achieved high-throughput analysis for half measurement time of conventional column (30 cm length) with the same or better resolution. Recently we employed multi-pore size distribution technology on polystyrene/divinylbenzene packing materials with 3 - 6 micron as particle size and packed in semi-micro columns (4.6 mm I.D. x 15 cm) and commercialized under the brand name of TSKgel SuperMultiporeHZ columns. This newly employed technology enables wide separation range of polystyrene calibration curves in tetrahydrofuran with good linearity on identical packing materials and no distortion on the chromatograms which are sometimes observed as mixed-bed type columns or series of individual columns are used.
Using this column reproducible SEC was achieved in addition to high-throughput and high resolution results.

Keywords: Size exclusion chromatography; Molecular mass; High-throughput analysis; Semi-micro SEC column.

Introduction

Size exclusion chromatography (SEC) is an analytical method for determination of molecular mass and molecular mass distribution of polymers, and widely used in various fields since it's a simple and precise method. Molecular mass determined by SEC is relative molecular mass, which is calculated by calibration curve which consists of retention time and molecular mass of reference standard. Therefore some problems have been pointed out that accuracy and precision of SEC measurement are affected by following facts; error between approximate calibration curve and real calibration curve, distorted chromatogram of polymer sample is observed when columns with different pore size are used in series. In addition, long measurement time has been problem of SEC for long time.
To overcome these problems, novel semi-micro SEC columns were developed and commercialized by Tosoh recently. The polystyrene/divinylbenzene packings which have broad pore size distribution in identical particle with small particle size, 3 - 6 um, are packed in semi-micro column, 4.6 mm I.D. x 15 cm and commercialized as 'TSKgel SuperMultiporeHZ' columns [1, 2].
Here we report the evaluation result of the columns for basic properties and some of applications to synthetic polymers.

Experimental

Instrumentation

SEC was carried out using HLC-8320GPC (EcoSEC), all-in-one equipment which consists of degass unit, dual pump, auto-sampler, column oven and refractive index (RI) detector (Tosoh, Japan). UV detector can be installed optionally (Tosoh, Japan).

Chemicals and Materials

Stabilized tetrahydrofuran (analytical grade) was used as the mobile phase without further treatment (Wako, Osaka). Synthetic polymers and other reference standards were obtained from Wako and other resin suppliers. The polystyrene narrow-molecular-mass distribution standards, PStQuick, were obtained from Tosoh (Tokyo, Japan).

SEC columns

TSKgel SuperMultiporeHZ (4.6 mm I.D. x 15 cm), TSKgel MultiporeHXL-M (7.8 mm I.D. x 30 cm) [3], TSKgel SuperHZ (4.6 mm I.D. x 15 cm) [4], TSKgel HXL (7.8 mm I.D. x 30 cm) were obtained from Tosoh (Tokyo, Japan).

Preparation of sample solutions

Synthetic polymers were dissolved in THF at concentrations of 3 g/L. Polystyrene standards were dissolved at concentrations of 0.2-1.0 g/L. Polystyrene standards were, occasionally, gently stirred for a period of 12-24 hrs prior to use.

Other SEC conditions

THF was used as mobile phase. Flow rates are 0.35 mL/min for semi-micro columns (4.6 mm I.D.) and 1.0 mL/min for conventional columns (7.8 mm I.D.). Detection was done by RI or UV 254 nm. Temperature of pumping unit, column oven and detector flow cell was controlled at 40 C.

Results and Discussion

Basic properties of TSKgel SuperMultiporeHZ columns

Physical properties of TSKgel SuperMultiporeHZ columns are shown in Table 1. The columns consist of three grades of columns which packing has different average pore size/exclusion limit as molecular mass, respectively, H grade has highest exclusion limit followed by M grade and N grade has lowest exclusion limit. Separation ranges of each column are $10^7 - 10^3$ for H grade, $10^6 - 5 \times 10^2$ for M grade and $5 \times 10^4 - 3 \times 10^2$ for N grade. Thus SuperMultiporeHZ columns have wide separation range and are applicable to various polymers with wide range of molecular mass. Fig. 1 shows SEM photograph of SuperMultiporeHZ-M column. The packing was found to be spherical mono-disperse particle, which has narrow particle size distribution.

The calibration curves of SuperMultiporeHZ columns are shown in Fig. 2. Each calibration curve was found to have different separation range as described in Table 1 and to show good linearity of >0.999 over separation range, respectively.

Fig. 3 shows errors of nominal weight average molecular mass (nominal Mw) of polystyrene std. from approximate calibration curve respectively. Compared to individual column system of TSKgel SuperHZ, which consist of columns with different individual pore size, SuperMultiporeHZ columns show smaller errors on average. This smaller error would contribute to more precise SEC measurement.

Resolution of TSKgel SuperMultiporeHZ as semi-micro SEC column

Chromatograms of polystyrene std. mixture on SuperMultiporeHZ-M columns, 4.6 mm I.D. x 15 cm x 2 were compared to TSKgel MultiporeHXL-M columns, 7.8 mm I.D. x 30 cm x 2. Both columns are packed by multipore-type packings with the same average pore size, but different particle size; 4 um for SuperMultiporeHZ-M, 6 um for MultiporeHXL-M. Moreover, former one has narrower particle size distribution, mono-disperse particle. Therefore, SuperMultiproeHZ-M as semi-micro columns were found to show the same or better resolution with half of analysis time compared to MultiporeHXL-M as conventional column, also solvent consumption of semi-micro column was 1/6th of conventional one.

Distortion of chromatogram of polymers with polydispersity

When polymers with polydispersity are measured by conventional column system, i.e. mixed-bed type columns in which packings with different pore size are packed or individual column system where series of columns, each of which is packed by packings with different pore size, distortion of chromatogram would be observed sometimes. The phenomena are caused by inflection points on total calibration curves which come from gap between calibration curves of different packings as pore size or from difference in slope between calibration curves of different packings. On the other hand, multipore-type packings have linear and seamless calibration curves. Therefore, smooth chromatograms of polymers would be observed on multipore-type columns. Fig. 5 shows comparison of chromatograms of acrylic resin between multipore-type column, SuperMultiporeHZ-M and individual columns, SuperHZ. Three distortions in red circles were observed in the chromatogram on SuperHZ columns where four individual columns with different pore size were connected in series, while smooth chromatogram was observed on SuperMultiporeHZ columns.

Precision of SEC measurement

Polystyrene and polycarbonate, which both have polydispersity were measured on multipore-type column and conventional columns, respectively and evaluated repeatability of Mw. The results are summarized in Table 2. TSKgel SuperMultiporeHZ-M column showed better repeatability for both polymers than conventional columns.

Applications to various polymers

TSKgel SuperMultiporeHZ-H column was applied to various polymers and chromatograms are shown in Fig. 6. Smooth chromatograms were obtained for all polymers measured.

Table 1 Physical properties of SuperMultiporeHZ columns.

Item	Properties		
	H grade	**M grade**	**N grade**
Packing material	Polystyrene/divinylbenzene		
Particle diameter (um)	6	4	3
	Mono-disperse		
Exclusion limit Mw (polystyrene)	-	2.0×10^6	1.2×10^5
Average pore diameter (nm)	-	14	8
Separation range (polystyrene)	$1.0 \times 10^7 - 1.0 \times 10^3$	$1.0 \times 10^6 - 5.0 \times 10^2$	$5.0 \times 10^4 - 3.0 \times 10^2$
Theoretical plates/15cm	11,000	16,000	20,000
Column size (mm I.D. x cm)	4.6 x 15		

Table 2 Repeatability of SEC measurement (Mw) for various polymers.

Column	Polystyrene		Polycarbonate	
	Ave. Mw	RSD (%)	Ave. Mw	RSD (%)
SuperMultipore	254,000	0.43	44,600	0.51
Individual	243,000	0.79	47,000	1.83
Mixed-bed	249,000	0.67	48,200	0.72

(n=5)

SuperMultipore; TSKgel SuperMultiporeHZ-M, 4.6 mm I.D. x 15 cm x 4, detection; RI, sample load; 3 g/L aech, 10 uL.

Figure 1 SEM photograph of SuperMultiporeHZ-M.

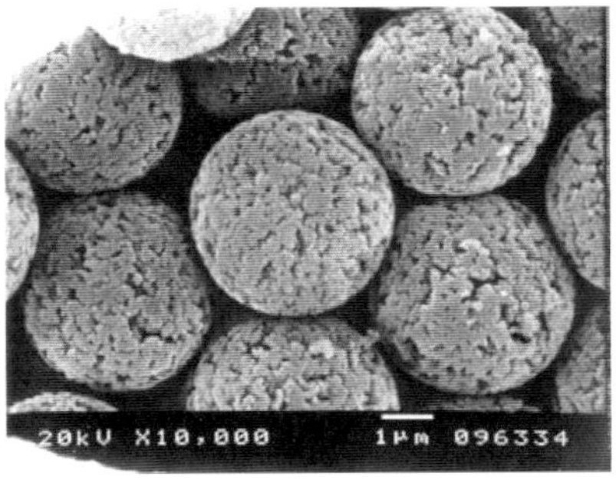

Figure 2 Polystyrene std. calibration curves of TSKgel SuperMultiporeHZ columns.
Column size; 4.6 mm I.D. x 15 cm, detection; UV 254 nm, column temp.; 25 C, sample; PStQuick, inj. volume; 5 uL.

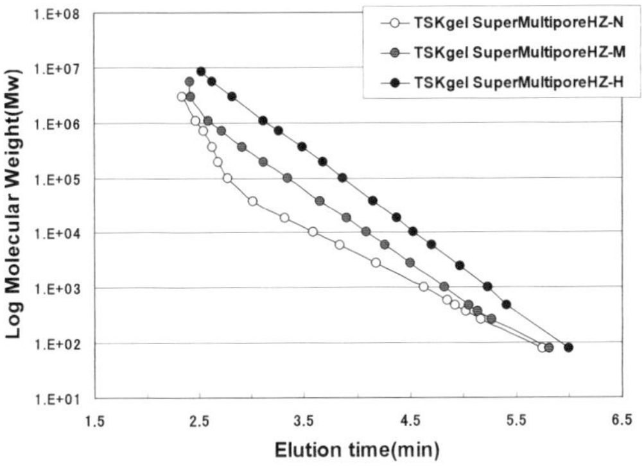

Column	Linearity	Slope
SuperMultiporeHZ-H	0.9995	-4.131
SuperMultiporeHZ-M	0.9998	-3.820
SuperMultiporeHZ-N	0.9996	-2.813

Figure 3 Errors of nominal Mw of polystyrene std. from calibration curve.
Column size for SuperHZ; 4.6 mm I.D. x 15 cm x 4, other conditions are the same in Fig.2.

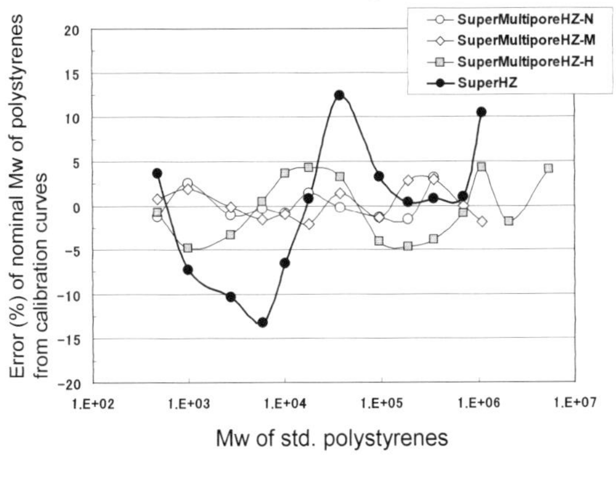

Figure 4 Comparison of chromatograms of polystyrene std. on semi-micro column (SuperMultiporeHZ-M) and conventional column (MultiporeHxl-M).
Column size; 7.8 mm I.D. x 30 cm x 2 (A), 4.6 mm I.D. x 15 cm x 2 (B), detection; RI, sample; PStQuick, inj. volume; 50 uL (A), 10 uL (B).

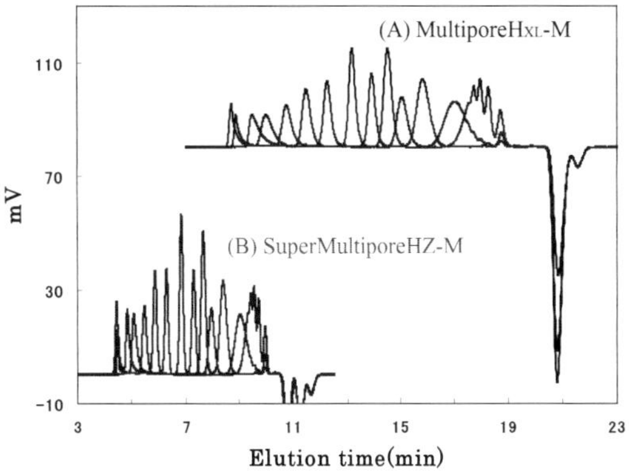

Figure 5 Comparison of chromatograms of acrylic resin on multipore-type column and individual column.
Column size; 4.6 mm I.D. x 15 cm x 4, detection; RI, sample; acrylic resin, 3 g/L, 10 uL.

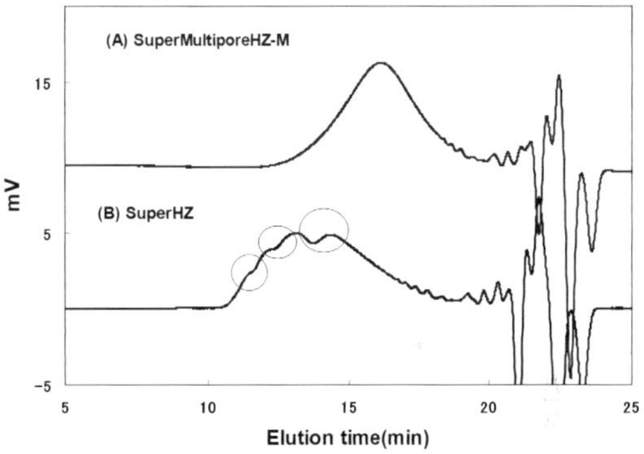

Figure 6 Chromatograms of various synthetic polymers on TSKgel SuperMultiporeHZ-H column.
Column size; 4.6 mm I.D. x 15 cm x 4, sample load; 3 g/L, 10 uL, samples; 1. polyisobutyrene, 2. acrylic resin-1, 3. polystyrene (SRM706), 4. polybutadiene, 5. PVB, 6. polycarbonate, 7. epoxy resin, 8. acrylic resin-2.

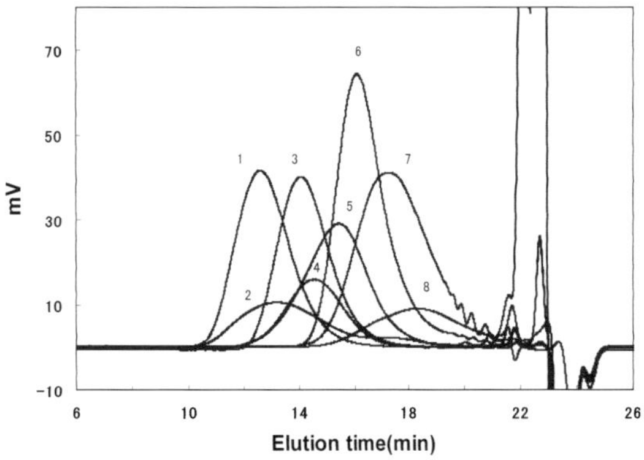

Conclusions

TSKgel SuperMultiporeHZ columns demonstrated a wide separation range according to their pore size, respectively, with a good linearity of the calibration curves for standard polystyrenes in THF as mobile phase.

Various polymers were analyzed on the SuperMultiporeHZ columns. It was confirmed that chromatograms were obtained in half the time required for conventional columns with the same or better resolution.

Solvent consumption was only 1/6th of that for conventional SEC columns.

Chromatograms of various polymers which have broad molecular mass distribution on SuperMultiporeHZ column were found to be very smooth without any distortion, although distortion points were observed in the chromatograms of other mixed-bed and individual columns.

References

1. K. Tokunaga, T. Tsuda, H. Moriyama, *TOSOH Res&Technol. Rev.*, 51 47 - 56 (2007).
2. *Separation Report.* Tosoh, No. 105 (2008).
3. M. Nagata, T. Kato, H. Furutani, *J. Liq. Chrom&Rel. Technol.*, 21 (10) 1471-1484 (1998).
4. *Separation Report.* Tosoh, No. 101 (2002).

MORPHOLOGY DEVELOPMENT IN FILLED NATURAL RUBBER BLENDS

Hai Hong Le[1,*], Sybill Ilisch[1,2], Gert Heinrich[3,4], Hans-Joachim Radusch[1]
[1]*Center of Engineering Sciences, Martin Luther University Halle-Wittenberg D-06099 Halle (Saale), Germany*
[2] *now: Styron Deutschland GmbH, Germany*
[3]*Leibniz-Institut für Polymerforschung Dresden e.V. (IPF), D-01069 Dresden, Germany*
[4]*Institut für Werkstoffwissenschaft, Technische Universität Dresden, D-01069 Dresden, Germany*

Abstract - The development of nanofiller localization (silica, carbon black, carbon nanotubes, nanoclay) and blend morphology of rubber blends consisting of natural rubber and different synthetic rubbers during mixing process was described comprehensively by combination of our developed methods. The dispersion kinetics of several fillers along the mixing time could be well characterized by the *method of online measured electrical conductance*. The phase selective filler localization in binary and ternary rubber blends as function of mixing time was experimentally determined by means of the *wetting concept*. For a quantitative prediction of filler localization in rubber blends the so-called *Z-model* based on thermodynamic data was proposed. Compared to synthetic blend partners like SBR, BR or EPDM, NR exhibits a high filler wetting speed and low filler-rubber interaction that leads to a complex behavior of filler localization in natural rubber blends. The effect of material and technological parameters on the localization kinetics of different fillers and its effect on the mechanical properties of blends are the focus of the present work.

Keywords: Rubber composites, rubber blends, filler localization, filler transfer, filler wetting.

Introduction

Tire tread compounds have been prepared from various rubber blends of two or more rubbers, such as styrene butadiene rubber (SBR), natural rubber (NR) and polybutadiene rubber (BR) and fillers like carbon black (CB) or/and silica. In passenger car sidewall mixtures NR is an essential blend partner. Especially, for truck tires the presence of NR is usually desired to obtain high mechanical performance, low heat build-up, good wear resistance, high flexibility at low temperature and high adhesion to steel cord. It is recognized that the desirable tire properties are not only dependent on the characteristics of rubber matrix and cross-linking agent but also on the type and loading of used filler as well as its dispersion and localization state in rubber blends. Assuming that filler transfer during masterbatch mixing is of minor importance, Hess et al. [1-3] demonstrated exemplarily the effect of the filler localization on hysteresis, tensile strength, fatigue and wear behavior of NR/BR and SBR/BR tread compounds. Waddell and Massie et al. [4,5] reported that the crack growth resistance of NR/BR blends with CB mainly dispersed in the NR phase has been significantly improved. Dynamic mechanical properties, hardness and fracture were found to be dependent on the filler localization in blends as well [6-8]. For the characterization of filler localization in low-filled blends microscopy methods like transmission electron microscopy (TEM) [1,9] and atomic force microscopy (AFM) [10-13] have been used. These methods are less suitable for highly filled rubber blends. In this regards several methods based on electrical resistivity measurements [14-17], nuclear magnetic resonance (NMR) [18,19], differential scanning calorimetry (DSC) [20,21], thermo-gravimetric analysis (TGA) [22,23] and dynamic mechanical analysis (DMA) [24-26] as well as *wetting concept* [7,27-30] have been developed for experimental determination of filler localization in highly filled rubber blends. Concerning the mechanisms of localization of filler in polymer blends different concepts have been also described in literature. Some authors attributed the observed particle arrangements to melt viscosity effects [31-33], but most frequently it is accepted that the phase selective localization of filler in rubber blends is governed by the polymer-filler interaction, which is in turn influenced by the nature of rubber components and filler [34-38]. In our previous works [7,27-30] it was found that the kinetics of filler localization is determined by the wetting process of filler by the matrix components in the early mixing period and in the subsequent period the rubber-filler interaction is essential for the re-localization of filler. Compared to other rubbers like SBR, BR or EPDM natural rubber can wet filler much faster on the one hand, i.e. more filler will be localized in the NR phase in the early state of mixing. However, the weak interaction between filler and NR surface due to its non-polar nature makes NR not favorable as a host for filler localization on the other hand. The competition between high wetting speed and low rubber-filler interaction leads to a complex behavior of filler localization in the rubber blends based on NR. In the present work the so-called *Z-model* was further developed for better understanding the wetting mechanism of filler by different rubber types. It was used for discussion of the tendency of filler transfer process during mixing. The influence of processing and material parameters on the wetting kinetics and localization behavior of different fillers in binary and ternary rubber blends has been the main focus of the present work.

Theoretical prediction of filler localization in rubber blends at an equilibrium state using the *Z-model*

The thermodynamic criterion for miscibility of polymer blends is a negative Gibbs free energy of mixing. The Gibbs free energy of mixing is given by the following equation:

$$\Delta G_m = \Delta H_m - T\Delta S_m \qquad (1)$$

where ΔH_m is the change in enthalpy of mixing, and ΔS_m the change in entropy of mixing. T is the absolute temperature. The change in enthalpy during mixing is the result of dipole interactions, van der Waals interactions, acid–base interactions, Coulombic interactions and the interaction energy between the different components arising from hydrogen bonds. Due to the constraints of segmental mobility of polymer chains the change in entropy is usually too small to compensate the change of mixing enthalpy. Thus, the Gibbs free energy of mixing is nearly similar to the change in enthalpy of mixing:

$$\Delta G_m \approx \Delta H_m \qquad (2)$$

Mixing a phase B into a phase A, the relationship between the change in enthalpy of mixing ΔH_m^{AB} of the phases A and B, and the interfacial tension γ_{AB} between two phases can be described according to Hildebrand and Scott by Eq. 3 [39]:

$$\frac{\Delta H_m^{AB}}{V_{AB}} = K\Phi_A\Phi_B(\gamma_{AB})^2 \qquad (3)$$

where V_{AB} is the average molar volume of the two components, Φ_A and Φ_B are the volume fractions of the components A and B, respectively, in the binary system AB. K is a constant and gets a value of 1 according to Scatchard [40,41]. From this equation it can be seen that blend with high interfacial tension γ_{AB} requires more energy for dispersion. If the interfacial tension is small enough, a potential for molecular miscibility might exist [42].

Mixing a component F to the system AB, we obtain a ternary system AFB consisting of two systems AF and BF. At the interface of AB the component F migrates between AF and BF in order to minimize the values of ΔH_m^{AF} and ΔH_m^{BF}. The filler transfer process goes on until the thermodynamic equilibrium state is reached. A thermodynamic equilibrium state is reached if

$$\frac{\Delta H_m^{AF}}{V_{AF}} = \frac{\Delta H_m^{BF}}{V_{BF}} \qquad (4)$$

According to Eq. 3 we can rewrite Eq. 4 as follows:

$$\frac{\Phi_A}{\Phi_A+\Phi_F^A}\frac{\Phi_F^A}{\Phi_A+\Phi_F^A}(\gamma_{AF})^2 = \frac{\Phi_B}{\Phi_B+\Phi_F^B}\frac{\Phi_F^B}{\Phi_B+\Phi_F^B}(\gamma_{BF})^2 \qquad (5)$$

where Φ_A and Φ_B are the volume fractions of the components A and B, respectively. Φ_F is the effective volume of the filler F in the ternary system AFB. Φ^A_F and Φ^B_F are the effective filler volume fraction of F in A and B at an equilibrium state with $\Phi_F = \Phi^A_F + \Phi^B_F$. γ_{AF} and γ_{BF} are the interfacial tension values between A and F as well B and F, respectively. Eq. 5 can be rewritten to:

$$\frac{\Phi_F^A}{\Phi_F^B} = \frac{\Phi_B}{\Phi_A}\left(\frac{\Phi_A+\Phi_F^A}{\Phi_B+\Phi_F^B}\right)^2\left(\frac{\gamma_{BF}}{\gamma_{AF}}\right)^2 \qquad (6)$$

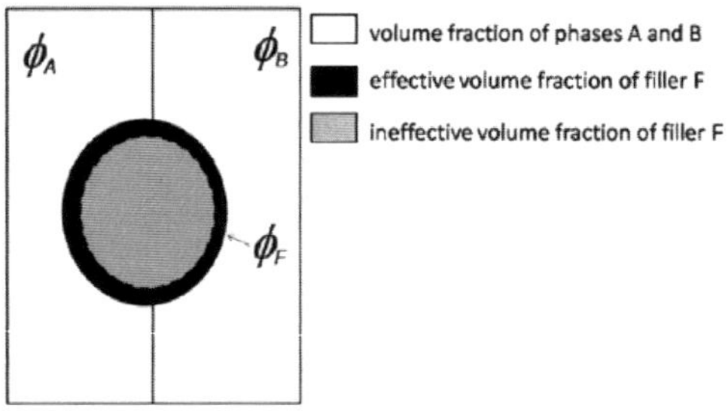

Figure 1. Filler localization in system AB

It is worthy of note that Eq. 3 was developed for a binary system of two polymers or solvents. As mentioned above, a molecular miscibility can be observed for this polymer-polymer system, if the interfacial tension becomes zero. However, in a polymer-filler system polymer molecules can get contact with filler only at its surface. The inner region (grey area, Figure 1) is considered as ineffective volume fraction of filler and it is not taken into consideration in Eq. 3. The effective volume fraction Φ_F (black area) has a thickness d, which corresponds to a diameter of a Kuhn segment of 0.5 nm [43]. For silica or CB with an average particle diameter of about 15 nm, the ratio of the effective volume fraction Φ_F to the total filler volume is 0.1. In a tire tread compound the total filler volume of F is usually in the range of 20 to 25 vol%. As a result, the effective filler volume Φ_F is 2 vol% and it is very small compared to that of the host blend phases A and B. Thus, we can neglect Φ^A_F and Φ^B_F from the term $(\Phi_A + \Phi^A_F)^2/(\Phi_B + \Phi^B_F)^2$ from Eq. 6 for simplification purpose. Eq. 6 can be expressed now as follows:

$$\frac{\varphi_F^A}{\varphi_F^B} = \frac{\Phi_F^A}{\Phi_F^B} \approx \frac{\Phi_A}{\Phi_B}\left(\frac{\gamma_{BF}}{\gamma_{AF}}\right)^2 = n_{A/B}\left(\frac{\gamma_{BF}}{\gamma_{AF}}\right)^2 \qquad (7)$$

φ^A_F and φ^B_F are the weight fractions of the component F in the phases A and B, respectively. $n_{A/B}$ is the blend ratio A to B. Using the Girifalco-Good equation [44] describing the relationship between surface tension and interfacial tension Eq. 26.8 can be written as follows:

$$\frac{\varphi_F^A}{\varphi_F^B} = n_{A/B}\left(\frac{\gamma_B + \gamma_F - 2\sqrt{\gamma_B\gamma_F}}{\gamma_A + \gamma_F - 2\sqrt{\gamma_A\gamma_F}}\right)^2 \quad (8)$$

γ_A, γ_B and γ_F are the surface tension values of the phases *A, B* and *F*, respectively. Setting $\varphi^B_F = 1 - \varphi^A_F$ into Eq. 8 the weight fraction φ^A_F can be calculated using Eq. 9.

$$\varphi_F^A = \frac{n_{A/B}\omega}{n_{A/B}\omega + 1} \quad (9)$$

with

$$\omega = \left(\frac{\gamma_B + \gamma_F - 2\sqrt{\gamma_B\gamma_F}}{\gamma_A + \gamma_F - 2\sqrt{\gamma_A\gamma_F}}\right)^2 \quad (10)$$

The validity of the *Z-model* was proved for different filled thermoplastic blends reported in literature [45]. It was also extended to ternary rubber blends in our previous work [30]. For example, the filler localization in SBR/BR/NR blends can be determined according to Eqs. 11 - 13.

$$\frac{\varphi_F^{BR}}{\varphi_F^{SBR}} = n_{BR/SBR}\left(\frac{\gamma_{SBR} + \gamma_F - 2\sqrt{\gamma_{SBR}\gamma_F}}{\gamma_{BR} + \gamma_F - 2\sqrt{\gamma_{BR}\gamma_F}}\right)^2 \quad (11)$$

$$\frac{\varphi_F^{BR}}{\varphi_F^{NR}} = n_{BR/NR}\left(\frac{\gamma_{NR} + \gamma_F - 2\sqrt{\gamma_{NR}\gamma_F}}{\gamma_{BR} + \gamma_F - 2\sqrt{\gamma_{BR}\gamma_F}}\right)^2 \quad (12)$$

$$\varphi_F^{SBR} + \varphi_F^{BR} + \varphi_F^{NR} = 1 \quad (13)$$

$n_{BR/SBR}$ and $n_{BR/NR}$ are the blend ratios of BR to SBR and BR to NR, respectively. γ_{SBR}, γ_{BR} γ_{NR} are the values of surface tension of SBR, BR and NR phase. By taking into consideration Eq. 13 the filler fraction φ_F^{SBR}, φ_F^{BR} and φ_F^{NR} localized in the SBR, BR and NR phase, respectively, can be calculated using Eqs. 14 - 18.

$$\varphi_F^{BR} = \frac{1}{1 + \frac{1}{n_{BR/SBR}\omega_{BR/SBR}} + \frac{1}{n_{BR/NR}\omega_{BR/NR}}} \quad (14)$$

$$\varphi_F^{SBR} = \frac{\varphi_F^{BR}}{n_{BR/SBR}\omega_{BR/SBR}} \quad (15)$$

$$\varphi_F^{NR} = \frac{\varphi_F^{BR}}{n_{BR/NR}\omega_{BR/NR}} \quad (16)$$

$$\omega_{BR/SBR} = \left(\frac{\gamma_{SBR} + \gamma_F - 2\sqrt{\gamma_{SBR}\gamma_F}}{\gamma_{BR} + \gamma_F - 2\sqrt{\gamma_{BR}\gamma_F}}\right)^2 \quad (17)$$

and

$$\omega_{BR/NR} = \left(\frac{\gamma_{NR} + \gamma_F - 2\sqrt{\gamma_{NR}\gamma_F}}{\gamma_{BR} + \gamma_F - 2\sqrt{\gamma_{BR}\gamma_F}}\right)^2 \quad (18)$$

Conclusions

In the present work, a comprehensive description of filler localization kinetics and development of blend morphology in rubber blends containing natural rubber could be done by combination of the *method of online measured electrical conductance* and the *wetting concept* as well as the *Z-model*. The development of dispersion of different fillers along the mixing time could be well characterized by the online conductance chart. The phase selective filler localization in binary and ternary rubber blends as a function of mixing time was experimentally determined by means of the *wetting concept*. For a quantitative prediction of filler localization in rubber blends a so-called *Z-model* based on thermodynamic data was proposed. Using this model, a master curve presenting a characteristic dependence of filler localization on the surface tension values of blend components was created. Compared to the synthetic blend partner like SBR, BR or EPDM natural rubber exhibits a high filler wetting speed and low filler-rubber interaction, which influence the kinetics of filler localization in rubber blends containing NR in an opposite manner. During the first mixing stage, filler localization is strongly affected by the high wetting speed of NR and consequently more filler is localized in the NR phase. In the subsequent mixing state after the wetting process is completed, the rubber filler interaction dominantly influences the localization kinetics. As a result, a re-localization of filler within blend phases takes place until the thermodynamic equilibrium state predicted by the *Z-model* is reached. It was found that the equilibrium state of silica filled blends can be reached within the experimental processing, while the filler localization of CB filled blends determined at the end of the mixing process is still far from the equilibrium state. That is related to the much higher transfer speed of silica compared to that of CB. Further investigation should be conducted for better understanding of this behavior. The strong effect of curing and processing additives as well as silane on the silica localization is sufficiently explained by the *Z-model* by taking into consideration the deactivation of the silanol groups on the silica surface by adsorbed additives and silane. Using the master curve the surface tension of filler affected by addition of additives and silane can be estimated, that may be useful for evaluation and comparison of the effect of different coupling agents.

Acknowledgements

The authors would like to thank the German Research Foundation (DFG) for the financial support.

References

1. W. M Hess, V. E. Chirico, *Rubber Chem. Technol.*, 1977, **50**, 301.
2. W. M. Hess, C. R. Herd, P. C. Vegvari, *Rubber Chem. Technol.*, 1993, **66,** 329.
3. W. M. Hess, *Rubber Chem. Technol.*, 1991, **64**, 386.
4. H. W. Waddell, *Rubber Chem. Technol.*, 1998, **71**, 590.
5. J. Massie, R. Hirst, A. Halasa, *Rubber Chem. Technol.*, 1993, **66**, 276.
6. J. Ziegler, R. H. Schuster, *Kautsch. Gummi Kunstst.*, 2003, **56**, 159.
7. H. H. Le, S. Ilisch, G. R. Kasaliwal, H.-J. Radusch, *Kautsch. Gummi Kunstst.*, 2007, **60**, 241.
8. G. Meier, M. Klüppel, H. Geisler, R. H. Schuster, *Kautsch. Gummi Kunstst.*, 2005, **58**, 587.
9. M. H. Walters, D. N. Keyte, *Rubber Chem. Technol.*, 1965, **38**, 62.
10. V. Herrmann, K. Unseld, H. B. Fuchs, *Kautsch. Gummi Kunstst.*, 2001, **54**, 453.
11. A. H. Tsou, W. H. Waddell, *Kautsch. Gummi Kunstst.*, 2002, **55**, 382.
12. I. H. Jeon, H. Kim, S. S. G. Kim, *Rubber Chem. Technol.*, 2003, **76**, 1.
13. C. C. Wang, J. B. Donnet, T. K.Wang, *Rubber Chem. Technol.*, 2005, **78**, 17.
14. B. G. Soares, F. Gubbels, R. Jerome, *Rubber Chem. Technol.*, 1997, **70**, 60.
15. B. G. Soares, F. Gubbels, R. Jerome, E. Vanlathem, R. Deltour, S. Blacher, F. Brouers, *Chem. Mater.*, 1998, **10**, 1227.
16. F. Gubbels, R. Jerome, P. H. Teyssib, E. Vanlathem, R. Deltour, A. Calderone, V. Parentb, J. L. Bredas, *Macromolecules,* 1994, **27**, 1972.
17. H. H. Le, Z. Qamer, S. Ilisch, H.-J. Radusch, *Rubber Chem. Technol.*, 2006, **79**, 621.
18. W. Hu, M. D. Ellul, A. H. Tsou, S. Datta, *Rubber Chem. Technol.*, 2007, **80**, 1.
19. M. Kotani, H. Dohi, H. Kimura, K. Muraoka, H. Kaji, *Macromolecules,* 2007, **40**, 9451.
20. A. Sircar, T. G. Lamond, *Rubber Chem. Technol.*, 1973, **46**, 178.
21. J. M. Massie, R. C. Hirst, A. F. Halasa, *Rubber Chem. Technol.*, 1992, **66**, 276.
22. G. R. Cotton, L. J. Murphy, *Kautsch. Gummi Kunstst.*, 1988, **41**, 54.
23. C. D. Woolard, B. J. McFadzean, Proceedings of the 28th Annual Conference on Thermal Analysis and Application, 2000 Orlando.
24. S. Maiti, S. K. De, A. K. Bhowmick, *Rubber Chem. Technol.*, 1992, **65**, 293.
25. M. Klüppel, R. H. Schuster, J. Schaper, *Gummi Fasern Kunstst.*, 1998, **51**, 508.
26. M. Klüppel, R. H. Schuster, J. Schaper, *Rubber Chem. Technol.*, 1999, **72**, 91.
27. H. H. Le, S. Ilisch, H.-J. Radusch , *Rubber Chem. Technol.*, 2008, **81**, 767.
28. H. H. Le, S. Ilisch, D. Heidenreich, A. Wutzler, H.-J. Radusch, *Polym. Comp.*, 2010, **31**, 1701.
29. H. H. Le, K. Osswald, S. Ilisch, D. Heidenreich, H.-J. Radusch, *Rubber Chem. Technol.*, 2011,**84**, 41.
30. H. H. Le, K. Oßwald, S. Ilisch, T. Pham, K. W. Stöckelhuber, G. Heinrich, H.-J. Radusch, *Macromol. Mater. Eng.* 2012, **297**, 464.
31. J. Y. Feng, C. M. Chan, J. X. Li, *Polym. Eng. Sci.*, 2003, **43**, 1058.
32. A. L. Persson, H. Bertilsson, *Polymer,* 1998, **39**, 5633.
33. R. Ibarra-Gomez, A. Marquez, L Valle, O. S. Rodriguez-Fernandez, *Rubber Chem. Technol.*, 2003, **76**, 969.
34. M Sumita, K. Sakata, S. Asai, K. Miyasaka, H. Nakagawa, *Polym. Bull.*, 1991, **25**, 265.
35. M. Sumita, K. Sakata, Y. Hayakawa, S. Asai, K. Miyasaka, M. Tanemura, *Colloid Polym. Sci.*, 1992, **270**, 134.
36. S. Wu, *Polymer Interface and Adhesion*, Marcel Decker, New York 1982.
37. L. Elias, F. Fenouillot, J. T. Majeste, G. Martin, G. Cassagnau, *J. Polym. Sci.: Part B: Polym. Phys.*, 2008, **46**, 1976.
38. F. Fenouillot, P. Cassagnau, J. C. Majeste, *Polymer*, 2009, **50**, 1333.
39. J. H. Hildebrand, R. L. Scott, *The Solubility of Nonelecrolytes*, Dover Publications, New York, 1964.
40. G. Scatchard, *Chem. Rev.*, 1931, **8**, 321.
41. G. Scatchard G, *Chem. Rev.*, 1949, **44**, 7.
42. D. R. Paul, S. Newman, *Polymer Blends*, Academic Press, New York, 1978.
43. N. Fatkullin, R. Kimmich, E. Fischer, C Mattea, U. Beginn, M. Kroutieva, *New Journal of Physics* 2004, **6(46),** DOI: 10.1088/1367-2630/6/1/046
44. L. A. Girifalco, R. J. Good, J. Phys. Chem., 1957, **61**, 904.
45. H.H. Le, K. Oßwald, M. Keller, S. Ilisch, T. Pham, H.-J. Radusch, *14. Problemseminar: Polymerblends*, Sep 14th -15th, 2011, Halle, Proceeding and CD p. 48, ISBN 978-3-86829-391-3

O-13-456

INFLUENCE OF AROMATIC CONTENT IN LOW-POLYAROMATIC HYDROCARBON OILS ON THE VISCOELASTIC BEHAVIOUR OF STYRENE-BUTADIENE RUBBER FOR TYRE TREAD APPLICATIONS

Y. Chokanandsombat[1]*and C. Sirisinha[1, 2]

[1]*Department of Chemistry and Center of Excellence for Innovation in Chemistry, Faculty of Science, Mahidol University, Bangkok, Thailand – yotwadee.cho@student.mahidol.ac.th*

[2] *Research and Development Centre for Thai Rubber Industry (RDCTRI), Faculty of Science, Mahidol University, Salaya Campus, Nakhon Pathom, Thailand – chakrit.sir@mahidol.ac.th*

Abstract - Nowadays, the most importance environmental challenge with regard to tyres is the reduction of tyre wear containing harmful compounds such as poly-aromatic hydrocarbons (PAHs) and polycyclic aromatic compounds (PCAs). These substances are identified as suspected carcinogens. As a consequence, in order to achieve both environmental friendliness and effective rubber compounding, the aromatic content in rubber processing oils (RPOs) must be optimised. In this work, an extensive study, focusing on the way that aromatic content in non-carcinogenic oils affects the properties of styrene butadiene rubbers (SBRs) compounds and vulcanisates, has been carried out. Results obtained indicate that the presence of RPOs leads to decreased compound viscosity, and thus enhanced processability. By adding an appropriate amount of RPOs into SBR compounds, mechanical properties including tensile strength, elongation at break and tear strength of SBR vulcanisates can be improved. Such improvement is more pronounced when using the RPO with high aromatic content, judged to be due to an increase in compatibility with the SBR matrix. Nonetheless, the influence of aromatic content on the crosslink density and viscosity of rubber compounds as well as hardness of vulcanisates is not significant.

Keywords: Rubber process oil; RPO; DAE; TDAE; Processing; Mechanical properties

Introduction

Processing oils are organic substances which are usually added to polymers for improving processability of such polymers. The further advantages of the processing oils are increments in desirable polymer properties, such as the softness, elongation and low-temperature flexibility. Furthermore, the processing oils also provide polymer compounds with lower intermolecular forces, and so lower glass transition temperatures (Tg) [1, 2]. The distillate aromatic extracts (DAEs) with high contents of poly-aromatic hydrocarbons (PAHs) and polycyclic aromatic compounds (PCAs) are widely used as rubber processing oils (RPOs) for enhancing processability of synthetic rubbers and natural rubber (NR) [3].

In 1994, the Swedish National Chemicals Inspectorate, KEMI, published a report which initiated the discussion about the use of the high aromatic extender oils in tyre treads. The oils with high content of PAH and PCA, which are identified as suspected carcinogens, caused more concern [4]. As a result, in 2005, the European Commission adopted the directive (EU Substance Directive 67/548/EEC). This legislation restricts the marketing and usability of certain PAHs and PCAs content in extender oils for tyre manufacture.

However, high aromatic extender oils still play a significant role in the fabrication of elastomers, especially those with aromatic groups such as styrene butadiene rubbers (SBRs) widely used in tyre manufacturing [5, 6, 7]. In this research an extensive study has been carried out on the effects of aromatic content (in non-carcinogenic oils) on styrene butadiene rubber (SBR) compounds. Cure characteristics, rheological properties and mechanical properties of SBR with these oils have been determined.

Experimental

Materials

All materials used in this work were used as-received. Styrene butadiene rubber (SBR 1502; styrene content of 23.5%) was manufactured by Bangkok Synthetic Co., Ltd. (Rayong, Thailand). Carbon Black (N330) as reinforcing filler was supplied by Thai Carbon Product Co., Ltd. (Bangkok, Thailand). Stearic acid was purchased from Petch Thai Chemical Co., Ltd. (Bangkok, Thailand). Zinc oxide (ZnO) and sulphur (S_8) were supplied by Chemmin Co. Ltd., (Samut Prakan, Thailand). N-t-butyl-2-benzothiazolesulfenamide (TBBS), as cure accelerator, was purchased from Kitpiboon chemical Ltd., Part. (Bangkok, Thailand). RPOs which have variable aromatic content were supplied by IRPC Plc. Co., Ltd.

Compound preparation

The compounding ingredients, as shown in **Table 1**, were mixed using a 2–roll mill (LabTech model LRM 150) with a two-stage mixing technique. In the first stage of mixing, ZnO, stearic acid, carbon black and RPO were incorporated sequentially into raw rubber with total mixing time of 14 minutes. Various types and loadings of RPOs were compounded in this step. The mix gained at the first stage is usually known as *rubber masterbatch*. In the second mixing stage, the rubber masterbatch was masticated for 2 minutes, and then sulphur and TBBS were charged into the rubber mix, and allowed 5 minutes to achieve good distribution and dispersion of ingredients. Finally the rubber compounds were sheeted and cooled.

Table 1 Compound recipe used in this work

Chemicals	Used as	Amount (phr*)
SBR (1502)	Raw rubber	100
Carbon black (N330)	Reinforcing filler	50
Zinc oxide	Cure activator	3
Stearic acid	Cure activator	1
RPO**	Processing aid/Softener	Variable: 5, 10, 15
Sulphenamide (TBBS)	Cure accelerator	1
Sulphur	Curing agent	1.75

* = parts per hundred of rubber

** = in this research work, the RPOs chosen to meet the research objective were prepared so as to have variable aromatic content. The percentage of aromatic carbon and saturated carbon, as stated by the supplier, as well as properties of each RPO are reported in Table 2.

Cured characteristics and viscoelatic properties

Cure characteristics were determined at the temperature of 145 °C with the use of moving die rheometer (MDR; TechPro, USA). Viscoelastic properties of rubber compounds were measured using the Rubber Process Analyser (Alpha Technology model RPA2000). Measurement of Mooney viscosity (ML1+4 @ 100°C) was carried out using the Mooney viscometer (TechPro viscTECH+).

Measurement of the mechanical properties

To prepare cured specimens, the rubber compounds were compression moulded under a clamping pressure of 12 MPa at 145 °C for the optimum cure time (t_{c90}) as pre-determined from the MDR.

Hardness of vulcanisates was measured using a durometer with Shore A scale (Cogenix Wallace, UK) as per ASTM D2240. Tensile properties were determined using a universal testing machine (Instron 5566, USA) according to ASTM D 412, Die C. Tear strength was determined according to ASTM D624.

Results and Discussion

RPO properties

Table 2 shows the properties of RPOs used in this work. Each RPO has variable aromatic content, as reported. One method to statistically relate to the dominant chemical structure of petroleum oil is to use ASTM D2501, the Viscosity-Gravity Constant (VGC) [7]. Generally a higher value of VGC indicates higher aromaticity in the RPOs [3, 5]. Evidently, the VGC and kinematic viscosity increases with increasing aromatic content in oils as expected.

Table 2 RPOs characteristics

Oil	Aromatic Carbon (%)*	Saturated Carbon (%)	Kinematic Viscosity at 100 °C (mm^2/sec)	Viscosity-Gravity Constant
Oil-1	60.09	39.01	14.745	0.8694
Oil-2	69.80	29.09	17.611	0.8966
Oil-3	80.28	19.03	23.543	0.9299

*Measured by IRPC Co. plc. Ltd.

Cure characteristics

Fig. 1 shows the plot between torque differences against aromatic content in RPOs. As reported elsewhere, the torque difference (M_H–M_L) could be used as an indication of the crosslink density of rubber compound [8]. Apparently, at any given oil loading, the crosslink density of SBR vulcanisates does not significantly change with aromatic content. With increasing RPO loading, the crosslink density decreases at a given aromatic content. In other words, the incorporation of RPO decreases cure efficiency in SBR compounds.

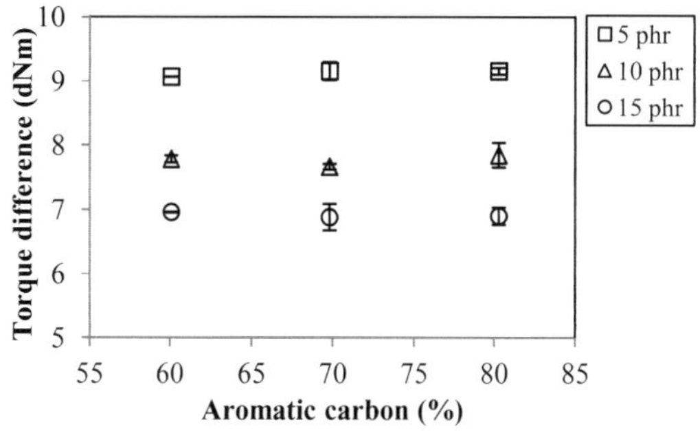

Figure 1 Cure torque difference (as an indication of crosslink density) in SBR compounds

Rheological behaviour

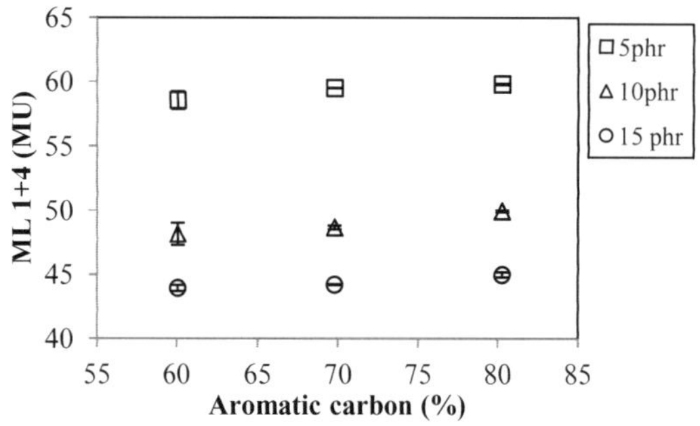

Figure 2 Mooney viscosity of SBR compounds with various RPOs

Fig. 2 demonstrates no significant alteration in Mooney viscosity with increasing aromatic content in

RPO. Also, as expected, the increase in RPO reduces clearly the Mooney viscosity, due to the plasticisation effect. This indicates the processability improvement as a result of RPO incorporation.

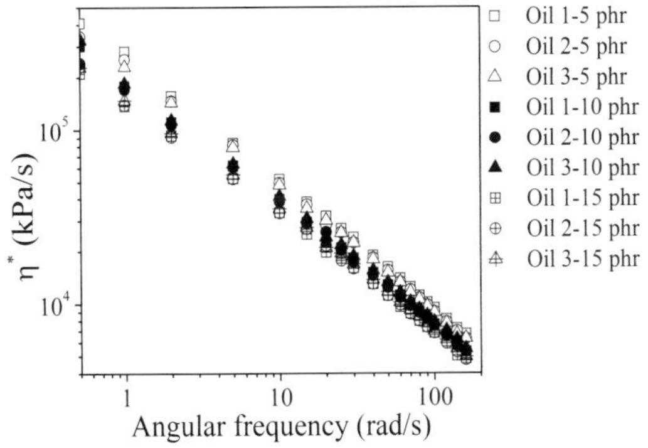

Figure 3 Complex viscosity of SBR compounds filled with various RPOs as a function of frequency

The logarithmic plots of complex viscosity (η^*) against angular frequency of SBR compounds is exhibited in **Fig. 3**. Results demonstrate clearly shear thinning behaviour or the so-called pseudoplastic behaviour of all SBR compounds [7, 9]. At any given oil loading, the η^* does not significantly change with aromatic content, but does affect with oil loading. With increasing RPO loading, a decrease in η^* could be observed, particularly at low test angular frequency. The results support the enhancement in processability of SBR compounds by of RPO loading rather than RPO type for a broad range of angular frequency (or shear rate).

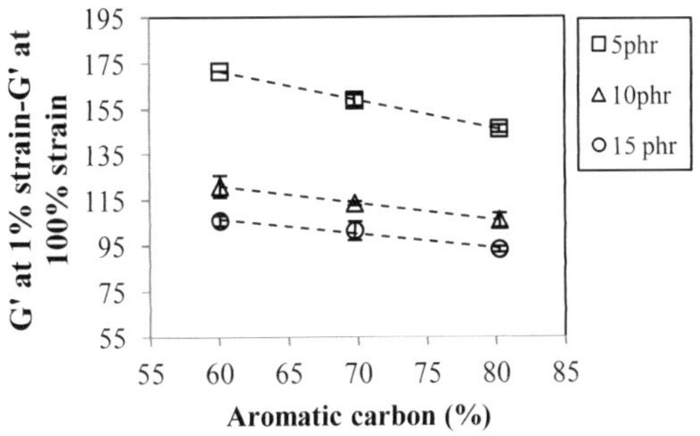

Figure 4 Payne effect of SBR compounds filled with various RPO

Apart from bulk viscosity, the degree of filler dispersion is a crucial factor controlling properties of rubber compounds and vulcanisates. Among various characterisation techniques for determining degree of filler dispersion in rubber matrix, the use of *Payne effect* has been proved to be an effective tool for achieving the purpose [1, 10, 11, 12]. The discrepancy of moduli at small and large strains is used to quantify the magnitude of Payne effect. The smaller the discrepancy, the lower the Payne effect magnitude and thus the greater the filler dispersion degree. **Fig. 4** exhibits the decrease in Payne effect with aromatic content and/or RPO loading, implying the enhancement in filler dispersion. It is believed that, at a given total mixing time, the decrease in bulk viscosity could facilitate the filler incorporation, i.e., a reduction in incorporation time. By this means, the mixing time remaining for allowing dispersion and distribution is increased, and thus the improvement in state-of-mix. Also, by increasing aromaticity of RPO, the increase in compatibility between RPO and rubber matrix is resulted via the interaction between aromatic carbon in RPO and aromatic rings on SBR backbone [5, 6]. This leads to the further increase in state-of-mix (i.e., the decrease in Payne effect.).

Mechanical properties

Fig. 5 shows hardness of cured SBR which does not significantly change with aromatic content in RPO at any given RPO loading. By contrast, the RPO loading plays strong role on hardness of SBR vulcanisates, as expected, as a result of plasticisation effect as mentioned previously. Additionally, referred to the results of crosslink density (see **Fig. 1**), the reduction in crosslink density with increasing RPO loading might give rise to the decrease in hardness.

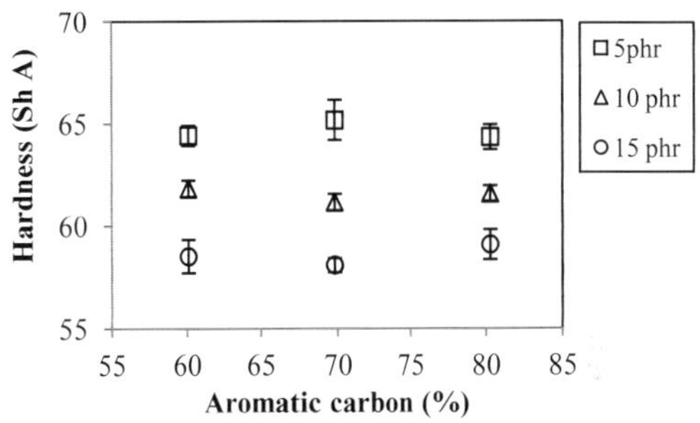

Figure 5 Hardness of SBR vulcanisates filled with RPOs having various aromatic contents.

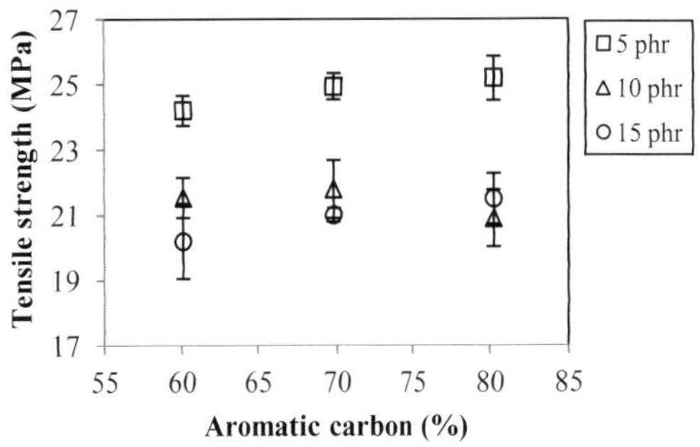

Figure 6 Tensile strength of SBR vulcanisates filled with RPOs having various aromatic contents

Tensile properties of the SBR vulcanisates as affected by RPOs are shown in **Figs. 6** and **7**. At any

RPO loading, tensile strength increases with increasing aromaticity of RPOs. This could be attributed to the enhancement in RPO-SBR compatibility and thus state-of-mix, as discussed previously. In addition, with increasing RPO loading, the tensile strength decreases due to the combined effect of plasticisation effect and decreased crosslink density. Likewise, elongation at break appears to increase with RPO loading and aromatic content, which could be explained by the enhanced oil compatibility, decresed crosslink density of SBR vulcanisates, increased SBR molecular mobility via the plasticisation effect and augmented state-of-mix.

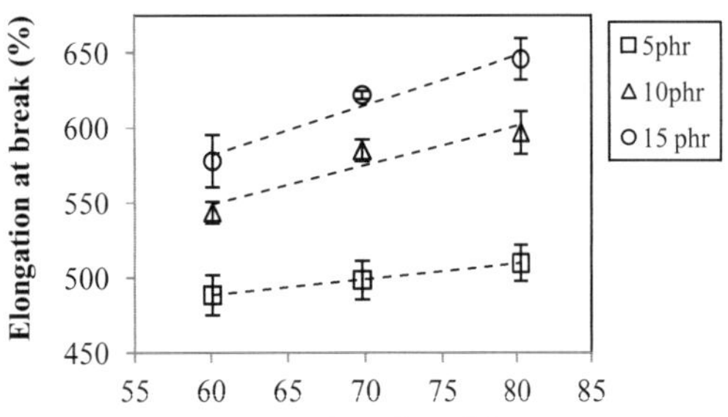

Figure 7 Elongation at break of SBR vulcanisates filled with RPOs having various aromatic contents

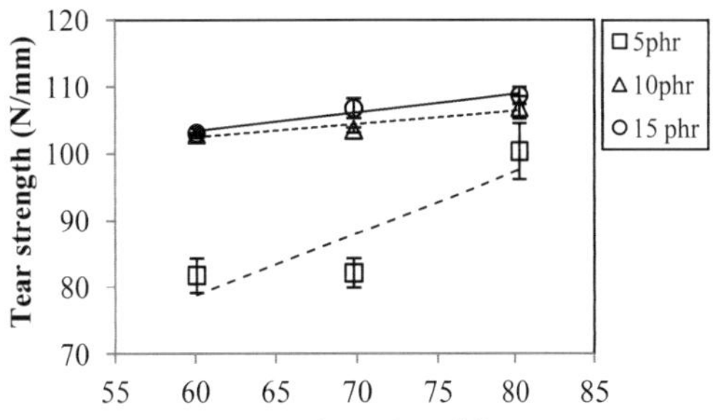

Figure 8 Tear strength of cured SBR filled with various RPOs

Fig. 8 exhibits tear strength of SBR vulcanisates with various RPOs, in which the tear strength tends to increase with RPO loading and aromatic content. The results could be explained by the enhanced RPO-SBR compatibility and the improvement in processability and, thus the state-of-mix in SBR vulcanisates.

Conclusions

The incorporation of RPOs into SBR compounds leads to decreased bulk viscosity, and thus enhanced processability in association with state-of-mix, but gives negative effect on some mechanical properties (i.e., hardness and tensile strength) via a plasticisation effect. The increases in elongation at break and tear strength with RPOs loading is believed to be the result of the processability improvement and thus state-of-mix. Aromatic contents in RPOs play no significant role on bulk viscosity, but affects strongly the state-of-mix and mechnical properties (i.e., enhancement in both state-of-mix and mechnical properties). The improvement in compatibility between RPO and SBR molecules is thought to be responsible for such enhancement.

Acknowledgements

The authors would like to acknowledge the IRPC Plc. Co., Ltd. for kindly supplying RPOs with various aromatic contents. The authors also gratefully acknowledge the Center of Excellence for Innovation in Chemistry, PERCH-CIC for the financial support of this research.

References

1. S. Dasgupta; S.L. Agrawal; S. Bandyopadhyay; R. Mukhopadhyay; R.K. Malkani; S.C. Ameta *Polym. Test.* 2009, 28, 251.
2. S. Dasgupta; S.L. Agrawal; S. Bandyopadhyay; S. Chakraborty; R. Mukhopadhyay; R.K. Malkani; S.C. Ameta *Polym. Test.* 2008, 27, 277.
3. S. Dasgupta; S.L. Agrawal; S. Bandyopadhyay; S. Chakraborty; R. Mukhopadhyay; R.K. Malkani; S.C. Ameta *Polym. Test.* 2007, 26, 489.
4. A. Hendriks *Rubber resources,* October 2006.
5. R.B. Simpson *Rubber basic*, Rapra Technology Ltd., United Kingdom, 2002.
6. B. Rodgers *Rubber compounding: Chemistry and Application,* Marcel Dekker Inc, United States of America, 2004.
7. J.S. Dick *Basic rubber testing : selecting methods for a rubber test program,* ASTM International, United States of America, 2003.
8. A. Neau; M. Rangstedt *Rubber World,* 21, November 2009.
9. J.S. Dick *Rubber technology: Compounding and testing for performance,*: Hanser publisher, Munich, 2001
10. N. Phewthongin; P. Saeoui; C. Sirisinha *J. Appl. Polym. Sci.* 2006, 100, 2565.
11. W. Dierkes; J.W.M. Noordermeer; M. Rinker; K.U. Kelting; P.C. Van De Pol; *Kaut. Gummi. Kunstst.* 2003, 56, 338.
12. P. Cassagnau *Polymer,* 2003, 44, 245.

O-13-573

MECHANICAL PROPERTIES OF FLAKE-SHAPED CELLULOSE PARTICLE REINFORCED NATURAL RUBBER COMPOSITES

Kunihiro Araki[1]*, Koki Matsumoto[1], Asahiro Nagatani[1,2], Tatsuya Tanaka[1], Yoshihiko Arao[1]

[1]*Applied Materials Engineering Laboratory, Faculty of Engineering, Doshisha University, 1-3, Tatara Miyakodani, Kyotanabe, Kyoto, 610-0321, Japan —tatanaka@mail.doshisha.ac.jp*

[2]*Hyogo Prefectural Institute of Technology, 3-1-12, Yukihira-cho, Suma-ku, Kobe, Hyogo, 654-0037, Japan*

Abstract - In this study, we produced green composites based on natural rubber (NR) and the flake-shaped cellulose particles (FSCP), as the biomass material. The FSCP were mechanically milled in a planetary ball mill with a grinding aid as a cellulose aggregation inhibitor. We used stearic acid, zinc oxide, sulfur, and an accelerant as vulcanizing ingredients of the rubber. The FSCP and NR were mixed with the vulcanizing ingredients and a silane coupling agent in an internal mixer. To improve interface affinity of the FSCP and NR, we used the silane coupling agent. The mechanical properties of the composites were evaluated by a tensile test, a tear test and a hardness test among others. As a result of tensile test, values of M_{100} (tensile modulus at 100[%] elongation) increased with an increased amount of the FSCP or of the silane coupling agent. The tear strength decreased when increasing the FSCP content, but the tear strength was improved by the addition of a silane coupling agent. The hardness of composites increased with an increased amount of the FSCP, but the amount of the silane coupling agent did not affect hardness.

Keywords: Natural rubber, Cellulose, Composites, Silane coupling agent, Mechanical properties

Introduction

In recent years, global warming and the depletion of fossil resources have become crucial problems. Moreover, waste disposal is a big problem due to our large consumption of plastic products that are made of fossil resources. One important way to address these problems is the replacement of fossil resources with "sustainable" biomass resources, which has been investigated in this research.

Cellulose is one of the most abundant biomass resources on earth[1]. We take notice of wood-based cellulose that make up most of the biomass because it has no involvement in food production. However, it is a fact that there is a large amount of biomass which is not yet used efficiently in Japan. For this reason, there is a great need to develop effective utilization and benefit from this enormous supply.

In this study, we produced green composites based on natural rubber (NR) and the flake-shaped cellulose particles (FSCP), as the biomass material. And, we surveyed the effects of amount of FSCP or of the silane coupling agent on mechanical properties of composites.

Experimental

Material

The FSCP were made from wood pulp (W400, Nippon Paper Chemicals Co., Ltd., Japan). The FSCP were mechanically milled using a planetary ball mill with a grinding aid as a cellulose aggregation inhibitor[2]. We used Ribbed Smoked Sheet Grade 1 NR. We used stearic acid, zinc oxide, sulfur, and an accelerant as vulcanizing ingredients of the rubber. The formulations used in this study are shown in Table 1. To improve interface affinity of the FSCP and NR, we used the silane coupling agent(0~15 [wt.%] of filler)[3-4].

Table 1 Compounding formulations

Ingredients	[phr]
NR	100
Zinc oxide	6
Stearic acid	3.5
Sulfur	0.5
Accelerant BBS	0.7
FSCP	0~50

Preparation of NR/FSCP composites

FSCP and NR were mixed with silane coupling agent in a melt mixer (10C100 Labo Plastomill, Toyo Seiki Seisaku-sho, Ltd., Japan). Table 2 shows the mixing parameters in this study. First, NR was masticated for 3 [min] alone in the mixer. Second, it was mixed with the vulcanizing ingredient for 1 [min]. Third, they were mixed with FSCP for 10 [min].

Table 2 Mixing parameters

	parameters
Mixture[cc]	70
Mixing temperature[℃]	60
Rotational speed[rpm]	30

Preparation of the sheet of composites

Sheets of composites were prepared by a compression molding method. The thickness was adjusted by using middle frame of 2 [mm] thickness. The composites were heat pressed using a compression molding machine (NF-50 Automatic Molding Press, Shinto Metal Ind., Ltd., Japan) at 150[℃] and 10 [MPa] for the respective optimal cure time obtained from the vulcanizing test.

Cure time measurements

The cure characteristics of the rubber compounds were measured on a curemeter (curelastmeter V, Orientec Co., Ltd., Japan) at 150[℃] according to JIS K 6300-2 (determination of cure characteristics with oscillating curemeters).

Tensile test, Tear test

Tensile properties and tear properties were measured using an Instron-typed tensile tester (Autograph AG-1000D, Shimadzu Co., Japan) according to JIS K 6251 (tensile testing methods for vulcanized rubber) and JIS K 6252 (tear testing methods for vulcanized rubber). The tests were carried out at room temperature (20± 2[℃]) with a crosshead speed of 500[mm/min.]

Hardness test

Hardness tests were carried out according to JIS K 6253 (hardness testing methods for vulcanized rubber) with a durometer-type tester (type A) at room temperature (20 ± 2[℃]). For measurement of hardness, three sheets of 2 [mm] thickness were stacked to attain a total of 6 [mm] thickness.

Observation of the shape of FSCP and the freeze-fractured surface

The shape of FSCP and the freeze-fractured surface of the composites were observed with a scanning electron microscope (SEM, JSM-7001FD, JEOL, Ltd., Japan). The acceleration voltage was set to 10 [kV] and all samples were 5 [nm] sputter-coated with gold to avoid charging.

Results and Discussion

Tensile properties of the composites

The tensile properties of the composites as a function of the FSCP content are presented graphically in Figs. 1, 2. Fig. 1 shows M_{100} (the tensile modulus at 100[%] elongation) of composites containing 0, 5, 10, 15, 20, 30, 40, and 50[phr] FSCP without silane coupling agent. Increasing the FSCP loading increased M_{100} because of the effect of FSCP's stiffness.

Fig. 2 shows M_{300} of above composites. Increasing the FSCP loading discontinuously increased M_{300}.

The tensile properties for the composites as a function of the silane coupling agent content are presented graphically in Figs. 3, 4. Fig. 3 shows M_{100} of composites containing 0, 1, 3, 5, 7, 10, and 15[wt.% of filler] silane coupling agent with 20 [phr] FSCP. Increasing the amount of the silane coupling agent increased M_{100} because of the interfacial adhesion between NR and the FSCP are improved.

Fig. 4 shows M_{300} of above composites. Increasing the silane coupling agent loading increased M_{300}. The results suggest that the interfacial adhesion between NR and the FSCP are improved.

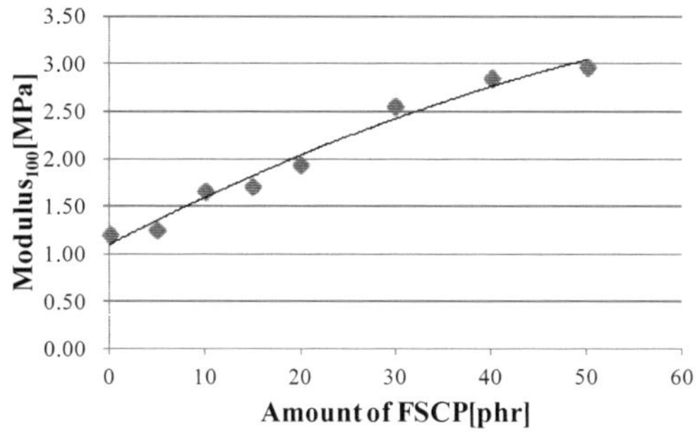

Figure 1 Effect of the FSCP content on M_{100} of composites without silane coupling agent

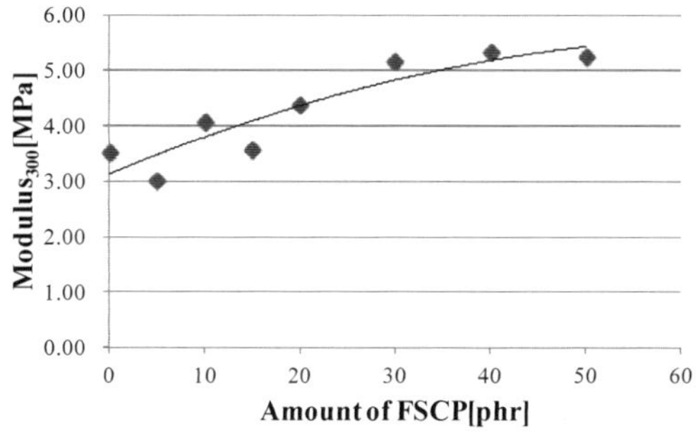

Figure 2 Effect of the FSCP content on M_{300} of composites without silane coupling agent

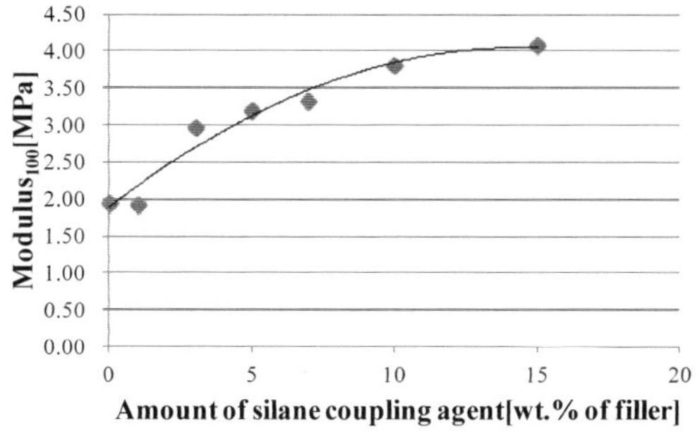

Figure 3 Effects of silane coupling agent contents on M_{100} of composites with 20[phr] FSCP

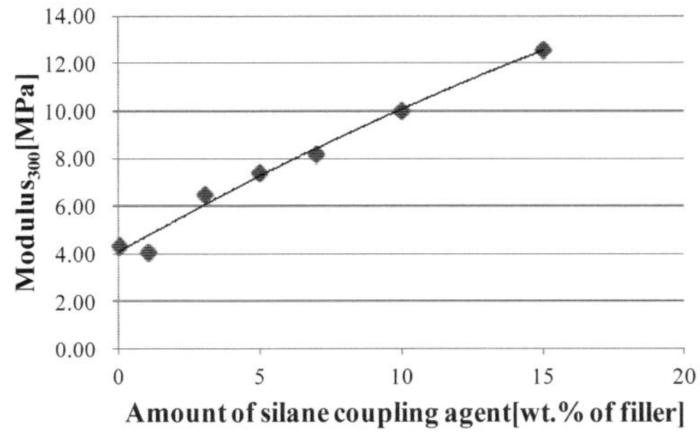

Figure 4 Effects of silane coupling agent contents on M_{300} of composites with 20[phr] FSCP

Tear properties of the composites

Fig. 5 shows tear strength of composites containing 0, 5, 10, 15, 20, 30, 40, and 50 [phr] FSCP without silane coupling agent. Increasing the FSCP loading discontinuously decreased tear strength because of the poor interfacial adhesion between NR and the FSCP.

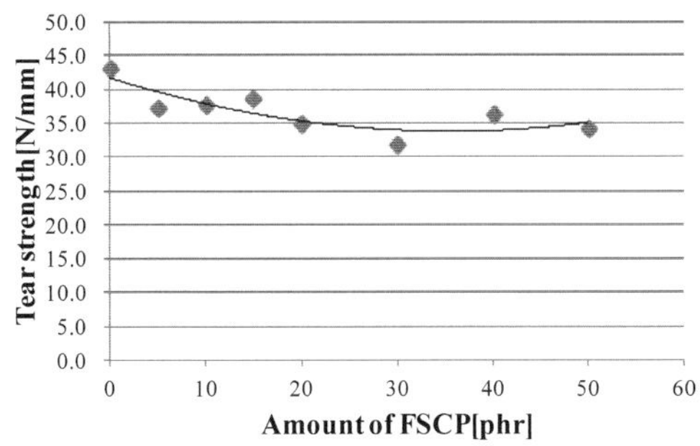

Figure 5 Effect of the FSCP content on tear strength of composites without silane coupling agent

Fig. 6 shows tear strength of composites containing 0, 1, 3, 5, 7, 10, and 15[wt.% of filler] silane coupling agent with 20 [phr] FSCP. Increasing amount of the silane coupling agent increased tear strength slightly because of the interfacial adhesion between NR and the FSCP are improved.

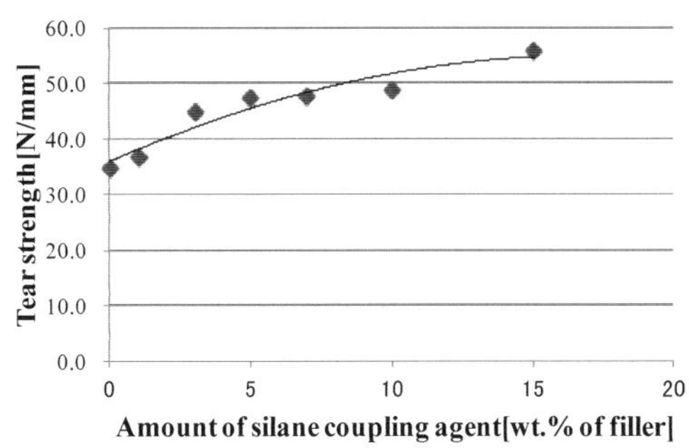

Figure 6 Effects of silane coupling agent contents on tear strength of composites with 20[phr] FSCP

Hardness of the composites

Fig. 7 shows hardness of composites containing 0, 5, 10, 15, 20, 30, 40, and 50 [phr] FSCP without silane coupling agent. Increasing the FSCP loading increased hardness because of the effect of hardness of FSCP.

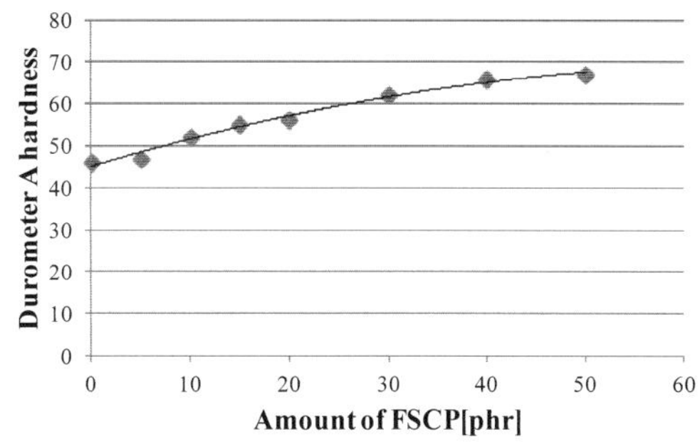

Figure 7 Comparison of hardness of composites with various FSCP content

Fig. 8 shows hardness of composites containing 0, 1, 3, 5, 7, 10, and 15[wt.% of filler] silane coupling agent with 20 [phr] FSCP. The amount of the silane coupling agent did not affect hardness.

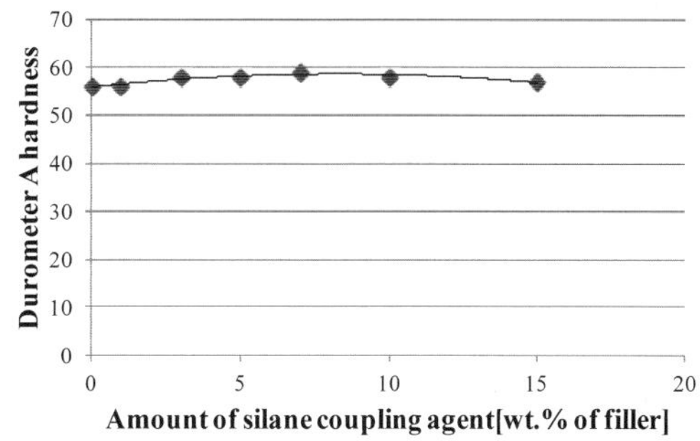

Figure 8 Comparison of hardness of composites with various the silane coupling agent

Observation of the shape of FSCP and the freeze-fractured surface

Fig. 9 shows an SEM picture of the FSCP. As seen in the image, the FSCP was fine flake shape, similar to talc, mica, etc., and the average particle size was about 20 [μm].

Fig. 10 shows an SEM picture of the freeze-fractured surface of the NR composite containing 20 [phr] the FSCP without silane coupling agent. The failure surface of the NR composite exhibits weak interfacial adhesion between the FSCP and the rubber matrix. Because, the failure surface shows that many holes (about 20 [μm] length) were left after the FSCP was pull out from the rubber matrix when stress was applied, and the failure occurred at the week FSCP and rubber matrix interface[5]. With the addition of 5 [wt.% of filler] silane coupling agent, there was less pullout of the FSCP from the NR matrix because of the strong interaction between the filler and rubber matrix (Fig. 11).

Conclusions

1. The M_{100} increases with increasing the FSCP loading, but the M_{300} discountly increases, because the poor filler-matrix interaction.
2. The presence of the silane coupling agent enhances the tensile properties and tear properties of the FSCP-filled NR composites.
3. SEM micrographs of freeze-fractured surfaces of the composites show that the presence of silane coupling agent causes a reduction in filler detachment and better interfacial interaction between the FSCP and NR.

Acknowledgements

This work was supported by JSPS KAKENHI Grant Number24560840.

References

1. A. Isogai, ed., *Cellulose no Kagaku* [Science of Cellulose], Asakura bookstore, 2003; 12-13
2. A. Nagatani, *et.al.*, "Preparation and Properties of Cellulose-Olefinic Thermoplastic Elastomer Composites," *Journal of Applied Polymer Science*, 2005; Vol. 95, 144-148.
3. N. Lopattananon, *et.al.*, "Hybridized Reinforcement of Natural Rubber with Silane-Modified Short Cellulose Fibers and Silica," *Journal of Applied Polymer Science*, 2011; Vol.120, 3242-3254.
4. M. Abdelmouleh, *et.al.*, "Short natural-fibre reinforced polyethylene and natural rubber composites: Effect of silane coupling agents and fibres loading," *Composites Science and Technology*, 2007; Vol. 67,1627-1639.
5. H. Ismail, and F. S. Haw, "Effects of Palm Ash Loading and Maleated Natural Rubber as a Coupling Agent on the Properties of Palm-Ash-Filled Natural Rubber Composites," *Journal of Applied Polymer Science*, 2008; Vol.110, 2867-2876.

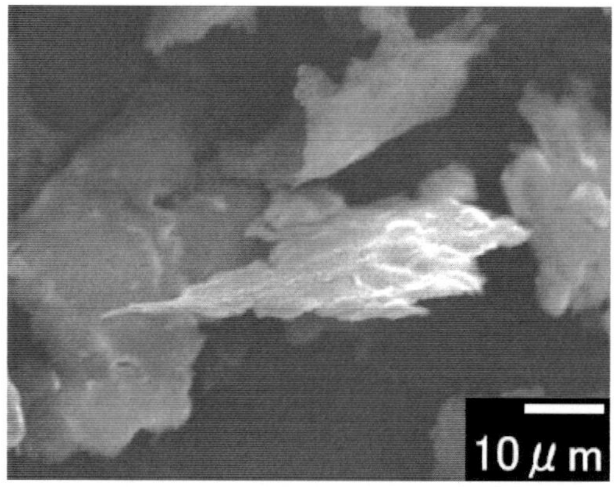

Figure 9 An SEM micrograph of the FSCP

Figure 10 An SEM micrograph of the freeze-fractured surface of the 20 [phr] FSCP-filled NR composites without silane coupling agent

Figure 11 An SEM micrograph of the freeze-fractured surface of the 20 [phr] FSCP-filled NR composites with 5[wt.% of filler] silane coupling agent

O-13-642

VISCOELASTIC PROPERTIES OF NATURAL RUBBER WITH SOFT NANOMATRIX STRUCTURE

Seiichi Kawahara[1], Kenichiro Kosugi[1] and Oraphin Chaikumpollert[1,2]

[1]*Department of Materials Science and Technology, Faculty of Engineering, Nagaoka University of Technology, Nagaoka, Niigata 940-2188, Japan – kawahara@mst.nagaokaut.ac.jp*

[2]*National Metal and Materials Technology Center, 114 Thailand Science Park, Paholyothin Road, Klong 1, Klong Luang, Pathumthani 12120, Thailand*

Abstract - Preparation of natural rubber (NR) with a soft nanomatrix structure was made by graft-copolymerization of butyl acrylate (BA) onto deproteinized natural rubber with tert-butyl hydroperoxide/tetraetylenepentamine in latex stage. The resulting graft-copolymer of deproteinized natural rubber and poly (butyl acrylate) (DPNR-*graft*-PBA) was characterized by FT-IR spectroscopy. Conversion and grafting efficiency of BA were dependent upon BA concentration, which were more than 90 mol% under a suitable condition of the graft-copolymerization. Morphology of DPNR-*graft*-PBA was observed by transmission electron microscopy after staining film specimens with I_2 vapor for 5 min. The NR particles of about 0.5 μm in diameter were dispersed in PBA matrix of about 15 nm in thickness. Storage modulus and loss tangent of DPNR-*graft*-PBA were measured and they were related with the soft nanomatrix structure. The tensile strength and elongation at break decreased, as monomer concentration increased.

Keywords: Nanomatrix structure, Natural rubber, Graft-copolymerization

Introduction

Soft nanomatrix may have a potential to provide various synergistic effects with rubbery dispersoid, since the nanomatrix and the dispersoid are similarly deformed by stretching or straining, due to the similar values of Poisson ratio, i.e., 0.5. Therefore, when we use natural rubber (NR) as the dispersoid of major component and functional rubbery polymer as the nanomatrix of minor component, we may prepare a novel functional rubbery material, since NR is well-known to be the most excellent rubbery polymer, which possesses outstanding tensile strength, tear strength, permanent set and so forth. It is expected to achieve not only the outstanding mechanical properties and functional properties but also the synergistic effect through the formation of the nanomatrix structure.

In the previous works [1, 2], we prepared NR with a hard nanomatrix structure through graft-copolymerization of styrene onto the NR particles in latex stage. High grafting efficiency and conversion were accomplished for the graft-copolymerization of styrene under a suitable condition, after purification of NR latex through deproteinization with urea in the presence of surfactant [3-5]. The nanomatrix structure, observed by transmission electron microtomography (TEMT), was found to be a hierarchical nano-phase separated structure, which consisted of NR particles of about 0.5 μm in diameter and polystyrene (PS) matrix of about 15 nm in thickness [6]. Tensile strength and dynamic mechanical properties of NR were significantly enhanced by forming the nanomatrix structure. However, it is quite difficult to explain the reason why the properties are enhanced with the nanomatrix of PS, since PS is the glassy polymer that is distinguished from the rubbery polymer, in which value of Poisson's ratio of PS (0.33) is quite smaller than that of NR (0.49). In order to elucidate the positive effect of the nanomatrix structure of PS on the properties of NR, it is necessary to form a soft nanomatrix structure of the rubbery polymer in NR and compare the properties of NR with the hard nanomatrix structure to those of NR with the soft nanomatrix structure.

Based on the previous works [1, 2], NR with the soft nanomatrix structure may be prepared by graft-copolymerization of a suitable monomer onto NR particles in latex stage, where the resulting matrix polymer has to be the rubbery polymer at ambient temperature to form the soft nanomatrix structure. Here, the necessarily requirement of the graft-copolymerization is phase separation after graft-polymerization of the monomer; that is, the resulting polymer has to be immiscible with NR. We, thus, selected butyl acrylate (BA) as a monomer, since glass transition temperature, T_g, of poly (butyl acrylate) is -55 °C. In this case, we have to investigate a condition of the graft-copolymerization, since a suitable initiator concentration depends on the monomer species: for instance, 3.3×10^{-2} mol/kg-rubber for 1.5 mol/kg-rubber of styrene, 20.0×10^{-2} mol/kg-rubber for 5.5 mol/kg-rubber of styrene and 6.6×10^{-2} mol/kg-rubber for 1.5 mol/kg-rubber of methyl methacrylate.

In the present work, an attempt to prepare NR with the soft nanomatrix structure is made by graft-copolymerization of BA onto deproteinized natural rubber (DPNR) particles with tert-butyl hydroperoxide /tetraethylenepentamine (TBHPO/TEPA) as an initiator in the latex stage. Effect of monomer concentration on conversion and grafting efficiency are investigated. Morphology of graft-copolymer of DPNR and PBA

(DPNR-*graft*-PBA) is observed by transmission electron microscopy (TEM) after staining the film specimens with I_2 vapor. The tensile properties of the DPNR-*graft*-PBA are measured to investigate the advantage of the soft nanomatrix structure. The relationship between viscoelastic properties and morphology of NR with the soft nanomatrix structure was investigated in comparison with those of NR with the hard nanomatrix structure.

Experimental

2.1 Materials

60 w/w% dry rubber content (DRC) of commercial high-ammonia natural rubber (HANR) latex was used in the present study (Golden Hope, Malaysia). TBHPO and TEPA were used as a redox initiator, which were purchased from Kishida Chemical Co., Ltd. BA monomer (99 %) was purchased from Nacalai Tesque Inc. Sodium dodecyl sulfate (SDS) (98 %) was purchased from Kishida Chemical Co., Ltd.

2.2 Deproteinization of natural rubber

Deproteinization of NR latex was carried out by incubation of HANR latex (DRC: 30 w/w%) with 0.1 w/w% urea in the presence of 1 w/w% SDS as a surfactant for 1 h at 30 °C. Resulting latex was subjected to centrifugation at 10,000 rpm for 30 min. The cream fraction was re-dispersed in 1 w/w% SDS solution and it was washed twice by centrifugation to prepare DPNR latex with 30 w/w% DRC. The resulting DPNR latex contained about 0.1 w/w% SDS.

2.3 Graft-copolymerization of butyl acrylate onto deproteinized natural rubber

Graft-copolymerization of BA onto DPNR was carried out with TBHPO/TEPA as an initiator. DPNR latex was charged with N_2 gas for 1 h at 30 °C and, then, 2 w/w% SDS was added into the latex. The initiator of 6.6×10^{-2} mol/kg-rubber and BA were added in turn to the latex. The reaction was carried out by stirring the latex at 400 rpm for 2 h at 30 °C. The unreacted BA was removed by using a rotary evaporator under reduced pressure. The gross polymer film was prepared by casting the latex into a petri dish followed by drying under reduced pressure at ambient temperature for more than a week. The gross polymer was extracted with acetone in a Soxhlet apparatus under nitrogen atmosphere in the dark for 24 h to remove free-PBA, which was isolated from NR.

2.4 Characterization

The presence of functional groups of the graft-copolymer was verified by Fourier-transform infrared (FT-IR) spectroscopy. The FT-IR spectra for DPNR-*graft*-PBA were recorded with a JASCO FT-IR 410 spectrometer in the range of 4000-400 cm^{-1} at a resolution of 4 cm^{-1}.

Differential scanning calorimetry (DSC) measurements were performed with a SII SSC/5200 DSC 220 at a heating rate of 10 °C/min.

The morphology for DPNR-graft-PBA was observed with a TEM, JEOL JEM-2100, at accelerating voltage of 200 kV. Reichert-Nissei FC S Ultra-microtrom was used to prepare an ultra thin section of the DPNR-graft-PBA at a temperature lower than T_g of NR.

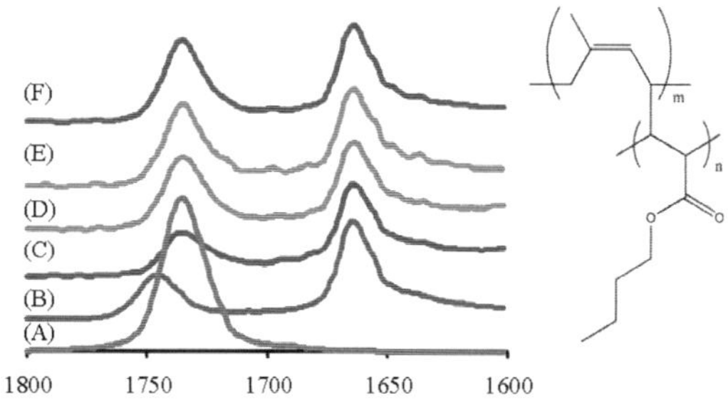

Fig. 1 FT-IR spectra for (A) PBA, (B) DPNR, (C) DPNR-*graft*-PBA 1.0 mol/kg-rubber, (D) DPNR-*graft*-PBA 1.5 mol/kg-rubber, (E) DPNR-*graft*-PBA 2.0 mol/kg-rubber, and (F) DPNR-*graft*-PBA 2.5 mol/kg-rubber.

The dynamic mechanical properties were measured with a TA Instrument ARES-G2 at frequencies between 0.1 and 50 Hz at 30 °C.

The tensile properties of the DPNR-*graft*-PBA were measured with an Instron universal tester machine, Instron 3365. The crosshead speed was 500 mm/min.

Results and Discussion

3.1 Characterization of graft-copolymer

Fig. 1 shows FT-IR spectra, ranging from 1600 to 1800 cm^{-1}, for PBA, DPNR and DPNR-*graft*-PBA prepared with various monomer concentrations. As for DPNR, infrared absorption peaks appeared at 1739 cm^{-1} and 1664 cm^{-1}, which were identified to ester linkage (C=O) of fatty acid esters and carbon-carbon double bond (C=C) of isoprene units, respectively. In contrast, the absorption peak appeared at 1735 cm^{-1} for PBA, which was identified to ester linkage (C=O) of BA units. As for DPNR-*graft*-PBA, intensity of the absorption peak at 1735 cm^{-1} was dependent upon the monomer concentration; that is, the higher the monomer concentration, the higher was the intensity of the absorption peak at 1735 cm^{-1}.

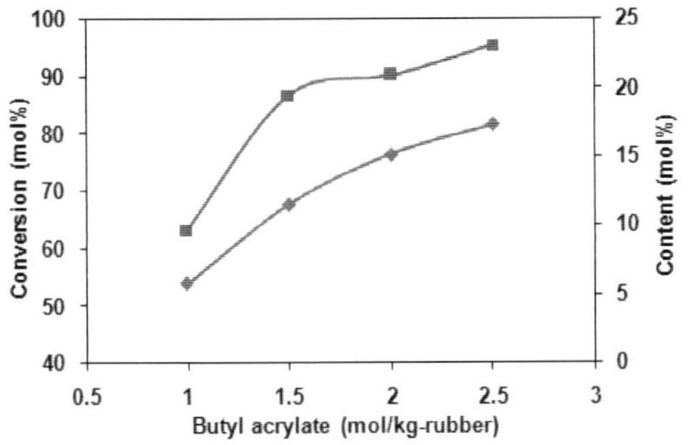

Fig. 2 Monomer concentrations dependence of (■), conversion and (♦), content of BA , which were determined by gravimetric method.

In order to investigate an effect of the monomer concentration on the graft-copolymerization, we estimated content of BA in DPNR-*graft*-PBA and conversion of BA from the intensity of the absorption peaks at 1735 cm^{-1} and 1664 cm^{-1}, using a calibration curve in a plot of the intensity versus BA content by weight, according to the following expressions:

$$\text{BA content (\%)} = \frac{\text{Weight of PBA in gross polymer}}{\text{Weight of gross polymer}} \times 100$$

$$\text{BA conversion (\%)} = \frac{\text{Weight of PBA in gross polymer}}{\text{Weight of monomer}} \times 100$$

where weight of PBA in gross polymer were determined by FT-IR measurement with the calibration curve, weight of gross polymer and weight of monomer were determined by gravimetric method. The weight of PBA in gross polymer was also estimated by gravimetric method. The estimated values of the conversion and content of BA determined by FT-IR measurement were consistent with those determined by gravimetric method.

The estimated values of the conversion and content of BA were shown in Fig. 2, which were determined by gravimetric method. The conversion and content of BA were dependent upon the BA concentration. Under a suitable condition for the graft-copolymerization, i.e., 2.5 mol/kg-rubber of monomer concentration, the conversion of BA was the highest, i.e., more than 95 mol%, suggesting that almost all BA was consumed for the graft-copolymerization.

It is important to evaluate how many the resulting PBA is chemically attached to the DPNR particles. We, thus, determined the grafting efficiency of BA for the graft-copolymerization. The grafting efficiency was estimated by the following expression:

$$\text{Grafting efficiency (\%)} = \frac{\text{Mole of PBA linked to natural rubber}}{\text{Mole of PBA produced during graft-copolymerization}} \times 100$$

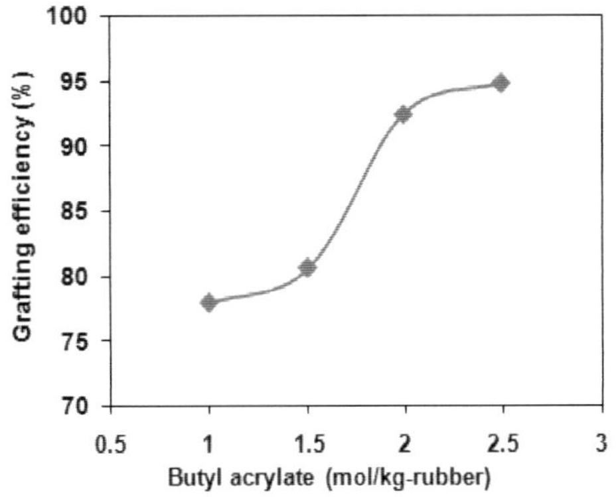

Fig. 3 Grafting efficiency of BA for DPNR-*graft*-PBA

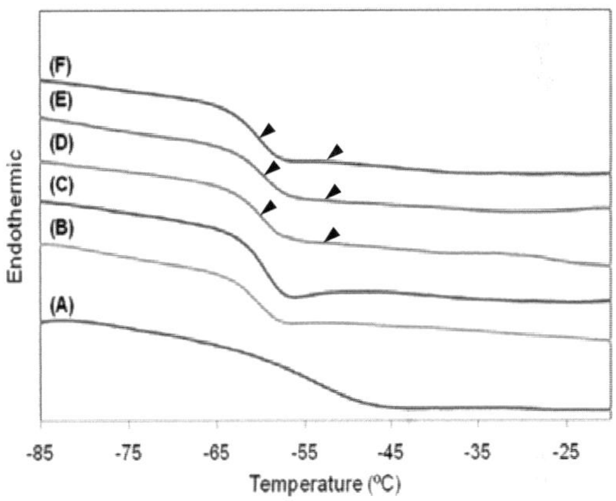

Fig. 4 DSC thermograms for (A) PBA, (B) DPNR, (C) DPNR-*graft*-PBA 1.0 mol/kg-rubber, (D) DPNR-*graft*-PBA 1.5 mol/kg-rubber, (E) 2.0 mol/kg-rubber, (F) DPNR-*graft*-PBA 2.5 mol/kg-rubber.

where the mole of PBA linked to DPNR was determined after removal of the free PBA, which did not link to the DPNR particles, through Soxhlet extraction with acetone for 24 h. The estimated value of the grafting efficiency of BA is shown in Fig. 3. The value of the grafting efficiency increased as the BA concentration increased. At 1.5, 2.0 and 2.5 mol/kg-rubber feed of BA, the value of the grafting efficiency was close to or higher than 90 mol%. This implies that almost all PBA was chemically attached to the DPNR particles. Particularly, at 2.5 mol/kg-rubber feed of BA, most BA was converted to the grafted PBA of DPNR-*graft*-PBA.

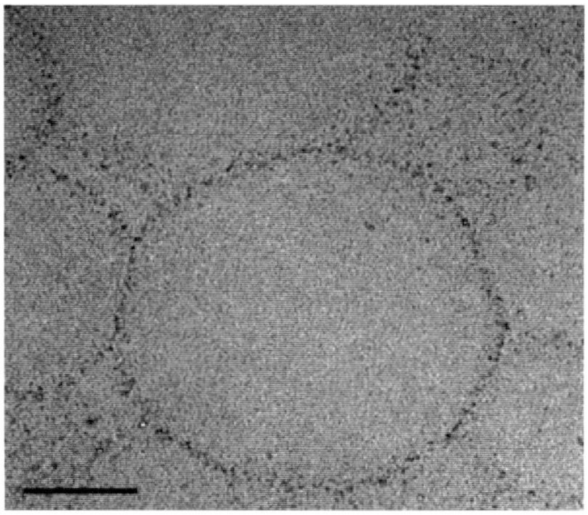

Fig. 5 TEM photograph for DPNR-*graft*-PBA 1.5 mol/kg
-rubber. The gloomy domains represent PBA and the bright domains represent DPNR. The bright DPNR particles of about 0.5 μm in average diameter were well dispersed in the gloomy PBA matrix of about 15 nm in thickness.

Fig. 4 shows the DSC thermograms for PBA, DPNR and DPNR-*graft*-PBA prepared with various monomer concentrations. Abrupt decrease in the heat capacity due to glass transition was shown at about -60 °C for DPNR and about -55 °C for PBA. Thus, T_g was determined as a point of inflection in DSC thermograms. The determined values of Tg was -62 °C for DPNR and -54 °C for PBA. As for DPNR-*graft*-PBA, two T_g's were shown in the thermograms, in which the values of T_g were -62 °C for DPNR and -54 °C. In the previous works, miscibility of the multi component systems was evaluated by T_g measurements. It was classified into the following three: one T_g for the miscible system, closed two T_g's for compatible system and distinct two T_g's for immiscible system. In the present work, values of two T_g's of the DPNR-*graft*-PBA were the same as those of DPNR and PBA, respectively. Thus, the DPNR and PBA in the DPNR-*graft*-PBA were found to be immiscible with each other.

3.2 Morphological observation

TEM images for DPNR-*graft*-PBA prepared with 1.5 mol/kg-rubber feed of BA are shown in Fig. 5, in which gloomy domains represent PBA and bright domains represent DPNR. The bright DPNR particles of about 0.5 μm in average diameter were well dispersed in the gloomy PBA matrix of about 15 nm in thickness. The morphology for DPNR-*graft*-PBA prepared with 1.5 mol/kg-rubber feed of BA was quite similar to that for DPNR-*graft*-PS prepared with 1.5 mol/kg-rubber feed of styrene reported elsewhere [1, 2].

3.3 Dynamic mechanical analysis

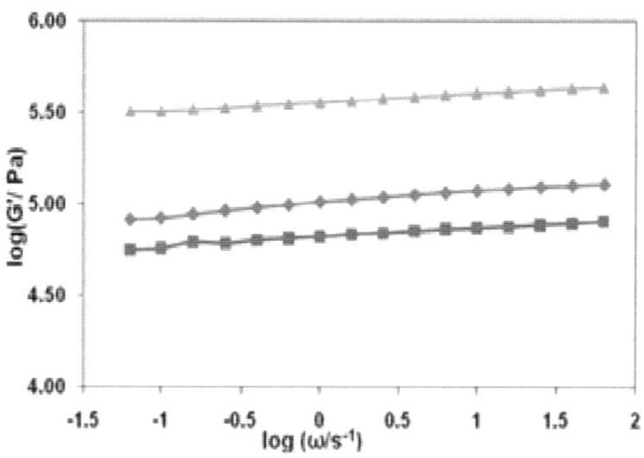

Fig. 6 Storage modulus versus frequency for (A) DPNR, (B) DPNR-*graft*-PS with hard nanomatrix structure and (C) DPNR-*graft*-PBA (PBA: ca. 11%) with soft nanomatrix structure.

Fig. 6 shows plot of storage modulus at plateau region (plateau modulus) versus frequency for DPNR, DPNR-graft-PBA prepared with 1.5 mol/kg-rubber feed of BA and DPNR-*graft*-PS prepared with 1.5 mol/kg-rubber feed of styrene. Value of the plateau modulus of DPNR was about 10^5 Pa, as reported in the literature [7]. After the soft nanomatrix structure was formed in DPNR, the plateau modulus of the rubber was 1.05 times as less as that of DPNR. This is different from the rubber with the hard nanomatrix structure, in which the plateau modulus of the rubber increased 3.5 times as high as that of DPNR. The decrease in the plateau modulus for DPNR-graft-PBA with the soft nanomatrix structure was found to be the same as that estimated from the moduli of NR and PBA by following equation [8]:

$$\sigma = E\gamma$$

$$E = \left\{ \frac{1-\lambda}{E_{\mathrm{PBA}}} + \frac{\lambda}{\phi E_{\mathrm{NR}} + (1-\phi) E_{\mathrm{PBA}}} \right\}^{-1}$$

where λ and ϕ mean volume fraction of the fragments in series and parallel models, respectively, and E_i is a storage modulus of i component, that is, $E_{NR} = 10^5$ Pa and $E_{PBA} = 10^4$ Pa.

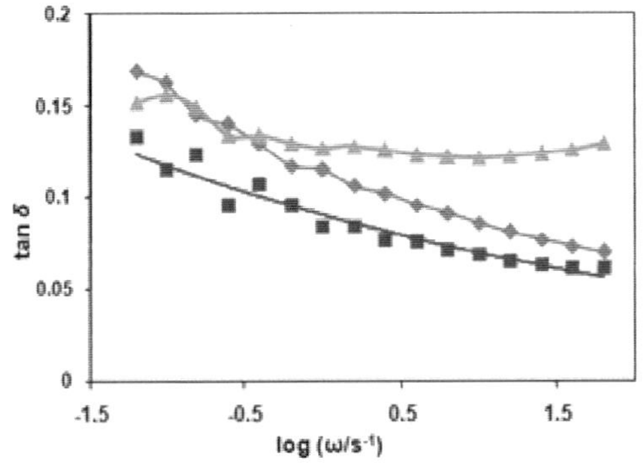

Fig. 7 Loss tangent versus frequency for (A) DPNR, (B) DPNR-*graft*-PS with hard nanomatrix structure and (C) DPNR-*graft*-PBA (PBA: ca. 11%) with soft nanomatrix structure.

Fig. 7 shows plot of loss tangent, tanδ, versus frequency for DPNR, DPNR-*graft*-PBA prepared with 1.5 mol/kg-rubber feed of BA and DPNR-*graft*-PS prepared with 1.5 mol/kg-rubber feed of styrene. Value of tanδ of DPNR was in a region of 0.1 to 0.15, as in the case of literature value. The value of tanδ decreased over the observed frequency range after the soft nanomatrix structure was formed in DPNR, whereas it increased after the hard nanomatrix structure was formed. The decrease in tanδ for the soft nanomatrix structure is not expected, because the value of tanδ of PBA is reported to be higher than that of DPNR [9,10]. This discrepancy will be investigated in our subsequent study.

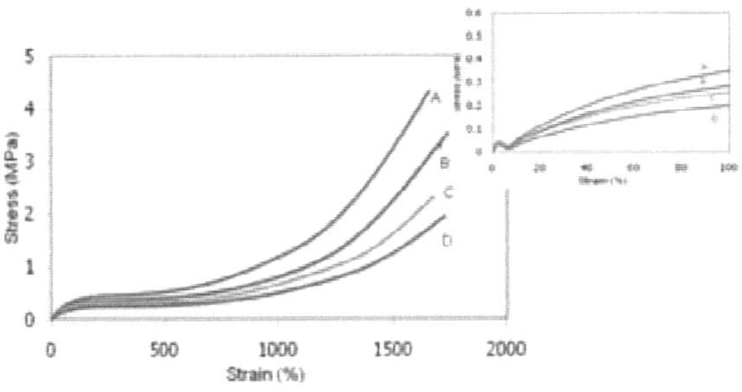

Fig. 8 Stress-strain curve for (A) DPNR-*graft*-PBA 1.0 mol/kg-rubber, (B) DPNR-*graft*-PBA 1.5 mol/kg-rubber, (C) DPNR-*graft*-PBA 2.0 mol/kg-rubber, (E) DPNR-*graft*-PBA 2.5 mol/kg-rubber.

Fig. 8 shows a plot of stress versus strain for the DPNR-*graft*-PBA prepared with various monomer concentrations. The highest tensile strength and elongation at break was shown for DPNR-*graft*-PBA prepared with 1.0 mol/kg-rubber feed of BA. The tensile strength of the DPNR-*graft*-PBA decreased gradually as BA content increased; that is, the higher the content of PBA, the lower is value of the stress at break of the DPNR-*graft*-PBA. The mechanical properties of the DPNR-*graft*-PBA could be significant rather than those of DPNR and DPNR-*graft*-PS, since the Poisson's ratio of PBA is similar to DPNR, which is distinguished from PS. The strain concentration of the DPNR-*graft*-PBA during streching was expected to be low, compared to DPNR-*graft*-PS. However, the tensile strength was reduced by forming the PBA nanomtrix into DPNR. This negative effect may be explained on the bassis of the empirical rule for multi component systems, in which the properties are governed by a matrix component. As is widly recognized, NR is the most superior rubbery polymer, which possesses the outstanding mechanical properties among rubbers, but PBA is not. Thus, when we form the PBA nanomatrix, the properties of the DPNR-*graft*-PBA must be lowered than DPNR. This is completely distinguished from the results obtained for DPNR-graft-PS. It is found that the hard nanomatrix results in remarkable reinforcement effects on the properties of NR, but the soft nanomatrix does not.

Conclusions

Graft-copolymerization of BA onto DPNR was carried out with TBHPO/TEPA as an initiator in latex stage, after deproteinization of NR latex with urea in the presence of SDS. The conversion and grafting efficiency of BA were dependent upon the feed of BA, which were more than 90 mol% under a suitable condition of the graft-copolymerization. At 1.5 mol/kg-rubber feed of BA, the rubber particles of about 0.5 μm in diameter were dispersed in the PBA matrix of about 15 nm in thickness, as observed by TEM. The values of storage modulus and stress at brake of DPNR were reduced by forming the PBA nanomatrix. The soft nanomatrix was found to be less effective to reinforce the rubber, which was distinguished from the hard nanomatrix reported in our previous papers.

Acknowledgements

This work was supported in part by a Grant-in-Aid (21655080) for Challenging Exploratory Research and Grant-in-Aid (223501000) for Scientific Research (B) from Japan Society for the Promotion of Science and JST-JICA SATREPS.

References

1. S. Kawahara; T. Kawazura; T, Sawada; Y. Isono *Polymer* 2003, *44*, 4527.
2. N. Pukkate; T. Kitai; Y. Yamamoto; T. Kawazura; J. Sakdapipanich; S. Kawahara *Eur. Polym. J.* 2007, *43*, 3208.
3. C. C. Ho; T. Kondo; N. Muramatsu; H. Ohshima *J. Colloid Interface Sci.* 1996, *178*, 442-445.
4. S. Kawahara; W. Klinklai; H. Kuroda; Y. Isono *Polym. Adv. Technol.* 2004, *15*, 181.
5 W. Klinklai; T. Saito; S. Kawahara; K. Tashiro; Y. Suzuki; T. T. Sakdapipanich; Y. Isono *J. Appl. Polym. Sci.* 2004, *93*, 555.
6. S. Kawahara; Y. Yamamoto; S. Fujii; Y. Isono; K. Niihara; H. Jinnai; H. Nishioka; A. Takaoka *Macromolecule* 2008, 41, 4510.

7. A. D. Roberts *Natural rubber science and technology*, Oxford University Press, Oxford, 1988
8. M. Takayanagi; S. Uemura; S. Minami *J. Polym. Sci. Part C, Polym. Sympo.* 1964, 5, 113.
9. C. Former; J. Castro; C. M. Fellows; R. I. Tanner; R. G. Gilbert *J. Polym. Sci. Part A Polym. Chem.* 2002, 40, 3335.
10. A. Zosel; G. Ley *Macromolecules* 1993, 26, 2222.

O-14-664

LASER CUT POLYMERIC STENTS WITH SHAPE MEMORY

Marek Kozlowski [1*], Andrzej Iwanczuk [1], Arkadiusz Antonczak [2], Bogusz Stepak [2]
1 – Faculty of Environmental Engineering, Wroclaw University of Technology, Wroclaw, Poland – marek.a.kozlowski@pwr.wroc.pl; andrzej.iwanczuk@pwr.wroc.pl
2 - Faculty of Electronics, Wroclaw University of Technology, Wroclaw, Poland – arkadiusz.antonczak@pwr.wroc.pl; bogusz.stepak@pwr.wroc.pl

Abstract - Polymers have been frequently used for a health care and medical devices, presenting a new field of applications in that group of materials. However, the medical grade polymers should be selected properly depending on their desired function. Those used in packaging should be different of that applied in the surgery or drug delivery. Therefore, the chemical structure of polymeric materials should be tailored specifically for the medical purposes. Selected application of the shape memory polymers in reconstructive surgery has been presented.
The copolymer based on the lactic acid has been used for manufacturing the polymeric implants. Influence of the material used for manufacturing of scaffolds on their properties has been discussed. The precise shaping of sophisticated implants was possible with a laser cutting. Throughout study on the technological parameters of a laser processing in conjunction with the material characteristics allowed for defining the polymer properties that are crucial for a precise laser cutting of a desired shape.
The study was performed for the commonly used RF-excited CO_2 laser (10.6μm) with the maximum average output power 25 W, beam quality factor $M^2 = 1.2$, pulse duration of the order of hundreds of μs, spot size of the laser beam 125 μm (@1.5" focusing lens) and pulse repetition rate *PRR* = 2–10 kHz. In order to achieve high precision of cutting (small kerf width) and reduce the heat affected zone *HAZ*, the polymer was cut on the high thermal conductivity substrate. Characteristics and comparison of tested copolymers have been presented and discussed.

Keywords: shape memory polymer, scaffold, laser cutting

Introduction

Smart materials which have been used in adaptive constructions are related to specific feature of a matter, which is the reversible phase transformation. That involve frequently the martensitic transformation in metals and glass transition in polymers. If the transformation occurs in a controlled manner within a defined range of external parameters, such materials may be implemented in functional devices. Geometry of such devices depends on pressure, temperature, or other activating parameters. Shape memory materials exhibit the above properties in a reversible manner, i.e. the material which has been deformed at lower temperature regains its original shape after heating to a higher temperature. The essential issue is a significant dependence of a selected material property to external parameter.
First objects of research in the area of shape memory materials have been metals, in particular alloys [1-2]. A few years ago shape memory polymers entered the market [3-7]. Such materials are thermoresponsive, with the glass transition temperature T_g and melting temperature T_m which constitute the material characteristic points between the glassy and rubbery state.
Shape memory polymers when processed with a conventional technology (injection moulding, press moulding, extrusion), then heated up above T_g, deformed and cooled down, keep a new shape. However, after heating again above T_g the materials recover to its original shape.

Experimental

The polymers used in this study were following:
- polylactid acid Ingeo 3052D (PLA), a NatureWorks product, designed for injection molding applications with high clarity and heat deflection temperature below 49°C;
- terpolymer of L-lactide-glycolide-trimethylene carbonate LP63 (PGA) delivered by the Centre of Polymer and Carbon Materials, Polish Academy of Sciences, Zabrze, Poland.

The polymers were characterised for their thermomechanical properties by means of DMA 2980 (bending mode), TA Instruments and for the heat flow using DSC Q20, TA Instruments.
The RF-excited CO_2 laser was used with the maximum average output power 25 W and the central wavelength of 10.6 μm. The laser beam with approximately Gaussian transverse intensity profile TEM_{00} (M^2<1.2) has been focused into spot with diameter of 125 μm by 1.5" FL lens.

Results and Discussion

Dynamic mechanical analysis has revealed how much a chemical structure influences properties of PLA and PLA-based terpolymer. PGA is stiffer than polylactide (storage modulus equals to 2.47 GPa and 1.66 GPa, respectively) but exhibits lower glass temperature (T_g) (maximum of a storage modulus at 66.66°C) in comparison to PLA (46.42°C) (**Fig. 1**). Such behaviour suggest the shape memory behaviour takes place in terpolymer at lower temperature, but a higher force of response should be expected.

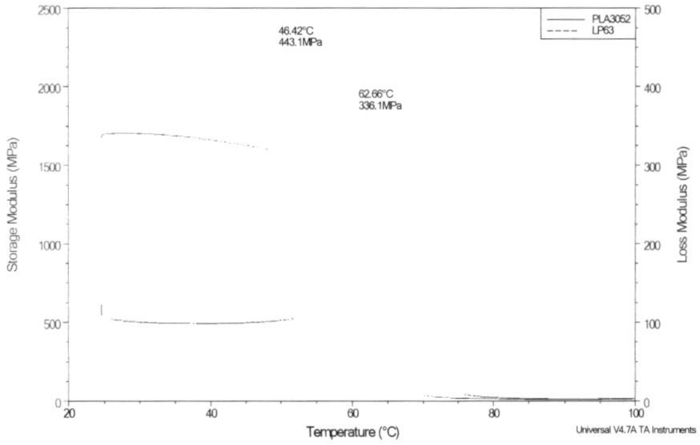

Figure 1. Storage and loss modulus for PLA and PGA

The heat flow tracks recorded at the differential scanning calorimetry measurements (**Fig. 2**) have confirmed that T_g value of terpolymer (49.35°C) is markedly lower than that of PLA (66.52°C). Moreover, PGA polymer started soften already at 44.83°C, which allows expecting first structural transformations in the terpolymer at that temperature.
Melting temperature of PGA corresponding to the maximum heat consumption was reported at 128.09°C, but melting of the terpolymer started around 108°C. PLA melting temperature was estimated to 148.42°C, whereas a heat consumption related to melting started at 118°C.

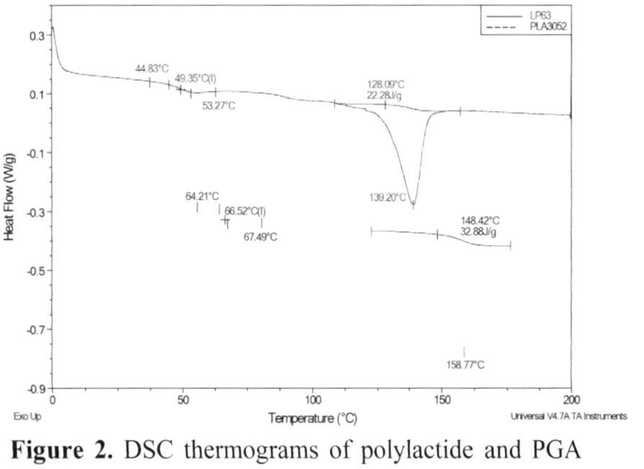

Figure 2. DSC thermograms of polylactide and PGA terpolymer

The research on the laser micromachining of the stents, were carried out in two stages. The first phase was aimed at determining the optimal process parameters for cutting biodegradable polymers using CO_2 laser. The second step concerned the creation of the complete, functional, stent prototype.
Because of a high energy density at the focal plane, the materials were rapidly heated and removal occured due to an explosive vaporization when a critical number of bonds were broken [8-10]. In this case, ablation of polymers occured mainly due to the photo-thermal phenomena, because of a long pulse duration, which lasted for ca. 50÷500 µs (depending on PRR).
Polymers were tested in a wide range of the process parameters: the average power, pulse repetition rate and pulse overlaping (scanning speed of the material). Response of a material to the laser processing was defined by measuring dimensions of channels created by the overlapped pulses. Each one has been measured taking into account both its depth and width – **Fig. 3**.
The impact of some parameters of the laser (as well as the type of a polymer) on the geometry of the cut profile (deep and width) has been shown in **Fig. 3** and **Fig. 4**, respectively. The measurements have been presented in the overlapping domain which indicates the percentage of overlapping for the following laser pulses.

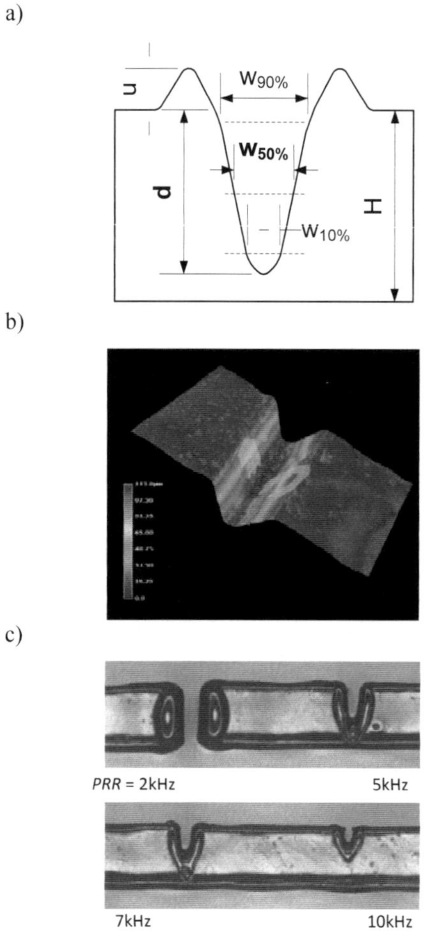

Figure 3. Cross-section of the cut material by a laser beam: a) definition with emphasized characteristic parameters: H – thickness of material, d – depth of cutting, $w_{10\%}$, $w_{50\%}$, and $w_{90\%}$ – the cutting widths at: 10 %, 50 % and 90 % of total depth, respectively; b) 3D profile of cut polymer measured by Keyence digital microscope; c) cross-section of cut channels for different pulse repetition rate *PRR*.

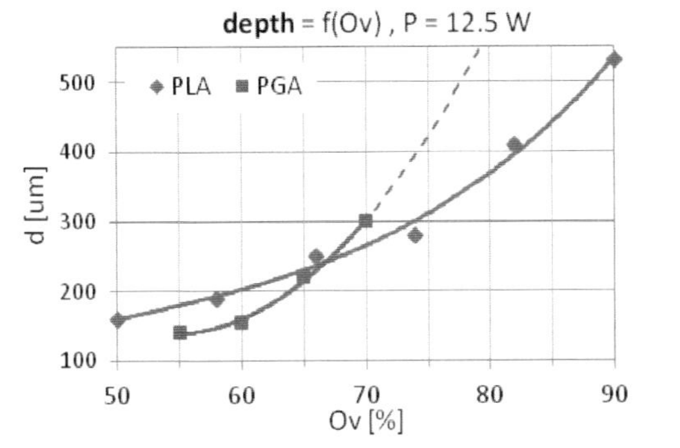

Figure 4. Depth of the channel formed by the laser as a function of the pulse overlapping.

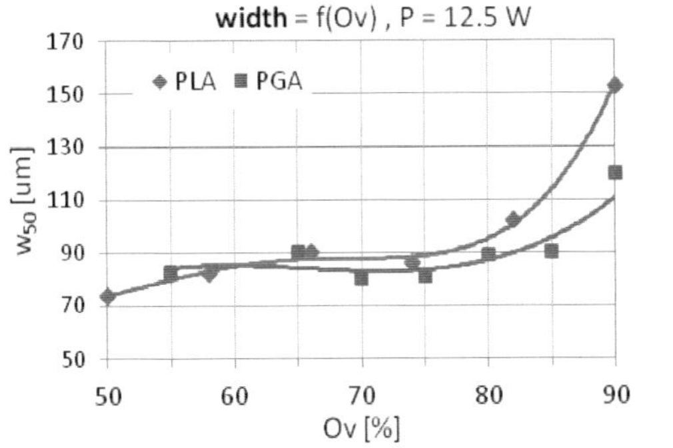

Figure 5. Width of the channel formed by the laser as a function of the pulse overlapping.

Comparing both grades of the cutting polymers (PGA and PLA), the difference in the processing of these materials has been noticed. It was assumed a priori, that optimal parameters should provide narrow cuts (small kerf width) and reduce the heat affected zone HAZ. Analyzing **Fig. 2** and **Fig. 3** in conjunction with **Fig. 6** one can conclude that PGA terpolymer allows for achieving narrower and deeper cuts, thus more precise material removal can be done than in case of PLA. The research performed led us to the manufacturing of a prototype stent (**Fig. 7**).

The best results have been obtained for the multiple (2÷3 times) laser passes. This allowed to achieve the highest precision and reduced a heat load to the material. The presented technology provides the ability to manufacture stents with structural dimensions lower than 300 μm.

a)

b)

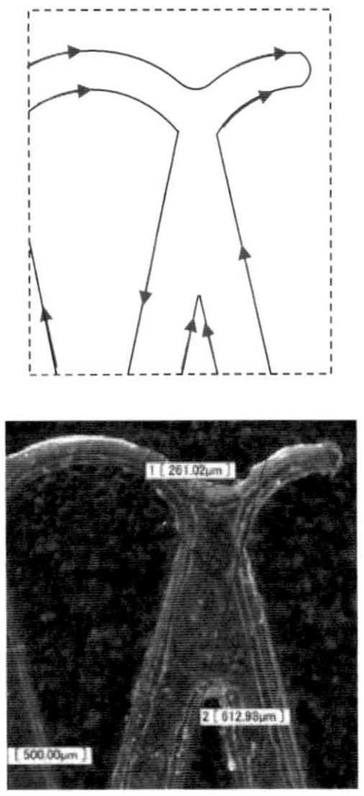

Figure 6. Manufacturing process of stent:
a) sequence of cutting (color of arrow designates different velocity); b) part of a cut stent;

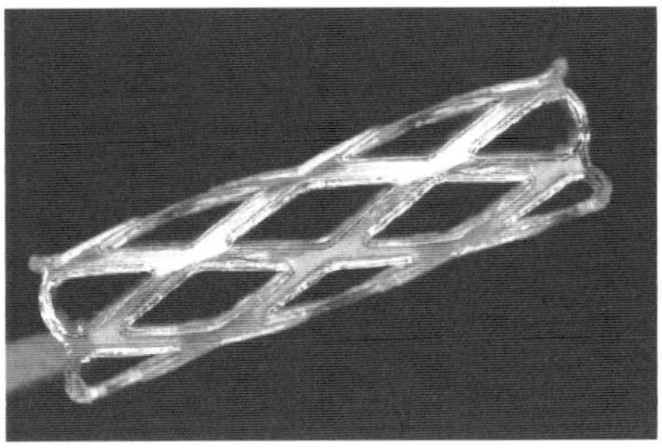

Figure 7. Complete 3D cylindrical stent.

Conclusions

1. Shape memory polymers represent a great potential for practical implementations in medical applications
2. Bioresorbable shape memory polymers are valuable materials for biomedical devices with a function of self-expansion after a temperature treatment.
3. Laser cutting allows precise shaping of a highly sophisticated geometry in polymeric matrices.
4. Selection of a proper polymer grade is of the first priority in designing the biomedical devices.

Acknowledgements

This research has been supported by the project "Polymeric resorbable surgery systems with shape memory" co-financed by the European Regional Development Fund and the Government of Poland.

References

1. A. Lendlein, S. Kelch, *Angew. Chem., Int. Ed.*, 2002,*41,* 2034.
2. N. Ma, G. Song, H-J. Lee, *Smart Mater. Struct.* 2004, *13,* 777.
3. M. Behl, A. Lendlein, *Materials Today*, 2007, *10,* 21.
4. S. Zhang, Z. Yu, T. Govender, H. Luo, B. Li, *Polymer*, 2008, *49,* 3205.
5. E. Zini, M. Scandola, P. Dobrzynski, J. Kasperczyk, M. Bero, *Biomacromolecules*, 2007, *8,* 3661.
6. T. Valimaa, S. Laasksovirta, *Biomaterials*, 2004, *25,* 1225.
7. C. Wischke, A.T. Neffe, S. Steuer, A. Lendlein, *J. Control. Rel.*, 2009, *138,* 143.
8. K. Zimmer, A. Braun in *Photo-Excited Processes, Diagnostics and Applications*, A. Peled, Ed., Springer, 2004, Chapter 11, 301.
9. R. Oritz, I. Quintana, J. Etxarri, *J. Appl. Phys.,* 2011, 110, 094902-094902-10.
10. S. V. Garnov, V. I. Konov, T. Kononenko, Laser Phys., 2004, *14,* 910.

O-14-449

DEVELOPMENT OF PRECISION INJECTION MOLDING BY COOLING RATE CONTROL AND STUDY OF SHRINKAGE CHARACTERISTICS OF CYLINDRICAL PARTS MOLDED WITH VARIOUS RESINS

M. Imaizumia[1]*, M. Ogawa[1], J. Kitayama[1], and H. Yokoi[2]

[1] *Mitsubishi Electric Corporation, 8-1-1 Tsukaguchi-Honmachi, Amagasaki, Hyogo 661-8661, Japan*

[2] *Institute of Industrial Science, The Univ. of Tokyo, 4-6-1 Komaba, Meguro-ku, Tokyo 153-8505, Japan*

Abstract - In this study, we proposed a precision injection molding by controlling cooling rate in which resins are injected into a mold which has been preheated to the appropriate temperature for the resin used (called initial mold temperature "T_M"), after which the core side of the mold is rapidly cooled to control the resin cooling rate and minimize the shrinkage of molded products. This molding method is already being practically applied to the molding of industrial products made of PPS. In order to clarify how this method can be applied to the molding of various resins, in this study, we attempted the injection molding of cylindrical pipe-shaped products with crystalline resins PPS, POM, and PBT, and non-crystalline resin PC. The shrinkage rates at the inner hole-circumference of the molded products were then measured and compared. The rapid cooling method did not show any effects for non-crystalline resins, but it was found to reduce shrinkage for crystalline resins only at initial mold temperatures above the glass transition temperature (T_g). On the other hand, with PBT, the longer the cooling time, the lower was the shrinkage rate. During cooling, stress reduced with the generation of thermal stress inside the molded product. Therefore, with PBT, as this stress decreases moderately, the effects of cooling time may have appeared more conspicuously. Based on these results, investigations were also conducted to clarify the types of resins and molding conditions for realizing better molding accuracy with this method, and to determine the resin shrinkage reduction mechanism with this method.

Keywords: Precision injection molding, Cooling rate control, Shrinkage characteristics, Cylindrical parts

Introduction

The dimensional accuracy of molded products is an important factor determining the functions of parts. However, resin shrinkage cannot be avoided during the injection molding process, posing as an obstacle in achieving high dimensional accuracy. In this study, the authors thus proposed the process "precision injection molding by cooling rate control" [1]. With this method, resins are injected into a mold which has been preheated to the appropriate temperature for the resin used, after which the part of the molded product requiring high accuracy is kept inside the mold and rapidly cooled. Generally, by setting the mold temperature low, the resin can be cooled rapidly to minimize shrinkage. But crystallization does not progress adequately in this case. One major advantage of the method proposed is that it realizes both crystallization and high dimensional accuracy which are opposing factors. This molding method is already being practically applied to the molding of high precision parts made of PPS such as valves and mobile phone chassis [2].

However, the mechanism by which high dimensional accuracy of parts is achieved with this method is still not known. In addition, there is a need to clarify for which resins this method is especially effective, and under what conditions. Thus in this study, we attempted the injection molding of cylindrical pipe-shaped products using a cylindrical mold and various resins, and investigated the shrinkage characteristics.

Experimental

Fig. 1 shows the mold cavity. To ensure even shrinkage of the test samples to facilitate the investigations, we used a cylindrical mold with a cavity thickness of 3mm. To minimize effects of shrinkage anisotropy due to the resin flow direction, we adopted a ring gate. Table 1 shows the resins used for the experiments. Three crystalline resins and one non-crystalline resin were selected focusing on T_g and its relation with room temperature. Table 1 also shows

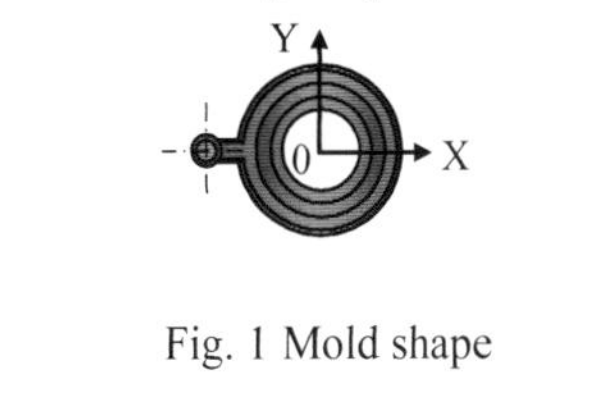

Fig. 1 Mold shape

Table 1 Molding materials

Resin		T_g		Resin Temp.
Crystalline resin	PPS	$T_g>T_r$	92°C	320°C
	PBT	$T_g=T_r$	22°C	260°C
	POM	$T_g<T_r$	−50°C	210°C
Non-crystalline resins	PC	$T_g>T_r$	151°C	300°C

T_r : Room temperature

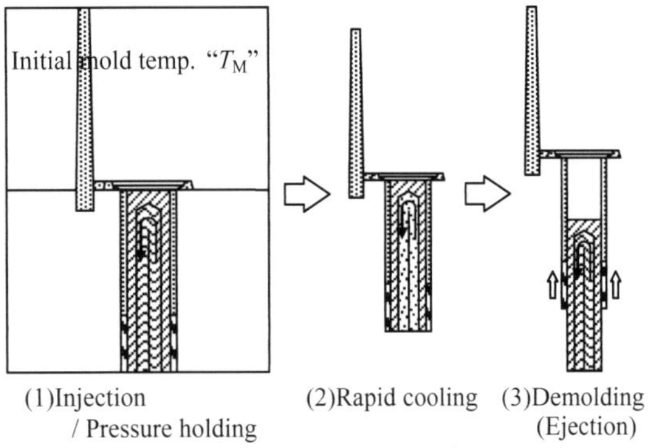

Fig. 2 Injection molding by cooling rate control

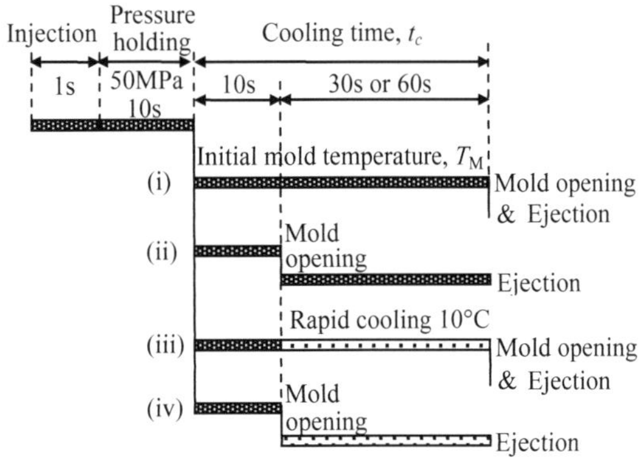

Fig. 3 Cooling chart of core pin

the resin temperatures during molding. Fig. 2 shows the outline of the molding process using the proposed method. Here, we assumed that high precision is required at the inner hole-circumference of the molded products. Thus, we first injected resin into a mold preheated to the appropriate temperature for that resin (Fig. 2 (1)). The mold temperature at this time is called initial mold temperature "T_M". With crystalline resins, crystallization can be promoted by setting this T_M to the appropriate temperature. After this, the test sample is rapidly cooled while keeping it inside the core pin (Fig. 2 (2)). After adequately cooling the test sample, it is removed from the core pin (Fig. 2 (3)). Fig. 3 shows the cooling method used in this study. The temperature of the core pin was controlled by either "normal cooling" which maintains the core pin temperature at T_M all the time, or "rapid cooling" which cools the core pin using 10°C water 10 seconds after the start of the cooling process. At this time, the mold for molding the outer circumference of the pipe-shaped is either closed or open. By combining the core pin cooling method and this open/closed state of the mold, four types of cooling methods were reviewed. To study the shrinkage characteristics of the test samples molded by each condition, the average inner diameter of the test samples was measured 40 mm deep from the gate using a 3-D coordinate measuring machine to calculate the shrinkage.

Results and Discussion

1. Effects of initial mold temperature on shrinkage characteristics

First, we investigated the shrinkage characteristics when T_M is varied. Figs. 4 (a) and 5 (a) show the results of crystalline PPS and PBT whose T_g is above room temperature. The cooling time was set at 60s. As for PPS, the shrinkage rates ((iii) and (iv)) for rapidly cooled test samples was constant regardless of T_M, while those of normally cooled test samples ((i) and (ii)) increased when T_M was above 90°C, widening the difference with the shrinkage rates for rapid cooling. However, when T_M was lower than 90°C, the shrinkage rate in normal cooling was more or less constant, and shrinkage control effects were no longer seen. In this way, the shrinkage characteristics of PPS vary significantly around 90°C, and this temperature was confirmed to match T_g (92°C). On the other hand, PBT showed the same shrinkage characteristics as PPS. At T_g (22°C), the shrinkage rates of normal cooling and rapidly cooling matched.

Next, Fig. 6 (a) shows the results of POM whose T_g (−50°C) is lower than room temperature. The shrinkage rates ((iii) and (iv)) during rapid cooling were more or less constant regardless of T_M. On the other hand, the shrinkage rates ((i) and (ii)) during normal cooling

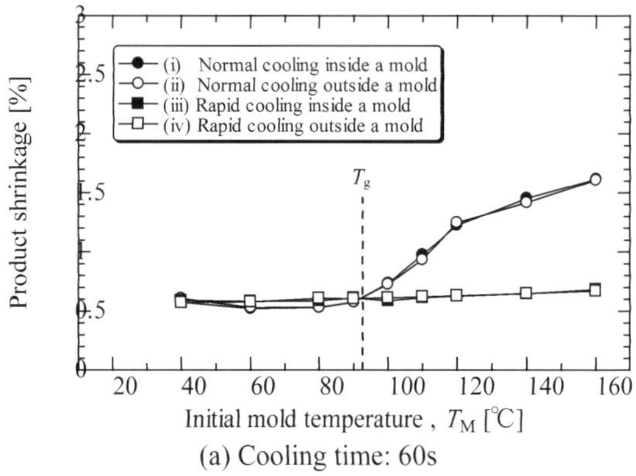

(a) Cooling time: 60s

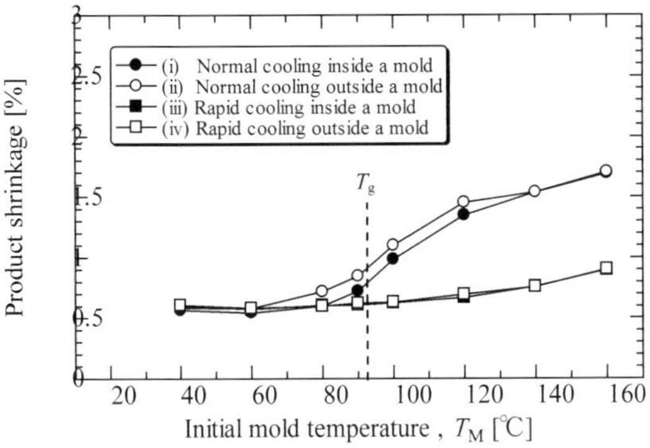

(b) Cooling time: 30s

Fig. 4 Effect of initial mold temperature on PPS product shrinkage

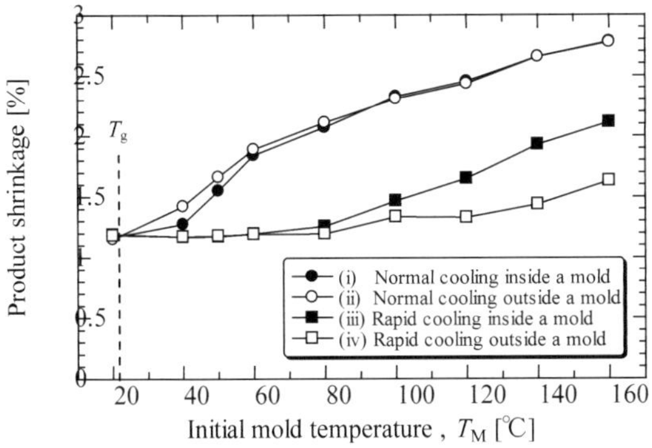

(a) Cooling time: 60s

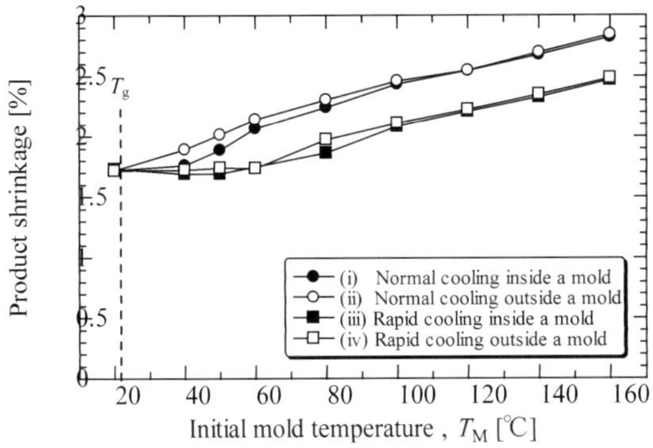

(b) Cooling time: 30s

Fig. 5 Effect of initial mold temperature on PBT product shrinkage

increased with increasing T_M to differ significantly from those of rapid cooling. However, the shrinkage rates of both cooling methods were the same at room temperature. These results also confirm that in the case of crystalline resins whose T_g is lower than room temperature, the temperature at which shrinkage control effects disappear does not always match T_g.

Finally, Fig. 7 shows the results for non-crystalline resin PC. Regardless of the value of T_M, there was no difference in the shrinkage rate between normal and rapid cooling.

Based on these results, the shrinkage mechanism is estimated to be as follows; when the test sample is rapidly cooling while being kept inside the mold, much of the thermal shrinkage that occurs when the temperature drops is absorbed by the non-crystalline part with fluidity. When the test sample is cooled to temperatures below T_g, the non-crystalline part loses its fluidity, and solidifies. For this reason, the thermal shrinkage from T_g to removal from the mold is replaced by thermal stress which acts on the test sample. Finally, this thermal stress is released after demolding, resulting in the elastic deformation of the test sample. The test sample also undergoes thermal shrinkage between the temperature at demolding to room temperature. Consequently, the thermal shrinkage from T_g to room temperature is not affected by the temperature at demolding, thus the shrinkage rate becomes stable. Even

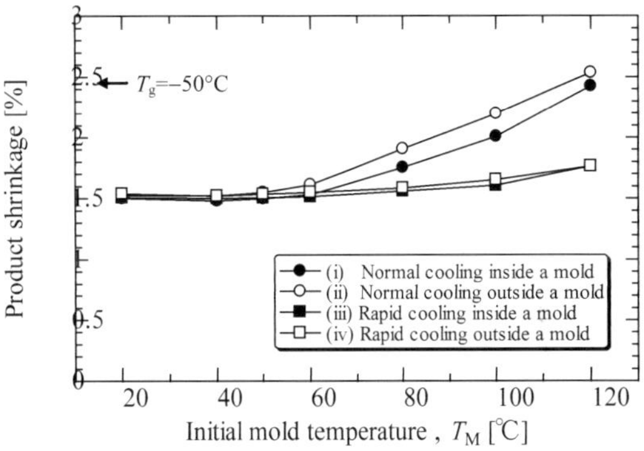

(a) Cooling time: 60s

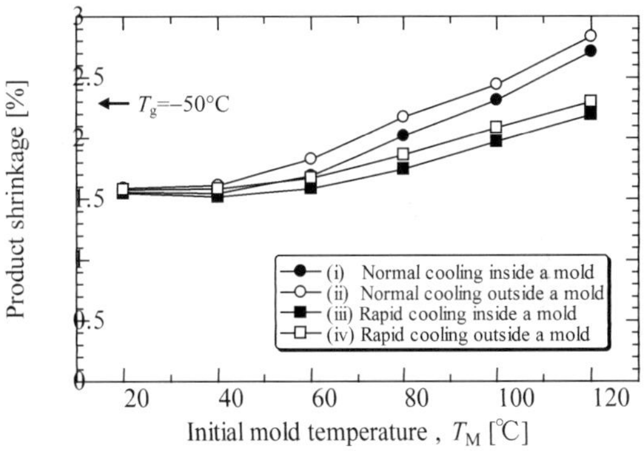

(b) Cooling time: 30s

Fig. 6 Effect of initial mold temperature on POM product shrinkage

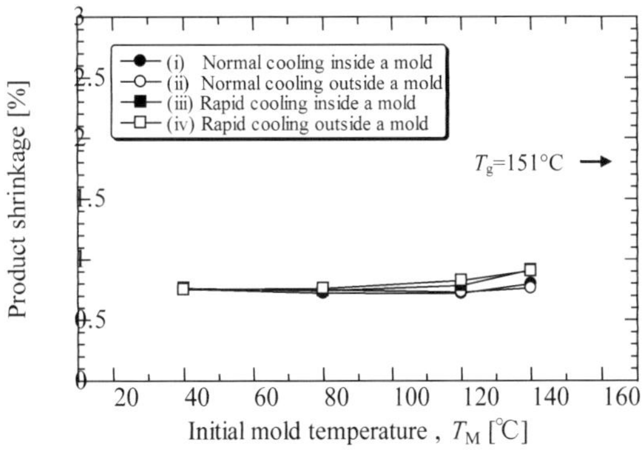

Fig. 7 Effect of initial mold temperature on PC product shrinkage (Cooling time: 60s)

in normal cooling, if T_M is below T_g, the shrinkage rate becomes constant owing to the same mechanism and the same value as that of rapid cooling. However for resins whose T_g is below the room temperature such as POM, even if rapid cooling is carried out with T_M set to around room temperature, no significant temperature difference is seen with normal cooling, and no shrinkage difference is seen around room temperature. In addition, as non-crystalline resins do not solidify above T_g, the mold temperature must be below T_g at least in order to carry out molding. For this reason, shrinkage control effects cannot be obtained by rapid cooling for non-crystalline resin.

2. Effects of cooling time on shrinkage characteristics

Shrinkage characteristics were compared at different cooling times of 60s and 30s. Figs. 4 (b), 5 (b), and 6 (b) show the results for crystalline resins PPS, PBT, and POM respectively when the cooling time is 30s.

When the cooling time is shortened, the shrinkage rate increases during rapid cooling with the increase in T_M, thereby reducing the difference in the shrinkage rate between rapid cooling and normal cooling. When the cooling time is 30s, the time to rapidly cool the core pin is 20s. This may be why the effects of rapid cooling can only be seen near the surface of the test sample. The higher T_M, the greater will the chance be for the test sample to be demolded without being cooled adequately, which increases the thermal shrinkage later. This shrinkage characteristic is thought to vary due to thermal and mechanical properties of various resins such as thermal conductivity, specific heat, and elastic modulus.

As for PBT, the following unique shrinkage characteristic was confirmed. Comparison of the results in Figs. 5 (a) and 5 (b) for the cooling times of 60s and 30s showed that the shrinkage rate is markedly different at the two cooling times when T_M is 20°C. On the whole, the shrinkage rate is higher at the cooling time of 30s. However, no resins except PBT showed any difference in shrinkage rate between the two cooling times. This is not only because of the difference in the test sample temperature due to the cooling time, but may be because of the following reason; while the test sample is inside the core pin, thermal stress occurs, but at the same time, stress is also relaxed according to the test sample temperature. Around the T_g temperature, as properties such as elastic modulus change significantly, the cooling rate around T_g has a major effect on the generation and relaxation of thermal stress. In this experiment, as rapid cooling was carried out using 10°C water, this may be why the effects of cooling time were conspicuous for PBT whose T_g is close to room temperature. However, further investigations are required for clarifying the mechanism which determines the shrinkage rate when the generation and relaxation of thermal stress are in good balance.

Although not shown in the figure, for the non-crystalline resin PC, even when the cooling time was shortened from 60s to 30s, no significant changes were seen.

3. Effects of cooling method on shrinkage characteristics

Finally, we compared the difference in shrinkage rate between inside-mold cooling and outside-mold cooling at the cooling time of 60s.

First, in normal cooling for PPS shown in Fig. 4 (a), no difference was seen in the shrinkage rate between inside-mold and outside-mold cooling. On the other hand, for the PBT and POM shown in Figs. 5 (a) and 6 (a), the shrinkage rate is slightly higher in outside-mold cooling compared to inside-mold. In outside-mold cooling, the cooling rate at the outer circumference of the pipe is slower than that at the inner circumference. In contrast, in inside-mold cooling, both the outer and inner circumferences are cooled evenly. With PBT and POM, as the non-crystalline portion is in the fluid state even at room temperature, the cooling rate along the wall thickness of the mold is thought to affect shrinkage characteristics.

Next, in rapid cooling, no conspicuous difference in the shrinkage rate was seen between inside-mold and outside-old cooling except for PBT. This may be because rapid cooling at the inner circumference of the pipe is more dominant, and as a result, the effects of the outer circumference become relatively smaller and difference in the shrinkage rate due to the cooling method does not appear clearly. However, when T_M becomes relatively higher than T_g, the temperature at the outer circumference of the pipe contacting the mold cannot be ignored. If the test sample is demolded at a large temperature gradient, the temperature at the inner circumference of the pipe is increased once again by the heat of the outer circumference to above T_g, therefore increasing shrinkage. With PBT, the shrinkage rate of inside-mold cooling is thought to be higher than that of outside-mold cooling.

Conclusions

In order to clarify the mechanism of shrinkage in the method proposed by the authors "precision injection molding by cooling rate control" which aims to control the shrinkage of molded products to enhance precision, and determine the appropriate molding conditions, in this study we investigate three factors thought to affect the shrinkage characteristics of various resins. As a result, the ideal molding conditions for the proposed method were found to be as follows; (i) using crystalline resins as the appropriate resin, (ii) set an initial mold temperature T_M which ensures adequate crystallization at a moderate cooling rate, and fill the resin into the mold, (iii) then leaving the mold open, complete crystallization with the molded product inside the core pin, and wait, (iv) after this, rapidly cool the molded product to control thermal shrinkage after demolding.

However, the shrinkage characteristics clarified in this study require both the generation and relaxation of thermal stress during the cooling process to be taken into account. These characteristics can therefore only be analyzed using viscoelastic models. Clarification of the detailed shrinkage mechanism using viscoelastic models needs to be undertaken in future studies.

References

1. M.Imaizumi, J.Kitayama, M.Tamura and K.Saito; *Mitsubishidenki-Giho*, 2007, *81*, 803.
2. M.Imaizumi, M.Ogawa, J.Kitayama, and H.Yokoi in Asian Workshop on Polymer Processing 2011 Proceedings, China, 2011, 205.

O-14-428

STRUCTURED POLYMER SURFACES TO INFLUENCE STEM CELL BEHAVIOR

M. Katschnig[a*], C. Holzer[a], S. Laske[a]
S. Haubenwallner[b], U. Schäfer[b], N. Andraschek[b], U. Fasching[b], S. Patz[b], M. Absenger[b]
W. Balika[c], M. Wagner[c]
[a*]*Department of Polymer Engineering and Science, Chair of Polymer Processing, Montanuniversitaet Leoben, Austria – matthias.katschnig@unileoben.ac.at*
[a]*Montanuniversitaet Leoben – clemens.holzer@unileoben.ac.at; stephan.laske@unileoben.ac.at*
[b]*Medical University of Graz – stefan.haubenwallner@medunigraz.at; ute.schaefer@medunigraz.at; natascha.andraschek@medunigraz.at; ulrike.fasching@medunigraz.at; silke.patz@medunigraz.at; markus.absenger@medunigraz.at*
[c]*Sony DADC Austria AG – werner.balika@sonydadc.com; Manuela.wagner@medunigraz.at*

Abstract - In vivo cells are influenced by a three-dimensional, dynamic extracellular matrix (ECM) at a micro- and nano-scale. Mimicking this control mechanism (e.g. to optimize implants) in vitro with structured polymeric cell culture scaffolds is the goal of this interdisciplinary project.
Basic requirement for cells to interact with a surface is sufficient adhesion. Hydrophilic surfaces are known to increase adhesion. So for basic information we examined at first selected unstructured polymers with various treatments regarding surface activation (surface energy), surface chemistry (XPS) and topography (AFM). It can be said that due to surface treatments surface energy increased, which is equivalent to an increase of hydrophilicity. XPS demonstrated that due to surface treatment polar functional groups (mostly containing oxygen and nitrogen) were implemented in the surface. These groups seem to be important for cell behavior. AFM investigations indicated a notable topography impact (roughening, characteristic formations) because of surface treatment, but there was no measurable impact on cell adhesion and viability.
Based on the investigations of unstructured polymer surfaces PMMA and COP culture plates were structured with cubes and ridges by injection molding and selected cell experiments were done. Cells aligned on ridges on small μm dimensions, showing an elongated dipolar phenotype. At the same time adhesion was enhanced. Furthermore gene expression controlling cellular structure characteristics was attenuated on ridges. Cell behavior on polymeric cubes at these dimensions was comparable to cells cultivated on unstructured substrates.

Keywords: polymer surfaces, microstructures, injection molding, cell behavior.

Introduction

There are distinctive environmental properties, to which cells respond. Physicochemical parameters like surface energy, surface charges and chemical composition are important [1, 2]. Cells attach to the surface by binding between integrins (protein receptor on cell surface) and surface adhesion proteins like fibronectin [3] or directly by cell membrane-surface connection. Based on the understanding of this biochemical-recognition process the first strategy employed in this work is to modulate and characterize the surface properties such as chemical composition, hydrophilicity (surface charges) and roughness to a state where cells can attach.
Cells can react to material stiffness [4, 5], which is not investigated in this work.
Micro- and nano-patterning, either topographical or chemical, can also manipulate cell behavior in terms of adhesion, migration, proliferation, shape and differentiation (gene expression) and is recently seen as one of the most important issues for cells [6-12]. This structure-recognition process leads to the second work package in this work where selected polymer microstructures were fabricated and cell impact was investigated.

Experimental

Polymer Selection

In the beginning some literature research was done to list often used polymers for cell culture dishes.
The outcome indicated hydrophobicity (low surface energy) and hydrophilicity (high surface energy) to be the initial parameter affecting protein adsorption and cell response. Hydrophobic materials for example were favored by Vogler [13] arguing that adsorption of proteins from aqueous solution is thermodynamically favored. On the other a hand it was demonstrated that hydrophilic surfaces are necessary to enhance adherence. Polar groups like carbonyls, hydroxyls or amines are needed therefore to immobilize cells via protein interaction on the formerly inert polymeric substrate.
To manage a wide range of wettability for testing COP/PP (hydrophobic), PS/PC (medium wettability) and PMMA/PLA (hydrophilic) were selected as base polymers. These polymers also managed sufficient

transparency, medical grade and appropriate processing properties. All polymers were injection molded to a microscopy standard format (L x W x H: 75 x 25 x 1°mm) for a clean and plane surface and easier handling
As commercial reference standard polystyrene cell culture dishes, that are commercially available, were used (CACD1, CADC2).

Structure Design
Based on literature study and technical considerations for COP and PMMA structuring cubes and rectangular walls were selected as basic design (Fig. 1). At first structures in the lower micrometer range were fabricated to collect information for future nano-structuring.

Figure 1: Basic design concept of structures: cubes (left) and rectangular walls (right).

Surface Activation by Plasma Treatment
In order to activate the polymer samples they were exposed to plasma, which either consists of either 100°% nitrogen (N_2) or oxygen (O_2). Through the high energy plasma bonds in the surface are broken, reactive links are generated and single atoms or molecule fragments can be integrated into the surface.
Plasma systems contain a vacuum chamber for the samples, a controlled process gas supply and a generator that ionizes the process gas in the chamber.
The plasma treatment was conducted with a low pressure plasma system with a vacuum chamber and a generator frequency of 40 kHz (Diener electronic GmbH + Co. KG, Germany).

Surface Activation by Corona Treatment
Corona treatment is comparable to plasma treatment. However, a treatment chamber is missing and surface activation takes place by means of electrode discharge under atmosphere ("air plasma"). The surface reactions are similar to plasma treatment. Without process gas the uncontrolled implementation of functional groups proves to be disadvantageous in this case [19].
Corona discharge systems are generally based on one or several high-voltage electrodes arranged at a distance ("discharge gap") of 1 to maximal 12 mm upon a grounded plate. When the polymeric microscopy slides are guided through the discharge gap, the surface becomes oxidized due to bond dissociations on the surface.
The corona treatment was carried out with a prototypic pre-treater based on an Ahlbrandt electrode system (Ahlbrandt System GmbH, Germany) combined with a controlled electric drive.

Contact Angle and Surface Energy Measurements
Water contact angles measure directly the overall wettability of a surface, while the surface energy more precisely differentiate between the polar part (caused by polar atomic elements) and disperse part (caused by van der Waal forces) of wettability effects.
Thus, a 2 µl distilled, deionized water droplet for contact angles and in addition a 2 µl ethylene glycol droplet for surface energies were positioned on the slide surface and images were captured to measure the angle formed at the liquid-solid interface.
Then contact angle measurements (static sessile drop method) were performed on all polymer surfaces using the Drop Shape Analysis System DSA100S (Kruess GmbH, Germany). Mean values were calculated from at least six individual measurements. Reproducibility was assured by a maximum standard deviation of ± 2°. Furthermore an untreated COP slide served as reference.

XPS Measurements
X-Ray photoelectron spectroscopy (XPS) is using the photoelectric effect for energy measurements of electrons emitted from the polymeric ground level. These energy spectra give information about the chemical composition of the surface.
XPS studies were performed with a Kα photoelectron spectrometer (Thermo Scientific, UK). For the radiation an aluminum Kα source was used. For the survey spot size of 400 µm, pass energy of 200 eV and step size energy of 1 eV were selected.
Again an untreated COP reference was used to verify the reproducibility.

AFM Measurements
The Atomic Force Microscope uses a cantilever with a sharp tip (radius of curvature on the order of nanometers) at its end to scan the specimen surface topography. When the tip is placed near the sample surface, forces between tip and sample lead to a deflection of the cantilever that is measured by a laser-photodiode-system.
A DimensionTM 3100 Series AFM (Digital Instruments Incorporated, USA) performed all topography measures in tapping mode. Scan area was 5° x 5 µm, evaluated area was 4 x 4 µm to avoid fringe effects. Tip velocity was 10 µm/s, amplitude set point was on average 1,5 V and drive amplitude was on average 250 mV. As cantilever tip a Veeco OTESPA Wafer silicon probe (k = 42 N/m, f_0 = 300 kHz; Asylum Research, USA) for general samples was used. Data receiving was done by NanoScope Version 5.12 (Digital Instruments Veeco Metrology Group, USA). Data processing was conducted by Gwyddion Version 2.26 (Czech Metrology Institute, Czech Rep.).
Untreated COP references were used to verify the reproducibility.

Cell Culturing
The NTERA-2 cl. D1 cells were grown in Dulbecco's modified Eagle's medium supplemented with 10 % fetal bovine serum (FBS) at 37 °C in 95 % air/5 % CO_2 to 70 - 90 % confluence. The cells were washed with 12 ml phosphate-buffered saline (PBS) before detachment with 1 ml Accutase for 10 min.

2 x 10^4 cells/cm² NTERA-2 cl.D1 were cultured in Dulbecco's modified Eagle's medium supplemented with 10 % FBS at 37°C in 95 % air/5 % CO_2 in 1 cm² wells on the sample slides through the usage of a system based on Millipore microscope slides for 2 hours, 48 hours and 3 days depending on the analytical test.

MTT Tests
In order to primarily analyse the effect of the native chemical surface composition on the viability (MTT) of neural stem cells initially 2 x 10^4 NTERA-2 cl. D1 cells, a neuronally committed human teratocarcinoma cell line, were seeded on the polymer samples and after 48 hours after wash steps with PBS fixated with 3 % paraformaldehyde for 10 min and permealized with 0,1 % Triton X100 for 5 min. Rhodamine phalloidin (540/565 nm; Invitrogen) is used to stain the F-actin bundles in the cells and DAPI counterstain (358/461 nm; Invitrogen) is used to show the nucleus of cells. The stained cells were analysed by fluorescence with an Inverse IX51 inverted light microscope (Olympus, PA, USA).

Results and Discussion

Surface Energy of Polymers
There was an increase of surface energy and decrease of contact angles (Fig. 2) due to corona treatment observed in all polymer samples. This was caused basically by increasing the polar fraction due to oxidation. All samples were responding to an increasing corona output with enhanced surface energy. Surface energy and its polar fraction were significantly increased in all treated samples as compared to CADC controls.
Plasma treatment with nitrogen or oxygen exhibited similar results. There was an increase of surface energy and decrease of contact angles in all treated polymer samples due to increased polar fractions. PMMA had 51 mN/m (d: 19 mN/m; p: 32 mN/m) treated with nitrogen and 56 mN/m (d: 9 mN/m; p: 47 mN/m) treated with oxygen. On the other hand COP had 52 mN/m (d: 18 mN/m; p: 34 mN/m) treated with nitrogen and 59 mN/m (d: 11 mN/m; p: 48 mN/m) treated with oxygen. Therefore plasma treatment did have a similar effect on both polymers regarding surface energy.

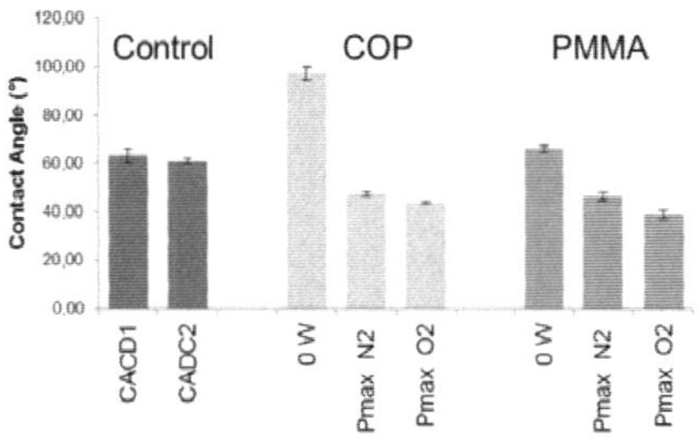

Figure 2: Contact angles (H_2O) of untreated and plasma-treated COP and PMMA compared to commercial references.

XPS of Base Polymers
Corona treatment implemented oxygen into less polar polymers (COP, PS) eliciting carbonyl groups, ether goups, hydroxyl groups etc.
Plasma results coincided with corona results, interestingly nitrogen process gas did not lead to enhanced nitrogen bonding on the surface.

AFM of Base Polymers
There were no significant topography differences between all measured untreated polymer samples. The average roughness height was in the range of 1,4 nm (PMMA) to 2,5 nm (PS). This was to be expected, because the profile of all untreated polymer samples (except commercial references) is a negative "copy" of the same polymer processing tool surface, which roughness is in the measured range. The controls were also fabricated by injection molding, so roughness values were in a similar scale.
Significant corona impact on average roughness was only observed on PMMA and PP. PP showed strong roughening (10 nm at maximum power) which is mainly due to the weak abrasiveness of the base polymer.
Plasma treatment had a smoothing effect due to abrasive etching effects to COP and PMMA. PMMA was even smoother than the commercial references then.

Surface Recognition of Cells (MTT-Tests)
As demonstrated in Fig. 3, strong correlations between surface oxygen content and cell viability, in particular carbonyl groups, was observed (r = 0,93). Furthermore a correlation between nitrogen and cell viability was demonstrated (r = 0,72).
Based on this information COP and PMMA were identified as best polymers regarding cell reaction and further structure fabrication. Therefore these polymers were treated with plasma and corona after structuring for all further cell investigations.

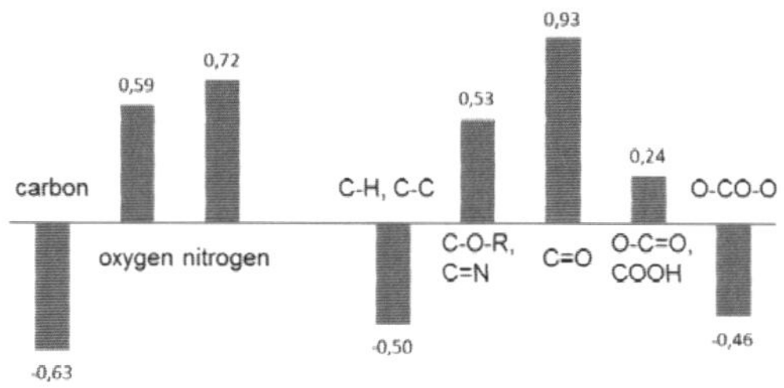

Figure 3: Correlation coefficients of surface chemistry and cell viability. Correlation was done with all commercial controls, untreated and treated polymers.

Contact Angles on Structures
One aim of this study was to analyse anisotropy of surface energy on structures. So the impact of treatment, structure and polymer was investigated by measuring water contact angles parallel (abbreviated: pa) and perpendicular (abbreviated: pe) to structure.
It could be demonstrated that untreated structuring leads to an increasing of hydrophobicity. Untreated

walls elicited this effect stronger perpendicularly than parallely ("spread"). COP showed spread (pe: 140°, pa: 112°) and PMMA also showed spread (pe: 122°, pa: 57°). Cubes indicated nearly isotropy (COP: 120°; PMMA: 83°) on both polymers. COP demonstrated generally significant higher hydrophobicity compared to blank control (control: 98°) and PMMA significant but less hydrophobicity regarding blank control (control: 67°).

Treated structures with corona or plasma, cubes and walls on COP, significantly changed their characteristics indicating more hydrophilicity then blank and treated controls (control: 42°). PMMA showed smaller but still present hydrophobicity compared to blank and treated samples (control: 39°). Spread tendency on treated walls remained the same like on untreated walls for both polymers (COP: 39° pe, 35° pa; PMMA: 42° pe, 35° pa). Cubes elicited isotropy again (COP: 35°; PMMA: 60°).

All in all it seems that contact angles are dependent from structure, base polymer, measuring direction and surface activation by plasma or corona treatment.

AFM of Structures

AFM profiles demonstrated high reproduction quality regarding injection molding. Additional structures possessed dimension stability over the entire structured area. Furthermore no corona or plasma treatment impact was determined on structure quality.

Surface Recognition of Cells (Morphology)

Microscopy pictures of adhered cells on treated structures demonstrated that cell morphology is strongly dependent on structure design.

Morphologies of cells cultured on cubes were comparable to morphologies observed on blank controls. On walls cells exhibited a neuronal-like, elongated morphology.

Conclusions

Cells showed on treated PMMA and COP better cell viability compared to commercial controls. Reason could be the basic polymeric chemistry in combination with enhanced nitrogen and oxygen groups implemented by surface treatment.

Cell morphology seemed to be strongly dependent on structure. A possible connection with anisotropic contact angle values is topic of further investigations.

It will be very interesting in future to explore the interdisciplinary of the polymer-cell interaction and more detailed the possible connection of bio-chemical recognition and structure-recognition of cells.

Next step will be the fabrication and characterization of nanostructures that possibly induce additive stem cell behavior.

Acknowledgements

We acknowledge the Chair of Synthesis of Special and Functional Polymers of Montanuniversitaet Leoben for the precious help in making the contact angle/XPS measurements and the Corona treatments.

Financial support was provided by the Austrian research funding association (FFG) under the scope of the BRIDGE program.

References

1. McNamara L. et al., Nanotopographical Control of Stem Cell Differentation, Journal of Tissue Engineering Volume 2010, Article ID 120623 (2010).
2. Ravichandran R et al., Effects of nanotopography on stem cell phenotypes, World Journal of Stem Cells 31, pp. 55-66 (2009).
3. Ma, Zuwei; Mao, Zhengwei; Gao, Changyou (2007): Surface modification and property analysis of biomedical polymers used for tissue engineering. In *Colloids and Surfaces B: Biointerfaces* 60 (2), pp. 137–157.
4. Yeung, Tony; Georges, Penelope C.; Flanagan, Lisa A.; Marg, Beatrice; Ortiz, Miguelina; Funaki, Makoto et al. (2005): Effects of substrate stiffness on cell morphology, cytoskeletal structure, and adhesion. Available online at http://www.uphs.upenn.edu/ime/janmey/Yeung.pdf, checked on 20/09/2012.
5. Leipzig, Nic D.; Shoichet, Molly S. (2009): The effect of substrate stiffness on adult neural stem cell behavior , Biomaterials 30, 6867–6878.
6. Y. Ito, G.P. Chen, Y. Imanishi, Micropatterned immobilization of epidermal growth factor to regulate cell function, Bioconjug. Chem. 9 (1998) 277–282.
7. Z.W. Mao, L. Ma, J. Zhou, C.Y. Gao, J.C. Shen, Bioactive thin film of acidic fibroblast growth factor fabricated by layer-by-layer assembly, Bioconjug. Chem. 16 (2005) 1316–1322.
8. C.S. Chen, M. Mrksich, S. Huang, G.M. Whitesides, D.E. Ingber, Geometric control of cell life and death, Science 276 (1997) 1425–1428.
9. R. Singhvi, A. Kumar, G.P. Lopez, G.N. Stephanopoulos, D.I.C. Wang, G.M.Whitesides, D.E. Ingber, Engineering cell-shape and function, Science 264 (1994) 696–698.
10. Z.W. Ma, S. Ramakrishna, Nanostructured Extracellular Matrix, Encyclopedia of Nanoscience and Nanotechnology, 7, American Scientific Publishers, California, USA, 2004, pp. 641–655.
11. A. Curtis, C. Wilkinson, Topographical control of cells, Biomaterials 18 (1997) 1573–1583.
12. R.G. Flemming, C.J. Murphy, G.A. Abrams, S.L. Goodman, P.F. Nealey, Effects of synthetic micro- and nano-structured surfaces on cell behavior, Biomaterials 20 (1999) 573–588.
13. E.A. Vogler, Structure and reactivity of water at biomaterial surfaces, Adv. Colloid Interface Sci. 74 (1998) 69–117.

ELECTRICALLY CONDUCTIVE LAYER FORMATION ON POLYMER SUBSTRATE USING NANO-SILVER-INK AND CO_2 LASER

T. Saito*, T. Kawaguchi, and I. Satoh

Department of Mechanical and Control Engineering, Tokyo Institute of Technology, O-okayama, Meguro-ku, Tokyo, Japan

tsaito@mep.titech.ac.jp, kawat@mep.titech.ac.jp, satohi@mep.titech.ac.jp

Abstract

As a technique for fabricating flexible and minute electrical wiring, use of nano-metal-ink is attracting people's attention. In this study, nano-silver-ink was applied on PMMA substrate to fabricate an electrical conductive part, and the optimum condition of CO_2 laser irradiation for the agglomeration (or sintering) was experimentally determined. The electrical resistance of the agglomerated part was measured by Four Probe method, and the electrical resistivity was calculated by the layer thickness obtained through SEM observation. As a result, the electrically conductive path of 20 mm length, 2 mm width, and 110 nm thickness was successfully generated under the condition as follows; the laser intensity and the scanning speed of the laser were 240 W/cm^2 and 20 mm/s, respectively. The obtained electrical resistivity was 36.1×10^{-8} Ωm, and this was 22.7 times higher than the electrical resistivity of bulk Ag (1.59×10^{-8} Ωm).

Keywords: Electrical resistivity, Polymer substrate, Nano-silver-ink, CO_2 laser

Introduction

"Nano-particles" of about 1-100 nm diameter show characteristic behavior different from that of the bulk material, i.e. optical, magnetic, mechanical, and thermal properties. For this reason, the nano-particles do not only attract the researcher's interest, but also promote the expectation of industrial application. There are several examples, such as a color filter using the plasmon resonance by metallic nano-particles [1] and a high-density magnetic recording media by magnetic nano-particles [2].

Because the metallic nano particles have high surface activity, the particles can easily agglomerate each other and form continuous phase at relatively low temperature. By using this feature, a new type of wiring technique that draws a pattern on the substrate using inkjet printing have been developed recently [3, 4]. To cause the electrical conduction to the printed ink, thermal treatment process of whole substrate is needed. The agglomeration of the particles occurring at a lower temperature is preferable to prevent the thermal damage of the substrate.

In this study, the effect of the thermal treatment on the electrical conductivity of the nano-silver-ink is investigated first, and then the nano-silver particles spin-coated on PMMA substrate is agglomerated by the scanning of the CO_2 laser irradiation. Through the investigation, the direction of the laser irradiation that can suppress the thermal damage of the substrate is discussed. From the consideration of the amount of energy brought by the laser irradiation, effective approach of the agglomeration control of the nano-silver particles by the laser irradiation is studied.

Test Sample and Experimental Procedure

-Nano-silver-ink-

Because the diameter of the nano-silver-particles is several tens nm, its surface area per volume is very large. For this reason, drop of the melting point arises compared with the bulk silver. Moreover, the high surface activity makes the nano-particles agglomerated easily. Therefore, the agglomeration control is needed to keep dispersing the nano-particles. There are two methods to prevent the agglomeration of the nano-particles; one is to apply the electrostatic repulsion and the other is to form a coat layer of polymeric material on the particle surface.

-Sample preparation and evaluation-

In the experiment, colloid liquid including nano-silver-particles (manufactured by Mitsubishi Materials Electronic Chemicals Co., Ltd.) was used, and its contents is as follows: Ag 10 wt.%, water 30 wt.%, ethanol 54 wt.%, methanol 6 wt.%. This colloid liquid utilizes a coat layer of polymeric material on the particle surface to prevent the agglomeration. Therefore, the agglomeration of the nano-silver-particles begins to be caused by removing the coat layer by the external excitation, and then, the electrical conduction arises. In the sample preparation procedure, nano-silver-ink was spread on a glass substrate or a PMMA substrate by a spin coater. The substrates used in this study had 20 mm length, 5 mm width, and 1 mm thickness. The applied ink was dried for 24 hours at the room temperature. The thermal treatment process to remove the surface coat of the nano-silver-particles was done by the electrical oven or the scanning of CO_2 laser (maximum power of 30 W).

-Differential scanning calorimetry (DSC) of the nano-silver-ink used-

The nano-silver-particles contained in the ink used in this study are coated with a polymeric material to prevent the agglomeration. To measure the thermal degradation temperature of the polymer layer, DSC of the used nano-silver-ink was performed from the room temperature to 300°C. Figure 1 shows the result, and the heating rate of this case was 10°C/min. The data showed that the thermal degradation temperature of the coat layer was approximately 180-190°C. The electrical conductivity of thermally processed nano-silver-layer was measured by the Four Probe method. The surface morphology of the nano-silver-layer was observed by the scanning electron microscopy (SEM).

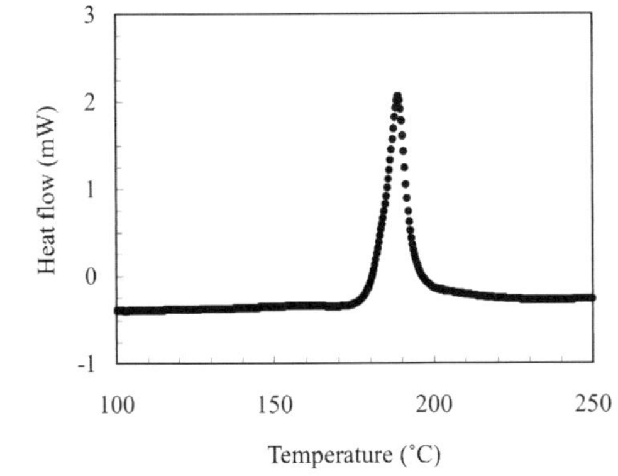

Figure 1 DCS result of the nano-silver-particles (The thermal degradation of the coat layer was exothermic phenomena.)

Results and Discussion

-Effect of thermal treatment time on electrical resistivity-

As an example of the effect of heat conditions, the electrical resistivity change with the thermal treatment time is examined. Figure 2 shows the result of which the sample was thermally processed at 190°C. Basically, the electrical resistivity decreased with the extension of the treatment time. In the temperature condition of 190°C, however, the value of the electrical resistivity did not reach to the electrical resistivity of bulk silver even if the treatment time was 50 min. In addition, the electrical conductivity did not arise at the initial condition (heat time 0 min.), thus the electrical resistivity of the initial condition was considered as infinite value.

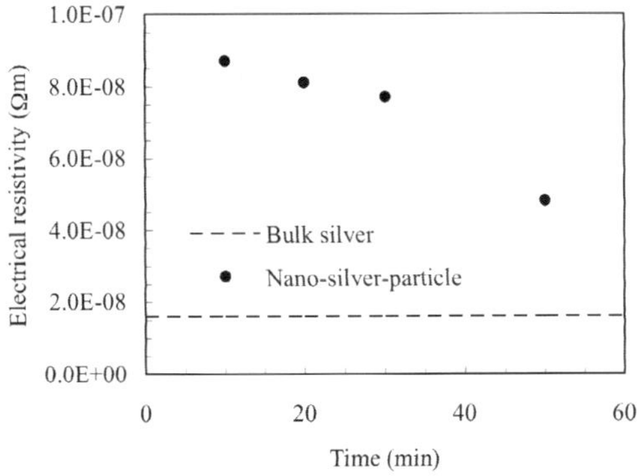

Figure 2 Electrical resistivity change with thermal treatment time

-Surface morphology observation by SEM-

The surface structure of the samples obtained in the previous section was observed by SEM. Figure 3 shows the obtained images at different process time. Although it was difficult to quantitatively connect the decrease of the electrical resistivity with the state of the agglomerate progress of the nano-silver-particles only from the photos in the figure, it seemed that there was a certain coincidence between them. Considerable decrease of the electrical resistivity was obtained even under the condition that the continuous phase of the nano-silver-particles was not formed enough because of the relatively short processed time (10 min.). This result was considered to be based on the good electrical conductivity of silver itself. On the other hand, the agglomeration size of the nano-silver-particles gradually increased after that. It was understood that this phenomena corresponded to the gradual decrease of the electrical resistivity shown in Fig. 2.

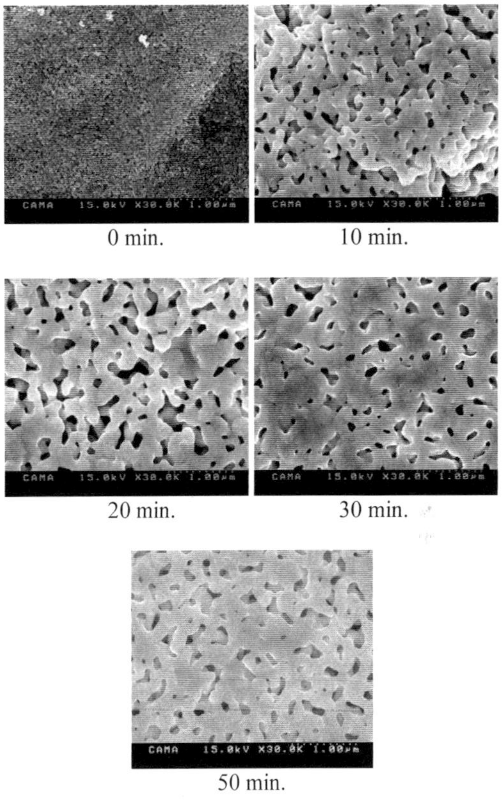

Figure 3 SEM images of the nano-particle surface (the morphology variation with the thermal treatment time)

-Occurrence of electrical conduction by irradiation of CO_2 laser-

When the CO_2 laser is irradiated to the substrate of polymeric material, the melting of the substrate will occur if the intensity of the laser is excessively high. The condition of the laser irradiation that realizes the agglomeration of the nano-silver-particles without the thermal damage of the substrate was explored in trial and error manner, and the obtained result was as follows; laser intensity 240 W/cm^2, and scan speed 2.0 cm/s. To discuss the effect of the condition of laser irradiation, the laser intensity I (W/cm^2) was changed together with the scanning speed v (cm/s) so that the amount of radiation energy given to a local point Q [J/cm^3] was kept constant.

$$Q = \frac{I}{v} \qquad (1)$$

Table 1 Thermal treatment condition by the laser irradiation

	A	B	C
Laser intensity (W/cm^2)	120	240	360
Scanning speed (cm/s)	10	20	30
Provided radiation energy (J/cm^3)		12	
Obtained electrical resistivity (x10^{-8} Ωm)	42.4	36.1	N/A

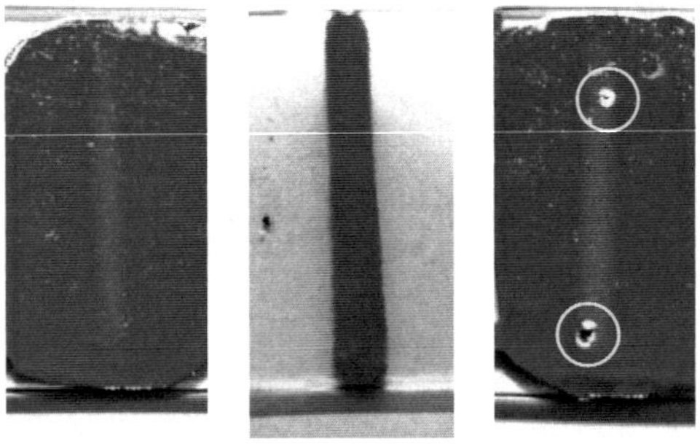

Figure 4 Optical micrograph of the samples after the thermal treatment by laser scanning; Left: Appearance of the sample obtained at case A, Center: Un-processed part was removed from the sample of case A, Right: Thermal damage was observed in case C.

The thermal process conditions by the laser irradiation, and the obtained results are summarized in Table 1. The state of the irradiated surface varied depending on the irradiation condition, though the amount of energy given by the laser irradiation at a local point was constant. In the case C (radiation intensity 360 W/cm^2), the thermal damage of the substrate was observed as shown in Fig. 4. As a result, it was experimentally concluded that the energy flux (or intensity) given by the laser was the dominant factor when the laser irradiation was used to control the agglomeration of the nano-silver-particles which was spread on the substrate of the polymeric material. The result also suggested that the agglomeration of the nano-silver-particles might be achieved without causing the thermal damage of the substrate, if the laser irradiation is controlled ideally.

Conclusions

In this study, the effect of the conditions of thermal treatment process on the agglomeration of nano-silver-particle was experimentally investigated to efficiently cause the electrical conduction. Based on the obtained results, the agglomeration of the nano-silver-particles spin-coated on the PMMA substrate was successfully formed by irradiating the CO_2 laser. The electrical resistivity obtained was 36.1x10^{-8} Ωm, and the value was 22.7 times higher than the electrical resistivity of bulk Ag (1.59x10^{-8} Ωm). By investigating the effect of the amount of thermal energy given by the laser irradiation, it was confirmed that the energy flux was the dominant factor of the present technique.

Acknowledgements

The authors thank Mr. Tomoya Komiyama and Mr. Satoru Kobayashi for their help with experiments. This work was partly supported by The Ministry of Education, Culture, Sports, Science and Technology of Japan (21686022, 24360077).

References

1. J. W. Oh, L. Choi, N. Kim, *Applied Optics* 2009, *48*, 3160-4.
2. X. Sun, Y. Huang, D. E. Nikles, *International Journal of Nanotechnology* 2004, *1*, 328.
3. Z. Liu, Y. Su, K. Varahramyan, *Thin Solid Films* 2005, *478*, 275.
4. K. Woo, D. Kim, J. S. Kim, S. Lim, J. Moon, *Langmuir* 2009, *25*, 429.

O-14-129

SUPERCRITICAL CARBON DIOXIDE-ASSISTED ELECTROLESS NICKEL PLATING ON POLYPROPYLENE SUBSTRATE—EFFECT OF INJECTION MOLDING ON HYDROPHILIC MODIFICATION WITH ANTISTATIC AGENTS—

S. Tengsuwan and M. Ohshima*

Department of Chemical Engineering, Kyoto University, Nishikyo-ku, Kyoto, 615-8510, Japan – pai@cheme.kyoto-u.ac.jp;, oshima@cheme.kyoto-u.ac.jp

Abstract - A supercritical carbon dioxide ($scCO_2$)-assisted electroless nickel–phosphorus (Ni-P) plating technique on a polymer substrate was developed for polypropylene (PP) substrates. The technique consists of three steps: the first step is the hydrophilic modification of PP, the second is the $scCO_2$-assisted impregnation of the substrate with $Pd(hfa)_2$, and the third step is the electroless plating reaction. However, the hydrophobic nature of PP makes the electroless plating reaction difficult because it hinders the diffusion of the aqueous plating solution in the polymer and reacting at the catalysts. It consequently results in a weak adhesion. In this study, we used an antistatic agent, which is a block copolymer between PP and polyethyleneoxide (PEO); PP-b-PEO, to improve hydrophilicity. We also prepared the PP/PP-b-PEO blends with three grades of PP with different viscosities to investigate the effect of viscosity ratio on blend morphology, concentration of PP-b-PEO and adhesiveness of metal layer to polymeric substrate. The blend morphology and the thickness of metal-polymer composite layer were observed by TEM. The adhesive strength of the metal layer to the polymeric surface was measured by peel-off testing. As the viscosity ratio of PP to PP-b-PEO approached to 1, the degree of elongation and orientation of PP-b-PEO domain near the surface at the skin layer increased, the metal-polymer composite layer, which was formed near the surface, increased and, consequently the adhesive strength of Ni-P layer increased. The uniform Ni-P metal film was successfully formed on a PP/PP-b-PEO blend substrates with an average adhesive strength of 8.8±1.8 N/cm to the polymeric substrate.

Keywords: Electroless plating, Polypropylene, Hydrophilic modification.

Introduction

In the automotive and electronics industries, plastic parts can be metalized for decorative or functional purposes. The electroless plating technique has been widely used for those purposes. However, the conventional electroless plating comprises a multistep process as illustrated in the left-hand side picture of Fig. 1, where the harmful oxidative acids are used to roughen the surface of polymer to create a strong adhesion of metal layer and a huge volume of wastewater is produced [1]. From the viewpoint of environmental concerns, we proposed a novel $scCO_2$-assisted process of electroless Ni-P metal plating on the polymeric substrates as an environmentally benign production process [2] (right-hand side picture of Fig. 1). In the developed process, $scCO_2$ is used as a carrier of $Pd(hfa)_2$ into polymers instead of using the strong and harmful acids for chemical etching and catalyst seeding. Our previous study found that a Ni-P metal-polymer composite layer (anchor) was formed at the interface between the metal layer and the polymer matrix during the electroless plating reaction and concluded that the thicker composite layer provided a stronger adhesion of the Ni-P metal layer to the polymeric substrate [2]. Moreover, we speculated that higher diffusion rate of the plating solution in polymer substrate can make the composite layer thicker and increase the adhesiveness. Because the major ingredient of the plating solution is water, the mass transfer rate of the plating solution in the polymer substrate was strongly affected by the hydrophilicity of the polymer. Unfortunately, most polymers are highly hydrophobic and thus our electroless plating method has been limited to polymers of which water absorption rate is fairly high, such as poly(methyl methacrylate) (PMMA), polyimide (PI) and polyamide 6 (PA6).

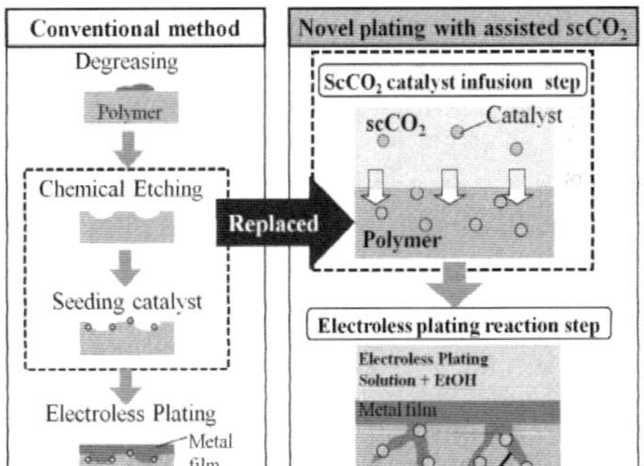

Fig. 1 Schematics of developed $scCO_2$-assisted electroless plating process

PP and acrylonitrile-butadiene-styrene (ABS) are two of the most commonly used polymers for automotive parts. Because of this higher degree of hydrophobicity, the PP and ABS substrates are barely metalized by our previous electroless plating method. The plating reaction occurred only on the surface of the PP and ABS substrates, and the composite layer did not form inside the substrate. Thus, a lesser anchor effect was provided and the adhesiveness was poor [3]. Therefore, the surface of the PP and ABS substrates must be modified to be hydrophilic to promote a stronger adhesion.

In this study, the hydrophilic property of PP was modified and enhanced with polypropylene(PP) polyethyleneoxide(PEO) block copolymer (PP-b-PEO). Namely, after dry-blending PP-b-PEO with PP, the blended polymer was injection-molded to prepare a polymeric substrate. The electroless Ni-P metal plating method was then conducted by simply immersing the modified PP into electroless plating solution after the impregnation of $Pd(hfa)_2$ by using $scCO_2$ infusion. The blend morphology of the molded substrate would strongly influence the sorption rate of the $Pd(hfa)_2$ as well as the plating solution. Thus, the blend morphology and the interface of the electroless plated metal and polymer surface were observed by TEM to confirm the location of the PP-b-PEO domains and the thickness of the metal-polymer composite layer respectively. The increase in the mass transfer of plating solution within the hydrophilic-modified polymer substrates were confirmed by the sorption test. The adhesive strength of the Ni-P metal layer to the polymeric surface was measured by peel-off testing after the metal layer was thickened by Cu electroplating. Furthermore, the effects of the process variables and process conditions on the adhesive strength of the Ni-P metal film to the polymer substrate were investigated.

Experimental

Materials

Three different PP i.e., two isotactic polypropylene (i-PP) with different tacticity 97% and 90% (Prime Polymer, F-133A and F-300SP) and random PP (r-PP) (Prime Polymer, E-330GV) were used to prepare the polymeric substrate. PP-b-PEO (Sanyo Chemical Ind., Pelestat® 300), which was developed as antistatic agents for PP, was used to enhance the hydrophilicity of PP substrates. Pd-complex ($Pd(hfa)_2$, Sigma-Aldrich) was used as the catalyst precursor. The electroless Ni-P plating standard solutions was prepared by mixing DK-M (10%) and DK-1 (5%) (Okuno Chem. Ind.) with distilled water (85%).

Substrate preparation

PP-b-PEO copolymer was dry-blended with PP at different weight ratios of PP to the copolymer, 95/5, 90/10, 85/15 and 75/25 and injection-molded into a flat plate. Injection molding was conducted by a 35-tonage clamping force injection molding machine (J35ELIII-F, Japan Steel Works Ltd.) at 210°C of barrel temperature with three different mold cavity temperatures, 40, 70 and 100°C. The reference substrates were also prepared from plain PA6 and PP under the same molding conditions.

Rheological measurement

Viscosity data of molten state sample polymers were measured by rheometer (ARES, Rheometric Scientific) equipped with parallel plate geometry. Dynamic temperature ramp tests starting from 220°C to 100°C at a cooling rate 2 °C/min with frequency 1 rad/s, strain rate 0.1% were performed.

Supercritical CO_2 impregnation of $Pd(hfa)_2$

Polymer substrates were placed in an autoclave with $Pd(hfa)_2$ at different loading amounts (1-25 mg) under the same $scCO_2$ condition (80°C, 10 MPa). The infusion of $Pd(hfa)_2$ into substrate were conducted for 120 min. Then, the reduction of $Pd(hfa)_2$ was performed thermally at 120°C for 45 min.

Electroless Ni-P plating reaction

A Pd-infused polymer substrate was immersed in the electroless Ni-P plating solution at 65°C, which is a mixture of the standard plating solution (60%) and ethanol (40%) [2]. Plating reaction time was 60 min.

Results and discussion

Part A: Hydrophilic modification and blend morphology

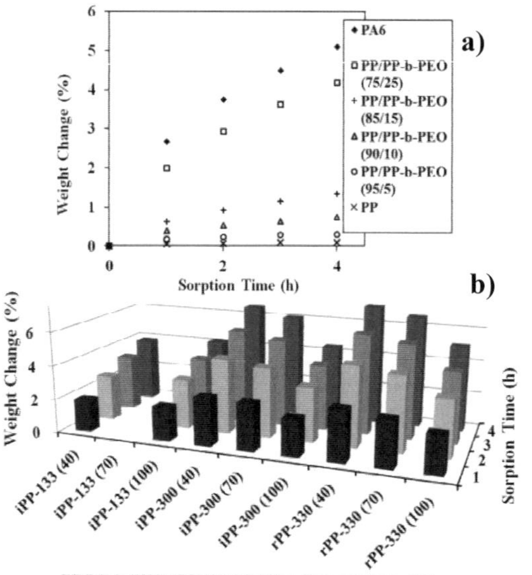

Fig. 2 Intake of plating solution in substrates

Sorption of plating solution to polymer substrates

Without infusing the $Pd(hfa)_2$, the polymer substrates were immersed in the plating solutions at 65°C to measure the solubility and diffusivity of the plating solution in polymer from the weight difference of the substrates before and after immersion. In Fig. 2(a) shows that the mass transfer of the plating solution in the polymer blend substrates increased with the increase of copolymer content. iPP(F133A) blended with PP-b-PEO at weight ratio, 25%, showed the hydrophilicity closer to PA6 level [3].

There was no large difference in hydrophilicity among neat PP substrates (F-133A, F-300SP and E-330GV). The weight gains of three grades of PP in the 24 hrs sorption test were in the range from 0.21 to 0.26%, which were very low. However, once they were blended with PP-b-PEO at the 75/25 weight ratio by injection-molding, the blends showed the different sorption ability of plating solution. The amounts of absorbed plating solution in PP/PP-b-PEO blends were different from the base PP grade and the mold temperature of injection machine as shown in Fig. 2(b).

Rheology data of the PP and PP-b-PEO copolymer

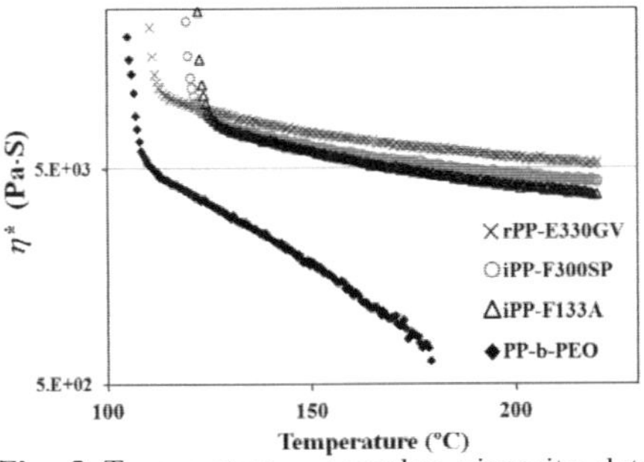

Fig. 3 Temperature - complex viscosity data of three PP and PP-b-PEO

Fig. 3 shows a complex viscosity, η^*, of three PP grades and PP-b-PEO samples by plotting η^* versus temperature. Fig. 3 clearly shows that the viscosities of three PP grades were different and extremely higher than the viscosity of PP-b-PEO.

Morphology of the PP/PP-b-PEO blend substrates

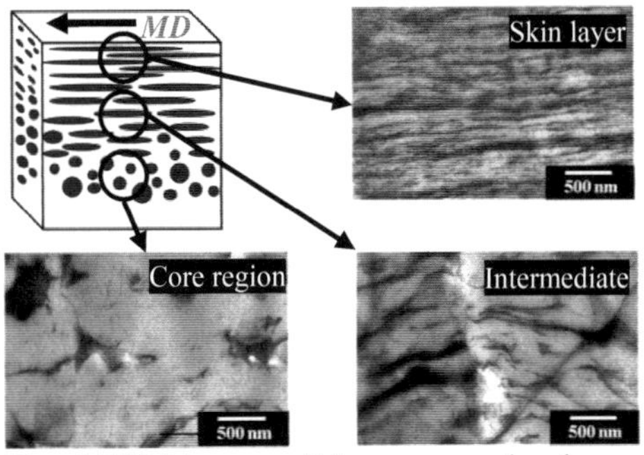

Fig. 4 TEM images of the cross-sectional area of iPP(F133A)/PP-b-PEO (75/25) blend substrates after staining. The dark regions represent PP-b-PEO domains and the white regions represent PP matrix.

Fig. 4 shows the blend morphology of iPP(F133A)/PP-b-PEO blend at different points of thickness direction (TD direction). When the polymer blends were injection-molded, the degree of orientation of copolymer (PP-b-PEO) domain (black) and the copolymer content were increased as the area was closer to the surface of the substrate. These morphology and concentration distribution could be caused by the difference in viscosity between the matrix polymer (iPP) and the PP-b-PEO domains, and also fountain flow behavior of injection molding.

Fig. 5 shows the blend morphologies of PP/PP-b-PEO (75/25) blended with three different PP grades and at two different temperatures of mold cavity. As shown in TEM images, the morphology was changed by PP grades and mold temperatures. As illustrated in Fig. 3, the complex viscosity of iPP(F133A) was closer to PP-b-PEO than other PP grades at temperatures higher than the melting points and the viscosity ratio of PP to PP-b-PEO became larger and apart from 1 as the temperature increases. Thus, the degree of elongation and orientation of PP-b-PEO domains in the skin layer of iPP(F133A)/PP-b-PEO is higher than in the other PP grades' blends. Therefore, it could be considered that the viscosity ratio of PP to PP-b-PEO could be a key of determining the morphology.

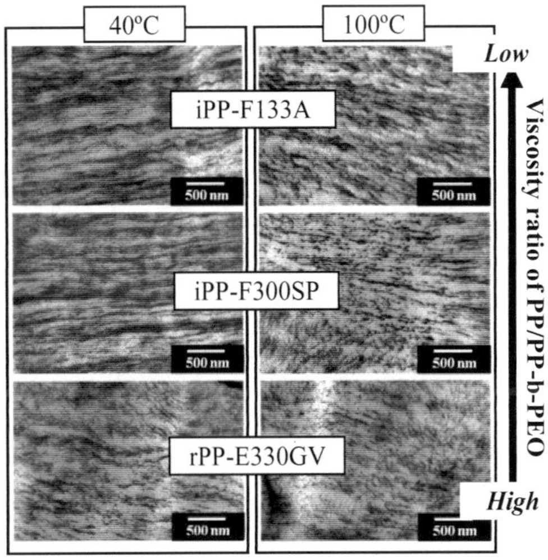

Fig. 5 TEM images of the cross-sectional area in the skin layer of the stained PP/PP-b-PEO (75/25) blend substrates (left column; mold cavity temperature 40°C, right column; 100°C)

Part B: Electroless plating of the plain polymer and the hydrophilicity-modified polymer substrates

Fig. 6 Image of a digital camera and optical microscope of surface of (a) Ni-P plated plain PP substrates and (b) Ni-P plated PP blend substrate

Because of poor hydrophilicity of PP alone, the metal layer electroless-plated on the neat PP substrates has many cracks and the plated metal layer was easily peeled off as shown in Fig. 6(a). On the other hand, a substantially-uniform Ni-P layer could be formed on PP/PP-b-PEO blend substrates (Fig. 6(b)).

Even though the Ni-P layer could be uniformly formed on the surface of iPP(F133A)/PP-b-PEO substrates with 5% PP-b-PEO blending, the adhesive strength of Ni-P metal to the substrates was weak and it was dramatically increased by blending 25% PP-b-PEO with PP [3]. Thus, in this study, adhesive strength of metal layer was measured for PP/PP-b-PEO with three PP grades at 75% weight percentage to PP-b-PEO.

Adhesive strength of Ni-P metal to polymer substrates

Fig. 7 shows the average peeling forces for the PP/PP-b-PEO (75/25) blend substrates, which were prepared by injection-molding PP-b-PEO with three different PP grades at two different mold cavity temperatures, 40 and 100°C. The stronger adhesive strength of the Ni-P metal layer to the polymeric substrates is the higher peeling force becomes. iPP(F300SP)/PP-b-PEO with blend ratio 75/25 injection-molded at mold cavity temperature 40°C shows the strongest adhesiveness of the metal to polymer substrate while iPP(F133A)/PP-b-PEO with blend ratio 75/25 stably showed the high adhesiveness against the change in mold temperature. The peeling forces of rPP(E330GV)/PP-b-PEO blend substrates were not so high as other PP grades' blends showed. Furthermore, PP/PP-b-PEO blend substrates prepared at 40 °C mold temperature showed higher peeling force than the substrates prepared at 100°C. Namely, as the degree of elongation and orientation of PP-b-PEO domain in the skin layer increased, the adhesiveness increased. It can be said that the oriented PP-b-PEO disperse domain morphology and the concentration near the surface could affect the adhesiveness between metal layer and polymeric substrate besides the overall hydrophilicity of the substrates.

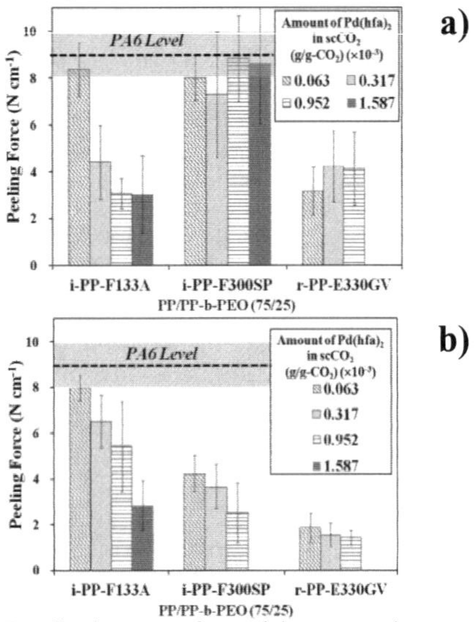

Fig. 7 Adhesiveness of metal layer to substrates, injection-molded at **(a)** 40°C and **(b)** 100°C molding temperatures. The blend substrates were treated under $scCO_2$ with different amounts of loaded $Pd(hfa)_2$ in infusion process

The higher amount of Pd infused in polymer substrate is not always the case that produces the higher adhesive strength of the metal to polymer as shown in Fig. 7. As can be seen in the peeling test results of PP/PP-b-PEO blend substrates prepared at 100°C mold temperature, an optimal Pd concentration exists to achieve the higher peeling force. For example, the concentration range of Pd loaded in high pressure autoclave on the infusion stage should be optimal within the range from 6.35×10^{-5} to 1.59×10^{-3} g-$Pd(hfa)_2$/g-CO_2 for the iPP(F300SP)/PP-b-PEO (75/25) blend substrate molded at 40°C.

Metal-polymer composite layer

The metal-polymer composite layers were formed at the interface between the electroless-plated metal and polymer substrate and they were observed by TEM. Fig. 8 shows TEM images of the cross-sectional area at the interface between the Ni-P metal layer and iPP(F300SP)/PP-b-PEO and rPP(E330GV)/PP-b-PEO (75/25) blend substrates which were injection-molded at 40°C molding temperature. The thickness of the composite layer was changed by the Pd concentration, PP grades, and mold temperature. As our previous study indicated, the thickness of the composite layer corresponds to the adhesive strength of Ni-P metal layer to the polymer substrate [3].

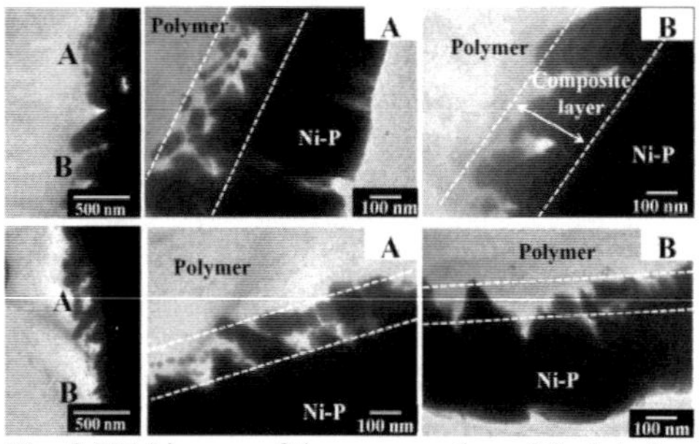

Fig. 8 TEM images of the cross-section of the interface between the Ni-P film and **(upper)** iPP(F300)/PP-b-PEO and **(lower)** rPP(E330)/PP-b-PEO (75/25) blends using concentrations of $Pd(hfa)_2$ of 6.35×10^{-5} g-$Pd(hfa)_2$/g-CO_2

Conclusions

The hydrophilicity of PP was modified by blending the PP-b-PEO and the $scCO_2$ assisted electroless Ni-P plating technique was applied. The modification could provide uniform metal layer with stronger adhesiveness of metal to PP substrates. The adhesiveness of Ni-P metal layer to PP/PP-b-PEO substrate was dominated by the thickness of metal-polymer composite layer. The thickness of the composite layer could be controlled by hydrophilicity modification and amount of infused Pd. In addition, the viscosity ratio of PP to PP-b-PEO and the molding conditions could affect the characteristic of the composite layer.

Acknowledgements

We appreciated Sanyo Chemical Industry for providing Pelestat® 300. We thank Dr. Taki for helping the TEM observation.

References

1. G.O. Mallory in *Electroless Plating : Fundamentals and Applications*, G. O. Mallory; J. B. Hadju, Ed.; AESF, Orlando, 1990; 1-56.
2. H. Adachi; K. Taki; S. Nagamine; A. Yusa; M. Ohshima *J. Supercrit. Fluids*, 2009, *49*, 265-270.
3. S. Tengsuwan; M. Ohshima *J. Supercrit. Fluids*, 2012, *69*, 117-123.

KN-MK-01

IMPROVEMENT OF BARRIER PROPERTIES OF POLYESTERS USING NANOCLAYS

P.J. Carreau*, A. Ghanbari, and M.C. Heuzey
CREPEC, Chemical Engineering Department, Ecole Polytechnique, Montreal, QC, Canada
**Author to whom correspondence should be addressed: pcarreau@polymtl.ca*

Abstract - In this presentation we summarize key results obtained on the development of PET and PLA nanocomposites using organo-modified clay, as a part of a major project funded three years ago by Natural Science and Engineering Council of Canada (NSERC) entitled: "Polyester Nanocomposites for Greener Transportation, Construction and Packaging Applications". Improvements up to 50 % of the barrier properties of PET using 4 wt% of an organo-modified clay has been achieved. Many challenges still need to be solved, in particular the control of the thermal degradation of both matrices during processing. Promising results with that respect using chain extenders are presented.

Keywords: Nanocomposites, polyethylene terephthalate, polylactide, organo-modified clay

Introduction

Polyethylene terephthalate (PET) is a semi crystalline engineering polymer extensively used for packaging applications. In the recent years, a biopolymer, poly(lactic acid) (PLA), has been intensively studied as a suitable, more environnementally acceptable substitute for PET. However, gas permeability for both polymers is a major issue for applications such as soft drink, beer and wine bottles.

In collaboration with Professor Musa R. Kamal, we have completed successful investigations on PET nanocomposites based on organo-modified clays. The most significant results are the following. An experimental grade high viscosity PET Selar PTX 295 (DuPont) and a low viscosity general purpose PET 9921 (Eastman Chemical Co.) were blended at a ratio of 4:1 (low: high viscosity) [1, 2]. A commercial ammonium modified clay (Cloisite® 30B (C30B), Southern Clay Products Inc.) was compounded with the PET blend at 3 wt% nominal level of clay. Melt compounding was carried out in an intermeshing co-rotating twin-screw extruder (TSE) (Leistritz, Germany). X-ray diffraction (XRD) and transmission electron microscopy (TEM) graphs showed a partially exfoliated/intercalated structure plus dispersed tactoids. The relative permeability to oxygen was found to be equal to K=0.77. This value is quite close to predicted values from theoretical models using aspect ratios of clay obtained by image analysis of TEM graphs [1].

In parallel, we have investigated the properties of nanocomposites based on poly(L-lactide) (PLA) supplied by NatureWorks (PLA 4032D). C30B was mixed to the PLA using a small (18 mm) twin-screw extruder or an internal mixer (Brabender). Three different chain extenders were used in this study. Polycarbodiimide (PCDI), a carboxyl-reactive chain extender, and tris (nonylphenyl) phosphite (TNPP)were obtained from Sigma Aldrich. Joncryl® ADR 4368-F (Joncryl), supplied by BASF, is a modified acrylic copolymer with epoxy functions and it was found to be the most effective chain extender [3, 4].

Typical thermal stability tests were carried out in rheometry in small amplitude oscillatory shear flow for the neat PLA and the PLA containing up to 6 wt% of C30B in presence or not of the chain extender, Joncryl. A rapid and significant degradation of the neat PLA with time during rheological tests carried out at 180 °C. The initial value of the complex viscosity was considerably reduced by the processing, indicating a serious degradation of the PLA when prepared using an internal mixer and more so using a twin-screw extruder. Note that in the presence of 3 wt% C30B, the thermal degradation was more rapid. However, using a chain extender, especially 0.5 wt% Joncryl, the complex viscosity was stabilized, indicating that the thermal degradation was stopped by this chain extender. Moreover the molecular weight of the PLA was increased by the reaction of the chain extender with the PLA. We noted also that in presence of a chain extender the behavior becomes more shear-thinning. This suggests stronger particle-particle interactions and a better dispersion of the clay. A refined analysis of the rheological data of PLA nanocomposites suggests that Joncryl reacts with PLA chains to form long branches [3].

TEM morphology analysis of two PLA/2 wt% C30B nanocomposites based on the work of Luo and Koo [5] revealed partial exfoliation for both samples. However, but the $D_{0.1}$ value was found to be 6.84 and 7.86 % for the nanocomposite without Joncryl and containing 1 wt% Joncryl, respectively. Note that a $D_{0.1}$ value of 8.0 % is considered to represent a totally exfoliated system [5]. By adding the Joncryl after the master-batch was prepared, the considerably enhanced viscosity due to the reaction of Joncryl with PLA was a major factor that favored delamination of tactoids and exfoliation.

The barrier properties of PLA using C30B improved by 37 % compared to the neat PLA when using 3 wt% clay. Adding 6 wt% Cloisite® 30B reduced the permeability by 48%. However, the use of 0.5 wt% Joncryl increased slightly the permeability of the 6 wt% nanocomposite. We have observed a significant increase of the permeability when Joncryl was added to the neat PLA. This is explained by a slight decrease of the crystallinity in the presence of Joncryl and possibly to the increase of the free volume of the PLA/Joncryl system. Finally, we note that the mechanical properties of PLA nanocomposites were considerably enhanced by using Joncryl.

In the following we will report some recent results on PET nanocomposites obtained using Joncryl as a chain extender.

Experimental

A commercial grade PET (PET 9921), with an intrinsic viscosity of 0.8 dL/g and a melting point of 243 °C, was obtained from Eastman Chemical Company. Two types of organically modified montmorillonites were employed: C30B from Southern Clay Products Inc. and Nanomer® I.28E (N28E) from Nanocor Inc. C30B is modified with methyl, tallow, bis-2-hydroxyethyl, quaternary ammonium (MT2EtOH), while N28E is modified with octadecyl ammonium (ODA). Joncryl® ADR-4368F (Joncryl), which is a styrene-acrylic multifunctional epoxy-based chain extender, was kindly supplied from BASF.

The nanocomposites were prepared by a master-batch approach. First, the PET, organoclays, and Joncryl were vacuum dried at 90 °C for 24 h before extrusion. Then the PET granules were ground into a powder and dry-mixed with 6 wt% C30B or N28E and processed using a co-rotating twin-screw Leistritz extruder (screw diameter = 18 mm and L/D = 40). Finally, the prepared master-batches were diluted with the appropriate amount of the neat PET and Joncryl using the same twin-screw extruder to obtain nanocomposites containing 4 wt% the organoclays (nominal value) and 1 wt% Joncryl. The temperature profile from the feed to the die was set between 240 and 265 °C and the extrusion was performed at 200 rpm and feeding rate of 1.8 kg/h. A 20 cm wide slit die with a 1 mm die gap was used to prepare films. Calender rolls were employed to stretch the extrudate and an air knife was mounted on both sides of the film. The thickness of the films was 50 μm corresponding to a draw ration around 21. The same extrusion procedure was followed to prepare pellets using a circular die of 2 mm diameter. A bath containing a mixture of water and ice was used to cool the extrudate. The extruded samples were granulated and vacuum dried at 90 °C for 24 h before compression molding. Disk-shape samples of approximately 2 mm thickness and 25 mm diameter were prepared using a laboratory press at 265 °C for 9 min at a pressure of 20 MPa under a purge of nitrogen, followed by quenching in another laboratory press for 5 min. The compression molded samples were vacuum dried at 90 °C for 24 h before being subjected to rheological testing.

Results and Discussion

Morphology

XRD patterns showed that the gallery spacing of both C30B and N28E increased from 1.85 to 3.4 nm and from 2.4 to 3.4 nm, respectively after melt blending with PET. Hence, C30B showed a higher level of intercalation in comparison to N28E, and the XRD pattern of the PET/4 C30B nanocomposite exhibited a shoulder instead of a defined peak, corresponding to a disordered intercalation, while the XRD pattern of sample PET/4 N28E showed a well-defined peak, characteristic of an ordered intercalation.

Fig. 1 presents TEM images of nanocomposites containing 4 wt% C30B, with and without Joncryl, at two magnifications. Both delaminated platelets and tactoids are seen in the micrographs. Comparing the low magnification micrographs (Figs. 1a and c) we observe that the addition of Joncryl improves the distribution of clay particles in the PET matrix, displaying a higher clay density (number of particles per unit area) and a more uniform distribution of clay particles in comparison to the nanocomposite prepared without Joncryl. The high magnification images (Figs. 1b and d) reveal that the presence of Joncryl reduces significantly the tactoid size. Clay tactoids are broken down to more clay particles with smaller numbers of platelets per particle. Similar observations were made for the nanocomposites containing N28E.

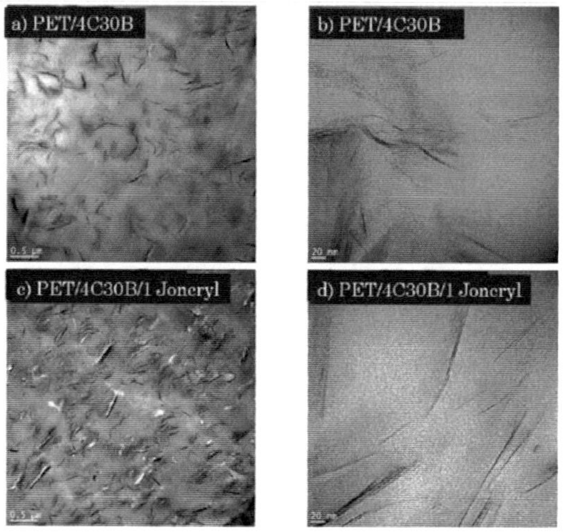

Fig. 1 TEM images of PET/4 C30B (a, b) and PET/4 C30B/1 Joncryl (c, d) at two magnifications.

Following the Luo and Koo [5] analysis the TEM images were used to quantify the level of clay dispersion within the PET matrix. A dimensionless parameter called the dispersion value, $D_{0.1}$, was calculated based on the distribution of the free-path spacing distances between the silicate nanolayers. It has been proposed that $D_{0.1}$ below 4% characterizes immiscible systems or microcomposites, and over 8% is a characteristic of exfoliated nanocomposites. The dispersion value is between 4 and 8% for intercalated nanocomposites. In this work around 600 measurements were carried out for each sample to ensure the validity of the values reported for $D_{0.1}$. The values obtained are 5.0 and 6.9 % for PET/4 C30B and PET/4 C30B/1 Joncryl, respectively; $D_{0.1}$ is 4.1 and 5.6 % for PET/4 N28E and PET/4 N28E/1 Joncryl, respectively. These results suggest a higher degree of clay exfoliation for nanocomposites based on C30B and those containing Joncryl. It is worth mentioning that a value of 7.5 for $D_{0.1}$ was obtained for a PET containing 2 wt % C30B in drawn films prepared via a larger twin-screw extruder [2].

Rheological Properties

The viscoelastic properties was found to be time-dependent due to possible thermal degradation of the PET matrix and/or chemical reactions between PET, clay organomodifiers, and Joncryl. However, less than 10% variations took place during the rheological measurements of all samples up to 5 min. Hence, we believe that the following frequency data are meaningful.

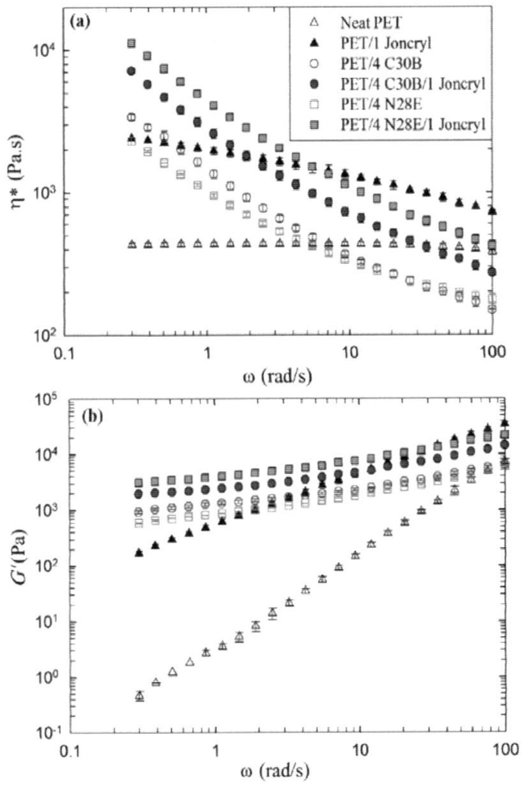

Fig. 2 Complex viscosity (a) and storage modulus (b) of the neat PET and PET-based nanocomposites with and without Joncryl as functions of frequency.

Fig. 2 present the complex viscosity and the storage modulus of the neat PET and PET-based nanocomposites, with and without Joncryl, as functions of frequency. The neat PET exhibits a pseudo-Newtonian behavior, while all PET-based nanocomposites show a marked shear-thinning behavior, characteristic of nanocomposites. At low frequencies the complex viscosity and the storage modulus of PET/4 C30B are larger than those of PET/4 N28E, which is due to the higher degree of C30B exfoliation over N28E revealed by XRD and TEM results. At high frequencies, PET/4 C30B shows a lower complex viscosity than PET/4 N28E. The better dispersion of C30B increases the exposure surface of this organoclay to the PET matrix and consequently causes a higher level of degradation. Contrary to the organic modifier of N28E, that of C30B contains an unsaturated tallow with double bonds as well as hydroxyl groups, which causes more degradation of the PET matrix. On the other hand, the introduction of the chain extender increases the viscoelastic properties of the neat PET and its corresponding nanocomposites. The epoxy groups of Joncryl may link the functional terminal groups of degraded PET chains and rebuild the molecular weight and viscoelastic properties [6]. The higher level of exfoliation observed for nanocomposites containing Joncryl, according to XRD and TEM results, may be explained by the larger viscosity of the matrix, which generates larger shear stresses that break down large clay tactoids into individual layers and double layer particles. At high frequencies, the viscosity of the nanocomposites is lower than that of the matrix. This confirms the severe degradation of the polymer matrix in the presence of nanoclay. Since the behavior of the neat PLA is almost Newtonian, the frequency data at 100 rad/s can be used to assess the polymer degradation using the Maron-Pierce equation (Eq. 1):

$$\frac{\eta^*}{\eta_M^*} = (1 - \frac{\varphi}{\varphi_m})^{-2} \quad (1)$$

where η^* and η_M^* are the complex viscosity of the nanocomposite and the matrix, respectively, φ is the volume fraction of the clay and φ_m is the maximum packing volume fraction which is set here as $\varphi_m = 3.55/p$, where p is the clay aspect ratio [7]. Using a value of 46 as determined by Ghasemi et al [1] for C30B aspect ratio, the viscosity was calculated from Eq. 1 and the molecular weight estimated from Eq. 2:

$$\eta^*_{M(100rad/s)} \approx \eta_0 = kM_w^{3.4} \quad (2)$$

The results for the nanocomposites based on the PET matix, presented in Table 1, show that the molecular weight of the PET is reduced by almost half when mixed to 4 wt% nanoclay. The same analysis could not be carried out for the nanocomposites based on the PET/ 1 % Joncryl since the behavior of the modified PET is no longer Newtonian as shown in Fig. 2a.

Table 1 Apparent complex (zero -shear) viscosity and apparent molecular weight for the neat PET and the PET matrix in nanocomposites calculated from Eqs.1 and 2, respectively.

Sample	Matrix apparent complex viscosity at 100 rad/s	Matrix apparent molecular weight (kg/mol)
Neat PET	380	65
PET/4 C30B	59.4	37.6
PET/4 N28E	65.4	38.7

The elastic modulus of the PET-based nanocomposites (Fig. 2b) exhibits a plateau at low frequencies, corresponding to the formation of a percolated three-dimensional network of clay particles, which acts like a weak solid. The introduction of Joncryl increases the viscoelastic properties of the PET nanocomposite containing N28E more than that containing C30B. Joncryl functional groups connect shorter chains in the case of PET-based nanocomposites containing C30B in comparison to those containing N28E, due to the more severe chain scission induced by C30B. Besides, the organo-modifier of C30B contains hydroxyl groups that might also react with the epoxy groups of Joncryl and, consequently, reduce the amount of Joncryl functional groups available for chain extension.

Barrier Properties

Fig. 3 reports the measured oxygen permeability values of the neat PET and PET-based nanocomposite films, with and without Joncryl. All nanocomposite films exhibit higher barrier properties than the neat PET due to the presence of clay layers and larger crystallinity. The barrier properties of the films containing Joncryl are better than those without Joncryl, which can be attributed to the better clay dispersion and/or larger degree of crystallinity. Overall, a 46% improvement in barrier properties was obtained for the PET nanocomposites containing C30B and Joncryl in comparison to the neat PET films. The corresponding improvement for the PET/4 N28E/1 Joncryl was 40%. According to XRD, TEM, and DSC results (not shown here), the PET/4 C30B/1 Joncryl films show the highest level of exfoliation and contain the highest amount of crystals, which explains the lowest oxygen permeability. It is worth noting that barrier properties of the PET/4 C30B films are better than those of the PET/4 N28E films because of the larger crystallinity and better clay dispersion.

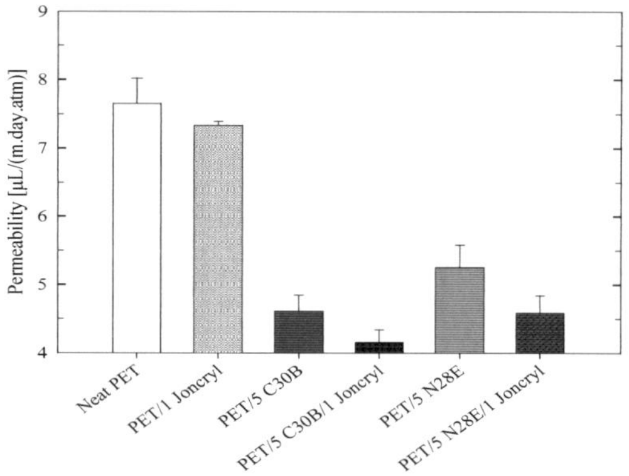

Fig. 3 Oxygen permeability of the neat PET and PET-based nanocomposites with and without Joncryl.

Conclusion Remarks

The addition of an organo-modified clay in a small amount (2 - 6 wt%) leads to significant decreases of the permeability of PET and PLA. In this work we have shown improvements of the barrier properties of about 50 % for PET using 4 wt% C30B and up to 58 % for PLA using 6 wt% C30B. Better improvements could be obtained if one could achieve complete exfoliation of the nanoclay. We have investigated quite a variety of commercially available nanoclays and found so far that C30B is the best for PET and PLA.

We have shown that Joncryl (or other chain-extenders) in PET and PLA nanocomposites can control the thermal degradation of the matrix. Moreover, it increases the molecular weight of the polymer chains during melt processing and helps delamination (exfoliation).

Acknowledgements

Financial support received from the Natural Science and Engineering Research Council of Canada (NSERC) is gratefully acknowledged

References

1. H. Ghasemi; P.J. Carreau; M.R. Kamal; S.H. Tabatabaei *Polym. Eng. Sci.* 2011, *51*, 1178.
2. H. Ghasemi; P.J. Carreau; M.R. Kamal; N. Chapleau *Intern, Polym. Proc.* 2011, *26*, 219.
3. N. Najafi C.; M.C. Heuzey; P.J. Carreau; P.M. Wood-Adam *Polym. Degrad. Stab.* 2012, *97*, 554.
4. N. Najafi C.; M.C. Heuzey; P.J. Carreau *Composites Sci. Techn.* 2012, *72*, 608.
5 Z.P. Luo; J.H. Koo *Polymer* 2008, *49*,1841.
6. M. Villalobos; A. Awojulu; T. Greeley; G. Turco; G. Deeter *Energy* 2006, *31*, 3227.
7. L. Sun;W.J. Boo; J. Liu; A. Clearfield; H.J. Sue; N.E. Verghese; H.Q. Pham; J. Bicerano *Macromol. Mater. Engin.* 2009, *294*, 103.

O-MK-481

MICROMECHANICAL MODELS FOR POLYMER COMPOSITES CONTAINING ROD-LIKE NANOFILLERS

M. R. Loos[1] and I. Manas-Zloczower[2]*

[1]*Department of Macromolecular Science and Engineering, Case Western Reserve University, USA-mrl75@case.edu*
[2]*Department of Macromolecular Science and Engineering, Case Western Reserve University, USA-ixm@case.edu*

Abstract – In this work we derive different micromechanical models which account for the presence of dispersed and agglomerated nanofillers, specifically carbon nanotubes (CNTs) and nanowhiskers, in a polymeric matrix and consider the effect of a percolation threshold on the elastic modulus of the composite. We demonstrate that the critical filler volume fraction where a percolating network is forming marks a 'turning point' in the reinforcement efficiency. The suitability of these models is verified by comparing simulated values with experimental data from literature. The results show that the models are able to predict mechanical properties over a wide range of testing conditions.

Keywords: Carbon Nanotubes, Nanowhiskers, Composites, Mechanical Properties

Introduction

Numerous studies in literature suggest that the mechanical properties of polymer composites containing carbon nanotubes (CNTs) show a 'turning point' in the reinforcement efficiency above a critical filler concentration: the properties either drop or the effectiveness decreases considerably. It is well known that carbon nanotubes tend to aggregate and above a critical concentration will form a 3D network inside the polymeric matrix. The volume fraction of CNTs at which this 3D network forms is called the percolation threshold. The 'turning point' in the reinforcement efficiency usually takes place at volume fractions coincident with the percolation threshold [1-2]. On the other hand, the outstanding reinforcement observed for polymer/cellulose nanowhisker composites is attributed also to the formation of a percolated architecture of rigid nanowhisker network in which stress transfer is facilitated by hydrogen-bonding among the nanowhiskers [3-4].

In this work, we analyze the mechanical percolation effect of CNTs and cellulose nanowhiskers, on the elastic modulus of polymer based composites. The effect of a percolating network is taken into consideration by using the Takayanagi models I and II. The change in reinforcement efficiency and network formation at temperatures below and above the glass transition temperature (T_g) is also explored.

Composites containing cellulose nanowhiskers

When used as reinforcement in polymer matrices, nanowhiskers can have a remarkable effect on the mechanical properties of the system. Let us consider different polymers. Generally, for temperatures below the T_g, nanowhiskers addition does not affect the composite modulus to a significant extent since the elastic moduli of the whiskers, E_w, and of the matrix, E_m, are not that different ($E_w/E_m \geq 1$). When the temperature is above the T_g, the reinforcement efficiency of the whiskers changes. In this case the ratio whisker-to-matrix modulus is extremely high ($E_w >> E_m$) and the reinforcement effect becomes apparent. At such temperatures and for volume fractions above the percolation threshold, the reinforcement effect of whiskers increases drastically with whisker concentration.

This behavior is attributed to the formation of a stiff cellulose whisker network held together by strong hydrogen bonds. Simulations of such behavior using classical models like Halpin-Kardos predict moduli lower than observed experimentally [5]. These observations suggest that interactions between the fillers induce the mechanical percolation of the whiskers. This effect on the elastic moduli of composites can be calculated following the method of Ouali et al. in their adaptation of the percolation theory to the parallel-series model of Takayanagi.

Takayanagi and co-authors developed a two-phase mechanical model to predict the modulus of a crystalline polymer, taking into consideration the crystalline and non-crystalline phases of a polymer and the fact that those different parts of the crystalline phase can undergo different deformation under the application of stress [6]. The *parallel-series* model of Takayanagi has been extended by Ouali and co-authors by introducing a percolating concept [7] which enabled the application of the model to predict mechanical properties for polymer composites with cellulose nanowhiskers [8-10]. Recently we have extended the *series-parallel* model of Takayanagi to include the percolating concept and use it to predict mechanical properties of carbon nanotube reinforced composites [10].

A schematics for the series-parallel (model I) and parallel-series (model II) models is shown in **Figure 1a** and **1b**. According to the model, parameters λ and ϕ or their combinations reflect volume fractions of the system components [10]. We define $\lambda\phi$ and $1-\lambda$ as the volume fraction of the matrix and dispersed phases, respectively. $V_F=1-\lambda(1-\phi)$ is the total volume fraction of the filler (F) [10].

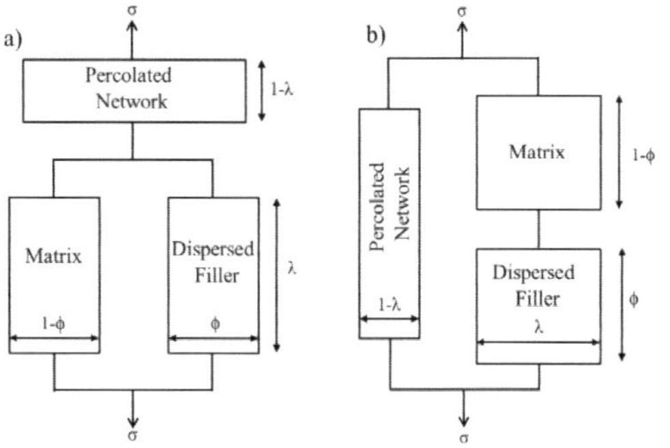

Figure 1 - Schematic representation of Takayanagi model I (a) and II (b). ϕ is a function of the volume fraction of the parallel element and λ of the series element. Models I and II are also known as the series-parallel and parallel-series model, respectively.

Following the concept of a "switching function", f_s, employed by Chatterjee [11], the percolation behavior in the vicinity of the percolation threshold can be emulated by the function f_s, defined as:

$$f_s(V_F)=0 \qquad V_F<V_c$$
$$f_s(V_F)=1-e^{-A(V_F/V_c)-1)^{0.474}} \qquad V_F \geq V_c \qquad (1)$$

where A is a fitting parameter which modulates the width of the transition [11]. The percolation threshold, V_c, can be measured or calculated as a function of the filler aspect ratio ($\eta=l/d$) and is estimated to be $V_c=0.5/\eta$ [12].

According to the Takayanagi model I shown in **Fig. 1a**, the modulus of the composite E_c is given by [9]:

$$E_c=\frac{(1-V_F)E_mE_F^{agg}+(V_F+\lambda-1)E_F^{dis}E_F^{agg}}{(1-\lambda)(1-V_F)E_m+(1-\lambda)(\lambda+V_F-1)E_F^{dis}+\lambda^2E_F^{agg}} \qquad (2)$$

where E_F^{agg}, E_F^{dis} and E_m are the Young's moduli of the percolated filler network, dispersed fillers and matrix phase, respectively and V_F is the filler volume fraction.

For the model II shown in **Fig. 1b**, the elastic modulus of the composite can be written as [9]:

$$E_c=\frac{(1-\lambda)(1-V_F)E_F^{dis}E_F^{agg}+(1-\lambda)(V_F+\lambda-1)E_mE_F^{agg}+\lambda^2E_mE_F^{dis}}{(1-V_F)E_F^{dis}+(V_F+\lambda-1)E_m} \qquad (3)$$

If $V_F<V_c$, $\lambda=1$ and Eq. (2) and (3) reduces to the upper bound and lower bound, respectively:

$$E_c=(1-V_F)E_m+V_FE_F^{dis}$$

$$E_c=\frac{E_mE_F^{dis}}{(1-V_F)E_F^{dis}+V_FE_m}$$

We analyzed the mechanical behavior of polymers reinforced with cellulose whiskers above the T_g using Takayanagi model II. For $V<V_c$, Takayanagi model II assumes that the system consist of only dispersed nanowhiskers. The matrix and whiskers phases are in a series arrangement and the composite modulus is described by a lower bound. This fits the physical picture of a soft matrix above the glass transition temperature and high modulus filler at relatively low concentrations. When the concentration of whiskers increases above V_c, a 3D network is formed throughout the matrix. At this point, three phases are present in the system: matrix, dispersed nanowhiskers and the percolated network. According to Takayanagi model II, the network of whiskers will elongate as much as the combined matrix and dispersed whiskers phases. Due to the hydrogen bonds linking adjacent whiskers, the strong 3D network will support a higher load than the matrix or the dispersed whiskers. If the nature of the cohesive forces among whiskers changes, the stiffness of the network is affected. For instance, when water is present in the system, water-nanowhiskers interactions prevail over filler-filler interactions and the network stiffness is tremendously reduced [4]. **Figure 2** shows two examples of the application of the Takayanagi model II for fitting experimental data from literature. The elastic moduli of composites comprising rubbery ethylene oxide–epichlorohydrin copolymer (EO-EPI) and tunicates-cellulose nanowhiskers [4] were measured by Capadona et al. Individual tunicates whiskers have a modulus of ~ 143 GPa while films prepared from the same material have a modulus of 4 GPa [4]. As shown in **Figure 2** the experimental results are well fitted by the Takayanagi model II for both dry EO-EPI/whisker nanocomposites as well as the ones conditioned by equilibrium swelling in de-ionized water.

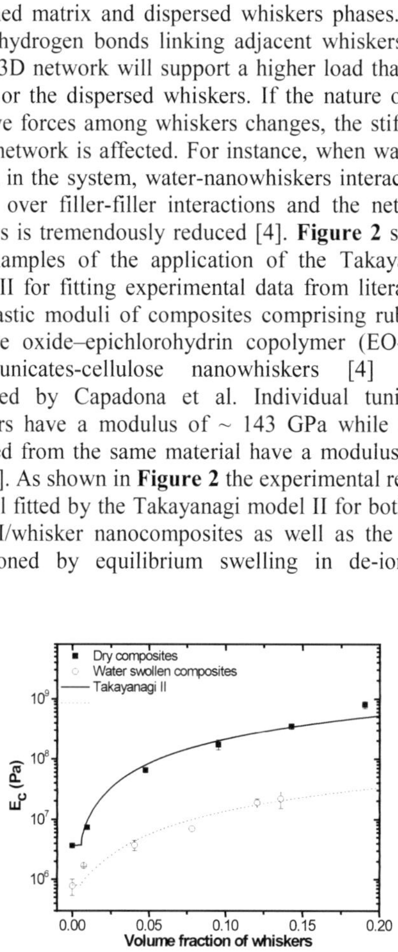

Figure 2 - Tensile storage moduli of EO-EPI/whisker nanocomposites as function of volume fraction of cellulose whiskers. Dry composites (■,———) [4] reported whiskers and matrix material parameters: $l=2.2\mu m$, $d=26nm$, $E_m=3.66MPa$; $E_{whi}^{dis}=143GPa$, $E_{whi}^{agg}=4GPa$, fitting parameter: $A=0.2$. Water swollen composites (○,--------) [4] reported whiskers and matrix material parameters: $l=2.2\mu m$, $d=26nm$, $E_m=0.78MPa$; $E_{whi}^{dis}=143GPa$, fitting parameters: $E_{whi}^{agg}=300MPa$, $A=0.15$.

Composites containing CNTs

Since Iijima's report in 1991, carbon nanotubes have been the focus of considerable research [13]. During the last decade, a great deal of effort has been given toward maximizing the potential of CNTs as reinforcing agents in polymer matrix composites. Despite this effort, the full potential of CNT-reinforced composites has not been realized due to current technological limitations in obtaining homogeneous dispersions of CNTs and sometimes less than favorable matrix/nanotube interfacial adhesion.

According to different micromechanical models such as rule of mixtures or Halpin-Tsai, the addition of CNTs should render outstanding improvements in mechanical properties such as elastic modulus. However, numerous experimental results suggest that above a certain filler concentration - the percolation threshold - the reinforcement efficiency decreases [14]. There are many experimental evidences that at low fraction of CNTs the composite modulus increase linearly with filler concentration whereas after certain concentration, increasing the amount of CNTs leads to a less significant improvement in properties [14,15].

When added to a polymer matrix, CNTs tend to aggregate as a consequence of the van der Waals interactions and large surface area. After reaching a critical concentration (coincident with the percolation threshold, V_c) CNTs form a percolating network throughout the matrix. The 3D network formed by CNTs is not held together by strong hydrogen bonds, as in the case of nanowhiskers. As a matter of fact the often weak CNT assemblies can have a detrimental result on the elastic properties. Recently we formulated and compared different micromechanical models for the prediction of elastic moduli of CNT reinforced polymers, assuming that only part of the CNTs are dispersed, and also considering the percolation of the fillers. For volume fractions below the critical percolation threshold and assuming temperatures below the glass transition, the system is comprised of a relatively stiff matrix and the dispersed carbon nanotubes. These two phases are in a parallel arrangement in the mechanical model (**Fig. 1a**) and the system properties will define an upper bound. Once the percolation threshold is achieved, a 3D network which can support a higher load than the matrix or the dispersed carbon nanotubes is formed [9,16,17]. As this network is not strengthened by hydrogen bonds, it cannot reinforce the composite to the same extent as individually dispersed CNTs and this explains the 'negative' turning point observed in the experimental results.

As shown in **Fig. 3**, the mechanical behavior described previously is well fitted by Takayanagi model I. The Takayanagi model I captures well the change in CNTs reinforcing efficiency once the percolation concentration is reached (**Fig. 3a**). Moreover, the result presented in **Fig. 3b** shows that up to the critical percolation threshold, the CNTs reinforce the polymer matrix but at higher volume fractions of CNTs the percolated network has a detrimental effect on the modulus. The composite modulus drops to values close to the modulus of the neat resin and this behavior is well fitted by the model.

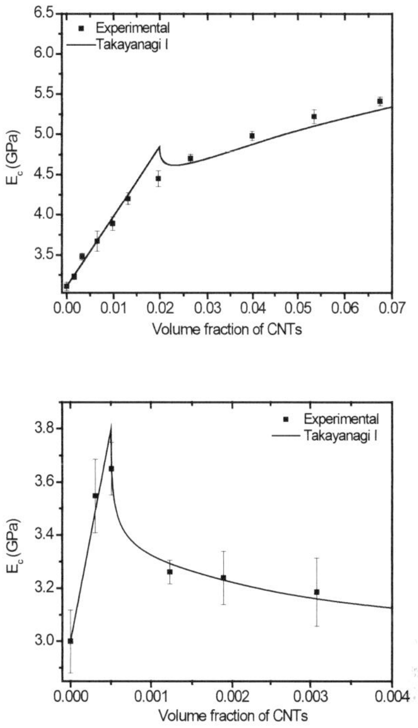

Figure 3 – Comparison of experimental and simulated elastic modulus of MWCNT/epoxy composites. (a) Reported [16] MWCNTs and matrix material parameters: $l = 1\mu m$, $d = 40nm$, E_m=3.11GPa; fitting parameters: $E_{NT}^{dis} = 90GPa$, $E_{NT}^{agg} = 40GPa$, $A = 0.75$. (b) Reported [17] MWCNTs and matrix material parameters: $l = 10\mu m$, $d = 10nm$, E_m=3GPa; fitting parameters: $E_{NT}^{dis} = 1600GPa$, $E_{NT}^{agg} = 100GPa$, $A = 1.6$.

Conclusions

Due to their differences in network characteristics, CNTs and whiskers have distinct effects on composites. For the case of whiskers, at temperatures above the T_g and volume fractions above the threshold, whiskers form a strong network and the change observed in reinforcement efficiency above V_c is well captured by the Takayanagi model II with a percolation concept. Since there is a drastic increase in the composite modulus, the threshold marks a *positive* "turning point" in the reinforcement efficiency of the fillers. Takayanagi model II assumes that the filler network will support much higher loads than the matrix, a reasonable assumption for systems at temperatures above the matrix glass transition temperature.

On the other hand, for CNTs, the percolated network formed is not linked by strong hydrogen bonds and its effect on the elastic modulus can be detrimental. In this case the threshold marks a *negative* "turning point" in reinforcement efficiency of the fillers. The Takayanagi model I differentiates between the deformation of the dispersed and percolated CNTs and captures well such behavior. According to this model the dispersed CNTs will support a larger portion of the load applied to the composite than the matrix and also elongate as much as the matrix. The model also assumes that the filler network will support more load than the matrix or the dispersed filler and, as the network is not sufficiently strong, the reinforcement efficiency is altered.

Acknowledgements

This material is based upon work supported by the Department of Energy and Bayer MaterialScience LLC. under Award Number DE-EE0001361.

Disclaimer

This report was prepared as an account of work sponsored by an agency of the United States Government. Neither the United States Government nor any agency thereof, nor any of their employees, makes any warranty, express or implied, or assumes any legal liability or responsibility for the accuracy, completeness, or usefulness of any information, apparatus, product, or process disclosed, or represents that its use would not infringe privately owned rights. Reference herein to any specific commercial product, process, or service by trade name, trademark, manufacturer, or otherwise does not necessarily constitute or imply its endorsement, recommendation, or favoring by the United States Government or any agency thereof. The views and opinions of authors expressed herein do not necessarily state or reflect those of the United States Government or any agency thereof.

References

1. S.-Y. Wu; S.-M. Yuen; C.-C. M. Ma; C.-L. Chiang; Y.-L. Huang; H.-H. Wu; C.-C. Teng; C.-C. Yang; M.-H. Wei *J. Appl. Polym. Sci.* 2010, *115*, 3481.
2. W. Jeong; M. R. Kessler *Carbon* 2009, 47, 2406.
3. J.R. Capadona; O. Van Den Berg; L.A. Capadona; M. Schroeter; S.J. Rowan; D.J. Tyler; C. Weder *Nat. Nanotechnol.* 2007, *2*, 765.
4. J.R. Capadona; K. Shanmuganathan; D.J. Tyler; S.J. Rowan; C. Weder *Science* 2008, *319*, 1370.
5. N. Ljungberg; C. Bonini; F. Bortolussi; C. Boisson; L. Heux; J.Y. Cavaille *Biomacromolecules* 2005, *6*, 2732.
6. M. Takayanagi; S. Uemura; S.J. Minami *J. Polym. Sci. C*, 1964, *5*, 113.
7. N. Ouali; J.Y. Cavaillé; J. Pérez *Plast. Rubber. Comp. Process. Appl.* 1991, *16*, 55.
8. V. Favier; H. Chanzy; J.Y.Cavaille *Macromolecules* 1996, *28*, 6365.
9. M.R. Loos; I. Manas-Zloczower *Macromol Theory Simul.* 2012, 21, 130.
10. M.R. Loos; I. Manas-Zloczower *Polym. Eng. Sci.* 2012, *in press*.
11. A.P. Chatterjee *J. Appl. Phys,* 2006, *100*, 054302-1.
12. A. Celzard; E. McRae; C. Deleuze; M. Dufort; G. Furdin; J.F. Marêché *Phys. Rev. B* **53**, 6209 (1996).
13. M. R. Loos; K. Schulte *Macromol. Theor. Simul.* 2011, *20*, 350
14. S.-Y. Wu; S.-M. Yuen; C.-C. M. Ma; C.-L. Chiang; Y.-L. Huang; H.-H. Wu; C.-C. Teng; C.-C. Yang; M.-H. Wei *J. Appl. Polym. Sci.* 2010, *115*, 3481.
15. S.-M. Yuen; C.-C. M. Ma; Y.-Y. Lin; H.-C. Kuan, *Compos. Sci. Technol*. 2007, *67*, 2564.
16. A. Martone; C. Formicola; M. Giordano; M. Zarrelli *Compos. Sci. Technol.* 2010, *49*, 1154.
17. M. Omidi; D. T. Rokni; A. S. Milani; R. J. Seethaler; R. Arasteh *Carbon* 2010, *48*, 3218.

O-MK-473

AN OVERVIEW OF THE CALENDERING PROCESS

J.F.Agassant[1]* and R.Magnier[1]
[1]*Mines ParisTech, CEMEF, UMR CNRS 7635, 06560 Valbonne, France – jean-francois.agassant@mines-paristech.fr*

Abstract - Calendering was historically dedicated to PVC formulations. The processing parameters (rolls velocities, gap between the rolls, temperature) and the rheology of the polymer influence the roll separating force and so the film profile. In the same way, defects may be observed in the calendering sheet (mattness for rigid PVC, or bubbles, or V-shaped defects in more general formulations) and these defects may be linked to flow parameters. Numerical modelling has been developed, especially when high friction coefficient is imposed between the two rolls velocities, and this allows predicting a priori the roll deflexion and the conditions of appearance of defects.

Introduction

Calendering is one of the most traditional polymer forming process. Developed initially for the rubber industry it has subsequently been then intensively used for PVC processing (for example for flooring or packaging production) and more recently for polyolefin formulations.

Due to the high viscosity of formulations, the roll separating force is important and so the roller elastic deformation can lead to a non-uniform sheet or film thickness. Several thickness correction methods may be used (rolls with non-uniform diameters, roll bending, cross-axis) but they require important trial and error adjustments before obtaining an uniform thickness at the highest possible calendering rate. Modelling the calendering process allows prediction a priori of the pressure within the calender nip, the roll separating force and so the roll deflection and correction methods to obtain a sheet of uniform thickness can be made.

In addition defects may be observed in the final calendered product (mattness, air bubbles, V-shaped defects) and the origin of these defects is clearly linked to the processing parameters (roll velocity, minimum gap between the cylinders, temperature) and the rheology of the polymer formulation. Comparison between the process conditions for the occurrence of these defects and some relevant thermo-mechanical parameters allows optimization a priori of the calendering conditions.

Kinematics analysis

A calender is generally constructed from several rollers that rotate at around the same velocity. Upstream two successive rolls, there is an important polymer reservoir (called the bank) where the polymer encounters complex 3D kinematics which contributes to increasing the width of the sheet, and also dissipating energy in order to improve melting and homogenization of the polymer formulation (figure 1).

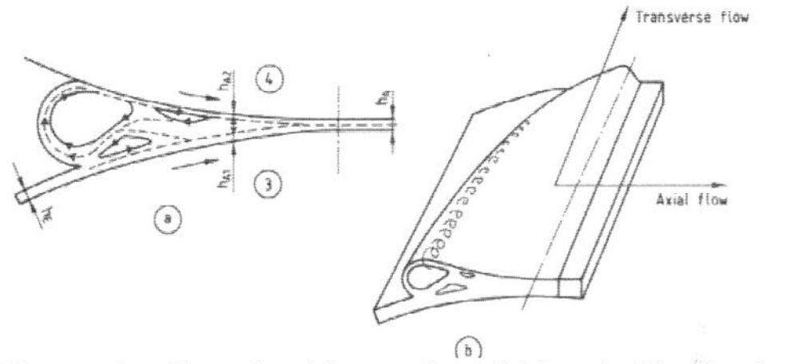

Figure 1 – Complex kinematics within a half-calender bank. [1]

Most of the authors neglect the component of the velocity in the sheet width direction. Using finite element methods, Agassant et al [2] and Mitsoulis et al [3] were able to account for the complex kinematics in the bank.

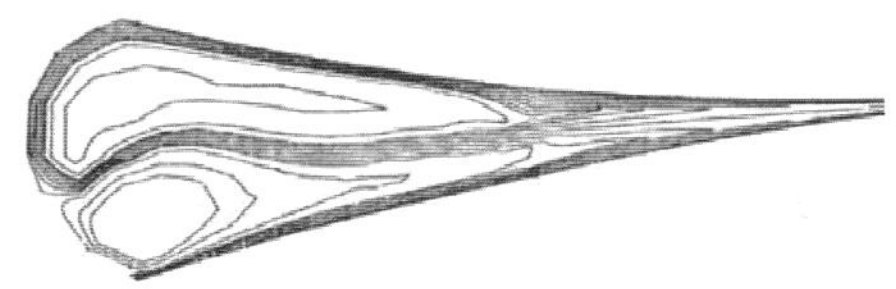

Figure 2 – 2D modelling of the flow kinematics within the calender bank. [2]

As pointed out on figure 2, the flow kinematics becomes symmetric within the nip which justify using the so called "lubrication approximations" [4,5].

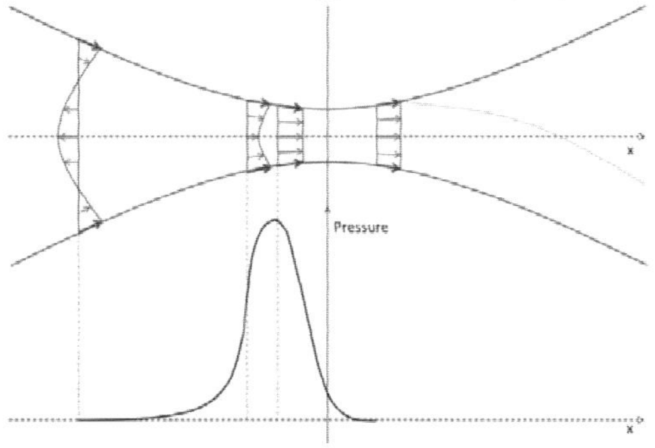

Figure 3 – Flow kinematics within the calender bank resulting from the lubrication approximations

Most of the authors assume that the polymer sheet leaves the contact between the two rolls with a thickness which is greater than the minimum gap (this is called the spread height, see figure 3).
The existence of this spread height has been recently demonstrated by Serrat et al. [6] who used a finite element Lagrangian computation by feeding progressively the gap between the two rolls until stabilization of the process was achieved (bank dimension, pressure trace and spread height) (figure 4).

Figure 4 – Pressure distribution between the two rolls (the bank is small)

With the development of highly filled PVC or PE formulations, it is necessary to impose a friction coefficient (the ratio between the rotation velocity of the last roll as compared to the velocity of the previous one). In these circumstances, the spread height hypothesis is no longer valid, as the final sheet thickness, governed by the velocity of the last roll, may be far less significant than the gap between the two rolls, and the sheet will leave the contact between the rolls at the minimum gap.

The last kinematic hypothesis concerns the polymer velocity at the roll surface. Most of the authors assume a sticky contact, but especially PVC formulations may slip along metallic tools. Slippage has been accounted for by Agassant [7] and Ray [8]. The difficult point is that the roll may slip on the polymer in the upstream region (see figure 3) where the shear stress is positive, and the polymer may slip along the roll in the downstream region where the shear stress is negative. In-between, a sticky contact can be assumed.

Calendering Modelling
Several models have been proposed, based on different kinematics hypothesis and different rheology constitutive equations. The finite element models which account for the complex flow in the bank [2] show that pressure and stress distributions are equivalent to those resulting from a more simple "lubrication approximation" approach. Simple power law constitutive equations have been generally considered [4, 5, 9-11]. Kiparissides and Vlachopoulos [12] compared the lubrication approximation with symmetric finite element solution.

When considering the symmetric case (the same rotation velocity for both rolls), mass and force balance equations result in a generalized Reynolds equation which can be easily integrated. In the following, a non-symmetric case is considered, where the two roll velocities are significantly different. This leads to more complex itegration steps. Figure 5 shows that the computed roll separating force as a function of the friction coefficient between the two rolls is able to capture quantitatively the experiment carried out on a PVC formulation for flooring applications.

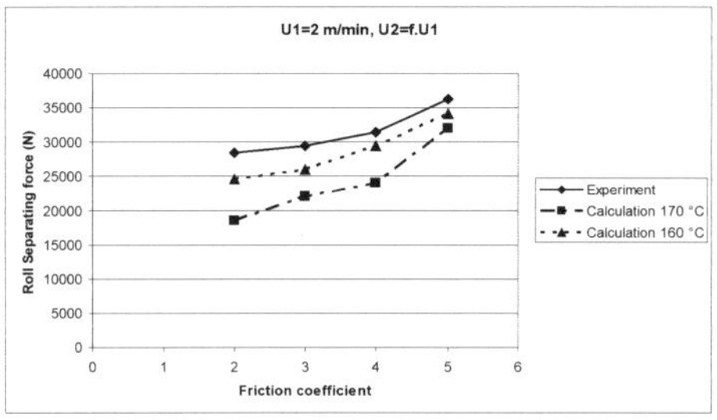

Figure 5 – roll separating force as a function of the friction between the two rolls (the velocity of the first roll is kept constant

Due to the high viscosity of polymer calendering formulations, the viscous dissipation may lead to a significant temperature increase as well as large temperature gradients within the gap (figure 6).

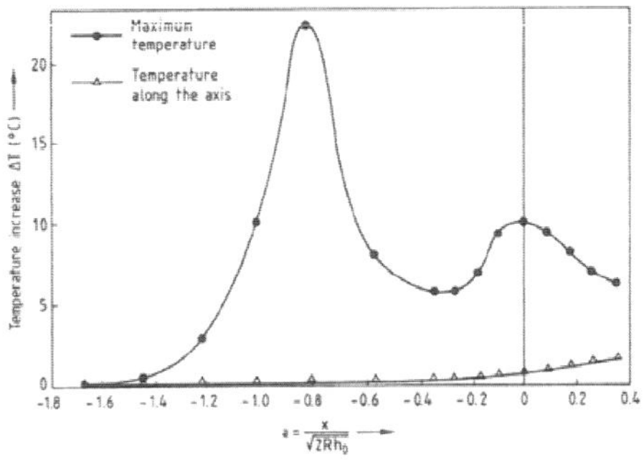

Figure 6 – Temperature profile along the symmetry axis and maximum temperature [13]

Calendering defects
The most common calendering defect is a non-uniform thickness through the width of the sheet. Starting from the rheology of a new polymer formulation, it is now possible to compute the roll separating force for target processing parameters, to deduce the roll deflexion and so to fix a priori the necessary corrections (roll-bending, cross-axis...) which will achieve a uniform thickness without tedious trial and error tests.

When calendering rigid PVC formulations for packaging applications, a so called "Mattness" defect can occur which corresponds to a rough surface of the sheet and so to a bad transparency. If the roll temperature, the roll velocity and the calender gap are varied, as shown in (figure 7), a discriminating line can be established that splits the calendering conditions.

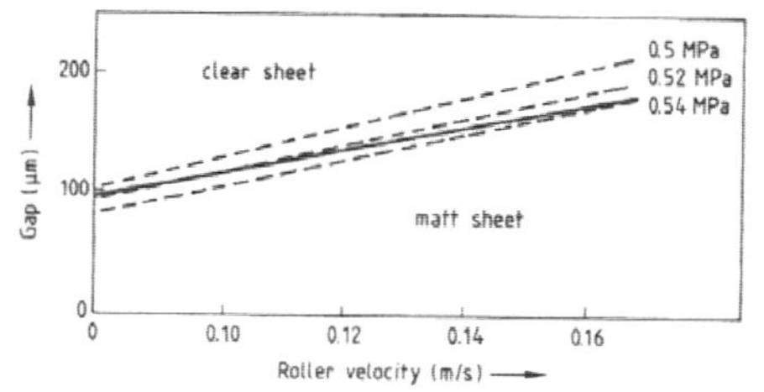

Figure 7 – Quality of the sheet as a function of the calendering parameters, and iso-values of the maximum shear stress [14]

It can be seen that this boundary line corresponds to an iso-value of the maximum shear stress at the roll surface within the contact: when the shear stress exceeds this critical value the sheet will present the mattness defect.

For packaging applications, the sheet may be subjected to subsequent thermoforming and the presence of bubbles in the calendered sheet is undesirable . When measuring the bubble density as a function of the calendering parameters, one may again discriminate processing conditions with acceptable bubble density (figure 8).

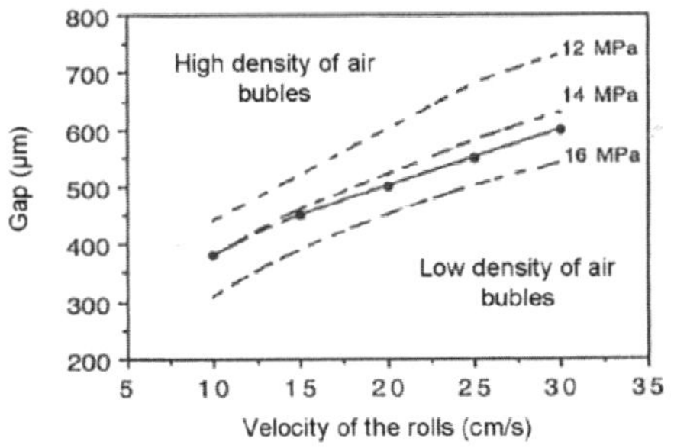

Figure 8 – Bubble density as a function of the processing parameters and iso-values of the maxumum pressure in the contact [14]

The critical parameter for the occurrence of the defect is now the maximum pressure in the contact: when this pressure is too low, air entrapped at the bank inlet will flow through the roll gap and induce bubbles. On the contrary, if the pressure is high enough, it will block air bubbles within the bank where they will be progressively dissolved.

Conclusions

A calendering model with relevant kinematics and rheology assumptions enables the physical understanding of the process and then in some cases the understanding of certain defects. Some defects, such as V-shaped defects which correspond to periodic polymer discharge from the upstream bank to the sheet, necessitate the development of full 3D models as already proposed by Luther and Meves [15] which in addition have to include viscoelastic constitutive equations.

References

1. W. Unkruer *Kunststoffe* 1972, *62*, 7.
2. J. F. Agassant; M. Espy *Polym. Eng. Sci.* 1985, *25*, 118-121.
3. E. Mitsoulis; J. Vlachopoulos; F. A. Mirza *Polym. Eng. Sci.* 1985, *25*, 6-18.
4. J. M. McKelvey *Polymer Processing*, John Wiley & Sons, New York, 1962.
5. J. F. Agassant; P. Avenas *J. Macromol. Sci. Phys.* 1977, *B14*, 345-365.
6. M. C. Serrat; J. F. Agassant; J. Bikard; S. Devisme *Int. Polym. Proc.* 2012, *27*, 318-327.
7. J. F. Agassant, Thèse de Doctorat d'Etat, Paris 6, 1980.
8. A. Ray; A. V. Shenoy *J. Appl. Polym. Sci.* 1985, *30*, 1-18.
9. J. R. A. Pearson *Mechanical Principles of Polymer Melt Processing*, Pergamon Press, Oxford, 1966.
10. J. S. Chong *J. Appl. Polym. Sci.* 1968, *12*, 191-212.
11. I. Brazinsky; H. F. Cosway; C. F. Valle; R. Clark Jones; V. Story *J. Appl. Polym. Sci.* 1970, *14*, 2771-2784.
12. C. Kiparissides; J. Vlachopoulos *Polym. Eng. Sci.* 1976, *16*, 712-719.
13. C. Kiparissides; J. Vlachopoulos *Polym. Eng. Sci.* 1978, *18*, 210-214.
14. J. F. Agassant; A. Philippe in *Interrelation between Processing, Structure and Properties of Polymeric Materials*, J.C. Siferis; P.S. Theocaris Eds.; Elsevier, Amsterdam, 1984; 231-240.
15. S. Luther; D. Meves *Polym. Eng. Sci.* 2004, *44*, 1642-1647.

O-MK-258

STRUCTURE DEVELOPMENT IN MICRO INJECTION MOLDING OF POLYOXYMETHYLENE

Abdessalem Derdouri [a,*], **Musa R. Kamal** [b] **and Rabie El Otmani** [b]
[a]*Industrial Materials Institute, National Research Council, Boucherville, QC, Canada*
[b]*Department of Chemical Engineering, McGill University, Montreal, QC, Canada*

Abstract - The structure development of polyoxymethylene (POM) under severe strain rates and high thermal gradients provided by the micro injection molding process was investigated using a three zones stepped plaque cavity. Two levels (low and high) of mold temperature and injection speed were tested. The morphology developed in each zone was examined in the flow and transverse planes using polarized light microscopy (PLM). Compared to conventional injection molding, more complex morphological gradient are obtained, caused by local flow conditions and thermal gradients. The morphology depends strongly on the local thickness of the molding and thickness changes. Differential scanning calorimetry (DSC) was used to assess locally the developed crystallinity.

Keywords: Morphology, Micro Injection Molding, Polyoxymethylene.

Introduction

The increasing use of the micro-injection molding process for the production of complex parts with micro scale dimensions makes it important to study resin-process-microstructure-property relations for this process. Although, micro-injection molding is basically like conventional injection molding, the very small dimensions of the parts present some challenges: early freezing-off, filling difficulty with high aspect ratio part features, high injection velocities which may result in excessive shearing and viscous heating. Despite the fast growth of the process, little effort has been devoted to the study of microstructure development. Earlier work in a simple rectangular mold showed that micro injection molded parts exhibit complex stratified morphology, which varies substantially in both the injection and transverse directions. This variability was reflected in the mechanical and thermal properties of the moldings [1-4].

Injection molded poly(oxy-methylene) exhibits a well developed processing dependent skin-core morphology, which greatly affects its mechanical performance [2,6]. Similar morphology has been observed in several semi-crystalline polymer systems [2, 5-8].

The present work reports some observations regarding the character and distribution of crystalline morphology in microinjection molded POM. It also provides insight into how changes in thickness and process conditions influence the morphology and properties of POM micro parts.

Experimental

Commercial grade POM (DuPont Delrin 900) was used, in conjunction with an end-gated, microinjection mold, to make rectangular plaques with stepwise reduced thickness (0.8 mm, 0.45 mm, 0.2 mm, in that order). The nominal planar dimensions of the stepwise cavity were 3 mm×15 mm. A Battenfeld® Microsystem 50 injection-molding machine having a clamping force of about 50 KN was used for that purpose. The melt temperature at the nozzle was set at 215°C. the mold temperature at 40 or 90 °C, while the injection speed was 200 or 800 mm/s. In all cases, the nominal cooling time after filling was set to 5s..The morphology was investigated with polarized light microscopy using an Olympus BX50 optical microscope system, equipped with a Sony Digital HD camera. For this purpose, thin cross-sections (~5μm) were cut in the flow direction (FD) and transverse direction (TD) at the mid-plane of each of the part thickness zones, using a Leica Reichert Ultracut microtome Differential scanning calorimetry (DSC) was used to assess locally the developed crystallinity.

Results and discussion

The evolution of morphology, observed for typical microtomed thin sections along the injection direction and the transverse direction, from the thickest step to the thinnest one, is depicted in Figure 1. Although the morphology observed can be more complex, the polarized light microscopy revealed the presence of three distinct morphological zones with various microstructures: the skin layer, the shear layer and the core.

Fig.1 POM morphology along the flow direction (*xz* plane) and the transverse direction (y-z plane). a, b and c refer to the thick step near the gate, the middle step and the thin step respectively.

Considering the top micrographs of Fig. 1 (xz-plane), in the vicinity of the entrance to the thick step (a), a parabolic aggregate of small spherulites is observed, probably formed during packing of the cavity. Under the fast conditions of micro-injection molding, the nucleation rate is high and thus large numbers of growth-restricted small spherulites are produced. However, a distinctive feature of the observed morphology is the presence of oblate spherulites that grow randomly in the thermal gradient direction (z-axis) from a single nucleus and that are mostly located in the shear layer. This is well illustrated in Figure 2 where a probable shish-kebab structure is also seen. These microstructures are also observed in the transverse direction. They are representative of oriented crystallization, due to the existing local thermal gradient producing conditions that favor spherulitic growth primarily in one direction [9, 10].

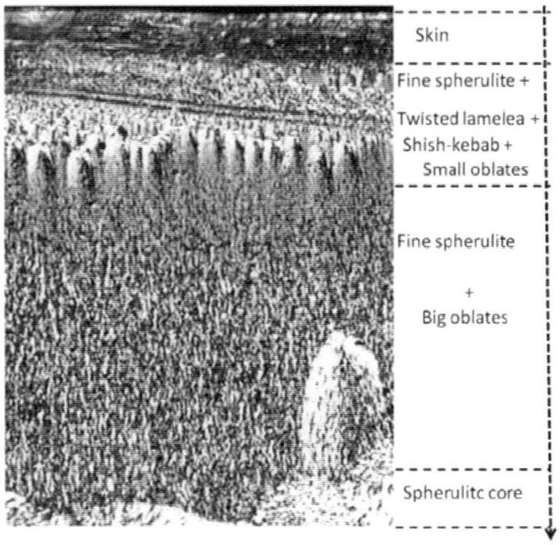

Fig. 2 Micro-injected POM: distinctive morphological zones with various crystalline features (Magnified view from Fig.1a, xz plane, thickest step near the gate)

Radical morphology changes, induced by secondary flows during cavity filling, due to abrupt thickness changes, were observed near thickness transitions in the sample, as shown in Figure 3. This seems to strongly suggest that crystallization may occur during filling at locations where flow channels are drastically reduced in size.

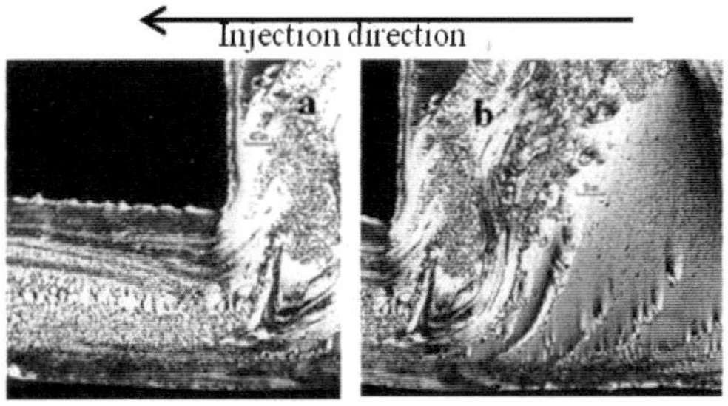

Fig. 3 Morphology at small step thickness ((a) at a magnification of 4X, (b) at a magnification of 10X

Figure 4 shows the thickness of each layer as a percent of the total thickness at different positions in the injection direction (case 1).

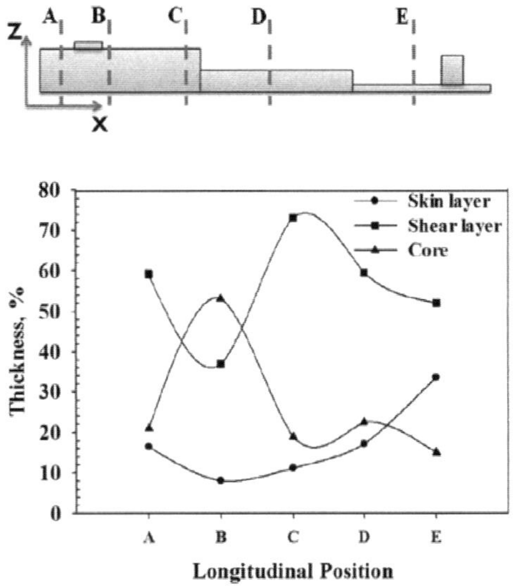

Fig.4 Percent of total thickness developed by skin, shear and core layers in the flow direction (rpm=200; T(mold)=40°C)

These results demonstrate the influence of microscale phenomena, especially the dependence of the size of the formed layers on the local thickness. The skin layer for instance tends to increase in size as the thickness of the step zones is decreased, while the shear layer and the core, after a more complex variation in the thickest step at the entrance of the part, evolve in the opposite direction. The effect of the processing conditions used in this study on thickness layer is shown in Figure 5 for the shear layer. The results indicate that this effect is more important when the zone thickness is highest (step closest to gate, positions A, B, C).

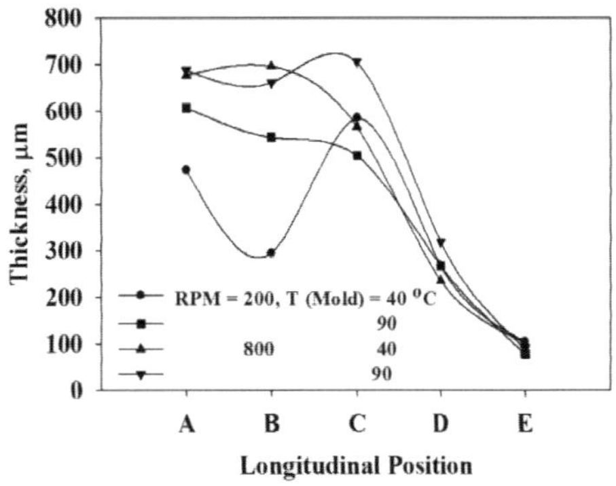

Fig. 5 Shear layer total thickness, flow direction, all processing conditions

The results presented so far provide insight into how changes in cavity thickness and process conditions influence the morphology and properties of POM micro-parts. More understanding is gained when we consider the developed local crystallinity which is displayed in Figure 6 for all four processing conditions. Overall, the proportion of crystalline material is highest in the thickness zone closest to the gate (thick step), where the morphology displayed a higher degree of diversity. This zone is also the zone that is packed and stays hot longer which may enhance crystallization more than in the middle step and the thin step zones. The processing conditions of case 3 (high injection speed and low mold temperature) resulted in the highest degree of crystallinity due to the high shear rates and thermal gradients generated.

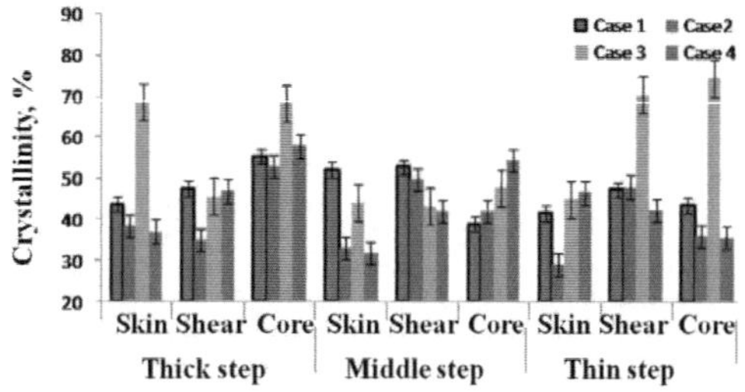

Fig. 6 Crystallinity rates for each thickness zone and all processing conditions

Figure 7 shows the resulting shrinkage in the thickness direction (Z-axis) from the previous crystallinity distribution results.

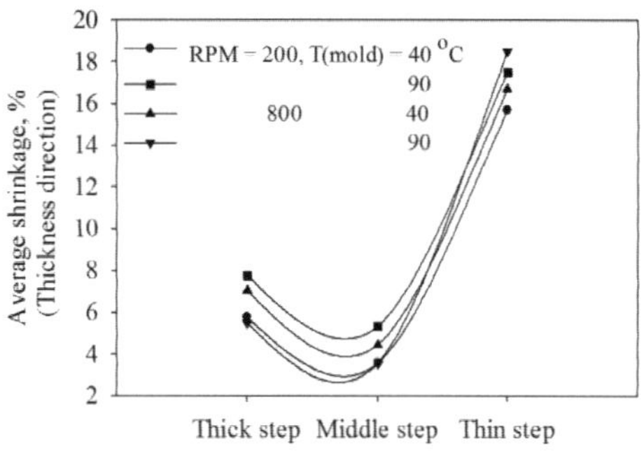

Fig.7 Average shrinkage for each thickness zone and all processing conditions

The molding conditions affected only slightly the shrinkage, but it is obvious that the local part thickness has a preponderant effect in close relation with the local cristallinity shown in Figure 6.

Conclusion

This study has shown that the morphological study in microinjection molding needs more investigations. The microstructure development observed by polarized light microscopy is very sensitive to the evolution of local thermal and flow conditions, which are likely to change significantly in time and space. Within the different zones, regular spherulites (core layer), oriented spherulites, twisted lamellae, shish-kebab and fine spherulites (skin and shear layers) may be found. Abrupt changes in part thickness induce severe flow patterns resulting in complex structures at these transition regions. The size and the nature of the crystalline features depend on the local thickness and, to a lesser extent, on the processing conditions particularly in thin flow channels. Owing to the complex coupling between the shear flow and the thermal gradient in the micro part during the microinjection cycle, it is difficult at this point to judge the dispersed morphological zones along the thickness of the micro part.

Acknowledgement

Financial support from NSERC (Natural Science and Engineering Research Council of Canada) and NRC (National Research Council Canada) is gratefully acknowledged.

Table 1: Processing conditions

Parameters	Case 1	Case 2	Case 3	Case 4
Melt T, °C	215	215	215	215
Mold T, °C	40	90	40	90
Injection Speed, mm/s	200	200	800	800

References

[1] J. Chu, M. R. Kamal, S. Derdouri, and A. N. Hrymak, *Polymer Engineering and Science*, ***50,*** issue 6, page 1214-1225(2010)

[2] M. R. Kamal, J. Chu, S. Derdouri, and A. N. Hrymak,, *SPE ANTEC,* 1323-1327 (2009)

[3] J. Chu, M. R. Kamal, S. Derdouri, and A. N. Hrymak, *Polymer Engineering and Science*, (2011), DOI: 10. 1002/pen

[4] M. R. Kamal, J. Chu, S. Derdouri, and A. N. Hrymak, *Plastics, Rubber and Composites: Macromolecular Engineering*, **39**, number 7 (2010)

[5] E.S.Clark, *Applied Polymer Symposia, l,*. **20**, p. 325-332(1973)

[6] E.S. Clark, *Applied Polymer Symposia, ,* **24**: p. 45-53(1974)

[7] G.W.M. Peters, J. W. Housmans, M. Gahleitner, H. E. H. Meijer. *Polymer,* **50**(10): p. 2304-2319(2009)

[8] S.S. Katti and J.M. Schultz, *Polymer Engineering and Science*, **22**(16): p. 1001-1017(1982)

[9] D. Hansen, A.N. Taskar, O. Casale, J. Polym. Sci. , Part A-2, **10**, 1625-1619 (1972)

[10] A.J. Lovinger, C.C. Gryte, Macromolecules, **9**(2), 247-253 (1976)

O-MK-257

SEALABILITY AND SEAL CHARACTERISTICS OF BLENDS OF PE/EVA AND PLA/PCL

Zahra Najarzadeh, Ramin Yousefzadeh Tabasi, , A. Ajji
3SPack, CREPEC, Chemical Engineering Department
Ecole Polytechnique de Montreal, Montreal, Quebec, Canada

Abstract
Seal strength behavior of low density polyethylene and ethylene vinyl acetate copolymer (PE/EVA) blends as well as that of blends of a seal grade PLA with aliphatic polyester (PCL) was studied. Incorporation of EVA in polyethylene improves its sealability in terms of decrease in plateau initiation temperature and broadness of sealability plateau. Formation of fine crystals, decrease in melting point and presence of vinyl acetate as a polar group were considered as the main explanations through differential scanning calorimetry (DSC) and Fourier transform infrared spectroscopy (FTIR) analysis. For the PLA/PCL system, the dispersed phase was stretched into elongated ellipsoidal domains. This type of morphology affected the mechanical and seal properties of the blends. As a result of blending, both Hottack initiation temperature and strength as well as seal initiation temperature were enhanced. The enhancement in properties was significant when the concentration of the dispersed phase exceeded 20wt% of the blend. Hottack strength of up to twice the initial strength of pure PLA was achieved through blending. The final Hottack strength (1300 g/25mm) at 40% dispersed content compared advantageously to commercially available polyolefin based sealant resins. The Seal and Hottack initiation temperature were shifted to lower temperatures by as much as 30 °C, which can allow faster and more energy efficient sealing process.

Introduction

Heat sealing is an important operation in packaging. Modern machinery for vertical and horizontal form fill seal (VFFS and HFFS) have a need for operating at higher and higher speeds. This requirement and the fact that the final package integrity is ultimately dependent on the results of the sealing process requires that the polymers used for the seal layer have superior hot tack, low sealing initiation temperature and wider sealing temperatures range. Although resin suppliers provide a wide range of resins for the seal layer, with a variety of performances and physical properties, materials optimization is still a challenge.
Coextrusion of low density polyethylene and ethylene vinyl acetate copolymer has been used in industry to tailor a sealant. However studies on correlation of physical properties and final seal performance of LDPE/EVA blends in open literature are limited. Studies on PE/EVA blends in a multilayer structure showed the effect of incorporating a low content vinyl acetate EVA (9%) in seal layer and tie layer. [1–3]
In this work we investigate the effect of EVA incorporation in monolayer PE sealant for several compositions produced by the film blowing process. We explain the consequences of this blending on microstructure and physical properties of the films and correlate them to the final seal properties.

On the other hand, many efforts and energy is used to overcome the problems caused by plastic wastes, particularly those produced every year by the packaging industry. Increasing costs for the removal, incineration and landfill of municipal wastes urge the packaging industry to shift their productions to more sustainable packaging. [4]. In this study, we also consider the applicability of biodegradable resins for multilayer packaging, particularly for seal layer.

Some bio-based resins such as Polylactides (PLA) and Polyhydroxyalkonoate (PHA) have good moisture barrier but low flexibility, while others such as Polyvinylalcohol (PVA, PVOH) have significantly high oxygen barrier, but highly sensitive to humidity. Others like PCL have low oxygen and water vapor barrier but a very good flexibility [5]. What all these resin come short of is the capability to be heat sealed at an acceptable dwell time and low enough temperature not to deteriorate other properties of the package. This hermetic seal property is a guaranty of the efficiency of barrier layers against oxygen permeation, odor loss and water vapor transmission. Available bio-based, compostable polymers to be used as a hermetic sealant layer are Polybutylene succinate (PBS) or amorphous PLA grade. PBS lacks seal strength and PLA has low flexibility and high seal initiation temperature. In this study the focus is on improvement of PLA seal properties to make it more plausible for packaging application.

Experiments

Low density polyethylene (LDPE, MFI=2.3) and ethylene vinyl acetate (EVA containing 18%VA, MFI=8) were obtained from Dow Chemicals and Dupont respectively. The blends studied contained EVA weight content of 100%, 60%, 40% and 20% and are designated as EVA100, EVA60, EVA40 and EVA20 respectively. Monolayer films (40μ thickness) were produced in a 45mm Killion single screw

extruder accompanied by a laboratory scale film blowing line which is equipped with a helical blown film die. The cooling air was at room temperature.
The thermal properties were analyzed using differential scanning calorimetry (DSC) TA Instruments Q1000.

The seal grade PLA (polylactic acid) was obtained from Natureworks is copolymer of D,L lactic acid with D-monomer content of higher than 10%. Rubbery aliphatic polyester has been used to improve both flexibility and the seal initiation temperature of the PLA sealant resin. All samples have been processed in twin-screw extruder (Leistritz model ZSE 18 HP co-rotating) and were made into film using a cast die and all the film have been drawn to a thickness of 50 μ m for further analysis. The temperature profile of the extruder along the barrel was set at 170/180/190/190 and the die was set to the temperature of 190°C. Different blend composition have been used to prepare the samples. The number following the sample name is the content of the dispersed phase.

Seal strength experiments were done using SL100 Lako-Tool hot-tack machine based on the ASTM F88 and F2029 experimental methods. Heat seals were made with two metallic jaws (19.1mm×25.4mm) covered by Teflon on film strips in (1in width×13in length) dimension. Biaxially oriented polypropylene tape was used to provide the support for sealant. All the sealing experiments were performed in a range of jaw temperatures at a constant pressure of 0.5 (N/mm^2) and dwell time of 0.5(s).

Results and discussion for PE/EVA system:

Differential Scanning Calorimetry (DSC) results

Through the DSC results, the composition of PE/EVA blend determines amount of crystallinity and the melting temperature range. An increase in the amount of EVA in the blend caused less crystallinity and broader range of crystal sizes. (Figure 1) Presence of vinyl acetate groups at crystal growth front can prevent crystal growth of polyethylene chains in PE/EVA blends, which would reduce the lamellar thickness and crystal perfection. It also decreases the enthalpy of melting and shifts the melting point to lower temperatures. By increasing the EVA content, another melting peak appears at lower temperatures due to the crystal formation of EVA chains, which contain sequences of ethylene groups capable of forming fine lamellas. [6]

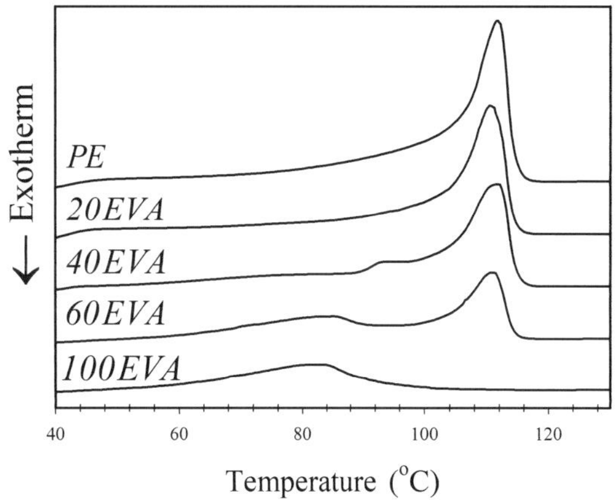

Figure 1: DSC thermographs of melting of films

Seal Strength Experiments

Turning to the heat sealing behavior of the blends, it is useful to compare the sealing curves to see how the seal initiation temperature (T_{si}), plateau initiation temperatures (T_{pi}), sealability temperature window and plateau seal strengths vary for the various blends. Polyethylene show T_{si}=108°C and T_{pi}=115°C, while in neat EVA has T_{si}=74°C and T_{pi}=80°C. Blends of these two resins are located in between. (Figure 2) Incorporating EVA in polyethylene cause a decrease in T_{si} and T_{pi} but the trend of seal temperature dependence remains the same for low EVA content blends (20% and 40% EVA). For the 60% EVA blend, the behavior of seal strength converts to a trend similar to that of neat EVA. The large gap between 40% and 60% EVA blends might be due to the dominance of EVA in 60% film. Also Vinyl acetate group polarity create attraction force between two film surfaces as well as finer crystals make EVA rich blends capable of making stronger seals at low temperatures. [7], [8]
Lower plateau seal strength for 60% EVA and neat EVA is most likely due to the decrease in their tensile yield strength, as can be observed in Figure 3. [9]

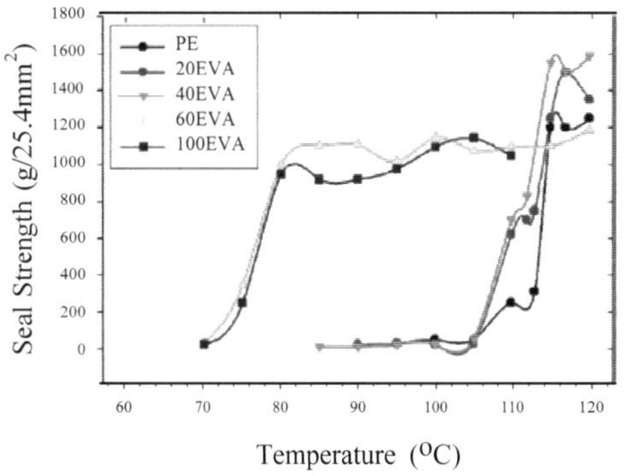

Figure 2: Seal Strength Temperature dependence of PE/EVA blends films

In failure mode analysis three temperature zones were observed for each film, representing three different failure modes. As illustrated in (**Erreur ! Source du renvoi introuvable.**), peeling failure zone is the region in which the seal area peels apart at the film interface, because of the insufficient heat for chain interdiffusion. Plateau zone, in which delamination and elongation failure occur, is the range of temperatures capable of melting crystals and providing enough energy for interdiffusion and entanglements such that the seal area is strong. Breakage zone is where high temperature causes merging the combination of the two film surfaces make adhesive debonding impossible.

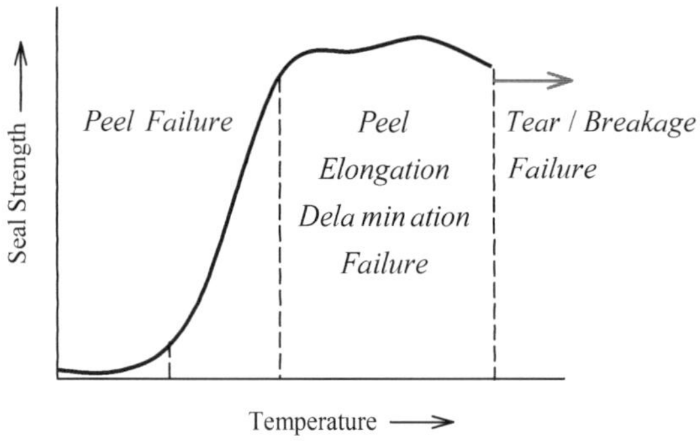

Figure 3 Failure modes observed in monolayer blown PE/EVA blend films

Results and discussion for PLA/PCL system:

Viscoelastic behavior analysis

Figure 5 illustrates complex viscosity of pure PLA and the modifier. In shear rates of actual process (about 30-50 s^{-1} in cast die and 200-300 s-1 in the extruder) in the Die the viscosity ratio ($\lambda=\eta_d/\eta_m$) of the melts are about 2.3, with the Seal modifier having a bit higher viscosity which prevent the coalescence of the droplets in lower shear rates than the extruder and hence the droplet volume of the dispersed phase would not change dramatically in the die. In this condition the very low interfacial tension of PLA and the dispersed phase make it plausible to expect laminar morphing in the cast films.

Thermal Characterization

Samples have been analyzed by differential scanning calorimetry instrument to investigate the thermal behavior of the prepared blends. The DSC results are shown in Figure 6. The endothermic peak at around 60°C is associated with the melting point of dispersed phase. Slight shift in the melting point of the latter in the blend has been detected, which indicate a partial miscibility between PLA and the seal modifier.

In blend samples of higher than 80% PLA content the rigid PLA chains, would lower the chain mobility of modifier molecules and hence lower the spatial ordering thereof. The latter would lead to significant decrease in crystalline content of dispersed phase in blends with higher PLA concentration as shown in Figure 6.

Seal and Hot-tack

The change in the hot tack and seal initiation temperature of pure PLA and the blends with 20% and 40% of seal modifier is shown in Figure 7 the significant shift in initiation temperature after blending is observed, which indicates the capability of selected modifier resin to enhance the mobility and molecular diffusion when blended with PLA. As shown in Figure3 the hot tack strength of the blend increases about twice as much as pure PLA strength when the Modifier content reaches 40%. This can be attributed to the linear structure of the modifier with no side chains in the backbone of the molecules. Researchers have investigated the latter in commercial seal grade polyethylene resins where the metallocene resins with few side chains and lower melting point show better seal strength and lower seal initiation temperature than LDPE resin.

The sealability can be affected by flexibility of polymer chains at the test temperature. The raise in temperature enhances this mobility and the strength of the sealed area would reach a maximum. By increasing the temperature afterward the high mobility of the softened polymeric chains would lead to flow under the pressure exerted by machine, while sealing happens. This phenomenon will decrease the thickness of seal area at very high temperatures and will cause the seal strength to drop.

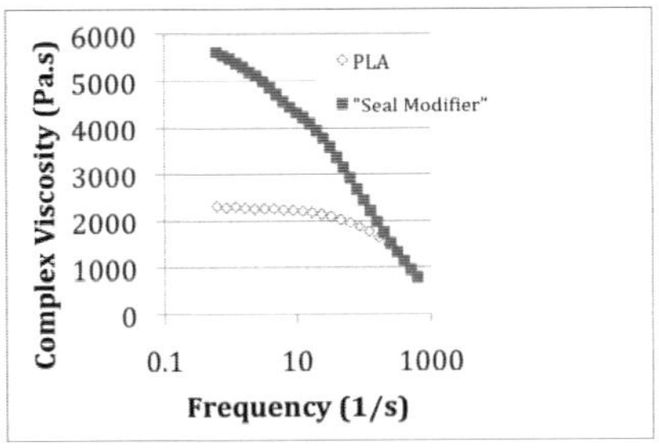

Figure 5: the Comparison of viscoelastic behavior of PLA and Modifier

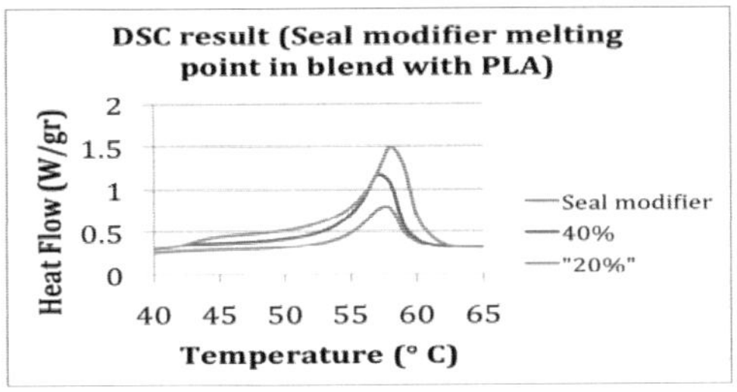

Figure 6: DSC results shows the change in the melting point of seal modifier in blend

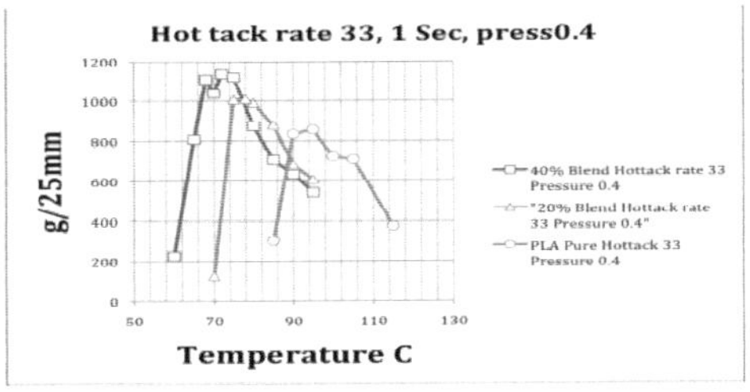

Figure 7: Seal initiation and Hottack Strength of PLA and blends of 20% &40%

Conclusion:

The results of PE/EVA blend blown film studies have demonstrated their sealing performance can approach, and surpass the performance of pure EVA film. This was shown by seal strength experiments. Reduction in crystallinity and melting temperature observed through DSC results as well as presence of vinyl acetate are the main reasons.
The failure analysis revealed that at low temperatures, the seal fails by peeling, for intermediate temperatures, the failure occur by delamination and elongation, and at high temperatures by tearing.
Deformation of EVA droplets at the surface of the films observed in SEM micrographs can be an explanation for the effect of EVA incorporation in seal performance of polyethylene.

For PLA/PCL system, addition of seal modifier polyester through melt extrusion to PLA increased both Hottack strength and seal initiation temperature. This behavior have been related to linear structure and low melting point of dispersed phase. This mobility have also lead to toughening of rigid PLA resin which will be addressed in future work and was out of the focus of this paper.

References:

[1] C. Poisson, V. Hervais, M. F. Lacrampe, and P. Krawczak, *Journal of Applied Polymer Science*, vol. 99, no. 3, pp. 974–985, Feb. 2006.

[2] C. Poisson, V. Hervais, M. F. Lacrampe, and P. Krawczak, *Journal of Applied Polymer Science*, vol. 101, no. 1, pp. 118–127, Jul. 2006.

[3] J. Zhang, E. Manias, G. Polizos, J.-Y. Huh, A. Ophir, P. Songtipya, and M. Jimenez-Gasco, *Journal of Adhesion Science and Technology*, vol. 23, no. 5, pp. 709–737, Dec. 2011.

[4] Chang, A. C.; Chum, S. P.; Hiltner, A.; Baer, E. J Appl Polym Sci 2002, 86, 3625.

[5] M.F. Koening, S.J. Huang; Polymer 36, 1877 (1995).

[6] M. Brogly, M. Nardin, and J. Schultz, *Journal of Applied Polymer Science*, vol. 64, no. 10, pp. 1903–1912, Jun. 1997.

[7] K. L. Mittal and A. Pizzi, *Handbook of Sealant Technology*, 1st ed. CRC Press, 2009.

[8] E. M. Petrie, *Handbook of adhesives and sealants*. McGraw-Hill Professional, 2000.

[9] R. W. Halle and B. P. Center, "Plastomer-mVLDPE Blends for High Performance Heat Sealing Applications."

O-MK-254

THE ROLE OF CHAIN DYNAMICS ON COLD CRYSTALLIZATION OF PLA-CLAY NANOCOMPOSITES

S.Karami and P.G.Lafleur [1]*

[1] *Génie Chimiquel, École Polytechnique de Montréal, CREPEC, Canada – Pierre.lafleur@polymtl.ca;*

Abstract - this work deals with the study of the effect of chain dynamics on isothermal and non-isothermal crystallization of PLA nanocomposites. Commercially available organically modified MMT clay (Cloisite 15A), at different concentrations, has been employed, to produce an intercalated nanocomposite. Differential scanning calorimetry and rheological measurements were utilized to study the molecular dynamics during the cold crystallization process and in the molten state, respectively. Solid-like behavior was observed in the low frequency region within the rheological studies, when the nanoparticles content was increased. The cold crystallization temperature shifted to a higher value when nanoparticles were added to a PLA, and it decreased as the experiment time-scale (inversely related to heating rate), for a certain composition, below the percolation threshold. The overall crystallization growth rate decreased upon the addition of nanoparticles during the isothermal process, while it was enhanced by the increase of annealing temperature above a critical value. This study demonstrated that nanoparticles act as heterogeneous nucleating sites at a combination of a time-scale and an annealing temperature such that the trapped chains can be devitrified along the interphase.

Keywords: PLA Nanocomposite, Chain dynamics, Isothermal and Non-isothermal crystallization, Morphological constraints

Introduction

Polymer nanocomposites show sizeable improvement of physical and mechanical properties compared to conventional composites. The excellent properties of the nanocomposites are attributed to the formation of an extensive polymer-particle interphase. The interphase layer is extended along the phase border of the polymer and filler due to the attractive interaction between them and a large surface area provided by the dispersion of the nanoparticle. The trapped chains show different molecular dynamics within the confined environment of the interphase imposed by the strong coupling, not participating in large-amplitude cooperative conformational motion of the bulk-like chains at the apparent glass transition temperature [1, 2].

The microstructure of semi-crystalline polymers depends on crystallization parameters such as the degree of crystallinity, crystallization kinetics and the morphology of the crystallites [3]. The addition of the inorganic phase influences the crystallization of the matrix [4-7]. The nanoparticles can act as heterogeneous nucleating sites at low particle contents [7]. Nevertheless, the nucleation might be hindered by the shielding effect of the organically modified nanoparticles [6]. Moreover, the crystallization kinetics reduced at higher particle loading due to the topological constraints. Diffusion of crystallizing segments into the growth front is impeded by the confined segmental motions following the formation of a percolated structure [4, 5].

In this work, we used PLA as a model polymer. This semi-crystalline polyester is recognized as a popular biopolymer produced from renewable resources. Polylactide exhibits promising physical and mechanical properties compared to conventional petroleum-based polymers. It is, recently, turning into a desired material in the packaging industry due to its biocompatibility and biodegradability. PLA crystallization has been investigated extensively over the past few years [8, 9].

In this paper, the kinetics of cold crystallization during the non-isothermal and isothermal processes are investigated in order to ascertain the influence of the addition of organically modified nanoparticles on chain dynamics. Finally, a relationship between devitrification of the polymer-particle interphase and the crystallization kinetics is illustrated.

Experimental

Materials. Poly(L,D-lactic acid) used in this study is a commercial grade of PLA(4043D) with approximately 2 mol% D-content, provided by NatureWorks LLC. Cloisite 15A, employed as the nanoparticles, was purchased from Southern Clay Product Inc. The organoclay was modified through a surface treatment process of montmorillonite in the presence of a quaternary ammonium salts. Hereinafter, the hybrid of PLA/Cloisite 15A is designated PLAAi where A represents Cloisite 15A and i stands for the particle loading (wt%) of the nanoparticles.

Sample Preparation. PLA and the nanoparticles were dried prior to any processing. An overnight drying procedure was applied at 70°C under reduced pressure. The PLA granules and the nanoparticles were manually mixed by tumbling in a sealed plastic zip-lock bag, and subsequently fed into an extruder. The mixture was melt-compounded into pellets via a Leistritz co-rotating twin-screw extruder (TSE). The outgoing melt was fed into a water bath and cut into pellets using a pelletizer. The pure sample also went through the same procedure as the nanocomposites. The obtained pellets were compression molded into thin sheets and discs at 200°C under a Nitrogen atmosphere. The samples were held for an additional 3 min at the molding temperature after removing the pressure to eliminate any thermal and deformation history. Then, the samples were quenched in cold water at 0°C. The quenched samples are designated as PLAAi(Q).

Rheology. The rheological measurements were done using a plate-plate geometry (R=25m, gap=1mm) on a Gemini stress-controlled rheometer, at 190°C under a Nitrogen atmosphere. All dynamic measurements were carried out in the linear viscoelastic region. A strain sweep experiment was performed over a strain range of 0.1–100% at frequencies of 0.1 rad/s and 10 rad/s to determine the linear domain. Thermal stability of the samples during frequency sweep measurements was verified by a time sweep experiment within the linear region. The frequency sweep experiments were carried out over a frequency range of 0.1-150 rad/s to determine the storage and loss moduli.

Thermal Analysis. Non-isothermal and isothermal studies were performed on a TA-Instruments DSC model Q1000. Approximately 7 mg of the sample was weighed and sealed in an aluminum pan and lid. During the non-isothermal experiments, the samples were equilibrated at 30°C and then heated to 200°C at rates of 3°C/min and 10°C/min. The isothermal measurements were performed at several annealing temperatures, in the range of 90°C to 130°C, during 1 hr. The annealed samples are designated PLAAi(Q1h)T, where T represents the annealing temperature.

Results and Discussion

Rheology. The storage modulus becomes nearly frequency-independent with the increase of particle loading from (i =) 1.2wt% to 3.6wt%, within the low frequency region, as shown in Fig. 1. The microstructure is not notably disturbed by the small-amplitude oscillatory deformation in the linear viscoelastic region. A percolation threshold occurred at a composition of less than 3.6wt% in the Intercalated nanocomposites, resulting in the formation of a network structure [10]. The appearance of the solid-like behavior suggests a strong polymer-particle interaction, as well as, the formation of morphological restrictions. The trapped chains in an interphase might entangle to molecules from another interphase rather than with those of the bulk-like chains, increasing the time required for disentanglement [1].

Non-Isothermal Crystallization. The cold crystallization temperature (T_c) of the quenched samples shifted to higher temperatures upon the addition of nanoparticles at a heating rate of 10°C/min. Meanwhile, for the compositions above the percolation threshold, the cold crystallization is barely noticeable for the same experiment time-scale (reversely proportional to heating rate), as revealed in Fig. 2a. The cold crystallization temperature shifted to a lower temperature for PLAA1.2(Q) compared to PLA(Q) following a decrease of the heating rate to 3°C/min as indicated in Fig. 2b.

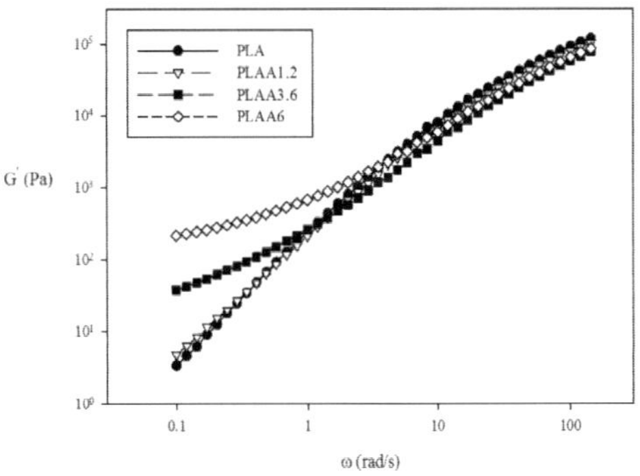

Figure 1 – Storage modulus as function of frequency (0.1-150 rad/s).

Basically, a non-equilibrium glassy state is obtained by quenching the polymer melt to a temperature far below the glass transition. The re-adjustment of chain conformation occurs during an increased time scale of the experiment when the metastable structure reaches a more energy-favorable state.

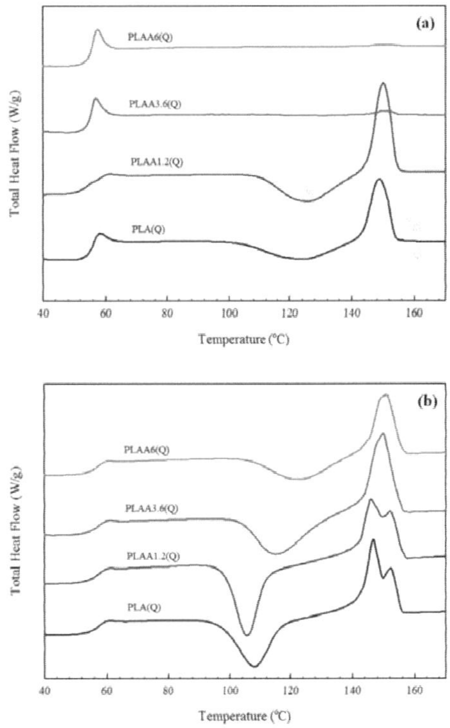

Figure 2 –Total heat flow by the standard DSC at heating rate of (a) 10°C/min and (b) 3°C/min.

Isothermal Crystallization. To study isothermal cold crystallization, the relative crystallinity of the quenched samples was obtained, and the crystallization half-life was estimated, during several annealing temperaures, as can be seen in Fig. 3 & Fig. 4. The crystallization half-life ($t_{1/2}$) is defined as the time required to reach half of the final degree of the ultimate crystallinity. In cold crystallization, the overall crystallization growth rate G, is proportional to $1/t_{1/2}$ [9].

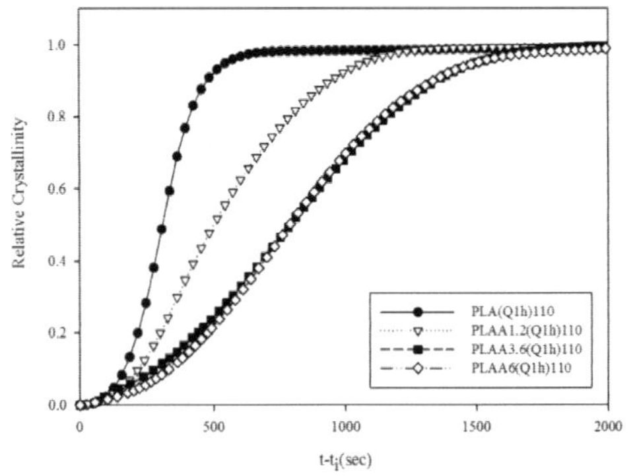

Figure 3 – Development of crystallinity of samples annealed at 110 °C. ti represents the induction time of crystallization.

The fastest relative crystallization and overall growth rate occur for PLA(Q1h)110 as illustrated in Fig. 3 and Fig. 5, respectively. Attractive interaction along the interphase and topological constraints reduce the chain mobility and subsequently affect the short distance diffusion of the crystallizing segments across the phase boundary. This influences both nucleation and crystallization growth rate due to an increase in the free enthalpy of activation, as well as, the free energy of crystallization of the chain segments. The self diffusion coefficient levels off at a certain particle loading, which corresponds to the percolation threshold. Thus, the overall growth rate is considerably lower for PLAA3.6(Q1h)T and PLAA6(Q1h)T due to the formation of a network structure, while the growth rate does not show any noticeable dependence on particle loadings above the percolation threshold, as can be seen in Fig 3 and Fig. 4.

It was found that, generally, the overall growth rate increases as a function of annealing temperatures. However, the entropy barrier of the uncrystallizing segments becomes significant by the increase of annealing temperature. At that point, this rate starts diminishing. The maximum overall growth rate of PLA(Q1h)T occurred at the annealing temperature of 110°C as shown in Fig. 5. Surprisingly, A larger overall growth rate have been observed for PLAA1.2(Q1h)T compared to PLA(Q1h)T at annealing temperatures above 120°C, as shown in Fig. 5. Similarly, the relative crystallization of PLAA1.2(Q1h)130 also goes faster than PLA(Q1h)130, as illustrated in Fig. 4.

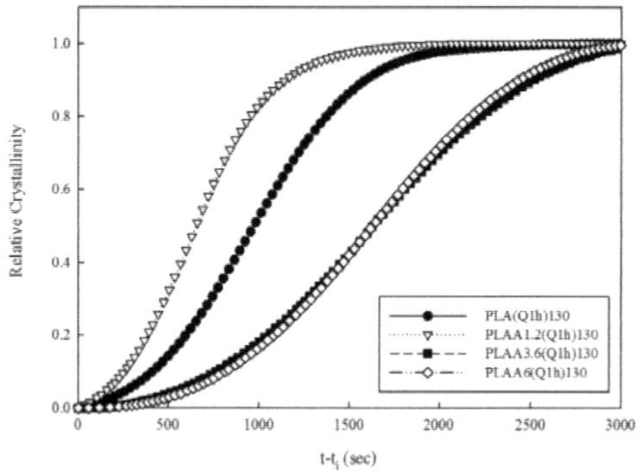

Figure 4 – Development of crystallinity of samples annealed at 130 °C. t_i represents the induction time of crystallization.

Above the critical annealing temperature, the enhanced overall growth rate can be attributed to the heterogeneous nucleating role of the nanoparticles at this composition. A portion of amorphous molecules trapped in the polymer-particle interphase shows restricted dynamics compare to bulk-like chains. It is assumed that the interphase layer is devitrified at 120°C, improving the nucleation rate while the crystal growth rate is fairly high at the annealing temperature. The devitrified trapped chains are not as mobile as bulk-like chains, due to the strong polymer-particle interaction, due to the entanglement of chain segments to the absorbed layer.

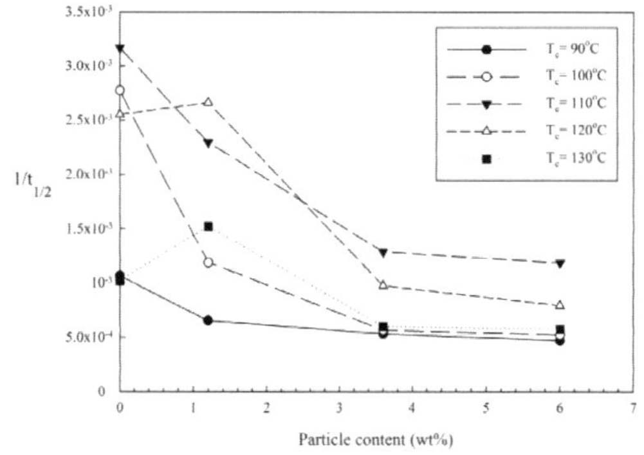

Figure 6 – Overall crystallization growth rate of the samples as a function of annealing temperature.

Beyond a certain experiment time-scale, the appearance of the solid-like behavior leads to a long disentanglement time even in the molten state Therefore, the constrained interphase cannot be devitrified at these annealing temperatures.

Conclusions

The formation of network structure was illustrated by the investigation of dynamic viscoelastic properties within the linear region, resulting in a confined topology, above the percolation threshold. Restricted chain mobility leads to higher energy of crystallization of chain segments, and larger free enthalpy of activation.

From the non-isothermal experiments, it was deduced that increasing the time scale of the measurement (reducing the heating rate) results in a structural rearrangement of the interphase layer. Beyond the percolation threshold such structural rearrangements do not take place due to the longer disentanglement time.

The overall growth rate of crystallization decreases following the appearance of topological constraints during the annealing process, however, devitrification of the trapped molecules at the critical annealing temperature improves the overall growth rate compared to the pure sample. Moreover, the short distance diffusion reduces non-monotonically corresponding to the particle content.

It was concluded that nanoparticles show heterogeneous nucleating effect at a certain time scale and annealing temperature where the trapped chain segments in the interphase can be devitrified.

Acknowledgements

This research was funded by CREPEC and the Natural Sciences and Engineering Research Council of Canada (NSERC).

References

1. Rao, Y.Q. and J.M. Pochan, *Macromolecules*, 2007. **40**(2): p. 290-296.
2. Ciprari, D., K. Jacob, and R. Tannenbaum, *Macromolecules*, 2006. **39**(19): p. 6565-6573.
3. Lu, X.F. and J.N. Hay, *Polymer*, 2001. **42**(23): p. 9423-9431.
4. Fornes, T.D. and D.R. Paul, *Polymer*, 2003. **44**(14): p. 3945-3961.
5. Somwangthanaroj, and M.J. Solomon, *Macromolecules*, 2003. **36**(7): p. 2333-2342.
6. Nam, P.H., et al., *Polymer Engineering and Science*, 2006. **46**(1): p. 39-46.
7. Maiti, P., et al., *Macromolecules*, 2002. **35**(6): p. 2042-2049.
8. Pan, P., et al., *Macromolecules*, 2007. **40**(19): p. 6898-6905.
9. Kawai, T., et al., *Macromolecules*, 2007. **40**(26): p. 9463-9469.
10. Vermant, J., et al., *Journal of Rheology*, 2007. **51**(3): p. 429-450.

O-MK-191

APPLICATION OF ADVANCED CONTROL ALGORITHMS IN INJECTION MOLDING

Feng Zhou[b], Shengyong Mo[b], Yi Yang[a], Ke Yao[b], and Furong Gao[a,b,*]
[a] *Department of Control Science and Engineering, Zhejiang University, Hangzhou, China*
[b] *Centre for Polymer Processing and Systems, Fok Ying Tung Graduate School, The Hong Kong University of Science and Technology, Hong Kong*

Control system plays a crucial role in injection molding product quality consistency. In terms of the volume, China is ranked as the number one IMM manufactory country in the world. These machines however are mainly low-end products, with very low profit margin. Technology and product upgrade is the key to business success. An advanced control system has been designed by the Center for Polymer Processing and Systems (CPPS) at Hong Kong University of Science and Technology (HKUST) to work directly with the low-end injection molding machines with minimal machine modification. This paper shows the control system architecture and advanced features. It has been shown experimentally that with this controller fitted to a local Hong Kong machine, the performance can be compared to that of the high cost and high performance European or Japanese machines, in terms of repeatability and accuracy. The applications of such a controller to the Asian machines can be a cost effective solution for the manufactures to upgrade their machines to a higher grade at minimal costs, with much improved performance.

Keywords: injection molding, model predictive control, iterative learning control, barrel temperature, injection velocity

***Corresponding Author**: Prof. Furong Gao, kefgao@ust.hk

1 Introduction

Injection molding, a major polymer processing process, transforms plastics into products of various shapes and types. It plays an important role in plastics industry, due to its high production rates, cost effectiveness, and ability to produce complex articles with high precision. In recent years, due to the rapid development of plastics industry, injection molding industry expands rapidly, especially in China. In terms of the output, China is ranked as the number one for the amount of injection molding machines produced. However, these machines are of low grade and very low profit margins, and poor market competitiveness. Lack of advanced control technology may be the main reason behind it.

A typical reciprocating screw injection molding machine comprises of an injection unit, a clamping unit, a power unit and a control unit as schematically shown in Figure 1. The task of the injection unit is mainly to melt the solid plastic, accumulate the melt in the screw chamber, inject the melt into the mold cavity and maintain the holding pressure during the packing phase. Mold operations including opening, closing and part ejection are accomplished by the clamping unit. The power unit supplies and distributes either hydraulic or electric power to the corresponding machine components. The control unit is the brain of an IMM. It is responsible for machine sequence control, key process variables control and supervisory functions, for example, process monitoring statistical analysis and fault detection/diagnosis.

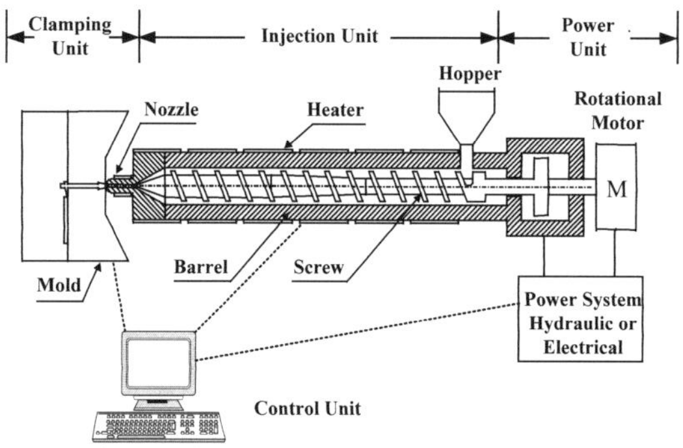

Fig. 1: Injection molding machine: Structure

As a typical stage-based batch process, injection molding includes mold open/close, filling, packing, plastication, and cooling, as illustrated in Figure 2. The injection cycle starts when the mold is closed. The filling phase starts after this during which the screw moves forward, pushing the melt through the runner and the gate, and finally into the mold cavity. When the mold is completely filled, the process switches to the packing-holding phase. In the packing-holding phase, additional materials are packed into the cavity at high pressure to compensate for the material shrinkage due to cooling. This phase continues until the gate freezes off, isolating the material in the mold from that in the injection unit. After this, the melt can no longer be pushed into the mold, and the process enters the cooling phase. In cooling phase, the polymer inside the mold continues to cool down. At the same time, the polymer in the barrel undergoes plastication to prepare the melt for next cycle. During plastication, the melting polymer is conveyed to the screw tips through the solid conveying, melting, and metering zones due to the screw rotation. It accumulates in front of the barrel, building up the pressure at the nozzle, which pushes the screw to retract. This process continues until a preset plastication stroke is reached. When the part in the mold is sufficiently solidified and becomes rigid enough, the mold opens and the part is ejected. The process repeats itself to produce more products.

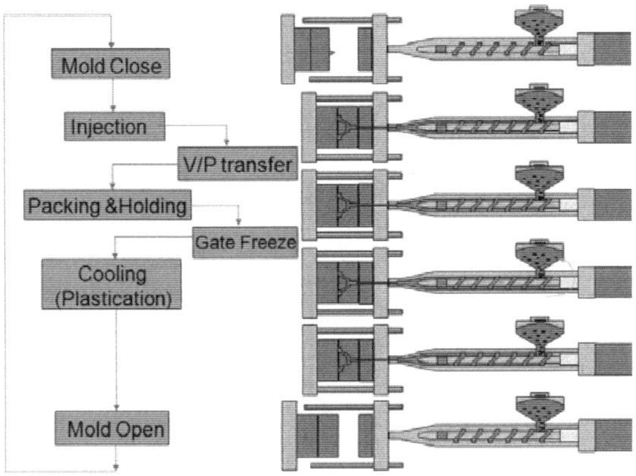

Fig. 2 Illustration of injection molding process

Although it appears to be simple, injection molding is a complex nonlinear dynamic process, during which the machine parameters, material properties and process variables interact with each other. The quality of the molded part, characterized in terms of its dimensions, appearance and mechanical properties, is a strong function of the processing conditions. A good control of the key process variables is necessary to ensure the repeatability and reliability of the product quality.

This paper shows the advanced injection molding control system developed by Center for Polymer Processing and Systems (CPPS). This advanced control system works directly with the low-end local Hong Kong injection molding machines with minimal machine modification. It is proved that the performance of machine equipped with this advanced control system can be compared to that of high-cost and high-performance European or Japanese machines, in terms of repeatability and accuracy.

2 Structure of Control System

2.1 Hardware Structure

The CPPS control system consists of a hardware base and a set of real-time control programs. The controller hardware, which is independently designed by CPPS specifically for injection molding machine, is constructed base on the modular design. The two basic modules are a central processing board and a signal input/output board, interconnected with CAN bus. Each sub-module can be independently designed and manufactured to reduce the cost and the effort of customization test.

With high-performance X86 based structure, the central processing module provides great compatibility, full support of network and other communication functions and easy storage of data, and reduces the workload of software system development in the meantime. It is also powerful carrier, in which the real-time programs are embedded for supporting graphical user interface (GUI), controlling the machine operation sequence and all important process variables during injection molding, such as barrel temperature, packing pressure and so on.

The signal input/output can be further divided into four sub-modules, including a digital input module measuring the limit switches and proximity sensors, a digital output module controlling the pump operation, solenoid valves and robot operations in certain cases, an analog input module acquiring the measurements of process variable sensors, and an analog output module controlling the proportional valves. The digital input signals are opto-isolated with the capability of high-speed pulse measurement specially designed for rotary encoders. The digital outputs are opto-isolated outputs. The analog input module uses 16bit isolated A/D conversion, which is sufficient for injection molding process measurement. It also includes special design for temperature measurement with thermocouples. The analog output module uses 14bit direct voltage/current output instead of the common pulse width modulation output, to control proportional valves with much higher accuracy and reduce response time.

2.2 Software Structure

The control software system is built based on a real-time Linux platform, which supports multi-task, priority-driven preemptive scheduling, and fast context switching, with accurate timers easily assigned for different processes. After custom, the operating system is more stable and more robust, therefore provides high precision and safety for the molding process control. Furthermore, the whole operating system was built from scratch and incorporated with all the key elements of real-time systems. Therefore, it is ideal for real-time control implementation.

The controller software system is constructed base on the modular design, as shown in Figure 3.It consists of multiple independent modules, in which each module performs a certain function. The input and output of digital signals and analog signals are handled by digital signal manager and analog signal manager respectively. The motion and sequence controller modular controls the operation sequence of injection molding machine, temperature controllers modular controls barrel and mold temperatures, and each sub-phases of molding process is controlled by the corresponding control modular. All the process data can be recorded by a data logger task, and interchanged via shared memories. The message-based inter-process communication technology is also adopted for communication between modules.

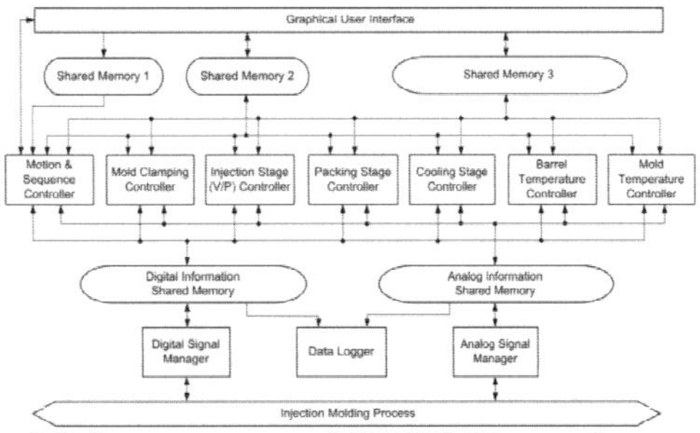

Fig. 3 Software structure of CPPS IMM control system

3 Test Setup

The control system is tested in the local Hong Kong injection molding machines, Chen Hsong MJ55, with minimal machine modification. The hydraulic pressure and flow rate of all test machines are controlled by proportional valves.

4 Test Results and Discussions

4.1 Barrel Temperature Control

Melt temperature is a major concern for the production of consistent quality products in injection molding. However, because melt temperature is difficult to be measured or controlled, the industry expects the barrel temperatures to be controlled within a small variation of their set-point. Compared to other key process variables, barrel temperature has its specific properties that could lead to control difficulties. It exhibits integrator-like dynamics in the relevant operation range. It also has significant time delays due to the sensor position and heat transfer characteristics. Moreover, barrel temperature normally has significant nonlinearities due to heat radiation and shear heating caused by melt flow. Therefore, during the operation, it is required that temperature is controlled as consistent as possible to guarantee the final product quality consistency.

The control system has a series of control approaches that can achieve very accurate temperature control during operation. Figure 4 shows the barrel temperatures operation control results with CPPS advance control system. As can be seen, temperature can be controlled within 0.4℃ of set-point. Therefore, this control system greatly reduces temperature fluctuation and thus enhances the material melt flow, thermal stability and final product quality consistency.

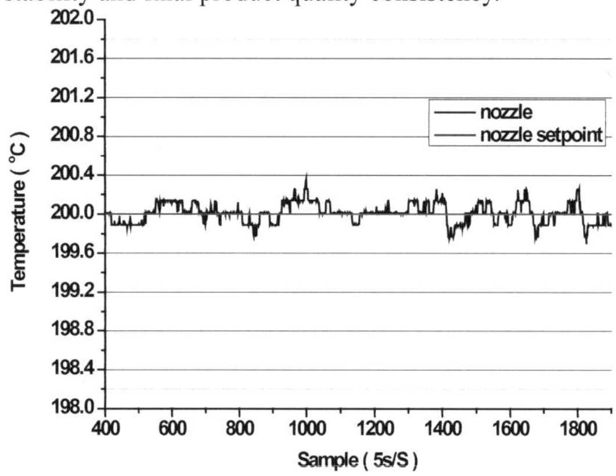

a. temperature operation control result of nozzle

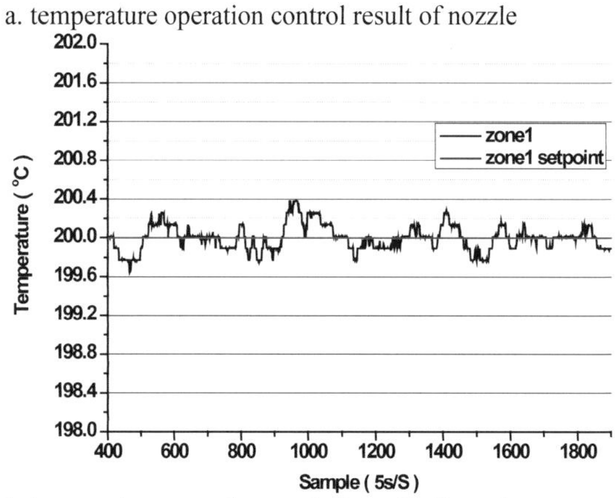

b. temperature operation control result of zone1

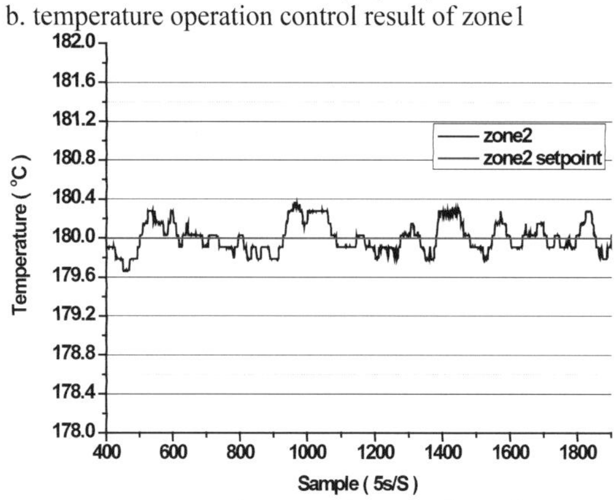

c. temperature operation control result of zone2

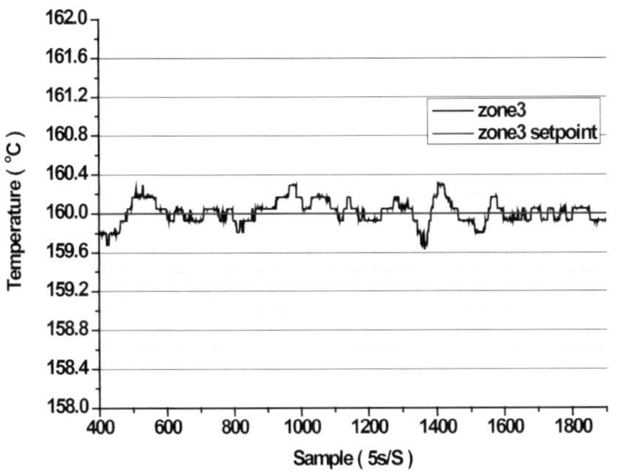

d. temperature operation control result of zone3

Fig. 5 Temperature control result of CPPS control system

4.2 Injection Velocity Control

In the filling phase, polymer melt is forced into a mold cavity by screw forward motion. The nature of the polymer flow entering the cavity strongly influences the quality of the mold part, especially the mechanical properties, such as tensile strength and impact strength. However, because melt flow rate is difficult to be measured or controlled, the industry expects filling rate is approximated by injection velocity. A proper setting of injection velocity can achieve a more evenly distributed melt flow pattern, and avoid over-pressurization, high thermal stress and high residual flow stress. Consequently, it is extremely necessary to control injection velocity accurately. Due to the technical requirement of product, the injection velocity has to be in the scenario of 'fast-slow'. However, the original controller has no velocity measurement, and completely open-loop control. Figure 6 shows the injection velocity control result of original controller, which exhibits poor performance. Using existing electrical scale of original machine, the CPPS advanced control system can make accurate measurement of velocity, and realize close-loop control. The control result is illustrated in Figure 7. It is obvious that the injection velocity accurately follows and rapidly tracks the step change profile in the set point, showing the great performance of this control system.

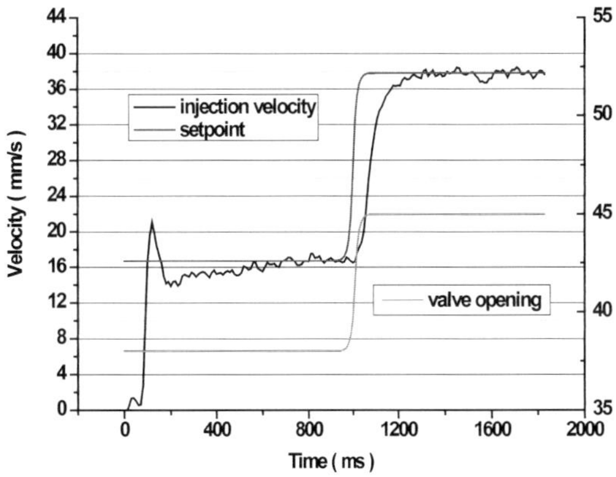

Fig. 6 Injection velocity control result of original controller

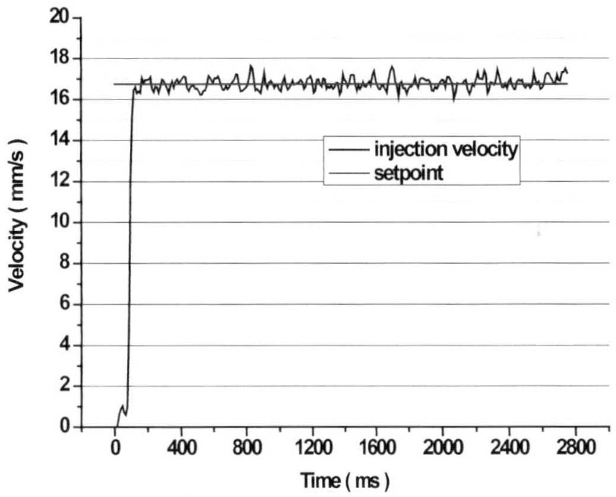

a. constant set-point

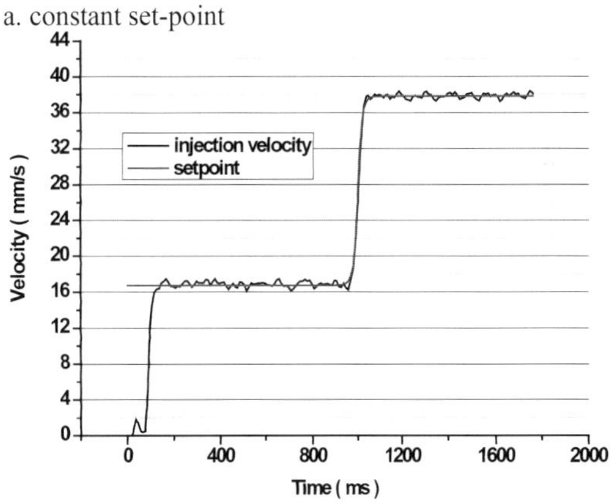

b. step change set-point

Fig. 7 Injection velocity control result of CPPS controller

4.3 Packing Pressure Control

After mold is completely filled, the process switches to the packing-holding phase. In the packing-holding phase, additional materials are packed into the cavity at high pressure to compensate for the material shrinkage due to cooling. Therefore, in this phase, proper packing pressure control is exceptionally important for producing part with good mechanical properties, dimensional stability, and satisfactory surface characteristics. However, the original controller completely adopts open-loop control. Figure 8 shows the packing pressure control result of original controller, which exhibits poor performance. The pressure control result of CPPS control system is illustrated in Figure 9. It is obvious that the control system accurately control the pressure, showing the great performance of this control system control performance.

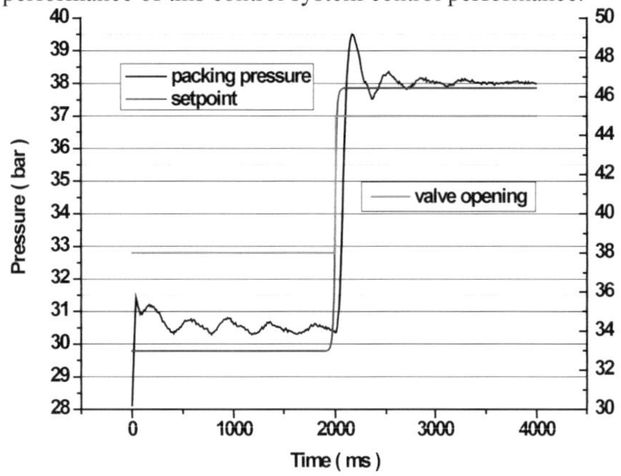

Fig. 8 Packing pressure control results of original controller

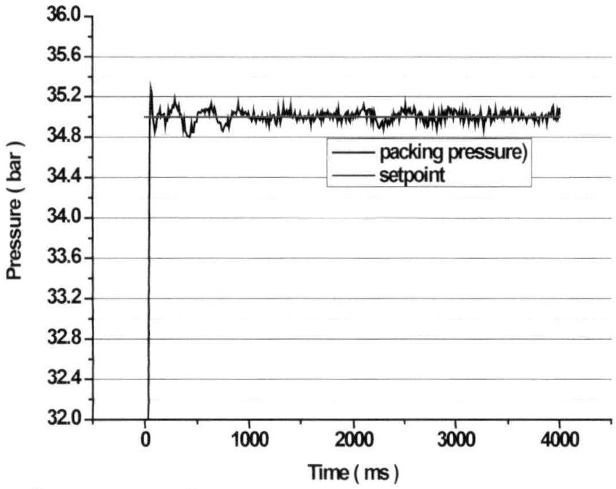

a. Constant set-point

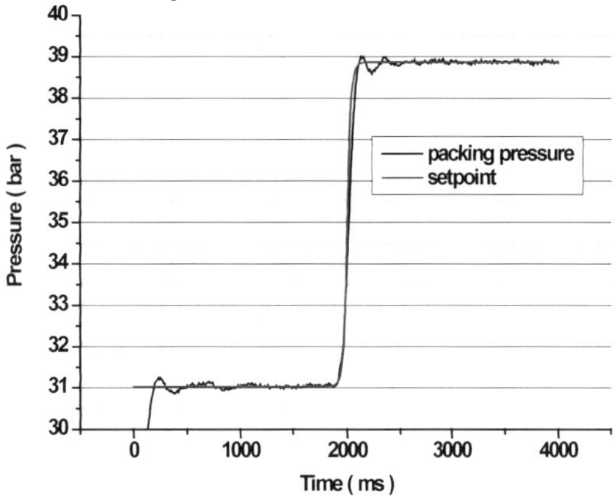

b. Step change set-point

Fig. 9 Packing pressure control results of CPPS control system

4.4 Mold Open Position Control

As the demand of automation production rises, robot arms are more widely used to take the molded part out of the machine in modern molding industry. The application of robot arms requires precise control of mold open position. Without adding dedicated proportional valve on the machine, the CPPS advanced control system successfully controls the mold open position with high precision. As shown in Figure 11, the mold open position can be controlled within the range of ±0.2mm.

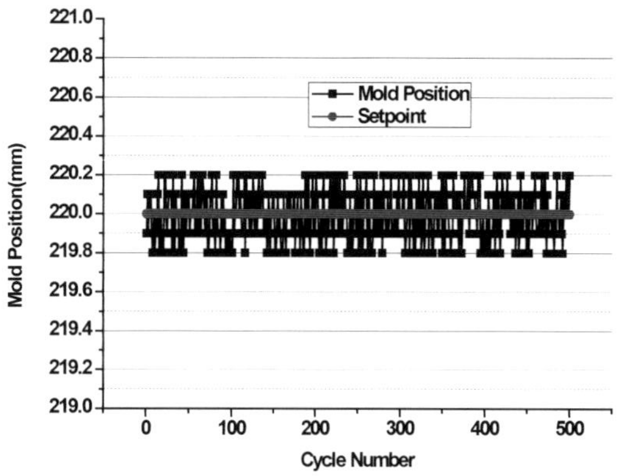

Fig. 10 Mold open position control result of CPPS control system

5 Conclusions

As the brain of IMM, the control system plays a crucial role in injection molding product quality consistency. A poor control system severely lowers the integrated performance of injection molding machine. Conversely, excellent control system can guarantee quality consistency of final product, upgrade the quality of machine, improve the market competitiveness and raise the value of machine. In this paper, an advanced IMM control system, with superior hardware and software structures, designed by CPPS, has been presented. It can work directly with low-end injection molding machines with minimal machine modification. The results show that the control system greatly improves the control accuracy of key process variables. The performance can be compared to that of the high cost and high performance European or Japanese machines, in terms of repeatability and accuracy.

O-MK-151

THE DEPENDENCY OF PLA'S CRYSTALLIZATION KINETICS ON DISSOLVED CO_2 PRESSURE

M. Nofar and C. B. Park*

Microcellular Plastics Manufacturing Laboratory, Department of Mechanical and Industrial Engineering, University of Toronto, 5 King's College Road, Toronto, Ontario, Canada- 1-mnofar@mie.utoronto.ca, 2- park@mie.utoronto.ca

Abstract - This study investigated the crystallization kinetics of polylactide (PLA) at various CO_2 pressures using a high-pressure and a regular differential scanning calorimeter. The isothermal and nonisothermal melt crystallization results showed that increasing the CO_2 pressure increased the crystallization rate. However, a very high crystallinity was achieved at 15 bar CO_2 pressure, during isothermal and low-cooling-rate non-isothermal crystallization. This was by facilitating more perfect crystal formation with the plasticization effect of CO_2 while limiting the crystal nucleation rate. At higher CO_2 pressures, a larger number of less close-packed crystals were formed due to chain entanglement, and consequently, the final crystallinity was decreased. The non-isothermal results showed that the crystallization temperature and glass transition temperature decreased at increased pressures through the CO_2's plasticization effect.

Keywords: Polylactide, crystallinity, high pressure DSC

Introduction

Due to its origin in renewable sources and its biodegradability, polylactide (PLA) is one of the polymers that have received much attention [1]. PLA is a promising candidate for a range of applications, and it could replace other non-biodegradable polymers such as polystyrene in the areas such as packaging, cushioning, and plastic utensils [2-3].

The crystallization control for PLA is crucial to enhance its heat deflection temperature and mechanical properties since the utility of PLA has been limited because of the low glass transition temperature (T_g) (and a low crystallinity) [4]. Currently, PLA's slow crystallization behavior is one of its most challenging issues. This structural parameter is also paramount in such polymer processes as foaming. It was found that the increase in melt strength of the connected molecules by crystallization improved the foamability of PLA significantly [5, 6]. On the other hand, too high crystallinity will decrease the foam's expansion ability and cell density because of too high stiffness and less gas dissolution. Furthermore, according to heterogeneous cell nucleation theory, the local pressure variations around the formed crystals can promote cell nucleation [6, 7]. Therefore, the crystallization control can significantly affect the foaming ability while improving the heat-deflection/service temperature [2] and mechanical properties of the final products.

During the foaming processes, the crystallization behavior of PLA will be more challenging in presence of dissolved gas. First of all, the dissolved CO_2 causes swelling of polymer [8, 9] and thereby affects the crystallization kinetics. Also, the expansion and cell growth itself has the effect of increasing the final crystallinity [5, 10]. While measuring the crystallization kinetics during the biaxial stretching that occurs in the last stage of foaming would not be easy, at least the initial crystallization of the PLA melt can be investigated accurately in the presence of dissolved gas while no foaming occurs.

Without having the effects of foaming, this study investigated the crystallization behaviors of a linear PLA in the presence of dissolved CO_2. A high-pressure differential scanning calorimeter (HP-DSC) was used. The maximum CO_2 pressure in the HP-DSC was 60 bar (870psi). Due to the equipment's limitations, pressures beyond 60 bar were not explored. A regular DSC was also used to measure the PLAs' crystallization at atmospheric pressure.

Experimental

1- Materials:

The PLA supplied by NatureWorks® LLC was the commercially available Ingeo™ 8051D linear PLA with D-lactide molar content of 4.6 mol % and molecular weight of 90,111 g/mol. To decrease their moisture level, the PLA pellets were oven-dried at 65°C for at least 6 hours. Talc of Mistron Vapor-R grade from Luzenac Corp. was also melt blended at 0.5%wt for better crystal nucleating in PLA. The melt blending was processed in a Minilab twin-screw extruder (Haake Minilab) at 180°C for 4 min. with a screw rpm of 40. CO_2 with a 99.98% purity produced by Linde Gas was used as the saturation gas for the HP-DSC.

2- Thermal behavior analysis:

To analyze the isothermal and non-isothermal melt crystallization behaviors of the PLA samples at atmospheric pressure and at various CO_2 pressures, DSC experiments were implemented using a DSC2000 (TA Instruments, New Castle, DE) and an HP-DSC (NETZSCH DSC 204 HP, Germany).

3- Crystallization morphology analysis using WAXD:

X-ray diffraction analysis was used to analyze the effect of CO_2 gas pressure on the crystals' structure and their morphology after various isothermal and non-isothermal treatments. The X-ray diffraction was run on a Siemens D5000 Diffractometer System operating at 50 kV/35 mA.

Results and Discussion

1- Isothermal melt crystallization analysis

Fig. 1 (a-c) shows the effect of different isothermal temperatures on the melt crystallization of PLA samples at atmospheric pressure (1 bar) and at 15 bar and 45 bar CO_2 pressures. For each case, the crystallization half-time ($t_{1/2}$) and the final crystallinity are listed in Table 1.

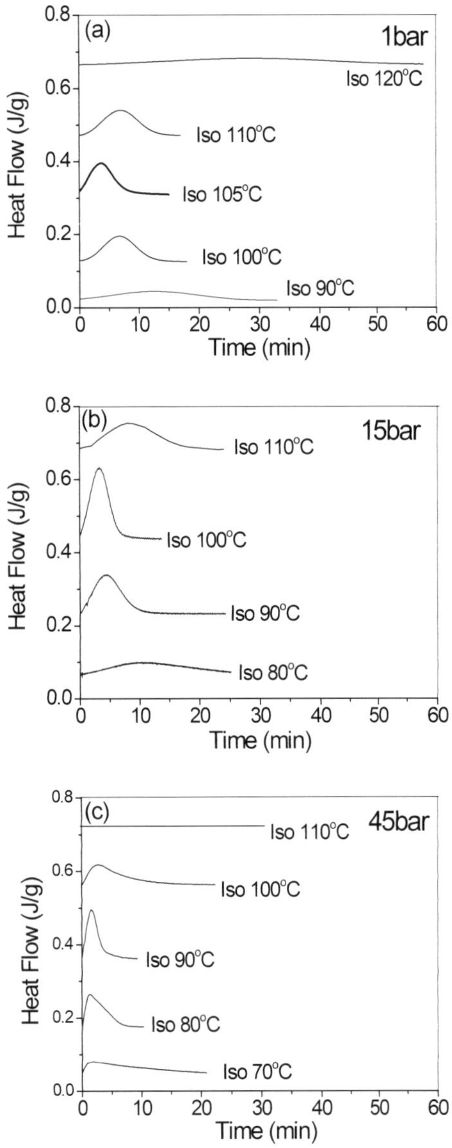

Figure 1 – Melt crystallization at various isothermal temperatures and atmospheric pressures (1 bar), 15 bar CO_2 pressure, and 45 bar CO_2 pressure

As Fig. 1 and Table 1 show, the crystallization half-time was decreased as the amount of CO_2 dissolved in the PLA increased with a higher pressure. It was also observed that the crystallization half-time became less sensitive to the temperature by the dissolved CO_2. In other words, the crystallization rate became high for a wide range of temperature as the CO_2 pressure increased. This clearly indicates that the dissolved CO_2 molecules increased the crystallization rate through its plasticization effect on the molecular mobility [11-14], i.e., the reduction of the dissipation energy required for crystallization [13]. This plasticization effect of CO_2 was also demonstrated by the decrease of the critical isothermal temperature, $T_{critical}$, at which the crystallization rate is the highest, as the pressure increased. The $T_{critical}$ decreased from 105°C at 1 bar to 90°C at 45 bar CO_2 pressure.

Despite the decreased crystallization half-time with the dissolved CO_2, the final crystallinity was not increased proportionally to the CO_2 pressure. The final crystallinity increased significantly at 15 bar, but decreased dramatically at 45 bar. This indicates that the crystallinity was governed by different mechanisms in these two cases. At low pressure CO_2, with limited number of crystal nuclei, the CO_2's plasticization increases the molecules mobility and facilitates the crystal growth with more close-packed structure. Therefore, without causing much entanglement, the final crystallinity increased. On the other hand, at increased CO_2 pressure, although the CO_2's plasticization effect increased the molecules mobility, a large number of crystals may also increase the molecules entanglement and thereby retard the crystal growth as a crystal may affect the chain mobility of neighboring crystals. Consequently, the crystals may become less perfect (less close-packed) due to the CO_2's lubrication effect and the swelling and the final crystallinity may be decreased due to the hindered molecules regularity despite the enhanced nucleation rate in the early stage of crystallization.

Table 1 – Crystallization half-time and the final crystallinity of PLA at various isothermal temperatures and various CO_2 pressures

Pressure	Iso. Temp.	$t_{1/2}$ (min)	*Crystallinity (%)*
1 bar (Atm. P)	90	14.3	21.39
	100	5.9	28.17
	105	4.8	30
	110	6.5	31.53
	120	33	36.17
15 bar CO_2	80	11.5	28.42
	90	4.5	36
	100	3.5	42.7
	110	10.5	54.4
45 bar CO_2	70	2.6	15.6
	80	1.4	21.36
	90	1.7	21.77
	100	2.7	22.6

110	-	--

Wide-Angle X-ray Diffraction (WAXD) measurements were also taken to explore the crystallization morphology after isothermal treatment at $T_{critical}$ under the three pressures. Fig. 2 shows that all PLA samples exhibited diffraction peaks at $2\theta = 14.8$, 16.7, 19.0, and 22.3° corresponding to α crystal form of PLA [14-16]. As shown, the presence of CO_2 dissolved gas did not induce a new crystalline phase in the PLA samples. When isothermal treatment was done under 15 bar CO_2 pressure, the relative intensity of the characteristic crystalline peak at $2\theta = 16.7$ increased. This reflects that the PLA's crystallinity was enhanced with larger crystals and more close-packed lamellas [17-18]. When the samples were exposed to 45 bar CO_2 pressure, the relative intensity at $2\theta = 16.7$, however, decreased, and the peak's broadness increased. This reveals that the crystal size decreased (i.e., the number of crystals increased), and that the crystals were less close-packed [17-18]. At the increased CO_2 pressure of 45 bar, the molecular mobility did not remain constant because of the growing entanglements through the larger number of crystals.

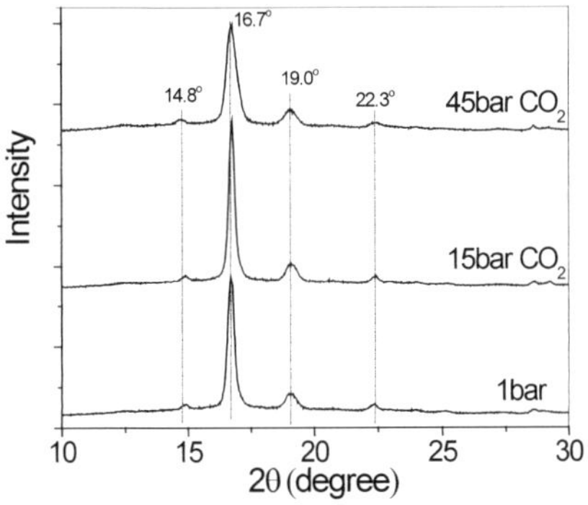

Figure 2 – WAXD patterns of PLA samples after isothermal melt crystallization under atmospheric pressure (100°C), 15 bar CO_2 pressure (100°C), and 45 bar CO_2 pressure (90°C)

2- Non-isothermal melt crystallization analysis

Fig. 3 shows the cooling graphs of PLA at atmospheric pressure and at 15, 30, 45, and 60 bar CO_2 pressures at a cooling rate of 2°C/min. Fig. 4 shows the degree of crystallinity of the cooled PLA samples versus the CO_2 pressure. Similar to the isothermal melt crystallization, the maximum crystallinity during cooling 2°C/min was achieved when the PLA samples were exposed to 15 bar CO_2 pressure. As the pressure increased from 15 bar to 30 bar, 45 bar, and 60 bar, the crystallinity for all PLA samples decreased and fluctuated in a range of 15-25%. In other words, at a cooling rate of 2°C/min the PLAs' crystallinity was retarded at higher CO_2 pressures compared with 15 bar CO_2, most likely for the same reason discussed above.

These results are comparable to the isothermal melt crystallization results, since at a slow cooling rate there is a long annealing time. At 15 bar CO_2 pressure, the final crystallinity was improved along with the reduced crystallization half-time. As discussed, this was due to the facilitated movement of the PLA molecules to form more close-packed crystals by the plasticizing CO_2 without causing much entanglement. On the other hand, at an increased CO_2 pressure over 15 bar, despite the decreased crystallization half-time, the final crystallinity was depressed due to the less close-packed crystals formed by the increased number of crystals and too rapid crystallization.

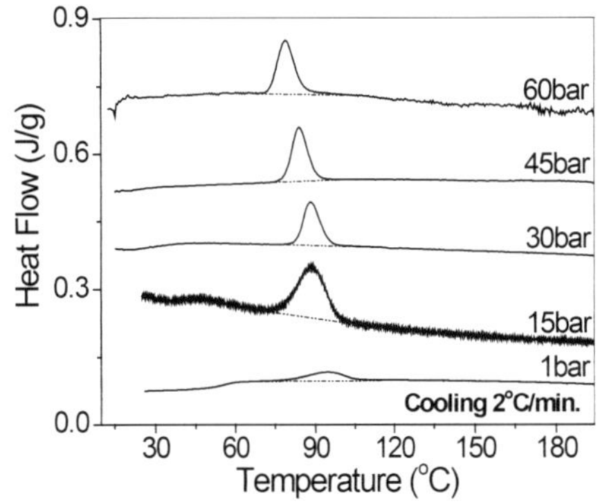

Figure 3 – Cooling graphs of PLA samples at atmospheric and various CO_2 pressures measured in HP-DSC

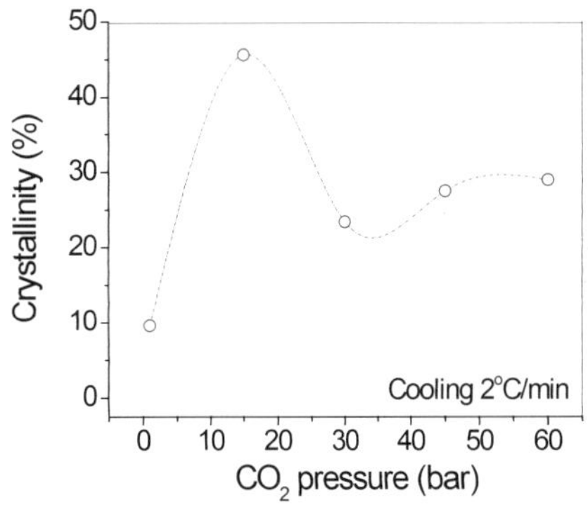

Figure 4 – Final crystallinity vs. CO_2 pressure for PLA samples after cooling 2°C/min

To verify the crystallization behavior of the linear PLA samples cooled at a rate of 2°C/min. under various pressures, wide-angle X-ray diffraction (WAXD) measurements were also conducted (Fig. 5). Very similar results were obtained as in the case of isothermal crystallization. As in Fig. 2, only α crystal phase of PLA was detected and the presence of CO_2 did not induce a new phase [14-16]. For the 15 bar CO_2 case, the relative intensity of the characteristic crystalline peak at $2\theta = 16.7$ increased compared to the other pressures, indicating the bigger-sized crystals with higher close-packed structure [17-18] as discussed earlier. As the CO_2 pressure increased to 60 bar in stages, the peak's relative intensity was decreasing with a broader shape, which corresponds to the smaller sized crystals with less close-packed crystals as in the case of isothermal crystallization.

The effects of the CO_2 pressure, on the crystallization temperature (T_c) and the glass transition temperature (T_g) were also investigated for the PLA samples and the results showed that the CO_2's plasticization reduced the T_c and T_g linearly.

At a cooling rate of 2°C/min, the difference in T_c for the samples at atmospheric pressure (1 bar) and at the highest CO_2 pressure (60 bar) was close to 15°C. Also, the T_g decreased from around 60°C to 20°C when the CO_2 pressure increased to 60 bar.

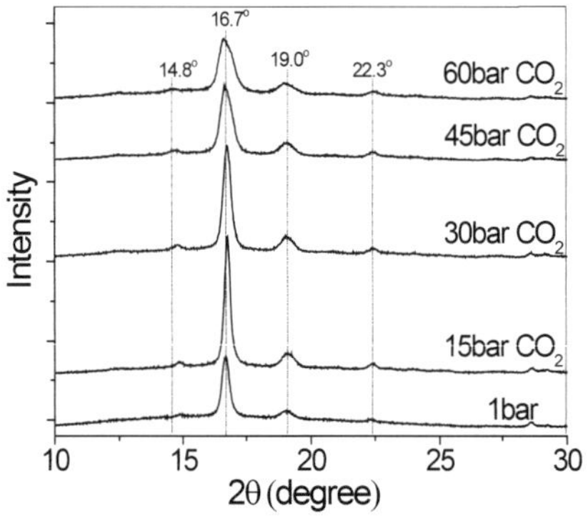

Figure 5 – WAXD patterns of PLA samples after cooling at 2°C/min under atmospheric pressure and various CO_2 pressures

Conclusions

In this study, the crystallization behavior of a linear PLA was investigated at atmospheric pressure in a DSC and under various CO_2 pressures in an HP-DSC. Both isothermal and non-isothermal melt crystallization results showed that increasing the CO_2 pressure decreased the crystallization half-time. The isothermal melt crystallization results and Wide-Angle X-ray Diffraction (WAXD) measurements showed that the maximum crystallinity was achieved at a low CO_2 pressure of 15 bar by facilitating the chain movement to create more close-packed crystals with the lubricating gas molecules dissolved in the PLA matrix, while maintaining a small number of crystals compared to the higher pressure case. At a high CO_2 pressure of 45 bar, the final crystallinity decreased because a larger number of less close-packed crystals were formed. Although the higher plasticization effect of CO_2 at an increased pressure increased the movement of the molecules in the beginning, the chain mobility was hindered in the later stage and dominated by the large number of nucleated crystals and the molecular entanglement through the nucleated crystals' network.

The non-isothermal melt crystallization at slow cooling rates showed a very similar behavior to the isothermal results. A very high crystallinity was observed at 15 bar CO_2 pressure and the crystallinity decreased at elevated pressures.

As the CO_2 pressure increased, both of T_c and T_g decreased consistently due to the plasticization effect of the dissolved CO_2 gas.

Acknowledgements

The financial support from the NSERC and the Consortium for Cellular and Microcellular Plastics is greatly appreciated. The donation of the PLA materials from NatureWorks and Jed Randall's assistance are also greatly acknowledged.

References

1. DW. Grijpma; AJ. Pennings; Macromol. Chem. Phys., 1994, 195, 1649
2. R. Auras; B. Harte; S. Selke; Macromol. Biosci., 2004, 4, 835–864
3. B. Gupta; N. Revagade; J. Hilborn; Prog. Polym. Sci., 2007, 32, 455–482
4. M. Nofar; W. Zhu; C.B. Park; J. Randall; Ind. Eng. Chem. Res. 2011, 50, 13789-13798
5. M. Mihai; M.A. Huneault; B.D. Favis; J. Appl. Polym. Sci., 2009, 113, 2920-2932
6. PAM. Lips; I.W. Velthoen; P.J. Dijkstra; M. Wessling; J. Feijen; Polymer, 46, 2005, 9396–9403
7. N.S. Leung; A. Wong, Q. Guo, C.B. Park; J.H. Zong; Chem. Eng. Sci., 64, 2009, 4899-4907
8. Y.G. Li; C.B. Park; H.B. Li; J. Wang; Fluid Phase Equilibr, 2008, 270, 15–22
9. Y.G. Li; C.B. Park; Ind Eng Chem Res, 2009, 48 6633–6640
10. W.T. Zhai; T. Kuboki; L. Wang; C.B. Park; E.K. Lee; H. Naguib; Ind. Eng. Chem. Res., 2010, 49, 9834-9845
11. L. Yu; H. Liu; K. Dean; Polym. Int., 2009, 58, 368-372
12. M. Takada; S. Hasegawa; M. Ohshima; Polym. Eng. Sci., 2004, 44 (1), 186-196
13. D. Li; T. Liu; L. Zhao; X. Lian; W. Yuan; Ind. Eng. Chem. Res. 2011, 50, 1997–2007
14. M. Nofar; W. Zhu; C.B. Park; Polymer, 2012, 53, 3341-3353
15. ML. Di Lorenzo; M. Cocca; M. Malinconico; Thermochim Acta, 2011, 522, 110-117
16. S. Huang, H. Li, S. Jiang, X. Chen, L. An, Polymer, 2011, 52, 3478-3487
17. W. Xu; G. Liang; W. Wang; S. Tang; P. He; W. Pan; J. of Appl. Polym. Sci., 2003, 88, 3225–3231
18. X. Zhang; M. Yang; Y. Zhao; S. Zhang; X. Dong; X. Liu; D. Wang; D. Xu; J. of Appl. Polym. Sci., 2004, 92, 552–558

O-MK-109

A CRITICAL REVIEW OF MATHEMATICAL MODELS AND COMPUTER SIMULATIONS IN BLOWN FILM EXTRUSION

John Vlachopoulos
MMRI/CAPPA-D, Chemical Engineering, McMaster University, Hamilton, ON, Canada L8S 4L7
vlachopj@mcmaster.ca

Abstract - Mathematical models and computer software have been developed for die design purposes during the last forty years. Most dies used for the production of blown film are of the spiral variety, either with a cylindrical mandrel in the center or stackable disk dies, for co-extrusion. Computer simulations are used for routine die design purposes, with the objective of obtaining uniform outflow from the die lips and for minimization of slow flow or stagnation regions. Mathematical models have also been developed for the bubble formation process. Such models have been aimed at determining the bubble shape. Despite their mathematical sophistication these models are not used in industry and it is hard to identify if they have had any tangible impact on blown film technology. It is argued that the failure of these models is due to the fact that they focused on determining the bubble shape. However, blown film bubbles are manipulated by mechanical means and the aerodynamics of impinging jets for cooling. It is argued that modeling ought to focus on the stress development under the externally imposed mechanical and aerodynamic constraints.

Keywords: die design, bubble formation, computer simulation

Introduction

The ever increasing demand for high production rates and high quality films has led to significant developments in materials and equipment which continue to widen the range of applications and markets for blown film. The blow-up ratio (BUR) is defined as the ratio of the final bubble diameter to the die diameter and is a good indicator of the amount of expansion in the radial direction. In typical installations BUR ranges from 1.2 to 4. The amount of expansion in the machine direction is characterized by a different dimensionless number, the take-up ratio (TUR), which is defined as the ratio of the final film velocity over the velocity at the die lips. In the literature the TUR is often called draw-down ratio (DDR).

The hot melt is cooled externally (and sometimes internally) by annular streams of high-speed air from air-rings. The maximum achievable output rate is often limited by the air-cooling capability of the line. The die specific output rate for lines equipped with dual-lip air-rings varies from 0.45 to 1.1 kg/hr per mm-of-die-diameter. Internal bubble cooling (IBC) is finding continuously increasing applications, increasing throughput rates from 0.7 to 1.7 kg/hr per mm-of-die-diameter.

Die Design

In blown film extrusion a thin tubular film is formed as the melt flows through the die lips. The die lip gap width usually ranges from 0.76mm to 3mm and die diameter from a few centimeters for laboratory lines to more than one meter for industrial installations producing more than one ton of film per hour. The annular flow is formed in the gap between the inner mandrel and the outer die body. Several types of dies have been proposed and build involving mandrels supported by spider legs, screen packs or a breaker plate. Also, side-fed dies are sometimes used. The problem with these designs is that they result in the formation of weldlines in the machine direction, along which the polymer is poorly bonded. By far the most common die geometry for blown film production is the spiral one. The polymer is fed by a number of melt tubes ending with a "port" at the start of each spiral. It flows both along the spirals and in the gap between the mandrel and the die body. The flow rate becomes progressively uniform around the circumference towards the die exit.

Above the die lips the outgoing polymer stream must have uniform thickness. Uneven thickness tends to be exaggerated by the subsequent inflation and stretching operations resulting in unacceptable film production. Polymer melts tend to flow preferentially directly above the ports resulting in periodic thick and thin sections. The tendency for uneven thickness distribution becomes greater as the power-law exponent in the viscosity function decreases, i.e. shear thinning increases. Thickness variation of more than 4% above the die lips is usually unacceptable. Another problem is the poor flow distribution along the spirals and between mandrel and die body. Many dies are currently in operation in industrial installations capable of producing a film of acceptable uniformity but having part of the spiral channel with a stagnant flow region. Stagnant flow regions may result in polymer degradation and other production problems.

Computer modeling is necessary for die designs producing uniform thickness without any stagnant flow regions. A good practice is to determine the local shear rates on the mandrel surface and avoid designs that have regions with wall shear rates less than 8 s^{-1}. Spiral die design issues are dealt with in great detail in a recent book chapter [1]. Computer modeling is used extensively for die design purposes and it has had a major impact on technology of the film blowing process.

Bubble Forming and Modeling

As the molten polymer emerges from the annular die lips, the thin tube is undergoing non-uniform biaxial stretching. The deformation effectively ceases at the frost line, because the stresses are not large enough to deform the solid film. For typical film production rates and thicknesses, it is reasonable to neglect inertia forces, surface tension, air drag and gravity on the highly viscous melt. Perhaps the neglect of gravity is the weakest assumption when high production rates are involved.

For an axisymmetric thin tubular film (of thickness h) with an embedded Cartesian system having unit vectors in the direction of flow m (meridional), in the tangential t and normal n directions as shown in Fig. 1, mass conservation gives

$$Q = 2\pi R h V_m = \text{constant} \qquad (1)$$

where Q is the volumetric flow rate, R is the bubble radius, h the bubble thickness and V_m the velocity in the meridional direction.

The force balance on any plane z gives

$$-\pi R^2 \Delta P + 2\pi R h \sigma_m \cos\theta = F_T = \text{constant} \qquad (2)$$

where F_T is the total force exerted on the thin membrane.

If F_T and ΔP are known, the stress σ_m can be directly determined from this equation and subsequently the stress in the tangential direction σ_t from the equilibrium of forces for a thin membrane

$$\frac{\Delta P}{h} = \frac{\sigma_m}{\rho_m} + \frac{\sigma_t}{\rho_t} \qquad (3)$$

where ρ_m and ρ_t are the radii of curvature in the m and t directions respectively.

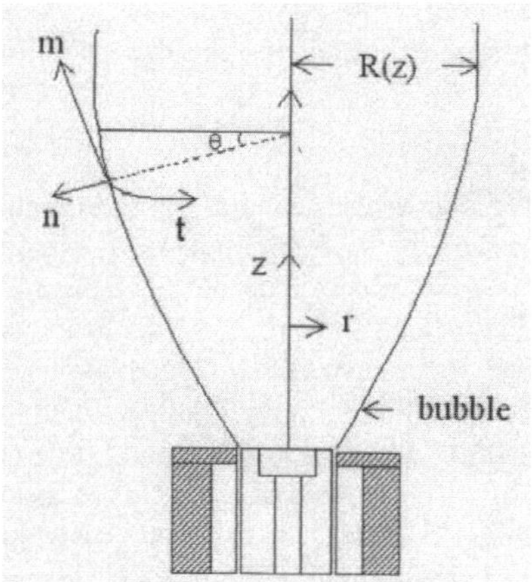

Figure 1. Blown film bubble and intrinsic coordinate system (n,m,t).

On the basis of the above equations Pearson and Petrie [2,3] developed a method for determining the shape of the bubble for Newtonian fluids. This method has be used by numerous other investigators, including [4-6], for improved solutions. Some investigators have also considered viscoelastic models and temperature effects. Despite their mathematical sophistication these models have had virtually no impact on blown film technology beyond providing a framework for understanding. The reason is that prediction of the bubble shape per se is of little interest. The film producer is interested in increasing output rate, avoiding instabilities, reducing film defects and improving the end-use properties. Some attempts have been made to correlate end-use properties to the stresses at the freeze- line, but with limited success thus far. There-in lies the author's proposal: Instead of using the Pearson and Petrie approach to determine the bubble shape, why not follow Kurz [7] who suggested to use the bubble shape as a parameter and analyze the steady state film blowing process in terms of the ratio of axial to hoop stress.

The industrial practice is manipulation of the bubble by confining in a "cage", using collapsing frames and adjusting the aerodynamics of the cooling jets [8]. Consequently, the approach ought to be aimed at stress determination for bubble shape which can be changed at will, following the practice of bubble manipulation at the factory floor.

Concluding Remarks

Mathematical modeling and computer software for polymer flow in spiral channels have had an enormous impact on blown film die technology and improved significantly the quality and production rates. The bubble formation models have had little or no technological impact thus far. It is argued that the models ought to shift the focus from aiming at determining the bubble shape to determination of stresses for mechanically and aerodynamically manipulated bubbles.

References

1. J. Vlachopoulos; R. Castillo, N. Polychronopoulos; S. Tanifuji in *Design of Extrusion Forming Tools*, O.S. Carneiro; M. Nobrega Eds.; Smithers, Rapra, London, 2012 (to be published).
2. J. R. A. Pearson; C. J. S. Petrie *J. Fluid Mech.* 1970, *40*, 1.
3. J. R. A. Pearson; C. J. S. Petrie *J. Fluid Mech.* 1970, *42*, 609.
4. I.A. Muslet; M.R. Kamal *J. Rheol* 2004, *48*, 525.
5. M. Beaulne; E. Mitsoulis *J. Appl. Pol. Sci.* 2007, *105*, 2098.
6. K.D. Housiadas *Polym. Eng. Sci.* 2011, *51*, 1301.
7. S.J. Kurtz *Intern. Pol. Proc.* 1995, *10*, 148.
8. V. Sidiropoulos; J. Vlachopoulos *Polym. Eng. Sci.* 2000, *40*, 1611.

O-MK-077

STRUCTURING OF INJECTION MOLDING PRODUCTS CONTAINING FIBER AND NANO REINFORCED POLYMERS

S.Kenig

Shenkar College of Engineering & Design, Ramat Gan, Israel 52526-samkenig@shenkar.ac.il

Abstract

Forty years ago K&K (Kamal & Kenig) presented a comprehensive model for the injection molding process taking into account the simultaneous flow and heat transfer that occur during the filling, compression and cooling stages. Since the combination of melt flow and heat transfer, has the decisive effect on the structuring and properties of molded products, an holistic investigation into the structuring mechanisms was pursued.

The structuring and derived properties of injection molded reinforced products are governed by the orientation development during flow of the polymers themselves and the high aspect ratio micro-fibers and nano-fillers with simultaneous cooling.

Theoretical modeling of the orientation patterns was the starting stage for studying the evolving structure followed by experimental investigation comprising optical and electronic microscopy and supplemented by mechanical properties characterization.

The modeling and experimental results have shown that the elongational and shear flows with simultaneous cooling cause the formation of a layered structure. The relative thickness of the resulting layers is determined by the heat transfer process that takes place during flow in the mold cavity.

The paper covers the structuring of short fiber glass and carbon polymer composites in addition to nanoclay and carbon nanotubes containing polymers.

Keywords: injection molding, structuring, fiber reinforced, nano reinforced

Introduction

The injection molding process comprises flow of a hot melt into a cold cavity, followed by pressurization with simultaneous cooling. Kamal and Kenig [1,2] were the first to present a comprehensive model taking into account the concurrent melt flow, pressurization and heat transfer that occur during the filling, compression and cooling stages of the injection molding process.

The combination of flow and heat transfer has the decisive effect on the structuring, residual stresses and derived mechanical properties of injection molded products [3, 4].

As shown by Menges et al [5] structuring of neat polymers is governed by the orientation development of the polymer which depends on the location of the polymer with respect to its distance from the gate and the distance from the cold cavity wall. Kaliske [6] have demonstrated the when fibers are incorporated into the polymer melt a distinct layered structure is formed due to the orientation of the elongated rigid fibers. The orientation of the fibers is determined by the flow pattern which once again is dependent on the position of the specific particle/fiber with respect to the distance from the gate and the wall cavity.

In the present paper a first order theoretical model for the orientation patterns and evolving structure is proposed followed by experimental investigation comprising optical microscopy and supplemented by mechanical properties characterization.

Experimental

Materials

The following materials were used in the investigation:

Nylon6/30% glass fibers (Polyram, Israel)

Nylon6/30% carbon fibers (Polyram, Israel)

Nanoclays (NCs) (30B, Southern Clay, USA), were compounded with Nylon6 at 5% wt. loading.

Carbon nanotubes (CNTs) (CNT 7000 Nanocyl, Belgium) were compounded with Nylon6 at 2 and 5% wt. loading

Mold

Plaques of 10cm x 10cm having various thicknesses were injection molded. The gate had a wide slit shape (flash gate) to control the melt front linear shape and eliminate the radial spreading flow near the gate.

Testing

Mechanical testing (tensile properties, flexural properties, impact), electrical conductivity and optical microscopy (OM), were performed.

Results and Discussion

To study the structuring development in fiber and nanoparticle reinforced polymer systems the investigation commenced with first order modeling of the orientation development of elongated particles

Modeling

Jeffery [7] was the first to relate the rate of orientation of a rigid elongated particle suspended in a Newtonian fluid. In 2D Poiseulle flow (see Fig. 1)

Figure 1- Schematic representation of shear and elongational flow using Euler's notation

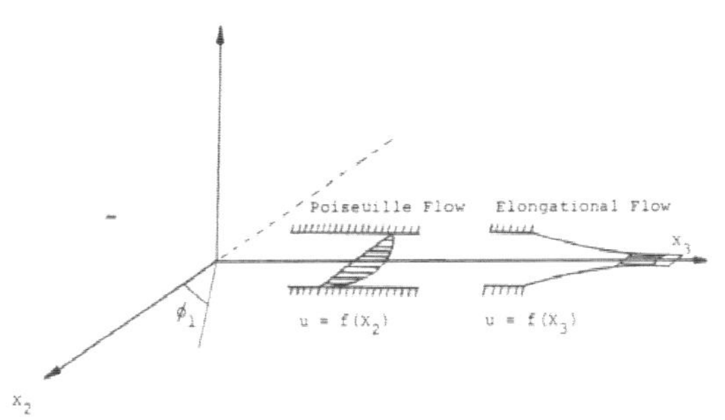

For simple shear flow Jeffery's Equation may be reduced to Eq. 1:

$$\Phi'_1 = \dot{\gamma}/2\,(1 + B \cos 2\Phi_1) \quad (1)$$

Integration of Eq. 1 results in:

$$\tan \Phi_1 = (1 + B)/(1 - B^2)^{1/2} \tan\{(1 - B^2)^{1/2} / 2\,\dot{\gamma}\,t\} \quad (2)$$

Where:

Φ'_1 - rate of angle change
Φ_1 – the orientation angle
$\dot{\gamma}$ - shear rate
$B = (r_p^2 - 1) / (r_p^2 + 1)$
r_p – aspect ratio of the elongated particle
t – time of flow

As can be concluded from Eq. 2, in the case of shear flow the elongated particle rotates at a non uniform angular velocity fastest at 90° and slowest at 0°. The net effect – particles favor alignment in the direction of flow. Moreover, the driving force is the product of shear rate and time.

In the case of simple elongational flow Jeffery's equation can be reduced to Eq. 3:

$$\Phi'_1 = B/2 \times (\dot{\varepsilon} \sin 2\Phi_1) \quad (3)$$

Where

$\dot{\varepsilon}$ - elongation rate

As can be concluded from Eq. 3 the rotation depends on the elongation rate and the sinus of the instantaneous angle that the elongated particle makes with the flow direction. Thus, at 90° the elongated particle will tend to align with the flow direction while at 0° the particle does not tend to rotate. The net effect is the alignment of the particles with the flow direction.

Structuring Analysis

Kenig [8] used the abovementioned first order modeling to analyze the structuring origins of polymer injection molding containing fibers.

As can be observed in Fig. 2, as the melt enters a rectangular mold from a centered narrow gate the melt front assumes a "spreading radial flow" until the radius of the front R equals half width of the mold –h.

In Region 1, substituting Eq. 3 into the momentum equation describing spreading flow (8) leads to Eq. 4

$$\tan \Phi_r = C\, r^{\lambda K} \quad (4)$$

Where:

Φ_r – the angle that the elongated particle makes with the radius r
C and K - integration constants
λ – orientability parameter (replaces B)

Figure 2- Schematic representation of different regions in filling of a rectangular mold

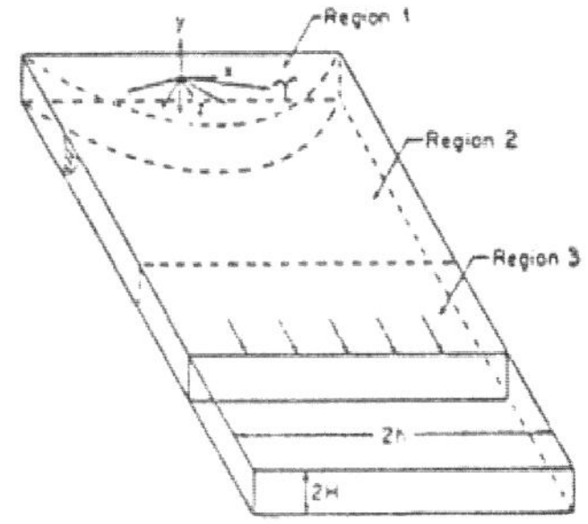

As can be concluded from Eq. 3, as the radius r increases, the angle Φ_r increases i.e. the elongated particles are oriented gradually perpendicular to r.

In Region 3 – the advancing front region, a "fountain flow" takes place. Basically it is an elongational flow, where fluid elements from the core region are elongated and frozen as they contact the cold cavity walls. Consequently, the fibers are aligned in the flow direction and freeze on the surface forming the "skin layer".

In Region 4 – down-stream from the melt front, shear flow takes place. In the case of filling of a cold wall, it was shown that the shear rate assumes a maximum at some distance from the wall [8]. According to Eq.2 the elongated particles will tend to be oriented in the shear direction (flow direction).

Superposition of the flow patterns in region 1, 3 and 4 governs the overall particles orientation and gives rise to a layered structure as shown experimentally by Kaliske et al [5].

Structuring and Mechanical Properties of Injection molding of Polymer Fiber Composites

Based on the above modeling and analysis the resultant mechanical properties of injection molded rectangular plaques having a narrow - wide gate were studied by Kenig et al [9]. In the latter study polyamide 66 having 33% glass fibers and a blend of polystyrene and poly (phenylene oxide) containing 22% glass fibers were injection molded in a variety of thicknesses. It was concluded that the fountain flow is the major structuring mechanism due to the orientation development of the fibers. Since the fountain flow affects mainly the skin area, the thickness of the skin was studied as function of melt velocity. Experimental results indicated that the slower the melt velocity the thicker is the skin thickness due to the enhanced heat transfer between the cold cavity and the hot flowing melt. The increase of the skin layer resulted in increased elastic modulus and strength of flow direction specimens.

Following the same rationale a Design of Experiment (DOE) was carried out to investigate the effect of processing conditions on the thickness of the skin layer for PA6 containing 30% glass fibers (GF) and 30% carbon fibers (CF).

Table 1-The effect of fiber type on skin thickness and modulus of injection molded fiber reinforced PA6

Composition	Mold thickness (mm)	Skin		Modulus (GPa)
		Thick (mm)	Fract. (%)	
PA6 neat		-	-	2.20
PA6/30% GF	3.4	1.63	48	3.14
PA6/30% GF	1.3	0.84	64	4.88
PA6/30% CF	3.4	2.12	62	4.50
PA6/30% CF	1.3	0.91	70	5.19

As can be seen in Tab. 1 as the thickness of the mold is decreased the relative skin thickness and modulus of the fiber reinforced PA6 increases. Furthermore, the effect of carbon fiber on skin thickness is more pronounced due to the higher thermal conductivity and enhanced heat transfer of the carbon fibers.

Fig. 3 depicts an image processed optical micrograph of the fiber's orientation of injected molded PA 6/30% glass fibers in the skin and core regions.

Figure 3 – Optical micrograph of fracture surface following etching and image analysis showing fiber orientation of PA3/30% glass fibers

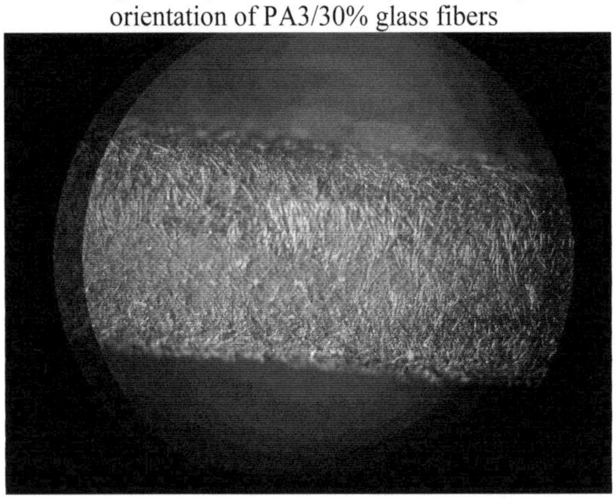

Structuring and Mechanical Properties of Nanoclay Reinforced Polymers

With the development of commercial nanoclays (NC) and related technologies, the investigation of the structuring and orientation of NC in polymers was pursued. Among the systems studied were Nylon6 and 5% hydroxyl treated nanoclays (30B by Southern Clay). Tab. 2 summarizes the mechanical properties of Nylon6/5%NC in the flow direction of a 3mm thick plaque. As evident from Tab. 2 the properties of the molded plaque depends on the position in the mold (Position 1-close to the gate, position 4-midway the mold, position 7- close to the mold end) and highly affected by mold temperature and filling velocity of the melt. Transmission Electron Microscopy supported the macro mechanical properties and demonstrated a structured morphology of the NCs which are parallel in the skin region, as was the case in the short fiber PA6 specimens.

Structuring and Mechanical Properties of Carbon Nantubes Reinforced Polymers

The structure development during injection molding of carbon nanotubes (CNTs) containing polymers was investigated among other polymers in PA6. Fig. 4 summarizes the properties of PA6 containing 2 and 5% CNTs incorporated using a commercial masterbatch (denoted by MB) and a self prepared compounds (denoted by N).

Figure 4 – Properties of injection molded PA6 plaques containing 2% (2N;2MB) and 5% (5N;5MB) CNTs

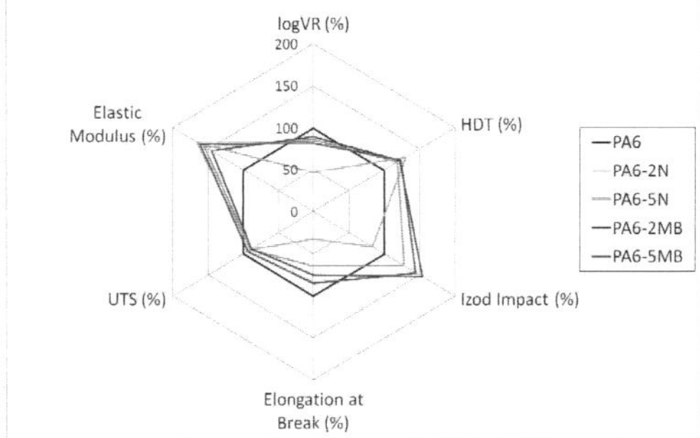

Fig. 4 indicates that both the elastic modulus and the izod impact with decreased volume resistivity were obtained when 2% and 5% CNTs were incorporated in PA6 compared to the neat PA6. This is due to the orientation of the CNTs in the skin area as were the cases for the short fibers and NCs containing polymers.

Table 2 – Elastic modulus and strength of PA6/5% NC Vs. processing conditions in 3mm thick plaque.

Melt temp. (°C)	Mold temp. (°C)	Inj. vel. (mm/sec)	Pos	σ brk (MPa)	Modulus (MPa)
230	60	60	1 4 7	31 21 32	3157 3411 3642
310	60	60	1 4 7	12 9 12	3656 3792 3584
270	60	60	1 4 7	20 12 19	4205 5097 3431
270	60	60	1 4 7	32 15 35	4183 4014 4656
270	20	20	1 4 7	32 26 30	3591 3775 4175
270	100	100	1 4 7	34 25 31	4205 3996 4783
270	60	60	1 4 7	39 30 32	4439 3818 4132

PA6/30% GF

Conclusions

The structuring and derived properties of injection molded polymers containing GFs and CFs, NCs and CNTs are governed by the orientation development of the high aspect ratio micro-fibers and nano-fillers.

The propsed theoretical modeling and analysis of the orientation development of fibers and elongated nanoparticles were shown to describe the evolving structure in the injection molding process. Experimental investigation comprising optical and electronic microscopy have indicated that the orientation of the fibers and elongated particles in the skin layer and its thickness are the governning parameters affecting the resultant mechanical properties. A DOE study indicated that the mold temperature and injection speed are the decisive process parametres that determine the skin thickness. As the temperature of the mold and the injection velocity are decreased the thickness increases. Mechanical properties characterization confirmed that the thickness of the skin affects directly the elastic modulus and strength in the flow direction.

Acknowledgements

The Author wishes to thank Dr. Fabian Rios and Moshe Omer from the Israel Plastics and Rubber Center and to the Shenkar College Graduate Students: Yonatam Amit, Ron Ricardo, Ido Offenbach and Meir Carmi for carrying out the experimental work and analysis of the results.

References

1. M.R. Kamal; S. Kenig, *Poly. Eng. Sci. 1972, 12, 294.*
2. M.R. Kamal; S. Kenig, *Poly. Eng. Sci. 1972, 12, 302.*
3. A. Siegmann; S. Kenig, *Poly. Eng. Sci. 1981, 21, 997.*
4. A. Siegmann ; S. Kenig, *Poly.Eng. Sci.,1982, 22, 560.*
5. G. Menges; W. Wubken, *in SPE ANTEC Proceedingd, Montreal, 1973, 519.*
6. G. Kaliske; H. Selfert, *Plaste Kauschuk, 22, 1973, 837.*
7. G.B. Jeffery, *Proc. Roy. Soc. 1922, a102, 1612.*
8. S. Kenig, *Poly. Comp. 1986, 7, 50.*
9. S.Kenig; B.Trattner;H.Anderman, *Sci. Eng. Comp. Mat. 1994, 1, 51.*

O-MK-053

MODELING OF MIXING AND SILICA AGGLOMERATE BREAKUP IN MODULAR COROTOTATING TWIN SCREW EXTRUDER

Sug Hun Bumm, James L. White and Avraam I. Isayev*
Department of Polymer Engineering, The University of Akron, Akron, OH 44325-0301, USA
Correspondence to: aisayev@uakron.edu

Abstract - Extensive experimental studies on silica agglomerate breakup during compounding with polymer melts of various viscosities in a modular corotating twin screw extruder were conducted. To avoid a subjectivity of the result, due to small size particles involved, silica agglomerates were characterized by measuring their mass average values. Increasing the screw speed, melt viscosity and silica concentration was found to increase the silica agglomerate breakage. The effect of these parameters on agglomerate breakup was ranked as follows: silica concentration > polymer viscosity ≈screw rpm. A correlation between silica agglomerate breakage and power input was also found. Based on the experimental data and dispersion process, a composite modular kinetic model for evaluating silica agglomerate breakup during compounding in a co-rotating twin screw extruder was developed. The kinetic constants of breakup and re-agglomeration of silica agglomerates were calculated based on the stresses applied to the agglomerates and their cohesive strength. These constants for silica agglomerates were found to be not significantly different. The latter was in contrast to experimental data available in literature on compounding of calcium carbonate with polypropylene where the re-agglomeration kinetic constant of calcium carbonate played a major role in the agglomerate breakup. Comparison of experimental and calculated results on the silica agglomerate size evolution during compounding with polymer melts indicated a reasonable agreement between them, as shown in Figure 1 for mixtures of polyamide 12 with silica.

Keywords: Polymer compounding, silica agglomerate, breakup, kinetic model, flow simulation.

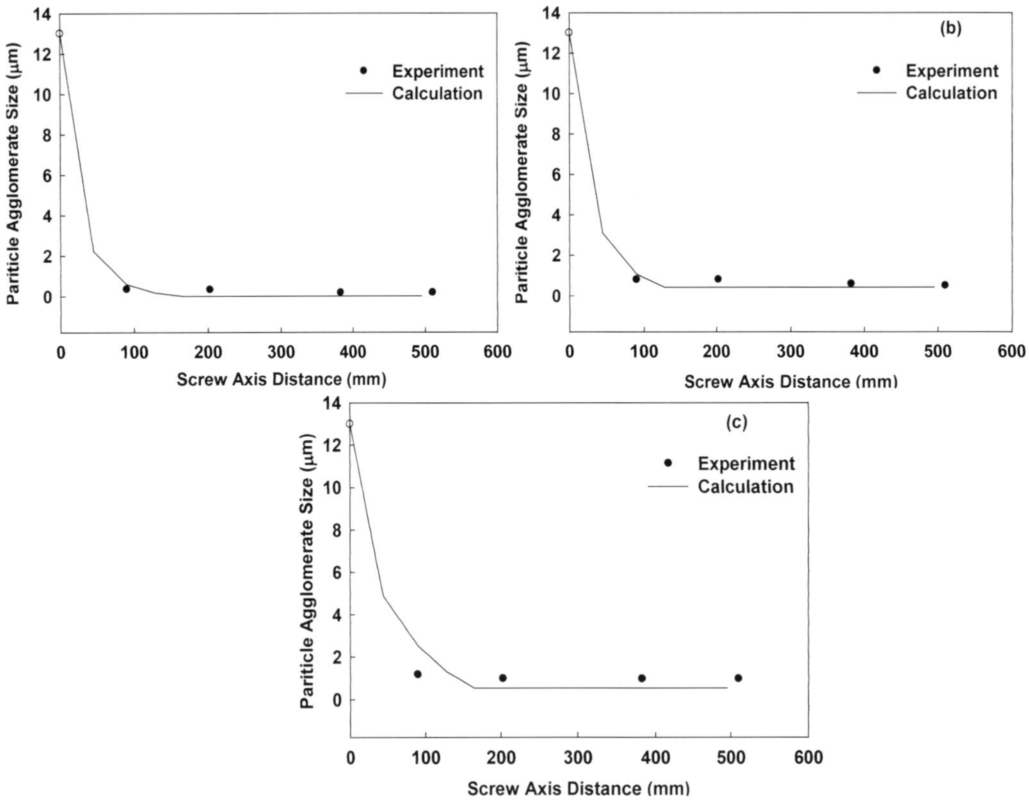

Figure 1. Comparison of calculated and measured agglomerate particle size for PA12 at silica concentration of 1 vol%

(a), 10 vol% (b) and 20 vol% (c) at a screw speed of 200 rpm

O-TM-455

INVESTIGATION OF A MODIFIED MOLECULAR CONSTITUTIVE EQUATION FOR PREDICTION OF NON-LINEAR RHEOLOGICAL PROPERTIES OF LONG CHAIN BRANCHED POLYMERS

Mahdi Abbasi[1], Nadereh Golshan Ebrahimi[1*], Masood Khabazian Esfahani[1], Mahdi Nadali[1],

Corresponding Author Address: Chemical Engineering Department/Polymer Group Tarbiat Modares University, P.O.Box 14115-114, Tehran, I.R.Iran

Corresponding Author E-mail: ebrahimn@modares.ac.ir

Abstract:

Molecular constitutive equations in polymer melts have been used not only to produce the stress tensor in fluid dynamics, but also for prediction of different molecular structures. Molecular stress function (MSF) theory have been successful to predict the rheological behavior of linear and long chain branched (LCB) polymer melts in transient shear, exponential shear, medium amplitude oscillatory shear (MAOS), large amplitude oscillatory shear (LAOS) flows as well as transient uniaxial, biaxial and planar extensional flows. Recently Abbasi et al. (2012) modified this model by a new evolution equation to predict the extensional behavior of various types of LDPEs by three nonlinear parameters β (LCB content), f_{max} (maximum stretching) and τ_a (branch stretch relaxation time). It should be mentioned that this constitutive equation should be able to predict the various deformation fields by the same nonlinear material parameter. Based on this idea, the effect of different parameters of MSF theory on the transient shear and uniaxial extensional flows were investigated theoretically.

Introduction:

The idea of impeded lateral motion of polymer chains in three dimensional network , reptation in confined tube, goes back to de Gennes [1]. Based on de Gennes work, Doi and Edwards improved a constitutive equation to describe molecular viscoelasticity of entangled linear polymer liquid. Doi and Edwards theory showed better predictions than the network theory of Green and Tobolsky, yet there are shortcomings to this theory [2].
However, recently the DE model has received a lot of attention and many authors have tried to improve the theory, double reptation, contour length fluctuation, convective constrain release and dynamic dilutionare among them [3].
One manifestation of the DE theory is that the chain tube diameter does not change during the deformation which consequently results in showing no strain hardening in elongational flows. The idea of tube diameter decrease was first suggested by Marrucci [4]. Based on this assumption Wagner [5]introduced the molecular stress function (MSF), f, into the DE theory which is the ratio of tube diameter before and after the deformation. The last version of the MSF theory for long chain branched polymers [6] includes two nonlinear fitting parameters β and f_{max}. Where β represents the ratio of molar mass of the branched polymer to the molar mass of the effective backbone and determines the slope of strain hardening in transient extensional flows. f_{max} is related to the maximum stretch that the polymer chain can be subjected to in flow before chains slip past one another without further stretch. In this version the total strain energy of a branched section is divided into two parts: backbone stretching and side chain compression.

Theory:

Molecular stress function (MSF) theory is a successful development of Doi-Edwards (DE) model for prediction of strain hardening phenomena in extensional flows. According to this theory, the extra stress tensor is given by [7]:

$$\boldsymbol{\sigma}(t) = \int_{-\infty}^{t} m(t-t') f^2(t,t') \boldsymbol{S}_{DE}^{IA}(t,t') dt' \tag{1}$$

with memory function $m(t\text{-}t')=\partial G(t\text{-}t')/\partial t'$ which would be obtained through either direct measurement of linear viscoelastic data or molecular theories. The relative tension of a Gaussian chain in a tube segment of diameter a at time t is $3kT/a(t,t')$ times of its equilibrium state value $3kT/a_0$ at time t'. With an assumption of tension-stretch proportionality, molecular stress function f is defined by:

$$f(t,t') = \frac{a_0}{a(t,t')} \tag{2}$$

where, $f(t,t')$ depends on the deformation history of the tube segments and the stretch will vary along the tube. $\boldsymbol{S}_{DE}^{IA}(t,t')$ is the strain measure of DE with independent alignment assumption:

$$\boldsymbol{S}_{DE}^{IA}(t,t') = 5\left\langle \frac{(\boldsymbol{F}_{t,t'}^{-1}.\boldsymbol{u}(t'))(\boldsymbol{F}_{t,t'}^{-1}.\boldsymbol{u}(t'))}{\left|\boldsymbol{F}_{t,t'}^{-1}.\boldsymbol{u}(t')\right|^2} \right\rangle = 5\boldsymbol{S}(t,t') \tag{1}$$

where $\boldsymbol{u}$ is the tube segment orientation vector at time t' and $\boldsymbol{F}_{t,t'}^{-1}$ is displacement gradient tensor between t and t'. $\boldsymbol{S}$ is the orientation tensor and <···> denotes an ensemble average over surface of a unit sphere. The orientation tensor $\boldsymbol{S}$ can be approximated using Currie's formula[8]:

$$\boldsymbol{S} = \left(\frac{1}{J-1}\right)\boldsymbol{B} - \left(\frac{1}{(J-1)(I_2+\frac{13}{4})^{0.5}}\right)\boldsymbol{C} \tag{2}$$

$$J = I_1 + 2(I_2 + \frac{13}{4})^{0.5} \tag{3}$$

Where $\boldsymbol{C}$ is Cauchy tensor defined by $\boldsymbol{C}=\boldsymbol{B}^{-1}$. I_1 and I_2 are the traces of the $\boldsymbol{B}$ and $\boldsymbol{C}$, respectively.

To use the Eq. 1 as an integral constitutive equation, another equation is required for molecular stress function f. Different types of empirical and theoretical equations have been proposed for such purpose [6, 7, 9]. Wagner et al. [6] proposed a strain energy function for branched polymer melts. They assumed that a tube section of a LCB polymer consists of β segments in which one segment is stretched in the direction of the tube (backbone) and β-1 segments (side chins) are compressed perpendicular to the tube. The hypothesis leads to the evolution equation of:

$$\frac{\partial f^2}{\partial t} = \frac{\beta f^2}{1+\frac{\beta-1}{f^4}}(\boldsymbol{\kappa}:\boldsymbol{S} - \frac{CR}{f^2-1}) \tag{6}$$

Recently, we [10] presented a new strain energy function that is based on MSF framework with both backbone stretching and side chain stretching, however these stretchings are different. Stretching in side chains is modelled by the interchain pressure concept of Marucci and Ianniruberto [11]. Therefore, a couple of evolution equations for long chain branch polymers proposed: [10]:

$$\frac{\partial f_s}{\partial t} = f_s\left((\boldsymbol{\kappa}:\boldsymbol{S}) - \frac{f_s(f_s^{\,3}-1)}{\tau_a}\right) \tag{7a}$$

$$\frac{\partial f^2}{\partial t} = \beta f^2\left((\boldsymbol{\kappa}:\boldsymbol{S}) - \frac{CR}{f^2-1}\right) - (\beta-1)\frac{\partial f_s^2}{\partial t} \tag{7b}$$

Where f_s represents side chain stretching and τ_a is the relevant stretch relaxation time of side chains. *CR* denotes constraint release effects that for constant strain rate flows can be expressed as [7]:

$$CR = \frac{(f^2-1)^2}{f_{\max}^2-1}\sqrt{\boldsymbol{D}^2:\boldsymbol{S}} + a_2(f^2-1)^2\sqrt{|\boldsymbol{W}\boldsymbol{D}:\boldsymbol{S}|} \qquad (8)$$

Where $\boldsymbol{\kappa}$, $\boldsymbol{D}$ and $\boldsymbol{W}$ are velocity gradient, rate of deformation and rate of rotation tensors, respectively. f_{max} is related to the maximum stretch that the polymer chain can be subjected before chains slip past one another without further stretch [12], and governs the steady state value of viscosity in extensional flows. a_2 is a constant that determines the overshoot in transient shear viscosity. In the next section, the effect of different parameters on the molecular stress function, f, is explained in shear and extensional flows.

Shear:

The shear counterpart of new MSF model Eqs. (7a and b) are

$$\frac{\partial f_s}{\partial \gamma} = f_s\left(S_{12} - \frac{f_s(f_s^3-1)}{\dot{\gamma}\tau_a}\right) \qquad (9a)$$

$$\frac{\partial f^2}{\partial \gamma} = \beta f^2\left(S_{12} - \frac{(f^2-1)}{2\left(f_{\max}^2-1\right)}\sqrt{S_{11}+S_{22}} - \frac{a_2(f^2-1)}{2}\sqrt{|S_{11}-S_{22}|}\right) - (\beta-1)\frac{\partial f_s^2}{\partial \gamma} \qquad (9b)$$

And for MSF model Eq. 6 is

$$\frac{\partial f^2}{\partial \gamma} = \frac{\beta f^2}{1+\frac{\beta-1}{f^4}}\left(S_{12} - \frac{(f^2-1)}{2\left(f_{\max}^2-1\right)}\sqrt{S_{11}+S_{22}} - \frac{a_2(f^2-1)}{2}\sqrt{|S_{11}-S_{22}|}\right) \qquad (10)$$

Figure 1 shows the numerical solution of Eq. 9a in shear flows. As seen f_s has a maximum around shear strain of 3.5. At large strains, f_s gets back to the initial value of one. This means the molecule is simply oriented in the flow direction and, due to dissipative constraint release, the tube diameter returns to its equilibrium value a_0. Also, increasing of *De* has no pronounce effect on f_s in shear flows.

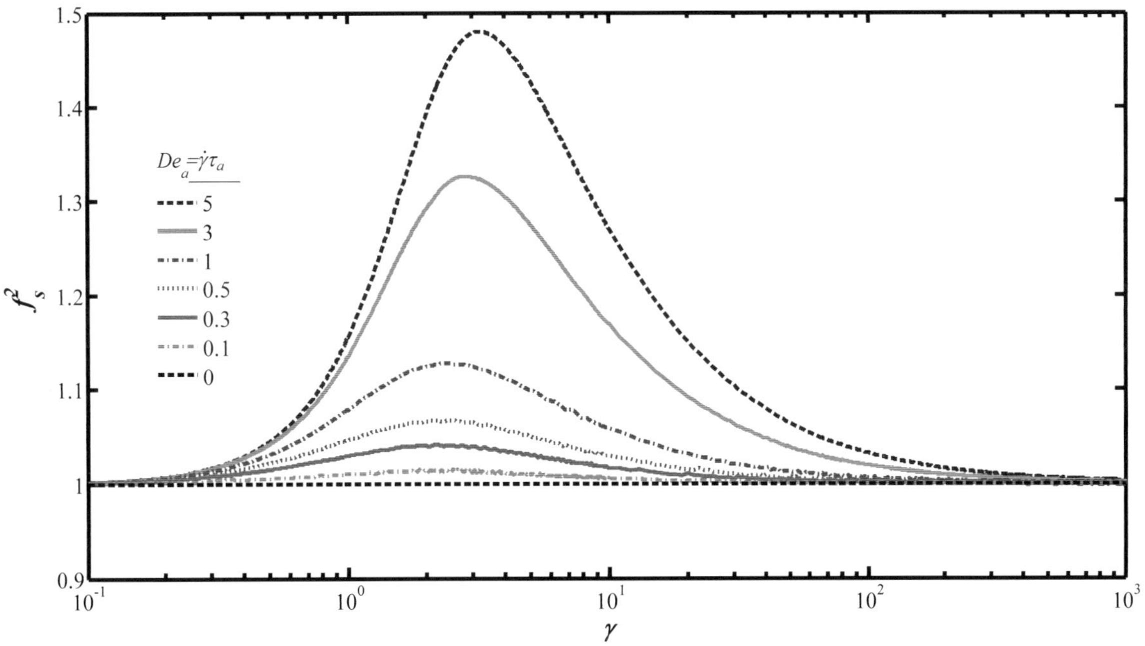

Fig. 1 The effect of shear strain on f_s^2 in shearing deformation

In order to investigate the effects of Debora number, β and a_2 on f for new MSF model, Eqs. 9a and b should be solved simultaneously. In this regard, f_{max} is set at 80 (a normal value for LDPEs). The results are shown in Figs. 2, 3 and 4. Equation 10 is used to plot f for the MSF model.

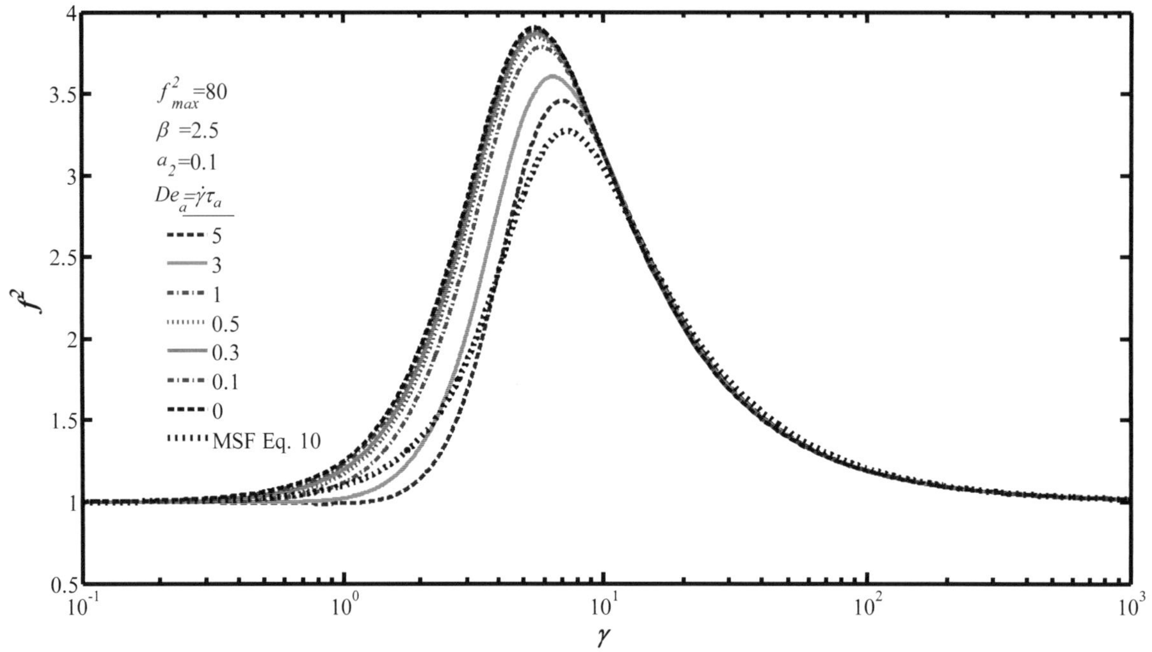

Fig. 2 The effect of shear strain on f^2 in shearing deformation

Figures 2 and 3 illustrate the effect of *De* and β on *f*. *f* decreases with Debora number increase and increases with β increase. However, these variations are little as *f* never gets to its maximum value f_{max}. This means that in shear deformations the effect of side chain stretching f_s is negligible. Also the effect of backbone stretching on stress relaxation is little. From the other point of view, orientational relaxation is dominant relaxation mechanism and chains do not tend to stretch These results are in agreement with strain softening observed in shear deformations [13].

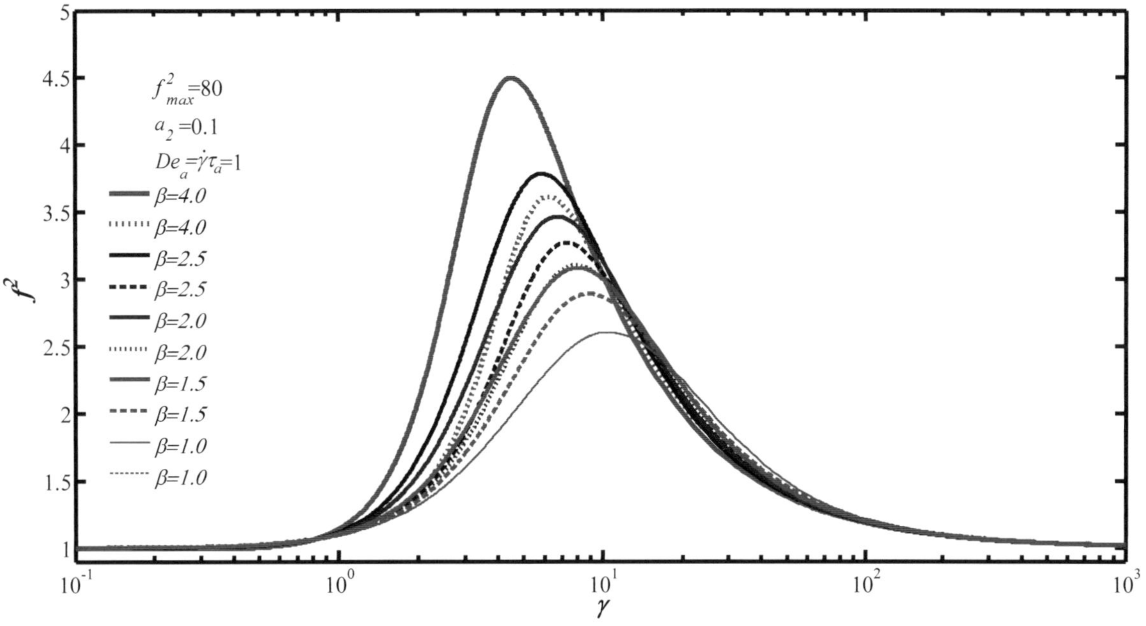

Fig. 3 The effect of β parameter on f^2 in shearing deformation. The new MSF model (solid lines) the MSF model (dash lines)

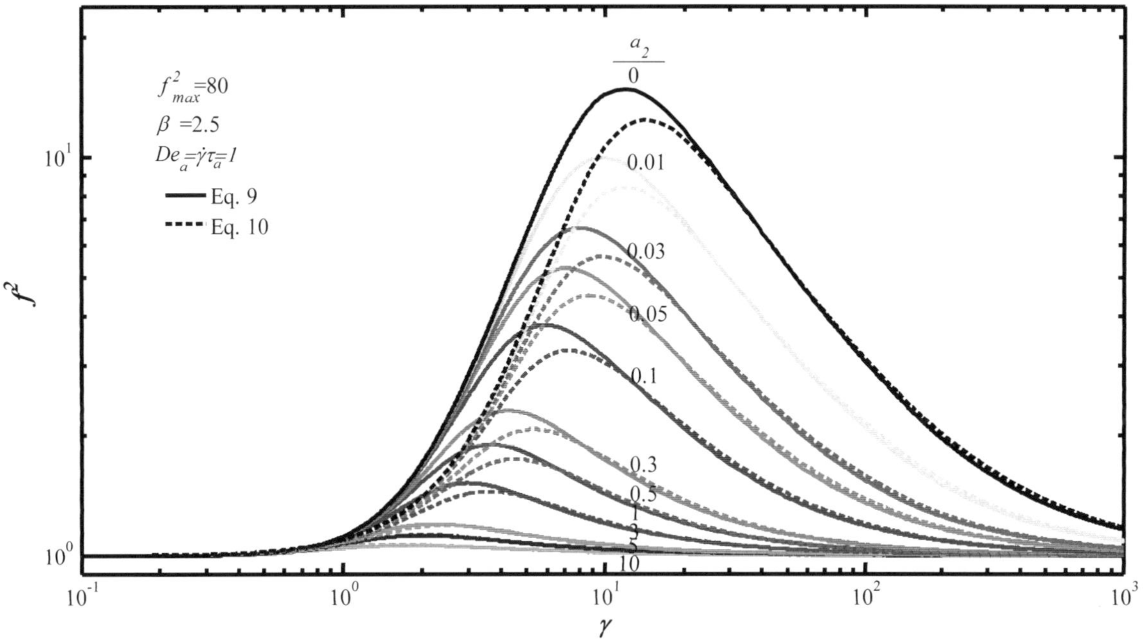

Fig. 4 The effect of a_2 on f^2 in shearing deformation. The new MSF model (solid lines) the MSF model (dash lines).

Figure 4 shows the effect of a_2 on f. It is observed that as a_2 approaches zero not only f increases but also, the strain at which maximum stretching takes place, goes to higher strain. The molecular stress function, f, of new MSF model (solid lines) is broader than that of the MSF model (dash lines).

Uniaxial:

Equations 7a and b can be rewritten for uniaxial extension as:

$$\frac{\partial f_s}{\partial \varepsilon} = f_s\left(S_{11} - S_{33} - \frac{f_s(f_s^{\ 3}-1)}{\dot{\varepsilon}\tau_a}\right) \tag{11a}$$

$$\frac{\partial f^2}{\partial \varepsilon} = \beta f^2\left(S_{11} - S_{33} - \frac{f^2-1}{f_{\max}^2-1}\sqrt{S_{11}+\frac{1}{2}S_{33}}\right) - (\beta-1)\frac{\partial f_s^2}{\partial \varepsilon} \tag{11b}$$

Figure 5 illustrates the numerical solution of Eq. 11a at different Debora numbers. As it is conspicuous, f_s^2 grows monotonically and reaches a constant value of $f^2_{s,max}$ for Debora numbers greater than 0. At De_a=0 no stretching is imposed to the side chains f_s=1.

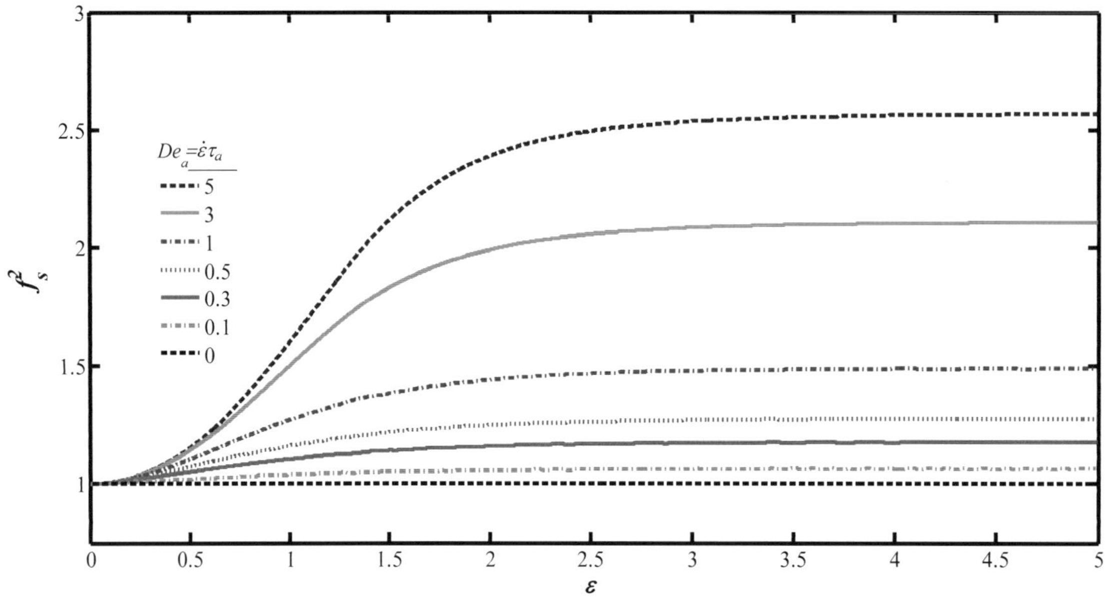

Fig. 5 f_s^2 versus uniaxial strain at different Debora numbers

Since the term S_{11}-S_{33} has the upper-limit value of 1, for $f_{s,max}$>>1 Eq. 11a can be simplified as

$$f_{s_{max}}(f_{s_{max}}^{3} - 1) = \dot{\varepsilon}\tau_a \tag{12}$$

Figure 6 represents the effect of Debora number on maximum molecular stress function of the side chains, $f^2_{s,max}$,. At $f_{s,max}$>>1, $f^2_{s,max}$ becomes proportional to $De_a^{0.5}$.

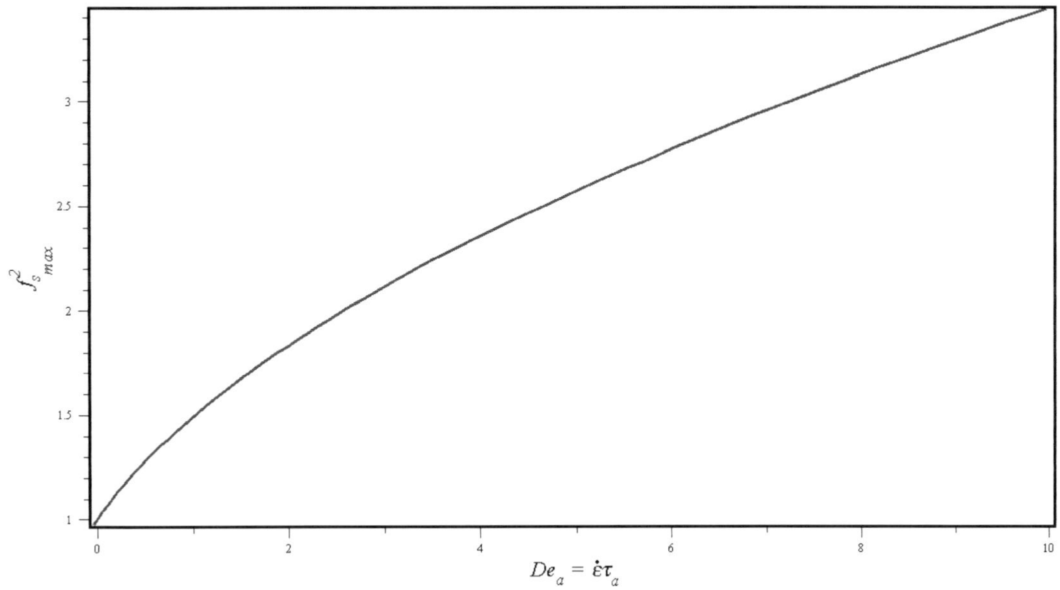

Fig. 6 $f^2_{s,max}$ versus Debora number

To investigate the effect of Debora number on f, Eqs. 11a and b should be solved simultaneously. Also, equation 6 can be rewritten in its strain form as:

$$\frac{\partial f^2}{\partial \varepsilon} = \frac{\beta f^2}{1 + \frac{\beta - 1}{f^4}} \left(S_{11} - S_{33} - \frac{(f^2 - 1)}{(f^2_{\max} - 1)} \sqrt{S_{11} + \frac{1}{2} S_{33}} \right) \tag{13}$$

According to Eq. 13 the MSF model depends solely on strain rather than strain rate. Whereas in accordance to Eq. 11, the new MSF model is strain and strain rate dependent. Consequently the MSF model is time-strain separable while the new one is not.

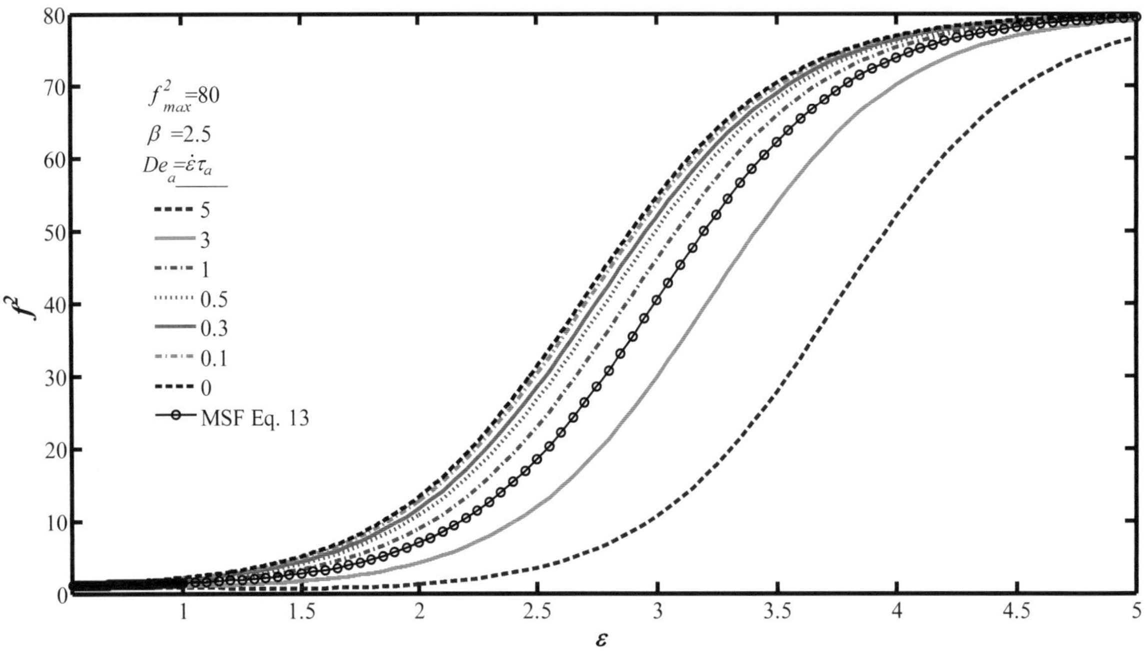

Fig. 7 The effect of Debora number on f^2 in uniaxial elongational flow

Figure 7 shows the result of Eq. 11 for new MSF model and Eq. 13 for MSF model at various values of Debora number. As discussed in shear flows, f_{max}=80 and β=2.5 were used. It is clear that for new MSF model, f decreases as Debora number increases and strain-hardening is postponed to later times (higher strain values). This means that the chain stretching of a branched polymer in uniaxial flow at constant strain rate decreases with increasing the relaxation time of the side chains. It is worth noting that for the new MSF model at De_a<1 the effect of Debora number on f is negligible. Even though, in such a case the difference is conspicuous. In Fig. 8 the effect of β parameter on the predictions of new MSF model (Eq. 11) and the MSF model (Eq. 13) is investigated. At β=1 both models give the same graph. As β increases the predicted f value of new MSF model gets higher values than the MSF model. However as the strain increases both predictions fall on the same value of f_{max}.

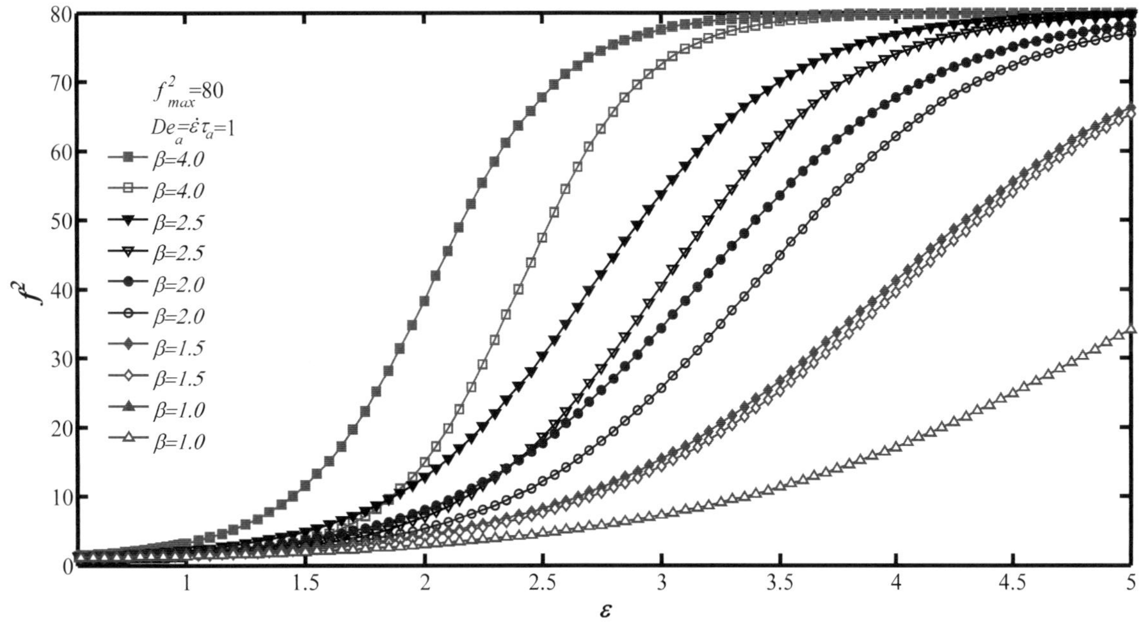

Fig. 8 The effect of the parameter β on f^2 for MSF model (empty symbols) and new MSF model (solid symbols)

Conclusion:

The effect of the MSF model parameters on f^2_s and f^2 in shear and uniaxial elongational flows is investigated. In shear flows f^2_s and f^2 undergo a maximum as the shear strain increases. At large strains the values of f^2_s and f^2 get back to the original value, 1 which shows that no stretching is imposed to the backbone and side chains. Having a maximum at intermediate strains shows that during shearing deformation the backbone and the side chains tension deviate from the equilibrium state. For uniaxial elongational flow, on the other hand, the values of f^2_s and f^2 increase as the strain increases and reaches to a maximum value at large starins. The effect of Debora number on f^2_{smax} becomes linear at high Debora number values. As the β value increases f^2 increases more rapidly with strain and at lower strains it reaches to $f^2_{\max}$.

References:

1. P. G. de Gennes *J. Chem. Phys.*, 1971, *55*,572
2. M. Doi; S. F. Edwards, *The Theory of Polymer Dynamics,* Clarendon Press, Oxford, 1986
3. H. Watanabe *Prog. Polym. Sci.*, 1999, *24*, 1253
4. G. Marrucci; N. Grizzuti *Gazz. Chim. Ital.*,1988, *118*,179
5. M. H. Wagner; J. Schaeffer *J. Rheol.*, 1993, *37(4)*,643
6. M. H. Wagner; M.Yamaguchi; M.Takahashi *J. Rheol.*, 2003, *47(3)*,779–793
7. M. H. Wagner; P.Rubio; H. Bastian *J. Rheol.*,2001, *45*, 1387
8. **C.**Pattamaprom; J. J. Driscroll; R. G. Larson *Macromol. Symp.,* 2000, *158*,1
9. V. H. Rolón-Garrido; M. H. Wagner *Rheol. Acta.*,2007,*46(5)*,583

10. M. Abbasi; N. Golshan Ebrahimi; M. Nadali; M. Khabazian Esfahani *Rheol. Acta.,* 2012, *51*, 163
11. G. Marrucci; G. Ianniruberto *Macromol.,* 2004, *37*,3934

12. V. H. Rolón-Garrido; R. Pivokonsky; P. Filip; M. Zatloukal; M. H. Wagner *Rheol. Acta.* 2009, *48*, 691
13. **J.** M. Dealy; K. F. Wissbrun *Melt rheology and its role in plastics processing*, Kluwer Academic Publishers, Amsterdam, 1999

O-TM-307

STRUCTURAL INTERPRETATION FOR TENSILE NECKING BEHAVIOR OF HIGH DENSITY POLYETHYLENE SOLIDS

K.-H. Nitta*, M. Kuriyagawa

Division of Material Science, Graduate School of Natural Science and Technology, Kanazawa University, Kakuma, Kanazawa, 920-1192, Japan –nitta@t.kanazawa-u.ac.jp

Abstract – The drawabilities of polymer materials are majorly determined by the necking deformation, therefore the natural draw ratio corresponding to elongation of necking is a dominant quantity for determining their drawabilities. Tensile tests were performed at various crosshead speeds using metallocene catalyzed high density polyethylene specimens with different molecular weights and different cross-sectional shapes, in order to investigate the dependences of crosshead speed, molecular weights, and cross-sectional shape on the natural draw ratio. The values of the natural draw ratio was estimated from the dimension of the necked specimens, because conventional natural draw ratio which is estimated from the strain at the terminal point of the neck propagation region includes the elastic components such as residual strain and elastic aftereffect were eliminated. It was found that the perfect plastic deformation took place below a critical crosshead speed, which points dependent on the molecular weight of samples and above their critical points the void formation was observed by polarized optical microscopy. On the basis of the SEM images, we proposed a simple structural model for explaining the necking formation, and then according to our model the molecular weight and the cross-sectional shape dependences of the natural draw ratio were discussed.

Keywords: natural draw ratio, high density polyethylene, necking deformation

Introduction

Most polymer solids show necking behavior during tensile deformation below their glass transition temperature or their melting temperature. These necking behavior are unique deformation for polymer materials which cannot be observed for ceramic and metal materials. In general, the stress-strain curves for polymer solids show distinct yield points as maximums, and beyond these yield points a concave construction suddenly appears in the middle of the specimens. The stress keeps constant during the neck propagate along gauge length, and the strain at the terminal point of neck propagation region is generally called the natural draw ratio. The crystal structure transforms from spherlitic to fibril in the necking deformation [1].

Many researchers have studied tensile necking deformation since Carothers and Hill termed it 'cold drawing' in 1932 [2]. The cross-sectional area of a necked specimen is locally decreased compared to the initial area, thus it is difficult to understand necking behavior using nominal stress-strain curves. Therefore they have analyzed these curves using various geometrical methods such as Considère construction and Maxwell rules [3-5]. It has long been considered that the uniaxial extension is converted into local adiabatic heating, leading to heterogeneous deformations in necking. However, Brauwer et al. [6] and Lazurkin [7] found that necking behavior was observed under quasi-isothermal conditions in such a way that a specimen is elongated at a very slow crosshead speed.

Peterlin et al. [8-10] investigated the necking deformation from a structural point of view, and concluded that a transformation from lamellar crystals to a fibril structure resulted from the large-scale rearrangement of crystal blocks occur during necking deformation. However, the molecular and structural mechanism underlying the neck drawing process is currently not well clarified, as suggested in prominent review by Séguéla [11].

When the necking behavior is investigated, the structural understanding of the deformation at the yield point is important because the structural units organized at a yield point act as the dominant units in the necking process. Nitta-Takayanagi proposed that at the yield point lamellar cluster fragmented into the lamellar cluster units, which is thought to be formed at crystallizing process by considering the exclusion of chain ends [12-14].

In this work, in order to explore the structural factors that govern the neck drawing mechanism and to provide physical insight into the natural draw ratio on the basis of a novel structural concept, we examined the intrinsic natural draw ratio which does not include the elastic component and void formation using a series of metallocene catalyzed high density polyethylene (PE) with different molecular weight.

Experimental

Materials and sample preparation

Metallocene-catalyzed high density polyethylenes (HDPEs) with different molecular weights of 74, 96, 120, and 150×10^3, which have narrow molecular weight distribution of about 2.5, were used. Pellet-shape sample were compression-molded for forming sample sheets at 483K and 10 MPa for 5 minutes in a laboratory hot press, and then quenched at 273K in iced water. These sheets were annealed in a vacuum oven at 373K for 5 hours for the reason that thermodynamically stable structures is obtained. The sheet thicknesses varied from about 0.1 to 1 mm for measurements of natural draw ratio. The end numeral of the sample code indicates the weight-average molecular weight in the 10^{-3} order of magnitude, i.e. HDPE74 have the weight-average molecular weight of 150×10^3.

Characterization

The crystallinities of sample sheets were estimated from the Archimedes method using a METTLER electrobalance XS205, with ethanol as the solvent. We use densities of the crystalline and amorphous regions of 1000 and 856 kg/m^3, respectively [15] in order to calculate crystallinity.

In order to characterizing the long period, we performed Small angle X-ray scattering (SAXS) measurements in 10C beam line in the Photon Factory, Japan. Four HDPE sample sheets were irradiated with monochromatized X-ray synchrotron radiation (wavelength, λ=0.1488 nm). Scattered X-rays were detected using a one-dimensional position sensitive proportional counter.

Using a Perkin Elmer Diamond instrument, differential scanning calorimetry (DSC) was carried out to obtain the melting point of samples. Samples of about 3 mg were sealed in aluminum pans, which were cut out from sheets with thickness of 0.3 mm, and were heated from 323 to 473 K at 20 K/min.

The characteristics of four HDPE sample sheets are listed in Table 1.

Table 1. Characteristics of samples.

Sample code	χ_v/vol%	L_p/nm	T_m/K
HDPE74	63	25.7	405
HDPE96	61	27.5	406
HDPE120	59	28.3	407
HDPE150	57	30.0	405

χ_v:crystallinity in volume fraction
L_p:long period
T_m:melting temperature

Tensile tests

Specimens were elongated up to an extension ratio of 4.4, which is in the necking region, and the necked specimen was held for 10 minutes in order to examine the necking deformation. Specimens were then released from the clamps and kept under unloading for one day at room temperature. The values of natural draw ratio were calculated from the dimension of the necked specimens.

Firstly, in order to examine the temperature dependence on the necking behavior, the tensile tests were performed with HDPE96 using the specimens with a gauge length of 10 mm, width of 4 mm, and thickness of about 0.3 mm (cross-section area of 1.2 ±0.08 mm^2), varying temperatures from 298 to 408 K.

Secondly, the crosshead speed dependence for HDPE96 was investigated at 298 K using the same type specimens with the case of temperature dependence. At the measurement temperature of 298 K, the crystalline lamellae are rigid bodies and the amorphous phases between adjacent crystalline lamellae are rubbery. Therefore, the necking deformation at 298 K is dominated by the plastic deformation of crystalline phases.

Thirdly, a series of the same shape specimens with the above cases were prepared for all HDPE samples for the purpose of investigating the molecular weight dependence of natural draw ratio. These specimens were elongated at 298 K at a fixed crosshead speed of 0.16 mm/s.

Finally, the series of specimens with a gauge length of 10 mm and a fixed cross-section area (1.2 ±0.08 mm^2) with different widths and thicknesses, were prepared for HDPE120. The influence of the shape of cross-section on natural draw ratio was investigated using this series.

Scanning electron microscopy (SEM)

In order to observe the morphological change caused by crosshead speeds and temperatures and the structural change in the shoulder region on the specimens, SEM images were recorded of the surfaces of necked specimens, which were first acid etched and sputter coated with gold using an E-1030 Ionsputter (HITACHI) under low vacuum (60 kg cm^{-2}). Acid etching was carried out similarly to previous reports [16, 17] using potassium permanganate, concentrated phosphoric acid and concentrated sulfuric acid ($KMnO_4/H_3PO_4/H_2SO_4$). The SEM observations were made using an S-4500 (HITACHI) microscope at 15.0 kV.

Results and Discussion

Stress-strain curve

Two yield points were observed on the stress-strain curves of all samples shown in Fig. 1, as known that the stress-strain curves for PE materials shows double yield points [18, 19]. Beyond the first yield point, concave construction appeared on the specimen, and neck propagation regions appear on the stress-strain curves, where the stress maintains constant values. The terminal extension ratio at the neck propagation region is known as the natural draw ratio, and a measure of material drawability.

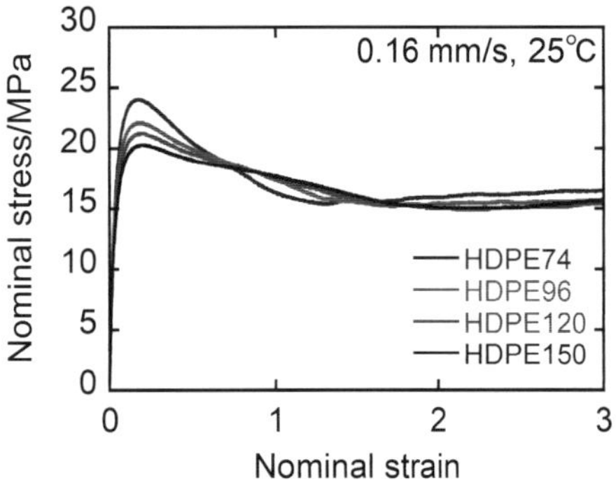

Fig.1 Stress-strain curves for four HDPE samples with different weight-average molecular weights measured at 0.16 mm/min and 25°C.

The experimental values of Poisson's ratio in the necked region for all samples were calculated from the width, thickness, and length of the necked portion. It was found that the values of Poisson's ratio were almost 0.5 being independent of crosshead speeds, molecular weight, and shape of cross-section area. This indicates that no volume change takes place during neck propagation.

Assuming that the neck propagation region of specimen consists of unnecked and necked portions, where the extension ratio achieves a fixed natural draw ratio, the natural draw ratio was estimated from the ratio of cross-section area of the necked portion and the original cross-section area using the following equation because there is no volume changes upon necking.

$$\lambda_N = \frac{L_N}{L_0} = \frac{A_0}{A_N} \qquad (1)$$

where L_N is the length of specimen at the natural draw ratio, L_0 is the initial gauge length of specimen, and A_0 and A_N are the cross-sectional areas before and after necking, respectively.

Crosshead speed dependence on NDR

Fig.2 shows crosshead speed dependence of natural draw ratio, and it was found that there are critical crosshead speeds at around 1 mm/s. On our SEM images [20], it was observed that both specimen which elongated above and below critical crosshead speed also have similar granular units of thickness of a few dozen nm. In addition, only the specimen has voids. Considering the fact that perfect plastic deformation takes place under incompressible state and is independent of elongation speeds, we focused only on the data of the natural draw ratio using necked specimens prepared below the critical crosshead speed.

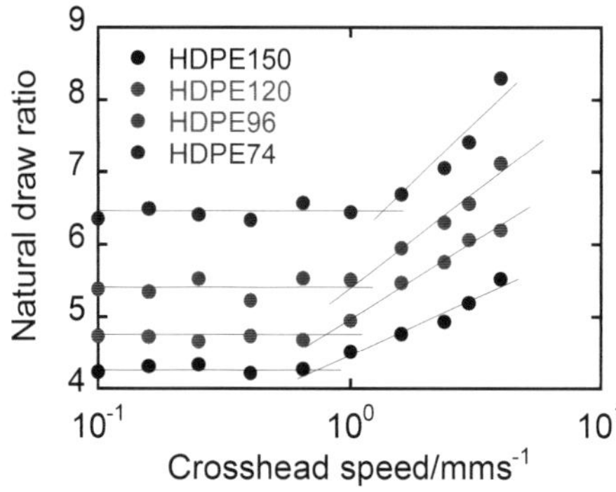

Fig.2 Crosshead speed dependence of natural draw ratio for all samples.

Morphology on the necking deformation

Fig. 3 shows SEM images of a surface structure from unnecked to necked regions. In the unnecked region, ribbonds consisting of granular units of 30-40 nm size are isotropically assembled. In the necking transition region, which corresponds to the neck shoulder region of the specimen, granular units gradually became closely-packed. Finally in the necked portion, rectangular blocks of thickness about 100-200 nm appeared. The closely packed state may appear at the inflection point of the neck shoulder.

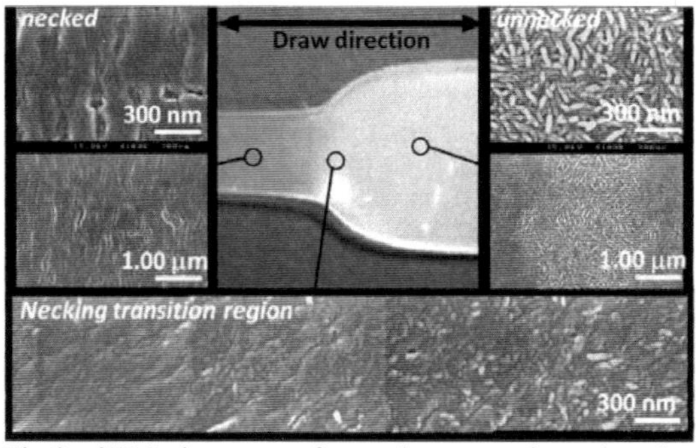

Fig.3 SEM images of specimen surface in the unnecked, necking transition, and necked regions.

Nitta-Takayanagi [12-14] showed that the deformation units which govern the yielding behavior are not individual lamellar crystals but are the lamellar clusters composed of a few stacked lamellae. The lamellar clusters are fragmented into the cluster units with random chain size at the yield point by the bundle of tie molecules. According to their theory, the size of lamellar cluster units for HDPE96 was estimated as 38.9 nm, and this saze is accordance with the granular unit size observed from SEM images (as seeing in Fig.3, unnecked region). This indicates that the granular units can be considered to be cluster units, and that necking is caused by the rearrangement of these cluster units.

Mechanical model for necking

Here we consider the structural model for necking.Considering the mass balance, the spatial changes along a path of necking trace corresponds to the change of the necking area at a fixed position on the specimen. Therefore SEM images shown in Fig.3 can be regarded as transient neck phenomena at a temporal deformation. As assumed that there are layers structure in the specimen perpendicular to the stretching direction, which called 'necking layer', as shown in Fig. 4. Focusing on a single necking layer, in unnecked region, the necking layer has about one cubic block length, and the blocks randomly assembled in the layer. In the neck shoulder region, the cubic blocks become closely-packed, and then the concave construction appear on the specimen, where the blocks are imaged as the lamellar cluster units. This close-packed necking layer is called frontier wall. Frontier wall is collapsed by sliding the blocks through the intercluster tie links. The more these slides of cluster units develops, the concave construction becomes more distinct. These cluster units within the frontier wall are rearranged to a thermodynamically stable object. The continual collapse from one frontier wall to another corresponds to the neck propagation.

Assuming that cubic blocks in Fig.4 are lamellar cluster units, we examined the validity of the deformation model of necking. It is notice that the necking layer just before necking correspond to the frontier wall. The surface free energy of the object after necking is expressed as;

$$F=2\gamma_B L_w^2 n_2 n_3+2\gamma_L L_w^2 n_1(n_2+n_3) \tag{2}$$

where γ_B is the basal surface free energy, γ_L is the lateral surface free energy of the cuboid object, n_1 is the number of cluster units on the lateral line, n_2 on width line, and n_3 on thickness line, and L_w is the size of lamellar cluster.

Assuming that the rearrangement of cluster units occurs in such a way that surface free energy becomes minimum, and the volume keep constant and the Poisson constriction in the thickness and width direction is the same, and there is a one-to-one correspondence of cluster units within each layer, we have

$$\lambda_N=\left(\frac{\gamma_B}{\gamma_L}\right)^{2/3}\left(\frac{2\sqrt{W_0D_0V_f^*}}{W_0+D_0}\right)^{2/3}\sqrt{\frac{V_f^*}{W_0D_0}}L_w^{-3/2} \tag{3}$$

where W_0 and D_0 is initial width and thickness, respectively, and V_f^* is the volume of frontier wall. As a result, the natural draw ratio is obtained like this, and the natural draw ratio is proportional to the weight-average cluster unit size to -3/2 powers.

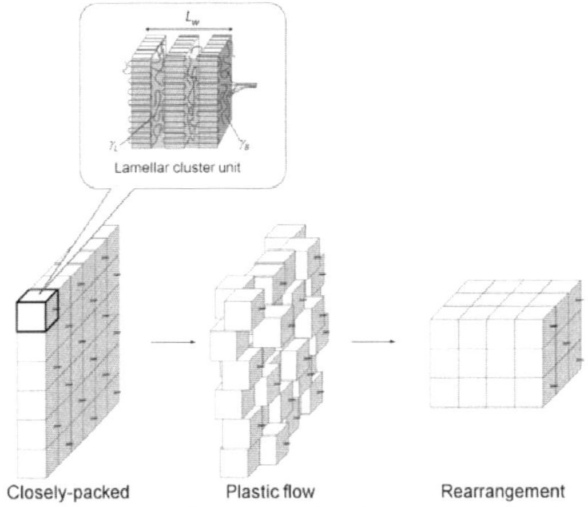

Fig.4 Shematic illustration of deformation model for necking region.

According to Eq. 3, the molecular weight and cross-sectional shape dependence of natural draw ratio expressed, and it was found that experimental data surprisingly fall on the theoretical curves.

Final Remarks

The present study showed that the dimensions of the specimen as well as molecular constitution and structural organization strongly effect on the nature of natural draw ratio. Thus, it will be an important forthcoming task to investigate the dependence of molecular parameters such as molecular species crystallinity, superstructure, and molecular distribution on the natural draw ratio.

Acknowledgements

We deeply thank to Dr. T. Kawamura for experimental help for SAXS measurements.

References

1. I. M. Ward; E. H. Immergut; E. A. Grulke; Abe; D. R. Bloch editors *Polymer Handbook. 4th ed.* Wiley New York, 1999
2. Carothers WH, Hill JW. J Am Chem Soc 1932;54:1579e96.
3. W. J. McTegart *Elements of Mechanical Metallurgy*, Chapter 1, Macmillan Co., New York and London, (1967
4. B. Crist, C. Metaxas, *J Polym Sci Part B Polym Phys.* **42**, 2081 (2004)
5. J. L. Ericsen, *J Elasticity*, **5**, 191 (1975)
6. Brauwer P, Müller FH. Kolloid A 1954;65:135
7. Lazurkin YS. J Polym Sci 1958;30:595e603
8. G. Meinel, A. Peterlin, *J. Polym. Sci. Part B, Polym Phys.*, **9**, 67 (1971)
9. A. Peterlin, G. Meinel, *Makromol Chem.*, **142**, 227 (1971)
10. A. Peterlin, *Colloid Polym Sci.*, **253**, 809 (1975)

11. Séguéla R. Macromol Mater Eng 2007;292:235e44.
12. Takayanagi M, Nitta K-H, Kojima O. J Macromol Sci Phys 2003;B42(5):1049e59
13. Nitta K-H, Takayanagi M. J Macromol Sci Phys 2002;42B(1):107e26
14. Nitta K-H, Asuka K, Liu B, Terano M. Polymer 2006;47(18):6463e87
15. L. C. Sawyer, D. T. Grubb, *Polymer Microscopy 2nd ed.* CHAPMAN & HALL, London, UK (1966)
16. L. E. Malvern, *Introduction to the Mechanics of a Continuous Medium*, Chapter 5, 6, Prentice-Hall, Inc. New Jersey (1969)
17. L. C. Sawyer, D. T. Grubb, *Polymer Microscopy 2nd ed.* CHAPMAN & HALL, London, UK (1966)
18. N. W. Brooks, R. A. Duckett, I. M. Ward, *Polymer Volume*, **33**, 1872 (1992)
19. R. Seguela, F. Rietsch, *Journal of Materials Science Letters*, **9**, 46 (1990)
20. Mizue Kuriyagawa, Koh-hei Nitta Polymer 52 (2011) 3469e3477

O-TM-262

PROPERTIES OF POLYMER/MULTI-WALLED CARBON NANOTUBE NANOCOMPOSITES

Sung Chul Kim[1*], Heonjoo Ha[2] and KiRyong Ha[3]

[1]Department of Chemical & Biomolecular Engineering, Korea Advanced Institute of Science and Technology, 335 Gwahangno, Yuseong-gu, Daejeon, 305-701, Republic of Korea; kimsc@kaist.ac.kr

[2] Automotive Material Development Group, Samsung Cheil Industries Co., Gocheon-dong, Uiwang-si, Gyeonggi-do 437-711, Republic of Korea; heonjoo.ha@samsung.com

[3] Department of Chemical Engineering, Keimyung University, 1095 Dalgubeoldaero, Dalseo-gu, Daegu 704-701, Republic of Korea; ryongi@kmu.ac.kr

Abstract - In this study, three issues are discussed. First, the effect of molecular weight of the polymer matrix on the dispersion of multi-walled carbon nanotubes was investigated for the high-density polyethylene and polycarbonate composites. The dispersion of carbon nanotubes in the polymer matrix showed strong correlations with the melt viscosity of the polymer matrix, in other words, with the molecular weight of the polymer material. The effect of the Van der Waals force interactions between carbon nanotubes and polymer matrix was investigated using three different kinds of polyamides: polyamide 6, polyamide 11, and polyamide 12. By comparing the degree of dispersion of carbon nanotubes and comparing the physical properties of each composite, it was concluded that polyamides with longer hydrophobic methylene groups in the repeating unit had better dispersion of carbon nanotubes and correspondingly lower surface resistivity. Finally, an empirical equation has been derived to fit to electrical measurements obtained by several experiments. The proposed empirical equation fit all of the experimental data very well in the concentration range above the percolation threshold, and it was shown that each parameter of the empirical equation had a strong correlation with the state of dispersion of the carbon nanotubes. Furthermore, the Hansen solubility parameter was used to predict the compatibility between the carbon nanotube and the thermoplastic polymer, and the predicted result agreed well with the data obtained in several other reports.

Keywords: Carbon Nanotubes, Electrical Properties, Hansen Solubility Parameter

Introduction

The desire and demand to find novel materials with superior property generated numerous scientific breakthroughs for the past few centuries. Scientists and engineers have long ago figured out that by simply mixing two or more different materials, they can accomplish the desire to make materials having superior property. By mixing two or more materials with completely different or similar physical and chemical properties, composite materials are prepared, and one of the most active and successful field for the past decade has been, without a doubt, the polymer/carbon nanotube nanocomposite area[1-6].

In this study, industrial grade multi-walled carbon nanotubes (MWCNTs) were incorporated in various polymers by simple melt mixing process. The main interest of this research is to investigate the effect of the molecular weight of the polymer matrix and the effect of interaction between MWCNTs and polymers on the dispersion of the MWCNTs and on their properties, such as electrical, rheological, thermal, and mechanical properties, *etc*. Another focus is to come up with a universal empirical equation which can simply, but also accurately predict the electrical property of the polymer/MWCNT nanocomposites.

Experimental

Materials

Ninety percent pure MWCNT materials, Nanocyl 7000 (Nanocyl S.A., Belgium), were directly incorporated into the following thermoplastic polymers by simple melt mixing: polyamide 6 (PA6) (Amtai Nylon TP-4407, Amtai's Group Inc., Taiwan), polyamide 11 (PA11) (Rilsan BESNO TL, Arkema, France), polyamide 12 (PA12) (Vestamid L1600 nf, Evonik Degussa AG, Germany), low molecular weight high-density polyethylene (HDPE) (2600J, Honam Petrochemical Corp., Korea), medium molecular weight HDPE (2200J, Honam Petrochemical Corp., Korea), high molecular weight HDPE (7000F, Honam Petrochemical Corp., Korea), polypropylene (PP) (5016S, Korea Petrochemical Ind. Co., Korea), low molecular weight polycarbonate (PC) (SC1620, Samsung Cheil Ind., Korea), medium molecular weight PC (SC1190, Samsung Cheil Ind., Korea), and high molecular weight PC (PDHP, Samsung Cheil Ind., Korea). Especially for HDPE and PC, the notations L, M, and H are used to represent low, medium and high molecular weight. The complex viscosity of PA6, PA11 and PA12 was 810, 3570, 960 Pa·s, respectively.

Sample Preparation

All materials, such as nanotubes and polymer pellets, were properly dried in a vacuum oven for 12 hours at 100 °C before mixing. The mixing was conducted in a Brabender mixer Lab station+W50 EHT (Brabender, Germany), equipped with a roller blade.

Samples collected from the Brabender mixer were placed in a mold consisting of two metal plates and a spacer. The spacer used to obtain JIS K 7194 standard samples had a thickness of 1 mm.

Electrical Properties

In the low resistivity region (resistance less than 10^7 Ω), 4-point probe resistivity meter (Loresta-GP MCP-T600, Mitsubishi Chemical, Japan) was used. Samples with dimensions of 50×80×1 mm (JIS K 7194) were conditioned at 23 °C and 50% RH for 48 hours (ASTM D 618) before measurement, and calibration using a standardized probe checker was conducted prior to each test. Volume resistivity (Ω·cm) was measured by multiplying the surface resistivity (Ω/sq) with the thickness of the material in centimeter

Results and Discussion

Effect of Molecular Weight of Polymer Matrix

Fig. 1 shows the surface resistivity versus the content of MWCNT for the composites having different molecular weight of HDPE and PC. As shown in both graphs, we can conclude that with increasing the molecular weight of the polymer matrix, the degree of dispersion of MWCNTs in the system is lowered since the percolation threshold concentration gradually increases with increasing molecular weight of the matrix polymer.

Primarily, due to high melt viscosity of the high molecular weight polymer matrix, during mixing process the MWCNTs experience extremely low mobility, which eventually lowers the dispersion of MWCNTs. In addition to that, high molecular weight polymers possess high entanglement density, which disables the penetration of the polymer chains to separate the bundled MWCNT to individual ones. In summary, not only polymers with high molecular weight are not capable of separating agglomerates of MWCNTs, but also due to the high melt viscosity of the polymer melt, the mobility of MWCNTs in the polymer melt is significantly low.

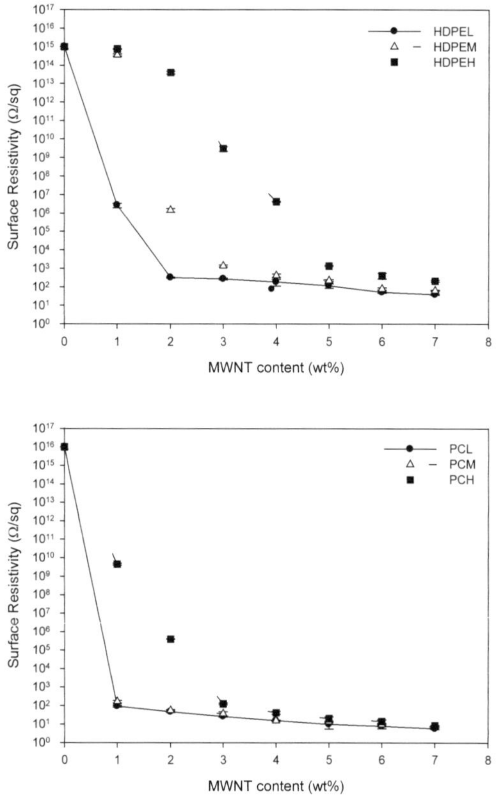

Figure 1 – Surface resistivity for HDPEs and PCs having different molecular weight: samples conditioned at 23 °C and 50% RH.

Effect of Interaction between MWCNT and Polymer Matrix

Fig. 2 shows the surface resistivity as a function of the MWCNT content for various PA composites. PA with longer methylene group in the repeating unit showed lower surface resistivity, which indicated that PA12 was more compatible with MWCNT.

In comparing PA11 and PA12, it is apparent that both composites showed similar electrical property in the low resistivity region, while significant difference was observed in the high resistivity region. It is assumed that the similarity in the low resistivity region is due to the resemblance in the chemical structure of PA11 and PA12. But PA12 showed a lower resistivity in the low resistivity region, indicating slightly improved interaction with MWCNT than PA11.

In the high resistivity region, PA6 without MWCNT showed lower surface resistivity than PA11 and PA12, due to the high moisture absorption which is originated from higher concentration of polar amide groups. In comparing PA11 and PA12, PA11 has more than three times higher melt viscosity than PA12. With this high melt viscosity, the MWCNTs experience significantly low mobility in the polymer melt during mixing process, which eventually prohibited the dispersion of MWCNTs. In addition, high molecular weight PA11 possesses high entanglement density, which disables the penetration of the polymer chains to separate the bundled form MWCNT into individual ones. So PA11 shows high surface resistivity than PA12, when the composite has lower MWCNT content than the percolation threshold.

When considering the melt viscosity of PA6, PA6 having the lowest melt viscosity should show lower surface resistivity than PA12. But actually, as shown in Fig. 2, the surface resistivity of PA6 is higher than PA12. This is presumably due to the poor dispersion of MWCNT in PA6 matrix, since comparably polar PA6 has lower compatibility with MWCNT.

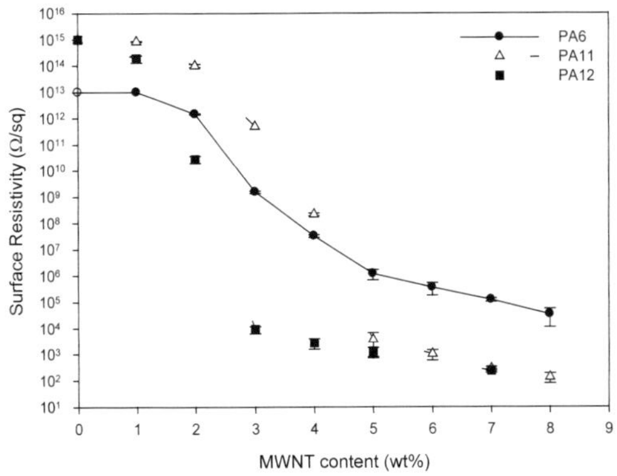

Figure 2 – Surface resistivity curves for various PAs: samples conditioned at 23 °C and 50% RH.

For the case of PA6, although we have considerably low melt viscosity which provides good mobility of MWCNT, strong hydrogen bonding formed between polymer-polymer chains retards the dispersion of MWCNTs. The hydrogen bonding formed in the melt state acts as a physical entanglement which impedes the penetration of polymer chains to separate bundled MWCNTs into individual ones. Furthermore, due to the limited space for MWCNTs to be dispersed, addition of MWCNTs will lead to an overlapping or reformation of bundled form MWCNTs, rather than maintaining the individual nanotube. But for the case of PA12, not only the system has more pronounced van der Waals force interaction between polymer and MWCNT, the hydrogen bonding sites are much less compared to PA6, providing better dispersion of MWCNTs. In summary, interaction is more important than the effect of the molecular weight of the polymer matrix in dispersing MWCNTs.

The Empirical Equation

The proposed empirical equation is expressed in a simple exponential form, where *R* represents the surface resistivity in Ω/sq, *a* and *b* are empirical constants and *x* is the weight percent of MWCNTs.

$$R = a \cdot e^{-bx} \qquad \text{Eq.1}$$

In addition, when a natural logarithm is applied to both sides of the equation, the equation can be transformed into a linear equation as follows:

$$\ln R = \ln a - bx \qquad \text{Eq.2}$$

All of the polymeric materials examined in this research have been fitted to the empirical equation Eq.2. The surface resistivity as a function of MWCNT content is also shown in Fig. 3.

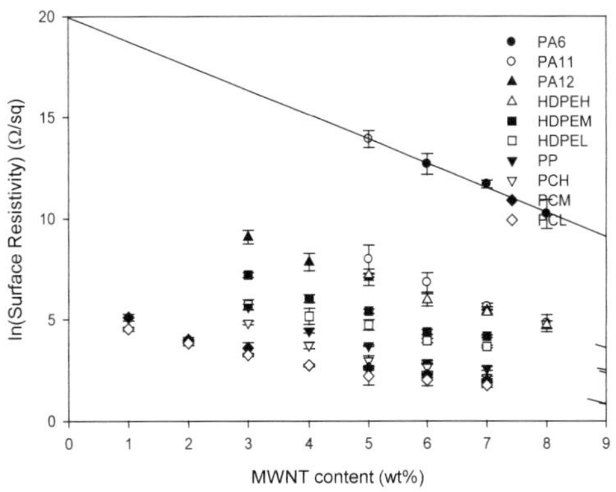

Figure 3 – ln (surface resistivity) curve for various polymer/MWCNT nanocomposites: samples were conditioned at 23 °C and 50% RH.

Since the *ln a* and *b* values have a strong correlation with each other, a regression line of *ln a* versus *b* is shown in Fig. 4. Surprisingly, *ln a* and *b* values show a linear correlation ($R^2 = 0.9002$) for most of the polymer/MWCNT nanocomposite system as shown in Fig. 4. This indicates that a strong relationship is present for all polymer/MWCNT nanocomposites, and it can be extended to predict surface resistivity values when information on the reduction in surface resistivity per addition of MWCNT (*b*) is provided. Since *ln a* is a function of *b*, the equation can be simplified as:

$$\ln R = b(18-x)-4 \qquad \text{Eq.3}$$

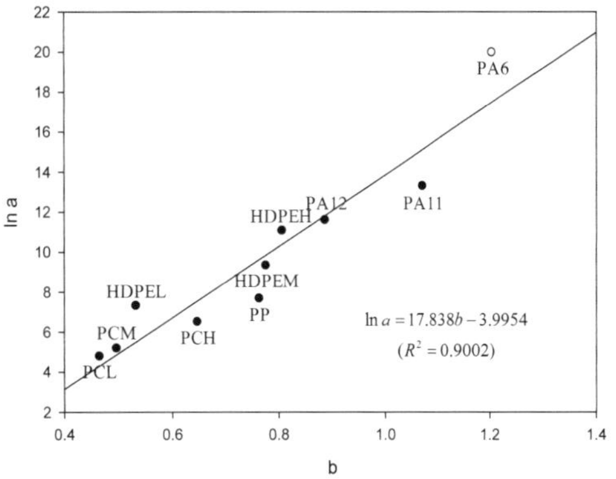

Figure 4 – Regression plot of *ln a* as a function of *b* for various polymer/MWCNT nanocomposites.

The above equation indicates that for all polymers, when wt% of MWCNT reaches 18 wt%, the surface resistivity is constant at e^{-4} Ω/sq. This *R* value is comparable to the volume resistivity data of single SWNT rope that has been reported as 10^{-4} Ω·cm [7]. This also indicates that the equation is applicable up to 18 wt%.

Hansen Solubility Parameter (HSP)

Previously, we have shown that the surface resistivity has a strong relationship with the agglomerates formed in the composite material. Instead of conducting additional experiments to find the origin of these phenomena, using a well-known theory can be more effective and persuasive. So, a well-established theory such as the Hansen solubility parameter (HSP) has been investigated [8], for the first time, to see whether its theoretical prediction and the experimental data correspond well.

The HSP data of PA6, PA12, HDPE, and PC were obtained from literature [8] and internet website[9]. The solubility parameter of the same MWCNTs used in this experiment has been published recently [10], and the data were used to construct the three dimensional solubility graph as shown in Fig. 5.

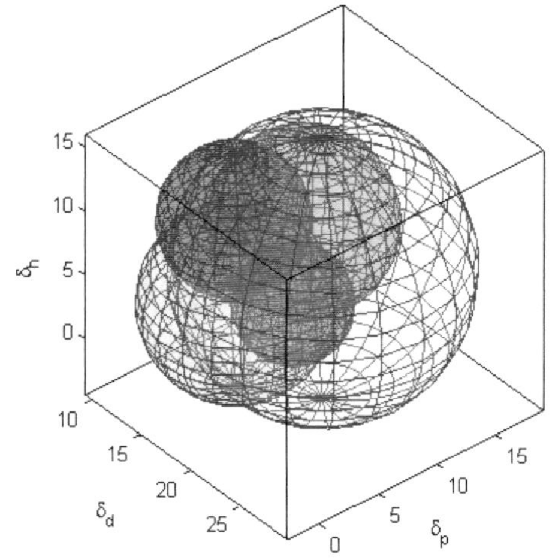

Figure 5 – Three-dimensional graphical presentation of Hansen solubility parameter of PA6, PA12, HDPE, PC, and MWCNT. The largest white sphere is PC, the green sphere is PA12, the blue sphere is PA6, the yellow sphere is HDPE and the red sphere is MWCNT.

The result shows that the whole volume of the MWCNT sphere is included inside the PC sphere. On the other hand, other polymers are partially intersected with the MWCNT sphere. The volume ratio of PA6, PA12, HDPE and PC sphere intersected with the MWCNT sphere are 6.38 %, 54.2 %, 89.0 % and 100 %, respectively.

In conclusion, MWCNTs should be most compatible to PC, because the intersected volume is the largest among all of the other polymers. The order of compatibility with respect to MWCNTs is PC, HDPE, PA12, and PA6, which is a perfect match with the surface resistivity data.

Conclusions

We have confirmed that the proposed empirical equation sufficiently explains the surface resistivity versus MWCNT content behavior in the low resistivity region. Although the equation is applicable in a limited range, the equation is simple, effective, and suitable for commercial applications.

HSP was used, for the first time, to gain better insight into the proposed equation. Additionally, it was also used to prove whether the state of dispersion of MWCNTs can be theoretically examined without doing any troublesome experiments. Eventually, the HSP method showed promising results and it has been revealed that the experimental data obtained in various reports had strong correlations with the results obtained by the theoretical prediction.

References

1. R. H. Baughman; A. A. Zakhidov; W. A. de Heer; *Science*, 2002, *297*, 787.
2. X. L. Xie; Y. W. Mai; X. P. Zhou; *Mater. Sci. Eng. R*, 2005, *49*, 89.
3. H. Ha; K. Ha; S. C. Kim; *Carbon*, 2010, *48*, 1939.

4. H. Ha; K. Ha; S. C. Kim; *Macromol. Res.*, 2010, *18*, 512
5. H. Ha; K. Ha; S. C. Kim; *Macromol. Res.*, 2010, *18*, 660
6. H. Ha; K. Ha; S. C. Kim; *Macromol. Res.*, 2010, *18*, 674
7. A. Thress; R. Lee; P. Nikolaev; H. Dai; P. Petit; J. Robert; C. Xu; Y. H. Lee; S. G. Kim; A. G. Rinzler; D. T. Colbert; G. E. Scuseria; D. Tománek; J. E. Fischer; R. E. Smalley; *Science*, 1996, *273*, 483.
8. C. M. Hansen, *Hansen Solubility Parameter: A User's Handbook*, 2nd Edition, CRC Press, Taylor & Francis Group, 2007.
9. www.accudynetest.com (accessed 10.26.09).
10. S. Detriche; J. B. Nagy; Z. Mekhalif; J. Delhalle; *J. Nanosci. Nanotech.*, 2009, 6015.

O-TM-176

RHEOLOGICAL MODIFICATION AND ITS APPLICATION TO POLYMER PROCESSING

M. Yamaguchi[1]*

[1]*School of Materilas Science, Japan Advanced Institute of Science and Technology, Japan – m_yama@jaist.ac.jp*

Abstract - Precise control of viscous and elastic properties is considerably important for good processability at various processing operations for conventional polymer melts. In this presentation, new methods to improve the elastic nature are shown, which will be applicable to the processing operations requiring high melt elasticity such as foaming and blow-molding. One is to add a weak gel just beyond sol-gel transition point. When the chain segments of a weak gel are miscible with matrix polymer chains, the strain-hardening behavior in elongational viscosity is enhanced to a great extent by a small amount of the gel, whereas the shear viscosity increases slightly. The other method is to add flexible nanofibers. Interdigitation of nanofibers provides long time relaxation mechanism. Furthermore, the elastic deformation of the network composed of the interdigitated fibers, which is presumably owing to bending energy of deformed fibers, is responsible for marked elastic properties such as normal stress difference, drawdown force, and Barus effect. The shear viscosity for the blend with flexible nanofibers is not greatly changed.

Keywords: rheology; melt elasticity; polymer processing; critical gel; nanofiber

Introduction

Rheological properties and processability strongly depend on the molecular structure such as molecular weight distribution and long-chain branches. The rheological behavior of a simple polymer melt including miscible polymer blends [1] is easily predicted these days. For example, low-density polyethylene produced by radical polymerization under high pressure shows strong melt elasticity and thus is preferably employed for foaming and extrusion coating rather than linear low-density polyethylene [2]. Similarly, high-density polyethylene by chromium catalyst is often used for blow-molding rather than that by Zieglar-Natta catalyst.

Because melt elasticity has a great impact on the processability at some processing operations, intensive efforts have been carried out to incorporate long-chain branches for various polymeric materials. Although the rheological properties of long-chain branched polymers are not clarified in detail because of the complicated phenomenon, called shear modification [2-4], new type of long-chain branched polymers will be developed by advanced technology of polymer synthesis and/or catalyst. However, it would be difficult to apply the technology to industries, because it often requires a huge investment to produce a new polymer in a commercial scale. Therefore, a processing modifier that enhances the melt elasticity has been desired for a long time.

Results and Discussion

Mixing Critical Gel

It has been found that blending a small amount of a weak gel near the sol-gel transition point, in which the molecular segments in the gel is miscible with a matrix polymer, enhances melt elasticity drastically, whereas it has less/no influence on the steady-state shear viscosity [5-7].

Figure 1 exemplifies the growth curves of uniaxial elongational viscosity at various strain rates for blends of linear isotactic polypropylene (PP) and gEHDM that is the gel fraction of a crosslinked terpolymer composed of ethylene, 1-hexene, and ethylidene norbornene with a low density of crosslink points. The precursor of gEHDM, *i.e.*, uncrosslinked terpolymer (EHDM), is miscible with PP in a molten state because of the high 1-hexene content, 67.5 mol% [1]. The strain-hardening behavior in the elongational viscosity, which hardly takes place for linear PP, is clearly detected in the curves for the blends, whereas the linear region in the growth curves is the same as that for the pure PP. The steady-state shear viscosity of the blend is, on the other hand, almost similar to that of the pure PP. These results prove that the gEHDM acts as a modifier of PP to enhance the elastic properties remaining the shear viscosity.

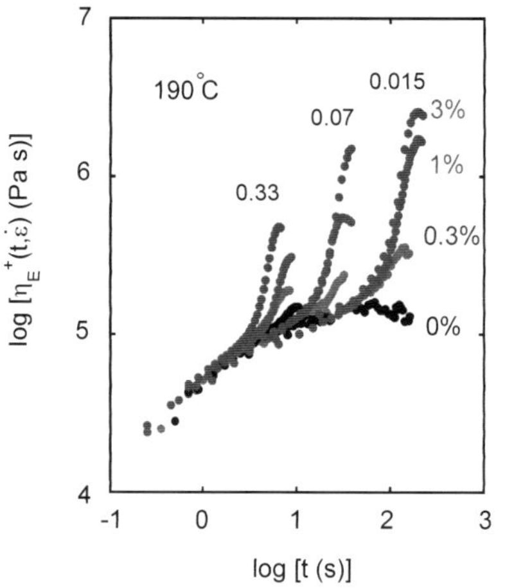

Figure 1 Growth curves of elongational viscosity for PP/gEHDM blends at 190 °C

Mixing Nanofibers

Recently, surface modified poly(tetrafluoro ethylene) PTFE is commercialized as a processing modifier for PP. Although PTFE is in the solid state at mixing temperature, it is deformed into fibrous shape with a diameter of approximately two hundred nanometers because of the complicated structure growth due to weak cohesive strength in PTFE crystals [8].

Addition of flexible PTFE nanofibers affects the rheological properties of PP to a great extent [9]. Figure 2 shows the drawdown force, defined as the force required for stretching an extruded strand from a capillary rheometer, of the blends prepared by a twin-screw extruder. As seen in the figure, the drawdown force increases rapidly with the PTFE content. On the contrary, steady-state shear viscosity does not change greatly. The enhanced drawdown force, shown in Figure 2, is found to be attributed to the marked strain-hardening in elongational viscosity.

This phenomenon is significantly different from a polymer composite with rigid fibers. As revealed by Mewis and Mezner [10], elongational viscosity is greatly enhanced irrespective of strain/time by addition of rigid fibers with large aspect ratio. However, the high level of elongational stress is ascribed to the localized shear deformation between fibers during elongational flow [11]. As a result, strain-hardening behavior is not detected for the blend systems, indicating that localized deformation at the elongational deformation cannot be improved by the addition of rigid fibers.

The mechanism of the enhanced elongational stress of the present system should be different from that of a polymer composite with rigid fibers. Since the fibers are flexible, network structure appears in the molten PP owing to interdigitation of fibers. The deformation of the network during flow with keeping the interparticle interaction provides the elastic response by the bending force of each fiber and frictional force between fibers, which will be responsible for the marked elastic properties of the blend with flexible nanofibers as similar to pulp-dispersion system [12-15]. It has been demonstrated by our research group that similar phenomenon is observed for other composite systems with nanofibers such as poly(lactic acid) containing poly(butylene terephthalate) PBT nanofibers [16] and PP containing PBT nanofibers [17].

Moreover, distribution state of PTFE fibers depends on the applied flow history. Therefore, the mixing condition and processing history affect the rheological properties and thus the processaibility of the blend system. This phenomenon is observed also for the blends with the critical gel and long-chain branched polymers, which is known as shear modification [2-4]. Because of the shear modification, actual processability does not always correspond with the rheological properties evaluated by original pellets.

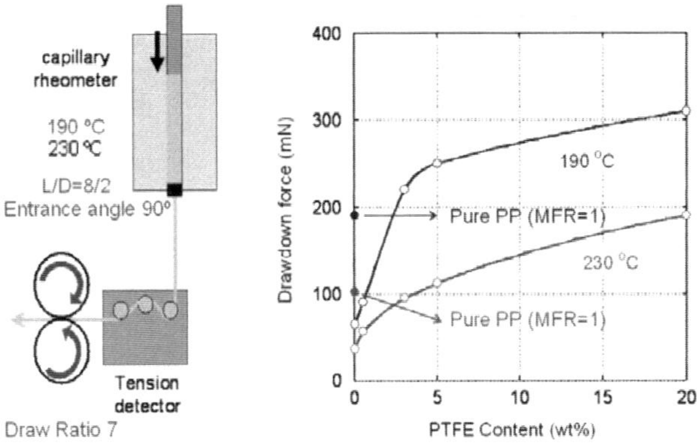

Figure 2 Apparatus of the measurement of drawdown force and the results plotted against PTFE contents

Conclusions

Two novel methods to enhance the melt elasticity of a polymer melt are proposed in the paper. A critical gel whose chain segments are miscible with the matrix polymer is effective to enhance the strain-hardening in elongational viscosity. Further, blending flexible nanofibers is another effective method to improve the melt elasticity. Since both methods do not change the shear viscosity greatly, viscoelastic balance of the polymer melts can be controlled by the methods. Including long-chain branched polymers, applied processing history affects the rheological properties to some degree for the polymer melts with enhanced melt elasticity. Therefore, processing conditions such as temperature, out-put rate, screw segments, and die geometry have to be controlled carefully to adjust the rheological properties for the aimed processing operation.

Acknowledgements

The authors would like to express their gratitude to the late Prof. Toshiro Masuda for his valuable and thoughtful advice to accomplish this work.

A part of this study was supported by Grant-in-Aid for Scientific Research (B) No. 22350102.

The authors acknowledge Mitsubishi Rayon Co. Ltd. for their kind supply of the samples.

References

1. M. Yamaguchi, K. Nitta, H. Miyata, T. Masuda, *J. Appl. Polym. Sci.*, 1997, *63*, 467.
2. M. Yamaguchi, M. Takahashi, *Polymer*, 2001, *42*, 8663.
3. M. Yamaguchi, D. B. Todd, C. G. Gogos, *Adv. Polym. Technol.*, 2003, *22*, 179.
4. M. Yamaguchi, M. H. Wagner, *Polymer*, 2006, *47*, 3629.
5. M. Yamaguchi, H. Miyata, *Polym. J.*, 2000, *32*, 164.

6. K. Arakawa, T. Yokohara, M. Yamaguchi, *J. Appl. Polym. Sci.*, 2007, *107*, 1320.
7. M. Yamaguchi, H. Miyata, *Macromolecules*, 1999, *38*, 5911.
8. B. M. A, Mohd Amran, Nobukawa S, Yamaguchi M, *Pure Appl Chem,* 2011, *83*, 1817.
9. B. M. A, Mohd Amran, K. Okamoto, M. Yamaguchi, A. Koshirai, T. Kasai, *J. Polym. Sci. Polym. Phys. Ed.*, 2009, *47*, 2008.
10. J. Mewis, A. B. Metzner, *J. Fluid Mech.*, 1974, *62*, 593.
11. M. H. Laun, *Colloid Polym. Sci.*, 1984, *262*, 257.
12. O. L. Forgacs, S. G. Mason, *J. Colloid Interface Sci*, 1959, *14*, 457.
13. R. Meyer, D. Wahren, *Sven. Papperstidn*, 1964, *67*, 432.
14. L. H. Switzer, D. J. Klingenberg, *J. Rheol.*, 2003, *47*, 759.
15. M. Keshtkar, M. C. Heuzey, P. J. Carreau, *Polym. Eng. Sci.*, 2010, *31*, 1474.
16. T. Yokohara, S. Nobukawa, M. Yamaguchi, *J. Rheol.*, 2011, *55*, 1205.
17. M. Yamaguchi, K. Fukuda, B. M. A. Mohd Amran, S. Nobukawa, *Macromol. Mat. Eng.*, 2012, *197*, 654.

O-TM-137

A MECHANISM OF THE LONG TIME RELAXATION OBSERVED IN HETEROGENEOUS POLYMER LIQUIDS

Takashi Yamamoto, Toshikazu Takigawa* and Toshiro Masuda
Department of Material Chemistry, Kyoto University, Japan – takigawa.toshikazu.7u@kyoto-u.ac.jp

Abstract – Heterogeneous polymer systems often show a new mechanical relaxation at long times. This relaxation occurs after the entanglement relaxation if polymers in the system have chains long enough to form entanglements. This long time relaxation is due to the diffusion of the cores of molecules.

Keywords: Long time relaxation, Second plateau, Diffusion, Heterogeneous polymer liquid

Introduction

Star-shaped polymers with large arm numbers and highly branched polymers have heterogeniety inside a molecule: the central part of the molecule becomes high and the marginal does low in density. We call melts or concentrated solution of these polyemers heterogeneous polymer liquids. The word "heterogeneous polymer liquids" can be extended to involve the conventional dispersions such as particle-filled polymers. Masuda et al first pointed out the possibility that a new mechanical relaxation at long times for the heterogeneous polymer liquids originates from diffusion of the hard cores of molecules in a viscous medium.[1, 2]

The aim of this study is to check if the diffusional motion can really explain the mechanical relaxation, i.e., the second plateau, for the heterogeneous polymer liquids.

Experimental

samples used were 4 types of multi-arm star polystyrenes (LM23, LM24, LM26 and LM27). The average number of arms (f) ranged from 15.6 to 36.9. Detailes of the sample characteristicss were described elsewhere.[3] The solvent used was tricresylphosphate (TCP). Dynamic storage and loss moduli (G' and G'', respectively) were measured as a function of angular frequency (ω) by using an ARES rheometer.

Results and Discussion

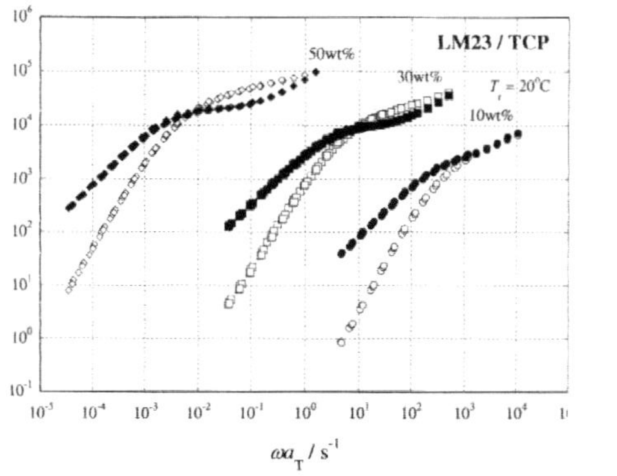

Figure 1 ω dispersion curves of G' and G'' for the TCP solutions of LM23.

Figure 1 shows the master curves of G' and G'' for the solutions of LM23. Here, T_r stands for the reference temperature and a_T is the shift factor. The time-temperature superposition hold well to produce nice master curves. On the curves the rubbury plateau and flow zone are clearly obserbed for all samples. Using the G' and G'', the relaxation spectrum ($H(\tau)$, τ being the relaxation time) was calculated for the solutions of LM23 (Figure 2).

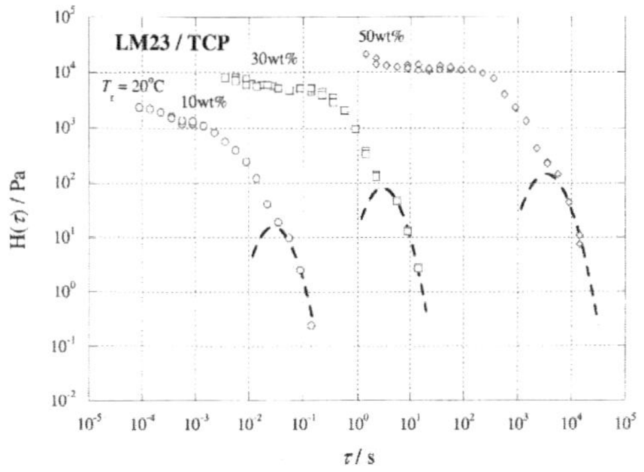

Figure 2 Reraxation Spectra for the TCP solutions of LM23.

Every Spectrum has a clear shoulder at longer relaxation times, although a trace of new relaxation was hardly detected on the G' and G''curves. The shoulder corresponds to a new mechanical relaxation for the solutions. Each relaxation spectrum was successfully decomposed into two components. The long time component after decomposition is shown in a convex curve for each solution, and τ at which the curve has a maximum corresponds to the average relaxation time of the new relaxation, originating from the diffusion of the cores.

References

1. T. Masuda; Y. Ohta; T. Yamauchi; S. Onogi *Polym. J.*, 1984, *16*, 273.
2. T. Masuda; Y. Ohta; S. Onogi *Macromol.*, 1986, *19*, 2524.
3. T. Yamamoto, Y. Ohta, T. Takigawa, T. Masuda *NIHON REOROJI GAKKAISHI*, 2001, *29*, 211.

EFFECT OF ASPECT RATIO OF MULTIWALLED CARBON NANOTUBES ON THE PERCOLATION AND GEL-LIKE BEHAVIOR OF MWCNT/PP COMPOSITES

Yongzheng Pan and Lin Li*
School of Mechanical and Aerospace Engineering, Nanyang Technological University
50 Nanyang Avenue, Singapore 639798 – mlli@ntu.edu.sg

Abstract - We present our recent study on the effect of aspect ratio of multiwalled carbon nanotubes (MWCNTs) on the rheological, percolation and gel-like properties of MWCNT/polypropylene (PP) composites. In our study, we found that with increasing MWCNT content, the composites show a typical transition from a liquid-like to a solid-like viscoelastic behavior. Utilizing the scaling laws (including the Winter-Chambon method), the rheological percolation thresholds and gel points of different composites could be obtained, respectively. It was further revealed that the composites containing MWCNTs with the largest aspect ratio had the lowest values of percolation threshold and gel point. On the other hand, the electrical conductivities were examined and the electrical percolation thresholds were obtained for the same composites. Based on these results, the differences among gel point, rheological and electrical percolation thresholds will be discussed and explained.

Keywords: nanocomposite, carbon nanotube, aspect ratio, gel-like behavior, percolation

Introduction

Carbon nanotubes (CNTs) such as multiwalled carbon nanotubes (MWCNTs) have been extensively used as effective nanofillers to produce CNT/polymer nanocomposites because CNTs are potentially able to significantly contribute both mechanical and electrical properties to a matrix polymer. The effect of CNTs on the enhancement of mechanical and electrical properties of a polymer depends on many factors such as aspect ratio, content and dispersion of CNTs in the polymer matrix, as well as the interfacial adhesion between CNTs and the polymer. To achieve high electrical conductivity at a low content of CNTs, the dispersion and the connection of CNTs into a network are two interdependent key factors.

With adding CNTs to a non-conducting polymer, the CNT/polymer composite will show a sudden increase in electrical conductivity at a critical content of CNTs, which is generally referred to as the electrical percolation threshold. More and more efforts have been recently made on fabricating CNT/polymer composites with lower percolation thresholds in a large scale via melt compounding for practical applications [1, 2]. The percolation threshold can be influenced by various factors including i) properties of the matrix polymer, ii) properties of CNTs, and iii) dispersion of CNTs in the polymer matrix. Any factor that takes effect on the dispersion of CNTs and the formation of CNT network can influence the electrical percolation threshold.

In addition to electrical characterization of a CNT/polymer composite, dynamic viscoelastic measurements can also be employed to detect the formation of a CNT network in CNT/polymer composites. For example, with increasing MWCNT content in a composite, the composite melt undergoes a typical liquid-to-solid transition at low frequencies, indicating the emergence of a CNT network or a CNT-polymer network. The CNT content at which this transition occurs is defined as the rheological percolation threshold.

The liquid-to-solid transition in a composite melt can also be considered as a physical gelling process [3] where a rheological percolation threshold is defined. Recently, some researchers studied the physical gelling-like behavior of CNT/polymer composites using the Winter-Chambon method [4-6], which is a classical method used to explore the gelation characteristic of physical and chemical gels. The gel point for a CNT/polymer composite is supposed to be relevant to the rheological and electrical percolation thresholds, which may not be exactly the same.

In this study, our interest focuses on the effect of the aspect ratio of MWCNTs on the rheological percolation and gel-like behavior for MWCNT/ polypropylene (PP) composites. We have applied the power scaling and Winter-Chambon method to determine the rheological percolation thresholds and gel points of the composites respectively, and then compared them with the electrical percolation thresholds.

Experimental

Polypropylene (PP) with a molecular weight of 190,000 was purchased from Sigma Aldrich. Three different types of multiwalled carbon nanotubes (MWCNTs) with different aspect ratios were obtained from three suppliers, and their outer diameters and lengthes were (i) 8-15 nm and 10-50 μm (Cheap Tubes Inc., USA), (ii) 10-15 nm and 10-20 μm (Iljin Nano Tech., Korea), and (iii) 5-20 nm and 1-10 μm (Bayer Material Science AG, Germany) respectively. A counter-rotating twin-screw micro-compounder (Haake

MiniLab) was used to mix MWCNTs with PP at 200 °C. The mixed MWCNT/PP composites with various contents of MWCNTs were injection molded at 200 °C by a mini injection machine (Haake MiniJet), into test samples. The through-plane electrical conductivity of the composites was measured in accordance with ASTM D4496. Rheological measurements were carried out using a rotational rheometer (ARES) with the molded samples (Φ25 mm x 1mm). A parallel plate geometry (25mm diameter) with a gap of 1 mm was used.

Results and Discussion

1. Rheological Percolation

MWCNTs may undergo a significant reduction in length during melt mixing using a twin-screw micro-compounder or a similar compounder. Therefore, the real values of aspect ratios of MWCNTs used in this study had to be measured. After melt mixing, three aspect ratios were about 50, 80, and 170, which are within a good range for the study of aspect ratio effect.

Fig.1 shows the storage modulus (G') and loss modulus (G'') as a function of angular frequency (ω) for the MWCNT/PP composites at 200 °C, where the average aspect ratio of MWCNTs was 80. As shown in Fig.1, in the low ω zone, neat PP exhibited a typical terminal behavior, in which G' and G'' nearly followed the scaling laws: $G' \sim \omega^1$ and $G'' \sim \omega^2$, due to the full relaxation of PP chains. With addition of MWCNTs, both G' and G'' gradually increased in magnitude and showed a deviation from the terminal behavior, which gave rise to the appearance of a G' plateaus in the low ω zone, indicating the increased elastic response of the composites. As the content of MWCNTs was further increased, both G' and G'' became almost independent of ω at low ω, implying a transition from a liquid-like to solid-like viscoelastic behavior.

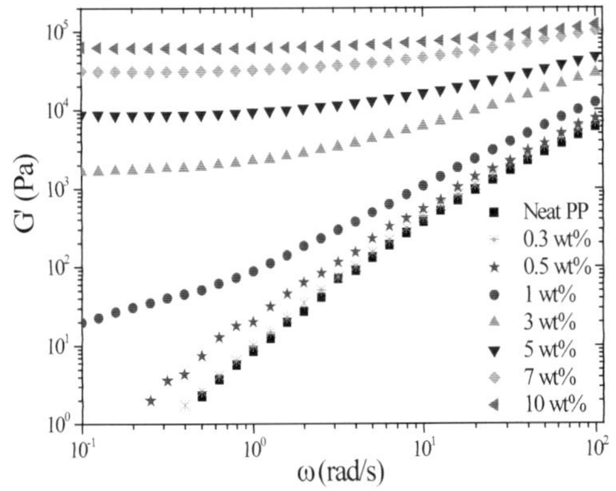

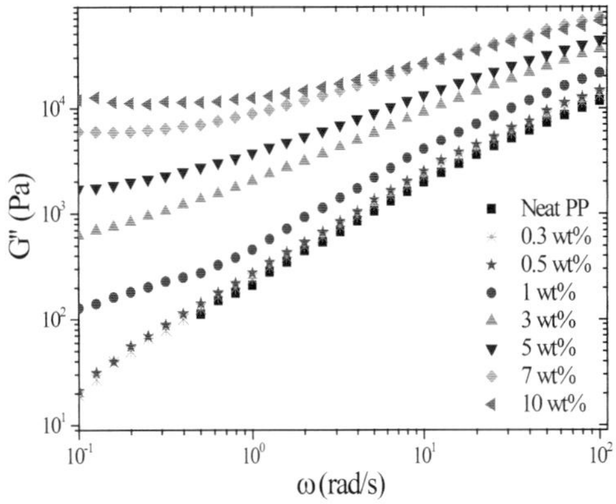

Figure 1. Storage modulus G and loss modulus G" as a function of angular frequency ω for MWCNT/PP composites at 200 °C. The average aspect ratio of MWCNTs was 80.

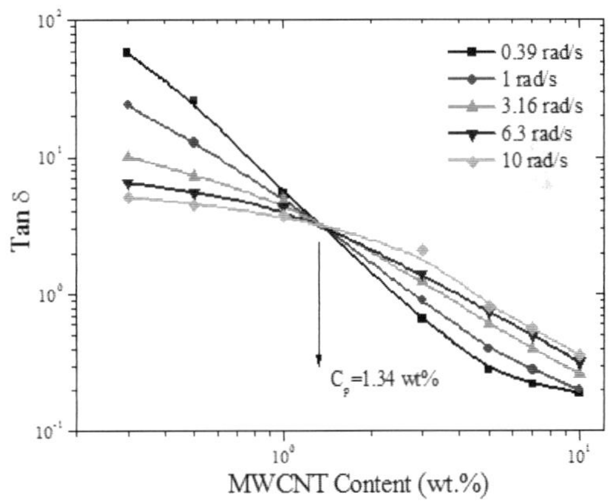

Figure 2. Loss tangent (tan δ) as a function of MWCNT content for MWCNT/PP composites at 200 °C. C_p is the MWCNT content at the gel point. The average aspect ratio of MWCNTs was 80.

To study the gel-like behavior of MWCNT/PP composites and the influence of MWCNT aspect ratio, we attempted to apply the well-established Winter and Chambon method to analyze the rheological data and to derive the gel content (c_g), relaxation exponent (n) and gel strength (S_g) for each composite. To determine the gel point, we need to observe a frequency-independent point of tan δ from a multifrequency plot of tan δ versus MWCNT content. As shown in Fig. 2, all the tan δ curves pass through a common point at a MWCNT content, which is defined as the gel point. The similar results were obtained for the other MWCNT/PP composites with the MWCNT aspect ratios of 50 and 170. Therefore, the accurate determination of the gel point for the MWCNT/PP composites suggests that the Winter-Chambon method is applicable to heterogeneous composite systems, in addition to its validity for many gel systems. The gel point, c_g, could be read directly from the convergence point. For the MWCNT/PP composites shown in Fig. 2, the gel point was determined to be 1.34 wt%. For the MWCNT/PP composites with the three aspect ratios (50, 80, and 170) of MWCNTs, their corresponding gel points are 1.92wt%, 1.34wt% and 0.7wt% respectively. Thus, it is clearly known that the gel point decreases with increasing the aspect ratio significantly.

2. **Electrical Percolation**

In order to understand how aspect ratio of MWCNTs further affects electrical percolation of MWCNT/PP composites, we measured electrical conductivities of MWCNT/PP composites as a function of MWCNT content. Our previous studies have shown that electrical conductivities of MWCNT/PP composites directly prepared by melt mixing and injection molding were relatively low due to the disruption of MWCNT connections. Annealing above melt temperature of PP for a certain time could significantly increase electrical conductivities [7]. Therefore, before electricity measurements, all the MWCNT/PP composites were annealed at 200 °C for 30 min in an oven with N_2 as purge gas.

Fig. 3 shows the electrical conductivities as a function of MWCNT content for three composites after annealing at 200 °C for 30 min. The electrical conductivities of the composites strongly depend on MWCNT content and aspect ratio. The MWCNT/PP composites with the largest aspect ratio of 170 have electrical conductivities almost 60 times higher than those of the MWCNT/PP composites with the smallest aspect ratio of 50.

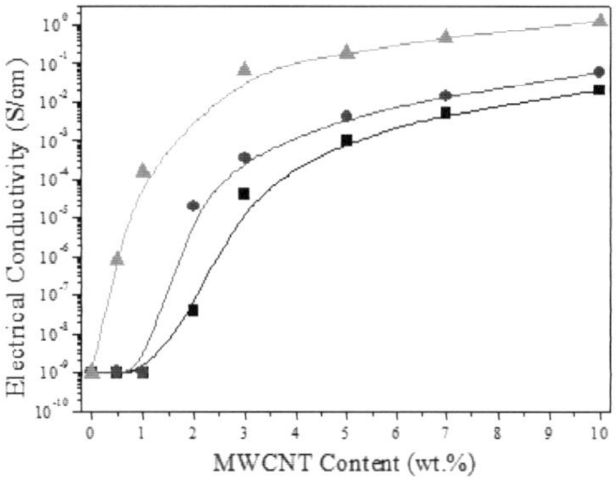

Figure 3. Electrical conductivity of MWCNT/PP composites as a function of MWCNT content. The black, red and green symbols indicate the average aspect ratios of 50, 80 and 170 respcectively.

From the electrical conditivity curves given in Fig. 3, we are able to determine the electrical percolation parameters for the MWNCT/PP composites, assuming that the conductivity σ follows a simple scaling law: $\sigma = (c - c_p)^\beta$, where σ is the electrical conductivity, *c* is the MWCNT content, c_p is the percolation threshold of electrical conductivity, and *β* is the critical exponent. The percolation threshold values we obtained are 2.11wt%, 1.51wt% and 0.44wt% for the MWCNT/PP composites with the average respect ratios of 50, 80, and 170, respectively.

The rheological results and the electrical conductivity results show that the electrical percolation thresholds (2.11, 1.51, and 0.44wt%) are similar to or silghtly higher than the corresponding gel points (1.92, 1.34, and 0.70wt%) obtained using the Winter-Chambon method. The differences among these values should originate from the different stages of nanotube network formation. It is considered that the rheological percolation occurs due to the formation of not only a network of nanotubes but also a network of polymer-MWCNTs. When the distance between two nanotubes is smaller than the gyration radius of a polymer chain, the nanotubes may be connected by random coils of polymer chains and restrict the polymer chain mobility [8]. As a result, the rheological percolation can be achieved early when a nanotube network has not been formed for the electric percolation.

Conclusions

The MWCNT/PP composites were prepared by melt mixing from three kinds of multiwalled carbon nanotubes (MWCNTs) with different aspect ratios and polypropylene (PP). The effects of MWCNT aspect ratio on the rheological and electrical properties of the composites have been studied. With a particular interest, we examined how the aspect ratio of MWCNTs could influence the rheological percolation ("gelation") and the electrical percolation differently. We found that (i) as the aspect ratio increases the percolation occurs early at a lower content of MWCNTs and (ii) the rheological percolation may occur earlier than the electrical percolation.

References

1. I. Alig; P. Potschke; D. Lellinger; T. Skipa; S. Pegel; G.R. Kasaliwal; T. Villmow *Polymer* 2012, 53, 4.
2. W. Bauhofer; J.Z. Kovacs *Compos. Sci. Technol.* 2007, 69, 1486.
3. P. Potschke; M. Abdel-Goad; I. Alig; S. Dudkin; D. Lellinger *Polymer* 2004, 45, 8863.
4. S-D. Park; D-H, Han; D. Teng; Y. Kwon *Curr. Appl. Phys.* 2008, 8, 482.
5. C.Y. Liu; J. Zhang; J.S. He; G.H. Hu *Polymer* 2003, 44, 7529.
6. S.L. Huang; Z.Y. Liu; C.L. Yin; Y. Wang; Y.J. Gao; C. Chen; M.B. Yang *Macromol. Mater. Eng.* 2012, 297,51.
7. Y. Pan; H.K.F. Cheng; L. Li; S.H. Chan; J. Zhao; Y. K. Juay *J. Polym. Sci., Part B: Polym. Phys.* 2010, 48, 2238.
8. G. Hu; C. Zhao; S. Zhang; M. Yang; Z. Wang *Polymer* 2006, 47, 480.

O-TM-098

A STUDY ON THE MECHANICAL PROPERTIES OF POLY(BUTYLENE TEREPHTHALETE) BY ADDING SEVERAL KINDS OF PBT ELASTOMER

K. Tokumitsu*, K. Nishizaka, Y. Yamashita and K. Kikuchi

Department of material Science, School of Engineering, The University of Shiga Prefecture, 2500 Hassaka-cho, Hikone, 522-8533, Shiga, Japan

Abstract - In order to improve the mechanical properties, especially for the impact strength, of poly(butylene terephthalate) (PBT), the effect of adding two kinds of PBT-derived elastomers, PELPRENE(P-40H having much butyl-ether structure and P-70B having much PBT structure), with/without a reactive compatibilizer into PBT was investigated. Although the impact strength of neat PBT was 3.5 kJ/m^2, when adding P-40H with 3.2 phr of the compatibilizers into PBT, the impact strength can be increased to 19.0 kJ/m^{-2} at 223 K. Furthermore, on the basis of Takayanagi's model for complex modulus of the blends, it was found that the volume fraction of interaction phase between PBT and PBT-derived elastomers can also be enhanced by adding the compatibilizer. When adding compatibilizer into PBT/PEL blends, the compatibility between PBT and PEL was improved.

Keywords: PBT, elastomer, polymer blend, compatibilizer, and mechanical properties.

Introduction

Poly (butylene terephthalate) (PBT) is an engineering plastic and widely used for many industrial application because of its superior chemical and physical properties, such as high heat resistance, chemical resistance, electrical insulation, moldability, dimensional stability, and so on. Impact strength of PBT, however, is much poorer than the other engineering plastics, and it has been necessary to improve the impact strength of PBT for achieving wider application.

On the other hand, a thermoplastic elastomer having PBT structure and butyl-ether structure as shown in Fig.1 has been commercialized. Then, it is expected that this kind of elastomer is one of the most promising candidate to improve the impact strength of PBT. Moreover, some kinds of reactive regents are expected to enhance the compatibility of PBT and the elastomer. In this study, the mechanical properties and morphological change of PBT and PBT-derived elastomer; blends with/without the reactive compatibilizer were investigated. The reactive regent having acryl-grycidy moiety in its chemical structure was used as a compatibilizer.

Experimental

Materials

As a PBT sample, VYLOPET (TOYOBO Co., LTD; melting point = 498 K, specific gravity = 1.32) was used. As thermoplastic elastomers, two kinds of PELPRENE (PEL, TOYOBO Co., LTD) were used. One of PEL consists of much butyl-ether structure, namely P-40H, and as a result, the melting temperature is 445 K, and the specific gravity is 1.12. Another one of PEL, namely P-70B, has much PBT structure, and the melting temperature is 473 K, and the specific gravity is 1.17. As a reactive compatibilizer, three kinds of regent consist of acrylic-glycidyl structure were used; v-1: glycidyl content 2,100eq/t, v-2: glycidyl content 1,800eq/t, v-3: glycidyl content 1,400eq/t. Before blending PBT and PEL, they were dried in a vacuum oven more than 24 h at 353 K.

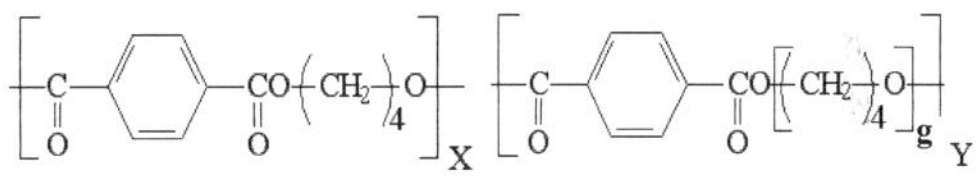

Fig.1 Chemical structure of PELPRENE

Sample preparation condition

PBT (70 wt%)/PEL (30 wt%)/ compatibilizer blends (0- 3.2 phr) were prepared by twin screw extruder (TECHNOVEL Co., LTD. L/D = 45) at 513 K, rotating speed at 200 rpm. All samples were fed through a main feeder at the same time.

Injection molding was carried out to obtain an impact test specimen (80 × 10 × 4 mm^3).

Film specimens (100 × 100 × 0.3 mm^3) were prepared by hot-pressing, at the temperature 513 K, and the pressure 7.0 MPa.

Impact strength measurement

The impact test was carried out by using Charpy impact tester (YASUDA SEIKI SEISAKUSHO Co., LTD. No.141-IS) at ambient temperature and 223 K. The energy of the impact hummer was 2J, and single notched specimen (Type A) was used according to JIS K 7111.

Dynamic mechanical measurement

Dynamic mechanical measurement was carried out with temperatures from 123 K to 493 K by using Rheogel-E400 (UBM Co. LTD). It was carried out in a tensile mode at a constant heating rate of 3 K/min, the frequency of 3 Hz, and the sample deformation 3μm.

SEM observation

The morphology of the blend samples was observed by SEM (HITACHI Co. LTD. S-3000N). The morphological change of the cryofractured surface of the samples was estimated.

Mechanical property measurement

Mechanical property at tensile mode was measured by using EZ-Test (SHIMADZU Co. LTD). The tensile rate was 30 mm/min, and the dimension of the specimen was 30 mm in length (initial length), 5 mm in width, and 0.3 mm in thickness.

Rheometer measurement

Rheometer measurement was carried out by using CVO-100 (Bohlin Instruments Co. LTD). It was carried out at 513 K and frequency range of 0.06~126 rad/s. The dimension of the specimen was 25 mm in diameter and 0.3 mm in thickness

Results and Discussion

Impact strength of blend materials

Fig.2 shows the result of Charpy impact test of neat PBT, and PBT/PEL(P-40H) /compatibilizer (0-3.2 phr) blends at 223 K. It was found that the neat PBT is brittle, and the impact strength of neat PBT was only 3.5 kJ/m^2. When adding P-40H into PBT, the impact strength of PBT/P-40H(70/30w/w%) increased up to 12.8 kJ/m^2. Furthermore, by adding compatibilizer (v-1, v-2, v-3) from 0.8 phr to 3.2 phr into PBT/P-40H blend, the impact strength was getting better than that of the neat PBT and PBT/P-40H. Particularly, the v-1 compativilizer having high glycidyl content had a remarkable effect on the impact strength even with small amount of 0.8 phr, but its impact strength increased slightly with high amount of v-1, compatibilizer up to 3.2 phr. On the contrary, the v-2, and v-3 compativilizers having lower glycidyl content had little effect on the impact strength with small amount of 0.8 phr, but its impact strength increased remarkably with high amount of these compatibilizers up to 3.2 phr. These results indicated that the amount of the reactive site of the compatibilizer should be controlled to highly improve the impact strength of PBT/P-40H because it is necessary to achieve a moderate cross-linking structure formed in the blend samples. On the other hand, when Charpy impact test was carried out at room temperature, the impact strength of PBT/P-40H was almost the same as the blend sample with 0.8 phr of each compatibilizer. When adding more than 1.6 phr of each compatibilizer into PBT/P-40H, the specimens cannot be broken completely, and so that, the impact strength of the blends exceeded the limit of the apparatus. From this result, it should be noted that PBT/P-40H added more than 1.6 phr of the reactive compatibilizer can be modified not to break easily at room temperature. The similar phenomena can be observed for PBT/P-70B with/without the compatibilizers. Hereinafter, the results obtained from PBT/P-40H with/without compatibilizer were discussed.

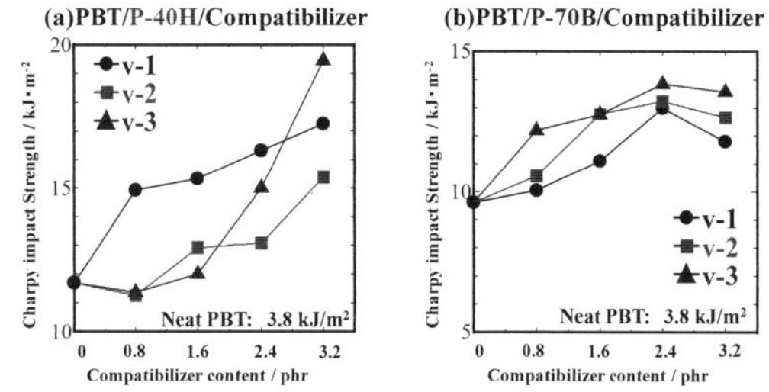

Fig.2 Charpy impact strength of (a)PBT/P-40H, (b)PBT/P-70B with compatibilizer (v-1, v-2, and v-3) at 223 K.

Viscoelastic properties of blends

Wada *et al.* reported that the ratio of dynamic modulus E' was proportional to the impact strength as shown Eq.(1):[1]

$$I \cong K'\tau^2 \frac{\left(E'_{T=100K} - E'_{T=300K(\omega=1/t_0)}\right)^2}{E'_{T=100K}} \quad (1)$$

where, I is the impact strength, τ is the relaxation time, E' is the dynamic modulus at a given temperature and K' is the proportional constant. As the relaxation time increased, the impact strength increased. In this study, assuming the relaxation time of PBT, PBT/PEL and PBT/PEL/compatibilizer blends were almost the same, the impact strength can be related to the ratio of dynamic modulus, $(E'_{T=100K} - E'_{T=300K})^2 / E'_{T=100K}$.

In the meantime, the E' at different temperature was described following Eq.(2):

$$E'_{100K} - E'_{300K} = \frac{2\Delta H^*}{\pi R}\int_{100K}^{300K} E''_{(\omega)}\, d\left(\frac{1}{T}\right) \quad (2)$$

where E'' is the loss modulus, ΔH^* is activation energy, π is the circular constant, R is the gas constant, and T is temperature, respectively. Therefore, as ΔH^* and the integrated area of the loss modulus, E'' in the tempearture of 100 K to 300 K were obtained, the ratio of dynamic modulus, $(E'_{T=100K} - E'_{T=300K})^2 / E'_{T=100K}$ can be calculated.

Figure 3 shows dynamic modulus (E'), loss modulus (E'') and loss tangent (tanδ) of neat PBT , PBT/P-40H blend and PBT/P-40H/v-1 (1.6 and 3.2 phr) blends. The temperature of the α relaxation of neat PBT was around 60°C and the β relaxation temperature was around -60°C. Here, the α relaxation corresponds to the glass transition temperature (T_g) of PBT. The β relaxation of PBT/P-40H blend was shifted to a higher temperature than that of neat PBT.

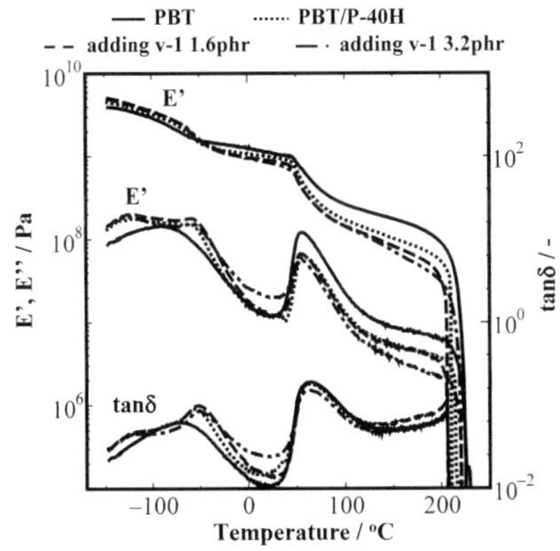

Fig.3 The variation of E', E'' and tand with temperature for PBT/P-40H/v-1 blend materials.

Figure 4 shows the relationship between the impact strength and the activation energy, ΔH^*, which was obtained by the dispersion map concerning the β relaxation for PBT/P-40H with various compatibilizer content: v-1=0, 0.8, 1.6, 2.4, and 3.2 phr. As can be seen in this figure, the activation energy increased gradually with an increase of the compatibilizer content. Hence, it was deduced that the compatibilizer reacted with PBT and/or PEL, and so that, the activation energy of molecular motions concerning PBT and PEL was getting higher.

Figure 5 shows the relationship between Charpy impact strength and dynamic modulus rations, $(E'_{T=100K}- E'_{T=300K})^2/ E'_{T=100K}$ calculated by the Eq.(1). As can be seen in this figure, Charpy impact strength has a relatively good correlation with dynamic modulus rations.

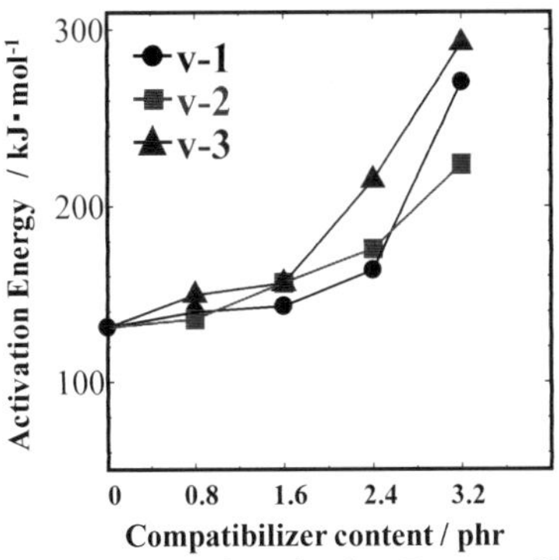

Fig.4 The variation of Activation Energy of PBT/P-40H with various compatibilizer content: v-1=0, 0.8, 1.6, 2.4, and 3.2 phr.

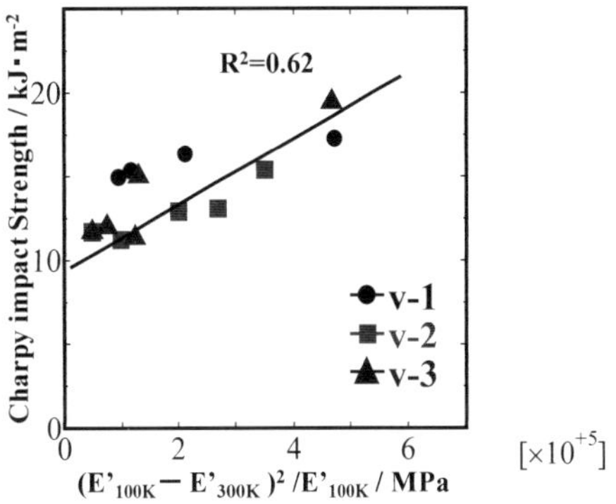

Fig.5 Relationship between Sharpy impact strength and dynamic modulus rations, $(E'_{T=100K}- E'_{T=300K})^2/ E'_{T=100K}$.

Morphology of blend materials

Figure 6 shows SEM images at the cryofractured surface of PBT/P-40H with different amount of compatibilizer (0- 3.2 phr). Although the brittle fracture patterns can be observed for the neat PBT, PBT/PEL blends can be changed to ductile property. Especially, when adding the compatibilizer more than 0.8 phr, the morphological change of the blend materials can be observed clearly, and the impact property can also be changed more ductility. These results indicated that interfacial adhesion between PBT and PEL was improved by adding the reactive compatibilizer, and as a result, the impact strength of PBT/PEL/compatibilizer increased as well as the mechanical property of them can be changed from brittleness to ductility.

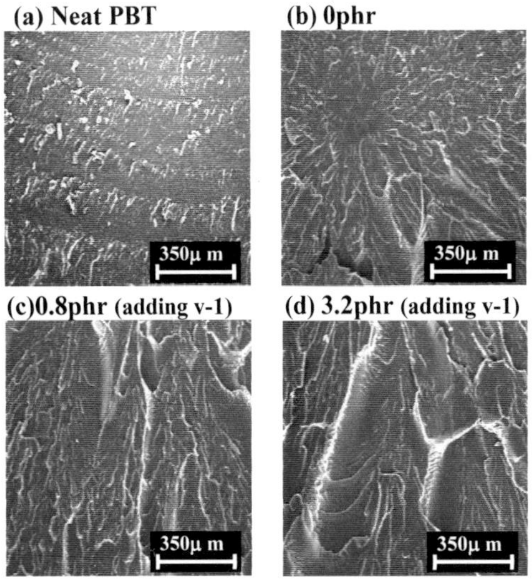

Fig.6 SEM images of (a)Neat PBT and PBT/ P-40H with different compatibilizer content; (b)0phr, (c)0.8phr, and (d)3.2phr.

Viscoelastisic properties of blend materials

Figure 7 shows storage modulus (G') on each frequency for neat PBT, PBT/P-40H blend and PBT/P-40H/v-1 (1.6 and 3.2 phr) blends at 240 °C. The slope of G' for general polymer at flow region was 2. It was shown Newtonian fluid such as cross-linked structure or non-entanglement structure. In this study, the slope of G' for neat PBT and neat PEL was about 2. Then, the value of G' for PBT/P-40H/v-1 blends was increased from neat PBT and neat PEL, as compatibilizer content increased, the value of G' was increased. Furthermore, in PBT/PEL/Compatibilizer blends, rubber like plateau region was shown at low frequency region. It is considered that three-dimensional network structure was formed in the PBT/PEL blend materils.

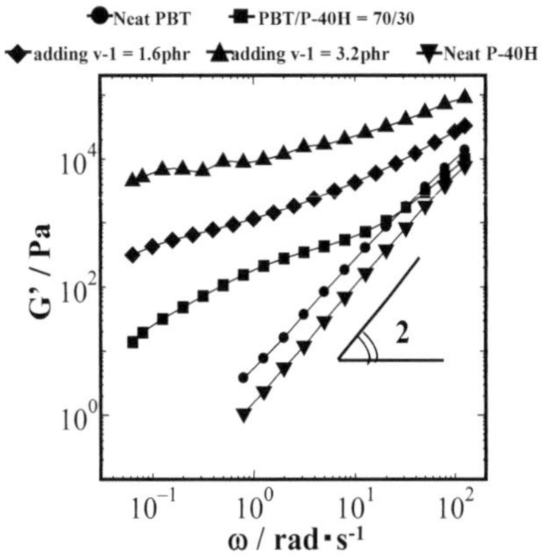

Fig.7 Dynamic modulus of PBT/P-40H with various amount of Compatibilizer(v-1).

Then, the molecular weight of cross-linked chains (*Mc*) was estimated by the value of G' at rubbery plateau region as shown Eq.(3)

$$Mc = \frac{\rho RT}{G'} \tag{3}$$

where ρ is a density, *R* is the gas constant, and T is a temperature.

Figure 8 shows cross-linked molecular weight of PBT/P-40H/v-1 blends. As compatibilizer content increased, cross-linked molecular weight was decreased. The three dimension cross-linked structure between PBT and PEL was formed by adding compatibilizer. It was concerned that cross-linked structure was compatible phase between PBT and PEL. That is, when adding compatibilizer into PBT/PEL blend, the compatibility between PBT and PEL was improved.

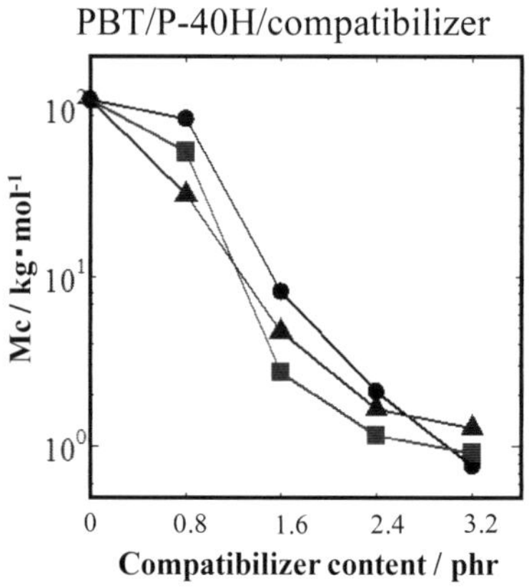

Fig.8 Relationship between Compatibilizer content and cross-linked molecular weight of PBT/P-40H/ Compatibilizer.

Conclusions

Mechanical properties and morphological change of PBT/PBT-derived elastomers (P-40H, P-70B; PELPLENE) with/without a reactive compatibilizer were investigated. The impact strength of PBT increased by adding PELPRENE, and the impact strength of PBT/ PEL/compatibilizer much increased with compatibilizer content. Moreover, the integrated *E''* from 100K to 300K increased, indicating the property of PBT can be changed from brittleness to ductility. Then, the modulus was decreased by adding PEL and compatibilizer into PBT. As compatibilizer content increased, the modulus of PBT/PEL/ Compatibilizer blends was decreased. It was found that the volume fraction of interaction phase PBT', between PBT and PEL, can be improved with the amount of compatibilizer content. Finally, as compatibilizer content increased, cross-linked molecular weight was decreased. The three dimension cross-linked structure between PBT and PEL was formed by adding compatibilizer. It was concerned that cross-linked structure was compatible phase between PBT and PEL. That is, when adding compatibilizer into PBT/PEL blend, the compatibility between PBT and PEL was improved. The impact strength was increased as a result of improving compatibility between PBT and PEL.

References

1. Y. Wada, T. Kasahara, *Journal of Applied Polymer Science,* Vol. 11, pp.1661-1665 (1967).

PL-02

EFFECT OF STRUCTURE FORMATION ON PROCESSING BEHAVIOR IN FIBER AND FILM PROCESSIING

T. Kikutani
Department of Organic and Polymeric Materials, Graduate School of Science and Engineering, Tokyo Institute of Technology – kikutani.t.aa@m.titech.ac.jp
2-12-1-S8-32, O-okayama, Meguroku, Tokyo 152-8552, Japan

Abstract – In polymer processing, processing conditions and structure formation have close correlation, however structure formation is not merely the results caused by the processing conditions. To emphasize the important concept of the mutual interaction of processing behavior and structure formation behavior, various experimental results of fibers and films processing in which structure development causes the alteration of processing behavior are introduced. One of the typical examples matching to such concept is the crystallization. Crystallization has strong influences to the processing behavior through the evolution of latent heat and change of the flowability.

Keywords: structure formation, polymer processing, crystallization, orientation, solidification

Introduction

In polymer processing, structure development or structural change of polymers proceeds during the process, either in the melt processing or in the solid state processing. Development of higher-order structure such as molecular orientation and crystallization would be affected by the processing conditions and accordingly properties of final products would vary. Temperature and stress histories applied to the material are the key factors for determining the structure formation behavior.

It should be emphasized here that the processing behavior would be affected by the structural change of material. Typical example is the variation of rheological behavior of polymer melt depending on the flow history. Another important example is the effect of crystallization. If crystallization occurs, there can be a change in the flowability of polymer melt. The polymer melt would solidify at a relatively low level of crystallinity. On the other hand, crystallization behavior would be affected significantly by the polymer flow. Evolution of latent heat through the crystallization also affects the cooling behavior.

Important point is that the crystallization behavior itself is affected by the flow field and temperature history. In other words, "processing condition and structure formation" does not hold the "cause and effect" relationship. These two factors affect each other during the polymer processing. To emphasize this point, structure formation in the fibers and films processing will be reviewed in this paper from the view point opposite to ordinary direction, that is: the "effect of structure formation on polymer processing". Polymer structure is known to have hierarchical nature, and therefore there can be various types of structure formation which have an influence on processing behavior.

Orientation–induced Crystallization in High-speed Melt Spinning Process

Variation of birefringence of poly(phenylene sulfide) (PPS) fibers is shown in Fig.1[1]. Birefringence increases continuously with the increase of take-up velocity however there is a hesitation of increase at around 5 km/min. At low speed region, as-spun fiber is amorphous, and orientation-induced crystallization starts to occur at around 5 km/min. It is well known that birefringence is governed by the tensile stress applied to the spin-line at the solidification point. Solidification of polymer melt occurs either by glass transition or crystallization. On the other hand, factors dominating the spin-line stress in the high-speed melt spinning process are inertia stress and air-friction stress. When crystallization starts, the solidification point shifts to the upstream of the spin-line because crystallization proceeds at temperatures higher than the glass transition temperature (Tg). Eventually, increase of air friction stress is suppressed, and therefore increase of birefringence can be suppressed. In other words, crystallization shifts the solidification temperature and reduces the spin-line stress at the solidification point.

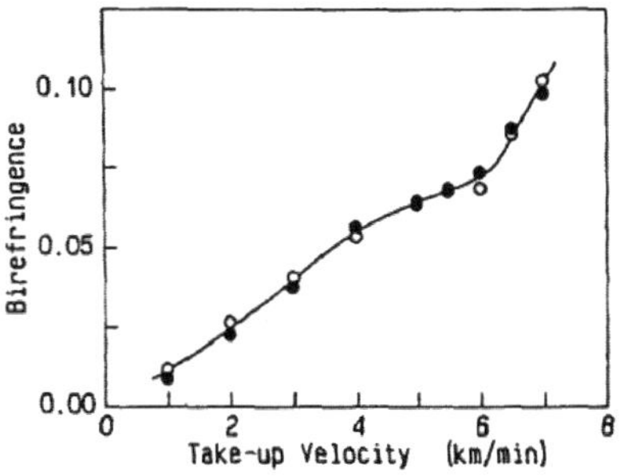

Fig.1 – Variation of birefringence of poly(phenylene sulfide) fibers with take-up velocity.

On the other hand, if liquid isothermal bath (LIB) is introduced to the spin-line [2], crystallization of the spin-line in the liquid causes a significant increase of solidification stress because of the higher friction stress in the liquid, and eventually the birefringence of resultant fibers also increases.

Effect of Variation of Structure Development Behavior in Fiber Cross-section

Suppression of birefringence increment with take-up velocity in the high-speed melt spinning of extremely high speed region explained in the previous section also can be originated from the evolution of temperature distribution in the fiber cross-section. Because of rapid cooling in the high-speed melt spinning process (>10000 K/s), material near the surface can be cooled faster than that in the core. This causes the enhancement of molecular orientation, crystallization and solidification near the surface. If a certain amount of material solidifies near the surface, then the further thinning of the spin-line would be prevented. As the temperature of the core part is still higher than the Tg, there can be a relaxation of molecular orientation in the core. In this manner, structural variation in the cross-section of high-speed spun fibers can be enhanced even though the fiber is extremely thin, and eventually increase of mean birefringence would be suppressed [3].

Similar effect of structural variation in fiber cross-section can be observed in the high-speed melt spinning of bicomponent fibers. In case of poly(ethylene terephthalate) (PET) / polystyrene (PS) bicomponent fibers, in comparison with the single component spinning of PET, significant suppression of orientation development occurs for the PET component in the bicomponent fibers. This is originated from the fact that Tg of PS is higher than that of PET. In this case, even though there is no temperature distribution in the fiber cross-section, thinning of spin-line stops at the Tg of PS (=90 °C) and before the spin-line temperature reaches the Tg of PET (=70 °C) through the further cooling, orientation relaxation of PET component proceeds [4].

Structure development behavior becomes more complicated when both components are highly crystalline. Fig.2 shows the wide-angle X-ray diffraction (WAXD) patterns of the high-speed spun polypropylene (PP) / polyethylene (PE) bicomponent fibers. At low take-up velocities, orthorhombic crystals of PE show c-axis orientation along the fiber axis whereas PP only shows smectic structure of no orientation. At high speeds, PE shows lower orientation with its (200) reflection on the meridian whereas PP exhibits highly oriented α-form crystals. Drastic structural change occurs at a certain take-up velocity, which becomes higher when extrusion temperature is increased.

From the view point of the effect of structure formation on processing behavior, these phenomena can be explained as follows. At low speeds, crystallization of PE proceeds prior to that of PP because PE has higher crystallization rate than PP. If crystallization of PE prevents the further thinning of the spin-line, orientation relaxation occurs in the PP component. On the other hand, at high speeds, tensile stress applied to the spin-line increases and the effect of stress induced crystallization can be enhanced for both components. It should be noted here that the melting temperature (Tm) of PP is higher than that of PE. Therefore, PP can start to crystallize at a temperature where crystallization rate of PE is still slow or even at a temperature higher than the Tm of PE, then orientation of PE may relax before its crystallization. Increase of the take-up velocity where drastic structural change occurs with the extrusion temperature is due to the reduction of tensile stress applied to the spin-line [5].

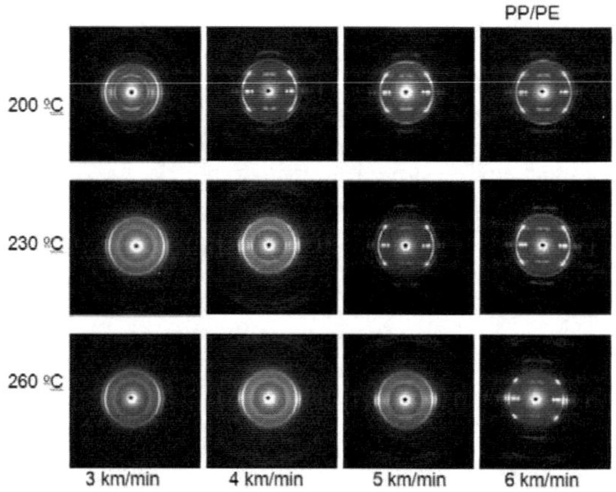

Fig.2 – WAXD patterns of high-speed spun PP/PE bicomponent fibers prepared at three different extrusion temperatures.

Effect of Pre-annealing of Polyester Films on Stretching Behavior

Sequential biaxial stretching behavior of PET films is shown in Fig.3. Original film has amorphous structure with negligibly small birefringence. Pre-annealed film was prepared by annealing the original film at a temperature higher than Tg. Slightly opaque appearance of the pre-annealed film indicates that the film is partially crystallized. It is obvious that the stretching behavior of the pre-annealed film is stable both during the first stretching in the machine direction (MD) and the second stretching in the transverse direction (TD), whereas the original film shows significant heterogeneity of the local draw ratio especially at the late stage of TD stretching.

We speculated that this behavior is originated from the transition of the structural characteristics of PET, that is, starting of spontaneous increase of orientation with the crystallization when the degree of orientation exceeds a certain level. If a region in the film happens to exceed this threshold, that part of film can be stretched easily or tends to stretch by itself whereas other parts try to shrink because a certain level of orientation is already imposed in the film.

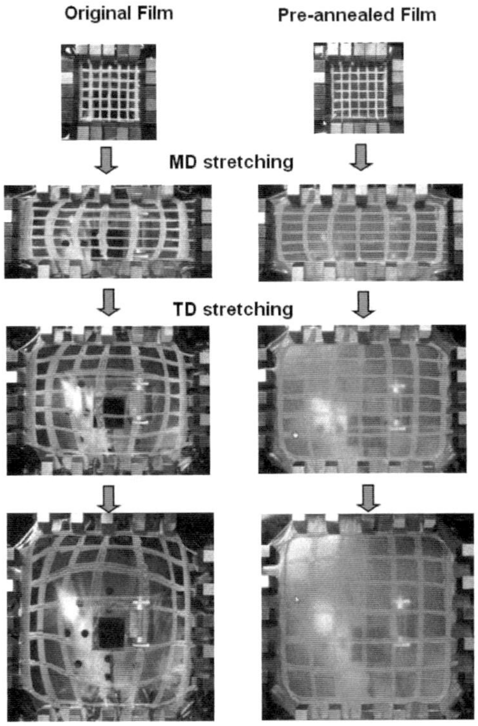

Fig.3 – Observed stretching behavior of PET films during sequential biaxial stretching. Original film is amorphous with no orientation whereas low degree of crystallinity is introduced for the pre-annealed film.

Effect of the spontaneous orientation increment accompanied by the stress or orientation induced crystallization can be the origin of the change in processing behavior in various fields of polymer processing. Development of POY (Partially oriented yarn) -DTY (Draw textured yarn) process to produce draw-textured yarn directly from the melt-spun fibers without applying conventional drawing process is one of the typical examples. There has been a development of fibers which exhibits spontaneous elongation during annealing. PE/PP bicomponent fibers with such characteristics are now widely used for the production of non-woven fabrics of soft-touch, which are applied for the surface of hygiene products such as diaper [6].

Processing with Sequential Steps

As described in the previous section, in cases of the polymer processing consisting of sequential steps, structure development in an earlier step naturally causes the variation of processing behavior in subsequent steps. Melt spinning – drawing & annealing process for the production of fibers and sequential biaxial stretching for the production of biaxially oriented films are typical examples. It should be noted that the concept for the formation of structure is not always clear. For example, in the course of developing high-strength PET fibers, we found that the high-strength and high toughness fibers can be produced by applying the drawing & annealing process to the as-spun fibers, which is prepared keeping the Deborah number at a low level in the melt spinning process [7]. This result suggests that the structure of as-spun amorphous PET fibers with low orientation can be manipulated even though difference may not be detectable through ordinary structural analysis technique.

Typical example of the changes of MD and TD stresses and MD-TD birefringence in the sequential biaxial stretching of cyclo-olefin copolymer (COC) film is shown in Fig.4. In the MD stretching and relaxation steps, as expected, MD stress is higher than the TD stress, and birefringence increases and relaxes in the stretching and relaxation processes, respectively. In the second stretching after a certain period of relaxation, TD stress increases steeply while birefringence decreases steeply and reaches a negative value. More surprisingly, birefringence comes back to a positive value during the relaxation process after the TD stretching. These results suggests that there is a formation of "structure" of various relaxation time in the MD stretching, and the TD stretching and subsequent relaxation processes are strongly affected by the formation of such "structure" [8].

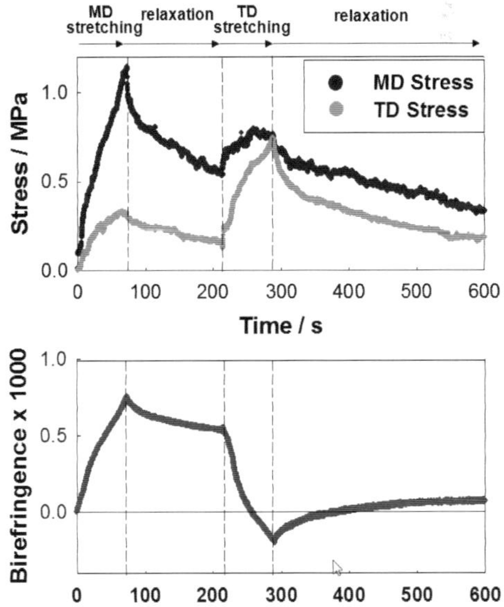

Fig.4 – Variations of MD and TD stresses and MD-TD birefringence in the sequential biaxial stretching of COC film.

Concluding Remark

Properties of polymer products vary significantly by changing the higher-order structure. Therefore, deep understanding on the structure development behavior in the polymer processing is an important subject. In this article, various experimental results which are useful for explaining the basic concept on the "effect of structure development on processing behavior" as shown in Fig. 5 is introduced.

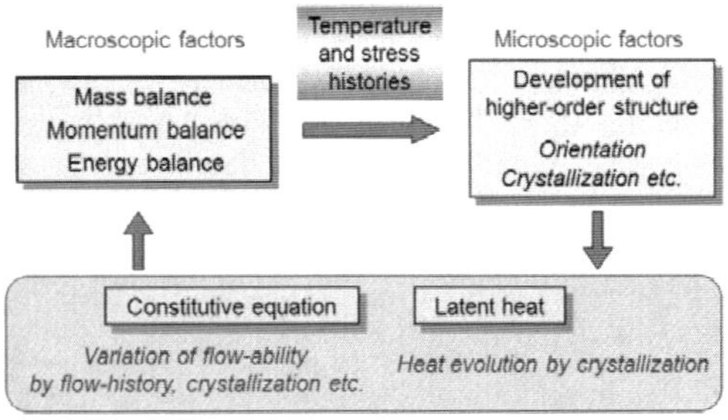

Fig. 5 – A schematic for explaining the effect of structure on the processing behavior.

References

1. T. Kikutani, K. Wakayama, M. Sato, A. Takaku, *Sen'i Gakkaishi*, 1992, **48**, 549.
2. T.Kikutani, J.Radhakrishnan, M. Sato, N. Okui, A. Takaku, *Int. Polym. Process.*, 1996, **11**, 42.
3. J. Shimizu, N. Okui, T. Kikutani, *High-speed Fiber Spinning*, John Wiley & Sons, New York, 173.
4. T. Kikutani, J. Radhakrishnan, S. Arikawa, A. Takaku, N. Okui, X. Jin, F. Niwa, Y. kudo, *J. Appl. Polym. Sci.*, 1996, **62**, 1913.
5. Y. Shimomura, H. Ito, T. Kikutani, *in PPS Europe/Africa Proceedings,* Athens, Greece, 2003
6. M. Matsui, W. Saka, K. Nagashima, Y. Nakano, *Fiber Preprints, Japan*, 2012, **67**, 12.
7. M. Masuda, W. Takarada, T. Kikutani, Int. Polym. Process., 2010, 25, 159.
8. K. Ikeda, W. Takarada, T. Kikutani, *in PPS Europe/Africa Proceedings*, Larnaca, Cyprus, 2009.

PL-03

Making light and ultra-strong tapes for high performance sails in America's Cup boat racing

Han Meijer[1*], Paul Smith[2], Kirill Feldmann[2], Theo Tervoort[2], Jérôme Lefèvre[2], Jan Giesbrecht[2]
[1] *TU/e, Materials Technology, P.O. Box 513. 5600 MB Eindhoven, The Netherland; h.e.h.meijer@tue.nl*
[2] *ETHZ, Department of Materials, Wolfgang Pauli Str. 10, 8093 Zurich Switzerland; www.polytech.mat.ethz.ch*

Abstract - The principal objective of the work presented here is to develop a processing route to produce high-performance tapes that can be glued together with minimal use of matrix materials into isotropic, and anisotropic, light weight composite products. The approach is based on the general experience that in building walls, rectangular bricks are to be preferred over cylindrical ones, since the better packing potential of the first avoids the excessive use of cement. We use a thermotropic liquid crystalline copolyester, Vectra®, given its convenient processing possibilities and still impressive mechanical properties. Vectra® tapes can be glued together via transesterification and the resulting high performance products are 100% recyclable. Sails for fast racing are an appealing first potential application.

Keywords: High performance, tapes, liquid crystalline polymers, HBA-HNA copolyesters, recyclability.

Introduction

Fibers.

High-performance polymer fibers nowadays are used in a broad spectrum of applications. The common feature of these fibers is that they all exhibit a high degree of orientation and extension of their constituent covalently-bonded macromolecular chains. Unfortunately, the response of such highly anisotropic structures is strongly dependent on the direction of applied load or strain, which has major consequences for the design and ultimate characteristics of the products in which they are utilized. Traditionally, fibers are most often applied in twisted form, such as in yarns and ropes, are woven, as in fabrics and cloth, or are used in cut form, e.g. in non-wovens and in composites.

Films.

Attempts have been made to avoid the adversary effects associated with weaving or with the use of short-fiber-reinforced composites, and to manufacture high-performance, biaxially oriented polymer films. Disappointingly, their characteristics are far inferior when compared to those of the corresponding uniaxially oriented fibers. In order to employ high-performance fibers more efficiently and to better exploit their ultimate potential, laminates of multi-filament strands and of spread multi-filaments, have been developed. Unfortunately, there are considerable difficulties in manufacturing such products with a high fiber content. In practical applications, that fraction appears to be limited to only about 55-65%, due to their approximately circular cross sections and incorporated voids. As the stiffness of the matrix materials used generally is low in comparison with that of the high performance fibers, the mechanical properties of the resulting composite are modest.

Tapes.

The principal objective of the work presented here is to develop processing routes to produce high-performance tapes that can be glued together with minimal use of matrix materials into isotropic, and anisotropic, light weight composite products. The approach is based on the general experience that in building walls rectangular bricks are to be preferred over cylindrical ones, since the better packing potential of the first avoids the excessive use of cement. The second inspiration originates from the so-called *all-PP composites* concept, where oriented polypropylene (PP) fibers or tapes, with their (if constrained) high melting point, are glued together with unoriented PP (or PP co-polymers) with a (considerably) lower melting point. Although the ultimate properties, and especially the modulus, of PP fibers and tapes are not that impressive, given their helical molecular configuration, the extra interest in the concept is in its 100% recyclability.

Experimental

LCP polymer.

In this study we focus on thermotropic liquid crystalline polymers, LCP's, and chose for a HBA-HNA copolyester, sold under the trade name Vectra® by Ticona (Germany), since it provides an adequate compromise between convenient processability and – compared to lyotropic LCP's inevitably more modest-mechanical properties due to the semi-rigid nature of the macromolecules. It has been demonstrated that high-performance fibers can be directly spun from their thermotropic melt at temperatures of 250-350 ^{0}C by applying relatively high wind-up speeds (draw down) that induce elongational flow fields causing the polymer molecules to orient in the direction of flow.

Advantageously, relaxation times of such an oriented melt are relatively long and, therefore, macromolecular orientation can largely be retained during solidification, yielding the desired high-performance fibers, with elastic moduli of typical 70 GPa.

Die Design.
The secret of high-performance tape making is basically in the design of the die that combines a fiber forming part with a fiber fusing part such that every individual fiber made upstream can be guided (by applying sufficient axial stress) to its end position in the tape without crossing any other neighboring fiber, see Figure 1. Basically this means that the d = 100μm diameter fiber forming holes are staggered such that they project next to each other into the central line downstream that forms the tape. In order to conserve mass without the need of fiber acceleration or deceleration, the tape thickness at the die exit is set at $t = \pi/4*d = 75\mu m$.

Air gap extrusion.
Preserving the molecular orientation created inside the upstream die requires a certain axial stress in the melt that is realized by applying a draw down λ after exiting the die. Indeed the modulus of the tape increases, however this is at the cost of the tape width, expressed in the ratio between the width of the tape relatively to that of the die, see the grey symbols in the two grey-shaded areas in Figures 2 and 3. Orientation lost and width reduction can be prevented by applying a sufficiently high stress via air gap extrusion and rapid efficient cooling, see the colored symbols in Figures 2 and 3.

Results and Discussion

Based on these base tapes, a number of different foils, films and products have been made. Interestingly, no glue is needed in these systems, since at high temperature and sufficient pressure trans-esterification proves to give sufficient adhesion between the tapes. This implies not only an almost factor two improvement in stiffness-to-weight but also guarantees 100% recyclability. Moreover, as in the case of fibers, a simple heat treatment gives post-condensation at elevated temperatures yielding enhanced tape properties upto 80 GPa modulus and 2 GPa strength. Most remarkable could, however, be the barrier properties of these, basically infinite, single crystals and oxygen barrier is the best of all polymers, not just in dry conditions, but also in 100% humidity. As a consequence, other applications than those of high performance sails for the America's Cup are explored.

Conclusions

A new route to produce high performance tapes based on, in this example of the Vectra polymer used thermotropic, liquid crystalline polymers is developed. Tapes can be stacked and glued to form high performance composite structures. The process is clean, fast, and te product is 100% recyclable.

References

Jérôme Lefèvre, *Ultra-high-performance polymer foils,* thesis ETH Zurich, no 17603 (2008)

Figure 1

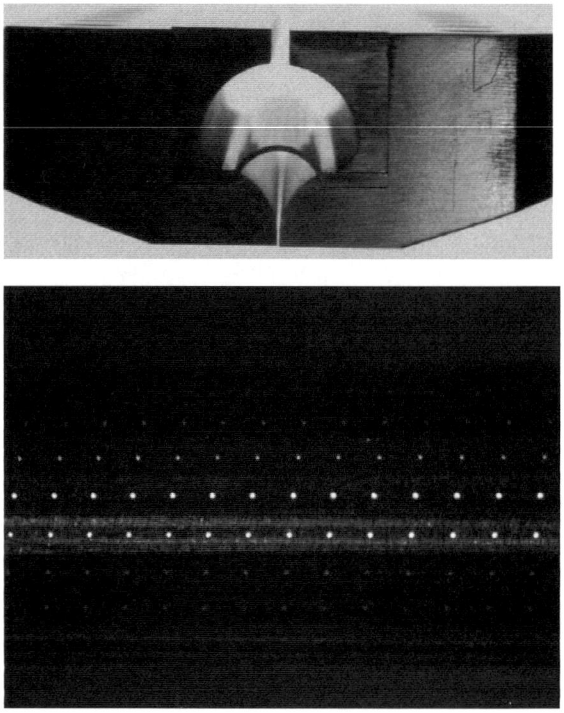

Top: cross section of the die with in the middle the fiber forming part in the semi-circle and the fiber fusing part below. Bottom: the hierarchy in the position of the nozzles that, projected on the midline constitute exactly one row that forms the tape.

Figure 2

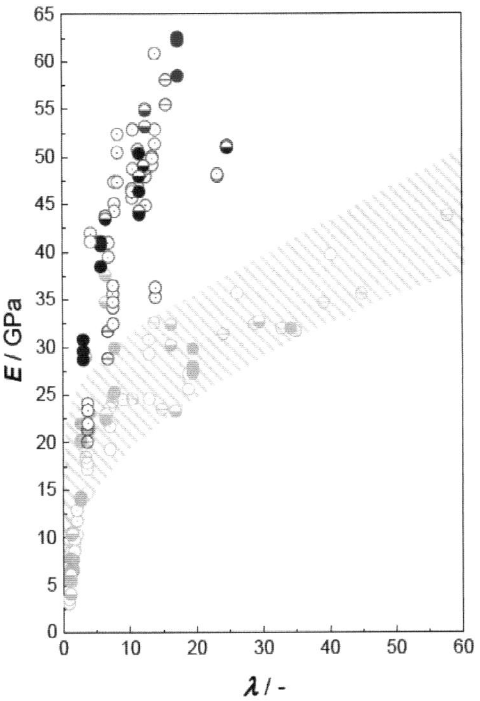

Tape modulus as a function of draw down. The grey symbols represent standard extrusion, the colored symbols relate to the high stress confined air gap extrusion.

Figure 3

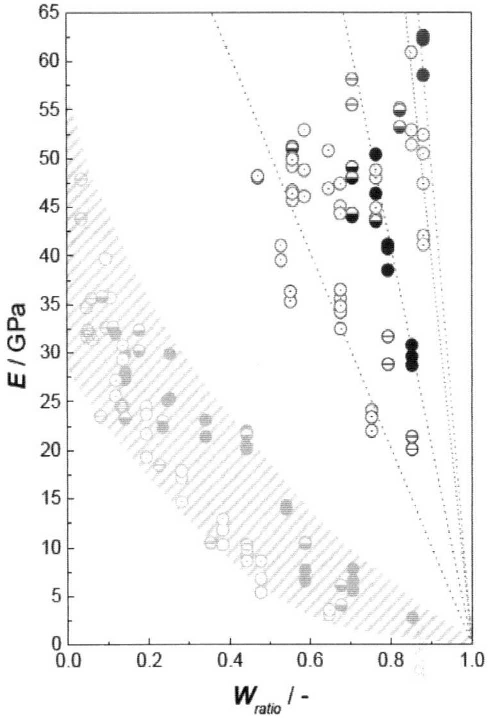

The resulting reduction in width. The grey symbols represent standard extrusion, the colored symbols relate to the high stress confined air gap extrusion.

PL-06

APPROACHING PROCESSING CONDITIONS WITH CLASSICAL RHEOMETRICAL FLOWS

P. Van Puyvelde
Department of Chemical Engineering, KU Leuven, Belgium - peter.vanpuyvelde@cit.kuleuven.be

Abstract - During processing, polymers are subjected to rather severe thermo-mechanical histories. In this paper, examples will be demonstrated on how classical rheometrical flows can be used to obtain information on the processing behaviour. As an example, flow-induced crystallization of particle-filled poly-propylene will be investigated using a sliding plate rheometer that allows to cover shear rates and shear strains that are closely related to processing conditions.

Keywords: processing flow, flow-induced crystallization, rheometry

Introduction (Times New Roman 10, bold)

During the past decades, many different rheometers have been developed, each of them with their own advantages and drawbacks. Despite the many differences amongst all these machines, the main goal has always been to characterize the material as good as possible and at conditions as close as possible to real-life processing conditions. For instance, in the field of flow-induced crystallization, one is of course interested in obtaining insight in the relation between flow conditions, the kinetics of crystallization and the final resulting morphology. Since the seminal work of Lagasse and Maxwell[1], many studies have been devoted to this topic[2]. Many of them use real processing flows to study this problem. However, whereas these flows are close to reality, it remains difficult to isolate the effect of flow because of the occurrence of thermal and flow gradients. On the other hand, many flow-induced crystallization studies use well-defined rheometrical flows with controlled thermal and flow conditions. However, the drawback of this approach is that it becomes difficult to attain strains and strain rates that are characteristic of a processing operation. In this work, an attempt is made to bridge the gap by using rheometrical flows to attain high flow rates. As an example, results on the flow-induced crystallization of particle-filled polypropylene will be demonstrated.

Experimental (Times New Roman 10, bold)

To demonstrate the effect of particle shape on the flow-induced crystallization of polypropylene, home-made hematite particles with different anisotropies have been used. Hematite has been chosen since the surface properties, important for the interaction with the polymer, can be controlled accurately. Hence, particles with different shapes but with identical interactions with the polymer can be synthesized.

The flow cell used in this work consists of a parallel plate geometry, equipped with a robust temperature control unit. The use of such a cell ensures a simple shear flow and by changing the gap thickness, shear rates and shear strains can be varied up to values of respectively 1000 s^{-1}. The crystallization is followed on-line by means of birefringence measurements. This in-situ and time-resolved technique is able to characterize the amount of orientation in the material and to obtain information about the kinetics of the crystallization process. Details of the cell can be found elsewhere[3,4].

The experimental protocol used in this work is the so-called 'short term shearing protocol' originally introduced by the group of Janeschitz-Kriegl[5]. It consists of bringing the material above the melting temperature in order to erase the thermo-mechanical history. Subsequently, the material is cooled to a certain crystallization temperature. When this temperature is reached, a shear step is applied using the shear rate and the shear time as variables. By using this approach, temperature and flow effects can be separated.

Results and Discussion

The optical methodology used in this work has the advantage that it can follow the molecular dynamics during and after the application of flow. A typical example for the unfilled polypropylene is shown in figure 1. In this figure, different shear rates have been applied but the total shear strain has been kept constant. For instance, when applying a shear rate of 94 s^{-1}, it is seen that the birefringence increases towards a plateau value, characteristic of the induction of molecular orientation. When flow is removed, this orientation relaxes. However, when the shear rate becomes sufficiently high (e.g. 255 s^{-1}), it is seen that even during flow, an upturn is observed in the birefringence. This upturn has been shown to result from oriented structures that finally will lead to a shish-kebab morphology[4,6,7].

In figure 2, the birefringence for the same experiment is shown but now for times after cessation of flow. From figure 2, interesting conclusions can be derived.

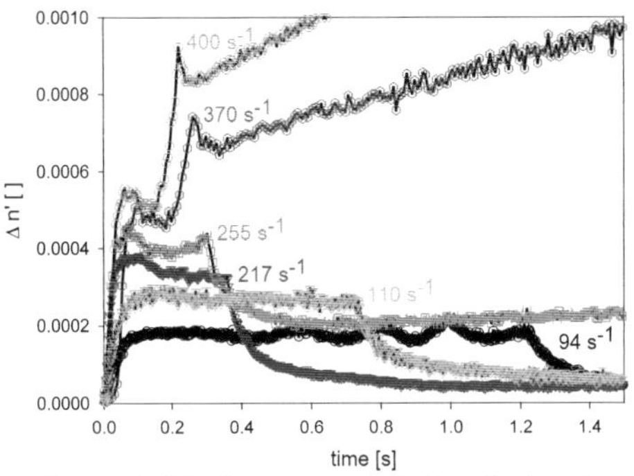

Figure 1: Birefringence evolution during the application of flow (different shear rates, strain=100, 145°C)

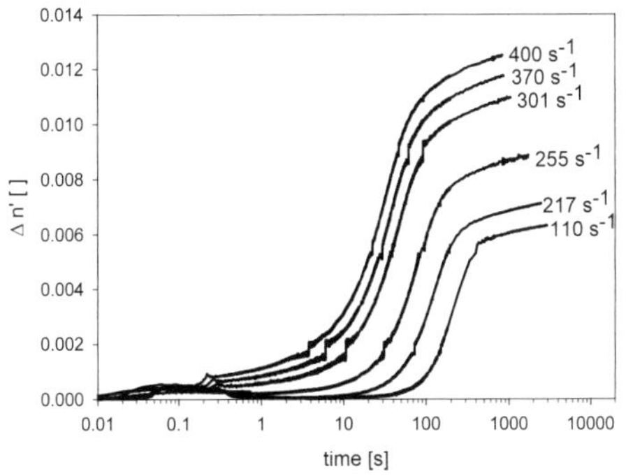

Figure 2: evolution of the birefringence after cessation of flow (different shear rates, strain=100, 145°C)

First, it can be seen that, as expected, the kinetics of the crystallization process is significantly increased when the shear rates becomes larger. Secondly, when the flow intensity is increased, the final value of the birefringence also increases which points to a higher degree of orientation. From the birefringence curves of figure 2, characteristic crystallization times can be defined. Here, the inflection point of the curves is taken as a characteristic value.

The effect of particle anisotropy on the flow-induced crystallization can be summarized in figure 3. Here, the characteristic crystallization times as defined above are plotted as a function of the applied shear rate. In this graph, three different particle shapes have been used: spherical hematite particles (AR 1) and two other anisotropies (1.8 and 3.6).

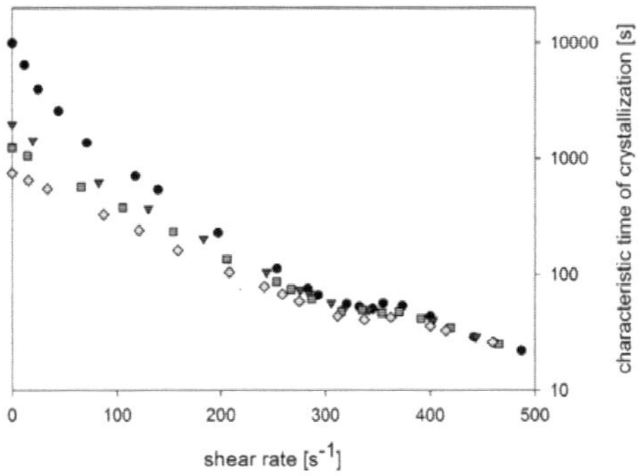

figure 3: Characteristic times as a function of shear rate (circles: neat polymer, triangles particles with aspect ratio 1, squares aspect ratio 1.8 and diamonds aspect ratio 3.6).

From the graph, it is clear that at low shear rates, the particles have a clear nucleating ability. For instance, when comparing the neat polymer with the polymer containing hematite particles with aspect ratio of 3.6, it is clear that the characteristic crystallization time differs with almost one decade. However, when shear rate is increased, the difference in crystallization kinetics between the different materials becomes insignificant and flow starts to dominate the nucleation efficiency. The latter observation was also made by for instance D'Haese et al.[8], in which the effect of particles size on flow-induced crystallization was discussed. At low values of the shear rate, particle size greatly influenced crystallization dynamics whereas at high values of the flow intensity, the applied flow field dominated the crystallization kinetics. From these data, it is clear that orientation effects, induced by flow, are indeed a very effective route to nucleate polymers, a method that is proven to be even more efficient that heterogeneities present in the material.

ACKNOWLEDGMENTS

PVP is indebted to the Onderzoeksfonds K.U. Leuven (GOA 02/009) for their financial support.

REFERENCES

1. Lagasse, R.R. and Maxwell, G. (1976), "Experimental study of kinetics of polymer crystallization during shear flow", *Polym. Eng. Sci.*, **16**, 189-199.
2. Kumaraswamy, G. (2005) "Crystal-lization of polymers from stressed melts", *J. Macromol. Sci. Polym. Rev.*, **C45**, 375-397.
3. Langouche, F. (2006) "Orientation development during shear flow-induced crystallization of i-PP", *Macromol.*, **39**, 2568-2573.
4. Baert, J., Van Puyvelde, P., Langouche, F. (2006) "Flow-induced crystallization of PB-1: from the low shear rate region up to processing rates", *Macromol.*, **39**, 9215-9222.

5. Jerschow, P., Janeschitz-Kriegl, H. (1997) "The role of long molecules and nucleating agents in shear-induced crystallization of isotactic polypropylene", *Int. Polym. Proc.*, **12**, 72-77.
6. Kornfield, J.A., Kumaraswamy, G., Issaian, A.M. (2002) "Recent advances in understanding flow effects on polymer crystallization", *Ind.&Eng. Chem. Res.*, **41**, 6383-6392.
7. D'Haese, M., Mykhaylyk, O.O., Van Puyvelde, P. (2011) "On the onset of oriented structures in flow-induced crystallization of polymers: a comparison of experimental techniques", *Macromol.*, **44**, 1783-1787.
8. D'Haese, M., Van Puyvelde, P., Langouche, F. (2010) "Effects of particles on the flow-induced crystallization of polypropylene at processing speeds", *Macromol.*, **43**, 2933-2941.

AUTHOR INDEX

AUTHOR INDEX

AUTHOR INDEX

AUTHOR INDEX

AUTHOR INDEX

AUTHOR INDEX

AUTHOR INDEX

AUTHOR INDEX